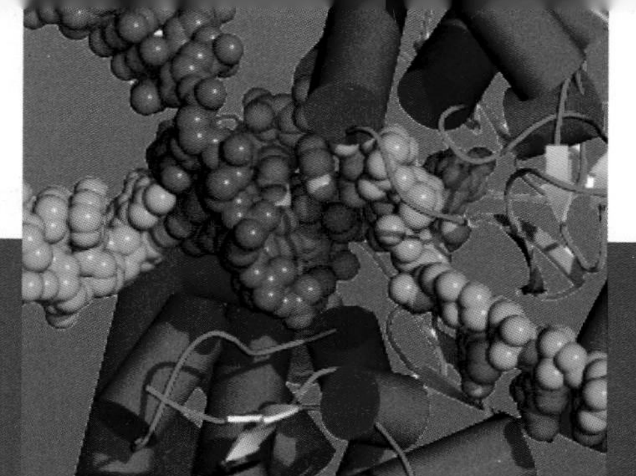

Textbook of BIOCHEMISTRY With Clinical Correlations

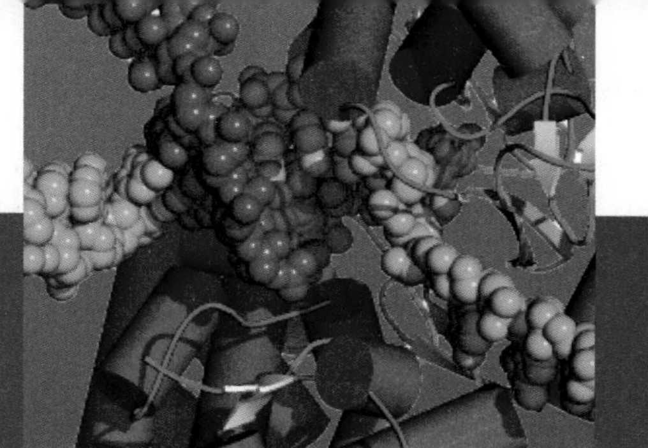

Textbook of BIOCHEMISTRY With Clinical Correlations

EDITED BY

Thomas M. Devlin

Professor Emeritus
Department of Biochemistry and Molecular Biology
College of Medicine
Drexel University

John Wiley & Sons, Inc.

IMAGE ON FRONT COVER

Model of a partial telomerase elongation complex. The enzyme telomerase is a ribonucleoprotein reverse transcriptase responsible for maintaining the length and integrity of the ends of linear chromosomes, termed telomeres, present in eukaryotes. The figure depicts the partial structure of the telomerase elongation complex at the end of a chromosome (blue spheres). The catalytic subunit of telomerase (red rods) uses an integral RNA template (green spheres) to add multiple identical repeats of deoxyribonucleotides to the 3' end of the DNA strand of the chromosome. Telomerase and telomere dysfunction are considered to contribute to replicative senescence and programmed cell aging. Activation of telomerase is associated with the uncontrollable proliferation of cells found in about 85 percent of human cancers. A discussion of telomerase is presented in chapter 4. Gillis, A, J., Schuller, A. P., Skordalakes, E. Structure of the *Tribolium castaneum* telomerase catalytic subunit TERT. *Nature* 455:633, 2008. Figure generously supplied by Dr. Emmanuel Skordalakes, The Wistar Institute, Philadelphia, PA 19104, USA

PUBLISHER	KAYE PACE
ASSOCIATE PUBLISHER	PETRA RECTER
ACQUISITIONS EDITOR	JOAN KALKUT
EDITORIAL ASSISTANT	YELENA ZOLOTOREVSKAYA
SENIOR PRODUCTION AND MANUFACTURING MANAGER	MICHELINE FREDERICK
SENIOR PRODUCTION EDITOR	KERRY WEINSTEIN
MARKETING MANAGER	KRISTINE RUFF
SENIOR DESIGNER	KEVIN MURPHY
INTERIOR DESIGNER	LAURA IERARDI
COVER DESIGNER	M77 DESIGN
SENIOR ILLUSTRATION EDITOR	SANDRA RIGBY
EXECUTIVE MEDIA EDITOR	TOM KULESA
MEDIA EDITOR	MARC WEZDECKI

This book was set in India by MPS Limited, A Macmillan Company and printed and bound by Worldcolor. The cover was printed by Worldcolor.

Library of Congress Cataloging in Publication Data:

Textbook of biochemistry : with clinical correlations / edited by Thomas M. Devlin. — 7th ed.

p. ; cm.

Includes bibliographical references and index.

ISBN 978-0-470-28173-4 (cloth)

BRV ISBN 978-0-470-60152-5

1. Biochemistry. 2. Clinical biochemistry.

I. Devlin, Thomas M. II. Title: Biochemistry.

[DNLM: 1. Biochemical Phenomena. QU 4 T355 2010]

QP514.2.T4 2010

612'.015—dc22 2009046304

Printed in the United States of America

10 9 8 7 6 5 4 3 2 1

• TO MY FAMILY

Steve, Bonnie, Mark, Cathy, Kate, Matt, Ryan, and Laura
who have been constant sources of pride and love

and

• TO MARJORIE

who has been with me through seven editions
for her enduring encouragement, support, and love

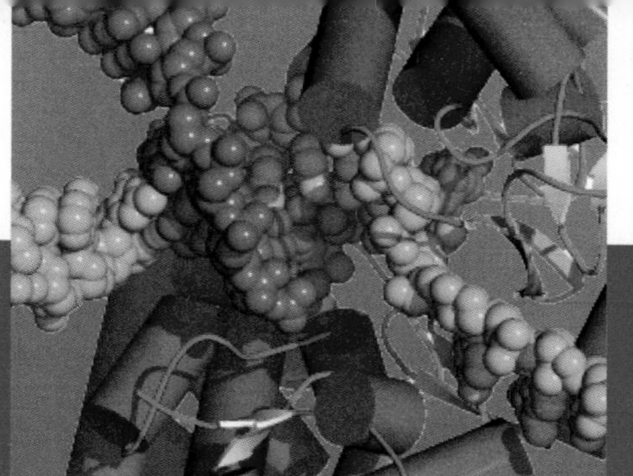

Contents in Brief

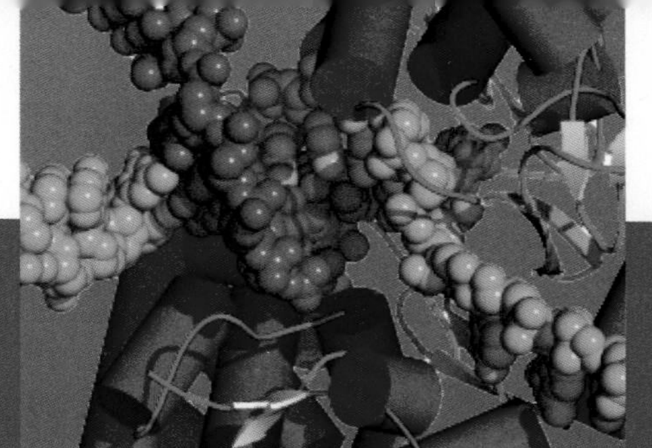

Contents

Part I

STRUCTURE OF MACROMOLECULES

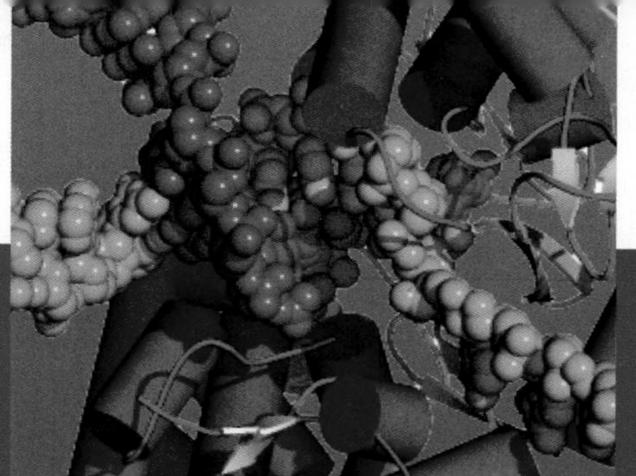

Clinical Correlations

8 Regulation of Gene Expression

9 Proteins II: Structure–Function Relationships in Protein Families

10 Enzymes: Classification, Kinetics, and Control

11 The Cytochromes P450 and Nitric Oxide Synthases

12 Biological Membranes: Structure, Receptors, and Solute Transport

13 Fundamentals of Signal Transduction

14 Bioenergetics, Mitochondria, and Oxidative Metabolism

15 Carbohydrate Metabolism I: Major Metabolic Pathways and Their Control

16 Carbohydrate Metabolism II: Special Pathways and Glycoconjugates

17 Lipid Metabolism I: Synthesis, Storage, and Utilization of Fatty Acids and Triacylglycerols

18 Lipid Metabolism II: Pathways of Metabolism of Special Lipids

19 Amino Acid and Heme Metabolism

20 Purine and Pyrimidine Nucleotide Metabolism

21 Metabolic Interrelationships

22 Biochemistry of Hormones

23 Molecular Cell Biology

24 Cell Cycle, Programmed Cell Death and Cancer

25 Digestion and Absorption of Basic Nutritional Constituents

26 Vitamins and Minerals: Requirements and Function

27 Macronutrients: Metabolic Effects and Health Implications

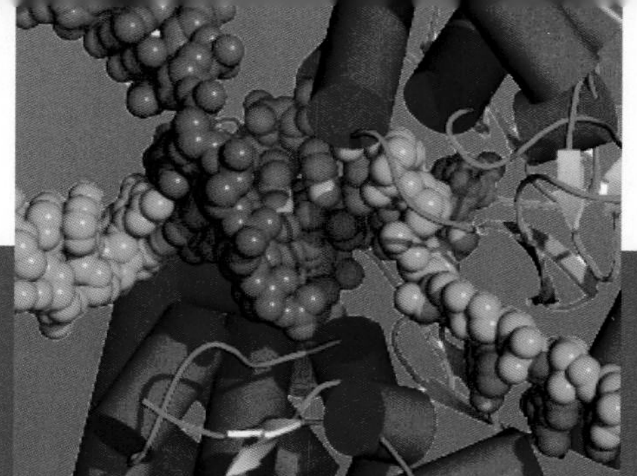

A Closer Look

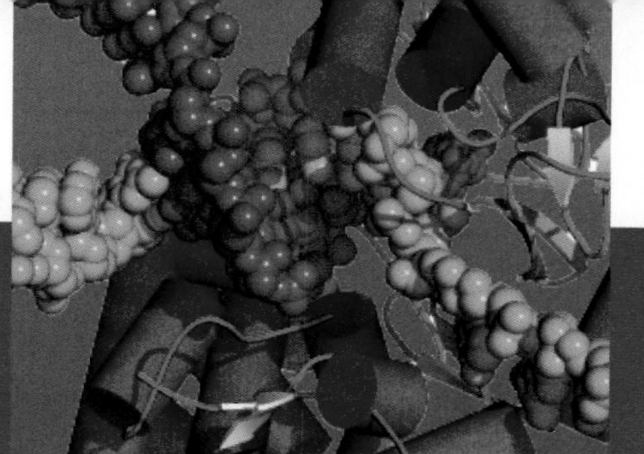

Preface

In the early 1980s, we were motivated to prepare the First Edition of the **Textbook of Biochemistry with Clinical Correlations** because at that time no biochemistry text emphasized the astonishing advances in the second half of the twentieth century regarding our knowledge about normal and abnormal mammalian cell biochemistry, specifically that of humans. Thus the decision was made that in this textbook the focus of the presentation would be on the basic cellular biochemistry of eukaryotes, with an emphasis on mammalian cells and tissues. The depth and scope of the presentation is still that of a basic biochemistry text. In the preparation of the First Edition we also wanted students to catch a glimpse of how biochemical research has lead to an understanding of the causes of many human diseases. This was accomplished by presenting descriptions of the biochemistry of selected diseases in separate **Clinical Correlation** boxes. These correlations provide significance to the biochemistry, thus facilitating the learning process. The popularity of the correlations with students encouraged us to add new ones in all subsequent editions.

Why a Seventh Edition?

The ever-increasing depth of our knowledge of the biochemistry and control mechanisms of normal cells and tissues has been the primary incentive to update the content of the textbook.

Results of recent research have lead to an understanding of the molecular events of many cellular and physiological processes, which until now were poorly understood, such as the extracellular matrix proteins, programmed cell death, molecular motors, and cell signaling. Thus, it is appropriate to include these and other such topics in this textbook. In addition, many of these topics are now presented in biochemistry courses.

Finally it was also deemed necessary to reorganize some topics as research has uncovered the complex relationships between many cellular and tissue processes. Along with these changes the scope and number of Clinical Correlations has increased, covering topics as diverse as HIV infections, hypercholesteremia, diabetes, and Parkinson disease.

Objectives and Scope of Presentation for the Seventh Edition

The following objectives were established for the First Edition and have been retained in all subsequent editions. They are

- To present a clear and precise discussion of the biochemistry of eukaryotic cells, with an emphasis on those of mammalian tissues
- To have the scope and depth of presentation meet the requirements of upper-level undergraduate, graduate level, and professional school courses in biochemistry
- To relate biochemical events at the cellular level to physiological processes in the whole animal;
- To cite examples of abnormal biochemical processes in human disease

The textbook is organized and written such that any sequence of topics considered most appropriate by an instructor can be presented. Throughout this text, results from research of nonmammalian organisms are presented if the information is more advanced than that of similar studies with mammalian cells. Thus the book presents a current picture of our biochemical knowledge of eukaryotic cells.

Significant Changes in the Seventh Edition

Every chapter was updated with inclusion of new information and condensation of some material. Some of the changes are

- An expanded discussion of microRNAs
- An in-depth presentation of the basal lamina protein complex, molecular motors, programmed cell death, and cancer
- A presentation of membrane transport mechanisms that conforms to current nomenclature and research directions
- A discussion of unstructured proteins
- A reorganized discussion of amino acid metabolism that separates the synthesis, degradation, and roles of amino acids;
- A discussion of heme metabolism included with the amino acids, which is the more usual placement in a teaching program
- A comprehensive presentation of the absorption and transport of iron
- An inclusive discussion of the function of vitamins concentrated in one chapter
- An updating of chapter bibliographies, with selection of references from readily available sources, many accessible on the Web. References are generally to review articles and seminal publications; carefully selected websites are also listed
- Approximately one half of the questions and annotated answers are new to this edition; they are similar to those used in graduate and professional school admission examinations, some are based on data presented in a clinical vignette, and each set of questions has several problem-solving questions.

In addition, in response to reviewer recommendations, sections in many chapters have been reorganized for a better flow of information. Related content that in previous editions was presented in several different chapters has been consolidated into single presentations. Illustrations were updated and new figures added. The adage "A picture is worth a thousand words" is appropriate, and the reader is encouraged to study the illustrations because they are meant to clarify confusing aspects of a topic. As in previous editions, there is frequent cross-referencing between chapters.

New Features in the Seventh Edition

To facilitate the students' learning, we have added several new features to the text.

- **Key Concepts:** A list of Key Concepts appears at the beginning of each chapter for students to use as a guide while studying the chapter and as a reference for self-evaluation at completion of study.
- **A Closer Look:** These boxes contain supplementary information on the topic under discussion.
- **Key Terms:** A list of Key Terms appears at the end of each chapter;
- **Online Mendelian Inheritance in Man (OMIM):** Accession numbers associated with a disease or enzyme in the Online Mendelian Inheritance in Man (OMIM) database are indicated at appropriate places in the text. The OMIM database (www.ncbi.nlm.nih.gov/Omim) catalogues known diseases with a genetic component.

Content and Organization of the Seventh Edition

The content of the seventh edition is divided into five major parts.

- **Part I, Structure of Macromolecules**, contains an introductory chapter on eukaryotic cell structure (Chapter 1), followed by two chapters on the chemistry and structure of nucleic acids (Chapter 2) and protein (Chapter 3).
- **Part II, Transmission of Information**, begins with separate chapters on the synthesis of the major cellular macromolecules, that is, DNA (Chapter 4), RNA (Chapter 5), and proteins (Chapter 6). A chapter on recombinant DNA and biotechnology is included because the knowledge and techniques in this area have had and continue to have a profound impact on research in nearly all facets of biochemistry (Chapter 7). Part II concludes with a chapter on the Regulation of Gene Expression in which mechanisms of both prokaryotes and eukaryotes are presented (Chapter 8).
- **Part III, Functions of Proteins**, opens with a presentation of the structure-function relationship of four major families of proteins, that is, antibody molecules, serine proteases, hemoglobin, and basal lamina proteins (Chapter 9). This is followed by a detailed discussion of enzyme function and kinetics (Chapter 10) and a separate chapter on the cytochromes P450, a unique and important family of enzymes (Chapter 11). A chapter on membrane structure and the essentials of transmembrane transport mechanisms (Chapter 12), and a chapter on the basic mechanisms of cellular signal transduction conclude Part III (Chapter 14). These chapters present the fundamentals of these topics and subsequent chapters present their role in specific cellular processes.
- **Part IV, Metabolic Pathways and Their Control**, opens with a chapter on bioenergetics and oxidative metabolism (Chapter 14). Separate chapters describe the major metabolic pathways of carbohydrates (Chapter 15), and special pathways of carbohydrates and glycoconjugates (Chapter 16). This is followed by a chapter covering the synthesis, storage, and utilization of energy in lipid form (Chapter 17), then one describing the metabolism of phospholipids, sphingolipids, cholesterol, and prostaglandins (Chapter 18). Amino acid and heme metabolism are covered in Chapter 19, followed by the synthesis and degradation of the purine and pyrimidine nucleotides (Chapter 20). A chapter on the integration of these metabolic pathways in humans completes this part (Chapter 21). A major emphasis throughout Part IV is on the controls of each pathway or process.
- **Part V, Physiological Processes**, covers those areas unique to mammalian cells and tissues beginning with a chapter on hormones that emphasizes their biochemical functions as messengers (Chapter 22) and a chapter on molecular cell biology containing discussions of four major physiological signal-transducing systems: the nervous system, the eye, muscle contraction and molecular motors, and blood coagulation (Chapter 23). A discussion of the cell cycle, programmed cell death, and cancer, three closely related topics, is presented (Chapter 24). A chapter on the complex and integrated biochemistry of digestion and absorption of basic nutritional constituents (Chapter 25) is followed by one on the functions and nutritional requirements of vitamins and minerals in metabolism (Chapter 26). The last chapter covers the general principles of human nutrition for proteins, carbohydrates, and fats (Chapter 27).

A **Glossary** with precise definitions serves as a ready reference to the most common words in the ever-expanding language of the biochemical sciences. New terms have been added to enrich this section.

A **Review of Organic Chemistry**, as an Appendix, is designed as a ready reference for the nomenclature and structures of important organic molecules encountered in biochemistry; it is not intended as a comprehensive review. The reader should become familiar with the content of the Appendix so that it can be used when necessary while reading the main text.

Clinical Correlations boxes in every chapter describe examples of human diseases where the ramifications of deviant biochemical processes are well established. There are 260 Clinical Correlations presenting the aberrant biochemistry of very common to relatively rare medical conditions, and in some cases their treatment based on the biochemical knowledge of the condition. The presentations are discussions of the altered biochemistry rather than a medical case study. In some instances, the same clinical condition is presented in different chapters, but each time it is based on the biochemistry being presented. In the case of several major diseases, as example diabetes, a single Clinical Correlation is designated as the primary discussion and other Correlations on the same disease cite the primary one for general background information. An understanding of the material in the main text does not require a reading of the Clinical Correlations. References are included in the Correlations to facilitate exploration of the topic in more detail. In a few cases, clinical conditions are discussed as part of the primary text because studies of the medical condition have lead to an understanding of a basic biochemical process.

Supplements

The *Textbook of Biochemistry with Clinical Correlations*, 7e offers a variety of innovative resources to support both students and instructors:

For students

Guided Explorations. 50 self-contained presentations, many with narration, employing extensive animated computer graphics to enhance student understanding of key topics.

Interactive Exercises. 22 molecular structures that have been rendered in Jmol, a browser-independent interface for manipulating structures in three dimensions, and paired with questions to facilitate comprehension of concepts.

Animated Figures. 25 figures illustrating various concepts, techniques, and processes; presented as brief animations that serve as helpful learning tools.

For instructors

PowerPoint Slides: Includes all of the figures in the text.

Image Gallery: All of the figures in the text in JPEG format.

Test Bank: Over 2,700 multiple choice questions, many derived from question banks compiled by the Association of Medical and Graduate Departments of Biochemistry.

In Conclusion

As in previous editions, this work is a multicontributor textbook. Each contributor holds a senior academic rank, and they are all members of the faculties of different universities. All of the contributors have been involved actively in teaching biochemistry in a graduate and/or medical program and each has an active research interest in the subject presented in the chapter that he or she has written. They prepared their chapters from the perspective of the classroom instructor, with the experience to select the topics and determine the emphasis required for students in a general biochemistry course.

Every contributor brings to the book an individual approach, leading to some differences in presentation. Every chapter, however, was edited to have a consistent writing style and to eliminate unnecessary repetitions and redundancies. A few topics are discussed in two different places in the book in order to make the individual discussions complete and self-contained. This repetition should facilitate the learning process.

The textbook is not intended as a compendium of biochemical facts or a review of the current literature, but each chapter contains sufficient detail on the subject to make it useful as a resource. Contributors were requested not to reference individual researchers and not to dwell on the historical aspects of their topic; our apologies to the many scientists who deserve recognition for their outstanding research contributions that have made this book possible.

One person must accept the responsibility for the final product in any project. The decisions concerning the selection of topics and format and reviewing the drafts and the responsibility for the final checking of the book were entirely mine. I accept full responsibility for these decisions. I welcome comments, criticisms, and suggestions from students, faculty, and professionals. It is our hope that this work will be of value to those embarking on the exciting experience of learning about the chemistry of life for the first time, as well as to those returning to a topic in which the information is expanding so rapidly.

Thomas M. Devlin
Berwyn, Pennsylvania
September 2009

Contributors

Carol N. Angstadt, Ph.D.
Professor Emerita
School of Nursing and Health Professions
Drexel University
490 S. Old Middletown Road
Media, PA 19063
Email: cnang@verizon.net

Diana S. Beattie, Ph.D.
Professor and Former Chair
Department of Biochemistry
West Virginia University School of Medicine
PO Box 9142
Morgantown, WV 26506
E-mail: dbeattie@hsc.wvu.edu

Stephen G. Chaney, Ph.D.
Professor
Departments of Biochemistry and Biophysics and of Nutrition
Genetic Medicine Building
School of Medicine CB# 7260
University of North Carolina at Chapel Hill
Chapel Hill, NC 27599-7260
Email: stephen_chaney@med.unc.edu

Marguerite W. Coomes, Ph.D.
Associate Professor
Department of Biochemistry and Molecular Biology
Howard University College of Medicine
3411 Murdock Road,
Kensington, MD 20895-1630
Email: mcoomes@howard.edu

Ann H. Cory, M.S.
Research Associate
Department of Biochemistry and Molecular Biology
Brody School of Medicine
East Carolina University
Greenville, NC 27834-4354
Email: corya@ecu.edu

Joseph G. Cory, Ph.D.
Professor and Former Chair
Department of Biochemistry and Molecular Biology
Brody School of Medicine
East Carolina University
Greenville, NC 27858-4354
Email: coryjo@ecu.edu

David W. Crabb, M.D.
John B. Hickam Professor and Chair
Departments of Medicine
Professor, Department of Biochemistry and Molecular Biology
Emerson Hall 317
Indiana University School of Medicine
545 Barnhill Drive
Indianapolis, IN 46202-5124
Email: dcrabb@iupui.edu

Thomas M. Devlin, Ph.D.
Professor Emeritus and Former Chair
Department of Biochemistry and Molecular Biology
Drexel University College of Medicine
159 Greenville Court
Berwyn, PA 19312-2071
Email: tdevlin@drexelmed.edu

John E. Donelson, Ph.D.
Professor and Former Chair
Department of Biochemistry
Carver College of Medicine
University of Iowa
Bowen Science Building
Iowa City, IA 52242-0001
Email: john-donelson@uiowa.edu

George R. Dubyak, Ph.D.
Professor
Department of Physiology and Biophysics
Case School of Medicine
Case Western Reserve University
2109 Adelbert Road
Cleveland, OH 44106
Email: george.dubyak@case.edu

Howard J. Edenberg, Ph.D.
Chancellor's Professor, Professor of Biochemistry and Molecular Biology and of Medical and Molecular Genetics
Department of Biochemistry and Molecular Biology
Indiana University School of Medicine
635 Barnhill Drive, Med. Sci. 4063
Indianapolis, IN 46202-5122
Email: edenberg@iupui.edu

Robert H. Glew, Ph.D.
Professor Emeritus
Department of Biochemistry and Molecular Biology
MSC08 4670
University of New Mexico
Albuquerque, NM 87131
Email: rglew@salud.unm.edu

Dohn G. Glitz, Ph.D.
Professor Emeritus
Department of Biological Chemistry
UCLA School of Medicine
11260 Barnett Valley Road
Sebastopol, CA 95472
E-mail: dglitz@mednet.ucla.edu

Richard W. Hanson, Ph.D.
Leonard & Jean Skeggs Professor of Biochemistry
Department of Biochemistry
Room W414
Case School of Medicine
Case Western Reserve University
Cleveland, OH 44106-4935
E-mail: rwh@cwru.edu

Robert A. Harris, Ph.D.
Distinguished Professor Emeritus and Showalter Professor of Biochemistry Emeritus
Department of Biochemistry and Molecular Biology
Indiana University School of Medicine
Richard Roudebush VA Medical Center
Research 151; Room D-3034
1481 West Tenth Street
Indianapolis, IN 46202
Email: raharris@iupui.edu

Ulrich Hopfer, M.D., Ph.D.
Professor of Physiology and Biophysics and Medicine
Department of Physiology and Biophysics
Case School of Medicine
Case Western Reserve University
109000 Euclid Ave.
Cleveland, OH 44106-4970
Email: ulrich.hopfer@case.edu

Bettie Sue Siler Masters, Ph.D. D., Sc., M.D. (Hon.)
The Robert A. Welch Distinguished Professor of Chemistry
Department of Biochemistry, MSC 7760
University of Texas Health Science Center at San Antonio
7703 Floyd Curl Drive
San Antonio, TX 78229-3900
Email: masters@uthscsa.edu

Linda J. Roman, Ph.D.
Associate Professor
Department of Biochemistry-MSC 7760
University of Texas Health Science Center at San Antonio
7703 Floyd Curl Dr.
San Antonio, TX 78229-3900
Email: roman@uthscsa.edu

Francis J. Schmidt, Ph.D.
Professor
Department of Biochemistry
117 Schweitzer Hall
University of Missouri
Columbia, MO 65211
Email: schmidtf@missouri.edu

Thomas J. Schmidt, Ph.D.
Professor
Department of Molecular Physiology and Biophysics
6-452 Bowen Science Building
Carver College of Medicine
University of Iowa
Iowa City, IA 52242-1109
Email: thomas-schmidt@uiowa.edu

Richard M. Schultz, Ph.D.
Professor
Programs in Biochemistry and Molecular Biology
Department of Microbiology and Immunology, Room 102-6652
Stritch School of Medicine
Loyola University of Chicago
2160 South First Avenue,
Maywood, IL 60153
Email: rschult@lumc.edu

Nancy B. Schwartz, Ph.D.
Professor
Departments of Pediatrics and of Biochemistry and Molecular Biology
University of Chicago, MC 5058
5841 S. Maryland Ave.
Chicago, IL 60637-1463
Email: n-schwartz@uchicago.edu

David R. Setzer, Ph.D.
Professor
Division of Biological Sciences
410 Tucker Hall
University of Missouri
Columbia, MO 65211
Email: setzerd@missouri.edu

Thomas E. Smith, Ph.D.
Professor and Former Chair
Department of Biochemistry and Molecular Biology
College of Medicine
Howard University
520 W Street, N.W.
Washington, DC 20059
Email: tsmith@howard.edu

Martin D. Snider, Ph.D.
Associate Professor
Department of Biochemistry, Room W433
Case School of Medicine
Case Western Reserve University
109000 Euclid Ave
Cleveland, Ohio 44106-4935
Email: martin.snider@case.edu

Gerald Soslau, Ph.D.
Professor and Senior Associate Dean
Department of Biochemistry and Molecular Biology
M.S. 344, Room 4104 NCB
Drexel University College of Medicine
245 North 15th Street
Philadelphia, PA 19102-1192
Email: gerald.soslau@drexelmed.edu

Francis Vella, Ph.D.
Professor (Retired)
Department of Biochemistry
University of Saskatchewan
18 Leyden Crescent
Saskatoon, Saskatchewan
SK S7J 2S4, Canada
Email: f.vella@sasktel.net

Daniel L. Weeks, Ph.D.
Professor
Department of Biochemistry
Bowen Science Building
Carver College of Medicine
University of Iowa
Iowa City, IA 52242
Email: daniel_weeks@uiowa.edu

Henry Weiner, Ph.D.
Professor
Department of Biochemistry
Purdue University
175 S. University Street
West Lafayette IN 47907-2063
Email: hweiner@purdue.edu

Stephen A. Woski, Ph.D.
Associate Professor
Department of Chemistry
Box 870336
University of Alabama
Tuscaloosa, AL 35487-0336
Email: swoski@bama.ua.edu

Acknowledgments

The seventh edition of the *Textbook of Biochemistry with Clinical Correlations* was made possible by the efforts and encouragement of many people, and I extend to everyone my most sincere thank you. I am very indebted to each of the contributors for accepting the challenge of preparing the chapters, sharing their ideas to improve the book, readily accepting suggestions to modify their contributions, and cooperating throughout the preparation. To each I extend my deepest appreciation for a job well done. The contributors and I extend a special acknowledgement to our former teachers, colleagues, and students for their support and inspiration that made this text possible.

In the preparation of this edition, chapters from the 6th edition were critically reviewed by the following: Dr. David J. Edwards, University of Pittsburgh, Dr. Kevin Gaston, University of Bristol, UK, Professor James J. A. Heffron, University College Cork, Ireland, Dr. Thomas E. Smith, Howard University, Dr. Frank Vella, University of Saskatchewan, and Dr. Edward J.Wood, University of Leeds, UK. We, the contributors and I, are deeply indebted to them; their excellent comments and suggestions were the basis for major changes in the seventh edition. With regrets I note the death of Dr. Edward J. Wood in December, 2008; we have lost a dedicated and inspiring colleague. Our gratitude is extended to Dr. Emmanuel Skordalakes, The Wistar Institute, Philadelphia, PA, for supplying a model of a partial telomerase elongation complex for the cover.

I extend my sincerest appreciation and thanks to the members of the staff of the Higher Education Division of John Wiley & Sons for their participation in the production of this edition. It has been a pleasure to work with an extremely intelligent, professional, and encouraging group of individuals. My deepest gratitude is extended to Joan Kalkut, Biochemistry Editor, who patiently and conscientiously guided me during the production of this edition, and coordinated my activities. Joan made many valuable suggestions and was always available to answer my queries. I extend my appreciation to Kaye Pace, Vice President, and Executive Publisher for Sciences, for her commitment to the project. I am in debt to Petra Recter, Associate Publisher of Chemistry and Physics, Micheline Frederick, Production Manager and Kerry Weinstein, Senior Production Editor, for their efforts; each demonstrated the highest standards of professionalism. A special acknowledgement to Hilary Newman, Manager, Photo Department and many thanks to Yelena Zolotorevskaya, Editorial Program Assistant, who handled efficiently the administrative details. I extend my deepest appreciation to Marc Wezdecki, Media Editor, and Kevin Murphy, Senior Designer, and Hope Miller, designer, who created the cover. The attractive design of the pages was created by Laura Ireardi, to whom I offer special thanks. Marketing had an important role in the design of the book and I extend my deepest appreciation to Kristine Ruff for her very valuable input.

The copyediting and compositing of this edition were the responsibility of MPS Content Services, Macmillan Publishing Solutions. I want to express my special thanks to John Sollami, Vice President, Onshore Content Services, for his support and counsel. John and I worked together on the fourth edition, and it was my good fortune to be able to interact with him again. We wish to acknowledge Pat O'Maley, Operations Director, who was responsible for overseeing this phase of the project. The person responsible for the day by day activities of production was Edward Dionne, Project Manager, who patiently and meticulously oversaw the transformation of our manuscripts to pages. Ed kept me well informed, managed the many details involved, acted promptly to my suggestions and concerns, and kept us on schedule. It has been a pleasure to work with an efficient, knowledgeable, and conscientious professional, as well as a very pleasant individual; I extend my heartfelt thanks to him. The excellent copy-editing of the manuscripts was completed by Carol A. Loomis, and the Index was prepared by Diana Witt; to both my sincerest thank you.

I would be amiss if I did not acknowledge the members of the STM Division of John Wiley who guided the preparation of the manuscripts and gave me invaluable assistance and support. My special thanks to Michael Forster, Vice President and Associate Publishing Director, Physical Sciences, Wiley-Blackwell, for his continuing support, and Darla P. Henderson, Senior Acquisitions Editor, Anita Lekhwani Senior Acquisitions Editor, and Rebekah Amos, Senior Editorial Assistant, who were a constant support and had an important and valuable role in the completion of the seventh edition.

Finally, a very special note of gratitude to my wife, Marjorie, who had the foresight many years ago to encourage me to undertake the preparation of a textbook, who supported me during the days of intensive work, and who created an environment in which I could devote the many hours required for the preparation of this textbook. To Marjorie, my deepest and sincerest thank you.

THOMAS M. DEVLIN

PART I

STRUCTURE OF MACROMOLECULES

1

Eukaryotic Cell Structure

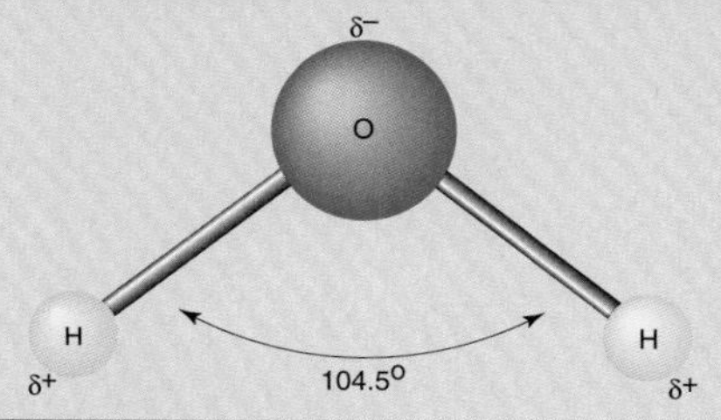

Thomas M. Devlin

Professor Emeritus, Drexel University College of Medicine

CLINICAL CORRELATIONS

Key Concepts

- All living organisms consist of individual cells, delineated by an outer lipid membrane. Cells can have a variety of intracellular structures.
- Living cells include archae, eubacteria, and eukaryotes. Multicellular organisms can have a variety of specialized cells.
- Living cells control their internal environment, extract material from the environment for energy and synthesis of required molecules, and can replicate themselves.
- Water is an essential component of living cells. Water molecules form hydrogen bonds with each other and other molecules.
- Weak acids and acidic groups on macromolecules permit cells to control their intracellular hydrogen ion concentration (pH).
- Mammalian cells are compartmentalized into a variety of intracellular structures each with specialized functions.
- Cellular functions are integrated and can be controlled by both intracellular mechanisms and extracellular influences.

1.1 • CELLS ARE THE BASIS OF LIVING ORGANISMS

The basis of all living organisms, from the simplest to the most complex, is a unit of space enclosed by a membrane. This unit of space is termed a cell. The space enclosed can have a diameter of from 10 to 10,000 microns and varying degrees of internal structural complexity depending on the organism. Living cells exist as single independent cells, such as bacteria, and as multicellular organisms, such as humans with over 100 trillion cells. Regardless of the complexity of the organism, individual biological cells have the following characteristics in common:

- Many of the same inorganic ions and organic molecules, including carbohydrates and lipids, and the macromolecules, proteins, and nucleic acids.
- An external cellular membrane, the **plasma membrane**, composed of **amphipathic** phospholipids and protein. This delineates the space occupied and separates a variable and potentially hostile environment outside from a relatively constant intracellular milieu.
- Systems that link the interior to the external milieux.
- The ability to transform external energy sources into utilizable energy to power endergonic reactions in the cell. These extracellular sources include light, organic molecules, and concentration gradients of molecules across membranes.
- The ability to convert ingested nutrients into required cellular constituents and to eliminate degraded and toxic materials. The reactions of these processes are catalyzed by protein catalysts, that is, enzymes.
- The ability to synthesize macromolecules such as nucleic acids and proteins.
- A genome of **deoxyribonucleic acid (DNA)** containing the instructions for all cell functions.
- The ability to replicate, transferring the cells hereditary genomic information to progeny.

Approximately 3.7 billion years ago under conditions not entirely clear and in a time span difficult to comprehend, the elements hydrogen, oxygen, nitrogen, sulfur, and phosphate formed simple chemical compounds. These combined, dispersed, and recombined to form a variety of larger molecules until structures were achieved that were capable of replication. There is no conclusive record, however, of the environmental conditions that fostered these reactions. With continued formation of ever more complex molecules, these self-replicating molecules became encapsulated by a membrane to form cells. With the passage of time these simple life forms evolved into a cell that would become the "last common ancestor" of all the millions of species and subspecies on Earth, including humans. The "last common ancestor" of all humans appeared about 450 million years ago. The evolution of organisms continues. The vast diversity of life today, from the simplest bacteria to complex multicellular organisms such as plants and animals, is the product of these evolutionary changes.

Classification of Living Cells

All organisms are grouped into one of three major domains: archaea, eubacteria, and eukaryotes. The evidence is that all three were derived from an unknown common precursor cell. Archaea, which may be the most primitive of the domains, and eubacteria, that is, common bacteria, are classified as **prokaryotes** because they have many common structural features, including the lack of a defined nucleus or internal membrane structures. They are usually unicellular (Figure 1.1*a*) but in some cases form colonies or filaments. Prokaryotes have a variety of shapes and sizes and can live under a variety of conditions, some very extreme. The plasma membrane is often invaginated. DNA of prokaryotes is a single circular strand and often segregated into a discrete mass in the cell, the **nucleoid** region, which is not surrounded by a membrane or envelope. Even without defined membrane compartments, the intracellular milieu of prokaryotes is organized into functional compartments.

Eukaryotes include single-cell organisms such as yeasts, fungi, and multicellular plants and animals. Their cell volume is 1000 to 10,000 times larger than most prokaryotic cells.

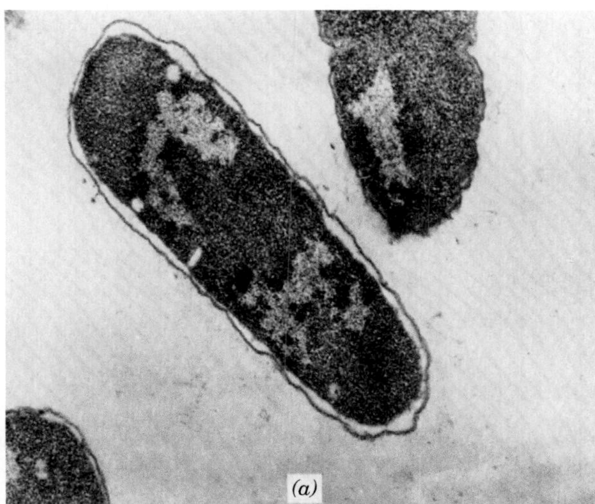

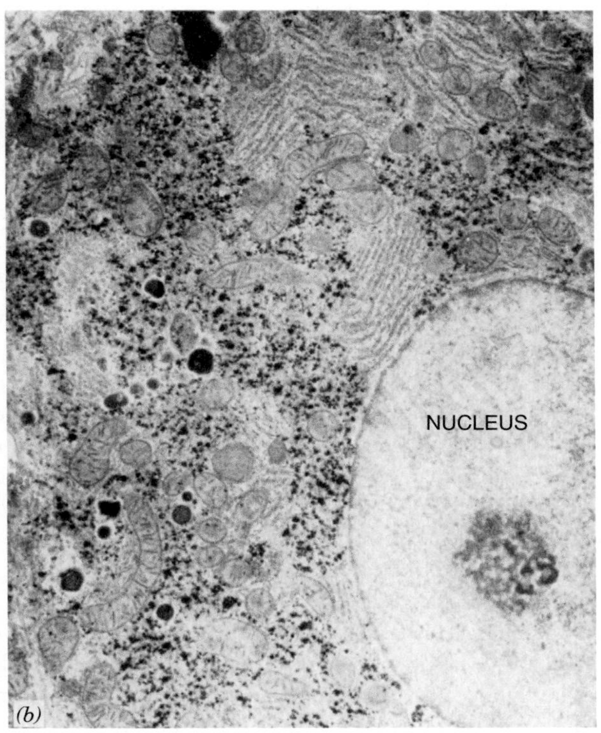

Figure 1.1 Cellular organization of prokaryotic and eukaryotic cells. (*a*) Electron micrograph of *Escherichia coli,* a representative prokaryote; approximate magnification ×30,000. There is some apparent intracellular organization but no membrane enclosed organelles. Chromatin is condensed in a nucleoid but not surrounded by a membrane. Prokaryotic cells are much smaller than eukaryotic cells. (*b*) Electron micrograph of a thin section of a liver cell (rat hepatocyte), a representative eukaryotic cell; approximate magnification ×7500. Note the distinct nuclear membrane, different membrane-bound organelles or vesicles, and extensive membrane systems.
Photograph (*a*) generously supplied by Dr. M. E. Bayer, Fox Chase Cancer Institute, Philadelphia, PA. Photograph (*b*) reprinted with permission of Dr. K. R. Porter, from Porter, K. R., and Bonneville, M. A. In *Fine Structure of Cells and Tissues.* Philadelphia: Lea & Febiger, 1972.

Eukaryotes have a defined **nucleus** with a well-defined membrane that contains the bulk of the cell's DNA. They also have extensive membrane systems and intracellular **organelles** surrounded by membranes (Figure 1.1*b*). These intracellular membrane systems establish distinct cellular compartments (p. 14), permitting a unique degree of subcellular organization. By compartmentalization, different chemical reactions that require different environments can occur simultaneously.

The basic chemical components and fundamental chemical reactions of all living cells, whether prokaryotes or eukaryotes, are very similar. There are many noteworthy differences, however, in chemical composition and biochemical activities between eukaryotes and prokaryotes. As an example, prokaryotes do not contain histones, a highly conserved class of proteins in all eukaryotes that complex with DNA (p. 54). There are also differences in enzyme content and in the ribonucleic acid–protein complexes, called ribosomes, involved in biosynthesis of proteins. The universality of many biochemical phenomena permits many extrapolations from prokaryotes to eukaryotes, including humans.

Emphasis throughout this book is on the biochemistry of eukaryotes, particularly mammals, but much of our knowledge of the biochemistry of living cells has come from studies of prokaryotic and non-mammalian eukaryotic cells.

1.2 • THE ENVIRONMENT OF CELLS: WATER AND SOLUTES

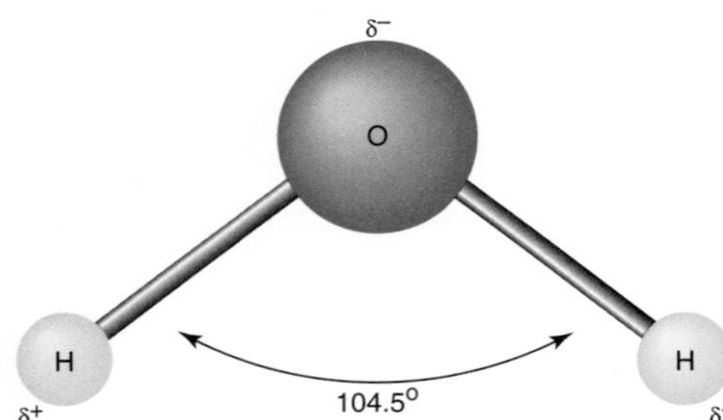

Figure 1.2 Structure of a water molecule. The H—O—H bond angle is 104.5°. Both hydrogen atoms carry a partial positive charge and the oxygen atom a partial negative charge, creating a dipole.

Life as we know it exists because of **water**, a common component of all biological cells and their extracellular environment.

Hydrogen Bonds Form between Water Molecules

The dynamic interaction of individual water molecules leads to a dynamic "water structure" in aqueous systems. A water molecule is formed when two hydrogen atoms share their electrons with an unshared pair of electrons of an oxygen atom (Figure 1.2). Water is a polar molecule because the oxygen nucleus has a stronger attraction for shared electrons than hydrogen, and positively charged hydrogen nuclei are left with an unequal share of electrons. This creates a partial positive charge on both hydrogens and a partial negative charge on oxygen. The bond angle between hydrogens and oxygen is 104.5°, making the molecule electrically asymmetric, and thus producing an electric dipole.

Water molecules interact with each other because positively charged hydrogen atoms on one molecule are attracted to negatively charged oxygen atoms on other water molecules. A weak bond between two water molecules is formed (Figure 1.3*a*). This bond, indicated by a dashed line, is a **hydrogen bond**. Recent studies suggest that the bond between two water molecules is partially covalent. Hydrogen bonds are relatively weak compared to covalent bonds but their large number is the reason for the stability of liquid water. A detailed discussion of noncovalent interactions, including electrostatic, van der Waals, and hydrophobic, between molecules is presented on page 116.

Five molecules of water form a **tetrahedral structure** by hydrogen bonding (Figure 1.3*b*), with each oxygen sharing its electrons with four hydrogen atoms and each hydrogen with another oxygen. This tetrahedral lattice structure is responsible for the crystalline structure of ice. In the transition from ice to liquid water, only some hydrogen bonds are broken. Liquid water has a rapidly changing structure as hydrogen bonds break and new bonds form; the half-life of hydrogen bonds in water is less than 1×10^{-11} s. Thus the structure of liquid water is constantly fluctuating with a variety of structures containing many water molecules being constantly formed and changed. Water at the molecular level is heterogeneous with dynamic structures containing clusters of water molecules of varying size involving hundreds of molecules. Even at 100°C, liquid water contains a significant number of hydrogen bonds, which accounts for its high heat of vaporization. In the transformation from liquid to vapor state, hydrogen bonds are disrupted. Many models for the structure of liquid water have been proposed but none adequately explains all its properties.

The structure of pure water is altered when its atoms hydrogen-bond to other ions and chemical structures. Interaction of water molecules with other molecules leads to substantial changes in the orientation of water. As an example, water molecules are more ordered near the surface of membranes because of the amphiphilic nature of the membrane phospholipids (p. 465). Water present on and within molecules of proteins and nucleic acids stabilizes these macromolecules. Water structure may be different in the extracellular aqueous environment from that of the intracellular because of differences in ionic composition.

Substances required for the existence of cells are dissolved or suspended in an aqueous medium and their activities are influenced by the organization of the water molecules. Microenvironments with different water structures form in and on the surface of macromolecules and on the surface of lipid membranes because of the interaction of water with groups on these molecules. The presence of these microenvironments can lead to variations in activity of ions and molecules within different sites in a cell.

Hydrogen bonding also occurs between molecules other than water, wherever electronegative oxygen or nitrogen atoms come in close proximity with hydrogen covalently bonded

(*a*)

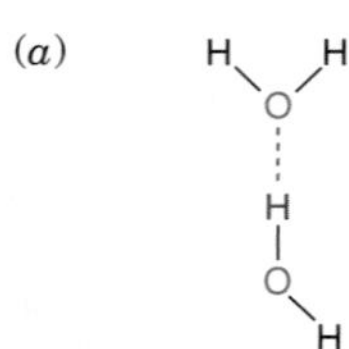

(*b*)

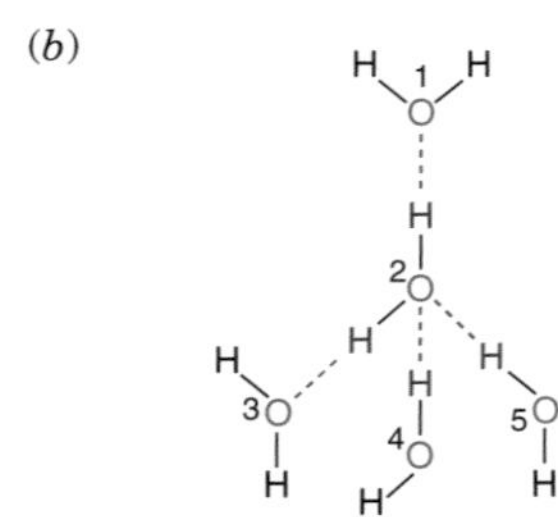

Figure 1.3 Hydrogen bonding in water. (a) Hydrogen bonding, indicated by dashed lines, between two water molecules. (b) Tetrahedral hydrogen bonding of five water molecules. Water molecules 1, 2, and 3 are in the plane of the page, 4 is below, and 5 is above.

to another electronegative atom. Representative hydrogen bonds are presented in Figure 1.4. Intramolecular hydrogen bonding occurs extensively in large macromolecules such as proteins and nucleic acids; it is partially responsible for their structural stability.

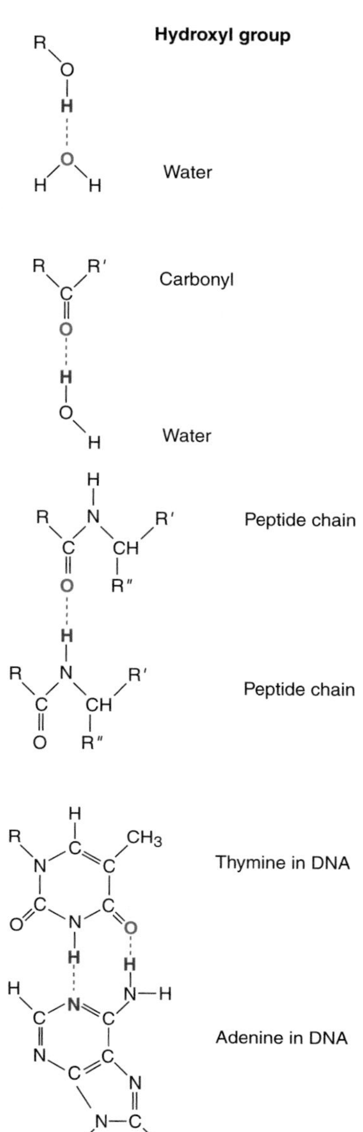

Figure 1.4 Representative hydrogen bonds of importance in biological systems.

Water Has Unique Solvent Properties

The polar nature of water and its ability to form hydrogen bonds are the basis for the ability of water to dissolve a variety of inorganic and organic molecules. Polar molecules such as salts are dispersed readily in water. The crystal lattice of **salts** is held together by attraction of positively and negatively charged atoms or groups. They dissolve in water because electrostatic forces in the crystal can be overcome by the attraction of the charged components to the dipole of water. The attraction of the charged Na^+ and Cl^- atoms in NaCl is overcome by interaction of Na^+ with the negative charge on oxygen atoms of water, and Cl^- with the positive charge on hydrogen atoms. In solution, the individual ions are surrounded by a shell of water. The number of weak charge-charge interactions between water and Na^+ and Cl^- ions is sufficient to maintain the physical separation of the charged ions.

Nonionic organic molecules containing weakly polar groups are also soluble in water because of attraction of the polar groups to molecules of water. Sugars and alcohols are readily soluble for this reason. Compounds that contain both polar and nonpolar groups, that is, **amphipathic** molecules, disperse in water if attraction of the polar group for water can overcome hydrophobic interactions of nonpolar portions of the molecules. Very hydrophobic molecules, such as lipids containing long hydrocarbon chains do not readily disperse as single molecules in water. They interact with one another to exclude the polar water molecules (p. 463).

Electrolytes: Dissociation of Molecules in Water

Molecules that dissociate in water form **cations** (positively charged ions) and **anions** (negatively charged ions). These are classified as **electrolytes** because the ions facilitate conductance of an electrical current. Sugars or alcohols are **nonelectrolytes** because they dissolve readily in water but do not carry a charge or dissociate into charged species.

Salts of alkali metals (e.g., Li, Na, and K) and acids such as hydrochloric and sulfuric at low concentrations dissociate completely when dissolved in water but not necessarily at high concentrations. In biological systems, such compounds as well as salts of organic acids are considered to be dissociated totally because of their low concentrations. If a solution contains a number of different salts (e.g., NaCl and K_2SO_4), these molecules do not exist as such in solution, only the dissociated ions (e.g., Na^+, K^+, and SO_4^{2-}) are present. Salts that dissociate completely are referred to as **strong electrolytes**. In water, the dissociated anions of organic salts react to some degree with free protons (H^+) from dissociation of water to form the undissociated acid (Figure 1.5).

In contrast to salts, many acids when dissolved in water do not dissociate totally but establish equilibrium between undissociated and dissociated components. Thus lactic acid, an important metabolic intermediate, dissociates partially into a lactate anion and a proton as follows:

$$CH_3{-}CHOH{-}COOH \rightleftharpoons CH_3{-}CHOH{-}COO^- + H^+$$

A dynamic equilibrium is established in which the products of the reaction reform the undissociated reactant while other molecules dissociate. The degree of dissociation of such an electrolyte depends on the affinity of the anion for an H^+. There will be more dissociation if the weak dipole forces of water that interact with the anion and cation are stronger than the electrostatic forces between the anion and H^+. On a molar basis such compounds, termed **weak electrolytes**, have a lower capacity to carry an electrical charge in comparison to those that dissociate totally.

In partial dissociation of a weak electrolyte, represented by HA, the concentration of various species can be determined from the equilibrium equation

$$K'_{eq} = \frac{[H^+][A^-]}{[HA]} \quad (1.1)$$

(1) $CH_3{-}CHOH{-}CHOONa \longrightarrow$
Na lactate

$Na^+ + CH_3{-}CHOH{-}COO^-$
Lactate ion

(2) $CH_3{-}CHOH{-}COO^- + H^+ \rightleftharpoons$
Lactate ion

$CH_3{-}CHOH{-}COOH$
Lactic acid

Figure 1.5 Reactions that occur when sodium lactate is dissolved in water.

where K'_{eq} is a physical constant, A^- represents the dissociated anion, and square brackets indicate the concentration of each component in units such as moles per liter (mol/L or M) or millimoles per liter (millimol/L or mM). The **activity** of each species rather than concentration should be employed in the equilibrium equation, but since most compounds of interest in biological systems are present in low concentration, the value for the activity approaches that of concentration. The equilibrium constant, however, is notated as $\boldsymbol{K'_{eq}}$ to indicate that it is an apparent constant based on concentrations. Since the dissociation of an acid increases with increasing temperatures, the K'_{eq} will also increase.

From the dissociation equation, it is apparent that K'_{eq} will be a small number if the degree of dissociation of a substance is small (large denominator in Eq. 1.1) but large if the degree of dissociation is large (small denominator). A K'_{eq} cannot be determined for strong electrolytes because at equilibrium there is no remaining undissociated solute.

Water Is a Weak Electrolyte

Water dissociates as follows:

$$HOH \rightleftharpoons H^+ + OH^-$$

Protons that dissociate interact with oxygens of other water molecules to form clusters of water molecules, that is, $H^+(H_2O)_n$, where n has been determined to be from 6 to 27. This hydration of H^+ is often presented as $\mathbf{H_3O^+}$, the **hydronium ion**. It is a generally accepted practice, and one that will be employed in this book, to present the proton as H^+ rather than H_3O^+, while recognizing that $H^+(H_2O)_n$ is the actual chemical species. At 25°C the value of K'_{eq} for dissociation of water is about 1.8×10^{-16}.

$$K'_{eq} = 1.8 \times 10^{-16} = \frac{[H^+][OH^-]}{[H_2O]} \tag{1.2}$$

With such a small K'_{eq} an extremely small number of water molecules actually dissociate relative to the number of undissociated molecules. The concentration of water, which is 55.5 M, is essentially unchanged by the very small dissociation and is a constant. Therefore, Eq. 1.2 can be rewritten as follows:

$$K'_{eq} \times [H_2O] = [H^+][OH] \tag{1.3}$$

$K'_{eq} \times [55.5]$ is a constant and is termed the **ion product of water**. Its value at 25°C is 1×10^{-14}. In pure water the concentration of H^+ equals OH^-, and by substituting $[H^+]$ for $[OH^-]$ in Eq. 1.3, $[H^+] = 1 \times 10^{-7}$ M. Similarly, $[OH^-] = 1 \times 10^{-7}$ M. The equilibrium reaction of H_2O, H^+, and OH^- always exists in aqueous solutions regardless of the presence of dissolved substances. If the $[H^+]$ is increased, as occurs on addition of an acid, a decrease of $[OH^-]$ must occur in order to satisfy the equilibrium relationship of water. Similarly, if the $[OH^-]$ is increased, $[H^+]$ will decrease. Using the equation for ion product, $[H^+]$ or $[OH^-]$ can be calculated if the concentration of one ion is known. The importance of hydrogen ions in biological systems will become apparent in discussions of enzyme activity (p. 387) and metabolism.

TABLE 1.1 • Relationships Between $[H^+]$, pH, $[OH^-]$, and pOH

$[H^+]$ (M)	pH	$[OH^-]$ (M)	pOH
1.0	0	1×10^{-14}	14
0.1 (1×10^{-1})	1	1×10^{-13}	13
1×10^{-2}	2	1×10^{-12}	12
1×10^{-3}	3	1×10^{-11}	11
1×10^{-4}	4	1×10^{-10}	10
1×10^{-5}	5	1×10^{-9}	9
1×10^{-6}	6	1×10^{-8}	8
1×10^{-7}	7	1×10^{-7}	7
1×10^{-8}	8	1×10^{-6}	6
1×10^{-9}	9	1×10^{-5}	5
1×10^{-10}	10	1×10^{-4}	4
1×10^{-11}	11	1×10^{-3}	3
1×10^{-12}	12	1×10^{-2}	2
1×10^{-13}	13	0.1 (1×10^{-1})	1
1×10^{-14}	14	1.0	0

1.3 • pH, WEAK ACIDS, AND THEIR CONJUGATE BASES

For convenience $[H^+]$ is usually expressed in terms of pH, defined as

$$\mathbf{pH = \log \frac{1}{[H^+]}} \tag{1.4}$$

In pure water $[H^+]$ and $[OH^-]$ are both 1×10^{-7} M, and pH = 7.0. $[OH^-]$ can also be expressed as pOH and has a value of 7. For the equation describing dissociation of water, $1 \times 10^{-14} = [H^+][OH^-]$; taking negative logarithms of both sides, the equation becomes $14 = pH + pOH$. Table 1.1 presents the relationship between pH and $[H^+]$.

CLINICAL CORRELATION 1.1

Abnormal Medical Conditions as Reflected in the pH of Blood

In mammals, the different pHs of the various intra- and extracellular aqueous environments are in a dynamic steady state, with changes in one leading to changes in other compartments. Blood pH reflects changes in pH in tissues, and values above or below the normal range of pH of 7.35 to 7.45 indicates a potential pathological condition. Blood pH values below 7.0 (H^+ = 0.0000001 M) or above 7.8 (H^+ = 0.000000016 M) are life threatening, and medical intervention is necessary. Many medical conditions can cause significant changes of blood pH from normal. If the blood pH falls below 7.35, the condition is referred to as an **acidosis**, and above 7.45, as an **alkalosis**. Conditions of acidosis or alkalosis are further defined on the basis of the source of the increase of acid or base in the body, that is, whether it is due to a metabolic or respiratory change.

A **metabolic acidosis** can be due to excess production of organic acids (e.g., lactic acid or ketone bodies [p. 697]) or loss of HCO_3^- from the body. Excess production of acids can occur in diabetes, hypoxemia (e.g., excess lactic acid production as occurs in long-distance runners), and metabolism of xenobiotics that form acids. Loss of HCO_3^-, which changes the balance of base and acid, happens in severe diarrhea, uremia, and chronic renal diseases. A **respiratory acidosis** occurs when there is retention of CO_2, the anhydride of H_2CO_3, and is caused by conditions restricting the exhaling of CO_2 from the lungs such as when there is an accumulation of fluids in the lungs from conditions such as emphysema or asthma, restriction in breathing as in trauma, poliomyelitis, and severe obesity.

The major causes of **metabolic alkalosis** are retention of HCO_3^- and ingestion of bases. A **respiratory alkalosis** occurs by hyperventilation due to hysteria or tenseness, overdose of some drugs (e.g., salicylate), and fever.

Measurement of blood pH to monitor the acid–base status is a routine in many diseases because an uncontrolled decrease or increase in blood pH can lead to rapid and severe consequences.

Preston, R. A. *Acid-Base, Fluids, and Electrolytes Made Ridiculously Simple.* Miami, FL: Medmaster, 2002.

pH values of different biological fluids are presented in Table 1.2. In blood plasma, $[H^+]$ is 0.00000004 M or a pH of 7.4. Other cations such as Na^+ and K^+ are between 0.001 and 0.10 M, well over 10,000 times higher than $[H^+]$. Clin. Corr. 1.1 describes the significance of changes in the pH of blood.

TABLE 1.2 • pH of Some Biological Fluids

Fluid	*pH*
Blood plasma	7.4
Interstitial fluid	7.4
Intracellular fluid	
Cytosol (liver)	6.9
Lysosomal matrix	Below 5.0
Gastric juice	1.5–3.0
Pancreatic juice	7.8–8.0
Human milk	7.4
Saliva	6.4–7.0
Urine	5.0–8.0

The definitions of an acid as a **proton donor** and of a base as a **proton acceptor**, proposed by Lowry and Brønsted, are convenient in considering biological systems. HCl and H_2SO_4 are defined as strong acids because they dissociate totally, releasing protons. OH^- is a strong base because it readily associates with available protons to form H_2O. Addition of either an acid or base to water will lead to the establishment of a new equilibrium of $OH^- + H^+ \rightleftharpoons H_2O$. When a strong acid and OH^- are combined, H^+ from the acid and OH^- interact essentially totally and neutralize each other. Anions produced when strong acids dissociate, such as Cl^- from HCl, are not bases because they do not reassociate with protons in dilute solution.

Most organic acids found in biological systems dissociate partially and are classified as **weak acids**. They establish equilibrium between HA (proton donor), an anion (A^-) of the dissociated acid, and an H^+ as follows:

$$HA \rightleftharpoons A^- + H^+$$

The anion formed in this dissociation is a base because it can accept a proton to reform the acid. A weak acid and its base (anion) formed on dissociation are referred to as a **conjugate pair**. Some biologically important conjugate pairs are presented in Table 1.3. The ammonium ion (NH_4^+) is a weak acid because it dissociates to yield H^+ and uncharged ammonia (NH_3), a conjugate base. Phosphoric acid (H_3PO_4) is an acid and PO_4^{3-} is a base, but $H_2PO_4^-$ and HPO_4^{2-} are either a base or acid depending on whether the phosphate group is accepting or donating a proton.

The tendency of a conjugate acid to release H^+ can be assessed from the K'_{eq} (Eq. 1.1). The smaller the value of K'_{eq}, the less the tendency to give up a proton and the weaker the acid. The larger the value of K'_{eq} the greater the tendency to dissociate and the stronger the acid. Water is a very weak acid with a K'_{eq} of 1.8×10^{-16} at 25°C.

A convenient method of stating the K'_{eq} is in the form of pK', defined as

$$\mathbf{p}K' = \log \frac{1}{K'_{eq}} \tag{1.5}$$

TABLE 1.3 • Some Conjugate Acid–Base Pairs of Importance in Biological Systems

Proton Donor (Acid)		*Proton Acceptor (Base)*
$CH_3—CHOH—COOH$ (lactic acid)	$\rightleftharpoons$	$H^+ + CH_3—CHOH—COO^-$ (lactate)
$CH_3—CO—COOH$ (pyruvic acid)	$\rightleftharpoons$	$H^+ + CH_3—CO—COO^-$ (pyruvate)
$HOOC—CH_2—CH_2—COOH$ (succinic acid)	$\rightleftharpoons$	$2H^+ + {}^-OOC—CH_2—CH_2—COO^-$ (Succinate)
${}^+H_3NCH_2—COOH$ (glycine)	$\rightleftharpoons$	$H^+ + {}^+H_3N—CH_2—COO^-$ (glycinate)
H_3PO_4	$\rightleftharpoons$	$H^+ + H_2PO_4^-$
$H_2PO_4^-$	$\rightleftharpoons$	$H^+ + HPO_4^{2-}$
HPO_4^{2-}	$\rightleftharpoons$	$H^+ + PO_4^{3-}$
Glucose 6-PO_3H^-	$\rightleftharpoons$	H^+ + glucose 6-PO_3^{2-}
H_2CO_3	$\rightleftharpoons$	$H^+ + HCO_3^-$
NH_4^+	$\rightleftharpoons$	$H^+ + NH_3$
H_2O	$\rightleftharpoons$	$H^+ + OH^-$

Note the similarity of this definition with that of pH; as with pH and $[H^+]$, the relationship between pK' and K'_{eq} is an inverse one, and the smaller K'_{eq}, the larger pK'. Representative values of K'_{eq} and of pK' for conjugate acids of importance in biological systems are presented in Table 1.4. Details of the **carbonic acid**/bicarbonate system are presented in A Closer Look 1.1.

TABLE 1.4 • Apparent Dissociation Constant and pK' of Some Compounds of Importance in Biochemistry

Compound	*Structures*	$K'_{eq}(M)$	*pK'*
Acetic acid	$CH_3—COOH$	1.74×10^{-5}	4.76
Alanine	$CH_3—CH(NH_3^+)—COOH$	4.57×10^{-3} 2.04×10^{-10}	2.34 (COOH) 9.69 (NH_3^+)
Citric acid	$HOOC—CH_2—COH(COOH)—CH_2—COOH$	8.12×10^{-4} 1.77×10^{-5} 3.89×10^{-6}	3.09 3.74 5.41
Glutamic acid	$HOOC—CH_2—CH_2—CH(NH_3^+)—COOH$	6.45×10^{-3} 5.62×10^{-5} 2.14×10^{-10}	2.19 (COOH) 4.25 (COOH) 9.67 (NH_3^+)
Glycine	$CH_2(NH_3^+)—COOH$	4.57×10^{-3} 2.51×10^{-10}	2.34 (COOH) 9.60 (NH_3^+)
Lactic acid	$CH_3—CHOH—COOH$	1.38×10^{-5}	3.86
Pyruvic acid	$CH_3—CO—COOH$	3.16×10^{-3}	2.50
Succinic acid	$HOOC—CH_2—CH_2—COOH$	6.46×10^{-5} 3.31×10^{-6}	4.19 5.48
Glucose 6-PO_3H^-	$C_{12}H_{11}O_5\ PO_3H^-$ H_3PO_4 $H_2PO_4^-$	7.76×10^{-7} 1×10^{-2} 2.0×10^{-7}	6.11 2.0 6.7
HPO_4^{2-}		3.4×10^{-13}	12.5
H_2CO_3		1.70×10^{-4}	3.77
NH_4^+		5.62×10^{-10}	9.25
H_2O		1.8×10^{-16}	15.74

A CLOSER LOOK 1.1

Carbonic Acid (H_2CO_3) Is a Weak Acid Important in the Homeostasis of Animals

Carbonic acid (H_2CO_3), a weak acid, is very important in control of pH in mammals. CO_2 is constantly produced in catabolic reactions in tissues and is removed by the lungs in expired air. Measurements of blood CO_2 are important in determining the acid /base status of a patient. Medical conditions restricting the release of CO_2 from the body lead to an accumulation of H_2CO_3 and a condition of respiratory acidosis (see Clin. Corr. 1.1, p. 7).

Carbon dioxide when dissolved in aqueous systems is involved in the following equilibrium reactions:

$$CO_2 + H_2O \overset{K_2'}{\rightleftharpoons} H_2CO_3 \overset{K_1'}{\rightleftharpoons} H^+ + HCO_3^-$$

Carbonic acid has a pK_1 of 3.77, which is comparable to organic acids such as lactic acid. The equilibrium equation for this reaction is

$$K_1' = \frac{[H^+][HCO_3^-]}{[H_2CO_3]} \quad (1)$$

H_2CO_3 is, however, in equilibrium with dissolved CO_2 and the equilibrium equation for this reaction is

$$K_2' = \frac{[H_2CO_3]}{[CO_2][H_2O]} \quad (2)$$

By solving Eq. 2 for H_2CO_3 and substituting for the H_2CO_3 in Eq. 1, the two equilibrium reactions can be combined into one equation.

$$K_1' = \frac{[H_1^+][HCO_3^-]}{K_2'[CO_2][H_2O]} \quad (3)$$

Rearranging to combine constants, including the concentration of H_2O (55.5 M), simplifies the equation and yields a new combined constant 3′ as follows:

$$K_1'K_2'\,[H_2O] = K_3' = \frac{[H^+][HCO_3^-]}{[CO_2]} \quad (4)$$

It is common practice in medicine to refer to dissolved CO_2 as the conjugate acid; it is the acid anhydride of H_2CO_3. The term K_3' has a value of 7.95×10^{-7} and $PK_3' = 6.1$. If the aqueous system is in contact with an air phase, dissolved CO_2 will also be in equilibrium with CO_2 in the air phase. Decrease or increase of one component—CO_2 (air), CO_2 (dissolved), H_2CO_3, H^+, or HCO_3^-—causes a change in every component. The CO_2/HCO_3^- system is extremely important for maintaining pH homeostasis in animals. CO_2 is constantly produced in catabolic reactions in tissues and is removed by the lungs in expired air. Respiratory acidosis can be either an acute (holding one's breath) or a chronic condition (pneumonia).

Henderson–Hasselbalch Equation Defines the Relationship between pH and Concentrations of Conjugate Acid and Base

A change in concentration of any component in an equilibrium reaction necessitates a concomitant change in every component. For example, an increase in $[H^+]$ decreases the concentration of conjugate base (e.g., lactate ion) with an equivalent increase in the **conjugate acid** (e.g., lactic acid). This relationship is conveniently expressed by rearranging the equilibrium equation and solving for H^+, as shown for the following dissociation:

$$\text{Conjugate acid} \rightleftharpoons \text{conjugate base} + H^+$$

$$K_{eq}' = \frac{[H^+][\text{conjugate base}]}{[\text{conjugate acid}]} \quad (1.6)$$

Dividing Eq. 1.6 by both $[H^+]$ and K_{eq}' leads to

$$\frac{1}{[H^+]} = \frac{1}{\mathrm{l}K_{eq}'} \cdot \frac{[\text{conjugate base}]}{[\text{conjugate acid}]} \quad (1.7)$$

Taking the logarithm of both sides gives

$$\log\frac{1}{[H^+]} = \log\frac{1}{1K_{eq}'} + \log\frac{[\text{conjugate base}]}{[\text{conjugate acid}]} \quad (1.8)$$

Since $pH = \log 1/[H^+]$ and $pK' = \log 1/K_{eq}'$, Eq. 1.8 becomes

$$\mathbf{pH = pK' + \log\frac{[conjugate\ base]}{[conjugate\ acid]}} \quad (1.9)$$

Equation 1.9, referred to as the **Henderson–Hasselbalch equation**, is a convenient way of viewing the relationship between pH of a solution and relative amounts of individual conjugate base and acid pairs present. Analysis of Eq. 1.9 demonstrates that when the ratio of [base] to [acid] is 1:1, the pH equals the pK' of the acid because $\log 1 = 0$. If pH is

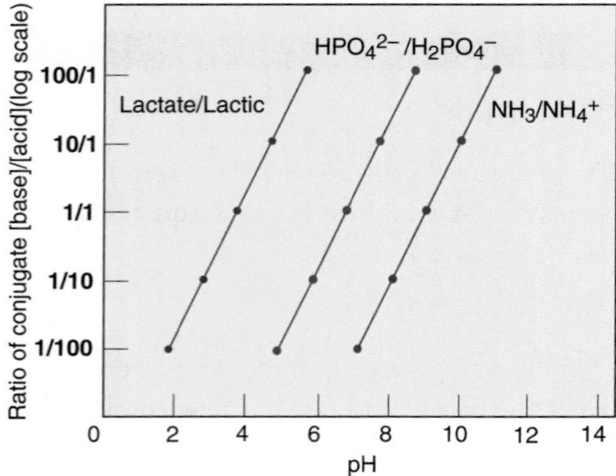

Figure 1.6 Ratio of conjugate [base]/[acid] as a function of pH. When the ratio of [base] to [acid] is 1, pH equals pK' of weak acid.

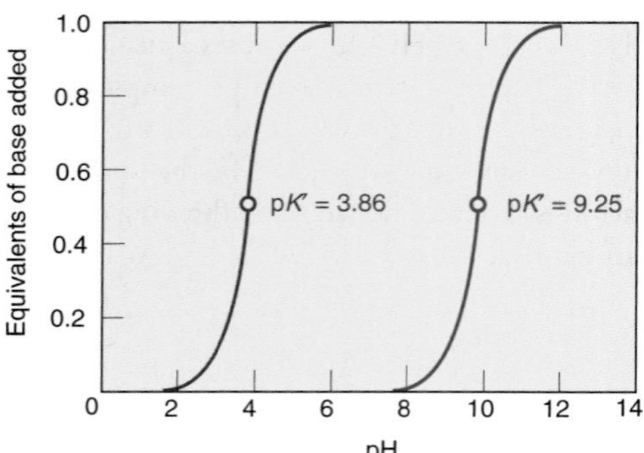

Figure 1.7 Acid–base titration curves for lactic acid (pK' 3.86) and NH_4^+ (pK' 9.25). At pH equal to respective pK' values, there will be an equal amount of acid and base for each conjugate pair.

one unit less than pK', the [base]/[acid] ratio is 1:10, and if pH is one unit above pK', the [base]/[acid] ratio is 10:1. Figure 1.6 is a plot of ratios of conjugate base to conjugate acid versus pH of several weak acids; note that ratios are presented on a logarithmic scale.

Buffering Is Important to Control pH

When NaOH is added to a solution of a weak acid, NaOH dissociates totally and the OH^- formed is neutralized by H^+ from the partially dissociated acid to form H_2O. The decrease in $[H^+]$ causes further dissociation of the weak acid to comply with requirements of its equilibrium reaction. The amount of weak acid dissociated will be so nearly equal to the amount of OH^- added that it is considered equal. Thus, the decrease in the conjugate acid concentration is equal to the amount of conjugate base that is formed and the ratio of [conjugate base]/[conjugate acid] of the weak acid changes. The events are represented in titration curves of two weak acids presented in Figure 1.7. When 0.5 equivalent (equiv) of OH^- is added, 50% of the weak acid is dissociated and the [acid]/[base] ratio is 1.0; pH at this point is equal to pK' of the acid. Shapes of individual titration curves are similar but displaced due to differences in pK' values. There is a steep rise in pH when only 0.1 equiv of OH^- is added, but between 0.1 and 0.9 equiv of added OH^-, the pH change is only ~2. Thus, a large amount of OH^- is added with a relatively small change in pH. This is called **buffering** and is defined as the ability of a solution to resist a change in pH when an acid or base is added. If weak acid is not present, a small amount of OH^- would lead to a change to a high pH because there would be no significant source of H^+ to neutralize the OH^-.

The best buffering range for a conjugate pair is in the pH range near the pK' of the weak acid. Starting from a pH one unit below and going to a pH one unit above pK', about 82% of a weak acid in solution will dissociate, and therefore an amount of base equivalent to about 82% of original acid can be neutralized with a change in pH of 2. Maximum buffering ranges for conjugate pairs are considered to be between 1 pH unit below and 1 pH unit above pK'. The conjugate pair lactate ion/lactic acid with a pK' = 3.86 is an effective buffer between pH 3 and pH 5, with no significant buffering capacity at pH = 7.0. The $HPO_4^{2-}/H_2PO_4^-$ pair with pK' = 6.7 is an effective buffer at pH = 7.0. Thus at the pH of the cell's cytosol (~ 7.0), the lactate–lactic acid pair is not an effective buffer but $HPO_4^{2-}/H_2PO_4^-$ is.

Buffering capacity also depends on concentrations of conjugate acid and base. The higher the concentration of conjugate base, the more added H^+ with which it can react. The more conjugate acid the more added OH^- can be neutralized by dissociation of the acid. A case in point is blood plasma at pH 7.4. The pK' for $HPO_4^{2-}/H_2PO_4^-$ of 6.7 suggests that this conjugate pair would be a significant buffer in plasma. The concentration of the phosphate pair, however, is low compared to that of the HCO_3^-/CO_2 system with a

A CLOSER LOOK 1.2

pH and Buffering Problems

1. Calculate the ratio of HPO_4^{2-} to $H_2PO_4^-$ (pK' = 6.7) at pH 5.7, 6.7, and 8.7.

Solution:
$$\mathbf{pH = p\mathit{K}' + \log \frac{[HPO_4^{2-}]}{[H_2PO_4^-]}}$$

$$5.7 = 6.7 + \log \text{ of ratio}$$

Rearranging, $5.7 - 6.7 = -1 = \log$ of ratio

The antilog of $-1 = 0.1$, or 1/10. Thus, $HPO_4^{2-}/H_2PO_4^- =$ 1/10 at pH 5.7. Using the same procedure, the ratio of HPO_4^{2-} to $H_2PO_4^-$ at pH 6.7 = 1:1 and at pH 8.7 = 100:1.

2. If the pH of blood is 7.1 and the HCO_3^- concentration is 8 mM, what is the concentration of CO_2 in blood (pK' for HCO_3^-/CO_2 = 6.1)?

Solution:
$$\mathbf{pH = p\mathit{K}' + \log \frac{[HCO_3^-]}{[CO_2]}}$$

$$7.1 = 6.1 + \log \frac{8\text{ mM}}{[CO_2]}$$

Rearranging, $7.1 - 6.1 = 1 = \log \frac{8\text{ mM}}{[CO_2]}$

The antilog of 1 = 10. Thus, 10 = 8 mM/$[CO_2]$. Rearranging,

$$[CO_2] = \frac{8\text{ mM}}{10} = 0.8\text{ mM}$$

CLINICAL CORRELATION 1.2

Importance of Blood HCO_3^- in Metabolic Acidosis

Buffers in blood maintain the pH at about 7.40. Cellular production of acids leads to an acidification of blood where the H^+ is buffered by several different bases including HCO_3^-, hemoglobin, and HPO_4^{2-}. The pH of blood will decrease as the bases are consumed in the buffering process. If all the bases are used, there will be no further buffering capacity and the pH will decrease precipitously. The concentrations of HCO_3^- and CO_2 and pH of blood are measured in clinical practice to determine the acid/base status of patients. In a metabolic acidosis where the blood pH is less than 7.35 (Clin. Corr. 1.1, p. 7), it is important to monitor the pH as well as to evaluate the amount of HCO_3^- remaining to buffer against a continuing production of acid. Using the Henderson–Hasselbalch equation, with knowledge of two variables, the third can be calculated.

Consider the blood values of a patient with a metabolic acidosis were pH = 7.03, and $[CO_2]$ = 1.10 mM. Normal values are pH = 7.40, $[HCO_3^-]$ = 24.0 mM, and $[CO_2]$ = 1.20 mM. What is the patient's blood $[HCO_3^-]$, and how much of the normal $[HCO_3^-]$ has been used in buffering the acid causing the condition?

1. The Henderson–Hasselbalch equation is

$$\text{pH} = \text{p}K' + \log\frac{[HCO_3^-]}{[CO_2]}$$

$\text{p}K'$ for $[HCO_3^-]/[CO_2]$ is 6.10.

2. Substitute given values in the equation.

$$7.03 = 6.10 + \log\frac{[HCO_3^-]}{1.10\text{ mM}}$$

or $$7.03 - 6.10 = 0.93 = \log\frac{[HCO_3^-]}{1.10\text{ mM}}$$

The antilog of 0.93 = 8.5; thus,

$$8.5 = \frac{[HCO_3^-]}{1.10\text{ mM}}$$

or $[HCO_3^-]$ remaining in blood = 9.4 mM

3. Thus, there has been a decrease of 14.6 mmol of HCO_3^-/L of blood. If much more HCO_3^- is lost, this important buffer would be unavailable to buffer any additional acid and the pH would drop rapidly, very likely leading to death of the patient.

pK' of 6.1, which is present at a 20-fold higher concentration. The HCO_3^-/CO_2 system accounts for most of the buffering capacity even though its pK' is 1.3 pH units lower than plasma pH. Both pK' and the concentration of a conjugate pair must be taken into account when considering buffering capacity. Most organic acids are relatively unimportant as buffers in cellular fluids because their pK' values are more than several pH units lower than the pH of the cell, and their concentrations are too low in comparison to buffer systems such as $HPO_4^{2-}/H_2PO_4^-$ and HCO_3^-/CO_2.

A typical problem using the Henderson–Hasselbalch equation is presented in A Closer Look 1.2. Control of pH and buffering in humans cannot be over emphasized; Clin. Corr. 1.2 is a representative problem in clinical practice.

1.4 • EUKARYOTES: MAMMALIAN CELLS AND TISSUES

Eukaryotic cells contain subcellular structures, organelles, such as a **nucleus**, **mitochondria**, **lysosomes**, and **peroxisomes**, each enclosing a subcellular compartment that is delineated by a membrane (Figure 1.8). Membranes also form a tubular network throughout the cell, the **endoplasmic reticulum** and **Golgi apparatus**, enclosing an interconnecting space or cisternae respectively. The lipid–protein nature of all cellular membranes (p. 465) prevents rapid exchange of many molecules, including water, from one compartment to another. Specific translocation mechanisms in membranes control the concentrations of small and large, charged and uncharged molecules in the various subcellular compartments. Thus the composition of inorganic ions, organic molecules, proteins, and nucleic acids is different in the diverse organelles. Sequestering of specific substrates, cofactors, and enzymes in membrane-enclosed spaces increases metabolic efficiency and permits a greater degree of control of biological processes.

The chemical composition, activities, and functions of the various cellular compartments have been studied by using a variety of techniques. With intact cells, histochemical, immunological, and fluorescent staining methods have been employed. Methods for observations in real time of intracellular events in viable cells have been employed. Changes of

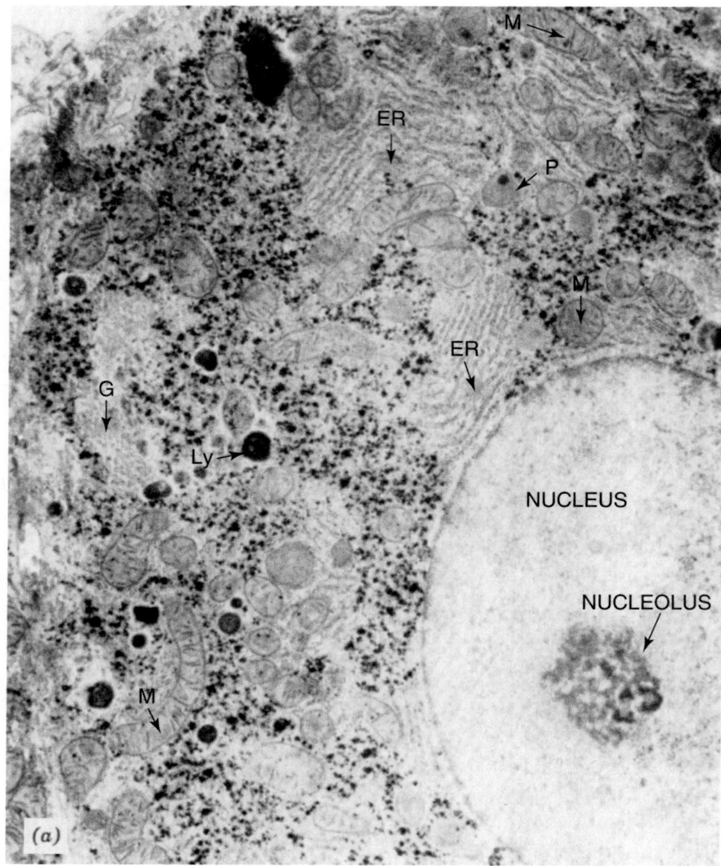

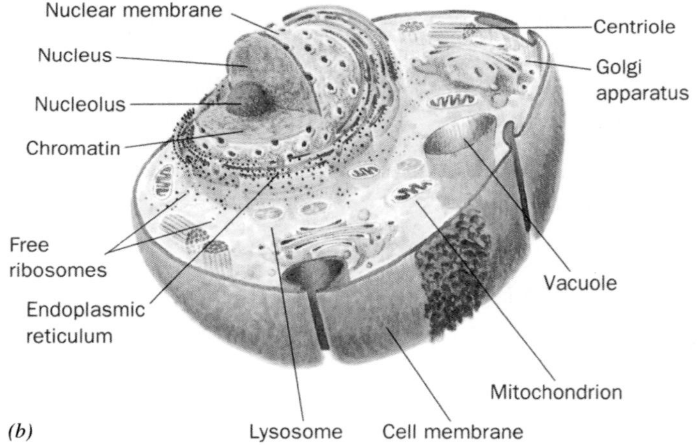

Figure 1.8 The Mammalian Cell. (*a*) Electron micrograph of a rat liver cell labeled to indicate the major structural components of eukaryotic cells. (*b*) A schematic drawing of an animal cell. Note the number and variety of subcellular organelles and the network of interconnecting membranes enclosing channels or cisternae. All eukaryotic cells are not as complex in their appearance, but most contain the major structures shown: ER, endoplasmic reticulum; G, Golgi apparatus; Ly, lysosome; P, peroxisome; and M, mitochondrion.
Photograph (*a*) reprinted with permission of Dr. K. R. Porter, from Porter, K. R., and Bonneville, M. A. In *Fine Structure of Cells and Tissues.* Philadelphia: Lea & Febiger, 1972. Schematic (*b*) reprinted with permission from Voet, D., and Voet, J. G. *Biochemistry,* 2d ed. New York: Wiley, 1995.

cytosolic pH and ionic calcium concentration are studied by use of ion-specific indicators. Cellular localization of specific proteins is determined by fluorescent antibodies. Individual organelles, membranes, and components of the cytosol can be isolated and analyzed after disruption of the plasma membrane. Techniques for disrupting the plasma membrane include use of osmotic shock, detergents, or homogenization of tissues. In the latter technique, shearing forces disrupt the plasma membrane. Differential centrifugation permits separation of cell components because of differences in their size and density. In addition, components of organelles, such as mitochondria and peroxisomes, can be isolated after disruption of the organelle membrane.

Most isolated structures and cellular fractions appear to retain the chemical and biochemical characteristics of the structure *in situ.* But biological membrane systems are subject to damage even under very mild conditions and changes in composition do occur during isolation. Damage to a membrane alters its permeability properties and allows transfer of substances that would normally be excluded by the membrane barrier. In addition, many proteins are associated loosely with membranes and dissociate easily (p. 470).

Mammalian Cells

Mammals are one of the most complex multicellular organisms; there are over 5400 species within this class of vertebrates. They are not only multicellular, but the cells in each of the organ systems (liver, kidney, brain, etc.) are specialized to carry out specific functions. Many of the individual organs consist of a number of different cell types, each with a specialized role. There are over 200 different cell types in the human as determined by their physical and chemical attributes. These range from single cells such as leukocytes in blood to cells grouped together in organs such as neurons in the nervous system. With the exception of mature erythrocytes, which contain no DNA, all cells in an individual contain the identical genetic information in their DNA. Only the necessary information for the functions of an individual cell is expressed. In the postfertilization period of an egg, some cells, referred to as **embryonic stem cells**, have the potential to develop into every cell of every organ in the adult organism. These are **pluripotent cells**. Differentiated tissues contain **multipotent cells**, which can differentiate into a limited number of different cell types. Recent studies have demonstrated that the genome of differentiated cells can be reprogrammed to form multipotent or pluripotent cells. Thus the genetic information for development in the differentiated cell is not lost irreversibly but only suppressed.

All mammalian cells, regardless of the tissue of origin and degree of specialization, have many physical, chemical, and biochemical properties in common. Activities and components of cells of an organ of one species of mammals will be similar, if not identical, to those of the same organ of some other species. Within one mammal, including humans, the same protein or enzyme is present in cells of many different cell types (e.g., muscle and liver) but there are frequently marked differences in the quantity (e.g., concentration). A particular metabolic pathway may be very important in one cell type but less so in another. These differences are due to variations in the needs of individual specialized cells; these differences are controlled by expression of the genome in the diverse cell types.

All components of cells are in a dynamic equilibrium of being synthesized and degraded, creating a steady state of cellular concentration of these substances. This dynamic state varies between mammalian cell types. These **catabolic** (degradative) and **anabolic** (synthetic) pathways are regulated to maintain the proper balance of chemical intermediates and products. In many cases control is exerted by chemical signals generated intracellularly. Mammals also have complex and integrated chemical signaling systems for information transmission between cell types (p. 504). These pathways include synthesis and release of signal molecules from one cell type, transport to cells in another organ system, and specific recognition of the signal by target cells. In most cases, the signal molecule does not affect all cell types. An example is the role of hormones (e.g., insulin) synthesized by cells in the endocrine system altering the metabolism of glucose in the muscular system.

As indicated, a major difference between prokaryotes and eukaryotes is the presence of intracellular membrane systems and organelles in eukaryotes. This permits another level for facilitation and control of enzyme catalyzed chemical reactions. As an example, by having a metabolic pathway within the smaller space of an organelle (e.g., mitochondria), the concentration of chemical intermediates can be maximized to facilitate the output of the pathway or reaction. If this were not the case, the total amount of reactants in a cell would have to be higher to achieve the same outcome.

TABLE 1.5 • Chemical Components of Biological Cells

Component	*Range of Molecular Weights*
H_2O	18
Inorganic ions Na^+, K^+, Cl^-, SO_4^{2-}, HPO_4^{2-} HCO_3^- Ca^{2+}, Mg^{2+}, etc.	23–100
Small organic molecules Carbohydrates, organic acids, lipids, nucleotides, peptides	100–1200
Macromolecules Proteins, polysaccharides, nucleic acids	6000–1,000,000

Chemical Composition of Mammalian Cells

The basic chemical components of biological cells are presented in Table 1.5. In different types of mammalian cells, the concentrations of intracellular inorganic ions are essentially similar but very different from those of the extracellular milieu. An approximation of the ionic composition of intracellular fluid, considered to represent primarily the **cytosol**, compared

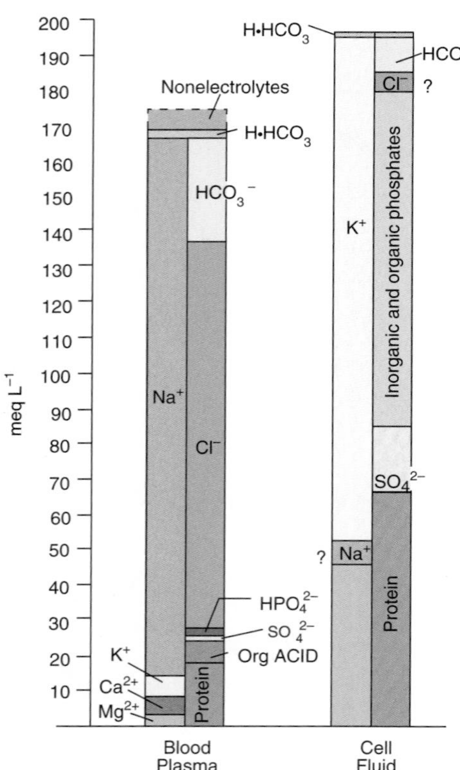

Figure 1.9 Major chemical constituents of blood plasma and cell fluid. Height of left half of each column indicates total concentration of cations; that of right half, concentration of anions. Both are expressed in milliequivalents per liter (meq/L) of fluid. Note that chloride and sodium values in cell fluid are questioned. It is probable that, at least in muscle, the cytosol contains some sodium but no chloride.
Adapted from Gregersen, M. I. In P. Bard (Ed.), *Medical Physiology*, 11th ed. St. Louis, MO: Mosby, 1961, 307.

to blood plasma is presented in Figure 1.9. The volume and degree of hydration of cells are dynamic, and thus, these concentrations are only estimates of the actual concentrations. Na^+ is the major extracellular cation, with a concentration of ~140 meq/L (mM); very little Na^+ is present in intracellular fluid. K^+ is the major intracellular cation. Mg^{2+} is present in both extra- and intracellular compartments at concentrations much lower than Na^+ and K^+. The major extracellular anions are Cl^- and HCO_3^- with smaller amounts of phosphate and sulfate. Most proteins have a negative charge at pH 7.4 (p. 85), being anions at the pH of tissue fluids. Major intracellular anions are inorganic phosphate, organic phosphates, and proteins. Other inorganic and organic anions and cations are present in concentrations well below the meq/L level. Except for very small differences created by membranes and leading to the development of membrane potentials, total cation equals total anion concentration in the different fluids.

Intracellular concentrations of most small molecular weight organic molecules, such as sugars, organic acids, amino acids, and phosphorylated intermediates, are in the range of 0.01–1.0 meq/L, but some are significantly lower. Coenzymes, organic molecules required for activity of some enzymes, are in the same range of concentration. The overall cellular concentrations of substrates for enzymes are relatively low in comparison to inorganic ions; localization in specific organelles or cellular microenvironments, however, can increase their concentrations significantly.

It is not very meaningful to determine the molar concentration of individual proteins in cells. In many cases, they are localized in specific structures or in combination with other proteins to create functional units. It is in a restricted compartment that individual proteins carry out their role, whether structural, catalytic, or regulatory. It is of interest that blood plasma contains thousands of distinct proteins, ranging in concentration from about 10^{-3} M (albumin) to 10^{-13} M (parathyroid hormone), with some proteins that have even lower concentrations.

1.5 • FUNCTIONS OF SUBCELLULAR ORGANELLES AND MEMBRANE SYSTEMS IN EUKARYOTE CELLS

Organelles have very specific functions and their enzymatic activities serve as markers for the organelle during isolation. Some metabolic pathways are located in a single cellular compartment. As examples, the tricarboxylic acid cycle (p. 552) is entirely in mitochondria, glycolysis in cytosol (p. 592), and DNA replication only in nuclei and mitochondria. Some pathways are divided between two locations, with the pathway intermediates diffusing or being translocated from one compartment to another. The enzymes for heme synthesis (p. 791) are split between mitochondria and cytosol.

A summary of functions of the major structures of eukaryotic cells is presented in Table 1.6 and the structures are identified in Figure 1.8. With the ability to selectively inactivate or delete specific genes in animals (p. 292), pathological conditions and diseases have been identified involving most of the cellular organelles and structures. In many cases, however, the biochemical details that lead to the consequences of a modified gene have not as yet been determined.

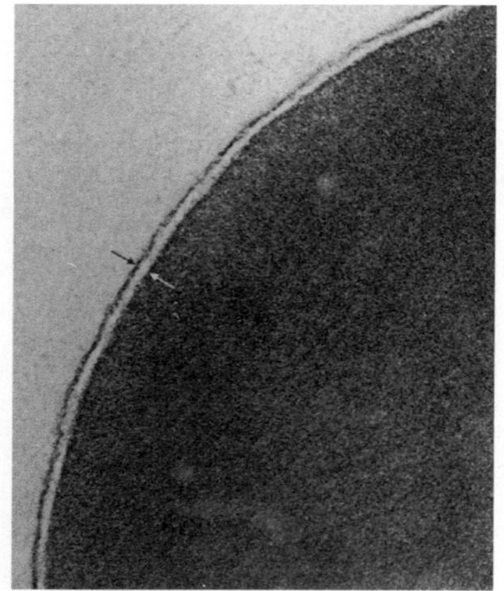

Figure 1.10 Plasma membrane of a mammalian cell. Electron micrograph of the erythrocyte plasma membrane.
Courtesy of J. D. Robertson, Duke University, Durham, NC.

Plasma Membrane Is Limiting Boundary of a Cell

The outer surface of **plasma membranes** (Figure 1.10) is in contact with a variable external environment and the inner surface with a relatively constant environment provided by the cell's cytoplasm. The protein composition of plasma membranes varies between cell types but there is a commonality in the lipid composition. Proteins and lipids are not randomly distributed throughout the membrane. The two sides of mammalian plasma membranes have a different lipid composition (p. 467).

The outer surface makes adhesive interactions with the extracellular matrix and other cells through **integrins**. These transmembrane proteins bind to the cytoskeleton inside the cell and participate in bidirectional signaling across the membrane. Through **cytoskeletal elements** the plasma membrane is involved in determining cell shape and movement. The

TABLE 1.6 • Summary of Eukaryotic Cell Compartments and Their Major Functions

Compartment	*Major Functions*
Plasma membrane	Transport of ions and small molecules Exo- and Endo-Cytosis Recognition Cell-Cell Receptors for small and large molecules Cell morphology and movement
Nucleus	DNA synthesis and repair RNA synthesis
Nucleolus	RNA processing and ribosome synthesis
Endoplasmic reticulum	Membrane synthesis Synthesis of proteins and lipids for some organelles and for export Lipid and steroid synthesis Detoxification reactions Ca^{2+} signaling
Golgi apparatus	Modification and sorting of proteins for incorporation into membranes and organelles, and for export Export of proteins
Mitochondria	Production of ATP Cellular respiration Oxidation of pyruvate, amino acids, and fatty acids Urea and heme synthesis
Lysosomes	Apoptosis Cellular digestion: hydrolysis of proteins, carbohydrates, lipids, and nucleic acids
Peroxisomes	Lipid oxidation Oxidative reactions involving O_2 Utilization of H_2O_2
Microtubules, intermediate filaments, and microfilaments	Cell cytoskeleton Cell morphology Cell motility Intracellular movements
Cytosol	Metabolism of carbohydrates, amino acids, and nucleotides; synthesis of fatty acids Protein synthesis

lipid nature of membranes excludes many substances, but transport mechanisms or pores allow transmembrane movement of ions, organic molecules, and water. Intercellular communication occurs by the presence of specific **protein receptors** on the outer surface of the plasma membrane for binding extracellular signals, such as hormones and neurotransmitters. The plasma membrane interacts with a variety of cytoplasmic substances to carry out endo- and exocytosis. Besides its role in the uptake of extracellular fluid, **endocytosis** has a significant role in programmed cell death, termed apoptosis (p. 1010). Details of membrane structure and biochemistry are presented in Chapter 12.

Nucleus Is Site of DNA and RNA Synthesis

The **nucleus** (Figure 1.11) is surrounded by two membranes, termed the **nuclear envelope**, with the outer membrane being continuous with membranes of the endoplasmic reticulum. The space between the membranes, the **perinuclear space**, is continuous with the lumen of the rough endoplasmic reticulum. The nuclear envelope contains numerous multi protein complexes, termed **nuclear pore** complexes, which cross the envelope and have a pore with an average diameter of about 90 Å. The complex permits controlled movement of particles and large molecules between the nuclear matrix and the cytoplasm. Transport of macromolecules requires expenditure of energy, and it is facilitated by a diverse series of cargo proteins and soluble transport factors, which cycle between the two compartments. It is now considered that the nuclear envelope plays important roles other than just as a barrier between the nuclear matrix and the

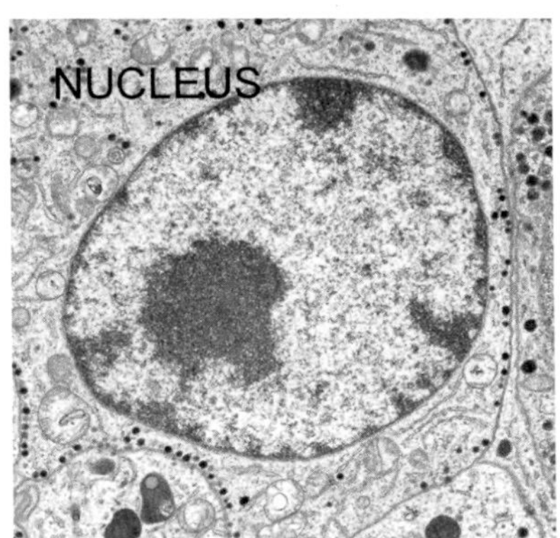

Figure 1.11 **Nucleus of a mammalian cell.** Electron micrograph generously supplied by Dr. John A. McNulty, Department of Cell Biology, Neurobiology and Anatomy, Stritch School of Medicine, Loyola University, Chicago.

CLINICAL CORRELATION 1.3

Accelerated Aging and the Cell Nucleus

Hutchinson–Gilford progeria syndrome (HGPS) (OMIM 17660) is a very rare condition where patients experience a rapid rate of aging beginning at birth. It is estimated that the syndrome occurs in about 1 in 4 to 8 million births. It is detected early in life with limited growth, loss of hair, distinctive facial characteristics, small fragile bodies, wrinkled skin, atherosclerosis, and cardiovascular problems, all of which are common in the elderly. Patients do not, however, have symptoms of neurodegenerative diseases. They rarely live past the early teens. The aging process in these individuals is 6 to 8 times the normal rate of aging.

A genetic change from normal has been identified in patients with HGPS. The nuclear envelope of patients appears distorted. Lamina A, a structural protein in nuclei, is involved in nuclear synthesis of DNA and RNA. It is formed from a precursor, attached to the nuclear membrane, by an enzyme-catalyzed reaction. HGPS patients have a deficit in this conversion, leading to accumulation of an abnormal lamina A in normal cells with aging. The presence of this abnormal protein causes blebbing of the nucleus, an alteration in gene expression, and interference with mitosis. At this time there is no established therapy for the condition, but by defining the causative changes, approaches to treatment may be possible. Results of studies of HGPS should also lead to unraveling details of the normal aging process.

Mattout, A., Dechat, T., Adam, A. A., Goldman, R. D., and Gruenbaum, Y. Nuclear lamins, disease and aging. *Curr. Opinion in Cell Biol.* 18:335, 2006; Korf, B. Hutchinson-Gilford progeria syndrome, aging, and the nuclear lamina. *N. Engl. J. Med.* 358: 552, 2008; and Merideth, M. A., Gordon, L. B., Clauss, S., Sachdev, V., et al., Phenotype and course of Hutchinson-Gilford progeria syndrome. *N. Engl. J. Med.* 358: 592, 2008.

cytoplasm (Clin. Corr. 1.3). There are also intranuclear subcompartments, the major one being the **nucleolus**. A meshwork structure underlying the inner nuclear membrane, the nuclear lamina, supports the structure of the nucleus.

Deoxyribonucleic acid (DNA), the repository of genetic information, is located in the nucleus as a DNA–protein complex, **chromatin**, which is organized into chromosomes. The nucleus contains the proteins and enzymes for replication of DNA and for repair of DNA that has been damaged (p. 164). Transcription of the genetic information in DNA into a form that can be translated into cell proteins (p. 187) involves synthesis and processing of a variety of forms of **ribonucleic acid (RNA)**. The processing of RNA for assembly of ribosomes, required for protein synthesis in the cytosol, occurs in the nucleolus.

Endoplasmic Reticulum Has a Role in Protein Synthesis and Many Synthetic Pathways

The cytoplasm of eukaryotic cells contains an extensive network of interconnecting membranes that extend from the nuclear envelope to the plasma membrane, termed **endoplasmic reticulum (ER)**. It consists of membranes with a smooth (smooth ER or SER) appearance in some parts and rough (rough ER or RER) in other places (Figures 1.12 and 1.13). This dynamic organelle changes size and shape depending on the needs of the cell. The ER encloses the ER lumen, where newly synthesized proteins are modified. The rough appearance is due to the presence of **ribonucleoprotein particles**, called **ribosomes**, attached on its cytosolic side. During cell fractionation the endoplasmic reticulum network is disrupted and the membrane reseals to form small vesicles, termed **microsomes**. These vesicles are not present in cells.

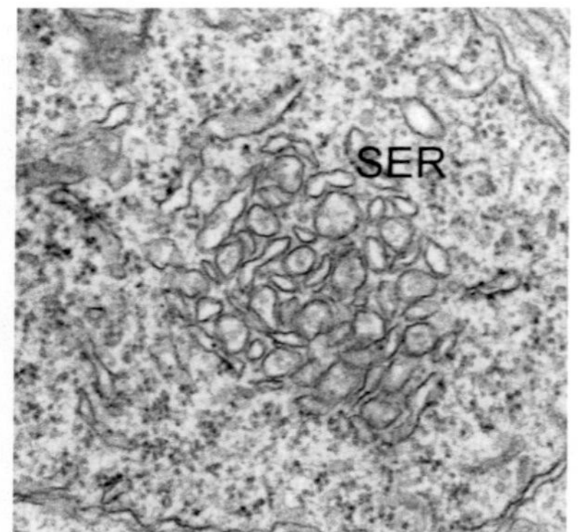

Figure 1.12 Smooth endoplasmic reticulum of a mammalian cell. SER, smooth endoplasmic reticulum.
Electron micrograph generously supplied by Dr. John A. McNulty, Department of Cell Biology, Neurobiology and Anatomy, Stritch School of Medicine, Loyola University, Chicago.

A major function of ribosomes on rough endoplasmic reticulum is biosynthesis of proteins for incorporation into membranes and cellular organelles, and for export to the outside of the cell. The ER is involved in the folding of proteins; if the folding system is overwhelmed and unfolded proteins accumulate, a series of processes are induced to reduce the stress on the ER. There is some evidence that ER stress is involved in several diseases. Smooth endoplasmic reticulum is involved in lipid synthesis and contains enzymes termed cytochromes P450 (p. 426), which catalyze hydroxylation of a variety of endogenous and exogenous compounds. These enzymes are important in the biosynthesis of steroid hormones, removal of toxic substances, and drug metabolism. The endoplasmic reticulum and Golgi apparatus are also involved in formation of other cellular organelles such as lysosomes and peroxisomes and in Ca^{2+} signaling (p. 532).

Golgi Apparatus Is Involved in Secretion of Proteins

The **Golgi apparatus** (also referred to as **Golgi complex**) (Figure 1.14) is a network of flattened smooth membrane stacks—cisternae—and vesicles. It works in consort with the endoplasmic reticulum where proteins for specific destinations are synthesized. Enzymes in Golgi membranes catalyze transfer of carbohydrate units to proteins to form glycoproteins, a process that is important in determining the proteins' eventual destination. The Golgi is the major site of new membrane synthesis and participates in formation of lysosomes and peroxisomes. Membrane vesicles shuttle proteins between the cisternae; they are pinched off one cisterna and fuse with another with the aid of a family of proteins, termed **SNARE** proteins. Vesicles originating from the Golgi apparatus transport proteins such as hormones, blood plasma proteins, and digestive enzymes to the plasma membrane for secretion. The Golgi apparatus also has a role in movement of lipids in the cell.

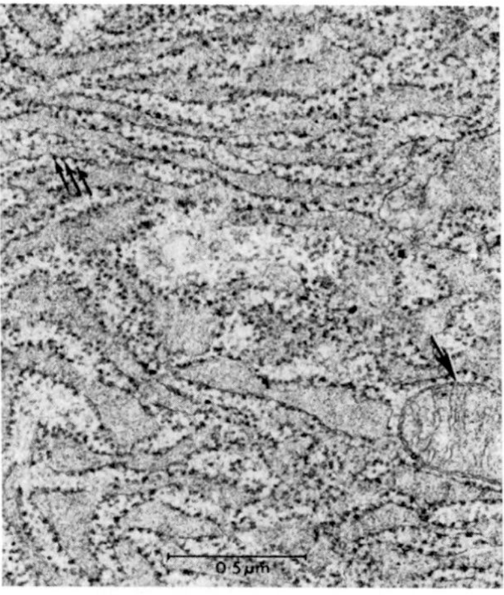

Figure 1.13 Rough endoplasmic reticulum of a mammalian cell. Three parallel arrows indicate three ribosomes among the many attached to the membranes. Single arrow indicates a mitochondrion for comparison.
Courtesy of Dr. U. Jarlfors, University of Miami, Miami, FL.

Mitochondria Supply Most of the Cellular Need for ATP

Mitochondria (Figure 1.15) are well established as the powerhouse of the cell, responsible for synthesis of over 90% of the required adenosine triphosphate (ATP) of cells (p. 572). They also have important roles in diverse cellular functions, including apoptosis (p. 1010), formation of reactive oxygen species (p. 583), cell signaling, and various metabolic processes. In electron micrographs, mitochondria appear as rods, spheres, or filamentous bodies, usually about 0.5–1 μm in diameter and up to 7 μm in length; some of these differences in shape may be due to the preparation of the tissue or cell for microscopic evaluation. The internal matrix of the organelle, the **mitosol**, is surrounded by two membranes, distinctly different in appearance, composition, and biochemical function. The inner membrane convolutes into the matrix to form **cristae** and contains numerous small spheres attached by stalks extending into the mitosol, which are responsible for ATP synthesis. The structure of cristae varies from tubular to lamellar depending on the tissue and functional state of the mitochondria. Mitochondria can divide by fission and fuse together; the proteins required for these processes have been identified, but many details of the actual mechanisms are not known. Exercise promotes an increase in the number of mitochondria in skeletal muscle. Components of the electron transport system and oxidative phosphorylation are part of the inner membrane (p. 559). Metabolic pathways for oxidation of pyruvate produced by glycolysis, fatty acids, and amino acids and some reactions in biosynthesis of urea and heme are located in the mitosol. The outer and inner membranes both contain mechanisms for translocation of specific proteins. There are a variety of transmembrane systems in the inner membrane for translocation of various metabolites. Mitochondria have a key role in aging; cytochrome c, a component of the mitochondrial electron transport system, is an initiator of apoptosis.

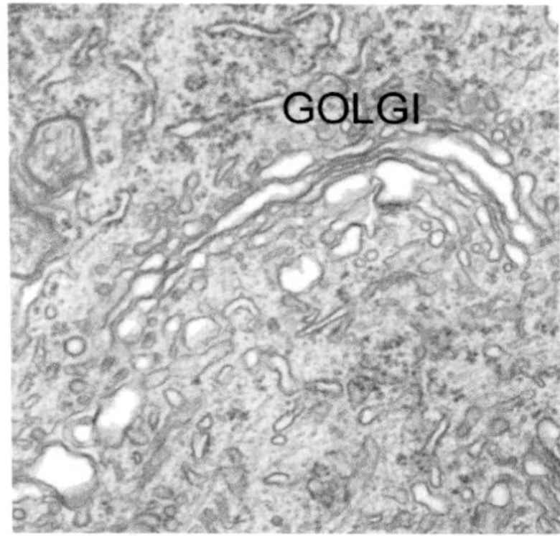

Figure 1.14 Golgi apparatus of a mammalian cell.
Electron micrograph generously supplied by Dr. John A. McNulty, Department of Cell Biology, Neurobiology and Anatomy, Stritch School of Medicine, Loyola University, Chicago.

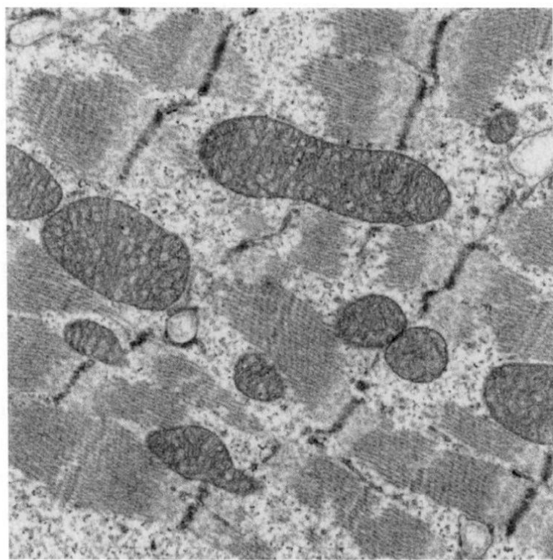

Figure 1.15 Mitochondria of a mammalian cell. Electron micrograph of mitochondria in muscle fibers from rabbit heart (×39,600).
Courtesy of Dr. W. B. Winborn, Department of Anatomy, University of Texas Health Science Center at San Antonio, and the Electron Microscopy Laboratory, Department of Pathology, University of Texas Health Science Center at San Antonio.

CLINICAL CORRELATION 1.4

Mitochondrial Diseases

The first disease specifically involving mitochondrial energy transduction (Lufts disease) (OMIM 238800) was reported in 1962. A 30-year-old patient had general weakness, excessive perspiration, a high caloric intake without increase in body weight, and an excessively elevated basal metabolic rate (a measure of oxygen utilization). She had a defect in the mechanism that controls mitochondrial oxygen utilization (p. 573). Since then several hundred genetic abnormalities have been identified that lead to alterations in mitochondrial enzymes, ribonucleic acids, electron transport components, and membrane transport systems. Some conditions result from mutations of mitochondrial DNA and others to mutations of nuclear DNA coding for mitochondrial proteins.

The first disease to be identified as due to a mutation of mtDNA was Leber Hereditary Optic Neuropathy (OMIM 535000), which leads to sudden blindness in early adulthood. Many mitochondrial diseases involve skeletal muscle and central nervous systems. Mitochondrial DNA damage may occur due to free radicals (superoxides) formed in the mitochondria. Abnormalities in mitochondria have been implicated in the pathophysiology of schizophrenia, bipolar disorder, and in age-related degenerative diseases, such as Parkinson disease, Alzheimer disease, and cardiomyopathies. Recently it has been suggested that a single mutation for a mitochondrial t-RNA leads to a constellation of symptoms including hypertension, high blood cholesterol, and decreased levels of plasma Mg^{2+}. The diseases caused by mutations of mtDNA are transmitted by a female to her offspring because mitochondria in the fertilized egg are derived from the ovum.

See Clin. Corrs. 6.6, page 229; 14.4, page 583; 14.5, page 583; 14.6, page 584; and 14.7, page 585, for details of specific mitochondrial diseases.

Luft, R. The development of mitochondrial medicine. *Proc. Natl. Acad. Sci. USA* 91:8731, 1994; Lin, M. T., and Beal, M. F. Mitochondrial dysfunction and oxidative stress in neurodegenerative diseases. *Nature* 443:787, 2006; Schapira, A. H. Mitochondrial disease. *Lancet* 368:70, 2006; and Pieczenik, S. R., and Neustadt, J. Mitochondrial dysfunction and molecular pathways of disease. *Exp. Mol. Path.* 83:84, 2007.

Mitochondria have a role in their own replication. They contain a number of copies of a circular DNA (**mitochondrial DNA [mtDNA]**), with genetic information for 13 mitochondrial proteins and some RNAs. The DNA is similar in size (16,569 base pairs) to that of bacterial chromosomes. The presence of an independent genome and the similarity to bacteria have lead to the widely accepted hypothesis that these organelles were bacteria that were taken up by a more advanced cell, and instead of being destroyed formed an endosymbiotic relationship. It is estimated that this event might have occurred some 3 billion years ago. The inheritance of mitochondria is by maternal transmission, and it has been possible to study the global movement of humans by evaluating variations in mtDNA. Mitochondria also have the requisite machinery (p. 582) to catalyze protein synthesis. Most mitochondrial proteins (over 90%) are derived from genes present in nuclear DNA and synthesized in the cytosol, then imported into the organelle. There are several hundred genetic diseases of mitochondrial function (Clin. Corr. 1.4).

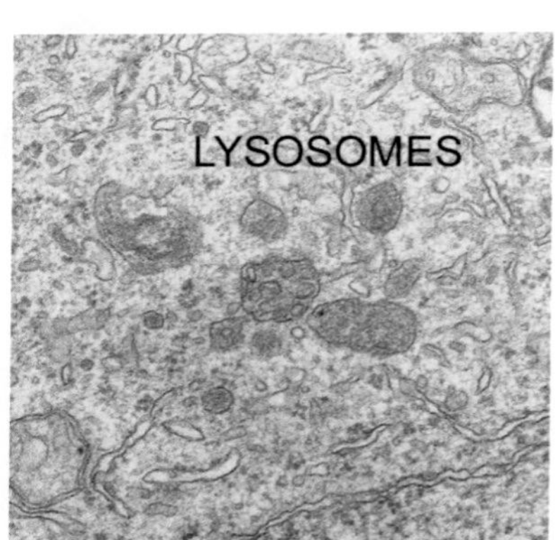

Figure 1.16 Lysosomes of a mammalian cell. Electron micrograph generously supplied by Dr. John A. McNulty, Department of Cell Biology, Neurobiology and Anatomy, Strich School of Medicine, Loyola University, Chicago.

Lysosomes Are Required for Intracellular Digestion

Lysosomes (Figure 1.16) are responsible for intracellular digestion of both extracellular and intracellular substances. With a single limiting membrane, they maintain an acidic intralysosomal pH of about 5. Encapsulated in these organelles is a class of glycoprotein enzymes, **hydrolases**, which catalyze hydrolytic cleavage of carbon–oxygen, carbon–nitrogen, carbon–sulfur, and oxygen–phosphorus bonds in proteins, lipids, carbohydrates, and nucleic acids. A partial list of lysosomal enzymes is presented in Table 1.7. The enzyme content of lysosomes varies in different tissues and depends on specific tissue functions. Lysosomal hydrolases are most active at an acid pH and split complex molecules into simple low molecular weight compounds that can be reutilized. The relationship between pH and enzyme activity is discussed on page 387. Disruption of the lysosomal membrane within cells leads to cellular digestion. Various pathological conditions have been attributed to release of lysosomal enzymes, including arthritis, allergic responses, several muscle diseases, and drug-induced tissue destruction (Clin. Corr. 1.5).

Cells ingest foreign particles, such as microorganisms and viruses, by phagocytosis, fluids by **pinocytosis**, and specific proteins by **endocytosis**, a receptor mediated process. In each case material from the exterior of a cell is encapsulated in plasma membrane bound vesicles (Figure 1.17). The vesicles containing external material fuse with primary lysosomes to form secondary lysosomes, or **digestive vacuoles**, which contain material to be digested

CLINICAL CORRELATION 1.5

Lysosomal Enzymes and Gout

Catabolism of purines (p. 820), nitrogen containing heterocyclic compounds found in nucleic acids, leads to formation of uric acid, which is excreted in urine. Gout is an abnormality in which uric acid is produced in excess, leading to an increase of uric acid in blood and deposition of urate crystals in joints. Clinical manifestations include inflammation, pain, swelling, and increased warmth of some joints, particularly the big toe. Uric acid is rather insoluble, and some of the clinical symptoms of gout can be attributed to damage done by urate crystals. Crystals are phagocytized by cells in the joint, and they accumulate in intracellular digestive vacuoles that contain lysosomal enzymes. Crystals cause physical damage to the vacuoles, releasing lysosomal hydrolases into the cytosol. Even though the pH optimum of lysosomal enzymes is lower than the pH of the cytosol, they have some hydrolytic activity at the higher pH, which causes digestion of cellular components and cellular autolysis.

Weissmann, G. Crystals, lysosomes, and gout. *Adv. Intern. Med.* 19:239, 1974; and Burt, H. M., Kalkman, P. H., and Mauldin, D. Membranolytic effects of crystalline monosodium urate monohydrate. *J. Rheum.* 10:440, 1983.

and lysosomal hydrolases to carry out the digestion. They are identified microscopically by their size and often by the presence of partially digested structures. Primary lysosomes are those which have not fused with vesicles containing material to be degraded.

Cellular proteins, nucleic acids, lipid, and organelles, such as mitochondria, are in a dynamic state of synthesis and degradation. Lysosomes are responsible for hydrolysis of these cellular components by a highly regulated process termed autophagy. Substances destined to be degraded are identified and taken up by lysosomes or are first encapsulated

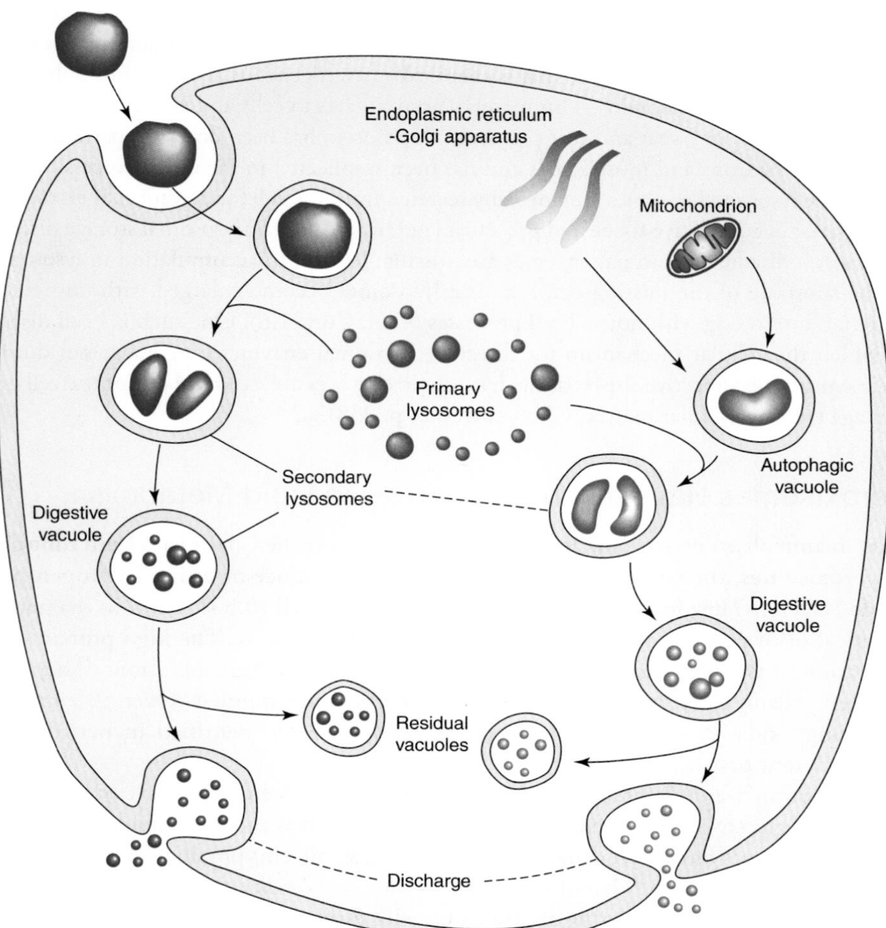

Figure 1.17 Diagrammatic representation of the role of lysosomes in intracellular digestion of substances internalized by phagocytosis (heterophagy) and of cellular components (autophagy). In both processes substances to be digested are enclosed in a membrane vesicle, followed by fusion with a primary lysosome to form a secondary lysosome where digestion occurs.

TABLE 1.7 • Representative Lysosomal Enzymes and Their Substrates

Type of Substrate and Enzyme	*Specific Substrate*
POLYSACCHARIDE HYDROLYZING ENZYMES	
α-Glucosidase	Glycogen
α-Flucosidase	Membrane fucose
β-Galactosidase	Galactosides
α-Mannosidase	Mannosides
β-Glucuronidase	Glucuronides
Hyaluronidase	Hyaluronic acid and chondroitin sulfates
Arylsulfatase	Organic sulfates
Lysozyme	Bacterial cell walls
PROTEIN-HYDROLYZING ENZYMES	
Cathepsins	Proteins
Collagenase	Collagen
Elastase	Elastin
Peptidases	Peptides
NUCLEIC ACID-HYDROLYZING ENZYMES	
Ribonuclease	RNA
Deoxyribonuclease	DNA
LIPID-HYDROLYZING ENZYMES	
Lipases	Triacylglycerol and cholesterol esters
Esterase	Fatty acid esters
Phospholipase	Phospholipids
PHOSPHATASES	
Phosphatase	Phosphomonoesters
Phosphodiesterase	Phosphodiesters
SULFATASES	
Chondroitin sulfatase	Heparan sulfate
Arylsulfatase B	Dermatan sulfate

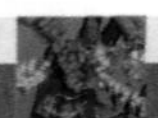

CLINICAL CORRELATION 1.6

Lysosomal Acid Lipase Deficiency

Human lysosomal acid lipase (hLAL) hydrolyzes triacylglycerol to free fatty acids and glycerol, and cholesteryl esters to cholesterol and fatty acids. It is a critical enzyme in cholesterol metabolism, serving to make cholesterol available for the needs of cells. Cholesteryl ester storage disease (CESD) and Wolman disease (OMIM 278000) are two distinct phenotypic forms of a genetic deficiency of hLAL; both are rare autosomal recessive diseases. CESD is usually diagnosed in adulthood and is evidenced by hypercholesterolemia, hepatomegaly, and early onset of severe atherosclerosis. Affected individuals have hLAL activity but at a very low level (< 5% of normal). Apparently, this level is sufficient to hydrolyze triacylglycerol but not cholesteryl esters. Patients are homozygous for a mutation at a splice junction on both alleles leading to a catalytically defective and unstable enzyme.

Wolman disease is manifested in infants and is usually fatal by age one. There is no detectable activity of hLAL. Both triacylglycerol and cholesteryl esters accumulate in tissues. Patients are homozygous for a mutation that leads to an absence of active enzyme.

Hegele, R. A., Little, J. A., Vezina, C., et.al. Hepatic lipase deficiency: Clinical, biochemical, and molecular genetic characteristics. *Atherosclerosis and Thrombosis* 13:720, 1993; Anderson, R. A., Bryson, G. M. and Parks, J. S. Lysosomal acid lipase mutations that determine phenotype in Wolman and cholesteryl ester storage disease. *Mol. Genetics & Met.* 68:333, 1999; and Zschenker, O., Illies, T., and Ameis, D. Overexpression of lysosomal acid lipase and other proteins in atherosclerosis. *J. Biochem.* 140:23, 2006.

within membrane vesicles that fuse with primary lysosomes (Figure 1.17). Autophagy is important not only in stressed but also in normal mammalian cells; starvation and specific hormones induce autophagy.

Products of lysosomal digestion are released from lysosomes and reutilized by the cell. Indigestible material accumulates in vesicles called residual bodies whose contents are normally removed from the cell by exocytosis. Some residual bodies contain high concentrations of a pigmented substance that is chemically heterogeneous and contains polyunsaturated fatty acids and proteins. This material accumulates in cells and is termed **lipofuscin** or "age pigment" or "wear and tear pigment." Lipofuscin has been observed particularly in postmitotic neurons and muscle cells and has been implicated in the aging process.

Some lysosomal enzymes are normally secreted from the cell for digestion of extracellular material in connective tissue and prostate gland. In the genetic lysosomal storage diseases (p. 236), individual lysosomal enzymes are missing, leading to accumulation in lysosomes of the substrate of the missing enzyme. The lysosomes become enlarged with undigested material, interfering with normal cell processes (Clin. Corr. 1.6). One such is I cell disease in which the cellular mechanism for directing lysosomal enzymes into lysosomes during their synthesis is defective. Instead, the hydrolytic enzymes are exported out of the cell and damage the extracellular matrix (Clin. Corr. 6.8, p. 238).

Peroxisomes Have an Important Role in Lipid Metabolism

Most mammalian, protozoan, and plant cells have organelles, designated **microbodies** or **peroxisomes**, the latter because of their ability to produce or utilize **hydrogen peroxide** (H_2O_2). They have a single membrane and are small (0.3–1.5 μm in diameter), spherical or oval with a fine network of tubules in their matrix. The high protein concentration in the matrix of some peroxisomes leads to crystalline inclusions. They vary between cells in their enzyme composition, function, and number. Over 50 enzymes, catalyzing oxidative and biosynthetic reactions, have been identified in peroxisomes from different tissues.

Peroxisomes are essential for oxidation of very long-chain fatty acids (p. 701) and synthesis of glycerolipids, glycerol ether lipids (plasmalogens), and isoprenoids (p. 720). In mice, peroxisomes have been demonstrated to have a significant physiological role in lipid synthesis and degradation. They also contain enzymes that oxidize D-amino acids, uric acid, and various 2-hydroxy acids using molecular O_2 with formation of H_2O_2. **Catalase,** a heme enzyme (p. 587) present in peroxisomes, catalyzes the conversion of H_2O_2 to water and oxygen and oxidation of various compounds by H_2O_2 (Figure 1.18). Cells protect themselves from the toxicity of H_2O_2 by having both peroxide-producing and peroxide-utilizing enzymes in one compartment.

$$(1)\ 2H_2O_2 \longrightarrow 2H_2O + O_2$$

$$(2)\ RH_2 + H_2O_2 \longrightarrow R + 2H_2O$$

Figure 1.18 Reactions catalyzed by catalase.

Proteins responsible for the biogenesis of peroxisomes have been identified. Several classes of **xenobiotics**, that is, foreign substances, including aspirin, lead to proliferation of peroxisomes in liver. Over 25 genetic diseases involve peroxisomes; they are grouped together as disorders of peroxisome biogenesis (Clin. Corr. 1.7).

Cytoskeleton Organizes the Intracellular Contents

Eukaryotic cells contain microtubules, intermediate filaments, and actin filaments (microfilaments) as parts of a cytoskeletal network, or scaffolding in the cytoplasm (p. 980). This dynamic structure has a role in maintenance of cellular morphology, intracellular transport of vesicles and organelles, cell motility, and cell division. The cytoskeleton includes **microtubules** consisting of multimers of **tubulin**, a protein that assembles rapidly into tubular structures and disassembles depending on the needs of cells. Actin and myosin form very important cellular filaments in striated muscle where they are responsible for muscular contraction (p. 969). Three separate super families of mechanochemical proteins—myosin, dynein, and kinesin—convert chemical energy into mechanical energy for movement of cellular components (p. 979). These molecular motors are associated with the cytoskeleton. Dynein is involved in ciliary and flagellar movement, whereas kinesin is a driving force for the movement of vesicles and organelles along microtubules, especially in neuronal axons. These motor proteins have a significant role in the pathogenesis of a variety of diseases.

Cytosol Contains Soluble Cellular Components

The least complex in structure, but not in chemistry, is the organelle-free cell sap, or **cytosol**. It is here that many reactions and pathways of metabolism (e.g., glycolysis, glycogenesis, glycogenolysis, and fatty acid synthesis) occur. Synthesis of proteins on both free ribosomes often in a polysome form and those bound to endoplasmic reticulum occurs in the cytosol, which contains all the required intermediates. Although there is no apparent structure to the cytosol, the high protein content precludes it from being a truly homogeneous mixture of soluble components, and it is believed that there are functional compartments throughout the cytosol. Many reactions are localized in selected areas where substrate availability is more favorable. The actual physicochemical state of the cytosol is poorly understood. Another complication of evaluating the composition of the cytoplasm is that some enzymes and proteins occur as soluble proteins when the cytosol is isolated, but in the intact cell may actually be loosely attached to membrane structures or to cytoskeletal components.

1.6 • INTEGRATION AND CONTROL OF CELLULAR FUNCTIONS

A eukaryotic cell is a complex structure that maintains an intracellular environment that permits many complex reactions and processes to occur simultaneously. This requires a high degree of control and integration. Cells of multicellular organisms also participate in maintaining the well-being of the whole organism by exerting influences on each other to maintain the balance between tissue and cellular activities. Very few functions operate totally independently; changes in one function can exert an influence, positive or negative, on other functions. As will be described throughout this book, the control of function is mediated at many levels, from the expression of a gene, to altering the concentration of an enzyme or a protein effector, to changes in substrate or coenzyme levels to adjust the rate of a specific enzyme reaction. The integration of many cellular processes is controlled by proteins that serve as activators or inhibitors that maintain cellular homeostasis. Many cellular processes are programmed to occur under specific conditions; for example, cell division in normal cells occurs only when the processes required for cell division are activated (p. 1004). Then and only then is there an orderly and integrated series of reactions culminating in the division of a cell into two daughter cells. Programmed cell death

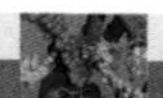

CLINICAL CORRELATION 1.7

Peroxisome Biogenesis Disorders (PBD)

Peroxisomes are responsible for a number of important metabolic reactions, including synthesis of glycerol ethers, shortening very long-chain fatty acids so that mitochondria can completely oxidize them, and oxidation of the side chain of cholesterol needed for bile acid synthesis. Peroxisome Biogenesis Disorders (PBD) (OMIM 601539) comprise more than 25 genetically and phenotypically related disorders that involve enzymatic activities of peroxisomes. They are rare autosomal recessive diseases characterized by decreased levels of glycerol–ether lipids (plasmalogens), and increased levels of very long-chain fatty acids (C_{24} and C_{26}) and cholestanoic acid derivatives (precursors of bile acids). The disease can affect liver, kidney, brain, and skeletal systems. The most severe is Zellweger syndrome, which is due to the absence of functional peroxisomes; death frequently occurs by age 6 months. In this condition, the genetic defect is in the mechanism for importing enzymes into the matrix of peroxisomes. Some PBD conditions are caused by donor splice or missense mutations (p. 166); in some, there is the absence of a single metabolic enzyme or defect in a membrane transport component. In some instances the disease can be diagnosed prenatally by assay of peroxisomal enzymes or fatty acids in cells of amniotic fluid.

FitzPatrick, D. R. Zellweger syndrome and associated phenotypes. *J. Med. Genetics.* 33:863, 1996; Warren, D. S., Wolfe, B. D., and Gould, S. J. Phenotype-genotype relationships in PEX10-deficient peroxisome biogenesis disorder patients. *Human Mutation* 15:509, 2000; and Steinberg, S. J., Dodt, G., Raymond, G. V., Braverman, N.E., et al. Peroxisome biogenesis disorders. *Biochim. Biophys. Acta.* 1763:1733, 2006.

(p. 1010) is a highly regulated process occurring in all mammalian cells but individual steps in the process vary from tissue to tissue. Many diseases are due to a failure in specific control mechanisms. As we continue to comprehend the complexity of biological cells, we are amazed that there are not many more errors occurring and many more individuals with abnormal conditions.

Thus as we proceed to study the separate chemical components and activities of cells in subsequent chapters, it is important to remember the concurrent and surrounding activities, constraints, and influences of the environment. Only by bringing together all the separate parts and activities of a cell, that is, reassembling the puzzle, will we appreciate the wonder of living cells.

Bibliography

Water and Electrolytes

Bukowski, R., Szalewicz, K., Groenenboom, G. C., and van der Avoird, A. Predictions of the properties of water from first principles. *Science* 315:1249, 2007.

Chaplin, M. Water Structure and Science. www.isbu.ac.uk/water, 2008.

Gestein, M., and Levitt, M. Simulating water and the molecules of life. *Sci. Amer.* 279:100, 1998.

Klein, M. L. Water on the move. *Science* 291:2106, 2001.

Miyazaki, M., Fujii, A., Ebata, T., and Mikami, N. Infrared spectroscopic evidence for protonated water clusters forming nanoscale cages. *Science* 304:1134, 2004.

Pollack, G. H., Cameron, I. L., and Wheatley, D. N. *Water and the Cell.* Dordrecht: Springer, 2006.

Westof, E. *Water and Biological Macromolecules.* Boca Raton, FL: CRC Press, 1993.

Wiggins, P. M. Role of water in some biological processes. *Microbiol. Rev.* 54:432, 1990.

Cell Structure

Alberts, B., Johnson, A., Lewis, J., Raff, M., Roberts, K., and Walter, P. *Molecular Biology of the Cell,* 4th ed. New York: Garland, 2002.

Becker, W. M., and Deamer, D. W. *The World of the Living Cell,* 2d ed. Redwood City, CA: Benjamin, 1991.

De Duve, C. *Guided Tour of the Living Cell,* Vols. 1 and 2. New York: Scientific American, 1984.

Dingle, J. T., Dean, R. T., and Sly, W. S. (Eds.). *Lysosomes in Biology and Pathology.* Amsterdam: Elsevier (a serial publication covering all aspects of lysosomes).

Holtzman, E., and Novikoff, A. B. *Cells and Organelles,* 3d ed. New York: Holt, Rinehart & Winston, 1984.

Hoppert, M., and Mayer, F. Principles of macromolecular organization and cell function in bacteria and archaea. *Cell Biochem. & Biophys.* 31:247, 1999.

Misteli, T. Protein dynamics: Implications for nuclear architecture and gene expression. *Science* 291:843, 2001.

Porter, K. R., and Bonneville, M. A. *Fine Structure of Cells and Tissues.* Philadelphia: Lea & Febiger, 1972.

Vale, R. D. Intracellular transport using microtubule-based motors. *Annu. Rev. Cell. Biol.* 3:347, 1987.

Cell Organelles

Baumann, O., and Walz, B. Endoplasmic reticulum of animal cells and its organization into structural and functional domains. *Inter. Rev. Cytology* 205:149, 2001.

Benedetti, A. *Endoplasmic Reticulum: A Metabolic Compartment.* New York: Ios, 2005.

Brocard, J. B., Rintoul, G. L., and Reynolds, I. J. New perspectives on mitochondrial morphology in cell function. *Biol. Cell.* 95:239, 2003.

Cuervo, A. M. Autophagy: in sickness and in health. *Trends Cell Biol.* 14:70, 2004.

De Duve, C. The Birth of Complex Cells. *Sci. Amer.* April:50, 1996.

Eckert, J. H., and Erdmann, R. Peroxisome biogenesis. *Rev. Physiol. Biochem. & Pharm.*147:75, 2003.

Enns, G. M. The contribution of mitochondria to common disorders. *Molecular Genetics Metab.* 80:11, 2003.

Fujiki, Y. Peroxisome biogenesis and peroxisome biogenesis disorders. *Fed. Euro. Biochem. Letters* 476:42, 2000.

Hirokawa, N., and Takemura, R. Biochemical and molecular characterization of diseases linked to motor proteins. *Trends Biochem. Sciences* 28:558, 2003.

Holtzman, E. *Lysosomes.* New York: Plenum, 1989.

Jaeken L. A new list of functions of the cytoskeleton. *IUBMB Life* 59:127, 2007.

Latruffe, N., and Bugaut, M. *Peroxisomes.* New York: Springer-Verlag, 1994.

Lemasters, J. J., and Nieminen, A. *Mitochondria in Pathogenesis.* New York: Kluwer, 2001.

Lin, J. J., Eppinga, R. D., Warren, K. S., and McCrae, K. R. Human tropomyosin isoforms in the regulation of cytoskeleton functions. *Adv. Exp. Med. Biol.* 644:201, 2008.

Luzio, J. P., Poupon, V., Lindsay, M. R., Mullock, B. M., Piper, R. C., and Pryor, P. R.. Membrane dynamics and the biogenesis of lysosomes. *Molecular Memb. Biol.* 20:141, 2003.

Moser, H. W. Molecular genetics of peroxisomal disorders. *Frontiers in Bioscience* 5:D298, 2000.

Osteryoung, K. W., and Nunnari, J. The division of endosymbiotic organelles. *Science* 302:1698, 2003.

Pavelka, M. *Functional Morphology of the Golgi Apparatus.* New York: Springer-Verlag, 1987.

Preston, T. M., King, C. A., and Hyams, J. S. *The Cytoskeleton and Cell Motility.* New York: Chapman & Hall, 1990.

Rothman, J. E., and Orci, L. Budding vesicles in living cells. *Sci. Amer.* 274:70, 1996.

Scheffler, I. E. *Mitochondria.* New York: Wiley, 1999.

Strauss, P. R., and Wilson, S. H. *The Eukaryotic Nucleus: Molecular Biochemistry and Macromolecular Assemblies.* Caldwell, NJ: Telford, 1990.

Van der Klei, I. J., and Veenhuis, M. Peroxisomes: Flexible and dynamic organelles. *Curr. Opinions in Cell Biol.* 14:500, 2002.

McBride, H. M., Neuspiel, M.,and Wasiak, S. Mitochondria: More than just a powerhouse. *Curr. Biol.* 16:R551, 2006.

Key Terms

DNA
eukaryotes
nucleoid
prokaryotes
organelles
water
hydrogen bonds
amphipathic
anion
cation
nonelectrolytes
salts
strong electrolytes
k'_{eq}
hydronium ion
weak electrolytes
Brønsted acid
conjugate base
ion product of water
pH
proton acceptor
proton donor
weak acid
pK'
conjugate pair
carbonic acid
buffering
buffering capacity
Henderson-Hasselbalch equation
ionic composition
plasma membrane
cytoskeleton
nuclear envelope
nucleus
nucleolus
endoplasmic reticulum
Golgi apparatus
ribosomes
mitochondria
mitosol
lysosomes
digestive vacuoles
endocytosis
hydrolase
microbodies
peroxisomes
catalase
microtubules
cytosol
xenobiotic

Questions • CAROL N. ANGSTADT

Multiple-Choice Questions

1. Both prokaryotic and eukaryotic cells have all of the following except
A. a plasma membrane.
B. a genome of DNA.
C. subcellular organelles defined by membranes.
D. the ability to replicate.
E. the ability to transform external sources of energy into utilizable energy.

2. Factors responsible for a water molecule being a dipole include the
A. similarity in electron affinity of hydrogen and oxygen.
B. tetrahedral structure of liquid water.
C. magnitude of the H—O—H bond angle.
D. ability of water to hydrogen bond to various chemical structures.
E. difference in bond strength between hydrogen bonds and covalent bonds.

3. Hydrogen bonds can be expected to form only between electronegative atoms such as oxygen or nitrogen and a hydrogen atom bonded to
A. carbon.
B. an electronegative atom.
C. hydrogen.
D. iodine.
E. sulfur.

4. Which of the following is both a Brønsted acid and a Brønsted base in water?
A. $H_2PO_4^-$
B. H_2CO_3
C. NH_3
D. NH_4^+
E. Cl^-

5. Biological membranes are associated with all of the following except
A. prevention of free diffusion of ionic solutes.
B. release of proteins when damaged.
C. specific systems for the transport of uncharged molecules.
D. sites for biochemical reactions.
E. free movement of proteins and nucleic acids across the membrane.

6. Analysis of the composition of the major fluid compartments of the body shows that
A. the major blood plasma cation is K^+.
B. the major cell fluid cation is Na^+.
C. one of the major intracellular anions is Cl^-.
D. one of the major intracellular anions is phosphate.
E. plasma and cell fluid are all very similar in ionic composition.

Questions 7 and 8: A patient with Luft disease presented with general weakness, excessive perspiration, a high caloric intake without increase in body weight, and an excessively elevated basal metabolic rate. Luft disease was the first disease involving a defect in mitochondria to be described. It is a defect in the mechanism that controls oxygen utilization in mitochondria.

7. Components of the electron transport system and oxidative phosphorylation are found associated with the
A. outer mitochondrial membrane.
B. inner mitochondrial membrane.
C. mitosol.
D. peroxisomes.
E. protein synthetic components.

8. Which of the following is a characteristic of mitochondria?
A. The inner membrane forms cristae and contains small spheres attached by stalks on the inner surface.
B. Only the outer membrane has transmembrane systems for translocation of metabolites.
C. The mitosol is relatively inert metabolically.
D. Mitochondrial DNA is similar to nuclear DNA in size and shape.
E. Mitochondria have no role in apoptosis.

Questions 9 and 10: Gout is a condition in which excessive production of uric acid leads to deposition of urate crystals in joints. Clinical manifestations include inflammation, pain, and swelling of joints, especially the joint of the big toe. Crystals are phagocytosed by cells in the joint and accumulate in digestive vacuoles that contain lysosomal enzymes. Crystals cause physical damage to the vacuoles, releasing lysosomal enzymes into the cytosol.

9. Lysosomal enzymes
A. are hydrolases.
B. usually operate at acidic pH.
C. are normally isolated from their substrates by the lysosomal membrane.
D. can lead to cellular digestion if the lysosomal membrane is disrupted.
E. all of the above.

10. Individual lysosomal enzymes are missing in a number of genetic diseases referred to as Lysosomal Storage diseases. In these diseases
A. the defect is an inability to direct enzymes to the lysosome after synthesis.
B. lysosomes of affected cells become enlarged with undigested materials.
C. lipofuscin, or "wear and tear pigment," accumulates in cells.
D. any material taken into the lysosome will accumulate.
E. residual bodies contain the products of digestion.

11 and 12: Zellweger syndrome is one of a class called Peroxisome Biogenesis Disorders (PBD). PBDs are characterized by abnormalities of the liver, kidney, brain, and skeletal system. Zellweger syndrome is particularly severe and death usually occurs by age 6 months. There is an absence of functional peroxisomes.

11. Peroxisomes have a role in all of the following except
A. oxidation of very long chain fatty acids.
B. synthesis of glycerolipids.
C. hydrolysis of cholesteryl esters.
D. oxidation of D-amino acids.
E. oxidation of uric acid.

12. One of the roles of peroxisomes is to render H_2O_2 nontoxic by
A. oxidizing amino acids with O_2.
B. the action of the enzyme catalase.
C. transporting H_2O_2 from blood into peroxisomes.
D. the action of cathepsins.
E. shortening of a fatty acid chain.

Problems

13. If a weak acid is 91% neutralized at pH 5.7, what is the pK' of the acid?

14. If normal plasma pH is 7.4 and normal $[CO_2]$ is 1.2 mM, what is the normal $[HCO_3^-]$? The pK' for this system is 6.1.

Answers

1. C The intracellular milieu of prokaryotes is organized into functional compartments, but they are not defined by membranes. A, B, D, and E: These are features that prokaryotes and eukaryotes have in common.

2. C Water is a polar molecule because the bonding electrons are attracted more strongly to oxygen than to hydrogen. The bond angle gives rise to asymmetry of the charge distribution; if water were linear, it would not be a dipole. A: Hydrogen and oxygen have very different electron affinity. B and D are consequences of water's structure, not factors responsible for it.

3. B Only hydrogen atoms bonded to one of the electronegative elements (O, N, or F) can form hydrogen bonds. A hydrogen participating in hydrogen bonding must have an electronegative element on both sides of it.

4. A $H_2PO_4^-$ can donate a proton to become HPO_4^{2-}. It can also accept a proton to become H_3PO_4. B and D are Brønsted acids; C is a Brønsted base. The Cl^- ion in water is neither.

5. E These molecules are too large to cross the membrane freely unless the membrane is damaged (B). A: Ionic solutes do not cross the lipid membrane readily. C: Most substances require transport across the membrane. D: Different membranes have different reactions.

6. D Phosphate and protein are the major intracellular anions. A, B, and E: plasma and cell fluid are strikingly different. The Na^+ ion is the major cation of plasma. C: Most chloride is extracellular.

7. B The electron transport and oxidative phosphorylation proteins are part of the inner membrane. D: Peroxisomes are separate organelles, not part of mitochondria. E: Mitochondria produce some of their own proteins but this is a separate process and not part of electron transport or oxidative phosphorylation.

8. A The cristae are the site of ATP synthesis. B: The outer membrane is relatively freely permeable; it is the inner membrane that is not and has multiple transport systems. C: The mitosol has numerous metabolic pathways–oxidation of pyruvate, fatty acids, and amino acids, for example. D: Mitochondrial DNA (mtDNA) is small and circular. E: Cytochrome c is an initiator of apoptosis.

9. E A: There are hydrolases for all classes of macromolecules. B and D: Lysosomal pH is usually about 5 but the enzymes retain some activity at higher pH. C: Substrates and enzymes are brought together by phagocytosis or pinocytosis.

10. B Engorgement interferes with normal cell processes. A: This is I cell disease and affects all of the lysosomal enzymes exported out of the cell. C: This is a normal process. Lipofuscin is a chemically heterogeneous pigmented substance and not caused by a missing lysosomal enzyme. D: Each Lysosomal Storage disease affects a single enzyme so only substrates of that enzyme would be affected. E: Residual bodies contain indigestible material.

11. C This occurs in lysosomes.

12. B H_2O_2 is converted to O_2 and H_2O. A and E: These produce H_2O_2. C: H_2O_2 is produced and utilized within peroxisomes. D: These are enzymes of lysosomes.

13. If a weak acid is 91% neutralized, 91 parts are present as conjugate base and 9 parts remain as the undissociated acid. Thus the conjugate base/acid ratio is 10:1. Substituting this ratio into the Henderson–Hasselbalch equation gives 5.7 = pK' + log(10/1). Solving the equation for pK' gives an answer of 4.7. The acid could be β-hydroxybutyric acid, an important physiological acid, which has this pK'.

14. In this system CO_2 is the conjugate acid and HCO_3^- the conjugate base. Substituting the values in the Henderson–Hasselbalch equation would give

$$7.4 = 6.1 + \log(x/1.2)$$

The antilog of 1.3 = 20. Therefore, $[HCO_3^-] = 20 \times 1.2 = 24$ mM. The ability to solve for any value in this equation given three of the terms is very important in assessing the acid–base status in the blood.

2

DNA and RNA: Composition and Structure

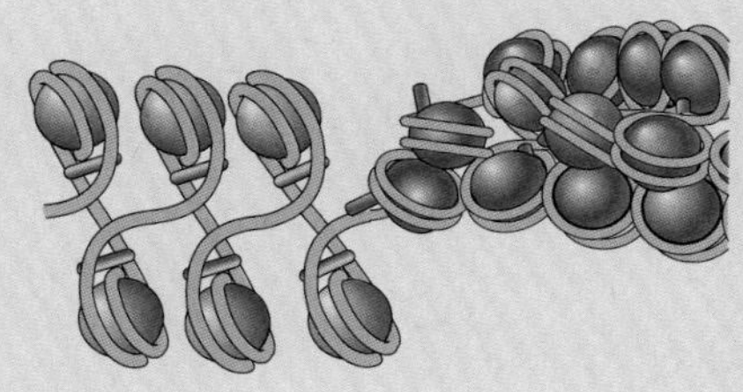

Stephen A. Woski
Associate Professor, University of Alabama

Francis J. Schmidt
Professor, University of Missouri

CLINICAL CORRELATIONS

Key Concepts

- The central dogma of molecular biology holds that biological information specified by nucleic acid sequences in DNA and RNA determines protein sequences, and therefore, protein properties.
- The structures of nucleic acids are determined by the chemical properties of their constituent parts: phosphate, sugar, and nucleobase.
- Cellular DNAs are typically double-stranded polymers of 2′-deoxyribonucleoside 5′ monophosphates.
- Complementarity between the bases in the double helix allows DNA to function as a cellular genetic material.
- The length of a DNA molecule is much greater than the dimensions of a cell; therefore, DNA adopts higher-order structures that greatly compact the molecule into chromatin.
- The nucleotide sequence of DNA determines its function.
- Cellular RNAs are single-stranded polymers of ribonucleoside 5′ monophosphates.
- RNA tertiary structure is determined by intramolecular hydrogen bonding.
- Cellular RNAs have roles in information transfer, and catalytic and regulatory roles.

2.1 • NUCLEIC ACIDS AND BIOLOGICAL INFORMATION

Central Dogma of Molecular Biology

One of the hallmarks of living organisms is their ability to reproduce. The information that makes each individual life form unique must be preserved and passed on to progeny. All life on Earth uses **nucleic acids** for the storage of genetic information. With the exception of some viruses, the biomolecule utilized for information storage is **deoxyribonucleic acid (DNA)**. This molecule is remarkably well suited for its task because of its chemical stability and its ability to encode vast amounts of information using a simple four-letter code. However, DNA is only a part of the core architecture of life. **Ribonucleic acid (RNA)** and protein play equally important roles. The interrelationship of these three classes of molecules constitutes the "**central dogma of molecular biology**" (Figure 2.1), which holds that DNA stores information that determines the sequence of RNA which in turn determines the structure of protein. Much of the structure and biochemistry of cells is due to the properties of their constituent proteins. These properties are determined by portions of the sequence of DNA. Information, however, cannot flow directly from DNA to protein, but depends on RNA to transport the information. Genetic information is transmitted from DNA to RNA by **transcription.** The sequence of RNA is then **translated** into a protein sequence at the ribosome. DNA also plays an essential role in heredity by serving as the template for its own **replication.** Several discoveries have blurred the distinct roles of each of these biomolecules in the central dogma. For example, RNA can act as a catalyst **(ribozyme)** in biochemical reactions and constitutes the active component of ribosomes, the protein synthesis machinery of the cell. RNA can serve as a template for DNA synthesis, reversing the normal flow of information **(reverse transcription).** Some disease can be transmitted by proteins in the absence of nucleic acids. Nevertheless, nucleic acids play central roles in the storage and transmission of biological information.

DNA Can Transform Cells

Prior to the 1950s, the general view held that nucleic acids were substances of limited cellular importance. The first evidence that DNA is the genetic material was found during the 1920s, but the definitive demonstration of DNA's role was not accomplished until 1944. The key experiments involved two strains of pneumococcus, a bacterium that causes a form of pneumonia. When cultured, one strain formed smooth colonies and the other formed rough colonies; these were labeled the S- and R-forms, respectively. The S-form is virulent while the R-form is nonvirulent, and the two forms are genetically distinct and cannot interconvert spontaneously. Treatment of R-form bacteria with pure DNA extracted from the S-form resulted in their **transformation** into the S-form. The transformation was inherited permanently by subsequent generations. This demonstrated that DNA was the transforming agent and the material responsible for transmitting genetic information from one generation to the next. Clin. Corr. 2.1 describes vaccines based on transformation of mammalian cells with DNA.

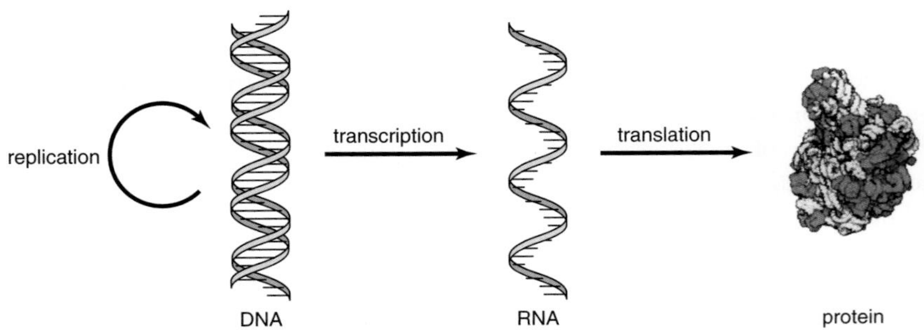

Figure 2.1 Central dogma of molecular biology.

Information Capacity of DNA Is Enormous

A striking characteristic of DNA is its ability to encode an enormous quantity of biological information. For example, a human cell contains information for the synthesis of about 20,000 to 25,000 proteins. This information is stored in the cell nucleus, a package roughly 0.00001 m in diameter. Despite this compactness, information in DNA is readily accessed and duplicated on command. The ability to store large amounts of information on molecules and to access it readily is still far beyond modern information technology. The capacity of nucleic acids to maintain and transmit the archived information efficiently arises directly from their chemical structure.

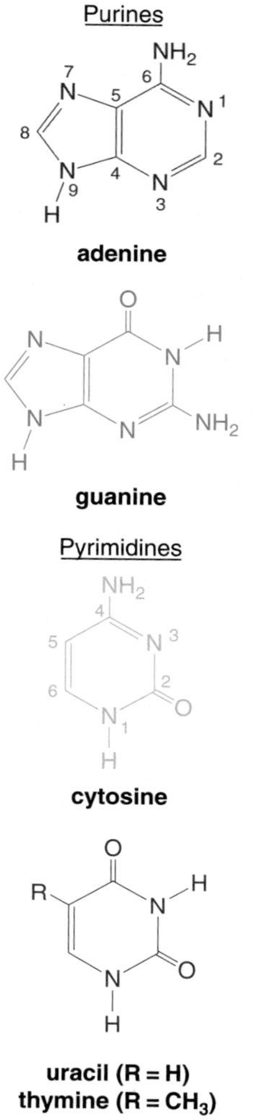

2.2 • STRUCTURAL COMPONENTS OF NUCLEIC ACIDS: NUCLEOBASES, NUCLEOSIDES, AND NUCLEOTIDES

Nucleic acids are linear polymers of **nucleotide** units. Each nucleotide consists of a phosphate ester, a pentose sugar, and a heterocyclic nucleobase. In RNA, the sugar is **D-ribose**; in DNA, it is the closely related sugar, **2-deoxy-D-ribose**. In either case, the base is attached to the 1-position of the sugar through a β-*N*-glycosidic bond. Two classes of major nucleobases are found in nucleic acids: **purines** and **pyrimidines** (Figure 2.2). The major purines are **guanine** and **adenine**, which occur in both DNA and RNA and are attached to the sugar at N-9. The three major pyrimidine nucleobases are **cytosine**, **uracil**, and **thymine**. Cytosine is present in both DNA and RNA. However, uracil is generally found only in RNA, and thymine is generally found only in DNA. Each pyrimidine is linked to the sugar through the N1-position. A nucleobase glycosylated with either pentose sugar is a **nucleoside.** Nucleosides that contain ribose are **ribonucleosides** (Figure 2.3) whereas those with **deoxyribose** are **deoxyribonucleosides** (Figure 2.4). Four ribonucleosides are commonly found in RNA—**adenosine (A), guanosine (G), cytidine (C),** and **uridine (U)**—and four deoxyribonucleosides in DNA—**deoxyadenosine (dA), deoxyguanosine (dG), deoxycytidine (dC), and deoxythymidine (dT)**. In addition, more than 80 minor nucleosides have been found in naturally occurring nucleic acids.

Nucleotides are phosphate esters of nucleosides (Figure 2.5). Any of the hydroxyl groups on their sugars can be phosphorylated, but the bases are not. Nucleotides that contain a phosphate monoester are nucleoside monophosphates. For example, the 5′-nucleotide of deoxycytidine is deoxycytidine 5′-monophosphate (5′-dCMP). More than one phosphate

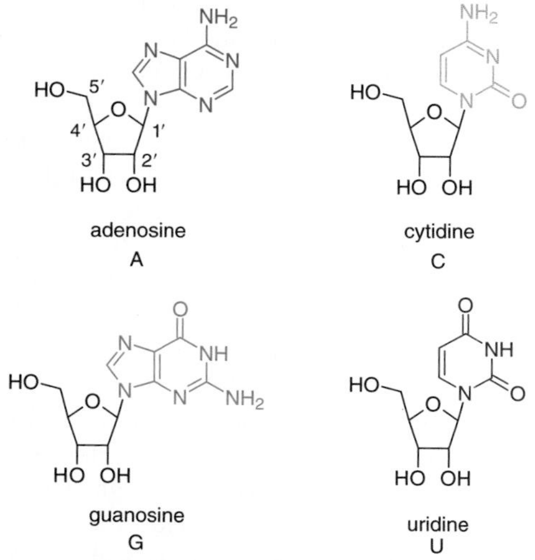

Purines

adenine

guanine

Pyrimidines

cytosine

uracil (R = H)
thymine (R = CH_3)

Figure 2.2 Major purine and pyrimidine bases found in DNA and RNA.

adenosine
A

cytidine
C

guanosine
G

uridine
U

Figure 2.3 Structures of ribonucleosides. Shown are one-letter abbreviations for each compound. These abbreviations are also used for the corresponding bases and nucleotides and, in some instances, for the deoxyribonucleosides.

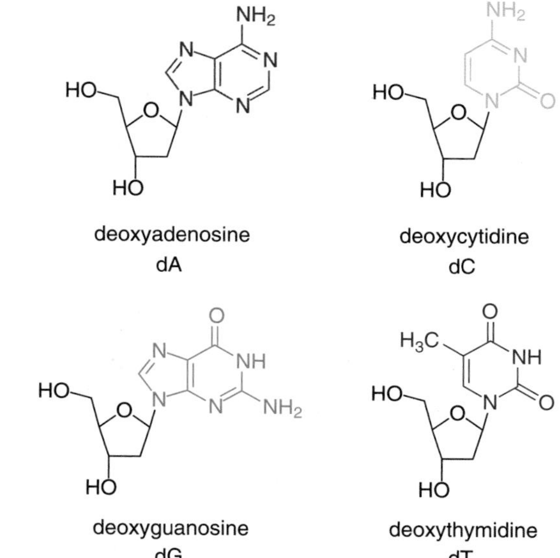

Figure 2.4 Structures of deoxyribonucleosides. Presence of 2-deoxyribose is denoted by a "d" preceding the one-letter notation. Note that thymidine (T) is often used interchangeably with deoxythymidine (dT).

CLINICAL CORRELATION 2.1

DNA Vaccines

Traditional procedures of vaccination have used components of infectious organisms, dead organisms, or attenuated intact cells to elicit production of specific antibodies that can provide individuals with active immunity. Such vaccines have successfully provided protection against diseases such as polio, smallpox, whooping cough, typhoid fever, and diphtheria. However, the development of immunizations against infectious pathogens such as HIV and malaria has proven to be difficult. The enormous worldwide impact of these pathogens, the potential bioterror uses of others, and the worsening problems of antibiotic resistance have prompted the efforts to produce new vaccines.

A recent approach to immunization has used DNA containing a sequence of the pathogen's genome. This DNA is typically a bacterial plasmid engineered to include the sequence of an antigenic protein from the pathogen. This DNA can enter a number of cell types, and it can be expressed there using cellular transcription and translation machinery. In this respect, DNA vaccines act much like viruses. However, these DNAs contain only a very limited amount of genetic information and cannot become infectious. The mechanisms of uptake and induction of the immune response are not yet clear. However, promising results have been observed against viruses, bacteria, and parasitic microorganisms. This approach holds great promise for the development of effective vaccines against intractable diseases including HIV/AIDS, tuberculosis, and malaria.

DNA vaccines may also be useful for vaccination against cancers. Although the antigens presented by tumor cells are only weakly immunogenic, model studies using plasmid DNA have shown promising results. Mice vaccinated by oral delivery of plasmids grown within attenuated *Salmonella* bacteria were able to slow or completely stop the growth of a lethal dose of carcinoma cells. Death of the bacteria presumably releases large numbers of the plasmids that are taken up by antigen-presenting cells of the immune system. Extension of this work to humans is currently under investigation.

Kutzler, M. A., and Weiner, D. B. DNA vaccines: Ready for prime time? *Nature Rev. Genet.* 9:776, 2008; Belakova, J., Horynova, M., Krupka, M., Weigl, E., and Raska, M. *Arch. Immol. Ther. Exp.* 55:387, 2007; Cui, Z. DNA vaccine. *Adv. Genet.* 54:257, 2005; and Chiarella, P., Massi, E., Robertis, M., Fazio, V. M., and Signori, E. Strategies for effective naked-DNA vaccination against infectious diseases. *Recent Pat. Anti-Infect. Drug Discovery* 3:93, 2008.

can be linked by an anhydride linkage, resulting in the corresponding di-, tri-, and tetraphosphate esters. The 5′-triphosphate of guanosine is guanosine 5′-triphosphate (GTP). **Phosphate diesters** are also possible including the important second messengers—adenosine-3′, 5′-cyclic monophosphate (cAMP) and guanosine-3′, 5′-cyclic monophosphate (cGMP).

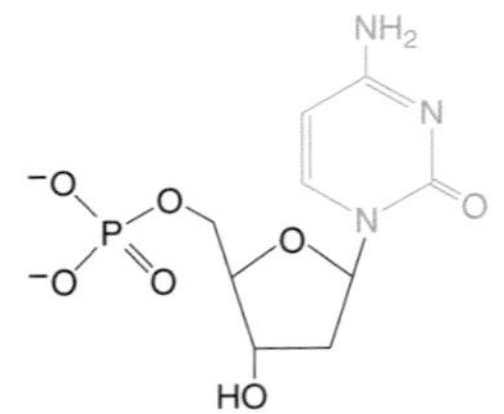

deoxycytidine 5′-monophosphate
dCMP

guanosine 5′-triphosphate
GTP

adenosine 3′,5′-cyclic monophosphate
cAMP

Figure 2.5 **Structures of some representative nucleotides.**

Physical Properties of Nucleosides and Nucleotides

To varying degrees, nucleobases, nucleosides, and nucleotides are soluble in water over a wide range of pHs. At biologically relevant pHs (pH 6–8), nucleobases and nucleosides are neutral species. The first p*Ka*s of phosphate mono- and diesters are ~1; thus, nucleotides carry a negative charge at physiological pHs. The second p*Ka* of a phosphate monoester is ~6.5, and an equilibrium exists between the monoanion and the dianion at neutral pHs. Nucleotides with di- and triphosphates carry multiple negative charges. The presence of charged phosphate groups on nucleotides provides sites for electrostatic interactions with positively charged sites on proteins and metal ions.

Nucleosides and nucleotides are stable over a wide range of pHs. Under strongly basic conditions, hydrolysis of phosphate esters slowly occurs. *N*-Glycosidic bonds of nucleosides and nucleotides are stable under these conditions, but under acidic conditions they are considerably more labile. At elevated temperatures, protonation of purine bases (G and A) results in rapid scission of the bond between the sugar and the base. Pyrimidine (C, T, and U) nucleosides and nucleotides are much more resistant to acid treatment. Conditions that break the glycosidic bond (e.g., 60% perchloric acid at 100°C) also lead to destruction of the sugar.

Molecules that contain purine or pyrimidine bases strongly absorb UV light. Purines and purine-containing nucleosides and nucleotides have higher molar extinction coefficients (absorb light more strongly) than pyrimidine derivatives. The UV wavelength where maximum absorption occurs varies but is usually near 260 nm. Because of the high extinction coefficients of nucleobases and their high concentration in nucleic acids, the absorbance at 260 nm can be used to accurately quantitate the amount of nucleic acid (DNA or RNA) present in a sample. Further analysis and separation of nucleobases, nucleosides, nucleotides, and nucleic acids can be accomplished by use of a variety of techniques, including high-performance liquid chromatography (HPLC), thin-layer chromatography (TLC), paper chromatography, and electrophoresis.

Structural Properties of Nucleosides and Nucleotides

A striking feature of nucleosides and nucleotides is the considerable number of possible conformations that they can adopt. Unlike six-membered rings, five-membered rings are highly flexible. The five-ring atoms in a pentose sugar are not coplanar; typically one or two of the atoms of the ring twist out of plane. In cyclopentane rings, there are several envelope and half-chair conformations that rapidly interconvert. Because of the asymmetric substitution pattern of the pentose ring in nucleosides, two conformations are preferred (Figure 2.6). These modes of **sugar puckering** are defined by the displacement of the 2′- and 3′-carbons above the plane of the C1′–O4′–C4′ atoms. By convention, above is the direction in which the base and C5′ project from the ring and is termed the endo face of the pentose. If C2′ is displaced above the pentose ring, the conformation is C2′-endo. In the second common pucker, C3′ is displaced toward the endo face and is called C3′-endo. Notably, the groups attached to the sugar have very different orientations in each of these conformations. For example, 5′- and 3′-phosphate groups are much farther apart in the C2′-endo pucker than in the C3′-endo pucker. The orientation of the glycosidic bond also changes significantly in the two conformations. C2′-endo and C3′-endo conformations are in rapid equilibrium. An electronegative substituent at the 2′-position of the pentose favors the C3′-endo conformation. Thus, ribonucleosides in RNA prefer this sugar pucker because they have a 2′-hydroxyl group. The 2′-deoxynucleosides of DNA contain a hydrogen in place of the 2′-OH group, and the C2′-endo conformation is preferred.

The bases in nucleosides are planar. Although free rotation around the glycosidic bond is possible, two orientations of the base with respect to the sugar predominate (Figure 2.7). In purines, the **anti conformation** places H8 over the sugar, whereas the **syn conformation** positions this atom away from the sugar and the bulk of the bicyclic purine over the sugar. Purines rapidly interconvert between the two conformations but favor the anti orientation. However, guanine 5′-nucleotides are exceptions. In these cases, favorable interactions between the 2-NH_2 group and the 5′-phosphate group stabilize the syn conformation. 2′-Deoxyguanosine 5′-monophosphate (dGMP), for example, prefers the syn glycosidic conformation. This preference has also been observed in double-stranded DNA with sequences of alternating Gs and Cs. The syn conformation of the G residues in such DNA results in formation of an unusual left-handed helix, termed Z-DNA (p. 38). In pyrimidines, the H6 atom is above the pentose ring in the anti conformation, and the larger O2 atom is above the ring in the syn glycosidic conformation. Pyrimidines, therefore, show a large preference for the less sterically hindered anti conformation.

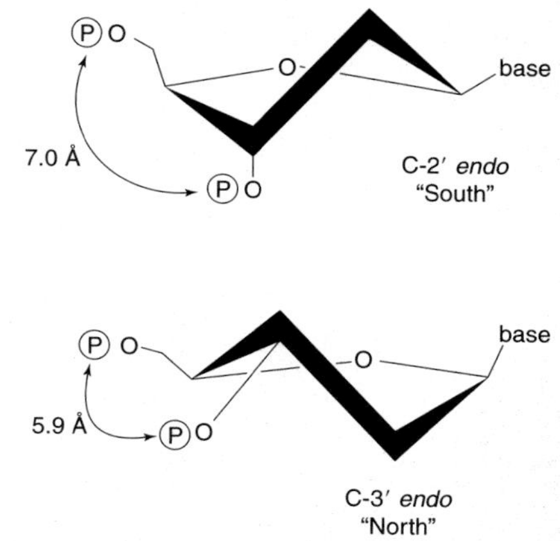

Figure 2.6 Preferred conformations of pentose sugars. The two conformations produce variations in relative orientation of the base (with respect to the sugar) and in the distance between 3′- and 5′-phosphate groups (P). Ultimately, these differences effect overall conformation of the double-helical complex.

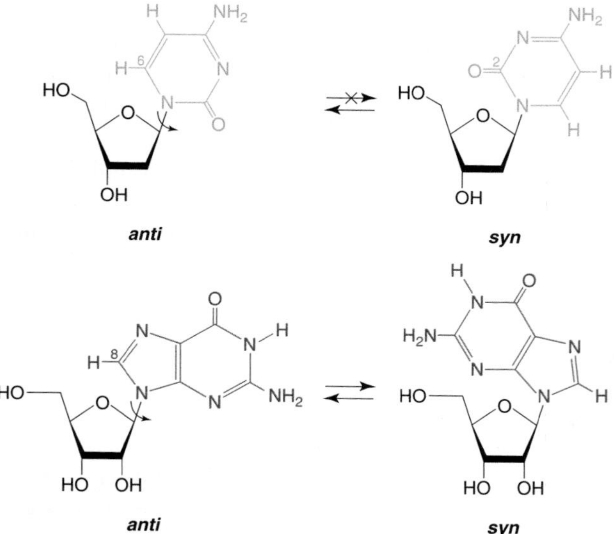

Figure 2.7 Glycosidic conformations of purines and pyrimidines. In pyrimidines, steric clashes between the sugar and the O2 of the base strongly disfavor the *syn* conformation. In purines, the *anti* and *syn* conformations readily interconvert, with *anti* being more stable in most cases. The *syn* conformation is stabilized in guanosine 5′-phosphates because of favorable interactions between the 2-NH_2 group and the phosphate oxygens.

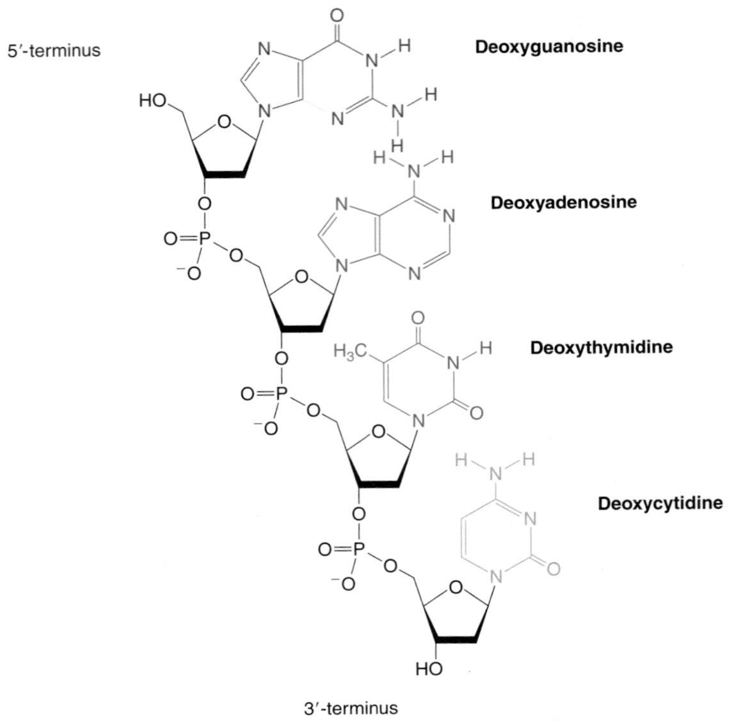

Figure 2.8 Structure of a DNA polynucleotide. Shown is a tetranucleotide. Generally, nucleic acids less than 50 nucleotides long are referred to as oligonucleotides. Longer nucleic acids are called polynucleotides.

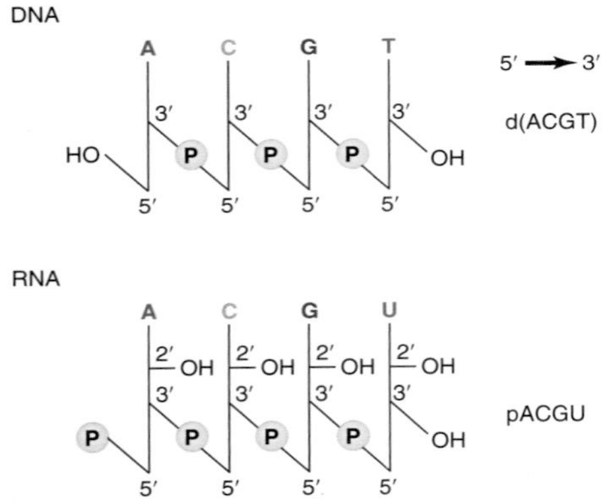

Figure 2.9 Shorthand notations for structures of oligonucleotides. The convention used in writing the structure of an oligo- or polynucleotide is a perpendicular bar represents the sugar moiety, with the 5′-OH position of the sugar located at the bottom of the bar and the 3′-OH (and 2′-OH, if present) at a midway position. Bars joining 3′- and 5′-positions represent the 3′→5′-phosphodiester bond, and the P on the left or right side of the perpendicular bar represents a 5′-phosphate or 3′-phosphate ester, respectively. The base is represented by its initial placed at the top of the bar. An alternative shorthand form is to use the one-letter initials for the bases written in the 5′→3′ direction from left to right. Internal phosphodiester groups are assumed, and terminal phosphates are denoted with a "p". Oligonucleotide sequences containing deoxyribose sugars are preceded by a "d".

2.3 • STRUCTURE OF DNA

Polynucleotide Structure and Properties

Nucleic acids are strands of nucleotides linked by **phosphodiester** bonds (Figure 2.8). The length of these strands varies considerably, ranging from two residues to hundreds of millions of residues. Typically, strands of nucleic acids containing 50 or fewer nucleotides are called **oligonucleotides,** whereas those that are longer are called **polynucleotides.** The phosphodiester bond links the 5′-hydroxyl group of one residue to the 3′-hydroxyl group of the next. Linkages between two 5′-OHs or two 3′-OHs are not seen in naturally occurring DNA. The **directionality** of this bond means that linear oligo- and polynucleotides have one end that terminates in a 5′-OH and another that terminates in a 3′-OH. These ends are the **5′-terminus** and **3′-terminus,** respectively. Circular polynucleotides have no free termini, and they are formed by joining the 5′-terminus of a linear polynucleotide with its own 3′-terminus through a phosphodiester bond. In both linear and circular polynucleotides, the sugar–phosphodiester backbone is highly uniform, and chemical diversity arises primarily from the bases. The **sequence** of bases provides each molecule with a unique chemical identity (Figure 2.9). This arrangement is analogous to that in polypeptides and proteins (p. 80) where variations in chemical structure primarily arise from the identity of the amino acid side chains.

Polynucleotide Conformations

The bases are largely responsible for the conformations of polynucleotides. The edges of the bases contain nitrogen and oxygen atoms (—NH_2, =N-, and =O groups) that can interact with other polar groups or surrounding water molecules. The faces of the rings, however, cannot participate in such interactions and tend to avoid contact with water. Instead they interact with one another, producing a stacked conformation. **Base stacking** reduces the hydrophobic surface area that must be solvated by polar water molecules. Release of these water molecules into the bulk solvent is entropically favorable. The stacked arrangement of bases is further stabilized by favorable electronic interactions (van der Waals forces). Like other aromatic systems, the bases possess highly delocalized π-orbitals above and below the planes of their rings. This electron density can be polarized by nearby dipoles.

The resulting induced dipole can then favorably interact with the polarizing group. These weak interactions are further enhanced by attractive London dispersion forces (induced dipole-induced dipole). Because the strength of these electronic interactions is strongly dependent on distance, no empty space remains between the stacked bases.

Polynucleotides adopt conformations that maximize favorable stacking interactions between neighboring bases. The constraints imposed by the structure of the sugar-phosphodiester backbone favor helical structures (Figure 2.10). The overall stability of these helical structures is dependent on factors that include sequence-dependent base stacking, pH, salt concentration, and temperature. For example, some synthetic polynucleotides, including poly(C) and poly(A), form right-handed helices in which the bases are highly stacked. Many other polynucleotides remain in a largely disordered **random-coil** conformation. Large polynucleotides may have both helical and disordered regions depending on the local sequence. The characteristics of the solution also determine whether polynucleotides adopt stacked conformations. For example, high concentrations of cations, especially polyvalent ions such as Mg^{2+}, stabilize the helical conformation by shielding the charges of the phosphodiester groups in the backbone. Without this shielding, the negative charges on the phosphates would destabilize the helix by electrostatic repulsion.

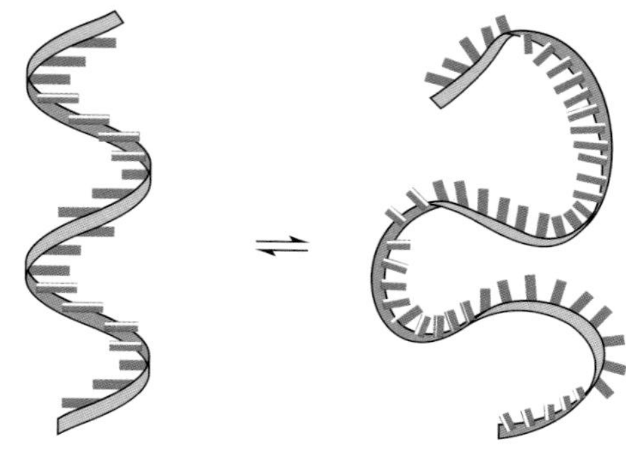

Figure 2.10 Stacked and random-coil conformations of a single-stranded polynucleotide. The helical band represents the sugar–phosphate backbone of the polynucleotide. Bases are shown in a side view as lines in contact with their neighbors. Unstacking of bases leads to a more flexible structure with bases oriented randomly.

Stability of the Polynucleotide Backbone

Polynucleotides are relatively stable in aqueous solutions near neutral pH. It has been estimated that the half-life for the spontaneous hydrolysis of phosphodiester linkages in DNA is about 200 million years. This high stability makes DNA suitable for the long-term storage of genetic information. In contrast, RNA is much more prone to hydrolysis. The presence of the 2′-OH group provides an internal nucleophile for transesterification of the 3′, 5′-phosphodiester linkage (Figure 2.11). The result is scission of the polynucleotide backbone, leaving a 2′, 3′-cyclic phosphate diester on one fragment and a free 5′-OH group on the other. The greater hydrolytic lability of RNA makes it less suitable as a genetic material than DNA.

While the DNA backbone is relatively stable to hydrolysis, numerous enzymes (**nucleases**) catalyze phosphodiester scission. These nucleases are characterized by the types of polynucleotides they hydrolyze and the specific bonds that are broken. Exonucleases cleave the last nucleotide residue at either the 5′- or 3′- terminus of a polynucleotide. Stepwise removal of individual nucleotides from one end of a polynucleotide can result in its complete degradation. Endonucleases cleave phosphodiester bonds located in the interior of polynucleotides. They do not require a free terminus; therefore, they can also cleave circular polynucleotides. Endonucleases such as DNase I and DNase II hydrolyze DNA with little sequence selectivity. The **restriction endonucleases** recognize and cleave very specific sequences. Nucleases also exhibit specificity with respect to the

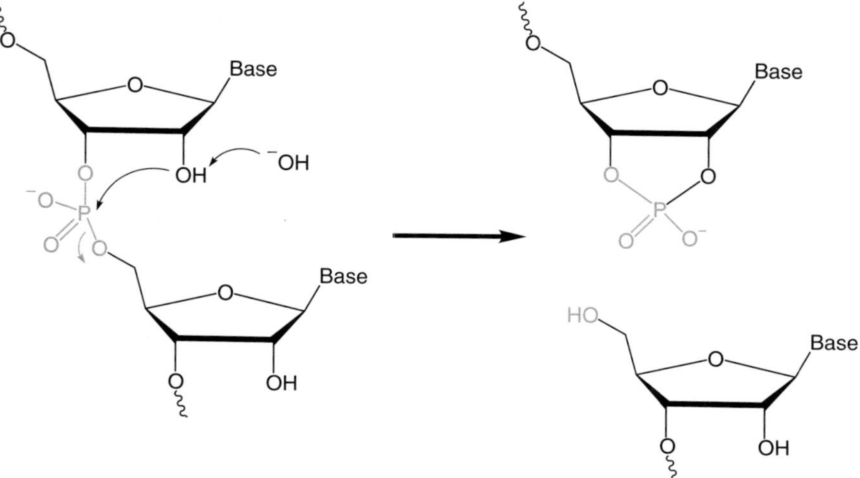

Figure 2.11 Hydrolysis of RNA is accelerated by participation of the 2′-OH group. Intramolecular nucleophilic attack by the 2′-OH group of ribose greatly increases the rate of hydrolytic cleavage of the phosphodiester backbone, especially under basic conditions. The resulting 2′, 3′-cyclic phosphodiester can subsequently react with water or hydroxide to a mixture of 2′- and 3′-phosphate monoesters.

overall structure of polynucleotides. For instance, some nucleases act on both single- and double-stranded polynucleotides, whereas others discriminate between these two structures. Some exclusively act on either DNA or RNA, whereas others are active toward only one type of polynucleotide.

Double-Helical DNA

With the recognition in the early- and mid-twentieth century that DNA served as the carrier of genetic information, work to establish the structural basis for information storage intensified.

Synthetic and X-ray diffraction experiments led to the acceptance of a linear polynucleotide structure for DNA. Three key pieces of information were necessary for deduction of the three-dimensional structure of DNA. The first was the determination that DNA did not contain equal amounts of the four nucleosides, but variable amounts in different organisms. However, it was found that the abundance of deoxyadenosine always equaled that of deoxythymidine and the abundance of deoxyguanosine always equaled that of deoxycytidine. This led to the consideration of structures with the nucleosides specifically paired together (dA with dT and dG with dC). The second was that X-ray diffraction data suggested that DNA contained **double-helical** structures, and symmetry suggested that the two polynucleotide strands were oriented **antiparallel** to each other. The third was the suggestion that the nucleobases were in the keto and amino **tautomeric** forms rather than enol and imino forms. Tautomers are isomers of a molecule that differ only in the position of a hydrogen atom. Each of the four nucleobases has two or more possible tautomeric forms that are in equilibrium (Figure 2.12). The incorrect assignment of the predominant tautomeric structures

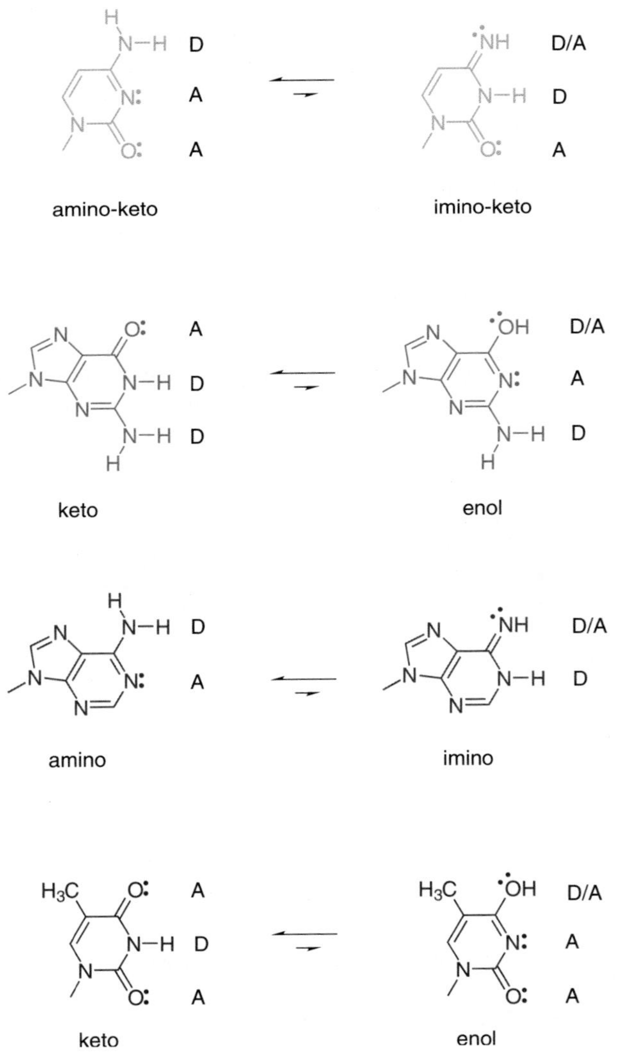

amino-keto D A A ⇌ imino-keto D/A D A

keto A D D ⇌ enol D/A A D

amino D A ⇌ imino D/A D

keto A D A ⇌ enol D/A A A

Figure 2.12 Tautomeric forms of nucleobases. The pattern of hydrogen bonding donating (D) and accepting (A) groups changes depending on the tautomer of the base that is present. Shown are predominant tautomeric forms (left) and one of the alternate forms (right) for each base. The hydrogen bonding pattern for each tautomer is also shown; ambiguity at the —OH and =NH positions of the minor tautomers arises from rotation or isomerism of these groups. The equilibrium ratio of the predominant form to all others is greater than 99:1.

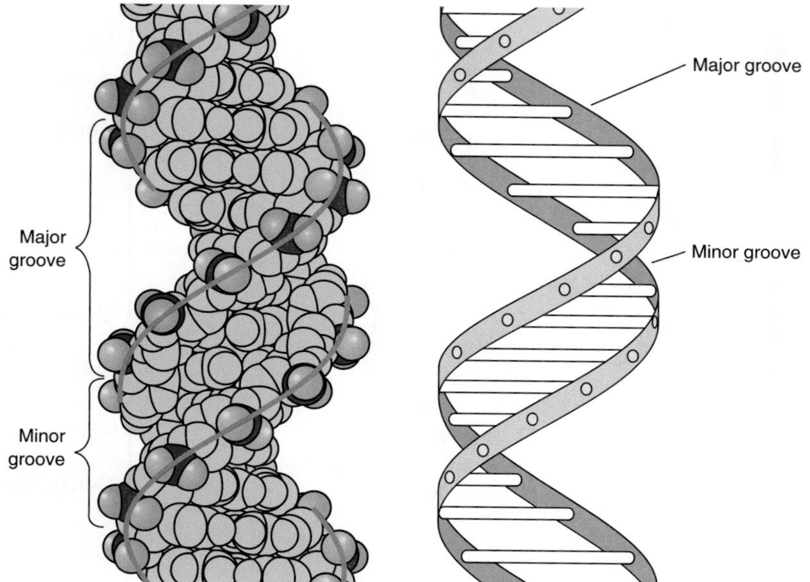

Figure 2.13 Watson–Crick model of DNA. On the left is a space-filling model of DNA; on the right is an idealized ribbon model. Bases are stacked in the interior of the helix, whereas hydrophilic sugar–phosphodiester backbone is located on the exterior.
Redrawn from Rich, A. J. *Biomol. Struct. Dyn.* 1:1, 1983.

meant that researchers were attempting to form base pairs using incorrect hydrogen-bonding patterns. With the correct tautomers, a model for DNA rapidly fell into place.

The structure that **Watson and Crick** proposed in 1953 for double helical DNA was attractive because of its simplicity and symmetry (Figure 2.13). Moreover, it explained all of the available structural data for DNA and immediately led to hypotheses regarding a mechanism for storage and replication of genetic information. The Watson–Crick double helix can be visualized as the interwinding of two right-handed helical polynucleotide strands around a common axis. The strands achieve contact through hydrogen bonds formed at the hydrophilic edges of the bases. The N—H groups of the nucleobases are good hydrogen bond donors, and the electron pairs on the sp^2-hybridized oxygens of the C=O groups and nitrogens of the =N– groups are good hydrogen bond acceptors. Pairing occurs when an acceptor and a donor are in a position to form a hydrogen bond. These bonds extend between purine residues in one strand and pyrimidine residues in the other, and the matching of hydrogen bond donors and acceptors results in two types of **base pairs:** adenine–thymine and guanine–cytosine (Figure 2.14). A direct consequence of these hydrogen-bonding specificities is that **double-stranded DNA (dsDNA)** must contain ratios of nucleosides that agree with experimental observations (dA = dT and dG = dC). Finally, the geometries of the dA/dT and dG/dC base pairs result in similar C1′–C1′ distances (~10.6 Å) and glycosidic bond orientations. This **structural isomorphism** means that any of the four possible base pairs (dA-dT, dT-dA, dG-dC, and dC-dG) can be placed into the double helix without significantly changing the structure of the backbone; note how this leads to the possible replacement of one base pair by any other, a key to understanding how DNA information can change, or **mutate.**

The relationship between bases on opposing strands within the double helix is described as **complementarity.** Bases are complementary because every nucleobase of one strand is matched by shape and hydrogen bonding to a complementary base on the other strand (Figure 2.15). For instance, for each adenine projecting toward the axis of the double helix, a thymine must be projected from the opposite strand to hydrogen-bond and exactly fill the space between the strands. Neither cytosine nor guanine fits precisely in the available space across from adenine in a manner that allows formation of hydrogen bonds across strands. These hydrogen-bonding specificities ensure that the entire base sequence of one strand is complementary to that of the other strand.

The exterior of the double helix consists of the sugar–phosphate backbones of its component strands. The two strands are aligned in opposite directions; if two adjacent nucleobases in the one strand, for example, thymine and cytosine, are connected in the 5′→3′ direction, their complementary nucleobases on the other strand, adenine and guanine, will

major groove
adenine
thymine
minor groove

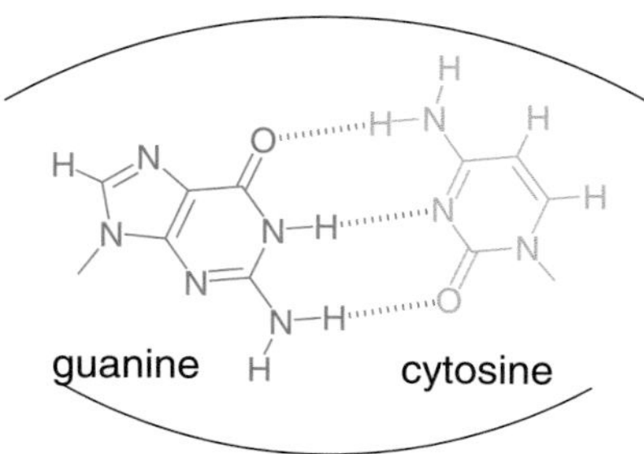

Figure 2.14 Watson–Crick base pairs. Selective base pairs are formed between adenine and thymine and between guanine and cytosine. Note the formation of two hydrogen bonds in an A-T base pair and three in a G-C base pair.

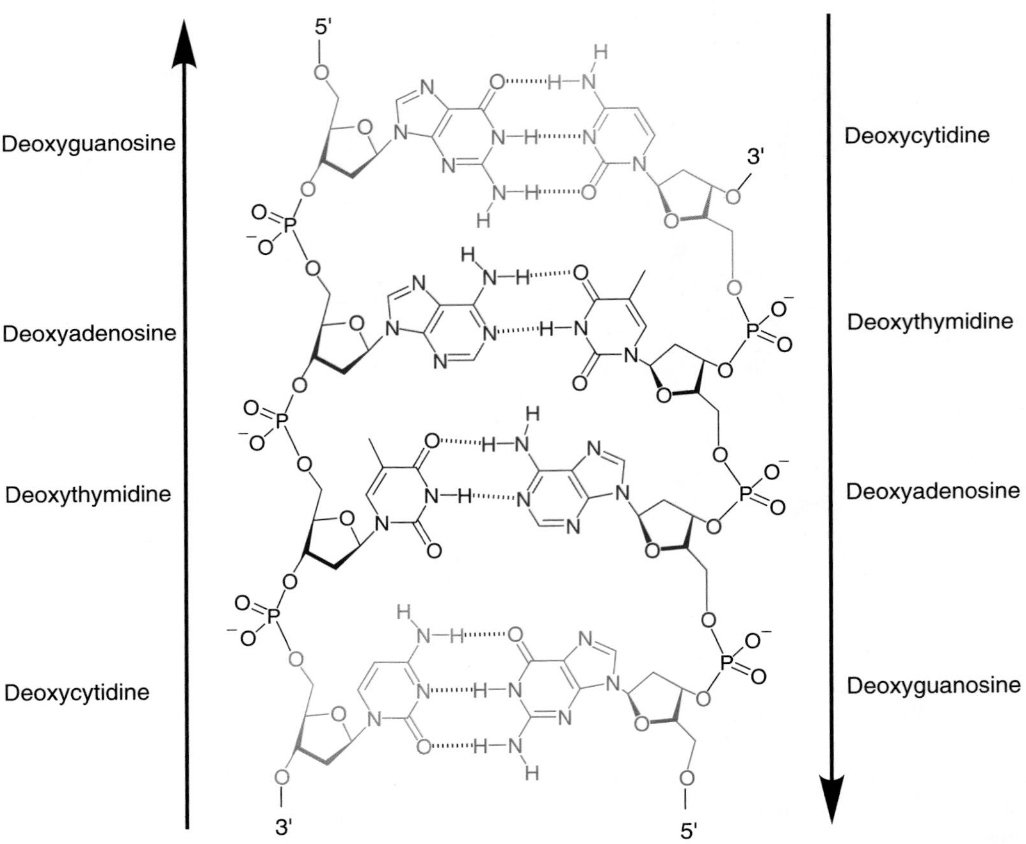

Figure 2.15 Formation of hydrogen bonds between complementary bases in double-stranded DNA. Interaction between polynucleotide strands is highly selective. Complementarity depends not only on the geometric factors that allow the proper fitting between the complementary bases of the two strands but also on the formation of specific hydrogen bonds between complementary bases. Note the antiparallel orientation of the strands of a double-stranded DNA. The geometry of the helices does not prevent a parallel alignment, but such an arrangement is not found in DNA.

be linked in the $3' \rightarrow 5'$ direction. This antiparallel alignment produces a stable association between strands to the exclusion of the alternate parallel arrangement.

Interwinding of the two antiparallel strands produces a structure that has two distinct helical grooves between the sugar–phosphate backbones (Figures 2.13 and 2.14). The **major groove** is much wider than the **minor groove**. This disparity arises from the geometry of the base pairs: The glycosidic bonds between the bases and the backbone pentose are not arranged directly opposite to each other but are displaced toward the minor groove. Significantly, the nucleotide sequence of DNA can be discerned without dissociating the double helix by looking inside these grooves. Each base always displays the same atoms into each of the grooves on the double helix. These atoms then constitute an important means of sequence-specific recognition of DNA by proteins and small molecules. For example, N7 of purines is always displayed in the major groove, and it can serve as a hydrogen bond acceptor in interactions with donor groups on proteins (Figure 2.14). Similarly, the exocyclic 2-NH_2 group of guanine always projects into the minor groove and can form a steric blockade to the binding of small molecules.

Factors That Stabilize Double-Helical DNA

Stacking interactions that stabilize the helical structures of single-stranded polynucleotides are also instrumental in stabilizing the double helix. The separation between the hydrophobic core of the stacked nucleobases and the hydrophilic exterior of the charged sugar–phosphate groups is even more pronounced in the double helix than in single-stranded helices. The stacking tendency of single-stranded polynucleotides can be viewed as resulting from a tendency of the bases to reduce their contact with water. The double-stranded helix is a more

TABLE 2.1 • Base-Pair Stacking Energies

Dinucleotide Base Pairs	*Stacking Energies per Pair*[a]	
	kJ mol^{-1}	*kcal mol*$^{-1}$
(GC)•(GC)	−61.0	−14.6
(AC)•(GT)	−44.0	−10.5
(TC)•(GA)	−41.1	−9.81
(CG)•(CG)	−40.5	−9.69
(GG)•(CC)	−34.6	−8.26
(AG)•(CT)	28.4	−6.78
(TG)•(CA)	−27.5	−6.57
(AT)•(AT)	−27.5	−6.57
(AA)•(TT)	−22.5	−5.37
(TA)•(TA)	−16.0	−3.82

[a]Data from Ornstein, R. L., Reim, R., Breen, D. L., and McElroy, R. D. *Biopolymers* 17: 2341, 1978.

favorable arrangement, essentially removing the nucleobases from the aqueous environment while permitting the hydrophilic phosphate backbone to be highly solvated by water.

Stacking interactions, a combination of hydrophobic forces and van der Waals interactions, are 16–61 kiloJoules per mole (kJ/mol; 4–15 kilocalories per mole [kcal/mol]) of stabilization energy for each adjacent base pair (Table 2.1). Additional stabilization of the double helix results from extensive networks of cooperative hydrogen bonds. Generally, these bonds are weak 10–30 kJ/mol (3–7 kcal/mol), and they are even weaker in DNA 8–12 kJ/mol (2–3 kcal/mol) because of geometric constraints within the double helix. It might appear that hydrogen bonds between nucleobases could cumulatively provide substantial stabilization for the double helix. However, the hydrogen bonds in a double helix merely replace energetically similar ones between the nucleobases and water in single-stranded polynucleotides. Therefore, hydrogen bonding is not the glue that maintains double-helical structures. Yet, in contrast to stacking forces, hydrogen bonds are highly directional and help discriminate between correct and incorrect base pairs. Because of their directionality, hydrogen bonds tend to orient the nucleobases to favor stacking. Therefore, the contribution of hydrogen bonds is indirectly vital for the stability of the double helix.

The relative importance of stacking interactions versus hydrogen bonding in stabilizing the double helix was not always appreciated. Experiments with reagents that reduce the stability of the double helix (denaturants) illustrate the greater relative importance of stacking interactions (Table 2.2). These results show that the destabilizing effect of a reagent is not related to its ability to break hydrogen bonds but is determined by the solubility of the

TABLE 2.2 • Effects of Various Reagents on the Stability of the Double Helix[a]

Reagent	*Adenine Solubility* $\times 10^{-3}$ *(in 1M Reagent)*	*Molarity Producing 50% Denaturation*
Ethylurea	22.5	0.60
Propionamide	22.5	0.62
Ethanol	17.7	1.2
Urea	17.7	1.0
Methanol	15.9	3.5
Formamide	15.4	1.9

Source: Data from Levine, L., Gordon, J., and Jencks, W. P. *Biochemistry* 2:168. 1963.

[a]The destabilizing effect of the reagents listed on the double helix is independent of the ability of these reagents to break hydrogen bonds. Rather, the destabilizing *effect* is determined by the solubility of adenine. Similar results would be expected if the solubilities of the other bases were examined.

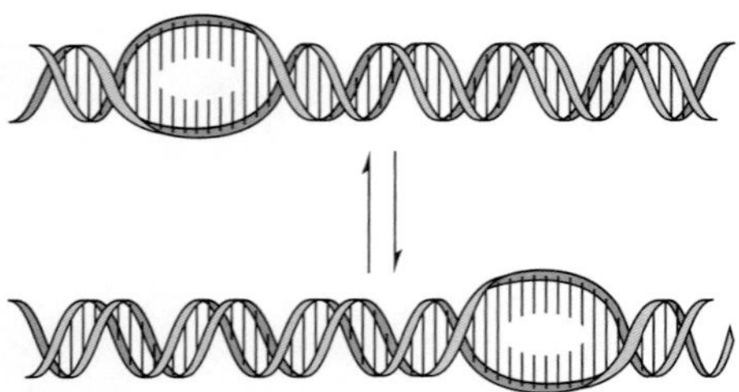

Figure 2.16 Migration of bubbles through double-helical DNA. DNA contains short open-stranded sections of DNA that can "move" along the helix.

free bases in the reagent. As the reagent becomes a better solvent for the bases, the driving force for stacking diminishes, and the double helix is destabilized.

Electrostatic forces also have an effect on the stability and conformation of the double helix. Phosphodiester groups are ionized at physiological pH; thus, the exterior of the double helix carries two negative charges per base pair. The interstrand electrostatic repulsion between negatively charged phosphates is destabilizing and tends to separate the complementary strands. In distilled water, DNA strands separate at room temperature. However, cations such as Mg^{2+}, spermine^{4+} (a tetramine), and basic amino acid side chains in proteins can shield the phosphate groups and decrease repulsive forces.

Denaturation and Renaturation

The double helix is disrupted during almost every important biological process in which DNA participates, including DNA replication, transcription, repair, and recombination. The forces that hold the two strands together are adequate for providing stability but weak enough to allow facile strand separation. The double helix is stabilized relative to the single strands by about 4 kJ/mol (1 kcal/mol) per base pair so that a relatively minor perturbation can produce a local disruption in a short section of the double helix. These base pairs can close up again, releasing free energy that can then cause the adjacent base pairs to unwind. In this manner, minor disruptions of the double helix can migrate along its length (Figure 2.16). Thus, at any particular moment, the large majority of bases remain hydrogen bonded, but all bases can transiently pass through the single-stranded state. The ability of the DNA double helix to breathe is an essential prerequisite for its biological functions.

Separation of DNA strands can be studied by increasing the temperature of a solution. At relatively low temperatures a few base pairs are disrupted, creating one or more open-stranded bubbles (Figure 2.17). These bubbles form initially in sections that contain relatively higher proportions of adenine–thymine pairs because of the lower stacking energies of such pairs. As the temperature is raised, the size of the bubbles increases, and the thermal motion of the polynucleotides eventually overcomes the forces that stabilize the double helix. At even higher temperatures, the strands separate and acquire a random-coil conformation. The process is most appropriately described as a helix-to-coil transition, but is commonly called denaturation or **melting**.

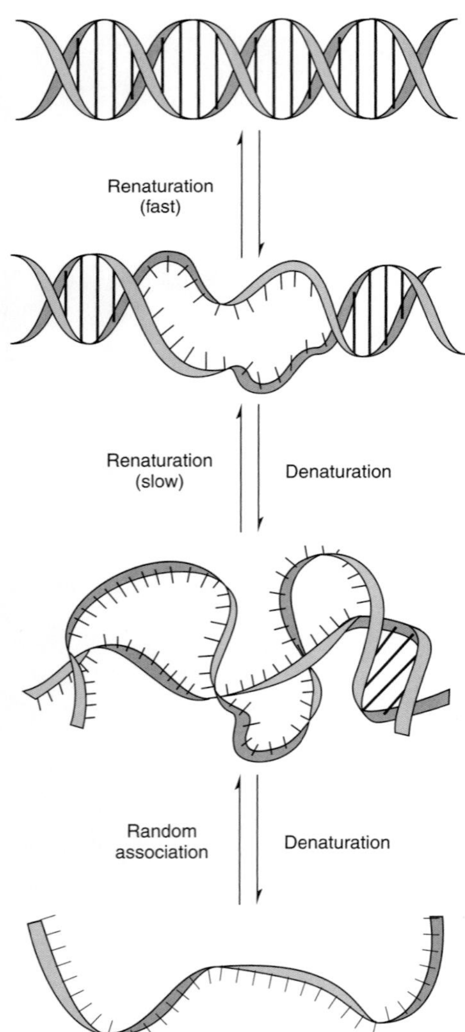

Figure 2.17 Denaturation of DNA. At high temperatures the double-stranded structure of DNA is completely disrupted, with eventual separation of strands and formation of single-stranded open coils. Denaturation also occurs at extreme pH ranges or at extreme ionic strengths.

Denaturation is accompanied by several physical changes in the DNA-containing solution including an increase in buoyant density, a reduction in viscosity, a change in ability to rotate polarized light, and changes in absorbance of UV light. The last are frequently used to follow the process of denaturation experimentally (Figure 2.18). Because of the strong absorbance by the purine and pyrimidine bases, DNA absorbs strongly in the UV region with a maximum near 260 nm. However, the absorbance by individual bases is reduced by electronic interactions between stacked bases. The total absorbance may be reduced by as much as 40% compared to that in an unstacked state. This reduction in the extinction coefficient of the aggregate bases is termed **hypochromicity**. Stacking interactions decrease gradually as the ordered structure of the double helix is disrupted by increasing temperatures. Therefore, a completely disordered polynucleotide approaches an absorbance comparable to the sum of the absorbances of its purine and pyrimidine constituents.

Measurement of the absorbance of a DNA complex at 260 nm while slowly increasing the temperature provides a means to observe denaturation. In a thermal denaturation experiment monitored using absorption spectroscopy; the polynucleotide absorbance typically changes very slowly at first, and then rises rapidly to a maximum value. Before the rise, the DNA is double stranded. In the rising section of the curve an increasing number of base pairs are interrupted. Complete strand separation occurs at temperatures corresponding to the upper plateau of the curve. The midpoint temperature of this transition, the melting temperature ($\mathbf{T_m}$), is characteristic of the base content of DNA under standard conditions of concentration and ionic strength. The higher the guanine–cytosine content, the higher the transition temperature between double-stranded helix and single strands. This difference in T_m values is attributed to the increased stability of guanine–cytosine pairs, which arises from more favorable stacking interactions.

DNA becomes denatured at pH >11.3 as the N–H groups on the bases become deprotonated, preventing them from participating in hydrogen bonding. Alkaline denaturation is often used to prevent damage to the DNA that can occur at a high temperature or low pH. Denaturation can also be induced at low ionic strengths, because of enhanced interstrand repulsion between negatively charged phosphates and by various denaturing agents (compounds that can effectively hydrogen bond to the bases while disrupting hydrophobic stacking interactions). A complete denaturation curve similar to that shown in Figure 2.18 is observed at a relatively low constant temperature by variation of the concentration of an added denaturant such as urea.

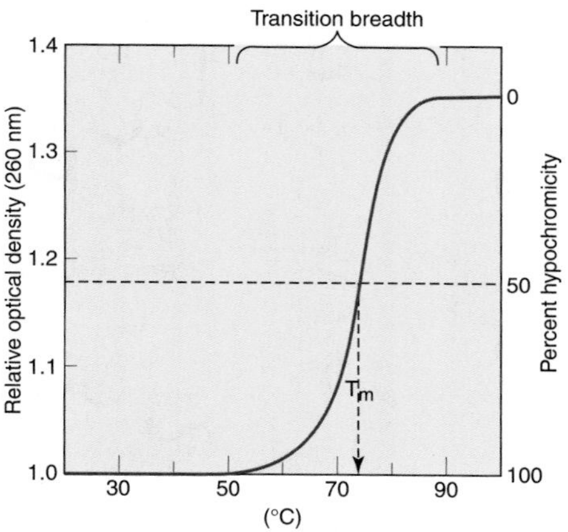

Figure 2.18 **Temperature-optical density profile for DNA.** When DNA is heated, the absorbance at 260 nm increases with rising temperature. A graph in which absorbance versus temperature is plotted is called a melting curve. Relative optical density is the ratio of the absorbance at any temperature to that at low temperatures. The temperature at which one-half of the maximum optical density is reached is the midpoint temperature T_m.

Complementary DNA strands, separated by denaturation, can reform a double helix if appropriately treated. This is called **renaturation** or **annealing**. If denaturation is not complete and a few nucleobases remain hydrogen bonded between the two strands, the helix-to-coil transition is rapidly reversible. Annealing is possible even after complementary strands have been completely separated. Under these conditions the renaturation process depends on the DNA strands meeting in a manner that can lead to reformation of the original structure. Not surprisingly, this is a slow, concentration-dependent process. As renaturation begins, some of the hydrogen bonds formed are extended between short tracts of polynucleotides that may have been distant in the original native structure. These randomly base-paired structures are short lived because the bases that surround the short complementary segments cannot pair and, thus, cannot form a stable, fully hydrogen bonded structure. Once the correct nucleobases begin to pair by chance, the double helix is rapidly reformed over the entire DNA molecule.

Sudden onset of denaturation or renaturation reveals the all-or-none nature of helix-to-coil or coil-to-helix transitions. The renaturation process requires formation of a short double-helical region to initiate formation of the double helix (Figure 2.19). This begins with the formation a single base pair that is rather unstable. However, the formation of a second neighboring base pair is enhanced because the process is now less entropically unfavorable. As new base pairs begin to form a stacked structure, the formation of subsequent base pairs is further facilitated. Unpaired nucleotides of each strand begin to stack on the growing helix, positioned optimally for base pairing. After formation of a double helix with four to five base pairs, a stable double helix forms, and the remainder of the complex will rapidly and spontaneously "zip up." After formation of the initial base pair, stacking and base pairing are not independent events but are influenced by the neighboring pairs. Such a process is called **cooperative.** The presence of a short double helix serves as a nucleation site for annealing by facilitating the formation of subsequent base pairs. Denaturation is the same process in reverse: A bubble serves as an initiation site for unstacking of the nucleobases and rapid unzipping of the helix.

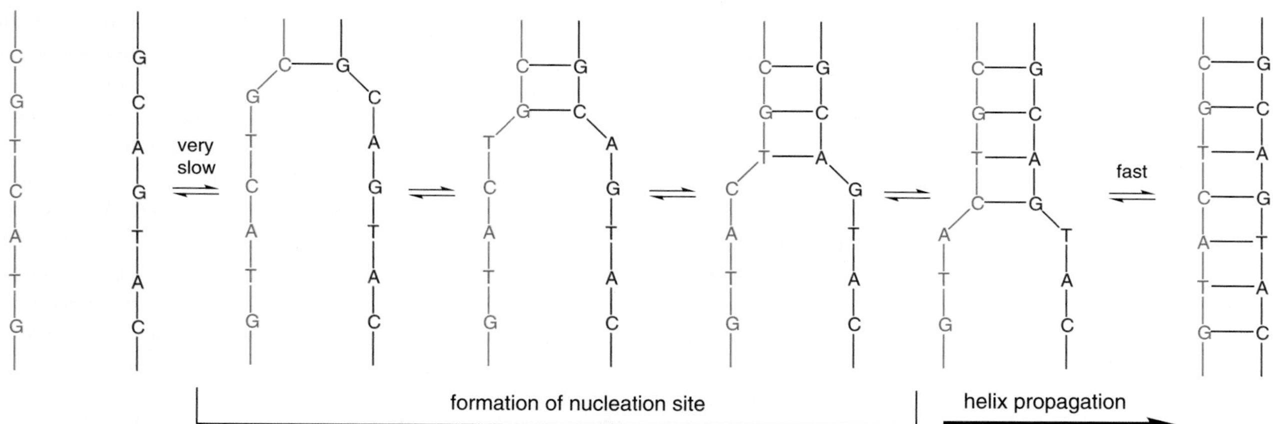

Figure 2.19 **Cooperativity of renaturation/denaturation of DNA double helices.** During renaturation, formation of the first correct base pair is very slow. Annealing of neighboring pairs is facilitated, especially after formation of a 3- to 5-bp nucleation site. Denaturation follows a similar course, but the order of steps is reversed.

Redrawn from Saenger, *W. Principles of Nucleic Acid Structure.* New York: Springer-Verlag, 1984, 141.

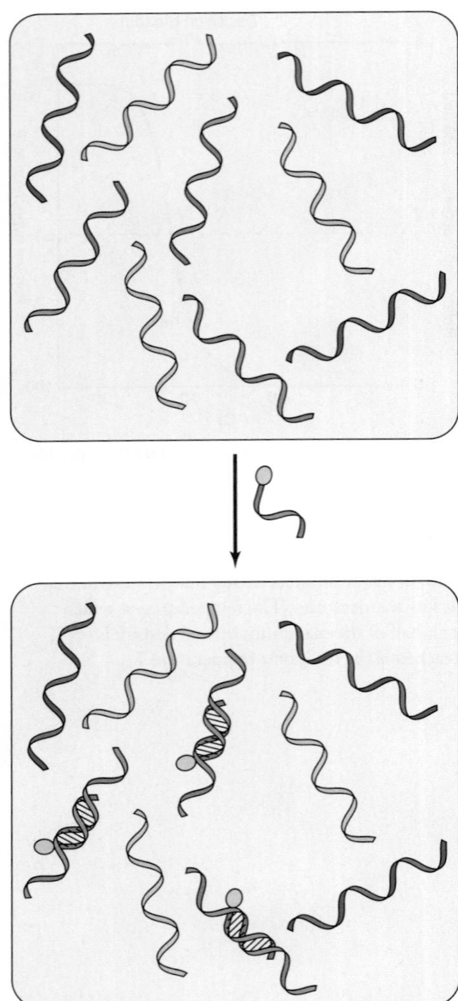

Figure 2.20 General representation of hybridization experiments. A mixture of denatured DNAs is treated with a DNA probe bearing a label. The probe can hybridize with those DNAs with complementary sequences. Detection of the double-helical complexes allows for detection and quantitation of DNA that contains the sequence of interest. Specific applications often feature steps to separate and immobilize the different DNAs in the mixture to be probed.

Hybridization

A technique based on the association of complementary polynucleotide strands, **hybridization** has been developed for the detection and quantitation of specific sequences of target nucleic acid. These are important basic tools in contemporary molecular biology and are being used for determining (a) whether a certain sequence occurs in the DNA of a particular organism, (b) a genetic or evolutionary relatedness between different organisms, (c) the number of genes transcribed in a particular mRNA, and (d) the location of any given DNA sequence. They are based on annealing a complementary polynucleotide, called a **probe,** which is appropriately tagged for easy detection of the hybrid double helix.

In a typical experiment, DNA or RNA to be tested for hybridization is denatured and immobilized by binding to a suitable insoluble matrix. Labeled DNA probes are then allowed to hybridize to complementary sequences bound to the matrix (Figure 2.20). Probes are short single-stranded RNA or DNA oligonucleotides that are complementary to specific sequences of interest. Under the proper conditions probes interact only with the segment of interest, indicating whether it is present in a particular sample of DNA. Probe molecules are generally >20 nucleotides long. Appropriate labels include radioactive elements, fluorescent chromophores, and biotin. Because the double-helical complex containing the hybridized probe is usually bound to an insoluble matrix, unhybridized probes can be washed away. Detection of bound labels allows direct quantitation of the sequence of interest. Determination of the maximum amount of DNA that can be hybridized can establish the degree of homology between the DNAs of different species since the base sequences in each organism are unique. The observed homologies serve as indices of evolutionary relatedness and have been particularly useful for defining phylogenesis in prokaryotes. Hybridization studies between DNA and RNA have provided very useful information about the biological role of DNA, particularly the mechanism of transcription. Arrays of probes are useful for definitive and rapid diagnosis of genetic disorders, infectious disease, and cancer, as described in Clin. Corr. 2.2.

Conformations of Double-Helical DNA

The early X-ray diffraction studies showed that there was more than one conformation of DNA (Figure 2.21). One of these, **A-DNA,** was found under conditions of low humidity and high salt concentration. Adding organic solvents such as ethanol reduced the humidity of these aqueous solutions. A second distinct form, B-DNA, appeared under conditions of high humidity and low salt concentration and was the basis of the Watson–Crick structure. Eleven distinct conformations of double-helical DNA have since been described. They vary in orientation of the bases relative to the helix and to each other and in other geometric parameters of the double helix. One form, Z-DNA, incorporates a left-handed helix rather than the usual right-handed variety.

The structural polymorphism of double-helical DNA depends on the base composition and on physical conditions. The local structure of DNA is sufficiently flexible to allow for changes in conformation that maximize stacking while minimizing unfavorable steric interactions. The stacking preferences of bases can favor one conformation over others. For example, consecutive guanines on one strand favor A-DNA–like conformations. The solution conditions also play a key role in determining the favored conformation. Water molecules interact differently with double helices in different conformations. For example, the phosphate groups in B-DNA are more accessible to water molecules than those in A-DNA. Also, polar groups on the bases are better hydrated in a B-DNA conformation. In fact, in AT-rich sequences, an ordered array of water molecules occupies the narrow minor groove of B-DNA (Figure 2.22*a*). With a decrease in humidity, the available water molecules solvate the highly polar phosphate groups in preference to the bases. The major groove narrows allowing water molecules to bridge the phosphates (Figure 2.22*b*), and thus stabilizing the A-DNA conformation.

The different conformations of DNA can be grouped into three families: A-DNA, **B-DNA,** and **Z-DNA.** The parameters for these conformations, listed in Table 2.3, have been determined by X-ray diffraction methods. It must be emphasized that the average overall structure of DNA in living organisms is believed to be B-DNA–like. Notably, the B-conformation, unlike the A- and Z-forms, is highly flexible. In native B-DNA, considerable local variation in conformation of individual nucleotides may occur. Such variations

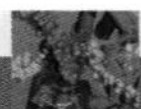

CLINICAL CORRELATION 2.2

Diagnostic Use of DNA Arrays in Medicine and Genetics

With completion of the Human Genome Project, a wealth of genetic information is rapidly becoming available. Application of this knowledge to medicine requires the development of new techniques to monitor gene expression and to analyze rapidly genes for mutations. Oligonucleotide arrays consist of a number of gene-specific oligonucleotide probes immobilized at specific sites on a solid matrix (chip). They can contain thousands of unique probe molecules, each within a spatially fixed address. Gene chips can then be treated with labeled target nucleic acids (DNA or RNA) derived from cells of an organism. Hybridization of the targets with complementary probe sequences allows for immobilization of the label at specific sites on the chip. The presence of specific sequences can be determined, and the amount of labeled target hybridized to a site can be quantitated.

Such techniques may lead to diagnostics for rapid screening of genomic DNA for disease-associated mutations. For example, high-density DNA arrays with thousands of oligonucleotide probes has been used to detect mutations leading to ataxia telangiectasia, a recessive disease characterized by neurological disorder, recurrent respiratory infection, and dilated blood vessels in the skin and eyes. Similar studies have examined mutations in the hereditary breast and ovarian cancer gene BRCA. Comparable assays can be used to identify accurately pathogens present in a clinical sample. The ability to quantitate the amounts of mRNAs in various cells allows for the profiling of gene expression. This may lead to techniques to evaluate illnesses such as cancer and select individualized treatments.

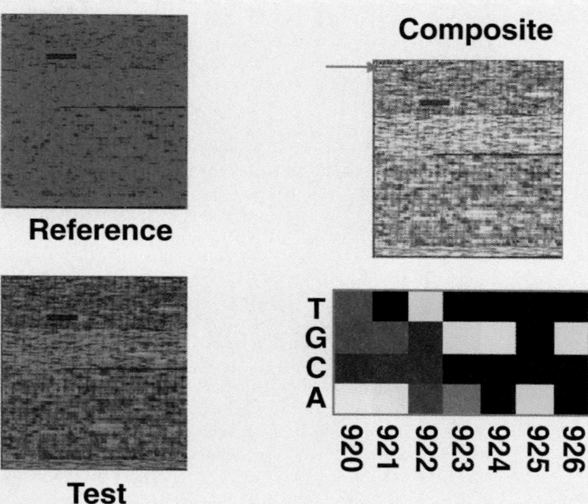

Freeman, W. M., Robertson, D. J., and Vrana, K. E. Fundamentals of DNA hybridization arrays for gene expression analysis. *BioTechniques* 29:1042, 2000; Stover, A. G., Jeffery, E., Xu, J., and Persing, D. H. Hybridization array technology. In Persing, D. H. (Ed), *Molecular Microbiology.*. Herndon, VA: ASM, 2004, 619; and Hacia, J. G., Brody, L. C., Chee, M. S., Fodor, S. P., and Collins, F. S. Detection of heterozygous mutations in BRCA1 using high density oligonucleotide arrays and two-colour fluorescence analysis. *Nat. Genet.* 14:441, 1996.

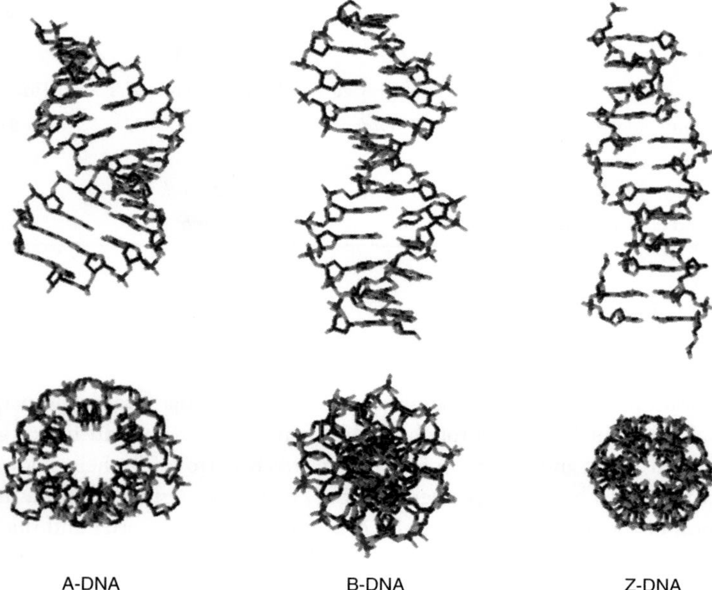

Figure 2.21 The varied geometries of double-helical DNA. Depending on conditions and base sequence, the double helix can acquire various distinct geometries. There are three main families of DNA conformations: A, B, and Z. The right-handed forms, B-DNA and A-DNA, differ in sugar pucker; this leads to differing helical structures. The A-form is underwound compared to the B-form, and the resulting helix is shorter and fatter. The Z-DNA structure is a left-handed helix with a zigzagging backbone. The sugar puckers and glycosidic conformations alternate from one residue to the next, producing a local reversal in chain direction.

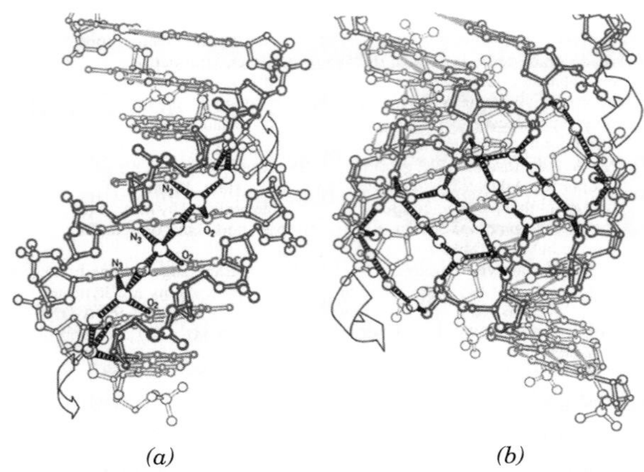

(*a*) (*b*)

Figure 2.22 Hydration of the grooves of DNA. (*a*) An organized spine of hydration fills the minor groove of B-form DNA. (*b*) The phosphates lining the major groove are spanned by a network of waters in A-form DNA.
Reproduced from Saenger, W. *Principles of Nucleic Acid Structure.* New York: Springer-Verlag, 1984, 379. With kind permission of Springer Science and Business Media.

TABLE 2.3 • Structural Features of A-, B-, and Z-DNA

Features	*A-DNA*	*B-DNA*	*Z-DNA*
Helix rotation	Right-handed	Right-handed	Left-handed
Base pairs per turn (crystal)	10.7	9.7	12
Base pairs per turn (fiber)	11	10	—
Base pairs per turn (solution)	—	10.5	—
Pitch per turn of helix	24.6 Å	33.2 Å	45.6 Å
Proportions	Short and broad	Longer and thinner	Elongated and thin
Helix packing diameter	25.5 Å	23.7 Å	18.4 Å
Rise per base pair (crystal)	2.3 Å	33 Å	3.7
Rise per base pair (fiber)	2.6	3.4 Å	—
Helix axis	Major groove	Through base pairs	Minor groove
Sugar ring conformation (crystal)	C3′ *endo*	Variable	Alternating
Sugar ring conformation (fiber)	C3′ *endo*	C2′ *endo*	—
Glycosyl bond conformation	*anti*	*anti*	*anti* at C, *syn* at G

may be important in regulation of gene expression, since they can influence the extent of DNA binding with various types of regulatory proteins.

DNA conformations in the B-family feature base pairs that are nearly perpendicular to the helical axis, which passes through the base pairs. The major and minor grooves are roughly the same depth, and the minor groove is relatively narrow. The helix is long and thin, with approximately 10 base pairs per helical turn. The rise per residue is 3.4 Å, the approximate thickness of the bases. In contrast, the A-DNA structure is shorter and thicker. There are about 11 base pairs per helical turn with a vertical rise of 2.56 Å per residue. In order to accommodate the thickness of the bases, the base pair is tilted 20° from the plane perpendicular to the helical axis. The helical axis is displaced to the major groove side of the base pairs. This results in a very deep major groove and a shallow minor groove and forms a hole ~3 Å in diameter that runs through the center of the helix. Low humidity favors the A-DNA structure that exposes more hydrophobic surface to the solvent than the B-DNA structure.

Z-DNA is a radically different left-handed, double-helical conformation for double-stranded DNA. It is generally observed in sequences of alternating purines and pyrimidines, particularly d(GC)*n*. The designation Z was chosen because the phosphodiester backbone assumes a zigzag arrangement compared to the smooth conformation that characterizes A-DNA and B-DNA. The Z-DNA structure is longer and much thinner than that of B-DNA and completes one turn in 12 base pairs. The minor groove is very deep and contains the helical axis. The base pairs are displaced so far into the major groove that a distinct channel no longer exists. These changes place the stacked nucleobases on the outer part of Z-DNA rather than in their conventional positions in the interior of the double helix.

Some evidence exists that suggests that Z-DNA influences gene expression and regulation. Apparently small stretches of DNA that contain alternating purines and pyrimidines are more commonly found at the 5′-ends of genes, regions that regulate transcriptional activities. Also, methylation of either guanine residues in the C8 and N7 positions or cytosine residues in the C5 position (Figure 2.23) stabilizes the Z-form. Sequences that are not strictly alternating purines and pyrimidines may also acquire the Z conformation because of methylation. The suggestion that Z-DNA may have a role in gene regulation is supported by modifications in methylation patterns that accompany the process of gene expression. However, Z-DNA has not yet been detected in DNA *in vivo*.

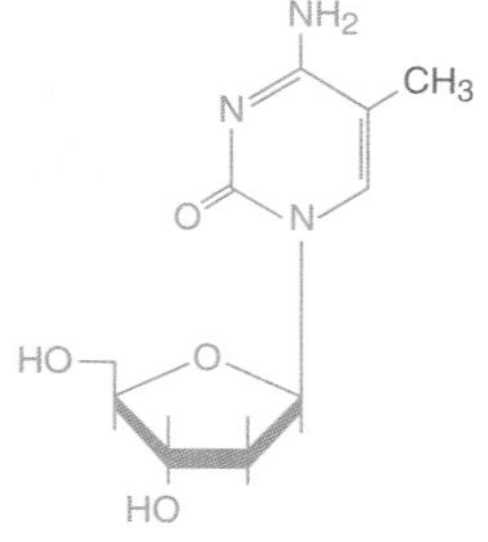

Figure 2.23 Structure of 5-methylcytidine.

Noncanonical DNA Structures

A-, B-, and Z-DNA are associated mainly with variation in the conformation of the nucleotide constituents of DNA. It is now recognized that even canonical B-DNA is not a straight, monotonous, and uniform structure. Instead, DNA bends and forms unusual structures such as cruciforms or triple-stranded arrangements as it interacts with certain proteins. Such variations in DNA conformation appear to be an important recurring theme in the process of molecular recognition of DNA by proteins and enzymes. Variations in DNA structure or conformation are favored by specific DNA sequence motifs such as inverted repeats, palindromes, mirror repeats, and direct repeats (Figure 2.24), as well as homopurine–homopyrimidine sequences, phased A tracts, and G-rich regions. AT-rich sequences, which are prone to easy strand separation, exist near the origins of DNA replication. The human genome is rich in homopurine–homopyrimidine sequences and alternating purine–pyrimidine tracts.

Bent DNA

DNA sequences with runs of 4 to 6 adenines separated by 10 base-pair (bp) spacers produce bent conformations. Structural studies have indicated that minor grooves of these sequences are compressed. However, it is not clear whether bending arises from this feature or from the boundary between this unusual conformation and normal B-DNA. DNA bending appears to be a fundamental element in the interaction between DNA sequences and proteins that catalyze central processes, such as replication, transcription, and site-specific recombination. Bending induced by interactions of DNA with enzymes and other proteins,

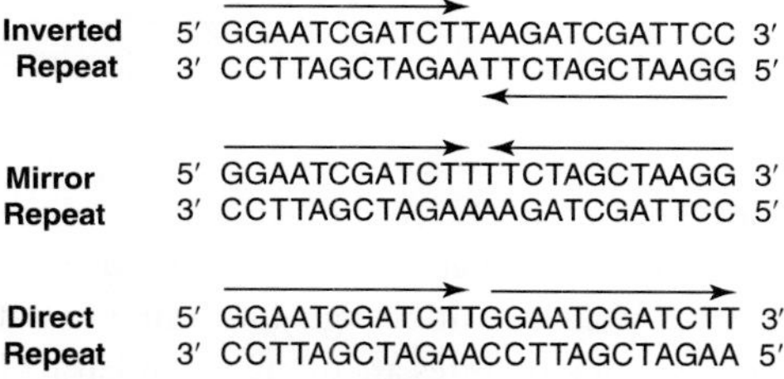

Figure 2.24 Symmetry elements in DNA sequences. Three types of symmetry elements for double-stranded DNA sequences are shown. Arrows illustrate the special relationship of these elements in each one of these sequences. In inverted repeats, or palindromes, each DNA strand is self-complementary within the inverted region that contains the symmetry elements. A mirror repeat is characterized by the presence of identical base pairs equidistant from a center of symmetry within the DNA segment. Direct repeats are regions of DNA in which a particular sequence is repeated. The repeats need not be adjacent to one another.

CLINICAL CORRELATION 2.3

Antitumor Antibiotics That Change the Shape of DNA

The local three-dimensional structure of DNA is important in interactions with proteins involved in repair, transcription, recombination, and chromatin condensation. It has been proposed that antibiotics can induce formation of DNA structures that can recruit these proteins with cytotoxic results. The best-studied example is the antitumor drug cisplatin, a tetracoordinate platinum complex (cis-$Pt(NH_2)_2Cl_2$). Cisplatin is used alone or in combination with other antitumor agents to treat a variety of tumors including testicular, ovarian, bone, and lung cancers. It forms inter- and intrastrand cross-links in double-stranded DNA with the latter adduct comprising 90% of DNA lesions. These bonds arise from displacement of chloride ligands on platinum by N7 atoms of two neighboring guanines. Structural studies on intrastrand cross-linked DNA adduct show that the double helix is strongly bent toward the major groove. In figure below, (a) Normal DNA, and (b) DNA adduct.

Bent structures of the cisplatin-DNA adduct are specifically recognized by several DNA-binding proteins such as nucleotide excision repair (NER) proteins and nonhistone DNA-binding proteins such as HMG-1. The cytotoxicity of cisplatin adducts is a complicated process mediated by specific interactions with these proteins. Cellular processes such as transcription and apoptosis are also affected by the formation of cisplatin–DNA adducts. The lesions themselves and the adduct–protein complexes are likely to interfere with transcription. NER proteins are recruited to repair the lesion, but excision repair is prone to produce DNA strand breaks. Accumulation of these breaks will ultimately induce apoptosis as the DNA becomes too damaged to function. Interactions of the cisplatin–DNA adduct with HMG proteins may also contribute to its cytotoxicity. Binding of HMG proteins may incorrectly signal that the damaged region of DNA is transcriptionally active and prevent condensation into folded chromatin structures. These complexes also perpetuate the lesion by shielding the cisplatin–DNA adducts from repair.

Zamble, D. B., and Lippard, S. J. The response of cellular proteins to cisplatin-damaged DNA. In Lippert, B. (Ed), *Cisplatin: Chemistry and Biochemistry of a Leading Anticancer Drug.* New York: Wiley-VCH, 1999, 73; Bhana, S., Hewer, A., Phillips, D. H., and Lloyd, D. R. Dependent global nucleotide excision repair of cisplatin-induced intrastrand cross links in human cells. *Mutagenesis* 23:131, 2008; Damsma, G. E., Alt, A., Brueckner, F., Carell, T, and Cramer, P. Mechanism of transcriptional stalling at cisplatin-damaged DNA. *Nat. Struct. Mol. Biol.* 14:1127, 2007; and Mukhopadhyay, R., Dubey, P., and Sarkar, S. Structural changes of DNA induced by mono- and binuclear cancer drugs. *J. Struct. Biol.* 150:277, 2005.

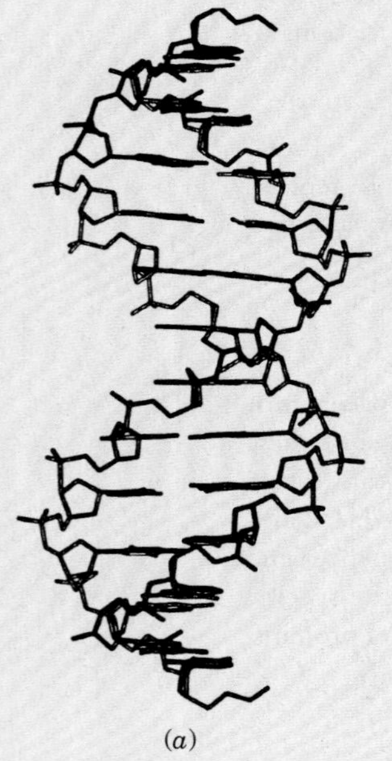

(*a*)

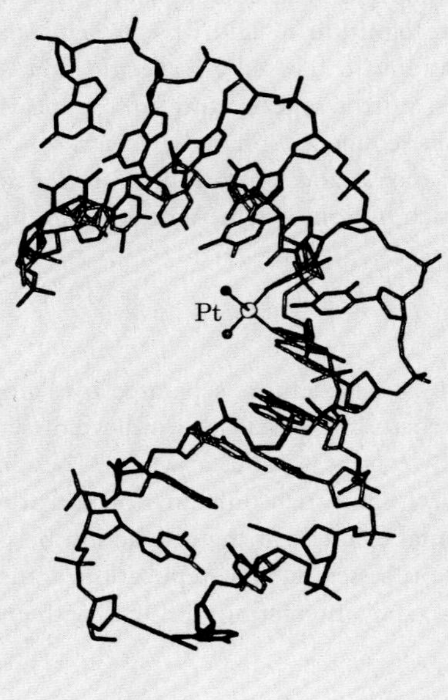

(*b*)

such as **histones**, does not require the nucleotide sequence conditions that are needed for bending of protein-free DNA. Bending also occurs because of photochemical damage or mispairing of bases and serves as a recognition signal for initiation of DNA repair. Antitumor antibiotics that produce bent structures are discussed in Clin. Corr. 2.3.

Cruciform DNA

Disruption of hydrogen bonds between the complementary strands and formation of intrastrand hydrogen bonds within the region of an inverted repeat produce a cruciform structure (Figure 2.25). The loops generated by cruciform formation require the unpairing of 3 to 4

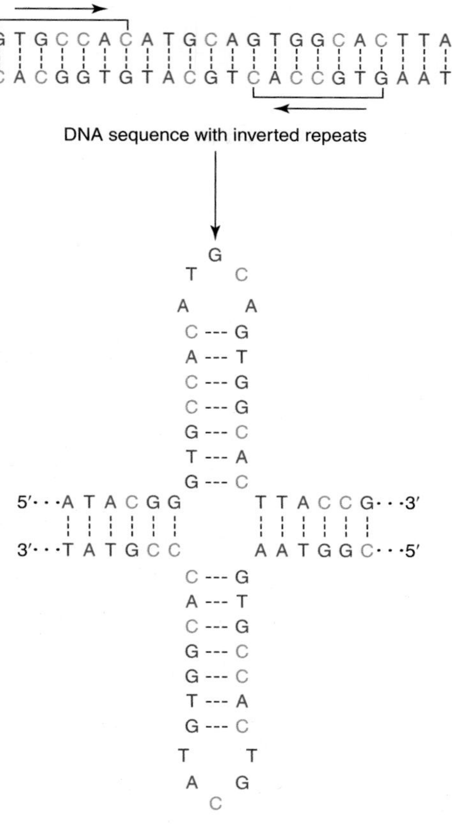

Figure 2.25 Formation of cruciform structures in DNA. The existence of inverted repeats in double-stranded DNA is a necessary but not a sufficient condition for the formation of cruciform structures. In relaxed DNA, cruciforms are not likely to form because the linear DNA accommodates more hydrogen-bonded stacked base pairs than the cruciform structure, making the formation of the latter thermodynamically unfavored. Unwinding is followed by intrastrand hydrogen bond formation between the two symmetrical parts of the repeat to produce the cruciform structure. Formation of cruciform structures is not favored at DNA regions that consist of mirror repeats because such cruciforms would be constructed from parallel rather than antiparallel DNA strands. Instead, certain mirror repeats tend to form triple helices.

bases at the end of the "hairpin." Depending on the sequence, these structures may be only slightly destabilizing because residues in the loop can remain stacked at the end of the helix.

It has been speculated that inverted repeats may function as molecular switches for replication and transcription. Inverted repeats are quite widespread within the human genome and often occur near putative control regions of genes or at origins of DNA replication. Cruciform structures at origins of DNA replication in mammalian cells have been shown to recruit cruciform-binding proteins that function during the initiation of DNA synthesis.

Triple-Stranded DNA

Some polynucleotides such as poly(dA) and poly(dT) combine to form triple-stranded complexes rather than the expected double helices. The stoichiometries of these complexes require that one strand contain a homopurine sequence while the other two have homopyrimidine sequences. Even when participating in Watson–Crick base pairing, purines possess two potential hydrogen-bonding sites in the major groove: N7 and O6 for guanine, and N7 and 6-NH_2 for adenine. Nucleobases appropriately positioned in the major groove can form specific **Hoogsteen** base triplets (Figure 2.26). For example, a thymine can selectively form two Hoogsteen hydrogen bonds to the adenine of an A-T pair. Likewise, a protonated cytosine can form two Hoogsteen H bonds with the guanine of a G-C pair, resulting in a base triplet isomorphous to the T-A-T. The pK_a' of cytidine is approximately 4.5, and triple helices containing C-G-C triplets show a strong dependence on pH. However, the templating effect of the triplet raises the apparent pK_a' of the cytosine, making it possible to form triple helices even in solutions that are only mildly acidic (pH 6). The unique Hoogsteen hydrogen-bonding patterns of guanine and adenine provide for specificity similar to the Watson–Crick pairs. Changing the orientation of the third-strand nucleobases in the major groove allows for the formation of **reverse-Hoogsteen** triplets (Figure 2.26). Selective triplets can be formed between an adenine and the A-T pair as well as between a guanine and a G-C pair. A reversed-Hoogsteen triplet can also be formed between thymine and A-T, but

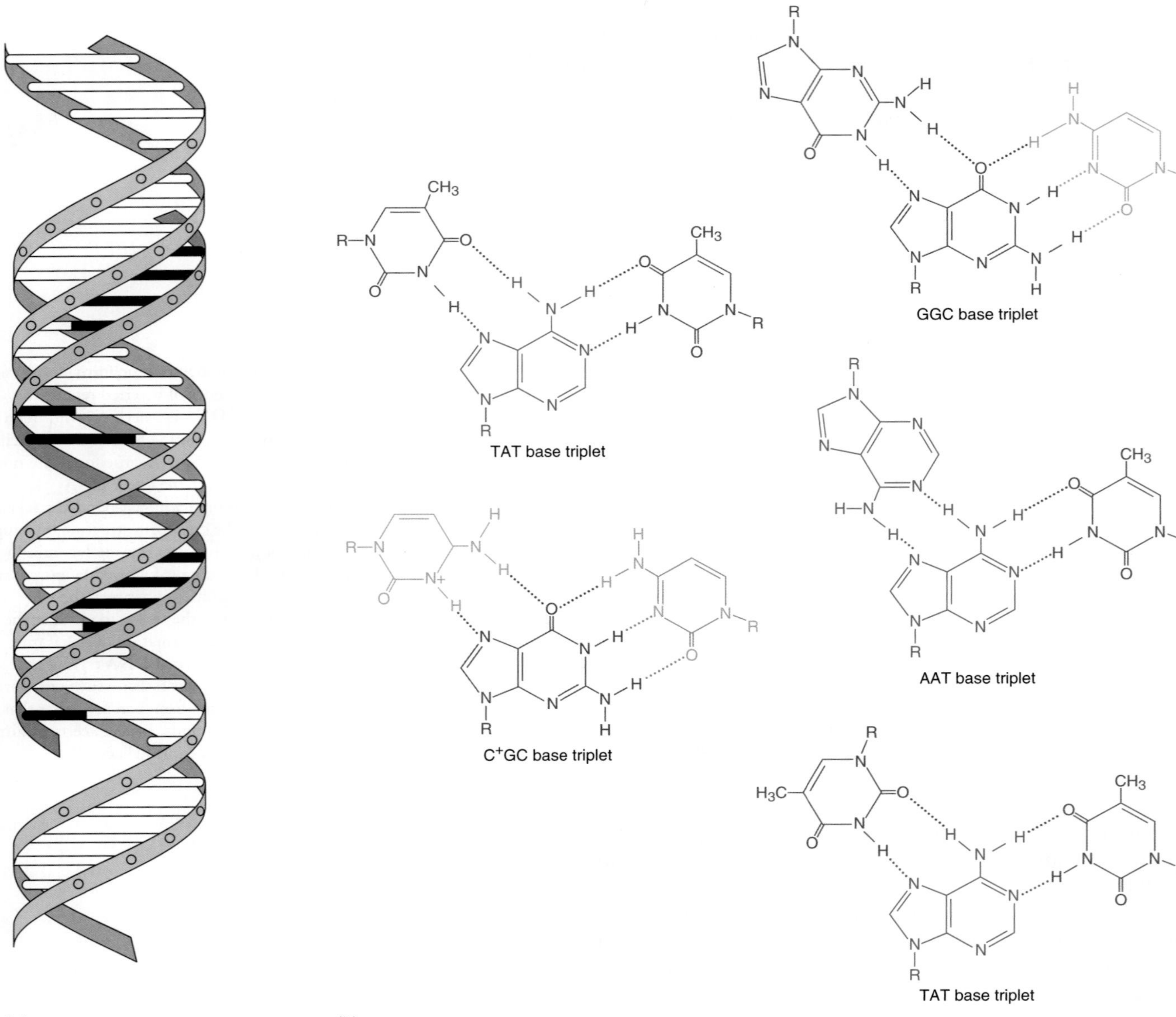

Figure 2.26 Triple helices. (*a*) Hoogsteen triple helix. Hoogsteen hydrogen bonds between a homopurine strand of a double helix and a parallel homopyrimidine strand in the major groove. The resulting isomorphous base triplets TAT and C+GC provide for sequence selective binding. (*b*) Reverse Hoogsteen triple helix. Triple helices can be formed by antiparallel binding of an oligonucleotide in the major groove to the homopurine strand of a Watson–Crick double helix. Reverse Hoogsteen hydrogen bonding produces three possible triplets: GGC, AAT, and TAT. The last triplet is not isomorphous with the others because the sugars (represented by R) are positioned differently with respect to the Watson–Crick base pair.

the resulting triplet is not isomorphous to the pu•pu•py triplets. However, the backbone is able to accommodate the distortions that result from incorporation of these triplets.

As in double-helical complexes, base stacking plays the key role in stabilizing the triplex structure. However, bringing three negatively charged backbone strands together increases electrostatic repulsion. Thus, the triple-helical complex is less stable than the associated Watson–Crick double helix. The presence of Mg^{2+} or other multivalent cations stabilizes the triple helix by shielding the phosphate charges.

Intramolecular triple helices can be formed by disruption of double-helical DNA with polypurine sequences in mirror repeats. A mirror repeat is a region such as AGGGGA that has the same base sequence when read in either direction from a central point. Refolding generates a triple-stranded region and a single-stranded loop in a structure called **H-DNA**

(Figure 2.27). Even though the formation of H-DNA is thermodynamically unfavorable because of a reduction of stacking interactions, intramolecular triple helices have been detected in cellular DNA when under **superhelical stress.** DNA **supercoiling** provides the energy to drive the unwinding of DNA that is necessary for the formation of the triple helix. Triple-strand formation produces a relaxation of negative supercoils. Also, the binding of proteins to the single-stranded DNA may further stabilize the H-DNA structure and prevent degradation of the loop by nucleases.

Many sequences in eukaryotic genomes have the potential to form triple-stranded DNA structures. Such regions occur with much higher frequency than expected from probability considerations alone. Polypurine tracts over 25 nucleotides long constitute as much as 0.5% of some eukaryotic genomes. These potential triple-helical regions are especially common near sequences involved in gene regulation. Because of this, it has been proposed that H-DNA may play a role in the control of RNA synthesis (Clin. Corr. 2.4). Other potential biological tasks have been proposed for triple-helical DNA, including possible roles in initiation and termination of replication and recombination. The abilities of triple-helical DNA to interfere with transcription have also led to efforts to use intermolecular triple helices to control artificially RNA synthesis and subsequent protein synthesis.

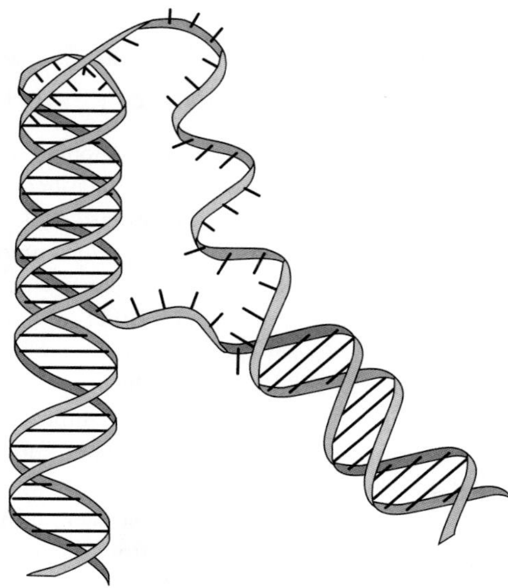

Figure 2.27 Intramolecular triple helices: H-DNA. Polypurine–polypyrimidine regions of DNA with a mirror repeat symmetry can form an intramolecular triple helix in which the third strand lies in the major groove, whereas its complementary strand acquires a single-stranded conformation.
Redrawn based on figure in Sinden, R. R. *DNA Structure and Function.* New York: Academic, 1994.

Four-Stranded DNA

Guanine nucleotides and highly G-rich polynucleotides form novel tetrameric structures called **G-quartets** that contain a planar array of guanines connected by Hoogsteen hydrogen bonds (Figure 2.28). Polynucleotides can interact to form **tetraplexes** where G-quartets stack on each other to form a multilayered structure (Figure 2.29). These structures are stabilized by Na^+ and K^+, which interact with guanine oxygens (O6) in the center of the quartet plane or between two adjacent planes. These cations stabilize the complex both electrostatically by balancing out the negative charges on the four polynucleotide strands and entropically by releasing a large number of water molecules into the bulk solution.

While G-tetraplexes have been observed by X-ray diffraction and NMR spectroscopy, their existence *in vivo* has not been proven. However, the ends of eukaryotic **chromosomes** (**telomeres**) contain repetitive G-rich sequences. Human telomeres contain 800–2400 copies of the hexameric repeat sequence $d(TTAGGG)_n$ and terminate in a single-stranded overhang that is roughly 150 nucleotides long. *In vitro,* oligonucleotides with this sequence can form tetraplex structures. The role tetraplexes play in telomere functions is unknown.

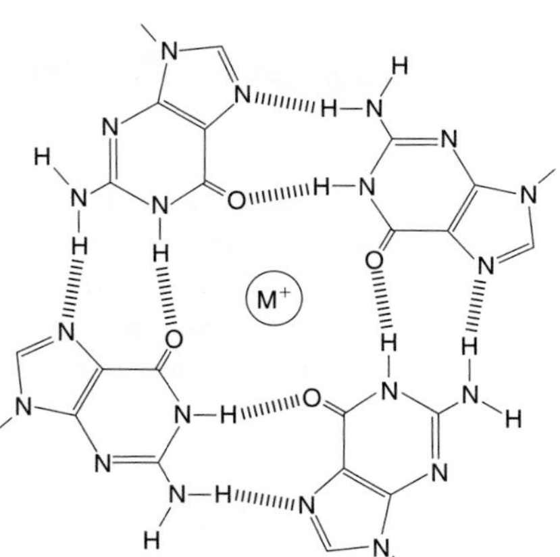

Figure 2.28 The structure of a G-quartet. The four coplanar guanines form a tetrameric structure by formation of Hoogsteen hydrogen bonds. The cavity in the center of the quartet can accommodate a sodium or potassium ion with coordination by the four O-6 oxygens.

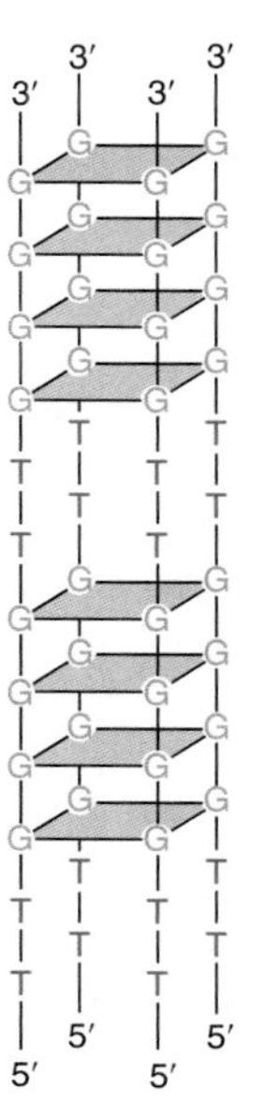

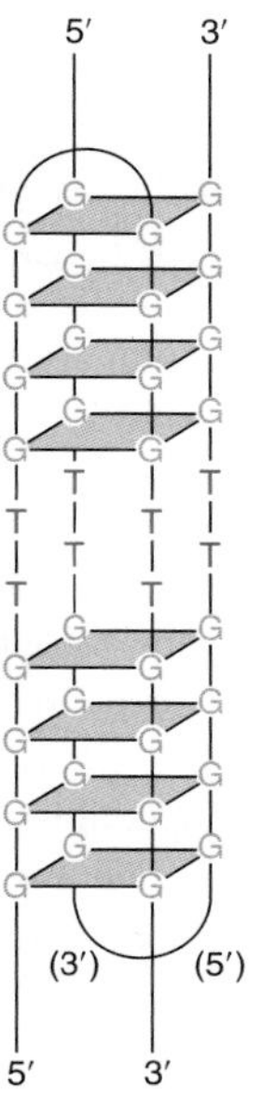

Figure 2.29 G-Quadruplex DNA. Four-stranded structures can arise from stacking of G-quartets. Quadruplex structures may be parallel or antiparallel (one possible arrangement shown). The latter may be formed by the G-rich sequences of telomeric DNA.
Redrawn based on figure in Sinden, R. R. *DNA Structure and Function.* New York: Academic, 1994.

CLINICAL CORRELATION 2.4

Hereditary Persistence of Fetal Hemoglobin

Hereditary persistence of fetal hemoglobin (HPFH) (OMIM 141749) is a group of conditions in which fetal hemoglobin synthesis is not terminated at birth but continues into adulthood. The homozygous form of the condition is extremely uncommon, being characterized by changes in red blood cells similar to those found in the genetic blood disorder β-thalassemia. HPFH, in either the homozygous or the heterozygous state, is associated with mild clinical or hematologic abnormalities. Mild musculoskeletal pains may occur infrequently, but HPFH patients are frequently asymptomatic.

The condition results from failure to stop transcription of human $^{G}\gamma$- and $^{A}\gamma$-globin genes, leading to elevated levels of fetal hemoglobin. The formation of an intramolecular DNA triple-helical structure located about 200 bp upstream from the initiation site for transcription of the γ-globin genes (between positions −194 and −215) acts as a brake for their expression. Hemoglobin genes of patients contain mutations in one or more positions in this region, decreasing the stability of the triple helix and reducing its ability to inhibit the protein synthesis.

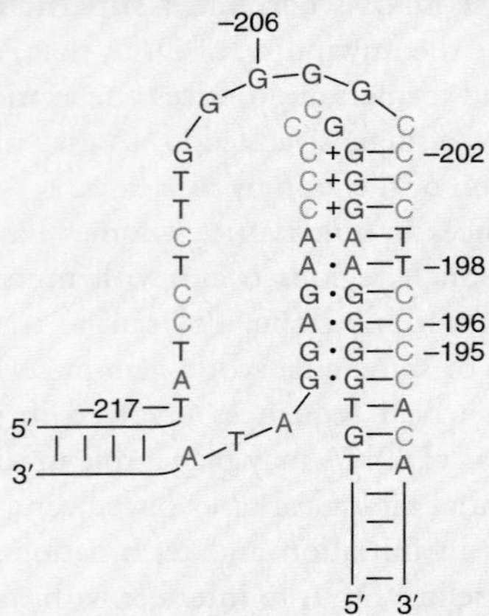

Ulrich, M. J., Gray, W. J., and Ley, T. J. An intramolecular DNA triplex is disrupted by point mutations associated with hereditary persistence of fetal hemoglobin. *J. Biol. Chem.* 267:18649, 1992; and Bacolla, A., Ulrich, M. J., Larson, J. E., Ley, T. J., and Wells, R. D. An intramolecular triplex in the human gamma-globin 5′-flanking region is altered by point mutations associated with hereditary persistence of fetal hemoglobin. *J. Biol. Chem.* 270:24556, 1995.

However, telomeres are attracting attention as targets for new anticancer chemotherapies (Clin. Corr. 2.5). In addition, G-tetraplexes have been implicated in recombination of immunoglobulin genes and in dimerization of double-stranded genomic RNA of the human immunodeficiency virus (HIV).

Slipped DNA

DNA regions with direct repeat symmetry can form slipped, mispaired DNA (SMP-DNA). Their formation involves unwinding of the double helix, realignment, and subsequent pairing of one copy of the direct repeat with an adjacent copy on the other strand. This generates two single-stranded loops. Two isomeric structures of an SMP-DNA are possible (Figure 2.30). Although SMP-DNA has not yet been identified *in vivo,* genetic evidence suggests that it is involved in spontaneous frameshift mutations that result in base addition or deletion within runs of single bases (Figure 2.31). Deletions and duplications of DNA segments

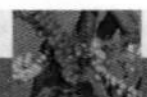

CLINICAL CORRELATION 2.5

Telomerase as a Target for Anticancer Agents

Telomeres, the ends of linear eukaryotic chromosomes, are critical for maintaining the stability of the genome. They are progressively shortened during each cycle of cell division and, on reaching a critical length, trigger apoptosis. Telomerase acts to maintain or lengthen the telomeres but is not active in normal somatic cells. Telomerase activity present in most tumor cell lines may be responsible for their immortalization and, when increased, correlates to poorer clinical prognosis.

Two approaches are being examined for selective inhibition of telomerase. The first involves targeting of the RNA-containing portion of the enzyme. This RNA serves as the template for extension of the telomeric repeat sequence. Nucleic acids with chemically modified sugar–phosphate backbones bind to telomerase RNA in immortal human cells, inhibit activity, and ultimately cause cell death. Modified nucleic acids were used to resist nuclease degradation and provide high affinity for forming double-helical complexes with RNA. A second approach involves drugs that bind to G-quadruplex DNA. Large aromatic molecules such as porphyrins and anthraquinones selectively bind and stabilize G-quadruplex DNA structures.

Herbert, B.-S., Pitts, A. E., Baker, S. I., Hamilton, S. E., et al. Inhibition of human telomerase in immortal human cells leads to progressive telomere shortening and cell death. *Proc. Natl. Acad. Sci. USA* 96:14276, 1999; Parkinson, E. K., and Minty, F. Anticancer therapy targeting telomeres and telomerase. *Biodrugs* 21:375, 2007; and Cuesta, J., Read, M. A., and Neidle, S. The design of G-quadruplex ligands as telomerase inhibitors. *Mini-Rev. Med. Chem.* 3:11, 2003.

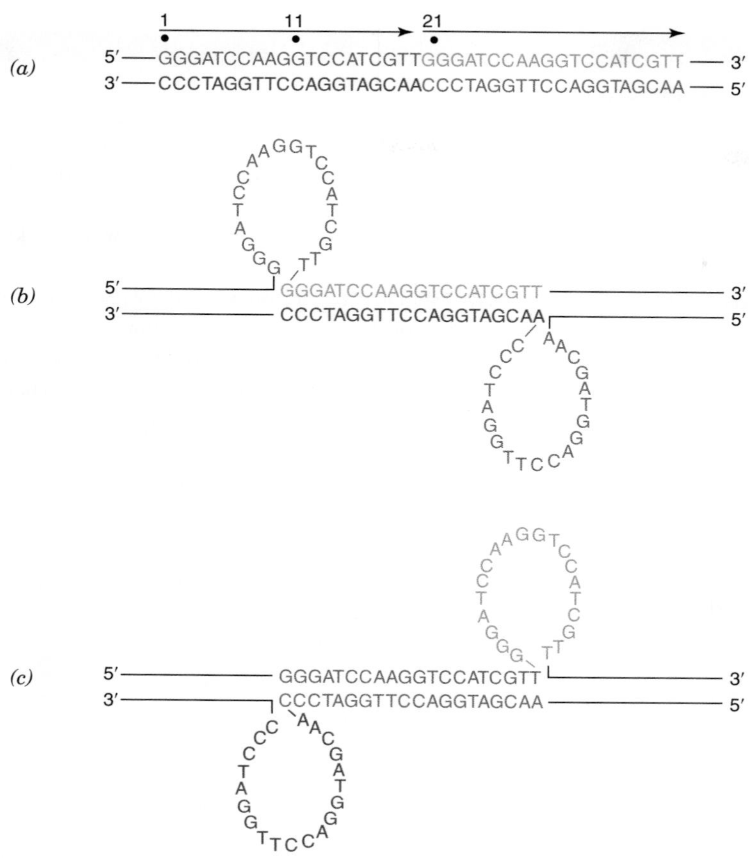

Figure 2.30 Slipped, mispaired DNA. (*a*) The presence of two adjacent tandem repeats can give rise to two isomers of slipped, mispaired DNA. (*b*) The second copy of the direct repeat in the top strand pairs with the first copy of repeat on the bottom strand. (*c*) The pairing of the first copy of the direct repeat in the top strand with the second copy of the direct repeat in the bottom strand. Two single-stranded loops are generated in each isomer.

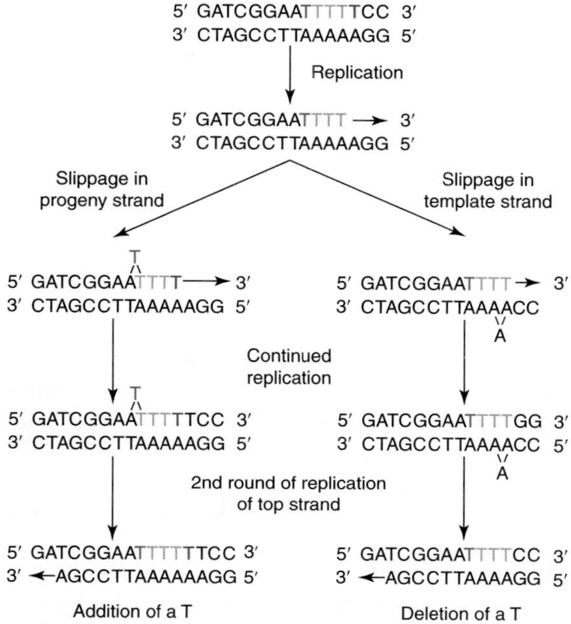

Figure 2.31 Frameshift mutagenesis by DNA slippage. DNA replication within a run of a single base can produce a single-base frameshift. In this example, a run of five As is replicated and, depending on whether a slippage occurs in progeny strand or template strand, a T may be added or deleted from the DNA.

CLINICAL CORRELATION 2.6

Expansion of DNA Triple Repeats and Human Disease

The presence of reiterated three-base-pair DNA sequences occurs in a number of human genetic diseases including fragile X syndrome (OMIM 300624), myotonic dystrophy (OMIM 160900), X-linked spinal and bulbar muscular atrophy (Kennedy syndrome), Friedrich ataxia, and Huntington disease (OMIM 143100). These diseases are associated with the expansion of nucleotide triplet repeats that appear within or near specific genes. Kennedy's disease was found to contain a CAG repeat in the first exon of the androgen receptor gene. Triplet repeats can also be found within untranslated regions of the gene: in Friedrich ataxia, a repeated GAA is found within an intron and, in myotonic dystrophy, a CTG repeat is found in the 3'-untranslated region of its gene. Fragile X syndrome, a leading cause of mental retardation, is characterized by expansion of a GCC triplet on the 5'-side of the FMR-1 gene. This repeat expands from 30 copies to thousands of copies. The disease develops when the normal expression of the FMR-1 gene is turned off by methylation of CpG sites. In all cases expansion of the triplet interferes with normal functioning of the related protein. Often, there is a loss of protein function, but occasionally, a gain of a deleterious function occurs. These diseases are characterized by an increase in severity of the disease with each successive generation, which is known as anticipation.

Triplet expansion may result from slipped mispairing during DNA synthesis. Because of massive amplification that characterizes these diseases, repeated or multiple slippages would have to be involved to explain the high degree of expansion. One possible mechanism involves slippage of nascent DNA during lagging strand synthesis. This process may be aided by formation of a stable hairpin structure by the slipped loop. Repetition of this process leads to accumulation of large numbers of triplet repeats.

Mirkin, S. Expandable DNA repeats and human disease. *Nature* 447:932, 2007; and Patel, P. I., and Isaya, G. Friedrich Ataxia: from GAA triplet-repeat expansion to frataxin deficiency. *Am. J. Hum. Genetics.* 69:15, 2001.

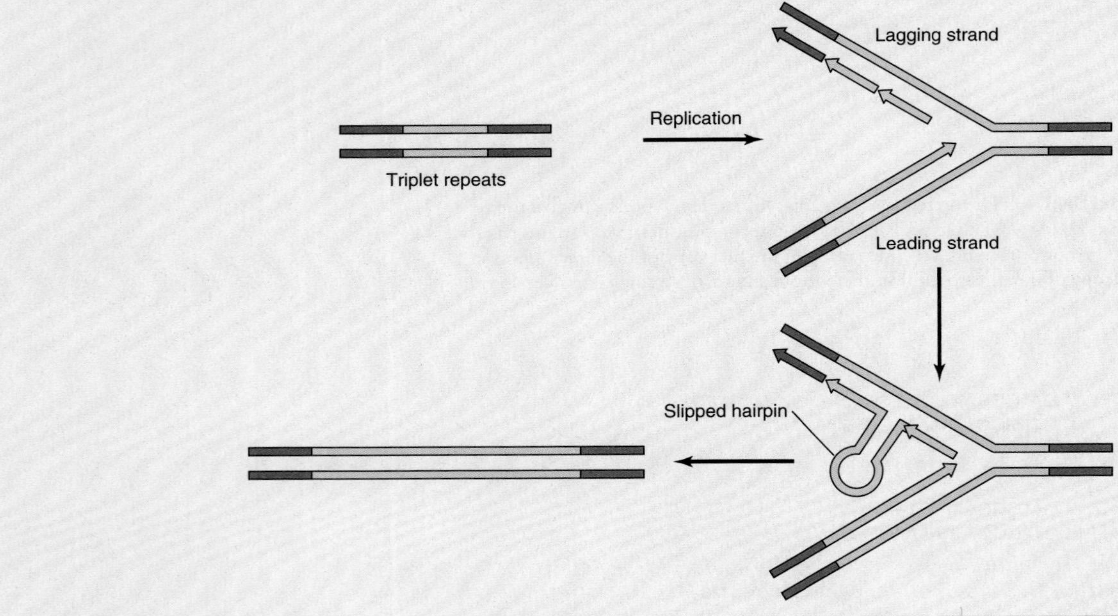

that are longer than a single base can occur during DNA replication between direct repeats, causing slipped-looped structures. Duplication of certain simple triplet repeats that are the basis of several human genetic diseases (Clin. Corr. 2.6) can also occur by this mechanism.

2.4 • HIGHER-ORDER STRUCTURE OF DNA

With the exception of RNA-containing viruses, all life on Earth uses DNA to store genetic information. The length of DNA varies from species to species and ranges from a few thousand base pairs for small viruses, to millions of base pairs in bacteria, and to billions of base pairs in plants and animals. In all organisms, the **contour length** (length of the DNA assuming a B-form double helix) of genomic DNA is usually much larger than the size of the cell that contains it. For example, a medium-sized virus, such as λ-phage, contains 4.8×10^4 bp of DNA which is 16.5 μm long. However, the viral particle is only 0.19 μm long. The common bacterium *Escherichia (E.) coli* is approximately 2 μm long and has a single chromosome with 4.6×10^6 bp with a contour length of 1.5 mm. A single diploid human cell contains 6×10^9 bp of chromosomal DNA packaged in 46 chromosomes. The contour length of this DNA approaches 2 m, which is packed within a nucleus about 10 μm in diameter. It is clear that the DNA of all organisms must be exquisitely packaged.

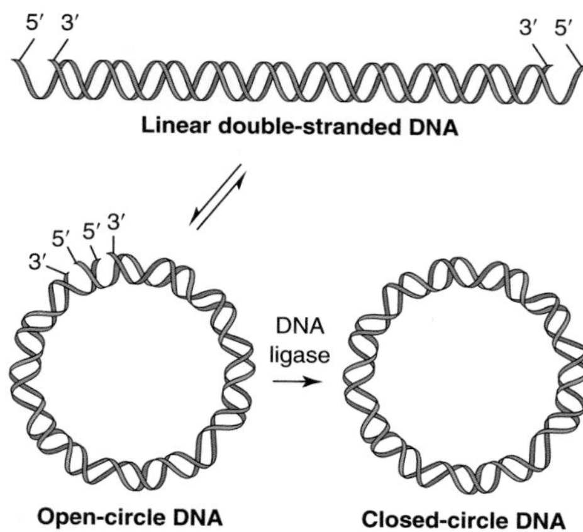

Figure 2.32 Circularization of λ DNA. DNA of bacteriophage λ exists in linear and circular forms, which are interconvertible. Circularization of λ DNA is possible because the 5′-overhangs of the linear form are complementary sequences.

Genomic DNA May Be Linear or Circular

With the exception of some small bacteriophages such as φX174 that can acquire a single-stranded form, most DNAs exist as double-helical complexes. Depending on the source, the complexes can be linear or circular. For example, DNAs of several small viruses are linear double-stranded helices. Some of these DNAs contain naturally occurring interior single-stranded breaks. The double-helical structure is maintained because the breaks in one strand are generally in different locations from those in the complementary strand. DNAs in most higher organisms are also linear. Each of the 46 **chromosomes** in a diploid human cell is a linear double-helical DNA complexed with proteins.

Circular DNA results from the formation of phosphodiester bonds between the 3′- and 5′-termini of linear polynucleotides. The circular nature of the single-stranded phage φX174 DNA was suspected from studies that showed that no ends were available for reaction with exonucleases. In addition, endonuclease cleavage at a single site yielded only one polynucleotide. These results were later confirmed by direct observation with electron microscopy.

Some organisms have DNAs that are either linear or circular at different points in their life cycles. Before entering *E. coli,* the λ-phage DNA is linear with single-stranded overhangs on the 5′-termini. These overhangs are approximately 20 nucleotides long and have complementary sequences. On infection, circularization occurs by hybridization of the ends to each other. Formation of phosphodiester bonds between the 3′- and 5′-ends of each strand by **DNA ligase** produces a covalently closed circle (Figure 2.32). The strands of a circular DNA cannot be irreversibly separated by denaturation because they exist as intertwined closed circles. The absence of 3′- or 5′-termini endows the circular DNA with complete resistance toward exonucleases, improving the longevity of DNA.

Most DNA in bacteria exists solely as closed circles. This includes the bacterial chromosomes and the smaller extrachromosomal **plasmid** DNAs. The latter are a few thousand base pairs long and encode accessory genes such as those for antibiotic resistance. Plasmids are maintained and replicated separately from chromosomal DNA and may number in the hundreds within a bacterium. Circular DNA exists in higher organisms too. Yeasts can also carry circular plasmids. Mitochondria and chloroplasts in higher eukaryotic cells contain circular DNAs 200–1500 $\times 10^3$ bp long which encode unique proteins used by the organelles (p. 582). The presence of an independent genome and a marked similarity to cyanobacteria has led to the hypothesis that these organelles arose billions of years ago from a symbiotic relationship between protoeukaryotic cells and bacteria.

DNA Is Superhelical

Circular double-stranded DNA formed by ligating the free termini of a linear DNA is **relaxed**; that is, it has the thermodynamically favored structure of B-DNA. This relaxed DNA has greatly reduced activity in replication, translation, and recombination. The biologically active form of DNA is **superhelical**, a topologically strained isomer created by underwinding or overwinding the double helix. Before circularization of a linear B-form DNA, the double helix contains about 10.5 base pairs per complete turn. If the DNA is untwisted before sealing the circle, the resulting structure will be strained (Figure 2.33). Untwisting reduces the total number of helical turns present in the circular structure. The underwinding of the helix can be accommodated by disrupting the base pairing over a small region to produce a pair of single-stranded loops in a relaxed circular structure. Loss of base stacking in this structure is energetically unfavorable. However, if all base stacking

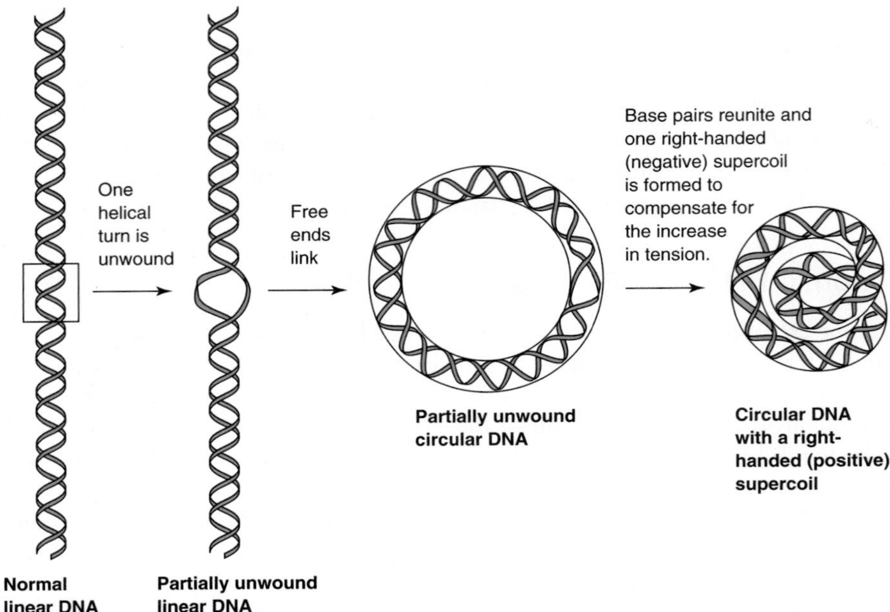

Figure 2.33 Negative DNA supercoiling. Right-handed supercoils (negatively supercoiled DNA) are formed if relaxed DNA is partially unwound. Unwinding may lead to a disruption of hydrogen bonds or may produce negative supercoils. The negative supercoils are formed to compensate for the increase in tension that is generated when disrupted base pairs are reformed.
Redrawn from Darnell, J., Lodish, H., and Baltimore, D. *Molecular Cell Biology.* New York: Freeman, 1986.

is maintained, the underwinding generates a torsional strain in the backbone of the double helix. Twisting the entire circular DNA in the direction opposite to the original rotation relieves the strain. This results in the formation of a coiled coil, better known as a superhelix. Underwound DNA is said to be negatively supercoiled, and the resulting superhelix is right handed (Figure 2.34). Conversely, overwound DNA is positively supercoiled and forms a left-handed superhelix.

Although the closed-circular form of DNA is ideal for acquiring a superhelical structure, any segment of double-stranded DNA that is immobilized at both ends can be superhelical. The DNA of eukaryotic cells, for instance, can acquire a superhelical form because its

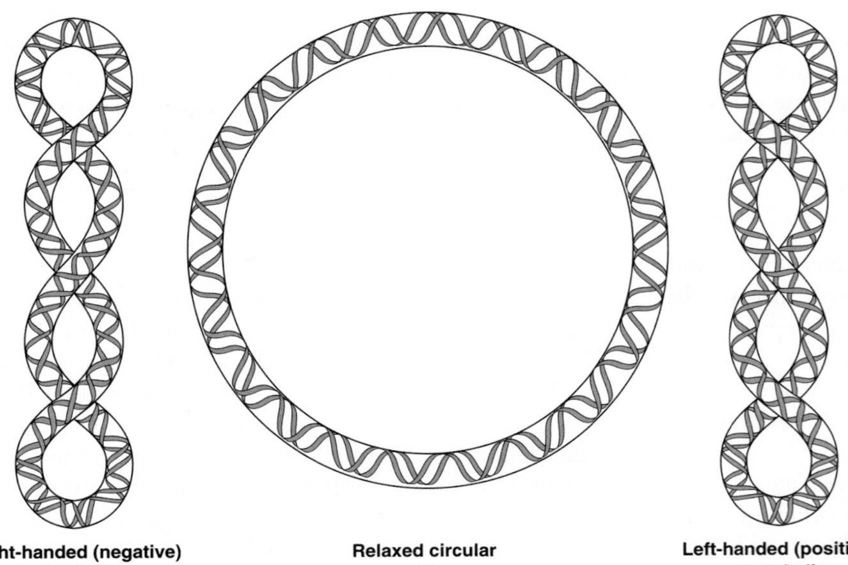

Figure 2.34 Relaxed and supercoiled DNA. Relaxed DNA can be converted to either right- or left-handed superhelical DNA. Right-handed DNA (negatively supercoiled DNA) is the form normally present in cells. Left-handed DNA may be transiently generated as DNA is subjected to enzyme-catalyzed transformations (replication, recombination, etc.) and is present in certain bacteria species. The distinctly different patterns of folding for right- and left-handed DNA superhelices are apparent in this representation.

anchoring by nuclear proteins creates numerous closed **topological domains**. A topological domain is defined as a DNA segment contained in a manner that restrains rotation of the double helix (Figure 2.35).

Overall, whether DNA is circular or linear, the existence of negative superhelicity appears to be an important feature. Supercoiling promotes packaging of DNA within the confines of the cell by facilitating formation of compact structures. Superhelices also possess a tendency to generate regions with disrupted hydrogen bonding (bubbles), which may be instrumental in facilitating the process of localized DNA strand separation during DNA repair, synthesis, and recombination.

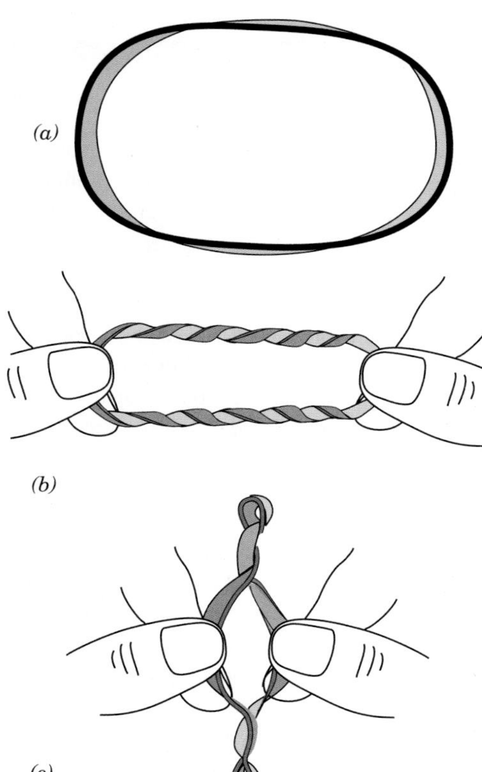

Figure 2.35 Superhelical model for DNA. A rubber band can demonstrate the topological properties of double-stranded circular DNA. (*a*) The relaxed form of the band. (*b*) The band can be twisted to generate two distinct topological domains, separated by the pair of "thumb-forefinger anchors". Left-handed (counterclockwise) turns are introduced into the upper section of the band, with compensating right-handed (clockwise) turns formed in the bottom section. (*c*) When the "anchors" are brought into close proximity with each other, the upper section that contained the left-handed turns forms a right-handed superhelix. The bottom section produces a left-handed superhelix. Superhelicity is not the property of a DNA molecule as a whole but rather a property of specific DNA domains.
Redrawn from Sinden, R. R., and Wells, R. D. DNA structure, mutations, and human genetic disease. *Curr. Opin. Biotech.* 3:612, 1992.

Topoisomerases

Topoisomerases regulate the formation of superhelices. These enzymes catalyze the concerted breakage and rejoining of DNA strands, producing a DNA that is more or less superhelical than the original (Table 2.4). The precise regulation of the cellular level of DNA superhelicity is important to facilitate protein interactions with DNA. The cellular ATP-to-ADP ratio may play a role in this process, because it influences the activity of some topoisomerases. Compounds that inhibit topoisomerases and gyrases are effective antibacterial and antitumor agents (Clin. Corr. 2.7).

Topoisomerases I make transient single-strand breaks in a negatively supercoiled DNA double helix; passage of the unbroken strand through the gap eliminates one supercoil from DNA (Figure 2.36). During the reaction, the enzyme remains bound to DNA by a covalent bond between a tyrosyl residue and a phosphoryl group at the incision site. This conserves the energy of the interrupted phosphodiester bond for the subsequent repair of the nick. The two subclasses of topoisomerase I differ in the formation of 5′-phosphotyrosine (class IA) or 3′-phosphotyrosine (class IB) bonds.

TABLE 2.4 • Properties of DNA Topoisomerases

Topoisomerase	*Type*[a]	*Functions*
E. coli topoisomerase I *(topA)*[b]	IA	Relaxes negatively supercoiled DNA
E. coli topoisomerase III *(topB)*	IA	Relaxes negatively supercoiled DNA, decatenation (unlinking) activity.
Eukaryotic topoisomerase III *(top3)*	IA	Possible role in recombination and/or chromosome decatenation?
Archaeal reverse DNA gyrase	IA	Introduces positive supercoils in DNA.
Eukaryotic DNA topoisomerase I *(top1)*	IB	Relaxes either negatively or positively supercoiled DNA.
Vaccinia (poxvirus) DNA topoisomerase	IB	Relaxes either negatively or positively supercoiled DNA; binds supercoiled DNA sequence specifically.
E. coli DNA gyrase *(gyrA, gyrB)*	II	Introduces negative supercoils in DNA, relaxes either negatively or positively supercoiled DNA.
E. coli topoisomerase IV *(parC, parE)*	II	Decatenation of chromosomes during replication; relaxes negatively supercoiled DNA.
Eukaryotic DNA topoisomerase II *(top2)*	II	Decatenation of linked chromosomes; relaxes either negatively or positively supercoiled DNA.

[a]Type I topoisomerases use Mg^{2+} as a cofactor, but do not require ATP. Type II topoisomerases require both Mg^{2+} and ATP.

[b]The name of the gene coding for the topoisomerase is shown in italics.

CLINICAL CORRELATION 2.7

Topoisomerases in the Treatment of Disease

Topoisomerases are important targets of antimicrobial and antineoplastic agents. These drugs do not inhibit overall activity of the enzyme, as is the case with most enzyme-targeting drugs. Instead, they trap the immediate complex between topoisomerase and DNA. This may result in degradation of DNA, introduction of mutations, or inhibition of translation and replication.

In the treatment of cancers, both topoisomerases I and II can be targeted. Camptothecin and its derivatives act on topoisomerase I. An excellent correlation has been noted between antitumor activity of various camptothecin derivatives on murine leukemia and their interference with topoisomerase activity. They may cause potentially lethal lesions by forming drug-stabilized covalent DNA cleavage complexes. Subsequent DNA replication may be a prerequisite for cell toxicity. The efficacy of camptothecin derivatives may be improved by increased levels of topoisomerase I in tumor cells such as advanced colon cancers. Two other potent antineoplastic agents, amsacrine and etoposide, act selectively on topoisomerases II and indicate that these clinically useful drugs stabilize covalent topoisomerase IIDNA cleavage complexes. Indirect evidence suggests that these drugs may stimulate formation of these complexes. Topoisomerase II-mediated DNA breaks can exert their cytotoxic effect in the absence of ongoing DNA synthesis, probably by inducing recombination and mutation at sites of formation of drug-induced topoisomerase II–DNA complexes. Many anticancer agents including anthracyclines (including adriamycin and doxorubicin), synthetic intercalators, ellipticines, and podophyllotoxins also target topoisomerases II. Hematologic neoplasms, such as lymphoid and nonlymphoid leukemias, high-grade non-Hodgkin lymphomas, and Hodgkin disease, are treated mostly with combinations of topoisomerase II inhibitors with or without additional cytotoxic agents.

Ciprofloxacin (Cipro®) gained considerable media attention for its use in the treatment of inhalation anthrax. It is a representative of the quinolones, a class of powerful broad-spectrum antibacterial activities. These antibiotics selectively inhibit bacterial DNA gyrase and topoisomerase IV. Quinolones bind to the complexes formed by these enzymes with DNA-inhibiting replication. Ultimately, cell death occurs due to the accumulation of lethal double-stranded breaks in the bacterial DNA. Because these enzymes are not found in eukaryotic cells, quinolone drugs are highly selective in their cytotoxic action. Fluoroquinolones such as ciprofloxacin show good activity against both Gram-positive and Gram-negative bacteria whereas newer drugs show even broader activities. For example, gemifloxacin (Factive®) is especially useful for the treatment of respiratory infections including those by multidrug-resistant *Staphylococcus pneumoniae, Mycoplasma, and Legionella.*

camptothecin

etoposide

ciprofloxacin

gemifloxacin

Potmesil, M., and Kohn, K. W. (Eds.). *DNA Topoisomerases in Cancer.* New York: Oxford University Press, 1991; Pommier, Y. Topoisomerase I inhibitors: camptothecins and beyond. *Nature Rev. Cancer* 6:789, 2006; Hooper, D. C. Clinical applications of quinolones. *Biochem. Biophys. Acta* 1400:45, 1998; Mitscher, L.A. Bacterial topoisomerase inhibitors: Quinolone and pyridine antibacterial agents. *Chem. Rev.* 105:559, 2005; and Oliphant, C. M., and Green, G. M. Quinolones: A comprehensive review. *Am. Fam. Physician* 65:455, 2002.

Topoisomerases II are dimeric proteins that bind to a double-helical DNA and cleave both strands (Figure 2.37). Passage of another double-helical DNA segment through the break removes or adds two supercoils. All eukaryotic and many prokaryotic topoisomerases II only relax supercoiled DNA. ATP hydrolysis is required for the turnover of the enzyme, but not actually for the relaxation reaction. The **gyrases**, a subset of type II topoisomerases, are the only enzymes that add negative supercoils into DNA. They occur only in bacteria; eukaryotes use the wrapping of DNA around chromosomal proteins for the introduction of negative supercoils. The gyrase reaction requires the hydrolysis of ATP as an energy source and can add negative supercoils at a rate of about 100 per minute. Gyrases bind to DNA in a conformation that introduces a positive and negative superhelical loop, leaving the topological state of the DNA unchanged. Double-strand breakage and passage of DNA through the gap introduces two negative supercoils.

Topoisomerases also catalyze other topological isomerizations. Bacterial type III topoisomerases have type I topoisomerase properties; that is, they relax supercoils without the

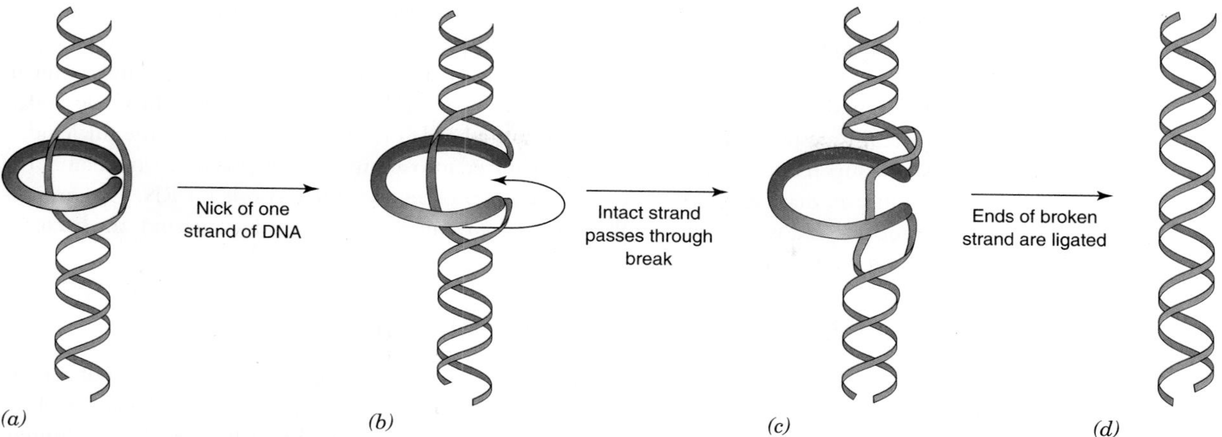

Figure 2.36 Mechanism of action of topoisomerases I. Topoisomerases I relax DNA by (*a*) binding and locally separating the complementary strands; then (*b*) nicking one strand and binding to the newly generated termini; and (*c*) passing the intact strand through the gap generated by the nick and closing the gap by restoring the phosphodiester bond. This gives rise to a relaxed structure (*d*).
Redrawn from Dean, F., et al. *Cold Spring Harbor Symp. Quant. Biol.* 47:773, 1982.

requirement of an energy source, such as ATP hydrolysis. These topoisomerases may specialize in the resolution of interlocking circular DNA products (catenates) that are generated just prior to the completion of DNA replication. Reverse gyrases, members of an unusual class of topoisomerases, were first isolated from archaebacteria. These gyrases introduce positive supercoils into DNA that may protect from the denaturing conditions of high temperature and acidity under which these bacteria exist.

Packaging of Prokaryotic DNA

In prokaryotic cells, the single chromosome is a supercoiled double-stranded circular DNA. Bacterial chromosomes are dynamic structures reflecting the need for rapid DNA synthesis, cell division, and transcription. However, in a common bacterium such as *E. coli,* the contour length of the DNA is some 750 times longer than the length of the cell. Thus, DNA must be packaged in a highly condensed form to fit inside the cell. Bacterial chromosomes are organized into compacted structures, called **nucleoids**, by interaction of **HU** and **H-NS** proteins and participation of various cations, **polyamines** (such as spermine, spermidine, putrescine, and cadaverine), RNA, and nonhistone proteins (Figure 2.38). HU is an 18-kDa heterodimer of two nearly identical subunits (HU-1 and HU-2). On binding, HU changes the shape and supercoiling of the DNA. HU binding *in vitro* compacts DNA

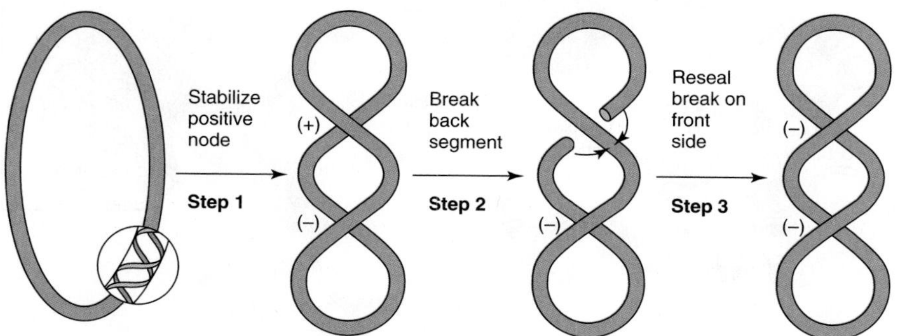

Figure 2.37 Mechanism of action of topoisomerases II. Topoisomerases II can relax or, in case of gyrases, introduce negative superhelices into DNA. The mechanism of action of gyrase is illustrated using the conversion of a relaxed DNA molecule to a molecule that contains first two supercoils, one positive and one negative (step 1). Passage of a DNA segment through the positive supercoil shown on the rightmost part of the figure (step 3) produces a molecule that contains two negative supercoils. This reaction requires the hydrolysis of ATP as an energy source.
Redrawn from Brown, P. O., and Cozzarelli, N. R. *Science* 206:1081, 1979.

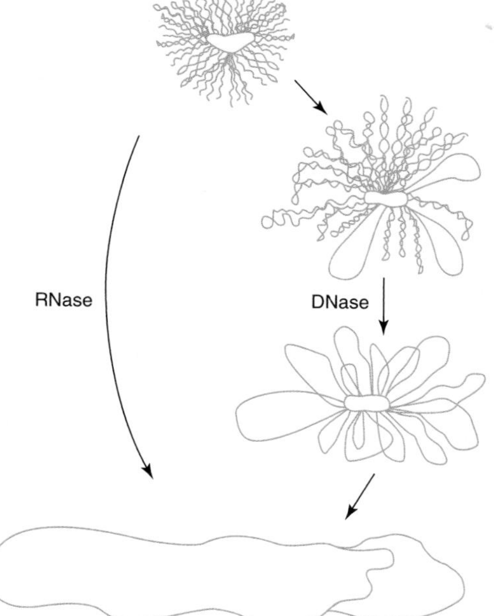

Figure 2.38 Bacterial chromosomes are packaged in nucleoids. The circular chromosome of a bacterium is compacted into about 40–50 loops of supercoiled DNA organized by a central RNA–protein scaffold. DNase relaxes the structure progressively by opening individual loops, one at a time. RNase completely unfolds the chromosome in a single step by disrupting the nucleoid core.
Redrawn from Worcel, A., and Burgi, E. *J. Mol. Biol.* 71:127, 1972.

and restrains supercoils in a concentration-dependent manner. At an equimolar ratio, the interaction of DNA with HU prevents topoisomerases I from relaxing negatively supercoiled DNA in the DNA–HU complex. Higher concentrations of HU do not result in additional supercoils. It appears that HU bends DNA sharply into a tight circle and is primarily responsible for the formation of a beaded nucleoid structure. In the nucleoid, a single supercoiled DNA molecule is organized into about 40 loops, each approximately 10^5 bp in size, that merge into a scaffold rich in protein and RNA. A role for RNA in chromosome packing is unique to prokaryotes. As a result of formation of a nucleoid, the bacterial genome can easily fit into a cell.

Organization of Eukaryotic Chromatin

The enormous length of the genome of most eukaryotes necessitates the division of genetic information into several independent chromosomes. Human cells contain 23 pairs of chromosomes with an average length of 1.3×10^8 bp, or approximately 43 mm. Each human chromosome consists of a DNA molecule varying in size from 263×10^6 bp for chromosome 1 to less than 50×10^6 bp for the Y chromosome. For this amount of DNA to fit within a cell nucleus with a diameter of approximately 10–20 μm requires a condensation ratio over five orders of magnitude (Figure 2.39).

DNA in eukaryotic cells is associated with numerous proteins to form **chromatin.** In nondividing (interphase) cells, chromatin is amorphous and dispersed throughout the nucleus. Just prior to cell division (metaphase), it becomes organized into highly compacted structures called **chromosomes.** Each chromosome is characterized by a **centromere,** which is the site for attachment to proteins that link the chromosome to the mitotic spindle. **Telomeres** define the termini of linear chromosomes. Chromosomes also contain sequences required for initiation of DNA replication (**origin of replication**).

Histones are the most numerous proteins in chromatin. There are five classes of these proteins: **histones H1, H2A, H2B, H3,** and **H4.** Because of their unusually high content of the basic amino acids lysine and arginine, histones are highly cationic and interact with the polyanionic phosphate backbone of DNA to produce uncharged nucleoproteins. The amino acid sequences of the histones are very highly conserved between species. Histones H4 from peas and cows differ by only two amino acids, although these species diverged more than a billion years ago. Histones H3 are also very highly conserved, whereas the nonbasic regions of histones H2A and H2B are less highly conserved. H1 is larger, more basic, and by far the most tissue-specific and species-specific histone. In some cell types of vertebrates, a sixth histone, H5, functions as a replacement for H1. A heterogeneous group of proteins with high species and organ specificity is also present in chromatin. These nonhistone proteins consist of several hundred members, most of which are present in trace

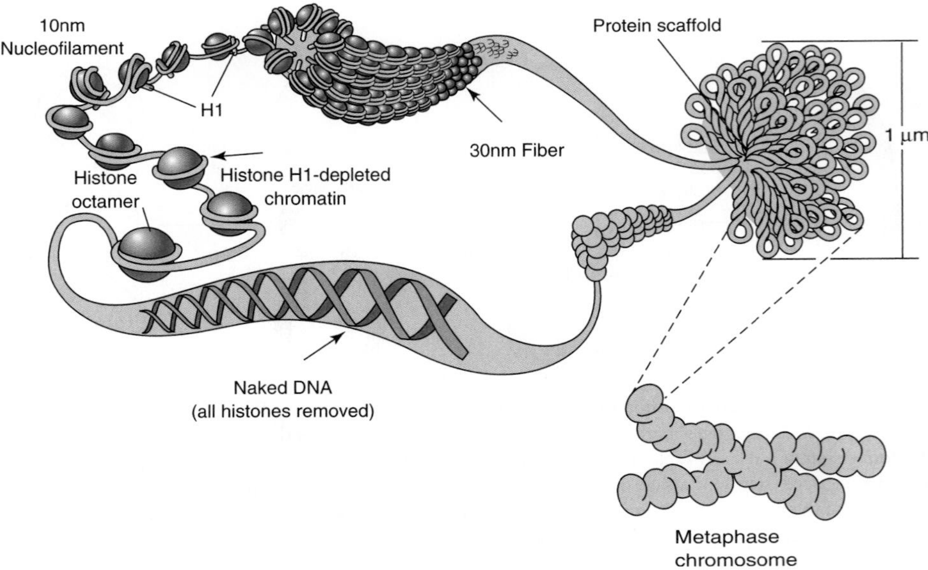

Figure 2.39 Organization of DNA into chromosomes. A drawing showing the stepwise condensation of DNA into chromatin. The DNA initially wraps around the histones of the nucleosome core. Condensation with histone H1 produces the 10-nm nucleofilament, which is subsequently packaged into a twisted, looped structure attached to a protein scaffold within the chromosome.

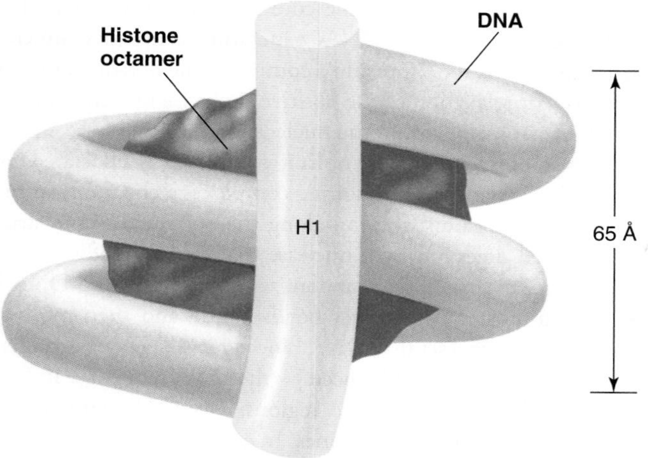

Figure 2.40 Models of the nucleosome complex. The nucleosome consists of approximately 146 bp of DNA corresponding to 1¾ superhelical turns wound around a histone octamer. The chromatosome consists of about 166 bp of DNA (two superhelical turns). The H1 histone is associated with the linker DNA. Reprinted with permission from Voet, D., and Voet, J. G. *Biochemistry.* 3d ed. New York: Wiley, 2004. © (2004) Donald and Judith Voet.

amounts. Many of them are associated with functions of chromosomes such as replication, gene expression, and chromosome organization.

Nucleosomes and Polynucleosomes

Histones interact with DNA to form a periodic beads-on-a-string structure, called a polynucleosome, in which the elementary unit is a **nucleosome** (Figure 2.40). Each nucleosome is disk shaped, about 11 nm in diameter and 6 nm in height, and consists of a DNA segment and a histone cluster composed of two molecules each of H2A, H2B, H3, and H4 histones. Each octameric cluster consists of the tetramer $(H3)_2$–$(H4)_2$ with two H2A–H2B dimers. Each histone is characterized by a central nonpolar domain, which forms a globular structure and is responsible for histone–histone interactions. The neighboring N-terminal regions contain most of the positively charged amino acid residues that provide favorable electrostatic interactions with DNA. These tail sequences appear to extend radially out from the histone core and may be involved in interactions between nucleosomes. The DNA is wrapped around the octamer with the $(H3)_2$-$(H4)_2$ core interacting with the central 70–80 bp of the surrounding DNA (Figure 2.41). Approximately 146 bp of DNA wraps around the histone octamer. The histones are in contact with the minor groove of DNA and leave the major groove available for interaction with proteins that regulate gene expression and other DNA functions. The structure of the nucleosome core explains why eukaryotic cells lack gyrases that can underwind relaxed DNA. Negative superhelicity is introduced because

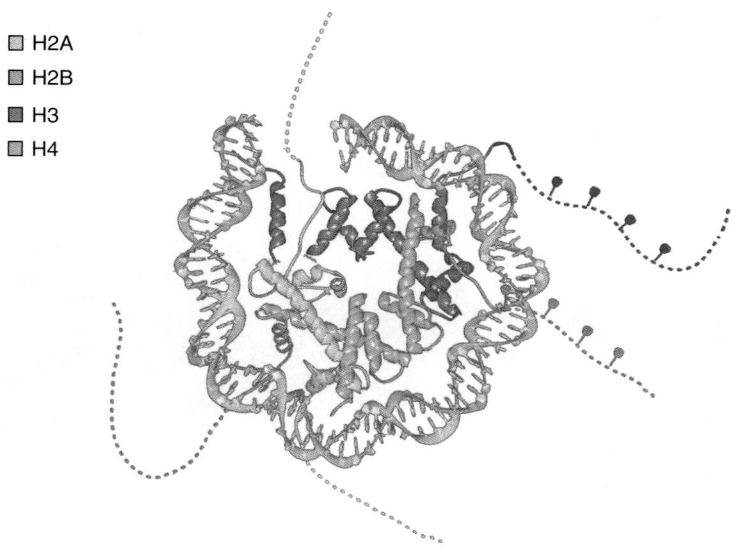

Figure 2.41 X-ray structure of the nucleosome core. The disk of the nucleosome core has been split in half to show one turn of DNA wrapped around four histone molecules. The dashed lines represent unstructured parts of the histone tails. Reproduced with permission from Kornberg, R. D., and Lorch, Y. *Cell* 98: 285, 1999.

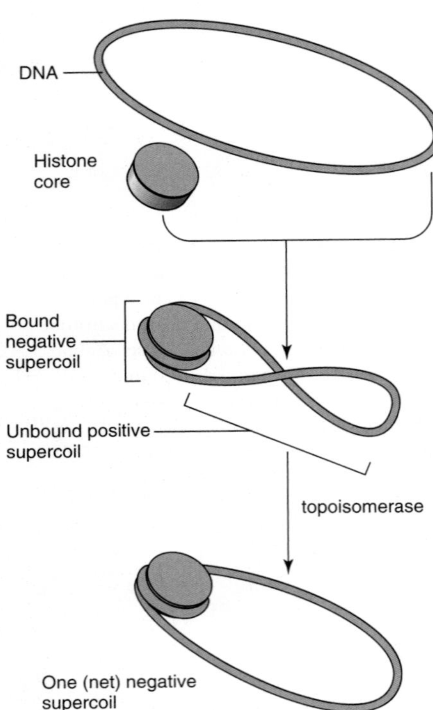

Figure 2.42 Generation of negative supercoiling in eukaryotic DNA. The binding of a histone octamer to a relaxed, closed-domain DNA forces the DNA to wrap around the octamer, generating a negative supercoil. In the absence of any strand breaks, the domain remains intact and a compensating positive supercoil must be generated elsewhere within the domain. The action of a eukaryotic topoisomerase subsequently relaxes the positive supercoil, leaving the closed domain with one net negative supercoil.

of DNA forming a coil around the histone core (Figure 2.42). Such wrapping removes approximately one helical turn from DNA, generating a negative supercoil within the region wrapped around the histone core and a compensating positive supercoil elsewhere in the molecule. Subsequent relaxation of the positive supercoil by eukaryotic topoisomerases leaves one net negative supercoil within the nucleosomal region.

Between nucleosomes is 20 to 90 bp of **linker DNA.** It is associated with histone H1 that locks the coiled DNA in place; the resulting complex is a **chromatosome** (Figure 2.40). The periodicity of nucleosome distribution along the polynucleosome structure has been determined by controlled digestion by a nuclease that preferentially attacks linker DNA. The distribution of nucleosomes is not random with respect to the DNA base sequence. DNA does not bend uniformly but rather bends gently and then more sharply around the histone octamers. This suggests that DNA binding is sequence dependent and that the positioning of nucleosomes may be influenced by the DNA sequence. In fact, nucleosomes tend to form preferentially in certain DNA regions. DNA that contains long A tracts or G-C repeats does not usually form nucleosomes. In contrast, certain bent DNA regions, for instance, periodically phased A tracts, associate strongly with histones. Histone octamers can migrate along the DNA strand. This mobility allows access to the DNA by polymerases and other proteins necessary for transcription and replication.

Polynucleosome Packing into Higher Structures

Wrapping of DNA around histones to form nucleosomes results in a 10-fold reduction of the apparent length of DNA and the formation of a **10-nm nucleofilament.** Condensation of 10-nm nucleofilaments into a **solenoid** arrangement involving six to seven chromatosomes per turn forms **30-nm fibers** (Figure 2.43). Histone H1 molecules bind to one another cooperatively, bringing the neighboring nucleosomes closer together in 30-nm fibers. At physiological salt concentrations, the 30-nm fibers form spontaneously, but at low ionic strength they dissociate into 10-nm nucleofilaments. Formation of polynucleosomes and their condensation into 30-nm fibers provides for a DNA compaction ratio of up to two orders of magnitude. The 30-nm fibers form only over selected regions of DNA that are characterized by the absence of binding with other sequence-specific nonhistone proteins. The presence of these and their effects on formation of 30-nm fibers may depend on the transcriptional status of the regions of DNA involved.

Models of the higher levels of packing of 30-nm fibers are based on indirect evidence obtained from the lampbrush chromosomes of vertebrate oocytes and polytene chromosomes of fruit fly giant secretory cells. These chromosomes are exceptional in that they maintain precisely defined higher-order structures in interphase, that is, when cells are in a resting (nondividing) state. By extrapolation, the structural features of interphase lampbrush chromosomes have led to the proposal that chromosomes in general are organized as a series of looped, condensed domains of 30-nm fibers of variable size for different organisms. These loops may contain 5,000 to 120,000 bp with an average of about 20,000. Thus, the haploid human genome would contain about 60,000 loops. A loop may contain a few linked genes. The loops are bound to a protein scaffold consisting of H1 histone and several nonhistone proteins, including two major scaffold proteins, Sc1 (a topoisomerase II)

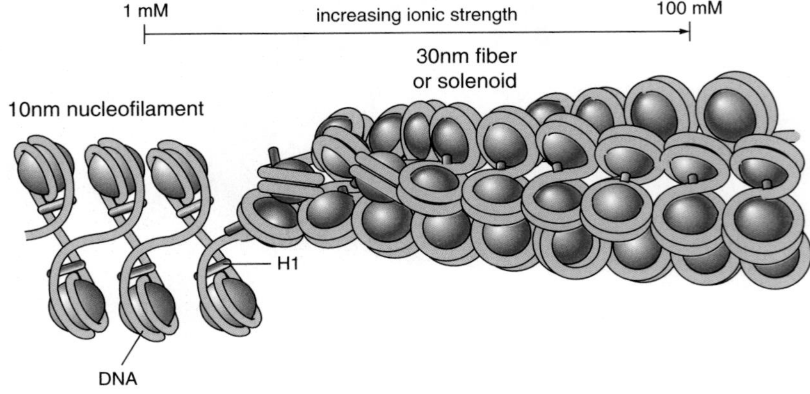

Figure 2.43 Nucleofilament structure. Histone H1 attached to the linker regions between nucleosomes results in condensation into 10-nm fibers. At higher ionic strengths the nucleofilament forms a very compact helical structure or a helical solenoid. The H1 histones interact strongly with one another in this structure.
Adapted from Kornberg, R. D., and Klug, A. *The Nucleosome.* San Diego, CA: Academic, 1989.

and Sc2. They are fixed at their bases and can therefore accumulate supercoils. Specific AT-rich regions known as scaffold attachment regions (SARs) are preferentially associated with the scaffold. SARs also contain topoisomerase II binding sites. The presence of the topoisomerases suggests that changes in supercoiling within these domains are biologically important. Formation of looped domains may account for an additional 200-fold condensation in the length of DNA and an overall packing ratio of more than four orders of magnitude. Each loop can be coiled and then supercoiled into 0.4 μm of a 30-nm fiber. Since a chromosome is about 1 μm in diameter, packing of the 30-nm fiber into a chromatid would require just one more order of folding. Further packing may be achieved by arranging the loops of 30-nm fiber into tightly stacked helical coils. **Chromatids,** the two linked arms of replicated DNA in the highly condensed metaphase chromosomes, may consist of helically packed loops of 30-nm fibers.

Packing changes at various stages of the cell cycle appear to be controlled in part by covalent modification of the core histones (**histone remodeling**). Reversible acetylation on the ε-amino groups of lysine and phosphorylation of serine and threonine residues are involved in regulating the activity of nucleosomal DNA. These modifications reduce the numbers of positive charges on the histones, causing them to bind less tightly to DNA. The resultant unraveling of 30-nm fibers and decondensation of chromatin produces loosely packed **euchromatin,** the transcriptionally active form of DNA. Methylation of basic amino acid side chains (lysine and arginine) promotes the opposite effect and, when accompanied by decreased acetylation, characterizes a highly condensed, inactive form of chromatin called **heterochromatin.** Throughout the cell cycle, permanently repressed genes and other untranscribed regions such as centromeres, telomeres, and " junk" DNA remain condensed.

The pattern of covalent modifications (the **histone code**) provides a highly tunable means of regulating gene expression (Clin. Corr. 2.8). While this code does not involve changes in the sequence of DNA (it is **epigenetic**), it is heritable due to coordination between histone modification and methylation of DNA at the 5-position of deoxycytidine (Figure 2.23; p. 41).

CLINICAL CORRELATION 2.8

Epigenetic Treatments for Cancer

Progression toward cancer can involve epigenetic alterations that affect processes including cell-cycle progression, repair of DNA damage, and programmed cell death (apoptosis). One such alteration, chromatin remodeling, has drawn attention as a target for cancer treatment. Histone deacetylases, enzymes responsible for the reversal of lysine and arginine acetylation, are overexpressed in a number of cancers including colon, prostate, breast, and cervical. Hydroxamic acids such as tricostatin A and suberoylanilide hydroxamic acid (SAHA) bind to a zinc ion in the enzyme-active site and block deacetylation of histone lysines and arginines. These compounds inhibit tumor cell growth and block cell-cycle progression. Notably, SAHA is currently approved for the treatment of cutaneous T cell lymphoma under the name Zolinza®. Other modifications of the histone code such as serine phosphorylation play roles in repair and chromosome stability and may also provide targets for antiproliferative drugs.

There is also a potential dietary component to the epigenetic control of cancer. Short-chain fatty acids such as butyrate are generated by bacterial flora of the large intestine and show a weak ability to regulate epithelial cell mitosis, differentiation, and apoptosis. Experiments have shown that butyrate increases the levels of histone acetylation, presumably from inhibition of histone deacetylase. These data suggest that the nutritional state of an individual can affect the histone code and its role in cancer.

Inche, A. G., and LaThangue, N. B. Chromatin control and cancer-drug discovery: Realizing the promise. *Drug Discov. Today* 11:97, 2006; Wang, G. G., Allis, C. D., and Chi, P. Chromatin remodeling and cancer, part I: Covalent histone modifications. *Trends Mol. Med.* 13:363, 2006; Marks, P. A., Richon, V. M., Breslow, R., and Rifkind, R. A. Histone deacetylase inhibitors as new cancer drugs. *Curr. Opin. Oncol.* 13:477, 2001; and Garfinkel, M. D., and Ruden, D. M. Chromatin effects in nutrition, cancer, and obesity. *Nutrition* 20:56, 2004.

tricostatin A

suberoylanilide hydroxamic acid (SAHA, Zolinza®)

2.5 • SEQUENCE AND FUNCTION OF DNA

The size and average base composition of DNA vary widely between species. What makes the DNA of a species unique is the nucleotide sequence. Until recently, direct determination of nucleotide sequences in genomic DNA was an intimidating undertaking. The technology developed in connection with the Human Genome Project has accelerated the rate at which DNA sequences are determined. DNA sequences of over 1000 species have already become available for analysis.

Restriction Endonucleases and Palindromes

One key event that enabled the development of methods to sequence genomic DNA was the discovery of bacterial **restriction endonucleases.** These cleave double-stranded DNA at a specific sequence by cutting each strand (Figure 2.44). Bacteria developed restriction enzymes as a defense against infection by phages. Cleavage exposes the viral DNA to eventual degradation by nonspecific bacterial exonucleases. Bacterial DNA can be protected from cleavage by sequence-specific methylation. The recognition sites for restriction methylases correspond to those of the endonucleases. Methylation of specific bases within the recognition site prevents cleavage by the cognate nuclease. The most common sites for base methylation are the 5-position of cytosine and the 6-NH_2 group of adenine. Notably, base methylation is also critical in gene regulation in higher organisms.

Hundreds of restriction endonucleases have been purified. With few exceptions, they recognize sequences four to six nucleotides long. Rare cutters, endonucleases with unusually long recognition sites, are valuable because of the relative infrequency of cleavage of very large DNAs such as eukaryotic chromosomes. Not I, for example, has an eight-nucleotide recognition sequence. The recognized sequences are often symmetrical inverted repeats or palindromes; the order of the bases is the same when the two complementary strands of the palindrome are read $5' \rightarrow 3'$. For EcoR1 from *E. coli,* the sequence is 5′-GAATTC-3′. Restriction endonucleases types I and III cut in the vicinity of the recognition site, whereas type II specifically cleaves DNA within the recognition sequence.

Restriction endonucleases recognize specific sequences that can have relatively low frequencie; they fragment DNA very selectively. For example, a typical bacterial DNA with 3×10^6 bp will be cleaved into a few hundred fragments. Plasmid DNAs may have few or no cutting sites at all for a particular restriction endonuclease (Figure 2.45). Thus, a particular restriction enzyme generates a unique family of fragments, or restriction digest, for any given DNA molecule. The availability of restriction enzymes and development of gel electrophoresis techniques for separating DNA fragments have made determination of sequences a simple matter (p. 259).

Most Prokaryotic DNA Codes for Specific Proteins

In prokaryotes a large percentage of the DNA codes for specific proteins. The entire *E. coli* genome consists of about 4.6×10^6 bp of DNA and contains ~4200 genes. Not all of the genes code for proteins. For example, 80 genes code for tRNA molecules, and some may

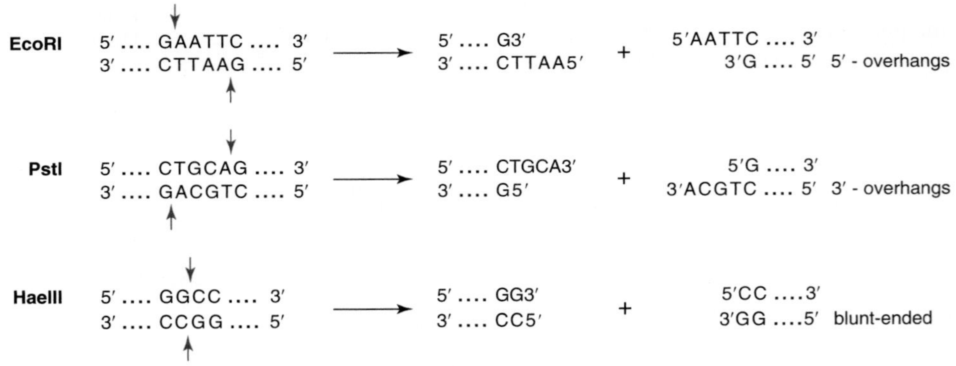

EcoRI 5′ .... GAATTC .... 3′ / 3′ .... CTTAAG .... 5′ → 5′ .... G3′ / 3′ .... CTTAA5′ + 5′AATTC .... 3′ / 3′G .... 5′ 5′ - overhangs

PstI 5′ .... CTGCAG .... 3′ / 3′ .... GACGTC .... 5′ → 5′ .... CTGCA3′ / 3′ .... G5′ + 5′G .... 3′ / 3′ACGTC .... 5′ 3′ - overhangs

HaeIII 5′ .... GGCC .... 3′ / 3′ .... CCGG .... 5′ → 5′ .... GG3′ / 3′ .... CC5′ + 5′CC ....3′ / 3′GG ....5′ blunt-ended

Figure 2.44 Types of products generated by type II restriction endonucleases. Enzymes exemplified by EcoRI and PstI nick on both sides of the center of symmetry of the palindrome, generating single-stranded overhangs. Many commonly used enzymes generate 5′-overhangs, although some produce ends with 3′-overhangs as shown for PstI. Other restriction nucleases cut across the center of symmetry of the recognition sequence, to produce flush or blunt ends, as exemplified by HaeIII.

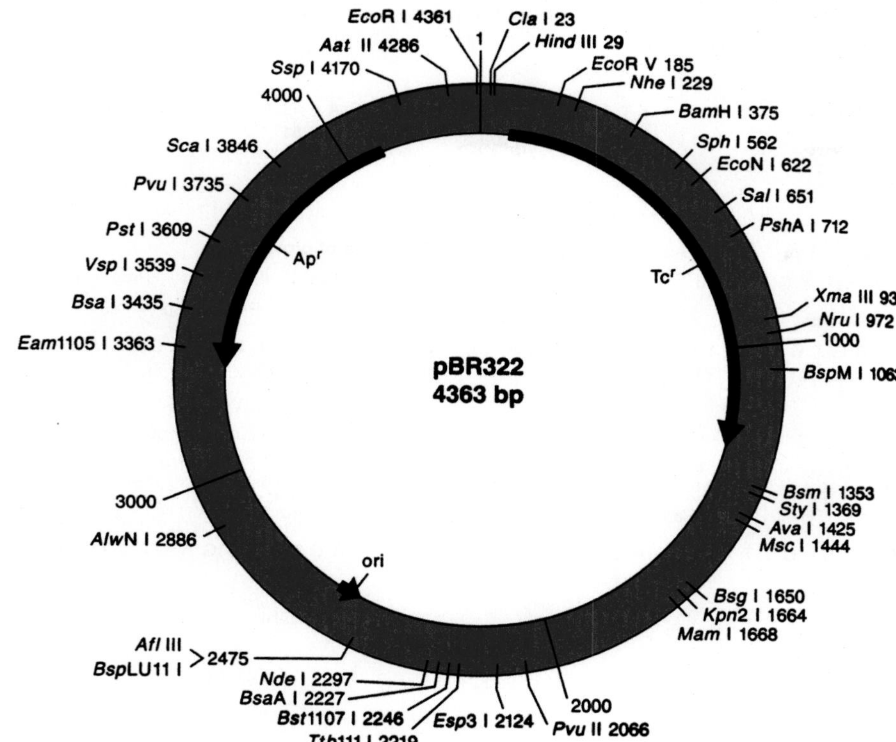

Figure 2.45 Restriction map for plasmid DNA. Bacterial plasmid pBR322 is a 4363-bp circular DNA containing an origin of replication (ori) and genes for resistance to the antibiotics ampicillin (Ap^r) and tetracycline (Tc^r). The locations of unique DNA sequences that are recognized by a number of restriction endonuclease are marked. Other endonucleases may cleave this plasmid many times or not at all. For example, the recognition sequence for Bgl I appears in three locations on this plasmid, whereas Bgl II does not cleave this plasmid at all.

not encode functional molecules at all. *E. coli* DNA is densely packed with sequence information, and there is little repetition of information within it. As much as 1% is composed of multiple copies of short **repetitive sequences** known as repeated extragenic palindromic (REP) elements. These are present at sites of DNA interaction with functional proteins, for example, in the region of initiation of DNA replication (referred to as OriC). At OriC, REP elements have a consensus sequence of 34 nucleotides and bind topoisomerase II. REP elements with the sequence GCTGGTGG (Chi sites) bind the enzyme RecBCD, which initiates DNA recombination. Chi sites are regularly spaced at intervals separated by about 4,000 bp. Genetic information is even more densely organized in smaller organisms, such as bacteriophages, where the primary sequence of DNA reveals that structural genes—nucleotide sequences coding for protein—do not always have exclusive physical locations. Rather, they frequently overlap with one another, as illustrated by the partial sequence of phage φX174 (Figure 2.46). The overlap makes the efficient and economic utilization of the limited DNA present and may control the order of gene expression.

Only a Small Percentage of Eukaryotic DNA Consists of Functional Genes

Eukaryotes have much larger genomes than prokaryotes, from about 1.5×10^7 bp for yeast to about 1.15×10^{11} bp for the haploid genome of the lily *Fritillaria assyrica.* Results from the Human Genome Project, however, indicate that the human genome codes for only 20,000–25,000 genes. As a result, genetic information need not be as densely packed as in bacteria. A typical mammalian DNA, with only 7 times as many genes as that of *E. coli,* contains 500 times more DNA than *E. coli.* In fact, only 2% to 4% of DNA in a mammalian cell may suffice for all its genes. Some of the remaining DNA, such as that in centromeres and telomeres, has a well-defined function. No specific function can be assigned to the majority of noncoding DNA, and it has been referred to as "junk DNA". However, there is increasing evidence that junk DNA may have a vital role in regulation of gene expression during development. Paradoxically, the amount of junk DNA correlates better to the complexity of the organism than the number of genes.

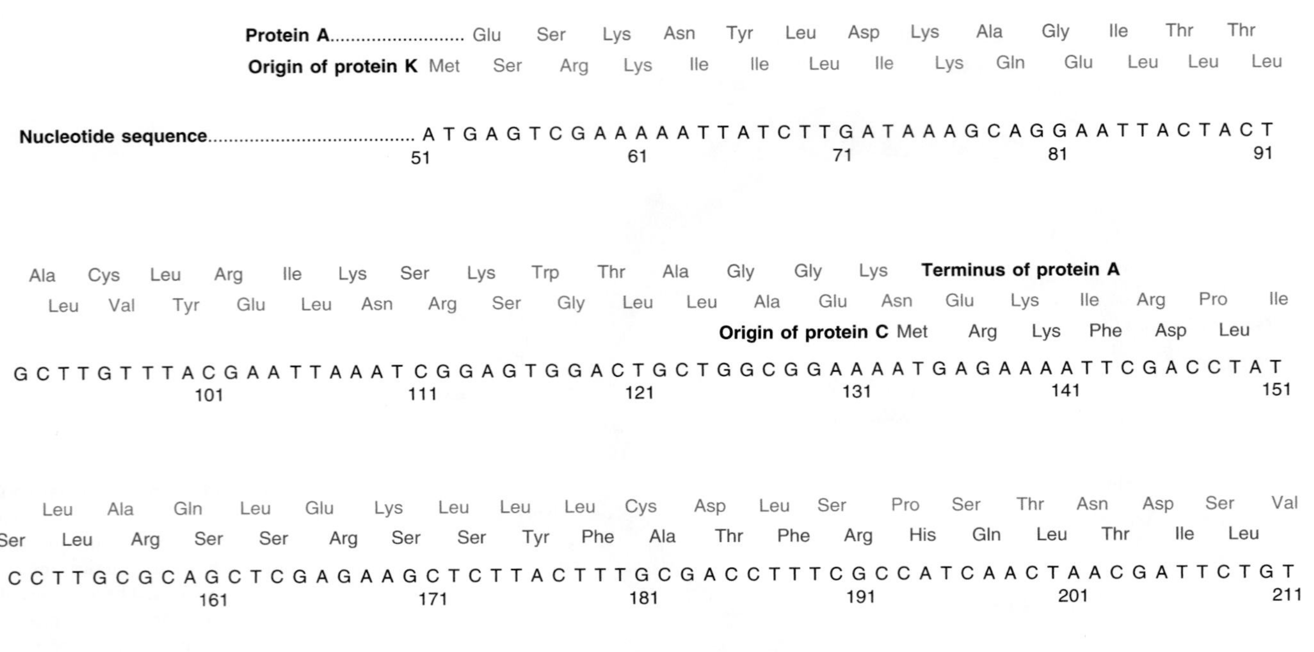

Figure 2.46 Partial nucleotide sequences of contiguous and overlapping genes of phage φX174. The sequence for protein K, nucleotides 51–219, is shown in this figure. This sequence codes. A part of this sequence, nucleotides 51–133, codes for part of protein A. Another part of the sequence, nucleotides 133–219, for part of protein C. Similar overlaps are noted between other genes of φX174.
Adapted from Smith, M. *Am. Sci.* 67:61, 1979.

Eukaryotic genes do not overlap and are spaced, on average, about 40,000 bp apart. However, some may be closer together in regions that contain genes expressed in a tightly coordinated manner (gene clusters). Most eukaryotic genes are interrupted by noncoding intervening nucleotide sequences, **introns** (Figure 2.47). The sequences in the gene that are expressed, either in the final RNA product (mature RNA) or as a protein, are termed **exons.** Introns are removed during the processing of the RNA transcript, and the exons are ligated together. This tailoring of the original transcript is called **splicing.** The sequence and size of introns vary greatly among species. Added together they may be 5 to 10 times the length of the exons they separate. Most genes are interrupted by introns at least once, whereas others are interrupted more frequently. Some genes, such as that for human interferon-α, contain no introns.

Introns are common in genes of vertebrates and flowering plants but infrequent in those of other species. The biological role of introns is not clear. They may have arisen relatively recently in evolution because of the migration of mobile DNA elements (transposons) from

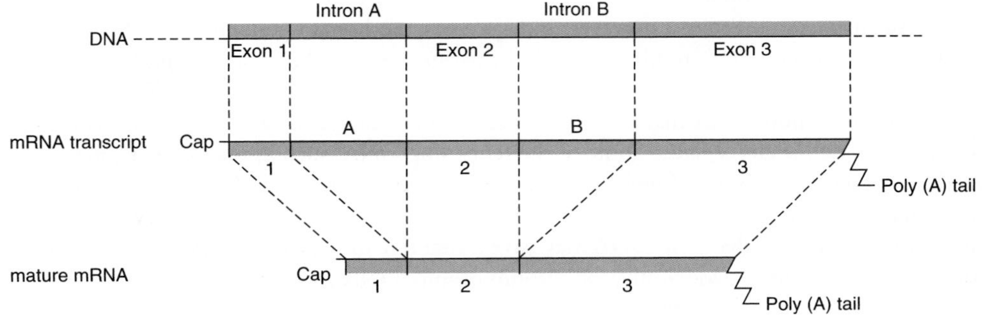

Figure 2.47 Schematic presentation of a eukaryotic gene. The top line represents a part of the DNA genome: the bottom line represents the mRNA produced by it. In this example the DNA consists of two introns and three exons. The intron sequences are transcribed but are removed during splicing; this produces a mature mRNA.
Redrawn from Crick, F. *Science* 204:264, 1979.

other parts of the genome and their insertion into protein-coding genes (p. 316). Through mutation, these inserts may have subsequently lost their mobility.

Repeated Sequences

While repetition of particular DNA sequences in prokaryotes is very limited, the DNA of eukaryotes contains sequences that are repeated a few times, for specified coding genes, to millions of times per genome, for certain simple, relatively short, sequences. Depending on the species, **repetitive DNA** may constitute 3% to 80% of total DNA. In mammalian genomes, including the human genome, 25% to 35% of the DNA is repetitive. Sequences can be classified as **single copy, moderately reiterated, highly reiterated,** and **inverted repeat.** The content of single-copy DNA varies among eukaryotes. The exact boundaries between the various types of reiterated DNAs are not strictly defined.

About one-half of the human genome consists of unique sequences, but only a small fraction of these encode proteins. Some DNA contains **pseudogenes,** sequences of DNA that have significant nucleotide homology to a functional gene but contain mutations that prevent gene expression. Pseudogenes significantly increase the size of eukaryotic genomes without contributing to their expressible genetic content. Additional single-copy DNA sequences serve as introns and as control regions that flank genes.

Moderately reiterated DNA consists of copies of identical or closely related sequences that are reiterated from a few to a thousand times. These sequences are relatively long, varying from a hundred to many thousands of nucleotides. About 20% of mouse DNA consists of sequences a few hundred base pairs long that are repeated more than a thousand times. About 15% of the human genome consists of moderately reiterated DNA. Some sequences involve genes for highly abundant proteins and for ribosomal RNAs (rRNAs). For example, the sea urchins carry hundreds of copies of histone genes. Eukaryotes generally possess hundreds of copies of the gene coding for rRNA. The function of other moderately repeated sequences is not clearly known.

Normally, single-copy and moderately reiterated sequences occur on a chromosome in an orderly pattern known as the **interspersion pattern,** which consists of alternating blocks of single-copy DNA and moderately reiterated DNA. **Long interspersed elements** (LINEs) consist of sequences several thousand nucleotides long with up to 1000 copies per genome. They are flanked on either side by sequences that are direct repeats. An example of a **short interspersed element** (SINE) is the Alu family, a sequence of unknown function that constitutes a substantial portion (about 5%) of the human genome. Alu sequences are approximately 300 bp long and are repeated over 500,000 times. The structures of the short interspersed repeats, including the Alu family, are reminiscent of transposons.

Approximately 1% to 15% of eukaryotic genomic DNA consists of sequences typically shorter than 20 nucleotides reiterated thousands or millions of times. Most highly reiterated sequences have a characteristic base composition and can be isolated by shearing the DNA into segments a few hundred nucleotides long and separating the fragments by density gradient centrifugation. These fragments are termed **satellite DNA** because they appear as satellites of the band of bulk DNA after centrifugation. Other highly reiterated sequences, which cannot be isolated by centrifugation, can be identified by their property of rapid reannealing. These highly reiterated DNAs are also called simple-sequence DNA. Simple sequences are typically present in the DNA of most, if not all, eukaryotes. In some species, one major sequence is present, while in others, several simple sequences are repeated up to 1 million times. Simple sequence DNA can often be isolated as satellite DNA. That found in the centromeres of higher eukaryotes consists of thousands of tandem copies of one or a few short sequences. Satellite sequences are only 5–10 bp long and are a constituent of telomeres where they have a well-defined role in DNA replication. Some longer simple-sequence DNA has been identified. For instance, in the genome of the African green monkey a 172-bp segment that contains some sequence repetitions is highly reiterated.

Inverted repeats are a structural motif of DNA. Short inverted repeats, consisting of up to six nucleotides long, for example, the palindromic sequence GAATTC, occur by chance about once for every 3000 nucleotides. Such short repeats cannot form a stable cruciform structure as is formed by longer palindromic sequences. Inverted repeat sequences that are long enough to form stable cruciforms are not likely to occur by chance and should be

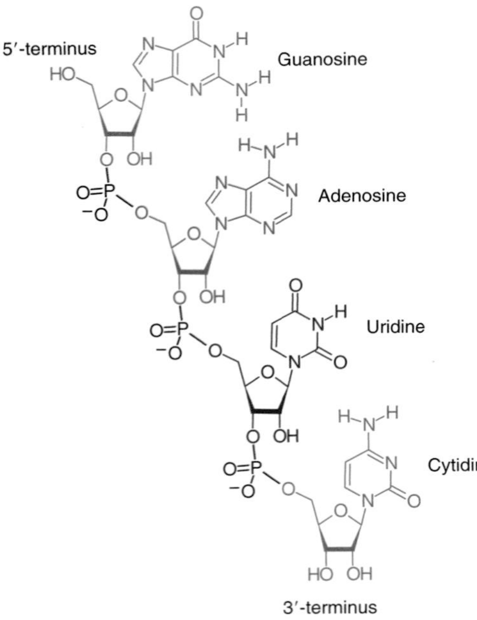

5′-terminus
Guanosine
Adenosine
Uridine
Cytidine
3′-terminus

Figure 2.48 Structure of the 3′, 5′-phosphodiester bonds between ribonucleotides forming a single strand of RNA. Phosphate joins the 3′ -OH group of one ribose with the 5′-OH group of the next ribose. This linkage produces a polyribonucleotide having a sugar–phosphate backbone. Purine and pyrimidine bases extend away from the axis of the backbone and may pair with complementary bases to form double-helical base-paired regions.

classified as a separate class of eukaryotic sequence. In human DNA about 2 million inverted repeats are present, with an average length of about 200 bp; however, inverted sequences longer than 1000 bp have been detected. Most inverted repeat sequences are repeated 1000 or more times per cell.

2.6 • STRUCTURE OF RNA

RNA Is a Polymer of Ribonucleoside 5′-Monophosphates

RNA is a linear polymer of ribonucleoside monophosphates. The purine bases in RNA are adenine and guanine; the pyrimidines are cytosine and uracil (Figure 2.2; p. 27). Except for uracil, which replaces thymine, these are the same bases found in DNA. A, C, G, and U nucleotides (Figure 2.3; p. 27) are incorporated into RNA during transcription. Many RNAs also contain modified nucleotides that are produced by **post-transcriptional processing.** Modified nucleotides are especially characteristic of stable RNA species (e.g., tRNA and rRNA); however, some methylated nucleotides are also present in eukaryotic mRNA. For the most part, modified nucleotides in RNA have fine tuning rather than indispensable roles in the cell.

The 3′,5′-phosphodiester bonds of RNA form a backbone from which the bases extend (Figure 2.48). Eukaryotic RNAs vary from approximately 20 nucleotides long to more than 200,000 nucleotides long. Each RNA is complementary to the base sequences of specific portions of only one strand of DNA. Thus, unlike the base composition of DNA, molar ratios of (A + U) and (G + C) in RNA are not equal. Cellular RNA is linear and single stranded. Stable, double-stranded RNA is not present in the cell. Rather, the presence of double-stranded RNA signals its destruction. This appears to be a part of the cellular response to virus infection, since some virus genomes are composed of double-stranded RNA, and double-stranded RNA is an intermediate in the replication of single-stranded RNA viruses.

Chemically, RNA is similar to DNA. Both contain negatively charged phosphodiester bonds, and the bases are very similar chemically. The differences between DNA and RNA are largely due to two factors. First, RNA contains ribose rather than 2′-deoxyribose as the nucleotide sugar component, and second, RNAs are generally single stranded rather than double stranded.

The 2′-hydroxyl group makes the phosphodiester bonds of an RNA molecule more susceptible to chemical hydrolysis, especially in alkaline solution, than those of DNA (Figure 2.11, p. 31). The chemical instability of RNA is reflected in its metabolic instability. Some RNAs, such as bacterial mRNA, are synthesized, used, and degraded within minutes. Others, such as human rRNA, are more stable metabolically, with a lifetime measured in days. Nevertheless, even the most stable RNAs are much less stable than DNA.

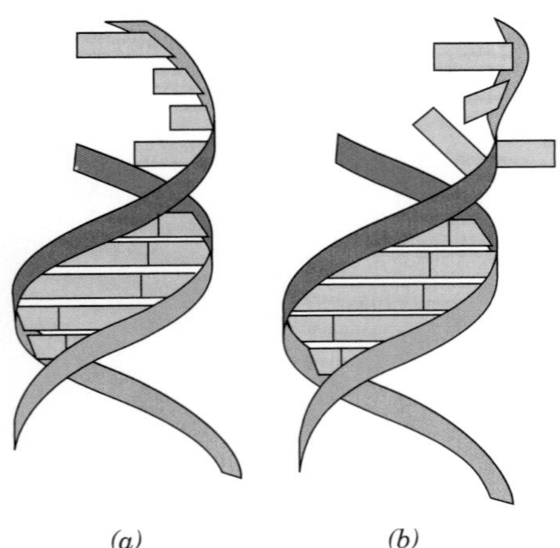

Figure 2.49 Helical structure of RNA Models indicating a helical structure due to (*a*) base stacking in the CCA terminus of tRNA and (*b*) lack of an ordered helix when no stacking occurs in this nonbase-paired region.
Redrawn from Sprinzl, M., and Cramer, F. *Prog. Nucl. Res. Mol. Biol.* 22:9, 1979.

Secondary Structure of RNA Involves Intramolecular Base Pairing

Since RNA molecules are single stranded, they do not usually form extensive double helices. Rather, the secondary structure of an RNA molecule results from relatively short regions of intramolecular base pairing. Even nonpaired sequences of single-stranded RNAs may contain considerable helical structure (Figure 2.49). Helices within RNA are usually the A-type with 11 nucleotides per turn.

Double-helical stem-loop regions in RNA often form "hairpins." There are considerable variations in the fine structural details of hairpin structures, including the length of base-paired regions and the size and number of unpaired loops (Figure 2.50). Transfer RNAs are excellent examples of base stacking and hydrogen bonding in a single-chain RNA molecule (Figure 2.51). About 60% of bases are paired in four double-helical stems, and two loops interact with each other, resulting in an L-shaped molecule. The anticodon region in tRNA is an unpaired, base-stacked loop of seven nucleotides. The partial helix caused by base stacking in this loop binds, by base pairing, to a complementary codon in mRNA so that translation can occur.

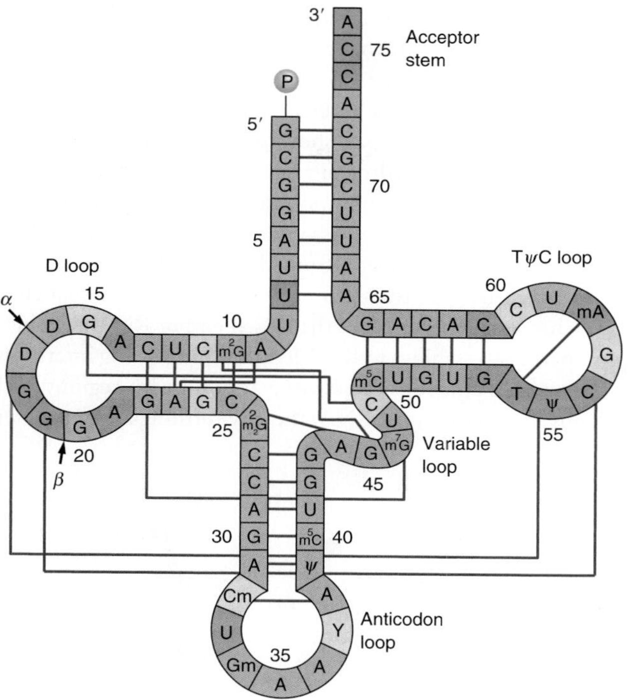

Figure 2.50 Cloverleaf structure of tRNA Cloverleaf diagram of the two-dimensional structure and nucleotide sequence of yeast tRNA[Phe]. Red lines connecting circled nucleotides indicate hydrogen-bonded bases. Solid squares indicate constant nucleotides; dashed squares indicate a constant purine or pyrimidine. Insertion of nucleotides in the D loop occurs at positions α and β for different tRNAs.
Redrawn from Quigley, G. J., and Rich, A. *Science* 194:797, 1976.

RNA Molecules Have Tertiary Structures

Structures of functional RNA molecules are more complex than is suggested by the base-stacked and hydrogen-bonded helices mentioned above. In particular, the bases in unpaired regions can form hydrogen bonds or otherwise interact with each other and with helical portions of the molecule. *In vivo* RNAs are dynamic molecules that can change their conformations during synthesis, processing, and functioning. Proteins associated with RNA molecules often lend stability to the RNA structure. In fact, cellular RNA usually functions as RNA–protein complexes, rather than as free RNA molecules.

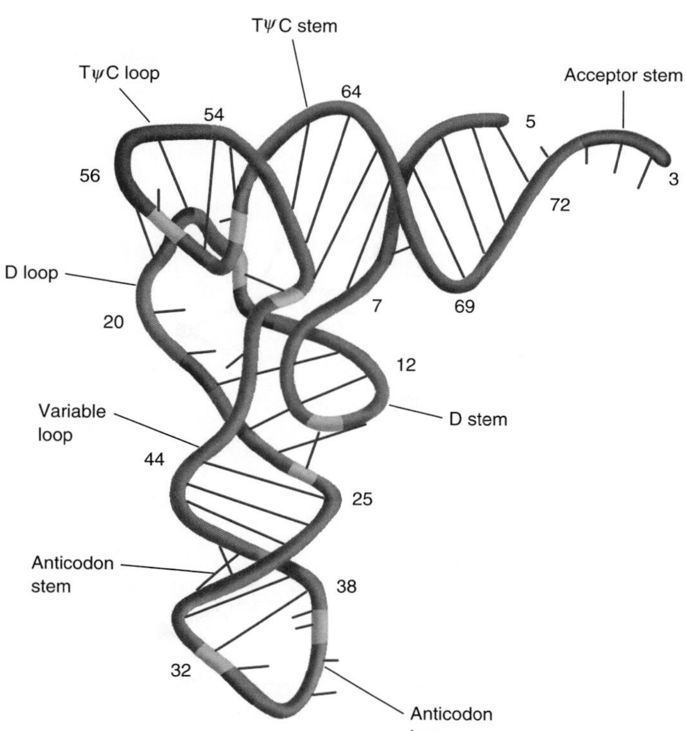

Figure 2.51 Tertiary structure on tRNA Tertiary folding of the cloverleaf structure of tRNA[Phe]. Hydrogen bonds are indicated by cross rungs. Compare the presentation with Figure 2.50.
Redrawn from Quigley, G. J., and Rich, A. *Science* 194:797, 1976.

The tertiary structure of RNA molecules results from base stacking and hydrogen bonding between different parts of the molecule. tRNA provides a number of examples. In solution, tRNA is folded into a compact "L-shaped" conformation (Figure 2.51), stabilized by Watson–Crick base pairing and base interactions involving more than two nucleotides. This is true for other RNAs. Figure 2.52*a* shows a well-defined base triplet from bacterial ribosomal RNA. Bases can donate hydrogen atoms to bond with the phosphodiester backbone. The 2′-OH of the ribose is also an important donor and acceptor for hydrogen-bond formation. All these interactions contribute to the folded shape of an RNA molecule.

Other tertiary features were deduced from the three-dimensional structure of RNAs in solution or in crystal form. The term *tetraloop* refers to specific sequences (for example, UUCG or GAGA) that are often found at the closed end of RNA hairpins (Figure 2.52*b*). These tetraloops impart extra stability to RNA hairpins by forming internal hydrogen bonds between bases and sugars. Tetraloops can hydrogen-bond with regions of the double helix or with internal loops in other parts of RNA, thereby forming a complex tertiary structure.

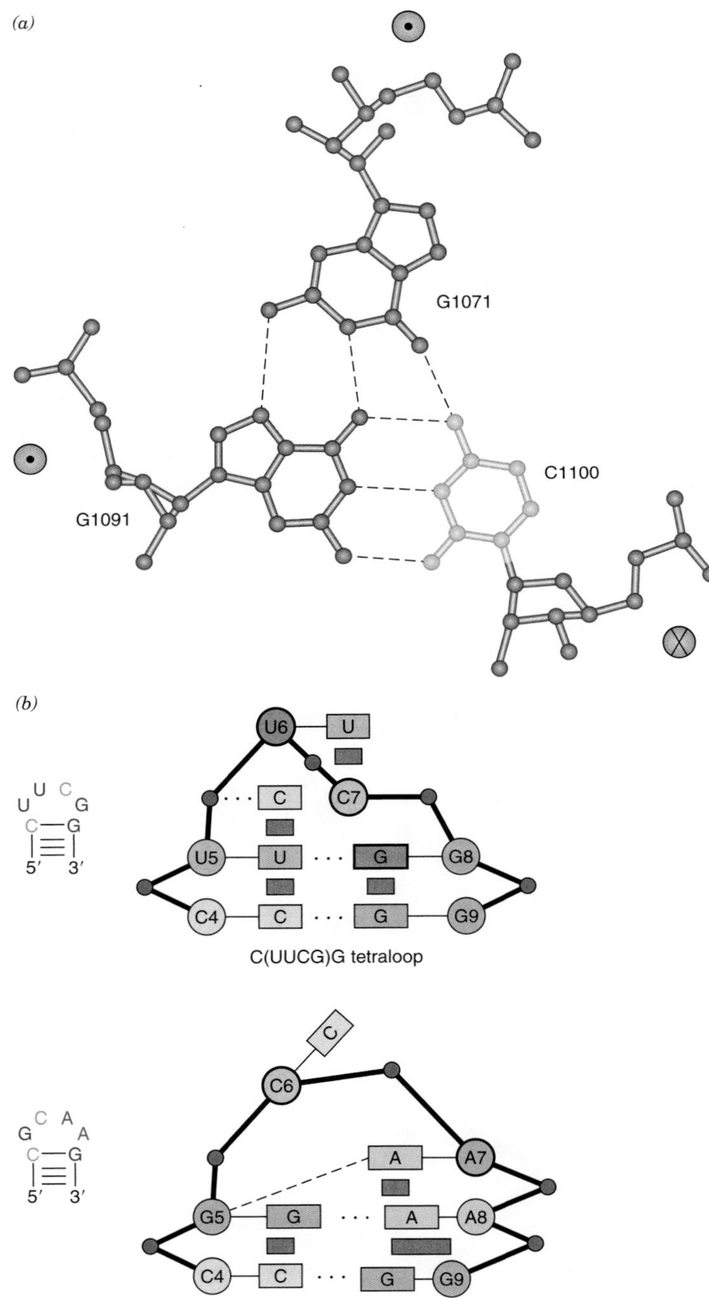

Figure 2.52 Hydrogen bonding in tRNA (*a*) A base triplet from 23S ribosomal RNA. In this case, the third base (G1071) forms hydrogen bonds with the top or Hoogsteen face of both G1091 and C1100. (*b*) Schematic diagram showing that two RNA tetraloops have similar structures. Both structures are stabilized by stacking of the loop bases, by non-Watson–Crick hydrogen bonds between bases, and by hydrogen bonding between sugar, phosphate, and ribose residues.
Redrawn with permission from Westhof and Fritsch (2000) RNA Folding: Beyond Watson-Crick base pairs. *Structure* 8:R55–R65. Redrawn from Wyatt, J., and Tinoco, I., Jr. RNA structure, 471. In Gesteland, R.F., and Atkins, J. F., (Eds.), *The RNA World.* 1st ed. Cold Spring Harbor, NY: Cold Spring Harbor, 1993.

2.7 • TYPES OF RNA

RNA molecules are traditionally classified as transfer, ribosomal, and messenger RNAs according to their usual function; however, RNA molecules perform a variety of other functions in a cell.

Transfer RNA Has Two Roles: Activating Amino Acids and Recognizing Codons in mRNA

About 15% of total cellular RNA is **tRNA**, an essential intermediary between DNA and protein information in the Central Dogma. The enzyme-catalyzed formation of aminoacyl-tRNA activates amino acids for protein synthesis so that formation of peptide bonds is energetically favored. Aminoacyl-tRNAs bind to ribosomes where the amino acids are transferred to growing peptide chains (hence tRNA's name). tRNA recognizes nucleotide sequence information in mRNA to ensure that the correct amino acid is incorporated into the growing peptide chain. Amino acid activation occurs at the 3′-OH terminal CCA sequence, to which specific amino acids are attached enzymatically. Codon recognition occurs by the binding of the anticodon triplet to a three-nucleotide codon in mRNA. Base pairing of the mRNA codon and tRNA anticodon triplets is the major mechanism to ensure accurate peptide bond formation.

Each tRNA can transfer only one amino acid. Although only 20 amino acids are used in protein synthesis, free-living organisms synthesize a larger set of tRNAs. This is a consequence of the genetic code being redundant (having more than one codon per amino acid). For example, the genomic sequence of the bacterium *Haemophilus influenzae* contains genes for 54 tRNA species. Mitochondria synthesize a much smaller number of tRNAs. tRNAs that accept the same amino acid are called isoacceptors. A tRNA that accepts phenylalanine would be written as $tRNA^{Phe}$, whereas one accepting tyrosine would be written $tRNA^{Tyr}$.

tRNAs are 65 to 110 nucleotides long, corresponding to a molecular weight range of 22,000 to 37,000. The sequences of all tRNA molecules (thousands are known) can be arranged into a common cloverleaf secondary structure by complementary Watson–Crick base pairing to form three stem-loop structures. The anticodon triplet sequence is at one "leaf" of the cloverleaf while the CCA acceptor stem is at the "stem". This arrangement is preserved in the tertiary structure of $tRNA^{Phe}$ (Figure 2.51), where the anticodon and acceptor stems are at each end of the L-shaped molecule. Additional non-Watson–Crick hydrogen bonds form the tertiary structure of the L-shaped molecule.

tRNAs contain modified nucleotides, for example, 7-methylguanine at position 46 in Figure 2.50. Modified nucleotides affect tRNA structure and stability but are not required for their basic functioning. For example, a modified base next to the anticodon makes codon recognition more efficient, but a tRNA without this modification can still be read correctly by the ribosome.

Many structural features are common to all tRNA cloverleafs. Seven base pairs form the amino acid acceptor stem, which terminates with the nucleotide triplet CCA. This CCA triplet is not base paired. The dihydrouracil- or D-stem has three or four base pairs, while the anticodon and T-stems have five base pairs each. Both the anticodon loop and T-loop contain seven nucleotides. Differences in the number of nucleotides in different tRNAs are accounted for by the variable loop. Most tRNAs have small variable loops of 4–5 nucleotides, while others have larger loops of 13–21 nucleotides. The positions of some nucleotides are constant in all tRNAs (see the dark orange boxes in Figure 2.50).

Ribosomal RNA Is Part of the Protein Synthesis Apparatus

Protein synthesis takes place on ribosomes. In eukaryotes, these complex assemblies are composed of four RNA molecules, representing about two-thirds of the particle mass, and 82 proteins. The smaller subunit, the 40S particle, contains one 18S RNA and 33 proteins. The larger subunit, the 60S particle, contains one copy each of three RNAs, 28S, 5.8S, and 5S rRNAs along with 49 proteins. The total assembly forms the 80S ribosome. Prokaryotic ribosomes are smaller; the 30S subunit contains a single 16S rRNA and 21 proteins while

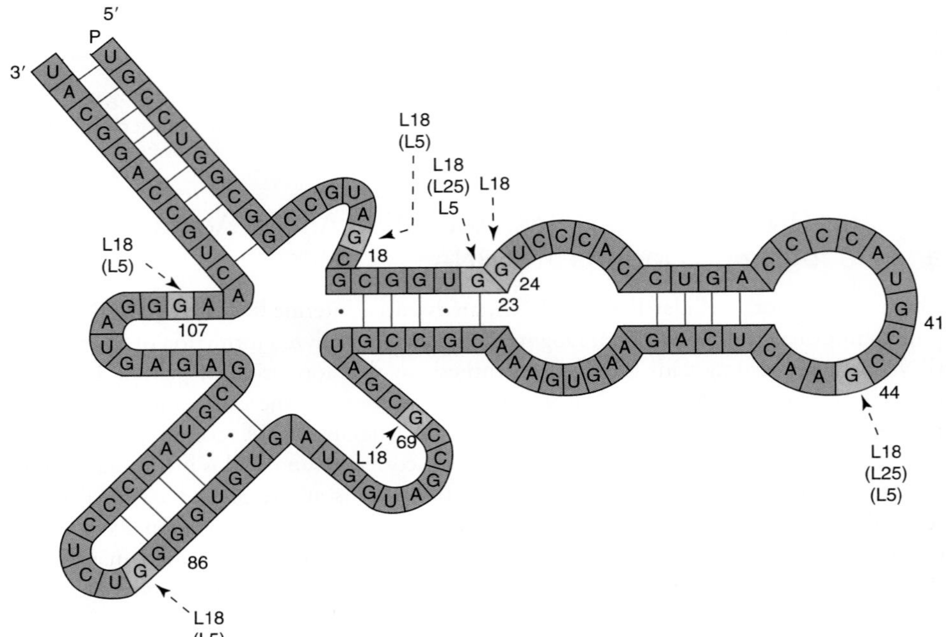

Figure 2.53 Secondary, base-paired, structure proposed for 5S rRNA. Arrows indicate regions protected by proteins in the large ribosomal subunit.
Combined information from Fox, G. E., and Woese, C. R. *Nature* (London) 256:505, 1975, and Gray, P.N., Bellemare, G., Monier, R., Garrett, R. A., and Stoffler, G. *J. Mol. Biol.* 77:133, 1973.

the larger 50S subunit contains one copy each of 5S and 23S rRNAs and 34 proteins (see p. 217 for a discussion of the structure of ribosomes).

rRNA accounts for 80% of cellular RNA and is metabolically stable. This stability is required for repeated functioning of the ribosome and is enhanced by association with ribosomal proteins. Eukaryotic 28S (4718 nucleotides), 18S (1874 nucleotides), and 5.8S (160 nucleotides) rRNAs are synthesized in the nucleolus. The 5S rRNA (120 nucleotides) is not transcribed in the nucleolus but rather from separate genes within the nucleoplasm (Figure 2.53). The three larger rRNAs are synthesized as part of one long polynucleotide chain that is then processed to yield the individual molecules.

Nucleotide modifications in rRNA are primarily methylations to form 2′-*O*-methylribose. Methylation of rRNA has been directly related to bacterial antibiotic resistance in a pathogenic species (Clin. Corr. 2.9). A small number of N^6-dimethyladenines is present in 18S rRNA. The 28S rRNA has about 45 methyl groups and the 18S rRNA has 30 methyl groups.

CLINICAL CORRELATION 2.9

Staphylococcal Resistance to Erythromycin

Bacteria exposed to antibiotics in clinical or agricultural settings often develop resistance to the drugs. This resistance can arise from a mutation in the target cell's DNA, which gives rise to resistant descendants. An alternative and clinically more serious mode of resistance arises when plasmids coding for antibiotic resistance proliferate through the bacterial population. These plasmids may carry multiple resistance determinants and render several antibiotics useless at the same time.

Erythromycin inhibits protein synthesis by binding to the large ribosomal subunit. *Staphylococcus aureus* can become resistant to erythromycin and similar antibiotics as a result of a plasmid-borne RNA methylase that converts a single adenosine in 23S rRNA to N^6-dimethyladenosine. Since the same ribosomal site binds lincomycin and clindamycin, the plasmid causes cross-resistance to these antibiotics as well. Synthesis of the methylase is induced by erythromycin.

A microorganism that produces an antibiotic must also be immune to it, or else it would be killed by its own toxic product. The producer of erythromycin, *Streptomyces erythraeus,* itself possesses an rRNA methylase that acts at the same ribosomal site as the one from *S. aureus.* Which came first? It is likely that many of the resistance genes in target organisms evolved from those of producer organisms. In several cases, DNA sequences from resistance genes of the same specificity have been conserved in producer and target organisms. We may therefore look on plasmid-borne antibiotic resistance as a case of "natural genetic engineering" whereby DNA from one organism (e.g., *Streptomyces* producer) is appropriated and expressed in another (e.g., the *Staphylococcus* target).

Cundliffe, E. How antibiotic-producing microorganisms avoid suicide. *Annu. Rev. Microbiol.* 43:207, 1989.

Ribosomal RNA is the catalytic component of the ribosome. The X-ray crystal structure of the large subunit of a bacterial ribosome species shows that the active site for peptide bond synthesis is composed of 23S RNA only; any potential catalytic amino acid side chains are too far away to participate directly in the reaction. The chemical reactions necessary for peptide bond synthesis are catalyzed by the large subunit rRNA (see p. 68 for a discussion of catalytic RNAs).

Messenger RNAs Carry the Information for the Primary Structure of Proteins

mRNAs are the direct carriers of genetic information from genome to ribosomes. Each eukaryotic mRNA is **monocistronic,** that is, contains information for only one polypeptide chain. In prokaryotes, mRNA species are often **polycistronic,** encoding more than one protein. A cell's phenotype and functional state are related directly to its mRNA content.

In the cytoplasm, mRNAs have relatively short life spans. Some are synthesized and stored in an inactive or dormant state in the cytoplasm, ready for a quick response when protein synthesis is required. In various animals, immediately on fertilization the egg undergoes rapid protein synthesis in the absence of transcription, indicating that preformed mRNA and ribosomes are present in its cytoplasm.

Eukaryotic mRNAs have unique structural features (Figure 2.54). Because the information within mRNA lies in the linear sequence of the nucleotides, the integrity of this sequence is extremely important. Any loss or change of nucleotides could alter the structure of the protein being translated. Translation of mRNA on the ribosomes must also begin and end at specific sequences. In eukaryotes, a "cap" structure at the 5′-end of an mRNA specifies where translation should begin. The caps of eukaryotic mRNAs contain an inverted phosphodiester bond, a methylated base attached to the mRNA via a 5′,5′-triphosphate diester rather than the usual 3′, 5′-phosphodiester linkage. The cap is attached to the first transcribed nucleotide, usually a 2′-*O*-methylated purine (Figure 2.55). The cap is

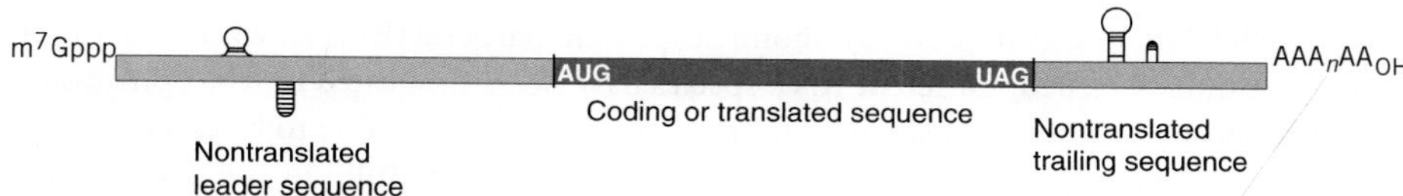

Figure 2.54 General structure for a eukaryotic mRNA. There is a blocked 5′-terminus cap, followed by an untranslated leader containing a promoter sequence. The coding region usually begins with the initiator codon AUG and continues to the translation termination sequence UAG, UAA, or UGA. This is followed by the untranslated trailer and a poly(A) tail on the 3′-end.

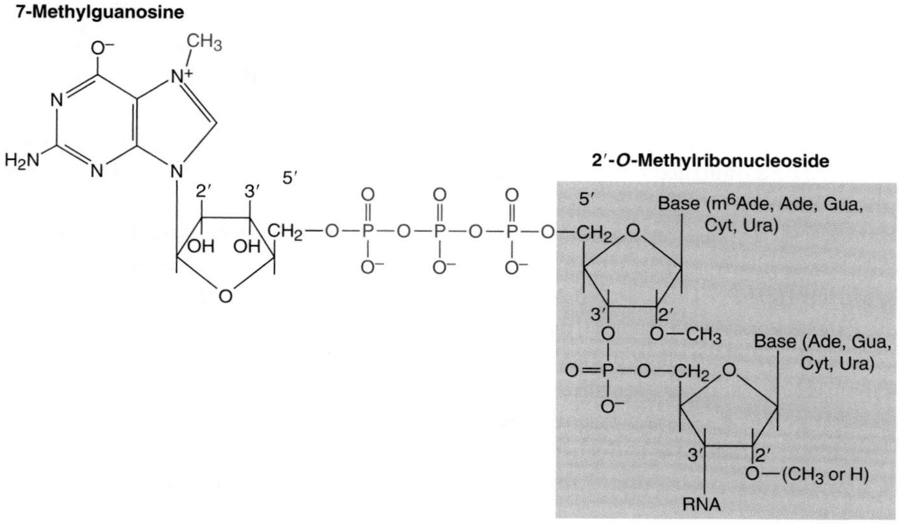

7-Methylguanosine

2′-*O*-Methylribonucleoside

Base (m^6Ade, Ade, Gua, Cyt, Ura)

Base (Ade, Gua, Cyt, Ura)

O—(CH_3 or H)

RNA

Figure 2.55 Diagram of "cap" structure or blocked 5′-terminus in mRNA. The 7-methylguanosine is inverted to form a 5′-phosphate to 5′-phosphate linkage with the first nucleotide of the mRNA. This nucleotide is often a methylated purine.

followed by an untranslated, or "leader" sequence. Translation starts at the initiation codon, most often AUG, and continues through the coding sequence. At the end of the coding sequence, a termination sequence signals termination of polypeptide formation and release from the ribosome. A second untranslated, or "trailer" sequence follows, terminated by a string of 20–200 adenine nucleotides, called a **poly(A) tail**, which makes up the 3′-terminus of the mRNA.

The 5′-cap has a positive effect on the initiation of message translation. During the initiation of translation of mRNA, the cap structure is recognized by a specific initiation factor (p. 221). The poly(A) sequence affects the stability of the mRNA; degradation of an mRNA generally begins with shortening of its poly(A) tail.

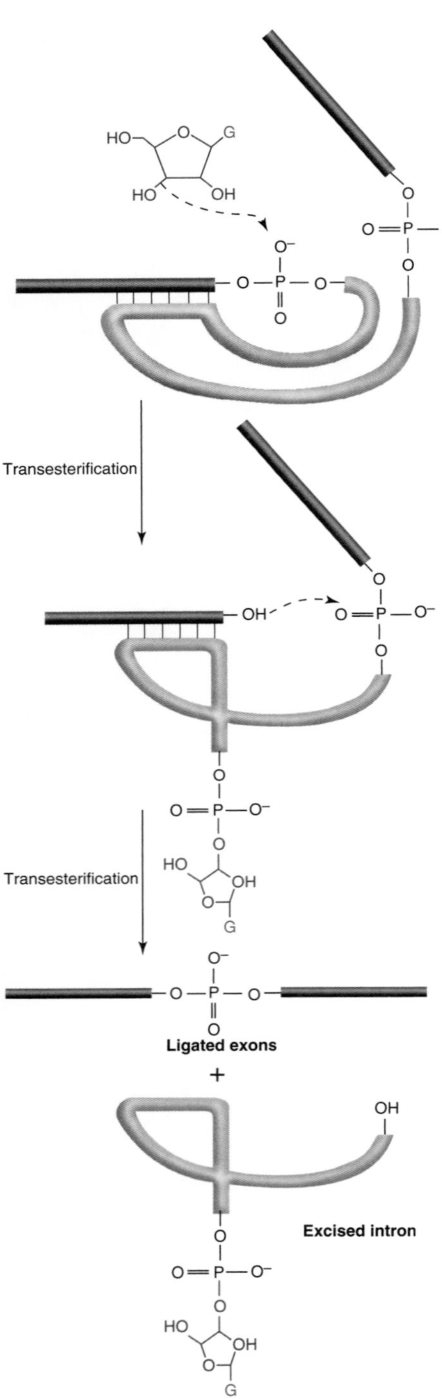

Figure 2.56 **Mechanism of self-splicing of the rRNA precursor of Tetrahymena.** Two exons of rRNA are denoted by white and shaded boxes, respectively. Catalytic functions reside in the intron, which is black. This splicing function requires an added guanine nucleoside or nucleotide. Reproduced with permission from Cech, T. R. *J. Am. Med. Assoc.* 260:308, 1988.

Mitochondria Contain Unique RNA Species

Mitochondria (mt) have their own protein synthesizing apparatus, including ribosomes, tRNAs, and mRNAs. Mitochondrial rRNAs, 12S and 16S, are transcribed from the mitochondrial DNA (**mtDNA**) as are (mtDNA) 22 specific tRNAs and 13 mRNAs, most of which encode proteins of the electron transport chain and ATP synthase. Note that there are fewer mitochondrial tRNAs than prokaryotic or cytosolic tRNA species; there is generally only one mitochondrial tRNA species per amino acid. The mtRNAs account for 4% of the total cellular RNA. They are transcribed by a mitochondrial-specific RNA polymerase and are processed from a pair of RNA precursors that contain tRNA and mRNA sequences. Expression of the nuclear and mitochondrial genomes is tightly coordinated. Most of the amino acylating enzymes for the mitochondrial tRNAs and all of the mitochondrial ribosomal proteins are specified by nuclear genes, translated in the cytosol and transported into mitochondria. The modified bases in mitochondrial tRNA species are synthesized by enzymes encoded in nuclear DNA.

RNA in Ribonucleoprotein Particles

Other small, stable RNA species can be found in the nucleus, cytosol, and mitochondria. These small RNA species function as ribonucleoprotein particles (RNPs), with one or more protein subunits attached. Different RNP species have been implicated in RNA processing, RNA transport, control of translation, and recognition of proteins due to be exported. The actual roles of these species, where known, are described more fully in the discussion of specific metabolic events.

Catalytic RNA: Ribozymes

RNA can be an enzyme. In several cases, the RNA component of a ribonucleoprotein particle is the catalytically active subunit of the enzyme. In other cases, catalytic reactions can be carried out *in vitro* by RNA in the absence of any protein. Enzymes whose RNA subunits carry out catalytic reactions are called ribozymes. There are five classes of ribozymes. Three of these RNA species carry out self-processing reactions while the others, ribonuclease P (RNase P) and rRNA, are true catalysts that act on separate substrates.

In the ciliated protozoan, *Tetrahymena thermophila,* an intron in the rRNA precursor is removed by a multistep reaction (Figure 2.56). A guanine nucleoside or nucleotide reacts with the intron–exon phosphodiester linkage to displace the donor exon from the intron. This reaction, a transesterification, is promoted by the folded intron itself. The free donor exon then similarly attacks the exon–intron phosphodiester bond at the acceptor end of the intron. Introns of this type (Group I introns) have been found in a variety of genes in fungal mitochondria, in bacteria, and the bacteriophage T4. Although these introns are not true enzymes *in vivo* as they only work for one reaction cycle, they can be made to carry out catalytic reactions under specialized conditions.

Group II self-splicing introns are present in the mitochondrial RNA precursors of yeasts and other fungi. Even though these RNAs self-splice, there are many parallels between this reaction and the removal of introns from mRNA precursors during processing.

Group III self-cleaving RNAs are found in the genomic RNAs of several viruses. These RNAs self-cleave during the generation of single genomic RNA molecules from larger precursors. The three-dimensional structure of the hammerhead ribozyme, a member of this third class, has been determined (Figure 2.57). Although the catalytic cycle is not completely determined, the amino group of a cytosine base plays an essential role in the cleavage reaction. The phosphate of the cleaved bond is left at the 3′-hydroxyl position of the RNA product. A self-cleaving RNA found in a small satellite virus, hepatitis delta virus, is implicated in severe human infectious hepatitis. All of the above self-processing RNAs can be made to act as true catalysts (i.e., exhibit multiple turnover) *in vitro* and *in vivo.*

The fourth type of ribozyme, Ribonuclease P, contains both a protein and an RNA component. It acts as a true enzyme in cells, cleaving tRNA precursors to generate the mature 5′-end of tRNA molecules. RNase P recognizes constant structures associated with tRNA precursors (e.g., the acceptor stem and CCA sequence) rather than using extensive base pairing to bind the substrate RNA to the ribozyme. The product of cleavage contains a 5′-phosphate in contrast to the products of hammerhead and similar RNAs. In all of these events, the structure of the catalytic RNA is essential for catalysis.

Finally, the X-ray crystallographic structure of a bacterial ribosome reveals that it, too, is a ribozyme. The active site of the ribosome is on the 50S subunit. There is no protein functional group close enough to catalyze peptide bond formation. Instead, residues within 23S RNA helps transfer a hydrogen ion during peptide bond synthesis.

Discovery of RNA catalysis has greatly altered our concepts of biochemical evolution and the range of allowable cellular chemistry. We now recognize that RNA can serve as both a catalyst and a carrier of genetic information. This has raised the possibility that the earliest living organisms were based entirely on RNA and that DNA and proteins evolved later. This model is sometimes called the "RNA world". Secondly, we know that many viruses, including human pathogens, use RNA genetic information; some of these RNAs have been shown to be catalytic. Thus, catalytic RNA presents opportunities for the discovery of RNA-based pharmaceuticals. Thirdly, many of the information-processing events in mRNA splicing require RNA components. These RNAs may also be acting as catalysts.

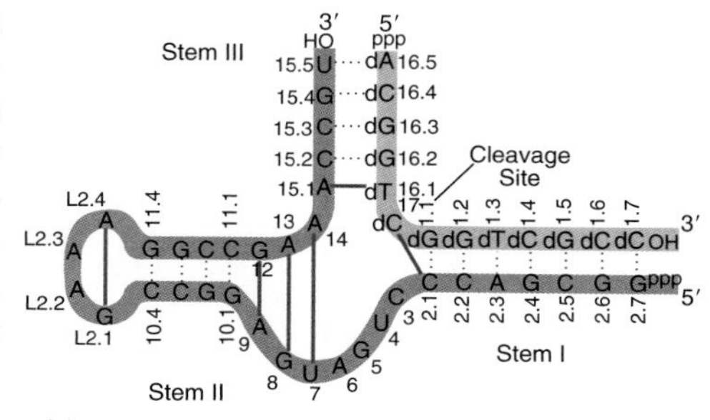

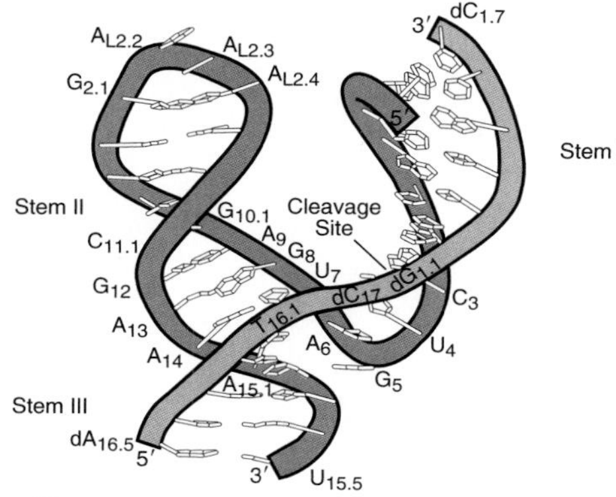

Figure 2.57 "Hammerhead" structure of viral RNA. (*a*) "Hammerhead" structure of a self-cleaving viral RNA. This artificial molecule is formed by the base pairing of two separate RNAs. Cleavage of the RNA sequence at the site indicated by the arrow in the top strand requires its base pairing with the sequence at the bottom of the molecule. The boxed nucleotides are a consensus sequence found in self-cleaving RNA of viral RNAs. (*b*) Three-dimensional folding of the hammerhead catalytic RNA. Star indicates the position of the cleaved bond while M indicates a binding site for a metal ion. Helices II and III stack to form an apparently continuous helix while non-Watson–Crick interactions position the noncomplementary bases in the hammerhead into a "uridine turn" structure identical to that found in tRNA.
(a) is redrawn from Sampson, J. R., Sullivan, F. X., Behlen, L. S., DiRenzo, A. B., and Uhlenbeck, O. C. *Cold Spring Harbor Sym. on Quan. Bio.* 52:267, 1987. *(b)* is redrawn from Pley, H. W., Flaherty, K. M., and McKay, D. B. *Nature* 372:68, 1994.

RNAs Can Bind Other Molecules

Consideration of the RNA world has led to a new type of "combinatorial biochemistry" based on the large number of potential sequences (4^N) that could be made if A, C, G, or U were inserted randomly in each of N positions in a nucleic acid. A set of chemically synthesized, randomized, nucleic acid molecules 25 nucleotides long would contain $4^{25} = 10^{15}$ potential members. Individual molecules within this large collection of RNAs would be expected to fold into a similarly large collection of shapes. The large number of molecular shapes implies that some member of this collection should be capable of strong, specific binding to ligands, as Group I introns bind guanine nucleotides specifically. These binding RNAs are termed **aptamers.** Though a single molecule would be too rare to study within the original population, the aptamer RNA can be selected and preferentially replicated *in vitro.* For example, an RNA capable of distinguishing theophylline from caffeine was selected from a complex population (Figure 2.58). Theophylline is used in the treatment of chronic asthma, but the level must be carefully controlled to avoid side effects. Monitoring of theophylline concentrations by conventional antibody-based assay is difficult because caffeine and theophylline differ only by a single methyl group. Therefore, anti-theophylline antibodies show considerable cross-reaction with caffeine. RNA aptamers have been found that bind theophylline 10,000-fold more tightly than caffeine. Other extensions of the technology have used selection procedures to identify new, synthetic ribozymes and potential therapeutic RNAs.

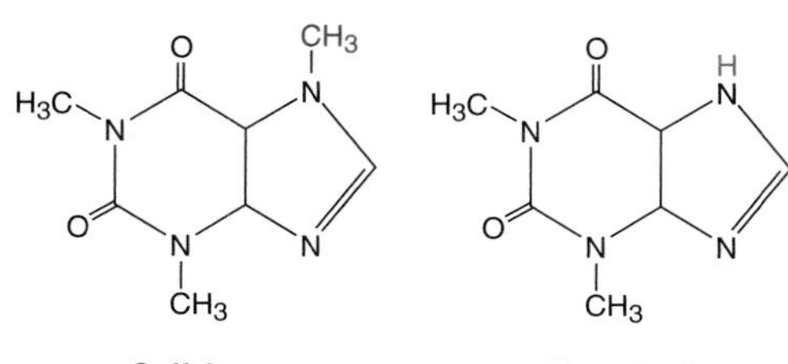

Figure 2.58 Structures of theophylline and caffeine. Although these compounds differ only by a single methyl group, a specific synthetic RNA can bind to theophylline 10,000-fold more tightly than to caffeine.

RNAs Control Translation

Translation of an mRNA requires that it be bound to and move relative to the ribosome. Controlling the efficiency of these processes allows for translational control. One way to do this is through aptamer sites contained in mRNA. Some bacterial mRNAs contain naturally occurring aptamer sites in the 5′-untranslated region of the mRNA molecule. These sites bind specific ligands such as amino acids, purine bases, glucosamine 6-phosphate, or enzyme cofactors that are products of the pathway whose enzymes are encoded by each mRNA. Binding of the ligand prevents translation of the mRNA, thereby controlling synthesis of the proteins that are responsible for the biosynthetic pathway for the ligand. This is a form of regulation in **cis** because control is exerted on the same molecule that contains the aptamer site.

Double-stranded RNA is a negative regulator of eukaryotic gene expression in **trans**, that is, on another molecule. For example, double-stranded RNA is a necessary intermediate in the replication of some RNA viruses. As a means of defense, the host processes the double-stranded RNA to produce short single-stranded RNAs about 23 nucleotides long. These RNAs, termed **microRNAs (miRNAs)** base-pair to specific mRNAs, and either stop their translation or trigger their degradation (p. 289). Further work on miRNAs and the closely related small inhibitory RNAs (siRNAs) has shown that these species are important regulators of many gene products, including those involved in malignancy (see Clin. Corr. 5.8, p. 203).

Bibliography

DNA

Ball, P. Pulling our strings. *Chem. World* 5:50, 2008.

Bates, A. D. DNA Topology. Oxford, UK: IRL Press, 1993.

Bell, J. I. The double helix in clinical practice. *Nature* 421: 414, 2003.

Blackburn, G. M., Gait, M. J., Loakes, D., and Williams, D. M. *Nucleic Acids in Chemistry and Biology,* 3d ed. Cambridge, UK: Royal Society of Chemistry, 2006.

Cech, T. R. Life at the end of the chromosome: Telomeres and telomerase. *Angew. Chem. Int. Ed.* 39: 34, 2000.

Champoux, J. J. DNA topoisomerases: Structure, function, and mechanism. *Annu. Rev. Biochem.* 70: 369, 2001.

De Bruijn, F. J., Lupski, J. R., and Weinstock, G. M. (Eds.). *Bacterial Genomes: Physical Structure and Analysis.* New York: Chapman & Hall, 1998.

Dong, K. C., and Berger, J. M. Structure and function of DNA topoisomerases. In Rice, P. A., and Correll, C. C. (Eds.), *Protein-Nucleic Acid Interactions.* Cambridge, UK: Royal Society of Chemistry, 2008.

Dorman, C. J. Nucleoid organization of bacterial chromosomes. In Lennarz, W. J., and Lane, M. D. (Eds.), *Encyclopedia of Biological Chemistry,* Vol. 3. New York: Elsevier, 2004, 115.

Felsenfeld, G., and Goudine, M. Controlling the double helix. *Nature* 421: 448, 2003.

Horn, P. J., and Peterson, C. L. Chromatin higher order folding: Wrapping up transcription. *Science* 297: 1824, 2002.

Hud, N. V., and Plavec, J. A unified model for the origin of DNA sequence-directed curvature. *Biopolymers* 69: 144, 2003.

Huppert, J. L. Four-stranded nucleic acids: Structure, function and targeting of G-quadruplexes. *Chem. Soc. Rev.* 37: 1375, 2008.

Jurka, J. Evolutionary impact of human Alu repetitive elements. *Curr. Opin. Genetics Dev.* 14: 603, 2004.

Kornberg, R. D., and Lorch, Y. Chromatin rules. *Nature Struct. Mol. Biol.* 14: 986, 2007.

Mirkin, S. M. Discovery of alternative DNA structures: A heroic decade (1979–1989). *Front. Bioscience* 13: 1064, 2008.

Pearson, C. E., Zorbas, H., Price, G. B., and Zannis-Hadjopoulos, M. J. Inverted repeats, stem-loops, and cruciforms: Significance for initiation of DNA replication. *Cell. Biochem.* 63: 1, 1998.

Saenger, W. *Principles of Nucleic Acid Structure.* New York: Springer-Verlag, 1984.

Tenover, F. C. *DNA Probes for Infectious Diseases.* Boca Raton, FL: CRC Press, 1989.

Wallrath, L. L., Danzer, J. R., Yazgan, O., and Geyer, P. K. Nuclear organization, chromatin structure, and gene silencing. In Lennarz, W. J., and Lane, M. D. (Eds.), *Encyclopedia of Biological Chemistry,* Vol. 3, 105. New York: Elsevier, 2004.

Watson, J. D., and Crick, F. H. C. Molecular structure of nucleic acids: A structure for deoxyribose nucleic acid. *Nature* 171: 737, 1953.

Wells, R. D. Non-B DNA conformations, mutagenesis and disease. *Trends Biochem. Sci.* 32: 271, 2007.

RNA

Cech, T. R. The ribosome is a ribozyme. *Science* 289: 878, 2000.

Draper D. E. A guide to ions and RNA structure. *RNA* 10: 335, 2004

Gesteland, R. F., Cech, T., and Atkins, J.F. (Eds.). *The RNA World,* 3d ed. Cold Spring Harbor, NY: Cold Spring Harbor Laboratory, 2006.

Gold, L., Polisky, B., Uhlenbeck, O., and Yarus, M. Diversity of oligonucleotide functions. *Annu. Rev. Biochem.* 64: 763, 1995.

Ibba, M., Becker, H. D., Stathopoulos, C., Tumbula, D. L., and Söll, D. The adaptor hypothesis revisited. *Trends Biochem. Sci.* 25: 311, 2000.

Leontis, N. B., and Westhof, E. Analysis of RNA motifs. *Curr. Opin. Struct. Biol.* 13: 300, 2003.

McKnight, K. L., and Heinz, B. A. RNA as a target for developing antivirals. *Antiviral Chem. Chemotherapy* 14: 61, 2003.

Pace, N. R., and Brown, J. W. Evolutionary perspective on the structure and function of ribonuclease P, a ribozyme. *J. Bacteriol.* 177: 1919, 1995.

Puerta-Fernandez, E., Romero-Lopez, C., Barroso-del Jesus, A., and Berzal-Herranz, A. Ribozymes: Recent advances in the development of RNA tools. *FEMS Microbiol. Rev.* 27: 75, 2003.

Tucker, B. J., and Breaker, R. R. Riboswitches as versatile gene control elements. *Curr. Opin. Struct. Biol.* 15: 342, 2005.

Westhof, E., and Fritsch, V. RNA folding: Beyond Watson-Crick pairs. *Structure* 8:R55, 2000.

Key Terms

central dogma
DNA
replication
transcription
adenine
cytosine
deoxyribose
guanine
nucleoside
nucleotide
thymine
uracil
phosphodiester
poly(A) tail
polynucleotide
base stacking
double helix
Watson–Crick base pair
complementarity
denaturation
renaturation
melting temperature
hybridization
A-DNA
B-DNA
Z-DNA
supercoiling
telomere
chromosome
topoisomerase
chromatin
centromere
histone
nucleosome
exon
intron
repetitive DNA
splicing
posttranscriptional
tRNA
rRNA
monocistronic
mRNA
polycistronic
mRNA cap
ribonucleoprotein particle
ribozyme
aptamer
micro RNA
small inhibitory RNA

Questions • CAROL N. ANGSTADT

Multiple-Choice Questions

1. The best definition of an endonuclease is an enzyme that hydrolyzes a
- **A.** nucleotide from only the 3′-end of an oligonucleotide.
- **B.** nucleotide from either terminal of an oligonucleotide.
- **C.** phosphodiester bond located in the interior of a polynucleotide.
- **D.** bond only in a specific sequence of nucleotides.
- **E.** bond that is distal to the base that occupies the 5′ position of the bond.

2. In the structure of a naturally occurring DNA,
- **A.** circular DNA is formed by linking the 3′-OHs at each end of a linear DNA.
- **B.** base stacking reduces the hydrophobic surface area that must be solvated by water and stabilizes the double helix.
- **C.** the two strands may be either parallel or antiparallel.
- **D.** which tautomeric form the nitrogen and oxygen functions assume is not relevant to their ability to form hydrogen bonds.
- **E.** interwinding of the two strands produces two helical grooves of nearly identical size.

3. The Z-DNA helix
- **A.** is the primary form in living organisms.
- **B.** is favored by an alternating G-C sequence.
- **C.** tends to be found at the 3′-end of genes.
- **D.** is inhibited by methylation of the bases.
- **E.** is a permanent conformation of DNA.

4. A nucleosome
- **A.** is a regularly repeating structure of DNA and histone proteins.
- **B.** has a core of DNA with proteins wrapped around the outside.
- **C.** uses only one type of histone per nucleosome.
- **D.** is separated from a second nucleosome by nonhistone proteins.
- **E.** has histones in contact with the major groove of the DNA.

5. RNA
- **A.** incorporates both modified and unmodified bases during transcription.
- **B.** does not exhibit any double-helical structure.
- **C.** structures exhibit base stacking and hydrogen-bonded base pairing.
- **D.** usually contains about 65–100 nucleotides.
- **E.** does not exhibit Watson–Crick base pairing.

6. Ribozymes
- **A.** are any ribonucleoprotein particles.
- **B.** are enzymes whose catalytic function resides in RNA subunits.
- **C.** carry out self-processing reactions but cannot be considered true catalysts.
- **D.** require a protein cofactor to form a peptide bond.
- **E.** function only in the processing of mRNA.

Questions 7 and 8: Nearly every process in which DNA participates requires that the DNA interact with proteins. Interaction with proteins depends on the local three-dimensional structure of DNA. One antitumor drug, cisplatin, is a tetracoordinate platinum complex. It forms intra-strand cross-links with DNA causing the double helix to strongly bend toward the major groove. Cellular processes such as transcription and programmed cell death are affected.

7. Bent DNA:
- **A.** occurs only in the presence of external agents like the antitumor drugs.
- **B.** may be a fundamental element in the interaction between DNA sequences and proteins.
- **C.** occurs primarily in the presence of triple-stranded DNA.
- **D.** requires the presence of inverted repeats.
- **E.** occurs only in DNA that is in the Z form.

8. Another unusual form of DNA is a triple-stranded complex. Triple-stranded DNA
- **A.** generally occurs in DNA in regions that play no role in transcription.
- **B.** involves Hoogsteen hydrogen bonding.
- **C.** is characterized by the presence of a string of alternating purine-pyrimidine bases.
- **D.** forms only intermolecularly.
- **E.** assumes a cruciform conformation.

Questions 9 and 10: Since topoisomerases play important roles in many activities of nucleic acid metabolism, they are targets of some antimicrobial and antineoplastic agents. They trap the immediate complex between topoisomerase and DNA, which may result in DNA degradation, introduction of mutations, or inhibition of translation and replication.

9. Topoisomerases
A. regulate the level of superhelicity of DNA in cells.
B. always break only one strand of DNA.
C. can create but not remove supercoils.
D. must hydrolyze ATP for their action.
E. of the subclass gyrases, introduce negative superhelices in eukaryotic DNA.

10. In addition to superhelicity, DNA must be still further packaged to fit into the bacteria or cell. Which of the following is correct?
A. Both prokaryotic and eukaryotic DNA form chromatin.
B. A bacterial chromosome contains a single DNA molecule, but a human chromosome contains more than one DNA.
C. Packaging of DNA into nucleosomes does not involve topoisomerases.
D. Polysomes are the highest level of packaging DNA undergoes.
E. Thirty-nm fibers (solenoids) can organize into a series of looped, condensed domains.

Questions 11 and 12: Eukaryotic DNA contains many sequences that can be repeated a few times, for certain coding genes, to millions of times, for some relatively short sequences. Abnormal expansion of reiterated three base-pair DNA sequences can lead to a number of diseases. Fragile X syndrome, which causes mental retardation, is characterized by expansion of a GCC triplet near the FMR-1 gene, from a normal 30 copies to thousands of copies. Diseases of this type generally increase in severity with each successive generation. One possible mechanism of expansion involves slippage of DNA during synthesis of the lagging strand.

11. Slipped, mispaired DNA occurs when the DNA region has
A. direct repeats.
B. homopurine–homopyrimidine sequences.
C. inverted repeats.
D. mirror repeats.
E. palindromes.

12. Normally, certain kinds of reiterated sequences occur in a chromosome as an interspersion pattern that is
A. highly repetitive DNA sequences.
B. the portion of DNA composed of single-copy DNA.
C. Alu sequences.
D. alternating blocks of single-copy DNA and moderately repetitive DNA.
E. alternating blocks of short interspersed repeats and long interspersed repeats.

Problems

13. Which conformation of DNA, totally double helix, minimally unwound or largely unwound, would have the highest relative optical density at 260 nm? Would a molecule of DNA having a higher content of guanine and cytosine than of adenine and thymine have a higher or lower T_m than one with the reverse composition?

14. An approach to reducing telomerase activity in tumor cells involves drugs that bind to G-quadroplex DNA since telomeres have the potential to form such structures. What is a G-quadroplex, and what about telomere structure might lead to it?

Answers

1. C Both A and B describe exonucleases. D refers specifically to a restriction endonuclease and is not a definition of the general type. E: Both endo- and exonucleases show specificity toward the bond hydrolyzed.

2. B This is a major contributor to the stability of the double helix. A: Joining of nucleotides is always by the 5′-OH of one to the 3′-OH of the other. C: The antiparallel arrangement permits opposing nucleotides to be in the proper position to form hydrogen bonds. D: Each pair in hydrogen bonding must be in the proper tautomeric form so that one is the donor and the other the acceptor. E: The grooves are different with the major groove wider than the minor groove.

3. B The alternating purine–pyrimidine sequence is important. A: This is the more flexible B-form. C: It is more likely to be found at the 5′-end, consistent with one of its proposed roles in transcriptional regulation. D: Methylation favors the Z-form in which the methyl is protected from water. E: B-to-Z transition is influenced by such things as methylation.

4. A The "beads-on-a-string" structure is called a polynucleosome. B: Histones form the core with DNA on the outside. C: All five types of histones are present, four in the core and one outside. D: The linker regions are DNA. E: Proteins that regulate gene expression and other activities bind to the major groove. Histones bind to the minor groove.

5. C Stacking stabilizes the single-strand helix. Folded portions of the structure have hydrogen-bonded base pairing. A: Only the four bases A, G, U, and C are incorporated during transcription. B: Although single stranded, RNA exhibits considerable secondary and tertiary structure. D: Only tRNA would be this small. E: This occurs in the intrachain helical regions.

6. B Ribozymes are a very specific type of particle. A: See previous comment. C: One of the four classes, RNase P, catalyzes a cleavage reaction. D: X-ray crystallography has shown there is no amino acid chain sufficiently close to catalyze peptide bond formation. E: Ribozymes have been implicated in the processing of ribosomal and tRNAs.

7. B The bent DNA may be a recognition site for specific proteins to bind. A and D: Bent DNA occurs naturally either in runs of 4 to 6 adenosines separated by spacers or is induced by interactions of DNA with certain proteins. C: Bent DNA is double stranded. E: Z-DNA does not have a prominent major groove and the bending is toward the major groove.

8. B Hoogsteen bonding in TAT and GGC triplets is responsible for holding the third strand in the major groove. A: They are found frequently in regions involved in gene regulation. C: The required sequence is a homopurine string. D: They can also form intramolecularly by unfolding and refolding of the DNA. E: A cruciform is an alternative conformation of DNA but does not involve a third strand.

9. A Topoisomerase I relaxes DNA, and topoisomerase II generates supercoils. B: Topoisomerase I breaks one strand, but topoisomerase II nicks both strands. C: Topoisomerase I removes supercoils. D: Topoisomerase II uses ATP, but Topoisomerase I does not. E: Gyrases are found only in bacteria.

10. E These loops are bound to a protein scaffold. A: Chromatin is found in eukaryotes, nucleoids in prokaryotes. B: A human chromosome contains a single DNA. C: There is adding and breaking of supercoils. D: Polysomes organize into 10-nm fibers and then 30-nm fibers, which in turn, further pack.

11. **A** When DNA unwinds and realigns one copy of a direct repeat on one-strand pairs with an adjacent copy on the other strand. B, C, D, and E: These are other types of structures important in DNA.
12. **D** A and B: These are two of several kinds of DNA but do not constitute patterns. C and E: Alu is a type of short interspersed repeat. Short and long interspersed repeats are the two classes of moderately repetitive DNA.
13. Unstacked purine and pyrimidine bases have a higher relative optical density than stacked bases so the most unwound form of DNA would have the highest value. T_m is the midpoint of the transition between double-stranded and separated DNA. Higher G-C content means the DNA is more stable so its T_m would be higher.
14. G-quartets contain a planar array of guanines connected by Hoogsteen hydrogen bonds, which can stack on each other to form a multilayered structure. They are stabilized by Na^+ and K^+. Telomeres contain 800–2400 copies of the required G-rich sequence d(TTAGGG)$_n$.

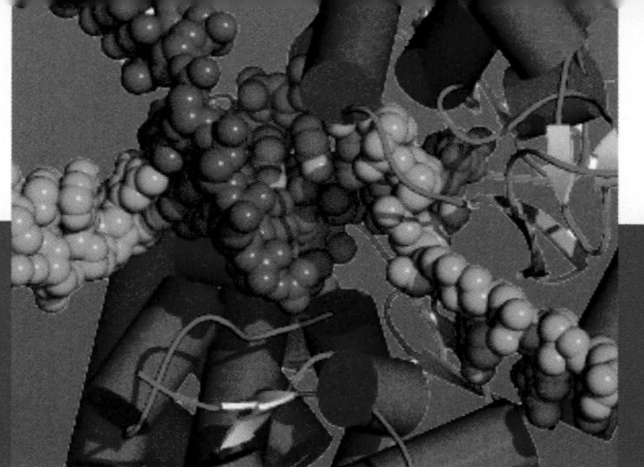

3

Proteins I: Composition and Structure

Richard M. Schultz

Professor, Stritch School of Medicine, Loyola University of Chicago

CLINICAL CORRELATIONS

Key Concepts

- Two types of amino acids compose polypeptides: common and derived amino acids. Common amino acids have a central carbon atom to which is attached an amino group, a carboxylic acid group, a hydrogen atom, and a side-chain group.
- Amino acids in polypeptide chains are joined by a peptide bond that connects the α-carboxylic acid group of one amino acid with the α-amino group of a second amino acid.
- Amino acids and proteins have acid-base and related charge properties that determine their biological activity and are used for characterization and purification.
- A protein is defined by its secondary, tertiary, and quaternary structure. α-Helical and β-structures are common stable secondary structures found in folded proteins.
- Globular proteins domains are formed from structural motifs and characteristic folds.
- Proteins in cells are present in complexes that are organized in networks and participate in interactomes.
- Some proteins are intrinsically unstructured, which is required for their biological functions.
- Nonglobular proteins include fibrous proteins such as collagen with characteristic repeating amino acid sequence and unusual helix and rod-shaped geometry.
- A protein folds to its conformation of lowest Gibbs free energy, within kinetic constraints, under the control of noncovalent forces.
- Folded protein structures are dynamic with atoms fluctuating and rotating around an average position.
- Proteins are purified and characterized by techniques that utilize charge and molecular weight. Protein structure is characterized by spectral, optical, and X-ray diffraction techniques.
- Proteomic techniques determine the expression of thousands of proteins within a cell or tissue in a single assay.

3.1 • FUNCTIONAL ROLES OF PROTEINS IN HUMANS

Proteins perform a surprising variety of essential dynamic and structural functions in mammalian organisms. Dynamic functions include catalysis of chemical transformations, transport, metabolic control, and contraction. In their structural roles, proteins provide the matrix for bone and connective tissue, giving structure and form to the human organism. The approximately 25,000 human genes through alternate splicing of RNA and post-translational modifications of translated proteins produce an estimated 1 million unique proteins. The number of proteins simultaneously expressed in a single human cell ranges between 20,000 and 50,000.

Enzymes are proteins that catalyze chemical reactions. Almost all of the thousands of chemical reactions in living organisms require a specific enzyme catalyst to ensure that these reactions occur at a rate compatible with life. The character of any cell is based on its particular chemistry, which is determined by its specific enzyme composition that establishes the cellular phenotype. Many genetic diseases result from altered levels of enzyme production or from specific alterations to their amino acid sequence. Approximately one-third of human genes code for enzyme proteins.

Transport is another major function of proteins. Examples discussed in detail in this text are hemoglobin and myoglobin, which transport oxygen in blood and store it in muscle, respectively. Proteins bind and carry lipids, metabolites, steroid hormones, vitamins, signaling molecules, and minerals in blood from their sites of synthesis or entry to their sites of action or elimination. Many drugs and toxic compounds are transported bound to proteins. Other proteins act intracellularly to transport compounds as well as other proteins to their sites of action. Proteins act to transmit compounds and signals across the plasma membrane of cells and organelle membranes. Proteins also participate in contractile mechanisms. Myosin and actin function in muscle contraction and are responsible for changes in cell shape.

Proteins with a protective role include immunoglobulins, complement proteins, and interferon that protect against bacterial or viral infection. The proteins of coagulation stop the loss of blood on injury to the vascular system.

Many hormones are proteins or peptides. Insulin, thyrotropin, somatotropin (growth hormone), prolactin, luteinizing hormone, and follicle-stimulating hormone are proteins. Many polypeptide hormones have a low molecular weight (<5 kDa) and are referred to as peptides. In general, the term **protein** is used for molecules that contain over 50 amino acids and **peptide** is used for those which contain less than 50 amino acids. Important peptide hormones include adrenocorticotropic hormone, antidiuretic hormone, glucagon, and calcitonin.

Proteins control and regulate gene transcription and translation. These include the histone proteins that are closely associated with DNA, repressor and enhancer transcription factors that control gene transcription, proteins that regulate DNA structure to promote or inhibit gene expressions, proteins that transcribe the DNA into RNA, and components of the heteronuclear RNA particles and ribosomes.

Structural proteins function in "brick-and-mortar" roles. They include collagen and elastin, which form the matrix of bone and ligaments and provide structural strength and elasticity to organs and the vascular system. Tubulin and actin are proteins that polymerize into fibers of microtubules and filaments that form the cytoskeleton of each cell.

An understanding of normal functioning and pathology of the mammalian organism requires a clear understanding of the properties of proteins.

3.2 • AMINO ACID COMPOSITION OF PROTEINS

All the different types of proteins are polymers of only 21 amino acids. The **common amino acids** are defined as the amino acids for which at least one codon exists in the genetic code. Transcription and translation of the DNA code result in the joining together of amino acids into a specific linear sequence characteristic of a protein (Figure 3.1). Many proteins also contain **derived amino acids**, which are usually formed by enzymatic modification of one of the common amino acids after it has been incorporated into a protein. Examples of derived amino acids are cystine (p. 79), desmosine, and isodesmosine found in elastin, hydroxyproline and hydroxylysine in collagen, γ-carboxyglutamate in prothrombin, and phosphoserine, phosphothreonine, and phosphotyrosine.

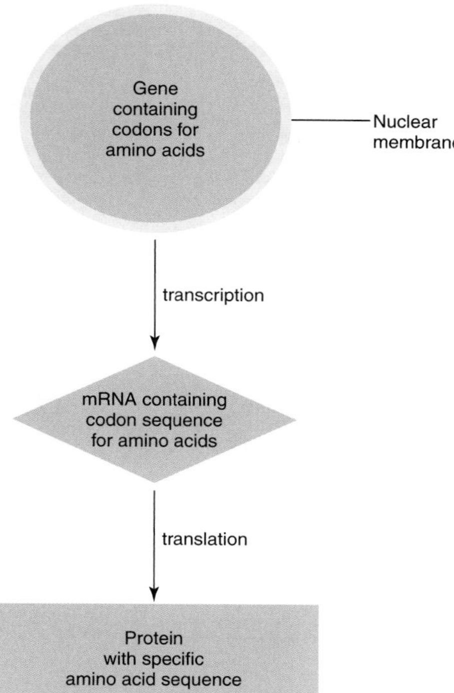

Figure 3.1 Genetic information is transcribed from a DNA sequence into mRNA and then translated to the amino acid sequence of a protein.

Common Amino Acids

Common amino acids have the general structure depicted in Figure 3.2. They contain in common a central alpha (α)-carbon atom to which a carboxylic acid group, an amino group, and a hydrogen atom are covalently bonded. In addition, the fourth valency of the α-carbon atom is linked to a specific chemical group, designated R and called the side chain that uniquely defines each of the common amino acids. Figure 3.2 depicts the ionized form of the common amino acids in solution at pH 7. Under physiological conditions, the α-amino group is protonated and in its ammonium ion form; the carboxylic acid group is in its unprotonated or carboxylate ion form.

Side Chains Define the Chemical Nature and Structures of α-Amino Acids

Structures of the common amino acids are shown in Figure 3.3. Amino acids that contain alkyl group side chains include glycine, alanine, valine, leucine, and isoleucine. **Glycine** has the simplest structure, with R = H. **Alanine** contains a methyl (CH_3—) group. **Valine** has an isopropyl R group (Figure 3.4, p. 79). The leucine and isoleucine R groups are isobutyl groups that are structural isomers of each other. In **leucine** the branching of the isobutyl side chain occurs on the gamma (γ)-carbon and in **isoleucine** it is branched on the beta (β)-carbon (Figure 3.4).

Phenylalanine, tyrosine, and tryptophan are aromatic amino acids. **Phenylalanine** contains a benzene ring, **tyrosine** a phenol group, and **tryptophan** the heterocyclic structure, indole. In each case the aromatic moiety is attached to the α-carbon through a methylene (—CH_2—) carbon (Figure 3.3).

Sulfur-containing amino acids are cysteine and methionine. The **cysteine** side-chain group is a thiolmethyl ($HSCH_2$—), and in **methionine** the side chain is a methyl ethyl thiol ether ($CH_3SCH_2CH_2$—).

The two hydroxy (alcohol)-containing amino acids are serine and threonine. In **serine** the side chain is a hydroxymethyl ($HOCH_2$—). In **threonine** an ethanol structure is connected to the α-carbon to produce a secondary alcohol (CH_3—CHOH—CH_α—).

Proline is unique in that it incorporates the α-amino group in its side chain and is more accurately classified as an α-imino acid, since its α-amine is a secondary amine with its α-nitrogen having two covalent bonds to carbon. Incorporation of the α-amino nitrogen into a five-member ring constrains the rotational freedom around the —N_α—C_α— bond in proline and thereby limits proline participation to particular polypeptide chain conformations.

The amino acids described so far contain side chains that are uncharged at physiological pH. The **dicarboxylic-monoamino acids** contain a carboxylic group in their side chain. In **aspartate**, this group is separated by a methylene carbon (—CH_2—) from the α-carbon (Figure 3.5, p. 79). In **glutamate** (Figure 3.5), the group is separated by two methy-lene (—CH_2—CH_2—) carbon atoms from the α-carbon (Figure 3.5). At physiological pH, these groups are unprotonated and negatively charged. The **dibasic-monocarboxylic** acids are lysine, arginine, and histidine (Figure 3.3). In these, the R group contains one or two nitrogen atoms that act as a base by binding a proton. In **lysine** the side chain is an *N*-butyl amine. In **arginine**, the side chain contains a guanidino group (Figure 3.6, p. 79) separated from the α-carbon by three methylene carbon atoms. Both the guanidino group of arginine and the ε-amino group of lysine are protonated at physiological pH (pH ~7) and positively charged. In **histidine** the side chain contains a five-member heterocyclic structure, the imidazole group

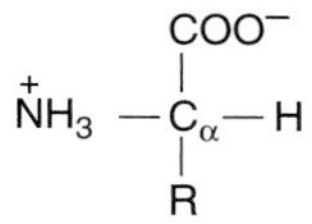

$$\begin{array}{c} COO^- \\ | \\ \overset{+}{N}H_3 - C_\alpha - H \\ | \\ R \end{array}$$

Figure 3.2 General structure of the common amino acids.

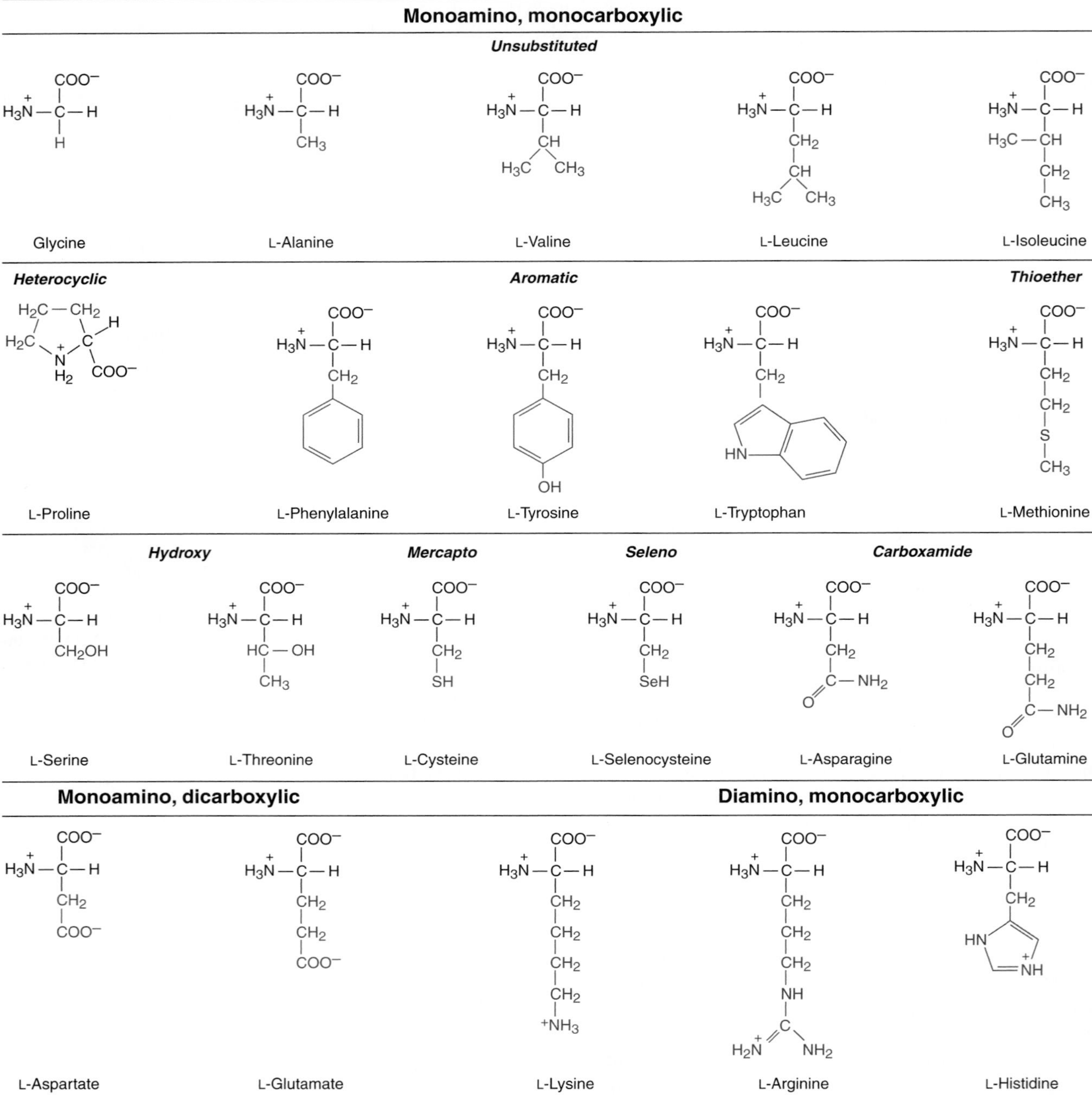

Figure 3.3 Structures of the common amino acids. Charge forms are those present at pH 7.0.

(Figure 3.6). The p$K'a$ of the imidazole group is approximately 6.0 in water; thus physiological solutions contain relatively high concentrations of both basic (imidazole) and acidic (imidazolium) forms of the histidine side chain.

Glutamine and **asparagine** are structural analogs of glutamic acid and aspartic acid with their carboxylic acid side-chain groups amidated. Unique DNA codons exist for glutamine and asparagine separate from those for glutamic acid and aspartic acid. The amide side chains of glutamine and asparagine cannot be protonated and are uncharged at physiological pH.

By the mid-1960s it was believed that the three-nucleotide DNA codons specifying the common amino acids found in proteins had been determined. The total number of common

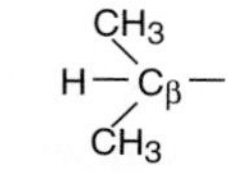

Isopropyl R group of valine

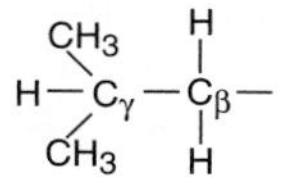

Isobutyl R group of leucine

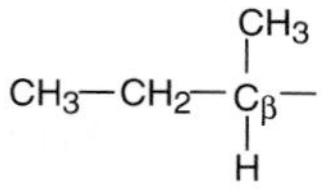

Isobutyl R group of isoleucine

Figure 3.4 Alkyl side chains of valine, leucine, and isoleucine.

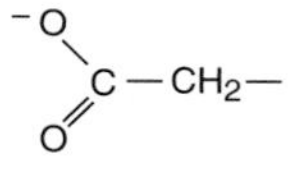

Aspartate R group

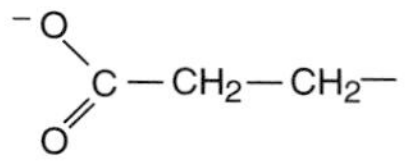

Glutamate R group

Figure 3.5 Side chains of aspartate and glutamate.

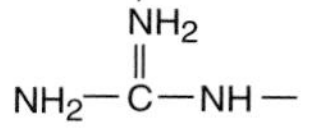

Guanidinium group (charged form) of arginine

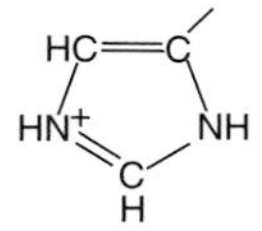

Imidazolium group of histidine

Figure 3.6 Guanidinium and imidazolium groups of arginine and histidine.

amino acids was 20 derived from 64 possible triplet codons with 3 codons assigned to terminate the translation of open reading frames (Table 6.1, p. 211). However, in 1988 a 21st common amino acid, **selenocysteine** (Figure 3.3), with a specific mRNA codon and tRNA that places the selenocysteine into polypeptide chains during mRNA translation was shown. The selenocysteine is structurally similar to cysteine, but with a selenium atom replacing the sulfur atom. A difficulty with the codon for selenocysteine is that it was identical to one of the termination codons that translates to a stop translation signal. Thus this codon, UGA, has two meanings. The codon codes a stop translation signal the great majority of the time, but within special RNA contexts it will code for selenocysteine incorporation into a polypeptide sequence. These instances are rare as only 25 genes have been identified within the human genome that code for proteins that incorporate selenocysteine. In addition, while selenocysteine proteins are found in all animals, the codon is not universal as it is not present in higher plants, yeast, or the majority of bacteria. In animal studies, knockout of the gene for the selenocysteine tRNA is lethal, inferring that selenocysteine is essential for life in animal species. Many of the selenoproteins have an antioxidant role in eliminating reactive oxygen species.

To represent sequences of amino acids in proteins, three-letter or one-letter abbreviations for the common amino acids are used (Table 3.1). These abbreviations are universally accepted and will be used throughout this book. The three-letter abbreviations of aspartic acid (Asp) and glutamic acid (Glu) should not be confused with those for asparagine (Asn) and glutamine (Gln). In determination of the amino acid composition of a protein by chemical procedures, one cannot easily differentiate between Asn and Asp, or between Gln and Glu, because the amide side-chain groups in Asn and Gln are often hydrolyzed in the procedure and generate Asp and Glu (Section 3.9). The symbols of Asx for Asp or Asn, and Glx for Glu or Gln indicate this ambiguity in the analysis. A similar scheme is used with the one-letter abbreviations for Asp or Asn, and Glu or Gln.

TABLE 3.1 • Abbreviations for the Amino Acids

	Abbreviation	
Amino Acid	*Three Letter*	*One Letter*
Alanine	Ala	A
Arginine	Arg	R
Asparagine	Asn	N
Aspartate	Asp	D
Asparagine or aspartate	Asx	B
Cysteine	Cys	C
Glycine	Gly	G
Glutamine	Gln	Q
Glutamate	Glu	E
Glutamine or glutamate	Glx	Z
Histidine	His	H
Isoleucine	Ile	I
Leucine	Leu	L
Lysine	Lys	K
Methionine	Met	M
Phenylalanine	Phe	F
Proline	Pro	P
Selenocysteine	Sec	U
Serine	Ser	S
Threonine	Thr	T
Tryptophan	Trp	W
Tyrosine	Tyr	Y
Valine	Val	V

Cystine Is a Derived Amino Acid Produced by Oxidation of Two Cysteine Residues

A derived amino acid found in many proteins is cystine. It is formed by the oxidation of two cysteine thiol side chains to form a covalent disulfide bond (Figure 3.7). Within proteins, disulfide links of cystine formed from cysteine residues, separated from each other within a polypeptide chain (intrachain) or between two polypeptide chains (interchain), have an important role in stabilizing the folded conformation of proteins (insulin structure; see Figure 3.23, p. 91).

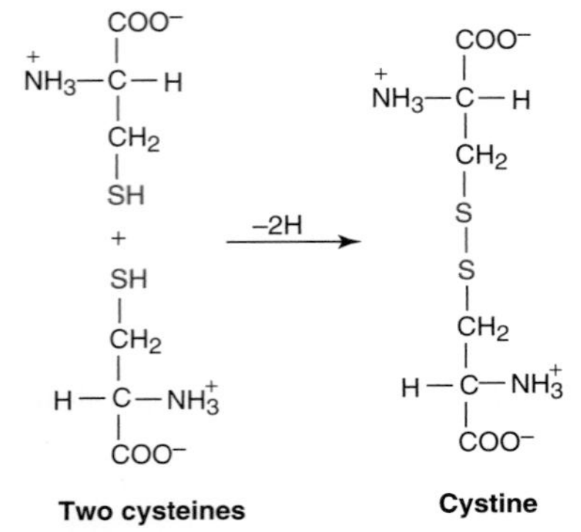

Two cysteines → (−2H) → Cystine

Figure 3.7 Cystine bond formation.

L-Configuration | D-Configuration

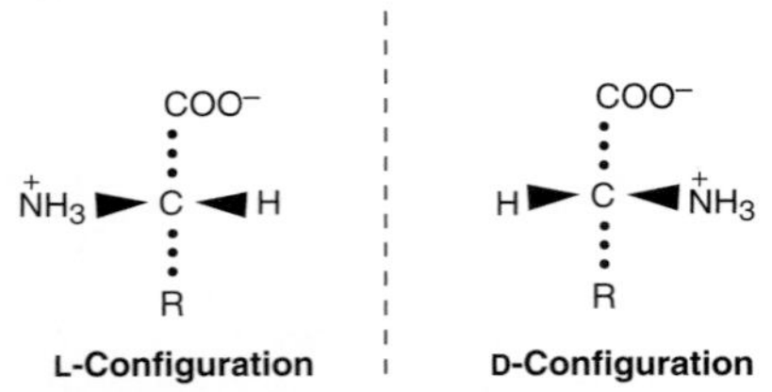

Figure 3.8 Absolute configuration of an amino acid.

α-Amino Acids Have an Asymmetric Center

The common amino acids (see Figure 3.2) have four substituents (R, H, COO^-, and NH_3^+) covalently bonded to an α-carbon atom. A carbon atom with four different substituents in a tetrahedral configuration is asymmetric and exists in two enantiomeric forms. Thus, all amino acids exhibit optical isomerism except glycine, in which R = H and so two of the four substituents on the α-carbon atom are hydrogen. The absolute configuration for an amino acid is depicted in Figure 3.8 using the Fischer projection to show the position in space of the tetrahedrally arranged α-carbon substituents. The α-COO^- group is directed up and behind the plane of the page, and the R group is directed down and behind the plane of the page. The α-H and α-NH_3^+ groups are directed toward the reader. An amino acid held in this way projects its α-NH_3^+ group to either the left or right of the α-carbon atom. By convention, if the α-NH_3^+ is projected to the left, the amino acid has an L absolute configuration. Its enantiomer, with α-NH_3^+ projected toward the right, has a D absolute configuration. Mammalian proteins contain amino acids of only L configuration. The L and D designations refer to the ability to rotate polarized light to the left (L, levo) or right (D, dextro) from its plane of polarization. As the amino acids in proteins are asymmetric, the proteins also exhibit asymmetric properties.

Amino Acids Are Joined into Peptides and Proteins

Linking of a selection of the common amino acids into polypeptide chains in cells is enzymatically catalyzed (p. 221). Chemically, this is a dehydration reaction (Figure 3.9). The α-carboxyl group of one amino acid forms a covalent **peptide bond** with the α-amino group of another amino acid by elimination of a molecule of water. The **dipeptide** (two amino acid residues joined by a single peptide bond) can then form a second peptide bond through its terminal carboxylic acid group and the α-amino of a third amino acid, to generate a tripeptide (Figure 3.9). Repetition of this process generates a **polypeptide** or protein of specific amino acid sequence. This is the **primary structure** of the protein, and it is predetermined by the nucleotide sequence of its gene. The unique primary structure enables a polypeptide chain to fold into a specific three-dimensional structure that gives the protein its chemical and physiological properties.

A peptide bond can be represented as two **resonance isomers** (Figure 3.10). In structure I, a double bond is located between the carbonyl carbon and carbonyl oxygen

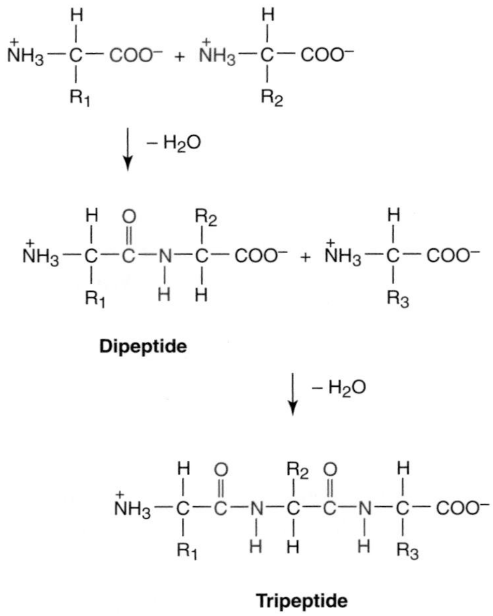

Dipeptide

Tripeptide

Figure 3.9 Peptide bond formation.

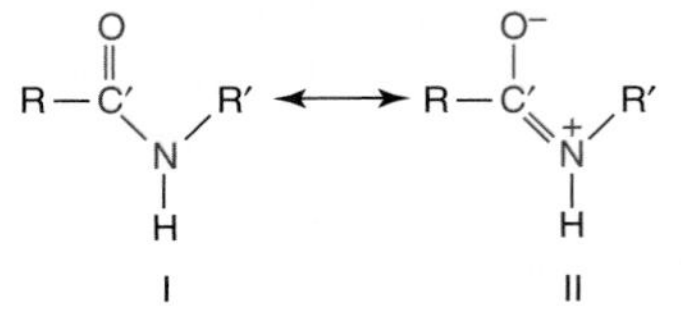

I ⟷ II

Figure 3.10 Electronic isomer structures of a peptide bond.

(C′=O), and the carbonyl carbon to nitrogen (C′—N) linkage is a single bond. In structure II, the carbon to oxygen bond ($C'—O^-$) is a single bond and the bond located between the carbon and nitrogen is a double bond (C′=N). In structure II there is a negative charge on the oxygen and a positive charge on the nitrogen. Actual peptide bonds are a resonance hybrid of these two electron isomer structures, the carbon to nitrogen bond having a 50% double-bond character. X-ray diffraction studies confirm this and show the carbonyl carbon to nitrogen bond length (1.33 Å) is approximately half way between that of a C′—N single bond (~1.45 Å) and a C′=N double bond (~1.25 Å).

A consequence of the partial double bond character is that rotation does not occur about the carbonyl carbon to nitrogen of a peptide bond at physiological temperatures. A second consequence of the double bond character is that all the atoms attached to C′ and N lie in a common plane.

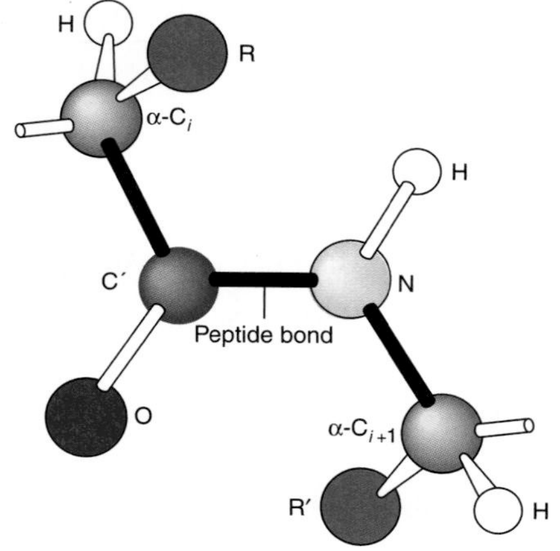

(a) *trans* configuration

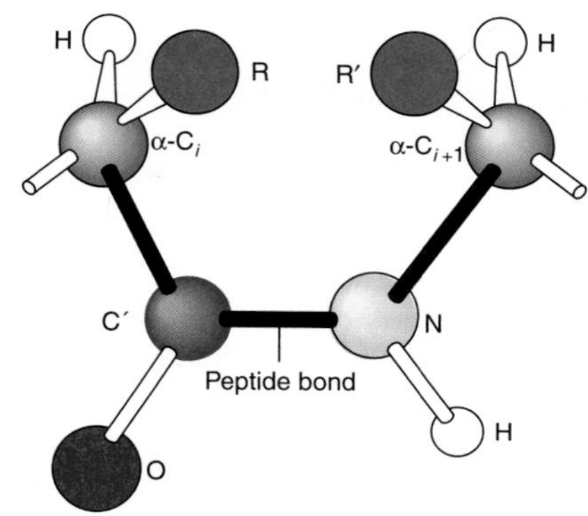

(b) *cis* configuration

Figure 3.12 (a) Trans peptide bond. (b) The rare *cis* peptide bond. C′–N bonds have a partial double-bond character.

Thus a polypeptide chain consists of peptide-bond planes interconnected at the α-carbon atoms. The α-carbon interconnects the peptide bonds through single bonds that allow rotation of adjacent peptide planes with respect to each other. Each **amino acid residue** contributes one α-carbon, two single bonds, and a peptide bond to the polypeptide chain (Figure 3.11). The term *residue* refers to the atoms contributed by an amino acid to a polypeptide chain including those of the side chain.

The peptide bond in Figure 3.12*a* shows a ***trans* configuration** of its oxygen (O) and hydrogen (H) atoms. This is the most stable configuration with the two side chains of the adjacent residues (R and R′) also being in *trans.* The ***cis* configuration** (Figure 3.12*b*) brings the two side-chain groups to the same side of the C′=N bond, which is unfavorable because of steric hindrance between the R groups. Accordingly, *trans* peptide bonds occur in proteins except where there are proline residues. In proline the side chain includes the α-amino group, and both the *cis* and *trans* peptide bond configurations with the proline α-imino group have unfavorable interactions with the α-carbon of the adjacent amino acid forming a peptide bond with the α-imino. Accordingly, the peptide bond of proline in an unstructured or denatured polypeptide chain will have significant amounts of its peptide bond in both *trans* and *cis* configurations. The configuration of the proline peptide bond in a folded protein depends on the forces on the proline bond generated by the three-dimensional structure of the particular protein molecule in which the proline is imbedded.

One of the largest polypeptides in humans is the muscle protein titin, which contains about 27,000 amino acid residues in a single polypeptide chain. Chain length alone, however, does not determine the function of a polypeptide. Many small peptides of less than 10 amino acids perform important biochemical and physiological functions in humans (Table 3.2). Primary structures are written in a standard convention and sequentially numbered from their NH_2-terminal end to their COOH-terminal end, consistent with the order of addition of the amino acid to the chain during biosynthesis. Accordingly, for thyrotropin-releasing hormone (Table 3.2) the glutamic acid residue on the left is the NH_2-terminal amino acid of the tripeptide and amino acid residue 1 in the sequence. The proline is the COOH-terminal amino acid and is residue 3. The defined direction of the polypeptide chain is from Glu to Pro (NH_2-terminal amino acid to COOH-terminal amino acid).

Figure 3.11 Amino acid residue within a polypeptide chain. Each residue of a polypeptide contributes two single bonds and one peptide bond to the chain. The single bonds are those between the C_α and carbonyl C′ atoms, and the C_α and N atoms. (See p. 92 for the definition of φ and ψ.)

TABLE 3.2 • Biologically Active Peptides

Amino Acid Sequence	*Name*	*Function*
1 3 pyroGlu-His-Pro(NH_2)[a]	Thyrotropin-releasing hormone	Secreted by hypothalamus; causes anterior pituitary gland to release thyrotropic hormone
1 9 H-Cys-Tyr-Phe-Gln-Asn-Cys-Pro-Arg-Gly(NH_2)[b,a] S———————S	Vasopressin (antidiuretic hormone)	Secreted by posterior pituitary gland; causes kidney to retain water from urine
1 5 H-Tyr-Gly-Gly-Phe-Met-OH	Methionine enkephalin	Opiate-like peptide found in brain that inhibits sense of pain
1 10 pyroGlu-Gly-Pro-Trp-Leu-Glu-Glu-Glu-Glu- 11 17 Ala-Tyr-Gly-Trp-Met-Asp-Phe(NH_2)[c,d] SO_3	Little gastrin (human)	Hormone secreted by mucosal cells in stomach; causes parietal cells of stomach to secrete acid
1 10 H-His-Ser-Gln-Gly-Thr-Phe-Thr-Ser-Asp-Tyr- 11 20 Ser-Lys-Tyr-Leu-Asp-Ser-Arg-Arg-Ala-Gln- 21 29 Asp-Phe-Val-Gln-Trp-Leu-Met-Asn-Thr-OH	Glucagon (bovine)	Pancreatic hormone involved in regulating glucose metabolism
1 8 H-Asp-Arg-Val-Tyr-Ile-His-Pro-Phe-OH	Angiotensin II (horse)	Pressor or hypertensive peptide; also stimulates release of aldosterone from adrenal cortex
1 9 H-Arg-Pro-Pro-Gly-Phe-Ser-Pro-Phe-Arg-OH	Plasma bradykinin (bovine)	Vasodilator peptide
1 10 H-Arg-Pro-Lys-Pro-Gln-Phe-Phe-Gly-Leu-Met(NH_2)[a]	Substance P	Neurotransmitter

[a]The NH_2-terminal Glu is in the pyro form in which its γ-COOH is covalently joined to its α-NH_2 via amide linkage; the COOH-terminal amino acid is amidated and thus also not free.

[b]Cysteine-1 and cysteine-6 are joined to form a disulfide bond structure within the nonapeptide.

[c]The Tyr 12 is sulfonated on its phenolic side-chain OH.

[d]The COOH-terminal amino acid is amidated.

3.3 • CHARGE AND CHEMICAL PROPERTIES OF AMINO ACIDS AND PROTEINS

Ionizable Groups of Amino Acids and Proteins Are Critical for Biological Function

Ionizable groups common to proteins and amino acids are shown in Table 3.3. The acid forms are on the left of the equilibrium sign and the base forms on the right. In forming its conjugate base, the acid form releases a proton. In reverse, the base form associates with a proton to form the respective acid. The dissociation of an acid is characterized by an acid dissociation constant (K_a') and its pK_a' value: p$K_a' = \log(1/K_a')$. Table 3.3 shows a range of pK_a' values for each acid group, as the actual pK_a' depends on the environment in which an acid group is placed. For example, when a positively charged ammonium group ($—NH_3^+$) is placed near a negatively charged group in a protein, the negative charge stabilizes the positively charged acid form, making it more difficult to dissociate its proton. The pK_a' of the $—NH_3^+$ will have a higher value than normal for an ammonium group in the absence of a nearby negative-charge stabilization. Other factors that affect the pK_a' include polarity of the environment, absence or presence of water, and the potential for hydrogen bond formation. In addition, acid groups (α-COOH or α-NH_3^+) at the ends of polypeptides

TABLE 3.3 • pK'_a Values for the Common Acid Groups in Proteins

Where Acid Group Is Found	*Acid Form*		*Base Form*	*Approximate pK'_a Range for Group*
NH_2-terminal or lysine side chain	$R{-}NH_3^+$ Ammonium	⇌	$R{-}NH_2 + H^+$ Amine	7.6–10.6
COOH-terminal or glutamate and aspartate side chains	R—COOH Carboxylic acid	⇌	$R{-}COO^- + H^+$ Carboxylate	3.0–5.5
Arginine side chain	$R{-}NH{-}\overset{+}{C}(NH_2){=}NH_2$ Guanidinium	⇌	$R{-}NH{-}C(NH_2){=}NH + H^+$ Guanidino	11.5–12.5
Cysteine side chain	R—SH Thiol	⇌	$R{-}S^- + H^+$ Thiolate	8.0–9.0
Histidine side chain	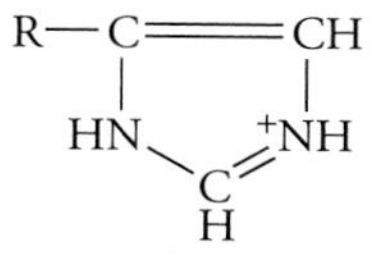 R—C=CH, HN, $^+$NH, C—H (ring) Imidazolium	⇌	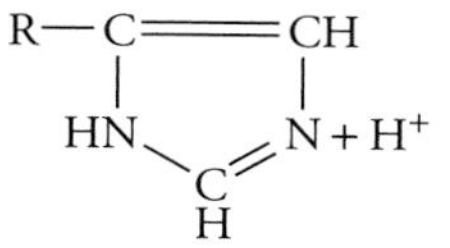 R—C=CH, HN, N, C—H (ring) $+ H^+$ Imidazole	6.0–7.0
Tyrosine side chain	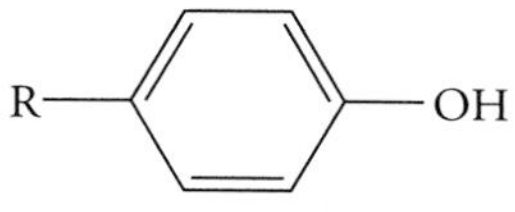 R—C_6H_4—OH Phenol	⇌	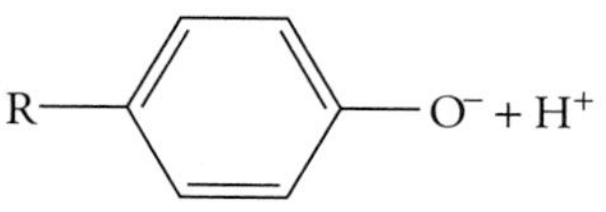 R—C_6H_4—$O^- + H^+$ Phenolate	9.5–10.5

TABLE 3.4 • pK'_a of Side Chain and Terminal Acid Groups in Ribonuclease

	—NH_3^+	*—COOH*
Side chain	Lysines ≃ 10.2	Glu and Asp ≃ 4.6
Chain end	N-terminal = 7.8	C-terminal = 3.8

$$pH = pKa + \log \frac{[\text{conjugate base}]}{[\text{conjugate acid}]}$$

or

$$pH - pKa = \log \frac{[\text{conjugate base}]}{[\text{conjugate acid}]}$$

Figure 3.13 Henderson–Hasselbalch equation. (For a more detailed discussion of this equation, see p. 9).

typically have a lower pK'_a value than the same types of acid groups in the side chains (Table 3.4). The amino acids whose R groups contain nitrogen atoms (Lys and Arg) are the **basic amino acids**, since their side chains have relatively high pK'_a values and function as bases at physiological pH. They are usually in their acid form and are positively charged at physiological pH. Amino acids whose side chains contain a carboxylic acid group have relatively low pK'_a values that easily lose their protons and are **acidic amino acids**. They are predominantly in their unprotonated forms and negatively charged at physiological pH. Proteins in which the ratio $(\Sigma\text{Lys} + \Sigma\text{Arg})/(\Sigma\text{Glu} + \Sigma\text{Asp})$ is greater than 1 are **basic proteins**. Proteins in which the ratio is less than 1 are **acidic proteins**.

Ionic Form of an Amino Acid or Protein Can Be Determined at a Given pH

From the pK'_a value for each acid group in an amino acid or protein and the Henderson–Hasselbalch equation (Figure 3.13), the ionic form of the molecule can be calculated at a given pH. This is an important equation as it shows the change in ionization state and charge of a molecule with pH. Physiological activities of a molecule differ with changes in pH and ionization state. For example, some enzymes require a histidine imidazole in its base form for catalytic activity. If the pK'_a of this histidine is 6.0, at pH 6.0 one-half of the enzyme molecules are in the active base (imidazole) form and one-half of the enzyme molecules are in the inactive acid (imidazolium) form. Accordingly, the enzyme exhibits 50% of its potential activity. At pH 7.0, the ratio of [imidazole]/[imidazolium] is 10:1 (Table 3.5) and the enzyme exhibits $10/(10 + 1) \times 100 = 91\%$ of its maximum potential activity. Thus a change in pH has a dramatic effect on the enzyme's activity.

TABLE 3.5 • Relationship between the Difference of pH and Acid pK'_a and the Ratio of the Concentrations of Base to Its Conjugate Acid

$pH - pK'_a$ (Difference Between pH and pK'_a)	*Ratio of Concentration of Base to Conjugate Acid*
0	1
1	10
2	100
3	1000
−1	0.1
−2	0.01
−3	0.001

NH_3^+—C(COOH)(H)—CH_2—CH(CH_3)—CH_3 $\underset{+H^+}{\overset{+OH^-}{\rightleftharpoons}}$ NH_3^+—C(COO^-)(H)—CH_2—CH(CH_3)—CH_3 $\underset{+H^+}{\overset{+OH^-}{\rightleftharpoons}}$ NH_2—C(COO^-)(H)—CH_2—CH(CH_3)—CH_3

	I	II	III
Charge	+1	0	−1
	pH < 2.4	2.4 < pH < 9.6	9.6 < pH

Figure 3.14 Ionic forms of leucine.

Titration of a Monoamino Monocarboxylic Acid: Determination of the Isoelectric pH

An understanding of a protein's acid and base forms and their relation to charge is made clearer by following the titration of the ionizable groups for a simple amino acid. In Figure 3.14, leucine has an α-COOH with $pK_a' = 2.4$ and an α-NH_3^+ group with $pK_a' = 9.6$. At pH 1.0 the predominant ionic form (form I) has a charge of +1 and migrates toward the cathode in an electrical field. Addition of 0.5 equivalent of base titrates half the α-COOH groups; that is, the ratio of $[COO^-]/[COOH]$ will equal 1. The Henderson–Hasselbalch equation, with the second term on the right side of the equation $\log_{10}[(\text{base})/(\text{acid})] = \log(1) = 0$ at a ratio of conjugate base to acid of 1:1, shows that the pH (when the α-COOH is half titrated) is directly equal to the $pK_{a\,(\alpha COOH)}$ (Figure 3.15).

Addition of 1 equiv of base at pH 6.0 completely titrates the α-COOH, but has no effect on the α-NH_3^+ group. In the resulting form (II), the negative and positive charges cancel each other and the net charge is zero. Form II is the **zwitterion** form, the ionic form in which the total of positive charges equals the total of negative charges. As the net charge on a zwitterion molecule is zero, it will not migrate in an electric field. Further addition of 0.5 equiv of base to the zwitterion form (total base added is 1.5 equiv) will half-titrate the α-NH_3^+ group. At this point, the ratio of $[NH_2]/[NH_3^+] = 1$, and the pH is equal to the value of the pK_a' for the α-NH_3^+ group (Figure 3.15). Addition of a further 0.5 equiv of base (total of two full equivalents of base added; Figure 3.15) completely titrates the α-NH_3^+ group to its base form (α-NH_2). The pH becomes greater than 11, and the predominant molecular species has a negative charge (form III).

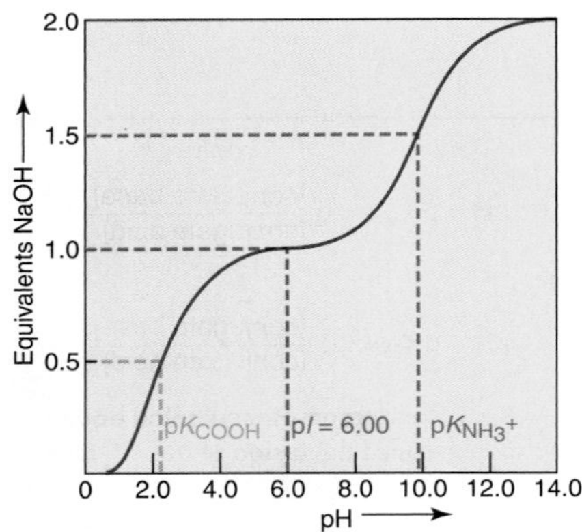

Figure 3.15 Titration curve of leucine.

It is useful to calculate the pH at which an amino acid is in its zwitterion form. This pH is the **isoelectric pH** for the molecule, represented as **pI**. The pI value is a constant for a compound at a particular ionic strength and temperature. For simple amino acid molecules, such as leucine, pI is calculated as the average of the two pK_a' values that regulate the boundaries of the zwitterion form. For leucine,

$$pI = \frac{pK_a'\text{COOH} + pK_a'\text{NH}_3^+}{2} = \frac{2.4 + 9.6}{2} = 6.0$$

At pH > 6.0, leucine has a partial negative charge that increases at high pH to a full negative charge of −1 (form III) (Figure 3.14). At pH < 6, it has a partial positive charge, until at a very low pH, it has a charge of +1 (form I). The charge at any pH can be calculated from the Henderson–Hasselbalch equation or from extrapolation from the titration curve of Figure 3.15.

Titration of a Monoamino-Dicarboxylic Acid

A more complicated example is provided by glutamic acid. As shown in Figures 3.16 and 3.17, in glutamic acid the α-COOH $pK_a' = 2.2$, the γ-COOH $pK_a' = 4.3$, and the α-NH_3^+ $pK_a' = 9.7$. The zwitterion form is generated after 1.0 equiv of base is added to

NH_3^+—C(COOH)(H)—CH_2—CH_2—COOH $\underset{+H^+}{\overset{+OH^-}{\rightleftharpoons}}$ NH_3^+—C(COO^-)(H)—CH_2—CH_2—COOH $\underset{+H^+}{\overset{+OH^-}{\rightleftharpoons}}$ NH_3^+—C(COO^-)(H)—CH_2—CH_2—COO^- $\underset{+H^+}{\overset{+OH^-}{\rightleftharpoons}}$ NH_2—C(COO^-)(H)—CH_2—CH_2—COO^-

Charge	+1	0	−1	−2
	pH < 2.2	2.2 < pH < 4.3	4.3 < pH < 9.7	9.7 < pH

Figure 3.16 Ionic forms of glutamic acid.

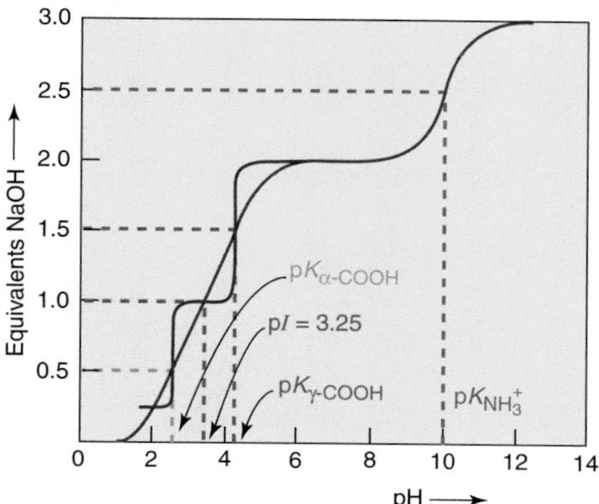

Figure 3.17 Titration curve of glutamic acid. Observed line (red) hides titrations of pK'_γ and pK'_α, which are too close to observe individually. Titrations of these groups are shown in overlapping blue line.

the low pH form, and the isoelectric pH (pI) is calculated from the average of the two pK'_a values that control the boundaries of the zwitterion form.

$$pI = \frac{2.2 + 4.3}{2} = 3.25$$

Accordingly, at values above pH 3.25 the molecule assumes a net negative charge, until at high pH it has a net charge of –2. At pH < 3.25 glutamic acid is positively charged, and at extremely low pH it has a net positive charge of +1.

TABLE 3.6 • pI Values for Some Proteins

Protein	*pI*
Pepsin	~1
Human serum albumin	5.9
α_1-Lipoprotein	5.5
Fibrinogen	5.8
Hemoglobin A	7.1
Ribonuclease	7.8
Cytochrome c	10.0
Thymohistone	10.6

General Relationship between Charge Properties of Amino Acids and Proteins, and pH

The relationship between pH and charge for leucine and glutamate is generally true for the other amino acids. At a pH less than pI, the molecule is positively charged. At a pH greater than pI, it is negatively charged. The degree of positive or negative charge is a function of the magnitude of the difference between pH and pI. As proteins are complex polyelectrolytes that contain many ionizable groups that regulate the zwitterion form, calculation of a protein's isoelectric pH from its multiple acid pK'_a values utilizing the Henderson–Hasselbalch relationship is difficult. Accordingly, pI values of proteins are experimentally measured by determining the pH value at which the protein does not move in an electric field. pI values for some proteins are given in Table 3.6.

As with amino acids, at a pH greater than its pI, a protein has a net negative charge. At a pH less than its pI, it has a net positive charge (Figure 3.18). The magnitude of the net charge increases as the difference between pH and pI increases. An example is human plasma albumin with 585 amino acid residues including 61 glutamates, 36 aspartates, 57 lysines, 24 arginines, and 16 histidines. The titration curve is shown in Figure 3.19. Albumin's pI = 5.9, at which pH its net charge is zero. At pH 7.5 the imidazolium groups of histidines are partially titrated and there is a negative charge of −10. At pH 8.6 the net charge is approximately −20, and at pH 11 it is approximately −60. At pH 3, the approximate net charge is +60.

pH > p*I*, then protein charge negative
pH < p*I*, then protein charge positive

Figure 3.18 Relationship between solution pH, protein pI, and protein charge.

Amino Acids and Proteins Can Be Separated on the Basis of Their Differences in pI

The techniques of electrophoresis, isoelectric focusing, and ion-exchange chromatography separate and characterize biological molecules based on their pI (p. 84). In clinical medicine, electrophoretic separation of **plasma proteins** has led to their

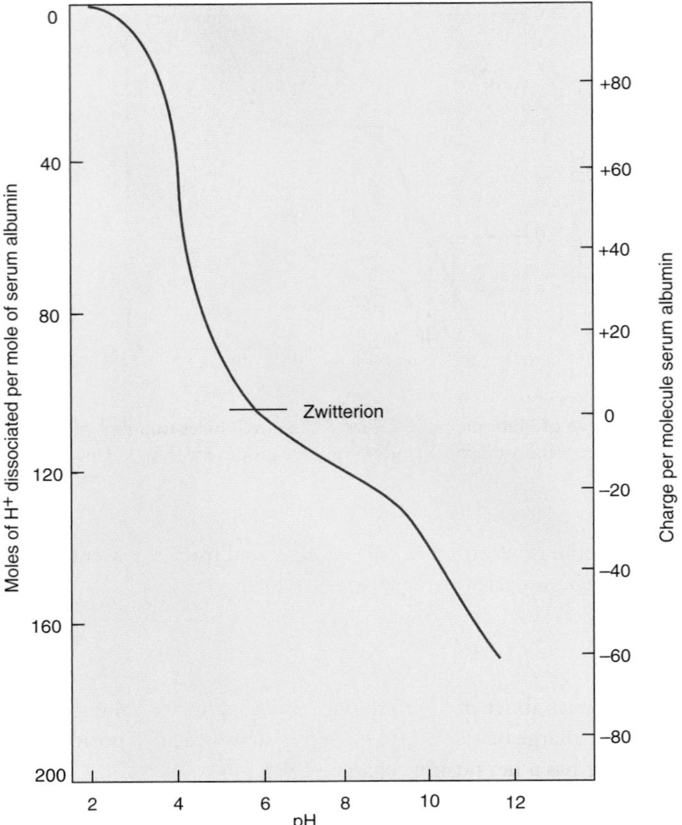

Figure 3.19 Titration curve of human serum albumin at 25 °C and an ionic strength of 0.150. Redrawn from Tanford, C., J. *Am. Chem. Soc.* 72:441, 1950.

classification being based on their relative electrophoretic mobility. The separation of the plasma proteins is commonly carried out at pH 8.6, which is higher than the pI values of the major proteins. Accordingly, the proteins are negatively charged and move toward the anode at a rate dependent on their net charge. Figure 3.20 shows an electrophoresis of plasma proteins in an agarose gel at pH 8.6 that separates the proteins into the five classical bands used to classify the plasma proteins. In order of their migration the major fractions are albumin, α1-, α2-, β-, and γ-globulins (Figure 3.20). Some of these bands actually represent tens to hundreds of different proteins that migrate similarly at pH 8.6. However, certain proteins predominate in each band, and variation in their relative amounts is characteristic of certain diseases (Figures 3.20 and 3.21, and Clin. Corr. 3.1).

Amino Acid Side Chains Have Polar and Apolar Properties

The **hydrophobicity** of amino acid side chains is critical for the folding of a protein to its native structure and for stability of the folded protein. Figure 3.22 (p. 88) plots the relative hydrophobicity of the amino acids on the basis of their tendency to partition in a mixture of water and a nonpolar solvent. The scale is based on a value of zero for glycine. Side chains that preferentially dissolve in the nonpolar solvent relative to glycine show a positive (+) hydrophobicity value, the more positive the greater the preference for the nonpolar solvent. In folded protein structures, the majority of hydrophobic amino acids are buried away from the water solvent that interacts with the surface of a soluble protein. However, the general correlation is not perfect due to the amphoteric nature of many of the hydrophobic amino acids that place the more polar portions of their side-chain structure near the surface to interact with water. In addition, some nonpolar side chains may be on the surface. However, when on the surface, the hydrophobic groups are generally dispersed among the polar side

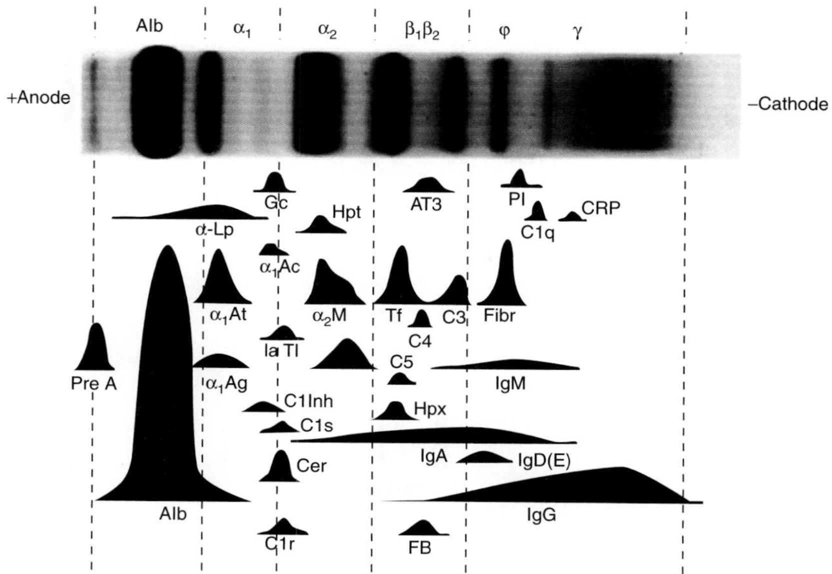

Figure 3.20 Electrophoresis pattern in agarose gel for plasma proteins at pH 8.6. Migration along horizontal axis from right to left (cathode toward anode) with proteins of highest mobility closest to anode. The intensity of the band along the vertical axis shows the protein concentration. Different major proteins are designated underneath their electrophoretic mobility peaks. ϕ Designates the fibrinogen peak, which is present in plasma but absent in serum. In some gels the γ-globin band is separated in γ_1 and γ_2 bands (not shown here).
Major protein peaks contained in the bands on the agarose gel are drawn in this figure. Area of drawn peaks show their relative concentration and position in the gel. Abbreviations: α_1Ac = α_1-antichymotrypsin; α_1Ag = α_1-acid glycoprotein; α_1At = α_1-antitrypsin; α_2-M = α_2-macroglobulin; α-Lp = α-lipoprotein; Alb = albumin; AT3 = antithrombin III; β-Lp = β-lipoprotein; complement components C1q, C1r, C1s, C3, C4, and C5 = as designated; C1Inh = C1 esterase inhibitor; Cer = ceruloplasmin; CRP = C-reactive protein; Gc = Gc-globulin (vitamin D–binding protein); FB = factor B; Fibr = fibrinogen; Hpt = haptoglobin; Hpx = hemopexin; immunoglobulins IgA, IgD, IgE, IgG, and IgM = as designated; IaTI = inter-α-trypsin inhibitor; Pl = plasminogen; PreA = prealbumin; and Tf = transferrin.
Redrawn from McPherson, R.A. In McPherson, R.A., and Pincus, M.R. (Eds.), *Henry's Clinical Diagnosis and Management by Laboratory Methods*, 21st ed., Chap. 19. Philadelphia: Saunders-Elsevier, 2007.

chains. When clustering of nonpolar side chains occurs on the surface, it is usually associated with a function, such as to provide a site for binding of substrate molecules through hydrophobic interactions.

Most charged side chains are on the surface of soluble globular proteins where they are stabilized by favorable energetic interactions with water. The rare positioning of a charged side chain in the interior usually implies an important functional role for that "buried"

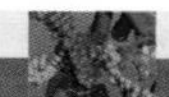

CLINICAL CORRELATION 3.1

Plasma Proteins in Diagnosis of Disease

Electrophoretic analysis of the plasma proteins is commonly used in diagnosis of disease. Electrophoresis of plasma buffered at pH 8.6 separates the major proteins as they migrate to the anode in the electric field into bands or peaks, on the basis of their charge differences (see text). Examples of abnormal electrophoresis patterns are shown in Figure 3.21. An immediate response that occurs with stress or inflammation caused by infection, injury, or surgical trauma is shown in pattern (*b*) in which haptoglobins in the α_2 mobility band are selectively increased. A delayed response shown in pattern (*c*) is associated with infection and shows an increase in the γ-globulin peaks due to an increase in immunoglobulins. An example of a hypo-γ-globulinemia due to an immunosuppressive disease is shown in pattern (*d*). In hepatic cirrhosis there is a broad elevation of the τ-globulins with reduction of albumin, as in pattern (*e*). Monoclonal gammopathies are due to the clonal synthesis of a unique immunoglobulin and give rise to a sharp γ-globulin band, as in pattern (*f*). Nephrotic syndrome shows a selective loss of lower molecular weight proteins from plasma, as in pattern (*g*). The pattern shows a decrease in albumin (65 kDa), but a retention of the bands composed of the large proteins α_2-macroglobulin (725 kDa) and β-lipoproteins (2000 kDa) in the α_2 band. Pattern (*h*) is from a patient with a protein-losing enteropathy, who is losing plasma by exudation in the intestinal tract. The slight increase in the α_2-band in pattern (*h*) is due to an immediate or late response from a stressful stimulus, as previously observed in patterns (*b*) and (*c*).

Ritzmann, S. E., and Daniels, J. C. Serum protein electrophoresis and total serum proteins. In Ritzmann, S. E., and Daniels, J. C. (Eds.), *Serum Protein Abnormalities, Diagnostic and Clinical Aspects.* Boston: Little, Brown, 1975, 3; McPherson, R. A. Specific proteins. In Henry, J. B. (Ed.), *Clinical Diagnosis and Management by Laboratory Methods,* 17th ed. Philadelphia: Saunders, 1984, 204; and Keren, D. F. *Protein Electrophoresis in Clinical Diagnosis.* London: Arnold/Hodder, 2003.

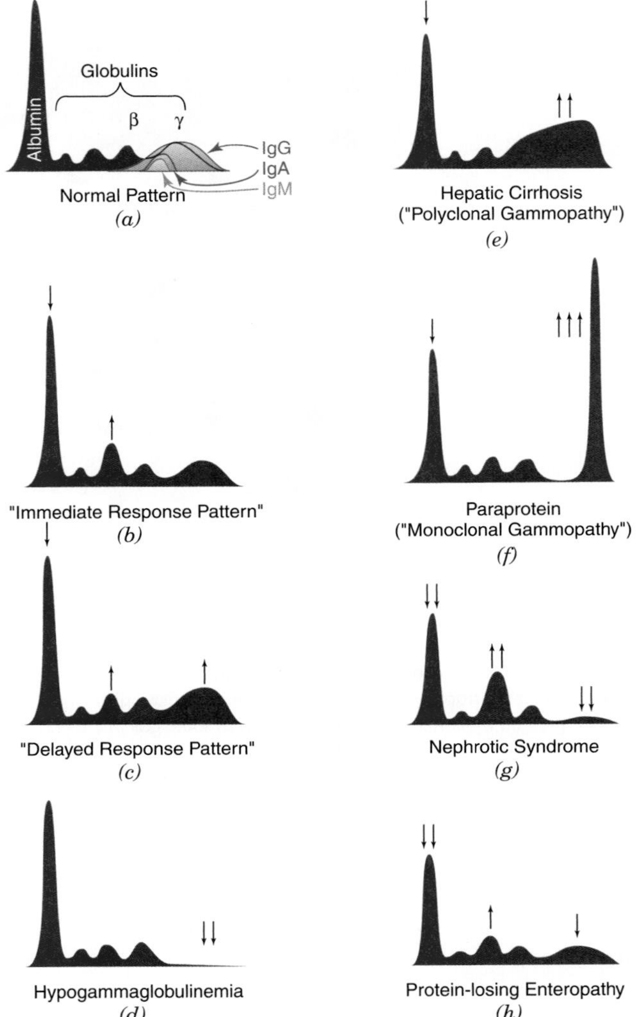

Figure 3.21 Electrophoretic mobility patterns observed for a normal individual and patients with abnormal concentrations of serum proteins, analyzed by agarose gel electrophoresis.
Redrawn from McPherson, R. A. Specific proteins. In Henry, J. B. (Ed.), *Clinical Diagnosis and Management*, 17th ed. Philadelphia: Saunders, 1984.

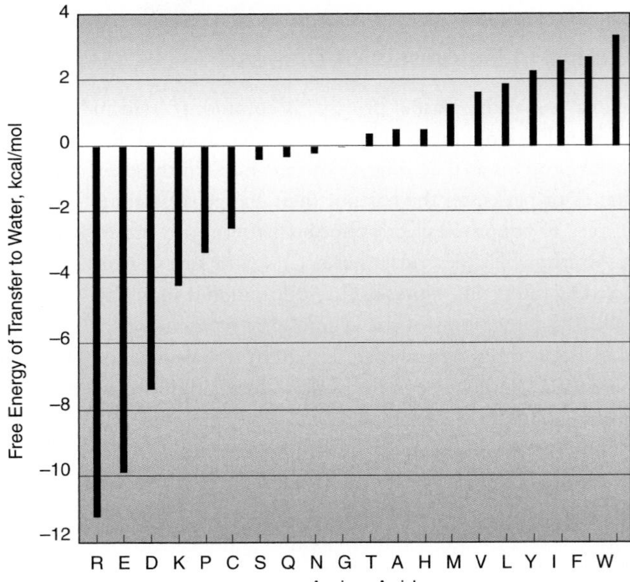

Figure 3.22 Relative hydrophobicity of amino acid side chains. Based on partition of amino acid between organic solvent and water. Negative values indicate preference for water, and positive values preference for nonpolar solvent (ethanol or dioxane) relative to glycine (see text).
Based on data from Von Heijne, G., and Blomberg, C., *Eur. J. Biochem.* 97:175, 1979; and from Nozaki, Y., and Tanford, C., *J. Biol. Chem.* 246:2211, 1971.

TABLE 3.7 • Chemical Reactions of Amino Acids

Reactive Group	*Reagent or Reaction*	*Product*
Amine (—NH_2) groups	Ninhydrin	Blue-colored product that absorbs at 540 nm[a]
	Fluorescamine	Product that fluoresces
Carboxylic acid groups	Alcohols	Ester products
	Amines	Amide products
	Carbodiimide	Activates for reaction with nucleophiles
—NH_2 of Lys	2,3,6-Trinitrobenzene sulfonate	Product that absorbs at 367nm
	Anhydrides	Acetylate amines
	Aldehydes	Form Shiff base adducts
Guanidino group of Arg	Sakaguchi reaction	Pink-red product that can be used to assay Arg
Phenol of Tyr	I_2	Iodination of positions ortho to hydroxyl group on aromatic ring
	Acetic anhydride	Acetylation of —OH
S atom of Met side chain	CH_3I	Methyl sulfonium product
	$[O^-]$ or H_2O_2	Methionine sulfoxide or methionine sulfone
—SH of Cys	Iodoacetate	Carboxymethyl thiol ether
	N-Ethylmaleimide	Addition product with S
	Organic mercurials	Mercurial adducts
	Performic acid	Cysteic acid (—SO_3H)
	Dithionitrobenzoic acid	Yellow product that can be used to quantitate —SH groups
Imidazole of His and phenol of Tyr	Pauly's reagent	Yellow to reddish product

[a]Proline imino group reacts with ninhydrin to form product that absorbs light at 440 nm (yellow color).

charge within the nonpolar interior in stabilizing protein conformation or participation in catalysis.

Transmembrane proteins reverse the positioning of their side-chain polarity from that of water-soluble globular proteins. Within the membrane, these proteins often position hydrophobic side chains on the outside and ionic groups on the inside to provide binding interactions and to form ion channels (p. 477).

Amino Acids Undergo a Variety of Chemical Reactions

Amino acids in proteins react with a variety of reagents that may be used to investigate the function of specific side chains. Some common chemical reactions are presented in Table 3.7. Reagents that modify acid side chains have been synthesized to bind to specific sites in a protein's structure, such as the substrate-binding site. The strategy is to model the structural features of the enzyme's natural substrate into the modifying reagent. The reagent binds to the active site like the natural substrate and reacts with a specific side chain. This identifies the modified amino acid as being located in the substrate-binding site and helps identify its role in catalysis.

A CLOSER LOOK 3.1

Proteolytic Cleavage of Proinsulin

Proinsulin is produced in pancreatic islet cells as a single polypeptide chain containing 86 amino acids and 3 intrachain cystine disulfide bonds (Figure 3.23). It is transformed into biologically active insulin by proteolytic cleavage prior to its secretion from islet cells. Proinsulin is cleaved, by proteases present in the islet cells, between residues 30 and 31, and 65 and 66. This releases two molecules, a 35-residue fragment (**C-peptide**) and insulin, which consists of two polypeptide chains (A and B) of 21 amino acids and 30 amino acids, respectively, covalently joined by the same disulfide bonds present in proinsulin. The C-peptide is further processed by proteases that hydrolyze a basic dipeptide from the N- and the C-terminal ends. The modified C-peptide is secreted with insulin.

3.4 • PRIMARY STRUCTURE OF PROTEINS

The **primary structure** of a protein refers to the covalent structure, which includes amino acid sequence and location of disulfide (cystine) bonds. The primary structure of a protein is required to understand its structure and mechanism of action, its biosynthesis including the posttranslational modifications of its structure, and its relationship to other proteins with similar physiological roles. The primary structure of several biologically active peptides was shown in Table 3.2 (p. 82). **Insulin** illustrates how knowledge of the primary structure leads to an understanding of the biosynthesis and the physiological forms of this bioactive hormone (Figure 3.23; A Closer Look 3.1 and Clin. Corr. 21.8). Insulin is initially synthesized as proinsulin, which is a single polypeptide chain of 86 amino acids and three intrachain cystine bonds (Figure 3.23). The hormone form consists of two polypeptide chains (A and B) covalently interconnected by two cystine bonds and chain A containing an intrachain cystine. This is the active insulin primary or covalent structure.

Besides providing information on its biosynthesis, the comparison of the primary structures of insulins from different animal species shows the residues essential and nonessential to its hormonal function. The aligned primary structures have identical residues in most amino acid positions, except for residues 8, 9, and 10 of the A chain and residue 30 of the B chain. Amino acids in these positions vary widely (Table 3.8) and apparently do not affect the biological properties (Clin. Corr. 3.2). Other residues are rarely substituted, suggesting that they have an essential role in function.

Sequence comparisons are commonly used to predict the similarity in structure and function between proteins. These comparisons typically require aligning of sequences to maximize the number of identical residues and minimize the number of insertions or deletions required to achieve this alignment. Two sequences are **homologous** when their sequences are highly alignable. Note that in its correct usage homology refers only to proteins that have evolved from the same gene. **Analogy** describes sequences from proteins that are structurally similar but for which no evolutionary relationship has been demonstrated. Substitution of an amino acid by one of similar polarity (i.e., Val for Ile in position 10 of insulin) is a **conservative** substitution and is commonly observed in sequences of the same protein from different animal species. If a particular amino acid is regularly found at the same position, it is an **invariant residue**. It can be assumed that such residues have an essential role in the structure or function of the protein. A **nonconservative** substitution involves replacement of an amino acid by one of different polarity. This may drastically change the properties of the protein or occur in regions that are apparently unimportant functionally (Clin. Corr. 6.2; p. 214). Polarity is only one physical property of amino acids that determines whether a substitution will significantly alter the protein's function. Other important properties are the molecular volume and surface area of the residue.

TABLE 3.8 • Variation in Positions A8, A9, A10, and B30 of Insulin

Species	*A8*	*A9*	*A10*	*B30*
Human	Thr	Ser	Ile	Thr
Cow	Ala	Ser	Val	Ala
Pig	Thr	Ser	Ile	Ala
Sheep	Ala	Gly	Val	Ala
Horse	Thr	Gly	Ile	Ala
Dog	Thr	Ser	Ile	Ala
Chicken[a]	His	Asn	Thr	Ala
Duck[a]	Glu	Asn	Pro	Thr

[a]Positions 1 and 2 of B chain are both Ala in chicken and duck; whereas in the other species in the table, position 1 is Phe and position 2 is Val in B chain.

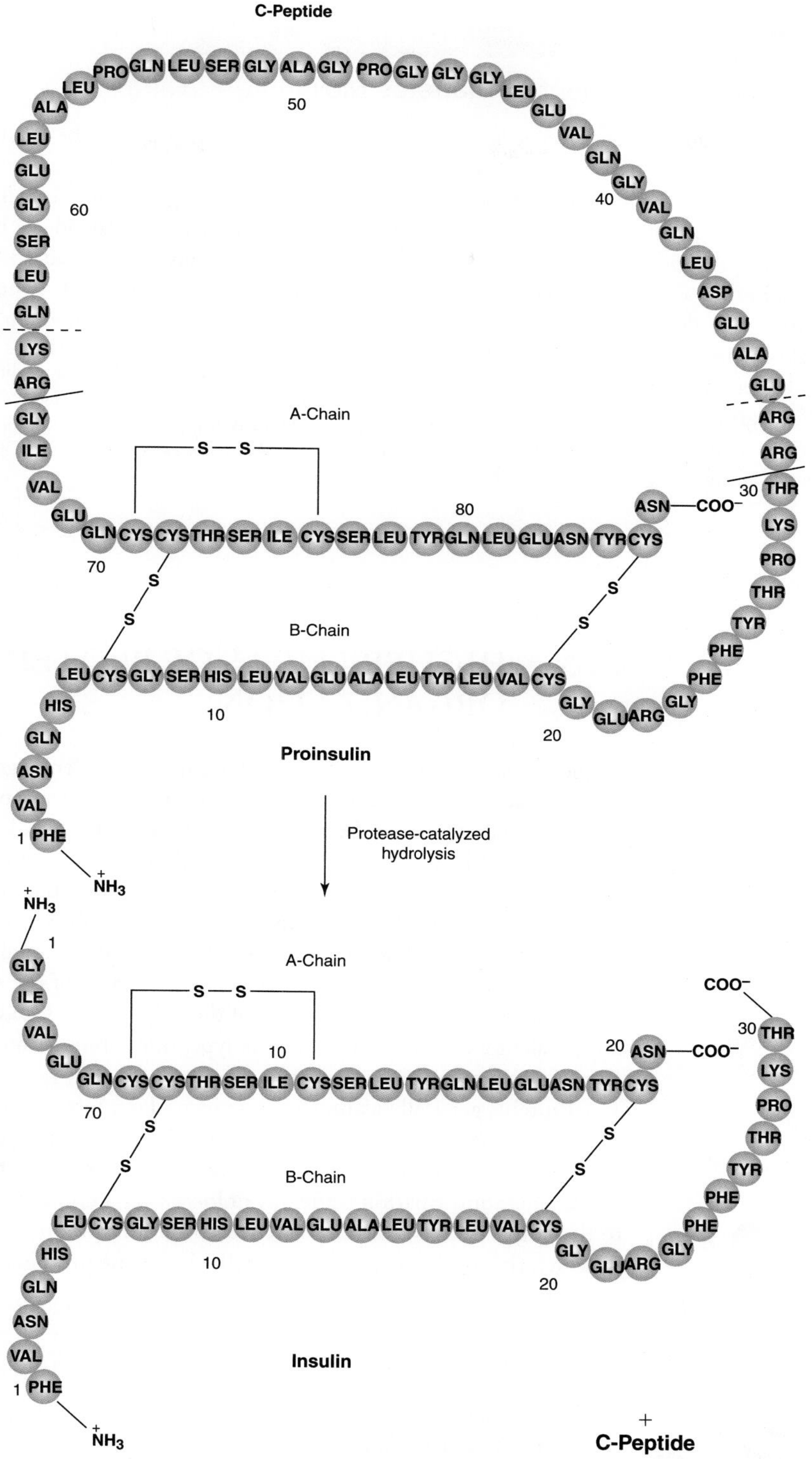

Figure 3.23 Primary structures of human proinsulin, insulin, and C-peptide. In the polypeptide sequence of proinsulin, the incipient B-chain peptide extends from Phe at position 1 to Thr at position 30, the C-peptide from Arg at position 31 to Arg at position 65, and the A-peptide from Gly at position 66 to Asn at position 86. Cystine bonds from positions 7 to 72, 19 to 85, and 71 to 76 are found in proinsulin.
Redrawn from Bell, G. I., Swain, W. F., Pictet, R., Cordell, et al. *Nature* 282:525, 1979.

CLINICAL CORRELATION 3.2

Differences in Insulins Used in Treatment of Diabetes Mellitus

Prior to the development of recombinant human insulin, both pig (porcine) and cow (bovine) insulins were used in the treatment of human diabetics. Insulin from animal sources is still used in treatment of diabetes in developing countries. Because of differences in sequence from human insulin, some diabetic individuals will have an initial allergic response to the injected insulin as their immunological system recognizes the insulin as foreign or develop an insulin resistance due to a high anti-insulin antibody titer at a later stage in treatment. However, the frequency of a deleterious immunological response to pig and cow insulins is small; the great majority of the population is able to utilize these insulins without complication. This is due to the small number and the conservative nature of the changes between the amino acid sequences and the fact that they do not significantly change the three-dimensional structure of the insulin relative to that of human insulin. Pig insulin is usually more acceptable than cow insulin in insulin-reactive individuals because it is more similar in sequence to human insulin (see Table 3.8). Human insulin is now the primary insulin used in developed countries, being made from genetically engineered bacteria or by modifying pig insulin.

Brogdon, R. N., and Heel, R. C. Human insulin: A review of its biological activity, pharmacokinetics, and therapeutic use. *Drugs* 34:350, 1987; and Richter, B., and Neises, G. Human insulin versus animal insulin in people with diabetes mellitus. *Cochrane Database of Systematic Reviews* 3, CD003816, 2002.

3.5 • HIGHER LEVELS OF PROTEIN ORGANIZATION

Higher levels of protein organization refer to noncovalently generated conformations of the polypeptide chain. These higher levels of protein organization are called secondary, tertiary, and quaternary structure. **Secondary structure** refers to the local folding of the polypeptide backbone into helical, pleated sheet, or random conformations. The polypeptide backbone is the covalently interconnected atoms of the peptide bonds and α-carbons that link the amino acid residues of the protein. Side chains are not considered at the level of secondary structure. **Tertiary structure** refers to the three-dimensional structure of the polypeptide. It includes the conformational relationships in space of the side chains and the geometric relationship between distant regions of the polypeptide. **Quaternary structure** refers to the noncovalent association of discrete polypeptide subunits into a multisubunit protein. Not all proteins have a quaternary structure.

Proteins generally assume unique secondary, tertiary, and quaternary conformations to produce the protein's **native conformation**. Folding of the primary structure into the native conformation occurs, in most cases, spontaneously through noncovalent interactions. This conformation is the one of lowest total Gibbs free energy kinetically accessible to the polypeptide(s) for the particular conditions of ionic strength, pH, and temperature in which the folding occurs. In the cell, chaperone proteins (p. 231) can facilitate protein folding.

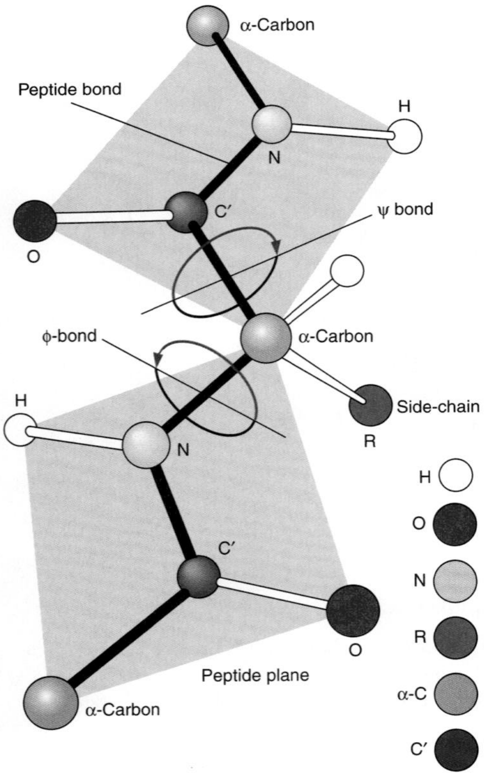

Secondary Structure

The conformation of a polypeptide backbone may be described by the rotational angles of covalent bonds that contribute to the chain. These are contributed by each of the amino acids and are between the nitrogen and α-carbon, and the α-carbon and carbonyl carbon. The first is designated the **phi (φ) bond** and the second the **psi (ψ) bond** (Figure 3.24). The third bond contributed by each amino acid to the polypeptide chain is the peptide bond. As previously discussed, due to the partial double-bond character of the C′—N bonds, there is a barrier to free rotation about this peptide bond.

Regular secondary structure occurs in segments of a polypeptide chain in which all φ bond angles are equal, and all ψ bond angles are equal. The rotational angles for φ and ψ bonds for common regular secondary structures are given in Table 3.9.

The α-helix and β-strand conformations of polypeptides are the most thermodynamically stable of the regular secondary structures. However, a particular sequence may have an

Figure 3.24 Polypeptide chain showing φ, ψ, and peptide bonds for residue R_i within a polypeptide chain. Redrawn with permission from Dickerson, R. E., and Geis, I. *The Structure and Action of Proteins.* Menlo Park, CA: Benjamin, 1969, 25.

TABLE 3.9 • Helix Parameters of Regular Secondary Structures

Structure	*Approximate Bond Angles (°)* ϕ	ψ	*Residues per turn, n*	*Helix Pitch,[a] p(Å)*
Right-handed α-helix [3.6_{13} -helix)	−57	−47	3.6	5.4
3_{10} -helix	+49	−26	3.0	6.0
Parallel β-strand	−119	+113	2.0	6.4
Antiparallel β-strand	−139	+135	2.0	6.8
Polyproline type II[b]	−78	+149	3.0	9.4

[a]Distance between repeating turns on a line drawn parallel to helix axis.
[b]Helix type found for polypeptide chains of collagen.

unordered, or random, secondary structure, in which neither the φ bond angles nor the ψ bond angles are equal. Proline interrupts α-helical conformation since its pyrrolidine side chain sterically interacts with the residue preceding it and prevents formation of an α-helical structure.

Helical structures are characterized by the number n of residues per turn of helix and the distance d between α-carbon atoms of adjacent amino acids measured parallel to the axis of the helix. The helix pitch p, the product of $n \times d$, then, measures the distance between repeating turns of the helix on a line drawn parallel to the helix axis (Figure 3.25).

α-Helical Structure

An amino acid sequence in a right-handed α-helical conformation is shown in Figure 3.26. Characteristic are 3.6 amino acid residues per 360° turn (n = 3.6). The peptide bond planes are parallel to the axis of the helix. In this geometry each peptide forms two hydrogen bonds, one to the peptide bond of the fourth residue above and the other to the peptide bond of the fourth amino acid below. Other parameters, such as the pitch p, are given in Table 3.9. In the hydrogen bonds between the peptide groups, the distance between the hydrogen-donor atom and the hydrogen-acceptor atom is 2.9 Å. In addition, the donor, acceptor, and hydrogen atoms are approximately in a straight line. This is an optimum geometry and distance for maximum hydrogen bond strength (Section 3.7, p. 112).

The side chains are on the outside of the spiral structure. Due to the characteristic 3.6 residues per turn, the first and every third and fourth R group of the amino acid sequence in the helix come close to each other. Helices often present polar and nonpolar faces if their amino acid sequences place polar or nonpolar R groups three or four residues apart. This gives rise to unique functional characteristics of the helix. However, if every third or fourth side chain that comes close to another has the same charge sign or is branched at its β-carbon (valine and isoleucine), its unfavorable ionic or steric interactions destabilize the helix structure. The α-helix may theoretically be left handed or right handed, giving it asymmetric properties and appropriate optical activity. A right-handed α-helix is more stable than a left-handed one.

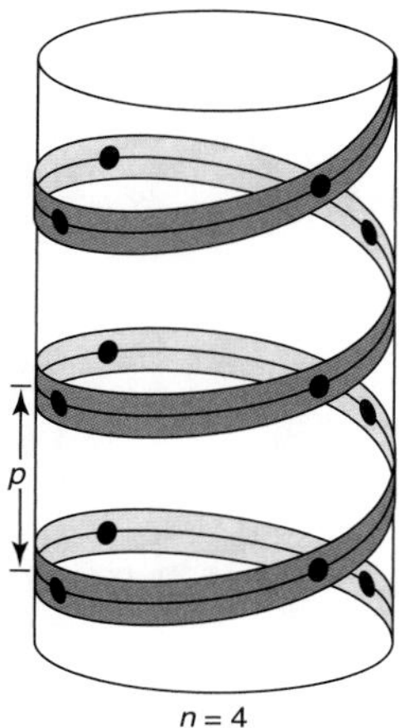

Figure 3.25 Helix pitch (*p*) for a helix with *n* = 4. Each circle on a line represents an α-carbon from an amino acid residue. The rise per residue would be p/n (see equation in text).
Reprinted with permission from Dickerson, R. E., and Geis, I. *The Structure and Action of Proteins.* Menlo Park, CA: Benjamin, 1969, 26.

β-Structure

A polypeptide chain in a β-structure (Figure 3.27) is hydrogen bonded to another similar polypeptide region aligned in a parallel or an antiparallel direction (Figure 3.28). Hydrogen-bonded β strands appear like a pleated sheet (Figure 3.29). The side chains project above and below the pleated sheet-like structure.

Structural Motifs and Protein Folds

Simple arrangements of secondary structure that occur in more than one protein are called **structural motifs**. They include the helix-turn-helix motif found in many DNA-binding proteins, the strand-turn-strand motif found in proteins with antiparallel β-structure, and the alternating strand-turn-helix-turn-strand motif found in many α/β-proteins. In these

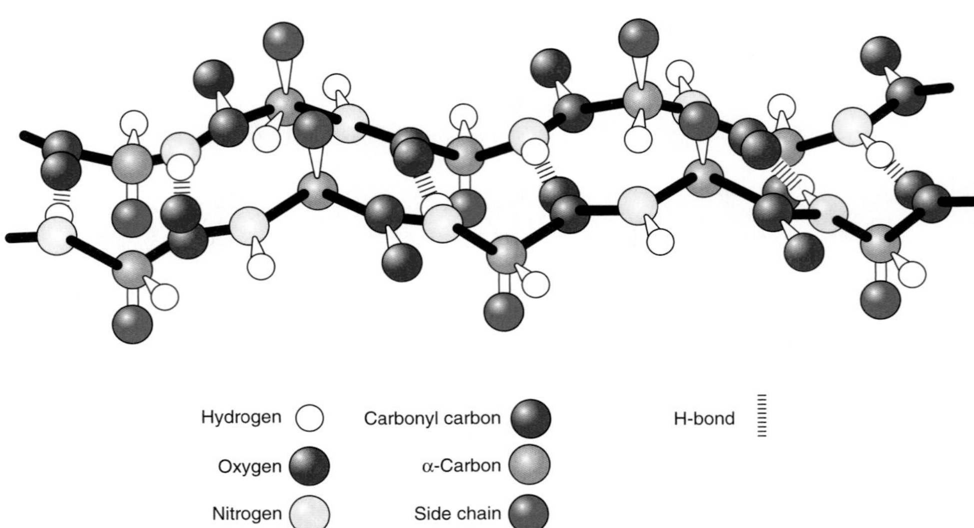

Figure 3.27 Two polypeptide chains in a β-sheet conformation. Polypeptide chains may be added to either edge of the two strands shown to generate a more extended structure.
Redrawn with permission from Fersht, A. *Enzyme Structure and Mechanism.* San Francisco: Freeman, 1977, 10.

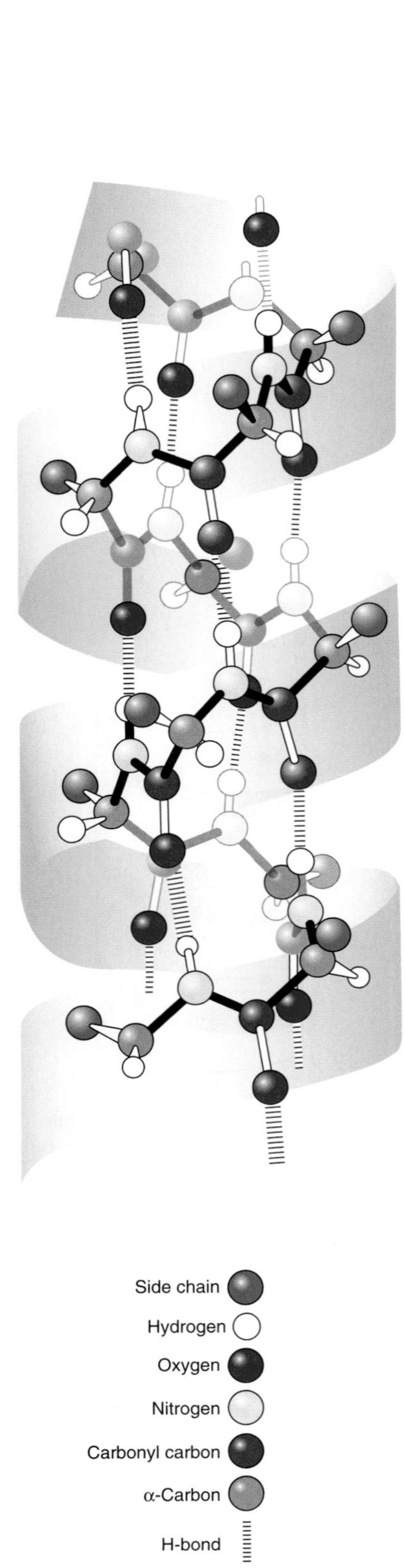

Figure 3.26 An α-helix. Redrawn with permission from Pauling, L. *The Nature of the Chemical Bond,* 3d ed., Ithaca, NY: Cornell University Press, 1960.

motifs a turn is a small segment of the polypeptide (approximately three or four residues) of nonregular secondary structure that connects regions of regular secondary structure, whereas a loop is a larger segment of connecting nonregular conformation. Combinations of motifs or more complex organizations of secondary structure may form a **fold**. A fold is the arrangement of secondary structure elements of a **domain**. A structural domain is a compact globular structural unit formed within the polypeptide with a hydrophobic core and hydrophilic surface and usually folds independently of other structural units within the polypeptide chain.

The calmodulin domain can serve as an example for defining the terms *structural motif, fold,* and *domain* (Figure 3.30). Calmodulin binds to target proteins where it acts to sense

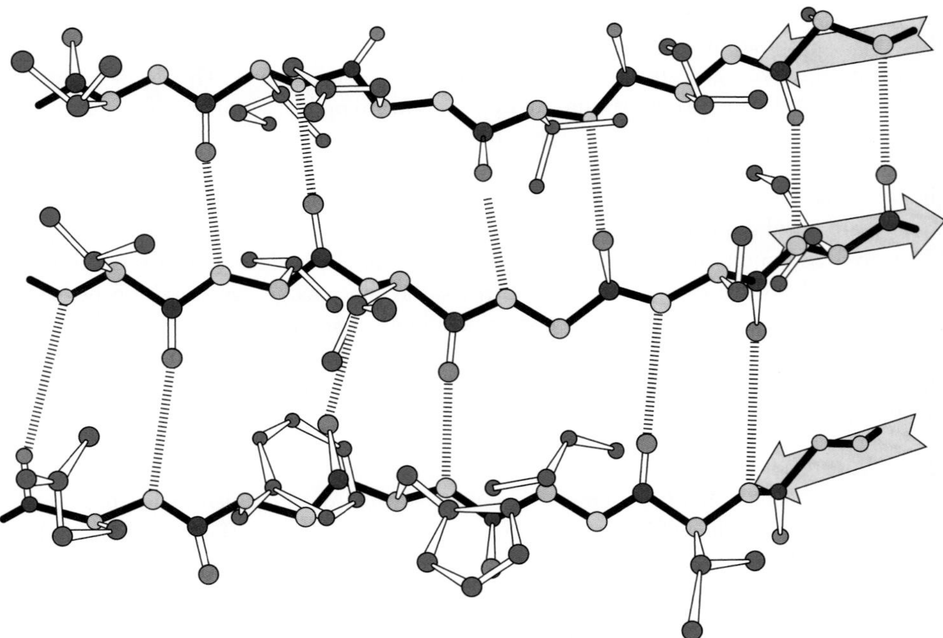

Figure 3.28 Example of antiparallel β-sheet (residues 93–98 [top], 28–33 [middle], and 16–21 [bottom] of Cu, Zn superoxide dismutase). Dashed line shows hydrogen bonds between carbonyl oxygen atoms (red) and peptide nitrogen atoms (grey); arrows show direction of polypeptide from N-terminal to C-terminal. In the characteristic antiparallel β-sheet, pairs of closely spaced interchain hydrogen bonds alternate with widely spaced hydrogen bond pairs.
Redrawn with permission from Richardson, J. S. *Adv. Protein Chem.* 34:168, 1981.

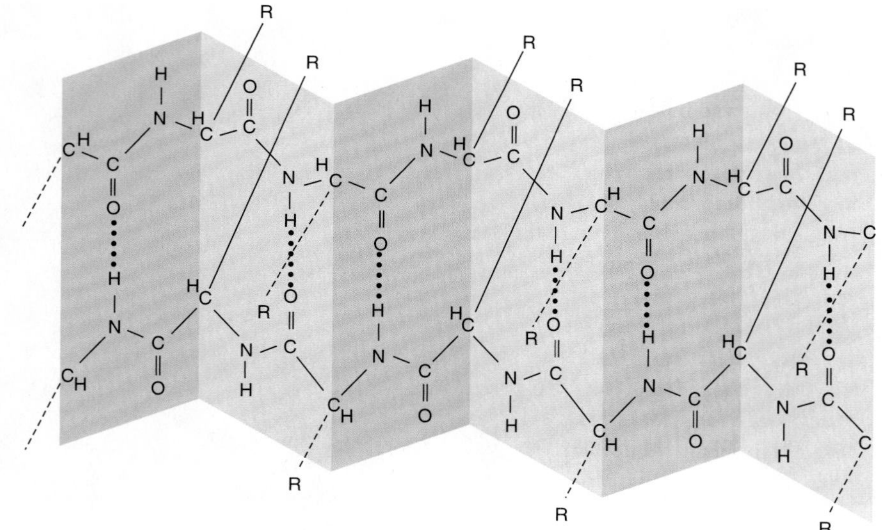

Figure 3.29 β-Sheet structure between two polypeptide chains showing planes of joined peptide bonds. Polypeptide chains may be added above and below to generate a more extended structure. Side chain groups along a polypeptide chain alternate their positions above and below the plane.

the levels of calcium in the cell. At increased cellular calcium levels, calcium activates the calmodulin to transmit a signal to its bound target protein. Target proteins of calmodulin participate in cell signaling, muscle contraction, fertilization, metabolism, programmed cell death, long- and short-term memory, nerve growth, immune response, and cell division. The calcium atom binds in calmodulin within the loop of a helix-turn-helix motif called an EF-hand (Figure 3.30*a* and see Figure 12.57, p. 493). The motif obtained its name from the E and F helices (helices 5 and 6 from the N-terminal end) in the muscle protein parvalbumin, in which this calcium-binding motif was first observed. The EF-hand motif is widely distributed in calcium-binding proteins and is found in over 70 different proteins with determined structures in the Protein Data Bank. The fold of the calmodulin domain contains two EF-hand motifs interconnected by an α-helical segment (Figure 3.30*b*). Addition of the side chain groups to the secondary structure traced by the polypeptide chain generates the complete tertiary structure of the domain (Figure 3.30*c*).

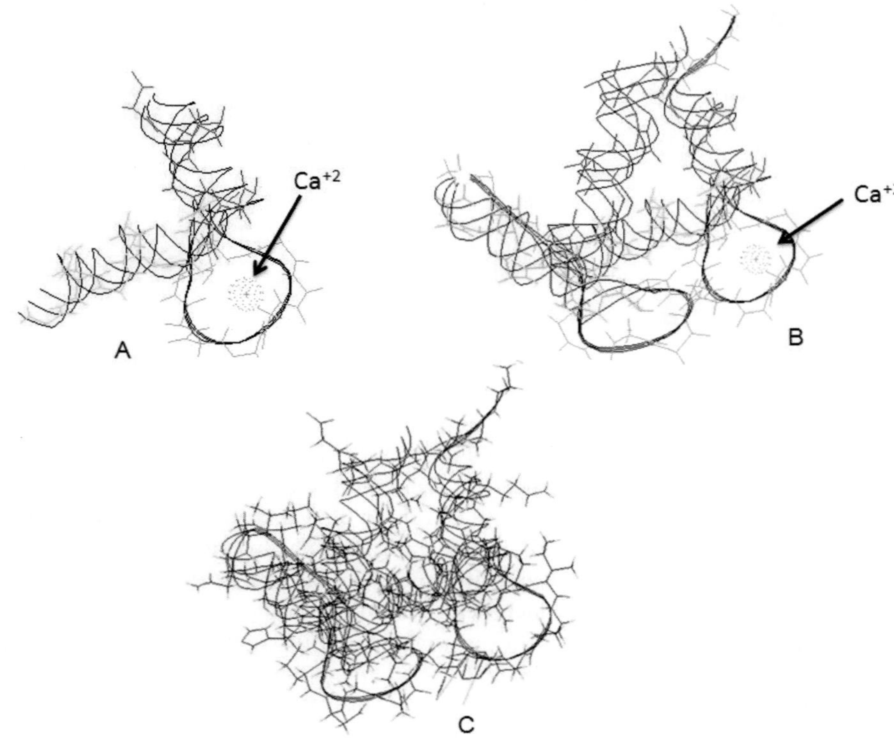

Figure 3.30 Motif and Fold Structures within Calmodulin Domain. (*a*) EF-Hand helix-turn-helix motif with Ca^{+2} bound. (*b*) Two EF-hand motifs combine in the fold of the calmodulin C-terminal domain. (*c*) The side-chain groups are added to generate the tertiary structure of the domain. Each α-helix is differently colored, with the polypeptide chain traced with a black ribbon to more clearly show the secondary structure of the polypeptide chain.
Based on structure PDB ID 1J7P, Chou, J. J. et al. *Nat. Struct. Biol.* 8:990, 2001. Pictures produced with Swiss PDB viewer, Guex, N., and Peitsch, M. C. *Electrophoresis* 18: 2714, 1997.

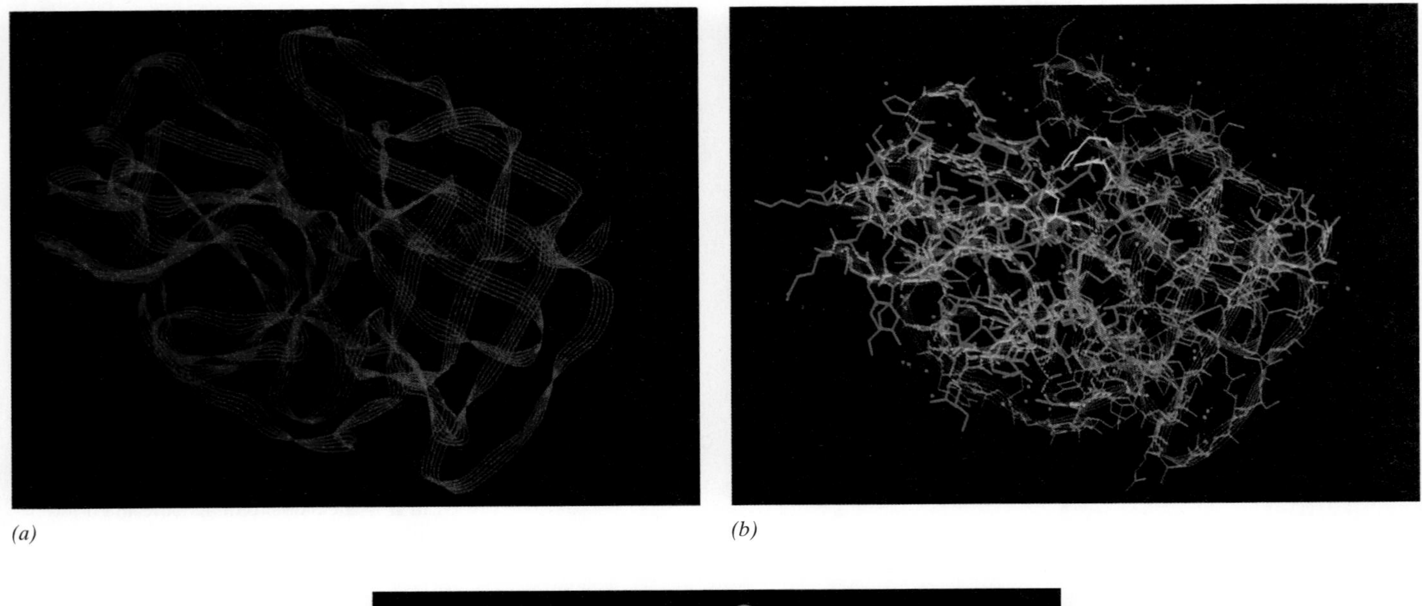

(a) (b)

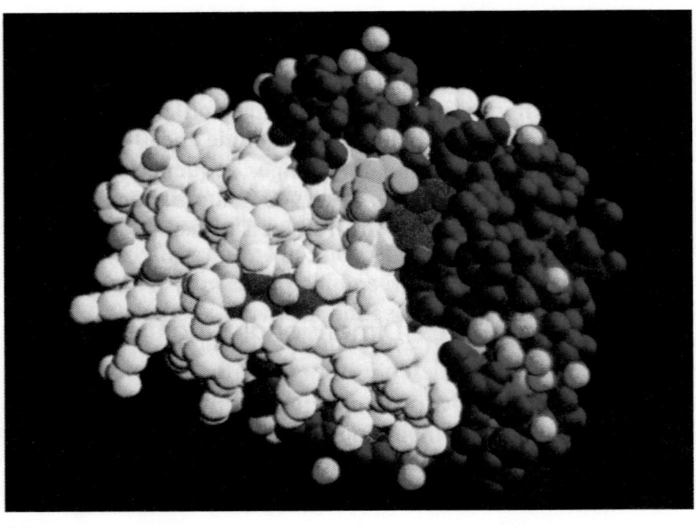

(c)

Figure 3.31 Tertiary structure of trypsin. (*a*) Ribbon structure outlines conformation of the polypeptide chain. (*b*) Structure shows side chains including active site residues (in yellow) with outline of polypeptide chain (ribbon) superimposed. (*c*) Space-filling structure in which each atom is depicted as the size of its van der Waals radius. In (*c*) the hydrogen atoms are not shown. Different domains are shown in dark blue and white. Active site residues are in yellow, and intrachain disulfide bonds of cystine in red. Light-blue spheres represent water molecules associated with the protein. This structure shows the density of packing within interior of the protein.

Tertiary Structure

The **tertiary structure** of a polypeptide depicts the location of each of its atoms in space. It includes the geometric relationship between distant segments of primary and secondary structures and the positional relationship of the side chains with one another. The tertiary structure of trypsin is shown in Figure 3.31. The ribbon structure (*a*) shows the conformation of the single polypeptide chain and the overall pattern of chain folding (fold structure). Figure 3.31*b* shows the position of side chains. Active site catalytic side chains are shown in yellow, which include the hydroxymethyl group of serine (residue 177), the imidazole of histidine (residue 40), and the side-chain carboxylate of aspartate (residue 85). Although these are widely separated in the primary structure, the tertiary structure brings them together to form the catalytic site. In Figure 3.31*c*, a space-filling model shows C, N, and O atoms represented by balls of radius proportional to their van der Waals radius. The structure conforms to the general rules for soluble proteins (Section 3.3).

Hydrophobic side chains are generally in the interior, away from the water interface. Ionized side chains are on the outside, where they are stabilized by the water of solvation. Within the protein structure (not shown) are buried water molecules often exhibiting specific stabilizing interactions. A large number of water molecules form a solvation shell around the outside of the protein.

A long polypeptide often folds into multiple compact semi-independent regions, or domains, each having a characteristic compact geometry with a hydrophobic core and polar surface. They typically contain 100–150 amino acids. Domains in a **multidomain protein** may be connected by a segment that lacks a regular secondary structure. Alternatively, the dense spherical folded regions may be separated by a cleft or region less dense in tertiary structure (Figure 3.32). Trypsin contains two domains with a cleft in between that contains the substrate-binding catalytic site. An active site within an interdomain interface is characteristic of many enzymes. Different domains within a protein can move with respect to each other. Hexokinase (Figure 3.33), which catalyzes phosphorylation of glucose by adenosine triphosphate (ATP) (p. 596), has a glucose-binding site in a region between two domains. When glucose binds in the active site, the surrounding domains move (Section 3.8) to enclose the substrate to trap it for phosphorylation (Figure 3.33). In enzymes with more than one substrate or allosteric effector sites (p. 413), the different sites may be located within different domains. In multifunctional proteins, each domain may perform a different task.

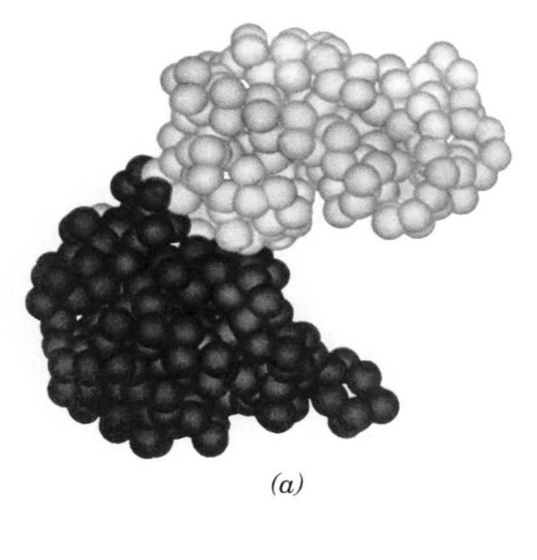

(*a*)

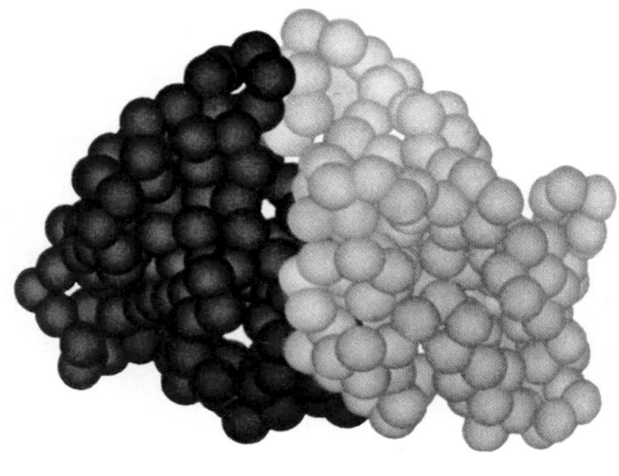

(*b*)

Figure 3.32 Globular domains within proteins. (*a*) Phosphoglycerate kinase has two domains with a relatively narrow neck in between. (*b*) Elastase has two tightly associated domains separated by a narrow cleft. Each sphere in the space-filling drawing represents the α-carbon position for an amino acid within the protein structure.
Reprinted with permission from Richardson, J. S. *Adv. Protein Chem.* 34: 168, 1981.

Quaternary Structure

Quaternary structure refers to the arrangement of polypeptide chains in a multichain protein. The subunits in a quaternary structure are associated noncovalently. α-Chymotrypsin contains three polypeptides covalently joined together by interchain disulfide bonds into a single covalent unit and therefore does not have a quaternary structure. Myoglobin consists of one polypeptide and has no quaternary structure. However, hemoglobin A contains four polypeptide subunits ($\alpha_2\beta_2$) held together noncovalently in a specific conformation as required for its function (p. 361). Thus, hemoglobin has a quaternary structure.

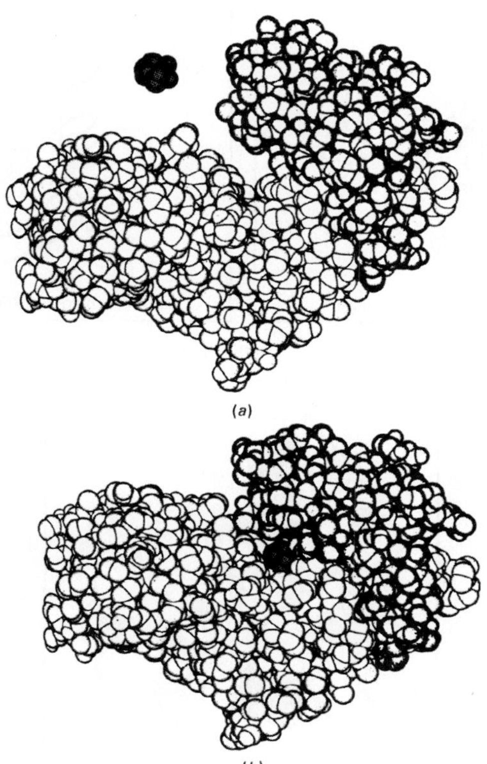

(*a*)

(*b*)

Figure 3.33 Drawings of (*a*) unliganded form of hexokinase and free glucose and (*b*) conformation of hexokinase with glucose bound. In this space-filling drawing, each circle represents the van der Waals radius of an atom in the structure. Glucose is black, and each domain is differently shaded.
Reprinted with permission from Bennett, W. S., and Huber, R. *CRC Rev. Biochem.* 15: 291, 1984.

Bacterial aspartate carbamoyl transferase (p. 822) has a quaternary structure comprised of 12 polypeptide subunits. The poliovirus coat protein contains 60 polypeptide subunits, and the tobacco mosaic virus protein contains 2120 noncovalently associated subunits.

Unstructured Proteins

Proteins described previously have defined folds and tertiary structures in which the average location in space of the atoms of the folded structures are known. In contrast, there are proteins that lack a stable folded structure. This may not be surprising, as folded proteins are typically only 21–42 kJ/mol (5–10 kcal/mol) more stable than their denatured conformation (Section 3.7, p. 112). It should not therefore be unexpected that there are proteins that do not have a stable folded conformation at physiological temperatures. Proteins with a nonfolded conformation are called **intrinsically unstructured proteins (IUPs)**. Other proteins may contain unstructured regions or domains that contain **partially unfolded conformations (PUFs)**. The IUPs and proteins with PUFs include scaffold proteins, hormones, activation domains of transcription factor, cyclin-dependent kinases and their inhibitors, proteins in cellular signal transduction, and the NH_2-terminal segments of histone proteins in nucleosomes.

The unfolded conformations are highly dynamic with the atoms showing mobility in displacement and rotation (see Section 3.8 for a discussion of the mobility of folded proteins). However, the unstructured conformations may contain regions of ordered secondary structure not completely randomized. These unstructured protein domains would resemble the molten globule intermediates present on protein-folding pathways (Section 3.5).

IUPs and PUFs often function by binding to other proteins or to DNA and RNA. The binding interaction induces a structure in the unfolded polypeptide (Figure 3.34). The induction of defined structure in the IUP or PUF is a negative entropic process that requires an unfavorable free energy. Thus, the binding strength of unstructured proteins to their binding partners is generally weaker than for the binding of a molecule to a preformed structure, which does not need to utilize a part of its favorable binding energy to drive an entropically unfavorable conformation change. This property of weak binding is often

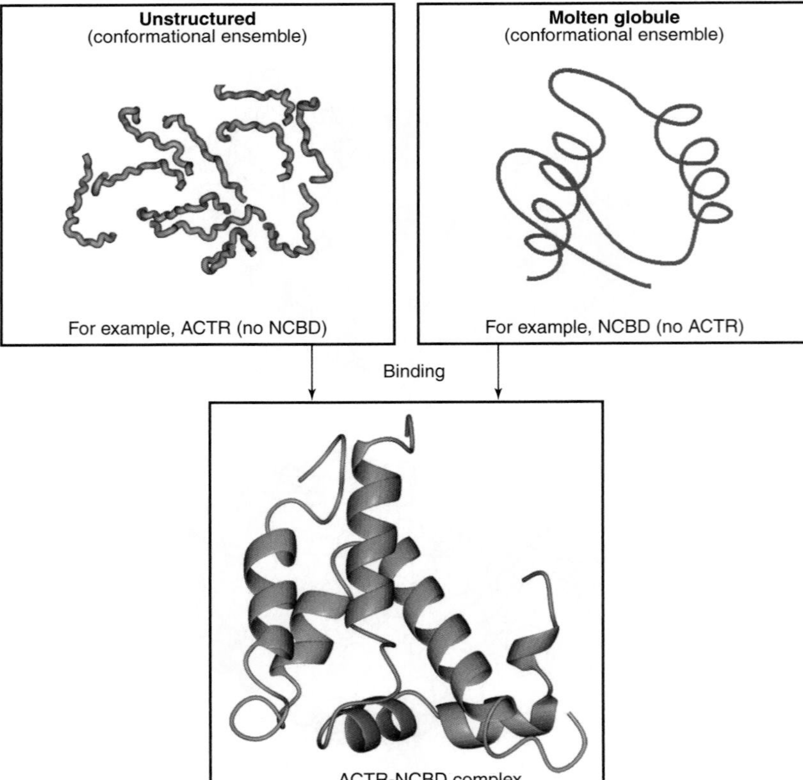

Figure 3.34 Structure Induced in Unstructured Domains on Binding. The interaction domain of activator for thyroid hormone and retinoid receptors (ACTR) is an ensemble of unstructured conformations (reddish brown, upper left panel) and the nuclear-receptor coactivator domain (NCBD) has a molten globule conformation with elements of regular secondary structure and an unstable tertiary structure (green, upper right panel). Association induces a stable secondary and tertiary structure in both proteins (lower panel). Reprinted by permission from Macmillan Publishers Ltd: from Dyson, H. J., and Wright, P. E. *Nature Rev. Mol. Cell Biol.* 6: 197, 2005.

advantageous, as weak and transient interactions are required for many biological processes. In addition, the unstructured domains show promiscuity in their binding partners, as their lack of a preformed structure gives them a plasticity to form complementary binding surfaces to many different binding partners. Thus the cyclin-dependent kinase inhibitor p21 has the ability to bind to different cyclin-dependent kinases and regulate cyclin-dependent kinases present in the different phases of the cell cycle. Scaffold proteins group many proteins into a functional complex. Their plasticity allows the binding of multiple partners into the complex and permits changes in the protein members according to physiological requirements.

Unstructured regions of proteins can be recognized from the amino acid sequence of their polypeptide. Disordered regions are enriched in polar and charged amino acids (glutamate, lysine, and glutamine) and in the amino acid proline and depleted in the aromatic and alkyl amino acids. Algorithms have been developed to search sequences and identify potential unstructured regions of polypeptide chains. The retention of the unstructured proteins and PUFs within specific proteins through evolution shows the importance of such regions to critical biological processes.

Protein Complexes, Networks, and Interactomes

Protein molecules in the cellular milieu are primarily present in protein complexes containing multiple protein subunits. The complexes are functionally essential to most cellular processes including cell signaling, gene transcription, RNA splicing and translation, apoptosis, and cellular metabolism. These complexes typically have 5–10 proteins, but may contain as many as 20–30 different proteins. The complexes communicate with each other through proteins present in two or more different complexes, which can move between the complexes to connect the complexes into **networks.** A complex that interconnects with more than three other complexes is a **hub** in the network, and an important target for drug therapies in disease. A functional network comprising interconnected protein complexes is an **interactome.**

Cellular protein complexes are characterized utilizing **tandem affinity purification (TAP)**, which involves insertion of a reporter gene in tandem with a gene for a protein of interest to produce a chimeric protein in which the target protein product is joined to a tagged protein. The tagged protein is isolated by elution of the cell lysate over an antitag protein antibody affinity column (Figure 3.35). The affinity resin binds the tagged protein and its associated proteins, which are eluted from the column, separated by electrophoresis, and identified by mass spectrometry (p. 127 for a discussion of techniques). Other methods for identification of the complex partners of a target protein include **co-immunoprecipitation** with an antibody directed against the target protein, and the **yeast two-hybrid assay** in which mammalian protein–protein binary interactions are assessed in a yeast cell reporter

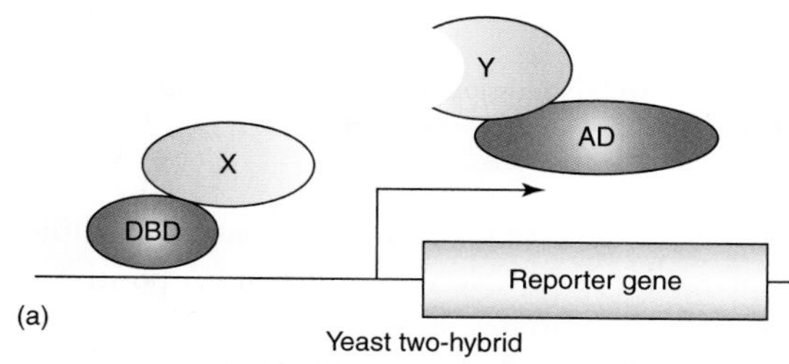

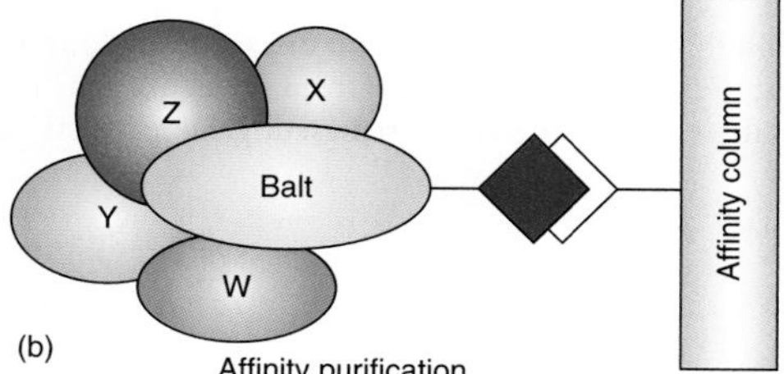

Figure 3.35 Two methods to determine protein–protein interactions. (*a*) The yeast two-hybrid technique determines binary protein–protein interactions in a yeast expression system. A transcription factor protein that binds to an upstream regulatory sequence of a yeast reporter gene such as *Gal4* consists of a DNA binding domain (DBD) and an activation domain (AD) required to activate reporter gene expression. The coding regions of the DBD and AD domains are separated and joined to two mammalian genes of interest (X and Y) and expressed in yeast cells to produce chimeric proteins DBD-X and Y-AD. If protein Y binds to protein X, the DBD and AD domains are brought together and upregulate reporter gene expression. The popular *Gal4* gene produces the enzyme β-galactosidase, which gives a blue color to cells on culture plates imbedded with the appropriate substrate.
(*b*) The tandem affinity purification (TAP) technique joins a tag protein (purple square) to a protein of interest (bait protein, orange) and the chimeric protein is expressed in a cell of interest. The cell lysate is passed over an affinity resin containing antibody to tag protein with the bait protein complex (proteins W–Z). Bound proteins are eluted from affinity column and indentified by mass spectrometry.
Redrawn based on figure from Aloy, P., and Russell, R. B. *Nature Reviews Mol. Cell Biol.* 7:188, 2006.

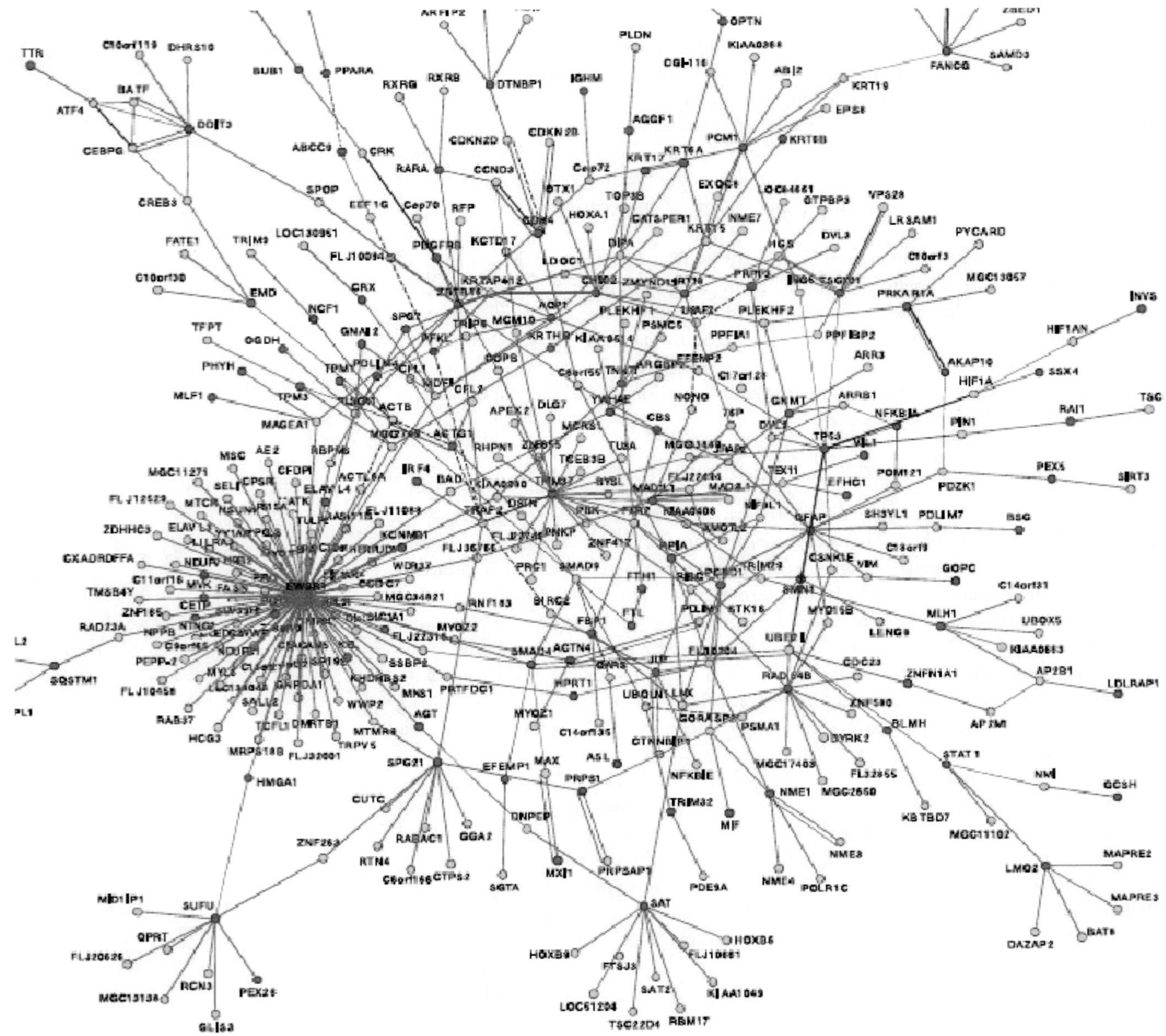

Figure 3.36 Preliminary disease-related interactome map for subset of 7200 human proteins. The yeast two-hybrid binary interactions were determined among a subset of 7200 gene products. The interactome shown is for the binding interactions of known disease-related proteins that are designated in the OMIM database. Green nodes are gene products in the OMIM database, and yellow nodes represent gene products without a current OMIM designation. Red edges show binary interactions identified by Rual et al., and solid blue edges interactions curated by the authors from the literature. Dashed edges show interactions through a third protein.

Reprinted with permission from Rual, J.-F., Venkatesan, K., Hao, T., Hirozane-Kishikawa, T., et al. *Nature* 437:1173, 2005.

system (Figure 3.35). Any one of these assay techniques alone is not sufficient to prove an *in vivo* protein–protein interaction, as artifacts are common. A positive result from more than one of these techniques demonstrates an interaction with high confidence. These techniques do not isolate transient complexes, membrane complexes, or complexes from physiological states or phenotypes not present in the cells analyzed.

Theoretical estimates indicate that a human cell has approximately 3000 core protein complexes, each with multiple isoforms (noncore exchangeable protein members generate different isoforms), and approximately 650,000 protein–protein binary interactions generating these protein complexes. Determining the characteristics of complexes and networks for which disease-associated proteins are present show the interdependencies of the multiple gene products affecting a disorder and their interrelationships in the disease-related pathways (Figure 3.36).

Bioinformatics Relates Structure and Function of Protein Gene Products

Over the last decade the nucleotide sequences of the genomes of multiple animal species have been obtained and the sequences of the RNA transcripts and polypeptide sequences produced from the genomic DNA. The amount of data is enormous and is stored in computer databases, which are freely accessible over the Internet. The number of polypeptide sequences in the UniProt Knowledge Base as of 2008 is over 6.4 million of which 75,000 are human. The Protein Data Bank contains 47,000 three-dimensional structures of folded proteins obtained by X-ray diffraction, nuclear magnetic resonance, and electron microscopy. **Bioinformatics** is a computationally based research area that focuses on the integration and analysis of the complex biological data with computer algorithms. A major emphasis of bioinformatics has been to identify patterns within nucleic acid or amino acid sequences that are **signatures** of structural features or motifs and of the protein family or class to which the gene product belongs. These homology-searching algorithms are used to identify and classify gene products on the basis of sequence similarity when the structure is known, homology searching may be based on structural similarity. The scoring for sequence similarity uses criteria from that of the absolute identity of residues at equivalent positions to similarity based on polarity, hydrophobicity, and size. The algorithms allow for insertion or deletion of segments of polypeptide chain or structure to give the maximum possible overlap between two proteins.

Protein domains are classified by class, fold, and family. The **class** is determined by the predominant type of secondary structure present in the protein. Some examples are mainly α-helix (all α), mainly β-strand (all β), and approximately equal amounts of α-helix and β-strand (alternating [α *or* β] and nonalternating [$\alpha + \beta$]). The **fold** is determined by the particular arrangement of secondary structure elements within the domain. The **family** is determined by the degree of sequence identity between the proteins. Proteins that are members of the same family have a common evolutionary relationship and are derived from the same primordial gene. Proteins of the same family have the same fold and often have similar functions.

Of interest in clinical medicine is the finding of mutations in the amino acid sequence, which can give rise to significant alterations in function. At the molecular level, changes in even one amino acid can be significant. An amino acid mutation may perturb the native conformation, the conformational flexibility; the energetics, or motion, of the molecule; and the selectivity of enzymes toward substrates and inhibitors. Examples of this are the hemoglobins for which there are extensive catalogs of mutations. Some hemoglobin mutations produce significant clinical symptoms, such as sickle-cell anemia (Clin. Corr. 6.2; p. 214). Certain positions in the amino acid sequence are variant among populations. These sequence positions, when they involve single changes in the base codon for that amino acid, are termed **single nucleotide polymorphisms** (SNPs) and can lead to an understanding of the differences in response to a disease or therapeutic treatment among human populations.

Homologous Fold Structures Are Often Formed from Nonhomologous Amino Acid Sequences

Although each native conformation is unique, comparison of tertiary structures of different proteins shows that similar arrangements of secondary structure motifs are often observed in the fold structures of domains. Folds of similar structure from proteins unrelated by function, sequence, or evolution are designated **superfolds**. They form because of the thermodynamic stability of their secondary structure arrangements or their kinetic accessibility.

A common all-α fold or domain is found in lysozyme and is designated the globin fold (Figure 3.37), as it was first reported in myoglobin and the subunits of hemoglobin (p. 355). These all-α structures have seven or eight sections of α-helix joined by smaller segments that allow the helices to fold back on themselves to form a characteristic globular shape. As this fold is generated by proteins from different sequence homology families, the globin fold is a superfold. Another common superfold is the ***α/β*-domain structure** of triose phosphate isomerase (Figure 3.38) in which the strands (designated by arrows) form a central ***β*-barrel** with each β-strand in the interior interconnected by α-helical

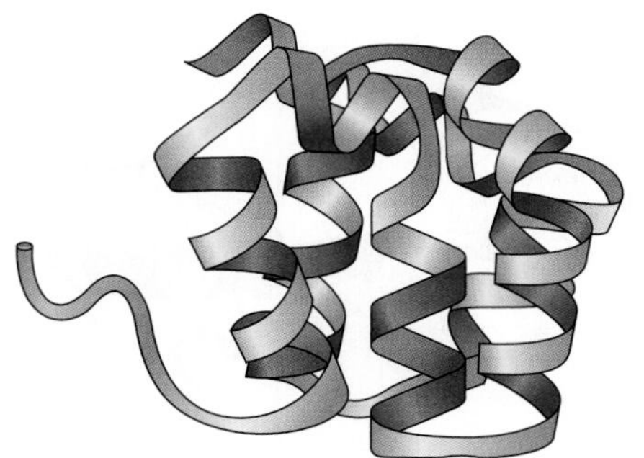

Figure 3.37 An example of an all α-domain globin fold in lysozyme. In this drawing and Figures 3.38, 3.39, and 3.40, only the outline of the polypeptide chain is shown. β-Strands are shown by arrows with direction of the arrow showing N—>C terminal direction; lightning bolts represent disulfide bonds, and circles represent metal ion cofactors (when present). All α domain is the globin fold.
Redrawn with permission from Richardson, J. S. *Adv. Protein Chem.* 34:168, 1981.

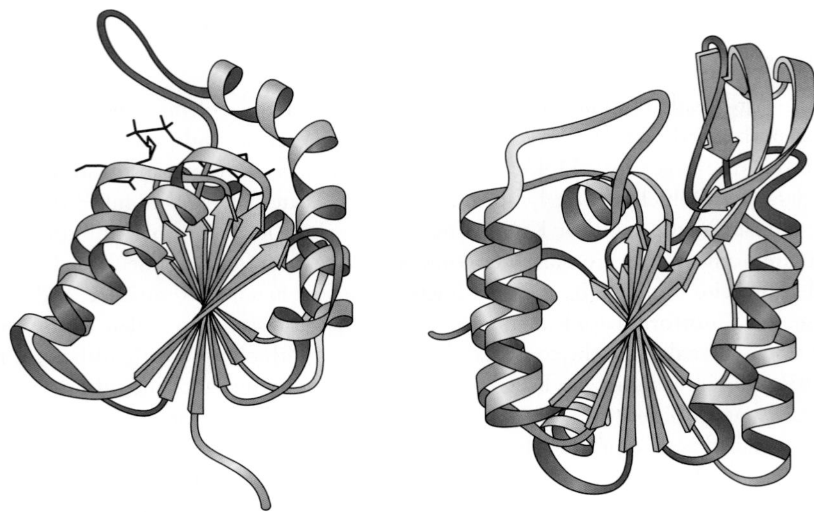

Figure 3.39 Example of an α/β-domain fold in which β-strands form a classical twisted β-sheet of lactate dehydrogenase and phosphoglycerate kinase. See the legend to Figure 3.37. As in previous α/β-domain fold, regions of α-helix alternate with regions of β-strand within polypeptide chain. β-Sheet structure is on the inside while the α-helical segments are on the outside. β-Strands are in parallel within the β-structure.
Redrawn with permission from Richardson, J. S. *Adv. Protein Chem.* 34:168, 1981.

regions located on the outside of the fold. A similar fold forms a domain of pyruvate kinase (Figure 3.38), which has no sequence or functional homology. A different type of α,β-superfold is present in the nonhomologous domain 1 of lactate dehydrogenase and domain 2 of phosphoglycerate kinase (Figure 3.39). In these, central sections participate in a **twisted β-sheet**. Again, the β-strand segments are joined by α-helical regions positioned on the outside to give the characteristic fold pattern. An **all-β-domain** superfold is present in Cu, Zn superoxide dismutase, in which the antiparallel β-sheet forms a Greek key β-barrel (Figure 3.40). A similar fold pattern occurs in each of the domains of the immunoglobulins. Concanavalin A (Figure 3.40) shows an all-β-domain superfold in which the antiparallel β-strands form a β-barrel fold called a "jelly roll". Proteins that are not water soluble may contain nonglobular fold patterns (Section 3.6).

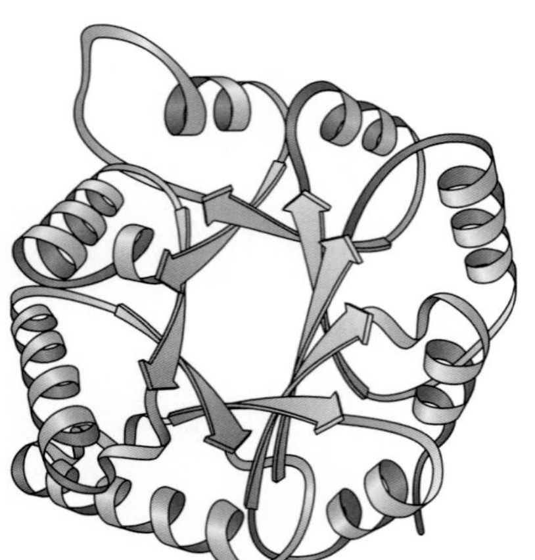

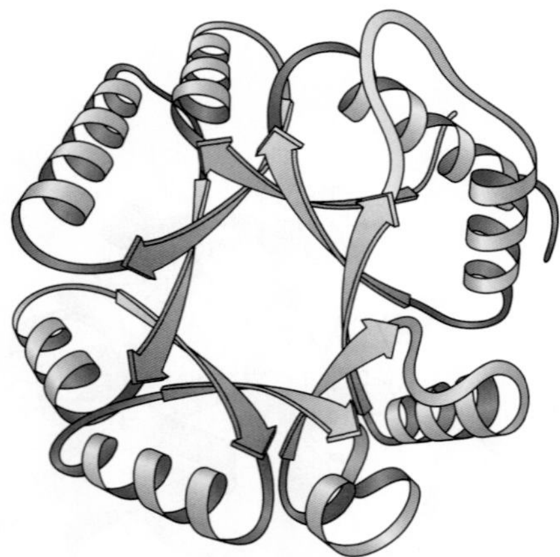

Figure 3.38 Example of an α,β-domain fold in triose phosphate isomerase and in pyruvate kinase domain 1. See the caption to Figure 3.37. In this commonly formed superfold, the β-strands form a β-barrel in the center of the domain while the α-helix segments are on the outside of the domain. The β-strands are in parallel directions. Regions of α-helix alternate with β-strands within the polypeptide chain.
Redrawn with permission from Richardson, J. S. *Adv. Protein Chem.* 34:168, 1981.

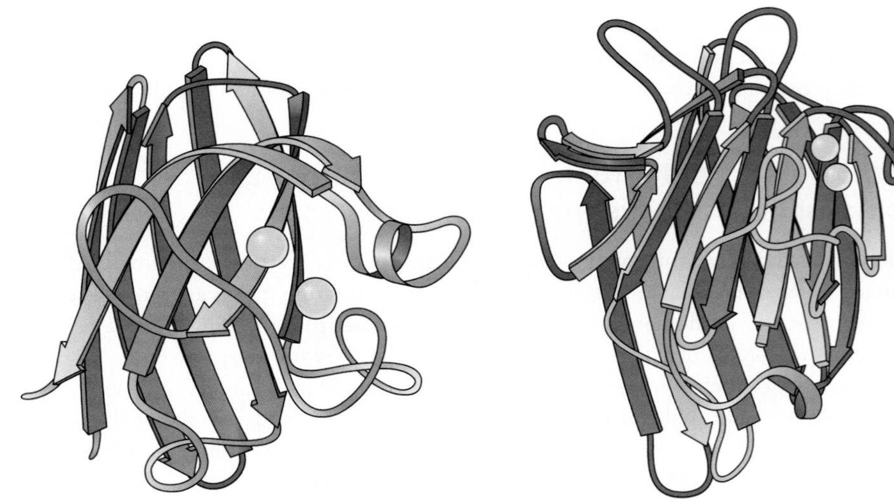

Figure 3.40 Examples of all β-domain superfolds: The Greek key barrel and jelly roll folds shown in superoxide dismutase and concanavalin A. See the legend to Figure 3.37. β-Strands are mostly antiparallel in all β-domain folds.
Redrawn with permission from Richardson, J. S. *Adv. Protein Chem.* 34:168, 1981.

3.6 • NONGLOBULAR STRUCTURED PROTEINS

The characteristics of protein structure already discussed are based on observations on globular and water-soluble proteins. **Globular proteins** have a spheroidal shape, vary in size, have relatively high water solubility, and function as catalysts, transporters, and regulators of metabolic pathways and gene expression. Proteins that do not conform to the globular soluble protein model are the fibrous proteins and membrane proteins. These are nonglobular and have low water solubility. In addition, lipoproteins and glycoproteins contain lipid and carbohydrate nonprotein components and may or may not have globular structures.

Fibrous proteins characteristically contain larger amounts of regular secondary structure, a long cylindrical (rodlike) shape, low solubility in water, and a structural rather than a dynamic role. Collagen, keratin, and tropomyosin are important fibrous proteins. Their multistrand structures have polypeptide chains with a high amount of regular secondary structure (polyproline type II helix or α-helix), which associate to form rod-shaped multichain supercoiled conformations. In each case, the amino acid sequences of the chains generate edges on the cylindrical surfaces that stabilize hydrophobic interactions between the chains in the supercoiled conformations. In turn, these molecular structures are aligned into multimolecular fibrils that are stabilized, in some cases, by covalent cross-links.

TABLE 3.10 • Percentage of Collagen in Representative Human Tissues and Organs

Tissue or Organ	*Percent Collagen by weight*
liver	4
lung	10
aorta	10–24
cartilage	50
cornea	64
whole cortical bone	23
skin	74

Collagen

Collagen is a family of extracellular proteins present in all tissues and organs, which provide the framework that gives the tissues their form and strength. It is the most prominent protein in humans (Table 3.10 and Clin. Corr. 6.13; p. 246).

Amino Acid Composition of Collagen

The amino acid composition of type I skin collagen and of the globular proteins ribonuclease and hemoglobin are given in Table 3.11. Skin collagen is rich in glycine (33% of its amino acids), proline (13%), and the derived amino acids 4-hydroxyproline (9%) and 5-hydroxylysine (0.6%) (Figure 3.41). Hydroxyproline is unique to collagens being formed enzymatically from proline. Most of the hydroxyproline has the hydroxyl group in the 4 position (γ carbon), although a small amount of 3-hydroxyproline is also formed (Table 3.11). Collagens are glycoproteins with carbohydrate joined to 5-hydroxylysine, by an *O*-glycosidic bond through the δ-carbon hydroxyl group.

Amino Acid Sequence of Collagen

The collagen family is made up of polypeptides derived from 40 known collagen chain genes which produce about 20 types of collagen. Each mature collagen or tropocollagen molecule contains three polypeptide chains. Some types of collagen contain three identical polypeptide chains. In type I (Table 3.12), there are two $\alpha 1$(I) chains and one $\alpha 2$(I). Type V collagen contains three different chains designated $\alpha 1$(V), $\alpha 2$(V), and $\alpha 3$(V). Collagens differ in amino acid sequence, but there are large regions of homologous sequence among all the different collagen types. In all the collagen types there are regions with the tripeptides **Gly-Pro-Y** and **Gly-X-Hyp** (where X and Y are any amino acid) repeated in tandem several hundred times. In type I collagen polypeptides, the triplet sequences encompass over 600 of approximately 1000 amino acids per polypeptide. The collagens differ in their carbohydrate component. Some characteristics of collagen types I–VI are summarized in Table 3.12.

Structure of Collagen

Synthetic polypeptide chains containing only proline (poly-Pro) can be prepared in the laboratory and assume in aqueous solutions a regular secondary helical structure different from an α-helix. The poly-Pro forms a tightly twisted extended helix with three residues per turn ($n = 3$). This helix with all *trans* peptide bonds is the **polyproline type II** helix (Figure 3.12, p. 81, for differences between *cis*- and *trans* peptide bonds). The poly-Pro helix closely resembles that found in collagen chains in regions that contain a proline or hydroxyproline at approximately every third position in the repeated tripeptide sequences, indicating that the thermodynamic forces leading to formation of the collagen helix are due to the properties of proline.

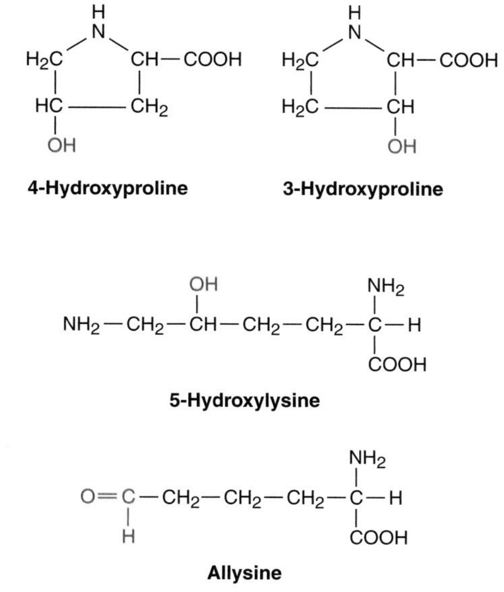

Figure 3.41 Derived amino acids in collagen. Carbohydrate is attached to 5-OH in 5-hydroxylysine by a type III glycosidic linkage (see Figure 3.49).

TABLE 3.11 • Comparison of Amino Acid Content of Human Skin Collagen (Type I) and Mature Elastin with That of Two Typical Globular Proteins[a]

Amino Acid	*Collagen (Human Skin)*	*Elastin (Mammalian)*	*Ribonuclease (Bovine)*	*Hemoglobin (Human)*
common amino acids		percent of total		
Ala	11	22	8	9
Arg	5	0.9	5	3
Asn			8	3
Asp	5	1	15	10
Cys	0	0	0	1
Glu	7	2	12	6
Gln			6	1
Gly	33	31	2	4
His	0.5	0.1	4	9
Ile	1	2	3	0
Leu	2	6	2	14
Lys	3	0.8	11	10
Met	0.6	0.2	4	1
Phe	1	3	4	7
Pro	13	11	4	5
Ser	4	1	11	4
Thr	2	1	9	5
Trp	2	1	9	2
Tyr	0.3	2	8	3
Val	2	12	8	10
derived amino acids				
Cystine	0	0	7	0
3-Hydroxyproline	0.1		0	0
4-Hydroxyproline	9	1	0	0
5-Hydroxylysine	0.6	0	0	0
Desmosine and isodesmosine	0	1	0	0

[a]Boxed numbers emphasize important differences in amino acid composition between the fibrous proteins (collagen and elastin) and typical globular proteins.

In the polyproline type II helix, the plane of each peptide bond is perpendicular to the axis of the helix. Thus the peptide bond oxygen and nitrogen atoms in a polyproline type II helix point toward neighboring polypeptide chains and form strong interchain hydrogen bonds. This contrasts with the α-helix, in which the plane of the peptide bond is parallel to the axis where the NH and carbonyl oxygen of the peptides form only intrachain hydrogen bonds.

The three chains of a collagen molecule, with each of the chains in a polyproline type II helix conformation, wind around each other to form a superhelical structure known as a superhelix (Figure 3.42, p. 106). The three-chain **superhelix** has a characteristic rise (d) and pitch (p) different from that found in each of the polypeptide chain polyproline helices. The superhelix is stabilized by the interchain hydrogen bonds between the three chains, because glycine occurs at every third position. With the polyproline type II helical conformation of each polypeptide chain with three residues per turn ($n = 3$), the glycines form an **apolar edge** along the length of each of the helices. The glycine edges then form nonpolar interactions between the chains that are critical to stabilize the three-chain superhelix. Any side

TABLE 3.12 • Classification of Collagen Types

Type	*Chain Designations*	*Tissue Found*	*Characteristics*
I	$[\alpha 1(\text{I})]_2\,\alpha 2(\text{I})$	Bone, skin, tendon, scar tissue, heart valve, intestinal and uterine wall	Low carbohydrate; <10 hydroxylysines per chain
II	$[\alpha 1(\text{II})]_3$	Cartilage, vitreous	10% carbohydrate; >20 hydroxylysines per chain
III	$[\alpha 1(\text{III})]_3$	Blood vessels, newborn skin, scar tissue, intestinal and uterine wall	Low carbohydrate; high hydroxyproline and Gly; contains Cys
IV	$[\alpha 1(\text{IV})]_3$ $[\alpha 2(\text{IV})]_3$	Basement membrane, lens capsule	High 3-hydroxyproline; >40 hydroxylysines per chain; low Ala and Arg; contains Cys; high carbohydrate (15%)
V	$[\alpha 1(\text{V})]_3\alpha 2(\text{V})$ $[\alpha 1(\text{V})]_3$ $\alpha 1(\text{V})\alpha 2(\text{V})\alpha 3(\text{V})$	Cell surfaces or extracellular matrix associated with cell; widely distributed in low amounts	High carbohydrate, relatively high glycine, and hydroxylysine
VI	$\alpha 1(\text{VI})\alpha 2(\text{VI})$ $\alpha 3(\text{VI})$	Aortic intima, placenta, kidney, and skin in low amounts	Relatively large globular domains in telopeptide region; high Cys and Tyr; molecular weight relatively low ($\sim$160 kDa); equimolar amounts of hydroxylysine and hydroxyproline

chain other than that of glycine along the apolar edge would impede the adjacent chains from coming together into the superhelix structure (Figure 3.42, Osteogenesis Imperfecta, and Clin. Corr. 6.13; p. 246).

In type I collagen the triple helix extends for most of the sequence and only the carboxyl-terminus and amino-terminal segments (known as the **telopeptides**) are not in a triple-helical conformation. The type I collagen molecule is 3000 Å long and only 15 Å wide, a very long cylindrical structure. In other collagen types, superhelical regions may be periodically broken by globular regions.

Formation of Covalent Cross-Links in Collagen

An extracellular enzyme acts on procollagen molecules (p. 107) to convert the ε-amino group of some lysine side chains to a δ-aldehyde (Figure 3.43). The derived amino acid is **allysine**. The newly formed aldehyde side chain spontaneously undergoes nucleophilic addition reactions with nonmodified lysine ε-amino groups and with the δ-carbon atoms of other allysine aldehydic groups. These covalent linkages can be between chains within the superhelical structure or between adjacent superhelical collagen molecules in a collagen fibril (p. 243 for a discussion of collagen biosynthesis).

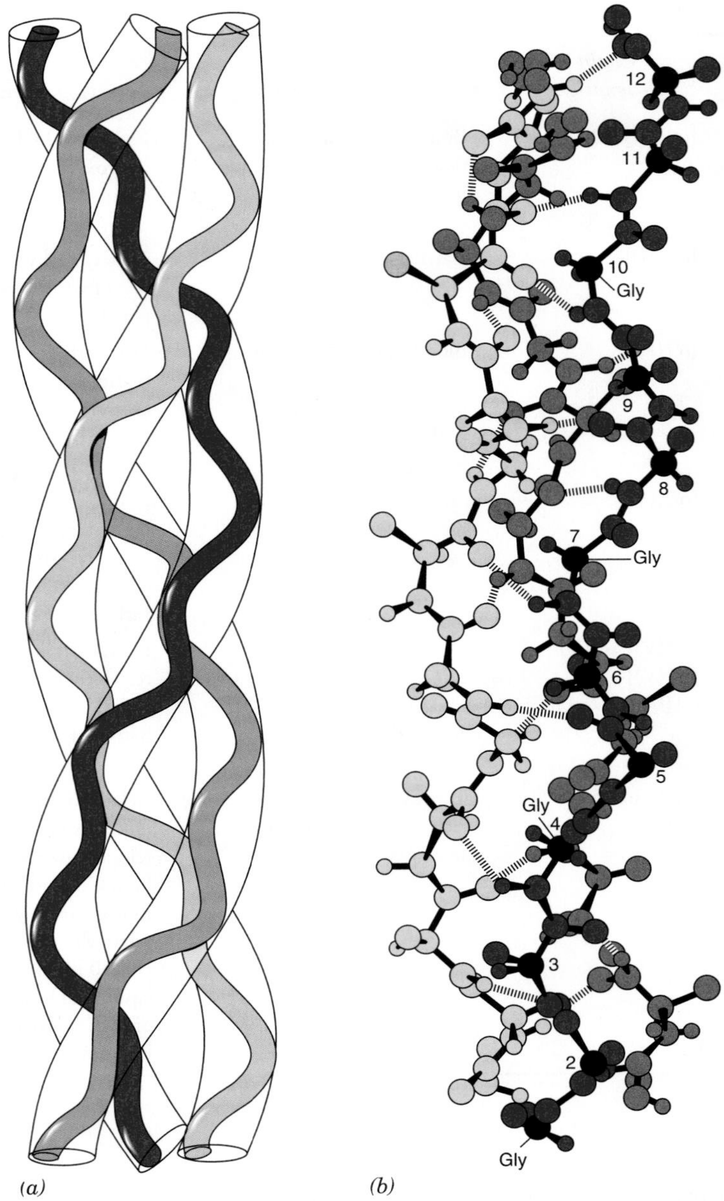

Figure 3.42 Diagram of collagen demonstrating the necessity for glycine in every third residue to allow different chains to be in close proximity in the structure. (*a*) Ribbon model for supercoiled structure of collagen with each individual chain in a polyproline type II helix. (*b*) More detailed model of supercoiled conformation. All α-carbon atoms are numbered, and proposed hydrogen bonds are shown by dashed lines.
Redrawn with permission from Dickerson, R. E., and Geis, I. *The Structure and Actions of Proteins.* Menlo Park, CA: Benjamin, 1969, 41, 42.

Elastin Is a Fibrous Protein with Allysine-Generated Cross-Links

Elastin gives tissues and organs the capacity to stretch without tearing. It is classed as a fibrous protein because of its structural function and relative insolubility in water; it is abundant in ligaments, lungs, walls of arteries, and skin. Elastin lacks a regular secondary structure. As in collagen, allysines form cross-links in **elastin**. An extracellular **lysine amino oxidase** converts lysine side chains in the sequence -Lys-Ala-Ala-Lys- and -Lys-Ala-Ala-Ala-Lys- to allysines. Three allysines and an unmodified lysine from different regions in the polypeptide chains react to form the heterocyclic structure of **desmosine** or **isodesmosine**, which cross-link the polypeptide chains in elastin networks (Figure 3.44).

Keratin and Tropomyosin

Keratin and tropomyosin are fibrous proteins in which each polypeptide is α-helical. Keratin is found in the epidermal layer of skin, in nails, and in hair. Tropomyosin is a component of the thin filament in muscle tissue. The sequences in both proteins show tandem

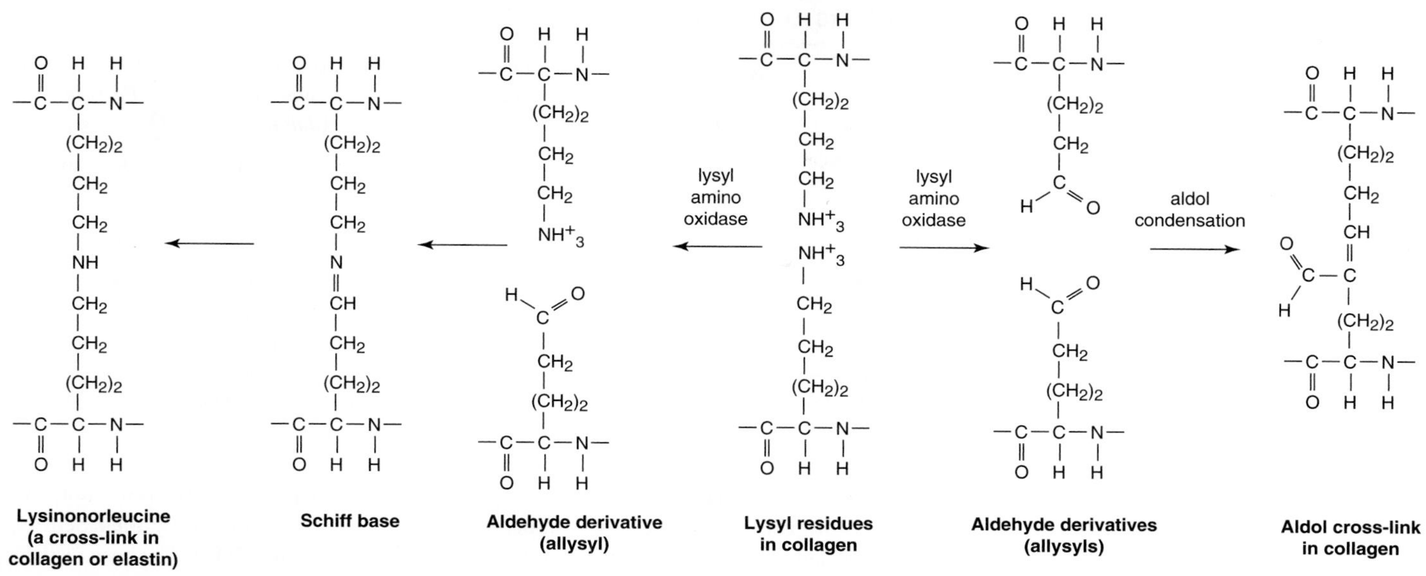

Lysinonorleucine (a cross-link in collagen or elastin) ← Schiff base ← Aldehyde derivative (allysyl) ← lysyl amino oxidase — Lysyl residues in collagen — lysyl amino oxidase → Aldehyde derivatives (allysyls) — aldol condensation → Aldol cross-link in collagen

Figure 3.43 Covalent cross-links formed in collagen through allysine intermediates. Formation of allysines is catalyzed by lysyl amino oxidase.

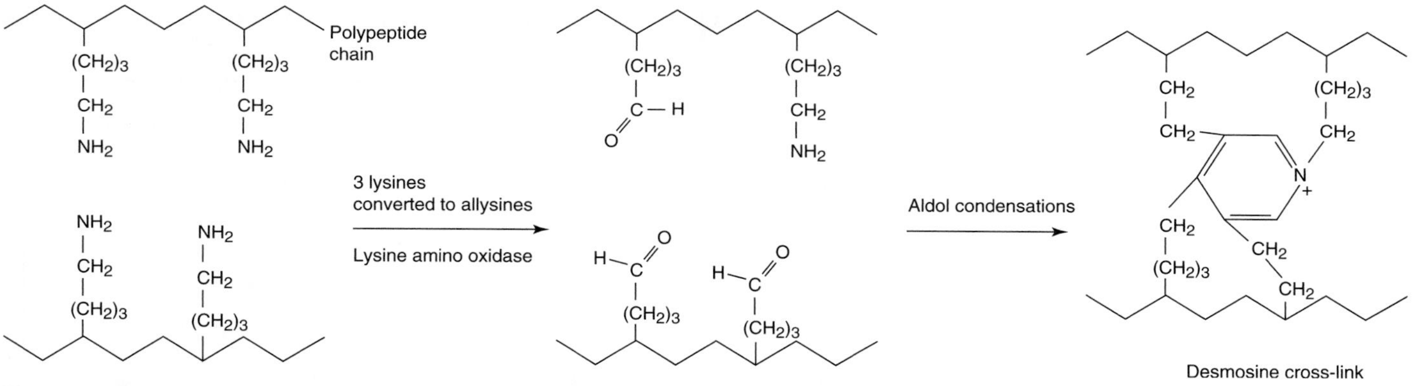

Polypeptide chain — 3 lysines converted to allysines — Lysine amino oxidase → — Aldol condensations → Desmosine cross-link

Figure 3.44 Formation of desmosine covalent cross-link in elastin from lysine and allysines. Polypeptide chain drawn schematically with intersections of lines representing placement of α-carbons.

repetition of seven residue segments (heptad), in which the first and fourth amino acids have hydrophobic side chains and the fifth and seventh polar side chains. The reiteration of hydrophobic and polar side chains in heptad segments is symbolically represented by the formulation $(\text{a-b-c-d-e-f-g})_i$, where a and d are hydrophobic amino acids, and e and g are polar or ionized side chain groups. Since a seven amino acid segment represents two complete turns of an α-helix ($n = 3.6$), the apolar residues at a and d align to form an apolar edge along one side of the α-helix (Figure 3.45). This apolar edge interacts with polypeptide apolar edges of other α-keratin chains to form a **coiled-coil** superhelical structure containing two or four polypeptide chains. Each polypeptide also contains a polar edge, due to residues e and g, that interacts with water on the outside of the superhelix and stabilizes the superhelix.

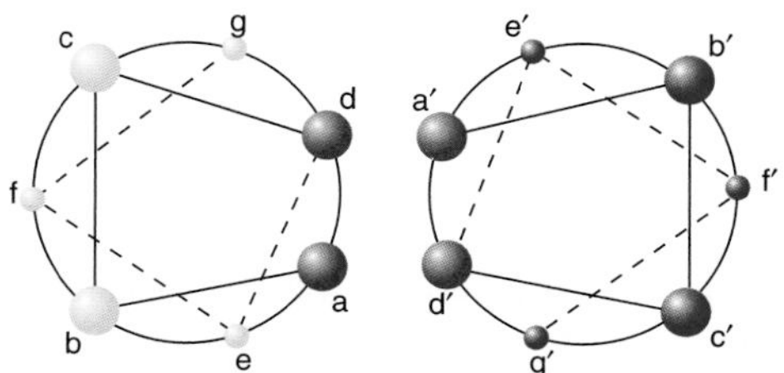

Figure 3.45 Interaction of apolar edges of two chains in α-helical conformation as in keratin and tropomyosin. Interaction of apolar d-a and a′-d residues of two α-helices aligned parallel in an NH_2-terminal (top) to COOH-terminal direction is presented.
Redrawn from McLachlan, A. D., and Stewart, M. *J. Mol. Biol.* 98: 293, 1975.

Plasma Lipoproteins Are Complexes of Lipids with Proteins

Plasma lipoproteins are complexes of proteins and lipids that form distinct aggregates with an approximate stoichiometry between protein and lipid components. Covalent bonds do not exist between lipid and protein molecules, rather the molecules of the complex are held in the particle structure through noncovalent interactions (p. 722). The lipoprotein particles function to transport lipids from tissue to tissue and participate in lipid metabolism (p. 724).

TABLE 3.13 • Hydrated Density Classes of Plasma Lipoproteins

Lipoprotein Fraction	*Density ($g\ mL^{-1}$)*	*Flotation Rate, S_f (Svedberg units)*	*Molecular Weight (daltons)*	*Particle Diameter (Å)*
HDL	1.063–1.210		HDL_2, 4×10^5	70–130
			HDL_3, 2×10^5	50–100
LDL (or LDL_2)	1.019–1.063	0–12	2×10^6	200–280
IDL (or LDL_1)	1.006–1.019	12–20	4.5×10^6	250
VLDL	0.95–1.006	20–400	$5 \times 10^6 - 10^7$	250–750
Chylomicrons	<0.95	>400	$10^9 - 10^{10}$	10^3–10^4

Source: Data from Soutar, A. K. and Myant, N. B. In: R. E. Offord (Ed.), *Chemistry of Macromolecules,* IIB. Baltimore, MD: University Park Press, 1979.

Four classes exist in plasma of normal fasting humans (Table 3.13) and in the postabsorptive period a fifth class, chylomicrons, is also present. These are distinguished by their density, as determined by ultracentrifugation and by electrophoresis (Figure 3.46). Changes in their relative concentrations are predictive of atherosclerosis, a major human disease (Clin. Corr. 3.3). Their protein components are termed **apolipoproteins**, and each class of lipoprotein having a characteristic apolipoprotein composition. The most prominent apolipoproteins (Table 3.14) are apolipoprotein **ApoA-I** in high-density lipoproteins (**HDLs**); **ApoB** in low-density lipoproteins (**LDLs**), intermediate-density lipoproteins (**IDLs**); and very-low-density

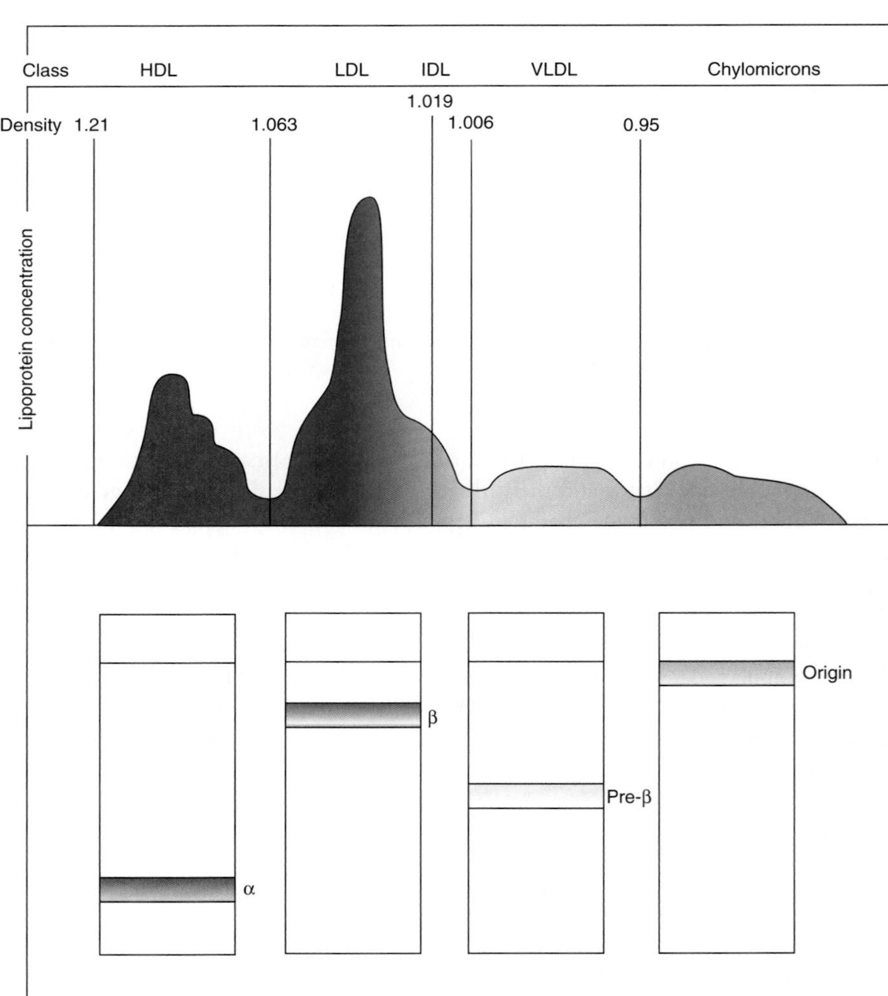

Figure 3.46 Correspondence of plasma lipoprotein density classes with electrophoretic mobility in a plasma electrophoresis. In the upper diagram, an ultracentrifugation Schlieren pattern is shown. At bottom, an electrophoresis on a paper support shows the mobilities of major plasma lipoprotein classes with respect to α- and β-globulin bands.
Reprinted with permission from Soutar, A. K., and Myant, N. B. In Offord, R. E. (Ed.), *Chemistry of Macromolecules, IIB.* Baltimore, MD: University Park Press, 1979.

CLINICAL CORRELATION 3.3

Hyperlipoproteinemias

Hyperlipoproteinemias (OMIM 608083) are disorders of the rates of synthesis or clearance of lipoproteins from the bloodstream. Usually they are detected by measuring plasma triacylglycerol and cholesterol and are classified on the basis of which class of lipoproteins is elevated.

Type I hyperlipoproteinemia is due to accumulation of chylomicrons. Two genetic forms are known: lipoprotein lipase deficiency and ApoCII deficiency. ApoCII is required by lipoprotein lipase for full activity. Patients with type I hyperlipidemia have exceedingly high plasma triacylglycerol concentrations (over 1000 mg/dL) and suffer from eruptive xanthomas (yellowish triacylglycerol deposits in the skin) and pancreatitis.

Type II hyperlipoproteinemia is characterized by elevated LDL levels. Most cases are due to genetic defects in the synthesis, processing, or function of the LDL receptor. Heterozygotes have elevated LDL levels; hence the trait is dominantly expressed. Homozygous patients have very high LDL levels and may suffer myocardial infarctions before age 20.

Type III hyperlipoproteinemia is due to abnormalities of ApoE, which interfere with the uptake of chylomicron and VLDL remnants. Hypothyroidism can produce a very similar hyperlipoproteinemia. These patients have an increased risk of atherosclerosis.

Type IV hyperlipoproteinemia is the commonest abnormality. The VLDL levels are increased, often due to obesity, alcohol abuse, or diabetes. Familial forms are also known.

Type V hyperlipoproteinemia is, like type I, associated with high chylomicron triacylglycerol levels, pancreatitis, and eruptive xanthomas.

Hypercholesterolemia also occurs in certain types of liver disease in which biliary excretion of cholesterol is reduced. An abnormal lipoprotein called lipoprotein X accumulates. This disorder is not associated with increased cardiovascular disease from atherosclerosis.

Havel, R. J., and Kane, J. P. Introduction: Structure and metabolism of plasma lipoproteins. In Scriver, C. R., Beaudet, A. L.. Sly, W. S., and Valle, D. (Eds.), *The Metabolic and Molecular Basis of Inherited Disease,* 8th ed. New York: McGraw Hill, 2001, Chap. 114; and Goldstein, J. L., Hobbs, H. H., and Brown, M. S. Familial hypercholesterolemia. In Scriver, C. R., Beaudet, A. L., Sly, W. S., and Valle, D. (Eds.), *The Metabolic and Molecular Bases of Inherited Disease,* 8th ed. New York: McGraw-Hill, 2001, Chap. 120.

lipoproteins (**VLDLs**), and **ApoC** in IDLs and VLDLs. Each apolipoprotein class is genetically and structurally distinct (Clin. Corr. 3.4). Apolipoproteins vary from 6 kDa (ApoC-I) to 550 kDa for ApoB-100. The latter is a long polypeptide (4536 amino acids) and occurs in truncated form (the N-terminal 2512 residues only) as ApoB -48 in chylomicrons.

A model for the structure of a VLDL particle is shown in Figure 3.47. On the inside are neutral lipids such as cholesteryl esters and triacylglycerols. Surrounding this inner core of neutral lipids, is a shell ~20 Å thick, in which reside the proteins and the charged amphoteric lipids such as unesterified cholesterol and phosphatidylcholines (p. 460). Amphoteric lipids and proteins in the outer shell place their hydrophobic apolar regions toward the inside of the particle and their charged groups toward the outside where they interact with each other and with water.

TABLE 3.14 • Apolipoproteins of Human Plasma Lipoproteins (Values in Percentage of Total Protein Present)[a]

Apolipoprotein	*HDL_2*	*HDL_3*	*LDL*	*IDL*	*VLDL*	*Chylomicrons*
ApoA-I	85	70–75	Trace	0	0–3	0–3
ApoA-II	5	20	Trace	0	0–0.5	0–1.5
ApoD	0	1–2			0	1
ApoB	0–2	0	95–100	50–60	40–50	20–22[b]
ApoC-I	1–2	1–2	0–5	<1	5	5–10
ApoC-II	1	1	0.5	2.5	10	15
ApoC-III	2–3	2–3	0–5	17	20–25	40
ApoE	Trace	0–5	0	15–20	5–10	5
ApoF	Trace	Trace				
ApoG	Trace	Trace				

Source: Data from Soutar, A. K, and Myant, N. B. In: R. E. Offord (Ed.), *Chemistry of Macromolecules,* IIB. Baltimore, MD: University Park Press, 1979; Kostner, G. M. *Adv. Lipid Res.* 20: 1, 1983.

[a]Values show variability from different laboratories.

[b]Primarily ApoB-48.

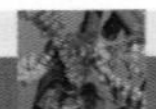

CLINICAL CORRELATION 3.4

Hypolipoproteinemias

Abetalipoproteinemia is a genetic disease that is characterized by absence of chylomicrons, VLDLs, and LDLs due to an inability to synthesize apolipoproteins apoB-100 and apoB-48. There is accumulation of lipid droplets in small intestinal cells, malabsorption of fat, acanthocytosis (spiny-shaped red cells), and neurological disease (retinitis pigmentosa, ataxia, and retardation).

Tangier disease, a α-lipoprotein deficiency, is a rare autosomal recessive disease in which the HDL level is 1–5% of its normal value. Clinical features are due to the accumulation of cholesterol in the lymphoreticular system, which may lead to hepatomegaly and splenomegaly. The plasma cholesterol and phospholipids are greatly reduced.

Deficiency of the enzyme lecithin:cholesterol acyltransferase is a rare disease that results in the production of lipoprotein X (see Clin. Corr. 3.5). Also characteristic is the decrease in the α-lipoprotein and pre-β-lipoprotein bands and increase in β-lipoprotein because of the presence of lipoprotein X on electrophoresis.

Kane, J. P., and Havel, R. J. Disorders of the biogenesis and secretion of lipoproteins containing the β apolipoproteins. In Scriver, C. R., Beaudet, A. L., Sly, W. S., and Valle, D. (Eds.), *The Metabolic and Molecular Basis of Inherited Disease,* 8th ed. New York: McGraw Hill, 2001, Chap. 115; Assmann, G., von Eckardstein, A., and Brewer, H. B., Jr. Familial high density lipoprotein deficiency: Tangier disease. In Scriver, C. R., Beaudet, A. L., Sly, W. S., and Valle, D. (Eds.), *The Metabolic and Molecular Basis of Inherited Disease,* 8th ed. New York: McGraw Hill, 2001, Chap. 122.

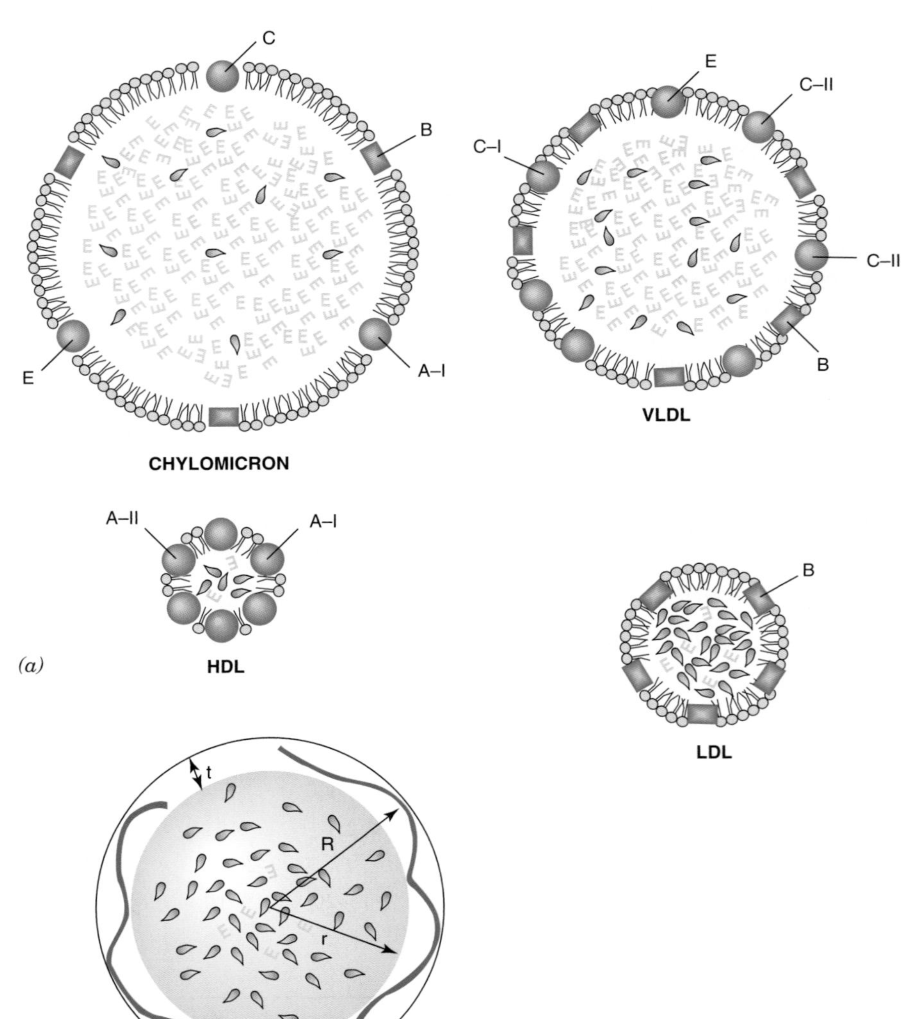

Figure 3.47 Generalized structure of plasma lipoproteins. (*a*) Spherical particle model consisting of a core of triacylglycerols (yellow Es) and cholesterol esters (orange drops) with a shell ~20 Å thick of apolipoproteins (lettered), phospholipids, and unesterified cholesterol. Apolipoproteins are embedded with their hydrophobic edges oriented toward the core and their hydrophilic edges toward the outside. (*b*) LDL particle showing ApoB-100 imbedded in the outer shell of particle.
Part (*a*) redrawn from Segrest, J. P., et al. *Adv. Protein Chem.* 45: 303, 1994; and part (*b*) redrawn from Schumaker, V. N., et al., *Protein Chem.* 45: 205, 1994.

TABLE 3.15 • Chemical Composition of Plasma Lipoprotein Classes

Total Lipoprotein Class	*Total Protein (%)*	*Lipid (%)*	*Percent Composition of Lipid Fraction*			
			Phospholipids	*Esterified Cholesterol*	*Unesterified Cholesterol*	*Triacylglycerols*
HDL_2[a]	40–45	55	35	12	4	5
HDL_3[a]	50–55	50	20–25	12	3–4	3
LDL	20–25	75–80	15–20	35–40	7–10	7–10
IDL	15–20	80–85	22	22	8	30
VLDL	5–10	90–95	15–20	10–15	5–10	50–65
Chylomicrons	1.5–2.5	97–99	7–9	3–5	1–3	84–89

Source: Data from Soutar, A. K. and Myant, N. B. In: R. E. Offord (Ed.), *Chemistry of Macromolecules,* IIB. Baltimore, MD: University Park Press, 1979.
[a]Subclasses of HDL.

This lipoprotein particle model applies to all classes of plasma lipoproteins. As the diameter of a particle decreases, the outer shell makes up a greater percentage of the total particle. Thus, the smaller particles have a higher percentage of the surface protein and amphoteric lipid molecules than the larger particles. The HDL particles, which are much smaller than VLDL particles, thus consist of 45% protein and 55% lipid, while the VLDL particles consist of 10% protein with 90% lipid (Table 3.15).

The apolipoproteins have high α-helical content when in association with lipid. Their helical regions are amphipathic since every third or fourth amino acid is charged. These residues form a polar edge that associates with the polar heads of phospholipids and the aqueous environment. The opposite sides of the helices have hydrophobic side chains that are oriented toward the nonpolar core of the lipoprotein particle. The α-helical structure of part of ApoC-I is shown in Figure 3.48. ApoB contains α-helical and β-strand segments. A single long polypeptide of ApoB-100 circles the circumference of the LDL particle like a belt weaving in and out of the phospholipid outer shell (Figure 3.47).

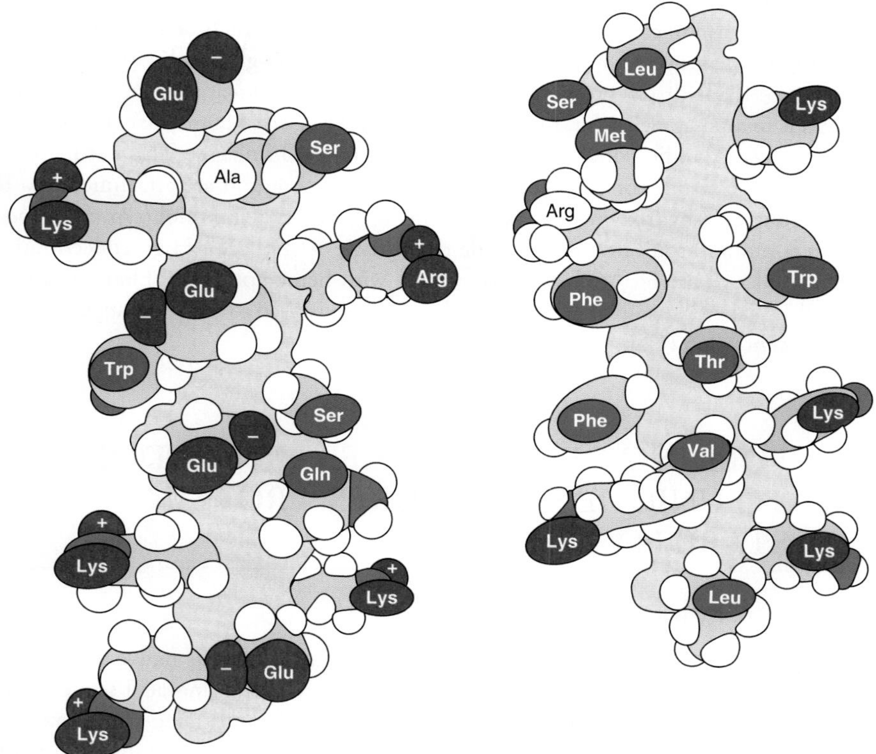

Figure 3.48 Side chains of a helical amphipathic segment of apolipoprotein C-1 between residues 32 and 53. The polar face shows acid residues in the center and basic residues at the edge. On the other side of the helix, hydrophobic residues form a nonpolar longitudinal face.
Reprinted with permission from Sparrow, J. T., and Gotto, A. M., Jr. *CRC Crit. Rev. Biochem.* 13: 87, 1983. Copyright (1983) CRC Press, Boca Raton, FL.

Glycoproteins Contain Covalently Bound Carbohydrate

Glycoproteins contain a single or multiple amino acids covalently bonded to carbohydrate (sugar) molecules. Glycoproteins participate in many normal and disease-related functions of clinical relevance. The best-characterized glycoproteins are those of the outer surface of plasma membranes, the extracellular matrix, and blood plasma. Those on plasma membranes provide information for identification of cells by other cells and for regulation of cell growth by contact inhibition. They also provide the antigenic determinants of the various blood group systems (e.g., ABO and Rh) on erythrocytes, the histocompatibility of tissue transplants, and the cellular receptors for hormones, neurotransmitters, and viruses. Tumorigenesis and malignant transformation are associated with changes in the glycoprotein composition of plasma membranes. In the extracellular matrix, many proteins (e.g., collagen and laminin) contain attached carbohydrate as do proteins of mucous secretions that lubricate and protect surfaces. The major plasma proteins (but not albumin) are glycoproteins, including blood-clotting proteins, immunoglobulins, complement proteins, and follicle-stimulating hormone, luteinizing hormone and thyroid-stimulating hormone.

The amount of carbohydrate in glycoproteins is variable. IgG contains 4% by weight, glycophorin of human erythrocyte membranes 60%, and human gastric glycoprotein 82%. The carbohydrate can be distributed evenly along the polypeptide chain or concentrated in defined regions. In proteins of plasma membrane, carbohydrate is attached only on the extracellular portion. The attached carbohydrate usually contains less than 15 sugar residues and in some cases only one. Glycoproteins with the same function from different animal species often have homologous amino acid sequences but vary in their carbohydrate. Heterogeneity in carbohydrate content is common in the same protein even within a single organism. For example, pancreatic ribonuclease A and B forms have an identical amino acid sequence but differ in their carbohydrate composition.

The incorporation of carbohydrate occurs in a series of enzyme-catalyzed reactions as the polypeptide chain is transported through the endoplasmic reticulum and Golgi network (p. 17).

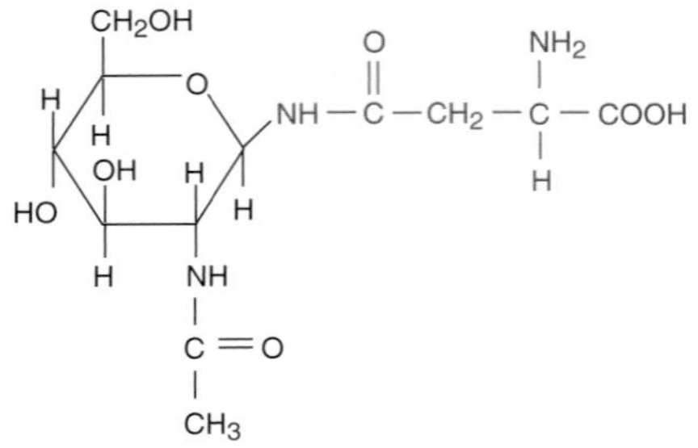

Carbohydrate–Protein Covalent Linkages

The two most common carbohydrate linkages are the ***N*-glycosidic linkage** (type I linkage) between the amide group of asparagine and a sugar, and the ***O*-glycosidic linkage** (type II linkage) between a hydroxyl group of serine or threonine and a sugar (Figure 3.49). In type I the bond is to asparagine within the sequence Asn-X-Thr(Ser). Mammalian glycoproteins also contain *O*-glycosidic bonds to 5-hydroxylysine (type III linkage). This occurs in collagens and in the serum complement protein C1q. Less common are attachments to the hydroxyl group of 4-hydroxyproline (type IV linkage), to a cysteine thiol (type V linkage), and to an α-amino group of a polypeptide chain (type VI linkage). High concentrations of type VI linkages are formed nonenzymatically with hemoglobin and blood glucose in uncontrolled diabetes. Assay of glycosylated hemoglobin is used to follow changes in blood glucose concentration (Clin. Corr. 3.5).

Type I *N*-Glycosyl linkage to asparagine

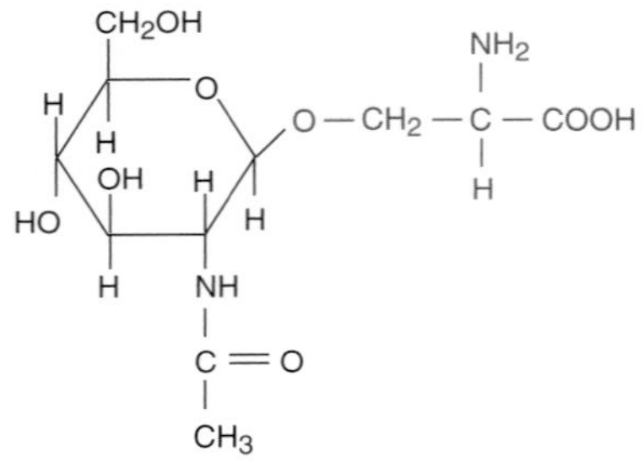

Type II *O*-Glycosyl linkage to serine

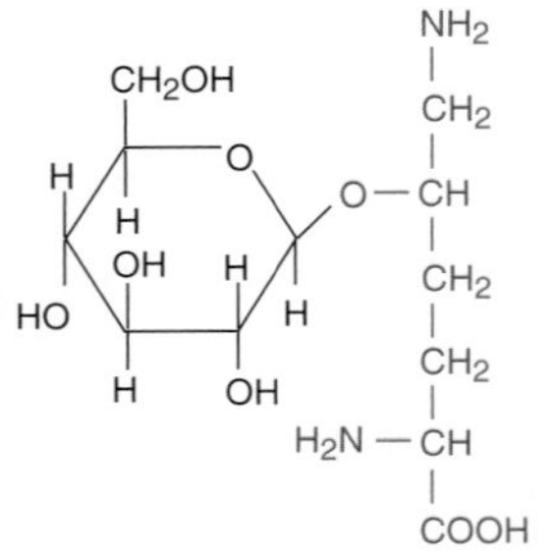

Type III *O*-Glycosyl linkage to 5-hydroxylysine

Figure 3.49 Examples of glycosidic linkages to amino acids in proteins. Type I is an *N*-glycosidic linkage through an amide nitrogen of Asn, type II is an *O*-glycosidic linkage through OH of Ser or Thr, and type III is an *O*-glycosidic linkage to 5-OH of 5-hydroxylysine.

3.7 • FOLDING OF PROTEINS FROM RANDOMIZED TO UNIQUE STRUCTURES: PROTEIN STABILITY

The Protein Folding Problem

The ability of a primary protein structure to fold spontaneously to its native secondary or tertiary conformation, without any information other than the amino acid sequence and the noncovalent forces that act on the sequence, has been demonstrated with many proteins. Thus pancreatic ribonuclease spontaneously refolds to its native conformation after being denatured and its disulfide bonds being reduced without the hydrolysis of peptide

CLINICAL CORRELATION 3.5

Glycosylated Hemoglobin, HbA_{1c}

A glycosylated hemoglobin, designated HbA_{1c}, is formed nonenzymatically in red blood cells by combination of the NH_2 terminal amino groups of the hemoglobin β chains and glucose. The aldehyde form of glucose first forms a Schiff base with the NH_2 terminal amino group,

$$-N{=}\underset{\displaystyle H}{\underset{|}{C}}-\overset{\displaystyle OH}{\overset{|}{\underset{\displaystyle H}{\underset{|}{C}}}}-$$

which then rearranges to a more stable amino ketone linkage,

$$-\underset{\displaystyle H}{\underset{|}{N}}-\overset{\displaystyle H}{\overset{|}{\underset{\displaystyle H}{\underset{|}{C}}}}-\overset{\displaystyle O}{\overset{\|}{C}}-$$

by a spontaneous reaction known as the Amadori rearrangement. The concentration of HbA_{1c} depends on the concentration of glucose in the blood and the duration of hyperglycemia. In prolonged hyperglycemia, the concentration may rise to 12% or more of the total hemoglobin. Patients with diabetes mellitus have high concentrations of blood glucose and therefore high amounts of HbA_{1c}. The changes in the concentration of HbA_{1c} in diabetic patients can be used to follow the effectiveness of treatment for the diabetes.

Bunn, H. F. Evaluation of glycosylated hemoglobin in diabetic patients. *Diabetes* 30:613, 1980; and Brown, S. B., and Bowes, M. A. Glycosylated hemoglobins and their role in management of diabetes mellitus. *Biochem. Educ.* 13:2, 1985.

bonds. Such observations led to the hypothesis that a polypeptide sequence contains the information for spontaneous folding to its unique active conformation under correct solvent conditions and in the presence of prosthetic groups that may be a part of its structure. As described in the following, chaperone proteins may facilitate protein folding. Quaternary structures also assemble spontaneously, after the tertiary structure of the subunits has formed.

It may appear surprising that a polypeptide chain folds into a single unique conformation given all the possible rotational conformations available around single bonds in its primary structure. For example, the α-chain of hemoglobin contains 141 amino acids and there are at least four single bonds per amino acid residue around which free rotation can occur. If each bond about which free rotation occurs has two or more stable rotamer conformations accessible to it, then there are a minimum of 4^{141} or 5×10^{86} possible conformations for the α-chain polypeptide.

The conformation of a protein is the one of lowest Gibbs free energy accessible to its sequence within a physiological time frame. Thus folding is under both thermodynamic and kinetic control. While a detailed knowledge of *de novo* folding of a polypeptide has at present been proposed for only a few examples, the involvement of certain processes appears reasonable in the majority of folding reactions. There is evidence that folding is initiated by short-range noncovalent interactions between a side chain and its nearest neighbors that form secondary structures in small regions of the polypeptide. Particular side chains have a propensity to promote the formation of α-helices, β-strands, and sharp turns or bends (β-turns) in the polypeptide. Such regions of polypeptide, called **initiation sites**, spontaneously assume a secondary structure. The partially folded structures then interact with each other to form a **molten-globule** state. This is a condensed intermediate on the folding pathway that contains much of the secondary structure elements of the native conformation, but many incorrect tertiary structure interactions. Regions of secondary structure in the molten-globule state are highly mobile relative to one another, and the molten globule is in rapid equilibrium with the fully unfolded state. The correct medium- and long-range interactions between different initiation sites are found by rearrangements within the molten globule, and the low free energy, native conformation of the polypeptide chain is formed. This is followed by formation of disulfide bonds (cystines). The rate-determining step for folding and unfolding of the native conformation lies between the molten globule and the native structure. For some proteins there may exist two or more thermodynamically stable folded conformations of low Gibbs free energy. Clin. Corr. 3.6 contains a discussion of proteins folding and prion infectious agents.

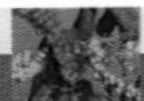

CLINICAL CORRELATION 3.6

Prion Diseases and Proteins as Infectious Agents: Human Transmissible Spongiform Encephalopathies (TSEs)

Prion proteins are able to act as infectious agents in the absence of DNA or RNA. Prion disease in humans can also appear spontaneously (sporadic disease) or due to the inheritance of a mutated prion protein gene. The disease is characterized by ataxia, dementia, and paralysis and is almost always fatal. Pathological examination of the brain shows amyloid plaques and spongiform (vacuolar) degeneration.

In prion disease, **PrP^{c}** refers to the highly soluble cellular conformation of the prion protein, and **PrP^{Sc}** to the insoluble toxic conformation originally isolated from sheep with scrapie (Sc) disease. The soluble PrP^{c} protein contains three α-helical and two small β-strand segments (Clin. Corr. Figure 1).

The conversion to the toxic PrP^{Sc} form is characterized by the conversion of an α-helical segment into a β-strand with the probable inducement of the addition of β-strands in the previously undefined N-terminal end of the polypeptide chain (Clin. Corr. Figure 1). The PrP^{Sc} conformation initiates other prion PcP^{c} molecules to convert to the PcP^{Sc} conformation and, through β-strand interactions between PrP^{Sc} molecules, polymerizes into amyloid fibers. Formation of the amyloid polymer is irreversible and in the brain causes neuronal cell death. In several other neurodegenerative diseases similar conformational equilibriums occur between a soluble predominantly α-helical protein conformation and a less soluble β-strand containing conformation of the protein, with the β-strand conformation proteins polymerizing into insoluble amyloid fibrils (Clin. Corr. Figure 2). The protein unit differs with the disease type, but the plaque formed all have a similar amyloid fibril structure (Clin. Corr. Figure 2). Prominent amyloid induced neuronal diseases include Alzheimer, Parkinson, Huntington, and amyotrophic lateral sclerosis (ALS, Lou Gehrig).

Clin. Corr. Figure 1. **Structure of Prion Forms** (*a*) Prp (PrP^{c}) fold structure. Approximately 120 amino acid residues at N-terminal end not observed. (*b*) Model of the toxic PrP (PrP^{Sc}) fold structure, in which an α-helical segment from the PrPc conformation forms a β-strand as well as β-strands now formed from loop region of PrPSc structure and the N-terminal of the segment of the polypeptide chain.
Figure from Web site of Dr. Fred Cohen, University of California, San Francisco.

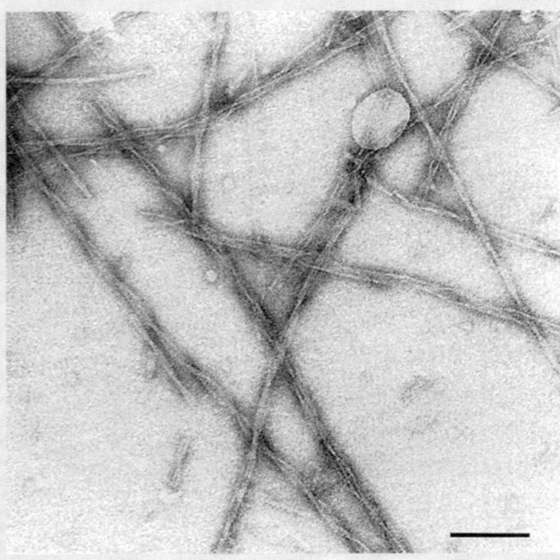

Clin. Corr. Figure 2. **Electron micrograph of amyloid fibril structure due β-strand polymerization of a SH3 domain from PI3 kinase.** Fibrils are similar to those observed in amyloid producing neuronal diseases. The scale bar is 100 nm.
Reprinted with permission from Guijarro, J. I., Sunde, M., Jones, J. A., Campbell, I. D., and Dobson, C. M. *Proc. Natl. Acad. Sci. USA* 95: 4224, 1998.

The prion protein is a glycoprotein containing covalently joined carbohydrate molecules and a phosphatidylinositol lipid. The lipid anchors the prion protein on the extracellular side of the plasma membrane. The PrP^{c} protein at the plasma membrane functions in conjugation with membrane receptor proteins in the transmission of extracellular signals to the cell.

Prion disease in humans is pleiotropic with the phenotype dependent on the cause of the disease (sporadic, inheritable, or infectious)

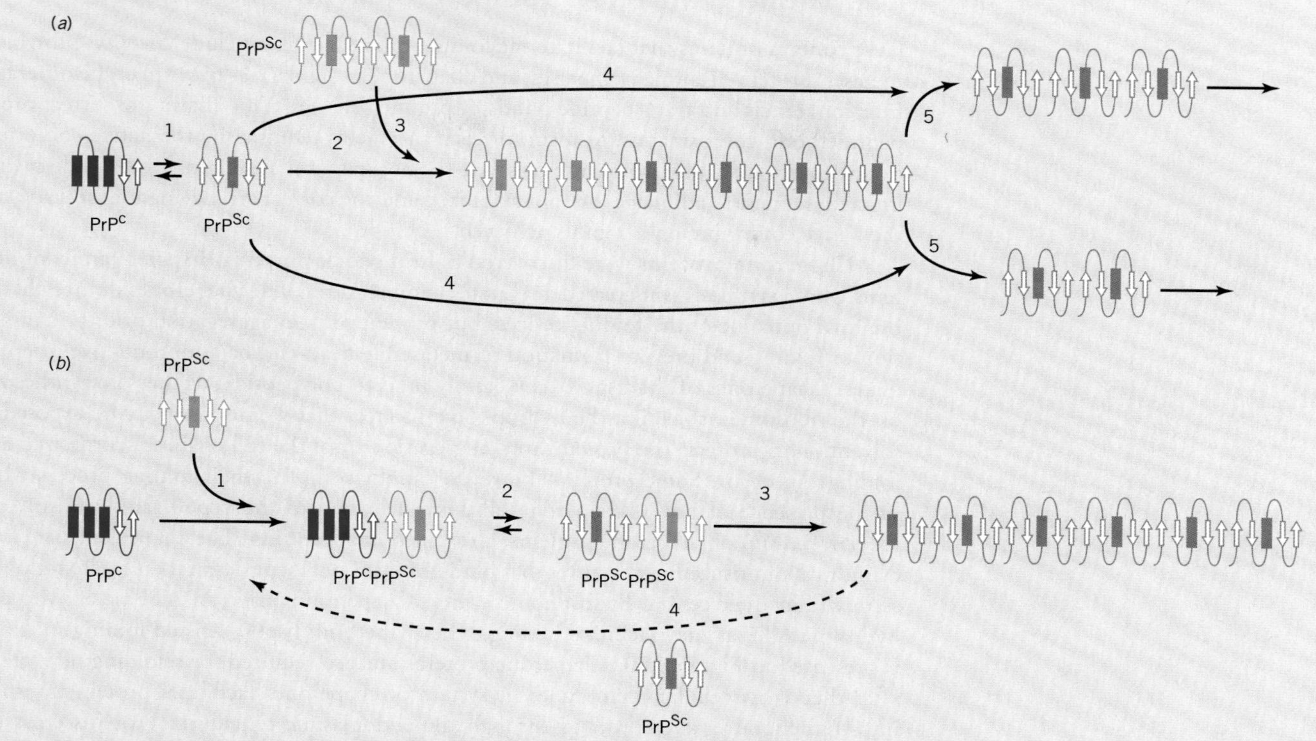

Clin. Corr. Figure 3. Mechanisms for amyloid fiber formation initiated by PrP^{Sc}.

and the type of polymorphisms presented in an individual's prion genes. There are multiple polymorphisms of the prion gene in the human population with over 50 reported mutations. Initiation of amyloid formation by a "seed" (initiator) PrP^{Sc} molecule can be through three different routes: (a) sporadic disease initiated by the spontaneous chance conformational transformation of a normal PrP^{c} molecule to the PrP^{Sc} conformation, (b) infective PrP^{Sc} proteins introduced through ingestion of PrP^{Sc} contaminated meat or the introduction of foreign PrP^{Sc} through other routes, and (c) inheritance of a mutated prion gene with ability of the gene product to more easily spontaneously fold into a PrP^{Sc} conformation. Inherited phenotypes include Gerstmann–Sträussler–Scheinker syndrome and fatal familial insomnia. Sporadic disease results in Creutzfeldt–Jakob Disease (CJD) and is typically fatal within 1 year of the onset of illness. The incidence of CJD is 1 case per 1 million per year worldwide, which translates into approximately 300 cases per year in the United States. Sporadic disease accounts for approximately 85%, inherited forms for 15%, and infectious disease less than 1% of the prion diseases.

Prion disease caused by extraneous infectious PrP^{Sc} protein is most interesting, as it shows the ability of a protein to act as an infectious agent in the absence of DNA or RNA. Humans can acquire prion disease through the ingestion of bovine PrP^{Sc} conformation prions in contaminated meat from cattle with bovine transmissible spongiform encephalitis (BSE; mad cow disease). The disease, variant Creutzfeldt–Jakob Disease (vCJD), has a different phenotype than the classic sporadic CJD. The median age of onset for vCJD is 28 years, and 68 years for sporadic CJD. The duration of the illness is approximately 14 months for vCJD and 5 months for sporadic CJD. The Centers for Disease Control (CDC) has identified a total number of 195 cases of vCJD as of August 2006, of which 162 were in the United Kingdom, 20 in France, and 2 in the United States. The incubation period from time of infection to onset of the disease is approximately 10 years. Infection by BSE appears to be age selective. As the peak of the United Kingdom outbreak occurred between 1994 and 1996 in patients with a median age of 28, the infection occurred around 1984–1986 at a median age of the infected individuals of 18.

A second significant infectious route is iatrogenic. Between 1985 and 1996 there were 136 vCJD cases reported worldwide in patients treated with human pituitary-derived growth hormone and 136 cases of patients receiving cadaveric *dura mater* transplants. A small number of iatrogenic CJDs have been caused with other procedures such as corneal transplants, blood transfusion, and treatment with human gonadotrophin.

Two mechanisms are suggested for amyloid plaque formation initiated by an infective prion (Clin. Corr Figure 3). In the Nucleation-Polymerization Mechanism (*a*), PrP^{c} (green) is in rapid equilibrium with the PrP^{Sc} conformation (brown) (step 1), but the polymerization into an amyloid fiber is slow (step 2) in the absence of an initiator molecule. The introduction of a PrP^{Sc} infective prion (orange) nucleates the polymerization process (step 3). The process is propagated by fragments from the newly generated polymer (step 5).

In the Template-Directed Mechanism (*b*), the prion, PrP^{Sc}, serves as a template to promote the change in conformation of endogenous PrP^{c} to PrP^{Sc} (steps 1 and 2). Once converted to the PrP^{Sc}, the PrP^{Sc} rapidly polymerize into insoluble fibrils.

Aguzzi, A., and Polymenidou, M. Mammalian prion biology: One century of evolving concepts. *Cell* 116:313, 2004; Collinge, J., and Clarke, A. R. A general model of prion strains and their pathogenicity. *Science* 318:930, 2007; and Soto, C., Estrada, L., and Castilla, J. Amyloids, prions and the inherent infectious nature of misfolded protein aggregates. *Trends Biochem. Sci.* 31:150, 2006.

Chaperone Proteins Assist the Protein Folding Process

Cells contain proteins that facilitate the folding process. These include *cis–trans*-prolyl isomerases, protein disulfide isomerases, and chaperone proteins. ***cis–trans*-prolyl isomerases** interconvert cis and trans peptide bonds of proline residues. This allows the correct prolyl peptide bond conformation to form for each proline as required by the native conformation. **Protein disulfide isomerases** catalyze the breakage and formation of disulfide cystine linkages so incorrect linkages are not stabilized and the correct arrangement of linkages for the native conformation is rapidly achieved.

Chaperone proteins were discovered as **heat shock proteins (hsps)**, a family of proteins whose synthesis is increased at elevated temperatures. The chaperones do not change the final outcome of the folding process but prevent protein aggregation prior to completion of folding and prevent formation of metastable dead-end or nonproductive intermediates. Chaperones of the hsp 70-kDa family bind to polypeptides as they are synthesized on the ribosomes, shielding the hydrophobic surfaces that would normally be exposed to solvent. This protects the protein from aggregation until the full chain is synthesized and folding can occur. Some proteins, however, cannot complete their folding process while in the presence of hsp70 chaperones and are delivered to the hsp60 family (GroEL in *Escherichia coli*) of chaperone proteins, also called **chaperonins**. The chaperonins are long cylindrical multisubunit structures that bind unfolded polypeptides in their molten-globule state within their central hydrophobic cavity. Chaperonins have an ATPase activity and hydrolyze ATP as they facilitate folding. The chaperonin hsp60-assisted folding in *E. coli* is presented in Figure 3.50. Chaperone proteins are also required for refolding of proteins after they cross cellular membranes. A system of chaperones facilitates protein transport into mitochondria and into and through the endoplasmic reticulum. Proteins cross the lipid bilayer of the mitochondrial and endoplasmic reticulum membranes in an unfolded conformation, and local chaperones are often required to facilitate their folding within the intracellular organelle.

Noncovalent Forces Lead to Protein Folding and Contribute to a Protein's Stability

Noncovalent forces cause a polypeptide to fold into a unique native conformation and then stabilize the native structure against denaturation. Noncovalent forces are weak bonding forces with strengths of 4–29 kJ/mol (1–7 kcal/mol). This may be compared to covalent bonds that have a bonding strength of at least 210 kJ/mol (50 kcal/mol) (Table 3.16). Even though individually weak, the large number of noncovalent contacts within a protein provides a large energy factor that promotes protein folding.

Hydrophobic Interaction Forces

Important noncovalent forces that cause a polypeptide to fold into its native conformation are the **hydrophobic interaction forces**. Their strength is not due to any intrinsic attraction between nonpolar groups but rather to the properties of water that surround the

TABLE 3.16 • Strength of Bonds Found in Protein Structures

	Bond Strength	
Bond Type	*(kJmol^{-1})*	*(kcal mol^{-1})*
Covalent	>210	>50
Noncovalent	2.5–29	0.6–7
Hydrophobic (i.e., two benzyl side chain groups of Phe)	8–13	2–3
Hydrogen	4–290	1–7
Ionic (low dielectric environment)	4–25	1–6
van der Waals	<4	<1

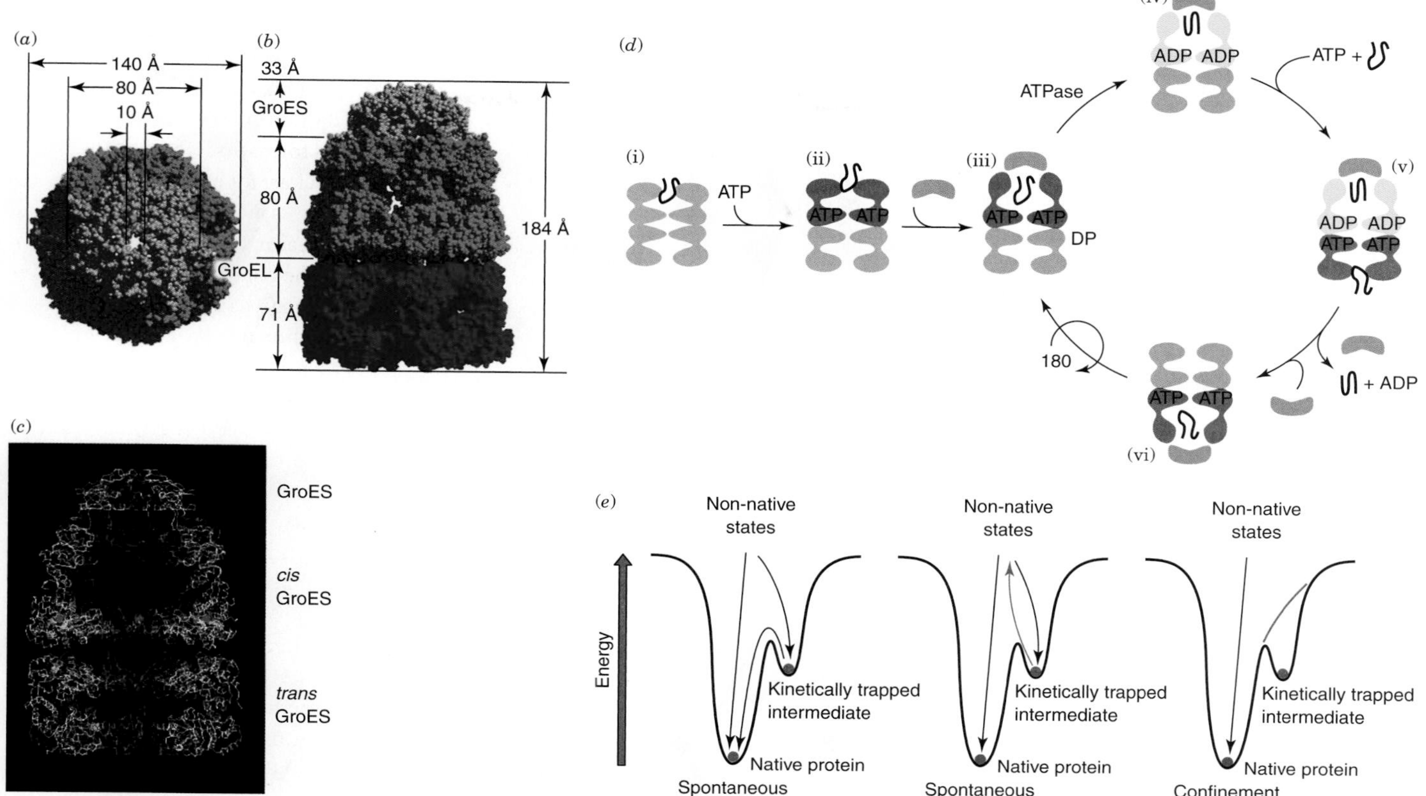

Figure 3.50 The GroEL-GroES Chaperonin system for protein folding. (*a*) Space-filling model of the entire chaperonin complex in a top view looking down from the GroES cap protein. The GroES is bound to the cis side of GroEL. GroES has 7 polypeptide subunits of 10 kD molecular weight and the GroEL cis and trans cylinder structures are each composed of 7 polypeptide subunits (total 14) each of 60 kD. (*b*) As (*a*) but in a side view, with the trans GroEL ring in red, the cis GroEL ring in green, and the GroES cap in gold. (*c*) Ribbons drawing of a slice through the center of the chaperonin complex showing internal cavities of the cis and trans cylindrical sections where denatured proteins enter and their folding occurs. (*d*) Simplified scheme for GroEL facilitated protein folding: (i) The denatured protein (black curved line) enters the GroEL and binds in the cavity with high affinity (black curved line), (ii) the ATP binds to one ring producing an altered conformation, (iii) the ATP-bound ring binds GroES to sequester the denatured protein in the folding chamber and a massive conformational change in the GroES-bound ring pulls open the denatured protein to untangle its nonproductive conformation and then releases the protein within the expanded cavity to fold spontaneously, (iv) the protein folds inside the chamber and ATP is hydrolyzed, (v) the ATP binds to the trans ring which releases GroES and the folded protein, and (vi) a new protein gets encapsulated. (*e*) Free-energy diagrams for protein folding. Left: Folding of protein in the absence of chaperonin results in a kinetically trapped intermediate. Middle: In chaperonin folding the pulling and release on chaperonin chamber expansion (annealing) untangles trapped intermediate and gives protein additional chances to fold without being trapped. Right: In addition the GroEL cavity smoothes the energy landscape to prevent the formation of certain kinetically trapped intermediates.
(*a*), (*b*), and (*c*) reproduced with permission from Xu, Z. H., Horwich, A. L., and Sigler, P. B. *Nature* 388: 741, 1997; (*d*) reproduced with permission from Saibil, H. R., and Ranson, N. A. *Trends Biochem.* Sci. 27: 627, 2002, and Ranson, N.A. et al. *Cell* 107, 869, 2001; (*e*) redrawn from Ulrich, F., Hartl, F., and Hayer-Hartl, M. *Science* 295: 1852, 2002. Copyright (2002) AAAS.

nonpolar groups. A nonpolar side chain or a region of a protein molecule dissolved in water induces a solvation shell of water in which water molecules are highly ordered. When two nonpolar side chains come together, the surface area exposed to solvent is reduced and some of the highly ordered water molecules in the solvation shell are released to bulk solvent. The entropy of the system (i.e., net disorder of the water molecules in the system) is increased. The increase in entropy is thermodynamically favorable and is the driving force causing nonpolar moieties to come together in aqueous solvent. A favorable free energy change of 8 kJ/mol (2 kcal/mol) for association of two phenylalanine side chains in water is due to this gain in entropy (Figure 3.51).

In transition from random to regular secondary conformation such as an α-helix or β-structure, approximately one-third of the water of solvation about the unfolded polypeptide is lost to bulk solvent. This approximates 2–4 kJ/mol (0.5–0.9 kcal/mol) for each peptide residue. Another one-third of the original solvation shell is lost when a polypeptide attains its native conformation. This brings different segments of folded polypeptide chain into close proximity with the release of water of solvation between these segments.

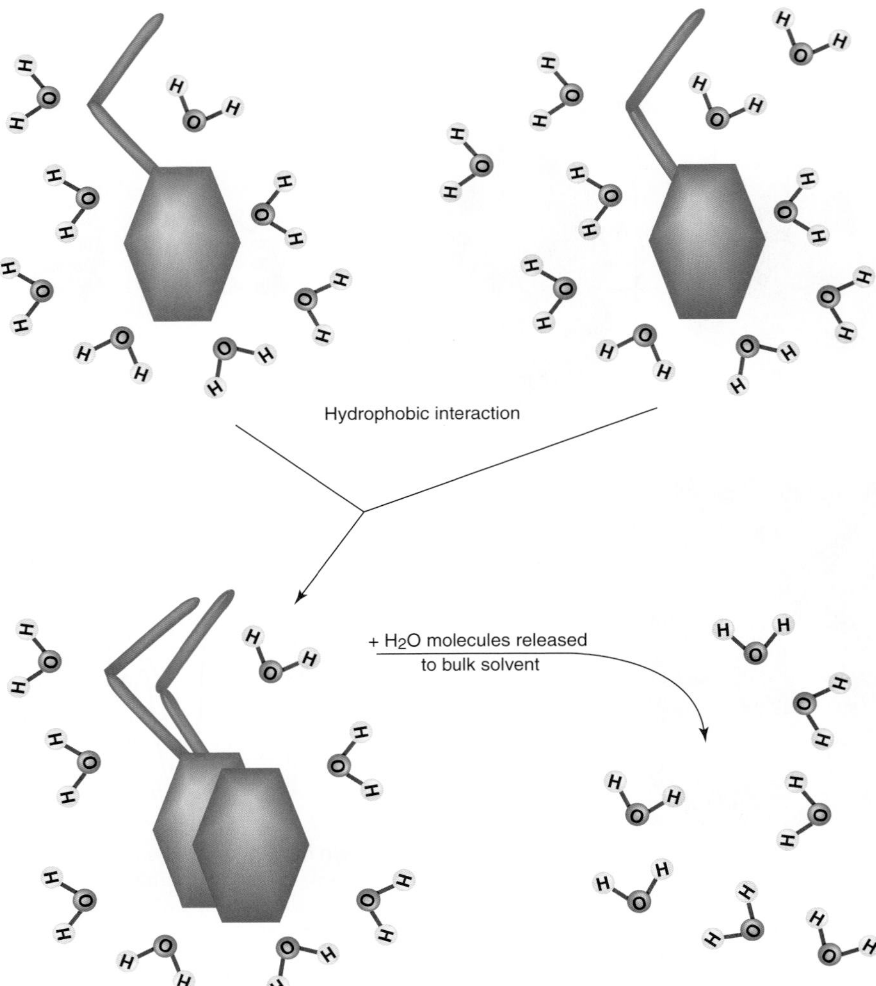

Figure 3.51 Formation of hydrophobic interaction between two phenylalanine side-chain groups.

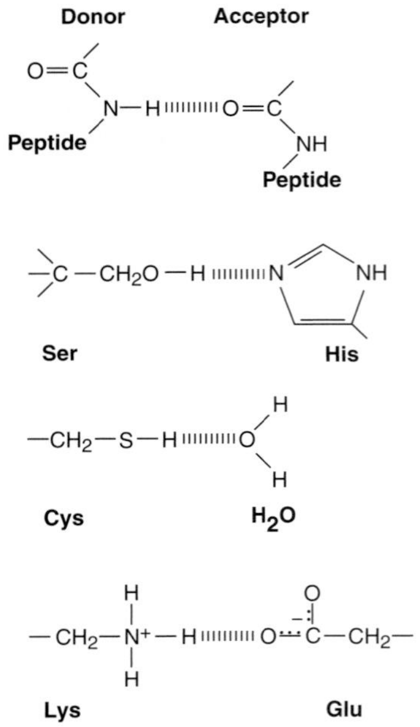

Donor Acceptor

O=C—N—H ········ O=C—NH

Peptide Peptide

—C—CH$_2$O—H ········ N (imidazole) NH

Ser His

—CH$_2$—S—H ········ O(H)(H)

Cys H_2O

—CH$_2$—N$^+$(H)(H)—H ········ O$\overset{-}{\cdots}$C(=O)—CH$_2$—

Lys Glu

Figure 3.52 Some common hydrogen bonds found in proteins.

Hydrogen Bonds

Hydrogen bonds are formed when a hydrogen atom covalently bonded to an electronegative atom is shared with a second electronegative atom. The atom to which the hydrogen atom is bonded is designated the hydrogen **donor atom**. The atom with which the hydrogen atom is shared is the hydrogen **acceptor atom**. Typical hydrogen bonds found in proteins are shown in Figure 3.52. α-Helical and β-strand conformations are extensively hydrogen bonded.

The strength of a hydrogen bond depends on the distance between donor and acceptor atoms. High bonding energies occur when this distance is between 2.7 and 3.1 Å. Of lesser importance, but not negligible, is the dependence of hydrogen-bond strength on geometry. Bonds of higher energy are geometrically collinear, with donor, hydrogen, and acceptor atoms lying in a straight line. The dielectric constant of the medium around the hydrogen bond may also be reflected in the bonding strength. Typical hydrogen bond strengths in proteins are 4–29 kJ/mol (1–7 kcal/mol). Although they contribute to thermodynamic stability of a protein's conformation, their formation may not be a major driving force for folding. This is because peptide bonds and other hydrogen-bonding groups form hydrogen bonds to the water solvent in the denatured state, and these bonds must be broken before the polypeptide can fold. In calculating the net contribution of hydrogen-bonding forces to the folding, the energy required to break the hydrogen bonds to water must be subtracted from that gained from the formation of new hydrogen bonds between atoms in the folded protein.

Electrostatic Interactions

Electrostatic interactions (**ionic** or **salt linkages**) between charged groups are important in the stabilization of protein structure and in binding of charged ligands and substrates to proteins. Electrostatic forces are repulsive or attractive depending on whether the interacting charges are of the same or opposite sign. The strength of an electrostatic force (E_{el}) is directly dependent on the charge (Z) of each ion and inversely dependent on the dielectric constant (D) of the solvent and the distance between the charges (r_{ab}) (Figure 3.53).

$$E_{el} \approx \frac{Z_A \cdot Z_B \cdot \varepsilon^2}{D \cdot r_{ab}}$$

Figure 3.53 Strength of electrostatic interactions.

Water has a high dielectric constant ($D = 80$), and charge interactions in water are relatively weak compared to those in the interior of a protein where the dielectric constant is low. However, most charged groups remain on the surface where they do not interact with other charged groups from the protein because of the high dielectric constant of water, but are stabilized by hydrogen bonding and polar interactions with water. These interactions generate the dominant forces that place most charged groups of a protein on the outside of the folded structures.

van der Waals Forces

van der Waals forces are the weakest of the noncovalent forces. They have an attractive term (A) inversely dependent on the sixth power of the distance between two interacting atoms (r_{ab}), and a repulsive term (B) inversely dependent on the 12th power of this distance (Figure 3.54). The A term contributes at its optimum distance an attractive force of less than 4 kJ/mol (1 kcal/mol) per atomic interaction due to the induction of complementary partial charges or dipoles in the electron density of adjacent atoms when the electron orbitals of the two atoms approach to a close distance. As the atoms come even closer, however, the repulsive component (term B) of the van der Waals force predominates as the electron orbitals of the adjacent atoms begin to overlap. The repulsive force is commonly called **steric hindrance**.

$$E_{VDW} = -\frac{A}{r_{ab}^6} + \frac{B}{r_{ab}^{12}}$$

Figure 3.54 Strength of van der Waals interactions.

The distance of maximum favorable interaction between two atoms is the **van der Waals contact distance**, which is the sum of van der Waals radii for the two atoms (Figure 3.55). The van der Waals radii for atoms found in proteins are given in Table 3.17.

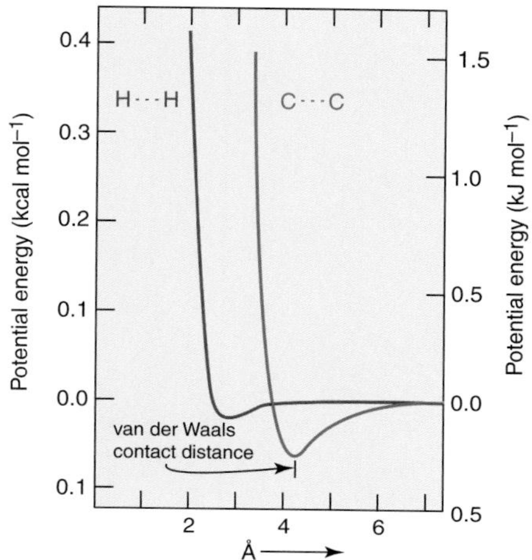

Figure 3.55 van der Waals–London dispersion interaction energies between two hydrogen atoms and two carbon (tetrahedral) atoms. Negative energies are favorable and positive energies unfavorable.
Redrawn from Fersht, A. *Enzyme Structure and Mechanism.* San Francisco: Freeman, 1977, 228.

The van der Waals repulsive forces between atoms of a peptide bond are weakest at the specific φ and ψ angles compatible with the α-helix and β-strand conformations. Thus, the absence of a van der Waals repulsive force is critical for secondary structure formation in proteins. In attaining the native conformation, the number of weak van der Waals interactions involved is in the thousands. The total contribution to the stability of a folded structure is substantial.

A special type of interaction (π-electron–π-electron) occurs when two aromatic rings approach each other with electrons of their aromatic rings interacting favorably (Figure 3.56). This interaction can result in attractive forces of up to 25 kJ/mol (6 kcal/mol).

TABLE 3.17 • Covalent Bond Radii and van der Waals Radii for Selected Atoms

Atom	*Covalent Radius (Å)*	*van der Waals Radius (Å)*[a]
Carbon (tetrahedral)	0.77	2.0
Carbon (aromatic)	0.69 along double bond	1.70
	0.73 along single bond	
Carbon (amide)	0.72 to amide N	1.50
	0.67 to oxygen	
	0.75 to chain C	
Hydrogen	0.33	1.0
Oxygen (—O—)	0.66	1.35
Oxygen (=O)	0.57	1.35
Nitrogen (amide)	0.60 ro amide C	1.45
	0.70 to hydrogen bond H	
	0.70 to chain C	
Sulfur, diagonal	1.04	1.70

Saurce: Fasman G. D. (Ed.). *CRC Handbook of Biochemistry and Molecular Biology*, 3rd ed., Sect. D, Vol. II. Boca Raton, FL CRC, Press, 1976, p. 221.

[a]The van der Waals contact distance is the sum of the rwo van der Waals radii for the two atoms in proximity.

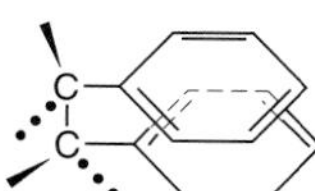

Figure 3.56 π-Electron–π-electron interactions between two aromatic rings.

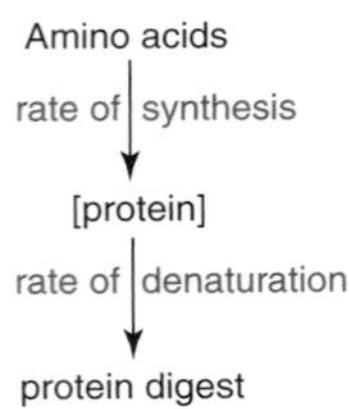

Figure 3.57 Steady-state concentration of a protein is due to its rates of synthesis and denaturation.

Denaturation Leads to Loss of Native Structure

Denaturation occurs when a protein loses its native secondary, tertiary, and/or quaternary structure. The peptide bonds are not broken by conformational denaturation. The **denatured state** is always correlated with loss of a protein's function. Loss of function is not necessarily synonymous with denaturation since small conformational changes can lead to loss of function without loss of the folded native structure. For example, a change in position of a single side chain in the active site of an enzyme or in protonation of a side chain may result in loss of enzyme activity but not to a loss of the native conformation.

Although conformational differences between denatured and native structures may be substantial, the free energy difference between such structures is typically as low as 20–40 kJ/mol (5–10 kcal/mol) (the energy of three or four noncovalent bonds). Thus, loss of a single hydrogen bond or electrostatic or hydrophobic interaction can lead to destabilization of a folded structure. Such a change in stability of a noncovalent bond can, in turn, be caused by a change in pH, ionic strength, or temperature. Presence of prosthetic groups, cofactors, and substrates also affects stability of its native conformation.

The statement that breaking of a single noncovalent bond can cause denaturation apparently conflicts with the observation that an amino acid sequence can often be extensively varied without loss of a protein's structure. The key to the resolution of this apparent conflict is the word "*essential*." Many noncovalent interactions are not essential for the structural stability of the native conformation of a protein. However, substitution or modification of an essential amino acid that provides a critical noncovalent interaction without a compensating stabilizing interaction dramatically affects the stability of a native protein conformation.

The cellular concentration of a protein is controlled by its rate of synthesis and of degradation (Figure 3.57). Under many circumstances the denaturation of a protein is the rate-controlling step in its degradation. Cellular enzymes and organelles that digest proteins "recognize" denatured proteins and eliminate them rapidly. In experimental situations, protein denaturation occurs on addition of urea, guanidine hydrochloride, or detergents (e.g., sodium dodecyl sulfate) that weaken hydrophobic bonding in proteins and stabilize the denatured state. Addition of a strong base, acid, or organic solvent or heating to temperatures above 60°C are also common ways to denature a protein.

3.8 • DYNAMIC ASPECTS OF PROTEIN STRUCTURE

Although high-resolution X-ray diffraction yields atomic coordinates for each atom in a protein, experimental evidence from NMR, fluorescence spectroscopy, and the temperature dependence of the X-ray diffraction reveals that the atoms in a protein molecule have a fluid-like dynamic motion and do not exist in one static position. Rather than an exact location, the atomic coordinates obtained by X-ray diffraction represent the time-averaged position for each atom. The time for position averaging is the duration of data collection, which may be up to several hours. Thus the active conformation may differ from the average conformation. An X-ray structure also shows small defects in packing of the folded structure, indicating the existence of holes in the structure that allow the protein space for flexibility. The concept that each atom in a protein is in constant motion, like molecules within a fluid, although constrained by its covalent bonds and the secondary and tertiary structure, is an important aspect of protein structure.

Calculations of theoretical **molecular dynamics** describe the changes in the coordinates of atoms in a protein structure and in the position of regions of the structure due to summation of the movements of atoms in that region. The computation is based on the solving of Newton's equations of motion simultaneously for all the atoms of the protein and the solvent that interacts with it. The energy functions used in the equation include representations of covalent and noncovalent bonding energies due to electrostatic forces, hydrogen bonding, and van der Waals forces. Individual atoms are randomly assigned a velocity from a theoretical distribution and Newton's equations are used to relax the system at a given temperature. The calculation is a computationally intensive activity, even when limited to less than several hundred picoseconds

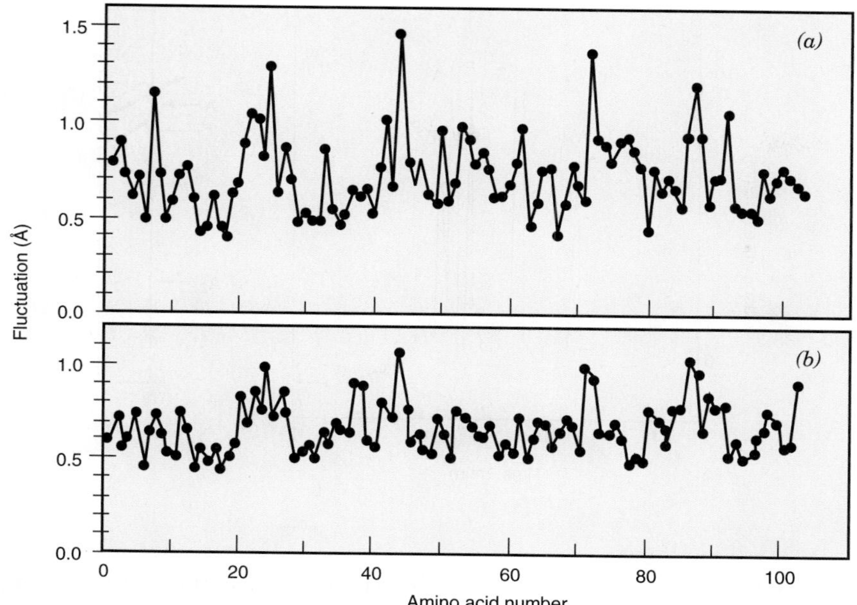

Figure 3.58 Fluctuation of structure of cytochrome c. (*a*) Calculated fluctuation on a picosecond time scale of α-carbons within each amino acid residue in folded structure of cytochrome c. (*b*) Experimentally observed fluctuation of each α-carbon of the amino acid residues determined from the temperature dependency of the X-ray diffraction pattern for the protein. Cytochrome-c has 103 amino acid residues. The *x*-axis plots amino acid residues in cytochrome-c from 1 to 103, and the *y*-axis the fluctuation distances in angstroms.
Redrawn from Karplus, M., and McCammon, J. A. *Annu. Rev. Biochem.* 53: 263, 1983.

(1 ps $= 10^{-12}$ s) of protein dynamic time. Such calculations indicate that the average atom in a protein oscillates over a distance of 0.7 Å on the picosecond scale. Some atoms or groups of atoms move smaller or larger distances than this calculated average (Figure 3.58).

Net movement of any segment of a polypeptide over time represents the sum of forces due to rapid atomic oscillations, and the local jiggling and elastic movements of covalently attached groups of atoms. These movements within the closely packed interior of a protein frequently allow the buried planar aromatic rings of tyrosines to flip. The small amplitude fluctuations provide the "lubricant" for large motions in proteins such as domain motions and quaternary structure changes, like those observed in hemoglobin on O_2 binding (p. 358). The dynamic behavior of proteins is the basis for conformational changes induced by substrate, inhibitor, or drug on binding to an enzyme and receptor, generation of allosteric effects in hemoglobins, electron transfer in cytochromes and in the formation of supramolecular assemblies such as viruses. The movements may also have a functional role in the catalytic action of enzymes.

3.9 • CHARACTERIZATION, PURIFICATION, AND DETERMINATION OF PROTEIN STRUCTURE AND ORGANIZATION

Separation of Proteins on Basis of Charge

In **electrophoresis**, protein dissolved in a buffer solution at a particular pH is placed in an electric field. Depending on the relationship of the buffer pH to the pI of the protein, the protein moves toward the cathode (−) or the anode (+) or remains stationary (pH = pI). Supports such as polymer gels (e.g., polyacrylamide, agarose, and cellulose acetate) are used. The inert supports are saturated with buffer solution, a protein sample is placed on the support as a line or spot, an electric field is applied across the support, and the charged proteins migrate as zones in the support toward the oppositely charged pole.

A technique of extremely high resolution is **isoelectric focusing**, in which mixtures of polyamino-polycarboxylic acid ampholytes with a defined range of pI values are used to establish a pH gradient across the applied electric field. A charged protein migrates through the pH gradient in the electric field until it reaches a pH region in the gradient equal to its pI value. At this point the protein becomes stationary and may be visualized (Figure 3.59). Proteins that differ by as little as 0.0025 in their pI values are separated on the appropriate pH gradient.

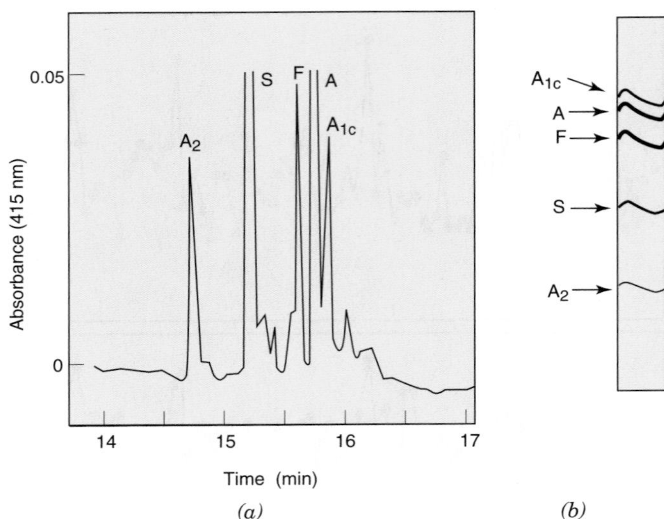

Figure 3.59 Isoelectric focusing of hemoglobins from a patient heterozygous for HbS and β-thalassemia. The figure shows the separation by isoelectric focusing of HbA_{1c} (HbA glycosylated on NH_2-end, see Clin. Corr. 3.5), normal adult HbA, fetal HbF, sickle-cell HbS (see Clin. Corr. 6.2; p. 214), and minor adult HbA_2. (*a*) Isoelectric focusing carried out by capillary electrophoresis with ampholyte pH range between 6.7 and 7.7 and detection of bands at 415 nm. (*b*) Isoelectric focusing carried out on gel with Pharmacia Phast System; ampholyte pH range is between 6.7 and 7.7.
Redrawn from Molteni, S., Frischknecht, H., and Thormann, W. *Electrophoresis* 15: 22, 1994.

Ion-exchange column chromatography is used for the preparative separation of proteins by charge. Ion-exchange resins consist of insoluble materials (agarose, polyacrylamide, cellulose, and glass) that contain charged groups (Figure 3.60). Negatively charged resins bind positively charged molecules and are **cation-exchange resins.** Positively charged resins bind anions strongly and are **anion-exchange resins.** The degree of retardation of a protein (or an amino acid) by a resin depends on the magnitude of the charge on the protein at the particular pH of the experiment. Molecules with the same charge as the resin are eluted first in a single band, followed by those with an opposite charge to that of the resin, in an order based on the protein's charge density (Figure 3.61). When it is difficult to remove a molecule from the resin because of the strength of the attractive interaction between the bound molecule and resin, systematic changes in pH or in ionic strength are used to weaken the interaction. For example, an increasing pH gradient through a cation-exchange resin reduces the difference between the solution pH and the pI of the bound protein. This decrease between pH and pI reduces the magnitude of the net charge on the protein and decreases the strength of the charge interaction between the protein and the resin. An increasing gradient of ionic strength also decreases the charge interactions, as the ions compete with the proteins for binding, and elutes tightly bound electrolytes from the resin.

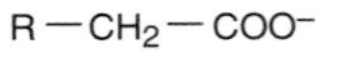

$R-CH_2-COO^-$

Negatively-charged ligand: carboxymethyl

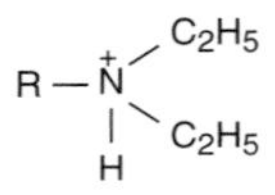

$R-\overset{+}{N}H(C_2H_5)_2$

Positively-charged ligand: diethylamino

Figure 3.60 Two examples of charged ligands used in ion-exchange chromatography.

Capillary Electrophoresis

Electrophoresis within a fused silica capillary tube has a high separation efficiency, utilizes very small samples, and requires only several minutes for an assay. A long capillary tube is filled with the electrophoretic medium, a sample is injected in a narrow band near the anode end of the tube, and the molecules of the sample are separated by their mobility toward the negatively charged pole. The fused silica wall of the capillary has a negative charge and an immobile cationic layer is fixed to it. An adjacent diffuse layer of cations moves toward the cathode in the applied electric field and causes a flow of solvent toward the cathode. This electro-osmotic flow creates a "current" that carries analyte molecules toward the cathode, irrespective of the charge of the analyte (Figure 3.62). Molecules with a high positive charge-to-mass ratio move with the current and have the highest mobility, followed by neutral molecules. Anionic molecules are repelled by the cathode and will move against the electro-osmotic flow. However, the electro-osmotic current toward the cathode overcomes any negative migration, and anions also migrate toward the cathode but at a slower rate than the cationic or neutral molecules.

In zone electrophoresis separations are made in the presence of a single buffer. Capillary electrophoresis may be performed in the presence of ampholytes to separate proteins by isoelectric focusing, in the presence of a porous gel to separate proteins by molecular weight or in the presence of a micellar component to separate by hydrophobicity. Detectors that utilize UV light, fluorescence, Raman spectroscopy, electrochemical detection, or mass spectroscopy make the capillary method sensitive and versatile.

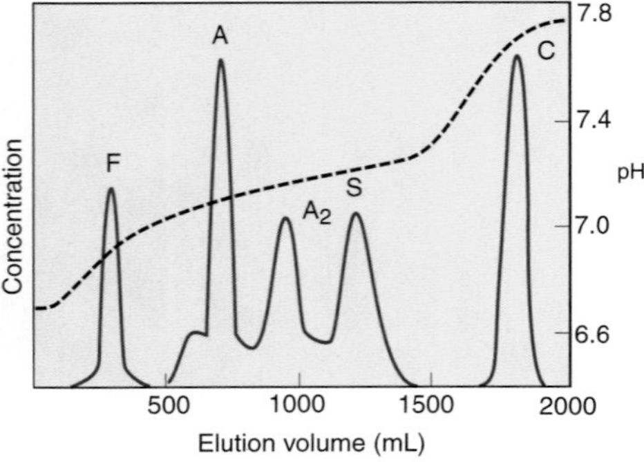

Figure 3.61 Example of ion-exchange chromatography. Elution diagram of an artificial mixture of hemoglobins F, A_1, A_2, S, and C on carboxymethyl-Sephadex C-50.
Redrawn from Dozy, A. M., and Huisman, T. H. J. *J. Chromatog.* 40: 62, 1969.

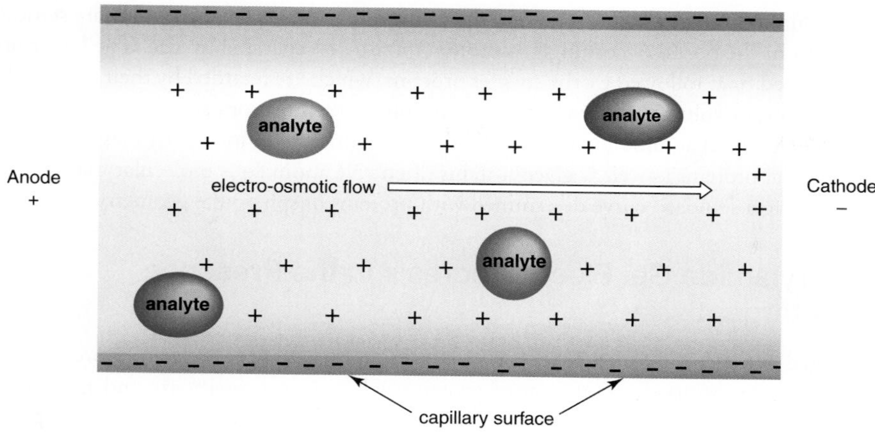

Figure 3.62 Generation of electro-osmotic flow toward the cathode in capillary electrophoresis.

Separation of Proteins Based on Molecular Mass or Size

Ultracentrifugation: Definition of Svedberg Coefficient

A protein subjected to centrifugal force moves in the direction of the force at a velocity dependent on its mass. The rate of movement is measured with an appropriate optical detection system, and from the rate, the sedimentation coefficient is calculated in **Svedberg units** (units of 10^{-13}s). In the equation (Figure 3.63), v is the measured velocity of protein movement, ω the angular velocity of the centrifuge rotor, and r the distance from the center of the tube in which the protein is placed to the center of rotation. Sedimentation coefficients between 1 and 200 Svedberg units (S) have been found for proteins (Table 3.18). Equations have been derived to relate the sedimentation coefficient to the molecular mass of a protein. One of the more simple equations is shown in Figure 3.64, in which R is the gas constant, T the temperature, s the sedimentation coefficient, D the diffusion coefficient of the protein, v the partial specific volume of the protein, and p the density of the solvent. The quantities D and v must be measured in independent experiments. The equation assumes a spheroidal geometry for the protein. Because this assumption may not be true and independent measurements of D and v are difficult, usually only the sedimentation coefficient for a molecule is reported. A protein's sedimentation coefficient is a qualitative measurement of its molecular mass.

$$s = \frac{v}{\omega^2 r}$$

Figure 3.63 Equation for calculation of the Svedberg coefficient.

$$\text{Molecular weight} = \frac{RTs}{D(1 - \bar{v}\rho)}$$

Figure 3.64 Equation relating the Svedberg coefficient to molecular weight.

Molecular Exclusion Chromatography

A porous gel in the form of small insoluble beads is commonly used to separate proteins by size in column chromatography. Small proteins penetrate the pores of the gel and have a larger

TABLE 3.18 • Svedberg Coefficients for Some Plasma Proteins

Protein	$s_{20,} \times 10^{-13}$ *(cms^{-1} dyn^{-1})*[a]	*Molecular Weight*
Lysozyme	2.19	15,000–16,000
Albumin	4.61	69,000
Immunoglobulin G	6.6–7.2	153,000
Fibrinogen	7.63	341,000
C1q (factor of complement)	11.1	410,000
α_2-Macroglobulin	19.6	820,000
Immunoglobulin M	18–20	1,000,000
Factor VIII of blood coagulation	23.7	1,120,000

Source: Fasman, G. D. (Ed.). *CRC Handbook of Biochemistry and Molecular Biology*, 3rd ed., Sect. A, Vol. II. Boca Raton, FL: CRC Press, 1976, p. 242.

[a] s_{20}, $\times 10^{-13}$, is sedimentation coefficient in Svedberg units, referred to water at 20°C, and extrapolated to zero concentration of protein.

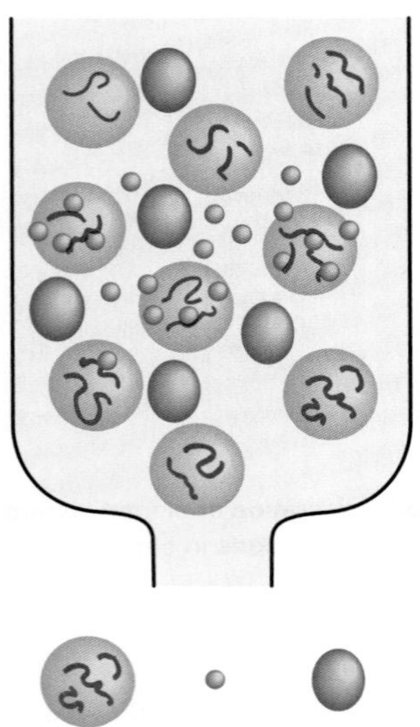

Figure 3.65 Molecular exclusion chromatography. A small protein can enter the porous gel particles and will be retarded on the column with respect to a larger protein that cannot enter the porous gel particles.

solvent volume through which to travel in the column than large proteins, which are sterically excluded from the pores. Accordingly, a protein mixture is separated by size. The larger proteins are eluted first, followed by the smaller proteins, which are retarded by their accessibility to a larger solvent volume (Figure 3.65). As with ultracentrifugation, an assumption is made as to the geometry of an unknown protein in the determination of molecular mass. Elongated nonspheroidal proteins as well as glycoproteins often give anomalous molecular masses when analyzed using a standard curve determined with proteins of spheroidal geometry.

Polyacrylamide Gel Electrophoresis in the Presence of a Detergent

If a charged detergent is added to a protein electrophoresis assay and electrophoresis occurs through a sieving support, separation of proteins is based on protein size and not charge. A common detergent is **sodium dodecyl sulfate (SDS)** and a common sieving support is **cross-linked polyacrylamide**. The dodecyl sulfates are amphiphilic C12 alkyl sulfates, which stabilize a denatured protein by forming a charged micellar SDS solvation shell around its polypeptide chain. The inherent charge of the polypeptide chain is obliterated by the negatively charged SDS molecules, and each protein-SDS solubilized aggregate has an identical charge per unit volume. Negatively charged particles move through the polyacrylamide gel toward the anode. Polyacrylamide acts as a molecular sieve and the protein–micelle complexes are separated by size; proteins of larger mass are retarded. A single band in an SDS polyacrylamide electrophoresis is often taken to demonstrate a pure protein. The conformation of the native structure is not a factor in the calculation of molecular mass, as molecular mass is determined by comparison to known standards that are similarly denatured. The detergent dissociates the quaternary structure of a multimeric protein and releases the constituent subunits. Only the molecular mass of the subunits of such a protein are determined by this method.

HPLC Techniques Separate Amino Acids, Peptides, and Proteins

In high-performance liquid chromatography (HPLC), a liquid solvent containing a mixture of molecules to be identified is passed through a column densely packed with a small-diameter insoluble beadlike resin. In column chromatography, the smaller and more tightly packed the resin beads, the greater the resolution of the separation technique. In HPLC, the resin is so tightly packed that the liquid must be pumped through the column at high pressure. Therefore HPLC uses precise high-pressure pumps with metal plumbing and columns rather than the glass and plastics used in gravity chromatography. Resin beads are coated with charged groups to separate compounds by ion exchange or with hydrophobic groups to retard hydrophobic nonpolar molecules. In hydrophobic chromatography, tightly associated nonpolar compounds are eluted from the hydrophobic beads in aqueous solvents containing various percentages of an organic reagent. The higher the percentage of organic solvent in the effluent, the faster the nonpolar component is eluted from the hydrophobic resin. This latter type of chromatography over nonpolar resin beads is called **reverse-phase HPLC** (Figure 3.66). HPLC separations have extremely high resolution and reproducibility.

Affinity Chromatography

Proteins have a high affinity for their substrates, prosthetic groups, membrane receptors, or specific noncovalent inhibitors and for antibodies made against them. These high-affinity compounds can be covalently attached to an insoluble resin, and the modified resin used to purify its conjugate protein in column chromatography. In a mixture of proteins eluted through the resin, the one of interest is selectively retarded.

General Approach to Protein Purification

A protein must be purified prior to determination of its chemical composition, structure, and function. As living cells contain thousands of genetically distinct proteins as well as lipids, polysaccharides, and polynucleotides, the purification of a specific protein from the other cellular constituents may be difficult. The first task is the development of a simple

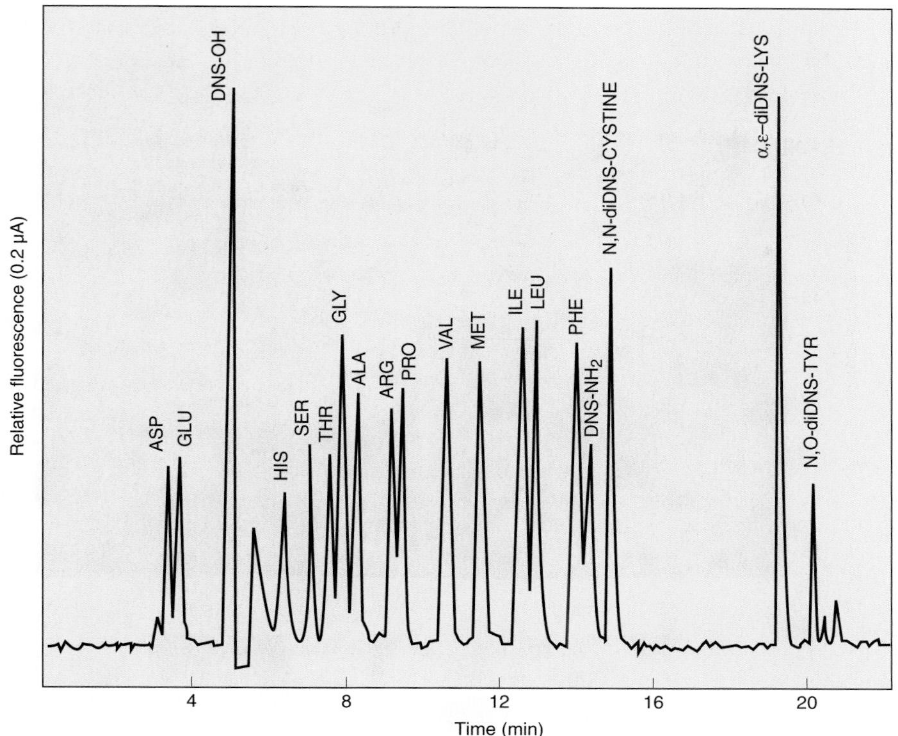

Figure 3.66 Separation of amino acids utilizing reverse-phase HPLC. The *x*-axis is time of elution from column. Amino acids are derivatized by reaction with dansyl chloride (DNS) so that they emit a fluorescence that is used to assay them as they are eluted from the column.
Reprinted with permission from Hunkapiller, M. W., Strickler, J. E., and Wilson, K. J. *Science* 226: 304, 1984. Reprinted with permission from AAAS.

assay for the protein. Whether it utilizes the rate of transformation of substrate to product, antibody–antigen reaction, or a physiological response in an animal assay system, a protein assay must give a quantitative measure of activity per unit of protein concentration. This quantity is known as the sample's **specific activity**. The aim of a purification procedure is to increase a sample's specific activity to the value expected for the pure protein. A typical protocol for purification of a soluble cellular protein involves disruption of the cell membranes, followed by differential centrifugation in a density gradient to isolate the protein from subcellular particles and high molecular weight aggregates. Further purification may utilize selective precipitation by inorganic salts such as ammonium sulfate (salting out) or by organic solvent. Final purification includes a combination of techniques that separate based on molecular charge, molecular size, and/or affinity.

Proteomic Techniques Determine All the Proteins Expressed in a Cell or Tissue in a Single Assay

The number of unique genes in the human genome is estimated to be about 25,000. Given that multiple protein products may be produced from a single gene by alternate splicing and posttranslational modifications generate additional forms, the number of unique proteins present in the human probably approaches 1 million. **Proteomics** is the science of determining exactly which proteins are produced in a cell or tissue, under a specific set of conditions.

Any cell or tissue may express thousands of different proteins simultaneously. To understand the properties of a cell or tissue, the type of proteins expressed and how the pattern of expressed proteins changes with development, differentiation, and disease is determined. Techniques have been developed to assay expressed mRNAs in cells and tissues in a single assay by hybridizing to DNA microarrays. More recently, techniques have been developed to analyze the active gene products, the proteins expressed by cells and tissues by use of **2-D electrophoresis.** In this, the proteins are first extracted from cells or tissues and then spotted onto a polyacrylamide gel in an electrophoresis apparatus. The proteins are separated in the first direction on the basis of their differences in pI. The gel is then turned by 90° and sodium dodecyl sulfate (SDS) added to buffer. The proteins are separated in the second

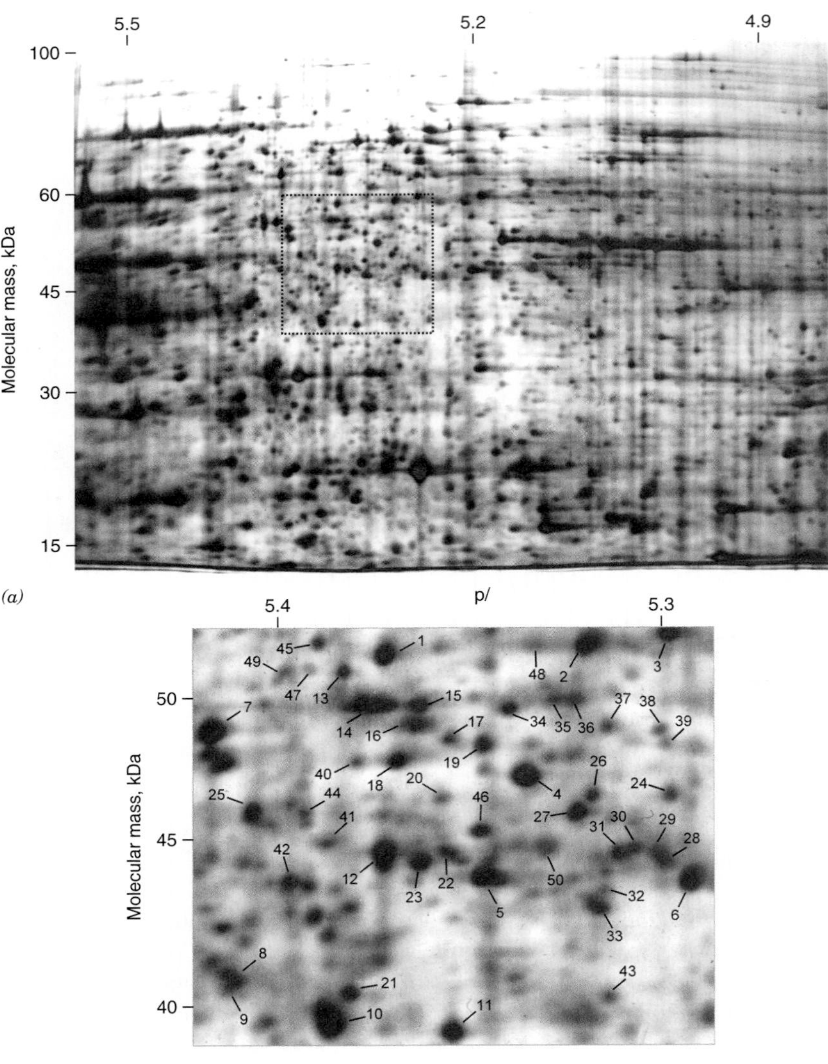

Figure 3.67 A two-dimensional (2-D) display of expressed proteins from cultured cells. (*a*) Soluble proteins from cellular extract (500 μg) loaded on the gel and separated by isoelectric focusing (between pH 4.9 and 5.5) in the horizontal direction and by molecular mass in the vertical direction (electrophoresed in the presence of SDS detergent). More than 1500 proteins are observed in the gel by silver staining. (*b*) A region from the gel expanded to show detail. Numbered proteins were analyzed by protease hydrolysis and mass spectrometry to determine their partial amino acid sequences, leading to their identification. Reproduced with permission from Gygi, S. P., Corthals, G. L., Zhang, Y., Rochon, Y., and Aebersold, R. *Proc. Natl. Acad. Sci. USA* 97: 9390, 2000.

direction on the basis off differences in their molecular mass (see Gel Electrophoresis in the Presence of Detergent). The resulting gel is stained for protein, and the intensity of each of the thousands of protein spots measured in order to determine whether a particular protein is expressed and its concentration (Figure 3.67).

Determining the identity of each of the protein spots in a 2-D gel is not a trivial task. The 2-D gel pattern, if carried out under standard conditions, may be compared to patterns obtained by reference laboratories that have determined the identity of the majority of spots in the 2-D pattern from a particular cell type. These reference 2-D patterns are available over the Internet. More definitively, a spot may be extracted from the gel and the protein partially hydrolyzed into smaller peptide fragments by proteolytic enzyme digestion (e.g., trypsin or chymotrypsin), and the peptide fragments subjected to mass spectroscopy. The mass spectroscopy rapidly determines the amino acid sequence of many of the small fragments. This technique is called peptide **mass fingerprinting**. Utilizing these sequences to search protein sequence or gene sequence databases leads to the identification of the protein extracted from the 2-D gel. Robotic instruments now perform each of the steps in protein spot extraction and identification. In this way, the thousands of expressed proteins may be identified.

The technique currently fails to identify low abundance proteins in cells or tissues. In addition, certain types of proteins are difficult to analyze due to low solubility, low molecular charge, or very low molecular mass. For example, integral membrane proteins are highly hydrophobic and are not soluble in the standard isoelectric focusing solvents.

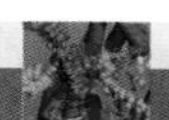

CLINICAL CORRELATION 3.7

Use of Amino Acid Analysis in Diagnosis of Disease

Elevated concentrations of amino acids are found in plasma or urine in a number of clinical disorders. An abnormally high concentration in urine is called an aminoaciduria.

Aminoacidurias in Humans

Amino Acid(s) Elevated	*Aminoaciduria*	*Clinical Correlation*
Phe	Phenylketonuria	19.7, p. 769
Cys, Lys, Arg, ornithine	Cystinuria	19.11, p. 774
Lys, Arg, ornithine	Lysinuric protein intolerance	19.15, p. 778
Neutral amino acids (mono-carboxylic mono-amino) Deficiency in Trp	Hartnup disorder	25.5, p. 1046
Pro, hydroxyproline, Gly	Iminoglycinuria	19.5, p. 766
Asp, Glu	Dicarboxylic aminoaciduria	

Bröer, S. Amino acid transport across mammalian intestinal and renal epithelia. *Physiology Rev.* 88:249, 2008; and Camargo, S. M., Bockenhauer, D., and Kleta, R. Aminoacidurias: Clinical and molecular aspects. *Kidney International* 73:918, 2008.

Determination of Amino Acid Composition of a Protein

Determination of the amino acid composition is an essential component in the study of a protein's structure and physiological properties. A protein is hydrolyzed to its constituent amino acids by heating it at 110°C in 6 N HCl for 18–36 h, in a sealed tube under a vacuum to prevent degradation of oxidation-sensitive side chains by oxygen in air. Tryptophan is destroyed in this method, and alternative procedures are used for its analysis. Side chain amides of asparagine and glutamine are hydrolyzed to aspartate and glutamate and free ammonia; they are included within the glutamic acid and aspartic acid content in the analysis.

Common procedures for amino acid identification use cation-exchange chromatography or reverse-phase HPLC to separate them and then react with ninhydrin, fluorescamine, dansyl chloride, or similar chromophoric or fluorophoric reagents for quantitation. With some types of derivatization, amino acids are identified at concentrations of 0.5×10^{-12} mol (pmol). Analysis of the amino acid composition of physiological fluids (i.e., blood and urine) is utilized in diagnosis of disease (Clin. Corr. 3.7).

Determination of Amino Acid Sequence

The ability to clone genes has led to the ability to determine the amino acid sequence of a protein from its DNA or mRNA sequence. This is often a much faster method for obtaining an amino acid sequence. Sequencing of a protein, however, is required for the determination of modifications to the protein structure that occur after its biosynthesis, to identify a part of the protein sequence in order that its gene can be cloned, and to identify a protein as the product of a particular gene. Determination of the primary structure of a protein requires a purified protein and determination of the number of chains in it. Individual chains are purified by the same techniques used in purification of the whole protein. If disulfide bonds join the chains, these bonds have to be broken (Figure 3.68).

Polypeptides are most commonly sequenced by the **Edman reaction** or by **mass spectroscopy**. In the Edman reaction, the polypeptide chain is reacted with phenylisothiocyanate, which reacts with the NH_2-terminal amino group. Acidic conditions catalyze intramolecular cyclization that cleaves the NH_2-terminal amino acid as a phenylthiohydantoin derivative (Figure 3.69). This amino acid derivative may be separated chromatographically and identified against standards. The remaining polypeptide is isolated, and the Edman reaction is repeated to identify the next NH_2-terminal amino acid. Theoretically, this can be repeated

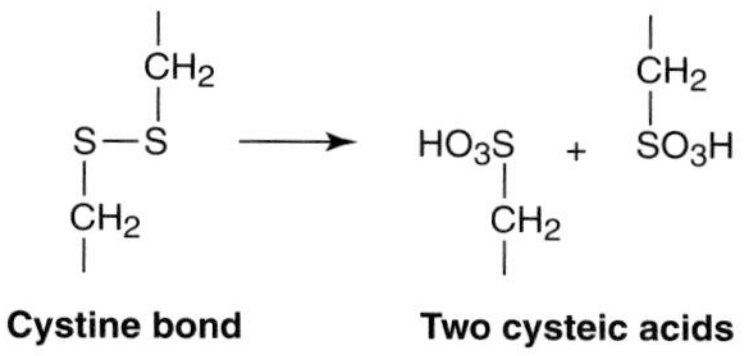

CH_2–S–S–CH_2 ⟶ HO_3S + SO_3H (each attached to CH_2)

Cystine bond — Two cysteic acids

Figure 3.68 **Breaking of disulfide bonds by oxidation to cysteic acids.**

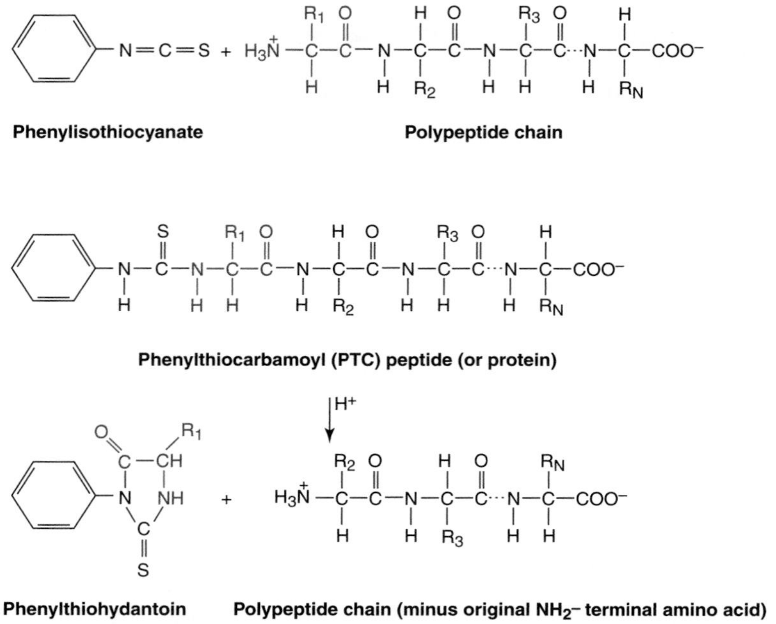

Figure 3.69 Edman reaction.

until the sequence of the entire polypeptide is determined but under favorable conditions can only be carried out for 30 or 40 amino acids into the polypeptide chain, when impurities generated from incomplete reactions in the reaction series make further **Edman cycles** infeasible. Polypeptides longer than 30 or 40 amino acids are hydrolyzed into smaller fragments and sequenced in sections. For sequencing by mass spectroscopy, it is also necessary to break long polypeptide chains into smaller fragments.

Enzymatic and chemical methods are used to break polypeptide chains into smaller fragments (Figure 3.70). **Trypsin** preferentially cleaves the peptide bond on the COOH-terminal side of lysine and arginine within polypeptide chains. **Chymotrypsin** cleaves the peptide bond on the COOH-terminal side of large apolar side chains. Other enzymes cleave polypeptide chains on the COOH-terminal side of glutamic and aspartic acid. **Cyanogen bromide** cleaves peptide bonds on the COOH-terminal side of methionine residues (Figure 3.70). After **partial hydrolysis**, the segments are separated, and the sequence of each is determined by the Edman reaction or mass spectroscopy. To place the sequenced peptides correctly into the complete sequence, another sample of the original polypeptide is subjected to partial hydrolysis by a different hydrolytic reagent from that used initially, and the fragments separated and sequenced. This generates overlapping sequences, which permit the determination of the complete sequence (Figure 3.71).

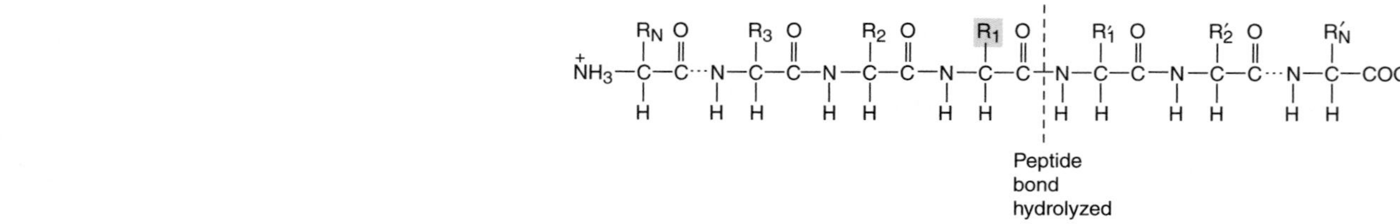

R_1	Reagent
Phe, Tyr, or Trp	Chymotrypsin
Arg, Lys	Trypsin
Met	Cyanogen bromide
Trp	*o*-Iodosobenzoic acid
Glu	*Staphylococcus aureus* endoprotease V8

Figure 3.70 Specificity of some polypeptide cleaving reagents.

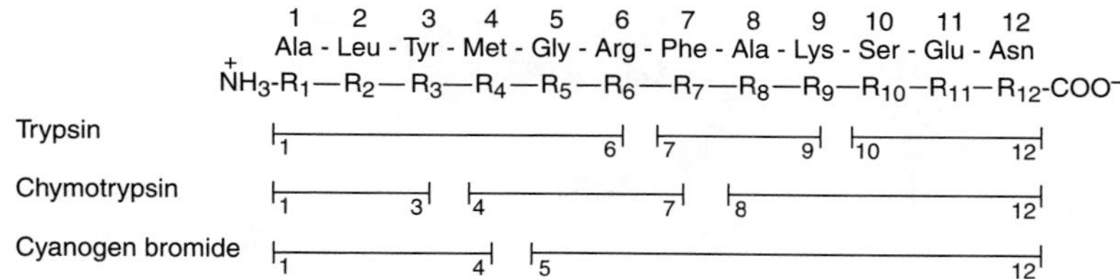

Figure 3.71 Ordering of peptide fragments from overlapping sequences produced by specific proteolysis of a peptide.

X-Ray Diffraction Techniques Are Used to Determine the Three-Dimensional Structure of Proteins

X-ray diffraction enables determination of the structure of proteins at near atomic resolution. The approach requires formation of a protein crystal, which contains solvent and is thus a concentrated solution, for use as the target. Our understanding of the detailed components of protein structure derived from this crystalline state correlate well with other physical measurements of protein structure in solution (e.g., by NMR spectroscopy; p. 133).

The generation of protein crystals can be the most time-consuming aspect of the process. Proteins have molecular dimensions at least an order of magnitude greater than a small molecule, and the packing of large protein molecules into the crystal lattice generates a crystal with large "holes" or solvent channels. A protein crystal typically contains 40% to 60% solvent, and it may be considered a concentrated solution rather than the hard crystalline solid obtained with most small molecules. The presence of solvent and unoccupied volume in the crystal permits the infusion of inhibitors and substrates into the protein molecules in the crystalline state.

A **dynamic flexibility** within regions of the protein structure may be seen as disorder in the X-ray diffraction pattern. Disorder describes the situation in which the observed electron density can be fitted by more than a single local conformation. Two explanations must be distinguished. The first involves the presence of two or more static molecular conformations, which are present in a stoichiometric relationship. The second involves the actual dynamic range of motion exhibited by atoms or groups of atoms in localized regions of the molecule. These explanations can be distinguished by lowering the temperature of the crystal to where dynamic disorder is "frozen out;" in contrast, the multiple static conformations are not temperature dependent and persist. Analysis of dynamic disorder by its temperature dependency using X-ray diffraction is an important method for studying protein dynamics (Section 3.8).

Crystallization techniques have so advanced that crystals are obtainable even from less abundant proteins. Interesting structures have been reported for proteins in which specific residues have been substituted, of antibody–antigen complexes, and of viral products such as the protease required for the infection of the human immunodeficiency virus (HIV) that causes acquired immunodeficiency syndrome (AIDS). Approximately 50,000 structures have been solved by X-ray diffraction and the details are stored in the **Protein Data Bank**, which is accessible on the Internet.

Diffraction of X-rays by a crystal occurs with incident radiation of a characteristic wavelength (e.g., copper, $K\alpha = 1.54$ Å). The X-ray beam is diffracted by the electrons that surround the atomic nuclei in the crystal, with an intensity proportional to the number of electrons around the nucleus. Thus, the technique establishes the **electron distribution** of the molecule and infers the nuclear distribution. Actual positions of atomic nuclei can be determined directly by diffraction with **neutron beam radiation**, an interesting but very expensive technique as it requires a nuclear reactor or particle accelerator. With the highest resolution now available for X-ray diffraction determination of protein structure, the diffraction from C, N, O, and S atoms can be observed. That from hydrogen atoms is not observed due to the low density of electrons, that is, a single electron around a hydrogen nucleus. The diffracted beam is typically detected on photographic film or electronic area detectors. This permits recording of the amplitude (intensity) of radiation diffracted in a defined orientation. Determination of the **phase**

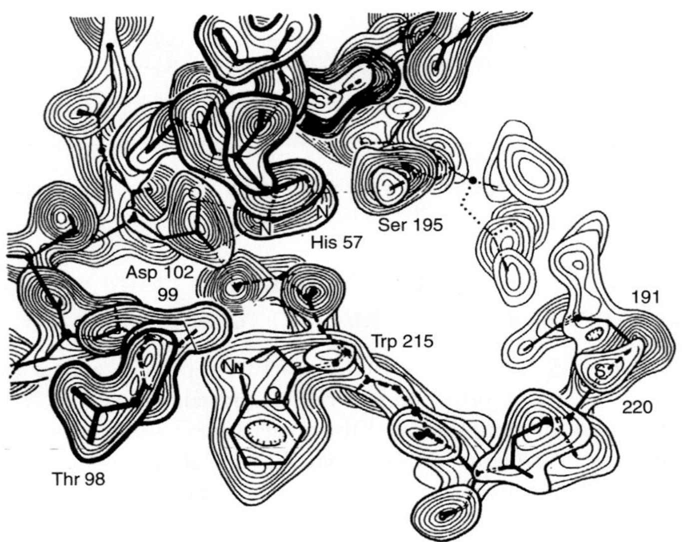

Figure 3.72 Electron-density map at 1.9-Å resolution of active site region of proenzyme form of trypsin. Active site amino acid residues are fitted onto an electron density map.
Reproduced with permission from Kossiakoff, A. A., Chambers, J. L., Kay, L. M., and Stroud, R. M. *Biochemistry* 16: 654, 1977. Copyright (1977) by American Chemical Society.

angles has historically required the placement of heavy atoms (such as iodine, mercury, or lead) in the protein molecule. Modern procedures can often solve the phase problem without use of heavy atoms.

It is convenient to compare X-ray crystallography with light microscopy to understand the processes involved. In light microscopy, incident radiation is reflected by an object under study and the reflected beam is recondensed by the objective lens to produce an image of the object. The analogy is appropriate to incident X-rays but no material exists that can serve as an objective lens for X-ray radiation. In place of the objective lens, amplitude and phase angle measurements of the diffracted radiation are mathematically reconstructed by **Fourier synthesis** to yield a three-dimensional **electron-density map** of the diffracted object. A few hundred reflections are required for a low-resolution electron-density map. For example, 400 reflections were used to obtain a 6-Å map of myoglobin. At this resolution one can clearly locate the molecule within the unit cell of the crystal and study the overall packing of the subunits in a multimeric protein. However, a trace of the conformation of the polypeptide chain is made with difficulty. Many more reflections are required to obtain higher-resolution maps. For myoglobin, where 400 reflections were utilized to obtain the 6-Å map, 10,000 reflections were needed for a 2-Å map, and 17,000 reflections for a 1.4-Å map. Much of the work is now partially automated using computers. A slice through a three-dimensional electron-density map of trypsinogen is shown in Figure 3.72. The known primary structure is fitted to the electron-density pattern by **refinement**, the computer-intensive process of aligning a sequence to the electron-density map until the best fit is obtained.

X-ray diffraction provides incomplete evidence of a protein's mechanism of action. Rather, it provides an average structure of a molecule whose atoms are normally undergoing rapid fluctuation in solution (Section 3.8). Indeed, the average structure may not be the active structure for the protein in solution. On the time scale of diffraction data collection, the structures of reactive enzyme–substrate complexes, intermediates, and reaction transition states cannot be observed. These structures must be inferred from the static pictures of an inactive form of the protein or from complexes with inactive analogs of its substrates (Figure 3.73).

Newer methods for X-ray diffraction use synchrotron radiation to generate an X-ray beam, at least 10,000 times more intense than that of standard X-ray generators, which enables the collection of diffraction data on a millisecond time scale. Application of these techniques will enable determination of short-lived structures and solve mechanistic and dynamic structural questions not addressable by the standard technology.

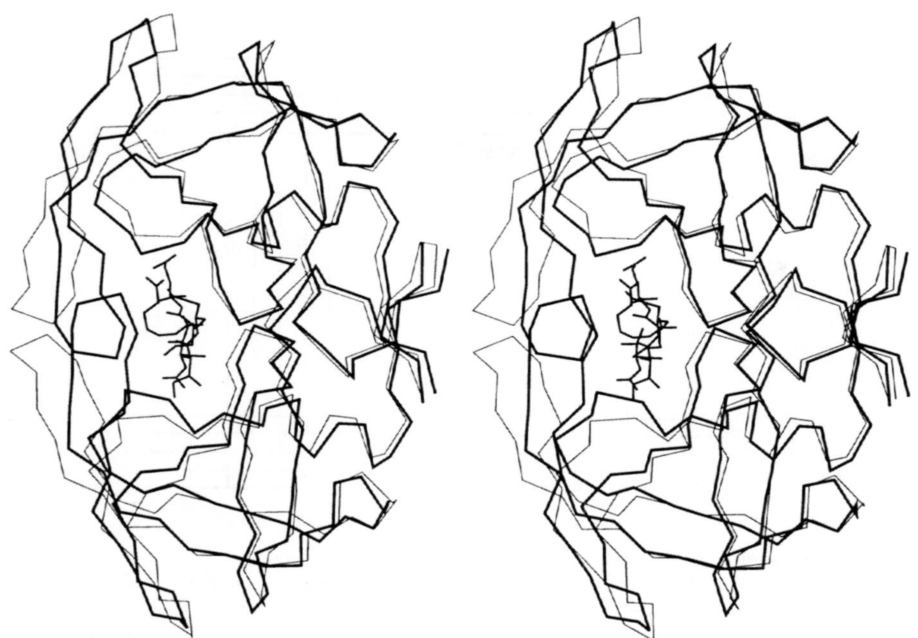

Figure 3.73 Stereo tracing of superimposed a-carbon backbone structure of HIV protease with inhibitor (thick lines) and that of the protease without inhibitor (thin lines). Redrawn with permission from Miller, M., Schneider, J., Sathyanarayana, B. K., Toth, M. V., et al. *Science* 246: 1149, 1989. Reprinted with permission from AAAS.

A CLOSER LOOK 3.2

Ultraviolet and Visible Light Absorbance Show a Linear Dependency on Concentration and Path Length

The Beer–Lambert law is an empirical equation that shows the linear relationship between the concentration of a chromophore and absorbance at a particular wavelength of UV or visible light. The equation is

$$A = \varepsilon_\lambda * l * c$$

where A is the light absorbance at wavelength λ, ε_λ is the molar absorptivity or extinction coefficient of the chromophore at wavelength λ, l is the path length of the light through the solution, and c is the concentration of chromophore. The extinction coefficient is a constant of the chromophore molecule at a set temperature and solution conditions. If ε_λ is in units of $M^{-1}cm^{-1}$, then l is in cm, c is in M, and A is in absorbance units.

Spectroscopic Methods for Evaluating Protein Structure and Function

Ultraviolet Light Spectroscopy

Side chains of tyrosine, phenylalanine, and tryptophan, and peptide bonds absorb ultraviolet (UV) light. The efficiency of absorption for each **chromophore** is related to its **molar extinction coefficient** (ε) (A Closer Look 3.2). A typical spectrum is shown in Figure 3.74. Absorbance between 260 and 300 nm is primarily due to aromatic side chains (Figure 3.75). When the tyrosine side chain is ionized at high pH (this R group has a p$K'_a \approx 10$), the absorbance is shifted to a higher wavelength (red shifted) and its molar absorptivity is increased (Figure 3.75). Peptide bonds absorb in the far-UV (180–230 nm). A peptide bond in α-helix conformation interacts with the electrons of peptide bonds above and below it to create an **exciton system** in which electrons are delocalized. The result is a shift of the absorption maximum for an isolated peptide bond to either a lower or a higher

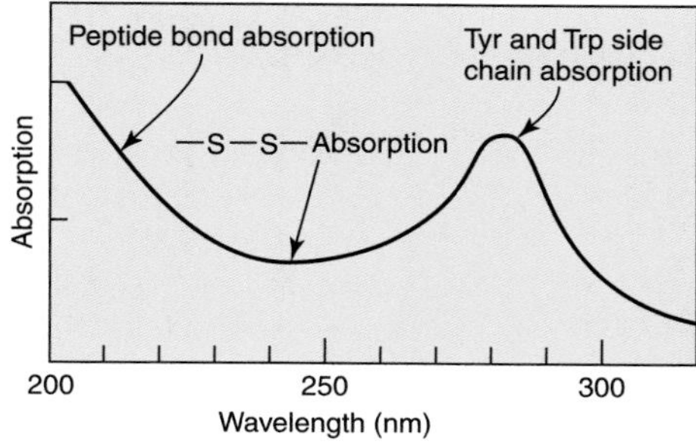

Figure 3.74 Ultraviolet absorption spectrum of the globular protein α-chymotrypsin.

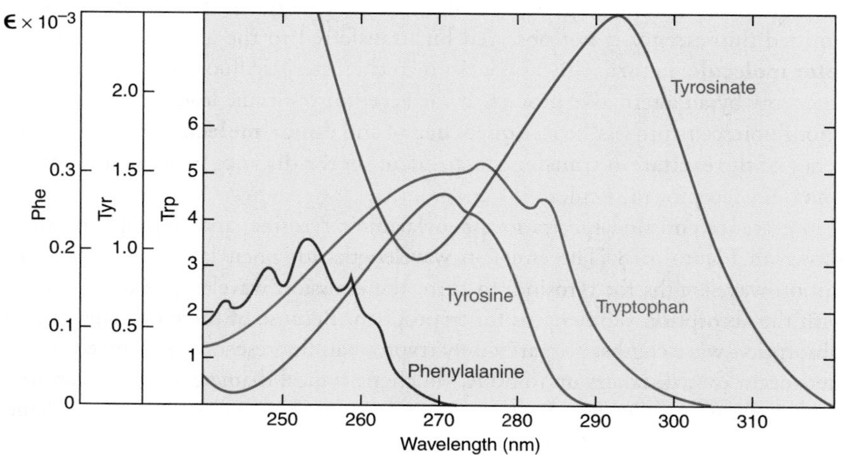

Figure 3.75 Ultraviolet absorption for chromophores of Phe, Tyr, Trp, and tyrosinate. Note differences in extinction coefficients on left axis for the different chromophores. Redrawn from d'Albis, A. and Gratzer, W. B. In Bull, A. T.. Lagnado, J. R.. Thomas, J. O., and Tipton. K. F. (Eds.), *Companion to Biochemistry.* London: Longmans, 1974, 170.

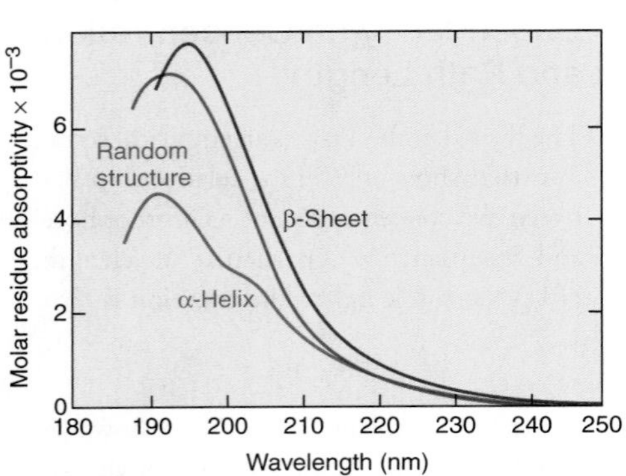

Figure 3.76 Ultraviolet absorption of the peptide bonds of a polypeptide chain in α-helix, random structure, and antiparallel β-sheet conformations.
Redrawn from d'Albis, A., and Gratzer, W. B. In Bull, A. T., Lagnado, J. R., Thomas J. O., and Tipton, K. F. (Eds.), *Companion to Biochemistry.* London: Longmans, 1974, 175.

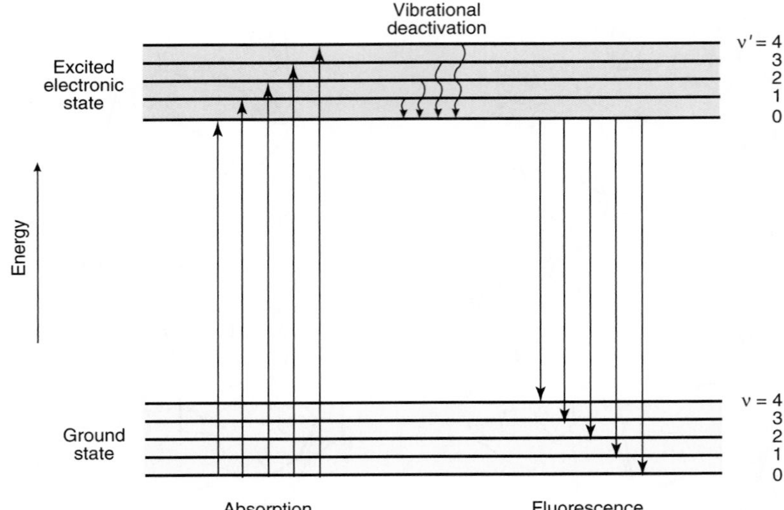

Figure 3.77 Absorption and fluorescence electronic transitions. Excitation on light absorption is from zero vibrational level in ground state to various higher vibrational levels in the excited state. Fluorescence emission is from the zero vibrational level in excited electronic state to various vibrational levels in the ground state. The fluorescence emission is at a higher wavelength (less energy) than the absorption wavelength.
Redrawn from d'Albis, A., and Gratzer, W. B. In Bull, A. T., Lagnado, J. R., Thomas, J. O., and Tipton, K. F. (Eds.), *Companion to Biochemistry.* London: Longmans, 1974, 166.

wavelength (Figure 3.76). Ultraviolet spectroscopy can provide information on a protein's secondary and tertiary structure. As a protein is denatured, differences in the absorption characteristics of the peptide bonds appear because of disruption of the exciton system. In addition, the absorption maximum for an aromatic chromophore is at a lower wavelength in an aqueous environment than in a nonpolar environment.

The **molar absorbancy** of a chromophoric substrate often changes on binding to a protein and can be used to measure its binding constant. Changes in chromophore extinction coefficients during enzyme catalysis are used to provide the kinetic parameters for the reaction.

Fluorescence Spectroscopy

The energy of an excited electron produced by light absorption is lost most commonly as thermal energy in a collision process. In some chromophores, the excitation energy is dissipated by fluorescence. The **fluorescent emission** is always at a longer wavelength of light (lower energy) than the absorption wavelength of the fluorophore as vibrational energy levels, formed during the light absorption (excitation), are lost prior to fluorescence (Figure 3.77). If another molecule is present to absorb light energy emitted by the **fluorophore**, the emitted fluorescence is not observed but transferred to the absorbing molecule. The **acceptor molecule**, in turn, either emits its own characteristic fluorescence or loses its excitation energy by an alternative process. If the acceptor molecule loses its excitation energy by a nonfluorescent process, it is a **quencher** of the **donor molecule's** fluorescence. The efficiency of the **excitation transfer** is dependent on the distance between and orientation of donor and acceptor molecules.

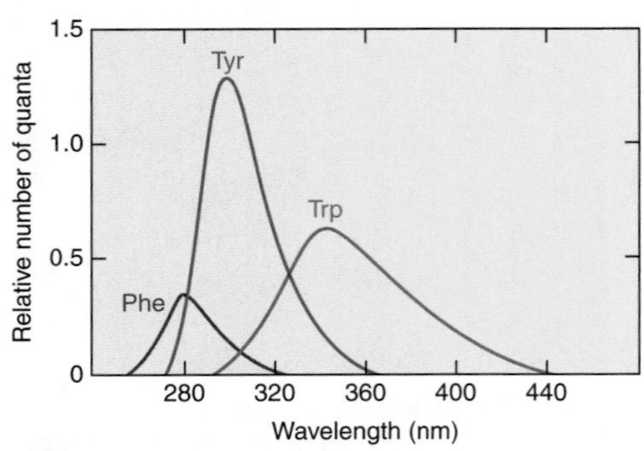

Figure 3.78 Characteristic fluorescence of aromatic groups in proteins. Redrawn from d'Albis, A., and Gratzer, W. B. In Bull, A. T., Lagnado, J. R., Thomas, J. O., and Tipton, K. F. (Eds.), *Companion to Biochemistry.* London: Longmans, 1974, 478.

Fluorescence emission spectra for phenylalanine, tyrosine, and tryptophan side chains are shown in Figure 3.78. The emission wavelengths for phenylalanine overlap with the absorption wavelengths for tyrosine. In turn, the emission wavelengths for tyrosine overlap with the absorption wavelengths for tryptophan. Because of these overlaps in emission and absorption wavelengths, primarily only tryptophan fluorescence is observed. Excitation transfers occur over distances up to 80 Å, which are typical diameters of globular domains. On denaturation, the distances between donor and acceptor groups increase and the efficiency of energy transfer to tryptophan decreases. Accordingly, an increase in tyrosine and/or

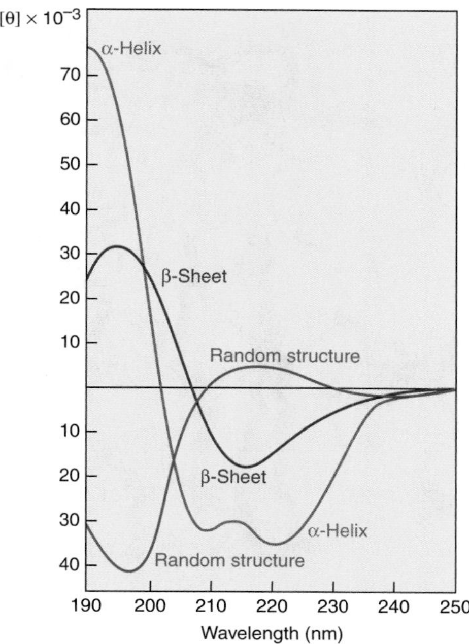

Figure 3.79 Circular dichroism spectra for polypeptide chains in α-helix, β-sheet and random conformations. Redrawn from d'Albis, A., and Gratzer, W. B. In Bull, A. T., Lagnado, J. R., Thomas, J. O., and Tipton, K. F. (Eds.), *Companion to Biochemistry*. London: Longmans, 1974,190.

phenylalanine and loss of tryptophan fluorescence is observed on denaturation of a protein. Since excitation transfer processes in proteins are distance and orientation dependent, the fluorescence yield is dependent on the conformation of the protein. This type of analysis detects changes due to conformational changes and binding interactions.

Circular Dichroism Spectroscopy

Circular dichroism (CD) is caused by differences in light absorption between the clockwise and counterclockwise component vectors of a beam of polarized light traveling through a solution of an optically active molecule such as an L-amino acid. A spectrum is generated when the circular dichroism is determined through a range of wavelengths. Aromatic amino acids in a protein and the polypeptide chain generate an optical rotation and CD spectrum (Figure 3.79). Because of the differences between spectra for α-helical, β-sheet, and random structures, circular dichroism has been a fairly sensitive assay for the amount and type of secondary structure and is commonly utilized to follow protein folding and denaturation.

Nuclear Magnetic Resonance

With **two-dimensional (2-D) NMR** and powerful NMR spectrometers the conformation in solution of small proteins of about 150 amino acids or less can be determined. Multidimensional NMR and triple resonance can extend the sensitivity to more than 250 residues.

Conventional NMR techniques use radiofrequency (rf) radiation to study the environment of atomic nuclei that are magnetic. The requirement for magnetic nuclei is absolute and based on an unpaired spin state in the nucleus. Thus, the naturally abundant carbon (^{12}C), nitrogen (^{14}N), and oxygen (^{16}O) do not absorb, while ^{13}C, ^{15}N, and ^{17}O do absorb. The information derived from absorption bands enables the determination of the identity and number of nearest-neighbor groups that can affect the response of absorbing species through bonded interactions, but it gives no information on through-space (nonbonded) interaction due to the native conformation of the protein. To determine through-space interactions and protein tertiary structure requires the use of **nuclear Overhauser effects (NOEs)** and the application of the two-dimensional technique.

The major difference between two-dimensional and one-dimensional (1-D) NMR is the addition of a second time delay rf pulse. The technique requires identification in the spectrum of proton absorbance from a particular position in the protein structure. Through-space

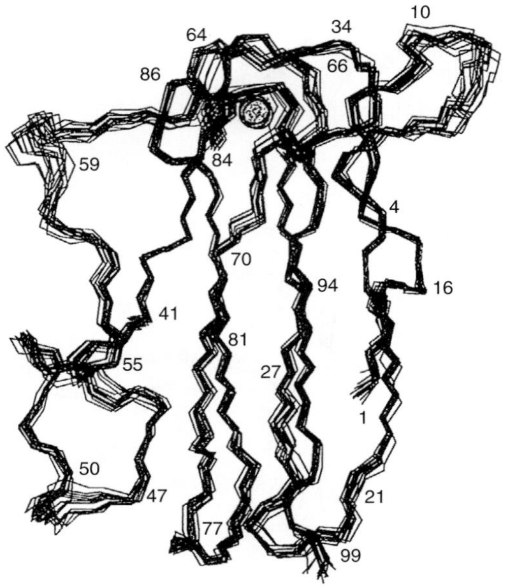

Figure 3.80 NMR structure of plastocyanin from the French bean. Structure shows superposition of eight structures of the polypeptide backbone for the protein, calculated from constraints of NMR spectrum.
From Moore, J. M., Lepre, C. A., Gippert, G. P., Chazin, W. J., et al. *J. Mol. Biol.* 221: 533, 1991. Figure generously supplied by P. E. Wright.

interactions can be observed at a maximum distance of about 5 Å. On the generation of distance information for interresidue pairs through the protein structure, protein conformations consistent with the spectra are generated. In this calculation, a distance matrix is constructed containing ranges of distances (minimum and maximum) for as many interresidue interactions as may be measured. Possible structures are generated from the data consistent with the constraints imposed by the NMR spectra. Computational refinements of the initially calculated structures can optimize covalent bond distances and bond angles. The method generates a family of structures, the variability showing either the imprecision of the technique or the dynamic "disorder" of the folded structure (Figure 3.80). Such computations have yielded structures that do not significantly differ from the time-averaged structure observed with X-ray diffraction methods.

Other enhancements of NMR, for determination of protein structure, include the ability to synthesize proteins that contain isotopically enriched (e.g., containing ^{13}C or ^{15}N) amino acids and the development of paramagnetic shift reagents to study localized environments on paramagnetic resonances, such as the lanthanide ion-reporting groups.

Bibliography

Bioinformatics and Proteomics Software Portals

The journal *Nucleic Acids Research* annually compiles and summarizes the Web-accessible biological databases (*Nucleic Acids Research* 36: January 2008) and web sites with tools for analysis of biological data (*Nucleic Acids Research* 36: July 1, 2008).

The National Institutes of Health Entrez site has access to sequence, genomic, and protein expression databases and analysis tools (http://www3.ncbi.nlm.nih.gov/Entrez).

The Expert Protein Analysis System (ExPASy) site of the Swiss Institute of Bioinformatics contains tools for protein study and links to other sites and to databases for the study and analysis of proteins (http:// http://ca.expasy.org).

The web site of the European Bioinformatics Institute also offers references to databases and a comprehensive package of analysis tools (http://www.ebi.ac.uk).

The Swiss PDB viewer (Deep View) can be downloaded and allows viewing of three-dimensional structures. The software and manual are available through the DeepView site (http://spdbv.vital-it.ch).

An excellent tutorial by Gale Rhodes can be accessed at http://spdbv.vital-it.ch/TheMolecularLevel/SPVTut/index.html.

Protein Explorer is available at http://www.umass.edu/microbio/chime/pe_beta/pe/protexpl.

Cn3D, is available at the National Center for Biotechnology Information (NCBI) at the NIH web site http://www.ncbi.nlm.nih.gov/Structure/CN3D/cn3d.shtml.

The molecular Viewers downloads protein structures stored at the Protein Data Bank (PDB) by the RCSB (http://www.rcsb.org/pdb/home/home.do).

Databases

Birney, E. et al. Databases and tools for browsing genomes. *Annu. Rev. Genomics Hum. Genet.* 3:293, 2002.

Day, R., Beck, D. A., Armen, R. S., and Daggett, V. A consensus view of fold space: Combining SCOP, CATH, and the Dali Domain Dictionary. *Protein Science* 12:2150, 2003.

Lee, D., Refern, O., and Orengo, C. Predicting protein function from sequence and structure. *Nature Reviews Mol. Cell Biol.* 8:995, 2007.

Sierk, M. L., and Kleywegt, G. J. Deja vu all over again: Finding and analyzing protein structure similarities. *Structure* 12:2103, 2004.

von Mering, C., Krause, R., Snel, B., Cornell, M., et al. Comparative assessment of large-scale data sets of protein-protein interactions. *Nature* 417:399, 2002.

Physical and Structural Properties of Proteins

Chothia, C. Principles that determine the structure of proteins. *Annu. Rev. Biochem.* 53:537, 1984.

Cremades, N., Sancho, J., and Freire, E. The native-state ensemble of proteins provides clues for folding, misfolding and function. *Trends Biochem. Sci.* 31:494, 2006.

Doolittle, R. F. The multiplicity of domains in proteins. *Annu. Rev. Biochem.* 64:287, 1995.

Dyson, H. H., and Wright, P. E. Intrinsically unstructured proteins and their functions. *Nature Reviews* 6:197, 2005.

Fasman, G. D. Protein conformational prediction. *Trends in Biochem. Sci.* 14:295, 1989.

Finkelstein, A. V., Gutun, A. M., and Badretdinov, A. Y . Why are the same protein folds used to perform different functions? *FEBS Lett.* 325:23, 1993.

Han, J. H., Batey, S., Nickson, A. A., Teichmann, S. A., and Clarke, J. The folding and evolution of multidomain proteins. *Nature Reviews Mol. Cell Biol.* 8:319, 2007.

Jones, D.T. Learning to speak the language of proteins. *Science* 302: 1347, 2003.

Laskowski, R. A., and Thornton, J. M. Understanding the molecular machinery of genetics through 3D structures. *Nature Reviews Genetics* 9:141, 2008.

Orengo, C. A., and Thornton, J. M. Protein families and their evolution: A structural perspective. *Annu. Rev. Biochem.* 74:867, 2005.

Richardson, J. S. The anatomy and taxonomy of protein structure. *Adv. Protein Chem.* 34:168, 1981.

Rose, G. D., and Wolfenden, R. Hydrogen bonding, hydrophobicity, packing, and protein folding. *Annu. Rev. Biophys. Biomol. Struct.* 22:381, 1993.

Rose, G. D., Fleming, P. J., Banavar, J. R., and Maritan, A. A backbone-based theory of protein folding. *Proc. Natl. Acad. Sci., USA* 103:16623, 2006.

Schueler-Furman, O., Wang, C., Bradley, P., Misura, K., and Baker, D. Progress in modeling of protein structures and interactions. *Science* 310:638, 2005.

Stevens, R. C. Long live structural biology. *Nature Structural Mol. Biol.* 11:293, 2004.

Tompa, P. Intrinsically unstructured proteins. *Trends Biochem. Sci.* 27:527, 2002.

Protein Folding

Fersht, A.R. From the first protein structures to our current knowledge of protein folding: Delights and skepticisms. *Nature Reviews Mol. Cell Biol.* 9:650, 2008.

Clark, P. L. Protein folding in the cell, reshaping the folding funnel. *Trends Biochem. Sci.* 29:527, 2004.

Dinner, A. R., Sali, A., Smith, L. J., Dobson, C. M., and Karplus, M. Understanding protein folding via free-energy surfaces from theory and experiment. *Trends Biochem. Sci.* 25:331, 2000.

Dobson, C. M., Protein folding and misfolding. *Nature* 426:884, 2003.

Hartl, F. U., and Hayer-Hartl, M. Molecular chaperones in the cytosol: From nascent chain to folded protein. *Science* 295:1852, 2002.

Saibil, H. R., and Ranson, N. A. The chaperonin folding machine. *Trends Biochem. Sci.* 27:627, 2002.

Selkoe, D. J. Cell biology of protein misfolding: the examples of Alzheimer's and Parkinson's diseases. *Nature Cell Biology* 6:1054, 2004.

Sharma, S., Chakraborty, K., Müller, B. K., Astola, N., et al. Monitoring protein conformation along the pathway of chaperonin-assisted folding. *Cell* 133:142, 2008.

Structural Genomics, Proteomics, and Protein Networks

Bray, D. Molecular networks: The top-down view. *Science* 301:1864, 2003.

Editorial, Looking ahead with structural genomics. *Nature Structural and Mol. Biol.* 14:1, 2007.

Eisenberg, D., Marcotte, E. M., Xenarios, I., and Yeates, T. O. Protein function in the post-genomic era. *Nature* 405:823, 2000.

Gavin, A. C., Bosche, M., Krause, R., et al. Functional organization of the yeast proteome by systematic analysis of protein complexes. *Nature* 415:141, 2002.

Ghavidel, A., Cagney, G., Emili, A. A skeleton of the human protein interactome. *Cell* 122: 830, 2005.

Hofmann, K. P., Spahn, C. M. T., Heinrich, R., and Heinemann, U. Building functional modules from molecular interactions. *Trends Biochem. Sci.* 31:497, 2006.

Lim, J., Hao, T., Shaw, C., Patel, A. J., et al. A protein–protein interaction network for human inherited ataxias and disorders of Purkinje cell degeneration. *Cell* 125:801, 2006.

Pandey, A., and Mann, M. Proteomics to study genes and genomes. *Nature* 405:837, 2000.

Pawson, T., and Nash, P. Assembly of cell regulatory systems through protein interaction domains. *Science* 300:445, 2003.

Taylor, I. W., and Wrana, J. L. SnapShot:The TGFβ pathway interactome. *Cell* 133:378, 2008.

Watson, J. D., Todd, A. E., Bray, J., Laskowski, R. A., et al. Target selection and determination of function in structural genomics. *IUBMB Life* 55:249, 2003.

Zhu, H., Bilgin, M., and Snyder, M. Proteomics. *Annu. Rev. Biochem.* 72:783, 2003.

Dynamics in Folded Proteins

Daggett, V., and Levitt, M. Realistic simulations of native-protein dynamics in solution and beyond. *Annu. Rev. Biophys. Biomol. Struct.* 22:353, 1993.

Joseph, D., Petsko, G. A., and Karplus, M. Anatomy of a conformational change: Hinged lid motion of the triosephosphate isomerase loop. *Science* 249:1425, 1990.

Karplus, M., and McCammon, J. A. Molecular dynamics simulations of biomolecules. *Nature Structural Biol.* 9:646, 2002.

Tompa, P., and Fuxreiter, M. Fuzzy complexes: Polymorphism and structural disorder in protein-protein interactions. *Trends Biochem. Sci.* 33:2, 2007.

Glycoproteins

Bellahcene, A., Castronovo, V., Ogbureke, K. U., Fisher, L. W., and Fedarko, N. S. Small integrin-binding glycoproteins (SIBLINGs): Multifunctional proteins in cancer. *Nature Reviews Cancer* 8:212, 2008.

Drickamer, K., and Taylor, M. E. Evolving views of protein glycosylation. *Trends Biochem. Sci.* 23:321, 1998.

Geyer, H., and Geyer, R. Strategies for analysis of glycoprotein glycosylation. *Biochim. Biophys. Acta* 1764:1853, 2006.

Grogan, M. J., Pratt, M. R., Marcaurelle, L. A., and Bertozzi, C. R. Homogeneous glycopeptides and glycoproteins for biological investigation. *Annu. Rev. Biochem.* 71:593, 2002.

Lis, H., and Sharon, N. Protein glycosylation: Structural and functional aspects. *Eur. J. Biochem.* 218:1, 1993.

Wassarman, P. M. Zona pellucida glycoproteins. *J. Biol. Chem.* 283:24285, 2008.

Lipoproteins

Jonas, A. In Vance, D. E., and Vance, J. E. (Eds), *Lipoprotein Structure in Biochemistry of Lipids, Lipoproteins, and Membranes,* 4th ed. Amsterdam: Elsevier, 2002, 483.

Khovidhunkit, W., Kim, M.-S., Memon, R. A., Shigenaga, J. K., et al. Effects of infection and inflammation on lipid and lipoprotein metabolism: Mechanisms and consequences to the host. *J. Lipid Research* 45:1169, 2004.

Myers, G. L., Cooper, G. R., and Sampson, E. J. Traditional lipoprotein profile: Clinical utility, performance requirement, and standardization. *Atherosclerosis* 108:S157, 1994.

Segrest, J. P., Garber, D. W., Brouillette, C. G., Harvey, S. C., and Anantharamaiah, G. M. The amphipathic α helix: A multifunctional structural motif in plasma apolipoproteins. *Adv. Protein Chem.* 45:303, 1994.

Collagen

Brodsky, B., and Baum, J. Modeling collagen diseases. *Nature* 453:998, 2008.

Brodsky, B., Thiagarajan, G., Madhan, B., and Kar, K. Triple-helical peptides: An approach to collagen conformation, stability, and self-association. *Biopolymers* 89:345, 2008.

Myllyharju, J., and Kivirikko, K. I., Collagens, modifying enzymes, and their mutations in humans, flies, and worms. *Trends in Genetics* 20:33, 2004.

Prockop, D. J., and Kivirikko, K. I. Collagens: molecular biology, diseases, and potentials for therapy. *Annu. Rev. Biochem.* 64:403, 1995.

Techniques for the Study of Proteins

Bax, A., and Grzesiek, S. Methodological advances in protein NMR. *Acc. Chem. Res.* 26:131, 1993.

Mann, M., Hendrickson, R. C., and Pandey, A. Analysis of proteins and proteomes by mass spectrometry. *Annu. Rev. Biochem.* 70:437, 2001.

Reif, O. W., Lausch, R., and Fritag, R. High-performance capillary electrophoresis of human serum and plasma proteins. *Adv. Chromatogr.* 34:1, 1994.

Rhodes, G. *Crystallography Made Crystal Clear,* 2d ed. San Diego: Academic, 2000.

Tugarinov, B., Hwang, P. M., and Kay, L. E. Nuclear magnetic resonance spectroscopy of high-molecular-weight proteins. *Annu. Rev. Biochem.* 73:107, 2004.

Key Terms

amino acids-common
amino acids-derived
amino acids-basic
selenocysteine
polypeptide
peptide
primary structure
amino acids-acidic
Henderson–Hasselbalch equation
isoelectric pH (pI)
p*K'a*
zwitterions
hydrophobic interaction
conservative substitution
homology
invariant residue
nonconservative substitution
secondary structure
native conformation
phi (φ) bond
psi (ψ) bond
quaternary structure
hydrogen bond
α-helix
protein fold
structural motif
β-structure
tertiary structure
intrinsically unstructured protein
globular protein
interactome
protein family
superfold
fibrous protein
polyproline type II helix
glycoprotein
apo-protein
lipoprotein
N-glycosidic linkage
O-glycosidic linkage
chaperone
molten globule
chaperonin
ionic interaction
van der Waals interaction
denaturation
protein dynamics

Questions • CAROL N. ANGSTADT

Multiple-Choice Questions

1. All of the following are correct about a peptide bond except
   A. it exhibits partial double bond character.
   B. it is more stable in the *cis* configuration than in the *trans* configuration.
   C. it has restricted rotation around the carbonyl carbon to nitrogen bond.
   D. it is planar.
   E. in proline, the nitrogen is attached to the side chain.

2. In an α-helix,
   A. side-chain groups can align to give a polar face.
   B. each peptide bond forms two hydrogen bonds.
   C. there are 3.6 amino acids per turn.
   D. all of the above.
   E. none of the above.

3. Chaperone proteins
   A. all require ATP to exert their effect.
   B. cleave incorrect disulfide bonds, allowing correct ones to subsequently form.
   C. guide the folding of polypeptide chains into patterns that would be thermodynamically unstable without the presence of chaperones.
   D. of the Hsp70 class are involved in transport of proteins across mitochondrial and endoplasmic reticulum membranes.
   E. act only on fully synthesized polypeptide chains.

4. Proteins may be separated according to size by
   A. isoelectric focusing.
   B. polyacrylamide gel electrophoresis.
   C. ion exchange chromatography.
   D. molecular exclusion chromatography.
   E. reverse-phase HPLC.

5. Similar arrangements of secondary structural motifs are often observed in the fold structure of protein domains. Which of the following is correct?
A. Superfolds are similar structures from proteins related by function or evolution from the same primordial gene.
B. Folds must be either all α or all β.
C. There is only one type of β domain.
D. If a protein has more than one domain, all domains are identical.
E. A common fold has a central β-barrel with the strands connected by α-helices around the outside.

6. Unstructured proteins
A. are those proteins that have been denatured by heat.
B. do not have any biological functions.
C. can be induced to have a defined structure by binding to other proteins or to DNA or RNA.
D. have no secondary or tertiary structure.
E. have regions that are very rich in aromatic amino acids.

Questions 7 and 8: Abnormalities in the synthesis or structure of collagen cause dysfunctions in cardiac organs, bone, skin, joints, and eyes. Problems may result from abnormal collagen genes, abnormal posttranslational modifications of collagen, or deficiency of cofactors needed by enzymes responsible for posttranslational modifications. Scurvy, a lack of vitamin C, is an example of the last type.

7. In collagen:
A. intrachain hydrogen bonding stabilizes the native structure.
B. three chains with polyproline type helical conformation can wind about one another to form a superhelix because of the structure of glycine.
C. the φ angles contributed by proline are free to rotate.
D. regions of superhelicity comprise the entire structure except for the N- and C-termini.
E. crosslinks between triple helices form after lysine is converted to allysine.

8. The formation of covalent cross-links in collagen
A. occurs during synthesis of the peptide chain.
B. uses hydroxyproline.
C. involves glycine residues.
D. requires conversion of some ε-amino groups of lysine to δ-aldehydes.
E. all of the above.

Questions 9 and 10: Glycoproteins, with carbohydrate molecules covalently bound to one or more amino acids, participate in many normal and disease-related functions, for example, antigenic determinants of blood groups, cell–cell recognition, and functioning of collagen. Usually the carbohydrate is added to the protein by enzymatic reactions. A glycosylated hemoglobin, HbA_{1c}, however, is formed nonenzymatically in red blood cells. In prolonged hyperglycemia as in patients with diabetes mellitus, the concentration of HbA_{1c} rises and it can be used to follow the effectiveness of treatment.

9. Glycoproteins
A. are found in cells but not in plasma.
B. in a plasma membrane typically have the carbohydrate portion on the cytosolic side.
C. may have the carbohydrate portion covalently linked to the protein at an asparagine.
D. that are carbohydrate to hydroxyl linked always have the linkage to hydroxylysine.
E. of a given type always have identical carbohydrate chains.

10. With glycoproteins,
A. the carbohydrate is added as the polypeptide chain is transported through the endoplasmic reticulum and Golgi network.
B. the amount of carbohydrate present is always a small percentage ($< 5\%$) of the weight of the protein.
C. the carbohydrate portion consists only of a single sugar on each amino acid.
D. the most common type of linkage is to an α-amino group of a polypeptide chain.
E. changes in composition would have no effect on biological function.

Questions 11 and 12: Many pathological hyperlipoproteinemias result from abnormalities in the rates of synthesis or clearance of lipoproteins in the blood. They are usually characterized by elevated levels of cholesterol and/or triacylglycerols in the blood. Type I has very high plasma triacylglycerol levels (> 1000 g/dL) because of an accumulation of chylomicrons. Type II (familial hypercholesterolemia) has elevated cholesterol, specifically in the form of LDL. Another abnormality of lipoproteins is hypolipoproteinemia in which lipoproteins are not formed because of the inability to make a particular apoprotein.

11. All lipoprotein particles in the blood have the same general architecture which includes
A. a neutral core of triacylglycerols and cholesteryl esters.
B. amphipathic lipids oriented with their polar head groups at the surface and their hydrophobic chains oriented toward the core.
C. most surface apoproteins containing amphipathic helices.
D. unesterified cholesterol associated with the outer shell.
E. all of the above.

12. In abetalipoproteinemia chylomicrons, VLDL and LDL are absent from the blood. Which of the following is correct?
A. In this disease, no apolipoproteins are synthesized.
B. If the blood from these individuals were centrifuged, the lipid bands would be found primarily in the upper half of the tube.
C. Failure to synthesize apolipoprotein ApoB-100 and ApoB-48 would account for the pattern shown in this disease.
D. Apolipoproteins are composed mostly of β structure.
E. All of the above.

Problems

13. In a plot of equivalents of OH^- versus pH, pH is $\sim$2 when 0.5 equiv has been used, pH is $\sim$6 for 1.5 equiv and pH is $\sim$9.5 for 2.5 eq. What amino acid has been titrated? In each case, the point indicated is the midpoint of a steep part of the curve.

14. After purification, the Edman reaction was used to sequence a dodecapeptide, yielding the following data: the C-terminal amino acid is isoleucine; the N-terminal amino acid is methionine; and the peptide fragments are Ala-Ala-Ile, Leu-Arg-Lys-Lys-Glu-Lys-Glu-Ala, Met-Gly-Leu, and Met-Phe-Pro-Met. What is the sequence of this peptide?

Answers

1. **B** This would put both side chains on the same side of the bond, and this is less favorable. C is a consequence of A. D: All atoms attached to the carbon and nitrogen of the bond are in a common plane. E: Proline has an imino group.
2. **D** A: They could also align to give a nonpolar face. B: One is to the fourth residue above the bond, and the other is to the fourth residue below. C: This is one of the characteristics of a right-handed α-helix.
3. **D** Proteins cross the membrane in an unfolded state and refold once they cross the membrane. A: The hsp60 family of chaperones is ATP linked, but the hsp70 family is not. B: Disulfide isomerases catalyze this reaction. C: The final product is thermodynamically stable; chaperones prevent unfavorable intermediate interactions. E: Hsp70 chaperones react with nascent polypeptide chains as they are synthesized by the ribosome. The protein may then be delivered to an hsp60 chaperone for facilitation of final folding.
4. **D** Another method that separates on the basis of size is SDS PAGE. A–C separate molecules on the basis of charge. E: Reverse-phase HPLC effects separations based on polarity.
5. **E** This α/β structure is found in pyruvate kinase, for example. A: This is called a family; superfolds are found in proteins unrelated in function or origin. B: See E. C: There are several different types. D: A protein may have multiple different domains, and the different domains may, in some cases, have different functions.
6. **C** Their unstructured domains allow them to bind with a variety of partners. A: These proteins naturally lack a stable folded conformation. B: They play a variety of roles such as scaffold proteins, cyclin-dependent kinases, and many others. D: They may have regions of secondary structure. E: They tend to be rich in polar and charged amino acids and low in aromatic ones.
7. **B** Close contacts in the interior of the triple helix are possible only when the R group at that position is very small, that is, hydrogen. A: Hydrogen bonding in collagen is interchain. C: The φ angle is part of the proline ring and is not free to rotate. D: Although the statement is true of type I collagen, the superhelical regions in other collagen types may be broken by regions of globular domains. E: The conversion and crosslinking are extracellular.
8. **D** These then react with each other or with unmodified ε-amino groups of lysine. A: It is posttranslational. B and C: This is not the role of hydroxyproline or glycine.
9. **C** In the N-linked type, asparagine is in the sequence Asn-X-Thr(Ser). A: Most plasma proteins, except albumin, are glycoproteins. B: Carbohydrate is on the outside of the cell for functions like cell–cell recognition. D: This link is in collagen only. Other O-linked carbohydrates have the linkage to serine or threonine. E: Carbohydrate structure is not determined by genes and is variable.
10. **A** The carbohydrate content is not directed by genes but is a co- or posttranslational event. B: The content is highly variable, for example, glycophorin in erythrocyte membrane has 60% carbohydrate. C: The attached carbohydrate can be a single carbohydrate or as many as 15 carbohydrate residues. D: This is a type VI linkage found in HbA_{1c} but is not the most common type. E: There could be a profound change in function.
11. **E** All lipoproteins share these characteristics. C: The polar face interacts with water and the hydrophobic face is oriented toward the core. D: Cholesterol's hydroxyl group is sufficiently polar that it orients toward the outer shell.
12. **C** Apo B-100 is a primary apolipoprotein of LDLs and VLDLs. Apo B-48, a truncated form of B-100, is found in chylomicrons. A: Since the various apolipoproteins come from different genes, one would not expect all genes to be affected. B: The missing lipoproteins are the ones found in the upper part of the tube on centrifugation. D: Apolipoproteins associated with lipid have a high content of amphipathic α-helices.
13. This is a titration curve for an acid with three dissociable groups. The points indicated represent half-titration of each of these groups, and the pHs are the pK values for each group. With these pK values, the amino acid must be histidine.
14. You must have the correct N- and C-terminal amino acids and the correct total number (12) of amino acids in the peptide. Using these values and overlapping the fragments, you should get the sequence Met-Phe-Pro-Met-Gly-Leu-Arg-Lys-Glu-Ala-Ala-Ile. Notice that the Lys-Glu fragment does not give any additional information for the sequencing.

PART II

TRANSMISSION OF INFORMATION

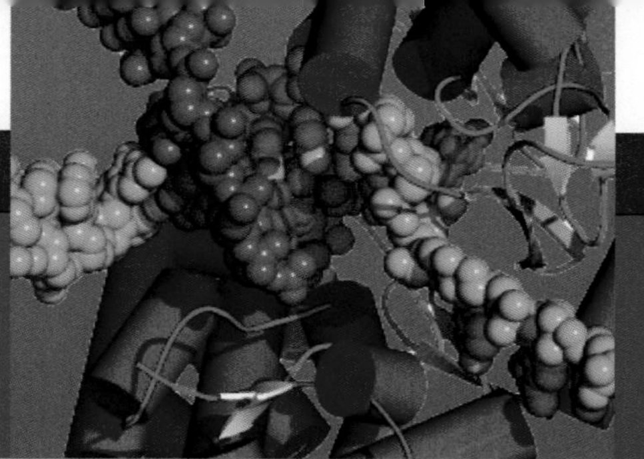

4

DNA Replication, Recombination, and Repair

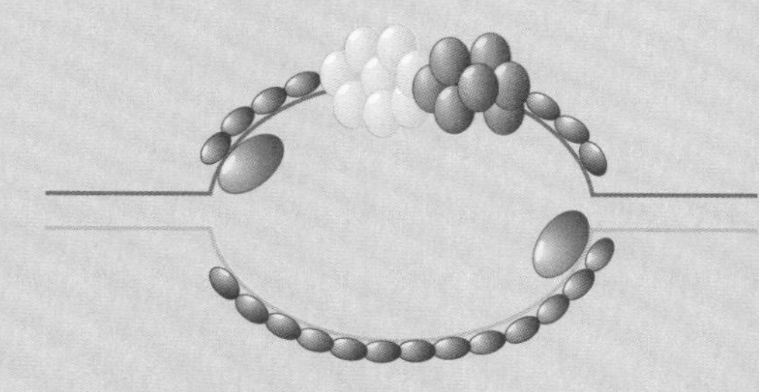

Howard J. Edenberg

Chancellor's Professor, Indiana University School of Medicine

Key Concepts

- DNA replication, recombination, and repair involve transesterification, the switching of partners across a phosphodiester bond. All three processes rely on complementary base pairing.
- DNA replication requires a template, a set of enzymes, and the proper precursors. It is semiconservative, with parental strands separating so that each can be a template for the synthesis of a complementary strand. The double helix opens at a replication fork followed by a priming step, the sequential addition of complementary nucleotides, and the sealing of adjacent fragments. The process is extremely accurate.
- Replication is very similar in bacteria and mammalian cells, but the differences allow antibiotics to selectively inhibit bacterial growth.
- Replication is highly regulated as part of the cell cycle. Telomerase preserves the ends of chromosomes.
- Recombination is essential for proper chromosome segregation into daughter cells. A key intermediate is a DNA structure called a Holliday Junction. Errors in recombination can lead to cancer.
- Changes in a single base pair of the 6 billion base pairs in the human genome can lead to disease. There are an estimated 10,000–50,000 damaging events, such as radiation and chemical reactions, to DNA per cell per day. There are several different mechanisms to repair damaged DNA. Base-excision repair and nucleotide excision repair are very accurate processes that remove damaged base(s). Defects in DNA repair mechanisms can cause cancer.
- Some repair mechanisms introduce errors into the DNA as a cost of allowing cells to deal with damage that blocks replication.

4.1 • COMMON FEATURES OF REPLICATION, RECOMBINATION, AND REPAIR

In addressing the "three R's" of DNA—replication, recombination, and repair—it is useful to consider some unifying similarities among them. The fundamental mechanisms of replication, recombination, and repair are conserved throughout evolution. Because these processes act on the same substrate, DNA chains, they share fundamental chemical mechanisms of making and breaking phosphodiester bonds. The chemistry of DNA is largely that of its phosphodiester bonds. The theme of a nucleophilic attack on a phosphodiester bond, leading to a switch in the partners across the bond, is common to all three processes. The double-stranded nature of cellular DNA and the complementarity of the two strands also play a major role in all three processes. Because the genetic information is present in two copies, each strand can serve as a template for synthesis or repair of the opposite strand. The antiparallel nature of the two strands is critical in all processes that act on DNA. The three processes share many enzymes. Requirements of accuracy and regulation, however, lead to specialization of groups of enzymes for a particular task. Recognizing similarities and differences will make understanding replication, recombination, and repair easier.

As the carrier of genetic information, DNA in a cell must be duplicated, maintained, and passed down to daughter cells with extraordinary accuracy. The scale of the problem is large. The common bacterium *Escherichia coli* (*E. coli*) has a genome containing 4.6 million base pairs (bp). The human genome consists of two copies of 23 chromosomes, totaling approximately 6×10^9 bp. Individual human chromosomes each contain a single DNA molecule, ranging from 34 million bp for the smallest chromosome to 263 million bp for the largest.

Overall accuracy of replication is extremely high: During a single cycle of replication, an average of only one error is introduced per billion bp. However, even with this extraordinary accuracy, a handful of errors are introduced into an average human cell at each cell division. These errors are mutations, heritable changes in the DNA sequence. Mutations are often harmful; they can affect cell viability or trigger the uncontrolled growth characteristic of cancer. Some serious diseases can result from the change of a single base pair in the 6×10^9 bp human genome. Eukaryotes have the additional task of ensuring that all parts of the genome are replicated only once during each cell cycle, to avoid chromosomal abnormalities and imbalances in gene expression.

Accurate replication, critical as it is, is not enough. DNA is subject to continuing attack by both chemical and physical agents including water, activated oxygen species that arise as by-products of normal cellular metabolism, chemicals in foods and in the environment, and radiation. There are systems that recognize and repair DNA damage, and thereby help maintain the integrity of the genome. In spite of all of this care, accumulation of mutations can lead to cell death, aging, and cancer.

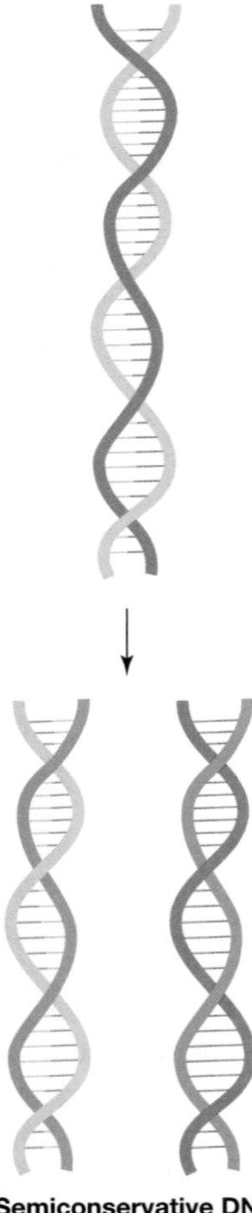

Figure 4.1 Semiconservative DNA synthesis. The two parental strands separate, and each is a template for synthesis of a complementary daughter strand. Therefore, the two daughter duplexes each have one parental strand and one newly synthesized strand.

4.2 • DNA REPLICATION: FUNDAMENTAL MECHANICS

The Basics

Replication leads to doubling of the DNA, preserving the genetic information that is carried as the sequence of bases. Replication requires a **template** to provide sequence information. The basic mechanism of DNA replication was obvious as soon as the complementary, double-stranded structure of DNA was recognized. Because each strand is the complement of the other, with an A in one strand always paired with a T in the other and likewise for C and G, the two strands can be separated and each can be used as a template for synthesis of a new complement. This is called **semiconservative replication**, because half of the parental DNA molecule (one strand) is conserved in each new double helix, paired with a newly synthesized complementary strand (Figure 4.1).

A template alone is not sufficient for DNA synthesis. **DNA polymerases** catalyze the addition of mononucleotides to a growing chain. DNA polymerases do not start a strand by

Figure 4.2 Template and primer. The substrate for addition of a nucleotide is a primer (light tan) hydrogen bonded to a template (dark tan). The template provides information about which nucleotide to add. The primer provides a free 3′-hydroxyl group to which nucleotides can be added. The two strands run antiparallel.

joining the first two nucleotides together; rather, they add nucleotides to an existing primer that provides a 3′-hydroxyl group (Figure 4.2, Table 4.1). The requirement for a primer is not due to the chemistry of phosphodiester bond formation, as demonstrated by the ability of **RNA polymerases** to start polynucleotide chains *de novo* (without primers). Rather, it probably evolved because it increases the overall accuracy of replication. Proofreading (see p. 143) during polymerization of the first few nucleotides is not likely to be effective because there are few base pairs to stabilize the double helix and allow recognition of correct base pairing. In most organisms, the initial primer is made of ribonucleotides by an enzyme called **primase**. The first few nucleotides are marked for eventual removal because they are ribonucleotides; their subsequent removal leaves a gap that can be resynthesized more accurately because the resynthesis is done by elongation of a longer strand of DNA. During growth of the DNA chain, the growing chain itself, base paired to the template, serves as a primer for addition of the next nucleotide.

TABLE 4.1 • Requirements for DNA Replication

Template	Provides sequence information
Primer	Provides free 3′-OH to which nucleotides are added
Precursors	5′-Deoxynucleoside triphososphates (5′-dNTPs)
Enzymes	DNA polymerases, sliding clamps, helicases, primases, single-stranded DNA-binding proteins, nucleases, ligases

Chemistry of Chain Elongation

The chemistry of DNA replication, repair, and recombination is largely that of the **phosphodiester bonds** that link neighboring nucleotides in a DNA chain. Figure 4.3 depicts the

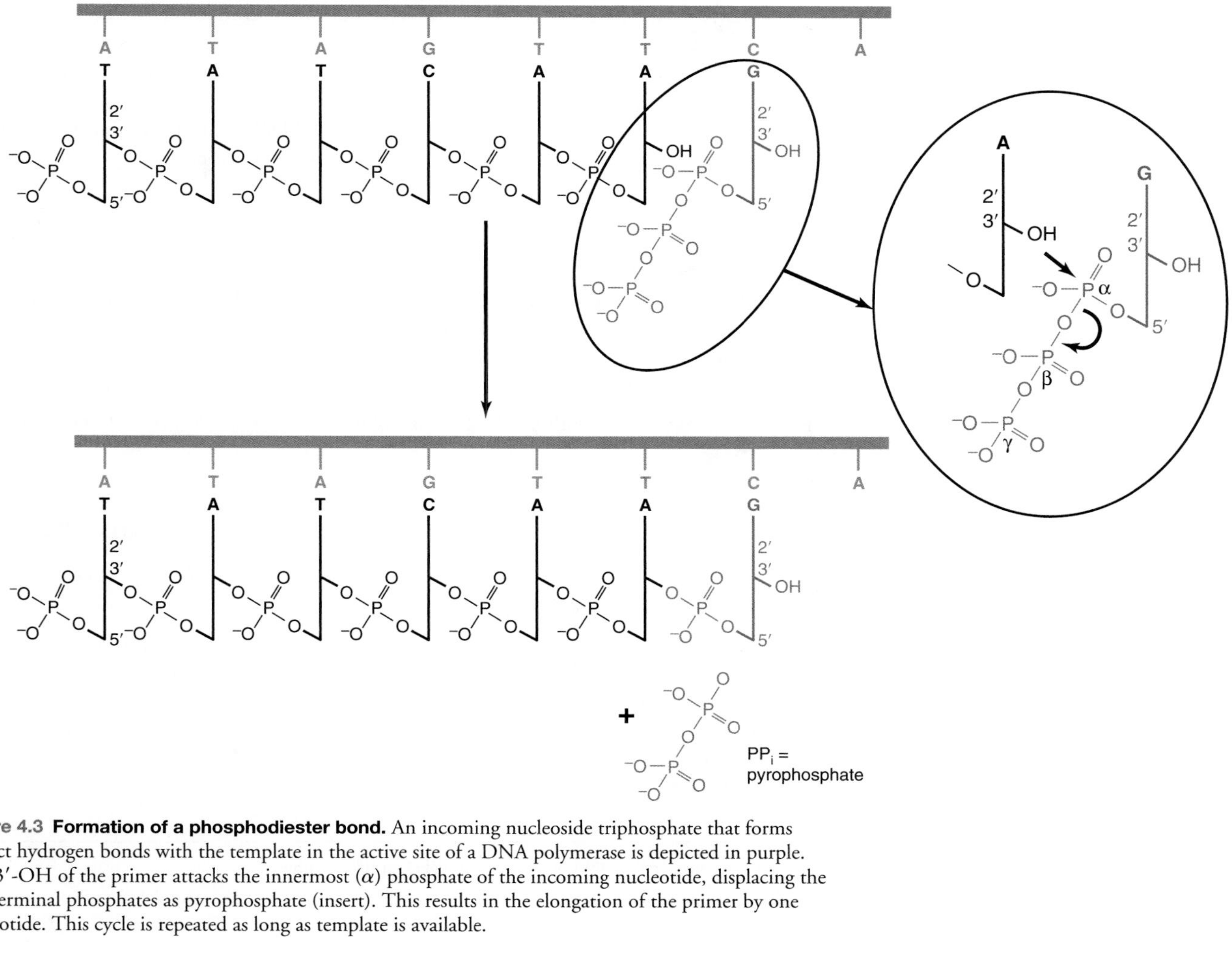

Figure 4.3 Formation of a phosphodiester bond. An incoming nucleoside triphosphate that forms correct hydrogen bonds with the template in the active site of a DNA polymerase is depicted in purple. The 3′-OH of the primer attacks the innermost (α) phosphate of the incoming nucleotide, displacing the two terminal phosphates as pyrophosphate (insert). This results in the elongation of the primer by one nucleotide. This cycle is repeated as long as template is available.

CLINICAL CORRELATION 4.1

Chemotherapy Can Target Precursors of DNA Synthesis

Because tumor cells replicate at higher rates than most normal tissues, much chemotherapy is directed at inhibiting DNA replication. Cells that are actively dividing are more sensitive to these drugs than are quiescent cells; this provides a therapeutic window. In addition to rapidly dividing cancer cells, the targets of chemotherapy, there are normal but rapidly dividing cells in the intestinal tract, bone marrow, and hair follicles; the sensitivity of these rapidly dividing normal cells explains many of the limiting side effects of chemotherapy.

One way to inhibit replication is to interfere with the pools of deoxyribonucleotides needed for DNA synthesis. Hydroxyurea is an antimetabolite that inhibits ribonucleotide reductase, the key enzyme needed to make deoxyribonucleotides from ribonucleotides. This decreases the concentrations of the dNTPs and thereby inhibits DNA synthesis, without dramatically altering the pool of rNTPs needed for transcription. Hydroxyurea is used primarily against melanoma and myeloproliferative diseases such as chronic myelogenous leukemia; it can be used in combination with radiation therapy that damages DNA, because by lowering pools of dNTPs it can also inhibit DNA repair. The main side effect is myelosuppression due to the inhibition of DNA synthesis in bone marrow, which leads to fewer red blood cells, white blood cells, and platelets.

formation of a phosphodiester bond between a short chain of DNA (a **primer**) and an incoming nucleotide. This process is repeated for each nucleotide added to the growing chain.

Addition of a nucleotide to a growing chain is not a spontaneous process, because the decrease in entropy is large. For this reason, nucleotide precursors of DNA are 5′-deoxyribonucleoside triphosphates (5′-dNTPs; Clin. Corr. 4.1). The phosphodiester bond connecting the first (α) phosphate (attached to the 5′ carbon of the sugar) and the outer two (β and γ) phosphates undergoes nucleophilic attack by the 3′-OH of the growing DNA chain (Figure 4.3). The bond is not hydrolyzed, but rather **transesterified;** the phosphodiester bond that had joined the α-phosphate to the β-phosphate now joins the α-phosphate to the 3′-end of the growing chain. The terminal two phosphates are released as inorganic pyrophosphate, which is hydrolyzed by phosphodiesterases present in cells. Cleavage of pyrophosphate renders the reaction essentially irreversible. Much of the chemistry of replication, repair, and recombination involves transesterification reactions.

DNA Polymerases

DNA polymerases catalyze the addition of nucleotides during chain elongation. They require both **templates** and **primers** and use 5′-dNTP precursors (Figure 4.3). The very powerful laboratory techniques of DNA sequencing and the polymerase chain reaction (p. 257) are based on the requirement for both template and primer to allow DNA polymerization. Cells contain several DNA polymerases that carry out specialized functions, including various aspects of replication and coping with damage to the template (Table 4.2). DNA polymerases

TABLE 4.2 • DNA Polymerases

Function	*E. coli Enzyme*	*Eukaryotic Enzyme*
Priming	DnaG primase[a]	Pol α/primase[b]
Synthesis of continuous (leading) strand	Pol III	Pol δ
Synthesis of initial part of Okazaki fragment		Pol α/primase
Synthesis of most of Okazaki fragment	Pol III	Pol δ
Gap filling after primer removal	Pol I	Pol δ
Damage bypass	Pol IV, Pol V	Pol ϵ, Pol ξ, Pol η, Pol ι
Damage bypass, DNA repair	Pol II	Pol ϵ?
Base excision repair	Pol I	Pol β
Mitochondria DNA replication	—	Pol γ

[a] DnaG is not considered a DNA polymerase; it makes an RNA primer.

[b] In eukaryotes, the primase component of the pol α/primase complex makes the initial RNA primer, which is extended 20–30 nt by the DNA polymerase α component using dNTPs.

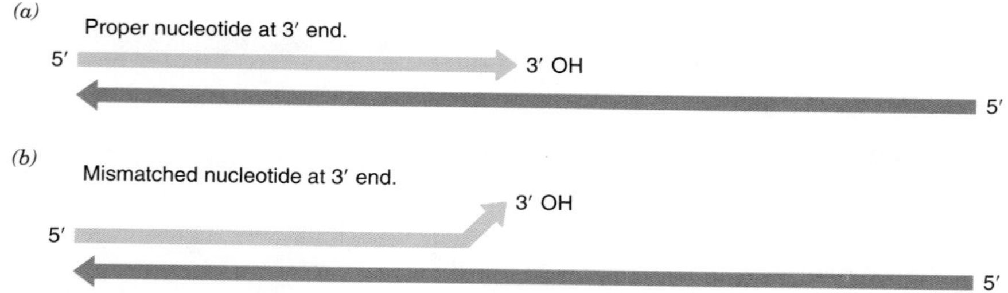

Figure 4.4 Proofreading. (*a*) When the proper nucleotide is added to the 3′-end of a primer, the result is a good substrate for addition of a nucleotide, but a poor substrate for the 3′ to 5′ exonuclease. (*b*) On the rare occasions when an incorrect nucleotide is added to the primer, the resulting primer with a mismatch at the 3′-end is a poor substrate for addition of a nucleotide, but a good substrate for the 3′ to 5′ proofreading exonuclease that removes the mismatched nucleotide. This leaves the primer shorter but ready for the correct nucleotide to be incorporated.

that participate in replication help to ensure accuracy in two ways: by the initial selection of the proper nucleotide to add and by enzymatic proofreading.

The initial selection is based on the fit of the incoming nucleotide into the active site of the polymerase. An incoming nucleotide that makes the proper hydrogen bonds with the nucleotide on the template strand can be added to the growing chain. An incoming nucleotide that does not make the correct hydrogen bonds is not aligned properly for catalysis. Conformational changes in the DNA polymerase itself are critical in assuring this accuracy, which leads to error rates in the range of 10^{-4} to 10^{-6}. Although good for most processes, this is not nearly as accurate as required for replication.

The second way DNA polymerases increase accuracy is by **proofreading**. This is carried out by a 3′- to 5′-**exonucleolytic activity** that removes mispaired nucleotides from the 3′ end of the growing chain. Proofreading is integral to many replicative DNA polymerases, but may also be carried out by an associated protein in the replication complex. A primer with a properly base-paired terminal nucleotide is a good substrate for the addition of the next nucleotide, but a poor substrate for the 3′- to 5′-exonuclease (Figure 4.4). A primer with a mismatched terminal nucleotide is a poor substrate for addition of a nucleotide but a good substrate for 3′- to 5′-exonuclease. This combination results in removal of virtually all mismatched nucleotides before the next nucleotide is added. It also leads to removal of some properly incorporated nucleotides, but that is a price that must be paid for the increased accuracy provided by proofreading. The ability to discriminate between properly and incorrectly incorporated nucleotides is compromised when the growing chain is very short, because the strength of base pairing between a short chain and template is weak. This is the most likely explanation for the evolution of DNA polymerases unable to start chains *de novo,* and for the use of ribonucleotides to prime DNA synthesis.

Some mutated DNA polymerases are actually more accurate than the normal polymerase. This is paradoxical if one assumes that there is an evolutionary advantage to ever-increasing accuracy. However, there are reasons to expect that beyond a certain level, higher accuracy is not advantageous and may in fact be disadvantageous. First, extremely accurate proofreading is energetically costly. To ensure that virtually all misincorporated bases are removed, a polymerase would also have to remove a large number of correctly incorporated bases that have moved slightly with respect to the template due to normal thermal fluctuations. This would greatly slow replication and utilize much additional energy in the form of dNTPs. Second, the ability of DNA repair systems to deal with residual errors reduces the advantages of ever more accurate polymerases. It has also been argued that a low level of mutations provides the raw material for evolution, producing a population with some genetic variation that is better able to survive changing conditions.

Despite their very high accuracy, polymerases can incorporate nucleotide analogs. Nucleoside analogs are often used in chemotherapy to kill rapidly growing cancer cells or viruses responsible for serious illnesses. Analogs that are phosphorylated to nucleotides can be incorporated into DNA, where they can inhibit further synthesis or lead to a high level of mutation. Differences between bacterial or viral polymerases and host polymerases in their ability to incorporate nucleotide analogs can provide a therapeutic window within which physicians can preferentially target infected cells (Clin. Corr. 4.2).

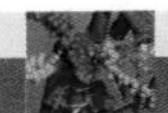

CLINICAL CORRELATION 4.2

Nucleoside Analogs and Drug Resistance in HIV Therapy

Acquired Immune Deficiency Syndrome (AIDS) is caused by infection with the retrovirus Human Immunodeficiency Virus (HIV). A key step in the HIV life cycle is synthesis of a DNA copy of the viral RNA genome, catalyzed by a reverse transcriptase. Reverse transcriptase is a major target of chemotherapy because it is not essential for normal cells.

Zidovudine (AZT; 3′-azido-2′,3′-dideoxythymidine) was the first drug approved for treatment of HIV infection. It is a nucleoside analog with an azido group on the sugar. It can be phosphorylated into the triphosphate form, which competes with dTTP for incorporation into the reverse transcript. Once incorporated, it terminates the growing chain of the transcript because the azido group on the 3′ carbon of the sugar is not a substrate for nucleotide addition. AZT is much less efficient at competing with the more accurate cellular DNA polymerases, providing a therapeutic window in which the effect is primarily on viral replication. Nevertheless, side effects include toxicity to bone marrow, which contains rapidly dividing cells, and myopathy that might be related to toxicity to mitochondria, which contain their own DNA polymerase, pol δ. Other nucleoside analogs have been used to treat HIV infection. Didanosine (DDI; 2′, 3′ dideoxyinosine) and zalcitabine (dideoxycytidine)also function as chain terminators after incorporation of their phosphorylated derivatives by the HIV reverse transcriptase.

Reverse transcriptases do not carry out proofreading, thus their error rate is much higher than that of cellular DNA polymerases. This high error rate complicates the treatment of AIDS, because the population of viruses carried by any one patient contains many mutants, and mutations continue to increase with time. Some of these mutants are likely to be resistant to any given therapeutic agent. Thus, many drugs initially reduce the viral load, but later become ineffective due to the selective growth of viruses in which the drug target is mutated to an insensitive form. Combination therapy, with multiple drugs that target different viral proteins, is an attempt to circumvent this problem.

There are interesting animations of HIV life cycle at www.galaxygoo.org/biochem/hiv/hiv_lifecycle.html and at http://www.roche-hiv.com/portal/eipf/pb/hiv/Roche-HIV/hivlifecycle#.

Separating Parental Strands: The Replication Fork

Both prokaryotes and eukaryotes solve problems of replication in fundamentally similar ways. Separation of the two parental strands of DNA allows each to serve as a template for a new complementary strand. The separation of parental strands creates a structure called a **replication fork** (Figure 4.5). This separation requires considerable input of energy. Melting double-stranded DNA into two single strands normally occurs only at elevated temperatures (over 90°C). To separate the parental strands at physiological temperatures, cells use enzymes called **helicases.** Helicases bind to single-stranded DNA and rachet along it in a fixed direction, with each step requiring hydrolysis of ATP. This "pushes apart" the parental DNA to form a replication fork (Figure 4.5).

In the absence of additional proteins, parental strands would quickly reanneal behind the helicase, because the complementary strands are close and aligned in proper register. Reannealing is prevented by **single-stranded DNA-binding proteins (SSBs)** (Figure 4.5). SSBs keep the strands apart, reduce potential secondary structure (hairpins that might impede polymerization), and align the template strands for rapid DNA synthesis. SSBs are important not only in replication, but also in recombination and repair.

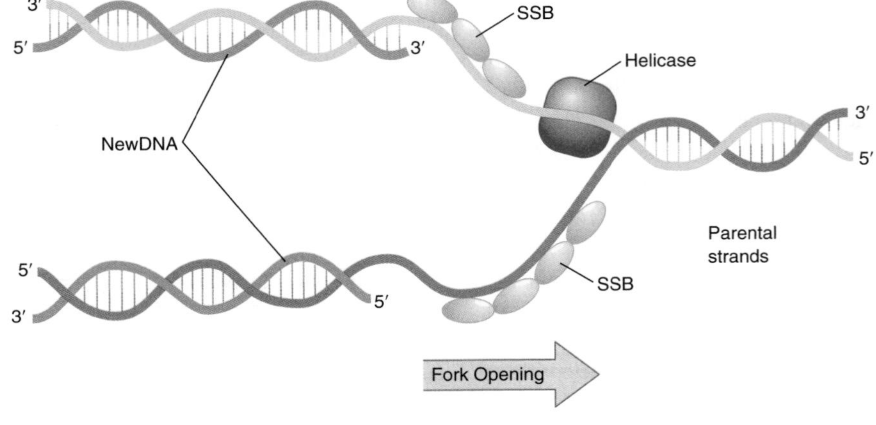

Figure 4.5 Separation of parental strands at a replication fork DNA synthesis occurs at a structure called a replication fork, at which parental strands are separated. A helicase is needed to separate parental strands and allow the fork to progress. Single-stranded DNA-binding proteins (SSBs) are needed to keep the strands apart. The parental strands twist around each other approximately every 10.5 base pairs and must untwist to separate; this creates topological problems. Note that one of the growing strands is oriented with its 3′-end toward the fork, while the other is oriented with its 5′-end toward the fork.

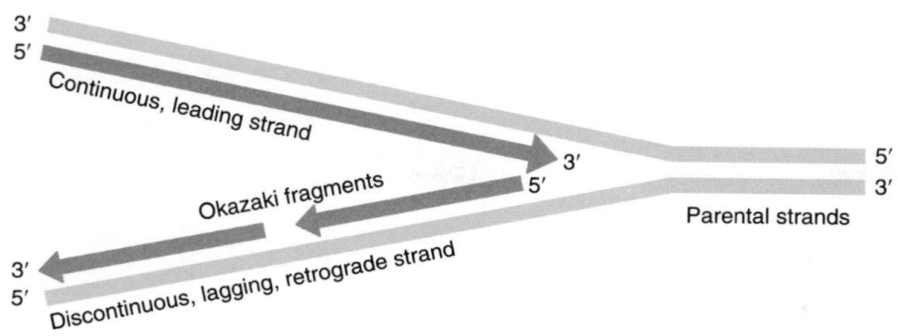

Figure 4.6 Solving the polarity problem: Semi-discontinuous DNA synthesis. Both growing strands are elongated at their 3′-ends. The one with its 3′-end toward the fork can grow continuously; it is called the continuous or leading strand. Because no DNA polymerase can add nucleotides to the 5′-end of a growing chain, the strand with its 5′-end toward the fork is made as a series of short pieces, called Okazaki fragments, each synthesized in 5′ to 3′ direction and later joined. This is called the discontinuous, retrograde, or lagging strand.

Solving the Polarity Problem: Semi-discontinuous DNA Synthesis

Antiparallel strands at a replication fork present an immediate problem for the process of replication; one daughter strand is oriented with the 3′-end toward the fork and the other strand has its 5′-end oriented toward the fork (Figure 4.5). The strand with its 3′-OH oriented toward the fork can be elongated simply by sequential addition of new nucleotides to this end (see Figure 4.3, p. 141); this strand is called the **continuous**, **leading**, or **antero-grade** strand. The other strand presents the problem: No DNA polymerase catalyzes the addition of nucleotides to the 5′-end of a growing chain. Yet when viewed on a larger scale, replication appears to proceed along both strands in the direction of fork movement. This problem is solved by synthesizing one strand as a series of short pieces, each made in the normal 5′ to 3′ direction, and later joining them (Figure 4.6). The strand made in pieces is called the **discontinuous**, **lagging**, or **retrograde** (backward-going) strand; the small pieces from which the discontinuous strand is made are called **Okazaki fragments** in honor of Reiji Okazaki, who first demonstrated this process. Okazaki fragments in human cells average about 130–200 nucleotides in length. In *E. coli*, they are about 10 times larger. The overall process is called **semi-discontinuous** synthesis, because one strand is synthesized continuously and the other discontinuously.

Replication Fork Movement

Priming

DNA polymerases require primers. The **continuous** (leading) **strand** needs only a single priming event, which occurs during replication initiation (p. 155). But each Okazaki fragment of the discontinuous (retrograde, lagging) strand requires a separate primer. These primers are short stretches of RNA, synthesized by a special enzyme called a **primase.** In eukaryotic cells, the RNA primers are about 8–10 nucleotides long. They provide a free 3′-OH to which the first deoxyribonucleotide can be covalently added (Figure 4.7). Therefore, each Okazaki fragment carries a short stretch of RNA at its 5′-end, covalently attached to the DNA.

Strand Elongation

DNA chains grow by repeated addition of nucleotides to the 3′-OH ends of the chains by the mechanism shown in Figure 4.3, page 142. This reaction is catalyzed by a DNA polymerase.

Primer Removal

The RNA primers are removed from the 5′-ends of Okazaki fragments by enzymes with **RNA Hybridase (RNase H)** activity. RNase H catalyzes hydrolysis of an RNA chain hydrogen bonded to a DNA chain (that is, an RNA–DNA hybrid). In eukaryotes, a special flap endonuclease (FEN1) also participates in this process; FEN1 cuts at a bend where the RNA portion has been peeled away from the template. All of the ribonucleotides are removed, leaving only DNA.

Gap Filling

Removal of RNA primers leaves a **gap** (Figure 4.7). Synthesis must fill the gap with deoxyribonucleotides, leaving only a **nick.** Note the terminology: A gap means that at least one nucleotide is missing. A nick is an interruption in the phosphodiester backbone with no missing nucleotides. The gap is filled with deoxyribonucleotides by a DNA polymerase,

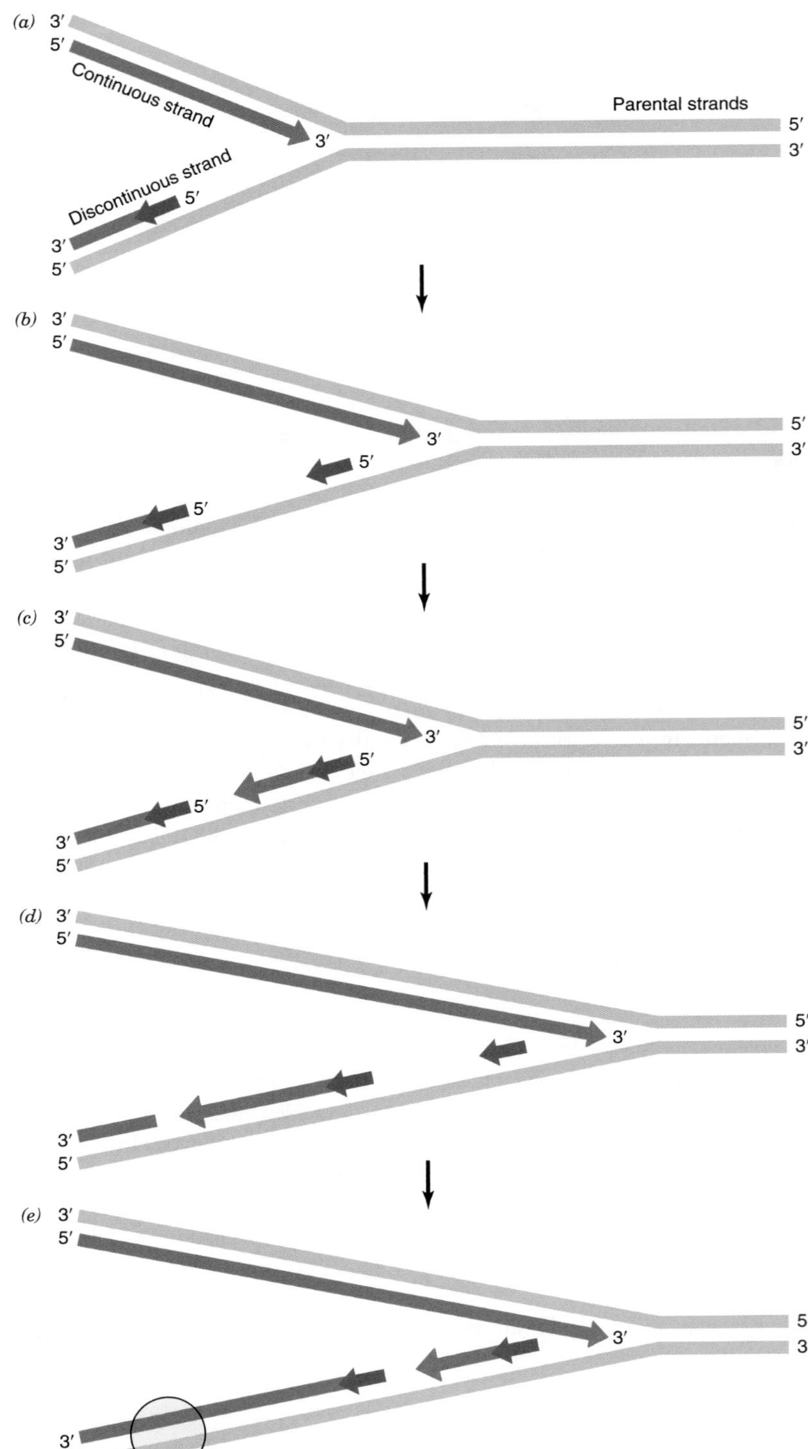

Figure 4.7 DNA synthesis at a replication fork. (*a*) The continuous strand is elongated at its 3′-end by repeated nucleotide additions as shown in Figure 4.2. A previously made Okazaki fragment with a short RNA primer is shown on the other side. (*b*) Using the parental strand as template, primase synthesizes an RNA primer (purple). (*c*) DNA polymerase covalently extends the 3′-end of the RNA primer, incorporating deoxyribonucleotides (tan) as shown in Figure 4.2. (*d*) As the growing Okazaki fragment approaches the previously synthesized fragment, the RNA primer on the older fragment is removed, and the gap is filled by a DNA polymerase elongating the newer Okazaki fragment. Meanwhile, the fork opens further and primase has synthesized a new primer. (*e*) When the gap has been filled and only a nick remains, a DNA ligase seals the two fragments together (within the circle). This process will be repeated, with the new primer elongated, the old primer removed, the gap filled, and the nick sealed.

using the more recently synthesized Okazaki fragment as primer. The Okazaki fragment provides a secure primer–template combination that allows accurate proofreading.

Ligation

The remaining nick is sealed by a **DNA ligase**. Sealing a nick requires formation of a new phosphodiester bond. It therefore requires energy, supplied by coupling the reaction to the splitting of ATP in eukaryotes or NAD^+ in prokaryotes. In both cases the first step is the covalent attachment of an AMP residue to the enzyme (Figure 4.8). This AMP is then

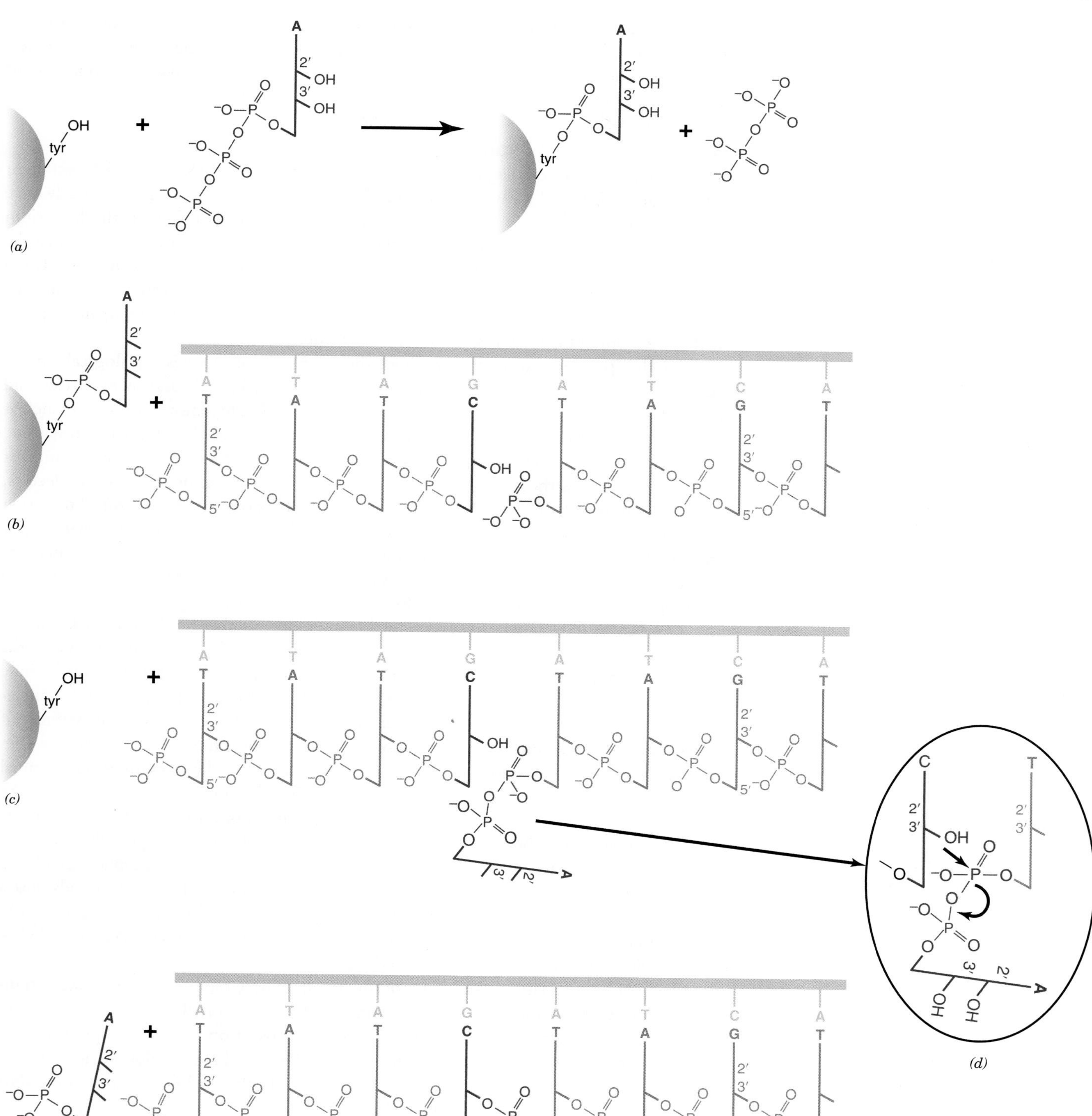

Figure 4.8 DNA ligase mechanism. (*a*) DNA ligase first catalyzes addition of an AMP residue to a tyrosine on the enzyme. In eukaryotes, the AMP donor is ATP (split to AMP + pyrophosphate, as shown here); in prokaryotes, it is NAD + (split to AMP + NMN). (*b*) The ligase-AMP complex binds to a nicked DNA duplex (broken backbone shown in red). (*c*) Ligase transfers the AMP onto the free 5′ phosphate of the nucleotide at the nick. (*d*) Inset: The 3′-OH at the other side of the nick attacks the α phosphate of the nucleotide at the nick to form a new phosphodiester bond, sealing the DNA backbone and releasing AMP. This attack is similar to that which occurs during DNA synthesis (compare Figure 4.2), but with an AMP as the group that leaves, rather than a pyrophosphate. (*e*) The result is a sealed phosphodiester backbone, at a net "cost" of an ATP split into an AMP + PPi (or an NAD split into AMP + NMN).

transferred to the 5′-end of the nick. The AMP serves as a good leaving group that is displaced by the 3′-OH at the other side of the nick; this last step is comparable to the reaction that adds a nucleotide to a growing chain but with an AMP as the leaving group instead of a pyrophosphate.

Untwisting Parental Strands

The above discussion and diagrams were simplified by depicting the DNA strands as straight lines (Figure 4.7, p. 146). But parental strands wrap around each other approximately once every 10.5 bp (Figure 4.5, p. 144). Cells cannot ignore this twisting. In the bacterium *E. coli,* for example, the genome can replicate in approximately 40 minutes. To untwist the 4.6 million bp DNA completely in 40 minutes requires rotation at approximately 11,000 revolutions per minute (rpm)! Clearly, the entire chromosome, compactly folded and carrying large replication and transcription complexes, cannot simply rotate at these speeds without whipping the contents of the cell into a froth.

Another problem caused by the double-helical nature of DNA is **topological**. In circular DNA (such as the *E. coli* chromosome, mitochondrial DNA, and many viruses and plasmids) or long linear DNA whose ends are not free to rotate around each other (such as in our chromosomes, bound at intervals to the nuclear matrix), the number of times one strand wraps around the other, called the **linking number**, is fixed. The linking number cannot be altered without breaking at least one of the two strands. In DNA molecules, the wrapping is of two types, the Watson–Crick **twisting** of one strand around the other approximately every 10.5 bp and the coiling of the double helix on its axis (**writhing** or **supercoiling**). Topologically, twisting and writhing are equivalent and can be interconverted; it is only the total number of times one strand wraps around the other, the linking number, that must remain constant as long as both strands are intact.

One way to allow untwisting and reduction in linking number would be to nick a parental strand ahead of the replication fork. This relieves the topological constraint and allows the parental strands to rotate around each other. A nick presents a serious danger, however, because if the replication fork reached the nick, the strand separation would transform it into a double-strand break in DNA. A double-strand break presents a serious threat to genome stability; if unrepaired, it is lethal. Enzymes called **topoisomerases** solve the unlinking problem in a safer manner by catalyzing changes in the linking number that allow the untwisting and eventual separation of parental strands. Topoisomerases act by forming a transient interruption in the DNA backbone and then resealing it. This interruption is not formed by hydrolysis of the sugar–phosphate bond, but by a transesterification reaction that creates a phosphate–enzyme bond as a transient intermediate (Figure 4.9). Rejoining of the backbone phosphodiester bond displaces the enzyme. Thus there is no net loss or creation of phosphodiester bonds, just a switch in partners across the bond. This means that "nicking" and "resealing" can occur without the need for coupling with a high-energy intermediate (e.g., hydrolysis of ATP or NAD^+). It also avoids the presence of free nicks in DNA.

There are two main classes of topoisomerases: Type I topoisomerases make a transient break in one strand (forming a protein–DNA bond) and allow the other strand to pass through. This changes the linking number in steps of one. Type II topoisomerases make transient breaks in both strands (slightly staggered) and allow a double helix to pass through, changing the linking number in steps of two. Both play important roles in DNA replication and are important targets of chemotherapy.

Sliding Clamps and Processivity

When an enzyme binds to a polymer, carries out a single step of a reaction (for example a polymerase adding a single nucleotide or an exonuclease removing a single nucleotide), and then dissociates from the substrate, the process is called **distributive**. In contrast, when an enzyme binds and carries out many additions (or excisions) before dissociating, the process is **processive**. Because it takes a finite time for an enzyme to dissociate from a template and then reassociate, distributive enzymes tend to work more slowly than processive enzymes.

For rapid synthesis of new DNA, processive polymerases are advantageous. To increase processivity, replicative DNA polymerases are associated with accessory proteins called **sliding clamps** that hold them in contact with the growing DNA chain. In mammalian cells,

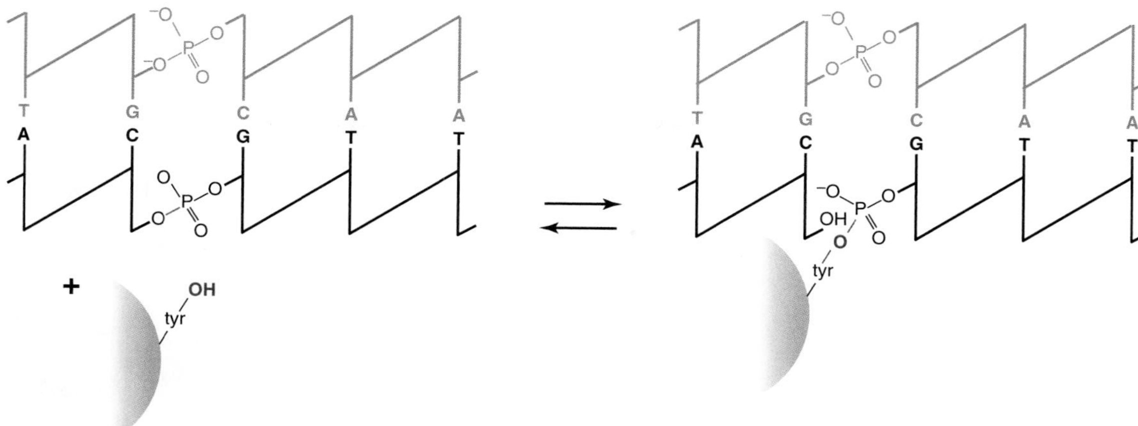

Figure 4.9 Topoisomerase mechanism. Topoisomerase I (mammalian nicking-closing enzyme is depicted here) catalyzes a transesterification reaction that results in a phosphodiester bond between the 3′ carbon of one residue on the DNA chain and a tyrosine residue on the topoisomerase (right panel). (Other topoisomerases form a phosphodiester bond with the 5′ carbon.) This breaks the continuity of the sugar–phosphate backbone and releases the other portion of DNA chain to rotate around (or pass through) its partner; this allows changes in linking number. Then the 5′-OH that was left at the other side of the interruption attacks the phosphodiester bond between nucleotide and protein, reforming the DNA backbone and freeing the protein (left panel). Note that because this is a series of transesterification reactions, no high-energy cofactor is needed.

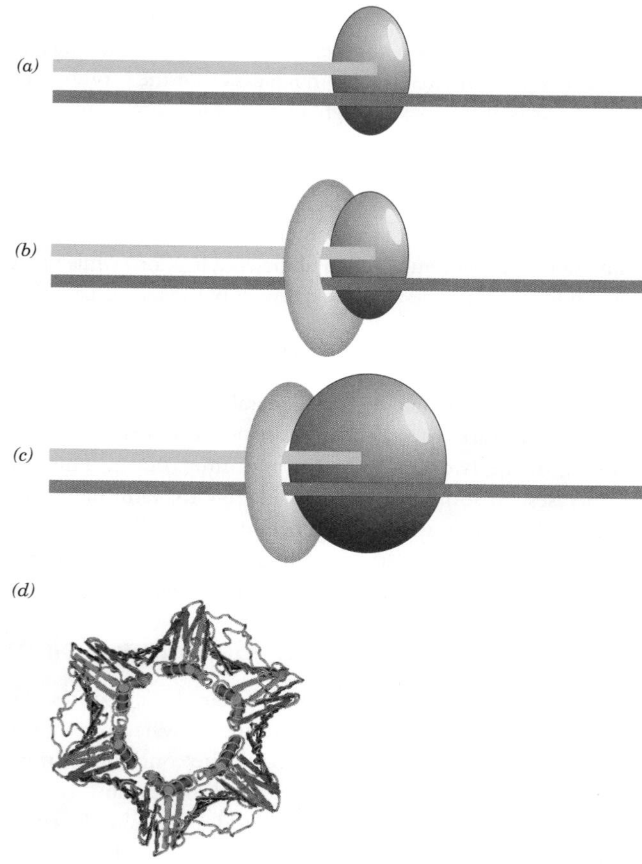

Figure 4.10 Sliding clamps. (*a*) A clamp-loading protein binds to DNA. (*b*) The clamp-loader assembles the sliding clamp from its subunits. (*c*) DNA polymerase associates with the assembled clamp and becomes processive. (*d*) The structure of the sliding clamp in mammalian cells, a trimer of PCNA subunits. DNA passes freely through the large hole in the center of the complex.

the sliding clamps are made of a protein called proliferating cell nuclear antigen (PCNA), with three PCNA molecules encircling the DNA double helix (Figure 4.10). Assembling this ring around the DNA requires accessory enzymes called **clamp-loading factors**. Binding to a sliding clamp makes a DNA polymerase more processive; this increases the speed of synthesis and its accuracy.

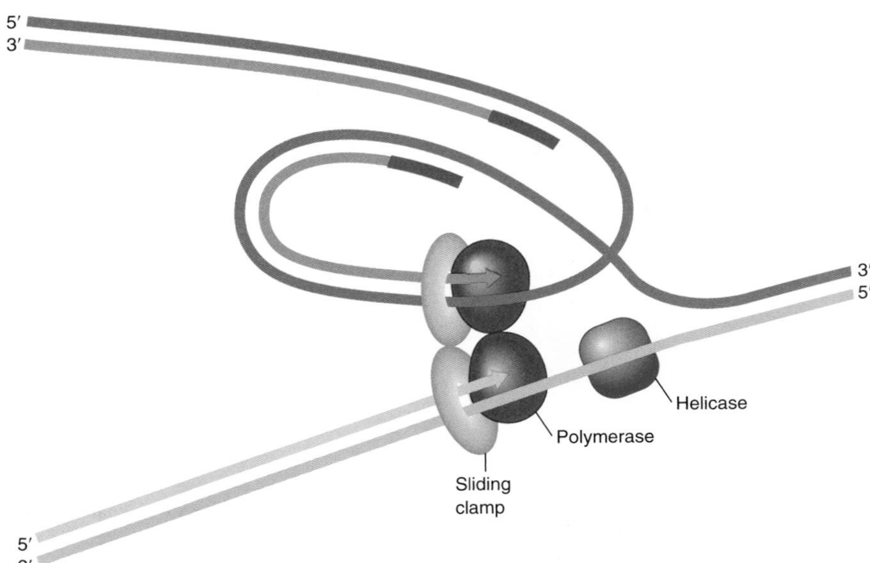

Figure 4.11 Replication of both strands in a replisome. By binding two replicative polymerases together and looping the discontinuous strand so that it can pass through the complex, both strands can be made in one place. DNA feeds through the complex, rotating as it passes through; the large protein complexes do not have to rotate around the DNA.

Choreography in Three Dimensions: The Replisome

The complex choreography described above occurs within the very confined volume of a bacterial cell or the nucleus of a eukaryotic cell, a volume that also contains many other molecules carrying out their functions. The idea of two huge replication complexes circling rapidly around the DNA at each replication fork is untenable. It is much easier for the very thin, cylindrical DNA to rotate around its own axis and move through a large replication complex. The polymerases working on the leading and lagging strands are assembled into a large complex called a **replisome** (Figure 4.11). As the DNA feeds through the replisome, both strands are synthesized. The continuous strand feeds straight through the replisome as it is being elongated at its 3′-end. The discontinuous strand binds to allow synthesis of an Okazaki fragment and is then released to allow the finishing steps (primer removal, gap filling, and ligation) while the template further along the molecule binds the replisome to allow synthesis of the next Okazaki fragment.

Termination of Replication in Circular Genomes

Termination of replication of a circular genome generally occurs 180° away from the origin. Two converging replication forks meet, and the last portion of the genome is synthesized. Topological unlinking of the two new chromosomes must occur. This is a key function of type II topoisomerases. In some small viruses like SV40, termination occurs wherever replication forks meet; there is no special sequence involved. The *E. coli* genome contains special termination sequences that constrain termination to occur within a defined region, by preventing replication forks from proceeding past the region.

Termination of Replication in Linear Genomes: Telomeres

Human cells, and eukaryotic cells in general, have linear chromosomes. There are special difficulties in replicating the ends of linear chromosomes. What exactly is the problem in replicating the ends of linear chromosomes? Although the continuous strand can theoretically be synthesized to the very end of its template, the discontinuous strand cannot. There is no place to synthesize a primer to which the nucleotides opposite the end of the template can be added (Figure 4.12*a*). Even were there a primase that could start at the very end of the template, removal of the RNA would leave a short gap. Although the failure to complete the lagging strand at the end of the chromosome would not be a problem in a single generation, over many cycles of replication chromosome ends would be shortened until essential genes were lost and the cell died. Therefore, it is essential to prevent continued loss of DNA at the ends of chromosomes.

A second problem that eukaryotes face is that the ends of DNA molecules tend to trigger recombination (discussed in the following). To avoid both problems, the ends of eukaryotic

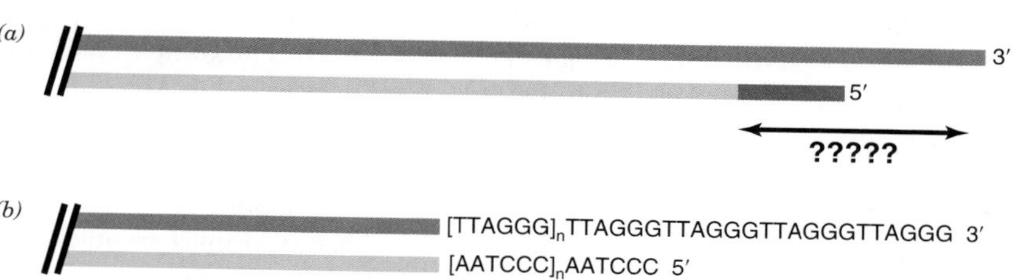

Figure 4.12 Human telomeres. (*a*) The telomere replication problem. The 3′-end of one parental strand, the template for the discontinuous (retrograde) daughter strand, is shown in dark tan, the daughter strand is shown in light tan, and an RNA primer is in purple. The problem is that there is no place to synthesize a primer that would allow the daughter strand to be completed (region shown with????), so the daughter strand will be shorter than the parental strand. Removal of the last RNA primer makes it even shorter. (*b*) A telomere consists of many tandem repeats of a 6 nt sequence, TTAGGG in humans, with the G-rich strand extending beyond the C-rich strand by about 12–18 nt.

linear chromosomes are special structures called **telomeres**, which contain many repeats of a six-nucleotide, G-rich repeated sequence. Human telomeres contain thousands of the repeat TTAGGG. The 3′-end of the chromosome extends about 18 nucleotides beyond the 5′-end (Figure 4.12*b*), leaving three repeats as an overhang. The overhanging 3′-end folds back on itself, forming non-Watson–Crick G-G hydrogen bonds, and binds proteins that define its length and protect the chromosome ends from undergoing recombination.

Epigenetics

An important epigenetic marking in mammalian DNA is the methylation of C residues on both strands of many CpG sequences. This can alter gene regulation and silence some genes (Chapter 8). Patterns of methylation can be inherited by daughter cells, but how? During replication, the new strand is made in unmethylated form, with unmodified C residues incorporated. Methylation is restored by DNA methyltransferases that recognize the hemimethylated DNA resulting from semiconservative replication (the C on the old strand remains methylated; that on the new strand is not) and methylate the newly incorporated C.

4.3 • DNA REPLICATION: ENZYMES AND REGULATION

Prokaryotic Enzymes of Replication

Enzymes that carry out the movement of a replication fork in *E. coli* are shown in Figure 4.13. The main replicative DNA polymerase in *E. coli* is **DNA polymerase III (pol III)**. Pol III is a multi-subunit complex that synthesizes the continuous strand and most of the lagging

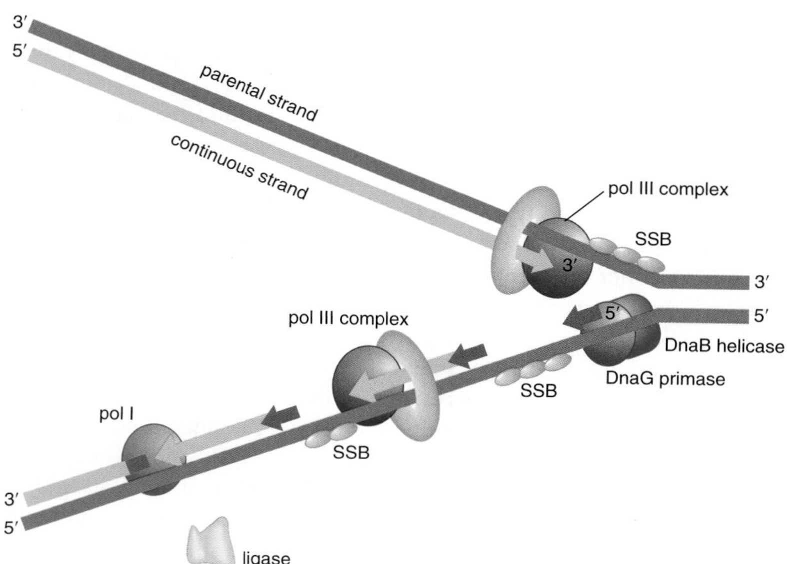

Figure 4.13 Replication enzymes in *E. coli*. The continuous strand (top) is elongated by pol III associated with a sliding clamp. The DNA helicase/primase complex (DnaB/DnaG) moves along the template for the discontinuous strand (bottom); it pauses to allow DnaG to synthesize an RNA primer every 1000 to 2000 nt. Pol III elongates the primer until it reaches the primer of the previously synthesized Okazaki fragment. At that time, pol III releases the DNA, and pol I binds to the end of the Okazaki fragment. Pol I removes the RNA primer one nucleotide at a time using its 5′ endonuclease activity and simultaneously fills the gap in a process called nick translation. When there is a remaining nick with deoxyribonucleotides on both sides, DNA ligase can seal the nick.

strand. The core polymerase contains a θ subunit that catalyzes phosphodiester bond formation, an β subunit that is the 3′ to 5′ proofreading exonucleases, and a β subunit. The sliding clamp in *E. coli* consists of two β subunits that are assembled onto the DNA by the γ complex (the clamp loader) in a step that requires the hydrolysis of ATP. The β subunits form a ring around the double helix that slides along it (Figure 4.10, p. 149), tethering the pol III complex to make the DNA synthesis a highly processive process. This complex can synthesize DNA until it runs out of template or hits certain types of damage in DNA.

On the lagging strand, the complex releases the DNA when it encounters the previous Okazaki fragment. Two molecules of pol III are held together by the τ subunit, allowing synthesis of both leading and lagging strands in one large replisome.

On the lagging strand the DNA primase (called DnaG) forms a complex with the DnaB helicase (DnaB). DnaB causes the complex to travel along the template for the discontinuous strand, pushing the parental strands apart in a movement that requires ATP hydrolysis. The single-stranded DNA that results from helicase action is coated with **single-stranded DNA-binding protein (SSB)** that prevents reannealing of parental strands and prevents hairpins and other secondary structures from forming in the single stranded DNA. The helicase/primase complex stops approximately once per 1000–2000 bp and the DnaG primase synthesizes a short RNA primer. The RNA primer is elongated into an Okazaki fragment of about 1000–2000 bp by DNA polymerase III with its associated sliding clamp. Elongation stops when the polymerase complex encounters the RNA primer from the previously synthesized Okazaki fragment, and the DNA dissociates from the pol III complex.

In *E. coli*, RNA primer removal and gap filling are both catalyzed by a single enzyme, **DNA polymerase I (pol I)**. Pol I contains a 5′- to 3′-exonuclease activity that removes the RNA primer from the 5′-end of the previously synthesized Okazaki fragment, thereby functioning as an RNase H. It also contains a DNA polymerase activity that catalyzes addition of deoxyribonucleotides to the 3′-end of the more recently made Okazaki fragment until the gap created by removal of the primer is filled (Figure 4.13). An intrinsic 3′-to 5′-proofreading exonuclease increases accuracy of gap filling. Pol I coordinates primer removal and gap filling by excising a ribonucleotide from the primer and elongating the newer Okazaki fragment with a deoxyribonucleotide. It repeats these steps until the primer is removed. Pol I is capable of continuing down the DNA, removing deoxyribonucleotides and replacing them, an activity called nick translation. This would be a wasteful procedure, but it is limited by the fact that pol I acts in a distributive manner, coming off the substrate frequently. As long as there is still at least one ribonucleotide remaining, DNA ligase cannot act on the substrate but pol I can bind again and replace the ribonucleotide. When the gap has been filled and only a nick remains, dissociation of pol I allows **DNA ligase** to bind to the nick and catalyze formation of a phosphodiester bond, sealing the Okazaki fragment to the growing chain.

As noted in the preceding, the circular *E. coli* chromosome is under **topological constraint**. Yet the parental double helix must be completely unwound for replication and the completed chromosomes to separate into daughter cells. Replication requires the removal of positive supercoils, because the linking number of the parental strands must be reduced from a large positive number (approximately 440,000, because they are twisted around each other once per 10.5 bp) to zero to allow the daughter chromosomes to separate. **Topoisomerases** are, therefore, crucial for DNA replication.

E. coli has both type I and type II topoisomerases. The *E. coli* type I topoisomerase is called the omega protein (ω). Omega acts unidirectionally, removing only negative supercoils. Therefore, ω is not sufficient to allow replication to occur. **DNA gyrase**, a type II topoisomerase, is essential for DNA replication. DNA gyrase acts as a "power swivel" to remove positive supercoils or introduce negative supercoils; the direction is identical. Gyrase is a heterotetramer with two "swivelase" subunits encoded by gyrA and two ATPase subunits encoded by gyrB. The swivelase subunits catalyze transesterification reactions that break and reform the phosphodiester backbone, creating and resealing a transient nick on both strands (a transient double-strand break). Hydrolysis of ATP is coupled with gyrase action not to form new phosphodiester bonds but rather to trigger the conformational changes that allow a double helix to pass through the transient double-strand break, resulting in unidirectional reduction in linking number. Antibiotics that target one or the other subunit of DNA gyrase

CLINICAL CORRELATION 4.3

Topoisomerases as Antibiotics

Antibiotics that target either subunit of DNA gyrase rapidly stop *E. coli* DNA replication, because preventing the reduction in linking number of the parental strands prevents the strands from untwisting. There are two modes of targeting topoisomerases. Topoisomerase inhibitors such as coumermycin A1 and novobiocin prevent catalytic activity; they target the ATPase subunits encoded by gyrB.

Topoisomerase poisons such as Nalidixic acid freeze the covalent DNA–protein links; these complexes are lethal if converted into double-strand breaks, as would happen during replication. Nalidixic acid is used against urinary tract infections; it targets the swivelase subunits encoded by gyrA. Ciprofloxacin, another topoisomerase poison, is one of the most effective oral antibiotics in clinical use today; it is used to prevent and treat anthrax and many other bacterial infections.

Froelich-Ammon, S. J., and Osheroff, N. Topoisomerase poisons: Harnessing the dark side of enzyme mechanism. *J. Biol. Chem.* 270:21429, 1995.

rapidly stop *E. coli* replication (Clin. Corr. 4.3). *E. coli* has a second type II topoisomerase activity, topo IV, which is important in chromosome segregation into daughter cells.

Eukaryotic Enzymes of Replication

Eukaryotes require the same types of enzymatic activities as do prokaryotes, because replication follows essentially the same pathway. But differences between bacterial and human enzymes in the details of their specificity and mechanisms are used in antibacterial therapy to target pathogen replication and spare the host cells.

The continuous strand at a replication fork is synthesized by **DNA polymerase δ (pol δ)**, associated with the **sliding clamp** called **proliferating cell nuclear antigen (PCNA)**. PCNA was first detected as an antigen in nuclei of replicating cells, hence its name. Three subunits of PCNA are assembled to form a ring around DNA (Figure 4.10, p. 149), to which pol δ attaches. Assembly of this ring requires a **clamp-loading factor** called Replication Factor C (RFC).

The situation on the lagging strand is slightly more complicated in eukaryotes than in *E. coli* (Figure 4.14). A helicase activity in eukaryotes separates the parental strands, and a single-strand DNA-binding protein called **replication protein A (RPA)** binds to the exposed single strands. In eukaryotes, the primase forms a complex with **DNA polymerase α (pol α)** that initiates Okazaki fragment synthesis. This **pol α/primase** complex synthesizes a primer of approximately 10 ribonucleotides, and then switches from primase to DNA polymerase activity and elongates the primer with approximately 15–30 deoxyribonucleotides. The product of this dual reaction is a short stretch of DNA covalently attached to the RNA primer. Once the Okazaki fragment has reached this length, the pol α/primase complex dissociates from DNA. RFC binds to this elongated primer and serves as a **clamp loader** to assemble the PCNA **sliding clamp**. Then pol β binds to the PCNA and completes the Okazaki fragments to a final length of about 130–200 bp.

Primer removal is carried out in two steps by RNase H and FEN1. **Rnase H** degrades the RNA primer, leaving a single ribonucleotide attached to the end of the Okazaki fragment. **Flap endonuclease 1 (FEN1)** removes the last ribonucleotide (and possibly some deoxyribonucleotides) by peeling back one or a few nucleotides to form a small "flap" and then cleaving the phosphodiester bond at the angle to release the flap. If there is a mismatch within the first few nucleotides of the Okazaki fragment, as a result of misincorporation by pol α, the mismatch would destabilize the 5′-end and create a larger flap that could be excised by FEN1. This increases accuracy of the replication process by removing errors introduced by pol α.

The gap that remains is filled by pol δ extending the 3′-end of the more recently synthesized Okazaki fragment (Figure 4.14). The pol δ/PCNA complex releases the DNA when it encounters dNTPs at the 5′-end of the previously synthesized Okazaki fragment. The remaining nick is sealed by DNA ligase. The pol δ/PCNA complex of the replisome must then reattach to the next partly synthesized Okazaki fragment to repeat this process. As in *E. coli*, two molecules of the main replicative polymerase, pol δ in this case, are held together in a replisome, or "replication factory" in which both strands are synthesized (Figure 4.11, p. 150).

Eukaryotic DNA is packaged into **nucleosomes** that contain approximately 200 bp of DNA. Dissociation of nucleosomes is required for replication and probably limits the rate of DNA synthesis. When a single nucleosome is dissociated, about 200 bp of parental DNA are available to be separated; primer synthesis can occur somewhere in the exposed single-stranded DNA. This would explain the limited size of the Okazaki fragments in human cells.

Humans have both type I and type II topoisomerases. The human type I topoisomerase, called nicking-closing enzyme (Figure 4.9, p. 149), can remove both positive and negative supercoils; it functions during DNA replication and transcription. Type II topoisomerase is critical at the termination step and for segregation of chromosomes, which would otherwise be tangled together as the many replication bubbles are completed. Topo II is an abundant protein that also plays a role in attaching DNA to special sites in the nuclear matrix during interphase. Human type II topoisomerase is not a gyrase, in that it does not introduce negative supercoiling. Cancer chemotherapy often targets topoisomerases, using poisons that lead to double-strand breaks during replication (see Clin. Corr. 2.7). Rapidly replicating cells are more sensitive to these drugs than quiescent cells.

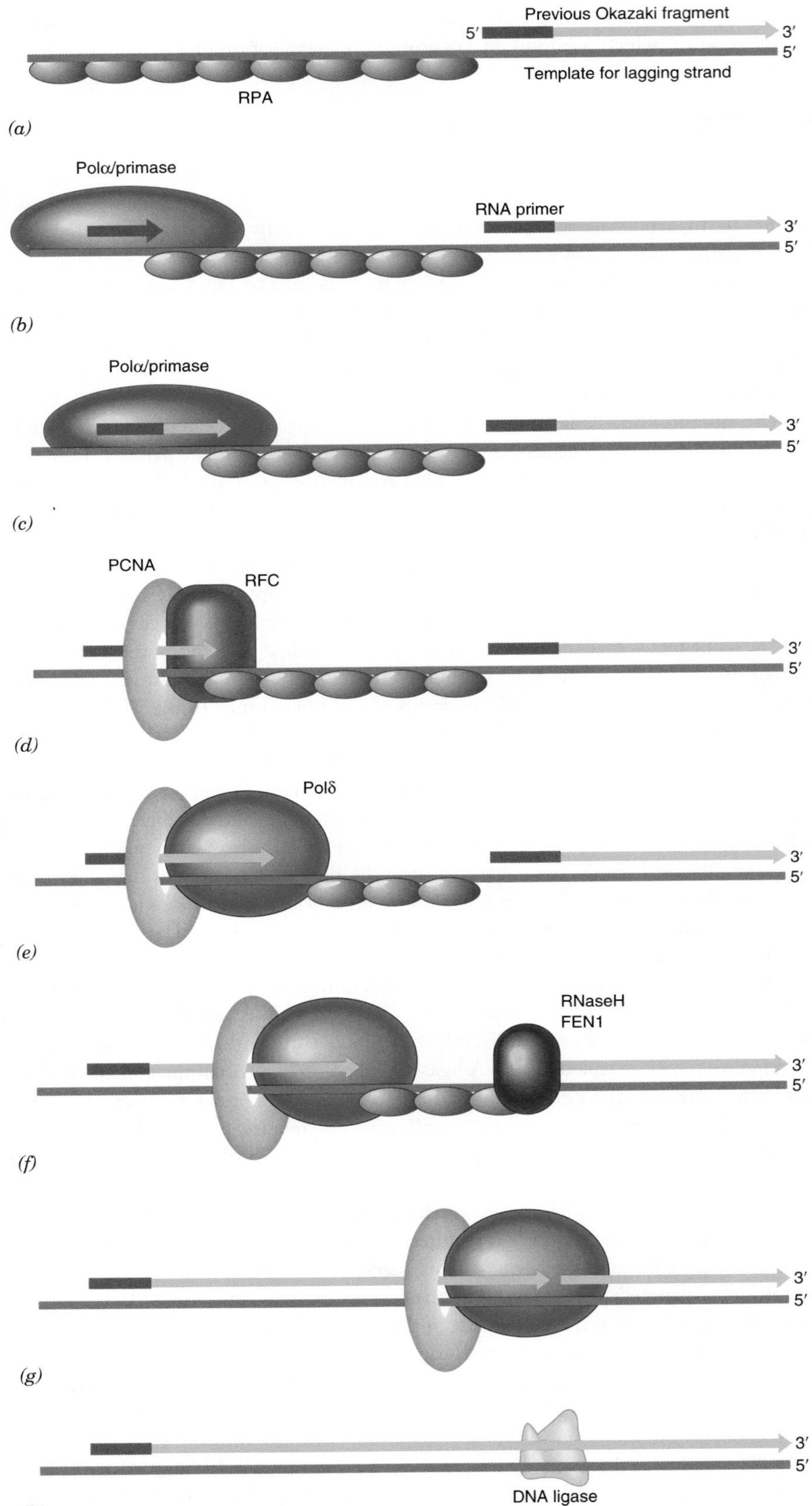

Figure 4.14 Enzymes replicating the lagging strand in eukaryotes (leading strand not shown). (*a*) RPA, an SSB, binds to the single-stranded template that has been opened up by the progressing replication fork. A previous Okazaki fragment is shown at the right, with the RNA primer in purple. (*b*) The primase activity of the polα/primase complex initiates synthesis of an RNA primer. (*c*) After the RNA primer reaches 10 nucleotides, the polymerase activity of the polα/primase complex takes over and elongates the primer with about 15 to 30 deoxyribonucleotides. Then the polα/primase dissociates. (*d*) RFC binds to the end of this partly completed Okazaki fragment and catalyzes assembly of a sliding clamp from three molecules of PCNA. (*e*) Pol δ binds to the sliding clamp and elongates the Okazaki fragment. (*f*) As the replication complex approaches the previous RNA primer, that primer is degraded by the combined action of RNase H and FEN-1. (*g*) The gap is filled by the continued elongation of the Okazaki fragment. (*h*) The remaining nick is sealed by DNA ligase.

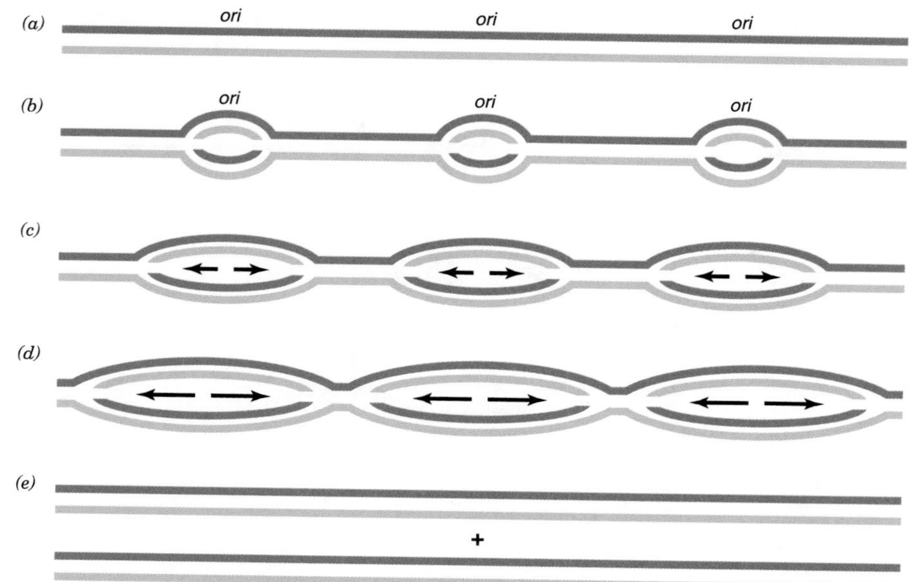

Figure 4.15 Tandem replicons and bidirectional replication in eukaryotes. (*a*) There are multiple origins of replication (ori) tandemly arrayed along each eukaryotic chromosome. In humans they are spaced at approximately 50,000- to 100,000-bp intervals. (*b*) Initiation occurs at each ori. Adjacent clusters of oris tend to function together. (*c*) Two diverging replication forks are established at each ori, so replication is bidirectional. The structures formed are called replication bubbles. (*d*) Replication bubbles enlarge as replication continues, until they are in close proximity. At that stage, termination of replication joins adjacent bubbles and unlinks parental DNA. Topoisomerase II is essential for termination of replication and segregation of the chromosomes. (*e*) The resulting duplicated chromosomes can then segregate into two daughter cells.

Initiation of Replication

In the previous discussion, we addressed progression of a replication fork. But, how does a replication fork get started? And how is initiation controlled so that the entire genome is copied once and only once? Replication begins from specific sites called **origins of replication.** Known origins contain multiple, short, repeated sequences that bind specific initiator proteins, and AT-rich regions at which the initial separation of parental strands occurs. The *E. coli* chromosome has a single origin of replication, origin of chromosomal replication (oriC), a region of approximately 245 bp. There are thousands of origins in eukaryotic cells (Figure 4.15), which allow those cells to replicate the large amount of DNA in the limited time of one cell cycle. In yeast the origins are termed **Autonomously Replicating Sequences (ARS)**. In humans, specific sequences that serve as origins have not been identified.

In *E. coli,* oriC is bound by an **initiator protein** called **DnaA. DnaC** then associates and acts like a "matchmaker" to allow **DnaB**, the helicase, to bind and begin separating the parental strands to create a replication fork. A replication bubble is formed at each origin and a pair of replication forks is established that move away from the origin, one in each direction (Figure 4.15). Thus, replication is **bidirectional**.

The **Origin Recognition Complex (ORC)** in eukaryotes assembles at multiple origins (Figure 4.16). Assembly of an ORC at an origin is necessary but not sufficient for initiation to occur. A second complex called **minichromosome maintenance proteins (MCM)**, which has a weak helicase activity, must also bind, forming a pre-replicative complex; Cdc6 (the protein encoded by *Cdc6*) is required for this binding. Once the ORC/MCM complex bound at an origin is activated, it catalyzes the initial separation of parental strands to form a small **replication bubble** (Figure 4.16). **SSBs** (the human SSB is RPA) bind and hold the separated strands apart. The **MCM** complex may then function as a helicase to allow replication fork unwinding. Activation of the ORC/MCM complex is regulated by **cyclins** and **cyclin-dependent protein kinases** (see p. 1005). Phosphorylation of Cdc6 by a **cyclin-dependent kinase** inactivates it and prevents reassembly of the complex; this is one of the ways in which reinitiation of replication, which would be disastrous for the cell, is prevented.

The initiation of DNA synthesis uses most of the same mechanisms and enzymes as the movement of a replication fork (compare Figure 4.14, p. 154, and Figure 4.16). In eukaryotes, the polα/primase complex initiates synthesis of an RNA primer and then switches to elongating it with dNTPs. This short polynucleotide is elongated by a DNA polymerase δ in the same way as on the discontinuous (retrograde) strand at a replication fork. The replisome elongating this first strand can continue synthesis and forms the leading strand of the fork on one side of the replication bubble (Figure 4.16). On that half of the bubble, synthesis of the first Okazaki fragment on the retrograde strand is initiated by the same mechanism as is used in moving the replication fork (compare Figure 4.13, p. 151). However, as this

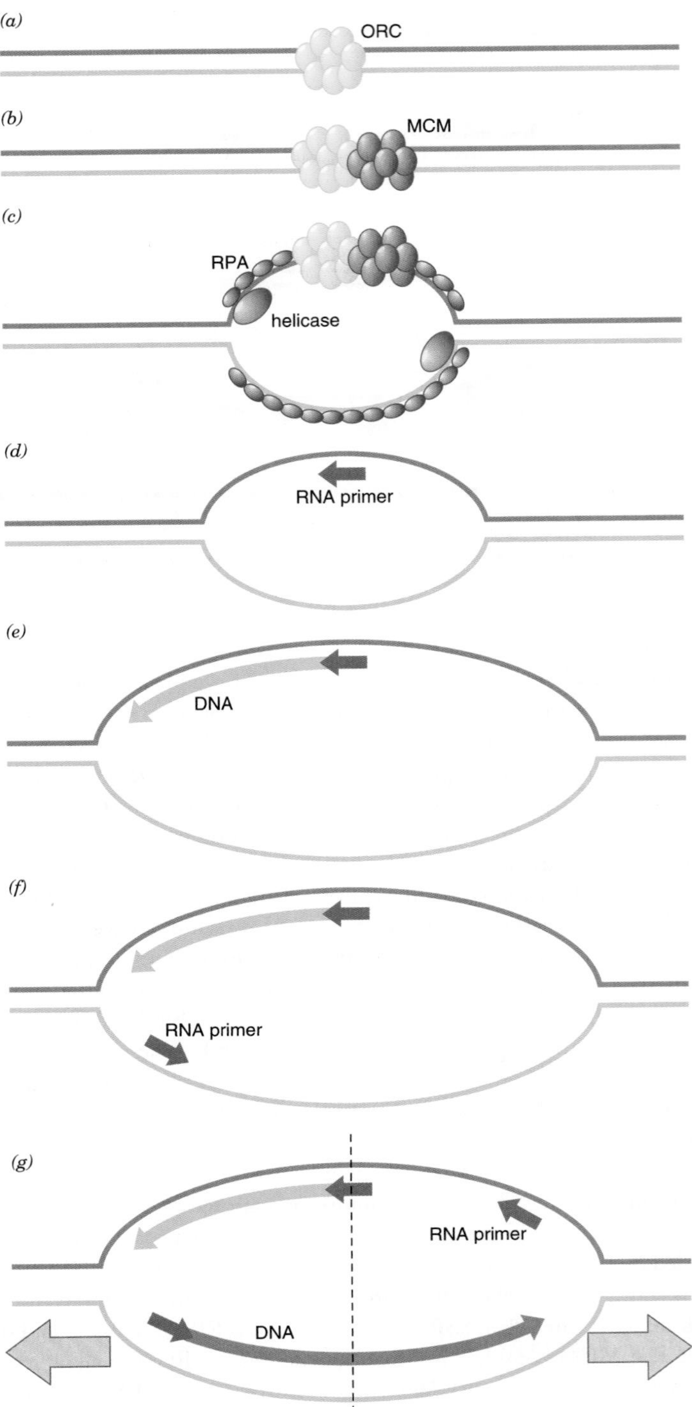

Figure 4.16 Initiation of replication at a eukaryotic replication origin. (*a*) The origin recognition complex (ORC) binds to an origin. (*b*) An MCM complex binds to this; cell division cycle 6 protein (cdc6) is important in this assembly. (*c*) The initiator complex is activated and the helicase activity opens the parental strands to form a very small bubble. SSB binds to the exposed single strands, helicases are loaded onto the DNA, and the bubble is enlarged. (*d*) Polα/primase synthesizes the first RNA primer and, after about 10 nt, switches to elongating it with deoxyribonucleotides. (In this and subsequent panels, the focus is on processes occurring on DNA, and proteins are not shown.) (*e*) After about 15–30 deoxyribonucleotides are added, polα/primase leaves and the chain is elongated by DNA polymerase δ, which incorporates deoxyribonucleotides. To this stage, the process is like that occurring on the discontinuous (retrograde) strand (compare Figure 4.14). However, the elongating strand will not encounter a previous Okazaki fragment; it can continue elongating, becoming the continuous (leading) strand on the leftward-moving replication fork. (*f*) An RNA primer is synthesized on the discontinuous side of this replication fork, as in normal fork progression. (*g*) This primer is elongated as previously described. However, this elongating strand will not encounter a previous Okazaki fragment and can continue elongating, becoming the continuous (leading) strand on the rightward-moving replication fork. An RNA primer can be synthesized on this fork by the normal mechanism (compare Figure 4.14). The result is two replication forks diverging from the origin. The process is symmetrical around the axis indicated by the dotted line, with mirror-image forks diverging from origin.

first Okazaki fragment is elongated, it will not run into a previous fragment so it continues to grow in the 5′ to 3′ direction; as it passes the origin, it becomes the leading strand of the replication fork moving in the opposite direction (Figure 4.16). Both leading strands continue to be elongated in a highly processive manner, and Okazaki fragments are synthesized on the opposite sides. The result is a pair of replication forks diverging from the origin.

Telomerase

Telomeres, the specialized structures at the end of linear chromosomes in eukaryotes (Figure 4.12, p. 151), are maintained by **telomerases**, enzymes that add new six-nucleotide repeats to the 3′-ends of the telomeres. Telomerases are ribonucleoprotein

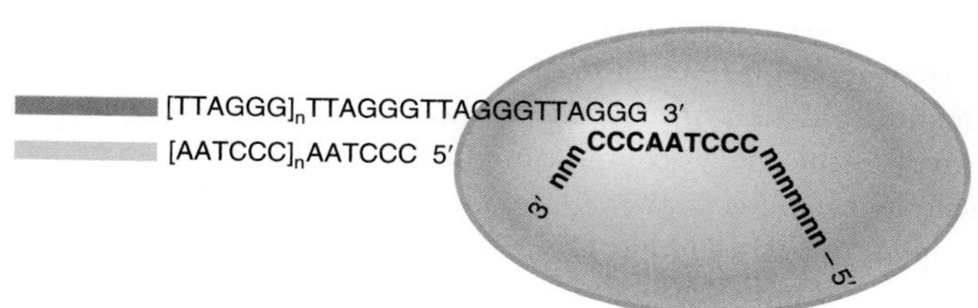

Figure 4.17 Telomerase. Telomerase is a ribonucleoprotein complex with a short RNA strand as an integral part; it catalyzes the addition of new 6-nt telomere repeats to the 3′-end of a DNA chain. Telomerase RNA partially base-pairs with the telomeric repeat and serves as the template for the reaction, while the protein component functions as a reverse transcriptase, synthesizing DNA using the RNA template. After a six-nucleotide repeat is added, the enzyme can dissociate and bind again, and add additional 6-nt repeats.

complexes containing a small RNA that serves as a template for addition of a new six-nucleotide repeat (Figure 4.17). A telomerase binds to the end of the 3′-strand with part of the telomerase RNA hydrogen bonded to the last few nucleotides of the chromosome. A six-nucleotide repeat is synthesized, using the RNA as a template. Then the telomerase can dissociate and reassociate to add another hexamer.

Telomeres do not have to remain exactly the same length; some shortening is not a problem, because the repeats do not encode proteins. Telomeres undergo cycles of shortening of the lagging strands due to the inability to complete synthesis (Figure 4.12*a*, p. 151) and lengthening by addition of new six-nucleotide repeats to the 3′-end by telomerase (Figure 4.17). Although the length of the telomere does not remain constant, progressive shortening is avoided by addition of repeats. Telomerases also reestablish the 3′-overhang characteristic of telomeres.

Cells that have differentiated and will divide only a limited number of times do not express telomerase. Thus, the telomeres shorten with each further division; this limits the number of times such cells can divide before the loss of telomeres triggers apoptosis, programmed cell death (see p. 1010). Telomerase expression is generally reactivated in tumor cells; this allows them to continue division indefinitely without chromosome shortening, and makes telomerase an attractive target for cancer chemotherapy. It should be noted that inactivating telomerase in a tumor would not lead to a rapid halt in tumor growth; the effect would be delayed by many cell cycles, until the chromosome ends shorten significantly. Thus, telomerase inhibitors are likely to be useful only in conjunction with other therapies.

Cell Cycle

In eukaryotes, there is a distinct and regulated pattern of activities that constitutes a **cell cycle** (Figure 4.18 and p. 157). The most striking morphological feature of the cell cycle is mitosis, the condensation, alignment, and separation of chromosomes followed by division of a cell into two cells. The phase of the cell cycle in which mitosis occurs is called **M-phase** (for mitotic phase). DNA replication takes place during the **S-phase** (for synthesis phase). **G1-phase** (Gap 1 phase) occurs between mitosis and S-phase, and **G2-phase** (Gap 2 phase) between S-phase and mitosis. G1-, S-, and G2-phases together are called interphase and constitute most of the total time of a cell cycle. A higher-than-normal fraction of cells in mitosis is a histological indicator of cancer (Clin. Corr. 4.4).

In eukaryotes, replication is regulated by determining whether to start DNA replication. Once started, cells generally proceed through the entire replication process. Initiation of replication at origins must be coordinated to ensure that all regions of all chromosomes are replicated in each cell cycle and that no region is replicated more than once. This is, obviously, a difficult challenge in a cell with thousands of origins. Failure to replicate the entire genome would lead to chromosome instability and loss of genetic information. If a portion of the chromosomal DNA is not replicated, the two progeny chromosomes cannot separate properly, or the chromosome could be broken during the attempt. Broken chromosomes are unstable and trigger recombination, which causes further problems. Either the lack of key genes in a cell missing a chromosome due to missegregation or the excessive expression of other genes in a cell with an extra copy of a chromosome can be disastrous. Trisomy 21, the presence of one extra copy of the smallest human chromosome, causes Down syndrome, and most other trisomies are lethal.

Different regions of the genome are replicated at different times during the S-phase of the cell cycle. Typically, regions that are actively transcribed are replicated early during

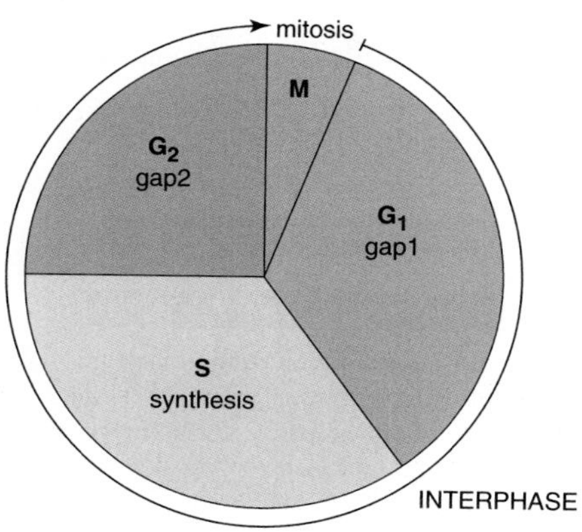

Figure 4.18 Cell cycle. In eukaryotic cells, DNA replication (S-phase) and mitosis (M-phase) are separated by two gaps, G1 and G2. The size of the segment represents the typical fraction of the cell division cycle time.

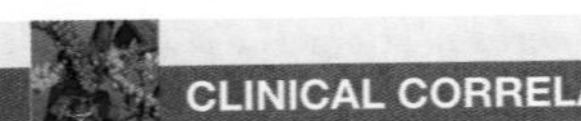

Cancer and the Cell Cycle

Cancer is defined as the unregulated, excessive division of cells. It is recognized pathologically by a higher fraction of cells actively in the cell cycle than is expected for the normal tissue from which it arose. This includes a higher fraction of cells in mitosis, recognizable by microscopy, and a higher fraction of cells in S-phase.

Many kinds of drugs are used to disrupt replication. Antimetabolites interfere with synthesis of precursors of 5′-NTPs. Methotrexate inhibits dihydrofolate reductase, needed to maintain reduced tetrahydrofolate required for nucleotide synthesis. 5-Fluorouracil inhibits thymidylate synthase; when metabolized into its triphosphate, it can be incorporated into DNA and lead to strand breaks. Vinca alkaloids inhibit microtubule assembly and thereby interfere with mitosis and chromosome segregation. Topoisomerase inhibitors slow or stop the process of replication by preventing untwisting of parental strands. Topoisomerase poisons inhibit resealing of the phosphodiester bond, leaving covalent protein–DNA junctions that are converted into strand breaks; this class of antineoplastic agents includes etoposide and doxorubicin.

S-phase. Conversely, heterochromatic regions (where the chromatin is condensed and not actively transcribed) are typically replicated late in S-phase.

Although the details of the process are not yet clear, some general principles can be stated. The current model is referred to as licensing. A licensing factor must be bound to the origins to allow initiation of replication at that origin. This occurs before the start of S-phase and involves the MCM complex binding to the ORC at the many origins. This binding is facilitated by "matchmaker" proteins, including cell division cycle protein 6 (Cdc6) and chromatin licensing and DNA replication factor 1 (Cdt1). Assembly of this pre-replication complex only occurs during G1-phase, before DNA synthesis occurs. Once initiation of replication has occurred, the licensing factor is inactivated by phosphorylation and dissociation of Cdc6 and of the MCM complex. Cyclin-dependent kinase (CDK) and Dbf4-dependent kinase (DDK) are involved in the initiation of replication, catalyzing the recruitment of the replisome to the origin, activating the helicase activity that unwinds the parental strands, and inactivating the licensing factors. For that origin to be used again, a new licensing factor must bind, but the assembly of the complex cannot occur until the next G1-phase when new Cdc6 protein is available. Another block to re-replication is the protein geminin, which binds and inhibits Cdt1. This prevents premature reinitiation during a single cell cycle.

For cell viability, it is important that DNA be completely replicated before mitosis. Cells have regulatory mechanisms that ensure coordination of the events of the cell cycle at several checkpoints. Cells are most sensitive to DNA damage during the process of replication, because replication forks can encounter the damage before it has been repaired (Clin. Corr. 4.4). Cells have evolved sensors that recognize DNA damage and arrest the cell cycle at either the G1/S boundary or the G2/M boundary. If the cell is already in S-phase, the blockage of replication forks prevents further initiation events during S-phase. The damage sensors also induce synthesis of proteins that aid in DNA repair. PCNA, RFC, RPA, and pol ϵ appear to play important roles in recognition of damage that blocks replication and coordination of replication and repair functions. The interaction of PCNA with $p21^{cip1}$, an inhibitor of cyclin-dependent kinase, appears to inhibit replication but not repair; it may thus allow repair to proceed before replication continues. Defects in several genes important in this checkpoint response, including ATM, TP53, and CHK2, increase genomic instability and the risk for cancer.

Replication of RNA Genomes

Some viruses have an RNA genome. Such genomes are replicated with much lower accuracy, and they can accumulate variations relatively rapidly. A particularly important example of this is the **Human Immunodeficiency Virus (HIV)**, which causes **Acquired Immune Deficiency Syndrome (AIDS)**. The HIV viral RNA is reverse transcribed into DNA and then the DNA integrates into the chromosome. The HIV reverse transcriptase is a target for antiviral chemotherapy (see Clin. Corr. 4.2, p. 144).

Reverse transcription of the RNA genome of HIV is far less accurate than DNA synthesis, and the reduced accuracy leads to the rapid generation of a collection of variant viruses within an individual. A small fraction of these variants is likely to be resistant to any single drug being used to treat the infection. That fraction can continue to replicate in the presence of the therapeutic agent to become the dominant variants; this leads to loss of efficacy of the drug. Current combination therapies are designed to reduce the probability that a virus will be simultaneously resistant to all drugs in the combination. Current combination therapies target both the HIV reverse transcriptase and the HIV protease.

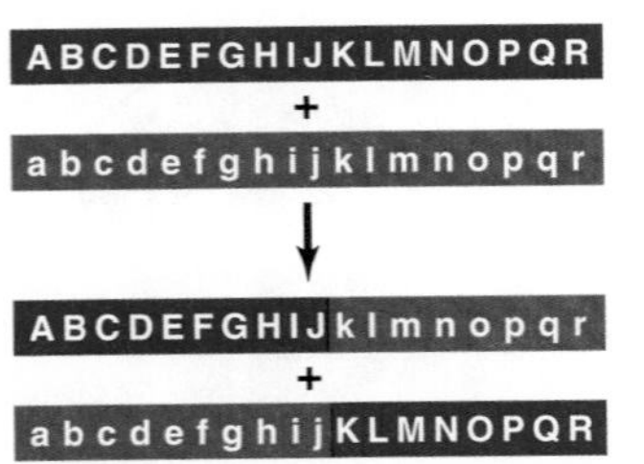

Figure 4.19 Homologous recombination. (*a*) The paternal chromosome is shown in blue with alleles shown in capital letters. The homologous maternal chromosome is shown in red with alleles shown in lower case. (*b*) After homologous recombination between genes j and k, both chromosomes contain DNA from both parents. There has been an equal, reciprocal exchange between them.

4.4 • RECOMBINATION

Recombination is the exchange of genetic information. There are two basic types: **homologous recombination** and **nonhomologous recombination**. Homologous recombination (also called **general recombination**) occurs between identical or nearly identical sequences, for example, between paternal and maternal chromosomes of a pair. Chromosomes are not passed down intact from generation to generation (Figure 4.19); rather, each chromosome you inherit from your father contains portions from both of his parents, and likewise for chromosomes inherited from your mother. Occasionally and often triggered by damage to a chromosome such as a double-strand break, recombination occurs between different chromosomes. This can lead to cancer (Clin. Corr. 4.5).

Homologous recombination is reciprocal, transferring part of one chromosome to the other and vice versa. It shuffles the combination of genes before they are passed to the next generation, generating genetic diversity. This is a normal part of the process of chromosome alignment and segregation necessary to ensure that each germ cell gets a single haploid set of chromosomes during meiosis. The probability that a recombination event will occur between any two points on a chromosome is roughly proportional to the physical distance between them. This is the basis for genetic mapping. A 1% frequency of recombination between two genes or markers is defined as a genetic distance of one **centimorgan (cM)**. In humans, 1 cM is approximately 1,000,000 base pairs along the chromosome.

CLINICAL CORRELATION 4.5

Recombination and Cancer

Errors in recombination can lead to gene duplications or deletions, triplet expansion, gene fusions, and chromosomal rearrangements. The frequency of recombinational events is increased by damage to DNA, in part due to disruptions of replication forks and in part due to breaks in the DNA. Recombination can occur between related but not identical sequences within a chromosome or between chromosomes. It can also occur by nonhomologous end joining following a double-strand break in DNA. Some forms of cancer are triggered by specific chromosomal rearrangements that result in gene fusions that form oncogenes. Both chronic myeloid leukemia (CML) and acute lymphocytic leukemia (ALL) can result from translocations between chromosomes 9 and 22 that create fusion proteins between the genes BCR and ABL. The BCR–ABL fusion protein is an active tyrosine kinase that functions as an oncogene.

Homologous Recombination

There are three major models for homologous recombination (Table 4.3). They are related in many ways, and have a common intermediate called a **Holliday junction.** The simplest is the Holliday model; understanding its key points will make it easy to see the distinctions of the others.

Holliday Model

The key features of the **Holliday model** are (a) homology, (b) symmetry of both breaks and strand invasion, and (c) presence of a four-stranded "Holliday junction" as a key intermediate (Table 4.3). In this model, recombination is initiated by a pair of single-strand breaks at homologous positions in the two aligned DNA duplexes (Figure 4.20). The strands of each duplex partly unwind, and each "invades" the opposite duplex. That is, a portion of one strand from each duplex base pairs with the opposite duplex in a reciprocal, symmetrical fashion. These strands can be covalently joined to the opposite duplex, creating a joint molecule in which two strands cross between the DNA molecules.

Branch migration then occurs (Figure 4.20*c*); branch migration is the simultaneous unwinding and rewinding of the two duplexes such that the total number of hydrogen bonds remains constant but the position of the crossover moves. Because the number of hydrogen bonds remains constant and the thin, cylindrical DNA molecule is easy to rotate around its axis, branch migration requires little energy. Branch migration creates a region of **heteroduplex**, a region where one strand comes from one original duplex and the other comes from the other duplex. In humans, two individuals differ by approximately one base pair per thousand; therefore, a region of heteroduplex often contains at least one mismatched base pair.

The crossed strands are covalently joined to the other duplex (Figure 4.20*c*) The two DNA duplexes, joined by a single crossover point, can rotate as shown in Figure 4.20*e* and *h* to create a four-stranded Holliday junction. Its resolution into two duplexes can occur

TABLE 4.3 • Homologous Recombination

	Holliday Model	*Meselson-Radding Model*	*Double-strand Break Model*
Homology	Yes	Yes	Yes
Initiation	2 single strand nicks at homologous positions	1 single strand nick	1 double strand break, ends resected
Strand invasion	1 strand of each duplex invades the other-reciprocal	Single elongated nick invades other duplex	1 or 2 strands invade other duplex
Branch migration	Yes	Yes	Yes
Key intermediate	Holliday Junction	Holliday Junction	2 Holliday Junctions
Heteroduplex formed	Yes-symmetrical	Yes-asymmetrical	Yes-asymmetrical and long

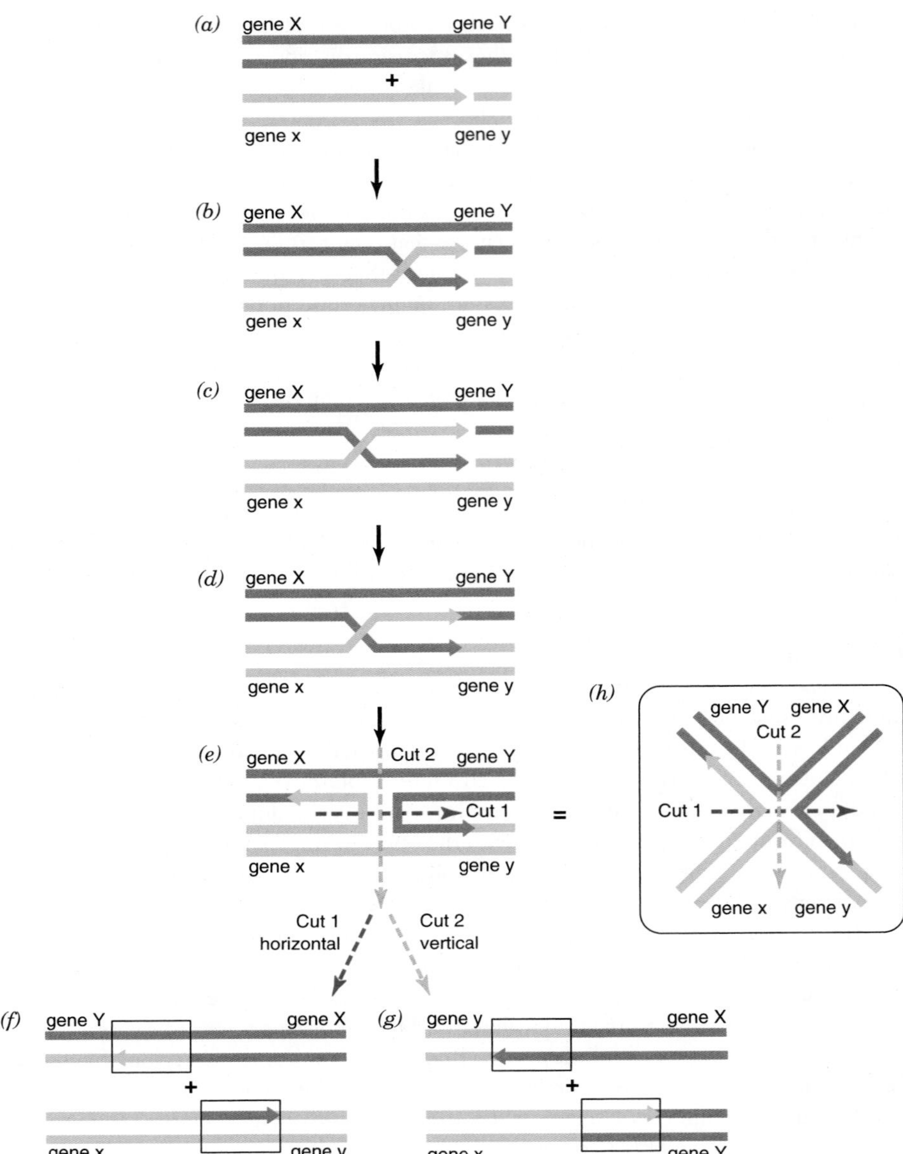

Figure 4.20 Holliday model of homologous recombination. (*a*) Synapsis (pairing): two homologous DNA molecules come together properly aligned. (X,x and Y,y represent different alleles of genes X and Y, one on either side of the crossover.) Single-strand breaks are made at homologous positions in strands of the same polarity in each duplex. (*b*) Strands partly unwind and reciprocally "invade" (base-pair with) the opposite molecule, forming a structure in which two intact strands are joined by two crossed strands. This is a symmetrical and reciprocal event. Regions in which one strand was from one duplex and the other from the homologous duplex are called heteroduplexes. (*c*) Branch migration occurs. (*d*) DNA ligase seals the nicks. Further branch migration can occur. All four strands are held together at one crossover point. (*e*) Rotation of this structure forms a Holliday intermediate shown in inset (*h*). (*f* and *g*) Resolution: Molecules are separated by a pair of symmetrical cuts in either of two directions (orange or green arrows). The ends are resealed by ligase. The direction of the cuts determines whether flanking regions are exchanged. (*f*) If crossed strands are cut (cut 1, horizontal), flanking genes remain as they were (XY/xy). You would not detect recombination between these loci. (*g*) If uncrossed parental strands are cut (cut 2, vertical), flanking genes or markers are exchanged (Xy/xY). This would be detected as a recombination event. Note that there is a region of heteroduplex in either case (boxed); mismatch repair in this region can lead to gene conversion.

in either of two ways, depicted as vertical or horizontal cuts in Figure 4.20*e*. If the crossed strands are cut (cut 1), markers flanking the region of heteroduplex remain linked together in the same phase as originally (X with Y and x with y; panel *f*); on a gross level, the chromosomes remain unchanged. If the intact strands are cut (cut 2), the flanking regions are exchanged between the two chromosomes, creating a new chromosome containing part from one parent and part from the other (X with y and X with y; Figure 4.20*g*). The heteroduplex may undergo mismatch repair (p. 170), converting one allele into another, a process called **gene conversion**.

Meselson–Radding Model

Once you understand the Holliday model, it is easy to understand the variant proposed by Meselson and Radding (Table 4.3). It is initiated by a single nick and a single invading strand, rather than a pair of nicks and reciprocal invasion (Figure 4.21*a*). The nicked strand serves as a template–primer combination that a DNA polymerase elongates, and this elongation displaces part of the invaded duplex (Figure 4.21*b*). The displaced strand invades the opposite, aligned duplex, forming a structure called a D-loop (Figure 4.21*c*). The single-stranded portion of the D-loop is degraded, and repair synthesis and resealing result

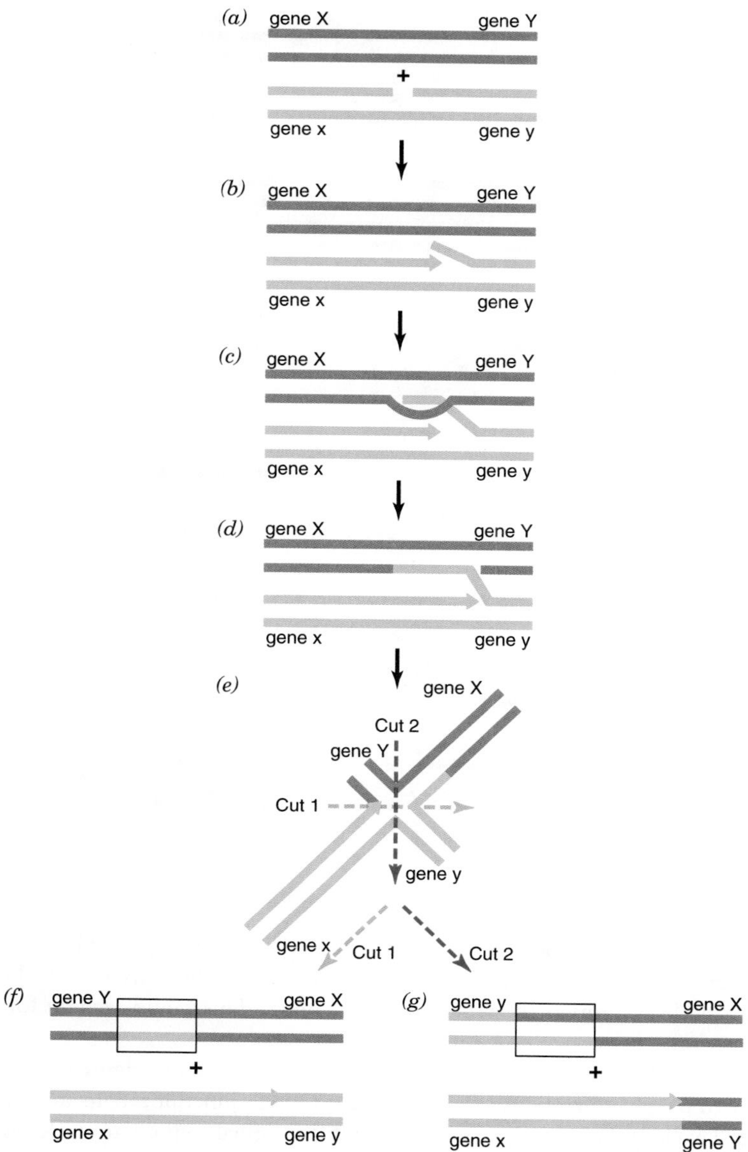

Figure 4.21 Meselson–Radding model of homologous recombination. This model is similar to the Holliday model, but initiation is by a single nick and a single invading strand. (*a*) Synapsis (pairing): Two homologous DNA molecules come together properly aligned. A single-strand break is made in one strand. (*b*) The 3′-OH-end is elongated by a DNA polymerase, displacing the 5′-end. (*c*) The free 5′-end invades the homologous duplex, forming a D-loop. (*d*) The D-loop is degraded, and repair synthesis and resealing by DNA ligase lead to a molecule in which one strand crosses between two duplexes. Branch migration occurs as one strand is elongated further (not shown). The elongating strand crosses over and is joined to the opposite duplex, leaving all four strands joined at a single crossover point, as in Figure 4.20. (*e*) Strands rotate to form a Holliday junction exactly as in Figure 4.20e and h. Resolution of the Holliday junction can occur in two ways, and the direction of the cut determines whether flanking regions are exchanged. Note that there is a region of heteroduplex in either case (boxed), but it is not reciprocal; it is only on one duplex. Mismatch repair in this region can lead to gene conversion.

in the crossing over of one strand. Branch migration occurs. The remaining nick is sealed and the strands can swivel to form a Holliday junction identical to that described for the Holliday model in the preceding (Figure 4.21*e*), which is resolved in exactly the same way, resulting in an exchange of flanking genes (cut 2: X with y and X with y; Figure 4.21*g*) or the original configuration of genes (cut 1: X with Y and x with y; Figure 4.21*f*). The key difference in the **Meselson–Radding model** is the asymmetrical heteroduplex that results from the invasion by one strand from one of the two duplexes. Mismatch repair can lead to gene conversion.

Double-Strand Break Model

The third model is the double-strand break model (Table 4.3). In this, recombination is initiated by a double-strand break in one duplex (Figure 4.22*a*). The ends are resected (Figure 4.22*b*), resulting in the loss of genetic information from one of the two duplexes. There are alternative forms of this model. In one, the two single-stranded segments of the broken duplex both invade the intact duplex, forming a joint molecule that has two Holliday junctions. The broken ends can be repaired by a polymerase, using the intact strand as a template, and a DNA ligase. Branch migration occurs as in the previous models. In an alternative model, the 3′-end of one of the two sides of the break invades the other

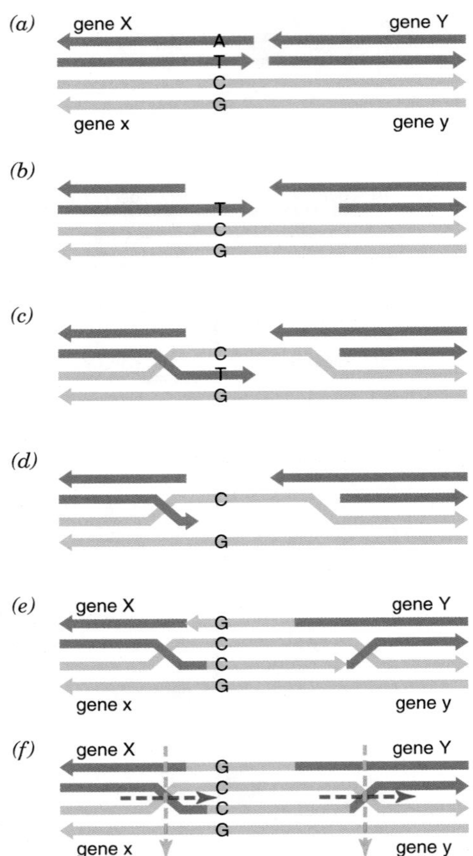

Figure 4.22 Double-strand break model of homologous recombination. This model is similar in many ways to the two models discussed previously, but initiation is by a double-strand break in one of the two DNAs. (*a*) Initiation is by a double-strand break in one DNA molecule. Two aligned duplexes are shown; flanking genes have different alleles (X,x;Y,y). Also shown is a position at which one duplex has an AT base pair and the other has a GC base pair. (*b*) DNA is resected from the break by an exonuclease that leaves 3′-OH-overhangs. This resection may remove one of the nucleotides at the site that differs between duplexes. (*c*) One or both of these overhangs can invade the intact homologous duplex. (*d*) Where there is a mismatch, the free end can be resected further by enzymes of mismatch repair. (*e*) The free end can be extended by a DNA polymerase, using the undegraded strand as a template and resulting in gene conversion. Remaining nicks are repaired, forming two Holliday junctions. (*f*) Both Holliday junctions are then resolved. Because there are two junctions and two ways to resolve each (as in Figure 4.20h), there are four possible results. Two of the four possible ways to resolve the Holliday junctions (vertical/vertical or horizontal/horizontal cuts) leave flanking genes as they were (XY/xy). The other two ways to resolve the Holliday junctions lead to exchange of flanking genes (Xy/xY). Resolution can also be carried out by topoisomerase.

strand and establishes a D-loop that allows copying of the intact duplex. This can result in a small replication bubble with synthesis on both strands for a short distance, and resolution when the bubble encounters the other side of the break. This would account for extensive gene conversion.

Resolution of the two Holliday junctions is by a pair of cuts (Figure 4.22*f*). Because there are two possible directions in which to resolve each junction, there are four possible outcomes. Two leave flanking genes together in their original configuration (X with Y and x with y), and two lead to the exchange of flanking genes (X with y and X with y). Resolution by a type II topoisomerase can also occur. Because one duplex has lost DNA in the region of the break, closure of the gap required the genetic information for both resulting chromosomes to come from the intact duplex; therefore if there were any sequence differences between the chromosomes in the gap, gene conversion will have occurred.

Key Proteins of Recombination in *E. Coli*

In *E. coli*, the key protein in the process of recombination is **RecA**, which binds in a cooperative manner to single-stranded DNA and forms a helical coil. This facilitates pairing of the single strand with a homologous duplex DNA to form a three-stranded D-loop structure. Strand switching can occur, leading to heteroduplex DNA. Eukaryotes have homologous proteins, the human version of which is called RAD51.

RecBCD is an *E. coli* complex with multiple activities; it binds at a double-strand break and processes it into a substrate for recombination by chewing back the DNA and, on encountering a special sequence called a χ site, leaving a protruding single strand with 3′-OH. RecA can bind to this single strand to initiate pairing with the homologous duplex. **RuvA** and **RuvB** form a complex with DNA helicase activity that catalyzes extensive branch migration, a key step in recombination. RuvC is an endonuclease that binds to this complex and catalyzes the resolution of Holliday junctions.

Nonhomologous Recombination

Site-Specific Recombination

In site-specific recombination, specialized enzymes catalyze the integration of a sequence into particular sites in the DNA. Integration of the bacteriophage λ into the *E. coli* chromosome is the best studied example. The **λ-integrase** catalyzes specific nicking of λ DNA and of a special sequence in the *E. coli* chromosome, and also catalyzes the resealing involved. This is a reversible process; the integrated λ-sequence can be excised from the chromosome. There are important examples of site-specific recombination that control processes such as mating type conversion in yeast, and phase change (alterations in major surface protein) in trypanosomes. Phase change in trypanosomes helps the parasite to evade the immune response.

Transposition

Transposition is the movement of specific pieces of DNA in the genome. There are two types of transposition events, simple transposition, in which a sequence is excised from one place in the genome and inserted elsewhere, and replicative transposition, in which a copy of the sequence is inserted elsewhere in the genome without the original being lost. **Transposases** are the enzymes that catalyze transposition. They recognize and act at **insertion sequences**. Some pieces of DNA, called **transposons**, encode transposases flanked by insertion sequences and can move nearly anywhere into a target chromosome. Transposons sometimes carry other genes with them when they transpose. Other pieces of DNA contain insertion sequences but do not encode their own transposase; these can move when a transposase encoded elsewhere is expressed.

Transposons can move to new places within a chromosome or integrate into other chromosomes. A significant fraction of the human genome has resulted from accumulation of transposons and insertion sequences. **Long interspersed elements (LINEs**, of which the L1 element is most common) are transposons present in about 50,000 copies in the human genome. Alu sequences are **short interspersed elements (SINE**s) present in about 500,000 copies per haploid human genome; they do not encode a reverse transcriptase. Some mutations that cause disease are due to insertion of a transposon into a gene, altering its coding or regulation.

Retrotransposons are sequences that were transcribed into RNA and then reverse-transcribed and integrated back into a random site in the genome. Retroviruses make DNA copies of their RNA genome (using **reverse transcriptase**, an enzyme that makes a DNA copy of an RNA) and insert them into the host chromosome by a transposition event catalyzed by a virus-encoded **integrase**.

Nonhomologous End Joining

In nonhomologous end joining (NHEJ; sometimes called illegitimate recombination) no homology is needed, nor are there special DNA sequences. Rather, broken ends of a DNA duplex can recombine with another duplex in the absence of homology. This can lead to the repair of a double-strand break in DNA, and therefore be of considerable survival value. It can also lead to chromosomal translocations or the insertion of a DNA fragment anywhere into the genome. In mammalian cells, a heterodimer called **Ku** binds avidly to DNA ends and can move along the DNA. Ku associates with a large catalytic subunit to form **DNA-PK**, a DNA-dependent serine/threonine protein kinase. This is thought to play a key role in nonhomologous end joining (NHEJ) by binding to and bringing together the two free ends of a double-strand break and protecting them from excessive nuclease action. It may unwind the ends until very short regions of "**microhomology**" (2–6 bp) can pair. The unpaired flaps could be removed by FEN1, the gaps filled by DNA polymerase, and the nicks ligated by DNA ligase. Defects in Ku lead to X-ray sensitivity, due to reduced ability to repair the double-strand breaks caused by X-rays. Defects in Ku can also lead to severe combined immunodeficiency disease (SCID) because another role of Ku is in V(D) J recombination, the site-specific cleavage and rejoining required to assemble functional immunoglobulin genes.

When genes are introduced into mammalian cells, they usually integrate randomly into the chromosome by nonhomologous end joining. This random integration of a fragment of DNA can disrupt genes and cause mutations or dysregulation. Therefore, it

CLINICAL CORRELATION 4.6

Recombination and Gene Therapy

Gene therapy is the introduction of new or altered genes into cells to correct a genetic defect or treat a disease. As we learn more about specific diseases, the possibilities for genetic intervention increase. The easiest type of disease to cure by gene therapy is one in which the supply of a circulating protein can be provided from cells engineered to produce it, with no requirement that those cells be in a specific location. Slightly more difficult would be the regulated production of a protein, for example insulin, that can be provided by cells located anywhere in the body. It will be much more difficult to replace a defective gene with a normal copy in particular cells in which it is expressed.

Several major limitations impede progress. One is the difficulty of introducing genes into the relevant cells; this is being addressed by more work on vectors for gene introduction and on stem cell biology. Another is the tendency of introduced genes to be incorporated randomly into the genome. DNA introduced into mammalian cells is most frequently integrated into the genome by nonhomologous end joining (NHEJ). This random insertion can result in the creation of a new mutation if the transgene inserts into either a coding or regulatory sequence of a gene. It can also create difficulties in sustaining gene expression over time, because the inserted gene will not be in its normal place within a chromosome and therefore may be inappropriately regulated by epigenetic mechanisms.

is possible that the attempt to correct one genetic problem by inserting a gene could cause another problem; this is a major limitation to gene therapy at present (Clin. Corr. 4.6).

Pseudogenes

The human genome contains many pseudogenes, nonfunctional copies of genes. These can result from retrotransposition of mRNAs, in which case they have no introns, or of partially processed transcripts, in which case they may contain some but not all of the introns. In other cases they appear to arise from recombination, such as unequal crossing over, and contain introns. Often these pseudogenes are nonfunctional because of missense or nonsense mutations (see p. 166) or because the sequences necessary for transcription initiation have been altered. In other cases, retrotransposition or unequal crossing over has provided the raw material for evolution, when accumulation of genetic changes leads to the gene copy taking on other functions.

4.5 • DNA DAMAGE AND MUTATIONS

DNA in cells is constantly being damaged by cellular constituents, including active oxygen species that are by-products of metabolism. Many environmental agents and chemicals in food attack and modify DNA. Damage to DNA can block essential processes including DNA replication and transcription and can lead to mutations.

In aqueous solution at 37°C, there is spontaneous **deamination** of C, A, and G bases in DNA. C deaminates to form U (Figure 4.23*a*), A to hypoxanthine, and G to xanthine. Spontaneous **depurination** due to cleavage of the glycosyl bond connecting purines to the backbone, leaving the backbone of the DNA intact (Figure 4.23*b*), occurs at a substantial rate. It has been estimated that between 2,000 and 10,000 purines are lost per mammalian cell in 24 hours. These depurinated sites are called abasic (lacking a base) or **AP sites** (originally meaning apurinic, lacking a purine, but since generalized to lacking any base).

Bases are oxidized at a substantial rate by reactive oxygen species that are by-products of oxidative metabolism; this creates altered bases, for example 8-hydroxyguanine (8-oxo G; Figure 4.23*c*). Oxidative damage tends to increase with age. Products of lipid peroxidation can form adducts with bases, particularly G residues.

Alkylating agents, including some used for chemotherapy, add methyl or other alkyl groups to the bases. S-adenosyl methionine, a carrier of methyl groups in normal metabolism, occasionally methylates a base in DNA.

Ultraviolet radiation (from sunlight or tanning lamps) covalently links adjacent pyrimidines along one strand of the DNA to form cis-syn cyclobutane **pyrimidine dimers** (Figure 4.23*d* and *e*) and other photoproducts, including pyrimidine 6–4 pyrimidones. Ionizing radiation, including X-rays and radioactive decay, creates strand breaks and produces reactive oxygen species that damage DNA.

Carcinogens and mutagens attack DNA. Some are direct acting and others are **pro-carcinogens**. Pro-carcinogens in their native form do not damage DNA, but they can be activated by metabolic processes (e.g. oxidized by cytochrome P450s) into carcinogens that damage DNA. This process is called **metabolic activation**. **Benzo(*a*)pyrene**, a constituent of coal tar, is an extremely potent pro-carcinogen that requires oxidation to the epoxide to attack DNA.

Many chemotherapeutic agents attack DNA; some cause bases to become alkylated, others, including cisplatin, cross-link the two strands (Clin. Corr. 4.7). The higher sensitivity of replicating cells to DNA damage means that rapidly dividing tumor cells are more sensitive to these damaging agents than are normal cells, but their damage to normal cells limits their dosing and use.

Mutations

Mutations are inheritable changes in the DNA sequence. They can result from replication errors, from damage to the DNA, or from errors introduced during repair of damage. Mutations that are changes of a single base pair are called **point mutations**.

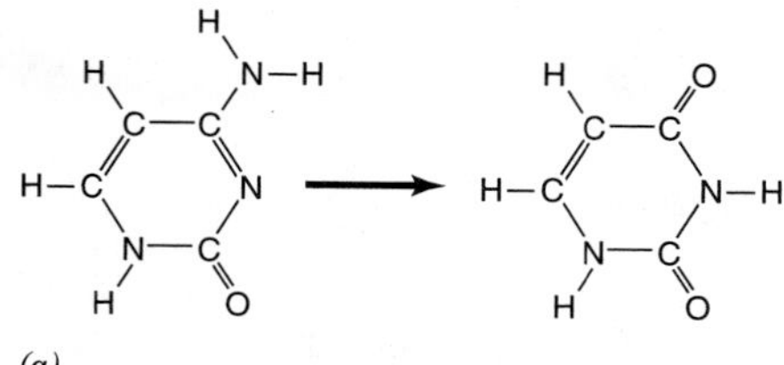

(a)

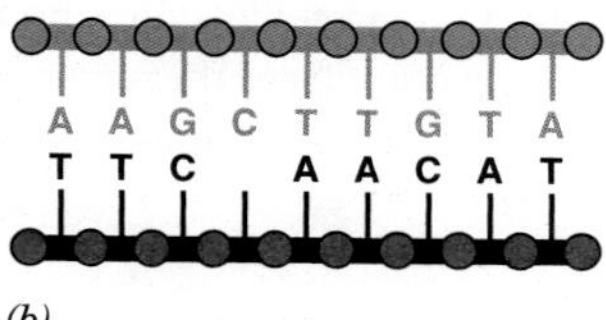

(b)

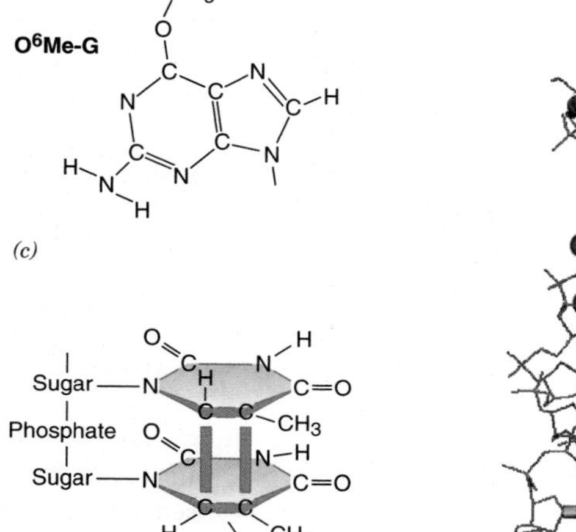

(c)

(d)

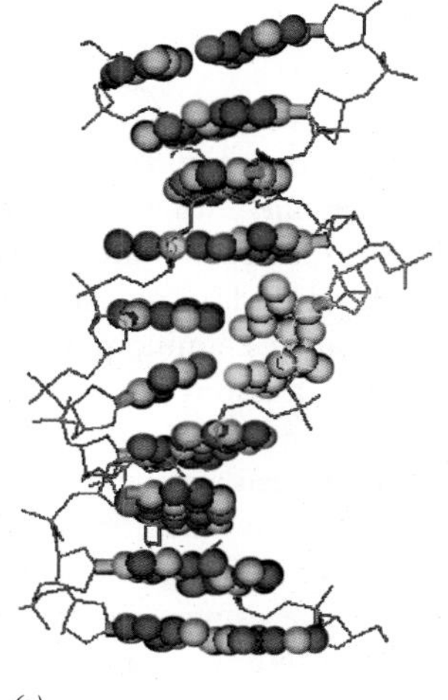

(e)

Figure 4.23 DNA damage. (*a*) Oxidative deamination converts C to U. (*b*) AP site. Depurination (or, less frequently, depyrimidination) is the cleavage of a glycosyl bond between the 1′ position of the sugar and the base. This leaves an abasic or AP site, but does not break the sugar–phosphate backbone. (*c*) O^6methyl guanosine is a highly mutagenic lesion. (*d*) Ultraviolet light leads to cross-linking of adjacent pyrimidines along one strand of DNA. A cis-syn cyclobutane thymine dimer is shown in schematic. (*e*) A cyclobutane dimer is shown within double-stranded DNA; the backbone is shown as a stick figure to reveal the distortion of the cross-linked bases (yellow) and complementary bases on the other strand.

CLINICAL CORRELATION 4.7

Cisplatin and the Tour de France

Many widely used chemotherapeutic drugs are alkylating agents, often ones with two functional groups that can create both intrastrand and interstrand cross-links in DNA. Cyclophosphamide, busulfan, and nitrosoureas are alkylating agents. Alkylating agents are not only cytotoxic but also mutagenic, sometimes resulting in secondary cancers such as leukemias.

Cisplatin (cis-diamminedichloroplatinum),

```
        NH3
         |
  Cl—Pt—NH3
         |
        Cl
```

used to treat metastatic testicular, ovarian, and a wide variety of other cancers, cross-links DNA. Among the serious side effects of cisplatin are bone marrow depletion, severe kidney impairment, and loss of hearing and balance. Disseminated testicular cancer used to be nearly always fatal (5% cure rate in 1973), until a combination chemotherapy, in which cisplatin was added to vinblastine and bleomycin, was developed by Drs. Larry Einhorn and John Donohue at the Indiana University School of Medicine; the new combination greatly increased survival. This regimen was later modified to a combination of cisplatin, etoposide, and bleomycin. Today, approximately 80% of patients survive. After treatment for advanced testicular cancer, world-leading cyclist Lance Armstrong went on to win the Tour de France a record seven consecutive times.

Einhorn, L. H. Curing metastatic testicular cancer. *Proc. Nat. Acad. Sci., USA* 99: 4592, 2002.

CLINICAL CORRELATION 4.8

Thiopurine Base Analogs as Drugs

6-Mercapto purine (6-MP) is an orally administered purine analog that is useful for chemotherapy of acute leukemias and for immunosuppression after organ transplantation. It acts through several mechanisms, including inhibiting purine biosynthesis and causing toxicity after incorporation into DNA. It is metabolized into the 6-MP riboside 5′-phosphate, which has a short half-life because it is degraded by xanthine oxidase. The rate of degradation is greatly reduced in patients being treated with allopurinol (an inhibitor of xanthine oxidase) for gout-related hyperuricemia, so the dose must be drastically reduced in such patients.

Another enzyme that metabolizes 6-MP (and the related antimetabolite 6-thioguanine) is thiopurine methyl transferase (TPMT). Some patients (about 10% of the population) are heterozygous for a polymorphism that inactivates the enzyme, and therefore have approximately 50% activity, and 1/300th of the people have no TPMT activity and are at extremely high risk for severe immunosuppression and death if treated with 6-MP. Conversely, people who metabolize the drugs more rapidly may not get a sufficient therapeutic dose. This pharmacogenetic difference, therefore, has serious implications for treatment with thiopurines.

Sanderson, J., Ansari, A., Marinaki, T., and Duley, J. Thiopurine methyltransferase: Should it be measured before commencing thiopurine drug therapy? *Annu. Clin. Biochem.* 41:294, 2004.

Point mutations can be categorized by the nature of the bases altered. **Transitions** are point mutations in which one purine is substituted for another (i.e., A for G or G for A) or one pyrimidine is substituted for another (i.e., T for C or C for T). Deamination of C to form U, if unrepaired, would lead to a transition. The frequency of transitions is increased by base analogs, including 2-amino purine and 6-mercapto purine (Clin. Corr. 4.8). **Transversions** are point mutations in which a purine is substituted for a pyrimidine or vice versa (e.g., A for C or C for A).

Point mutations can also be characterized by their effect on a coding sequence. **Missense mutations** are point mutations that change a single base pair in a codon such that the codon now encodes a different amino acid (Figure 4.24*a*). **Nonsense mutations** are point mutations that change a single base pair in a codon to a stop codon that terminates translation (Figure 4.24*b*). Nonsense mutations usually have more severe effects than missense mutations, because they lead to synthesis of truncated (and generally unstable) polypeptides. **Silent** or **synonymous** mutations do not alter the amino acid encoded; these include many changes in the third nucleotide of a codon. Some silent mutations may, however, have serious consequences if they alter the splicing pattern of the gene.

Point mutations

---CUGACGUAUUUUAAUGUCATG--- → ---CUGACGUCUUUUAAUGUCATG---
---LeuThrTyrPheAsnValMet--- → ---LeuThrSerPheAsnValMet---

(a) Missense mutation (A to C)

---CUGACGUAUUUUAAUGUCATG--- → ---CUGACGUAAUUUAAUGUCATG---
---LeuThrTyrPheAsnValMet--- → ---LeuThr***stop

(b) Nonsense mutation (U to A)

Insertions and deletions

---CUGACGUAUUUUAAUGUCATG--- → ---CUGAACGUCUUUUAAUGUCATG---
---LeuThrTyrPheAsnValMet--- → ---LeuAsnValPhe***stop

(c) Insertion (of A), changes reading frame and causes a Frameshift mutation

---CUGACGUAUUUUAAUGUCATG--- → ---UGACGUCUUUUAAUGUCATG---
---LeuThrTyrPheAsnValMet--- → ---***stop

(d) Deletion (of first C), changes reading frame and causes a Frameshift mutation

---CGGCGG [CGG]$_{45}$CGG--- → ---CGGCGG [CGG]$_{102}$CGG---

(e) Triplet expansion, can cause disease

Figure 4.24 Mutations. Point mutations. (*a*) A missense mutation changes a single amino acid in the encoded polypeptide. (*b*) A nonsense mutation changes a codon for an amino acid into a stop codon, terminating synthesis of the encoded polypeptide. Insertion/deletion mutations. (*c*) Insertion of a single nucleotide changes the reading frame of all codons beyond the point of insertion; this usually leads to formation of a new stop codon that terminates synthesis. (*d*) Deletion of a single nucleotide changes the reading frame of all codons beyond the point of insertion; this usually leads to formation of a new stop codon that terminates synthesis. (*e*) Triplet expansion is a great increase in the number of triplet repeats. Triplet expansion causes many diseases, including Huntington disease and Fragile X disease, by adding long stretches of a single amino acid to the encoded polypeptide or by disrupting regulation of a gene.

Insertions or deletions of one or more base pairs (if the number of base pairs is not a multiple of 3) lead to **frameshifts** that disrupt the coding of a protein (Figure 4.24 *c* and *d*). Translation of an mRNA does not have punctuation; rather, once the initiation codon is determined, successive triplets are read as codons. Therefore, whereas addition (or deletion) of a multiple of three base pairs in a coding region would add (or subtract) amino acids to (or from) a protein, the addition of other numbers of base pairs shifts the reading frame from that point onward. Frameshifts usually lead to premature termination (or more rarely elongation) of the encoded polypeptide chain, when stop codons are generated or removed by the frameshift. Some chemicals, including **acridines** and **proflavin**, intercalate into the DNA; that is, they insert between adjacent base pairs. This usually leads to insertions or deletions of a single base pair, and thus a frameshift. Mutations may also result from large-scale changes including the insertion of transposons. Although rare in any one generation, mutations have accumulated in populations over millions of years, such that two people differ by about 1 bp per 1000 along their genomes. Many of these genetic differences are without effect, but others affect our physiology, susceptibility to disease, and response to treatment (Clin. Corr. 4.9).

The genome contains many stretches of repeating 2, 3, 4, and 5 nucleotide sequences. These simple repeats have a higher-than-average mutation rate because they can occasionally misalign during replication and recombination, leading to the insertion or deletion of a number of repeats. Thus over many generations, different lengths of these sequences have accumulated in the human population; these are called simple sequence-length polymorphisms (also called microsatellites or simple sequence-repeat polymorphisms). There is a special kind of mutation that results from a great increase in the number of repeating triplets (e.g., CGGCGG[CGG]$_n$CGG), called **Triplet expansion** (Figure 4.24*e*). When the length of a triplet repeat gets much longer than normal, it can cause disease by encoding a much longer than normal stretch of the same amino acid residue in the encoded polypeptide, as in **Huntington disease** (Clin. Corr. 2.6, p. 48). Triplet expansion can also affect regulation of a gene, as in **Fragile X syndrome** (Clin. Corr. 2.6), in which the greatly lengthened triplet sequence in the promoter alters gene expression. For these triplet expansion diseases, there is an unusual phenomenon: The disease often gets more severe in succeeding generations, as a result of even further increases in triplet length, because the longer the repeating sequence is, the more likely there can be misalignment during replication or unequal crossing over to lead to dramatically longer repeats.

CLINICAL CORRELATION 4.9

Personalized Medicine

Individuals differ in the sequence of DNA in their genome in subtle ways, as a result of the very gradual accumulation of mutations in the population over many generations. The most common difference involves a single base pair; this is called a single nucleotide polymorphism, (SNP) (although single-base-pair polymorphism would be more accurate). There are nearly 7 million validated SNPs in the human genome, and many more reported; this averages 1 per 500 bp. Most polymorphisms do not affect cell functions, but others contribute to the many differences that we see. Some affect the susceptibility to specific diseases, some affect the metabolism of drugs and other compounds (pharmacokinetics), and still others affect the response to these drugs (pharmacodynamics).

As part of the human genome project, there is a major effort to find and catalog polymorphisms. As more is learned about the effects of these sequence differences on disease and treatment, medicine will become more tailored to the individual. Technology already exists to determine rapidly which polymorphisms are present at specific sites in the genome of an individual. When these technologies are made cheaper and more routine, determining the alleles present in key genes will become a routine part of diagnosis and planning of treatment. For example, there are significant differences between individuals in the rate at which they metabolize drugs (compare Clin. Corr. 4.7). A standard dose based on body weight may be too low to be effective in those who metabolize the drug rapidly, but toxic to those who metabolize it very slowly. Determination of the relevant polymorphisms and routine testing could lead to more effective therapies with fewer side effects.

4.6 • DNA REPAIR

Maintenance of the genetic information requires constant repair of DNA damage. All free-living organisms have several mechanisms to repair damage to DNA. Some lesions are directly repaired, but most are removed from DNA as part of the repair process. A key feature in most repair processes is the double-stranded nature of DNA, which allows restoration of the correct sequence on a damaged strand using the complementary genetic information on the other strand.

Excision Repair

Excision repair is a general mechanism that can very accurately repair many different kinds of damage. The defining characteristic of excision repair is removal of the damaged nucleotide(s), leaving a gap in the DNA, followed by resynthesis using the genetic information on the opposite strand, and then ligation to restore continuity of the DNA. There are two major modes of excision repair: **base excision repair (BER)** and **nucleotide excision repair (NER)** (Table 4.4).

Base Excision Repair

Base excision repair is an essential process that repairs many different types of damaged bases, including methylated bases, deaminated bases (e.g., U resulting from deamination of C), oxidized bases, and abasic (AP) sites. The initial step is removal of a single

TABLE 4.4 • Excision Repair

Basic steps
- Recognize damage
- Remove damage by excising part of one strand to leave a gap
- Resynthesize to fill gap; uses genetic information from other strand
- Ligate to restore continuity of DNA backbone

Base excision repair	*Nucleotide excision repair*
• Glycosylase removes base, leaves backbone intact • AP endonuclease cuts backbone, AP lyase removes sugar	• Double excision removes damage as part of an oligonucleotide (12–13 nt in *E. coli*, 27–29 nt in humans)
• DNA polymerase fills gap	• DNA polymerase fills gap
• DNA ligase seals nick	• DNA ligase seals nick

damaged base from the DNA backbone by a DNA **glycosylase**, an enzyme that cuts the N-glycosyl bond between the sugar and the base (Figure 4.25). This step does not break the sugar–phosphate backbone of the DNA; rather it leaves an abasic deoxyribose in the backbone, an AP site that must be removed. Two different activities are required to remove the remaining sugar: an **AP endonuclease** that cleaves the phosphodiester bond at the 5′-side but leaves the sugar still attached to the next nucleotide, and an **AP lyase** that cuts 3′ to the AP site to remove the sugar. The resulting single nucleotide gap has a free 3′-hydroxyl. The gap is filled by a DNA polymerase and ligated by DNA ligase (Figure 4.25).

There are at least eight different glycosylases in humans, each of which recognizes certain types of damaged bases. One of these, **uracil DNA glycosylase (UNG)**, is specialized to remove U from DNA. U is not a normal constituent of DNA but is formed when C is deaminated, and can occasionally be incorporated into DNA during replication. The frequent deamination of C to U noted in the preceding (Figure 4.23*a*, p. 165) would lead to a high rate of mutation; therefore, it is not surprising that a special repair mechanism evolved to remove U. But many C residues in mammalian DNA, in the sequence CG, are methylated at the 5 position ; this is important in the regulation of gene expression. Deamination of a 5-methyl C leaves a T, rather than a U, in the DNA. These T residues are not recognized by uracil DNA glycosylase, but there are other glycosylases that recognize the T:G mismatch and preferentially remove the T, leaving an AP site (Figure 4.23*b*, p. 165). Sometimes, however, the mismatch repair system removes the G instead of the T, creating a mutation. Over millions of years, this has led to a much lower than expected frequency of CG dinucleotides in mammalian DNA.

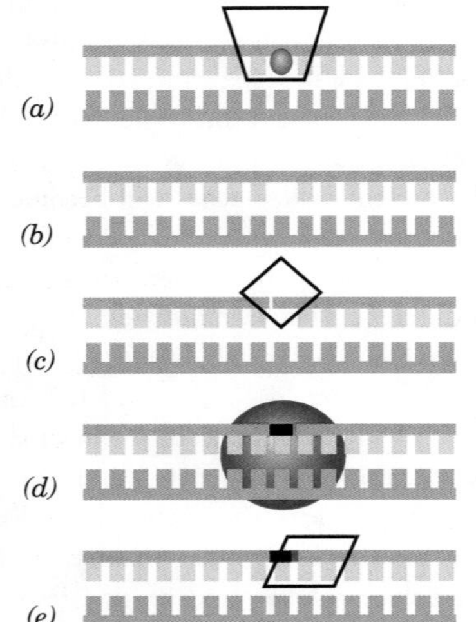

Figure 4.25 Base excision repair. (*a*) The damaged site (blue dot) is recognized by a DNA glycosylase. (*b*) The base is removed by cleavage of the glycosyl bond that connects it to the deoxyribose sugar; the sugar–phosphate backbone is not broken. This leaves an AP site. (*c*) The sugar–phosphate backbone is cleaved 5′ to the abasic site by an AP endonuclease. There is also a cleavage at the 3′-side of this site to remove the sugar residue; this can be done by some AP endonucleases or an AP lyase activity. (*d*) The single nucleotide gap is filled by a DNA polymerase. (*e*) The remaining nick is sealed by a DNA ligase.

Another common lesion is 8-oxo-G (Figure 4.23*c*, p. 165), formed by oxidative damage, which leads to a transversion. There is a special repair pathway designed to repair 8-oxo-G; OGG1 removes the lesion from DNA. If replication has already occurred, MUTYH removes the A that might have inappropriately been incorporated opposite the 8-oxo-G and allows the repair polymerase to put in the correct C.

The most important **AP endonuclease** in humans is APE1. After cleaving the backbone at the 5′-side of an AP site to leave a free 3′-OH, APE1 recruits DNA polymerase β (pol β) to the site. Pol β has **AP lyase** activity that removes the abasic sugar phosphate remaining at the site of the strand break, leaving a single-nucleotide gap with a free 3′-OH. Pol β also has DNA polymerase activity that can fill the gap. For many types of damage, the gap is filled with only one or two nucleotides; this is called **short patch repair**. In cases where AP lyase cannot readily remove the sugar phosphate, strand displacement can create a flap of several nucleotides and the endonuclease FEN1 (which also serves to help remove the RNA from Okazaki fragments as noted in the preceding) can cleave it. The somewhat larger gap is filled by a polymerase such as pol β, or possibly pol δ (or pol ϵ) stimulated by PCNA. In either case, the remaining nick is sealed by a DNA ligase. Protein–protein interactions coordinate this process, keeping the nick and the intermediates protected until the next enzyme has been recruited to the site.

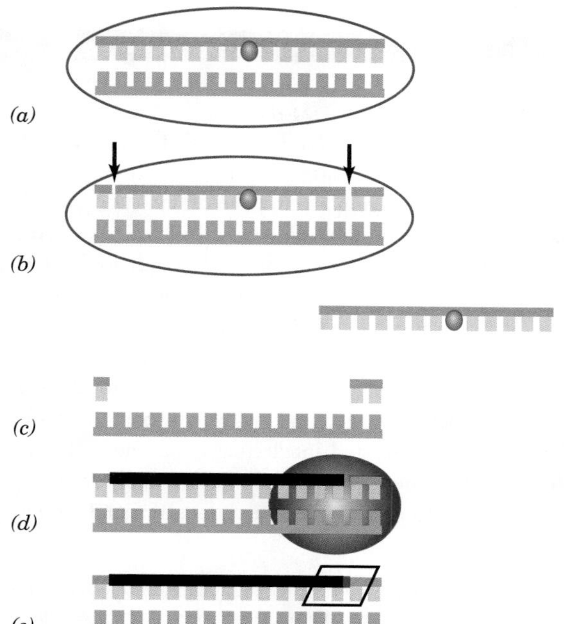

Figure 4.26 Nucleotide excision repair. (*a*) The damaged base (blue dot) is recognized by a DNA repair complex. (*b*) The segment around the damaged base is excised by an enzyme complex that makes two nicks in the damaged strand, one on either side of the damage. (*c*) An oligonucleotide (27–29 nt in humans, 12-13 nt in *E. coli*) is released. (*d*) Resynthesis: A DNA polymerase fills the gap, using the opposite strand as template. (*e*) DNA ligase seals the remaining nick.

Nucleotide Excision Repair

Nucleotide excision repair acts on a wide variety of damage, typically involving large adducts or distortion of the double helical structure of DNA. The best studied are pyrimidine dimers, the major products of ultraviolet damage (Figure 4.23*d* and *e*, p. 165). Chemical adducts with carcinogens such as **benzo(a)pyrene** and **aflatoxin**, and with chemotherapeutic agents such as **cisplatin**, are removed by this pathway, as are some mismatched bases and small loops in DNA. The damage is removed by an enzyme complex that cuts several nucleotides away on both sides of the damaged base(s), so the damage is released as part of an oligonucleotide (Figure 4.26).

In *E. coli*, a damage excision complex containing UvrA, UvrB, and UvrC recognizes the damage and makes two endonucleolytic cuts, one 3 to 5 nucleotides 3′ to the lesion and the other 8 nucleotides 5′ to the lesion. This excises the lesion as part of an oligonucleotide 12–13 nucleotides long. Damage recognition by UvrA appears to start the process, and opening of the strands by UvrB (a helicase) requires energy in the form of ATP hydrolysis. UvrC binds and catalyzes the dual cleavage, with one cut on either side of the damage. A DNA polymerase fills in the gap, followed by ligation of the nick.

In humans, several large enzyme complexes are involved in damage recognition, opening of DNA, and the two cleavages. The rare, autosomal recessive disease **Xeroderma pigmentosum (XP)**, which renders patients photosensitive and susceptible to skin cancers, results from mutation in any of seven genes involved in excision repair or in one gene involved in translesion DNA synthesis (Clin. Corr. 4.10). XPC is part of a complex that binds tightly to lesions that distort DNA structure. XPA is a key factor involved in damage recognition and assembly of the excision complex; mutations in the XPA gene render individuals essentially unable to remove UV-induced pyrimidine dimers. XPA recruits a large complex called **TFII-H**, a general transcription factor involved in initiation of transcription. TFII-H contains XPB and XPD subunits, both of which are ATPases with limited helicase activity. The DNA is partially unwound by the helicase activity of TFII-H, creating a bubble of about 25 base pairs. Then an excision repair cross-complementing 1 (XPF/ERCC1) heterodimer makes an endonucleolytic cut on the damaged strand approximately 22–24 nucleotides 5′ to the lesion, and XPG makes an endonucleolytic cleavage approximately 5 nucleotides 3′ to the lesion. These two cuts liberate a lesion-containing oligonucleotide approximately 27 to 29 nucleotides long. The single-stranded DNA-binding protein RPA is required for excision repair. The gap in DNA is bound by RPA and filled by pol δ (or pol ϵ) stimulated by PCNA. The remaining nick is sealed by a DNA ligase, probably LIG1. These latter steps resemble completion of an Okazaki fragment.

CLINICAL CORRELATION 4.10

Xeroderma Pigmentosum

Xeroderma pigmentosum (XP) (OMIM 194400) was the first disease demonstrated to be caused by defective DNA repair. Patients are photosensitive and highly susceptible to skin cancers in sun-exposed areas of the body; rates of skin cancer are 2000–5000-fold higher than average. Many patients also have neurological problems.

XP is a rare, autosomal recessive disease that can be caused by defects in any of eight different genes, reflecting the complexity of DNA repair of pyrimidine dimers, the most common damage introduced by exposure to ultraviolet light. Mutations in seven of the XP genes lead to defects in the initial incision step in nucleotide-excision repair of pyrimidine dimers. This explains the extreme photosensitivity that is characteristic of this disease. An additional class, called XP-V (for variant) is not detectably deficient in excision repair, but has a defect in translesion synthesis after UV damage. Cells from XP patients may also be hypersensitive to carcinogens in cigarette smoke.

XP-C is involved in initial binding to the damage in regions not being transcribed and recruits TFII-H, a general transcription factor. In transcribed regions, the stalled RNA polymerase recognizes the damage and recruits TFII-H. XPB and XPD encode proteins that function as subunits of TFII-H. TFII-H can function as a helicase, opening a region around the dimer. XPA is critical for the recognition of pyrimidine dimers and interacts with other repair proteins. ERCC1-XPF is a complex that binds to TFII-H and incises at the 5'-side of the dimer. XPG incises at the 3'-side.

It is notable that defects in XP-B, XP-D, or XP-G can also lead to Cockayne syndrome, with photosensitivity, growth defects, and mental deficiency among its symptoms. Defects in XP-B or XP-D can also lead to trichothiodystrophy, with symptoms including photosensitivity, short stature, and intellectual impairment. Thus different alterations in the same genes can lead to different syndromes, perhaps by differentially affecting the repair and transcription functions of these proteins.

van Brabant, A. J., Stan, R., and Ellis, N. A. DNA helicases, genomic instability, and human genetic disease. *Annu. Rev. of Gen. and Human Genet.* 1:409, 2000; and a good reference for genetic diseases is the Online Mendelian Inheritance in *Man, OMIM,* at http://www.ncbi.nlm.nih.gov/entrez/query.fcgi?db=OMIM.

Transcription-Coupled Repair

Many lesions block transcription in the same manner that they block replication. Thus cells face an immediate problem even if they are not synthesizing DNA: They must be able to make key proteins in order to survive. **Transcription-coupled repair** directs the immediate repair response to the template strand of transcribed regions, so that lesions there are repaired faster than lesions elsewhere and transcription can resume; the survival value of this is obvious. Transcription-coupled repair is a form of excision repair triggered by an RNA polymerase complex halted at a lesion. In eukaryotic cells, a complex containing CSA and CSB (mutations in which cause Cockayne syndrome, another DNA repair deficiency disease) recognizes the stalled RNA polymerase, causing it to back up away from the lesion, and recruits repair proteins including the TFII-H complex. Then, many kinds of damage that block transcription are repaired by the nucleotide excision repair pathway involving XPA, TFII-H, and other excision repair factors. Recent evidence suggests that transcription-coupled repair also acts on oxidative lesions, which are repaired by base excision repair after the transcription complex is backed away from the lesion. Once repair is complete, transcription can resume.

```
      m
      |
5' ---GATC--- 3'
3' ---CTAG--- 5'
        |
        m
```

(*a*) Methylated DNA in *E. coli*

```
      m
      |
5' ---GATC--- 3'
3' ---CTAG--- 5'
```

(*b*) Hemi-methylated DNA in *E. coli*

```
      m
      |
5' ---CG--- 5'
3' ---GC--- 3'
       |
       m
```

(*c*) Methylated DNA in human cells

```
      m
      |
5' ---CG---3'
3' ---GC---5'
```

(*d*) Hemi-methylated DNA in human cells

Figure 4.27 Methylated DNA (*a*) DNA methylation in *E. coli* is on the A within a GATC sequence. (*b*) Hemimethylated E. coli DNA. Parental strand methylated, daughter strand not yet methylated. (*c*) Methylation in human cells is on the C within a CG sequence. (*d*) Hemimethylated human DNA, daughter strand not yet methylated. Note that in both cases the sequence is a palindrome that reads the same (5′ to 3′) along both strands, and the residues in both strands are eventually methylated.

Mismatch Repair

Mismatch repair is a specialized form of nucleotide excision repair that removes replication errors. Mismatches are not like DNA damage: There is no damaged or modified base present, just the wrong one of the four bases. Thus, recognition of mismatches relies on the distortion of the double-helical structure. A major difference between repair of DNA damage and repair of mismatches is in the choice of which base to excise. Enzymes can recognize damaged bases specifically and remove them, either individually in base excision repair or as part of an oligonucleotide in nucleotide excision repair. But when a mismatch is recognized, both bases are normal; which one should be excised? In newly replicated DNA, removing the newly synthesized base would preserve the genetic information, whereas excising the base on the parental strand would permanently alter the DNA, producing a mutation. Randomly choosing one would lead to mutations half of the time, a rate that is unacceptably high. Therefore, the challenge is to recognize the newly synthesized strand and remove the mismatched base on that strand.

The mismatch repair system is best understood in *E. coli.* DNA in most organisms (but not all) is methylated at specific positions. In *E. coli,* the most frequent methylation is of A in a GATC sequence (Figure 4.27*a*), a palindromic sequence (meaning that the

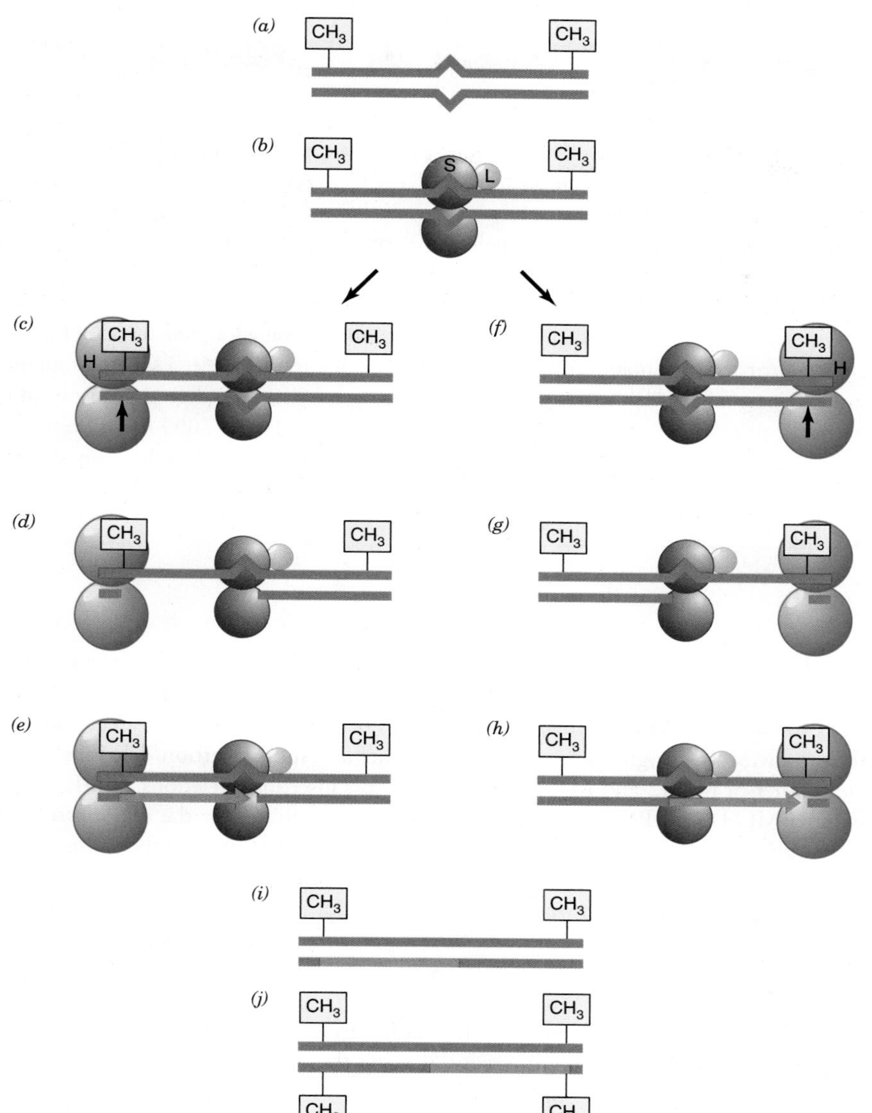

Figure 4.28 Mismatch repair in *E. coli*. (*a*) A mismatch in newly replicated DNA. The parental strand is methylated, but for the first few minutes after it is synthesized, the new strand is not yet methylated; such a site, with only one of the strands methylated, is called hemimethylated. (*b*) MutS and MutL bind to the mismatch. ATP is hydrolyzed as they extrude a short loop of DNA. (*c* and *f*) MutH binds at a hemimethylated site (which could be to either the left or right side of the mismatch). (*d* and *g*) The unmethylated strand is nicked and then excised, extending back past the site of the mismatch. This can be in either direction (*d* or *g*). (*e* and *h*) The resulting long gap is filled by a DNA polymerase, adding nucleotides to the 3′-end of the strand (red). (*i*) The remaining nick is sealed by a DNA ligase. (*j*) Hemimethylated sites are methylated.

opposite strand has the same sequence when read from 5′ to 3′); both strands are methylated. This specific methylation does not affect base pairing. The A residues are incorporated into the DNA in unmodified form, and later methylated by a **maintenance methylase** which recognizes **hemimethylated** sequences in DNA (sequences methylated on only one strand, the parental strand; Figure 4.27*b*) and methylates the appropriate base. During the very brief period in which the daughter strand is not yet methylated, the *E. coli* mismatch repair system can recognize the newly synthesized strand in the region of a mismatch and remove the mismatched nucleotide from the new strand. Mismatches are recognized by a MutS homodimer, which creates a loop that contains the mismatch (Figure 4.28). A MutL homodimer binds and coordinates the subsequent cleavage and excision. MutH nicks the unmethylated, newly synthesized strand on either side of the mismatch. A helicase (UvrD) unwinds the strand from the site of the nick through the site of the mismatch, and the nicked single strand is degraded by either a 5′ to 3′ exonuclease or a 3′ to 5′ exonuclease, depending on the free end. The long (hundreds-of-nucleotides) gap that is created is filled by the action of DNA polymerase, adding nucleotides to the 3′-end of the nicked strand. When the gap has been filled, the remaining nick is sealed by DNA ligase.

Humans have a similar mismatch repair system with homologous proteins, although the system is more complex and the mechanism by which it recognizes the new strand is not clear. Human DNA is methylated on the 5 position of C in CG sequences (Figure 4.27*c*), but it is not clear whether the hemimethylated CG sequences, transient single-strand breaks from

CLINICAL CORRELATION 4.11

Mismatch Repair and Cancer

Defects in mismatch repair cause hereditary nonpolyposis colon cancer (HNPCC) and are important in several other cancers. This was initially suggested by the finding that some colon tumors showed frequent mutations in microsatellites (simple sequence-length polymorphisms, short repeating sequences, particularly mono- and dinucleotides), a phenotype called microsatellite instability.

HNPCC is an autosomal dominant disease. A gene contributing to the risk for HNPCC was mapped to a region that contains a mismatch repair gene. This led to studies that demonstrated the primary defect to be a mutation in one of the mismatch repair genes. Defects in other mismatch repair genes have been found to cause HNPCC; the most common are MLH1 and MSH2.

Why does the inheritance of a single defective allele in a mismatch repair gene lead to HNPCC? Cells lining the intestinal tract actively divide throughout a person's life. There is a high probability that at some time in one's life, a somatic mutation will occur in one allele of a mismatch repair gene in at least one colon cell. Cells in which one of the alleles at any of these loci is inactive can still carry out mismatch repair, so this does not generally lead to a tumor. However, in an individual who inherits one inactive allele, a somatic mutation that inactivates the single active copy leads to defective mismatch repair. Mutation rates can be elevated 100- to 1000-fold in cells with inactivated mismatch repair. The result is the accumulation of mutations that eventually lead to tumor formation through mutations in growth-regulating genes.

unligated Okazaki fragments, or binding to PCNA aid in the recognition of the newly synthesized strand. There are several MutS homologs (MSH proteins) and MutL homologs (MLH proteins and PMS proteins), which work as heterodimers. MutSα, a heterodimer of MSH2 and MSH6, recognizes base–base mismatches and short insertion/deletions; MutSβ, a heterodimer of MSH2 and MSH3, recognizes short insertions/deletions. MutLα, dimers containing MLH1 and PMS2, are the main complexes required for the incision step. The enzymes responsible for excision of the nicked strand to a point beyond the mismatch have not been identified. As in *E. coli,* the excised region is hundreds of nucleotides long. PCNA, the sliding clamp, is required for mismatch repair. In addition to its role as a processivity factor for the DNA polymerase, it appears to function during the excision step. The resynthesis of the large gap requires DNA pol δ (or pol ϵ) and PCNA. The nick that remains is sealed by DNA ligase.

Mismatch repair can lead to gene conversion during recombination by removing one strand of a heteroduplex and resynthesizing it using the other strand as a template; it can also act to reduce the frequency of recombination.

Defects in mismatch repair cause **hereditary nonpolyposis colon cancer (HNPCC)** (Clin. Corr. 4.11). Between 60% and 70% of the cases of HNPCC are due to mutations in MLH1 or MSH2, with other cases due to mutations in other mismatch repair genes. Some sporadic cancers, particularly some colorectal, gastric, and endometrial carcinomas, also appear to be related to suppression or functional inactivation of the mismatch repair system.

Mismatch repair plays a paradoxical role in response of cells to cytotoxic agents such as methylating agents (*N*-nitrosourea) and cisplatin, which are used in cancer chemotherapy. The recognition of alkylation damage by mismatch repair proteins can trigger cell death by apoptosis; cells that are deficient in mismatch repair may escape this process, and thereby be resistant to chemotherapeutic agents. Mismatch repair-deficient cells have higher rates of mutation, and can thus evolve to tumors that are more aggressive.

Direct Demethylation

In addition to the normally methylated bases in DNA, some bases become alkylated inappropriately, by carcinogens or by interaction with the normal methyl carriers in cells (e.g., *S*-adenosyl methionine). Alkylation on positions that affect base pairing is highly mutagenic. For example, O^6methyl-guanine pairs with T rather than with C (Figure 4.29*a*), so its replication would lead to a mutation. Cells have glycosylases that recognize inappropriately methylated bases and trigger base excision repair, but they also have special proteins that recognize O^6-methyl guanine in DNA and directly remove the methyl group, leaving the DNA intact. **O^6-methyl guanine methyl transferase (MGMT)** recognizes O^6-methyl

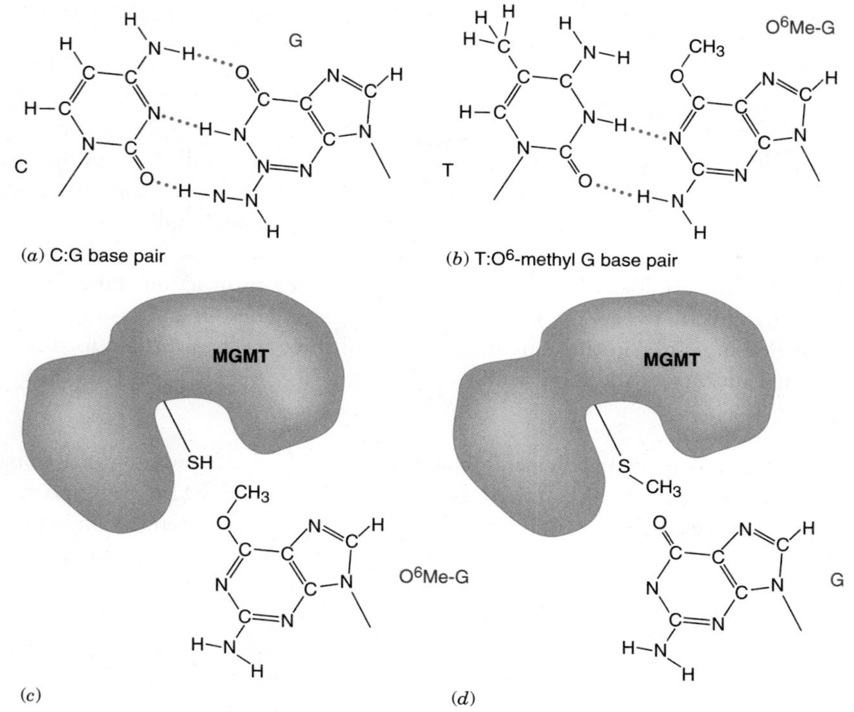

(*a*) C:G base pair

(*b*) T:O^6-methyl G base pair

(*c*)

(*d*)

Figure 4.29 Direct demethylation. (*a*) Normal C:G base pair. (*b*) T:O^6-methyl G base pair. This base pairing would allow incorporation of a T opposite O^6-methyl G during DNA replication and lead to a mutation in next generation. (*c*) Methyl guanine methyl transferase (MGMT) can bind to O^6-methyl G. (*d*) MGMT transfers the methyl group from the O^6 of guanine to a sulfhydryl group on the MGMT protein itself, leaving an intact G in DNA. The methylated MGMT protein is inactive, becomes ubiquitinated, and is then degraded.

guanine and transfers the methyl group from the guanine to a cysteine on the protein itself (Figure 4.29*b*). A single molecule of MGMT can remove only one methyl group, because methylation of the cysteine inactivates the protein and targets it for degradation through the ubiquitin proteolytic pathway. The cost to the cell of removing a single methyl group through this pathway is high, because cells must replicate and transcribe the gene for this enzyme, then process and translate the mRNA, and all of these steps require considerable energy and material. This reemphasizes how important it is to maintain the integrity of the genome, even at great metabolic cost.

Photoreactivation

Photoreactivation is a specific mechanism for repair of cyclobutane pyrimidine dimers (Figure 23*d* and *e*, p. 165), the major lesions produced by ultraviolet irradiation. It is a direct reversal of the damage. A specialized enzyme called **photolyase** binds to the damage (Figure 4.30). On absorbing light, photolyase catalyzes the reversal of the bonds between the adjacent pyrimidines, leaving the DNA exactly as it was before the dimer formed. This

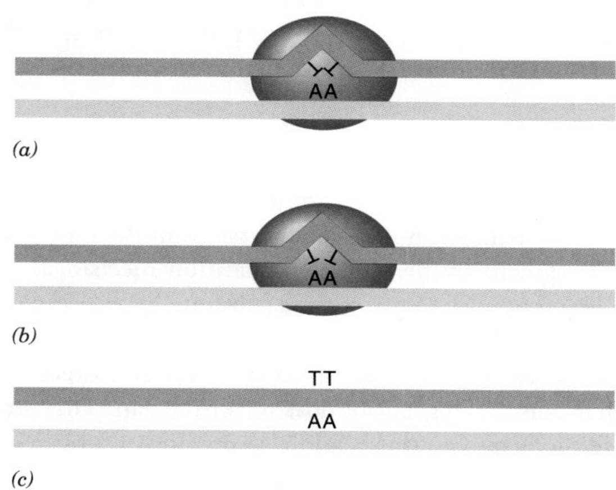

Figure 4.30 Photoreactivation. (*a*) Photolyase binds to a cyclobutane pyrimidine dimer; light is not needed for this step. (*b*) The complex absorbs light; this results in cleavage of the bonds linking adjacent pyrimidines. (*c*) The enzyme then dissociates from the DNA.

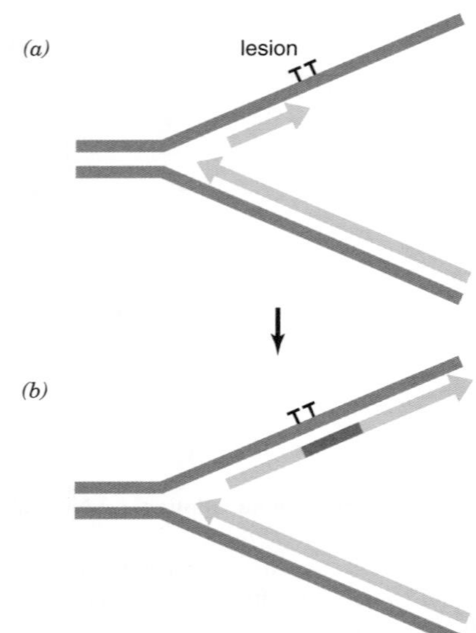

Figure 4.31 Bypass (translesion) synthesis. (*a*) Damage to the template generally halts replicative DNA polymerases, since they cannot add a nucleotide opposite a noncoding or strand-distorting lesion. (*b*) Specialized DNA polymerases with relaxed specificity (see Table 4.2) can insert nucleotides opposite lesions in the template, and thereby allow bypass of damage and synthesis of nongapped daughter strands. The relaxed specificity and lack of proofreading in these polymerases means that errors (potential mutations) are often introduced. These polymerases readily dissociate from the DNA, so the length of DNA they synthesize is usually short.

is a direct reversal of the damage. Photo-reactivation, while of great importance in most organisms, is not significant in mammals.

Lesions Can Block Replication

The accuracy and proofreading abilities of DNA polymerases generally prevent them from inserting nucleotides opposite damaged bases; they require very accurate base pairing before they insert a nucleotide and move on. Thus when a replication complex reaches a lesion in the template strand, it halts. There are three general ways for replication to proceed: the complex can either (a) disassemble or backup to allow repair of the lesion and then resume synthesis, (b) directly bypass the lesion, or (c) skip past the lesion and resume synthesis beyond it (leaving a gap in the daughter strand). All three mechanisms are used, and all present some advantages and some problems.

Bypass (Translesion) Synthesis

A very accurate, proofreading polymerase cannot insert nucleotides opposite damaged bases, and therefore would arrest synthesis (or leave a gap, as described in the following). A process called **bypass synthesis** or **translesion synthesis** can, however, allow replication to proceed past the lesion (Figure 4.31). There is a family of "error-prone" or "relaxed stringency" DNA polymerases that lack proofreading ability (Table 4.2). Due to their relaxed stringency, these polymerases can insert nucleotides opposite the damaged bases and thereby allow replication to continue. Their reduced accuracy increases the probability that the wrong nucleotide will be incorporated opposite the lesion, and in any other DNA made by these polymerases. The harmful consequences of this are evidently less than those of a long (or permanent) block to replication.

In *E. coli,* DNA polymerases IV (DinB) and V (UmuD$'_2$C complex) are much less accurate than the normal replicative DNA polymerases III and I. They have low processivity, generally synthesizing fewer than 10 nucleotides before dissociating from the DNA; this serves to limit the extent of mutations introduced. They are induced as part of the SOS response described in the following. This ensures that the error-prone bypass (translesion) synthesis mechanism functions only when there is sufficient damage to the DNA to trigger the SOS response, limiting the risk of mutations to situations in which the cell's viability is seriously threatened.

Eukaryotes also have error-prone polymerases that can carry out bypass (translesion) synthesis (Table 4.2). DNA polymerase η predominantly incorporates A opposite the thymine dimers introduced by UV irradiation; because A is the appropriate nucleotide to incorporate opposite the T of the dimer, the probability of mutagenesis at those lesions is reduced (but not eliminated). Polymerase η lacks intrinsic proofreading capability and incorporates one mismatch per 20 to 400 nucleotides synthesized, suggesting that its activity must be tightly controlled. DNA polymerase i incorporates nucleotides opposite other highly distorting lesions or lesions such as abasic sites that cannot normally serve as templates. These polymerases appear to function along with DNA polymerase ζ, which doesn't itself add nucleotides opposite lesions but can extend the structures formed by the other bypass polymerases until the normal replicative polymerase can take over. These bypass polymerases are distributive enzymes, so the region in which mismatches are likely to be introduced is typically small, limiting the introduction of mutations.

Daughter-Strand-Gap Repair

If DNA synthesis is blocked by a lesion but restarts beyond the lesion, a gap is left in the DNA (Figure 4.32). This situation is easiest to picture on the discontinuous strand where it can occur simply by dissociation of the polymerase from the blocked site and initiation of the next Okazaki fragment by the normal replication mechanism. Gaps in DNA are potentially lethal. A lesion opposite a gap is not a substrate for excision repair, because there is no intact strand opposite the lesion to provide coding information necessary for repair. **Daughter-strand gap repair** (previously called postreplication recombination) does not remove the lesion that caused the gap, but does repair the gap. This protects the cell from the potential damage that gaps can cause, and also provides a substrate for excision repair to act on the lesion itself.

Daughter-strand gap repair is by recombination. Single-stranded DNA, such as that opposite a gap, is recombinogenic; that is, it stimulates recombination. Recombination between the two newly formed duplexes allows transfer of a piece of the normal parental strand into the gap (Figure 4.32*a*; the parental strand on one side of the fork has the same polarity as the daughter strand on the other side). This leaves a gap in the parental strand that donated the patch (Figure 4.32*b*), but because that strand has a normal strand of DNA opposite, that gap can readily be filled by DNA polymerase and sealed by a DNA ligase (Figure 4.32*c*). The result is that the gap is repaired, although the lesion remains. As noted previously, in this configuration the lesion is a substrate for excision repair.

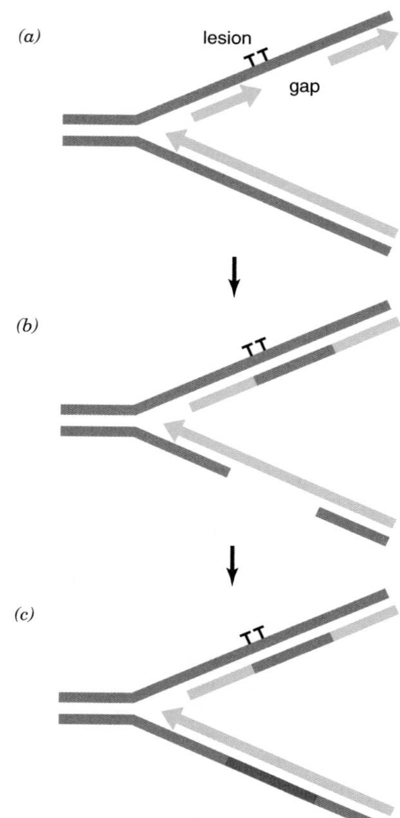

Figure 4.32 Daughter-strand gap repair. (*a*) Gaps are left in newly replicated DNA where replication forks are halted by lesions. Because the lesion does not have intact DNA opposite, it is not a substrate for excision repair. (*b*) Recombination allows the isopolar parental strand (dark tan) to fill the gap in the daughter strand (light tan). This leaves a gap in that parental strand. (*c*) The gap in the parental strand can be filled (purple) by a DNA polymerase, because there is an intact template opposite it. The result is repair of gap in daughter strand, but the lesion remains. Note that the lesion can now be repaired by excision repair, because it is now opposite an intact strand.

Rewinding and Repair of Replication Forks

It is easy for a replication complex that has stopped synthesis of the discontinuous strand at a lesion to resume synthesis to form the next Okazaki fragment, using the normal mechanisms of DNA replication (see p. 145). Reinitiating the continuous strand beyond a block is more difficult, because initiations are normally tightly controlled at replication origins. It appears that in this case, replication forks can regress by the annealing of the daughter strands, which allows repair of the lesion and restarting of replication. In *E. coli,* RecA binds to the single-stranded regions at blocked replication forks and maintains the integrity of the DNA. The replication fork can regress by branch migration while it is protected by the RecA, allowing NER or BER repair mechanisms to remove the lesion. Then, a reversal of the branch migration or degradation of the displaced strand could allow synthesis to resume. Details are still unclear, including whether the replication apparatus could remain with the displaced strand or whether the replication complex would need to be reassembled in an origin-independent manner.

Double-Strand Break Repair

Double-strand breaks (DSBs) in DNA are potentially lethal events. They stimulate genetic recombination that can lead to chromosomal translocations, and unrepaired DSBs lead to broken chromosomes and cell death. Double-strand breaks are caused by many agents, particularly ionizing radiation, reactive oxygens species, and chemotherapeutic agents that generate oxidative free radicals (e.g., bleomycin) or are topoisomerase poisons. When a replication fork encounters a single-strand break, that break is converted into a double-strand break. Cells have mechanisms, including checkpoint controls that slow or arrest replication in the presence of damage to allow time for repair and to minimize the likelihood that a single-strand break would be encountered. Nevertheless, it has been estimated that about 10 double-strand breaks are formed during a single cycle of replication in mammalian cells.

The major pathway for cell survival involves double-strand-break repair by homologous recombination (described previously; Figure 4.24, p. 166) or by nonhomologous end joining. In yeast, homologous recombination predominates; in mammalian cells, nonhomologous end joining predominates.

Regulation of DNA Repair: The SOS Regulon

In *E. coli,* damage to DNA triggers the **SOS response**, a coordinated change in gene expression that aids in recovery from damage. The SOS response is a coordinated induction of many genes whose transcription is at least partly regulated by a common repressor, LexA. A group of operons that are controlled by a common repressor is called a **regulon**. Among the genes in the SOS regulon are uvrA, uvrB, uvrC, recA and LexA itself. There is a low, constitutive level of many of the genes in the SOS regulon, because repression by LexA is not complete and some of these genes have alternative promoters not repressed by LexA.

The **SOS response** is triggered by long stretches of single-stranded DNA such as those in gaps left by blocked replication (Figure 4.33). When such a stretch is formed, the low, constitutive level of RecA protein can cooperatively polymerize along single-stranded DNA to form a nucleoprotein filament. In this state, it is activated and it binds LexA, increasing the rate at which LexA is cleaved into inactive fragments. This relieves the repression and allows rapid synthesis of the genes normally repressed by LexA. RecA participates in

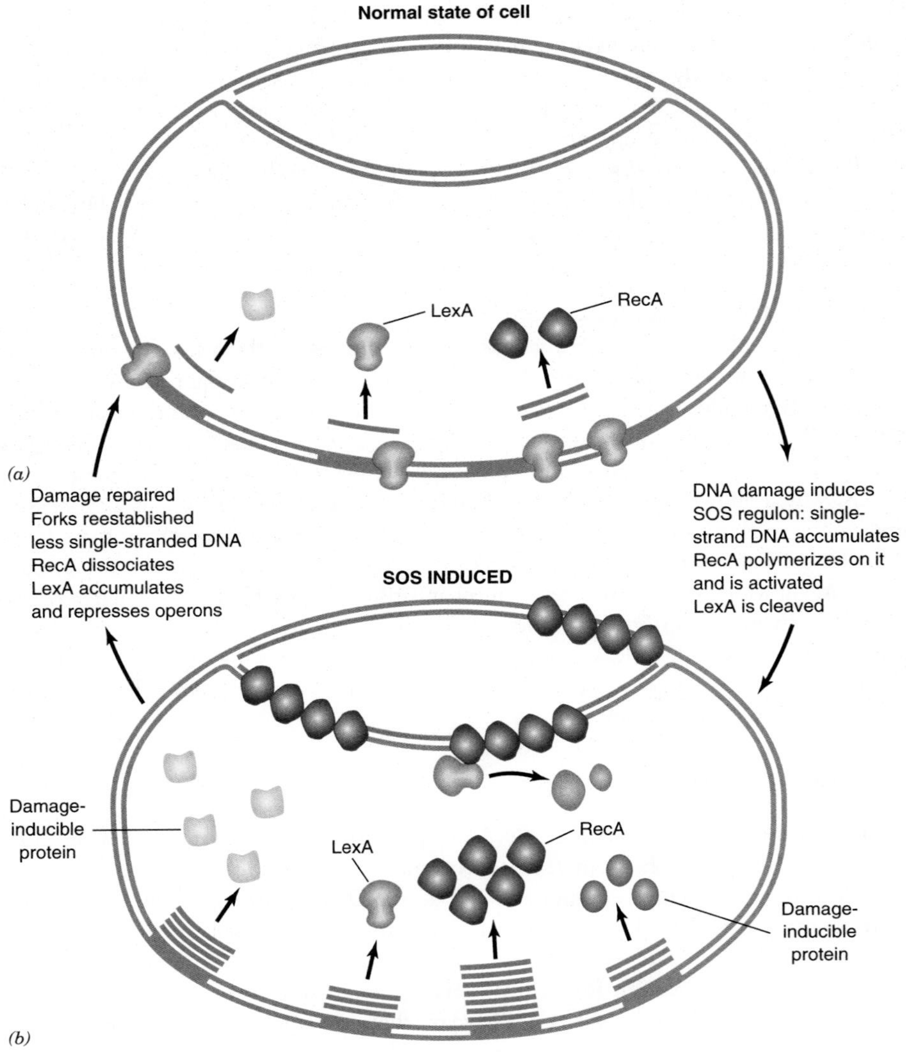

Figure 4.33 Regulation of DNA repair and recovery in *E. coli:* The SOS response. (*a*) In undamaged cells, a group of operons that constitute the SOS regulon is regulated by a common repressor, LexA. This regulon includes the lexA gene itself (autoregulation) and the recA gene, along with many genes encoding enzymes that act to repair damage. There is a low, constitutive transcription of many of these genes, including lexA and recA, even when LexA protein is present. (*b*) Damage blocks replication forks, and thereby leaves single-stranded DNA. RecA binds to single-stranded DNA and is activated. In that form, it aids in cleavage of LexA. The cleaved fragments of LexA cannot bind to operators, so the entire set of operons is induced. Much more RecA is made, along with other proteins including LexA. As long as the damage remains and RecA remains activated, LexA continues to be cleaved so the operons remain on. When damage is repaired, RecA is deactivated and no longer aids the cleavage of LexA, so active LexA builds up and shuts off the genes in this regulon.

daughter-strand gap repair and protection of the replication fork, as noted previously. The other damage-inducible genes (din genes) produce proteins that aid in DNA repair and cell recovery and inhibit initiation of new replication forks and cell division until repair is completed (Figure 4.33).

Autoregulation of LexA is based on a simple feedback loop that effectively allows a rapid but transient response to DNA damage. When LexA protein is cleaved, among the many genes of the SOS regulon that are induced is LexA itself, so the amount of LexA produced increases significantly. However, while DNA damage is still present and RecA protein is still activated, this LexA protein is cleaved and it cannot repress the many genes of the SOS regulon. As the damage is repaired, the amount of activated RecA drops and therefore cleavage of LexA decreases. The increasing levels of LexA bind to the operators of the din genes and gradually shut down the SOS response.

Bibliography

Baynton, K., and Fuchs, R. P. P. Lesions in DNA: Hurdles for polymerases. *Trends Biochem. Sci.* 25: 75, 2000.

Bell, S. P., and Dutta, A. DNA replication in eukaryotic cells. *Annu. Rev. Biochem.* 71: 333, 2002.

Benkovic, S. J., Valentine, A. M., and Salinas, F. Replisome-mediated DNA replication. *Annu. Rev. Biochem.* 70: 181, 2001

Berger, J. M., Gamblin, S. J., Harrison, S. C., and Wang, J. C. Structure and mechanism of DNA topoisomerase II. *Nature* 379: 225, 1996.

Blow, J. J., and Tada, S. A new check on issuing the license. *Nature* 404: 560, 2000.

Cooke, M. S., Evans, M. D., Dizdaroglu, M., and Lunec, J. Oxidative DNA damage: Mechanisms, mutation, and disease, *FASEB J.* 17: 1995, 2003.

Caldecott, K. W. Single-strand break repair and genetic disease. *Nat. Rev. Genet.* 9: 619, 2008.

Cox, M. M., Goodman, M. F., Kreuzer, K. N., Sherratt, D. J., et al. The importance of repairing stalled replication forks. *Nature* 404: 37, 2000.

David S. S., O'Shea, V. L., and Kundu, S. Base-excision repair of oxidative DNA damage. *Nature* 447: 941, 2007.

Evans, A. R., Limp-Foster, M., and Kelley, M. R. Going ape over Ref-1. *Mutation Res.* 461: 83, 2000.

Flores-Rozas, H., and Kolodner, R. D. Links between replication, recombination, and genome instability in eukaryotes. *Trends Biochem. Sci.* 25: 196, 2000.

Friedberg, E. C. DNA damage and repair. *Nature* 421: 436, 2003.

Friedberg, E. C. How nucleotide excision repair protects against cancer. *Nature Rev. Cancer* 1: 22, 2001.

Gerson, S. L. MGMT: Its role in cancer aetiology and cancer therapeutics. *Nature Rev. Cancer* 4: 296, 2004.

Haber, J. E. DNA recombination: the replication connection. *Trends Biochem. Sci.* 25: 271, 2000.

Hanawalt, P. C. Transcription-coupled repair and human disease. *Science* 266: 1957, 1994.

Hanawalt, P. C. The bases for Cockayne syndrome. *Nature* 405: 415, 2000.

Harrison, J. C., and Haber, J. E. Surviving the breakup: The DNA damage checkpoint. *Annu. Rev. Genet.* 40: 209. 2006.

Hubscher, U., Maga, G., and Spadari, S. Eukaryotic DNA polymerases. *Annu. Rev. Biochem.* 71: 133, 2002.

Kowalczykowski, S. C. Initiation of genetic recombination and recombination-dependent replication. *Trends Biochem. Sci.* 25: 156, 2000.

Lieber, M. R., Ma, Y., Pannicke, U., and Schwarz, K. Mechanism and regulation of human non-homologous DNA end joining. *Nature Rev. Mol. Cell Biol.* 4: 712, 2003.

Lindahl, T., and Wood, R. D. Quality control by DNA repair. *Science* 286: 1897, 1999.

Loeb, L. A., and Monnat, R. J., Jr. DNA polymerases and human disease. *Nat. Rev. Genet.* 9: 594, 2008.

Modrich, P. Mismatch repair, genetic stability, and cancer. *Science* 266: 1959, 1994.

Naktinis, V., Turner, J., and O'Donnell, M. A molecular switch in a replication machine defined by an internal competition for protein rings. *Cell* 84: 137, 1996.

Online Mendelian Inheritance in Man (OMIM): http://www.ncbi.nlm.nih.gov/entrez/query.fcgi?db=OMIM.

O'Driscoll, M., and Jeggo, P. A. The role of double-strand break repair: Insights from human genetics. *Nat. Rev. Genet.* 7: 45, 2006.

Rattray, A. J., and Strathern, J. N. Error-prone DNA polymerases: When making a mistake is the only way to get ahead. *Ann. Rev. Genet.* 37: 31, 2003.

Sancar, A. Excision repair in mammalian cells. *J. Biol. Chem.* 270: 15915, 1995.

Sclafani R. A., and Holzen, T. M. Cell cycle regulation of DNA replication. *Annu. Rev. Genet.* 41: 237, 2007.

Stahl, F. Meiotic recombination in yeast: Coronation of the double-strand -break repair model. *Cell* 87: 965, 1996.

Ulaner, G. A. Telomere maintenance in clinical medicine. *Am. J. Med.* 117: 262, 2004.

van Brabant, A. J., Stan, R., and Ellis, N. A. DNA helicases, genomic instability, and human genetic disease. *Annu. Rev. Gen. and Human Genet.* 1: 409, 2000.

Von Hippel, P. H., and Jing, D. H. Bit players in the trombone orchestra. *Science* 287: 2435, 2000.

Waga, S., and Stillman, B. Anatomy of a DNA replication fork revealed by reconstitution of SV40 replication in vitro. *Nature* 369: 207, 1994.

Watson, J. D., and Crick, F. H. C. Genetical implications of the structure of deoxyribonucleic acid. *Nature* 171: 964, 1953.

Wood, R. D., Mitchell, M., Sgouros, J., and Lindahl, T. Human DNA repair genes. *Science* 291: 1284, 2001.

Yu, Z., Chen, J., Ford, B. N., Brackley, M. E., and Glickman, B. W. Human DNA repair systems: An overview. *Environ. Molec. Mutagenesis* 33: 3, 1999.

Key Terms

antiparallel strands
template
DNA polymerases
primase
proofreading
RNA polymerase
semiconservative replication
transesterification
exonucleolytic activity
helicase
replication fork
semi-discontinuous synthesis
Okazaki fragment
RNA hybridase (RNase H)
DNA ligase
Double-strand break
supercoiling
topoisomerase
processive reaction
continuous strand
discontinuous strand
DNA gyrase
origin of replication
bidirectional replication
origin recognition complex (ORC)
telomerase
homologous recombination
non-homologous recombination
Holliday junction
branch migration
heteroduplex
gene conversion
insertion sequence
transposition
missense mutation
nonsense mutation
point mutation
frameshift mutation
DNA repair
excision repair
nucleotide excision repair
base excision repair
short patch repair
transcription-coupled repair
mismatch repair
photoreactivation
bypass synthesis
daughter-strand gap repair
SOS response
reverse transcriptase

Questions • CAROL N. ANGSTADT

Multiple-Choice Questions

1. Replication
   A. requires that a phosphodiester bond of the incoming dNTP be hydrolyzed in order to be added to the growing chain.
   B. uses 5′ to 3′ polymerase activity to synthesize one strand and 3′ to 5′ polymerase activity to synthesize the complementary strand.
   C. owes its accuracy, in part, to 3′ to 5′ exonucleolytic activity of DNA polymerases or associated proteins.
   D. begins with two dNTPs joining together.
   E. requires only proteins with DNA polymerase activity.

2. In eukaryotic DNA replication,
   A. only one replisome forms because there is a single origin of replication.
   B. separating the parental DNA strands to form the replication fork is an energy neutral process.
   C. Okazaki fragments form on the leading strand within a replication fork (bubble).
   D. topoisomerases catalyze changes in the linking number, facilitating untwisting of the parental DNA strands.
   E. the role of ligase is to replace the nucleotides in the gap left by the removal of the RNA primer.

3. All of the following statements about telomerase are correct except
   - A. the RNA component acts as a template for the synthesis of a segment of DNA.
   - B. it adds telomeric repeats to the 5′-ends of the DNA strands.
   - C. it provides a mechanism for replicating the ends of linear chromosomes.
   - D. it recognizes a G-rich single strand of DNA.
   - E. it is a reverse transcriptase.

4. A transition mutation
   - A. occurs when a purine is substituted for a pyrimidine or vice versa.
   - B. results from the insertion of one or two bases into the DNA chain.
   - C. is most frequently caused by chemicals (like acridine) that intercalate into DNA.
   - D. results from substitution of one purine for another or of one pyrimidine for another.
   - E. always is a missense mutation.

5. Homologous recombination
   - A. occurs only between two segments from the same DNA molecule.
   - B. requires that a specific DNA sequence be present.
   - C. requires one of the duplexes undergoing recombination be nicked in both strands.
   - D. involves a four-stranded intermediate (Holliday junction) which can be cut in either of two ways.
   - E. is catalyzed by transposases.

6. All of the following are true about transposons except
   - A. transposons move from one location to a different one within a chromosome.
   - B. they contain but do not code for transposases.
   - C. transposons have insertion sequences that are recognized by transposases.
   - D. the transposon may either be excised and then moved or be replicated with the replicated piece moving.
   - E. they may either activate or inactivate a gene.

Questions 7 and 8: Retroviruses, like HIV which causes AIDS, have their genetic information in the form of RNA. Reverse transcriptase synthesizes a DNA copy of the viral genome. One drug used in treating AIDS is AZT, an analog of deoxythymidine, which has an azido group at the 3′ position of the sugar. It can be phosphorylated and competes with dTTP for incorporation into the reverse transcript. Once incorporated, its presence terminates chain elongation.

7. The growing chain is terminated because
   - A. the analog can not hydrogen bond to RNA.
   - B. the presence of the AZT analog inhibits the proofreading ability of reverse transcriptase.
   - C. AZT does not have a free 3′-OH.
   - D. the analog causes distortion of the growing chain, inhibiting reverse transcriptase.
   - E. dTTP can no longer be added to the growing chain.

8. There is a window in which the effect is primarily on viral replication since AZT is much less effective at competing with dTTP for incorporation by cellular DNA polymerases because of the proofreading ability of DNA polymerases. Proofreading activity to maintain the fidelity of DNA synthesis
   - A. occurs after the synthesis has been completed.
   - B. is a function of 3′ to 5′ exonuclease activity intrinsic to or associated with DNA polymerases.
   - C. requires the presence of an enzyme separate from the DNA polymerases.
   - D. removes mismatched bases in the interior of the chain.
   - E. does not occur in prokaryotes.

Questions 9 and 10: Patients with the rare genetic disease xeroderma pigmentosum (XP) are very sensitive to light and are highly susceptible to skin cancers. The study of such patients has enhanced our knowledge of DNA repair because XP is caused by defective DNA repair–nucleotide excision repair. (A variant, XP-V, is deficient in postreplication repair.)

9. In nucleotide excision repair
   - A. removal of the damaged bases occurs on only one strand of the DNA.
   - B. only thymine dimers generated by UV light can be removed.
   - C. the excision nuclease is an exonuclease.
   - D. a single multifunctional enzyme carries out the repair process.
   - E. only the damaged nucleotides are removed.

10. Another type of DNA repair is base excision repair. Base excision repair
    - A. is used only for bases that have been deaminated.
    - B. uses enzymes called DNA glycosylases to generate an abasic sugar site.
    - C. removes about 10 to 15 nucleotides.
    - D. does not require an endonuclease.
    - E. recognizes a bulky lesion.

Questions 11 and 12: Interfering with topoisomerases is one way of inhibiting DNA replication. Certain antibiotics target DNA gyrase (type II topoisomerase) of *E. coli* inhibiting catalytic activity. Topoisomerase poisons prevent resealing of the phosphodiester bond, leaving covalent protein–DNA junctions. These compounds are used in treating infections and as chemotherapeutic agents.

11. DNA gyrase
    - A. ATP subunits hydrolyze ATP to form new phosphodiester bonds.
    - B. removes negative supercoils.
    - C. swivelase subunits create and reseal transient nicks on both strands.
    - D. increases the linking number.
    - E. occurs in eukaryotes as well as prokaryotes.

12. All of the following are correct about double-strand breaks in DNA *except* they
    - A. can lead to loss of genetic information.
    - B. are always involved in homologous recombination.
    - C. are involved in nonhomologous recombination.
    - D. are associated with a heterodimer (Ku) in mammals.
    - E. can lead to mutations or improper regulation of gene expression.

Problems

13. Mismatch repair removes replication errors by excising incorrect bases. There is no DNA damage or modified bases present. How does the cell distinguish the newly synthesized strand and preserve the correct parental DNA strand?

14. In the coding strand of DNA for the alpha gene of normal hemoglobin (HbA), the three bases that correspond to codon 142 of the mRNA are TAA and the alpha chain has 141 amino acids. In the coding strand of the gene for the alpha chain of Hemoglobin Constant Spring, the three bases are CAA and the chain contains 172 amino acids. Explain the mutation that has occurred.

Answers

1. **C** This can remove a mismatched terminal nucleotide, that is, proofreading. A: The reaction is a nucleophilic attack of the 3′-OH of the DNA chain on the $\alpha-\beta$ phosphate bond of the incoming dNTP—a transesterification. B: Replication involves Okazaki fragments because synthesis occurs only in the 5′ to 3′ direction. D: Replication begins with the formation of an RNA primer. E: Replication requires proteins like primase, ligase, helicase, and others.
2. **D** Since changing the linking number is a transesterification, it protects the integrity of the DNA and occurs without the need of additional energy. A: There are multiple initiation sites. B: Helicases opening the replication fork require the hydrolysis of ATP. C: The continuous strand has the correct orientation; it is the lagging strand that requires formation of Okazaki fragments because it has the 5′-end oriented toward the fork. E: Ligase simply forms the phosphodiester bond between adjacent nucleotides after the gap has been filled.
3. **B** Telomeres are at the 3′-end of each strand so that the 5′-ends can be replicated. A and C: Telomerase positions itself at the 3′-end of the DNA and provides the template for extending that end. D: This is a characteristic of the 3′-end. E: It is using an RNA template to synthesize DNA.
4. **D** This is the definition. A: This is a transversion. B: This is frame shift; transitions are point mutations. C: This would cause a frameshift mutation. E: It could be a missense mutation if the change coded for a different amino acid or a nonsense mutation if the code was changed to a stop signal.
5. **D** One of these cuts accounts for formation of a new chromosome in which part comes from one parent and part from the other. A and B: It may occur between two distinct DNA molecules if the two sequences are homologous. C: Nicks can be on a single strand. E: These are the enzymes of transpositional site-specific recombination.
6. **B** Transposons are DNA, not protein: They can code for a transposase. A: This is the definition. C: This is one of the key features. D: Both types of event are catalyzed by transposases. E: Insertion into the middle of a gene would inactivate it; insertion of a promoter next to a gene may activate it.
7. **C** The chemistry of nucleotide formation requires a free 3′-OH at the end of the chain for attachment of the next nucleotide. A: The presence of the azido group does not affect hydrogen bonding. B: Reverse transcriptase does not have proofreading properties. D: This does not happen. E: The AZT analog competes with dTTP; it does not eliminate it.
8. **B** This activity removes a newly added base if there is a mismatch with the template. A: This is called repair. C: Most polymerases are multifunctional and have proofreading ability. D: Only the terminal mismatched base is removed. E: pol I of *E. coli* has the intrinsic 3′ to 5′ exonuclease activity.
9. **A** The uncut strand serves as the template for repair. B: This repairs a variety of damage typically involving large adducts or disturbances to the DNA structure. C and E: Cuts are made several nucleotides on either side of the damaged bases so this is an endonuclease. D: Several enzyme complexes are involved including a polymerase, ligase, and helicase.
10. **B** These catalyze the first step of the process. A: Methylated and other chemically modified bases can also be removed. C and E: These are characteristics of a different repair system. D: The abasic sugar phosphate must be removed.
11. **C** This enables unwinding of the DNA. A: ATP hydrolysis triggers conformational changes. B: It removes positive supercoils and introduces negative ones. D: The linking number is reduced. E: DNA gyrase is specifically prokaryotic; eukaryotes have different topoisomerases that carry out the same activity.
12. **B** This can also occur with single-strand nicks. A: This can be repaired but covalent protein–DNA junctions caused by topoisomerase poisons can be lethal. C: Broken ends of a DNA duplex can recombine with another duplex. D: This plays a role in nonhomologous end joining. E: Random integration of a fragment of DNA by nonhomologous end joining can disrupt genes or disrupt promoters.
13. In *E. coli.* DNA is methylated on the A of a GATC sequence. Both strands are methylated. Methylation occurs after the unmethylated bases are incorporated into DNA. During synthesis, the DNA will be hemimethylated for a short period. The mismatch repair system recognizes the hemimethylated state and can remove the mismatch on the unmethylated (new) strand. Eukaryotes have a similar repair system although the details of recognizing the new strand are not yet known.
14. The coding strand of the gene has the same sequence as the mRNA (except U replaces T in the RNA). In HbA, the codon at position 142 of mRNA is a stop codon so the last amino acid added is 141. In Hb Constant Spring, a point mutation has mutated the DNA so that the mRNA codon at 142 now codes for an amino acid instead of stop. Translation continues until a stop codon appears at position 173 (so 172 amino acids).

5

RNA: Transcription and RNA Processing

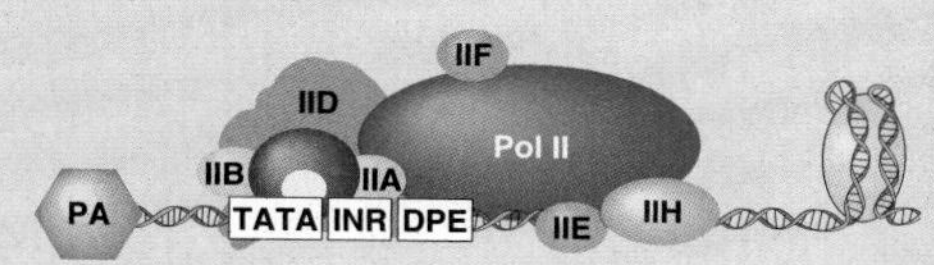

Frank J. Schmidt
Professor, University of Missouri

David R. Setzer
Professor, University of Missouri

Key Concepts

- Transcription is the synthesis of RNA whose sequence is complementary to that of a DNA template. Transcription begins at a promoter sequence and proceeds in the 5′ to 3′ direction.
- RNA polymerases, which initiate transcription, of all types are related by common ancestry.
- Bacterial RNA polymerase contains a special subunit (sigma); different sigma subunits recognize different promoter classes.
- Eukaryotic initiation operates by recruitment of RNA polymerase and accessory protein factors to promoters located in accessible regions of chromatin.
- RNA transcripts are modified by cleavage, additions, and base modification. Most eukaryotic mRNA transcripts are processed to synthesize a 5′ cap and 3′ poly(A) sequence, and to remove introns. Ribonucleoprotein particles participate in processing.
- Short RNAs can inhibit gene expression, allowing post-transcriptional gene regulation.
- There are no repair mechanisms for RNA so all classes of RNA molecules turnover. Exo- and endonucleases hydrolyse RNA to nucleotides for reutilization.
- The different RNAs turnover at different rates and by different mechanisms.

5.1 • INTRODUCTION

Synthesis of an RNA molecule involves copying one strand of a template sequence using Watson–Crick base pairing between nucleotides of the template (usually DNA) and the nucleotides that are being incorporated into the **transcript**. The initiation of transcription by RNA polymerase is perhaps the most important event in the control of gene expression. Transcription initiation requires specialized DNA sequences, called **promoters**, which signal where RNA synthesis should begin. The recognition of promoter sequences involves molecular contacts between DNA and protein **factors**. Factors bind both to RNA polymerase enzyme and to DNA nucleotides through hydrogen bonds and other contacts. During or after transcript elongation, **RNA processing** reactions remove, add, or modify nucleotides in the primary transcript. These processing reactions can occur co-transcriptionally. In other words, they occur on parts of the transcript while downstream sequences are still being transcribed.

5.2 • MECHANISMS OF TRANSCRIPTION

Initial Process of RNA Synthesis Is Transcription

Transcription is the process by which RNA chains are made from DNA templates. Transcription reactions take the following form:

$$\text{DNA template} + n\ (\text{NTP}) \rightarrow \text{pppN(pN)}_{n-1} + (n-1)\text{PP}_i + \text{DNA template}$$

Enzymes that catalyze this reaction are the RNA polymerases; it is important to recognize that, like DNA polymerases, they are absolutely template dependent. Unlike DNA polymerases, however, RNA polymerases do not require a primer molecule. The RNA polymerase reaction is driven forward by two factors: First, the 5′ α-nucleotide phosphate of the ribonucleoside triphosphate is converted from a phosphate anhydride to a phosphodiester bond with a change in free energy ($\Delta G'$) of approximately 12.5 kJ/mol (3 kcal/mol) under standard conditions. Second, the released pyrophosphate, PP_i, is cleaved to two phosphates by pyrophosphatase so that its concentration is low and phosphodiester bond formation is more favored relative to standard conditions (see p. 632).

Since a DNA template is required for RNA synthesis, eukaryotic transcription takes place in the nucleus or in the matrix of mitochondria and chloroplasts. Structural changes occur in DNA during its transcription. In polytene chromosomes of *Drosophila,* transcriptionally active genes are visualized in the light microscope as puffs distinct from the condensed, inactive chromatin. In fact, active genes exist in modified chromatin structures that are less compacted, as revealed by the accessibility of these genes to chemical reagents or enzymes. In both prokaryotes and eukaryotes, the DNA double helix is transiently opened (unwound) as the transcription complex proceeds along the DNA.

These openings and unwindings are necessary because DNA is a double helix and the two strands of DNA must be separated from one another to permit Watson–Crick base pairing between the template strand of DNA and nucleotides in the newly synthesized RNA. Local opening and unwinding of the DNA allows transcription to proceed without wrapping the RNA product around the DNA template. Type I and type II topoisomerases (p. 152) are involved in these changes in DNA helicity.

The process of transcription is generally divided into three parts: **Initiation** refers to the recognition of a specific DNA sequence by RNA polymerase and the beginning of the bond formation process. **Elongation** is the actual synthesis of the RNA chain, which is followed by chain **termination** and release.

DNA Sequence Information Signals RNA Synthesis

Much of the human genome is not transcribed into RNA. Even in bacteria, where almost all the DNA specifies gene products, not all genes are expressed at any one time. RNA polymerase molecules must, therefore, distinguish between genes and nongenes in the DNA, as well as between genes that need to be expressed and those that do not.

Initiation of transcription starts with recognition of a **promoter sequence** in the DNA template. Typically, promoters are located a short distance upstream of the nucleotide where transcription starts. Different promoters often contain similar sequences, permitting identification of a "consensus" promoter sequence (Figure 5.1), from which actual promoters vary to differing extents. Although prokaryotic and eukaryotic promoters differ from one another, conserved sequences are found within both categories of promoters. Some DNA sequences stimulate transcription but are located further away from the initiation site, and some of these can stimulate transcription independently of distance, location, or orientation relative to the transcriptional start site. Such regulatory sequences are called **enhancers.** Enhancers stimulate the synthesis of some prokaryotic RNAs and of most, if not all, eukaryotic mRNAs. Enhancers work by binding specific protein factors, called activators. When an activator binds to an enhancer, a structural change (often a looping or bending) in the DNA template allows interaction of the activator with other factors or with RNA polymerase. This interaction facilitates transcription by recruiting RNA polymerase and other factors to form an initiation complex.

RNA Polymerase Catalyzes the Transcription Process

RNA polymerases synthesize RNA in the $5' \rightarrow 3'$ direction using a DNA template. In this regard, they are similar to the template-dependent DNA polymerases discussed in Chapter 4. Unlike DNA polymerases, however, RNA polymerases initiate polymerization at promoter sequences without the need for a DNA or RNA primer.

Prokaryotic and eukaryotic RNA polymerases are large multisubunit enzymes. RNA polymerase from *Escherichia coli* consists of five subunits with an aggregate molecular weight of over 500,000 (Table 5.1). Two α-subunits, one β-subunit and one β'-subunit constitute the core enzyme, which is capable of faithful transcription but not of specific (i.e., promoter-initiated) RNA synthesis. Addition of a fifth protein subunit, designated σ, forms the

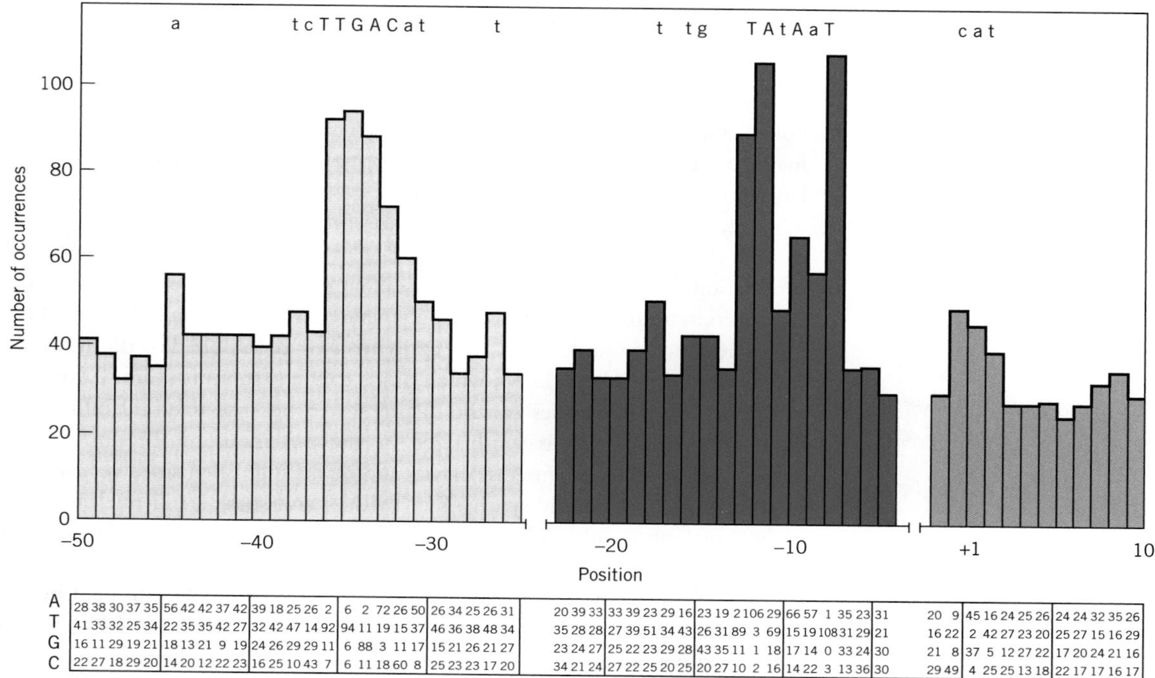

Figure 5.1 A consensus sequence for prokaryotic promoters. The sequences of a large number of promoters from *E. coli* were compiled and the number of times each base occurs at individual positions in the promoter was tabulated. The histogram displays the number of occurrences of the most frequent base at each position. This analysis permits the definition of a consensus sequence for *E. coli* promoters recognized by the form of RNA polymerase holoenzyme containing σ^{70}. This consensus sequence is shown at the top, with the most highly conserved bases in capital letters and the more weakly conserved bases in lowercase letters.

Redrawn from Hawley, D., and McClure, W. R. *Nucleic Acids Res.* 11: 2237, 1983.

TABLE 5.1 • Subunit Composition of Some RNA Polymerases[a]

Yeast Pol I	*Yeast Pol II*	*Yeast Pol III*	*E. coli RNA Polymerase Core*
A190	B220 (Rpb1)	C160	β'
A135	B150 (Rpb2)	C128	β
AC40	B45 (Rpb3)	AC40	α
AC19	B12.5 (Rpb11)	AC19	ω
ABC23	ABC23 (Rpb6)	ABC23	
ABC27	ABC27 (Rpb5)	ABC27	
ABC14.5	ABC14.5 (Rpb8)	ABC14.5	
ABC10α	ABC10α ((Rpb10)	ABC10α	
ABC10β	ABC10β (Rpb12)	ABC10β	
A12.2	B12.6 (Rpb9)	C11	
A14	B32 (Rpb4)	C17	
A43	B16 (Rpb7)	C25	
A49		C37	
A34.5		C53	
		C82	
		C34	
		C31	

[a]Homologous subunits are shown in the same row. Note that five subunits are shared between RNA polymerases I, II, and III from yeast, and that two additional subunits are shared between RNA polymerases I and III. In addition, RNA polymerase III contains three subunits that have no obvious homologs in the other polymerases. It is likely that A49/C37 and A34.5/C53 are homologous to the subunits of TFIIF, an initiation and elongation factor that associates with RNA polymerase II during transcription but which is not considered a part of the polymerase itself. Four subunits of each of the eukaryotic polymerases are homologous to the four different subunits of *E. coli* core RNA polymerase. The letters A, B, and C in the subunit names refer to the occurrence of that subunit in RNA polymerase I, II, or III, respectively, and the number in the subunit name refers to the subunit's molecular weight in kilodaltons. For RNA polymerase II, the gene encoding each of the subunits is given in parentheses.

holoenzyme that is capable of specific RNA synthesis *in vitro* and *in vivo.* Although one major σ factor is involved in most transcription, specific σ factors are made in response to large-scale environmental or developmental events. These σ factors recognize different classes of genes. Thus, a specific σ factor recognizes promoters for genes that are induced as a result of heat shock. In sporulating bacteria, specific σ factors recognize genes induced during sporulation. Some bacteriophages (viruses that infect bacteria) synthesize σ factors that allow the appropriation of the cell's RNA polymerase for transcription of the viral DNA.

The common prokaryotic RNA polymerases are inhibited by the antibiotic **rifampicin** (used in treating tuberculosis), which binds to the β subunit (Clin. Corr. 5.1). Eukaryotic nuclear RNA polymerases can be distinguished because they are inhibited differentially by **α-amanitin**, which is synthesized by the poisonous mushroom *Amanita phalloides.* Three nuclear RNA polymerase classes can be distinguished since very low concentrations of α-amanitin inhibit the synthesis of mRNA and some small nuclear RNAs (snRNAs); higher concentrations inhibit the synthesis of tRNA, 5S rRNA, and other snRNAs. rRNA synthesis (other than 5S rRNA synthesis) is not inhibited much even by very high concentrations of the toxin. The purified enzymes are differentially inhibited by α-amanitin, so it is possible to conclude that mRNA synthesis is the function of RNA polymerase II, the most sensitive of the purified RNA polymerase forms. Synthesis of tRNA, 5S rRNA, U6 snRNA, and other small cellular and viral RNAs is carried out by RNA polymerase III. rRNA genes are transcribed by RNA polymerase I, which is concentrated in the nucleolus. (The numbers refer to the order of elution of the enzymes from a chromatography column.) Each enzyme has a highly complex structure (Table 5.1).

An RNA polymerase in mitochondria is responsible for synthesis of mitochondrial mRNA, tRNA, and rRNA species. This enzyme, like bacterial RNA polymerase, is inhibited by rifampicin (Clin. Corr. 5.1).

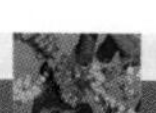

CLINICAL CORRELATION 5.1

Antibiotics and Toxins That Target RNA Polymerase

RNA polymerase is an essential enzyme for life since transcription is the first step of gene expression. Lack of RNA polymerase activity means no other proteins are made. Two natural products illustrate this principle; in both cases inhibition of RNA polymerase leads to death of the organism.

The *death cap* or *destroying angel* mushroom, *Amanita phalloides,* is highly poisonous and still causes several deaths each year despite widespread warnings to amateur mushroom hunters (it is reputed to taste delicious, incidentally). The most lethal toxin, α-amanitin, inhibits eukaryotic RNA polymerase II, thereby inhibiting mRNA synthesis. The poisoning starts with relatively mild gastrointestinal symptoms, followed about 48 h later by massive liver failure as essential mRNAs and their proteins are degraded but not replaced by newly synthesized molecules. The only therapy is supportive, including liver transplantation.

More benign (at least from the point of view of our own species) is the action of the antibiotic rifampicin to inhibit the RNA polymerases of a variety of bacteria, most notably in the treatment of tuberculosis. *Mycobacterium tuberculosis,* the causative agent, is insensitive to many commonly used antibiotics, but it is sensitive to rifampicin, the product of a soil streptomycete. Since mammalian RNA polymerase differs from the prokaryotic variety, inhibition of the latter enzyme is possible without great toxicity to the host. This implies a good therapeutic index for the drug, that is, the ability to treat a disease without causing undue harm to the patient. Together with improved public health measures, antibiotic therapy with rifampicin and isoniazid (an antimetabolite) has greatly reduced the morbidity due to tuberculosis in industrialized countries. Unfortunately the disease is still endemic in impoverished populations in the United States and in other countries. In increasing numbers, immunocompromised individuals, especially AIDS patients, have active tuberculosis.

Mitchel, D. H. *Amanita* mushroom poisoning. *Annu. Rev. Med.* 31:51, 1980; and Petri, W. A., Jr. Chemotherapy of tuberculosis, mycobacterium *avium* complex disease, and leprosy. In Brunton, L., Lazo, J., and Parker, K. *Goodman and Gilman's The Pharmacological Basis of Therapeutics,* 10th ed. New York: McGraw Hill, 2005, 1203.

Steps of Transcription in Prokaryotes

Chromosomal DNA is usually transcribed in only one direction. The DNA strand that serves as the template for RNA synthesis is complementary to the RNA transcript. Conventionally, the template strand is usually the "bottom" strand of a double-stranded DNA as written. The other strand, the nontemplate, or *top,* strand, has the same direction as the transcript when read in the 5′ to 3′ direction; this strand is sometimes called the coding strand. When only a single DNA sequence is given in this book, the coding strand is represented. Its sequence can be converted to the RNA transcript of a gene by simply substituting U (uracil) for T (thymine) bases.

Promoter Recognition

Prokaryotic transcription begins with binding of RNA polymerase to a gene's **promoter** (Figure 5.2). RNA polymerase holoenzyme binds to one face of the DNA extending 45 or so bp upstream and 10 bp downstream from the RNA initiation site. Two short sequences in this region are highly conserved (see Figure 5.1, p. 183). One sequence that is located about 10 bp upstream from the transcription start is the consensus sequence (sometimes called a -10 or *Pribnow* box):

T*A*TAAT*

The positions marked with an asterisk are the most conserved.

A second consensus sequence is located upstream from the –10 or Pribnow box. This –35 sequence

T*T*G*ACA

is centered about 35 bp upstream from the transcription start; the nucleotides with asterisks are most conserved. The spacing between the –35 and –10 sequences is crucial with 17 bp being highly conserved. The TTGACA and TATAAT sequences are asymmetrical; that is, they do not have the same sequence if the complementary sequence is read. Thus, the promoter sequence itself determines that transcription will proceed in only one direction.

What difference do these consensus sequences make to a gene? Measurements of RNA polymerase binding affinity and initiation efficiency on various promoter sequences show that the most active promoters fit the consensus sequences most closely. The degree of similarity of a particular promoter sequence to the consensus correlates well with the measured "strength" of a promoter, that is, its ability to initiate transcription with purified RNA polymerase.

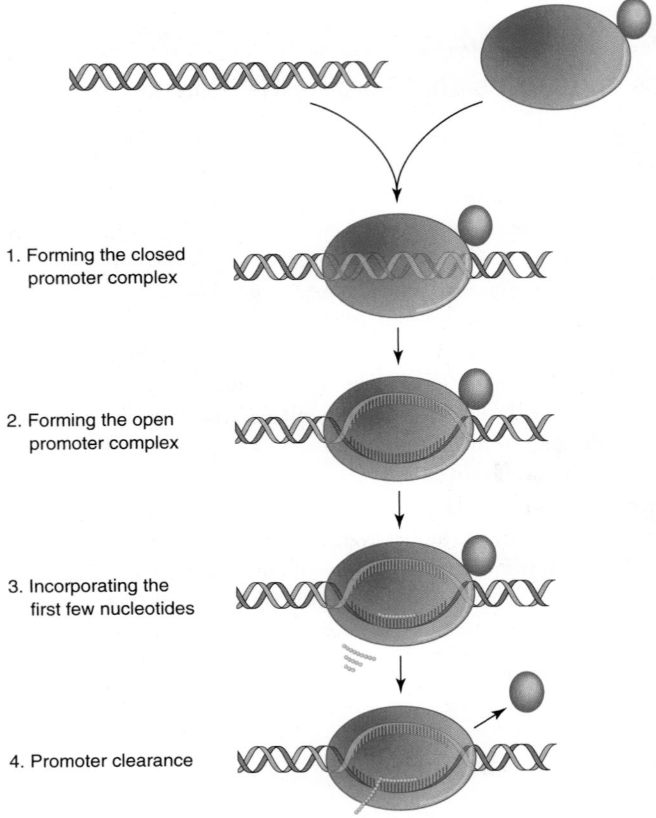

Figure 5.2 Early events in prokaryotic transcription. Promoter-containing DNA is shown in double-helical form, and *E. coli* RNA polymerase holoenzyme is shown as two ovals, the smaller of which represents the sigma factor and the larger of which represents the core. The nascent RNA is shown within the transcription bubble annealed to the DNA template strand, and abortive initiation products are shown as products released from the initiation complex. Although this figure depicts release of the sigma subunit as RNA polymerase enters productive elongation mode, the fate of the sigma factor at this stage of the transcription process remains somewhat controversial.

Bases flanking the –35 and –10 sequences, bases near the transcription start, and bases located near the –16 position are weakly conserved. In some of these weakly conserved regions, RNA polymerase may require that a particular nucleotide not be present or that local variations in DNA helical structure be present. Promoters for *E. coli* heat shock genes have different consensus sequences at the –35 and –10 regions. This is consistent with their being recognized by a different σ factor.

An RNA transcript usually starts with a purine riboside triphosphate; that is, pppG . . . or pppA . . . , but pyrimidine starts are also known (see Figure 5.1, p. 183). The position of transcription initiation differs slightly among various promoters, but usually is five to eight base pairs downstream from the final T of the Pribnow box.

Start of Synthesis

Two kinetically distinct steps are required for RNA polymerase to initiate synthesis of an RNA transcript (see Figure 5.2). In the first step, described above, RNA polymerase holoenzyme binds relatively weakly to the promoter DNA to form a *closed complex.* In the second step, the holoenzyme forms a more tightly bound *open complex,* characterized by a local opening of about 10 bp of the DNA double helix. Since the consensus Pribnow box is A-T rich, it can facilitate this local unwinding; its base pairs are more easily disrupted during opening. Opening 10 bp of DNA is topologically equivalent to the relaxation of a single negative supercoil. As might be predicted from this observation, the activity of some promoters depends on the superhelical state of the DNA template. Some promoters are more active on highly supercoiled DNA whereas others are more active when the superhelical density of the template is lower. The template strand of the unwound DNA base pairs with the initiating nucleoside triphosphate as well as with the second nucleotide of the RNA chain; RNA polymerase then catalyzes formation of the first phosphodiester bond. The enzyme translocates to the next position (this is the rifampicin-inhibited step) and continues synthesis. After several nucleotides have been added to the growing RNA chain, the enzyme enters an elongation mode characterized by a very stable association with the DNA template and does not disengage until specific sequence signals for transcription termination are encountered. Alteration of the association of the sigma subunit with the core polymerase may be associated with entry into this highly processive elongation state. Once the RNA polymerase has cleared the promoter region, other RNA polymerase molecules can bind and initiate at the promoter so that the gene can be transcribed by many polymerases at the same time (Figure 5.3).

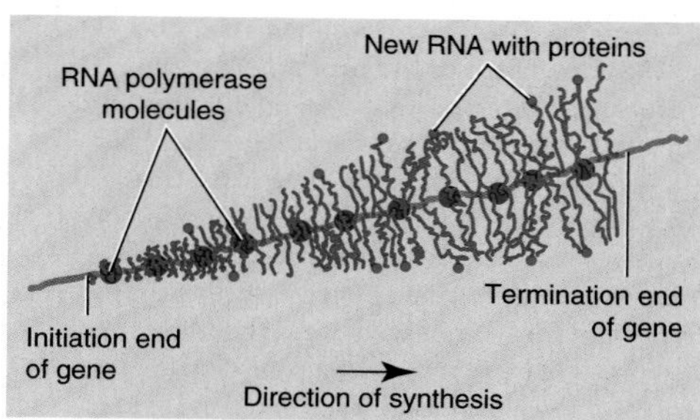

Figure 5.3 Simultaneous transcription of a gene by many RNA polymerases, depicting increasing length of nascent RNA molecules. Courtesy of Dr. O. L. Miller, University of Virginia. Reproduced with permission from O. L. Miller and B. R. Beatty, *J. Cell Physiol.* 74:225, 1969.

Elongation

RNA polymerase continues the nucleotide binding-bond formation-translocation cycle at an average rate of about 40 nucleotides per second. However, many examples are known where RNA polymerase pauses or slows down at particular sequences. As discussed in the following, these pauses can be linked to transcription termination.

As RNA polymerase moves along the double helix, it continues to separate the two strands of the DNA template. This process allows the template strand of the DNA to base-pair with the growing RNA chain. Thus a single mechanism of information transfer (Watson–Crick base pairing) serves several processes: DNA replication, DNA repair, and transcription of genetic information into RNA. Base pairing is also essential for translation (p. 21). The process of unwinding and restoring the DNA double helix leads to changes in superhelical density in the DNA, and the superhelical state of the DNA is controlled by the activity of DNA topoisomerases I and II.

Changes in the transcription complex during the elongation phase can affect subsequent termination events. These changes depend on the binding of other cellular proteins (NusA protein, for example) to core RNA polymerase. Failure to bind these protein factors can result in an increased frequency of termination and, consequently, a reduced level of gene expression.

Termination

The RNA polymerase complex also recognizes the ends of genes (Figure 5.4). Transcription termination can occur in either of two modes, depending on whether or not it is dependent on the protein factor ρ **(rho)**. Terminators are thus classified as **ρ-independent** or **ρ-dependent**; ρ-independent terminators are well characterized (Figure 5.4). A consensus-type sequence is involved: a G-C-rich palindrome (inverted repeat) which precedes a sequence of 6–7 U residues in the RNA chain. As a result the RNA chain forms a stem-and-loop structure just upstream of the oligoU residues. This secondary structure of the stem and loop is crucial for termination, as demonstrated by experiments showing that base change mutations that disrupt pairing in the RNA also reduce termination. The stem and loop left after termination stabilizes prokaryotic mRNA against nucleolytic degradation.

The ρ-dependent terminators are less well defined. The ρ factor is a hexameric protein that has an essential RNA-dependent ATPase activity. The sequences of ρ-dependent termination sites feature C-rich sequences located some distance upstream of the termination site, which is often a rather broad region within which termination occurs, rather than a specific site. The ρ factor is thought to translocate along the nascent RNA chain in an ATP-dependent fashion until it catches up with the elongating RNA polymerase. When that occurs, ρ in some way destabilizes the elongation complex, leading to release of both the template DNA and the completed RNA chain from each other and from the polymerase.

Prokaryotic ribosomes usually attach to nascent mRNA while it is being transcribed. This coupling between transcription and translation is important in gene control by attenuation (p. 313).

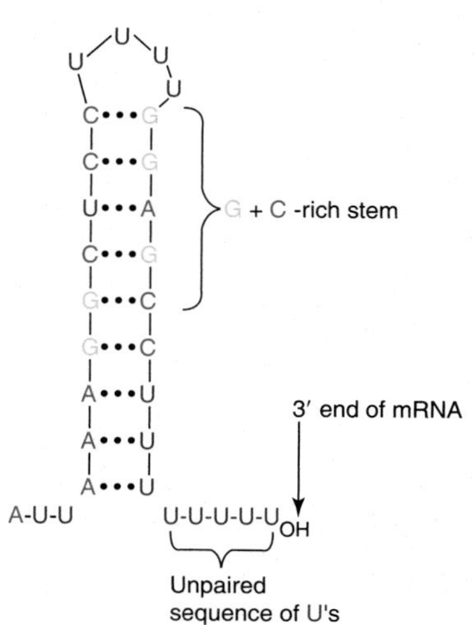

Figure 5.4 Stem-loop structure of RNA transcript that determines rho-independent transcriptional termination. Note the two components of the structure: the G + C-rich stem and loop, followed by a sequence of U residues.

5.3 • TRANSCRIPTION IN EUKARYOTES

Initiation of eukaryotic transcription differs substantially from its prokaryotic counterpart. Whereas the definition of a promoter is the same—DNA sequence information that specifies the start of transcription—the molecular events required for transcription initiation are more complex. First, chromatin containing the promoter sequence must be made accessible to the transcription machinery. Second, transcription factors distinct from RNA polymerase must bind to DNA sequences in the promoter region for a gene to be active. Third, enhancers and other cis-acting transcriptional control elements bind other protein factors (activators) to stimulate transcription.

Eukaryotic transcription factors bind to DNA and recruit RNA polymerase to the promoter. This contrasts with the action of bacterial σ factors, which do not bind DNA without

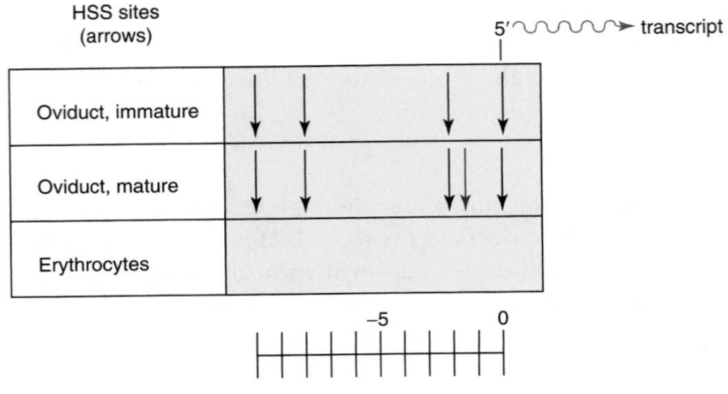

Figure 5.5 DNase-hypersensitive sites (HSS) upstream of the promoter for the chick lysozyme gene, a typical eukaryotic transcriptional unit. Hypersensitive sites, that is, sequences around the lysozyme gene where nucleosomes are not bound to DNA, are indicated by arrows. Note that some hypersensitive sites are found in the lysozyme promoter whether the oviduct is synthesizing or not synthesizing lysozyme; the synthesis of lysozyme is accompanied by the opening up of a new hypersensitive site in mature oviduct. In contrast, no hypersensitive sites are present in nucleated erythrocytes that never synthesize lysozyme.
Adapted from Elgin, S. C. R. *J. Biol. Chem.* 263:1925, 1988.

first binding to the RNA polymerase core enzyme. Eukaryotic RNA polymerase consists of three distinct enzyme forms, each specific form capable of transcribing different classes of cellular RNA. By contrast, all prokaryotic genes are transcribed by a single form of core RNA polymerase, although different σ factors may be involved in initiation on different genes.

Nature of Active Chromatin

The structural organization of eukaryotic chromosomes is discussed in Chapters 4 and 8. Although chromatin is organized into nucleosomes whether or not it is capable of being transcribed, an active gene has a generally "looser" conformation than transcriptionally inactive chromatin. This difference is most striking in the promoter sequences, parts of which are not organized into nucleosomes at all (Figure 5.5). The lack of nucleosomes is manifested experimentally by the enhanced sensitivity of promoter sequences to external reagents that cleave DNA, such as the enzyme DNase I. This enhanced accessibility of promoter sequences (termed DNase I hypersensitivity) ensures that transcription factors can bind to appropriate regulatory sequences. Although the transcribed sequences of a gene may be organized into nucleosomes, the nucleosomes are less tightly bound than those in an inactive gene. Various covalent modifications of the histones are associated with changes in the transcriptional state of genes organized into nucleosomes. One such modification that is usually associated with transcriptional activation is the acetylation of histones by histone acetyltransferases, a reaction that transfers acetyl groups from Acetyl CoA to histones, especially to the N-terminal regions of histones H4 and H3. Other histone modifications that influence gene activity include methylation, phosphorylation, and ubiquitination. Histone methylation is often associated with inactive chromatin, whereas the histones of actively transcribed genes are subject to the other modifications. Combinations of these modifications occurring at different specific positions in histones may constitute a "histone code" that couples histone modification, chromatin compaction, DNA modification, and gene activity (Clin. Corr. 5.2). An overall theme is that partially unfolded chromatin is necessary but not sufficient for transcription.

Transcription Activation Operates by Recruitment of RNA Polymerase

Eukaryotic protein factors, regardless of the sequence to which they bind, operate in a fundamentally different way than *E. coli* σ factor. Rather than first forming part of a protein complex and then seeking out the relevant DNA sequence, the factors bind to a specific site (sequence) on DNA and then bind to RNA polymerase (with or without involvement of intermediary factors). This mechanism is termed "recruitment", which is a minor means of gene activation in prokaryotes and the major mechanism in eukaryotes.

Enhancers

Enhancers increase expression of a gene manyfold. Enhancer-binding transcription factors are called activators. Activator proteins have at least two domains, one of which binds to the enhancer sequence, whereas the other binds to other protein factors or to RNA polymerase.

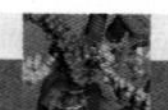

CLINICAL CORRELATION 5.2

Fragile X Syndrome: A Chromatin Disease?

Fragile X syndrome (OMIM 300624) is the single most common form of inherited mental retardation, affecting 1 out of 1250 males and 1 out of 2000 females. A variety of anatomical and neurological symptoms result from inactivation of the FMR1 gene, located on the X chromosome, which encodes the protein FMRP. The genetics of the syndrome are complex due to the molecular mechanism of the Fragile X mutation.

The Fragile X condition results from the expansion of a trinucleotide repeat sequence, CGG, found in the 5′ untranslated region of the FRM1 gene. Normally, this repeat is present in 30 copies, although normal individuals can have up to 200 copies of the repeat. In individuals with Fragile X syndrome, the FMR1 gene contains many more copies, from 200 to thousands, of the CGG repeat. The complex genetics of the disease result from the potential of the CGG repeat sequence to expand from generation to generation.

The presence of an abnormally high number of CGG repeats induces extensive DNA methylation of the entire promoter region of FMR1. The production of methylated DNA may be mechanistically linked to histone modifications that are associated with transcriptionally inactive chromatin. Certainly, heavily methylated DNA is transcriptionally inactive, and FMRP mRNA is not synthesized when large numbers of CGG repeats lead to FMR1 promoter methylation. The absence of FMRP leads to the pathology of the disease.

FMRP protein is an RNA-binding protein that can bind to cellular polyribosomal mRNAs. Association with FMRP negatively regulates translation of the target mRNAs, thus controlling gene expression in the dendrite. The action of FMRP is one of many recent examples that show how posttranslational mechanisms control gene expression and molecular phenotype.

Penagarikano, O., Mulle, J. G., and Warren, S. T. The pathophysiology of Fragile X syndrome. *Annu. Rev. Gen. Hum. Genet.* 8:109, 2007.

The most accepted model for these effects is that chromatin forms a loop that allows the enhancer and the promoter to be close together in space, even though they are separated by a relatively long sequence of DNA.

Transcription by RNA Polymerase II

RNA polymerase II is responsible for synthesis of mRNA in the nucleus. Several common themes have emerged from research on a large number of genes (Figure 5.6). (1) The DNA sequences that control transcription are complex; a single gene may be controlled by dozens of DNA sequence elements in addition to the promoter, which includes the site at which a multiprotein complex containing RNA polymerase assembles. This preinitiation complex is made up of a number of basal-level, or general, transcription factors (p. 320) whose function is to place RNA polymerase II at the correct site for initiating transcription, to assist

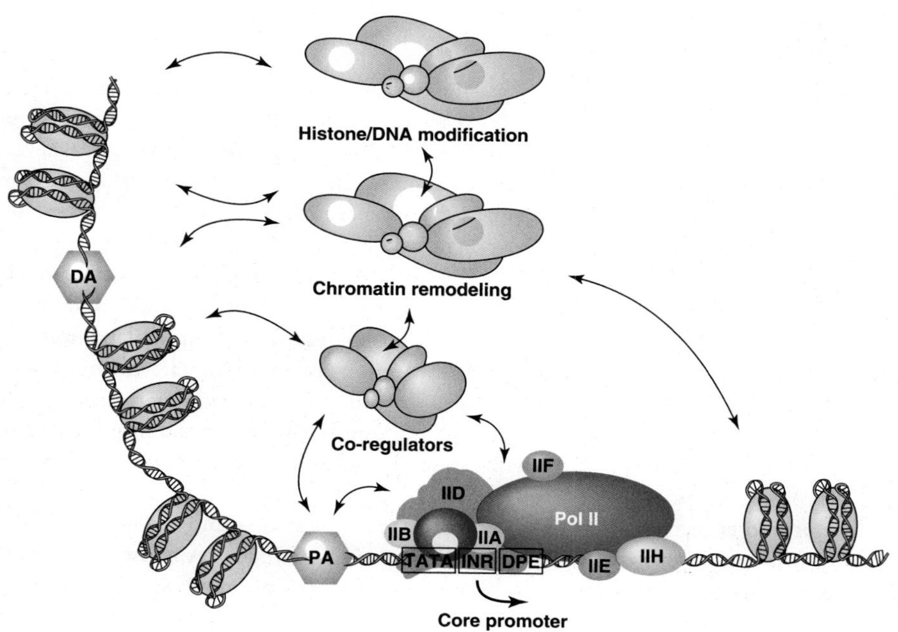

Figure 5.6 Interaction of transcription factors and chromatin-modifying factors with eukaryotic promoters. Cis-acting transcriptional control elements located near the transcription start site may include a TATA box, initiator element (INR), or a downstream promoter element (DPE) and are shown here interacting with the core factors required for specific transcription by RNA polymerase II. Promoters typically contain only a subset of these elements, not all three. The core factors include TFIIA, TFIIB, TFIID, TFIIE, TFIIF, and TFIIH, in addition to RNA polymerase II. Transcriptional activator proteins can bind to promoter-proximal (PA) or promoter-distal (DA) sequences and activate transcription through interactions with the core factors, coregulators, or chromatin-modifying complexes, as shown. Binding of either the core factors or transcriptional activators may require or be accompanied by modifications of chromatin structure, including loss of nucleosomes. Reproduced with permission from Hochheimer, A. and Tjian, R. *Genes and Dev.* 17:1309, 2003.

in separation of DNA strands at the start site, and to control the transition of polymerase from initiation to elongation mode. Without additional transcription control proteins interacting with enhancers and other cis-acting sequence elements, this process is generally quite inefficient, however. Controlling sequence elements function in combination to give a finely tuned pattern of control. (2) The effect of controlling sequences on transcription is mediated by binding of proteins to each sequence element. These transcription factors recognize the nucleotide sequence of the appropriate controlling sequence element. (3) Bound transcription factors interact with each other and ultimately act to recruit RNA polymerase, often acting through factors in the preinitiation complex or other intermediates. The DNA-binding and activation activities of the factors reside in separate domains of the proteins. (4) RNA polymerase II is modified during the transcription reaction. The modified polymerase recruits other nuclear components, including RNA processing enzymes, during the elongation phase of transcription.

Promoters for mRNA Synthesis

In contrast to prokaryotic RNA polymerase which recognizes only a single promoter sequence, transcription factors acting in conjunction with RNA polymerase II can recognize several classes of consensus sequences upstream from the mRNA start site. The first and most prominent of these, sometimes called the **TATA box**, has the sequence

TATA(A/T)(A/T) A

where the nucleotides in parentheses can be either an A or a T.

The TATA box is centered about 25 bp upstream from the transcription unit. Experiments in which it was deleted suggest that it is required for efficient transcription of many genes, but may only specify the correct transcriptional start site in others. Some promoters lack it entirely.

Other promoter-proximal control elements are often found in genes transcribed by RNA polymerase II, but are not universal. One such sequence is the **CAAT box**, with the sequence

GG(T/C) CAATCT

Other sequences, illustrated in Figure 5.6, may also promote transcription.

CAAT and TATA boxes, as well as other sequences shown in Figure 5.6, do not contact RNA polymerase II directly. Rather, they require the binding of specific transcription factors to function. Note how protein factors bind not only to their recognition sequences but also to each other and to RNA polymerase, itself a very large and complex enzyme. Mutated forms of some of these transcription factors are products of oncogenes (Clin. Corr. 5.3).

Transcription by RNA Polymerase I

Ribosomal RNAs are synthesized in a specific nuclear body, the nucleolus, and the rRNA genes are located in a specific chromosomal region termed the nucleolar organizer. Each transcriptional unit contains sequences for 28S, 5.8S, and 18S rRNAs, in that order. Several hundred copies of each transcriptional unit occur tandemly (one after another) in the chromosome. Transcriptional units are separated by spacer sequences. Spacer sequences include sequences that specify binding of RNA polymerase I and of Class I transcription factors, which promote RNA polymerase I activity. Figure 5.7 is a diagram of this arrangement. Each repeat unit is transcribed as a unit, yielding a primary transcript that contains one copy each of the 28S, 5.8S, and 18S sequences, ensuring synthesis of equimolar amounts of these three RNAs. The primary transcript is then processed by ribonucleases and modifying enzymes to the three mature rRNA species. Termination of transcription occurs within the nontranscribed spacer region before **RNA polymerase I** reaches the promoter of the next repeat unit. The mechanism of termination differs from those discussed previously for *E. coli* RNA polymerase.

The promoter recognized by RNA polymerase I is located within the nontranscribed spacer, from about positions -40 to $+10$ and from -150 to -110. A transcription factor binds to the promoter and thereby directs recognition of the promoter sequence by RNA polymerase I. In addition, an enhancer element is located about 250 bp upstream from the

CLINICAL CORRELATION 5.3

Involvement of Transcriptional Factors in Carcinogenesis

Conversion of a normally well-regulated cell into a cancerous one requires a number of independent steps whose end result is a transformed cell capable of uncontrolled growth and metastasis. Insights into this process have come from recombinant DNA studies of the genes, termed oncogenes, whose mutated or overexpressed products contribute to carcinogenesis. Oncogenes were first identified as constituents of the genomes of DNA or RNA tumor viruses, but those found in RNA tumor viruses have been derived from genes present in normal cells. The normal, nonmutated cellular analogs of oncogenes are termed protooncogenes. Their products are components of the many pathways that regulate growth and differentiation of a normal cell; mutation into an oncogenic form involves a change that makes the regulatory product less responsive to normal control.

Some protooncogene products are involved in transduction of hormonal signals or recognition of cellular growth factors and act cytoplasmically. Other protooncogenes have a nuclear site of action; their gene products are often associated with the transcriptional apparatus, and they are synthesized in response to growth stimuli. It is easy to visualize how overproduction or permanent activation of such a positive transcription factor could aid the transformation of a cell to malignancy: Genes normally transcribed at a low or controlled level would be overexpressed by such a deranged control mechanism.

A more subtle genetic effect predisposing to cancer is exemplified by the human tumor suppressor protein p53. This protein is the product of a member of another class of genes implicated in tumorigenesis, the tumor suppressor genes. In contrast to oncogenes, whose products positively regulate cell growth and division, the product of a tumor suppressor gene like p53 negatively regulates cell division. A single copy of the mutant p53 gene causes Li–Fraumeni syndrome, an inherited condition predisposing to carcinomas of the breast and adrenal cortex, sarcomas, leukemia and brain tumors.

Somatic mutations in p53 can be identified in about half of all human cancers. Mutations represent a loss of function, affecting either the stability or the DNA-binding ability of p53. The wild-type protein helps to control the checkpoint between the G1- and S-phases of the cell cycle, activates DNA repair, and in other circumstances, leads to programmed cell death (apoptosis). Thus the biochemical actions of p53 serve to keep cell growth regulated, maintain the information content of the genome, and finally, eliminate damaged cells. All of these functions would counteract the neoplastic transformation of a cell.

These varied roles are a function of the action of p53 as a transcription factor, inhibiting some genes and activating others. For example, p53 is a site-specific DNA-binding protein and promotes transcription of some genes, including those involved in DNA repair.

The three-dimensional structure of p53 has been determined. Mutations found in p53 from tumors affect the DNA-binding domain of the protein. For example, nearly 20% of all mutated residues involve mutations at two positions in p53. The crystal structure of the protein–DNA complex shows that these two amino acids, both arginines, form hydrogen bonds with DNA. Arginine 248 forms hydrogen bonds in the minor groove of the DNA helix with a thymine oxygen and with a ring nitrogen of adenine. Mutation disrupts this H-bonded network and therefore the ability of p53 to regulate transcription.

Weinberg, R. A. *The Biology of Cancer.* New York: Garland Science, 2007.

promoter in human ribosomal DNA. The size of nontranscribed spacer varies considerably among organisms, as does the position of the enhancer element.

Transcription of rRNA can be very rapid; this reflects the fact that synthesis of ribosomes is rate limiting for cell growth and the demand for rRNA can therefore be very high. In other situations, when growth is not so rapid, only some of the rDNA repeats are transcriptionally active.

Transcription by RNA Polymerase III

Many of the themes elaborated above for transcription of Class I and Class II promoters hold for the synthesis of 5S RNA, tRNA, and other small cellular and viral RNAs by **RNA polymerase III.** Transcription factors bind to DNA and direct the action of RNA polymerase. One unusual feature of RNA polymerase III action is the location of transcription factor-binding sequences; these can be located within the DNA sequence encoding the RNA

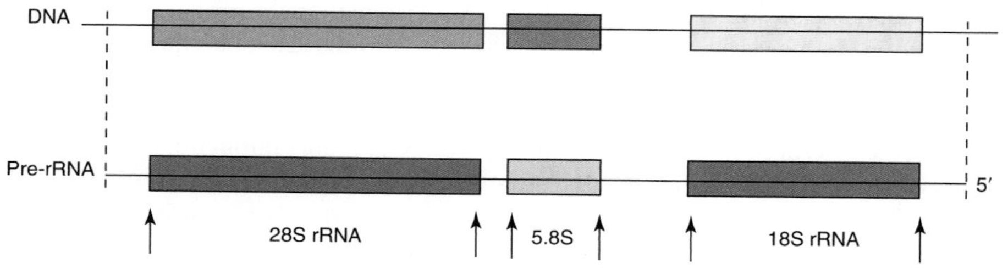

Figure 5.7 Structure of an rRNA transcription unit. Ribosomal RNA genes are arranged with many copies one after another. Each copy is transcribed separately, and each transcript is processed into three separate RNA species. Promoter and enhancer sequences are located in the nontranscribed regions of the tandemly repeated sequences.

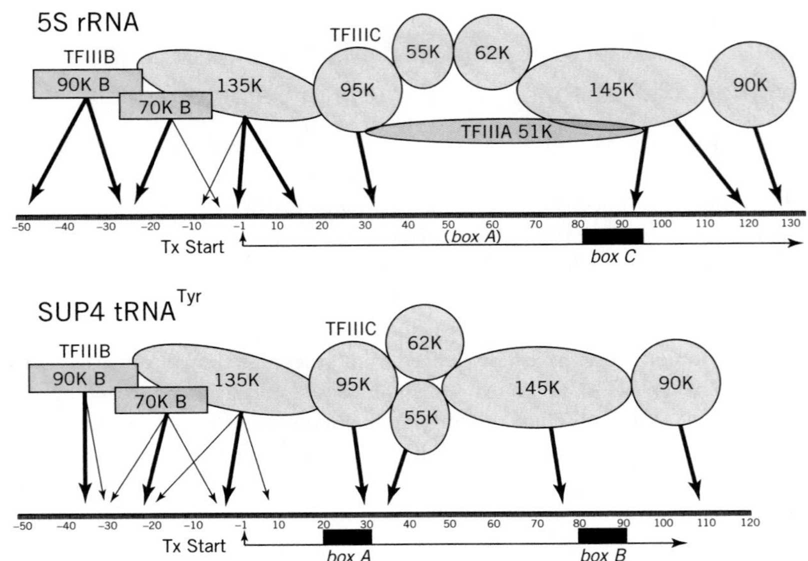

Figure 5.8 The sequences of the nontemplate strands of yeast genes encoding 5S rRNA and a tRNA are shown, along with schematic representations of the various subunits of TFIIIA, TFIIIB, and TFIIIC. The approximate locations of these subunits relative to the DNA sequences of the two genes are indicated by arrows. Note that TFIIIA is required to form a transcription complex on 5S rRNA genes but not on tRNA genes. TFIIIA is responsible for much of the sequence-specific recognition of the internal control region of the 5S rRNA gene, whereas TFIIIC performs this function for the tRNA gene. Nonetheless, both TFIIIC and TFIIIB are required for transcription of both 5S rRNA and tRNA genes. Redrawn from Braun, B. R., Bartholomew, B., Kassevetis, G. A., and Geiduschek, E. P. *J. Mol. Biol.* 228:1063, 1992.

and are often referred to as the internal control region. In some cases, the DNA in the region immediately 5′ to the transcribed region of the gene can be substituted by other sequences with relatively minor effects on transcription. In other cases, however, sequences upstream of the transcription start site can play a major role in establishing efficient transcription. The differences appear to result from the relative importance of various DNA–protein contacts in stabilizing the entire transcription complex made up of multiple proteins in addition to RNA polymerase III. The relationships of these various proteins to the underlying DNA sequence are shown for a 5S rRNA gene and for a tRNA gene in Figure 5.8.

The Common Enzymatic Basis for RNA Polymerase Action

Our understanding of the molecular basis of RNA polymerase action has grown recently, as the three-dimensional structures of a σ factor, a bacterial core RNA polymerase, the RNA polymerase holoenzyme, a eukaryotic RNA polymerase II, and even an RNA polymerase II engaged in transcription elongation have been determined by X-ray crystallography. Despite all the differences in subunit composition, size, and mechanism, RNA polymerases seem to interact with their DNA templates, nucleotide substrates, and nascent RNA chains in a similar fashion.

Remarkably, the shapes of the prokaryotic and eukaryotic enzymes are strikingly similar, particularly within the enzymes' catalytic core. Both bacterial core RNA polymerase and RNA polymerase II from yeast are shaped rather like the claw of a crab (Figure 5.9). DNA comes in at one end of the molecule, and moves through a deep cleft in the polymerase before making a 90-degree bend and exiting from the polymerase. The catalytic site, that is, the site at which a new nucleotide is added to the 3′-end of the growing RNA chain, is located at the end of the channel. The two strands of the double-helical DNA molecule are separated by the enzyme to form a "transcription bubble," and a transient nine-base pair RNA–DNA hybrid forms within the bubble. Nucleotides come in and RNA exits through two other, separate channels in the molecule. Despite the differences in sequence and subunit composition, the prokaryotic and eukaryotic RNA polymerases have retained similar structures to accomplish the same goals leading to synthesis of a new RNA chain. These include (1) threading the DNA template through a channel in the enzyme, (2) separating the DNA strands to form a bubble within which an RNA-DNA hybrid exists with the 3′-end of the RNA chain positioned in the active site, (3) bringing nucleotides into the active site through a pore in the enzyme, (4) using a metal ion to assist in catalyzing the formation of a new phosphate ester bond between the alpha phosphate of the incoming nucleoside triphosphate and the 3′-end of the nascent RNA chain, (5) extruding the RNA product through another pore, and (6) directing reannealing of the two DNA strands as they exit from the enzyme.

Recall that the σ-subunit of the prokaryotic holoenzyme is responsible for binding to the −10 and −35 boxes of promoters. Structural studies have shown that sigma is an

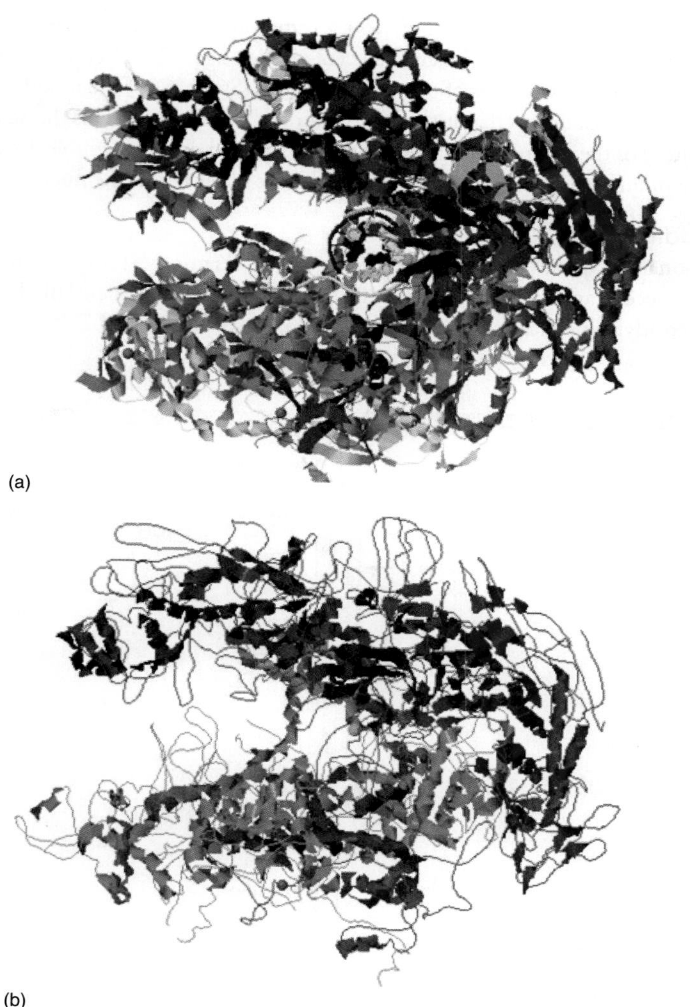

Figure 5.9 Comparison of the structures of eukaryotic and prokaryotic RNA polymerases. The same colors are used to represent homologous subunits in the two enzymes. Note the similarity in the overall shapes. (*a*) The three-dimensional structure of yeast RNA polymerase II engaged in transcription elongation. The various subunits are shown in different colors, with the largest subunit shown in blue and the second largest subunit in green. The DNA template strand is yellow and the nascent RNA is dark blue. This structure is missing two subunits of the RNA polymerase (products of the Rpb4 and Rpb7 genes), and the nontemplate strand of DNA was not resolved in the structural determination. (*b*) The structure of the RNA polymerase core from *Thermus aquaticus* is shown at the bottom. The β' subunit is shown in green, the β subunit in blue, the two α subunits in red, and the ω subunit in violet.
Figures drawn using the coordinates provided in the protein structure database (pdb), based on the following citations: (*a*) Gnatt, A. L., Cramer, P., Fu, J., Bushnell, D. A., and Kornberg, R. D. Structural basis of transcription: An RNA polymerase II elongation complex at 3.3 Å resolution. *Science* 292:1876, 2001; and (*b*) Minakhin, L., Bhagat, S., Brunning, A., Campbell, E. A., et al. Bacterial RNA polymerase subunit omega and eukaryotic RNA polymerase subunit RPB6 are sequence, structural, and functional homologs and promote RNA polymerase assembly. *Proc. Natl. Acad. Sci. USA* 98:892, 2001.

oblong molecule, composed of a bundle of α-helical residues, packed into an open V shape. One arm of the "V" contains critical residues for promoter recognition and core polymerase binding. One side of this arm contains an α-helix that binds to the -10 sequence and the other face binds to core polymerase through hydrophobic interactions.

5.4 • RNA PROCESSING

RNA copies of DNA sequences must be modified to mature, functional, molecules in prokaryotes and eukaryotes. The reactions of RNA processing can include: removal of extra nucleotides, base modification, addition of nucleotides, and separation of different RNA sequences by the action of specific nucleases. Processing reactions can occur either cotranscriptionally (while the RNA is still being synthesized by RNA polymerase) or posttranscriptionally (after the transcript is released by RNA polymerase). Finally, in eukaryotes, RNAs are exported from the nucleus.

Transfer RNA Is Modified by Cleavage, Addition, and Base Modification

Cleavage

The primary transcript of a tRNA gene contains extra nucleotide sequences both 5′ and 3′ to the tRNA sequence. These primary transcripts may also contain **introns** in the anticodon region. Posttranscriptional processing reactions occur in a closely defined but not necessarily rigid temporal order. First, the primary transcript is trimmed in a relatively nonspecific manner to yield a precursor molecule with shorter 5′- and 3′-extensions. Then ribonuclease P, a ribozyme (p. 68), removes the 5′-extension by endonucleolytic cleavage. The 3′-end is trimmed exonucleolytically, followed by synthesis of the CCA terminus. Synthesis of the modified nucleotides occurs in any order relative to the nucleolytic trimming. Intron removal is dictated by the secondary structure of the precursor (Figure 5.10) and is carried out by a soluble, two-component enzyme system; one enzyme removes the intron and the other reseals the nucleotide chain.

3′-End Addition

Each functional tRNA has the sequence CCA at its 3′-terminus. This sequence is essential for tRNA to accept amino acids. In most instances it is added sequentially by the enzyme tRNA nucleotidyltransferase. Nucleotidyltransferase uses ATP and CTP as substrates and

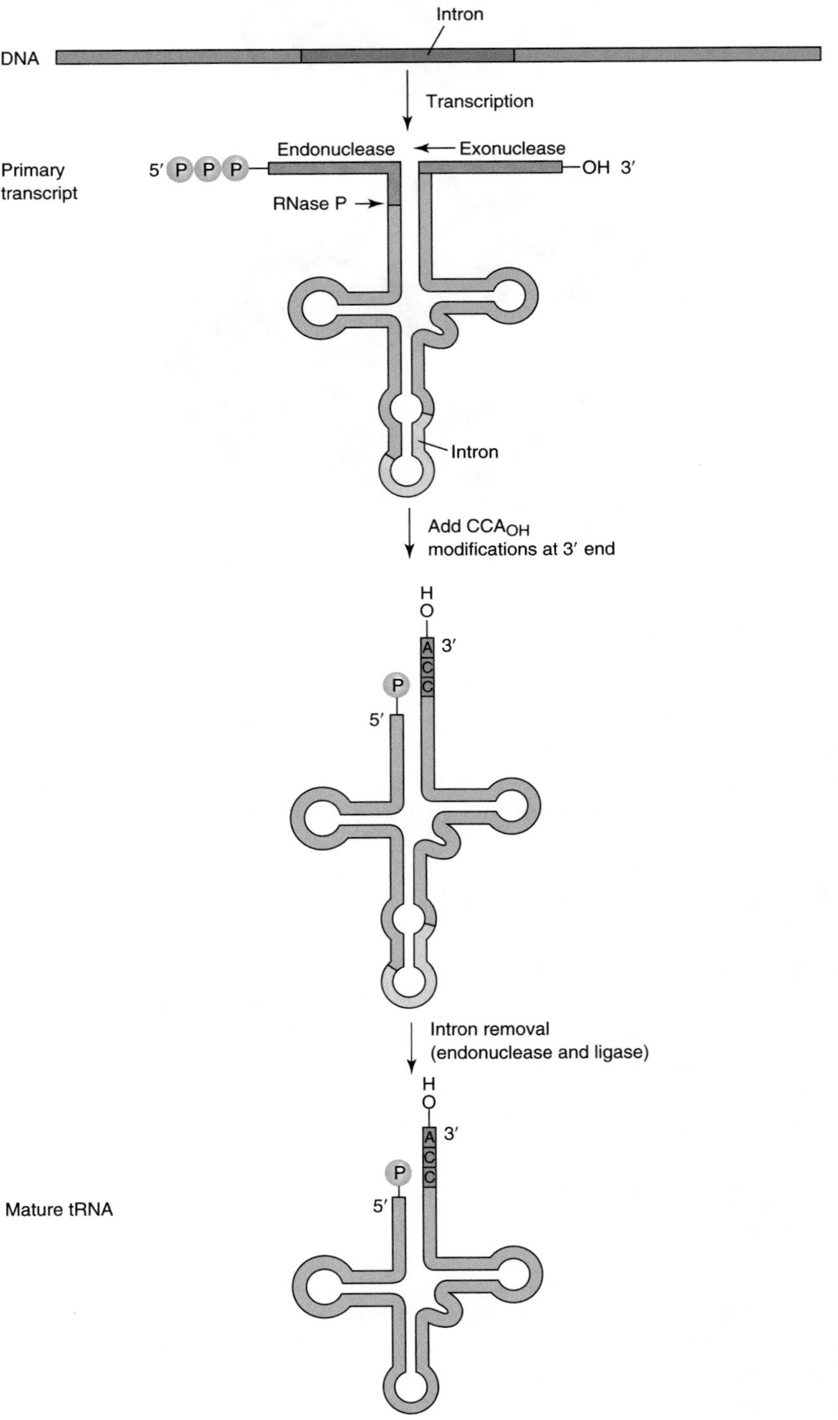

Figure 5.10 Scheme for processing a eukaryotic tRNA. Primary transcript is cleaved by RNase P and a 3′ exonuclease and the terminal CCA is synthesized by tRNA nucleotidyltransferase before the intron is removed, if necessary.

always incorporates them into tRNA at a ratio of 2C/1A. The CCA ends are found on both cytosolic and mitochondrial tRNAs.

Modified Nucleosides

Transfer RNA nucleotides are the most highly modified of all nucleic acids. More than 60 different modifications to the bases and ribose, requiring well over 100 different enzymatic reactions, have been found in tRNA. Many are simple, one-step methylations, but others involve multistep synthesis. Formation of some modified bases actually requires severing of the β-glycosidic bond between ribose and the base. Modifying enzymes produce the same specific modification in more than one species of tRNA; however, the modifying enzymes are location specific. Most modifications are completed before the tRNA precursors have been cleaved to the size of mature tRNA.

Ribosomal RNA Processing Releases Several RNAs from a Longer Precursor

The primary product of rRNA gene transcription is a long RNA, termed 45S RNA, which contains the sequences of 28S, 5.8S, and 18S rRNAs. Processing of 45S RNA occurs in the nucleolus and is carried out by large multisubunit ribonucleoprotein assemblies. Processing of the rRNAs follows a sequential order (Figure 5.11). Processing of pre-rRNA in prokaryotes also involves cleavage of high-molecular-weight precursors to smaller molecules. At an early stage of processing in eukaryotes, some bases in 35S rRNA are modified by methylation on the ring nitrogens of the bases and by the formation of pseudouridine. These reactions are specified by small nucleolar ribonuclear protein particles (snoRNPs) in the nucleolus. Each snoRNP contains a **guide RNA** which base-pairs to the rRNA transcript at the site of modification, specifying where a methylation or formation of pseudouridine should occur. A similar mechanism operates for modification of some other small RNAs.

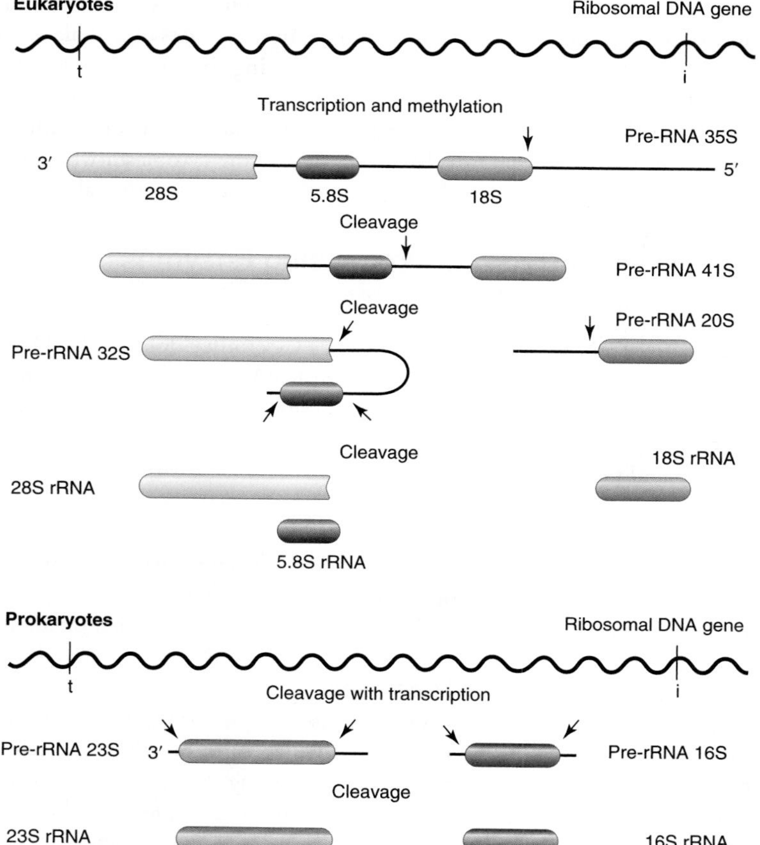

Figure 5.11 Schemes for transcription and processing of rRNAs.
Redrawn from Perry, R. *Annu. Rev. Biochem.* 45:611, 1976.

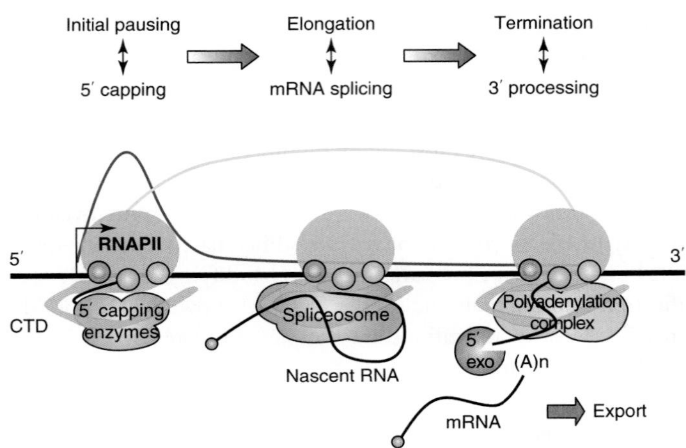

Figure 5.12 RNA processing occurs during transcriptional elongation. Phosphorylation of the C-terminal domain of RNA polymerase II (RNAPII) moves transcription into the elongation phase and results in the recruitment of processing factors. As the partially completed (nascent) mRNA precursor emerges from the transcription complex, the spliceosome removes introns. When the polyadenylation signal is reached, the polyadenylation complex cleaves the nascent mRNA and adds the poly(A) tail, $(A)_n$. The remaining nascent RNA is degraded by 5′ exonucleases, while the processed mRNA is exported for translation.
Adapted with permission from Pandit, S., Wang, D., and Fu, X.-D. *Curr. Opinion Cell Biol.* 20:260, 2008.

Messenger RNA Processing Ensures the Correct Coding Sequence

Most eukaryotic mRNAs have distinctive structural features added in the nucleus by enzyme systems other than RNA polymerase. These include the 3′-terminal poly(A) tail, methylated internal nucleotides, and the cap 5′-terminus. Mature mRNAs are shorter than their primary transcripts. Noncoding sequences present within pre-mRNA, but not present in mature mRNAs, are called intervening sequences or introns. The retained sequences are called **exons**. The general pattern for mRNA processing is depicted in Figure 5.12. Incompletely processed mRNAs make up a large part of the heterogeneous nuclear RNA (HnRNAs).

Processing of eukaryotic pre-mRNA involves a number of molecular reactions all of which must be carried out with exact fidelity. This is most clear in the removal of introns from an mRNA transcript. An extra nucleotide in the coding sequence of mature mRNA would cause the reading frame of that message to be shifted, and the resulting protein will almost certainly be nonfunctional. Indeed, mutations that interfere with intron removal are a major cause of human genetic diseases, for example, in some cases of β-thalassemia (Clin. Corr. 5.4). The task for cells becomes even more daunting since some important human genes consist of over 90% intron sequences. The complex reactions to remove introns are accomplished by multicomponent ribonucleoprotein enzyme systems in the nucleus; after these reactions are completed, the mRNA is exported to the cytoplasm where it interacts with ribosomes to initiate translation. Most processing is cotranscriptional.

RNA Polymerase II Recruits Processing Enzymes during Transcription in Eukaryotes

The largest subunit of RNA polymerase II contains a C-terminal domain (CTD) that functions to couple transcription and processing. When RNA polymerase II is in an initiation mode, the CTD is not modified. When the transcript is about 25 nucleotides long, the CTD is extensively phosphorylated. Phosphorylated CTD successively recruits the capping enzyme, splicing and polyadenylation complexes.

Capping

As the transcription complex moves along the DNA, the capping enzyme complex modifies the 5′-end of the nascent mRNA. Capping involves the synthesis of a 5′– 5′ triphosphate bond with the addition of a G residue. This structure is further modified by methylation.

Removal of Introns from mRNA Precursors

As pre-mRNA is extruded from the RNA polymerase complex, it is rapidly bound by **small nuclear ribonucleoproteins, snRNPs** ("snurps") which carry out the dual steps of RNA splicing: breaking the intron at the 5′ donor site and joining the upstream and downstream exon sequences together. Almost all introns begin with a GU sequence and end with AG; these are termed the donor and acceptor intron–exon junctions, respectively. Not all GU

CLINICAL CORRELATION 5.4

Thalassemia due to Defects in Messenger RNA Synthesis

Thalassemias are genetic defects in the coordinated synthesis of α- and β-globin peptide chains; a deficiency of β globin chains is termed β-thalassemia whereas a deficiency of α globin chains is termed α-thalassemia. Patients suffering from β-thalassemia present with anemia at about 6 months of age, at the time when HbF synthesis ceases and HbA synthesis becomes predominant. The severity of symptoms leads to the classification of the disease into either thalassemia major, where a severe deficiency of globin synthesis occurs, or thalassemia minor, where a less severe imbalance occurs. Occasionally an intermediate form is seen. Therapy for thalassemia major involves frequent blood transfusions, leading to a risk of complications from iron overload. Unless chelation therapy is successful, the deposition of iron in peripheral tissues, termed hemosiderosis, can lead to death before adulthood. Carriers of the disease are typically asymptomatic or nearly so. Ethnographically, the disease is common in persons of Mediterranean, Arabian, and East Asian descent. As in sickle-cell anemia (HbS) and glucose 6-phosphate dehydrogenase deficiency, the abnormality of the carriers' erythrocytes affords some protection from malaria. Maps of the regions where one or another of these diseases is frequent in the native population superimpose over the areas of the world where malaria is endemic.

The most common form of α-thalassemia is due to a genetic deletion which presumably results from unequal crossing over between adjacent, duplicated α-globin alleles. In contrast, β-thalassemia can result from a wide variety of mutations. Known events include mutations leading to frame shifts in the β-globin coding sequence, as well as mutations leading to premature termination of peptide synthesis. Many β-thalassemias result from mutations affecting the biosynthesis of β-globin mRNA. Genetic defects are known that affect the promoter of the gene, leading to inefficient transcription. Other mutations result in aberrant processing of the nascent transcript, either during splicing out of its two introns from the pre-mRNA or during polyadenylation of the mRNA precursor. Examples where the molecular defect illustrates a general principle of mRNA synthesis are discussed in the text.

Orkin, S. H. Disorders of hemoglobin synthesis: The thalassemias. In Stamatoyannopoulis, G., Nienhuis, A. W., Leder, P., and Majerus, P. W. (Eds.), *The Molecular Basis of Blood Diseases.* Philadelphia; Saunders, 1987; and Weatherall, D. J., Clegg, J. B., Higgs, D. R., and Wood, W. G., The hemoglobinopathies. In Scriver, C. R., Beaudet, A. L., Sly, W. S., and Valle, D. (Eds.), *The Metabolic Bases of Inherited Disease,* 7th ed. New York: McGraw Hill, 1995.

or AG sequences are spliced out of RNA, however. How does the cell know which GU sequences are in introns (and therefore must be removed) and which are destined to remain in mature mRNA? This discrimination is accomplished by formation of base pairs between U1 RNA and the sequence of the mRNA precursor surrounding the donor GU sequence (Figure 5.13; Clin. Corr. 5.5). Another snRNP, containing U2 RNA, recognizes important sequences at the 3′-acceptor end of the intron. Still other snRNP species, among them U5 and U6, then bind to the RNA precursor, forming a large complex termed a **spliceosome** (by analogy with the large ribonucleoprotein assembly involved in protein synthesis, the ribosome). The spliceosome uses ATP to carry out accurate removal of the intron. First, the phosphodiester bond between the exon and the donor GU sequence is broken, leaving a free 3′-OH group at the end of the first exon and a 5′-phosphate on the donor G of the intron. This pG is then used to form an unusual linkage with the 2′-OH group of an adenosine within the intron to form a branched, or **lariat**, RNA structure, as shown in Figure 5.14. After the lariat is formed, the second step of splicing occurs. The phosphodiester bond immediately following the AG is cleaved, and the two exon sequences are ligated together.

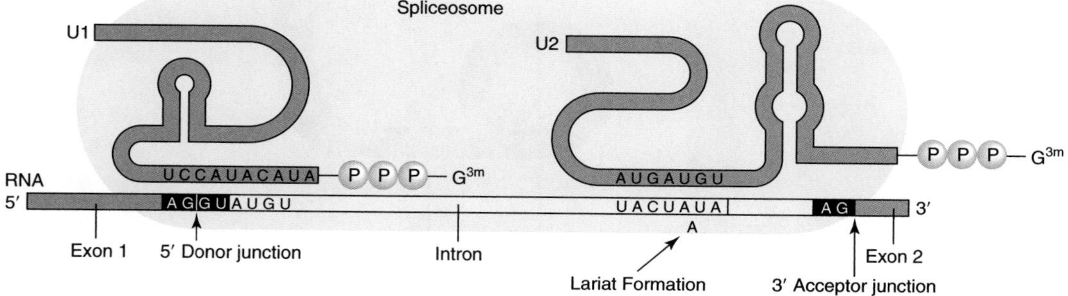

Figure 5 .13 Mechanism of splice junction recognition. Recognition of the 5′ splice junction involves base pairing between the intron–exon junction and the U1 RNA snRNP. This base pairing targets the intron for removal.
Adapted from Sharp, P. A. *J. Am. Med. Assoc.* 260:3035, 1988.

CLINICAL CORRELATION 5.5

Autoimmunity in Connective Tissue Disease

Humoral antibodies in sera of patients with various connective tissue diseases recognize a variety of ribonucleoprotein complexes. Patients with systemic lupus erythematosus exhibit a serum antibody activity designated Sm, and those with mixed connective tissue disease exhibit an antibody designated RNP. Each antibody recognizes a distinct site on the same RNA–protein (U1–RNP) complex that is involved in mRNA processing in mammalian cells. The U1–RNP complex contains U1 RNA, a 165 nucleotide sequence highly conserved among eukaryotes, that at its 5′-terminus includes a sequence complementary to intron–exon splice junctions. Addition of this antibody to *in vitro* splicing assays inhibits splicing, presumably by removal of the U1 RNP from the reaction. Sera from patients with other connective tissue diseases recognize different nuclear antigens, nucleolar proteins, and/or chromosomal centromere. Sera of patients with myositis have been shown to recognize cytoplasmic antigens such as aminoacyl-tRNA synthetases. Although humoral antibodies have been reported to enter cells that have Fc receptors, there is no evidence that this is part of the mechanism of autoimmune disease.

Polyadenylation

RNA polymerase continues transcribing the gene until a polyadenylation signal sequence is reached (Figure 5.15). This sequence, which has the consensus AAUAAA, appears in the mature mRNA but does not form part of its coding region. Rather, it signals cleavage of the nascent mRNA precursor about 20 or so nucleotides downstream. The poly(A) sequence is then added by a specialized polymerase to the free 3′-end produced by this cleavage. The C-terminal domain of RNA polymerase II recruits the cleavage and polyadenylation complex to the transcript. Termination of transcription by the elongating RNA polymerase occurs further downstream and is coupled to the cleavage and polyadenylation reaction, although the detailed mechanism is not clear. The end result of processing is a fully functional coding mRNA, all introns removed, and ready to direct protein synthesis.

Mutations in Splicing Signals Cause Human Diseases

Messenger RNA splicing is an intricate process dependent on many molecular events. If these events are not performed with precision, functional mRNA is not produced. This is illustrated in the human thalassemias, which affect the balanced synthesis of α- and β-globin chains (see Clin. Corr. 5.4, p. 197). Some mutations that lead to β-thalassemia interfere with splicing of β-globin mRNA precursors. For example, we know that all intron sequences begin with the dinucleotide GU. Mutation of the G in this sequence to an A means that the splicing machinery will no longer recognize this dinucleotide as a donor site. Splicing will skip over the correct exon–intron junction. This could lead to extra sequences appearing in the β-globin mRNA that would normally be spliced out, or alternatively, sequences could be deleted from the mRNA product (Figure 5.16). In either event, functional β-globin will be synthesized in reduced amounts and the anemia characteristic of the disease will result.

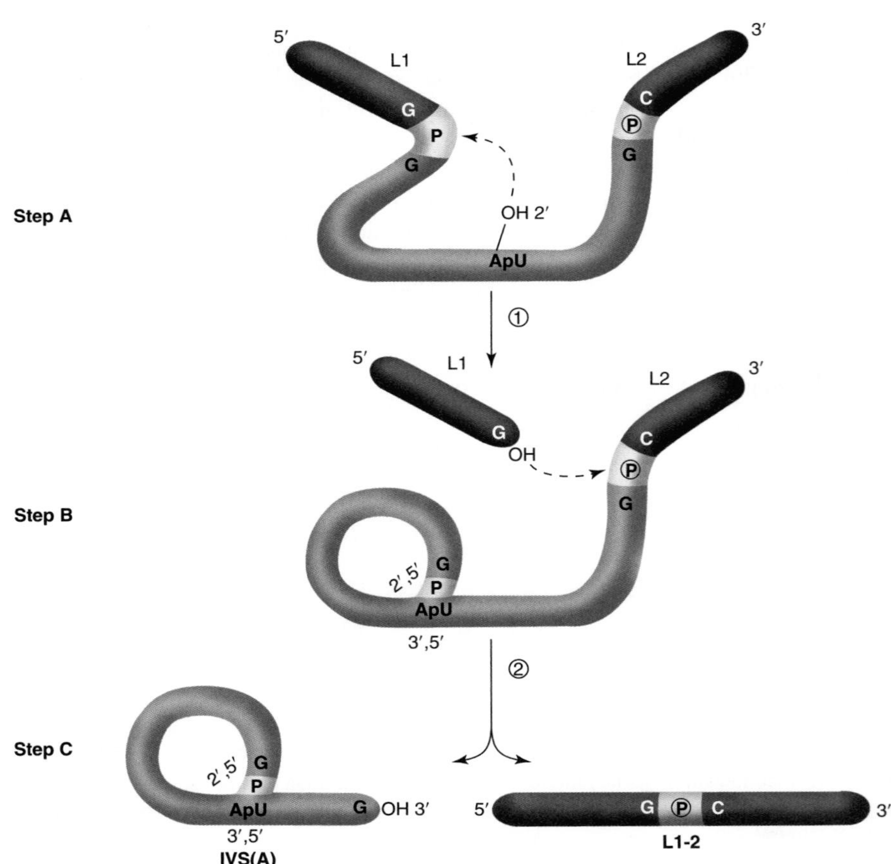

Figure 5.14 **Proposed scheme for mRNA splicing to include the lariat structure.** A messenger RNA is depicted with two exons (in blue) and an intervening intron (in orange). A 2′-OH group of the intron sequence reacts with 5′-phosphate of the intron's 5′-terminal nucleotide producing a 2′–5′ linkage and the lariat structure. Simultaneously, the exon 1–intron phosphodiester bond is broken leaving a 3′-OH terminus on this exon free to react with 5′-phosphate of the exon 2, displacing the intron and creating the spliced mRNA. The released intron lariat is subsequently digested by cellular nucleases.

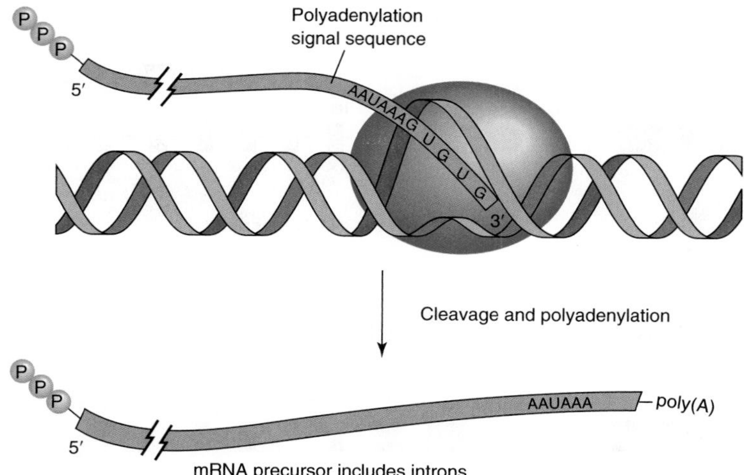

Figure 5.15 Cleavage and polyadenylation of eukaryotic mRNA precursors. The 3′-termini of eukaryotic mRNA species are derived by processing. The sequence AAUAAA in mRNA specifies cleavage of mRNA precursor. The free 3′-OH end of the mRNA is a primer for poly(A) synthesis.
Adapted from Proudfoot, N. J. *Trends Biochem. Science* 14:105, 1989.

Alternate Pre-mRNA Splicing Can Lead to the Synthesis of Multiple Protein Isoforms from a Single DNA Coding Sequence

The existence of intron sequences is paradoxical. Introns must be removed precisely so that the mRNA can accurately encode a protein. As described in Clin. Corr. 5.4, a single base mutation can drastically interfere with splicing and cause a serious disease. Further, the presence of intron sequences in a gene means that its overall sequence is much larger than is required to encode its protein product. A large gene is a target for more mutagenic events than is a small one. Indeed, common human genetic diseases like Duchenne muscular dystrophy occur in genes that encompass millions of base pairs of DNA information. Why has nature not removed introns completely over the long time scale of eukaryotic evolution? The answer seems to lie in the ability of differential processing to generate multiple proteins from a single transcript.

Tropomyosin proteins are essential components of the contractile apparatus in the three types of muscle (p. 970) and each contains a specific tropomyosin. This diversity arises from a single gene that is transcribed into a primary transcript. The transcript is then processed as diagramed in Figure 5.17. Each cell type processes the primary tropomyosin transcript in a characteristic fashion. The resulting mRNA species are translated to yield the tropomyosins characteristic of each cell type. This is a general theme in eukaryotic gene expression. There are only 20,000–25,000 protein-coding genes in the human genome; however, approximately two-thirds of these give rise to differentially spliced transcripts. Differentially spliced transcripts are translated into protein products that have different locations, molecular interactions, or enzymatic properties. Thus the existence of introns supplies the organism with a powerful method of generating protein diversity. The choice of a specific splicing pattern is often regulated by protein factors that bind to sequences within an exon. These proteins can either enhance or repress splicing at adjacent intron–exon junctions.

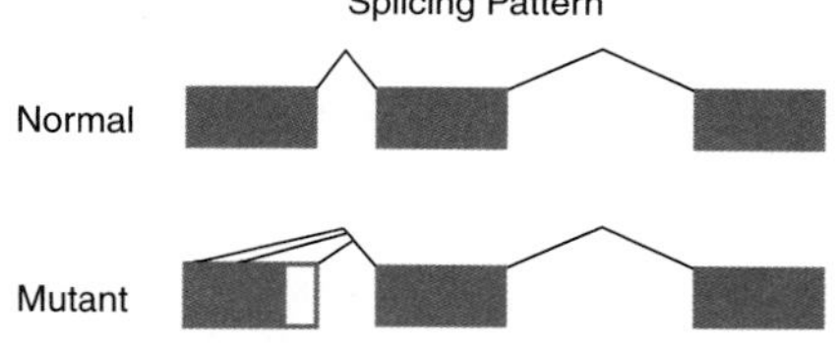

Figure 5.16 Nucleotide change at an intron–exon junction of a human β-globin gene, which leads to aberrant splicing and β-thalassemia. This figure shows the splicing pattern of a mutated transcript containing a change of G-U to A-U at the first two nucleotides of the first intron. Loss of this invariant sequence means that the correct splice junction cannot be used; therefore transcript sequences that base-pair with the U1 snRNA less well than the correct sequence junction are used as splice donors. The diagonal lines indicate the portions spliced together in mutant transcripts. Note that some of the mutant mRNA precursor molecules are spliced so that portions of the first intron (denoted as white boxes) appear in the processed product. In other instances the donor junction lies within the first exon and portions of the first exon are deleted. In no case is wild-type globin mRNA produced.
Adapted from Orkin, S. H. In Stamatoyannopoulis, G., et al. (Eds.), *The Molecular Basis of Blood Diseases.* Philadelphia: Saunders, 1987.

5.5 • RNA EXPORT AND QUALITY CONTROL

Messenger, transfer, and ribosomal RNAs function in the cytoplasm. Movement out of the nucleus is through the nuclear pore. The nuclear pore assembly is a cylinder that spans the membrane. Within the cylinder lie two stacked rings and a central "plug." Small ions can move through the pore between the plug and the rings, but RNAs are too big to pass through that space. RNA export occurs through a channel in the plug; this is a process that requires

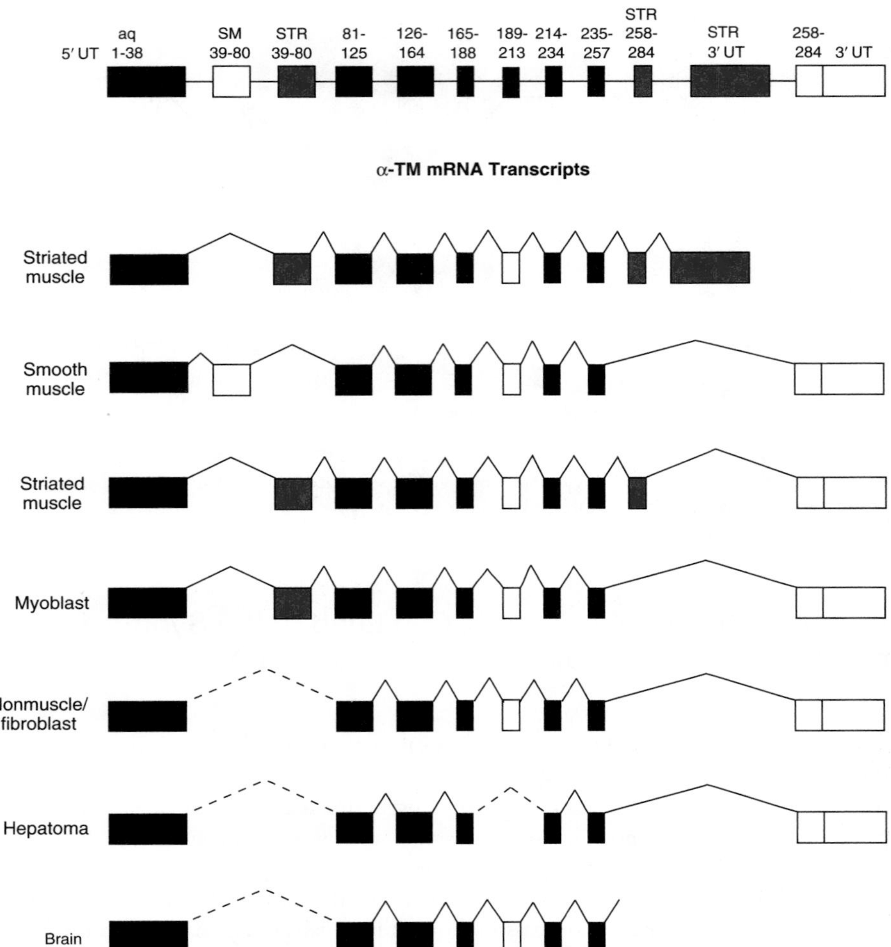

Figure 5.17 Alternate splicing of tropomyosin gene transcripts results in a family of tissue-specific tropomyosin proteins. Redrawn from Breitbart, R. E., Andreadis, A., and Nadal-Ginard, B. *Annu. Rev. Biochem.* 56:467, 1986.

that the RNAs be bound to proteins to form ribonucleoprotein particles. The movement of ribosomal RNAs occurs after they are assembled into ribosomes in the nucleolus. Transfer RNAs may be bound to aminoacyl-tRNA synthetases, since they can be aminoacylated in the nucleus prior to export. Messenger RNAs are exported in complexes that contain a cap-binding protein at the 5′-end and RNA-binding proteins along the rest of the sequence. The process requires metabolic energy in the form of GTP, specific signal sequences in the protein, and a G-protein GTPase.

RNA splicing and export ensure quality control of mRNA integrity. The chain termination codon (one of the three nonsense codons, UAA, UAG, or UGA) of an mRNA is usually located in the last exon. If mRNA were to contain an upstream nonsense codon, the protein coded would be shorter than the normal protein (Clin. Corr. 5.6). This shorter protein would likely be nonfunctional and could interfere with cellular function. The nonsense-mediated decay pathway prevents the export and translation of mRNAs whose **chain termination codon** is upstream of the last exon. A few spliceosomal proteins remain on the spliced mRNA at the junction of two exons. A nonsense codon in an upstream exon is followed by these junction proteins; these mRNAs are targeted for nonsense-mediated decay. Recognition of the upstream nonsense codon must involve translation. Apparently, the first translating ribosome removes the exon junction proteins. If the ribosome stops at a chain termination codon before all the exon junction proteins are removed, a specific set of nucleases degrades the mRNA, preventing its translation. Similarly, mRNA that lacks a normal chain-terminating nonsense codon is targeted for degradation because the mRNA most likely has been synthesized inaccurately or has been cleaved by mistake. Translation of this mRNA would likewise result in synthesis of a shortened protein.

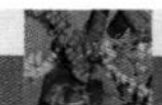

CLINICAL CORRELATION 5.6

Nucleic Acid Editing Enzymes: An Antiviral Defense as well as a Way of Altering Gene Expression

Apolipoprotein B, a key player in cholesterol metabolism, exists in two forms: one of M_r = 48,000 which is produced in the intestine and a second one with M_r = 100,000 which is made in the liver. The two RNAs are transcribed from a single gene, and the intestine-specific isoform results from an editing event. An intestinal cytidine deaminase enzyme, termed Apolipoprotein B mRNA editing enzyme, catalytic polypeptide 1 (APOBEC1) converts a specific C in ApoB mRNA to U, transforming a CAA (Gln) codon to a UAA (stop) codon. APOBEC enzymes form a family and different isoforms are expressed in different cell types.

Why such an unusual mechanism, when a similar result could be obtained by differential splicing of a single transcript or a duplicated gene? An answer may lie in the properties of the APOBEC family of enzymes. Primary T cells and macrophages are relatively resistant to HIV infection. Recall that HIV has an RNA genome, which must be reverse transcribed into cDNA (p. 257). One host-encoded isoform of the APOBEC family, APOBEC3G, is carried along in the viral particle as it buds off from an infected cell. APOBEC3G converts multiple C residues in the initial cDNA product to deoxy-U. The dUs then specify dAs during synthesis of the second-strand cDNA, which has the same sense as the viral RNA. The hypermutated viruses are defective in later phases of the life cycle.

APOBEC3G does not confer absolute resistance, however. In another example of how disease organisms evolve to evade host defenses, HIV encodes a protein termed Vif, which specifically targets APOBEC3G for degradation by the proteasome. Vif is related to ubiquitin–E3 ligase enzymes, although the exact mechanism by which it targets APOBEC3G is not yet clear.

Goila-Gaur, R., and Strebel, K. HIV-1, Vif, APOBEC, and intrinsic immunity. *Retrovirol.* 5:51, 2008; Harris, R., and Liddament, M. T. Retroviral restriction by APOBEC proteins. *Nature Rev. Immunol.* 4:868, 2004; and Wahl, S. M., Greenwill-Wild, T., and Vazquez, N. HIV accomplices and adversaries in macrophage infection. *J. Leukocyte Biol.* 80:973, 2006.

5.6 • RNA INTERFERENCE

Double-stranded regions in cellular mRNA are usually fairly short, less than 10 base pairs. Longer double-stranded sequences can result from replication of RNA viruses, from transcription of repetitive sequences, or from longer regions of some natural mRNAs. These double-stranded RNAs are processed into single-stranded, **inhibitory RNAs (RNAi)** which then act to silence expression from their target mRNAs.

The processing pathway for these RNAs is shown in Figure 5.18. A double-stranded RNA-specific RNase (a member of the RNase III family) first processes the double-stranded RNA to a smaller double-stranded species that is exported from the nucleus. Once it is in the cytoplasm, the double-stranded RNA is further processed by another RNase III termed Dicer to form a small double-stranded RNA about 21 base pairs long. Then the double-stranded RNA is unwound by a protein complex called RNA-induced silencing complex (RISC) or micro-RNA–protein (miRNP). In either case, one strand of the RNA duplex is selected, almost always the one that is complementary to the mRNA target. Then the protein-bound small RNA base pairs bind to the mRNA, causing one of two events. If there is perfect complementarity (the RNA is termed **small inhibitory RNA [siRNA]**), the enzymes of the RISC complex cleave the mRNA strand. If there is imperfect complementarity (the RNA is termed **micro-RNA** [**miRNA**]), the mi-RNP represses mRNA translation by base pairing of the miRNA with the mRNA. RNA interference is efficient because one RISC–siRNA complex can cleave many mRNA molecules, and translational repression may lead to mRNA degradation. Recent research suggests that miRNAs may regulate the expression of as many as one-third of all mammalian genes, and the involvement of noncoding RNAs in heterochromatin formation implies that they may regulate transcription as well as translation. Clin. Corr. 5.7 discusses the involvement of microRNAs in oncogenesis.

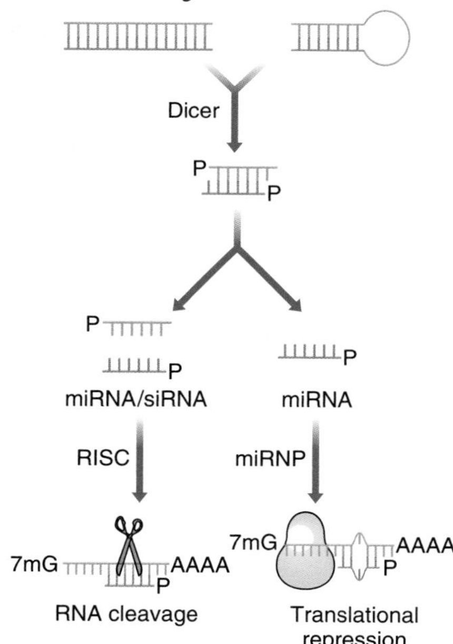

Figure 5.18 Effects of small inhibitory RNAs on eukaryotic mRNA metabolism. Double-stranded RNA, for example, from replication of RNA viruses (top left), or from hairpins in long RNA transcripts (top right) is processed by the Dicer enzyme into short double-stranded RNAs. One strand is complementary to mRNA and base-pairs with it, aided either by RISC (left pathway) or the miRNP complex (right pathway). RISC cleaves mRNA at the double-stranded region, whereas miRNP base pairing stops translation of the mRNA.
Redrawn from Meister, G., and Tuschl, T., *Nature* 431:343, 2004.

5.7 • TRANSCRIPTION-COUPLED DNA REPAIR

A chemical lesion in DNA affects gene expression only if the altered sequence is transcribed. Sequences that aren't expressed will not lead to an observed mutation. It makes sense, therefore, that the repair of DNA would be targeted to sequences that are actively being transcribed. This coupling of transcription and repair occurs because DNA lesions (such as

CLINICAL CORRELATION 5.7

Involvement of MicroRNAs in Oncogenesis

MicroRNAs (miRNAs) reduce expression of their target genes by binding mRNAs and thus either inhibiting their translation or triggering their destruction (Figure 5.18). Approximately 30% of all human mRNAs are potentially regulated by miRNAs. Since cancer is caused by deranged gene expression, it seems logical that the miRNA pathway would be involved in oncogenesis. This hypothesis has been supported by a number of observations. (1). Genes of miRNA families are often deleted in human cancer cell lines. (2) Conversely, some B-cell lymphomas are characterized by amplified DNA sequences that specify miRNAs. (3) The let-7 miRNA interacts with several oncogenes. Reduction of let-7 miRNA is correlated with poor prognosis of nonsmall cell lung cancer (NSCLC). Conversely, in a mouse model of NSCLC, functional let-7 transgenes reduce the oncogenic phenotype. (4) Several miRNAs inactivate the p53 pathway, which triggers apoptosis in response to genetic damage.

These and similar data have led to a working model of the roles of miRNAs in the development of cancers (see figure). Although the model is incomplete, it points to the possibility of therapies that would target this new regulatory mechanism.

Garzon, R., and Croce, C. M. MicroRNAs in normal and malignant hematopoeisis. *Curr. Opin. Hematol.* 15:352, 2008; Kumar, M. S., Erkeland, S. J., Pester, R. E., Chen, D. Y., et al. Suppression of non-small cell lung tumor development by the let-7 microRNA family. *Proc. Natl. Acad. Sci. USA* 105:3903, 2008; and Papagiannakopoulis, T., and Kosik, K. S. MicroRNAs: regulators of oncogenesis and stemness. *Biomedical Central Medicine* 6:15, 2008. (http://www.biomedcentral.com/1741-7015/6/15.)

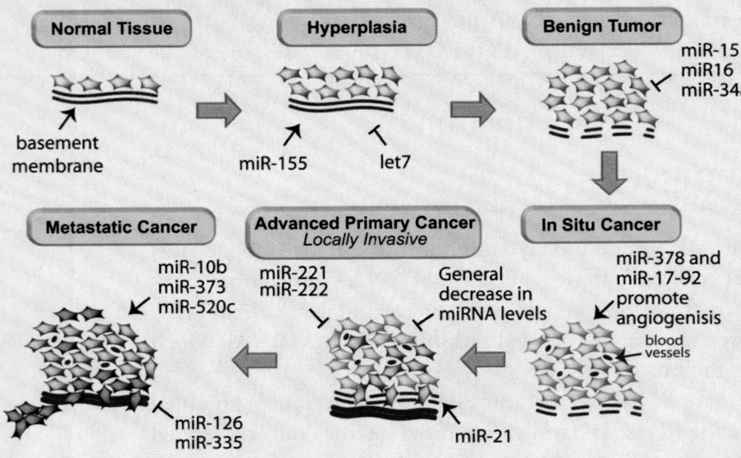

The progression of a normal cell to a metastatic cancer phenotype is regulated in part by specific microRNA species. An arrow means that a microRNA positively regulates the transition to a different stage, whereas a line ending in a bar means that the miRNA negatively regulates the transition. Figure generously supplied by Dr. Phillip Sharp, Koch Institute for Integrative Cancer Research, M.I.T.

cyclobutane dimers; p. 165) that occur on the template strand cause RNA polymerase II to stall. Stalled RNA polymerase provides a signal to the excision-repair systems to remove the altered sequences and resynthesize the correct one. Obviously, this cannot occur with the bulky RNA polymerase bound to the DNA, so the stalled transcript is sacrificed. A deficiency in transcription-coupled repair is a hallmark of Cockayne syndrome (Clin. Corr. 5.8).

5.8 • NUCLEASES AND RNA TURNOVER

The different roles of RNA and DNA in genetic expression are reflected in their metabolic fates. A cell's information store (DNA) must be preserved, thus the multiple DNA repair and editing systems in the nucleus. Although individual stretches of nucleotides in DNA may turn over, the molecule as a whole is metabolically inert when not replicating. RNA molecules, on the other hand, are individually dispensable and can be replaced by newly synthesized species of the same specificity. It is no surprise that RNA repair systems are not known. Instead, defective RNAs are removed from cells by degradation into nucleotides, which then are reused for new RNA species.

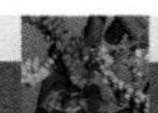

CLINICAL CORRELATION 5.8

Cockayne Syndrome

Cockayne syndrome (CS) (OMIM 216400) is a complex autosomal recessive disorder most commonly caused by a mutation in one of two genes. CS patients have developmental and neurological disorders, skeletal and retinal abnormalities, and a birdlike facial dysmorphia. Death usually results by the age of 20 years and is caused by neural degeneration. Most Cockayne syndrome patients also are photosensitive and predisposed to skin cancer. This photosensitivity points to a defect in DNA repair. For example, xeroderma pigmentosum (XP) results from mutations in a number of components of the DNA repair pathway. CS also is a DNA repair deficiency.

There are two genes whose mutation causes most cases of Cockayne syndrome; both are known to be involved in the repair of DNA damage. CSB encodes a DNA-dependent ATP helicase related to proteins involved in chromatin remodeling. CSA is an E3 ubiquitin ligase apparently involved in turnover of proteins. The CSA and CSB proteins cooperate with other DNA repair proteins to clear away the stalled RNA polymerase and transcripts and repair the damaged template. After repair, new transcription initiation can occur at the promoter, or other RNA polymerase II molecules that are stalled behind the lesion can proceed with accurate transcription.

Transcription-coupled repair occurs when RNA polymerase is stalled by encountering an altered base (for example, a thymine photodimer). Transcription is halted, the partial transcript is degraded, and the template DNA strand is repaired. Apparently, enhancement of transcription by the CSB protein also stimulates transcription-coupled DNA repair.

If this were all to the story, Cockayne syndrome would be a variant of xeroderma pigmentosum; however, XP patients are developmentally and neurologically normal while still being photosensitive. What causes the other features of CS? It is likely that these other symptoms are due to a primary deficiency in transcription elongation caused by the mutationally altered elongation factor CSB. This idea expands our understanding of the relationship between mutation and disease. Usually, genetic diseases are caused by a mutation in biochemical processes that are outside the central information pathways of cells. This makes sense because overall inhibition of DNA, RNA, or protein synthesis would be lethal at an early stage of development. The widespread defects of CS must be due to the mutation affecting the transcription of some genes more than of others.

Citterio, E., Vermeulen, W., and Hoeijmakers, J. H. J. Transcriptional healing. *Cell* 101:447, 2000; D'Errico, M., et al. The role of CSA in the response to oxidative DNA damage in human cells. *Oncogene* 26:4336, 2007; and Laine, J. P., and Egly, J. M. When transcription and repair meet: a complex system. *Trends Genet.* 22:430, 2006.

This is clearest for mRNA species, which are relatively unstable. However, even the so-called stable RNAs turn over; for example, the half-life of tRNA species in liver is about 5 days. A fairly long half-life for a mammalian mRNA is 30 h.

RNA degradation occurs both in the nucleus and in the cytoplasm. In the nucleus, degradation accomplishes two overlapping functions. First, it serves as quality control so that RNAs that result from incorrect transcription or processing are prevented from getting out of the nucleus. Secondly, RNAs in ribonucleoprotein particles turn over, especially during the cell cycle. RNAs are decapped by a nuclear enzyme and degraded in the 3′ to 5′ direction by the **exosome**, a multienzyme complex.

Cytoplasmic Messenger RNA Turnover Is Coupled to Translation

Removal of RNAs from the cytoplasm is accomplished by cellular ribonucleases. Messenger RNAs are initially degraded in the cytoplasm. The rates vary for different mRNA species, raising the possibility of control by differential degradation.

Two examples of the role of RNA stability in gene control illustrate how the stability of mRNA influences gene expression. Tubulin is the major component of the microtubules of the cytoskeleton of many cell types. When there is an excess of tubulin in the cell, the monomeric protein binds to and promotes degradation of tubulin mRNA, thereby reducing tubulin synthesis. A second example is provided by Herpes simplex viruses (HSVs), the agent causing cold sores and some genital infections. An early event in the establishment of HSV infection is the ability of the virus to destabilize all cellular mRNA molecules, thereby reducing the competition for free ribosomes. As a result, viral proteins are more efficiently translated.

Nucleases are of several types and specificities. The most useful distinction is between exonucleases, which degrade RNA from the 5′- or 3′-end, and endonucleases, which cleave phosphodiester bonds within a molecule. Products of RNase action contain either 3′- or 5′-terminal phosphates, and both endo- and exonucleases can be further characterized by the position (5′ or 3′) at which the monophosphate created by the cleavage is located.

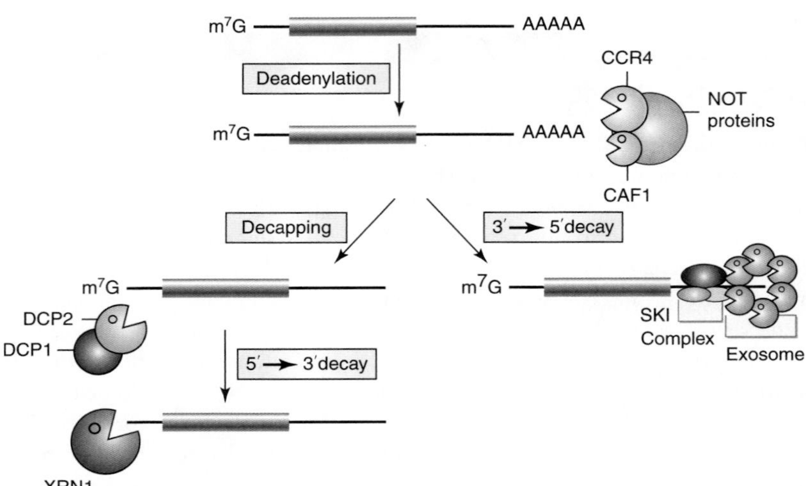

Figure 5.19 Two pathways for mRNA decay. Both pathways begin with the shortening of the poly(A) tail of the mRNA. The deadenylated mRNA is then degraded by either decapping followed by 5′–3′ decay (left) or the 3′–5′ pathway by the exosome and the SKI complex (right). Adapted from Eulalio, A., Behm-Ansmant, I., and Izaurralde, E. *Nature Rev. Mol. Cell Biol.* 8:9, 2007.

Another 3′–5′ exonuclease activity involved in RNA degradation is polynucleotide phosphorylase. Its mechanism for RNA degradation uses inorganic phosphate to displace the phosphodiester bond at the 3′-end of RNA, yielding a shorter RNA and a nucleotide diphosphate.

The structure of RNA can affect nuclease action. Most ribonucleases are less efficient on regions of highly ordered RNA structure. Thus, tRNAs are preferentially cleaved in unpaired regions of the sequence, and the exosome prefers unstructured or misfolded RNA substrates. On the other hand, many RNases involved in RNA processing require the substrate to have a defined three-dimensional structure for recognition.

Cytoplasmic mRNAs can be degraded by several overlapping pathways (Figure 5.19). The mRNA degradation machinery is commonly found in P-bodies, which are subcellular granules that contain a host of mRNA-metabolizing enzymes, including those for degradation. Most commonly, the 3′ poly(A) tail of mRNA is shortened primarily by exonuclease digestion. This deadenylation is inhibited by poly(A)-binding protein(s). Then the cap is removed by a specialized **decapping enzyme,** which is also inhibited by poly(A)-binding protein. Decapped RNA is the substrate for a 5′–3′ exonuclease that degrades the RNA to nucleotides. An alternative pathway involves deadenylation and 3′–5′ exonucleolytic degradation by cytoplasmic exosomes of mRNAs that still retain their cap (Figure 5.19).

Nonsense-mediated decay eliminates mRNAs that are incapable of making a complete protein. Nothing is perfect. Whereas transcription and processing use cellular energy to keep the frequency of mistakes low, they still occur. Since each mRNA is translated into multiple proteins, an incorrectly processed mRNA would both waste the energy required for translation and lead to misfolded proteins, which could trigger inflammation or more severe pathologies, as in the formation of amyloid plaques in Alzheimer disease.

A misprocessed or mistranscribed RNA would have a high probability of carrying chain-terminating (nonsense) codons. Premature nonsense codons trigger mRNA destruction, a process termed nonsense-mediated decay (NMD). During the first (pioneer) round of translation, any premature termination codon is recognized by termination factors. This complex recruits other factors which signal mRNA destruction and send the defective mRNA to the P body. The surveillance mechanism is able to distinguish between premature and correct nonsense codons because the former occur upstream of (5′ to) intron–exon junctions of the mRNA.

Bibliography

Bartel, D.P. MicroRNAs: Genomics, biogenesis, mechanism and function. *Cell* 116: 281, 2004.

Carmo-Fonseca, M., Mendes-Soares, L., and Campos, I. To be or not to be in the nucleolus. *Nature Cell Biol.* 2: E107, 2000.

Chung, Y-F., Imam, J. S., and Wilkinson, M. F. The nonsense-mediated decay RNA surveillance pathway. *Annu. Rev. Biochem.* 76: 51, 2007.

Citterio, E., Vermeulen, W., and Hoeijmakers, J. H. J. Transcriptional healing. *Cell* 101: 447, 2000.

Conaway, J. W., and Conaway, R. C. Transcription elongation and human disease. *Annu. Rev. Biochem.* 68: 301, 1999.

Cramer, P., Armache, K.-J., Baumli, S., Benkert, S., et al. Structure of eukaryotic RNA polymerases. *Annu. Rev. Biophys.* 37: 337, 2008.

Decatur, W. A., and Fournier, M. J. RNA-guided nucleotide modification of ribosomal and other RNAs. *J. Biol. Chem.* 278: 695, 2003.

Eulalio, A., Behm-Ansmant, I., and Izaurralde, E. P bodies: at the crossroads of post-transcriptional pathways. *Nature Revs. Mol. Cell Biol.* 8: 9, 2007.

Garzon, R., and Croce, C. M. MicroRNAs in normal and malignant hematopoiesis. *Curr. Opinion Hematol.* 15: 352, 2008.

Gesteland, R. F., Cech, T., and Atkins, J. F., (Eds.). *The RNA World,* 3d ed. Cold Spring Harbor, NY: Cold Spring Harbor Laboratory, 2006.

Goila-Guar, R., and Strebel, K. Hiv-I, Vif, APOBEC, and intrinsic immunity. *Retrovirol.* 5: 51, 2008. (http://www.retrovirology.com/content/5/1/51.)

Harris, R. S., and Liddament, M. T. Retroviral restriction by APOBEC proteins. *Nature Rev. Immunol.* 4: 868, 2004.

Hawley, D. K., and McClure, W. R. Compilation and analysis of *Escherichia coli* promoter DNA sequences. *Nucleic Acids Res.* 11, 2237, 1983.

Hochheimer, A., and Tjian, R. Diversified transcription initiation complexes expand promoter selectivity and tissue-specific gene expression. *Genes and Dev.* 17: 1309, 2003.

Houseley, J., LaCava, J., and Tollervey, D. RNA-quality control by the exosome. *Nature Rev. Mol. Cell. Biol.* 7: 529, 2006.

Kornberg, R. D. The molecular basis of eukaryotic transcription. *Proc. Natl. Acad. Sci. USA* 104: 12955, 2007.

Li, B., Carey, M., and Workman, J. L. The role of chromatin during transcription. *Cell* 128, 707–719, 2007.

Lindsay, M. A. MicroRNAs and the immune response. *Trends Immunol.* 29: 343, 2008.

Mooney, R. A., and Landick, R. RNA Polymerase Unveiled. *Cell* 98: 687, 1999.

Mooney, R. A., Darst, S. A., and Landick, R. Sigma and RNA polymerase: an on-again, off-again relationship? *Mol. Cell* 20: 335, 2005.

Orkin, S. H., Disorders of hemoglobin synthesis: The thalassemias. In Stamatoyannopoulis, G., Nienhuis, A. W., Leder, P., and Majerus, P. W. (Eds.), *The Molecular Basis of Blood Diseases.* Philadelphia: Saunders, 1987, 206.

Pandit, S., Wang, D., and Fu, X. D. Functional integration of transcriptional and RNA processing machineries. *Curr. Opin, Cell Biol.* 20: 260, 2008.

Parker, R., and Song, H. The enzymes and control of eukaryotic mRNA turnover. *Nature Struct. Mol. Biol.* 11: 121, 2004.

Paule, M. R., and White, R. J. Survey and summary: Transcription by RNA polymerases I and III. *Nucleic Acid Research* 28: 1283, 2000.

Penagarikano, O., Mulle, J. G., and Warren, S. T. The pathophysiology of Fragile X syndrome. *Annu. Rev. Gen. Hum. Genet.* 8: 109, 2007.

Proudfoot, N. Connecting transcription to messenger RNA processing. *Trends Biochem. Sci.* 25: 290, 2000.

Ptashne, M., and Gann, A. Transcriptional activation by recruitment. *Nature* 386: 569, 1997.

Shultzaberger, R. K., Chen, Z., Lewis, K. A., and Schneider, T. D. Anatomy of *Escherichia coli* sigma-70 promoters. *Nucleic Acids Res.* 35: 771, 2007.

Smith, C. W. J., and Valcárcel, J. Alternative pre-mRNA splicing: the logic of combinatorial control. *Trends Biochem. Sci.* 25: 381, 2000.

Turelli, P., and Trono, D. Editing at the crossroad of innate and adaptive immunity. *Science* 307: 1061, 2005.

Valencia-Sanchez, M. A., Lui, J., Hannon, G. J., and Parker, R. Control of translation and mRNA degradation by miRNAs and siRNAs. *Genes Dev.* 20: 515, 2006.

Wahl, S. M., Greenwell-Wild, T., and Vasquez, N. HIV accomplices and adversaries in macrophage infection. *J. Leukocyte Biol.* 80: 973, 2006.

Weatherall, D. J., Clegg, J. B., Higgs, D. R., and Wood, W. G. The hemoglobinopathies. In Scriver, C. R., Beaudet, A. L., Sly, W. S., and Valle, D. (Eds.), *The Metabolic Bases of Inherited Disease,* 7th ed. New York: McGraw Hill, 1995.

Key Terms

initiation
promoter
RNA processing
transcript
RNA polymerase
RNA polymerase I
RNA polymerase II
RNA polymerase III
template
elongation
termination
transcription factor
histone acetylation
enhancer
recruitment
cis-acting
RNA polymerase holoenzyme
sigma factor
intron
RNA processing
exon
capping
RNA polymerase C-terminal domain
snRNP
spliceosome
lariat
polyadenylation
alternative splicing
chain termination codon
microRNA (miRNA)
small inhibitory RNA (siRNA)
exosome
decapping

Questions • CAROL N. ANGSTADT

Multiple-Choice Questions

1. In eukaryotic transcription,
   - **A.** RNA polymerase does not require a template.
   - **B.** all RNA is synthesized in the nucleolus.
   - **C.** consensus sequences are the only known promoter elements.
   - **D.** phosphodiester bond formation is favored because there is pyrophosphate hydrolysis.
   - **E.** RNA polymerase requires a primer.

2. The sigma (σ) subunit of prokaryotic RNA polymerase
   - **A.** is part of the core enzyme.
   - **B.** binds the antibiotic rifampicin.
   - **C.** is inhibited by α-amanitin.
   - **D.** must be present for transcription to occur.
   - **E.** specifically recognizes promoter sites.

3. Termination of a prokaryotic transcript
   A. is a random process.
   B. requires the presence of the rho subunit of the holoenzyme.
   C. does not require rho factor if the end of the gene contains a G-C-rich palindrome.
   D. is most efficient if there is an A-T-rich segment at the end of the gene.
   E. requires an ATPase in addition to rho factor.

4. Eukaryotic transcription
   A. is independent of the presence of upstream consensus sequences.
   B. may involve a promoter located within the region transcribed rather than upstream.
   C. requires a separate promoter region for each of the three ribosomal RNAs transcribed.
   D. requires that the entire gene be in the nucleosome form of chromatin.
   E. is affected by enhancer sequences only if they are adjacent to the promoter.

5. Processing of transfer RNA (tRNA)
   A. involves cleaving out several different tRNAs from one primary transcript.
   B. involves only exonucleases.
   C. cleaves excess bases from the 3′-end after the sequence CCA.
   D. cleaves excess bases from either end of the primary transcript but not in the interior of the molecule.
   E. necessitates modification of some of the bases by methylation.

6. In the cellular turnover of RNA,
   A. repair is more active than degradation.
   B. regions of extensive base pairing are more susceptible to cleavage.
   C. a small double-stranded region of mRNA could lead to cleavage of the mRNA if one strand has perfect complementarity with a region of the mRNA.
   D. the products are always nucleotides with a phosphate at the 5′-OH group.
   E. all species except rRNA are cleaved.

Questions 7 and 8: Fragile X syndrome is a common form of inherited mental retardation. The mutation in the disease allows the increase of a CGG repeat in a particular gene from a normal of about 30 repeats to 200–1000 repeats. This repeat is normally found in the 5′ untranslated region of a gene for the protein FMR1. FMR1 might be involved in the translation of brain-specific mRNAs during brain development. The consequence of the very large number of CGG repeats in the DNA is extensive methylation of the entire promoter region of the FMR1 gene.

7. Methylation of bases in DNA usually
   A. facilitates the binding of transcription factors to the DNA.
   B. makes a difference in activity only if it occurs in an enhancer region.
   C. prevents chromatin from unwinding.
   D. inactivates DNA for transcription.
   E. results in increased production of the product of whatever gene is methylated.

8. Transcription of eukaryotic genes requires the presence of a promoter and usually the presence of enhancers. An enhancer
   A. is a consensus sequence in DNA located where RNA polymerase first binds.
   B. may be located in various places in different genes.
   C. may be on either strand of DNA in the region of the gene.
   D. functions by binding RNA polymerase.
   E. stimulates transcription in both prokaryotes and eukaryotes.

Questions 9 and 10: The synthesis of normal adult hemoglobin (HbA) requires the coordinated synthesis α-globin and β-globin. β-Thalassemia is a genetic disease leading to a deficiency of β-globin chains and an inability of the blood to deliver oxygen properly. β-Thalassemia can result from a wide variety of mutations.

9. All of the following could lead to lack of or nonfunctional β-globin except
   A. a frameshift mutation leading to premature termination of protein synthesis.
   B. mutation in the promoter region of the β-globin gene.
   C. mutations toward the 3′-end of the β-globin gene that codes for the polyadenylation site.
   D. mutations in the middle of an intron that is not at an A base.
   E. all of the above could lead to this result.

10. One mutation leading to β-thalassemia occurs at a splice junction. Which of the following statements about removing introns is correct?
   A. Small nuclear ribonucleoproteins (snRNP) are necessary for removing introns.
   B. The consensus sequences at the 5′- and 3′-ends of introns are identical.
   C. Removal of an intron does not require metabolic energy.
   D. The exon at one end of an intron must always be joined to the exon at its other end.
   E. The nucleoside at the end of the intron first released forms a bond with a 3′-OH group on one of the nucleotides within the intron.

Questions 11 and 12. Protooncogenes produce products that have specific roles in regulating growth and differentiation of normal cells. Mutations can turn these genes into oncogenes whose products are less responsive to normal control. Unmutated protein p53, a tumor suppressor, is a transcription factor, inhibiting some genes and activating others. P53 inhibits genes with TATA sequences and activates genes for DNA repair.

11. The TATA sequence
   A. occurs about 25 bp downstream from the start of transcription.
   B. binds directly to RNA polymerase.
   C. binds transcription factors which bind RNA polymerase.
   D. binds p53.
   E. is an enhancer sequence.

12. Transcription-coupled DNA repair
   A. occurs because a DNA lesion causes the RNA polymerase to stall during transcription.
   B. occurs when a DNA lesion forms in a region of DNA which is highly compacted.
   C. leads to a normal transcript because transcription continues after the repair is made.
   D. occurs because both the template and coding strands of the DNA have lesions.
   E. all of the above.

Problems

13. How does nonsense-mediated decay help prevent the export and translation of mRNA that would lead to a shorter, nonfunctional protein?

14. a. How could you experimentally determine whether a purified preparation of an RNA polymerase is from a prokaryotic or eukaryotic source?
b. A purified preparation of RNA polymerase is sensitive to inhibition by α-amanitin at a concentration of 10^{-8} M. The synthesis of what type or types of RNA is inhibited?

Answers

1. D This is an important mechanism for driving reactions. A and B: Transcription is directed by DNA, generating rRNA precursors in the nucleolus and mRNA and tRNA precursors in nucleoplasm. C: Eukaryotic transcription may have internal promoter regions as well as enhancers. E: This is a difference from DNA polymerase.

2. E A, D, and E: The sigma factor is required for correct initiation and dissociates from the core enzyme after the first bonds have been formed. The core enzyme can transcribe but cannot correctly initiate transcription. B and C: Rifampicin binds to the β-subunit, and α-amanitin is an inhibitor of eukaryotic polymerases.

3. C Rho-independent termination involves secondary structure, which is stabilized by high G-C content. A, B, and E: There is a rho-dependent as well as a rho-independent process. Rho is a separate protein from RNA polymerase and appears to possess ATPase activity. D: A G-C-rich region is required.

4. B RNA polymerase III uses an internal promoter. A: RNA polymerase II activity involves the TATA and CAAT boxes. C: RNA polymerase I produces one transcript, which is later processed to yield three rRNAs. D: Parts of the promoter are not in a nucleosome. E: Enhancers may be as much as 1000 or more bp away.

5. E Transfer RNAs are the most highly modified nucleic acids. A: This occurs with the precursor of rRNAs. B: Both endo- and exonucleases are required. C: CCA are added sequentially by nucleotidyltransferase and are not part of the primary transcript. D: There may be introns, for example, in the anticodon region, of the primary transcript.

6. C This strand is called siRNA. If the complementarity is imperfect (miRNA), translation is repressed. A: There are no RNA repair processes. B: Most degradative enzymes are less efficient on an ordered structure. D: It could be either 3′- or 5′ depending on which side the phosphodiester bond is cleaved. E: All RNAs turn over.

7. D Methylation-demethylation of DNA is one form of control, with the methylated DNA being inactive for transcription. A and E: These would indicate that methylation activates DNA. B: Altering the enhancer region could make a difference but alteration of the promoter region certainly would. C: Unfolding of chromatin in this region is necessary for transcription but methylation of DNA is not directly involved in the process.

8. B B and C: Enhancer sequences seem to work whether they are at the beginning or end of the gene, but they must be on the same DNA strand as the transcribed gene. A: RNA polymerase first binds at the promoter. D: They seem to function by binding proteins which themselves bind RNA polymerase.

9. D Mutation at either terminus of the intron would lead to a splicing error, but a mutation in the middle of the intron should not. The exception would be if the specific mutation site were at the adenosine that forms the branch point. A, B, and C: These are all correct.

10. A snRNPs are responsible both for cleaving the intron at the 5′-end and joining the exons together. B: The sequences at the two ends of an intron are different: GU at the 5′-end and AG at the 3′end. C: The spliceosome requires ATP to remove the intron. D: Nonconsecutive exons can be joined together, leading to multiple proteins from one gene. E: The lariat forms at a 2′-OH.

11. C B and C: RNA polymerase does not bind directly to TATA but to transcription factors that bind TATA. A and E: TATA is part of the promoter and so upstream from the transcription start. D: p53 probably binds to the complex formed between transcription factors and TATA, interfering with transcription.

12. A This signals the excision repair system to repair the lesion. B: Repair is targeted to actively transcribed genes where chromatin is more loosely organized. C: For repair to occur, RNA polymerase must dissociate and the transcript is sacrificed. D: A typical lesion is thymine dimers in the template strand. The coding strand is normal and acts as the template for repair.

13. A chain termination codon upstream of the last exon is abnormal. When translation stops at such a codon, some exon junction proteins still remain. This triggers a specific set of nucleases to degrade the mRNA, preventing its translation.

14. a. Test the preparation for inhibition by α-amanitin at a high concentration where eukaryotic RNA polymerases, except for I, are inhibited but prokaryotic enzyme is not. If the preparation is not inhibited by the amanitin, confirm that it is inhibited by rifampicin. It will not be possible to distinguish eukaryotic mitochondrial RNA polymerase from prokaryotic enzyme because they have the same sensitivity.
b. RNA polymerase II is sensitive to α-amanitin at a very low concentration so mRNA synthesis would be inhibited.

6

Protein Synthesis: Translation and Posttranslational Modifications

Dohn Glitz

Professor Emeritus, University of California at Los Angeles School of Medicine

Key Concepts

- Protein biosynthesis involves translation of the polynucleotide language of genes into the language of proteins.
- Three nucleotide sequences (codons) in mRNA specify a single amino acid, the starting point of translation, and the termination of the peptide chain. There are 64 possible codons and most amino acids are specified by more than one codon. The genetic code is termed degenerate and is almost universal.
- Translation is centered on RNA: mRNA transmits the genetic information, tRNA carries amino acids, and ribosomal RNA helps shape the ribosome and catalyzes peptide bond formation. Many proteins, enzymes, and factors are involved in protein synthesis.

- Synthesis is from the amino to the carboxyl terminus. At each step, (1) the appropriate amino acid is activated by attachment to a tRNA, (2) the aa-tRNA is positioned at the acceptor site of the ribosome, (3) the nascent peptide chain is coupled to the new amino acid by peptidyl transferase, and (4) the elongated peptidyl-tRNA is moved to the donor site of the ribosome.
- Translation is highly regulated, at both individual protein and global levels. Posttranslational and cotranslational modification of proteins is common and can affect protein shape, function, localization in the cell, export from the cell, and lifetime.
- Folding of a protein into its final conformation is spontaneous or assisted by protein chaperones.
- Mechanisms exist for degradation of proteins no longer needed, misfolded, damaged, or incomplete. The proteasome is an important but not exclusive degradation center.

6.1 • INTRODUCTION

Protein biosynthesis is called **translation** because it involves the biochemical translation of genetic information, stored and transmitted in the 4-letter alphabet and structural language of DNA, into the 20-letter alphabet and structural language of proteins. Translation is centered on multiple functions of RNA: messenger RNA transmits the genetic information, transfer RNA carries amino acids and decodes the message, ribosomal RNA catalyzes peptide bond formation and makes up the core of the assembly bench, and very small RNAs are important regulators of the process.

Cells vary in their need and ability to synthesize proteins. Terminally differentiated red blood cells lack the apparatus for translation and have a life span of only 120 days. Other cells synthesize proteins they need to maintain concentrations of enzymes and other proteins. Growing and dividing cells must synthesize much larger amounts of protein, and some cells also produce proteins for export. For example, liver cells are protein factories that synthesize enzymes used in multiple metabolic pathways plus proteins for export including serum albumin, the major protein of blood plasma.

6.2 • COMPONENTS OF THE TRANSLATIONAL APPARATUS

Messenger RNA Transmits Information Encoded in DNA

Genetic information is inherited and stored as nucleotide sequences of DNA. Selective expression of this information requires its transcription into mRNA that carries a specific and precise message from the nuclear "data bank" to the cytoplasmic site of protein synthesis.

In eukaryotes mRNAs are usually synthesized as larger precursor molecules that are processed prior to their export from the nucleus (p. 196). Eukaryotic mRNA is almost always **monocistronic;** that is, it encodes a single polypeptide. The 5′-end is **capped** with **7-methylguanosine** linked through a 5′-triphosphate bridge to the 5′-end of the mRNA (p. 196). A 5′-nontranslated region precedes the translation **initiation signal.** Usually, but not always, this is the first AUG sequence encountered as the message is read 5′ → 3′. Uninterrupted 3-nucleotide-long **codon** sequences that specify a unique polypeptide sequence follow the **initiation codon** until a translation termination codon is reached. This is followed by a 3′-untranslated sequence (or UTR), usually about 100 nucleotides long, before the mRNA terminates in a 100–200 nucleotide long polyadenylate tail.

Prokaryotic mRNA differs significantly. The 5′-end of the mRNA is not capped but it retains a terminal triphosphate from its synthesis by RNA polymerase. Most mRNAs are **polycistronic;** they encode several polypeptides and include more than one initiation AUG codon. A ribosome-positioning sequence is located about 10 nucleotides upstream of the AUG initiation codon. A UTR follows the termination signal but there is no polyadenylate tail.

Transfer RNA Is a Bilingual Translator Molecule

All **tRNA** molecules have several common structural characteristics including the 3′-terminal –CCA sequence to which an amino acid is bound, a conserved "cloverleaf" secondary structure, and an L-shaped tertiary structure (p. 63). However, each of the many tRNA species has a unique nucleotide sequence that allows great specificity in interactions with mRNA and with the aminoacyl tRNA synthetase that couples one specific amino acid to it.

The Genetic Code Uses a Four-Letter Alphabet of Nucleotides

Genetic information exists as linear sequences of nucleotides in DNA, analogous to the linear sequence of letters in these words. The DNA language uses a **four-letter alphabet** composed of two purines, A and G (adenine and guanine), and two pyrimidines, C and T (cytosine and thymine). Information in DNA is selectively transcribed into mRNA in a similar four-letter alphabet, but U (uracil) replaces T. The language of RNA is thus a dialect of the genetic language of DNA. Genetic information is **expressed** mostly in the form of proteins that derive their properties from their sequence of amino acids. In protein biosynthesis the 4-letter language of nucleic acids is **translated** into the 20-letter language of proteins. Implicit in the language analogy is the directionality of these sequences. By convention, nucleic acid sequences are written in a $5' \rightarrow 3'$ direction and protein sequences from the amino terminus to the carboxy terminus. These directions correspond in their reading and biosynthetic senses.

Codons in mRNA Are Three-Letter Words

A 1:1 correspondence of bases to amino acids would permit mRNA to encode only four different amino acids, while a 2:1 correspondence would encode $4^2 = 16$ amino acids. Neither is sufficient since 20 different amino acids occur in most proteins. The three-letter genetic code has $4^3 = 64$ permutations or words; this is also sufficient to encode start and stop signals, equivalent to punctuation. The triplets are called **codons** and are customarily shown in the form of Table 6.1. Two amino acids are designated by single codons: methionine by AUG and tryptophan by UGG. The rest are designated by two, three, four, or six codons. Multiple codons for a single amino acid represent **degeneracy** in the code. The genetic code is nearly **universal,** since the same code words are used in all organisms. An exception to universality occurs in mitochondria, in which a few codons have a different meaning than in the cytosol of the same organism (Table 6.2).

TABLE 6.2 • Nonuniversal Codon Usage in Mammalian Mitochondria

Codon	*Usual Code*	*Mitochondrial Code*
UGA	Termination	Tryptophan
AUA	Isoleucine	Methionine
AGA	Arginine	Termination
AGG	Arginine	Termination

Punctuation

Four codons act partially or fully as punctuation. The **start signal, AUG,** also specifies methionine. An AUG at an appropriate site in mRNA signifies methionine as the initial, amino-terminal residue. AUG codons later in the mRNA specify methionine residues

TABLE 6.1 • The Genetic Code[a]

5′ Base	U	C	A	G	3′ Base
U	UUU, UUC: Phe UUA, UUG: Leu	UCU, UCC, UCA, UCG: Ser	UAU, UAC: Tyr UAA, UAG: Stop	UGU, UGC: Cys UGA Stop UGG Trp	U C A G
C	CUU, CUC, CUA, CUG: Leu	CCU, CCC, CCA, CCG: Pro	CAU, CAC: His CAA, CAG: Gln	CGU, CGC, CGA, CGG: Arg	U C A G
A	AUU, AUC, AUA: Ile AUG Met	ACU, ACC, ACA, ACG: Thr	AAU, AAC: Asn AAA, AAG: Lys	AGU, AGC: Ser AGA, AGG: Arg	U C A G
G	GUU, GUC, GUA, GUG: Val	GCU, GCC, GCA, GCG: Ala	GAU, GAC: Asp GAA, GAG: Glu	GGU, GGC, GGA, GGG: Gly	U C A G

[a] The genetic code comprises 64 codons, which are permutations of four bases taken in threes. Note the importance of sequence: three bases, each used once per triplet codon, give six permutations: ACG, AGC, GAC, GCA, CAG, and CGA, for threonine, serine, aspartate, alanine, glutamine, and arginine, respectively.

A CLOSER LOOK 6.1

Aminoacyl–tRNA Synthetases

Eukaryotes and some bacteria have the expected 20 aminoacyl-tRNA synthetases, but many prokaryotes and archae lack synthetases for glutaminyl-, asparaginyl-, and sometimes cysteinyl-tRNA formation. Instead, these tRNAs are formed indirectly. The tRNAs are first aminoacylated with glutamate, aspartate, or *O*-phosphoserine by aminoacyl-tRNA synthetases with relaxed specificity for recognition of their cognate tRNA species. The bound amino acids are then converted by aminoacyl-tRNA-dependent modifying enzymes that recognize only the appropriate aminoacyl-tRNA.

Selenocysteine is formed from seryl-tRNA^{Sec} in a two-step process; the serine is phosphorylated by a kinase and then converted to selenocysteine by a synthase that uses selenophosphate. Many proteins that include selenocysteine have an antioxidant role in eliminating reactive oxygen species, and elimination of the seryl-tRNA^{Sec} gene is lethal.

Sheppard, K., Yuan, J., Hohn, M. J., Jester, B., et al. From one amino acid to another: tRNA-dependent amino acid biosynthesis. *Nucl. Acids Res.* 36:1813, 2008.

within the protein. Three codons, UAG, UAA, and UGA, are **stop signals** that specify no amino acid and are **termination codons** or, less appropriately, **nonsense codons.**

The UGA codon has an additional function, coding for **selenocysteine** (sec) in at least 25 human proteins. **Selenocysteinyl-tRNA** is synthesized from a specific seryl-tRNA$^{(sec)\ ser}$ (A Closer Look 6.1) and selenocysteine is incorporated only if the UGA codon occurs in a sec-insertion sequence, a defined stem-loop structure in mRNA. A specific elongation factor (eEFSec) and a selenocysteine insertion binding protein (SBP2) are also required.

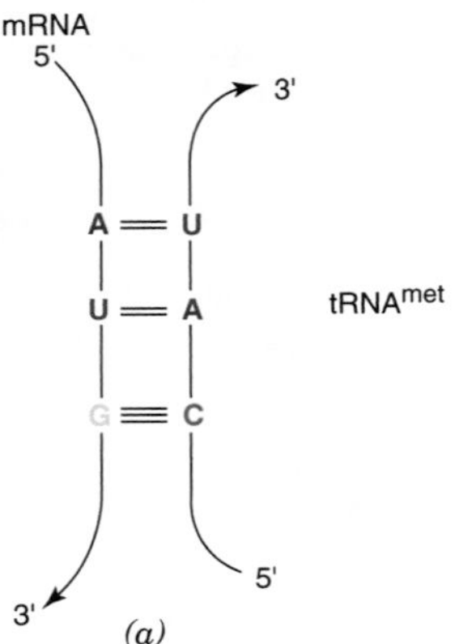

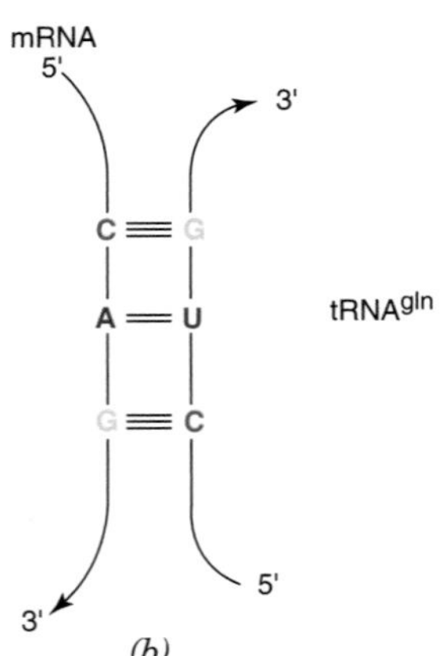

Figure 6.1 Codon–anticodon interactions. Interactions between (*a*) the AUG (methionine) codon and its CAU anticodon and (*b*) the CAG (glutamine) codon and a CUG anticodon show that base pairing of mRNA with tRNA is antiparallel.

Codon–Anticodon Interactions Permit Reading of mRNA

Translation of the codons of mRNA involves their interaction with complementary anticodon sequences in tRNA. Each tRNA species carries a unique amino acid, and each has a specific anticodon triplet. Like base pairing in DNA, codon-anticodon base pairing is **antiparallel,** as shown in Figure 6.1. Codons are read in a **sequential, non-overlapping reading frame.** Anticodon and amino acid-acceptor sites are at opposite ends of the L-shaped tRNA molecule (p. 63). The tRNA, conceptually and physically, bridges the gap between the nucleotide sequence of the ribosome-bound mRNA and the site of protein assembly on the ribosome.

Since 61 codons designate 20 amino acids, it might seem necessary to have exactly 61 different tRNA species. This is not the case. Many amino acids are carried by more than one tRNA species, but variation from standard base pairing is also common in codon–anticodon interactions. Many degenerate codons can be read by more than one tRNA (but always one carrying the correct amino acid). A frequent anticodon nucleotide is inosinic acid (I), the nucleotide of hypoxanthine, which base-pairs with U, C, or A. Modified nucleotides at or beside the first nucleotide of the anticodon in many tRNA species also modulate codon–anticodon interactions. The **wobble hypothesis** permits less stringent base pairing between the first position of the anticodon and the **degenerate** (third) position of the codon. Wobble base pairing rules are shown in Table 6.3. If the wobble rules are followed, the 61 nonpunctuation codons could be read by 31 tRNA molecules, but most cells have 50 or more tRNA species. Some codons are read more efficiently by one anticodon than another. Not all codons or codon pairs are used equally; some are used very rarely. Examination of many mRNA sequences shows that different organisms preferentially use different codons to generate similar polypeptide sequences.

"Breaking" the Genetic Code

The genetic code (see Table 6.1, p. 611) was determined before DNA or RNA sequencing was possible. Code-breaking experiments used simple synthetic mRNAs to provide insight into how proteins are encoded.

Polynucleotide phosphorylase catalyzes the template-independent reaction.

$$x\text{NDP} \leftrightarrow \text{polynucleotide (pN)}_x + x\text{P}_i$$

where NDP is any nucleoside 5′-diphosphate. With UDP a polymer of U, poly(U), is formed. *In vitro* protein synthesis with poly(U) as mRNA generates polyphenylalanine. Similarly, poly(A) encodes polylysine and poly(C) polyproline. An mRNA with a random sequence of only U and C produces polypeptides that contain not only proline and phenylalanine as predicted, but also serine (from UCU and UCC) and leucine (from CUU and CUC). Degeneracy in the code and the complexity of products made experiments with random sequence mRNAs difficult to interpret, but mRNAs of defined sequence can be transcribed from simple repeating synthetic DNAs. Thus poly(AU), transcribed from a repeating poly(dAT), produces only a repeating copolymer of Ile-Tyr-Ile-Tyr, read from successive triplets AUA UAU AUA UAU, and so on. Poly(CUG) contains possible codons CUG for Leu, UGC for Cys, and GCU for Ala, each repeating itself once the **reading frame** has been set. Initiation is random in these experiments, so three homopolypeptides are produced: polyleucine, polycysteine, and polyalanine. A perfect poly(CUCG) produces a polypeptide with the sequence (-Leu-Ala-Arg-Ser-) whatever the initiation point. These relationships, summarized in Table 6.4, show codons to be triplets read in exact sequence, without overlap or omission.

Other experiments used synthetic trinucleotide codons as minimal messages. No proteins were made, but ribosome binding of only one amino acid (as a tRNA conjugate) was stimulated by a given codon. The meaning of each possible codon was later verified by determination of mRNA sequences.

TABLE 6.3 • Wobble Base-Pairing Rules

3′ Codon Base	*5′ Anticodon Bases Possible*
A	U or I
C	G or I
G	C or U
U	A or G or I

Mutations

Understanding the genetic code and how it is read provides a basis for understanding the nature of mutations. A mutation is simply a heritable change in a gene. **Point mutations** involve a change in a single base pair in the DNA, and thus a single base in the corresponding mRNA. If this change occurs in the third position of a degenerate codon, there may be no change in the amino acid specified (e.g., UCC to UCA still codes for serine). Such **silent mutations** are commonly seen in comparison of genes for similar proteins, for example hemoglobins from different species, so they are usually without effect. (For an exception, see Clin. Corr. 6.1.)

Missense mutations arise from a base change that causes incorporation of a different amino acid in a protein (Clin. Corr. 6.2). Point mutations can also form or destroy a termination codon and thus change the length of a protein. Formation of a termination codon from one that encodes an amino acid is a **nonsense mutation;** it results in premature termination and a truncated protein (Clin. Corr. 6.3). Mutation of a termination codon to one for an amino acid allows the message to be read through until another stop codon is encountered. The result is a larger than normal protein. This phenomenon is the basis of several disorders (Table 6.5 and Corr. Clin. 6.4).

Insertion or deletion of a single nucleotide within the coding region of a gene results in a **frameshift mutation.** The reading frame is altered at that point and subsequent codons are read in the new context until a termination codon is reached. Table 6.6 illustrates this phenomenon. The significance of reading frame selection is underscored by a phenomenon in some very small viruses in which the size of the virus limits the amount of DNA it contains. As compensation, a single segment of DNA encodes different polypeptides that are translated using different reading frames. An example is the tumor-causing simian virus SV40 (Figure 6.2). HIV and some other viruses rely on frameshifting during translation to generate different proteins from a single message (Clin. Corr. 6.5).

TABLE 6.4 • Polypeptide Products of Synthetic mRNAs[a]

mRNA	*Codon Sequence*	*Products*
—(AU)$_n$—	—AUA UAU AUA UAU—	—(Ile-Tyr)$_{n/3}$—
—(CUG)$_n$—	—CUG CUG CUG CUG—	—(Leu)$_n$—
	—UGC UGC UGC UGC—	—(Cys)$_n$—
	—GCU GCU GCU GCU—	—(Ala)$_n$—
—(CUCG)$_n$—	CUC GCU CGC UCG	—(Leu-Ala-Arg-Ser)$_{n/3}$—

[a]The horizontal brackets accent the reading frame.

CLINICAL CORRELATION 6.1

Silent Mutations May Not Be Silent: (1) Hutchinson–Gilford Progeria Syndrome and (2) Multidrug Resistance 1 Gene

Single nucleotide polymorphisms, in which one nucleotide is changed from that in the normal gene, can be detected by DNA analysis. When a polymorphism in a coding region of the gene does not alter the amino acid encoded by the affected codon, it is considered a silent mutation. Nevertheless, there may be significant consequences.

Hutchinson–Gilford progeria syndrome is a very rare disorder characterized by extreme premature aging; children remain small and suffer hair and hearing loss, osteoporosis, cardiovascular disease, and other maladies of old age. Most die of myocardial infarctions in their early teens. In 90% of cases a change from GGC to GGT in codon 608 of the lamin A gene would be expected to be silent, since glycine is specified in each case. However, a cryptic splice site is activated, resulting in a Lamin A protein with an internal deletion of 50 amino acids and which is not properly processed. Lamin A is a component of the nuclear membrane; the mutant protein disrupts the membrane and somehow results in the disease. Since the affected children do not reach reproductive age, each case represents a new mutation at this precise spot in one allele of the gene.

Combinations of two and three silent mutations in the multidrug resistance 1 (MDR1) gene result in proteins with the normal sequence but a different conformation and altered drug and inhibitor interactions. The change from frequently used codons to rarely used codons slows or pauses translation in the affected region and results in changed cotranslational folding of the protein. Codon use preference extends to the frequency of use of adjacent codons that specify the same sequence. Synthetic genes that used disfavored codon pairs to encode poliovirus capsid protein were translated at decreased rates, even when the individual codons were favored in general.

Korf, B. Hutchinson-Gilford progeria syndrome, aging, and the nuclear lamina. *N. Engl. J. Med.* 358: 552, 2008; Kimchi-Sarfaty, C., Oh, J. M., Kim, I., Sauna, Z.E., et al. A "silent" polymorphism in the MDR1 gene changes substrate specificity. *Science* 315:525, 2007; and Coleman, J. R., Papamichail, D., Skiena, S., Futcher, et al. Virus attenuation by genome-scale changes in codon pair bias. *Science* 320:1784, 2008.

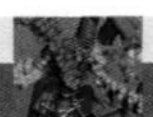

CLINICAL CORRELATION 6.2

Missense Mutations: Hemoglobinopathies

The most important missense mutation that affects hemoglobin structure occurs in sickle-cell anemia (see Clin. Corrs. 2.4, p. 46, and 9.6, p. 354). A change from A to U in either the GAA or GAG codon for glutamate gives a GUA or GUG codon for valine in the sixth position of the β-chain. Deoxygenated hemoglobin S (HbS) forms linear polymers that cause erythrocytes to adopt a sickle shape when oxygen tension is low. Hydroxyurea treatment stimulates synthesis of γ-chains and thus increases fetal hemoglobin levels; the HbF interferes in HbS polymerization. A second common hemoglobinopathy is hemoglobin C disease, in which a change from G to A in a GAA or GAG codon for glutamate results in an AAA or AAG codon for lysine in the sixth position of the β-globin. Hundreds of other hemoglobin missense mutations are known.

Platt, O.S. Hydroxyurea for the treatment of sickle cell anemia. *N. Engl. J. Med.* 358:1362, 2008.

CLINICAL CORRELATION 6.3

Mutations to Form a Terminator Codon

In Hemoglobin McKees Rocks (OMIM 141900) the UAU or UAC codon, normally designating tyrosine in position 145 of β-globin, is mutated to the terminator codon UAA or UAG. This shortens the β-globin from its normal 146 residues to 144 residues. Binding of 2, 3-bisphosphoglycerate is disrupted and the mutant hemoglobin has an unusually high oxygen affinity and decreased oxygen delivery. In response, erythropoietin is secreted by the kidney and increased red blood cell production produces a polycythemic phenotype.

Another terminator mutation causes a type of β thalassemia. Thalassemias are characterized by an imbalance in the stoichiometry of α- and β-globin synthesis. β-Globin is not synthesized in β^0 thalassemia. As a result, α-globin, unable to associate with β-globin to form hemoglobin, accumulates and precipitates in erythroid cells. This damages cell membranes, causing hemolytic anemia and stimulation of erythropoiesis. One variety of β^0 thalassemia, common in Southeast Asia, results from a termination mutation at codon 17 of β-globin; the normal AAG codon that designates a lysyl residue becomes a UAG stop codon; no useful β-globin sequence is synthesized, and β-globin is absent. β-Globin mRNA levels are also depressed, probably through the process called nonsense-mediated decay.

Winslow, R. M., Swenberg, M. L., Gross, E., Chervenick, P. A., et al. Hemoglobin McKees Rocks ($\alpha_2\beta_2{}^{145\ \mathrm{tyr} \rightarrow \mathrm{term}}$). A human nonsense mutation leading to a shortened β-chain. *J. Clin. Invest.* 57:772, 1976; Chang, J. C., and Kan, Y. W. β-Thalassemia: A nonsense mutation in man. *Proc. Natl. Acad. Sci. USA* 76:2886, 1979; and Rehwinkel, J., Raes, J., and Izaurralde, E. Nonsense-mediated mRNA decay: Target genes and functional diversification of effectors. *Trends Biochem. Sci.* 131:639, 2006.

TABLE 6.5 • "Read Through" Mutations in Termination Codons Produce Abnormally Long α-Globin Chains

Hemoglobin	*α-Codon 142*	*Amino Acid 142*	*α-Globin Length (Residues)*
A	UAA		141
Constant Spring	CAA	Glutamine	172
lcaria	AAA	Lysine	172
Seal Rock	GAA	Glutamate	172
Koya Dora	UCA	Serine	172

CLINICAL CORRELATION 6.4

α-Thalassemia (OMIM 604131)

There are two expressed α-globin genes on each chromosome 16. Many instances of α-thalassemia arise from deletion of two, three, or all four α-globin genes; clinical severity increases with the number of genes deleted. In contrast, the disorders summarized in Table 6.5 are forms of α-thalassemia that arise from abnormally long α-globin molecules. These replace normal α-globin and are present only in small amounts because of a decreased rate of synthesis or, more likely, from an increased rate of breakdown of the abnormal α-globin. The normal stop codon, UAA, mutates to any of four sense codons with resultant placement of four different amino acids at position 142. Normal α-globin is 141 residues long, but the four abnormal α-globins are 172 residues long because a triplet of nucleotides in the normally untranslated region of the mRNA becomes a termination codon at position 173. Elongated globin chains can also result from frameshift mutations or insertions.

Weatherall, D. J., and Clegg, J. B. The α-chain termination mutants and their relationship to the α-thalassemias. *Philos. Trans. R. Soc.* B [London] 271:411, 1975.

TABLE 6.6 • A Frameshift Mutation Produces Hemoglobin Wayne[a]

Position	*137*	*138*	*139*	*140*	*141*	*142*	*143*	*144*	*145*	*146*	*147*
Normal α-globin sequence	- Thr	- Ser	- Lys	- Tyr	- Arg						
Normal α-globin codon sequence	- ACP	- UC(U)	- AAA	- UAC	- CGU	- [UAA]	- GCU	- GGA	- GCC	- UCG	- GUA
Wayne α-globin codon sequence	- ACP	- UCA	- AAU	- ACC	- GUU	- AAG	- CUG	- GAG	- CCU	- CGG	- [UAG]
Wayne α-globin amino acid sequence	- Thr	- Ser	- Asn	- Thr	- Val	- Lys	- Leu	- Glu	- Pro	- Arg	

[a]The base deletion causing the frameshift is encircled. The stop codons are boxed.
P = A, G, U, or C.

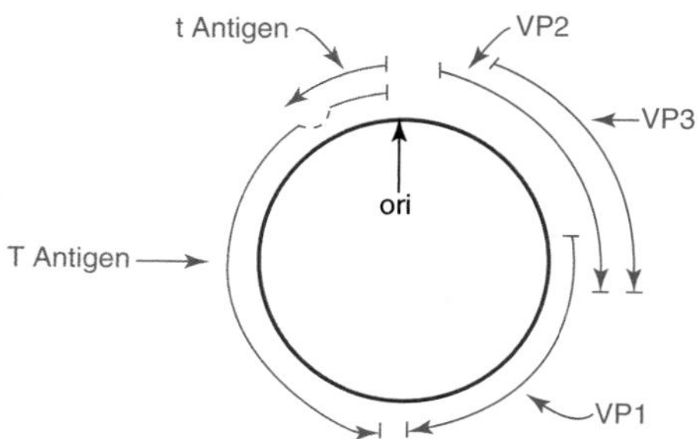

Figure 6.2 Genome of simian virus 40 (SV40). SV40 DNA, shown in red, is a double-stranded circle of slightly more than 5000 base pairs that encodes all information needed for viral survival and replication within a host cell. It shows extremely efficient use of the protein-coding potential of a genome. VP1, VP2, and VP3 are structural proteins of the virus; VP2 and VP3 are translated from different initiation points to the same carboxyl terminus. VP1 is translated in a different reading frame: Its N-terminal section overlaps the VP2/VP3 genes but its amino acid sequence in the overlapping segment is different from that of VP2 and VP3. Two proteins, the large T and small t tumor antigens, promote transformation of infected cells. They are encoded in a common mRNA precursor and have identical N-terminal sequences, but the C-terminal part of small t protein is encoded by a segment of mRNA that is spliced out of the large T message, and the carboxyl-terminal sequence of large T is encoded in RNA that follows termination of small t. The DNA replication origin (ori) is outside of coding regions.

CLINICAL CORRELATION 6.5

Programmed Frameshifting in Biosynthesis of HIV Proteins

Maintaining the reading frame during translation is central to fidelity of translation. Yet many retroviruses, including the AIDS virus HIV, use mRNA slippage and a change in reading frame to generate different proteins from the same messenger. A single mRNA, designated gag-pol, encodes two polyproteins that overlap by about 200 nucleotides and are in different reading frames. The gag polyprotein is translated from the initiation codon to an in-frame termination codon near the gag-pol junction; gag polyprotein is cleaved to generate several structural proteins of the virus. However, about 5% of the time a one-nucleotide frameshift occurs within the overlapping segment of mRNA and the termination codon is bypassed because it is no longer in the reading frame. A gag-pol fusion polyprotein is produced; proteolytic cleavage of the pol polyprotein produces the viral reverse transcriptase and other proteins needed in virus reproduction. (See Clin. Corrs. 4.2, p. 144, and 13.3, p. 523.)

Jacks, T., Power, M. D., Masiarz, F. R., Luciw, P. A., et al. Characterization of ribosomal frameshifting in HIV-1 gag-pol expression. *Nature* 331:280, 1988.

Aminoacylation of Transfer RNA Activates Amino Acids for Protein Synthesis

Amino acids must be *activated* by linkage to their appropriate tRNA carriers before they can be incorporated into proteins. This two-step process is catalyzed by a family of **aminoacyl-tRNA synthetases,** each of which is specific for a single amino acid and its appropriate tRNA species. The reactions are normally written as follows:

$$\mathrm{H{-}N(H){-}CH(R){-}C({=}O){-}OH + ATP + E \rightleftharpoons \left[H{-}N(H){-}CH(R){-}C({=}O) \sim AMP \cdot E\right] + PP_i} \quad (1)$$

$$\mathrm{\left[H{-}N(H){-}CH(R){-}C({=}O) \sim AMP \cdot E\right] + tRNA \rightleftharpoons H{-}N(H){-}CH(R){-}C({=}O){-}tRNA + AMP + E} \quad (2)$$

Sum:

$$\mathrm{H{-}N(H){-}CH(R){-}C({=}O){-}OH + ATP + tRNA \rightleftharpoons H{-}N(H){-}CH(R){-}C({=}O){-}tRNA + AMP + PP_i}$$

Brackets surrounding the aminoacyl-AMP–enzyme complex indicate that it is a transient, enzyme-bound intermediate. The aminoacyladenylate, a mixed acid anhydride with carboxyl and phosphoryl components, is a *high-energy* intermediate. The aminoacyl ester linkage in tRNA is lower in energy than in aminoacyl-adenylate, but still higher than in the carboxyl group of the free amino acid. Pyrophosphatases cleave the pyrophosphate that is produced and the equilibrium is strongly shifted toward formation of aminoacyl-tRNA. From the viewpoint of precision in translation, the amino acid, which had only its side chain (R group) to distinguish it, becomes linked to a large, complex, and easily recognized carrier.

Specificity and Fidelity of Aminoacylation Reactions

Almost all cells contain 20 different aminoacyl-tRNA synthetases, each specific for one amino acid, and at most a small family of carrier tRNAs for that amino acid (see A Closer Look 6.1 for exceptions). Since codon—anticodon interactions define the amino acid to be incorporated, a misacylated tRNA will cause an error in the protein. Correct selection of both tRNA and amino acid by the synthetase is therefore central to fidelity in translation. Aminoacyl-tRNA synthetases share a common mechanism and many occur within the cell in multiprotein complexes. Nevertheless, they are a diverse group of proteins that may contain one, two, or four identical subunits or pairs of dissimilar subunits. Separate structural domains can be involved in aminoacyl-adenylate formation, tRNA recognition, and, if it occurs, subunit interactions. Each enzyme is capable of almost error-free formation of correct aminoacyl-tRNA combinations. Some amino acids are easily recognized by their bulk (e.g., tryptophan) or lack of bulk (glycine) or by positive or negative charges on the side chains (e.g., lysine and glutamate). Others are much more difficult to discriminate. For example, recognition of valine rather than threonine or isoleucine by the valyl-tRNA synthetase is difficult since the side chains differ by either an added hydroxyl or single methylene group.

The amino acid recognition and activation sites of each enzyme have great specificity, but sometimes misrecognition does occur. An additional **proofreading** step increases discrimination by many (but not all) aminoacyl-tRNA synthetases. This most often occurs through hydrolysis of the aminoacyladenylate intermediate, with the release of amino acid and AMP. Valyl-tRNA synthetase efficiently hydrolyses threonyl-adenylate, and it hydrolyses isoleucyl-adenylate in the presence of bound (but not aminoacylated) $tRNA^{val}$. Alternatively, misacylated tRNA is recognized and deacylated. Valyl-, phenylalanyl-, and isoleucyl-tRNA synthetases deacylate tRNAs that have been mischarged with threonine, tyrosine, and valine, respectively. The net result is an average level of misacylation of one in 10^4 to 10^5.

Each synthetase must also correctly recognize one to several tRNA species that can carry the same amino acid, while rejecting incorrect tRNA species. This may be simpler than selection of a single amino acid. However, recall the conformational similarity and common sequence elements of all tRNAs (p. 216). Different synthetases recognize different elements of tRNA structure. Usually multiple structural elements contribute to recognition, but many of these elements are not absolute determinants. One expected element of tRNA recognition is the anticodon. In the case of $tRNA^{met}$, changing the anticodon alters recognition by the synthetase. In other instances, this is only partly true, and sometimes the anticodon is not a recognition determinant. This is shown by **suppressor mutations** that suppress the expression of classes of nonsense mutations. A point mutation in a glutamine (CAG) codon produces a termination (UAG) codon that causes premature termination of the encoded protein. A second suppressor mutation in the anticodon of a $tRNA^{tyr}$, in which the normal GUA anticodon is changed to CUA, allows "read through" of the termination codon. The initial mutation is suppressed and a nearly normal protein is made, with the affected glutamine replaced by tyrosine. Aminoacylation of the mutant $tRNA^{tyr}$ with tyrosine shows that in this case the anticodon does not determine synthetase specificity.

Other tRNA-identification features may include elements of the acceptor stem and parts of the variable loop or the D-stem–loop. In *E. coli* $tRNA^{ala}$, the primary recognition characteristic is a G3-U70 base pair in the acceptor stem; even if no other changes in the $tRNA^{ala}$ occur, any variation at this position destroys its acceptor ability with alanine-$tRNA^{ala}$ synthetase. The X-ray structure of glutaminyl synthetase-tRNA complex shown in Figure 6.3 shows binding at the concave tRNA surface, which is typical and compatible with the biochemical observations.

Ribosomes Are Machines for Protein Biosynthesis

Ribosomes are complex **ribonucleoprotein** particles that provide a workbench and catalytic basis for protein synthesis. They are made up of two dissimilar subunits, each of which contains RNA and many proteins. With one exception, each protein is present in a single

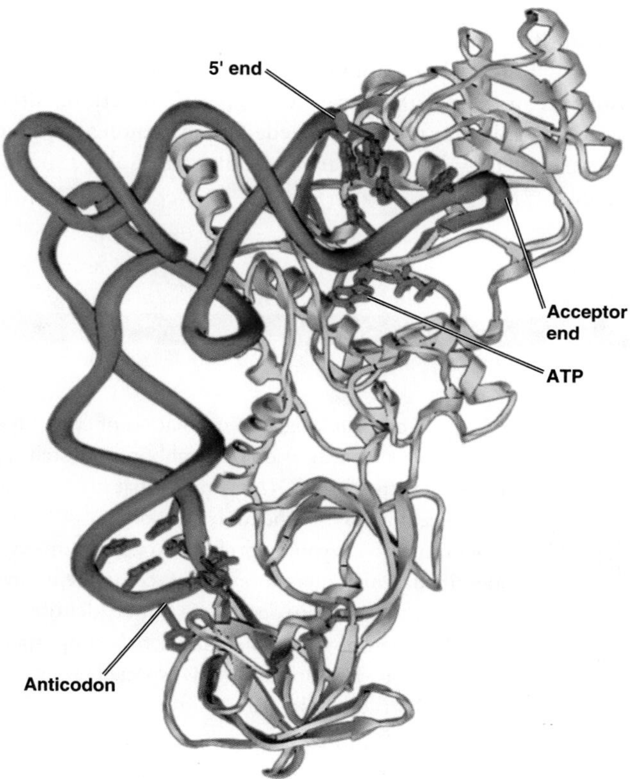

Figure 6.3 Interaction of a tRNA with its aminoacyl-tRNA synthetase. The sugar–phosphate backbone of *E. coli* glutaminyl-tRNA is shown in green and the peptide backbone of the glutaminyl-tRNA synthetase in multiple colors. The synthetase interacts with the partially unwound acceptor stem and the anticodon loop of tRNA. ATP, shown in red, is within a few angstroms of the 3′-end of tRNA. Space-filling models of the complex show both molecules as solid objects with several sites of direct contact. Adapted from Perona, J., Rould, M., and Steitz, T. *Biochem.* 32:8758, 1993.

TABLE 6.7 • Ribosome Classification and Composition

Ribosome Source	*Monomer Size*	*Subunits*	
		Small	*Large*
Eukaryotes			
Cytosol	80S	40S:	60S:
		34 proteins	50 proteins
		18S RNA	28S, 5.8S, 5S RNAs
Mitochondria			
Animals	55S–60S	30S–35S:	40–45S:
		12S RNA	16S RNA
		70–100 proteins	
Higher plants	77S–80S	40S:	60S:
		19S RNA	25S, 5S RNAs
		70–75 proteins	
Chloroplasts	70S	30S:	50S:
		20–24 proteins	34–38 proteins
		16S RNA	23S, 5S, 4.5S RNAs
Prokaryotes			
Escherichia coli	70S	30S:	50S:
		21 proteins	34 proteins
		16S RNA	23S, 5S RNAs

copy per ribosome, as is each RNA species. The composition of major ribosome types is shown in Table 6.7.

Ribosome architecture and function have been conserved in evolution, and similarities between ribosomes and subunits from different sources are more important than the differences. Structures of ribosomes and their subunits have been determined by X-ray crystallography of particles from microorganisms. Ribosomes and subunits from many sources have also been visualized by electron microscopy; their structure is highly conserved. Prokaryotic ribosomes can self-assemble; native structures can be reconstituted from mixtures of purified individual proteins and RNAs. Reconstitution of subunits from mixtures in which a single component is omitted or modified can show if it is important for assembly of the subunit or for some specific function. Although reconstitution of eukaryotic subunits has not yet succeeded, many studies show prokaryotic and eukaryotic ribosomes not only have similar morphology but also function in the same ways (A Closer Look 6.2).

A CLOSER LOOK 6.2

Ribosome Structure and Function

The complexity of ribosomes requires application of many techniques to decipher their architecture and functional elements, particularly if X-ray structures are not available. Major techniques include

1. Neutron scattering of reconstituted subunits in which two proteins are deuterated: This allows determination of the separation of those two proteins in the subunit.
2. Electron microscopy of complexes of particles with antibodies to a single component protein, RNA element, or functional probe: The Y-shaped antibody serves as a pointer to the location of the component.
3. Chemical cross linking: Determination of cross-linked components allows identification of neighboring proteins, proteins with RNA segments, and RNA elements that may be distant in the sequence but close in the folded structure.
4. Chemical and enzymatic probes: Exposed components can be identified, while internal or hidden components are resistant.
5. Sequence comparison: Sequences can be identified that are conserved in evolution and so are likely to be important in structure and function. This approach was especially important in establishing 2° structures in rRNA.

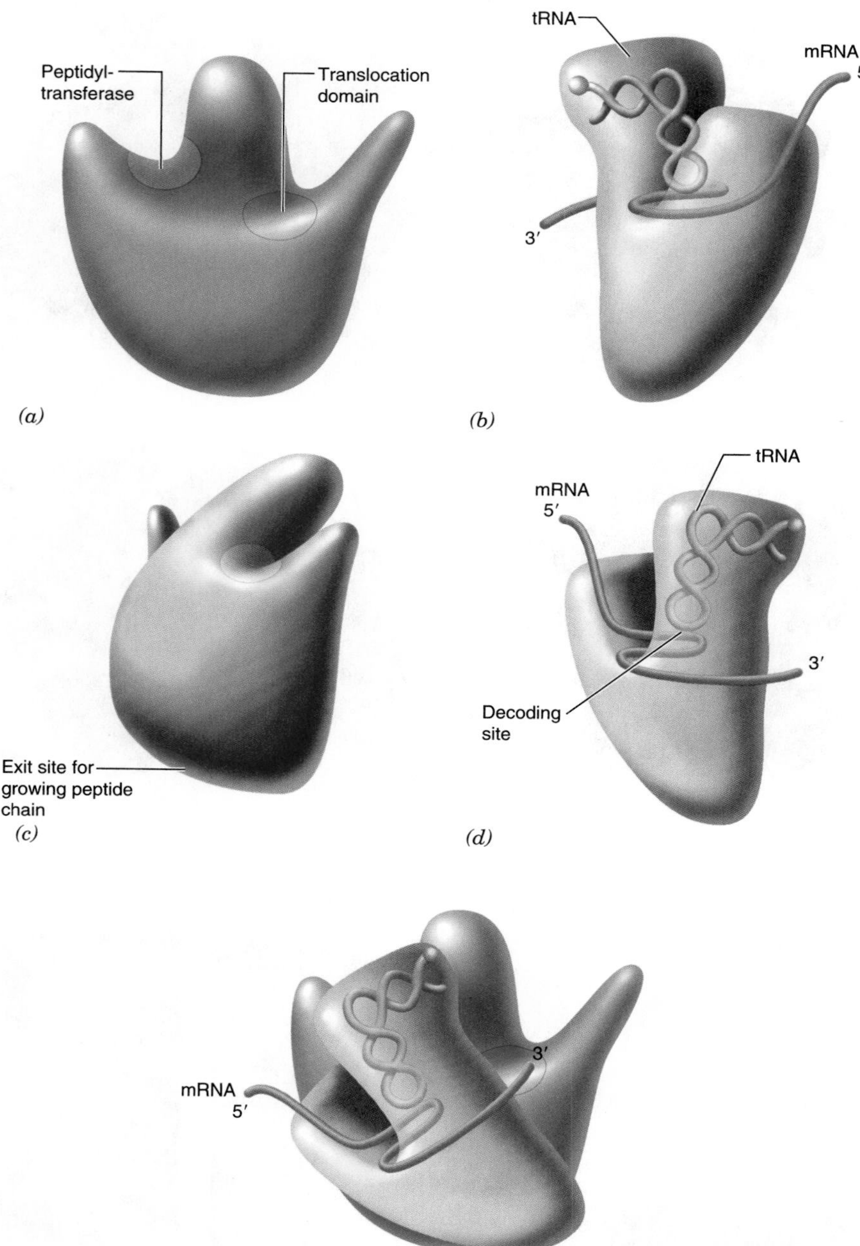

Figure 6.4 Model of ribosome structure and functional sites. Top row: The face of each subunit that interacts in the functional ribosome. In (*a*) the large subunit sites of peptide bond formation (peptidyl transferase) and of elongation factor binding (translocation domain) are seen on opposite sides of the central protuberance. The stalk projects to the right. In (*b*) the small subunit platform protrudes toward the reader. mRNA and tRNAs interact in the decoding site, in a cleft between the platform and the subunit body. In (*c*) the large subunit is rotated 90°, the stalk projects into the page, and the nascent peptide exit site is seen near the subunit base. This site contacts membrane in the bound ribosomes of rough endoplasmic reticulum. The site of peptide bond formation is distant from the exit site; the growing peptide passes through a groove or tunnel in the ribosome to reach the exit site. In (*d*) the small subunit has been rotated 90°; the platform projects into the page. In (*e*) subunits have been brought together to show their orientation in the ribosome. Aminoacyl- tRNA bound by the small subunit is oriented with the acceptor end near peptidyl transferase while the translocation domain is near the decoding site.

Structural and functional studies have led to ribosome models, features of which are described in Figure 6.4, and Figure 6.5 shows X-ray structures of ribosomes and their complexes with mRNA and tRNA. Ribosomes are organized in two more ways. Several ribosomes often translate a single mRNA molecule simultaneously. Electron micrographs of mRNA-linked **polysomes** are shown in Figure 6.6. In eukaryotic cells ribosomes occur free in the cytosol and bound to membranes of the rough **endoplasmic reticulum.** In general, **free ribosomes** synthesize proteins that remain within the cytosol or become targeted to the nucleus, mitochondria, or some other organelle. **Membrane-bound ribosomes** synthesize proteins that will be secreted from the cell or sequestered and function in the plasma membrane, endoplasmic reticulum, Golgi complex, or lysosomes. In cell homogenates, membrane fragments with bound ribosomes constitute the **microsome** fraction; detergents that disrupt membranes release these ribosomes.

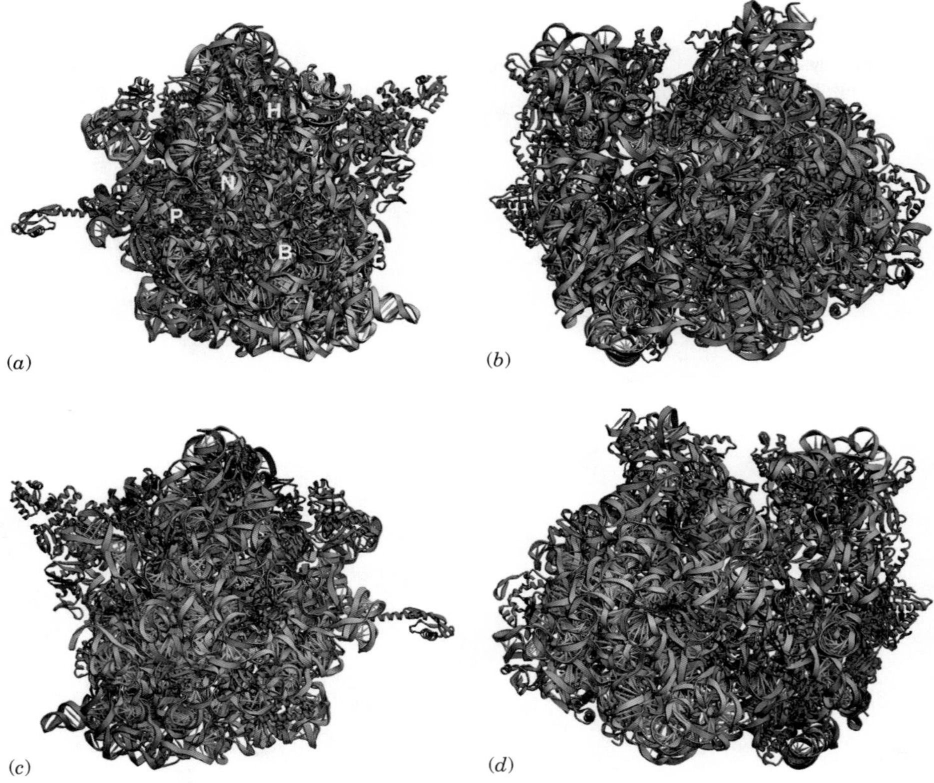

Figure 6.5 Crystallographic structure of a 70S ribosome. The structure of the *Thermus thermophilus* ribosome is shown at 5.5-Å resolution. Successive views are rotated 90° around the vertical axis. (*a*) The small subunit lies atop the large subunit, as in the model of Figure 6.4e. Small subunit features include the platform (P), head (H), body (B), and a connecting neck (N). The 16S RNA is colored cyan, and small subunit proteins are dark blue. (*b*) The large subunit is on the right; 23S RNA is grey, 5S RNA is light blue, and large subunit proteins are magenta. A tRNA molecule (gold) spans the subunits. (*c*) The large subunit lies on top with its stalk protruding to the left. (*d*) The large subunit is on the left and tRNA elements are visible in the subunit interface. Reprinted with permission from Yusupov, M. M., Yusupova, G. Z., Baucom, A., Lieberman, K., Earnest, T. N., Cate, J. H. D., and Noller, H. F. *Science* 292:883, 2001. Reprinted with permission from AAAS. Figure generously supplied by Drs. A. Baucom and H. Noller.

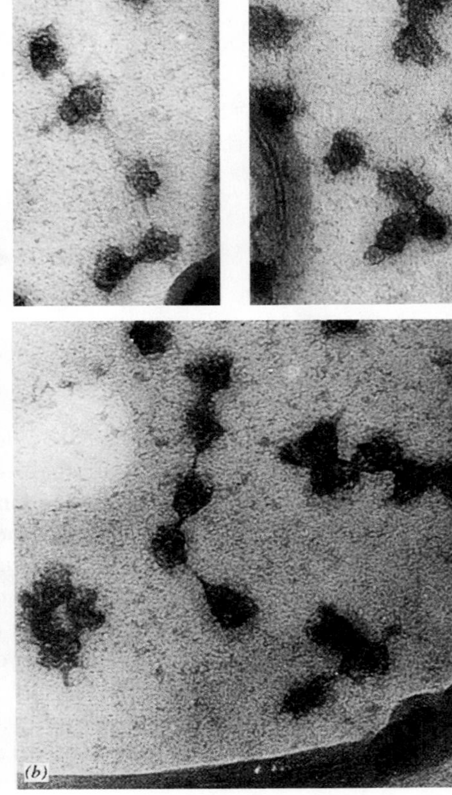

Figure 6.6 Electron micrographs of polysomes. (*a*) Reticulocyte polyribosomes shadowed with platinum are seen in clusters of three to six ribosomes, a number consistent with the size of mRNA for a globin chain. (*b*) Uranyl acetate staining and visualization at a higher magnification shows polysomes; parts of the mRNA are visible. Courtesy of Dr. Alex Rich, M.I.T.

6.3 • PROTEIN BIOSYNTHESIS

Translation Is Directional and Colinear with mRNA

Messenger RNA sequences are transcribed and written 5′ → 3′, and during translation they are read in the same way. Amino acid sequences are both written and synthesized from the amino-terminal residue to the carboxy terminus. A ribosome remains bound to an mRNA molecule: The messenger is moved along its length until a stop codon is reached. Comparison of mRNA sequences with sequences of the proteins they encode shows a perfect, colinear, nonoverlapping and gap-free correspondence of the mRNA coding sequence and that of the polypeptide synthesized. In fact, it is common to deduce the sequence of a protein solely from the nucleotide sequence of its mRNA or the DNA of its gene, although the final sequence of the protein may differ because of posttranslational modifications.

Initiation of Protein Synthesis Is a Complex Process

Initiation requires bringing together a small (40S) ribosomal subunit, an mRNA, and a tRNA complex of the amino-terminal amino acid, all in a proper orientation. This is followed by association of the large (60S) subunit to form a completed initiation complex on an 80S ribosome. This process requires several protein **initiation factors,** listed in Table 6.8, that act transiently and only in initiation. Figure 6.7 shows the initiation process. First, **eukaryotic initiation factor 2 (eIF2)** binds to GTP and the initiator tRNA, Met-$tRNA_i^{met}$ to form a ternary complex. No other aminoacyl-tRNA can replace the initiation-specific Met-$tRNA_i^{met}$ in this step. (Prokaryotes utilize a specific initiator tRNA whose methionine is modified by formylation of its amino group. Only fMet-$tRNA_i^{met}$ is recognized by prokaryotic IF2.)

The second step requires 40S ribosomal subunits associated with a very complex protein, **eIF3.** Mammalian eIF3 contains 13 different polypeptides; it binds to the 40S subunit surface that will contact the 60S subunit and physically blocks subunit association. Hence eIF3 is a ribosome **anti-association factor,** as is eIF6 which binds to 60S subunits. A complex that includes eIF2–GTP, Met-$tRNA_i^{met}$, eIF3–40S, and additional protein factors now forms. In the third step the **pre-initiation complex** is formed; mRNA, **eIF4F,** also

TABLE 6.8 • Major Mammalian Initiation Factors

Factor	*Subunits*	*Functions*
eIF 1	1	Aids pre-initiation complex formation
eIF1A	1	Aids pre-initiation complex formation, binds A-site
eIF2	3 (α, β, γ)	Delivers ternary complex
eIF2B		Recycles eIF2-GDP
eIF3	13 (a-m)	Organizer for other factors on 40S subunit, regulator binding, anti-association factor, other unknown roles
eIF4A	1	RNA helicase
eIF4B	1	Aids eIF4A
eIF4E	1	mRNA cap binding
eIF4F	3	Complex of eIF4A, eIF4E, eIF4G
eIF4G	1	mRNA recruitment, scanning, eIF3 binding
eIF5		GTPase for eIF2, aids 60S association
eIF5B		80S Ribosome formation
eIF6		Binds 60S (anti-association factor)
PAB		Binds polyA of mRNA

Source: Thornton, S., Anand, N., Purcell, D. and Lee, J. Not just for housekeeping: protein initiation and elongation factors in cell growth and tumorigenesis. *J. Mol. Med.* 81: 536, 2003. Noble, C. G. and Song, H. Structural studies of elongation and release factors. *Cell. Mol. Life Sci.* 65: 1335, 2008.

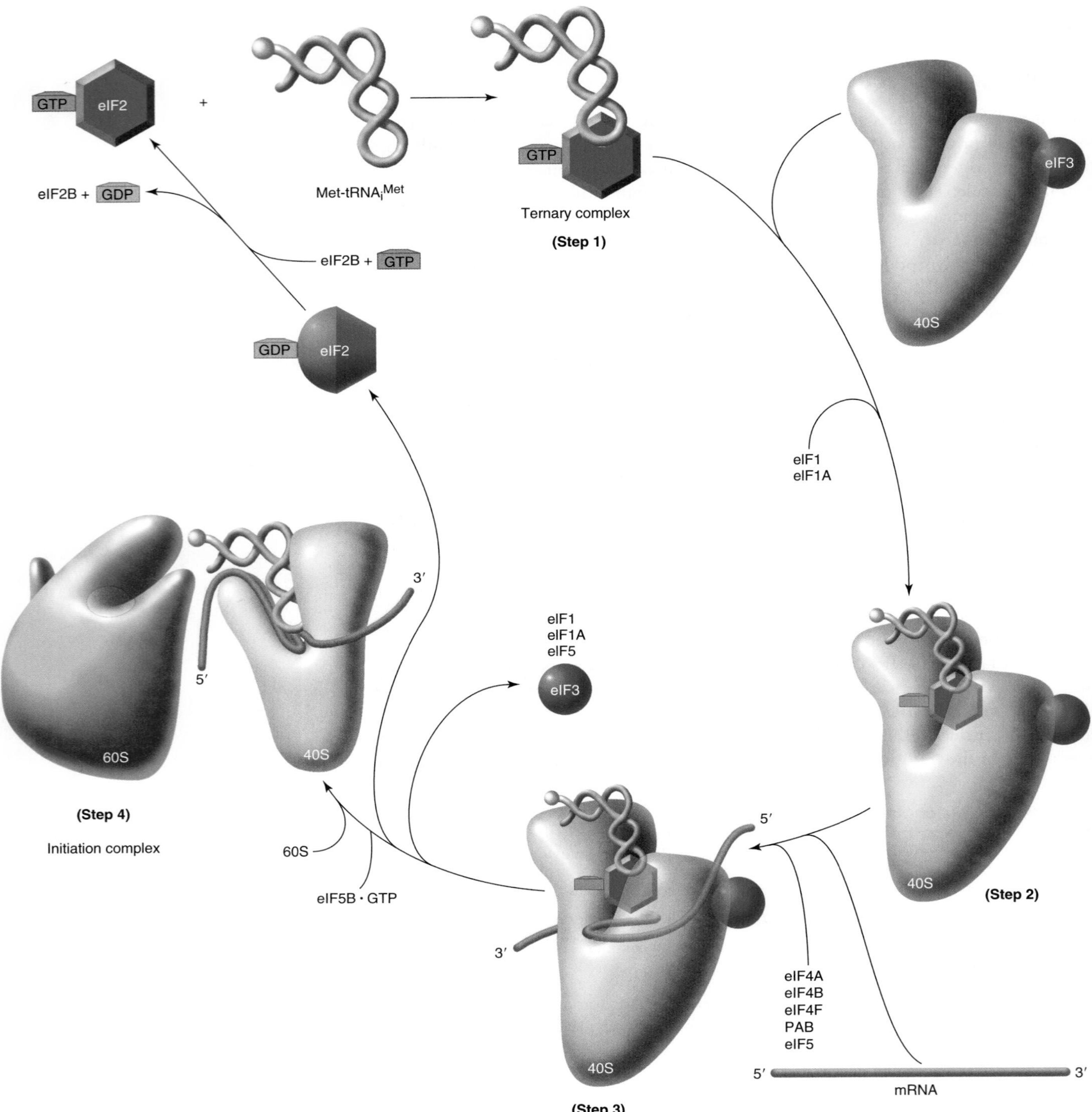

Figure 6.7 Initiation of translation in eukaryotes. Details in text. A ternary complex that includes the initiator tRNA (step 1) combines with a small ribosomal subunit (step 2). Interaction with mRNA forms a pre-initiation complex (step 3). Large subunit binding completes formation of the initiation complex (step 4). The different shape of eIF2 in complexes with GTP and GDP indicates that conformational changes occur on hydrolysis of triphosphate. After elongation has begun, additional small subunits will complex with the same mRNA to form polysomes.

called the **cap-binding complex, eIF4A,** an **RNA helicase** that unwinds secondary structure in the untranslated leader sequence of the mRNA, **PAB,** a polyA-binding protein that loops the 3′-end of the mRNA near to the 5′-cap, and several other proteins are needed. The mRNA is then scanned 5′ → 3′until the first AUG triplet is found. GTP from the ternary complex is hydrolyzed with the help of eIF5, and eIF2–GDP and other factors are released. The eIF2–GDP interacts with the guanine nucleotide exchange factor eIF2B and GTP to regenerate eIF2–GTP for another round of initiation. The final step requires binding of the pre-initiation complex by a 60S subunit and an additional factor, **eIF5B–GTP.** GTP is hydrolyzed, and eIF5B–GDP and other factors are released. The complete **initiation complex** is an 80S ribosome with the mRNA and initiator tRNA correctly positioned to begin translation.

Sometimes the first AUG is not used and a later AUG, characterized by its secondary structure within the mRNA, is selected for initiation. Such **internal ribosome entry sites** (IRES) were first found in viral mRNAs, but they also are common in messengers that encode proteins that protect cells from stress, when the translation of most mRNAs is repressed.

Prokaryotes use fewer initiation factors to form an initiation complex. Their 30S subunits, complexed with a simpler IF3, can first bind either mRNA or a ternary complex of IF2, fMet-$tRNA_i^{met}$, and GTP. Orientation of the mRNA relies in part on base pairing between a pyrimidine-rich sequence of eight nucleotides in 16S rRNA and a purine-rich sequence about 10 nucleotides upstream of the initiator AUG codon. Complementarity between rRNA and mRNA may include several mismatches, but greater complementarity usually leads to initiation that is more efficient. Initiation factor IF1also acts in formation of the preinitiation complex. Finally, a 50S subunit is bound, GTP is hydrolyzed, and the initiation factors are released. It is interesting that prokaryotes rely more on RNA–RNA interactions to position mRNA, whereas eukaryotes use many proteins to attain the same result.

Elongation Is the Stepwise Formation of Peptide Bonds

Protein elongation, as shown in Figure 6.8, occurs by stepwise lengthening of the polypeptide chain. At each step ribosomal **peptidyl transferase** transfers the growing peptide (or in the first step the initiating methionine residue) from its carrier tRNA to the α-amino group of the amino acid residue of the aminoacyl-tRNA specified by the following codon:

$$\cdots\text{NH–CH}(R_n)\text{–C(=O)–tRNA}_n + \text{H}_2\text{N–CH}(R_{n+1})\text{–C(=O)–tRNA}_{n+1} \longleftrightarrow \cdots\text{NH–CH}(R_n)\text{–C(=O)–NH–CH}(R_{n+1})\text{–C(=O)–tRNA}_{n+1}$$

peptidyl-tRNA (***n*** residue length) aminoacyl-tRNA peptidyl-tRNA (***n* + 1** residue length)

Efficiency and fidelity are enhanced by nonribosomal elongation factors that use energy released by GTP hydrolysis to ensure placement of the proper aminoacyl-tRNA species and to move the mRNA and associated tRNAs through the ribosome.

During elongation up to three tRNA molecules are bound at specific sites that span both ribosomal subunits. The initiating met-tRNA is positioned so that its methionyl residue can be transferred (or donated) to the free α-amino group of the incoming aminoacyl-tRNA; it thus occupies the **donor site,** also called the **peptidyl site** or **P site** of the ribosome. The aminoacyl-tRNA specified by the next codon of the mRNA is bound at the **acceptor site,** also called the **aminoacyl site** or **A site** of the ribosome. Selection of the correct aminoacyl-tRNA is enhanced by **elongation factor EF1.** A component of EF1, EF1α, first forms a ternary complex with aminoacyl-tRNA and GTP. The EF1α–aminoacyl-tRNA–GTP complex binds to the ribosome, and if codon–anticodon interactions are correct, the aminoacyl-tRNA is placed at the A site, GTP is hydrolyzed, and the EF1α–GDP complex dissociates. The initiating methionyl-tRNA and the incoming aminoacyl-tRNA are now juxtaposed on the ribosome. Their anticodons are paired with successive codons of the mRNA in the P and A sites of the small subunit, and their amino acids are beside one

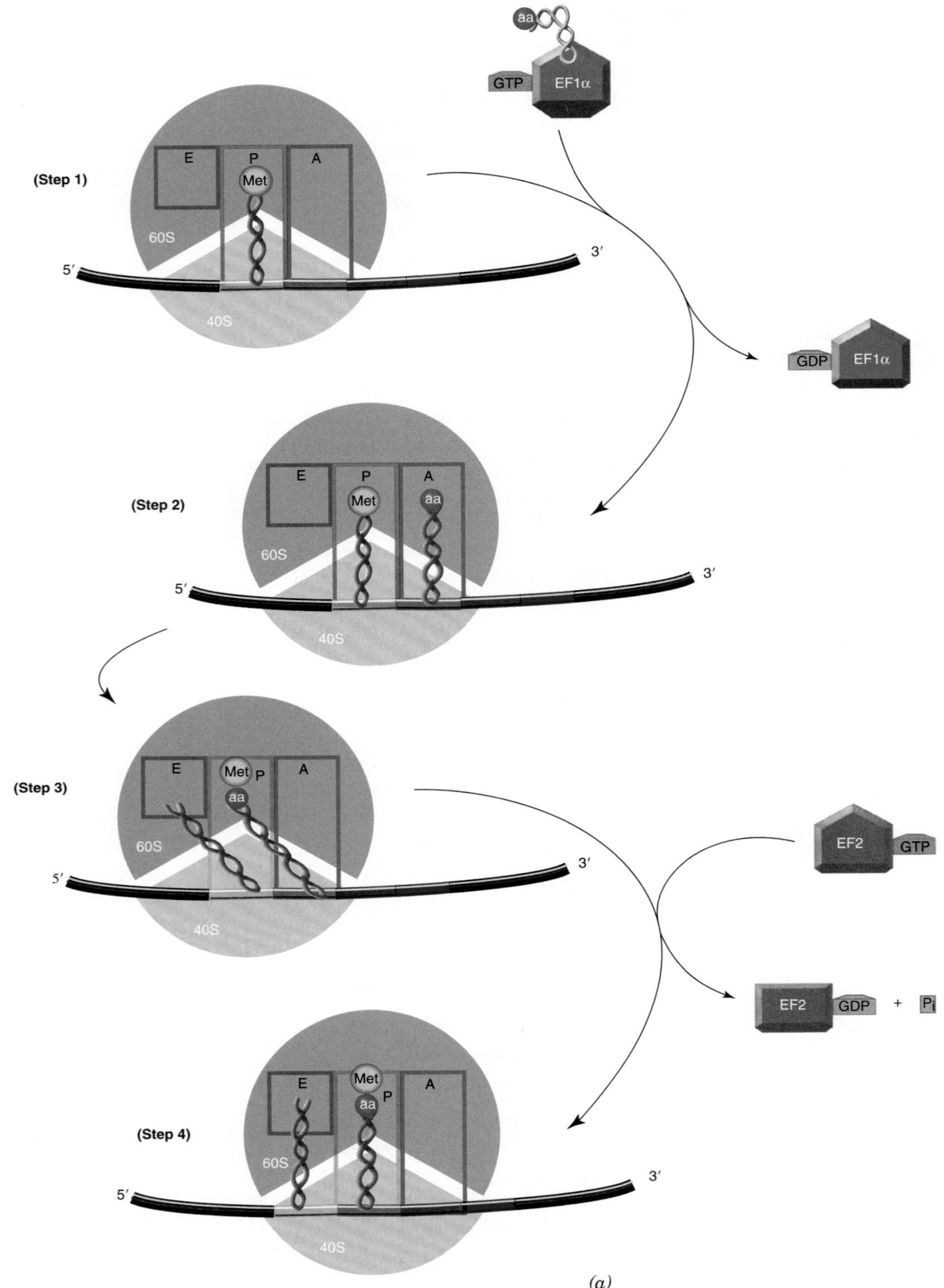

Figure 6.8 Elongation steps in protein synthesis.
(*a*) The first round of elongation is shown. Step 1: An initiation complex with methionyl tRNA$_i^{met}$ in 80S P site. Step 2: EF1α has placed an aminoacyl-tRNA in the A site. GTP hydrolysis results in a conformational change in EF1α. Step 3: A peptide bond has been formed; new peptidyl tRNA occupies a hybrid (A/P) site on the ribosome and the acceptor stem of deacylated tRNA$_i^{met}$ is in the large subunit E site. Step 4: The mRNA–peptidyl tRNA complex has been translocated to the P site while deacylated initiator tRNA is moved to the E site. (*b*) Further rounds of elongation. Step 5: Binding of aminoacyl-tRNA in the A site causes release of deacylated tRNA from the E site. Step 6: Formation of peptide bond results in the new peptidyl tRNA occupying a hybrid A/P site. Step 7: Translocation moves mRNA and new peptidyl tRNA in register into P site. Additional amino acids are added by successive repetitions of the cycle. Recycling of EF-1α (upper left) is detailed in Figure 6.9. For further details see text.

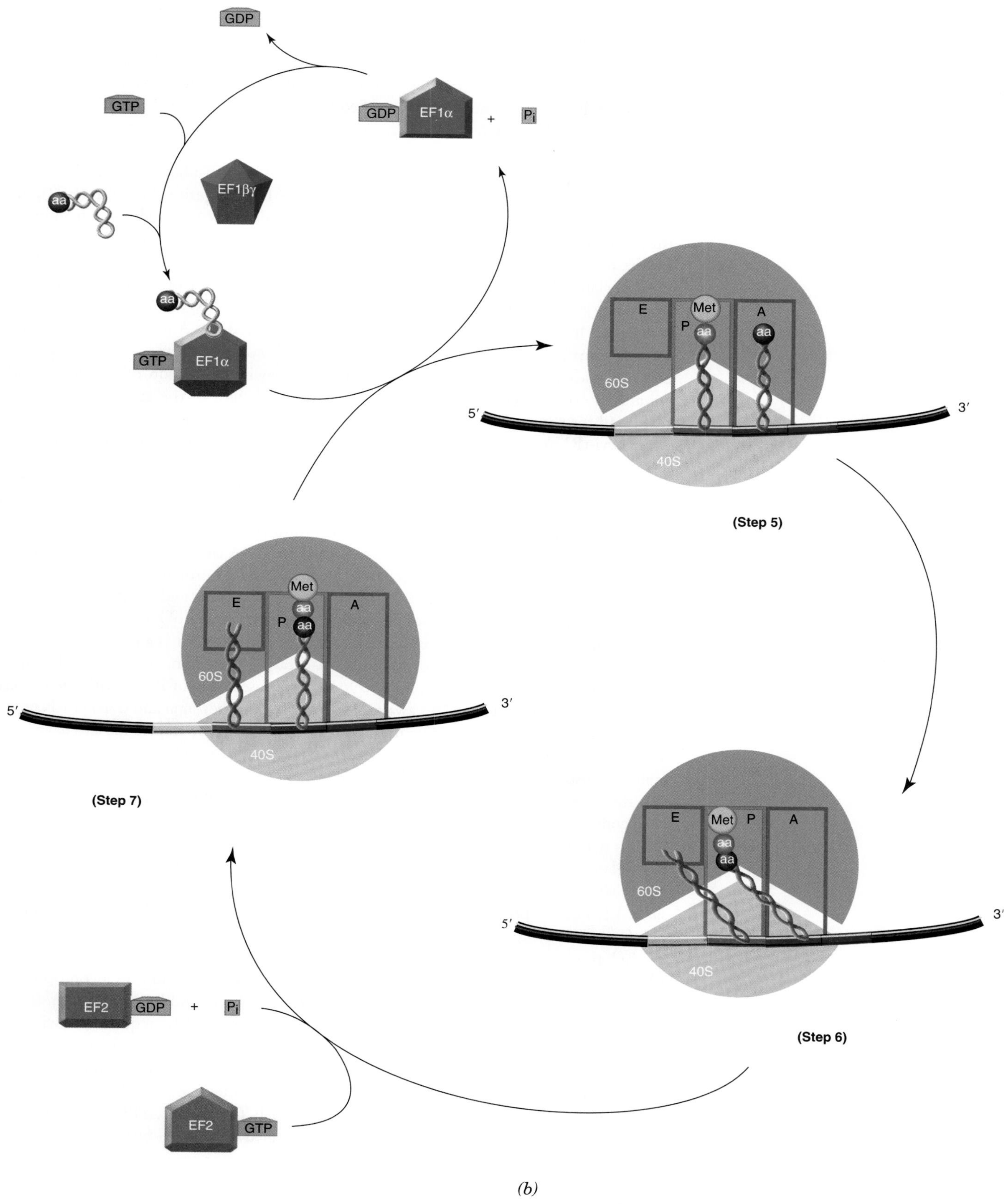

(b)

Figure 6.8
(continued)

another at the peptidyl transferase site of the large subunit. Peptidyl transferase catalyzes attack of the α-amino group of aminoacyl-tRNA on the carbonyl carbon of methionyl-tRNA. The result is transfer of methionine to the amino group of the aminoacyl-tRNA, which then occupies a "hybrid" position on the ribosome with the anticodon in the 40S A site, while the acceptor end and the attached peptide are in the 60S P site. The anticodon of the deacylated tRNA remains in the 40S P site; its acceptor end is located in the 60S **exit,** or **E, site.**

The mRNA and the dipeptidyl-tRNA at the 40S A site must be repositioned to permit another round of elongation to begin. This is done by **elongation factor 2 (EF2),** also called **translocase.** EF2 moves the mRNA and dipeptidyl-tRNA, in codon–anticodon register, from the 40S A site to the P site. In the process GTP is hydrolyzed, providing energy for the movement, and the A site is vacated. As the dipeptidyl-tRNA moves to the P site, the deacylated donor (methionine) tRNA also moves to the E site on the 60S subunit. The ribosome can now enter a new round. The next aminoacyl-tRNA specified by the mRNA is delivered by EF1α–GTP to the A site, and the deacylated tRNA in the E site is released. Peptide transfer again occurs. Successive rounds of binding of aminoacyl-tRNA, peptide bond formation, and translocation result in the stepwise elongation of the polypeptide toward its eventual carboxyl terminus. Whatever the length of the growing chain, peptide bond formation always occurs through attack of the α-amino group of the incoming aminoacyl-tRNA on the peptide carboxyl-tRNA linkage, hence the geometric arrangement of the reacting molecules at the peptidyl transferase site remains constant.

Peptide bond formation does not require an energy source such as ATP or GTP. The energy of the methionyl (or peptidyl) ester linkage to tRNA drives the reaction toward peptide bond formation. Stepwise elongation continues until a termination codon (UGA, UUA, or UAG) occurs in the mRNA. Note that the UGA codon has an additional function, coding for **selenocysteine** (sec) in a small number of proteins (p. 79). Ribosomal RNA is a major part of the tRNA binding sites of the ribosome, and isolated 60S subunits can catalyze peptidyl transferase activity without the involvement of any nonribosomal factors. Evidence from many sources indicates that the large subunit is a complex **ribozyme** in which peptide bond formation is an **RNA-catalyzed** reaction. X-ray crystallography of large subunits shows the transferase region is composed only of RNA, with the nearest amino acid side chain at least 20Å away. Elements of both ribosomal and transfer RNAs are involved in transferase action. The important nucleotides in the transferase site have been universally conserved in evolution, so the same mechanism is believed to occur in mammalian ribosomes, although crystallographic evidence is not yet available. It is speculated that the primordial ribosome was a naked RNA molecule and that transferase activity has been conserved in RNA.

As determined from prokaryotic models, the role of GTP in the action of EF1α and EF2 relates to conformational changes in these proteins. Crystallographic studies show a large rearrangement of domains, with movements of several angstroms on GTP hydrolysis in EF-Tu, the prokaryotic equivalent of EF1α. Both EF1α and EF2 bind ribosomes tightly as GTP complexes, while GDP complexes dissociate from the ribosome more easily. Viewed another way, GTP stabilizes a protein conformation that confers on EF1α high affinity toward aminoacyl-tRNA and the ribosome, while GDP stabilizes a conformation with lower affinity toward either, thus allowing tRNA delivery and EF dissociation. Restoration of the higher affinity GTP-associated conformation of EF1α requires participation EF1$\beta\gamma$, which causes release of GDP from EF1α and forms an EF1α–EF1$\beta\gamma$ complex (Figure 6.9). GTP then causes release of EF1$\beta\gamma$, forming an EF1α–GTP complex that can bind an aminoacyl-tRNA and then a ribosome. Prokaryotes use a similar mechanism in which EF-Tu binds GTP and aminoacyl-tRNA and EF-Ts replaces GDP. Prokaryotes also utilize a GTP-dependent translocase, like EF2 but called **EF-G** or **G factor.**

Termination of Polypeptide Synthesis Requires a Stop Codon

A chain-terminating UAG, UAA, or UGA codon in the A site promotes binding of **release factor (eRF1)** (Figure 6.10). The peptide-tRNA ester linkage is cleaved by peptidyl transferase, acting here as a hydrolase, and the completed polypeptide is released from its tRNA

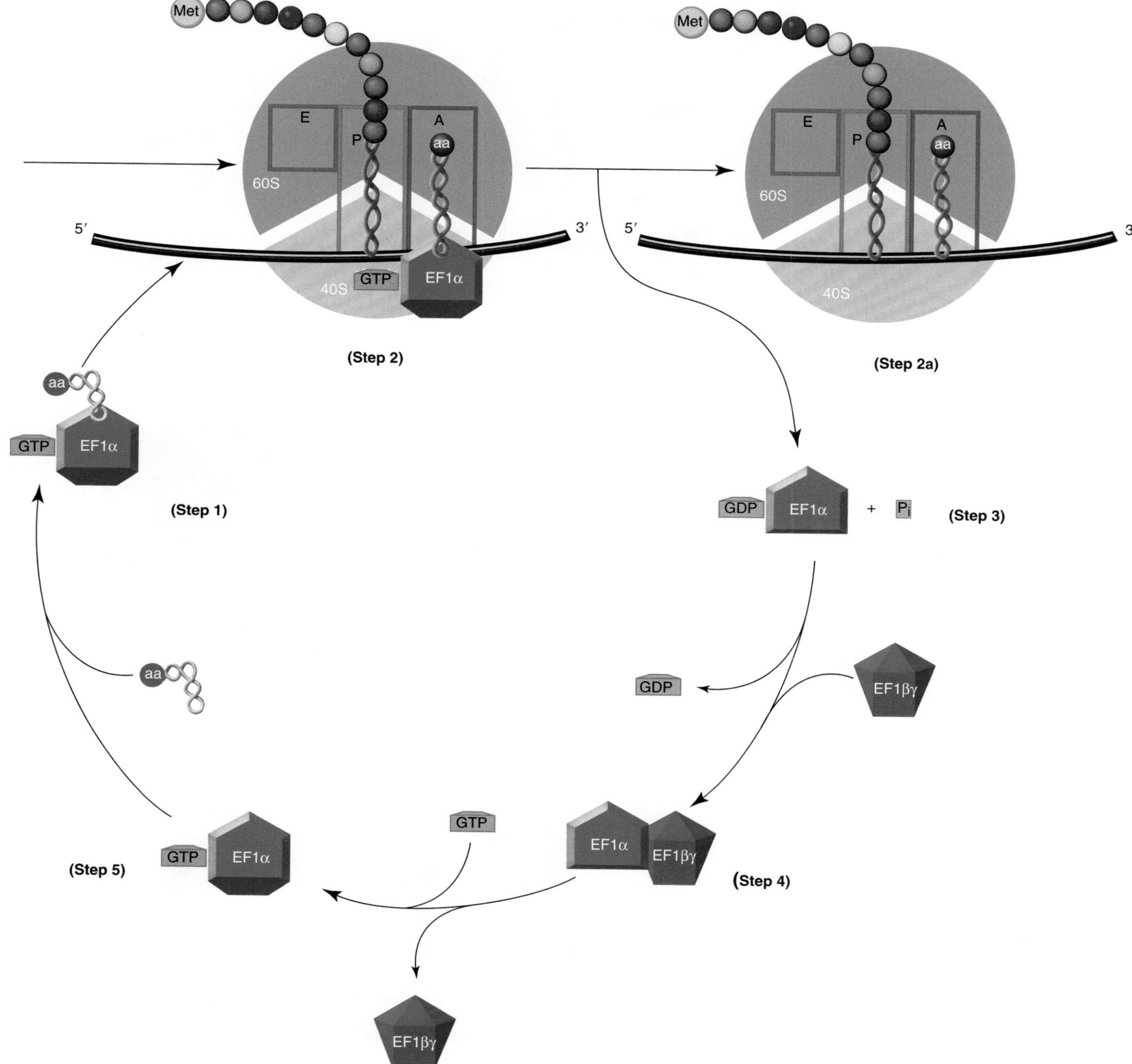

Figure 6.9 EF1 in the elongation cycle. Step 1: EF1α-GTP-aminoacyl-tRNA complex binds the ribosome. Step 2: Aminoacyl-tRNA is placed on the ribosome (step 2a) with hydrolysis of GTP, a change in conformation of EF1α, and its release from tRNA and the ribosome (step 3). Step 4: GDP is displaced from EF1α by EF1$\beta\gamma$. Binding of GTP then displaces EF1$\beta\gamma$ (step 5) and allows binding of an aminoacyl-tRNA by EF1α in its higher affinity conformation (step 1).

and the ribosome. Release of eRF1 from the ribosome follows hydrolysis of the bound GTP and frees the ribosome to dissociate into subunits and then reenter the protein synthesis process. A second protein, eRF3, participates in this process. Although human eRF1 is a protein, it is similar in size and shape to tRNA, and amino acids at one end of a protein domain recognize and interact with a termination codon. This "molecular mimicry" is also apparent in the structure of bacterial EF-G, in which a protein domain mimics the anticodon stem of tRNA and functions in elongation by displacing the tRNA from the A to the P subsite on the small subunit.

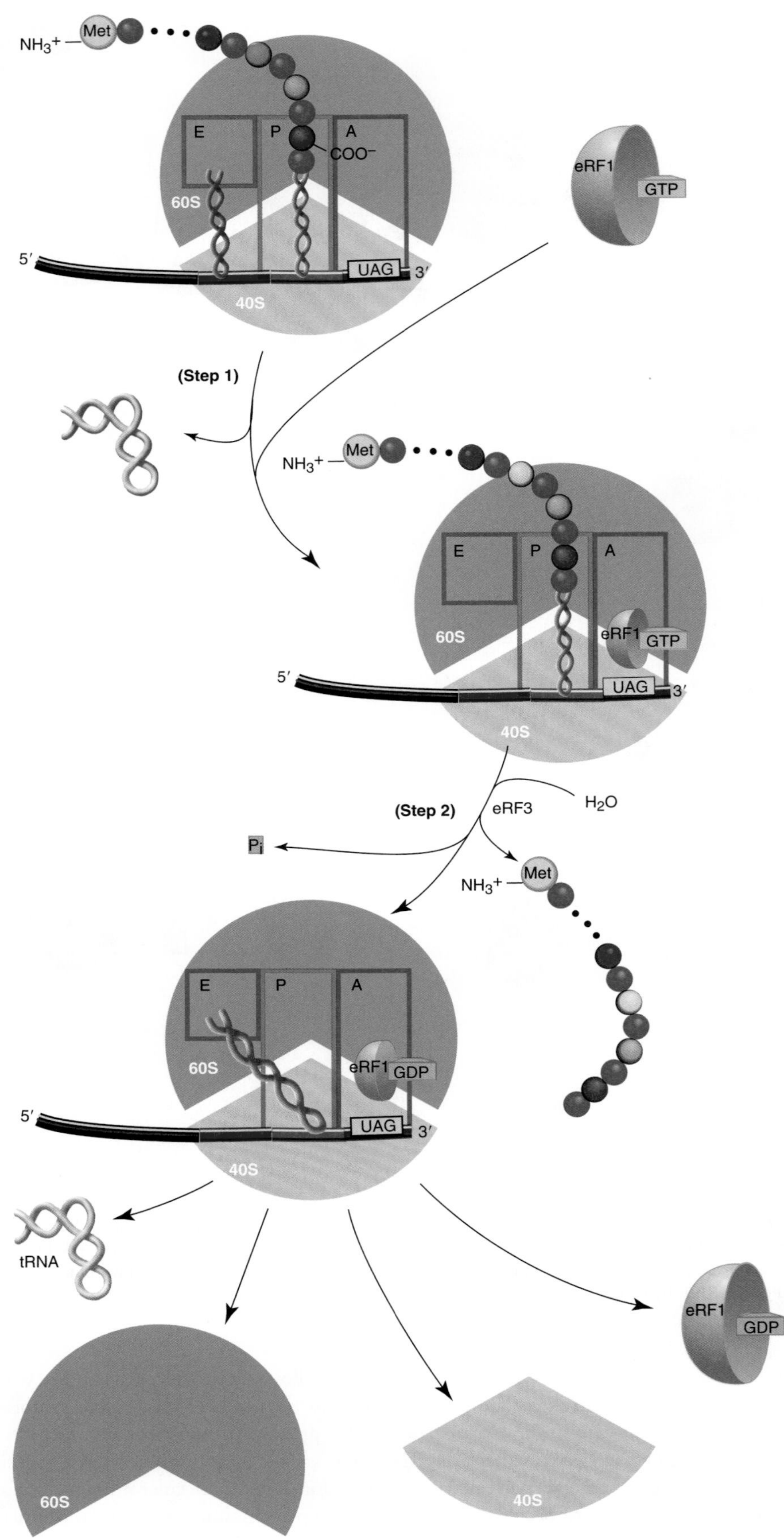

Figure 6.10 Termination of translation. When a termination codon in mRNA occupies the A site, binding of an RF1-GTP complex occurs (step 1) and deacylated tRNA is released. Step 2: Peptidyl transferase acts as a hydrolase; hydrolysis of the ester bond linking protein to tRNA releases the protein. The acceptor end of deacylated tRNA is probably displaced, GTP is hydrolyzed, and release factor-GDP dissociates. eRF3 promotes these events. Dissociated components can now enter additional rounds of protein synthesis.

Translation Has Significant Energy Cost

There is considerable use of energy in making a polypeptide. Amino acid activation converts an ATP to AMP and pyrophosphate, which is hydrolyzed to Pi; the net cost is two high-energy phosphates. Two more high-energy bonds are hydrolyzed in the action of EF1α and EF2, for a total of four per peptide bond formed. Initiation, termination, and posttranslational modifications add to the energy cost. Further energy is needed for biosynthesis of the multiuse mRNA, tRNAs, ribosomes, and protein factors, but these costs are distributed generally among the proteins formed during their lifetimes.

Protein Synthesis in Mitochondria Differs Slightly

Many characteristics of the mitochondrion suggest that it is a descendant of an aerobic prokaryote that formed a symbiotic relationship within a eukaryotic cell. Some independence and prokaryotic character are retained. Although most mitochondrial proteins are encoded in nuclear DNA, synthesized in the cytosol, and imported into the organelle, mitochondrial protein synthesis is important (Clin. Corr. 6.6). Human mitochondria have a circular DNA genome of 16,569 base pairs that encodes 13 proteins, 22 tRNA species, and mitochondrion-specific 12S and 16S rRNA species. Their apparatus for protein synthesis includes RNA polymerase, aminoacyl-tRNA synthetases, tRNAs, and ribosomes that are unique to the mitochondrion. Some details of translation differ: Mitochondrial ribosomes are smaller and the rRNAs are shorter than those of either the eukaryotic cytosol or of prokaryotes (see Table 6.7, p. 618). The genetic code is slightly different (see Table 6.2, p. 211), and like prokaryotes, the initiator Met-tRNA$_i^{met}$ is modified by a **transformylase** that uses N^{10}-formyl H_4-folate to produce **fMet-tRNA$_i^{met}$**. Cells must coordinate protein synthesis within mitochondria with the cytosolic synthesis of proteins destined for import into mitochondria.

Many Antibiotics and Toxins Target Protein Biosynthesis

Protein biosynthesis is central to the life and reproduction of cells. Because an organism can gain a biological advantage by interfering with the ability of its competitors to synthesize proteins, many antibiotics and toxins function in this way. Some are selective for prokaryotic protein synthesis and so are extremely useful in clinical practice. Examples showing several mechanisms of antibiotic action are listed in Table 6.9.

Many antibiotics bind the ribosome to inhibit an aspect of translation. **Streptomycin** binds the small subunit of prokaryotic ribosomes and causes misreading of mRNA (Clin. Corr. 6.6). Mutations in a ribosomal protein or the rRNA of the small subunit can confer streptomycin resistance or dependence. Other aminoglycoside antibiotics, such as

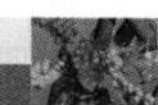

CLINICAL CORRELATION 6.6

Mutations in Mitochondrial rRNA and tRNA Result in Antibiotic-Induced Deafness

Instances of permanent deafness have been linked to use of normally safe and effective amounts of aminoglycoside antibiotics such as streptomycin, paromomycin, and gentamycin. The unusual sensitivity to aminoglycosides is maternally transmitted, suggesting a mitochondrial locus (sperm do not contribute mitochondria to the zygote). Aminoglycosides target bacterial ribosomes, so the prokaryotic-like mitochondrial ribosome is a logical place to look for a mutation site. A single A $\rightarrow$ G point mutation in mitochondrial DNA at nucleotide 1555 of the 12S rRNA gene has been identified in families with this aminoglycoside susceptibility. The mutation occurs in a highly conserved region of the rRNA sequence that is involved in aminoglycoside binding; some mutations in the same region confer resistance to the antibiotics, and the RNA region is part of the ribosomal decoding site. It is likely that the mutation increases affinity for aminoglycosides and their ability to inhibit protein synthesis. Reduced mitochondrial protein synthesis decreases levels of enzymes of the oxidative phosphorylation system and affected cells become starved of ATP. Many more mutations in mitochondrial rRNA and tRNA are associated with hearing loss and other problems.

Fischel-Ghodsian N., Prezant, T., Bu, X., and Öztas, S. Mitochondrial ribosomal RNA gene mutation in a patient with sporadic aminoglycoside ototoxicity. *Amer. J. Otolaryngology* 14:399, 1993; Xing, G., Chen, Z., and Cao, X. Mitochondrial rRNA and tRNA and hearing function. *Cell Res.* 17:227, 2007; and Wilson, F. H., Hariri, A., Farhi, A., Zhao, H., et al. A cluster of metabolic defects caused by mutation in a mitochondrial tRNA. *Science* 306:1190, 2004.

TABLE 6.9 • Some Antibiotic Inhibitors of Translation

Inhibitor Type	*Examples*	*Process Affected*	*Site of Action*
Aminoglycosides	Streptomycin Neomycin Gentamycins Paromomycin	Codon-anticodon Interaction fidelity	Small subunit Decoding center
Macrolides	Erythromycin Telithromycin Tylocin	Translocation	Large subunit peptidyl transferase
Oxazolidinones	Linezolid	Peptide bond formation	Large subunit A-site
Tetracyclins	Tetracycline Glycylcyclins	Aminoacyl-tRNA binding	Small subunit A-site
Thiopeptides	Thiostrepton	Elongation	Large subunit GTPase center
Aminonucleosides	Puromycin	Elongation	Large subunit A-site

Source: Hermann, T. Drugs targeting the ribosome. *Curt. Opinion Struct. Biol.* 15: 355 2005. Tenson, T. and Mankin, A. Antibiotics and the Ribosome. *Molec. Microbial.* 59: 1664, 2006.

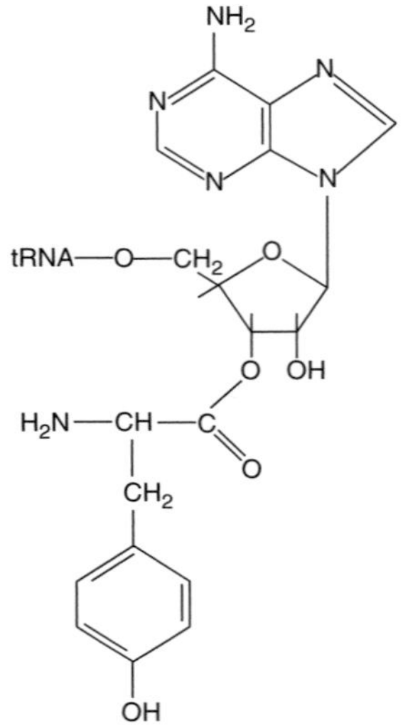

NH_2 N N N N tRNA—O—CH_2 O O OH H_2N—CH—C O CH_2 OH

3' end of tyrosyl-tRNA

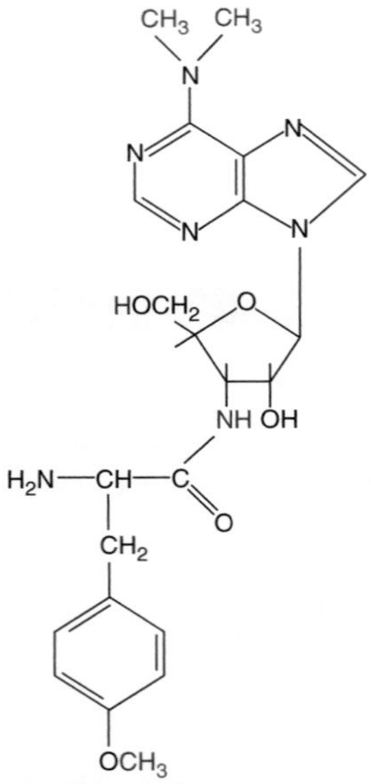

CH_3 CH_3 N N N N N $HOCH_2$ O NH OH H_2N—CH—C O CH_2 OCH_3

Puromycin

Figure 6.11 Puromycin (bottom) terminates translation by binding at the large subunit A site and acting as an analog of aminoacyl-tRNA (here tyrosyl-tRNA [top]). The α-amino group (arrow) is a peptidyltransferase acceptor, and peptidyl-puromycin dissociates from the ribosome.

the **neomycins** and **gentamycins,** interact with the small ribosomal subunit at a slightly different site but also cause mistranslation. **Erythromycin,** a macrolide antibiotic, interferes with translocation on prokaryotic ribosomes. The macrolides block passage of the elongating peptide through the ribosome. **Kasugamycin** inhibits the initiation of translation; sensitivity to kasugamycin depends on base methylation that normally occurs on two adjacent adenines of small subunit rRNA. **Tetracyclines** bind to ribosomes and interfere in aminoacyl-tRNA binding. **Puromycin** (Figure 6.11) resembles an aminoacyl-tRNA; it binds at the large subunit A site and acts as an acceptor of the nascent peptide in the peptidyl transferase reaction. Peptidyl-puromycin is not translocated, and it cannot serve as a peptide donor. Translation is prematurely terminated, and peptidyl-puromycin is released. **Chloramphenicol** inhibits peptidyl transferase by binding at the transferase center; no transfer occurs, and peptidyl-tRNA remains associated with the ribosome.

Some agents act directly on ribosomal RNA. **Ricin** (from castor beans) and related toxins are *N*-glycosidases that cleave a single adenine from the large subunit rRNA backbone. A fungal toxin, ***α*-sarcin,** cleaves large subunit rRNA at a single site; in both cases the ribosome is inactivated. Some *E. coli* strains make extracellular toxins that affect other bacteria. One of these, **colicin E3,** is a ribonuclease that cleaves 16S RNA near the mRNA-binding sequence and decoding site; it inactivates the small subunit and halts protein synthesis in competitors of the colicin-producing cell. Other agents affect translation factors: For example, eukaryotic translocation is inhibited by **diphtheria toxin,** a protein produced by *Corynebacterium diphtheriae.* The toxin binds at the cell membrane and a subunit enters the cytoplasm and catalyzes the ADP-ribosylation and inactivation of EF2, as represented in the reaction

$$\underset{\text{(active)}}{\text{EF-2}} + \text{NAD} \rightleftharpoons \underset{\text{(inactive)}}{\text{ADP-ribosyl EF-2}} + \text{nicotinamide} + \text{H}^+$$

ADP-ribose is attached to EF2 at a **diphthamide** residue that is itself a posttranslationally modified form of histidine (see Figure 6.20, p. 243). *Pseudomonas aeruginosa* produces a similar exotoxin.

6.4 • PROTEIN MATURATION: FOLDING, MODIFICATION, SECRETION, AND TARGETING

Some proteins emerge from the ribosome ready to function, whereas others undergo a variety of **posttranslational modifications.** The result may be conversion to a functional form, direction to a specific subcellular compartment, secretion from the cell, or an alteration

CLINICAL CORRELATION 6.7

Deletion of a Codon, Incorrect Folding, and Premature Degradation: Cystic Fibrosis (OMIM 219700)

Cystic fibrosis (CF) is the most common autosomal recessive disease in Caucasians, with a frequency of about 1 per 2000. The CF gene is 230 kb long and includes 27 exons that encode a protein of 1480 amino acids. The protein known as the cystic fibrosis transmembrane conductance regulator or CFTR (p. 494) is a cyclic AMP-regulated chloride channel. It contains two membrane-spanning domains, each having six transmembrane regions, two ATP-binding domains, and one regulatory domain, which includes several phosphorylation sites. Its complex folding pathway is affected by several chaperones.

CF epithelia are characterized by defective electrolyte transport. The organs most strongly affected include the lungs, pancreas, and liver, and the most life-threatening effects involve thick mucous secretions that lead to chronic obstructive lung disease and persistent lung infections. In about 70% of patients there is a deletion of the three nucleotides that encode phenylalanine 508, normally located in ATP-binding domain 1 on the cytoplasmic side of the plasma membrane. As with several other CF mutations, the Phe 508-deletion protein does not fold correctly in the ER and is not properly glycosylated or transported to the cell surface. Instead, it is degraded in the cytoplasm. Yet, in cultured cells at lower temperatures or with added chemical chaperones such as glycerol, the mutant protein folds, is transported, and functions properly. Drugs that foster chaperone interactions with mutant CFTR and/or aid in its folding and transport to the membrane are a potential therapeutic approach to this form of cystic fibrosis.

Cheung, J. C., and Deber, C. M. Misfolding of the cystic fibrosis transmembrane conductance regulator and disease. *Biochem.* 47:1465, 2008; Plemper, R. K., and Wolf, D. H. Retrograde protein translocation: ERADication of secretory proteins in health and disease. *Trends Biochem.* Sci. 24:266, 1999; and Egan, M. E., Pearson, M., Weiner, S. A., Rajendran, V., et al. Curcumin, a major constituent of turmeric, corrects cystic fibrosis defects. *Science* 304:600, 2004.

in activity or stability. Information that determines the posttranslational fate of a protein ultimately resides in its structure; the sequence and conformation of the protein determine whether it will be a substrate for a modifying enzyme and/or identify it for direction to a subcellular or extracellular location.

Chaperones Aid in Protein Folding

Many proteins can spontaneously generate their native conformation. For example, fully denatured pancreatic ribonuclease can, under appropriate laboratory conditions, refold and generate correct disulfide bridges and full activity. Many newly synthesized proteins fold correctly as they emerge from the ribosome. However, other proteins require the assistance of **chaperones** (p. 116) which reversibly bind hydrophobic regions of nascent proteins and proteins in an intermediate stage of folding. Their actions can help a protein attain its functional conformation. Chaperones can stabilize intermediates and promote correct folding, maintain proteins in an unfolded state to allow passage through membranes, help unfold misfolded segments, prevent formation of incorrectly folded intermediates, and prevent aggregation or other inappropriate interactions with other proteins. They can also assist in unfolding proteins destined for degradation. Some use energy from ATP hydrolysis to carry out their functions. Failure to fold correctly usually leads to rapid protein degradation (Clin. Corr. 6.7). Accumulation of misfolded proteins can result in protein aggregation and serious disease (see Clin. Corr. 6.14, p. 250).

Proteins for Export Follow the Secretory Pathway

Proteins destined for export and those destined for the plasma membrane, lysosomes, endosomes or the Golgi apparatus are synthesized on membrane-bound ribosomes of the **rough endoplasmic reticulum** (ER) (Figure 6.12). These proteins have a hydrophobic **signal peptide,** usually at or near the amino terminus. Signal peptides have a positively charged N-terminal region and a core of 8–12 hydrophobic amino acids in an α-helix, followed by a more polar C-terminal segment that eventually serves as a cleavage site for excision of the signal peptide. Translation begins on free cytosolic ribosomes. As the signal peptide of 15–30 amino acids emerges from the ribosome, it binds a **signal recognition particle (SRP)** (Figure 6.13). SRP is composed of six different proteins and a small (7S) RNA molecule. The signal peptide binds in a hydrophobic pocket of

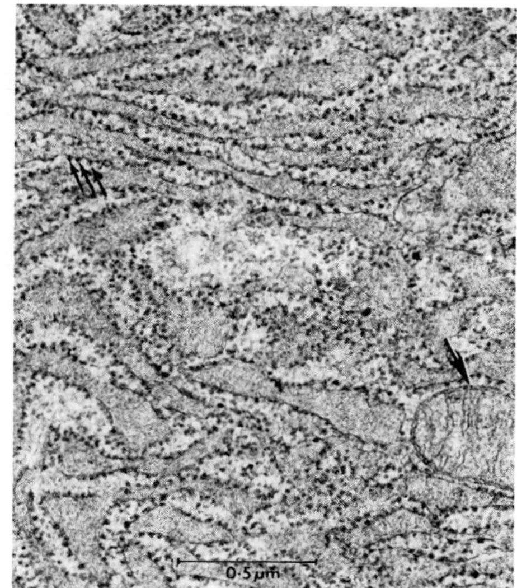

Figure 6.12 Rough endoplasmic reticulum. Three parallel arrows indicate three ribosomes among the many attached to the membranes. Single arrow indicates a mitochondrion for comparison.
Courtesy of Dr. U. Jarlfors, University of Miami.

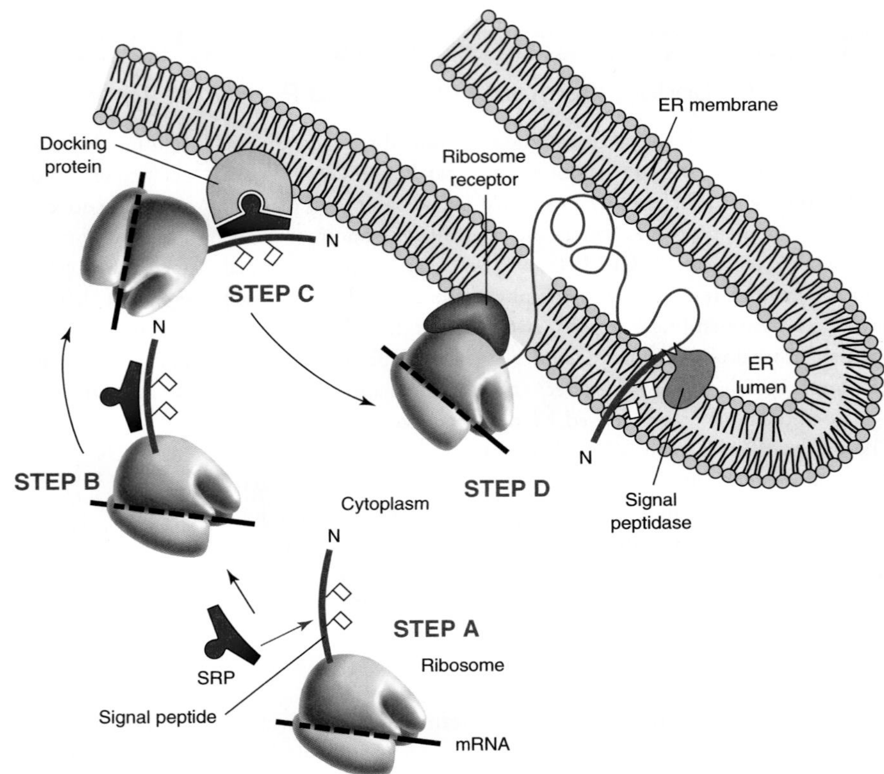

Figure 6.13 Secretory pathway: Signal peptide recognition. Step A: A hydrophobic signal peptide emerges from a free ribosome in the cytosol. Step B: Signal recognition particle (SRP) binds signal peptide and elongation is temporarily halted. Step C: The ribosome moves to ER membrane where docking protein binds SRP. Step D: Ribosome is transferred to a receptor (or translocon), elongation is resumed, and newly synthesized protein is extruded through the membrane into the ER lumen.

A CLOSER LOOK 6.3

Chaperones of the Endoplasmic Reticulum

Chaperone Family	Function
Hsp40	Assist Hsp 70
Hsp 70	General ER protein folding
Hsp90	General ER protein folding
Hsp100	Uncertain
GrpE-like	Assist Hsp 70
Lectins	Glycoprotein folding
Ribosome-tethered	Assist Hsp 70
PDI	Disulfide bond formation
PPI	Peptidyl-proline cis-trans isomerization
Other	Folding of specific proteins

Hebert, D. N., and Molinari, M. In and out of the ER: Protein folding, quality control, degradation, and related human diseases. *Physiol. Rev.* 87:1377, 2007.

the SRP with the positively charged N-terminal segment in contact with SRP RNA. Translation halts; the complex binds GTP and moves to the ER where the SRP binds to a GTP complex of an **SRP receptor** or **docking protein,** on the cytosolic surface of the ER membrane. The ribosome is transferred to a **translocon,** a ribosome receptor that crosses the membrane and serves as a passageway through the membrane. GTP is hydrolyzed and the SRP is released, relieving the translational block. The translocon passageway opens to allow transit of the nascent protein, the hydrophobic signal sequence is inserted, and translation and extrusion into or through the translocon are now coupled. Even very hydrophilic or charged segments are directed through the membrane into the ER lumen. Chaperones help the protein fold either co- or posttranslationally (A Closer Look 6.3). Disulfide bonds can form with the participation of thiol oxidoreductases, and components of multi-subunit proteins may assemble. The signal peptide is excised by **signal peptidase,** an integral membrane protein at the luminal surface of the ER. Other steps may include proteolytic processing and glycosylation that occur within the ER lumen and during transit of the protein through the Golgi apparatus and into secretory vesicles.

Glycosylation of Proteins Occurs in the Endoplasmic Reticulum and Golgi Apparatus

Glycosylation of proteins to form glycoproteins (p. 112) is important for many reasons. Glycosylation alters the properties of proteins, changing their stability, solubility, and physical bulk. The carbohydrate moieties can be recognition signals that direct protein interactions and targeting. Glycosylation sites are specified by the amino acid and its neighboring sequence in the protein. **Glycosyltransferases,** classes of which are summarized in Table 6.10, all carry out a similar basic reaction: A sugar is transferred from an activated donor substrate to an appropriate acceptor, usually another sugar residue that is part of an oligosaccharide under construction. The enzymes show specificity for the **monosaccharide** that is transferred, the structure and sequence of the **acceptor,** and the site and configuration of the **anomeric linkage** formed.

TABLE 6.10 • Glycosyltransferases in Eukaryotic Cells

Sugar Transferred	*Abbreviation*	*Donor*	*Glycosyltransferase*
Mannose	Man	GDP-Man	Mannosyltransferase
		Dolichol-Man	
Galactose	Gal	UDP-Gal	Galactosyltransferase
Glucose	Glc	UDP-Glc	Glucosyltransferase
		Dolichol-Glc	
Fucose	Fuc	GDP-Fuc	Fucosyltransferase
N-Acetylgalactosamine	GalNAc	UDP-GalNac	*N*-acetylgalactosaminyltransferase
N-Acetylglucosamine	GlcNAc	UDP-GlcNAc	*N*-acetylglucosaminyltransferase
N-Acetylneuraminic acid	NANA or NeuNAc	CMP-NANA	*N*-Acetylneuraminyltransferase
(or sialic acid)	SA	CMP-SA	(sialyltransferase)

In **N-linked glycosylation** an oligosaccharide is linked through the amide nitrogen of asparagine. Formation of N-linked oligosaccharides begins in the ER lumen and continues in the Golgi apparatus. A specific sequence, **Asn-X-Thr** (or **Ser**), in which X can be any amino acid except proline, is required. Not all Asn-X-Thr/Ser sequences are glycosylated because some may be unavailable for modification. The antibiotic tunicamycin prevents *N*-glycosylation.

N-linked glycosylation is initiated with a lipid-linked intermediate (Figure 6.14). **Dolichol phosphate** at the cytoplasmic surface of the ER membrane serves as glycosyl acceptor for *N*-acetylglucosamine. Stepwise glycosylation and formation of a branched $(Man)_5(GlcNAc)_2$-pyrophosphoryl-dolichol on the cytosolic side of the membrane follows. This intermediate is then **reoriented** to the luminal surface of the ER membrane, and four additional mannose and three glucose residues are added. A complex membrane bound **oligosaccharyltransferase** then transfers the complete oligosaccharide from its dolichol carrier to an asparagine residue of the polypeptide as it emerges into the ER lumen. Note that *N*-glycosylation is **cotranslational;** it occurs as the protein is being synthesized and it can affect protein folding.

Glycosidases and **glycosyltransferases** now modify the newly transferred oligosaccharide. Glucose residues, which were required for transfer of the oligosaccharide from the dolichol carrier, are removed. The folded glycoprotein is transported to an ER-Golgi intermediate compartment where sorting occurs, and then to the Golgi apparatus. Further trimming may now occur, or additional sugars can be added. The resulting N-linked oligosaccharides are diverse, but two classes are distinguishable. Each has a common core region ($GlcNAc_2Man_3$) linked to asparagine and originating from the dolichol-linked intermediate. The **high-mannose type** includes mannose residues in a variety of linkages and shows less processing from the dolichol-linked intermediate. The **complex type** is more highly processed and diverse, with a larger variety of sugars and linkages (Figure 6.15).

In **O-linked glycosylation,** oligosaccharides are bound through hydroxyl groups of serine or threonine in proteins (p. 112). O-linked glycosylation occurs only after the protein has reached the Golgi apparatus; hence *O*-glycosylation is **posttranslational** and occurs only on folded proteins. There is no specific amino acid sequence in which the serine or threonine must occur, but only residues on the protein surface serve as acceptors for the **GalNAc-transferase** that attaches *N*-acetyl galactosamine. Stepwise addition of sugars to the GalNAc acceptor follows. The oligosaccharides formed depend on the types and amounts of glycosyltransferases in a given cell. If an acceptor is a substrate for more than one transferase, the amounts of each transferase influence the competition between them. Oligosaccharide growth stops when structures are formed that are not acceptors for any of the glycosyltransferases present. This can result in different oligosaccharide structures on otherwise identical polypeptides, so **heterogeneity** in glycoproteins is common (Figure 6.16).

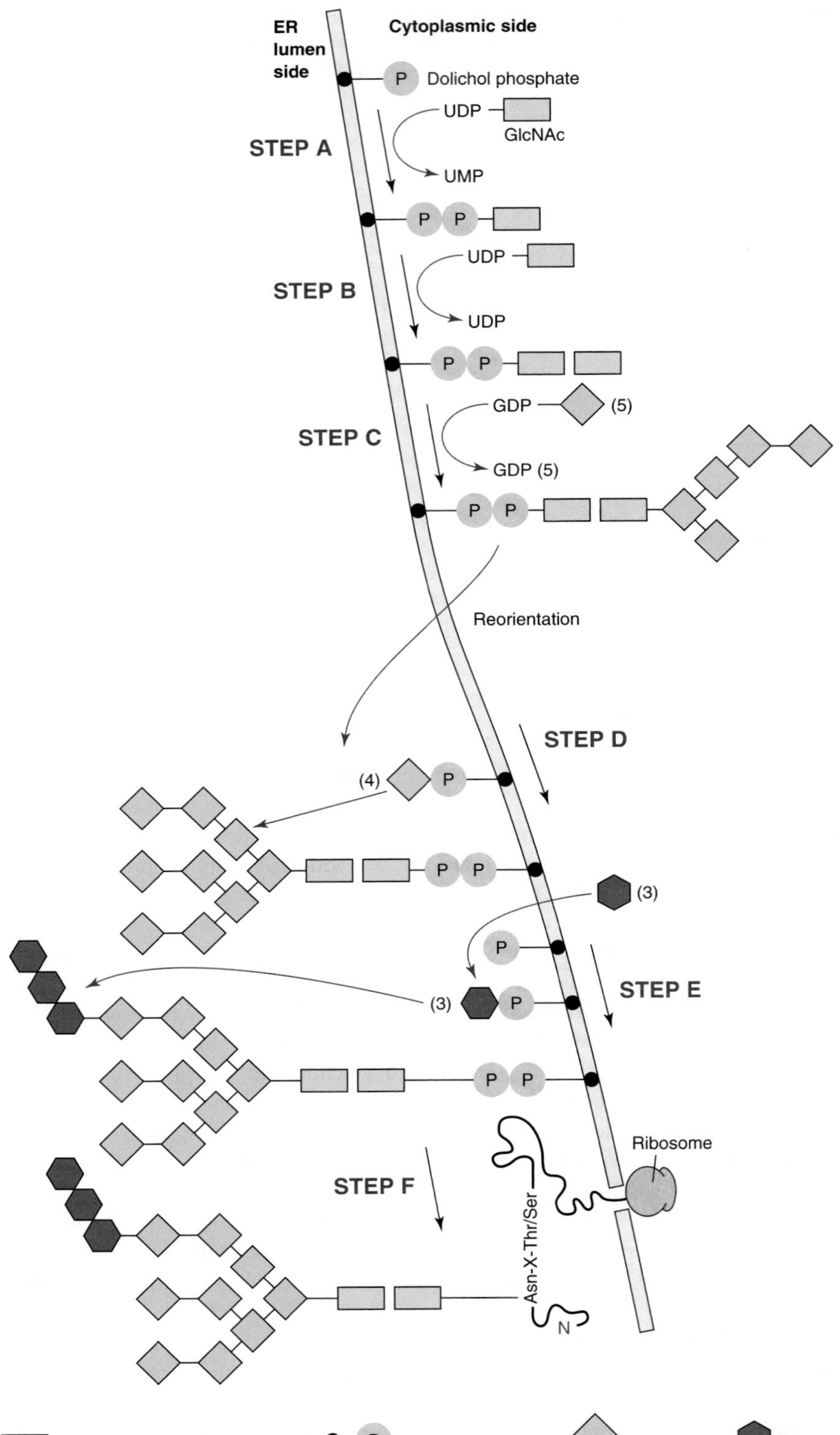

Figure 6.14 Biosynthesis of N-linked oligosaccharides at the endoplasmic reticulum. Step A: Synthesis begins on the cytoplasmic face of the ER membrane with transfer of *N*-acetylglucosamine phosphate to a dolichol acceptor. Step B: Formation of the first sugar-to-sugar glycosidic bond occurs on transfer of a residue of *N*-acetylglucosamine. Step C: Five residues of mannose (from GDP mannose) are added sequentially and the lipid-linked oligosaccharide is reoriented to the lumenal side of the membrane. Step D: Additional mannose and (step E) glucose residues are transferred from dolichol-linked intermediates. Dolichol-sugars are generated from cytosolic nucleoside diphosphate sugars. Step F: The completed oligosaccharide is transferred to a nascent polypeptide at the membrane surface; signal peptide may have already been cleaved at this point.

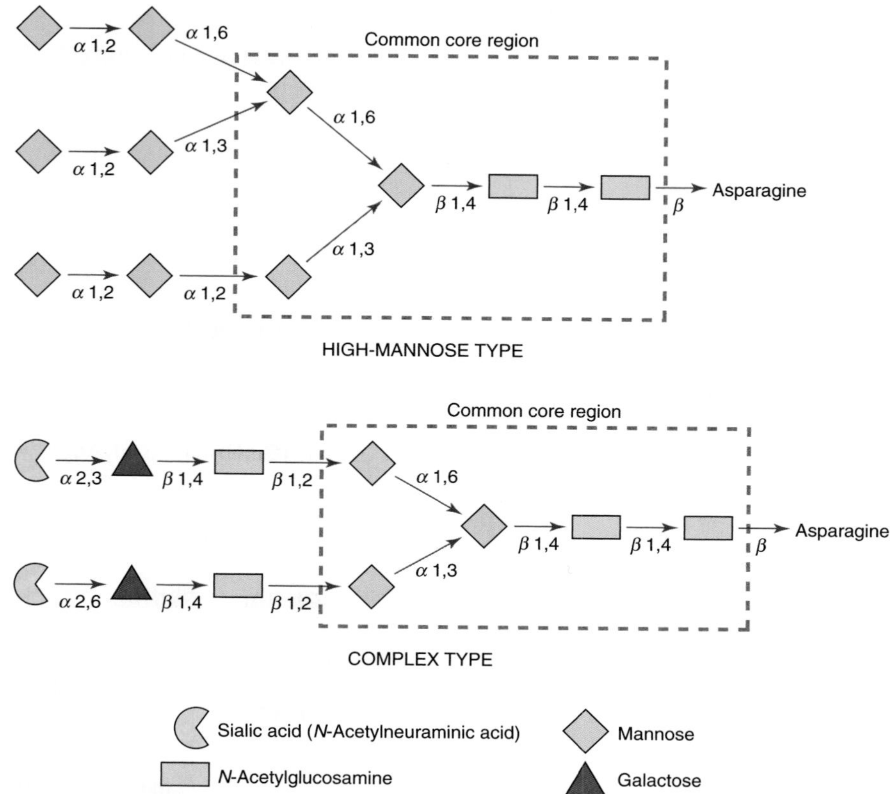

Figure 6.15 Structure of N-linked oligosaccharides.
Basic types of N-linked oligosaccharides are shown. Each structure is derived from the initial dolichol-linked oligosaccharide through action of glycosidases and glycosyltransferases. Note the variety of glycosidic linkages involved.

6.5 • MEMBRANE AND ORGANELLE TARGETING

Protein transport from the ER to and through the Golgi apparatus and beyond uses multiprotein carrier vesicles. Only correctly folded proteins are recognized as cargo for transport, and both general and protein-specific chaperones in the ER assist folding and foster correct disulfide formation. Chaperone networks also act at "checkpoints" in ER-Golgi transport to recognize misfolded or damaged proteins, which are returned to the cytosol for degradation in the process of **ER-associated degradation (ERAD).** Sorting of proteins for their ultimate destinations occurs in conjunction with their glycosylation and proteolytic trimming as they pass through the cis, medial, and trans elements of the Golgi apparatus. Families of vesicle and receptor proteins provide specificity for membrane targeting and fusion.

Sorting of Proteins in the Secretory Pathway

Targeting of specific glycoproteins to **lysosomes** is well understood. In the cis Golgi an aspect of tertiary structure allows lysosomal proteins to be recognized by a glycosyltransferase that attaches *N*-acetylglucosamine phosphate (GlcNAc-P) to high-mannose-type oligosaccharides. A glycosidase then removes the GlcNAc, forming an oligosaccharide that contains **mannose 6-phosphate** (Figure 6.17) that determines the compartmentation and vesicular transport of these proteins to lysosomes. Other oligosaccharide chains on the proteins may be further processed to form complex-type structures, but the mannose 6-phosphate defines their lysosomal destination. Patients with **I-cell disease** lack the GlcNAc-P glycosyltransferase, cannot mark lysosomal enzymes for their destination, and instead secrete them from the cell (Clin. Corr. 6.8). Genetic deficiencies of many lysosomal enzymes lead to diseases, some of which can be modulated by taking advantage of the targeting mechanism (Clin. Corrs. 6.9 and 18.5).

The secretory pathway directs proteins to lysosomes, to the plasma membrane, or for secretion from the cell. Proteins of the ER and Golgi apparatus are targeted through partial

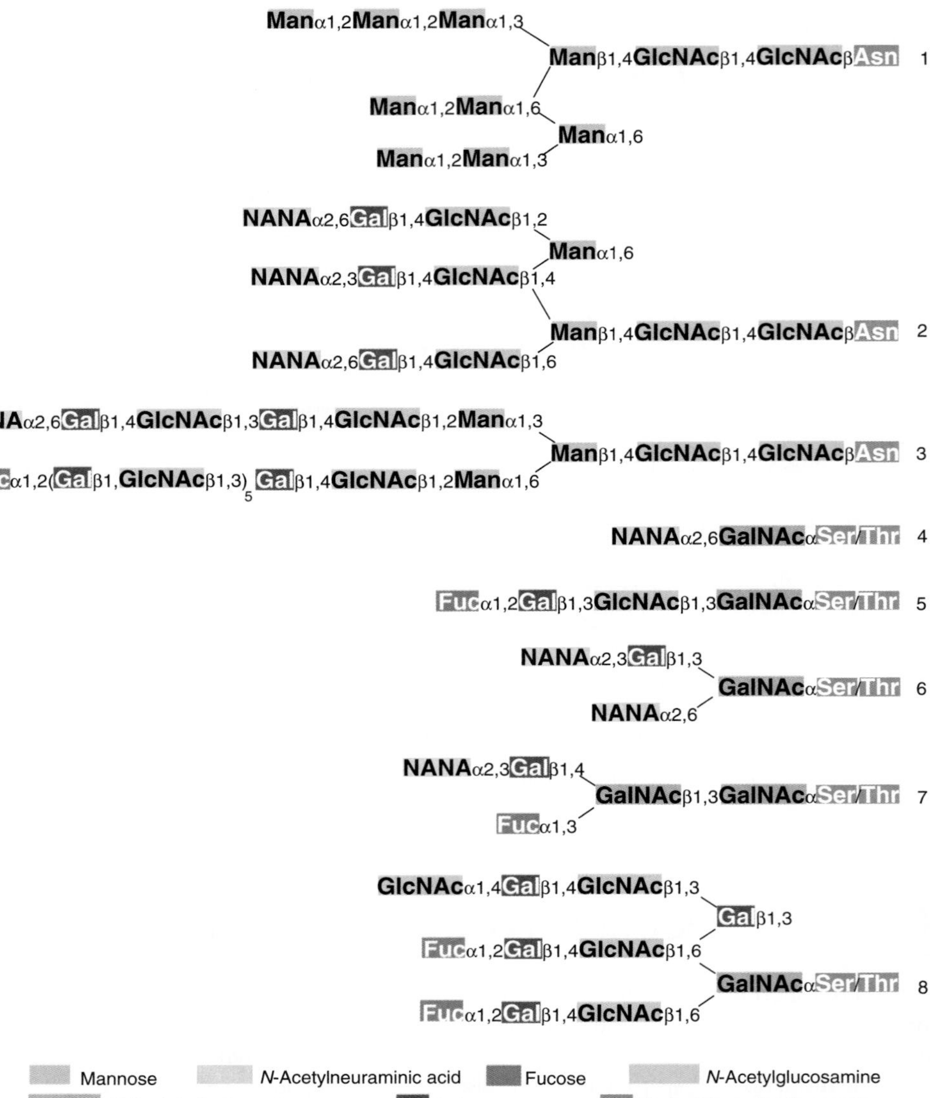

Figure 6.16 Examples of oligosaccharide structure. Structures 1–3 are typical N-linked oligosaccharides of high mannose (1) and complex types (2, 3); note the common core structure from the protein asparagine residue through the first branch point. Structures 4–8 are common O-linked oligosaccharides that may be simple or highly complex. Note that although the core structure (GalNAc-Ser/Thr) is unlike that of N-linked oligosaccharides, the termini can be quite similar (e.g., structures 2 and 6, 3, and 7). Abbreviations: Man = mannose, Gal = galactose, Fuc = fucose, GlcNAc = N-acetylglucosamine, GalNAc = N-acetylgalactosamine, and NANA = N-acetylneuraminic acid (sialic acid).
Adapted from Paulson, *J. Trends Biochem. Sci.* 14:272, 1989.

use of the pathway. For example, localization of proteins on either side of or spanning the ER membrane can utilize the signal recognition particle in different ways (Figure 6.18). Sometimes a signal sequence is not near the amino terminus of the protein but somewhere downstream where it is bound by SRP. An amino terminal domain remains on the cytoplasmic surface, and the signal sequence and C-terminal segment are passed into and through the translocon. Hydrophobic **anchoring sequences** in a protein can embed in the membrane yet allow much of the sequence to remain at the cytosolic surface or to be retained on the luminal surface of the ER membrane. Multiple anchoring sequences in a single polypeptide can cause it to span the membrane several times and thus retain it, largely buried in the membrane. Such hydrophobic sequences are separated by polar loops whose orientation is determined by positively charged flanking residues that predominate on the cytoplasmic side of the membrane.

Other sorting signals determine how proteins are directed (as cargo) to Golgi sub compartments, various storage and secretory granules, and specific elements of the plasma membrane. For example, soluble proteins are retained in the ER lumen in response to a C-terminal **KDEL** (Lys-Asp-Glu-Leu) sequence. A different sequence in an exposed C-terminus signals retention in the ER membrane. Some transmembrane domains result in retention in the Golgi. Polypeptide-specific glycosylation and sulfation of some glycoprotein hormones in the anterior pituitary mediates their sorting into storage granules.

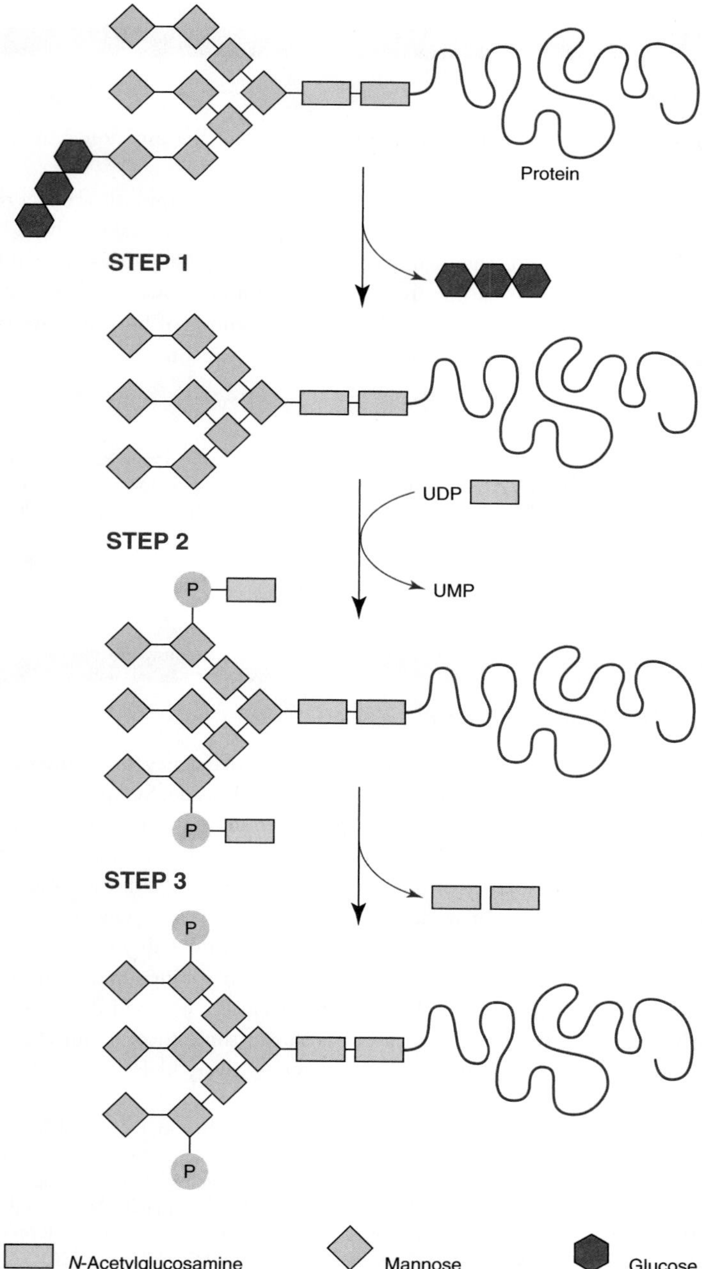

Figure 6.17 Targeting of proteins to lysosomes. Complete folded N-linked glycoprotein is released from the ER membrane and, prior to transport to the Golgi apparatus, glycosidases remove glucose residues (step 1). A mannose residue may also be removed. Step 2: In the Golgi apparatus a glycosyl transferase links one or sometimes two N-acetylglucosamine phosphate residues to the oligosaccharide. Step 3: A glycosidase removes N-acetylglucosamine, leaving one or two mannose 6-phosphate residues on the oligosaccharide. The protein is then recognized by a mannose 6-phosphate receptor and directed to vesicles that are targeted to lysosomes. Adapted from Kornfeld, R., and Kornfeld, S. *Annu. Rev. Biochem.* 54:631, 1985.

Polysialic acid modification of a neural cell adhesion protein is specific to the protein and regulated in development.

Import of Proteins by Mitochondria Is Complex

Mitochondria provide a special targeting problem, and defects in the system can lead to disease (Clin. Corr. 6.10). Most mitochondrial proteins are synthesized in the cytosol on free ribosomes as larger **preproteins.** N-terminal **presequences** mark the protein for the mitochondrion; the targeting signal is not a specific sequence, but rather a positively charged amphiphilic α-helix that is recognized by a mitochondrial receptor. The preproteins are transported with the aid of chaperones to the mitochondrion and translocated (unfolded) first across the outer membrane by the **translocase of the outer membrane (TOM) complex** and then into the mitochondrial matrix (or the inner membrane) by a **translocase of**

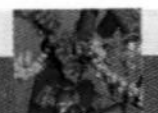

CLINICAL CORRELATION 6.8

Diseases of Lysosome Targeting: I-Cell Disease

I-cell disease (mucolipidosis II) arises from defects in lysosomal enzyme targeting because of a deficiency of **GlcNac-1-phosphotransferase,** the enzyme that transfers N-acetylglucosamine phosphate to the high-mannose-type oligosaccharides of proteins destined for the lysosome. The hexameric enzyme includes three types of subunit encoded in two genes, GNPTAB and GNPTG. Fibroblasts from affected individuals show dense inclusion bodies (hence I cells) and lack phosphotransferase activity; multiple lysosomal enzymes are secreted into the medium, and high levels of lysosomal enzymes are present in plasma and other body fluids of patients. The disease is characterized by severe psychomotor retardation, many skeletal abnormalities, coarse facial features, and restricted joint movement. Symptoms are usually present at birth and progress until death, usually by age 8. Almost all patients show frameshift or chain termination mutations in the GNPTAB gene.

Pseudo Hurler polydystrophy (mucolipidosis III) is a related but less severe disease in which phosphotransferase activity is reduced but not completely absent. Onset is usually delayed until the age of 2–4 years, the disease progresses more slowly, and patients can survive into adulthood. Prenatal diagnosis of both diseases is possible, but there is as yet no definitive treatment.

Kornfeld, S. Trafficking of lysosomal enzymes in normal and disease states. *J. Clin. Invest.* 77:1, 1986; and Braulke, T., Pohl, S. and Storch, S. Molecular analysis of the GlcNac-1-phosphotransferase. *J. Inherit. Metab.* Dis. 31:253, 2008.

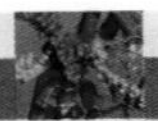

CLINICAL CORRELATION 6.9

Enzyme Replacement Therapy for Lysosomal Storage Diseases

Several diseases arise from a genetic deficiency of a single hydrolytic enzyme in lysosomes (see Clin. Corr. 18.5, p. 737), leading to accumulation of undegraded material. Several of these diseases are targets for **enzyme replacement therapy (ERT)** in which intravenous administration of purified human enzyme that is competent for lysosomal uptake can ameliorate symptoms (but not cure the underlying defect).

Mucopolysaccharidosis type I is caused by defects in α-L-iduronidase, a lysosomal enzyme that acts in the degradation of glycosaminoglycans. More than 70 mutations have been identified, and symptoms vary from severe mental retardation and early childhood death (Hurler syndrome) to milder effects including hearing loss and corneal clouding, but normal intelligence and lifespan. Enzyme replacement therapy using recombinant normal human enzyme was approved in 2003 and results in significant reduction in symptoms. Mucopolysaccharidosis type II (iduronate-2-sulfatase deficiency) and Mucopolysaccharidosis type VI (Maroteaux–Lamy syndrome, *N*-acetylgalactosamine deficiency) also are approved for ERT.

Gaucher disease results from deficiency of a glucocerebrosidase and the accumulation of glucocerebroside. Symptoms in type 1 Gaucher disease include hepatosplenomegaly and many bone problems; ERT, especially of young children, has been successful in reducing the severity of the disease. Infantile Pompe disease results from acid α-1, 4-glucosidase deficiency; it leads to glycogen accumulation in lysosomes and cardioskeletal myopathy. Fabry disease, caused by defective α-galactosidase-A, results in sphingolipid accumulation and organ damage that first appears at age 6–8 and death in middle age. ERT of these diseases has shown some success, but cost of purified enzymes for all of these therapies is very high.

Yogalingam, G., Guo, X.-H., Muller, V. J., Brooks, D. A., et al. Identification and molecular characterization of α-L-iduronidase mutations present in mucopolysaccharidosis type I patients undergoing enzyme replacement therapy. *Hum. Mutat.* 24:199, 2004; and Burrow, T. A., Hopkin, R. J., Leslie, N. D., Tinkle, B. T. and Grabowski, G. A. Enzyme reconstitution/replacement therapy for lysosomal storage diseases. *Curr. Opin. Pediatr.* 19:628, 2007.

the inner membrane (TIM) complex in energy-dependent reactions. Passage occurs at adhesion sites where inner and outer membranes are close together, and several intermembrane proteins are involved. Proteases remove the matrix targeting signal, but may leave other sequences that further sort the protein within the mitochondrion, back to the inner membrane, via a different TIM complex, or to the intermembrane space.

For example, in response to a hydrophobic sequence, a clipped precursor of cytochrome b_2 is moved from the matrix back across the inner membrane; further proteolysis frees it in the intermembrane space. In contrast, cytochrome c apoprotein (without heme) binds at the outer membrane and is passed into the intermembrane space. There it acquires its heme and undergoes a conformational change that assures retention. Outer membrane proteins are also targeted to the TOM complex, but anchoring sequences orient and retain them, either embedded in the membrane or with globular domains on the outer mitochondrial membrane surface.

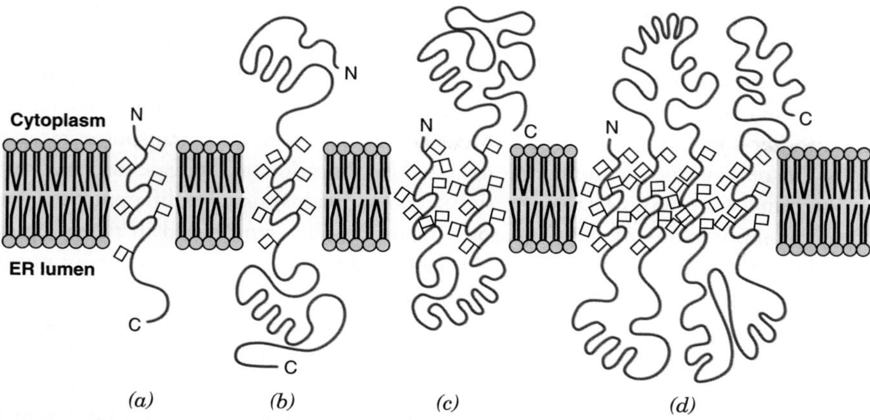

Figure 6.18 Topology of proteins at membranes of endoplasmic reticulum. Proteins are shown in several orientations with respect to membrane. In (*a*) the protein is anchored to lumenal surface of membrane by an uncleaved signal peptide. In (*b*) the signal sequence is not at the N-terminus; a domain of the protein was synthesized before emergence of signal peptide. Insertion of internal signal sequence, followed by completion of translation, resulted in a protein with a cytoplasmic N-terminal domain, a membrane-spanning central segment, and a C-terminal domain in the ER lumen. (*c*) A protein with the opposite orientation: an N-terminal signal sequence, which might also have been cleaved by signal peptidase, resulted in extrusion of a segment of protein into the ER lumen. A hydrophobic anchoring sequence remained membrane associated and prevented further passage of protein through the membrane, thus formatting a C-terminal cytoplasmic domain. In (*d*) several internal signal and anchoring sequences allow various segments of the protein to be oriented on each side of membrane.

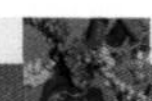

CLINICAL CORRELATION 6.10

Defective Mitochondrial Protein Import and Disease

Inefficient targeting to mitochondria can have major consequences. Lactic acidemia resulting from deficiencies in the pyruvate dehydrogenase (PDH) complex has been linked to mutations in several genes, most commonly the X-linked E1α subunit. A point mutation resulting in an Arg → Pro substitution in the mitochondrial targeting signal results in decreased importation of E1α, lowered PDH activity, and multiple neurological and other symptoms. Mistargeting to the mitochondrion occurs in some cases of primary hyperoxaluria type 1. Alanine/glyoxylate aminotransferase normally catalyzes the conversion of glyoxalate to glycine in peroxisomes. Separate point mutations that create a mitochondrial targeting signal and alter enzyme folding and dimerization result in localization of the enzyme in the mitochondria; reduced peroxisome function; oxalate deposition in the kidney, urinary tract, and elsewhere; and eventual kidney failure. Improper sorting and localization in the mitochondrion is a factor in some cases of severe alcoholic liver disease. Manganese superoxide dismutase (MnSOD) is normally located in the mitochondrial matrix. Valine (rather than alanine) at position 16 in the targeting sequence results in partial retention of the enzyme in the inner mitochondrial membrane, reduced MnSOD activity, and reduced ability to detoxify reactive oxygen species that are generated by alcohol consumption.

Defects in proteins of the import and sorting mechanism also can result in disease. Deafness dystonia syndrome has been linked to defects in a TIM component Tim 8 (see Clin. Corr. 6.6 for a different cause of deafness linked to mitochondria). Defects in Tim 14 have been linked to a rare cardiomyopathy with ataxia. Decreased levels of the mitochondrial matrix chaperone Hsp60 are linked to reduced numbers and atypical morphology of mitochondria and cases of spastic paraplegia and multisystem failure.

Robinson, B. H. Lactic acidemia and mitochondrial disease. *Mol. Genet. Metab.* 89:3, 2006; and MacKenzie, J. A., and Payne, R. M. Mitochondrial protein import and human health and disease. *Biochim. Biophys. Acta* 1772:509, 2007.

Targeting Signals Direct Proteins to Specific Organelles

The nucleus must import proteins that are synthesized in the cytosol. Nuclear proteins are targeted by localization signals that include two closely separated clusters of basic amino acids or sometimes a single basic sequence. They interact with carrier proteins (e.g., importins) that transport them through cylindrical **nuclear pore complexes** that span the nuclear membrane. Transport requires the GTPase Ran. A different class of regulatory proteins shuttles between cytoplasm and nucleus to transmit signals to activate transcription.

Phosphorylation of an SPS (Ser-Pro-Ser) or TPT (Thr-Pro-Thr) sequence induces nuclear uptake.

Peroxisomes import their matrix proteins from the cytosol, and may need to adjust their enzyme content to meet cellular needs. One targeting signal is a carboxy-terminal tripeptide, **Ser-Lys-Leu (SKL),** which interacts with a cargo-binding transporter for movement to the peroxisome membrane for docking, translocation, and release into the matrix. An N-terminal nonapeptide targeting signal also functions, using a different transporter complex. Some proteins of the peroxisome membrane are synthesized in the cytoplasm, and are localized to the membrane by internal targeting signals. A second class of proteins are synthesized in the ER and exported to the peroxisome membrane.

Some proteins reside in more than one subcellular compartment. A suboptimal localization signal can lead to inefficient targeting and a dual location, as in the partial secretion of an inhibitor of the plasminogen activator. A protein can contain two targeting signals, resulting in dual localization. Gene duplication and divergence can result in different targeting signals in closely related mature polypeptides. Alternative transcription initiation sites or alternative splicing of pre-mRNA can generate different messages from a single gene. Alternatively spliced mRNAs for calcium/calmodulin-dependent protein kinase differ with respect to the presence of an internal segment that encodes a nuclear localization signal. Without this segment, the protein remains in the cytosol.

Finally, the first initiation codon in an mRNA is sometimes bypassed. **Alternative translation initiation** leads to two forms of rat liver fumarase; one includes a mitochondrial targeting sequence, while the other does not and remains in the cytosol.

6.6 • FURTHER POSTTRANSLATIONAL MODIFICATIONS

Additional maturation events can modify proteins to generate their final, functional forms. Some are very common, while others are highly restricted. Many reversible modifications regulate protein activity.

Partial Proteolysis Releases Insulin and Activates Zymogens

Partial proteolysis is a common maturation step. Sequences can be removed from either end or from within the protein. Proteolysis in the ER and Golgi apparatus helps to mature the protein hormone insulin (Figure 6.19). Preproinsulin encoded by mRNA is inserted into the ER lumen. Signal peptidase cleaves the signal peptide to generate proinsulin, which folds to form the correct disulfide linkages. Proinsulin is transported to the Golgi apparatus where it is packaged into secretory granules. An internal **connecting peptide (C-peptide)** is removed by proteolysis, and mature insulin is secreted. In familial hyperproinsulinemia processing is incomplete (Clin. Corr. 6.11). The pathway for insulin biosynthesis ensures production of equal amounts of A and B chains. Moreover, proinsulin folds into a conformation in which the cysteine residues are positioned for correct disulfide bond formation. Reduced and denatured proinsulin can refold correctly, whereas renaturation of reduced and denatured insulin is inefficient and leads to incorrect disulfide linkages.

Cleavage of precursor proteins is a common means of enzyme activation. Digestive proteases are classic examples (p. 1033). Inactive **zymogen** precursors are packaged in storage granules and activated by proteolysis after secretion. Thus, **trypsinogen** is cleaved to give **trypsin** plus an amino terminal hexapeptide, and **chymotrypsinogen** is cleaved to form **chymotrypsin** and two peptides (p. 1044).

Amino Acids Can Be Modified after Incorporation into Proteins

Although 20 amino acids are encoded genetically, **posttranslational modification** allows formation of many amino acid derivatives in proteins. Modification may be permanent or easily reversible. While the number of modified amino acids in a protein may be small, they often

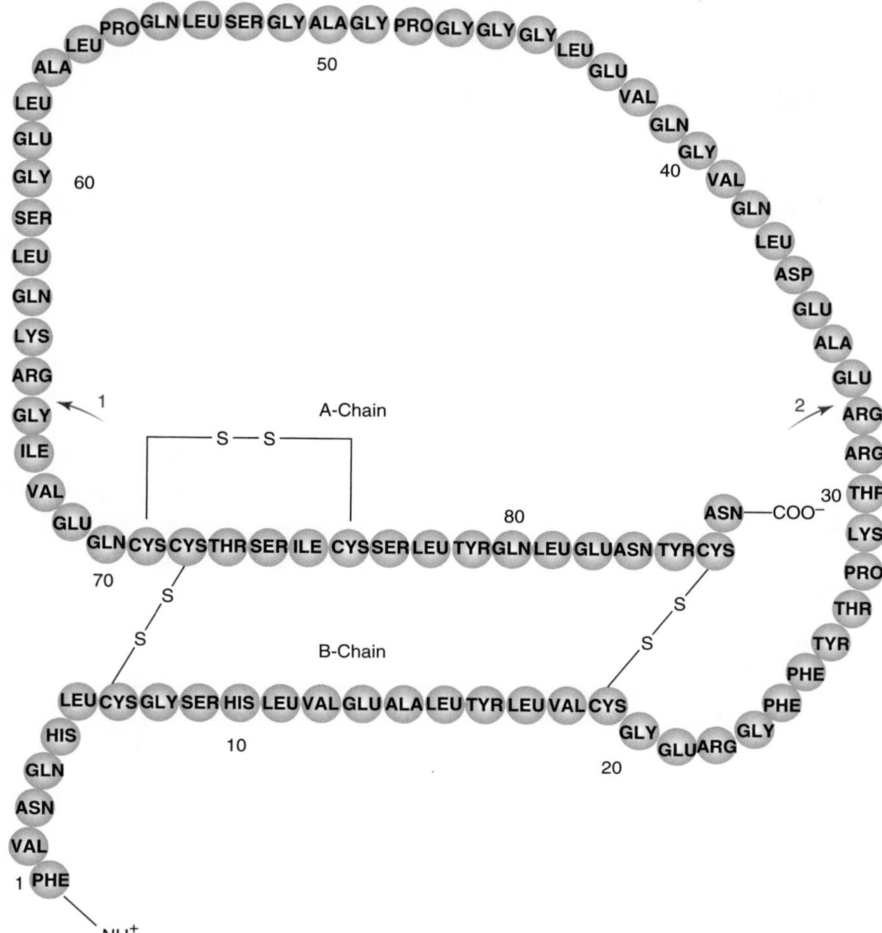

Figure 6.19 Maturation of human proinsulin. After cleavage at the two sites indicated by arrows, arginine residues 31, 32, and 65 and lysine residue 64 are removed to produce insulin and C-peptide. Redrawn from Bell, G. I., Swain, W. F., Pictet, R., Cordell, B., Goodman, H. M., and Rutter, W. J. *Nature* 282:525, 1979.

CLINICAL CORRELATION 6.11

Familial Hyperproinsulinemia (OMIM 176730)

Familial hyperproinsulinemia, an autosomal dominant condition, results in approximately equal amounts of insulin and an abnormally processed proinsulin being released into the circulation. Although affected individuals have high concentrations of proinsulin in their blood, they are apparently normal in terms of glucose metabolism, being neither diabetic nor hypoglycemic. The defect was originally thought to result from a deficiency of one of three proteases that process proinsulin: endopeptidases that cleave the Arg31-Arg32 and Lys64-Arg65 peptide bonds, and a carboxypeptidase. In several families the defect is the substitution of Arg65 by His or Leu which prevents cleavage between the C-peptide and the A chain of insulin resulting in secretion of a partially processed proinsulin. In one family a point mutation (His10 → Asp10) causes the hyperproinsulinemia, but how this mutation interferes with processing is not known.

Zhou, A., Webb, G., Zhu, X., and Steiner, D. F. Proteolytic processing in the secretory pathway. *J. Biol. Chem.* 274:20745, 1999.

have a major role in its function. Modified amino acids in proteins are listed in Table 6.11. Important examples include

- Amino-terminal modification: Protein synthesis is initiated using methionine, but proteolysis often removes one to several amino-terminal residues. The new amino-terminus is then often altered, for example, by acetylation. The α subunits of signal transducing G-proteins (p. 520) are derivatized with myristic or palmitic acid.

Amino-terminal glutamine often spontaneously cyclizes to form a pyroglutamyl residue. The amino terminus can also be lengthened by addition of an amino acid (see Section 6.8 on protein degradation).

- Interchain and intrachain disulfide bond formation: As catalyzed by **disulfide isomerase,** disulfide formation can prevent unfolding of proteins and their passage across membranes, so it also is a means of localization. As in the case of insulin, disulfide bonds can covalently link separate polypeptides and be necessary for function.
- Cysteine modification: *S*-palmitoylation aids membrane interactions and trafficking of some proteins. The γ-subunits of the heterotricyclic G proteins (p. 525) are modified by thioester linkage of an isoprenoid to a cysteine at or near the carboxy terminus. Multiple sulfatase deficiency arises from reduced ability to carry out a posttranslational modification of cysteine (Clin. Corr. 6.12).
- Lysine modification: Acetylation and methylation of lysine ϵ-amino groups in histones modulates their interactions with DNA. A fraction of the H2A histone is also modified through isopeptide linkage of a lysine ϵ-amino group to the C-terminal glycine of ubiquitin, a small protein with many regulatory functions. Biotin is also linked to a few carboxylases through amide linkages to lysine (p. 777).

TABLE 6.11 • Modified Amino Acids in Proteins[a]

Amino Acid	*Modifications Found*
Amino terminus	Formylation, acetylation, aminoacylation, myristoylation, glycosylation
Carboxy terminus	Methylation, ADP-ribosylation, glycosyl-phosphotidylinositol anchor formation
Arginine	N-Methylation, ADP-ribosylation
Asparagine	N-glycosylation, N-methylation, deamidation
Aspartic acid	Methylation, phosphorylation, hydroxylation, isomerization (iso Asp)
Cystine	Cystine formation, palmitoylation, heme linkage, S-glycosylation, prenylation, ADP-ribosylation, S-hydroxylation
Glutamine acid	Methylation, γ-carboxylation, ADP-ribosylation, polyglycination, polyglutamylation
Glutamine	Pyroglutamate formation, deamidation, crosslinking
Histidine	Methylation, phosphorylation, diphthamide formation, ADP-ribosylation
Lysine	N-Acetylation, N-methylation, hydroxylation, oxidation, ubiqutinylation, biotinylation, allysine formation, cross-linking
Methionine	Sulfoxide formation
Phenylalanine	β-Hydroxylation, glycosylation
Proline	Hydroxylation, glycosylation
Serine	Phosphorylation, glycosylation, acetylation, α-formylglycine formation, phosphopantetheinylation
Threonine	Phosphorylation, glycosylation,methylation
Tryptophan	β-Hydroxylation, dione formation, C-mannosylation
Tyrosine	Phosphorylation, iodination, adenylation, sulfonation, hydroxylation, nitration

Source: Krishna, R. G. and Wold, F. Post-translational modification of proteins. In: A. Meister (Ed.), *Advances in Enzymology,* Vol. 67, New York: Wiley InterScience, 1993, pp265–298. Walsh, C. T., Gameau-Tsodikova, S. and Gatto, G. J. Protein posttranslational modifications: the chemistry of proteome diversifications. *Angew. Chem. Int. Ed Engl.* 44: 7342, 2005.

[a]The listing is not comprehensive, and some modifications are rare. Note that no modifications of glycine alanine, leucine, isoleucine and valine have been reported in proteins.

- Serine and threonine hydroxyl groups: In addition to glycosylation, reversible phosphorylation by protein kinases and protein phosphatases is very common and functionally important. A classic example is phosphorylation of a serine residue of glycogen phosphorylase by phosphorylase kinase (p. 635). Tyrosine residues are phosphorylated by highly specific tyrosine kinases. Kinase activity is a property of many growth factor receptors in which growth factor binding stimulates cell division. Oncogenes, responsible in part for cell transformation and proliferation of tumor cells, often have tyrosine kinase activity and show strong homology with normal growth factor receptors. Dozens of other examples exist; together the protein kinases and protein phosphatases control the activity of many proteins that are central to normal and abnormal cellular development.
- ADP-ribosylation: **Diphtheria toxin** further modifies **diphthamide,** a modified histidine residue (Figure 6.20) of EF2, by ADP-ribosylation. EF2 activity and thus protein synthesis is inhibited. Physiological ADP-ribosylation, not mediated by bacterial toxins, usually involves reversible modification of arginine and cysteine residues and is important in cell signaling, and regulatory and surface proteins.
- **γ-carboxyglutamate** formation: Glutamic acid residues in several blood-clotting proteins including prothrombin and factors VII, IX, and X are modified. The γ-carboxyglutamate chelates calcium ion, which is required for blood clotting (p. 987). This modification requires vitamin K. It can be blocked by coumarin derivatives (p. 995), which antagonize vitamin K. Coumarin derivatives are therapeutic anticoagulants.

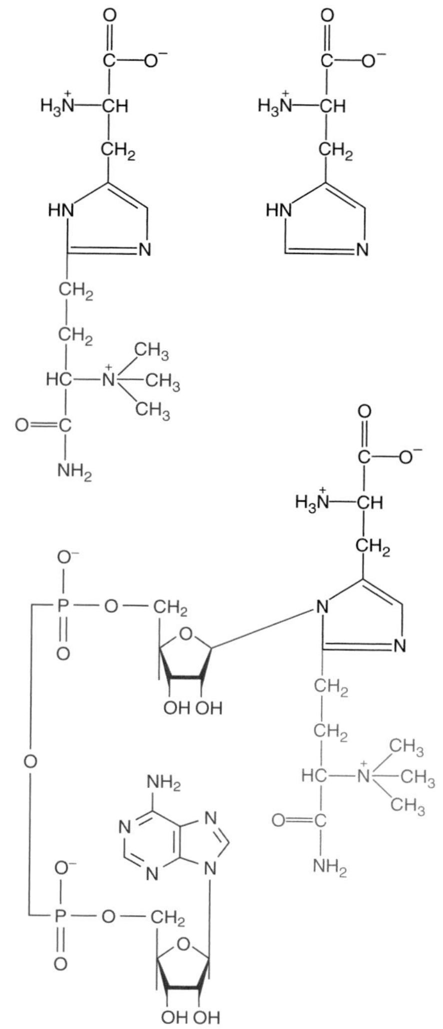

ADP-ribose

Figure 6.20 Diphthamide (top left) is a posttranslational modification of a specific residue of histidine (top right) in EF2. ADP-ribosyl-diphthamide is shown at the bottom.

Collagen Biosynthesis Requires Many Posttranslational Modifications

Collagens are the most abundant proteins in humans. The collagen family includes at least 28 different species of mostly fibrous proteins that provide the structural framework for tissues and organs, and additional collagen-like proteins share structural characteristics. Collagens undergo many posttranslational modifications that illustrate the complexity of these processes, their effect on structure and function, and modification defects that result in serious diseases. Different collagens, designated Type I, II, III, IV, and so on (see p. 105 for details of structure) are encoded on several chromosomes and expressed in different tissues. Their amino acid sequences differ, but a repeating Gly-X-Y sequence of about 1000 residues predominates. Every third residue is glycine, about one-third of the X positions are occupied by proline, and a similar number of Y positions are 4-hy-

CLINICAL CORRELATION 6.12

Absence of Posttranslational Modification: Multiple Sulfatase Deficiency (OMIM 272200)

Multiple sulfatase deficiency is a rare lysosomal storage disorder. Affected individuals develop slowly and from their second year of life lose the ability to stand, sit, or speak; physical deformities and neurological deficiencies develop and death before age 10 is usual. There is a severe lack of all sulfatase types. Degradation of sulfated molecules depends on the activity of several related sulfatases, most of which are lysosomal. Deficiencies in individual sulfatases are also known, and several distinct diseases are linked to single enzyme defects.

Multiple sulfatase deficiency arises from a defect in a posttranslational modification that is common to all sulfatase enzymes. In each case a cysteine residue of the enzyme is normally converted to C_α-formylglycine. Fibroblasts from individuals with multiple sulfatase deficiency catalyze this modification with significantly lowered efficiency, and the unmodified sulfatases are catalytically inactive.

Many biological molecules are sulfated, for example, glycosaminoglycans, steroids, and glycolipids. Poor sulfation is also responsible for diseases. For example, ineffective sulfation of the glycosaminoglycans chondroitin sulfate and keratan sulfate (p. 665) of cartilage results in major skeletal deformities.

Schmidt, B., Selmer, T., Ingendoh, A., and von Figura, K. A novel amino acid modification in sulfatases that is deficient in multiple sulfatase deficiency. *Cell* 82:271, 1995; and Dierks, T., Schmidt, B., Borissenko, L. V., Peng, J., et al. Multiple sulfatase deficiency is caused by mutations in the gene encoding the human C-formylglycine generating enzyme. *Cell* 113:435, 2003.

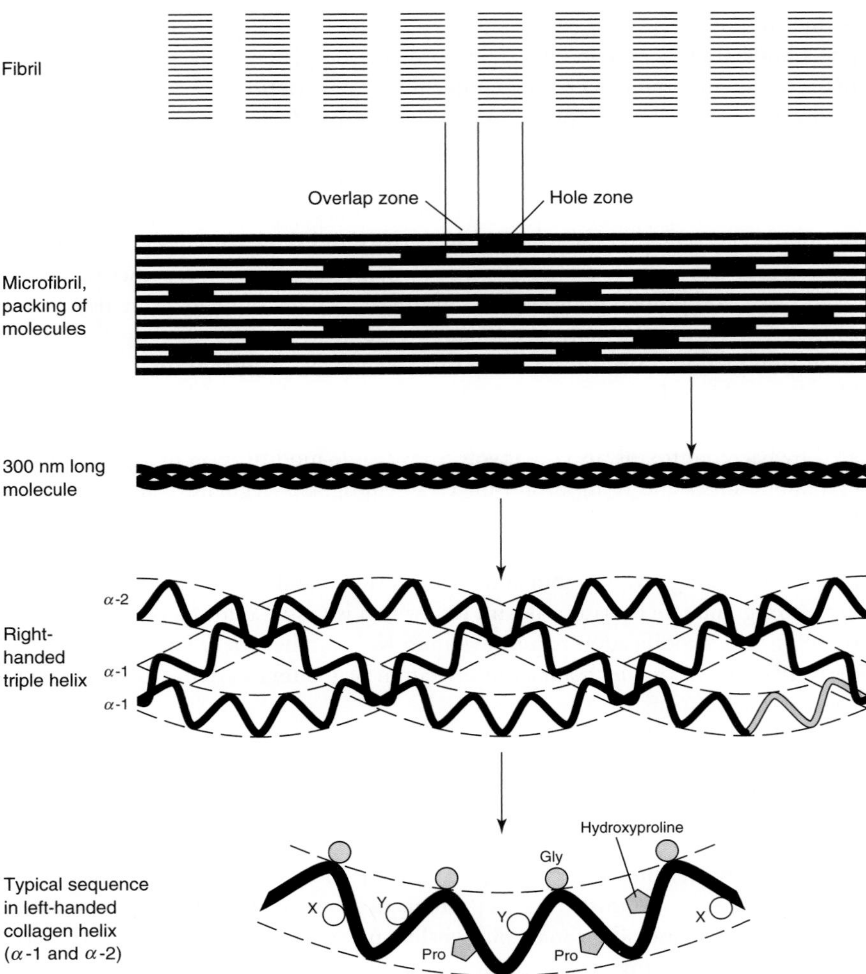

Figure 6.21 Collagen structure, illustrating (bottom to top) the regularity of primary sequence in a left-handed polyproline type II helix; the right-handed triple helix; the 300-nm molecule; and the organization of molecules in a typical fibril, within which collagen molecules are cross-linked.

droxyproline, a posttranslationally modified form of proline. Proline and hydroxyproline residues impart rigidity to their polyproline type II helix structure (Figure 6.21). Each collagen polypeptide is designated an α-chain; a collagen molecule has three α-chains intertwined in a collagen triple helix in which glycine residues occupy the center of the structure (p. 105).

Procollagen Formation in the Endoplasmic Reticulum and Golgi Apparatus

Collagen α-chain synthesis starts in the cytosol. Amino terminal signal sequences bind signal recognition particles and precursor forms, designated for example prepro α 1(I), are extruded into the ER lumen. Tissue-specific hydroxylation of proline and lysine residues occurs cotranslationally, before assembly of a triple helix. Prolyl 4-hydroxylase requires an -X-Pro-Gly- sequence (hence, 4-hydroxyproline is found only at Y positions in the -Gly-X-Y- sequence). Prolyl 3-hydroxylase modifies a smaller number of proline residues, and lysyl hydroxylase modifies some of the Y-position lysine residues. These hydroxylases require Fe^{2+} and ascorbic acid (vitamin C). Proline hydroxylation stabilizes collagen while lysine hydroxylation provides sites for interchain crosslinking and for glycosylation by specific glycosyltransferases of the ER. The extent of modification depends on the specific α-chain type and cell in which it occurs.

Chaperone-mediated **triple-helix assembly** occurs after the polypeptide chains have been completed and the signal peptides are cleaved. Carboxy-terminal globular proprotein domains fold and disulfide bonds are formed between them. This initiates winding of the triple helix from the carboxyl- toward the amino-termini. The completed triple helix, with globular proprotein domains at each end, moves to the Golgi apparatus where oligosaccharides

TABLE 6.12 • Selected Disorders in Collagen Biosynthesis and Structure

Disorder	*Collagen Defect*	*Clinical Manifestations*
Osteogenesis imperfecta 1	Decreased synthesis of type I	Long bone fractures prior to puberty
Osteogenesis imperfecta 2	Point mutations and exon rearrangements in triple helical regions	Perinatal lethality; malformed and soft, fragile bones
Ehlers—Danlos IV	Poor secretion, premature degradation of type III	Translucent skin, easy bruising, arterial and colon rupture
Ehlers—Dardos VI	Decreased hydroxylysine in types I and III	Hyperextensive skin, joint hypermobility
Ehlers—Danlos VII	Type I procollagen accumulation: N-terminal propeptide not cleaved	Joint hypermobility and dislocation
Cutis laxa (occipital horn syndrome)	Decreased hydroxylysine due to poor Cu distribution	Lax, soft skin; occipital horn formation in adolescents
Scurvy	Hydroxyproline decreased due to ascorbic acid deficiency	Poor bone growth, poor wound healing, severe bruising

are processed and matured. Sometimes tyrosine residues are modified by sulfation and some serines are phosphorylated. The completed procollagen is then released from the cell via secretory vesicles.

Collagen Maturation

Conversion of procollagen to collagen occurs extracellularly. In fibril-forming collagens (e.g., type I) the amino- and carboxyl-terminal propeptides are cleaved by separate proteases, while the carboxyl-terminal propeptides are retained in network-forming collagens (e.g., type IV). Concurrently, the triple helices assemble into fibrils by end-to-end and side-to-side aggregation, and the collagen is stabilized by extensive cross-linking. **Lysyl oxidase** converts some lysine or hydroxylysine to the reactive aldehydes, allysine, or hydroxyallysine. These residues condense with each other or with lysine or hydroxylysine residues in adjacent chains to form Schiff's base and aldol cross-links. Less well-characterized reactions can involve other residues including histidine and can link three α-chains. Defects at many of these steps are known and some of the best characterized are listed in Table 6.12 and described in Clin. Corr. 6.13.

6.7 • REGULATION OF TRANSLATION

The amount of protein in cells is regulated at the levels of transcription, translation, and degradation. Regulation of translation can be global, affecting protein synthesis in general, or it can be much more specific for individual proteins or groups of proteins. The most effective and common mechanisms for regulation of translation are at the initiation stage, and **phosphorylation** of initiation factors can serve, depending on circumstances, to either suppress or stimulate translation.

The best-understood mechanism of global regulation of translation involves reversible phosphorylation of serine 51 of the α-subunit of eIF2. Under conditions such as nutrient starvation, heat shock, and viral infection, eIF2 is phosphorylated. It remains functional, but phosphorylated eIF2–GDP binds very tightly to eIF2B, the guanine nucleotide exchange factor. eIF2B is present in limiting amounts and becomes unavailable for nucleotide exchange; hence no eIF2–GTP is available for initiation. Paradoxically, phosphorylated eIF2 stimulates translation of a transcription factor that that upregulates stress response genes.

CLINICAL CORRELATION 6.13

Defects in Collagen Synthesis

Osteogenesis Imperfecta (OMIM 166210)

Osteogenesis imperfecta is a group of at least seven clinically, genetically, or biochemically distinguishable disorders characterized by bone fragility, deformities, and low bone mass. Several variants result from mutations that produce modified α(I) chains, and the effect on severity of the symptoms varies widely. In the clearest example a deletion mutation causes absence of 84 amino acids in the α1(I) chain. The shortened α1(I) chains associate with normal α1(I) and α2(I) chains and prevent normal collagen triple helix formation, with resultant degradation of all of the chains. Three-fourths of all of the collagen molecules formed have at least one defective α1(I). Other forms of osteogenesis imperfecta result from point mutations that substitute another amino acid for a glycine. Since glycine has to fit into the interior of the collagen triple helix, these substitutions destabilize that helix.

Cheung, M. S., and Glorieux, F. H. Osteogenesis imperfecta: update on presentation and management. *Rev. Endocr. Metab. Disord.* 9:153, 2008; and Barsh, G. S., Roush, C. L., Bonadio, J., Byers, P. H., and Gelinas, R. E. Intron mediated recombination causes an α1(I) collagen deletion in a lethal form of osteogenesis imperfecta. *Proc. Natl. Acad. Sci. USA* 82:2870, 1985.

Ehlers–Danlos Syndrome, Type IV (OMIM 130050)

Ehlers–Danlos syndrome is a group of at least 10 disorders that are clinically, genetically, and biochemically distinguishable but that share manifestations of structural weaknesses in connective tissue. The usual problems are with fragility, hyperextensibility of skin, and hypermobility of joints. Type IV Ehlers–Danlos syndrome is caused by defects in type III collagen (p. 105), which is particularly important in skin, arteries, and hollow organs. Characteristics include thin, translucent skin through which veins are easily seen, marked bruising, and sometimes an appearance of aging in the hands and skin. Clinical problems arise from arterial rupture, intestinal perforation, and rupture of the uterus during pregnancy or labor. Surgical repair is difficult because of tissue fragility. Defects in type IV Ehlers–Danlos arise from changes in the structure of type III chains, due to point mutations that replace glycine residues and disrupt the collagen triple helix, and from exon skipping which shortens the polypeptide and can result in inefficient secretion, decreased stability of the collagen, and abnormal formation of type III collagen fibrils. In some cases, type III collagen accumulates in the rough ER and is degraded very slowly.

Germain, D. P. Ehlers–Danlos syndrome type IV. *Orphanet J. Rare Dis.* 2:32, 2007; and Superti-Furga, A., Gugler, E., Gitzelmann, R., and Steinmann, B. Ehlers–Danlos Syndrome type IV: A multi-exon deletion in one of the two COL 3A1 alleles affecting structure, stability, and processing of type III procollagen. *J. Biol. Chem.* 263:6226, 1988.

Deficiency of Lysylhydroxylase (OMIM 225400)

In type VI Ehlers–Danlos syndrome lysylhydroxylase is deficient. At least 20 different mutations are known that result in type I and III collagens (p. 105) with lowered hydroxylysine content and subsequently less stable cross linking of collagen fibrils. Some cross-links between lysine and allysine form but these are not as stable and do not mature as readily as hydroxylysine-containing cross links. Hydroxylysine is also a site of glycosylation, but the function of this modification is unclear. Clinical features include marked hyperextensibility of skin and joints, poor wound healing, and musculoskeletal deformities. Some patients have a mutant form of lysylhydroxylase with a higher Michaelis constant for ascorbic acid than the normal enzyme. Accordingly, they respond to high doses of ascorbic acid.

Pinnell, S. R., Krane, S. M., Kenzora, J. E., and Glimcher, M. J. A heritable disorder of connective tissue: hydroxylysine deficient collagen disease. *N. Engl. J. Med.* 286:1013, 1972; and Yeowell, H. N., and Walker, L. C. Mutations in the lysyl hydroxylase 1 gene that result in enzyme deficiency and the clinical phenotype of Ehlers–Danlos syndrome type VI. *Mol. Gen. Metab.* 71:212, 2000.

Ehlers–Danlos Syndrome, Type VII (OMIM 225410)

In this condition skin bruises easily and is hyperextensible, but the major manifestations are dislocations of major joints, such as hips and knees. Laxity of ligaments can be caused by incomplete removal of the amino-terminal propeptide of procollagen chains. One variant results from deficiency of procollagen *N*-protease known as ADAMTS2. A similar deficiency occurs in the autosomal recessive disease called dermatosparaxis of cattle, sheep, and cats, in which skin frailty is so extreme as to be lethal. In other variants the proα1(I) and proα2(I) chains lack amino acids at the cleavage site because of skipping of one exon in the genes. This prevents normal cleavage by procollagen *N*-protease.

Cole, W. G., Chan, D., Chambers, G. W., Walker, I. D., and Bateman, J. F. Deletion of 24 amino acids from the proα1(I) chain of type I procollagen in a patient with the Ehlers–Danlos syndrome type VII. *J. Biol. Chem.* 261:5496, 1986; and Tang, B. L. ADAMTS: A novel family of extracellular matrix proteases. *Int J. Biochem. Cell Biol.* 33:33, 2001.

Occipital Horn Syndrome (OMIM 304150)

Occipital horn syndrome is also known as type IX Ehlers–Danlos syndrome or cutis laxa and is related to Menkes' kinky hair syndrome. Defects in a gene encoding a trans-Golgi copper-transporting ATPase underlie these disorders, which appear as a deficiency in copper-dependent lysyl oxidase activity and thus cross-linking defects in collagen. In occipital horn syndrome characteristics include lax, soft skin and the appearance during adolescence of bony occipital horns. In Menkes' kinky hair syndrome, a variable but usually more severe disorder, patients show neurological problems such as seizures and developmental delay, failure to thrive, peculiar hair, and sometimes death in infancy or early childhood. A woman taking high doses of the copper-chelating drug, penicillamine, gave birth to an infant with an acquired Ehlers–Danlos-like syndrome, which subsequently improved. Side effects of d penicillamine therapy include poor wound healing and hyperextensible skin. Copper-deficient animals also show poor cross-linking of elastin and collagen, again because of the requirement for cuprous ion by lysyl oxidase.

Kaler, S. G. Metabolic and molecular bases of Menkes disease and occipital horn syndrome. *Pediatr. Dev. Pathol.* 1:85, 1998; and Peltonen, L., Kuivarniemie, H., Palotie, H., Horn, N., et al. Alterations of copper and collagen metabolism in the Menkes syndrome and a new subtype of Ehlers–Danlos syndrome. *Biochem.* 22:6156, 1983.

Continued on Page 247

Clinical Correlation 6.13 (continued)

Scurvy and Hydroxyproline Synthesis

Most animals, but not humans, can synthesize ascorbic acid (vitamin C). Scurvy is historically associated with long sea voyages and vitamin C deficiencies due to a lack of fresh fruit and vegetables. Clinical manifestations include poor wound healing, increased capillary fragility and severe bruising, gingival bleeding, and suppression of the orderly process of bone growth in children. Among other problems, ascorbic acid deficiency causes decreased hydroxyproline synthesis because prolyl hydroxylase requires ascorbic acid. Hydroxyproline provides additional hydrogen-bonding atoms that stabilize the collagen triple helix. Collagen containing insufficient hydroxyproline is significantly less stable than normal collagen at body temperature. Severe ascorbic acid deficiency leads secondarily to a decreased rate of procollagen synthesis.

Peterofsky, B. Ascorbate requirement for hydroxylation and secretion of procollagen: relationship to inhibition of collagen synthesis in scurvy. *Am. J. Clin. Med.* 54:1135S, 1991; and Olmedo, J. M., Yiannias, J. A., Windgassen, E. B., and Gornet, M. K. Scurvy: A disease almost forgotten. *Int. J. Dermatol.* 45:909, 2006.

Phosphorylation of eIF2 can be catalyzed by a **heme-regulated kinase** that is active in the absence of heme. This kinase is present in many cells, but has been best studied in reticulocytes that synthesize hemoglobin. A deficiency in energy supply or any heme precursor activates the kinase. A related **double-stranded RNA-dependent kinase** is activated by binding of ds-RNA that results from viral infection. Production of this kinase is also induced by interferon.

In contrast, initiation factor eIF4E is **activated** by phosphorylation of a serine residue in response to, for example, growth factors. Phosphorylation interferes with the binding of inhibitory eIF4E-binding proteins. Reversal by a protein phosphatase follows, for example, viral infection. These effects are mediated by eIF4G that is also activated by phosphorylation.

Regulation of translation of specific mRNAs is important in regulating the amounts and balance of proteins in a cell. Localization and packaging of mRNAs allows specific messages to be translated where they are needed or for mRNAs to be stored for future use. Localization within the cell relies on RNA sequences called **targeting elements** (or zip codes or localization elements), which interact with RNA binding proteins that concentrate the mRNAs at a subcellular site. Messages are sometimes packaged into storage granules during development or into **P bodies:** aggregates of translationally repressed mRNAs, regulatory proteins, and mRNA decay enzymes. These mRNAs may be degraded, but they can also be stored and returned to the cytoplasm for translation. P bodies help reprogram translation at times of stress.

A clear example of messenger-specific regulation is synthesis of the iron-binding protein ferritin. In the absence of iron a repressor protein binds to the **iron response element (IRE),** a stem–loop structure in the 5′-leader sequence of ferritin mRNA (p. 1089). This mRNA is sequestered for future use. δ-Aminolevulinic acid synthase, an enzyme of heme biosynthesis, is also regulated by a 5′-IRE in its mRNA. In contrast, more **ferritin receptor** is needed if iron is limited; its mRNA has IREs in its 3′-untranslated region. Binding of the repressor protein stabilizes the mRNA and prolongs its lifetime. Many growth-regulated mRNAs, including those for ribosomal proteins, have a polypyrimidine tract in their leader sequence. A polypyrimidine binding protein helps regulate their translation.

In addition to its roles in information transfer (mRNA) and translation (tRNA, rRNA), RNA is very important in regulation of gene expression in multiple biological functions, often in a tissue specific manner. Several classes of small RNAs bind specific groups of proteins and interact with mRNA to regulate transcription and translation in processes called **RNA silencing** and **RNA interference.** In humans almost 500 different **micro-RNA (miRNA)** genes exist. Larger hairpin RNAs are processed in the nucleus by a **Drosha** nuclease and then by a cytoplasmic endonuclease called **Dicer** to generate 21–23 nucleotide-long double-stranded RNAs. An RNA helicase separates the strands, one of which is bound by an **RNA-induced silencing complex (RISC),** the central component of which is an **Argonaute** protein. The RISC guides the miRNA to complementary sequences of mRNA, often in the 3′-untranslated region. Formation of an imperfect mRNA–miRNA duplex represses translation but does not immediately affect mRNA stability. Cooperative interaction of multiple miRNAs with an mRNA increases efficiency of inhibition. The miRNAmRNA complexes may be directed to cytoplasmic **P bodies** for storage or degradation. Dicer and the RISC also generate perfectly complementary duplexes of small RNA molecules with mRNA. These complexes result in cleavage and inactivation of the target mRNA by an RISC endonuclease called **slicer** and other nucleases that degrade the mRNA.

Several more classes of small RNA have been identified. **Small interfering RNAs (siRNAs)** are generated from double-stranded RNA; they can be endogenous or virally derived and function like miRNA. Other classes are transcribed from repetitive DNA sequences, derived from single-stranded transcripts, and may be larger than miRNAs. Chemically synthesized interfering **RNA**$_i$ mimics miRNA and is a valuable laboratory tool and potential therapeutic agent.

6.8 • PROTEIN DEGRADATION AND TURNOVER

Protein lifetimes vary widely. Lens cells are not replaced and their proteins are not recycled. Hemoglobin lasts the 120-day life of the red blood cell. Some blood-clotting proteins survive only a few days, so hemophiliacs are only protected for a short period by transfusion or injection of required factors. Diabetics require frequent insulin injections since hormone needs vary (see Clin. Corr. 21.8, p. 868). Metabolic enzymes vary quantitatively depending on need or changing situation; for example, the concentration of urea cycle enzymes changes in response to diet. Proteins are also damaged by oxidation, proteolysis, denaturation, or other irreversible modifications. Errors in translation and folding lead to nonfunctional proteins, and proteolytic processing generates nonfunctional peptides such as the C-peptide of proinsulin. Disposal mechanisms are needed; proteolysis reduces unwanted proteins to peptides and eventually amino acids. Most of these amino acids are recycled to synthesize new proteins, but some are metabolized and their degradation products excreted. Digestive proteases such as pepsin, trypsin, chymotrypsin, and elastase hydrolyze dietary protein and have no part in intracellular protein turnover, but the amino acids they generate contribute to the metabolic pool used in translation. This is particularly necessary for the essential amino acids (see Table 19.1, p. 753).

Figure 6.22 Model of the proteasome. A 20S central segment that includes 12–15 different polypeptides is made up of four stacked heptameric rings, with a hollow core that includes several proteases with different specificities. V-shaped segments at each end cap the cylinder and act in ATP-dependent substrate recognition, unfolding, and translocation into the core. The upper cap structure is shown displaced from the central segment in order to illustrate the hollow core of the cylinder. Adapted from Rubin, D., and Finley, D. *Cur. Biol.* 5:854, 1995; and Peters, J.-M. Trends *Biochem.* Sci. 19:377, 1994.

ATP-Dependent Proteolysis Occurs in Proteasomes

The best-described proteolytic pathway uses proteasomes, dumbbell-shaped structures that contain about 28 polypeptides (Figure 6.22). A cylindrical core is capped at each end by V-shaped complexes that help recognize and unfold polypeptides and transport them to the proteolytic core in an ATP-dependent mechanism. Targeting to proteasomes normally requires **ubiquitin,** a highly conserved 76 amino acid protein. Proteins are marked for degradation by **polyubiquitinylation,** as shown in Figure 6.23. Ubiquitin is activated by enzyme E1 to form a thioester; ATP is required and a transient AMP-ubiquitin complex forms. The ubiquitin is then passed to enzyme E2, and finally via one of a group of E3 multiprotein complexes to a targeted protein. Linkage of ubiquitin is through **isopeptide** bonds between ϵ-amino groups of lysine residues of the protein and the carboxyl-terminal glycine residues of ubiquitin. Several ubiquitin molecules are attached to the protein and to each other, and the polyubiquitinylated protein is brought to the proteasome and degraded; an **isopeptidase** releases intact ubiquitin for reuse. Damaged, defective, misfolded, and mutant proteins are rapidly degraded via the ubiquitin pathway, and accumulation of undegraded protein can lead to disease (Clin. Corr. 6.14). For example, deletion of one amino acid alters the stability of CFTR protein (see Clin. Corr. 6.7, p. 231) in cystic fibrosis. Selection of native proteins for degradation depends on E3 enzyme specificity; both conformation and amino acid sequence are important. Destabilizing PEST sequences (rich in Pro, Glu, Ser, and Thr) occur in several short-lived proteins, and an ubiquitin-interacting motif that binds ubiquitin and sometimes also promotes ubiquitinylation has been identified. Another determinant is the identity of the amino-terminal amino acid. In accord with the **N-end rule,** proteins with different amino-terminal residues are degraded at widely differing rates, and the lifetime of a protein can be modified by incorporation of an N-terminal destabilizing residue from an aminoacyl-tRNA donor. Misfolded proteins in the ER are usually degraded through the ubiquitin pathway. Chaperones and lectins in the ER keep misfolded N-linked glycoproteins from aggregating and target them for retrotranslocation back to the cytosol in the process called **ERAD** (**ER-associated degradation**).

Ubiquitin-dependent proteolysis plays a major role in the regulation of cellular events. Cyclins (p. 1016) and receptor protein tyrosine kinases (p. 509) are involved in control of cell division. Ubiquitin-dependent destruction of a cyclin allows cells to pass from the M phase into G1, while degradation of receptors prevents signal transduction and halts cell proliferation. Other proteins degraded by ubiquitin-dependent proteolysis include transcription factors, the p53 tumor suppressor and other oncoproteins, protein kinases, and immune system and other cell surface receptors.

Polyubiquitinylation is necessary to signal proteolysis; for example, polyubiquitinyl-p53 is degraded, while monoubiquitinyl-p53 is targeted for export from the nucleus. Defects in ubiquitin-dependent proteolysis lead to disease (Clin. Corr. 6.15). Other roles of ubiquitin are described in A Closer Look 6.4.

Like proteasomes, the complex prokaryotic proteases Lon and Clp and FtsH require ATP hydrolysis for their action, but ubiquitin is absent in prokaryotes. These proteases also are assembled into barrel-like structures and the ATPase powers unfolding of the substrate protein. An N-end rule applies through a smaller number of N-terminal destabilizing amino acids.

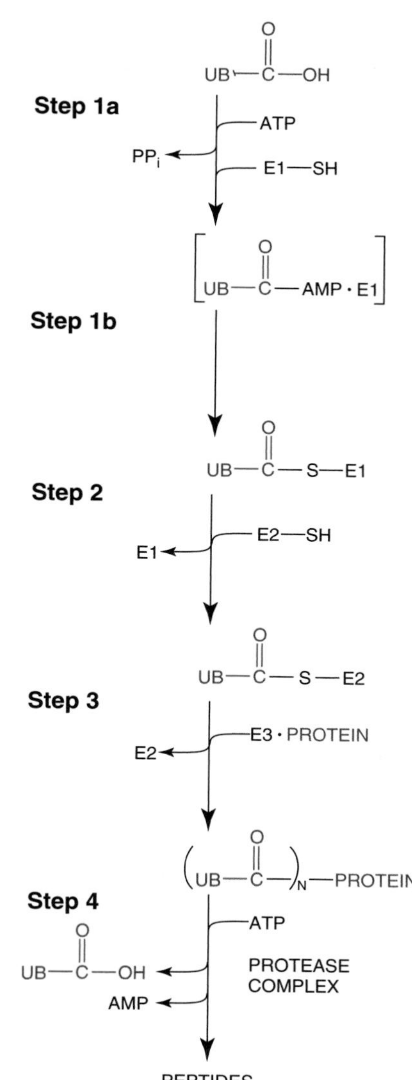

Figure 6.23 ATP and ubiquitin-dependent protein degradation. Ubiquitin is activated in a two-step reaction involving formation of a transient mixed anhydride of AMP and the carboxy-terminus of ubiquitin (step 1a), followed by generation of a thioester with enzyme E1 (step 1b). Enzyme E2 forms a thioester with ubiquitin (step 2) and serves as a donor in E3-catalyzed transfer of ubiquitin to a targeted protein (step 3). Several more ubiquitin molecules are attached to lysine residues of the ubiquitin and/or other lysines of the targeted protein at this stage. Polyubiquitinylated protein is degraded by ATP-dependent proteolysis (step 4); ubiquitin is not degraded, and can reenter the process at Step 1.

Intracellular Digestion of Some Proteins Occurs in Lysosomes

Intracellular digestion of proteins from the extracellular environment occurs within protease-rich **lysosomes.** Material that does not permeate the plasma membrane is imported by endocytosis. In **pinocytosis** large particles, molecular aggregates, or other material in the extracellular fluid are engulfed. Macrophages ingest bacteria and dead cells by this mechanism. **Receptor-mediated endocytosis** uses cell surface receptors to bind specific molecules at pits in the cell surface that are coated with the multisubunit protein clathrin. Invagination of the plasma membrane and the ligand-bound receptors forms intracellular clathrin-coated vesicles, one fate of which is fusion with a lysosome and degradation of the contents. Some intracellular protein turnover may also occur within lysosomes, and under some conditions significant amounts of cellular material can be degraded by lysosomes. For example, serum starvation of fibroblasts or starvation of rats leads to the lysosomal degradation of a subpopulation of cellular proteins. Recognition of a specific peptide sequence is involved, again indicating that the lifetime of a protein is ultimately encoded in its sequence.

Other Proteolytic Systems

Apoptosis (p. 1010), also called programmed cell death, follows on proteolytic activation of proteases known as **caspases** (cysteine aspartyl proteases). Stresses such as inflammation, cell damage, or a deficiency of growth factors can either directly activate caspases or cause mitochondria to release cytochrome c and other factors that activate a caspase cascade and result in the rapid degradation of cellular proteins. A caspase in the ER acts in ER stress-induced apoptosis, while other Golgi and ER proteases degrade misfolded proteins and peptide fragments that arise during maturation of proteins in the secretory pathway. Proteins are also exported from the ER for degradation by proteasomes.

Calcium-dependent thiol proteases, also called calpains, are present in most cells and function in, for example, cellular proliferation, differentiation, and progression through the cell cycle through selective and limited proteolysis. Calpains also function with proteasomes in protein turnover, and other roles are indicated but less well established. It appears that protein degradation is as complex and important a process as protein biosynthesis.

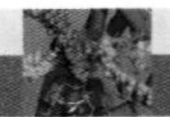

CLINICAL CORRELATION 6.14

Protein Misfolding and Aggregation: Huntington Disease and Alzheimer Disease

Misfolded proteins are usually degraded, as occurs with mutant CFTR protein in cystic fibrosis (see Clin. Corr. 6.7, p. 231). Several neurological diseases illustrate the result of cellular aggregation of misfolded proteins or their partial-degradation products.

Huntington disease (OMIM 143100) is an autosomal dominant neurodegenerative disorder in which patients show spasmodic movements in voluntary muscles, personality changes, and cognitive dysfunction; all grow more severe over time. Huntington disease is associated with accumulation of protein aggregates derived from **huntingtin,** a primarily cytoplasmic protein. Huntington is essential for embryonic development and is involved in transcription regulation, vesicle trafficking, and synapsis; mechanisms are not well understood. Variants of huntingtin are involved in Huntington disease; a CAG codon within the gene is expanded, resulting in polyglutamine repeats of 35 to 180 residues within huntingtin. Longer polyglutamine repeats are associated with more severe disease. Protein and peptides derived from huntingtin accumulate in inclusions in the nucleus, and transcription is affected. Mitochondrial function is also disturbed and reactive oxygen species increase, and the ubiquitin-proteasome pathway is impaired. Several other neurodegenerative disorders are also linked to polyglutamine tract expansion and protein deposition

Alzheimer disease (OMIM 104300) is characterized by progressive loss of memory and cognition, and it is most common in elderly people. Mitochondrial dysfunction in neural cells is prevalent. Alzheimer disease is characterized by accumulation of intraneuronal and extracellular bundles and filaments that form plaques. The major component of the plaque is β-amyloid, a 39–43 residue peptide derived from a larger amyloid precursor protein. Other components include the GM1 ganglioside, a microtubule-associated protein Tau, and other highly ubiquitinylated but protease-resistant components in compacted and tangled fibers that appear to overwhelm the proteasome degradative pathway. Disturbed calcium ion regulation may underlie neuronal death.

Amyotrophic lateral sclerosis (ALS or Lou Gehrig's disease) is a neurodegenerative disease characterized by progressive degeneration of motor neurons and accumulation of ubiquinated proteins in cellular inclusions. Superoxide dismutase 1 and TAR DNA-binding protein have both been identified in inclusions in both familial and sporadic forms of the disease.

Parkinson disease (see Clin. Corr. 19.19, p. 785) is also linked to accumulations of undegraded protein aggregates. Misfolding of the presynaptic protein alpha-synuclein (α-Syn) in the endoplasmic reticulum (ER) of neurons leads to the formation of cytoplasmic Lewy Bodies in which α-Syn is a major component. ER-associated degradation of proteins is impaired. In early onset Parkinson disease Lewy bodies are usually absent and mutations in parkin protein are common (Clin. Corr. 6.15). Prion diseases are also associated with protein aggregation (see Clin. Corr. 3.6, p. 114).

Walker, F. O. Huntington's disease. *Lancet* 369:218, 2007; Goedert, M. and Spillantini, M. G. A century of Alzheimer's disease. *Science* 314:777, 2006; Marx, J. A new take on Tau. *Science* 316:1416, 2007; Kabashi, E., and Durham, H. D. Failure of protein quality control in amyotrophic lateral sclerosis. *Biochim. Biophys. Acta* 1762:1038, 2006; Lozano, A. M., and Kalia, S. K. New movement in Parkinson's. *Sci. Am.* 293:68, 2005; and Cooper, A. A., Gitler, A. D., Cashikar, A., Haynes, C. M., et al. α-Synuclein blocks ER-Golgi traffic and Rab1 rescues neuron loss in Parkinson models. *Science* 313:324, 2006.

CLINICAL CORRELATION 6.15

Defects in the Ubiquitin-Proteasome System

The following are diseases linked to malfunctions of the ubiquitin-proteosome pathway.

Liddle syndrome (OMIM 177200): A mutation in a subunit of the renal epithelial sodium channel interferes in binding of Nedd4, an E3 ligase that mediates its degradation. Channel accumulation results in excessive Na^+ resorption and severe hypertension.

Autosomal recessive juvenile Parkinsonism (OMIM 605909): A small percentage of patients have an early onset familial form of Parkinson disease. Many have mutations in the parkin protein causing loss of E3 enzyme activity and accumulation of parkin substrates. A second cause of familial Parkinson disease is linked to a mutation in an ubiquitin-recycling enzyme, possibly resulting in a shortage of free ubiquitin.

Breast and ovarian cancer: Heterozygous mutations in BRCA1 (the breast cancer susceptibility locus) are seen in some cases of breast and ovarian cancer. BRCA1 protein is an E3 enzyme whose activity is abolished by mutations found in familial breast cancer.

HPV-induced cervical cancer: Infection with some strains of human papilloma virus can lead to invasive cervical cancer. A virus-encoded oncoprotein E6 interacts with the cellular p53 tumor suppressor protein; an E3 enzyme recognizes the complex, leading to p53 degradation via the ubiquitin-proteasome pathway.

Reinstein, E., and Ciechanover, A. Narrative review: Protein degradation and human diseases: The ubiquitin connection. *Ann. Intern. Med.* 145:676, 2006.

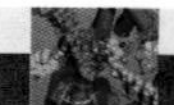
A CLOSER LOOK 6.4

Ubiquitin and SUMO

Ubiquitin has important roles beyond protein degradation. Reversible linkage of ubiquitin to histones H2A and H2B is unrelated to turnover since the proteins are stable, but modification affects chromatin structure and transcription. Ubiquitin participates in chromosome alignment and segregation, in DNA repair, and in regulating transcription in the inflammatory response. Ubiquitination is a signal for endocytosis and a sorting signal directing contents to lysosomes. Often these functions employ ubiquitin linkages that are different from those in the proteasome pathway. Small ubiquitin-related modifier (SUMO) proteins, whose conformation and ligation to target proteins are very similar to that of ubiquitin, also function in transcriptional regulation and other aspects of cellular regulation, sometimes with effects opposite to that of ubiquitin.

Schnell, J. D., and Hicke, L. Non-traditional functions of ubiquitin and ubiquitin-binding proteins. *J. Biol. Chem.* 278:35857, 2003; Mukhopadhyay, D., and Riezman, H. Proteasome-independent functions of ubiquitin in endocytosis and signaling. *Science* 315:201, 2007; and Liu, B., and Shuai, K. Regulation of the sumoylation system in gene expression. *Curr. Opin. Cell Biol.* 20:288, 2008.

Bibliography

Ribosomes and Transfer RNA

Daviter, T., Gromadski, K. B., and Rodnina, M. V. The ribosome's response to codon–anticodon mismatches. *Biochimie.* 88:1001, 2006.

Hale, S. P., Auld, D. S., Schmidt, E., and Schimmel, P. Discrete determinants in transfer RNA for editing and aminoacylation. *Science* 276:1250, 1997.

Hausmann, C. D., and Ibba, M. Aminoacyl-tRNA synthetase complexes: Molecular multitasking revealed. *FEMS Microbiol. Rev.* 32:705, 2008.

Hou, Y. M., Francklyn, C., and Schimmel, P. Molecular dissection of a transfer RNA and the basis for its identity. *Trends Biochem. Sci.* 14:233, 1989.

Korostelev, A., and Noller, H. F. The ribosome in focus: new structures bring new insights. *Trends Biochem. Sci.* 32:434, 2007.

Korostelev, A., Trakhanov, S., Laurberg, M., and Noller, H. F. Crystal structure of a 70S ribosome-tRNA complex reveals functional interactions and rearrangements. *Cell* 126:1065, 2006.

Lue, S. W., and Kelley, S. O. An aminoacyl-tRNA synthetase with a defunct editing site. *Biochem.* 44:3010, 2005.

Nomura, M. The role of RNA and protein in ribosome function: a review of early reconstitution studies and prospects for future studies. *Cold Spring Harbor Symp. Quant. Biol.* 52:653, 1987.

Rodnina, M. V., Beringer, M., and Wintermeyer, W. How ribosomes make peptide bonds. *Trends Biochem. Sci.* 32:20, 2007.

Silvian, L. F., Wang, J., and Steitz T. A. Insights into editing from an Ile-tRNA synthetase structure with tRNA and mupirocin. *Science* 285:1074, 1999.

Yusupova, G. Z., Yusopov, M. M., Cate, J. H. D., and Noller, H. F. The path of messenger RNA through the ribosome. *Cell* 106:233, 2001.

Protein Biosynthesis

Agarwal, R. K., Heagle, A. B., Penczek, P., Grassucci, R. A., and Frank, J. EF-G-dependent GTP hydrolysis induces translocation accompanied by large conformational changes in the 70S ribosome. *Nature: Struct. Biol.* 6:643, 1999.

Barrell, B., et al. Different pattern of codon recognition by mammalian mitochondrial tRNAs. *Proc. Natl. Acad. Sci. USA* 77:164, 1980.

Berchtold, H., Reshetnikova, L., Reiser, C. O., Shirmer, N. K., Sprinzl, M., and Hilgenfeld, R. Crystal structure of active elongation factor Tu reveals major domain rearrangements. *Nature* 365:126, 1993.

Beringer, M., and Rodnina, M. V. The ribosomal peptidyl transferase. *Molec. Cell* 26:311, 2007.

Burkhardt, N., Jünemann, R., Spahn, C. M. T., and Nierhaus, K. M. Ribosomal tRNA binding sites: Three site models of translation. *Crit. Revs. Biochem. Mol. Biol.* 33:95, 1998.

Dong, Z., and Zhang, J.-T. Initiation factor eIF3 and regulation of mRNA translation, cell growth, and cancer. *Crit. Revs. Oncol. Hematol.* 59:169, 2006.

Kapp, L., and Lorsch, J. R. The molecular mechanics of eukaryotic translation. *Annu. Rev. Biochem.* 73:657, 2004.

Hinnebusch, A. G. eIF3: A versatile scaffold for translation initiation complexes. *Trends Biochem. Sci.* 31:553, 2006.

Kieft, J. S. Viral IRES RNA structures and ribosome interactions. *Trends Biochem. Sci.* 33:274, 2008.

Komar, A. A., and Hatzoglou, M. Internal ribosome entry sites in cellular mRNAs: Mystery of their existence. *J. Biol. Chem.* 280:23425, 2005.

Lang, K., Erlacher, M., Wilson, D. N., Micura, R., and Polacek, N. The role of 23S ribosomal RNA residue A2451 in peptide bond synthesis revealed by atomic mutagenesis. *Chem. Biol.* 15:485, 2008.

Laurberg, M., Asahara, H., Korostelev, A., Zhu, J., Trakhanov, S., and Noller, H. F. Structural basis for translation termination on the 70S ribosome. *Nature* 454:852, 2008.

Sonenberg, N. eIF4E, the mRNA cap-binding protein: From basic discovery to translational research. *Biochem. Cell Biol.* 86:178, 2008.

Song, H., Mugnier, P., Das, A. K., Webb, H. M., Evans, D. R., Tuite, M. F., et al. The crystal structure of human eukaryotic release factor eRF1–mechanism of stop codon recognition and peptidyl-tRNA hydrolysis. *Cell* 100:311, 2000.

Squires, J. E., and Berry, M. J. Eukaryotic selenoprotein synthesis: mechanistic insight incorporating new factors and new functions for old factors. *Life* 60:232, 2008.

Weinger, J. S., Parnell, K. M., Dorner, S., Green, R., and Strobel, S. R. Substrate-assisted catalysis of peptide bond formation by the ribosome. *Nat. Struct. Molec. Biol.* 11:1101, 2004.

Yates, S. P., Jørgensen, R., Andersen, G. R., and Merrill, A. R. Stealth and mimicry by deadly bacterial toxins. *Trends Biochem. Sci.* 31:123, 2006.

Ziff, E. B. Transcription and RNA processing by the DNA tumour viruses. *Nature* 287:491, 1980.

Protein Targeting and Posttranslational Modification

Baines, A. C., and Zhang, B. Receptor-mediated protein transport in the early secretory pathway. *Trends Biochem. Sci.* 32:381, 2007.

Bigotti, M. G., and Clarke, A. R. Chaperonins: the hunt for the group II mechanism. *Arch. Biochem. Biophys.* 474:331, 2008.

Buck, T. M., Wright, C. M., and Brodsky, J. L. The activities and function of molecular chaperones in the endoplasmic reticulum. *Semin. Cell Dev. Biol.* 18:751, 2007.

Byers, P. H. Collagens: building blocks at the end of the development line. *Clin. Genet.* 58:270, 2000.

Chuderland, D., Konson, A., and Seger, R. Identification and characterization of a general nuclear translocation signal in signaling proteins. *Molec. Cell* 31:850, 2008.

Corda, D., and Di Girolamo, M. Functional aspects of protein mono-ADP-ribosylation. *EMBO J.* 22:1953, 2003.

Egea, P. F., Stroud, R. M., and Walter, P. Targeting proteins to membranes: structure of the signal recognition particle. *Curr. Opin. Struct. Biol.* 15:213, 2005.

Dahms, N. M., Lobel, P., and Kornfeld, S. Mannose 6-phosphate receptors and lysosomal enzyme targeting. *J. Biol. Chem.* 264:12115, 1989.

Danpure, C. J. How can products of a single gene be localized to more than one subcellular compartment? *Trends Cell Biol.* 5:230, 1995.

Hartl, F. U., and Hayer-Hartl, M. Molecular chaperones in the cytosol: from nascent chain to folded protein. *Science* 295:1852, 2002.

Hassa, P. O., Haenni, S. S., Elser, M., and Hottiger, M. O. Nuclear ADP-ribosylation reactions in mammalian cells: Where are we today and where are we going? *Microbiol. Mol. Biol. Rev.* 70:789, 2006.

Jensen, R. E., and Dunn, C. D. Protein import into and across the mitochondrial inner membrane: role of the TIM23 and TIM22 translocons. *Biochim. Biophys. Acta* 1592:25, 2002.

Khoshnoodi, J., Cartailler, J.-P., Alvares, K., Veis, A., and Hudson, B. G. Molecular recognition in the assembly of collagens: Terminal noncollagenous domains are key recognition modules in the formation of triple helical protomers. *J. Biol. Chem.* 281:38117, 2006.

Nagradova, N. Enzymes catalyzing protein folding and their cellular functions. *Curr. Protein Pept. Sci.* 8:273, 2007.

Platta, H. W., and Erdmann, R. Peroxisomal dynamics. *Trends Cell Biol.* 17:474, 2007.

Rapaport, D. Finding the right organelle. Targeting signals in outer membrane proteins. *EMBO Rep.* 4:948, 2003.

Robins, S. P. Biochemistry and functional significance of collagen cross-linking. *Biochem. Soc. Trans.* 35:849, 2007.

Schlenstedt, G. Protein import into the nucleus. *FEBS Lett.* 389:75, 1996.

Wickner, W., and Schekman, R. Protein translocation across biological membranes. *Science* 310:1452, 2005.

Yan, A., and Lennarz, W. J. Unraveling the mechanism of protein N-glycosylation. *J. Biol. Chem.* 280:3121, 2005.

Zhou, A., Webb, G., Zhu, X., and Steiner, D. F. Proteolytic processing in the secretory pathway. *J. Biol. Chem.* 274:20745, 1999.

Regulation of Translation

Bartel, D. P. MicroRNAs: genomics, biogenesis, mechanism, and function. *Cell* 116:281, 2004.

Berkhout, B., and Jeang, K.-T. RISCy business: micro RNAs, pathogenesis, and viruses. *J. Biol. Chem.* 282:26641, 2007.

Czaplinsky, K., and Singer, R. H. Pathways for mRNA localization in the cytoplasm. *Trends Biochem. Sci.* 31:687, 2006.

Farazi, T. A., Juranek, S. A., and Tuschl, T. The growing catalog of small RNAs and their association with distinct Argonaute/Piwi family members. *Development* 135:1201, 2008.

Parker, J. S., and Barford, D. Argonaute: a scaffold for the function of short regulatory RNAs. *Trends Biochem. Sci.* 31:622, 2006.

Parker, R., and Sheth, U. P bodies and the control of mRNA translation and degradation. *Molec. Cell.* 25:635, 2007.

Pierrat, O. A., Mikitova, V., Bush, M. S., Browning, K. S., and Doonan, J. H. Control of protein translation by phosphorylation of the mRNA 5′-cap-binding complex. *Biochem. Soc. Trans.* 35:1634, 2007.

Sonenberg, N., and Hinnebusch, A. G. New modes of translational control in development, behavior, and disease. *Mol. Cell* 28:721, 2007.

Yamasaki, S., and Anderson, P. Reprogramming mRNA translation during stress. *Curr. Opin. Cell Biol.* 20:222, 2008.

Protein Turnover and Proteasomes

Anelli, T., and Sitia, R. Protein quality control in the early secretory pathway. *EMBO J.* 27:315, 2008.

Ciechanover, A. The ubiquitin-proteasome proteolytic pathway. *Cell* 79:13, 1994.

DeMartino, G. N., and Slaughter, C. A. The proteasome, a novel protease regulated by multiple mechanisms. *J. Biol. Chem.* 274:22123, 1999.

Fesik, S. W., and Shi, Y. Controlling the caspases. *Science* 294:1477, 2001.

Koppen, M., and Langer, T. Protein degradation within mitochondria: Versatile activities of AAA proteases and other peptidases. *Crit. Rev. Biochem. Mol. Biol.* 42:221, 2007.

Licht, S., and Lee, I. Resolving individual steps in the operation of ATP-dependent proteolytic molecular machines: from conformational changes to substrate translocation and processivity. *Biochem.* 47:3595, 2008.

Löwe, J., Stock, D., Jap, B., Zwickl, P., Baumeister, W., and Huber, R. Crystal structure of the 20S proteasome from the archaeon T. acidophilum at 3.4Å resolution. *Science* 268:33, 1995.

Mogk, A., Schmidt, R., and Bukau, B. The N-end rule pathway for regulated proteolysis: prokaryotic and eukaryotic strategies. *Trends Cell Biol.* 17:165, 2007.

Mukhopadhyay, D., and Riezman, H. Proteasome-independent functions of ubiquitin in endocytosis and signaling. *Science* 315:201, 2007.

Nakatsukasa, K., and Brodsky, J. L. The recognition and retrotranslocation of misfolded proteins from the endoplasmic reticulum. *Traffic* 9:861, 2008.

Perrin, B. J., and Huttenlocher, A. Calpain. *Int. J. Biochem. Cell Biol.* 34:722, 2002.

Perry, J. J. P., Tainer, J. A. and Boddy, M. N. A SIM-ultaneous role for SUMO and ubiquitin. *Trends Biochem. Sci.* 33:201, 2008.

Rogers, S., Wells, R., and Rechsteiner, M. Amino acid sequences common to rapidly degraded proteins: The PEST hypothesis. *Science* 234:364, 1986.

Tamura, T., Cormier, J. H., and Hebert, D. N. Sweet bays of ERAD. *Trends Biochem. Sci.* 33:298, 2008.

Taylor, R. C., Cullen, S. P., and Martin, S. J. Apoptosis: controlled demolition at the cellular level. *Nat. Rev. Mol. Cell Biol.* 9:231, 2008.

Key Terms

translation
monocistronic
polycistronic
codons
anticodons
initiation codons
termination codons
genetic code
initiation factors
elongation factors
termination factors
aminoacyl–tRNA synthetase
point mutations
silent mutations
frameshift mutations
suppressor mutations
termination mutations
ribosome
polysome.
peptidyl transferase

posttranslational modification
chaperone
endoplasmic reticulum
signal peptidase
signal peptide
signal recognition particle
translocon
N-linked glycosylation
O-linked glycosylation
complex-type oligosaccharide
glycosyl transferase
high-mannose oligosaccharide
anchoring sequence
endoplasmic reticulum associated degradation
protein targeting
collagen triple helix
translational regulation
RNA interference
RNA silencing
RNA-induced silencing complex
proteasome
ubiquitin
N-end rule

Questions • CAROL N. ANGSTADT

Multiple-Choice Questions

1. Degeneracy of the genetic code denotes the existence of
   A. multiple codons for a single amino acid.
   B. codons consisting of only two bases.
   C. base triplets that do not code for any amino acid.
   D. different systems in which a given triplet codes for different amino acids.
   E. codons that include one or more of the unusual bases.

2. In the formation of an aminoacyl-tRNA,
   A. ADP and Pi are products of the reaction.
   B. aminoacyl adenylate appears in solution as a free intermediate.
   C. aminoacyl-tRNA synthetase is believed to recognize and hydrolyze incorrect aminoacyl-tRNAs it may have produced.
   D. separate aminoacyl-tRNA synthetases exist for every amino acid in the functional protein.
   E. there is a separate aminoacyl-tRNA synthetase for every tRNA species.

3. During initiation of protein synthesis,
   A. methionyl-tRNA appears at the A site of the 80S initiation complex.
   B. eIF3 and the 40S ribosomal subunit participate in forming a preinitiation complex.
   C. eIF2 is phosphorylated by GTP.
   D. the same methionyl-tRNA is used as is used during elongation.
   E. a complex of mRNA, 60S ribosomal subunit, and certain initiation factors is formed.

4. During the elongation stage of eukaryotic protein synthesis,
   A. the incoming aminoacyl-tRNA binds to the P site.
   B. a new peptide bond synthesized by peptidyl transferase requires GTP hydrolysis.
   C. the peptidyl- tRNA is translocated to a different site on the ribosome.
   D. streptomycin can cause premature release of the incomplete peptide.
   E. peptide bond formation occurs by the attack of the carboxyl group of the incoming amino acyl tRNA on the amino group of the growing peptide chain.

5. Formation of mature insulin includes all of the following except
   A. removal of a signal peptide.
   B. folding into a three-dimensional structure.
   C. disulfide bond formation.
   D. removal of a peptide from an internal region.
   E. γ-carboxylation of glutamate residues.

6. Streptomycin binds the small subunit of prokaryotic ribosomes and
   A. causes premature release of the incomplete peptide.
   B. prevents binding of the 40S and 60S subunits.
   C. interferes with initiation of protein synthesis.
   D. inhibits peptidyl transferase activity.
   E. acts as an N-glycosidase.

Questions 7 and 8: Cystic fibrosis is a frequent genetic disease of Caucasians. The CF gene codes for a protein called the cystic fibrosis transmembrane conductance regulator (CFTR) which functions as a cAMP-regulated chloride channel. The protein has two membrane-spanning domains, two domains that interact with ATP, and one regulatory domain. The most common defect is in the gene for one of the ATP binding domains. The result is a protein that does not fold correctly in the endoplasmic reticulum, is not properly glycosylated, and is not transported to the cell surface. Rather, it is degraded in the cytosol within proteasomes. Drugs that foster chaperone interaction with the mutant protein are a potential therapeutic approach.

7. Chaperones
   A. are always required to direct the folding of proteins.
   B. when bound to protein increase the rate of protein degradation.
   C. usually bind to strongly hydrophilic regions of unfolded proteins.
   D. sometimes maintain proteins in an unfolded state to allow passage through membranes.
   E. foster aggregation of proteins into plaques.

8. Targeting a protein to be degraded within proteasomes usually requires ubiquitin. In the function of ubiquitin all of the following are true except
   A. ATP is required for activation of ubiquitin.
   B. a peptide bond forms between the carboxyl terminal of ubiquitin and an ϵ-amino group of a lysine.
   C. linkage of a protein to ubiquitin does not always mark it for degradation.
   D. the N-terminal amino acid is one determinant of selection for degradation.
   E. ATP is required by the enzyme that transfers the ubiquitin to the protein to be degraded.

Questions 9 and 10: Collagen is unusual in its amino acid composition and requires a wide variety of posttranslational modifications to convert it to a functional molecule. Because of the complexity of collagen synthesis, there are many diseases, resulting in structural weaknesses in connective tissue, caused by defects in the process. Scurvy leads to a less stable collagen lacking sufficient hydroxyproline.

9. 4-Hydroxylation of specific prolyl residues during collagen synthesis requires all of the following except
   A. Fe^{2+}.
   B. a specific amino acid sequence at the site of hydroxylation.
   C. ascorbic acid.
   D. co-hydroxylation of lysine.
   E. individual α chains, not yet assembled into a triple helix.

10. Much of procollagen formation occurs in the endoplasmic reticulum and Golgi apparatus which requires signal peptide. All of the following statements about targeting a protein for the ER are true except
   A. signal peptide usually has a positively charged N-terminus and a stretch of hydrophobic amino acids.
   B. signal peptide emerging from a free ribosome binds signal recognition particle (SRP).
   C. signal peptide is usually cleaved from the protein before the protein is inserted into the ER membrane.
   D. docking protein is actually an SRP receptor and serves to bind the SRP to the ER.
   E. SRP and docking protein do not enter the ER lumen but are recycled.

Questions 11 and 12: A number of diseases occur because of defects in targeting signals that direct proteins to particular organelles. I-Cell disease (mucolipidosis II), characterized by severe psychomotor retardation and skeletal abnormalities, arises from a defect in targeting proteins to lysosomes. Mistargeting can also cause problems. In primary hyperoxaluria type I, an enzyme that normally functions in peroxisomes, because of mutations that create a mitochondrial targeting signal, is localized in the mitochondria instead.

11. I-Cell patients lack the enzyme that transfers N-acetylglucosamine phosphate to high mannose type oligosaccharides on the protein. This *N*-linked glycosylation
   A. occurs only after the protein has been completely translated and folded.
   B. does not require any specific amino acid sequence.
   C. occurs by transfer of an oligosaccharide chain from a dolichol phosphate carrier to the protein.
   D. has the oligosaccharide chain attached to serine or threonine.
   E. contains only mannoses in the chain.

12. The targeting signal for mitochondria is
   A. a hydrophobic anchoring sequence.
   B. a positively charged amphipathic α-helix.
   C. mannose 6-phosphate.
   D. one or two clusters of basic amino acids.
   E. a carboxy-terminal Ser-Lys-Leu (SKL) sequence.

Problems

13. The most common mutation in cystic fibrosis is a deletion of three consecutive bases in the gene for one of the ATP binding domains. How would the protein from this gene most likely be related to the normal protein?

14. Some proteins may be degraded in lysosomes. How does this differ from protein degradation within proteasomes?

Answers

1. **A** This is the definition of degeneracy. B and E are not known to occur, although sometimes tRNA reads only the first two bases of a triplet (wobble), and sometimes unusual bases occur in anticodons. C denotes the stop (nonsense) codons. D is a deviation from universality of the code, as found in mitochondria.
2. **C** Bonds between a tRNA and an incorrect smaller amino acid may form but are rapidly hydrolyzed. A and B: ATP and the amino acid react to form an enzyme-bound aminoacyl adenylate; PPi is released into the medium. D: Some amino acids, such as hydroxyproline and hydroxylysine, arise by co- or posttranslational modification. E: An aminoacyl-tRNA synthetase may recognize any of several tRNAs specific for a given amino acid.
3. **B** This then binds the mRNA. A: Methionyl-$tRNA_i^{met}$ appears at the P site. C: Phosphorylation of eIF-2 inhibits initiation. D: Methionyl-$tRNA_c^{met}$ is used internally. E: mRNA associates first with the 40S subunit.
4. **C** This is necessary to free the A site for the next incoming tRNA. A: Incoming amino acyl tRNA binds to the A site. B: Peptide bond formation requires no energy source other than the aminoacyl-tRNA. D: Streptomycin inhibits formation of the prokaryotic 70S initiation complex and causes misreading of the genetic code. E: The electron pair of the amino group carries out a nucleophilic attack on the carbonyl carbon.
5. **E** γ-Carboxylation is of special importance in several blood-clotting proteins but not in insulin formation. A: Preproinsulin is inserted into the ER. B: All proteins, except fibrous ones, have to fold into a three-dimensional structure. C: A proinsulin folds and forms disulfide bonds before the chain is cleaved. D: This is called the C-peptide.
6. **C** By altering interactions of tRNA, mRNA, and ribosomal subunit, it interferes with initiation and also causes misreading. A: Puromycin, resembling aminoacyl-tRNA, does this. B: These are eukaryotic subunits. D: Chloramphenicol does this. E: This is the action of ricin and related toxins.
7. **D** This is only one of many functions chaperones serve. A: Many proteins spontaneously fold correctly. B: Misfolded proteins are recognized and rapidly degraded. C: Chaperones bind to hydrophobic regions of unfolded proteins. E: Misfolded proteins are more likely to do this.
8. **E** ATP is required in the ubiquitin activation and the protease steps but not here. B and D: These are both correct. C: Linkage to histones does not result in their degradation.
9. **D** Lysine is hydroxylated but by a different enzyme. A and C: Prolyl hydroxylase requires both Fe^{+2} and ascorbic acid. B: The sequence is -X-Pro-Gly-. E: Hydroxylation is a cotranslational event.
10. **C** Signal peptidase is located on the luminal surface of the ER. A: These are common features along with a polar segment that signal peptidase recognizes. B, D, and E: These are all essential features of the process.
11. **C** The chain is linked to the amide of asparagine in an Asn-X-Thr/Ser sequence. A, B, and D: These are characteristics of the O-linked type. E: *N*-Acetylglucosamine is also present.
12. **B** This is recognized by a mitochondrial receptor. A: This would embed the protein in the ER membrane. C: This is the signal for the lysosome. D: This targets the nucleus. E: This is the signal for peroxisomes.
13. Deletion of an entire codon means that particular amino acid is not present in the protein. It is not usual for the deletion of three nucleotides to span two codons so you would not expect a frameshift mutation.
14. Degradation in proteasomes usually requires ubiquitin adding to the protein. Proteins are taken into lysosomes by endocytosis, and enzymes present in the lysosomes degrade them. Specific extracellular proteins bind to cell surface receptors in clathrin-coated pits. Invagination of the membrane and ligand-bound receptors forms vesicles which may fuse with lysosomes. Again the proteins are degraded by the lysosomal enzymes.

7

Recombinant DNA and Biotechnology

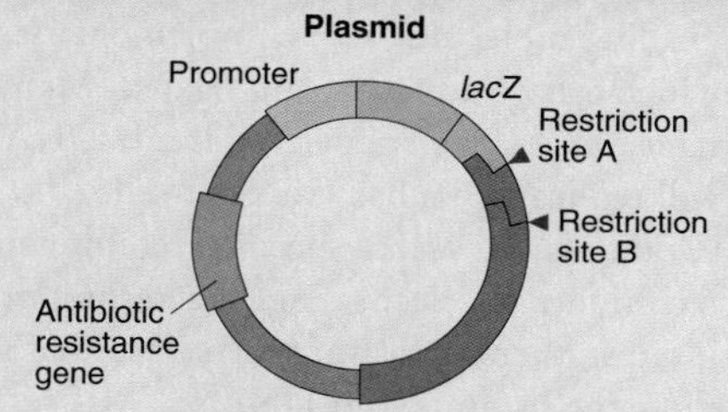

Gerald Soslau

Professor, Drexel University College of Medicine

CLINICAL CORRELATIONS

Key Concepts

- The polymerase chain reaction (PCR) is used to determine nucleic acid sequences of genes, their regulatory sites, and their products.
- Restriction endonucleases (RE) generate DNA molecules for synthesis of recombinant DNA molecules. DNA molecules from two different species can be cleaved with the same RE-yielding molecules with complementary single-stranded ends, which can be annealed and ligated. Many copies of the recombinant DNA can be generated if one DNA species is a vector that will replicate in a bacterium or cell.
- Cloning a cell with a specific recombinant molecule requires techniques to select a single bacterium or cell with the DNA of interest, then replicating the cell so that all daughter cells have the same recombinant DNA.
- Genomic libraries are generated with the multitude of recombinant DNA molecules that have pieces of DNA from a whole genome linked to a vector and transformed into bacteria or cells. Specific clones can be selected.
- Nucleic acid hybridization techniques with labeled DNA or RNA probes can detect polymorphic DNA patterns of restriction endonuclease digested genomic DNA from different individuals, specific RNA species, and cell clones carrying a specific recombinant DNA.
- Reverse transcription generates cDNA from cellular mRNA. A cDNA library consists of all the cDNAs produced from total cellular mRNA.
- Different vectors can be recombined with foreign DNA with the number of base pairs ranging from a few thousand to hundreds of thousands. Small DNA inserts can represent cDNA of an expressed gene while much larger inserts are required to carry a genomic gene with introns and exons along with upstream and downstream regulatory sites.
- Expression vectors are DNA species that combine with functional foreign "genes" such that when introduced into a bacterium or cell the recombinant "gene" is transcribed and translated.
- Site-directed mutagenesis with cloned recombinant DNA can result in the selective loss of a DNA region, the addition or loss of one or a few nucleotides, or the selective change of a single nucleotide. Altered "genes" are utilized to study the structure and function of the encoded protein.
- Regulation of gene expression in diseased or altered metabolic states can be studied by recombinant DNA technologies, including antisense nucleic acid technologies, introduction of normal genes for altered genes, and gene destruction in whole animals (knockout animals).

7.1 • INTRODUCTION

By 1970, the stage was set for modern molecular biology based on the studies of numerous scientists in the previous 30 years during which ignorance of what biochemical entity orchestrated the replication of life forms with such fidelity gave way to a state where sequencing and manipulation of the expression of genes would become commonplace. The relentless march toward a full understanding of gene regulation under normal and pathological conditions has moved with increasing rapidity since then. Identification, purification, and characterization of restriction endonucleases permitted the development of recombinant DNA methodologies. Development of DNA sequencing revealed the secrets within the organization of diverse biological genomes with the complete sequencing of numerous prokaryotes and eukaryotes including the human genome. The delineation of all of the human genes and their regulatory sequences should greatly enhance our understanding of many genetic diseases. This knowledge along with our technological advances to manipulated genes and gene expression should also open new avenues to regulate or cure these diseases. Hundreds of clinical trials in gene therapy involving many patients have been initiated. Genetic diseases now identified and to be identified should eventually be curable by gene replacement therapy when the technical roadblocks are surmounted. If one looks at the enormous advances made in molecular biology in just the past three decades, it is reasonable to believe the when will not be that far off. The old challenge confronting scientists was how to sequence the human genome; the new challenge is how to effectively manipulate that knowledge to benefit mankind.

7.2 • POLYMERASE CHAIN REACTION

Prior to 1987 the analysis of the sequence and function of a selected region of DNA required relatively large amounts of the purified DNA segment. The rapid production of large quantities of a specific DNA sequence took a leap forward with the development of the **polymerase chain reaction (PCR)**. This requires two nucleotide oligomers that hybridize to the complementary DNA strands in a region of interest. The oligomers serve as primers for a DNA polymerase that extends each strand. Repeated cycling of the PCR yields large amounts of each DNA molecule of interest in a matter of hours as opposed to days and weeks required for cloning techniques.

The amplification of a specific DNA sequence by PCR is accomplished with purified DNA or with a complex mixture of DNA. The principles of the reaction are shown in Figure 7.1. The nucleotide sequence of the DNA to be amplified must be known, or it must be cloned in a vector (p. 277) for which the sequence of the flanking DNA has been established. The product of PCR is a double-stranded DNA (dsDNA) molecule and the reaction is completed in each cycle when all of the template molecules are copied. In order to initiate a new round of replication, the sample is heated to melt the dsDNA and, in the presence of excess oligonucleotide primers, cooled to permit hybridization of the single-stranded template with free oligomers. DNA replication will start in the presence of DNA polymerase and all four dNTPs. Heating to about 95°C as required for melting DNA inactivates most DNA polymerases, but a heat-stable polymerase, termed Taq DNA polymerase isolated from *Thermus aquaticus,* is now employed and obviates the need for fresh polymerase being added after each cycle. This has permitted the automation of PCR with each DNA molecule capable of being amplified one million-fold.

A unique temperature stable DNA polymerase—Phusion™ High-Fidelity DNA polymerase—was commercially developed for PCR under routine and more demanding PCR applications. This enzyme was constructed by the fusion of a unique dsDNA-binding domain to a *Pyrococcus*-like proofreading polymerase. The processivity of this fused enzyme is about twice the Taq DNA polymerase and an error rate 50-fold lower. Also, unlike Taq DNA polymerase the enzyme is not inhibited by any quantity of blood. Therefore, genomic DNA can be amplified directly from blood drops without the costly and time-consuming purification steps required with the Taq enzyme.

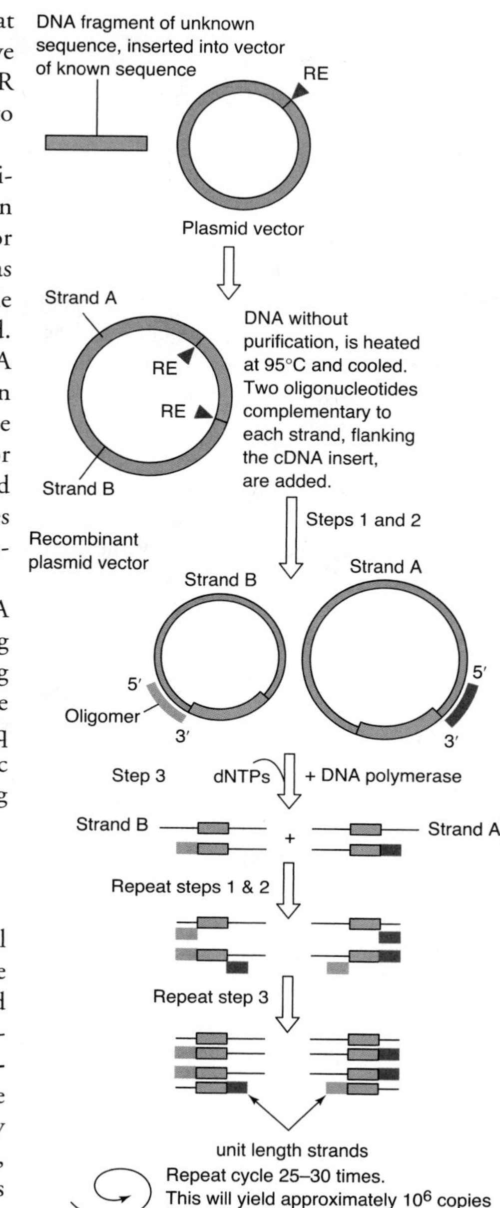

Figure 7.1 Polymerase chain reaction (PCR). A DNA fragment of unknown sequence is inserted into a vector of known sequence by normal recombinant methodology. The recombinant DNA of interest does not need to be purified. The DNA is heated to 95°C to dissociate the double strands and cooled in the presence of excess amounts of two different complementary oligomers that hybridize to the known vector DNA sequences flanking the foreign DNA insert. Only recombinant single-stranded DNA species can serve as templates for DNA replication yielding double-stranded DNA fragments of foreign DNA bounded by the oligomer DNA sequences. The heating-replication cycle is repeated many times to rapidly produce large amounts of the original foreign DNA. The DNA fragment of interest can be purified from the PCR mixture by cleaving it with the original restriction endonuclease (RE), electrophoresing the DNA mixture through an agarose gel, and eluting the band of interest from the gel.

Nested PCR

When the DNA to be amplified is present in very low concentrations relative to the total DNA in the sample, the DNA region of interest along with other spurious sequences can be amplified. In this situation, the specificity of the amplification can be enhanced by nested PCR. After conducting the first PCR with one set of primers for 10–20 cycles, a small aliquot is removed for a second PCR. However, this is conducted with primers that are **complementary** to the template DNA just downstream of the first primers or "nested" between the original set of primers. This process amplifies the DNA region of interest twice with a greatly enhanced specificity. PCR has many applications in gene diagnosis; forensic investigations, for which only a drop of dried blood or a single hair is available; and evolutionary studies on preserved biological material (Clin. Corr. 7.1). In 2008 a new technology, touch DNA/PCR, was applied to clear the parents of the murder of their young daughter, Jon Benet Ramsey. This methodology could take one or a few cells adhering to any surface touched by a person (in this case Jon Benet's pants touched by the murderer) as a sufficient source of DNA to amplify by PCR and analyze by DNA fingerprinting. After more than a decade of doubt, the results finally cleared the parents in this horrific and highly publicized case.

Quantitative Real-Time PCR

Quantitative real-time PCR (qRT-PCR) couples the traditional PCR amplification reaction with the simultaneous quantification of the DNA synthesized after each cycle. This technique allows one to calculate the number of copies of a DNA sequence within a genome. One can also use the method to determine the relative level of expression of a gene in a cell by initiating the reaction with RT-PCR (reverse transcriptase–PCR) producing cDNA as the template for PCR. In each case the quantification of DNA product must be compared to an internal housekeeping gene whose level of expression is nearly equal in all cells being studied. There are currently a few variations to the techniques employed, some being more rigorous in the calculated quantification than others, but only the most straightforward process will be described here.

CLINICAL CORRELATION 7.1

The Polymerase Chain Reaction

A. PCR in Screening for Human Immunodeficiency Virus

The use of PCR to amplify minute quantities of DNA has revolutionized the detection and analysis of DNA species. With PCR it is possible to synthesize sufficient DNA for analysis. Conventional detection and identification of the human immunodeficiency virus (HIV), for example by Southern blot–DNA hybridization and antigen analysis, is labor intensive and expensive and has low sensitivity. An infected individual, with no sign of AIDS (acquired immunodeficiency syndrome), may test false negative for HIV by these procedures. Early detection of HIV infections is crucial to initiate treatment or monitor the progression of the disease. In addition, a sensitive method is required to ensure that blood contributed by donors does not contain HIV. PCR amplification of potential HIV DNA sequences within DNA isolated from an individual's white blood cells permitted the detection of infection prior to appearance of antibodies, the so-called seronegative state. With the advent of new combination antiviral (protease inhibitors) treatments for HIV the potential for viral mutations leading to drug resistance has been reduced. However, some patients do present with multidrug resistance. A common viral genome mutation, L9OM, leads to protease inhibitor resistance. The viral gag-pro region was amplified and sequenced using L9OM allele specific PCR with samples from 15 patients early in their diseased state. Results were then compared to samples analyzed after the patients failed drug treatment. The L9OM viruses were a minority of the total viruses in the early samples. However, these mutant species became the dominant species in the plasma of drug resistant patients. It appears that the emergence of the dominant L9OM containing viruses occurred on multiple occasions in many of the patients as determined by Bayesian evolutionary analysis. Early detection of the conversion of drug sensitive to drug resistant viruses by PCR analysis could help determine clinical treatment of these patients.

B. Nested PCR to Detect Microchimerism

Donor leukocytes transferred to patients during blood transfusion have been shown to survive in the recipient's peripheral blood. The significance, if any, of this microchimeric population of leukocytes remains to be resolved. One of the best ways to detect donor-derived cells in the recipient's blood is to use PCR detection of polymorphisms in the HLA–DR region of the major histocompatibility complex (MHC). A nested PCR assay was at least 100-fold more sensitive than a standard PCR assay. However, because of the increased sensitivity, nonspecific products may appear due to mispriming events that are generally associated with pseudogenes. As such it is essential to establish a baseline pattern with pre-transfusion blood samples. Once potential false positives are excluded with these baseline patterns, the detection of donor leukocytes is greatly enhanced.

Kwok, S., and Sninsky, J. J. Application of PCR to the detection of human infectious diseases, in *PCR technology*, 235. In Erlich, H. A. (Ed.), *PCR: Technology Principles and Applications for DNA Amplification,* 2d ed. New York: Stockton, 1989; and Kapoor, A., Shapiro, B., Shafer, R. W., Shulman, N., et al. Multiple independent origins of a protease inhibitor resistance mutation in salvage therapy patients. *Retrovirology* 5:7, 2008.

Carter, A. S., Cerundolo, L., Bunce, M., Koo, D. D. H., et al. Nested polymerase chain reaction with sequence-specific primers typing for HLA-A, -B, and -C alleles: Detection of microchimerism in DR-matched individuals. *Blood* 94:1471, 1999.

Although RT-PCR/PCR usually uses forward and reverse primers that generate DNA products in the 300–600-bp range, qRT-PCR more often employs primers that are only about 100 bases apart. The method described here takes advantage of the fact that the chromofluor, SYBR green, only binds to double-stranded DNA and not to single-stranded DNA. When it is bound to double-stranded DNA, it fluoresces many times more brightly than when it is free in solution. DNA is amplified from either DNA or cDNA templates in the presence of SYBR green in a PCR instrument that measures the fluorescence of the SYBR green–DNA complex after each cycle. Therefore, the increase in fluorescence with each cycle is proportional to the real time synthesis of the amplified DNA. A plot of the log of the fluorescence units versus cycle number will yield a linear relationship during the logarithmic amplification phase of PCR. The linear part of the curve is compared to a known internal housekeeping standard for quantification of the DNA/cDNA product. This technique is employed to identify the differential expression of specific genes in normal versus diseased tissues where the gene product may have a role in the development of abnormal cells (Clin. Corr. 7.2).

7.3 • RESTRICTION ENDONUCLEASE, RESTRICTION MAPS, AND DNA SEQUENCING

Restriction Endonucleases Selectively Hydrolyze DNA

Restriction endonucleases are capable of selectively dissecting DNA molecules of many sizes and origins into smaller fragments. They confer some protection on bacteria against invading viruses (bacteriophage). Bacterial DNA sequences normally recognized by a restriction endonuclease are protected from cleavage in host cells by methylation of bases within the palindrome. The unmethylated viral DNA is recognized as foreign and is hydrolyzed. Numerous Type II restriction endonucleases are now commercially available (see p. 58 for discussion of restriction endonuclease activities).

Restriction endonucleases permit construction of a **restriction map**, in which the site of cleavage within the DNA is identified. Purified DNA species that contain susceptible sequences are subjected to restriction endonuclease cleavage. By regulating the time of exposure of the purified DNA for cleavage, a population of DNA fragments of different sizes is generated. Separation of these fragments by agarose gel electrophoresis allows for the construction of a restriction map; an example is presented in Figure 7.2. The sequential use of different restriction endonucleases has permitted a detailed restriction map of numerous circular DNA species including bacterial plasmids, viruses, and mitochondrial DNA. The method is equally amenable to linear DNA fragments that have been purified to homogeneity.

Restriction Maps Permit Routine Preparation of Defined Segments of DNA

Restriction maps may yield little information as to the genes or regulatory elements within the various DNA fragments. They are used to demonstrate sequence diversity of organelle DNA, such as mitochondrial DNA, within species (Clin. Corr. 7.3). They are

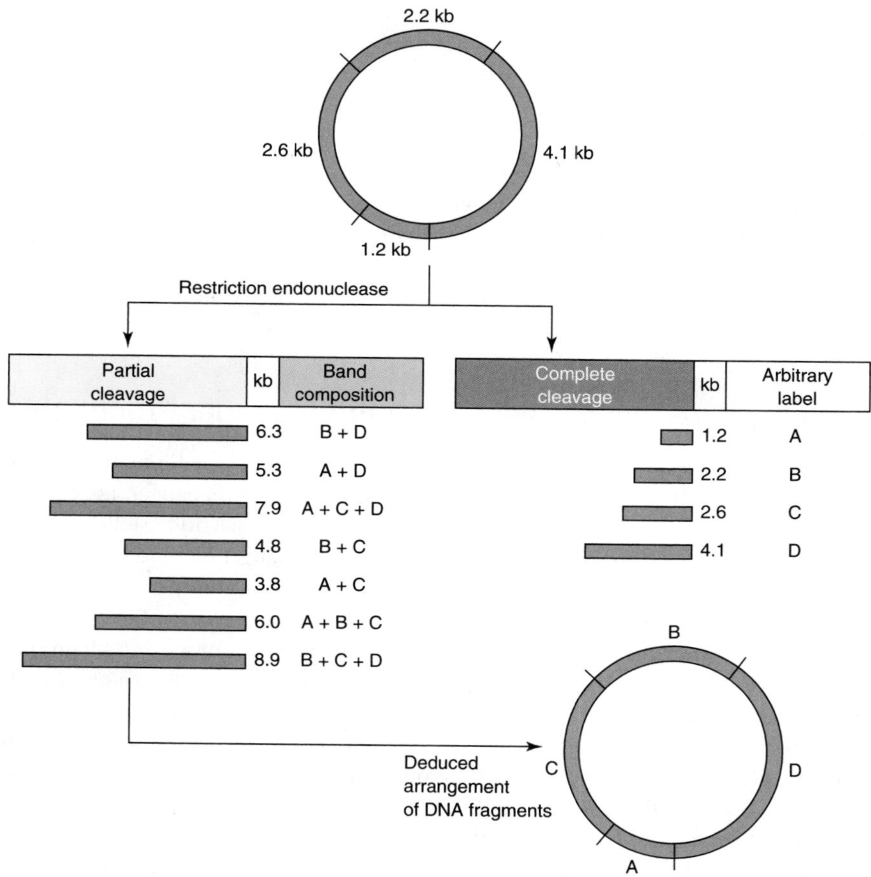

Figure 7.2 Restriction endonuclease mapping of DNA. Purified DNA is subjected to restriction endonuclease digestion for varying times which generates partially to fully cleaved DNA fragments. The fragments are separated by agarose gel electrophoresis and stained with ethidium bromide. The bands are visualized with a UV light source and photographed. The size of the fragments is determined by the relative migration through the gel as compared to co-electrophoresed DNA standards. The relative arrangement of each fragment within the DNA molecule can be deduced from the size of the incompletely hydrolyzed fragments.

CLINICAL CORRELATION 7.2

Quantitative Real-Time PCR (qRT–PCR in the Analysis of a Gene Associated with Prostate Cancer

Prostate cancer is the predominant malignancy found in North American males. There is no known cure for prostate cancer once it has metastasized. Therefore, it is very important to identify biomarkers which are associated with the metastatic disease and which may serve as new therapeutic targets. Two cell lines were developed from a prostate cancer patient that when transplanted into immunodeficient mice, either had metastatic potential or were nonmetastatic. The differential analysis of genes expressed in the two cell lines detected one gene, ASAP1, in the metastatic cell line that had not previously been associated with the metastatic process in prostate cancer. qRT-PCR confirmed that ASAP1 (encodes an ADP-ribosylation factor GTPase-activating protein) was up-regulated in the metastatic cell line. Furthermore, the inhibition of expression of ASAP1 in prostate cancer cells by small interfering RNA (siRNA; see p. 201) greatly reduced the metastatic potential of these cells by *in-vitro* analysis. This gene may be an important target for the clinical treatment of metastatic prostate cancer.

Lin, D., Watahiki, A., Bayani, J., Zhang, F., et al. ASAP1, a gene at 8q24, is associated with prostate cancer metastasis. *Cancer Res.* 68:4352, 2008.

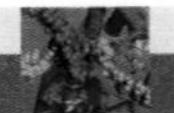

CLINICAL CORRELATION 7.3

Restriction Mapping and Evolution

In the past, evolutionary studies of species have depended solely on anatomical changes observed in fossil records and on carbon dating. These studies are being supported by the molecular analysis of the sequence and size of selected genes or whole DNA molecules. Alterations in a selected DNA molecule from different species can be rapidly assessed by restriction mapping which requires a pure preparation of DNA. Mammalian mitochondria contain a closed circular DNA molecule of 16,569 bps that can rapidly be purified from cells. The mitochondrial DNA (mtDNA) can be employed directly for the study of evolutionary changes in DNA without the need of cloning a specific gene.

mtDNA has been purified from the Guinea baboon, rhesus macaque, guenon, and human and cleaved with 11 different restriction endonucleases. Restriction maps were aligned relative to the direction and the nucleotide site where DNA replication is initiated. A comparison of restriction endonuclease sites permitted the calculation of the degree of divergence in nucleotide sequence between species. The rate of base substitution (calculated from the degree of divergence versus the time of divergence) has been about 10-fold greater than in the nuclear genome. This high rate of mutation of the readily purified mtDNA molecule makes this an excellent model to study evolutionary relationships between species.

Analysis of mtDNA can be used to study the migration patterns of people who settled in diverse geographic regions but possess the same restriction enzyme-generated patterns (restriction fragment length polymorphism [RFLP]). Native Americans have been grouped into four major haplogroups based upon their mtDNA restriction patterns. These groups are thought to be descended from the Clovis people (Paleoindian) who migrated from Asia and Siberia and entered North America, just before the major ice age period. The timing of this migratory event is quite variable from study to study. However, analysis of mtDNA, amplified from fossil feces (human coprolites) in Oregon, appears to prove that humans populated the Americas before the so-called Clovis people. The mtDNA sequences represent the Native American founding haplogroups A_2 and B_2 and the samples predate the Clovis remains by at least 1000 years.

Brown, W. M., George, M., Jr., and Wilson, A. C. Rapid evolution of animal mitochondrial DNA. *Proc. Natl. Acad. Sci. USA* 76:1967, 1979; Schurr, J. G. Mitochondrial DNA and the people of the new world. *Amer. Sci.* 88:246, 2000; Gilbert, M. T., Jenkins, D. L., Gotherstrom, A., Naveran, N., et al. DNA from pre-Clovis human coprolites in Oregon, North America. *Science* 320:786, 2008; and Balter, M. DNA from fossil feces breaks Clovis barrier. *Science* 320:37, 2008.

also used to detect deletion mutations in which a DNA fragment from the parental strain migrates as a smaller fragment in the mutated strain. Most importantly, the restriction endonucleases cut DNA into defined homogeneous fragments that can be readily purified. These maps are crucial for cloning and for sequencing genes and their flanking DNA regions.

Interrupted Enzymatic Cleavage Method: Sanger Procedure

In the late 1970s two different sequencing techniques were developed, one by A. Maxam and W. Gilbert, the chemical-cleavage approach, and the other by F. Sanger, the enzymatic approach. Both procedures employed labeling of a terminal nucleotide, followed by the separation and detection of generated oligonucleotides. The Sanger method has become the method of choice due to the relative ease of the procedure and its ability to sequence longer stretches of DNA (400 bases) relative to the Maxam and Gilbert method (250 bases).

The **Sanger procedure** is based on the random termination of a DNA chain during enzymatic synthesis. The technique depends on the dideoxynucleotide analog of each of the four normal nucleotides (Figure 7.3) being incorporated into a growing DNA chain by DNA polymerase and blocking further elongation. The ribose of the **dideoxynucleoside triphosphate (ddNTP)** lacks the OH group at both the 2′ and 3′ positions whereas dNTP lacks only the OH group at the 2′ position. Thus, the ddNTP incorporated into the growing chain cannot form a phosphodiester bond with another dNTP because the 3′ position of the ribose does not contain a OH group. The growing DNA molecule is terminated at random points by including in the reaction system the normal dNTP and the ddNTP (e.g., dATP and ddATP) at concentrations such that the two nucleotides compete for incorporation. Identification of DNA fragments requires labeling of the 5′-end of the DNA molecules or the incorporation of labeled nucleotides during synthesis. The technique, outlined in Figure 7.4, is best conducted with pure single-stranded DNA; however, denatured double-stranded DNA can also be used. The DNA to be sequenced is

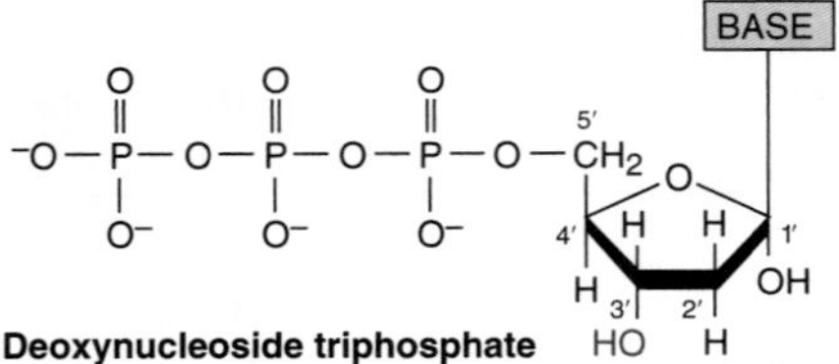

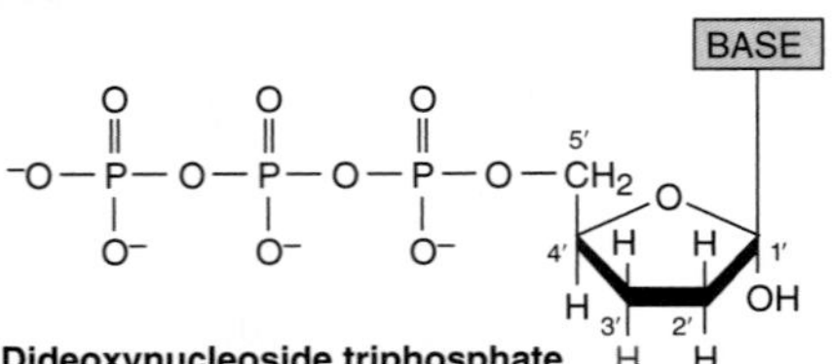

Figure 7.3 Structure of deoxynucleoside triphosphate and dideoxynucleoside triphosphate. The 3′-OH group is lacking on the ribose component of the dideoxynucleoside triphosphate (ddNTP). This molecule can be incorporated into a growing DNA molecule through a phosphodiester bond with its 5′-phosphates. Once incorporated the ddNTP blocks further synthesis of the DNA molecule since it lacks the 3′-OH acceptor group for an incoming nucleotide.

frequently isolated from a recombinant single-stranded bacteriophage (p. 276) in which a region that flanks the DNA of interest contains a sequence that is complementary to a universal primer. The primer can be labeled with either ^{32}P or ^{35}S nucleotide. Primer extension is accomplished with one of several DNA polymerases; one with great versatility is a genetically engineered form of the bacteriophage T 7 DNA polymerase. The reaction mixture, composed of the target DNA, labeled primer, and all four dNTPs, is divided into four tubes, each containing a different ddNTP. ddNTPs are randomly incorporated during the enzymatic synthesis of DNA and cause chain termination. Since the ddNTP is present in the reaction tube at a low level relative to the corresponding dNTP, termination of DNA synthesis occurs randomly at all possible complementary sites of the DNA template. This yields DNA molecules of varying sizes, labeled at the 5′-end that can be separated by polyacrylamide gel electrophoresis. The labeled species are detected by X-ray autoradiography, and the sequence is read.

Initially, this method required a single-stranded DNA template, production of a specific complementary oligonucleotide primer, and a relatively pure preparation of the Klenow fragment of *E. coli* DNA polymerase I. These difficulties have been overcome and modifications have simplified the procedure.

The PCR and the Sanger methods are combined for **direct sequencing** of small DNA regions of interest. The double-stranded PCR product is employed directly as template. Conditions are set so that one strand of melted DNA (template) anneals with the primer in preference to reannealing with the complementary strand. Sequencing then follows the standard dideoxy chain termination reaction (typically with Sequenase replacing the Klenow polymerase) with synthesis of random length chains occurring as extensions of the PCR primer. This method has been successfully employed for diagnosis of genetic disorders (Clin. Corr. 7.4).

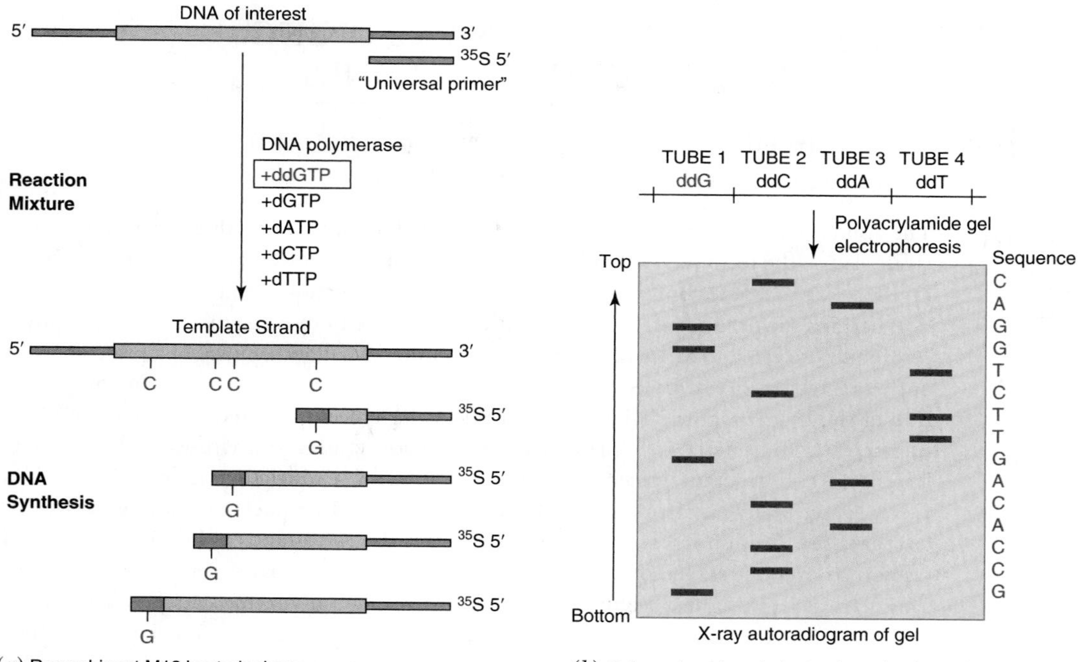

(*a*) Recombinant M13 bacteriophage

(*b*) Polyacrylamide gel electrophoresis of reaction mixture

Figure 7.4 **The Sanger dideoxynucleoside triphosphate method to sequence DNA.** The DNA region of interest is inserted into bacteriophage DNA molecule. Replicating bacteriophage produces a single-stranded recombinant DNA molecule that is readily purified. The known sequence of the bacteriophage DNA downstream of the DNA insert serves as a hybridization site for an end-labeled oligomer with a complementary sequence, a universal primer. Extension of this primer is catalyzed with a DNA polymerase and all four dNTPs plus one ddNTP, for example ddGTP. Synthesis stops whenever a ddNTP is incorporated into the growing molecule. Note that the ddNTP competes for incorporation with the dNTP. This generates end-labeled DNA fragments of all possible lengths that are separated by electrophoresis. The DNA sequence can then be determined from the electrophoretic patterns.

CLINICAL CORRELATION 7.4

Direct Sequencing of DNA for the Diagnosis of Bloom Syndrome: A Genetic Disorder

Early methodologies to determine the sequence of DNA regions or genes that may be associated with genetic disorders have been costly, time consuming, and all too often inaccurate. Direct sequencing of PCR-amplified genomic DNA circumvents these shortcomings. Bloom syndrome (BS) (OMIM 210900) is a rarely encountered autosomal recessive disease. Patients with BS present with all types of cancer far more often than the general population. Cells from BS patients are characterized by a high level of genetic instability. The current diagnosis of BS is based on the detection of a high level (~10-fold) of sister chromatid exchanges in cells of the patient. Direct genomic DNA sequencing was employed to screen for mutations in the BLM (Bloom syndrome) gene of suspected BS patients. This method detected four mutations in the BLM gene. All of the mutations (three were frameshift and one was nonsense mutation) resulted in the introduction of a stop codon. It is postulated that this methodological approach will allow for a direct correlation of mutations to the clinical status of the patient.

Amor-Gueret, M., Dubois-d'Enghien, C., Lauge, A., Onciercq-Delic, R., et al. Three new BLM gene mutations associated with Bloom syndrome. *Genet. Testing* 12:257, 2008.

Dye-Terminator Sequencing

The development of the dye-terminator sequencing method has greatly simplified, sped up, and reduced the cost of DNA sequencing. The basic principles of the Sanger enzymatic method are employed; however, each ddNTP is fluorescently labeled with a different colored fluor. Since each colored ddNTP terminates the reaction yielding different-sized oligonucleotides with a different fluoresent color, it is not necessary to run four different reaction tubes. A single reaction is run with the products separated by capillary electrophoresis and sequenced by an automated, high-throughput DNA sequencer. A sample scan of a chromatogram and derived sequence is shown in Figure 7.5.

The dye-terminator sequencing method is, like the Sanger method, able to sequence only about 400 bases at a time. New methods are being developed and explored that would greatly lower the cost and concominantly greatly increase the size of DNA chains being sequenced. The ultimate goal of these improved methods is the ability to rapidly sequence the entire genome of individuals at costs below $1000. Achieving this goal would allow clinicians and scientists to correlate altered or mutated genes with genetic diseases, predict clinical progression of diseased states, and develop drug or genetic therapies for each individual.

The ability to routinely sequence the individual's genome is likely to be achieved several years from now. However, other technologies are currently available to detect human genetic variations that give clues to an individual's potential to develop specific diseased states. It is postulated that as many as 15 million sites within the human genome may vary from one individual to another as single-nucleotide polymorphisms (SNP). Using array technologies (p. 295), one can search for as many as 500,000 SNPs at a time in hundreds to thousands of individuals. By October 2007 a human haplotype map (HapMap) of over 3.1 million SNPs had been generated. The appearance of specific SNPs or combinations of SNPs is being correlated to diseases such as type 1 diabetes and autism.

7.4 • RECOMBINANT DNA, CLONING, AND SELECTION OF CLONES

DNA from Different Sources Can Be Ligated to Form a New DNA Species: Recombinant DNA

The ability to cleave a population of DNA molecules selectively with restriction endonucleases led to the technique for joining two different DNA molecules termed **recombinant DNA**. This procedure combined with techniques for replication, separation, and identification permits the production of large quantities of purified DNA fragments. The combined techniques, referred to as recombinant DNA technology, allow removal of a piece of DNA out of a larger complex molecule, such as the genome of a virus or human, and amplification of the DNA fragment. Recombinant DNAs have been prepared that combine DNA fragments from bacteria with fragments from humans, viruses with viruses, and so on. The joining of two different pieces of DNA is achieved by a restriction endonuclease and a DNA ligase. Many restriction endonucleases, varying in their nucleotide sequence specificity, are used (Section 7.3, p. 258). Some hydrolyze the two strands of DNA in a staggered fashion to produce sticky, or cohesive, ends (Figure 7.6) whereas others cut both strands symmetrically to produce a blunt end. A specific restriction enzyme cuts DNA at exactly the same sequence site regardless of source of DNA (bacteria, plant, mammal, etc.). A DNA molecule may have none or numerous recognition sites for a particular restriction endonuclease. The staggered cut results in DNA fragments with single-stranded ends. When different DNA fragments generated by the same restriction endonuclease are mixed, their single-stranded ends can hybridize or anneal together. DNA ligase joins the two fragments to produce a recombinant DNA molecule.

The DNA fragments that contain blunt ends can also be ligated but with much less efficiency. The efficiency is increased by enzymatically adding poly(dA) tails to one species of DNA and poly(dT) tails to the second species. The fragments with complementary tails can then be annealed and ligated.

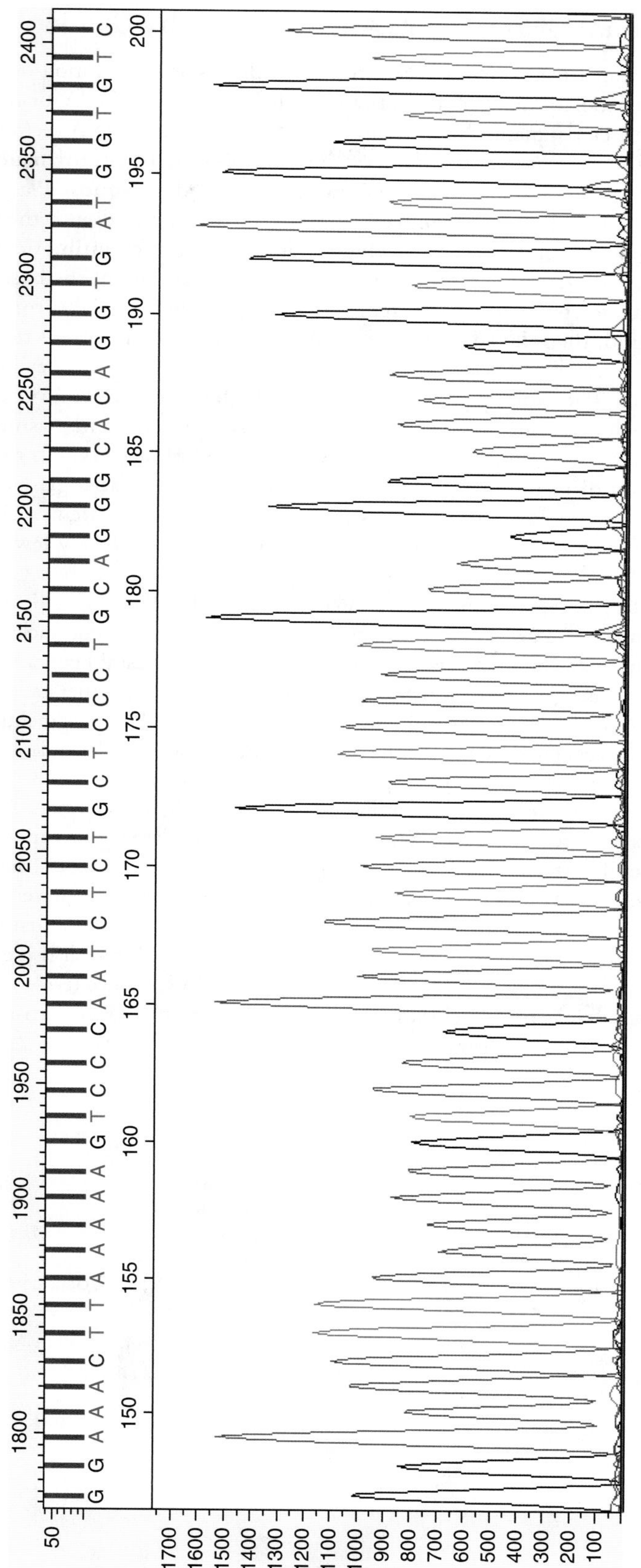

Figure 7.5 Chromatogram of DNA sequence generated by the dye-terminator sequencing method. The partial sequence of PCR amplified DNA of the human granulocyte matrix metalloproteinase 2 (MMP2) cDNA is depicted. Each ddNTP is tagged with a different colored fluor. The color of each peak corresponds to the colored nucleotide listed across the top of the figure.
Data from the laboratory of G. Soslau, Drexel Universty College of Medicine, Philadelphia, PA.

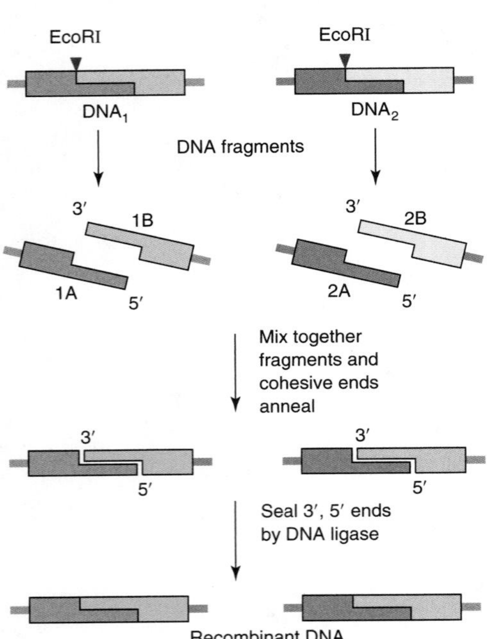

Figure 7.6 The formation of recombinant DNA from restriction endonuclease-generated fragments that contain cohesive ends. Many restriction endonucleases hydrolyze DNA in a staggered fashion yielding fragments with single-stranded regions at their 5′- and 3′-ends. DNA fragments generated from different molecules with the same restriction endonuclease have complementary single-stranded ends that can be annealed and covalently linked together with a DNA ligase. All different combinations are possible in a mixture. When two DNA fragments of different origin combine, a recombinant DNA molecule results.

Recombinant DNA Vectors Are Produced by Cloning

By incorporating a recombinant DNA into a cell that allows replication of recombinant DNA, amplification of DNA of interest can be achieved. A carrier DNA, or cloning vector, is employed. Bacterial plasmids are ideally suited as recombinant DNA vectors.

Many bacteria contain a single circular chromosome of approximately 4×10^6 base pairs (bp) and minicircular DNA molecules called **plasmids**. Plasmids usually contain only a few thousand base pairs and are rarely associated with the larger chromosome. Genes within a plasmid include those which confer antibiotic resistance to the bacterium, an attribute useful in selecting specific colonies of the bacterium. Plasmids replicate independently of the replication of the main bacterial chromosome. One type of plasmid, the **relaxed-control** plasmids, may be present in tens to hundreds of copies per bacterium, and its replication depends solely on host enzymes that have long half-lives. Therefore, replication of "**relaxed**" **plasmids** occurs in the presence of a protein synthesis inhibitor. Bacteria accumulate several thousand plasmid copies per cell under these conditions. Because of this, the relaxed plasmid type is routinely used for production of recombinant DNA. In plasmids subject to **stringent control**, replication depends on continued synthesis of plasmid-encoded proteins. They replicate at about the same rate as the large bacterial chromosome, and only few copies occur per cell.

The first practical recombinant DNA contained foreign DNA ligated with the vector *E. coli* **plasmid pSC101**, which contains one EcoRI restriction endonuclease site and a gene that confers antibiotic resistance to the bacterium. The plasmid contains an origin of replication and associated DNA regulatory sequences that are together called a **replicon**. However, the single restriction endonuclease site limits the DNA fragments that can be cloned to those generated by EcoRI; the one antibiotic-resistance selectable marker reduces the convenience in selection, and the plasmid replicates poorly.

Plasmid vectors with broad versatility are constructed by recombinant DNA technology. The desirable features of a plasmid vector include a relatively low molecular weight (3–5 kilobases [kb]) to accommodate larger fragments, several different restriction sites useful in cloning a variety of fragments, multiple selectable markers to aid in selecting bacteria with recombinant DNA molecules, and a high rate of replication. The first plasmid constructed (Figure 7.7) to satisfy these requirements was **pBR322**, which is used to generate many vectors in use today. These contain an inserted DNA sequence termed a **polylinker**, **restriction site bank**, or **polycloning site**, which contains numerous restriction endonuclease sites unique to the plasmid.

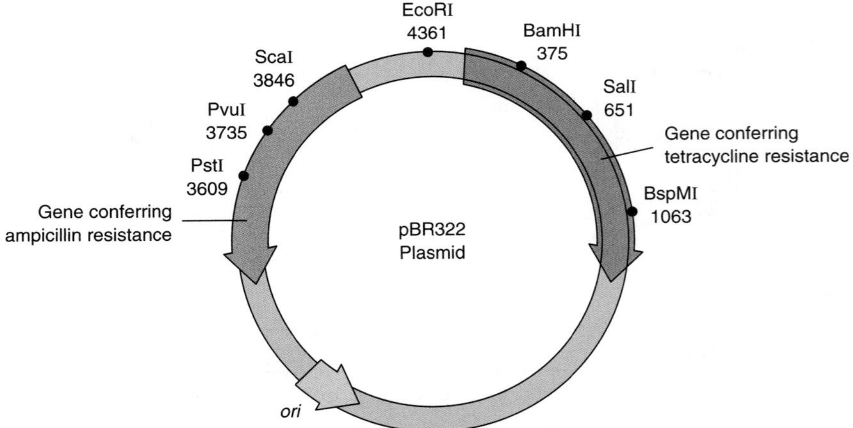

Figure 7.7 The pBR322 plasmid constructed in the laboratory to contain features that facilitate cloning of foreign DNA fragments. By convention the numbering of the nucleotides begins with the first T in the unique EcoRI recognition sequence (GAATTC) and the positions on the map refer to the 5′-base of the various restriction endonuclease-recognition sites. Only a few of these within the antibiotic resistance genes and none of the sites where an enzyme cuts more than once are shown.

Directional Cloning: DNA Inserted into Vector DNA in a Specific Direction

Directional cloning reduces the number of variable recombinants and increases the probability of selection of the desired recombinant. Insertion of foreign DNA, with a defined polarity, into a plasmid vector without the plasmid resealing itself can be accomplished by use of two restriction endonucleases (Figure 7.8); vectors with polylinkers are ideally suited for this purpose. The use of two enzymes yields DNA fragments and linearized plasmids with different sticky ends. Under these conditions the plasmid is unable to reanneal with itself. In addition, the foreign DNA is inserted into the vector in only one orientation. This is extremely important when one clones a potentially functional gene downstream from the promoter-regulatory elements in expression vectors.

Bacteria Transformed with Recombinant DNA and the Need for a Selection Process

The artificial introduction of DNA into bacteria is referred to as **transformation**. It is accomplished by briefly exposing the cells to divalent cations that make them transiently permeable to small DNA molecules. Recombinant plasmid molecules, containing foreign DNA, are then introduced into bacteria and replicate normally.

Once a plasmid has been introduced into a bacterium, both can replicate. During the transformation process, some bacteria may not take up a plasmid or may be transformed with a vector not carrying foreign DNA; in preparation some vectors may reanneal without inclusion of the foreign DNA. In some experimental conditions DNA fragments are generated that are readily purified from small, highly purified DNA species, for example, some DNA viruses. More typically, however, restriction endonuclease generates numerous DNA fragments whose number depends on the size and complexity of DNA being used, and individual fragments cannot be isolated from these samples. Restriction endonucleases

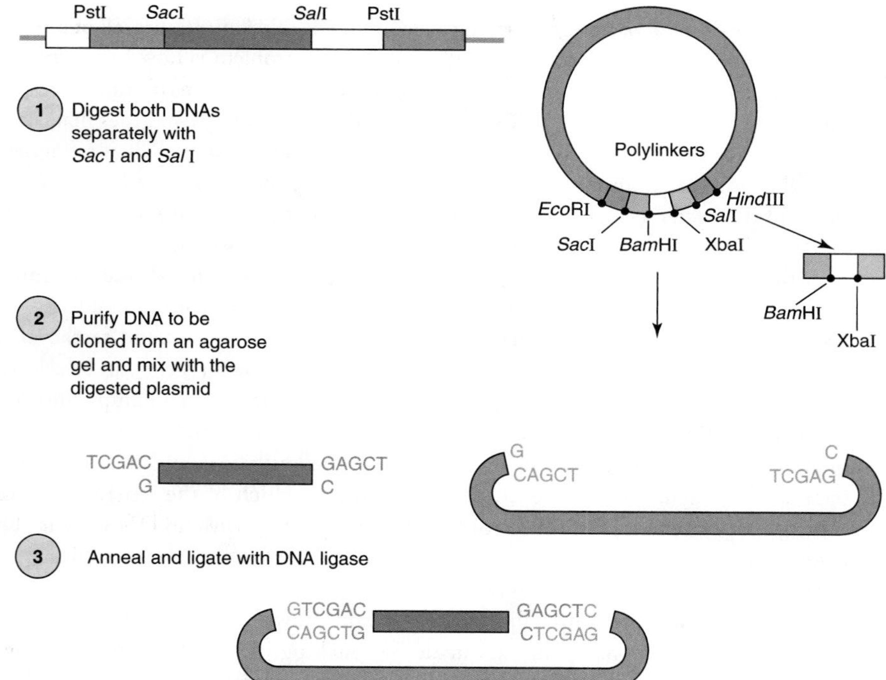

Figure 7.8 Directional cloning of foreign DNA into vectors with a specified orientation. Insertion of a foreign DNA fragment into a vector with a specified orientation requires two different annealing sequences at each end of the fragment and the corresponding complementary sequence at the two ends generated in the vector. Two different annealing sequences are generated by specific restriction endonucleases that hydrolyze both the foreign DNA fragment and vector. A polylinker with numerous unique restriction endonuclease sites within the vector facilitates directional cloning. Knowledge of the restriction map for the DNA of interest allows for the selection of appropriate restriction endonucleases.

do not necessarily produce fragments that contain intact genes. A fragment may contain an entire gene but lack the required flanking regulatory sequences, such as the promoter region. If the foreign gene is of mammalian origin, its regulatory sequences are not recognized by the bacterial synthetic machinery. The primary gene transcript (premRNA) may contain introns that cannot be processed by the bacteria. Methods have been developed to select bacteria that contain the desired DNA.

Recombinant DNA Molecules in a Gene Library

When a complex mixture of thousands of different genes, located on different chromosomes as in the human genome, is subjected to cleavage by a single restriction endonuclease, thousands of DNA fragments are generated. These DNA fragments can be annealed with a plasmid vector that has been cleaved to produce a linear molecule by the same restriction endonuclease. By adjusting the ratio of plasmid to foreign DNA, the probability of joining at least one copy of each DNA fragment within a circularized recombinant-plasmid DNA approaches 1. Usually, only one DNA fragment is inserted into each plasmid vector. Bacteria are transformed with the recombinant molecules so that one plasmid is taken up by one bacterium. Each recombinant molecule is then replicated within the bacterium; this produces progeny, each carrying multiple copies of the recombinant DNA. The resulting total bacterial population will contain fragments of DNA that may represent the entire human genome. This is termed a **gene library**. As in any library containing thousands of volumes, a selection system must be available to retrieve the book or gene of interest.

Plasmids are commonly employed to clone DNA fragments generated from DNA of limited size and complexity, such as viruses, and to subclone large DNA fragments previously cloned in other vectors. Genomic DNA fragments are usually cloned from vectors that can carry larger foreign DNA fragments than plasmids (p. 277).

Selection of Transformed Bacteria by Loss of Antibiotic Resistance

When a single transformed bacterium carrying a recombinant DNA multiplies, all of its progeny will carry copies of the same recombinant plasmid. The problem is how to identify the colony that contains the desired plasmid in a field of thousands to millions of different bacterial colonies. The plasmid construct pBR322 and its modifications carry two genes that confer antibiotic resistance and are sensitive to several restriction endonucleases. When a fragment of foreign DNA is inserted into a restriction site within one of them, the gene becomes nonfunctional and the bacteria carrying this recombinant plasmid are sensitive to the antibiotic (Figure 7.10). The second antibiotic resistance gene within the plasmid, however, remains intact, and the bacteria will be resistant to this antibiotic. This **insertional inactivation** of plasmid gene products permits selection of bacteria that carry recombinant plasmids.

pBR322 contains genes that confer resistance to **ampicillin** (amp^r) and **tetracycline** (tet^r). A gene library of cellular DNA fragments inserted within the tet^r gene can be selected and screened in two stages (Figure 7.9). First, the bacteria are grown in an ampicillin containing growth medium. Bacteria that are not transformed because they lack a normal or recombinant plasmid will not grow in the presence of the antibiotic, and this population of bacteria is eliminated. This, however, does not indicate which of the remaining viable bacteria carry a recombinant plasmid vector rather than a plasmid with no DNA insert. The second step is to identify bacteria that carry recombinant vectors with nonfunctional tet^r genes, which are, therefore, sensitive to tetracycline.

The bacteria resistant to ampicillin are plated and grown on agar plates containing ampicillin (Figure 7.9). Replica plates are made by touching the colonies on the original agar plate with a filter and then touching additional sterile plates with the filter. All the plates will contain portions of each original colony at identifiable positions on the plates. The replica plate can contain tetracycline, which will kill bacteria harboring a disrupted tet^r gene. Comparison of replica plates with and without tetracycline indicates which colonies on the original ampicillin plate contain recombinant plasmids.

Either DNA or RNA probes (pp. 268 and 270) are used to identify the DNA of interest. Ampicillin resistant bacterial colonies on agar can be plated onto a nitrocellulose filter

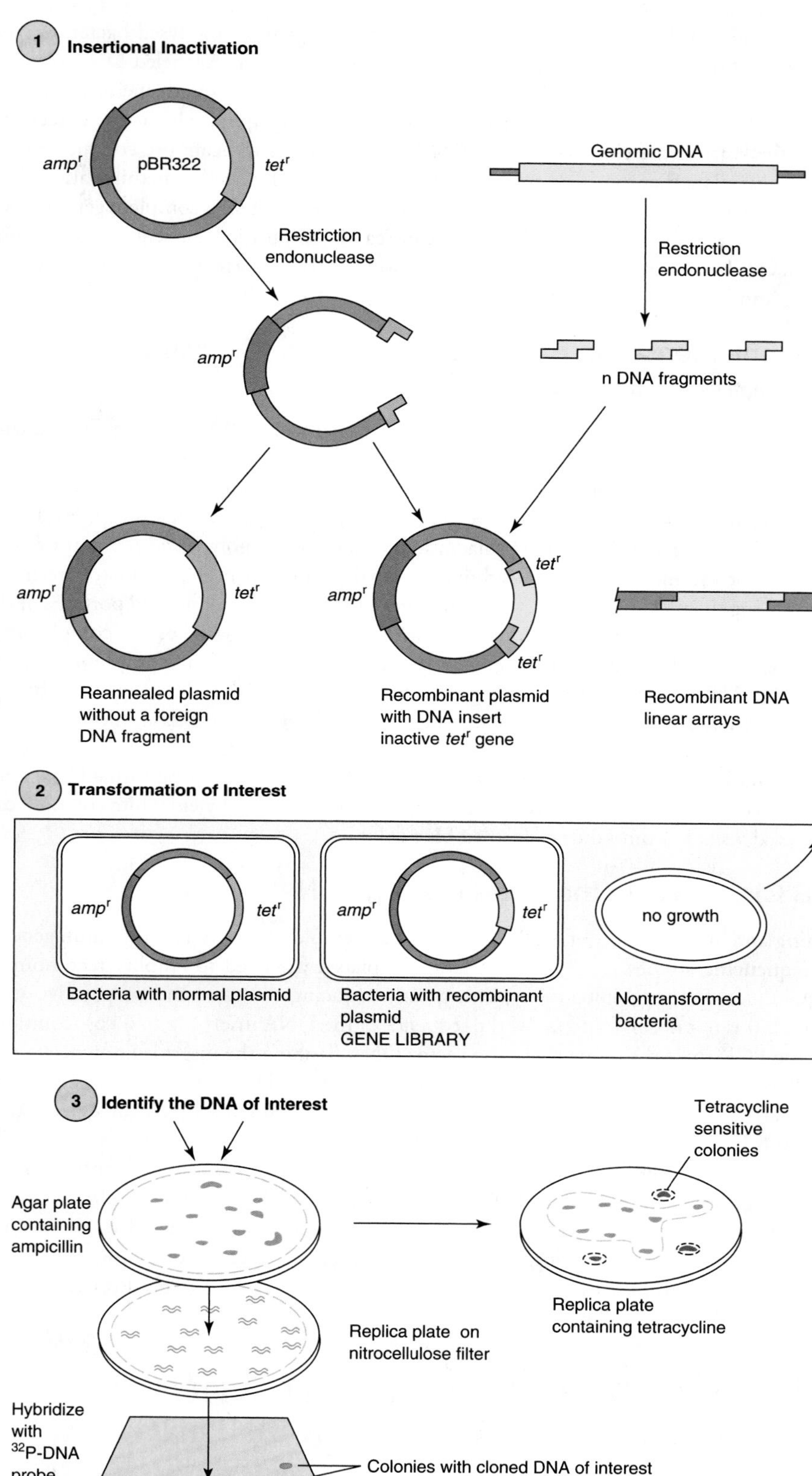

Figure 7.9 Insertional inactivation of recombinant plasmids and detection of transformed bacteria carrying a cloned DNA of interest. When the insertion of a foreign DNA fragment into a vector disrupts a functional gene sequence, the resulting recombinant DNA does not express the gene. The gene that codes for resistance to tetracycline (*Tet^r*) is destroyed by DNA insertion while the ampicillin resistance gene (*Amp^r*) remains functional. Destruction of one resistance gene and the retention of a second antibiotic resistance gene permits the detection of bacterial colonies carrying the foreign DNA of interest within the replicating recombinant vector.

CLINICAL CORRELATION 7.5

Multiplex PCR Analysis of the HGPRTase Gene Defects in the Lesch–Nyhan Syndrome

Lesch–Nyhan syndrome (OMIM 300322) results from a deficiency in hypoxanthine-guanine phosphoribosyltransferase (HG-PRTase) activity (see Clin. Corr. 20.2, p. 811). Several variant forms of HGPRTase defects have been detected. Multiplex PCR amplification of the HGPRT gene locus has been employed to analyze this gene in cells derived from Lesch–Nyhan patients and results account for the variability of the HG-PRTase. The gene, composed of nine exons, can be multiplex amplified using 16 different primers in a single PCR reaction. The products can be separated by agarose gel electrophoresis. Analysis of the HGPRT gene locus by multiplex amplification of DNA derived from cells of several patients detected great variations ranging from deletions of regions of exons to total absence of the exons.

Rossiter, B. J. F., et al. PCR, A Practical Approach, Vol. 1. In McPherson, M.J., et al. (Eds)., Oxford: Oxford University Press, 1994, 67.

and adhering cells lysed with NaOH (Figure 7.9). DNA within the lysed bacteria is also denatured by the NaOH and becomes firmly bound to the filter. A labeled DNA or RNA probe that is complementary to the DNA of interest will hybridize with the nitrocellulose bound DNA. The filter is then tested by X-ray autoradiography, which will detect any colony that carries the cloned DNA of interest. These spots indicate the colonies on the original agar plate that can be grown in a large-scale culture for further manipulation.

Cloned and amplified DNA fragments usually do not contain a complete gene and are not expressed. The DNA inserts, however, can readily be purified for sequencing or used as probes to detect genes within a mixture of genomic DNA, to assay transcription levels of mRNA, and to detect disease producing mutations.

α-Complementation for Selecting Bacteria Carrying Recombinant Plasmids

Vectors have been constructed (the pUC series) such that selected bacteria transformed with these vectors carrying foreign DNA inserts can be identified visually (Figure 7.11). The pUC plasmids contain the regulatory sequences and part of the 5′-coding sequence (N-terminal 146 amino acids) for the β-galactosidase gene (lac Z gene) of the lac operon (p. 305). The translated N-terminal fragment of β-galactosidase is an inactive polypeptide. Mutant *E. coli,* that code for the missing inactive carboxy-terminal portion of β-galactosidase, are transformed using the pUC plasmids. The translation of the host cell and plasmid portions of the β-galactosidase in response to an inducer, **isopropyl thio-β-D-galactoside**, complements each other to yield an active enzyme. The process is referred to as **α-complementation**. When these transformed bacteria are grown in the presence of a chromogenic substrate (5-bromo-4-chloro-3-indolyl-β-D-galactoside [X-gal]) for β-galactosidase, they form blue colonies. If, however, a foreign DNA fragment has been inserted into the sequence for the N-terminal portion of β-galactosidase, the active enzyme cannot be formed. Bacteria, transformed with these recombinant plasmids and grown on X-gal yield white colonies, can be selected visually from nontransformed blue colonies.

PCR Circumvents the Need to Clone DNA

Cloning and amplification of a DNA fragment are employed in subcloning, mutagenesis, and sequencing studies. The PCR has largely replaced the need to amplify recombinant DNA in a replicating biological system, greatly reducing the time and preparative steps required. It is not necessary to know the sequence of the DNA insert (up to 6 kb) to amplify it by the PCR since the sequence of the vector DNA flanking the insert is known.

PCR may completely circumvent the need to clone the DNA of interest. For instance, a gene of known sequence can readily be analyzed in a patient's DNA for detection of mutations by a **multiplex PCR strategy**. DNA is isolated from the patient's blood cells, and multiple pairs of oligonucleotide primers are synthesized to amplify the entire gene or selected regions within the gene (Figure 7.10). Analysis of the amplified DNA fragments by agarose gel electrophoresis reveals any deletion mutation as compared to the normal gene products. Direct sequencing of multiple PCR products detects point mutations in the patient's gene. Multiplex PCR has been used to detect defects in the HGPRTase gene in Lesch–Nyhan patients (Clin. Corr. 7.5).

7.5 • DETECTION AND IDENTIFICATION OF NUCLEIC ACIDS AND DNA-BINDING PROTEINS

Nucleic Acids as Probes for Specific DNA or RNA Sequences

DNA and RNA probes are used for selection of bacteria that harbor recombinant DNA of interest, for analysis of mRNA expressed in a cell or for identification of DNA sequences within a genome. They contain sequences complementary to the target nucleic acid and

Step 1

Amplification of the DNA region of interest by the PCR

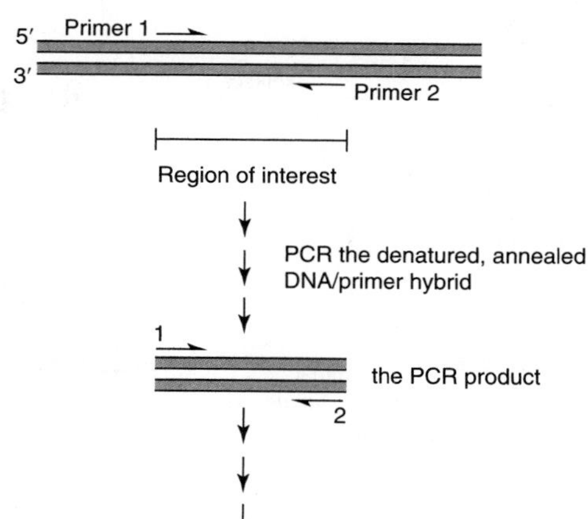

Step 2a

Employ multiple primers (1′ – 6′) simultaneously with the isolated PCR product to yield secondary PCR products of varying sizes

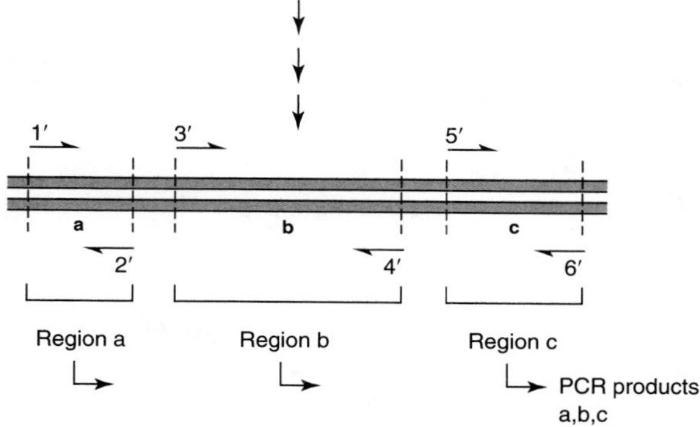

Step 2b

Repeat steps 1 and 2a with products from patient material containing a potentially deleted region b

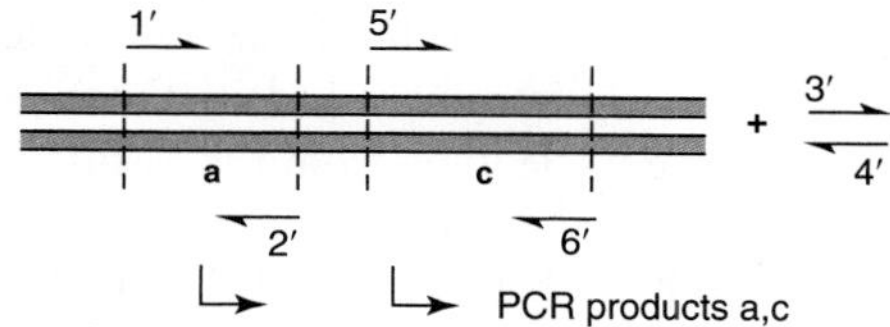

Step 3

Separate and detect the PCR products from control and patient samples by agarose gel electrophoresis and ethidium bromide staining to demonstrate the deletion of region b in the patient DNA

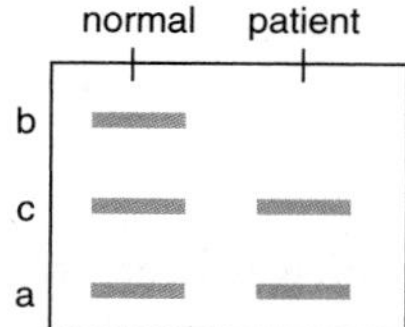

Figure 7.10 **Multiplex PCR to analyze a DNA region of interest for mutations.** A region of DNA within a complex DNA molecule is amplified by the PCR with primers that are complementary to sequences flanking the DNA region of interest (step 1). The PCR product is then used as a template simultaneously for multiple pairs of primers (step 2a) that are complementary throughout the DNA (here they cover three segments, *a*, *b*, and *c*). This procedure requires prior knowledge of the sequence of the normal DNA or gene. Step 2a is repeated for DNA derived from a patient with potential mutation(s) in the DNA region of interest (step 2b). The amplified DNA products from the multiplex PCR step (step 2a and b) are then analyzed by agarose gel electrophoresis to ascertain if the patient sample contains a mutation (step 3).

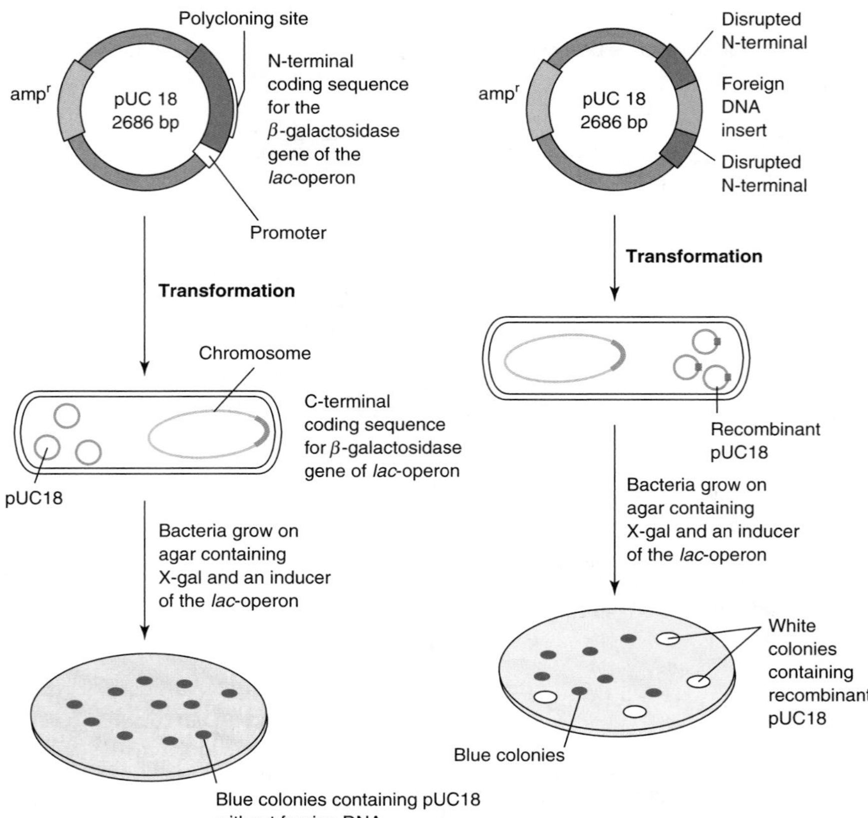

Figure 7.11 α-Complementation for detection of transformed bacteria. A constructed vector (pUC 18) expresses the N-terminal coding sequence for β-galactosidase of the lac operon. Bacterial mutants coding for the C-terminal portion of β-galactosidase are transformed with pUC 18. These transformed bacteria grown in the presence of a special substrate for the intact enzyme (X-gal) result in blue colonies because they contain the enzyme to react with substrate. The functional N-terminal and C-terminal coding sequences complement each other to yield a functional enzyme. If, however, a foreign DNA fragment insert disrupts the pUC 18 N-terminal coding sequence for β-galactosidase, bacteria transformed with this recombinant molecule will not produce a functional enzyme. Bacterial colonies carrying these recombinant vectors can then be visually detected as white colonies.

hybridize with the nucleic acid of interest. The degree of complementarity determines the tightness of binding of the probe. The probe does not need to contain the entire complementary sequence of the DNA. The probe can be labeled, usually with ^{32}P or with nonradioactive labels that depend on enzyme substrates coupled to nucleotides, which when incorporated into the nucleic acid is detected by an enzyme-catalyzed reaction.

Labeled probes are produced by **nick translation** of double-stranded DNA. Nick translation (Figure 7.12) involves the random cleavage of a phosphodiester bond in the backbone of a DNA strand by DNase I; the breaks are called nicks. *E. coli* DNA polymerase I, with its $5' \rightarrow 3'$ exonucleolytic activity and its DNA polymerase activity, creates single-strand gaps by hydrolyzing nucleotides from the 5′-side of the nick and then filling in the gaps with its polymerase activity. The reaction is usually carried out in the presence of an α-^{32}P-labeled dNTP and the other three unlabeled dNTPs. The DNA employed in this method is usually purified and is derived from cloned DNA, viral DNA, or cDNA.

Random primer labeling of DNA has distinct advantages over the nick translation method. The random primer method typically requires only 25 ng of DNA as opposed to 1–2 μg of DNA for nick translation and results in labeled probes with a specific activity ($>10^9$ cmp/ug) approximately 10 times higher. This method generally produces longer labeled DNA probes. The double-stranded probe is melted and hybridized with a mixture of random hexanucleotides containing all possible sequences (ACTCGG, ACTCGA, ACTCGC, etc.). Hybridized hexanucleotides serve as primers for DNA synthesis with a DNA polymerase, such as the Klenow enzyme, in the presence of one or more radioactively labeled dNTPs.

Labeled RNA probes have advantages over DNA probes. For one, relatively large amounts of RNA can be transcribed from a template that may be available in very limited quantities. A double-stranded DNA (dsDNA) probe must be denatured prior to hybridization with the target DNA and rehybridization with itself competes for hybridization with the DNA of interest. No similar competition occurs with single-stranded RNA probes that hybridize with complementary DNA or RNA molecules. Synthesis of an RNA probe requires DNA as a template. To be transcribed the template must be covalently linked to an upstream promoter that can be recognized by a DNA dependent RNA polymerase. Vectors have been constructed that are well suited for this technique.

A labeled DNA or RNA probe is hybridized to nitrocellulose-bound nucleic acids and identified by the detection of labeled probe. Nucleic acids of interest are transferred to nitrocellulose from bacterial colonies grown on agar or from agarose gels on which they are electrophoretically separated by size.

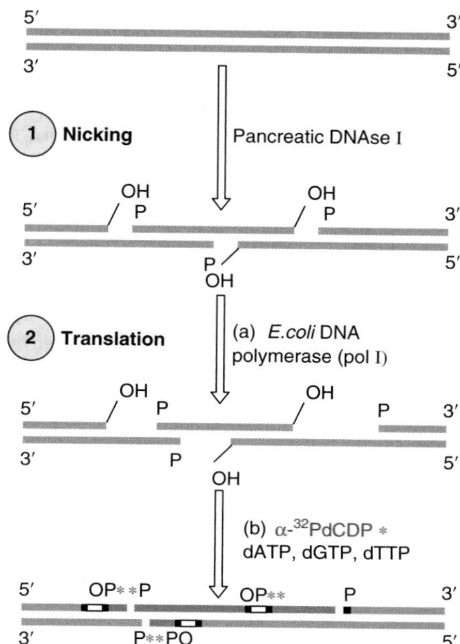

Figure 7.12 Nick translation to label DNA probes. Purified DNA molecules can be radioactively labeled and used to detect, by hybridization, the presence of complementary RNA or DNA in experimental samples. (1) Nicking step: Introduces random single-stranded breaks in DNA; (2) Translation step: *E. coli* DNA polymerase (pol I) has both 5′–3′ exonuclease activity that removes nucleotides from the 5′-end of the nick and polymerase activity that simultaneously fills in the single-stranded gap with radioactively labeled nucleotides using the 3′-end as a primer.

Southern Blot Technique for Identifying DNA Fragments

The transfer of DNA species separated by agarose gel electrophoresis to a filter for analysis was developed in the 1970s by E. M. Southern and is an indispensable tool. The method is called the **Southern blot technique** (Figure 7.13). A mixture of restriction endonuclease-generated fragments can be separated according to size by electrophoresis through an agarose gel. The DNA is denatured by soaking the gel in alkali. The gel is then placed on absorbent paper, and a nitrocellulose filter is placed directly on top of the gel. Several layers of absorbent paper are placed on top of the nitrocellulose filter. The absorbent paper under the gel is kept wet with a concentrated salt solution that is pulled up through the gel, the nitrocellulose, and into the absorbent paper layers above by capillary action. The DNA is eluted from the gel by the upward movement of the high-salt solution onto the nitrocellulose filter directly above, where it becomes bound. The position of the DNA bound to the nitrocellulose filter is the same as that present in the agarose gel. Single-stranded membrane-bound DNA can be analyzed with labeled probes.

The Southern blot technique is invaluable for detection and determination of the number of copies of particular sequences in complex genomic DNA, for confirmation of DNA cloning results, and demonstration of polymorphic DNA in the human genome that correspond to pathological states. An example of the use of Southern blots is shown in Figure 7.13. Whole human genomic DNA, isolated from three individuals, was digested separately with a restriction endonuclease generating thousands of fragments. These were distributed throughout the agarose gel according to size in an electric field. The DNA was transferred (blotted) to a nitrocellulose filter and hybridized with a ^{32}P-labeled DNA or RNA probe for a gene of interest. The probe detected two bands in all three individuals indicating that the gene of interest was cleaved at one site within its sequence. Individuals A and B presented a normal pattern while patient C had one normal band and one lower-molecular-weight band. This is an example of detecting altered restriction endonuclease-generated DNA fragments from different individuals within a single species, **restriction fragment length polymorphism (RFLP).** The detection of a reduced molecular weight fragment in patient C implies that deletion of a segment of the gene occurred and may be associated with a pathological state. The gene from this patient is cloned, sequenced, and fully analyzed to characterize the altered nature of the DNA (Clin. Corr. 7.6).

Other techniques that employ the principles of Southern blot are the transfer of RNA (Northern blots), as described below and of proteins (Western blots) to nitrocellulose filters or nylon membranes.

Single-Strand Conformation Polymorphism

Southern blot analysis and detection of base changes in DNA from different individuals by RFLP analysis is dependent on alteration of a restriction endonuclease site. A base substitution, deletion or insertion (**a single-nucleotide polymorphism [SNP]**) rarely occurs within a restriction endonuclease site. However, these modifications are detected at a specific site by **single-strand conformation polymorphism (SSCP).** This technique takes advantage of the fact that single-stranded DNA, smaller than 400 bases long, subjected to electrophoresis through a polyacrylamide gel migrates with a mobility partially dependent on its conformation. A single base alteration usually modifies the DNA conformation sufficiently to be detected as a mobility shift on electrophoresis through a non-denaturing polyacrylamide gel.

The analysis of a small region of genomic DNA or cDNA for SSCP is accomplished by PCR-amplification of the region of interest. Sense and antisense oligonucleotide primers are synthesized that flank the region of interest, and this DNA is amplified by PCR in the presence of radiolabeled nucleotide(s). The purified radiolabeled double-stranded PCR product is then heat denatured in 80% formamide and immediately loaded onto a nondenaturing polyacrylamide gel. The mobilities of control products are compared to samples from experimental studies or patients. Detection of mutations in samples from patients identify genetic lesions. These procedures were successfully applied to the analysis of genes associated with the long-QT syndrome that has been implicated in the sudden infant death syndrome (SIDS) (Clin. Corr. 7.7). The method depends on prior knowledge of the sequence of the gene or gene fragment of interest whereas analysis by RFLP requires only restriction map analysis of DNA.

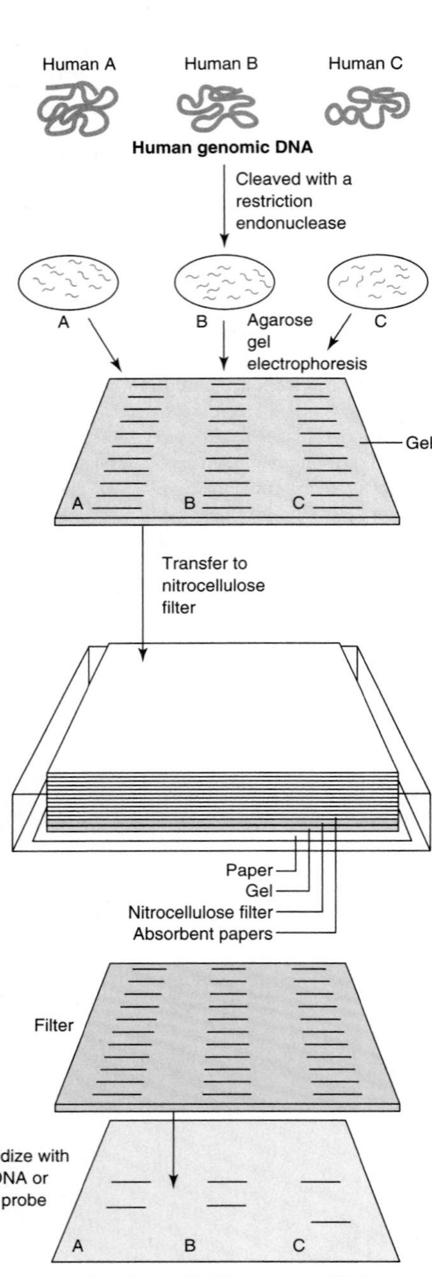

Figure 7.13 Southern blot to transfer DNA from agarose gels to nitrocellulose. Transfer of DNA to nitrocellulose, as single-stranded molecules, allows for the detection of specific DNA sequences within a complex mixture of DNA. Hybridization with nick translated labeled probes can demonstrate if a DNA sequence of interest is present in the same or different regions of the genome.

Detection of mRNA

Analysis of the amount of mRNA species in tissue or total RNA preparations is often critical for our understanding gene regulation of cell growth and tissue differentiation. We can assess timing, level, and site within a tissue of gene expression through the analysis of specific mRNA species. Several techniques are available to assay the amount of a specific mRNA species in an RNA preparation. (a) **Northern blot analysis** requires the electrophoretic separation of RNA species, by size, on an agarose gel followed by the transfer and cross-linking to a membrane as in the Southern blot technique. The fixed RNA species are then hybridized with a labeled probe specific for the mRNA species of interest. If the probe is radiolabeled, hybrids are detected by autoradiography. Nonradioactive probes are also available. Northern blot analysis allows ready determination of the size of the mRNA and identification of potential alternatively spliced transcripts and/or the presence of multigene family transcripts. (b) **RT-PCR** (Section 7.8, p. 281), unlike the Northern blot, allows us to measure the amount of an mRNA species present in an RNA sample. Measurement is based on comparison of the relative amount of the PCR product of an internal control transcript to the amount of a specific mRNA in the same sample. Typical internal control transcripts include GAPDH (glyceraldehyde 3-phosphate dehydrogenase) and β-actin. (c) The **nuclease protection assay** (NPA) does not give any information about the size of the mRNA of interest, but is ideally suited for the simultaneous analysis of multiple mRNA species. Radiolabeled or nonisotopic probe(s) are hybridized with the RNA sample in solution. Hybridization in solution is more efficient than hybridization to membrane-fixed RNA, as used in Northern blots. Any remaining unhybridized probe and RNA are then hydrolyzed with nucleases. The mixture of hybridized species is then separated through a low percentage (6%) acrylamide gel. The size of each hybridized species is determined by the size of the probe employed. NPA is commonly used to detect mRNA transcripts in different tissues (Figure 7.14). (d) ***In situ* hybridization** is the only method described in this section to define which cells in a sample of formalin-fixed tissue express a specific gene. In this method, RNA is not isolated or subjected to electrophoretic separations. Rather tissues of interest are fixed, thin sectioned, and mounted onto microscope slides. Proteins are digested with protease K to increase accessibility of the labeled probe to hybridize with the cellular mRNA of interest. This method supports findings of the nuclease protection assays and also defines which cell types within the tissues are expressing the mRNA of interest.

Detection of Sequence-Specific DNA-Binding Proteins

Regulatory proteins bind to specific DNA sequences that flank genes that up- or down-regulate gene expression. The **electrophoretic mobility shift assay** (**EMSA**), or gel retardation, method has been employed extensively to analyze sequence-specific DNA-binding proteins along with the DNA sequence required for binding. Proteins with potential DNA-binding characteristics are prepared from whole cell extracts, nuclear extracts, purified protein preparations, or recombinant proteins from genetically engineered expression systems. The DNA probe is radio-labeled. The DNA employed may be DNA fragments, double-stranded synthetic oligonucleotides, or cloned DNA containing a known protein-binding site(s) or potential protein-binding site(s). The DNA must be double stranded since single-stranded DNA would bind nonspecifc single-strand binding proteins (with cell or nuclear extracts) that would interfere with the interpretation of the results. The purified labeled DNA probe

CLINICAL CORRELATION 7.6

Restriction Fragment Length Polymorphisms Determine the Clonal Origin of Tumors

It is generally assumed that most tumors are monoclonal in origin; that is, a rare event alters a single somatic cell genome such that the cells grow abnormally into a tumor mass with all-daughter cells carrying the identically altered genome. Proof that a tumor is monoclonal versus polyclonal can help to distinguish hyperplasia (increased production and growth of normal cells) from neoplasia (growth of new or tumor cells). The detection of restriction fragment length polymorphisms (RFLPs) in Southern blotted DNA samples allows one to define the clonal origin of human tumors. If tumor cells were collectively derived from different parental cells, they should contain a mixture of DNA markers characteristic of each cell of origin. However, an identical DNA marker in all tumor cells would indicate a monoclonal origin. The analysis is limited to females to take advantage of the fact that each cell carries only one active X chromosome of either paternal or maternal origin with the second X chromosome being inactivated. Activation occurs randomly during embryogenesis and is faithfully maintained in all daughter cells with one-half the cells carrying an activated maternal X chromosome and the other one-half an activated paternal X chromosome.

Analysis of the clonal nature of a human tumor depends on the fact that activation of an X chromosome involves changes in the methylation of selected cytosine (C) residues within the DNA molecule. Several restriction endonucleases, such as Hha I which cleaves DNA at GCGC sites, do not cleave DNA at their recognition sequences if a C is methylated within this site. Therefore, the methylated state (activated versus inactivated) of the X chromosome can be probed with restriction endonucleases. Also, the paternal X chromosome can be distinguished from the maternal X chromosome in a significant number of individuals by differences in the electrophoretic migration of restriction endonuclease generated fragments derived from selected regions of the chromosome. These fragments are identified on a Southern blot by hybridization with a DNA probe that is complementary to this region of the X chromosome. An X-linked gene that is amenable to these studies is the hypoxanthine guanine phosphoribosyltransferase (HGPRTase) gene. The HGPRTase gene consistently has two Bam HI restriction endonuclease sites (B1 and B3 in figure), but in some individuals a third site (B2) is also present.

The presence of site B2 in only one parental X chromosome HGPRT permits detection of RFLPs. Therefore, a female cell may carry one X chromosome with the HGPRT gene possessing 2Bam HI sites (results in one DNA fragment of 24 kb) or 3Bam HI sites (results in a single detectable DNA fragment of 12 kb). This figure depicts the expected results for the analysis of tumor cell DNA to determine its monoclonal or polyclonal origin. As expected, three human tumors examined by this method were shown to be of monoclonal origin.

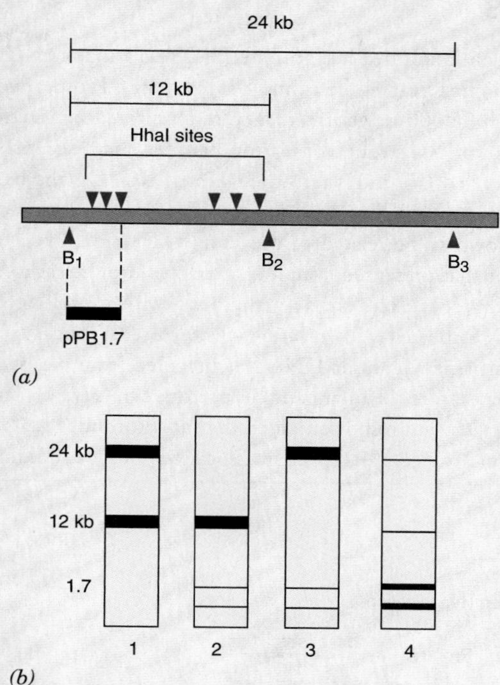

Analysis of genomic DNA to determine the clonal origin of tumors. (*a*) The X chromosome-linked hypoxanthine guanine phosphoribosyltransferase (HGPRTase) gene contains two invariant Bam HI restriction endonuclease sites (B1 and B3) whereas in some individuals a third site, B2, is also present. The gene also contains several Hha I sites; however, all of these sites, except H1, are usually methylated in the active X chromosome. Therefore, only the H1 site would be available for cleavage by Hha I in the active X chromosome. A cloned, labeled probe, pPB1.7, is used to determine which form of the HGPRTase gene is present in a tumor and if it is present on an active X chromosome. (*b*) Restriction endonuclease patterns predicted for monoclonal versus polyclonal tumors are as follows: (1) Cleaved with Bam HI alone. 24-kb fragment derived from a gene containing only B1 and B3 sites and 12-kb fragment derived from a gene containing extra B2 site. Pattern is characteristic for heterozygous individual. (2) Cleaved with Bam HI plus Hha I, monoclonal tumor with the 12 kb derived from an active X chromosome (methylated). (3) Cleaved with Bam HI plus Hha I; monoclonal tumor with the 24-kb fragment derived from an active X chromosome (methylated). (4) Cleaved with Bam HI plus Hha I, polyclonal tumor. All tumors studied displayed patterns as in Lane 2 or Lane 3.

Vogelstein, B., Fearon, E. R., Hamilton, S. R., and Feinberg, A. B. Use of restriction fragment length polymorphism to determine the clonal origin of tumors. *Science* 227:642, 1985.

and protein sample are preincubated to form a stable protein-DNA complex prior to the electrophoretic analysis of the sample.

Figure 7.15 schematically depicts the expected results where the radio-labeled DNA complexes with a protein resulting in its retarded movement through the gel relative to the unreacted DNA. If an antibody to the DNA-binding protein of interest is added to the preincubation tube, one of two possible reactions occurs. If the antibody binds to an epitope that does not impede protein binding to the DNA, the complex size is increased resulting in a **supershift** (greater retardation) in the gel migration. Conversely, if the antibody blocks

CLINICAL CORRELATION 7.7

Single-Strand Conformational Polymorphism for Detection of Spontaneous Mutations that May Lead to SIDS

The sudden infant death syndrome (SIDS) is a major cause of death during the first year of life in the United States. Prospective study of more then 34,000 newborns that were monitored by electrocardiography indicated a strong correlation between increased risk of SIDS and prolonged QT interval in their heart EKG. On the basis of this study, it was decided to look for a mutation in one or more of the genes known to be related to the Long QT syndrome in a 44-day-old infant that presented cyanotic, apneic, and pulseless to a hospital emergency room. The child's arrhythmia with a prolonged QT interval was stabilized with multiple electrical DC shocks followed by drug treatment. Genomic DNA was prepared from peripheral blood lymphocytes from the infant and his parents. One gene associated with the Long QT syndrome contained the substitution of AAC for TCC at position 2971 to 2972 (protein associated with the sodium channel) child's but not the parents' DNA sample. The mutation replaced a serine residue with an asparagine in a highly conserved region of the protein that is presumed to participate in the function of the sodium channel. The mutation was not detected in 200 control subjects. The conclusion was that the child had a spontaneous mutation in a gene that was associated with a prolonged QT interval and that this contributed to a SIDS-like event. After treatment, the child was symptom free at nearly 5 years of age. This study points to the potential value of neonatal electrocardiographic screening to reduce infant mortality due to SIDS.

Schwartz, P. J., Priori, S. G., Dumaine, R., Napolitano, C., et al. A molecular link between the sudden infant death syndrome and the Long-QT syndrome. *N. Engl. J. Med.* 343: 262, 2000.

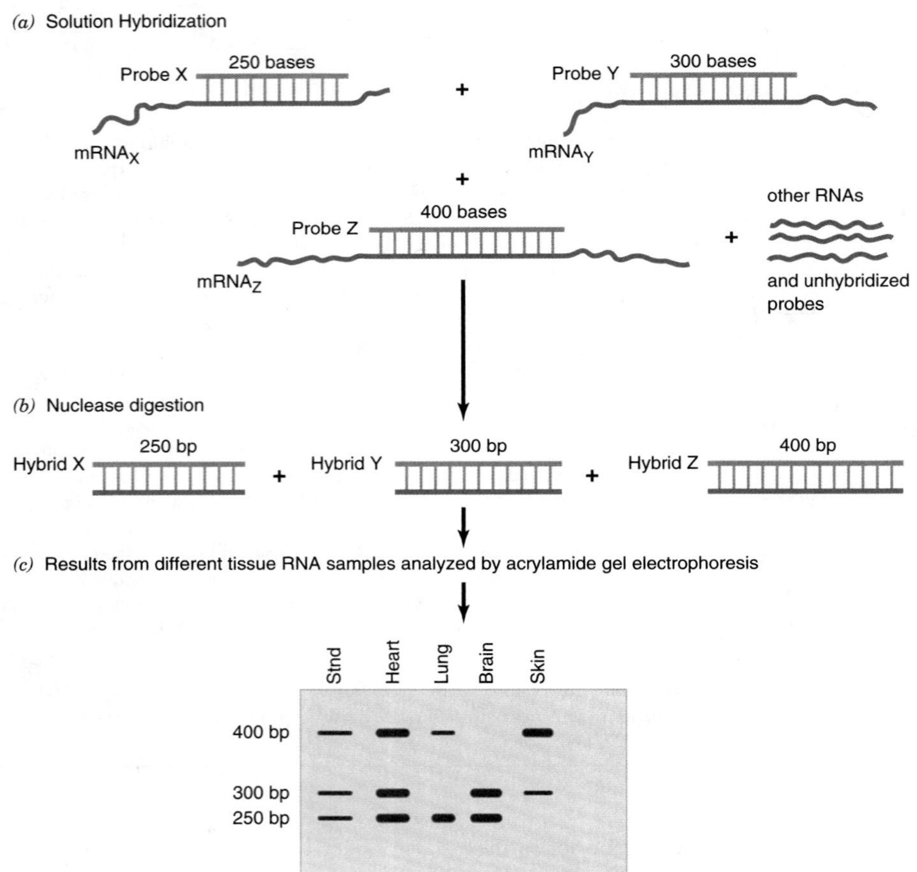

Figure 7.14 Nuclease protection assay. Total cellular mRNA can be isolated from different tissues. Single-stranded DNA probes that are complementary to known sequences of different gene transcripts (mRNAx, mRNAy, mRNAz) are hybridized with the RNA mixture. Nuclease digestion with a ribonuclease will hydrolyze single-stranded RNA regions of mRNA not hybridized with the DNA probe and all nonhybridized RNA species. Only the DNA–RNA hybrids protected against nuclease will remain for analysis by acrylamide gel electrophoresis. Differential expression of genes in specific tissues is then readily observed.

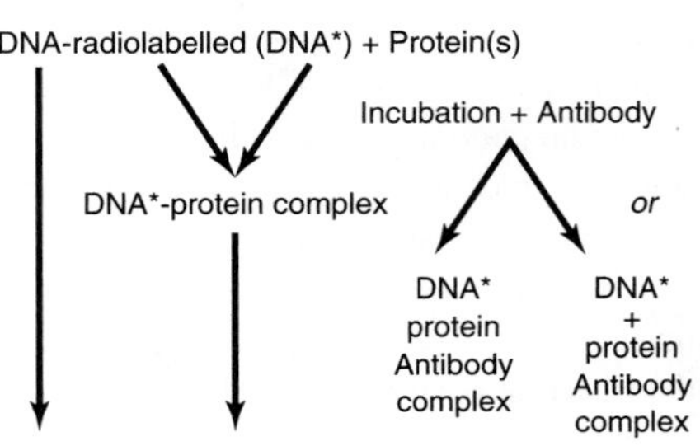

Figure 7.15 Electrophoretic mobility shift assay. Purified DNA migrates through a gel when subjected to an electric field based on its charge and molecular mass. If a protein(s) is(are) added to this, purified DNA that complexes with the DNA the total apparent molecular mass of the DNA is increased. This would slow its migration through the gel—a mobility shift. The addition of an antibody that reacts with the protein complexed with the DNA, but does not interfere with the DNA–protein interaction, will further increase the size of the DNA–protein complex and slow its migration. If the antibody reacts with the DNA-binding region of the protein a DNA–protein complex will not form. In both cases the antibody helps to identify the DNA-binding protein.

the protein's DNA-binding site, a protein–antibody complex will form and the labeled DNA probe will be unmodified in its migration through the gel. In either case the result with the antibody would confirm the identity of the DNA-binding protein.

7.6 • COMPLEMENTARY DNA AND COMPLEMENTARY DNA LIBRARIES

Insertion of specific functional eukaryotic genes into vectors that can be expressed in a prokaryotic cell could produce large amounts of genetically engineered proteins with significant medical, agricultural, and experimental potential. Hormones and enzymes, including insulin, erythropoietin, thrombopoietin, interleukins, interferons, and tissue plasminogen activator, are currently produced by these methods. It is not possible to clone functional genes from genomic DNA, except in rare instances. One reason is that most genes within the mammalian genome yield transcripts that contain introns that must be spliced out of the primary mRNA transcript. Prokaryotic systems cannot splice out the introns to yield functional mRNA transcripts. This problem is circumvented by synthesizing **complementary DNA (cDNA)** from functional eukaryotic mRNA.

mRNA as Template for DNA Synthesis Using Reverse Transcriptase

Messenger RNA can be reverse-transcribed to cDNA, and the cDNA inserted into a vector for amplification, identification, and expression. Mammalian cells normally contain 10,000–30,000 different species of mRNA molecules at any time during the cell cycle. In some cases, however, a specific mRNA species may approach 90% of the total mRNA, such as mRNA for globin in reticulocytes. Many mRNAs are normally present at only a few (1–14) copies per cell. A cDNA library can be constructed from the total cellular mRNA, but if only a few copies per cell of mRNA of interest are present, the cDNA may be very difficult to identify. Methods that enrich the population of mRNAs or their corresponding cDNAs permit reduction in number of different cDNA species within a cDNA library and greatly enhance the probability of identifying the clone of interest.

Desired mRNA Is Enriched by Separation Techniques

Messenger RNA can be separated by size by gel electrophoresis or centrifugation. Isolation of mRNA in a specific molecular size range enriches severalfold an mRNA of interest. Knowledge of the molecular weight of the protein encoded by the gene of interest gives a clue to the approximate size of the mRNA transcript or its cDNA; variability in the predicted size, however, arises from differences in the length of the untranslated regions of the mRNAs.

Enrichment of a specific mRNA molecule is also accomplished by immunological procedures, but it requires the availability of antibodies against the encoded protein by the gene of interest. Antibodies added to an *in vitro* protein synthesis mixture react with the growing polypeptide chain associated with polysomes and precipitate it. The mRNA is then purified from the immunoprecipitated polysomal fraction.

Complementary DNA Synthesis

An mRNA mixture is used as a template to synthesize complementary strands of DNA using RNA-dependent DNA polymerase, reverse transcriptase (Figure 7.16). A primer is required; advantage is taken of the poly(A) tail at the 3′-terminus of eukaryotic mRNA. An oligo(dT) with 12–18 bases is employed as the primer that hybridizes with the poly(A) sequence. After cDNA synthesis, the hybrid is denatured or the mRNA hydrolyzed in alkali to obtain the single-stranded cDNA. The 3′-termini of single-stranded cDNAs form a hairpin loop that serves as a primer for the synthesis of the second strand of the cDNA by the Klenow fragment or a reverse transcriptase. The resulting double-stranded cDNA contains a single-stranded loop that is recognized and digested by S1 nuclease. The ends of the cDNA must be modified prior to cloning in a vector. One method involves incubating

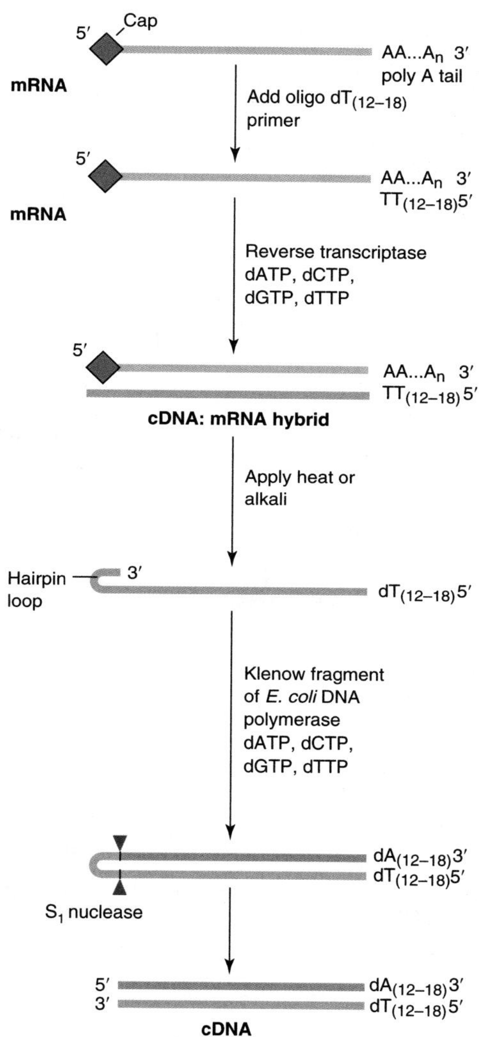

Figure 7.16 Synthesis of cDNA from mRNA. The 3′ poly(A) tail of mRNA is hybridized with an oligomer of dT (oligo$[dT]_{12\text{-}18}$) that serves as a primer for reverse transcriptase which catalyzes the synthesis of the complementary DNA (cDNA) strand in the presence of all four deoxynucleotide triphosphates (dNTPs). The resulting cDNA:mRNA hybrid is separated into single-stranded cDNA by melting with heat or hydrolyzing the mRNA with alkali. The 3′-end of the cDNA molecule forms a hairpin loop that serves as a primer for the synthesis of the second DNA strand catalyzed by the Klenow fragment of *E. coli* DNA polymerase. The single-stranded unpaired DNA loop is hydrolyzed by S_1 nuclease to yield a double-stranded DNA molecule.

blunt-ended cDNA molecules with linker molecules and a bacteriophage T4 DNA ligase that catalyzes the ligation of blunt-ended molecules (Figure 7.17). The synthetic linker molecules contain restriction endonuclease sites that are now cleaved with the appropriate enzyme for insertion of the cDNA into a vector cleaved with the same endonuclease.

Bacteriophage DNA (p. 281) is the most convenient and efficient vector to create cDNA libraries because they can readily be amplified and stored indefinitely. Two bacteriophage vectors, **λgt10** and **λgt11,** and their newer constructs are employed to produce cDNA libraries. The cDNA libraries in λgt10 are screened only with labeled nucleic acid probes when the sequence of the DNA is unknown, whereas, those in λgt11, an **expression vector,** can also be screened with antibody for the production of the protein or antigen of interest. If the sequence of the desired cDNA is known, then PCR is used to screen the recombinant bacteriophage.

Total Cellular RNA as Template for DNA Synthesis Using RT–PCR

Alternative methods to construct cDNA libraries employ a **reverse transcriptase-PCR (RT–PCR)** technique and obviate the need to purify mRNA. One such strategy is depicted in Figure 7.18 and begins with the reverse transcriptase production of a DNA-mRNA hybrid. The method uses terminal transferase to add a dG homopolymer tail to the 3′-end followed by hydrolysis of the mRNA. PCR primers are synthesized to hybridize with the dG and dA tails and terminate with two different restriction endonuclease sequences. The resulting PCR-amplified cDNA is then hydrolyzed with two different restriction endonucleases for directional cloning (p. 265) into an appropriate vector.

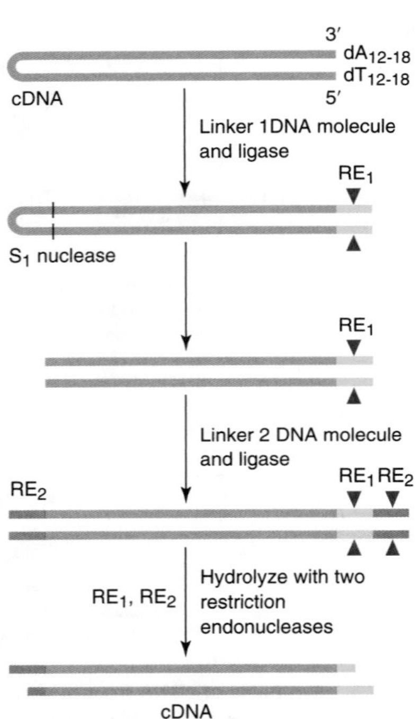

Figure 7.17 Modification of cDNA for cloning. The procedure begins with double-stranded DNA that contains a hairpin loop. A linker DNA that contains a restriction endonuclease site (RE_1) is added to the free end of the cDNA by blunt end ligation. The single-stranded hairpin loop is next hydrolyzed with S_1 nuclease. A second linker with a different restriction endonuclease site within (RE_2) is blunt end ligated to the newly created free cDNA. The second linker will probably bind to both ends but will not interfere with the first restriction endonuclease site. The modified DNA is hydrolyzed with the two restriction endonucleases and can be inserted into a plasmid or bacteriophage DNA by directional cloning.

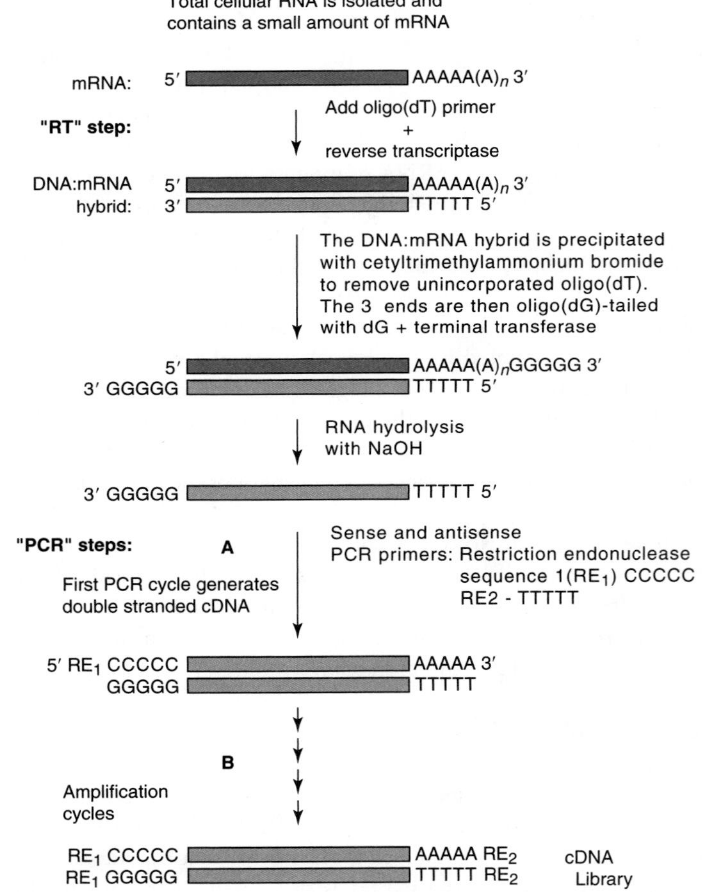

Figure 7.18 Generation of cDNA by reverse transcriptase–PCR (RT–PCR). Total cellular RNA or mRNA can be used to generate cDNA by RT–PCR. The mRNA with an oligo rA tail is reverse transcribed with an oligo dT primer. An oligo dG tail is added to the 3′-ends of the RNA and DNA strands, and the RNA strand is subsequently hydrolyzed with NaOH. Sense and antisense primers, modified with restriction site sequences, are then employed to amplify the cDNA by the PCR. The products can be hydrolyzed with the specific restriction endonucleases (RE and RE_2) for cloning and subsequent studies.

7.7 • BACTERIOPHAGE, COSMID, AND YEAST CLONING VECTORS AND ANALYSIS OF LONG STRETCHES OF DNA

Detection of noncoding sequences (i.e., introns) in most eukaryotic genes and distant regulatory regions that flank the genes necessitated cloning strategies to package larger DNA fragments than can be cloned in plasmids. Plasmids accommodate foreign DNA inserts 5–10 kb long. Recombinant DNA fragments larger than this are randomly deleted during replication of the plasmid within the bacterium. Thus, alternate vectors have been developed.

Bacteriophage as Cloning Vectors

Bacteriophage λ (λ phage), a virus that infects and replicates in bacteria, is an ideal vector for DNA inserts about 15 kb long. The λ phage selectively infects bacteria and replicates by a lytic or nonlytic (lysogenic) pathway. The λ phage contains a self-complementary 12-base single-stranded tail (cohesive termini) at both ends of its 50-kb double-stranded DNA. On infection of the bacteria, the cohesive termini (cos sites) of a single λ phage DNA self-anneal and the ends are covalently linked by the host cell DNA ligase. The circular DNA serves as a template for transcription and replication. λ phage, with restriction endonuclease-generated fragments representing a cell's whole genomic DNA inserted into it, are used to infect bacteria. Recombinant bacteriophage, released from the lysed cells, are collected and constitute a genomic library in λ phage. The phage library can be screened more rapidly than a plasmid library because of the increased size of the DNA inserts.

Numerous λ phage vectors have been constructed for different cloning strategies. Only a generic λ phage vector will be described. A 15- to 25-kb segment of the phage DNA can be replaced without impairing its replication in *E. coli* (Figure 7.19). The length

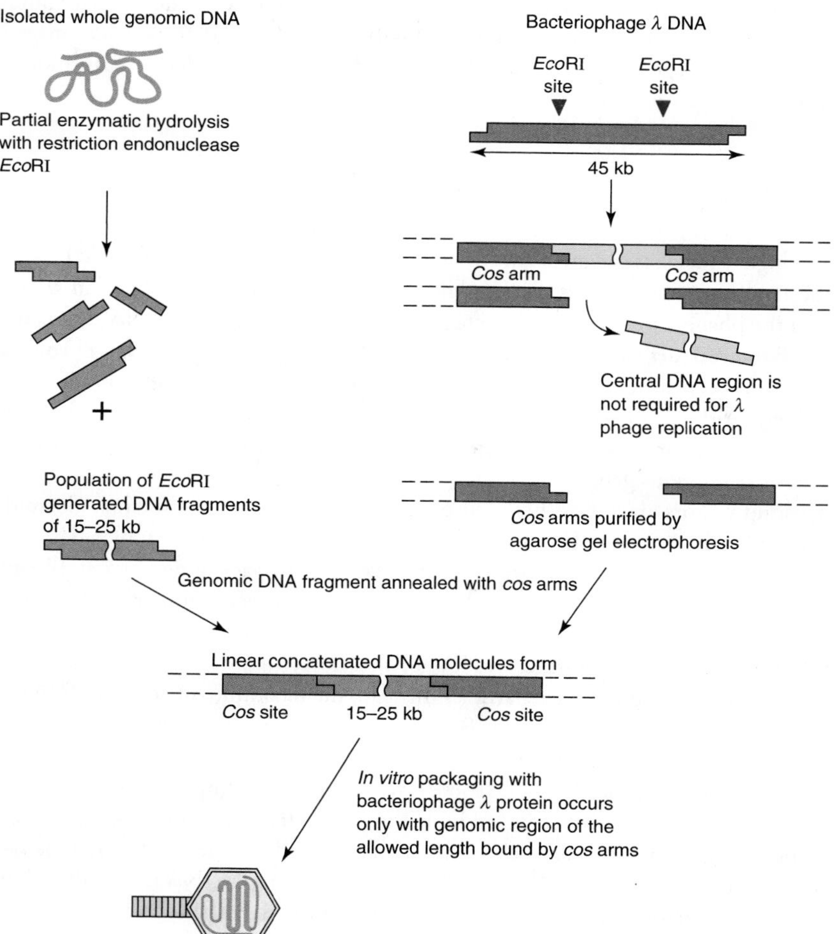

Figure 7.19 Cloning genomic DNA in bacteriophage λ. Whole genomic DNA is incompletely digested with a restriction endonuclease (e.g., EcoR1). This results in DNA of random size fragments with single-stranded sticky ends. DNA fragments called Cos arms, are generated with the same restriction endonuclease from bacteriophage DNA. The purified Cos arms carry sequence signals required for packaging DNA into a bacteriophage virion. The genomic fragments are mixed with the Cos arms, annealed and ligated, forming linear concatenated DNA arrays. The *in vitro* packaging with bacteriophage λ proteins occurs only with genomic DNA fragments of allowed lengths (15–25 kb) bounded by Cos arms.

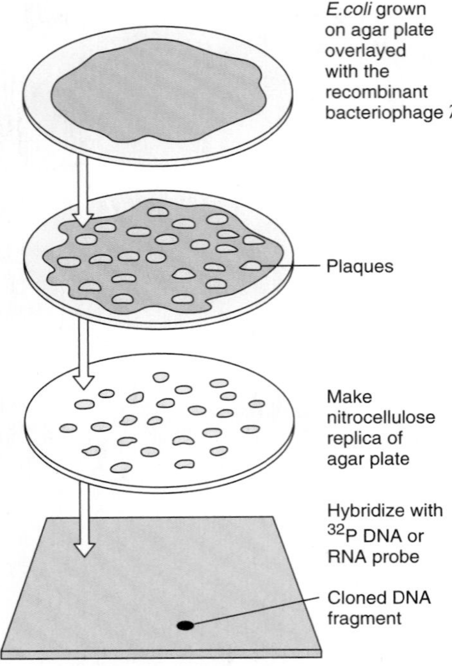

Figure 7.20 Screening genomic libraries in bacteriophage λ. Competent *E. coli* are grown to confluence on an agar plate and then overlaid with the recombinant bacteriophage. Plaques develop where bacteria are infected and subsequently lysed by the phage λ. Replicas of the plate can be made by touching the plate with a nitrocellulose filter. The DNA is denatured and fixed to the nitrocellulose with NaOH. The fixed DNA is hybridized with a ^{32}P-labeled probe and exposed to X-ray film. The autoradiograph identifies the plaque(s) with recombinant DNA of interest.

of phage DNA to be packaged into the virus particle must be about 50 kb long. The linear λ phage DNA is digested with a specific restriction endonuclease that generates small terminal fragments with their cos sites (arms). These are then separated from larger intervening fragments. Cellular genomic DNA is partially digested by appropriate restriction enzymes to permit annealing and ligation with the phage arms. DNA fragments smaller or larger than 15–25 kb hybridize with the cos arms but are excluded from being packaged into infectious bacteriophage particles. All the information required for phage infection and replication in bacteria is carried within the cos arms. The recombinant phage DNA is mixed with λ phage proteins *in vitro,* which assemble into infectious virions. These are then propagated in an appropriate *E. coli* strain to yield a λ phage library. Many *E. coli* strains have been genetically altered to sustain replication of specific recombinant virions.

Screening Bacteriophage Libraries

A bacteriophage library can be screened by plating the virus on a continuous layer of bacteria (a bacterial lawn) grown on agar plates (Figure 7.20). Individual phage will infect, replicate, and lyse one cell. The progeny virions will then infect and subsequently lyse bacteria immediately adjacent to the site of the first infected cell creating a clear region or plaque in the opaque bacterial lawn. Phage, within each plaque, is picked up on a nitrocellulose filter (as for replica plating) and the DNA fixed to the filter with NaOH. The location of cloned DNA fragments of interest are located by hybridizing the filter-bound DNA with a labeled DNA or RNA probe followed by autoradiography. The bacteriophage in the plaque corresponding to the labeled filter-bound hybrid are picked up and amplified in bacteria for further analysis. PCR is also employed if the full or partial sequence is known of the desired cDNA. Here one takes portions of several plaques, and if the DNA of interest is present in one or more, a region can be amplified with an appropriate primer pair by PCR. The PCR product is then detected by gel electrophoresis. Complementary DNA libraries in bacteriophage are also constructed that contain the phage cos arms. If the cDNA is recombined with phage DNA that permits expression of the gene, such as λgt 11, plaques can be screened immunologically with antibodies specific for the antigen of interest.

Cloning DNA Fragments into Cosmid and Artificial Chromosome Vectors

Even though λ phage is the most commonly used vector to construct genomic DNA libraries, the length of many genes exceeds the maximum size of DNA that can be inserted between the phage arms. A **cosmid vector** can accommodate foreign DNA inserts of about 45 kb. **Bacterial artificial chromosomes (BACs)** and **yeast artificial chromosomes (YACs)** have been developed to clone DNA fragments 100–200 and 200–500 kb long, respectively. Although cosmid and YAC vectors are difficult to work with, their libraries permit the cloning of large genes with their flanking regulatory sequences, as well as families of genes or contiguous genes.

Cosmid vectors are a cross between plasmid and bacteriophage vectors. Cosmids contain an antibiotic resistance gene for selection of recombinant DNA molecules, an origin of replication for propagation in bacteria, and a cos site for packaging of recombinant molecules in bacteriophage particles. The bacteriophage with recombinant cosmid DNA can infect *E. coli* and inject its DNA into the cell. Cosmid vectors contain only approximately 5 kb of the 50-kb bacteriophage DNA and, therefore, cannot direct its replication and assembly of new infectious phage particles. Instead, the recombinant cosmid DNA circularizes and replicates as a large plasmid. Bacterial colonies with recombinants of interest are selected and amplified by methods similar to those described for plasmids.

Standard cloning procedures and some novel methods are employed to construct YACs. Very large foreign DNA fragments are joined to yeast DNA sequences, one that functions as telomeres (distal extremities of chromosome arms) and another that functions as a centromere and as an origin of replication. The recombinant YAC DNA is introduced into the yeast by transformation. The YAC constructs are designed so that yeast transformed with

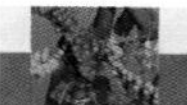

CLINICAL CORRELATION 7.8

The Use of Yeast Artificial Chromosome (YAC) Transgenic Mice to Study Huntington Disease

It is often difficult or impossible to study the role of a mutant protein's role in the genesis of the diseased state in human disorders. The development of animal models is an invaluable tool to study human diseases such as the adult onset neurodegenerative disorder, Huntington disease (HD) (OMIM 143100). HD is caused by the expansion of the CAG trinucleotide in exon1 of the HD gene resulting in polyglutamine stretchs in the coded Huntington (htt) protein. The resulting polyglutamine tracts appear to be toxic to selected regions of the brain. A transgenic mouse model was generated with a YAC transgene that contained the full-length human HD gene plus 25-kb upstream and 120-kb downstream sequences to ensure the inclusion of endogenous regulatory regions. YAC mice containing 46, 72, and 128 CAG repeats were created. The YAC 128 transgenic model most mirrored the human disease. These mice presented with motor and cognitive deficits that were similar to what was observed in humans. The wild-type htt protein is mainly found in the cytoplasm while the mutant protein localizes to the nucleus. Anti-htt antibody evaluation of brain tissues from wild-type (WT) mice and Y128 mice at 1 mon and 2 mon of age demonstrated that the normal and mutant htt proteins were not present in the nuclei of WT mice or 1-month-old Y128 mice; the mutant htt protein was detected in the nucleus only in the striatum of the 2-month-old Y128 mice; and as the Y128 mice aged further, although some mutant htt was detected in nuclei in other brain regions, it increased and remained greatest in the striatum. The striatum is the predominant region affected in both human HD and 2-month-old Y128 mice, and the accumulation of htt also corresponded with the age when motor and cognitive defects were first observed in these mice. It appears that the onset of the nuclear localization of the mutant htt protein in the striatum is the major contributor to the selective degeneration observed in HD. The mechanism of how the mutant htt protein exerts its toxic effects when localized to the nucleus is unknown but is thought to involve modification of gene expression and possibly activation of apoptosis.

Van Raamsdonk, J. M., Warby, S. C., and Hayden, M. R. Selective degeneration in YAC mouse models of Huntington Disease. *Brain Res. Bull.* 72: 124, 2007.

recombinant chromosomes grow as visually distinguishable colonies. This facilitates the selection and analysis of cloned DNA fragments. **Transgenic animals** can also be generated by the introduction of recombinant YAC DNA into their genome. If these animals express a defective human gene, they can be used as a model to study the corresponding human disease (Clin. Corr. 7.8).

Subcloning Permits Definition of Large Segments of DNA

Complete analysis of functional elements in a cloned DNA fragment requires sequencing of the entire molecule. Current techniques can sequence 200–400 bases, yet cloned DNA inserts are frequently much larger. Restriction maps of the initial DNA clone are essential for cleaving the DNA into smaller pieces to be **subcloned** for further analysis. The sequence of each subcloned DNA fragment can be determined. Overlapping regions of the subcloned DNA properly align and confirm the entire sequence of the original DNA clone.

Sequencing is often accomplished without subcloning. Antisense primers are used that are complementary to the initially sequenced 3′-ends of the cloned DNA. This process is repeated until all of the cloned DNA has been sequenced. This method obviates the need to prepare subclones but requires synthesis or purchase of numerous primers. On the other hand, the subcloned DNA is always inserted back into the same region of the plasmid. Therefore, one set of primers complementary to the plasmid DNA sequences flanking the inserted DNA is used for all of the sequencing reactions with subcloned DNA.

Chromosome Walking Defines Gene Arrangement in Long Stretches of DNA

Knowledge of how genes and their regulatory elements are arranged in a chromosome should lead to an understanding of how sets of genes may be coordinately regulated. It is difficult to clone DNA fragments large enough to identify contiguous genes. The combination of several techniques allows for analysis of stretches of DNA 50–100 kb long. The method, **chromosome walking**, is possible because λ phage or cosmid libraries contain partially cleaved genomic DNA cut at specific restriction endonuclease sites. The cloned fragments will contain overlapping sequences with other cloned fragments. Overlapping

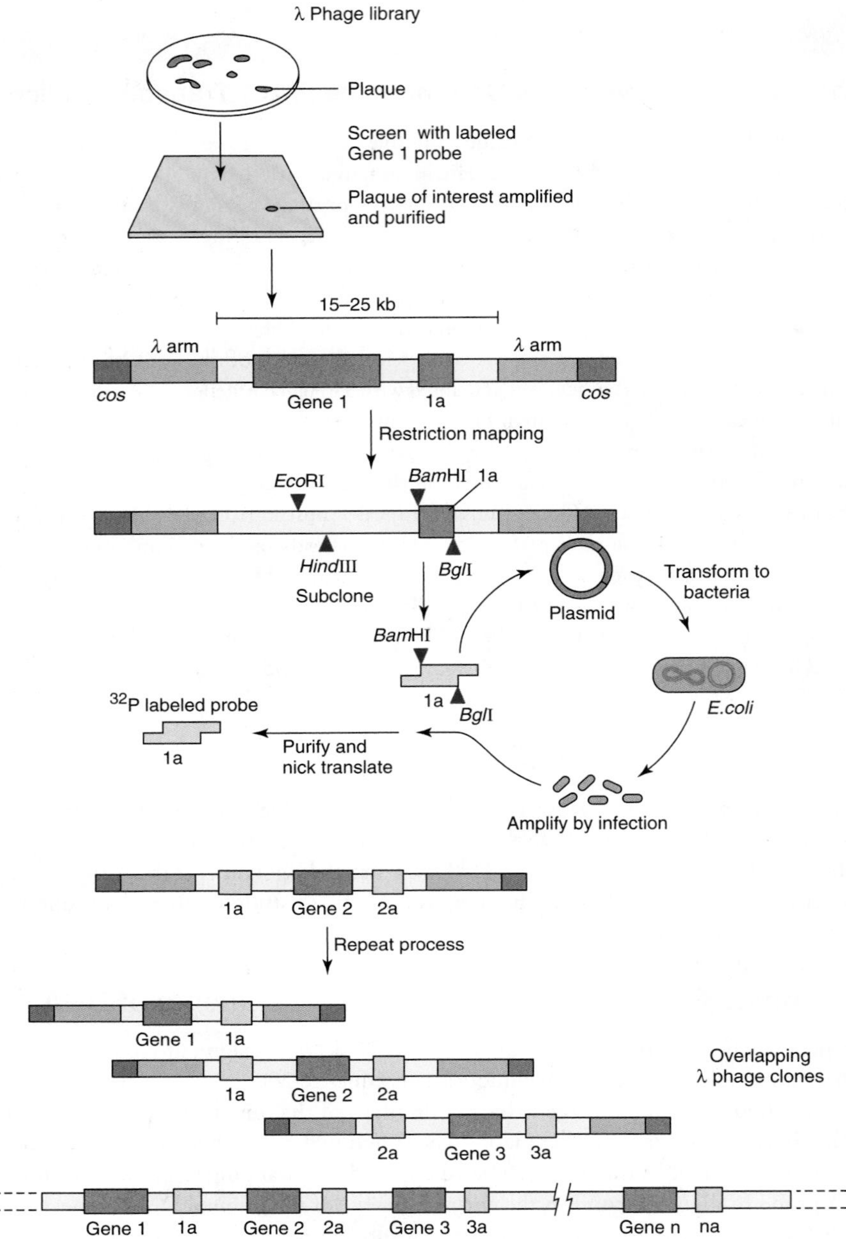

Figure 7.21 Chromosome walking to analyze contiguous DNA segments in a genome. Initially, a DNA fragment is labeled by nick translation to screen a library for recombinant λ phage carrying a gene of interest. The amplified DNA is mapped with several restriction endonucleases to select a new region (1a) within the original cloned DNA that can be recloned (subcloned). The subcloned DNA (1a) is used to identify other DNA fragments within the original library that would overlap the initially amplified DNA region. The process can be repeated many times to identify contiguous DNA regions upstream and downstream of the initial DNA (gene 1) of interest.

regions are identified by restriction mapping, subcloning, screening λ phage or cosmid libraries, and sequencing procedures.

The procedure of chromosome walking is shown in Figure 7.21. A λ phage library is screened with a DNA or RNA probe for a sequence of interest. The selected DNA is cloned and restriction mapped, and a small segment is subcloned in a plasmid, amplified, purified, and labeled by nick translation. This labeled probe is used to rescreen the λ phage library for complementary sequences and these are then cloned. The newly identified overlapping cloned DNA is then treated in the same fashion as the initial DNA clone to search for other overlapping sequences. In higher eukaryotic genomes caution must be taken so that subcloned DNA does not contain one of the large numbers of repeating DNA sequences. If a subcloned DNA probe contains a repeat sequence, it hybridizes to numerous bacteriophage plaques and prevents identification of a specific overlapping clone. These procedures in conjunction with chromosome jumping were employed to identify the human Cystic Fibrosis gene (Clin. Corr. 7.9).

CLINICAL CORRELATION 7.9

The Use of Chromosome Walking and Jumping to Identify the Cystic Fibrosis Gene

Cystic fibrosis (CF) (OMIM 602421) is a commonly encountered severe autosomal recessive disorder (1 in 2000 births) resulting from mutations in the CF transmembrane conductance regulator (CFTR). These mutations result in defective transport in the conducting pulmonary and exocrine glands. The primary clinical symptom associated with CF is chronic pulmonary disease. The determination of the CF locus was determined by linkage analysis of numerous polymorphic DNA markers to the long arm of chromosome 7, band q31. Approximately a 500-kb region of chromosome 7 was further analyzed by chromosome walking and jumping to identify the CF gene. Chromosome jumping allowed the researchers to initiate a new bidirectional walk from the end point of each jump. A restriction map was constructed from the phage and cosmid clones derived from chromosome walking and jumping and ultimately led to the identification of the CF gene (CFTR gene), which spans about 250 kb. The identification of the CFTR gene and its major mutation has allowed for the accurate prenatal and clinical diagnosis of the disease. In the ensuing years since the identification of the CFTR gene many studies have been directed toward the correction of the disease and the technological aspects required for the delivery of genes to target cells. However, success has been elusive. It has been recently reported that a new BAC vector has been constructed with a 250.3-kb insert that contains both the whole genomic CFTR gene and its upstream and downstream regulatory sites. The human CFTR gene was correctly expressed in a cultured mouse cell line from integrated copies of the BAC vector. These results point to the potential application for gene therapy of CF.

Rommens, J. M., Iannuzzi, M. C., Kerem, B., Drumm, M. L., et al. Identification of the Cystic Fibrosis gene: Chromosome walking and jumping. *Science* 245: 1059, 1989.

7.8 • EXPRESSION VECTORS AND FUSION PROTEINS

The recombinant DNA methodology described to this point has dealt primarily with screening, amplification, and purification of cloned DNA species. An important goal of recombinant DNA studies is to have a foreign gene expressed in bacteria with the product in a biologically active form. Sequencing of many bacterial genes and their flanking regions has identified the spatial arrangement of regulatory sequences required for expression of genes. A promoter and other regulatory elements upstream of the gene are required to transcribe a gene (p. 305). The mRNA transcript of a recombinant eukaryotic gene, however, is not translated in a bacterial system because it lacks the bacterial recognition sequence, the Shine–Dalgarno sequence, required to orient it properly with a functional bacterial ribosome. **Expression vectors** facilitate the functional transcription of DNA inserts. A foreign gene can be inserted into such a vector downstream of a regulated promoter but within a bacterial gene, commonly the lacZ gene. The mRNA transcript of the recombinant DNA contains the lacZ Shine–Dalgarno sequence, codons for a portion of the 5′-end of the lacZ protein, followed by the codons of the complete foreign gene of interest. The protein product is a **fusion protein** that contains a few N-terminal amino acids of the lacZ protein and the complete sequence of the foreign protein product.

Foreign Genes Expressed in Bacteria Allow Synthesis of Their Encoded Proteins

Many plasmid and bacteriophage vectors permit the expression of eukaryotic genes in bacterial cells. Rapidly replicating bacteria become a biological factory to produce large amounts of specific proteins, which have research, clinical, and commercial value. As an example, recombinant technologies have produced human protein hormones for treatment of patients with aberrant or missing hormone production. Figure 7.22 depicts a generalized plasmid vector for the expression of a mammalian gene. Recall that the inserted gene must be in the form of cDNA from its corresponding mRNA since the bacterial system cannot remove the introns in the pre- mRNA transcript. The DNA must be inserted in register with the codons of the 3′-terminal region of the bacterial protein when creating a fusion protein. That is, insertion must occur after a triplet codon of the bacterial protein and at the beginning of one for the eukaryotic protein to ensure proper translation. Finally, to yield a

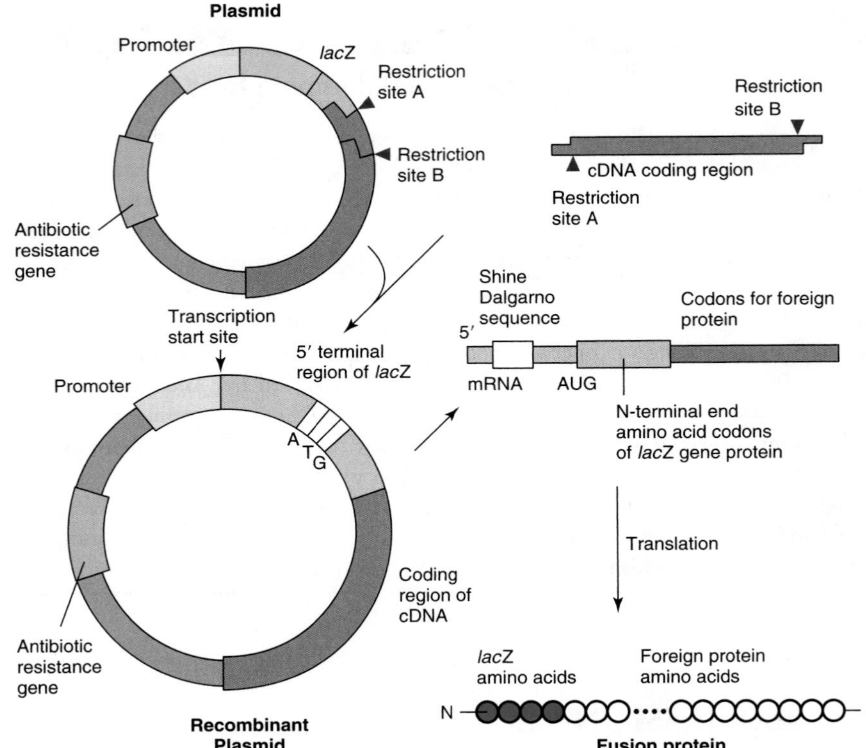

Figure 7.22 Construction of a bacterial expression vector. A cDNA for a protein of interest is inserted downstream of bacterial regulatory sequences (promoter [P]) for the lacZ gene, the coding sequence for the mRNA Shine–Dalgarno sequence, the AUG codon, and a few codons for the N-terminal amino acids of the LacZ protein. The mRNA produced from this expression vector directs synthesis in the bacterium of a fusion protein with a few of its N-terminal residues identical to those of the LacZ protein.

functional transcript, the foreign gene must be inserted in the proper orientation relative to the promoter. This is achieved by directional cloning.

Eukaryotic proteins synthesized within bacteria are often unstable and are degraded by intracellular proteases. Fusion protein products, however, are usually stable. The amino acids encoded by the prokaryotic genome may be cleaved from the purified fusion protein by enzymatic or chemical procedures. A strategy to circumvent the intracellular instability of some proteins is to produce a foreign protein that is secreted. This requires cloning the foreign gene in a vector so that the fusion protein synthesized contains a signal peptide that is recognized by the bacterial signal peptidase that properly processes the protein for secretion.

Expression Vectors in Eukaryotic Cells

Genetic diseases in mammals result from missing or defective intracellular proteins. To use recombinant techniques to treat these diseases, vectors have to be incorporated into mammalian cells. They also require some targeting mechanism such that the vectors are selective for the affected tissue or cells containing the aberrant protein. Numerous vectors permit the expression of foreign DNA genes in mammalian cells grown in tissue culture. They are used extensively for elucidation of the posttranslational processing and synthesis of proteins in cultured eukaryotic cells. The targeting of foreign genes to specific tissues or at specific developmental stages has met with very limited success.

Expression vectors have been developed that allow replication, transcription, and translation of foreign genes in cultured eukaryotic cells. They include both RNA and DNA viral vectors. Shuttle vectors contain both bacterial and eukaryotic replication signals thus permitting replication of the vector in both bacteria and mammalian cells. Shuttle vectors allow a gene to be cloned and purified in large quantities from a bacterial system and then transferred for expression in mammalian cells. Some expression vectors become integrated into the host cell genome and are stably expressed in the daughter cells. Other expression vectors remain as extrachromosonal or episomal DNA and permit only transient expression of their foreign gene prior to cell death.

Introduction of foreign DNA, such as viral expression vectors, into cultured eukaryotic cells is called **transfection**, a process that is analogous to transformation of DNA into bacterial cells. Two common methods involve the formation of a complex of DNA with calcium phosphate or diethylaminoethyl (DEAE)-dextran, which is then endocytosed and transferred to the nucleus, where it is replicated and expressed. Both methods are employed to establish transiently expressed vectors while the calcium phosphate procedure is also used for permanent expression of foreign genes. Typically, 10%–20% of the cells in culture are transfected by these procedures.

DNA Elements Required for Expression of Vectors in Mammalian Cells

Expression of recombinant genes in mammalian cells requires the presence of appropriate controlling elements within the vector. To be expressed, the cloned gene is inserted in the vector in the proper orientation relative to control elements which include a promoter, polyadenylation signals, and an enhancer. Expression may be improved by the inclusion of an intron. Some or all of these DNA elements may be present in the recombinant gene if whole genomic DNA is used for cloning. A particular cloned fragment generated by restriction endonuclease cleavage, however, may not contain these elements. A cDNA would not contain these required DNA elements. It is, therefore, necessary that the expression vector to be used in mammalian cells be constructed so that it contains all of the required controlling elements.

The controlling elements can be inserted into the vector by recombinant technologies. Enhancer and promoter elements should be recognized by a broad spectrum of cells in culture for the greatest applicability of the vector. Controlling elements derived from viruses with a broad host range, for example, **papovavirus**, **simian virus 40 (SV40)**, **Rous sarcoma virus**, and **human cytomegalovirus** are used. The vector must contain a viral origin of replication (ori) sequence. Specific protein factors, encoded by genes inserted in the vector or previously introduced into the host genome, recognize and interact with the ori sequence to initiate DNA replication.

Selection of Transfected Eukaryotic Cells by Utilizing Mutant Cells with Specific Nutrient Requirements

It is important to grow the transfected cells selectively since they often represent only 10%–20% of the cell population. A gene can be incorporated into a cell that confers resistance to a drug or confers selective growth capability. This is achieved by **cotransfection** in which two different vectors are efficiently taken up by cells, one that carries the selectable marker and the other that carries the gene of interest. In most cases, more than 90% of transfected cells carry both vectors. Two commonly used selectable markers are the thymidine kinase (tk) and the dihydrofolate reductase (dhfr) genes. Thymidine kinase is expressed in most mammalian cells where it participates in the salvage pathway of thymidine. Several mutant cell lines are available that lack a functional thymidine kinase gene (tk–) and do not survive in growth medium that contains hypoxanthine, aminopterin, and thymidine. Only those tk–mutant cells cotransfected with a vector carrying a tk gene, usually from herpes simplex virus, will grow in the medium. In most instances, these cells have been cotransfected with the gene of interest.

Dihydrofolate reductase maintains cellular concentrations of tetrahydrofolate for nucleotide biosynthesis (p. 813). Cells lacking this enzyme survive only in media that contains thymidine, glycine, and purines. Mutant cells (dhfr–), transfected with the dhfr gene, are, therefore, selectively grown in a medium that lacks these supplements. Expression of foreign genes in mutant cells, cotransfected with selectable markers, is limited to cell types with the required gene defect that can be isolated.

Normal cells, transfected with a vector that carries the dhfr gene, are resistant to methotrexate, an inhibitor of dihydrofolate reductase and are selected for by growth in methotrexate. Similarly, normal cells transfected with a bacterial gene coding for aminoglycoside 3′-phosphotransferase (APH) are resistant to aminoglycoside antibiotics such as neomycin and kanamycin; this inhibits protein synthesis in both prokaryotes and eukaryotes. Vectors that carry an APH gene are used as a selectable marker in both bacterial and mammalian cells.

Foreign Genes Expressed in Virus-Transformed Eukaryotic Cells

Figure 7.23 depicts the transient expression of a transfected gene in COS cells, a commonly used system to express foreign eukaryotic genes. COS cells are cultured simian cells, which were transformed with a SV40 genome with a defective origin. The defective viral genome has become integrated into the host cell genome and constantly expresses viral proteins. Infectious

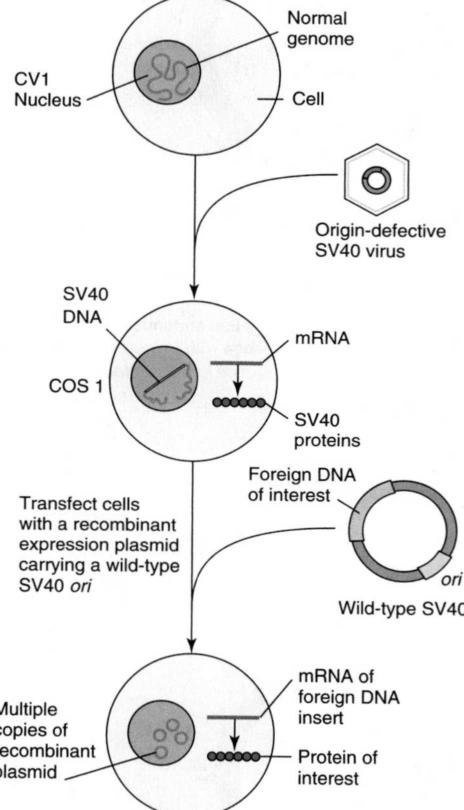

Figure 7.23 Expression of foreign genes in eukaryotic COS cells. CV1, an established tissue culture cell line of simian origin, can be infected and supports the lytic replication of the simian DNA virus, SV40. Cells are infected with an origin(ori)-defective mutant of SV40 whose DNA permanently integrates into the host CV-1 cell genome. The defective viral DNA continuously codes for proteins that can associate with a normal SV40 ori to regulate replication. Due to its defective ori, the integrated viral DNA will not produce viruses. The SV40 proteins synthesized in the permanently altered CV-1 cell line, COS-1, can, however, induce the replication of recombinant plasmids carrying a wild-type SV40 ori to a high copy number (as high as 10^5 molecules per cell). The foreign protein synthesized in the transfected cells may be detected immunologically or enzymatically.

viruses that are normally lytic to infected cells are not produced because the viral origin of replication is defective. The SV40 proteins expressed by the transformed COS cell interact with a normal SV40 ori carried in a transfected vector and therefore, promote the repeated replication of the vector that may reach a copy number in excess of 10^5 molecules per cell. Transfected COS cells die after 3–4 days probably from toxic overload of the episomal vector DNA.

7.9 • SITE-DIRECTED MUTAGENESIS

By mutating selected regions or single nucleotides within cloned DNA, it is possible to define the role of DNA sequences in gene regulation and amino acid sequences in protein function. **Site-directed mutagenesis** is the controlled alteration of selected regions of a DNA molecule. It may involve insertion or deletion of selected DNA sequences or replacement of a specific nucleotide with a different base. A variety of chemical methods mutate DNA usually at random sites within the molecule.

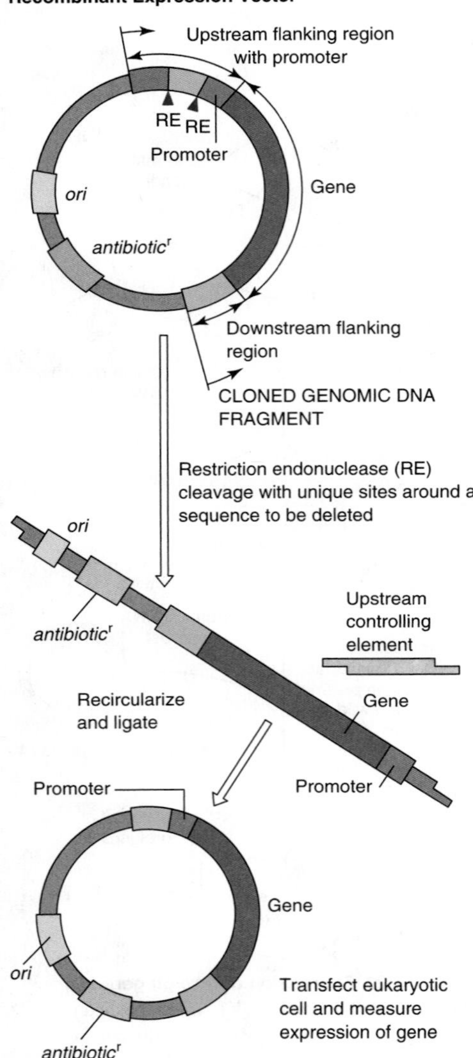

Figure 7.24 Use of expression vectors to study DNA regulatory sequences. The gene of interest with upstream and/or downstream DNA flanking regions is inserted and cloned in an expression vector, and the base-line gene expression is determined in an appropriate cell. Defined regions of potential regulatory sequences can be removed by restriction endonuclease cleavage, and the truncated recombinant DNA vector can be recircularized, ligated, and transfected into an appropriate host cell. The level of gene expression in the absence of the potential regulator is determined and compared to controls to ascertain the regulatory role of the deleted flanking DNA sequence.

Role of DNA Flanking Regions Evaluated by Deletion and Insertion Mutations

Site-directed mutagenesis is carried out in various regions of a DNA sequence including the gene itself or the flanking regions. Figure 7.24 depicts a simple deletion mutation strategy where the sequence of interest is selectively cleaved with restriction endonuclease, the specific sequences removed, and the altered recombinant vector recircularized with DNA ligase. The role of the deleted sequence can be determined by comparing the level of expression (translation) of the gene product, measured immunologically or enzymatically, to the unaltered recombinant expression vector. A similar technique is used to insert new sequences at the site of cleavage. Deletion of a DNA sequence within the flanking region of a cloned gene can indicate its regulatory role in gene expression. The spatial arrangement of regulatory elements relative to one another, to the gene, and to its promoter may also be important in the regulation of gene expression (p. 307).

Analysis of a potential regulatory sequence involves inserting the sequence upstream of a reporter gene in an expression vector. A reporter gene, usually of prokaryotic origin, encodes a protein that can be distinguished from proteins which are normally present in the nontransfected cell and which can be conveniently and rapidly assayed. A commonly used reporter gene is the chloramphenicol acetyltransferase (CAT) gene of bacteria. This enzyme acetylates and inactivates chloramphenicol, an inhibitor of protein synthesis in prokaryotic cells. The effect of a regulatory element on expression of the CAT gene then can be determined. The regulatory element can be mutated prior to insertion into the vector carrying the reporter gene to determine the element's spatial and sequence requirements.

Regulatory elements may lack restriction endonuclease sites. In such a case deletion mutations are made by linearizing cloned DNA by restriction endonuclease cleavage downstream of the regulatory sequence. The DNA is then systematically truncated with an exonuclease that removes nucleotides from the free end of both strands of the linearized DNA. Longer digestion generates smaller DNA fragments. Figure 7.25 demonstrates how larger deletion mutations (yielding smaller fragments) can be tested for functional activity. Hydrolysis of the DNA occurs at both ends of the linearized vector and destroys the original restriction endonuclease site (RE2). A unique restriction endonuclease site is reestablished to recircularize the truncated DNA molecule for further manipulations to evaluate the function of the deleted sequence. This is accomplished by ligating the blunt ends with a linker DNA, a synthetic oligonucleotide containing one or more restriction endonuclease sites. The ligated linkers are cut with the appropriate enzyme permitting recircularization and ligation of the DNA.

Site-Directed Mutagenesis of a Single Nucleotide

The procedures discussed previously can elucidate the functional roles of DNA sequences. Frequently, however, one wants to evaluate the role of one nucleotide at a selected site within the DNA so as to evaluate the role of specific amino acids in a protein (Clin. Corr. 7.10).

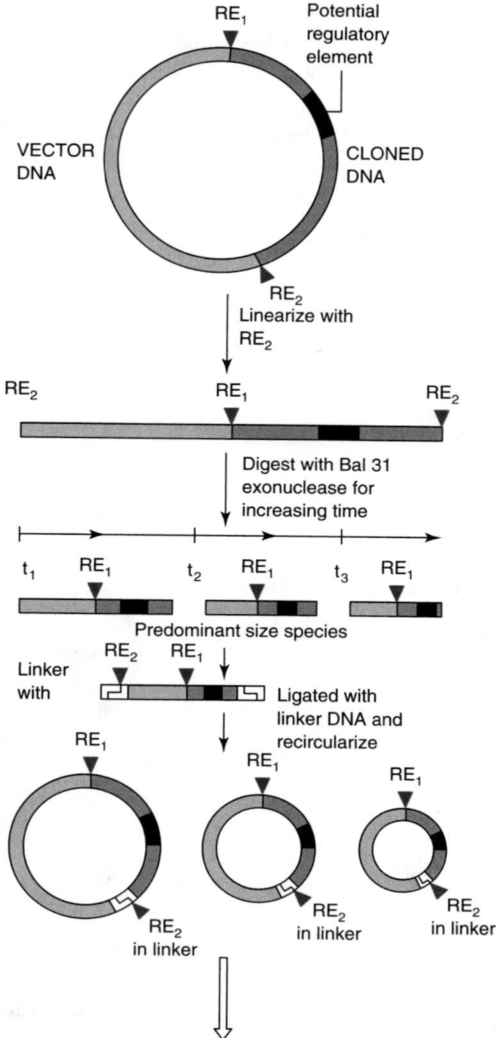

Figure 7.25 Enzymatic modification of potential DNA regulatory sequences. A purified recombinant DNA molecule with a suspected gene regulatory element within flanking DNA regions is cleaved with a restriction endonuclease (RE_2). The linearized recombinant DNA is digested for varying periods with the exonuclease, Bal31, reducing the size of the DNA flanking the potential regulatory element. The resulting recombinant DNA molecules of varying reduced sizes have small DNA oligomers (linkers) containing a restriction endonuclease sequence for RE_2 ligated to their ends. The linker modified DNA is hydrolyzed with RE_2 to create complementary single-stranded sticky ends that permit recircularization. The potential regulatory element, bounded by various reduced-sized flanking DNA sequences, can be amplified, purified, sequenced, and inserted upstream of a competent gene in an expression vector. Modification of the expression of the gene in an appropriate transfected cell can then be monitored to evaluate the role of the potential regulatory element placed at varying distances from the gene.

This method also permits the creation or destruction of a restriction endonuclease site at specific locations. Site-directed mutagenesis of a specific nucleotide is a multistep process that begins with cloning the normal gene in a bacteriophage (Figure 7.26). The M13 series of recombinant bacteriophage vectors are commonly employed. This filamentous bacteriophage specifically infects male *E. coli* that express sex pili (F factor) coded on a plasmid. The M13 bacteriophage contains DNA in a single-stranded or replicative form, which is replicated to double-stranded DNA within an infected cell. The double-stranded form is isolated from infected cells and used for cloning the gene to be mutated. The plaques of interest are visually identified by α-complementation (p. 268).

The recombinant M13 is used to infect susceptible *E. coli.* The progeny bacteriophage released into the growth medium contains single-stranded DNA. In **oligonucleotide-directed mutagenesis** an oligonucleotide (18–30 nucleotides long) is synthesized that is complementary to a region of interest except for the nucleotide to be mutated. This oligomer, with one mismatched base, hybridizes to the single-stranded gene and serves as a primer. Primer extension by bacteriophage T4 DNA polymerase results in double-stranded DNA that is transferred into susceptible *E. coli* where the mutated DNA strand serves as a template for new (+) strands that carry the mutated nucleotide.

The bacteriophage plaques, that contain mutated DNA, are screened by hybridization with a labeled probe of the original oligonucleotide carrying the altered nucleotide. By adjusting the wash temperature of the hybridized probe, only the perfectly matched hybrid remains complexed with the mutated DNA whereas the oligomer with mismatched

CLINICAL CORRELATION 7.10

Site-Directed Mutagenesis of Type VII Collagen (C7)

Dystrophic epidermolysis bullosa (DEB) (OMIM 226600) is a genetic disease resulting from a mutation in the COL7A1 gene that encodes for type VII collagen (C7). The C7 homotrimer molecule is an important component of the anchoring fibrils that link the basement membrane of the skin to the underlying papillary dermis. Mutated COL7A1 proteins result in diminutive or reduced numbers of anchoring fibrils. More than 300 distinct mutations have been identified within the COL7A1 gene with little correlation of genotype with the diseased phenotype. Patients with DEB present with mild to severe intensities of the diseased state depending on which mutation they have. All patients present with some degree of blistering, scarring, and skin involvement. Site-directed mutagenesis was used to generate four different single amino acid mutations. The purified recombinant C7 mutant proteins were then employed to analyze functional properties of the proteins *in vitro* that appeared to correspond to clinical phenotypes of the DEB patients.

Woodley, D. T., Hou, Y., Martin, S., Li, W., and Chen, M. Characterization of molecular mechanisms in underlying mutations in dystrophic epidermolysis bullosa using site-directed mutagenesis. *J. Biol. Chem.* 283:17838, 2008.

nucleotide will dissociate from the wild-type DNA. The M13 that carries the mutated gene is then replicated in bacteria, the DNA purified, and the mutated region of the gene sequenced to confirm the identity of the mutation. A modified method to selectively replicate the mutated strand has been developed to improve the efficiency of site-directed mutagenesis. The M13 bacteriophage, replicated in a mutant *E. coli,* incorporates some uracil residues in place of thymine because of a metabolic defect in synthesis of dTTP from dUTP and lack of an enzyme that normally removes uracil from DNA. The purified

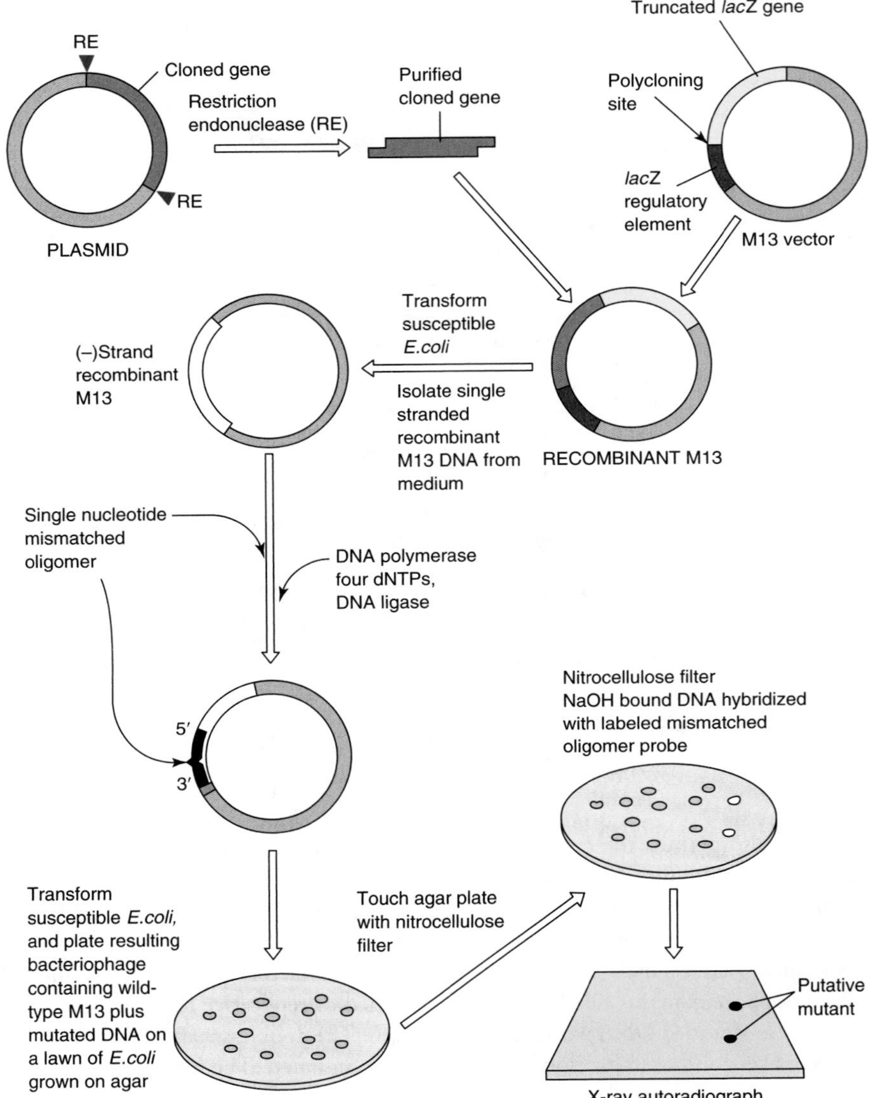

Figure 7.26 Site-directed mutagenesis of a single nucleotide and detection of the mutated DNA. The figure is a simplified overview of the method. This process involves the insertion of an amplified pure DNA fragment into a modified bacteriophage vector, M13. Susceptible *E. coli*, transformed with the recombinant M13 DNA, synthesize the (+) strand DNA packaged within the virion bacteriophage proteins. The bacteriophage are isolated from the growth medium and the single-stranded recombinant M13 DNA is purified. The recombinant M13 DNA serves as a template for DNA replication by DNA polymerase, deoxynucleoside triphosphates (dNTPs), DNA ligase, and a special primer. The DNA primer (mismatched oligomer) is synthesized to be exactly complementary to a region of the DNA (gene) of interest except for the one base intended to be altered (mutated). The newly synthesized M13 DNA, therefore, contains a specifically mutated base which when reintroduced into susceptible *E. coli* will be faithfully replicated. The transformed *E. coli* are grown on agar plates with replicas of the resulting colonies picked up on a nitrocellulose filter. DNA associated with each colony is denatured and fixed to the filter with NaOH and the filter-bound DNA is hybridized with a ^{32}P-labeled mismatched DNA oligomer probe. The putative mutants are then identified by exposing the filter to X-ray film.

single-stranded M13 uracil-containing DNA is hybridized with a complementary oligomer containing a mismatched base at the nucleotide to be mutated. The oligomer serves as the primer for DNA replication *in vitro* with the template (+) strand containing uracils and the new (–) strand containing thymines. When the double-stranded M13 DNA is transformed into a wild-type *E. coli* the uracil-containing strand is destroyed and the mutated (−) strand serves as the template for the progeny bacteriophage, most of which will carry the mutation of interest.

The PCR is also employed for site-directed mutagenesis. Strategies have been developed to incorporate a mismatched base into one of the oligonucleotides that primes the PCR. Some of these procedures employ M13 bacteriophage and follow the principles described in Figure 7.27. A variation of these PCR methods, **inverse PCR mutagenesis,** has been applied to small recombinant plasmids (4–5 kb). The method is very rapid with 50%–100% of the generated colonies containing the mutant sequence. The two primers are synthesized so that they anneal back-to-back with one primer carrying the mismatched base.

A method for site-directed mutagenesis that obviates the need for ss DNA and laborious selection steps has been commercially developed by Stratagene. The process employs oligonucleotide-directed mutagensis and PCR with any plasmid DNA vector isolated from

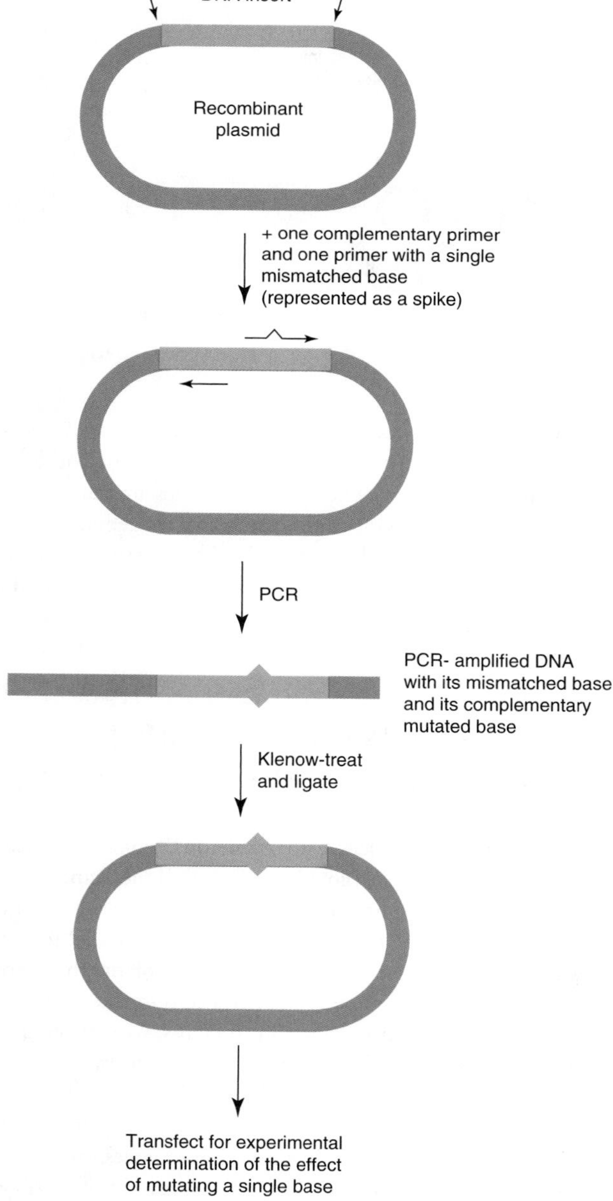

Figure 7.27 Inverse PCR mutagenesis. A single base can be mutated in recombinant DNA plasmids by inverse PCR. Two primers are synthesized with their antiparallel 5′-ends complementary to adjacent bases on the two strands of DNA. One of the two primers carries a specific mismatched base that is faithfully copied during the PCR amplification steps yielding ultimately a recombinant plasmid with a single mutated base.

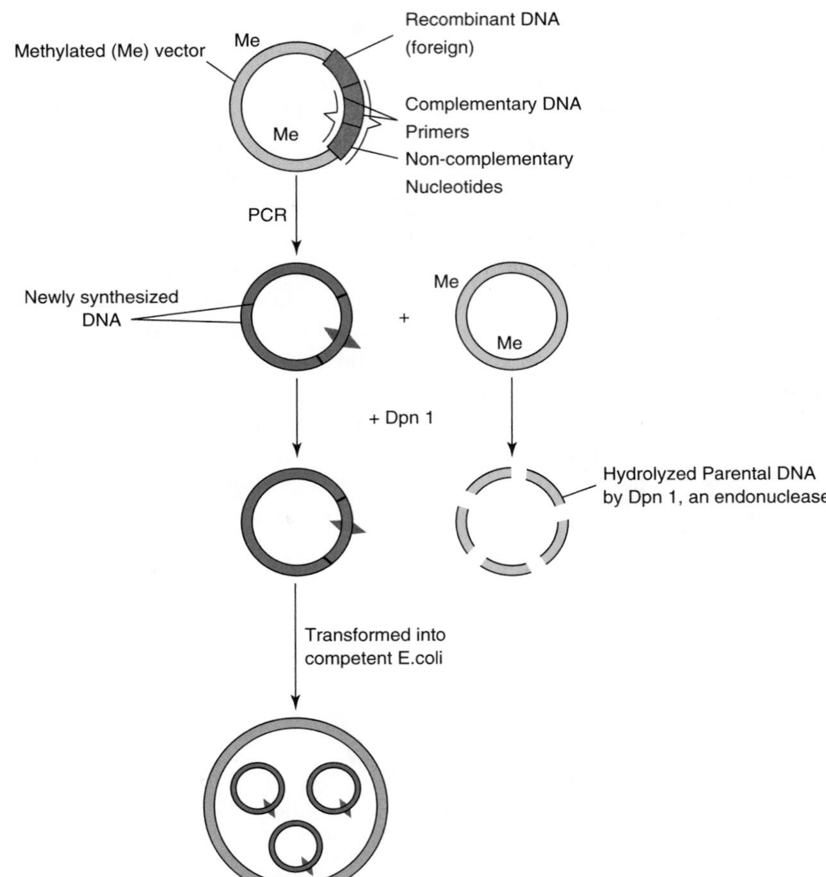

Figure 7.28 Site-directed mutagenesis of double-stranded DNA. A recombinant DNA molecule can be mutated at a selected site by synthesizing complementary primers that each contains a nucleotide that is not complementary to the parental DNA strand but are complementary to each other. The primers are extended by PCR leaving nicks (X) in each strand which is ligated upon transformation into competent E. coli. The methylated (Me) parental DNA is removed by hydrolysis with the endonuclease, Dpn I, prior to transformation.

DNA methylating competent *E. coli* strains. As depicted in Figure 7.28 one anneals two complementary oligonucleotides, carrying the desired altered nucleotide, to a region of a cloned DNA inserted within the vector. The oglionucleotides are then PCR extended with a Phusion-type DNA polymerase without primer displacement, generating a mutated recombinant plasmid molecule with nicks in both strands. The reaction mixture is then treated with an endonuclease, *Dpn*-I, which selectively digests the methylated parental (nonmutated) DNA. The remaining nicked mutated DNA plasmids (nonmethylated) are transformed into competent cells that ligate the nick and replicate the mutated recombinant plasmids.

7.10 • APPLICATIONS OF RECOMBINANT DNA TECHNOLOGIES

Recombinant DNA methods are applicable to numerous biological disciplines including agriculture, studies of evolution, forensic biology, and clinical medicine. Genetic engineering can introduce new or altered proteins into crops (e.g., corn), so that they contain amino acids essential to humans but often lacking in plant proteins. Toxins lethal to specific insects but harmless to humans can be introduced into crops to protect plants thus avoiding the use of environmentally destructive pesticides. The DNA from cells in the amniotic fluid of a pregnant woman can be analyzed for genetic defects in the fetus. Minuscule quantities of DNA from biological samples preserved in ancient tar pits or frozen tundra have been amplified and sequenced for evolutionary studies. DNA from a single hair, a drop of blood, or sperm from a rape victim can be isolated, amplified, and mapped for forensic purposes. Improvement in current technologies should permit selective introduction of genes into defective cells and of nucleic acids into specific cells to inhibit selectively the expression of detrimental genes.

Antisense Nucleic Acids in Research and Therapy

Antisense nucleic acids (RNA or DNA) are used to study the intracellular expression and function of specific proteins. Natural and synthetic antisense nucleic acids that are complementary to mRNA inactivate it and block translation. There are hundreds, perhaps thousands, of different micro RNAs (antisense RNA) encoded by the human genome. Each cell type expresses different combinations of these **microRNAs** that selectively regulate gene expression at the RNA level. Both natural and synthetic antisense RNAs are processed to or associated with a protein complex, RISC (RNA-induced silencing complex), that presents the approximately 21 oligomer (21 mer) antisense RNA to the sense RNA (usually mRNA) and the complementary strands hybridize. The sense/antisense RNA is then recognized for destruction/inhibition **(RNA interference [RNAi])**. Natural or synthetic antisense RNAs that are larger than the 21 mer must be hydrolyzed to the appropriate size by an enzyme called Dicer resulting in the **small-interfering RNA (siRNA)**. The stable introduction of siRNA into cells can permanently reduce or knockdown the expression of selected genes. RNAi technologies are currently being used in numerous laboratories as a research tool to elucidate specific gene function and by many pharmaceutical and biotechnology companies to develop RNAi-based therapeutics. The introduction of antisense nucleic acids into cells has opened new avenues to explore how selectively repressed proteins function within that cell. This method also holds great promise in control of viral infections. Antisense technology and site-directed mutagenesis are part of reverse genetics (from gene to phenotype), which selectively modifies a gene to evaluate its function, while classical genetics depends on the isolation and analysis of cells carrying random mutations that can be identified. A second use of the term **reverse genetics** refers to the mapping and ultimate cloning of a human gene associated with a disease for which the molecular agents are still unknown.

Antisense RNA is introduced into cells by common cloning techniques. Figure 7.29 demonstrates a gene cloned in an expression vector downstream of a promoter but in the opposite direction to normal. This places the complementary or antisense strand of the DNA under the control of the promoter with expression- or transcription-yielding antisense RNA. Transfection of cells with the antisense expression vector introduces antisense RNA that hybridizes with normal mRNA. The mRNA–antisense RNA complex is not translated due to a number of reasons such as its inability to bind to ribosomes, blockage of normal processing, and rapid enzymatic degradation.

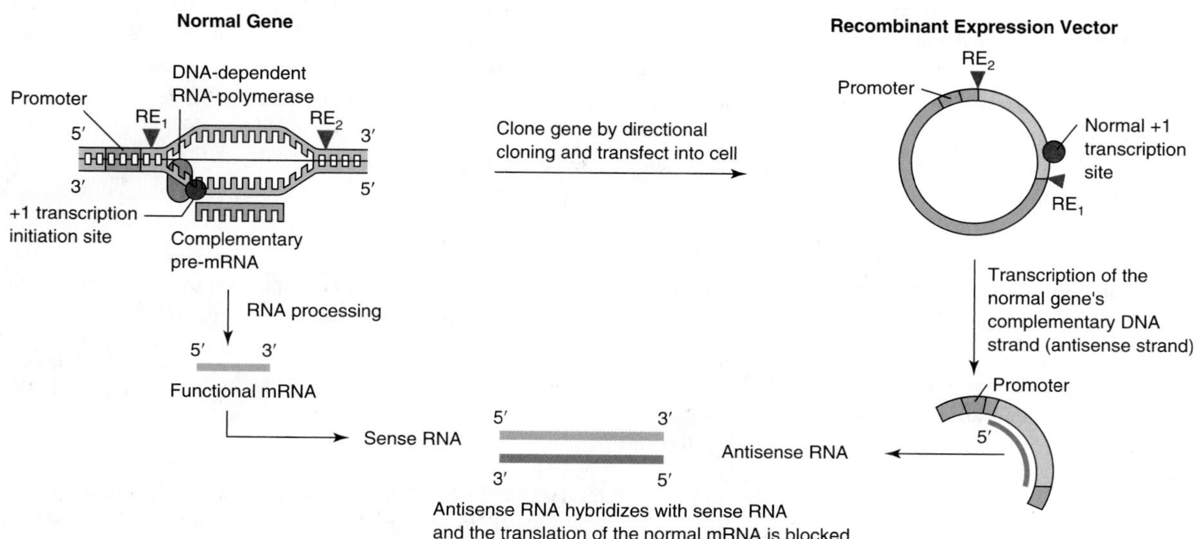

Figure 7.29 Production of antisense RNA. A gene, or a portion of it, is inserted into a vector downstream of a promoter by directional cloning and in the reverse orientation to that normally found. Transfection of this recombinant DNA into the parental cell that carries the normal gene results in the transcription of RNA (antisense RNA) from the cloned reversed-polarity DNA along with a normal cellular mRNA (sense RNA) transcript. The two antiparallel complementary RNAs hybridize within the cell resulting in blocked expression (translation) of the normal mRNA transcript.

Synthetic antisense DNA oligonucleotides have been produced that inhibit viral infection including that by the human immunodeficiency virus (HIV). It is conceivable that antisense HIV nucleic acids will be introduced into bone marrow cells removed from AIDS patients. These protected cells will then be reintroduced into the patient and replace those cells normally destroyed by the virus. Progress is being made with antisense nucleic acids that regulate the expression of oncogenes, genes involved in cancer formation. Harnessing antisense technologies holds great promise for the treatment of human diseases (Clin. Corr. 7.11). Although many studies have been conducted with siRNA that are 21 bases long to trigger RNAi, it has been recently demonstrated that the use of slightly longer double-stranded antisense RNAs, which are then substrates for Dicer, are more effective inhibitors of gene expression. This is thought to result from the participation of Dicer with the formation of the RISC complex (Clin. Corr. 7.11B).

Molecular Techniques Applied to the Whole Animal

As described, many *in vitro* and intracellular systems facilitate the purification, sequencing, and modification of genes or their cDNAs and DNA flanking regions of genes with potential regulatory functions. Introduction of foreign genes into a whole animal, deletion of a gene from the whole animal genome, and cloning of a whole animal has been achieved. While clear bioethical issues and hurdles must be addressed when using some of these methodologies, the methods themselves hold great promise for future gene therapy.

Gene Therapy: Normal Genes Can Be Introduced into Cells with a Defective Gene

Individuals who possess a defective gene resulting in a debilitating or fatal condition can theoretically be treated by supplying their cells with a normal gene. **Gene therapy**, the transfer of a normal gene to humans, has been accomplished using retroviral vectors (Clin. Corr. 7.12). The success of gene transfer depends, in part, on the integration of the gene into the host

CLINICAL CORRELATION 7.11

siRNA-Mediated Regulation of Gene Expression

A. Aptamer–siRNA for HIV-1 Therapy

The use of short-interfering RNA, siRNA, (antisense RNA) to regulate gene expression of HIV-1 proteins in diseased cells requires the efficient delivery of siRNA to target cells. An aptamer has been engineered to recognize and selectively bind to gp120 (glycoprotein 120). The gp120 is a HIV-1 protein that is expressed on the membrane of infected cells. An aptamer is a nuleic acid molecule capable of selective binding of target molecules such as specific proteins. Chimeric molecules of the anti-gp120 aptamer and anti-tat/rev siRNA were produced by covalently linking the two species. The siRNA was a 27-mer that serves as a Dicer substrate. The aptamer-siRNA was selectively taken up by HIV-1 infected cells expressing gp120. The 27-mer siRNA was then processed intracellularly by Dicer yielding an inhibitory 21-mer anti-tat/rev siRNA. The knockdown of expression of the HIV-1 tat/rev genes along with the antiviral effect of the anti-gp120 aptamer had a combined inhibitory effect on HIV-1 replication. The use of aptamer–siRNA chimeras holds great promise for the systemic treatment of HIV infections and other diseases that present potential surface target markers.

B. Use of Dicer-Substrate siRNA for Pain Research

The majority of drugs sold today that act on the Central Nervous System (CNS) are directed at the treatment of pain. The two main classes of analgesic drugs are opioids and nonsteroidal anti-inflammatory drugs. All too often, these drugs are ineffective, or lose their potency in the treatment of chronic pain. It has been shown in a model that G-protein-coupled receptor, NTS2, in the CNS is involved in the transmission of pain. RNAi (RNA interference) was employed to knock down the expression of the NTS2 gene in mice. It was found that use of a 27-mer (antisense RNA to the NTS2 mRNA) that would serve as a Dicer-substrate siRNA (DsiRNA) efficiently suppressed the expression of NTS2. The DsiRNA was suspended in a cationic lipid, i-Fect, and administered into the rat spinal cord by intrathecal injection. The administration of a low dose of DsiRNA ameliorated pain for up to 3–4 days which corresponded to a knockdown of the NTS2 receptor protein over the same time period. This type of treatment with DsiRNA may ultimately help reduce intractable chronic pain in humans.

Zhou, J. I., Li, H., Li, S., Zaia, J., and Rossi, J. J. Novel dual inhibitory function aptamer-siRNA delivery system for HIV-1 therapy. *Mol. Therapy* 16:1358, 2008.
Dore-Savard, L., Roussy, G., Dansereau, M. A., Collingwood, M. A., et al. Central delivery of Dicer-substrate siRNA: A direct application for pain research. *Mol. Therapy* 16:1331, 2008.

CLINICAL CORRELATION 7.12

Gene Therapy: Normal Genes Can Be Introduced into Cells with Defective Genes

A. Gene therapy to treat severe combined immune deficiency disease (SCID) (OMIM 102700)

More than 4000 genetic diseases are known, many of which are debilitating or fatal. Most are currently incurable. With the advent of new technologies in molecular biology, the clinical application of gene transfer and gene therapy is becoming a reality. Diseases that result from a deficiency in adenosine deaminase (ADA) or a mutation in the gene that encodes a subunit of several cytokine receptors, γc, are but two of many genetic diseases that may be readily cured by gene therapy.

ADA is important in purine salvage, catalyzing the conversion of adenosine to inosine or deoxyadenosine to deoxyinosine. It contains 363 amino acids and has highest activity in thymus and other lymphoid tissues. Over 30 mutations in the ADA gene are associated with severe combined immune deficiency disease (SCID), an autosomal recessive disorder. These immune-compromised children usually die in the first few years of life from overwhelming infections. The first authorized gene therapy in humans began on September 14, 1990, with the treatment of a 4-year-old girl with ADA deficiency. The patient's peripheral blood T cells were cultured with appropriate growth factors. The ADA gene was introduced within these cells by retroviral-mediated gene transfer. A modified retrovirus was constructed to contain the human ADA gene such that it would be expressed in human cells without virus replication. The modified T cells carrying a normal ADA gene were then reintroduced to the patient by autologous transfusion. Levels of ADA as low as 10% of normal are sufficient to normalize the patient. The patient, 10 years later at age 13, was alive and well and maintained circulating gene-corrected T cells at a level of 20%–25% of total T cells.

The X-linked inherited disorder, SCID-X1, results in the blocked differentiation of T and natural killer (NK) lymphocytes. SCID-X1 results from a mutation in the cytokine receptor subunit, γc, common to interleukin-2, -4, -7, -9, and -15 receptors. The methodology of treatment for an 8- and an 11-month-old patient was similar to that employed for the ADA-deficient SCID patients; this tends to be a less severe disease than the SCID-X1. The SCID-X1 studies transduced CD34+ stem cells with a retroviral vector carrying the normal γc cDNA. This resulted in a much higher level of gene transduction than observed in earlier studies. Both infants had their immune systems normalized sufficiently after 3 months to leave the protective isolation of the hospital.

A moratorium on gene therapy for SCID-X1 was triggered in 2003 by the induction of a leukemia-like syndrome in 2 of 14 European children treated for this disease. The T-cell leukemia was caused by retroviral insertion and up regulation of the LMO2 gene that encodes a transcription factor required for hematopoiesis. It is anticipated that future modifications in the vector used for gene therapy should prevent the apparent insertional preference of the retrovirus near active genes.

B. Gene Therapy to Treat Leber Congenital Amaurosis

An exciting use of genetic therapy has recently been reported to correct visual function in Leber congenital amaurosis which represents multiple forms of severe, recessively inherited infantile-onset rod-cone dystrophies. A 65-kD protein is a key component of the visual cycle and is encoded for by the RPE65 gene in the retinal pigment epithelium. A nonfunctional RPE65 gene causes a loss of 11-cis retinal, which is required for red photoreceptor cells to respond to light. It appears that children possess an alternate pathway for the availability of 11-cis retinal which disappears as they age. Children with this disease have visual function that diminishes with age with the progressive degeneration of cone photoreceptor cells. Three patients (ages 17–23) had a recombinant adeno-associated virus vector containing the human RPE65 coding sequence surgically administered to the subretinal space of one eye. At the time of surgery each patient retained some visual function in good lighting but limited, or no vision in low lighting. The results of this study indicate that patients with advanced degeneration of the cone photoreceptor cells achieved modest improvement in visual function with RPE65 gene therapy. It is postulated that such gene therapy given to younger patients would be more beneficial than with adults.

Blaese, R. M., Progress toward gene therapy. *Clin. Immun. Immunopath* 61:574, 1991; Mitani, K., Wakamiya, M., and Caskey, C. T. Long-term expression of retroviral-transduced adenosine deaminase in human primitive hematopoietic progenitors. *Human Gene Therapy* 4:9, 1993; Anderson, W. F. The best of times, the worst of times. *Science* 288:627, 2000; Cavazzana-Calvo, M., Hacein-Bey, S., de Saint Basile, G., Gross, F., et al. Gene therapy of human severe combined imunodeficiency(SCID)-X1 disease. *Science* 288:669, 2000; Hacein-Bey, S., VonKalle C., Schmidt, M., McCormack, N. P., et al. LMO2- associated clonal T-cell proliferation in two patients after gene therapy for SCID-X1. *Science* 302:415, 2003; Bainbridge J. W., Smith, A. J., Barker, S. S., Robbie, S., et al. Effect of gene therapy on visual function in Leber congenital amaurosis. *N. Eng. J. Med.* 358:2231, 2008; and Maguire, A. M., Simonelli, F., Pierce, E. A., Pugh, E. N., Jr., et al. Safety and efficacy of gene transfer for Leber's congenital amaurosis. *N. Engl. J. Med.* 358:2240, 2008.

genome, which is directed by the retroviral integration system. Integration, however, is normally a random event that could have deleterious consequences. Studies indicate that the viral integration machinery is selectively tethered to specific target sequences within the host DNA by protein–protein interactions to obviate these potential problems.

Transgenic Animals

In order to investigate the role of a selected gene product in growth and development of a whole animal, the gene must be introduced into the fertilized egg. Foreign genes can be inserted into the genome of a fertilized egg. Animals that develop from such a fertilized egg carry the inserted gene in every cell and are referred to as **transgenic animals**.

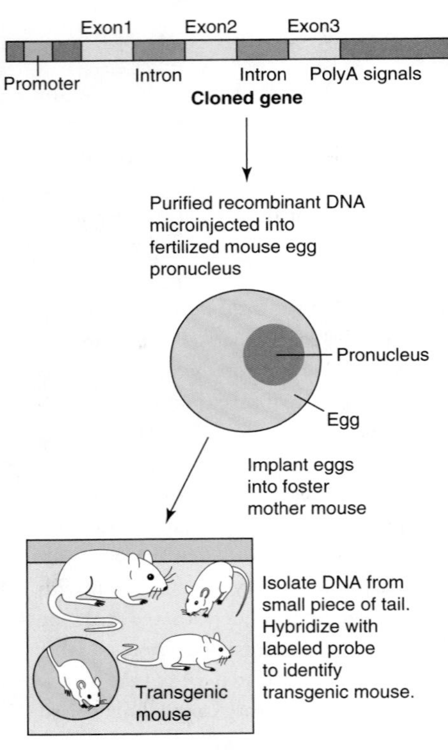

Figure 7.30 Production of transgenic animals. Cloned, amplified, and purified functional genes are microinjected into several fertilized mouse egg pronuclei *in vitro*. The eggs are implanted into a foster mother. DNA is isolated from a small piece of each offspring pup's tail and hybridized with a labeled probe to identify animals carrying the foreign gene (transgenic mouse). The transgenic mice can be mated to establish a new strain of mice. Cell lines can also be established from tissues of transgenic mice to study gene regulation and the structure or function of the foreign gene product.

Figure 7.30 outlines a popular method to create transgenic animals. The gene of interest is usually a cloned recombinant DNA molecule with its own promoter or is cloned with a different promoter that can be selectively regulated. Multiple copies of the gene are microinjected into the pronucleus of the fertilized egg. The foreign DNA inserts randomly within the chromosomal DNA. If an insert disrupts a critical cellular gene, the embryo will die. Usually, nonlethal mutagenic events result from the insertion of foreign DNA into the chromosome.

Transgenic animals are currently being used to study DNA regulatory elements, expression of proteins during differentiation, tissue specificity, and the role of oncogene products on growth, differentiation, and induction of tumorigenesis. Such technologies should permit replacing defective genes in the developing embryo (Clin. Corr. 7.13).

Knockout Mice

The creation of an animal with a selected gene destroyed in every cell allows researchers to define the biological role of the gene if its loss is nonlethal. The basic principles behind creating a **null**, or **knockout, mouse** involves inactivating a recombinant purified gene, introducing this altered gene into an embryonic stem (ES) cell such that it replaces the normal gene, injecting the modified ES cell into a developing blastocyst, implanting the blastocyst into a foster mother, and selecting offspring lacking a normal gene. A common approach to inactivate a selected gene is to insert an antibiotic resistance gene, known as neo, into the cloned gene of interest. This both inactivates the gene of interest and allows identification of ES cells that have been successfully transfected. The normal gene in a very low percentage of ES cells may be replaced by the neo-disrupted recombinant gene by **homologous recombination**. The nucleotide sequences of the altered or disrupted gene align with the homologous sequence of the normal gene and swap places, or may insert within the normal gene. In either case the net result is to destroy or knockout the normal ES cellular gene. Selection of altered ES cells, microinjection into a mouse blastocyst, birth of chimeric offspring, and breeding of offspring homozygous for the altered gene (Figure 7.31) is an extremely labor-intensive process. The process has been applied in numerous experiments to evaluate the role of selected genes (Clin. Corr. 7.14).

Dolly, a Lamb Cloned from an Adult Cell

The cloning of animals raises numerous ethical questions such as: Is it man's purview to select the genetic composition of animals and to propagate these animals? Clearly, there are many experimental advantages to developing animals with identical genetic information. However, the application of whole animal cloning techniques to humans raises ethical issues that most consider unacceptable.

Cloning viable offspring from adult mammalian cells is: very labor intensive. The method that produced Dolly, the cloned lamb, began with the culturing of mammary gland cells from a 6-year-old Finn Dorset ewe. Cells were arrested in G_o by being maintained in

CLINICAL CORRELATION 7.13

Transgenic Animal Models

Transgenic animal models hold promise to correct genetic diseases early in fetal development. They are used to study the regulation of expression and the function of specific gene products in an animal and may lead to creation of new breeds of commercially valuable animals. Transgenic mice have been produced from fertilized mouse eggs with rat growth hormone (GH) genes microinjected into their male pronuclei (p. 292). Approximately 600 copies of the gene, fused to the mouse metallothionein-I (MT-I) promoter region were introduced into each egg, which was then inserted into the reproductive tract of a foster mother mouse. The resulting transgenic mouse carried the rat GH gene within its genome. The diet of the animals was supplemented with $ZnSO_4$ at 33 days postparturition to activate the mouse MT-I promoter and initiate transcription of the rat GH gene. Continuous overexpression of rat GH in some transgenic animals produced mice nearly twice the size of littermates that did not carry the rat GH gene. A transgenic mouse transmitted the rat GH gene to one-half of its offspring, indicating that it had stably integrated into the germ cell genome and that new breeds of animals can be created.

Palmiter, R. D., Brinster, R. L., Hammer, R. E., Trumbauer, M. E., et al. Dramatic growth of mice that develop from eggs microinjected with metallothionein-growth hormone fusion genes. *Nature* 300:611, 1982.

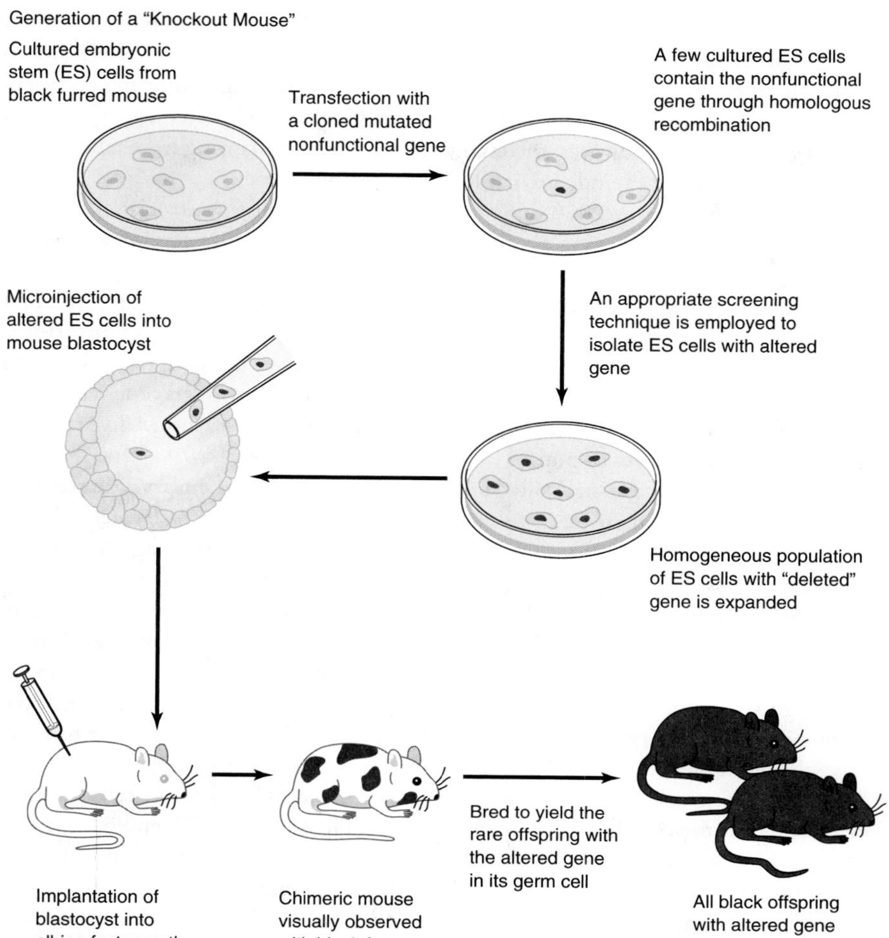

Figure 7.31 Generation of a knockout mouse. Cultured embryonic cells can be manipulated through recombinant technology to carry a defective or "deleted" gene. The altered cells can be introduced into a blastocyst that is then implanted into a foster mother. Offspring can then be selected that have the nonfunctional "deleted" gene in all of their cells—the knockout animal.

CLINICAL CORRELATION 7.14

Knockout Mice to Define a Role for the P2Y$_1$ Purinoceptor

Platelets play a central role in hemostasis and thrombosis. An important physiologic agonist that induces platelet aggregation is ADP. Studies in the 1970s established the dogma that ADP, bound to a G protein-coupled receptor, induced aggregation while extracellular ATP, acting at the same receptor site, competitively inhibits aggregation. Sometimes rules, based on limited evidence, become firmly entrenched and take years to be reworked. In the 1990s it became clear that platelets possess at least three different purinoceptors with different binding specificities for ATP versus ADP. An understanding of how these potentially interactive receptors function is very important clinically. ATP and ADP are released into the cardiovascular system/blood under a variety of normal and pathological conditions. The balance between the agonist action of ADP and the antagonist action of ATP may play a significant role in the regulation of platelet responses. Defining the mechanisms of action may lead to the production of new drugs to combat thrombosis. The recent development of a knockout (null) mouse model allowed for the assessment of the role of the purine receptor, P2Y$_1$.

Although the P2Y$_1$ receptor is found in many tissues other than platelets, the knockout mouse had no apparent abnormalities. However, normal levels of ADP failed to induce their platelets to aggregate, whereas high concentrations of ADP did. Aggregated platelets did not show the usual change in shape. The P2Y$_1$-null mice were resistant to thromboembolism, and thus this receptor may be a good target for antithrombotic drugs.

Soslau, G., and Youngprapakorn, D. A possible dual physiological role of extracellular ATP in the modulation of platelet aggregation. *Biochim. Biophys. Acta* 1355:131, 1997; and Leon, C., Hechler, G., Freund, M., Eckly, A., et al. Defective platelet aggregation and increased resistance to thrombosis in purinergic P2Y$_1$ receptor-null mice. *J. Clin. Invest.* 104:1731, 1999.

culture medium with a low serum content. The nuclei from quiescent diploid donor cells were then transferred to anucleated oocytes derived from a Scottish Blackface ewe. The artificially fertilized eggs developed to the morula or blastocyst stage in culture and then transferred to a recipient ewe that carried the fetus to term. Only a small percentage of implanted embryos survive to term. However, the methodology to clone offspring from adult somatic cells has been successfully extended to cattle, mice, and monkeys.

Dolly was euthanized in 2004 due to a progressive lung disease associated with what appeared to be a premature aging process. It is not clear if these defects were a result of the cloning process and/or environmental factors. However, cloning techniques continue to be explored with the intent that cloning will have great therapeutic value. Recently a female mouse was created from two female eggs by bypassing a process called **imprinting**. Imprinting dictates that some active genes must be contributed by the mother and others from the father. Normally mammalian embryos engineered with two sets of female or male chromosomes will not develop. The female mouse created from two mothers lived to adulthood and bore her own litter of pups.

With the success of creating Dolly and the isolation of **human embryonic stem cells (hESC)**, scientists began their quest to combine the two techniques to create embryonic stem cells genetically tailored for a particular patient, **therapeutic cloning**. A major step in this direction has been accomplished by the Herculean effort that created a single hESC line derived from a cloned blastula. The cloned blastula was created by inserting the nucleus of a human cumulus cell (cells surrounding an egg in an ovary) into an anucleated human egg and subjecting it to the appropriate chemical environment to induce replication. The hESC line grows normally in culture and should be able to differentiate into all cell types of the body. These cells could be induced to form normal cells of any tissues to replace defective tissues in patients.

Embryonic stem cells are **pluripotential cells** that can develop into any cell type within an organism given the correct environment and signaling molecules. Much hope and ethical concerns have swirled around the development and use of embryonic stem cells for the treatment of many human diseases. Recent studies demonstrate that human somatic cells can be reprogrammed to pluripotential, embryonic-like stem cells. The stable introduction of four transcription factors, Oct 4, Sox2, Klf4, and Myc, into mouse or human somatic cells by viral vectors resulted in the formation of **induced-pluripotent stem (iPS) cells**. Although much work remains to demonstrate the safety of using these iPS cells for clinical treatment of diseases, their potential is enormous. Furthermore, iPS cells can be patient specific to obviate any immune response.

Recombinant DNA in Agriculture Has Commercial Impact

Perhaps the greatest gain to humanity would be the practical use of recombinant technology to improve agricultural crops. Genes must be identified and isolated that code for properties such as higher crop yield, rapid plant growth, resistance to adverse conditions such as arid or cold periods, and plant size. New genes, not common to plants, may be engineered into plants so as to confer resistance to insects, fungi, or bacteria. Finally, genes for structural proteins can be modified to contain essential amino acids not normally present, without modifying the function of the protein. Production of plants with new genetic properties requires introduction of genes into plant cells that can differentiate into whole plants.

New genetic information carried in crown gall plasmids can be introduced into plants infected with soil bacteria known as agrobacteria. Agrobacteria naturally contain a crown gall or Ti (tumor-inducing) plasmid whose genes integrate into an infected cell's chromosome. The plasmid genes direct the host plant cell to produce new amino acid species called opines (such as the amino acid derivatives of arginine and pyruvate, octopine, or arginine and α-ketoglutarate, nopaline) that are required for bacterial growth. A crown gall, or tumor mass, of undifferentiated plant cells develops at the site of bacterial infection. New genes engineered into the Ti plasmid can be introduced into plant cells on infection with the Agrobacteria. The cells can then be grown in culture and induced to redifferentiate into whole plants. Every cell would contain the new genetic information and would represent a transgenic plant.

Some limitations in producing plants with improved genetic properties must be overcome before significant advances in our world food supply can be realized. Clearly, genes for desired characteristics must be identified and isolated. Also, important crops such as corn and wheat cannot be transformed by Ti plasmids; therefore, other vectors must be identified. Significant success has been achieved in designing crop plants resistant to insects and viruses. Recently genetic engineering inserted a foreign gene into pea plants to produce a protein that inhibits the feeding of weevil larvae on the pea seeds. This will permit storage of peas and other legume seeds without the need of protective chemical fumigants (currently Brazilian farmers lose 20%–40% of their stored beans to pests).

7.11 • GENOMICS, PROTEOMICS, AND MICROARRAY ANALYSIS

Genomics is the study of the molecular characteristics of the whole genome. This includes the sequence of the genome, the identification of genes and their regulatory sequences, and the pattern of gene expression. Dozens of prokaryotic and eukaryotic genomes, as well as over one thousand viruses, have been sequenced (Table 7.1). The sequencing of numerous genomes of marine microbes has recently added approximately one million new genes to the 180,000 documented genes. It is estimated that there are more than 10 billion genes in the Earth's repertoire. A singe gene may give rise to multiple protein species due to splice choice of RNA transcript and variable posttranslational processing of the protein product. The complement of proteins within a cell is referred to as its **proteome**. Proteomics is the study of proteins expressed in tissues under various conditions, protein interactions, and protein modifications. Clin. Corr. 7.15 describes how proteomics has been used to diagnose early stages of cancer by identifying tumor-associated biomarkers.

TABLE 7.1 • A Partial Listing of Prokaryotic and Eukaryotic Species Whose Genomes Have Been Sequenced

	Species	*Genome Size*
Prokaryotes (over 100 microbes)	*Hemophilus influenzae*	1.83 Mb
	Chlamydia pneumoniae	1.23 Mb
	Neisseria meningitides	2.27 Mb
	Vibrio cholerae	4.0 Mb
		Number of Chromosomes
Simple Eukaryotes	*Saccharomyces cerevisiae* (yeast)	16
	Plasmodium falciparum (malarial protozoan)	14
	Anopheles gambiae (mosquito)	3
	Drosophila melanogaster (fruit fly)	5
Plants	*Avena sativa* (oat)	7
	Hordeum vulgare (barley)	7
	Oryza sativa (rice)	12
	Zea mays (corn)	10
Vertebrates	*Danio rerio* (zebra fish)	25
	Gallus gallus (chicken)	39
	Mus musulus (mouse)	21
	Rattus norvegicus (rat)	22
	Homo sapiens (human)	24

CLINICAL CORRELATION 7.15

Microarray Techniques to Detect and Treat Diseases

A. Nucleic Acid Microarray Analysis of Breast Cancer

The treatment modality of human breast cancer patients depends, in large part, on the estrogen receptor (ER) status of the tumor cells. Estrogen receptor-positive tumors generally respond well to adjuvant hormone or drug therapy whereas ER-negative tumors usually do not. Patients with ER-negative tumors have variable responses to chemotherapy, and there are no known indicators to predict postoperative prognoses. The amplified cellular RNA derived from 10 ER-negative patients who died within 5 years of breast cancer and 10 ER-negative patients who were alive, disease free, was analyzed on a cDNA microarray consisting of 25,344 genes. The analysis detected 71 genes with higher expression in the group that died than those who lived and 15 genes expressed at higher levels in the survivors versus those who died. A scoring system was developed based on gene markers to classify ER-negative breast cancer patients into a poor prognosis versus a good prognosis group. The scoring system appears to have a high degree of accuracy and may help direct treatment protocols and define new gene targets for drug treatment.

B. Antibody Microarray Technology for Treatment of Brain Tumors

Glioblastoma multiforme (GBM) is the most common central nervous system tumor found in human adults and, unfortunately, one of the most lethal cancers. Abnormal activation of phosphatidylinositol-3-kinase (PI3K) pathway components is a hallmark for GBM, as well as other cancers. Receptor tyrosine kinases (RTKs) activate PI3K. Some examples of RTKs include the growth factor receptors EGFR, PDGFR, and VEGFR. It was found that multiple RTKs activated PI3K in GBM by analyzing tumor cell extracts incubated with RTK antibody arrays. Here antibodies, specific to numerous RTKs, are fixed to a solid support and will bind the cellular RTKs. The activated RTK present in the cell extract is phosphorylated at tyrosine residues which are readily detected in the RTK–antibody complex by a second horseradish peroxidase (HRP) conjugated antibody against phospho-tyrosines. A substrate (fluorescent or colored) for HRP is then used to identify which active RTKs were present in a tumor cell extract. On the identification of which RTKs were involved in the activation of PI3K in a particular tumor cell line, specific RTK inhibitors were tested for their ability to reduce colony formation and cell viability. A single RTK inhibitor had little effect, and dual RTK inhibitors were more effective. However, use of three RTK inhibitors was most effective. Assays of untreated primary human GBM tumors all contained multiple phosphorylated RTKs while tissue from normal brain specimens had no detectable RTK activation. These studies demonstrate the ability to rapidly profile the complement of activated RTKs present in GBM samples from patients such that clinicians can personalize their drug treatment.

Nagahata, T., Onda, M., Emi, M., Nagai, H. K., et al. Expression profiling to predict postoperative prognosis for estrogen receptor-negative breast cancer by analysis of 25,344 genes on a cDNA microarray. *Cancer Sci.* 95: 218, 2004.

Stommel, J. M., Kimmelman, A. C., Ying, H., Nabioullin, R., Ponugoti, A. H., et al. Coactivation of receptor tyrosine kinase affects the response of tumor cells to targeted therapies. *Science* 318: 287, 2007.

Microarray Analysis

Analysis of genomes and proteomes has been greatly facilitated by the development of microarray technologies developed first for analyzing nucleic acids. Nucleic acid hybridization and antigen–antibody complex formation are employed extensively in techniques described in preceding sections for the analysis of one or a few macromolecular species. However, techniques with a very high throughput are required to analyze simultaneously the hundreds to thousands of genes expressed in a cell. Microarray techniques depend on these same macromolecular interactions described previously such as in Southern, Northern, and Western blots. The fixation of DNA probes to solid impermeable surfaces reduces the solution volumes and times required for hybridization. DNA microarrays are produced by computer-controlled additions of microdrops of cloned or PCR synthesized DNA sequences to a glass slide with a poly-L-lysine substrate to which these DNA probes are covalently cross-linked. DNA or RNA to be analyzed is isotopically or fluorescently labeled and then hybridized with the fixed probes on a glass slide (Figure 7.32). Positive reactions are detected and quantified by high-resolution imaging and computer analysis. Many DNA microarrays are available that focus on selected genes involved in discreet processes or pathways such as signal transduction and apoptosis. A specific example of altered gene expression in malaria-infected cells is depicted in Figure 7.33.

DNA oligonucleotides (usually 20 or more bases) are also synthesized on micromatrices fixed to a glass slide. The size of the micromatrix varies from $1 \times 10 \times 5$ μm to $100 \times 100 \times 20$ μm. Thousands of different oligonucleotides, each complementary to a specific gene, are immobilized on what is termed a **DNA chip**. A single–chip microarray for the human genome, which will be commercially available, contains 1 million oligonucleotide

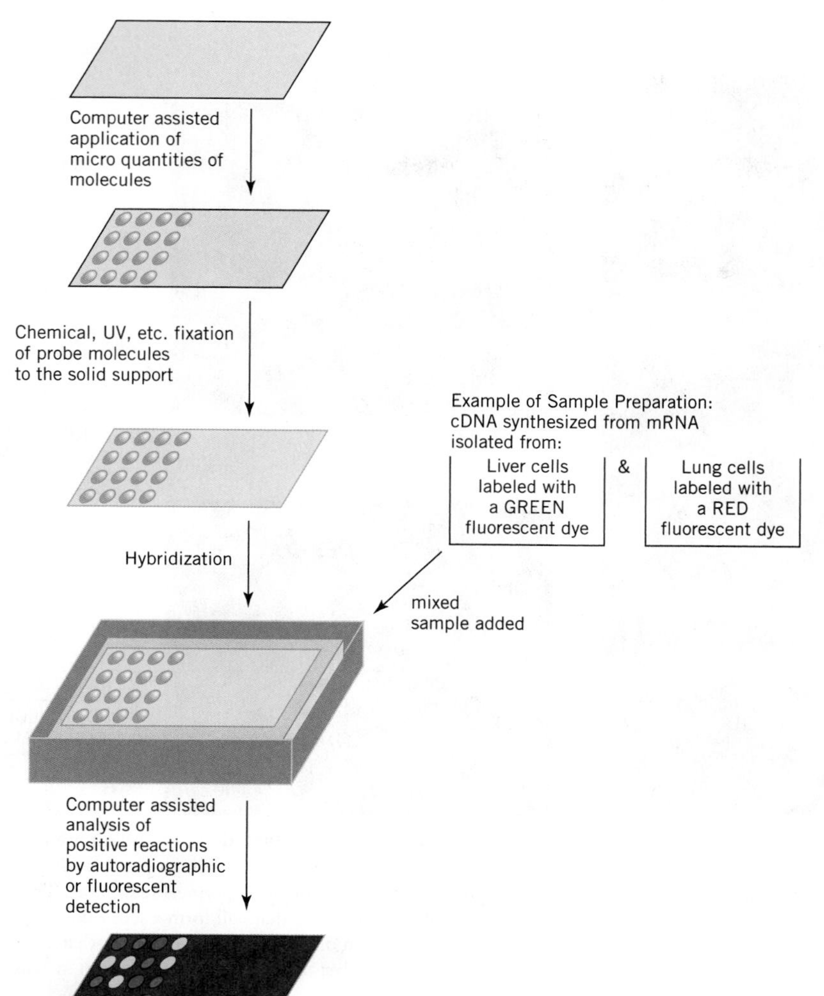

Figure 7.32 DNA microarray analysis. The surface of an impermeable support, typically glass or polypropylene, is prepared to apply multiple samples of DNA or oligonucleotides. Slide with oligonucleotides or nucleic acids representing 100s to 1,000s of genes is incubated in a chamber under appropriate conditions for hybridization followed by appropriate washes to remove nonhybridized cDNA. Green fluorescent spots would indicate genes expressed in liver cells, red fluorescent spots for genes expressed in lung cells, and yellow fluorescent spots for genes expressed in both tissues.

probes. This DNA chip will detect about 50,000 RNA transcripts from approximately 30,000 human genes. These biological microchips are compared to electronic microchips since both are capable of performing multiple reactions in parallel in a high-throughput fashion.

Microarrays allow for the analysis of expression of the whole genome or genes in a selected pathway. Genes expressed in a specific tissue or under a selected metabolic or diseased state could give clues as to which genes are active in the specific condition. Unfortunately, a single microarray plate gives a static snapshot of genomic genes expressed under a particular condition. Regulatory genes involved in the final altered expression cannot be readily deduced from this single picture since many genes may be turned on or off by downstream pathways during the transition to the final cellular state. Coregulatory genes involved in the switch between different metabolic or disease states could be determined by microarray analysis at multiple time points between the initial and final stage of altered gene expression. Computer analysis of microarrays generated over the time course can cluster genes that appear to be coordinately expressed. **Cluster analysis** allows identification of potential genes that coregulate the alteration of the cell from one state to another.

Protein microarrays are produced with antibodies or nonantibody proteins fixed to slides or filters that can capture target proteins in a sample in a fashion similar to those used for DNA microarrays. Other protein probes are being developed for protein microarrays such as short stretches of synthesized DNA or RNA (**aptamers**) that selectively bind specific proteins. Protein arrays are generally produced with recombinant proteins that may be generated from tissue-specific cDNA expression libraries. These microarrays are readily analyzed with specific antibodies that are fluorescently labeled. This approach has been

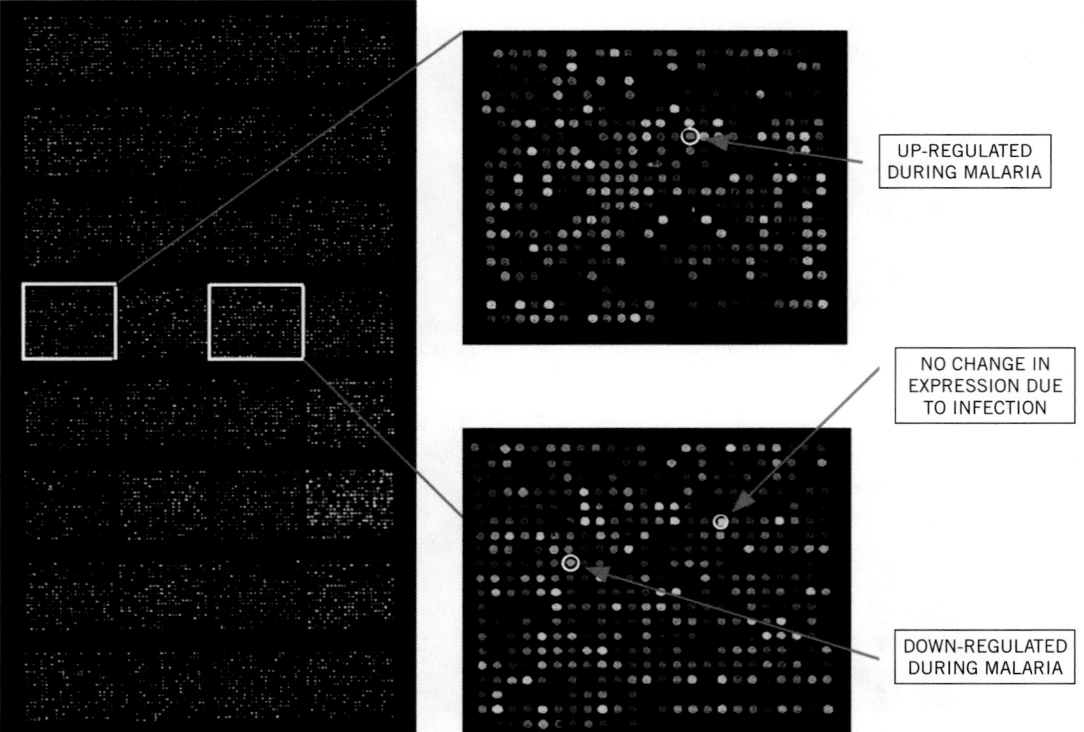

Figure 7.33 **Microarray analysis of gene expression in mice infected with a rodent malaria parasite.** DNA, RNA, or oligonucleotides, representing part or all of the cellular genome, or proteins or antibodies, representing part of the cellular proteome, are fixed on an impermeable support. The fixed probes are reacted with appropriately labeled cellular components that will form a stable complex through hybridization or protein–protein interactions. Identification of nucleic acids or proteins that associate with the fixed samples in the microarrays allows for the analysis of cellular gene expression in cells maintained under a variety of conditions.
Mouse DNA microarrays were produced for the mouse genome which contains 13,443 synthetic oligonucleotides representing well- characterized mouse genes. Each oligonucleotide is 70 bases in length derived from the 3′-end of each gene, optimized to minimize cross-hybridization to all known mouse genes. The array includes a comprehensive set of immune system related genes encoding a lymphocyte subset and activation markers, cytokines, chemokines, their receptors, adhesion molecules, costimulatory molecules, immunoregulatory factors, transcription factors and signal transduction proteins, and others.
Samples: Splenocyte RNA from uninfected mice was used to synthesize cDNA fluorescently labeled with Cy3 (green). Splenocyte RNA from mice on day 8 of infection with a rodent malaria parasite was used to synthesize cDNA fluorescently labeled with Cy5 (red). The mixed fluorescently labeled cDNA probes are hybridized with the target oligonucleotide array. Genes that are up-regulated in mice infected with malaria are red and those that are down-regulated in these animals are green. Genes that are unaffected by the malarial parasite are yellow.
Data are unpublished work from the laboratory of Dr. James Burns, Department of Microbiology, Drexel University College of Medicine.

successfully employed to detect the antibody repertoire in human plasmas associated with some autoimmune diseases.

Microarray techniques are critical for the analysis of genomic expression and the proteomic state of cells in normal cells as compared to the patterns in cells that are altered due to changes such as a modified metabolic environment, differentiation, response to drugs, response to growth factors, or a diseased state.

Human Genome

By 1996 the genome of yeast, a eukaryote, consisting of approximately 14×10^6 bps distributed among 16 chromosomes, was sequenced. The genome of the nematode, *C. elegans,* sequenced by 1998, contains approximately 18,000 genes as compared to the 6,000 genes of yeast. In 2000, a **shotgun sequencing** approach was employed to sequence essentially the whole 180-$\times$ 10^6-bps genome of the fruit fly *Drosophila melanogaster*. This method involves the breakage of the whole genome into small pieces, sequencing the pieces by a rapid automated procedure, and assembling the sequenced fragments into the proper order assisted by overlapping sequences and a super computer. The availability of previously cloned sequences facilitated the sequencing.

The shotgun sequencing approach was employed to speed up the sequencing of the human genome. Several human chromosomes had already been mapped and thousands of cDNA clones had been sequenced or were ready to be sequenced, which would ultimately provide landmarks of the huge human genetic map. In 2000, J. C. Venter and F. S. Collins jointly announced that their groups had sequenced 97% of the human genome.

The determination of a genomic sequence is only a part of the genome puzzle. The start and stop site of each gene must be determined along with what the gene codes for a process referred to as **annotation**. Eukaryotic genomes usually contain permanently condensed or heterochromatic regions around the centromeres. These regions contain repetitive DNA that is not cloned and sequenced by any of the current sequencing methods. It may be decades or more before the complete determination of all the genes and regulatory elements are resolved.

The completion of the human genome project generated enormous amounts of data. The achievement of the project is only the beginning of our understanding of the molecular underpinnings of how a single cell functions, let alone an organism. It is now considered that the human genome encodes about 30,000 genes, far fewer genes than anticipated, and only about 20% more than a plant genome and about 40% more than *C. elegans* (roundworm). However, the RNA transcripts of the "static" genome are relatively "dynamic" as are their translated protein products. Splice choice of the primary RNA transcript results in different protein products from the same gene (p. 295). Protein products may also undergo numerous chemical modifications and association with other peptide subunits or proteins. It is estimated that the 30,000 genes ultimately give rise to 200,000 to 2 million protein products whose expression is dependent, in part, on tissue specificity, stage of development, and metabolic conditions. The task before us now is to be able to analyze transcripts and their final translation products at the structure or function level.

The delineation of all of the human genes and their regulatory sequences should greatly enhance our understanding of many genetic diseases. Technological advances to manipulate genes and gene expression should open new avenues to regulate or cure many diseases. Genetic diseases should eventually be curable by gene replacement therapy; limited clinical trials in gene therapy have been initiated. Considering the enormous advances made in molecular biology in just the past three decades, it is reasonable to believe the when will not be that far off. The old challenge confronting scientists was how to sequence the human genome; the new challenge is how to manipulate effectively that knowledge to benefit humankind.

Bibliography

Adams, D. M., Celniker, S. E., and Holt, R. A., et al. The genome sequence of Drosophila melanogaster. *Science* 287: 2185, 2000.

Arany, Z. P. High-throughput quantitative real-time PCR. *Curr. Protocols Human Genet.*, 2008, Chap. 11.

Brown, W. M., George, M., Jr., and Wilson, A. C. Rapid evolution of animal mitochondrial DNA. *Proc. Natl. Acad. Sci. USA* 76: 1967, 1979.

Bushman, F. Targeting retroviral integration. *Science* 267: 1443, 1995.

Capecchi, M. R. Altering the genome by homologous recombination. *Science* 244: 1288, 1989.

C. elegans Sequencing Consortium. Genome sequence of the nematode *C. elegans:* A platform for investigating biology. *Science* 282: 2012, 1998.

Chilton, M. D. A vector for introducing new genes into plants. *Sci. Amer.* 51: 36, June 1983.

Church, G. M. Genomes for all. *Sci. Amer.* 294: 47, 2006.

Davis, L. G., Kuehl, W. M., and Battey, J. F. Basic methods in molecular biology. In Erlich, H. A. (Ed.), *PCR Technology. Principles and Applications for DNA Amplification,* 2d ed. New York: Stockton, 1989.

Feinberg, A. and Vogelstein, B. Addendum: A technique for radiolabeling DNA restriction endonuclease fragments to high specific activity. *Anal. Biochem.* 137: 266, 1984.

Fields, S. Proteomics in genome land. *Science* 291: 1221, 2001.

Goffeau, A., Barrell, B. G., Bussey, H., Davis, R. W., et al. Life with 6000 genes. *Science* 274: 546, 1996.

Hall, N. Advanced sequencing technologies and their wider impact in microbiology. *Exptl. Biol.* 209: 1518, 2007.

International HapMap Consortium. A second generation of human haplotype map of over 3.1 million SNPs. *Nature* 449: 851, 2007.

International Human Genome Sequencing Consortium. Initial sequencing of the human genome. *Nature* 409: 860, 2001.

Jaenisch, R. Transgenic animals. *Science* 240: 1468, 1988.

Kerr, L. D. Electrophoretic mobility shift assay. *Methods Enzymol.* 254: 619, 1995.

Kreeger, K.Y. Influential consortium's cDNA clones praised as genome research time-saver. *Scientist* 9: 1, 1995.

Kunkel, T. A. Rapid and efficient site-specific mutagenesis without phenotypic selection. *Proc. Natl. Acad. Sci. USA* 82: 488, 1985.

McPherson, M. J., Quirke, P., and Taylor, G. R. (Eds). *PCR. A Practical Approach,* Vol. 1. Oxford: Oxford University Press, 1994.

Marshall, E. Gene therapy's growing pains. *Science* 269: 1050, 1995.

Maxam, A. M., and Gilbert, W. A new method of sequencing DNA. *Proc. Natl. Acad. Sci. USA* 74: 560, 1977.

Mulligan, R. C. The basic science of gene therapy. *Science* 260: 926, 1993.

Palmiter, R. D., Brinster, R. L., Hammer, R. E., Trumbauer, M. E., et al. Dramatic growth of mice that develop from eggs microinjected with metallothionein-growth hormone fusion genes. *Nature* 300: 611, 1982.

Park, I. H., Zhao, R, West, J. A., Yabuuchi, A., et al. Reprogramming of human somatic cells to pluripotency with defined factors. *Nature* 451: 141, 2008.

Pennis, E. Breakthrough of the year: Human genetic variation. *Science* 318: 1842, 2007.

Reddy, K. S., and Perrotta, P. L. Proteomics in transfusion medicine. *Transfusion* 44: 601, 2004.

Rigby, P. W. J., Dieckmann, M., Rhodes, C., and Berg, P. Labelled deoxyribonucleic acid to high specific activity in vitro by nick translation with DNA polymerase I. *J. Mol. Biol.* 113: 237, 1977.

Sambrook, J., Fritsch, E. F., and Maniatis, T. Molecular Cloning. *A Laboratory Manual,* 2d ed. New York: Cold Spring Harbor Laboratory, 1989.

Sanger, F., Nicklen, S., and Coulson, A. R. DNA sequencing with chain-terminating inhibitors. *Proc. Natl. Acad. Sci. USA* 74: 5463, 1977.

Simon, R., Mirlacher, M., and Sauter, G. Tissue microarrays. *Bio-Techniques* 36: 98, 2004.

Smith, L. M., Sanders, J. Z., Kaiser, R..J., Hughes, P., et al. Fluorescence detection in automated DNA sequence analysis. *Nature* 321: 674, 1986.

Southern, E. M. Detection of specific sequences among DNA fragments separated by gel electrophoresis. *J. Mol. Biol.* 98: 503, 1975.

Southern, E. M.. Microarrys. In Rampal, J. B. (Ed.), *DNA Arrays, Methods and Protocols. Methods in Molec. Biol.* 170: 1, 2001.

Venter, J. C., Adams, M. D., Myers, E. W., Li, P. W., et al. The sequence of the human genome. *Science* 291: 1304, 2001.

Venter, J. C., Remington, K, Heidelberg, J. F., Halpern, A. L., et al. Environmental genome shotgun sequencing of the Sargasso Sea. *Science* 304: 66, 2004.

Vogelstein, B., Fearon, E. R., Hamilton, S. R., and Feinberg, A. P. Use of restriction fragment length polymorphism to determine the clonal origin of human tumors. *Science* 227: 642, 1985.

Watson, J. D., Tooze, J., and Kurtz, D. T. *Recombinant DNA a Short Course.* San Francisco: Scientific American Books, Freeman, 1983.

Weintraub, H. M. Antisense RNA and DNA. *Sci. Amer.* 262: 40, 1990.

Wilmut, I., Schnieke, A. E, McWhir, J., Kind, A. J., and Campbell, K. H. S. Viable offspring derived from fetal and adult mammalian cells. *Nature* 385: 810, 1997.

Zhang, Y., and Yunis, J. J. Improved blood RNA extraction microtechnique for RT-PCR. *BioTechniques* 18: 788, 1998.

Key Terms

polymerase chain reaction (PCR)
nested PCR
quantitative real time PCR
restriction endonuclease
restriction maps
dideoxynucleoside triphosphate
direct sequencing
dye-terminator sequencing
recombinant DNA
plasmids
polycloning site
multiplex PCR
insertional inactivation
α complementation
complementary DNA (cDNA)
Northern blot
restriction fragment length polymorphism (RFLP)
Southern blot
single-strand conformation polymorphism (SSCP)
in situ hybridization
nuclease protection assay
electrophoretic mobility shift assay (EMSA)
bacteriophage libraries
cosmid vector
bacterial artificial chromosomes (BACS)
yeast artificial chromosomes (YACS)
chromosome walking
transgenic animals
expression vectors
fusion protein
transfection
site-directed mutagenesis
inverse PCR mutagenesis
antisense nucleic acids
microRNAs
RNA interference, RNAi
small-interfering RNA (siRNA)
gene therapy
knockout mice
pluripotential cells
genomics
microarray analysis
proteomics
shotgun sequencing

Questions • CAROL N. ANGSTADT

Multiple-Choice Questions

1. Construction of a restriction map of DNA requires all of the following except
   A. partial hydrolysis of DNA.
   B. complete hydrolysis of DNA.
   C. electrophoretic separation of fragments on a gel.
   D. staining of an electrophoretic gel to locate DNA.
   E. cyclic heating and cooling of the reaction mixture.

2. Preparation of recombinant DNA requires
   A. restriction endonucleases that cut in a staggered fashion.
   B. restriction endonucleases that cleave to yield blunt-ended fragments.
   C. poly(dT).
   D. DNA ligase.
   E. cDNA.

3. In the selection of bacterial colonies that carry cloned DNA in plasmids, such as pBR322, that contain two antibiotic resistance genes,
   A. one antibiotic resistance gene is nonfunctional in the desired bacterial colonies.
   B. untransformed bacteria are antibiotic resistant.
   C. both antibiotic resistance genes are functional in the desired bacterial colonies.
   D. radiolabeled DNA or RNA probes play a role.
   E. none of the above.

4. Complementary DNA (cDNA)
   A. cannot be expressed by prokaryotic systems.
   B. is reverse transcribed from functional eukaryotic mRNA.
   C. has all of the elements of the original gene for the mRNA of interest.
   D. synthesis requires the prior formation of an RNA primer.
   E. remains as a single-stranded molecule.

5. The best vector to use to carry a segment of DNA that is 350 kb in size would be
   A. bacteriophage.
   B. bacterial artificial chromosome (BAC).
   C. cosmid.
   D. plasmid.
   E. yeast artificial chromosome (YAC).

6. Antisense nucleic acids
   A. complementary to mRNA would enhance translation.
   B. could result if a gene is inserted downstream of a promoter but in opposite direction to normal.
   C. can have no clinical uses.
   D. react only with DNA.
   E. are necessary for recombinant DNA technology.

Questions 7 and 8: In the United States, a major cause of death of babies during the first year is sudden infant death syndrome (SIDS). One study showed a strong correlation for an increased risk of SIDS with a prolonged QT interval in their electrocardiograms. In one child, a gene associated with the Long QT syndrome had a substitution of AAC for TCC. This gene codes for a protein associated with the sodium channel. The mutation was detected by single-strand conformation polymorphism (SSCP).

7. Which of the following statement(s) about SSCP is/are correct?
   A. The electrophoretic mobility on polyacrylamide gel of small, single-stranded DNA during the SSCP technique depends partly on the secondary conformation.
   B. There must be a restriction endonuclease site in the region studied for SSCP to work.
   C. Radiolabeling is not used in this technique.
   D. It is not necessary to know the sequence of the DNA to be studied.
   E. All of the above.

8. In order to get enough DNA to analyze, DNA is amplified by a polymerase chain reaction (PCR). In a PCR,
   A. the nucleotide sequence of the DNA to be amplified must be known.
   B. the sample is protected from heat which would denature the DNA.
   C. the role of oligonucleotides in the reaction mixture is to act as primers for the synthesis of new DNA.
   D. the final product is single-stranded DNA.
   E. the original DNA can be amplified only about 10-fold.

Questions 9 and 10: Recombinant DNA technology has uses of interest in the medical field. One is gene therapy to introduce a normal gene into cells containing a defective gene. The first authorized human gene therapy was given to a 4-year-old girl with severe combined immune deficiency (SCID) who had a defective gene for adenosine deaminase (ADA). A modified retrovirus was constructed to contain the human ADA gene which could be expressed in human cells without replication of the virus. Another clinical use for recombinant DNA technology is to have rapidly replicating bacteria produce large amounts of specific proteins, for example, human proteins such as hormones.

9. Expression of recombinant genes in mammalian cells
   A. will not occur if the gene contains an intron.
   B. occurs most efficiently if cDNA is used in the vector.
   C. does not require directional insertion of the gene into the vector.
   D. requires that the vector have enhancer and promoter elements engineered into the vector.
   E. does not require that the vector have an origin of replication (ori).

10. Expression of a eukaryotic gene in prokaryotes involves
   A. a Shine–Delgarno (SD) sequence in mRNA.
   B. absence of introns.
   C. regulatory elements upstream of the gene.
   D. a fusion protein.
   E. all of the above.

Questions 11 and 12: Whole genomic DNA was isolated from three individuals, digested separately with a restriction endonuclease to fragments and the fragments separated on agarose gel in an electric field. The gene of interest was isolated and analyzed using Southern blot technique. Each individual sample showed two bands. The bands were identical for two of the individuals. For the third, one band was identical to one band of the other two but the second band was of lower molecular weight than the others. This is an example of restriction fragment length polymorphism (RFLP).

11. A reasonable explanation for this RFLP is that the gene in the third individual
   A. lacked the recognition sequence for the restriction endonuclease.
   B. had a mutation at the cleavage site.
   C. had an additional recognition site for the endonuclease.
   D. had a deletion of a segment of the gene.
   E. underwent a random cleavage.

12. The Southern blot technique
   A. transfers DNA fragments from agarose gel to a nitrocellulose filter.
   B. requires that the DNA fragments remain double stranded.
   C. requires that the DNA is radiolabeled prior to addition to the agarose gel.
   D. alters the position of the DNA fragments during the process.
   E. amplifies the amount of DNA material.

Problems

13. The X-ray autoradiogram of one strand of a fragment of DNA sequenced by the Sanger method was obtained. The 5′-end of the nucleotide was labeled with ^{32}P. Numbering from bottom to top of the autoradiogram, the lane from the ddG tube showed bands at positions 4 and 11; ddC tube had bands at 1, 2, 7, 8, 9, and 13; ddA at 5, 10, and 15; and ddT at 3, 6, 12, and 14. What is the sequence of the fragment? Construct the autoradiogram pattern of the complementary sequence.

14. What makes the Taq DNA polymerase frequently used for a polymerase chain reaction (PCR) different from other DNA polymerases, and why is it preferable?

Answers

1. **E** Cyclic heating and cooling is part of the PCR process, not of restriction mapping. A and B: Restriction mapping involves all degrees of hydrolysis. Partial hydrolysis gives fragments of varying sizes, and complete hydrolysis gives the smallest possible fragments. C and D: Fragments are electrophoretically separated by size on agarose gel, which is stained to reveal the DNA.
2. **D** DNA ligase covalently connects fragments held together by interaction of cohesive ends. A: This is the most desirable type of restriction endonuclease to use, but it is not essential. B: Restriction nucleases that make blunt cuts can also be used if necessary. C: This is used in conjunction with poly(dA) if restriction endonucleases that make blunt cuts are employed, but it is not essential to all of recombinant DNA preparation. E: cDNA is only one type that can be used.
3. **A** The foreign DNA is inserted into one antibiotic resistance gene, thus destroying it. B: Resistance is due to the plasmids. C: See the comment for A above. D: Radiolabeling detects the DNA of interest, not the colonies that contain cloned DNA.
4. **B** This uses an RNA-dependent DNA polymerase (reverse transcriptase). A and C: Prokaryotic systems cannot splice out introns to yield functional mRNA transcripts—thus cDNA which does not contain the sequences for these elements. D. The poly(A) tail at the 3′-end of mRNA is used as the primer to make the strand complementary to the mRNA template. E. Once the first strand is synthesized, it becomes the template for the complementary strand. cDNA is double stranded.
5. **E** It can accept DNA from 200–500 kb. A: Bacteriophage is very good for DNA inserts about 15 kb long. B: BAC can accept DNA from about 100–200 kb. C: Cosmids accept inserts of about 45 kb. D: Plasmid vectors can be very useful but can accept only small pieces of DNA.
6. **B** This would put the antisense strand of DNA under control of the promoter with subsequent transcription of antisense mRNA. A: Antisense nucleic acid binds mRNA blocking translation. C: If not now, in the future. Antisense DNA nucleotides have been produced that inhibit viral infection. D: React with both DNA and RNA.
7. **A** A single base substitution usually modifies the conformation enough to shift the mobility as detected by SSCP. B: SSCP is the method of choice if there is no restriction endonuclease site. C: DNA is amplified by PCR in the presence of radiolabeled nucleotides. There has to be a detection method for the bands. D: SSCP requires prior knowledge of the sequence.
8. **C** The oligonucleotides hybridize with the single-stranded templates in the region of interest so that replication can begin. A: DNA of unknown sequence can be used but must be inserted into a vector with known flanking regions. B: The sample must be heated to 90° to separate the DNA strands. D: PCR produces a double-stranded DNA (dsDNA). E: With automation, amplification can be a millionfold.
9. **D** Eukaryotic systems require controlling elements, which are not necessary in bacterial systems. A: Expression may be improved if an intron is present. B: cDNA does not possess the required controlling elements. C: The gene must be inserted in the proper orientation relative to the control elements. E: The vector must replicate so it needs the sequence to promote replication.
10. **E** A: The SD sequence is necessary for the bacterial ribosome to recognize the mRNA. B: Bacteria do not have the intracellular machinery to remove introns from mRNA. C: Appropriate regulatory elements are necessary to allow the DNA to be transcribed. D: A fusion protein may be a product of the reaction.
11. **D** This would account for one band being of lower molecular weight. A, B, and F: The cleavage site has a specific recognition sequence. C: This would yield more than two bands.
12. **A** This occurs by elution from the gel by a high-salt solution. B: DNA is hydrolyzed by alkali and single-strand DNA binds to the filter. C: The probe to identify the bands on the filter is radiolabeled. D: The positions on the filter are identical to the gel. E: Doesn't happen.
13. The sequence is C C T G A T C C C A G T C T A reading 5′ to 3′. The autoradiograph of the complementary strand should have bands for ddG at 1, 2, 7, 8, 9, and 13; ddC at 4 and 11; ddA at 3, 6, 12, and 14; and ddT at 5, 10, and 15.
14. PCR requires cycling between low temperatures, where hybridization of template DNA and oligomeric primers occurs, and high temperatures, where DNA melts. The Taq DNA polymerase, isolated from a thermophilic organism discovered in a hot spring, is stable at high temperatures and makes the cycling possible with no addition of fresh polymerase after each cycle. Phusion^{TM} High-Fidelity DNA polymerase, which has been commercially developed, has even more desirable properties.

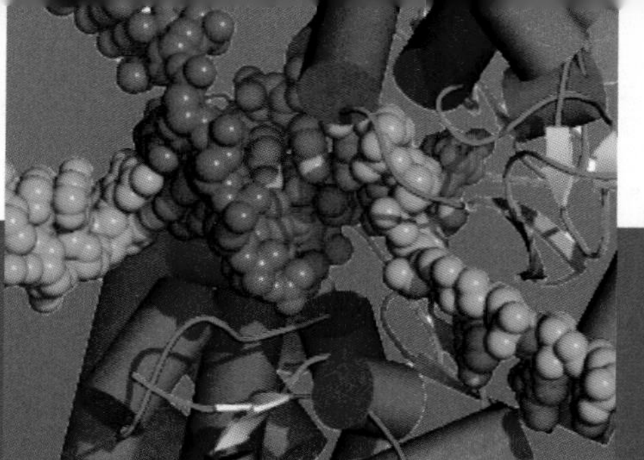

8

Regulation of Gene Expression

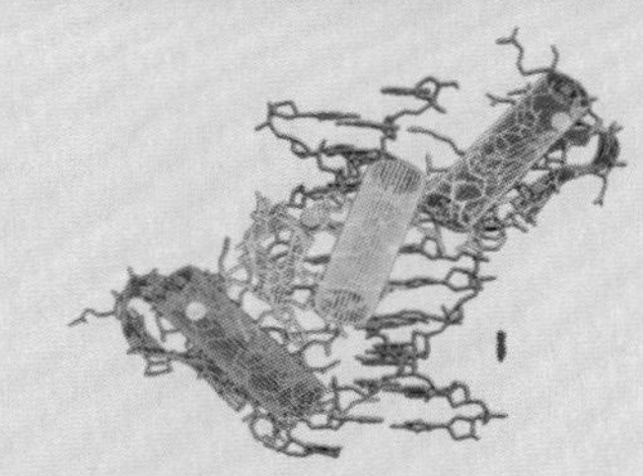

Daniel L. Weeks

Professor, Carver College of Medicine, University of Iowa

John E. Donelson

Professor, Carver College of Medicine, University of Iowa

CLINICAL CORRELATIONS

Key Concepts

- Transcription of genes provides the link between the DNA reservoir of genetic information and the expression of RNA and ultimately protein.
- Genes generally include a nontranscribed promoter that is "upstream" of the protein-encoding portion of the gene.
- The transcripts encoded by prokaryotic organisms are often polycistronic, grouping genes encoding proteins with interrelated functions together under common control. In eukaryotic organisms most genes are monocistronic.
- Transcription can be activated or repressed in response to nutritional, environmental, or other cellular signals. Regulatory responses aim to provide appropriate gene expression for existing conditions, often by incorporating small molecules indicative of cellular conditions as coregulators.
- Control of transcription often depends on sequence-specific binding proteins. The amino acid motifs of sequence-specific binding proteins have been identified, allowing transcription factors to be grouped by structural similarities.
- The identification of DNA sequences that bind to transcription factors may help identify genes that are coordinately regulated. Eukaryotic promoters often have multiple transcription factor–binding sites and utilize combinations of transcription factors to regulate gene expression.

- Modifications of DNA and chromatin play a significant role in eukaryotic gene regulation. DNA methylation and histone methylation, acetylation, and phosphorylation influence the ability to transcribe a gene.
- Transcription factors are often modular, with DNA-binding domains, protein interaction domains, activation domains, and regulatory domains.
- Transcription factors may affect the assembly of the RNA polymerase complex, the recruitment of other transcription factors, or the recruitment of modification factors.

8.1 • INTRODUCTION

To survive, a living cell must respond to changes in its environment. One way cells adjust to these changes is to alter expression of specific genes; this affects the number of the corresponding protein molecules in the cell. This chapter focuses on some of the molecular mechanisms that determine when a given gene is expressed and to what extent. Understanding how expression of genes is regulated is one of the most active areas of biochemical research today.

E. coli cells contain genes for about 4300 different proteins, but *E. coli* does not need to synthesize all of them at the same time. The classic illustration of bacterial gene regulation is that of the gene for **β-galactosidase**, which converts the disaccharide lactose to glucose and galactose. When *E. coli* grows in a medium containing glucose as the carbon source, only about five molecules of the enzyme are present in a cell. When lactose is the sole carbon source, however, there are 5000 or more molecules. Thus, when bacteria need to metabolize lactose, they increase synthesis of β-galactosidase. If lactose is removed from the medium, synthesis of this enzyme stops as rapidly as it began.

Eukaryotic cells have more extensive mechanisms of gene regulation than prokaryotic cells. Differentiated cells of higher organisms have a highly complicated structure and often a specialized biological function that is determined by expression of their genes. For example, **preproinsulin** is synthesized in β-cells of the pancreas but not in kidney cells even though the nuclei of all cells of the body contain the gene for preproinsulin. During development of the organism, the presence of proteins in specific cell types is tightly controlled with respect to timing and order of developmental events.

Much more is understood about gene regulation in prokaryotes than in eukaryotes. Furthermore, studies on prokaryotes often provide exciting new ideas that can be tested in eukaryotic cells and are important for developing rational therapies for diseases caused by bacteria.

Several well-studied examples of gene regulation in bacteria will be discussed, followed by examples of the many protein transcription factors regulating expression of genes in the human genome.

8.2 • UNIT OF TRANSCRIPTION IN BACTERIA: THE OPERON

The *E. coli* chromosome is a circular double-stranded DNA molecule of 4.6 million base pairs whose entire sequence has been determined. Most of its approximately 4300 genes are not distributed randomly; instead, genes encoding enzymes of a specific metabolic pathway are clustered in one region of the DNA. Similarly, genes for associated structural proteins, such as the 70 or so proteins that compose the **ribosome**, are frequently adjacent to one another. A set of clustered genes is usually coordinately regulated; that is, the genes are transcribed together into a *polycistronic* mRNA species that contains the coding sequences for the various proteins. The term **operon** describes the complete regulatory unit of a set of clustered genes. It includes the adjacent **structural genes** for the related enzymes or associated proteins, a **regulatory gene** or genes encoding regulator protein(s), and **control elements** or sites on the DNA near the structural genes at which regulator proteins act. Figure 8.1 shows a partial genetic map of the *E. coli* genome that gives locations of structural genes of some *E. coli* operons.

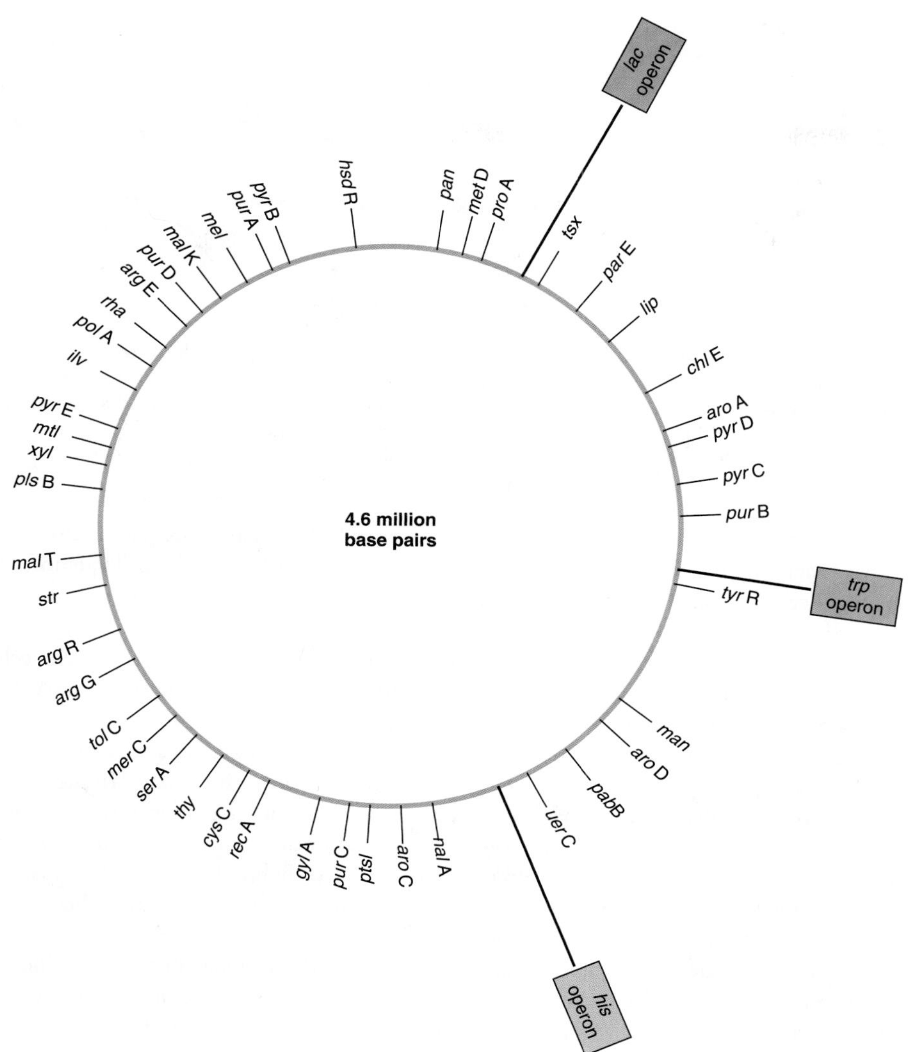

Figure 8.1 Partial genetic map of E. *coli*. Locations of a few of the 4300 genes in the circular *E. coli* genome are shown. Three operons discussed in this chapter are indicated.
Reproduced with permission from Stent, G. S., and Calendar, R. *Molecular Genetics: An Introductory Narrative.* San Francisco: Freeman, 1978, 289; modified from Bachmann, B. J., Low, K. B., and Taylor, A. L. *Bacteriol. Rev.*40:116, 1976.

When transcription of the structural genes of an operon increases in response to a specific substrate in the medium, the effect is known as **induction**. An example is the increase in transcription of the β-galactosidase gene when lactose is the sole carbon source. Bacteria also respond to nutritional changes by quickly turning off synthesis of enzymes no longer needed. For example, *E. coli* synthesizes tryptophan as the end product of a specific biosynthetic pathway. However, if tryptophan is present in the medium, synthesis of enzymes for this pathway stops. The process is called **repression**, and it prevents bacteria from using their energy to make unnecessary proteins.

Induction and repression are manifestations of the same phenomenon. The signal for each is the small molecule that is a substrate for the metabolic pathway or a product of the pathway, respectively. These small molecules are called **inducers** when they stimulate induction and **corepressors** when they cause repression.

8.3 • LACTOSE OPERON OF *E. COLI*

The lactose operon contains three adjacent structural genes (Figure 8.2). *LacZ* codes for β-galactosidase, which is composed of four identical subunits of 1021 amino acids each. *LacY* codes for a permease, a 275-amino acid protein of the cell membrane that participates in the transport of sugars including lactose, across the membrane. *LacA* codes for thiogalactoside transacetylase, a 275-amino acid enzyme that transfers an acetyl group from

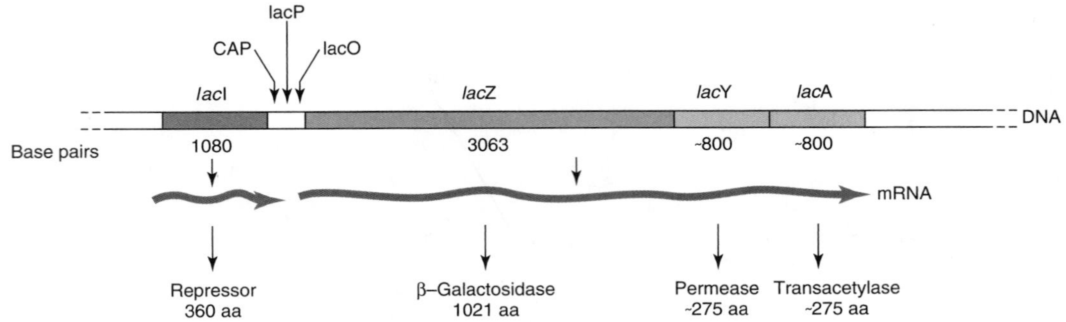

Figure 8.2 Lactose operon of *E. coli*. Lactose operon is composed of *lac*I gene, which codes for a repressor; control elements of CAP, *lac*P, and *lac*O; and three structural genes, *lac*Z, *lac*Y, and *lac*A, which code for β-galactosidase, a permease, and a transacetylase, respectively. The *lac*I gene is transcribed from its own promoter. The three structural genes are transcribed from the promoter, *lac*P, to form a polycistronic mRNA from which the three proteins are translated.

acetyl CoA to β-galactosides. Of the three proteins, only β-galactosidase participates in a known metabolic pathway. The permease helps lactose move across the cell membrane. The transacetylase may be associated with detoxification and excretion reactions of nonmetabolized analogs of β-galactosides.

Mutations in *lac*Z or *lac*Y that inactivate β-galactosidase or permease prevent cells from cleaving lactose or acquiring it from the medium, respectively. Mutations in *lac*A that inactivate transacetylase do not seem to affect cell growth and division.

A single mRNA species containing the coding sequences of all three structural genes is transcribed from a **promoter**, called *lac*P, located just upstream from *lac*Z. Induction of these genes occurs during initiation of their transcription. Without the inducer, transcription occurs at a very low level. In presence of inducer, transcription begins at *lac*P and proceeds through all three genes to a transcription terminator located slightly beyond *lac*A. Thus, the genes are **coordinately expressed**; all are transcribed in unison or none is transcribed.

The presence of three coding sequences on the same mRNA molecule suggests that the relative amount of the three proteins is always the same under varying conditions of induction. The inducer is a molecular switch that influences synthesis of one mRNA species for all three genes. However, the amount of each of the three proteins is often different because of differences in rates of translation or of degradation of the proteins themselves.

The mRNA induced by lactose has a half-life of about 3 min. Therefore, expression of the operon can be altered very quickly. Transcription ceases as soon as inducer is no longer present, existing mRNA molecules disappear within minutes, and cells stop making the proteins.

Repressor of Lactose Operon Is a Diffusible Protein

The regulatory gene of the lactose operon, *lac*I, codes for the lactose **repressor** whose only function is to control initiation of transcription of the *lac* structural genes. *Lac*I is located just upstream of the controlling elements for the *lac*ZYA cluster. However, a regulatory gene need not be close to the gene cluster it regulates. Transcription of *lac*I is not regulated; this gene is transcribed from its own promoter at a low rate that is relatively independent of the cell's status.

Lactose repressor is synthesized as a monomer of 360 amino acids, which forms an active homotetramer. Usually there are about 10 tetramers per cell. It has strong affinity for its main DNA-binding site, called the **operator** or *lac*O, which lies between *lac*P and *lac*Z. *lac*O partly overlaps *lac*P so bound repressor prevents **RNA polymerase** from binding to *lac*P and initiating transcription.

Besides recognizing and binding to *lac*O, the repressor has strong affinity for inducer molecules of the lactose operon. Each monomer binds to an inducer molecule, which causes a **conformational change** that greatly lowers the repressor's affinity for *lac*O (Figure 8.3). Thus, when inducer is present, the repressor no longer binds to *lac*O and RNA polymerase begins transcription from the promoter.

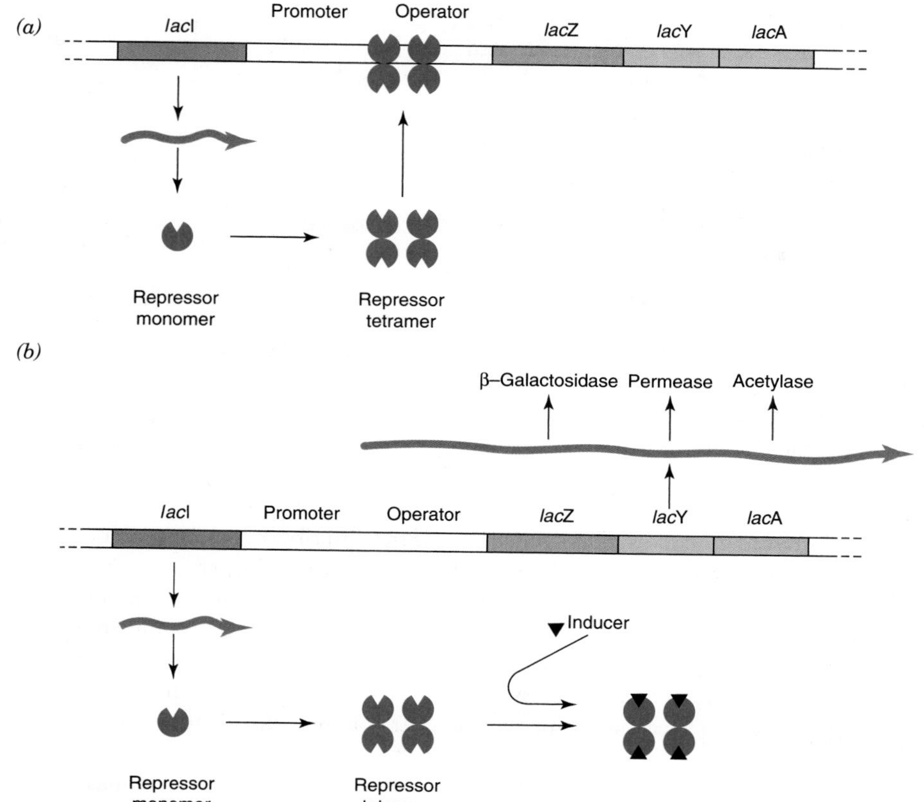

Figure 8.3 Control of lactose operon. (*a*) Repressor tetramer binds to operator and prevents transcription of structural genes. (*b*) Inducer binds to repressor tetramer, which prevents repressor from binding to operator. Transcription of the three structural genes can occur from the promoter.

Study of the lactose operon was greatly facilitated by the discovery that some small molecules, such as isopropylthiogalactoside (**IPTG**), fortuitously serve as inducers but are not metabolized by β-galactosidase. These **gratuitous inducers** bind like inducers to the repressor molecule.

The repressor protein acts in trans; that is, it can diffuse to its site of action. Some *lac*I mutations change or delete amino acids of the repressor that are part of the binding site for the inducer. These changes do not affect the affinity of repressor for the operator, so that the repressor is always bound to the operator, even in the presence of inducer, and the *lac*ZYA genes are never transcribed above a very low basal level. Other *lac*I mutations change the amino acids in the operator-binding site and lessen the affinity of repressor for operator. Thus, the repressor does not bind the operator and the *lac*ZYA genes are continuously transcribed. Such mutations are **repressor-constitutive mutations** because the *lac* genes are permanently expressed. A few rare *lac*I mutants increase the affinity of repressor for operator. In these cases inducer molecules can bind to repressor, but they are less effective in releasing repressor from the operator.

Repressor-constitutive mutants illustrate the features of a negative control system. An active repressor, in the absence of an inducer, shuts off expression of the *lac* structural genes. An inactive repressor results in the constitutive, unregulated, expression of these genes. Using recombinant DNA techniques described in Chapter 7, a plasmid-containing wild-type *lac*I (but not the rest of the lactose operon) can be introduced into constitutive *lac*I mutant cells. Such cells synthesize both active and inactive repressor molecules. Under these conditions, normal wild-type regulation of the lactose operon occurs. Thus, in genetic terms, wild-type induction is dominant over mutant constitutivity.

Operator Sequence of Lactose Operon Is Contiguous with a Promoter and Three Structural Genes

The control elements upstream of the structural genes of the lactose operon are the operator and the promoter. The operator was identified, like *lac*I, by mutations that affect transcription of the *lac*ZYA region. Some of these mutations result in the constitutive synthesis of

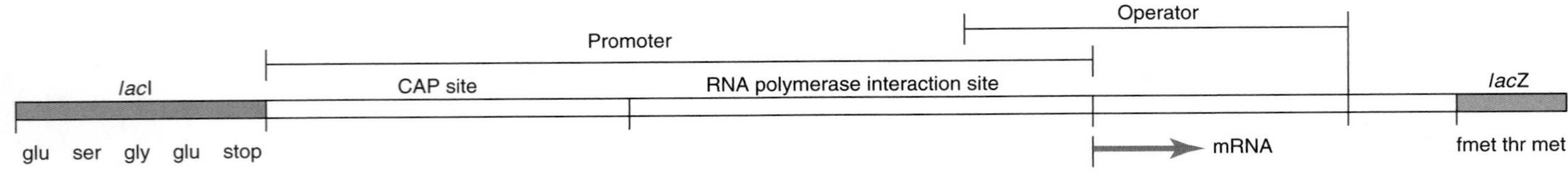

5′ GGAAAGCGGGCAGTGAGCGCAACGCAATTAATGTGAGTTAGCTCACTCATTAGGCACCCCAGGCTTTACACTTTATGCTTCCGGCTCGTATGTTGTGTGGAATTGTGAGCGGATAACAATTTCACACAGGAAACAGCTATGACCATG 3′
3′ CCTTTCGCCCGTCACTCGCGTTGCGTTAATTACACTCAATCGAGTGAGTAATCCGTGGGGTCCGAAATGTGAAATACGAAGGCCGAGCATACAACACACCTTAACACTCGCCTATTGTTAAAGTGTGTCCTTTGTCGATACTGGTAC 5′

Figure 8.4 Nucleotide sequence of control elements of lactose operon. The end of lacI gene (coding for the lactose repressor) and beginning of lacZ gene (coding for β-galactosidase) are also shown. Lines above and below the sequence indicate symmetrical sequences within the CAP site and operator.

lac mRNA; that is, they are **operator-constitutive mutations**. In these cases the operator DNA sequence has one or more base pair changes so the repressor no longer binds as tightly to the sequence. Thus, the repressor is less effective in preventing RNA polymerase from initiating transcription.

In contrast to mutations in *lac*I which encode the diffusible repressor, mutations in the operator do not change a diffusible protein. They exert their influence on transcription of only the three *lac* genes immediately downstream of the operator on the same DNA molecule. If a second lactose operon is introduced into a bacterium on a recombinant plasmid, the operator of one operon has no influence action on the other operon. Thus, an operon with a wild-type operator will be repressed under the usual conditions, whereas in the same engineered bacterium a second operon with an operator-constitutive mutation will be continuously transcribed.

Operator mutations are frequently referred to as **cis-dominant** to emphasize they affect only adjacent genes on the same DNA molecule. Cis-dominant mutations occur in DNA sequences that are recognized by proteins rather than in DNA sequences that code for the diffusible proteins. **Trans-dominant** mutations occur in genes that specify diffusible products. Therefore, cis-dominant mutations also occur in promoter and transcription termination sequences, whereas trans-dominant mutations also occur in genes for subunit proteins of RNA polymerase, ribosomal proteins, and so on.

Figure 8.4 shows the sequence of the *lac* operator and promoter. The operator sequence has an axis of **dyad symmetry**. The sequence of the upper strand on the left side of the operator is nearly identical to that of the lower strand on the right side; only three differences occur. This symmetry in the DNA recognition sequence reflects symmetry in the tetrameric repressor and facilitates tight binding of the repressor subunits to the operator. Dyad symmetry in the double-stranded DNA sequence is a common feature of many protein-binding or recognition sites, including most recognition sites for restriction enzymes.

The *lac* operator of about 30 bp is an extremely small fraction of the total *E. coli* genome and occupies an even smaller fraction of the volume of the cell. However, the approximately 10 tetrameric repressors are also confined to a small fraction of the cell volume. Since the repressor gene is very close to the *lac* operator, the repressor does not have far to diffuse if its translation begins before its mRNA is fully synthesized. More importantly, the repressor has a low general affinity for all DNA sequences. When the inducer binds to the repressor, its affinity for the operator is reduced about 1000-fold, but its low affinity for random DNA sequences is unaltered. Therefore, all of the lactose repressors in the cell are in loose association with DNA. When binding of inducer releases a repressor molecule from the operator, it quickly binds a nearby DNA region. Therefore, induction redistributes the repressor on the DNA rather than generates freely diffusing repressor molecules.

How does lactose enter a cell in the first place if the *lac*Y gene encoding the permease is repressed, yet permease is required for lactose transport across the cell membrane? Even in the repressed state, there is a very low basal level of transcription of the *lac* operon that provides five or six molecules of permease per cell. Perhaps this is just enough to get a few molecules of lactose inside the cell and begin the process.

Another curious observation is that lactose is not the natural inducer of the lactose operon. When the repressor is isolated from fully induced cells, the small molecule bound to each repressor monomer is **allolactose**, not lactose. Allolactose, like lactose, is composed of galactose and glucose, but the linkage between the two sugars is different. A side reaction of β-galactosidase (which normally breaks down lactose to galactose and glucose) converts these two products to allolactose.

Therefore, a few molecules of lactose are taken up and converted by β-galactosidase to allolactose, which binds to the repressor and induces the operon. Further confirmation that lactose is not the real inducer comes from experiments indicating that binding of lactose to purified repressor actually increases the repressor's affinity for the operator. Therefore, in the induced state a small amount of allolactose must be present to overcome this *anti-inducer* effect of lactose.

RNA Polymerase and a Regulator Protein Recognize Promoter Sequence of Lactose Operon

Immediately upstream of the *lac* operator is the promoter. This sequence contains binding sites for RNA polymerase and **catabolite activator protein (CAP;** also called **cAMP receptor protein** or **CRP)** (Figure 8.4). The site to which RNA polymerase binds has been identified by genetic and biochemical approaches. Point mutations and deletions (or insertions) in this region dramatically affect RNA polymerase binding. The end points of the sequence to which RNA polymerase binds were identified by DNase protection experiments. Purified RNA polymerase was bound to the *lac* promoter region cloned in a bacteriophage DNA or a plasmid, and this protein–DNA complex was digested with DNase I. The DNA segment protected from degradation by DNase was recovered, and its sequence determined. The ends of this protected segment varied slightly with different DNA molecules but corresponded closely to the boundaries of the RNA polymerase-binding site shown in Figure 8.4.

The RNA polymerase-binding site does not have symmetrical elements like the operator sequence. This is not surprising since RNA polymerase must associate with DNA in an asymmetrical fashion for transcription to be initiated in only one direction. However, the part of the promoter sequence recognized by CAP does contain some symmetry.

Catabolite Activator Protein Binds Lactose Promoter

E. coli prefers glucose over other sugars as a carbon source. For example, if both glucose and lactose are in the medium, the bacteria selectively metabolize glucose. Figure 8.5 shows that β-galactosidase does not appear until the glucose is depleted. This indicates that glucose interferes with induction of the lactose operon, a process called **catabolite repression** since it occurs during the catabolism of glucose. An identical effect is seen with other inducible operons, including the arabinose and galactose operons. This is probably a general coordinating system for turning off synthesis of unwanted enzymes when glucose is present.

Catabolite repression occurs because glucose inhibits adenylate cyclase, which synthesizes **cyclic AMP (cAMP)**, leading to a lower intracellular concentration of cAMP. cAMP forms a complex with **Catabolite Activator Protein (CAP)**, which then binds the CAP regulatory site at promoters of lactose (and other) operons (Figure 8.6). The CAP–cAMP complex exerts a positive control on transcription by causing the DNA helix to bend, or kink, about 90° at the site of the binding. This DNA bend and the interaction between CAP–cAMP and RNA polymerase activate transcription initiation. If the CAP site is not occupied, RNA polymerase has difficulty binding to the promoter and transcription initiation is much less efficient.

The lactose operon demonstrates how bacteria can coordinate a general response to a metabolic condition (the need to use a sugar as an energy or carbon source) and a specific response to that condition (the need to utilize lactose as the sugar).

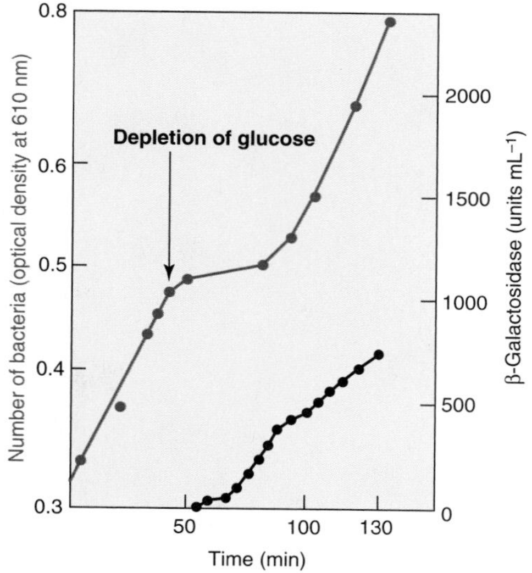

Figure 8.5 Lack of synthesis of β-galactosidase in *E. coli* when glucose is present. Bacteria are growing in a medium containing initially 0.4 mg glucose/mL and 2 mg lactose/mL. Left-hand ordinate indicates optical density of growing culture, an indicator of the number of bacterial cells. Right-hand ordinate indicates units of β-galactosidase per milliliter. Note that appearance of β-galactosidase is delayed until the glucose is depleted. Redrawn from Epstein, W., Naono, S., and Gros, F. *Biochem. Biophys. Res. Commun.* 24: 588, 1966.

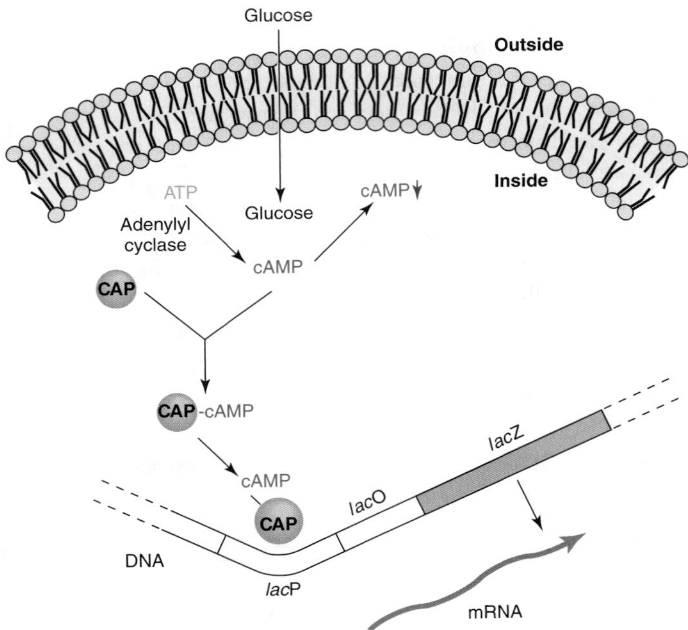

Figure 8.6 Control of *lac*P by cAMP. A CAP–cAMP complex binds to CAP site and enhances transcription at *lac*P. Catabolite repression occurs when glucose lowers intracellular concentration of cAMP. This reduces the amount of the CAP–cAMP complex and decreases transcription from *lac*P and from promoters of several other operons.

8.4 • TRYPTOPHAN OPERON OF *E. COLI*

Bacteria must have the proper amount and relative balance of the 20 amino acids that make up proteins. **Tryptophan**, for example, is required for the synthesis of all proteins that contain it. Therefore, if tryptophan is not present in sufficient quantity in the medium, the bacterial cell has to make it. In contrast, lactose is not absolutely required for the cell's growth; many other sugars can substitute for it. As a result, synthesis of biosynthetic enzymes for tryptophan is regulated differently than synthesis of proteins encoded by the lactose operon.

Tryptophan Operon Is Controlled by a Repressor Protein

In *E. coli* tryptophan is synthesized from chorismic acid in a five-step pathway catalyzed by three enzymes (Figure 8.7). The **tryptophan operon** contains five structural genes that code for these three enzymes (two of which contain two different subunits). Upstream from this gene cluster is a promoter where transcription begins and an operator to which a repressor protein encoded by a separate *trp*R gene binds. Transcription of the lactose operon is generally *turned off*, or repressed, unless induced by a small molecule inducer. The tryptophan operon, on the other hand, is always active, or derepressed, unless it is repressed by a small molecule **corepressor**, which for the tryptophan operon is tryptophan itself.

Tryptophan biosynthesis is regulated by both the synthesis and the activity of enzymes that catalyze the pathway. For example, anthranilate synthetase, which catalyzes the first step, contains two subunits encoded by *trp*E and *trp*D and its enzyme activity is regulated by **feedback inhibition**. This is a common short-term means of regulating the first committed step in a metabolic pathway. Tryptophan can bind to an allosteric site on anthranilate synthetase and prevent its enzymatic activity. Thus, as the concentration of tryptophan builds up, it inhibits anthranilate synthetase. Tryptophan also is a corepressor that shuts down transcription of the tryptophan operon. Feedback inhibition is a short-term control with an immediate effect on the pathway, whereas repression takes longer but has the more permanent effect of repressing transcription.

The tryptophan repressor is a homodimer whose subunits have 108 amino acids each. Under normal conditions about 20 molecules of the repressor dimer are present. The repressor must be complexed with two molecules of tryptophan to bind to the operator. Recall that the lactose repressor binds its operator only in the absence of its inducer.

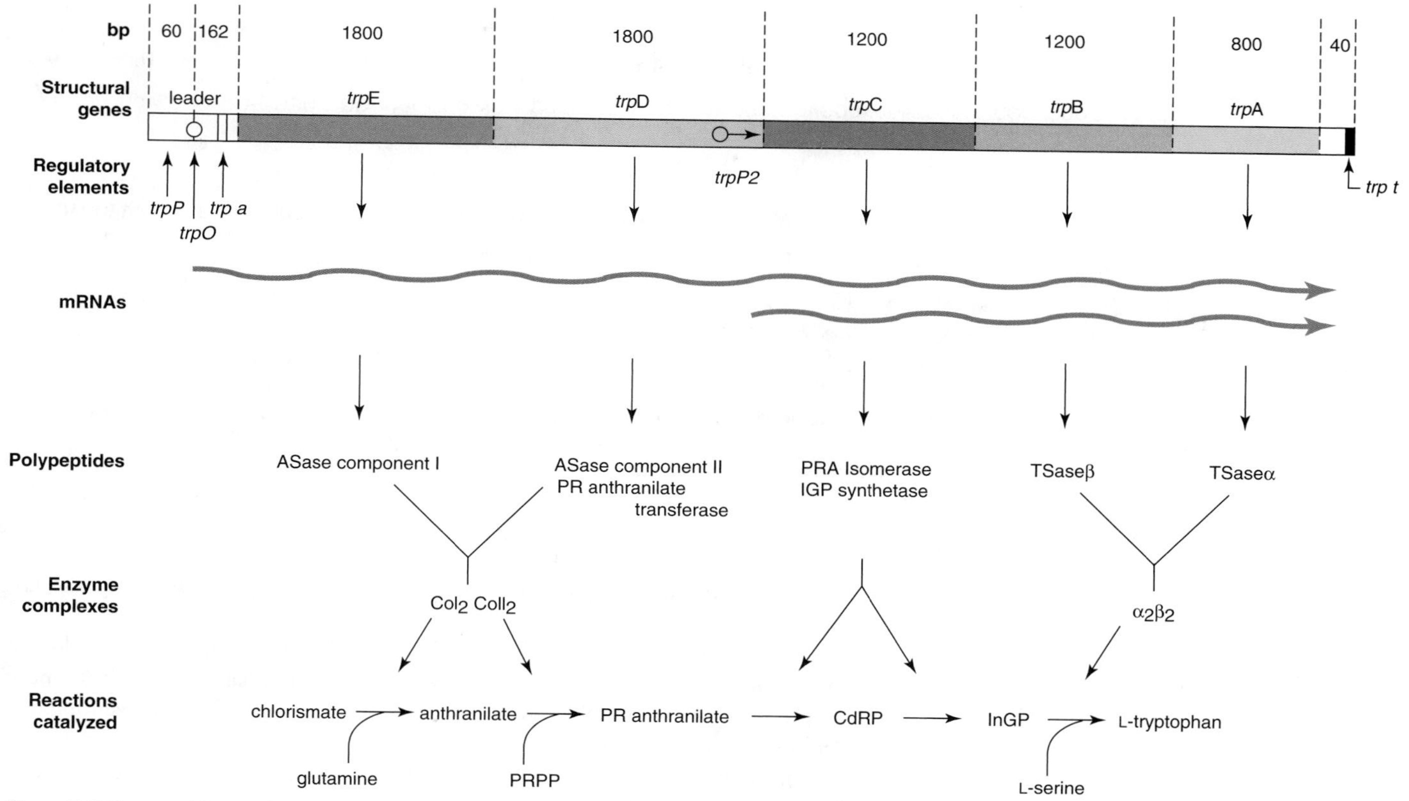

Figure 8.7 Genes of tryptophan operon of E. *coli*. Regulatory elements are the primary promoter (*trp*P), operator (*trp*O), attenuator (*trp*a), secondary internal promoter (*trp*P2), and terminator (*trp*t). Direction of mRNA synthesis is indicated by wavy arrows representing mRNAs. CoI_2 and $CoII_2$ signify components I and II, respectively, of the anthranilate synthetase (ASase) complex; PR-anthranilate is *N*-5′-phosphoribosyl-anthranilate; CdRP is 1-(o-carboxy-phenylamino)-1-deoxyribulose-5-phosphate; InGP is indole-3-glycerol phosphate; PRPP is 5-phosphoribosyl-1-pyrophosphate; and TSase is tryptophan synthetase.
Described in Yanofsky, C. *Trends in Genet.* 20: 367, 2004.

The tryptophan repressor also regulates transcription of *trp*R, its own gene. As tryptophan accumulates in cells, the repressor–tryptophan complex binds to a region upstream of *trp*R, turning off its transcription and maintaining the equilibrium of 20 repressors per cell. The *trp* operator occurs entirely within the *trp* promoter rather than adjacent to it (Figure 8.8). The operator is a region of dyad symmetry, and the mechanism for preventing transcription is the same as in the lactose operon; that is, binding of the repressor–co-repressor complex to the operator prevents binding of RNA polymerase.

Repression decreases the rate of transcription initiation at the *trp* promoter by about 80-fold. (The basal level of lactose operon gene products is about 1000-fold lower than the induced level.) However, the tryptophan operon contains additional regulatory elements. A secondary promoter designated *trp*P2 is located within the coding sequence of *trp*D (Figure 8.7) and is not regulated by the tryptophan repressor. Transcription from it occurs constitutively at a relatively low rate to generate an mRNA that contains the coding sequences for *trp*CBA. Thus, two polycistronic mRNAs are derived from the tryptophan

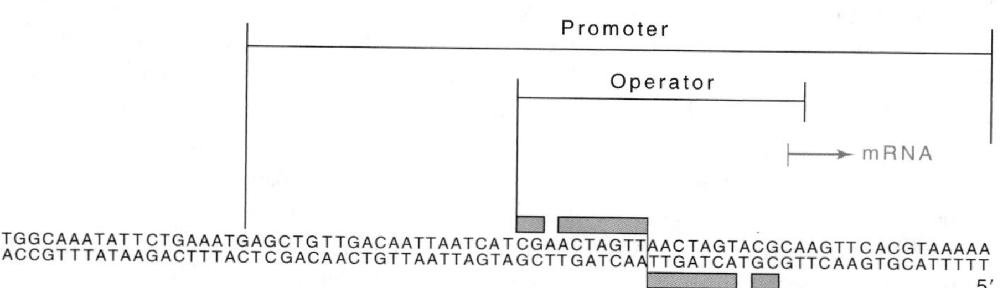

Figure 8.8 Nucleotide sequence of control elements of tryptophan operon. Boxes above and below sequence indicate symmetrical sequences within operator.

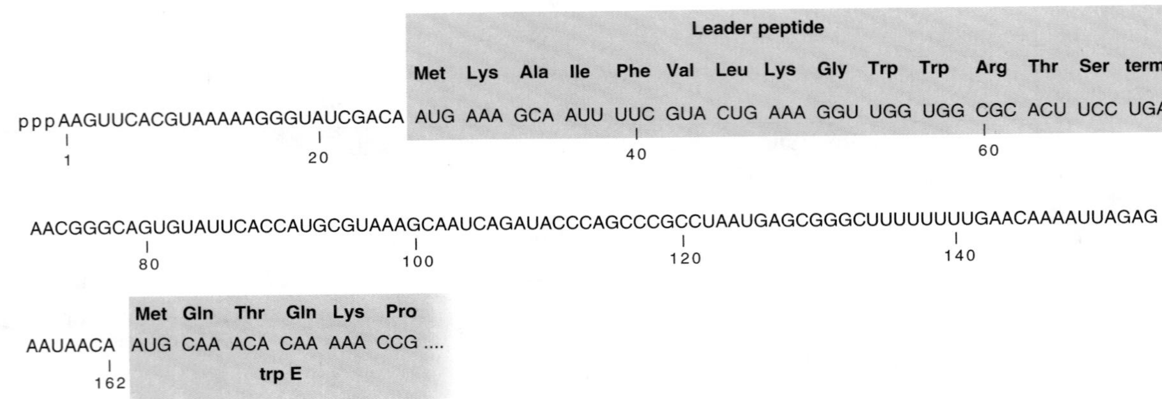

Figure 8.9 Nucleotide sequence of leader RNA from tryptophan operon. The 14 amino acids of the putative leader peptide are indicated over their codons.
Redrawn with permission from Oxender, D.L., Zurawski, G., and Yanofsky, C. *Proc. Natl. Acad. Sci. USA* 76: 5524, 1979.

operon, one containing all five structural genes and the other possessing only the last three.

A second internal promoter may be needed because three of the five genes do not contain a tryptophan codon; only *trp*B and *trp*C contain a tryptophan codon. Thus, under extreme tryptophan starvation, these two proteins are not synthesized, preventing the pathway from being activated. However, since both of these genes lie downstream of the unregulated second promoter, their protein products will always be present at the basal level necessary to maintain the pathway.

Attenuator Region of Tryptophan Operon

Another important control element of the tryptophan operon not present in the lactose operon is the **attenuator** region (Figure 8.9), which lies within the 162 nucleotides between the promoter and the initiator methionine codon of *trp*E. Its existence was first deduced from mutations that mapped in this region and increased transcription of all five structural genes. Within this region is the **leader sequence** of 14 adjacent codons that begins with a methionine codon and ends with an in-phase termination codon. These codons are preceded by a canonical ribosome-binding site and could potentially specify a 14-residue leader peptide. This peptide has not been detected in bacterial cells, perhaps because it is degraded very rapidly.

The attenuator region provides RNA polymerase with a second chance to regulate transcription of the tryptophan operon. In the presence of tryptophan, transcription begins at the promoter but it is prematurely terminated at the end of the attenuator region. This produces a short 140-nucleotide transcript. In the absence of tryptophan, the attenuator region has no effect on transcription, and the entire **polycistronic mRNA** of the five structural genes is synthesized. Therefore, at both the operator and attenuator, tryptophan exerts the same general influence. At the operator, it participates in repressing transcription, and at the attenuator it terminates transcription by those RNA polymerases that have escaped repression. Attenuation has about a 10-fold effect on transcription of the tryptophan structural genes. Derepression at the operator has an 80-fold effect, so transcription of the tryptophan operon is regulated over an 800-fold range.

Transcription is terminated at the attenuator site by cooperative interaction between transcription and translation. The tryptophan leader sequence has two adjacent tryptophan codons at positions 10 and 11(Figure 8.9). This is unusual since tryptophan is a relatively rare amino acid in *E. coli.* These codons provided an early clue that $tRNA^{trp}$ is involved in attenuation. If tryptophan in the cell is low, the amount of charged tryptophanyl-$tRNA^{trp}$ will also be low. The ribosomes will be unable to translate through the two tryptophan codons and will stall in the leader RNA sequence.

The RNA sequence of the attenuator region can adopt several secondary structures (Figure 8.10). The position of the ribosome on the leader sequence determines which

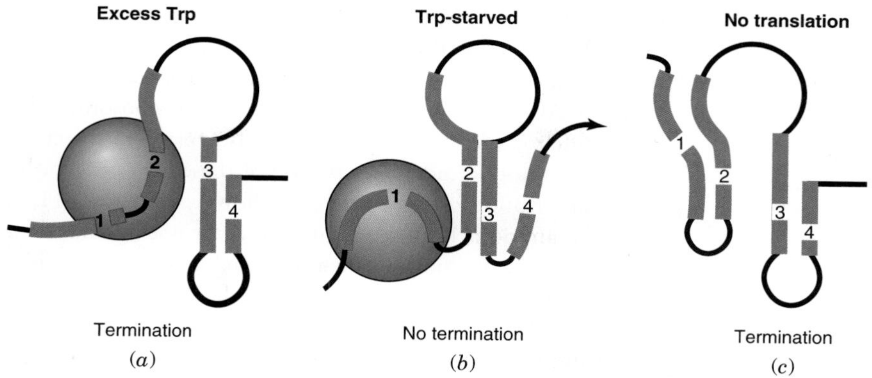

Figure 8.10 Schematic model for attenuation in the tryptophan operon of *E. coli.* (*a*) Under conditions of excess tryptophan, the ribosome (green sphere) translating the newly transcribed leader RNA synthesizes the complete leader peptide. During this synthesis the ribosome binds to regions 1 and 2 of the RNA and prevents formation of stem and loop 1–2 or 2–3. Stem and loop 3–4 will be free to form and signal the RNA polymerase molecule (not shown) to terminate transcription. (*b*) Under conditions of tryptophan starvation, tryptophanyl-tRNA^{Trp} will be limiting, and the ribosome will stall at the adjacent tryptophan codons at the beginning of region 1 in the leader peptide-coding region. Because region 1 is bound to the ribosomes, stem and loop 2–3 will form, excluding formation of stem and loop 3–4, which is required as the signal for transcription termination. Therefore RNA polymerase continues transcription into the structural genes. (*c*) Under conditions in which the leader peptide is not translated, stem and loop 1–2 forms, preventing formation of stem and loop 2–3, and permitting formation of stem and loop 3–4. This signals transcription termination.
Reproduced with permission from Oxender, D. L., Zurawski, G., and Yanofsky, C. *Proc. Natl. Acad. Sci. USA* 76: 5524, 1979.

one occurs. The secondary structure, in turn, is recognized by the RNA polymerase that has transcribed through this region. The structure that forms when a ribosome is not stalled at the tryptophan codons is a termination signal for the RNA polymerase, so that transcription stops after the synthesis of a 140-nucleotide transcript, which is quickly degraded. The secondary structure that forms when the ribosomes are stalled at the tryptophan codons is not recognized as a termination signal, and transcription continues into the *trp*E gene.

Transcription Attenuation Controls Other Operons for Amino Acid Biosynthesis

Transcription attenuation does not occur in eukaryotic organisms because transcription (in the nucleus) and translation (in the cytoplasm) take place in different cellular compartments. However, attenuation is common in bacterial gene expression and occurs in at least six other operons for enzymes catalyzing amino acid biosynthetic pathways. Figure 8.11 shows the **leader peptide** sequences specified by these operons. In each, the leader peptide has several copies of the amino acid that is the end product of the pathway. The 16-residue leader peptide encoded in the histidine operon contains seven contiguous histidines. Starvation for histidine decreases the amount of histidinyl-tRNA^{his} and increases transcription of the *his* operon. The nucleotide sequence of the attenuator region suggests that ribosome stalling at these codons influences the formation of alternate hairpin loops, one of which resembles a termination hairpin followed

Operon	Leader peptide sequence	Regulatory amino acids
his	Met-Thr-Arg-Val-Gln-Phe-Lys-His-His-His-His-His-His-His-Pro-Asp	His
*phe*A	Met-Lys-His-Ile-Pro-Phe-Phe-Phe-Ala-Phe-Phe-Phe-Thr-Phe-Pro	Phe
thr	Met-Lys-Arg-Ile-Ser-Thr-Thr-Ile-Thr-Thr-Thr-Ile-Thr-Ile-Thr-Thr-Gly-Asn-Gly-Ala-Gly	Thr Ile
leu	Met-Ser-His-Ile-Val-Arg-Phe-Thr-Gly-Leu-Leu-Leu-Leu-Asn-Ala-Phe-Ile-Val-Arg-Gly-Arg-Pro-Val-Gly-Gly-Ile-Gln-His	Leu
ilv	Met-Thr-Ala-Leu-Leu-Arg-Val-Ile-Ser-Leu-Val-Val-Ile-Ser-Val-Val-Val-Ile-Ile-Ile-Pro-Pro-Cys-Gly-Ala-Ala-Leu-Gly-Arg-Gly-Lys-Ala	Leu, Val, Ile

Figure 8.11 Leader peptide sequences specified by biosynthetic operons of *E. coli.* All of the leader peptide sequences contain multiple copies of amino acid(s) synthesized by enzymes encoded by that operon.

by several U residues. Unlike the *trp* operon, transcription of the *his* operon is regulated primarily by attenuation since it does not possess an operator recognized by a repressor protein. Instead, the ribosome acts rather like a positive regulator protein, similar to the cAMP–CAP complex in the *lac* operon. If the ribosome is bound to (that is, stalled at) the attenuator site, transcription of the downstream structural genes is enhanced. If the ribosome is not bound, transcription of these genes is greatly reduced.

Transcription of some operons, shown in Figure 8.11, can be attenuated by more than one amino acid. For example, the *thr* operon is attenuated by threonine or isoleucine; the *ilv* operon by leucine, valine, or isoleucine. This effect can be explained in each case by stalling of the ribosome at the corresponding codon, which interferes with formation of a termination hairpin. It is possible that in longer leader sequences, stalling at more than one codon is necessary to achieve maximal transcription through the attenuation region.

8.5 • OTHER BACTERIAL OPERONS

Synthesis of Ribosomal Proteins Is Regulated in a Coordinated Manner

Many bacterial operons possess the same general regulatory mechanisms as the *lac*, *trp*, and *his* operons. However, each operon has its own distinctive characteristics. One example is the structural genes for the 70 or more proteins that compose a ribosome (Figure 8.12). Each ribosome contains one copy of each **ribosomal protein** (except for protein L7–L12, which is probably present in four copies). Therefore, all 70 proteins are required in equimolar amounts, and it makes sense that their synthesis is regulated in a coordinated fashion. Six different operons, containing about one-half of the ribosomal protein genes, occur in two major gene clusters. One cluster contains four adjacent operons (*Spc*, *S10*, *str*, and *a*), and the other contains two operons (*L11* and *rif*) located elsewhere in the *E. coli* chromosome. There is no obvious pattern in the distribution of genes among these operons. Some operons code for proteins of one ribosomal subunit; others code for proteins of both subunits. These operons also contain genes for other (related) proteins. For example, the *str* operon contains genes for soluble translation elongation factors, EF-Tu and EF-G, and genes for some proteins in the 30S ribosomal subunit. The α operon contains genes for proteins of both ribosomal subunits and a gene for the α subunit of RNA polymerase. The *rif* operon has genes for the β and β' subunits of RNA polymerase and for ribosomal proteins.

A common feature of the six operons for ribosomal proteins is that their expression is regulated by one of their own structural gene products; that is, they are **self-regulated**. In some cases, regulation occurs at the level of translation, not transcription as discussed for

Operon	Regulator protein	Proteins specified by the operon
Spc	S8	L14-L24-L5-S14-S8-L6-L18-S5-L15-L30
S10	L4	S10-L3-L2-L4-L23-S19-L22-S3-S17-L16-L29
str	S7	S12-S7-EF.G-EF.Tu
α	S4	S13-S11-S4-α-L17
L11	L1	L11-L1
rif	L10	L10-L7-β–β'

Figure 8.12 Operons containing genes for ribosomal proteins *E. coli*. Genes for protein components of the small (S) and large (L) ribosomal subunits of *E. coli* are clustered on several operons. Some of these operons also contain genes for RNA polymerase subunits α, β, and β', and protein synthesis factors EF.G and EF.Tu. At least one protein product of each operon usually regulates expression of that operon (p. 306).

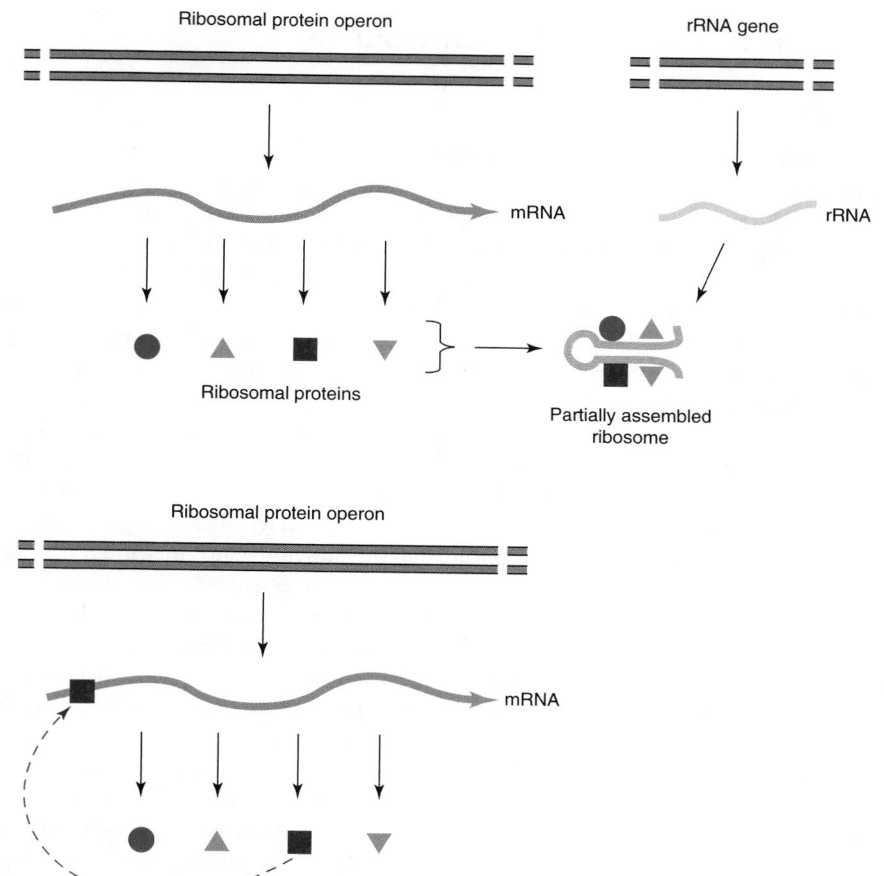

Figure 8.13 Self-regulation of ribosomal protein synthesis. If free rRNA is not available for assembly of new ribosomal subunits, individual ribosomal proteins bind to polycistronic mRNA from their own operon, blocking further translation.

the *lac* and *trp* operons. After the polycistronic mRNA is made, the regulatory ribosomal protein binds to this mRNA and determines which regions, if any, are translated. In general, the ribosomal protein that regulates expression of its own operon associates with ribosomal RNA (rRNA) in the ribosome. This protein has a high affinity for rRNA and a lower affinity for one or more regions of its own mRNA. Therefore, competition occurs between rRNA and the operon's mRNA for binding with the protein. As the protein accumulates to a higher level than free rRNA, it binds to its own mRNA and prevents protein synthesis at one or more of the coding sequences on this mRNA (Figure 8.13). As more ribosomes are formed, the excess of this protein is used up and translation of its mRNA can begin again.

Stringent Response Controls Synthesis of rRNAs and tRNAs

Bacteria respond in several ways to **extreme general stress**. One such situation is when there are insufficient amino acids to maintain protein synthesis. Under this condition the cell invokes the **stringent response** that reduces synthesis of rRNAs and tRNAs about 20-fold. Synthesis of mRNAs also decreases about 3-fold.

The stringent response is triggered by the presence of an uncharged tRNA in the A site of the ribosome, which occurs when the concentration of the corresponding charged tRNA is very low. The first result is that further peptide elongation by the ribosome stops. This stoppage causes a protein called the **stringent factor**, the product of the relA gene, to synthesize **guanosine tetraphosphate** (ppGpp) and **guanosine pentaphosphate** (pppGpp) from ATP and GTP or GDP as shown in Figure 8.14. Stringent factor is loosely associated with some ribosomes of the cell. Perhaps a conformational change in the ribosome is induced by the presence of an uncharged tRNA in the A site, which, in turn, activates the associated stringent factor. The exact functions of ppGpp and pppGpp are not known. However, they inhibit transcription initiation of rRNA and tRNA genes and they affect transcription of some operons more than others.

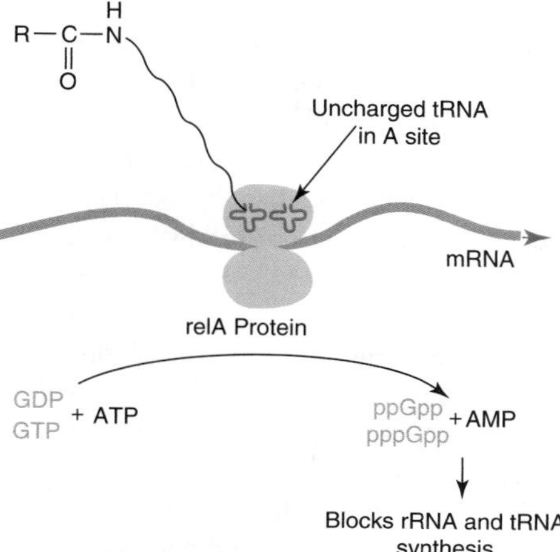

Figure 8.14 Stringent control of protein synthesis in *E. coli*. During extreme amino acid starvation, an uncharged tRNA in the A site of the ribosome activates relA protein to synthesize ppGpp and pppGpp, which are involved in decreasing transcription of the genes coding for rRNAs and tRNAs.

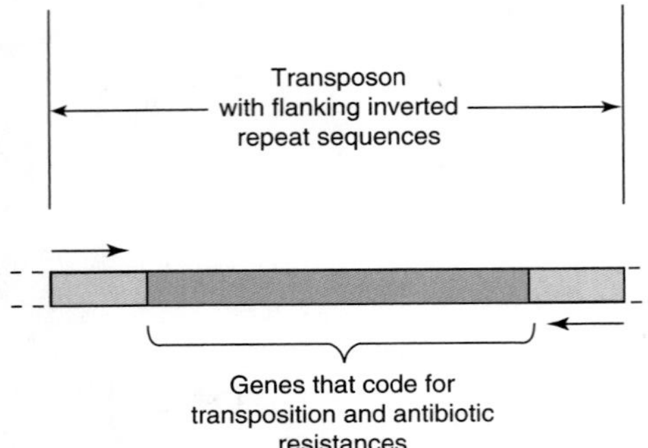

Figure 8.15 General structure of transposons. Transposons are relatively rare mobile segments of DNA that contain genes coding for their own rearrangement and often have genes that specify resistance to various antibiotics.

8.6 • BACTERIAL TRANSPOSONS

Transposons Are Mobile Segments of DNA

The vast majority of bacterial genes have fixed locations in the chromosome. In fact, the genetic maps of *E. coli* and *Salmonella typhimurium* are very similar, indicating the lack of much evolutionary movement of most genes within bacterial chromosomes. There is a small class of bacterial genes, however, in which newly duplicated gene copies "jump" from one genomic site to another with a frequency of about 10^{-7} per generation, the same rate as spontaneous point mutations. These mobile segments of DNA are called **transposable elements** or **transposons** (Figure 8.15). Some genes within bacterial transposons control the presence and transposition of the transposon itself, whereas others, usually antibiotic-resistance genes, provide the bacterium with a selective advantage against other bacteria.

Transposons were first detected as rare insertions of foreign DNA into structural genes of bacterial operons. Usually, these insertions interfere with expression of the affected structural gene and all downstream genes of the operon. This is not surprising since insertions can destroy the translation-reading frame, introduce transcription termination signals, affect the mRNA stability, and so on. Many transposons and the sites into which they insert have been isolated using recombinant DNA techniques and have been extensively characterized.

Some transposons consist of a few thousand base pairs and contain two or three genes; others are much longer and contain many genes. Sometimes, small transposons can occur within a large transposon. All active transposons contain at least one gene that codes for a **transposase**, which is required for the transposition, or "jumping" event. Often they contain genes that code for resistance to antibiotics or heavy metals. Most transpositions involve generation of an additional copy of the transposon and insertion of this copy into another location. The original copy is unaffected by insertion of its duplicate into a new target site. Transposons contain short inverted **terminal repeat sequences** that are essential for the insertion mechanism, and are often used to define the two boundaries of a transposon. Most target sites are fairly random in sequence, although some transposons have a propensity for insertion at specific "hot spots." The duplicated transposon can be located in a different DNA molecule than its donor. Frequently, transposons are found on plasmids that pass from one bacterial strain to another and are the source of a suddenly acquired resistance to one or more antibiotics by a bacterium (Clin. Corr. 8.1).

CLINICAL CORRELATION 8.1

Transmissible Multiple-Drug Resistance

An alarming trend is that pathogenic bacteria are becoming increasingly resistant to a large number of antibiotics. Many cases have been documented in which a bacterial strain in a patient being treated with one antibiotic suddenly became resistant to that antibiotic and, simultaneously, to several other antibiotics even though the bacterial strain had never been previously exposed to these other antibiotics. This occurs when a bacterium suddenly acquires from another bacterial strain a plasmid that contains several different transposons, each containing one or more antibiotic-resistance genes. Examples include the genes encoding β-lactamase, which inactivates penicillins and cephalosporins; chloramphenicol acetyltransferase, which inactivates chloramphenicol; and phosphotransferases, which modify aminoglycosides such as neomycin and gentamycin.

Neu, H. C. The crisis in antibiotic resistance. *Science* 257:1064, 1992.

Transposon *Tn*3 Contains Three Structural Genes

An example of a transposon is **transposon *Tn*3**, which contains 4957 bp including inverted repeats of 38 bp at each end (Figure 8.16). Three genes are present in *Tn*3. One codes for β-lactamase, which hydrolyzes ampicillin and renders the cell resistant to this antibiotic. The others, *tnp*A and *tnp*R, code for a transposase and a repressor protein, respectively. The transposase contains 1021 amino acids and binds to single-stranded DNA. It recognizes the repetitive ends of the transposon and participates in cleavage of the target site into which the new transposon copy inserts. The repressor protein of 185 amino acids controls

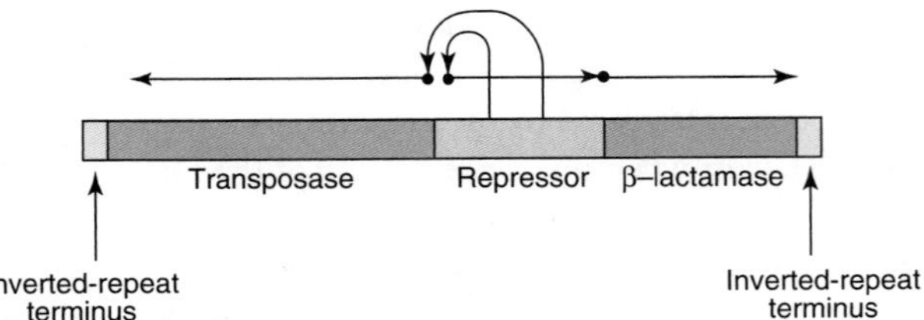

Figure 8.16 Functional components of the transposon *Tn*3. Genetic and DNA sequence analyses show there are four distinct regions: the 38-bp inverted repeat termini; a gene for the enzyme β-lactamase, which confers resistance to ampicillin and related antibiotics; a gene encoding an enzyme required for transposition (transposase); and a gene for a repressor protein that controls transcription of the genes for the transposase and the repressor itself. Horizontal arrows indicate directions of transcription.
Redrawn from Cohen, S. N., and Shapiro, J. A. *Sci. Amer.* 242:40, 1980. W. H. Freeman and Company, Copyright © 1980.

transcription of both the transposase gene and its own gene. *tnp*A and *tnp*R are transcribed divergently from a 163-bp control region located between them that binds the repressor. The repressor also participates in the insertion of the new transposon, but does not affect transcription of the ampicillin-resistance gene.

Mutations in *tnp*A that inactivate the transposase decrease the frequency of *Tn*3 transposition. Those in *tnp*R that inactivate the repressor increase the frequency of transposition. These mutations de-repress *tnp*A resulting in more molecules of the transposase, which enhances the formation of more duplicated transposon copies. They also derepress *tnp*R, but since the repressor is inactive, the depression has no effect.

Transposons located on bacterial plasmids are of increasing importance in clinical uses of antibiotics. Bacterial plasmids that have not been altered for experimental use usually contain genes that facilitate their transfer from one bacterium to another. As these plasmids transfer among different infecting bacterial strains, their transposons containing **antibiotic-resistance genes** are moved into new bacterial strains. Once inside a new bacterium, the transposon can be duplicated onto the chromosome and become permanently established in that cell's lineage. The result is that more and more pathogenic bacterial strains have become resistant to an increasing number of antibiotics (Clin. Corr. 8.2).

CLINICAL CORRELATION 8.2

The Bacterial Pathogen of the Stomach: *Helicobacter pylori*

Helicobacter pylori (*H. pylori*) is a gram-negative bacterium that infects the stomach and duodenum of more than 50% of the world's human population. The organism's helical shape enhances its motility in the mucus overlying the gastric mucosa. Although most infected people do not exhibit any symptoms, *H. pylori* is known to cause many cases of peptic ulcers (lesions in the lining of the stomach or duodenum), gastritis, duodentis and stomach cancers. The bacterium survives in the highly acidic environment of the stomach by synthesizing a large amount of urease enzyme, which metabolizes stomach urea to carbon dioxide and ammonia that neutralizes gastric acid. Acid pH increases the expression of urease and a proton-gated urea channel protein of *H. pylori* at the level of both transcription and mRNA stability. *H. pylori* is typically treated with a therapy of three or more antibiotics, but some isolates of *H. pylori* have been found that exhibit drug resistance. Two Australian physicians, Robin Warren and Barry Marshall, received the 2005 Nobel Prize in Medicine for their work on *H. pylori*.

Scott, D. R., Marcus, E. A., Wen, Y., Oh, J., and Sachs, G. Gene expression *in vivo* shows that *Helicobacter pylori* colonizes an acidic niche on the gastric surface. *Proc. Natl. Acad. Sci. USA*104:7235, 2007; and Blaser, M. J. The endangered species in the stomach. *Sci. Amer.* 292:38, 2005.

8.7 • GENE EXPRESSION IN EUKARYOTES

Gene transcription in eukaryotic organisms is also regulated to provide the appropriate response to biological needs. In addition to modulating gene expression in response to nutritional or environmental conditions, multicellular organisms regulate the expression of specialized genes that drive cellular differentiation and typify specific cell types. Some genes, referred to as housekeeping genes, are expressed in most cells, other genes are activated on demand, and still other genes are rendered permanently inactive in all but a few cell types. In eukaryotic cells, the nuclear membrane serves as a barrier that selectively allows some proteins access to DNA while keeping others in the cytosol. In bacteria one RNA polymerase is responsible for transcription of all RNAs (tRNA, rRNA, and mRNA). In eukaryotic organisms, three different RNA polymerases are used (p. 189). RNA polymerase I transcribes the rRNA genes, RNA polymerase II the protein-encoding genes whose transcripts become mRNAs, and RNA polymerase III the genes for tRNAs and for most other small RNAs. Although some principles for eukaryotic gene activation and control are applicable for all three RNA polymerases, in this section the focus will be on transcription by RNA polymerase II. RNA polymerase II is composed of at least 10 different subunits, ranging in size from 10 to 220 kDa. Some of the subunits are also part of the RNA polymerase I and III complexes; others are unique to RNA polymerase II. The largest subunit of RNA polymerase II has, depending on the species, as many as 52 repeats of the amino acid motif PTSPSYS in its C-terminal domain (CTD). A distinguishing feature of these repeats is that threonine (T), serine (S) and tyrosine (Y) can be phosphorylated.

To better understand the regulation of transcription in eukaryotes, it is useful to recall the organization of DNA into **chromatin** and the role of DNA modification, specifically **methylation** of cytosine, on gene activation. In addition, we will consider how RNA polymerase II is positioned at the appropriate spot in the promoter of a gene to transcribe that gene by the formation of a preinitiation complex that involves assembly of **general transcription factors (TFs)** with RNA polymerase II. Next, we will look at how specific gene activity can be regulated through the use of enhancers, **transcription factor–binding sites** and **RNA polymerase assembly sites.** Finally, we will discuss the activation of transcription by specific transcription factors, some of their general characteristics and how they are regulated.

Eukaryotic DNA Is Bound by Histones to Form Chromatin

Segments of eukaryotic DNA are wrapped around an octamer of histone proteins that contain two molecules each of histone H2A, H2B, H3, and H4 to form **nucleosomes** (p. 55). Histones associate with DNA through electrostatic interactions. Histones have a large number of positively charged lysine residues that interact with the negatively charged phosphodiester backbone of DNA. In most cells histone H1 or H5 binds when the octamer–DNA association is established. About 200 bp of DNA make up a single nucleosomal unit, with about 130–160 bp in direct contact with the octamer core. Some of the remaining DNA binds histones H1 and H5, and the rest is the **linker** DNA between nucleosomes. Although the histone octamer–DNA interaction is not sequence specific, some sequence-dependent patterns of association have been noted. When, for instance, the double-stranded DNA helix bends because of the presence of an AT-rich region, the minor groove generally faces the nucleosome core particle. In GC-rich regions, the minor groove faces away from the histone octamer core. Despite these tendencies, there is not a clear way to predict the sequences within a DNA duplex that will compose the major and minor grooves on the outside of the nucleosome complex. This positioning is important because assembly of the transcription initiator complex and binding of other proteins that influence gene expression need access to the grooves in the DNA duplex. Sequence-specific interactions occur via hydrogen bonds formed with the edges of the base pairs in the major or minor groove. When DNA is wrapped around a histone octamer, the specific DNA sequences that bind transcription factors may be occluded. The complete removal of DNA from the nucleosome is not required to allow access to specific sequences. By rolling the DNA helix relative to the histone octamer, different major and minor groove contacts can be made available on the outer surface of the nucleosome (Figure 8.17).

Genes that are not transcribed within a particular cell form highly condensed heterochromatin. In contrast, transcriptionally active regions of DNA have a less condensed, more open structure. Part of this difference is due to posttranslational modification of histones. The acetylation of the ε-amino group of the lysine residues near the N-termini of histones reduces net positive charge and thus weakens the electrostatic attraction between the histones and the DNA. In general, acetylation of histones leads to activation of gene expression, whereas deacetylation reverses the effect. Acetylation of histones, and the subsequent repositioning of the nucleosome by the destabilization of histone DNA interactions, is important in providing sequence-specific access to DNA (Figure 8.18). Modification of histones is thought to be critical in epigenetic control of gene expression as will be discussed later in this chapter.

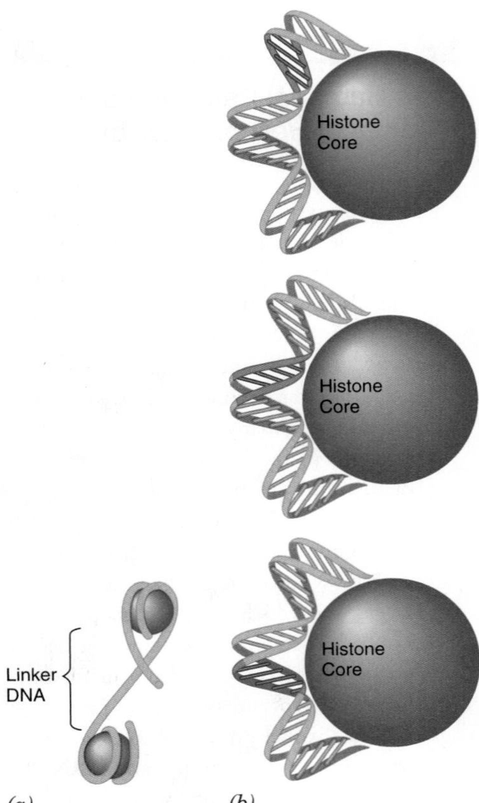

Figure 8.17 Access to specific sequences in major and minor groove of the DNA depends on nucleosome positioning. (*a*) DNA wound around the core histone octamer (two molecules each of histone H2A, H2B, H3 and H4) is called the nucleosome core particle. Under most conditions histone H1 or H5 (not shown) associates with DNA where the winding around the histone octamer begins. DNA between nucleosome core particles is called linker DNA. When exposed to agents like micrococcal nuclease, linker regions are more accessible than DNA wound around the histone core particle. (*b*) Within the nucleosome, the side of the helix facing away from the histone core is accessible, but the side facing in is not. If we rotate the helix with respect to the histone core of a nucleosome and follow the position of the same five base pairs, the access to the major and minor groove to other interactions is changed depending on whether the base pair is on the inside or the outside of the wrapped nucleosome core particle.
Redrawn from Wolffe, A. P. *Chromatin: Structure and Function,* 3d ed.: New York: Academic Press, 1998.

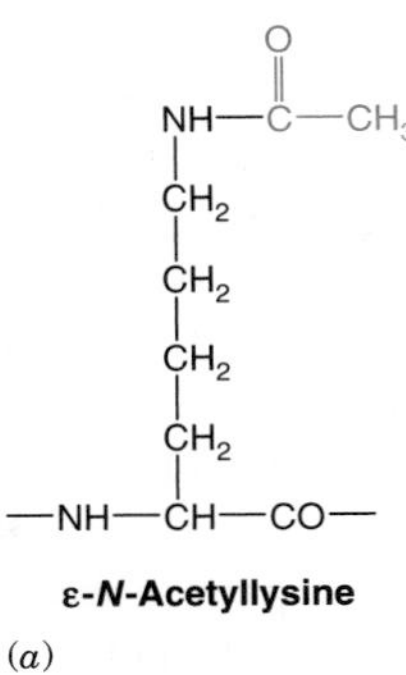

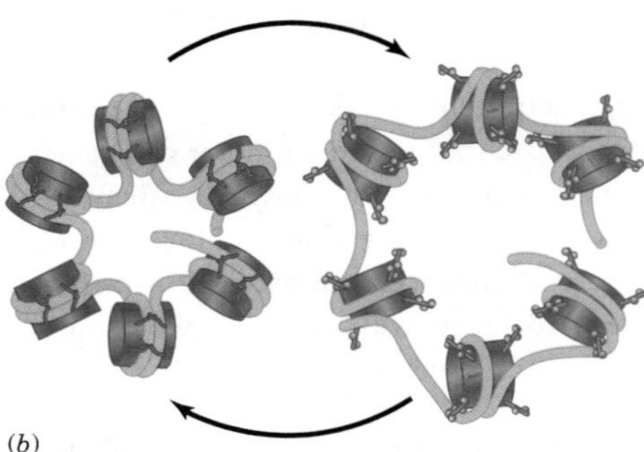

Figure 8.18 Strength of histone DNA association is modified by acetylation of lysine residues in the N-terminus of histone proteins. Histones that make up the core octamer contain lysine residues that, by virtue of their positively charged side group can promote interaction with the negatively charged phosphodiester linkages of DNA. Modification of the lysines by acetylation replaces the positive charge with a neutral acetyl group and weakens the electrostatic interaction between the octamer core and the DNA. This process is reversible, and the acetylation and deacetylation of histones provides a way to loosen or tighten chromatin structure.
Redrawn from Wolffe, A. P. The cancer-chromatin connection. *Sci. and Med.* 6:28, 1999.

The relative position of a nucleosome can also be influenced by the SWI–SNF complex, first identified and best characterized in the yeast *S. cerevisiae* and named after mating type switching and sucrose non-fermenting strains. This complex contains about 10 proteins and interacts with the C-terminal domain of the large subunit of RNA polymerase II disrupting nucleosomal arrays in an ATP-dependent manner. The result is to open regions of DNA for interaction with transcription factors, and thus facilitate gene activation. There are fewer SWI/SNF complexes in the cell than genes being transcribed, suggesting that the SWI /SNF complex acts in a catalytic manner (Clin. Corr. 8.3).

Methylation of DNA Correlates with Gene Inactivation

Methylation of human DNA involves formation of 5-methylcytosine (Figure 8.19). The reaction is sequence specific with 5′CpG3′ as a substrate. The result is that both strands of the duplex are methylated. About 70% of CpG sequences in human DNA are methylated. Methylation is implicated in the **imprinting** of genomic DNA. In imprinting, the methylation patterns of DNA inherited from the sperm or egg correlate with choice of allelic expression. Methylation patterns are conserved after replication by a hemimethylase (which recognizes methylation in the parental strand and methylates the newly replicated strand). More recently it has been suggested that methylation of DNA that results in gene silencing can be induced by the action of siRNA (short interfering RNA) that are complementary to the sequence that is methylated.

A terminology often encountered in the DNA methylation literature is *CpG islands* or *CpG-rich regions.* The sequence CpG is under represented in the genome. This is likely due to the accumulation, over many generations, of C to T transitions caused by the deamination of 5-methylcytosine. Maintenance of CpGs is presumably due to selective pressure to maintain a regulatory region. In fact CpG islands are most commonly found in gene promoter regions. Methylation of DNA often correlates with lack of transcriptional activity. This is thought to be mediated by proteins that recognize and bind methylated DNA, which prevents binding of transcription factors. Methylation of DNA correlates with deacetylation of histones, providing two different means of repression of transcription at a specific location (Figure 8.20).

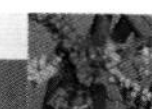

CLINICAL CORRELATION 8.3

Rubinstein–Taybi Syndrome

Regulation of histone acetylation influences activation and inactivation of gene expression. Two major enzymes responsible for acetylation of histones in mammalian cells are the acetyltransferases p300 and CBP. p300 was named for its molecular mass and CBP stands for CREB binding protein, with CREB (a cAMP regulatory-element binding protein) being a transcription factor. These acetyltransferases allow activation of appropriate genes in association with CREB and other transcription factors, leading to the opening of chromatin structure through weakening of histone–DNA interactions.

Rubinstein–Taybi syndrome (OMIM 180849), characterized by mental retardation and other developmental abnormalities, is caused by point mutations, small deletions, and rearrangements in the CBP gene. More severe developmental defects correspond to substantially more mutations. The complete loss of the CBP gene product is probably lethal.

Petrij, F., Giles, R. H., Dauwerse, H. G., Saris, J. J., et al. Rubinstein–Taybi Syndrome caused by mutations in the transcriptional co-activator CBP. *Nature* 376:348, 1995; and Wolffe, A. P. The cancer chromatin connection. *Sci. and Med.* 6:28, 1999.

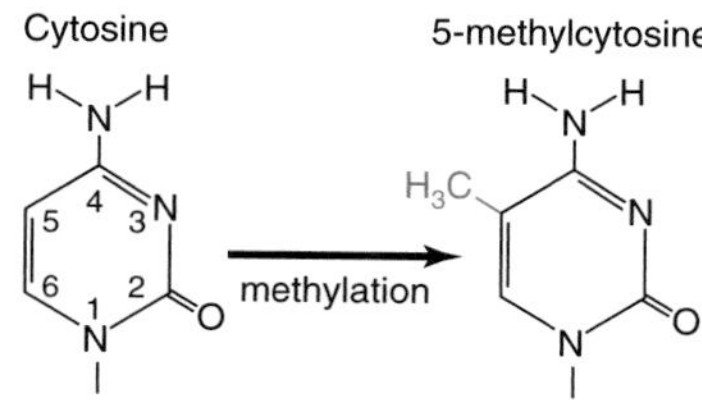

Figure 8.19 The most common methylated base in humans is 5-methylcytosine.

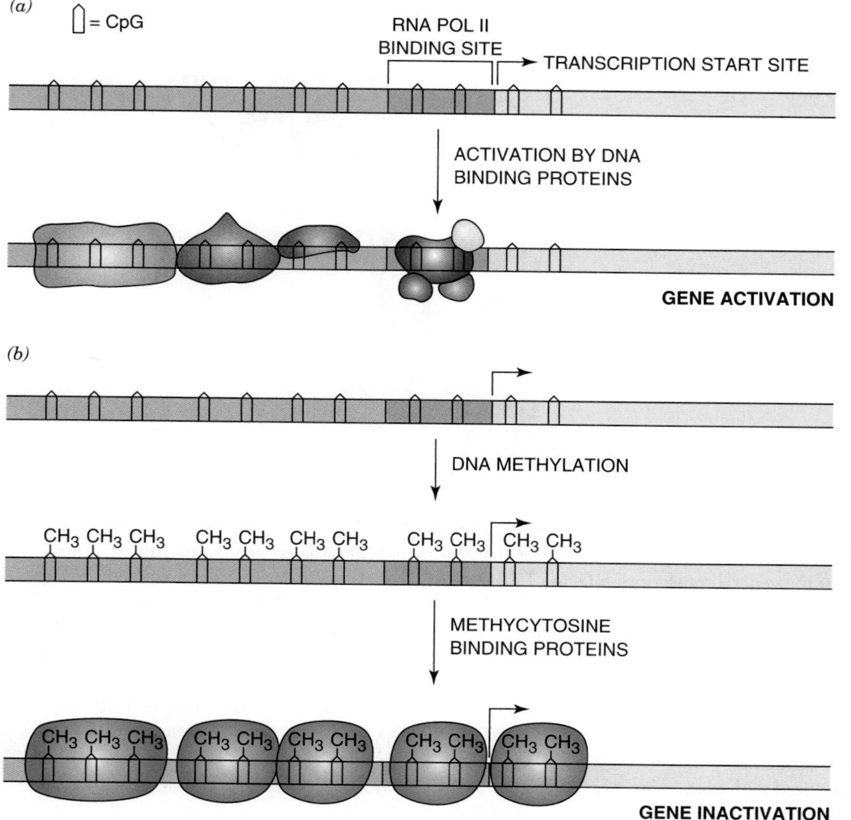

Figure 8.20 Methylation of DNA leads to altered gene activity. For maximal transcription of most genes transcription factors must recognize and bind to specific sequences of DNA in the promoter region. Their interaction with the DNA and the general transcription factors in the RNA polymerase II initiation complex leads to the expression of a gene. Methylation of DNA, specifically the formation of 5-methyl cytosine, provides a new target for protein DNA interaction. The association of 5-meC DNA binding proteins with methylated DNA may block the ability of other transcription factors to bind to DNA. This inhibition is usually not through the sequence specific competition for DNA binding, but rather by steric hindrance.
Redrawn after Alberts, B., Bray, D., Lewis, J., Raff, M., Roberts, K., and Watson, J. *Molecular Biology of the Cell.* New York: Garland, 1994.

TABLE 8.1 • General Transcription Factors Found in Eukaryotes

Factor	*Number of Subunits*	*Mass (kDa)*
TFIID		
TBP	1	38
TAFs	12	15–250
TFIIA	3	12, 19, 35
TFIIB	1	15
TFIIE	2	34, 57
TFIIF	2	30, 74
TFIIH	9	35–89

Source: Roeder, R. G. *Trends Biochem. Sci.* 21:329, 1996.

Hypermethylation of DNA is a common feature in cells of cancerous tissue. There is increasing evidence that methylation occurs in genes encoding proteins that would direct abnormally dividing cells into programmed cell death (apoptosis). Conversely, in mammalian totipotent and pluripotent cells, such as a fertilized egg and cells of the very early embryo, DNA undergoes fairly global demethylation.

8.8 • PREINITIATION COMPLEX IN EUKARYOTES: TRANSCRIPTION FACTORS, RNA POLYMERASE II, AND DNA

Unlike bacterial RNA polymerase, eukaryotic RNA polymerase II does not undergo sequence-specific binding to eukaryotic promoters. Rather, an initiation complex is formed through the initial contact of the promoter with the general transcription factor TFIID. TFIID is one of at least six general transcription factors (TFIIA, TFIIB, TFIID, TFIIE, TFIIF, and TFIIH) required for basal transcription by RNA polymerase II. The nomenclature TFII reflects that they are transcription factors involved in the assembly of the RNA polymerase II preinitiation complex (Table 8.1).

TFIID is a multi-subunit complex that contains the TATA binding protein (TBP) and a number of different TAFs (TBP-associated factors). TFIID binds in the minor groove of DNA at a consensus sequence called the TATA box, located about 27 bp upstream of the transcription start site. The TATA box must be accessible to TFIID, so it cannot be on the inner face of a nucleosome core particle if the preinitiation complex is to form. Interaction of TBP with the DNA causes a large distortion in the DNA duplex (Figure 8.21). TBP was once thought to be essential for transcription using any RNA polymerase, but the recent identification of additional TLFs (TBP-like factors) suggests that some genes use an alternate protein. After binding to the TATA box, TBP directs assembly of the preinitiation complex by ordered addition of several general transcription factors and RNA polymerase II (Figure 8.22). Recall that the large subunit of RNA polymerase II has a C-terminal domain (CTD) containing many repeats of the amino acid sequence PTSPSYS.

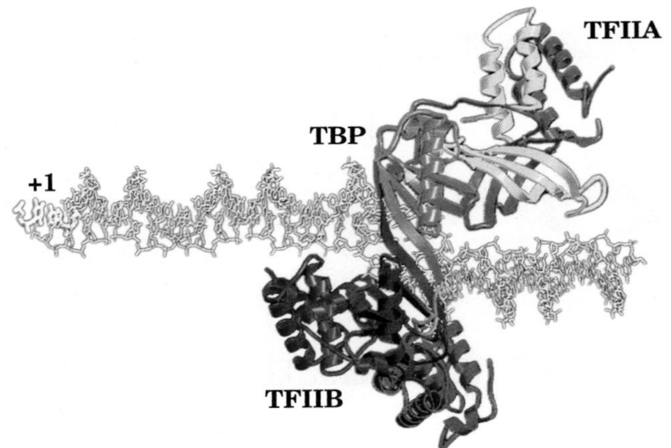

Figure 8.21 The TATA binding protein (TBP) has been cocrystallized with DNA The first step in forming the transcription complex that allows RNA polymerase II mediated gene transcription is the association of the general transcription factor TFIID with DNA. In most cases this is mediated through one of the proteins of the TFIID protein complex, TBP, which binds to DNA through contacts with the sequence TATAA. The cocrystallization of TBP with DNA allowed a closer look at the interaction and indicated that binding of TBP introduced a significant bend in the DNA. Included in the figure are two other general transcription factors, TFIIA and TFIIB, neither of which are involved in the initial contact with DNA. TFIIA binds to TBP and stabilizes the TBP–DNA interaction. TFIIB binds to TBP, leads to the recruitment of RNA polymerase II to the initiation complex, and is involved in identification of the transcription start site.

Figure reproduced with permission from Voet, D., Voet, J., and Pratt, C. W. *Fundamentals of Biochemistry.* New York: Wiley, 1999. © (1999) John Wiley & Sons, Inc.

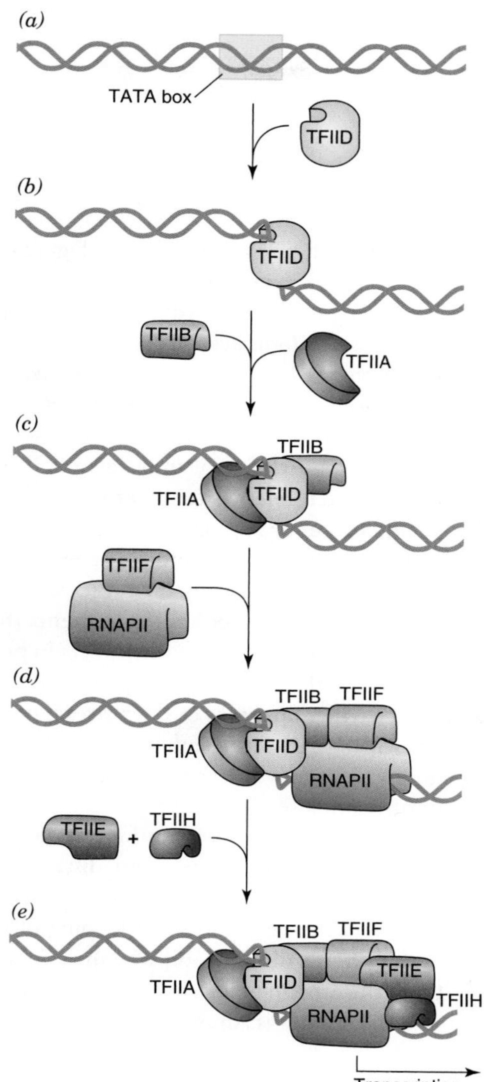

Figure 8.22 Formation of the initiation complex for genes transcribed by RNA polymerase II is ordered. The current models of initiation complex assembly include the following steps: (*a*) The promoter region of a gene with a TATA box before TFIID binding. (*b*) TFIID, which includes TBP binds at the TATA box and bends the DNA. (*c*) The DNA–TFIID complex serves as a coordination site for binding other general transcription factors. The TFIID complex associates with TFIIA and TFIIB. TFIIA serves to stabilize the TFIID DNA interaction, and TFIIB provides an appropriate interaction site for the binding of RNA polymerase II. (*d*) TFIIF and RNA polymerase II join the complex. At this stage the carboxy-terminal domain (CTD) of RNA polymerase II is unphosphorylated. TFIIF is thought to destabilize nonspecific RNA polymerase II DNA interactions, thus targeting the RNA polymerase II to the growing initiation complex. (*e*) Two additional general transcription factors join the complex. TFIIE recruits TFIIH to the complex. TFIIH has helicase, ATPase, and kinase activity. TFIIH is thought to mediate transient opening of the DNA duplex and also the phosphorylation of the CTD of RNA polymerase II allowing the polymerase to leave the initiation complex and start transcription.
Redrawn from Voet, D., Voet, J., and Pratt, C. W. *Fundamentals of Biochemistry.* New York: Wiley, 1999. © (1999) Wiley.

The entry of RNA polymerase II into the preinitiation complex occurs when its CTD is not phosphorylated. However, the movement of RNA polymerase II away from the assembly site is correlated with phosphorylation of the CTD. Thus, the assembly of the preinitiation complex does not ensure transcription; further signaling, such as phosphorylation, must occur for transcription to begin.

Formation of the preinitiation complex allows the ATP-dependent helicase activity associated with TFIIH to open the two strands of the DNA and provides a template for the subsequent elongation phase of RNA synthesis. TFIID remains behind, still bound to the TATA box, to promote the assembly of additional preinitiation complexes. TFIIF moves with RNA polymerase II during the elongation phase, but the other general transcription factors dissociate from the elongation complex.

Eukaryotic Promoters and Other Sequences That Influence Transcription

Promoters of eukaryotic genes transcribed by RNA polymerase II are operationally defined as those sequences that influence the initiation of gene transcription. The hallmark of eukaryotic promoters is their utilization of multiple transcription factor–binding sites to regulate gene activity. In general, these binding sites are relatively close to the TATA box that marks the site of assembly of the preinitiation complex. Other consensus sequences often found in

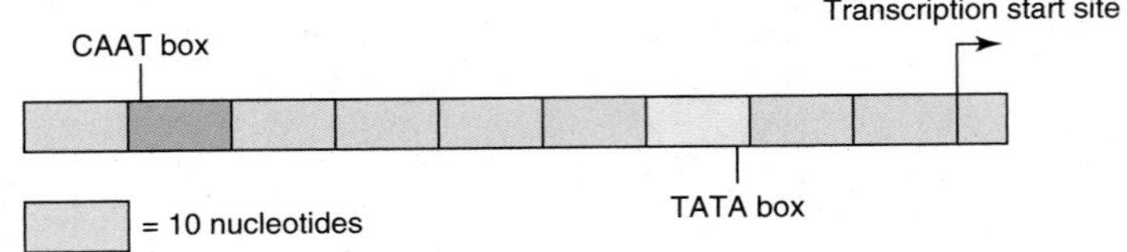

Figure 8.23 DNA sequences in genes transcribed in eukaryotes typically contain a TATA box and may contain a CAAT box, GC-rich regions (GC boxes), and multiple transcription factor–binding sites.

eukaryotic promoters include the CAAT box and GC boxes (Figure 8.23). The exact position of the CAAT box and the position, orientation, and number of GC boxes varies among promoters. The CAAT box is a binding site for several different transcription factors, including NF1. The presence of a CAAT box usually indicates a strong promoter. The GC box provides a binding site for the transcription factor SP1 and is characteristic of many housekeeping genes.

Binding of transcription factors to other sequences in promoters may be influenced by a signaling molecule, such as hormones associated with steroid response transcription factors, or by modification, such as phosphorylation of amino acids. DNA sequences that bind regulated transcription factors are often referred to as **response elements**. Transcription factors may be unique to specific tissues or appear at developmentally specific times. The relative effectiveness of regulatory sequences in promoters depends upon their orientation or directionality, and may be diminished by altering their distance from the TATA box.

Many cellular and viral genomes also have **enhancer elements** that bind proteins that activate transcription. Enhancers differ from promoter sequences in two important respects. First, enhancers can be located many thousands of base pairs away from the preinitiation complex assembly site and can be either upstream or downstream of the transcription start site. Secondly, the enhancer element can act in either orientation. Transcription factors that bind to response elements close to the transcription initiation site may act directly with the initiation complex or through the use of proteins that make a bridge to the initiator complex (see Figure 8.30, p. 328). It is quite common for proteins influencing transcription to serve this bridging function. However, the linear distance between an enhancer and the preinitiation complex may be very large. Transcription factors bound to enhancers are brought together with the preinitiation complex by looping of the DNA. This helps to explain how the enhancer can act at a distance and why the orientation of the enhancer sequence does not matter. Once the protein binds to the enhancer, the proper alignment of the protein with the preinitiation complex is adjusted as the DNA loops back.

Modular Design of Eukaryotic Transcription Factors

Eukaryotic transcription factors have multiple domains that carry out specific interactions. The DNA recognition domains contribute site-specific binding, and the activation domains contact general transcription factors, RNA polymerase II, or other regulators of transcription. Many transcription factors also have dimerization domains that promote the formation of homodimers or heterodimers, protein interaction domains that allow association with proteins like histone acetylase or domains that bind to coactivators. Coactivators include steroid hormones and cAMP that change the ability of the transcription factor to bind DNA or serve as an activator. The domains involved in such activities can be grouped into a few characteristic motifs.

Common Motifs in Proteins That Bind DNA and Regulate Transcription.

Several amino acid motifs are commonly found in transcription factors (transacting regulators). They include **helix–turn–helix (HTH)**, **zinc finger**, **helix–loop–helix (HLH)**, and basic region-leucine zipper (bZIP) proteins (p. 324). These motifs are present in about 80% of known sequence-specific binding proteins.

The lactose repressor of *E. coli* (see Section 8.3) and a group of developmentally important transcription factors called homeodomain proteins are among the proteins with the **h**elix–**t**urn–**h**elix (HTH) motif. The homeodomain is the DNA recognition portion of these transcription factors. Homeodomain proteins were first identified in the fruit fly

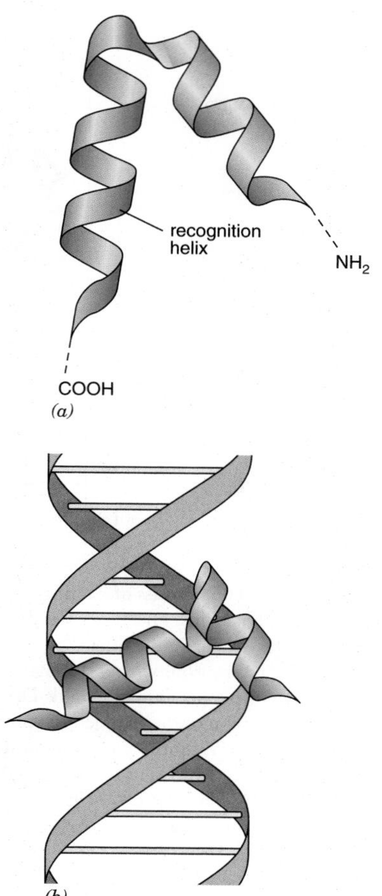

Figure 8.24 Helix–turn–helix proteins use one helix to bind in the major groove while the other supports that binding through hydrophobic interaction. (*a*) The HTH domain of a protein typically contains about 20 amino acids that form two α-helices, joined by a nonhelical turn. (*b*) The dimensions of the α helix of the protein allow it to fit into the major groove of the DNA helix. The helix that interacts directly with the DNA (the recognition helix) includes amino acids with side chains capable of forming hydrogen bonds with specific bases that are exposed in the major groove. The second helix does not directly interact with DNA, but stabilizes the binding of the recognition helix through hydrophobic interactions. Thus, both helices include amino acids like valine or leucine that allow these hydrophobic interactions to occur. Redrawn from Alberts, B., Bray, D., Lewis, J., Raff, M., Roberts, K., and Watson, J. *Molecular Biology of the Cell.* New York: Garland, 1994.

when mutations in genes' encoding them led to homeosis, or the replacement of one part of the body for another. For instance, one mutation in a homeodomain protein can cause legs to form where antennae should be. Homeodomain proteins are key regulators in mammalian development as well. The HTH motif is about 20 amino acids and forms a relatively small part of a much larger protein. Within the motif the first seven amino acids form one helix followed by a four-amino acid turn and then a nine-amino acid helix. The second helix is the *recognition helix* that binds in a sequence-specific manner in the major groove, while hydrophobic interactions between the first and second helix stabilize the structure. The rest of the protein may have allosteric sites that bind to regulatory molecules and regions that allow other protein–protein interactions (Figure 8.24).

TFIIIA, a general transcription factor that binds to promoters of RNA polymerase III-transcribed genes, SP1 (which gives 10- to 20-fold stimulation of all genes with GC boxes), Gal4 in yeast, and the steroid hormone receptor superfamily are all examples of zinc finger proteins. Different subclasses of zinc finger proteins are defined by the specific amino acids that coordinate Zn binding. For instance, in TFIIIA two cysteines and two histidines coordinate Zn binding and are in the C2H2 class, whereas the steroid hormone receptor transcription factors use four Cys for each Zn and are of the C_4 class. The zinc finger motif binds in the major groove of the DNA in a sequence-specific manner, mediated by an alpha helix formed on one side of the finger region. As was seen with HLH proteins, the α-helix fits in the major groove and amino acid side chains form hydrogen bonds with the bases for sequence-specific binding (p. 322).

The two classes of zinc finger proteins have binding sites characteristic of the way each class positions itself on DNA. For TFIIIA, sequential zinc fingers follow the major groove, each forming hydrogen bonds with specific bases. Zinc finger proteins of the C_4 class, such as the steroid hormone receptor transcription factors, use one zinc finger to bind to DNA and a second zinc finger to stabilize DNA binding by the first. The steroid hormone

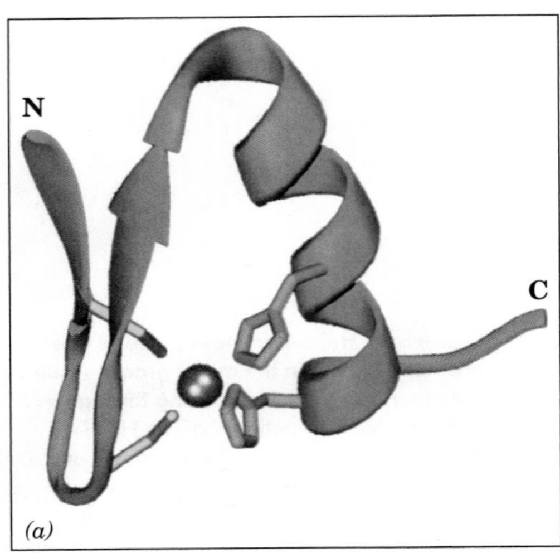

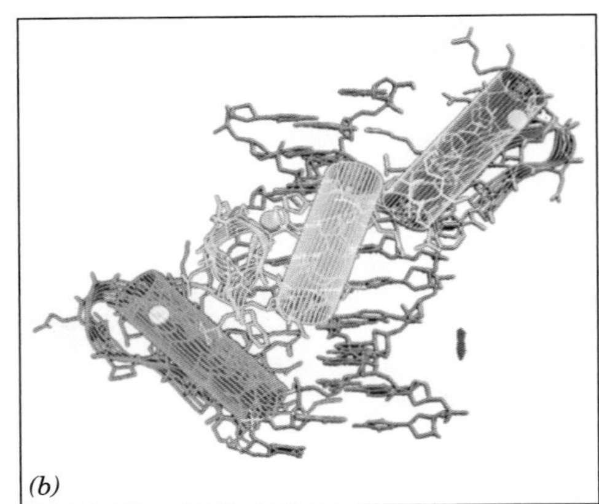

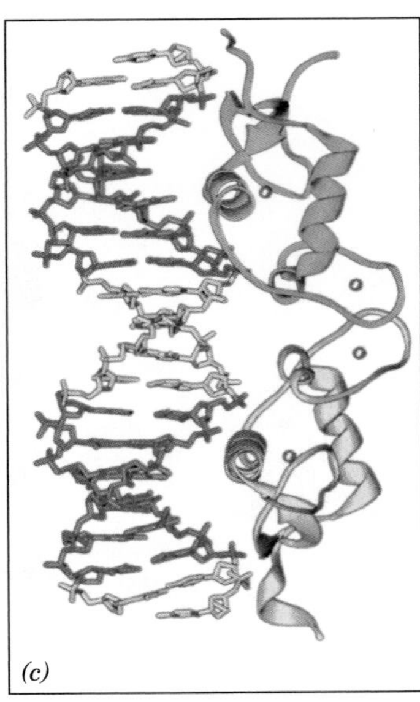

Figure 8.25 Two different Zn finger motifs are found in transcription factors. Although both Zn finger proteins use the coordinate binding of a Zn molecule to assume their final structure, there are recognizable differences in the two major motifs that have been identified. Both classes take advantage of the formation of α-helical structure to form domains that bind in the major groove of DNA. (*a*) The C2H2 class includes proteins that may have many Zn finger domains. Each α-helix from a Zn finger has the potential to bind in a sequence-specific manner to sites along the major groove. The result may be a procession of protein DNA interactions each dependent on the particular array of amino acid side chains found in each α-helical domain of the Zn finger protein. (*b*) The Cx class of Zn finger proteins commonly has two Zn finger domains. The α-helix from one Zn finger binds the DNA in the major groove, while the α-helix in the other Zn finger supports that interaction by hydrophobic interactions with the domain binding to the DNA. The Cx class of Zn finger proteins normally binds DNA by forming dimers. Shown in (*c*) is the interaction of the estrogen receptor dimer, with each monomer contacting the DNA. Zn molecules are indicated by spheres.
(*a*) Reproduced with permission from Voet, D., and Voet, J. G. *Biochemistry,* 2d ed. New York: Wiley, 1995. © (1995) John Wiley & Sons, Inc. Part (*b*) and (*c*) generously supplied by C. Pabo, M.I.T.

receptor transcription factors associate with DNA as dimers. Their binding sites consist of two palindromic "half-sites" spaced to accommodate the two DNA-binding fingers of the dimer (Figure 8.25).

Basic region–leucine zipper (bZIP) proteins include fos, jun, and CREB. bZIP proteins contain leucine residues at every seventh position in an α-helix. These leucines form hydrophobic interactions with a second protein that also has this pattern of leucine repeats to form either homo- or a heterodimers (Figure 8.26).

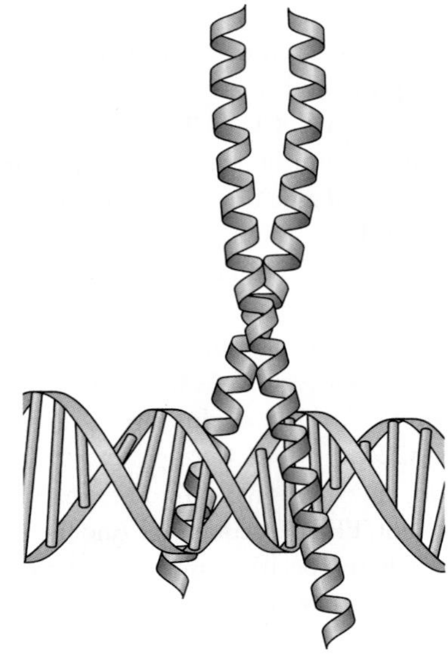

Figure 8.26 Leucine zipper proteins bind to DNA as dimers. Leucine zipper proteins form dimers by virtue of a leucine every seventh residue along an α-helix. These leucines form a hydrophobic face that interact with the hydrophobic face of a similar α-helix. The protein–protein interaction domains are in blue. The α helical regions may continue beyond the **protein–protein** interaction domain, allowing each monomer to bind to the major groove of DNA (green). For homodimers, the DNA-binding site is characterized by two recognizable and symmetric half-sites. Modified from Alberts, B., Bray, D., Lewis, J., Raff, M., Roberts, K., and Watson, J. *Molecular Biology of the Cell.* New York: Garland, 1994.

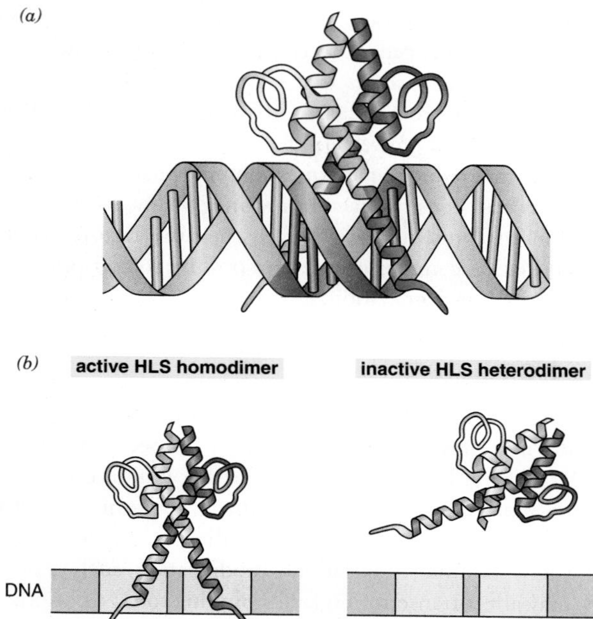

Figure 8.27 Transcription factor dimer formation is mediated through helix–loop–helix interactions. The helix–loop–helix motif brings together two monomers to form a dimer that binds to DNA. (*a*) Each monomer has two helices joined by a loop. One helix is used for protein–protein interaction; the other is used to bind the major groove of DNA. Thus, the dimer consists of a four-helix bundle. If the dimer is formed by two identical monomers, then the DNA-binding sites are expected to be very similar or identical; however, if the monomers are different proteins (forming heterodimers), then the DNA-binding sites may be unrelated. (*b*) When transcription factors bind as dimers, the presence of a truncated monomer can prevent DNA binding even in the presence of full-length monomers. For example, if the protein dimerization helix is made without the DNA-binding domain, the dimerization with a full-length monomer produces a product unable to bind effectively to DNA.
Modified from Alberts, B., Bray, D., Lewis, J., Raff, M., Roberts, K., and Watson, J. *Molecular Biology of the Cell.* New York: Garland, 1994.

bZip proteins have a DNA-binding domain composed of a basic region, defined by the presence of arginine and lysine, located seven amino acids before the first leucine and α-helix. The basic amino acids stabilize the DNA–protein association through electrostatic interactions with the negatively charged DNA backbone in addition to forming hydrogen bonds in the major groove. Homodimer-binding sites have dyad symmetry, whereas this symmetry is not found in heterodimer-binding sites.

The helix–loop–helix class of transcription factors includes myoD, myc, and max. Two amphipathic α-helical segments separated by an intervening loop characterize helix–loop–helix proteins. The helices are not responsible for DNA-binding, as in the zinc finger proteins, but for dimerization with another protein. As was described for the bZip proteins, the dimers formed can be homo- or heterodimers. The DNA-binding domain is an extension of one of the α helices that form the four-helix bundle generated by dimerization (Figure 8.27).

8.9 • REGULATION OF EUKARYOTIC GENE EXPRESSION

As indicated above, a relatively small region of a regulatory protein may be dedicated to sequence-specific binding of DNA, whereas other domains are involved in protein–protein or ligand interactions. Several characteristic activation domains have been identified, including acidic domains (high concentration of amino acids with acidic side chains), glutamine-rich domains, and proline-rich domains. Experimentally, overproduction of any of these domains by recombinant techniques, even without their corresponding DNA-binding domain, can lead to inappropriate activation of transcription from a variety of genes. They appear to activate transcription by increasing the rate of assembly of the preinitiation complex. Some interact directly with TFIID to enhance binding to the TATA box, whereas others interact with TFIIB or TAFs that are part of the TFIID complex. When multiple transcription factors bind to a promoter, they can have a combinatorial effect on the binding and assembly of the preinitiation complex. Activation domains of many transcription factors target the same proteins in the preinitiation complex. Thus, transcription regulation is linked to the placement of DNA-binding sites and not to protein–protein interactions unique to a specific gene.

Transcription factors can also regulate gene expression by recruitment of other proteins to the promoter area. These proteins may not bind DNA but form a bridge between the DNA-bound transcription factor and the initiation complex. Alternatively, the interacting

CLINICAL CORRELATION 8.4

Tamoxifen and Targeting of Estrogen Receptor

Tamoxifen, a drug used to treat breast cancer, is a competitive inhibitor of the estrogen receptor (ER). Breast cancer cells respond to normal estrogen by increasing their proliferation rate. Estrogen activates the estrogen receptor, whereas tamoxifen prevents normal activation and reduces transcription from genes regulated by the estrogen receptor, and hence reduces growth of breast cancer cells. The National Cancer Institute in 2004 stated, "The benefits of tamoxifen as a treatment for breast cancer are firmly established and far outweigh the potential risks." However, tamoxifen treatment can increase the risk of uterine cancer. The apparent cause lies in the presence of two estrogen receptor subtypes (α and β). Both subtypes exist in breast tissue but estrogen receptor-α predominates in uterine tissue. In uterine tissue tamoxifen apparently activates receptor-α, rather than inhibiting it. The result is stimulation of transcription in conjunction with transcription factors fos and jun.

Paech, K., Webb, P., Kuiper, G., Nilsson, S., et al. Differential ligand activation of estrogen receptors ERα and ERβ at AP1 sites. *Science* 277:1508, 1997.

protein may bring chromatin modification enzymes, such as histone acetylase (for example, the CBP–p300 complex) to specific genes. Acetylation of histones loosens chromatin structure. Some regulated transcription factors that recruit the CBP–p300 complex, and the molecules or events to which they respond, are CREB and cAMP, SREBP and cholesterol, NFkB and cytokines, and p53 and growth arrest.

Transcription factors that reduce gene expression may preemptively bind to the DNA, either to the same sequence as a positive factor or more globally, such as the negative effect of methylated DNA-binding proteins. Alternatively, negative factors may inhibit binding or assembly of the preinitiation complex or bind to a positive transcription factor to prevent its binding to DNA or speed its degradation.

Regulating the Regulators

Transcription factors are regulated in a variety of ways. Probably the simplest examples are those which are synthesized only in specialized cells or at specific times in development. Some are regulated by cofactor binding, which may either inhibit or stimulate their binding to DNA. For transcription factors that form dimers, the formation of nonproductive homo- or heterodimer complexes can alter DNA binding or the protein–protein interactions that allow gene activation. Posttranslational modification, such as protein processing or phosphorylation, may affect transcription factor binding to DNA or other proteins, and their ability to move from the cytoplasm into the nucleus (Clin. Corrs. 8.4 and 8.5).

Since enhancers can act over long distances and in either orientation, there needs to be a way to ensure they are acting on the correct gene. This is accomplished by the use of insulator sequences that set the boundaries of influence of an enhancer. Insulators are DNA sequences that when bound to the Zn finger protein CCTF (CCCTC-binding factor) and positioned between an enhancer and the TATA box, prevent enhancer-mediated gene activation. An additional activity of insulators is to prevent the spread of heterochromatin that would lead to gene silencing.

Activation of Transcription of the LDL Receptor Gene Illustrates Many Features Found in Eukaryotic Gene Regulation.

The description of the regulation of expression of specific eukaryotic genes can rapidly turn into an alphabet soup of transcription factors and regulators of transcription factors that makes sense to researchers studying that gene, but is confusing to most others. However, many features of transcription factors described above are illustrated by the transcriptional control of the gene for the low-density lipoprotein (LDL) receptor. This gene is transcribed in response to the lack of cellular cholesterol. Increased transcription leads to an increased amount of the LDL receptor and enhanced uptake of LDLs and their cholesterol in the blood (p. 724).

The promoter of the LDL receptor gene has a TATA box and binding sites for a variety of transcription factors (Figure 8.28). The TATA box provides the site for TFIID binding

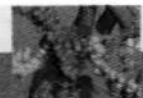

CLINICAL CORRELATION 8.5

Transcription Factors and Cardiovascular Disease

Among genes recently identified as causing human cardiovascular disease are two that encode transcription factors. One, Nkx2–5, is a homeodomain-containing protein. Homeodomain proteins often are involved in regulating cell fate in embryonic cells and Nkx2–5 regulates genes involved in heart formation. Mutations in one allele of Nkx2–5 are linked to atrial septal defects and conduction defects. The null mutation in the mouse is embryonic lethal. Mutations in the other gene, Tbx, are responsible for Holt–Oram syndrome. Holt–Oram syndrome often includes atrial septal defects (ASD), ventricular septal defects (VSD) conduction defects and defects in hand and arm development.

Barinaga, M. Tracking down mutations that can stop the heart. *Science* 128:32, 1998; Schott, J. J., Benson, D. W., Basson, C. T., Pease, W., et al. Congenital heart disease caused by mutations in the transcription factor NKX2–5. *Science* 281:108, 1998; and Li, Q. Y., Newbury-Ecob, R. A., Terrett, J. A., et al. Holt–Oram syndrome is caused by mutations in TBX5, a member of the Brachyury (T) gene family. *Nature Genetics* 15:21, 1997.

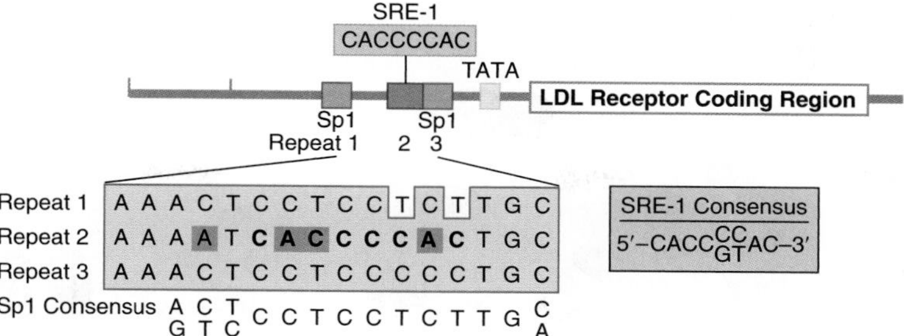

Figure 8.28 A schematic of the LDL receptor gene promoter. The LDL receptor gene promoter, like most eukaryotic genes, has several different transcription factor–binding sites. Shown here are the principal sites involved in regulating the LDL receptor gene in response to cholesterol levels. They include the TATA box, just upstream from the transcription initiation site, several GC boxes (Sp1-binding sites), and the response element (SRE) where SREBP binds. Modified from Goldstein, J. L., and Brown, M. S. Regulation of the mevalonate pathway. *Nature* 343:425, 1990.

and the formation of the preinitiation complex with RNA polymerase II. There are three consensus binding sites for the zinc finger containing transcription factor Sp1 (recall that Sp1 binds to GC boxes in promoters). Sp1 has a glutamine-rich activation domain, which is thought to help recruit TFIID to the TATA box. This recruitment needs an additional factor called the **c**ofactor **r**equired for **Sp**1 activation (CRSP). However, the binding of Sp1 is not sufficient to activate transcription of the gene. Activation requires the participation of a second transcription factor, **s**terol **r**esponsive **e**lement-**b**inding **p**rotein 1a (SREBP-1a). SREBP-1a is a helix–loop–helix–leucine zipper protein that binds to the sterol response element between the Sp1-binding sites in the LDL receptor gene.

The capacity to respond to the concentration of cholesterol is controlled by the release of SREBP from a membrane-bound state that prevents it from being transported into the nucleus. SREBP-1a contains two domains that loop through the membrane of the endoplasmic reticulum (ER), leaving the domains capable of serving as a transcription factor exposed to the cytoplasm but tethered to the ER. In order for SRCBP-1a to move to the nucleus, it must be cleaved by two proteolytic steps, the first of which is carried out by a protease found in an active form in a post-ER compartment called the cis-Golgi. Transport of SRCBP-1a to the cis-Golgi requires **S**REBP **c**leavage **a**ctivating **p**rotein (**SCAP**), which also is partially embedded in the ER membrane. SCAP has a cholesterol-sensing region, and when membrane cholesterol levels are low, the SCAP–SRCBP-1a complex is transported to the cis-Golgi, where proteolysis of SRCBP-1a occurs, releasing the cytoplasmic domain, the transcription factor, for entry into the nucleus (Figure 8.29). Once there, it binds the **s**teroid **r**esponse **e**lement (SRE)

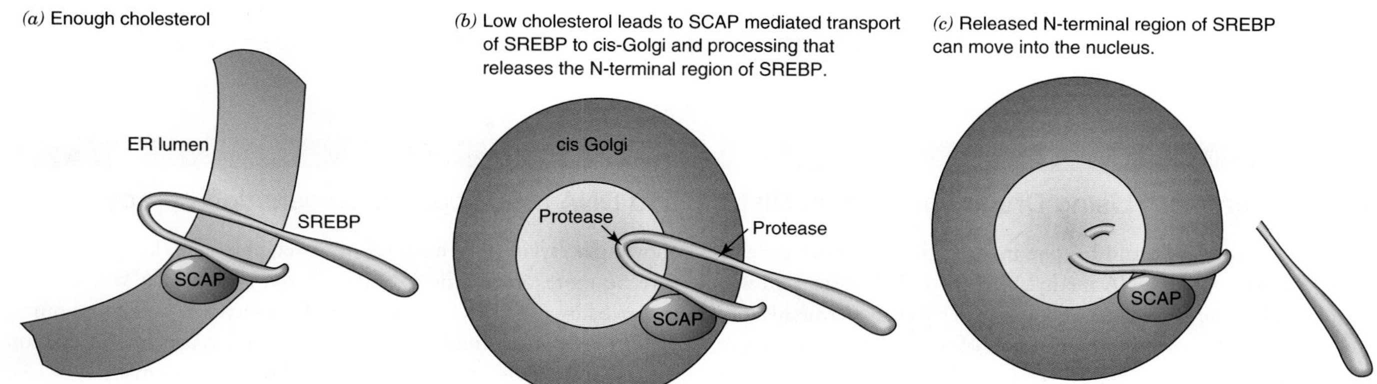

Figure 8.29 SREBP is released from membrane-bound precursor by protease action. (*a*) The sterol responsive element-binding protein (SREBP) is synthesized as a precursor protein that must be proteolytically processed before it acts as a transcription factor. The precursor is tethered to the membrane of the ER by two membrane-spanning regions. It is in close association with SREBP cleavage-activating protein (SCAP), but remains unprocessed when cholesterol levels in the ER membrane are normal. (*b*) SCAP senses low levels of cholesterol and moves with SREBP to the cis-Golgi where two different proteases cleave SREBP, (*c*) allowing the cytoplasmic domain to move to the nucleus. In this way the SREBP does not activate the LDL receptor gene unless cholesterol levels are low. Modified from Brown, M. S., Ye, J., Rawson, R. B., and Goldstein, J. L. Regulated intermembrane proteolysis: A control mechanism conserved from bacteria to humans. *Cell* 100:391, 2000.

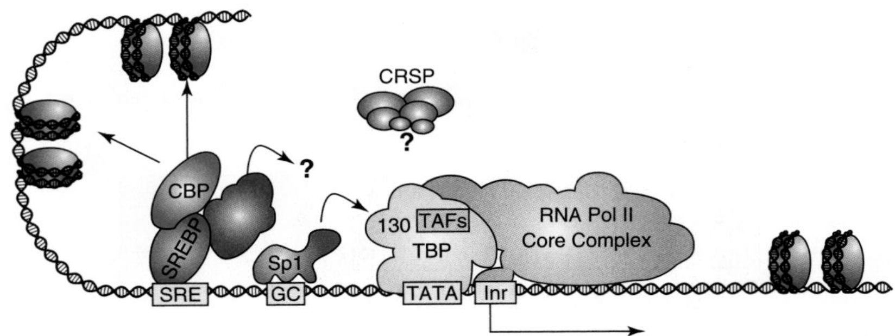

Figure 8.30 LDL receptor gene is activated through coordinate effect of several transcription factors. Activation of the LDL receptor gene requires several factors. Once SREBP, Sp1, and the cofactor required for Sp1 activation (CRSP) bind, the histone CBP is recruited to the promoter region. This combination of factors provides the positive signal needed to enhance binding of the initiation complex to the TATA box, and through recruitment of CBP, with its histone acetyl transferase activity, also affects nearby chromatin structure.
Modified from Naar, A. M., Ryu, S., and Tijian, R. Cofactor requirements for transcriptional activation by Sp1. *Cold Spring Harbor Sym. Quant. Biol.* 63:189, 1998.

on the LDL promoter and recruits a histone acetyl transferase called CBP and other proteins to the promoter area. Then, acting with SP1 and CRSP, the LDL receptor gene is transcribed (Figure 8.30). This illustrates the composite nature of eukaryotic gene regulation, in which activation of transcription factors, coordination of multiple transcription factors, and recruitment of chromatin remodeling enzymes work in concert to regulate gene expression.

Epigenetic Control of Gene Expression

The transcription factors discussed above regulate gene expression via interaction with specific DNA sequences. However, as noted earlier in this chapter, access to DNA can be an equally powerful regulator of gene expression. Modulating chromatin structure via histone modification, nucleosome remodeling, or modification of DNA via methylation can all be viewed as non-sequence-dependent regulation of gene expression. Non-sequence-dependent regulation of gene expression is often called epigenetic (Clin. Corr. 8.6).

Many posttranslational modifications have been identified on histones. Lysines can be acetylated, methylated (mono-, di-, and trimethylated), ubiquinated, and sumoylated. Arginines can be methylated (mono- or dimethylated) or deiminated (forming citrulline), serine and threonine can be phosphorylated, glutamate can be ADP-ribosylated, and proline can be isomerized. Over 60 enzymes that add or remove histone modifications have been identified, and some but not all of the modifications have been linked to gene activation or repression. Specific modifications can alter association of histones with DNA, other histones, other nucleosomes, and transcription factors. There is emerging evidence that specific small RNAs participate in the stable inheritance of histone modification patterns once a pattern is established.

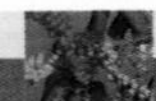

CLINICAL CORRELATION 8.6

Cancer Treatment Using Drugs That Target Histone and DNA Modification: Epigenetic Therapy

Both DNA methylation and histone modification influence gene expression. For example, DNA methylation is normally associated with a reduction in gene expression, and in some cancers hypermethylation of DNA is found in the promoter regions of tumor suppression genes. DNA methylation mediated by DNA methyltransferase (DNMT) can be inhibited by 5-azacytidine and related nucleoside analogs. Several of these compounds have been approved by the FDA for treatment of myodysplastic syndrome and leukemia. Other nucleoside analogs and nonnucleoside analogs that inhibit DNMTs are being sought.

Inhibitors of histone deacetylase (HDAC) are also being sought, with the aim of restimulating tumor suppressor gene expression by allowing acetylation of histones to be retained and chromatin structure be more open. The HDAC inhibitor vorinostat was recently approved by the FDA for treatment of cutaneous T-cell lymphoma.

The use of these inhibitors, individually and as part of combination treatments with other cancer therapies, show great promise.

Gore, S. D. Combination therapy with DNA methyltransferase inhibitors in hematologic malignancies. *Nature Clinical Practice Oncology* 2:S30, 2005; Lyko, F. and Brown, R. DNA methyltransferase and the development of epigenetic cancer therapies. *J. Nat. Cancer Inst.* 97:1498, 2005: and Schmidt, K. Lamarckism Revisited: Epigenetics and Its implications for Modern Health Care. Sequenom White Paper, 2007.

Bibliography

Prokaryotic Gene Expression

Bennett, P. M. Plasmid encoded antibiotic resistance: Acquisition and transfer of antibiotic resistance genes in bacteria. *Brit. J. Pharm.* 153:S347, 2008.

Blattner, F. R., Plunkett, G., Bloch, C. A., Perna, N. T., et al. The complete genome sequence of *Escherichia coli* K–12. *Science* 277:1453, 1997.

Cohen, S. N., and Shapiro, J. A. Transposable genetic elements. *Sci. Amer.* 242:40, 1980.

Miller, J.H. The lac gene: Its role in lac operon control and its use as a genetic system. In Miller, J. H., and Resnikoff, W. S. (Eds.), *The Operon.* Cold Spring Harbor, NY: Cold Spring Harbor Laboratory, 1978, 31.

Schultz, S. C., Shields, G. C., and Steitz, T. A. Crystal structure of a CAP–DNA complex: The DNA is bent by 90 degrees. *Science* 253:1001, 1991.

Yanofsky, C. The different roles of tryptophan transfer RNA in regulating trp operon expression in *E. coli* versus *B. subtilis. Trends Genet.* 20:367, 2004.

Eukaryotic Gene Expression

Alberts, B., Johnson, A., Lewis, J., Raff, M., Roberts, K., and Walter, P. *Molecular Biology of the Cell.* New York: Garland Science, 2002.

Allis, C. David, Jenuwein, T., and Reinberg, T. (Eds.), Caparros, M. L. (Assoc. Ed.). *Epigenetics.* Cold Spring Harbor, NY: Cold Spring Harbor Laboratory, 2007.

Brown, M. S., Ye, J., Rawson, R. B., and Goldstein, J. L. Regulated intramembrane proteolysis: A control mechanism conserved from bacteria to humans. *Cell* 100:391, 2000.

Brown, C. E., Lechner, T., Howe, L., and Workman, J. L. The many HATs of transcription coactivators. *Trends Biochem. Sci.* 25:15, 2000.

Burley, S. K., and Roeder, R. Biochemistry and structural biology of transcription factor IID (TFIID). *Annu. Rev. Biochem.* 65:769, 1996.

Dantonel, J. C., Wurtz, J. M., Poch, J. M., Moras, D., and Tora, L. The TBP-like factor: An alternative transcription factor in Metazoa. *Trends Biochem. Sci.* 24:335, 1999.

Goldstein, J. L., and Brown, M. S. Regulation of the mevalonate pathway. *Nature* 343:425, 1990.

Gore, S.D. Combination therapy with DNA methyltransferase inhibitors in hematologic malignancies. *Nature Clinical Practice Oncology* 2:S30, 2005.

Kawasaki, H., and Taira, K. Induction of DNA methylation and gene silencing by short interfering RNAs in human cells. *Nature* 431:211, 2004.

Kornberg, R. D. Mechanism and regulation of yeast RNA polymerase II transcription. *Cold Spring Harbor Sym. Quant. Biol.* LXIII:229, 1998.

Kouzarides, T. Chromatin modifications and their function. *Cell* 128:693, 2007.

Latchman, D. S. *Eukaryotic Transcription Factors,* 3d ed. New York: Academic Press, 1998.

Lemon, B., and Tjian, R. Orchestrated response: A symphony of transcription factors for gene control. *Genes and Dev.* 14:2551, 2000.

Lewin, B. *Genes VII.* New York: Oxford University Press, 2000.

Lyko, F., and Brown, R. DNA methyltransferase and the development of epigenetic cancer therapies. *J. Nat. Cancer Inst.* 97:1498, 2005.

Naar, A. M., Ryu, S., and Tijian, R. Cofactor requirements for transcriptional activation by Sp1. *Cold Spring Harbor Symp. Quant. Biol.* LXIII:189, 1998.

Ng, H., and Bird, A. Histone deacetylases: silencers for hire. *Trends Biochem. Sci.* 25:121, 2000.

Pruss, D., Hayes, J. J., and Wolffe, A. Nucleosomal anatomy—where are the histones? *BioEssays* 17:161, 1995.

Schmidt, K. Lamarckism Revisited: Epigenetics and Its Implications for Modern Health Care. Sequenom White Paper, 2007.

Stewart, S., and Crabtree, G. Regulating the regulators. *Nature* 408:46, 2000.

West, A. G., Gaszner, M., and Felsenfeld, G. Insulators: Many functions, many mechanisms. *Genes Dev.* 16:271, 2002.

Wolffe, A. P. The cancer-chromatin connection. *Sci. and Med.* 6:28, 1999.

Wolffe, A. P. *Chromatin: Structure and Function,* 3d ed. New York: Academic Press, 1998.

Woychik, N., and Young, R. RNA polymerase II: Subunit structure and function. *Trends Biochem Science* 15:347, 1990.

Key Terms

operon
inducer
repressor
co-repressor
control element
transcription
RNA polymerase
attenuation
polycistronic
chromatin
nucleosomes
linker region
promoter
transcription factor
methylation
acetylation
phosphorylation
major groove
minor groove
enhancer
histone
DNA-binding motifs
epigenetic
response elements
leader sequence
leucine zipper
zinc-finger
helix–loop–helix
helix–turn–helix

Questions • CAROL N. ANGSTADT

Multiple-Choice Questions

1. In an operon,
   - **A.** each gene of the operon is regulated independently.
   - **B.** control may be exerted via induction or via repression.
   - **C.** operator and promoter may be *trans* to the genes they regulate.
   - **D.** the structural genes are either not expressed at all or fully expressed.
   - **E.** control of gene expression consists exclusively of induction and repression.

2. The *E. coli lac*ZYA region will be upregulated if
   - **A.** there is a defect in binding of the inducer to the product of the *lacI* gene.
   - **B.** glucose and lactose are both present, but the cell cannot bind the CAP protein.
   - **C.** glucose and lactose are both readily available in the growth medium.
   - **D.** the operator has mutated so it can no longer bind the repressor.
   - **E.** the *lac* corepressor is not present.

3. All of the following describe an operon except
   - A. it is a control mechanism for eukaryotic genes.
   - B. it includes structural genes.
   - C. it is expected to code for polycistronic mRNA.
   - D. it contains control sequences such as an operator.
   - E. it can have multiple promoters.

4. In bacteria, amino acid starvation is associated with the production of guanosine tetraphosphate and guanosine pentaphosphate. This situation is referred to as
   - A. attenuation.
   - B. corepression.
   - C. repression.
   - D. self-regulation.
   - E. stringent response.

5. A number of specific sequences have been found in many promoter and operator regions. The sequence that is very closely associated with methylation of DNA is
   - A. CG island.
   - B. CAAT box.
   - C. TATA box.
   - D. CAP site.
   - E. CTD.

6. In eukaryotic transcription by RNA polymerase II, formation of a preinitiation complex
   - A. begins with the binding of a protein (TBP) to the TATA box of the promoter.
   - B. involves the ordered addition of several transcription factors and the RNA polymerase.
   - C. allows an ATP-dependent opening of the two strands of DNA.
   - D. requires that the C-terminal domain of RNA polymerase II not be phosphorylated.
   - E. all of the above.

Questions 7 and 8: Since the initiation of eukaryotic transcription involves the interaction of a multitude of transcription factors, there must be regulation of these. Estrogen normally binds to its nuclear receptor and the complex binds to the sterol response element to regulate transcription. Tamoxifen, a drug used to treat breast cancer, competes for the estrogen receptor and reduces transcription of the genes it regulates. Another type of alteration is seen in Holt–Oram syndrome in which there is a mutation in a gene for a transcription factor (Tbx), leading to defects in formation of the heart.

7. Response elements
   - A. are proteins that bind to the promoter region.
   - B. are more effective the closer they are to the TATA box.
   - C. may be thousands of base pairs away from the preinitiation complex assembly.
   - D. can be either upstream or downstream of the transcription start.
   - E. bind transcription factors that interact with the initiation complex only upon looping of the DNA.

8. Transcription factors are frequently sequence-specific binding proteins and most likely have one of several specific structural motifs. The helix–turn–helix motif
   - A. coordinates zinc between cysteines and histidines.
   - B. joins two proteins via hydrophobic interactions between leucines.
   - C. forms dimers held together by interaction of a helix on each monomer.
   - D. has one helix that recognizes and binds in the major groove of DNA while hydrophobic interactions with a second helix stabilizes the structure.
   - E. is the only motif that binds in the major groove of DNA.

Questions 9 and 10: The problem of pathogenic bacteria becoming resistant to a large number of antibiotics is a serious public health concern. A bacterial strain in a patient being treated with one antibiotic may suddenly become resistant not only to that antibiotic but to others as well even though it has not been exposed to the other antibiotics. This occurs when the bacteria acquire a plasmid from another strain that contains several different transposons.

9. All of the following phrases describe transposons except
   - A. a means for the permanent incorporation of antibiotic resistance into the bacterial chromosome.
   - B. contain short inverted terminal repeat sequences.
   - C. code for an enzyme that synthesizes guanosine tetraphosphate and guanosine pentaphosphate, which inhibit further transposition.
   - D. include at least one gene that codes for a transposase.
   - E. contain varying numbers of genes.

10. In the operation of transposons,
   - A. typically the transposon moves from its original site and relocates to a different site.
   - B. a duplicated transposon must be inserted into the same DNA molecule as the original.
   - C. all transposons are approximately the same size.
   - D. the insertion sites must be in a consensus sequence.
   - E. the transposase may recognize the repetitive ends of the transposon and participate in the cleavage of the recipient site.

Questions 11 and 12: Genes present in a region of DNA which is in the highly condensed heterochromatin form cannot be transcribed. To be transcribed, this region of DNA must change to a more open structure; this may occur by acetylation of histones by an acetyltransferase like CBP, CREB-binding protein. CREB is a transcription factor. Rubinstein–Taybi syndrome is caused by mutations of the CBP gene. Rubinstein–Taybi patients are mentally retarded and have other developmental abnormalities, the severity of which depends on the extent of mutation.

11. In chromatin,
   - A. a nucleosome consists of four molecules of histones surrounding a DNA core.
   - B. DNA positioned so that the major and minor grooves are on the outside of the nucleosome is more accessible for transcription than if the grooves face the interior.
   - C. DNA must be completely removed from the nucleosome structure for transcription.
   - D. linker DNA is the only DNA capable of binding transcription factors.
   - E. the histone octamer consists of eight different kinds of histone proteins.

12. Acetylation of histones can lead to a more open DNA structure by
    A. weakening the electrostatic attraction between histones and DNA.
    B. causing histones to interact with the C-terminal domain (CTD) of RNA polymerase.
    C. causing electrostatic repulsion between histones and DNA.
    D. facilitating methylation of DNA.
    E. attracting transcription factors to DNA.

Problems

13. What will be the status of transcription of the *lac* operon in (a) the presence of glucose and (b) the absence of glucose, in each case with lactose present, if there is a mutation that produces an inactive adenylate cyclase?

14. In an operon for synthesis of an amino acid that is controlled wholly or in part by attenuation, why does the presence of the amino acid prevent transcription of the whole operon while the absence of the amino acid permits it?

Answers

1. **B** Induction and repression are among the mechanisms used to control operons. A: The structural genes are under coordinate control. C: The operator and promoter are elements of the same strand of DNA as the operon they control, not diffusible. D: Typically, regulation of operators is somewhat leaky. E: Another mechanism for regulation of an operon is attenuation.
2. **D** If the operator is unable to bind the repressor, the rate of transcription is greater than the basal level. A: The product of the *lacI* gene is the repressor protein. When this protein binds an inducer, it no longer binds to the operator and transcription increases. Failure to bind an inducer prevents this sequence. B and C: In the presence of glucose, catabolite repression occurs. Glucose lowers the intracellular level of cAMP so there is no CAP–cAMP complex to activate transcription. E: The *lac* operon does not involve corepression.
3. **A** Operons are prokaryotic mechanisms. B and D: An operon is the complete regulatory unit of a set of clustered genes, including the structural genes, regulatory genes, and control elements, such as the operator. E: An operon may have more than one promoter, as does the tryptophan operon of *E. coli.*
4. **E** The mechanism is unclear but transcription initiation of rRNA is inhibited.
5. **A** Methylation is on the cytosine. B: This is a binding site for several transcription factors. C: This is the site where RNA pol II binds. D: This is the site in the *lac operon,* which binds the catabolite activator protein. E: This stands for the C-terminal domain of a protein and is not a DNA sequence.
6. **E** A: TBP is part of the general transcription factor TFIID. B: Many proteins are involved. C: The helicase activity is associated with one of the transcription factors (TFIIH). D: The CTD must be in the unphosphorylated form for RNA polymerase to enter the preinitiation complex.
7. **B** They are part of the promoter region. A: They are DNA sequences, not proteins. C, D, and E: These are characteristics of enhancers, which are also DNA sequences but not usually part of the promoter.
8. **D** The turn is only four amino acids so the two helices are very close. A: This is the zinc finger. B: This is called the leucine zipper (bZIP). C: This is characteristic of the helix–loop–helix. E: All of the motifs have some part that binds in the major group.
9. **C** These guanosine phosphates are synthesized by the product of the relA gene; they inhibit initiation of transcription of the rRNA and tRNA genes, shutting off protein synthesis in general. This is the stringent response.
10. **E** This has been demonstrated for transposon Tn3. A: Most transpositions involve generation of an additional copy of the transposon which is then inserted somewhere else in DNA. B: The copy could be inserted into a different DNA molecule. C: Length varies considerably, depending on the number of genes incorporated into the transposon. D: Most target sites seem to be fairly random in sequence although some transposons tend to insert at specific *hot spots.*
11. **B** Assembly of the transcription initiation complex and other proteins involved in gene expression need access to the grooves of DNA. A and E: The core of the nucleosome is an octamer of two molecules each of four different histones. C: A five-base pair shift in the association of DNA with the histone core is sufficient to change the orientation of the grooves and allow access. D: It seems to be the DNA associated with the nucleosome that is involved in binding transcription factors and other proteins.
12. **A** Acetylation of an ε-amino group of lysines near the N-terminus of histones changes a positive charge to a neutral species so there is less attraction to the negative phosphates of DNA. B: Histones do not interact directly with CTD of RNA polymerase II. C: The change is from positive to neutral, not negative, so there is no electrostatic repulsion. D: Methylation of DNA does affect transcription but is a separate phenomenon from acetylation of histones. E: The nucleosome undergoes repositioning.
13. Normally, glucose lowers the intracellular level of cAMP. There is then little or no CAP–cAMP complex to activate transcription so transcription of the operon is low. The same situation would exist in this mutation. Normally, in the absence of glucose, cAMP is high, positive control is exerted, and the *lac* operon is transcribed. In this mutation, cAMP cannot form since the enzyme for its formation is defective. There is no positive control and the *lac* operon will not be transcribed effectively.
14. Such operons will code for a leader peptide that has one or more codons for the amino acid in question. Once the RNA for the leader peptide has been synthesized, it can form different secondary structures depending on whether or not the ribosome is stalled at this region. If the ribosome does not stall because there is enough amino acid (and thus charged tRNA), the secondary structure is a termination signal and synthesis stops. If the ribosome stalls because of insufficient charged tRNA, the secondary structure that forms is not recognized as a termination signal and transcription through the operon continues.

PART III

FUNCTIONS OF PROTEINS

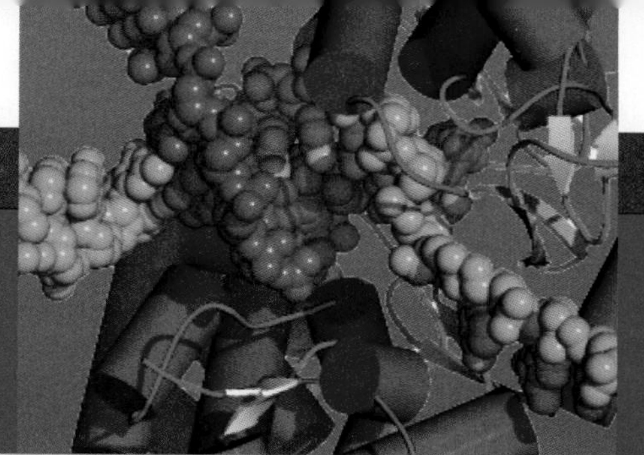

9

Proteins II: Structure–Function Relationships in Protein Families

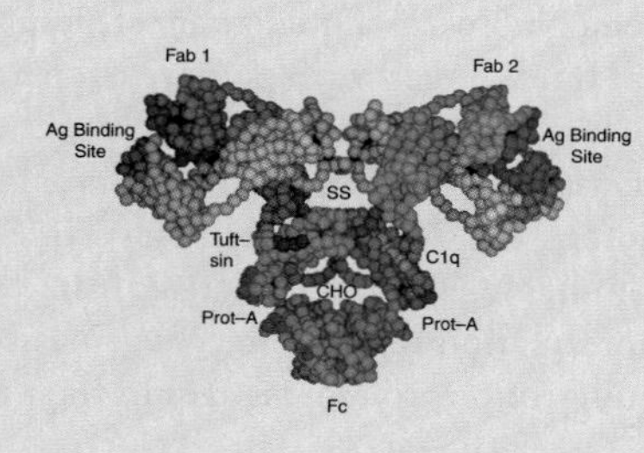

Richard M. Schultz

Professor, Stritch School of Medicine, Loyola University of Chicago

CLINICAL CORRELATIONS

Key Concepts

IMMUNOGLOBULINS

- Immunoglobulin molecules contain repeating homologous sequences that produce multiple immunoglobulin folds within the light (L) and heavy (H) chains of the basic four polypeptide $(LH)_2$ structure.
- The immunoglobin folds from different chains associate to form domains that function to bind antigen, activate complement, and promote polymerization in particular antibody classes.
- The N-terminal domains from the L and H chains (the V_H—V_L domains) have a hypervariable loop region that binds to the targeted antigen specific for that antibody.
- The immunoglobulin fold is a superfold, formed in a wide variety of molecules with non-homologous sequence and function.

SERINE PROTEASES

- Serine proteases are defined by a common catalytic mechanism with identical catalytically essential amino acid residues.
- Trypsin-like serine proteases have two β-barrel folds comprising the catalytic unit with the active site residues in the interdomain crevice. Through convergent evolution, other families of serine proteases have evolved with a nonhomologous tertiary structure, but with an identical catalytic mechanism to the proteases in the trypsin-like family.
- Substrate specificity for different physiological substrates among trypsin-family proteases is determined by the loop regions of the fold structures that provide secondary site interactions with the polypeptide substrates.

- Proteases catalyze reactions by stabilizing the tetrahedral structure and oxyanion of the hydrolytic transition state.
- Serine proteases are often synthesized as zymogens, activated by an upstream protease in an enzyme cascade, and inhibited by protein inhibitors known as serpins.

HEMOGLOBIN AND MYOGLOBIN

- Hemoglobin and myoglobin chains have identical fold structures.
- A heme prosthetic group is part of the O_2-binding site in hemoglobin and myoglobin globin chains.
- Hemoglobin has a quaternary structure with four heme units that bind oxygen with positive cooperativity through conformational changes in the subunit transferred from one heme site to other heme sites within the quaternary structure.
- Protons and 2, 3-bisphosphoglycerate molecules that dissociate from hemoglobin as it changes from the deoxy- to the oxy-conformation regulate the oxygen association–dissociation equilibrium through changes in their concentrations.
- Hemoglobin sequesters NO to deliver to the capillary wall on conversion to the deoxy-conformation.

BASAL LAMINA STRUCTURE

- The basal lamina is an extracellular molecularly organized structural complex composed of multiple protein subunits with an interacting core consisting of type IV collagen, laminin, nidogen/entactin, and perlecan heparan sulfate.
- Proteins of basal lamina are present in different isoforms and the isoforms expressed in a particular tissue are characteristic of the tissue that synthesizes its associated basement membrane.
- Laminin and type IV collagen self-polymerize to form defined molecular networks with binding sites for coproteins and for cell receptor proteins.
- The laminin of the basal lamina is interconnected to the intracellular cytoskeleton of bordering cells at focal adhesion receptor sites in the cellular plasma membranes.

9.1 • INTRODUCTION

The fundamentals of protein architecture were presented in Chapter 3. This chapter discusses the relationships between structure and function in three **protein families**: immunoglobulins, serine proteases, and hemoglobins by examination of amino acid sequence, structural organization, and biological function. The organization of proteins forming the basement membrane of the extracellular matrix that surrounds cells and tissues is also described.

Immunoglobulins provide examples of multidomain architecture that supports recognition and binding to foreign molecules and leads to their sequestration. Diversity among family members is the source of specific recognition of molecular and individual binding capabilities.

Serine proteases provide examples of enzymes that appear to have diverged through evolution to perform unique physiological functions, frequently within highly organized enzyme cascades. Similarities in their catalytic mechanisms and three-dimensional structures form a common link.

Hemoglobins offer examples of highly fine-tuned proteins that simultaneously perform multiple highly regulated physiological functions. They can accommodate small substitutions or mutations, many of which have clinical implications, and still retain their physiological functions. They reveal the diversity of amino acid sequence substitutions that can be structurally tolerated and allow the protein to function. Hemoglobin demonstrates on the atomic level how a protein performs multiple physiological functions in a regulated manner.

The **basement membrane** is an extracellular protein complex formed by cell secreted proteins that bind with each other to form a molecular network in the extracellular space that provides the boundaries and regulatory environment of the cells and tissues. It provides an example of a complex network composed of multiple different protein subunits that spontaneously associate into a specific molecular arrangement. Such protein networks provide essential intracellular as well as extracellular functions.

9.2 • ANTIBODY MOLECULES: THE IMMUNOGLOBULIN SUPERFAMILY OF PROTEINS

Antibody molecules are produced by lymphocytes in response to foreign compounds that can be proteins, complex carbohydrate polymers, or nucleic acid. They noncovalently associate with such foreign compounds present in the protein coat of invading viruses and bacterial cells, initiating a process by which the invading organism is eliminated. Molecules that induce antibody production are called **antigens** and may contain multiple antigenic determinants or epitopes, small regions of the antigen molecule that elicit the production of a specific antibody to which the antigen binds. In proteins for example, an antigenic determinant may comprise only six or seven amino acid residues and a single protein may have many antigenic determinants.

A **hapten** is a small molecule that cannot alone elicit production of specific antibodies but when covalently attached to a larger molecule acts as an antigenic determinant and induces antibody synthesis. Although haptens need attachment to a larger molecule to elicit antibody synthesis, in their free state they bind strongly to antibody.

Each human has the potential to produce about 1×10^8 different antibody structures. All antibodies, however, have a similar overall structure. This has been determined from studies of their amino acid sequence and X-ray diffraction of the antibody molecule alone or in complex with antigen. Structural studies require pure homogeneous preparations of proteins. Antibodies are extremely difficult to purify from plasma because of the many thousands of different antibody molecules present. Homogeneous antibodies can be obtained, however, by the monoclonal hybridoma technique in which mouse myeloma cells are fused with mouse antibody-producing B lymphocytes to construct immortalized hybridoma cells that express a single antibody.

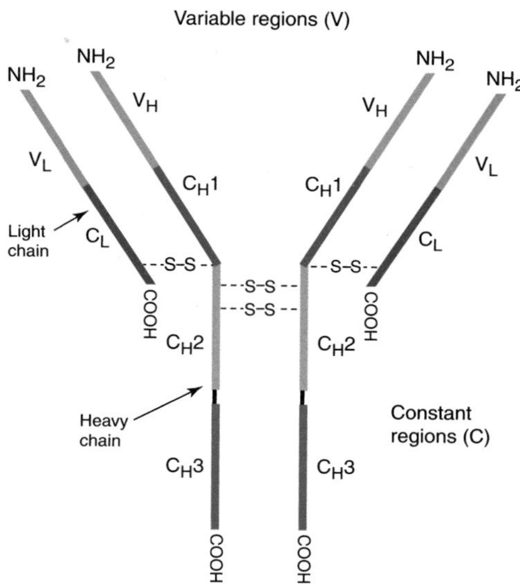

Figure 9.1 Linear representation of IgG antibody molecule. Two heavy (H) chains and two light (L) chains are co-oriented parallel to each other. Interchain disulfide bonds link H chains to each other, and L chains to H chains. Ig fold repeats in the constant (C) region of the H chain are C_H1, C_H2, and C_H3. The constant region of L chain is designated C_L, and variable (V) regions are V_H and V_L of H and L chains, respectively. Based on figure by Burton, D. R. In Calabi, F., and Neuberger, M. S. (Eds.), *Molecular Genetics of Immunoglobulin.* Amsterdam: Elsevier, 1987, 1.

Antibody Molecules Contain Four Polypeptide Chains

Antibody molecules are glycoproteins composed of two light chains (L) of identical sequence combined with two identical heavy chains (H) to form the structure $(LH)_2$. In the most common immunoglobulin class, IgG, the H chains contain approximately 440 amino acids (50 kDa). The L chains contain about 220 amino acids (25 kDa). The four chains are covalently interconnected by disulfide bonds (Figures 9.1 and 9.2). Each H chain is associated

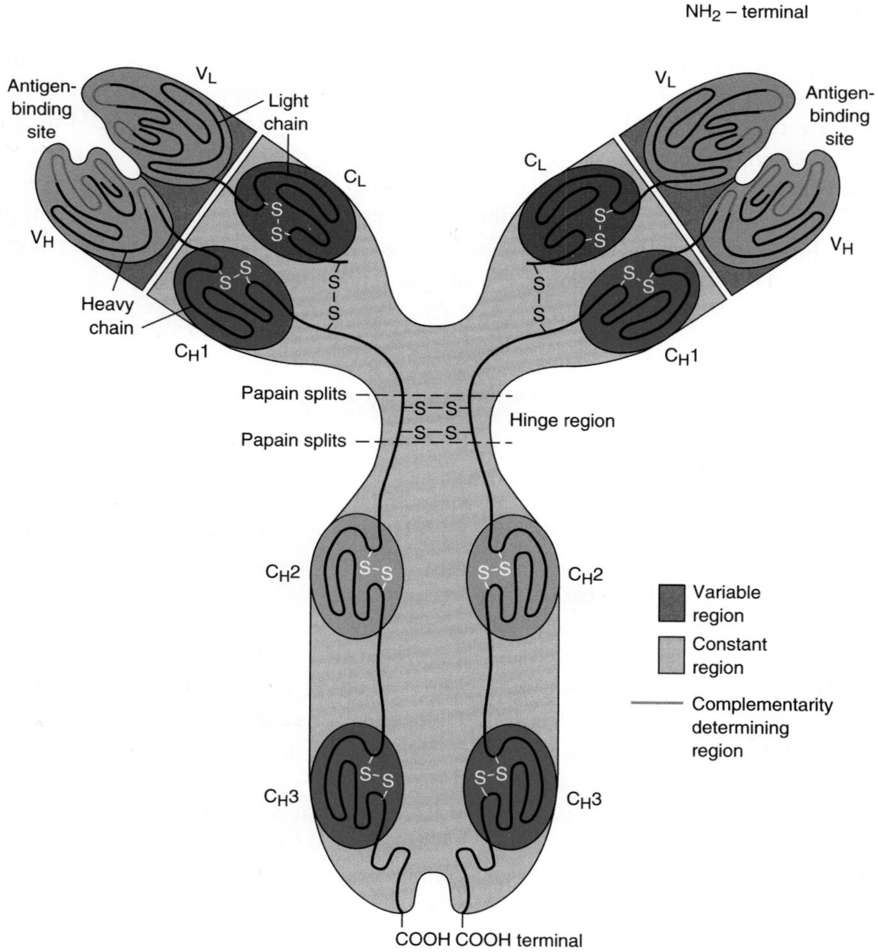

Figure 9.2 Diagrammatic structure of IgG. Light chains (L) are divided into Ig folds V_L (variable amino acid sequence) and C_L (constant amino acid sequence). Heavy chains (H) are divided into V_H and C_H1, C_H2, and C_{H3}. Antigen binding sites are V_H—V_L. "Hinge" polypeptides interconnect domains. Positions of inter- and intrachain cystine bonds are shown. From Cantor, C. R., and Schimmel, P. R. *Biophysical Chemistry,* Part I. San Francisco: Freeman, 1980. Reproduced with permission of Mr. Irving Geis, New York.

TABLE 9.1 • Immunoglobulin Classes

Classes of Immuno-globulin	*Approximate Molecular Mass*	*H-Chain Isotype*	*Carbohydrate by Weight (%)*	*Concentration in Serum (mg 100 mL^{-1})*
IgG	150,000	γ, 53,000	2–3	600–1800
IgA	170,000–720,000[a]	α, 64,000	7–12	90–420
IgD	160,000	δ, 58,000		0.3–40
IgE	190,000	ϵ, 75,000	10–12	0.01–0.10
IgM	950,000[a]	μ, 70,000	10–12	50–190

[a]Forms polymer structures of basic structural unit.

with an L chain such that the NH_2-terminal ends of both chains are near each other. The L chain is half the size of the H chain and only the NH_2-terminal half of the H chain is associated with the L chain.

In the other classes of immunoglobulins (Table 9.1) the H chains are longer than in the IgG class. A variable amount of carbohydrate (2%–12%, depending on class) is attached to the H chain.

Constant and Variable Regions of Primary Structure

Comparison of amino acid sequences of antibody molecules elicited by different antigens shows regions of sequence homology and others of sequence variability. In particular, sequences of the NH_2-terminal half of L chains and the NH_2-terminal quarter of H chains are highly variable. These are the **variable (V) regions** designated V_H and V_L of H and L chains, respectively. Within these V regions certain subregions are hypervariable. Three **hypervariable regions** of between 5 and 7 residues in the V_L and three or four hypervariable regions of between 6 and 17 residues in the V_H domain are common. They are the **complementarity-determining regions (CDRs)** as they form the three-dimensional antigen-binding site complementary to the topology of the antigen.

In contrast, the COOH-terminal three-quarters of H chains and the COOH-terminal half of L chains are homologous in sequence with other H or L chains of the same class. These **constant (C) regions** are designated C_H and C_L in the H and L chains, respectively.

The C_H regions determine the antibody class, provide for binding of complement proteins (Clin. Corr. 9.1), and contain the site necessary for antibodies to cross the placental membrane. The V regions determine the antigen specificity of the antibody.

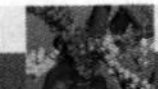

CLINICAL CORRELATION 9.1

The Complement Proteins

The complement system consists of at least 30 distinct complement proteins in plasma (see p. 983 for their designations). These proteins participate in a biochemical cascade that acts to eliminate pathogens and foreign materials from an organism. Three distinct pathways (classical, lectin, and alternative pathways, see figure) utilize different initial signals to activate the complement system. The three pathways converge on factor C3 to form protein fragments C3a and C3b, which cause further activation events.

Many of the complement proteins are inactive precursors or zymogens of serine protease enzymes. On activation, they activate a succeeding protein of the pathway by cleavage of a specific peptide bond in the substrate protein, causing it to become an active serine protease. Activation by hydrolysis of a specific peptide bond is an important method for activation of extracellular proteases. For example, the serine protease enzymes that catalyze blood clot formation, fibrinolysis of blood clots, and digestion of dietary proteins in the gut are mostly activated by specific upstream proteases (pp. 984 and 1033). These pathways of successive protease activations constitute cascades in which amplification of the trigger event occurs.

The classical pathway is initiated by IgG or IgM antibody binding to antigens on the outer surface of invading bacterial cells, protozoa, or tumor cells. On binding to a cellular antigen the complement binding site in the antibody's F_c region becomes exposed as a result of a conformational change. This allows binding of the C1 complement proteins to the cell-surface-bound antibody. The C1 complex is composed of the proteins Clq, Clr, and Cls. Proteins Clr and Cls in the complex undergo a conformational change and become active serine protease enzymes on the cell surface. The activated C1 complex (Cla) hydrolyzes a peptide bond in C2 and C4, which then also associates on the cell surface. The now proteolytically active C2a—C4b complex is the C3 convertase of the classical pathway (figure, complex 5) and hydrolyzes a peptide bond in C3 to produce products C3a and C3b. C3a is an anaphylatoxin-causing platelet and neutrophil activation, mast cell degranulation, and increased vascular permeability to facilitate inflammatory cell recruitment. C3b binds to the target cell surface and the activated C2—C4—C3b complex proteolytically hydrolyzes C5 to products C5a and C5b. Activated C5b associates with C6, C7, C8, and multiple molecules of C9, which bind to the bacterial cell surface and initiate membrane lysis.

$$\text{IgG or IgM} \xrightarrow{\text{Clq, Clr, Cls}} \overline{\text{Cla}} \xrightarrow{\text{C2, C4}}$$

$$\text{C4b} \cdot \overline{\text{C2a}} \xrightarrow{\text{C3}}$$

$$\text{C4b} \cdot \overline{\text{C2a}} \cdot \text{C3b} \xrightarrow{\text{C5, C6, C7, C8, C9}} \text{lysis}$$

(Line above designates active protease.) In addition, target cell membrane bound C5b has opsonic and immune adherence activities, promoting destruction of the target cell by phagocytic cells. Product C5a with C3a contribute to the inflammation.

The alternative pathway can be activated by polysaccharides of invading microorganisms, in the absence of a specific antibody, or by aggregates of IgA. The alternative pathway involves properdin and factor D. Factor D proteolytically cleaves B into products Ba and Bb, which on the target cell surface binds C3, and C3 in this complex autocatalytically hydrolyzes a peptide bond to produce C3a and C3b. C3b combines with Bb to produce a C3 convertase, different from the convertase formed in the classical pathway, that further acts on C3 molecules in blood to produce increased amounts of C3a and C3b (see figure, pathway 3).

The lectin pathway is activated by the opsonin, mannose-binding lectin (MBL), which binds mannose polysaccharides on the invading microorganism surface. The MBL on the cell surface then binds the MBL-associated serine proteases MASP-1 and MASP-2. This complex is similar to the C1 complex in the classical pathway and cleaves C4 in the C2—C4 complex to produce C4b that converts C3 to C3a and C3b (figure, pathway 2).

Extrinsic proteases released from leukocytes or activated by other pathways can also activate the complement system independently of the convertase complexes (figure, pathway 4).

Abnormalities in the complement system can result in inflammatory disorders, repeated pathogen infections, autoimmune diseases including rheumatoid arthritis, glomerular injury, and ocular diseases such as uveitis and macular degeneration. Complement has a role in ischemia and cardiovascular diseases, and in organ and tissue transplantation.

Markiewski, M. M., and Lambris, J. D. The role of complement in inflammatory diseases from behind the scenes into the spotlight. *Am. J. Pathology* 171:715, 2007; Okroj, M., Heinegård, D., Holmdahl, R., and Blom, A.M. Rheumatoid arthritis and the complement system. *Ann. Med.* 39:517, 2007; and Zipfel, P.F., Mihlan, M., and Skerka, C. The alternative pathway of complement: a pattern recognition system. *Adv. Exp. Med. Biol.* 598:80, 2007.

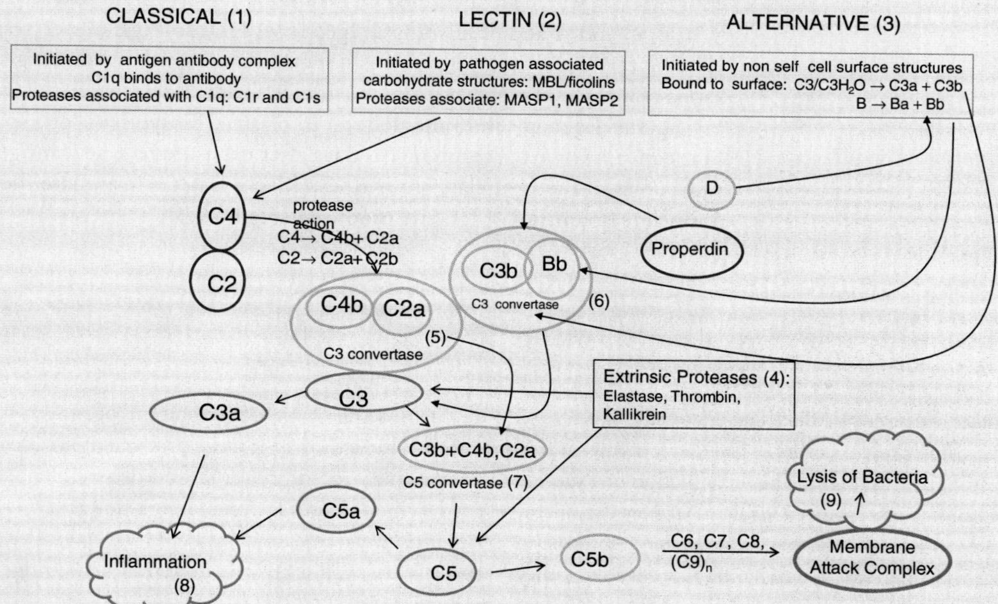

Pathways of Complement Activation The three pathways are the classical (1), lectin (2), and alternative (3). In addition, extrinsic proteases (4) can activate the complement system. The classical C3 convertase (5) is generated by both the classical and lectin pathways, while a different C3 convertase complex (6) is formed in the alternative pathway. The C3b product participates in the C5 convertase complex (7). Outcomes of the convertase pathways are inflammation (8) and lysis of invading bacteria ((9). Color scheme: outcomes, red; activators, brown; proteolytic enzymes, orange; and inert protein substrates, blue.
Redrawn from Markiewski, M. M., and Lambris, *J. D. Am. J. Pathol.* 171:715, 2007.

CLINICAL CORRELATION 9.2

Functions of Different Antibody Classes

The IgA class of immunoglobulins is found primarily in mucosal secretions (bronchial, nasal, and intestinal mucous secretions; tears; milk; and colostrum). They are the initial defense against invading viral and bacterial pathogens prior to their entry into plasma or other internal space.

The IgM class is found only in plasma. They are the first antibodies elicited in significant quantity on exposure to a new foreign antigen. They promote phagocytosis of microorganisms by macrophage and polymorphonuclear leukocytes and are also potent activators of complement (see Clin. Corr. 3.1). They occur in many external secretions but at levels lower than those of IgA.

The IgG class occurs in high concentration in plasma. Their synthesis in response to foreign antigens takes a longer period of time than that of IgM. At maximum concentration they are present in significantly higher concentration than IgM. Like IgM, IgG antibodies promote phagocytosis in plasma and activate complement.

The normal biological functions of the IgD and IgE classes are not known; however, the IgE play an important role in allergic responses such as anaphylactic shock, hay fever, and asthma.

Immunoglobulin deficiency usually causes increased susceptibility to infection. X-linked agammaglobulinemia and common variable immunodeficiency are examples. The commonest disorder is selective IgA deficiency, which results in recurrent infection of sinuses and the respiratory tract.

Cerutti, A., Qiao, X., and He, B. Plasmocytoid dendritic cells and the regulation of immunoglobulin heavy chain class switching. *Immunology and Cell Biol.* 83:554, 2005; Rosen, F. S., Cooper, M. D., and Wedgewood, R. J. P. The primary immunodeficiencies. *N. Engl. J. Med.* 333:431, 1995; and Wines, B. D., and Hogarth, P. M. IgA receptors in health and disease. *Tissue Antigens* 68:103, 2006.

Immunoglobulins in a Class Contain Homologous C_H Regions

Differences in sequence of the C_H regions between immunoglobulin classes determine the characteristics of each class. In some classes, the C_H sequence promotes polymerization of molecules of the basic structure $(LH)_2$. Thus antibodies of the IgA class are often present as dimer structures $[(LH)_2]_2$. Similarly, IgM molecules are commonly pentamers $[(LH)_2]_5$. The different H chains, which are identified by their homologous amino acid sequences, are designated γ, α, δ, ϵ, and μ and occur in IgG, IgA, IgM, IgD, and IgE classes, respectively (Table 9.1; Clin. Corr. 9.2). Two types of L chain sequences are synthesized, designated lambda (λ) and kappa (κ), either of which are found combined with each of the five classes of H chains.

IgG is the major immunoglobulin in plasma. Biosynthesis of a specific IgG in significant concentrations takes about 10 days after exposure to a new antigen (Clin. Corr. 9.3). Antibodies of the IgM class are synthesized at a faster rate and are the first line of defense until large quantities of IgG are produced (Figure 9.3).

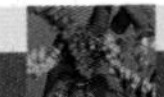

CLINICAL CORRELATION 9.3

Immunization

An immunizing vaccine consists of killed bacterial cells, inactivated viruses, killed parasites, nonvirulent forms of live bacteria, denatured bacterial toxins, or recombinant protein. The introduction of a vaccine into a human leads to protection against a virulent form of the microorganisms or toxic agents that contain the same antigen. Antigens in nonvirulent material cause differentiation of lymphoid cells so that they produce antibody toward the foreign antigen and differentiation of some lymphoid cells into memory cells. Memory cells do not secrete antibody but place antibodies to the antigen onto their outer surface, where they act as future sensors for the antigen. These memory cells are like long-standing radar for the potentially virulent antigen. On reintroduction of the antigen later, binding of the antigen to the cell surface antibody in the memory cells stimulates the memory cell to divide into antibody-producing cells and new memory cells. Once introduction of an antigen occurs, this reduces the time required for antibody production and increases the initial concentrations of antigen-specific antibody produced. This is the basis for the protection provided by immunization.

Vaccines recently introduced for adults include pneumococcal vaccine (to prevent pneumonia due to *Diplococcus pneumoniae),* hepatitis B vaccine, and influenza vaccine. The composition of the latter is changed each year to account for antigenic variation in the influenza virus.

Barouch, D.H. Challenges in the development of a HIV-1 vaccine. *Nature* 455:613, 2008; Benkö, S., Magyarics, M. Szabó, A., and Rajnavölgyi, E. Dendritic cell subtypes as primary targets of vaccines: The emerging role and cross-talk of pattern recognition receptors. *Biol. Chem.* 389:469, 2008; Moscicki, A. HPV vaccines: Today and in the future. *J. Adolescent Health* 43:S26, 2008; Pejawar-Gaddy, S., and Finn, O.J. Cancer vaccines: accomplishments and challenges. Critical Reviews in *Oncology-Hematology* 67:93, 2008; and Webby, R. J. and Sandbulte, M.R. Influenza vaccines. *Frontiers in Bioscience* 13:4912, 2008.

Repeating Sequences and Homologous Three-Dimensional Domains within an Antibody

Within each chain of an antibody molecule is a repeating pattern of amino acid sequences. For the IgG class, the repetitive pattern contains approximately 110 amino acids within both L and H chains. This homology is far from exact, but clearly a number of amino acids match identically on alignment of these segments. Other residues are matched in the sequence by having similar nonpolar or polar side chains. The repetition occurs four times in the H chains, in the V_H region and three C_H regions (designated C_H1, C_H2, and C_H3) (Figures 9.1 and 9.2). The L chain contains one V_L and one C_L region. Each repeat contains an intrachain disulfide bond (Figure 9.2).

Each 110 amino acid segment has a similar arrangement of antiparallel β-strands known as an **immunoglobulin fold** (Figure 9.4). The fold consists of seven to nine polypeptide strands in two antiparallel β-sheets aligned face-to-face. The association of two immunoglobulin folds from different chains forms globular domains, V_L—V_H, and C_L—C_H1 between the H and L chains, and in the C-terminal half of the H chains the domains C_H2—C_H2 and C_H3—C_H3 (Figure 9.2). A "hinge" region interconnects the two V_H—C_H1 N-terminal domains with the C_H2—C_H2 domain of the H chains. Thus the basic antibody structure exhibits six domains, each domain formed from the interaction of two immunoglobulin folds (Figures 9.2 and 9.5). The NH_2-terminal V_L—V_H domains contain a shallow crevice in the center of a hydrophobic core that binds the antigen. The binding site of the crevice is composed primarily of the **hypervariable sequences** of the V domain that are within loops that come close together and are the complementarity binding site (complementarity

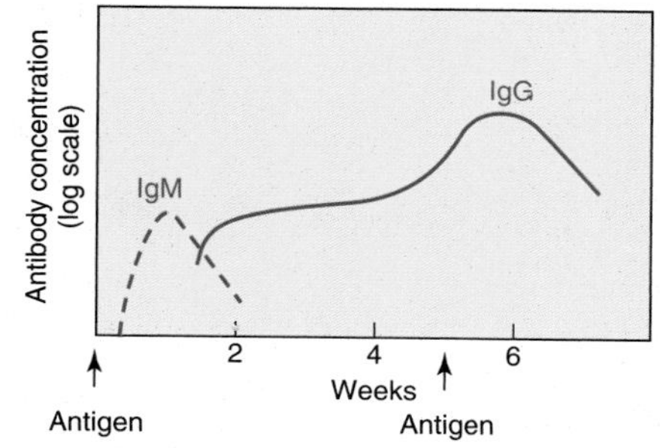

Figure 9.3 Time course of specific antibody IgM and IgG response to added antigen. Based on a figure in Stryer, L. *Biochemistry.* San Francisco: Freeman, 1988, 890.

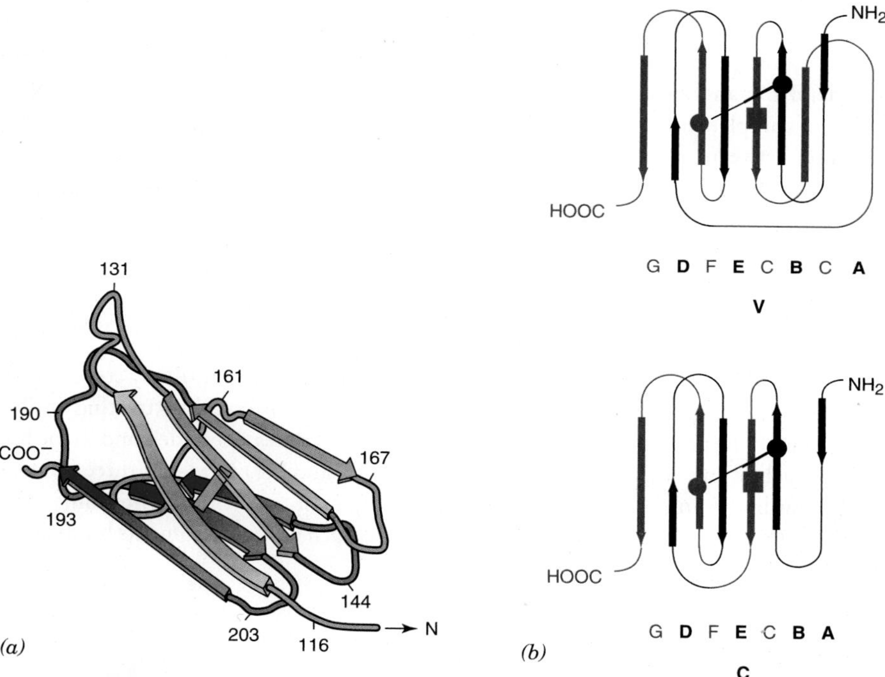

Figure 9.4 Immunoglobulin fold. (*a*) Schematic diagram of folding of a C_L domain, showing β-pleated sheet structure. Arrows show strands of β-sheet, and bar (blue) shows position of cystine bond. Pink arrows are for β-strands in plane above, and dark red arrows are β-strands in plane below. (*b*) Diagrammatic outline of arrangement (topology) of β-strands in immunoglobulin fold motif. Examples are for IgG variable (upper diagram) and constant (lower diagram) regions. Thick arrows indicate β strands, and thin lines loops that connect the β strands. Circles indicate cysteines that form intradomain disulfide bond. Squares show positions of tryptophan residues that are an invariant component of the core of the immunoglobulin fold. Boldface black letters indicate strands that form one plane of the sheet, whereas other letters form the parallel plane behind the first plane.
(*a*) Reproduced with permission from Edmundson, A. B., Ely, K. R., Abola, E. E., Schiffer, M., and Pavagiotopoulos, N. *Biochem.* 14:3953, 1975. Copyright © 1975 by American Chemical Society.
(*b*) Redrawn from Calabi, F. In Calabi, F., and Neuberger, M. S. (Eds.), *Molecular Genetics of Immunoglobulin.* Amsterdam: Elsevier, 1987, 203.

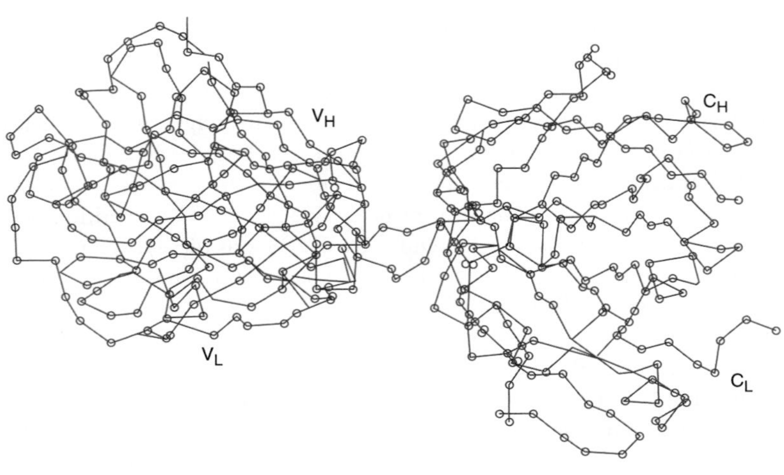

Figure 9.5 α-Carbon (O) structure of, an N-terminal fragment of IgG KOL showing V_L-V_H and C_L-C_H1 domains, each domain formed from two immunoglobulin folds.
Redrawn with permission from Huber, R., Deisenhofer, J., Coleman, P. M., Matsushima, M., and Palm, W. In *The Immune System*, 27th Mosbach Colloquium. Berlin: Springer-Verlag, 1976, 26.

determining regions, CDRs) for the antigen (Figures 9.6 and 9.7). The hypervariable sequences of the domain loops provide a unique three-dimensional conformation for each antibody that makes it specific for its antigenic specificity. The binding of antigen into the V_L–V_H domains may induce small conformational changes in the CDRs inducing functionally important conformational changes between the V_L—V_H domains and the other domains of the antibody molecule leading to the activation of effector sites, such as the binding of complement to the C_H2—C_H2 domain; see C1q site in Figure 9.6. The strength of binding between antibody and antigen is due to noncovalent forces (p. 116). Complementarity of the structures of the antigenic determinant and antigen-binding site results in extremely high equilibrium affinity constants, between 10^5 and 10^{10}/M (strength of 30–60 kJ/mol or 7–14 kcal/mol) for this noncovalent association.

There Are Two Antigen-Binding Sites per Antibody Molecule

There are two NH_2-terminal V_L—V_H domains in the $(LH)_2$ basic antibody structure. The existence of an antigen-binding site in each LH pair is demonstrated by treating antibody molecules with the proteolytic enzyme papain, which hydrolyzes a peptide bond in the hinge region of each of the H chains (see Figures 9.2, p. 335, and 9.8) to release three fragments from the antibody molecule. Two are identical, each consisting of the NH_2-terminal half of an H chain (V_H—C_H1) associated with a full L chain. Each of these fragments binds antigen

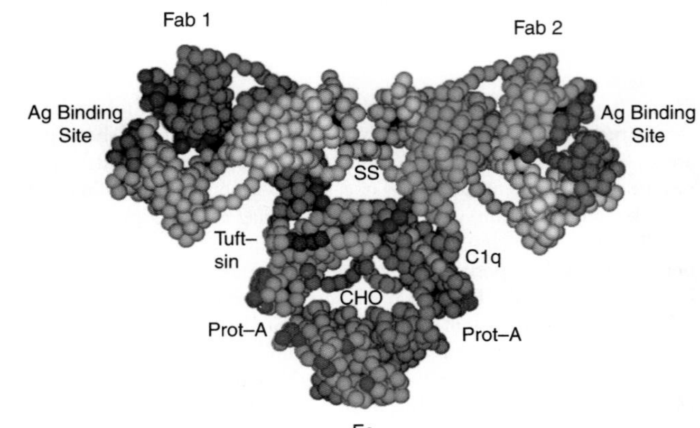

Figure 9.6 Model of an IgG antibody molecule. Only the α-carbons of the structure appear. The two L chains are represented by light gray spheres, and H chains by lavender and blue spheres. Carbohydrates attached to the two C_H2 domains are colored light green and orange. The CDR regions of the V_H—V_L domains are dark red in the H chains and pink in L chains. The interchain disulfide bond between the L and H chains is a magenta ballandstick representation (partially hidden). The heptapeptide hinge between C_H1 and C_H2 domains are dark red. The center of the C1q binding site in the C_H2 domains is yellow, the protein A docking sites at the junction of C_H2 and C_H3 are magenta, and the tuftsin binding site in C_H2 is gray. Tuftsin is a natural tetrapeptide that induces phagocytosis by macrophages and may be transported bound to an immunoglobulin. Protein A is a bacterial protein with a high affinity for immunoglobulins.
Photograph generously supplied by Dr. Allen B. Edmundson, from Guddat, L. W., Shan, L., Fan, ZC., et al. *FASEB J.* 9:101, 1995.

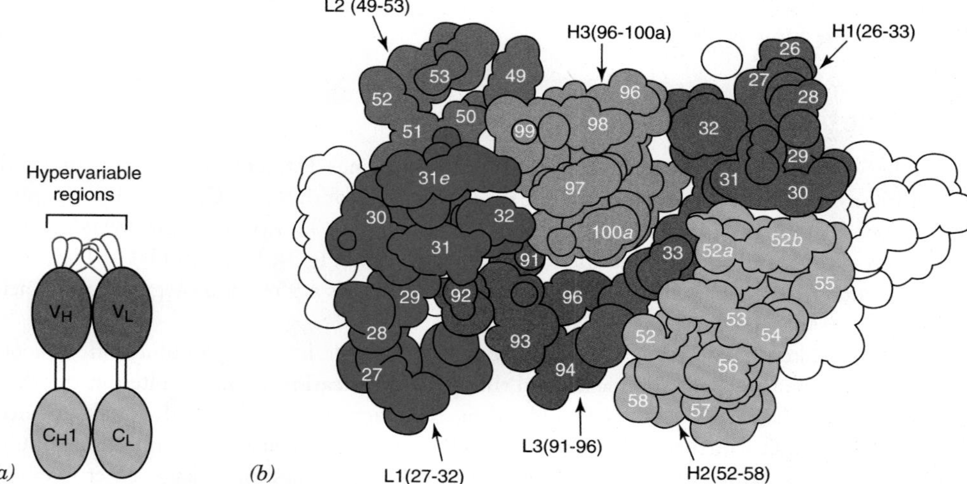

Figure 9.7 Hypervariable loops in immunoglobin. (*a*) Schematic diagram showing hypervariable loops (CDRs) in V_L—V_H domain that form the antigen binding site. (*b*) A cut through an antigen binding site showing contributions of different CDRs using CPK space-filling models of the atoms. Numbers refer to amino acid residue numbers in the sequence of the hypervariable loop regions of the light (L) and heavy (H) chains.

(*a*) Redrawn from Branden, C., and Tooze, J. *Introduction to Protein Structure.* New York: Garland Publishing, 1991, 187. (*b*) Redrawn from Branden, C., and Tooze, J. *Introduction to Protein Structure.* New York: Garland, 1991, 189, and attributed to Chothia, C., and Lesk, A. J. *Mol. Biol.* 196:914, 1987.

with a similar affinity to the intact molecule and is designated a $\mathbf{F_{ab}}$ **(antigen binding fragment)**. The other fragment is the cystine-linked COOH-terminal halves of the two H chains $(C_H2—C_H3)_2$. This is the $\mathbf{F_c}$ **(crystallizable fragment)**, which does not bind the antigen. The V_L—C_L chain can be dissociated from its V_H–C_H1 chain partner within the F_{ab} fragment by reduction of linking disulfide bonds; this eliminates antigen binding. Accordingly, each antigen-binding site must be formed from components of both the L chain and the H chain domains acting together.

The major features of antibody structure and antibody–antigen interactions are: (1) Repeating homologous sequences in the light (L) chains generate two immunoglobulin folds and in the heavy (H) chains four immunoglobulins folds. (2) Immunoglobulin folds on separate chains associate to form the six domains of the basic immunoglobulin structure. (3) The antigen-binding site is generated by hypervariable loops (CDRs) in each V_L—V_H domain, which form a continuous surface topology complementary to the antigenic determinant. (4) The strong interactions between antigen and antibody CDRs are non-covalent and include van der Waals, hydrogen bonding, hydrophobic interactions, and to a smaller extent ionic salt bridges. (5) Small conformational changes occur in the V_L—V_H domain on binding of antigen, indicating an induced fit which induces conformational changes in distant domains of the antibody that alter the binding affinity of effector sites in the

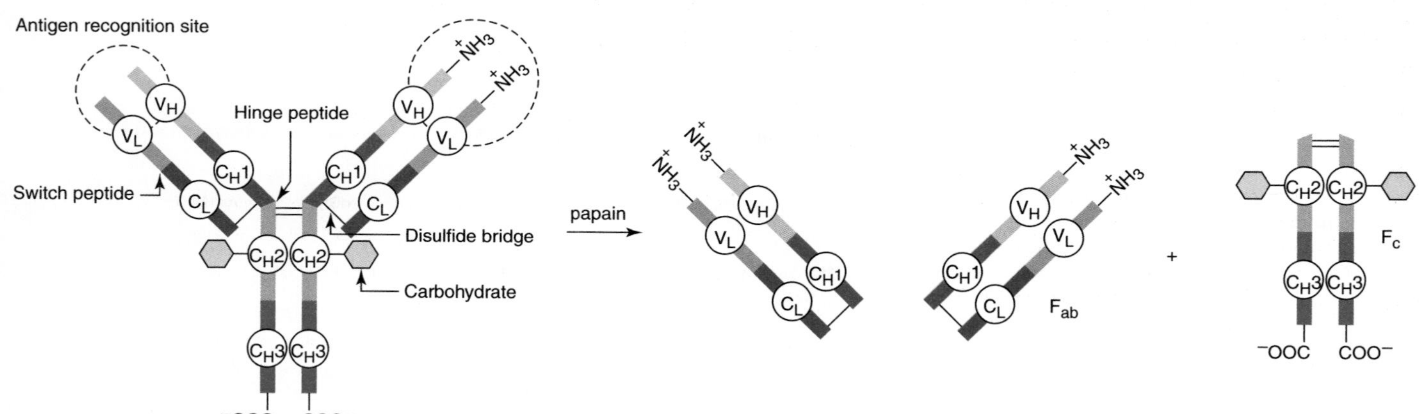

Figure 9.8 Hydrolysis of IgG into two F_{ab} and one F_c fragments by papain, a proteolytic enzyme.

constant domains such as that for binding of complement protein C1q to the $\mathrm{C_H2}$—$\mathrm{C_H2}$ domain (Clin. Corr. 9.1, p. 336).

Genetics of the Immunoglobulins

The V region of the L and H chains are specified by genes distinct from the genes that code for the C region. There are four unique genes that code for the $\mathrm{C_H1}$–$\mathrm{C_H2}$–$\mathrm{C_H3}$ polypeptide segment of the H chain in the IgG antibody class. They are known as gamma (γ) genes, that is, γ_1, γ_2, γ_3, and γ_4, that give rise to **IgG isotypes** $\mathrm{IgG_1}$, $\mathrm{IgG_2}$, $\mathrm{IgG_3}$, and $\mathrm{IgG_4}$. Figure 9.9 presents the sequences of three γ-gene proteins. There is a 95% homology in amino acid sequence among the genes.

It is likely that a primordial gene coded for a single immunoglobulin fold sequence of approximately 110 amino acids and that **gene duplication** events resulted in the three repeating units within the same IgG H γ gene. Mutations modified the individual sequences so that an exact sequence correspondence no longer exists. However, each immunoglobulin fold has a similar length and folding pattern stabilized by a cystine linkage. Later in evolution further gene duplications led to the multiple genes (γ_1, γ_2, γ_3, and γ_4) that code for the constant regions of the IgG classes.

```
Constant Region C1:
Cγ1: AlaSerThrLysGlyProSerValPheProLeuAlaProSerSerLysSerThrSerGlyGlyThrAlaAlaLeuGly
Cγ2                                         C     R           E  S
Cγ4

Cγ1  CysLeuValLysAspTyrPheProGluProValThrValSerTrpAsnSerGlyAlaLeuThrSerGlyValHisThr
Cγ2
Cγ4

Cγ1  PheProAlaValLeuGlnSerSerGlyLeuTyrSerLeuSerSerValValThrValProSerSerSerLeuGly
Cγ2                                                                    N  F
Cγ4

Cγ1  ThrGlnThrTyrIleCysAsnValAsnHisLysProSerAsnThrLysValAspLysLysVal
Cγ2              T           D                                T
Cγ4     K                                                     R

Hinge Region H:
Cγ1  GluProLysSerCysAspLysThrHisThrCysProProCysPro
Cγ2     R     C     V  E  C  P  P        –  –  –
Cγ4     S     Y  G  P  P        S        –  –  –

Constant Region C2:
Cγ1  AlaPro   GluLeuLeuGlyGlyProSerValPheLeuPheProProLysProLysAspThrLeuMetIleSerArg
Cγ2           –  P  V  A
Cγ4         E F     G

Cγ1  ThrProGluValThrCysValValValAspValSerHisGluAspProGluValLysPheAsnTrpTyrValAspGly
Cγ2                                                        Q
Cγ4                                      Q

Cγ1  ValGluValHisAsnAlaLysThrLysProArgGluGluGlnTyrAsnSerThrThrArgValValSerValLeuThr
Cγ2                                                F        F
Cγ4                                                         Y

Cγ1  ValLeuHisGlnAspTrpLeuAsnGlyLysGluTyrLysCysLysValSerAsnLysAlaLeuProAlaProIleGlu
Cγ2     V                                                     G
Cγ4                                                                    S  S

Cγ1  LysThrIleSerLysAlaLys
Cγ2                 T
Cγ4

Constant Region C3:
Cγ1  GlyGlnProArgGluProGlnValTyrThrLeuProProSerArgAspGluLeuThrLysAsnGlnValSerLeuThr
Cγ2                                                E     M
Cγ4                                             Q

Cγ1  CysLeuValLysGlyPheTyrProSerAspIleAlaValGluTrpGluSerAsnGlyGlnProGluAsnAsnTyrLys
Cγ2
Cγ4

Cγ1  ThrThrProProValLeuAspSerAspGlySerPhePheLeuTyrSerLysLeuThrValAspLysSerArgTrpGln
Cγ2              M
Cγ4                                                     R

Cγ1  GlnGlyAsnValPheSerCysSerValMetHisGluAlaLeuHisAsnHisTyrThrGlnLysSerLeuSerLeuSer
Cγ2
Cγ4  E

Cγ1  ProGlyLysStop
Cγ2
Cγ4  L
```

Figure 9.9 Amino acid sequence of heavy chain constant regions of IgG heavy chain γ_1, γ_2, and γ_4 genes. Sequences of $\mathrm{C_H1}$, hinge region H, $\mathrm{C_H2}$, and $\mathrm{C_H3}$ regions are presented. Sequence for γ_1 is complete and differences in γ_2 and γ_4 from γ_1 sequence are shown using single-letter amino acid abbreviations. Dashed line (—) indicates absence of an amino acid in position correlated with γ_1, in order to better align sequences to show maximum homology. Sequence of γ_1 chain from Ellison, J. W., Berson, B. J., and Hood, L. E. *Nucleic Acid Res.* 10:4071, 1982; and sequences of the 1/2 and 1/4 genes from Ellison, J., and Hood, L. *Proc. Natl. Acad. Sci. USA* 79:1984, 1982.

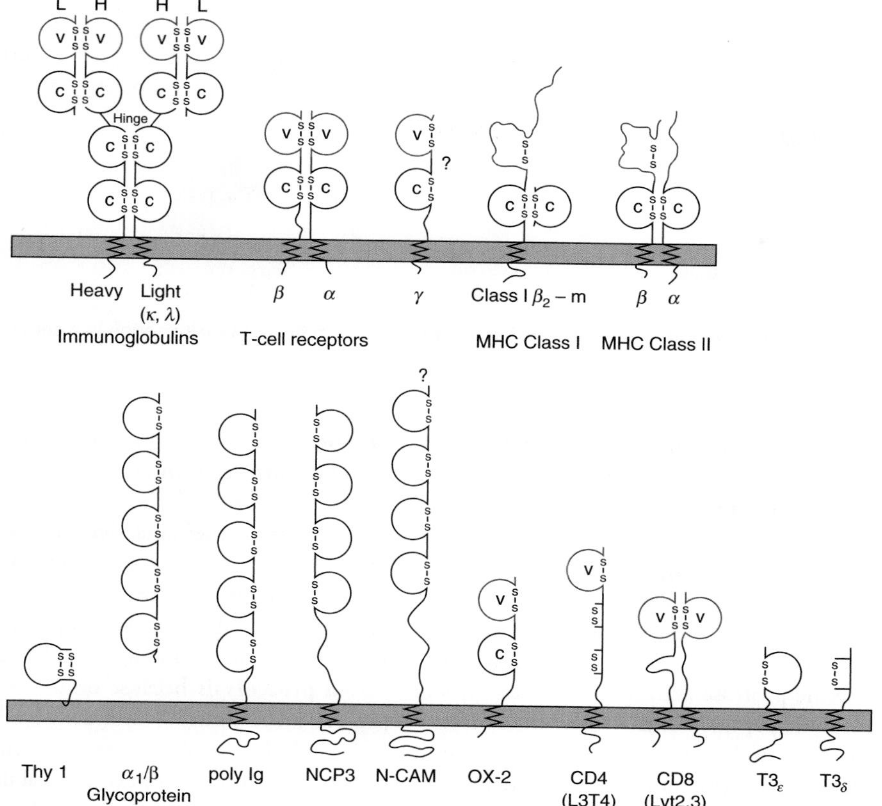

Figure 9.10 Diagrammatic representation of immunoglobulin fold structure found in different proteins of immunoglobulin gene superfamily. Proteins presented include heavy and light chains of immunoglobulins, T-cell receptors, major histocompatibility complex (MHC) Class I and Class II proteins, T-cell accessory proteins involved in Class I (CD8) and Class II (CD4) MHC recognition and possible ion channel formation, a receptor responsible for transporting certain classes of immunoglobulin across mucosal membranes (poly-Ig), β_2-microglobulin which associates with class I molecules, a human plasma protein with unknown function (α_1/β-glycoprotein), two molecules of unknown function with a tissue distribution that includes lymphocytes and neurons (Thy-1, OX-2), and two brain-specific molecules, neuronal cell-adhesion molecule (N-CAM) and neurocytoplasmic protein 3 (NCP3).
Redrawn from Hunkapiller, T., and Hood, L. *Nature* 323:15, 1986.

Immunoglobulin Fold Is Found in a Large Family of Proteins with Different Functional Roles

The immunoglobulin fold is present in many nonimmunological proteins, which exhibit widely different functions. On the basis of their structural homology they are grouped into a **protein superfamily** (Figure 9.10). The Class I major histocompatibility complex proteins are in this superfamily; they have immunoglobulin folds like those in antibody molecules consisting of two stacked antiparallel β-sheets enclosing an internal space filled mainly by hydrophobic amino acids. Two cysteines in the structure form a disulfide bond linking the facing β-sheets. Transcription factors Nf-κB and p53 also contain an immunoglobulin fold motif. It has been speculated that gene duplication during evolution led to distribution of the immunoglobulin fold motif in functionally diverse proteins.

9.3 • PROTEINS WITH A COMMON CATALYTIC MECHANISM: SERINE PROTEASES

Serine proteases are a family of enzymes that use a uniquely activated serine residue in their substrate-binding site to catalytically hydrolyze peptide bonds. This serine is characterized by the irreversible reaction of its side chain hydroxyl group with diisopropylfluorophosphate (DFP) (Figure 9.11). Of all the serines in the protein, DFP reacts only with the catalytically active serine to form a phosphate ester.

Proteolytic Enzymes Are Classified by Their Catalytic Mechanism

Proteolytic enzymes are classified according to their catalytic mechanism. Besides serine proteases, other classes utilize cysteine (**cysteine proteases**), aspartate (**aspartate proteases**), or metal ions (**metalloproteases**) to perform their catalytic function. Those which hydrolyze

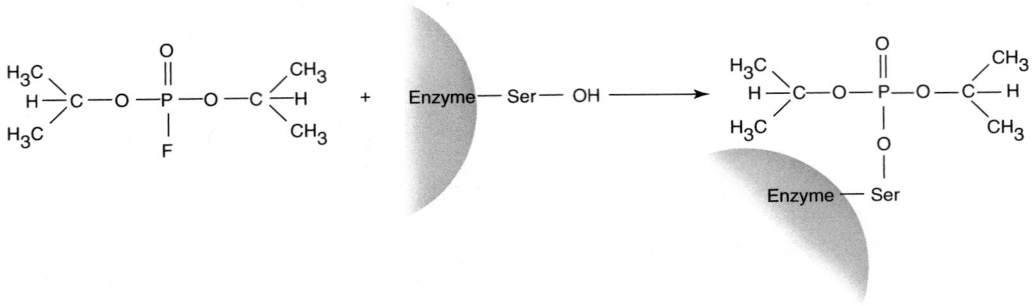

Figure 9.11 Reaction of diisopropylfluorophosphate (DFP) with the active-site serine in a serine protease.

peptide bonds in the interior of a polypeptide are **endopeptidases**, and those which cleave the peptide bond of either the COOH- or NH_2-terminal amino acid in polypeptide substrates are **exopeptidases**.

Serine proteases often activate other serine proteases from their inactive precursor form, termed a **zymogen**, by cleavage of a specific peptide bond. This mechanism of zymogen activation allows serine proteases to participate in carefully controlled physiological processes such as blood coagulation (Clin. Corr. 9.4), fibrinolysis, complement activation (see Clin. Corr. 9.1, p. 336), fertilization, and hormone production (Table 9.2). The serine protease activations are examples of **limited proteolysis** because only one or two specific peptide bonds of the hundreds in the protein substrate are hydrolyzed. Under denaturing conditions, however, these same proteases hydrolyze multiple peptide bonds that lead to digestion of peptides, proteins, and even self-digestion (autolysis). Several diseases, such as emphysema, arthritis, thrombosis, cancer metastasis (Clin. Corr. 9.5), and some forms of hemophilia, are thought to result from the lack of regulation of a specific serine protease.

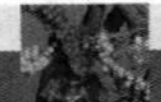

CLINICAL CORRELATION 9.4

Fibrin Formation in a Myocardial Infarct and Use of Recombinant Tissue Plasminogen Activator (rt-PA)

Blood coagulation results from an enzyme cascade process in which inactive precursor serine proteases (zymogens) are activated by other serine proteases in a stepwise manner (the coagulation pathway is described in Chapter 22). These activation events generate catalytic products producing a dramatic amplification of the initial signal of the pathway. The end product of the coagulation pathway is a cross-linked fibrin clot. The zymogen components include factor II (prothrombin), factor VII (proconvertin), factor IX (Christmas factor), factor X (Stuart factor), factor XI (plasma thromboplastin antecedent), and factor XII (Hageman factor). The Roman numeral designation indicates the order of their discovery and not their order of action within the pathway. On activation, the enzymes are denoted with the suffix "a." Thus, prothrombin is denoted as factor II, and catalytically active thrombin factor as IIa.

The main function of coagulation is to maintain the integrity of the closed circulatory system after blood vessel injury. The process, however, can become dangerously activated in a myocardial infarction, resulting in a decrease blood flow to heart muscle. About 1.2 million individuals suffer heart attacks each year in the United States, which result in approximately 500,000 deaths.

A fibrinolysis pathway in plasma degrades the fibrin clots once formed. The counteracting pathway also utilizes serine proteases activated from zymogens. The end reaction of the fibrinolysis pathway is formation of plasmin, a serine protease, which acts directly on fibrin to degrade the fibrin clot. Tissue plasminogen activator (t-PA) is a plasminogen (plasmin zymogen) activator. Recombinant t-PA (rt-PA) is produced by gene cloning technology (p. 262). Clinical studies show that the administration of rt-PA shortly after a myocardial infarct significantly enhances recovery. Plasminogen activators such as urokinase and streptokinase are also effective.

The GUSTO investigators (authors). An international randomized trial comparing four thrombolytic strategies for acute myocardial infarction. *N. Engl. J. Med.* 329:673, 1993; Grotta, J., and Marler, J. Intravenous rt-PA: A tenth anniversary reflection. *Surgical Neurology* 68:S12, 2007; Khaja, A. M., and Grotta, J. C. Established treatments for acute ischaemic stroke. *Lancet* 369:319, 2007; and Simpson, D., Siddiqui, M. A., Scott, L. J., and Hilleman, D. E. Reteplase: A review of its use in the management of thrombotic occlusion disorders. *Am. J. Cardiovascular Drugs* 6:265, 2006.

TABLE 9.2 • Some Serine Proteases and Their Biochemical and Physiological Roles

Protease	*Action*	*Possible Disease Due to Deficiency or Malfunction*
Plasma kallikrein Factor XIIa Factor XIa Factor IXa Factor VIIa Factor Xa Factor IIa (thrombin) Activated protein C	Coagulation (see Clin. Corr. 9.4)	Cerebral infarction (stroke), coronary infarction, thrombosis, bleeding disorders
Factor Clr Factor Cls Factor D Factor B C3 convertase	Complement (see Clin. Corr. 9.1)	Pathogen infections, inflammation, rheumatoid arthritis, autoimmune disease
Trypsin Chymotrypsin Elastase (pancreatic) Enteropeptidase	Digestion	Pancreatitis, digestive disorders
Urokinase plasminogen activator Tissue plasminogen activator Plasmin	Fibrinolysis, cell migration, embryogenesis, menstruation	Clotting disorders, tumor metastasis (see Clin. Corr. 9.5)
Tissue kallikreins	Hormone activation	
Acrosin	Fertilization	Infertility
α-Subunit of nerve growth factor γ-Subunit of nerve growth factor	Growth factor activation	
Granulocyte elastase Cathepsin G Mast cell chymases Mast cell tryptases	Extracellular protein and peptide degradation, mast cell function	Inflammation, allergic response

CLINICAL CORRELATION 9.5

Involvement of Serine Proteases in Tumor Cell Metastasis

The serine protease urokinase plasminogen activator is believed to be required for metastasis of cancer cells. Metastasis is the process by which a cancer cell leaves a primary tumor and migrates through the blood or lymph systems to another tissue or organ, where a secondary tumor grows. Increased synthesis of urokinase-type plasminogen activator (uPA) has been correlated with an increased ability to metastasize in many cancers. uPA activates plasminogen to form plasmin. Plasminogen is ubiquitously located in the extracellular space and its activation to plasmin can cause degradation of proteins in the extracellular matrix through which the metastasizing tumor cells migrate. Plasmin can also convert procollagenase to collagenase, which promotes degradation of collagen in the basement membrane surrounding the capillaries and lymph system. This promotion of proteolytic activity by uPA secreted by tumor cells or stromal cells in the environment of tumor cells allows tumor cells to degrade the extracellular matrix proteins and invade through the matrix to form secondary tumor sites.

Annecke, K., Schmitt, M., Euler, U., Zerm M., et al. uPA and uPAI-1 in breast cancer: Review of their clinical utility and current validation in the prospective NNBC-3 trial. *Adv. Clinical Chemistry* 45:31, 2008; Harbeck, N., Kates, R. E., Gauger, K., Willems, A., et al. Urokinase-type plasminogen activator (uPA) and its inhibitor PAI-I: novel tumor-derived factors with a high prognostic and predictive impact in breast cancer. *Thrombosis and Hemostasis* 91:450, 2004; and Hildenbrand, R., and Schaaf, A. The urokinase-system in tumor tissue stroma of breast and breast cancer cell invasion. *Int. J. Oncology* 34:15, 2009.

Serine Proteases Exhibit Remarkable Specificity in Peptide Bond Hydrolysis

Many serine proteases prefer to hydrolyze peptide bonds next to a particular type of amino acid. Thus trypsin preferentially cleaves on the carboxyl side of the basic amino acids arginine and lysine, and chymotrypsin cleaves on the carboxyl side of large hydrophobic side-chain amino acid residues such as tryptophan, phenylalanine, tyrosine, and leucine. Elastase cleaves on the carboxyl side of small hydrophobic residues such as alanine. A serine protease is trypsin-like, chymotrypsin-like, or elastase-like if it has specificity similar to the named enzymes. The specificity for a certain type of amino acid only indicates an enzyme's relative preference. Trypsin also cleaves peptide bonds following hydrophobic amino acids, but at a much slower rate than the basic amino acids. Thus specificity for hydrolysis of a peptide bond may not be absolute, but may be more accurately described as a range of most likely targets. Each of the identical amino acid hydrolysis sites within a protein substrate is not equally susceptible. Trypsin hydrolyzes each of the arginine peptide bonds in a particular protein at a different catalytic rate, and some may require a conformational change to make them accessible.

The specificity of serine proteases for particular peptide bonds has been determined with synthetic substrates with fewer than 10 amino acids (Table 9.3). Because these are significantly smaller than the natural ones, they interact only with the catalytic site (primary binding site S_1, p. 347) and are said to be **active-site directed**. Studies with small substrates and inhibitors indicate that the site of hydrolysis in polypeptide substrates is flanked by approxi-

TABLE 9.3 • Reactivity of α-Chymotrypsin and Elastase Toward Substrates of Various Structures

Structure	*Variation of Side Chain Group in S_1 Site (Chymotrypsin)*	*Relative Reactivity*[a]
Glycyl	H—	1
Leucyl	$(H_3C)_2CH-CH_2-$	1.6×10^4
Methionyl	$CH_3-S-CH_2-CH_2-$	2.4×10^4
Phenylalanyl	(benzene ring)—CH_2-	4.3×10^6
Hexahydrophenylalanyl	(cyclohexane ring, S)—CH_2-	8.2×10^6
Tyrosyl	HO—(benzene ring)—CH_2-	3.7×10^7
Tryptophanyl	(indole ring, N H)—CH_2-	4.3×10^7
Variation in chain length (elastase hydrolysis of Ala N-terminal amide)[b]		
Ac-Ala-NH_2		1
Ac-Pro-Ala-NH_2		1.4×10^1
Ac-Ala-Pro-Ala-NH_2		4.2×10^3
Ac-Pro-Ala-Pro-Ala-NH_2		4.4×10^5
Ac-Ala-Pro-Ala-Pro-Ala-NH_2		2.7×10^5

[a]Calculated from values of k_{cat}/K_m found for *N*-acetyl amino acid methyl esters in chymotrypsin substrates.

[b]Calculated from values of k_{cat}/K_m in Thompson, R. C. and Blout, E. R. *Biochemistry* 12:57, 1973.

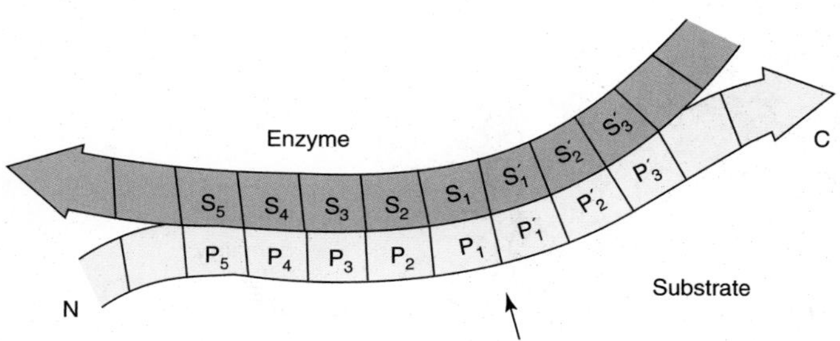

Figure 9.12 Schematic diagram of binding of a polypeptide substrate to binding site in a proteolytic enzyme. P_5—P'_5 are amino acid residues in the substrate that bind to subsites S_5—S'_5 in the enzyme with peptide hydrolysis occurring between P_1—P'_1 (arrow). NH_2-terminal direction of substrate polypeptide chain is indicated by N, and COOH-terminal direction by C. Redrawn from Polgar, L. In Neuberger, A. and Brocklehurst, K. (Eds.), *Hydrolytic Enzymes.* Amsterdam: Elsevier, 1987, 174.

mately four residues in both directions that bind to the enzyme and impact on the reactivity of the bond hydrolyzed. The two residues in the substrate that contribute the hydrolyzable bond are designated $\mathbf{P_1}$**—**$\mathbf{P'_1}$. Thus in trypsin substrates the P_1 residue will be lysine or arginine, and in chymotrypsin substrates P_1 will be a hydrophobic amino acid. The P_1 residue interaction with the S_1 site provides the primary interaction between substrate and enzyme. The other interacting amino acid residues of the protein substrate are labeled P_4—P_2 and P'_2—P'_4 on either end of the scissile bond, respectively. The complementary regions in the enzyme that bind these residues in the substrate are designated subsites $\mathbf{S_4}$**—**$\mathbf{S'_4}$ (Figure 9.12). It is the **secondary interactions** with the substrate outside S_1—S'_1 that ultimately determines a protease's specificity toward a particular protein substrate. Thus, the coagulation serine protease factor Xa cleaves only at a particular arginine in prothrombin to produce thrombin. It is the secondary interaction that allows factor Xa to recognize the particular arginine it cleaves in prothrombin. The interactions of P_4—P'_4 with the subsites S_4–S'_4 are noncovalent. The extended binding site in the substrate interacts with the binding site in the enzyme to produce a β-sheet structure between the enzyme and the substrate, which places the scissile peptide bond of the substrate in the S_1–S'_1 position (Figure 9.13).

Serine Proteases Are Synthesized as Zymogens and in Multiple Domain Proteins

Serine proteases are synthesized as **zymogens** that require limited proteolysis to produce the active enzyme. The coagulation zymogens are synthesized in liver cells and are secreted into the blood for subsequent activation following vascular injury by other serine proteases. Zymogens are usually designated by adding the suffix -ogen to the enzyme name; as in trypsin*ogen* and chymotrypsin*ogen*. In some cases the zymogen is referred to as a **proenzyme**; for example, the zymogen of thrombin is prothrombin.

Several plasma serine proteases have zymogen forms that contain **multiple nonsimilar domains**. Protein C, involved in a fibrinolysis pathway in blood, has four distinct domains (Figure 9.14). The NH_2-terminal domain contains the derived amino acid, **γ-carboxyglutamic acid** (Figure 9.15), which is formed by carboxylation of glutamic acid

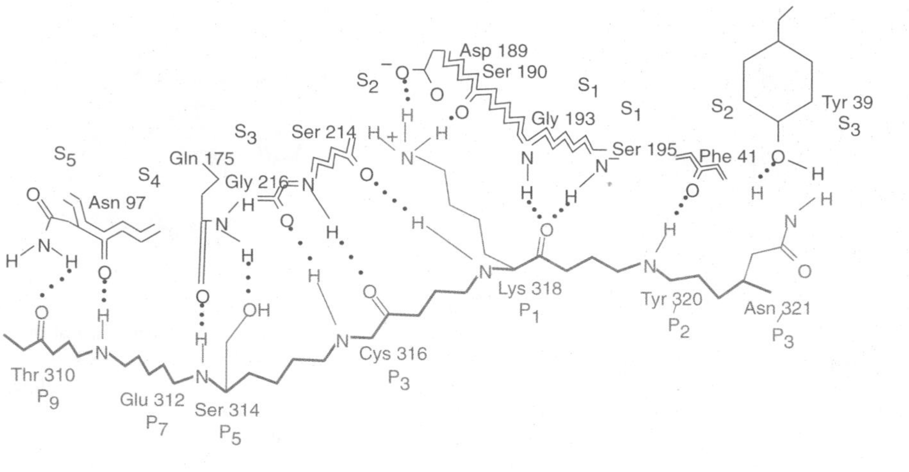

Figure 9.13 Schematic drawing of binding of pancreatic trypsin inhibitor to trypsin. Binding site region of trypsin (blue) shown in complex with region of pancreatic trypsin inhibitor (brown), which models the binding of a polypeptide substrate. Dotted lines are hydrogen bonds. The inhibitor has an extended conformation so that amino acids P_9, P_7, P_5, P_3, and P_1—P'_3 interact with binding subsites S_5—S'_3 in trypsin. Potentially hydrolyzable bond in inhibitor is between P_1—P'_1 (amino acids 318 and 319 of which only Lys318 is visible).

The diagram is based on X-ray diffraction of the complex of pancreatic trypsin inhibitor with trypsinogen. The binding region of trypsinogen in the complex assumes a conformation like that of active trypsin bound to a protein substrate. Redrawn from Bolognesi, M., Gatti, B., Menegatti, E., Guarneri, M., Papamokos, E., and Huber, R. J. *Mol. Biol.* 162:839, 1983.

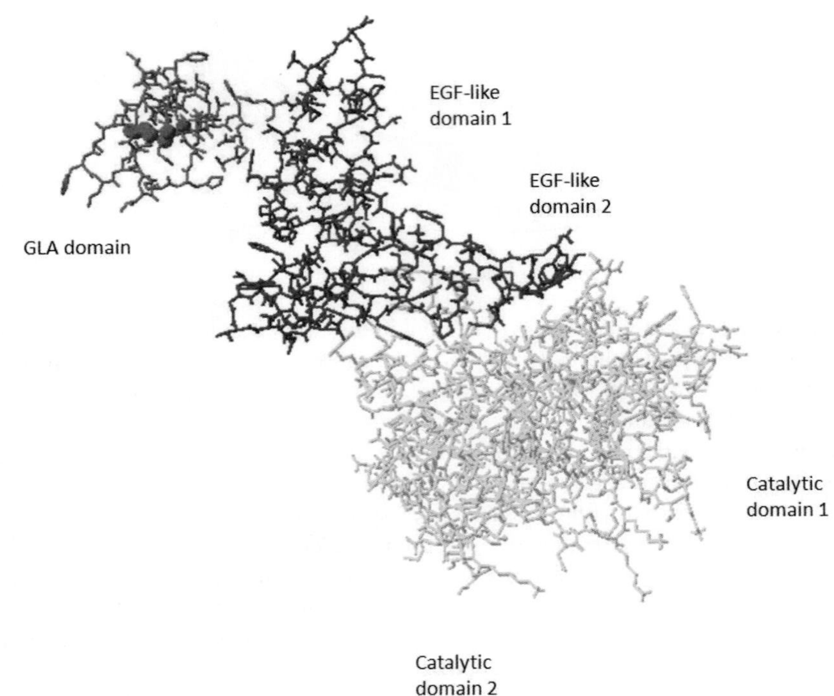

Figure 9.14 Schematic of domain structure for activated protein C showing multidomain structure. GLA refers to the derived amino acid, γ-carboxy-glutamic, residues present in the NH_2-terminal GLA domain (red) where they function to bind calcium ions (red balls). EGF-like domains (blue) contain fold present in epidermal growth factor, and the two catalytic domains (yellow) comprise the protease enzyme at the COOH-terminal end of the protein.
Diagram prepared by combining the GLA domain of human activated protein C (PDB 1LQV) with GLA-domainless structure of activated protein C (PDB 1AUT). Structure references: Oganesyan, V., et al. *J. Biol. Chem.* 277:24851. 2002; and Mather, T. et al. *EMBO J.* 15:6822, 1996. Structure generated using SPDBV, Guex, N., and Peitsch, M.C. *Electrophoresis* 18:2714, 1997.

residues in a vitamin K–dependent reaction (pp. 987 and 994). The γ-carboxyglutamates chelate $Ca^{2+}$and form part of a site that binds cell membranes. The COOH-terminal region contains the catalytic domains. Activation from the zymogen form requires peptide bond cleavage outside the catalytic domains and is controlled by the binding of the γ-carboxyglutamate-containing NH_2-terminal end through calcium ions to a membrane site.

Tertiary Structures of the Catalytic Domains of Serine Proteases Are Similar

Trypsin-like serine proteases constitute a superfamily of proteins with two distinct closed β-barrel domains of the same fold and equal size (Figure 9.16), which do not penetrate one another. Loop regions protrude from the barrel structures, being almost symmetrically presented by each of the two folded domains. These loops combine to form a surface region that extends outward, above the catalytic site, and are functionally similar to the CDRs of immunoglobulins. Structure determinations by X-ray crystallography have been carried out

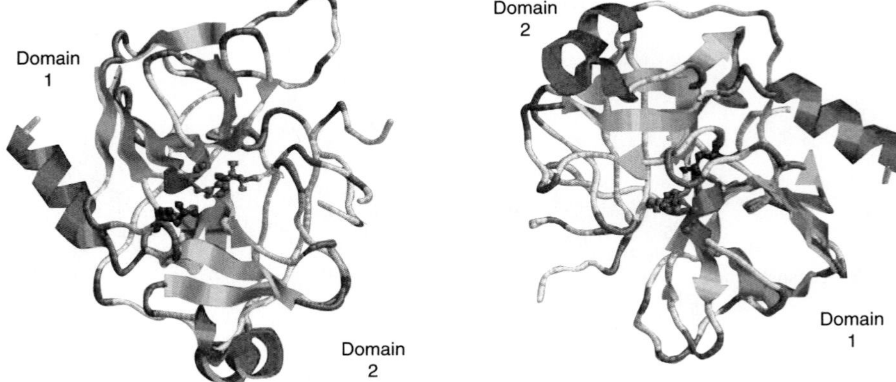

Figure 9.16 Two views of trypsin-like catalytic domains showing β-barrel fold structure of the domains with active site residues at the domain interface. β-strands are yellow arrows, α-helices are magenta, turns are pale blue, and all other residues are white. Active site serine-195 (orange), histidine-57 (blue), and aspartate-102 (reddish-brown) are shown.
B and C chains from human α-chymotrypsin, PDB 6CHA, are visualized with RasMol.

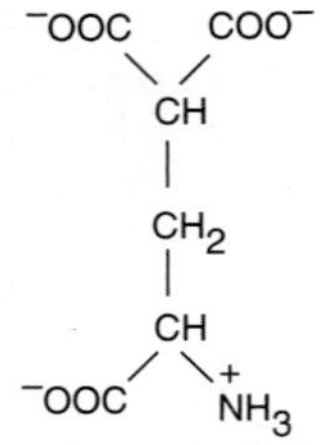

$^{-}OOC \quad COO^{-}$
CH
|
CH_2
|
CH
$^{-}OOC \quad \overset{+}{N}H_3$

Figure 9.15 Structure of the derived amino acid γ-carboxyglutamic acid (abbreviation Gla), found in NH_2-terminal domain of many coagulation proteins.

TABLE 9.4 • Serine Protease Structures Determined by X-Ray Crystallography

Enzyme	*Species Source*	*Inhibitors Present*	*Resolution (Å)*
Chymotrypsin[a]	Bovine	Yes[b]	1.67[c]
Chymotrypsinogen	Bovine	No	2.5
Elastase	Porcine	Yes	2.5
Kallikrein	Porcine	Yes	2.05
Proteinase A	*S. griseus*	No	1.5
Proteinase B	*S. griseus*	Yes	1.8
Proteinase II	Rat	No	1.9
Trypsin[a]	Bovine	Yes[b]	1.4[c]
Trypsinogen[a]	Bovine	Yes[b]	1.65[c]

[a]Structure of this enzyme molecule independently determined by two or more investigators.

[b]Structure obtained with no inhibitor present (native structure) and with inhibitors. Inhibitors used include low molecular weight inhibitors (i.e., benzamidine, DFP, and tosyl) and protein inhibitors (i.e., bovine pancreatic trypsin inhibitor).

[c]Highest resolution for this molecule of the multiple determinations.

on many members of the trypsin-like serine protease protein family (Table 9.4), including active enzyme forms, zymogens, the same enzyme in multiple species, enzyme-inhibitor complexes, and a particular enzyme at different temperatures and in different solvents.

Alignment of three-dimensional structures can be performed on serine proteases. Table 9.5 shows for sets of two proteases the number of amino acids that are superimposable in the 3-dimensional structures and the number that are chemically identical in comparison of the sequences. Structurally superimposable amino acids, even if they differ chemically, cannot be distinguished from one another at the resolution of the X-ray diffraction. The proteins are more homologous in their structures than in sequence. The regions of greatest difference are in the CDR-like loop regions. Altering these loops changes the **macromolecular-binding specificity** of the protease. The structure of the loop in factor Xa, for example, allows it to bind specifically to prothrombin. Protease inhibitors interact with different proteases through their affinity for the loop structures. Bacterial proteases related to the eukaryotic serine protease family contain the same two domains but lack most of the loops. This agrees with their lack of a requirement for complex interactions and the observation that bacterial proteases are not synthesized as zymogens.

Ser-195 in chymotrypsin reacts with the inhibitor diisopropylfluorophosphate (DFP), with a 1:1 enzyme:DFP stoichiometry (p. 344 and Figure 9.11). The three-dimensional structure of chymotrypsin reveals that the Ser-195 is situated within an internal pocket, with access to the solvent interface. His-57 and Asp-102 are oriented so that they participate with the Ser-195 in the catalytic mechanism. Thus the serine protease family constitutes

TABLE 9.5 • Structural Superposition of Selected Trypsin Family Serine Proteases and the Resultant Amino Acid Sequence Comparison

	Number of Amino Acids in Sequence			
Comparison	*Protease 1*	*Protease 2*	*Number of Structurally Equivalent Residues*	*Number of Chemically Identical Residues*
Trypsin–elastase	223	240	188	81
Trypsin–chymotrypsin	223	241	185	93
Trypsin–mast cell protease	223	224	188	69
Trypsin–prekallikrein	223	232	194	84
Trypsin–*S. griseus* protease	223	180	121	25

a structurally related series of proteins that use a catalytically active serine. During evolution, the basic two-domain structure and the catalytically essential residues have been maintained, but the regions of the secondary interactions (loop regions) have changed to provide for different specificities toward substrates, activators, and inhibitors, related to their diverse physiological roles.

Serine Proteases Have Similar Structure–Function Relationships

The relationships between structure and physiological function of the serine proteases are summarized as follows: (1) Only one serine residue is catalytically active and participates in peptide bond cleavage. Bovine trypsin contains 34 serine residues with only one catalytically active or able to react with the inhibitor DFP (Figure 9.11, p. 344). (2) X-ray diffraction and sequence homology studies demonstrate that a histidine and an aspartate are always associated with the activated serine in the catalytic site. On the basis of their positions in chymotrypsinogen, these three invariant residues are named Ser 195, His 57, and Asp 102 and called the *catalytic triad.* This numbering, based on their sequence number in chymotrypsinogen, is used to identify these residues irrespective of their exact position in any serine protease. (3) Eukaryotic serine proteases exhibit high sequence and structural similarity. (4) Genes that code for serine proteases are organized similarly (Figure 9.17). Their **exon–intron patterns** show that each of the catalytically essential residues (Ser-195, His-57, and Asp-102) is on a different exon. The essential histidine and serine are all almost adjacent to their exon boundary. The cross-species homology in gene structure supports the concept that the serine proteases evolved from a common primordial gene. (5) The catalytic portion of serine proteases comprises two domains, of approximately equal size. The catalytic site is within their interface or the crevice between the two domains. The three-dimensional structure of the catalytic site in serine proteases is complementary to structure of the reaction transition state and contains an oxyanion site to stabilize the transition state of the peptide hydrolysis reaction. (7) Serine proteases that interact with membranes typically have an additional domain to provide for this specific function. (8) Natural protein substrates and inhibitors of serine proteases bind through an extended specificity site. (9) Specificity for natural protein inhibitors is marked by extremely tight binding. The binding constant for trypsin to pancreatic trypsin inhibitor is about 10^{13}/M, reflecting a binding free energy of approximately 75 kJ/mol (18 kcal/mol). (10) Natural protein inhibitors are usually poor substrates with strong inhibition requiring hydrolysis of a peptide bond in the inhibitor by the protease. (11) Serine proteases in eukaryotes are synthesized as zymogens to permit their transport in an inactive state to their sites of action. (12) Zymogen activation frequently involves hydrolysis by another serine protease. (13) Several serine proteases undergo **autolysis** or self-hydrolysis.

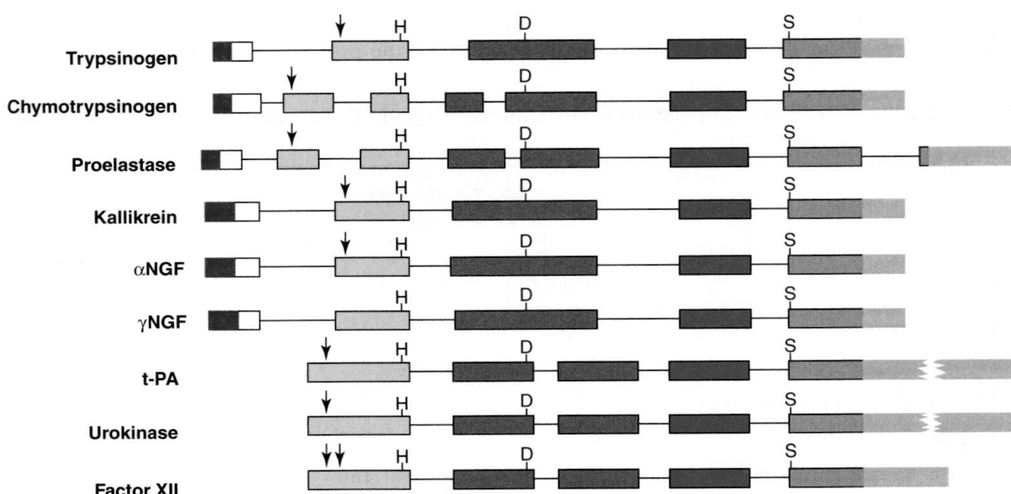

Figure 9.17 Organization of exons and introns in genes for serine proteases. t-PA is tissue plasminogen activator and NGF is nerve growth factor which contain a structure homologous to a serine protease. Exons are shown by boxes and introns by connecting lines. Position of the codons for active-site serine, histidine, and aspartate are denoted by S, H, and D, respectively. Red boxes, on left, show regions that code for NH_2-terminal signal peptide that is cleaved before secretion. Light-colored boxes, on right, represent part of gene sequence transcribed into messenger RNA (mRNA), but not translated into protein. Arrows show codons for residues at which proteolytic activation of zymogen occurs. Redrawn and modified from Irwin, D. M., Roberts, K. A., and MacGillivray, R. T. J. *Mol. Biol.* 200:31, 1988.

Sometimes this leads to specific peptide bond cleavage and further activation of the catalytic activity. At other times autolysis leads to inactivation of the protease.

Sequence Homology in Serine Proteases

Much early knowledge of the serine protease family came from trypsin and chymotrypsin purified from bovine materials. This has yielded a useful but nonintuitive nomenclature, which uses a sequence alignment against the amino acid sequence of chymotrypsin, to name and number residues of other serine proteases. As mentioned previously, the catalytically essential residues are Ser-195, His-57, and Asp-102. Insertions and deletions of the amino acids in another serine protease are compared to the numbering of residues in chymotrypsin. Alignment is made by algorithms that maximize sequence homology, with exact alignment of the essential serine, histidine, and aspartate residues. These three residues are invariant in all serine proteases and the sequences surrounding them are invariant among the serine proteases of the trypsin family (Table 9.6).

Members of the trypsin family also occur in prokaryotes. Thus, bacterial serine proteases from *Streptomyces griseus* and *Myxobacteria* 450 are homologous with trypsin. However, a separate class of serine proteases first discovered in bacteria has no structural homology to the mammalian trypsin family. The serine protease subtilisin, isolated from *Bacillus subtilis,* hydrolyzes peptide bonds and contains an activated serine with a histidine and aspartate in its active site and an active site structure complementary to the hydrolytic reaction transition state, but its active site arises from structural regions of the protein that have no sequence or structural homology with the trypsin family. Serine proteases of the subtilisin family are also found in human cells. This is an example of **convergent evolution** of an enzyme catalytic mechanism. Apparently a completely different gene evolved the same catalytic mechanism for peptide hydrolysis.

TABLE 9.6 • Invariant Sequences Around Catalytically Essential Serine (S) and Histidine (H)

Enzyme	*Sequence (Residues Identical to Chymotrypsin Are in Bold)*																								
	Residues Around Catalytically Essential Histidine																								
Chymotrypsin A	F	**H**	**F**	**C**	**G**	**G**	**S**	**L**	**I**	**N**	**E**	**N**	**W**	**V**	**V**	**T**	**A**	**A**	**H***	**C**	G	V	T	T	S	D
Trypsin	Y	**H**	**F**	**C**	**G**	**G**	**S**	**L**	**I**	**N**	S	Q	**W**	**V**	**V**	S	**A**	**A**	**H**	**C**	Y	K	S	G	**I**	Q
Pancreatic elastase	A	**H**	T	**C**	**G**	**G**	T	**L**	**I**	R	Q	**N**	**W**	**V**	M	**T**	**A**	**A**	**H**	**C**	V	D	R	**E**	L	T
Thrombin	**E**	L	L	**C**	**G**	A	**S**	**L**	**I**	S	D	R	**W**	**V**	L	**T**	**A**	**A**	**H**	**C**	**L**	L	Y	P	P	W
Factor X	**E**	G	**F**	**C**	**G**	**G**	T	**I**	L	**N**	**E**	F	Y	**V**	L	**T**	**A**	**A**	**H**	**C**	**L**	H	Q	A	K	R
Plasmin	**M**	**H**	**F**	**C**	**G**	**G**	T	**L**	**I**	S	P	**E**	**W**	**V**	**L**	**T**	**A**	**A**	**H**	**C**	**L**	**E**	**K**	S	P	R
Plasma kallikrein	S	F	Q	**C**	**G**	**G**	V	**L**	V	**N**	P	**K**	**W**	**V**	**L**	**T**	**A**	**A**	**H**	**C**	K	N	D	N	Y	E
Streptomyces trypsin	—	—	—	**C**	**G**	**G**	A	**L**	Y	A	Q	D	I	**V**	**L**	**T**	**A**	**A**	**H**	**C**	V	S	G	S	G	N
Subtilisin	V	G	G	A	S	F	V	A	G	**E**	A	Y	N	T	D	G	N	G	H	G	T	H	V	A	G	T
	Residues Around Catalytically Essential Serine																								
Chymotrypsin A	**C**	**A**	**G**	—	—	—	A	S	**G**	**V**	—	—	**S**	**S**	**C**	**M**	**G**	**D**	**S***	**G**	**G**	**P**	**L**	**V**		
Trypsin	**C**	**A**	**G**	Y	—	—	L	E	**G**	G	K	—	D	**S**	**C**	Q	***G***	***D***	***S***	***G***	***G***	***P***	V	**V**		
Pancreatic elastase	**C**	**A**	**G**	—	—	—	G	N	**G**	**V**	**R**	—	**S**	G	**C**	Q	**G**	**D**	**S**	**G**	**G**	**P**	L	**H**		
Thrombin	**C**	**A**	**G**	Y	K	P	G	E	**G**	K	**R**	G	D	A	**C**	E	**G**	**D**	**S**	**G**	**G**	**P**	F	**V**		
Factor X	**C**	**A**	**G**	Y	—	—	D	T	**Q**	**P**	E	—	D	A	**C**	Q	**G**	**D**	**S**	**G**	**G**	**P**	H	**V**		
Plasmin	**C**	**A**	**G**	H	—	—	L	A	**G**	G	T	—	D	**S**	**C**	Q	**G**	**D**	**S**	**G**	**G**	**P**	L	**V**		
Plasma kallikrein	**C**	**A**	**G**	Y	—	—	L	P	**G**	G	K	—	D	T	**C**	**M**	**G**	**D**	**S**	**G**	**G**	**P**	L	I		
Streptomyces trypsin	**C**	**A**	**G**	Y	—	P	D	T	**G**	G	V	—	D	T	**C**	Q	**G**	**D**	**S**	**G**	**G**	**P**	M	F		
Subtilisin	A	G	V	Y	S	T	Y	P	**T**	N	T	Y	A	T	L	N	**G**	T	**S**	M	A	S	P	H		

Source: Barrett, A. J. In: A. J. Barrett and G. Salvesen (Eds.), *Proteinase Inhibitors.* Amsterdam: Elsevier, 1986, p. 7.

Mechanism of Serine Protease Catalysis

All trypsin-like serine proteases catalyze peptide bond hydrolysis by an identical catalytic mechanism with the coordinate actions of an active site Ser-195, His-57, and Asp-102. These residues are in different domains of the two domains that make up the catalytic unit. In the folded protein the three-dimensional structures bring these residues together at the domain interface that comprises the primary site for catalysis (Figure 9.18). The imidazole nitrogens of His-57s form hydrogen bonds to the Ser-195 $^{\gamma}$OH and to the Asp-102 side-chain carboxylate. In protease-catalyzed hydrolysis, the histidine acts as a general base catalyst and the $^{\gamma}$O of the serine forms a transient covalent ester intermediate with a part of the substrate. These mechanisms involving the Ser-195 and His-57 facilitate the breaking of the scissile peptide bond of the substrate bound to the active site.

In addition to the three catalytically essential amino acids of the active site, all active sites have features that stabilize the transition-state structure of the reaction. The transition state is the highest energy structure on the reaction pathway to which the substrate is elevated. The rate of the reaction is dependent on the energy required to reach the structure of the transition state, and enzymes accelerate reaction rates by decreasing the energy required to reach the transition-state structure. One way of decreasing this energy is to stabilize the hydrolytic transition-state structure. All serine protease active sites contain two hydrogen-bond donating groups oriented and poised to donate hydrogen bonds that stabilize the transition-state structure (Figure 9.18). The exact placement of these hydrogen-bond donating groups within the active site is a catalytically critical structural feature of the protease enzymes.

Specific Protein Inhibitors of Serine Proteases

Evolution of the serine protease family for participation in physiological processes promoted a parallel evolution of inhibitors. Specific proteins inhibit the activity of serine proteases when their physiological role has ended (Table 9.7). Thus, coagulation is limited to the site of vascular injury and complementation activation leads to lysis only of cells exhibiting foreign antigens. Inability to control these proteases, such as by a deficiency of a specific inhibitor, can lead to undesirable consequences, such as thrombus formation initiating myocardial infarction and stroke or uncontrolled reactions of complement in autoimmune disease. These natural inhibitors of serine proteases are termed **serpins** for **ser**ine **p**rotease **in**hibitors. They occur in animals that produce the proteases, but surprisingly they are also found in plants that lack proteases.

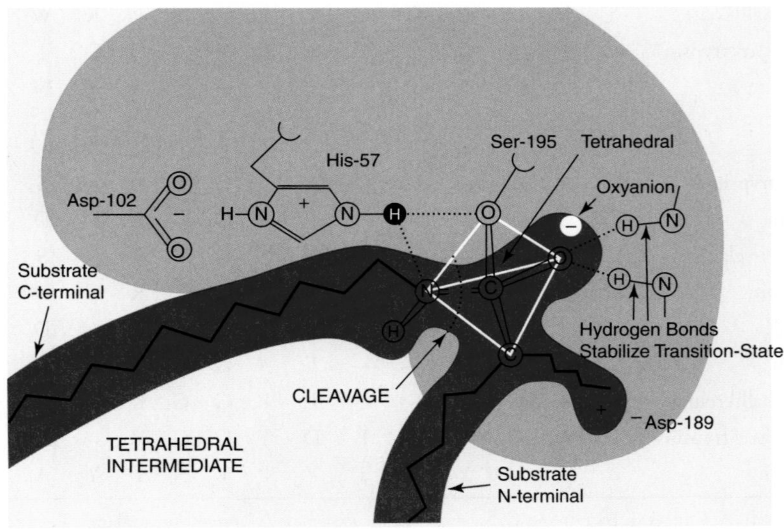

Figure 9.18 **Stabilization of peptide hydrolysis reaction transition-state Structure by trypsin.** Polypeptide substrate is shown in red with the atoms of the α-carbon, side chain (lysine), and labile peptide bond of the P_1 residue in the substrate. The labile peptide bond is in an intermediate structure like that of the proposed reaction transition state, in which the carbonyl carbon to nitrogen peptide bond is being broken while the bond between the Ser-195 $^{\gamma}$O and carbonyl carbon is being made. In this structure the carbonyl carbon is transiently bonded to four different atoms and the carbonyl oxygen is negatively charged. The enzyme active is in blue and contains the catalytically essential Ser-195, His-57, and Asp-102. The polypeptide chain of the enzyme active site contributes two hydrogen bonds that stabilize the transition-state structure. Modified from figure by Stroud, R. M. *Sci. Am,* 231:74, 1974.

TABLE 9.7 • Some Human Proteins that Inhibit Serine Proteases

Inhibitor	*Action*
α_1-Proteinase inhibitor	Inhibits tissue proteases including neutrophil elastase; deficiency leads to pulmonary emphysema
α_1-Antichymotrypsin	Inhibits proteases of chymotrypsin-like specificity from neutrophils, basophils, and mast cells including cathepsin G and chymase
Inter-α-trypsin inhibitor	Inhibits broad range of serine protease activities in plasma
α_2-Antiplasmin	Inhibits plasmin
Antithrombin III	Inhibits thrombin and other coagulation proteases
C_1 Inhibitor	Inhibits complement reaction
α_2-Macroglobulin	General protease inhibitor
Protease nexin I	Inhibits thrombin, urokinase plasminogen activator, and plasmin
Protease nexin II	Inhibits growth factor-associated serine proteases, identical to NH_2-terminal domain of amyloid protein secreted in Alzheimer's disease
Plasminogen activator inhibitor I	Inhibits plasminogen activators
Plasminogen activator inhibitor II	Inhibits urokinase plasminogen activator

9.4 • HEMOGLOBIN AND MYOGLOBIN

Hemoglobins are globular proteins, present in high concentrations in red blood cells that bind oxygen in the lungs and transport it to cells of the tissues. They also transport CO_2 and H^+ from the tissues to the lungs, and carry and release nitric oxide (NO) in the blood vessels of the tissues. NO is a potent vasodilator and inhibitor of platelet aggregation. The molecular structure and chemistry of the hemoglobins beautifully fit these complex and interrelated physiological roles.

Human Hemoglobin Occurs in Several Forms

A hemoglobin molecule consists of four polypeptide chains, two each of two different sequences. Each chain contains a heme prosthetic group that binds oxygen. On the binding of O_2 to a hemoglobin molecule to form $Hb(O_2)_4$, the four O_2-binding sites act in a cooperative manner. The major form of human adult hemoglobin, **HbA_1**, consists of two α and two β chains ($\alpha_2\beta_2$). The α-polypeptide contains 141, and the β-polypeptide 146 amino acids. The fetal form (**HbF**) (Figure 9.19) contains the same α chains as HbA_1, and γ chains that differ in sequence from the β chain (Table 9.8). Two other hemoglobin forms appear in the first months after conception (embryonic) in which the α chains are substituted by zeta (ζ) chains of different sequence and the ε chains serve as the β chains. A minor adult hemoglobin, **HbA_2**, composes about 2% of normal adult hemoglobin and contains

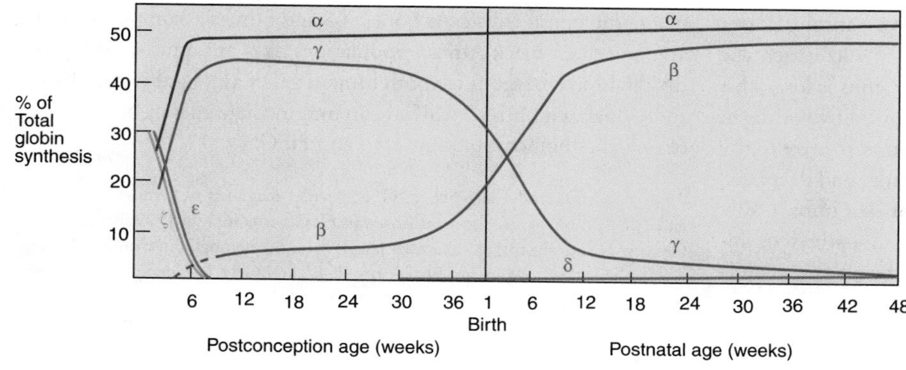

Figure 9.19 Changes in globin chain production during development. Redrawn from Nienhuis, A. W., and Maniatis, T. In Stamatoyannopoulos, G., Nienhuis, A.W., Leder, P., and Majerus, P. W. (Eds.), *The Molecular Basis of Blood Diseases.* Philadelphia: Saunders, 1987, 68, in which reference of Weatherall, D. J., and Clegg, J. B., *The Thalassemia Syndromes,* 3d ed. (Oxford: Blackwell Scientific, 1981), is acknowledged.

TABLE 9.8 • Chains of Human Hemoglobin

Developmental Stage	*Symbol*	*Chain Composition*
Adult	HbA_1	$\alpha_2\beta_2$
Adult	HbA_2	$\alpha_2\delta_2$
Fetus	HbF	$\alpha_2\gamma_2$
Embryo	Hb Gower-1	$\zeta_2\epsilon_2$
Embryo	Hb Portland	$\zeta_2\gamma_2$

two α chains and two chains designated delta (δ) (Table 9.8). A discussion of abnormal hemoglobins is presented in Clin. Corr. 9.6.

Myoglobin: A Single Polypeptide with One O_2-Binding Site

Myoglobin (Mb) is an O_2 binding protein that captures and releases O_2 with changes in the O_2 concentration in the sarcoplasm of skeletal muscle cells. In contrast to hemoglobin, which has four chains and four O_2-binding sites, myoglobin contains only a single polypeptide chain and one O_2-binding site. Myoglobin is a model for what occurs when a single protomer molecule acts alone without the interactions exhibited among the four O_2-binding sites in the more complex tetramer molecule of hemoglobin.

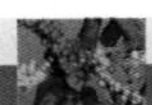

CLINICAL CORRELATION 9.6

Hemoglobinopathies

There are well over 800 known mutant human hemoglobins. Mutations cause instability in hemoglobin structure, increased or decreased oxygen affinity, or an increase in the rate of oxidation of the heme ferrous iron (Fe^{+2}) to the ferric state (Fe^{+3}). A nonfunctional ferric hemoglobin is designated methemoglobin and is symbolized HbM.

Unstable hemoglobin can arise by the substitution of proline for an amino acid within an α-helical region of the globin fold. Prolines do not participate in α-helical structure and the breaking of an α-helical segment causes an unstable hemoglobin (examples are $Hb_{Saki}{}^{\beta Leu14\rightarrow Pro}$ and $Hb_{Genova}{}^{\beta 28Leu\rightarrow Pro}$). Other unstable hemoglobins arise from substitution by an amino acid that is either too large or too small to make appropriate contacts or by placement of a charged or polar group on the inside of a domain. Unstable hemoglobins easily denature and may precipitate to form Heinz bodies, which damage the erythrocyte membrane. Patients with unstable hemoglobins develop anemia, reticulocytosis, splenomegaly, and urobilinuria.

Some mutant hemoglobins have an increased oxygen affinity (lower P_{50}). An interesting example is $Hb_{Cowtown}{}^{\beta His146\rightarrow Leu}$, in which the histidine that dissociates 50% of the Bohr-effect protons is lost. This mutation impedes the regulation of oxygen dissociation by hydrogen ion concentration and destabilizes the T conformation relative to the R conformation, causing an increase in oxygen affinity and decreased release of O_2 to the tissues. Hemoglobin mutations that impair BPG binding also increase oxygen affinity. Hemoglobinopathies of increased oxygen affinity are often characterized by hemolytic anemia and Heinz body formation.

Mutant hemoglobins that form methemoglobin include $HbM_{Iwate}{}^{\alpha 87His\rightarrow Tyr}$ and $HbM_{Hyde\ Park}{}^{\beta 92His\rightarrow Tyr}$ where the proximal histidines F8 in the α and β chains, respectively, are mutated. In $HbM_{Boston}{}^{\alpha 58His\rightarrow Tyr}$ and $HbM_{Saskatoon}{}^{\beta 63His\rightarrow Tyr}$ the distal E7 histidines are mutated. Mutations of amino acids that pack against the heme or form the oxygen-binding site often lead to methemoglobinemia. Patients with high concentrations of methemoglobin show cyanosis (bluish color of the skin).

Two of the most prevalent mutations occur at the same amino acid position, the β6Glu. When this glutamate is replaced by valine, the result is $HbS^{\beta 6Glu\rightarrow Val}$, whereas replacement by lysine yields $HbC^{\beta 6Glu\rightarrow Lys}$. Homozygotes with HbS have sickle-cell anemia in which the hemoglobin molecules precipitate as tactoids or long arrays, which produce the sickle shape of the erythrocyte. HbC forms a different aggregate structure consisting of blunt-ended crystalloids. This reduces the survival time of the red blood cells but produces less hemolysis than HbS. This form of hemoglobinopathy exhibits more limited pathological effects. As both HbS and HbC are commonly found among certain black African populations, it is not unusual to find individuals heterozygous for both mutant genes among this population. Individuals with HbSC will have an intermediate anemia between that observed of homozygotes for HbS and HbC.

Arcasoy, M. O., and Gallagher, P. G. Molecular diagnosis of hemoglobinopathies and other red blood cell disorders. *Seminars Hematology* 36:328, 1999: Benz, E. J., Jr. Genotypes and phenotypes: Another lesson from the hemoglobinopathies. *N. Engl. J. Med.* 351:1490, 2004; and Dickerson, R. E., and Geis, I. *Hemoglobin: Structure, Function, Evolution, and Pathology.* Menlo Park, CA: Benjamin/Cummings, 1983.

The Heme Prosthetic Group Is Site of O_2 Binding

The four globin subunits in hemoglobin and the one of myoglobin each contain a heme prosthetic group. A **prosthetic group** is a nonpolypeptide moiety that forms a functional part of a protein. Without its prosthetic group, a protein is designated an **apoprotein**. With its prosthetic group it is a **holoprotein**.

In hemoglobins and myoglobin the heme is protoporphyrin IX (p. 791) with an iron atom in its center (Figure 9.20). The iron is in the ferrous (+2 charge) state in functional hemoglobin and myoglobin where it can form five or six covalent bonds, depending on whether or not O_2 is bound. Four bonds are to the pyrrole nitrogen atoms of the porphyrin. As the pyrrole rings and connecting carbons are part of the same aromatic system, these atoms lie in a common plane. The porphyrin bonded Fe^{+2} will tend to lie in this same porphyrin plane. The fifth and the potentially sixth bonds to Fe^{+2} are directed along an axis perpendicular to the plane of the porphyrin ring (Figure 9.21). The fifth bond is to a nitrogen of a histidine imidazole. This is designated the **proximal histidine** in globin structures (Figures 9.21 and 9.22). In oxy-globins, O_2 forms the sixth bond, the O_2 being between the ferrous atom and a second histidine imidazole, designated the **distal histidine**. In deoxy-globins, the sixth position is unoccupied.

The heme is positioned within a hydrophobic pocket of each globin subunit. In the pocket about 18 amino acid residues, mostly between apolar side chains and the apolar regions of the porphyrin, provide approximately 80 interactions. The driving force for the interaction is the expulsion of water of solvation on association of the hydrophobic heme with the apolar side chains in the heme pocket (see hydrophobic forces, p. 116). Additional interactions in myoglobin are made between the negatively charged propionate groups of the heme and positively charged arginine and histidine side chains. In hemoglobin chains a difference in the amino acid sequence in this region of the heme binding site leads to stabilization of the porphyrin propionates by interaction with an uncharged histidine imidazole and with water molecules of solvent toward the outer surface of the molecule.

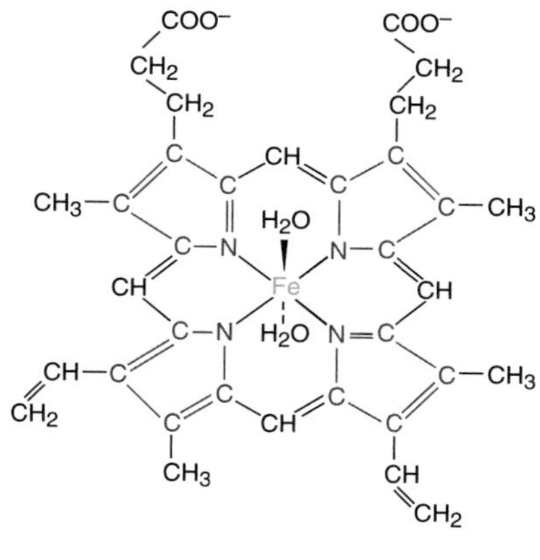

Figure 9.20 Structure of heme. Iron (Fe^{+2}) has four bonds to the pyrrole nitrogens in planar porphyrin. The fifth and six bonds are perpendicular to porphyrin plane. Porphyrin structure is protoporphyrin type IX.

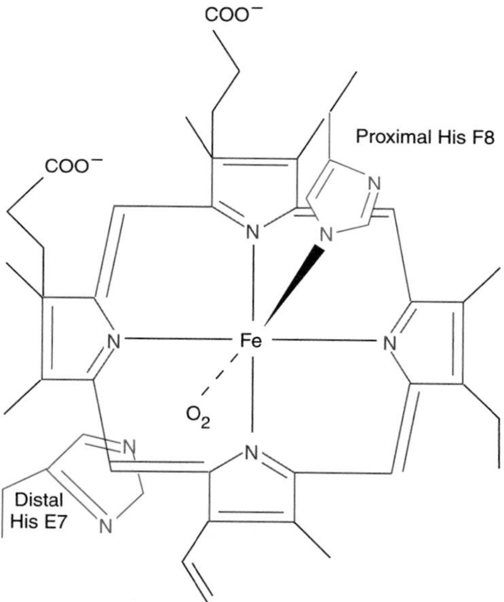

Figure 9.21 Ligand bonds to ferrous atom in oxyhemoglobin.

X-Ray Crystallography Has Defined the Structures of Hemoglobin and Myoglobin

The structures of deoxy and oxy forms of hemoglobin and myoglobin have been resolved by X-ray crystallography. In fact, sperm whale myoglobin was the first protein whose structure was determined by this technique, followed by horse hemoglobin. These structures show that

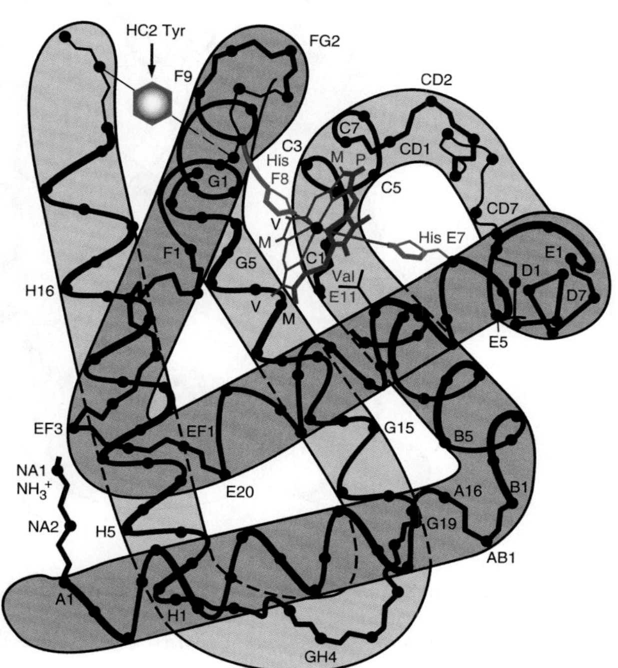

Figure 9.22 Secondary and tertiary structure characteristics of globins of hemoglobin. Proximal His F8, distal His E7, Tyr HC2, and Val E11 side chains are shown. Other amino acids of polypeptide chain are represented by α-carbon positions only; the letters M, V, and P refer to the methyl, vinyl, and propionate side chains of the heme.

Reprinted with permission from Perutz, M. *Br. Med. Bull.* 32:195, 1976.

each globin consists of multiple α-helical regions connected by turns that allow folding into a spheroidal shape characteristic of the globin fold (Figure 9.22 and Figure 3.37, p. 101).

Primary, Secondary, and Tertiary Structures of Myoglobin and Hemoglobin Chains

Comparison of the amino acid sequences of myoglobin from many animal species shows all myoglobin chains contain 153 amino acids, of which 83 are invariant. Only 15 invariant residues are identical to the invariant residues of the globin chains in mammalian hemoglobins. While there is surprising variability in sequences, the secondary and tertiary structures are almost identical (Figure 9.23). In addition, all the globin chains function to reversibly bind oxygen. However, when comparing the myoglobin sequences to the sequences of the globin chains in hemoglobin, the changes are mostly conservative and preserve the physical properties of the residues (Table 9.9). Changes in the sequence between the myoglobin and hemoglobin sequences are partially explained by myoglobin acting as a monomer; hence many of its surface positions interact with water and prevent another molecule of myoglobin from associating. In contrast, surface residues of the individual subunits in hemoglobin provide hydrogen bonds and nonpolar contacts with other subunits in the quaternary structure. The proximal and distal histidines to the heme iron are preserved in all the globin chains, as are the hydrophobic heme pockets that form essential nonpolar contacts with the heme.

Any significant differences in the physiological properties between α, β, γ, and δ chains of hemoglobins and the chain of myoglobin are due to rather small specific changes in their structures. The similarity in tertiary structure shows that the same fold structure of a polypeptide can be arrived at by different sequences.

Approximately 70% of the residues participate in the α-helices of the globin fold, generating seven α-helices in the α chain and eight in the β chain. These latter eight helical regions are commonly labeled A–H, starting from the NH_2-terminal end. The interhelical regions are designated as AB, BC, CD, . . . , GH, respectively. The nonhelical region at the NH_2-terminal end is designated NA, and that at the COOH-terminal end is HC (Figure 9.22). Amino

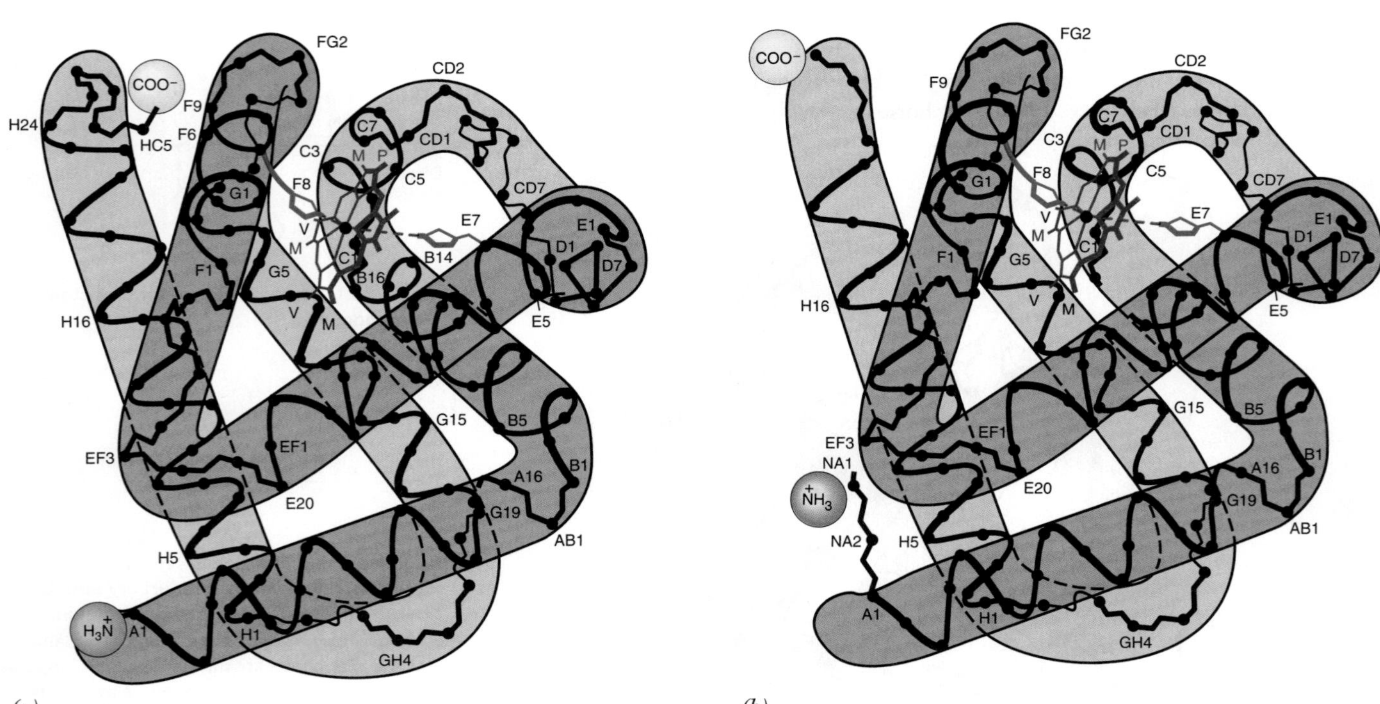

Figure 9.23 Comparison of conformation of (*a*) myoglobin and (*b*) β chain of HbA_1. Overall structures are very similar, except at NH_2-terminal and COOH-terminal ends.
Reprinted with permission from Fersht, A. *Enzyme Structure and Mechanism.* San Francisco: Freeman, 1977, 12 and 13.

TABLE 9.9 • Amino Acid Sequences of Hemoglobin Chains and Myoglobin[a]

	NA 1	2	3	A 1	2	3	4	5	6	7	8	9	10	11	12	13	14	15	A 16	AB 1	B 1	2	3	4	5	6
MYOGLOBIN	Val	. . .	Leu	Ser	Glu	Gly	Glu	Trp	Gln	Leu	Val	Leu	His	Val	Trp	Ala	Lys	Val	Glu	Ala	Asp	Val	Ala	Gly	His	Gly
HEMOGLOBIN Horse α	Val	. . .	Leu	Ser	Ala	Ala	Asp	Lys	Thr	Asn	Val	Lys	Ala	Ala	Trp	Ser	Lys	Val	Gly	Gly	His	Ala	Gly	Glu	Tyr	Gly
β	Val	Gln	Leu	Ser	Gly	Glu	Glu	Lys	Ala	Ala	Val	Leu	Ala	Leu	Trp	Asp	Lys	Val	Asn	. . .	. . .	Glu	Glu	Glu	Val	Gly
Human α	Val	. . .	Leu	Ser	Pro	Ala	Asp	Lys	Thr	Asn	Val	Lys	Ala	Ala	Trp	Gly	Lys	Val	Gly	Ala	His	Ala	Gly	Glu	Tyr	Gly
β	Val	His	Leu	Thr	Pro	Glu	Glu	Lys	Ser	Ala	Val	Thr	Ala	Leu	Trp	Gly	Lys	Val	Asn	. . .	. . .	Val	Asp	Glu	Val	Gly
γ	Gly	His	Phe	Thr	Glu	Glu	Asp	Lys	Ala	Thr	Ilu	Thr	Ser	Leu	Trp	Gly	Lys	Val	Asn	. . .	. . .	Val	Glu	Asp	Ala	Gly
δ	Val	His	Leu	Thr	Pro	Glu	Glu	Lys	Thr	Ala	Val	Asn	Ala	Leu	Trp	Gly	Lys	Val	Asn	. . .	. . .	Val	Asp	Ala	Val	Gly

	7	8	9	10	11	12	13	14	15	16	C1	2	3	4	5	6	7	CD1	2	3	4	5	6	7	8	D1
MYOGLOBIN	Gln	Asp	Ilu	Leu	Ilu	Arg	Leu	Phe	Lys	Ser	His	Pro	Glu	Thr	Leu	Glu	Lys	Phe	Asp	Arg	Phe	Lys	His	Leu	Lys	Thr
HEMOGLOBIN Horse α	Ala	Glu	Ala	Leu	Glu	Arg	Met	Phe	Leu	Gly	Phe	Pro	Thr	Thr	Lys	Thr	Tyr	Phe	Pro	His	Phe	. . .	Asp	Leu	Ser	His
β	Gly	Glu	Ala	Leu	Gly	Arg	Leu	Leu	Val	Val	Tyr	Pro	Trp	Thr	Gln	Arg	Phe	Phe	Asp	Ser	Phe	Gly	Asp	Leu	Ser	Gly
Human α	Ala	Glu	Ala	Leu	Glu	Arg	Met	Phe	Leu	Ser	Phe	Pro	Thr	Thr	Lys	Thr	Tyr	Phe	Pro	His	Phe	. . .	Asp	Leu	Ser	His
β	Gly	Glu	Ala	Leu	Gly	Arg	Leu	Leu	Val	Val	Tyr	Pro	Trp	Thr	Gln	Arg	Phe	Phe	Glu	Ser	Phe	Gly	Asp	Leu	Ser	Thr
γ	Gly	Glu	Thr	Leu	Gly	Arg	Leu	Leu	Val	Val	Tyr	Pro	Trp	Thr	Gln	Arg	Phe	Phe	Asp	Ser	Phe	Gly	Asn	Leu	Ser	Ser
δ	Gly	Glu	Ala	Leu	Gly	Arg	Leu	Leu	Val	Val	Tyr	Pro	Trp	Thr	Gln	Arg	Phe	Phe	Glu	Ser	Phe	Gly	Asp	Leu	Ser	Ser

	2	3	4	5	6	7	E 1	2	3	4	5	6	7	8	9	10	11	12	13	14	E 15	16	17	18	19	20
MYOGLOBIN	Glu	Ala	Glu	Met	Lys	Ala	Ser	Glu	Asp	Leu	Lys	Lys	His	Gly	Val	Thr	Val	Leu	Thr	Ala	Leu	Gly	Ala	Ilu	Leu	Lys
HEMOGLOBIN Horse α	. . .	. . .	. . .	. . .	. . .	Gly	Ser	Ala	Gln	Val	Lys	Ala	His	Gly	Lys	Lys	Val	Ala	Asp	Gly	Leu	Thr	Leu	Ala	Val	Gly
β	Pro	Asp	Ala	Val	Met	Gly	Asn	Pro	Lys	Val	Lys	Ala	His	Gly	Lys	Lys	Val	Leu	His	Ser	Phe	Gly	Glu	Gly	Val	His
Human α	. . .	. . .	. . .	. . .	. . .	Gly	Ser	Ala	Gln	Val	Lys	Gly	His	Gly	Lys	Lys	Val	Ala	Asp	Ala	Leu	Thr	Asn	Ala	Val	Ala
β	Pro	Asp	Ala	Val	Met	Gly	Asn	Pro	Lys	Val	Lys	Ala	His	Gly	Lys	Lys	Val	Leu	Gly	Ala	Phe	Ser	Asp	Gly	Leu	Ala
γ	Ala	Der	Ala	Ilu	Met	Gly	Asn	Pro	Lys	Val	Lys	Ala	His	Gly	Lys	Lys	Val	Leu	Thr	Ser	Leu	Gly	Asp	Ala	Ilu	Lys
δ	Pro	Asp	Ala	Val	Met	Gly	Asn	Pro	Lys	Val	Lys	Ala	His	Gly	Lys	Lys	Val	Leu	Gly	Ala	Phe	Ser	Asp	Gly	Leu	Ala

	EF1	2	3	4	5	6	7	8	F1	2	3	4	F5	6	7	8	9	FG1	2	3	4	5	G1	2	3	4
MYOGLOBIN	Lys	Lys	Gly	His	His	Glu	Ala	Glu	Leu	Lys	Pro	Leu	Ala	Gln	Ser	His	Ala	Thr	Lys	His	Lys	Ilu	Pro	Ilu	Lys	Tyr
HEMOGLOBIN Horse α	His	Leu	Asp	Asp	Leu	Pro	Gly	Ala	Leu	Ser	Asp	Leu	Ser	Asn	Leu	His	Ala	His	Lys	Leu	Arg	Val	Asp	Pro	Val	Asn
β	His	Leu	Asp	Asn	Leu	Lys	Gly	Thr	Phe	Ala	Ala	Leu	Ser	Glu	Leu	His	Cys	Asp	Lys	Leu	His	Val	Asp	Pro	Glu	Asn
Human α	His	Val	Asp	Asp	Met	Pro	Asn	Ala	Leu	Ser	Ala	Leu	Ser	Asp	Leu	His	Ala	His	Lys	Leu	Arg	Val	Asp	Pro	Val	Asn
β	His	Leu	Asp	Asn	Leu	Lys	Gly	Thr	Phe	Ala	Thr	Leu	Ser	Glu	Leu	His	Cys	Asp	Lys	Leu	His	Val	Asp	Pro	Glu	Asn
γ	His	Leu	Asp	Asp	Leu	Lys	Gly	Thr	Phe	Ala	Gln	Leu	Ser	Glu	Leu	His	Cys	Asp	Lys	Leu	His	Val	Asp	Pro	Glu	Asn
δ	His	Leu	Asn	Asp	Leu	Lys	Gly	Thr	Phe	Ser	Gln	Leu	Ser	Glu	Leu	His	Cys	Asp	Lys	Leu	His	Val	Asp	Pro	Glu	Asn

	5	6	7	8	G 9	10	11	12	13	14	15	16	17	18	19	GH 1	2	3	4	5	6	H 1	2	H 3	4	5
MYOGLOBIN	Leu	Glu	Phe	Ilu	Ser	Glu	Ala	Ilu	Ilu	His	Val	Leu	His	Ser	Arg	His	Pro	Gly	Asn	Phe	Gly	Ala	Asp	Ala	Gln	Gly
HEMOGLOBIN Horse α	Phe	Lys	Leu	Leu	Ser	His	Cys	Leu	Leu	Ser	Thr	Leu	Ala	Val	His	Leu	Pro	Asn	Asp	Phe	Thr	Pro	Ala	Val	His	Ala
β	Phe	Arg	Leu	Leu	Gly	Asn	Val	Leu	Ala	Leu	Val	Val	Ala	Arg	His	Phe	Gly	Lys	Asp	Phe	Thr	Pro	Glu	Leu	Gln	Ala
Human α	Phe	Lys	Leu	Leu	Ser	His	Cys	Leu	Leu	Val	Thr	Leu	Ala	Ala	His	Leu	Pro	Ala	Glu	Phe	Thr	Pro	Ala	Val	His	Ala
β	Phe	Srg	Leu	Leu	Gly	Asn	Val	Leu	Val	Cys	Val	Leu	Ala	His	His	Phe	Gly	Lys	Glu	Phe	Thr	Pro	Pro	Val	Gln	Ala
γ	Phe	Lys	Leu	Leu	Gly	Asn	Val	Leu	Val	Thr	Val	Leu	Ala	Ilu	His	Phe	Gly	Lys	Glu	Phe	Thr	Pro	Glu	Val	Gln	Ala
δ	Phe	Arg	Leu	Leu	Gly	Asn	Val	Leu	Val	Cys	Val	Leu	Ala	Arg	Asn	Phe	Gly	Lys	Glu	Phe	Thr	Pro	Gln	Met	Gln	Ala

	6	7	8	9	10	11	12	13	14	15	16	17	18	19	20	H 21	22	23	2	HC 1	2	3	4	5
MYOGLOBIN	Ala	Met	Asn	Lys	Ala	Leu	Glu	Leu	Phe	Arg	Lys	Asp	Ilu	Ala	Ala	Lys	Tyr	Lys	Glu	Leu	Gly	Tyr	Gln	Gly
HEMOGLOBIN Horse α	Ser	Leu	Asp	Lys	Phe	Leu	Ser	Ser	Val	Ser	Thr	Val	Leu	Thr	Ser	Lys	Tyr	Arg						
β	Ser	Tyr	Gln	Lys	Val	Val	Ala	Gly	Val	Ala	Asn	Ala	Leu	Ala	His	Lys	Tyr	His						
Human α	Ser	Leu	Asp	Lys	Phe	Leu	Ala	Ser	Val	Ser	Thr	Val	Leu	Thr	Ser	Lys	Tyr	Arg						
β	Ala	Tyr	Gln	Lys	Val	Val	Ala	Gly	Val	Ala	Asn	Ala	Leu	Ala	His	Lys	Tyr	His						
γ	Ser	Trp	Gln	Lys	Met	Val	Thr	Gly	Val	Ala	Ser	Ala	Leu	Ser	Ser	Arg	Tyr	His						
δ	Ala	Tyr	Gln	Lys	Val	Val	Ala	Gly	Val	Ala	Asn	Ala	Leu	Ala	His	Lys	Tyr	His						

Source: Based on diagram in Dickerson, R. E. and Geis, I. *The Structure and Function of Proteins.* New York: Harper & Row, 1969, p. 52.

[a]Myoglobin from sperm whale. Residues that are identical are enclosed in box. A, B, C, . . . designate different helices of tertiary structure (see text).

acid residues are sequentially numbered from 1 in each of the lettered helices and interhelical regions. This numbering system allows discussion of particular regions and residues that have similar functional and structural roles in hemoglobin and myoglobin.

A Simple Equilibrium Defines O_2 Binding to Myoglobin

The association of oxygen with myoglobin is characterized by a simple equilibrium constant (Eqs, 9.1 and 9.2). [MbO_2] is the solution concentration of oxymyoglobin, [Mb] is deoxymyoglobin, and [O_2] is oxygen, all in units of moles/liter. The equilibrium constant,

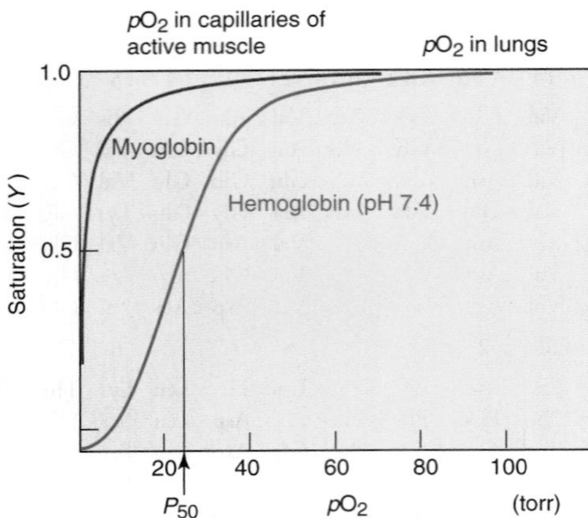

Figure 9.24 Oxygen-binding curves for myoglobin and hemoglobin. The plots place the myoglobin curve near the zero pO_2 value in this coordinate system. If the pO_2-axis was expanded around the region of the myoglobin P_{50}, the lower sensitivity of the change in Y in myoglobin to fold-change in oxygen concentration to that of hemoglobin (see Cooperativity Index, Table 9.10) would be demonstrated. The index, based on the observed Hill coefficients, shows that myoglobin changes from $Y = 0.1$ to $Y = 0.9$ over a 81-fold change in oxygen concentration while hemoglobin takes only a 4.8-fold change in oxygen concentration.

K_{eq}, has units of moles/liter. The value of K_{eq} is dependent on pH, ionic strength, and temperature.

$$\mathrm{Mb} + \mathrm{O_2} \underset{}{\overset{K_{eq}}{\rightleftharpoons}} \mathrm{MbO_2} \tag{9.1}$$

$$K_{eq} = \frac{[\mathrm{Mb}][\mathrm{O_2}]}{[\mathrm{MbO_2}]} \tag{9.2}$$

Since oxygen is a gas, it is convenient to express its concentration as the pressure of oxygen in torr (1 torr equals the pressure of 1 mm Hg at 0°C and standard gravity). In Eq. 9.3 this conversion of units has been made as shown by the change in symbols: P_{50}, the equilibrium constant, and pO_2, the concentration of oxygen, being expressed in torr.

$$P_{50} = \frac{[\mathrm{Mb}] \cdot pO_2}{[\mathrm{MbO_2}]} \tag{9.3}$$

In an oxygen binding or saturation curve the fraction of oxygen-binding sites that contain oxygen (Y, Eq. 9.4) is plotted on the ordinate *versus* pO_2 (oxygen concentration) on the abscissa. The Y-value is defined for myoglobin by Eq. 9.5. Substitution into Eq. 9.5 of the value of $[\mathrm{MbO_2}]$ obtained from Eq. 9.3, and dividing through by [Mb] results in Eq, 9.6, which shows the dependence of Y on the value of the equilibrium constant, P_{50}, and on pO_2. From Eqs. 9.3 and 9.6, the value of P_{50} is equal to pO_2 when $Y = 0.5$ (50% of the available sites occupied), hence the designation of the equilibrium constant as P_{50}.

$$Y = \frac{\text{number of binding sites occupied}}{\text{total number of binding sites in solution}} \tag{9.4}$$

$$Y = \frac{[\mathrm{MbO_2}]}{[\mathrm{Mb}] + [\mathrm{MbO_2}]} \tag{9.5}$$

$$Y = \frac{pO_2}{P_{50} + pO_2} \tag{9.6}$$

A plot of Eq. 9.6 for Y versus pO_2 generates an oxygen-saturation curve for myoglobin and has the form of a rectangular hyperbola (Figure 9.24).

A simple algebraic manipulation of Eq. 9.6 leads to Eq. 9.7. Taking the logarithm of both sides of Eq. 9.7 results in Eq. 9.8, the **Hill equation**. A plot of $\log[Y/(1 - Y)]$ versus $\log pO_2$, according to Eq. 9.8, yields a straight line with a slope equal to 1 for myoglobin (Figure 9.25). This is the Hill plot, and the slope (n_H) is the **Hill coefficient** (Eq. 9.9).

$$\frac{Y}{1 - Y} = \frac{pO_2}{P_{50}} \tag{9.7}$$

$$\log \frac{Y}{1 - Y} = \log pO_2 - \log P_{50} \tag{9.8}$$

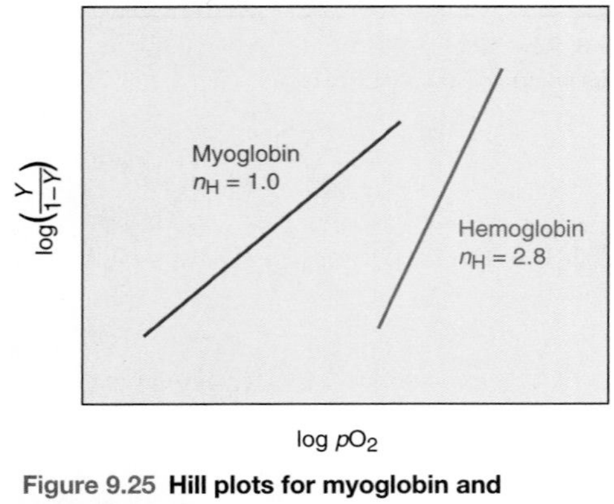

Myoglobin $n_H = 1.0$

Hemoglobin $n_H = 2.8$

$\log\left(\frac{Y}{1-Y}\right)$

$\log pO_2$

Figure 9.25 Hill plots for myoglobin and hemoglobin A_1.

Binding of O_2 to Hemoglobin Involves Cooperativity between Subunits

Binding of four O_2 to hemoglobin manifests as **positive cooperativity**, since binding of the first O_2 facilitates binding of O_2 to the other subunits. Conversely, dissociation of the first O_2 from $\mathrm{Hb(O_2)_4}$ will make dissociation of O_2 from the other subunits easier.

Because of this cooperativity, the oxygen-saturation curve for hemoglobin differs from that for myoglobin. A plot of Y versus pO_2 for hemoglobin is sigmoidal, indicating cooperativity in oxygen association (Figure 9.24). A plot of the Hill equation (Eq. 9.9) gives a slope (n_H) equal to 2.8 (Figure 9.25). Values of $n_H = 1.0$ show no cooperativity in substrate binding to multiactive site proteins. Values of $n_H > 1.0$ show positive cooperativity with higher values correlating with greater positive cooperativity in binding. Values of n_H greater than 0 and less than 1.0 show negative cooperativity, with values closer to 0 having higher negative cooperativity.

$$\log \frac{Y}{1-Y} = n_{\mathrm{H}} \log pO_2 - \text{constant} \tag{9.9}$$

The meaning of the Hill coefficient to cooperative O_2 association is evaluated quantitatively as presented in Table 9.10. A **cooperativity index**, *Rx*, is calculated, which shows the fold change of pO_2 required to change Y from a value of $Y = 0.1$ (10% of sites filled) to a value of $Y = 0.9$ (90% of sites filled) for Hill coefficient values found experimentally. For myoglobin $n_{\mathrm{H}} = 1$ and an 81-fold change in oxygen concentration is required to change from $Y = 0.1$ to $Y = 0.9$. For hemoglobin, $n_{\mathrm{H}} = 2.8$, and only a 4.8-fold change in oxygen concentration is required for the same change in Y.

TABLE 9.10 • Relationship between Hill Coefficient (n_{H}) and Cooperativity Index (R_x)

n_{H}	R_x	*Observation*
0.5	6560	Negative substrate cooperativity
0.6	1520	
0.7	533	
0.8	243	
0.9	132	
1.0	81.0	Noncooperativity
1.5	18.7	Positive substrate cooperativity
2.0	9.0	
2.8	4.8	
3.5	3.5	
6.0	2.1	
10.0	1.6	
20.0	1.3	

Source: Based on Table 7.1 in Cornish-Bowden, A. *Principles of Enzyme Kinetics*. London: Butterworths Scientific Publishers, 1976.

Molecular Mechanism of Cooperativity in O_2 Binding

X-ray diffraction data on deoxyhemoglobin show that the ferrous atoms actually sit out of the plane of their porphyrins by about 0.4–0.6 Å. The electronic configuration of the five-coordinated ferrous atom in deoxyhemoglobin has a slightly larger radius than the distance from the center of the porphyrin to each of the pyrrole nitrogen atoms. Accordingly, the iron can be placed in the center of the porphyrin only with some distortion of the porphyrin conformation. Probably more important is that if the iron atom sits in the plane of the porphyrin, the proximal His F8 imidazole will interact unfavorably with the porphyrin. This unfavorable steric interaction is due, in part, to conformational constraints on the His F8 and the porphyrin in the deoxyhemoglobin conformation that forces the approach of the His F8 toward the porphyrin to a particular path (Figure 9.26). These constraints become less significant in oxyhemoglobin. With the iron atom out of the plane of the porphyrin, the conformation is unstrained and energetically favored for the five-coordinated ferrous atom. When O_2 forms the sixth bond of the iron, however, this conformation becomes strained. A more energetically favorable conformation for the O_2-bonded iron is that in which the iron atom is within the plane of the porphyrin.

The binding energy of O_2 overcomes the repulsive interaction between His F8 and porphyrin, and the ferrous atom moves into the plane of the porphyrin. This is the most thermodynamically stable position for the oxy-heme, the six-bonded iron atom having one axial ligand on either side of the plane of the porphyrin, and the steric repulsion of one axial ligand is balanced by the repulsion of the second axial ligand on the opposite side when the ferrous atom is in the center. Also, the radius of the iron atom with six bonds is reduced

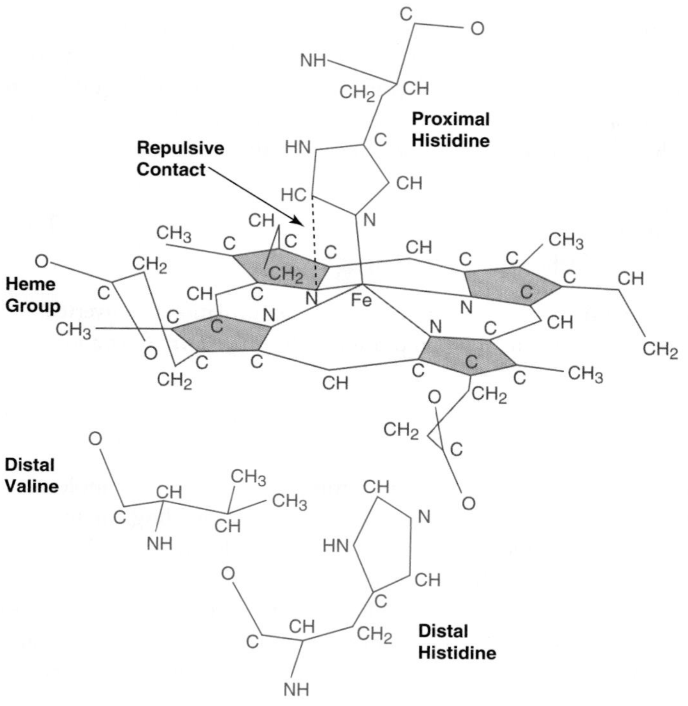

Figure 9.26 Steric hindrance between proximal histidine and porphyrin in deoxyhemoglobin. Redrawn from Perutz, M. *Sci. Am.* 239:92, 1978.

so that it just fits into the center of the porphyrin without distortion of the porphyrin conformation.

Since steric repulsion between porphyrin and His F8 in the deoxy-conformation must be overcome on O_2 association, binding of the first O_2 is characterized by a relatively low affinity constant. However, on binding of O_2 to the first heme, the movement of the iron into the center of the porphyrin triggers a conformational change in other subunits of the hemoglobin molecule. This change results in a greater affinity of O_2 to other empty heme sites.

The conformation change involves the following interconnected sequential events: (1) The binding of O_2 pulls the Fe^{2+} into the porphyrin plane. (2) The movement of the Fe^{+2} moves the bonded proximal His F8 toward the porphyrin. (3) The position of the covalently connected F-helix of which the His F8 is a part consequently changes. (4) The interconnected FG corner of the subunit changes position; this destabilizes the FG corner's noncovalent interaction with the C-helix of the adjacent subunit at $\alpha_1\beta_2$ or $\alpha_2\beta_1$ subunit interfaces (Figures 9.27 and 9.28). (5) The adjacent globin structure changes to a conformation that allows O_2 to bind to its heme with high affinity.

Thus, the FG to C inter-subunit contacts act as a *switch* between two different arrangements between the FG corner of one subunit and the C-helix of the adjacent subunit. The new conformation allows the His F8 residues in the deoxy globin chains to approach their porphyrins on O_2 association with less steric repulsion (Figure 9.28). O_2 binds to the empty hemes in the modified conformation with higher affinity. In addition, ionic interactions that stabilize the deoxy-conformation (Figure 9.29) are broken by the conformation change induced by the binding of O_2 to one of the hemes, allowing the oxy-conformation in other subunits to more easily form.

The deoxy-conformation of hemoglobin is the *tense*, or **T conformational state**. The oxy form with higher O_2 affinity is the *relaxed*, or **R conformational state**. The allosteric mechanism describes how initial binding of the O_2 to one of the heme subunits of the tetrameric molecule pushes the molecular conformation from the T to R conformational state. The affinity constant for O_2 is greater in the R state by a factor of 150–300, depending on the solution conditions.

Hemoglobin Facilitates Transport of CO_2 and NO

The survival of cells relies on the ability of hemoglobin in red blood cells to deliver O_2 for cellular metabolism and to facilitate the transport of CO_2 away from the cells to the lung. Hemoglobin also binds the vasodilator, NO, and delivers it to the blood vessel walls in the tissues. The T to R conformational equilibrium of the hemoglobin molecule controls the delivery of O_2, CO_2, and NO to their appropriate sites. Due to changes in the p*Ka* of side-chain acid groups between the T and R conformations, the T $\leftrightarrows$ R equilibrium is regulated by the hydrogen ion concentration. The changes in pH link the delivery of O_2 and NO to the tissues with the transport of CO_2 away from the tissues.

Decrease in pK_a of Acid Groups with Change from T to R Conformation Releases Protons

Equation 9.10 shows the release of protons as the T conformation is converted to the R conformation. This release of protons at blood pH is the alkaline **Bohr effect** and is of physiological importance.

$$\underset{T}{Hb} + 4O_2 \rightleftharpoons \underset{R}{Hb(O_2)_4} + nH^+ \tag{9.10}$$

The value of n in Eq. 9.10 (equivalents of protons released when one molecule of deoxy-Hb is converted to oxy-Hb) varies between 1.2 and 2.7 depending on solution conditions and the concentration of chloride ion and 2,3-bisphosphoglycerate. According to the law of mass action, increasing the hydrogen ion concentration on the right side of Eq. 9.10 forces the equilibrium to the left, toward increasing concentrations of deoxy-Hb and free O_2. Increased cellular metabolic activity increases release of both carbonic acid from CO_2 and lactic acid (p. 601), and thereby increases the hydrogen ion concentration (acidity) of

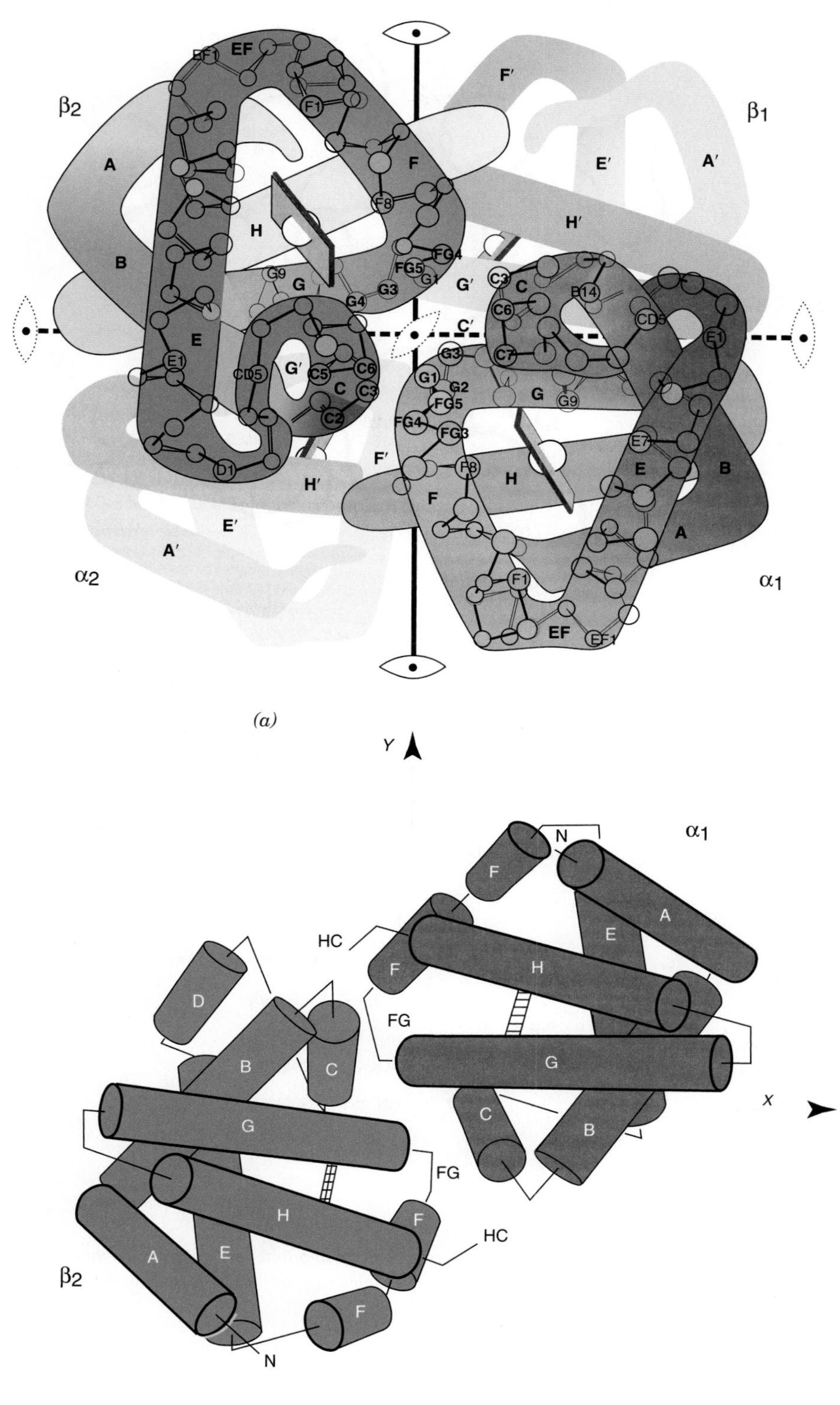

Figure 9.27 Quaternary structure of hemoglobin showing FG corner C helix interactions across α_1–β_2 interface. (*a*) α_1–β_2 interface contacts between FG corners and C helix are shown. (*b*) Cylinder representation of α_1 and β_2 subunits in hemoglobin showing α_1–β_2 interface contacts between FG corner and C helix, viewed from opposite side of *xy*-plane from (*a*).

(*a*) reproduced with permission from Dickerson, R. E., and Geis, I. *The Structure and Action of Proteins.* Menlo Park, CA: Benjamin, 1969, 56. (*b*) Reproduced with permission from Baldwin, J., and Chothia, C. J. *Mol. Biol.* 129:175, 1979. Copyright (1979) by Academic Press. Inc. [London] Ltd.

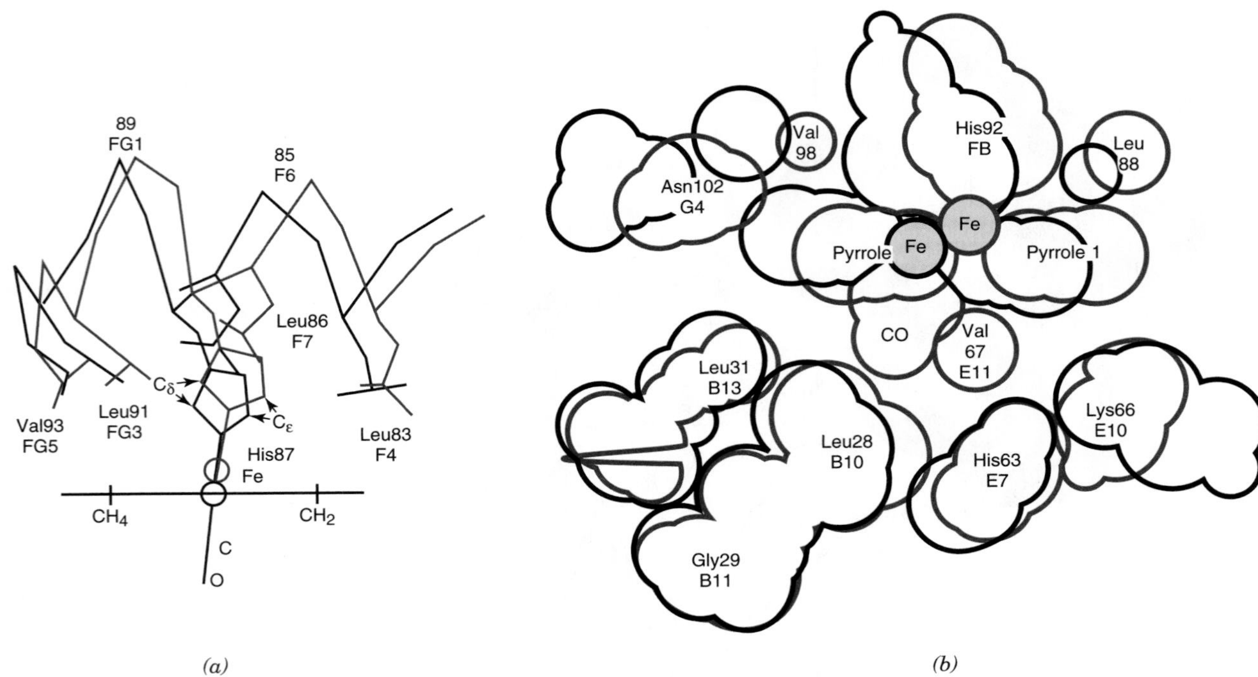

Figure 9.28 Stick and space-filling diagrams drawn by computer graphics showing movements of residues in heme environment on transition from deoxyhemoglobin into oxyhemoglobin.
(*a*) Black line outlines position of polypeptide chain and His F8 in carbon monoxide hemoglobin, a model for oxyhemoglobin. Red line outlines the positions of the same regions for deoxyhemoglobin. Position of iron atom shown by circle. Movements are for an α subunit. (*b*) Similar movements in amino acid residue position in a β subunit using space-filling diagram shown. Residue labels centered in density for the deoxy conformation.
Redrawn with permission from Baldwin, J., and Chothia, C. J. *Mol. Biol.* 129:175, 1979. Copyright © 1979 by Academic Press, Inc. [London] Ltd.

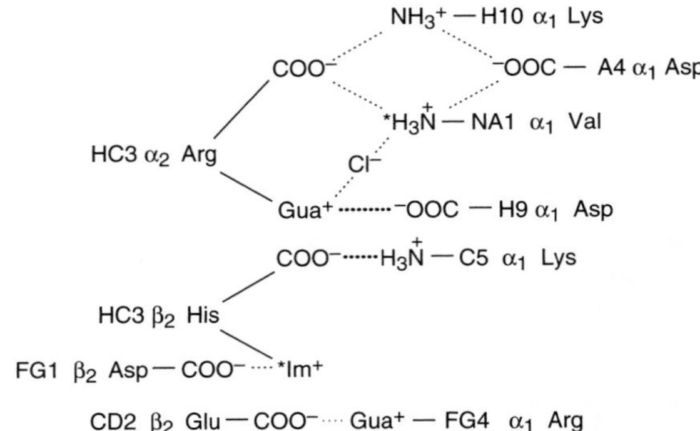

Figure 9.29 Salt bridges between subunits in deoxyhemoglobin that are broken in oxyhemoglobin.
Im^+ is imidazolium, Gua^+ is guanidinium, and starred residues account for approximately 60% of alkaline Bohr effect.
Redrawn from Perutz, M. *Br. Med. Bull.* 32:195, 1976.

the blood in the cellular environment. The increase in acidity forces oxy-Hb to deoxy-Hb with the dissociation of O_2, which is delivered to the metabolizing cells. This is a feedback loop in which the cellular metabolic acid waste signals an increase in the dissociation of O_2 required for the continuation of cellular metabolism. The increase in hydrogen ion concentration induces O_2 delivery through its effect on the equilibrium between the R and T conformations of the hemoglobin molecule.

Why do protons dissociate from hemoglobin as the T conformation converts to the R conformation? Protons dissociate when acid side chains of hemoglobin are more acidic in the R conformation than in the T conformation. This requires the pK_a of these groups to be lower in the R conformation.

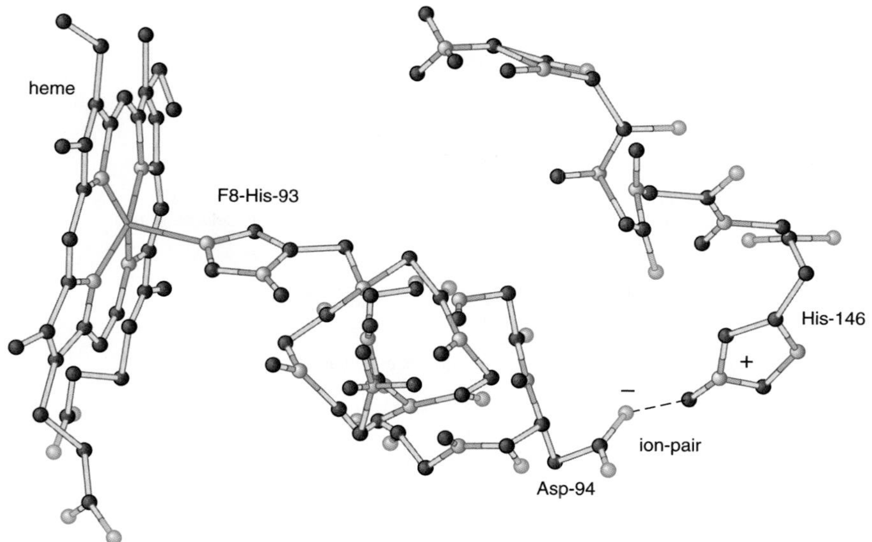

Figure 9.30 Ion pair between the βHis-146 imidazolium and the βAsp-94 carboxylate side-chain groups in the deoxy(T)-conformation of hemoglobin. A partial structure of the β-chain showing the backbone polypeptide chain for amino acids 87 through 95 and 142 through 146 of the β-chain in the deoxy-conformation. Only the side chains of Asp-94, His-146, and the F8 His-92 liganded to the β-subunit heme are shown. Oxygen atoms are colored red, nitrogen atoms are blue, and carbon and hydrogen atoms are black. The hydrogen bond (broken line) is shown between the positively charged imidazolium N-H of βHis-146 and negatively charged carboxylate oxygen of βAsp-94.

Drawing made with the Swiss-PdbViewer using PDB structure 1A3N.

His-146(β) is the major contributor to this proton dissociation on conversion of the T to R conformation. In the T conformation, the positive charge on the His-146(β) side-chain imidazolium is in an ion pair with the negative charge of the carboxylate side chain of the Asp-94(β) (Figure 9.30). The negatively charged carboxylate stabilizes the positively charged imidazolium. The stabilized imidazolium tends to retain its positive charge and has a higher than normal pK_a (approximately 7.7). Conversion of hemoglobin to the R conformation breaks this ion pair, placing the Asp and His side chains at new positions in the R conformation where the oppositely charged groups no longer strongly interact, and this loss of opposite charge interaction reverts the His-146(β) imidazolium to a more normal pK_a of approximately 7.3. The change in the His-146(β) pK_a to a more acidic value on conversion to the R conformation results in the dissociation of protons at the pH of blood (7.4). Approximately 50% of the protons that dissociate come from the β-His-146 midazolium. The rest come from other acid groups that similarly change their p*Ka* from higher to lower values as the conformation changes from T to R.

Figure 9.31 shows the plot of the fractional saturation of O_2 binding sites in hemoglobin versus O_2 concentration for Hb at different values of pH. At more acidic pH (higher [H^+]), hemoglobin dissociates its O_2 more easily (Eq. 9.10) and the curve is shifted to the right. Accordingly, the P_{50} value at more acidic pH is higher. The higher P_{50} represents a poorer O_2 affinity as oxygen-binding sites are 50% saturated at a higher $p(O_2)$ concentration. At higher pH, the curve is shifted to the left and the P_{50} value decreases reflecting a higher affinity of O_2.

CO_2 and O_2 Transport Are Linked by Bohr-Effect Protons

Metabolizing cells utilize O_2 and produce CO_2. The CO_2 produced by the cells diffuses into the surrounding blood and enters the red blood cell. There the CO_2 is rapidly converted by the action of the enzyme carbonic anhydrase to carbonic acid by a hydration reaction (Eq. 9.11).

$$CO_2 + H_2O \rightleftharpoons H_2CO_3 \tag{9.11}$$

$$H_2CO_3 \rightleftharpoons HCO_3^- + H^+ \tag{9.12}$$

Subsequently, the carbonic acid spontaneously dissociates to HCO_3^- and a H^+ (Eq. 9.12). The HCO_3^- diffuses out of the red blood cell and is carried to the lung by the plasma (Figure 9.32). This transport of CO_2 from tissues to lung in plasma as bicarbonate (HCO_3^-) is known as **isohydric transport**. Approximately 70%–80% of CO_2 produced by cells is transported to the lung by this mechanism.

The hydrogen ions that dissociate from the carbonic acid bind to hemoglobin, according to Eq. 9.10, and force $Hb(O_2)_4$ to dissociate its O_2, which diffuses out of the red blood cell to the cells of the tissue. Hemoglobin thus acts as a buffer of blood pH by binding the hydrogen ions produced by the metabolizing cells and preventing the blood pH from

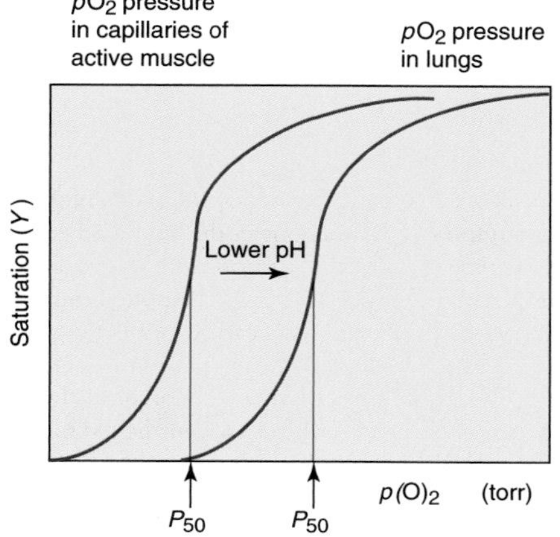

Figure 9.31 Change in oxygen–hemoglobin saturation curve to higher P_{50} value with decrease in pH (increase in [H^+]).

(a) Red Blood Cell in Capillaries of Tissues

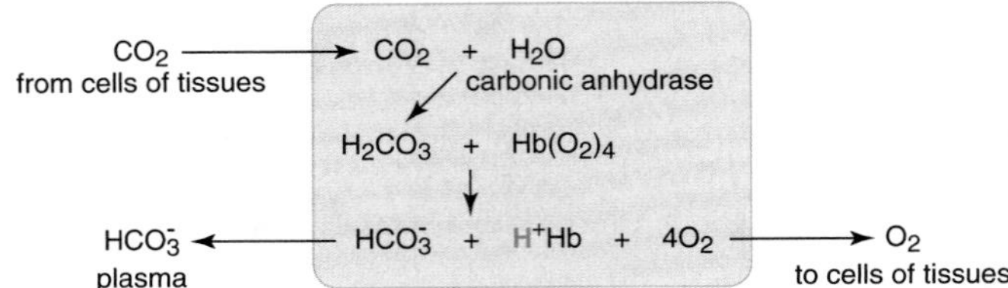

(b) Red Blood Cell in Capillaries of Lung

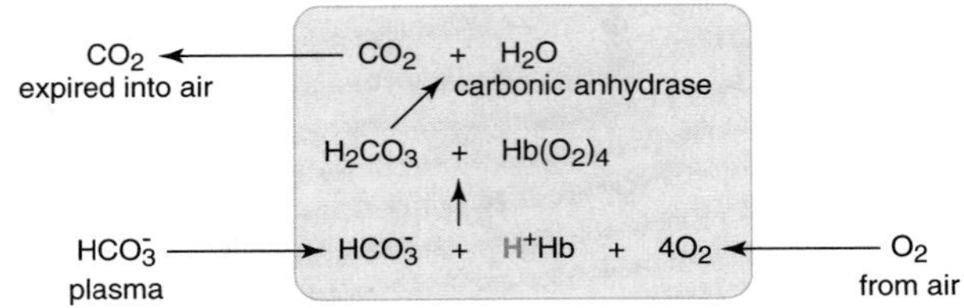

Figure 9.32 The isohydric transport of CO_2 as bicarbonate. *(a)* Reactions in the red blood cell at the tissues. CO_2 diffuses into red blood cell from tissues and is converted into carbonic acid by carbon anhydrase (Eq. 9.11). The carbonic acid spontaneously dissociates into H^+ and HCO_3^- (Eq. 9.12). The proton that dissociates from carbonic acid (boldface H^+) binds to deoxy-Hb forcing the O_2–Hb equilibrium from oxy-hemoglobin to deoxy-hemoglobin (Eq. 9.10) with the dissociation of O_2, which diffuses out of the red blood cell to the tissues. The HCO_3^- diffuses out of the red blood cell and is transported in plasma to the lung (outside of cell). *(b)* Reactions in the red blood at the level of the lung. In the lung, the high O_2 pressure forces the reactions in the opposite direction. Reactions are the reverse of those in the capillaries.

becoming too acidic. The protons bind side chains such as the imidazole of the His-146(β) in the two β chains. The deoxy $H^+{\cdot}Hb$ is transported to the lungs in the red blood cell.

In the lungs the same reactions that occurred at the tissues are now forced in the reverse direction by the high concentration of O_2 in the lung. The high pO_2 concentration converts $H^+Hb(T)$ to $Hb(O_2)_4(R)$ conformation, the protons dissociate (Eq. 9.10), and combine with HCO_3^- molecules that diffuse back into the red blood cells from the plasma to form carbonic acid (H_2CO_3, reverse reaction of Eq. 9.12). The carbonic anhydrase of the red blood cell converts H_2CO_3 to CO_2 and H_2O (reverse of Eq. 9.11). The CO_2 diffuses into the lung alveoli and is expired into air (Figure 9.32).

A second mechanism of CO_2 transport, which accounts for 15%–20% of the CO_2 transported from tissues to lungs, is through **carbamino-hemoglobin**, formed by the nonenzymatic reaction of CO_2 with the NH_2-terminal amino groups of Hb polypeptide chains (Eq. 9.13). This reaction producing carbamino-Hb also produces a H^+.

$$HbNH_2 + CO_2 \rightleftharpoons HbNHCO_2^- + H^+ \qquad (9.13)$$

The excess H^+ produced by carbamino-Hb formation then binds to hemoglobin and stabilizes the deoxy form, promoting the release and delivery of O_2 to the actively metabolizing cells (Eq. 9.10 and Figure 9.33).

In the lung, the high pO_2 concentration generates $Hb(O_2)_4$ with dissociation of H^+ (Eq. 9.10). This increase in H^+ forces dissociation of the carbamino group with release of CO_2 (reverse of Eq. 9.13), which is expired from the lung (Figure 9.33).

(a) Red Blood Cell in Capillaries of Tissues

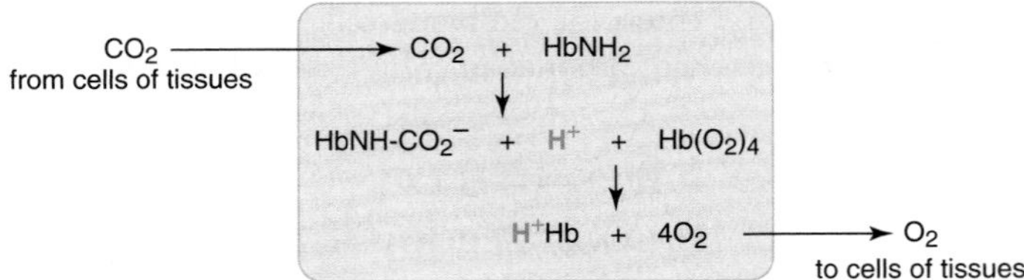

(b) Red Blood Cell in Capillaries of Lung

CO2 to air
CO_2 + $HbNH_2$
$HbNH\text{-}CO_2^-$ + H^+ + $Hb(O_2)_4$
H^+Hb + $4O_2$
O2 from air

Figure 9.33 Transport of CO_2 as carbamino-hemoglobin. *(a)* Reactions in the red blood cell at the tissues. CO_2 diffuses into red blood cell and reacts with the NH_2-terminal amino group of hemoglobin chains to form carbamino-hemoglobin (Eq. 9.13). The reaction releases a proton (boldface H^+), which promotes the dissociation of O_2 from hemoglobin. The O_2 diffuses out of the red blood cell to the tissues. *(b)* Reactions in the red blood cell at the level of the lung. In the lung, the high O_2 pressure forces the reactions in the opposite direction leading to the expiration of CO_2 from the lung. Reactions are the reverse of those in the capillaries.

2,3-Bisphosphoglycerate (BPG) in Red Blood Cells Modulates Oxygen Release from Hemoglobin

An important modulator of the hemoglobin equilibrium is **2,3-bisphosphoglycerate (BPG)** or **2,3-diphosphoglycerate (DPG)** (Figure 9.34). BPG is formed in a minor pathway for glucose metabolism and is present in small amounts in all cells. However, in the red blood cell this pathway is highly active, and BPG concentrations are approximately equimolar to that of hemoglobin. BPG binds to the deoxy (T) but not the oxy (R) conformation. Binding of BPG to Hb stabilizes the T conformation and increases its concentration relative to the R conformation. BPG dissociates as deoxy-Hb is converted to oxy-Hb (Eq. 9.14).

$$H^+BPG \cdot Hb + 4O_2 \rightleftharpoons Hb(C_2)_4 + BPG + nH^+ \tag{9.14}$$

Increased concentrations of BPG force the equilibrium (Eq. 9.14) to the left, and correspondingly shift the saturation plot to the right with an increase in P_{50} (Figure 9.35). In contrast, lowered concentrations of BPG force this equilibrium to the right, and correspondingly shift the saturation plot to the left with a lower P_{50}.

A single BPG molecule binds within a pocket formed at the β_1–β_2-subunit interface in the T conformation. This site contains eight positively charged groups [the His-143(β), Lys-82(β), His-2(β), and NH_2-terminal ammonium residues from both β chains] (Figure 9.36). The BPG has a charge of minus 5 (Figure 9.34) and is strongly attracted to the positive charges of the β–β-interface binding site. In the R conformation, the size of the binding pocket is decreased making the BPG unable to fit easily into the binding pocket. Hence the conversion of T to R leads to the dissociation of bound BPG.

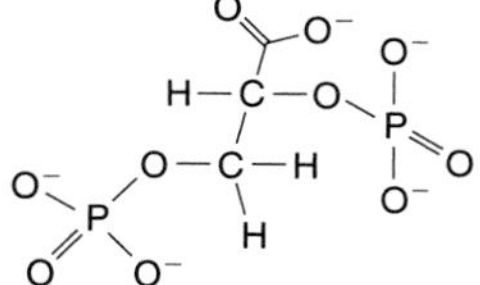

Figure 9.34 Structure of 2,3-bisphosphoglycerate (BPG). Molecule has a charge of –5 at pH 7.4.

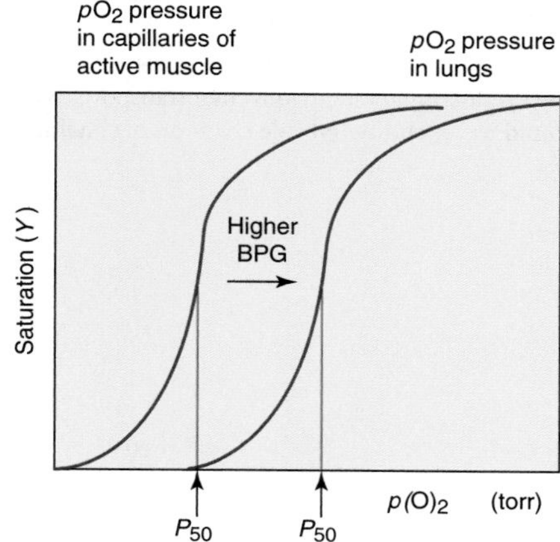

Figure 9.35 Change in oxygen–hemoglobin saturation curve to higher P_{50} value with increase in BPG concentration.

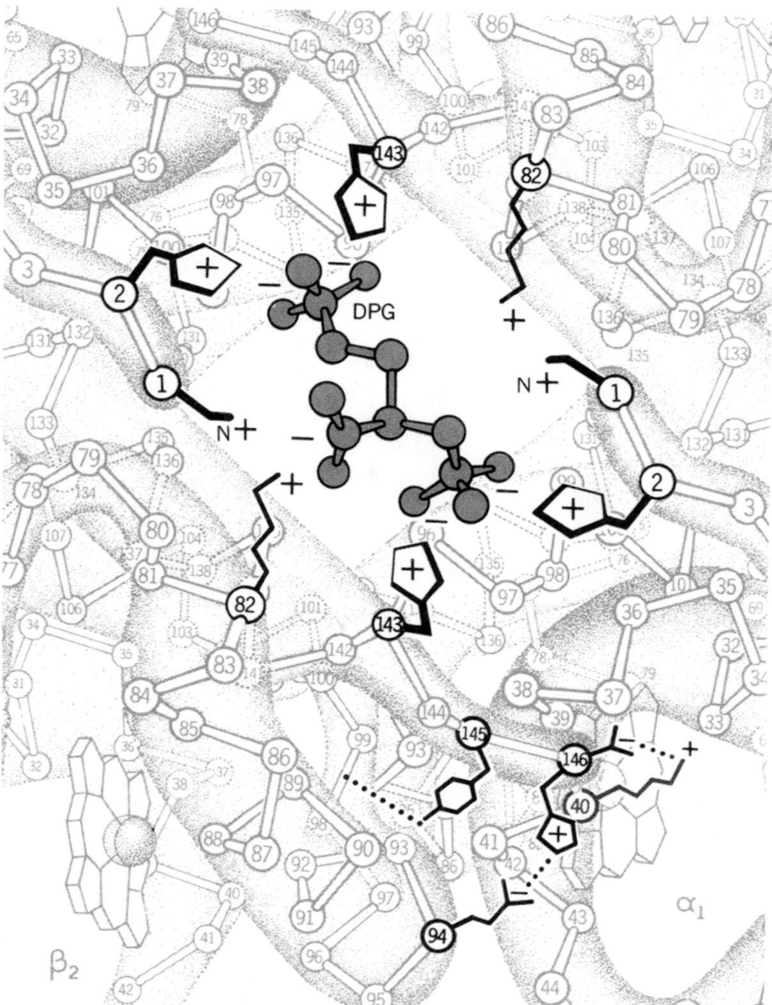

Figure 9.36 2,3- Bisphosphoglycerate binding site at the β–β interface of deoxy-hemoglobin. Shown are the positively charged side chains of two βVal-1 amino terminus ammonium, βHis-2 imidazolium, βLys-82 ε-ammonium, and β143-His imidazolium. The negatively charged 2,3-bisphosphoglycerate binds in the middle of the ring of positively charged groups.
Reproduced with permission from Dickerson, R. E., and Geis, I. *Hemoglobin: Structure, Function, Evolution, and Pathology.* Menlo Park, CA: Benjamin/Cummings, 1983. Illustration by Irving Geis. Image from Irving Geis Collection, Howard Hughes Medical Institute. Rights owned by HHMI. Not to be reproduced without permission.

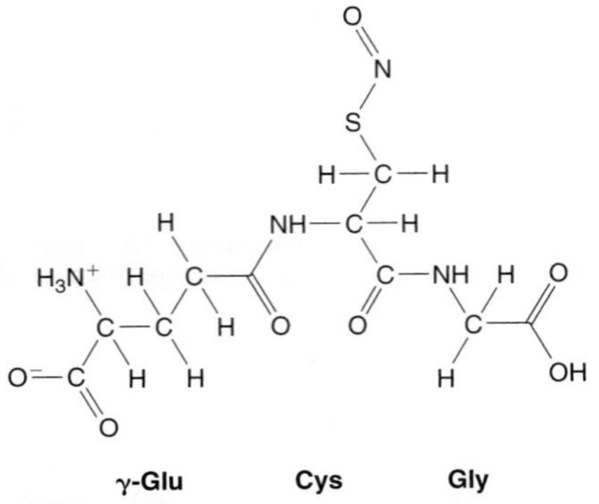

γ-Glu Cys Gly

Figure 9.37 The X-S-NO Transporter. Glutathione (γ-glutamyl-cysteinyl-glycine) transports NO bound to the sulfhydryl side chain of its cysteine.

Conditions that cause hypoxia (deficiency of oxygen) such as anemia, smoking, and high altitude increase BPG levels in the red blood cells. In turn, conditions leading to hyperoxia result in lower levels of BPG. Changes in red blood cell levels of BPG are slow and occur over hours and days to compensate for chronic changes in pO_2 levels.

Hemoglobin Delivers Nitric Oxide (NO) to the Capillary Wall of Tissues Where It Promotes O_2 Delivery

Hemoglobin reversibly binds nitric oxide (NO) a potent vasodilator, with a very short lifetime in blood. By binding NO, hemoglobin sequesters it from rapid destruction. Hb releases NO by transferring it to small sulfhydryl (X-SH) molecules as hemoglobin changes from oxy(R)- to the deoxy(T)-conformation. A common X-SH molecule is glutathione (Figure 9.37). The X-S-NO compound continues to stabilize NO against degradation and allows efficient delivery of a bioactive NO equivalent (X-S-NO) to the NO receptors in cells of the blood vessel wall, promoting relaxation of the vascular wall. The relaxation facilitates the transfer of gases between the blood and the cells of the tissues. The molar concentration of NO in blood is 1/70th the concentration of Hb. However, this low concentration is physiologically significant because of the potent biological activity of X-S-NO.

NO is first captured by the Fe^{+2} at the heme and then transferred to the sulfhydryl of the β-chain cysteine-93 (βCys^{93}) residues. The heme iron preferentially binds NO when in the T conformation. The NO is transferred to βCys^{93} when hemoglobin is in the R conformation. Then when the R changes again to the T conformation to deliver O_2 to the tissues, the NO is transferred from the βCys^{93} to small S-XH molecules such as glutathione (Figures 9.37 and 9.38). The net effect is the conversion by hemoglobin of unstable free plasma NO to stable X-S-NO.

Figure 9.39 shows the region of hemoglobin that contains βCys^{93}-S-NO. In the T conformation the S-NO is on the outside and accessible to glutathione molecules. This allows *trans*-nitrosylation to occur. In contrast, in the R conformation the βCys^{93}-SH is pointed toward the heme iron and inaccessible to the outside solvent. The distance in the R conformation between the heme iron and βCys^{93} is optimal for transfer of NO from the heme to the βCys^{93}-SH.

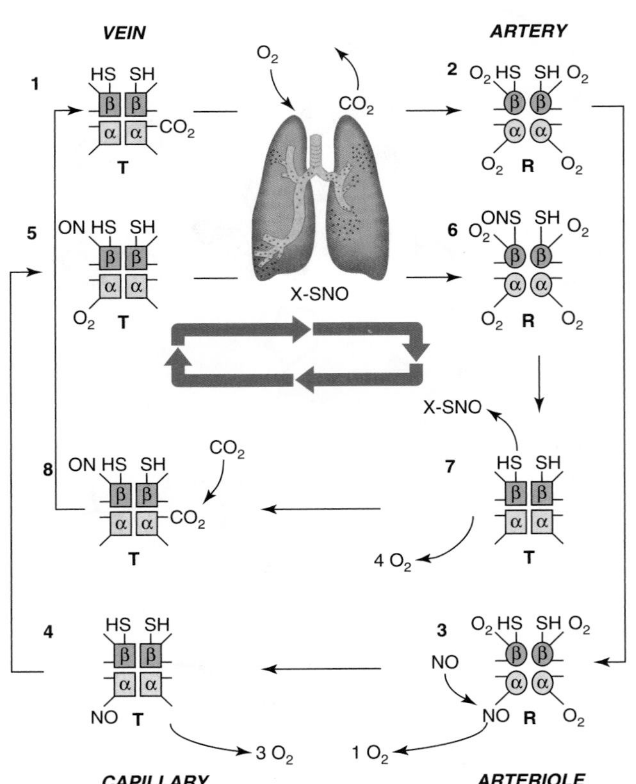

Figure 9.38 Binding and release of NO by hemoglobin during the respiratory cycle. The model shows the binding and dissociation of NO, O_2, and CO_2 as a hemoglobin molecule makes two complete cycles through the circulation. The first cycle involves intermediates 1–4, and the second cycle intermediates 5–8. The T and R conformations are shown, and the SH groups are from the side chain of βCys^{93}. The NO is either directly bound to a heme iron or to the βCys^{93} SH.

The key steps in the NO transport are its initial binding to a heme in intermediate 3 and transfer from a β-subunit heme to βCys^{93} in intermediate 6 (R conformation), and its transfer to a small molecular thiol X-SH in intermediate 7 (T conformation) when hemoglobin is converted from R to T. The hemoglobin molecule depicted may be only 1 in 1000 hemoglobin molecules that are cycling, due to the low relative molar concentration of NO in blood.

Redrawn from Gross, S. S., and Lane, P. *Proc. Natl. Acad. Sci. USA* 96: 9967, 1999.

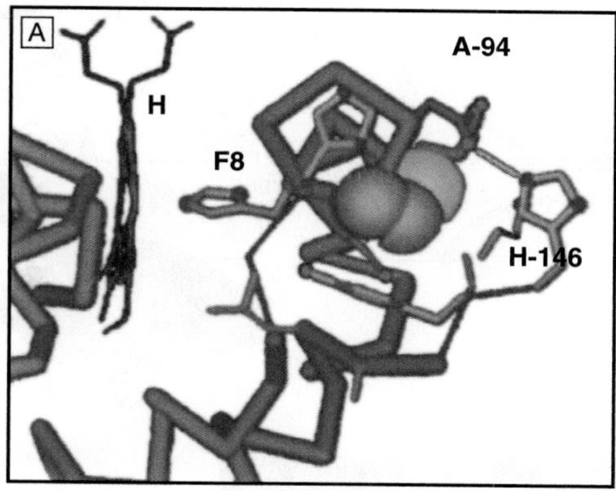

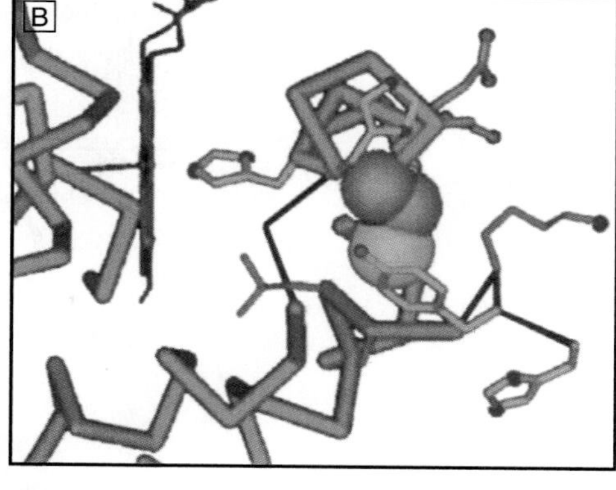

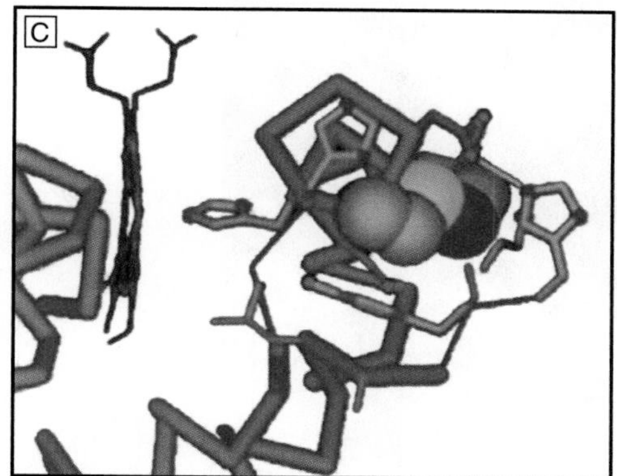

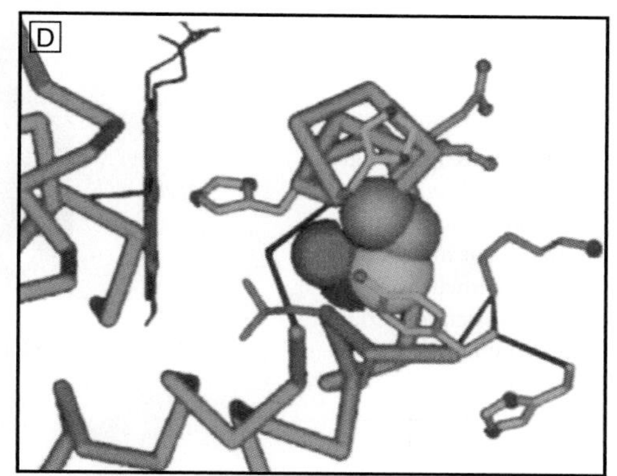

Figure 9.39 Structure of βCys93 and βCys93-S-NO in the T and R conformations. In all structures the heme (H, red) edge is pointed toward reader and is shown to be bonded to proximal histidine imidazole (F8). The two carbons (grey) and sulfur (yellow-green) of the Cys-93 side chain are shown with space filling models. In panels C and D, NO is bonded to the sulfur of βCys^{93} with the N atom colored blue and the O atom colored red.

A. The deoxy(T) conformation with the βCys^{93} side chain ($—CH_2—SH$) is on the surface of the molecule away from the heme. The cysteine side chain is prevented from entering the heme binding site by the βHis^{146}-imidazolium to βAsp^{94}-carboxylate hydrogen bonded ion pair (H bond between groups shown in yellow) on the upper right and behind the βCys^{93}-SH side chain.

B. The oxy(R) conformation with the βCys^{93} side chain pointed toward the heme and away from the solvent on the outside of the molecule. The βHis^{146} to βAsp^{94} salt bridge is broken in the R conformation, which allows the folding of the βCys-SH toward the heme pocket.

C. Model of βCys^{93}-SNO in the deoxy(T) conformation. SNO is positioned on the outside accessible to react with X-SH small molecules in the solvent. As in A, the cysteine side chain is prevented from entering heme site by the βHis^{146}–βAsp^{94} ion pair.

D. Model of βCys^{93}-SNO in oxy(R) conformation. SNO is buried near the heme and away from the outside solvent. This conformation facilitates the transfer of NO from the heme iron to the cysteine SH and prevents the reaction of the βCys^{93}-SNO with X-SH molecules in the solvent.

From Stamler, J. S., Jia, L., Eu, J. P., McMahon, T. J., et al. *Science* 276: 2034, 1997. Reprinted with permission from AAAS.

Thus the conformational changes in the hemoglobin molecule, induced by changes in oxygen pressure between the lungs and tissues, regulate NO uptake and release from the βCys^{93}. The net effect is delivery of bioactive NO to the tissue capillaries under low oxygen tension.

9.5 • THE BASAL LAMINA PROTEIN COMPLEX

A **basal lamina** is a highly structured complex of extracellular matrix proteins observed microscopically as an amorphous densely packed region of about 50–100 nm thickness surrounding tissues or cells (Figure 9.40). The term **basement membrane** is used to describe both the basal lamina and connecting fibrillar collagens attached to its outside. The basement membrane gives support to tissues and regulates access of cells to the interstitial stroma. It also participates in determining the properties of the cells that are attached to it, including the critical processes of cell division, death (apoptosis), differentiation, and migration. All cells produce constituents of basement membrane, and each basement membrane has characteristics of the cell type from which it is synthesized. A basement membrane underlies sheets of epithelial and endothelial cells and surrounds other cell types (Figure 9.40). A basement membrane separates two sheets of cells in the kidney glomerulus, where it acts as a selective filter.

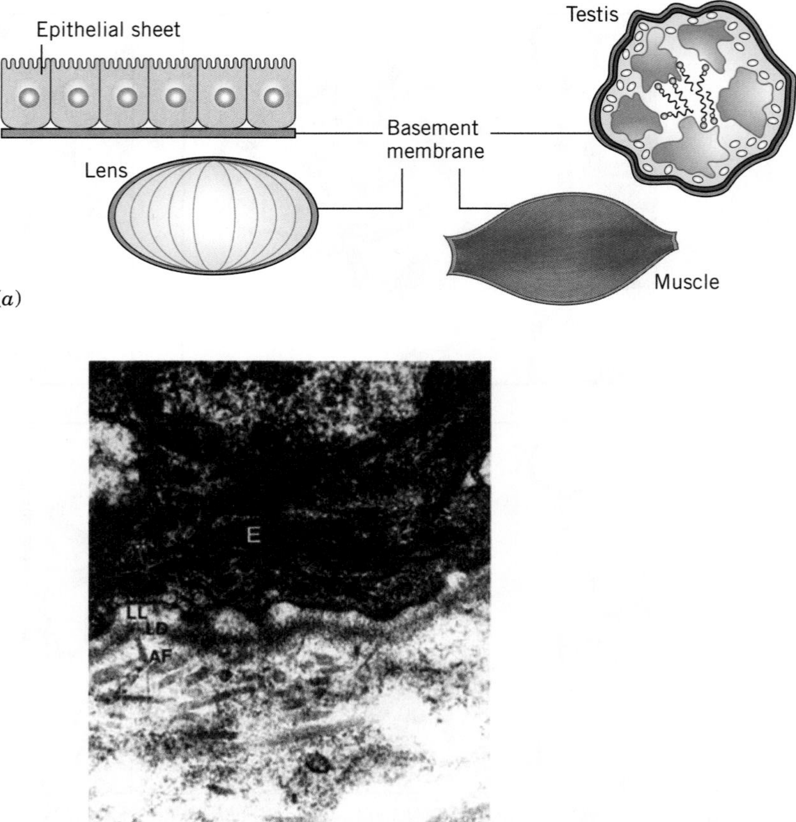

Figure 9.40 Basement membranes. (*a*). Diagram of the basement membranes surrounding various tissues and cell types. (*b*) Electron micrograph showing ultrastructure of extracellular matrix with basement membrane adjacent to epithelial cell (E). The lamina lucida (LL) and lamina densa (LD), or basal lamina of the basement membrane is shown. Below lamina densa is stroma of extracellular matrix. Bar represents 100 nm. (*a*) Reproduced with permission from Kalluri, R. *Nature Rev. Cancer* 3:422–433, 2003. (*b*) Reproduced with permission from Bosman, F. T., and Stamenkovic, I. J. *Pathol.* 200:423, 2003.

The basal lamina is formed by the non-covalent associations between specific sites located in binding domains of the associated proteins. Many basement membrane proteins also bind to cells through cellular binding domains in the proteins and cell receptor proteins in the outer cell membrane. The structure of many of the proteins is composed of module units with sequence and fold homology to common superfolds, such as the immunoglobulin (Ig) and the epidermal growth factor (EGF) folds. These folds are found repetitively and are the building blocks of extracellular matrix proteins. While the EGF modules within these proteins do not appear to have a growth factor function, growth factor and cytokine proteins are found within the basal lamina, particularly in association with the carbohydrate parts of the proteoglycan components. During turnover of the basement membrane induced by proteases and heparanases, these growth factors and cytokines are freed to act on nearby cells. In addition, many basal lamina proteins hide cryptic activities that are activated when cleaved out of the full sequence by action of proteases (see endostatin, p. 1020). The pro-protease plasminogen is ubiquitously present within extracellular matrix and is activated by cellular secretions of plasminogen activators (p. 994).

Protein Composition of the Basal Lamina

The basal lamina is composed of type IV collagen, laminin, nidogen (also referred to as entactin), and perlecan, a heparin sulfate proteoglycan. In addition, minor amounts of perhaps 50 other proteins may be present including osteopontin (also referred to as BM-40 or SPARC), fibulin, type XV collagen, type XVIII collagen, and the proteoglycan agrin. The diversity and tissue specificity of a basement membrane is determined by the isoforms of type IV collagen and laminin and the types of minor proteins present. Type IV collagen isoforms are produced by seven different type IV collagen genes. These isoforms share domain structure homology, but differ by 30%–50% in their amino acid sequences. There are at least 12 different laminin isoforms. The isoforms of type IV collagen and laminin expressed are characteristic of the cell type and tissue that synthesizes the associated basement membrane.

The Molecular Structure of Basal Lamina Is Formed from Networks of Laminin and Type IV Collagen

The structure of the basal lamina protein complex is produced by the connecting of planar networks of laminin and of type IV collagen. Nidogen/entactin and perlecan proteoglycan molecules link these networks. The laminin polymer structure initiates formation of the basement membrane, facilitated by its attachment to cellular receptors (Figure 9.41).

Laminin Network

Laminin is composed of three-polypeptide chains (α, β, and γ) cross-linked by disulfide bonds. Each chain contains about 1500 amino acids, and the molecular weight of the laminin-1 isoform is approximately 800 kD. The overall structure appears as an asymmetric cross (Figure 9.42). The cruciform structure contains three short arms, each formed by a different chain, and a long arm composed of all three chains. There are five distinct genes that code for an α chain and three each for the β and γ chains, which combine to form at least 12 distinct $\alpha\beta\gamma$ isoforms. Different domains in the structure bind to perlecan and nidogen, and at least two domains contain binding sites for cell surface receptors. Near the cell surface, the laminin molecules self-associate to form a sheetlike network structure through interactions between binding sites in the domains of the short arms of the cruciform structure (Figure 9.43).

Type IV Collagen Network

Collagen molecules contain three polypeptide chains that initially associate through noncovalent interactions (p. 103). Fibrillar forming collagens, such as type I collagen, contain long

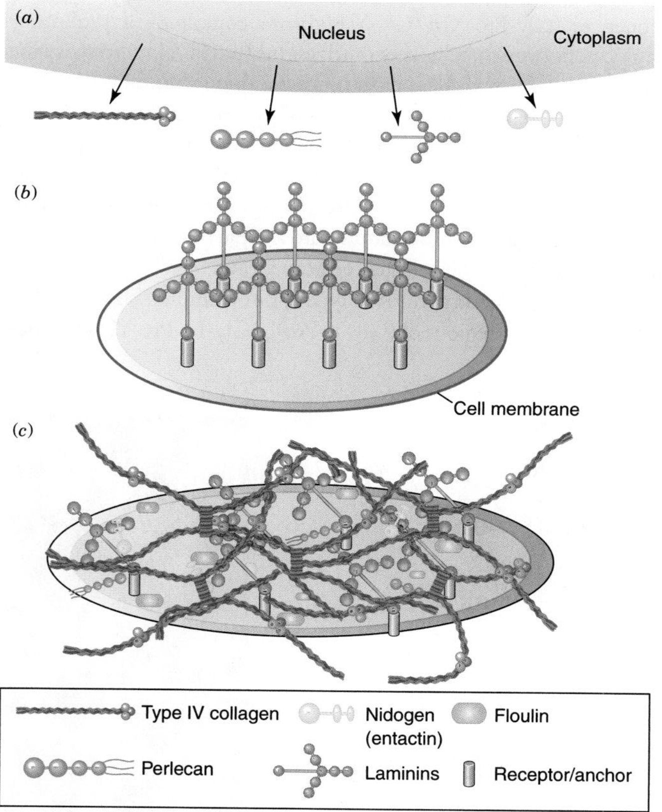

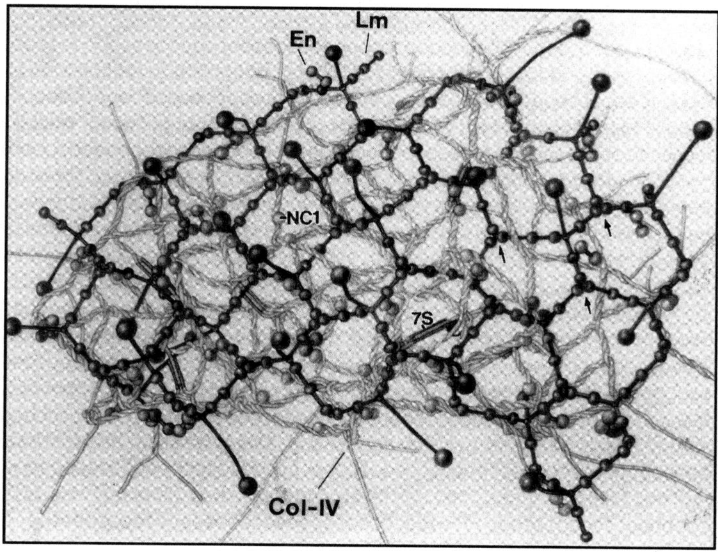

Figure 9.41 Molecular structure of basal lamina. The laminin and Type IV collagen molecules self-associate to form sheet-like network structures, which are linked by the proteins nidogen/entactin and by the heparan sulfate proteoglycan perlecan. (*a*) Synthesis of basement membrane proteins by cell. (*b*) Assembly of laminin network initiated by binding of laminin to cell surface receptor. (*c*) Assembly of Type IV collagen network and association of proteins linking the laminin and Type IV collagen scaffolds. (*d*) Type IV collagen network joined through 7S and NC1 domain interactions and with entactin/nidogen (En) bridging the collagen and laminin (Lm) networks. (*a*)–(*c*) reproduced with permission from Kalluri, R. *Nature Rev. Cancer* 3:422, 2003. Copyright (2003) *Nature*. (*d*) reproduced with permission from Yurchenco, P. D. Assembly of basement membrane networks. In Yurchenco, P. D., Birk, D. E., and Mecham, R. P. (Eds.), *Extracellular Matrix Assembly and Structure.* New York: Academic Press, 1994, 351. Copyright (1994) Elsevier.

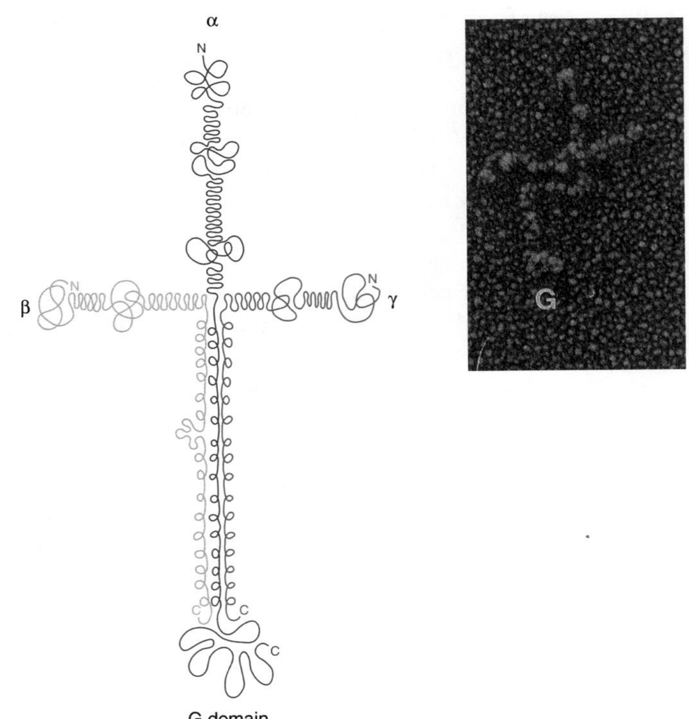

Figure 9.42 Structure of laminin (isoform 1). (*a*) Diagrammatic structure of laminin-1. The α, β, and γ laminin chains each form a short arm of the cruciform structure. The long arm is formed by a coiled-coil structure of α-helical regions of all three chains with the α laminin chain extending out to form the COOH-terminal globular G domain. The short arms are composed of globular domains, which are separated by epidermal growth factor-like (EGF-like) repeats. (*b*) Contrast-reversed glycerol rotary shadowed replica of laminin molecule. α-Chain NH_2-terminal and α-chain globular (G) domain indicated.
(*a*) Redrawn based on figure from Yurchenco, P. D. Assembly of basement membrane networks. In Yurchenco, P. D., Birk, D. E., and Mecham, R. P. (Eds.), *Extracellular Matrix Assembly and Structure.* New York: Academic Press, 1994, 351. (*b*) Reproduced with permission from Yurchenco, P. D. Assembly of basement membrane networks. In Yurchenco, P. D., Birk, D. E., and Mecham, R. P. (Eds.), *Extracellular Matrix Assembly and Structure.* New York: Academic Press, 1994, 351.Copyright © (1994) Elsevier.

stretches of the repeating sequences (Gly-Pro-X)*n*/(Gly-Y-HyPro)*n* containing a proline or hydroxyproline (HyPro) and a glycine approximately every third residue (p. 106). Hydroxyproline is a derived amino acid formed by proline hydroxylase in the endoplasmic reticulum on proline residues within the nascent procollagen chain. The high content of proline and hydroxyproline generates a helical conformation for this sequence known as the polyproline type II helix (p. 106). This helix is characterized by three residues per helix turn ($n = 3$), with a Gly every third residue forming a longitudinal glycine edge along one side of the helix that promotes self-association of three polypeptides, each in a polyproline helical conformation, into a triple helical or superhelical structure. This three-chain molecule composes the collagen protomer unit.

Type IV collagen does not form fibrillar structures as type I, but a laminar sheetlike network. Type IV collagen polypeptides contain regions of $(\text{Gly-X-Pro})_n/(\text{Gly-HyPro-Y})_n$ sequence

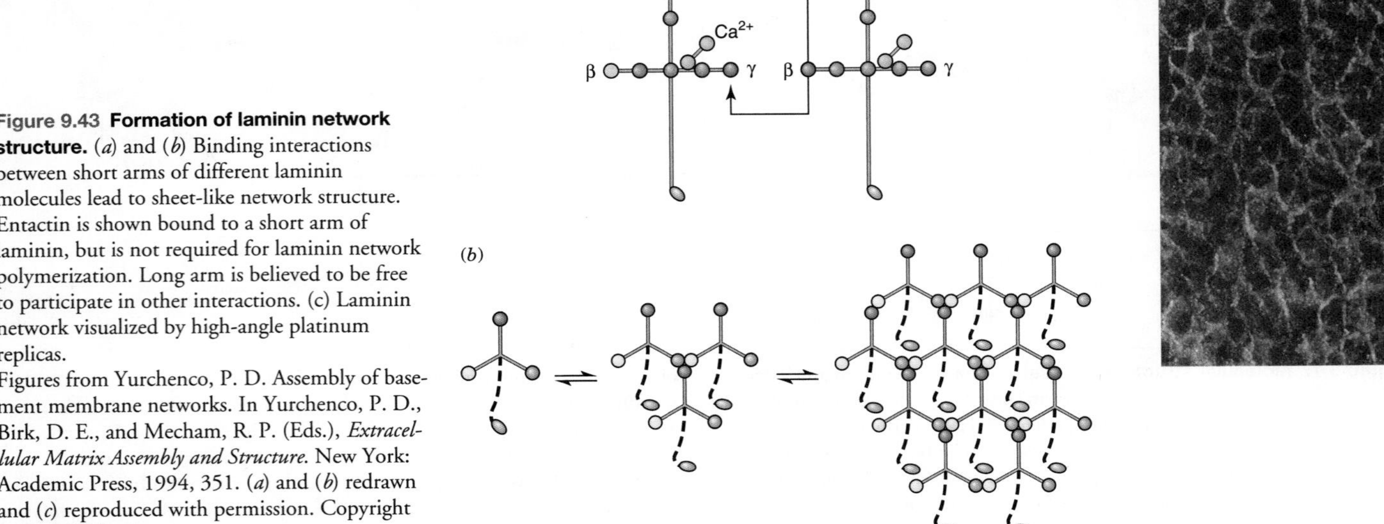

Figure 9.43 Formation of laminin network structure. (*a*) and (*b*) Binding interactions between short arms of different laminin molecules lead to sheet-like network structure. Entactin is shown bound to a short arm of laminin, but is not required for laminin network polymerization. Long arm is believed to be free to participate in other interactions. (c) Laminin network visualized by high-angle platinum replicas.
Figures from Yurchenco, P. D. Assembly of basement membrane networks. In Yurchenco, P. D., Birk, D. E., and Mecham, R. P. (Eds.), *Extracellular Matrix Assembly and Structure.* New York: Academic Press, 1994, 351. (*a*) and (*b*) redrawn and (*c*) reproduced with permission. Copyright © (1994) Elsevier.

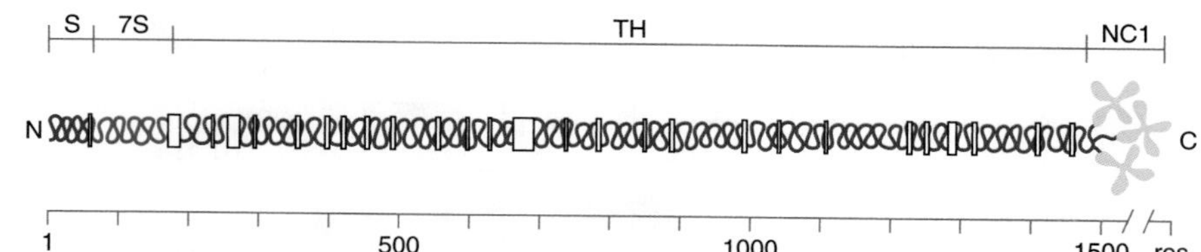

Figure 9.44 Diagram of the type IV collagen protomer composed of three polypeptide chains. Top line shows the location in the primary structure of the NC1 region, the central triple-helical (TH) region with a superhelical structure, and the N-terminal 7S region. Black bars, lines, and boxes show locations of the multiple interruptions of the Gly-X-Y sequence in the chains of type IV collagen.
Redrawn from Yurchenco, P. D. Assembly of basement membrane networks. In Yurchenco, P. D., Birk, D. E., and Mecham, R. P. (Eds.), *Extracellular Matrix Assembly and Structure.* New York: Academic Press, 1994, 351. Copyright (1994) Elsevier.

that form rod-shaped triple-helical regions, but different from type I these regions are interspersed by linker and globular domains that break up the superhelical regions (Figure 9.44). The globular domain at the C-terminal end of the protomer is called the NC1 (*n*on-*c*ollagenous 1) domain. This C-terminal domain is preceded by a triple-helical (TH) central region and a hinge connecting to a smaller triple-helical (7S) region at the N-terminal end.

In assembling the type IV collagen network, each protomer binds to another through a NC1 to NC1 head-to-head interaction (Figure 9.45). Then the dimers interact through their 7S regions to form tetramers, which further self-aggregate to form the planar network. The 7S interactions are the nodes of the collagen network structure.

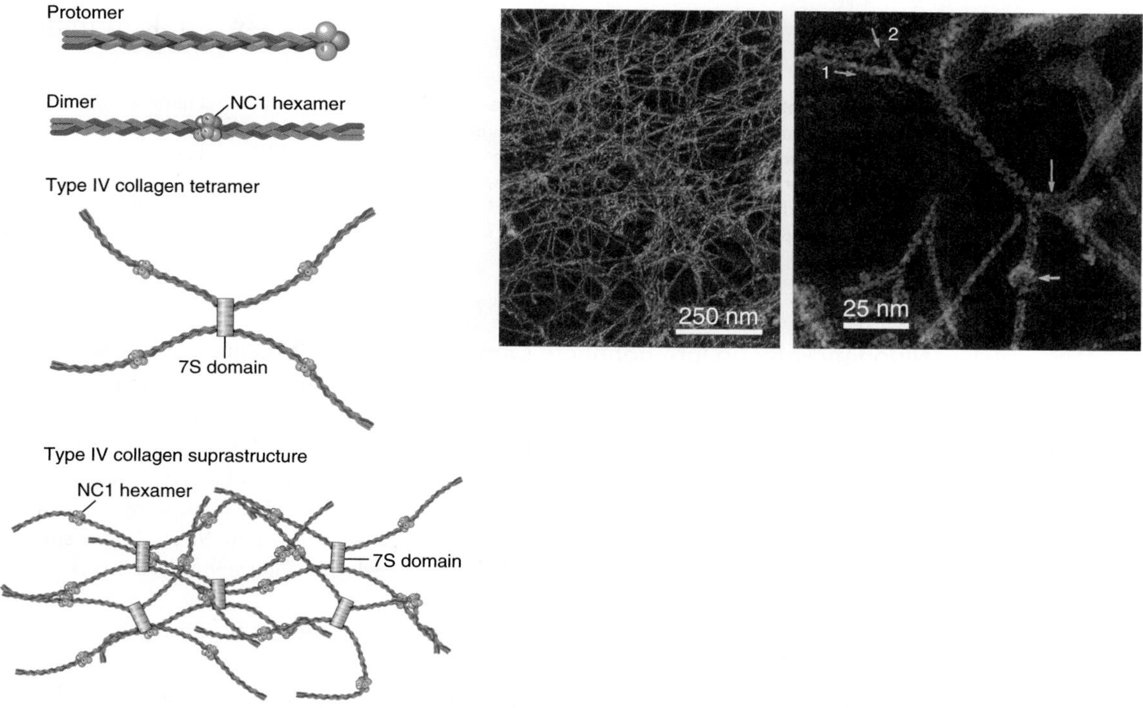

Figure 9.45 Formation of type IV collagen network structure. A. Diagrammatic depiction of network formation. (*a*) Protomer type IV collagen molecule is formed from three polypeptide chains (red, purple, and blue) in the endoplasmic reticulum and golgi of cell. Collagen chains are aligned with their NH_2-terminal 7S regions on one end and COOH-terminal NC1 regions at other end. (*b*) Two protomers are joined by their NC1 domains to form a dimer of protomers (each protomer has three chains so that interaction is between NC1 domains of six chains). (*c*) Two protomer dimers bind through 7S regions. (*d*) Polymerization of tetramers of protomers forms type IV collagen network. B. Type IV collagen networks viewed by a high-angle replicate of amniotic basement membrane observed *in situ.*
A redrawn from Kalluri, R. *Nature Rev. Cancer* 3:422, 2003. B reprinted with permission from Yurchenco, P. D., and Ruben, J. *Cell Biol.* 105:2559, 1987. Modified from Yurchenco, P. D. Assembly of basement membrane networks. In Yurchenco, P. D., Birk, D. E., and Mecham, R. P. (Eds.), *Extracellular Matrix Assembly and Structure.* New York: Academic Press, 1994, 351. Copyright (1994) Elsevier.

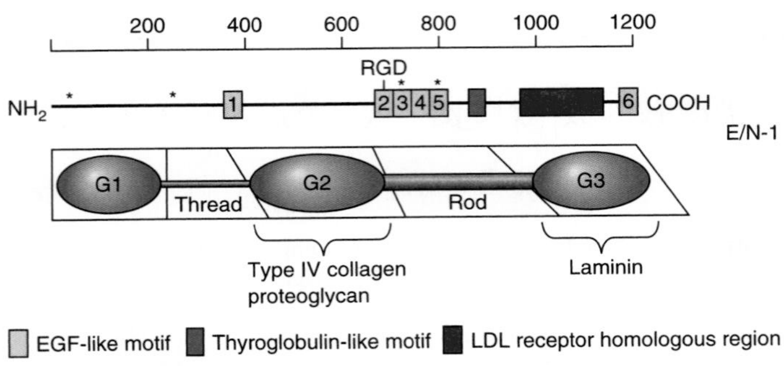

Figure 9.46 Structure of nidogen. Schematic structure of the entactin/nidogen-1 molecule. The potential calcium binding sites are marked by stars. Binding sites for type IV collagen and proteoglycan are in G2 domain, and the binding site to laminin in the G3 domain. The RGD (Arg-Gly-Asp) sequence binds to integrin type cell receptors. Redrawn from Erickson, A. C., and Couchman, J. R. *J. Histochem Cytochem.* 48:1291, 2000.

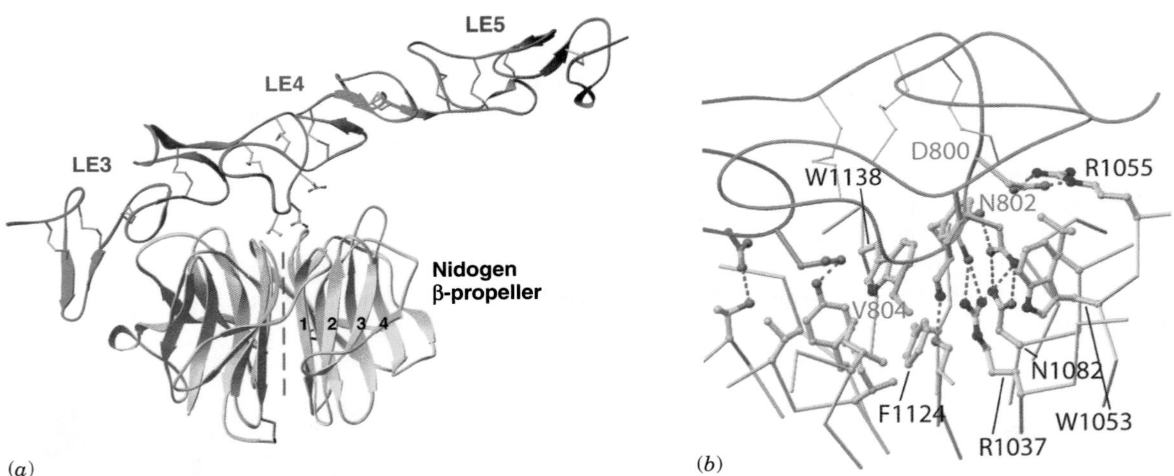

Figure 9.47 Molecular details of molecular interactions between nidogen and laminin. (*a*) Ribbon diagram of the nidogen β-propeller complex of domain G3 with laminin modules LE3-5. β-Strands are numbered on the propeller β-sheet. (*b*) View of the nidogen interaction with laminin domain LE4 and the adjacent portion of laminin domain LE3. Portions of the nidogen backbone as C_α-trace and side chains forming the interaction are shown in gold. Green dashed lines show hydrogen bonds. Reprinted with permission from Takagi, I., Yang, y., Lu, J., Wang, H. and Springer, T. A. *Nature* 424, 969, 2003.

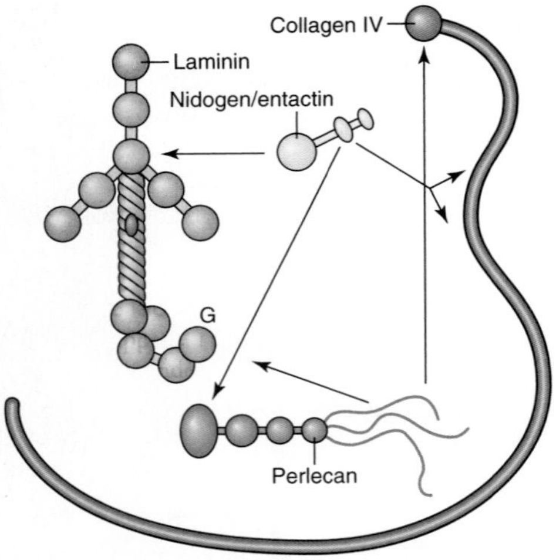

Figure 9.48 Interactions of nidogen and the proteoglycan percleran. Arrows show domain–domain binding interactions between the major proteins in basement membrane superstructure. Redrawn from Kalluri, R. *Nature Rev. Cancer* 3: 422, 2003.

Nidogen/Entactin Interconnects Laminin and Type IV Collagen Networks.

The two isoforms, nidogen-1 and nidogen-2, have a 46% homology. Both forms are composed of three globular domains (G1, G2, and G3) separated by a link region between G1 and G2 and a longer rod region between G2 and G3 (Figure 9.46). Domain G3 is a β propeller fold composed of antiparallel β strands that appear as blades of an airplane propeller. It is similar to a fold of the LDL receptor protein. Nidogens bind to laminin by interaction of the β-propeller interface with laminin domains III and IV (Figure 9.47). Other sites in nidogen form complexes with type IV collagen and perlecan (Figure 9.48).

Heparan Sulfate Proteoglycan Perlecan Interconnects Laminin and Type IV Collagen Networks

Perlecan (MW 600 kDa) is a major proteoglycan of the *basal lamina.* Proteoglycans have a polypeptide core attached by serine side chains through a tetrasaccharide to a glycosaminoglycan (GAG) chain, composed of a disaccharide of an amino sugar (*N*-acetylglucosamine or *N*-acetylgalactosamine) and a uronic acid (glucuronic or iduronic). In perlecan the GAG is the heavily sulfated heparin sulfate (Figure 9.49). Each perlecan contains 2–15 heparin sulfate (**GAG**) chains (Figure 9.50). The sulfated GAG chains are highly negatively charged and bind cations and water to form gels. This provides a swelling pressure that enables the basement membrane to withstand compressive forces. Perlecan interacts with the other three major components of basement membrane through either its core protein to type IV collagen or through its heparan sulfate chains to bind laminin. Other sites in the heparin sulfate bind to cellular receptors and still others to growth factors such as TGF-β.

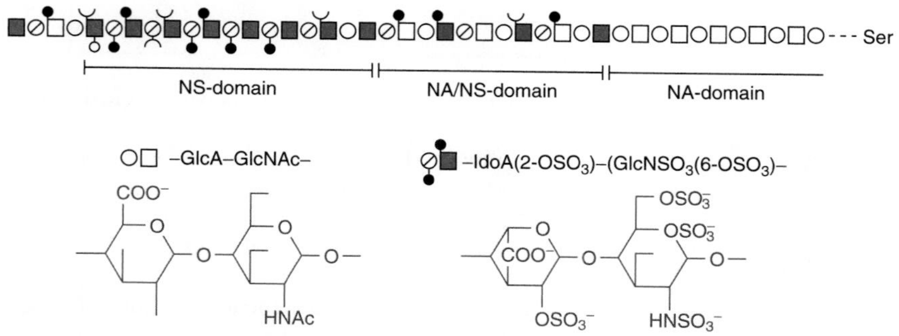

Figure 9.49 Structure of a heparan sulfate chain. The GAG disaccharide polymers contain alternating hexuronic acid (D-glucuronic acid (GlcA) or L-iduronic acid (IdoA)) and D-glucosamine (GlcN) units. The GlcN is *N*-acetylated in regions of the chain (NA region). In further processed regions of the heparan polymer the *N*-acetyl groups have been partially replaced by *N*-sulfate (NA/NS regions) or the *N*-acetyl groups have been completely replaced by *N*-sulfate (NS regions). Open circle is 3-*O*-sulfate.

The polymers are further modified by C-5 epimerization of GlcA to IdoA residues, and, finally incorporation of *O*-sulfate groups at various positions. The heparan chain is joined to the protein core by forming a bond to a serine side chain.

Redrawn from Lindahl, U., Kusche-Gullberg, M., and Kjellen L. *J. Biol. Chem.* 273:24979, 1998.

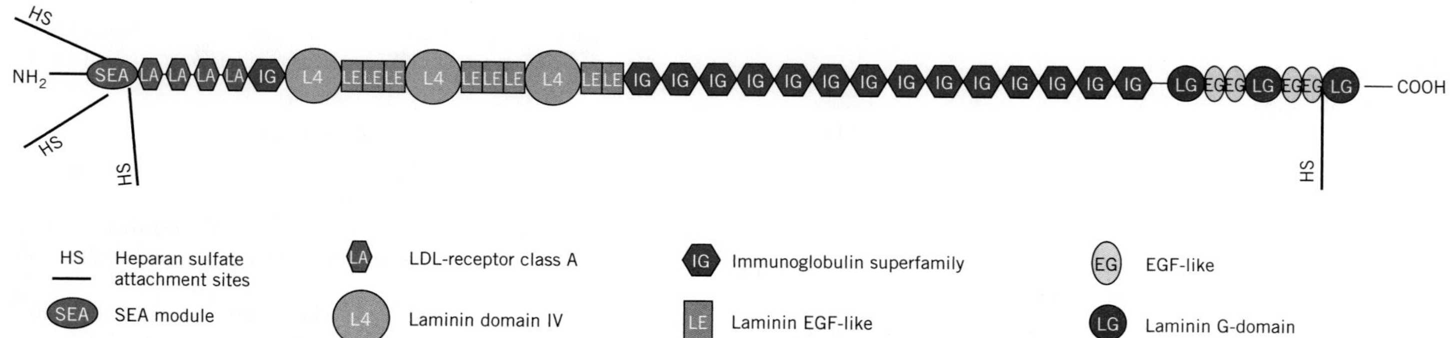

Figure 9.50 Module organization of domains in the protein core of perlecan. Domain organization of mouse perlecan. The attachment of the heparan sulfate (HS) chains is indicated. The domains present are SEA (homologous to the domain found in sea urchin sperm protein, enterokinase, and agrin); LA (homologous to LDL receptor type A domain); L4 (homologous to laminin domain IV), IG (immunoglobulin-like domain); LE (homologous to the laminin type epidermal growth factor-like domain); EG (epidermal growth factor-like), and LG (homologous to laminin G-like domain).
Redrawn from Hopf , M., Göhring, W., Kohfeldt, E., Yamada, Y., and Timpl, R. *Eur. J. Biochem.* 259, 917-925 (1999); and Kvansakul, M., Hopf, M., Ries, A., Timpl, R., and Hohenester, E. *EMBO J.* 20: 5342, 2001.

Focal Contact in Cell Membrane Interconnects Extracellular Matrix with Cytoskeleton

The basement membrane is an extracellular protein complex that is connected to the intracellular cytoskeletal protein complex, as both complexes are connected to the same receptor proteins present in the focal contact regions of cellular membranes. These protein–protein binding interactions connect the extracellular matrix to the cellular cytoskeleton and are critical for the regulation of intracellular processes, such as cell motility, cell division, and cell death. In most cancer cells, this binding of the membrane receptor protein complex to both the extracellular *basal lamina* and the cytoskeletal protein complexes is broken.

Bibliography

Immunoglobulins

Almagro, J. C. and Fransson, J. Humanization of antibodies. *Frontiers in Biosci.* 13:1619, 2008.

Alzari, P. M., Lascombe, M. B., and Poljak, R. J. Structure of antibodies. *Annu. Rev. Immunol.* 6:555, 1988.

Barclay, A. N., Ig-like domains: evolution from simple interaction molecules to sophisticated antigen recognition. *Proc. Nat. Acad. Sci. USA* 96:14672, 1999.

Chothia, C., Lesk, A. M., Tramontano, A., Levitt, M., et al. Antibody therapeutics, antibody engineering, and the merits of protein stability. *Curr. Opin. Drug Discovery & Development* 11:675, 2008.

Guddat, L. W., Shan, L., Fan, Z.-C., Andersen, K. N., et al. Diversity of the immunoglobulin gene superfamily. *Adv. Immunol.* 44:1, 1989.

James, L. C., Roversi, P., and Tawfik, D. S. Antibody multispecificity mediated by conformational diversity. *Science* 299:1362, 2003.

Padlan, E. A. Anatomy of the antibody molecule. *Mol. Immunol.* 31:169, 1994.

Rini, J. M., Schultze-Gahmen, U., and Wilson, I. A. Structural evidence for induced fit as a mechanism for antibody-antigen recognition. *Science* 255:959, 1992.

Serine Proteases

Bobofchak, K. M., Pineda, A. O., Mathews, F. S. and Di Cera, E. Energetic and structural consequences of perturbing Gly-193 in the oxyanion hole of serine proteases. *J. Biol. Chem.* 280:25644, 2005.

Liebman, M. N. Structural organization in the serine proteases. *Enzyme* 36:115, 1986.

López-Otin, C. and Bond, J. S. Proteases: multifunctional enzymes in life and disease. *J. Biol. Chem.* 283:30433, 2008.

Neurath, H. Proteolytic processing and physiological regulation. *Trends in Biochem. Sci.* 14:268, 1989.

Perona, J. J., and Craik, C. S., Structural basis of substrate specificity in the serine proteases. *Protein Sci.* 4:337, 1995.

Polgar, L. Structure and function of serine proteases. In Neuberger, A. and Brocklehurst, K. (Eds.), *Hydrolytic Enzymes,* series in *New Comprehensive Biochemistry,* Vol. 16. Amsterdam: Elsevier, 1987, 159.

Hemoglobin and Myoglobin

The Globin Gene Server Database of Sequence Alignments and Experimental Results for the β-like Globin Gene Cluster of Mammals. http://globin.cse.psu.edu/.

Baldwin, J., and Chothia, C. Haemoglobin: the structural changes related to ligand binding and its allosteric mechanism. *J. Mol. Biol.* 129:175, 1979.

Benz, E. J., Jr. Genotypes and phenotypes: Another lesson from the hemoglobinopathies. *N. Engl. J. Med.* 351:1490, 2004.

Cossins, A., and Berenbrink, M. Myoglobin's new clothes. *Nature* 454:416, 2008.

Cui, Q., and Karplus, M. Allostery and cooperativity revisited. *Protein Science* 17:1295, 2008.

Dickerson, R. E., and Geis, I. *Hemoglobin: Structure, Function, Evolution and Pathology.* Menlo Park, CA: Benjamin/Cummings, 1983.

Gross, S. S, and Lane, P., Physiological reactions of nitric oxide and hemoglobin: A radical rethink. *Proc. Natl. Acad. Sci. USA* 96:9967, 1999.

Hsia, C. C. W. Mechanisms of disease: respiratory function of hemoglobin. *N. Engl. J. Med.* 338: 239, 1998.

Perutz, M. Hemoglobin structure and respiratory transport. *Sci. Am.* 239:92, 1978.

Perutz, M. F., Fermi, G., and Shih, T.-B. Structure of deoxy Cowtown [His HC3(146)beta to Leu]: Origin of the alkaline Bohr effect. *Proc. Natl. Acad. Sci. USA* 81:4781, 1984.

Perutz, M. F., Wilkinson, A. J., Paoli, M., and Dodson, G. G. The stereochemical mechanism of the cooperative effects in hemoglobin revisted. *Annu. Rev. Biophysics Biomolecular Structure* 27:1, 1998.

Takashi, Y. T., Park, S., Tsuneshige, A., Imai, K., and Kanaori, K. Global allostery model of hemoglobin: Modulation of O_2 affinity, cooperativity, and Bohr effect by heterotropic allosteric effectors. *J. Biol. Chem.* 277:34508, 2002.

Veeramachaneni, N. K., Harken, A. H., and Cairns, C. B. Clinical implications of hemoglobin as a nitric oxide carrier. *Arch. of Surg.* 134:434, 1999.

Basal Lamina and Basement Membrane

Annes, J. P., Munger, J. S., and Rifkin, D. B. Making sense of latent TGFβ activation. *J. Cell Science* 116:217, 2003.

Brooke, B. S., Karnik, S. K., and Li, D. Y. Extracellular matrix in vascular morphogenesis and disease: structure versus signal. *Trends in Cell Biology* 13:51, 2003.

Dityatev, A., and Schachner, M. Extracellular matrix molecules and synaptic plasticity. *Nature Reviews Neuroscience* 4:456, 2003.

Erickson, A. C., and Couchman, J. R. Still more complexity in mammalian basement membranes. *J. Histochem. Cytochem.* 48:1291, 2000.

Heino, J. The collagen family members as cell adhesion proteins. *BioEssays* 29:1001, 2007.

Kalluri, R. Basement membranes: structure, assembly and role in tumour angiogenesis. *Nature Cancer Reviews* 3:422, 2003.

LeBleu, V., MacDonald, B., and Kalluri, R. Structure and function of basement membranes. *Exp. Biol. and Med.* 232:1121, 2007.

Sundaramoorthy, M., Meiyappan, M., and Hudson, B.G. Crystal structure of NC1 domains: Structural basis for type IV collagen assembly in basement membranes. *J. Biol. Chem.* 277:31142, 2002.

Nyberg, P., Salo, T., and Kalluri, R. Tumor microenvironment and angiogenesis. *Frontiers in Bioscience* 13:6537, 2008.

Yurchenco, P. D. Assembly of basement membrane networks. In Yurchenco, P. D., Birk, D. E., and Mecham, R. P. (Eds.), *Extracellular Matrix Assembly and Structure.*)New York: Academic Press, 1994, 351.

Key Terms

ANTIBODY MOLECULES

constant and variable regions
F_{ab} fragment
F_c fragment
H chains
hapten
hypervariable region
immunoglobulin class
immunoglobulin fold
L chains

SERINE PROTEASES

active-site directed
β-barrel domain
catalytic residues
convergent evolution
endopeptidase
exopeptidase
γ-carboxyglutamic acid
proenzyme

reaction transition-state stabilization
sequence homology
serine proteases
serpin
zymogen

HEMOGLOBIN

2,3-bisphosphoglycerate
allosteric mechanism
Bohr effect
carbamino-hemoglobin
CO_2 transport
distal histidine
HbA_1
HbF
Hill coefficient
isohydric effect
myoglobin
NO transport
O_2 transport
P_{50}
pK_a perturbation
positive cooperativity
proximal histidine
R and T conformations

BASAL LAMINA

basement membrane
collagen network
focal contact
laminin network
nidogen
perlecan

Questions • CAROL N. ANGSTADT

Multiple-Choice Questions

1. Haptens
- **A.** can function as antigens.
- **B.** strongly bind to antibodies specific for them.
- **C.** may be macromolecules.
- **D.** never act as antigenic determinants.
- **E.** can directly elicit the production of specific antibodies.

2. In the three-dimensional structure of immunoglobulins,
- **A.** β-sheets align edge to edge.
- **B.** in each chain (H and L) the C and V regions fold onto one another, forming CV associations.
- **C.** C_L–V_L associations form the complementary sites for binding antigens.
- **D.** free-SH groups are preserved to function in forming tight covalent bonds to antigens.
- **E.** hinge domains connect globular domains.

3. In immunoglobulins all of the following are true except
- **A.** there are four polypeptide chains.
- **B.** there are two copies of each type of chain.
- **C.** all chains are linked by disulfide bonds.
- **D.** carbohydrate is covalently bound to the protein.
- **E.** immunoglobulin class is determined by the C_L regions.

4. Hemoglobin and myoglobin both have, are, can, or do all of the following except
- **A.** subunits that provide hydrogen bonds to and nonpolar interaction with other subunits.
- **B.** highly α helical.
- **C.** bind one molecule of heme per globin chain.
- **D.** bind heme in a hydrophobic pocket.
- **E.** bind one O_2 per heme.

5. Isohydric transport of carbon dioxide from tissues to lung refers to
- **A.** free CO_2 dissolved in plasma.
- **B.** HCO_3^- in plasma.
- **C.** carbamino compounds.
- **D.** the Bohr effect.
- **E.** the presence of carbonic anhydrase in plasma.

6. Myoglobin has a Hill coefficient of 1.0 and hemoglobin has a Hill coefficient of 2.8. This indicates that
- **A.** an oxygen saturation curve for myoglobin would be sigmoidal.
- **B.** hemoglobin exhibits a simple equilibrium binding of oxygen.
- **C.** myoglobin shows negative cooperativity in its binding of oxygen.
- **D.** hemoglobin shows positive cooperativity in its binding of oxygen.
- **E.** only myoglobin would yield a straight line in a Hill plot.

Questions 7 and 8: A serine protease is believed to be required for metastasis of cancer cells. Tumor cells secrete urokinase-type plasminogen activator (uPA), a serine protease, which activates plasminogen to plasmin. Plasmin causes hydrolysis of proteins in the extracellular matrix. It also activates procollagenase to collagenase; this degrades collagen. The destruction of extracellular matrix proteins allows tumor cells to invade through the matrix to form secondary tumor sites.

7. Basal lamina structure is produced by connecting planar networks of laminins and type IV collagen. Which of the following statements about these is/are correct?
- **A.** Both laminin and type IV collagen are composed of three polypeptide chains.
- **B.** Laminin has a cruciform structure.
- **C.** Laminin and type IV collagen are interconnected by a heparin sulfate proteoglycan.
- **D.** Type IV collagen contains repeating sequences of $(\text{Gly-HyPro-Y})_n$.
- **E.** All of the above.

8. In addition to serine, the other two essential residues at the catalytic site of the serine proteases are
- **A.** proline and hydroxyproline.
- **B.** histidine and aspartate.
- **C.** proximal and distal histidines.
- **D.** the C-terminal and N-terminal amino acids.
- **E.** aspartate and glutamate.

Questions 9 and 10: When a blood vessel is injured, the coagulation process is initiated to prevent loss of blood and maintain the integrity of the circulatory system. Coagulation is a process in which zymogens are converted to active serine proteases in a stepwise, cascade process with the final result the production of a cross-linked fibrin clot. In a myocardial infarction, the process can be activated to such an extent that blood flow to the heart muscle is decreased. The fibrinolysis pathway to dissolve fibrin clots also involves activating zymogens to active serine proteases, the final step being the activation of plasminogen to plasmin, which acts directly on the fibrin clot. One of the current treatments for myocardial infarctions is rapid administration of t-PA (tissue plasminogen activator). Actually recombinant t-PA, produced by gene cloning technology, is used.

9. Serine proteases
- **A.** hydrolyze peptide bonds involving the carboxyl groups of serine residues.
- **B.** are characterized by having several active sites per molecule, each containing a serine residue.
- **C.** are inactivated by reacting with one molecule of diisopropylfluorophosphate per molecule of protein.
- **D.** are exopeptidases.
- **E.** recognize only the amino acids that contribute to the bond to be broken.

10. All of the following are characteristic of serine proteases as a class except
   A. only one serine residue is catalytically active.
   B. natural protein substrates and inhibitors bind very tightly to the protease.
   C. the genes that code for serine proteases are organized in a similar fashion.
   D. catalytic units exhibit two structural domains of dramatically different size.
   E. conversion of zymogen to active enzyme usually involves one or more hydrolytic reactions.

Questions 11 and 12: There are over 800 mutant human hemoglobins. Mutations may cause instability leading to rapid degradation, alterations in oxygen affinity, or more rapid oxidation of heme Fe^{2+} to Fe^{3+}. All of these alter hemoglobin's ability to carry out its physiological functions. In the mutation $Hb_{Cowtown}$, histidine 146, which contributes to 50% of the Bohr effect, is lost. Regulation of oxygen dissociation by protons is impeded, the T confirmation is destabilized relative to the R conformation, and oxygen affinity is increased.

11. All of the following are believed to contribute to the stability of the deoxy or T conformation of hemoglobin except
   A. a larger ionic radius of six-coordinated ferrous ion as compared to five-coordinated ion.
   B. unstrained steric interaction of His F8 with the porphyrin ring when iron is above the plane.
   C. interactions between the FG corner of one subunit and the C helix of the adjacent subunit.
   D. ionic interactions.

12. When hemoglobin is converted from the deoxy (T) form to oxyhemoglobin (R),
   A. it becomes more acidic and releases protons.
   B. carbamino formation is promoted.
   C. binding of 2,3-bisphosphoglycerate (BPG) is favored.
   D. bound NO is transferred to glutathione.
   E. all of the above.

Problems

13. One of the adaptations to high altitude is an increase in the concentration of BPG. What effect does this have on a saturation versus pO_2 curve? Why does increasing [BPG] increase the delivery of O_2 to tissues?

14. What would be the likely order of appearance of various immunoglobulins in response to a viral pathogen introduced via the nasal passages and subsequently into the blood?

Answers

1. **B** Haptens are small molecules and cannot alone elicit antibody production; thus they are not antigens. They can act as antigenic determinants if covalently bound to a larger molecule, and free haptens may bind strongly to the antibodies thereby produced.
2. **E** See Figure 9.2. A: The β-sheets align face-to-face. B: The V and C regions are adjacent to each other. C: The complementarity regions are the variable regions of both the heavy and light chains (V(H)—V(L). D: Antigen binding is noncovalent.
3. **E** The C_H regions determine class. A: There are two copies of each of two types of polypeptide chain.
4. **A** Hemoglobin has four chains and four oxygen-binding sites, whereas myoglobin has one chain and one oxygen-binding site. Each oxygen-binding site is a heme.
5. **B** Protons released as H_2CO_3 dissociates to HCO_3^- and H^+ bind to oxyhemoglobin, facilitating release of O_2 to tissues. A: Very little CO_2 is in this form. C: This is another way to carry CO_2. D: Bohr effect releases H^+ as deoxy Hb picks up O_2. E: Carbonic anhydrase is in red cells.
6. **D** A Hill coefficient greater than 1 indicates positive cooperativity which hemoglobin has. A: A coefficient of 1.0 indicates non-cooperativity; this would give a hyperbolic oxygen saturation curve. B: This coefficient indicates a simple equilibrium, which is what myoglobin shows. C: The coefficient is between 0 and 1 for negative cooperativity. Myoglobin is a single polypeptide and does not show cooperativity. E: The Hill plot is the log of each value for a normal saturation curve and gives a straight line. The slopes, which are the Hill coefficients, differ.
7. **E** All of these contribute to the basil lamina, which with connecting fibrillar collagen attached to the outside, is called the basement membrane.
8. **B** They are designated Ser-195, His-57, and Asp-102, on the basis of the numbering of chymotrypsin, regardless of the actual numbers in the individual proteases. A: These are found in collagen. C: These are important residues in hemoglobin. D and E: These are not involved.
9. **C** This is the distinguishing characteristic of the serine proteases, and of the serine hydrolases in general. A: They have various specificities. B: There is only one active site per molecule. D: They are all endopeptidases. E: An "extended active site" containing the hydrolyzable bond and about four amino acids on either side is responsible for specificity.
10. **D** The domains are of about equal size.
11. **A** Six-coordinated ferrous ion has a smaller ionic radius than the five-coordinated species and just fits into the center of the porphyrin ring without distortion.
12. **A** Positively charged histidine is no longer stabilized by close proximity to an aspartate carboxylate group. This is the Bohr effect. B: Increased H^+ favors dissociation of carbamino groups; this is what happens in the lung. C: DPG does not bind effectively to oxyhemoglobin because the binding pocket is too small. D: In oxyhemoglobin, NO is bound to cysteine and it gets transferred to glutathione when O_2 is released (R form reverts to T form).
13. Binding of BPG to hemoglobin stabilizes the T conformation, so for any given amount of O_2, less O_2 will be bound to hemoglobin. This shifts the saturation curve to the right and increases P_{50}. The equilibrium shown illustrates that increasing BPG causes the release of O_2, which must happen in order for O_2 to enter tissues.

$$H^+BPG{\cdot}Hb + 4O_2 \rightleftarrows Hb(O_2)_4 + BPG + nH^+ \text{ (Eq. 9.14 from text)}$$

14. The initial defense against pathogens in the nasal mucus is immunoglobulins of the IgA class. If the organism invades the plasma, IgM immunoglobulins are the first antibodies elicited. With time, the IgG class antibodies are synthesized and will reach higher concentrations than IgM antibodies.

10

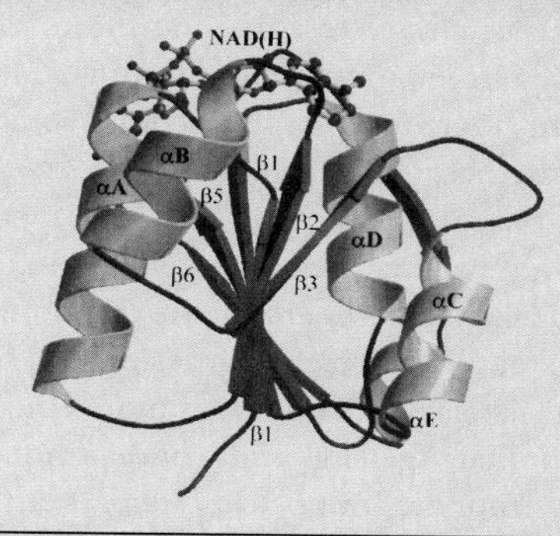

Enzymes: Classification, Kinetics, and Control

Henry Weiner
Professor, Purdue University

CLINICAL CORRELATIONS

Key Concepts

- Enzymes are proteins that catalyze chemical reactions by increasing the rate of conversion of a specific-substrate to a specific product.
- Enzymes lower-the energy of activation of reactions by stabilizing the substrate in a form that allows formation of product. This is referred to as the transition state.
- Coenzymes, most derived from vitamins, are essential participants in many enzyme-reactions; some are chemically altered and others unchanged during the reaction. Some are covalently bound to the enzyme. Metal ions are required for the activity of some enzymes; they are often bound to the protein or a cofactor.
- The chemistry of enzymatic transformations is related to general chemical principles such as acid-base catalysis and nucleophilic attack. Enzyme catalysis is more efficient because groups that protonate (general acids) or deprotonate (general bases) and nucleophilic amino acids are part of the enzyme.
- Rates of enzyme catalyzed reactions are related to concentration of enzyme and substrate and are described by the Michaelis–Menten equation, in which one term, K_m, is related to how well the substrate binds to the enzyme. Another kinetic term, k_{cat}, indicates how fast the enzyme converts the substrate to product.
- Specific chemicals inhibit the binding of substrate and the rate of catalysis. Many drugs are enzyme inhibitors by binding to the substrate-binding site or to a different domain of the protein.
- In addition to the substrate- and coenzyme-binding sites on the enzyme, some enzymes possess additional binding sites (allosteric sites) such that when a compound (ligand) binds, it regulates the catalytic activity. These ligands (allosteric modifiers or effectors) usually induce a change in the structure of the enzyme. The catalytic properties of some enzymes are modified by covalent modification.
- Tissue-specific enzymes released into blood from damaged cells are measured to determine the clinical condition. Enzymes are also used in treatment of some medical conditions.

10.1 • INTRODUCTION

Enzymes are specialized proteins that catalyze biological reactions. Virtually every reaction that occurs within a cell requires the action of an enzyme because most reactions will not occur at a detectible rate under physiological conditions (pH, temperature, and ionic milieu) of the cell. Enzymes are efficient catalysts not only increasing the rate of conversion of substrate to product but also recognizing a specific chemical structure in the presence of similar structures to produce a specific product.

Not all biological catalysts, however, are enzymes. Catalytic RNAs (p. 68) have been identified that participate in processing introns and tRNA, while others perform a self-processing step by hydrolyzing a phosphodiester bond within its own nucleotide chain. Artificial enzymes, termed **abzymes**, have been synthesized by making antibodies against chemicals that are transition-state analogs. Abzymes have been designed to catalyze over 100 different chemical reactions.

10.2 • CLASSIFICATIONS OF ENZYMES

With the advent of genomic sequences, many enzymes have been identified. In mid-2004, information was available for over 83,000 different enzymes from 9800 different organisms; by 2008 the number of entries doubled. All enzymes are classified as belonging to one of six classes, defined by the chemical reaction they catalyze. A numbering system, consisting of four numbers for each enzyme, has been developed by the International Union of Biochemistry and Molecular Biology (IUBMB) to characterize each enzyme. The first number defines the type of reaction that is catalyzed, followed by numbers to define details of the reaction. It is interesting that the entire chemistry of life can be reduced to six different types of chemical reactions. The classes of enzymes are listed in Table 10.1.

The systematic names for enzymes include the substrate and the type of reaction; the enzyme name often ends in "ase." Alcohol dehydrogenase, an enzyme that oxidizes an alcohol to an aldehyde, has the IUBMB number 1.1.1.1. This indicates (1) that the enzyme is involved in an oxidation–reduction reaction (first number), (2) it removes hydrogen as a hydride ion with NAD^+ as the electron acceptor (second number), and (3) substrates for the enzyme can be most primary alcohols (third number). (4) The last number is reserved

TABLE 10.1 • Summary of the Enzyme Classes and Major Subclasses

1. Oxidoreductases Dehydrogenases Oxidases Reductases Peroxidases Catalase Oxygenases Hydroxylases	2. Transferases Transaldolase and transketolase Acyl, methyl, glucosyl, and phosphoryltransferases Kinases Phosphomutases
3. Hydrolases Esterases Glycosidases Peptidases Phosphatases Thiolases Phospholipases Amidases Deaminases Ribonucleases	4. Lyases Decarboxylases Aldolases Hydratases Dehydratases Synthases Lyases 6. Ligases Synthetases Carboxylases
5. Isomerases Racemases Epimerases Isomerases Mutases (not all)	

to differentiate each enzyme that catalyzes the same overall reaction but with different substrates. Lactate dehydrogenase (1.1.1.27) catalyzes an identical reaction as alcohol dehydrogenase, but the substrate is lactate. Enzymes are often named for the direction of the reaction of importance in the cell.

Arbitrary names are frequently given to enzymes; for example, aldolase was the name given and still used for the enzyme that catalyzes an aldol-condensation reaction between glyceraldehyde-3-phosphate and dihydroxyacetone phosphate (p. 597), and lysozyme for an enzyme that breaks down bacterial cell walls. There has been an effort to drop the trivial in favor of the systematic name of many enzymes but the literature continues to contain references to both names for many enzymes.

Isozymes are different structural forms of a protein that catalyze the same reaction. They are products of different genes, possess different degrees of sequence identity, and can have different kinetic properties. There are two isozymes of lactate dehydrogenase, a heart form (H) and a muscle form (M), which have slightly different amino acid sequences. Lactate dehydrogenase is a tetrameric protein, consisting of the M and H monomers, and can have five different tetrameric structures (Table 10.2).

Class 1: Oxidoreductases

Class 1 enzymes catalyze oxidation–reduction reactions and are referred to as oxidoreductase. Oxidation means the loss of electrons, and reduction means the addition of electrons.

TABLE 10.2 • Lactate Dehydrogenase Isozymes

Type	*Composition*	*Location*
LDH_1	HHHH	Myocardium and RBC
LDH_2	HHHM	Myocardium and RBC
LDH_3	HHMM	Brain and kidney
LDH_4	HMMM	
LDH_5	MMMM	Liver and skeletal muscle

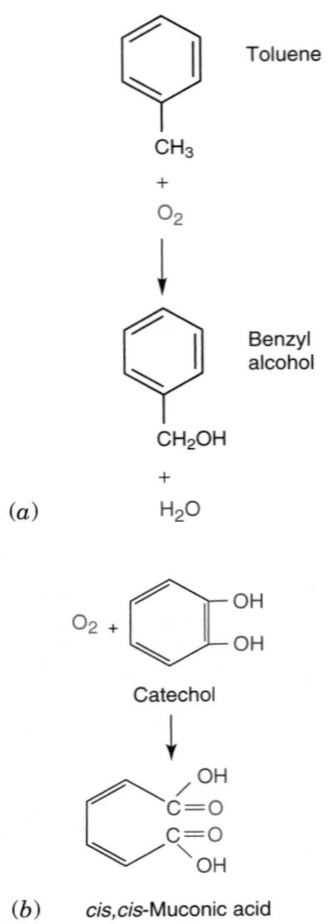

Electrons are removed from one substrate (donor or reductant) that is oxidized and added to a second substrate (acceptor or oxidant) that is reduced. Many different electron acceptors are used in biological systems (p. 561). Dehydrogenases are typically named in the oxidizing direction. Examples are lactate dehydrogenase rather than pyruvate reductase and glyceraldehyde-3-phoshpate dehydrogenase rather than 1,3-bisphosphoglyerate reductase.

Oxidations and reduction reactions include the following pairs of donors–acceptors: saturated–unsaturated carbon–carbon bonds, alcohols–aldehydes, aldehydes–acids, and amines–imines. Molecular oxygen (O_2) as an electron acceptor is involved in a variety of irreversible oxidation–reduction reactions. Monooxygenases catalyze the insertion of a hydroxy group into the substrate with the other atom of oxygen ending up in water (Figure 10.1) whereas dioxygenases incorporate both atoms of O_2 into a substrate. Cytochromes P450 (p. 426) are an important group of enzymes that use oxygen in the metabolism of xenobiotics such as drugs and toxins; these enzymes convert a saturated compound into an alcohol, a hydroxylation reaction using one of the oxygen atoms in O_2 for the hydroxyl group.

Toluene

CH_3 + O_2 →

Benzyl alcohol

CH_2OH + H_2O

(*a*)

O_2 + OH OH

Catechol

↓

OH C=O C=O OH

(*b*) *cis,cis*-Muconic acid

Figure 10.1 Enzyme-catalyzed oxidation reactions. Hydroxylation of toluene, an industrial solvent, and oxygenation of catechol by an oxygenase.

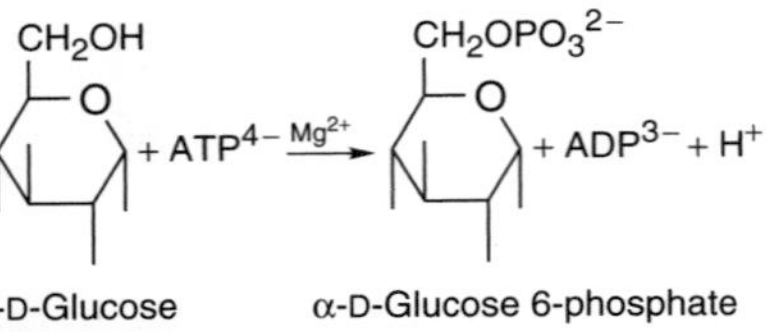

Class 2: Transferases

Members of this large family transfer a chemical group from one molecule to another; thus they have two substrates and produce two products. Hexokinase transfers a phosphate group from ATP to an acceptor such as glucose (Figure 10.2). An enzyme that transfers phosphate to another molecule using a nucleotide triphosphate as the donor is referred to as a **kinase**; the acceptor can be a small molecule or a protein. Another example of a transferase is acyltransferase that catalyzes the transfer of an acetyl group from acetylCoA to an acceptor such as lysine in a histone or a serine in other proteins (Figure 10.3).

An important amine transfer reaction occurs between an amino acid and a carbonyl-containing compound. These enzymes, **aminotransferases**, transfer the amino group and convert the amino acid to a keto acid; the accepting substrate, a keto acid, is transformed in to an amino acid (Figure 10.4). Aminotransferases require pyridoxal phosphate, derived from vitamin B_6, as a coenzyme (p. 755).

CH_2OH + ATP^{4-} $\xrightarrow{Mg^{2+}}$ $CH_2OPO_3^{2-}$ + ADP^{3-} + H^+

α-D-Glucose α-D-Glucose 6-phosphate

Figure 10.2 Phosphorylation reaction. Phosphorylation of glucose by ATP catalyzed by hexokinase.

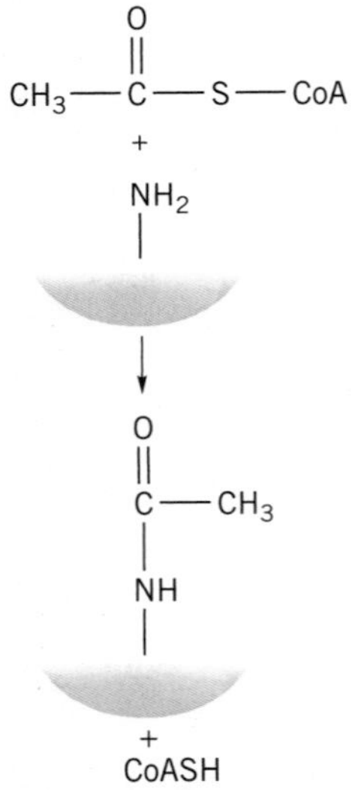

Class 3: Hydrolases

Hydrolases catalyze hydrolysis reactions, which are the addition of water to a chemical bond (Figure 10.5). This is essentially a transfer of an –OH from water to the substrate, but the enzymes are not classified as transferases. Hydrolysis is essentially an irreversible reaction. The substrate is typically an ester or an amide, for example, an ester between an alcohol and a carboxylic acid. Such a reaction is the hydrolysis of a cholesterol ester to produce cholesterol and a fatty acid. The acid need not be a carboxylic acid; sulfuric acid, phosphoric acid, or any acid can form an ester bond. RNases and DNases are enzymes that hydrolyze the phospho–ester bond. The clinical use of a hydrolase is presented in Clin. Corr. 10.1.

CH_3—C(=O)—S—CoA + NH_2 → C(=O)—CH_3 NH + CoASH

Figure 10.3 Acetylation of amino acid residues in a protein.

Class 4: Lyases

Lyase implies a breaking apart. These enzymes usually catalyze a carbon–carbon bond cleavage. Other enzymes in this class can form or break a carbon–nitrogen bond or release a CO_2 from a β-keto acid. Some of the reactions are reversible so a carbon–carbon bond can form, as illustrated by the reaction catalyzed by aldolase (Figure 10.6). Some lyases require pyridoxal phosphate as the cofactor. Important neurochemical reactions including the formation of dopamine (Figure 10.7) and serotonin are examples of this reaction. If a nucleotide triphosphate is not involved in forming the new bond, the enzyme is called

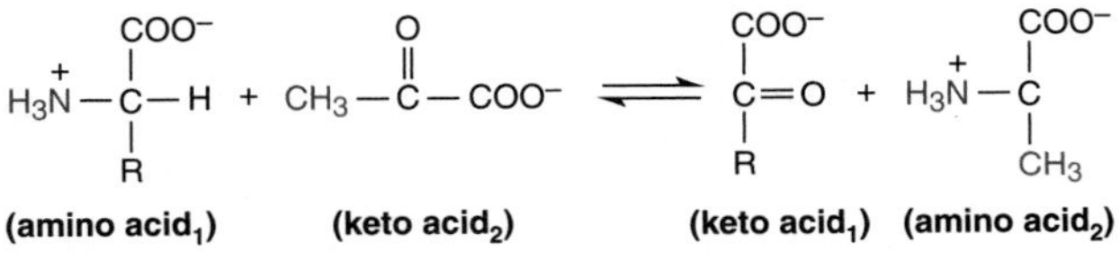

$$H_3\overset{+}{N}-\underset{R}{\overset{COO^-}{C}}-H + CH_3-\overset{O}{\overset{\|}{C}}-COO^- \rightleftharpoons \underset{R}{\overset{COO^-}{C}}=O + H_3\overset{+}{N}-\underset{CH_3}{\overset{COO^-}{C}}$$

(amino acid$_1$) (keto acid$_2$) (keto acid$_1$) (amino acid$_2$)

Figure 10.4 Example of a reaction catalyzed by an aminotransferase. Aminotransferase reactions require pyridoxal phosphate (PLP).

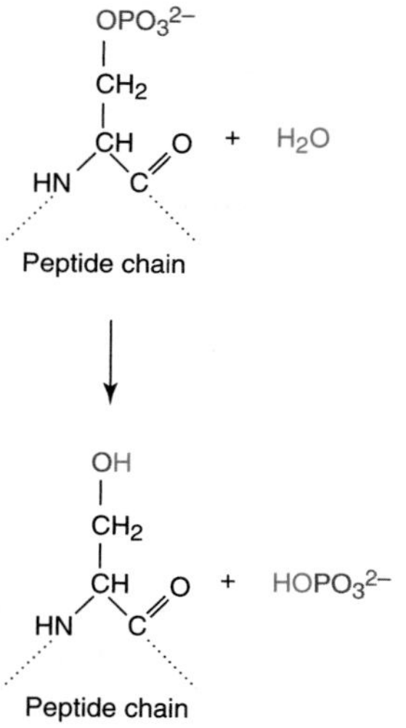

Peptide chain + H_2O → Peptide chain + $HOPO_3^{2-}$

Figure 10.5 Hydrolysis of a phosphorylated protein by a protein phosphatase.

a **synthase**, but if the energy released when ATP is hydrolyzed is required, the enzyme is referred to as a **synthetase** (see Class 6: Ligases).

Class 5: Isomerases

The enzymes in this class are involved with moving a group or a double bond within the same molecule. These include exchanging the position of a hydroxyl and a carbonyl, such as found when glucose-6-phosphate is converted to fructose-6-phosphate (p. 596), and moving of a double bond from one position to the adjacent one, as found in fatty acid metabolism (p. 695). The enzyme is called a mutase when a phosphate is moved from one carbon to another within the same molecule such as occurs with phosphoglycerate mutase that converts 2-phosphoglycerate to 3-phosphoglycerate (Figure 10.8). Isomerases and epim-erases may change the stereochemistry at a carbon atom. The conversion of D-lactate to L-lactate is an example of an isomerase and D-xylulose 5-phosphate to D-ribulose 5-phosphate an epimerase (Figure 10.9). The UDP-glucose 4-epimerase-catalyzed reaction requires the involvement of NAD, so it is assumed that an oxidation–reduction reaction occurs during the reaction but there is no net oxidation–reduction of the coenzyme.

CLINICAL CORRELATION 10.1

Hydrolysis of Asparaginase and Leukemia

Some forms of juvenile leukemia require the nonessential amino acid asparagine. Early investigators immobilized bacterially derived asparaginase, a hydrolase enzyme that converts asparagine into aspartic acid, and passed the patient's blood over a column of immobilized enzyme. This placed the patient afflicted with acute lymphoblastic leukemia into remission. Later investigators found that they could inject the enzyme into the patient where it would hydrolyze the substrate circulating in blood. More recently the enzyme was encoded in polyethlenglycol as some patients had a reaction to the bacterial enzyme even though it was purified to complete homogeneity. The encoded enzyme functioned well in children and helped clear their blood of asparagine. This form of the enzyme is commercially available and is approved for human treatment.

Appel, I. M., Kazemier, K. M., Boos, J., Lanvers, C., et al. Pharmacokinetic, pharmacodynamic, and intracellular effects of PEG-asparaginase in newly diagnosed childhood acute lymphoblastic leukemia: Results from a single agent window study. *Leukemia* 22:1665, 2008.

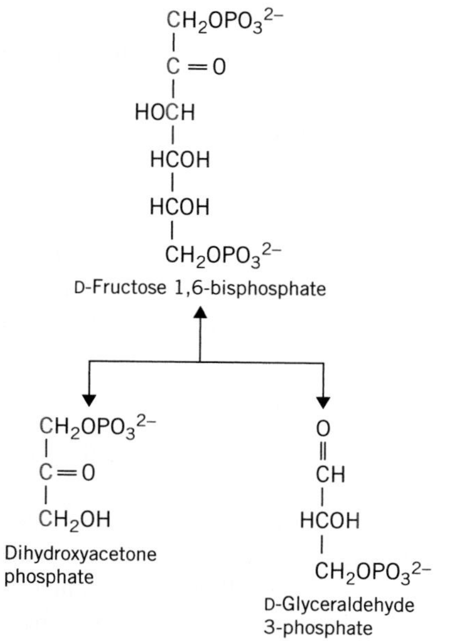

D-Fructose 1,6-bisphosphate ⇄ Dihydroxyacetone phosphate + D-Glyceraldehyde 3-phosphate

Figure 10.6 Lyase reaction catalyzed by aldolase.

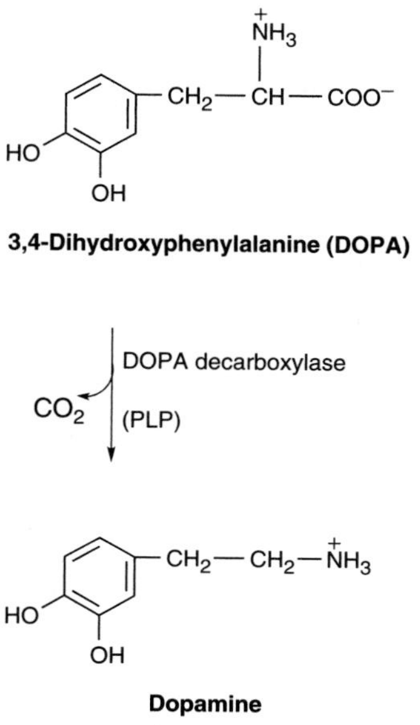

3,4-Dihydroxyphenylalanine (DOPA) → (DOPA decarboxylase, PLP; CO_2) → Dopamine

Figure 10.7 Dopamine synthesis involves a lyase reaction. Synthesis of dopamine requires pyridoxal phosphate (PLP).

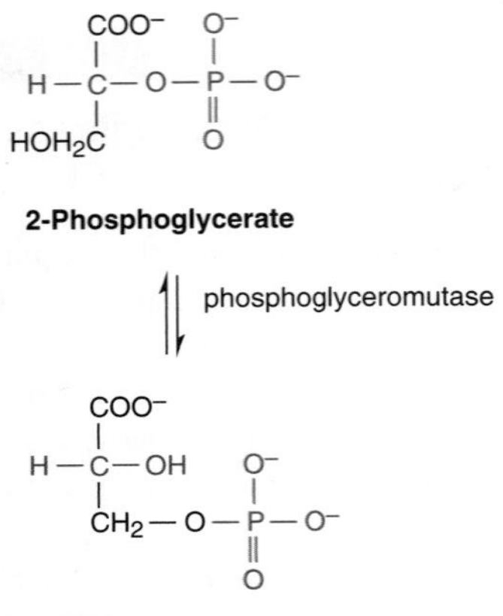

2-Phosphoglycerate

phosphoglyceromutase

3-Phosphoglycerate

Figure 10.8 Interconversion of 2- and 3-phosphoglycerates.

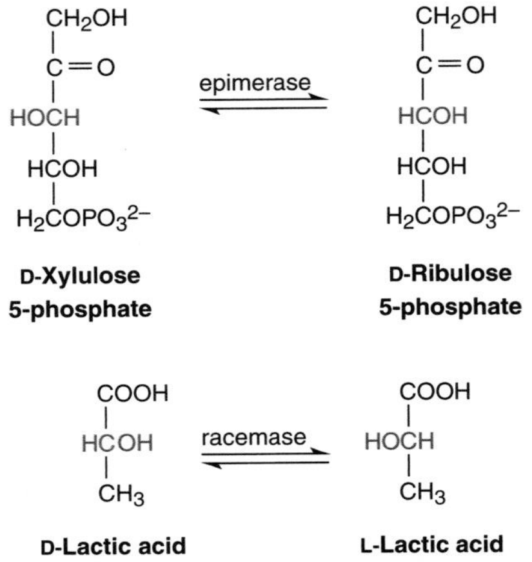

D-Xylulose 5-phosphate ⇌ (epimerase) D-Ribulose 5-phosphate

D-Lactic acid ⇌ (racemase) L-Lactic acid

Figure 10.9 Examples of reactions catalyzed by an epimerase and a racemase.

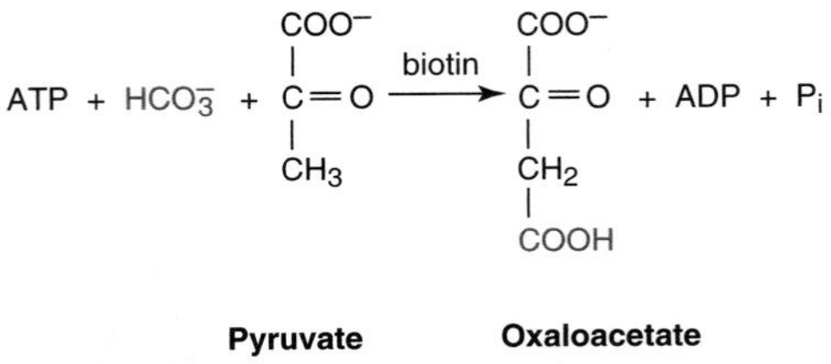

ATP + HCO_3^- + Pyruvate —(biotin)→ Oxaloacetate + ADP + P_i

Figure 10.10 Pyruvate carboxylase reaction. See page 622 for details of biotins role.

Class 6: Ligases

Ligases join carbon atoms together but unlike lyases, Class 4, they require energy for the reaction. They are referred to as **synthetases**. Typically, the energy comes from the involvement of ATP. For example, to add CO_2 to pyruvate, CO_2 is incorporated into the coenzyme biotin, which requires hydrolysis of ATP (Figure 10.10). The CO_2 is transferred to pyruvate by the same enzyme. The addition of an amino acid to tRNA is another example of an enzyme in this class (p. 216). There must be some interaction between the components of the reaction to utilize the energy released from ATP hydrolysis for synthesis of the new bond.

10.3 • GENERAL CONCEPTS OF ENZYME MECHANISMS

Thermodynamic Considerations

Energy is required to form the starting material of a reaction and to form the product of a reaction. This energy is referred to as the **heats of formation**. There is a **thermodynamic relationship** between the difference in energy needed to form the starting material (reactants) and to form the products. The energy difference is ΔG_o, and the units can be expressed as either kilojoules/mole (kJ/mol) or kilocalories/mole (kcal/mol) (A Closer Look 10.1). When the equilibrium of the reaction is established, the concentration of each is related as indicated in Eq. 10.1.

$$\Delta G_0 = -2.3RT \times \log[\text{Products}]/[\text{Reactants}] \qquad (10.1)$$

R is the universal gas constant (8.31 joule/mole/°K, or 1.99cal/mole/°K) and T is the temperature in °K.

When added to a reaction, a catalyst increases the rate of reaching equilibrium between the starting material and the product. The ratio of [Products]/[Reactants] at equilibrium is the equilibrium constant, K_{eq}. This implies that the difference in the free energy (heats of formation) of the products and reactants will determine the amount of each that is present when equilibrium is reached. It does not dictate the rate at which equilibrium is reached. A catalyst increases the rate of the reaction to reach equilibrium but does not change the equilibrium. K_{eq} is governed by the heats of formation, which is a constant for any reaction.

Many reactions are considered irreversible; it is not possible for products of irreversible reactions to be reconverted to starting material even in the presence of an enzyme because essentially 100% of the starting material will become product. This statement is not completely valid, for even with essentially irreversible reactions, the ratio of starting material to products might be 1:200,000 or even 1:1,000,000. The reaction reaches equilibrium but

A CLOSER LOOK 10.1

Units of Energy

Two different conventions exist for describing the units of energy released or required for a chemical reaction. One is calories and the other is joules with the latter now being used in the scientific literature. Calories are still used to describe how much energy is available in food. For example, a typical glass of beer contains approximately 125 Cal (125 kilocal) of energy when metabolized. To convert calories to joules, one multiplies the caloric value by 4.184. Thus, the universal gas constant (R) in Eq. 10.1 is 1.987 cal/mol/°K, or 8.31 J/mol/°K. This book will use kilojoules (*kJ*) to describe the energetics of biological transformations, but it will also present the values in kilocalories (kcal or Cal). Nutritional discussions will use calories (Cal).

the amount of starting material present at equilibrium is lower than the level of detection, so the reaction is considered essentially irreversible.

Binding of Substrate by an Enzyme

For an enzyme to function as a catalyst it must first bind the substrate(s), and then lower the energy of activation so that the reaction will proceed at a faster rate than it would in the absence of the catalyst. Two prevalent theories exist to explain how substrate binds to the enzyme. The classical one is the **lock-and-key theory**, which proposes that the binding site for substrate on the enzyme is a rigid entity and only a compound with a particular shape will fit, analogous to how a lock (enzyme) allows only one key (substrate) to make the proper contact. An alternative theory is the **induced-fit theory**. This assumes that the enzyme is flexible, and after substrates bind to the enzyme, the conformation of the protein changes so that a stable binary complex forms. Both of the theories were presented prior to knowledge of the actual structure of any protein. Now that three-dimensional structures of many enzymes are known, it is found that substrate binding causes a small movement of the peptide bonds supporting the induced-fit hypothesis. A large conformational change is illustrated in Figure 3.33, p. 97. Figure 10.11 illustrates a change in conformation.

The altered structure of an enzyme after a substrate binds allows for complementary interactions between substrate and enzyme to take place. That is, a hydrophilic portion of the substrate will lie near amino acid residues that allow a hydrogen bond to be shared, and a charged portion of the substrate will interact such that either its charge is neutralized by an opposite charge or it is stabilized by forming a hydrogen bond. Hydrophobic portions of the substrate bind such that they are in a hydrophobic portion of the protein, referred to as a **hydrophobic pocket**.

Transition State

The theory to describe how a chemical reaction occurs is based on the transition-state model. The basic concept is that the starting material (substrate) has certain structural and chemical properties, and a product has others. The transition state is a structure where the bond(s) undergoing a transformation are such that they are not like either the starting material or the product.

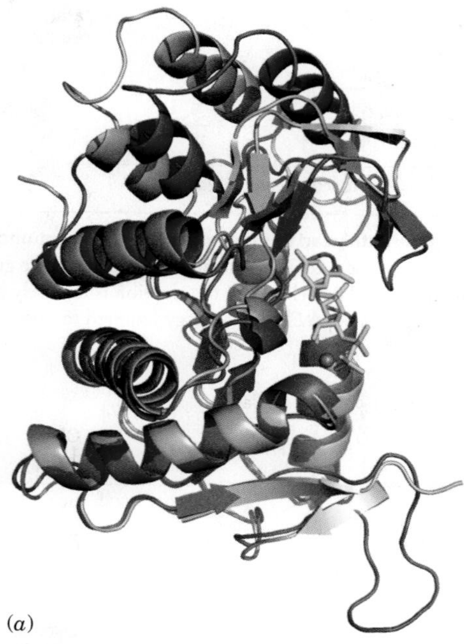

(*a*)

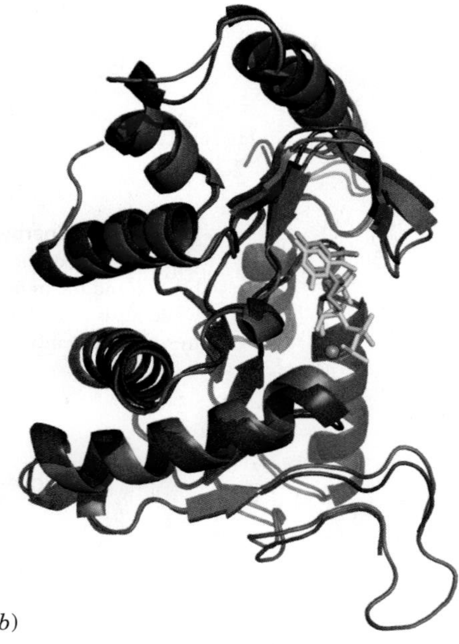

(*b*)

Figure 10.11 Orotate phosphoribosyltransferase. Orotate phosphoribosyltransferase (OPRTase) catalyzes the $Mg2^+$-dependent condensation of orotic acid (OA) with PRPP (5-α-D-phosphorylribose 1-diphosphate) to yield pyrophosphate (PP_i) and the nucleotide OMP (orotidine 5′-monophosphate). (*a*) Free enzyme in cyan and the ternary complex in blue. (*b*) Ternary complex in red and binary complex in blue. See page. 821 for reactions involved in pyrimidine biosynthesis.
Reprinted with permission from Gonzlaes-Segurea, L, Whitte, J. F., McClard, R. W., and Hurley, T. D. *Biochemistry* 46:14075, 2007. Copyright © (1977) by American Chemical Society.

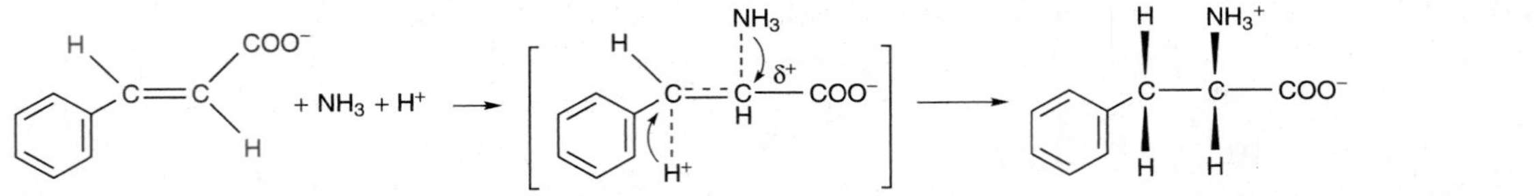

Figure 10.12 Transition state. The transition-state structure shows electrons in the double bond moving to accommodate the incoming proton, which makes the other carbon positive so the pair of electrons on the nitrogen can attack.

This can be illustrated in the reaction of ammonia adding to *trans*-cinnamic acid to produce phenylalanine (Figure 10.12), where the double bond in the starting material is planar. The product has the amine attached so the carbon that was part of the double bond (planar) is now tetrahedral in structure. The transition state is where the bond is starting to form and the geometry is changing. It takes energy to change the shape of the substrate so it will resemble the transition state, and the velocity of a reaction is inversely proportional to the energy needed to reach the transition state. A reaction energy plot is presented in Figure 10.13 and used to depict the energy change necessary to convert a substrate to the transition state. At the transition state, an entity exists that has never been observed or isolated. The transition-state entity can proceed to produce product, or it can return to the starting material. A catalyst simply lowers the energy of the transition state (ΔE^*), thus speeding up the rate of its formation from the starting material.

Enzymes stabilize the transition state by lowering the energy of activation, hence speeding up the reaction. This is accomplished by having residues in the enzyme interact with the substrate while in the transition state.

The hydrolysis of a peptide bond to produce a carboxylic acid and a free amino group illustrates how it might be possible for the enzyme to stabilize the transition state (Figure 10.14). As water adds to the peptide bond, the carbonyl group becomes tetrahedral and hence has to change its bond angle and shape. The transition state is not the tetrahedral intermediate but is the conformation that would lead to it should the covalent

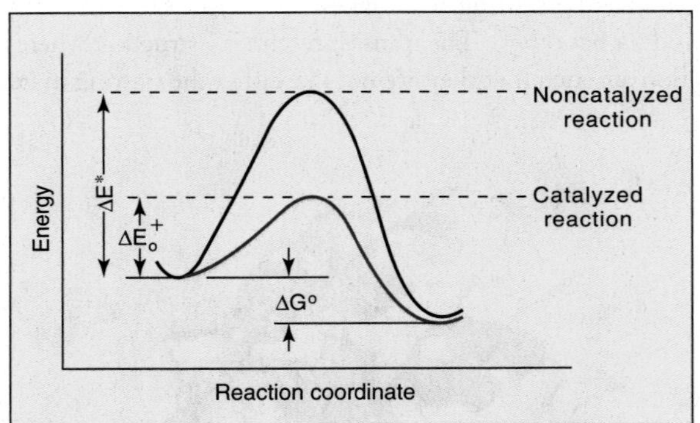

Figure 10.13 Energy diagrams for catalyzed versus noncatalyzed reactions. Diagrammatic representation of the energy differences between the reactant and product. The ground state is indicated by ΔG_0 and the transition state by ΔE^*. The velocity of the reaction is inversely proportional to ΔE^* while K_{eq} is related to the difference in ΔG_0. ΔE^* is the energy required to raise the ground state to the hypothetical transition state.

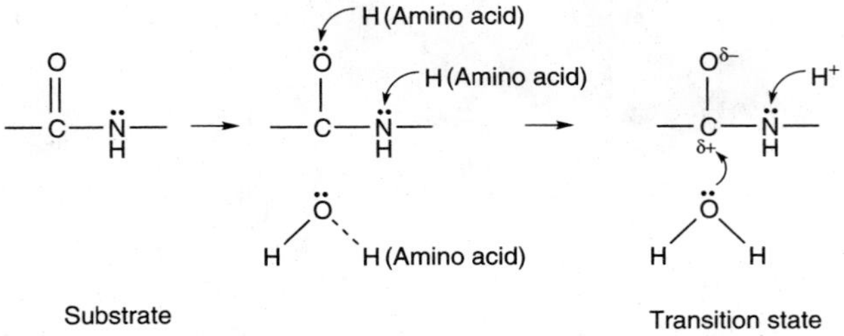

Substrate

Transition state

Figure 10.14 Transition state for the hydrolysis of a peptide bond. The carbonyl group in the planar peptide bond becomes tetrahedral as water (or OH^-) attacks and the nitrogen is protonated, followed by hydrolysis of the peptide bond. Amino acids in the active site furnish an H^+ to stabilize the developing negative charge on the carbonyl oxygen atom and protonate the nitrogen so it will leave as an amine. A different amino acid in the protein removes a proton from water, producing the more reactive OH^-.

bond actually form. During the reaction not only does the carbon–oxygen bond of the carbonyl group on the substrate change from being planar to being tetrahedral, but a negative charge develops on the oxygen that needs to be stabilized.

Tight Binding in the Transition State

An enzyme binds tightly the transition-state structure but not the substrate or final product. If it binds substrate tightly, then it takes more energy to reach the transition state. Similarly, the enzyme cannot bind the product too tightly, for if it did, it would become too difficult for the product to dissociate from the enzyme. The enzyme needs to release product so it can again interact with another molecule of substrate. These concepts are illustrated in Figure 10.15.

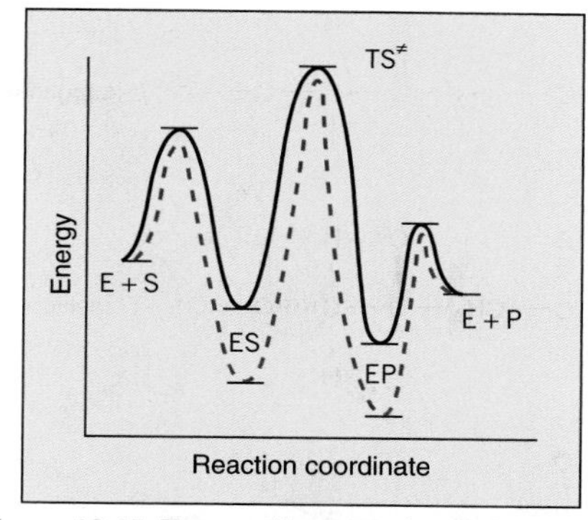

Figure 10.15 Energy diagrams for transition state of tightly bound substrate. Red line indicates tight binding of substrate in comparison to a less tightly bound substrate. More energy is required if the substrate is tightly bound, hence a slower reaction to reach the transition state. A similar condition occurs if the product is bound too tightly; more energy would be required to reach the transition state for releasing product. There is a transition state for each step in the reaction.

Ionic Reactions Need Not Involve Ions

The majority of reactions found in nature can be explained using ionic models. That is, most cellular reactions require positive and negative charges within the molecules to allow the components to interact. The enzyme can add or remove protons from the substrate, changing its charge, or a neutral group of the enzyme becomes ionic. The removal of a proton makes a group more nucleophilic, which is a group (nucleophile) that will attack a positive center. For example, OH^- is more nucleophilic than is water and ionized cysteine is more nucleophilic than is the unionized sulfur atom. Similarly, enzymes can protonate a group to make it positive so it will be more susceptible to nucleophilic attack. In the hydrolysis of a peptide bond the more nucleophilic $-OH^-$ group attacks the bond rather than water. In addition, the attack will be on a polarized carbonyl group, where the carbon atom would have a partial positive charge, rather than on an unpolarized carbonyl group (Figure 10.14). The addition and removal of protons is referred to as **general acid** and **general base catalysis**, respectively. The advantage of enzymes using general acid and general base catalysis is that at a physiological neutral pH, they can catalyse a reaction even though the concentration of OH^- and H^+ is very low.

Partially Charged Bonds

A common way for forming a carbon–carbon bond is to have a carbanion attack a carbonyl group, as illustrated with the formation of fructose-1,6-bisphosphate from two three-carbon compounds (Figure 10.16*a*). The presence of the carbonyl group in dihydroxyacetone phosphate allows the hydrogen atoms on the carbon adjacent to be more acidic. The enzyme removes a proton, leaving a negative charge on the carbon. Electrons of a carbonyl group reside closer to the oxygen; thus the carbon carries a partial positive charge. This stabilizes the carbanion since the electrons are delocalized and not just associated with one carbon. The carbanion then can attack the carbonyl group to form the new bond. The enzyme both holds the two substrates near to each other increasing the likelihood of favorable collision and is involved in polarizing the carbonyl bond and removing a proton.

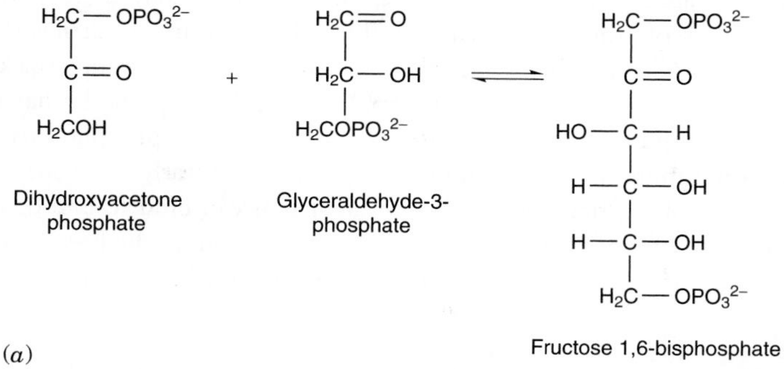

Dihydroxyacetone phosphate + Glyceraldehyde-3-phosphate ⇌ Fructose 1,6-bisphosphate

(*a*)

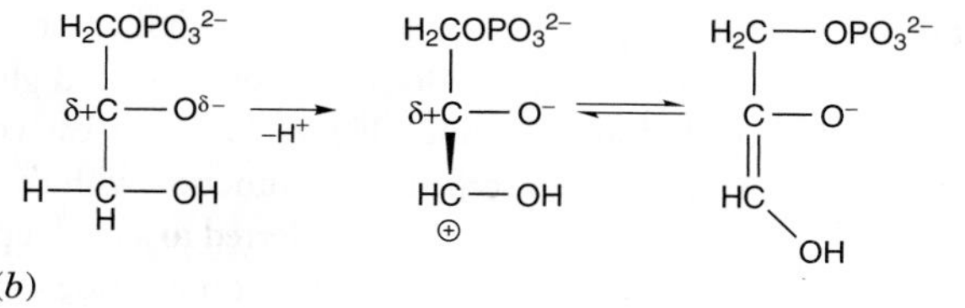

(*b*)

Figure 10.16 A carbonyl group stabilizes a negative charge on the adjacent carbon. (*a*) Reaction of dihydroxyacetone phosphate and glyceraldehydes-3-phosphate to form fructose 1,6-bisphosphate. (*b*) Dihydroxyacetone phosphate transition states. The partial positive charge on the carbon can be attacked by a carbanion to form a new carbon–carbon bond. The presence of the carbonyl group will allow a proton to be removed from the adjacent carbon atom so the carbanion can be produced.

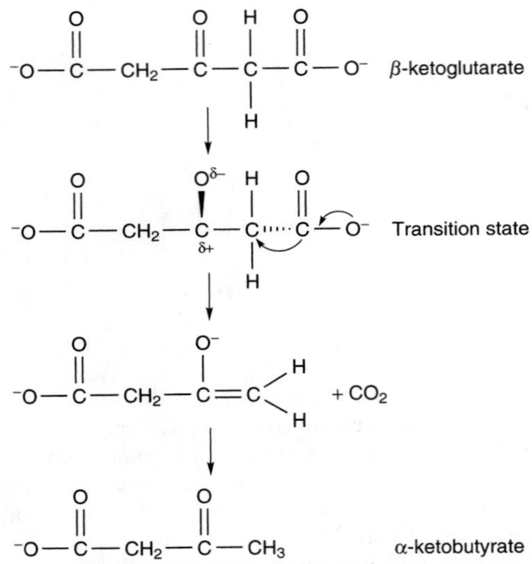

Figure 10.17 **Decarboxylation of β-ketoglutarate.** The decarboxylation of β-ketoglutarate occurs because the partial positive charge on the carbon of the carbonyl group stabilizes the developing negative charge that forms as the bond to the carboxyl group starts to break. A carbonyl adjacent to the leaving CO_2 cannot stabilize the charge.

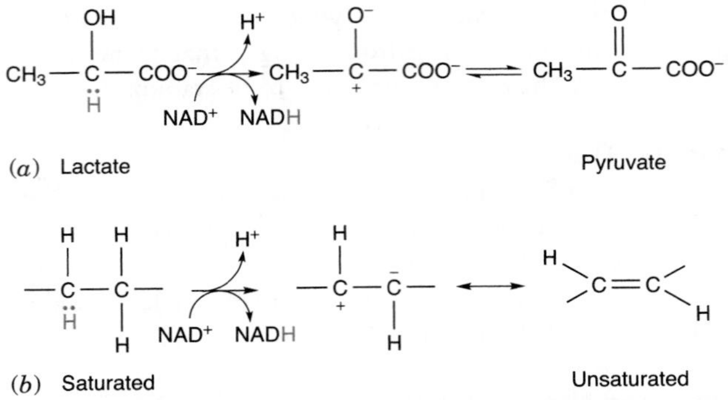

Figure 10.18 **Oxidation of lactate and a saturated carbon bond.** (*a*) Lactate to pyruvate. (*b*) Saturated bond to an unsaturated bond. These reactions undergo an ionic-type reaction even though a hydride ion (H^-) is involved.

Importance of the Carbonyl Group

Many metabolic transformations are based on the special nature of the carbonyl group and its ability to stabilize a negative charge on the carbon adjacent to it. For example, β-keto acids such as β-ketoglutarate (Figure 10.17) can be decarboxylated but it is energetically more difficult for an α-keto acid such as pyruvate to be decarboxylated. As the —CH_2—COO^- bond breaks, the electrons in the bond are left behind on the β-carbon. The developing negative charge on what was the β-carbon can be stabilized by the partial positive charge on the carbonyl carbon. The enzyme stabilizes the transition state by having a proton interact with the carbonyl oxygen that is partially negative due to the resonance of the carbonyl carbon and the developing negative charge on the β-carbon.

In contrast, pyruvate cannot undergo a simple decarboxylation reaction, as there would be no way to stabilize the negative charge that would result from the pair of electrons being left as the CO_2 group departs. For decarboxylation of α-keto acids, coenzymes are required.

Oxidations

Even oxidation reactions have an ionic component. During the oxidation of an alcohol to an aldehyde/ketone or a carbon–carbon saturated bond to an unsaturated bond, the electrons are moved not as free electrons but as a hydride ion, that is, a proton with two electrons carrying a negative charge (Figure 10.18). The hydride ion does not go into the solvent as does a proton but to an acceptor such as NAD(P) or FAD.

Addition and Removal of Protons

Most enzymes efficiently add and remove protons but often in a unique way. Proteases can polarize the carbonyl group of the peptide bond so that the bond is more easily hydrolyzed, catalyzing general base catalysis, or acid catalysis, or metal ion catalysis (p. 393).

Phosphorylation of proteins, especially on the hydroxyl group of a serine, threonine, or tyrosine residues by ATP or GTP, proceeds by having the —OH group attack phosphate (Figure 10.19). It is necessary to make the —OH more nucleophilic by having it lose a proton while the phosphate–oxygen bond is being polarized so the phosphorous atom is partially positive and is more susceptible to nucleophilic attack. Nearly every enzyme (kinases) catalyzing this type of reaction uses Mg^{2+}. Divalent ions can bind to two or three of the phosphates. The presence of the positive ion causes the electrons in the P═O bond to reside closer to the oxygen atom thus leaving the phosphate ion with a slightly positive charge.

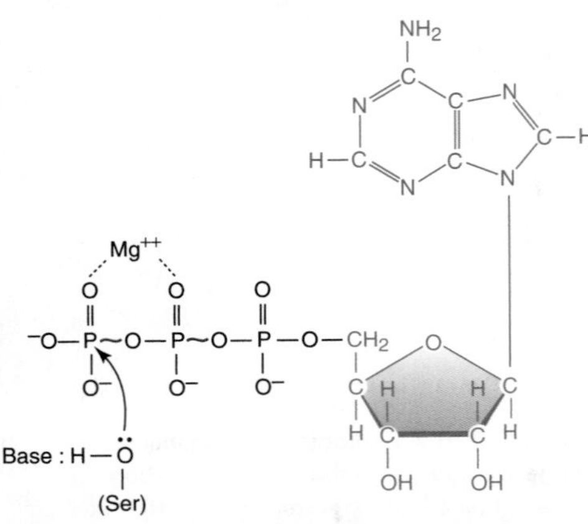

Figure 10.19 **Phosphorylation is an attack of a nucleophile on a nucleoside triphosphate (NTP).** ATP or another NTP is attacked by the pair of electrons on the oxygen of an alcohol such as serine, threonine, or tyrosine. A Mg^{2+} ion when bound by the NTP serves to pull electrons away from the phosphorous group making it more susceptible to attack by the nucleophile.

Covalent Binding of Substrate to Enzyme

Enzymes not only function in performing general acid and base catalysis, but in some cases the substrate forms a covalent bond with the enzyme. Both chymotrypsin and glyceraldehyde 3-phosphate dehydrogenase are examples of such enzymes (Figure 10.20). In each case, a nucleophilic amino acid attacks a carbonyl group. Even some kinases function with covalent catalysis; ATP initially binds to the enzyme, and the phosphate is transferred to a nucleophilic residue forming a phosphorylated enzyme; in this case a hydroxyl group on a serine or threonine is

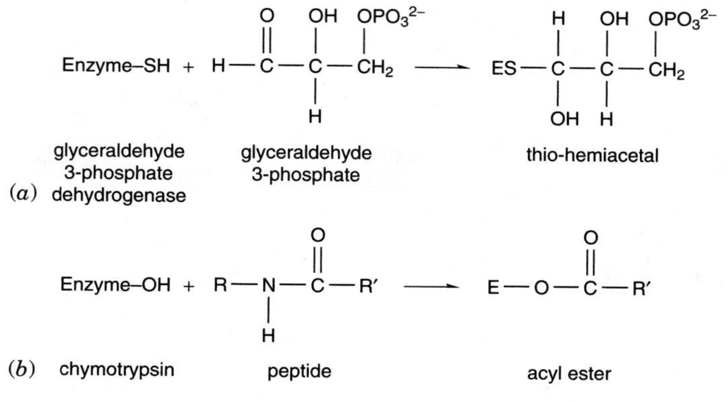

$$\text{Enzyme–SH} + \text{H—}\overset{\text{O}}{\overset{\|}{\text{C}}}\text{—}\underset{\text{H}}{\overset{\text{OH}}{\text{C}}}\text{—}\overset{\text{OPO}_3^{2-}}{\text{CH}_2} \longrightarrow \text{ES—}\underset{\text{OH}}{\overset{\text{H}}{\text{C}}}\text{—}\underset{\text{H}}{\overset{\text{OH}}{\text{C}}}\text{—}\overset{\text{OPO}_3^{2-}}{\text{CH}_2}$$

(*a*) glyceraldehyde 3-phosphate dehydrogenase; glyceraldehyde 3-phosphate; thio-hemiacetal

$$\text{Enzyme–OH} + \text{R—}\underset{\text{H}}{\text{N}}\text{—}\overset{\text{O}}{\overset{\|}{\text{C}}}\text{—R}' \longrightarrow \text{E—O—}\overset{\text{O}}{\overset{\|}{\text{C}}}\text{—R}'$$

(*b*) chymotrypsin; peptide; acyl ester

Figure 10.20 Some substrates form a covalent bond with an enzyme. (*a*) Glyceraldehyde-3-phosphate dehydrogenase and (*b*) chymotrypsin function by first forming a covalent intermediate with substrate.

activated and does the initial attacking. Substrate then enters, and it is activated by the general base on the enzyme; the base attacks the phosphate group forming the final product.

pH Alters a Reaction by Affecting General Acids and Bases

Nearly all enzymes involve general acid/base catalysis, and it is not unexpected that reaction rates would be affected by pH. An enzyme requiring a histidine to donate a proton in a reaction will not proceed at pHs above 8, where most of the imidazole groups of histidine are unprotonated. The pK_a of the imidazole is near 7 (Figure 10.21). All enzymes have a pH versus rate profile (p. 405), which in some cases gives an insight as to the groups involved in the active site. Since the overall structure of a protein is important in the catalytic activity, if a change in pH alters the overall protein conformation, then the activity will also be changed. In some disease states, the intracellular pH is not normal; this will affect cellular enzyme activity.

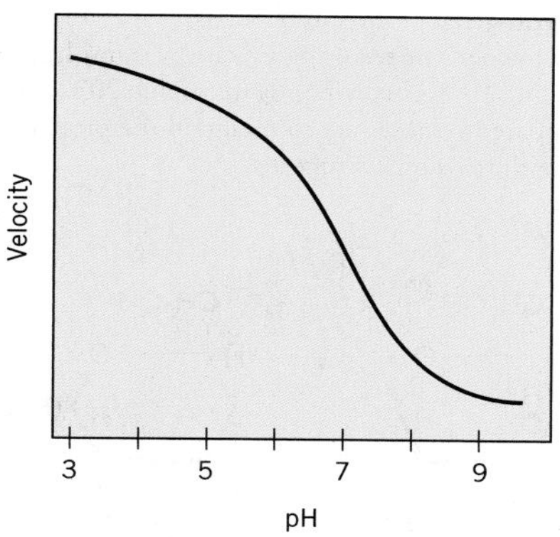

Figure 10.21 Velocity of an enzyme-catalyzed reaction can be affected by pH if a group dissociates a H^+. If an imidazole of histidine needs to be protonated for catalytic activity of an enzyme, the velocity of the reaction will decrease as the pH increases above the pK_a (assume 7) of the residue. If the group needed to be unprotonated for the enzyme to function, then the curve would have the reverse shape, with maximum activity above pH8. Half maximal activity would be found at the pK_a of the group because 50% of the molecules will be ionized and 50% unionized.

10.4 • ACTIVE SITE OF AN ENZYME

The active site of an enzyme is small compared to the overall size of the protein molecule. Most of the amino acid residues are not in contact with the substrate (Figure 10.22). The distances and angles between the catalytic residues of the enzyme and the substrate must be exact to permit catalysis to occur. Most of the amino acids in the protein serve as a large scaffold to allow for the proper alignment of the functional groups of the substrates. A mutation of a residue distant from the active site can cause the enzyme to have altered activity. This is illustrated in Figure 10.23 for a point mutation in human liver aldehyde dehydrogenase.

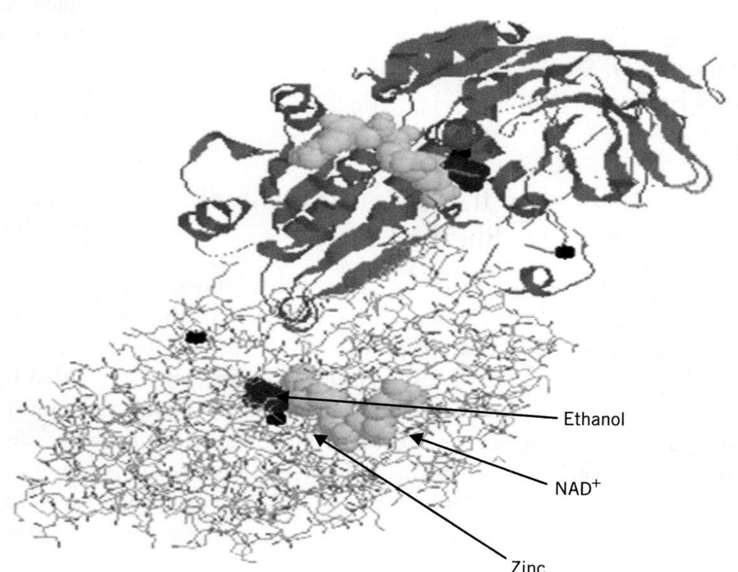

Figure 10.22 Enzymes are much larger than their substrates. Alcohol dehydrogenase with ethanol bound to a Zn^{2+} and an NAD^+ at the active site. Most of the amino acids of the protein are not in contact with the substrate. One subunit of the dimer is shown as ribbons and the other as atoms.

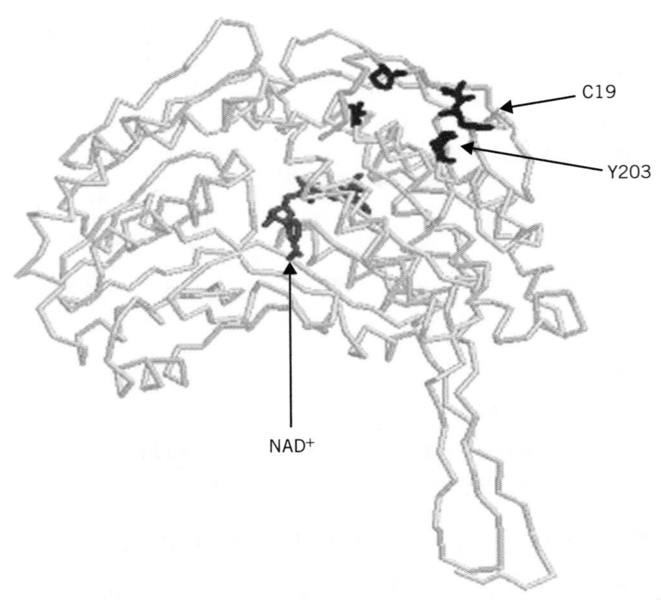

Figure 10.23 Amino acids at a distance from the active site are critical. An amino acid residue on the surface of an enzyme might be necessary to keep the enzyme in its proper shape. Residue 19, a cysteine, of a subunit of aldehyde dehydrogenase (500 amino acids) was changed to tyrosine. The resulting mutant was insoluble. Residue 19 normally bonds to residue 203, which is apparently necessary to maintain the proper three-dimensional structure.

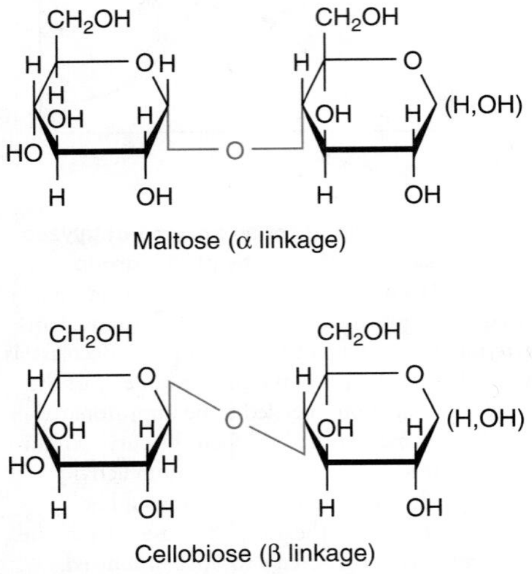

Maltose (α linkage)

Cellobiose (β linkage)

Figure 10.24 Enzymes can differentiate between α and β linkages. An enzyme that hydrolyzes maltose does not hydrolyze cellobiose.

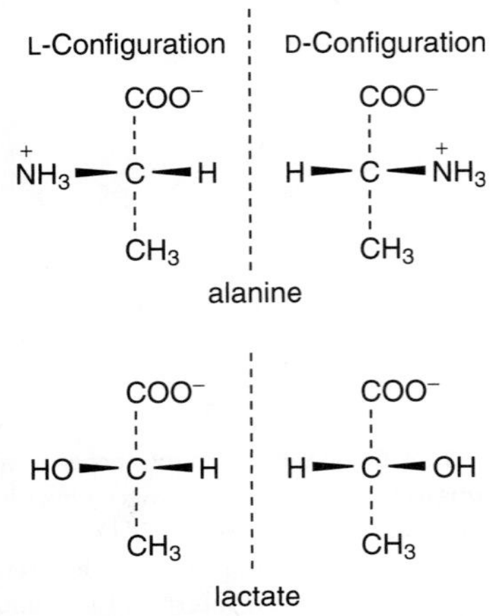

L-Configuration | D-Configuration

alanine

lactate

Figure 10.25 Enzymes can differentiate between optical isomers.

Stereochemistry of Active Site

The active site is specific for the substrate. As an example, subtle geometric constraints prevent an enzyme from hydrolyzing a β-glycosidic linkage between two glucose molecules while allowing the enzyme to hydrolyze the α-linkage (Figure 10.24). Enzymes also differentiate between optical isomers; though both the D and L forms of many compounds exist, mammalian cells utilize only one stereoisomer. D-Lactate is hardly recognized by mammalian lactate dehydrogenase, which is specific for the L-isomer (Figure 10.25). The binding can also react with pro-chiral centers to produce an optically active compound. A chiral carbon has four different groups attached and is optically active; a **pro-chiral center** is a carbon that is not optically active but becomes optically active on a stereo-specific addition of a group. The non-optically active compound binds to the enzyme in just one orientation, forcing the chemical reaction to produce an optically pure product. Glycerol, a non-optically active compound, is phosphorylated by glycerol kinase; only the L-product is produced. The enzyme binds the compound such that just one of the primary alcohols is near ATP. This occurs because the surface of the enzyme is asymmetric since the amino acids are all L configuration and the helices are all right handed (p. 93). Though perhaps an over simplification, it is often stated that if there were a three-point attachment of the substrate to the enzyme, then an optically pure product would be formed from a pro-chiral center (Figure 10.26). See A Closer Look 10.2 for a discussion on how stereochemistry explains some enzyme mechanisms.

Influence of Groups on the Substrate Distal to the Bond Being Modified

The chemistry at the active site in some cases involves a limited portion of the substrate, but adjoining portions of the substrate can be necessary for the reaction to occur. For example, the mitochondrial-processing protein peptidase requires at least 12 residues before the site of hydrolysis in order to function. Lysozyme, an enzyme that hydrolyzes bacterial cell walls, involves sugar residues located on both sides of the bond to be split (Figure 10.27). The binding of adjacent sugar moieties helps create a change in the conformation of the sugar at the bond being broken. Thus, the active site is more than just the amino acid residues involved in the bond breakage or formation; it is a complex region on the enzyme that can adjust its structure to accommodate the proper substrate and stabilize the transition state.

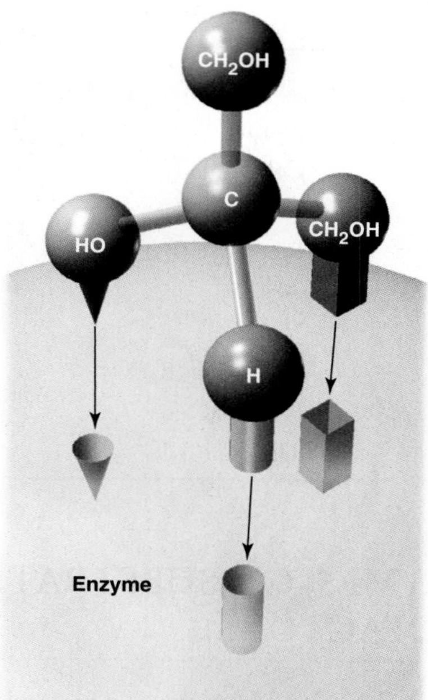

Figure 10.26 Three-point attachment of a symmetrical substrate to an asymmetric substrate-binding site. Glycerol kinase by virtue of dissimilar binding sites for the —H and —OH group of glycerol binds only the α'-hydroxymethyl group to the active site. One stereoisomer results from the kinase reaction, L-glycerol 3-phosphate. Active site is box on the right.

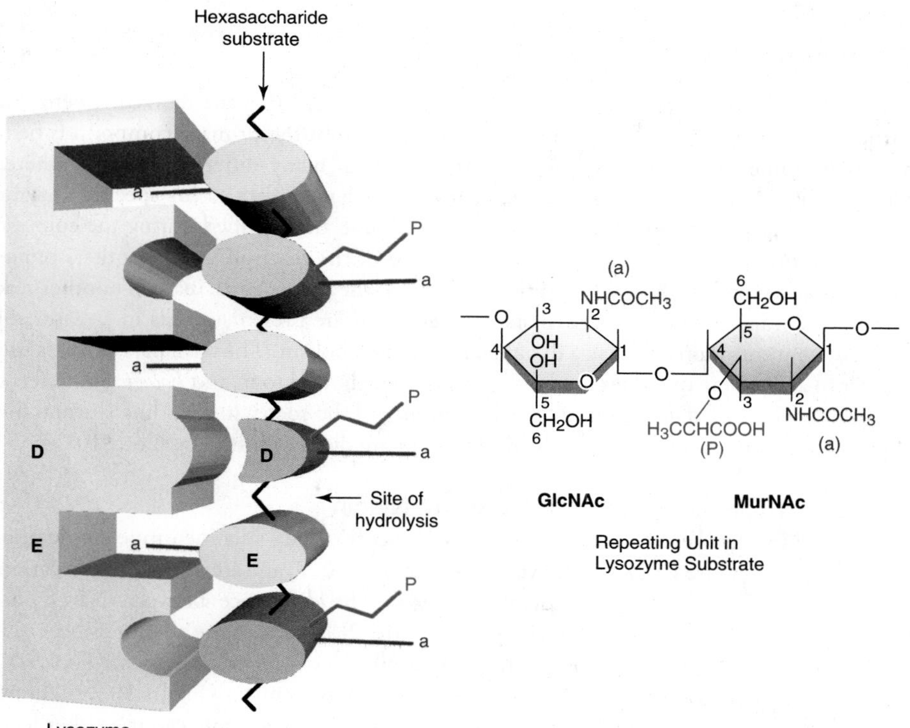

Figure 10.27 Hexasaccharide binding at active site of lysozyme. In the model substrate pictured, the ovals represent individual pyranose rings of the repeating units of the lysozyme substrate shown to the right. Ring D is strained by the enzyme to the half-chair conformation, and hydrolysis occurs between the D and E rings. Six subsites on the enzyme bind the substrate. Alternative sites are specific for acetamido groups (*a*) but are unable to accept the lactyl (P) side chains, which occur on the *N*-acetylmuramic acid residues. Thus the substrate can bind to the enzyme in only one orientation.

Redrawn based on model proposed by Imoto, T., et al. In Boyer, P. (Ed.), *The Enzymes*, 3d ed., Vol. 7. New York: Academic Press, 1972, 713.

A CLOSER LOOK 10.2

Stereochemistry Helps Explain the Mechanism of Enzyme-Catalyzed Reactions

Many naturally occurring substrates are optically active. The stereochemistry of the product can give a clue to the mechanism the enzyme employs. For example, if the reaction produces a carbonium ion, one would expect the product to be racemic. However, an enzyme-catalyzed carbonium ion reaction will produce a product with retention of stereochemical configuration. The carbonium ion would not be free in solution but would be associated with the enzyme, so the incoming nucleophilic attack could come only from the same direction as the leaving group. Retention of configuration can also be achieved by an enzyme that functions with covalent catalysis. Reactions catalyzed by a ping-pong mechanism (p. 405) involve two inversion steps that produce a product with retention of configuration. As an example, the phosphorolysis of a disaccharide starts with the sugar in the β conformation and ends with a phosphorylated product that is also β. The attachment of the substrate to the enzyme causes inversion to an α configuration, but phosphate attacks to give another inversion resulting in retention of configuration. If the reaction occurred by a direct attack of the phosphate on the substrate, then the product would have been inverted. Inversion of configuration at an optically active carbon usually occurs when there is a ternary complex and the nucleophile attacks the substrate.

TABLE 10.3 • Coenzymes

Coenzyme	*Vitamin*	*Reaction Mediated*
Biotin	Biotin	Carboxylation
Cobalamin (B_{12})	Cobalamin (B_{12})	Alkylation
Coenzyme A	Pantothenate	Acyl transfer
Flavin coenzymes	Riboflavin (B_2)	Oxidation–reduction
Lipoic acid		Acyl transfer
Niacin coenzymes	Niacin	Oxidation–reduction
Pyridoxal phosphate	Pyridoxine (B_6)	Amino group transfer
Tetrahydrofolate	Folic acid	One-carbon group transfer
Thiamin pyrophosphate	Thiamin (B_1)	Carbonyl transfer

10.5 • COENZYMES, CO-SUBSTRATES, AND COFACTORS

Many enzymes require the participation of a nonprotein molecule during the enzymatic reaction. **Coenzymes** are small organic molecules; most are derivatives of vitamins (Chapter 26) Coenzymes are involved in many different reactions and may or may not be modified in the reaction. Those that are altered are also referred to as **co-substrates**. For some reactions the energy from hydrolysis of ATP is required without incorporation of phosphate in the product; in this case ATP is a co-substrate. Metal ions are required for many enzyme reactions and are termed **cofactors**.

Coenzymes

Table 10.3 lists the coenzymes and the vitamins from which they are derived. Coenzymes participate in enzyme-catalyzed reactions, but they are not the primary compounds being modified. Some coenzymes participate in the action of many different enzymes whereas others only in a limited number of reactions. They have affinities for specific enzymes and may be tightly bound or covalently attached. Some are modified during the course of a reaction but are in their original state at the end of the reaction whereas others remain modified at the end. If they are modified at the end, they must participate in another reaction to be returned to their original state. Coenzymes are present in cells in a reasonably constant concentration and play a dynamic role in metabolism. The term *apoenzyme* is used to refer to an enzyme that does not have the coenzyme bound to it, and *holoenzyme* refers to an enzyme with its cofactor bound. The importance of the coenzyme-binding site and how alterations in this site cause metabolic dysfunction are discussed in Clin. Corr. 10.2.

NAD and NADP Are Coenzyme Forms of Niacin

Nicotinamide adenine dinucleotide (NAD) and nicotinamide adenine dinucleotide phosphate (NADP) (Figure 10.28) are derived from niacin (p. 1077) and are involved in oxidation–reduction reactions catalyzed by dehydrogenases. The difference between NAD^+ and $NADP^+$ is the presence of a phosphate group on the 2′ carbon of the adenine ribose. The nomenclature for the abbreviations for these coenzymes is described in A Closer Look 10.3. A generalized dehydrogenase reaction involving NAD^+ is presented in Figure 10.29. Though most dehydrogenases are specific for one or the other coenzyme, some can use either form. Most dehydrogenases catalyze reversible reactions.

The functional portion of the oxidized coenzyme is the nicotinamide ring that has a nitrogen atom at position 1 and carries a positive charge due to the attachment to a ribose ring. The positive charge makes it possible for the nicotinamide to accept a hydride ion, transferring two electrons (Figure 10.30). A ternary complex between enzyme, coenzyme, and substrate forms and a hydride ion is transferred from substrate to $NAD(P)^+$. A proton from the substrate is released to the medium. For the reverse

CLINICAL CORRELATION 10.2

Mutation of a Coenzyme-Binding Site Results in Clinical Disease

Cystathioninuria is a genetic disease in which γ-cystathionase is either deficient or inactive. Cystathionase catalyzes the reaction

$$\text{Cystathionine} \rightarrow \text{cysteine} + \alpha\text{-ketobutyrate}$$

Deficiency of the enzyme leads to accumulation of cystathionine in the plasma. Since cystathionase is a pyridoxal phosphate-dependent enzyme, vitamin B_6 was administered to patients whose fibroblasts contained material that cross-reacted with antibodies against cystathionase. Many responded to B_6 therapy with a fall in plasma levels of cystathionine. These patients produce the apoenzyme that reacted with the antibody. In one patient the enzyme activity was undetectable in fibroblast homogenates but increased to 31% of normal with the addition of 1 mM of pyridoxal phosphate to the assay mixture. It is thought that the K_m for pyridoxal phosphate binding to the enzyme was increased because of a mutation in the binding site. Activity is partially restored by increasing the concentration of coenzyme. Apparently these patients require a higher steady-state concentration of coenzyme to maintain γ-cystathionase activity.

Pascal, T. A., Gaull, G. E., Beratis, N. G., Gillam, B. M., et al. Vitamin B_6-responsive and unresponsive cystathionuria: two variant molecular forms. *Science* 190:1209, 1975.

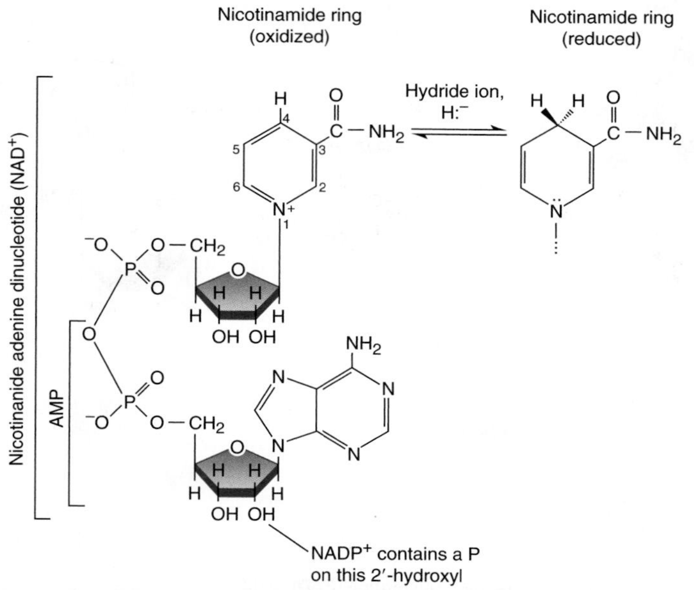

Figure 10.28 Nicotinamide adenine dinucleotide (NAD) and nicotinamide adenine dinucleotide phosphate (NADP).

A CLOSER LOOK 10.3

Abbreviations for NAD and NADP

The notation NAD and NADP are used generically to indicate the coenzymes whether oxidized or reduced. NAD^+ and $NADP^+$ specify the oxidized forms, and NADH and NADPH the reduced forms. Some enzymes use either NAD or NADP, so they are referred to as NAD(P)-dependent enzymes.

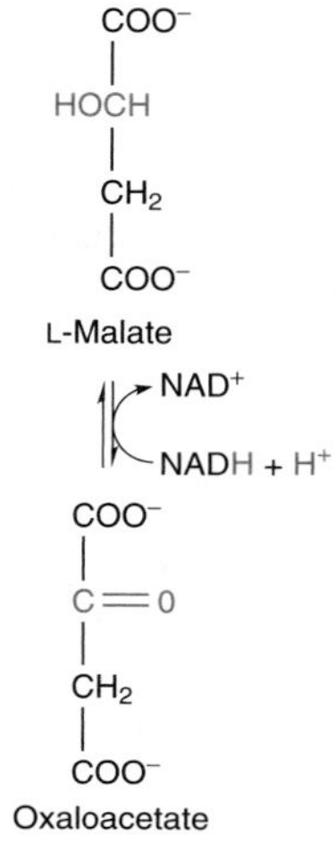

reaction, the reduced nicotinamide ring is involved and the ring does not have the positive charge until the hydride is transferred to an oxidized substrate. The reactions in which NAD(P) participate can be considered as ionic though the ion they transfer is a hydride ion that carries a negative charge. Hydride ions do not exist in the free state; the ion is transferred directly between substrate and coenzyme. After accepting or donating the hydride ion, the coenzyme dissociates from the enzyme and enters the cellular pools of NAD^+ or $NADP^+$ and is available for other reactions. The ratios of NAD^+/NADH and $NADP^+$/ NADPH are independent of a specific dehydrogenase but are measures of the oxidation–reduction state of the cellular compartment (p. 561). NAD-dependent dehydrogenases are more often involved in catabolism whereas NADP-dependent ones in anabolism.

Nearly all NAD(P)-dependent dehydrogenases possess a similar structural motif (**Rossmann fold**) involved in the coenzyme binding domain; the motif for alcohol dehydrogenase is presented in Figure 10.31.

Figure 10.29 Reaction catalyzed by malate dehydrogenase. The enzyme forms a ternary complex by binding both NAD^+ and malate before the hydride ion is passed to NAD^+.

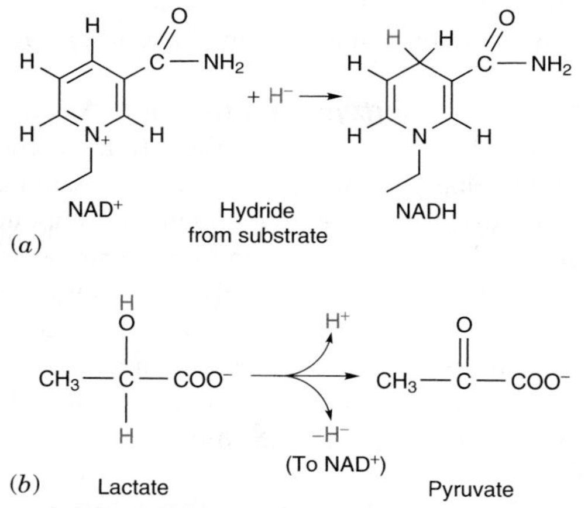

Figure 10.30 NAD is involved in oxidation–reduction reactions. (*a*) Nicotinamide group in NAD^+ accepts a hydride ion (H^-) while oxidizing substrate. (*b*) Lactate donates the hydride ion when oxidized to pyruvate.

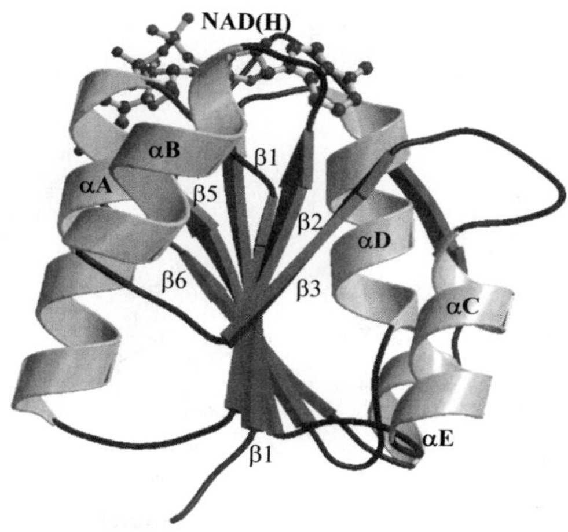

Figure 10.31 The Rossmann fold in liver alcohol dehydrogenase. NAD(H) binds to the Rossmann fold in most dehydrogenases.

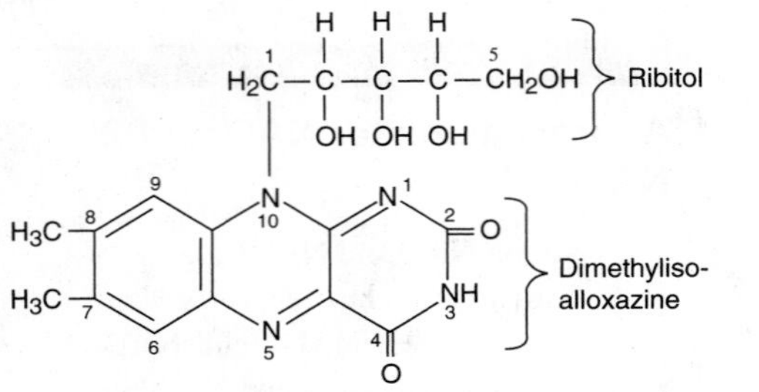

Riboflavin

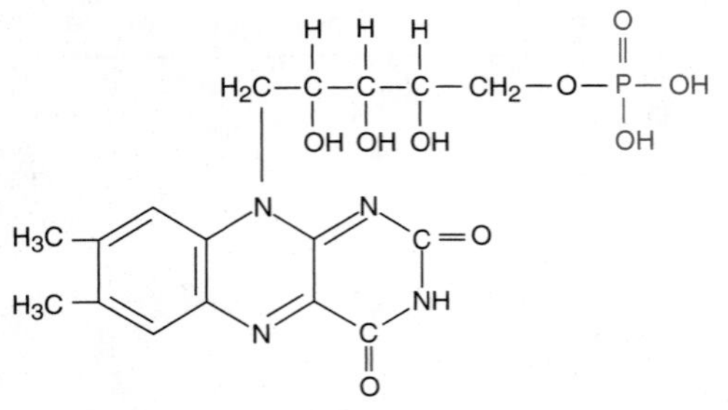

Flavin mononucleotide (FMN)

Figure 10.32 Riboflavin and flavin mononucleotide.

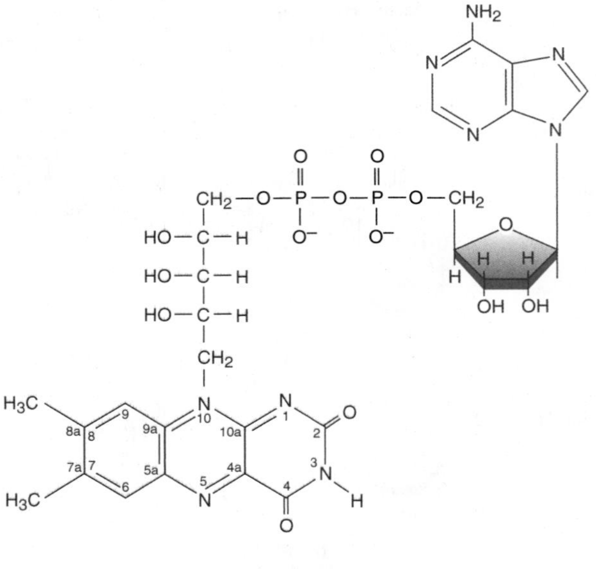

Flavin adenine dinucleotide (FAD) (oxidized)

Figure 10.33 Flavin adenine dinucleotide (FAD).

FMN and FAD Are Coenzyme Forms of Riboflavin

The two coenzyme forms of riboflavin are **FMN** (**flavin mononucleotide**) and **FAD** (**flavin adenine dinucleotide**). The vitamin riboflavin consists of the heterocyclic ring, isoalloxazine (flavin) connected through N-10 to the alcohol ribitol (Figure 10.32). FMN has a phosphate esterified to the 5′-OH group of ribitol. FAD is structurally analogous to NAD in having adenosine linked by a pyrophosphate linkage to sugar of a flavin mononucleoside (Figure 10.33). Both FAD and FMN function in oxidoreduction reactions by accepting and donating $2e^-$ in the isoalloxazine ring in the form of a hydride. In some cases, these coenzymes are $1e^-$ acceptors, which lead to flavin semiquinone formation (a free radical). Flavin coenzymes are not found free in solution but are bound either very tightly or covalently. They are considered a **prosthetic group** in the latter case. This requires that an enzyme involved in an oxidation reaction generating reduced bound flavin must also carry out the reduction of another substrate so that oxidized flavin will be regenerated.

The oxidation of a saturated carbon–carbon bonds, such as in oxidation of succinic acid to fumaric acid (Figure 10.34) or a saturated to an unsaturated fatty acid, involves flavin-dependent dehydrogenase reactions. In the case of monoamine oxidase, a flavin-containing enzyme, a hydride ion is removed from the amine but the terminal acceptor of electrons is molecular oxygen, leading to formation of hydrogen peroxide.

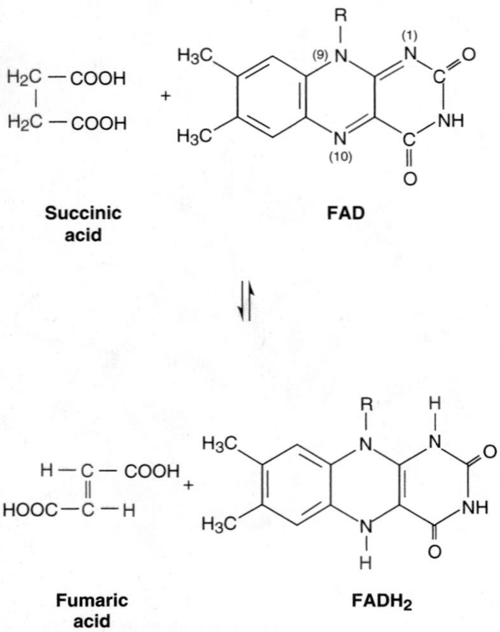

Succinic acid + FAD ⇌ Fumaric acid + $FADH_2$

Figure 10.34 FAD is the coenzyme in the succinic dehydrogenase reaction.

Pyridoxal Phosphate Is the Coenzyme Form of Pyridoxal

Pyridoxal phosphate (Figure10.35) is involved primarily in reactions involving amino acids. A major reaction utilizing pyridoxal phosphate is transamination, a reaction (see Figure 10.4, p. 380) that introduces an amino group into a compound. The acceptor is a carbonyl group such as an α-keto acid and the donor is an amino acid such as glutamate or aspartate (see Figure 19.3, p. 753). The amino acid bound to pyridoxal phosphate transfers the amino group to it making an enzyme-bound pyridoxamine. After the substrate which donated the amino group leaves the enzyme, a different α-keto acid enters and there is a transfer of the amine from pyridoxamine. Pyridoxal is thus regenerated. This is an example of the modification of a coenzyme during the reaction but returning to its original state at the end.

Resonance in the pyridine ring of the coenzyme stabilizes a negative charge at what was the α-carbon. The basis of the reaction is that the aldehyde in the pyridoxal moiety binds with the amino acid making a Schiff base. It is on this covalent intermediate that the enzyme catalyzes the removal of the group. A pair of electrons is left on the carbon, and their charge is stabilized by resonance with the pyridoxal ring. The formation of dopamine and serotonin, important neurotransmitters, involves a pyridoxal phosphate enzyme.

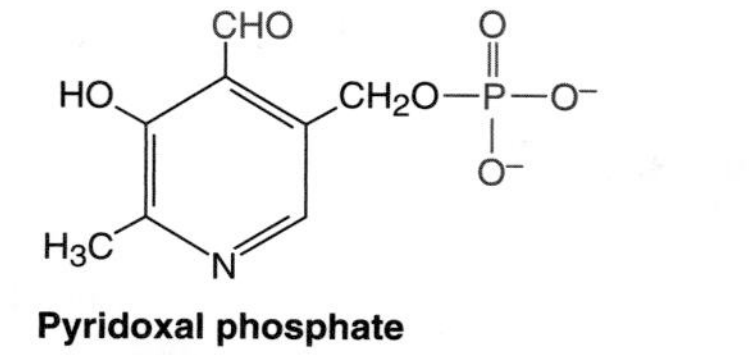

Pyridoxal phosphate

Figure 10.35 **Pyridoxal phosphate.** Pyridoxal phosphate binds to the amino group of an amino acid and assists in the removal of the R group, the H or the CO_2 of the amino acid.

Adenosine Triphosphate May Be a Second Substrate or a Modulator of Activity

Adenosine triphosphate (ATP) (Figure 10.36) functions as co-substrate, but can also serve as a cofactor in modulation of the activity of some enzymes. This compound is central in biochemistry and is synthesized in all mammalian cells. The biochemically functional end is the reactive triphosphate. The terminal phosphate–oxygen bond has a high free energy of hydrolysis, which means that the phosphate can be transferred from ATP to other acceptor groups. For example, as a co-substrate ATP is utilized by the kinases for the transfer of the terminal phosphate to various acceptors. A typical example is the reaction catalyzed by hexokinase (see Figure 15.4; p. 594).

ATP also serves as a modulator of the activity of some enzymes. These enzymes have binding sites for ATP, occupancy of which changes the affinity or reactivity of the enzyme toward its substrates. In these cases, ATP in these cases is an **allosteric effector** (p. 413).

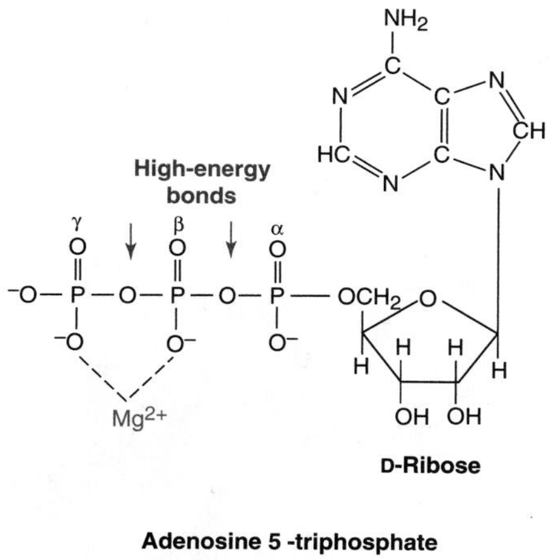

High-energy bonds

D-Ribose

Adenosine 5 -triphosphate

Figure 10.36 **Structure of Mg^{2+} –ATP Orotate phosphoribosyltransferase.**

Metal Ion Cofactors

Many enzymes require the presence of a metal ion for activity. Metal ions serve a structural role or function as a Lewis acid (a positive ion that can bind to unpaired electrons) or as an electron acceptor/donor in oxidation–reduction reactions. They can bind directly to an enzyme or to the substrate. The metal-ion-bound substrate is the actual substrate for the reaction. A very common reaction involves a Mg^{2+} ion bound to a nucleotide triphosphate; the metal ion binds to the phosphates making the phosphate ion more positive and susceptible to nucleophilic attack (Figure 10.37).

Metalloenzymes contain a tightly bound transition metal, such as Zn^{2+} or Fe^{2+}, which facilitates binding of substrate by forming a metal-bridged complex. Metals function either by binding substrate and promoting electrophilic (nucleophilic) catalysis at the site of bond cleavage or by stabilizing intermediates in the reaction pathway. In alcohol dehydrogenase, an enzyme-bound Zn^{2+} interacts with the oxygen in the substrate, changing the properties of the carbon to which the oxygen is attached (Figure 10.38). Zn^{2+} functions in carboxypeptidase and thermolysin, proteases with identical active sites, to

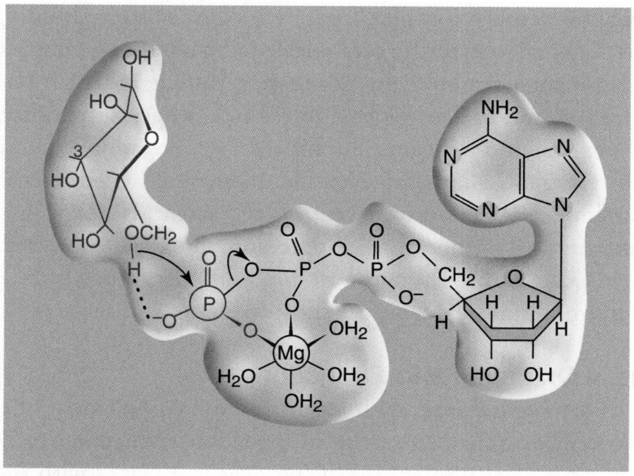

Figure 10.37 **Role of Mg^{2+} as a substrate-bridged complex in the active site of the kinases.** In hexokinase the terminal phosphate of ATP is transferred to glucose, yielding glucose 6-phosphate. Mg^{2+} coordinates with the ATP to form the true substrate and may labilize the terminal P—O bond of ATP to facilitate transfer of the phosphate to glucose. There are specific binding sites (light blue) on the enzyme (darker blue) for glucose (upper left in red) as well as the adenine and ribose moieties of ATP (black).

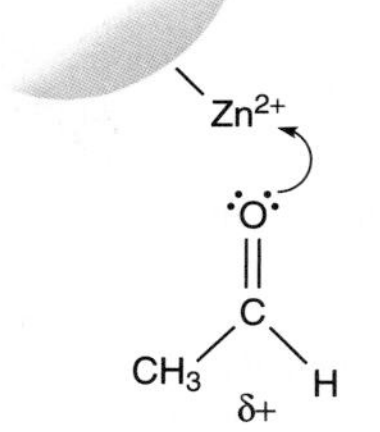

Figure 10.38 **Binding of a carbonyl to Zn^{2+}.** Zn^{2+} binding of the carbonyl of an aldehyde results in making the carbon atom partially positive and more receptive to accepting a hydride ion from NADH.

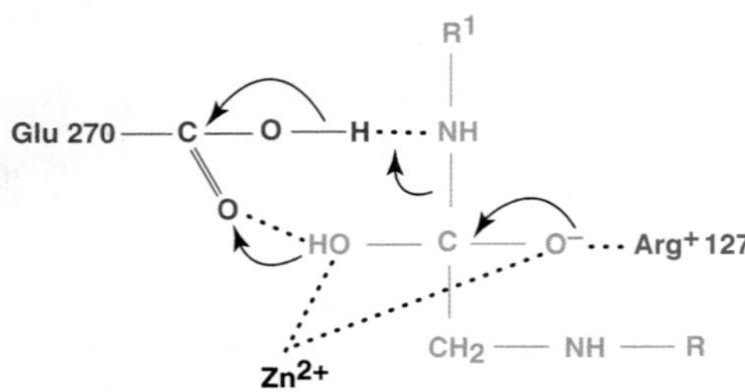

Figure 10.39 Stabilization of the transition state of the tetrahedral intermediate by Zn^{2+}. Positive charge on the Zn^{2+} stabilizes the negative charge that develops on the oxygens of the tetrahedral carbon in the transition state. The tetrahedral intermediate then collapses as indicated by the arrows, resulting in breakage of the peptide bond.

Figure 10.40 Zn^{2+} in the mechanism of reaction of carboxypeptidase A. Enzyme-bound Zn^{2+} generates a hydroxyl nucleophile from bound water, which attacks the carbonyl of the peptide bond as indicated by the arrow. Glu 270 assists by pulling the proton from the zinc-bound water.
Redrawn from Lipscomb, W. N., and Robert, A. *Welch Found. Conf. Chem. Res.* 15:140, 1971.

generate a hydroxyl group from water and to stabilize the transition state resulting from attack of the hydroxyl on the peptide bond. Figure 10.39 depicts the generation of the active-site hydroxyl by Zn^{2+}. Glu 270 functions as a base necessary to remove the proton from water. Stabilization of the tetrahedral transition state by Zn^{2+} is shown in Figure 10.40. The Zn^{2+} provides a counterion to stabilize the negative oxygen on the tetrahedral carbon.

In addition to the role of binding enzyme and substrate, metals also bind directly to the enzyme to stabilize it in the active conformation or perhaps to induce the formation of an active site. Mn^{2+} has such a role as do weakly bound alkali metals (Na^+ or K^+). In pyruvate kinase, K^+ induces an initial conformational change, which is necessary, but not sufficient, for ternary complex formation. On substrate binding, K^+ induces a second conformational change to the catalytically active ternary complex as indicated in Figure 10.41. Mammalian alcohol dehydrogenase is unique in that it possesses two molecules of Zn^{2+} per subunit; one is involved in catalysis whereas the other has a structural role. Each metal ion is bound to ligands; in this case they are amino acid side chains of the protein. For example, Zn^{2+} binds four ligands. In alcohol dehydrogenase the structural Zn^{2+} has four amino acids bound to it whereas the active site Zn^{2+} has just three with water or substrate serving as the fourth. Some compounds such as azide ion (N_3^-) or carbon monoxide (CO) bind to the substrate-binding site on metals, thus preventing the enzyme or protein from binding its proper substrate.

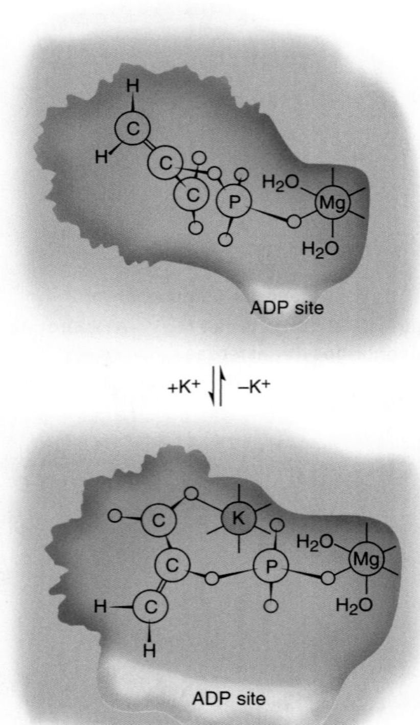

Figure 10.41 Model of the role of K^+ in the active site of pyruvate kinase. Pyruvate kinase catalyzes the reaction: phosphoenolpyruvate + ADP → ATP + pyruvate. Initial binding of K^+ induces conformational changes in the kinase, which result in increased affinity for phosphoenolpyruvate. In addition, K^+ orients the phosphoenolpyruvate in the correct position for transfer of its phosphate to ADP (not shown), the second substrate. Mg^{2+} coordinates the substrate to the enzyme active site. More recent results show the precise role of K^+ ions is not really known, but it appears that the ion allows for the proper orientation of the active site to permit catalysis to occur.
Modified with permission from Mildvan, A. S. *Annu. Rev. Biochem.* 43:365, 1974. Copyright © (1974) Annual Reviews, Inc.; and Oria-J., Cabrera, N., Oerez-Montfort, R. and Ranirez-Silva, L. *J. Biol. Chem.* 280:37924, 2005.

Role of Metals in Oxidation and Reduction

Metal ions such as copper and iron, with less than a full complement of electrons in their outer shell, can accept or donate electrons in oxidation–reduction reactions, but monovalent ions (Na^+ and K^+) or divalent ions (Zn^{2+}, Mg^{2+}, or Ca^{2+}) with their outer shell of electrons complete do not.

Iron–sulfur proteins, often referred to as nonheme iron proteins, are a unique class of metalloenzymes in which the active center consists of one or more clusters of sulfur-bridged iron chelates. In some cases the sulfur atoms come only from cysteine, and in

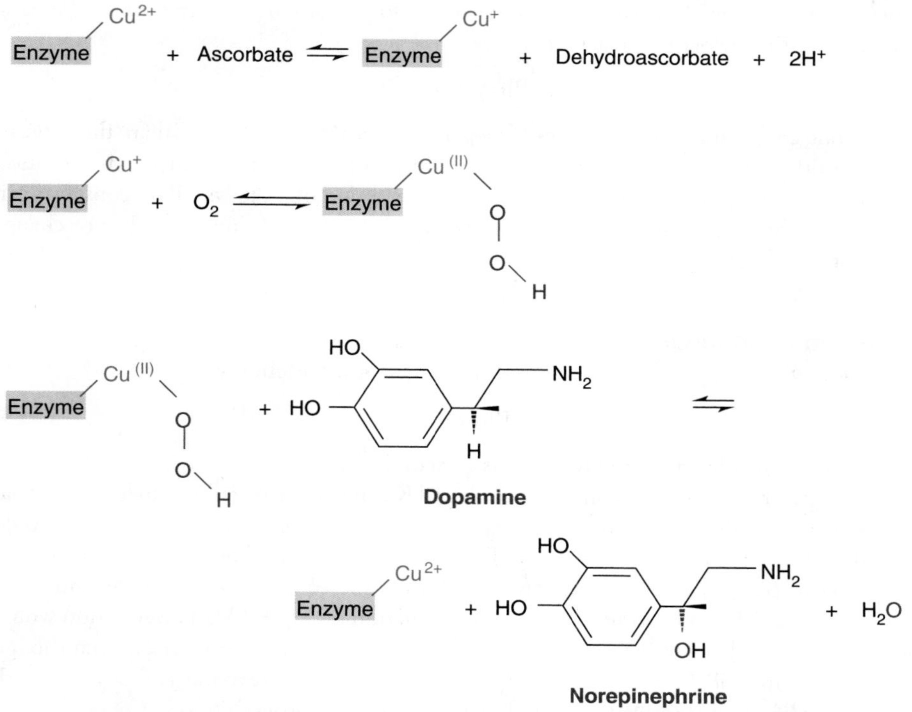

Figure 10.42 Role of copper in activation of molecular oxygen by dopamine hydroxylase. The normal cupric form of the enzyme is not reactive with oxygen but on reduction by the cosubstrate, ascorbate, generates a reactive enzyme-copper-bound oxygen radical that then reacts with dopamine to form norepinephrine and an inactive cupric enzyme.

others they come from both cysteine and free ionic sulfur. These nonheme iron enzymes have reasonably low reducing potentials (E_0') and function in electron-transfer reactions (p. 562). **Cytochromes** are heme iron proteins where the iron undergoes reversible one electron transfers. In some cases heme is bound to the apoprotein by coordination of amino acid side chains to the iron of heme. Thus, the metal serves not only a structural role but also participates in the chemical event. Copper and iron also have a role in activation of molecular oxygen. Copper is an active participant in several oxidases and hydroxylases. For example, dopamine β-hydroxylase catalyzes the introduction of one oxygen atom from O_2 into dopamine to form norepinephrine (Figure 10.42). The active enzyme contains one atom of cuprous ion that reacts with oxygen to form an activated oxygen–copper complex. The copper–hydroperoxide complex is thought to be converted to a copper (Il) $-O^-$ species that serves as the *active oxygen* in the hydroxylation of DOPA. Other species of active oxygen are generated in metalloenzymes and used for hydroxylation.

10.6 • KINETICS OF CHEMICAL REACTIONS

Rate of Product Formation

The rates of all chemical reactions can be expressed in mathematical terms. In developing the rate equations, either the formation of product or the disappearance of initial reactants can be determined; in most cases product formation is measured. From the mathematical relationships between the rate(s) of product(s) formation (termed **velocity**) as a function of substrate(s) concentration(s), some aspects of the mechanism of enzyme catalysis can be determined. An equation to describe the rate of product formation in any reaction contains a rate constant (k) and the concentration of all the substrates that change during the course of the reaction. In the reaction $A + B \rightarrow P$, the change in concentration of product as a function of time is d[P]/dt where [P] is concentration of product at any time t. The basic equation relating velocity to reactants is

$$d[\mathrm{P}]/dt = k[\mathrm{A}][\mathrm{B}] \tag{10.2}$$

k is the rate constant and is a constant for a particular reaction. [A] and [B] are the concentrations of the substrates at any time. The rate constant is related to the energy of activation to reach the transition state and is changed by the addition of a catalyst. Rate constants can be altered only by changing the reaction conditions, that is, addition of catalyst or change of pH or ion composition.

First- and Second-Order Reactions

A reaction with only one substrate (A → P) is a first-order reaction with

$$d[\mathrm{P}]/dt = k[\mathrm{A}] \tag{10.3}$$

If there are two substrates, the reaction is a second-order reaction (Eq. 10.2). The units of k are different in the equations for a first- and second-order reaction. Since $d[\mathrm{P}]/dt$ has the units moles/minute (or any other unit of concentration or time), k in Eq. 10.2 would have the unit of 1/moles × min while in Eq. 10.3, the unit would be 1/min. If it takes two molecules of compound A to form product P, then the equation would be second order because the velocity of making P is related to a reaction of [A] + [A]. The equation would be $d[\mathrm{P}]/dt = k[\mathrm{A}][\mathrm{A}]$ or $k[\mathrm{A}]^2$, which is identical to a second-order reaction even though only one compound is involved. For $d[\mathrm{P}]/dt = k[\mathrm{A}][\mathrm{B}]$, the reaction is first order with respect to both A and B, but the overall reaction is second order.

The concentration terms in the equation are only for those substrates whose concentrations change during the course of the reaction. In reactions involving water, the concentration of water is not altered (or not measurably). Thus if B is water in the reaction A + B → C, the value of B is a constant and the reaction is first order (Eq. 10.3).

Rate of Substrate Disappearance

For the simple reaction A → P, the velocity of the reaction can also be given by disappearance of substrate as a function of time, $-d[\mathrm{A}]/dt$. The minus sign is necessary since the concentration of A is decreasing with time. The change in the concentration of product P with time is $d[\mathrm{P}]/dt$, which is equal to the loss of substrate or $-d[\mathrm{A}]/dt$.

$$d[\mathrm{P}]/dt = -d[\mathrm{A}]/dt = k[\mathrm{A}] \tag{10.4}$$

The equation for the loss of substrate as a function of time is

$$d[\mathrm{A}]/dt = -k[\mathrm{A}] \tag{10.5}$$

and the integral of dA/A is ln A leading to

$$\ln[\mathrm{A}] = 2.3\log[\mathrm{A}] = -kt + \text{constant} \tag{10.6}$$

Eq. 10.6 is the mathematical relationship for the loss of substrate as a function of time. At time of zero, [A] is the initial concentration [A_0], so the equation can be rewritten as

$$\ln[\mathrm{A}] = -kt + [\mathrm{A}]_0$$

Thus

$$\ln[\mathrm{A}/\mathrm{A}_0] = -kt \tag{10.7}$$

or

$$2.3\log[\mathrm{A}/\mathrm{A}_0] = kt$$

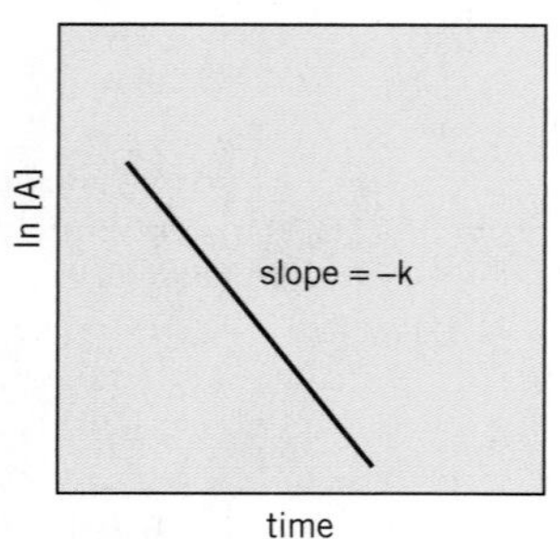

Figure 10.43 A plot of the ln[A] versus time for a first-order reaction.

A plot of the ln[A] versus time is a straight line with a slope of $-k$ (Figure 10.43).

For comparison purposes, the term *half-life* is often used. The half-life ($t_{1/2}$ or $t_{50\%}$) is the time it would take for 50% of the substrate to be converted to product. Since the value [A/A_0] would be 0.5 and the ln(0.5) = −0.69; this leads to

$$-0.69 = -kt_{50\%} \tag{10.8}$$

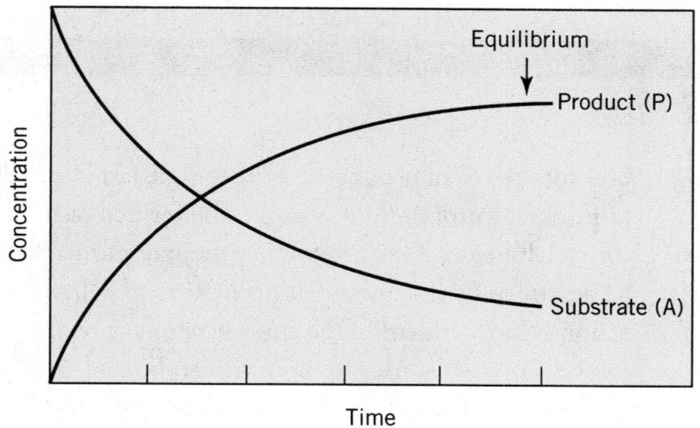

Figure 10.44 Plot of substrate disappearance and product formation. For a reversible reaction, no net change in substrate or product concentration occurs at equilibrium.

or half-life is equal to $0.69/k$. Thus, the larger the rate constant, the faster the reaction will deplete half the substrate. Radioactive decay is an example of a first-order reaction as are enzyme-catalyzed reactions under certain conditions.

Reversible Reactions

Many reactions are reversible. That is,

$$A \underset{k_2}{\overset{k_1}{\leftrightharpoons}} P$$

One can write a velocity equation for the formation of product P or of starting material A.

$$d[\mathrm{P}]/dt = k_1[\mathrm{A}] - k_2[\mathrm{P}] \tag{10.9}$$

$$d[\mathrm{A}]/dt = k_2[\mathrm{P}] - k_1[\mathrm{A}] \tag{10.10}$$

At equilibrium, where there is no net change in concentrations of components, the rate of formation of one component is equal to the rate of formation of the other. Figure 10.44 is a plot of substrate disappearance and product formation. The velocity equation for product formation reflects that the material is being produced (k_1[A]) but is also being lost as it goes back to starting material ($-k_2$[P]). The negative sign shows that the material is being removed. A similar equation is written for the change is starting material. Since at equilibrium

$$d[\mathrm{P}]/dt = d[\mathrm{A}]/dt \tag{10.11}$$

then

$$k_1[\mathrm{A}] = k_2[\mathrm{P}] \quad \text{so} \quad k_1/k_2 = [\mathrm{P}]/[\mathrm{A}] \tag{10.12}$$

The ratio k_1/k_2 is the equilibrium constant, K_{eq}, and is equal to [P]/[A]. Even though the individual rate constants for a reversible reaction are independent of the equilibrium constant, the ratio of the constants is constrained by the equilibrium constant. The latter is related to the difference in free energy needed to make A and P (called heats of formation). Kinetics of complex reactions is presented in A Closer Look 10.4.

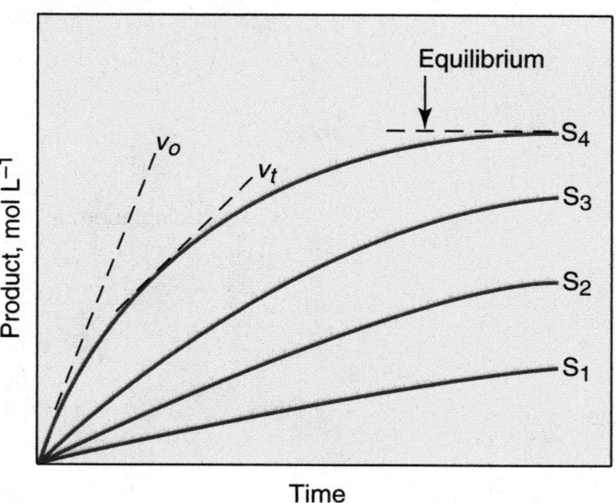

Figure 10.45 Progress curves for an enzyme-catalyzed reaction at different substrate concentrations and a constant concentration of enzyme. The initial velocity (v_0) of the reaction is determined from the slope of the progress curve at the beginning of the reaction. The initial velocity increases with increasing substrate concentration (S1–S4) but reaches a limiting value characteristic of each enzyme. The velocity at any time t is denoted as v.

10.7 • ENZYME KINETICS OF ONE-SUBSTRATE REACTIONS

An enzyme-catalyzed reaction is like a second-order reaction. Hence, with constant enzyme concentration, as substrate concentration is increased, the rate of reaction increases as shown in Figure 10.45. The velocity slows down during the course of the

A CLOSER LOOK 10.4

Complex Reactions

Nucleophilic displacement is a common organic reaction occurring in biochemical reactions. A general form of the reaction is $R-X + Y \rightarrow Y-R + X$ (where R represents part of an organic compound and X or Y represents a component such as an alcohol or amine group). The reaction can occur in two different ways, each leading to a different rate equation.

The first possibility is that the compound R–X dissociates, forming $R^+ + X^-$. Next Y can attack R^+ to form the final product, R–Y. The second possibility is Y will attack R–X directly, displacing X to form the same final product. The kinetic equations for the two reactions are different.

The velocity for the first step in the reaction of R–X going to $R^+ + X^-$ is

$$d[\text{R–X}]/dt = -k[\text{RX}]$$

For the second half of the reaction, $R^+ + Y$, the velocity is given by $k'[R^+][Y]$. If the first step is very slow compared to the second step, then the overall velocity will be governed by only the first step. In kinetics, only the slow steps are measured. To illustrate, assume that R–X goes to $R^+ + X^-$ at a rate of 10 μM/min and the attack of Y on R^+ proceeds at 1000 μM/min. Thus, as a molecule of R^+ is formed, it very rapidly becomes the final product. Increasing the fast step will not alter the rate of product formation for it cannot be formed faster than the velocity of the slowest step. This means that the rate of formation of product will be described by the rate equation $d[R-(-X)]/dt = -k_1[RX]$. This reaction is independent of the concentration of component Y. That is, the reaction is a first-order reaction. It can be said that the reaction is **zero order** with respect to [Y], meaning that the reaction is independent of the concentration of Y. (Recall that $[Y]^0 = 1$.) Increasing the concentration of Y will have no affect on the velocity of the reaction.

In the reaction of Y attacking R–X, the velocity term contains concentrations of both substrates. That is, $d[\text{R–X}]/dt = -k_2[\text{R–X}][Y]$, a traditional second-order reaction. The major difference between the two reactions pathways is that the slowest step in one involves the transition state for R–X going to $R^+ + X^-$. In the other, the transition step involves Y attacking R–X (Figure 10.45, p. 397). To reiterate, kinetics tells something about the slower steps that occur during a reaction. The slowest step that occurs during an enzyme-catalyzed reaction is called the rate-limiting step.

Complex reaction kinetics occurs when monitoring the flow of compounds through a metabolic pathway. The velocity of the slowest step is what will govern the rate of product formation.

$CH_3-CH_2-CH(O-PO(O^-)-O-PO(O^-)-O^-)-CH_3 + NH_3 \rightarrow CH_3CH_2CH(NH_3^+)-CH_3 + {}^-O-PO(O^-)-O-PO(O^-)-O^-$

(*a*) ionization

$CH_3-CH_2-CH(OPP)-CH_3 \xrightarrow{\text{slow}} CH_3-CH_2-\overset{+}{C}H-CH_3 + {}^-O-PO(O^-)-O-PO(O^-)-O^- \xrightarrow[\text{fast}]{NH_3} CH_3-CH_2-C(NH_3^+)(H)-CH_3$

v = k [substrate]

(*b*) direct displacement

$CH_3-CH_2-CH(OPP)-CH_3 + NH_3 \xrightarrow{\text{slow}} CH_3-CH_2-C(H)(NH_3^+)-CH_3 + {}^-O-PO(O^-)-O-PO(O^-)-O^-$

v = k′ [substrate] [NH$_3$]

Nucleophilic displacement reactions. (*a*) Ionization mechanism. (*b*) Direct displacement mechanism. Note that in the ionization mechanism, the velocity of the reaction is a function of the concentration of only one substrate, whereas in the direct displacement the velocity depends on both substrate concentrations.

reaction because the concentration of substrate is becoming depleted or product is starting to accumulate. Assuming the reaction is reversible, it will reach equilibrium. By calculating the initial velocity (v_0) (i.e., velocity before product starts to accumulate) of the reaction at different substrate concentrations, one can obtain the curve presented in Figure 10.46. The velocity becomes zero order with respect to substrate at high concentrations of substrate. That is, the velocity no longer increases with an increase in [S]. The reason is that the enzyme and substrate form a complex, and the rate of product formation is proportional to the concentration of the complex. All of the enzyme will be in the complex at high substrate concentrations. The enzyme will be released from the complex by reversibly dissociating the substrate or proceed to form products. The overall reaction is written as

$$E + S \underset{k_2}{\overset{k_1}{\rightleftharpoons}} ES \xrightarrow{k_3} E + P \tag{10.13}$$

Rate constants k will be numbered such that the ones leading to products will be odd numbers (k_1, k_3, and k_5), while the reversible steps will be even ones (k_2, k_4, and k_6); some are first-order and others second-order constants. *The units of first- and second-order constants are different, meaning that they cannot be added to each other but can be multiplied or divided.*

The rate of formation of product is the term leading to product times the rate constant.

$$d[P]/dt = k_3[ES] \tag{10.14}$$

To determine the rate of product formation, it is necessary to find the value of ES as a function of the concentration of enzyme and substrate. Eq. 10.14 can be solved for the rate of the reaction (see A Closer Look 10.5 for derivation of the equation). The final form is

$$v = \frac{k[E][S]}{K + [S]} \tag{10.15}$$

where k is a rate constant and K is a constant which includes the individual rate constants relating to the substrate's interaction with the enzyme.

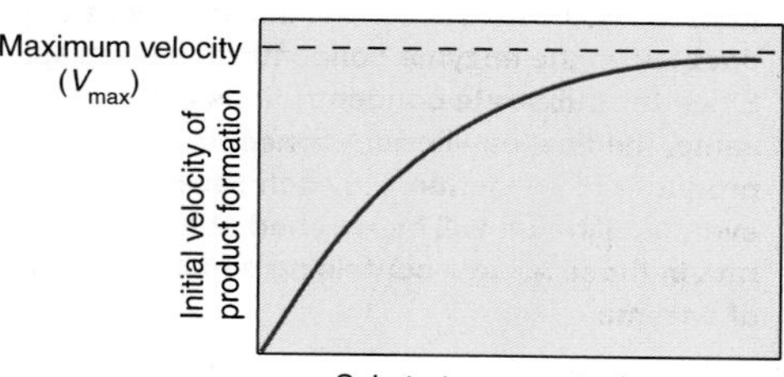

Figure 10.46 Plot of velocity versus substrate concentration for an enzyme-catalyzed reaction. Initial velocities are plotted against the substrate concentration at which they were determined. The curve is a rectangular hyperbola, which asymptotically approaches the maximum velocity possible with a given amount of enzyme.

A CLOSER LOOK 10.5

Derivation of the Michaelis–Menten Equation

In order to derive the Michaelis–Menten equation, it is necessary to determine the concentration of ES. The rate of formation of ES is given by $k_1[E_f][S]$, where $[E_f]$ is the concentration of free enzyme. The utilization of ES is governed by two reactions: ES returning to starting material, given by k_2[ES], and ES going to products, k_3[ES]. Combining these terms leads to an equation describing the change in concentration of ES.

$$d[ES]/dt = k_1[E_f][S] - k_2[ES] - k_3[ES]$$

Assuming that the concentration of ES is constant except for the first few milliseconds after contact of substrate and enzyme, $d[ES]/dt = 0$. This is referred to as the *steady-state assumption,* meaning the concentration of ES during the initial time period of the enzyme-catalyzed reaction (period for measuring the initial rate). The validity of the steady-state condition can be proven mathematically. Substituting $[E_f]$ with $[E_t] - [ES]$, where E_t is the total enzyme concentration and rearranging, leads to

$$k_1([E_t] - [ES])[S] = (k_2 + k_3)\,[ES]$$

Solving for [ES], one obtains

$$[ES] = \frac{k_1[E_t][S]}{(k_2 + k_3) + k_1[S]}$$

Since the rate of product formation (v) of the reaction is k_3[ES], multiplying the above equation by k_3 gives

$$v = \frac{k_3\, k_1[E_t][S]}{(k_2 + k_3) + k_1[S]}$$

To convert this equation into the Michaelis–Menten equation, the numerator and denominator are divided by k_1.

$$v = \frac{k_3\,[E_t][S]}{(k_2 + k_3)/k_1 + [S]} \quad \text{or} \quad v = \frac{V_m[S]}{K_m + [S]}$$

k_3E_t is the maximum velocity of the reaction (V_m), with all the enzyme in the ES form, and therefore k_3 is k_{cat}. The constants in the denominator are defined as K_m. It is apparent that K_m is not a simple dissociation constant. If k_3 were very small compared to k_2, then the resulting ratio would be k_2/k_1, which is K_d, the dissociation constant.

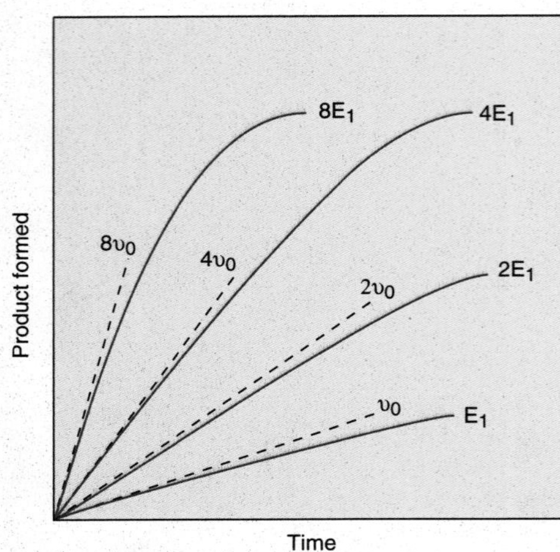

Figure 10.47 Progress curves at variable concentrations of enzyme and saturating concentrations of substrate. The initial velocity (v_0) doubles as the enzyme concentration doubles. Since the substrate concentrations are the same, the final equilibrium concentrations of product will be identical in each case; however, equilibrium will be reached at a slower rate in those assays containing small amounts of enzyme.

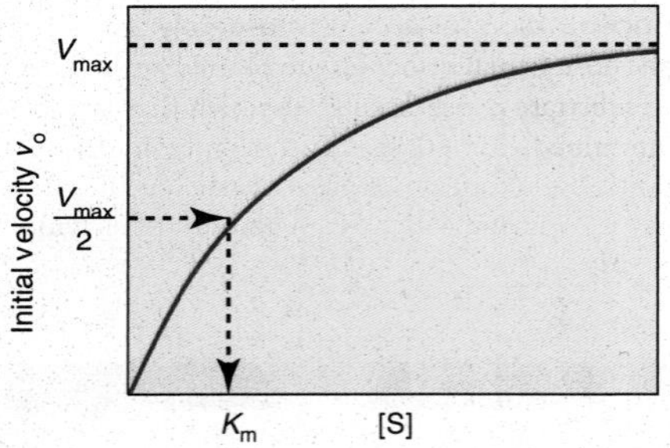

Figure 10.48 Graphic estimation of K_m from the v_0 versus [S] plot. K_m is the substrate concentration at which the enzyme has half-maximal activity.

Michaelis–Menten Equation

The simplest enzyme reaction where one substrate goes to one product is

$$\mathrm{E} + \mathrm{S} \underset{k_2}{\overset{k_1}{\rightleftharpoons}} \mathrm{ES} \underset{k_4}{\overset{k_3}{\rightleftharpoons}} \mathrm{E} + \mathrm{P} \tag{10.16}$$

If the reaction in Eq. 10.16 is essentially irreversible ($k_4 = 0$ or [P] is not present as would be found when enzyme and S are first mixed), the equation for the rate product formation is

$$v = \frac{k_3[\mathrm{E}][\mathrm{S}]}{(k_3 + k_2)/k_1) + [\mathrm{S}]} \tag{10.17}$$

Combining constants in the denominator into a term called K_m and if the reaction ES → E + P in Eq. 10.16 is the rate-limiting step, then k_3 defines the catalytic activity of the enzyme (k_{cat}), and the equation becomes

$$v = \frac{k_{cat}[\mathrm{E_t}][\mathrm{S}]}{(K_m + [\mathrm{S}])} \tag{10.18}$$

Note the similarity of Eqs. 10.15 and 10.18. Since k_{cat} (the rate constant) and [E_t] (the total enzyme concentration) are constants, $k_{cat}[E_t] = V_{max}$. The equation can be written as

$$v = \frac{V_m[\mathrm{S}]}{(K_m + [\mathrm{S}])} \tag{10.19}$$

and is referred to as the **Michaelis–Menten equation**. For reactions that are more complicated, k_{cat} is composed of a collection of rate constants. V_m in Eq. 10.19 is the **maximum velocity** of the reaction at a defined enzyme concentration and has the unit of moles of product per unit time. Thus V_m (that is, $k_3[E_t]$ or $k_{cat}[E_t]$) is a constant for a fixed concentration of enzyme. If the enzyme concentration is increased or decreased, V_{max} will increase or decrease respectively (Figure 10.47). K_m is composed of individual rate constants; however, from Eq. 10.18, it can be stated that *K_m is the concentration of substrate where the velocity of the reaction (v) is one-half of the maximum velocity (V_m)*, that is, $v/V_m = 0.5$ (Figure 10.48) when K_m = [S]. When v is equal to ½V_m, one-half the concentration of total enzyme is in a complex with substrate and one-half is free. In contrast, when [S] ≫ K_m essentially 100 percent of the enzyme is in the ES complex. Under this condition the reaction reaches V_m since the velocity would be $v = V_m$. It is usually considered that V_m is approached when [S] is close to 20[K_m], since under these conditions $v = V_m \times 20/21$, or $0.95\,V_m$. The biphasic nature of enzyme mechanisms, that is binding followed by modification of substrate, is reinforced in Clin. Corr. 10.3.

CLINICAL CORRELATION 10.3

A Case of Gout Demonstrates Two Phases in the Mechanism of Enzyme Action

The two phases of the Michaelis–Menten model of enzyme action, binding followed by modification of substrate, are illustrated by studies on a family with gout. The patient excreted three times the normal amount of uric acid per day and had markedly increased levels of 5′-phosphoribosyl-α-pyrophosphate (PRPP) in his red blood cells. PRPP is an intermediate in the biosynthesis of AMP and GMP, which are converted to ATP and GTP. Uric acid is a product of the degradation of AMP and GMP. Assays *in vitro* revealed that the patient's red cell PRPP synthetase activity was increased threefold. The pH optimum and the K_m of the enzyme for ATP and ribose 5-phosphate were normal, but V_{max} was increased threefold! This increase was not due to an increase in the amount of enzyme; immunologic testing with an antibody specific to the enzyme revealed similar quantities of the enzyme protein as in normal red cells. This finding demonstrates that the binding of substrate as reflected by K_m and the subsequent chemical event in catalysis, which is reflected in V_{max}, are separate phases of the overall catalytic process. This situation holds only for those enzyme mechanisms in which $k_3 \gg k_2$.

Becker, M. A., Kostel, P. J., Meyer, L. J., and Seegmiller, J. E. Human phosphoribosyl pyrophosphate synthetase: Increased enzyme specific activity in a family with gout and excessive purine synthesis. *Proc. Natl. Acad. Sci. USA* 70:2749, 1973.

If the reaction is more complex such that

$$E + S \underset{k_2}{\overset{k_1}{\rightleftharpoons}} ES \xrightarrow{k_3} ES' \xrightarrow{k_5} E + P \qquad (10.20)$$

where ES′ represents a modified but independent form of the complex, then the velocity equation becomes

$$v = \frac{k_3k_5/(k_3 + k_5)[E_t][S]}{k_5(k_2 + k_3)/k_1(k_3 + k_5) + [S]} \qquad (10.21)$$

Inspection of Eqs. 10.17 and 10.21 reveals that the generalized form of the equation is

$$v = \frac{\text{Constant}[E_t][S]}{\text{Constant} + [S]} \qquad (10.22)$$

This is the same general equation as Eq. 10.18. The numerator rate constants are grouped to become the k_{cat} term while the constants in the denominator are defined as K_m. Thus, both k_{cat} and K_m are typically composed of a complex set of individual rate constants. The specific constants that make up the terms are related to the actual kinetic mechanism and magnitude of the individual rate constants. As an example, if the value of k_3 is much larger than k_5 in Eq. 10.21, k_{cat} would reduce to k_5 because the sum of k_3 and k_5 would be essentially the value of k_3. If, though, the two rate constants had similar values, then k_{cat} would be composed of many terms, as would K_m whose units are concentration.

Concentration of Free Enzyme

The derivation of the Michaelis–Menten equation requires a careful account of all forms of the enzyme. It is assumed that the concentration of substrate is greater than that of the enzyme, so during the initial course of the reaction being measured the value of [S] does not change. Obviously, various values of [S] can be used, but for each it will remain constant during the reaction. For the simple reaction $E + S \rightleftharpoons ES \rightarrow E + P$, the concentration of enzyme is always the sum of the free enzyme and the substrate-bound enzyme; that is

$$[E_t] = [E_f] + [ES] \qquad (10.23)$$

$[E_t]$ is the total concentration of enzyme and $[E_f]$ is the concentration of free enzyme. It is E_f that is necessary in a reaction to interact with substrate. The concentration of free enzyme is not directly measured but rather the total enzyme concentration. This means

$$[E_t] - [ES] = [E_f] \qquad (10.24)$$

If there were other forms of the enzyme such as a form with an inhibitor bound (p. 407), then Eq. 10.24 would have to be expanded to include each form of enzyme.

Meaning of K_m

K_m consists of a number of rate constants and is not a derived dissociation constant (K_d). Its value, however, is used often to indicate how well a substrate interacts with an enzyme. A K_m of 10^{-7} M indicates that the substrate has a greater affinity for the enzyme than if the K_m is 10^{-5} M. The smaller the value of K_m, the tighter is the interaction between substrate and enzyme. Even though K_m is not a dissociation constant, the smaller value means it takes less substrate to bind 50% of the enzyme (Clin. Corr. 10.4).

An example of the importance of K_m is the physiological utilization of glucose. Glucose can be phosphorylated by two different kinases to form glucose 6-phosphate. Liver contains both hexokinase and glucokinase that catalyze the identical reaction of glucose + ATP → glucose 6-phosphate + ADP. For hexokinase, the K_m for glucose is 0.1 mM whereas for glucokinase it is 5 mM. When the concentration of blood sugar is low, as occurs in the fasted state, hexokinase is used to phosphorylate glucose, but when blood glucose increases after feeding, the high K_m enzyme also functions.

CLINICAL CORRELATION 10.4

Physiological Effect of Changes in Enzyme K_m Values

The unusual sensitivity of some Asians to alcoholic beverages has a biochemical basis. In some Japanese and Chinese, much less ethanol is required to produce vasodilation, resulting in facial flushing and rapid heart rate, than is required to achieve the same effect in Europeans. The physiological effects are due to acetaldehyde generated by liver alcohol dehydrogenase. Acetaldehyde is converted to acetate by aldehyde dehydrogenase (ALDH). One amino acid at position 487 in the 500–amino acid subunit of the tetrameric enzyme is changed in some of the affected individuals. The active enzyme has a glutamate while the inactive variant has a lysine. The Asian variant was found to have a very low activity, and the K_m for NAD^+ increased from 30 μM to 7000 μM. Even though the enzyme was active, it would have very little activity in the liver as the K_m was so high and k_{cat} was low.

In addition affected individuals were heterozygotic, having genes for both the active glutamate enzyme and the essentially inactive lysine enzyme. It was expected that their enzyme would be 50% active, yet it had very low activity. ALDH forms tetramers with the two monomers (E_4, E_3K, E_2K_2, EK_3, and K_4) where E_4 and K_4 are homotetrameric forms of the glutamate- and lysine-containing subunits, respectively and the others are the heterotetramers. E_3K had 50% of the total activity, not 75%, while EK_3 had essentially no activity, not the 25% one could expect from the one active subunit. The K subunit was dominant in that it could inactivate the subunit to which it was paired as the residue at position 487 interacted with an arginine at position 475. When a glutamate was at position 487 a stable salt bond formed, but when a lysine was at position 487, it caused the arginine to move. The movement disrupted the NAD binding pocket even though residue 487 was not in contact with the Rossmann fold. This example illustrates the fact that a point mutation on the enzyme can affect the active site even though the residue is not in contact with the region. It further demonstrates that one subunit can be dominant over another.

Crabb, D. W., Edenberg, H. J., Bosron, W. F., and Li, T.-K. Genotypes for aldehyde dehydrogenase deficiency and alcohol sensitivity: The ALDH22 allele is dominant. *J. Clin. Invest.* 83:314, 1989; and Zhou, J., and Weiner, H. Basis for half-of-the-site reactivity and the dominance of the K487 oriental subunit over the E487 subunit in heterotetrameric human liver mitochondrial aldehyde dehydrogenase. *Biochemistry* 39:12019, 2000.

Turnover Number (k_{cat})

In Eq. 10.18, k_{cat} in the numerator has units of 1/time. This constant is also referred to as the **turnover number** of the enzyme, that is, the number of molecules of substrate converted to product per unit time per molecule of enzyme. The larger the value of k_{cat} for an enzyme, the faster the reaction will be. In general, this term represents the slowest step(s) in the overall reaction.

Significance of k_{cat} in the Michaelis–Menten Equation

When Substrate Concentration Is Much Larger than K_m

Eq. 10.18 is in the form of a general hyperbolic equation. When the value of [S] is much larger than the value of K_m, the value of the denominator is essentially the value of [S], and the equation can be approximated by

$$v = \frac{k_{cat}[E][S]}{[S]} = k_{cat}[E] \tag{10.25}$$

That is, the velocity becomes independent of the concentration of S; the reaction becomes zero order with respect to S. This is found in the portion of curve where the velocity essentially levels off and is not increased as the concentration of [S] increases (slope = 0) (see Figure 10.46, p. 399). The reaction is proceeding at its maximum velocity under these conditions because essentially 100 percent of the enzyme is in the ES complex. The instant a molecule of product is made it leaves the enzyme and another molecule of S binds enzyme, always keeping the enzyme saturated with S. Under these conditions the rate is governed strictly by the terms that govern the reaction of ES going to product ($ES \rightarrow E + P$).

When Substrate Concentration Is Much Smaller than K_m

When [S] is very small compared to K_m, Eq. 10.17 can be approximated as

$$v = \frac{k_{cat}[E_t][S]}{K_m} \quad \text{or} \quad v = \frac{V_{max}[S]}{K_m} \tag{10.26}$$

A plot of v versus [S] is linear only when [S] $\ll K_m$ value. Since for any chemical reaction the velocity of product formation is a rate constant times the concentration of the reactants, the term k_{cat}/K_m can be considered as a second-order rate constant since there are two concentration terms in the rate equation ($[E_t]$ and [S]). The fact that this ratio is equivalent to a second-order rate constant puts certain constraints on the ratio. For example, a rate constant cannot be faster than the rate two molecules diffuse in solution. Even if no energy of activation is required, the two molecules have to collide with each other in order for a reaction to occur. The rate constant for a second-order diffusion reaction is around 10^9/mole/sec. Thus, if k_{cat} is very large, K_m cannot be very small or the ratio would exceed the diffusion rate. As an example, catalase, an enzyme that converts hydrogen peroxide to oxygen and water, has a k_{cat} of 4×10^7/sec. The K_m for hydrogen peroxide, however, is 1 M, a very high value. The k_{cat}/K_m ratio is less by a factor of 25 (10^9 versus 4×10^7) of the theoretical maximum velocity that second-order reactions can attain. Catalase is one of the fastest known enzymes when the term for k_{cat}/K_m is measured.

k_{cat} can be a very low number only (such as 1 to 10/sec) if the K_m for substrate is very low (as an example, 1×10^{-9} M). This slow rate means that one molecule of enzyme would produce just 1 to 10 molecules of product per second. The reaction is so slow because the substrate binds very tightly to the enzyme. A great deal of energy is required to convert the bound substrate to the transition state. An extremely small K_m (high affinity) for substrate requires the chemical reaction to be slow, whereas a large K_m (low affinity) means that k_{cat} can be high. The values of k_{cat}/K_m for enzymes range from 10 to 10^7/sec.

Reversible Reactions

Another limitation to the ratio of k_{cat}/K_m occurs when the reaction is reversible.

$$E + S \underset{k_r}{\overset{k_f}{\rightleftharpoons}} E + P \tag{10.27}$$

k_f and k_r are the forward and reverse rate constants (the terms ES and EP are ignored because they are the same for both directions). It can be derived that k_f is $(k_{cat}/K_m)_f$ and k_r is $(k_{cat}/K_m)_r$. For a reversible reaction, the equilibrium constant K_{eq} = ([P]/[S]) is k_f/k_r or

$$K_{eq} = \frac{(k_{cat}/K_m)_f}{(k_{cat}/K_m)_r} \tag{10.28}$$

The values for k_{cat} and K_m cannot change randomly since their ratio has to be equal to K_{eq}. This implies that if a mutant of an enzyme has a greater value of k_{cat} than the original enzyme, some other constant must change to satisfy the equilibrium situation. These equilibrium-type relationships are called **Haldane** relationships.

Low K_m versus High k_{cat}

Since the ratio of k_{cat}/K_m has an upper limit, the question can be raised as to whether it would be preferable for an enzyme to have a very low K_m or a very high k_{cat}. Frequently it is considered that a low K_m is most desirable, and compounds with the lowest K_m are reported as the best substrates. Actually, the compound with the highest k_{cat}/K_m ratio is the best substrate. At a constant k_{cat}/K_m, it is better to have a high k_{cat} rather than a low K_m. An example is presented in Table 10.4. At a fixed concentration of substrate, 1 mM in the example, the overall velocity is always faster when K_m is high as that forces k_{cat} to be larger to satisfy the k_{cat}/K_m ratio. Thus, *when evaluating different substrates for an enzyme, the substrates with the highest k_{cat}/K_m value should be considered the better substrate for the enzyme not the one with the lowest K_m.*

In a cell it is not just the k_{cat} and K_m terms that are important, but one must take into account the total concentration of enzyme. The velocity term is first order with respect to enzyme, so a few picomoles of an enzyme with a k_{cat}/K_m of 10^7/sec/M in the cell would be less efficient at converting a substrate to product than would a nanomole of an enzyme with a k_{cat}/K_m of 10^5/sec/M.

TABLE 10.4 • K_m and k_{cat} Relationships[a]

K_m (*M*)	k_{cat} (s^{-1})	*Rate* (s^{-1})
10^{-6}	1	1
10^{-5}	10	9
10^{-4}	10^2	90
10^{-3}	10^3	500
10^{-2}	10^4	909
10^{-1}	10^5	990
1	10^6	999

[a]Reaction conditions: $k_{cat}/K_m = 10^6 M^{-1} s^{-1}$ and [S] = 10^3M.

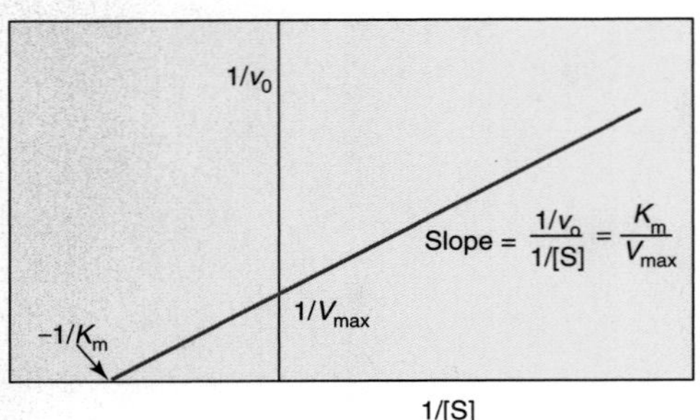

Figure 10.49 Determination of K_m and V_{max} from the Lineweaver–Burk double-reciprocal plot. Plots of the reciprocal of the initial velocity versus the reciprocal of the substrate concentration used to determine the initial velocity yield a line whose x-intercept is $-1/K_m$, because at $1/v_0 = 0$, $1/[S] = 1/Km$ in Eq. 10.29.

Calculating the Constants

The K_m can be approximated from plots of v versus [S]. Graphical presentations allow for the calculation of the kinetic constants. The most common graphical presentation is based on the **Lineweaver–Burk equation**, which is the reciprocal of the Michaelis–Menten equation. The reciprocal of Eq. 10.19 is

$$\frac{1}{v} = \frac{1}{V_m} + \frac{K_m}{V_m} \times \frac{1}{[S]} \quad (10.29)$$

A plot of $1/v_0$ versus $1/[S]$ is a straight line (Figure 10.49); the intercept on the y-axis is $1/V_m$, the intercept on the x-axis is $-1/[S]$, which is equivalent to $-1/K_m$; the slope of the line is K_m/V_m. Thus from the graph of $1/v_0$ at different concentrations of $1/[S]$, two key parameters of the enzyme-catalyzed reaction can be determined. Alternatively, computer programs can calculate the k_{cat} and K_m from the velocity and substrate data.

Substrate and Product Bind to the Same Site

The presence of product will act as an inhibitor of the reaction since its presence can prevent substrate from binding. It will be a competitive inhibitor since both S and P bind to E_f. The equation relating the initial velocity of the reaction in the presence of product is

$$v = \frac{V_m[S]}{K_m(1 + [P]/K_{mp}) + [S]} \quad (10.30)$$

where K_{mp} is the K_m for product, assuming the reaction could be reversible. Product is initially not present when running an assay *in vitro* but starts to accumulate with time. A graph of product formation as a function of time will initially be linear, but the slope will decrease with time as product accumulates and acts as an inhibitor (see Figure 10.44, p. 397).

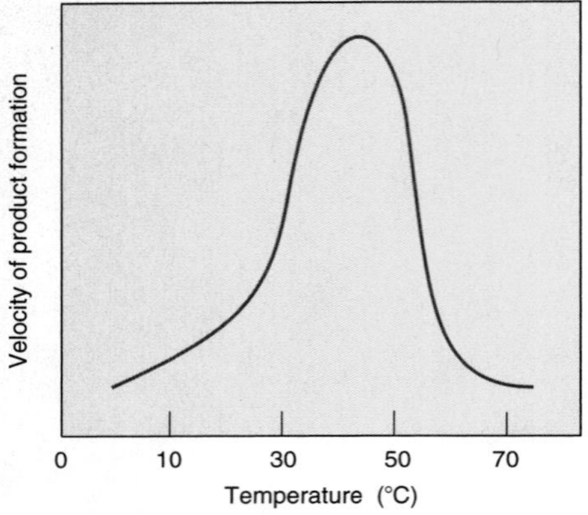

Figure 10.50 Temperature dependence of a typical mammalian enzyme. To the left of the optimum, the rate is low because the environmental temperature is too low to provide enough kinetic energy to overcome the energy of activation. To the right of the optimum, the enzyme is inactivated by heat denaturation.

Effect of Assay Conditions

When running a reaction under V_m conditions, essentially 100% of the enzyme is in the ES complex. In contrast, when a reaction is studied where $[S] \ll K_m$, the effect being measured would reflect what is occurring with both free and substrate-bound enzyme. Thus to study the effect of an environmental change such as pH, the assay is usually conducted under V_{max} conditions. The effect of experimental conditions under V_{max} conditions gives information about the k_{cat} term for all the enzyme is bound to substrate. To evaluate the effect of pH on the free enzyme, then k_{cat}/K_m as a function of pH would be determined.

External conditions, such as pH, temperature, and salt concentration, affect enzyme activity. These effects are very important when running enzyme assays *in vitro.*

Temperature

A plot of velocity versus temperature for most mammalian enzymes is a bell-shaped curve with an optimum between 40°C and 45°C (Figure 10.50). Above this temperature, heat denaturation of the enzyme occurs; some enzymes will denature at lower temperatures. Typically between 0°C and 40°C, enzymes show a two-fold increase in activity for every 10°C increase. Mutation of an enzyme to a thermolabile form can have serious consequences (Clin. Corr. 10.5).

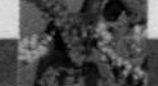

CLINICAL CORRELATION 10.5

Thermal Lability of Glucose-6-Phosphate Dehydrogenase Results in Hemolytic Anemia

In red cells, glucose-6-phosphate dehydrogenase (G6PD) is an important enzyme for the maintenance of the membrane integrity. A deficiency or inactivity of this enzyme leads to a hemolytic anemia. In other cases, a variant enzyme is present that normally has sufficient activity to maintain the membrane but fails under conditions of oxidative stress. A mutation of this enzyme leads to a protein with normal kinetic constants but a decreased thermal stability. This condition is especially critical to the red cell, since it is devoid of protein-synthesizing capacity and cannot renew enzymes as they denature. The end result is a greatly decreased lifetime for those red cells that have an unstable G6PD. Those red cells are also susceptible to drug induced hemolysis.

Luzzato, L., and Mehta, A., Glucose-6-phosphate dehydrogenase deficiency. In Scriver, C. R., Beaudet, A. L., Sly, W. S., and Valle, D. (Eds.), *The Metabolic and Molecular Bases of Inherited Disease,* 7th ed. New York: McGraw Hill, 1995, 3369.

pH

Nearly all enzymes show a bell-shaped pH-velocity profile, but the maximum (**pH optimum**) varies greatly with different enzymes. Alkaline and acid phosphatases with very different pH optima are both found in humans (Figure 10.51). The bell-shaped curve and its position on the *x*-axis are dependent on the particular ionized state of the substrate that will be optimally bound to the enzyme. This in turn is related to the ionization of specific amino acid residues that constitute the substrate-binding site. In addition, amino acid residues involved in catalyzing the reaction must be in the correct charge state to be functional. Clin. Corr. 10.6 describes the effect of a mutation leading to a change in the pH optimum of a physiologically important enzyme.

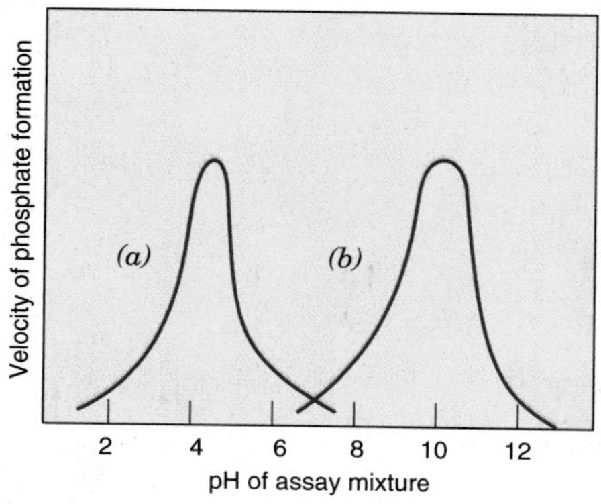

Figure 10.51 The pH dependence of (*a*) acid and (*b*) alkaline phosphatase reactions. In each case the optimum represents the ideal ionic state for binding of enzyme and substrate and the correct ionic state for the amino acids side chains involved in the catalytic event.

10.8 • KINETICS OF TWO-SUBSTRATE REACTIONS

Sequential Mechanism

There are two mechanisms of interactions for a two-substrate reaction as illustrated in Figure 10.52. In the **sequential mechanism**, both substrates bind to the enzyme in a sequential manner to form a ternary complex. With substrates A and B, the ternary complex would be E-A-B (A, B, C, . . . are substrates and P, Q, R, . . . are products). If substrate A binds prior to B, this is called an **ordered-sequential reaction** whereas if either A or B bind first, the reaction is called a **random-sequential reaction**. The equation for the sequential reaction contains K_m terms for both substrates A and B (K_a and K_b) and a term K_{ia}, the dissociation constant for substrate A. To find the various constants, a double reciprocal plot of $1/v$ versus 1/[A] is plotted for the reaction with different fixed concentrations of substrate B (Figure 10.53). The resulting plot shows a series of lines that intersect with each other. The individual constants can be found by plotting the slopes and intercepts as a function of [B].

Ping-Pong Mechanism

The second mechanism for a two-substrate reaction is called a **ping-pong mechanism**. Substrate A binds to the enzyme, and it is converted to product P, leaving a modified enzyme. The modified enzyme binds substrate B, which is then converted to product Q. Enzymes that exhibit ping-pong kinetics are modified by the substrate. Kinases that follow a ping-pong mechanism do so by having ATP or another nucleotide triphosphate first bind to and then phosphorylate the enzyme. ADP dissociates, and the second substrate

CLINICAL CORRELATION 10.6

Alcohol Dehydrogenase Isoenzymes with Different pH Optima

In addition to the change in aldehyde dehydrogenase isoenzyme composition in some Asians, different alcohol dehydrogenase (ADH) isoenzymes are also observed. ADH is encoded by three genes, which produce three different polypeptides, α, β, and γ. Three alleles are found for the β-gene that differ in a single nucleotide base; this causes substitutions for arginine. The substitutions are shown here.

	Residue 47	*Residue 369*
β_1	Arg	Arg
β_2	His	Arg
β_3	Arg	Cys

The liver β_3 form has ADH activity with a pH optimum near 7, compared with 10 for β_1, and 8.5 for β_2. The rate-determining step in alcohol dehydrogenase is the release of NADH. NADH is held on the enzyme by ionic bonds between the phosphates of the coenzyme and the arginines at positions 47 and 369. In the β_1 isozyme this ionic interaction is not broken until the pH is quite alkaline and the guanidinium group of arginine starts to dissociate H^+. Substitution of amino acids with lower pK values, as in β_2 and β_3, weakens the interaction and lowers the pH optimum. Since the release of NADH is facilitated, the V_{max} values for β_2 and β_3 are also higher than for β_1.

Burnell, J. C., Carr, L. G., Dwulet, F. E., Edenberg, H. J., Li, TK., and Bosron, W. F. The human β_3 alcohol dehydrogenase subunit differs from β_1 by a cys- for arg-369 substitution which decreases NAD(H) binding. *Biochem. Biophys. Res. Commun.* 146: 1227, 1987.

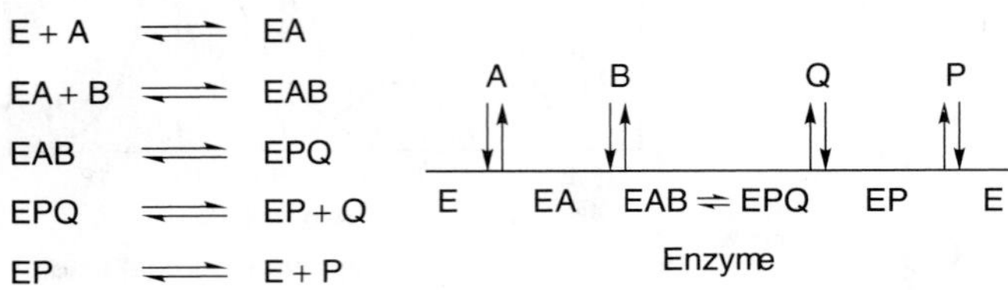

Figure 10.52 Mechanisms of interaction for two substrate reactions. (*a*) Sequential mechanism. (*b*) Ping-pong mechanism.

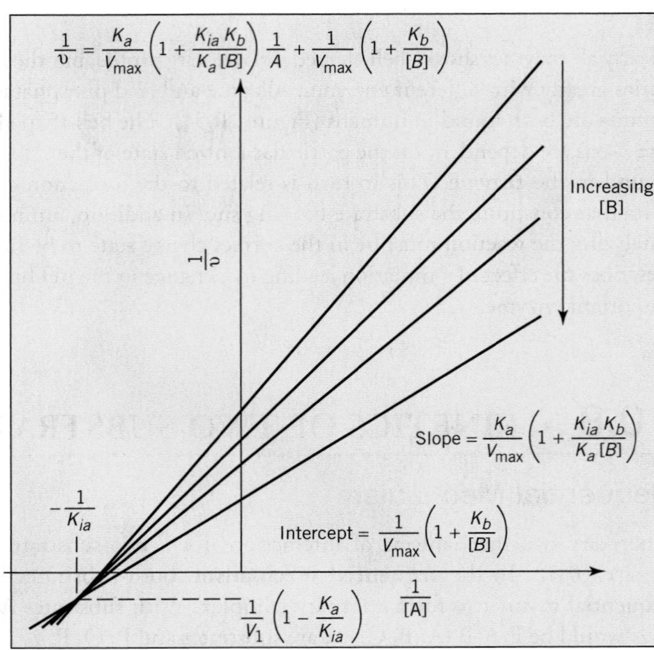

Figure 10.53 Double-reciprocal plot of initial velocity for a sequential reaction. Substrate B is fixed at specified concentrations, and substrate A varied. The rate equation is

$$\frac{1}{v} = \frac{K_a}{V_{max}}\left(1 + \frac{K_{ia}K_b}{K_a B}\right)\frac{1}{A} + \frac{1}{V_{max}}\left(1 + \frac{K_b}{B}\right)$$

binds to the phosphorylated enzyme followed by phosphate transfer to produce the phosphorylated acceptor. Many enzymes that use coenzymes that do not dissociate employ ping-pong reactions.

The equation relating the velocity to the concentrations of A and B in the ping-pong mechanism is similar to the equation for the sequential reaction, but most significantly it does not have the K_{ia} term as found in the sequential reaction. Double reciprocal plots for this reaction result in a series of parallel lines (Figure 10.54). Again, the individual K_m and V_m values can be obtained from the slopes and intercepts after plotting the values obtained at each concentration of B.

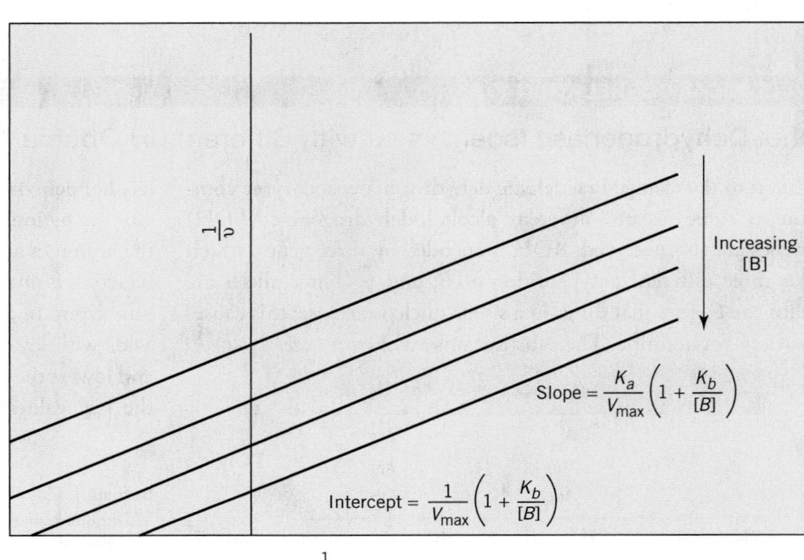

Figure 10.54 Double-reciprocal plot of initial velocity for a ping-pong reaction. Substrate B is fixed at specified concentrations, and substrate A varied. The rate equation is

$$\frac{1}{v} = \frac{K_a}{V_{max}}\left(1 + \frac{K_b}{B}\right) + \frac{1}{V_{max}}\left(1 + \frac{K_b}{B}\right)$$

K_{mapp}

Often the K_m for just one of the substrates is desired. Performing the reaction in the presence of very high concentrations of substrate A and varying the concentration of B permits determination of a K_{mapp} for substrate B. This is possible because when $[A] \gg K_a$, the velocity equation reduces to a simple Michaelis–Menten equation.

10.9 • INHIBITORS

Much of our basic knowledge of metabolic pathways was obtained by using inhibitors of specific enzymes. Pharmaceutical compounds are now designed as inhibitors of specific enzymes, and many naturally occurring compounds are inhibitors (Clin. Corr. 10.7). Inhibitory compounds can bind to the free enzyme or to the ES complex and affect the velocity of the reaction. With an inhibitor, EI and/or ESI (where I is an inhibitor) can exist. An equation relating the total concentration of enzyme to all its possible complexes results in a new Michaelis–Menten equation containing the term $1 + [I]/K_i$ where K_i is the dissociation constant of EI or ESI. k_{cat} in the presence of the inhibitor is an apparent k_{cat} (k_{catapp}), and K_m will be an apparent K_m (K_{mapp}); that is, a value will be measured, but it will be different from the value that would be found in the absence of inhibitor. The inhibitor concentration does not change during the course of the reaction, so [I] is a constant. Several classes of inhibitors are defined on the basis of these reaction kinetics.

CLINICAL CORRELATION 10.7

Inhibitors of Xanthine Oxidase Isolated from Plants

Gout is an inflammatory disease caused by the overproduction or undersecretion of uric acid, the end product of purine catabolism (p. 820). The terminal step in purine metabolism is catalyzed by xanthine oxidase. A common treatment for gout is the use of allopurinol, a structural analog of xanthine and an inhibitor of the enzyme. Many older societies used medicinal plants and have maintained records of which are effective. Records of Vietnamese traditional medicine indicate that 96 plants, including *Chrysanthemum sinense,* have been used to treat gout. A few extracts from these plants were found to inhibit mammalian xanthine oxidase, some nearly as effective as allopurinol. Their structures are quite different from that of allopurinol and xanthine. The results indicate that study of compounds from traditional medicine may lead to new pharmaceuticals.

Nguyen, M. T. T., Awale, S., Tezuka, Y., Tran, Q. L., et al. Xanthine oxidase inhibitory activity of Vietnamese medicinal plants. *Biol. Pharm. Bull.* 27:1414, 2004.

Competitive Inhibition

Competitive inhibition occurs when an inhibitor competes with the substrate for binding to the free enzyme. Some **competitive inhibitors** are structurally similar to the substrate or product, whereas others are structurally quite different. For example, malonate is a competitive inhibitor of succinic dehydrogenase (Figure 10.55 and p. 555), binding presumably to the substrate-binding site. Pyrazole is a competitive inhibitor of alcohol dehydrogenase, but its structure is quite different from the substrate. If an inhibitor binds to the active site or to a different site and prevents S from binding, the reaction would be as in Figure 10.56. The inhibitor competes with substrate for binding to the enzyme, so when there is a very large concentration of substrate, the action of the inhibitor will be overcome. The amount of enzyme in the ES complex will be a function of [S], [I], K_m, and K_i (k_5/k_6). Since $E_t = E_f + ES + EI$, Eq. 10.31 is obtained.

$$v = \frac{k_3 k_1 [E_t][S]}{(k_2 + k_3)(1 + [I]/K_i) + k_1[S]} \tag{10.31}$$

Dividing all terms by k_1 yields Eq. 10.32 since $(k_2 + k_3)/k_1$ is K_m in the absence of I.

$$v = \frac{k_3 [E][S]}{K_m(1 + [I]/[K_i]) + [S]} \tag{10.32}$$

Note the similarity of Eqs. 10.32 and 10.16. A number of constants are found in the denominator, and those can be group to be called K_{mapp} since the [I] is constant, the term $1 + [I]/K_i$ is a constant. The K_m measured in the presence of inhibitor will be a K_{mapp}, but it is the true K_m times $(1 + [I]/K_i)$. The presence of the inhibitor will cause the reaction to become slower at lower substrate concentrations because the constants in the denominator are larger than if no inhibitor is present. Only when [S] is much greater than K_{mapp} would V_{max} be obtained as the denominator would be approximately just [S]. Under this condition, virtually no inhibitor will bind to enzyme (Clin. Corr. 10.8). Both the substrate and the inhibitor compete for binding, thus increasing the concentration of the substrate will overcome the action of the inhibitor. Therapeutically, competitive inhibitors are useful only if the value of $[I]/K_i$ is high so the inhibitor can function in the presence of even high concentrations of substrate. Low values for K_i (tight binding) are desired so large amounts of inhibitor need not be required to inhibit the enzyme.

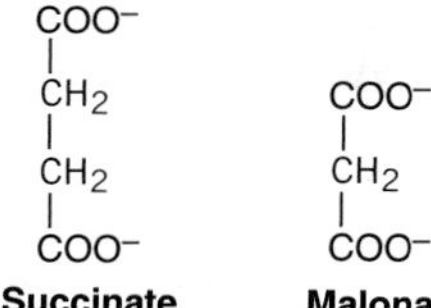

Figure 10.55 Substrate and inhibitor of succinate dehydrogenase.

$$E + S \underset{k_2}{\overset{k_1}{\rightleftharpoons}} ES \xrightarrow{k_3} E + P$$
$$+ \; I \;\; \underset{k_6}{\overset{k_5}{\rightleftharpoons}} \;\; EI$$

Figure 10.56 Reaction of a competitive inhibitor reacting with free enzyme.

CLINICAL CORRELATION 10.8

Design of a Selective Inhibitor

Prostaglandins are a very important class of paracrine hormones derived from long-chain unsaturated fatty acids (p. 737). One of the early steps in their synthesis involves an enzyme, cyclooxygenase (COX). Among the many physiological effects of the prostaglandins are the pain and inflammation associated with arthritis. Traditionally, arthritis has been treated with aspirin and other nonsteroidal anti-inflammatory drugs that have been shown to work by inhibiting the cyclooxygenases, thus decreasing the levels of "bad" prostaglandins causing the symptoms. A serious problem with these drugs is the development of gastrointestinal bleeding and perforation caused by a concomitant decrease in the "good" prostaglandins which are protective of the gastrointestinal mucosa.

There are two different cyclooxygenases, COX-1 and COX-2, which have different tissue distributions. The COX-2 enzyme is inducible by mediators from the arthritic condition, whereas COX-1 constitutively synthesizes the mucosa-sparing prostaglandins. The only significant difference in the active sites of the two enzymes is that COX-2 has a valine rather than isoleucine at residue 523. The smaller valine allows for preferential binding of custom designed inhibitors resulting in selective inhibition of COX-2. Such a custom-designed drug, VIOXX® is shown modeled into the COX-1 active site.

Key amino acid side chains forming the substrate-binding pocket are shown in red, and the heme, which participates in prostaglandin synthesis, is shown in blue. It has been shown that aspirin acetylates serine 530 in both COX-1 and -2 whereas VIOXX® selectively binds to and inhibits COX-2, thus preserving the production of "good" prostaglandins by COX-1. The phenyl group of aspirin occupies the identical location in the active site as the phenyl group of VIOXX®. Initially VIOXX®-like drugs appeared to be competitive inhibitors. With COX-2, but not with COX-1, however, a time-dependent irreversible inhibition occurs that does not involve a covalent linkage. VIOXX® came on the market in 2000 and made a tremendous difference in the quality of life for arthritic patients. Unfortunately, significant side effects involving the heart have developed with some patients, and the drug was withdrawn from the market.

Tyr 385
Ser 530
Val 523
Arg 120
Leu 357

Gierse, J. K., McDonald, J. J., Hauser, S. D., Rangwala, S. H., Koboldt, C. M., and Seibert K. A single amino acid difference between cyclooxygenase-1 (COX-1) and -2 (COX-2) reverses the selectivity of COX-2 specific inhibitors. *J. Biol. Chem.* 271: 15810, 1996.

Uncompetitive Inhibition

An inhibitor can bind to the ES complex rather than the free enzyme, as indicated in Figure 10.57. It is not binding to the substrate-binding site but to an alternative binding site that does not prevent substrate from binding. The Michaelis–Menten equation will be

$$v = \frac{V_{max}\,[S]/(1 + [I]/K_i)}{(K_m + [S])/(1 + [I]/K_i)} \qquad (10.33)$$

k_{cat} is now decreased, so the V_{max} will be decreased compared to the value in the absence of inhibitor and the K_{mapp} term is also changed. This situation is called **uncompetitive inhibition**. In contrast to competitive inhibition, increasing the concentration of substrate will not overcome the effect of the inhibitor because [I] is binding to ES complex and not free enzyme.

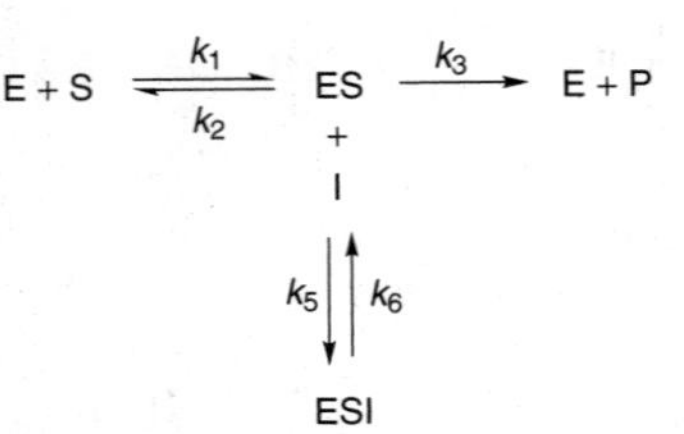

Figure 10.57 Reaction of an uncompetitive inhibitor reacting with the enzyme–substrate complex.

Noncompetitive Inhibition

A more complicated situation occurs if a reversible inhibitor binds to both the free enzyme and the ES complex. A variety of situations can occur depending on how the presence of I affects the binding of substrate and the ability of the enzyme to catalyze the reaction (Figure 10. 58).

The Michaelis–Menten equation is very complex as both ES and ES-I can yield products. If I binds with the same affinity to both E and ES and k'_3 is zero, then the following equation can be derived:

$$v = \frac{V_{max}[S]/(1 + [I]/K_i)}{K_m + [S]} \tag{10.34}$$

This mode of binding leads to an inhibition of activity in the presence of even high concentrations of [S]. If k'_3 is not zero, then a different situation arises since the ES-I complex will produce product.

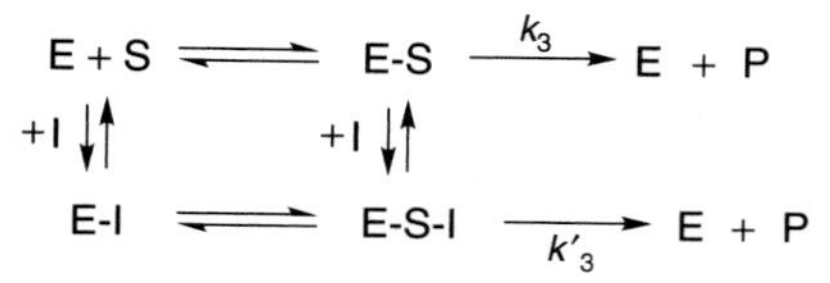

$$\begin{array}{ccccc} E + S & \rightleftharpoons & E\text{-}S & \xrightarrow{k_3} & E + P \\ +I \updownarrow & & +I \updownarrow & & \\ E\text{-}I & \rightleftharpoons & E\text{-}S\text{-}I & \xrightarrow[k'_3]{} & E + P \end{array}$$

Figure 10.58 Reaction of a noncompetitive inhibitor reacting with an enzyme.

Lineweaver–Burk Plots in the Presence of Inhibitors

Competitive and Noncompetitive Inhibitors

The Michaelis–Menten equation for an enzyme reaction performed in the presence of the three different types of reversible inhibitors leads to different double-reciprocal plots (Figure 10.59). The reason being that the term $1 + [I]/K_i$ is found associated with the K_m or k_{cat} term depending on what type of inhibitor is being used. In the case of a competitive inhibitor (Figure 10.59*a*), V_{max} is not changed at high substrate concentrations, so the velocity lines in the absence and presences of different inhibitor concentrations intersect on the *y*-axis. With uncompetitive inhibition as the inhibitor concentration is changed, a series of parallel lines forms since the slope of the line is V/K and both V and K terms contain the $(1 + [I]/K_i)$ term (Figure 10.59*b*). With noncompetitive inhibition, the inhibitor binds to both E and ES and the Lineweaver–Burk plot is a family of lines that intersect on the *x*-axis.

The double-reciprocal plots can be used to determine the mode of inhibitor binding. From the intercepts on the *x*- and *y*-axis, V_{mapp} and K_{mapp} can be calculated. If the K_m for a substrate is known, then the K_i for the inhibitors can be calculated.

Noncompetitive inhibitors effectively reduce the concentration of free enzyme, and thus velocity is decreased (Figure 10.59*c*). Even if substrate were saturating ($[S] \gg K_m$), the observed maximum velocity would be lower than it would be in the absence of the inhibitor. Compounds that are these types of inhibitors are more useful as drugs since they inhibit the enzyme independent of the concentration of substrate.

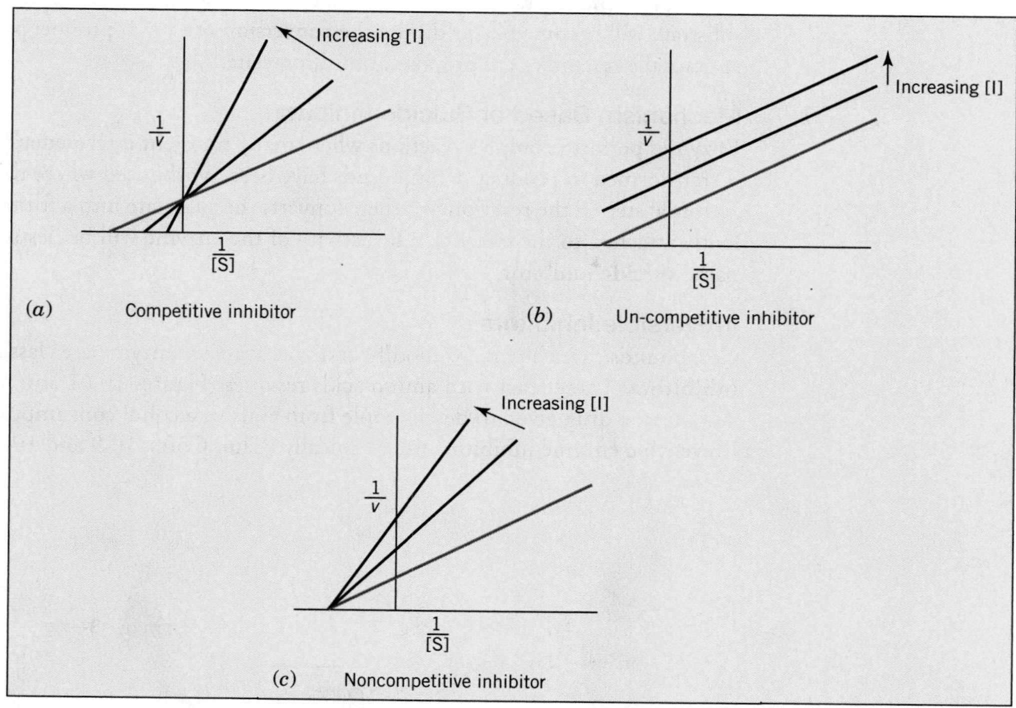

Figure 10.59 Double-reciprocal plots for competitive, uncompetitive, and reversible noncompetitive inhibition. A competitive inhibitor binds at the substrate- binding site and effectively increases the K_m for the substrate. An uncompetitive inhibitor causes an equivalent shift in both V_{max} and K_m, resulting in a line parallel to that given by the uninhibited enzyme. A reversible noncompetitive inhibitor binds with both E and ES, presumably at a site other than the substrate- binding site; therefore the effective K_m does not change, but the apparent V_{max} decreases. Red lines are in the absence of inhibitor.

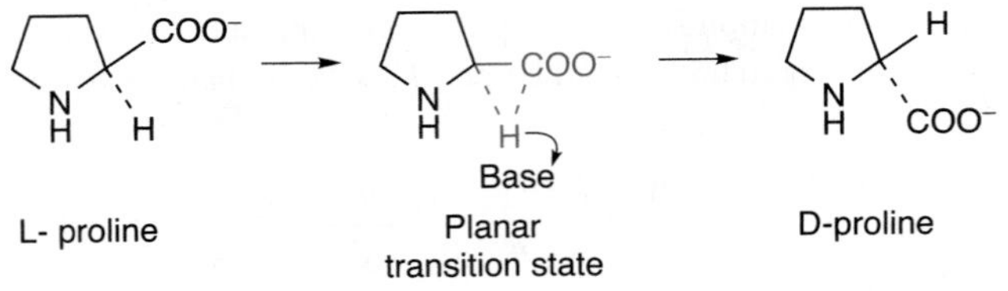

L- proline → Planar transition state → D-proline

(*a*) Proline racemase reaction

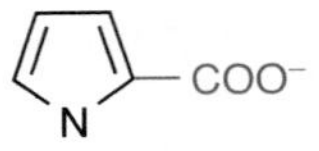

(*b*) Pyrrole-2-carboxylate is planar as in transition state.

Figure 10.60 Pyrrole-2-carboxylate mimics the planar transition state and is an excellent inhibitor of proline racemase.

Inhibitors of Two-Substrate Reactions

A competitive inhibitor of a two-substrate reaction usually inhibits only one of the two substrates. If the compound binds to only free enzyme, it will be competitive against substrate A. If the inhibitor binds to the E-A complex, then it will be competitive against substrate B. The K_i value for inhibitors of two-substrate reactions can be determined by varying the concentration of one substrate in the presences of an excess of the other substrate. The reactions for noncompetitive inhibitors would be the same as with a single substrate.

Other Inhibitors

Transition-State Inhibitors

Enzymes lower the energy of activation by stabilizing the transition state not the binding of substrate. Thus, if a compound resembled the transition state, it might bind more tightly to the enzyme than a compound resembling the substrate. These inhibitors have a K_i value much lower than the K_m for substrate. No one has ever seen a transition state, so proposed structures are based on chemical intuition. An example of a transition-state analog is an inhibitor of proline racemase (Figure 10.60). During the enzyme reaction, the tetrahedral substrate will become planar during the conversion of L- to D-proline; pyrrole 2-carboxylate structurally resembles the proposed transition state.

Mechanism Based or Suicide Inhibitors

Enzymes perform complex reactions where many transient intermediates occur as substrate is transformed to product. Compounds have been synthesized where the enzyme performs an initial step in the reaction but then converts the substrate into a form that remains covalently attached to the enzyme. The activity of the enzyme will be destroyed and hence the name **suicide** inhibitor.

Irreversible Inhibitors

Compounds that chemically modify and inactivate an enzyme are classified as **irreversible inhibitors**. These react with amino acids residues (Figures 10.61 and 10.62). Aspirin and Antabuse, a drug given to deter people from abusive alcohol consumption, are examples of irreversible enzyme inhibitors used clinically (Clin. Corrs. 10.9 and 10.10).

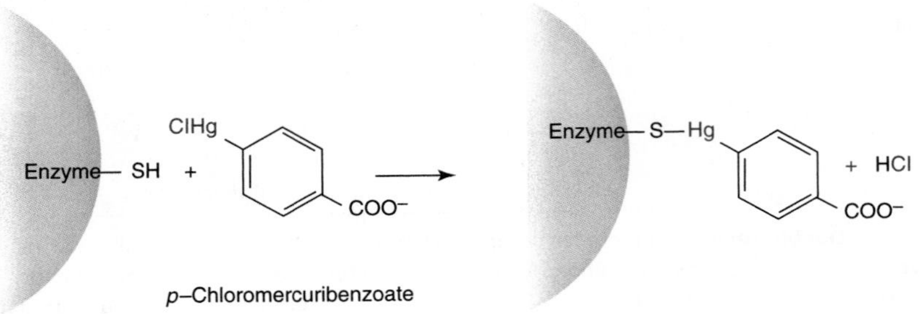

Enzyme—SH + ClHg–C₆H₄–COO⁻ → Enzyme—S—Hg–C₆H₄–COO⁻ + HCl

p–Chloromercuribenzoate

Figure 10.61 Enzyme inhibition by covalent modification of an active site cysteine.

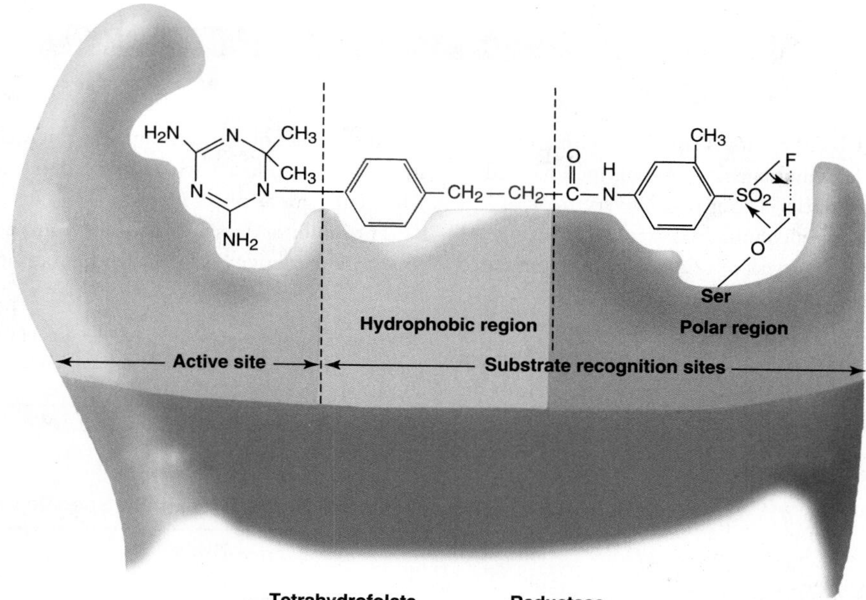

Figure 10.62 Site-directed inactivation of tetrahydrofolate reductase. The irreversible inhibitor, a substituted dihydrotriazine, structurally resembles dihydrofolate and binds specifically to the dihydrofolate site on dihydrofolate reductase. The triazine portion of the inhibitor resembles the pterin moiety and therefore binds to the active site. The ethylbenzene group (in red) binds to the hydrophobic site normally occupied by the *p*-aminobenzoyl group. The reactive end of the inhibitor contains a reactive sulfonyl fluoride that forms a covalent linkage with a serine hydroxyl on the enzyme surface. Thus, this inhibitor irreversibly inhibits the enzyme by blocking access of dihydrofolate to the active site. This is not a suicide inhibitor but an example of an inhibitor that binds to the active site and has a reactive group.

Enzyme Inhibitors as Drugs

New pharmacologically active compounds are screened for their affect on a variety of enzymes. Drugs designed to inhibit enzymes unique to a microorganism will produce fewer side effects in patients. A classic example is sulfa drugs because bacteria convert *p*-aminobenzoic acid to folic acid whereas humans do not have the enzyme for this reaction (Figure 10.63). Subtle differences that exist between isozymes of mammalian enzymes are being exploited in order to produce specific inhibitors. An example is **sildenafil** (trade name **Viagra**), a drug for erectile dysfunction, which inhibits a specific

CLINICAL CORRELATION 10.9

A Case of Poisoning

Emergency room personnel encounter many instances of pesticide poisoning and must be equipped to recognize and treat these cases. Many of the common insecticides are organophosphate compounds that irreversibly inhibit the action of acetylcholine esterase (AChE) in the postsynaptic fibers of the cholinergic neurons (p. 948) by forming stable phosphate esters with a specific serine in the active site of the esterase. Inhibition of AChE prevents the hydrolysis of acetylcholine in the synapse resulting in constant stimulation of the end organs of these neurons. The most prominent effects of pesticide poisoning in humans are paralysis of the respiratory muscles and pulmonary edema. If given early enough, a drug like pralidoxime can displace the alkyl phosphate from the pesticide bound to the active site serine and regenerate an active AchE.

Main, R. In Hodgson, E., and Guthrie, F. E. (Eds.), *Introduction to Biochemical Toxicology.* New York: Elsevier, 1980, 193.

CH=N—O⁻ → P(=O)(OR)(OR)—O—seryl (enzyme) → O—H seryl (enzyme) + CH=N—O—P(=O)(OR)—OR

Pralidoxime | Inhibited AChE | Active AChE | Inactive inhibitor

CLINICAL CORRELATION 10.10

Mushrooms and Alcohol Metabolism

Coprine, a toxin produced by the Inky Cap mushroom, forms a covalent bond with aldehyde dehydrogenase, an enzyme involved in the metabolism of alcohol. Coprine is not usually very toxic except in individuals who consume an alcoholic beverage while eating the mushrooms. That person gets a toxic reaction to the accumulated acetaldehyde formed in the oxidation of ethanol.

Disulfuram, a drug used to deter abusive consumption of alcohol is sold under the name Antabuse®, which is also a covalent inhibitor of aldehyde dehydrogenase. The drug makes the patient ill if he or she consumes alcohol. Disulfuram was originally used as an antioxidant in the rubber industry where it was observed that workers became ill when having a drink after work.

The inhibitor does not dissociate from the enzyme and the only way to restore the activity is by biosynthesis of new enzyme in the tissue.

Wiseman, J. S., and Abeles, R. H. Mechanism of inhibition of aldehyde dehydrogenase by cyclopropanone hydrate and the mushroom toxin coprine. *Biochemistry* 18: 427, 1979.

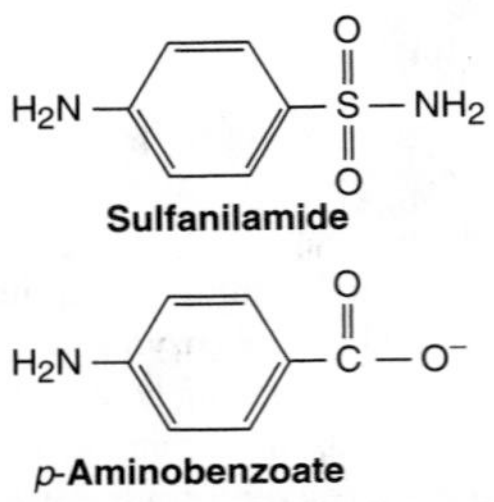

H_2N–C_6H_4–SO_2–NH_2

Sulfanilamide

H_2N–C_6H_4–COO^-

***p*-Aminobenzoate**

Figure 10.63 Structure of *p*-aminobenzoate and sulfanilamide, a competitive inhibitor of a bacterial enzyme involved in the synthesis of folic acid.

TABLE 10.5 • Representative Drugs That Inhibit Specific Enzymes

Drug	*Enzyme Target*	*Disease*
Amrubicin®	Topoisomerase 2	Cancer chemotherapy
Antabuse®	Aldehyde Dehydrogenase	Alcoholism
Captopril®	Angiotensin-converting enzyme	Hypertension
Celebrex®	Cyclooxygenase-2	Arthritis
Digoxin®	Na^+-K^+-Exchanging ATPase	Heart problems
Hycamtin®	Topoisomerase 1	Cancer chemotherapy
Lipitor®	3-hydroxy-3-methylglutaryl CoA	High cholesterol
Viagra®	Phosphodiesterase 5	Erectile dysfunction

isozyme of phosphodiesterase (p. 447). A compound may be found to be a specific for an enzyme *in vitro,* but it may also inhibit enzymes not tested in the screening process. Representative drugs that are inhibitors of enzymes are presented in Table 10.5. Clin. Corrs. 10.11 and 10.12 describe two very specific examples. A Closer Look 10.6 describes a method to rapidly identify new enzyme inhibitors and activators.

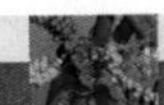

CLINICAL CORRELATION 10.11

Testosterone and Prostate Cancer

Prostate cancer is one of the most common forms of cancers and affects only men. During an early stage of the disease the tumor requires testosterone. With time, the tumor cells mutate so they no longer need testosterone. The tumor cells are referred to as hormone refractive cells. Some investigators feel that even in the early stages of disease, there are a few cells that are already hormone refractive.

Investigators consider that the cells they thought were hormone refractive actually respond to a very low level of testosterone which might be found even after castration (surgically or chemically). A new approach to combat hormone refractive disease is to use an inhibitor of a cytochrome P450 (Cyp17A1), an enzyme involved in the biosynthesis of the hormone. Abiraterone has been tested as an *in vivo* inhibitor of the nontestis testosterone pathway.

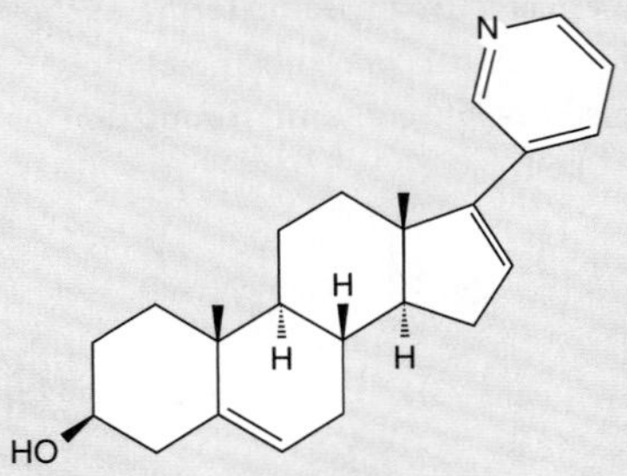

Structure of Abiraterone.

CLINICAL CORRELATION 10.12

Natural Products as Inhibitors of Enzymes

Some fruits contain compounds that affect the metabolism of specific drugs. Individuals consuming the food and the drug can have an adverse effect. Perhaps the best-studied interaction is with grapefruit. This fruit contains a class of organic compounds called furanocoumarins, and it was found that some members inhibit a cytochrome P450 (CYP3A4) in the intestine but not in the liver. A number of statin drugs used to treat high cholesterol and some blood pressure medications are partially metabolized in the intestines and require the involvement of the CYP3A4. Consuming grapefruit while taking the drug alters the pharmokinetics of a variety of drugs and may present dire consequences. Once the drug is in the circulating system, its metabolism is no longer inhibited.

In spite of the effect of grapefruit juice on CYP3A4, studies have shown that it has many known good effects on the body because of the antioxidants it possesses. The grapefruit effect was discovered during a study on the interaction of ethanol with a blood pressure medication called felodipine (brand name Plendil®). Grapefruit juice was used as the vehicle for the ethanol, and ultimately the investigators realized that the inhibition of drug metabolism was due to the juice and not ethanol.

Kiani, J., and Imam, S. Z. Medicinal importance of grapefruit juice and its interaction with various drugs. *Nutr. J.* 6: 33, 2007.

A CLOSER LOOK 10.6

High-Throughput Screening to Find Inhibitors and Activators of Enzymes

Inhibitors of enzymes have been found from natural compounds occurring in plants, bacteria, and mold. A chemical-library screening procedure is used for the search of new enzyme inhibitors, especially by the pharmaceutical industry. Tens of thousands of compounds and extracts of natural product are rapidly tested for their ability to inhibit an enzyme or affect cell function. Different combinations of test materials are robotically added to the test system. Those which affect the enzyme are then analyzed so that new biologically active chemicals can be identified. The advantage of this approach is that compounds with a variety of different structures can be tested, without a preconceived idea of what should be the structure of the effector. Depending on the assay, it is possible to screen for any type of effector.

Assay Guidance Manual 5.0. Eli Lilly and Company and NIH Chemical Genomics Center, 2008. Available online at http://www.ncgc.nih.gov/guidance/manual_toc.html.

10.10 • REGULATION OF ENZYME ACTIVITY

Covalent Modification

Cells have the ability to regulate the activity of key enzymes by covalent modification of the protein. There are many protein kinases that catalyze phosphorylation of serine, threonine, or tyrosine residues on specific enzymes and protein phosphorylases that remove the phosphate by hydrolysis. Depending on the enzyme, phosphorylation can activate an inactive enzyme or inactivate an active enzyme. Groups such as sulfate and acetate can also be added to alter enzymatic activity. Since both phosphorylation and dephosphorylation are performed by enzymes, this reversible covalent modification is a rapid and effective way of controlling enzyme activity.

Allosteric Control of Enzyme Activity

The activity of many enzymes can be modulated by ligands binding to the enzyme at sites other than the substrate-binding site. These sites are referred to as allosteric sites (Figure 10.64). **Allosteric** is derived from the Greek root *allo,* meaning "the other." An allosteric site is a unique region of the enzyme quite different from the substrate-binding site. The ligands that bind at the allosteric site are called allosteric effectors or modulators and are not modified during the course of the enzyme reaction. Binding of an allosteric effector causes a conformational change of the enzyme so that the affinity for the substrate or other ligands also changes. Positive (+) allosteric effectors increase the enzyme affinity for

CLINICAL CORRELATION 10.13

A Case of Gout Demonstrates the Difference between an Allosteric- and Substrate-Binding Site

The realization that allosteric inhibitory sites are separate from allosteric activator sites as well as from the substrate binding and catalytic sites is illustrated by a study of a patient with gout whose red blood cell PRPP level was increased. It was found that the patient's PRPP synthetase had normal K_m and V_{max} values and sensitivity to activation by phosphate. The increased PRPP levels and hyperuricemia arose because the end products of the pathway (ATP, GTP) were not able to inhibit the synthetase through the allosteric inhibitory site. It was suggested that a mutation in the inhibitory site or in the coupling mechanism between the inhibitory and catalytic site led to failure of the feedback control mechanism.

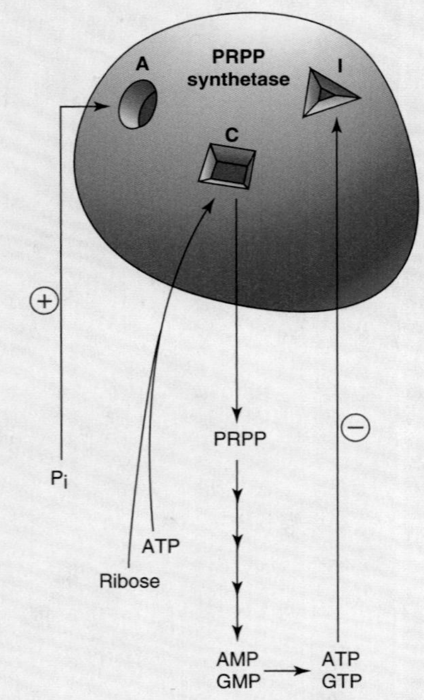

Sperling, O., Persky-Brosh, S., Boen, P., and DeVries, A. Human erythrocyte phosphoribosyl pyrophosphate synthetase mutationally altered in regulatory proprties. *Biochem. Med.* 7: 389, 1973.

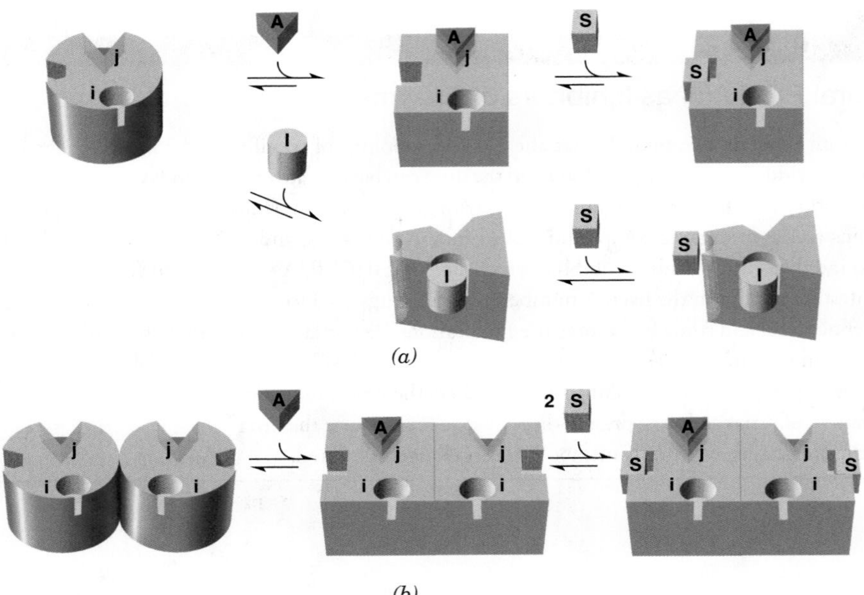

Figure 10.64 Models of allosteric enzyme systems. (*a*) Model of a monomeric enzyme. Binding of a positive allosteric effector, A (green), to the activator site, j, induces a new conformation to the enzyme, one that has a greater affinity for the substrate. Binding of a negative allosteric effector (purple) to the inhibitor site, i, results in an enzyme conformation having a decreased affinity for substrate (orange). (*b*) A model of a polymeric allosteric enzyme. Binding of the positive allosteric effector, A, at the j site causes an allosteric change in the conformation of the protomer to which the effector binds. This change in the conformation is transmitted to the second protomer through cooperative protomer–protomer interactions. The affinity for the substrate is increased in both protomers. A negative effector decreases the affinity for substrate of both protomers.

substrate. The reverse is true for negative (–) allosteric effectors. The allosteric site at which the positive effector binds is referred to as an activator site; the negative effector binds at an inhibitory site. The existence of allosteric sites is illustrated in Clin. Corr. 10.13.

Allosteric enzymes are divided into two classes on the basis of the effect of the allosteric effector on the K_m or V_{max}. In the **K class** the effector alters the K_m but not V_{max}, whereas in the **V class** the effector alters V_{max} but not K_m. In the K class, a negative effector binding at an allosteric site increases the K_m for the substrate. Similarly in V class enzymes, positive and negative allosteric modifiers increase or decrease the rate of breakdown of the ES complex to products. There are a few enzymes in which both K_m and V_{max} are affected by the modifier.

In theory, a monomeric enzyme can undergo an allosteric transition in response to a modulating ligand (Figure 10.64*a*). However, only a few monomeric allosteric enzymes are known including ribonucleoside diphosphate reductase and pyruvate-UDP-N-acetylglucosamine transferase. Most allosteric enzymes are oligomeric consisting of several subunits (Figure 10.64*b*). Binding of ligand to one protomer can affect the binding of ligands on the other protomers in the oligomer. If the same ligand binds at the allosteric site and is also the substrate for the enzyme, the effect is referred to as a **homotropic interaction**. Homotropic interactions are almost always positive. A **heterotropic interaction** is the effect of one ligand binding at the allosteric site on the binding of a different ligand at the active site. Heterotropic interactions can be positive or negative. Examples are the effect of a negative effector on the binding of substrate or the positive effect of an allosteric activator on substrate binding. Heterotropic and homotropic effects in oligomeric enzymes are mediated by cooperativity between subunits.

Most of the enzymes subject to allosteric modulation are rate-determining enzymes in metabolic pathways or are located at a junction where the substrate can be used for more than one pathway.

Multi-subunit Enzymes: Cooperativity

Many enzymes are multimers consisting of either the same or different monomeric subunits. Subunits of some multimeric enzymes act independently of each other, with

CLINICAL CORRELATION 10.14

Ambiguity in the Assay of Mutated Enzymes

Structural gene mutations leading to production of enzymes with increases or decreases in K_m are frequently observed. A case in point is a patient with hyperuricemia and gout, whose red blood cell hypoxanthine-guanine-phosphoribosyltransferase (HGPRT) showed little activity in assays *in vitro*. This enzyme is involved in the salvage of purine bases and catalyzes the reaction.

$$\text{Hypoxanthine} + \text{PRPP} \rightarrow \text{inosine monophosphate} + \text{PPi}$$

where PRPP is phosphoribosyl pyrophosphate.

The absence of HGPRT activity results in a severe neurological disorder known as Lesch–Nyhan syndrome (p. 816), yet this patient did not have the usual clinical signs of this disorder. Immunological testing with a specific antibody to the enzyme revealed as much cross-reacting material in the patient's red blood cells as in normal controls, indicating that the enzyme was being synthesized but was inactive in the assay *in vitro*. Increasing substrate concentration in the assay of the patient's red cell hemolysate restored full activity. This anomaly is apparently due to a mutation in the substrate-binding site of HGPRT, leading to an increased K_m. The substrate concentration in the standard assay and in the red blood cells was not high enough to bind to the enzyme. This case reinforces the point that an accurate enzyme determination is dependent on zero-order kinetics, that is, the enzyme being saturated with substrate.

Sorenson, L., and Benke, P. J. Biochemical evidence for a distinct type of primary gout. *Nature* 213: 1122, 1967.

events taking place in one subunit not influencing the activity of the other subunits. For these enzymes k_{cat} or V_{max} values can be divided by the number of subunits to obtain the actual turnover number per subunit. In other multi-subunit enzymes, an event in one subunit affects the ability of the adjacent subunits to function. In these situations, the K_m for substrate is usually changed but in some cases k_{cat} is altered (Clin. Corr. 10.14).

One subunit in a multimer can affect the properties of an adjacent subunit by changing the conformation of the subunits after substrate binds to it (Figure 10.64*a*). The altered monomer could bind substrate more easily and hence is more saturated without the substrate concentration increasing. If the binding becomes easier (decrease in K_m), then the enzyme is said to exhibit **positive cooperativity**; this is analogous to hemoglobin binding oxygen (p. 358). When the structural change makes binding of substrate to the second subunit more difficult (increase in K_m), then the enzyme is said to exhibit **negative cooperativity**.

The test for an enzyme exhibiting cooperativity is a plot of velocity versus [S]. The curve will be sigmoidal rather than hyperbolic (Figure 10.65*a*) and a double-reciprocal plot will not be linear (Figure 10.65*b*). This occurs because the K_m will change as the concentration of substrate changes.

It is very difficult to calculate precise K_m values for an enzyme exhibiting either positive or negative cooperativity. The value, however, is estimated from the concentration of substrate that allows v to be 50% of V_{max}; it is referred to as $K_{0.5}$. The Hill equation, employed to explain the cooperative binding of oxygen to hemoglobin (p. 358), is used to calculate the Hill coefficient that estimates the degree of cooperativity between interacting subunits of an enzyme. Values of less than 1 demonstrate negative cooperativity, and values greater than 1 positive cooperativity. Representative examples of positive and negative cooperativity are illustrated in Figure 10.66.

Models to Explain Cooperativity

There are two models to explain cooperativity. Both use the premise that there are interactions between the subunits, and the conformation of the subunit binding a ligand is different from that of the subunit in the absence of ligand. The major difference between the models relates to whether the two conformations are preexisting or they occur only after ligand binds.

The **sequential model** proposes that when ligand binds to one subunit (A), there is a change in the conformation of the subunit, which then induces a conformational change in a contiguous subunit (B). The effect of ligand binding is sequentially transmitted through the interface between subunits producing increased or decreased affinity for the ligand by B (Figure 10.67*a*). In this model numerous hybrid states occur, giving rise to cooperativity and

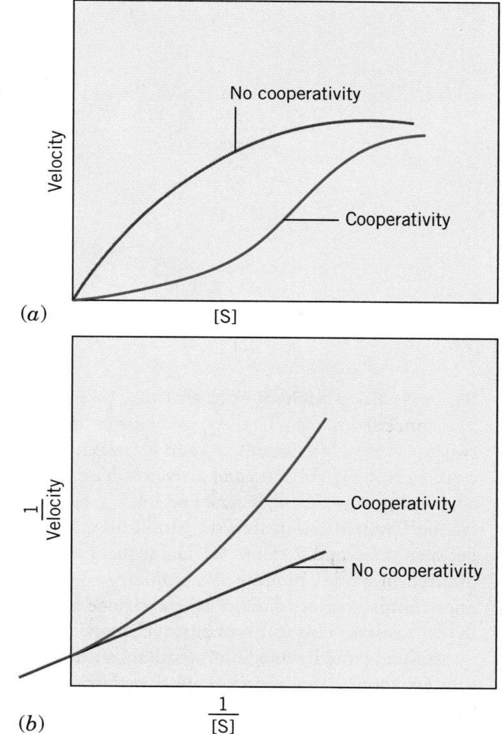

Figure 10.65 Plots for enzymes that exhibit cooperativity are atypical. (*a*) Velocity versus substrate plot and (*b*) double-reciprocal plot. Cooperative curves are for positive cooperativity.

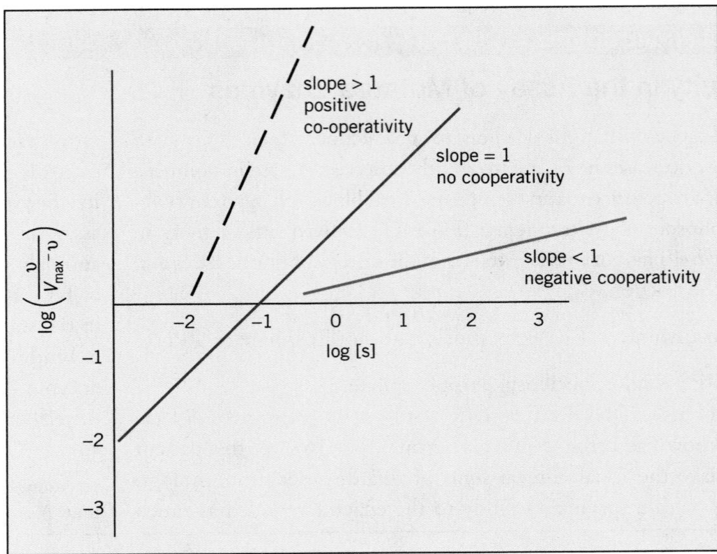

Figure 10.66 Hill plot for an allosteric enzyme.

sigmoid plots of velocity versus [S]. Both positive and negative cooperativity can be accommodated by the model. Positive allosteric modulators can induce a conformation in the subunit that has an increased affinity for the substrate, and a negative modulator induces a different conformation in the subunit resulting in decreased affinity for substrate. A similar argument can be made for enzymes where k_{cat} is changed.

The **concerted model** proposes that the two conformations are in equilibrium in the absence of ligand and the equilibrium is driven toward the ligand-bound conformation as ligand binds. The two states in the concerted model are called the T (tense or taut) and the R (relaxed) (Figure 10.67*b*). Though both states can bind ligand, one state binds it more tightly. Activators and substrates bind to the R state and shift the preexisting equilibrium toward the R state. With more of the protein in the R state, it is easier for the next molecule of substrate or activator to bind. Conversely, inhibitors favor the T state. In the presence of

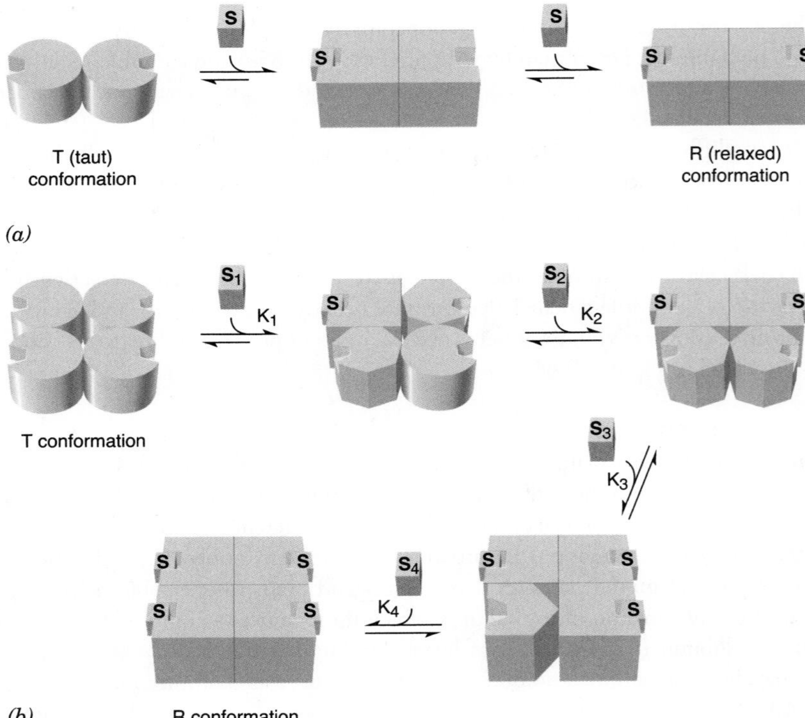

Figure 10.67 Models of cooperativity. (*a*) The concerted model. The enzyme exists in only two states, the T (tense or taut) and R (relaxed) conformations. Substrates and activators have a greater affinity for the R state and inhibitors for the T state. Ligands shift the equilibrium between the T and R states. (*b*) The sequential induced-fit model. Binding of a ligand to any one subunit induces a conformational change in that subunit. This conformational change is transmitted partially to adjoining subunits through subunit–subunit interaction. Thus the effect of the first ligand bound is transmitted cooperatively and sequentially to the other subunits (protomers) in the oligomer, resulting in a sequential increase or decrease in ligand affinity of the other protomers. The cooperativity may be either positive or negative, depending on the ligand.

an inhibitor, it is more difficult for the substrate to bind to the T conformation. Although this model accounts for the kinetic behavior of many enzymes, it cannot easily account for negative cooperativity.

Regulatory Subunits Modulate the Activity of Catalytic Subunits

The allosteric-binding site, as discussed, is considered to reside on the same subunit as the catalytic site, and all subunits of the enzyme are identical. In several enzymes a distinct regulatory protein subunit exists. These regulatory subunits (R) have no catalytic function, but their binding with the catalytic subunit (C) modulates the activity of the catalytic subunit through an induced conformational change. Protein kinase A is regulated by this mechanism (Figure 10.68). In the absence of cAMP, the R subunit binds to the C subunit causing inhibition of the activity. When cAMP binds to the R subunit, there is a conformational change of R leading to a dissociation of R from C. The C subunit is now catalytically active.

Calmodulin (p. 493), a Ca^{2+}-binding protein, is a regulatory subunit for some enzymes that use Ca^{2+} as a modulator of their activity. Binding of Ca^{2+} to calmodulin causes a conformational change, allowing it to bind to the Ca^{2+}-dependent enzyme.

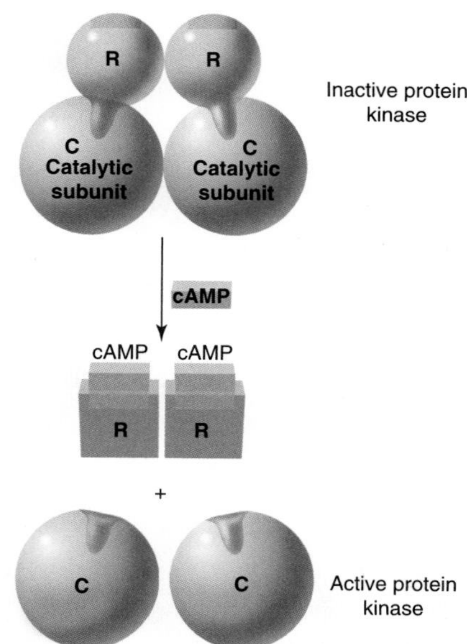

Figure 10.68 Model of allosteric enzyme with separate catalytic (C) and regulatory (R) subunits. The regulatory subunit of protein kinase A contains a pseudo substrate region in its primary sequence that binds to the substrate site of the catalytic subunit. In the presence of cAMP the conformation of the R subunit changes so that the pseudo substrate region can no longer bind, resulting in release of active C subunits.

10.11 • REGULATION OF METABOLIC PATHWAYS

The physiological integration of many enzymes into a metabolic pathway as well as the interrelationship of the products of one pathway with the activity of other pathways is controlled. It is not necessary to regulate the activity of every enzyme in a pathway. Rather modulation of the activity of one or more key enzymes of the pathway is sufficient. One enzyme-catalyzed reaction is usually slower than the others. This is the **rate-limiting step** for the pathway. Thus, the rate of flux of substrates in the sequence is dependent on one enzyme in the entire pathway. Altering the kinetic properties of this one enzyme has a large effect of the rate of formation of the final product.

Another site to regulate a metabolic pathway is at the first irreversible reaction unique to the pathway; this step is referred to as the **commitment step**. The rate-limiting enzyme is not necessarily the one catalyzing the committed step. In addition, in some pathways one intermediate substrate serves as an intermediate for two or more metabolic transformations each leading to different products. Regulating the enzymes at these branch points would control the flux of metabolites through the two pathways. The activity of the enzyme associated with the committed step or with the rate-limiting step can be regulated in a number of ways.

Cellular Enzyme Concentration

Many inducible enzymes have half-lives of less than 1 h, whereas many constitutive enzymes have half-lives of days. Cells regulate enzyme concentration by changing the rate of de novo synthesis at either the transcriptional or translational level. For example, glucose represses the de novo synthesis of pyruvate carboxykinase, the rate-limiting enzyme in the conversion of pyruvate to glucose. If sufficient glucose is available in the blood, there is a low level of pyruvate carboxykinase in the liver. Low blood levels of glucose lead to an increase in the enzyme and increase in glucose synthesis.

Activators, Inhibitors, and Covalent Modification

Short-term regulation of an enzyme occurs through modulation of the activity by activators, inhibitors, and covalent modification. When the cellular concentration of deoxyribonucleotides increases such that the cell has sufficient amounts for synthesis of DNA, the key enzyme of the synthetic pathway is inhibited by the end products of the pathway, resulting in a slowdown of the synthetic pathway. This is referred to as **feedback inhibition** (Figure 10.69*a*). In addition to feedback within a pathway, feedback inhibition of other pathways can also occur. This is referred to as **cross-regulation**. Here a product of one pathway serves as an inhibitor or activator of an enzyme occurring early in another pathway (Figure 10.69*b*). A good example is the cross-regulation of the production of the four deoxyribonucleotides for DNA synthesis.

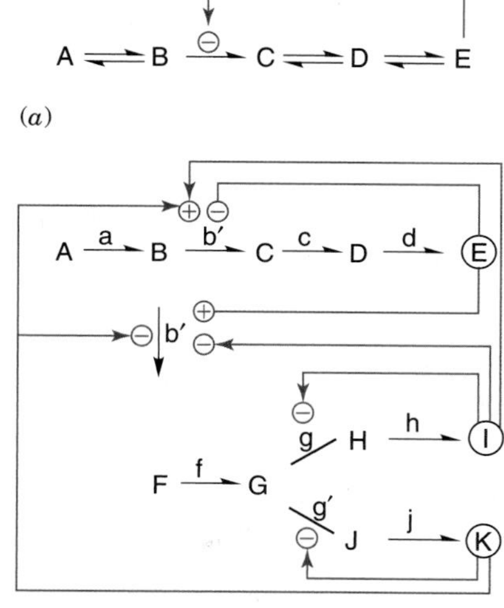

Figure 10.69 Control of metabolic pathways. (*a*) Feedback inhibition. Product E could inhibit the enzyme that catalyzes reaction of B to C. This may be an irreversible or rate-limiting reaction. (*b*) Regulation of a hypothetical complex pathway. Product E could inhibit enzyme b while activating enzyme b'. Similarly, product I and K could inhibit their synthesis by inhibiting enzymes b', g (and g') while activating enzyme b.

Compartmentation

Finally, the activity of a pathway can be regulated by physically partitioning reactions in specific cellular compartments separated by impermeable membranes. The transport of substrates, products, and cofactors across the membrane controls their concentration in the compartment. This regulates their availability and can limit enzyme activity. This is referred to as **compartmentation**. For example, acetyl CoA is formed in the mitochondria and must be transported out to the cytosol for use in fatty acid synthesis. Acetyl CoA carboxylase, the first step in fatty acid synthesis, requires acetyl CoA. The substrate cannot diffuse into the cytosol so a specific transport mechanism is present in the membrane (p. 684). Anabolic and catabolic pathways are often segregated in different organelles in order to better control each pathway. There would be no point for oxidation of fatty acids occurring at the same time and in the same compartment as biosynthesis of fatty acids. By maintaining fatty acid biosynthesis in the cytoplasm and oxidation in the mitochondria, control can be exerted by regulating transport of common intermediates across the mitochondrial membrane.

10.12 • CLINICAL APPLICATIONS OF ENZYMES

Because of their diverse properties enzymes find many uses in therapeutics, clinical analysis, and disease diagnostics. For example, hydrolytic enzymes, such as the proteases pronase and RNases, are used in debridement of wounds. **Streptokinase**, an enzyme mixture prepared from a streptococcus, is useful in clearing blood clots that occur in myocardial infarcts and in the lower extremities. It activates the fibrinolytic preenzyme **plasminogen** (p. 994) that is normally present in plasma. The activated enzyme is **plasmin**, a serine protease that cleaves the insoluble fibrin in blood clots into several soluble components. Another serine protease, human tissue plasminogen activator t-PA, is commercially produced by bioengineered *E. coli* and used in dissolving blood clots in patients suffering myocardial infarction. t-PA also functions by activating plasminogen.

Measurement of Plasma Enzymes

Enzymes found in plasma are of two types: those which are normally present and have a functional role in plasma and those which are released from tissues. The latter have no functional role in the plasma but are used for diagnostic purposes. Normally, the plasma content of intracellular enzymes is low or absent, but when cells are damaged, the contents can appear in blood. A disease process may cause changes in cell membrane permeability or increased cell death, resulting in release of intracellular enzymes. Lower molecular weight components can also diffuse out of cells when there is an abnormal increase in their cellular concentration. Analyzing blood for cellular components represents a convenient way for clinicians to assess damage occurring to specific tissues. In the diagnosis of specific organ involvement in a disease process it would be ideal if there were an enzyme unique to each organ that could be measured in plasma. Unfortunately, identical enzymes are found in most tissues. In some instances, however, different isozymes (p. 379) occur in different organs; these are useful for diagnostic purposes if released from a diseased tissue. For example, an isozyme of alcohol dehydrogenase that is liver specific and an acid phosphatase form found primarily in prostate are useful for specific identification of diseases in these organs. For some enzymes the tissue specific isozymes possess different kinetic properties, so determination of K_m and k_{cat} values can be used to determine from what tissue they arose.

Evaluation of the timing of appearance and disappearance of particular enzymes in plasma permits a diagnosis of specific organ involvement. Figure 10.70 illustrates the time-dependent changes in plasma activities of enzymes from the myocardium following a heart attack. Such profiles allow one to establish when the attack occurred and whether treatment is effective. Clin. Corr. 10.15 describes how diagnosis of a specific enzyme defect led to a rational clinical treatment that restored the patient to health. Another example of how isozymes are used to diagnose diseases is the isozyme pattern for lactate dehydrogenase as a function of time after a myocardial infarction (Figure 10.71, p. 420, and Table 10.2, p. 379). Not only does the total

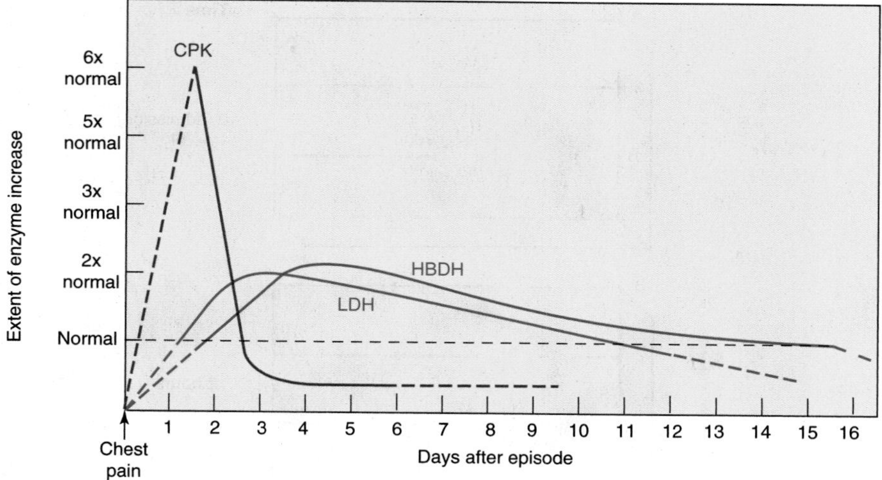

Figure 10.70 Kinetics of release of cardiac enzymes into serum following a myocardial infarction. CPK, creatine kinase; LDH, lactic dehydrogenase; HBDH, α-hydroxybutyric dehydrogenase. Such kinetic profiles allow one to determine where the patient is with respect to the infarct and recovery. *Note:* CPK rises sharply but briefly; HBDH rises slowly but persists. Reproduced with permission from Coodley, E. L. *Diagnostic Enzymes.* Philadelphia: Lea & Febiger, 1970, 61.

CLINICAL CORRELATION 10.15

Identification and Treatment of an Enzyme Deficiency

Enzyme deficiencies usually lead to increased accumulation of specific intermediary metabolites in plasma and hence in urine. Recognition of the intermediates that accumulate in biological fluids is useful in pinpointing possible enzyme defects. After the enzyme deficiency is established, metabolites that normally occur in the pathway but are distal to the block may be supplied exogenously in order to overcome the metabolic effects of the enzyme deficiency. In hereditary orotic aciduria, there is a double enzyme deficiency in the pyrimidine biosynthetic pathway leading to accumulation of orotic acid. Both orotate phosphoribosyltransferase and orotidine 5′-phosphate decarboxylase are deficient, causing decreased *in vivo* levels of CTP and TTP. The two activities are deficient because they reside in separate domains of a bifunctional polypeptide of 480 amino acids. dCTP and dTTP, which arise from CTP and TTP, are required for DNA synthesis. In these enzyme deficiency diseases, the patients are pale, weak, and fail to thrive. Administration of the missing pyrimidines as uridine or cytidine promotes growth and general well-being and also decreases orotic acid excretion. The latter occurs because the TTP and CTP formed from the supplied uridine and cytidine repress carbamoylphosphate synthetase, the committed step, by feedback inhibition, resulting in a decrease in orotate production.

Webster, D. R., Becroft, D. M. O., and Suttie, D. P. Hereditary orotic aciduria and other diseases of pyrimidine metabolism. In Scriver, C. R., Beaudet, A. L., Sly, W. S., and Valle, D. (Eds.), *The Metabolic and Molecular Bases of Inherited Disease,* 7th ed. New York: McGraw Hill, 1995, 1799.

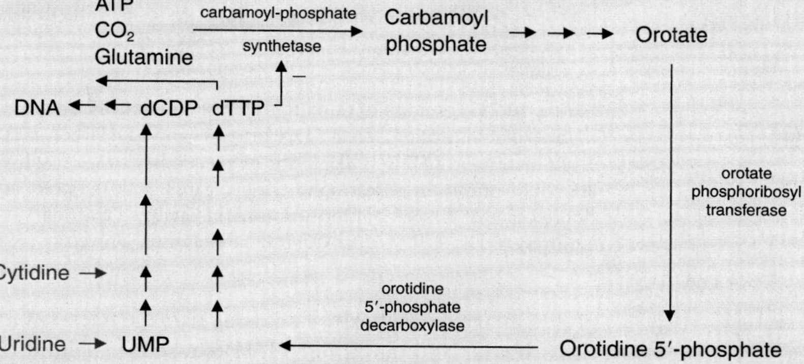

enzymatic activity increase, but o there is a change in the isozyme pattern. More LDH_1, found primarily in myocardium and red blood cells, is found. In this example, LDH_5 is increased, indicative of liver damage to the patient. Thus secondary complications of heart failure can be monitored. After two weeks the isozyme pattern returns to that of a healthy patient.

Kinetic analyses of both creatine kinase and lactate dehydrogenase are useful for diagnostic purposes. CPK and LDH isozymes are plotted in Figure 10.72 as a function of time

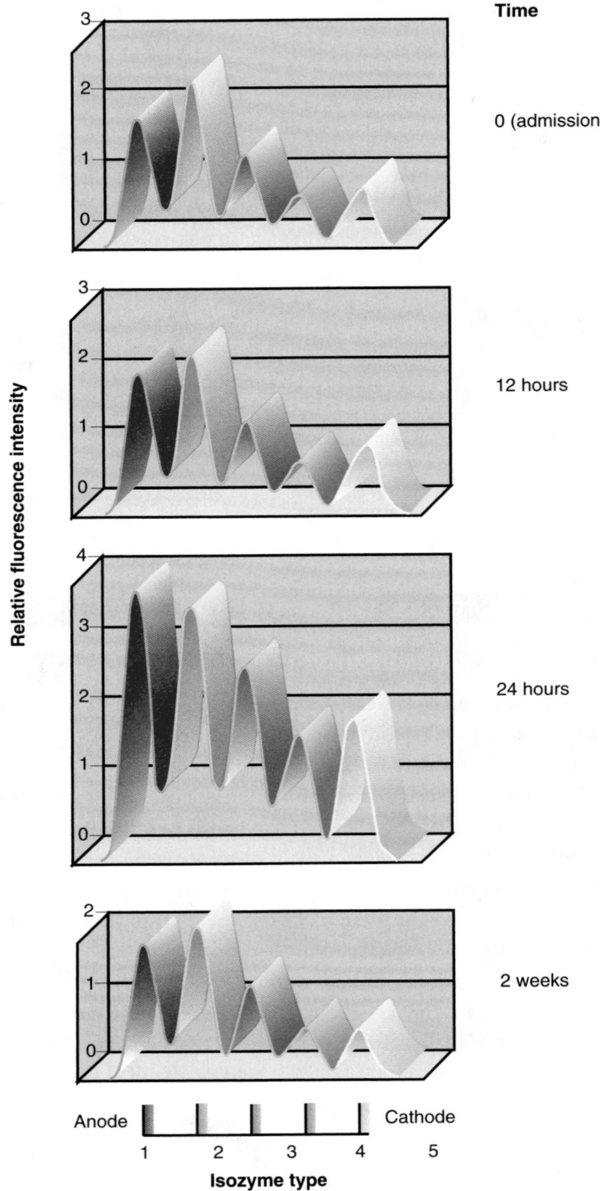

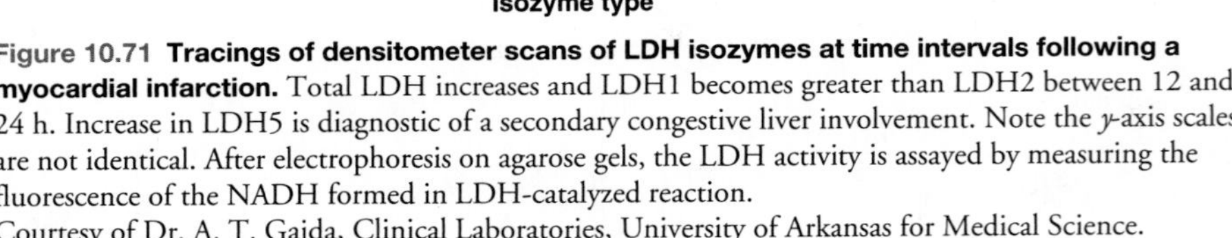

Figure 10.71 Tracings of densitometer scans of LDH isozymes at time intervals following a myocardial infarction. Total LDH increases and LDH1 becomes greater than LDH2 between 12 and 24 h. Increase in LDH5 is diagnostic of a secondary congestive liver involvement. Note the *y*-axis scales are not identical. After electrophoresis on agarose gels, the LDH activity is assayed by measuring the fluorescence of the NADH formed in LDH-catalyzed reaction.
Courtesy of Dr. A. T. Gajda, Clinical Laboratories, University of Arkansas for Medical Science.

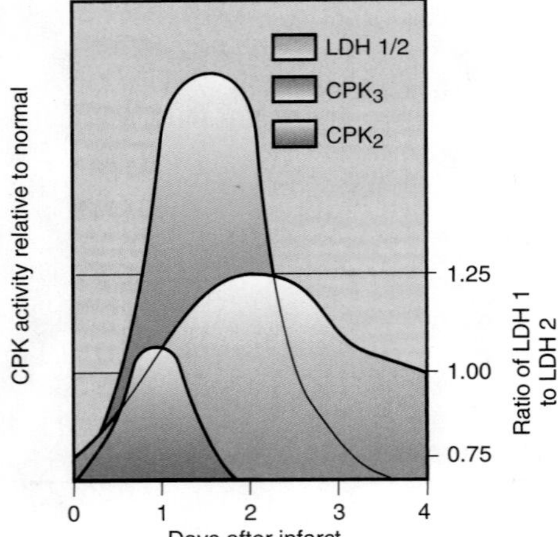

Figure 10.72 Characteristic changes in serum CPK and LDH isozymes following a myocardial infarction. CPK2 (MB) isozyme increases to a maximum within 1 day of the infarction. CPK3 lags behind CPK2 by about 1 day. Total LDH level increases more slowly. The increase of LDH1 and LDH2 within 12–24 h coupled with an increase in CPK2 is diagnostic of myocardial infarction.

after damage to the heart. The tissue damage releases CPK_2 into the blood within the first 6–18 h after an infarct, but LDH release lags behind the appearance of CPK_2 by 1 to 2 days. Normally, the activity of the LDH_2 isozyme is higher than that of LDH_1; however, in the case of infarction, the activity of LDH_1 becomes greater than LDH_2, at about the time CPK_2 levels are back to baseline (48–60 h). The LDH isozyme "switch" coupled with increased CPK_2 is diagnostic of myocardial infarct (MI) in virtually 100% of the cases.

The use of enzymes in the clinical laboratory is described in A Closer Look 10.7 and 10.8.

Metabolomics and Proteomics

With mass spectrophotometers becoming relatively common instruments, many investigators are using them with the hope of finding specific metabolites, proteins, or

A CLOSER LOOK 10.7

Clinical Laboratory Assays Employing Enzymes

Typically enzymes found in organs are not present in serum unless there is damage to the tissue. Hence measurement of serum levels of enzymes can be diagnostic of a diseased state. Assays for measurement-of plasma enzymes-are based on control of the assay system, including temperature, pH, and concentration of substrates, co-substrates, and cofactors. The K_m for the enzyme to be determined must be known. Typically, assays are performed under V_{max} conditions though they need not be. Assay conditions are optimized for the properties of the normal enzyme, which may not correctly measure levels of a variant enzyme because pH optima and/or the K_m for substrate and cofactors may be different from that of the normal enzyme. Enzyme assays measure the concentration of a product formed or substrate remaining as a function of time. In some reactions, the products or substrates absorb light; change in light absorption as a function of time is measured.

Many substrates have been synthesized where they or their products are chromophoric compounds (absorb light or emit fluorescent light). The coenzymes NAD(P) and FAD are easily measured because their reduced forms absorb light in the visible range. Many enzymes that do not directly reduce NAD(P) or FAD generate products that are substrates for NAD(P) or FAD-linked dehydrogenases. By coupling two enzyme reactions, the activity of the enzyme of interest can be determined.

Enzymes, as reagents, are used to measure metabolites in automated clinical analyzers. Glucose oxidase, an enzyme that oxidizes glucose, is used clinically to measure the concentration of glucose in body fluids. Assays for cholesterol employ cholesterol oxidase, and for triacylglycerols lipase is employed. The tests can be completed in a few minutes using small amounts of plasma. The enzymes can be immobilized in a solid bilayer with cofactors and indicator reagents, which become chromophores when reduced. In the case of cholesterol oxidase, hydrogen peroxide is a product of the oxidase reaction and the peroxide formed oxidizes a colorless dye to a colored product mediated by a peroxidase, another enzyme.

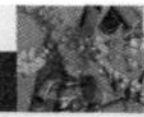

A CLOSER LOOK 10.8

Enzyme-Linked Immunoassays Employ Enzymes as Indicators

Modern clinical chemistry has benefited from the merging of enzyme chemistry and immunology. Antibodies specific to a protein antigen are coupled to an indicator enzyme such as horseradish peroxidase to generate a very specific and sensitive assay. After binding of the peroxidase-coupled antibody to the antigen, the peroxidase is used to generate a colored product that is measurable and whose concentration is related to the amount of antigen in a sample. This assay has been given the acronym **ELISA** for **e**nzyme-**l**inked **i**mmunoad**s**orbent **a**ssay. Monoclonal antibodies generate antibodies that are very specific for an antigen, As an example, antibodies to a specific isozyme or to one member of a family of very similar enzymes have been produced allowing their determination by **ELISA**.

Horseradish peroxidase is used as an indicator in ELISA to test for the presence of human immunodeficiency virus (HIV) coat protein antigens. Peroxidase generates many molecules of product per antigen as time progresses. Such amplified enzyme assays allow the measurement of remarkably small amounts of antigens.

enzymes released from tumors. These metabolites and proteins could be employed as markers of specific types of tumors. It is anticipated that unique enzymes and proteins associated with various disease states will be identified in plasma. These enzymes and proteins will have to be purified and characterized to determine their effects on normal cellular activity.

Bibliography

Bialit, S., and Kimchi, A. The death-associated protein kinases: Structure, function, and beyond. *Annu. Rev. Biochem*, 75: 189, 2006.

Blackburn, G. M., Kang, A. S., Kingsbury, G. A., and Burton, D. R. Review of abzymes. *Biochem. J.* 262: 381, 1989.

Bugg, T. *An Introduction to Enzyme and Coenzyme Chemistry.* London: Blackwell, 1997.

Chen, C-H, Budas, G. R., Churchill, E. N., Disatnik, M-H., et al. Activation of aldehyde dehydrogenase-2 reduces ischemic damage to the heart. *Science* 321: 1493. 2008.

Cornish-Bowden, A. *Fundamentals of Enzyme Kinetics.* London: Portland, 1995.

Eliot, A. C., and Kirsch, J. F. Pyridoxal phosphate enzymes: Mechanistic, structural, and evolutionary considerations. *Annu. Rev. of Biochem.* 73: 383, 2004.

Fenton, A. W. Allostery: An illustrated definition for the 'second secret of life.' *Trends in Biochem. Sci.* 33: 420, 2008.

Fersht, A. *Enzyme Structure and Mechanism in Protein Science: A Guide to Enzyme Catalysis and Protein Folding.* New York: Freeman, 1999.

Hammes-Schiffer, S., and Benkovic, S, J. Relating protein motion to catalysis. *Annu. Review Biochem.* 75: 519, 2006.

Holden, H. M., Rayment, I., and Thoden, J. B. Structure and function of enzymes of the Leloir pathway for galactose metabolism. *J. Biol. Chem.* 278: 43885, 2003.

Imlay, J. A. Cellular defenses against superoxide and hydrogen peroxide. *Annu. Rev. Biochem.* 77: 755, 2008.

Johnson, D. C., Dean, D. R, Smith, A. D., and Johnson, M. K. Structure, function, and formation of iron sulfur complexes. *Annu. Rev. Biochem.* 74: 247, 2005.

Knowles, J. R., and Alberty, W. J. Evolution of enzyme function and the development of catalytic efficiency. *Biochemistry* 15: 5631, 1976.

Koshland, D. E., Jr.. A new model for protein stereospecificity. *Nature* 403: 614, 2000.

Koshland, D. E., Jr., and Hamadani, K. Proteomics and models for enzyme cooperativity. *J. Biol. Chem.* 277: 46841, 2002.

Kraut, D. A., Carroll, K. S., and Herschlag, D. Challenges in enzyme mechanism and energetics. *Annu. Rev. Biochem.* 72: 517, 2003.

Kyte, J. *Mechanism in Protein Chemistry.* New York: Garland, 1995.

Lerner, R. A., Benkovic, S. J., and Schultz. P. G. At the crossroads of chemistry and immunology: Catalytic antibodies. *Science* 252: 659, 1991.

Lilley, D. M. C. The origins of RNA catalysis in ribozymes. *Trends in Biochem. Sci.* 28: 495, 2003.

McMurry, J., and Begley, J. *The Organic Chemistry of Biological Pathways.* Greenwood Village, COL: Roberts, 2005.

Nevinsky, G. A., and Buneva, V. N. Catalytic antibodies in healthy humans and patients with autoimmune and viral diseases. *J. Cell Mol. Med.* 7: 265, 2003.

Plowman, K. M. *Enzyme Kinetics.* New York: McGraw Hill, 1972.

Silverman, R. B. *The Organic Chemistry of Enzyme-Catalyzed Reactions.* San Diego, CA: Academic Press, 2000.

Smith, T. J., and Stanley C. A. Untangling the glutamate dehydrogenase allosteric nightmare. *Trends in Biochem. Sci.* 33: 557, 2008.

Valentine J. S., Doucette, A., and Potter, S. Z. Copper–zinc superoxide dismutase and amyotrophic lateral sclerosis. *Annu. Rev. Biochem.* 74: 563. 2005.

Wackett, L. P. Evolution of enzymes for the metabolism of new chemical inputs into the environment. *J. Biol. Chem.* 279: 41259, 2004.

Key Terms

enzymes
cofactors
coenzymes
vitamins
metals
energy of activation
transition state
stabilization of transition state
thermodynamic considerations
equilibrium
tight binding
induced fit
active site
allosteric site
covalent catalysis
general acid and base catalysis
inhibitors
activators
Michaelis–Menten equation
double-reciprocal plots
maximum velocity
K_m
K_m(app)
K_i
two-substrate reactions
ping-pong reaction
sequential reaction
positive cooperativity
negative cooperativity
Hill plot
competitive inhibitors
noncompetitive inhibitors
uncompetitive inhibitors
inhibitors as drugs
redox reactions
kinases
phosphatases
reversible reactions
irreversible reactions
regulation of a pathway
therapeutic use of enzymes
isozymes and disease

Questions • CAROL N. ANGSTADT

Multiple-Choice Questions

1. In all enzymes the active site
- **A.** contains the substrate binding site.
- **B.** is contiguous with the substrate binding site in the primary sequence.
- **C.** lies in a region of the primary sequence distant from the substrate binding site.
- **D.** contains a metal ion as a prosthetic group.
- **E.** contains the amino acid side chains involved in catalyzing the reaction.

2. When added to a reaction, a catalyst
- **A.** supplies the heats of formation.
- **B.** alters the equilibrium constant K_{eq} to favor the formation of products.
- **C.** increases the rate at which equilibrium is reached.
- **D.** changes the order of the reaction (e.g., first order to second order).
- **E.** alters the ΔG_0 of the reaction.

3. Metal cations may do all of the following except
- **A.** donate electron pairs to functional groups found in the primary structure of the enzyme protein.
- **B.** serve as Lewis acids in enzymes.
- **C.** participate in oxidation-reduction processes.
- **D.** stabilize the active conformation of an enzyme.
- **E.** form chelates with the substrate, with the chelate being the true substrate.

4. An enzyme can facilitate the rate of a reaction by
- **A.** stabilizing the transition state.
- **B.** binding very tightly to the substrate.
- **C.** binding very tightly to the product.
- **D.** preventing the substrate from changing its ionic state.
- **E.** preventing the reaction from proceeding in the reverse direction.

5. In the reaction sequence, below the best point for controlling production of Compound 6 is reaction

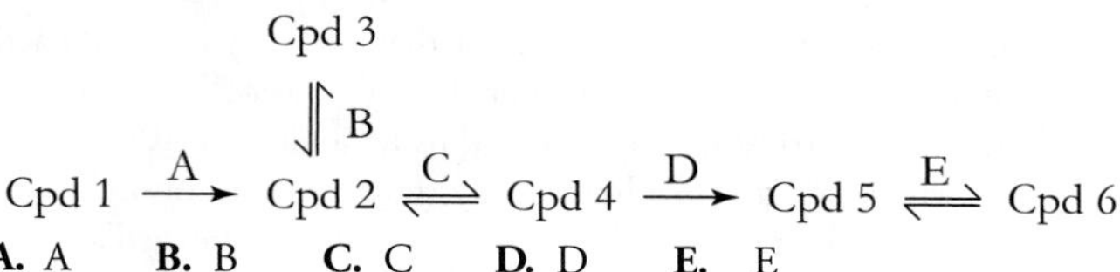

A. A B. B C. C D. D E. E

6. Turnover number (k_{cat})
   A. is a ratio of the rate constants for the formation of ES and of product.
   B. has units of 1/time.
   C. is inversely related to how fast the reaction is.
   D. for a mutant enzyme can change without any change in the K_m of the reaction.
   E. has units of substrate concentration.

Questions 7 and 8: A man of Japanese ancestry found himself to be experiencing severe flushing and a very rapid heart rate after consuming one alcoholic beverage. His companion, a Caucasian male, did not have the same symptoms even though he had finished his second drink. These physiological effects are related to the presence of acetaldehyde (CH_3CHO) generated from the alcohol. Acetaldehyde is normally removed by the reaction of mitochondrial aldehyde dehydrogenase which catalyzes the reaction

$$CH_3CHO + NAD^+ \rightleftharpoons CH_3COO^- + NADH + H^+$$

7. Acetaldehyde dehydrogenase is a(n)
   A. oxidoreductase.
   B. transferase.
   C. hydrolase.
   D. lyase.
   E. ligase.

8. The explanation for the difference in physiological effects is that the Japanese man was missing the normal mitochondrial aldehyde dehydrogenase and had only a cytosolic isozyme. The cytosolic isozyme
   A. does not react with CH_3CHO.
   B. activates the enzyme that produces CH_3CHO.
   C. differs from the mitochondrial enzyme in that it has a higher K_m for CH_3CHO.
   D. would be expected to have a greater affinity for the substrate than the mitochondrial enzyme.
   E. produces a low steady-state level of acetaldehyde following alcohol consumption.

Question 9 and 10: A research technician who is working with organophosphate compounds is required to have a weekly blood test for acetylcholine esterase activity. Typically, esterase activity remains relatively constant for some time and then abruptly drops to zero. If this happens, the technician must immediately stop working with the organophosphate compounds. The organophosphate compounds form stable esters with a critical serine hydroxyl group in the esterase.

9. In the esterase, serine transfers a proton to a histidine residue. Which of the following is correct?
   A. Serine is acting as a general acid.
   B. Histidine is acting as a general acid.
   C. Serine and histidine form a covalent intermediate.
   D. The enzyme would be relatively insensitive to pH changes.
   E. Serine is acting as a transition stabilization catalyst.

10. Organophosphate compounds inactivate the esterase by
   A. competitive inhibition.
   B. uncompetitive inhibition.
   C. noncompetitive inhibition.
   D. suicide inhibition.
   E. irreversible inhibition.

Questions 11 and 12: Gout is a disease in which uric acid is high in blood and urine. One patient who excreted three times normal uric acid had very high blood levels of PRPP, an intermediate in biosynthesis of AMP and GMP, which are precursors of ATP and GTP. Degradation of these products produces uric acid. The patient's PRPP synthetase had normal K_m and V_{max} values but was insensitive to regulation by the end products of the pathway (ATP, GTP). These are negative allosteric modifiers of PRPP synthetase.

11. All of the following statements about allosteric effectors are correct except they
   A. may increase the enzyme's affinity for its substrate.
   B. may decrease the enzyme's affinity for its substrate.
   C. bind at the substrate binding site.
   D. cause a conformational change in the enzyme.
   E. can change either the K_m or the V_{max} of the reaction.

12. Most allosteric enzymes
   A. are monomers.
   B. have more then one subunit.
   C. exhibit only homotropic interactions.
   D. exhibit only heterotropic interactions.
   E. bind the allosteric effector with no effect on binding other ligands.

Problems

13. An experiment measuring velocity versus substrate concentration was run, first in the absence of substance A and then in the presence of substance A. The following data was obtained:

[S] *μM*	*Velocity in absence of A* *μ mol/min*	*Velocity in presence of A* *μ mol/min*
2.5	0.32	0.20
3.3	0.40	0.26
5.0	0.52	0.36
10.0	0.69	0.56

Is substance A an activator or an inhibitor? If it is an inhibitor, what kind of inhibitor is it?

14. For the experiment in problem 13, calculate the K_m and V_{max} both in the absence and in the presence of substance A. Are these results consistent with your answer for problem 13?

Answers

1. **E** The active site contains all of the machinery, including the amino acid side chains, involved in catalyzing the reaction. A–D are all possible, but none is necessarily true.
2. **C** A catalyst lowers the energy of the transition state, thus speeding the rate of its formation. A: Heat of formation is the energy to form reactants or products. B: K_{eq} is governed by heats of formation which are constant in a reaction. D: This is determined by the number of reactants. E: ΔG_0 is related to the difference in heats of formation of reactants and products—a constant.
3. **A** Metal cations are electron deficient and may accept electron pairs, serving as Lewis acids, but they do not donate electrons to other functional groups. C: On the contrary, they sometimes accept electron pairs from groups in amino acid side chains. D: In doing so, they may become chelated that may stabilize the appropriate structure. E: Sometimes they are chelated by the substrate, with the chelate being the true substrate.
4. **A** Residues in the enzyme interact with the substrate while it is in the transition state, lowering the energy of activation. B: This would require more energy to reach the transition state. C: It would be difficult for the product to dissociate. D: A key to many reactions is for the enzyme to add or remove protons from the substrate to make the substrate more or less nucleophilic. E: Catalysis is reversible although thermodynamic considerations may make some reactions appear irreversible.
5. **D** Reaction D is irreversible; if it were not controlled, Cpd 5 might build up to toxic levels. A: Control of reaction A would control production of both Cpds 3 and 6. B: Reaction B is not on the direct route. C and E: Reactions C and E are freely reversible, so do not need to be controlled.
6. **B** This is the number of molecules of substrate converted per unit time per molecule of enzyme. A and E: These refer to the K_m. C: The higher the k_{cat}, the faster the reaction. D: k_{cat}/K_m is equal to K_{eq}, so they must both change to maintain the equilibrium.
7. **A** Aldehyde-to-acid conversion is an oxidation. This is of the subclass dehydrogenases as indicated by the presence of the NAD^+.
8. **C** The lower affinity (higher K_m) makes it more difficult for the enzyme to remove CH_3CHO. A: Isozymes, by definition, catalyze the same reaction. B: It is highly unlikely that one enzyme would activate another enzyme. D: With a greater affinity (lower K_m), the reaction would be occurring at a rapid rate, thus removing CH_3CHO. E: These effects are due to a high steady-state concentration.
9. **A** Acids donate protons. General acids are weakly ionized at physiological pH. The environment of the serine in the protein facilitates its ability to transfer a proton and thus act as a nucleophile. B: Bases accept protons. C: No covalent bonds are formed. D: The ability of histidine to accept a proton is very pH dependent. E: This is not likely.
10. **E** The phosphate ester does not dissociate and the enzyme is not restored. A, B, and C: These types of inhibitors bind to the enzyme and/or enzyme substrate complex reversibly. D: A suicide inhibitor is formed from the substrate and remains covalently attached to the enzyme. Organophosphates are not substrates for the esterase.
11. **C** "*Allo*" means other. A and B: Allosteric effectors may be either positive or negative. D: This alters the affinity for the substrate. E: K class alters K_m and V class alters V_{max}.
12. **B** Most are oligomers. A: Only two monomeric allosteric enzymes are known. C and D: Both homotropic and heterotropic interactions are common. E: The principle of allosterism is that binding one ligand affects binding of other ligands.
13. Take the reciprocals of both [S] and v and construct a Lineweaver–Burke plot. You should find that the two curves cross the y-axis at the same point but the curve in the presence of A crosses the x-axis closer to the origin. This pattern indicates that A is a competitive inhibitor.
14.

	$-1/K_m$	K_m	$1/V_{max}$	V_{max}
Absence of A	–0.14	7.1	0.8	1.25
Presence of A	–0.08	12.5	0.8	1.25

With a competitive inhibitor, the V_{max} remains constant (be sure you understand why) but the apparent K_m is larger. It takes more substrate to reach a given velocity because the substrate has to compete with the inhibitor.

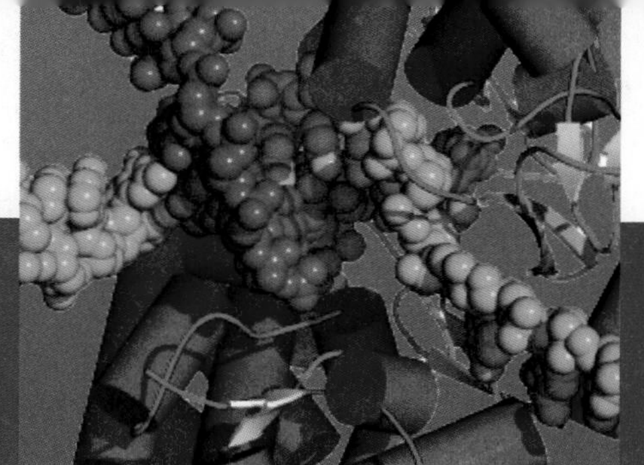

11

The Cytochromes P450 and Nitric Oxide Synthases

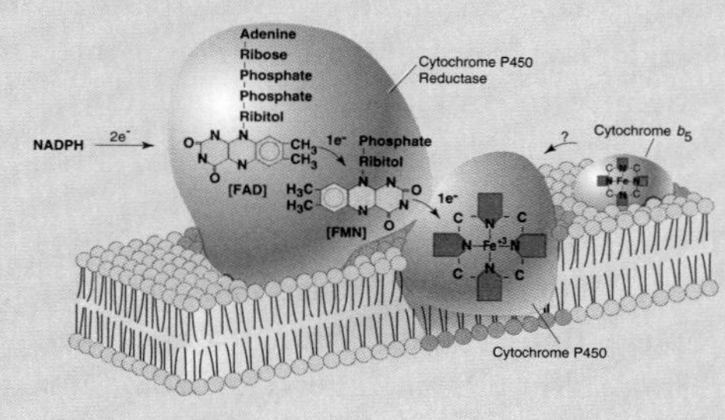

Linda J. Roman
Associate Professor, University of Texas Health Science Center at San Antonio

Bettie Sue Siler Masters
Robert A. Welch Distinguished Professor in Chemistry, University of Texas Health Science Center at San Antonio

CLINICAL CORRELATIONS

Key Concepts

- Cytochromes P450 catalyze incorporation of an oxygen atom into both natural substrates and foreign compounds. They contain heme (iron porphoryrin IX); the iron is reduced by electrons donated by a reductase containing FAD or FMN. Their spectrum derives from the fifth axial ligand of the heme iron to a cysteine residue in each of the proteins.

- The oxygenation of steroids produces regiospecific and stereospecific products that are essential for sexual differentiation (androgens and estrogens) and for metabolic and electrolyte balance (corticosteroids and mineralocorticoids).
- Some cytochromes P450 have less substrate specificity and metabolize exogenous compounds, to produce more polar products for elimination. Some products are more toxic than the original substrate but subsequent metabolism negates these effects. Most drugs are metabolized by several specific cytochromes P450 in the endoplasmic reticulum. Drug–drug interactions occur frequently.
- Some cytochromes P450 can be induced while others are constitutive. There are 57 human cytochromes P450; 50 occur in the endoplasmic reticulum and require NADPH-cytochrome P450 reductase.
- Mitochondrial hydroxylation of steroids requires electrons supplied from NADPH through a FAD-containing reductase to an iron-sulfur protein, adrenodoxin, and finally to cytochrome P450.
- Genetic polymorphisms have been demonstrated in the human population in both cytochromes P450 and NADPH-cytochrome P450 reductase. Mutations can result in reduced metabolism of drugs and abnormal sex steroid metabolism.
- Nitric oxide (NO) is a signaling molecule produced through monooxygenation catalyzed by nitric oxide synthases (NOSs), which contain FAD, FMN, and heme bound in a single molecule that functions as a homodimer. NO is derived from the guanidino group of L-arginine and *N*-hydroxy-L-arginine is an intermediate in the reaction.
- NOSs are composed of two distinct domains, one containing FAD and FMN, and the other heme, connected by a calmodulin binding sequence. Binding of Ca^{+2}-calmodulin is required for electron transfer between the domains. Tetrahydrobiopterin and Zn are also required.
- NOSs are encoded by three genes producing NOSI (neuronal NOS), NOSII (inducible NOS) and NOSIII (endothelial NOS). NOSI is constitutively expressed primarily in skeletal muscle and neurons of the central and peripheral nervous system. It is activated by binding of calmodulin and the NO produced acts as a neurotransmitter and modulator of muscular contractility.
- NOSII is primarily in neutrophils, macrophages, astrocytes, and hepatocytes. It is induced by cytokines; calmodulin is bound under basal physiological conditions. NOSII functions in the early immune response, producing NO as a cytotoxic agent against pathogens.
- NOS III is primarily in vascular endothelial cells, attached to the caveolar membrane through myristoyl and palmitoyl residues. It is constitutively expressed and activated by Ca^{2+}. NOSIII produces NO as a vasodilator, which acts on guanylate cyclase in adjacent smooth muscle cells, ultimately leading to vasorelaxation.

11.1 • INTRODUCTION

Cytochromes P450 constitute a unique family of heme proteins, present in bacteria, fungi, insects, plants, fish, mammals, and primates that catalyze the monooxygenation (i.e., insertion of one atom of molecular oxygen) of a large variety of structurally diverse compounds. Substrates for this enzyme system include both endogenously synthesized compounds, such as cholesterol, steroid hormones, and fatty acids, and exogenous compounds, such as drugs, food additives, components of cigarette smoke, pesticides, and chemicals, that are ingested, inhaled, or absorbed through the skin. The cytochromes P450 are involved in (1) production of steroid hormones; (2) metabolism of fatty acids, prostaglandins, leukotrienes, and retinoids; (3) inactivation or activation of therapeutic agents; (4) conversion of chemicals to highly reactive molecules that may produce unwanted cellular damage; and (5) enzyme inhibition or induction resulting in drug–drug interactions and adverse effects.

The nitric oxide synthases are bidomain enzymes, consisting of a heme-containing (or oxygenase) domain and a flavin-containing (or reductase) domain. These enzymes catalyze the formation of nitric oxide, a highly reactive gaseous free radical molecule that functions in neurotransmission, hemodynamic regulation, or in the immune response, depending on the tissue in which it is produced. Like the cytochromes P450, nitric oxide synthases are heme-containing enzymes that catalyze the monooxygenation of their substrates using similar catalytic mechanisms. The following will discuss both the cytochromes P450 and the nitric oxide synthases in detail, addressing both the similarities and differences between the two systems.

11.2 • CYTOCHROMES P450: PROPERTIES AND FUNCTION

Cytochromes P450 are integral membrane proteins containing a single iron protoporphyrin IX (heme) prosthetic group (p. 791). The heme iron of cytochromes P450 is capable of forming six bonds, four to each of the four pyrrole nitrogen atoms of the porphyrin ring and two to axial ligands. One of the axial (so-called proximal) ligands is a sulfhydryl group

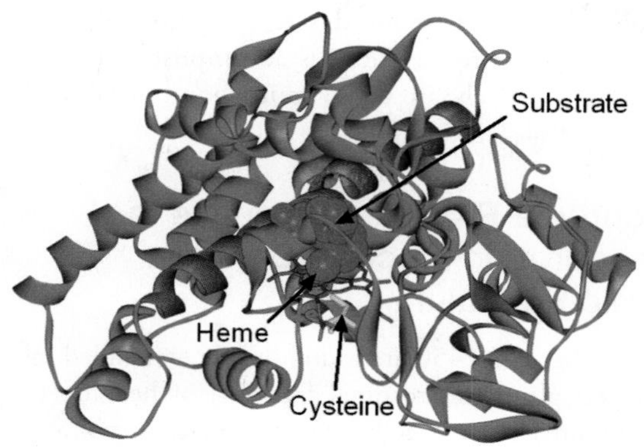

Figure 11.1 Model of the active site of mammalian CYP2C5 showing the protoporphyrin IX prosthetic group (red) with the cysteine thiolate ligand (yellow) attached to the heme iron. The substrate diclofenac, a nonsteroidal anti-inflammatory drug (green), is in the active site of the P450. The I-helix (purple) spans the molecule and is one of the most easily identifiable structural characteristics of cytochromes P450.
Generated from Protein Data Bank file 1NR6, deposited by M. R. Wester, E. F. Johnson, and C. D. Stout, using WebLab Viewer Lite (Molecular Simulations, Inc.).

from a cysteine residue located toward the carboxyl end of the cytochrome P450 molecule (Figure 11.1) and the other (distal) ligand-binding site can be open (producing a penta-coordinated, high-spin heme) or occupied by oxygen, CO, NO, water, or other ligand (producing a hexa-coordinated, low-spin heme). The spin state of the heme is a description of the occupancy of the *d*-electronic shells within the iron atom determined by the ligand state; these can be distinguished spectroscopically.

All cytochromes P450 are triangle-shaped molecules, closely resembling that shown in Figure 11.1. Half of the enzyme consists of α-helical structure and the other half is β-sheet and other structures; a long I-helix runs throughout the enzyme behind the heme. This basic scaffolding is conserved among all known cytochromes P450, although relative placement may differ and some of the minor elements may be missing. The mammalian proteins contain an N-terminal membrane anchor sequence and internal membrane association sequences that result in these proteins being deeply buried in the membrane.

Cytochromes P450 are so called because of the unique absorbance spectrum that is produced when CO is bound to the reduced, ferrous form of the heme. The spectrum exhibits a peak at approximately 450 nm (Figure 11.2); thus the name P450 for a **p**igment with an absorbance at **450** nm. This spectrum is characteristic of a heme thiolate-liganded protein. Much as with the heme iron of hemoglobin, CO binds to the heme iron of cytochromes P450 with a much higher binding affinity than oxygen and thereby is a potent inhibitor of its function.

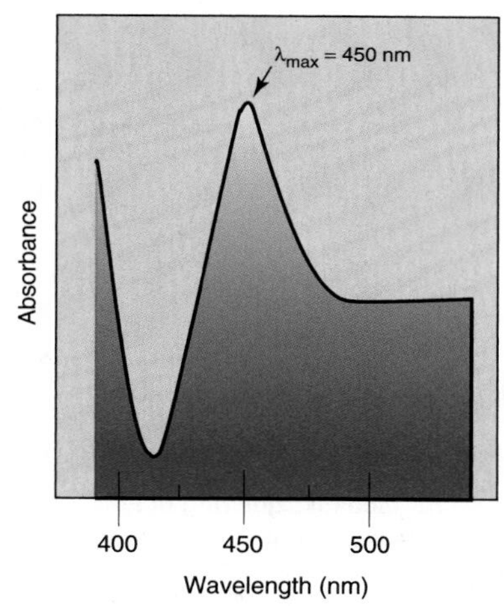

Figure 11.2 The absorbance spectrum of the carbon monoxide–bound cytochrome P450. The reduced form of cytochrome P450 binds carbon monoxide to produce a maximum absorbance at approximately 450 nm.

The general reaction catalyzed by a cytochrome P450 is

$$\text{NADPH} + \text{H}^+ + \text{O}_2 + \text{SH} \rightarrow \text{NADP}^+ + \text{H}_2\text{O} + \text{SOH}$$

where NADPH is a two-electron donor and the substrate (S) may be a steroid, fatty acid, drug, or other chemical that has an alkane, alkene, aromatic ring, or heterocyclic ring substituent that can serve as a site for oxygenation. The reaction is a monooxygenation, and the enzyme is a monooxygenase as it incorporates only one of the two oxygen atoms into the substrate.

11.3 • CYTOCHROMES P450 ELECTRON TRANSPORT SYSTEMS

The two electrons required for the overall monooxygenase reaction are provided by NADPH, but a basic mechanistic problem is created in that NADPH is a two-electron donor (p. 562). Cytochrome P450, with its single heme iron, can accept only one electron at a time. This problem is solved by the presence of a NADPH-dependent flavoprotein reductase, which accepts two electrons from NADPH simultaneously but transfers the electrons individually, either to an intermediate iron–sulfur protein or directly to cytochrome P450. The active redox group of the flavoprotein is the isoalloxazine ring, which is uniquely suited to perform this chemical task since it can exist in fully oxidized or one- or two-electron reduced states (Figure 11.3). The transfer of electrons from NADPH to cytochromes P450 is accomplished by distinct electron transport systems present exclusively in either mitochondria or endoplasmic reticulum (microsomes).

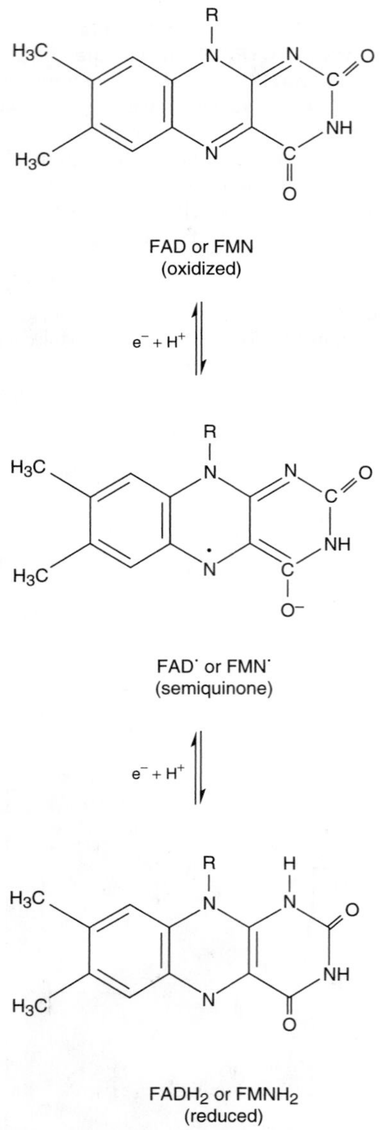

NADPH-Cytochrome P450 Reductase Is the Obligatory Flavoprotein Electron Donor in the Endoplasmic Reticulum

Most mammalian cytochromes P450 are found deeply embedded in the cytoplasmic side of the endoplasmic reticulum (microsomal fraction) of hepatocytes, renal and adrenal cortical cells, ovarian and testicular cells, and cells of the respiratory tract, among others. Fifty of the 57 human cytochrome P450 isoforms are of the microsomal type. These cytochromes P450 use a single NADPH-cytochrome P450 reductase, a peripheral membrane protein of 76.6 kD containing both flavin adenine dinucleotide (FAD) and flavin mononucleotide (FMN) as prosthetic groups, for electron delivery. The crystal structure of the human isoform is shown in Figure 11.4. Until the characterization of nitric oxide synthases, the P450 reductase was the only mammalian flavoprotein known to contain both FAD and FMN. NADPH donates electrons to the FAD moiety of P450 reductase and the FMN serves as the exit flavin, transferring two electrons, one at a time, to cytochrome P450. Because a single flavin molecule may exist as a one- or two-electron-reduced form and two flavin molecules are bound per reductase molecule, the enzyme may receive electrons from NADPH and store them between the two flavin molecules before transferring them to one of the numerous microsomal cytochromes P450 (Figure 11.5).

The overall mechanism by which cytochromes P450 catalyze the monooxygenation of their substrates is shown in Figure 11.6. The reaction cycle is initiated by the binding of substrate to the native, low-spin, hexa-coordinate, ferric form of the heme, converting it to a high-spin, penta-coordinate ferric heme. The first electron can be transferred to the heme, reducing it to the high-spin, penta-coordinate ferrous form only after substrate has been bound. Electrons cannot be transferred to the heme in the absence of substrate because the reduction potential of the heme (−300 V) in that state is more electronegative than that of the FMN moiety (−270 mV), making such a transfer thermodynamically unfavorable. Upon substrate binding, there is a conformational change of the protein structure surrounding the heme iron, and the reduction potential of the heme becomes more positive (−230 mV), thus allowing the P450 heme iron to be reduced (Figure 11.6).

FAD or FMN
(oxidized)

$e^- + H^+$

$FAD^{\cdot}$ or $FMN^{\cdot}$
(semiquinone)

$e^- + H^+$

$FADH_2$ or $FMNH_2$
(reduced)

Figure 11.3 Isoalloxazine ring of FMN or FAD in its oxidized, semiquinone (1e⁻-reduced form), or fully reduced (2e⁻-reduced) state.

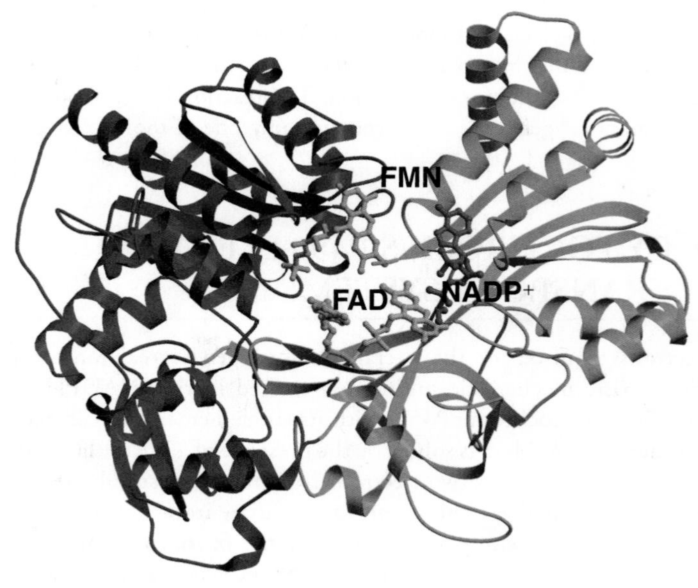

Figure 11.4 Ribbon diagram of human cytochrome P450 reductase based on its crystal structure. Three domains are present: the FMN-binding domain (blue), the FAD/NADPH-binding domain (green), and the intervening connecting domain (purple). Cofactors are shown in stick-and-ball models: blue, FMN; yellow, FAD; and red, $NADP^+$. The figure was rendered using the program Molscript (Kraulis, P. J. Molscript: A program to produce both detailed and schematic plots of protein structures. *J. Appl. Cryst.* 24:946, 1991). Generously supplied by Dr. Jung-Ja P. Kim, Professor of Biochemistry, Medical College of Wisconsin, before publication.

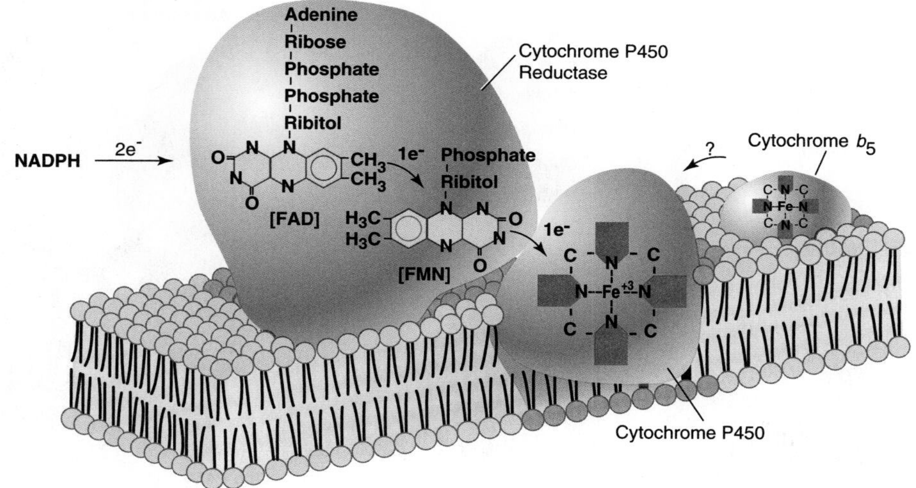

Figure 11.5 Components of the endoplasmic reticulum (microsomal) cytochrome P450 system. NADPH-cytochrome P450 reductase is bound by its hydrophobic tail to the membrane (a peripheral protein) whereas cytochrome P450 is deeply embedded in the membrane (an integral protein). Also shown is cytochrome b_5, which may participate in selected cytochrome P450-mediated reactions.

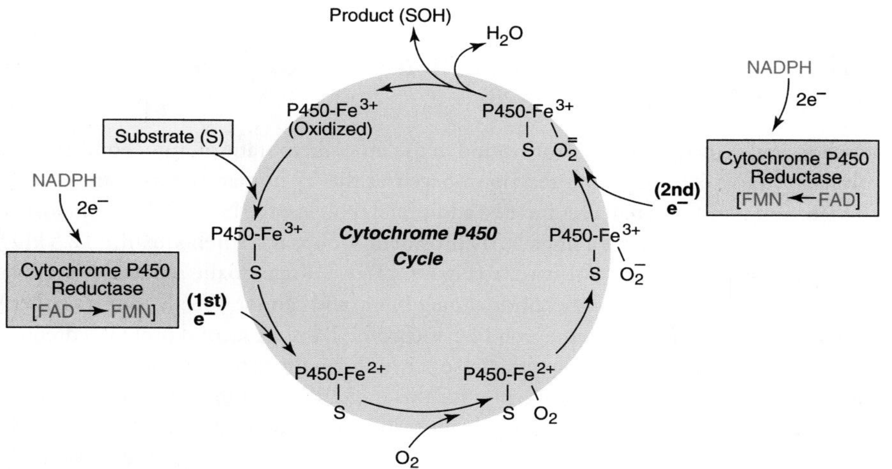

Redox Potentials (mV)
P450 from endoplasmic reticulum

NADPH	–324
FAD	–290
FMN	–270
P450 heme	–300
P450 heme + substrate	–230

Figure 11.6 Reaction cycle of cytochrome P450. Diagram shows the binding of substrate, transfer of the first and second electrons from NADPH-cytochrome P450 reductase, and binding of O_2. The reduction potentials of the various components are also shown.

Oxygen can now bind to the ferrous heme and the second electron transfer is facilitated. Subsequently, molecular oxygen is activated and cleaved; one atom is inserted into the substrate (monooxygenation), and the other combines with two protons and two electrons to form H_20.

The binding of the P450 and reductase to one another is electrostatic rather than hydrophobic, and P450 levels exceed those of reductase in liver by a 10:1 to 20:1 ratio. If a particular P450 has a higher affinity for association with reductase, electrons will flow preferentially to those P450s. Consequently, P450s with lower affinities must have mechanisms that allow them to receive electrons. In some cases, the binding of substrate increases both the affinity and the rate of association of the P450/reductase complex; this mechanism makes sense as an "empty" P450 monopolizing the reductase would be non-productive.

In some reactions, transfer of the second electron may not be directly from the P450 reductase, but may occur from cytochrome b_5, a small heme protein (15.2 kDa) that can be reduced by either the P450 reductase or another microsome-bound flavoprotein, NADH-cytochrome b_5 reductase. Some reactions catalyzed by specific cytochromes P450 require cytochrome b_5 for optimal enzymatic activity possibly because cytochrome b_5: (1) alters the affinity of the P450/reductase interaction, (2) alters the affinity of the P450 for a particular substrate, or (3) increases the rate of transfer of the second electron to the P450 heme.

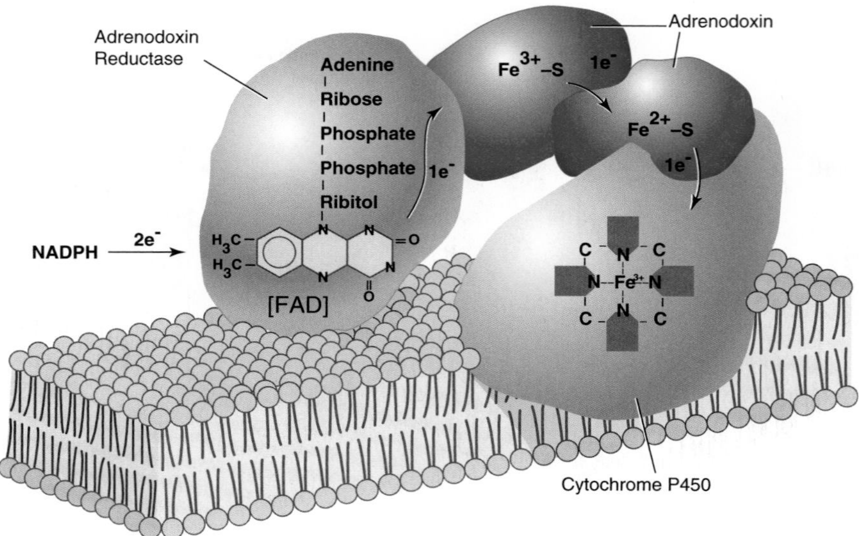

Figure 11.7 Components of mitochondrial cytochrome P450 system. Cytochrome P450 is an integral protein of the inner mitochondrial membrane. NADPH-adrenodoxin reductase and adrenodoxin are peripheral proteins, not embedded in the membrane.

NADPH-Adrenodoxin Reductase Is the Flavoprotein Electron Donor in Mitochondria

The mitochondrial cytochromes P450 are found in the inner membrane of mitochondria and are involved in steroid hydroxylation reactions. Seven of the 57 human isoforms are of this type. This electron transfer system requires two additional proteins, the NADPH-adrenodoxin reductase (51 kD), which contains a single FAD prosthetic group, and adrenodoxin, a 12.5 kD iron–sulfur protein, for product formation (Figure 11.7). Adrenodoxin reductase is only weakly associated with the inner mitochondrial membrane and cannot directly transfer either the first or second electron to the heme iron of cytochrome P450. A second protein, adrenodoxin, which is also weakly associated with the inner mitochondrial membrane, contains an iron–sulfur cluster that serves as a redox center for this molecule and functions as an electron shuttle between the FAD of adrenodoxin reductase and the heme iron of a mitochondrial cytochrome P450. One adrenodoxin molecule receives an electron from its mitochondrial flavoprotein reductase and interacts with a second adrenodoxin, which then transfers its electron to a specific cytochrome P450. Components of the mitochondrial cytochrome P450 system are synthesized in the cytosol as larger molecular weight precursors, transported into mitochondria, and processed by proteases into smaller molecular weight, mature proteins.

11.4 • CYTOCHROMES P450: NOMENCLATURE AND ISOFORMS

Because of the large number of cytochromes P450 that have been identified (over 7000, as of February 2008), a system for classification of the enzymes into functional groups and nomenclature had to be developed. The system chosen is based on classification according to the relative identity of the amino acid sequences of the enzymes. The superfamily of cytochromes P450 is thus divided initially into families in which the amino acid sequence identity of the members is greater than 40%. The family is designated by the prefix "CYP", for cytochrome P450, followed by an Arabic numeral (e.g., CYP1, CYP2, CYP3, etc.) The families are further divided into subfamilies in which the amino acid sequence identity of the members is greater than 55%. The subfamily is identified by an additional capital letter (e.g., CYP1A, CYP1B, CYP1C, etc.). The individual members of each subfamily are then numbered in the order in which they were identified (e.g., CYP1A1, CYP1A2, CYP1A3, etc.). Although cytochromes P450 are enzymes, the term *isoenzyme* or *isozymes* is generally not used to describe these proteins; rather, the term *form* or *isoform* is used.

Tables 11.1 and 11.2 list the known human P450 isoforms. There are 57 isoforms split into 18 families and 41 subfamilies. The human genome also encodes for 58 CYP pseudogenes

TABLE 11.1 • Human Cytochromes P450 Involved in Exogenous Metabolism

CYP Family	*Isoforms*	*Selected Substrate(s)*	*Selected Inhibitor(s)*	*Selected Inducer(s)*
1	1A1	Benzo(*a*)pyrene, diclofenac	Ketoconazole	Benzo(a)pyrene
	1A2	Benzo(*a*)pyrene, warfarin	Ciprofloxin	St. John's wort
	1B1	Benzo(*a*)pyrene, aflatoxin B1	Tamoxifen	NC[a]
2	2A6	Acetaminophen, nicotine	Cannabidol	Dexamethazone
	2A7	NC	NC	NC
	2A13	Hexamethylphosphoramide	NC	NC
	2B6	Diazepam, mephenytoin	Ketoconazole	Rifampicin
	2C8	Taxol, ibuprofen, verapamil	Quinine	Phenobarbital
	2C9	Amitriptyline, naproxen	Sulfaphenazole	Rifampicin
	2C18	Imipramine, methadone	NC	Rifampicin
	2C19	Diazepam, omeprazole	Isoniazid	Rifampicin
	2D6	Fluvastatin, codeine, risperidone	Quinidine	Dimethylsulfoxide
	2E1	Acetaminophen, halothane	Watercress	Isoniazid, ethanol
	2F1	Naphthalene, styrene	NC	NC
	2J2	Bufuralol	NC	NC
	2R1,2S1	NC	NC	NC
	2U1,2W1	NC	NC	NC
3	3A4	Erythromycin, nifedipine, codeine, warfarin, terfenadine	Troleandomycin, ketoconazole	Cortisol, rifampin, phenobarbital
	3A5	Verapamil, prevastatin	NC	Dexamethazone
	3A7	Retinoic acid, codeine, cortisol	DHEA	NC
	3A43	Testosterone	NC	NC

[a]NC, not well-characterized.

that do not form active protein. Table 11.1 lists the isoforms that utilize primarily exogenous compounds, that is, drugs and xenobiotics; these isoforms each metabolize a wide variety of substrates. Table 11.2 lists the isoforms involved in metabolism of endogenous compounds; these isoforms generally recognize only one or two specific substrates and these are listed in the table.

11.5 • CYTOCHROMES P450: SUBSTRATES AND PHYSIOLOGICAL FUNCTIONS

Cytochromes P450 metabolize a variety of lipophilic compounds of endogenous or exogenous origin. As shown in Figure 11.8 (p. 433), cytochromes P450 may catalyze hydroxylation of the carbon atom of a methyl group, the methylene carbon of an alkane, hydroxylation of an aromatic ring to form a phenol, or addition of an oxygen atom across a double bond to form an epoxide. They may also catalyze dealkylation reactions in which alkyl groups attached to oxygen, nitrogen, or sulfur atoms are removed. Oxidation of nitrogen, sulfur, and phosphorus atoms and dehalogenation reactions are also catalyzed by cytochrome P450 forms.

Both endogenous and exogenous substrates are metabolized by the cytochromes P450. Endogenous substrates include cholesterol, steroids, prostaglandins, and fatty acids. In these cases, the reactions are more stereo- and regiospecific and the cytochromes P450 bind only specific substrates. Exogenous substrates (*xenobiotics,* meaning "foreign to life") include many drugs, chemicals, environmental contaminants, and food additives. Cytochromes P450 in this group are less discriminating among substrates and oxidize these primarily lipophilic (hydrophobic) compounds to make them more water soluble (hydrophilic) for excretion.

TABLE 11.2 • Human Cytochromes P450 Involved in Endogenous Metabolism

CYP Family	*Isoforms*	*Substrate(s)*	*Activity*
4	4A11	Fatty acids	ω-Hydroxylase
	4B1	Arachidonic acid	12-Hydroxylase
	4F2	Arachidonic acid, leukotriene B_4	ω-Hydroxylase
	4F3	Leukotriene B_4	ω-Hydroxylase
	4F8	Arachidonic acid, prostaglandins	18-Hydroxylase
	4F12	Arachidonic acid	ω-Hydroxylase
	4A22,4F11,4F22,4V2,4X1,4Z1	Unknown	Unknown
5	5A1	Prostaglandin H_2	Thromboxin A_2 synthase
7	7A1	Cholesterol	7-α-Hydroxylase
	7B1	Pregnenolone, dehydroepiandrosterone (DHEA)	7-α-Hydroxylase
8	8A1	Prostaglandin H_2	Prostacyclin synthase
	8B1	Sterols	12-α-Hydroxylase
11	11A1	Cholesterol	Side-chain cleavage
	11B1	11-Deoxycortisol, 11-deoxycorticosterone	11-β-Hydroxylase
	11B2	Corticosterone	18-Hydroxylase
17	17A1	Pregnenolone, progesterone	17-α-Hydroxylase
		17-hydroxy pregnenolone, 17-hydroxy progesterone	17–20 Lyase
19	19A1	Androstenedione, testosterone	Aromatase
20	20A1	Unknown	Unknown
21	21A2	Progesterone, 17-hydroxy progesterone	21-Hydroxylase
24	24A1	25-Hydroxy vitamin D3	24-Hydroxylase
26	26A1	Retinoic acid	4-Hydroxylase
	26B1	Retinoic acid	Unknown
	26C1	Unknown	Unknown
27	27A1	Sterol	27-Hydroxylase
		Vitamin D3	25-Hydroxylase
	27B1	Vitamin D3	1-α-Hydroxylase
	27C1	Unknown	Unknown
39	39A1	24-Hydroxy cholesterol	7-Hydroxylase
46	46A1	Cholesterol	24-Hydroxylase
51	51A1	Lanosterol	14-α-Demethylase

Cytochromes P450 Participate in the Synthesis of Steroid Hormones and Oxygenation of Endogenous Compounds

The importance of cytochrome P450-catalyzed reactions with endogenous substrates is illustrated by the synthesis of steroid hormones from cholesterol in the adrenal cortex and sex organs. Both mitochondrial and microsomal (endoplasmic reticulum) cytochrome P450 systems are required to convert cholesterol to aldosterone and cortisol in adrenal cortex, testosterone in testes, and estradiol in ovaries.

Cytochromes P450 are responsible for several steps in the adrenal synthesis of aldosterone, the mineralocorticoid responsible for regulating salt and water balance, and cortisol, the glucocorticoid that governs protein, carbohydrate, and lipid metabolism. In addition, adrenal cytochromes P450 catalyze the synthesis of small quantities of the androgen androstenedione, a regulator of secondary sex characteristics and a precursor of both estrogens and testosterone. Figure 11.9 presents a summary of these pathways.

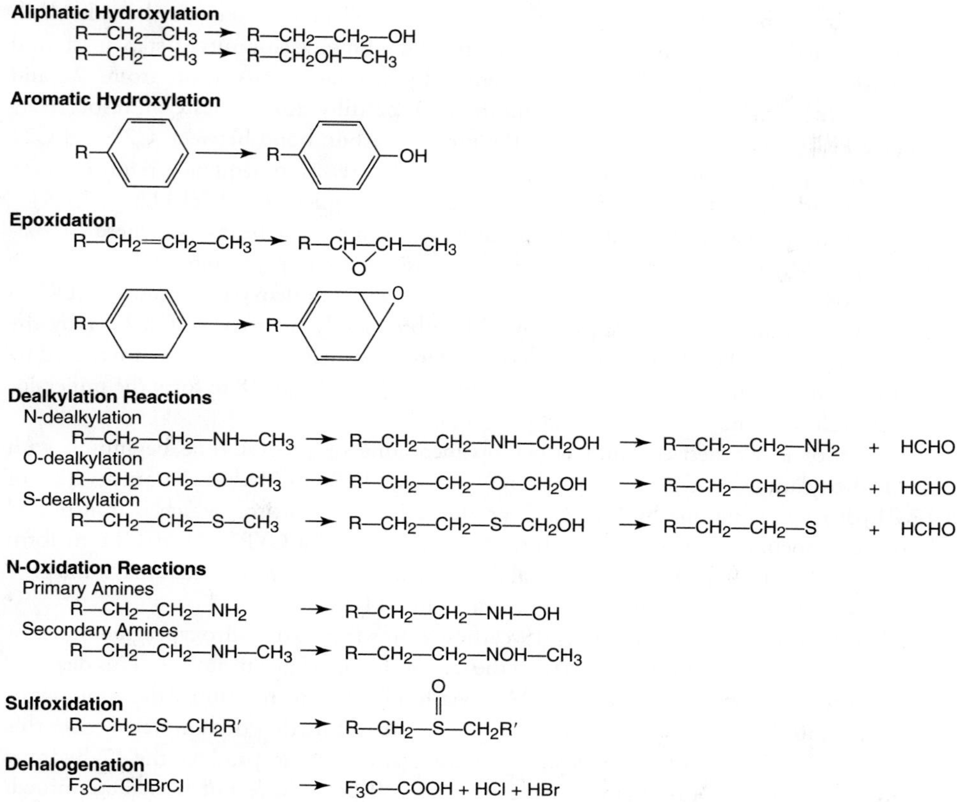

Aliphatic Hydroxylation

R—CH_2—CH_3 → R—CH_2—CH_2—OH
R—CH_2—CH_3 → R—CH_2OH—CH_3

Aromatic Hydroxylation

R—C_6H_5 → R—C_6H_4—OH

Epoxidation

R—CH_2=CH_2—CH_3 → R—CH—CH—CH_3 (epoxide, O bridging)

R—C_6H_5 → R—arene epoxide (O)

Dealkylation Reactions

N-dealkylation
R—CH_2—CH_2—NH—CH_3 → R—CH_2—CH_2—NH—CH_2OH → R—CH_2—CH_2—NH_2 + HCHO
O-dealkylation
R—CH_2—CH_2—O—CH_3 → R—CH_2—CH_2—O—CH_2OH → R—CH_2—CH_2—OH + HCHO
S-dealkylation
R—CH_2—CH_2—S—CH_3 → R—CH_2—CH_2—S—CH_2OH → R—CH_2—CH_2—S + HCHO

N-Oxidation Reactions

Primary Amines
R—CH_2—CH_2—NH_2 → R—CH_2—CH_2—NH—OH
Secondary Amines
R—CH_2—CH_2—NH—CH_3 → R—CH_2—CH_2—NOH—CH_3

Sulfoxidation

R—CH_2—S—CH_2R' → R—CH_2—S(=O)—CH_2R'

Dehalogenation

F_3C—CHBrCl → F_3C—COOH + HCl + HBr

Figure 11.8 Common reactions catalyzed by cytochromes P450.

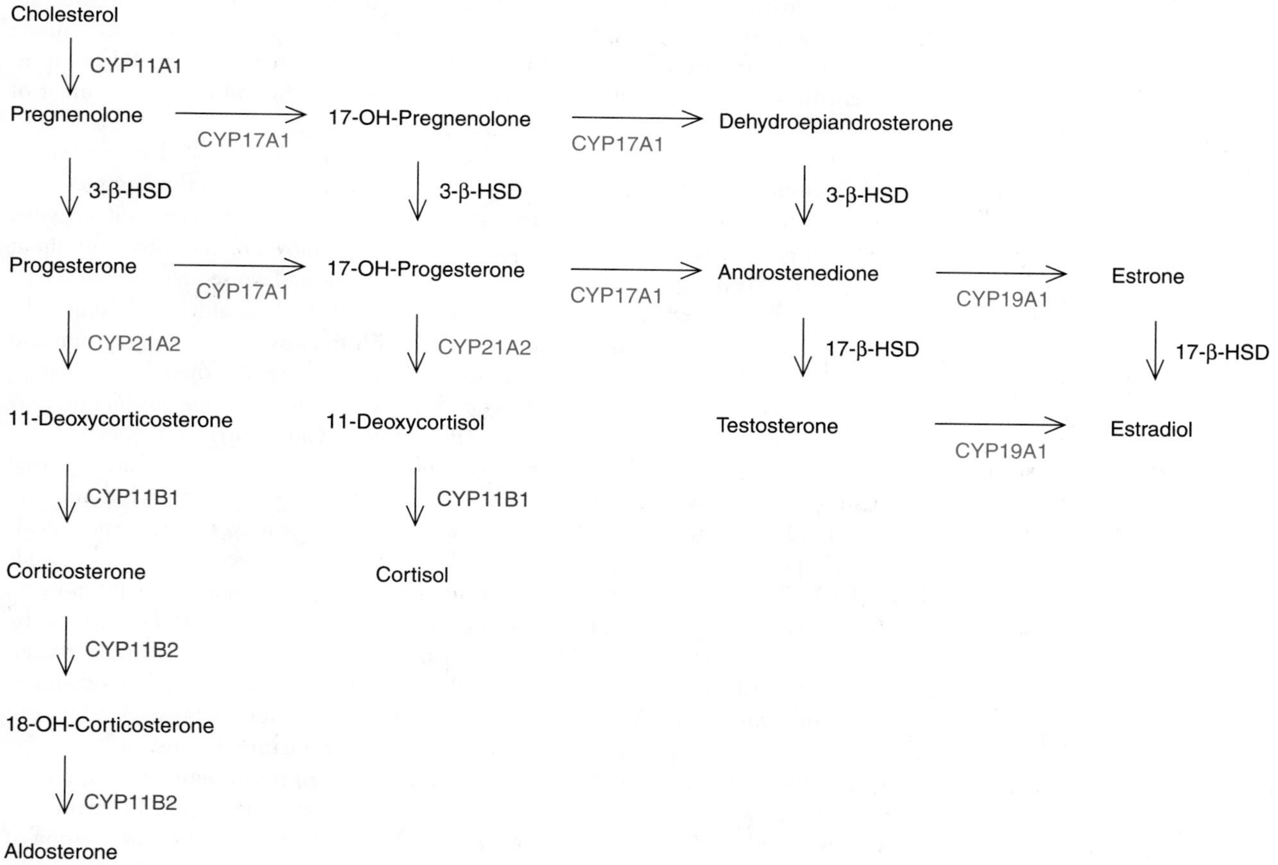

Cholesterol
↓ CYP11A1
Pregnenolone —CYP17A1→ 17-OH-Pregnenolone —CYP17A1→ Dehydroepiandrosterone

Pregnenolone ↓ 3-β-HSD → Progesterone
17-OH-Pregnenolone ↓ 3-β-HSD → 17-OH-Progesterone
Dehydroepiandrosterone ↓ 3-β-HSD → Androstenedione

Progesterone —CYP17A1→ 17-OH-Progesterone —CYP17A1→ Androstenedione —CYP19A1→ Estrone

Progesterone ↓ CYP21A2 → 11-Deoxycorticosterone
17-OH-Progesterone ↓ CYP21A2 → 11-Deoxycortisol
Androstenedione ↓ 17-β-HSD → Testosterone
Estrone ↓ 17-β-HSD → Estradiol

Testosterone —CYP19A1→ Estradiol

11-Deoxycorticosterone ↓ CYP11B1 → Corticosterone
11-Deoxycortisol ↓ CYP11B1 → Cortisol

Corticosterone
↓ CYP11B2
18-OH-Corticosterone
↓ CYP11B2
Aldosterone

Figure 11.9 Steroid hormone synthesis in the adrenal gland. The cytochromes P450 involved are CYPs 11A1, 11B1, and 11B2 in the mitochondria (red) and CYPs 17A1, 19A1, and 21A2 in the endoplasmic reticulum (blue). The other enzymes shown, 3-β-HSD and 17-β-HSD, are hydroxysteroid dehydrogenases.

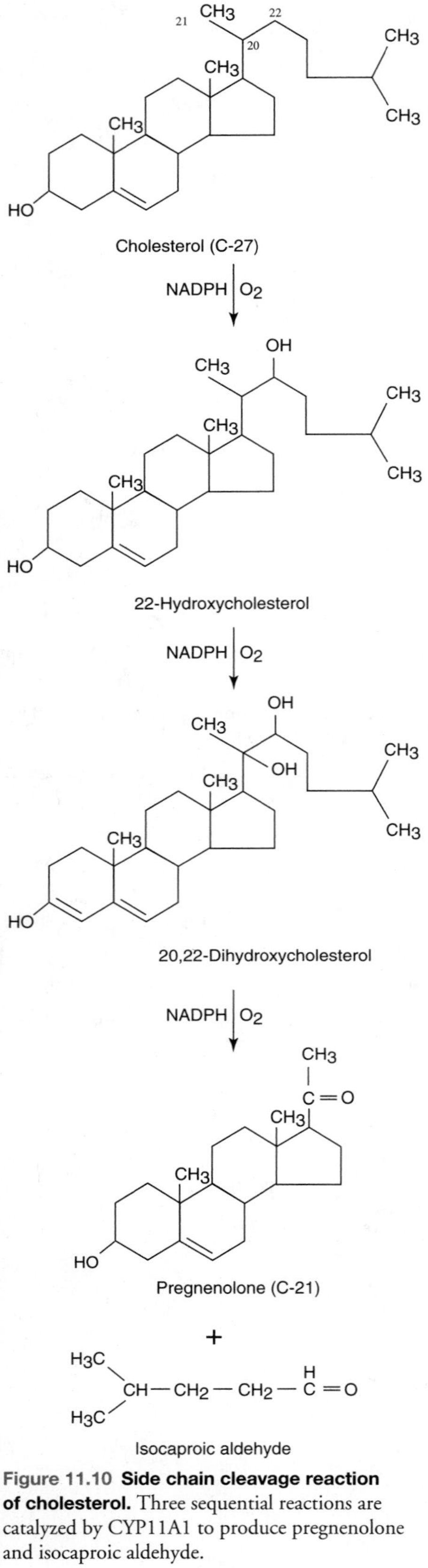

Figure 11.10 Side chain cleavage reaction of cholesterol. Three sequential reactions are catalyzed by CYP11A1 to produce pregnenolone and isocaproic aldehyde.

In adrenal mitochondria, a cytochrome P450 (CYP11A1) catalyzes the side chain cleavage reaction converting cholesterol to pregnenolone, a committed step in steroid biosynthesis. This reaction involves sequential hydroxylation at carbon atoms 22 and 20 to produce 22-hydroxycholesterol and then 20, 22-dihydroxycholesterol, respectively (Figure 11.10). CYP11A1 then cleaves the carbon–carbon bond between C20 and C22 to produce pregnenolone, a 21-carbon steroid. This reaction sequence requires three NADPH and three O_2 molecules, and it is catalyzed completely by CYP11A1.

Pregnenolone is produced in mitochondria, but is moved to the cytosol where it is oxidized by 3-β-hydroxysteroid dehydrogenase $\Delta^{4,5}$-isomerase to progesterone. Progesterone is hydroxylated at the C21 position of the steroid nucleus to deoxycorticosterone (DOC) by the microsomal P450 CYP21A2. DOC is subsequently hydroxylated at C11 by the mitochondrial P450 CYP11B1 to form corticosterone, which can then be hydroxylated by another mitochondrial cytochrome P450 enzyme, CYP11B2, at C18 to form the mineralocorticoid aldosterone (Figure 11.9 and p. 433).

Cortisol is synthesized from either pregnenolone or progesterone beginning with the hydroxylation of C17 by microsomal P450 CYP17A1. Hydroxylation at C21 of 17α-hydroxyprogesterone by CYP21A2 produces 11-deoxycortisol, which is transported into the mitochondrion where it is further hydroxylated by CYP11B1 at C11 to form cortisol (Figure 11.9, p. 433). Congenital deficiencies resulting from mutations in the gene that codes for CYP21A2 are discussed in Clin. Corr. 11.1.

Synthesis of steroids containing 19 carbon atoms from 7α-hydroxypregnenolone or 17α-hydroxyprogesterone proceeds with the loss of the acetyl group at C17. This deacetylation reaction is catalyzed by CYP17A1 (via a 17,20-lyase reaction), the same cytochrome P450 that hydroxylates steroids at C17. The factors that determine whether this cytochrome P450 performs only a single hydroxylation step to produce the 17-hydroxy product or proceeds further to cleave the C17–C20 bond have not been determined, although the presence of cytochrome b_5 in gonadal tissues appears to allow CYP17A1 to catalyze the cleavage reaction. The products formed after removal of the acetyl group are dehydroepiandrosterone (DHEA) from 17α-hydroxypregnenolone or androstenedione from 17α-hydroxyprogesterone. DHEA may be dehydrogenated at the 3-OH group to androstenedione, a potent androgenic steroid that serves as the immediate precursor of testosterone.

The estrogens are synthesized from the androgens in a reaction called aromatization because an aromatic ring is produced in ring A of the steroid product. In this complex reaction, multiple hydroxylation reactions are catalyzed by a single cytochrome P450 enzyme, CYP19A1 or aromatase, to form the aromatic ring with removal of the methyl group at C19. Figure 11.11, p. 436, details the aromatization reaction. Two sequential hydroxylation reactions at the methyl C19 result in the introduction of an aldehyde group. The final step may involve a peroxidative attack at C19 with the loss of the methyl group and elimination of the hydrogen atom to produce the aromatic ring. All three hydroxylation reactions are catalyzed by CYP19A1. The complexity of steroid hormone production during pregnancy and the role of P450 forms are illustrated in Clin. Corr. 11.2.

Cytochromes P450 also metabolize other endogenous compounds that are involved in physiological processes and thus regulate their activity. For example, arachidonic acid is metabolized by cytochromes P450 to form 19- and 20-hydroxyeicosatetraenoic acids (HETEs), epoxyeicosatetraenoic acids (EETs), and dihydroxyeicosatetraenoic acids (DiHETEs). These metabolites function in the regulation of vascular, renal, pulmonary, and cardiac systems and in inflammatory and growth responses. 20-HETE, formed by the metabolism of arachidonic acid by CYP4As and CYP4Fs in humans, is a potent vasoconstrictor. The EETs, formed by CYP1A, 2B, 2C, and 2J in humans, are endothelium-derived vasodilators. The CYP4As also catalyze the hydroxylation of prostaglandins and fatty acids, thus activating or inactivating these important regulatory factors.

Cytochromes P450 produce vital metabolites for maintenance of homeostasis in other systems as well. CYP7A1 catalyzes the first and rate-limiting step of bile acid synthesis in the liver (pp. 728 and 1053). In liver, CYP27B1 catalyzes the 1-hydroxylation of 25-hydroxy-vitamin D3, formed in kidney, to produce the 1,25-dihydroxy metabolite, which is the active form of this hormone, important in developing and maintaining bone

CLINICAL CORRELATION 11.1

Congenital Adrenal Hyperplasia: Deficiency of CYP21A2

The adrenal cortex, a major site of steroid hormone production during fetal and adult life, is metabolically more active in fetal life and may produce 100–200 mg of steroids per day in comparison to 20–30 mg in the nonstressed adult adrenal gland. Enzyme deficiencies have been reported at all steps of cortisol production. Diseases associated with insufficient cortisol production are referred to as congenital adrenal hyperplasias (CAH). The most common enzyme deficiency in CAH is in CYP21A2, a 21-hydroxylase, resulting in the failure to metabolize 17α-hydroxyprogesterone to 11-deoxycortisol, which is converted by CYP11B1 to cortisol. Because there is not enough cortisol, secretion of ACTH, the pituitary hormone that regulates production of cortisol, increases. Prolonged periods of elevated ACTH levels cause adrenal hyperplasia and an increased production of the androgenic hormones, dehydroepiandrosterone (DHEA) and androstenedione.

Clinical problems arise because the additional production of androgenic steroids causes virilization in females, precocious sex organ development in prepubertal males, or diseases related to salt imbalance because of decreased levels of aldosterone. Clinical consequences of severe 21-hydroxylase deficiency may be recognizable at birth, particularly in females, because the excessive buildup of androgenic steroids may cause obvious irregular development of their genitalia. In male newborns, a deficiency in 21-hydroxylase activity may be overlooked, because male genitalia appear normal, but there will be precocious masculinization and physical development later on. In late onset CAH, clinical symptoms may vary considerably from early development of pubic hair, early fusion of epiphyseal growth plates causing premature cessation of growth, or male baldness patterns in females.

Donohoue, P. A., Parker, K. and Migeon, C. J. Congenital adrenal hyperplasia. In Scriver, C. S.,, Beaudet, A. L., Sly, W. S., and Valle, D. (Eds.), *The Metabolic and Molecular Bases of Inherited Disease.* New York; McGraw Hill, 7th ed., Vol. 2, 1995, Chap. 94, 2929; and White, P. C. and Speiser, P. W. Congenital adrenal hyperplasia due to 21-hydroxylase deficiency. *Endocrine Reviews* 21: 245, 2000.

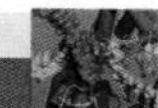

CLINICAL CORRELATION 11.2

Steroid Hormone Production during Pregnancy

Cytochromes P450 play a major role in estrogen synthesis. During pregnancy, a unique interaction among cytochromes P450 in different organs is needed to synthesize the large quantities that are required. Steroid hormone production increases dramatically and, at term, reaches 15–20 mg of estradiol, 50–100 mg of estriol, and approximately 250 mg of progesterone per day, amounts that are 1000 times greater than those produced by premenopausal, nonpregnant women.

The corpus luteum of the ovary is the major site for estrogen production in the first few weeks of pregnancy, but at approximately 4 weeks of gestation, the placenta begins synthesizing and secreting progesterone and estrogens. After 8 weeks of gestation, the placenta becomes the dominant source of progesterone. The human placenta lacks CYP17, which catalyzes both the 17α-hydroxylation reaction and the cleavage of the C17–C20 bond and thus cannot synthesize estrogens from cholesterol. The placenta, however, can catalyze the side-chain cleavage reaction to form pregnenolone from cholesterol and oxidizes pregnenolone to progesterone, which is secreted into the maternal circulation. How then does the placenta produce estrogens if it cannot synthesize DHEA or androstenedione from progesterone? The fetal adrenal gland, a highly active steroidogenic organ, catalyzes the synthesis of DHEA from cholesterol and releases it into the fetal circulation. A large proportion of fetal DHEA is metabolized by the fetal adrenal and liver to 16α-hydroxy-DHEA, which is converted by CYP19 in the placenta to the estrogen estriol. This is an elegant demonstration of the cooperativity of the cytochrome P450-mediated hydroxylating systems in the fetal and maternal organ systems leading to the progressive formation of estrogens during gestational development.

Cunningham, F. G., MacDonald, P. C. Gant, N. F., Leveno, K. J., et al. The placental hormones. In *Williams Obstetrics,* 20th ed. East Norwalk, CT: Appleton & Lange, 1997, Chapt.6, 125.

(p. 1067), while CYP24A1 is involved in degradation or inactivation of vitamin D metabolites. In the immune system, leukotriene B4 is a polymorphonuclear leukocyte chemoattractant that is hydroxylated by CYP4F3 to the less active 20-hydroxy-leukotriene B4, vital in the inactivation of the immune response. Retinoic acid is metabolized by CYP26A1 to the 4-hydroxy derivative, which may have greater activity than retinoic acid in certain organs, where it plays a role in cell cycling and signaling. Table 11.2, p. 432, lists the human cytochromes P450 that metabolize endogenous substrates and details their substrate specificities and functions.

Cytochromes P450 Oxidize Exogenous Lipophilic Substrates

Exogenous substrates metabolized by cytochromes P450 include therapeutic drugs, chemicals used in the workplace, industrial byproducts that become environmental contaminants, and food additives. The human cytochromes P450 involved in xenobiotic metabolism are listed in Table 11.1, p. 431. These cytochromes P450 exhibit broad substrate specificities, oxidizing a wide variety of xenobiotics, particularly lipophilic compounds. Addition of a hydroxyl group makes the compound more polar and, therefore, more soluble in the aqueous environment of the cell. Many exogenous compounds are highly lipophilic and will accumulate within cells over time, potentially interfering with cellular function, unless they are metabolized to more hydrophilic products.

It is estimated that CYP3A4 metabolizes approximately 50% of therapeutic drugs, CYP2D6 approximately 20%, CYP2C9 and 2C19 approximately 15%, and CYP1A2, CYP2A6, CYP2B6, and others metabolize the remaining 15% of the drugs. CYP3A4 is the major drug-metabolizing P450 in the human. It is present in the gastrointestinal tract as well as the liver and is responsible for the poor oral bioavailability of many drugs. Because cytochromes P450 have such broad substrate specificities, a compound may be metabolized by more than one cytochrome P450 isoform and/or at different sites, as shown in Figure 11.12 for the compound diazepam.

Drugs and xenobiotics are metabolized primarily via two pathways, Phase I (oxidative functionalization reactions) and Phase II (biosynthetic reactions). They may be metabolized by either or both pathways. In Phase I metabolism, a functional group, such as a hydroxyl group, is introduced into or exposed on the drug by cytochromes P450 (among other enzymes; Table 11.3). The enzymes in Phase II metabolism, in many cases, then connect this functional group to an endogenous species such as glucuronic acid, sulfate, glutathione, amino acids, or acetate. The net effect of either or both of these modifications is to make the foreign compound more water soluble and thus easier to excrete, generally through

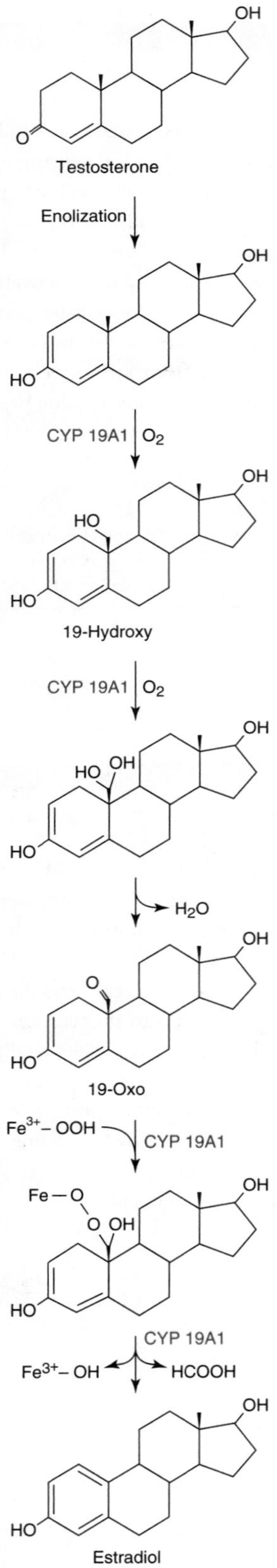

Figure 11.11 The sequence of reactions leading to the aromatization of androgens to estrogens. Adapted from Graham-Lorence, S., Amarneh, B., White R. E., Peterson, J. A., and Simpson E. R., A three dimensional model of aromatase cytochrome P450. *Protein Science* 4: 1065, 1995.

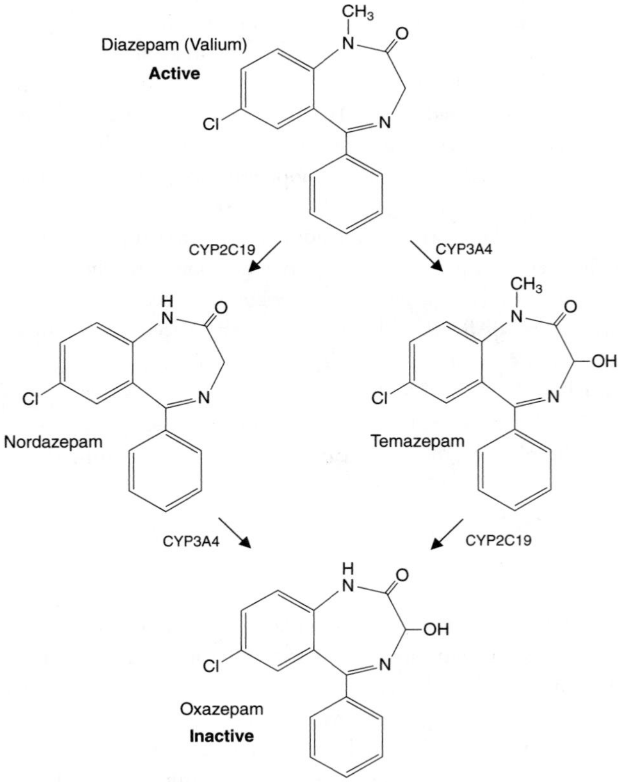

Figure 11.12 The metabolism of diazepam by CYPs 3A4 and 2C19. Diazepam may be metabolized by more than one P450. The metabolites formed from a compound will be affected by the amount of P450 forms that are expressed in tissues.

TABLE 11.3 • Xenobiotic-Metabolizing Reactions and Enzymes

Type of Reaction	*General Enzyme Name*	*General Enzymatic Reaction*
Oxidation	Dehydrogenases	Oxidation of alcohol or aldehyde groups
	Flavin-containing monooxygenases (FMO)	Oxidation of primary, secondary, and tertiary amines, nitrones, thioethers, thiocarbamates and phosphines.
	Monoamine oxidases	Removal of an amine by oxidative deamination
	Cytochrome P450	See Tables 11.1 and 11.2 and Fig. 11.7 for reactions catalyzed by P450 enzymes
Reduction	Aldehyde and ketone reductases	Reduction of carbonyl group to a hydroxyl group
	Azoreductases	Reduction of azo group to form primary amines
	Quinone reductases	Reduction of quinones to hydroquinones
Hydrolysis	Epoxide hydrolases	Addition of water to arene oxide or epoxide groups to form vicinal hydroxy groups
	Carboxylesterases	Addition of water to an ester bond
	Amidases	Addition of water to an amide bond
Conjugation	*N*-acetyltransferases	Addition of CoA from acetyl CoA to aromatic amines or compounds containing hydrazines
	UDP-glucuronosyl transferases	Addition of glucuronic acid from uridine-5′-diphosphate (UDP)–glucuronic acid to carboxyl, phenolic, aliphatic hydroxyl, aromatic or aliphatic amines, or sulfhydryl groups
	Sulfotransferases	Addition of sulfate group from phosphoadenosyl phosphosulfate to phenolic groups, aliphatic alcohols or aromatic amines
	Methyltransferases	Addition of methyl group from S-adenosyl methionine to compounds containing phenolic, catechol, aliphatic or aromatic amines, or sulfhydryl groups
	Glutathione *S*-transferase	Conjugation of compounds containing electrophilic heteroatoms, nitrenium ions, carbonium ions, free radicals, epoxide, or arene oxide groups

the kidneys, but also through the intestines via excretion into bile. Cytochromes P450 are involved in Phase I, but not Phase II, metabolism. General examples of xenobiotic-metabolizing enzymes and their reactions are shown in Table 11.3.

Metabolism of xenobiotics by cytochromes P450 can have one of three effects on the compound: **inactivation**, **activation**, or **formation of a toxic metabolite**. In inactivation,

CLINICAL CORRELATION 11.3

Cytochrome P450 Inhibition: Drug–Drug Interactions and Adverse Effects

The roles that cytochromes P450 play in drug metabolism and the serious consequences of drug–drug interactions were clearly shown for two drugs, terfenadine (Seldane) and cisapride (Propulsid), when their metabolism was inhibited by other drugs. A small percentage of users experienced severe adverse effects that forced their manufacturers and the Food and Drug Administration (FDA) to issue warnings that the drugs should not be taken with other drugs that inhibit CYP3A4. The seriousness of these adverse effects forced the FDA to have both drugs removed from the market.

The FDA approved terfenadine, a second-generation H1 antihistamine, in 1985 to treat seasonal allergies. Terfenadine is rapidly metabolized in the liver by CYP3A4 to fexofenadine, as shown in Figure 11.12, resulting in low levels of the parent drug soon after ingestion, but the therapeutic effects of terfenadine are actually affected by fexofenadine. Because many other drugs are substrates or inhibitors of CYP3A4, the metabolism of terfenadine is potentially inhibitable. Individuals who took terfenadine with the macrolide antibiotic, erythromycin, or the antifungal agent, ketoconazole, both strong inhibitors of CYP3A4, had significantly elevated plasma levels of terfenadine. In a small percentage of these terfenadine users, serious cardiac problems developed because the parent drug caused alteration in cardiac potassium channels and increased the risk of a rare ventricular tachycardia called Torsades de Pointes. Some individuals died from cardiac problems that developed after taking terfenadine with erythromycin or ketoconazole. Since the therapeutic properties of terfenadine were associated with its nontoxic metabolite, fexofenadine, the manufacturer tested fexofenadine as a new medication and sought FDA approval for this metabolite, now marketed as Allegra.

Cisapride was approved in 1993 for patients who suffer from nighttime heartburn as a result of gastroesophageal reflux disease or GERD. The elimination of this drug from the body is dependent on CYP3A4 metabolism, and when administered alone or with other drugs that do not inhibit CYP3A4, the parent drug does not accumulate in the plasma. However, when taken with drugs that are substrates or inhibitors of CYP3A4, cisapride metabolism is reduced, and it accumulates with subsequent intake. In some individuals, increased cisapride levels cause cardiac arrhythmias, and, by the end of 1999, heart rhythm abnormalities were reported in 341 patients taking cisapride, resulting in 80 deaths. Accordingly, the manufacturer of cisapride stopped marketing this drug in the United States after 2000.

Terfenadine: Proposal to withdraw approval of two new drug applications and one abbreviated new drug application. Fed. Reg. 62: 1889, 1997. This document may be accessed from the internet site for the Federal Register at http://www.access.gpo.gov/su_docs/aces/aces140.html; and Desta, Z., Soukhova, N., Mahal, S. K., and Flockhart, D. A. Interaction of cisapride with the human cytochrome P450 system: Metabolism and inhibition studies. *Drug Metab. Disp.* 28: 789, 2000.

the active form of a drug is converted to an inactive form, thus decreasing bioavailability or, in the case of harmful xenobiotics, decreasing potential harmful effects (e.g., phenobarbital). Diazepam (brand name Valium, prescribed as a sedative), for example, is metabolized by CYPs 2C19 (*N*-dealkylation) and 3A4 (hydroxylation) to the biologically inactive oxazepam (Figure 11.12), which then undergoes Phase II conjugation with glucuronic acid before elimination.

Other drugs are biologically inactive until metabolized by a cytochrome P450 to a biologically active form. The drug terfenadine (brand name Seldane, a histamine H1 receptor antagonist prescribed for seasonal allergies) is an inactive form of a drug, or a prodrug, that is activated by sequential CYP3A4 hydroxylation to fexofenadine, the functional, desired form (Figure 11.13). Problems associated with interference of terfenadine activation are detailed in Clin. Corr. 11.3.

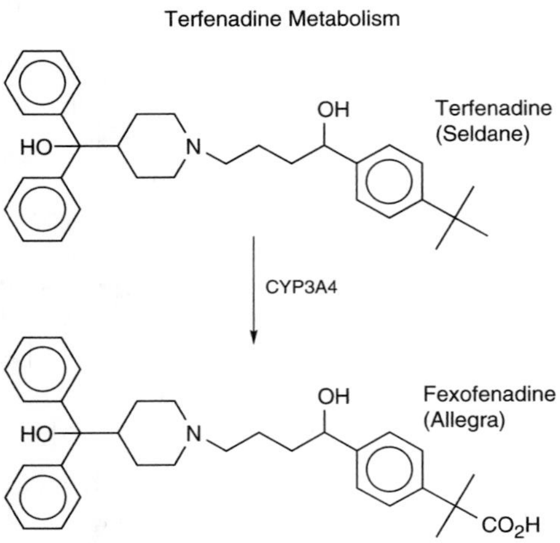

In addition to xenobiotic activation or inactivation, cytochrome P450 metabolism can inadvertently cause the formation of toxic metabolites. Benzo[a]pyrene is a weak carcinogen produced from the burning of coal, combustion of materials in tobacco products, food barbecued on charcoal briquettes, and industrial processing. It is converted by CYPs 1A1, 1A2, and 1B1 to benzo[a]pyrene-7,8-dihydrodiol-9,10-epoxide, a much more potent carcinogen (Figure 11.14). Benzo[*a*]pyrene is converted to the 7,8-epoxide form by CYP1A1. Subsequent hydrolysis by epoxide hydrolase forms the vicinal hydroxy derivative, benzo[*a*]pyrene-7,8-dihydrodiol, and a second epoxidation reaction by CYP1A1 or 3A4 forms benzo[*a*]pyrene-7,8-dihydrodiol-9,10-epoxide, which can form guanine adducts on DNA, causing disruption of gene function leading to mutations. The interaction of benzo[*a*]pyrene-7,8-dihydrodiol-9,10-epoxide with the p53 gene is a critical step in the mechanism by which benzo[*a*]pyrene is carcinogenic in humans. Benzo[a]pyrene also binds to the aryl hydrocarbon receptor (AhR) and induces these P450 forms, thus increasing its own metabolism.

Terfenadine Metabolism

Terfenadine (Seldane)

CYP3A4

Fexofenadine (Allegra)

Figure 11.13 The metabolism of terfenadine to fexofenadine, the bioactive compound.

Acetaminophen, commonly used as an analgesic and antipyretic, is converted by CYP2E1 to *N*-acetyl-*p*-benzoquinone imine (NAPQI), a highly reactive compound leading to protein adducts, oxidative stress, and toxicity. Normally, acetaminophen is primarily metabolized by glucuronidation and sulfation pathways to polar, inactive conjugates that are readily excreted (Figure 11.15, upper and lower pathways, respectively). Acetaminophen is also metabolized by CYP2E1 to a highly reactive NAPQI (Figure 11.15, middle pathway). The amount of acetaminophen that is metabolized by CYP2E1 is normally low in comparison to that of glucuronidation and sulfation. Whatever small amounts of NAPQI formed are rapidly conjugated by glutathione (p. 789) to a nontoxic metabolite. However, if glutathione levels are depleted, NAPQI will escape glutathione conjugation and react with liver cell components to cause cell damage. Normally, the levels of CYP2E1 are low in comparison to other P450 forms, and the production of NAPQI, when normal doses of acetaminophen are taken, is small. However, intake of large quantities of acetaminophen will cause an increase in the production of NAPQI that may cause severe liver damage. Another factor in acetaminophen-induced liver toxicity is that the consumption of alcoholic beverages induces CYP2E1, which will increase the production of NAPQI. However, alcohol is also a substrate for this cytochrome P450 and will inhibit the metabolism of other CYP2E1 substrates. The extent of liver damage thus depends on the timing and amount of acetaminophen taken. Acetaminophen overdose results in more calls to poison control centers in the United States than any other pharmacological substance. In addition, according to the American Liver Foundation, 35% of cases involving liver failure are caused by acetaminophen poisoning. A description of the effects of alcohol on acetaminophen metabolism is presented in Clin. Corr. 11.4.

As exemplified by benzo[*a*]pyrene and acetaminophen, the P450 system can produce toxic metabolites. Formation of toxic compounds by the cytochromes P450, however, does not mean that cell damage or a tumor will occur. Other cellular processes will determine the extent of cellular damage; for example, conjugation of the benzo[*a*]pyrene-7,8-dihydrodiol-9,10-epoxide by glutathione transferases plays a key protective role in inhibiting DNA adduct formation and reducing its carcinogenic activity. The actions of various detoxification

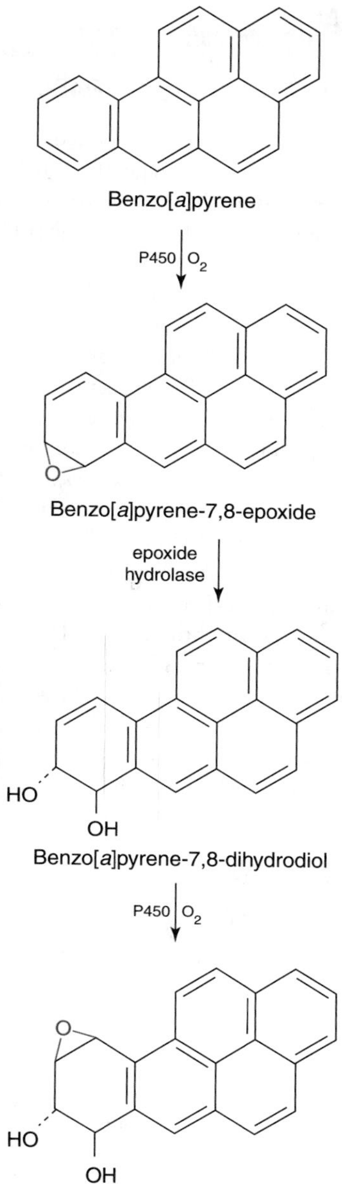

Benzo[*a*]pyrene

P450 O_2

Benzo[*a*]pyrene-7,8-epoxide

epoxide hydrolase

HO OH

Benzo[*a*]pyrene-7,8-dihydrodiol

P450 O_2

HO OH

Benzo[*a*]pyrene-7,8-dihydrodiol-9,10-epoxide

Figure 11.14 The metabolism of benzopyrene by cytochrome P450 and epoxide hydrolase to form benzopyrene-7,8-dihydrodiol-9,10-epoxide.

O-Glucuronide

Acetaminophen

CYP2E1

NAPQI

Protein

Cell Death

GSH

GSH

O-Sulfate

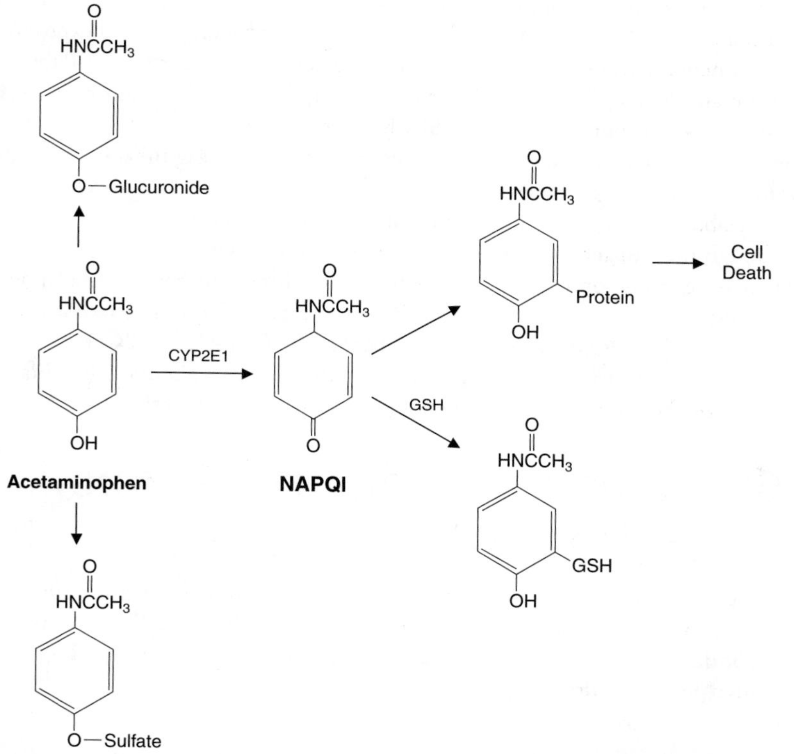

Figure 11.15 Acetaminophen may be metabolized by the phase II enzymes sulfotransferase and glucuronsyl transferase and by CYP2E1. Metabolism by CYP2E1 will lead to the formation of *N*-acetyl-*p*-benzoquinoneimine (NAPQI). When NAPQI formation is low, NAPQI can be conjugated by glutathione to a nontoxic metabolite; however, if NAPQI is formed in excessive amounts, it may escape glutathione conjugation and react with liver cell components to cause cell damage.

CLINICAL CORRELATION 11.4

Role of Cytochrome P450 2E1 in Acetaminophen-Induced Liver Toxicity

Acetaminophen is one of the most commonly used analgesic (pain-relieving) and antipyretic (fever-reducing) over-the-counter medications. It is available alone or as a component of more than 100 other nonprescription medications. The pathway of acetaminophen metabolism is shown in Figure 11.15. The consumption of alcohol, an inducer and substrate of CYP2E1, has a profound effect on acetaminophen metabolism. Alcohol consumption equivalent to drinking a 750-ml bottle of wine, six 12-oz cans of beer, or 9 oz of 80-proof liquor over a 6- to 7-hour period causes a 22% increase in CYP2E1-mediated metabolism of acetaminophen that manifests several hours post alcohol intake. Thus, the timing between the last consumption of alcohol and acetaminophen intake may be very critical in the development of acetaminophen-induced liver injury. In a person drinking alcohol, metabolism of acetaminophen taken concomitantly or shortly after cessation of drinking is delayed because ethanol, like acetaminophen, is a substrate for CYP2E1 and therefore competes for binding. In chronic alcohol consumption, CYP2E1 is induced and greater amounts of *N*-acetyl-*p*-benzoquinoneimine (NAPQI) are produced; however, alcohol may actually protect the liver from acetaminophen-induced injury if taken at the same time or shortly before. If acetaminophen is taken several hours post alcohol intake, the alcohol has already been metabolized and increased levels of CYP2E1 are then available to metabolize acetaminophen to NAPQI, increasing the risk of liver damage.

The length of time and amount of alcohol consumed daily affect the amount of CYP2E1 expressed in hepatocytes. The amount of acetaminophen consumed is also a major determining factor for risk, as normally recommended doses do not usually generate sufficient NAPQI to produce liver cell damage. Higher doses of acetaminophen increase the risk of liver cell injury as more drug will be metabolized through the CYP2E1 pathway. Both the sulfation and glucuronidation pathways are essential in determining the amount of acetaminophen that will be metabolized by CYP2E1. In addition, the levels of intracellular glutathione are critical as glutathione forms conjugates with NAPQI to protect the liver from this reactive metabolite. In the presence of high levels of NAPQI, glutathione is rapidly depleted and can no longer protect against damage.

Brunner, L. J., McGuinness, M. E., Meyer, M. M., and Munar, M.Y. Acute acetaminophen toxicity during chronic alcohol use. *U.S. Pharmacist* September:HS11–HS19, 1999. This article may be accessed at http://www.uspharmacist.com/; Thummel, K. E., Slattery, J. T., Ro, H., Chien, J. Y., et al. Ethanol and production of the hepatotoxic metabolite of acetaminophen in healthy adults. *Clin. Pharmacol. Ther.* 67: 591, 2000; and Slattery, J. T., Nelson, S. D., and Thummel, K. E. The complex interaction between ethanol and acetaminophen. *Clin. Pharmacol. Ther.* 60:241, 1996.

enzyme systems, including conjugation reactions such as glucuronidation and sulfation, the status of the immune system, nutritional state, genetic predisposition, and exposure to various environmental chemicals all play key roles in determining the extent of P450-mediated cell damage and toxicity. Although it seems counterproductive for mammalian systems to possess enzymes that potentially create highly toxic compounds, the purpose of the cytochrome P450 system is to add or expose functional groups making the molecule more polar and/or more susceptible to metabolism by other detoxification enzyme systems; the creation of toxic metabolites is an unintended consequence of this process.

Although the vast majority of the substrates for these cytochromes P450 (Table 11.1, p. 431) are exogenous compounds, some endogenous functions have also been postulated. CYP1A1 and 1A2 appear to be involved in heme catabolism and estradiol metabolism, respectively. There is also evidence that members of the CYP2A, 2B, 2C, and 3A subfamilies are involved in testosterone metabolism and those of the 2D and 2E subfamilies are involved in catecholamine metabolism and gluconeogenesis, respectively.

11.6 • CYTOCHROMES P450: INDUCTION AND INHIBITION

Clearly, the levels and activity of the cytochromes P450 in any given person will greatly influence the effect of a particular drug or xenobiotic on the system. Thus, compounds that inhibit or induce P450 activity, as well as polymorphisms occurring in a P450 gene, can produce unexpected results.

Drug–Drug Interactions

Many drugs induce the formation of the cytochrome P450 that metabolizes them. Other drugs cause inhibition of P450 activity. Altering the metabolism of a particular drug from its predicted rate may cause unexpected and adverse effects and is of particular concern for individuals who take a combination of drugs. For drugs that are dependent on P450 metabolism for elimination from the body, inhibition will lead to accumulation of the parent drug to concentrations that may be toxic. Conversely, induction of the P450 isoform required to metabolize a particular drug may lead to over-metabolism and sub-effective concentrations.

The unintended effects that can occur when levels of cytochromes P450 are induced or inhibited by other drugs are called drug–drug interactions. If these inducing or inhibiting drugs are given with other drugs that are normally metabolized by P450 enzymes, the lifetime of these other drugs will be altered. If these drugs affect critical systems, the result can be fatal. For example, CYP2E1 is induced by alcohol or isoniazid. It also metabolizes anesthetics, such as halothane and enflurane, to compounds that damage liver proteins. The immune system then attacks these abnormal proteins, producing a form of hepatitis. People with higher than normal CYP2E1 activity produced by alcohol intake are at higher risk for this sort of hepatitis reaction to an anesthetic.

Drug–drug interactions also occur with CYP3A4, the most important drug metabolizing P450 and the P450 most abundantly present in liver and small intestine. CYP3A4 levels are induced by rifampicin, anticonvulsants such as carbamazepine and phenytoin, and glucocorticoids such as dexamethasone and are inhibited by antifungals such as ketoconazole, HIV protease inhibitors, calcium channel blockers, cyclosporine, antibiotics such as erythromycin, SSRI antidepressants, and even grapefruit juice. A very few of the many compounds metabolized by this enzyme are listed in Table 11.1, p. 431.

It is readily apparent that mixing inducers and/or inhibitors of CYP3A4 with various drugs (which can themselves be inducers or inhibitors) may have unintended or even fatal effects. Taking rifampicin, a CYP3A4 inducer, with oral contraceptives, which are metabolized by CYP3A4, has resulted in lower efficacy of the contraceptives, leading to unintended pregnancies. Taking ketoconazole or even drinking grapefruit juice, both of which inhibit CYP3A4, along with the anticoagulant warfarin, could lead to excessive bleeding. Clinical consequences that relate to inhibition or induction of drug-metabolizing enzymes are presented in Clin. Corr. 11.3, p. 438, and Clin. Corr. 11.5.

The mechanisms of induction of cytochromes P450 are at either the transcriptional or posttranscriptional (e.g., stabilization of mRNA or a decrease in protein degradation) level. It is not possible to predict the mode of induction on the basis of the inducing compound. The complexity of the induction process is illustrated with CYP2E1 in the human versus rat liver. CYP2E1 protein is induced in both species by small organic molecules, such as ethanol, acetone, or pyrazole, or under fasting or diabetic conditions. In humans, the levels of mRNA for the protein are unaffected and the induction is likely due to stabilization of human CYP2E1 from proteolytic degradation. However, in diabetic rats the sixfold induction of CYP2E1 protein is accompanied by a 10-fold increase in mRNA without involving an increase in gene transcription, suggesting stabilization of the mRNA.

Specific cytosolic receptor proteins have been indicated for some inducing agents. One of the most extensively studied is the aryl hydrocarbon (or Ah) receptor, which interacts with 2,3,7,8-tetrachlorodibenzo-*p*-dioxin (TCDD) and causes induction of CYP1A1, CYP1A2, and CYP1B1. This is an example of a compound that persists in the lipid environments of the cell for extended periods of time and can induce carcinogenesis. Polycyclic aromatic hydrocarbons are ligands for the Ah receptor, and they produce a ligand-receptor complex that is translocated into the nucleus by the **A**h **r**eceptor **n**uclear **t**ranslocator (Arnt) where it binds to specific response elements in the upstream regulatory regions of cytochrome P450 genes. Figure 11.16, p. 442, illustrates this process.

Another receptor system of P450 genes is the pregnane X receptor (PXR) which regulates the CYP3A gene. Pregnane refers to 21-carbon steroids such as the progesterone metabolite 5α-pregnane-3, 20-dione, which is a ligand for PXR and a potent inducer of the human CYP3A4 gene. In addition to 21-carbon steroids, other structurally dissimilar compounds such as carbamazepine (a drug used for the treatment of convulsions) and rifampin (an antituberculosis drug), as well as St. John's wort (a nonprescription herbal agent used to treat mood disorders) are ligands for PXR. PXR binds the retinoid X receptor (RXR) to form a heterodimeric complex (PXR-RXR) that binds specific response elements in the 5′-flanking region of CYP3A genes. Recall that approximately 50% of prescription drugs are metabolized by this cytochrome P450 form.

CLINICAL CORRELATION 11.5

Cytochrome P450 Induction: Drug–Drug Interactions and Adverse Effects

Induction of specific cytochromes P450 may decrease the therapeutic effects of drugs, because increases in liver P450 levels increase rates of metabolism and, therefore, inactivation and/or excretion of drugs. Induction of the P450 system may also have adverse effects due to the increased formation of toxic metabolites (also Clin. Corr. 11.4).

Induction of CYP3A4 by, for example, the antituberculosis drug rifampin, may greatly increase elimination of the oral contraceptive ethinyl estradiol, causing subeffective plasma levels of the drug, which increases the risk of pregnancy. Treatment with the anticonvulsant drugs, phenytoin and carbamazepine, may also induce CYP3A4 and reduce the contraceptive potency of ethinyl estradiol. It is imperative that a woman taking ethinyl estradiol as an oral contraceptive simultaneously with a CYP3A4 inducer increase the dose of the contraceptive or use an alternative contraceptive method or drug.

The widely used herbal agent St. John's wort, which can be purchased without a prescription, also induces CYP3A4 activity, raising concerns that individuals who ingest the herb may not receive the therapeutic benefits of drugs that are CYP3A4 substrates. A reduction in plasma levels of oral contraceptives, HIV antiprotease drugs, the immunosuppressive drug cyclosporin, or certain statin drugs, used to lower cholesterol levels, can occur because each of these drugs is a substrate for CYP3A4. Because St. John's wort is considered a natural substance, rather than a drug, individuals may take this agent without notifying their physician or knowing its effects on the efficacy of other drugs they are taking.

Roby, C. A., Anderson, G. D., Kantor, E., Dryer, D. A., and Burstein, A. H. St. John's wort: Effect on CYP3A4 activity. *Clin. Pharmacol. Ther.* 67:451, 2000; and Shader, R. I. and Oesterheld, J. R. Contraceptive effectiveness: Cytochromes and induction. *J. Clin. Psychopharmacol.* 20:119, 2000.

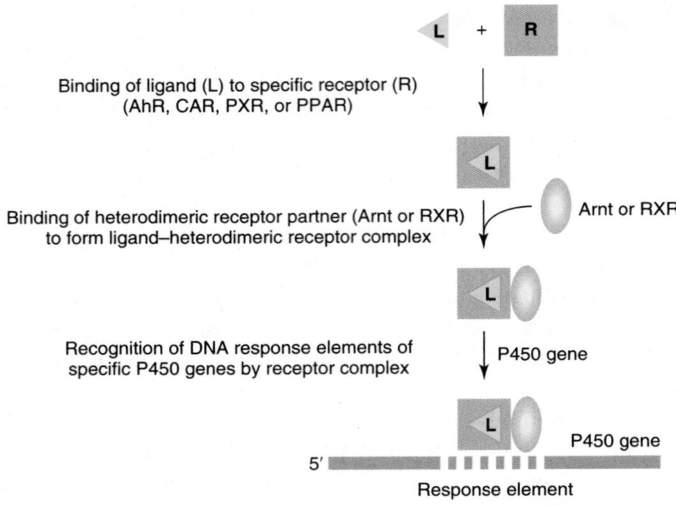

Figure 11.16 The interaction of a ligand with its receptor and receptor partner to form the heterodimeric receptor complex initiates induction of cytochrome P450 forms.

Another inducer of cytochrome P450 genes is phenobarbital. This induction is mediated by the **c**onstitutive **a**ndrostane **r**eceptor (CAR), which also forms a heterodimeric complex with RXR to interact with its specific response element in the 5′-flanking region of CYP2B genes. Two endogenous androstane steroids, androstanol and androstenol, are also ligands for CAR and, when either ligand is bound to CAR, they prevent the interaction of the CAR-RXR receptor complex with the protein, SRC-1. SRC-1 is a coactivator of nuclear receptors and is required for transcriptional activation by CAR. Administration of phenobarbital or other phenobarbital-like inducers displaces the androstane ligands, permitting SRC-1 binding and transcription of the CYP2B gene.

Cytochrome P450 Genetic Polymorphisms

In addition to exposure to various inducing agents that may change the cytochrome P450 expression pattern in the liver and other organs, individuals may differ in their metabolism rates for particular drugs because of unique cytochrome P450 genes or alleles they possess, *i.e.*, polymorphisms. A polymorphism is a difference in DNA sequence found at 1% or higher in a population. These differences in DNA sequence can lead to differences in drug metabolism. Since these are genetic variations, many of them are present in specific ethnic groups, which may, therefore, experience more frequent adverse effects with certain drugs. It is estimated that 40% of the different human cytochrome P450 forms that metabolize xenobiotics are polymorphic.

Genetic polymorphisms in CYP2D6 have been studied extensively and 65 different alleles have been identified. The population can be divided into *extensive metabolizers* or *poor metabolizers,* depending on the levels of CYP2D6 expression. If there is no other way to clear a particular drug from the system, poor metabolizers may be at risk for adverse drug reactions. Alternatively, extensive metabolizers may find many drugs to be ineffective.

Many persons of Ethiopian and Saudi Arabian origin exhibit high expression of CYP2D6. This isoform metabolizes a variety of drugs, making them ineffective; many antidepressants and neuroleptics are important examples. Conversely, prodrugs will be extensively activated; codeine will be converted very quickly to high amounts of morphine.

In contrast, many individuals lack functional 2D6. These patients will be predisposed to drug toxicity caused by antidepressants or neuroleptics, but will find codeine to be ineffective due to lack of activation. Perhexiline (a calcium channel blocker) was withdrawn from the market due to the neuropathy it caused in 2D6 inactive patients. Even beta-blocker removal may be impaired (for example, propranolol) in 2D6-deficient people. Clin. Corr. 11.6 discusses polymorphism effects.

Although genetic polymorphism of a particular cytochrome P450 may affect metabolism of a discrete set of compounds, genetic polymorphisms of the cytochrome P450 reductase, the sole and obligate electron donor to the 50 members of the microsomal class

CLINICAL CORRELATION 11.6

Genetic Polymorphisms of P450 Enzymes

Multiple alleles have been demonstrated for approximately 40% of human P450 genes. These genetic polymorphisms may produce defective P450 proteins. A description of the polymorphic CYP21 and its importance in the metabolism of an endogenous compound was given in Clin. Corr. 11.1. The absence of a cytochrome P450 activity may cause many problems due to the accumulation of therapeutic drugs that cannot be metabolized. The discovery of exaggerated hypotensive effects with the antihypertensive drug debrisoquine led to the characterization of individuals who inefficiently metabolized CYP2D6 substrates. Approximately 5% to 10% of the Caucasian, 8% of the African and African-American, and 1% of the Asian populations are deficient in catalytically active CYP2D6. In addition to debrisoquine, other drugs that are metabolized by CYP2D6 are sparteine, amitriptyline, dextromethorphan, and codeine. In the case of codeine, CYP2D6 in normal individuals catalyzes the *O*-demethylation of approximately 10% of the codeine to morphine. Individuals who lack functional CYP2D6 are unable to catalyze this reaction and so do not achieve the analgesic effects of codeine.

A genetic polymorphism associated with CYP2C19 has been demonstrated in individuals who are poor metabolizers of mephenytoin, a drug used in the treatment of epilepsy; CYP2C19 hydroxylates C4 of the S-enantiomer of mephenytoin to inactivate the physiological effect. Poor metabolizers of this drug suffer greater sedative effects at normal dosages. Approximately 14% to 22% of the Asian, 4% to 7% of the African and African-American, and 3% of the Caucasian populations lack the active form of CYP2C19.

CYP2C9 polymorphisms occur in less than 0.003% of the African or African-American, 0.08% of the Asian, and 0.36% of the Caucasian populations. The absence of a functional CYP2C9 may have serious consequences in the metabolism of S-warfarin, an orally administered drug used to inhibit blood coagulation in patients who have suffered a heart attack or a stroke to prevent the reoccurrence of life-threatening clots. Plasma levels of the drug must be maintained within a specific range because excessive amounts will cause uncontrolled bleeding and, possibly, death. Warfarin is metabolized by CYP2C9 for elimination. CYP2C9*3 is an allelic variant in which isoleucine is substituted for leucine at residue 359, resulting in substantial loss of enzymatic activity. Individuals deficient in CYP2C9 may require only 0.5 to 1 mg of warfarin per week in contrast to the 4 to 5 mg per day prescribed for an individual with normal CYP2C9. If a 5-mg dose of warfarin was taken by someone lacking functional CYP2C9, uncontrollable bleeding could result from a simple cut.

Ingelman-Sundberg., M., Oscarson, M., and McLellan, R. A. Polymorphic human cytochrome P450 enzymes: An opportunity for individualized drug treatment. *Trends Pharm. Sci.* 20:342, 1999.

CLINICAL CORRELATION 11.7

Genetic Polymorphisms of NADPH-Cytochrome P450 Reductase: Antley–Bixler Syndrome (OMIM 207410)

Mutations have been found in human NADPH-cytochrome P450 reductase that result in a patient phenotype of skeletal malformations characterized by craniosynostoses, radio-ulnar or radio-humeral synostoses, and/or bowed femora, a set of clinical features known as Antley–Bixler syndrome. These patients also exhibit disordered steroidogenesis and congenital adrenal hyperplasia (CAH). As described in Clin. Corr. 11.1, a defect in CYP21A2 is the most common cause of CAH. Because cytochrome P450 reductase deficiency potentially affects the activities of three steroidogenic cytochromes P450 (CYPs 19A and 17A, as well as CYP21A, Figure 11.8), variable and abnormal patterns of steroid hormones are observed in these cases, resulting in overvirilized females and undervirilized males. The pathways involved in embryonic skeletal development were also clearly affected in some cytochrome P450 reductase-deficient patients, and cytochrome P450-mediated drug metabolism was compromised in at least some of these patients. Because there is no redundancy in the electron-transferring portion of the microsomal cytochrome P450 system, defects in the cytochrome P450 reductase would be expected to have variable and pleotropic effects, depending on the severity of the effect on its activity. In addition, various cytochromes P450 could interact with different surface residues of the reductase, and thus be affected differentially by mutations in such residues. Therefore, depending on the locus of the mutation in the structure of the reductase (Figure 11.4), and the resultant alteration of enzymatic activities or protein–protein interactions, the mutations may vary greatly in resulting severity and phenotype.

Fluck, C. E., Tajima, T., Pandey, A. V., Arlt, W., et al. Mutant P450 oxidoreductase causes disordered steroidogenesis with and without Antley–Bixler syndrome. *Nat. Genet.* 36:228, 2004: and Scott, R. R. and Miller, W. L. Genetic and clinical features of P450 oxidoreductase deficiency. *Horm. Res.* 69:266, 2008.

of cytochromes P450, might be expected to have more global and dramatic effects on metabolism of both endogenous and xenobiotic substrates. Indeed, completely knocking out the cytochrome P450 reductase gene in an animal model is embryonically lethal to these animals. Recently, human mutations that alter activity in this gene have been described to have profound effects (Clin. Corr. 11.7).

Therapeutic Inhibition of Cytochrome P450

Because cytochromes P450 are present in almost all organisms and because they play such important roles in cellular homeostasis, selective inhibition of these enzymes is of clinical importance. Azole inhibitors (ketoconazole, itraconazole, and fluconazole), for example, are used to control yeast infections contracted by AIDS patients, cancer patients undergoing chemotherapy, and patients who are treated with immunosuppressive drugs. In yeast, CYP51 catalyzes the demethylation of sterol molecules in the synthesis of ergosterol, an important membrane lipid. In humans, a related P450 form, CYP51A1, catalyzes the demethylation of lanosterol, a reaction that is essential to cholesterol synthesis. Although the human and yeast CYP51 proteins are similar, sufficient differences exist that azole antifungal agents are effective inhibitors of ergosterol formation. As discussed earlier, however, these agents may also induce or inhibit native human cytochromes P450, possibly leading to drug–drug interaction phenomena.

Certain tumors, such as breast tumors, are dependent on estrogen for their growth. Thus, the prevention of estrogen synthesis without inhibiting production of other steroid hormones might serve as a selective chemotherapeutic tool to reduce and eliminate tumors. CYP19A1, the cytochrome P450 that catalyzes estradiol formation from androstenedione, is an attractive target for this type of intervention. A particularly selective type of inhibitor is a mechanism-based, or suicide, inhibitor. These compounds bear strong resemblance to a substrate of an enzyme but form an irreversible inhibition product during catalytic turnover with the enzyme prosthetic group or protein. Such inhibitors would be highly specific for a particular cytochrome P450 form and would, therefore, be a good tactical approach to drug design.

It is readily apparent that without an understanding of drug metabolism and drug interactions mediated by the cytochrome P450 system, medication cannot be safely prescribed, particularly when multiple medications are required, as in older patients. It is critical to determine how therapeutic drugs are metabolized and whether cytochrome P450 forms can be inhibited by different compounds. The Food and Drug Administration's (FDA's) drug approval process requires that extensive information be submitted on the metabolism of drugs before they are finally approved for general human use.

11.7 • NITRIC OXIDE SYNTHASES: PROPERTIES AND ENZYMATIC FUNCTION

The nitric oxide synthases (NOSs) are heme- and flavin-containing enzymes that catalyze the synthesis of nitric oxide (NO) through two serial monooxygenase reactions analogous to those of the cytochrome P450 systems. The NOSs are functional dimers and consist of two major domains, a heme-containing oxygenase domain, which is structurally unrelated to cytochrome P450s, and a flavin-containing reductase domain, which structurally resembles the P450 reductase (Figure 11.17). Electrons are transferred from NADPH into the reductase domain of one monomer, through the flavins FAD and FMN, to the heme iron of the other monomer where molecular oxygen is bound and activated. The domains are connected by a calmodulin-binding site, occupation of which by calmodulin is required for electron transfer from the reductase to the oxygenase domain.

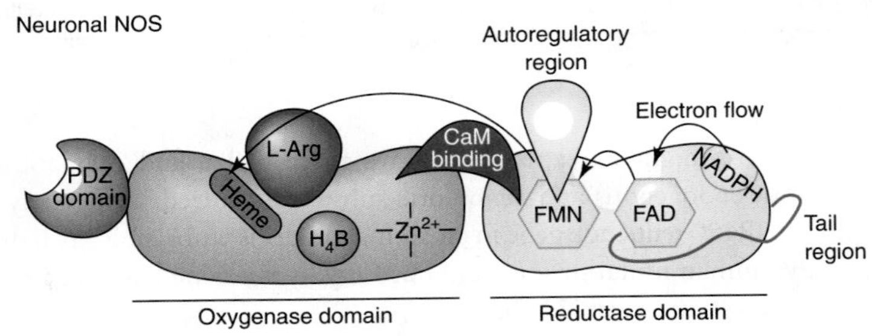

Figure 11.17 A modular structure of neuronal nitric oxide synthase showing approximate locations of prosthetic groups and cofactors.

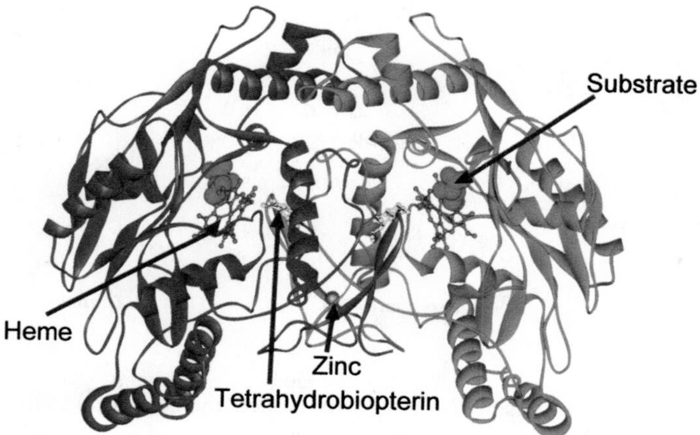

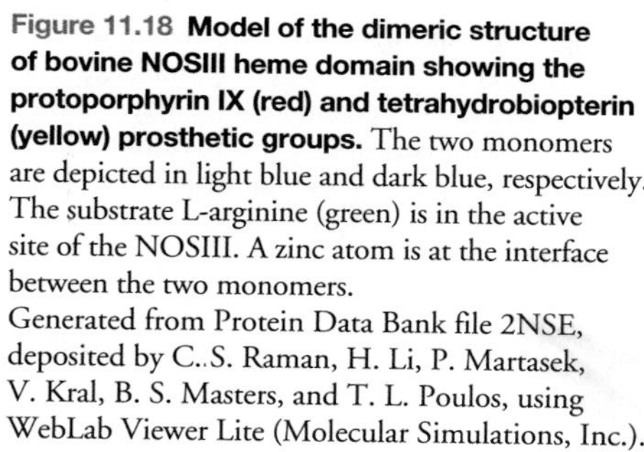
Figure 11.18 Model of the dimeric structure of bovine NOSIII heme domain showing the protoporphyrin IX (red) and tetrahydrobiopterin (yellow) prosthetic groups. The two monomers are depicted in light blue and dark blue, respectively. The substrate L-arginine (green) is in the active site of the NOSIII. A zinc atom is at the interface between the two monomers.
Generated from Protein Data Bank file 2NSE, deposited by C..S. Raman, H. Li, P. Martasek, V. Kral, B. S. Masters, and T. L. Poulos, using WebLab Viewer Lite (Molecular Simulations, Inc.).

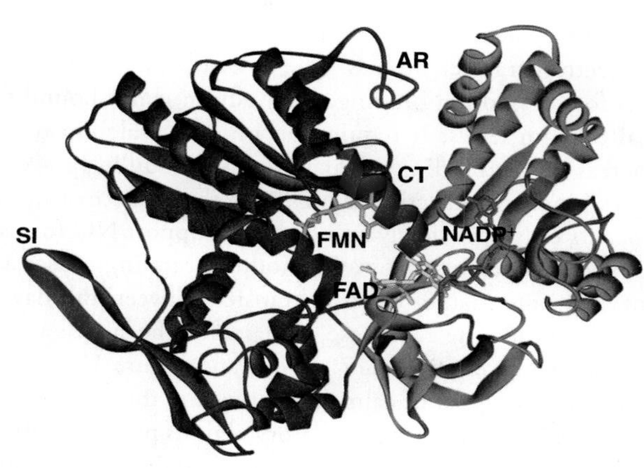

Figure 11.19 Ribbon diagram of nNOS reductase domain derived from its crystal structure. Similar to cytochrome P450 reductase, several subdomains are present—the FMN-binding domain (blue), the FAD/NADPH-binding domain (green), and the intervening connecting domain (purple). Unlike the P450 reductase, the nNOS reductase contains a calmodulin-binding site (not evident in the diagram), and regulatory regions (red) at the C-terminus (CT), in the FMN domain (AR), and in the connecting domain (SI). Cofactors are shown in stick models: blue, FMN; yellow, FAD; and red, $NADP^+$.
Generated from Protein Data Bank file 1TLL, deposited by E. D. Garcin, C. M. Bruns, S. J. Lloyd, D. J. Hosfield, et al., using WebLab Viewer Lite (Molecular Simulations, Inc.).

The heme moiety of the NOSs resembles that of cytochromes P450 in that it forms a thiolate ligand with a cysteine in the NOS protein, and it yields the same type of absorbance spectrum when CO is bound to the reduced, ferrous form of the heme, that is, an absorbance peak at about 450 nm. The NOS oxygenase domain is, however, structurally very different from the cytochromes P450 (compare the NOS oxygenase domain crystal structure in Figure 11.18 with that of the cytochrome P450 in Figure 11.1, p. 431). In addition to the heme moiety and substrate-binding site, the oxygenase domains of NOSs also contain a binding site for tetrahydrobiopterin, which apparently functions in electron transfer, and a tetra-coordinated zinc atom, which is important for structural integrity.

The NOS reductase domain is structurally very similar to that of cytochrome P450 reductase (compare the nNOS reductase domain crystal structure in Figure 11.19 with that of the cytochrome P450 reductase in Figure 11.4, p. 428). They share almost 60% sequence homology, and the overall architecture and subdomain structures, the FMN domain, the connecting domain, and the FAD/NADPH binding domain, closely resemble each other. The primary structural difference between the two is the inclusion in the NOS reductase domain of various regulatory elements, which are only partially visible in the crystal structure due to their extreme flexibility.

The general reaction catalyzed by the NOSs is sequential monooxygenation of the amino acid L-arginine to form NO and citrulline, as shown in Figure 11.20. The reaction cycle is initiated by the binding of calmodulin to "open the gate" between the reductase and oxygenase domains. The electron donor is NADPH, which donates two electrons to

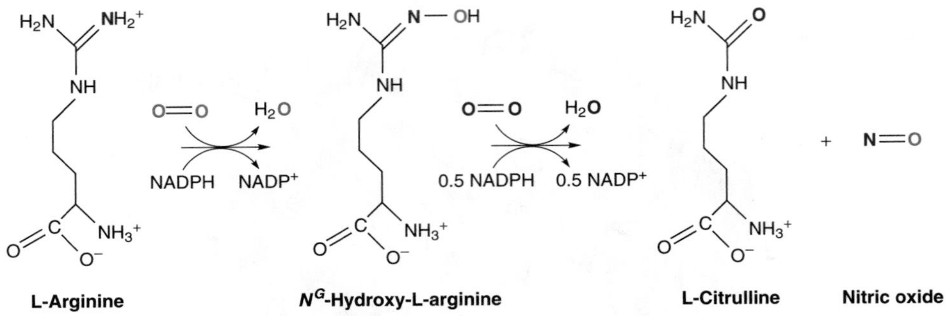

Figure 11.20 Production of nitric oxide. NOSs catalyze the formation of NO from L-arginine through two sequential monooxygenation steps which appear to be mechanistically similar to that of the microsomal cytochrome P450 system. The nitrogen atom (blue) of NO is derived from the guanidino group of the L-arginine side chain, and the oxygen atom (red) of NO is derived from molecular oxygen.

FAD, which, in turn, reduces FMN. The FMN reduces the heme iron to its ferrous form, to which oxygen can bind. In the first step, arginine is oxidized to the stable intermediate *N*-hydroxy-L-arginine. In the second step, *N*-hydroxy-L-arginine is oxidized to NO and citrulline. Note that the source of NO is the guanidino group of the substrate, arginine. This entire process requires 2 oxygen molecules and 1.5 molecules of NADPH and, like that of the cytochromes P450, is inhibited by carbon monoxide.

The overall electron transfer chain and mechanism by which the NOSs form NO is similar to that of the cytochromes P450, but there are some significant differences in regulation of the electron transfer process. The most obvious is the requirement of the constitutive NOSs for calmodulin binding. Calmodulin is not bound to the enzyme at intracellular, basal calcium levels. It requires an influx of calcium to raise the calcium concentration, increasing the binding of calcium to calmodulin (p. 493). The calcium–calmodulin complex binds to and activates NOSs. In the absence of calmodulin binding, heme reduction occurs only very slowly and does not support NO formation, and transfer between FAD and FMN is also slow. Calmodulin binding causes a conformational change in the enzyme, potentiating electron transfer between the flavins and allowing productive heme reduction. The reductase domains of the NOSs also contain several regulatory regions, shown in Figure 11.17, p. 444 and Figure 11.19, p. 445, which are involved in transducing the signal of calmodulin binding through the enzyme. In addition to the calmodulin-binding site, the regulatory regions present in the NOSs include a C-terminal extension of the FAD/NADPH binding domain, an insert of 45–50 residues in the FMN domain, and a small insertion in the connecting domain. All NOS isoforms include the C-terminus, but only the so-called constitutive isoforms contain the FMN and connecting domain inserts.

Another difference between the NOSs and P450s is that substrate binding is not required by the NOSs to initiate the reaction cycle, and the binding of substrate does not alter the reduction potential of the heme. If substrate is not present, the electron transferred to the heme can activate molecular oxygen, forming the oxy–ferrous complex, which can decay and be released as superoxide. If the second electron is transferred before the release occurs, the peroxy form of oxygen is generated, which will dissociate as hydrogen peroxide. Thus, if substrate is not present, but the NOS is active, potentially destructive reactive oxygen species may be formed. In addition, a tetrahydrobiopterin cofactor is required for NO synthesis by NOS, but is not required by any known cytochrome P450. Tetrahydrobiopterin serves as an electron donor to the oxy–ferrous complex (i.e., the one-electron reduced, oxygen-bound form of the heme) of NOS, delivering the second electron to the heme much faster than the relatively slow interdomain transfer of an electron from the NOS FMN. Thus, in the absence of tetrahydrobiopterin, NOS will produce superoxide due to the autoxidation of the oxy–ferrous complex. In the presence of tetrahydrobiopterin, but in the absence of substrate, NOS can also form hydrogen peroxide, due to the formation and subsequent decay of the peroxy–ferrous complex. The formation of these reduced oxygen species has been implicated in several inflammatory disease processes, such as endothelial dysfunction, and also in ischemia/reperfusion injury, where tissues are first depleted of oxygen and then reperfused with oxygen-rich blood. This type of injury occurs in cerebral thrombosis and in tissues removed for transplantation.

TABLE 11.4 • Properties of the NOS Isoforms

Property	*NOSI*	*NOSII*	*NOSIII*
Molecular mass	160 kD	130 kD	135 kD
Expression	Constitutive	Inducible	Constitutive
Cell fraction	Cytoplasmic	Cytoplasmic	Membrane-bound
Dependence on calcium influx	Dependent	Independent	Dependent
Physiological action	Neurotransmission	Cytotoxicity	Vasodilation

11.8 • NITRIC OXIDE SYNTHASE ISOFORMS AND PHYSIOLOGICAL FUNCTIONS

There are three major NOS isoforms, neuronal (NOSI or nNOS), inducible (NOSII or iNOS), and endothelial (NOSIII or eNOS), although variations of each of these types occur in many different organisms. Properties of these isoforms are summarized in Table 11.4.

Many of the physiological functions of NO are mediated through activation of soluble guanylate cyclase, a heterodimeric (α/β) heme-containing protein that converts GTP to cGMP. cGMP is a second messenger involved in many signal transduction cascades (p. 530). The inactive form of soluble guanylate cyclase has a penta-coordinate heme, bound by a histidine on the β-subunit. NO binds as the sixth ligand to form a hexa-coordinate heme and causes breakage of the heme–histidine bond, yielding the active, penta-coordinate nitrosyl complex of soluble guanylate cyclase (Figure 11.21). Activation of soluble guanylate cyclase causes up to a 400-fold increase in the rate of cGMP formation. The cGMP levels are regulated by a balance between guanylate cyclase activity and the phosphodiesterases (PDE), particularly PDE5, which is specific for cGMP, that hydrolyze cGMP to 5′-GMP, thus turning off the signal.

NOSI

NOSI is found primarily in skeletal muscle and in neurons of both the central and peripheral nervous systems. It is a soluble enzyme, although it is localized to the membrane via protein–protein interactions with a host of regulatory and localization proteins. It is constitutively expressed, meaning that it is not normally induced at the transcriptional level and is activated by the influx of calcium. In the case of NOSI (and NOSIII, p. 450), calmodulin (p. 493) is not bound to the enzyme at intracellular, basal calcium levels. It requires an influx of calcium to raise the calcium concentration intracellularly, allowing calmodulin to bind, and thus activating NO formation. The amount of NO synthesized in the activated state is very low, that is, picomolar. The NO produced serves as a neurotransmitter in the central and peripheral nervous systems. In skeletal muscle, NO also serves as a mediator of contractile force.

In the central nervous system, NO is produced generally by nNOS in the postsynaptic neuron, but diffuses back across the synapse to the presynaptic neuron. NO synthesis is regulated by the influx of calcium through receptor-dependent channels, that is, following postsynaptic stimulation of *N*-methyl-D-aspartate (NMDA) receptors by the excitatory neurotransmitter glutamate. Guanylate cyclase is activated by NO, producing cGMP, which

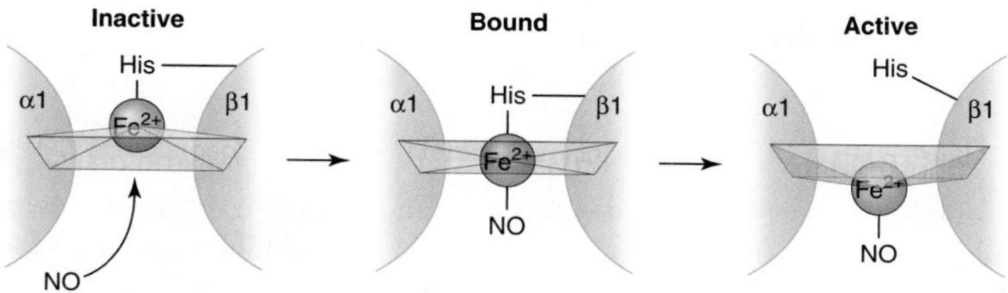

Figure 11.21 Activation of soluble guanylate cyclase by NO.
Redrawn based on figure in Bellamy, T. C. and Garthwaite, J. The receptor-like properties of nitric oxide-activated soluble guanylate cyclase in intact cells. *Mol. Cell. Biochem.* 230: 165, 2002.

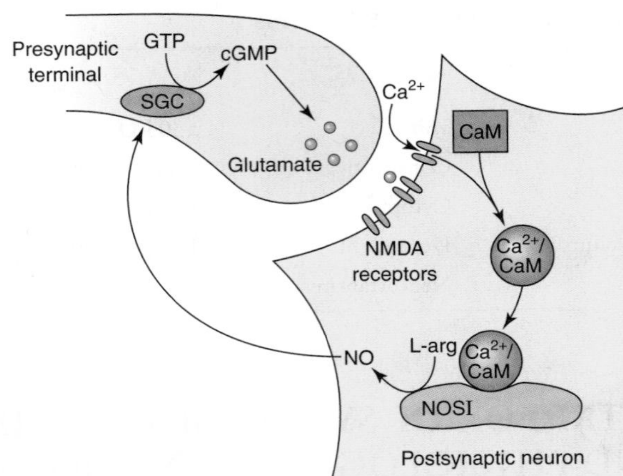

Figure 11.22 NO produced by NOSI in the central nervous system. In this instance, NO is produced by the postsynaptic cell and travels to the presynaptic cell where it activates sGC, creating cGMP that activates PKG. PKG phosphorylates proteins on the neurotransmitter vesicles, leading to more neurotransmitter release and potentiation of the signal.

regulates synthesis of the neurotransmitters norepinephrine and glutamate, thus enhancing NO production (Figure 11.22) in a positive feedback manner. NO has been implicated in neural signaling, neurotoxicity, synaptic plasticity, learning and memory, and perception of pain.

In addition to its oxygenase and reductase domains, NOSI has an *N*-terminal 300–amino acid PDZ domain, which mediates its interaction with other proteins. In the central nervous system, the PDZ domain of NOSI interacts with the postsynaptic density protein PSD-95. The NMDA receptor also interacts with PSD-95, thus bringing NOSI and the NMDA receptor into close proximity, so that NOSI is directly exposed to the calcium entering the ion channel of the activated NMDA receptor.

NO is very different from other known neurotransmitters (p. 942). NO is synthesized on demand and cannot be stored in lipid vesicles, as it freely diffuses across membranes. The action of NO is not terminated by degradation or reuptake but is removed by interaction with its target, which is an intracellular second messenger rather than a membrane-bound receptor.

In the peripheral nervous system, NO is produced by nNOS in myenteric neurons of the gastrointestinal, pulmonary, vascular, and urogenital systems. In the gastrointestinal system, NO is involved in gut motility and control of the pyloric sphincter. NO mediates the main bronchodilator pathway in the human pulmonary system and is involved in neural control of the vascular system, particularly important in cerebral blood flow. In the urogenital system, NO is involved in urethra and bladder control and in penile erection. During sexual excitement, NO is released from nerves adjacent to the blood vessels of the penis. Relaxation of these vessels causes blood to engorge the corpus cavernosum, producing erection. NOSI is activated in these systems by the influx of calcium through voltage-dependent calcium channels (p. 481). It diffuses to a neighboring smooth muscle cell causing activation of soluble guanylate cyclase. The smooth muscle relaxes via the mechanism described below for vascular smooth muscle. Clin. Corr. 11.8 describes the therapeutic manipulation of cGMP levels for treatment of erectile dysfunction.

High levels of NOSI are expressed in the skeletal muscle, where it is involved in mediation of contractile force, innervation of developing muscle, and perhaps glucose uptake. Skeletal muscle relaxation by NO also occurs through the cGMP pathway via the mechanism described in the following for vascular smooth muscle. NOSI is targeted to the sarcolemma of the muscle due to its association with α-syntrophin, a member of the muscle dystrophin complex. NOSI is activated by influx of calcium through voltage-dependent calcium channels, as well as from the sarcoplasmic reticulum.

NOSII

NOSII is found primarily in activated neutrophils and macrophages, astrocytes, and hepatocytes and is involved in the early immune response. It is a soluble enzyme and is induced at the transcriptional level by cytokines, such as interferon-γ, interleukin 1, and tumor

CLINICAL CORRELATION 11.8

Mechanism of Action of Sildenafil

The levels of cyclic nucleotides (cGMP and cAMP) present in the cell are dependent not only on their formation by adenylate or guanylate cyclases but also on their degradation by phosphodiesterases (PDEs), which hydrolyze them to 5′-nucleotides and so play a critical role in the modulation of these second messenger signaling pathways. The inhibition of PDE destruction of cyclic nucleotides would thus prolong the physiological response elicited by the cyclic nucleotides. Sildenafil (Viagra), a drug prescribed for erectile dysfunction in males, is a potent inhibitor of PDE5, which is specific for cGMP and is found primarily in the human corpus cavernosum and in vascular and gastrointestinal smooth muscle. In the penis, NO produces erection via a cGMP-dependent pathway, as discussed in the text. As the cGMP is degraded by PDE5, the vasodilatory signal attenuates and the erection subsides. Sildenafil inhibits PDE5, delaying cGMP degradation and thus prolonging erection.

Sildenafil is highly selective for PDE5 over PDE3, a cAMP-specific PDE involved in the regulation of cardiac contractility. This is important because inhibition of PDE3 increases the incidence of cardiac arrhythmias and decreases long-term mortality in heart failure patients. Interestingly, Sildenafil also inhibits, to a lesser extent, PDE6, which is localized to photoreceptors in the retina, and one of the side effects sometimes reported with Sildenafil use is a blue-green tint in the patient's vision.

Because PDE5 is present in vascular smooth muscle, Sildenafil causes a mild, clinically insignificant reduction in blood pressure due to vasodilation but, more importantly, it greatly exacerbates the effects of nitrates and external NO donors such as nitroglycerin, taken for angina, causing potentially fatal hypotension. These hypotensive effects occur regardless of the sequence and/or timing of the drugs. Sildenafil is therefore contraindicated in patients using nitrates in any form. In addition, Sildenafil is metabolized by both the CYP2C9 and 3A4 pathways. Inhibition of these pathways by other drugs may increase the plasma concentration of Sildenafil, intensifying side effects.

Cheitlin, M. D., Hutter, A. M., Brindis, R. G., Ganz, P., et al. Use of Sildenafil (Viagra) in patients with cardiovascular disease. *J. Am. Coll. Cardiol.* 33: 273, 1999; Corbin, J. D., Francis, S. H., and Webb, D. J. Phosphodiesterase type 5 as a pharmacologic target in erectile dysfunction. *Urology* 60: 4, 2002; and Seftel, A. D. Phosphodiesterase type 5 inhibitor differentiation based on selectivity, pharmacokinetic, and efficacy profiles. *Clin. Cardiol.* 27: I14, 2004.

necrosis factor-α, or endotoxins, such as lipopolysaccharides. Induction occurs over several hours via the inflammatory NFκB pathway. Calmodulin is bound to NOSII under basal physiological conditions, that is, independent of calcium influx, so the enzyme is always activated once it is synthesized. The amount of NO synthesized by NOSII is in the nanomolar range, approximately 1000 times greater than that of either NOSI or NOSIII.

The NO produced by NOSII is a potent cytotoxin, and its principal role is to destroy pathogens engulfed by neutrophils and macrophages. NOSII controls such intracellular pathogens as parasites, for example, plasmodia (malaria), schistosoma (schistosomiasis), leishmania (leishmaniasis), and toxoplasma (toxoplasmosis), as well as microbials, for example, bacteria and mycobacteria (tuberculosis and leprosy), fungi, and even tumor cells. The cellular targets for the NO produced by NOSII are metal-containing heme proteins, such as cytochromes P450, and iron-sulfur proteins, such as aconitase and mitochondrial complexes I and II, all of which are inhibited. NO also reacts with thiol groups and tyrosines, causing nitrosation or oxidation of proteins, and with oxygen or superoxide, forming highly reactive nitrogen oxide compounds, for example, peroxynitrite. These reactive intermediates cause a variety of DNA alterations including strand cleavage and deamination. These modifications thus affect many crucial metabolic enzymes, leading to cytostatic or cytotoxic effects against these pathogens.

Although NO is essential for the antimicrobial function of the macrophage, overproduction of NO by NOSII has been implicated in septic/cytokine-induced circulatory shock. Septic shock is characterized by a systemic inflammatory response to microbial infection in which blood pressure drops precipitously, mediated by the overproduction of NO, causing decreased tissue perfusion and oxygen delivery and, eventually, multiple organ failure. Hypotension in these patients is often refractory to conventional vasoconstrictor drugs. The role of NO in septic shock and therapeutic intervention by NOS inhibitors is discussed in Clin. Corr. 11.9.

In addition to being produced during acute inflammation, NO is generated by NOSII during chronic inflammation. NOSII is present and active in a number of chronic inflammatory conditions including rheumatoid arthritis, Crohn disease, and asthma, where it may potentially exacerbate some situations, although protective effects have also been reported.

CLINICAL CORRELATION 11.9

Nitric Oxide Overproduction in Septic Shock

Although NO is essential in tumoricidal and bactericidal functions of macrophages, overproduction of NO (by NOSII) has been implicated in septic/cytokine-induced circulatory shock in humans. In septic shock, production of too much NO leads to massive systemic vasodilation and a precipitous and often fatal drop in blood pressure, which causes an insufficient blood and oxygen supply to peripheral organs, eventually leading to multiple organ failure. Septic shock is associated with an extremely high mortality rate of 50%–70%; more people die annually from septic shock than from myocardial infarction, stroke, trauma, or lung or breast cancer. NO-induced hypotension in these patients occurs through the mechanism for smooth muscle relaxation, that is, activation of guanylate cyclase. Therapeutic intervention by NOS inhibitors has been investigated for treatment of septic shock. Interestingly, although NOS inhibition successfully and rapidly raises blood pressure and systemic vascular resistance, it also led to a progressive fall in cardiac output and exacerbated organ dysfunction, leading to increased mortality in clinical trials. Thus, NO plays a detrimental role in septic shock, due to its mediation of hypotension, but may also play a beneficial role in survival, perhaps due to localized vasodilation or the antiplatelet aggregation, antileukocyte adhesion, antiapoptotic or antioxidative properties of NO.

Assreuy, J. Nitric oxide and cardiovascular dysfunction in sepsis. *Endocr. Metab. Immune Disord.-Drug Targets* 6: 165, 2006; Cauwels, A. Nitric oxide in shock. *Kidney Int.* 72: 557, 2007; and Vallance, P. Nitric oxide: Therapeutic opportunities. *Fund. Clin. Pharmacol.* 17: 1, 2003.

NOSIII

NOSIII is found primarily in vascular endothelial cells, which line all blood vessels, and cardiac myocytes. It is localized to the membrane due to co-translational myristoylation and posttranslational palmitoylation of the *N*-terminus. Like NOSI, it is constitutively expressed and is activated by the influx of calcium, which binds to calmodulin facilitating its binding to both NOSI and NOSIII for activation. The amount of NO synthesized in the activated state is very low, that is, picomolar, and the NO produced serves as a vasodilator of vascular smooth muscle, both ligand mediated and flow dependent. The NO also serves antithrombotic and anti-inflammatory functions by inhibiting platelet and leukocyte adhesion and aggregation. In addition, NO also inhibits angiotensin II, a vasoconstrictor.

NO synthesis by NOSIII is activated in response to an increase in calcium following the binding of ligands such as acetylcholine, bradykinin, histamine, or insulin or following shear stress, the mechanical force of blood flow on the luminal surface of the vascular endothelium. Increased blood flow velocity thus stimulates calcium release and increased NOSIII activity, leading to vasodilation.

NO diffuses out of the endothelial cell and into adjacent smooth muscle cells where it binds to and activates soluble guanylate cyclase, which makes cGMP (Figure 11.23). cGMP activates protein kinase G, a cGMP-dependent kinase that phosphorylates a variety of channels, including L-type calcium channels, and receptors, all leading to inhibition of calcium influx into the smooth muscle cell. Vasoconstriction in vascular smooth muscle requires the influx of calcium, which binds to calmodulin that, in turn, activates muscle light chain kinase (MLCK). MLCK then phosphorylates myosin light chain (MLC), leading to cross-bridge formation between the myosin heads and the actin filament, resulting in contraction

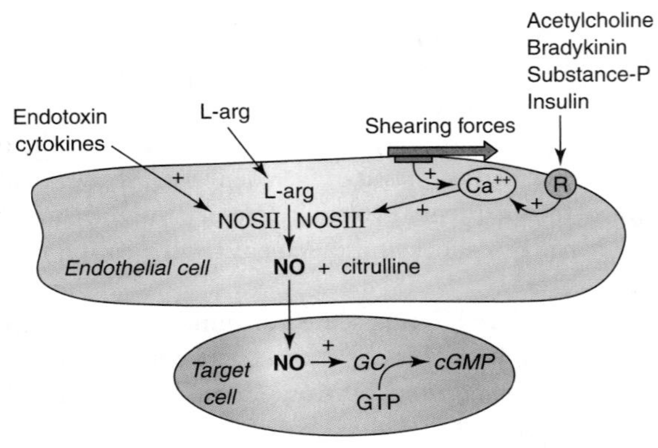

Figure 11.23 NO produced by NOSIII in the endothelial cell. Calcium concentration in the endothelial cell rises due to ligand binding or shear stress leading to activation of NO production by NOSIII. The NO diffuses across the membrane of the endothelial cell to a smooth muscle cell where it activates sGC.
Redrawn based on figure by Klabunde, R. E. Cardiovascular Physiology Concepts. This figure can be accessed at http://cvphysiogy.com/blood%20Flow/BFO11.htm.

CLINICAL CORRELATION 11.10

History and Biological Effects of Nitroglycerin

Nitroglycerin, more correctly called glyceryl trinitrate (GTN), was first synthesized by Ascanio Sobrero in 1846. Because GTN was extremely explosive and many fatal accidents occurred, Alfred Nobel developed a procedure in 1864 to stabilize GTN by mixing it with diatomaceous earth. The fortune he made from dynamite is the basis of the Nobel Prizes, awarded yearly in Sweden for outstanding contributions to the fields of physics, chemistry, physiology or medicine, literature, economics, and world peace. Shortly thereafter, in 1867, Sir Thomas Lauder Brunton first reported the use of inhaled amyl nitrate in his patients with chest pains, and in 1879, William Murrell published an article describing the use of the longer-lasting and more convenient compound GTN to better alleviate the pain of angina pectoris. GTN works by dilating the arteries of the heart, thus allowing more oxygenation of the tissue. Ironically, Murrell died in 1912 from untreated heart failure having never taken GTN, and Nobel refused to take it when prescribed by his doctor for heart disease.

It was not until 1987, over a century after Murrell's publication, that nitric oxide released from GTN metabolism was postulated to be the active agent in the dilation of the coronary blood vessels. It was not until 15 years later that the mechanism of nitric oxide release from nitroglycerin was elucidated. The mitochondrial aldehyde dehydrogenase (ALDH2) metabolizes GTN to 1,2-glycerol dinitrite and nitrite. The nitrite is further reduced to form NO. The mechanism of this reduction is unclear, but ALDH2 may also be involved in this step, and members of the mitochondrial electron transport chain have also been shown to reduce nitrite to NO. Alternatively, a major mechanism of nitrite reduction may be through direct reaction with deoxygenated hemoglobin or myoglobin.

When used over a long period of time for chronic conditions, GTN gradually loses its efficacy. Several mechanisms have been proposed to explain patient tolerance to this drug. Impaired biotransformation of GTN by ALDH2, which occurs on depletion of an ALDH2 reductant, or an increase in oxidative stress due to impaired mitochondrial metabolism are both possibilities.

Beretta, M, Gruber, K., Kollau, A., Russwurm, M., et al. Bioactivation of nitroglycerin by purified mitochondrial and cytosolic aldehyde dehydrogenases. *J. Biol. Chem.* 283: 17873, 2008; Chen, Z. Q., Zhang, J., and Stamler, J. S. Identification of the enzymatic mechanism of nitroglycerin bioactivation. *Proc. Natl. Acad. Sci., USA* 99: 8306, 2002; Marsh, N., and Marsh, A. A short history of nitroglycerine and nitric oxide in pharmacology and physiology. *Clin. Exp. Pharm. and Physio.* 27: 313, 2000; Napoli, C., and Ignarro, L. J. Nitric oxide-releasing drugs. *Annu. Rev. Pharmacol. Toxicol.* 43: 97, 2003; and Wells, W. From explosives to the gas that heals. Written for "Beyond Discovery: The Path from Research to Human Benefit," 2002. This article may be accessed at http://www.beyonddiscovery.org/content/view.article.asp?a=318.

(p. 978). Inhibition of calcium influx leads to decreased calmodulin stimulation of MLCK and decreased phosphorylation of MLC, causing diminished development of smooth muscle tension and, thus, vasodilation. Increased cGMP also directly causes MLC dephosphorylation by activation of MLC phosphatase. As with NOSI, cGMP levels are regulated by a balance between guanylate cyclase activity and the phosphodiesterases, which hydrolyze cGMP to 5′-GMP and thus turn off protein kinase G activity and vasodilation. Several therapeutic treatments are based on the administration of NO, either directly or in a precursor form. Clin. Corr. 11.10 discusses the history and use of nitroglycerin for vasodilation, and Clin. Corr. 11.11 discusses the use of direct NO administration as a therapeutic treatment.

NOSIII has a unique *N*-terminal sequence that undergoes myristoylation at one site and palmitoylation at two other sites within this sequence, localizing the enzyme to the membrane where it is targeted to plasmalemma caveolae. The caveolae are small invaginations of the plasma membrane, characterized by the presence of proteins called caveolins that serve to organize and attach signaling molecules such as receptors, *G*-proteins, and NOS to plasma membranes. At low cytoplasmic calcium concentrations, caveolin-1 binds to and inhibits NOSIII, maintaining it in an inactive state. On calcium influx, calmodulin competitively displaces the caveolin from NOSIII, resulting in activation of NOS synthesis.

Also localized to the caveolae are several other proteins important for NOSIII activity. The cationic amino acid transporter CAT-1, present in caveolae, is involved in the uptake of L-arginine, thus ensuring a supply of substrate for NO synthesis. NOSIII interacts with porin, a voltage-dependent anion/cation channel, which localizes NOSIII near a source of calcium influx. The bradykinin B2 receptor is also present, and it binds to and inactivates NOSIII. On stimulation with bradykinin, this complex dissociates, and NOSIII becomes activated. Localization to the caveolae is an important regulator of NOSIII activity; if palmitoylation is inhibited, NOSIII is not found in the caveolae, and synthesis of NO does not occur in those cells. Palmitoylation is a reversible process that is influenced by some agonists and is essential for membrane localization; cellular relocalization resulting in deactivation may occur on depalmitoylation.

CLINICAL CORRELATION 11.11

Therapeutic Uses of Inhaled Nitric Oxide

Systemically delivered nitric oxide causes systemic hypotension, severely limiting its usefulness as a therapeutic agent. On the other hand, nitric oxide that is inhaled would theoretically relax the pulmonary vasculature, but any excess NO that reached the bloodstream would be scavenged by oxy-hemoglobin, thereby preventing systemic vasodilation. Indeed, studies have shown that NO inhalation increased the systemic oxygenation of hypoxemic newborns with persistent pulmonary hypertension of the newborn (PPHN). Inhaled NO also decreased the incidence and mortality of bronchopulmonary dysplasia (BPD) in premature infants. In adults, chronic obstructive pulmonary disease (COPD) has been successfully treated with a combination of supplemental oxygen and NO; such a regimen reduced pulmonary arterial pressure and pulmonary vascular resistance and increased cardiac output in treated patients.

Many patients do not respond to inhaled NO therapy, however. In addition, the effects of NO are transient and disappear when treatment is discontinued, requiring continuous therapy. Approximately 70% of inhaled NO is excreted in the urine in the form of nitrate within 48 h. Several studies have described the use of Sildenafil, a drug used to treat erectile dysfunction, to augment or prolong the vasodilatory effects of inhaled NO. As discussed in Clin. Corr. 11.8, Sildenafil inhibits the breakdown of cGMP by inhibiting the cGMP-specific phosphodiesterase, PDE5, an isoform highly expressed in penile, retinal, and lung tissue. In patients with pulmonary hypertension, inhaled-NO-mediated pulmonary vasodilation was both augmented and prolonged by cotreatment with Sildenafil.

Interestingly, although inhaled NO has no effect on systemic blood pressure, other systemic effects have been reported, including platelet and leukocyte inhibition, which result in decreased thrombosis and reduced ischemia/reperfusion injury. It is believed that these systemic effects are mediated by blood-borne low- and high-molecular-weight thiols, which react with NO and transport it in a stable form throughout the body. *S*-nitrosylated hemoglobin and serum albumin, as well as other nitrosamines, iron-nitrosyls, and nitrated lipids are proposed mediators.

Bloch, K. D., Ichinose, F., Roberts, J. D., Jr., and Zapol, W. M. Inhaled NO as a therapeutic agent. *Cardiovascular Res.* 75: 339, 2007; and Griffiths, M. J. D. and Evans, T. W. Inhaled nitric oxide therapy in adults. *N. Engl. J. Med.* 353: 2683, 2005.

Many disease processes, such as hypertension, hypercholesterolemia, and diabetes lead to endothelial dysfunction, which is characterized by compromise of the normal functions of the endothelial cells, including vasodilation and mediation of platelet adhesion. One of the major hallmarks of endothelial dysfunction is the decreased bioavailability of NO, due to decreased function of NOSIII. Clin. Corr. 11.12 discusses some possible roles of NO in and mechanisms of endothelial dysfunction.

CLINICAL CORRELATION 11.12

The Role of eNOS in Endothelial Dysfunction

Nitric oxide derived from eNOS in the vascular endothelial cells is responsible for regulation of vascular tone, inhibition of platelet and leukocyte aggregation and adhesion, decreasing expression of proinflammatory genes, and limiting vascular smooth muscle cell proliferation. Impairment of endothelial function, endothelial dysfunction, is characterized by reduced NO synthesis leading to increased vasospasm, vascular inflammation, and thrombosis and is very often seen concomitant with other cardiovascular risk factors, such as diabetes, hypertension, atherosclerosis, or hypercholesteremia. It is believed that vascular oxidative stress, due to increased production of radical oxygen species, particularly superoxide, contributes to endothelial dysfunction due, in part, to reaction of superoxide with NO to form peroxynitrite. Not only is less NO available for its protective biological function, but peroxynitrite is a powerful oxidant and nitrating agent, which can damage protein, DNA, and lipid molecules. Covalent nitration of tyrosine residues in proteins by reactive oxygen species is often used as a marker of oxidative stress. Peroxynitrite is also responsible for depletion of tetrahydrobiopterin, which leads to uncoupling of eNOS electron transfer, causing eNOS to produce superoxide rather than nitric oxide, further exacerbating oxidative stress. In the absence of NO and in the presence of oxidative stress, several proinflammatory and proatherosclerotic pathways are activated; this is the primary molecular basis of endothelial dysfunction.

Treatment with antioxidant drugs, such as vitamin C or E, does not restore bioavailable NO levels, nor does it improve cardiovascular outcome, perhaps because the vitamins do not scavenge the superoxide before it reacts with NO. One of the most effective class of drugs is the statins, which function both by decreasing oxidative stress, via inhibition of NADPH oxidase activity, a prominent source of superoxide production, and by increasing NO production by eNOS, via both protein expression and phosphorylation. Angiotensin-converting-enzyme (ACE) inhibitors and angiotensin II receptor agonists also effectively improve cardiovascular outcome. These drugs decrease the formation/action of angiotensin II, a protein that activates the NADPH oxidases, thus reducing oxidative stress.

Forstermann, U. Oxidative stress in vascular disease: Causes, defense mechanisms, and potential therapies. *Nat. Clin. Pract. Cardiovasc. Med.*, 5, 338, 2008; and Liao, J. K. Beyond lipid lowering: The role of statins in vascular protection. *Int. J. Cardiol.* 86: 5, 2002.

All of the NOSs produce NO, which serves such diverse functions as neurotransmission, vasodilation, muscle relaxation, and cytotoxicity. The consequences of NO can be beneficial, as in maintenance of vascular tone, acquisition of memory and learning, and protection against pathogen invasion, as well as detrimental, as with septic shock or chronic inflammatory and mutagenic responses. The net effect of the NO produced is dependent on where it is being produced, regulation of its production, how much is being produced, and which other reactive species are present.

Bibliography

General References on Drug Metabolism and Cytochrome P450

Belle, D. J., and Singh, H. Genetic factors in drug metabolism. *Am. Fam. Physician* 77: 1553, 2008.

Coon, M. J. Multiple oxidants and multiple mechanisms in cytochrome P450 catalysis. *Biochem Biophys. Res. Commun.* 312: 163, 2003.

Guengerich, F. P. Cytochrome P450 and chemical toxicology. *Chem. Res. Toxicol.* 21: 70, 2008.

Ingelman-Sundberg, M., Sim, S. C., Gomez, A., and Rodriguez-Antona, C. Influence of cytochrome P450 polymorphisms on drug therapies: pharmacogenetic, pharmacoepigenetic, and clinical aspects. *Pharmacol. Ther.* 116: 496, 2007.

Iyanagi, T. Molecular mechanism of phase I and phase II drug-metabolizing enzymes: Implications for detoxification. Int. *Rev. Cytol.* 260: 35, 2007.

Lewis, D. F. V. *Guide to Cytochromes P450 Structure and Function.* New York: Taylor & Francis, 2001.

Lewis, D. F. 57 varieties: the human cytochromes P450. *Pharmacogenomics* 5: 305. 2004.

Nelson, D. R. Website: http://drnelson.utmem.edu/CytochromeP450.html.

Zanger, U. M., Turpeinen, M., Klein, K., and Schwab M. Functional pharmacogenetics/genomics of human cytochromes P450 involved in drug biotransformation. *Anal. Bioanal. Chem.* 392:1093, 2008.

Substrate Specificity of Cytochromes P450: Physiological Functions

Cunningham, F. G., MacDonald, P. C., Gant, N. F., Leveno, K. J., et al. The placental hormones. In *Williams Obstetrics.* Stamford, CT: Appleton & Lange, 20th ed., 1997, Chap. 6, 125.

Estabrook, R. W., Cooper, D. Y., and Rosenthal, O. The light reversible carbon monoxide inhibition of the steroid C21-hydroxylase system of the adrenal cortex. *Biochem. Zeit.* 338: 741, 1963.

Hsu, M. H., Savas, U., Griffin, K. J., and Johnson, E. F. Human cytochrome P450 family 4 enzymes: Function, genetic variation and regulation. *Drug Metab. Rev.* 39: 515, 2007.

Kroetz, D. L.. and Zeldin, D. C. Cytochrome P450 pathways of arachidonic acid metabolism. *Curr. Opin. Lipidology* 13: 273, 2002.

Masters, B. S. S., Muerhoff, A. S., and Okita, R. T. Enzymology of extrahepatic cytochromes P450. In Guengerich, F. P. (Ed.), *Mammalian Cytochromes P450.* Boca Raton, FL: CRC Press, 1987, Chap. 3, 107.

Miller, W. L. Steroidogenic enzymes. *Endocr. Dev.* 13: 1, 2008.

New, M. I. Inborn errors of adrenal steroidogenesis. *Mol. Cell Endocrinol.* 211: 75, 2003.

Rendic, S. Summary of information on human CYP enzymes: Human P450 metabolism data. *Drug. Met. Rev.* 34: 83, 2002.

Stratakis, C. A., and Bossis, I. Genetics of the adrenal gland. *Rev. Endocr. Metab. Disord.* 5: 53, 2004.

Cytochrome P450 Induction and Inhibition

Cascorbi, I. Pharmacogenetics of cytochrome P450 2D6: Genetic background and clinical implication. *Eur. J. Clin. Invest.* 33: 17, 2003.

Guengerich, F. P. Cytochromes P450, drugs, and diseases. *Mol. Interv.* 3: 194, 2003.

Handschin, C., and Meyer, U.A. Induction of drug metabolism: the role of nuclear receptors. *Pharmacol. Rev.* 55: 649. 2003.

Ingelman-Sundberg, M. Pharmacogenetics of cytochrome P450 and its applications in drug therapy: The past, present and future. *Trends Pharmacol. Sci.* 25: 193, 2004.

Schwarz, U. I. Clinical relevance of genetic polymorphisms in the human CYP2C9 gene. *Eur. J. Clin. Invest.* 33: 23, 2003.

Shimada, T., and Fujii-Kuriyama, Y. Metabolic activation of polycyclic aromatic hydrocarbons to carcinogens by cytochromes P450 1A1 and 1B1. *Cancer Sci.* 95: 1, 2004.

Schuster, I., and Bernhardt, R. Inhibition of cytochromes p450: Existing and new promising therapeutic targets. *Drug Metab. Rev.* 39: 481, 2007.

Tompkins, L. M., and Wallace, A. D. Mechanisms of cytochrome P450 induction. *J. Biochem. Mol. Toxicol.* 21: 176, 2007.

Waxman, D. J. P450 gene induction by structurally diverse xenochemicals: Central role of nuclear receptors CAR, RXR, PPAR. *Arch Biochem. Biophys.* 369: 11, 1999.

Zanger, U. M., Raimundo, S., and Eichelbaum, M. Cytochrome P450 2D6: overview and update on pharmacology, genetics, and biochemistry. *Naunyn Schmiedebergs Arch. Pharmacol.* 369: 23, 2004.

Zhou, S., Chan, E., Pan, S. Q., Huang, M., and Lee, E. J. Pharmacokinetic interactions of drugs with St John's wort. *J. Psychopharmacol.* 18: 262, 2004.

Cytochrome P450 Reductase

Fluck, C. E., Tajima, T., Pandey, A. V., Arlt, W., et al. Mutant P450 oxidoreductase causes disordered steroidogenesis with and without Antley–Bixler syndrome. *Nat. Genet.* 36: 228, 2004.

Hart, S. N., and Zhong, X. B. P450 oxidoreductase: genetic polymorphisms and implications for drug metabolism and toxicity. *Expert Opin. Drug Metab. Toxicol.* 4: 439, 2008.

Masters, B. S., and Marohnic, C. C. Cytochromes P450: A family of proteins and scientists understanding their relationships. *Drug Metab. Rev.* 38: 209, 2006.

Scott, R. R., and Miller, W. L. Genetic and clinical features of P450 oxidoreductase deficiency. *Horm. Res.* 69: 266, 2008.

Biochemistry and Physiology of Nitric Oxide Formation

Alderton, W. K., Cooper, C. E., and Knowles, R. G. Nitric oxide synthases: structure, function and inhibition. *Biochem. J.* 357: 593, 2001.

Bredt, D. S. Nitric oxide signaling specificity the heart of the problem. *J. Cell Sci.* 116: 9, 2003.

Coggins, M. P., and Bloch, K. D. Nitric oxide in the pulmonary vasculature. Arterioscler. *Thromb. Vasc. Biol.* 27: 1877, 2007.

Dudzinski, D. M., and Michel, T. Life history of eNOS: Partners and pathways. *Cardiovasc. Res.* 75: 247, 2007.

Esplugues, J. V. NO as a signalling molecule in the nervous system. *Brit. J. Pharmacol.* 135: 1079, 2002.

Gorren, A. C., and Mayer, B. Nitric-oxide synthase: A cytochrome P450 family foster child. *Biochim. Biophys. Acta* 1770: 432, 2007.

Ignarro, L. J., Buga, G. M., Wood, K. S., Byrns, R. E., and Chaudhuri, G. Endothelium-derived relaxing factor produced and released from artery and vein is nitric oxide. *Proc. Natl. Acad. Sci. USA* 84: 9265, 1987.

Khan, M. T., and Furchgott, R. F. Additional evidence that endothelium-derived relaxing factor is nitric oxide. In Rand, M. J., and Raper, C. (Eds.), *Pharmacology*. Amsterdam: Elsevier, 1987, 341.

Kleinert, H., Schwarz, P. M., and Forstermann, U. Regulation of the expression of inducible nitric oxide synthase. *Biol. Chem.* 384: 1343, 2003.

Liaudet, L., Soriano, F. G., Yaffe, M. B., and Fink, M. P. Biology of nitric oxide signaling. *Crit. Care Med.* 28: N37, 2000.

Marin, E., and Sessa, W. C. Role of endothelial-derived nitric oxide in hypertension and renal disease. *Curr. Opin. Nephrol. Hypertens.* 16: 105, 2007.

Masters, B. S. S. Nitric oxide synthases: Why so complex? *Annu. Rev. Nutrition.* 14: 131, 1994.

Murphy, S., and Gibson, C. L. Nitric oxide, ischaemia and brain inflammation. *Biochem. Soc. Trans.* 35: 1133, 2007.

Palmer, R. M. J., Ferrige, A. G., and Moncada, S. Nitric oxide release accounts for the biological activity of endothelium-derived relaxing factor. *Nature* 327: 524, 1987.

Raman, C. S., Martasek, P., and Masters, B. S. S. Structural themes determining function in nitric oxide synthases. In *The Porphyrin Handbook*, Vol. 4. San Diego, CA: Academic Press, , 2000, 293.

Roman, L. J., Martasek, P., and Masters, B. S. S. Intrinsic and extrinsic modulation of nitric oxide synthase activity. *Chem. Rev.* 102: 1179, 2002.

Schulz, E., Jansen, T., Wenzel, P., Daiber, A. and Münzel T. Nitric oxide, tetrahydrobiopterin, oxidative stress, and endothelial dysfunction in hypertension. *Antioxid. Redox Signal.* 10: 1115, 2008.

Stuehr, D. J., Santolini, J., Wang, Z. Q., Wei, C. C., and Adak, S. Update on mechanism and catalytic regulation in the NO synthases. *J. Biol. Chem.* 279: 36167, 2004.

Vallance, P. Nitric oxide: Therapeutic opportunities. *Fund. Clin. Pharmacol.* 17:1, 2003.

Walford, G., and Loscalzo, J. Nitric oxide in vascular biology. *J. Thromb. Haemostasis* 1: 2112, 2003.

Website: http://metallo.scripps.edu/PROMISE/NOS.html

Key Terms

cytochrome P450
heme
porphyrin
integral membrane protein
oxygen metabolism
cytochrome P450 reductase
flavoprotein
heme protein
monooxygenase
NADPH
endoplasmic reticulum
flavin
electron transfer
steroid synthesis
cortisol
fatty acid metabolism
testosterone
hydrophobic
DHEA
immune response
polymorphisms
drug metabolism
xenobiotic metabolism
carcinogen metabolism
drug–drug interactions
gaseous signaling
nitric oxide
nitric oxide synthase
arginine
citrulline
tetrahydrobiopterin
calcium
calmodulin
cGMP
guanylate cyclase
protein kinase G
vasodilation
phosphorylation
neurotransmission

Questions • CAROL N. ANGSTADT

Multiple-Choice Questions

1. All of the following are correct about a molecule designated as a cytochrome P450 except it
   A. contains a heme as a prosthetic group.
   B. catalyzes the hydroxylation of a hydrophobic substrate.
   C. may accept electrons from a substance such as NADPH.
   D. undergoes a change in the heme iron on binding a substrate.
   E. comes from the same gene family as all other cytochromes P450.

2. Flavoproteins are usually intermediates in the transfer of electrons from NADPH to cytochrome P450 because
   A. NADPH cannot enter the membrane.
   B. flavoproteins can accept two electrons from NADPH and donate them one at a time to cytochrome P450.
   C. they have a more negative reduction potential than NADPH so accept electrons more readily.
   D. production of NADPH in the cytosol is the only source of electrons.
   E. they contain iron–sulfur centers.

3. NADPH-cytochrome P450 reductase
   A. uses both FAD and FMN as prosthetic groups.
   B. binds to cytochrome P450 by strong hydrophobic interactions.
   C. requires an iron–sulfur center for activity.
   D. always passes its electrons to cytochrome b_5.
   E. can use NADH as readily as NADPH.

4. Reactions after the 21-hydroxylation leading to cortisol and androgens occur in mitochondria which require NADPH-adrenodoxin reductase and adrenodoxin; NADPH-adrenodoxin reductase
   A. contains both FAD and FMN.
   B. passes its electrons to a protein with iron–sulfur centers.
   C. is an integral protein of the membrane.
   D. reacts directly with cytochrome P450.
   E. reacts directly with cytochrome b_5.

5. Genetic polymorphism in genes for cytochromes P450 could cause
A. an individual to metabolize poorly certain drugs.
B. an individual to metabolize certain drugs more rapidly than normal.
C. a specific ethnic group to experience more effects with certain drugs.
D. all of the above.
E. none of the above.

6. In the presence of active nitric oxide synthase but absence of substrate,
A. electrons cannot be transferred to heme.
B. calmodulin cannot bind Ca^{2+}.
C. the oxygenase and reductase domains dissociate.
D. cGMP production is increased.
E. superoxide and/or hydrogen peroxide will be formed.

Questions 7 and 8: Acetaminophen, an analgesic and antipyretic, is available alone or as a component of many nonprescription medications. Acetaminophen overdose, which can lead to liver damage, is a significant health problem. Consumption of alcoholic beverages can exacerbate the problem, especially if the acetaminophen is taken several hours post-alcohol intake. Acetaminophen, in normal doses, is nontoxic because it is metabolized using normal mechanisms for increasing water solubility.

7. Normally, acetaminophen is primarily metabolized by
A. oxidation by cytochrome P450 (CYP2E1).
B. conjugation with glucuronide or sulfate.
C. conjugation with glutathione.
D. addition of an –OH to the ring.
E. reduction with NADPH.

8. Alcohol is both an inducer and substrate for CYP2E1. Alcohol taken at the same time as acetaminophen may protect the liver from injury. Which of the following is correct?
A. CYP2E1 metabolizes alcohol but not acetaminophen.
B. CYP2E1 is increased and converts acetaminophen to a nontoxic form.
C. Taken at the same time, alcohol and acetaminophen compete for CYP2E1 so less toxic metabolite is formed.
D. With the longer time gap, only alcohol reacts with CPY2E1.
E. None of the above.

Questions 9 and 10: Because herbal remedies are not prescription drugs, many patients fail to inform their physicians that they are using such products. A patient was told by his physician that his blood cholesterol levels were not responding to the usually effective statin drug he was taking. The patient was also taking St. John's wort, an herbal agent, to improve his mood. St. John's wort induces a cytochrome P450 (CYP3A4) activity. Statins are one of the many drugs that are metabolized by CYP3A4.

9. The induction of cytochromes P450
A. occurs only by exogenous compounds.
B. occurs only at the transcriptional level.
C. necessarily results from increased transcription of the appropriate mRNA.
D. necessitates the formation of an inducer-receptor protein complex.
E. may occur by posttranscriptional processes.

10. Statins could be considered xenobiotics (exogenous substances metabolized by the body). Many xenobiotics are oxidized by cytochromes P450 in order to
A. make them carcinogenic.
B. increase their solubility in an aqueous environment.
C. enhance their deposition in adipose tissue.
D. increase their pharmacological activity.
E. all of the above.

Questions 11 and 12: Some patients show profound hypotension after abdominal surgery complicated by bacterial infections that produce endotoxins. Such hypotension is often refractory to treatment with conventional vasoconstrictor drugs. This hypotension may be caused by an overproduction of nitric oxide by the induced form of nitric oxide synthase (NOS). Administration of NOS inhibitors specific to this form might be an appropriate treatment for such patients.

11. Nitric oxide
A. is formed spontaneously by a reduction of NO_2.
B. is synthesized only in macrophages.
C. is synthesized from arginine.
D. acts as a potent vasoconstrictor.
E. has three isoforms.

12. Nitric oxide synthase
A. catalyzes a dioxygenase reaction.
B. is similar to cytochromes P450 in binding zinc and tetrahydrobiopterin.
C. accepts electrons from NADH.
D. uses a flow of electrons from NADPH to FAD to FMN to heme-iron.
E. is inhibited by Ca^{+2}.

Problems

13. Phenobarbital is a potent inducer of cytochrome P450. Warfarin, an anticoagulant, is a substrate for cytochrome P450 so the drug is metabolized more rapidly than normal. If phenobarbital is given to a patient, with no change in warfarin dosage, what would happen? What would happen if the warfarin dosage were adjusted for a proper response, and then phenobarbital withdrawn without adjusting the warfarin dosage?

14. Congenital adrenal hyperplasia (CAH) occurs because of a deficiency in CYP21A2, a 21-hydroxylase of endoplasmic reticulum. The disease results in decreased cortisol, increased ACTH, increased androgenic hormones, and problems related to salt balance. Explain why these should be so.

Answers

1. **E** Several gene families are known. A: All cytochromes are heme proteins. B: The types of substrates are hydrophobic. It is classified as a monooxygenase. C: This is the usual electron donor. D: The change from hexa- to penta-coordinated gives the compound a more positive reduction potential.
2. **B** Heme can accept only one electron at a time whereas NADPH always donates two at a time. A: NADPH passes only electrons; it does not have to enter the membrane. C: If this were true, the flow of electrons would not occur in the way it does. D: NADPH is required for both endoplasmic reticulum and mitochondrial mediated cytochrome P450 reactions. E: Iron–sulfur centers play a role in some, but not all, systems.
3. **A** This enzyme is one of two mammalian proteins known to do so. B: The binding is electrostatic. C: Some reductases do so but not this one. D: Only certain reactions catalyzed by the enzyme do. E: There are NADH-dependent reductases, but they are different enzymes.
4. **B** The iron–sulfur protein is adrenodoxin. A: It has only FAD. C: It is weakly associated with the membrane. D and E: Adrenodoxin acts as an electron shuttle between the reductase and heme of cytochrome P450.
5. **D** A: This occurs if the allelic variants code for inactive or less active proteins. B: If there is gene duplication, an individual may have larger amounts of active protein. C: There may be greater expression of specific allelic variants in specific ethnic populations.
6. **E** O_2 is activated leading to these toxic products. A: Superoxide results from the transfer. B and C: These do not happen. D: NO increases cGMP, and NO is not formed in absence of substrate.
7. **B** These are common mechanisms for increasing water solubility. A: This occurs but normally in small amounts; the product, NAPQ1, is the toxic metabolite. C: Glutathione is conjugated with NAPQ1 and renders it nontoxic. D and E: These do not occur.
8. **C** The competition lowers the production of NAPQ1, the toxic metabolite. A: Both are substrates. B: CYP2E1 converts acetaminophen to the toxic metabolite. D: Alcohol is metabolized more rapidly so the CYP2E1 it has induced is available to metabolize acetaminophen.
9. **E** There may be a stabilization of mRNA or decrease in the degradation of the protein. A: Both endogenous and exogenous substances can induce cytochromes P450. B and C: Transcriptional modification is only one of the mechanisms of induction. D: This has been shown with induction of some compounds but not all.
10. **B** Xenobiotics oxidized by cytochrome P450 are usually highly lipophilic but must be excreted in the aqueous urine or bile. A: This may happen but is certainly not the purpose. C: They do that prior to oxidation. D: Oxidation tends to reduce pharmacological activity.
11. **C** The other product is citrulline. A: It is formed from arginine. B: One of the isoforms of NO synthase has been found in macrophages, but neuronal and endothelial isoforms also exist. D: Nitric oxide is a vasodilator. E: Three isoforms of nitric oxide synthase have been identified but nitric oxide is a specific compound.
12. **D** This is one of two mammalian enzymes that use both FAD and FMN. A: The reaction is a monooxygenation. B: Cytochromes do not bind these. C: The donor is NADPH. E: The system requires Ca^{+2}-calmodulin, at least the neuronal and endothelial isoforms.
13. If warfarin is metabolized and cleared more rapidly by cytochrome P450, its therapeutic efficiency is decreased. At the same dosage, it will be less effective as an anticoagulant. If the phenobarbital is withdrawn, the cytochrome P450 levels will decrease with time to the uninduced level. With no change in the warfarin level, eventually there will be an increased possibility of hemorrhaging.
14. This enzyme converts 17-OH-progesterone to 11-deoxycortisol, the precursor of cortisol. It also converts progesterone to 11-deoxycorticosterone, a precursor of aldosterone which regulates salt balance. The decreased cortisol leads to an increased secretion of ACTH because of a decrease in feedback control. Prolonged elevated ACTH causes adrenal hyperplasia and an increased production of the androgenic hormones. Abnormalities in sexual development are common in infants with CAH.

12

Biological Membranes: Structure, Receptors, and Solute Transport

Thomas M. Devlin

Professor Emeritus, Drexel University College of Medicine

Key Concepts

- Major biological membrane lipids are glycerophospholipids, sphingolipids, and cholesterol. Specific lipid composition varies with the source of the membrane.
- In aqueous medium, amphipathic lipids can form spheres (micelles) and vesicles (liposomes), which have a membrane consisting of a bilayer of randomly distributed lipids. Polar groups are in contact with the aqueous medium. Biological membranes have a similar structure.
- Membranes proteins are classified as integral (tightly bound) or peripheral (easily dissociated). Integral proteins have a single segment or multiple amino acid domains spanning the lipid bilayer. Some form homo- or hetero-oligomeric structures. Peripheral proteins have various means of attachment to the surface of the bilayer. Protein composition varies depending on function of the membrane.
- Lipids and proteins readily diffuse laterally in the bilayer but not transversely. Biological membranes have a chemical asymmetry. There are lipid–protein microdomains in membranes with specific roles.
- Membrane protein receptors interact with growth factors, hormones, and other small molecules initiating a cellular response. Some receptors are ion channels, while others initiate enzymatic reactions or intracellular protein–protein interactions.
- Specific protein structures in membranes facilitate translocation across the membrane of non-diffusable ions and compounds. They are classified as protein channels/pores or transporters. Transporters facilitate translocation without input of metabolic energy. Others require energy from hydrolysis of ATP or a $[Na^+]$ or $[H^+]$ gradient for translocation of a substrate.
- Some toxins form pores in membranes. Ionophores facilitate movement of ions across membranes.

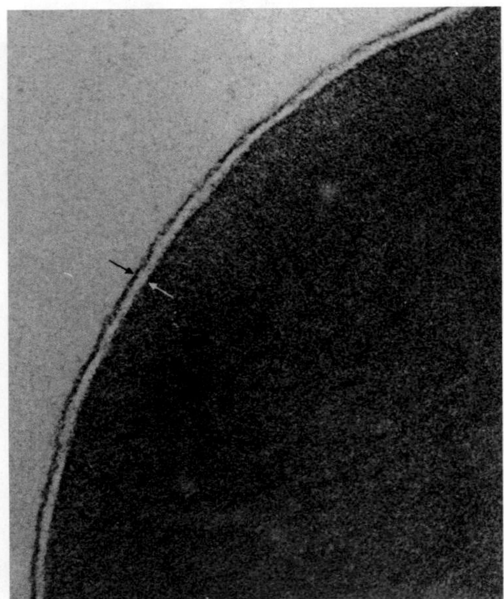

Figure 12.1 Electron micrograph of the erythrocyte plasma membrane showing the trilaminar appearance. A clear space separates the two electron-dense lines. Electron microscopy has demonstrated that the inner dense line is frequently thicker than the outer line. Magnification about 150,000×.
Courtesy of J. D. Robertson, Duke University, Durham, NC.

12.1 • INTRODUCTION

Membranes of eukaryotic or prokaryotic cells have the same classes of chemical components, are similar in structural organization, and have many properties in common. They have a trilaminar appearance when viewed by electron microscopy (Figure 12.1), with two dark bands on each side of a light band. Most mammalian membranes have a width of 7–10 nm and a dense inner layer often thicker than an outer dense layer. Intracellular membranes are usually thinner than plasma membranes. Membranes have a chemical asymmetry and their surfaces are dotted with protruding globular proteins. They are dynamic structures that permit cells and subcellular structures to adjust their shape and to change position. The lipid is in a fluid state, in which both proteins and lipids are able to diffuse and interact.

Cellular membranes control the composition of space they enclose by excluding a variety of ions and molecules and by the presence of selective transport systems for transmembrane movement of substances. Concentrations of substances in cellular compartments are modulated by controlling translocation of substrates, cofactors, and ions across membranes. Extracellular regulators of cellular function bind to protein receptors on plasma membranes and their signal is transmitted from the membrane receptors to the appropriate metabolic pathway. Plasma membranes have a role in cell–cell recognition, in maintenance of cell shape, and in cell locomotion.

12.2 • CHEMICAL COMPOSITION OF MEMBRANES

Lipids and proteins are the major components of all membranes. Their concentrations vary greatly between membranes (Figure 12.2). Proteins are responsible for the function of membranes, and depending on the source, different membranes have very specific complements of proteins. A small amount of various polysaccharides, as either glycoproteins or glycolipids, are present in membranes but there is no free carbohydrate.

Lipids Are Major Components of Membranes

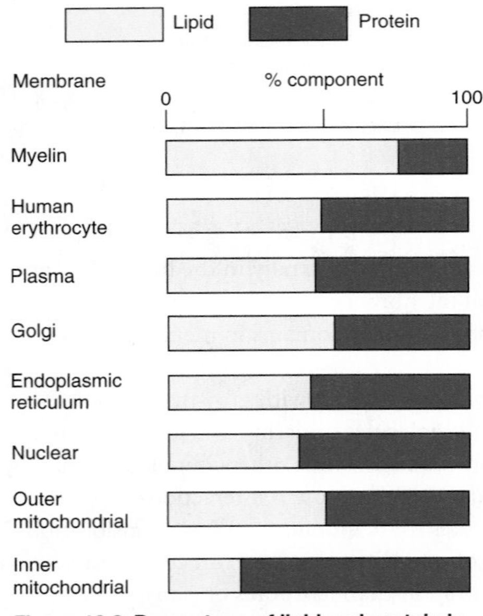

Figure 12.2 Percentage of lipid and protein in various cellular membranes. Values are for rat liver, except for the myelin and human erythrocyte plasma membrane. Values for liver from other species, including human, indicate a similar pattern.

Glycerophospholipids, sphingolipids, and cholesterol are the three major lipid components of eukaryotic membranes. Glycerophospholipids and sphingomyelin, a sphingolipid that contains phosphate, are classified as **phospholipids**. Bacteria and blue-green algae contain glycerolipids in which a carbohydrate is attached directly to the glycerol. Some membranes contain small quantities of other lipids, such as triacylglycerol and diol derivatives (p. 687 for structures), as well as lipids covalently linked to protein. Lipids have a structural role, but many also have other cellular functions.

Glycerophospholipids Are the Most Abundant Lipids of Membranes

Glycerophospholipids (phosphoglycerides) have a glycerol molecule with a phosphate esterified at the α-carbon (Figure 12.3) and two long-chain fatty acids esterified to the remaining carbon atoms (Figure 12.4). Glycerol does not contain an asymmetric carbon, but the α-carbon atoms are not stereochemically identical. Esterification of a phosphate to an α-carbon makes the molecule asymmetric. Naturally occurring glycerophospholipids are designated by the stereospecific numbering (sn) system (Figure 12.3) discussed on p. 708.

1,2-Diacylglycerol 3-phosphate, or **phosphatidic acid**, is the parent compound of the glycerophospholipids. **A phosphodiester bridge to glycerol attaches choline, ethanolamine, serine, glycerol, and inositol** (Figure 12.5). **Phosphatidylethanolamine (ethanolamine glycerophospholipids** or **cephalin)** and **phosphatidylcholine (choline glycerophospholipid** or **lecithin)** are the most common glycerophospholipids in membranes (Figure 12.6). **Phosphatidylglycerol phosphoglyceride** (Figure 12.7) **(diphos-**

phatidylglycerol or **cardiolipin**) contains two phosphatidic acids linked by a glycerol and is found nearly exclusively in mitochondrial inner membranes and bacterial membranes. Inositol, a hexahydroxy alcohol, is esterified to phosphate in **phosphatidylinositol** (Figure 12.8). Different phosphorylated derivatives are localized in different membranes; phosphatidylinositol 4-phosphate predominates in Golgi, phosphatidylinositol 4,5-bisphosphate in plasma membranes, and phosphatidylinositol 3-phosphate and phosphatidylinositol 3,5-bisphosphate in endosomes (A Closer Look 12.1).

Two fatty acids are esterified to C1 and C2 of glycerol in glycerophospholipids; the major ones are presented in Table 12.1. A saturated fatty acid is usually on C1 of glycerol and an unsaturated fatty acid on C2. There are 500 to 1000 molecular species of lipids in membranes principally because of the number of possible combinations of fatty acids in the different phospho- and sphingolipids. Designation of the different classes of phospholipids does not specify which fatty acids they contain. Phosphatidylcholine usually contains palmitic or stearic on C1 and a C_{18} unsaturated fatty acid, oleic, linoleic, or linolenic, on C2.

Phosphatidylethanolamine contains palmitic or oleic on C1 but a long-chain polyunsaturated fatty acid, such as arachidonic, on the C2 position. Each type of cell membrane has a distinctive composition of fatty acyl groups in its glycerophospholipids. The fatty acid composition of lipids has a significant impact on membrane function. Fatty acyl groups of

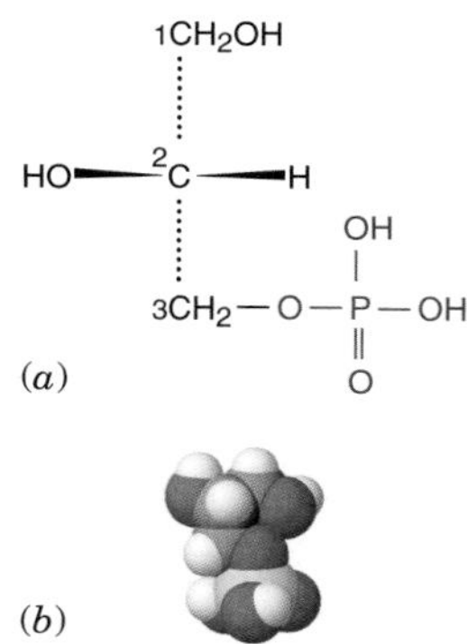

Figure 12.3 L-glycerol 3-phosphate. (*a*) Stereochemical configuration of L-glycerol 3-phosphate (sn-glycerol 3-phosphate). The H and OH attached to C2 are above and C1 and C3 are below the plane of the page. (*b*) Space-filling model of L-glycerol 3-phosphate. Color scheme: H white, C gray, O red, N green, and P yellow.
(*b*) courtesy of Dr. Daniel Predecki, Shippensburg University, PA.

Polar head

Hydrophobic tails

(*a*) (*b*)

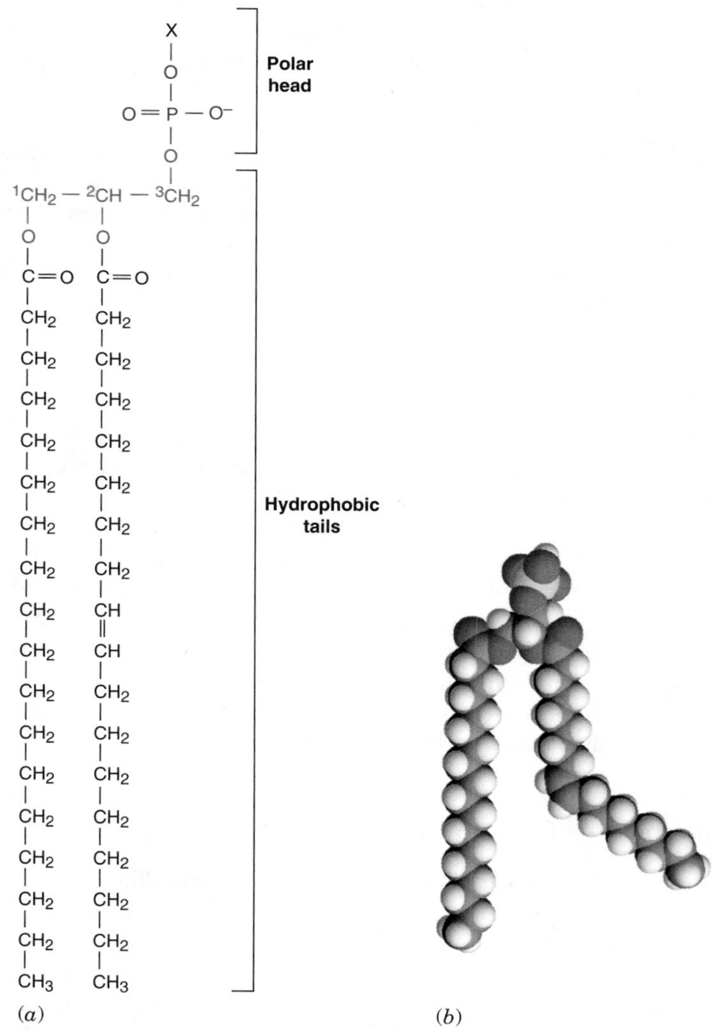

Figure 12.4 Structure of glycerophospholipid. (*a*) Long-chain fatty acids are esterified at C1 and C2 of the L-glycerol 3-phosphate. X can be a H (phosphatidic acid) or one of several alcohols presented in Figure 12.5. (*b*) Space-filling model. See Figure 12.3 for color scheme.
(*b*) Courtesy of Dr. Daniel Predecki, Shippensburg University, PA.

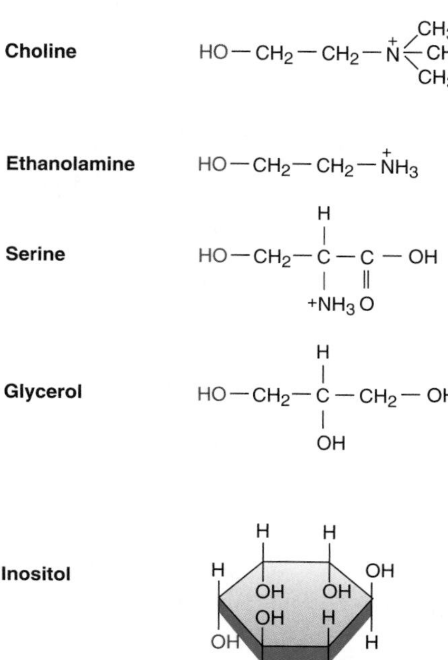

Figure 12.5 Structures of major alcohols esterified to phosphatidic acid to form the glycerophospholipid.

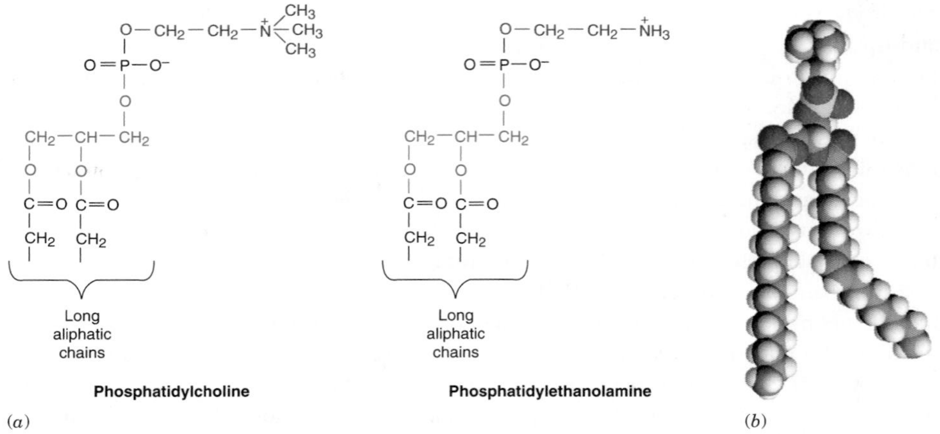

Long aliphatic chains

Phosphatidylcholine

Long aliphatic chains

Phosphatidylethanolamine

(*a*) (*b*)

Figure 12.6 Structures of the two most common glycerophospholipids. (*a*) Phosphatidylcholine and phosphatidylethanolamine. (*b*) Space-filling model of phosphatidylcholine. See Figure 12.3 for color scheme. (*b*) Courtesy of Dr. Daniel Predecki, Shippensburg University, PA.

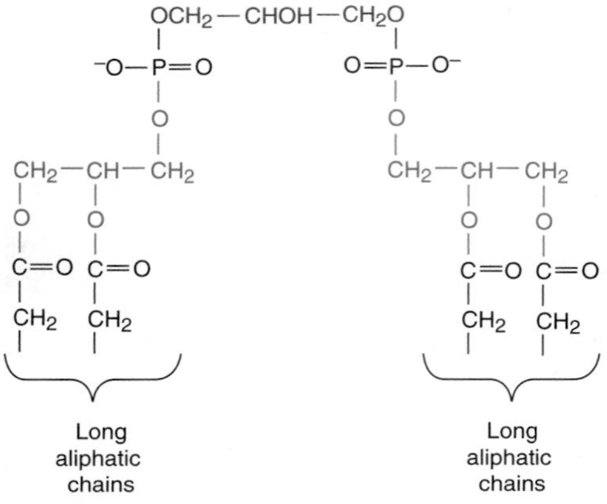

the same tissue in different species are very similar. Acyl group composition can change under different nutritional and pathophysiological conditions.

A saturated fatty acid is a straight chain, as is a *trans* unsaturated fatty acid. *Cis* double bonds, which occur in most naturally occurring fatty acids, create a kink in the hydrocarbon chain (Figure 12.9). The presence of unsaturated fatty acids in the phospholipid has a marked effect on the fluidity of the membrane (p. 472).

Glycerol ether phospholipids contain a long aliphatic chain in ether linkage to the glycerol at the C1 position (Figure 12.10). Ether phospholipids contain an alkyl group (alkyl acyl glycerophospholipid) or an α,β-unsaturated ether, termed a **plasmalogen**. Plasmalogens containing ethanolamine (**ethanolamine plasmalogen**) or choline (**choline plasmalogen**) esterified to the phosphate are abundant in nervous tissue and heart but not in liver. More than 50% of the ethanolamine glycerophospholipids in human heart are plasmalogens. High levels of ether-linked lipids in plasma membranes of metastatic cancer cells have been reported, suggesting a role for the lipids in the invasive properties of these cells.

Long aliphatic chains

Figure 12.7 Phosphatidylglycerol phosphoglyceride (cardiolipin).

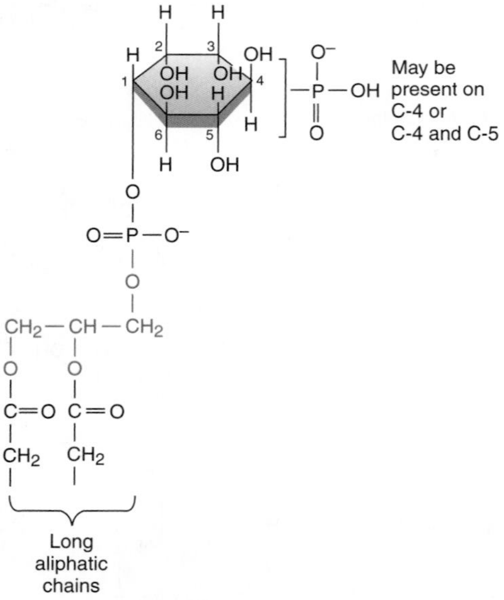

Glycerophospholipids Are Amphipathic

Glycerophospholipids are **amphipathic**, containing a polar and nonpolar end. The charge on the polar end (head group) consists of a charged phosphate ($pK \sim 2$), which is negatively charged at pH 7.0, and any charge of the groups (bases) on the phosphate (Table 12.2). Choline and ethanolamine glycerophospholipids are **zwitterions** at pH 7.0, with a negative charge on phosphate and a positive charge on nitrogen. Phosphatidylserine has a positive charge on the α-amino group of serine and two negative charges, one on phosphate and one on the carboxyl group of serine, with a net charge of −1. In contrast, glycerophospholipids containing inositol and glycerol have only one negative charge on phosphate. The 4-phosphoinositol and 4,5-bisphosphoinositol derivatives are very polar compounds with negative charges on phosphate. The nonpolar end of glycerophospholipids consists of the hydrophobic hydrocarbon chains of the fatty acyl groups.

May be present on C-4 or C-4 and C-5

Long aliphatic chains

Figure 12.8 Phosphatidylinositol. Phosphate groups are also found on C4 or C4 and C5 of the inositol. The additional phosphate groups increase the charge on the polar head of this glycerophospholipid.

Sphingolipids Are Present in Membranes

The amino alcohols **sphingosine** (D-4-sphingenine) and **dihydrosphingosine** (Figure 12.11) are the basis for the **sphingolipids**. A saturated or unsaturated long-chain fatty acyl group in amide linkage with the amino group of sphingosine (Figure 12.12) is a **ceramide**, which is similar in structure to diacylglycerol. Various substitutions are found on the hydroxyl group at C1 of sphingosine. The **sphingomyelins**, the most abundant sphingolipids in mammalian tissues, have phosphorylcholine esterified at C1 (Figure 12.13) and are classified as phospholipids. Sphingomyelin and choline glycerophospholipid are similar in structure and have many properties in common; both are amphipathic. The fatty acid composition of sphingomyelin varies from tissue to tissue; myelin sphingomyelin contains predominantly longer-chain fatty acids (C_{24}).

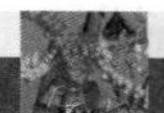

A CLOSER LOOK 12.1

Phosphatidylinositols Have Many Functions

Phosphatidylinositols in cells have a variety of functions. They are involved in membranes as the source of second messengers (p. 534), in activation of membrane ion channels (p. 477), in activation of enzymes, involvement in endo- and exocytosis, and as anchors of proteins (p. 471). Phosphatidylinositol 4,5-bisphosphate is the source of the second messengers inositol trisphosphate and diacylglycerol (p. 535) that are involved in the action of some hormones. Phosphatidylinositol 3-phosphate and phosphatidylinositol 3,5-bisphosphate are presumed to be involved in control of membrane traffic in endosomes. They are also involved in the cytoskeleton and control of cell shape and motility and have been implicated in cytosolic and nuclear functions. Enzymes metabolizing phosphatidylinositols have been linked to human diseases.

Di Paolo, G., and De Camilli, P. Phosphoinositides in cell regulation and membrane dynamics. *Nature* 443: 651, 2006.

TABLE 12.1 • Major Fatty Acids in Glycerophospholipids

Common Name	*Systematic Name*	*Structural Formula*
Myristic acid	Tetradecanoic	$CH_3—(CH_2)_{12}—COOH$
Palmitic acid	Hexadecanoic	$CH_3—(CH_2)_{14}—COOH$
Palmitoleic acid	*cis*-9-Hexadecenoic	$CH_3—(CH_2)_5—CH{=}CH—(CH_2)_7—COOH$
Stearic acid	Octadecanoic	$CH_3—(CH_2)_{16}—COOH$
Oleic acid	*cis*-9-Octadecenoic acid	$CH_3—(CH_2)_7—CH{=}CH—(CH_2)_7—COOH$
Linoleic acid	*cis,cis*-9,12-Octadecadienoic	$CH_3—(CH_2)_3—(CH_2—CH{=}CH)_2—(CH_2)_7—COOH$
Linolenic acid	*cis,cis,cis*-9,12,15-Octadecatrienoic	$CH_3—(CH_2—CH{=}CH)_3—(CH_2)_7—COOH$
Arachidonic acid	*cis,cis,cis, cis*-5,8, 11,14-Icosatetraenoic	$CH_3—(CH_2)_3—(CH_2—CH{=}CH)_4—(CH_2)_3—COOH$

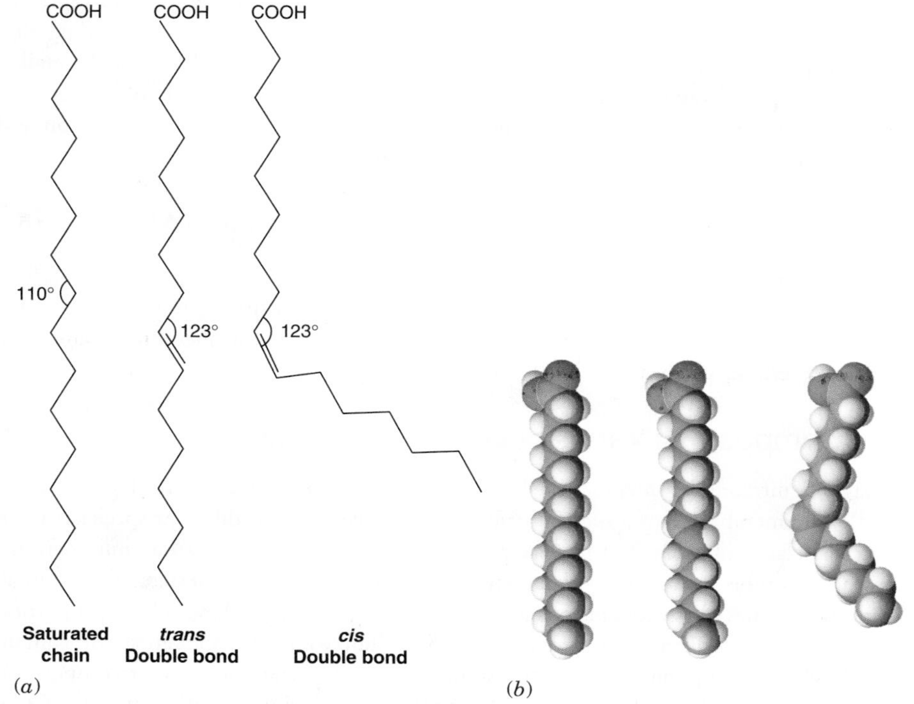

Figure 12.9 **Conformation of fatty acyl groups in phospholipids.** (*a*) Saturated (palmitic) and unsaturated fatty acids (palmitoleic) with *trans* double bonds are straight chains in their minimum energy conformation, whereas a chain (palmitoleic) with a *cis* double bond has a bend. The *trans* double bond is rare in naturally occurring fatty acids. (*b*) Space-filling models. See Figure 12.3 for color scheme.
(*b*) Courtesy of Dr. Daniel Predecki, Shippensburg University, PA.

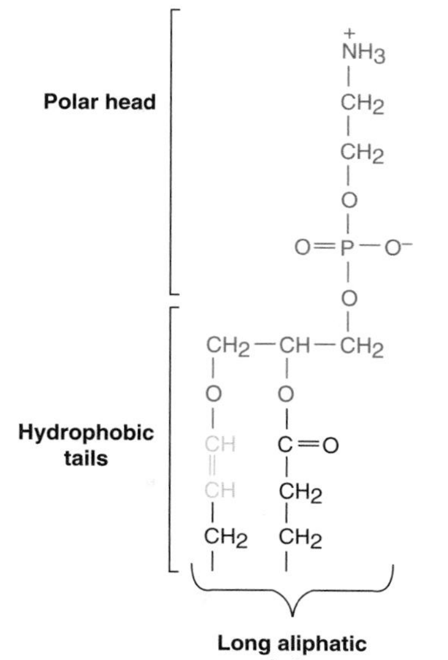

Figure 12.10 **Ethanolamine plasmalogen.** Note the ether linkage of the aliphatic chain on C1 of glycerol.

TABLE 12.2 • Predominant Charge on Glycerophospholipids and Sphingomyelin at pH 7.0

Lipid	*Phosphate Group*	*Base*	*Net Charge*
Phosphatidylcholine	−1	+1	0
Phosphatidylethanolamine	−1	+1	0
Phosphatidylserine	−1	+1, −1	−1
Phosphatidylglycerol	−1	0	−1
Diphosphatidylglycerol (cardiolipin)	−2	0	−2
Phosphatidylinositol	−1	0	−1
Sphingomyelin	−1	+1	0

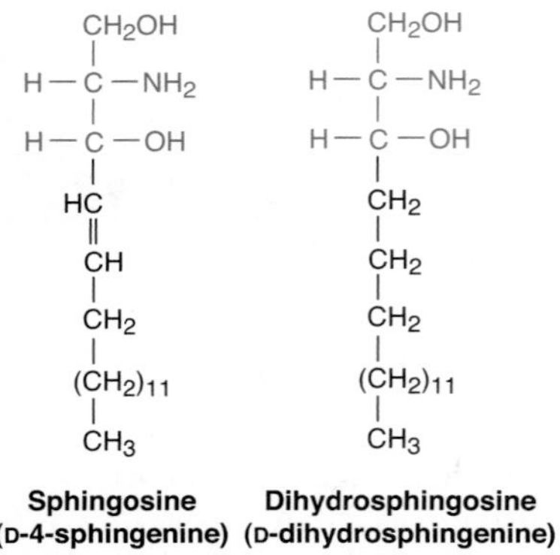

Sphingosine (D-4-sphingenine) **Dihydrosphingosine (D-dihydrosphingenine)**

Figure 12.11 Structures of sphingosine and dihydrosphingosine.

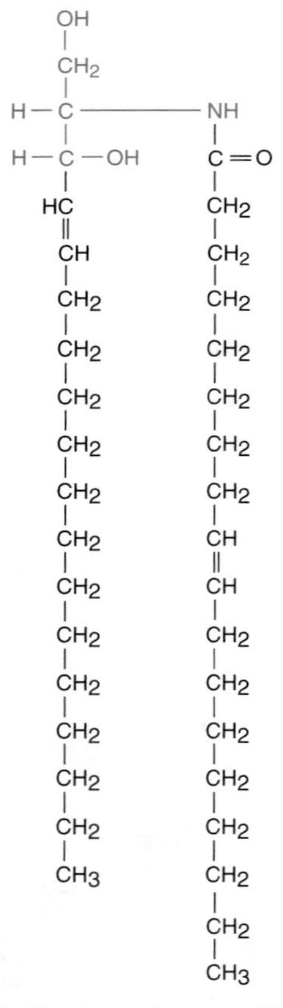

Figure 12.12 Structure of a ceramide.

Glycosphingolipids contain a sugar linked by a β-glycosidic bond to C1 OH group of a ceramide; they lack phosphate and are uncharged. A subgroup is the **cerebrosides**, which contain either glucose (**glucocerebrosides**) or galactose (**galactocerebrosides**). **Phrenosine** (Figure 12.14, p. 464) is a galactocerebroside that contains a C_{24} fatty acid in amide linkage. Galactocerebrosides predominate in brain and nervous tissue, whereas glucocerebrosides occur in small quantities in nonneural tissues. Galactocere-broside containing a sulfate group esterified on the C3 of the sugar are classified as **sulfatides** (Figure 12.15, p. 464). Cerebrosides and sulfatides usually contain fatty acids with 22–26 carbons. Neutral glycosphingolipids often have 2 (dihexosides), 3 (trihexosides), or 4 (tetrahexosides) sugar residues attached to the 1-OH group of ceramide. The usual sugar components are diglucose, digalactose, ***N*-acetyldiglucosamine**, and ***N*-acetyldigalactosamine**.

Gangliosides, the most complex glycosphingolipids, contain an oligosaccharide group with one or more residues of ***N*-acetylneuraminic acid (sialic acid)** (see p. 657 for structure); they are amphipathic compounds with a negative charge at pH 7.0 and represent 5%–8% of the total lipids in brain. Some 20 different types have been identified that differ in the number and relative position of the hexose and sialic acid residues. A detailed description of gangliosides is presented on p. 733. The carbohydrate moiety of gangliosides that extends beyond the surface of the membrane is involved in cell–cell recognition and binding of hormones and bacteria toxins, such as **cholera toxin** (p. 527).

Cholesterol Is an Important Component of Plasma Membranes

Cholesterol is a compact, rigid, hydrophobic molecule (Figure 12.16, p. 464), with four fused rings and a C8-branched hydrocarbon chain attached to the D ring at position 17. It has a polar hydroxyl group at C3. The presence of cholesterol in membranes alters the fluidity and reduces permeability.

Lipid Composition Varies between Membranes

Plasma and intracellular membranes have distinctive lipid compositions (Figure 12.17, p. 464). The membranes of a specific tissue, for example, liver, in different species contain very similar classes of lipids. The plasma membrane exhibits the greatest variation in percentage composition because cholesterol content is affected by the nutritional state of the animal. Myelin membranes of neuronal axons are rich in sphingolipids and have a high proportion of glycosphingolipids. Intracellular membranes, for example endoplasmic reticulum, contain primarily glycerophospholipids and little sphingolipids. The membrane lipid composition of mitochondria, nuclei, and rough endoplasmic reticulum is similar, with that of the Golgi complex being somewhere between that of intracellular and plasma membranes. Cardiolipin is high in inner mitochondrial membrane and low in outer membrane with essentially none in other membranes. The choline containing lipids, phosphatidylcholine and sphingomyelin, are most common, followed by phosphatidylethanolamine. The constancy of composition of various membranes suggests a relationship between lipids and the specific functions of those membranes.

Membrane Proteins

Membrane proteins have a variety of functions, including a structural role to maintain the shape of the cell, as receptors for the binding of hormones and growth factors, as facilitators of transmembrane movement of solute and ions, and as enzymes involved in transduction of signals (p. 507). About 30% of genes in mammalian cells encode membrane proteins.

Membranes proteins are classified on the basis of the ease of separation of the protein from the lipid of the membrane. Isolation of **integral (intrinsic) membrane proteins** requires disruption of the membrane by detergents or organic solvents, and when isolated, they usually contain tightly bound lipid. These proteins are immersed in the lipid milieu of the membrane (p. 468). They contain sequences of hydrophobic amino acids, which constitute the segments of the protein in contact with the lipids of the membrane. The hydrophobic nature of these proteins, thus their water insolubility, has made it difficult to determine their three-dimensional structure and to study their functional properties.

Peripheral (extrinsic) membrane proteins are loosely attached to the surface of the lipid membrane. Treatment with salt solutions of different ionic strength, acidic or basic pHs of the medium, or cleavage of covalently bound lipid releases these proteins. Peripheral membrane proteins are subclassified on the basis of their mode of attachment to a membrane (p. 468). Many loosely attached peripheral proteins are typical water-soluble globular proteins.

Proteolipids are hydrophobic lipoproteins soluble in chloroform and methanol but insoluble in aqueous systems. They are present in many membranes but particularly in myelin, where they represent about 50% of the protein component. **Lipophilin**, a major lipoprotein of brain myelin, contains over 65% hydrophobic amino acids and covalently bound fatty acids. (See p. 470 for a discussion of localization and attachment of membrane proteins.)

Membrane Carbohydrates Are Part of Glycoproteins or Glycolipids

Carbohydrates are present in membranes as oligosaccharides covalently attached to proteins (glycoproteins) and to lipids (glycolipids). The sugars in the oligosaccharides include glucose, galactose, mannose, fucose, *N*-acetylgalactosamine, *N*-acetylglucosamine, and ***N*-acetylneuraminic acid** (sialic acid) (Figure 12.18 and Appendix). Structures of glycoproteins and glycolipids are presented on pages 659 and 732, respectively. The carbohydrate is on the extracellular surface of plasma membrane and luminal surface of endoplasmic reticulum. The roles for protein-bound carbohydrates include cell–cell recognition and adhesion, and receptor action. There is little or no free carbohydrate in membranes.

12.3 • MICELLES, LIPID BILAYERS, AND LIPOSOMES

Lipids Form Vesicular Structures

The basic structural characteristic of membranes is due to the physicochemical properties of the glycerophospholipids and sphingolipids. These amphipathic compounds with a hydrophilic head and a hydrophobic tail (Figure 12.19*a*) interact in aqueous systems to form spheres, termed **micelles** (Figure 12.19*b*). The charged polar head groups are on the outside of the sphere while the hydrophobic tails interact to exclude water. Micelles have only one polar surface, which is the side presented to the aqueous phase. Micelles that contain a single lipid or a mixture of lipids can be prepared. The concentration of lipid required for

(*a*)

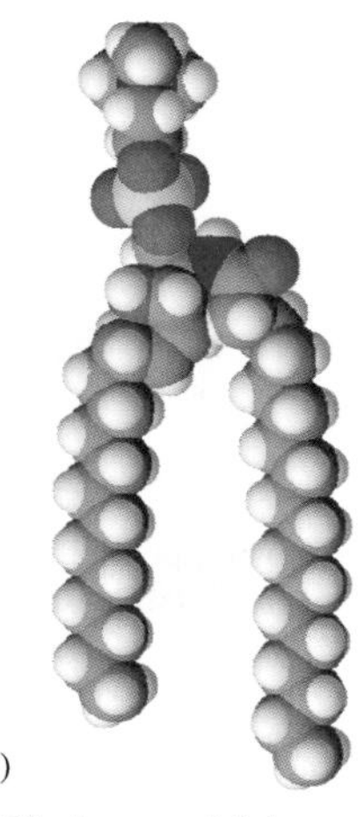

(*b*)

Figure 12.13 Choline-containing sphingomyelin. (*a*) Structure of choline containing sphingomyelin. (*b*) Space-filling model. See Figure 12.3 for color scheme.
(*b*) Courtesy of Dr. Daniel Predecki, Shippensburg University, PA.

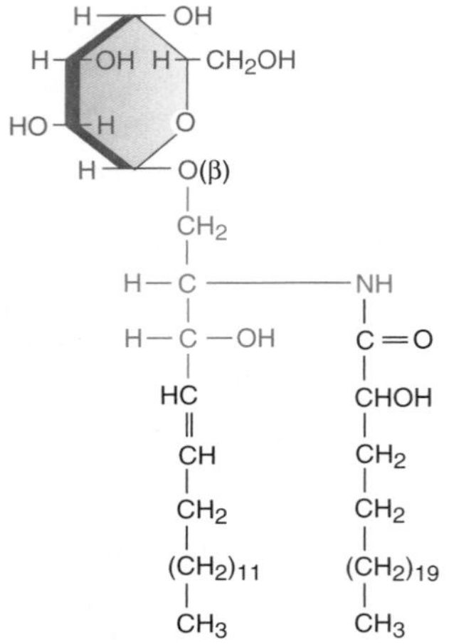

Figure 12.14 Structure of a galactocerebroside containing a C_{24} fatty acid.

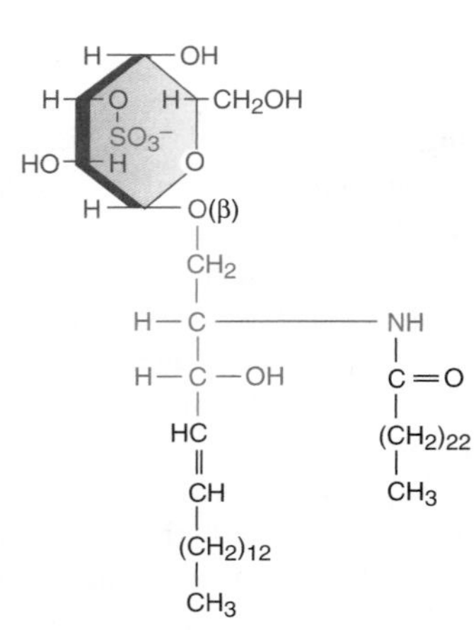

Figure 12.15 Structure of a sulfatide.

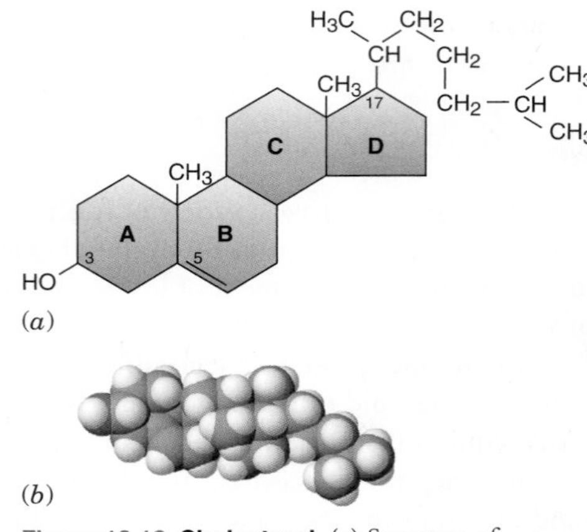

Figure 12.16 Cholesterol. (*a*) Structure of cholesterol. (*b*) Space-filling model. See Figure 12.3 for color scheme.
(*b*) Courtesy of Dr. Daniel Predecki, Shippensburg University, PA.

micelle formation is the **critical micelle concentration** (p. 1058). Formation of micelles also depends on the temperature and, if a mixture of lipids is present, on the ratio of concentrations of the different lipids. The structure of a micelle is very stable because of hydrophobic interactions between hydrocarbon chains and attraction of the polar head groups to water. Micelles are important in intestinal digestion and absorption of lipids (p. 1058).

Synthetic Lipid Bilayers and Liposomes

Amphipathic lipids, such as glycerophospholipids, can also form a bilayer structure with two layers of lipid. In this structure there is minimal contact of hydrocarbon chains with water. Polar head groups are at the interface between aqueous medium and lipid, and hydrophobic tails interact creating an interior hydrophobic environment that excludes water (Figure 12.19*c*). This bilayer configuration is the basic lipid structure of biological membranes. Lipid bilayers are extremely stable. They are held together by hydrophobic forces of the hydrocarbon chains and ionic interactions of charged head groups with water. Lipid bilayers self-seal if disrupted.

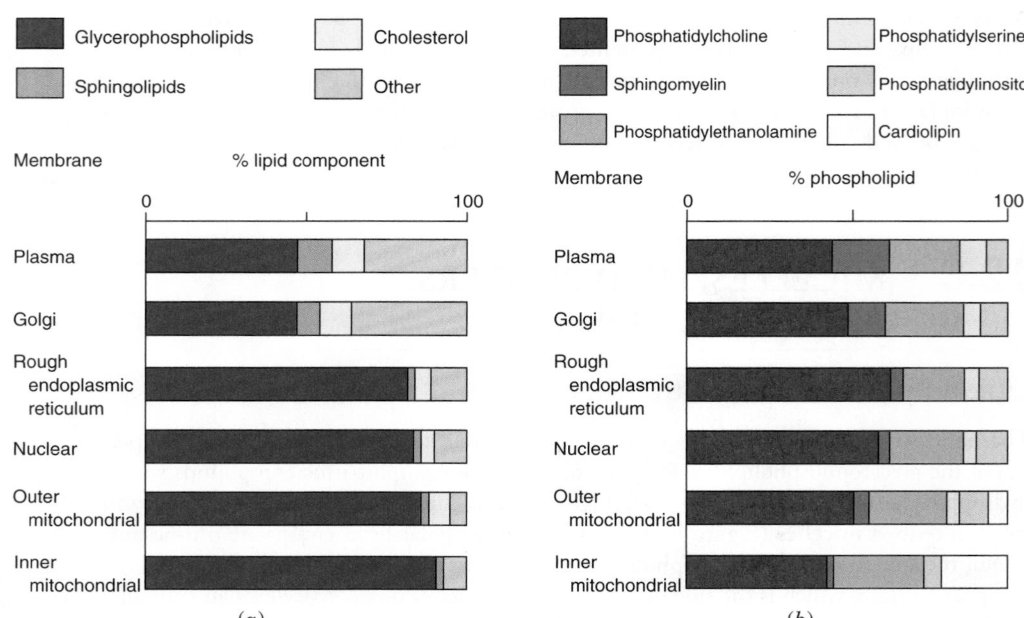

Figure 12.17 Lipid composition of cellular membranes isolated from rat liver. (*a*) Amount of major lipid components as percentage of total lipid. The area labeled "Other" includes mono-, di-, and triacylglycerol, fatty acids and cholesterol esters.
(*b*) Phospholipid composition as a percentage of total phospholipid.
Values from Harrison, R., and Lunt, G. G. *Biological Membranes.* New York: Wiley, 1975.

A lipid bilayer will form a spherical vesicle, separating the external environment from an internal aqueous compartment. Such vesicles, termed **liposomes** (Figure 12.19*d*), have been prepared using purified lipids and mixtures of lipids from biological membranes. Depending on the procedure, unilamellar and multilamellar (vesicles within vesicles) vesicles of various sizes (20 to 1000 nm in diameter) are prepared. The ability of amphipathic lipids to self-assemble into bilayers is an important property for formation of cell membranes.

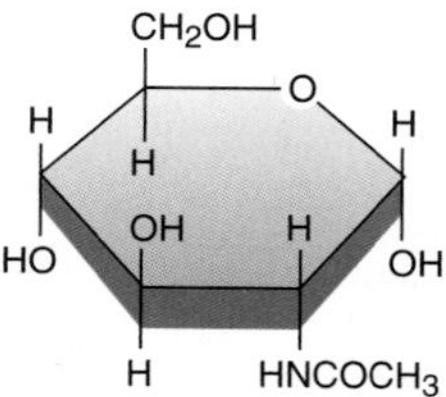

N-Acetyl-α-D-glucosamine

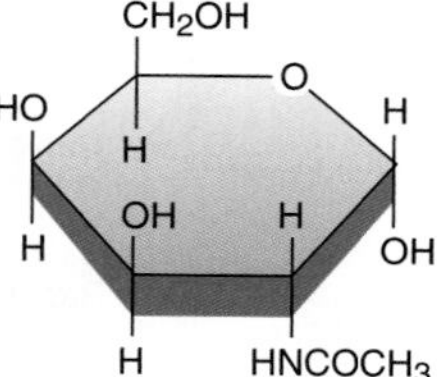

N-Acetyl-α-D-galactosamine

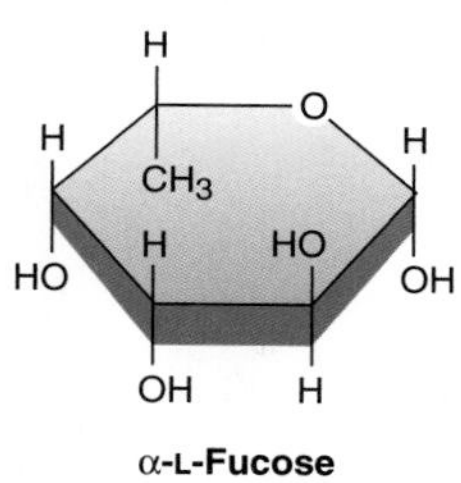

α-L-Fucose

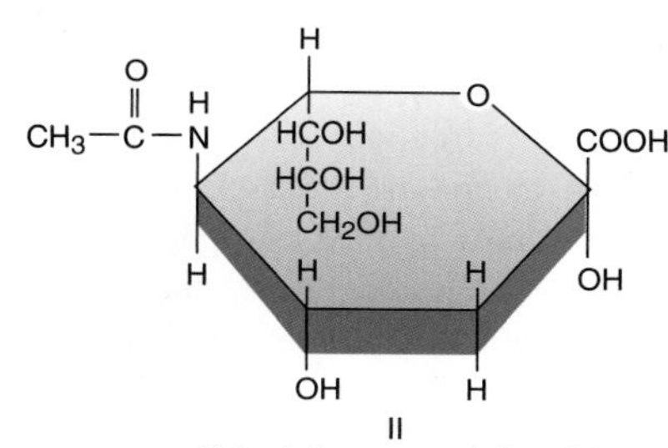

N-Acetyl-D-neuraminic acid

Figure 12.18 Structures of some membrane carbohydrates.

General Properties of Lipid Bilayers

Phospholipid molecules readily exchange places with neighboring molecules in the separate monolayers, leading to rapid lateral diffusion in the plane of the membrane (Figure 12.20). In the preparation with different lipids of artificial membranes, lipids distribute randomly between monolayers. Rotation occurs around the carbon–carbon bonds in fatty acyl chains; in fact, there is a greater degree of rotation nearer the methyl end, leading to greater motion at the center than at the surface of the lipid bilayer. Individual lipid molecules do not flip-flop readily from one monolayer to the other because of thermodynamic constraints on movement of a charged head group through the lipophilic core. In addition, lipids do not readily escape from the bilayer. Lipid bilayers, therefore, have an inherent stability in which individual molecules move rapidly in their own monolayer but do not readily exchange with an adjoining monolayer.

Interaction of lipids in a bilayer is very different from that illustrated in Figure 12.19*c*. The interior of membrane bilayers is very fluid and in constant motion. Acyl chains of glycerophospholipids and sphingolipids have a random motion, intercalating with chains in the opposing monolayer. The viscosity is higher closer to the lipid head groups. Cholesterol in the bilayer reduces the fluidity near the surface of the membrane because cholesterol does not extend into the layer as far as the acyl chains of phospholipids. Figure 12.21 is a computer representation of the arrangement of phospholipids in a bilayer at a moment in time. Several layers of ordered water molecules on the surface of membranes influence the environment of the membrane.

The lipid bilayer is essentially impermeable to nonlipid (as an example, carbohydrates) and charged molecules but not to neutral hydrophobic molecules. Inorganic ions (Na^+, K^+, Cl^-, etc.), charged organic molecules, and macromolecules outside liposomes do not have access to the interior of liposomes. During liposome formation, nonpermeable solutes, however, can be trapped in the aqueous interior. Thus, both the external and internal environments of liposomes can be manipulated, and properties, including ability to exclude molecules, interaction with various substances, and stability under different conditions can be evaluated. Liposomes are an important tool in studying aspects of natural membrane structure and function. Proteins isolated from biological membranes have been incorporated into liposome membranes for study of their function. With some limited success, liposomes of varying lipid and protein composition have been tested in both animals and humans as a replacement for red cell, to deliver encapsulated drugs, and as vectors for gene therapy (Clin. Corr. 12.1).

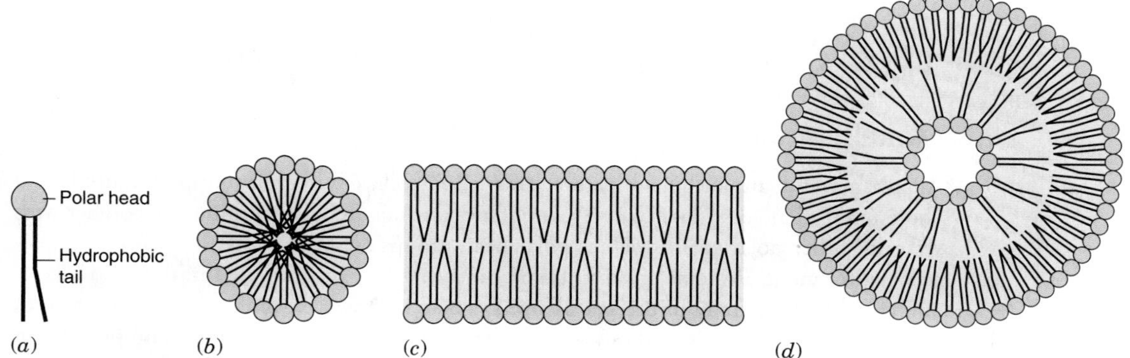

Figure 12.19 Interactions of phospholipids in an aqueous medium. (*a*) Representation of an amphipathic lipid. (*b*) Cross-sectional view of the structure of a micelle. (*c*) Cross-sectional view of the structure of lipid bilayer. (*d*) Cross section of a liposome. Each structure has an inherent stability due to the hydrocarbon chains and the attraction of the polar head groups to water.

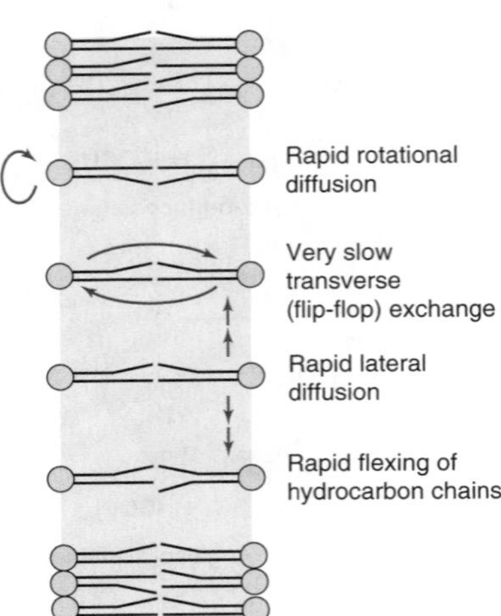

Figure 12.20 Mobility of lipid components in membranes.

Figure 12.21 Model of a lipid bilayer stopped at a moment in time. An artificial bilayer consisting of dipalmitoyl phosphatidylcholine surrounded by water as modeled by computer. Atom colors are chain C gray (except terminal methyl C yellow and glycerol C brown), ester O red, P and O green, choline C and N pale violet, water O dark blue, and water H light blue. Lipid hydrogens have been omitted.
Figure generously supplied by Richard Pastor and Richard Venable, FDA, Bethesda, MD.

CLINICAL CORRELATION 12.1

Liposomes as Carriers of Drugs, Proteins, and Nucleic Acids

Liposomes have been used for administration of a variety of therapeutic substances in order to target specific organs or tissues. Both hydrophilic and hydrophobic compounds can be carried, the former in the aqueous interior and the latter in the liposome membrane. Liposomes prepared from purified phospholipids are nontoxic and biodegradable. Cationic liposomes have been evaluated that will interact with the cell membrane surface charge, and others with specific proteins to bind specific cellular membrane receptors. Uptake of liposomes can be by fusion with the cell membrane or by endocytosis. Liposomes have found a use with drugs that are metabolized very rapidly. Incorporation of the drug in liposomes permits release over a longer time period, thus increasing their effectiveness. Antibiotic, antineoplastic, antimalarial, antiviral, antifungal, and antiinflammatory agents are effective when administered in liposomes. It may be possible to prepare liposomes with a high degree of tissue specificity so that drugs and perhaps even enzyme replacement may be possible with this technique. Liposomes have been tested as a vehicle to introduce nucleic acids into cells for gene therapy and treatment with small RNAs. In another use, liposomes containing concentrated hemoglobin have been tested with some success as a transfusion alternative in place of red blood cells.

Torchilin, V. P. Recent approaches to intracellular delivery of drugs and DNA, and organelle targeting. *Annu. Rev. Biomed. Engineering* 8: 343, 2006; Dass, C. R., and Choong, P. F. Selective gene therapy for cancer therapy using cationic liposomes: In vivo proof of applicability. *J. Controlled Release* 113: 155, 2006; Tsuchida, E., Sakai, H., Horinouchi, H., and Kobayashi, K. Hemoglobin-vesicles as a transfusion alternative. Artificial cells, blood substitutes. *Immobilization Biotech.* 34: 581, 2006; and Sofou, S., and Sgouros, G. Antibody-targeted liposomes in cancer therapy and imaging. *Expert Opin. Drug Delivery* 5: 189, 2008.

12.4 • STRUCTURE OF BIOLOGICAL MEMBRANES

Mosaic Model of Biological Membranes

All plasma and intracellular membranes are lipid bilayers, with amphipathic lipids and cholesterol oriented so that their hydrophobic portions interact to minimize contact with water or other polar groups. Their polar head groups are at the interface with the aqueous environment. Proteins are distributed on the surface or immersed in the lipid bilayer and can diffuse laterally in the **sea of lipid**. There is, however, molecular organization of biological membranes, including constraints on random diffusion of lipids and proteins in some domains. Figure 12.22 is a model of a biological membrane, indicating a degree of organization. The general characteristics of lipid bilayers explain many of the properties of cellular membranes, including fluidity, flexibility that permits change of shape and form, ability to self-seal, and impermeability.

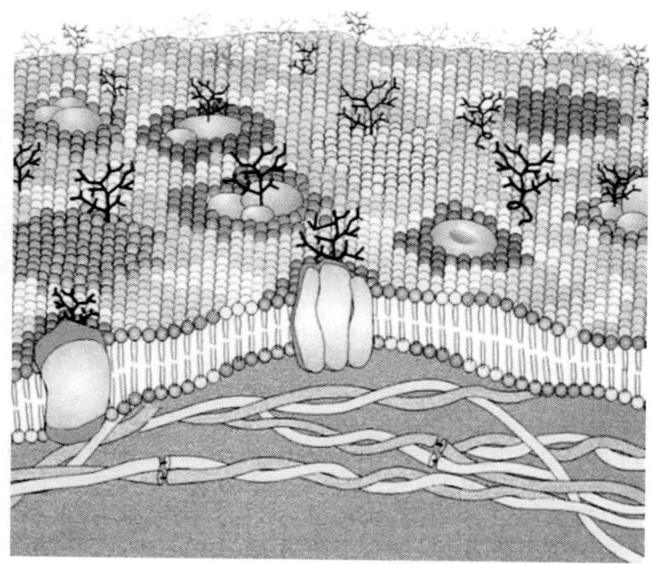

Figure 12.22 Model of a mammalian plasma membrane. The various lipids are different colored spheres; they would be more random than indicated. Proteins are represented by globular structures, and carbohydrates are attached to proteins by the black trees on the external surface of the membrane. The fiber network underneath the membrane represents the cytoskeleton and its association with the membrane.
Figure very generously supplied by Professor P. Kinnunen, University of Helsinki, Finland.

Lipids Are Asymmetrically Distributed in Membranes

Lipids are distributed asymmetrically in the two bilayers of biological membranes. Each layer has a different composition of glycerophospholipids and sphingolipids. The asymmetric distribution of lipids of the plasma membrane of the human erythrocyte is presented in Figure 12.23. Sphingomyelin is predominant in the outer layer, whereas phosphatidylethanolamine is predominant in the inner lipid layer. Phosphatidylinositols are not shown in Figure 12.23 because of their low amounts but are located primarily in the inner layer; the amount varies depending on the tissue. The transbilayer distribution of cholesterol is not known. Lipids in intracellular membranes are also distributed asymmetrically but not with

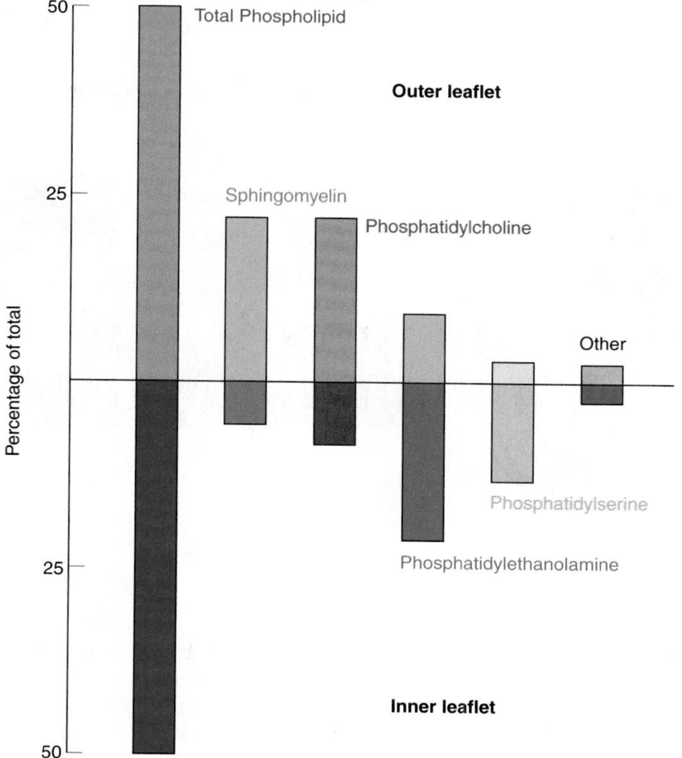

Figure 12.23 Distribution of phospholipids between inner and outer layers of the human erythrocyte membrane. Values are percentage of each phospholipid in the membrane. Others include phosphatidylinositol, phosphatidylinositol 4-phosphate, phosphatidylinositol 4,5-phosphate, and phosphatidic acid.
Data from Verkeij, A. J., Zwaal, R. F. A., Roelofsen, B. Comfurius, P. et al. *Biochim. Biophys. Acta* 323: 178, 1973; and Zachowski, A. *Biochem. J.* 294: 1, 1993.

the same pattern as in the plasma membrane. The asymmetry accounts for some of the functional differences in the two layers.

The asymmetry of lipids is established during biosynthesis of membranes on the endoplasmic reticulum. Uncatalyzed transverse movement (flip-flop) between bilayers of membrane lipids is slow but does occur. Asymmetry of lipids is maintained by **lipid transporters**, which catalyze unidirectional movement of specific lipids from one layer to the other sometimes against their concentration gradient. A subclass of ATP-dependent lipid transporters, **aminophospholipid translocases** or **flippases**, specific for phosphatidylserine and phosphatidylethanolamine catalyze the transport of these aminoglycerolipids from the extracellular to the cytoplasmic layer. They are responsible for maintaining low concentrations of these phospholipids in the outer layer of plasma membranes. An outward-directed ATP-dependent transporter, termed **floppase**, is nonspecific for phospholipids. A third type of lipid transporter, **phospholipid scramblase**, facilitates bidirectional mixing of phospholipids between the two leaflets, thus randomizing the phospholipids between the layers. It is nonspecific with respect to phospholipids, not ATP dependent, and is stimulated by increases in intracellular Ca^{2+}. Phospholipid scramblase 1 (PLSCR 1) can be reversibly palmitoylated-depalmitoylated and contains a protein kinase C phosphorylation site for control of activity (p. 535). Scramblases have an important role in the cell-mediated coagulation cascade (p. 984), and recognition by the reticuloendothelial system of dying cells (p. 1010). In both responses, the movement of phosphatidylserine from the inner to the outer leaflet leads to exposure of phosphatidylserine on the extracellular surface and a change in charge on the surface.

Integral Membrane Proteins

Interactions of proteins to biological membranes are illustrated in Figure 12.24. **Integral membrane proteins** are embedded in the membrane with a defined orientation determined by their primary structure. Their **transmembrane segments** contain sequences rich in hydrophobic amino acids such as leucine, isoleucine, valine, and phenylalanine, which are in contact with the lipid milieu. Some have large globular domains in contact with the aqueous environment on one or both sides of the membrane, and some, particularly those in the plasma membrane, are glycoproteins.

Depending on the specific protein, the amino and carboxyl termini can be on opposite sides or both on one side of the membrane. In the former case, the carboxyl terminus

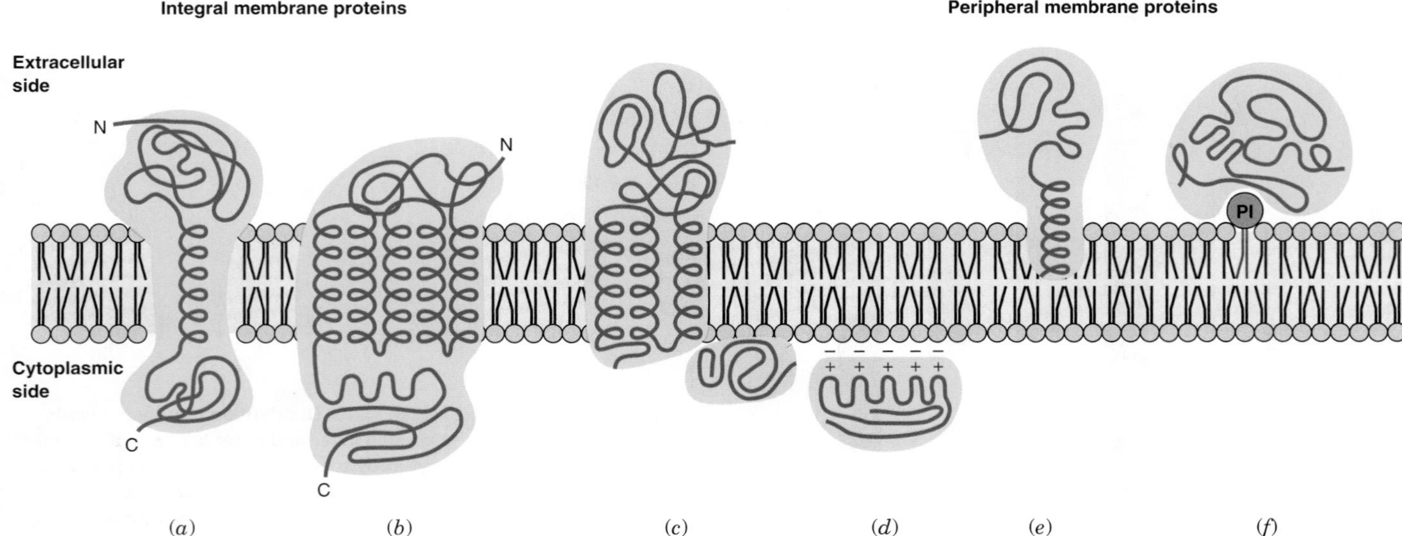

Figure 12.24 Interactions of membrane proteins with the lipid bilayer. Diagram illustrates the multiple types of binding of proteins in or to the lipid bilayer. (*a*) A single transmembrane segment. (*b*) Multiple transmembrane segments. (*c*) Bound to an integral protein. (*d*) Bound electrostatically to the lipid bilayer. (*e*) Attached by a short terminal hydrophobic sequence of amino acids. (*f*) Non-covalent binding to a phosphatidylinositol (PI) in the membrane.

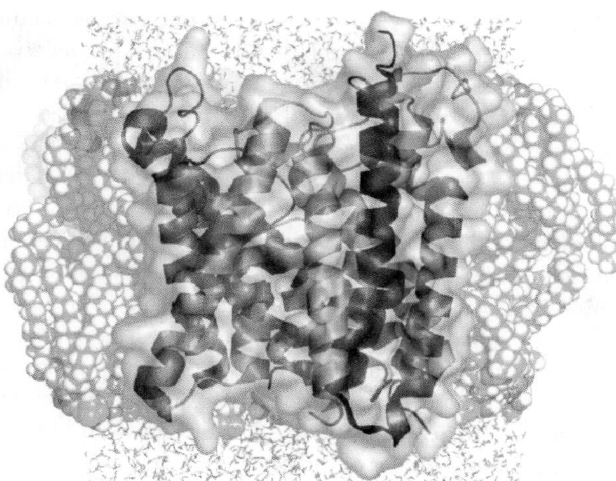

Figure 12.25 Integral membrane protein. The illustration shows one subunit of the trimeric ammonium transporter protein AmtB from *E. coli* embedded in a lipid bilayer membrane. Note the lipid–protein interaction, and the aqueous periplasmic (top) and intracellular (bottom) compartments on both sides of the membrane. Figure very generously supplied by Professor Fritz K. Winkler and Dr. Peter Hasler, Paul Scherrer Institute, Switzerland.

is often found on the cytoplasmic face. Many proteins of the plasma and endoplasmic reticulum membranes have a fourfold abundance of positively charged lysine and arginine residues on the cytoplasmic surface compared to the extracellular or luminal surface. Transverse (flip-flop) movement of integral proteins is prevented by their bulkiness and thermodynamic restrictions.

The transmembrane segments of most integral membrane proteins are α-helical but β-barrel motifs are also found. Some proteins contain a single transmembrane α-helical segment, as illustrated in Figure 12.24*a*. An example is **glycophorin**, with 131 residues, present in the plasma membrane of human erythrocytes; the transmembrane sequence consists of amino acid residues 73–91. The amino-terminus is exterior to the cell and contains various oligosaccharides including the ABO and MN blood group determinants; glycophorin contains 60% by weight carbohydrate. The **integrins** also have a single transmembrane segment; this ubiquitous family of proteins links the cytoplasm to the extracellular matrix (p. 367) and mediates transmembrane signaling in both directions.

Other integral proteins have multiple transmembrane sequences that loop back and forth across the lipid bilayer (Figure 12.24*b* and *c*). Figure 12.25 illustrates such a protein; note the interaction of the lipid with the protein and the aqueous compartments on both sides of the membrane. The number of putative transmembrane segments is determined by plotting the degree of hydrophobicity of each amino acid residue in the protein (p. 86). Those sequences with a high proportion of hydrophobic amino acids suggest a transmembrane segment. The hydropathy plot for aquaporin, an integral membrane protein, is presented in Figure 12.26; the plot indicates the presence of six

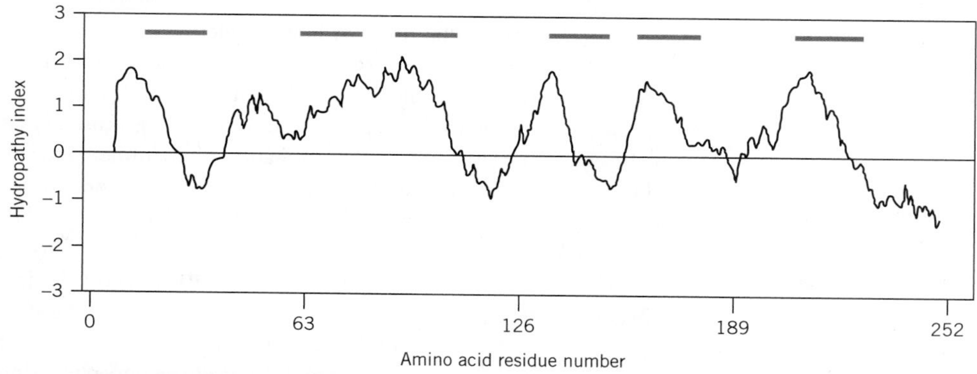

Figure 12.26 Hydropathy plot of aquaporin1 from human erythrocytes. Hydropathy index for an amino acid is a value indicating their hydrophobic and hydrophilic tendency. The more positive the value, the more hydrophobic the amino acid; a positive value is a good predictor of whether the amino acid will be in contact with an aqueous environment. Segments of the protein containing a high proportion of hydrophobic amino acids are considered to be putative transmembrane segments.

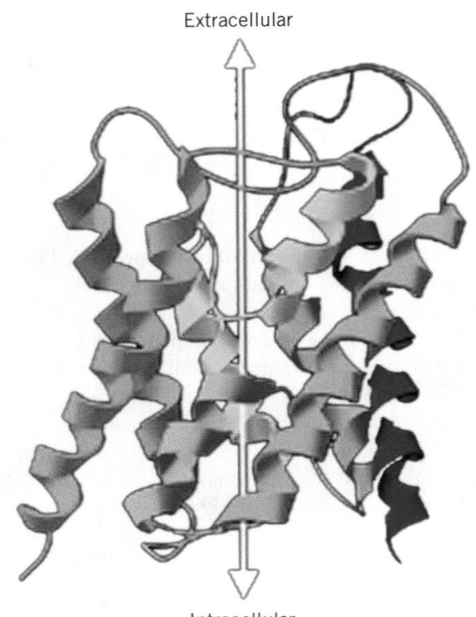

Figure 12.27 Multiple α-helical transmembrane segments. Ribbon model of human red cell AQP1 monomer. Multiple α-helical transmembrane segments form a tubular structure. The amino-terminus is blue and the carboxy-terminus is red. Six membrane-spanning tilted helices surround two hemipores (loops with short helices—cyan and orange) that meet in the center of the bilayer. The white arrow illustrates the aqueous channel through the protein.
Figure reproduced with permission from Kozono, D., Yasui, M., King, L. S., and Agre, P. Aquaporin water channels: atomic structure and molecular dynamics meet clinical medicine. *J. Clin. Invest.* 109: 1395, 2002. Copyright (2002) *Amer. Soc. Clin. Invest.* via Copyright Clearance Center.

putative transmembrane sequences. Proteins with putative transmembrane sequences from 2 to 24 have been identified on the basis of amino acid analysis of the protein or by gene analysis. Multiple α helical transmembrane segments of an integral protein are often organized to form a tubular structure, which can serve as a passageway for movement of molecules through the membrane (Figure 12.27). The **anion channel** of human erythrocytes forms a tubular channel (p. 490) that mediates the exchange of Cl^- and HCO_3^- across the membrane. It has two major domains: a hydrophilic amino-terminus domain on the cytoplasmic side of the membrane with binding sites for **ankyrin**, a protein that anchors the cytoskeleton and other cytoplasmic proteins, and a domain of 509 amino acids with 12 transmembrane sequences. With an even number of transmembrane segments, the carboxyl terminus is also on the cytoplasmic side. A common motif of integral proteins is oligomerization of the same monomers (homomeric complexes) and oligomerization of different monomers (heteromeric complexes).

Integral membrane proteins contain specific domains for ligand binding, for catalytic or transport activity, and for attachment of carbohydrate. The presence of bound lipids and specific sites for water molecules suggest a high degree of structural and functional variability. They are not rigid structures, and their shape in the membrane can be altered by interaction with other proteins and small molecules. The modulating effects of lipids on a protein may reflect a change in ordering and fluidity of the membrane as well as a direct influence on the activity of the protein. Many integral proteins bind specific lipids by non-covalent interactions between the charged head groups as well as the hydrophobic tail. Examples of the influence of lipids on protein activity are the requirement of D-β-hydroxybutyrate dehydrogenase (p. 698) for phosphatidylcholine for activity, and the stabilization of respiratory chain complexes by cardiolipin (p. 460). Maximum activity of diacylglycerol kinase occurs with phosphatidylcholine with C_{18} chains, and cholesterol has been implicated in the activity of various membrane ion pumps and acetylcholine receptors.

Peripheral Membrane Proteins: Lipid Anchors

Peripheral membrane proteins are on the surface of membranes and can be easily removed without disrupting the lipid bilayer. Many have a transient association with the membrane, with cyclic attachment and reattachment. This controls their activity. Many are involved in intracellular signal transduction (p. 509).

Their different methods of attachment are illustrated in Figure 12.24*c–f* (p. 468). Some bind to integral membrane proteins, such as ankyrin binding to the anion channel in erythrocytes (Figure 12.24*c*). Negatively charged phospholipids of membranes interact with positively charged regions of proteins and produce electrostatic binding (Figure 12.24*d*); in some cases Ca^{2+} mediates the binding. Several peripheral proteins have short sequences of hydrophobic amino acids at one end that serve as a membrane anchor (Figure 12.24*e*). Other peripheral membrane proteins have a specific conserved domain that binds non-covalently to the inositol 3-phosphate head group of phosphatidylinositol fixed in the membrane (Figure 12.24 *f*). Some proteins have several different forms of attachment.

A number of peripheral proteins have covalently linked lipid that is inserted into the lipid membrane, thus anchoring the protein to the membrane. The various types of lipid anchors are presented in Table 12.3 and Figure 12.28. One such anchor involves phosphatidylinositol (see Figure 12.8, p. 460), attached to a **glycan** consisting of ethanolamine, phosphate, mannose, mannose, mannose, and glucosamine (p. 659). The glycan is covalently bound to the carboxyl terminus of a protein by ethanolamine (Figure 12.28*a*), and the glucosamine is linked covalently to phosphatidylinositol. The fatty acyl groups of phosphatidylinositol are inserted into the lipid membrane. This form of attachment is referred to as a **glycosylphosphatidylinositol** (**GPI**) anchor; the fatty acyl groups appear to be specific for different proteins. Release and reattachment of the protein to the anchor is controlled, allowing regulation of the activity of the protein. A specific phosphatidylinositol-specific phospholipase C catalyzes the hydrolysis of the phosphate–inositol bond leading to release of the protein The GPI anchor also controls the localization of the protein on the membrane. Over 50 proteins including enzymes, antigens, and cell adhesion proteins are so attached (Table 12.3). The GPI anchor has been conserved through evolution.

TABLE 12.3 • Types of Lipid Anchors

Type of Anchor	*Lipid Involved in Attachment*	*Attachment*	*Representative Proteins*
Phosphatidyl inositol (GPI)	Phosphatidylinositol	Glycan (ethanolamine, phosphate, mannose, mannose, mannose, and glucosamine)	Acetylcholine esterase Alkaline phosphodiesterase Carbonic anhydrase Cell–cell adhesion molecules Cell surface hydrolases Lipoprotein lipase Scrapie prion protein Surface antigens
Myristoyl	Myristic acid (C_{14})	Amide linkage at N-terminal glycine	β-Adrenergic receptor c-CAMP protein kinase Insulin receptor α-Subunit of G protein
Thioester and hydroxy ester	Myristic acid (C_{14}) Palmitic acid (C_{16}) Stearic acid (C_{18}) Oleic acid (C_{18})	—SH or —OH of Cysteine Serine Threonine	G-protein-coupled receptors Transferrin receptor
Thioether-linked	Isoprenoid lipid Farnesyl (C_{15}) Geranylgeranyl (C_{20}) Dolichol	Cysteine	Many GTP-binding proteins Nuclear lamins Protein kinases Protein phosphatases

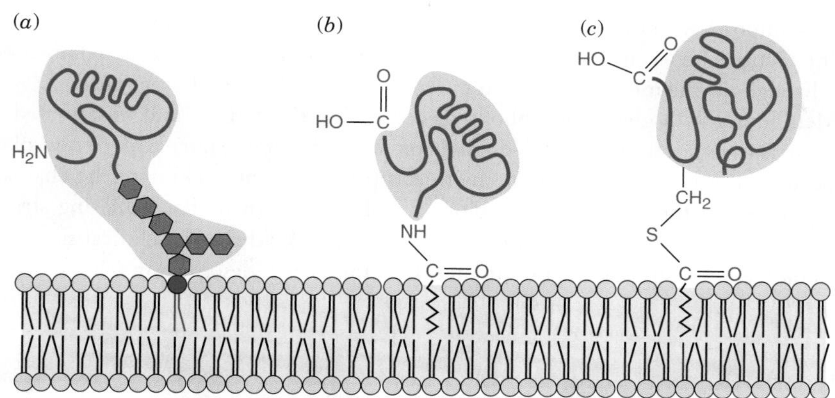

Figure 12.28 Types of lipid anchors for attachment of membrane proteins. (*a*) Glycosylphosphatidylinositol (GPI) anchor. (*b*) Myristoyl anchor. (*c*) Thioester anchor.

Thio-ester-linked acyl anchors (Table 12.3) involve myristic acid (C_{14}), palmitic acid (C_{16}), stearic acid (C_{18}), and oleic acid (C_{18}) covalently linked to a cysteine. The acyl groups serve as membrane anchors (Figure 12.28*c*) and acylation affects the protein in many ways including function, subcellular localization, stability, and interaction with other proteins. The acyl group is often found close to a myristic acid in amide linkage to an N-terminal glycine or a prenylated cysteine at the C-terminus. The function of the acylated protein is controlled by a dynamic cycling of acylation and deacylation. These acylated proteins serve as signaling molecules and are involved in vesicular transport and the neuronal synapse.

Another form of lipid attachment of a peripheral protein involves a thioether linkage between the protein and the isoprenoid lipids farnesyl (C_{15}) or geranylgeranyl (C_{20}) (Figure 12.29). Attachments are referred to as **CAAX prenylated proteins** because the isoprenoid is attached to a cysteine residue in the sequence Cys-Aliphatic-Aliphatic-Any residue close to the carboxyl terminus. The X apparently directs which isoprenoid is linked; glutamine, methionine, or serine is required for attachment of farnesyl, and leucine for geranylgeranyl. CAAX prenylated proteins are usually attached on the cytoplasmic surface.

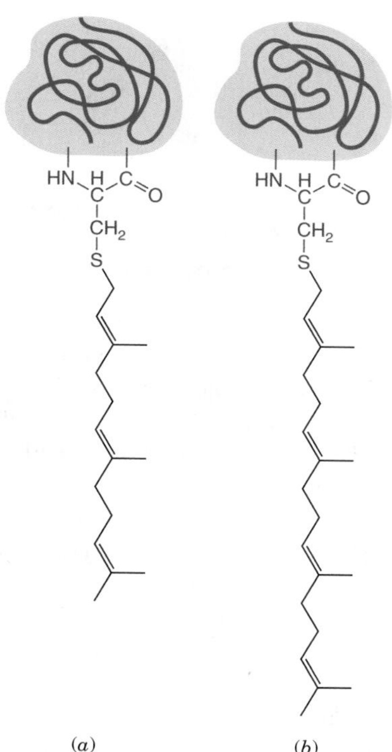

Figure 12.29 Isoprenoid thioether anchor for attachment of membrane proteins. (*a*) Farnesyl anchor. (*b*) Geranylgeranyl anchor.

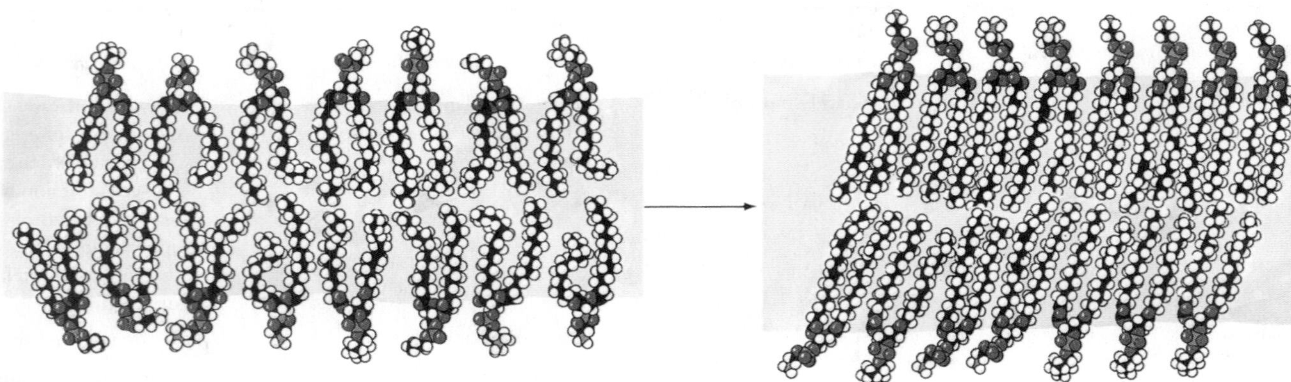

Figure 12.30 Structure of lipid bilayer above (disordered state) and below (ordered state) transition temperature. Figure reproduced with permission from Voet, D., and Voet, J. *Biochemistry,* 2d ed. New York: Wiley, 1995. Copyright (1995) John Wiley & Sons.

Lipids and Proteins Diffuse in Membrane Leaflets

Interactions among different lipids and between lipids and proteins are very complex and dynamic. There is fluidity in the lipid portion of membranes, and individual lipids and proteins move rapidly across the surface of membranes. With artificial lipid membranes, the degree of fluidity is dependent on the temperature. At low temperatures, lipids are in a gel-crystalline state, restricting their mobility; as temperature increases a phase transition occurs, with an increase in fluidity (Figure 12.30). In biological membranes there is a mixture of liquid-ordered and liquid-disordered phases presumably depending on the lipid and protein composition, rather than a transition gel-crystalline to liquid-crystalline state. In fact, interactions between lipids and proteins lead to variations in the ordered–disordered states throughout the membrane and to differences in fluidity in different areas.

Membranes with glycerophospholipids containing short-chain and unsaturated fatty acyl groups have greater fluidity. *Cis*-double bonds in unsaturated fatty acids of phospholipids lead to kinks in the hydrocarbon chain; this prevents tight packing of the chains and creates pockets in the hydrophobic regions. Cholesterol with its flat, stiff ring structure reduces the coiling of the fatty acid chain and decreases fluidity. The clinical significance of high blood cholesterol on the fluidity of cell membranes is described in Clin. Corr.12.2.

CLINICAL CORRELATION 12.2

Abnormalities of Cell Membrane Fluidity in Disease

Membrane fluidity can control the activity of membrane-bound enzymes and functions such as phagocytosis, and cell growth and death. A major factor in the fluidity of the plasma membrane in higher organisms and mammals is the presence of cholesterol. With increasing cholesterol content, the lipid bilayers become less fluid on their outer surface but more fluid in the hydrophobic core. Erythrocyte membranes of individuals with spur cell anemia have increased cholesterol content, a spiny shape, and the cells are destroyed prematurely in the spleen. This condition occurs in severe liver disease such as alcoholic cirrhosis. Cholesterol content is increased 25%–65%, and the fluidity of the membrane is decreased. Erythrocyte membranes require a high degree of fluidity to pass through the capillaries. Increased plasma membrane cholesterol in other cells leads to an increase in intracellular membrane cholesterol, which also affects their fluidity. The intoxicating effect of ethanol on the nervous system is probably due to modification of membrane fluidity, altering membrane receptors and ion channels. Individuals with abetalipoproteinemia have an increase in sphingomyelin content and a decrease in phosphatidylcholine in cellular membranes, with a decrease in membrane fluidity.

Changes in membrane fluidity have been suggested as a factor in lecithin:cholesterol acyltransferase deficiency, hypertension, and Alzheimer disease. As techniques for measurement and evaluation of cellular membrane fluidity improve, some of the pathological manifestations in disease may be explained based on changes in membrane structure and function.

Cooper, R. A. Abnormalities of cell membrane fluidity in the pathogenesis of disease. *N. Engl. J. Med.* 297: 371, 1977; Muller, W. E., Kirsch, C., and Eckert, G. P. Membrane-disordering effects of beta-amyloid peptides. *Biochem. Soc. Trans.* 29: 617, 2001; and Tsudo, K., and Nishio, I. Membrane fluidity and hypertension. *Amer. J. Hypertension* 16: 259, 2003.

Ca^{2+} ion decreases fluidity of membranes because of its interaction with negatively charged head groups of the phospholipids, reducing repulsion between polar groups and increasing packing of lipid molecules. This causes aggregation of lipids into clusters and reduces fluidity. The asymmetric distribution of phospho- and sphingolipids leads to a difference in fluidity of the two leaflets. The hydrocarbon chains of the lipids are flexible, producing a greater degree of fluidity in the hydrophobic core than closer to the surface, where there are more constraints from stiffer portions of the hydrocarbon chains.

The composition of fatty acyl chains of phospholipids and of cholesterol in membranes is altered by diet, changes in cellular Ca^{2+}, free radicals, and lipid peroxidation, leading to a change in membrane fluidity. Anesthetics increase membrane fluidity *in vitro* and may act *in vivo* through their effect on membrane fluidity of specific cells. Anesthesia is induced by structurally unrelated compounds with common lipid solubility. Changes in fluidity of membranes in hypercholesteremia, hypertension, diabetes mellitus, obesity, alcoholism, schizophrenia, and Alzheimer disease have been reported.

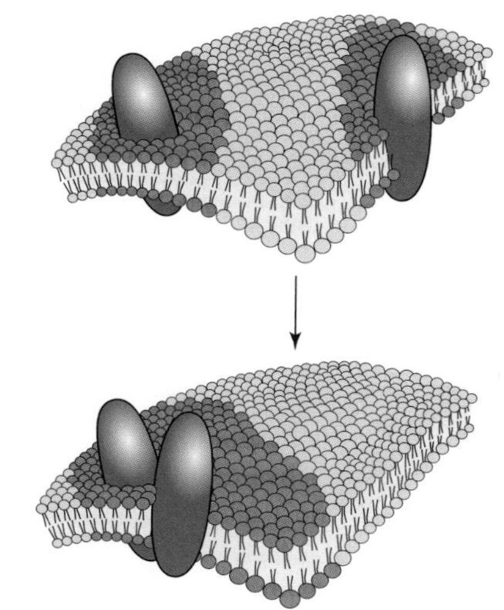

Figure 12.31 Schematic representation of lipid rafts of membrane. The diagram indicates the movement of lipid rafts with merging of rafts and association of two proteins.

Microdomains of Lipid–Protein Complexes Are Present in Membranes

There is a heterogeneous rather than homogeneous distribution of lipids and proteins in membranes. **Microdomains** containing specific lipids and proteins are present. These domains, referred to as **lipid rafts**, are organized bilayer structures with a different distribution of lipids than the surrounding membrane. Thus, biological membranes are actually a fluid mosaic of different microdomains (Figure 12.31). In plasma membranes they constitute a significant percentage of the total area. The function of lipid rafts is to segregate and concentrate specific proteins to facilitate their activity. Smaller rafts assemble and disassemble constantly and diffuse freely in the membrane. Plasma membranes and membranes destined to interact with the plasma membrane, such as the *trans* side of the Golgi complex, contain an abundance of these microenvironments but they may also occur in other cellular membranes. Membrane rafts are defined as small (10–200 nm), heterogeneous, dynamic microenvironments enriched with cholesterol and sphingolipids. The lipids in all rafts, however, are not necessarily in the same physical/chemical state. The higher concentration of cholesterol makes them less fluid than the surrounding lipid environment. Cholesterol is important in the assembly of the lipid rafts in the Golgi complex and in maintaining their integrity and function. The domains may play a role in cholesterol transport.

The outer leaflets of rafts have a higher concentration of ceramides and glycosphingolipids (p. 460), which usually contain longer chain fatty acids. Their accumulation in the outer leaflet leads to areas of increased membrane thickness. The inner leaflet contains phospholipids with more highly saturated acyl chains than in the nonraft areas, which allows close packing of the chains. Thus, lipids in the rafts are more ordered and tightly packed than the surrounding bilayer. In addition, electrostatic interactions of polar head groups, hydrophobic interactions of cholesterol with selected phospholipids or glycolipids, and protein–lipid interactions constrain movement of lipids. Through protein–protein and protein–lipid interaction smaller rafts cluster to form larger, ordered subdomains in the membrane. Some of these are located in specific regions of the plasma membrane.

Some membrane proteins are associated mainly with rafts, some move in and out, and others are present only in the nonraft portions of the membrane.

Glycosylphosphatidylinositol (GPI)-anchored proteins (p. 470) partition into lipid rafts and CAAX prenylated anchored proteins (p. 471) are in rafts on the inner leaflet. In some cells the raft and the sequestered protein are attached to the cytoskeleton of the cell. Certain proteins involved in cell signaling are maintained in the inactive state in individual rafts; activation occurs when several rafts cluster together, forming a larger raft and bringing together the inactive proteins (Figure 12.31). Other proteins have a weak affinity for rafts in the unliganded state but after binding to a ligand undergo a conformational change or oligomerize increasing their affinity for a specific raft. Some peripheral membrane proteins are palmitoylated leading to association with a raft; they dissociate on depalmitoylation.

Plasma membranes of cells from some tissues contain **Caveolae**, small flasklike invaginations whose lipid composition is similar to the lipid rafts; they have a high content of cholesterol and glycosphingolipids. Caveolae contain the cholesterol-binding proteins **caveolin-1, -2,** or **-3** that may be responsible for stabilizing the structure.

Lipid rafts are involved in many cellular processes, including membrane sorting and trafficking, and cell polarization. Growth factor and G protein receptor signaling also involve membrane rafts. The roles of lipid rafts in the pathogenesis of conditions such as Alzheimer, Parkinson, cardiovascular diseases, systemic lupus erythematosus, and HIV have been reported. Bacteria, prions (Clin Corr. 3.6, p. 114), viruses, and parasites utilize lipid rafts of cells for their infectivity.

Dynamic Environment of Membranes

Biological membranes (see Figure 12.22, p. 467) are in a state of constant change, with movement of proteins and lipids laterally, with association and dissociation of proteins in the membrane, with proteins and lipids moving into and out, and with modification (e.g., acylation and phosphorylation) of proteins. Even the thickness of the lipid bilayer varies due to the changing composition of lipids and proteins. Microenvironments are constantly being formed and dispersed. A membrane is a very crowded environment considering the estimated number of enzymes, receptors, channels, and transporters present. Mammalian plasma membrane components interact with the cytoskeleton of the cytoplasm, extracellular components, and other cells. Neither words nor illustrations can capture the complexity and dynamic state of biological membranes or their extensive role in so many different biological processes.

12.5 • MEMBRANE RECEPTORS

Biological cells have a large number of specific protein **receptors** that bind other molecules or react to specific stimuli to initiate a cellular response. Receptor proteins are either bound to plasma or intracellular membranes, or are present as soluble proteins in the cytoplasm or nucleus. The molecules binding to a receptor, referred to as **ligands**, are a communication link, or **signal**. Signal molecules can be generated in one compartment of a cell to stimulate a change in an activity present in another part of the same cell, or the signal may be generated by one cell to alter the activity of a different cell. Intercellular signal molecules, that is, ligands generated by one cell to change the activity of another cell, react either with receptors on the plasma membrane or with intracellular receptors. Ligands such as charged molecules, peptides, and proteins cannot transverse the membrane because of their charge or size and react with plasma membrane receptors on the extracellular face of the membrane. Molecules that can diffuse through the plasma membrane, such as steroids, bind to intracellular soluble receptors. Intracellular signals can either bind to soluble receptors or ones on the cytoplasmic side of the membrane. There are a variety of initial responses to the ligand receptor interaction including activation of an enzyme, release of an intermediate molecule that is recognized by a subsequent intermediate in a signal pathway, or opening of a channel in the membrane to permit the flux of ions. The major signal transduction pathways are presented in Chapter 13 (p. 503), and tissue responses to signal molecules will be described in presentations of individual cellular processes.

TABLE 12.4 • Cellular Responses to Ligand-Receptor Interaction

Stimulation of growth and differentiation
Modulation of the peripheral and central nervous system
Control of reproductive system
Regulation of metabolic processes
Stimulation of immunological functions

Signaling between cells allows individual cells to respond to the requirements of the whole organism; representative responses to ligand receptor binding are listed in Table 12.4. There are hundreds of different plasma membrane receptors; Table 12.5 lists three major classes. The largest and most diverse group of membrane receptors is the G protein-coupled receptors (p. 522) that occur in both prokaryotic and eukaryotic cells.

TABLE 12.5 • Classification of Plasma Membrane Receptors

Classification of Receptor	*Examples of Receptor*
G-Protein coupled	Glucagon Acetylcholine
Catalytic: tyrosine kinase	Insulin Erythropoietin
Ion transport	γ-Aminobutyric acid Glutamate

Membrane receptors are classified on the basis of homology of structure of the protein and mechanism; representative receptors with their ligands/activity are listed in Table 12.6. Receptors have a high degree of specificity for ligands and some receptors respond to stimuli such as light and temperature (Table 12.7). There are also plasma membrane receptors, such as the Toll-like receptors (TLR) that induce a response when bacteria or viruses bind; binding stimulates a host defense against the invading pathogens. Transport of proteins destined for secretion between endoplasmic reticulum and Golgi involve membrane receptors that recognize specific cargo-encoded sorting signals on the proteins.

TABLE 12.6 • Representative Plasma Membrane Receptors

Receptor	*Number of Receptors*	*Representative Ligand or Activity*
Cytokine type 1 receptors	43	Leptin
Cytokine type 2 receptors	11	Interferon
GPI-anchored	4	Neurotrophic factor
Guanylyl cyclase receptors	7	Retinyl guanylate cyclase
Interleukin-17 receptors	8	Interleukin
Integrins	28	Laminin
Low-density lipoprotein (LDL) receptor	14	Low-density lipoprotein
Other non-7 TM* receptors	156	Transferrin
Receptor-like protein tyrosine phosphatases	24	Protein tyrosine phosphatase
Receptor tyrosine kinases	87	Insulin receptor precursor
Seven transmembrane (7TM) receptors	754	Dopamine; oxytocin; light
TGF-beta serine/threonine kinase receptors	13	Serine/threonine kinase R2
Tetraspanins	36	Cell adhesion
TNF/NGF	33	Tumor necrosis factor precursor

*TM, transmembran; Source: Human Plasma Membrane Receptors, www.receptome.org.

TABLE 12.7 • Receptor Ligands

Ions	Proteins
Amino Acids	Tastants
Lipids	Toxins
Nucleotides	Viruses
Polypeptides	Odorants
Hormones	Pheromones
Neurotransmitters	Photons

Armbruster, B. N. and Roth, B. L. *J. Biol. Chem.* 280: 5129, 2005.

Binding of a **signal molecule** to its receptor is an equilibrium process (p. 512). This is similar to the binding of substrate to an enzyme, but in most cases of receptor binding there is no chemical modification of the bound ligand. As the concentration of free ligand in the aqueous phase increases or decreases, the ligand associates or dissociates from the receptor; dissociation ends the signaling process. Most receptors have very high affinities for their ligands, and some receptors react with ligands of dissimilar structure. Receptors are excellent targets for action of pharmaceuticals.

It has been estimated that about 5% of proteins coded in the human genome are plasma membrane receptors. The high degree of tissue specificity of action of signal molecules is because cells in different tissues express different receptors. As an example, insulin, a signal molecule, alters the metabolism of glucose in muscle and liver but not in most other tissues. Cellular response to a ligand is modulated by the number of receptor molecules in the plasma membrane. An increase in number of receptors, termed **upregulation** of receptor, the greater the cellular response; a decrease, or **downregulation**, leads to a decrease in response. Incorporation into and removal from the membrane of receptors is a controlled process.

Membrane receptors that span the membrane can have a single transmembrane or multiple transmembrane amino acid domains. The G protein-coupled receptors that occur in bacteria and eukaryotic cells (p. 522) have seven discrete transmembrane domains (7TM receptors). The human 7TM receptors are classified into six families on the basis of the sequence similarities of their transmembrane domains. Some receptors require formation of homomeric or heteromeric oligomers for activity. Binding of a ligand alters the tertiary or quaternary structure, leading to initiation of the cellular response. In one class of receptors a surface receptor binds to a ligand and then binds to second receptor (signal receptor) that initiates the cellular reaction. In this case, the ligand binding receptor is referred to as a **coreceptor**.

12.6 • TRANSLOCATION OF MOLECULES THROUGH MEMBRANES

Some Molecules Diffuse through Cell Membranes

Diffusion of a chemical species through a membrane involves (1) leaving the aqueous environment on one side and entering the membrane, (2) traversing the lipid milieu of the

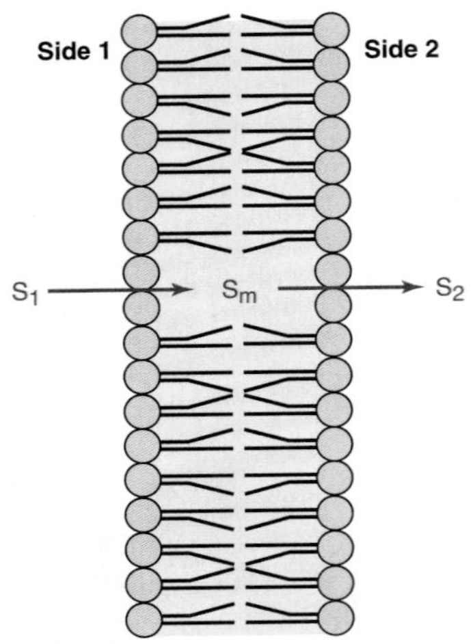

Figure 12.32 Diffusion of a solute molecule through a membrane. S_1 and S_2 are solute on each side of membrane, and S_m is solute in membrane.

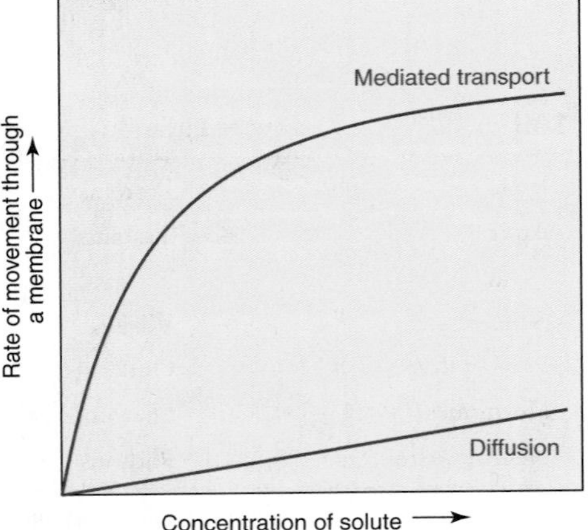

Figure 12.33 Kinetics of movement of a solute molecule through a membrane. Initial rate of diffusion is directly proportional to the concentration of solute. In mediated transport, rate will reach a V_{max} when carrier is saturated.

membrane, and (3) leaving the membrane and entering the aqueous phase on the opposite side (Figure 12.32). The flow can only be down the concentration gradient of the substance. Thermodynamic and kinetic constraints control the concentration equilibrium on two sides of a membrane and the rate at which it attains equilibrium. The rate of diffusion of a chemical species in a membrane is directly proportional to its solubility and diffusion coefficient in lipids; the latter is a function of the size and shape of the substance. Movement of solutes by diffusion is from a higher to a lower concentration, and the rate is described by Fick's first law of diffusion (Eq. 12.1).

$$J = -D\left(\frac{\delta c}{\delta x}\right) \tag{12.1}$$

where J = net amount of substance moved per time, D = the diffusion coefficient, and $\delta c/\delta x$ = chemical gradient of substance. As the concentration of solute on one side of the membrane is increased, there is an increasing *initial rate* of diffusion as illustrated in Figure 12.33. Net movement of molecules from one side to another will continue until the concentration in each is at chemical equilibrium. Continued exchange of solute molecules from one side to another occurs after equilibrium is attained but there is no net accumulation on either side as this would recreate a concentration gradient.

Gases, such as O_2, N_2, CO_2, NO, CO, and H_2S, and lipophilic molecules diffuse through the hydrophobic regions of a membrane. Water diffuses under osmotic forces, but in most cells the rate does not meet the needs for rapid equilibration of the aqueous phases on each side of the plasma membrane; a family of proteins, the aquaporins, facilitates transmembrane movement of water (p. 579). The shell of water surrounding many solutes must be stripped away before the solute enters the lipid milieu. Inorganic ions and charged organic molecules do not diffuse at a significant rate because of their attraction to water and the exclusion of charged species by the hydrophobic environment of membranes. Uncharged organic molecules, such as sugars, are essentially excluded, as are peptides, proteins, and nucleic acids because of their size and charge. If diffusion does occur, the rate is too slow to accommodate cellular needs.

Protein-Based Mechanisms for Translocation

Several hundred protein-mediated systems, each with a different set of proteins, have evolved to facilitate movement of water, ions, nutrients, metabolic waste products, and macromolecules across different cellular membranes. The proteins involved in each system from different organisms and even between organs in a species are very similar but not identical. A functional classification system of the membrane transport proteins, recommended by the **International Union of Biochemistry and Molecular Biology (IUBMB)** (Table 12.8), divides them into classes, subclasses, families, and subfamilies on the basis of mechanism of transport and their properties. Many of the proteins are present only in prokaryotes; many others in both prokaryotes and eukaryotes, including mammals. There is a structural homology of the proteins in many of the subfamilies. The following discussion will focus on transport systems in mammalian cells and specifically on the classes designated **pore/channel** (1.A., 1.B, and 1.C) and **transporter** (2.A and 3.A) (Table 12.8). A variety of other terms for describing transport systems appear in the scientific literature, but the current recommended nomenclature will be used here.

Pores and channels involve the presence of an aqueous pore in the protein structure through which molecules diffuse; movement is controlled by a gating mechanism. Transporter proteins are similar to enzymes, binding the solute on one side of the membrane, then changing conformation, and moving the substrate to the environment on the opposite side of the membrane. This occurs within the membrane portion of transport protein. Most pores/channels and transporters have a high degree of solute specificity. The rate of substrate translocation by channels is higher than by specific transporters; for channels the rate is about 10^7 ions/s, which is the rate of diffusion of a substance in water, whereas with transporters it is about 10^2–10^3 molecules/s.

A variety of substances, termed **substrates**, are transported, including inorganic anions and cations, water, metabolic intermediates, organic molecules, drugs, and macromolecules. In eukaryotic cells, substances translocated are unchanged during the process. Some bacteria have a class of transporters termed group translocation where the substrate is chemically

TABLE 12.8 • Classification of Membrane Transport Proteins

Class	*Subclass (# of families)*
1. Pores and channels	
	1.A α-Helical channels (36)
	1.B β-Strand porins (34)
	1.C Pore-forming toxins (58)
	1.D Non-ribosomally synthesized channels (10)
2. Electrochemical-potential-driven transporters	2.A Transporters or carriers (uni-, sym- and antiporters) (80)
	2.B Non-ribosomally synthesized transporters (6)
3. Primary active transporters	
	3.A P-P-bond-hydrolysis-driven transporters (14)
	3.B Decarboxylation-driven transporters (1)
	3.C Methyl-transfer-driven transporters (1)
	3.D Oxidoreduction-driven transporters (9)
	3.E Light-driven transporters (2)
4. Group translocators	
	4.A Phosphotransferases (6)
5. Transmembrane electron carriers	
	5.A Two electron carriers (2)
	5.B One electron carriers (3)

Source: Moss, G. P. Membrane Transport Proteins. http://www.chem.qmul.ac.uk/iubmb/mtp/

modified during transport (A Closer Look 12.2). Some substrates are transported by two different mechanisms in different cells and in some cases in the same cell. Subfamilies of membrane transport proteins have many isoforms, all with the same function. As an example, the kidney nephron has at least seven different isoforms of aquaporins, all responsible for translocation of water. In addition, many transporters are controlled by intra- or extracellular signal molecules and are also referred to as receptors for the specific ligand (Section 12.5; p. 474).

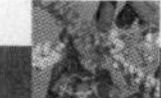

A CLOSER LOOK 12.2

Group Translocation and the γ-Glutamyl Cycle

Some bacteria have a mechanism for uptake of sugars, termed **group translocation**, where the sugar to be transported is chemically modified during translocation. The sugar is phosphorylated after binding to but before release from the transporter. The phosphorylated sugar is released into the cytoplasm and not the original unphosphorylated sugar. No group-translocation system has been found in mammalian cells. The γ-glutamyl cycle (p. 790), an important system for amino acid uptake by some mammalian cells, involves the modification of an amino acid by γ-glutamyl transferase on the external surface of the cell membrane. The modified amino acid is then transported to the cytoplasm by an amino acid transporter. This cycle is not classified as group translocation because separate proteins catalyze the chemical modification and the transport.

12.7 • MEMBRANE CHANNELS AND PORES

Membrane **channels** and **pores** are ubiquitous in living organisms, being present in all types of organisms, from bacteria to higher eukaryotes. Channels and pores permit diffusion in aqueous channels created by the transporter proteins. They do not sequester molecules or ions in transit, and their specificity is based on the size and charge of the substance. Diffusion in channels and pores is (1) down the concentration gradient of the substance crossing the membrane, (2) active without utilization of metabolic energy for the actual movement of the substance, (3) regulated by specific mechanisms that open or shut the passageway, and (4) inhibited by various compounds. Pores consist of a number of different proteins and, in comparison to channels, have a very complex structure. A major difference between channels and pores is the specificity in what the system will permit to be translocated. Pores have relatively large openings, permitting a variety of small and large molecules to tranverse the membrane, whereas channels have a high degree of solute specificity.

Structures of Membrane Channels

Channel proteins have one of two major structures: α-type channels and β-barrel channels. **α-Type channels** are homo- or hetero-oligomeric proteins, in some cases with several dissimilar subunits, and can have 2 to 22 transmembrane **α-helical segments** (see Figure 12.27, p. 470). A common physical characteristic of channels is a structure formed by

Figure 12.34 Structure of Ompf porin in a tetragonal crystal form. The transmembrane segments of the porin are β-helices. PDB ID: 1OPF. Cowan, S. W., Garavito, R. M., Jansonius, J. N., Jenkins, J. A. et al. The structure of Ompf porin in a tetragonal crystal form. *Structure* 3: 1041, 1995.

amphipathic α-helices from different transmembrane segments within a single polypeptide chain or of different protein subunits to create a central aqueous channel. In addition to the domains containing the actual pore, channel proteins contain domains for ligand binding or for sensing physical changes, such as changes in the membrane potential, which control the opening and closing of the channel. α-Type channels in mammals function as cellular sensors, responding to temperature changes, touch, pain, osmolarity, pheromones, taste, and other stimuli.

β-Barrel channels have a transmembrane sequence consisting of **β-strands** that form a β-barrel structure (p. 93) with a water filled channel from 0.6 to 3 nm in diameter (Figure 12.34). They are present in the outer membranes of mitochondria (porins), some bacteria, and plastids. Several bacterial porins consist of 16-stranded antiparallel β-barrel structures with hydrogen bonding between nearest neighbors along the chain. Polypeptide loops line the inner barrel wall limiting channel width and restricting size of molecules diffusing through. This subclass of channels has a wide range of substrate selectivity from inorganic ions to proteins. Porins can either be nonselective or exhibit some degree of selectivity, and some allow passage of any molecule smaller than a specific cutoff size.

Control and Selectivity of Membrane Channels

The two most studied classes of channels are those that respond to a change in the transmembrane potential, referred to as **voltage-gated channels**, and those responding to binding of a ligand, the **ligand-regulated channels.** Voltage-gated channels have a specific sensor domain in the protein that detects membrane potential changes and transfers the energy to the channel domain to control its gate. As examples, depolarization of the plasma membrane leads to opening of Na^+ channels. The mitochondrial outer membrane contains a voltage-dependent anion channel consisting of β-sheets, which permits the passage of substances of up to 10 kDa and has a role in apoptosis (p. 1010). Ligand-regulated channels respond to the binding of specific extracellular or intracellular chemical ligands (also referred to as **agonists)** and, depending on the channel, leads to either an opening or closing of the channel. Some have several ligand-binding domains, which permit cooperativity in binding. Opening and shutting is achieved in different ways including (1) with α-helical transmembrane segments that line the pore moving as a rigid body by tilting, rotating, or bending and (2) an attached terminal end that swings into the channel and physically plugs it.

Binding of acetylcholine to the **nicotinic-acetylcholine receptor** opens the channel, allowing flow of Na^+ into the cell. This is important in neuronal electrical signal transmission (p. 942). Other channels are controlled by binding neurotransmitters (p. 514), cAMP (p. 530), inositol 1,4,5-trisphosphate, diacylglycerol (p. 534), and G proteins (p. 522). Opening of most channels is very fast, permitting bursts of ion flow of 10^7 ions per second through the membrane. This rate is necessary because many of these channels are involved in nerve conduction and muscle contraction. The surface of a nerve terminal can contain several hundred different channels, including voltage-dependent channels for Ca^{2+} and K^+, Ca^{2+}-gated K^+ channels, Cl^- channels, ligand-gated channels, and stretch-activated channels. Some channels are controlled not only by ligand binding but also by phosphorylation-dephosphorylation. A number of pharmacological agents are used therapeutically to modulate channels; those that inhibit are referred to as **antagonists**.

Different classes of channels are very selective with respect to the inorganic ion, small organic molecule (e.g., urea), or protein allowed to move through the channel. The **Na^+ channel** of plasma membranes of eukaryotic cells permits movement of Na^+ at a rate more than 10 times greater than that for K^+, and some K^+ channels are 100–1,000 times more permeable to K^+ than Na^+. Other channels are less specific. Some members of the family of aquaporins are selective only for water while others permit translocation of both water and small, neutral organic molecules (p. 479). There are also differences from tissue to tissue in channels transporting the same solute. Over 15 different K^+ specific voltage-dependent channels have been reported in humans. One method of selectivity involves rings of positively charged residues in the channel, which exclude cations and facilitate flow of anions; negatively charged residues facilitate flux of cations.

TABLE 12.9 • Site of Expression of Selected Aquaporins in Humans

Tissue	*Location of Expression*	*Aquaporin*	*Function*
Blood cells	Erythrocytes	AQP1	Osmotic protection
	Leukocytes	AQP9[a]	?
Brain	Choroid plexus	AQP1	Cerebral spinal fluid production
	Ependymal cells	AQP4	Cerebral spinal fluid balance
	Hypothalamus	AQP4	Osmolarity sensing (?)
Eye	Lens fiber cells	AQP0	Fluid balance in lens
	Ciliary epithelium	AQP1	Production of aqueous humor
Kidney	Proximal tubules	AQP1	Concentration of urine
	Collecting ducts	AQP2	Mediates ADH[b] activity
	Collecting ducts	AQP3[a]	Reabsorption of water into blood
	Collecting ducts	AQP4	Reabsorption of water
	Collecting ducts	AQP6	?
Lung	Alveolar epithelial cells	AQP1	Alveolar hydration
	Bronchial epithelium	AQP4	Bronchial fluid secretion
Mouth and	Trachea epithelial cells	AQP3[a]	Secretion of water into trachea
throat	Salivary glands	AQP5	Production of saliva

[a] An aquaglyceroporin.

[b] Antidiuretic hormone.

Source: King, L. S., Kazono,D., and Agre, P. From structure to disease: The evolving tale of aquaporin biology. *Nature Rev.: Mol. Cell Biol.* 5: 687, 2004.

Representative Membrane Channels and Pores

Aquaporins and Aquaglyceroporins

It is essential for all cells to control the distribution of water across the plasma and internal membranes because osmotic pressure changes can cause expansion or contraction of enclosed compartments. Nonfacilitated diffusion of water across biological membranes occurs, but it is too slow to account for the required rates of water movement. Over 200 different channel proteins in plants, bacteria, insects, and vertebrates permit the translocation of water and other small molecules and as a group are referred to as **aquaporins (AQP)**. Twelve mammalian aquaporins have been identified, which are divided by amino acid sequence homology and functional characteristics into channels that are selective only for water (aquaporins) and those that permit translocation of water and small solutes (**aquaglyceroporins**). The aquaporins permit movement of H_2O but not hydronium ions (H_3O^+), thereby maintaining the pH gradient across the cell membrane. Humans have at least 10 aquaporins and aquaglyceroporins (AQP0 to AQP9); sites of expression of selected aquaporins in humans are presented in Table 12.9. Some tissues have several different isoforms with different functions; the kidney has at least seven isoforms located at distinct sites along the nephron and collecting duct. Clin. Corr. 12.3 describes the distribution of aquaporins in kidneys and examples of pathophysiological changes. Plant aquaporins contribute to water uptake by rootlets, transpiration, and seed desiccation.

Aquaporins are small, hydrophobic, intrinsic membrane proteins with several highly conserved amino acid sequences. They assemble in membranes as homotetramers (Figure 12.35). Each monomer, consisting of six transmembrane α-helical domains, contains a distinct water pore. AQP1 of the human erythrocyte contains a highly conserved sequence motif consisting of an asparagine–proline–alanine in two separate loops. The loops fold into the lipid bilayer in an hourglass fashion forming an aqueous channel. The selectivity is due to a constriction of the channel to about 2.8 Å, which limits the size of molecules that can pass through. The aqueous pathway of AQP1 (Figure 12.36, p. 481) is lined with conserved hydrophobic residues and a few hydrophilic residues that attract H_2O but not charged molecules such as H_3O^+. Water molecules apparently flow through the channel in single file.

CLINICAL CORRELATION 12.3

The Mammalian Kidney and Aquaporins

The primary role of the kidney is to regulate the pH of blood, eliminate toxic products of metabolism, and regulate water balance of the body. Each kidney contains about one million nephrons, the multicellular functional units of the tissue. Each nephron has several distinct regions (see diagram), each with very specific functions in the processing of the blood filtrate, which occurs in the glomerulus. A large quantity of blood is filtered (about 150 L/day) and the water must be reabsorbed in the nephron and collecting ducts except for about 1.5 L that is excreted as urine. Aquaporins are responsible for the recovery of water from the filtrate as it flows through the lumen of the nephron. As indicated in the diagram, at least seven different isoforms of aquaporins, located at different sites along the nephron and collecting ducts, are responsible for water reabsorption.

There are a number of renal disorders where the reabsorption of water is abnormal and where the expression and function of the aquaporins has been investigated. Several animal models of these conditions have been invaluable in determining the cause of some of these conditions. Low levels of AQP2 and polyuria (excessive excretion of urine) are found in acquired nephrogenic diabetes insipidus (NDI), acquired hypokalemia (low blood K^+), and hypercalcemia (increased blood Ca^{2+}). In many cases, NDI is caused by the inability of the kidney to respond to vasopressin (p. 892); it is considered that this leads to a decreased expression and /or incorporation of AQP2 in the membrane. In other cases of NDI, there is a defect in the aquaporin gene; in some cases, this defect leads to an inability of the monomers to form normal tetramer structures. Levels of AQP1, AQP2, and AQP3 in animal models are reduced in tissue ischemia. In some conditions, such as congestive heart failure, liver cirrhosis, and pregnancy, there is an increase in the amount of AQP2 in the kidney, leading to an expansion of the extracellular fluid volume. Humans lacking AQP1 activity of the kidney apparently do not have any measurable problems under normal but may under stress conditions (dehydration). It is expected that additional clinical conditions will be attributed to changes in the other aquaporins.

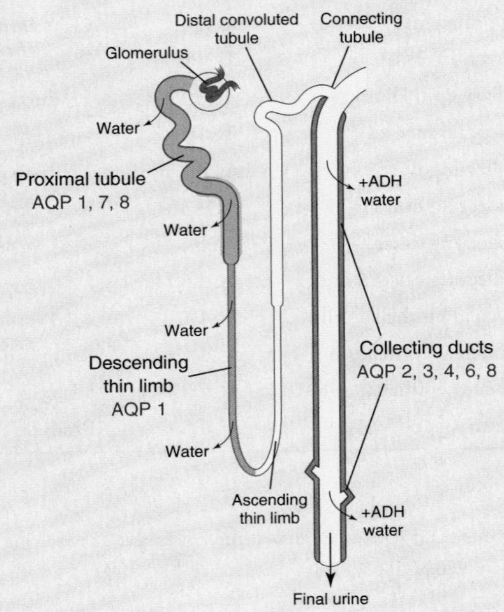

Localization of aquaporins in each segment of the nephron and collecting duct of the kidney. Modified from figures in the following citations.

King, L. S., and Yasui, M. Aquaporins and disease: Lessons from mice to humans. *Trends in Endocrinology & Metabolism*13: 355, 2002; Nielsen, S., Frokiaer, J., Marples, D., Kwon, T-H., et al. Aquaporins in the kidney: From molecules to medicine. *Physiol. Rev.* 82: 205, 2002; and King, L. S., Kozono, K., and Agre, P. From structure to disease: The evolving tale of Aquaporin biology. *Nature Reviews: Molecular Cell Biology* 5: 687, 2004.

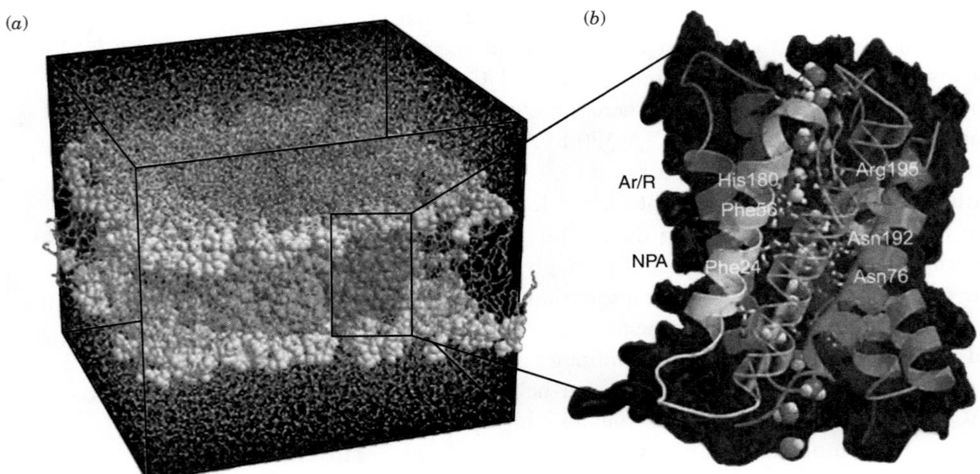

Figure 12.35 Simulation of water permeation through AQP1. (*a*) The simulation of the AQP1 tetramer (orange/magenta/blue/cyan) embedded in a palmitoyl oleoyl phosphatidylethanolamine bilayer (yellow/green) surrounded by water (red/white) consisted of approximately 100,000 atoms. (*b*) Simulation of the pathway of one of the monomers of APQ1 for water permeation observed through the pore.
Figure reproduced with permission from Fujiyoshi, Y., Mitsuoka, K., de Groot, B. L., Philippsen, A. et al. Structure and function of water channels. *Curr. Opin. Struct. Biol.* 12: 509, 2002. © Elsevier.

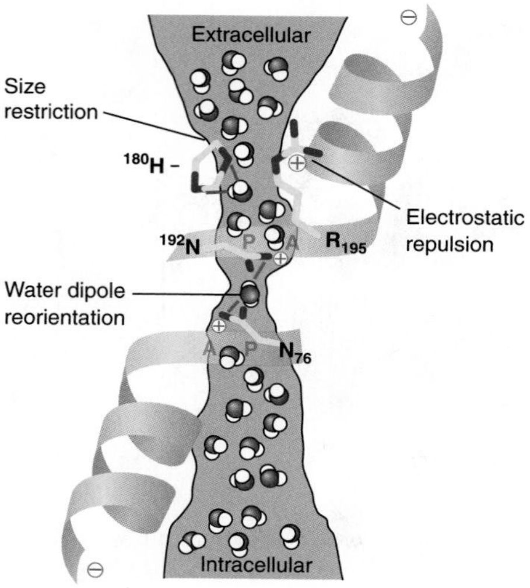

Figure 12.36 Water channel within an AQP1 subunit. The shape of the aqueous pore (blue) is derived from calculations based on the structure of bovine AQP1. Three features of the channel specify selectivity for water: (*a*) Size restriction. Eight Å above the midpoint of the channel, the pore narrows to a diameter of 2.8 Å (approximately the diameter of a water molecule). (*b*) Electrostatic repulsion. A conserved residue (R, Arg-195) at the narrowest constriction of the pore imposes a barrier to cations, including protonated water (H_3O^+). (*c*) Water dipole reorientation. Two partial helices meet at the midpoint of the channel, providing positively charged dipoles that reorient a water molecule as it traverses this point. Disrupting hydrogen bonding in the single-file chain water molecules prevents the formation of a proton conductance.
Reproduced with permission from Kozono, D., Yasui, M., King, L.S., and Agre, P. Aquaporin water channels: atomic structure and molecular dynamics meet clinical medicine. *J. Clin. Invest.* 109: 1395, 2002. © Elsevier.

In addition to water, AQP1 permits translocation of CO_2, AQP3 of glycerol, and AQP9 of glycerol, urea, polyols, purines, pyrimidines, nucleosides, and monocarboxylates. AQP1 is reversibly inhibited by $HgCl_2$. AQP2 in kidneys of humans is under hormonal control and at least three others may be controlled by pH.

TABLE 12.10 • Subunit Composition of Voltage-Gated Cation Channels

Cation Channel	*Subunits*
Na^+	α, β_1, and β_2
K^+	4α, β
Ca^{2+}	α_1, $\alpha_2\delta$, and β

Na^+, Ca^{2+}, K^+, and Cl^- Voltage-Gated Ion Channels

The individual **voltage-gated ion channels (VIC)** for Na^+, Ca^{2+}, and K^+ are controlled by the transmembrane electrical potential. They open and close remarkably fast (milliseconds), and in mammalian cells are responsible for the generation of conducted electrical signals in neurons and other excitable cells. These ion channels are members of a large superfamily found in bacteria, archaea, viruses, and eukaryotes. In humans, there are six types of Ca^{2+} channels and at least 10 types of K^+ channels responding to different stimuli.

Each ion channel consists of specific glycoprotein subunits as homo- or hetero-oligomeric structures (Table 12.10). The large α-subunits (about 260 kDa) contain the ion-conducting channel. They have four homologous domains each with six transmembrane segments, connected by both intracellular and extracellular loops, for a total of 24 transmembrane segments (Figure 12.37). Accessory subunits (β, γ, or δ) do not share a common structure; some have several transmembrane segments and others are peripheral proteins located entirely intra- or extracellularly. They are usually involved in regulation of the channel; β-subunits may interact with cytoskeletal and extracellular matrix proteins. A model of the Na^+ channel is

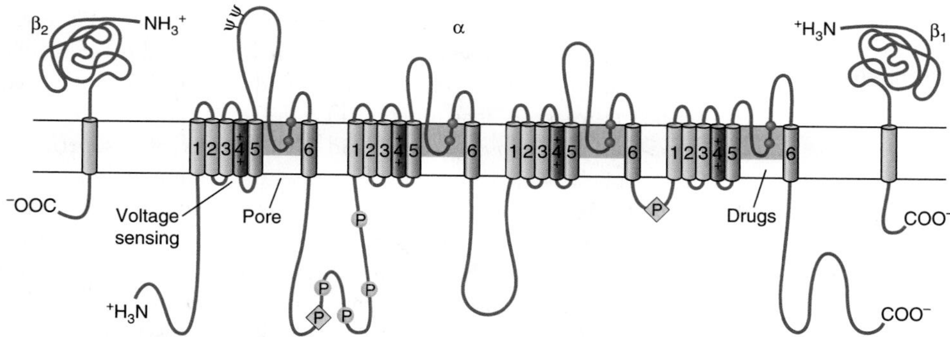

Figure 12.37 Subunit structure of the voltage-gated Na⁺ channel. The primary structures of the subunits of the voltage-gated Na^+ channel are illustrated as a transmembrane folding diagram. Cylinders represent probable α-helical segments. The red cylinders represent the voltage sensor and the blue cylinders the segments in the pore (see Figure 12.38). Bold lines represent the polypeptide chains of each subunit, with length approximately proportional to the number of amino acid residues in the brain Na^+ channel subtypes. The extracellular domains of the β_1- and β_2-subunits are shown as globular structures. Protein phosphorylation sites are indicated.
Redrawn from Catterall, W. A., Goldin, A. I., and Waxman, S. G. Internat. Union Pharm. XXXIX. Compendium of Voltage-Gated Ion Channels: Sodium Channels. *Pharmacol. Rev.* 55: 575, 2003.

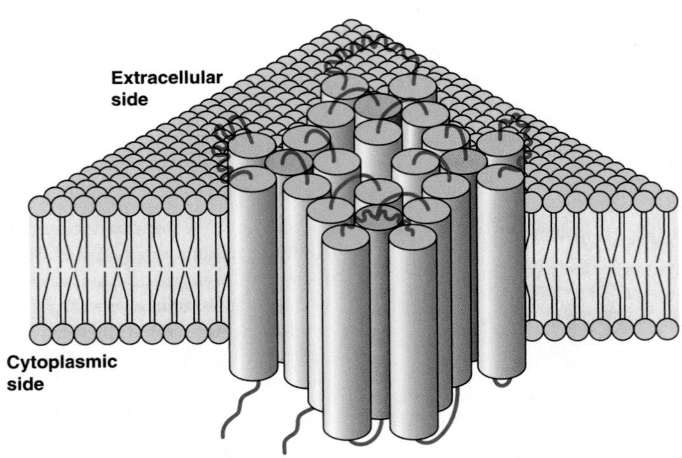

Figure 12.38 Model of the Na^+ channel. The single peptide consists of four repeating units with each unit folding into six transmembrane helices. The red transmembrane segments represent the voltage sensor. Redrawn from Noda, M., et al. *Nature* 320: 188, 1986.

presented in Figure 12.38. One transmembrane segment has a positively charged residue at every third position and may be the voltage-sensor domain; mechanical shift of this segment may lead to a conformational change in the protein, resulting in the opening of the channel. Two very different mechanisms have been suggested, but neither proven, for how the channel and the voltage-sensor domains are coupled to initiate the opening of the channel. The membrane protein responsible for the translocation of protons contains domains for sensing membrane voltage changes, but a channel domain has not been described.

Voltage-gated ion channels can exist in one of three conformational and functional states: closed (resting), open (active), and closed (inactive). In response to membrane depolarization, there is a transition from the closed to an open state. Transition between resting and inactive states occurs on binding of ligands or by protein phosphorylation.

In contrast to the cation channels, not much is known about the structure of the voltage-gated Cl^- channels (p. 943); they apparently have 12 transmembrane helical segments. Cl^- channels regulate cell volume, stabilize membrane potential, are involved in signal transduction, and regulate transepithelial transport. The cystic fibrosis transmembrane regulator is a Cl^- channel.

The voltage-gated ion channels have a distinctive expression pattern in different tissues. In humans, there are a considerable number of inherited ion channel diseases, termed channelopathies, caused by mutations in the anion and cation channels (see Clin. Corr. 3.8, p. 971). Clin. Corr. 3.12, p. 978, describes changes in cation voltage-gated channels in myotonic muscle disorders.

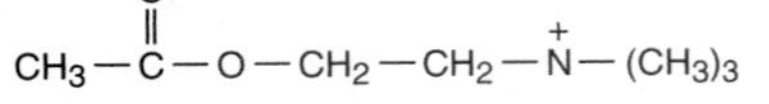

$$CH_3-\overset{\overset{\displaystyle O}{\|}}{C}-O-CH_2-CH_2-\overset{+}{N}-(CH_3)_3$$

Figure 12.39 Structure of acetylcholine.

Nicotinic-Acetylcholine Channel (nAChR)

The **nicotinic-acetylcholine channel**, also referred to as the **acetylcholine receptor**, is an example of a ligand-regulated channel, in which binding of acetylcholine (Figure 12.39) opens the channel. The dual name is used to differentiate this receptor from other acetylcholine receptors that function in a different manner. The channel is expressed at the nerve–muscle synapse (p. 948) and is a member of a superfamily of transmitter-gated ion channels, which includes the serotonin, γ-aminobutyric-acid, and glycine receptors (p. 942). Acetylcholine diffuses to the skeletal muscle membrane where it interacts with the acetylcholine receptor, opening the channel and allowing positively charged ions and small nonelectrolytes but not negatively charged ions to flow through the channel. The channel complex consists of a pentameric complex of two α- and one each of β-, γ-, and δ-subunits (Figure 12.40). Each α-subunit is glycosylated, and two other subunits contain covalently bound lipid. Each subunit has an extracellular N- and C-terminus, four membrane-spanning domains (M1, M2, M3, and M4), with a long cytoplasmic loop between M3 and M4 that contains consensus sites for phosphorylation by protein kinases. Five juxtaposed M2 domains (one from each protein subunit) form the ion-conducting pore. The M2 domains of cation-selective receptor channels, such as the nicotinic-acetylcholine receptor,

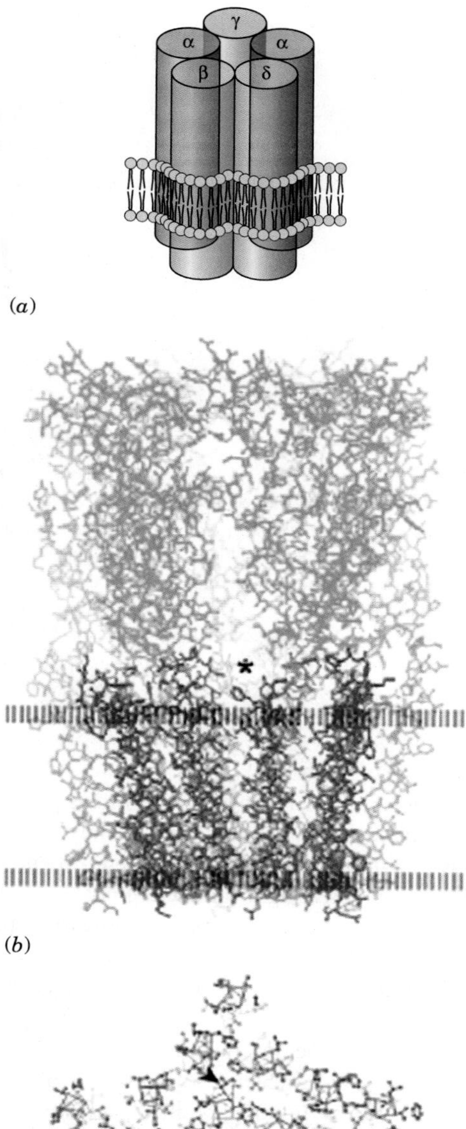

Figure 12.40 Model of acetylcholine (ACh) receptor. (*a*) The arrangement of the ACh receptor subunits and their common topology. (*b*) AcH receptor showing the α-helical pore structure (blue, pore-facing; red, lipid-facing helices) in relation to the membrane surfaces and the β-sheet structure (green) composing the ligand-binding domain; the asterisk denotes open space at a subunit interface. (*c*) Cross-sectional view through the pentamer at the middle of the membrane, showing partitioning of the structure into pore- and lipid-facing parts, with intervening spaces.
(*a*) Redrawn from Wilson, G. G., and Karlin, A. Acetylcholine receptor channel structure in the resting, open, and desensitized states probed with the substituted-cysteine accessibility method. *Proc. Nat.l Acad, Sci.USA* 98: 1241, 2001. (*b*) Reproduced with permission from Miyazawa, A., Fujiyoshi, Y., and Unwin, N. Structure and gating mechanism of the acetylcholine receptor pore. *Nature* 423: 949, 2003. Copyright (2003) *Nature.*

are enriched in negatively charged amino acids at the intracellular "mouth" of the pore. This region acts to attract cations like Na^+ and repel anions like Cl^-. Binding of the neurotransmitter to extracellular sites on the complex induces very small, subtle rearrangements of the M2 domains. This near-instantaneous rearrangement removes the energetic barriers to ionic flow through the water-lined pore. Phosphorylation of the α-subunit is also required for activity. Closure of the channel occurs within a millisecond due to rapid hydrolysis of acetylcholine and dissociation of the products, acetate and choline, from the protein.

Several deadly neurotoxins, including ***d*-tubocurarine**, the active ingredient of **curare**, and several toxins from snakes, including **α-bungarotoxin**, **erabutoxin**, and **cobratoxin**, inhibit the nicotinic-acetylcholine receptor. **Succinylcholine**, a muscle relaxant, opens the channel leading to depolarization of the membrane; succinyl choline is used in surgical procedures because its activity is reversible due to the rapid hydrolysis of the compound after cessation of administration.

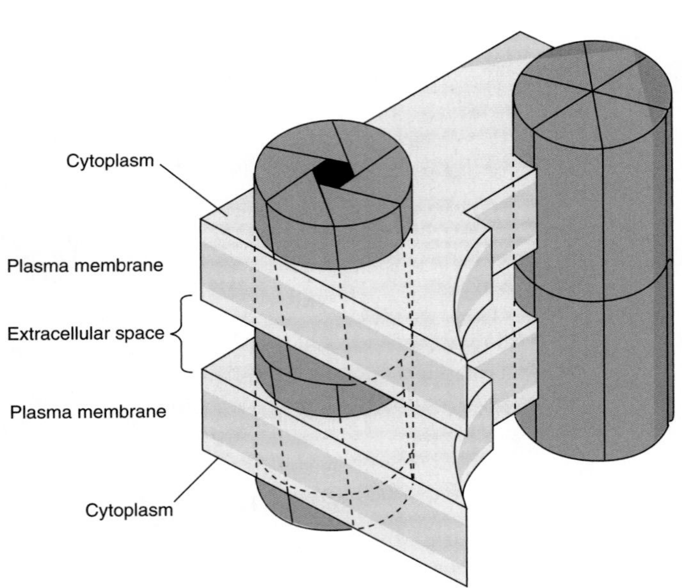

Figure 12.41 Model for a channel in the gap junction.

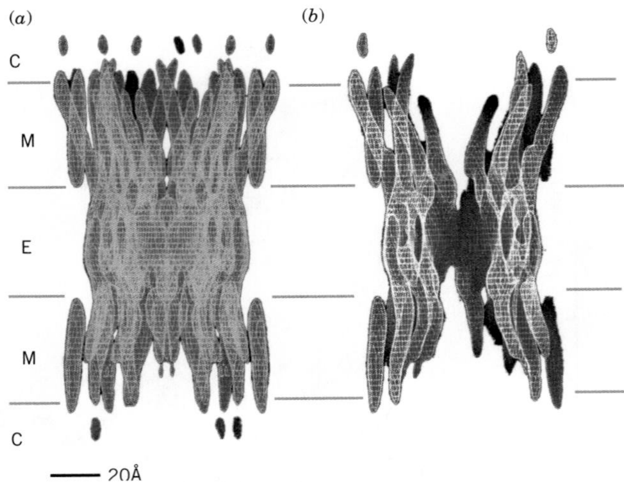

Figure 12.42 Molecular organization of recombinant cardiac gap junction. Model developed from electron crystallography. (*a*) A full side view is shown. (*b*) The density has been cropped to show the channel interior. M, membrane bilayer; E, the extracellular space; and C, the cytoplasmic space.

Reprinted from Unger, V. M., Kumar, N. M., Gilula, N. B., and Yeager, M. *Science* 283: 1176, 1999. With permission from AAAS.

Gap Junctions and Nuclear Pores

Gap junctions and nuclear pores are relatively large multimeric protein channels. **Gap junctions** found in mammalian plasma membranes consist of transmembrane protein channels, referred to as **connexons**. Connexons in one cell dock end-to-end with connexons in the membrane of a neighboring cell, creating a connecting channel (Figure 12.41). Connexons consist of homo- or hetero-hexameric proteins of **connexins (gap junction proteins)**. Twelve subunits, six in each cell forming a hexameric structure in each membrane. Over 15 isoforms of connexin subunit are known; each has four putative α-helical transmembrane sequences. The molecular organization of a recombinant gap junction is presented in Figure 12.42. The diameter of the opening ranges from 1.2 to 2 nm. The aqueous channel mediates electrical coupling, and exchange between cells of ions and metabolites but not macromolecules. Gap junctions composed of different connexins may exhibit differing specificities for solutes. The channels are normally open. Closure occurs with increases in cytoplasmic Ca^{2+}, a change in metabolism, a drop in transmembrane potential, or acidification of the cytoplasm. Control may also involve specific growth factors and phosphorylation of the connexins. When the channel is open the subunits appear to be slightly tilted but when closed they appear to be more perpendicular in the membrane, suggesting that subunits slide over each other.

Nuclear pore complexes are dynamic macromolecular assemblies that span the two membranes of the nuclear envelope. They create a channel for passage of proteins, RNA, and small molecules in both directions. The pores have a flexible diameter of about 9 to 20 nm and are 45 nm in length. The nuclear pore complex is composed of 50 to 100 proteins, called nucleoporins. Small molecules diffuse passively in both directions through the channel but macromolecules do not. Proteins, RNA, and ribonucleoprotein particles are carried in both directions by specific carrier proteins. Nuclear receptors, **karyopherins**, and cytosolic factors are involved in nuclear import and export of proteins. Karyopherin β mediates export of macromolecules bound to nucleoporins and a GTPase protein, **Ran**. Proteins destined for the nucleus from the cytoplasm contain short amino acid sequences, termed **nuclear localization sequences**, which target them for import. Transport through the nuclear pore complex occurs by facilitated diffusion of the soluble carrier proteins and carrier-macromolecule complexes. The activity is regulated.

The least understood of the pores is one involved in the mitochondrial permeability transition. This opening occurs in the inner membrane in response to oxidative stress, loss of adenine nucleotides, and increases in mitosolic Ca^{2+}. The structure has not been conclusively defined. Opening of the pore can lead to loss of mitochondrial matrix components and cytochrome c, and initiation of apoptosis (p. 1010). This pore may have an important role in tissue perfusion/reperfusion injury following tissue ischemia such as occurs in a stroke.

12.8 • MEMBRANE TRANSPORT PROTEINS

Membrane transport proteins facilitate movement of ions and molecules through a membrane by binding and physically moving them to access the environment on the opposite side of the membrane. It should be noted that physically moving the substrate does not require the protein to actually move the substrate across the entire width of a membrane. As will be discussed (p. 485), a small distance can move the substrate between two different environments in the protein. For most substances the rate of mediated transport is significantly greater than that of simple diffusion through a lipid membrane.

If S_1 is the solute on side 1 and S_2 on side 2, then the transporter promotes the establishment of equilibrium as follows:

$$[S_1] \rightleftharpoons [S_2]$$

Brackets represent the concentration of solute. If the transporter T is included, the reaction becomes

$$[S_1] + T \rightleftharpoons [S\text{–}T] \rightleftharpoons [S_2] + T$$

If no energy from another source is expended, the concentration on both sides of the membrane will be equal at equilibrium, but if there is an expenditure of energy, a concentration gradient can be established. As with an enzyme-catalyzed reaction, the transporter increases the rate but does not determine the final equilibrium. The transporter protein is unchanged at the end of the reaction. Its activity can be evaluated in the same kinetic terms as an enzyme-catalyzed reaction, because it has specificity for the substance to be transported, defined kinetics, catalyzes a vectorial transport, and is affected by both competitive and noncompetitive inhibitors. The transporter demonstrates saturation kinetics (Figure 12.33, p. 476), whereas simple diffusion does not. Constants such as V_{max} and K_m are determined for transporters. Theoretically a transporter can facilitate movement of a solute in both directions across the membrane but in most situations there is a unidirectional movement because of a large $\Delta G^{o\prime}$ for the reaction.

Most transporters have a high degree of structural and stereospecificity for substrate. In mediated transport of D-glucose in erythrocytes, the K_m for D-galactose is 10 times larger and for L-glucose 1000 times larger than for D-glucose, demonstrating a much lower affinity for D-galactose and L-glucose. The transporter has essentially no activity with D-fructose or disaccharides. Structural analogs of the substrate for a transporter inhibit competitively and reagents that react with specific groups on proteins are noncompetitive inhibitors.

Four Steps in the Transport of Solute Molecules

There are four steps of mediated transport (Figure 12.43): (1) recognition by transporter of appropriate solute from amongst a variety of solutes in the aqueous environment, (2) translocation of solute, (3) release of solute by transporter, and (4) recovery of transporter to its original condition.

The first step, **recognition** of a specific substrate by the transporter, is the same as recognition of a substrate by an enzyme (p. 387). Specific binding sites on the protein recognize the correct structure of the solute. The second step, **translocation**, requires a conformational change in the transporter after binding of the substrate (Figure 12.44), which moves the substrate a short distance, perhaps only 2 or 3 Å to the environment of the opposite side of the membrane. It is not necessary for the transporter to move the molecule the entire distance across the membrane (Figure 12.45). The conformational change of the protein is sufficient to change the environment. Step 3, **release** of

Recognition:	$S_1 + T_1$	$\rightleftharpoons$	$S - T_1$
Transport:	$S - T_1$	$\rightleftharpoons$	$S - T_2$
Release:	$S - T_2$	$\rightleftharpoons$	$T_2 + S_2$
Recovery:	T_2	$\rightleftharpoons$	T_1

Figure 12.43 Steps involved in mediated transport across a biological membrane. S_1 and S_2 are solutes on side 1 and 2 of the membrane, respectively; T_1 and T_2 are binding sites on the transporter on sides 1 and 2, respectively.

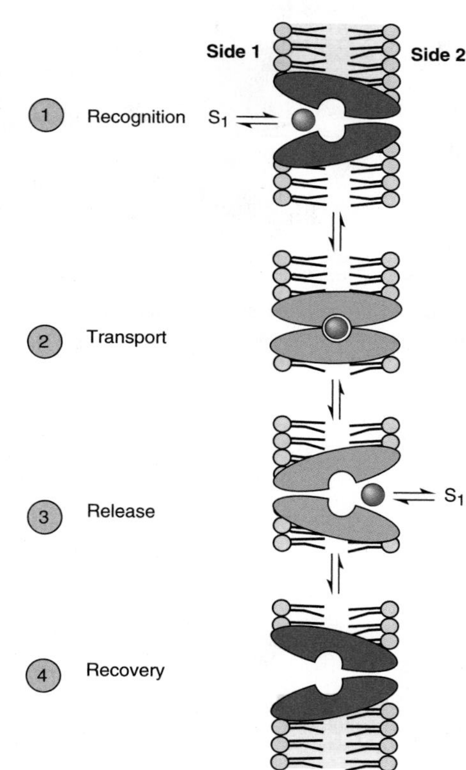

Figure 12.44 Model for a mediated transport system in a biological membrane. Model is based on the concept of specific sites for binding of substrate and a conformational change in the transporter to move bound solute a short distance but into the environment of the other side of the membrane. Once moved, solute is released from the transporter.

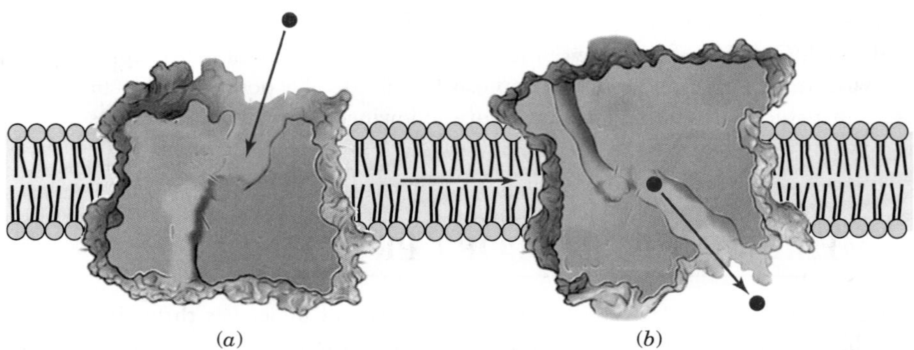

Figure 12.45 Transport of a substrate through a membrane. After binding of substrate on one side of the membrane, a conformational change in the protein occurs moving the substrate to an environment on the opposite side.
Model of protein redrawn from Arkin, I. T., et al. Mechanism of Na^+/H^+ antiporting. *Science* 317: 799, 2007.

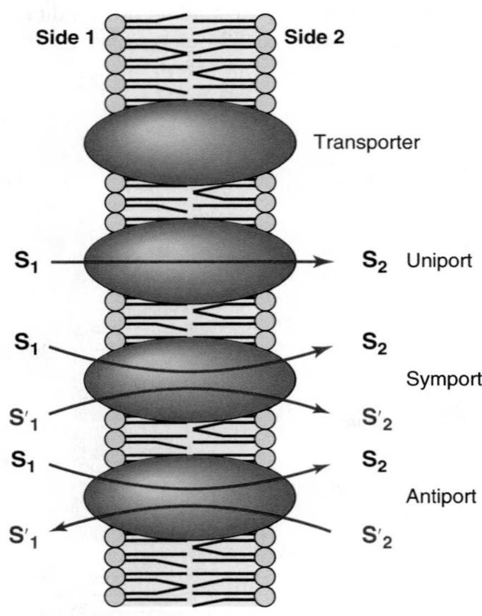

Figure 12.46 Uniport, symport, and antiport mechanisms for translocation of substances. S and S' represent different molecules.

substrate, depends on the affinity of the transporter for the substrate. Without a change in affinity (K_{eq}) for substrate, the substrate will be released in the new environment if the concentration of substrate is lower in the new environment than in the initial environment because of a shift in the equilibrium between bound and free substrate. For a transporter that moves a substrate against a concentration gradient, release of the substrate at the higher concentration requires the transporter to change so that it has a lower affinity (higher K_m) for the substrate. A conformational change of transporter can lead to a decrease in affinity. Finally in **recovery**, step 4, the transporter returns to its original conformation on release of substrate.

This discussion centered on movement of a single substrate by a transporter (**uniport** mechanism). There are systems that move two substrates simultaneously in one direction (**symport**) and two substrates simultaneously in opposite directions (**antiport**) (Figure 12.46). When a charged substance is translocated and no ion of the opposite charge is moved or two molecules with different charges are translocated by an antiport mechanism, a charge separation occurs across the membrane. This mechanism is termed **electrogenic transport** and leads to development of a membrane potential. If an oppositely charged ion is moved to balance the charge or two molecules of the same charge are moved in opposite directions across the membrane, the mechanism is termed **neutral** or **electrically silent transport**.

Energetics of Membrane Transport Systems

Eq. 12.2 gives the change in free energy when an uncharged molecule moves from concentration C1 to concentration C2 on the other side of a membrane.

$$\Delta G' = 2.3RT\log\left(\frac{C_2}{C_1}\right) \tag{12.2}$$

When $\Delta G'$ is negative—that is, there is release of free energy—movement of solute occurs without the need for a driving force. When $\Delta G'$ is positive, as would be the case if C_2 is larger than C_1, then there needs to be input of energy to drive the transport. For a charged molecule (e.g., Na^+) both the electrical potential and concentrations of solute are involved in calculating the change in free energy as indicated in Eq. 12.3

$$\Delta G' = 2.3RT\log\left(\frac{C_2}{C_1}\right) + Z\mathcal{F}\Delta\Psi \tag{12.3}$$

where $\Delta G'$ is electrochemical potential, Z is charge of the transported species, $\mathcal{F}$ is Faraday constant (96.49 kJ/V/mol, 23.06 kcal/V/mol), and $\Delta\Psi$ is electrical potential difference in volts across the membrane. The magnitude and sign of both Z and $\Delta\Psi$ influence $\Delta G'$. When $\Delta G'$ is negative movement of solute occurs spontaneously; this is often referred to as passive transport. When $\Delta G'$ is positive, input of energy from some source is required and the process is called active transport.

Transport in mammalian cells requiring energy input is driven by either hydrolysis of nucleoside triphosphate (e.g., ATP $\rightarrow$ ADP + phosphate) or utilization of the electrochemical gradient of Na^+ (**sodium motive force, SMF**) or H^+ (**proton motive force, PMF**). Prokaryotes also have transporters driven by decarboxylation and methyl transfer reactions, and plants have light-absorption-driven transporters. The SMF is generated at the expense of ATP hydrolysis and the PMF by oxidation–reduction reactions (p. 572). Inhibition of ATP synthesis or collapse of the PMF will dissipate the Na^+ and/or H^+ gradient leading to a cessation of transport.

In the simplest transport mechanism, that is, the transport of a single solute by a uniport mechanism, molecules are transported down their chemical gradient if they have no charge or down an electrochemical gradient if charged; this is often referred to as **protein-mediated diffusion** or **facilitated diffusion.** These transporters cannot create a concentration gradient of the substance across the membrane. In a symport (two molecules being moved simultaneously in the same direction) or antiport (two molecules being moved simultaneously in opposite directions), movement down a concentration gradient of one of the solutes can drive the translocation of the other molecule. These have been referred to as **secondary active transporters**, to distinguish them from primary active transporters, which utilize energy derived from cellular metabolism directly. The large Na^+ concentration gradient across the plasma membrane (high extracellular and low intracellular concentration; Figure 1.9, p. 14) is used to translocate sugars and amino acids by a symport mechanism into cells.

Some transporters maintain very large concentration gradients, such as the plasma membrane transporter that maintains the Na^+ and K^+ gradient (p. 491). Another striking example is an active transporter in parietal cells of gastric glands, which is responsible for secretion of HCl into the lumen of the stomach (p. 1042). The pH of plasma is about 7.4 (4×10^{-8} M H^+), and the luminal pH of the stomach can reach a pH of 0.8 (0.15 M H^+). H^+ is transported against a concentration gradient of greater than 1×10^6. Assuming there is no electrical component, the energy for H^+ secretion under these conditions is calculated from Eq. 12.1 and is 38 kJ mol/L (9.1 kcal mol/L) of HCl.

Mammalian Membrane Transporters

The two major protein classes of transporters in mammalian cells (see Table 12.8, p. 477) are

- **electrochemical-potential-driven transporters**, which move their substrate across a membrane by utilizing the concentration gradient of the substrate or another solute, and
- **primary active transporters**, which require the input of metabolic energy to move the substrate.

Mammalian cells have a wide variety of transporters and in many instances use different subfamilies of transporters to translocate the same substrate depending on the specific requirements of the cell or tissue. Transport systems are present in plasma as well as intracellular membranes, including mitochondria, lysosomes, peroxisomes, and endosomes. The number of active transporters present in a membrane at a given time establishes the maximum rate of uptake by a cell or organelle. Thus, in many cases, the synthesis of the protein transporter and incorporation into a membrane controls the concentration of a particular substrate in a cell or cellular compartment. All of these translocation systems are regulated, permitting moment-by-moment control of transport. Many genetic diseases are attributable to defects in the various transport systems (Clin Corr. 12.4).

12.9 • ELECTROCHEMICAL-POTENTIAL-DRIVEN TRANSPORTERS

Electrochemical-potential-driven transporters utilize

1. a uniport mechanism where a single solute is transported either by mediated diffusion or in a membrane-potential-dependent manner if the solute is charged;

CLINICAL CORRELATION 12.4

Diseases due to Loss of Membrane Transport Systems

A number of pathological conditions are due to an alteration in a transport system for specific cellular components. Individuals with a decrease in glucose uptake from the intestinal tract lack the specific sodium-coupled glucose–galactose transporter. Fructose malabsorption syndromes are caused by an alteration in the activity of the transport system for fructose. In Hartnup disease there is a decrease in the transport of neutral amino acids in the epithelial cells of the intestine and renal tubules.

In cystinuria, renal reabsorption of cystine and the basic amino acids lysine, arginine, and ornithine is abnormal, resulting in formation of renal cystine stones. In hypophosphatemic, vitamin D–resistant rickets, renal absorption of phosphate is abnormal. Recently there have been reports about genetic-based diseases involving the substrate transporters for dopamine and creatine, and the mitochondrial aspartate glutamate transporter. The various ATP-requiring active transporters have been implicated in a number of pathophysiological conditions.

Evans, L., Grasset, E., Heyman, M., Dumontier, A. M., et. al. Congenital selective malabsorption of glucose and galactose. *J. Pediatr. Gastroenterol. Nutr.* 4: 878, 1985; Mueckler, M. Facilitative glucose transporters. *Eur. J. Biochem.* 219: 713, 1994; deGrauw, T. J., Cecil, K. M., Byars, A. W., Salomons., G. S., et al. The clinical syndrome of creatine transporter deficiency. *Mol. Cell. Biochem.* 244: 45, 2003; and Pederson, P. L. Transport ATPases in biological systems and relationship to human disease: A brief overview. *J. Bioenergetics Biomembranes* 34: 327, 2002.

2. an antiport mechanism where two or more solutes are transported in opposite directions in a tightly coupled process not utilizing any form of energy other than the electrochemical-potential gradient; the gradient across the membrane of one solute can drive the movement of the other solute
3. a symport mechanism where two or more species are transported together in the same direction in a coupled process not utilizing any form of energy other than the electrochemical-potential gradient of one substrate; the gradient of one solute drives the movement of the other solute

Hundreds of such transporters have been reported, many exclusively in bacteria, with specificity for inorganic ions, sugars, amino acids, and other metabolic intermediates. Table 12.11 illustrates the variety of solutes transported in eukaryotic cells by this mechanism. Many of the transporters have 2 to 24 putative α-helical transmembrane sequences. Electrochemical-potential-driven transporters demonstrate saturation kinetics, specificity of substrate moving across the membrane, and specificity for inhibitors.

Representative Electrochemical-Potential-Driven Transporters

Glucose is Translocated by a Passive Uniport Mechanism

A family of **glucose transporters** (**GLUT**) facilitates transport of D-glucose across the plasma membranes of mammalian cells by a uniport mechanism. The genes for the GLUT

TABLE 12.11 • Representative Substrates for Membrane Transport Proteins in Eukaryotes

Inorganic cations	H^+, K^+, Na^+, NH_4^+, Ca^{2+}, Co^{2+}, Cu^{2+}, Fe^{2+}, Mg^{2+}, Zn^{2+}
Inorganic anions	$C1^-$, HCO_3^-, SO_4^{2-}, iodide, phosphate, arsenite
Sugars	Fructose, glucose, hexoses, lactose, maltose, myo-inositol
Organic acids	Acetate, bile acids, bilirubin, citrate, glucose 6-phosphate, α-ketoglutarate, lactate, malate, oxalacetate, prostaglandins
Amino acids	Acidic, basic, neutral, branched chain
Peptides/proteins	Some
Nucleosides/nucleotides	ATP, ADP, all others
Vitamins/cofactors	Ascorbate, biotin, folate, lipoate, thiamin
Drugs	Multiple drugs

Source: Moss, G. P. Membrane Transport Proteins. http://www.chem.qmul.ac.uk/iubmb/mtp/

TABLE 12.12 • Major Isoforms of Glucose Transporter

Isoform	*Tissue*	*Cellular Localization*	*Comments*
GLUT1	Muscle Heart Blood-brain barrier Glia cells Placenta	Plasma membrane	Activity in muscle increased by insulin, hypoxia, and diet
GLUT2	Liver Pancreas Intestine Kidney	Plasma membrane	
GLUT3	Brain Kidney Placenta	Plasma membrane	
GLUT4	Muscle Adipose Heart Blastocysts	Plasma membrane	Activity in muscle and adipose tissue stimulated by insulin, and in muscle by hypoxia and diet
GLUT5	Muscle Spermatozoa	Sarcolemmal vesicles	Fructose as substrate

family are expressed in a tissue specific manner. The glucose transporters, designated as GLUT1, GLUT2, and so on (Table 12.12), all have 12 transmembrane segments, with a significant amino acid similarity. Most are in the plasma membrane where direction of movement of glucose is usually out to in. Depending on the concentration gradient, however, erythrocyte **GLUT1** facilitates transport in both directions demonstrating the reversibility of the system. The three high-affinity transporters (GLUT1, **GLUT3**, and **GLUT4**) function *in vivo* at rates close to maximal velocity because their K_ms are below the normal blood glucose concentration. D-Galactose, D-mannose, D-arabinose, and several other D-sugars as well as glycerol are translocated by some isoforms. **GLUT5** of sarcolemmal membranes of skeletal muscle transports fructose preferentially. GLUT1 is important for supplying erythrocytes and brain cells with glucose. The rate of glucose uptake by the low-affinity GLUT2 increases in parallel with the rise in blood glucose concentration and has a major role in intestinal absorption of sugars (p. 1049). In liver cells, **GLUT2** catalyzes both glucose influx and efflux, having responsibility for glucose export (p. 596); in pancreatic β-cells, GLUT2 it is involved in sensing blood glucose levels. GLUT4 is an insulin-responsive transporter (p. 595) in adipose tissue, heart muscle, and skeletal muscle. Activity of some glucose transporters in muscle is stimulated by exercise and hypoxia apparently because there is a need for increased glucose utilization by the tissue. Several sugar analogs as well as phloretin and 2,4,6-trihydroxyacetophenone (Figure 12.47) are competitive inhibitors.

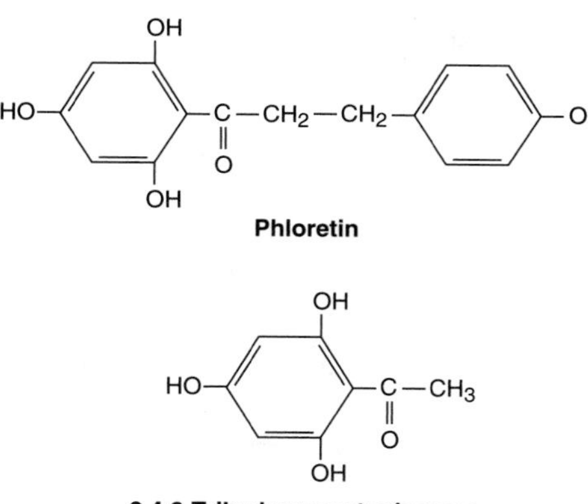

Figure 12.47 Inhibitors of passive transport of D-glucose in erythrocytes.

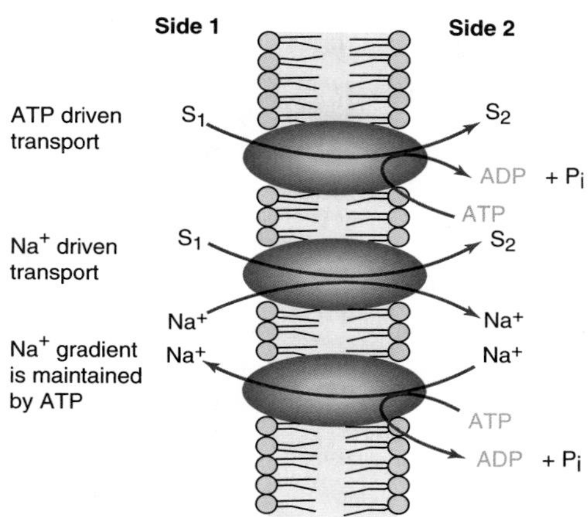

Figure 12.48 Involvement of metabolic energy (ATP) in active mediated transport systems. Chemical energy released on hydrolysis of ATP to ADP and inorganic phosphate is used to drive transport of various substances, including Na^+. The transmembrane concentration gradient of Na^+ is used for active transport of substances by symport and antiport (not shown) mechanisms.

Glucose and Amino Acids Are Translocated with Na+ by a Symport Mechanism

The Na^+ electrochemical gradient (**sodium motive force, SMF**) across the plasma membrane (p. 14) is an energy source for symport movement of Na^+ with sugars, amino acids, ions, and other small molecules. There are over 400 expressed transporters in this family of proteins. As described previously (p. 487), the SMF is generated by transport of Na^+ coupled to hydrolysis of ATP (Figure 12.48); inhibition of ATP synthesis will lead to dissipation of the Na^+ gradient and cessation of transport.

The general mechanism of the **sodium/glucose cotransporter** (**SGLT**) is presented in Figure 12.49. The diagram represents transport of D-glucose driven by the movement of Na^+. In the process two Na^+ move into the cell down the Na^+ electrochemical gradient by passive facilitated transport, and glucose is carried along against its concentration gradient.

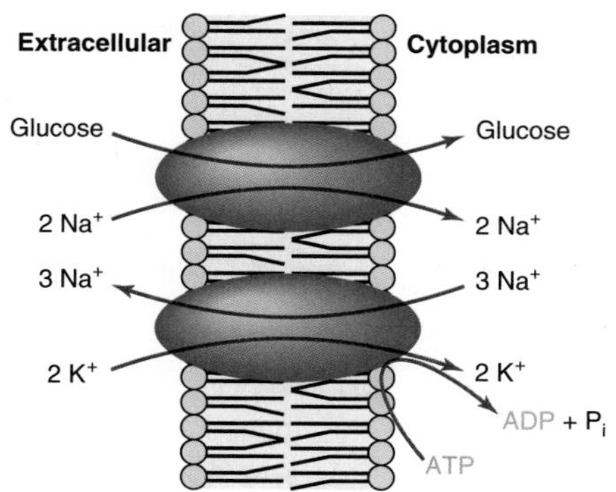

Figure 12.49 Na^+-dependent transport of glucose across the plasma membrane by a symport mechanism transport.

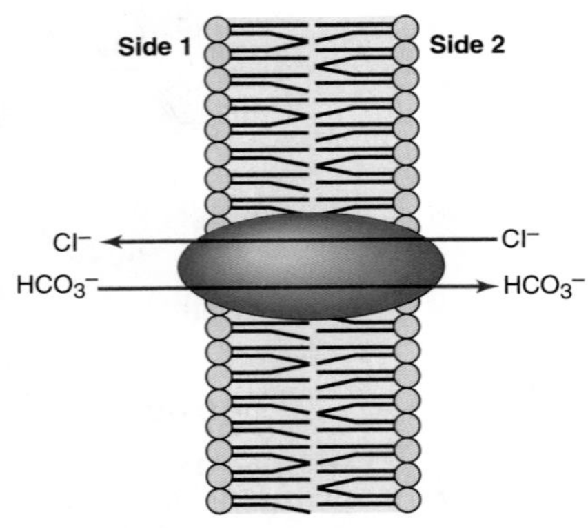

Figure 12.50 Anion antiport mechanism for movement of Cl^- and HCO_3^- across the erythrocyte plasma membrane.

The Na^+ gradient dissipates in the process but the Na^+/K^+-exchanging ATPase continuously reestablishes it (p. 491). Thus, metabolic energy in the form of ATP is indirectly involved in transport because it is required to maintain the Na^+ gradient. Inhibition of the Na^+/K^+-exchanging ATPase leads to a decrease in the Na^+ gradient, thus preventing glucose uptake by this mechanism.

Over 37 members of the SGLT gene family have been identified in animal cells, yeast, and bacteria; four are expressed in humans. Some isoforms are expressed in proximal renal tubules and small intestinal epithelium (p. 1042). All SGLTs have a common core structure of 13 transmembrane helices, but different isoforms have additional transmembrane segments, including human SGLT1 which has 14. SGLTs can catalyze a uniport movement of Na^+ and water in the absence of glucose. Na^+ binds in an extracellular cavity of the transporter, inducing a conformational alteration that increases the affinity of the transporter for glucose. The binding of glucose, forming a ternary complex, induces another structural change that exposes Na^+ and substrate to a cavity in the protein contiguous with the intracellular environment on the other side of the membrane. Na^+ dissociates from the transporter because of the low intracellular Na^+ concentration leading to a return of the transporter to its original conformation, which has a decreased affinity of glucose. Thus glucose is released. The empty transporter is reorientated in the membrane allowing the cycle to begin again.

Amino acids are translocated across intestinal epithelial cells by **Na^+/amino acid transporters** by a symport mechanism; at least seven transporters have been identified for different classes of amino acids (p. 78). The Na^+ gradient across the plasma membrane is also utilized to drive the transport of other ions, including a symport mechanism for uptake of Cl^- and an antiport mechanism for excretion of Ca^{2+}.

Cl^- and HCO_3^+ Are Transported by an Antiport Mechanism

An anion transporter in erythrocytes and kidneys mediates the electroneutral antiport movement of Cl^- and HCO_3^- (Figure 12.50). The transporter is referred to as the **Na^+ independent Cl^-–HCO_3^- exchanger** or **anion exchange protein**, and in erythrocytes as band 3, the latter because of its position in SDS polyacrylamide gel electrophoresis of erythrocyte membrane proteins. The direction of ion flow depends on the concentration gradients of the ions across the membrane. The transporter is important in adjusting the erythrocyte HCO_3^- concentration in arterial and venous blood. The kidney transporter is a truncated form of the erythrocyte protein and is responsible for base (HCO_3^-) efflux to balance ATP driven H^+ efflux.

Na^+/H^+ Exchanger Regulates Cytoplasmic pH

The nine mammalian isoforms of the **Na^+/H^+ transporter** are members of a large family of proteins in bacteria, archaea, yeast, plants, and animals. They vary in size from 541 to 894 residues and have 10–12 predicted transmembrane α-helical segments. A major physiological function is to extrude H^+ generated during metabolism, thus regulating cytoplasmic pH and preventing intracellular acidification. In the kidney, isoform-3 serves to recover Na^+ from the nephron lumen. Isoform-1 in heart muscle is activated by decreased pH and by phosphorylation. This isoform may have a significant role in the pathogenesis of a myocardial infarct and tissue hypertrophy.

Mitochondria Have a Number of Electrochemical-Potential-Driven Transporters

The inner mitochondrial membrane contains transporters for exchange of anions between cytoplasm and mitochondrial matrix, including (1) an antiporter for exchange of ADP and ATP, (2) a symporter for transport of phosphate and H^+, (3) a dicarboxylate transporter, and (4) an antiporter for glutamate-aspartate (see p. 578 for a detailed discussion). These transporters are important in energy conservation and metabolism. In the absence of an input of energy, these transporters can mediate a passive exchange of metabolites down their concentration gradient to achieve a thermodynamic equilibrium. In some cases, the transporter mediates the antiport movement of an equal number of electrical charges on the substrates with the mitochondrial membrane potential influencing the equilibrium.

12.10 • PRIMARY ACTIVE TRANSPORTERS

Primary active transporters in mammalian cells include transport systems where

1. the free energy of hydrolysis of a pyrophosphate bond (p. 543) is utilized to drive the movement of substrate against their chemical or electrochemical gradient; and
2. the transport protein may be transiently phosphorylated during the transport process, but the substrate is not phosphorylated or modified.

This family of transporters catalyzes the translocation of a solute against its concentration gradient, leading to creation of a chemical or electrochemical gradient. Like all transporters, they are characterized by saturation kinetics, specificity for one or a few related substrates, and an ability to be inhibited. They are found in all living organisms. Mammalian cells utilize ATP or another nucleoside triphosphate as the energy source. Transporters utilizing ATP are also referred to as **ATPases**, indicating their enzyme activity in hydrolyzing ATP during translocation. Decarboxylation, methyl transfer, oxidoreduction, and light absorption serve to drive transport in other organisms.

Three subfamilies of primary active transporters catalyze the translocation of inorganic cations (Na^+, K^+, Ca^{2+}, or H^+) and are classified as P-, V-, or F-type transporters. **P-type transporters** are phosphorylated and dephosphorylated during transport; there are over 300 members of this subfamily. **V-type transporters** (V for vacuole) are proton pumps, responsible for acidification of the interior of lysosomes, endosomes, Golgi vesicles, and secretory vesicles. **F-type transporters**, present in mitochondria, chloroplasts, and bacteria, translocate protons at the expense of ATP hydrolysis, but synthesize ATP when functioning in the reverse direction, that is, when protons are transported down their concentration gradient (p. 572).

The ATP-binding cassette (ABC) transporters are a very large and unique superfamily of ATP dependent primary active transporters, required for transport of a variety of inorganic and organic molecules.

Representative Primary Active Transporters

Antiport Translocation of Na^+ and K^+ is by Primary Active Transport

The external or plasma membrane of cells from all forms of life contain a P-type Na^+/K^+ antiporter, which utilizes the energy from hydrolysis of ATP to ADP+ phosphate to drive translocation. The transporter is termed the **Na^+/K^+-exchanging ATPase**, but it is frequently referred to as **Na^+/K^+-ATPase** or **Na^+ pump**. This superfamily of multi-subunit proteins found in bacteria, archaea, and eukaryotes have a large subunit that catalyzes both the ATPase activity and ion translocation. Members of the superfamily have been described in bacteria for uptake and/or efflux for K^+, Mg^{2+}, Ca^{2+}, Ag^+, Zn^{2+}, Co^{2+}, Pb^{2+}, Ni^{2+}, Cd^{2+}, and Cu^{2+}.

The plasma membranes of all mammalian cells catalyze the reaction

$$\text{ATP} \xrightarrow[\text{Mg}^{2+}]{\text{Na}^+ + \text{K}^+} \text{ADP} + \text{Pi} \tag{12.4}$$

with an absolute requirement for Na^+ and K^+ ions, and for Mg^{2+}, a cofactor for ATP-requiring reactions. The transporter is responsible for maintaining the high K^+ and low Na^+ concentrations in the cytoplasm of mammalian cells (p. 14). Excitable tissues, such as muscle and nerve, and cells actively involved in movement of Na^+ ion, such as those in the salivary gland and kidney cortex, have high activities of the Na^+/K^+-exchanging ATPase. It has been estimated that this transporter uses about 60%–70 % of the ATP synthesized by nerve and muscle cells and about 35% of ATP generated in a resting individual.

The transporter consists of an α-subunit (110 kDa) and a β-subunit (55 kDa); there is some question whether the actual transporter may be a tetramer with two α- and two β-subunits. The α-subunit has 10 transmembrane α-helical segments (Figure 12.51), contains both the ATPase and translocation activity, and is phosphorylated and dephosphorylated on a specific aspartic acid residue to form a β-aspartyl phosphate during ion translocation. The cation and ATP-binding sites are on different domains of the α-subunit. The β-subunit

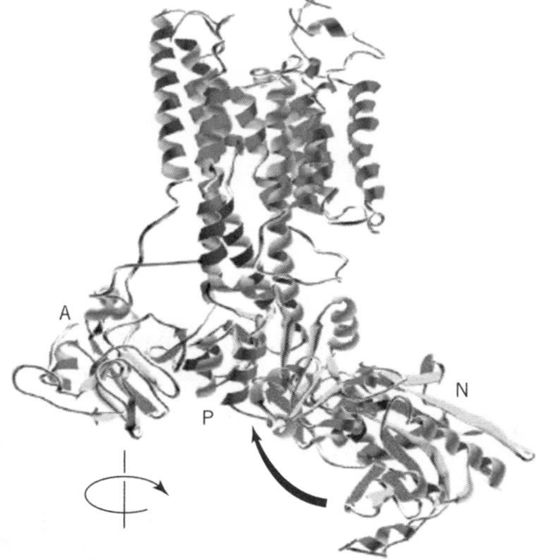

Figure 12.51 Structure of the α-subunit of Na^+/K^+-exchanging ATPase. Model of the sheep Na^+/K^+-exchanging ATPase α-subunit. The predicted domains are the nucleotide binding (*N*) and the phosphorylation domain (*P*). α-Helices are shown in shades of blue and green, β-sheets are shown in shades of red and yellow, and coils are shown in gray.
Reproduced with permission from Kaplan, J. H. Biochemistry of Na^+, K^+-ATPase. *Annu. Rev. Biochem.* 71: 511, 2002. Copyright (2002) Annual Reviews; www.annualreviews.org

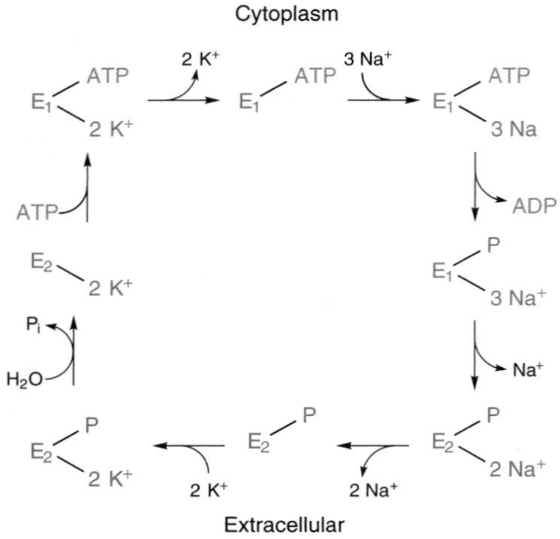

Figure 12.52 Proposed sequence of reactions and intermediates in hydrolysis of ATP by the Na^+/K^+-exchanging ATPase. E_1 and E_2 are different conformations of the enzyme. Phosphorylation of the enzyme requires Na^+ and Mg^{2+} and dephosphorylation involves K^+.

is heavily glycosylated, facilitates insertion of the α-subunit into the membrane after synthesis, and stabilizes the complex; it may also function actively in the transport activity. The transporter from the kidney contains a γ-subunit, which may influence kinetic parameters of the activity. Phosphorylation of the α subunit requires Na^+ binding and Mg^{2+} but not K^+, whereas dephosphorylation requires K^+ binding but not Na^+ or Mg^{2+} (Figure 12.52). The protein complex has two distinguishable conformations and is classified as an **E1-E2-type transporter**, to indicate the change in conformation.

Movement of Na^+ and K^+ is an electrogenic antiport process, with three Na^+ ions moving out and two K^+ ions into the cell for each ATP molecule hydrolyzed. This leads to an increase in external positive charge and is part of the mechanism for maintaining the transmembrane potential in cells. ATP hydrolysis by the transporter occurs only if Na^+ and K^+ are translocated. A model for movement of Na^+ and K^+ is presented in Figure 12.53. The protein goes through conformational changes during which the Na^+ and K^+ move short distances within the protein. This conformational transition decreases the affinity of the binding site for the cations, resulting in the release of the cation into a milieu where the concentration is higher than that from which it was transported.

Various isoforms of the transporter are regulated by different mechanisms including substrate concentration, the cytoskeleton, circulating inhibitors, and a variety of hormones. The activity has a requirement for phospholipids; the transporter has binding sites for about 30 lipid molecules and levels of docosahexaenoic acid (long-chain polyunsaturated fatty acid) in brain correlate with the ATPase activity. **Cardiotonic steroids**, such as **digitalis**, are inhibitors of the Na^+/K^+-exchanging ATPase and increase the force of contraction of heart muscle by altering the excitability of the tissue. **Ouabain** (Figure 12.54) is one of the most active inhibitors of Na^+/K^+-exchanging ATPase; it binds on the small subunit on the external surface of the membrane.

Ca^{2+} Translocation Is Catalyzed by a Primary Active Transporter

Calcium is an important intracellular messenger that regulates many cellular processes from carbohydrate metabolism to muscle contraction. Cytosolic Ca^{2+} is about 0.10 μM, over 10,000 times less than that of extracellular Ca^{2+}. Ca^{2+} in the cytosol increases rapidly by release of Ca^{2+} from the endoplasmic or sarcoplasmic reticulum or transient opening of Ca^{2+} channels in the plasma membrane permitting flow of Ca^{2+} down the large concentration gradient (p. 481). To reestablish low cytosolic levels, Ca^{2+} is actively transported by two different **Ca^{2+}-transporting ATPases (Ca^{2+}-ATPase)**, one transporting Ca^{2+} back into the lumen of the endoplasmic or sarcoplasmic reticulum, and the other out of the cell across the plasma membrane. The Ca^{2+}-ATPases are P-type transporters, phosphorylated by ATP on an aspartyl residue; they exist in two conformational states, thus are E1-E2-type transporters.

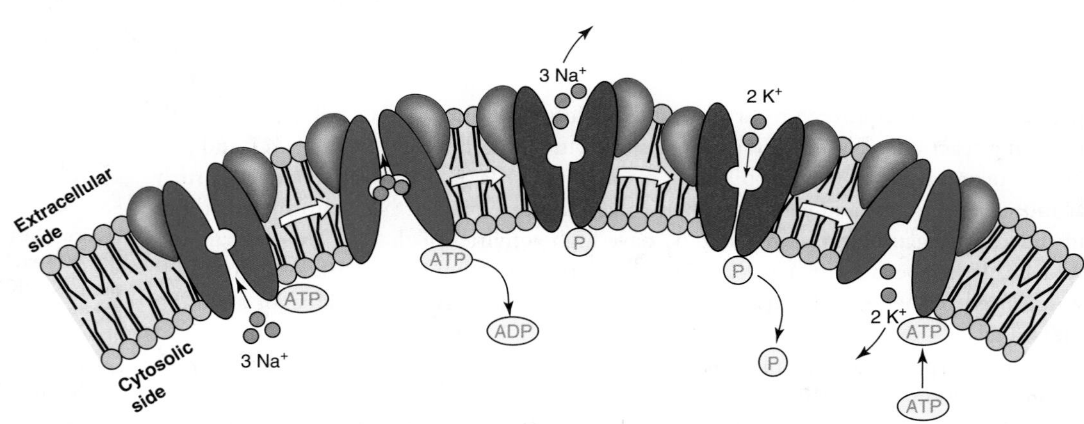

Figure 12.53 Model for translocation of Na^+ and K^+ across plasma membrane by the Na^+/K^+-exchanging ATPase. (1) Transporter in conformation 1 binds Na^+. (2) Transporter in conformation 2 translocates and releases Na^+. (3) Transporter in conformation 2 binds K^+. (4) Transporter in conformation 1 translocates and releases K^+.

The Ca^{2+}-ATPase of muscle sarcoplasmic reticulum (SERCA) is involved in the contraction–relaxation cycles of muscle, representing 80% of the integral membrane protein of the sarcoplasmic reticulum and occupying one-third of the surface area (p. 975). It consists of a single polypeptide chain with ten transmembrane helices (Figure 12.55). Two Ca^{2+}s are translocated from the cytosol to the lumen against a large concentration gradient (10^4), in an antiport exchange for 2 H^+ per ATP hydrolyzed. In the E1 state (Figure 12.56), the binding sites have a high affinity for Ca^{2+} and are accessible from the cytosol, whereas in the E2 state, the Ca^{2+}-binding sites have low affinity and are open to the lumen. The Ca^{2+}-free form shows large conformational differences from the Ca^{2+}-bound form (Figure 12.56). The two Ca^{2+}s are bound side by side, surrounded by four transmembrane helices. It has been proposed that the sarcoplasmic reticulum Ca^{2+}-ATPase may also play a role in **thermogenesis** by binding Ca^{2+}, followed by hydrolysis of ATP, but releasing the Ca^{2+} back into the cytosol without translocation of the cation. The energy of the ATP hydrolysis is released as heat. A small peptide, **sarcolipin**, increases heat production.

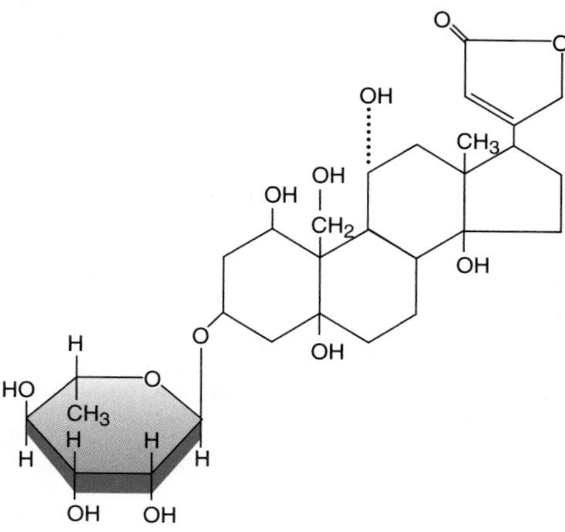

Figure 12.54 Structure of ouabain, a cardiotonic steroid, a potent inhibitor of the Na^+/K^+-exchanging ATPase.

Calmodulin Controls the Ca^{2+}-Transporting ATPase of Plasma Membranes

The Ca^{2+}-transporting ATPase of plasma membranes (PMCA) resembles that of the sarcoplasmic reticulum. Thirty isoforms of PMCA are known. In contrast to the SERCA, it transports only one Ca^{2+} per ATP hydrolysed. In eukaryotic cells, the transporter is regulated by cytoplasmic Ca^{2+} levels through **calmodulin,** a Ca^{2+}–binding protein (Figure 12.57). As cytosolic Ca^{2+} levels increase, Ca^{2+} binds to calmodulin, which has a dissociation constant of ~ 1 μM for Ca^{2+}. The Ca^{2+}–calmodulin complex binds to the Ca^{2+}

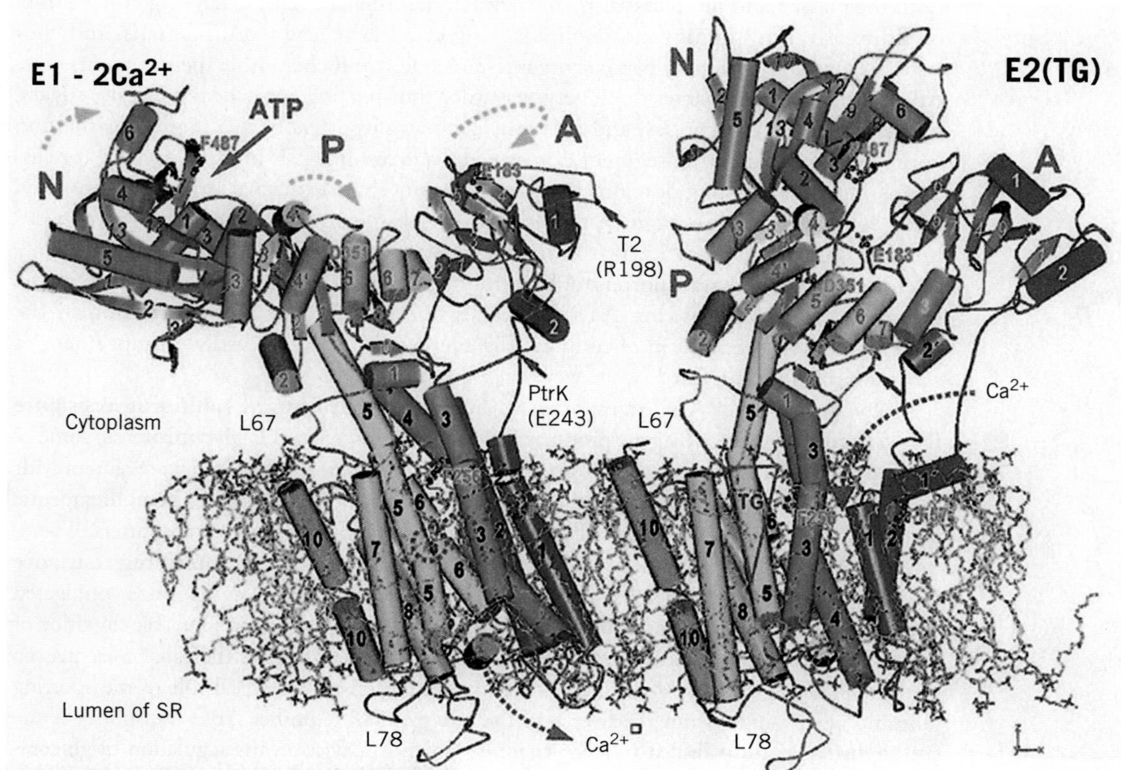

Figure 12.55 Structure of Ca^{2+}-transporting ATPase. Conformational changes of the Ca^{2+}-transporting ATPase of the sarcoplasmic reticulum. Points of entry and exit of Ca^{2+} are shown by the purple arrow. Figure on the left is with bound Ca^{2+}, and on the right without Ca^{2+}.
Reproduced with permission from Toyoshima, C., and Inesi, G. Structural basis of ion pumping by Ca^{2+}–ATPase of the sarcoplasmic reticulum. *Annu. Rev. Biochem.* 73: 269, 2004.

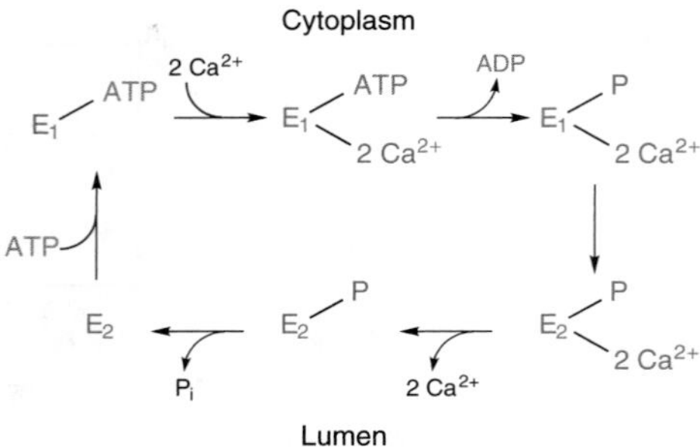

Figure 12.56 Proposed sequence of reactions and intermediates in translocating Ca^{2+} by Ca^{2+}-transporting ATPase. E_1 and E_2 are different conformations of enzyme.

transporter, lowering the K_m for Ca^{2+} from about 20 to 0.5 μM and increasing Ca^{2+} transport. Increased activity reduces cytosolic Ca^{2+} to its normal resting level ($\sim$ 0.10 μM) at which concentration the Ca^{2+}–calmodulin complex dissociates and the rate of Ca^{2+} transport returns to its basal value.

Other cellular processes are also regulated by the Ca^{2+}–calmodulin complex. Calmodulin is one of several Ca^{2+}-binding proteins, including **parvalbumin** and **troponin C**, which have very similar structures. Calmodulin (17 kDa) has the shape of a dumbbell with two globular ends connected by a seven-turn α-helix; it contains four Ca^{2+}–binding sites, two of high affinity on one end and two of low affinity on the other. Binding of Ca^{2+} to the lower-affinity binding sites causes a conformational change which reveals a hydrophobic region that interacts with a protein it controls. Each Ca^{2+}–binding site consists of a helix–loop–helix secondary structure (Figure 12.57) with Ca^{2+} bound in the loop connecting the helices. The motif is called the EF hand, on the basis of studies with parvalbumin where the Ca^{2+} binds between helices E and F of the protein.

The two families of Ca^{2+} transporters share some properties; for example, the high affinity for Ca^{2+}, the membrane topography, and the organization of the catalytic domain. They are different in structural and functional characteristics, however, principally in their regulation.

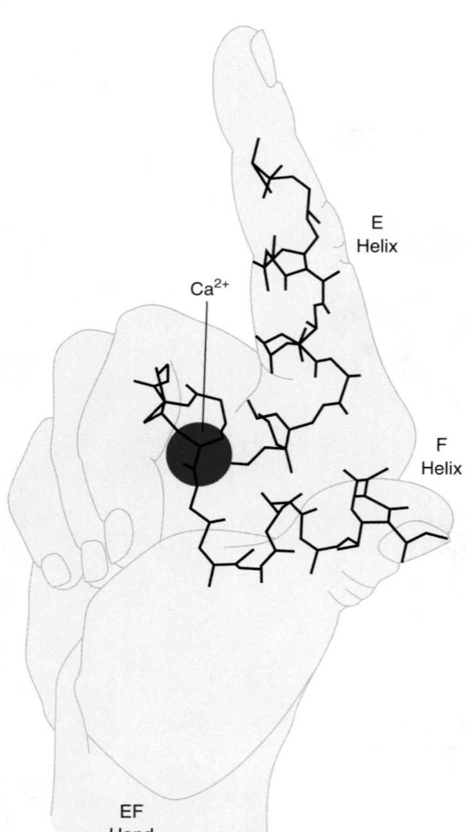

Figure 12.57 Binding site for Ca^{2+} in calmodulin. Calmodulin contains four Ca^{2+}-binding sites, each with a helix–loop–helix motif. Ca^{2+} ion is bound in the loop that connects two helices. This motif occurs in various Ca^{2+}-binding proteins and is referred to as the EF hand.

ATP-Binding Cassette (ABC) Transporters

Membranes from prokaryotes to eukaryotes have transporters that belong to the superfamily of **ATP-binding cassette (ABC) transporters** that catalyze an ATP-dependent vectorial movement of diverse substances. There are dozens of subfamilies within the superfamily and over 1000 genes, mostly from bacteria, for ABC transporters have been cloned; the human genome contains at least 49 genes. ABC transporters are present in both the plasma and intracellular membranes of mammalian cells, catalyzing either influx or efflux of various lipids (phospholipids, long-chain fatty acyl CoA, bile salts, and cholesterol), peptides, a variety of toxic organic molecules, and chemotherapeutic agents. The ABC transporters in bacteria are very versatile, transporting ions, heavy metals, sugars, drugs, amino acids, peptides, and proteins. These transporters have a modular structure, with a six-transmembrane-segment domain, and a cytosolic ATP-binding domain (or cassette). The ATP-binding domains in the superfamily share extensive sequence homology. Most subfamilies have a tandem repeat of two transmembrane domains and two ATP-binding domains (Figure 12.58). Individual ABC transporters function mechanistically as either channels or transporters. ATP binding induces a conformational change in the transmembrane domains, but ATP hydrolysis does not lead to phosphorylation of the transporter. The mechanism of coupling the energy of ATP hydrolysis with movement of substrate is still not defined.

One subfamily of ABC transporters in humans is coded by the **multidrug resistance (Mdr) family** of genes. The gene products are glycoproteins, termed **P-glycoproteins**, some of which are responsible for extrusion from cells of a variety of xenobiotics. Overexpression of Mdr genes by tumor cells leads to increased resistance to a variety of drugs because chemotherapeutic agents are rapidly transported out of cells with the increased numbers of transporters. Therefore, the drug does not attain effective therapeutic concentrations. Other **multidrug resistance associated proteins (MRP)** are responsible for the efflux from cells of compounds conjugated to glutathione, glucuronate, and sulfate. P-glycoprotein transporters are responsible for efflux of lipids such as phosphatidylcholine, cholesterol, and of bile acids across the canalicular plasma membrane of hepatocytes. One member has the important physiological role of transporting bilirubin glucuronides from the liver into the bile (p. 800). Another ABC transporter is the **sulfonylurea receptor** linked to a K^+-channel, which functions in the regulation of glucose-induced insulin secretion.

The **c**ystic **f**ibrosis **t**ransmembrane conductance **r**egulator (CFTR) is an ABC transporter but mechanistically is both a conductance regulator and a Cl^- channel. The channel mediates transepithelial salt and liquid movement in the apical plasma membrane of epithelial cells. CFTR consists of five domains, two domains each of six transmembrane helices, two ATP-binding domains, and a regulatory globular domain (Figure 12.59). Phosphorylation

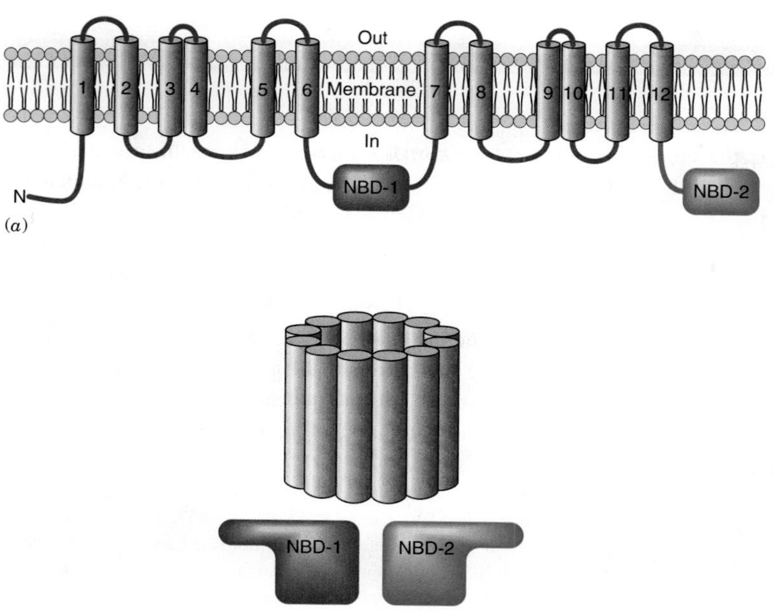

Figure 12.58 Model of a P-glycoprotein, an ATP-binding cassette (ABC) transporter. NBD-1 and NBD-2 are nucleotide-binding domains.

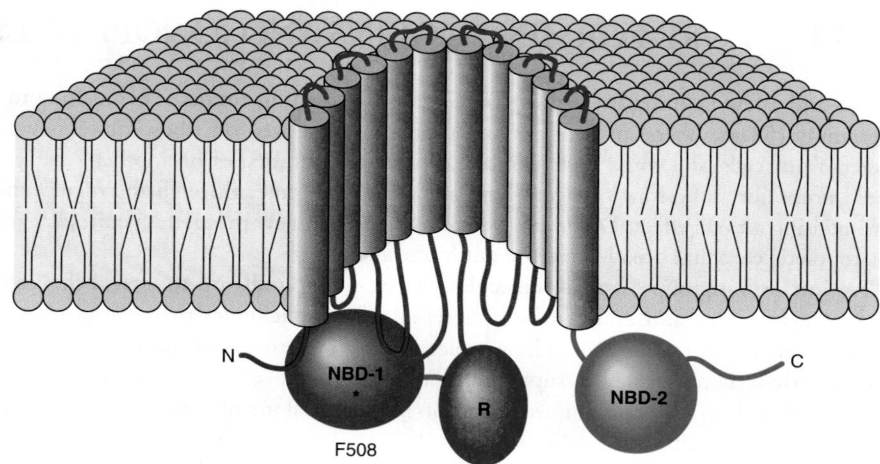

Figure 12.59 Schematic diagram of the cystic fibrosis transmembrane conductance regulator (CFTR), a member of the ABC transporter family. NBD-1 and NBD-2 are ATP-binding subunits and R is a regulatory subunit. F508 is the site of deletion of phenylalanine which occurs in about 70% of patients with cystic fibrosis.
Redrawn based on a figure in Ko, Y. H., and Pedersen, P. L. Frontiers in research on cystic fibrosis: Understanding its molecular and chemical basis and relationship to the pathogenesis of the disease. *Bioenerg. Biomemb.* 29: 417, 1997.

of the regulatory domain regulates channel activity whereas ATP hydrolysis controls gating of the channel. A dysfunction of CFTR causes the genetic disease cystic fibrosis. A deletion of phenylalanine residue F508 in the ATP-binding loop NBD-1 (nucleotide-binding domain-1) is present in about 70% of the cases of cystic fibrosis. This deletion prevents the proper folding of NBD-1 (Clin. Corr. 12.5, and Clin. Corr. 6.7, p. 231).

Other genetic diseases attributed to defects in specific ABC transport proteins include adrenoleukodystrophy, Stargardt macular dystrophy, a hereditary degenerative disease of the macula lutea leading to rapid loss of visual acuity, and Dubin–Johnson syndrome (Clin. Corr. 12.6).

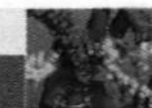

CLINICAL CORRELATION 12.5

Cystic Fibrosis and the Cl^- Channel

Cystic fibrosis (CF), an autosomal recessive disease, is the commonest, serious, inherited disease of Caucasians, occurring with a frequency of 1 in 2000 live births. It is a multiorgan disease, with pulmonary obstruction as a principal manifestation; thick mucous secretions obstruct the small airways allowing recurrent bacterial infections. Exocrine pancreatic dysfunction occurs early and leads to steatorrhea (fatty stool); see page 1032 for a discussion of the role of the pancreas in fat digestion and absorption. CF patients have reduced Cl^- permeability that impairs fluid and electrolyte secretion, leading to luminal dehydration. Diagnosis of CF is confirmed by a significant increase of Cl^- content of sweat of affected in comparison to normal individuals.

The gene responsible for CF was identified in 1989 and over 800 mutations leading to CF have been found. The most common mutation affects about 70% of the patients and is a deletion of a single phenylalanine at position 508 on the protein, but missense, nonsense, frameshift, and splice-junction mutations (p. 197) have been reported. Many mutations lead to a change in protein folding. The CF gene product is the **c**ystic **f**ibrosis **t**ransmembrane conductance **r**egulator (**CFTR**). CFTR is a cAMP-dependent Cl^- channel, which may regulate other ion channels; it is expressed in epithelial tissues. Phosphorylation of a cytoplasmic regulatory domain by protein kinase A activates the channel. CFTR is a polypeptide of 1480 amino acids with structural homology to the superfamily of ATP-binding cassette (ABC) transporters. The gene has been cloned, and a major effort is under way to treat the disease by gene therapy, using both viral and nonviral vectors including liposomes (see Clin. Corr. 12.1).

Kunzelmann, K., and Mall, M. Pharmacotherapy of the ion transport defect in cystic fibrosis. *Clin. Exp. Pharm. & Physiol.* 28: 857, 2001; Zeitlin, P. L. Therapies directed at the basic defect in cystic fibrosis. *Clinics in Chest Med.* 19: 515, 1998; Frizzell R. A., Functions of the cystic fibrosis transmembrane conductance regulator protein. *Am. J. Respir. Crit. Care Med.* 151: 54, 1995; and Naren, A. P., Cormet-Boyaka, E., Fu, J., et al. CFTR chloride channel regulation by an interdomain interaction. *Science* 286: 544, 1999.

12.11 • PORE-FORMING TOXINS AND IONOPHORES

Bacteria and a few mammalian tissues synthesize **pore-forming toxins**. They belong to a large family (58 members) of peptides and proteins synthesized on ribosomes. They are secreted from cells and form transmembrane pores in membranes of another cell. The created pores allow the flow of electrolytes and small molecules through the membrane, thus initiating a toxic effect. **Defensin**, a pore-forming toxin synthesized by epithelial and hematopoietic cells, has broad-spectrum antibiotic activity.

An interesting group of small molecular weight compounds (up to several thousand Daltons), synthesized and excreted by bacteria and fungi, facilitates the translocation of inorganic ions across membranes. These molecules, referred to as **ionophores**, have antibacterial activity because they disrupt the ionic balance of a cell. They are also valuable experimental tools in studies of ion translocation in biological membranes and for manipulation of the ionic compositions of cells.

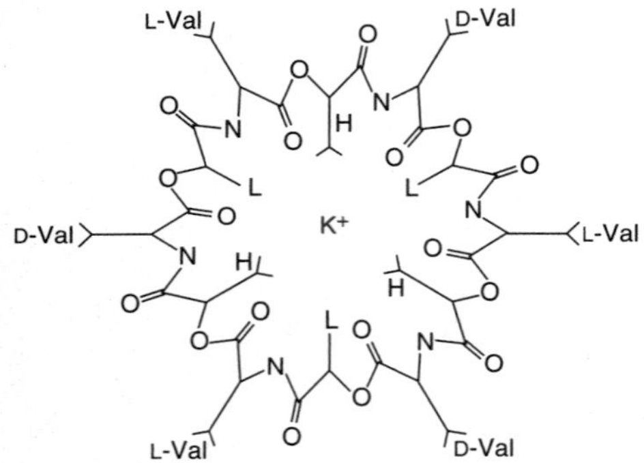

Figure 12.60 Structure of valinomycin-K^+ complex. Abbreviations: D-Val, D-valine; L-Val, L-valine; L, L-lactate; and H, D-hydroxyisovalerate.

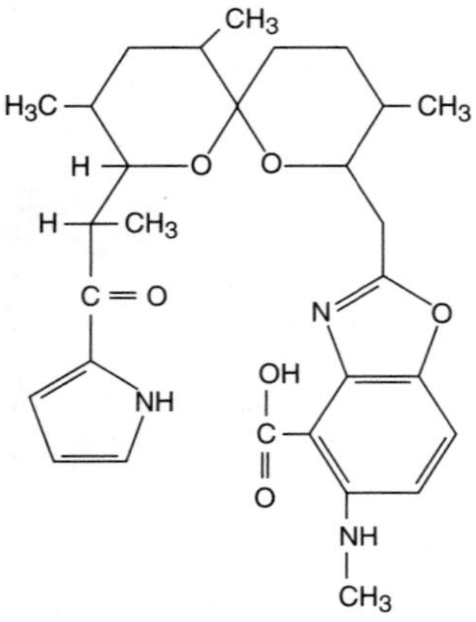

Ionophores are classified as nonribosomally synthesized channels (p. 477); there are two major subgroups: mobile transporters that bind an ion and readily diffuse in a membrane, and channel formers. Some major ionophores are listed in Table 12.13.

Each mobile transporter has an ion specificity. **Valinomycin** (Figure 12.60) has an affinity for K^+ 1000 times greater than for Na^+, and **A23187** (Figure 12.61) has an affin-

Figure 12.61 Structure of A23187, a Ca^{2+} ionophore.

TABLE 12.13 • Major Ionophores

Compound	*Major Cations Transported*	*Action*
Valinomycin	K^+ or Rb^+	Uniport, electrogenic
Nonactin	NH_4^+, K^+	Uniport, electrogenic
A23187	$Ca^{2+}/2\ H^+$	Antiport, electroneutral
Nigericin	K^+/H^+	Antiport, electroneutral
Monensin	Na^+/H^+	Antiport, electroneutral
Gramicidin	H^+, Na^+ K^+, Rb^+	Forms channels
Alamethicin	K^+, Rb^+	Forms channels

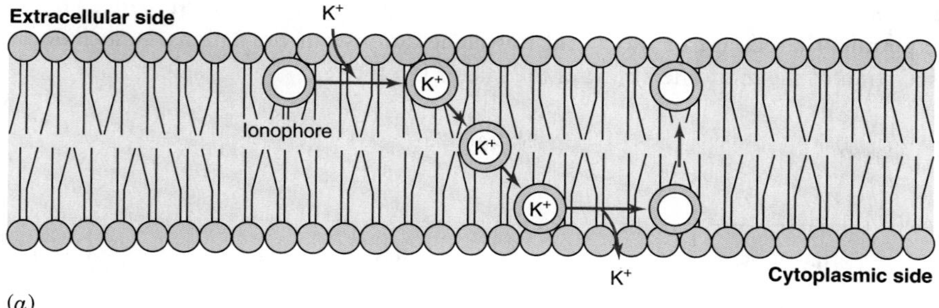

(*a*)

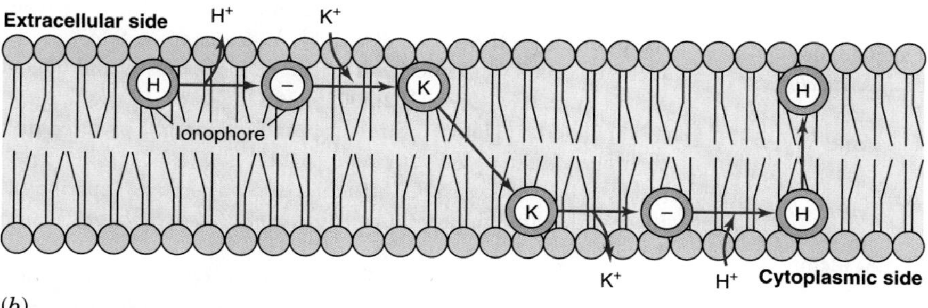

(*b*)

Figure 12.62 Mechanism for ionophoretic activities of valinomycin and nigericin. (*a*) Transport by valinomycin. (*b*) Transport by nigericin. I represents ionophore. The valinomycin-K^+ complex is positively charged and translocation of K^+ is electrogenic leading to creation of a charge separation across membrane. Nigericin translocates K^+ in exchange for a H^+ across membrane and the mechanism is electrically neutral.
Diagram adopted from Pressman, B. C, *Annu. Rev. Biochem.* 45: 501, 1976.

ity for Ca^{2+} 10 times greater than Mg^{2+}. Several of the mobile transporters have a cyclic structure and the ion coordinates to oxygen atoms in the core of the structure; the periphery of the molecule consists of hydrophobic groups. When an ion is chelated by the ionophore, its water shell is stripped away and the ion encompassed by the hydrophobic shell. The ionophore–ion complex freely diffuses across the membrane. Since interaction of ion and ionophore is an equilibrium reaction, a steady-state concentration of the ion develops on both sides of the membrane. Valinomycin transports K^+ by an electrogenic uniport mechanism that creates an electrochemical gradient across a membrane as it transports a positively charged K^+ (Figure 12.62*a*). **Nigericin** is an electrically neutral antiporter; its carboxyl group when dissociated binds a positive ion, such as K^+, leading to a neutral complex that moves across a membrane. It transports a proton back on diffusion through the membrane, leading to an exchange of K^+ for H^+ (Figure 12.62*b*).

Gramicidin A is a 15-residue peptide with alternating D- and L-amino acids. In membranes it forms a β-helix and dimerizes, forming a long (25 Å) and narrow (5 Å) diameter) transmembrane channel (Figure 12.63). Polar residues line the channel and hydrophobic

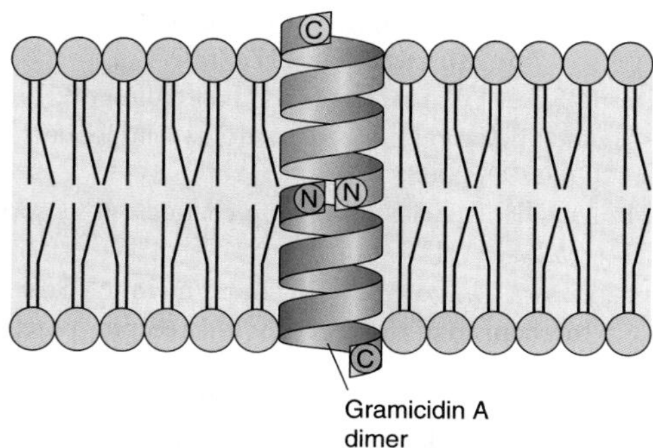

Figure 12.63 Action of gramicidin A Two molecules of gramicidin A form a dimer by hydrogen binding at the amino ends of the peptides creating a channel in a membrane.

CLINICAL CORRELATION 12.6

Diseases Involving the Superfamily of ABC Transporters

ATP binding cassette (ABC) transporters are a superfamily of proteins with a number of subfamilies that all contain a conserved ATP-binding domain and diverse transmembrane domains. The members of the superfamily catalyze the ATP-dependent transport of a wide variety of substances and are present in many different tissues. A number are found in mutant form in various diseases. The following lists selected ABC transporters, their substrates, and related diseases:

Transporter	*Substrate*	*Disease*
ABCA1	Phospholipid	Tangier disease
ABCG5	Plant sterols	Sitosterolemia
ABCR	Phosphatidyl-ethanolamine	Juvenile macular degeneration Retinitis pigmentosa
ABC7	Iron?	Sideroblastic anemia with ataxia
BSEP	Bile salts	Progressive familial intrahepatic cholestasis
MRP1	Xenobiotics	Multidrug resistance
MRP2	Bilirubin glucuronides	Dubin–Johnson syndrome
PGY3	Long-chain phosphatidyl-choline	Cholestasis of pregnancy
TAP1	Peptide transporter	Ankylosing spondylitis Insulin-dependent diabetes mellitus Celiac disease

Note that genetic diseases involving the ABC transporters involve a variety of pathophysiological conditions, involving a number of different tissues, and various biochemical systems.

For a detailed list of the ABC transporters and their diseases, see http://nutrigene.4t.com/humanabc.htm.

Borst, P., and Elferink, R. O. Mammalian ABC transporters in health and disease. *Annu. Rev. Biochem.* 71: 537, 2002.

groups are on the periphery of the channel interacting with the lipid membrane. The structure permits the passage of water and univalent cations but not anions. Association and dissociation of the monomers control the rate of ion flux.

Bibliography

General

Chin, D., and Means, A. R. Calmodulin: A prototypical calcium sensor. *Trends Cell Biol.* 10:322, 2000.

Dobler, M. *Ionophores and Their Structure.* New York: Wiley Interscience, 1981.

Gil, T., Ipsen, J. H., Mouritsen, O. G., Sabra, M. C., et al. The theoretical analysis of protein organization in lipid membranes. *Biochim. Biophys. Acta* 1376:245, 1998.

Jones, M. N. *Micelles, Monolayers, and Biomembranes.* New York: Wiley-Liss, 1995.

Scott, H. L. Modeling the lipid components of membranes. *Curr. Opinion in Structural Biology* 12:495, 2002.

Van Winkle, L. J. *Biomembrane Transport.* San Diego, CA: Academic Press, 1999.

Membrane Structure

Daleke, D.L. Phospholipid flippases. *J. Biol. Chem.* 282:821, 2007.

Daleke, L. D., and Lyles, J. V. Identification and purification of aminophospholipid flippase. *Biochim. Biophys. Acta* 1486:108, 2000.

Gregoriadis, G. Engineering liposomes for drug delivery: progress and problems. *Trends Biotechnol.* 13:527, 1995.

Hanzal-Bayer, M. F., and Hancock, J. F. Lipid rafts and membrane traffic. *FEBS Letters* 581:2098, 2007.

Ikezawa, H. Glycosylphosphatidylinositol (GPI)-anchored proteins. *Biol. Pharm. Bull.* 25:409, 2002.

Kinnunen, P. K. J. On the functional ordering in biological membranes. *Chem. Phys. Lipids* 57:375, 1991.

Lee, A. G. How lipids affect the activities of integral membrane proteins. *Biochim. Biophys. Acta* 1666:62, 2004.

Lenoir, G., Williamson, P., and Holthius, J. C. On the origin of lipid asymmetry: the flip side of ion transport. *Curr. Opin. Chem. Biol.* 11:654, 2007.

Liang, J., Adamian, L., and Jackups, R. The membrane-water interface region of membrane proteins:structural bias and the anti-snorkeling effect. *Trends Biochem. Sci.* 30:355, 2005.

Maxfield, F. R. Plasma membrane microdomains. *Curr. Opinion in Cell Biology* 14:483, 2002.

McMurchie, E. J. Dietary lipids and the regulation of membrane fluidity and function. In Aloia R, C., Curtain, C. C., and Gordon, L.M, (Eds.), *Physiological Regulation of Membrane Fluidity.* New York: Liss, 1988, 189.

Michel, V., and Bakovic, M. Lipid rafts in health and disease. *Biol. of Cell.* 99:129, 2007.

Pike, L. J. Lipid rafts:heterogeneity on the high seas. *Biochem. J.* 378:281, 2004.

Roth, M. G. Phosphoinositides in constitutive membrane traffic. *Physiological Revs.* 84:699, 2004.

Rothman, J. E., and Lenard, J. Membrane asymmetry. *Science* 195:743, 1977.

Simons, K., and Ehehalt, R. Cholesterol, lipid rafts, and disease. *J. Clin. Invest.* 110:597, 2002.

Receptors

Allen, S. J., Crown, S. E., and Handel, T. M. Chemokine: Receptor structure, interactions, and antagonism. *Annu. Rev. Immun.* 25:787, 2007.

Armbruster, B. N., and Roth, B. L. Mining the receptorome. *J. Biol. Chem.* 280:5129, 2005.

Cascio, M. Structure and function of the glycine receptor and related nicotinicoid receptors. *J. Biol. Chem.* 279:19383, 2004.

Catalog of human plasma membrane receptors. Accessed at http://www.receptome.org/HPMR/.de Weerd, N. A., Samarajiwa, S. A., and Hertzog, P. J. Type I interferon receptors: biochemistry and biological functions. *J. Biol. Chem.* 282:20053, 2007.

Gay, N. J., and Gangloff, M. Structure and function of Toll receptors and their ligands. *Annu. Rev. Biochem.* 76:141, 2007.

Jeon, H., and Blacklow, S. C. Structure and physiologic function of the low-density lipoprotein receptor. *Annu. Rev. Biochem.* 74:535, 2005.

Niswender, C. M, and Conn, P. J. Metabotropic Glutamate receptors: Physiology, pharmacology, and disease. *Annu. Rev. Neuroscience* July 2009.

Yip, C. C., and Ottensmeyer, P. Three dimensional structural interactions of insulin and its receptor. *J. Biol. Chem.* 278:27239, 2003.

Channels and Pores

Borgnia, M., Nielsen, S., Engel, A, and Agre, P. Cellular and molecular biology of the aquaporin water channels. *Annu. Rev. Biochem.* 68:425, 1999.

Carbrey, J. M., and Agre, P. Discovery of the aquaporins and development of the field. *Handbook Exp. Pharm.*190:3, 2009.

Chapman, D. E. TRP channels as cellular sensors. *Nature* 426:517, 2003.

Clark, K., Middelbeek, J., and van Leeuwen, F. N. Interplay between TRP channels and the cytoskeleton in health and disease. *Eur. J. Cell Biol.* 87:631, 2008.

DeFelice, L.J., and Goswami, T. Transporters as channels. *Annu. Rev. Physiol.* 69:87, 2007.

Dutzler, R., Campbell, E. B., and MacKinnon. Gating the selectivity filter in ClC chloride channels. *Science* 300:108, 2003.

Elinder, F., Nilsson, J., and Arhem, P. On the opening of voltage-gated ion channels. *Physiol. Behavior.* 92:1, 2007.

Evans, W. H., De Vuyst, E., and Leybaert, L. The gap junction cellular internet: Connexin hemichannels enter the signalling limelight. *Biochem. J.* 397:1, 2006.

Gonen, T., and Walz, T. The structure of aquaporins. *Quarterly Reviews of Biophysics* 39:361, 2006.

Hatta, S., Sakamoto, J., and Horio, Y. Ion channels and diseases. *Medical Electron Microscopy.* 35:117, 2002.

Lim, R. Y., Ullman, K. S., and Fahrenkrog, B. Biology and biophysics of the nuclear pore complex and its components. *Rev. Cell Molec. Biol.* 267:299, 2008.

Lingueglia, E. Acid-sensing ion channels in sensory perception. *J. Biol. Chem.* 282:17325, 2007.

Miyazawa, A., Fujiyoshi, Y., and Unwin, N. Structure and gating mechanism of the acetylcholine receptor pore. *Nature* 423:949, 2003.

Murata, K., Mitsuoka, T., Hirai, T., Walz, P., Agre, J. B., et al. Structural determinants of water permeation through aquaporin-1. *Science* 407:599, 2000.

Nielsen, S., Frøkiær, J., Marples, D., Kwon, T., Agre, P., and Knepper, M. A. Aquaporins in the kidney: From molecules to medicine. *Physiol. Rev.* 82:205, 2002.

Peracchia, C. (Ed.). *Gap Junctions: Molecular Basis of Cell Communication in Health and Disease.* San Diego, CA: Academic Press, 2000.

Venkatachalam, K., and Montell, C. TRP channels. *Annu. Rev. Biochem.* 76:387, 2007.

Verkmann, A. S. Mammalian aquaporins: Diverse physiological roles and potential clinical significance. *Expert Reviews in Molecular Medicine* 10:e13, 2008.

Electrochemical-Potential-Driven Transporters

Cordat, E., and Casey, J. R. Bicarbonate transport in cell physiology and disease. *Biochem. J.* 417:423, 2009.

Jung, H. The sodium/substrate symporter family: structural and functional features. *FEBS Letters* 529:73, 2002.

Kakuda, D. K., and MacLeod, C. L. Na^+-independent transport (uniport) of amino acids and glucose in mammalian cells. *J. Exp. Biol.* 196:93, 1994.

Klingenberg, M. The ADP and ATP transport in mitochondria and its carrier. *Biochim. Biophys. Acta* 1778:1978, 2008.

Poole, R. C., and Halestrap, A. P. Transport of lactate and other monocarboxylates across mammalian plasma membranes. *Am. J. Physiol.* 264:C761, 1993.

Saier, M. H. A functional-phylogenetic classification system for transmembrane solute transporters. *Micro. Molecular Biol. Rev.* 64:354, 2000.

Simpson, I. A., Dwyer, D., Malide, D., Moley, K. H., et al. The facilitative glucose transporter GLUT3: 20 years of distinction. *Amer. J. Physiol., Endocrinol. Met.* 295:E242, 2008.

Slepkov, E. R., Rainey, J. K., Sykes, B. D., and Fliegel, L. Structural and functional analysis of the Na^+/H^+ exchanger. *Biochem. J.* 401:623, 2007.

Wright, E. M., and Turk, E. The sodium/glucose cotransport family SLC5. *Eur. J. Physiol.* 447:510, 2004.

Primary Active Transporters

Szakacs, G., Varadi, A., Ozvegy-Laczka, C., and Sarkadi, B. The role of ABC transporters in drug absorption, distribution, metabolism, excretion, and toxicity. *Drug Discovery Today* 13:379, 2008.

Borst, P., and Elferink, R. O. Mammalian ABC transporters in health and disease. *Annu. Rev. Biochem.* 71:537, 2002.

Carafoli, E., and Brini, M. Calcium pumps: structural basis for and mechanism of calcium transmembrane transport. *Curr. Opinion Chem. Biol.* 4:152, 2000.

Esmann, M., and Marsh, D. Lipid-protein interaction with the Na,K-ATPase. *Chem. Phys. Lipids* 141:94, 2006.

Facciotti, M. T., Rouhani-Manshadi, S., and Glaeser, R. M. Energy transduction in transmembrane ion pumps. *Trends Biochem. Sci.* 29:445, 2004.

Holland, I. B., Kuchler, K., Higgins, C, and Cole, S. (Eds.). *ABC Proteins: From Bacteria to Man.* London: Academic Press, 2002.

Kaplan, J. H. Biochemistry of Na, K-ATPase. *Annu. Rev. Biochem.* 71:511, 2002.

Lee, A. G. A calcium pump made visible. *Curr. Opin. Struct. Biol.* 12:547, 2002.

Leo, T. W., and Clarke, D. M. Molecular dissection of the human multidrug resistance P-glycoprotein. *Biochem. and Cell Biol.* 77:11, 1999.

Rees, D. C., Johnson, E., and Lewinson, O. ABC transporters: The power to change. *Nature Rev. Mole. Cell Biol.* 10:218, 2009.

Riordan, J. R. CFTR function and properties for therapy. *Annu. Rev. Biochem.* 77:701, 2008.

Scarborough, G. A. Structure and function of the P-type ATPases. *Curr. Opinion. Cell Biol.* 11:517, 1999.

Sharom, F. J. ABC multidrug transporters: Structure, function and role in chemoresistance. *Pharmacogenomics* 9:105, 2008.

Toyoshima, C. Structural aspects of ion pumping by Ca^{2+}-ATPase of sarcoplasmic reticulum. *Arch. Biochem. Biophys.* 476:3, 2008.

Key Terms

glycerophospholipids
phosphatidic acid
cardiolipin
phosphatidylinositol
plasmalogen
amphipathic lipids
sphingolipids
ceramide
sphingomyelin
glycosphingolipids
cholesterol
integral proteins
peripheral proteins
micelles
liposomes
membrane fluidity
membrane asymmetry
microdomains
mosaic model
glycosylphosphatidylinositol anchor
receptors
signal molecules
up- and down-regulation
channels
voltage-gated ion channels
ligand-regulated channels
nicotinic-acetylcholine channel (receptor)
Na^+ channel
aquaporins
gap junctions
nuclear pore complexes
uniport mechanism
symport mechanism
antiport mechanism
electrically silent transport
sodium motive force
proton motive force
protein-mediated diffusion
electrochemical-potential-driven transport
sodium/glucose cotransporter
mitochondrial transporters
primary active transport
pyrophosphate bond
P-, V-, and F-type transporters
Na^+/K^+-exchanging ATPase
Ca^{2+}-transporting ATPases
calmodulin
ATP-binding cassette (ABC) transporters
pore-forming toxins
ionophores

Questions • CAROL N. ANGSTADT

Multiple-Choice Questions

1. All of the following are ways in which peripheral proteins bind to membranes except
   A. binding to an integral protein.
   B. electrostatic binding between phospholipids and positive groups on the protein.
   C. by a short hydrophobic group at one end of the protein.
   D. attached by the charged carboxyl group at the carboxyl terminus of the protein.
   E. binding non-covalently to membrane phosphatidylinositol.

2. Characteristics of a mediated transport system include
   A. nonspecific binding of solute to transporter.
   B. release of the transporter from the membrane following transport.
   C. a rate of transport directly proportional to the concentration of solute.
   D. release of the solute only if the concentration on the new side is lower than that on the original side.
   E. a mechanism for translocating the solute from one side of the membrane to the other.

3. The translocation of Ca^{2+} across a membrane involves all of the following except
   A. active transport by Ca^{2+}-transporting ATPases.
   B. maintaining $[Ca^{2+}]$ very much higher in the cell than in extracellular fluid.
   C. phosphorylation of the transporter.
   D. regulation by the binding of a Ca^{2+}-calmodulin complex to the transporter in eukaryotes.
   E. different transporters for transport into the lumen or out across the plasma membrane

4. Cell membranes typically
   A. are about 90% phospholipid.
   B. have both integral and peripheral proteins.
   C. contain cholesteryl esters.
   D. contain free carbohydrate such as glucose.
   E. contain large amounts of triacylglycerols.

5. All of the following are correct about an ionophore except it
   A. requires the input of metabolic energy for mediated transport of an ion.
   B. may diffuse back and forth across a membrane.
   C. may form a channel across a membrane through which an ion may diffuse.
   D. may catalyze electrogenic-mediated transport of an ion.
   E. will have specificity for the ion it moves.

6. Cells control the distribution of water across membranes by aquaporins and aquaglyceroporins because nonfacilitated diffusion is too slow. Aquaporins (AQP)
   A. permit translocation of water and small solutes.
   B. reduce the pH gradient because they transport H_3O^+.
   C. are peripheral proteins of the membrane.
   D. form channels through which water flows.
   E. have no specific controls to open the channel.

Questions 7 and 8: Two problems encountered with oral or intravenous administration of drugs are the lack of tissue specificity in the action of the drug and rapid metabolism, and therefore limited period of effectiveness, of some drugs. One attempt to circumvent these problems is the use of liposomes to encapsulate the drugs. Some drugs have a longer period of effectiveness when administered this way. Liposomes can be prepared with specific proteins to bind specific cellular membrane receptors. Liposomes are also useful as a research tool to study the properties of biological membranes since they have a similar structure and properties. Much of our understanding of biological membranes has been obtained using liposomes.

7. Plasma membrane receptors
   A. usually have as ligands molecules like steroids.
   B. are always coupled to G-proteins.
   C. are fixed in number for a given cell.
   D. often span the membrane with one or more transmembrane domains.
   E. when bound to their ligand, always result in the release of a small molecule (second messenger) into the cell.

8. Which of the following statements concerning membranes is correct?
   A. Microdomains, called lipid rafts, are fixed in position in membranes.
   B. Lipid composition of the two layers of the membrane equilibrate.
   C. As demonstrated with liposomes, the membrane is most fluid at the surfaces.
   D. An increase in the cholesterol content of a membrane increases membrane fluidity.
   E. Lipid transporters catalyze unidirectional movement of specific lipids from one layer to the other.

Questions 9 and 10: Cystic fibrosis is a relatively common (1 in 2000 live births) genetic disease of Caucasians. Although it affects many organs, pulmonary obstruction is a major problem. CF patients have reduced Cl^- permeability and the disease can be diagnosed by elevated $[Cl^-]$ in sweat. Genetic mutations lead to defects in cystic fibrosis transmembrane conductance regulator (CFTR) protein which is a cAMP-dependent Cl^- channel. CFTR has structural homology to the superfamily of ATP-binding cassette (ABC) transporters.

9. Membrane channels
   A. have a large aqueous area in the protein structure so are not very selective.
   B. commonly contain amphipathic α-helices.
   C. are opened or closed only as a result of a change in the transmembrane potential.
   D. are the same as gap junctions.
   E. allow substrates to flow only from outside to inside of the cell.

10. ATP-binding cassette (ABC) transporters
    A. have both a membrane-spanning domain that recognizes the substrate and an ATP-binding domain.
    B. all effect translocation by forming channels.
    C. are found only in eukaryotes.
    D. all have two functions: forming a channel and conductance regulation.
    E. are all P-glycoproteins.

Questions 11 and 12: Alterations in membrane transport systems for specific components lead to a number of diseases. In Hartnup disease there is a decrease in transport of neutral amino acids by intestine and renal tubules. Individuals with a decreased glucose uptake from the intestinal tract lack a specific glucose–galactose transporter. In these diseases the transport systems are Na^+/(amino acid) or (glucose) co-transporters.

11. This type of transport system
   A. moves Na^+ and the amino acid or glucose in opposite directions across the membrane.
   B. uses the energy of the Na^+ gradient (SMF) to concentrate the other substance against its gradient.
   C. results in the hydrolysis of ATP during the transport.
   D. is the same as coded for by the multidrug resistance (Mdr) family of genes.
   E. is the only type of system used to transport glucose across membranes.

12. A different type of transport system that maintains the Na^+ and K^+ gradients across the plasma membrane of cells
   A. involves an enzyme that is an ATPase.
   B. is a symport system.
   C. moves Na^+ either into or out of the cell.
   D. is an electrically neutral system.
   E. in the membrane hydrolyzes ATP independently of the movement of Na^+ and K^+.

Problems

13. Draw a curve illustrating the rate of movement of a solute through a membrane as a function of the concentration of the solute for (a) O_2 and (b) uptake of glucose into the erythrocyte.
14. What kind of transporter is the acetylcholine receptor of skeletal muscle membrane, and how is it controlled?

Answers

1. **D** A single electrostatic charge probably would not be sufficient. A: An example is ankyrin binding to the anion channel in erythrocytes. B: The negative charge of the phospholipids is involved. C: This would insert into the lipid bilayer. E: A specific domain binds to the inositol 3-phosphate head group.
2. **E** This is essential to having the solute transported across the membrane. A: Specificity of binding is an integral part of the process. B: Recovery of the transporter to its original condition is one of the characteristics of mediated transport. C: Only at low concentrations of solute; transporters show saturation kinetics. D: Active transport, movement against a gradient, is also mediated transport.
3. **B** Extracellular Ca^{2+} is about 10,000 times higher than intracellular. A: This is a primary active transport. C: These are P-type transporters (phosphorylation is on an aspartate). D: This binding lowers the K_m for Ca^{2+} and increases Ca^{2+} transport. E: They are known as CERCA and PMCA.
4. **B** Some proteins are embedded in the membrane whereas others are at the surface. A: This is more than the total lipid. C: Cholesterol in membranes is unesterified. D: All carbohydrate in membranes is in the form of glycoproteins and glycolipids. E: This is a minor component.
5. **A** Ionophores transport by passive mediated mechanisms. B and C: These are the two major types of ionophores. D: Valinomycin transports K^+ by a uniport mechanism. There are also antiport systems that are electroneutral. E: Valinomycin has an affinity for K^+ 1000 times greater than for Na^+.
6. **D** They assemble as homotetramers with each monomer forming a specific pore. A: This is a description of aquaglyceroporins. B: They do not transport H_3O^+. C: They are integral proteins that span the membrane. E: Channels need controls.
7. **D** This is a common mechanism for anchoring the protein in the membrane. A: Steroids can diffuse through the plasma membrane and have intracellular receptors. B: This is just one of the three major classes of plasma membrane receptors. C: Receptors can be upregulated or downregulated to modify the cellular response to a ligand. E: There are a variety of initial responses of the receptor- ligand binding, including opening a channel.
8. **E** Asymmetry is maintained by ATP-dependent flippases and floppases. A: They diffuse freely and can cluster into larger domains. B: Moving a charged head group through the lipid core of the membrane is thermodynamically unfavorable. C: The area of greatest fluidity is the center of the membrane. D: Cholesterol is a rigid molecule which decreases fluidity.
9. **B** The helices typically form the channel with hydrophilic side chains lining the channel and hydrophobic ones facing the lipid core of the membrane. A: This describes a pore; channels are quite specific. C: Voltage gated channels, like that for Na^+, are controlled this way but others, like the nicotinic acetylcholine channel, are chemically regulated. D: Clusters of membrane channels work together to form a gap junction. E: Substances may move in either direction as dictated by the concentration gradient.
10. **A** These are the common features. B: They operate by a variety of mechanisms, including transporters and receptors. C: Both prokaryotes and eukaryotes have them. D: CFTR is unusual in having both of these functions. E: Some, such as those coded for by the Mdr gene family or the transporter for cholesterol out of cells are P-glycoproteins, but others are not.
11. **B** These are Na^+-dependent secondary active transporters. A: These are symports, moving both substances in the same direction. C: ATP is used to maintain the Na^+ gradient, not in the transport of the amino acid or glucose. D: Mdr are ABC transporters. E: Glucose can also be transported by a passive uniport system.
12. **A** The Na^+, K^+-transporter is the Na^+/K^+-exchanging ATPase. B, C, and D: It is an antiport, vectorial (Na^+ out), electrogenic ($3Na^+$, $2K^+$) system. E: ATP hydrolysis is not useless.
13. See Figure 12.32. (a) O_2 crosses membranes by simple diffusion; thus, rate increases with concentration as illustrated by the lower curve on the figure. (b) Uptake of glucose by erythrocytes is a passive mediated transport so should show saturation kinetics as illustrated by the upper curve in the figure.
14. This is also called the nicotinic-acetylcholine channel and is an example of an agonist regulated channel. Acetylcholine interacts with the acetylcholine receptor on the membrane causing a conformational change in the protein which opens the channel and allows cations (but not anions) to flow through. Hydrolysis of acetylcholine with release of the products closes the channel.

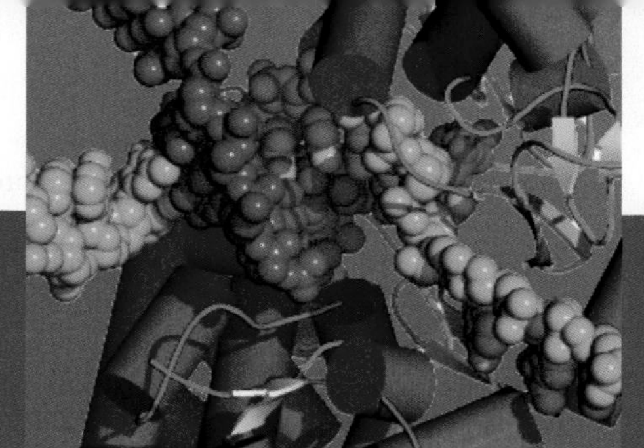

13

Fundamentals of Signal Transduction

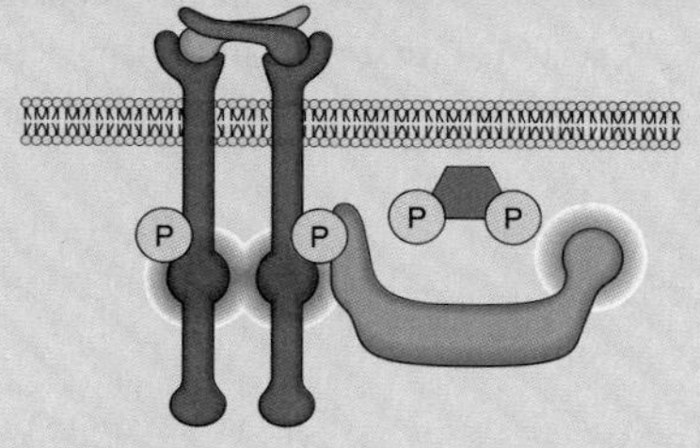

George R. Dubyak

Professor, Case Western Reserve University School of Medicine

CLINICAL CORRELATIONS

Key Concepts

- Communication between cells involves intercellular signaling via secreted molecules (hormones, neurotransmitters, and other mediators) and intracellular signaling cascades regulated by receptors that bind the secreted molecules. Receptors include: (1) cell surface proteins that bind hydrophilic or large secreted molecules, and (2) intracellular proteins that bind hydrophobic secreted molecules.
- Cell surface receptors initiate reactions that rapidly change the intracellular concentrations of small second-messenger molecules, such as cyclic nucleotides, Ca^{2+} ions, or lipid metabolites.
- Two major types of molecular switches control the transmission of intracellular signals: (1) the phosphorylation status of proteins; and (2) G proteins that when bound to GTP, switch between conformational states that turn on or off the activity of other proteins.
- Some protein kinases and phosphatases are intrinsic activities of cell surface receptor proteins that bind growth factors; others are separate proteins that are regulated by transient association with cell surface proteins or by binding small second messengers.
- Protein kinases and phosphatases regulate intracellular signaling by changing the phosphorylation status and function of: (1) enzymes, including other kinases and phosphatases; (2) ion channels and ion transporters; (3) transcription factors; and (4) structural proteins that regulate cell shape, size, and motility.
- G proteins are regulated by their interaction with other proteins that increase the rate of GTP/GDP exchange by the G proteins;

these include the G protein-coupled receptors (GPCR) that constitute the largest superfamily of cell surface receptors and include the sensory receptors involved in vision, smell, and taste.

- G proteins regulate the activity of other proteins including enzymes that synthesize second messengers, ion channels that control membrane potential or the flux of regulatory ions, and other protein kinases.
- Combinations of different intercellular and intracellular signaling reactions constitute signaling networks that allow cells in one organ or tissue to respond to changes in the function of cells in other organs or tissues; these signaling networks ensure homeostasis within the entire organism.
- Aberrant expression or function of signaling proteins, including secreted molecules, receptors, G proteins, and protein kinases, is the underlying cause of common human diseases including many cancers, heart failure, diabetes, and some behavioral disorders; about 50% of the most widely used therapeutic drugs target signaling proteins.

13.1 • SIGNAL TRANSDUCTION BETWEEN CELLS

Communication or signal transduction between cells and tissues consists of two phases. Intercellular signal transduction characterizes the passage of a signal from one cell to target cells via the extracellular environment. Intracellular signal transduction comprises the biochemical decoding of that signal on receipt by the target cells, a process that involves stepwise regulation of intracellular signaling proteins that ultimately results in altered function of proteins involved in metabolism, gene regulation, membrane transport, and cell motility (Figure 13.1). The combination of intercellular and intracellular signaling allows cells in one organ or tissue to respond to changes in the function of cells in other organs or tissues to ensure homeostasis within the entire organism.

13.2 • INTERCELLULAR SIGNAL TRANSDUCTION

Two Fundamental Modes of Intercellular Signal Transduction

Intercellular signaling describes the mechanisms by which one cell (the "sender" cell) sends a message to change the function of another cell (the "target" or recipient cell). Two cells can communicate (i.e., transmit information to each other) in several different ways (Figure 13.2). One requires that they be in direct physical contact; this mode is **juxtacrine** or **contact dependent**. A second and more common mode is contact independent and involves release of **secreted molecules** from the sender cells. The secreted molecules diffuse through extracellular space to the target cells where they interact with receptor proteins resulting in an activation of intracellular signaling and altered function. This latter mode permits intercellular communication

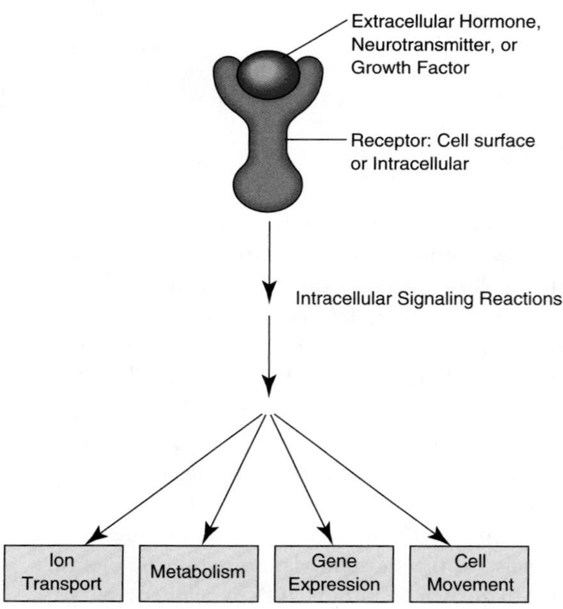

Figure 13.1 Basic elements of a signal transduction pathway at the cellular level.

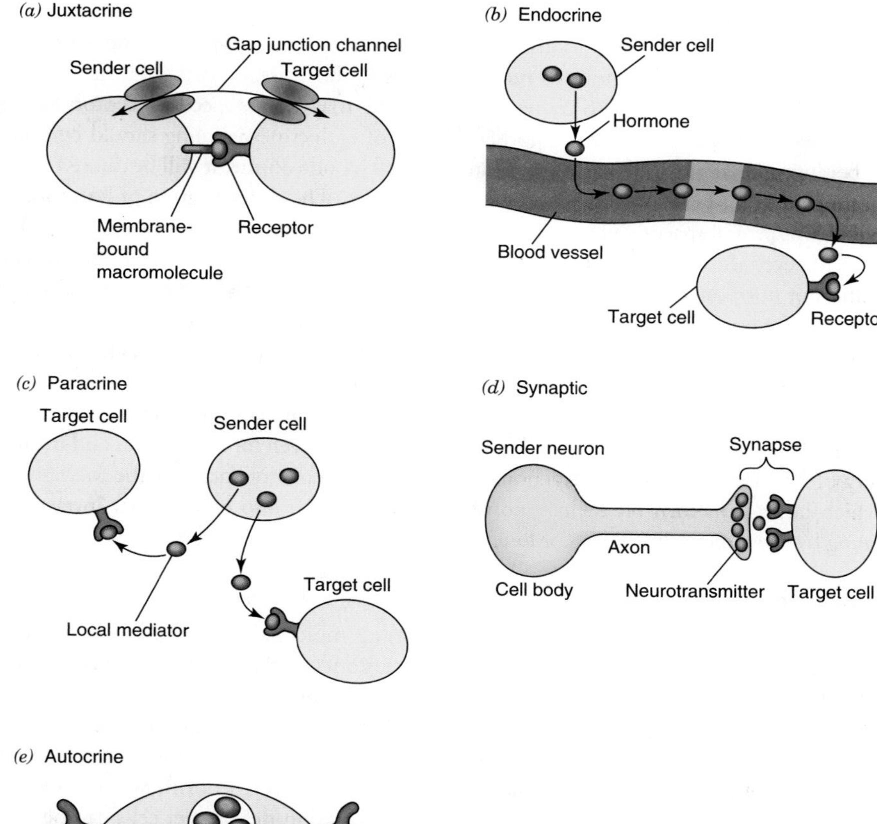

Figure 13.2 Five major modes of intercellular signal transduction.
Redrawn based on figures from Alberts, B., et al., *Essential Cell Biology,* 2d ed. New York: Garland Science, 2004; and Lodish, H., et al., *Molecular Cell Biology,* 4th ed. New York: Freeman, 2000.

between cells separated by large distances, for example, cells in different tissues or organs or between cells separated by short distances. Contact-independent signaling by secreted molecules includes four modes: **endocrine**, **paracrine**, **synaptic** or neuronal, and **autocrine**. These modes are distinguished by: (1) the actual distance between the sender and target cells; and (2) the biochemical nature of the secreted molecules.

Juxtacrine, or Contact-Dependent, Signaling

In juxtacrine signaling, cells communicate by either of two mechanisms. One is through gap junction channels that allow direct transfer of small ions, metabolites, and second messenger molecules between neighboring cells. The structure and function of these channels is described in detail on page 477 and in Figures 12.41, page 484, and 12.42, page 484. The second mechanism involves the interaction of a protein expressed on the surface of the sender cell with a receptor protein expressed on the surface of the target cell (Figure 13.2*a*). This interaction changes the conformation of the receptor protein and triggers a cascade of intracellular signaling events in the target cell that results in altered function of that cell. Important elements of juxtacrine signaling include a variety of proteins, such as integrins, involved in cell–cell adhesion and regulation of cell shape and motility.

Endocrine Signaling

This mode involves signaling between cells separated by short or long distances. Specialized sender cells in glandular tissues synthesize, package, and secrete molecules called hormones

(Figure 13.2*b*). By definition, hormones are molecules secreted into the blood for long-distance transport to the target cells in various tissues. This is a broadcast type of remote signaling because blood carries the hormone to most cell types in most tissues. Although most cells in the body will be exposed to the hormone, only those cells that express specific receptors for that hormone will be target cells. Two important aspects of endocrine signaling should be noted. First, because a hormone is being secreted into the total blood volume, it will be diluted manyfold by the time it reaches target cells in peripheral tissues. The concentration of hormones in blood and interstitial spaces is always low (picomolar to nanomolar concentrations, i. e., 10^{-12} to 10^{-9} M) even during maximal rates of secretion from the source gland. Given the low concentration of hormones, the target cell receptors must have a very high affinity and selectivity for a particular hormone. One consequence of this high affinity is that once the hormone is bound, it cannot easily dissociate from its receptor. Second, it takes a considerable amount of time to increase and or to decrease the concentration of a hormone in blood because of the large volume or capacity of the blood. Following secretion, hormone concentrations in the blood and interstitial fluid can remain elevated for many minutes or even for hours. This and the high affinity of hormone—receptor interactions have important consequences for the mechanisms by which target cells turn off their response to a hormone. It also ensures that intracellular signaling by the target cell will last for long periods.

Paracrine Signaling

In this mode, a sender cell does not secrete signaling molecules into the blood for long-distance transport, but only secretes such molecules into its immediate or local environment (Figure 13.2*c*). This greatly restricts the distances over which the molecules can travel (by passive diffusion) and thus limits responses only to target cells in the immediate vicinity of the sender cell. Given their localized action, the secreted molecules used for paracrine signaling are generically termed *local mediators*. Because, unlike secreted hormones, the secreted local mediators are not massively diluted, their concentration near target cells can be fairly high (nanomolar to micromolar range, i. e., 10^{-9} to 10^{-6} M). Thus, target cell receptors for local mediators usually have lower affinity for these secreted molecules. Lower affinity means that the local mediator can rapidly dissociate from the receptor when the local, extracellular concentration of the mediator is reduced. Moreover, because the mediator is secreted locally into a small extracellular volume, its concentration at the target cell will be only transiently elevated due to rapid diffusion and dilution into the larger tissue space. Thus, paracrine signaling is used for rapid and localized communication between cells.

Synaptic or Neuronal Signaling

This type of paracrine signaling is specific to nerve cells that send out specialized cellular extensions (axons) to the immediate vicinity of a target cell that is usually another neuron or a muscle cell (Figure 13.2*d*). The synapses between a neuron and its target cells constitute a restricted extracellular space into which the neuron secretes signaling molecules called neurotransmitters. Given this special synaptic geometry, neurotransmitters travel only very short distances to the target cell. Because of this short intercellular distance and the restricted synaptic volume, the local concentration of neurotransmitter reaching the target cell can be very high (micromolar to millimolar range, i. e., 10^{-6} to 10^{-3} M). Thus, receptors on the target cell can have relatively low affinity for the neurotransmitter. This low affinity ensures that neurotransmitters can rapidly dissociate from their receptors following decreases in their local concentration. This is essential for the very rapid (millisecond) termination of neurotransmission from the presynaptic neuron to its postsynaptic targets.

Autocrine Signaling

This type of paracrine signaling occurs when one cell type is both the sender cell and the target cell (Figure 13.2*e*). Autocrine signaling is often used by organisms during tissue growth, organ development, and immune and inflammatory responses. The general features of autocrine signaling are the same as for nonsynaptic paracrine signaling except that the sender cell provides both the secreted molecule (often a large polypeptide growth factor) and a receptor for that molecule. For example, early tissue development often involves increased expression of a growth factor and the receptor for that growth by a particular progenitor or stem cell. The released growth factor induces autocrine activation of its target receptor to initiate replication of the original stem cell and its progeny cells.

Secreted Signaling Molecules

Many types of biological molecules, including proteins, glycoproteins, small peptides, amino acids, amines, lipids (fatty acids or steroids), nucleotides, nucleosides, and gases (e.g., nitric oxide), are used as secreted signaling molecules by various cells. The major differences between the signaling molecules are in their size and relative water solubility. Hormones used for endocrine signaling are amino acid products (large proteins, glycoproteins, polypeptides, or small peptides) or steroids. Neurotransmitters used for synaptic signaling are small, simple molecules such as amino acids (e.g., glutamate, glycine), substituted amines (e.g., epinephrine, acetylcholine, γ-amino-butyric acid [GABA]), nucleotides (e.g., ATP), or small peptides. Local mediators used for paracrine (nonsynaptic) and autocrine signaling are very diverse and include proteins, small peptides, simple organic molecules (e.g., histamine, ATP, and adenosine), or fatty acid derivatives (e.g., prostaglandins).

13.3 • RECEPTORS FOR SECRETED MOLECULES

Receptors for secreted molecules can be divided into two major classes: **intracellular receptors** and **cell surface receptors** (Figure 13.3). Hydrophobic molecules readily pass through the plasma membrane of target cells and bind to **intracellular receptor** proteins.

Intracellular receptors are specific for secreted steroid hormones, derivatives of vitamin D_3, retinoic acid, and thyroid hormone (Figure 13.3*a*). They compose a highly homologous group of structurally related proteins. Their structures and functions (see Figures 22.46,

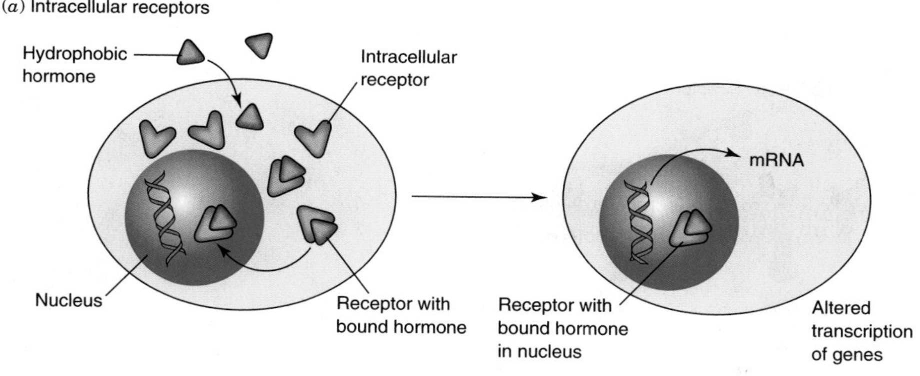

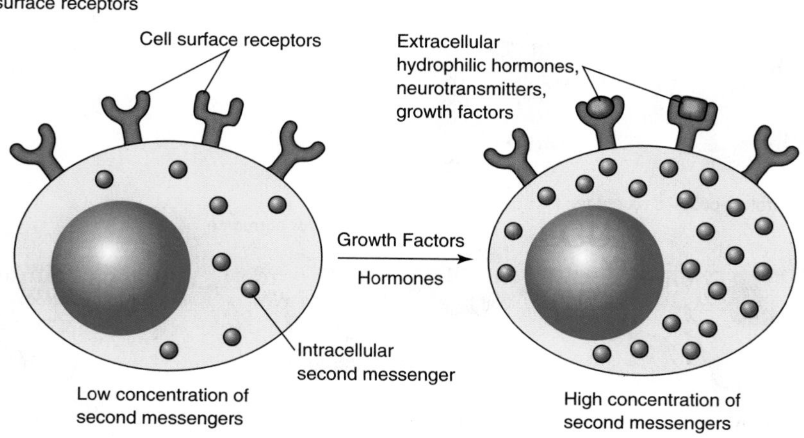

Figure 13.3 **Basic properties of intracellular receptors versus cell surface receptors.**

p. 927, and 22. 47, p. 929) are described in detail in Chapter 22. Some are cytosolic proteins that traffic to the nucleus only when bound to their cognate hormone; others enter the nucleus even in the absence of bound hormone. In all cases, the hormone-bound receptors function as dimeric complexes (homomers or heteromers) that bind to specific DNA sequences within the upstream regulatory elements of various genes to enhance or repress transcription and thus alter expression of proteins encoded by these genes. Because of their effects on gene expression, activation of intracellular receptors produces long-lasting changes (hours to days) in the function of target cells.

Cell surface receptors are recognition sites for the vast majority of signaling molecules that are either too big (protein or polypeptide hormones) or too hydrophilic to rapidly cross the target cell plasma membrane. Such signaling molecules interact with the target cell by binding to cell surface receptors that are integral membrane proteins (Figure 13.3*b*). Surface receptor proteins are coupled to a variety of intracellular biochemical reactions termed **signal transduction cascades** or **pathways**. The earliest signal transduction event triggered by binding of a signal molecule to such a receptor is a conformational change in the receptor that results in: (1) the generation of intracellular signaling molecules known as **second messengers** (the extracellular secreted molecule is the first messenger) or (2) a change in plasma membrane electrical potential or (3) activation of enzymatic cascades involving protein kinases, protein phosphatases, or proteases. Four major superfamilies of structurally and functionally related cell surface receptors that mediate the vast majority of intercellular communication pathways (Figure 13.4) are **(1) ligand-gated ion channel receptors**, **(2) enzyme-linked or catalytic receptors**, **(3) cytokine family receptors**, and **(4) G protein-coupled receptors** or **GPCR**.

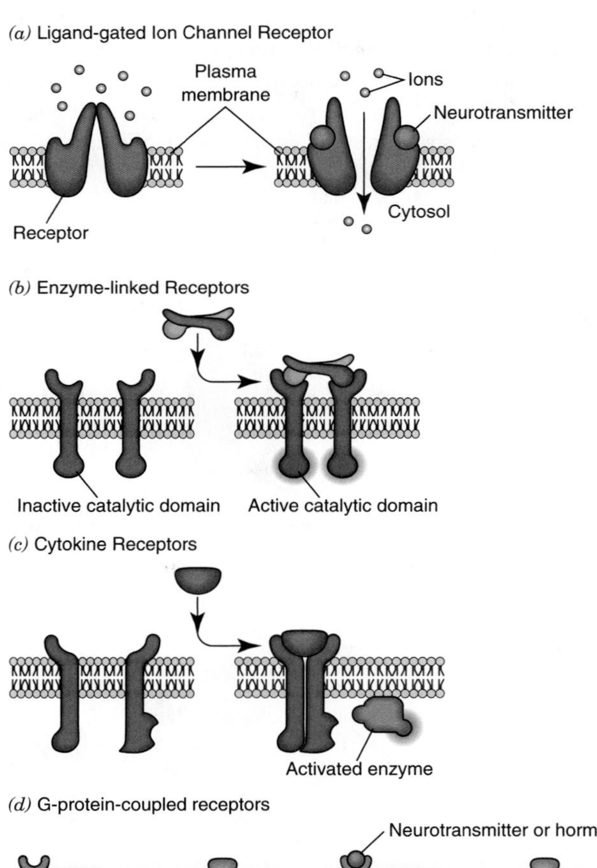

Figure 13.4 Major classes of cell surface receptors for secreted signaling molecules. Redrawn based on figure from Alberts, B., et al., *Essential Cell Biology,* 2d ed. New York: Garland Science, 2004.

13.4 • INTRACELLULAR SIGNAL TRANSDUCTION BY CELL SURFACE RECEPTORS

Ligands, Receptors, and Receptor–Ligand Interactions

A **ligand** is any molecule that binds to a receptor protein. An **agonist** is a ligand that on binding, activates signal transduction, whereas an **antagonist** is a ligand that prevents signal transduction when it binds to the receptor. A **physiological agonist** or **antagonist** is a naturally occurring molecule (like a hormone or neurotransmitter) that acts as a receptor ligand. A **pharmacological agonist** or **antagonist** is a synthetic molecule that acts as a receptor ligand. Many therapeutic drugs are agonists or antagonists of receptors.

Certain physiological agonists stimulate multiple types of receptors that are termed **receptor subtypes**. This means that the same extracellular signaling molecule may bind to different receptor proteins that are the products of different genes. For example, the neurotransmitter acetylcholine interacts with a ligand-gated ion channel receptor that causes contraction of skeletal muscle, or with a G protein-coupled receptor that causes relaxation of cardiac muscle (Figure 13.5). Each of the neurotransmitters, for example, acetylcholine, serotonin, and dopamine, recognizes a dozen or more receptor subtypes. Each receptor subtype can be differentially expressed in various tissues or/cells or can be coexpressed in the same tissue or cell. Moreover, receptor subtypes often activate opposing signal transduction responses. This greatly complicates the design and testing of synthetic agonists and antagonists and is a major reason why most drugs aimed at receptors have dangerous side effects when used improperly.

Relationships between Receptors, Effectors, and Second Messengers

The first step in signaling by cell surface receptors is the binding of the extracellular agonist to the receptor protein that changes the intrinsic energy and structural conformation of the receptor protein. In this conformation, the receptor interacts with **effector proteins** that are signaling proteins (enzymes or ion channels) distal to, but activated by, the agonist-bound receptor (Figure 13.4). In some cases, the receptor protein itself is also the effector. Activation of an effector leads to generation of **second messengers** or to changes in membrane potential. Some receptors interact directly with effector proteins whereas others require **coupling** or **transducing proteins** to interact with the effector. GTP-binding regulatory proteins (G proteins) are a very important class of such coupling proteins.

Enzyme effectors may be intrinsic components of a receptor protein (e.g., receptor tyrosine kinases) or distinct proteins (e.g., nonreceptor tyrosine kinases). They can be integral membrane proteins (e.g., adenylate cyclase) or cytosolic proteins (e.g., guanylate cyclase)

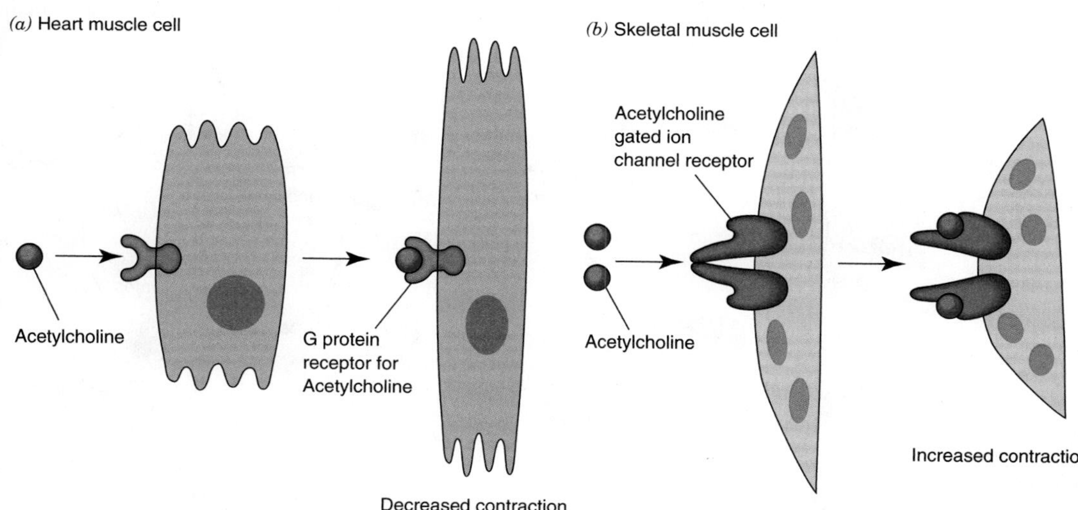

Figure 13.5 Basic characteristics of receptor subtypes as functionally distinct receptor proteins that bind a common extracellular signaling molecule.

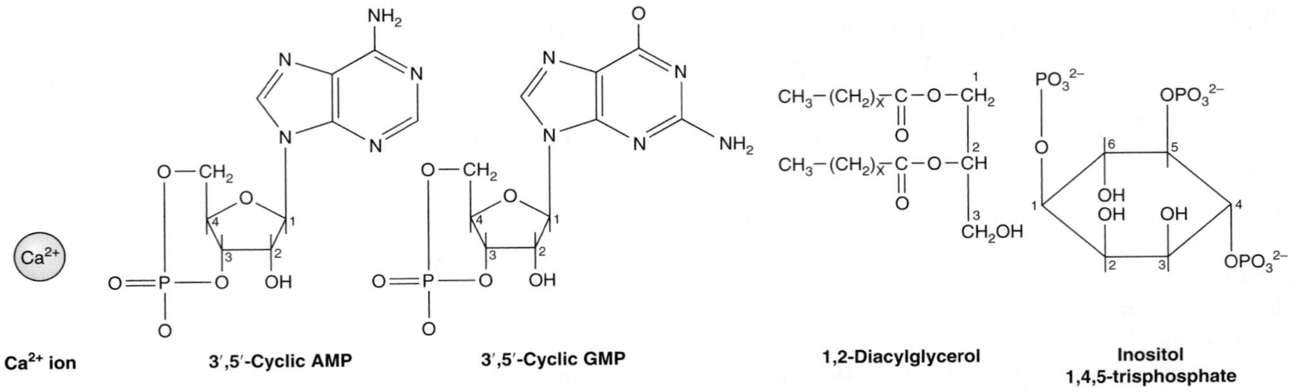

Ca^{2+} ion | 3′,5′-Cyclic AMP | 3′,5′-Cyclic GMP | 1,2-Diacylglycerol | Inositol 1,4,5-trisphosphate

Figure 13.6 Structures of four second-messenger molecules.

and include nucleotide cyclases, phospholipases, protein kinases, and protein phosphatases. These will be discussed in later sections. Ion channel effectors are an intrinsic component of some receptors (ligand-gated ion channels) whereas others are indirectly regulated by receptors via G proteins or second messengers.

Second messengers are small intracellular molecules that transmit and amplify the initial signal from agonist-activated receptors (Figure 13.6). They can be inorganic ions or organic products of enzyme-catalyzed reactions.

Ions passing through receptor-regulated channels function as second messengers in two ways. If the ionic flux is sufficiently great, the cytosolic concentrations of certain regulatory ions can change appreciably; Ca^{2+} is the major intracellular regulatory ion. If the particular ionic flux is the predominant current across the plasma membrane, the membrane potential will be driven to the equilibrium potential for that ion. This will result in depolarization or hyperpolarization of the membrane.

Effector enzymes often catalyze the rapid production of **water-soluble second-messenger molecules**, for exmple, cyclic AMP (cAMP), cyclic GMP (cGMP), and inositol trisphosphate (IP_3) (Figure 13.6).

Other effector proteins, such as phospholipases or lipid kinases, catalyze rapid production of membrane-associated **lipid second messengers**, for example, diacylglycerol (DAG) or phosphatidylinositol 3,4,5-trisphosphate (PIP_3) (Figure 13.6).

Protein Phosphorylation in Signal Transduction

Most pathways of intracellular signal transduction involve changes in the phosphorylation status of certain proteins. Phosphorylation of proteins can change their secondary, tertiary, or quaternary structure and result in altered function. Such phosphorylation involves the regulated activity of a variety of **protein kinases** and **protein phosphatases** (Figure 13.7*a*). Protein kinases are classified on the basis of their selectivity for particular amino acids. Most protein

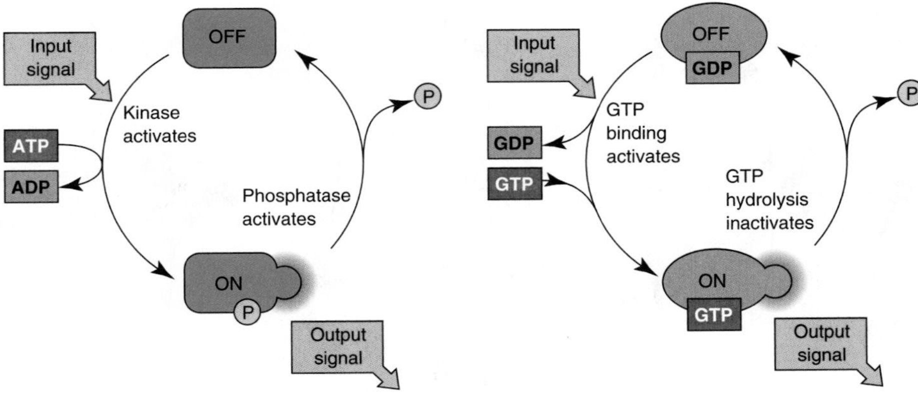

(*a*) Signaling by protein phosphorylation

(*b*) Signaling by GTP-binding regulatory protein

Figure 13.7 Protein phosphorylation versus guanine nucleotide binding as two major types of molecular switches for changing protein function during intracellular signaling.

kinases activated by second messengers phosphorylate specific proteins on serine or threonine residues. Many growth factors and cytokines will also stimulate kinases that phosphorylate proteins on tyrosine residues. These protein tyrosine kinases play important roles in regulation of cell growth and differentiation. Protein substrates for either serine/threonine kinases or tyrosine kinases are very diverse and include metabolic enzymes, ion channels, transcriptional factors, or other protein kinases. Protein phosphatases also exhibit selectivity for serine/threonine phosphate or tyrosine phosphate residues. The ability of some protein kinases to phosphorylate and activate target proteins, which are themselves protein kinases, underscores the involvement of **protein kinase cascades** in the regulation of complex cellular responses. These cascades are often critical elements in the **amplification** process that characterizes the intracellular signaling reactions initiated by binding of ligands to cell surface receptors.

Protein kinases can be regulated by receptors in several ways. Some tyrosine kinases are intrinsic components of the receptor protein structure (e.g., receptor tyrosine kinases) whereas others are cytosolic proteins that transiently associate with activated receptors (e.g., nonreceptor tyrosine kinases that bind to cytokine receptors). Most serine/threonine kinases are soluble proteins that are regulated by an increase in second-messenger concentration (e.g.. cAMP-dependent protein kinase). Likewise, there are multiple types of phosphatases, which reverse the signals generated by receptor- or second-messenger activated kinases.

GTP-Binding Regulatory Proteins in Signal Transduction

The G proteins that are functionally coupled to the G protein-coupled receptors belong to a larger family of proteins that bind and hydrolyze GTP as part of their specific cellular function (Figure 13.7*b*). These **GTPases** have an important **molecular switch** role, being involved in critical biological functions such as initiation, elongation, and termination of protein synthesis; microtubule assembly; regulation of enzyme activity; and membrane trafficking. Many of these "switch" GTP-binding proteins are monomeric proteins; this distinguishes them from the trimeric, receptor-coupled G proteins. All GTP-binding proteins (monomeric and trimeric molecules) have different conformations and activities depending on whether they have guanosine-5′-diphosphate (GDP) or GTP bound. The GDP-bound form interacts with another protein (a **guanine nucleotide exchange factor [GEF]**) that catalyzes the exchange of GDP for GTP. The GTP-bound form interacts with, and changes the activity of, downstream effector protein. The GTPase activity converts the GTP-bound form to a GDP-bound form, which has decreased ability to interact with effector proteins and must interact with a GEF in order to resume its active conformation. A major function of many cell-surface receptors is to trigger, directly or indirectly, the GDPGTP exchange reactions that result in G protein activation.

Other Components of Receptor-Mediated Signaling Complexes and Cascades

In addition to effector proteins, second messengers, trimeric G proteins, and protein kinases, a wide variety of intracellular proteins may be involved in the cell-surface receptor-mediated intracellular signaling pathways. By physically colocalizing multiple signaling proteins within a given signaling pathway, **scaffold** and **adaptor proteins** increase the fidelity and speed of a complex signaling cascade. Proteins termed **anchoring proteins** localize and concentrate soluble signaling proteins at particular subcellular locations. For example, a wide variety of A-Kinase Anchoring Proteins (AKAPs) localize cAMP-regulated protein kinases (protein kinase A [PKA]) at the plasma membrane, the cytoskeleton, or within the nucleus. The assembly of receptors, adaptor proteins, effector proteins, and other signaling components into signaling complexes often involves the specific interactions of particular domains within these proteins (Figure 13.8). **SH2** (**S**rc-**H**omology type **2**) **domains** and **PTB** (**p**hospho**t**yrosine **b**inding) **domains** recognize and bind to specific phosphorylated tyrosine residues on a variety of signaling proteins. **SH3** (**S**rc-**H**omology type **3**) **domains** bind to proline-rich domains that characterize certain signaling proteins. **PH** (**P**leckstrin **H**omology) **domains** specifically recognize and bind to certain inositol phospholipids that have important roles as either lipid second messengers or lipid substrates for phospholipase C.

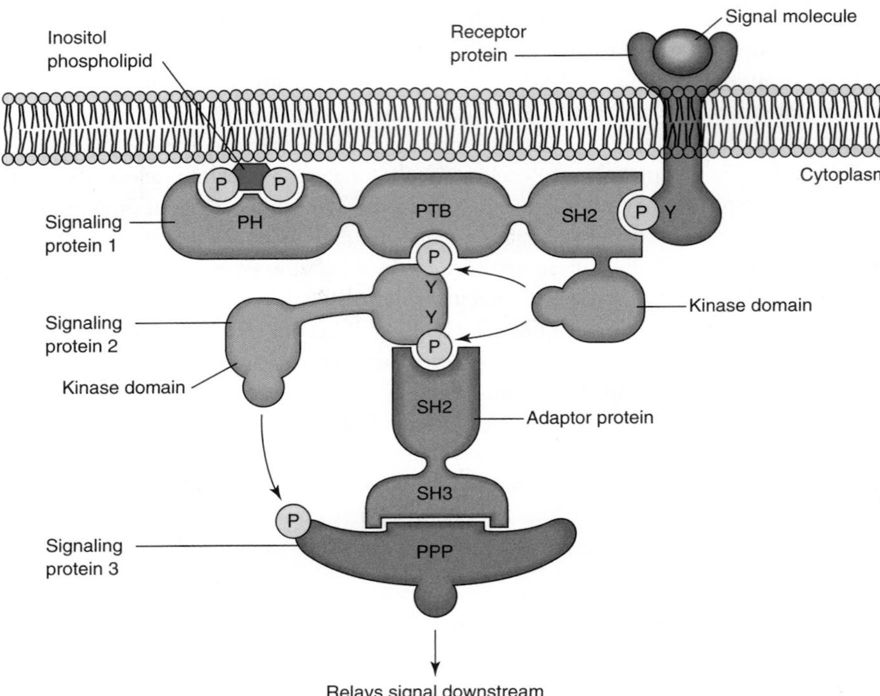

Figure 13.8 Roles of adapter proteins and protein–protein interaction domains in the assembly of intracellular signaling complexes. Redrawn based on figure from Alberts, B., et al., *Molecular Biology of the Cell,* 4th ed. New York: Garland Science, 2002.

Ligand–Receptor Interaction and Downstream Signaling Events

Binding of an **agonistic ligand** to a receptor is characterized by the receptor's binding **affinity** and **capacity**. Affinity describes the relationship between agonist concentration and formation of agonist–receptor complexes. Affinity is inversely related to the **dissociation constant (K_D)** (measured in ligand Molarity) that characterizes the agonist–receptor complex (Figure 13.9). The K_D is equal to the concentration of free agonist that permits binding of 50% of the total number of binding sites on the receptor (half-maximal occupancy). **Potency** is a term used to measure the relative affinity of an agonist for its receptor. An agonist with high potency binds even when present at very low concentrations (high affinity, low K_D). The capacity of the interaction refers to the absolute number of receptor sites. Because receptor proteins are usually expressed in relatively small amounts per cell, agonist–receptor binding curves show saturation. In addition to directly measuring the binding of an agonist to its receptor, one can indirectly measure the agonist–receptor interaction by assaying the various functional responses triggered by the agonist-bound receptor.

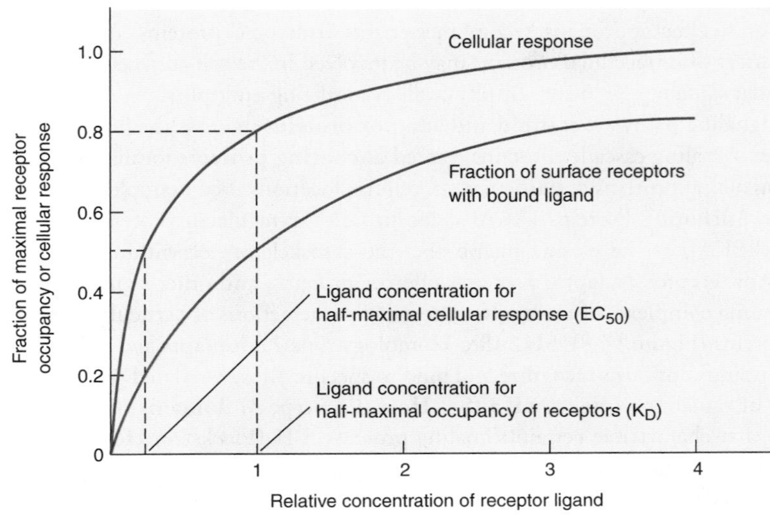

Figure 13.9 Relationship between K_d, as a measure of ligand-receptor binding affinity, and EC_{50}, as a measure of functional response to the formation of ligand–receptor complexes.

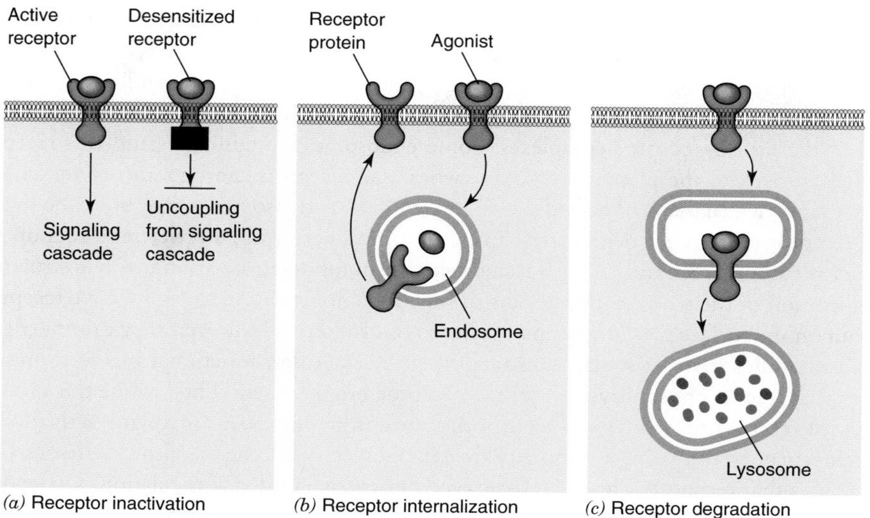

Figure 13.10 Major mechanisms for the termination of receptor-dependent signal transduction.

In general, the magnitude of these responses is proportional to the amount of agonist-bound receptor complex. The **dose–response curve** measures the relationship between agonist concentration and biological response.

The concentration of agonist that causes a half-maximal response is the half-maximal effective concentration or $\mathbf{EC_{50}}$. This is directly proportional to the K_D of the agonist–receptor complex (Figure 13.9). However, EC_{50} is usually much lower than the K_D; that is, the agonist concentration required for half-maximal activation of the cell's biological response is lower than the agonist concentration required for half-maximal occupancy of the cell's total number of receptors. This means that a cell usually expresses more receptors than it needs for effective biological responses. These extra, or spare, receptors permit the cell to respond to agonist even after the inactivation or degradation of receptors activated by previous agonist occupation. As described below, receptor inactivation or degradation plays an important role in terminating signal transduction by receptors.

Termination of Signal Transduction by Cell Surface Receptors

Cells express multiple mechanisms that permit them to terminate or attenuate their responses to various hormones, neurotransmitters, or local mediators. These mechanisms allow cells to return to their basal, prestimulus state and prevent overactivation or inappropriate activation of the functions controlled by a particular receptor (Figure 13.10). They include: (1) reducing agonist availability in the extracellular vicinity of a target cell, (2) internalizing and degrading the agonist-bound receptor complex, and (3) rapidly modifying the receptor (e.g., by phosphorylation) so that it becomes inactive or desensitized.

The simplest mechanism for terminating signal transduction is to reduce the extracellular concentration of the secreted agonist. The effectiveness and rapidity of this mechanism is determined by (1) the stability of the secreted agonist, (2) the size or capacity of the extracellular compartment into which the agonist is secreted, and (3) the proximity of sender cells to target cells. It is difficult to terminate endocrine signaling by reducing agonist availability rapidly because hormones are secreted into a large-capacity extracellular environment. In contrast, rapid reduction in agonist availability plays a significant role in terminating synaptic signaling; neurotransmitters are locally secreted into a very low-capacity space, the synaptic cleft, and diffuse away rapidly. Some cells also actively reduce the concentration of nearby agonist by reaccumulating the released agonist (e.g., synaptic reuptake of neurotransmitters like serotonin and dopamine) or by metabolizing the extracellular agonist. Extracellular metabolism can be catalyzed by specific enzymes (e.g., the acetylcholinesterases) present on the surface of target cells. Certain neurotoxins act as inhibitors of neurotransmitter breakdown.

Cells can also terminate signal transduction by regulating the functional availability of receptors, a process known as **adaptation** or **desensitization**. If repeatedly exposed to a particular agonist, a cell adapts its sensitivity to that agonist. Over time, a cell will become less and less sensitive to a concentration of agonist that normally causes maximal activation. This desensitization results from several processes. First, many receptors are rapidly internalized when bound by their agonistic ligand. These receptor–ligand complexes are incorporated in endosomes and subsequent acidification of endosomes causes dissociation of the ligand–receptor complexes. Some endosomes containing ligand-free receptor are routed back to the plasma membrane where the receptors can be reincorporated into the surface membrane. Other endosomes are routed to lysosomes where both the ligand and receptor protein are degraded. Thus, over time, there will be fewer receptors on the surface membrane. Second, a much faster mechanism for desensitization involves structural modification of the receptor that results in functional inactivation. Such inactive receptors remain on the surface membrane and may even bind agonist. However, they cannot trigger the intracellular steps of signal transduction. In many cases, functional inactivation of a receptor is due to phosphorylation of the receptor protein itself. Thus, while the agonist-occupied receptor stimulates a cascade of signal transduction events involving activation of protein kinases, some of these protein kinases phosphorylate the receptor itself; this then inactivates that receptor. This is a classic type of negative feedback regulation. Depending on the receptor, this functional inactivation can be reversed rapidly or slowly.

13.5 • LIGAND-GATED ION CHANNEL RECEPTORS

Very rapid (milliseconds) signaling is required at nerve–nerve or nerve–muscle synapses. Thus, neurotransmitter receptors need to change cell responses via a minimum number of enzymatic steps because some biochemical reactions are relatively slow (seconds–minutes). Many (but not all) neurotransmitters bind to receptors that are **ligand-gated ion channels** (see Figure 13.4*a*, p. 507). The extracellular ligands for these receptors are invariably very small organic molecules like amino acids, substituted amines, or nucleotides. These are packaged in vesicles and released via rapid exocytosis at synapses. Because these ligands are so small, they diffuse rapidly (Figure 13.11). The ligands bind to their receptors with low affinity (micromolar to millimolar K_D). This permits rapid dissociation of the ligand–receptor complex on reduction of the extracellular concentration of the neurotransmitter.

The binding of neurotransmitter to extracellular ligand-binding sites of such receptors triggers a near-instantaneous change in receptor conformation that permits ions to permeate through the receptor protein complex. The flux of these ions rapidly changes the

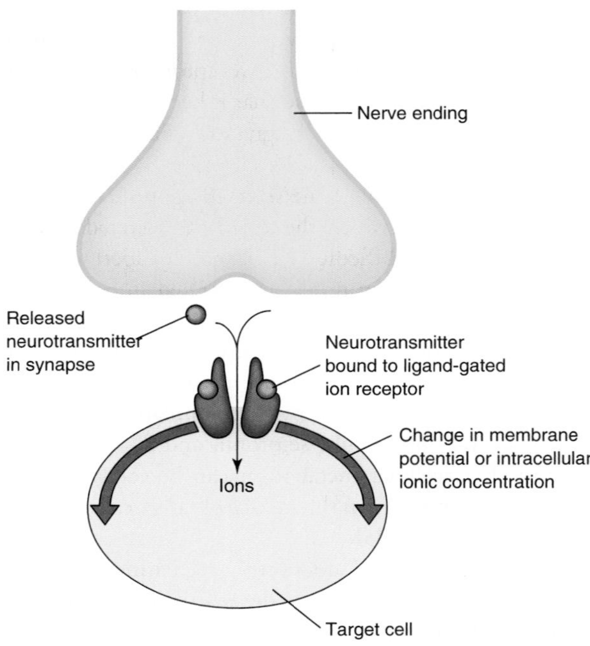

Figure 13.11 **Neurotransmitter-gated ion channel receptors as major signal transduction elements at neuronal synapses.**

cell's membrane potential. If a cation like Na^+ passes through the receptor channel, the cell becomes depolarized; if an anion like Cl^- is the permeating species, the cell becomes hyperpolarized (Figure 13.12). Cation-selective receptors are gated by the **excitatory neurotransmitters** (acetylcholine, glutamate, serotonin, and ATP) whereas anion-selective receptors are gated by the **inhibitory neurotransmitters**, (γ-amino butyric acid [GABA] and glycine). Activation of these neurotransmitter-gated ion channels changes the cell's plasma membrane potential and results in secondary regulation (activation or inhibition) of many types of voltage-gated channels, including voltage-gated channels for Ca^{2+}. Changes in the activity of these channels will alter the cytosolic concentration of Ca^{2+}, a key second messenger. In turn, increased cytosolic Ca^{2+} drives acute responses in target cells, such as exocytotic release of neurotransmitter-containing vesicles from neurons or contraction of muscle cells, or long-term responses, such as activation of Ca^{2+}-sensitive gene expression. Most of the neurotransmitters that activate ligand-gated ion channel receptors can also interact with other receptor subtypes that are G protein-coupled receptors. Thus, receptors for acetylcholine are either **nicotinic receptors** (because they also bind the drug nicotine) that are ligand-gated channels, or **muscarinic receptors** (because they also bind the drug muscarine) that are G protein coupled.

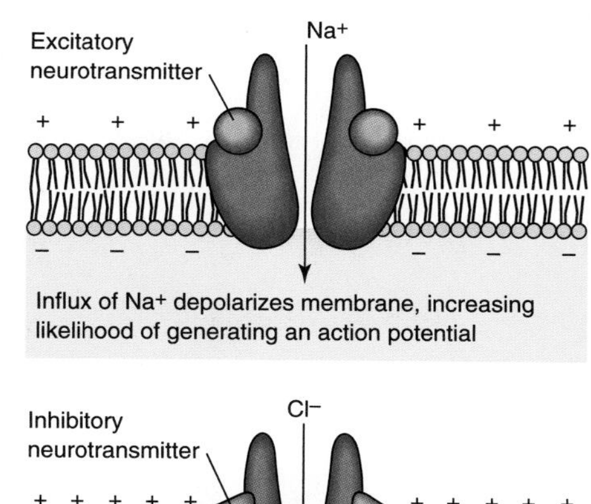

Figure 13.12 Excitatory versus inhibitory neurotransmitters as agonists for ligand-gated ion channel receptors.

Ion Channel Receptors

All ion channel receptors are oligomeric complexes of three to five protein subunits. Some are homomeric complexes that contain several copies of the same protein subunit (i.e., the product of the same gene). Others are heteromeric complexes of different protein subunits (i.e., each subunit is the product of a different gene). Each individual subunit is an integral membrane protein with multiple transmembrane domains. **Nicotinic acetylcholine receptors** provide a good example of the important structural features that characterize ligand-gated ion channel receptors. These are pentameric complex of subunits, each of which has an extracellular N-and C-terminus and four membrane-spanning domains (M1, M2, M3, and M4). The five juxtaposed M2 domains, one from each protein subunit, form the ion-conducting pore. Binding of the neurotransmitter to extracellular sites on the receptor complex causes very small and subtle rearrangements of how the M2 domains pack together. This near-instantaneous rearrangement removes the energetic barriers to ionic flow through the water-lined pore. In this state, the receptor is ligand bound and activated. However, the activated state of ligand-bound receptor channels persists for only a few seconds or minutes (it varies with different types of ligand-gated receptor channel) before a second rearrangement of the subunits occurs. That causes the channel to assume a nonconducting conformation even though the neurotransmitter remains bound (i.e., the receptor is ligand bound but inactive). This rapid inactivation is a non-covalent form of receptor desensitization (Figure 13.13). Reactivation of the receptor channel occurs only after the neurotransmitter dissociates from the extracellular binding sites. Thus, these receptors cycle between three major functional states (no ligand/inactive > ligand-bound/active > ligand-bound/inactive).

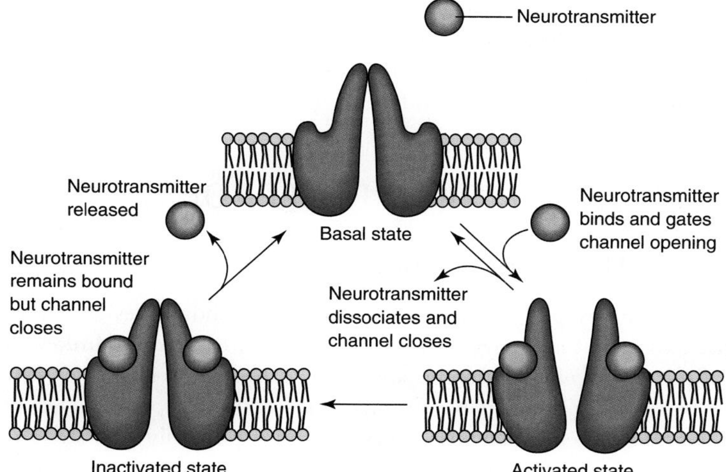

Figure 13.13 Conformational changes in ligand-gated ion channels during activation and inactivation of receptor function. Redrawn based on figure from Alberts, B., et al., *Molecular Biology of the Cell,* 4th ed. New York: Garland Science, 2002.

Termination of Signaling by Ion Channel Receptors

Several mechanisms exist for termination of signaling by ion channel receptors. First, the ligands for these receptors are secreted very transiently in very small, defined extracellular spaces (e.g., the synapse) permitting them to diffuse away from the receptor rapidly. Second, the extracellular concentration of the neurotransmitter is reduced by rapid breakdown by enzymes (like acetylcholinesterase) on the outer surface of the plasma membrane or by rapid re-uptake into the secreting neuron. Finally, the formation of the ligand-bound, inactive state ensures a very brief period of signal transduction by these neurotransmitter receptors.

Although rapid gating of ion fluxes is their direct function, these receptors also associate with a large number of other cellular proteins. Some of these associated proteins act as adapter proteins to co-localize efficiently these receptors with other downstream signaling proteins. They form concentrated **signaling complexes** at particular subcellular domains such as the plasma membrane at synaptic contacts. For example, the intracellular C-terminal tail of the glutamate-gated ion channel receptor associates with an **adapter protein** called PSD-95 at the **P**ost-**S**ynaptic **D**ensity under the synaptic contact site. This interaction is facilitated by a domain on PSD-95 called the PDZ domain. (PDZ is an acronym combining the first letters of three **p**roteins–**p**ostsynaptic **d**ensity protein [PSD95], **D**rosophila disc large tumor suppressor [DlgA], and **z**onula occludens-1 protein [Zo-1]–that were first found to share the domain). PSD-95 contains multiple PDZ domains (in addition to the specific PDZ site that binds to the glutamate-gated channels) that permit the co-localization of the receptor with various kinases, phosphatases, and cytoskeletal proteins.

Other proteins that associate with neurotransmitter-gated channels act as chaperones to ensure the localization and clustering of these receptors at the synaptic membrane. For example, nicotinic acetylcholine receptors are highly concentrated at the neuromuscular synapse via their physical interaction with the protein rapsyn, a 43 kDa membrane-associated protein that tightly associates with the intracellular domain of nicotinic receptors. When rapsyn is knocked out in mice, nicotinic receptors are not appropriately clustered at the neuromuscular junction, and this results in myasthenia or muscle weakness.

Other Ligands of Ion Channel Receptors

Receptors other than the neurotransmitter-gated ion channels utilize ion channels for intracellular signal transduction. Certain types of channels are activated by direct protein–protein interaction with subunits from activated trimeric G proteins, which in turn are activated when upstream G protein-coupled receptors are activated by agonistic ligands. Other types of ion channels are activated on binding with intracellular second-messenger ligands such as cAMP, cGMP, or Ca^{2+}. These second messengers accumulate on activation of many types of cell surface receptors, including G protein-coupled receptors, catalytic receptors for certain growth factors, and some cytokine receptors. Thus, these ligand-gated ion channels are indirectly regulated by cell surface receptors. Very important second-messenger-gated ion channels are the Ca^{2+} channels localized in the endoplasmic reticulum. Finally, the activity of yet other ion channels can be increased (or decreased) on phosphorylation by various second-messenger-regulated protein kinases. This is another example of indirect regulation of channel function by cell surface receptors.

13.6 • ENZYME-LINKED RECEPTORS

Physiological Roles and Extracellular Ligands

The receptors for many polypeptide growth factors and hormones are transmembrane proteins that have intrinsic catalytic activity (Figure 13.14). **Enzyme-linked** or **catalytic receptors** of this *superfamily* of proteins are distinguished by several major structural features. Their intracellular domains have protein kinase, protein phosphatase, protease, or nucleotide phosphodiesterase activities. These receptors primarily regulate long-term cell functions on a time scale of minutes to hours by initiating intracellular signaling cascades

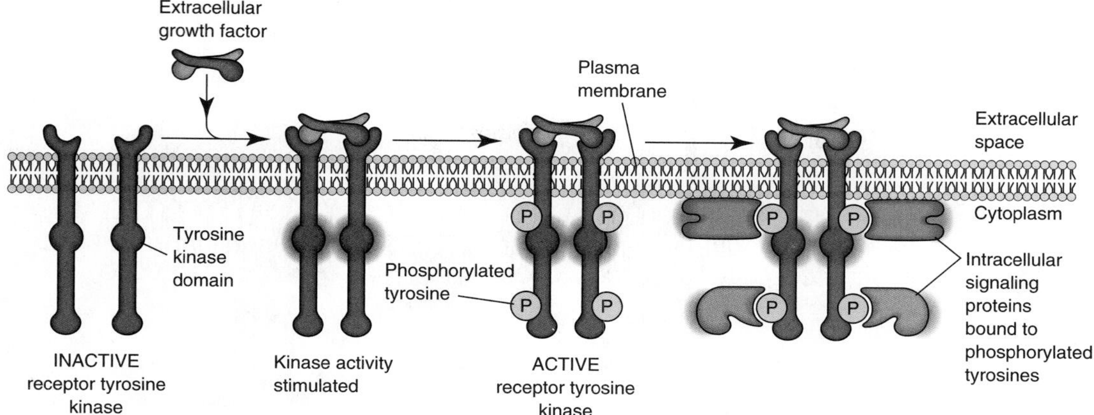

Figure 13.14 Conformational and functional changes in a receptor tyrosine kinase during activation by growth factor binding. Redrawn based on figure from Alberts, B., et al., *Essential Cell Biology,* 2d ed. New York: Garland Science, 2004.

that culminate in activation or inhibition of gene expression. In turn, these changes in gene expression direct very fundamental pathways of integrated cellular response, such as **cell division**, **programmed cell death**, or **cell differentiation**. Their agonists are usually large, secreted proteins that function as paracrine/endocrine growth factors or differentiation factors. These ligands generally bind to their cognate receptors with very high affinity (picomolar K_D). However, the agonistic ligands for some types of catalytic receptors are macromolecules expressed on the extracellular surfaces of adjacent cells (juxtacrine signaling). The most common catalytic receptors have tyrosine kinase activity. These **receptor tyrosine kinases (RTK)** include the receptors for epidermal growth factor (EGF), platelet-derived growth factor (PDGF), insulin, and many other polypeptide **growth factors**. Some catalytic receptors have **serine/threonine kinase** activity. Finally, certain hormones involved in blood pressure regulation bind to receptors with **guanylate cyclase** activity (see Figure 22.33, p. 915).

Receptor Tyrosine Kinases (RTK)

Most RTKs are single subunit receptors but some (like the insulin receptor) exist as multimeric complexes (see Figure 22.29, p. 911). Each monomer has a single transmembrane (TM) spanning domain that comprises 25–28 amino acids. The NH_2-terminal ends of such receptors are extracellular and compose a very large domain for binding the growth factor or hormone. The COOH-termini are intracellular and contain the domains responsible for the catalytic activities of the receptors.

When a growth factor binds to the extracellular domain of an RTK, it triggers dimerization with adjacent RTK subunits (Figure 13.14); this leads to rapid activation of the cytoplasmic kinase domain. The first protein substrate for this activated tyrosine kinase is the receptor itself; the intracellular domain of the receptor becomes autophosphorylated on multiple tyrosine residues. The phosphorylated tyrosine residues then act as recognition or anchoring sites for other proteins that are substrates for the RTK. These protein substrates usually contain a SH2-type protein–protein interaction domain which recognizes the phosphorylated tyrosines on the receptor. This permits tight binding to the receptor and subsequent phosphorylation of tyrosines on the substrate proteins. Thus, activation of RTK results in accumulation of many types of tyrosine-phosphorylated proteins. These "downstream" substrates act as effectors for the RTK and include signaling proteins such as other protein kinases, regulators of small GTPases, and enzymes that modify phospholipid synthesis and breakdown. When phosphorylated on tyrosine residues, these effector proteins become active links in signaling cascades that ultimately result in changes in the location or functional activity of transcription factors or other proteins involved in cell division or differentiation.

Mutation of certain RTK genes results in expression of receptors that assume activated conformations in the absence of binding of growth factor (Figure 13.15). Mutated RTK genes can act as **oncogenes** (see p. 1016 and Table 24.2) that contribute to the initiation or progression of cancer (Clin. Corr. 13.1).

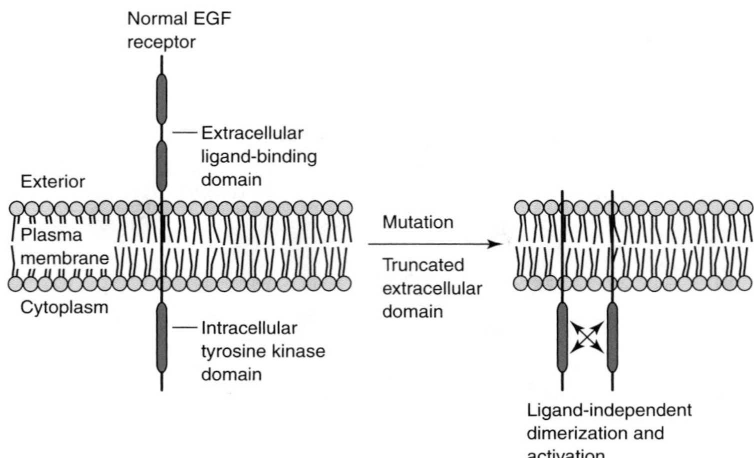

Figure 13.15 Mutated forms of receptor tyrosine kinases as the products of cancer-causing oncogenes.

Receptor dimerization also triggers an endocytic internalization of the growth factor–RTK complex. Acidification of these endosomes leads to dissociation of the growth factor from the receptor and degradation of the internalized growth factor by lysosomes. Although some of the internalized receptors are also degraded, most are rerouted back to the plasma membrane. This internalization of receptor–ligand complex, followed by dissociation of the ligand is the major mechanism for terminating signal transduction by these receptors.

Ras GTPase and MAP Kinase

Ras is a small GTPase or monomeric GTP-binding regulatory protein that is a critical regulator in cell proliferation. About 30% of all human tumors involve cells that express mutated **Ras oncogenes**. Proteins that directly bind to phosphotyrosine sites of an activated RTK include adapter molecules that recruit and stimulate Ras-activating proteins (Figure 13.16). The stimulated Ras-activating proteins act as GEFs to enhance the exchange of GTP for GDP on Ras itself; this results in accumulation of Ras molecules with an active,

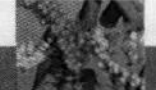

CLINICAL CORRELATION 13.1

ErbB/HER Receptor Tyrosine Kinases as Targets for Cancer Chemotherapy

Receptors belonging to the ErbB/HER family are linked to many different types of human cancers. Overexpression of the human ErbB1 gene, which encodes the EGF receptor (HER1), is common in bladder, breast, kidney, prostrate, and non-small-cell lung cancers. A mutant ErbB1 gene produces a receptor that lacks the extracellular EGF-binding domain, and this is highly expressed in glioblastomas that account for 25% of the brain tumors in adult humans. Human ErbB3 and ErbB4 encode the HER3 and HER4 receptors that bind extracellular proteins of the NRG (neuregulin/heregulin/neu) family of growth and differentiation factors. ErbB3 is often overexpressed in breast, colon, prostate, and stomach cancers, whereas ErbB4 overexpression occurs in ovarian granulosa cell tumors. ErbB2 encodes the HER2 protein that does not bind any known extracellular growth factor. However, HER2 can form homodimers or heterodimers with growth-factor-bound HER1, HER3, or HER4 receptors. Because dimerization activates the intrinsic tyrosine kinase activity of this protein, even modest overexpression of HER2 can alter normal cell growth regulation. Significantly, ErbB2 gene expression is amplified by up to two orders of magnitude in 20%–30% of human subjects with invasive breast cancer. The aberrant expression of ErbB/HER genes in multiple human cancers has prompted the development of several drug therapies that target these receptors. One group consists of monoclonal antibodies that bind to functionally significant extracellular domains of different HER subtypes. Trastuzumab (Herceptin® from Genentech), an anti-HER2 antibody, has been used in the treatment of those breast cancers characterized by overexpression of ErbB2/HER2. Its mechanism of action involves recruitment of immune factors that kill the tumor cells and also suppression of constitutive HER2 dimerization. Cetuximab (IMC-C225 or Erbitux® from ImClone), an antibody against the EGF-binding domain of the ErbB1/HER1 protein, prevents ligand-induced activation of the receptor. It is being tested in patients who have either squamous cell cancers of the head and neck or non-small-cell cancers of the lung. A variety of small molecule drugs have been designed to target the intracellular tyrosine kinase domains of the ErbB/HER proteins. These drugs are competitive inhibitors of the ATP-binding sites of the kinases, particularly the ErbB1/HER1 subtype. They are being tested in patients suffering from non-small-cell lung cancers that fail to respond to other chemotherapies.

Roskowski, R., Jr. The ErbB/HER receptor protein-tyrosine kinases and cancer. *Biochem. Biophys. Res. Commun.* 319:1, 2004.

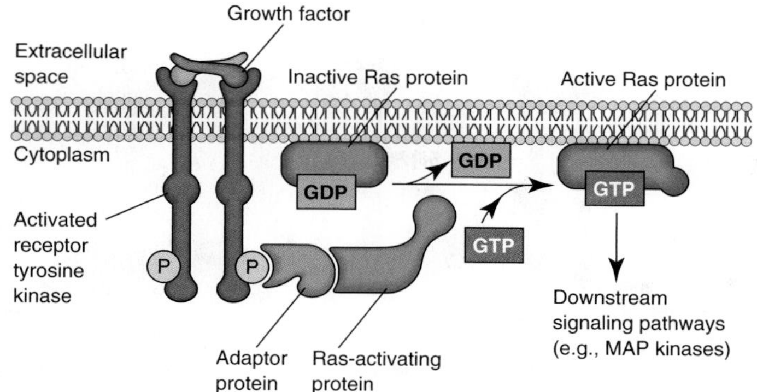

Figure 13.16 Role of the ras GTPase during intracellular signal transduction by an activated receptor tyrosine kinase.

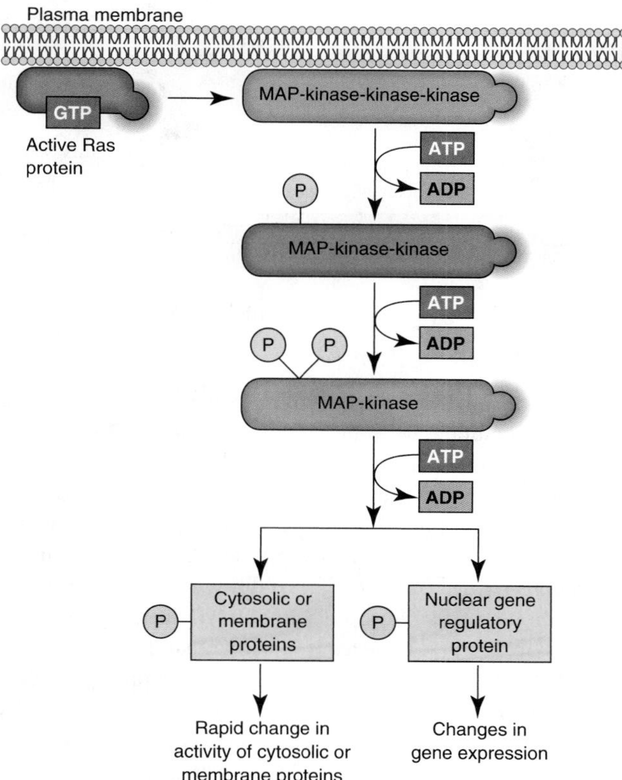

Figure 13.17 Role of the MAP kinase cascade during intracellular signal transduction by an activated receptor tyrosine kinase.
Redrawn based on figure from Alberts, B., et al., *Essential Cell Biology,* 2d ed. New York: Garland Science, 2004.

GTP-bound conformation. In turn, these active Ras proteins transiently bind to and stimulate a family of serine/threonine protein kinases that trigger the **m**itogen-**a**ctivated **p**rotein kinase cascade, or **MAP kinase cascade** (Figure 13.17). This amplification cascade involves the serial actions of three protein kinases; the initial Ras-activated kinase (or MAP kinase kinase kinase) activates an intermediary set of MAP kinase kinases, which activate the terminal **MAP kinase** effectors. Activated terminal MAP kinases phosphorylate multiple target proteins in both the cytosol and the nucleus, including transcription factors that regulate the expression of genes required for cell division, cell survival, or phenotypic differentiation.

Receptor Serine/Threonine Kinases

The growth regulatory factors **transforming growth factor-β (TGF-β)** and **bone morphogenetic proteins (BMP)** have receptors with intrinsic serine/threonine kinase activity. These growth factors have important signaling roles in tissue development at the fetal and neonatal stages and in the maintenance of a differentiated tissue phenotype in adult organisms. As in the case of receptor tyrosine kinases, mutations in TGF-β receptors and other serine/threonine kinase receptors contribute to the initiation or progression of certain cancers.

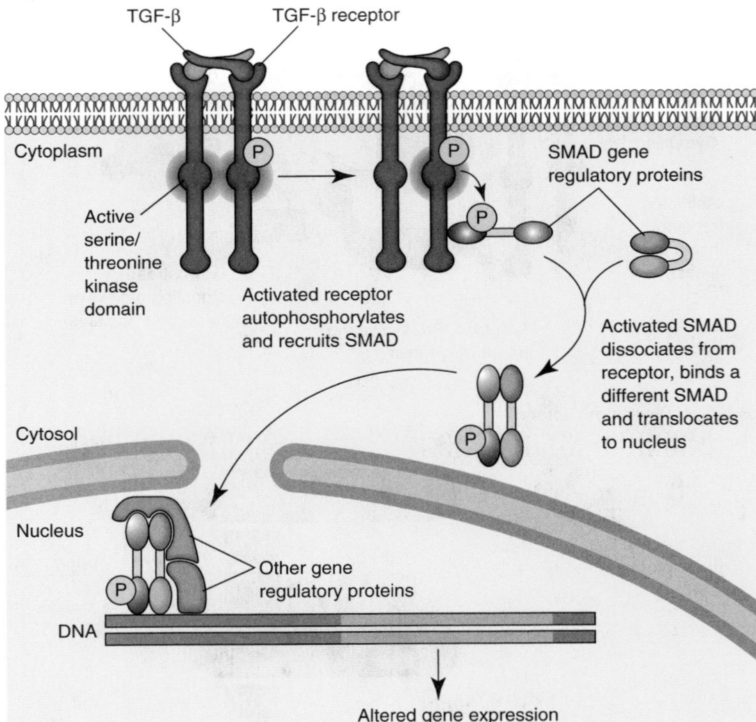

Figure 13.18 Intracellular signaling cascades triggered by a receptor serine/threonine kinase during activation by TGFβ (transforming growth factor β).
Redrawn based on figure from Alberts, B., et al., *Essential Cell Biology,* 2d ed. New York: Garland Science, 2004.

Similarly, these receptors compose plasma membrane proteins with an extracellular amino terminus, a single transmembrane-spanning region, and an intracellular carboxy terminus that contains the catalytic domain. They also form homodimeric complexes when their agonists bind to the tandem extracellular domains (Figure 13.18). The resulting conformational change results in one catalytic subunit phosphorylating its partner subunit on serine/threonine residues. This initial phosphorylation allows the partner subunit to recruit and to phosphorylate cytoplasmic proteins termed **SMADs** that comprise a family of gene regulatory factors. (SMAD proteins are homologs of a Drosophila protein "mothers against decapentaplegic," or MAD, and a C. elegans protein SMA; the name is a combination of the two). In their dephosphorylated state, SMADs adopt a folded conformation, which prevents their interaction with other SMADs and also maintains their localization as cytosolic proteins. However, phosphorylation by activated receptor kinases allows certain SMAD subtypes to unfold and to form dimeric complexes with other SMAD subtypes. This dimerization also exposes **nuclear localization sequences (NLS)** on the SMADs that result in translocation of the cytosolic complex into the nucleus where the SMAD dimers interact with other gene regulatory proteins to modulate the transcription of genes involved in organ development or tissue differentiation.

13.7 • CYTOKINE RECEPTORS

Cytokines represent another group of secreted polypeptides that act as autocrine/paracrine regulators of growth and differentiation. Many cytokines control the growth and differentiation of **hematopoietic** (blood-forming) cells including the various types of white blood cells, or leukocytes. For this reason, some cytokines are known as **interleukins** because they regulate information transfer among different types of leukocytes during various stages of immune or inflammatory responses (Clin. Corr. 13.2). Other cytokines are called **interferons** for their ability to interfere in the changes in function induced in cells and tissues infected with viral or bacterial **pathogens**. Like the growth factors that bind to catalytic receptors, cytokines bind to their cognate receptors with high affinity and trigger the rapid activation of protein kinase cascades and accumulation of phosphorylated signaling proteins that ultimately direct long-lasting changes in gene expression. However, the plasma membrane receptors for cytokines lack intrinsic tyrosine kinase or serine/threonine kinase activity (see Figure 13.4*c*, p. 508).

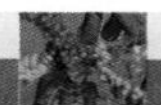

CLINICAL CORRELATION 13.2

Endogenous Antagonists of Interleukin-1 Receptors as Therapy for Inflammatory Diseases

Interleukin-1 (IL-1) is the composite designation for two related cytokines (IL-1α and IL-1β) that bind to the same multi-subunit IL-1 receptors (IL-1R). The broad expression of IL-1R underlies the central role of IL-1 in many immune, inflammatory, metabolic, and hematopoietic responses. These include fever, increased expression of leukocyte adhesion proteins on endothelial cells, induction of acute phase proteins by the liver, and stimulated production of other cytokines and colony-stimulating factors. IL-1β is a major proinflammatory cytokine whose local and circulating levels are tightly regulated to prevent aberrant activation of pathways that can lead to chronic inflammatory diseases. IL-1β accumulates as a biologically inactive procytokine in the cytoplasm of monocytes and macrophages. Its activation requires cleavage to a mature, secreted form by the protease, caspase-1. Activation of caspase-1 involves its oligomerization via assembly of multiprotein complexes termed inflammasomes. Under normal conditions the assembly of inflammasomes is regulated by adapter proteins that become active only in response in bacterial infection or tissue damage. However, some people express mutated versions of these adapter proteins, and this causes activation of inflammasomes and hyperproduction of IL-1β in the absence of microbial infection. The resulting overproduction of IL-1β has been linked to multiple inflammatory diseases including the periodic fever syndromes, Crohn disease, rheumatoid arthritis, and gout. Therapeutic agents, such as IL-1 receptor antagonist (Anakinra or Kineret®), are highly effective for the treatment of autoinflammatory syndromes. Notably, Anakinra is a recombinant version of a naturally occurring protein, human IL-1Ra (Ra for receptor antagonist), which is produced by the same cells that secrete IL-1α and IL-1β. Endogenous IL-1Ra is an important natural anti-inflammatory protein in arthritis, colitis, and other inflammatory diseases; maintenance of a balance between IL-1 and IL-1Ra production prevents the development or progression of such diseases. However, when endogenous IL-1 is overproduced in individuals with mutations in inflammasome regulatory proteins, the injection of recombinant IL-1RA acts to restore this balance.

Mariathasan, S., and Monack , D. M. Inflammasome adaptors and sensors: Intracellular regulators of infection and inflammation. *Nature Rev. Immunol.* 7:31, 2007: and Church, L. D., Cook, G.P., and McDermott, M. F. Primer: inflammasomes and interleukin 1beta in inflammatory disorders. *Nat Clin Pract Rheumatol.* 4:34, 2008.

Cytokine Receptors: Structure and Function

Cytokine receptors are multimeric receptors. Each subunit has a single transmembrane domain, a large extracellular amino-terminal domain for binding the cytokine, and an intracellular carboxy-terminal domain that contains different types of protein–protein interaction motifs but lacks intrinsic enzyme activity. Functional cytokine receptors exist as a stable **hetero-oligomeric complex** of two or more different receptor subunits. High-affinity binding of the cytokine usually requires interaction with at least two different receptor subunits. This coordinated binding of extracellular cytokine ligands triggers conformational changes in the intracellular domains of the receptor subunits that facilitate their rapid association with other intracellular signal-transducing enzymes. Cytokine receptors exhibit significant diversity in the structure of their individual subunits, as well as in their assembled multimeric complexes. Thus, different subfamilies recruit a broad range of intracellular signaling proteins. Among the most important of these associated proteins are **nonreceptor tyrosine kinases**, such as the src-family kinases and the **Janus kinases (JAKs)** (Figure 13.19). These tyrosine kinases are not integral membrane proteins, but exist as soluble or weakly membrane-associated proteins that become activated only when they associate with ligand-occupied cytokine receptors. They often phosphorylate the other subunits of the cytokine receptor on tyrosine residues which then act as recognition sites for the binding of other signaling proteins with SH2-interaction domains. Thus, several of the interleukin-type cytokine receptors recruit JAKs, which phosphorylate the receptors to generate binding sites for gene regulatory proteins known as **STATs** (**signal transducers and activators of transcription**). Like SMAD proteins, unphosphorylated STAT proteins are cytosolic monomers. Subsequent tyrosine phosphorylation by activated JAKs induces dimerization which results in translocation of the complex to the nucleus and association with other gene regulatory proteins.

Other cytokine receptors recruit adapter proteins that initiate the assembly of protein kinase complexes that culminate in the cytosol-to-nucleus translocation of the **NFκB** (nuclear factor of activated B cells) complex, an important transcription factor for regulation of expression of genes involved in multiple types of immune and inflammatory responses. Many cytokine receptors also trigger activation of the MAP kinase cascades.

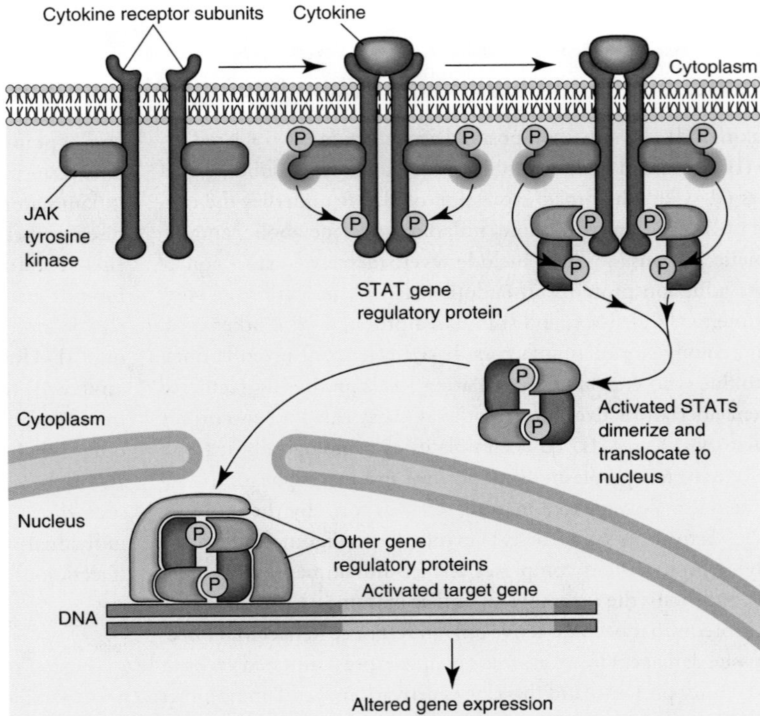

Figure 13.19 Intracellular signaling cascades triggered by prototypical noncatalytic cytokine receptor.

13.8 • G PROTEIN-COUPLED RECEPTORS

Physiological Roles and Extracellular Ligands

The superfamily of **G protein-coupled receptors** (GPCR) is targeted by an extraordinarily diverse range of agonistic ligands that include proteins, peptides, amino acid derivatives, catecholamines, lipids, nucleotides, and nucleosides (Figure 13.20). These ligands include hormones, neurotransmitters, and local mediators. The G protein-coupled receptors play important roles in endocrine, synaptic, paracrine, or autocrine signaling in virtually all

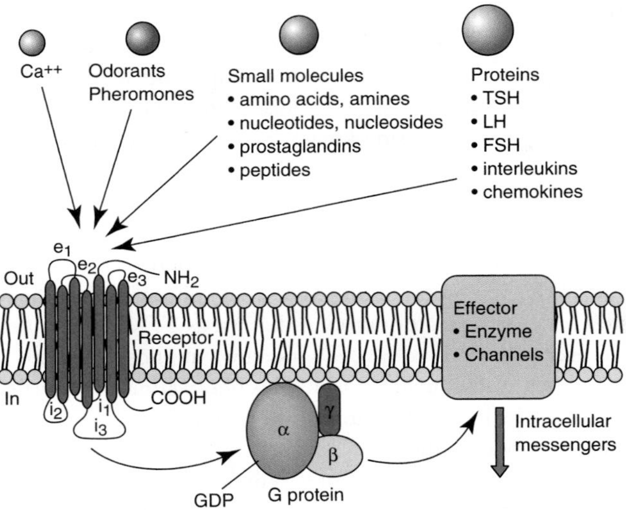

Figure 13.20 Major elements of signal transduction initiated by the 7-transmembrane domain G protein-coupled receptors.
Redrawn based on figure from Bockaert, J., and Pin, J-P. Molecular tinkering of G protein-coupled receptors: an evolutionary success. *EMBO J.* 18:1723, 1999.

CLINICAL CORRELATION 13.3

G Protein-Coupled Chemokine Receptors as Targets for the Human Immunodeficiency Virus (HIV)

Among the 1000 or so human genes that encode G protein-coupled receptors is the subgroup of chemokine receptors that are highly expressed in leukocytes (white blood cells). Chemokines compose approximately 50 polypeptide factors that are secreted by multiple cell types (*e.g.*, epithelial cells and stromal cells) within various tissues. They are agonistic ligands for 19 types of chemokine receptors that are differentially expressed in the various classes of leukocytes (*e.g.*, monocytes, T-lymphocytes, B-lymphocytes, natural killer cells), which execute critical immune and inflammatory responses within tissues invaded by foreign pathogens (protozoans, bacteria, or viruses). Chemokines function as: (1) chemotactic agents that recruit leukocytes to the infected tissue location,; (2) regulators of G protein-dependent signaling pathways involved in the rapid killing or sequestering of pathogens, and (3) inducers of gene expression that contribute to long-term adaptive immunity to pathogens. Not surprisingly, the disruption or disabling of chemokine receptor signaling can contribute to a wide range of infectious diseases. Of particular significance is the finding that human immunodeficiency virus (HIV) utilizes endogenous chemokine GPCR to infect and kill immune effector leukocytes.

HIV entry into leukocytes requires the direct interaction of the viral envelope glycoprotein (gp120) with extracellular surface proteins of the target host cell. The CD4 membrane protein, a critical immune receptor of helper T-lymphocytes, is a major target for gp120 binding. This explains why helper T cells are drastically reduced in most HIV-infected patients who develop AIDS (acquired immune deficiency syndrome). However, different strains of HIV preferentially infect T-lymphocytes (T-trophic virus) or monocyte-macrophages (M-trophic virus), and epidemiological analyses have revealed individuals who are highly resistant to infection by M-trophic strains of HIV. This resistance is due to mutations in the gene encoding CCR5, a chemokine receptor that is highly expressed in monocyte-macrophages; these mutations result in low or no expression of cell surface CCR5 protein. Thus, M-trophic strains of HIV require the presence of both the CCR5 and CD4, as co-receptors, to efficiently infect monocyte-macrophages.

CXCR4, a chemokine receptor that is highly expressed in T-lymphocytes, is a coreceptor with CD4 for binding the gp120 of T-trophic strains of HIV. Significantly, the ability of CCR5 and CXCR4 to bind gp120 does not require activation of the G_i-family proteins to which these receptors are normally coupled. Although G protein activation by CCR5 and CXCR4 is not obligatory for HIV entry, gp120 binding to these receptors does induce functional activation of the downstream G protein signaling cascades. It seems that activation of these signaling pathways within the HIV-infected cells may facilitate the chemo-attraction of naive uninfected leukocytes and so aid in propagation and dissemination of the virus within the infected subject. The identification of CCR5 and CXCR4 as HIV coreceptors has elicited intense interest in the development of antagonistic ligands that selectively attenuate gp120 binding and retard the colonization and spread of this devastating pathogen within infected subjects.

Sodhi, A., Montaner, S., and Gutkind, J. S. Viral hijacking of G protein-coupled receptor signaling networks. *Nat. Rev. Mol. Cell Biol.* 5:998, 2004.

tissues and cell types. It is estimated that several percent of the genes in the human genome encode GPCR. Ligands bind to GPCRs with affinities in the nanomolar to micromolar range, which is intermediate between the high affinity of enzyme-linked receptors and the low affinity of the ligand-gated channel receptors. Many important sensory proteins are G protein-coupled receptors including rhodopsin (see p. 958 and Figure 23.21) and the many receptors for perception of taste and smell, the actual ligands for which are largely unknown. Many GPCR are involved in the acute regulation of critical physiological responses such as cardiac contractility, metabolism, and complex behavior. These receptors or their downstream signaling pathways are the targets for many of the drugs most widely used in the treatment of human disease (Clin. Corr. 13.3 and 13.5).

Structure of G Protein-Coupled Receptors

When occupied by agonistic ligands, GPCRs act as GEFs for **heterotrimeric GTP-binding regulatory proteins** (**G proteins**) (see Figure 13.20), that interact with downstream effector proteins which are usually enzymes or ion channels. The GPCRs do not directly regulate effector proteins. Thus, even the earliest signaling events require the participation of at least five proteins: the receptor itself, the intermediary G protein (with its three distinct protein subunits), and the effector protein. The existence of multiple G protein types that can differentially couple to various receptors and effectors provides for considerable molecular diversity in the types of signal transduction pathways controlled by these receptors.

Despite the extreme diversity of the ligands for GPCR, the receptors themselves show a high degree of structural but not sequence similarity (see Figure 13.20). The crystal structure of bovine rhodopsin provides the essential topographical features of the GPCR family (see Figure 23.21). Each receptor is a single subunit membrane protein with seven α-helical transmembrane spanning domains. (GPCR have also been termed **heptahelical**, or **7-transmembrane, domain receptors**). Each has an extracellular amino-terminus, three extracellular loops, three intracellular loops, and an intracellular carboxy-terminus. The extracellular region provides sites for glycosylation and the extracellular loops contain essential cysteine residues for formation of interloop disulfide bridges. The transmembrane(TM)-spanning domains are homologous, but not identical among different GPCR. These form a pocket which often contains the critical amino acid residues for ligand binding. The intracellular loops, particularly the third loop, contain critical sequences for selective interaction with G proteins. The intracellular tail varies greatly in length and sequence among the different receptors, and contains sites that are involved in recognition or activation of G proteins. It also contains serine and threonine sites for phosphorylation by protein kinases. Phosphorylation of these residues plays important roles in the termination of signal transduction by activated GPCR.

The tertiary structure of GPCR within biological membranes involves a clustering of the seven transmembrane helices that controls the conformations and apposition of the intracellular loops and the carboxy-terminus. Large ligands such as polypeptide hormones bind to the extracellular loops whereas small ligands, such as acetylcholine and epinephrine, bind to the pocket formed by the membrane-spanning segments. In either case, binding causes a rearrangement of the membrane-spanning helices and changes the conformations of the intracellular loops and carboxy-terminus to expose recognition sites for particular G proteins. Thus, a GPCR occupied by an agonist can transiently form a complex with a G protein. An antagonist can bind to the receptor but does not cause appropriate rearrangement of the membrane-spanning domains or exposure of the G protein recognition sites on the intracellular loops and tail.

The transient interaction between the agonist-occupied receptor and its G protein affects the function of both. The functional changes in the G protein are described in the following section. The major change in the receptor following its binding to a G protein is that its affinity for the agonist is significantly reduced, and this usually leads to dissociation

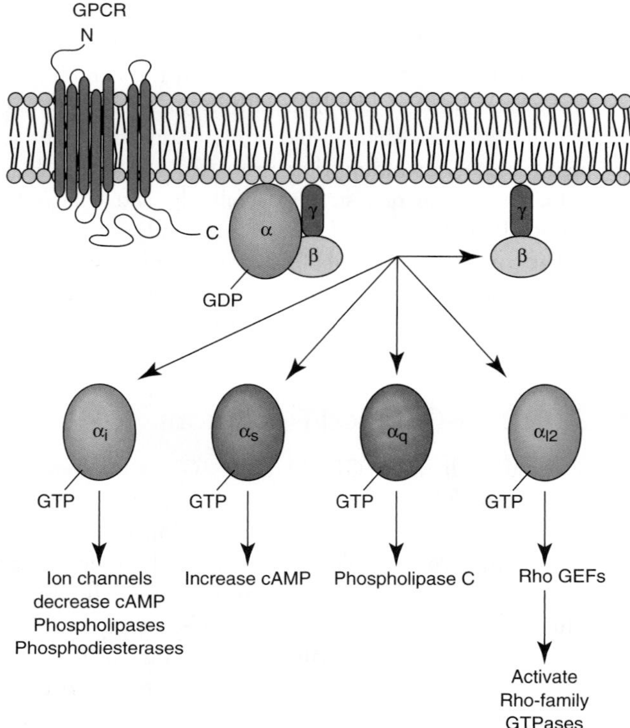

Figure 13.21 **Major subclasses of heterotrimeric G proteins that are activated by G by the 7-transmembrane domain G protein-coupled receptors.**

of the agonist. The agonist-free receptor assumes its inactive conformation with loss of its ability to interact with the G protein. The G protein dissociates from the receptor–G protein complex and returns the receptor to its inactive, ligand-free state.

Heterotrimeric G Proteins

G proteins that are functionally coupled to GPCR consist of heterotrimeric complexes of **α-, β-, and γ-subunits** (Figure 13.21). The α-subunit, the largest, is a relatively hydrophilic protein in the 39–46 kDa range. It contains the guanine nucleotide-binding site, the GTPase activity that is critical for proper regulation of the receptor–G protein interaction, and domains that interact with various effector proteins. Twenty distinct α-subunits are grouped into several subfamilies on the basis of both structural features and type of coupling to particular effector proteins and second-messenger cascades. The **α_s**-subunits (stimulatory α-subunit) of the **Gs** family stimulate **adenylate cyclase** and result in an increase in intracellular **cAMP** (Clin. Corr. 13.4). In contrast, the **α_i**- subunits (inhibitory α-subunits) of the **G_i** family inhibit adenylate cyclase and lower the cellular levels of cAMP. The **α_o**- and **α_z**-subunits, which also belong to the Gi family, predominantly modulate voltage-sensitive ion channels in excitable tissues. The G_i family also includes the **α_t**-subunits (t for transducin), the G proteins that transduce light perception by the rhodopsin-type GPCR that stimulate **cGMP phosphodiesterase** and rapidly decrease intracellular **cGMP**. The **Gq** family includes the α_q-, α_{11}-, α_{15}-, α_{16}-subunits that are coupled to **phospholipase C** and trigger increases in three second messengers: IP_3 (inositol trisphosphate), DAG (diacylglycerol), and Ca^{2+}. Finally, the α_{12}- and α_{13}-subunits of the **$G_{12/13}$** family stimulate **GEFs** for a variety of Rho-family small GTPases involved in cell migration, cell shape regulation, and cell division (Clin. Corr. 13.4).

CLINICAL CORRELATION 13.4

G Protein Mutations in Pituitary Gland Tumors and Endocrine Diseases

The pituitary gland plays a central role in regulating the endocrine status of mammals by secreting a variety of hormones that regulate the function of other organs including the liver, thyroid, adrenal, and reproductive glands. Release of these hormones is regulated by neurotransmitters, neuropeptides, and other hormones released by the hypothalamic region of the brain. For example, the hypothalamus and pituitary cells known as somatotrophs act in concert to control the release of Growth Hormone (GH), which regulates metabolism and cell growth in many peripheral tissues. The hypothalamus-derived Growth Hormone Releasing Hormone (GHRH) is an agonist for a G_s protein-coupled receptor which is expressed by pituitary somatotroph cells. Increased cAMP triggered by this receptor activates CREB-dependent transcription of the GH gene and results in increased synthesis and release of GH. Certain pituitary tumors cause the hypersecretion of GH, which in adults causes acromegaly, a condition characterized by abnormal growth patterns in many tissues. Such tumors are generally non-metastatic and believed to arise from the replication of single, transformed pituitary cells that have acquired a growth or survival advantage due to spontaneous mutations in key regulatory genes.

30%–40% of the GH-secreting pituitary tumors in humans express a mutated version of the GNAS1 gene that encodes the α_s-subunit of G_s protein. The mutated gene, termed *gsp*, results from single amino acid substitutions in residues that play critical roles in the binding and hydrolysis of GTP. Such mutations greatly diminish GTPase activity and thereby cause the constitutive activation of G_s and adenylate cyclase. The increase in adenylate cyclase activity markedly increases cAMP generation, protein kinase A (PKA) activation, the PKA-dependent phosphorylation of CREB and other cAMP-sensitive gene regulatory proteins, and, as a consequence, the expression of cAMP-inducible genes such as that for GH. Similar *gsp* mutations have been found in pituitary corticotroph tumors that hypersecrete ACTH (adrenocorticotropic hormone) with consequent overproduction of glucocorticoid hormones by the adrenal cortex.

The two amino acids altered in these naturally occurring *gsp* mutants were identified by *in vitro* biochemical studies as significant for the regulation of GTPase activity. Arginine-201, which is replaced by cysteine in the most common *gsp* mutant, is the acceptor site for the covalent ADP-ribosylation induced by cholera toxin. Glutamine-227, which is replaced by alanine or other amino acids in some *gsp* mutations, is a conserved glutamine located in the switch-II domain that undergoes large conformational movements depending on whether GTP or GDP is bound to the α-subunit. Substitution of this glutamine residue within all known α-subunits cause decreased GTPase activity, and it is widely used to generate constitutively active versions of the various G proteins for transgenic animal studies.

Lania, A., Mantovani, G., and Spada, A. Genetics of pituitary tumors: Focus on G protein mutations. *Exp. Biol. Med.* 228:1004, 2003.

The β(~35 kDa)- and γ (10 kDa) **subunits** are very hydrophobic proteins and, under non-denaturing conditions, are isolated as a **dimeric complex**. Purified $\beta\gamma$ complexes isolated from one G protein can often be recombined with the unique, purified α-subunit of a second G protein to produce a heterotrimer, which is functionally indistinguishable from the native heterotrimer of the second G protein. The $\beta\gamma$ complex is necessary for the interaction of the α subunit with receptors. In addition, the $\beta\gamma$ complexes from some activated G proteins also directly interact with, and activate, a variety of downstream effector proteins. For example, G_i-derived $\beta\gamma$ dimers control the opening of certain K^+ channels in the heart, thereby inducing cardiac myocyte hyperpolarization and a slowing of heart rate.

The G Protein Cycle

In their basal state, the $\alpha\beta\gamma$ heterotrimeric complex has GDP at the nucleotide-binding site of the α-subunit (Figure 13.22). The rate of release of GDP in the presence of unoccupied receptors (basal conditions) is low but not zero. The intrinsic GTPase activity of the α-subunit converts GTP into GDP and P_i. While the Pi is readily released, the GDP remains bound for a much longer time. This sequential release makes the GTPase reaction irreversible. The α-GDP displays high affinity for the $\beta\gamma$ dimer, and the heterotrimer is therefore regenerated. Consequently, in the unstimulated steady state, only a small fraction of the G protein molecules is in the active conformation (GTP bound). On interaction of a heterotrimeric complex with agonist-occupied GPCR, the rate of release of GDP is increased by over an order of magnitude. This allows GTP, which is much more abundant than GDP in the cytosol, to bind to the α-subunit and to trigger dissociation of α-GTP from the $\beta\gamma$ dimer. The released α-subunits and $\beta\gamma$ dimers then interact with appropriate

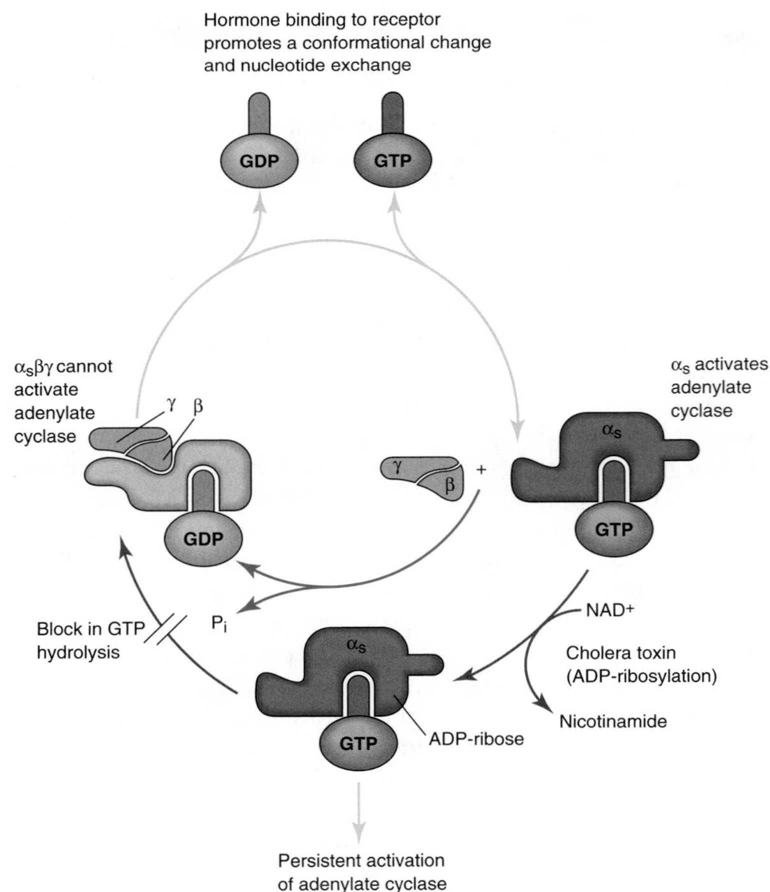

Figure 13.22 Activation and inactivation cycle of a heterotrimeric G protein.
Redrawn based on figure from Lodish, H., et al. *Molecular Cell Biology,* 4th ed. New York: Freeman, 2000.

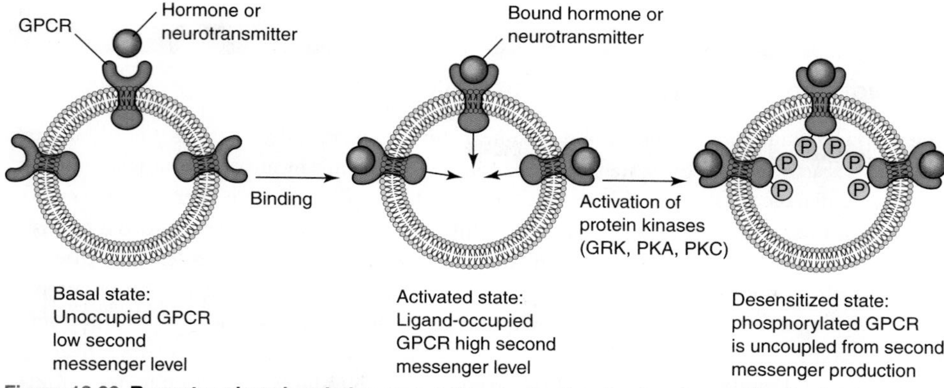

Figure 13.23 Receptor phosphorylation as a major mechanism for the desensitization/inactivation of G protein-coupled receptors.

downstream effector enzymes to change second-messenger levels. Activation of G proteins by agonist-occupied GPCR is catalytic rather than stoichiometric. Thus, depending on the type of GPCR, one molecule of agonist-occupied receptor may serially interact with 10–100 molecules of trimeric G protein; this results in major signal amplification at the earliest stages of GPCR activation. Although the intrinsic GTPase activity of the released α-subunits ensures that the dissociation of $\alpha\beta\gamma$ heterotrimeric complex will be transient, the overall repetition rate of the cycle will be increased as long as there are agonist-bound GPCR in the cell's plasma membrane.

Termination of Signaling by G protein-Coupled Receptors

Many of the small ligands for GPCR can be rapidly metabolized or inactivated by extracellular enzymes and most receptors exhibit rapid **desensitization** or adaptation. Although some of this adaptation is due to receptor internalization and down-regulation, the primary mechanism for rapid GPCR desensitization involves phosphorylation of the receptor that results in loss of function (Figure 13.23). The phosphorylated receptors cannot efficiently activate G proteins. Two families of protein kinases phosphorylate GPCR and induce desensitization. Second-messenger-dependent protein kinases, such as cAMP-dependent **protein kinase A (PKA)**, phosphorylate many types of GPCR regardless of whether the receptor is in an agonist-occupied or agonist-free conformation. GPCR-specific protein kinases (**GRKs**) are also known as **β ARKs** (βAR kinases) because the β-adrenergic receptor (βAR) was their first identified substrate (Clin. Corr. 13.5). In contrast to PKA and other second-messenger-regulated kinases, β ARKs only phosphorylate GPCR when they are occupied by agonistic ligands.

Effects of Bacterial Toxins on Heterotrimeric G Proteins

Some α-subunits of G proteins contain arginine or cysteine residues that can be ADP-ribosylated by exotoxins (cholera or pertussis toxins) secreted from some bacterial pathogens, such as *Vibrio cholerae* or *Bordetella pertussis* (see Figure 13.22, p. 526). These toxins possess NAD glycohydrolase activity that catalyzes breakdown of NAD^+ into nicotinamide and ADP-ribose, and ADP-ribosyl transferase activity that catalyzes transfer of the ADP-ribose moiety to the G protein α-subunit. These toxin-catalyzed ribosylation reactions interfere with the normal function of the G protein and cause abnormal signal transduction and regulation of cell function. For example, ADP-ribosylation of the Gs α-subunits by cholera toxin represses the intrinsic GTPase activity thereby causing constitutive dissociation of the GTP-bound α-subunits from the $\beta\gamma$ dimers even when GPCR is not occupied by agonist. This chronic increase in GTP-bound α_s induces continuous activation of the adenylate cyclase and cAMP production. In the large intestine, this increase in cAMP results in a sustained PKA-mediated phosphorylation of chloride channels that normally regulate salt and water transport; the hyperactivation of these channels severely disrupts salt and water transport and results in life-threatening diarrhea.

CLINICAL CORRELATION 13.5

Alterations in β-Adrenergic Receptor Signaling in Congestive Heart Failure

In congestive heart failure the heart fails to pump blood in amounts sufficient for adequate perfusion of peripheral tissues. Heart failure results from adaptations of cardiac myocytes for the maintenance of cardiac output following myocardial injury or from excessive hemodynamic burdens, such as high blood pressure. These adaptations include myocardial hypertrophy and activation of sympathetic nerve output, especially the release of norepinephrine and epinephrine from cardiac adrenergic nerves, which augments myocardial contractility. These adaptive mechanisms ultimately become inadequate and result in a progressive decrease in cardiac performance.

This shift from adaptive to maladaptive response is particularly evident in changes with the β-adrenergic receptor signaling system. In normal cardiac myocytes, epinephrine activates the G_s-coupled $\beta 1$- and $\beta 2$-adrenergic receptors that simulate adenylate cyclase, increase cAMP accumulation, and activate protein kinase A (PKA). PKA phosphorylates L-type dihydropyridine-sensitive Ca^{2+} channels and phospholamban, an allosteric regulator of cardiac sarcoplasmic reticulum (SR) Ca^{2+}-translocating ATPase. Phosphorylation of former increases the time these channels stay open in response to depolarization and results in enhanced Ca^{2+} influx. Unphosphorylated phospholamban represses the function of the SR Ca^{2+}ATPase pump. When phospholamban is phosphorylated by PKA, this inhibitory effect is relieved and results in an increased rate and extent of Ca^{2+} pumping by the SR. The combined action of enhanced plasma membrane Ca^{2+} influx (through the L-type channels) and enhanced Ca^{2+} sequestration within the SR results in greater Ca^{2+}-induced Ca^{2+} release from the SR during subsequent cardiac action potentials. Thus, during adrenergic stimulation, more Ca^{2+} is released during each action potential resulting in a stronger contraction (positive inotropic effect). However, the released Ca^{2+} is also more rapidly sequestered by the SR resulting in a more rapid cycle of contraction/ relaxation (positive chronotropic effect). Thus, at the whole organ level, β-adrenergic stimulation of heart results in both stronger contractions and more blood pumped per heart contraction (stroke volume) and a more rapid contraction/relaxation cycle (heart rate). Cardiac output (stroke volume $\times$ heart rate) is increased to deliver more blood to the systemic circulation. However, this means the heart will do more work (higher-energy output) and will require a higher delivery of oxygen and metabolites (fatty acids and glucose) to generate that energy.

Because of reduced myocardial function due to chronic disease, heart failure triggers an increased secretion of epinephrine/norepinephrine to provide a compensatory hyperactivation of the inotropic signaling responses described. While this initially increases activation of the $\beta 1$ and $\beta 2$ receptors, it also induces their chronic desensitization and down-regulation. Moreover, these β receptors increasingly activate other signaling cascades. For example, $\beta 1$ receptors stimulate cascades that are predominantly maladaptive, such as decreased expression of survival genes and increased activation of proapoptotic signals. During development of heart failure there is a decrease in the ratio of $\beta 1$-subtype receptors to $\beta 2$-subtype receptors. Heart failure also increases activity of the GRK2/βAR-kinase that phosphorylates the receptors and uncouples them from activation of G protein-dependent signaling pathways.

Although it may seem paradoxical, the treatment of heart failure is based on β-adrenergic antagonists (β-blockers), most commonly **metoprolol** or bisoprolol, which are highly selective for the $\beta 1$ receptors. The effectiveness of β-blocker therapy seems to involve repression of the maladaptive signaling pathways activated by $\beta 1$-receptors. Recent clinical studies have reported the usefulness of **carvedilol**, which nonselectively blocks both $\beta 1$- and $\beta 2$-adrenergic receptors, but also antagonizes $\alpha 1$-adrenergic receptors that are highly expressed in vascular smooth muscle cells. These receptors activate the Gq/phospholipase C cascade and release of IP_3-sensitive Ca^{2+} pools to cause contraction of smooth muscle with consequent constriction of blood vessels. Antagonism of these latter responses decreases the resistance to blood flow and the amount of work required by the heart to deliver effective cardiac output.

Lohse, M. J., Englehardt, S., and Eschenhagen, T. What is the role of β-adrenergic signaling in heart failure? *Circ. Res.* 93:896, 2003.

13.9 • CYCLIC AMP-BASED SIGNAL TRANSDUCTION

Regulation of Cyclic AMP Synthesis and Degradation

Many metabolic and behavioral responses to different hormones and neurotransmitters are mediated by increase in intracellular cAMP. The use of cAMP as a second messenger is largely limited to GPCRs. Cells actively regulate both the synthesis and degradation of this second messenger. cAMP synthesis is catalyzed by **adenylate cyclases** that use ATP as a substrate to produce cAMP and pyrophosphate (PPi) (Figure 13.24). The normal basal level of cAMP in cytosol is $\sim 10^{-7}$ to 10^{-6} M. During maximal hormonal stimulation, the cAMP concentration can increase 2- to 100-fold depending on cell type. At least six genes that encode different subtypes of adenylate cyclase have been identified. All are integral membrane proteins with 12 membrane-spanning domains, one catalytic domain within the intracellular segment connecting transmembrane segments six and seven, and a second catalytic domain at the intracellular carboxy-terminus. The various subtypes are expressed in tissue-specific patterns and can be differentially regulated by particular signaling proteins. All adenylate cyclase isoforms are positively regulated by the α_s-subunits of G_s proteins. However, the $\beta\gamma$-subunit complex from various G proteins (including G_s) can stimulate some adenylate cyclase isoforms while inhibiting others. The α_i-subunits from G_i proteins can directly inhibit certain subtypes of adenylate cyclase (Figure 13.25).

The breakdown of cAMP is catalyzed by **cyclic nucleotide phosphodiesterases (PDEs)** that hydrolyze the cyclic diester bond to produce AMP (Figure 13.24). The activity of these PDEs can be regulated by both hormones and certain drugs. Cellular cAMP can be increased by inhibition of the PDEs. Xanthine derivatives, such as theophylline and caffeine, inhibit PDEs and result in increased cAMP levels in the absence of hormonal stimulation.

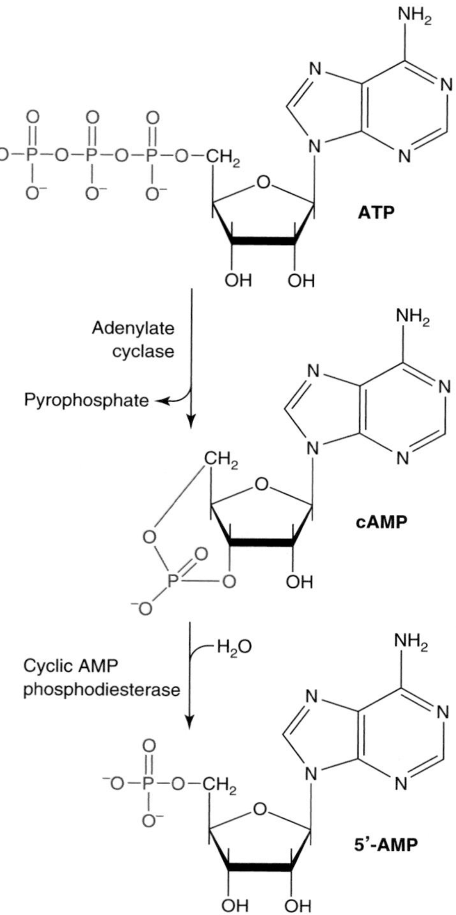

Figure 13.24 Synthesis and degradation of cyclic AMP by adenylate cyclases and cyclic nucleotide phosphodiesterases.

Intracellular Signaling Mechanisms of Cyclic AMP

cAMP-dependent **protein kinase A (PKA)** mediates most of the effects of cAMP (Figure 13.26, p. 530). PKA is a tetramer composed of two regulatory subunits (R-subunits), which contain the cAMP-binding sites, and two catalytic subunits (C-subunits), which contain the kinase activity. The tetrameric form cannot catalyze phosphorylation of target proteins. On the cooperative binding of several cAMP molecules to the regulatory subunits, the tetramer dissociates and the free catalytic subunits become active. Many of the protein substrates of the cAMP-dependent kinase are themselves kinases with specific target enzymes. This underscores the cascading nature of this (and most other) receptor-linked signaling systems. Specific phosphatases dephosphorylate the phosphorylated proteins and thus inactivate the cAMP-mediated signaling cascade. These phosphatases are subject to receptor-linked regulation.

cAMP profoundly alters cellular metabolism by altering both the activity and expression of many catabolic enzymes, including enzymes involved in lipid and in carbohydrate metabolism. Metabolic regulation by cAMP often involves the PKA-mediated regulation of *secondary* kinases or phosphatases by changing the phosphorylation states of the latter enzymes. PKA can also phosphorylate several types of plasma membrane ion channels or ion transporters and lead to either altered membrane potential or the activation/inhibition of the influx of regulatory ions such as Ca^{2+}. Because neurotransmitters, such as serotonin, dopamine, and epinephrine, interact with GPCR that regulate G_s or G_i, the ability of cAMP and PKA to modulate ion channel activity is a critical component of regulation in the central nervous system.

Increased cAMP in eukaryotic cells rapidly alters the transcription and expression of the genes coding for important metabolic enzymes, polypeptide hormones, and ion transport proteins. The dissociation of catalytic subunits from the regulatory subunits exposes nuclear localization sequences on the catalytic subunits and facilitates their

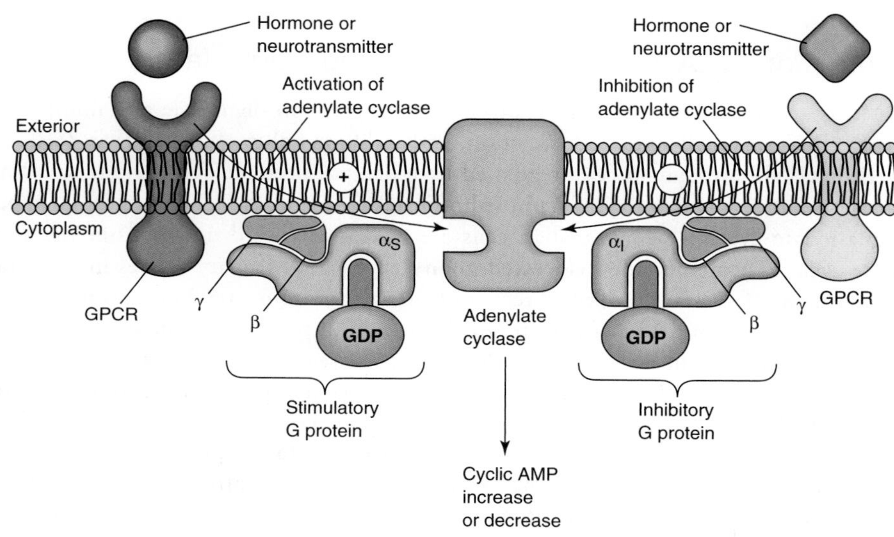

Figure 13.25 Positive and negative regulation of adenylate cyclase effector enzymes by G_s- and G_i-family heterotrimeric G proteins.

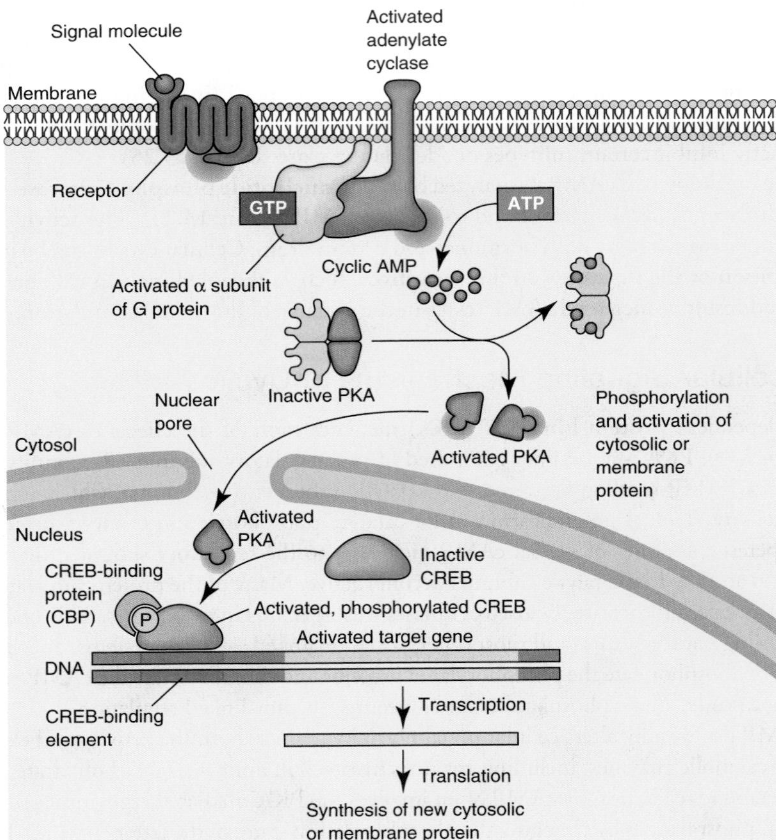

Figure 13.26 Role of protein kinase A in the intracellular signaling cascades regulated by cyclic AMP.
Redrawn based on figure from Alberts, B., et al. *Molecular Biology of the Cell,* 4th ed. New York: Garland Science, 2002.

cytosol-to-nucleus translocation (Figure 13.26). The intranuclear PKA induces phosphorylation and activation of cAMP-regulated proteins called **CREBs** to control the expression of genes that contain cAMP-sensitive regulatory elements (**CRE**) within the 5′-flanking regions of many cyclic AMP-responsive genes. In this way, the transient increases in cAMP triggered by GPCR at the cell surface can be translated into long-lasting changes in nuclear gene expression and cell function.

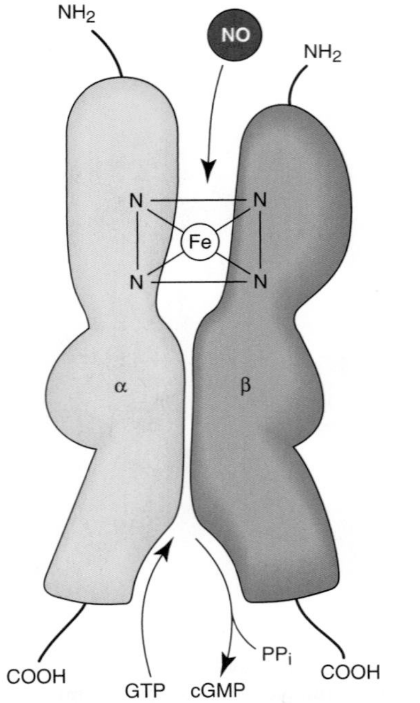

13.10 • CYCLIC GMP-BASED SIGNAL TRANSDUCTION

Regulation of Cyclic GMP Synthesis and Degradation

cGMP is an important second messenger in the regulation of muscle and nonmuscle contractility, in visual signal transduction, and in blood volume homeostasis. Cellular levels of cGMP are dynamically regulated by a balance between synthetic (**guanylate cyclases**) and degradative (**cGMP phosphodiesterases**) enzymes. The two major types of gua-nylate cyclase in mammalian cells are fundamentally different from the adenylate cyclase. A **membrane-associated guanylate cyclase** activity resides in the catalytic domains of the enzyme-linked receptors for the hormone **atrial natriuretic factor (ANF)** secreted from cardiac myocytes of the atrium in response to increases in circulating blood volume and/or pressure. ANF receptors are expressed primarily in vascular smooth muscle cells and in renal cells. This receptor has a large extracellular domain for binding ANF, a single transmembrane domain, and a relatively short intracellular catalytic domain. ANF binding stimulates catalytic activity via receptor dimerization and binding of regulatory proteins to the hormone-occupied receptor (see Figures 22.32, p. 914, and 22.33, p. 915). In contrast to the tissue-restricted expression of the ANF receptor–guanylate cyclase, a cytosolic form of guanylate cyclase is present in most cell

Figure 13.27 NO binding to soluble guanylate cyclase.

CLINICAL CORRELATION 13.6

Nitric Oxide/cGMP Signaling as Therapeutic Targets in Cardiac and Vascular Disorders

The endothelial cells that line the internal surfaces of all blood vessels are major sources of the nitric oxide used for the relaxation of vascular smooth muscle cells. In these myocytes, cGMP generation by the soluble guanylate cyclase is stimulated by nitric oxide generated by adjacent endothelial cells. cGMP induces myocyte relaxation and vasodilation of blood vessels with increased blood flow to nearby tissue. Similar changes can be produced by drugs that increase nitric oxide production. cGMP levels are also regulated by phosphodiesterases (PDE) that convert cGMP to GMP. Thus, drugs that inhibit vascular isoforms of PDE improve blood flow to particular tissues.

Angina pectoris is the formal name for chest pain suffered by patients with different types of cardiac insufficiency. Nitroglycerin is commonly prescribed and rapidly alleviates angina (see Clin. Corr. 11.10, p. 451). It is metabolized to nitric oxide and also binds directly to nitric oxide–sensitive proteins (like soluble guanylate cyclase). Thus, it can induce smooth muscle relaxation and vasodilation. How nitroglycerin alleviates angina is not completely understood. The coronary arteries and cardiac microvasculature during coronary insufficiency are likely maximally dilated due to local ischemia. Thus, nitroglycerin probably alleviates angina by decreasing cardiac workload and myocardial oxygen requirement via vasodilatory effects on the veins that return blood to the heart (preload), and by decreasing peripheral blood pressure (afterload) through vasodilatory effects on systemic arteries.

Viagra® is the trade name for Sildenafil citrate that is used to treat erectile dysfunction. It inhibits selectively the type 5 cyclic nucleotide phosphodiesterase (PDE5) that is highly expressed in vascular smooth muscle. It is 80- to 4000-fold less potent as an inhibitor of other cyclic nucleotide PDE isoforms, including PDE3 which is predominantly expressed in cardiac muscle. It is only 10-fold more selective for PDE5 versus PDE6, which is the predominant PDE in the retina. When blood vessels in erectile tissue are vasodilated in response to released NO and activation of guanylate cyclase, the inhibition of cGMP breakdown by Viagra leads to higher and more sustained levels of cGMP accumulation and PKG activation. This increases the vasodilation in erectile tissue. High doses of Viagra can lead to mild alterations in color vision due to side effects on the retinal PDE6.

Ignarro, L. J., Napoli, C., and Lascalzo, J. Nitric oxide donors and cardiovascular agents modulating the bioactivity of nitric oxide, *Circ. Res.* 90:21, 2001.

types (Figure 13.27). This **soluble guanylate cyclase** is a dimeric complex of 82-kDa and 70-kDa subunits. Each dimer constitutively binds a molecule of **heme**. In the absence of gaseous ligands that bind to this heme group, soluble guanylate cyclase has a very low rate of catalysis. However, binding of **nitric oxide (NO)** to the heme group induces conformational changes that greatly increase activity of the guanylate cyclase. Thus, a major stimulus for cGMP accumulation in most cells is an elevated level of NO. NO is generated by **nitric oxide synthases (NOS)** (p. 447). Because NO readily permeates biological membranes, it can be produced in one type of cell, for example, a vascular endothelial cell, and rapidly diffuse into neighboring cell types, for example, vascular smooth muscle cells, where it activates soluble guanylate cyclase. Accumulation of cGMP in smooth muscle triggers rapid and sustained relaxation of the contractile apparatus (Clin. Corrs. 13.6).

cGMP degradation is catalyzed by several soluble or weakly membrane-associated **cGMP phosphodiesterases** (PDE). Although some appear to be passive catabolic enzymes (i.e., not directly regulated), the cGMP-dependent PDE in the light-sensing cells of the retina is an effector enzyme for the trimeric G protein transducin (p. 963).

Intracellular Signaling Mechanisms of Cyclic GMP

Cyclic GMP exerts most of its second messenger effects by activating **cGMP-sensitive protein kinases (PKG)** that are widely expressed in many tissues. Depending on cell type, PKG phosphorylates a variety of important targets, including enzymes involved in metabolism, structural proteins, and ion transport proteins. For example, stimulation of PKG in smooth muscle and nonmuscle cells results in the coordinated phosphorylation of many proteins that act to inhibit Ca^{2+}-dependent contraction or motility. These include: (1) proteins that attenuate Ca^{2+} release channels of the endoplasmic reticulum, (2) myosin light chain phosphatases that repress myosin interaction with actin, and (3) K^+ channels that induce plasma membrane hyperpolarization and thereby repress voltage-gated Ca^{2+} influx. cGMP also acts as a direct (i.e., PKG-independent) regulator of cyclic nucleotide-gated cation channels expressed in certain sensory tissues such as the retina.

13.11 • CALCIUM-BASED SIGNAL TRANSDUCTION

Regulation of Cytosolic Ca^{2+} Concentration

Ca^{2+} is an important intracellular regulator of many fundamental cell functions including: contraction in all types of muscle, secretion of hormones and neurotransmitters, cell division, directed migration of nonmuscle cells, and regulation of gene expression. Cells respond to many types of hormones and neurotransmitters with rapid and transient increases in cytosolic Ca^{2+} that then binds to a variety of Ca^{2+}-dependent regulatory proteins. Increases in cytosolic Ca^{2+} are elicited by opening of either Ca^{2+} channels in the plasma membrane or Ca^{2+} channels in the endoplasmic reticulum (ER). This released Ca^{2+} can be rapidly reaccumulated by the ER to terminate the response to the hormone/neurotransmitter.

The basal concentration of free Ca^{2+} in the cytosol of most cell types is maintained at approximately 10^{-7} M (Figure 13.28). Small increases in cytosolic free Ca^{2+} to 10^{-6} M rapidly and maximally activate the various Ca^{2+}-regulated cell functions. Because extracellular Ca^{2+} concentration is 10^{-3} M, there is a 10,000-fold concentration gradient for Ca^{2+} across the plasma membrane and the driving force for rapid Ca^{2+} influx is further increased by the inside-negative membrane potential. In excitable cells, such as neurons and muscle cells, Ca^{2+} influx involves voltage-gated Ca^{2+} channels (see Table 12.10, p. 481). This type of Ca^{2+} signaling pathway is generally initiated by the ion channel receptors for excitatory neurotransmitters, such as acetylcholine, glutamate, serotonin, and ATP. The plasma membrane also contains two transport systems, the Na^+/Ca^{2+} antiport transporter and the Ca^{2+}-transporting ATPase that efficiently pump the excess Ca^{2+} out of the cytosol (Figure 13.28).

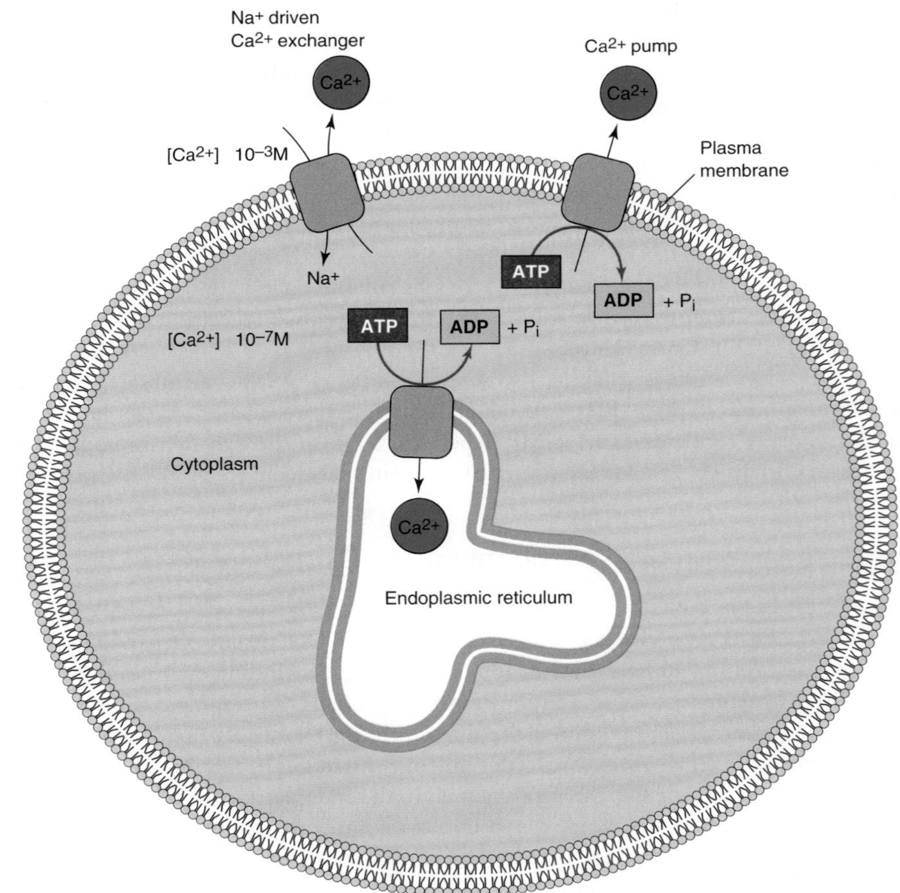

Figure 13.28 Role of plasma membrane and organellar Ca^{2+} transport proteins in the regulation and compartmentation of intracellular Ca^{2+} pools.

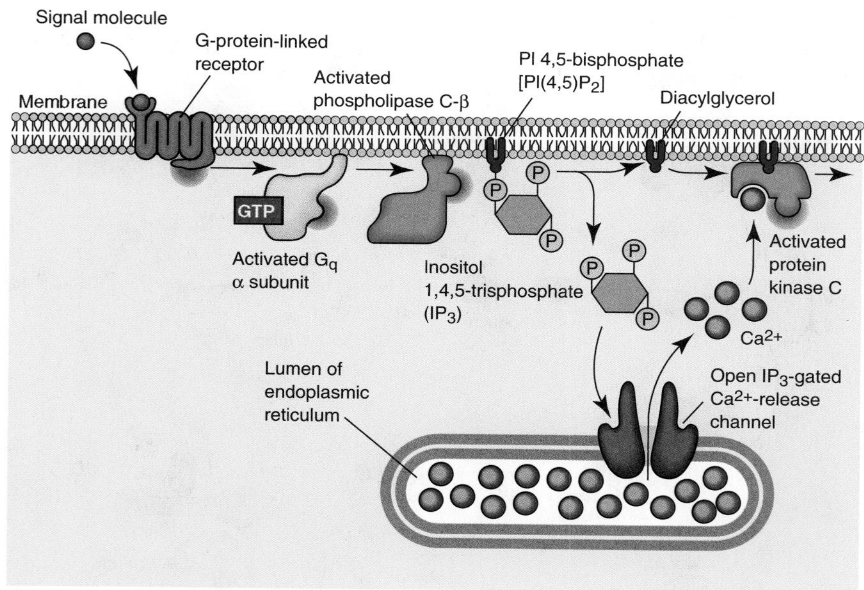

Figure 13.29 Mobilization of endoplasmic reticulum Ca^{2+} pools and activation of protein kinase C by receptors that stimulate the phospholipase C-mediated hydrolysis of inositol phospholipids.

Ca^{2+} ions are efficiently concentrated within the ER (termed sarcoplasmic reticulum or SR in muscle cells) by Ca^{2+}-transporting ATPase pumps. This ATPase permits Ca^{2+} to accumulate to a concentration of about 1 mM (10^{-3} M) within the lumen of the ER/SR; this is equivalent to the concentration of Ca^{2+} in the extracellular space. This accumulated Ca^{2+} can be released into the cytosol via two types of ER/SR-localized Ca^{2+} channels that are regulated by signals generated by many types of cell surface receptors. The first is a very widely expressed channel known as the **1,4,5-inositol trisphosphate (IP_3) receptor (IP3R)** or **IP_3-gated Ca^{2+} channel** (Figure 13.29). IP_3, a second messenger, is rapidly generated when phospholipase C is stimulated by different hormones or neurotransmitters. A different ligand-gated Ca^{2+} release channel is expressed in the ER/SR of skeletal and cardiac muscle. This is the **ryanodine receptor** (RyR) or ryanodine-sensitive Ca^{2+} release channel because it binds the drug ryanodine (Figure 13.30). In cardiac muscle, the RyR channels are gated by the Ca^{2+} that enters through plasma membrane voltage-gated Ca^{2+} channels sensitive to channel-blocking drugs known as dihydropyridines. This interplay of multiple Ca^{2+} channels to regulate cardiac muscle function is termed Ca^{2+}-induced Ca^{2+} release. In skeletal muscle, RyR channels open via direct protein–protein interactions with the voltage- and dihydropyridine-sensitive Ca^{2+} channels; this regulation is possible because SR in skeletal muscle forms direct contacts with specialized invaginations of the plasma membrane.

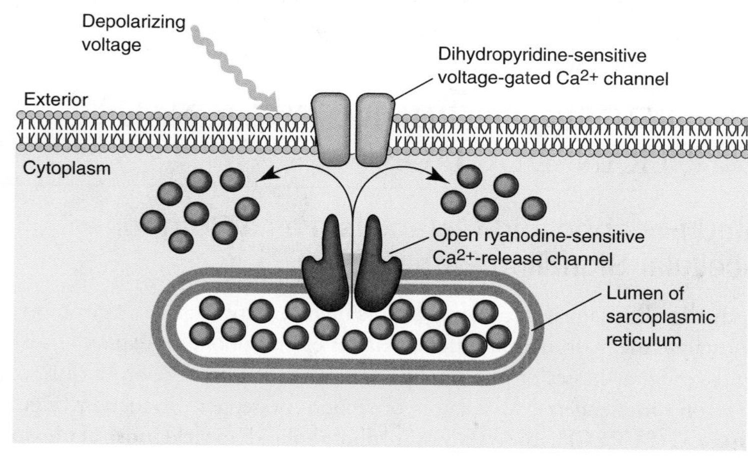

Figure 13.30 Mobilization of sarcoplasmic reticulum Ca^{2+} pools in striated muscle cells by depolarization-induced activation of ryanodine-sensitive Ca^{2+} release channels.

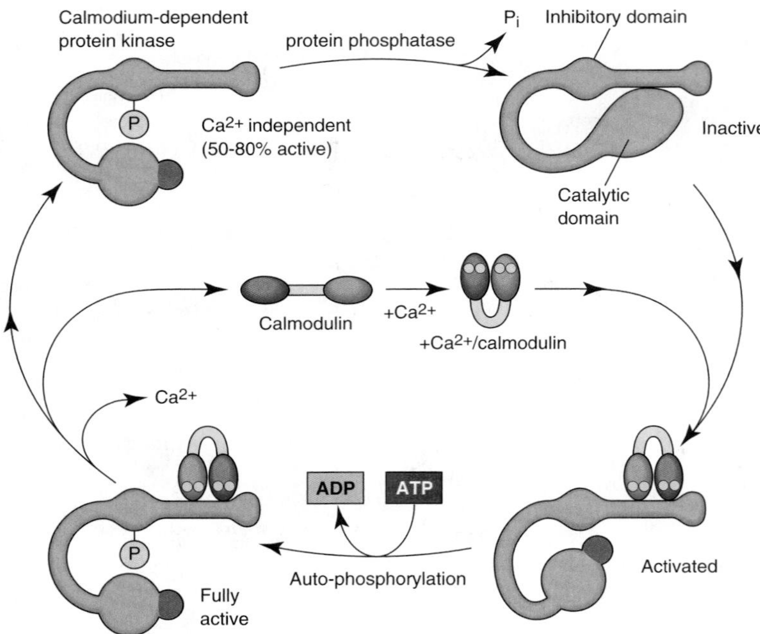

Figure 13.31 Regulation of calmodulin and calmodulin-regulated protein kinases by Ca^{2+}. Redrawn based on figure from Alberts, B., et al. *Molecular Biology of the Cell,* 4th ed. New York: Garland Science, 2002.

Calcium Activation of Calmodulin-Dependent Protein Kinases and Phosphatases

Increased cytosolic Ca^{2+} results in activation of various calcium-binding regulatory proteins. Some, such as troponin-C, which regulates contraction in striated muscles, are highly specialized and expressed in only a few tissues (see Figure 23.35, p. 976). In contrast, **calmodulin (CaM)** (p. 493) is a ubiquitously expressed protein that mediates the regulatory actions of Ca^{2+} on a broad range of cellular functions via its ability to activate Ca^{2+}-sensitive effector proteins that include serine/threonine-specific protein kinases (Figure 13.31), phosphatases, and nitric oxide synthases (see Figure 11.17, p. 444). On binding of Ca^{2+}, calmodulin undergoes conformational changes that facilitate its interaction with the downstream effector proteins. Some calmodulin-dependent kinases, such as myosin light chain kinase , target only specific proteins whereas other forms, such as the multifunctional CaM Kinase II, can phosphorylate a broad range of protein substrates. Altering the activity of these proteins is the primary mechanism by which increased Ca^{2+} alters cellular behavior. CaM-kinase II also becomes autophosphorylated when bound to calmodulin, and this autophosphorylation allows CaMKII to remain active long after cytosolic Ca^{2+} has returned to a resting level. CaM-kinases can phosphorylate and modulate a wide range of enzymes, ion channels, contractile proteins, and gene regulatory proteins.

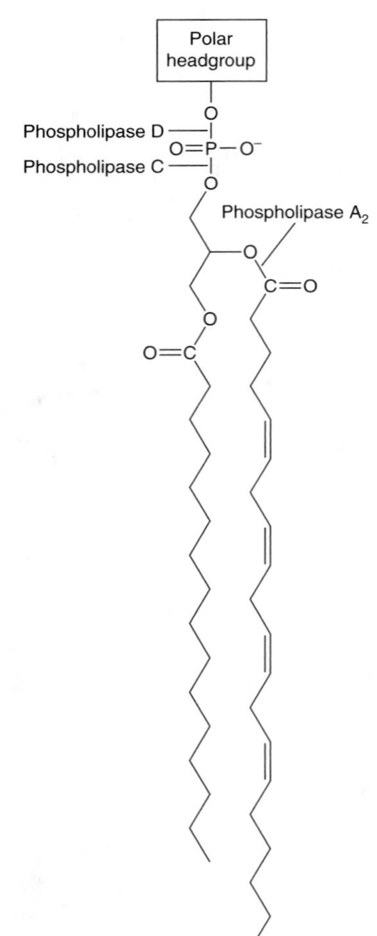

Figure 13.32 Major classes of phospholipases used during receptor-mediated signal transduction.

13.12 • PHOSPHOLIPID-BASED SIGNAL TRANSDUCTION

Regulated Phospholipid Metabolism as a Component of Intracellular Signaling Pathways

Signal transduction by many cell surface receptors involves activation of one or more phospholipases that catalyze the hydrolysis of different classes of phospholipids (e.g., inositol or choline phospholipids). Several classes of phospholipases catalyze the production of distinct products that act as second messengers or regulators of second-messenger production (Figure 13.32). **Phospholipase C (PI-PLC)** hydrolyses inositol phospholipids to yield **inositol phosphates** and

diacylglycerols as second messengers. Inositol phospholipids can be further phosphorylated by **phosphoinositide-3-kinase (PI-3K)** to produce **phosphatidylinositol-3,4,5-trisphosphate (PIP_3)**, another second messenger whose function is described in the following. **Phospholipase D (PLD)** predominantly hydrolyses choline or ethanolamine phospholipids to produce **phosphatidic acid** that is subsequently hydrolysed by phosphatidic acid phosphohydrolyases (PAP) to yield diacylglycerol. **Phospholipase A2 (PLA2)** attacks various phospholipids to produce free fatty acids, such as arachidonic acid, and lyso-phospholipids.

Regulation of Phospholipase C and Phospholipase D

As previously described, many G protein-coupled receptors and receptor tyrosine kinases stimulate the release of Ca^{2+} from the endoplasmic reticulum by activating the production of **1,4,5-inositol trisphosphate (IP_3)**. This water-soluble second messenger is derived from the hydrolysis of **phosphatidylinositol-4,5-bisphosphate (PIP_2)**, a relatively minor membrane phospholipid that is generated by phosphorylation of two hydroxyls on the inositol residue of phosphatidylinositol. Hydrolysis of PIP_2 is catalyzed by C-type phospholipases (**PLC**) with high selectivity for inositol phospholipids (Figure 13.32) and generates two second messengers: (1) the water soluble product IP_3 and (2) **diacylglycerol (DAG)**, a hydrophobic product. **PI-PLCβ** enzymes are activated by interaction with either the α-subunits of $G_{q/11}$-family G proteins or the $\beta\gamma$-subunits of G_i-family. In contrast, the **PI-PLCγ** enzymes are activated by tyrosine phosphorylation and association with receptor tyrosine kinases. Some nonreceptor tyrosine kinases also phosphorylate PLC enzymes.

In addition to IP_3, the activation of PI-PLCβ and PI-PLCγ results in accumulation of DAG. However, the magnitude of receptor-induced production of DAG in many cell types cannot be ascribed solely to the breakdown of PIP_2. Although PIP_2 hydrolysis yields small amounts of DAG over short periods, the hydrolysis of more abundant phospholipids, such as phosphatidylcholine (PC), underlies most of the sustained production of DAG triggered in response to receptor activation. Consequently, PC hydrolysis by PLC and **phospholipase D (PLD)** mediates the regulation of cellular processes that require prolonged elevation of DAG. PC-PLD and PC-PLC are indirectly activated during receptor stimulation by a complex network of primary signals that include Ca^{2+}, small GTPases, and protein kinase C.

Diacylglycerol and Protein Kinase C

Diacylglycerol (DAG), which results from breakdown of PIP_2 or PC, binds to and activates members of the **protein kinase C (PKC)** family (see Figure 13.29, p. 533). Inactive PKCs exist as soluble, cytosolic proteins in which the N-terminus (containing the DAG-binding sites) occludes the C-terminus (containing binding sites for substrate proteins and ATP). When DAG is generated in membranes, PKC binds to it, and this results in unfolding and exposure of the substrate- and ATP-binding domains of the kinase. Other N-terminal sites on PKC enzymes bind to acidic phospholipids and facilitate association of PKC with the membrane. Some isoforms of PKC contain Ca^{2+}-binding sites in their N-terminal domains so that increased cytosolic Ca^{2+} potentiates PKC binding to membranes. Activation of PKC enzymes results in serine/threonine phosphorylation of many substrate proteins, which in turn, elicit a number of cellular responses. Like cAMP-dependent protein kinases and calmodulin-dependent protein kinases, PKC isozymes phosphorylate: (1) transcription factors involved in regulation of gene expression, (2) ion channels and transporters and (3) other types of protein kinases. Some PKC isozymes also trigger the various MAP kinase cascades that control multiple gene regulatory proteins. The ability of PKC enzymes to regulate directly or indirectly the phosphorylation of transcription factors underlies the critical role of PKC in regulating the growth, differentiation, or death of many cells and tissues.

PIP_3, Phosphatidylinositol 3-Kinases, and Protein Kinase B

Inositol phospholipids are substrates for inositol lipid kinases that specifically phosphorylate inositol at the 3′-OH . These **phosphatidylinositol 3-kinases (PI3-kinases)** compose a family of signaling proteins. Some associate with, and are activated by, tyrosine phosphorylated growth factor receptors or activated cytokine receptors via SH-2 domain interactions. Others are activated by the $\beta\gamma$ dimers derived from activated G proteins. PI-3 kinases predominantly phosphorylate phosphatidylinositol-4,5-bisphosphate to produce **phosphatidylinositol-3,4,5-trisphosphate (PIP_3)**, which does not act as a substrate for PLC, but has critical roles as a second messenger in the regulation of membrane trafficking, cell motility, and activation of cell survival signaling pathways (Figure 13.33). The **Akt/protein kinase B (PKB)** family are major mediators of PIP_3 action. Active PKB potentiates cell survival by repressing the activity of certain cell death signaling pathways and regulates other proteins and kinases involved in glucose transport and glycogen metabolism.

Phospholipase A_2 and Generation of Arachidonic Acid Metabolites

Occupancy of cell surface receptors in many cell types triggers the release of **arachidonic acid (AA)** from membrane phospholipids. The release of AA is catalyzed by members of the family of phospholipase A_2 (**PLA_2**) enzymes (see Figure 13.32, p. 534). The regulation of PLA_2 is complex but includes Ca^{2+}, which seems to control the translocation of some PLA_2 isoforms from cytosol to membranes, and kinase-mediated phosphorylation. Arachidonic acid can be rapidly metabolized to multiple oxidation products collectively known as **eicosanoids** (p. 737); these include prostaglandins, thromboxanes, and leukotrienes. In contrast to other second messengers, such as cyclic nucleotides or Ca^{2+}, eicosanoids are generally secreted from the cells in which they are produced to act as extracellular agonists for cell surface receptors in neighboring cells. The eicosanoids are formed by virtually all cell types and play fundamental paracrine/autocrine roles in processes such as inflammation, blood clotting, control of vascular tone, renal function, and others.

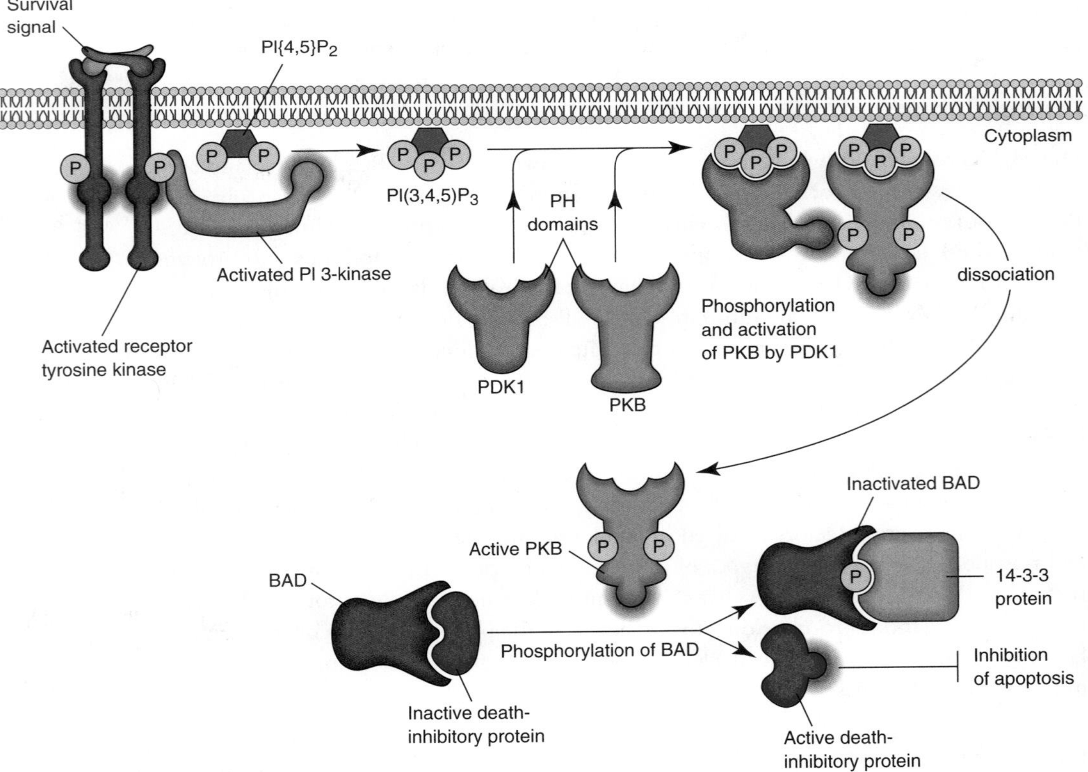

Figure 13.33 Role of phosphatidylinositol-3-kinase and protein kinase B in the intracellular signaling cascades regulated by 3,4,5-phosphatidylinositol trisphosphate.
Redrawn based on figure from Alberts, B., et al. *Molecular Biology of the Cell,* 4th ed. New York: Garland Science, 2002.

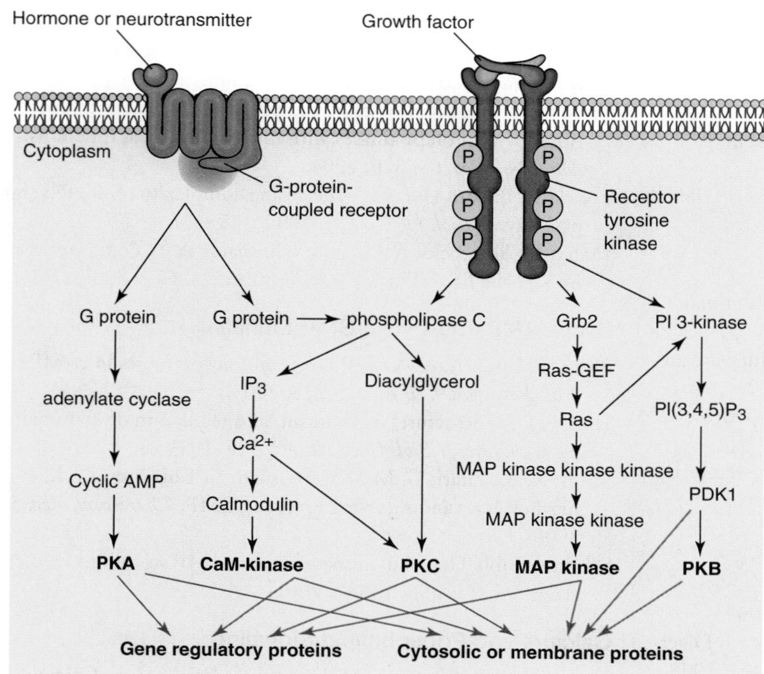

Figure 13.34 Cross-talk and integration among the major intracellular signaling cascades regulated by different cell surface receptors. Redrawn based on figure from Alberts, B., et al. *Molecular Biology of the Cell,* 4th ed. New York: Garland Science, 2002.

13.13 • INTEGRATION OF SIGNAL TRANSDUCTION PATHWAYS INTO SIGNAL TRANSDUCTION NETWORKS

Depending on their tissue localization and degree of specialization, individual cells may be genetically programmed to respond to many distinct extracellular signaling molecules. Many important intracellular signaling cascades can be regulated synergistically by receptors of various types and families (Figure 13.34). For example, the MAP kinase pathways can be regulated by G protein-coupled receptors and growth factor receptor kinases. This "cross-talk" provides cells the ability to integrate signals from multiple extracellular stimuli into unique and particular patterns of altered cellular response. In other cases, different second-messenger-activated kinases may phosphorylate distinct residues on a common target protein, such as a transcription factor, to produce either divergent or convergent changes in gene expression patterns. Thus, the integrated response of a cell to activation of a particular cell-surface receptor will often vary depending on what other cell surface receptors are activated simultaneously. For example, a receptor that normally induces specialized gene expression in a particular cell type may trigger the programmed death of that same cell if other cell surface receptors are being simultaneously activated or, in some cases, are not being activated. Although modern cell biology has made significant advances in defining the basic components and sequential steps of individual signaling pathways, a major challenge remains in understanding how these individual pathways are integrated into signal transduction networks.

Bibliography

Ligand-Gated Ion Channel Receptors

Lester, H. A., Dibas, M. I., Dahan, D. S., Leite, J. F., and Dougherty, D. A. Cys-loop receptors: New twists and turns. *Trends Neurosci.* 27: 329, 2004.

Wollmuth, L. P., and Sobolevsky, A. I. Structure and gating of the glutamate receptor ion channel. *Trends Neurosci.* 27: 32, 2004.

Vial, C., Roberts, J. A., and Evans, R. J. Molecular properties of ATP-gated P2X receptor ion channels. *Trends Pharmacol. Sci.* 25: 487, 2004.

Receptor and Nonreceptor Tyrosine Kinases

Linggi, B., and Carpenter, G. ErbB receptors: new insights on mechanisms and biology. *Trends Cell Biol.* 16: 649, 2006.

Parsons, S. J., and Parsons, J. T. Src family kinases, key regulators of signal transduction. *Oncogene* 23: 7906, 2004.

Schlessinger, J. Ligand-induced, receptor-mediated dimerization and activation of the EGF receptor. *Cell* 110: 669, 2002.

Yaffe, M. B. Phosphotyrosine-binding domains in signal transduction. *Nat. Rev. Mol. Cell Biol.* 3: 177, 2002.

Wiley, H. S. Trafficking of the ErbB receptors and its influence on signaling. *Exp Cell Res.* 284: 78, 2003.

Cytokine Receptors

Aaronson, D. S., and Horvath, C. M. A road map for those who don't know JAK-STAT. *Science* 296: 1653, 2002.

Derynck, R., and Zhang, Y. E. Smad-dependent and Smad-independent pathways in TGFα family signaling. *Nature* 425: 577, 2003.

Chen, G., and Goeddel, D.V. TNF-R1 signaling: A beautiful pathway. *Science* 296: 1634, 2002.

O'Neill, L. A., and Dinarello, C. A. The IL-1 receptor/toll-like receptor superfamily. *Immunol. Today* 21: 206, 2000.

G Protein-Coupled Receptors

Bockaert, J., and Pin, J-P. Molecular tinkering of G protein-coupled receptors: An evolutionary success. *EMBO J.* 18: 1723, 1999.

Ji, T. H., Grossman, M., and Ji, I. G protein-coupled receptors: I. Diversity of receptor-ligand interactions. *J. Biol. Chem.* 273: 17299, 1998.

Lefkowitz, R. J and Shenoy, S. K. Transduction of receptor signals by α-arrestins. *Science* 308:512, 2005.

Sodhi, A., Montaner, S., and Gutkind, J. S. Viral hijacking of G protein-coupled receptor signaling networks. *Nat. Rev. Mol. Cell Biol.* 5: 998, 2004.

Trimeric G Proteins

Berman, D. M., and Gilman, A. G. Mammalian RGS proteins: Barbarians at the gate. *J. Biol. Chem.* 273: 1269, 1998.

Oldham, W. M., and Hamm, H. E. Heterotrimeric G protein activation by G protein-coupled receptors. *Nat. Rev. Mol. Cell Biol.* 9: 60, 2008.

Preminger, A. M., and Hamm, H. E. G protein signaling: Insights from new structures. *Science-STKE* 218: 1, 2004.

Small GTPases and MAP Kinase Cascades

Jaffe, A. B., and Hall, A. 2005. Rho GTPases: Biochemistry and biology. *Annu. Rev. Cell Dev. Biol.* 21: 247, 2005.

Roux, P. P., and Blenis J. ERK and p38 MAPK-activated protein kinases: A family of protein kinases with diverse biological functions. *Micro Mol. Biol. Rev.* 68: 320, 2004.

Vojtek, A. B., and Der, C. J. Increasing complexity of the Ras signaling pathway. *J. Biol. Chem.* 273: 19925, 1998.

Yang S-H, Sharrocks, A. D., and Whitmarsh A. J. Transcriptional regulation by the MAP kinase signaling cascades. *Gene* 320:3, 2003.

Cyclic AMP and Cyclic GMP Signaling

Bos, J. L. Epac: A new cAMP target and new avenues in cAMP research. *Nat. Rev. Mol. Cell Biol.* 4: 733, 2003.

Hurley, J. H. Structure, mechanism, and regulation of mammalian adenylyl cyclase. *J. Biol. Chem.* 274: 7599, 1999

Lucas, K. A., Pitari, G. M., Kazerounian, S., Ruiz-Stewart, I., et al. Guanylyl cyclases and signaling by cyclic GMP. *Pharmacol. Rev.* 52: 375, 2000.

Schwartz J. H. The many dimensions of cAMP signaling. *Proc. Natl. Acad. Sci., USA* 98: 13482, 2001.

Calcium and Phospholipid Signaling

Berridge, M. J., Bootman, M. D., and Roderick, H. L. Calcium signalling: dynamics, homeostasis and remodeling. *Mol. Cell Biol.* 4: 517, 2003.

Brazil, D. P., Park, J., and Hemmings, B. A. PKB binding proteins: Getting in on the Akt. *Cell* 111: 293, 2002.

Oude Weernink, P. A., Han, L., Jakobs, K. H., and Schmidt, M. Dynamic phospholipid signaling by G protein-coupled receptors. *Biochim. Biophys. Acta* 1768: 888, 2007.

Ron, D., and Kazanietz, M. G. New insights into the regulation of protein kinase C and novel phorbol ester receptors. *FASEB J.* 13: 1658, 1999.

Suh, P. G., Park, J. I., Manzoli, L., Cocco, L., et al. Multiple roles of phosphoinositide-specific phospholipase C isozymes. *BMB Rep.* 2008 41: 415, 2008.

Toker, A. Protein kinases as mediators of phosphoinositide 3-kinase signaling. *Mole. Pharm.* 57: 652, 2000.

Key Terms

juxtacrine signaling
receptor
endocrine signaling
hormone
local mediator
paracrine signaling
autocrine signaling
neurotransmitter
growth factor
synaptic signaling
second messenger
agonist
antagonist
cytokine
G protein-coupled receptor
ligand
effector protein
phospholipase
phosphatase
potency
ligand-gated ion channel receptor
nuclear receptor
receptor desensitization
adapter protein
enzyme-linked receptor
guanylate cyclase
serine/threonine kinase
tyrosine kinase
MAP kinase
small GTPase
heterotrimeric G protein
adenylate cyclase
receptor internalization
cyclic nucleotide
cyclic nucleotide phosphodiesterase
Ca^+ influx channel
Ca^+ release channel
calmodulin
eicosanoid
second-messenger-regulated protein kinase

Questions • CAROL N. ANGSTADT

Multiple-Choice Questions

1. The type of intercellular signaling in which one cell can communicate with another over long distances is called
   A. autocrine.
   B. endocrine.
   C. juxtacrine.
   D. paracrine.
   E. synaptic.

2. Intracellular receptors
   A. usually bind hydrophobic ligands.
   B. may be located either in the cytosol or nucleus in unbound state.
   C. when bound to their ligand, regulate gene transcription.
   D. when bound to their ligand, function as dimeric complexes binding to specific DNA sequences.
   E. all of the above.

3. A cell surface receptor
   A. reacts only with molecules too large to cross the plasma membrane.
   B. when bound to its ligand, could result in activation of an enzymatic cascade.
   C. always opens an ion channel when bound to its ligand.
   D. must produce a second messenger when it binds to its ligand.
   E. is usually also called GPCR.

4. Cells can terminate signal transduction by cell surface receptors by
   A. reducing agonist availability in the vicinity of the target cell.
   B. internalizing and degrading the receptor–agonist complex.
   C. modifying the receptor so that it is inactive or desensitized.
   D. all of the above.
   E. none of the above.

5. Calmodulin is
   A. a nonspecific kinase.
   B. a protein that binds Ca^{2+}.
   C. a second messenger.
   D. an activator of nitric oxide synthase.
   E. a protein channel that facilitates the influx of Ca^{2+}.

6. The guanylate cyclase that responds to NO
   A. is in the catalytic domain of a membrane receptor.
   B. is also activated when atrial natriuretic factor (ANF) binds to its receptor.
   C. increases activity because of a conformational change when NO binds to its heme.
   D. is a monomeric enzyme.
   E. is found only in smooth muscle cells.

Questions 7 and 8: Manic depression may be caused by overactivity of certain central nervous system cells, perhaps caused by abnormally high levels of hormones or neurotransmitters which stimulate phospholipid-based signal transduction (e.g., from phosphatidylinositol [PI]). Lithium has been used for many years to treat manic depression. In the presence of Li^{+}, the PI system is slowed despite continued stimulation and cells become less sensitive to these stimuli. Li^{+} may have two functions, inhibition of the phosphatase that dephosphorylates inositol trisphosphate and direct interference with the function of G proteins.

7. The PI system begins with activation of phospholipase C, which initiates a sequence of events including all of the following except
   A. activation of IP_3 by action of a phosphatase.
   B. increase in intracellular Ca^{2+} concentration.
   C. release of diacylglycerol (DAG) from a phospholipid.
   D. activation of protein kinase C.
   E. phosphorylation of certain cytoplasmic proteins.

8. Which of the following statements concerning G proteins is correct?
   A. G proteins bind the appropriate hormone at the cell surface.
   B. GTP is bound to G protein in the resting state.
   C. α-Subunit may be either stimulatory or inhibitory because it has two forms.
   D. Adenylate cyclase can be activated only if α- and β-subunits of G protein are associated with each other.
   E. Hydrolysis of GTP is necessary for G protein subunits to separate.

Questions 9 and 10: Growth hormone releasing hormone (GHRH) produced by the hypothalamus binds to its pituitary receptor and leads to the production of growth hormone (GH) because of increase in cyclic AMP. Certain pituitary tumors result in hypersecretion of GH because of a mutation that produces a G_s-α protein with a greatly diminished GTPase activity.

9. Elements leading to increased cyclic AMP in response to GHRH binding to its receptor include
   A. activation of a monomeric G protein.
   B. activation of adenylate cyclase by α_s-subunit of a G_s protein.
   C. activation of cyclic nucleotide phosphodiesterase.
   D. activation of protein kinase A.
   E. all of the above.

10. Low GTPase activity in the mutated protein results in consitutive activation of G_s and adenylate cyclase because
   A. GTP-bound α-subunit does not reform the $\alpha\beta\gamma$ trimer.
   B. GTP-bound G protein binds more strongly to the membrane receptor.
   C. GTP reacts directly with adenylate cyclase to activate it.
   D. the trimeric form of the G protein is stabilized.
   E. adenylate cyclase is phosphorylated more readily.

Questions 11 and 12: The ErbB/HER family of receptor tyrosine kinase genes are linked to many different types of human cancers. Overexpression of any of these genes would lead to an increase in the various receptors. ErbB2 encodes the HER2 protein. This receptor does not bind any known extracellular growth factor, but it does form dimers with other growth-factor-bound HER receptors. Even modest overexpression of HER2 can alter normal cell growth regulation.

11. Which of the following statements about receptor tyrosine kinases is correct?
   A. The catalytic domain is on the N-terminal end.
   B. Activation of the kinase requires ATP.
   C. Growth factor binding to the receptor triggers dimerization which activates the kinase activity.
   D. Active tyrosine kinase can phosphorylate other proteins but not itself.
   E. All of the above.

12. Ras protein is a critical regulator in cell proliferation, and its activity is enhanced by activated tyrosine kinase. Elements of its action include all of the following except
   A. formation of cyclic GMP.
   B. adaptor proteins binding to phosphorylated tyrosines on receptor tyrosine kinase.
   C. recruitment and stimulation of Ras-activating protein.
   D. exchange of GDP for GTP on the Ras protein.
   E. initiation of a cascade in which several kinases are activated sequentially by phosphorylation.

Problems

13. How does elevation of cyclic AMP in eukaryotic cells lead to altered transcription of certain genes?

14. How do excitatory and inhibitory neurotransmitters differ in their effects on ligand-gated ion channels?

Answers

1. **B** A hormone is released into the blood and travels to the target tissue. A: One cell type is both the sender and target. C: This is contact-dependent signaling. D: The cells are in the immediate vicinity of each other. E: This is a type of paracrine signaling used by nerve cells.
2. **E** A: Things like steroid hormones. B: Once bound to ligand, cytosolic receptors translocate to the nucleus. C and D: They bind to upstream regulatory elements and either enhance or repress transcription.
3. **B** This is one of the ways that signal transduction occurs. A: They also react with very hydrophilic signaling molecules. C and D: These are additional ways in which cell surface receptors can act. E: This is true only if the receptor is coupled to a G protein.
4. **D** These are all possible.
5. **B** This is a ubiquitous Ca^{2+}-binding protein. A: Ca^{2+}-calmodulin regulates some kinases but not this one. C: Second messengers are usually small molecules or ions. D: It is the Ca^{2+}–calmodulin complex that regulates. E: Calmodulin binds the Ca^{2+} that enters but is not part of the channel to allow it to enter.
6. **C** The soluble enzyme has low activity when heme has not bound a gaseous molecule. A and B: The membrane-bound enzyme is stimulated when ANF binds its receptor but not by NO. D: It is a dimer. E: The soluble enzyme is present in most cell types.
7. **A** IP_3 is the active form. Action of phosphatase on it renders it inactive as it is converted to inositol. B: IP_3 causes release of calcium. C and D: This is the other second messenger, and it activates protein kinase C. E: This is what protein kinase C does.
8. **C** Which form is released depends on the specific hormone and receptor that have interacted. A: Receptor binds the hormone and the complex interacts with G protein. B and E: GDP is bound to the resting enzyme. Its replacement by GTP causes α-subunit to dissociate. Actually hydrolysis of GTP allows subunits to reassociate. D: α-subunit, when dissociated from the other two, interacts with the enzyme to either activate or inhibit it depending on whether it is an α_s or an α_i.
9. **B** Adenylate cyclase then converts ATP to cAMP. A: G protein-coupled receptors are all heterotrimeric. C: This would decrease cAMP by hydrolyzing it to AMP. D: Cyclic AMP activates protein kinase, not the other way.
10. **A** α-GTP dissociates from the $\beta\gamma$ dimer. B: G proteins do not bind to the receptor. C: It is the α-GTP that activates the enzyme. D: See A. E: Activation of adenylate cyclase is a conformational change, not phosphorylation.
11. **C** This is why HER2 alters growth even though it does not bind a growth factor. A: The N-terminal end is extracellular and site of binding of the ligand; the catalytic site must be on an intracellular domain. B: Dimerization causes the activation; once activated it uses ATP to phosphorylate proteins. D: The first protein phosphorylated is the receptor itself, which then attracts other proteins to be phosphorylated.
12. **A** This is not a mechanism that produces a small second messenger. B to E: These are the sequential steps.
13. Cyclic AMP binding to protein kinase A causes dissociation of catalytic from regulatory subunits, exposing nuclear localization sequences on catalytic subunits. These can translocate to the nucleus where they can phosphorylate and activate cAMP-regulated gene regulatory proteins (CREBs), which control genes containing cAMP-sensitive regulatory elements (CREs).
14. Excitatory neurotransmitters (e.g., acetylcholine, glutamate) bind to cation-selective receptors and allow ions like Na^+ to enter, depolarizing the membrane. Inhibitory neurotransmitters (e.g., GABA, glycine) bind to anion-selective receptors and allow anions like Cl^- to enter, hyperpolarizing the membrane.

14

Bioenergetics, Mitochondria, and Oxidative Metabolism

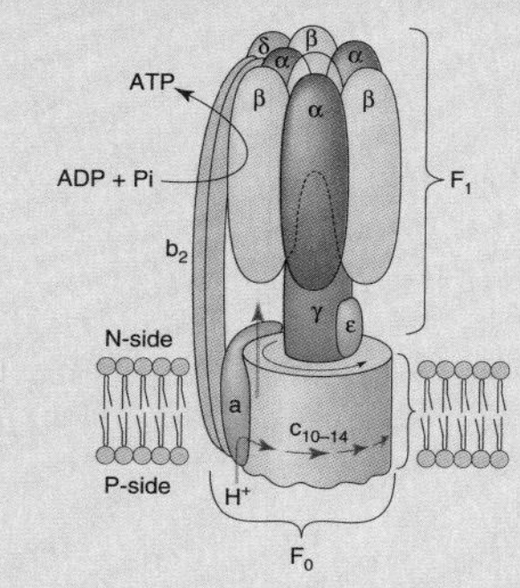

Diana S. Beattie
Professor, West Virginia University School of Medicine

CLINICAL CORRELATIONS

Key Concepts

- Energy is conserved in living organisms by oxidation of metabolic fuels and stored as the *high-energy* bond of ATP to provide energy for biosynthetic reactions, muscle contraction, and active ion transport.
- Gibbs free-energy predicts the direction of an enzymatic reaction and is related to the equilibrium constant.
- Pyruvate, the end product of glucose metabolism, is oxidatively decarboxylated to acetylCoA by the regulated pyruvate dehydrogenase complex.
- AcetylCoA, the end product of carbohydrate, lipid, and protein catabolism, is oxidized by the tricarboxylic acid cycle located in the mitochondrial matrix, producing NADH and $FADH_2$.
- NADH and $FADH_2$ are oxidized by the electron transport chain, organized in four large enzyme complexes in the inner mitochondrial membrane, by transferring electrons in a series of steps to the terminal electron acceptor oxygen.
- The energy released during the oxidative reactions of the electron transport chain is conserved as a proton and charge gradient across the inner mitochondrial membrane. The gradient drives the synthesis of ATP catalyzed by ATP synthase. This process is called oxidative phosphorylation.
- Transport systems located in the inner mitochondrial membrane facilitate movement of substrates and intermediates into and out of the mitochondrial matrix.

- Mammalian mitochondria have a unique circular DNA that encodes proteins of the electron transport chain, ATP synthase, and mitochondrial ribosomal RNA. Diseases result from mutations in mitochondrial genes.
- Stepwise transfer of electrons to oxygen results in formation of superoxide anions (O_2^-), hydrogen peroxide (H_2O_2), and hydrogen free radicals (OH·), which cause cell damage by lipid peroxidation, protein oxidation, and DNA mutations.

14.1 • ENERGY-PRODUCING AND ENERGY-UTILIZING SYSTEMS

Living cells depend on a complex, intricately regulated system of energy-producing and energy-utilizing chemical reactions called metabolism. Metabolism consists of two contrasting processes, **catabolism** and **anabolism**, which together constitute the chemical changes that convert foodstuffs into usable forms of energy and into complex biological molecules. Catabolism is responsible for degradation of ingested foodstuffs or stored fuels such as carbohydrate, lipid, and protein into either usable or storable forms of energy. Catabolic reactions generally result in conversion of large complex molecules to smaller molecules (ultimately CO_2 and H_2O), and in mammals often require consumption of O_2. Energy-utilizing reactions perform various necessary, and in many instances tissue-specific functions, for example, nerve impulse conduction, muscle contraction, growth, and cell division. Catabolic reactions are generally exergonic with the released energy generally trapped in the formation of ATP. The oxidative reactions of catabolism transfer reducing equivalents to the coenzymes NAD^+ and $NADP^+$ to form NADH and NADPH. Anabolic pathways are responsible for biosynthesis of large molecules from smaller precursors and require input of energy either in the form of ATP or reducing equivalents of NADPH (see Figure 10.28, p. 391).

ATP Links Energy-Producing and Energy-Utilizing Systems

The relationship between energy-producing and energy-utilizing functions of cells is illustrated in Figure 14.1. Energy is derived from oxidation of metabolic fuels utilized by the organism usually as carbohydrate, lipid, and protein. The proportion of each fuel utilized as an energy source depends on the tissue and the dietary and hormonal state of the organism. For example, mature erythrocytes and adult brain in the fed state use only carbohydrate as a source of energy, whereas the liver of a diabetic or fasted individual metabolizes primarily lipid to meet the energy demands. Energy may be consumed during performance of various energy-linked (work) functions, some of which are indicated in Figure 14.1. Note that liver and pancreas are primarily involved in biosynthetic and secretory work functions, whereas cardiac and skeletal muscles convert metabolic energy into mechanical energy during muscle contraction.

The essential link between energy-producing and energy-utilizing pathways is the nucleoside triphosphate **adenosine 5′-triphosphate (ATP)** (Figure 14.2). ATP is a purine (adenine) nucleotide in which adenine is linked by a glycosidic bond to D-ribose. Three phosphoryl

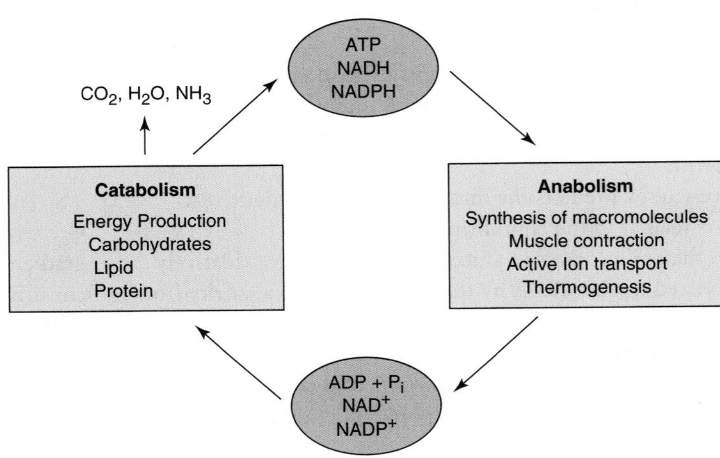

Figure 14.1 Energy relationships between energy production (catabolism) and energy utilization (anabolism). Oxidative breakdown of foodstuffs is an exergonic process releasing free energy and reducing power that are trapped as ATP and NADH or NADPH respectively. Anabolic processes are endergonic and use chemical energy stored as ATP and NADPH.

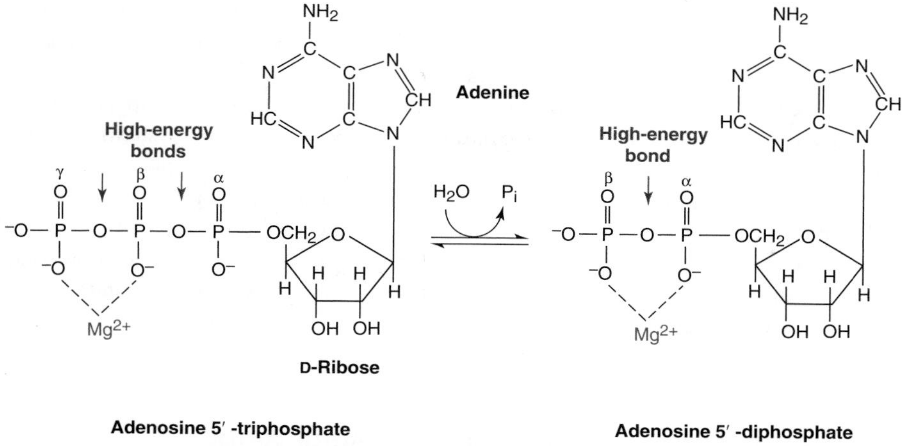

Adenosine 5′-triphosphate   Adenosine 5′-diphosphate

Figure 14.2 Structure of ATP and ADP complexed with Mg^{2+}. High-energy bonds are highlighted.

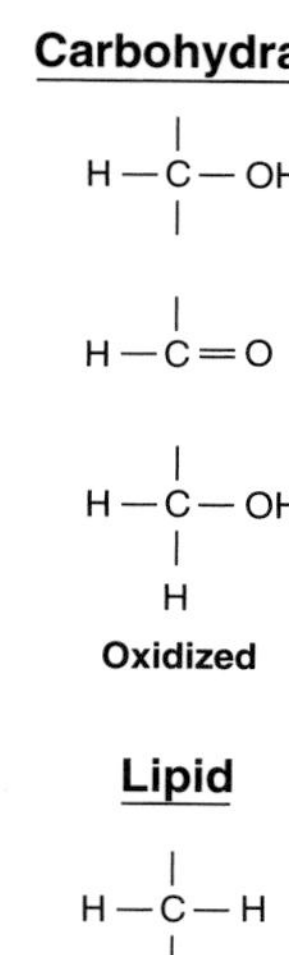

Figure 14.3 Oxidation states of typical carbon atoms in carbohydrates and lipids.

groups are esterified to the 5 position of the ribose moiety. The two terminal phosphoryl groups (i.e., β and γ) are energy-rich phospho-anhydride bonds or **high-energy bonds**. Synthesis of ATP as a result of a catabolic process or consuming ATP in an energy-linked process involves formation and either hydrolysis or transfer of the terminal phosphate group of ATP. This nucleotide is chelated with a divalent metal cation such as magnesium under physiological conditions.

NAD^+ and NADPH in Catabolism and Anabolism

Many catabolic processes are oxidative in nature because the carbons in the substrates, carbohydrates, fats and proteins, are in a partially or highly reduced state (Figure 14.3). **Reducing equivalents** are released from substrates as protons and electrons, which are transferred to **nicotinamide adenine dinucleotide (NAD^+)** by enzymes called dehydrogenases with formation of NADH (see Figure 10.28, p. 391). The reducing equivalents of NADH are transported into mitochondria and transferred by the electron transport chain to O_2 as the ultimate electron acceptor (p. 562). The oxidative reactions in mitochondria are exergonic; the energy produced is used for synthesis of ATP in a process called **oxidative phosphorylation**. The reductive and oxidative reactions of the NAD^+−NADH cycle are central in conversion of the chemical energy of carbon compounds in foodstuffs into that of phosphoanhydride bonds of ATP. This process, called energy transduction, will be discussed in detail later in this chapter.

Anabolism, by contrast, is largely a reductive process as small, more highly oxidized molecules are converted into large complex molecules (Figure 14.4). The reducing power used in biosynthesis of highly reduced compounds, such as fatty acids, is provided by NADPH (p. 391), a 3′ phosphorylated NADH.

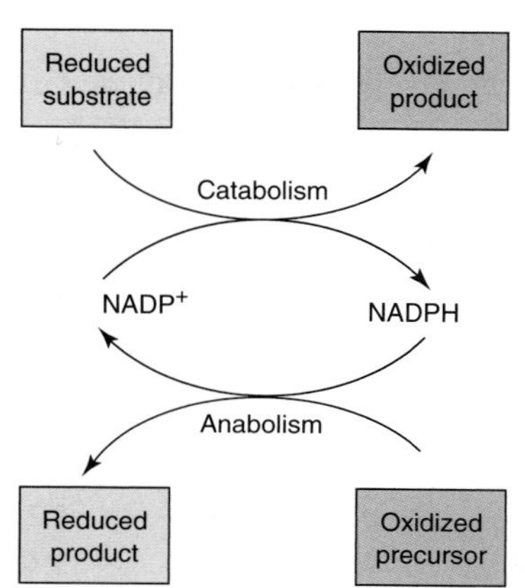

Figure 14.4 Transfer of reducing equivalents during catabolism and anabolism using NADPH and NADH.

14.2 • THERMODYNAMIC RELATIONSHIPS AND ENERGY-RICH COMPONENTS

Living cells interconvert different forms of energy and also exchange energy with their surroundings. The principles of **thermodynamics** govern reactions of this type. Knowledge of these principles facilitates a perception of how energy-producing and energy-utilizing reactions occur within the same cell and how an organism is able to accomplish various work functions.

The **first law of thermodynamics** states that energy can neither be created nor destroyed. This law of energy conservation stipulates that, although energy may be converted from one form to another, the total energy in a system remains constant. For example, chemical energy available in a metabolic fuel such as glucose is converted in glycolysis to the chemical energy of ATP. In skeletal muscle, chemical energy involved in energy-rich phosphate bonds of ATP is converted to mechanical energy during muscle contraction. The energy of an osmotic electropotential gradient of protons across the mitochondrial membrane is converted to chemical energy during ATP synthesis.

The **second law of thermodynamics** is concerned with **entropy**. Entropy, denoted by *S*, is a measure or indicator of the degree of disorder or randomness in a system. Entropy is viewed as the energy in a system that is unavailable to perform useful work. All processes, whether chemical or biological, tend to progress toward a situation of maximum entropy. Hence, living systems that are highly ordered are never at equilibrium with their surroundings as equilibrium in a system results when the randomness or disorder (entropy) is at a maximum. In biological systems, therefore, it is nearly impossible to quantitate entropy changes as such systems are rarely at equilibrium. For simplicity and because of its inherent utility in these considerations, a quantity termed **free energy** is employed.

Free Energy Is Energy Available for Useful Work

The free energy (denoted by G or Gibbs free energy) of a system is that portion of the total energy that is available for useful work. It is defined by

$$\Delta G = \Delta H - T\Delta S$$

In this expression for a system proceeding toward equilibrium at a constant temperature and pressure, ΔG is the change in free energy, ΔH is the change in enthalpy or the heat content, T is the absolute temperature, and ΔS is the change in entropy. If ΔG of a reaction is equal to zero, the process is at equilibrium and there is no net flow in either direction. Furthermore, any process that exhibits a negative ΔG (free-energy change) proceeds spontaneously toward equilibrium in the direction written, in part, due to an increase in entropy or disorder in the system. Such a process releases energy and is **exergonic**. A process that exhibits a positive ΔG will proceed spontaneously in the reverse direction written. Energy from some other source must be applied to allow it to proceed toward equilibrium. This process is termed **endergonic**. Note that the sign and value of ΔG do not predict how fast the reaction will go.

The rate of a given reaction depends on the free energy of activation but not on the magnitude of ΔG. In addition, the change in free energy in a biochemical process is the same regardless of the path or mechanism used to attain the final state. The change in free energy for a chemical reaction is related to the equilibrium constant. For example, a reaction may be described as

$$A + B \rightleftharpoons C + D$$

with the equilibrium constant expressed as

$$K_{eq} = [C][D]/[A][B]$$

Under standard conditions, when reactants and products are initially present at 1-M concentrations, at 1-Atm pressure and 1 M $[H^+]$ or pH 0, the standard free-energy change is defined as ΔG^0. Biochemists have modified this expression so that the standard free energy is calculated at pH 7.0 ($[H^+] = 10^{-7}$ M) where biological reactions generally occur. Under these conditions the change in free energy is expressed as $\Delta G^{0'}$ and K'_{eq}. Since the value of $\Delta G^{0'}$ is zero at equilibrium, the following relationship is described:

$$\Delta G^{0'} = -RT \ln K'_{eq}$$

where R is the gas constant, which is 1.987 cal/mol × °K, or 8.134 J/mol × °K, depending on whether the resultant free-energy change is expressed in calories (cal) or joules (J) per mole, and T is the absolute temperature in degrees Kelvin (K).

Hence, if the **equilibrium constant** is determined, the standard free-energy change ($\Delta G^{0'}$) also can be calculated. The relationship between $\Delta G^{0'}$ and K'_{eq} is illustrated in Table 14.1. When the equilibrium constant is less than 1, the reaction is endergonic and $\Delta G^{0'}$ is positive. When the equilibrium constant is greater than 1, the reaction is exergonic and $\Delta G^{0'}$ is negative.

As discussed, $\Delta G^{0'}$ of a reaction represents the free energy available in a reaction when substrates and products are present at 1 M concentrations. This situation does not occur in

TABLE 14.1 • Values of K_{eq} and $\Delta G^{0'}$

K_{eq}	$\Delta G^{0'}$ *(kcal/mol)*	$\Delta G^{0'}$ *(kJ/mol)*
10^{-4}	5.46	22.8
10^{-3}	4.09	17.1
10^{-2}	2.73	11.4
10^{-1}	1.36	5.7
1	0	0
10	−1.36	−5.7
10^{2}	−2.73	−11.4
10^{3}	−4.09	−17.1
10^{4}	−5.46	−22.8

cells, as biomolecules are rarely present at 1 M concentration. Hence, an expression related to actual intracellular concentrations of substrates and products provides insight into work available in a reaction. The expression for ΔG at any concentration of substrate or product includes the energy change for a 1 M concentration of substrate and product to reach equilibrium ($\Delta G^{0'}$) and the energy change to reach a 1 M concentration of substrates and products.

$$\Delta G = \Delta G^{0'} + RT\ln([C][D]/[A][B])$$

For example, in a muscle cell the concentration of ATP = 8.1 mM, of ADP = 0.93 mM, and of P_i = 8.1 mM. If $\Delta G^{0'}$ for the reaction: ATP + HOH $\leftrightarrow$ ADP + P_i is 32.1 kJ/mol at 37°C, pH 7.4, then the overall ΔG for the reaction is

$$\Delta G = \Delta G^{0'} + RT\ln([ADP][P_i]/[ATP])$$

$$\Delta G = \Delta G^{0'} + RT\ln(0.93 \times 10^{-3})$$

$$\Delta G = -32.1 \text{ kJ/mol} + (-17.5 \text{ kJ/mol}) = -49.6 \text{ kJ/mol}$$

These calculations demonstrate that considerably more free energy is available to perform work in a muscle cell than indicated by the value of $\Delta G^{0'}$. Moreover, synthesis of ATP in muscle cells under these conditions, the reverse reaction, would require +50.1 kJ/mol of energy.

In energy-producing and energy-utilizing metabolic pathways, free-energy changes of individual enzymatic reactions are additive, for example,

$$A \rightarrow B \rightarrow C \rightarrow D$$

$$\Delta G^{0'}_{A\rightarrow D} = \Delta G^{0'}_{A\rightarrow B} + \Delta G^{0'}_{B\rightarrow C} + \Delta G^{0'}_{C\rightarrow D}$$

Although any given enzymatic reaction in a sequence may have a positive free-energy change, as long as the sum of all free-energy changes is negative, the pathway will proceed. Another way of expressing this principle is that enzymatic reactions with positive free-energy changes may be coupled to or driven by reactions with negative free-energy changes. In a metabolic pathway such as glycolysis, various reactions either have positive $\Delta G^{0'}$ values or $\Delta G^{0'}$ values close to zero, whereas other reactions have large and negative $\Delta G^{0'}$ values, which drive the entire pathway. The crucial consideration is that the sum of $\Delta G^{0'}$ values of all reactions in a pathway must be negative in order for such a pathway to be thermodynamically feasible. As for all chemical reactions, individual enzymatic reactions in a metabolic pathway or the pathway as a whole would be facilitated if the concentrations of reactants (substrates) exceed those of products.

Caloric Value of Dietary Components

During complete stepwise oxidation of glucose, a primary metabolic fuel in cells, a large quantity of energy is made available. This is illustrated in the equation

$$C_6H_{12}O_6 + 6\ O_2 \rightarrow 6\ CO_2 + 6\ H_2O;\ \Delta G^{0'} = -2864 \text{ kJ/mol } (-684 \text{ kcal/mol})$$

When this process occurs under aerobic conditions in most cells, it is possible to conserve approximately one-half of this "available" energy as 38 molecules of ATP. The caloric values for oxidation of other metabolic fuels are listed in Table 14.2. Carbohydrates and proteins (amino acids) have a caloric value of 12–16 kJ/g, while lipid (i.e., palmitate, a long-chain

TABLE 14.2 • Free-Energy Changes and Caloric Values for Total Metabolism of Various Metabolic Fuels

Compound	*Molecular Weight*	$\Delta G^{0'}$ *kJ/mol*	*kcal/mol*	*Caloric Values kJ/g*	*kcal/g*
Glucose	180	−2,864	−684	15.9	3.81
Lactate	90	−1,361	−325	15.1	3.62
Palmitate	256	−9,987	−2,393	38.8	9.30
Tripalmitin	809	−31,772	−7,612	38.8	9.30
Glycine	75	−976	−233	13.0	3.12

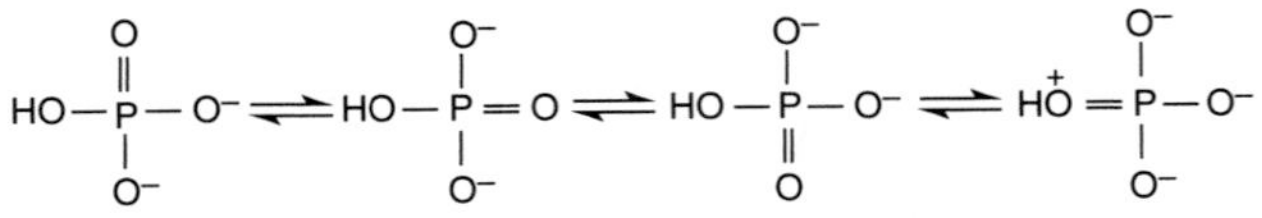

(*a*) **Resonance forms of phosphate**

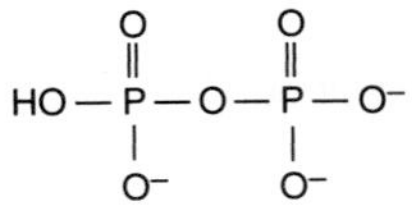

(*b*) **Pyrophosphate**

Figure 14.5 **(*a*) Resonance forms of phosphate. (*b*) Structure of pyrophosphate.**

fatty acid, or a triacylglycerol) has a value nearly three times greater. The reason that more energy is derived from lipid than from carbohydrate or protein relates to the average oxidation state of carbon atoms in these substances. Carbon atoms in carbohydrates are considerably more oxidized (or less reduced) than those in lipids (see Figure 14.3). Hence during sequential breakdown of lipid many more reducing equivalents (a reducing equivalent is defined as a proton plus an electron, i.e., $H^+ + e^-$) can be extracted than from carbohydrate.

Compounds Are Classified on the Basis of Energy Released on Hydrolysis of Specific Groups

The two terminal phosphoryl groups of ATP are high-energy bonds, since free energy of hydrolysis of a phosphoanhydride bond is much greater than that of a simple phosphate ester. High-energy is not synonymous with stability of the chemical bond in question, nor does it refer to the energy required to break such a bond. The concept of high-energy compounds implies that products of their hydrolytic cleavage are in more stable forms than the original compound. As a rule, phosphate esters (low-energy compounds) exhibit negative $\Delta G^{0\prime}$ values of hydrolysis of 42 kJ/mol, whereas high-energy bonds have negative $\Delta G^{0\prime}$ values of 21−63 kJ/mol. Phosphate esters such as glucose 6-phosphate and glycerol 3-phosphate are low-energy compounds.

There are various reasons why certain compounds or bonding arrangements are energy rich. First, products of hydrolysis of an energy-rich bond may exist in more **resonance forms** than the precursor molecule. The more possible resonance forms in which a molecule can exist stabilize that molecule. The resonance forms for inorganic phosphate (P_i) are indicated in Figure 14.5. Fewer resonance forms can be written for ATP or pyrophosphate (PP_i) than for phosphate (P_i).

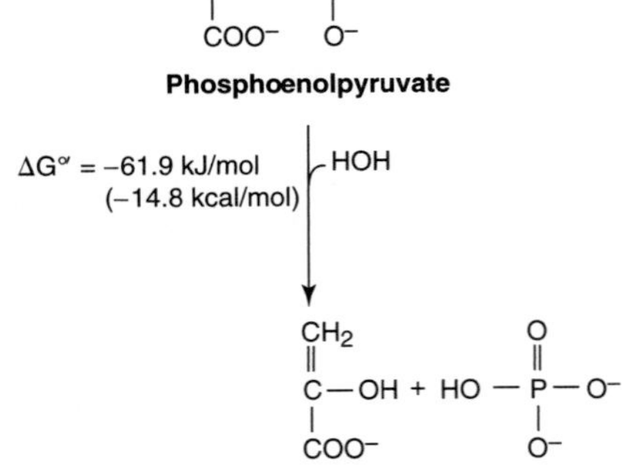

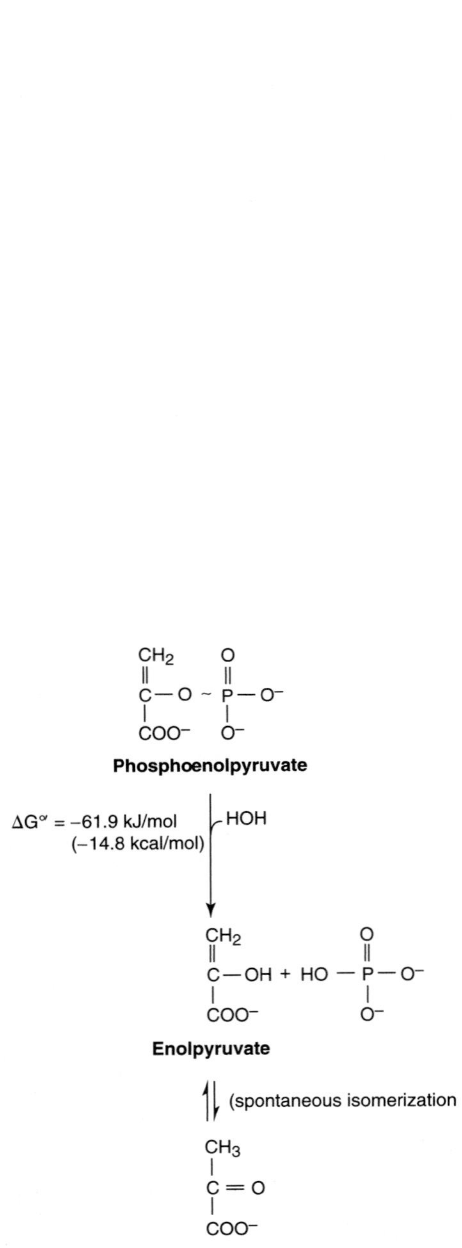

Pyruvate (stable form)

Figure 14.6 **Hydrolysis of phosphoenolpyruvate indicating the free energy released.**

Second, many high-energy bonding arrangements contain groups of similar electrostatic charges located in close proximity to each other. Like charges repel one another, and consequently hydrolysis of highly charged energy-rich bonds alleviates this situation and lends stability to products of hydrolysis. Third, hydrolysis of certain energy-rich bonds results in formation of an unstable compound, which may isomerize spontaneously to form a more stable compound. Hydrolysis of phosphoenolpyruvate is an example of this type of compound (Figure 14.6). The $\Delta G^{0\prime}$ is considerable for isomerization, and the final product, in this case pyruvate, is much more stable. Finally, if a product of hydrolysis of a high-energy bond is an undissociated acid, dissociation of the proton and its subsequent buffering may contribute to the overall $\Delta G^{0\prime}$ of the hydrolytic reaction. In general, any property or process that stabilizes products of hydrolysis tends to confer a high-energy character to that compound. The high-energy character of **3′,5′-cyclic adenosine monophosphate (cAMP)** has been attributed to the fact that its phosphodiester bond is strained as it bridges the 3′ and 5′ positions on ribose. The energy-rich character of thiol ester compounds such as **acetyl CoA** or succinyl CoA results from the relatively acidic character of the thiol group. Hence, the thioester bond of acetyl CoA is nearly equivalent in energy to a phosphoanhydride bond.

Free-Energy Changes Can Be Determined from Coupled Enzyme Reactions

The $\Delta G^{0'}$ value of hydrolysis of the terminal phosphate of ATP is difficult to determine simply because the K_{eq} of the reaction is far to the right.

$$\text{ATP} + \text{HOH} \rightleftharpoons \text{ADP} + \text{P}_i + \text{H}^+$$

However, $\Delta G^{0'}$ of hydrolysis of ATP is determined indirectly because of the additive nature of free-energy changes discussed above. Hence, free energy of hydrolysis of ATP is determined by adding $\Delta G^{0'}$ of an ATP-utilizing reaction such as hexokinase to $\Delta G^{0'}$ of a reaction that cleaves phosphate from the product of the hexokinase reaction, glucose 6-phosphate (G6P), as indicated here.

$$\text{Glucose} + \text{ATP} \xrightarrow{\text{hexokinase}} \text{G6P} + \text{ADP} + \text{H}^+ \quad \Delta G^{0'} = -16 \text{ kJ/mol}$$

$$\text{G6P} + \text{HOH} \xrightarrow{\text{glucose 6-phosphatase}} \text{ADP} + \text{P}_i + \text{H}^+ \quad \Delta G^{0'} = -14 \text{ kJ/mol}$$

Free energies of hydrolysis for other energy-rich compounds are determined in a similar fashion.

High-Energy Bond Energies of Various Groups Can Be Transferred from One Compound to Another

Energy-rich compounds can transfer various groups to an acceptor compound in a thermodynamically feasible fashion in the presence of an appropriate enzyme. The energy-rich intermediates of glycolysis, 1,3-bisphosphoglycerate and phosphoenolpyruvate, can transfer their high-energy phosphate moieties to ADP in the phosphoglycerate kinase and pyruvate kinase reactions, respectively (Figure 14.7). The $\Delta G^{0'}$ values of these reactions are -18.8 and -31.3 kJ/mol, respectively, and hence transfer of "high-energy" phosphate is thermodynamically possible, and ATP synthesis is the result. ATP can transfer its terminal phosphoryl group to form a compound of relatively similar high-energy character (i.e., creatine phosphate in the creatine kinase reaction [Figure 14.7]) or compounds of considerably lower energy, such

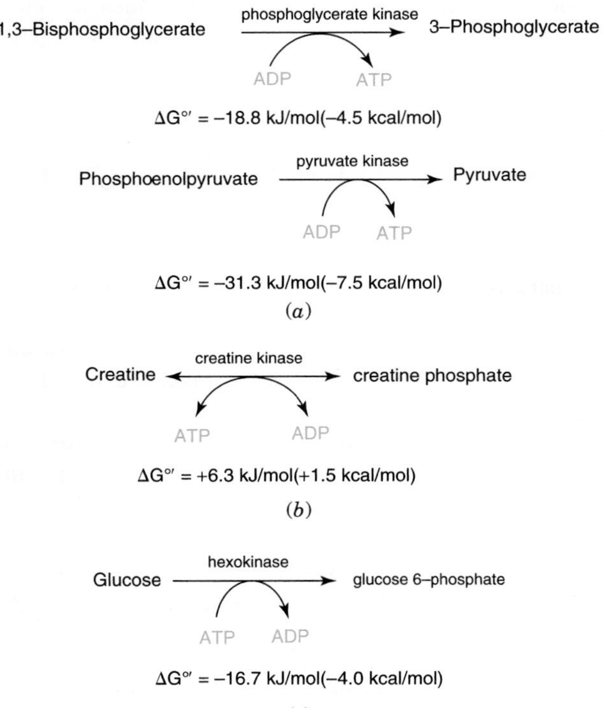

Figure 14.7 **Examples of reactions involved in transfer of "high-energy" phosphate.**

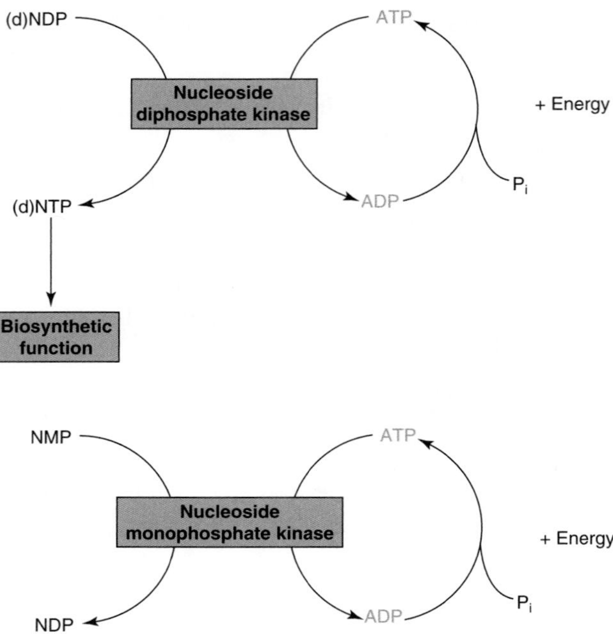

Figure 14.8 Nucleoside diphosphate kinase and nucleoside monophosphate kinase reactions. *N* represents any purine or pyrimidine base; (d) indicates a deoxyribonucleotide.

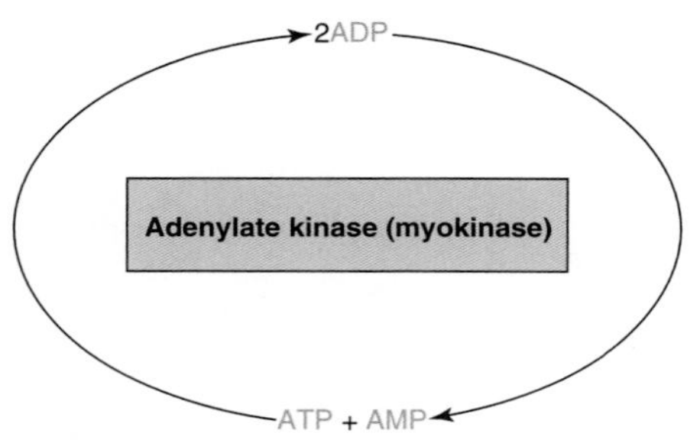

Figure 14.9 Adenylate kinase (myokinase) reaction.

as glucose 6-phosphate in the hexokinase reaction (Figure 14.7). Such transfers are crucial in linking energy-producing and energy-utilizing metabolic pathways in living cells.

Although adenine nucleotides are mainly involved in energy generation or conservation, various nucleoside triphosphates, including ATP, are involved in energy transfer in biosynthetic pathways. The guanine nucleotide GTP is the source of energy in gluconeogenesis and protein synthesis, whereas UTP (uracil) and CTP (cytosine) are utilized in glycogen and lipid synthesis, respectively (see p. 27 for structures). The energy in the phosphate bonds of ATP may be transferred to other nucleotides by nucleoside diphosphate kinase or nucleoside monophosphate kinase (Figure 14.8). Two nucleoside diphosphates are converted to a nucleoside triphosphate and a nucleoside monophosphate in various nucleoside monophosphate kinase reactions, such as that of adenylate kinase (Figure 14.9). Thus, the terminal energy-rich phosphate bond of ATP may be transferred to appropriate nucleotides and utilized in a variety of biosynthetic processes.

14.3 • SOURCES AND FATES OF ACETYL COENZYME A

The acetate group that serves as the source of fuel for the **tricarboxylic acid (TCA)** cycle is derived from the major energy-generating metabolic pathways of cells. These include oxidation of long-chain fatty acids by β-oxidation, breakdown of ingested or stored carbohydrate by **glycolysis**, oxidation of the **ketone bodies (acetoacetate** and **β-hydroxybutyrate)**, oxidation of ethanol, and oxidative breakdown of certain amino acids (Figure 14.10). All of these eventually result in production of the two-carbon unit **acetyl coenzyme A**. Coenzyme A (CoA or CoASH), consists of β-mercaptoethylamine, the vitamin **pantothenic acid**, and the adenine nucleotide, adenosine 3′-phosphate 5′-diphosphate (Figure 14.11). In cells, coenzyme A exists as the reduced thiol (CoASH), which forms high-energy thioester bonds with acyl groups and is involved in acyl group transfer reactions in which CoA serves as acceptor, then donor, of the acyl group. Various metabolic pathways involve only acyl CoA derivatives, for example, β-oxidation of fatty acids and branched-chain amino acid degradation. Because CoA is a large, hydrophilic molecule, it and its derivatives such as acetyl CoA, are not freely transported across cellular membranes. This has necessitated evolution of certain transport or shuttle mechanisms by which various intermediates or groups are

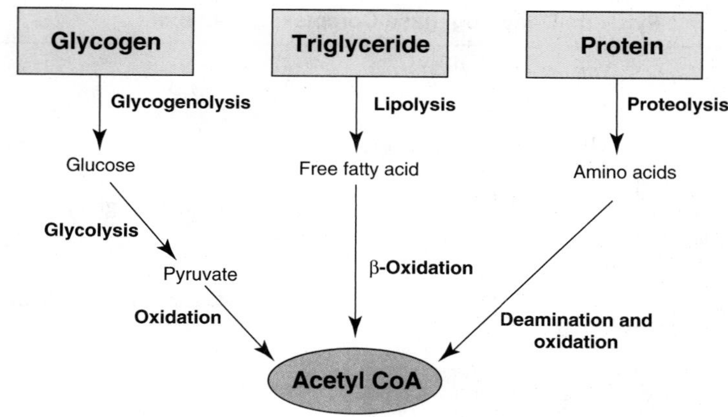

Figure 14.10 General precursors of acetyl CoA. Carbohydrates, lipids, and proteins are broken down to form acetyl CoA.

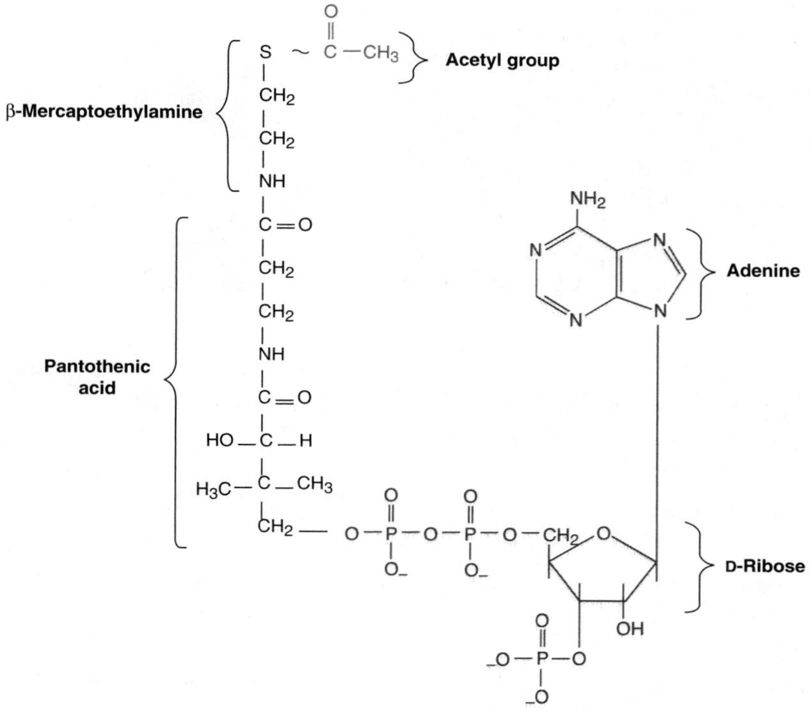

Figure 14.11 Structure of acetyl CoA. Note the presence of pantothenic acid, one of the B vitamins essential for humans.

transferred across membranes. Such acyl transferase reactions for acetyl groups and long-chain acyl groups will be discussed in Chapter 16. Since the thiol ester bond in acyl CoA derivatives is an energy-rich bond, these compounds are effective donors of acyl groups in acyl transferase reactions. To synthesize an acyl CoA derivative, two high-energy bonds of ATP must be expended, such as in the **acetate thiokinase** reaction

$$\text{Acetate} + \text{CoASH} + \text{ATP} \xrightarrow{\text{acetate kinase}} \text{acetyl CoA} + \text{AMP} + \text{PP}_i$$

Metabolic Sources and Fates of Pyruvate

During aerobic glycolysis (p. 592), glucose or other hexoses are converted to pyruvate, the end product of this cytosolic pathway. Pyruvate is also formed in degradation of amino acids such as alanine or serine and has several fates depending on the tissue and its metabolic state. The fates of pyruvate and types of reactions in which it participates are indicated in Figure 14.12. The oxidative decarboxylation of pyruvate in the **pyruvate dehydrogenase** reaction is discussed next; see page 753 for a discussion of other reactions involving pyruvate.

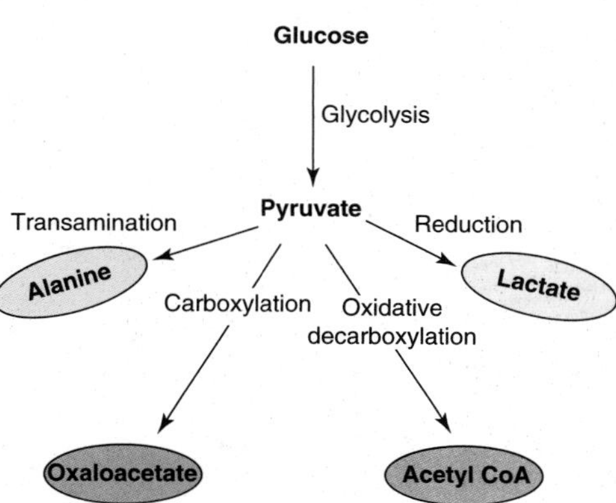

Figure 14.12 Metabolic fates of pyruvate. Pyruvate is at a crossroads of metabolism. It can be converted to lactate, alanine, oxaloacetate, or acetyl CoA depending on the needs of the cell.

TABLE 14.3 • Pyruvate Dehydrogenase Complex of Mammals

Enzyme	*Number of Subunits*	*Prosthetic Group*	*Reaction Catalyzed*
Pyruvate dehydrogenase	20 or 30	TPP	Oxidative decarboxylation of pyruvate
Dihydrolipoyl transacetylase	60	Lipoamide	Transfer of the acetyl group to CoA
Dihydrolipoyl dehydrogenase	6	FAD	Regeneration of the oxidized form of lipoamide and transfer of electrons to NAD

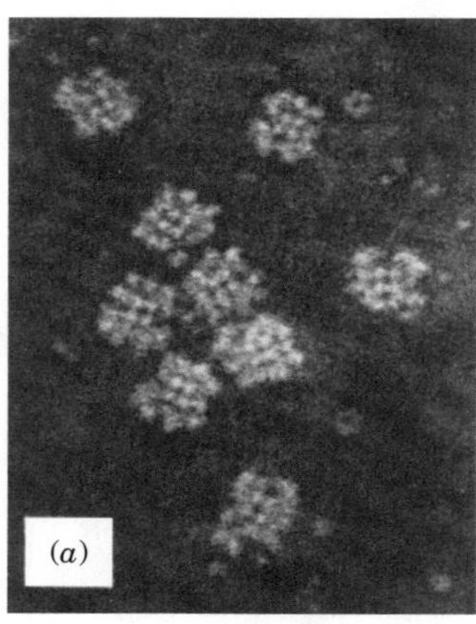

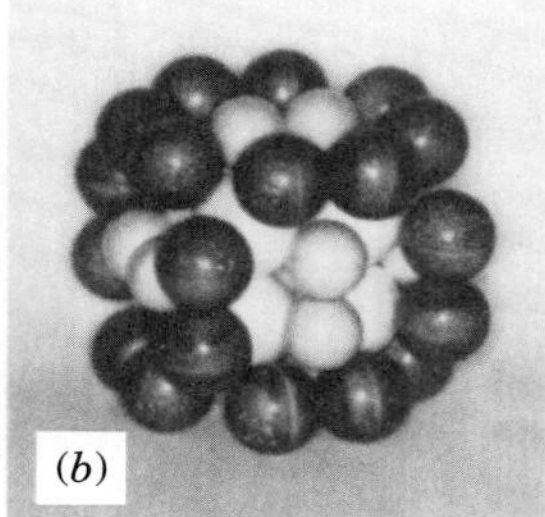

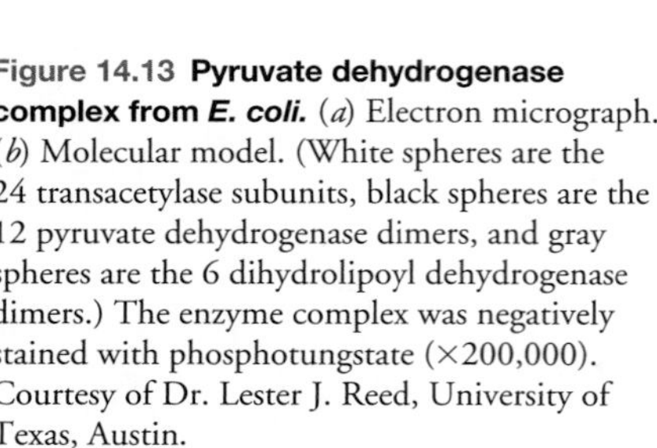

Figure 14.13 Pyruvate dehydrogenase complex from *E. coli*. (*a*) Electron micrograph. (*b*) Molecular model. (White spheres are the 24 transacetylase subunits, black spheres are the 12 pyruvate dehydrogenase dimers, and gray spheres are the 6 dihydrolipoyl dehydrogenase dimers.) The enzyme complex was negatively stained with phosphotungstate (×200,000). Courtesy of Dr. Lester J. Reed, University of Texas, Austin.

Pyruvate Dehydrogenase Is a Multienzyme Complex

Pyruvate is converted to acetyl CoA by the pyruvate dehydrogenase multienzyme complex

$$\text{Pyruvate} + \text{NAD}^+ + \text{CoASH} \longrightarrow \text{acetyl CoA} + \text{CO}_2 + \text{NADH} + \text{H}^+$$

$$\Delta G^{0\prime} = -33.4 \text{ kJ/mol}$$

The mechanism of this reaction is more complex than might be inferred from the overall stoichiometry. Three of the cofactors, **thiamin pyrophosphate** (TPP), **lipoamide**, and **flavin adenine dinucleotide** (FAD) are bound to subunits of the complex. The reaction has a $\Delta G^{0\prime}$ of -33.4 kJ/mol and hence is irreversible under physiological conditions. The mammalian pyruvate dehydrogenase complex contains three types of catalytic subunits associated in a multienzyme complex of mass 7 to 8.5×10^6 kDa for the complex from kidney, heart, or liver. The catalytic subunits and their associated cofactors are presented in Table 14.3.The arrangement of the catalytic subunits of pyruvate dehydrogenase (Figure 14.13) in a multiprotein complex provides for greater efficiency in the overall reaction as the intermediates are tightly bound to the subunits and are not released into the surrounding medium.

The mechanism of the reaction is illustrated in Figure 14.14. The functional group of TPP participates in the formation of a covalent intermediate. Lipoic acid is linked by an amide bond with a lysine on each transacetylase subunit and is called lipoamide, while FAD is tightly bound to each dihydrolipoyl dehydrogenase subunit. The structure of FAD is presented in Figure 10.33 (p. 392).

Pyruvate Dehydrogenase Is Strictly Regulated

The pyruvate dehydrogenase complex is regulated in two ways. First, two products of the reaction, acetyl CoA and NADH, inhibit the complex in a competitive fashion, as feedback inhibitors. Second, the pyruvate dehydrogenase complex undergoes phosphorylation and dephosphorylation. The complex is active when dephosphorylated and inactive when phosphorylated. Inactivation is accomplished by a Mg^{2+}-ATP-dependent **protein kinase**, which is tightly associated with the complex. Activation is accomplished by a **phosphoprotein phosphatase**, which is also associated with the complex and functions in a Mg^{2+}- and

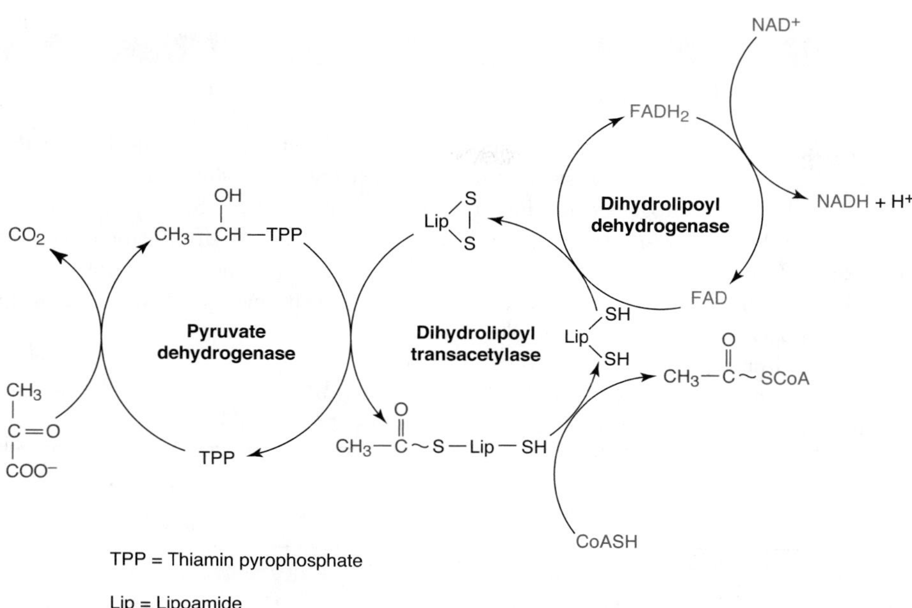

Figure 14.14 Mechanism of the pyruvate dehydrogenase multienzyme complex. Pyruvate dehydrogenase catalyzes oxidative decarboxylation of pyruvate and transfer of the acetyl group first to lipoamide. Dihydrolipoyl transacetylase transfers the acetyl group from lipoamide to coenzyme A. Dihydrolipoyl dehydrogenase oxidizes reduced lipoamide.

Ca^{2+}-dependent manner. The differential regulation of the pyruvate dehydrogenase kinase and phosphatase is the key to the overall regulation of the complex. Products of the enzyme stimulate the protein kinase reaction, leading to an inactivation of the complex (Figure 14.15). Activity of the complex is stimulated by Mg^{2+} and Ca^{2+}, a potent activator of the protein phosphatase. These effects of Ca^{2+} may play an important role in skeletal muscle where the release of Ca^{2+} during contraction should activate the protein phosphatase stimulating the oxidation of pyruvate and hence energy production. Finally, administration of insulin activates pyruvate dehydrogenase in adipose tissue, and catecholamines, such as epinephrine, activate pyruvate dehydrogenase in cardiac tissue (Clin. Corr. 14.1).

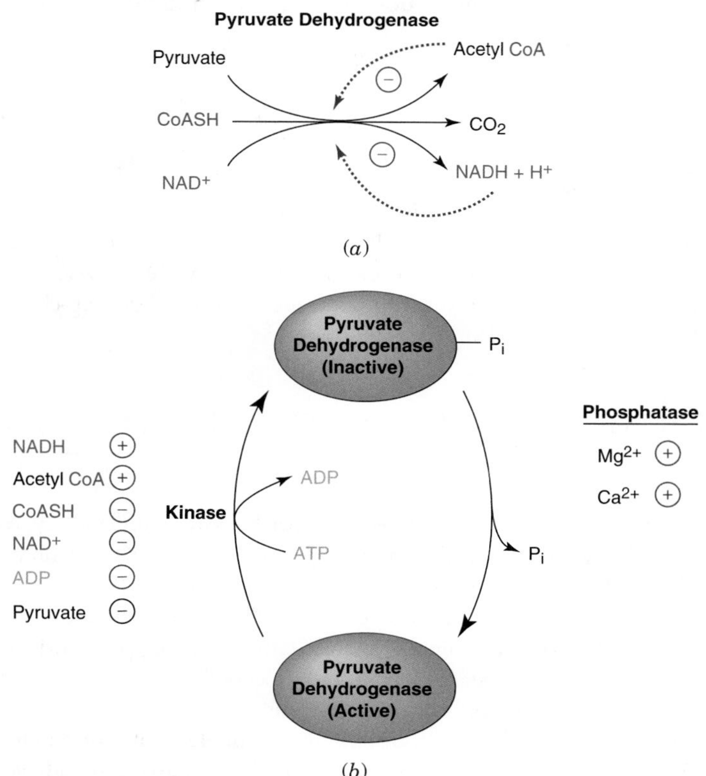

Figure 14.15 Regulation of the pyruvate dehydrogenase multienzyme complex. (*a*) Pyruvate dehydrogenase is inhibited by its products, acetylCoA, and NADH. (*b*) Pyruvate dehydrogenase is also inactivated by phosphorylation and activated by dephosphorylation. The phosphatase is stimulated by Mg^{2+} and Ca^{2+} ions. The kinase is stimulated by ATP, NADH, and acetylCoA and inhibited by CoASH, NAD^+, and ADP.

CLINICAL CORRELATION 14.1

Pyruvate Dehydrogenase Deficiency

A variety of disorders of pyruvate metabolism have been detected in children. Some involve deficiency of the catalytic or regulatory subunits of the pyruvate dehydrogenase complex. Children with pyruvate dehydrogenase deficiency usually exhibit elevated serum levels of lactate, pyruvate, and alanine, which produce a chronic lactic acidosis. They frequently exhibit severe neurological defects, which generally result in death. The diagnosis of pyruvate dehydrogenase deficiency is usually made by assaying the enzyme complex and/or its enzymatic subunits in cultures of skin fibroblasts taken from the patient. Some patients respond to dietary management in which a diet low in carbohydrates is administered. Patients may be in shock from lactic acidosis since decreased delivery of O_2 inhibits pyruvate dehydrogenase and increases anaerobic metabolism. Some patients have been treated with dichloroacetate, an inhibitor of the protein kinase subunit of the pyruvate dehydrogenase complex. Complete inhibition of the kinase, which inhibits the enzyme, will therefore activate the enzyme complex.

Patel, M. S., and Harris, R. A. Mammalian α-keto acid dehydrogenase complexes: Gene regulation and genetic defects. *FASEB J.* 9:1164, 1995.

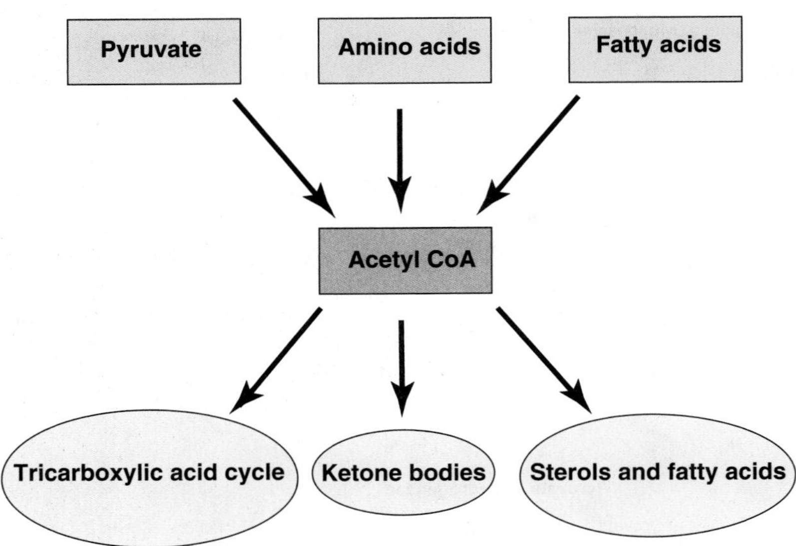

Figure 14.16 Sources and fates of acetylCoA.

Acetyl CoA Is Used in Several Different Pathways

Fates of acetyl CoA generated in the mitochondrial matrix include (1) complete oxidation of the acetyl group in the TCA cycle for energy generation; (2) conversion of excess acetyl CoA into the **ketone bodies**, acetoacetate and β-hydroxybutyrate in **liver**; and (3) transfer of acetyl units as citrate to the cytosol with subsequent synthesis of long-chain fatty acids (p. 680) and sterols (Figure 14.16).

14.4 • THE TRICARBOXYLIC ACID CYCLE

Acetyl CoA produced in the energy-generating catabolic pathways of most cells is completely oxidized to CO_2 in a cycle of reactions termed the **tricarboxylic acid (TCA) cycle**. This cycle is also called the **citric acid cycle** or the **Krebs cycle** after Sir Hans Krebs who postulated its essential features in 1937. The primary location of enzymes of the TCA cycle is in mitochondria, although isozymes of some enzymes are present in the cytosol. This location is appropriate as the pyruvate dehydrogenase complex and the fatty acid β-oxidation sequence, the two primary sources of acetyl CoA, are also located in mitochondria. Four reactions of the TCA cycle transfer electrons to either NAD^+ or FAD. The resulting NADH or $FADH_2$ is then oxidized by the mitochondrial **electron transport chain** (the **electron transfer chain** or **respiratory chain**) to generate energy that is used to form ATP by **oxidative phosphorylation**

(p. 572). The enzymes of the electron transport chain and those involved in ATP synthesis are exclusively localized in mitochondria. Figure 14.17 is an overview of the reactions of the TCA cycle. In the first step, the acetyl moiety of acetyl CoA is condensed with oxaloacetate (a 4-carbon dicarboxylic acid) to form citrate (a 6-carbon tricarboxylic acid). After rearrangement of the carbons of citrate, two oxidative decarboxylation reactions produce two molecules of CO_2, two molecules of NADH + H^+, one molecule of succinate (a 4-carbon dicarboxylic acid), and a high-energy bond in GTP. Two more oxidations produce another NADH + H^+ plus one $FADH_2$ and regenerate oxaloacetate.

In summary, the substrate of the TCA cycle is the two-carbon unit acetyl CoA and the products of a complete turn of the cycle are two CO_2, one high-energy phosphate bond (as GTP), three NADH, and one $FADH_2$. The NADH and $FADH_2$ are subsequently oxidized by the electron transport chain with the production of nine ATPs (see p. 573 for yields of ATP during oxidation of the coenzymes). Hence, 10 ATPs or its equivalent (GTP) are produced during oxidation of one acetate by the TCA cycle.

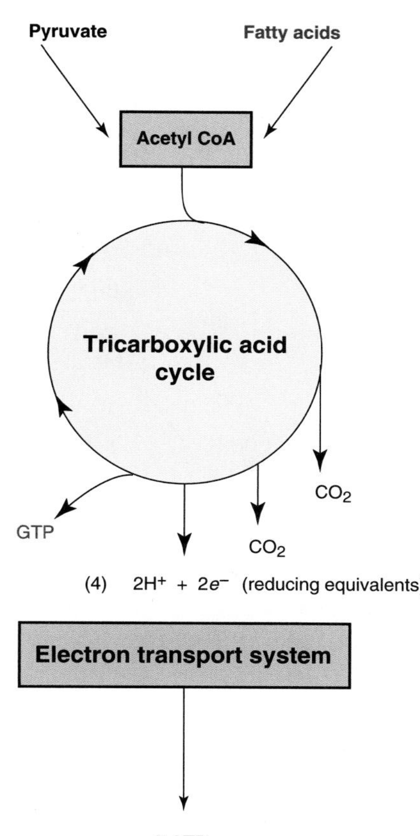

Figure 14.17 General description of oxidation of foodstuffs to provide energy for ATP synthesis within mitochondria. Acetyl CoA produced by oxidation of pyruvate and fatty acids is metabolized by the tricarboxylic acid cycle to reducing equivalents, which are oxidized by the electron transport system. Energy released during the oxidative process is used to drive synthesis of ATP.

Reactions of the Tricarboxylic Acid Cycle

The individual reactions of the TCA cycle are presented in Figure 14.18. The initial step of the cycle is catalyzed by **citrate synthase**, in the mitochondrial matrix. This highly exergonic reaction commits acetyl groups to **citrate** formation and complete oxidation in the cycle. As shown below citrate synthase condenses an acetyl moiety to the α-keto carbon of the dicarboxylic acid **oxaloacetate**. The citroyl-SCoA intermediate remains bound to the catalytic site on citrate synthase.

$$\underset{\text{Acetyl CoA}}{{}^{*}CH_3{}^{*}C(=O)-SCoA} + \underset{\text{Oxaloacetate}}{O=C(COO^-)-CH_2COO^-} \longrightarrow \underset{\text{Enzyme-bound citroyl-SCoA}}{\left[{}^{*}C(=O)SCoA-{}^{*}CH_2-C(OH)(COO^-)-CH_2-COO^-\right]} \underset{}{\overset{H_2O}{\rightleftharpoons}} \underset{\text{Citrate}}{{}^{*}COO^- -{}^{*}CH_2-C(OH)(COO^-)-CH_2-COO^-} + \text{CoASH}$$

CITRATE SYNTHASE

The equilibrium of this reaction is far toward citrate formation with a $\Delta G^{0'}$ near -38 kJ/mol. Note that the intramitochondrial concentration of oxaloacetate is very low (less than 1 μM); however, the strong, negative $\Delta G^{0'}$ drives the reaction forward. The low concentration of oxaloacetate, which is below the K_m of the reaction, may also be a major factor controlling this reaction.

Citrate is converted to **isocitrate** in a reversible reaction catalyzed by **aconitase** in which the hydroxyl group of citrate is exchanged for a H atom on an adjacent carbon. The hydroxyl group is thus located next to the carboxyl group of isocitrate where oxidative decarboxylation can occur. Conversion of citrate to isocitrate occurs on aconitase without release of the intermediate **cis-aconitate.** Aconitase contains a nonheme iron–sulfur cluster that is involved in the catalytic mechanism. The overall equilibrium of the reaction favors the formation of citrate.

Fluoroacetate is a potent inhibitor of the cycle, although it does not inhibit directly any of the enzymes of the cycle. Fluoroacetate is converted to fluorocitrate, which is a potent inhibitor of aconitase. Fluoroacetyl CoA is formed by acetyl CoA synthetase and converted to fluorocitrate by citrate synthase. Fluoroacetate is lethal in small doses and has been used as a rat poison; the LD_{50}, the lethal dose for 50% of animals consuming it, is 0.2 mg per kilogram of body weight.

Isocitrate dehydrogenase converts isocitrate to α-ketoglutarate in an oxidative decarboxylation reaction with concomitant reduction of NAD^+ to NADH + H^+. The isocitrate dehydrogenase of mammalian mitochondria requires NAD^+ as acceptor of reducing equivalents, has a mass of 380 kDa, and consists of eight identical subunits. The reaction requires a divalent metal cation (e.g., Mn^{2+} or Mg^{2+}) for removal of the β carboxylate of oxalosuccinate.

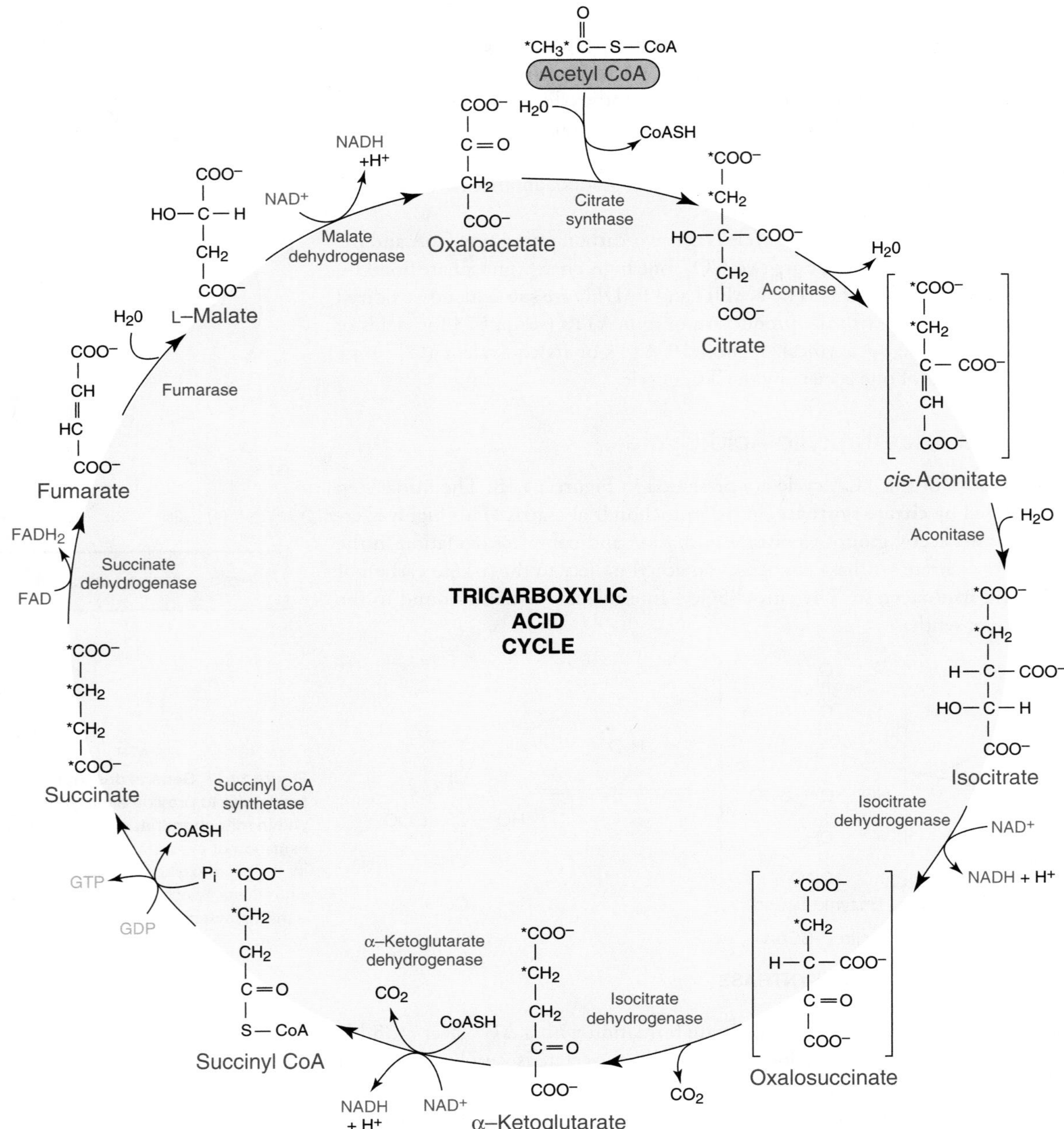

Figure 14.18 The tricarboxylic acid cycle. Asterisked carbons indicate fate of the carbons of the acetyl group.

The equilibrium of this reaction lies strongly toward α-ketoglutarate formation with a $\Delta G^{0\prime}$ of nearly -21 kJ/mol.

Mitochondria also contain an isocitrate dehydrogenase that requires $NADP^+$. The $NADP^+$-linked enzyme is also found in the cytosol, where it provides reducing equivalents for cytosolic reductive processes.

Conversion of α-ketoglutarate to succinyl CoA is catalyzed by the **α-ketoglutarate dehydrogenase** complex, which is nearly identical to the pyruvate dehydrogenase complex in terms of the individual reactions catalyzed and its structural features. Again**, thiamin pyrophosphate**, **lipoic acid**, CoASH, **FAD**, and NAD^+ participate in the catalytic mechanism. The complex consists of subunits of α-ketoglutarate dehydrogenase, **dihydrolipoyl transsuccinylase**, and dihydrolipoyl dehydrogenase. The equilibrium of the reaction lies strongly toward succinyl CoA formation with a $\Delta G^{0\prime}$ of -33 kJ/mol. In this reaction the second molecule of CO_2 and the second reducing equivalents (i.e., NADH + H^+) of the cycle are produced. The product of this reaction, **succinyl CoA**, is an energy-rich thiol ester similar to acetyl CoA.

The energy-rich character of the thiol ester linkage of succinyl CoA is conserved by **substrate-level phosphorylation** in the next step of the cycle. **Succinyl-CoA synthetase** (or **succinate thiokinase**) converts succinyl CoA to succinate and, in mammalian tissues, results in the phosphorylation of GDP to GTP. This reaction is freely reversible with a $\Delta G^{0'} = -3.0$ kJ/mol and the catalytic mechanism involves an enzyme-succinyl phosphate intermediate.

$$\text{Succinyl CoA} + P_i + \text{Enz} \rightarrow \text{Enz-succinyl phosphate} + \text{CoASH}$$

$$\text{Enz-succinyl phosphate} \rightarrow \text{Enz-phosphate} + \text{succinate}$$

$$\text{Enz-phosphate} + \text{GDP} \rightarrow \text{Enz} + \text{GTP}$$

The enzyme is phosphorylated on the 3 position of a histidine residue during the reaction; this conserves the energy of the thioester for formation of GTP. The GTP is used for the mitochondrial synthesis of protein, RNA and DNA.

Succinate is oxidized to fumarate by **succinate dehydrogenase**, a complex enzyme tightly bound to the inner mitochondrial membrane. Succinate dehydrogenase is composed of a 70-kDa subunit that contains the substrate binding site (FAD covalently bound to a histidine residue), a 30-kDa subunit that contains three **iron–sulfur centers** (**nonheme iron**), and two small hydrophobic proteins. The enzyme is a typical flavoprotein in which electrons and protons are transferred from the substrate through covalently bound FAD and the iron–sulfur centers in which the nonheme iron undergoes oxidation and reduction. The electrons are then transferred to coenzyme Q for further transport through the electron transfer chain, as will be discussed in Section 14.6. Succinate dehydrogenase is strongly inhibited by malonate and oxaloacetate and is activated by ATP, P_i, and succinate. Malonate inhibits succinate dehydrogenase competitively with respect to succinate because of the very close structural similarity between malonate and succinate (Figure 14.19).

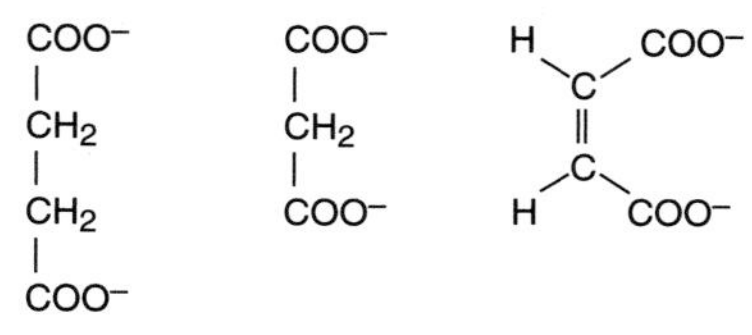

Figure 14.19 Structures of succinate, a TCA cycle intermediate; malonate, an inhibitor of succinate dehydrogenase, and the cycle; and maleate, a compound not involved in the cycle.

Fumarate is then hydrated to form L-malate by fumarase. **Fumarase** is a homo-tetramer (200 kDa) and is stereospecific for the trans form of substrate (the cis form, maleate, is not a substrate; (Figure 14.19). The reaction is freely reversible under physiological conditions. Clin. Corr. 14.2 describes a genetic deficiency of fumarase.

The final reaction in the cycle is catalyzed by malate dehydrogenase in which the reducing equivalents are transferred to NAD^+ to form NADH + H^+. The equilibrium of the reaction lies far toward L-malate formation with a $\Delta G^{0'} = +29$ kJ/mol. This endergonic reaction is pulled in the forward direction by the action of citrate synthase and other reactions, which remove oxaloacetate.

The NADH produced by the three NAD^+-linked dehydrogenases in the TCA cycle is oxidized rapidly to NAD^+ by the respiratory chain, thus favoring the forward direction of malate dehydrogenase.

Conversion of the Acetyl Group of Acetyl CoA to CO_2 and H_2O Conserves Energy

The TCA cycle (see Figure 14.18) is the terminal oxidative pathway for most metabolic fuels. Two-carbon moieties in the form of acetyl CoA are oxidized completely to CO_2 and H_2O, and four oxidative steps result in the formation of 3 NADH + H^+ and 1 $FADH_2$, which are used subsequently for ATP generation. Oxidation of each NADH + H^+ results in formation of 2.5 ATP by oxidative phosphorylation, whereas oxidation of $FADH_2$ formed in the succinate dehydrogenase reaction yields 1.5 ATPs. A high-energy bond is formed as GTP in the succinyl-CoA synthetase reaction. Hence the net yield of ATP or its equivalent (i.e., GTP) for the complete oxidation of an acetyl group in the TCA cycle is 10.

Tricarboxylic Acid Cycle Is a Source of Biosynthetic Intermediates

The previous discussion of the TCA cycle has concentrated on its role in the oxidative breakdown of acetyl groups to CO_2 and H_2O, formation of reduced coenzymes, and synthesis of ATP. Overall, the TCA cycle is the final common pathway for breakdown of

CLINICAL CORRELATION 14.2

Fumarase Deficiency (OMIM 606812)

Deficiency of enzymes of the TCA cycle is rare indicating the importance of this pathway for survival. Several cases, however, have been reported of severe deficiency of fumarase in mitochondria and cytosol of tissues (e.g., blood lymphocytes). The disease is characterized by severe neurological impairment, encephalomyopathy, and dystonia developing soon after birth. Urine contains abnormally high amounts of fumarate and elevated levels of succinate, α-ketoglutarate, citrate, and malate. The mitochondrial and cytosolic isozymes of fumarase are derived from a single gene. In affected patients, both parents had half-normal levels of enzyme activity but were clinically normal, as expected for an autosomal recessive disorder. The first mutation characterized in the gene for fumarase contains a glutamine substituted for a glutamate residue 319.

Bourgeron, T., Chretien, D., Poggi-Bach, J., et al. Mutation of the fumarase gene in two siblings with progressive encephalopathy and fumarase deficiency. *J. Clin. Invest.* 93:2514, 1994.

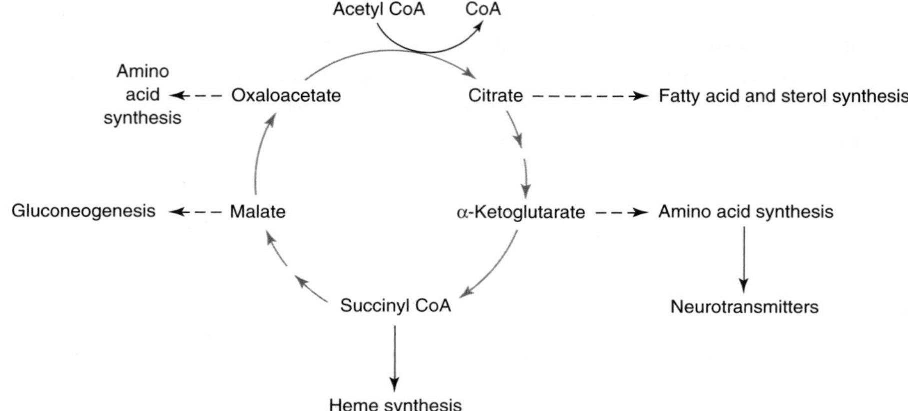

Figure 14.20 **The TCA cycle is a source of precursors for amino acid, fatty acid, and glucose synthesis.**

foodstuffs; however, as summarized in Figure 14.20, the four-, five- and six-carbon intermediates generated in the reactions of the TCA cycle are important intermediates in **biosynthetic processes**. Succinyl CoA, malate, oxaloacetate, α-ketoglutarate, and citrate are all precursors in the biosynthesis of important cellular compounds.

Transamination converts α-ketoglutarate to glutamate, which can leave mitochondria and be converted into several other amino acids. In nervous tissue, α-ketoglutarate is converted to the neurotransmitters, glutamate and γ-aminobutyric acid (GABA). Glutamate is also produced from α-ketoglutarate by the mitochondrial enzyme glutamate dehydrogenase in the presence of NADH or NADPH and ammonia. The amino group incorporated into glutamate can then be transferred to form various amino acids by different aminotransferases. These enzymes and the relevance of the incorporation or release of ammonia into or from α-keto acids are discussed in Chapter 19.

Succinyl CoA represents a metabolic branch point (Figure 14.21) as it may be formed either from α-ketoglutarate in the TCA cycle or from methylmalonyl CoA in the final steps in the breakdown of odd-chain-length fatty acids or the branched-chain amino acids valine

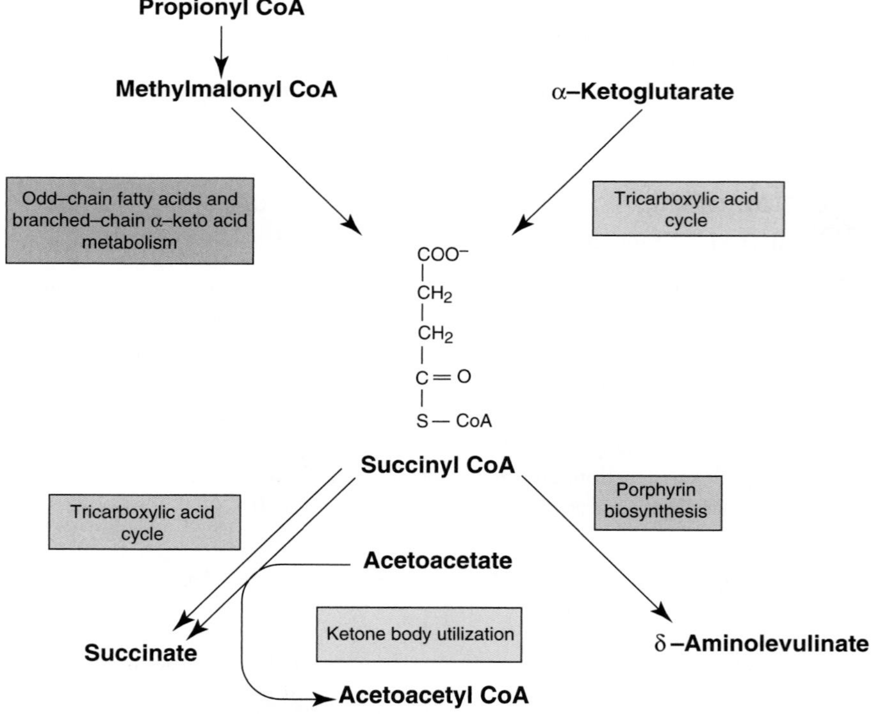

Figure 14.21 **Sources and fates of succinyl CoA.**

and isoleucine. Succinyl CoA may also be converted to succinate or condensed with glycine to form Δ-aminolevulinate, the initial reaction in porphyrin biosynthesis (p. 791).

Oxaloacetate is transaminated to aspartate the precursor of asparagine and of the pyrimidines cytosine, uracil, and thymine. Oxaloacetate is converted to **phosphoenolpyruvate (PEP)**, a key intermediate in **gluconeogenesis** (p. 619). Oxaloacetate cannot cross the inner mitochondrial membrane but is converted to malate, which is transported on a specific carrier out of mitochondria and oxidized to oxaloacetate, which is then converted to PEP (p. 580).

Citrate is exported from mitochondria into cytosol. **Citrate lyase** converts it to oxaloacetate and acetyl CoA, a precursor for synthesis of long-chain fatty acids and sterols. The oxaloacetate is rapidly reduced to malate, which is converted by **malic enzyme** to pyruvate and NADPH, a source of reducing equivalents for biosynthetic processes in cytosol. In addition, citrate is a regulatory effector of other metabolic pathways (p. 684).

Anaplerotic Reactions Replenish Intermediates of the Tricarboxylic Acid Cycle

In its role in catabolism, the TCA cycle oxidizes acetyl CoA with release of two CO_2 molecules. Indeed, oxaloacetate, the acceptor of the acetate group, is regenerated during the cycle. However, metabolic pathways in all tissues remove intermediates of the cycle for biosynthetic pathways. Hence, in order to maintain a functional cycle, a source of four-carbon acids is required to replenish the loss of oxaloacetate. The reactions that supply four- or five-carbon intermediates to the cycle are called **anaplerotic** (meaning "filling up") reactions (Figure 14.22). The most important is catalyzed by **pyruvate carboxylase**, which converts pyruvate and CO_2 to oxaloacetate (Figure 14.23). The enzyme contains a biotin molecule linked to the ε amino group of a lysine residue by an amide bond. The biotin binds CO_2 in the presence of ATP and Mg^{2+} ions and then transfers it to pyruvate as a carboxyl group (p. 1079). Levels of pyruvate carboxylase are high in both liver and nervous tissues, because these tissues experience a constant efflux of intermediates from the TCA cycle, which are used for gluconeogenesis in liver and neurotransmitter synthesis in nervous tissues.

Some amino acids are sources of four- or five-carbon intermediates. Glutamate is converted by glutamate dehydrogenase to α-ketoglutarate in mitochondria. Aspartate is converted to oxaloacetate by transamination, whereas valine and isoleucine are broken down to propionyl CoA that enters the TCA cycle as succinyl CoA. Amino acids derived from muscle breakdown become an important source of malate for gluconeogenesis during fasting (p. 845).

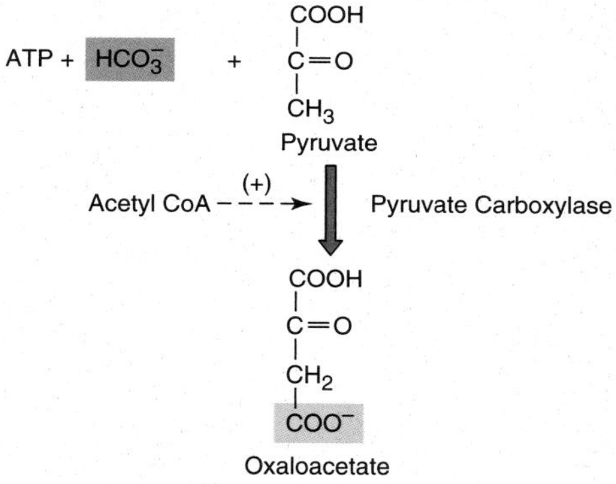

ATP + HCO_3^- + COOH–C=O–CH_3 (Pyruvate)

Acetyl CoA --(+)--> Pyruvate Carboxylase

COOH–C=O–CH_2–COO^- (Oxaloacetate)

Figure 14.23 Pyruvate carboxylase reaction. Acetyl CoA is an essential activator of pyruvate carboxylase.

Activity of Tricarboxylic Acid Cycle Is Carefully Regulated

Various factors regulate the TCA cycle. First, the supply of acetyl units, whether derived from pyruvate (by glycolysis) or fatty acids (by β-oxidation), is crucial in determining the

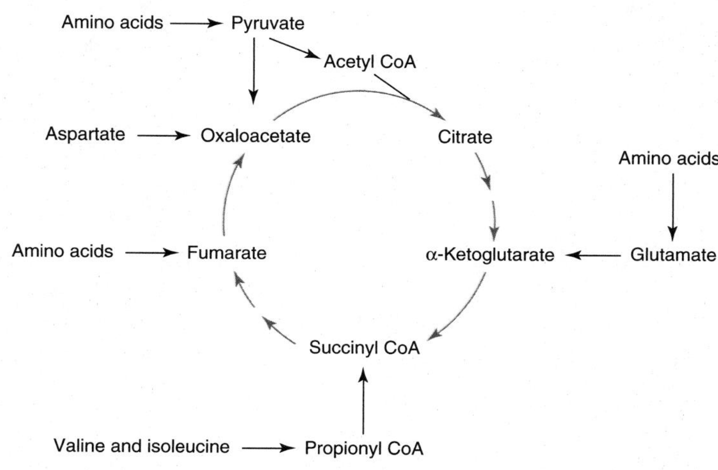

Figure 14.22 Anaplerotic reactions replenish intermediates of the TCA cycle.

rate of the cycle. Regulation of the pyruvate dehydrogenase complex, the transport of fatty acids into mitochondria, and β-oxidation of fatty acids are effective determinants of cycle activity. Second, because the dehydrogenases of the cycle are dependent on a continuous supply of NAD^+ and FAD, their activities are very stringently controlled by the respiratory chain that oxidizes NADH and $FADH_2$. As discussed in Section 14.7, the activity of the respiratory chain is coupled obligatorily to generation of ATP in reactions of oxidative phosphorylation, a regulatory process called **respiratory control**. Consequently, activity of the TCA cycle is very dependent on the rate of ATP synthesis (and hence the rate of electron transport), which is strongly affected by availability of ADP, phosphate, and O_2. Hence, an inhibitory agent or any metabolic condition that interrupts the supply of O_2, the continuous supply of ADP, or the source of reducing equivalents (e.g., substrate for the cycle) results in decreased activity of the TCA cycle. In general, these control mechanisms of the TCA cycle provide a coarse control of the cycle.

A variety of effector-mediated regulatory interactions between various intermediates or nucleotides and individual enzymes of the cycle have been postulated to exert a fine control of the cycle. Some of these effectors are shown in Figure 14.24. Note that the physiological relevance of many of these regulatory interactions has not been established.

Purified citrate synthase is inhibited by ATP, NADH, succinyl CoA, and long-chain acyl CoA derivatives; however, these effects have not been demonstrated under physiological conditions. The most probable means for regulating the citrate synthase reaction is the availability of the substrates, acetyl CoA and oxaloacetate. As discussed, very low concentrations of oxaloacetate (lower than the K_m for oxaloacetate on citrate synthase) are present in mitochondria.

The NAD^+-linked isocitrate dehydrogenase, often considered to be the key regulatory enzyme of the TCA cycle, is stimulated by Ca^{2+} ions, ADP, and AMP and is inhibited by ATP and NADH. Hence, under high-energy conditions (i.e., high ATP/ADP + P_i and high $NADH/NAD^+$ ratios), the activity of this dehydrogenase is inhibited. By contrast, during periods of low energy, the activity of this enzyme and consequently of the TCA is stimulated. Respiratory control by the electron transport chain coupled to ATP synthesis thus regulates the TCA cycle at the NAD^+-linked isocitrate dehydrogenase step by affecting levels of ADP and NAD^+.

The α-ketoglutarate dehydrogenase complex is inhibited by ATP and GTP, NADH, and succinyl CoA, whereas Ca^{2+} activates the complex in certain tissues. Unlike the pyruvate

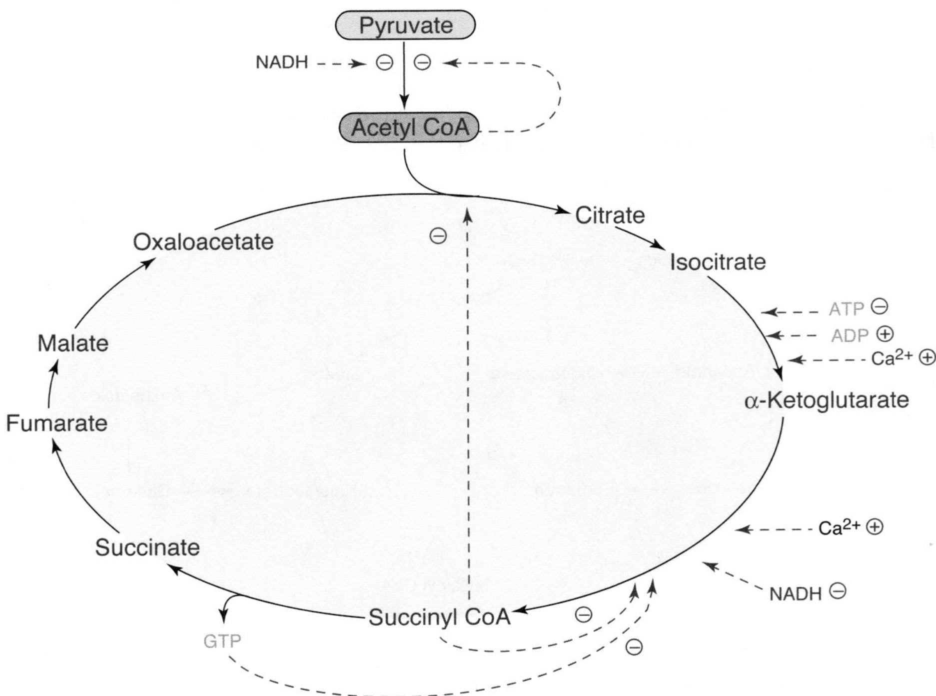

Figure 14.24 Examples of regulatory interactions in the TCA cycle.

dehydrogenase complex, the α-ketoglutarate dehydrogenase complex is not regulated by a protein-kinase-mediated phosphorylation.

The stimulation of both isocitrate and α-ketoglutarate dehydrogenases by Ca^{2+} occurs at concentrations that initiate muscle contraction and activate phosphorylase b during gycogenolysis. These effects of Ca^{2+} ensure that tension development and energy supply in muscle tissue are integrated following nerve stimulation.

14.5 • STRUCTURE AND COMPARTMENTATION BY MITOCHONDRIAL MEMBRANES

The final steps in breakdown of carbohydrates and fatty acids are located in mitochondria where energy released during oxidation of NADH and $FADH_2$ is transduced into chemical energy of ATP by the process of **oxidative phosphorylation**. Consequently, mitochondria are often called the powerhouse of the cell. The role of a tissue in aerobic metabolic functions and its need for energy is reflected in the number and activity of its mitochondria (Figure 14.25). Cardiac muscle is highly aerobic, needing a constant supply of ATP. Approximately one-half of the cytoplasmic volume of cardiac cells consists of mitochondria, which contain numerous invaginations of the inner membrane called **cristae** and, consequently, a high concentration of the enzyme complexes of the electron transport chain. The liver is also highly aerobic with each mammalian hepatocyte containing 800–2000 mitochondria. By contrast, erythrocytes contain no mitochondria and obtain energy only from glycolysis.

Mitochondria have different shapes, depending on the cell type. In Figure 14.25, mitochondria from liver are nearly spherical, whereas those found in cardiac muscle are oblong or cylindrical and contain more numerous cristae than liver mitochondria.

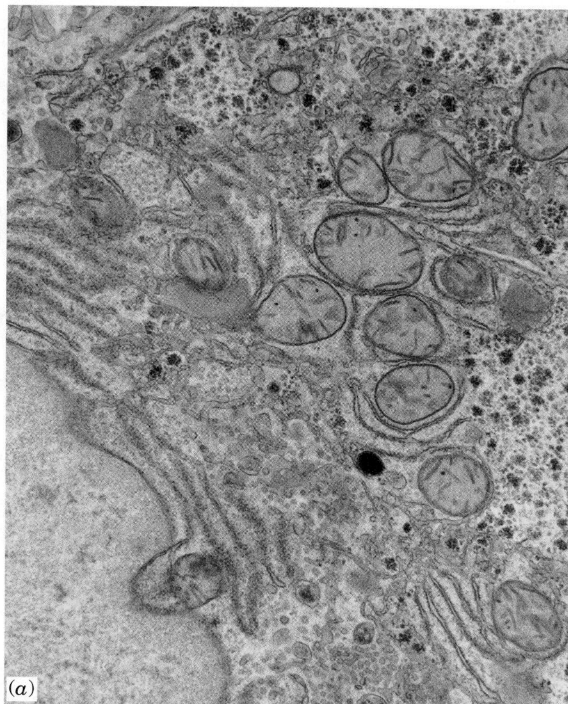

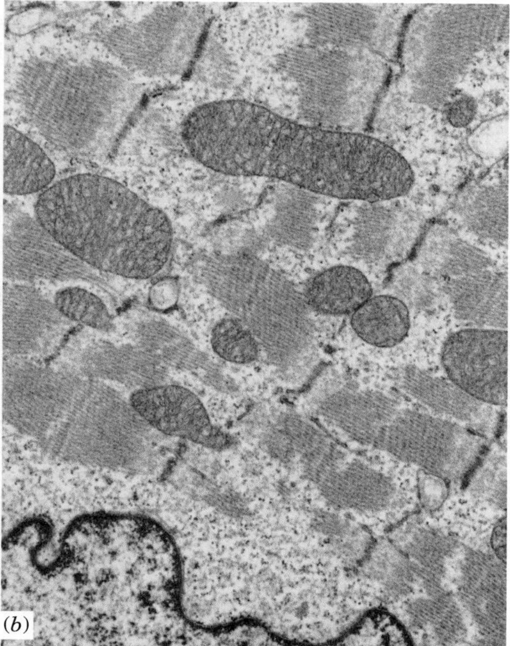

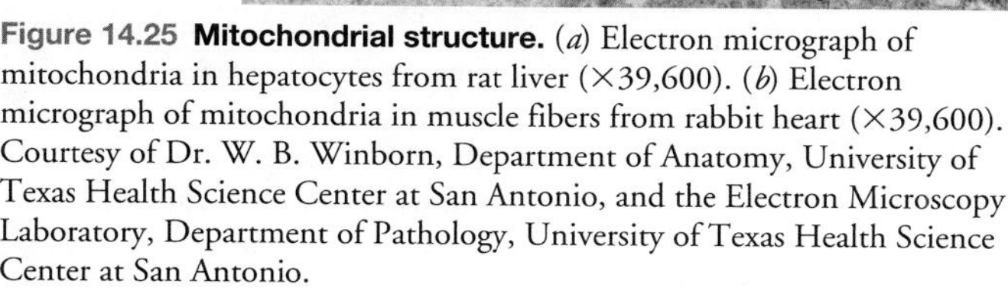
Figure 14.25 Mitochondrial structure. (*a*) Electron micrograph of mitochondria in hepatocytes from rat liver (×39,600). (*b*) Electron micrograph of mitochondria in muscle fibers from rabbit heart (×39,600). Courtesy of Dr. W. B. Winborn, Department of Anatomy, University of Texas Health Science Center at San Antonio, and the Electron Microscopy Laboratory, Department of Pathology, University of Texas Health Science Center at San Antonio.

Courtesy of Dr. W. B. Winborn, Department of Anatomy, University of Texas Health Science Center at San Antonio, and the Electron Microscopy Laboratory, Department of Pathology, University of Texas Health Science Center at San Antonio.

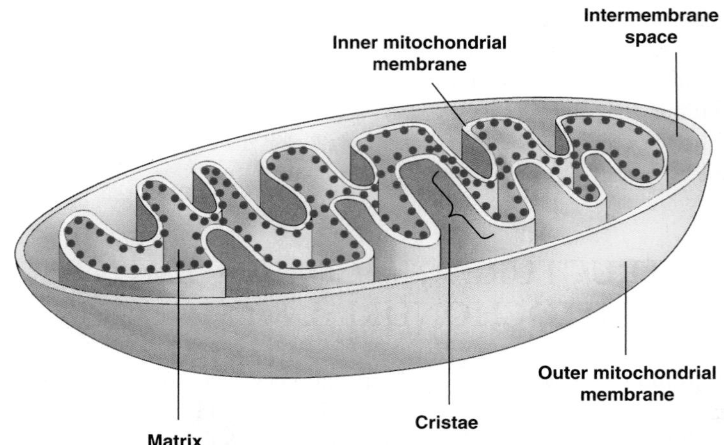

Figure 14.26 Diagram of submitochondrial compartments. Green spheres represent localization of the F_1 portion of ATP synthase on the inner mitochondrial membrane.

Inner and Outer Mitochondrial Membranes Have Different Compositions and Functions

Mitochondria contain an **outer membrane** and a structurally and functionally more complex **inner membrane** (Figure 14.26); the space between the membranes is the **intermembrane space**. Enzymes involved in transfer of energy from the γ-phosphoryl bond of ATP, such as adenylate kinase, creatine kinase, and nucleoside diphosphate kinase, are located in the intermembrane space (Table 14.4). The outer membrane consists of about 30%−40% lipid and 60%−70% protein, with relatively few enzymatic or transport proteins. It is rich in the integral membrane protein called **porin** (or **VDAC**, voltage-dependent anion channel) consisting of β-sheets, which form a channel that permits the passage through the membrane of particles of up to 10 kDa. Monoamine oxidase and kynurenine hydroxylase, of importance in nervous tissues for removal of neurotransmitters, are located on the outer surface of the outer membrane.

The inner membrane consists of 80% protein and is rich in unsaturated fatty acids. In addition, **cardiolipin** (diphosphatidylglycerol) is present in high concentrations. The enzyme complexes of electron transport and oxidative phosphorylation are located in this membrane as are various dehydrogenases and several transport systems involved in transferring substrates, metabolic intermediates, and adenine nucleotides between cytosol and matrix. The inner membrane appears to be invaginated into folds or **cristae**, which increase the surface area (Figure 14.26).

TABLE 14.4 • Enzymes of Mitochondrial Subcompartments

Outer Membrane	*Intermembrane Space*	*Inner Membrane*	*Matrix*
Monoamine oxidase Kynurenine hydroxylase Nucleoside diphosphate kinase Phospholipase A	Adenylate kinase Nucleoside diphosphate kinase Creatine kinase	Succinate dehydrogenase F_1F_0 ATP synthase NADH dehydrogenase β-Hydroxybutyrate dehydrogenase	Pyruvate dehydrogenase complex Citrate synthase Isocitrate dehydrogenase α-Ketoglutarate dehydrogenase complex
Fatty acyl CoA synthetases NADH: cytochrome *c* reductase (rotenone-insensitive)		Cytochromes b, c_1, c, a, a_3 Carnitine: acyl CoA transferase	Aconitase Fumarase Succinyl CoA synthetase
Choline phosphotransferase		Adenine nucleotide translocase Mono-, di-, and tricarboxylate transporters Glutamate–aspartate transporters Glycerol 3-phosphate dehydrogenase	Malate dehydrogenase Fatty acid β-oxidation system Glutamate dehydrogenase Glutamate–oxaloacetate transaminase Ornithine transcarbamoylase Carbamoyl phosphate synthetase I Heme synthesis enzymes

The space inside the inner membrane, the **matrix**, contains the enzymes of the TCA cycle with the exception of succinate dehydrogenase which is bound to the inner membrane, enzymes for fatty acid oxidation, and some enzymes of porphyrin (p. 791) and urea synthesis (p. 759). In addition, **mitochondrial DNA (mtDNA)**, ribosomes, and proteins necessary for transcription of mtDNA and translation of mRNA are located in the matrix.

14.6 • ELECTRON TRANSPORT CHAIN

During the reactions of fatty acid oxidation and the TCA cycle, reducing equivalents derived from oxidation of substrates are transferred to NAD^+ and FAD (forming NADH and $FADH_2$) which are then oxidized by the **electron transport chain**, a system of electron carriers located in the inner membrane (Figure 14.27). In the presence of O_2, electron transfer chain converts reducing equivalents into utilizable energy, as ATP, by oxidative phosphorylation. The complete oxidation of NADH and $FADH_2$ by the electron transport chain produces approximately 2.5 and 1.5 mol of ATP per mole of reducing equivalent transferred to O_2, respectively.

Oxidation–Reduction Reactions

The mitochondrial electron transport consists of a sequence of linked oxidation–reduction reactions. Such reactions transfer electrons from a suitable electron donor (**reductant**) to a suitable electron acceptor (**oxidant**). In some oxidation–reduction reactions only electrons are transferred from reductant to oxidant (e.g., electron transfer between cytochromes).

$$\text{Cytochrome } c\,(Fe^{2+}) + \text{cytochrome } a\,(Fe^{3+}) \longrightarrow \text{cytochrome } c\,(Fe^{3+}) + \text{cytochrome } a\,(Fe^{2+})$$

whereas in others, both electrons and protons (hydrogen atoms) are transferred (e.g., electron transfer between NADH and FAD).

$$NADH + H^+ + FAD \longrightarrow NAD^+ + FADH_2$$

An oxidant and its reductant form a **redox couple**, or pair. The ease with which an electron donor (reductant) gives up its electrons to an electron acceptor (oxidant) is expressed as the **oxidation–reduction potential** of the system. This is measured in volts as an **electromotive force** (emf) of a half-cell made up of an oxidation-reduction couple when compared to a standard reference half-cell (usually the hydrogen electrode reaction). The potential of the standard hydrogen electrode is set by convention at 0.0 V at pH 0.0; however, in biological systems where pH is 7.0, the reference hydrogen potential is −0.42 V. The potentials for a variety of important biochemical reactions are tabulated in Table 14.5. To interpret the

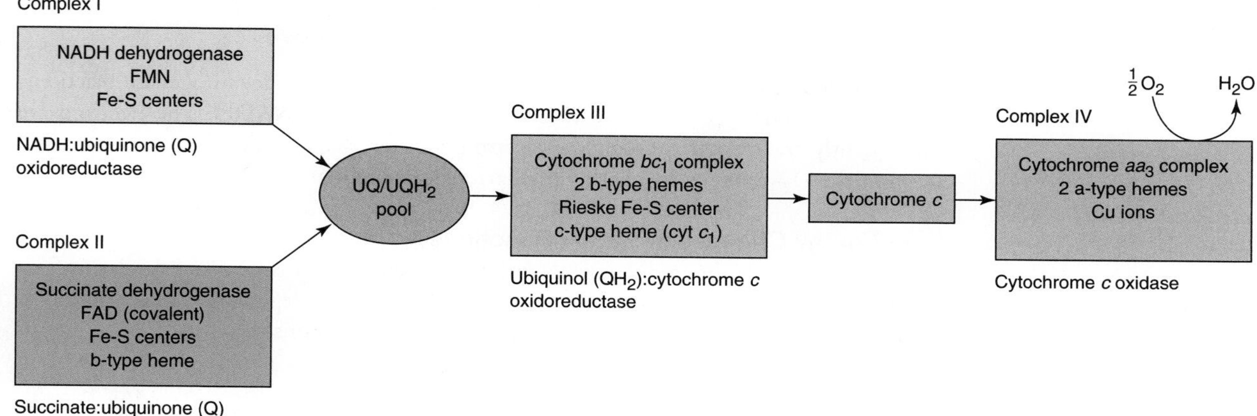

Figure 14.27 Overview of the complexes and pathways of electron transfer in mitochondrial electron transport chain.

TABLE 14.5 • Standard Oxidation–Reduction Potentials for Various Biochemical Reactions

Oxidation–Reduction System	*Standard Oxidation–Reduction Potential E_0' (V)*
Acetate + $2H^+$ + $2e^-$ $\rightleftharpoons$ acetaldehyde	−0.60
$2H^+ + 2e^- \rightleftharpoons H_2$	−0.42
Acetoacetate + $2H^+$ + $2e^-$ $\rightleftharpoons$ β-hydroxybutyrate	−0.35
NAD^+ + $2H^+$ + $2e^-$ $\rightleftharpoons$ NADH + H^+	−0.32
Acetaldehyde + $2H^+$ + $2e^-$ $\rightleftharpoons$ ethanol	−0.20
Pyruvate + $2H^+$ + $2e^-$ $\rightleftharpoons$ lactate	−0.19
Oxaloacetate + $2H^+$ + $2e^-$ $\rightleftharpoons$ malate	−0.17
Coenzyme Q_{ox} + $2e^-$ $\rightleftharpoons$ coenzyme Q_{red}	+0.10
Cytochrome *b* (Fe^{3+}) + e^- $\rightleftharpoons$ cytochrome*b* (Fe^{2+})	+0.12
Cytochrome *c* (Fe^{3+}) + e^- $\rightleftharpoons$ cytochrome*c* (Fe^{2+})	+0.22
Cytochrome *a* (Fe^{3+}) + e^- $\rightleftharpoons$ cytochrome*a* (Fe^{2+})	+0.29
½O_2 + $2H^+$ + $2e^-$ $\rightleftharpoons$ H_2O	+0.82

data in the table, recall that the reductant of a redox pair with a large negative potential will give up its electrons more readily than redox pairs with smaller negative or positive potentials. Compounds with large negative potentials are strong reducing agents. By contrast, a strong oxidant (e.g., one characterized by a large positive potential) has a very high affinity for electrons and acts to oxidize compounds with more negative standard potentials.

The **Nernst equation** characterizes the relationship between standard oxidation–reduction potential of a redox pair (E_0'), observed potential (E), and ratio of concentrations of oxidant and reductant in the system.

$$E = E_0' + 2.3(RT/nf)\,\log([\text{oxidant}]/[\text{reductant}])$$

where E equals the observed potential and E_0' is the standard potential when all of reactants are present under standard conditions. R is gas constant of 8.3 J/mol × °K, T is absolute temperature in kelvin units (K), n is number of electrons being transferred, and f is the Faraday constant of 96,500 J/V.

From standard oxidation–reduction potentials of a diverse variety of biochemical reactions, one can predict direction of electron flow or transfer when more than two redox pairs are linked together by the appropriate enzyme. For example, Table 14.5 shows that the NAD^+−NADH pair has a standard potential of −0.32 V, and the pyruvate–lactate pair has a standard potential of −0.19V. This means that electrons will flow from NAD^+− NADH to pyruvate–lactate as long as lactate dehydrogenase is present as indicated here.

$$\text{Pyruvate} + \text{NADH} + H^+ \rightarrow \text{lactate} + NAD^+$$

Reducing equivalents are produced in NAD^+-and FAD-linked dehydrogenase reactions, which have standard potentials at or close to that of NAD^+−NADH. The electrons are subsequently transferred through the electron transfer chain, which has as its terminal acceptor the O_2–water couple with a standard redox potential of +0.82 V.

Free-Energy Changes in Redox Reactions

Differences in oxidation–reduction potentials between two redox pairs are similar to free-energy changes in chemical reactions, in that both quantities depend on concentration of reactants and products of the reaction and the following relationship exists:

$$\Delta G^{0'} = -nf\,\Delta E_0'$$

Using this expression, free-energy change for electron transfer reactions can be calculated if potential difference between two oxidation-reduction pairs is known. Hence, for the mitochondrial electron transfer chain in which electrons are transferred between the

NAD^+-NADH couple ($E_0' = -0.32$ V) and the $1/2O_2-H_2O$ couple ($E_0' = +0.82$ V), the free-energy change for this process can be calculated.

$$\Delta G^{0'} = -nf\,\Delta E_0' = -2 \times 96.5 \text{ kJ/V} \times 1.14 \text{ V}$$

$$\Delta G^{0'} = -219 \text{ kJ/mol}$$

where 96.5 is the Faraday constant in kJ/V and n is the number of electrons transferred; for example, in the case of NADH $\rightarrow O_2$, $n = 2$. The free energy available from the potential span between NADH and O_2 in the electron transfer chain is capable of generating more than enough energy to synthesize three molecules of ATP per two reducing equivalents or two electrons transported to O_2. In addition, because of the negative sign of the free energy available in the electron transfer, this process is exergonic and proceeds, if the necessary enzymes are present.

Mitochondrial Electron Transport Is a Multi-component System

The final steps in the overall oxidation of foodstuffs—carbohydrates, fats and amino acids-result in formation of NADH and $FADH_2$ in the matrix. The electron transport chain oxidizes these reduced cofactors by transferring electrons in a series of steps to O_2, the terminal electron acceptor, while capturing the free energy of the reactions to drive the synthesis of ATP (Figure 14.27, p. 561). During removal of electrons from the coenzymes, protons are pumped from the matrix to the intermembrane space to form an electrochemical gradient across the inner membrane, which provides energy for synthesis of ATP. The carriers that transfer electrons from NADH to O_2 have standard redox potentials that span the range from -0.32 V that of the most electronegative electron donor NADH to $+0.82$ V the most electropositive electron acceptor O_2 (Figure 14.28). The mitochondrial electron carriers, however, are not organized in a linear arrangement but are grouped into four large complexes (complexes I–IV) that catalyze different partial reactions of the electron transport chain (see Figure 14.27).

Complex I, NADH-ubiquinone oxidoreductase, catalyzes the transfer of electrons from NADH to **ubiquinone (UQ)** or **coenzyme Q (CoQ)**; **complex II**, succinate-ubiquinone oxidoreductase or succinate dehydrogenase, transfers electrons from succinate to coenzyme Q; **complex III**, or the **cytochrome** bc_1 **complex**, ubiquinol-cytochrome *c* reductase, transfers electrons from ubiquinol (reduced form of ubiquinone denoted as **$CoQH_2$** or **UQH_2**) to **cytochrome *c***; and **complex IV**, cytochrome *c* oxidase, transfers electrons from cytochrome *c* to O_2 (see Figure 14.27). Another complex, the **ATP synthase**, or **complex V**, uses energy of the electrochemical gradient for synthesis of ATP. Complexes I–IV consist of electron carriers that include **flavoproteins**, which contain tightly bound FMN or FAD and can transfer one or two electrons, the heme-containing proteins **cytochromes**, (cytochromes b, c_1, c, a, and a_3) that transfer one electron from Fe^{2+} of heme, **iron–sulfur proteins** that

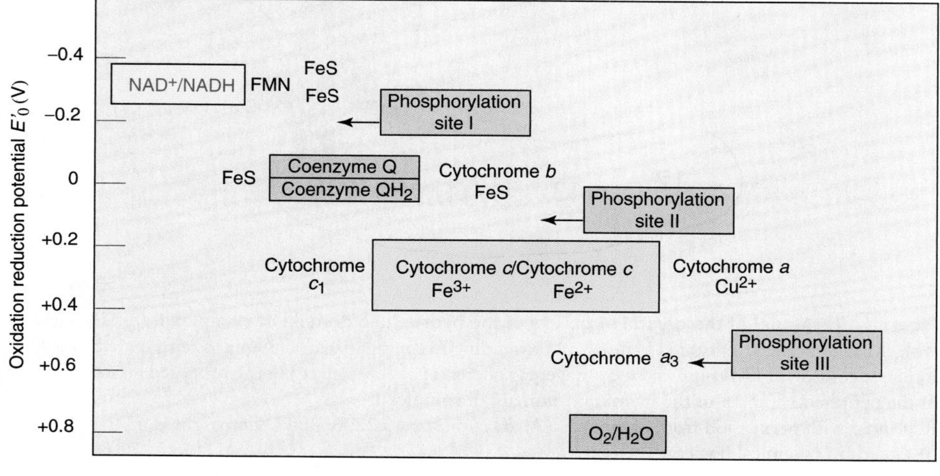

Figure 14.28 Oxidation–reduction potentials of the mitochondrial electron transport chain carriers listed from the most negative ($NAD^+/NADH$) to the most positive (O_2/H_2O).

contain bound inorganic Fe and S and transfer one electron, and **copper** in complex IV (cytochrome *c* oxidase) that transfers one electron. UQ participates in one or two electron transfer reactions.

Complex I: NADH–Ubiquinone Oxidoreductase

Complex I, the most complicated found in mammalian mitochondria, contains a minimum of 40 different polypeptides with a total mass of approximately 1 MDa. A simpler complex I containing 14 subunits, which catalyzes similar electron transfer and proton pumping reactions, is present in bacterial membranes where structure and enzymatic activity have been extensively investigated. Complex I transfers electrons from NADH to ubiquinone (coenzyme Q) coupled with the transport of four protons across the membrane, thus contributing to the proton-motive force required for the synthesis of ATP. Both the mammalian and bacterial forms of complex I have a characteristic L-shaped structure with a long hydrophobic arm localized in the membrane and a peripheral hydrophilic arm protruding into the mitochondrial matrix (Figure 14.29). Electrons are transferred from NADH to

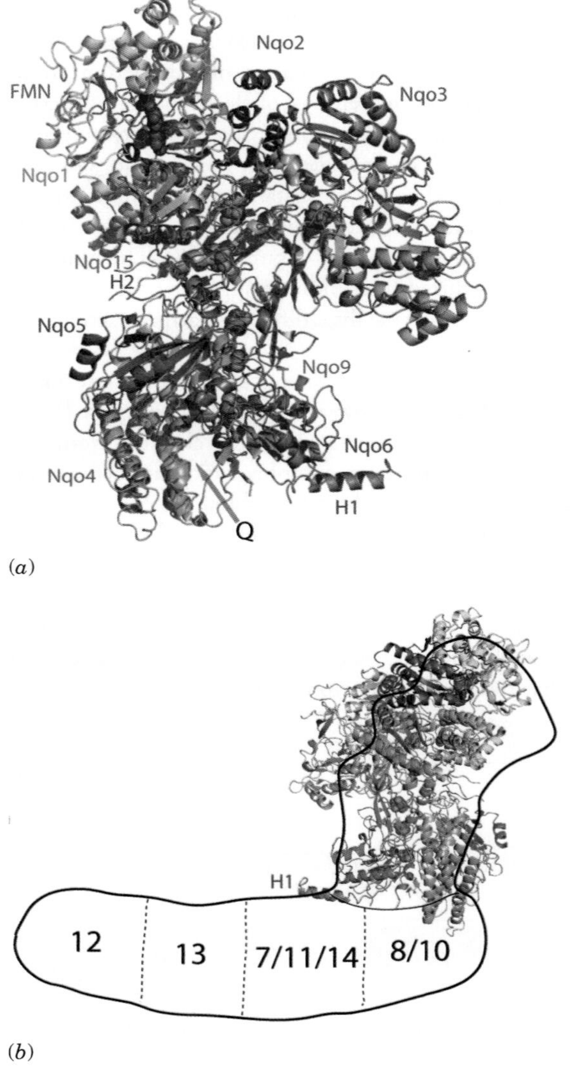

Figure 14.29 Model of the crystal structure of the hydrophilic domain of complex I. (*a*) Side view with the membrane arm located beneath and extending to the right. Each subunit is represented by a different color with FMN indicated by magenta spheres. (*b*) View indicating the proposed attachment of the peripheral domain to the membrane domain of complex I.
Reprinted with permission from Sazanov, L. A. *Biochemistry* 46:2276, 2007. Copyright (1977) by American Chemical Society.

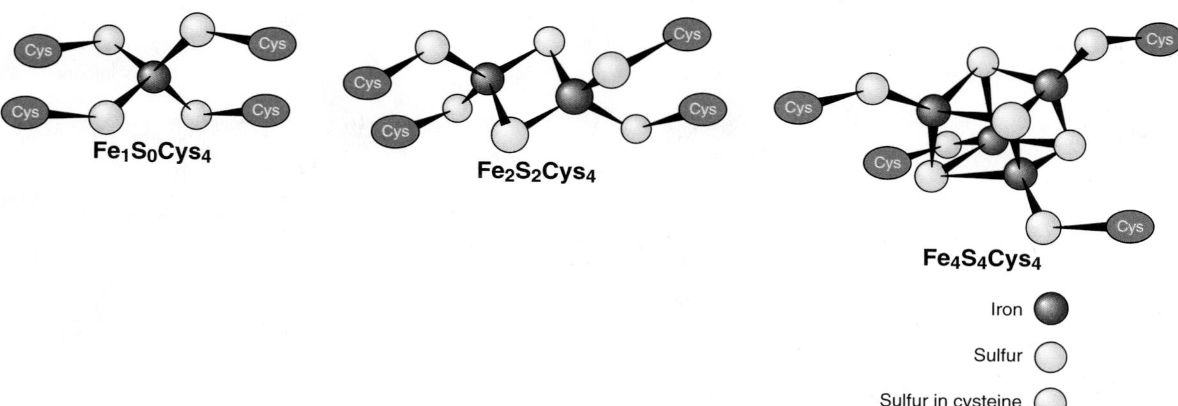

Figure 14.30 Structures of iron–sulfur centers. Yellow, inorganic sulfur; gray, sulfur in cysteine; and red, iron.

FMN, flavin mononucleotide (see Figure 10.32, p. 566), which is tightly bound to a subunit in the hydrophilic arm of complex I.

$$\text{NADH} + \text{H}^+ + \text{FMN} \rightarrow \text{NAD}^+ + \text{FMNH}_2$$

The electrons are then transferred one at a time via a series of FeS centers, of both the 2Fe2S and 4Fe4S types (Figure 14.30), located in different subunits of the hydrophilic arm of complex I. These iron–sulfur clusters subsequently reduce a membrane-embedded ubiquinone to form ubiquinol (Figure 14.31). During transfer of two electrons to ubiquinone by complex I, four protons are also translocated across the inner membrane to the intermembrane space by a mechanism suggested to involve proteins located in the hydrophobic membrane arm of complex I.

Mutations in subunits of complex I have been shown to lead to a number of neurodegenerative diseases. In addition, complex I has been shown to be a major source of reactive oxygen species (ROS) that may damage mitochondrial DNA and may be a cause of aging (see Sections 14.9 and 14.10).

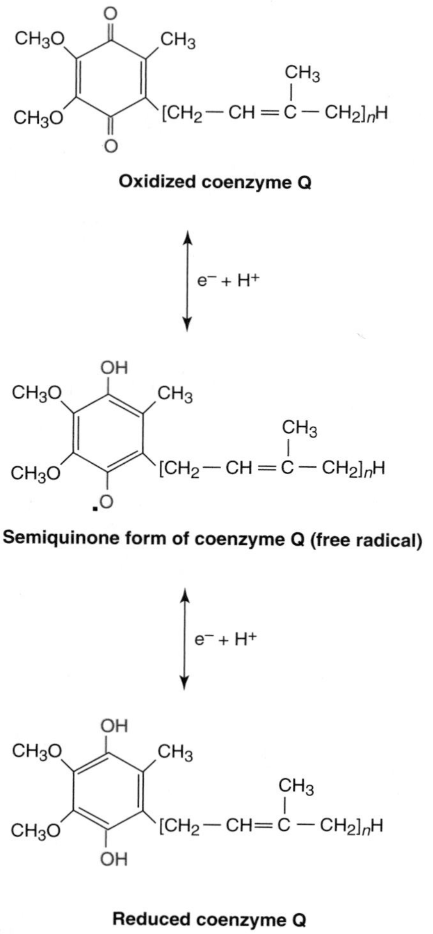

Complex II: Succinate–Ubiquinone Oxidoreductase

Complex II, better known as succinate dehydrogenase, consists of a 70-kDa subunit that contains FAD covalently bound to a histidine residue, a 30-kDa subunit that contains three iron–sulfur centers, and two small hydrophobic proteins. In the oxidation of succinate to fumarate, two electrons and two protons are transferred to FAD (Figure 14.32). The $FADH_2$ transfers electrons to ubiquinone via FeS centers of complex II in the reactions.

$$\text{Succinate} \rightarrow \text{fumarate} + 2\text{H}^+ + 2\text{e}^-$$

$$\text{UQ} + 2\text{H}^+ + 2\text{e} \rightarrow \text{UQH}_2$$

$$\text{Overall} \quad \text{Succinate} + \text{UQ} \rightarrow \text{fumarate} + \text{UQH2}$$

$$\Delta E^{0\prime} = 0.029 \text{ V} \qquad \Delta G^{0\prime} = -5.6 \text{ kJ/mol}$$

The amount of free energy liberated in these reactions is insufficient for proton pumping across the membrane so no gain in free energy is accomplished in complex II. Figure 14.32 provides a schematic representation of these events.

Other Mitochondrial Flavoprotein Dehydrogenases

Other mitochondrial dehydrogenases feed electrons into the electron transport chain at the level of ubiquinone. Glycerol 3-phosphate, formed from glycerol released by hydrolysis of triacylglycerols or by reduction of dihydroxyacetone phosphate produced during glycolysis, is oxidized by **glycerol 3-phosphate dehydrogenase** (Figure 14.32).

$$\text{Glycerol 3-phosphate} + \text{FAD} \rightarrow \text{dihydroxyacetone phosphate} + \text{FADH}_2$$

Oxidized coenzyme Q

$e^- + H^+$

Semiquinone form of coenzyme Q (free radical)

$e^- + H^+$

Reduced coenzyme Q

Figure 14.31 Oxidation–reduction of ubiquinone (coenzyme Q). Note that ubiquinone can accept one electron at a time forming an intermediate semiquinone.

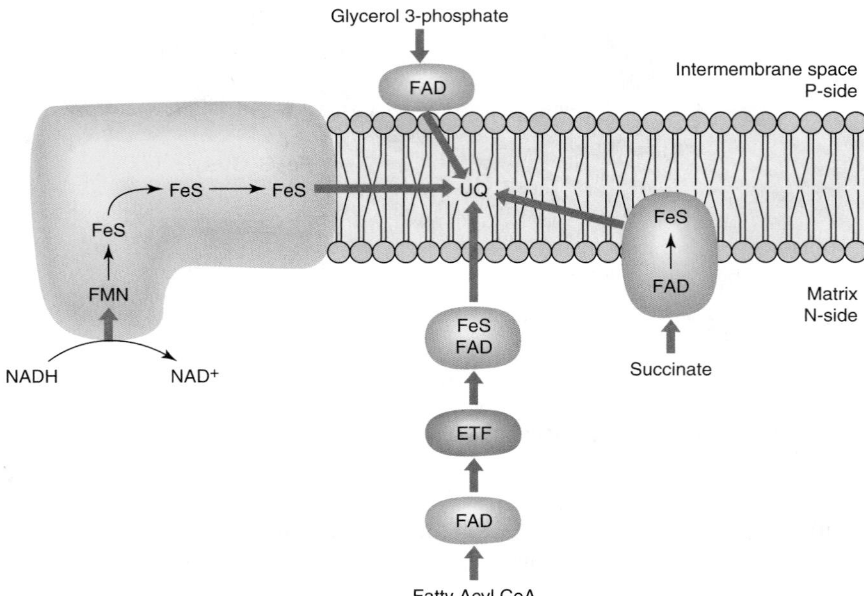

Figure 14.32 Reduction of ubiquinone (UQ) in the mitochondrial inner membrane by the flavoproteins NADH, succinate, glycerol 3-phosphate, and fatty acyl CoA dehydrogenases.

This flavoprotein, a single polypeptide chain, is localized on the outer face of the inner membrane and transfers electrons directly to ubiquinone in the membrane. The importance of glycerol 3-phosphate dehydrogenase in shuttling of reducing equivalents from NADH in the cytosol to the mitochondrial electron transport chain will be discussed in Section14.8 (p. 477).

Acyl-CoA dehydrogenase, a flavoprotein that catalyzes the first step in β-oxidation of fatty acids, transfers electrons from fatty acyl CoA to FAD to form $FADH_2$, which then transfers electrons to **e**lectron **t**ransferring **f**lavoprotein (**ETF**). Electrons are then transferred from ETF to **ETF–ubiquinone oxidoreductase** which transfers electrons directly to ubiquinone in the inner membrane. Figure 14.32 illustrates the reduction of the ubiquinone pool by complex I, complex II, glycerol 3-phosphate dehydrogenase, and ETF–ubiquinone oxidoreductase. Ubiquinol is subsequently oxidized by complex III.

Complex III: Ubiquinol–Cytochrome *c* Oxidoreductase

Complex III, or cytochrome bc_1 complex, catalyzes transfer of two electrons from ubiquinol to cytochrome *c* with translocation of four protons across the membrane. In mammals this enzyme complex consists of 11 subunits of which 3 have prosthetic groups that serve as redox centers. These are **cytochrome *b***, which contains two hemes, b562 and b566; **cytochrome c_1**, which contains one heme group; and the **Rieske iron–sulfur protein**, which contains a 2Fe2S cluster. Resolution of the complete structure of complex III by X-ray crystallography was achieved recently (Figure 14.33). The dimeric complex (250 kDa for each monomer) is pear shaped and has a large domain protruding 75 Å into the mitochondrial matrix and a smaller domain containing the head groups of the Rieske iron–sulfur protein and cytochrome c_1 protruding into the intermembrane space. The transmembrane domain of each monomer of complex III consists of eight α-helices of the hydrophobic protein, cytochrome *b*, plus membrane-anchoring helices of the Rieske iron–sulfur protein, cytochrome c_1, and other subunits of each monomer. Oxidation of ubiquinol occurs at the Q_O site located at the P side of the mitochondrial membrane facing the intermembrane space with the transfer of one electron to the iron–sulfur protein and the second electron to b_L and b_H with the release of two protons to the intermembrane space. The proposed mechanism for the transfer of electrons and protons in complex III, termed the Q cycle, is described in Figure 14.34 and A Closer Look 14.1, p. 568.

Cytochromes

The cytochromes are proteins that contain a heme group tightly bound to the protein (p. 791). Unlike the heme in hemoglobin or myoglobin in which heme iron remains in the Fe^{2+} state during oxygen transport, iron in heme of a cytochrome is alternately oxidized (Fe^{3+}) or reduced (Fe^{2+}) as it functions in the transport of electrons. The cytochromes of mammalian

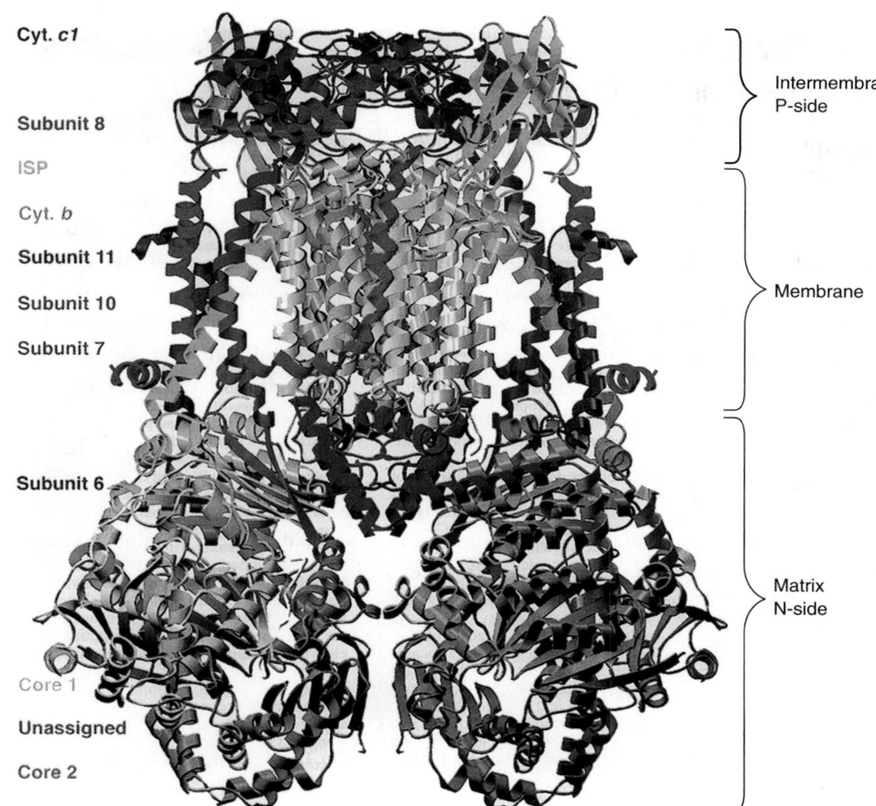

Figure 14.33 Model of the crystal structure of the dimeric cytochrome bc_1 complex. The α-helices of cytochrome *b* (pale green) form the transmembrane domain of the complex. The complex protrudes 75 Å into the matrix and 38 Å into the intermembrane space. Colors identifying the subunits are shown on the left. ISP is iron–sulfur protein.
Reproduced with permission from Kim H., Xia, D., Yu, C.-A., Xia, J.-Z., Kachurin, A. M., Zhang, L., Yu. L., and Deisenhofer, J. *Proc. Natl. Acad. Sci., USA* 95:8026, 1998. Copyright 1998, National Academy of Sciences, USA. Figure generously supplied by Dr. J. Deisenhofer.

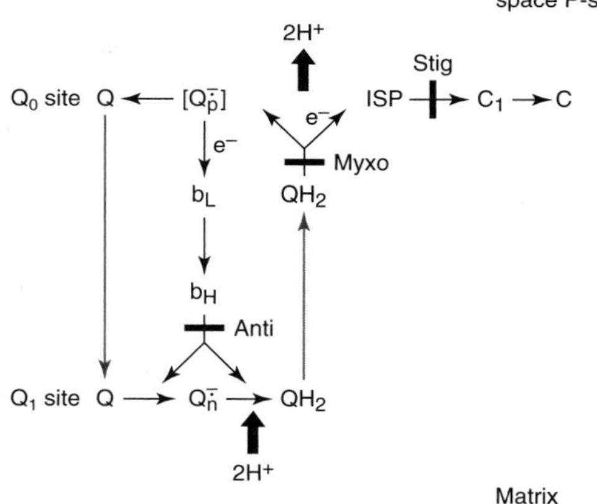

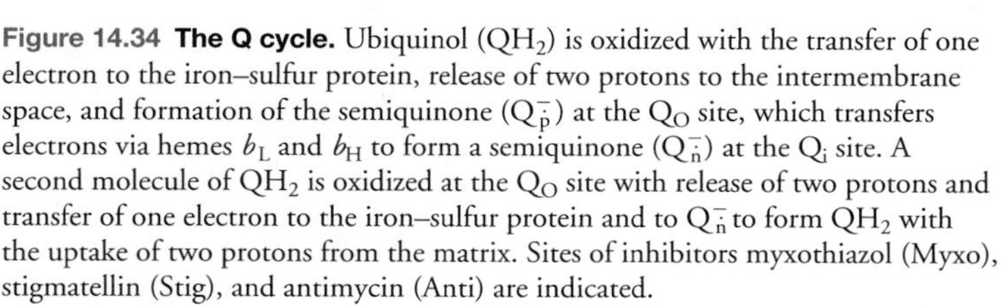

Figure 14.34 The Q cycle. Ubiquinol (QH_2) is oxidized with the transfer of one electron to the iron–sulfur protein, release of two protons to the intermembrane space, and formation of the semiquinone (Q_p^-) at the Q_O site, which transfers electrons via hemes b_L and b_H to form a semiquinone (Q_n^-) at the Q_i site. A second molecule of QH_2 is oxidized at the Q_O site with release of two protons and transfer of one electron to the iron–sulfur protein and to Q_n^- to form QH_2 with the uptake of two protons from the matrix. Sites of inhibitors myxothiazol (Myxo), stigmatellin (Stig), and antimycin (Anti) are indicated.

mitochondria are designated as *a*, *b*, and *c* on the basis of the α band of their absorption spectrum and the type of heme group attached to the protein (Figure 14.35). The absorption band and standard redox potential depend on the structure of the heme and its environment in the protein. Cytochrome *b* and other ***b*-type cytochromes** contain the same iron–protoporphyrin IX (Figure 14.35) found in hemoglobin and myoglobin; however, these hemes are buried in the membrane and cannot bind O_2. The ***c*-type cytochromes** contain heme *c* that is covalently bound to two cysteine residues of the protein via thioether linkages involving vinyl side chains of protoporphyrin IX. The ***a*-type cytochromes**, contain heme *a*, a modified form of protoporphyrin IX (p. 793) in which a formyl group and an isoprenoid side chain have been added. Two forms of cytochrome *a* are present in cytochrome *c* oxidase, complex IV.

Cytochrome *c* Is a Mobile Electron Carrier

Electrons are transferred through complex III to cytochrome *c*, a globular hydrophilic protein of 13 kDa. The planar heme group is located in the middle of the protein surrounded

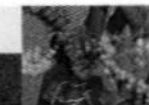

A CLOSER LOOK 14.1

The Q Cycle, for Electron Transfer and Proton Pumping in Complex III

Complex III contains two *b*-type hemes, a high-potential b_L (+0.50 V) and a low-potential b_H (0.10 V) plus cytochrome c_1. The transfer of electrons through this complex is best explained by the Q cycle in which four protons are translocated across the membrane for every two electrons transferred from ubiquinol to cytochrome *c* (Figure 14.38). For electron transfer to continue, two separate ubiquinone- or ubiquinol-binding sites are required in the bc_1 complex: a ubiquinol oxidizing site (Q_O) involving the low potential heme *b* (heme b_L) located at the positive (P) side of the membrane, where inhibitors such as myxothiazol and stigmatellin bind; and a ubiquinone-reducing site (Q_i), involving the high-potential heme *b* (heme b_H) located at the negative (N) side of the membrane, where inhibitors such as antimycin bind. Oxidation of ubiquinol at the Q_O site results in the transfer of one electron to the 2Fe2S cluster of the iron–sulfur protein with the release of two protons to the intermembrane space. The iron–sulfur protein then transfers an electron to the heme of cytochrome c_1. The strongly reducing ubisemiquinone anion formed at the Q_O site after the transfer of the first electron from ubiquinol rapidly transfers an electron to cytochrome b_L which then transfers an electron to high-potential heme of cytochrome b_H at the Q_i site. The reduced cytochrome b_H then transfers this electron to ubiquinone at the Q_i site to form a stable ubisemiquinone. To complete the Q cycle, a second molecule of ubiquinol is oxidized at the Q_O site with the release of another two protons and transfer of one electron to the iron–sulfur protein and the second electron to heme b_L. Heme b_L transfers an electron to b_H and eventually to the ubisemiquinone at the Q_i site to form ubiquinol with the uptake of two protons from the matrix. The Q cycle explains how during the oxidation of two ubiquinols at the Q_O site, four protons are released to the intermembrane space, while two protons are taken up from the matrix side to reduce ubiquinone at the Q_i site, a net of two protons released per ubiquinol oxidized.

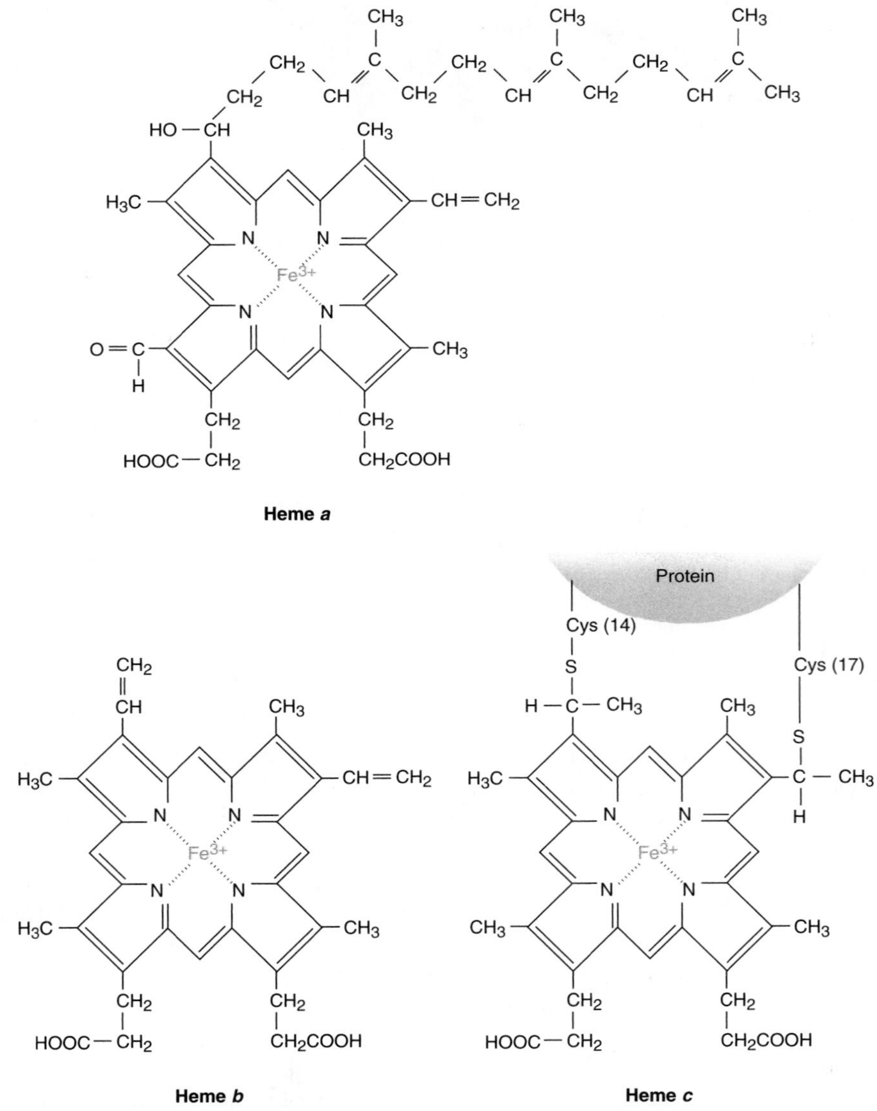

Heme *a*

Heme *b*

Heme *c*

Figure 14.35 Structures of heme *a*, heme *b*, and heme *c*.

by hydrophobic residues and covalently bound to two conserved cysteine residues through vinyl ether linkages (Figure 14.35). The heme iron is coordinated to a nitrogen of a histidine and a sulfur atom of a methionine, thus preventing the interaction of the heme with O_2 (Figure 14.36).

Cytochrome *c*, like ubiquinone, is a mobile electron carrier. It is held loosely by electrostatic forces to the outer face of the inner membrane where it binds to cytochrome c_1 of complex III and accepts electrons from it. The reduced cytochrome *c* then moves along the membrane surface to interact with subunit II of cytochrome *c* oxidase through electrostatic linkages and donates electrons to the Cu_A site.

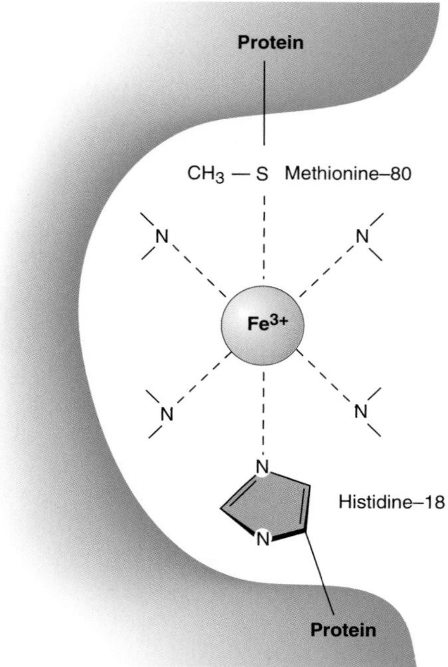

Figure 14.36 Six coordination positions of cytochrome *c*.

Complex IV: Cytochrome *c* Oxidase

Complex IV transfers electrons from cytochrome *c* to O_2, the terminal electron acceptor, to form water coupled to translocation of protons across the membrane. The mammalian complex consists of 13 subunits with a total mass of 200 kDa and contains two cytochromes, ***a*** and $\boldsymbol{a_3}$, and two copper centers, known as $\mathbf{Cu_A}$ and $\mathbf{Cu_B}$. A simpler cytochrome *c* oxidase containing only three or four subunits, which catalyzes similar electron transfer and proton pumping reactions is present in bacterial membranes. The three subunits are homologous to the three largest subunits of mammalian cytochrome *c* oxidase, which are encoded in mitochondrial DNA (mtDNA). The remaining subunits of complex IV are encoded in nuclear DNA and may be regulatory subunits or function in assembly of the complex.

The crystal structure of a bacterial cytochrome *c* oxidase and complex IV from beef heart mitochondria have been solved (Figure 14.37). Subunit I, the largest, contains twelve transmembrane helices but lacks any significant extramembranous domains. Two heme groups, *a* and a_3, are bound to subunit I with the heme coordinated by nitrogen atoms of conserved histidine residues. The planes of both hemes lie perpendicular to the membrane. Subunit I also contains a copper atom (Cu_B) that with heme a_3 forms a binuclear center involved in transfer of electrons from heme a to O_2 (Figure 14.38). Subunit II has a large domain protruding from the cytosolic face of the inner membrane, where reduced cytochrome *c* binds, and contains two atoms of copper bound through sulfhydryl groups to two cysteine moieties (called Cu_A). Subunit III contains seven transmembrane helices with negligible extramembranous domains but does not have any redox carriers. Subunits II and III are localized on opposite sides of

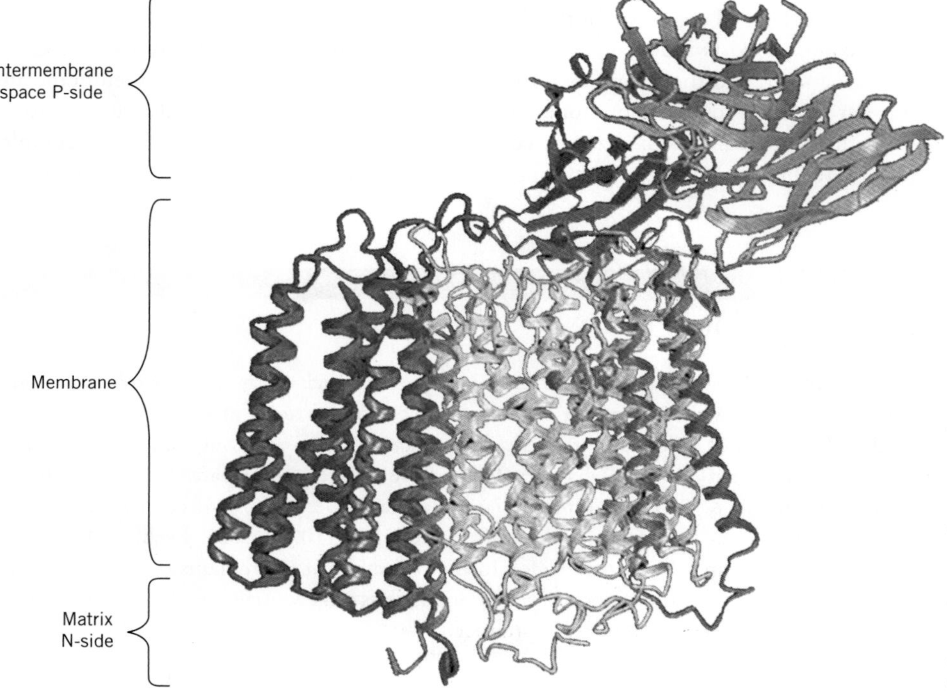

Figure 14.37 Model of crystal structure of cytochrome *c* oxidase from bacterium *Paracoccus denitrificans*. Subunit I (12 transmembrane helices) is yellow, subunit II (two transmembrane helices) is purple, and subunit III (7 transmembrane helices) is blue with an embedded phospholipid in pink. The antibody fragment used to drive crystallization is cyan. Reproduced with permission from Iwata, S., Ostermeier, C., Ludwig, B., Michel, H., et al. *Nature* 376:660, 1995. Copyright © (1995), Macmillan Magazines Limited. Figure generously supplied by Professor S. Iwata.

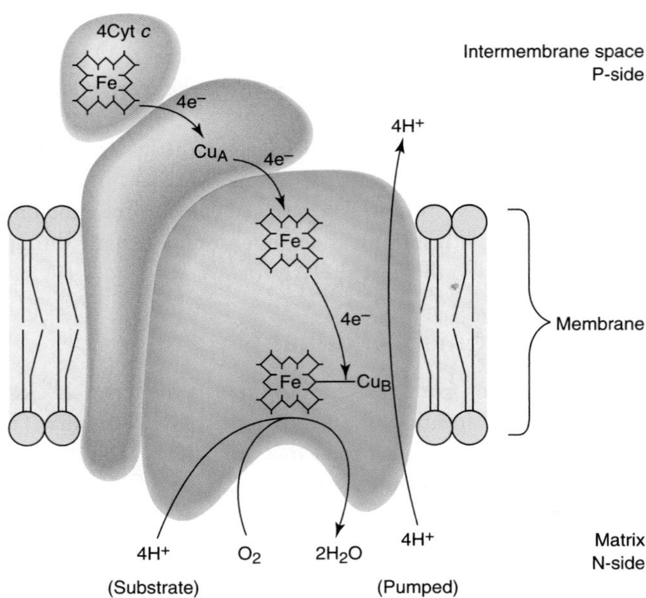

Figure 14.39 Pathways of electron and proton transfer through cytochrome c oxidase. Cytochrome *c* binds on the surface of subunit II and transfers electrons to CuA. Electrons are transferred from CuA to heme *a* and then to the binuclear center (heme a_3 and CuB) where oxygen is reduced to water. Four protons are transferred to the binuclear center for reduction of oxygen and four protons are pumped across the membrane by a different channel.

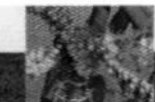

Figure 14.38 Binuclear center of cytochrome c oxidase indicating heme *a*3 and CuB. L is an unknown but proposed ligand.

subunit I; the role of subunit III is unclear. Electrons are transferred from reduced cytochrome *c* first to Cu_A, then to heme *a*, and finally to the binuclear center containing Cu_B and heme a_3 where the transfer of four electrons to oxygen occurs (Figure 14.39 and A Closer Look 14.2). The transfer of four electrons to form water results in the uptake of four protons from the matrix to reduce oxygen and the translocation of four protons across the mitochondrial membrane to contribute to the electrochemical gradient (Figure 14.39).

Electron Transport Chain Inhibitors

A dynamic picture of the electron transport chain has developed with increasing knowledge of the detailed chemistry of the different respiratory chain complexes (Figure 14.40). Each complex exists independently in the inner membrane and is freely mobile. Complexes I and II, and the other flavoprotein dehydrogenases, diffuse in the membrane and transfer electrons to the ubiquinone pool in the membrane. Ubiquinol also diffuses freely in the membrane and is oxidized by complex III. Electrons are transferred from complex III to cytochrome *c* which moves along the surface of the membrane to complex IV where

A CLOSER LOOK 14.2

Pathways of Electron Transfer through Complex IV

Electrons are transferred from reduced cytochrome *c* to the Cu_A site on subunit II and then to heme *a* on subunit I of complex IV (Figure 14.43). The Cu_A and heme *a* are localized within 1.5 Å of each other, permitting rapid electron transfer. Electrons are then transferred to the binuclear center consisting of Cu_B and heme a_3 where final transfer of electrons to O_2 occurs. Initially, two electrons are transferred to an O_2 tightly bound to the binuclear center to form a peroxy derivative of oxygen (O_2^{2-}). Two additional electrons are transferred with uptake of four protons from the matrix to form water. Since each of the redox carriers present in complex IV is a one-electron carrier and the reduction of O_2 to water requires four electrons, the reactions catalyzed by complex IV have evolved to prevent release of partially reduced toxic oxygen intermediates such as superoxide, hydrogen peroxide, or hydroxyl radicals (see Section 14.10, p. 583). Each of the intermediates formed in the reduction of O_2 remains tightly bound to the binuclear center and is thus largely prevented from dissociating until water is produced.

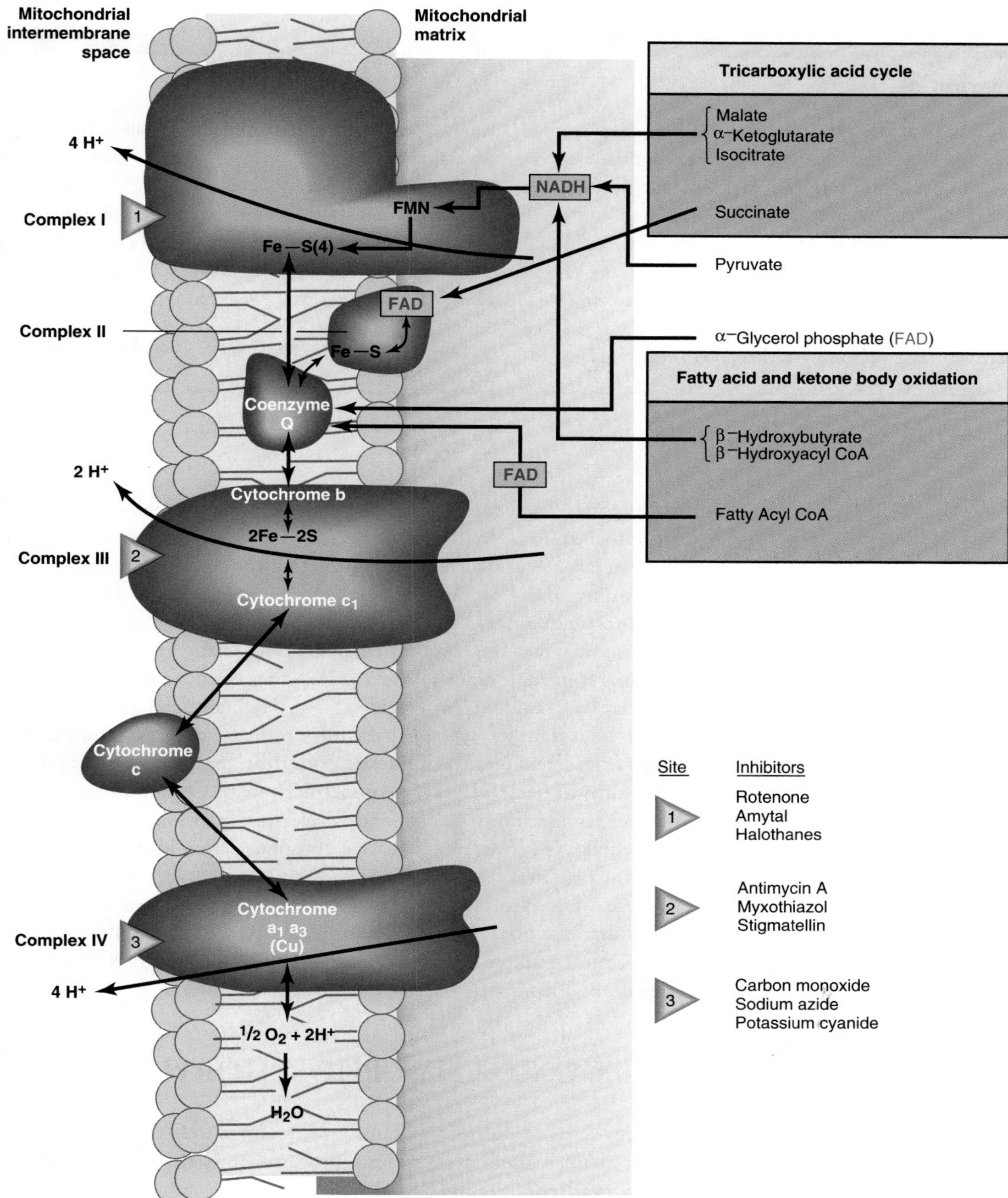

Figure 14.40 Overview of mitochondrial electron transport chain indicating locations of complexes I–IV, ubiquinone (CoQ) and cytochrome *c* in the inner membrane, pathways of electron transfer, and sites of proton pumping. The binding sites for specific inhibitors on the complexes are indicated at complex I (rotenone, amytal, and halothanes), at complex III (antimycin A, myxothiazole, and stigmatellin), and at complex IV (carbon monoxide, sodium azide, and potassium cyanide).

its electrons are transferred to O_2. Currently, it is accepted that the transfer of two electrons from NADH to O_2 results in the translocation of 10 protons across the membrane, 4 each from complexes I and IV and 2 from complex III. The electrochemical gradient, thus formed, provides the potential energy that is used to drive ATP synthesis by the ATP synthase (p. 572).

Figure 14.40 indicates the sites where specific inhibitors bind and block electron flow. Rotenone, a commonly used insecticide, binds stoichiometrically to complex I and prevents the reduction of ubiquinone. Piericidin, Amytal, and other barbiturates including halothanes used in anesthetics also inhibit complex I by preventing the transfer of electrons

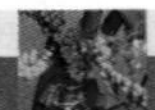

CLINICAL CORRELATION 14.3

Cyanide Poisoning

Inhalation of hydrogen cyanide gas or ingestion of potassium cyanide causes a rapid and extensive inhibition of the mitochondrial electron transport chain at the cytochrome oxidase step. Cyanide is one of the most potent and rapidly acting poisons known. It binds to the Fe^{3+} of heme a_3 in cytochrome *c* oxidase, which catalyzes the terminal step of the electron transport chain. Mitochondrial respiration and energy production cease, and cell death occurs rapidly. Death occurs from tissue anoxia, most notably of the central nervous system. An antidote to cyanide poisoning, if the poisoning is diagnosed rapidly, is the administration of various nitrites that convert oxyhemoglobin to methemoglobin by oxidizing Fe^{2+} of hemoglobin to Fe^{3+}. Methemoglobin (Fe^{3+}) competes with cytochrome a_3 (Fe^{3+}) for cyanide by forming a methemoglobin–cyanide complex. Administration of thiosulfate causes the cyanide to react with the enzyme rhodanese, forming the nontoxic thiocyanate. Cytochrome *c* oxidase is also inhibited by carbon monoxide (CO) that binds to the reduced form of heme a_3 and by H_2S.

Holland, M. A., and Kozlowski, L. M. Clinical features and management of cyanide poisoning. *Clin. Pharmacol.* 5:737, 1986.

from the iron–sulfur centers to ubiquinone. Complex II is inhibited by carboxin and thenoyltrifluoroacetone as well as by malonate that acts as a competitive inhibitor with the substrate succinate. Antimycin, an antibiotic, inhibits electron transfer through complex III by binding to the Q_i site and blocking the transfer of electrons from the cytochrome heme b_H to ubiquinone. Other antibiotics, such as myxothiazol and stigmatellin, inhibit electron transfer through complex III by binding to the Q_O site and blocking the transfer of electrons from ubiquinol to the 2Fe2S center of the iron–sulfur protein. Complex IV is inhibited by cyanide (CN^-), azide (N_3^-), H_2S, and carbon monoxide (CO). Cyanide and azide bind tightly to the oxidized form of heme a_3 (Fe^{3+}) and prevent the transfer of electrons from heme *a* to the binuclear center. By contrast, carbon monoxide binds to the reduced form of heme a_3 (Fe^{2+}) competitively with O_2 and prevents electron transfer to O_2. Hence, inhibition of mitochondrial electron transport results in impairment of the energy-generating function of oxidative phosphorylation leading to the death of the organism (Clin. Corr. 14.3).

Figure 14.40 also indicates the three sites where protons are translocated across the mitochondrial membrane during electron transport to contribute to the formation of the electrochemical gradient used for ATP synthesis. Four protons are translocated by electron transfer through complexes I and IV, while two protons are translocated by complex III.

14.7 • OXIDATIVE PHOSPHORYLATION

The energy released during the transfer of electrons to O_2 via the mitochondrial electron transport chain is used to translocate protons across the inner membrane and establish a **proton gradient** (Figure 14.41). This makes the intermembrane space more acidic and the matrix space more alkaline. Simultaneously, the external face of the membrane becomes more positively charged and the matrix face becomes more negatively charged establishing a **charge gradient**, since there is no compensating translocation of a negatively charged ion.

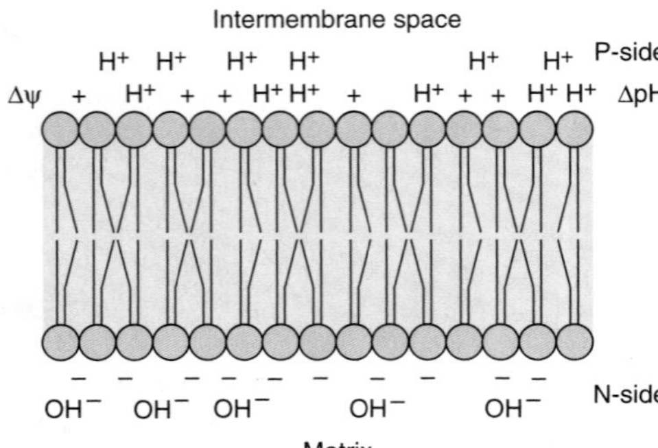

Figure 14.41 The electrochemical gradient consists of a gradient of charges (Δψ) and proton concentration (ΔpH) across inner mitochondrial membrane.

During the transfer of two electrons from NADH to O_2, approximately 10 protons are pumped across the membrane to establish the electrochemical gradient (see Figure 14.40). The total free energy derived from the translocation of protons and the charge distribution across the membrane can be calculated by the following equation in which Z is the absolute value of the charge, f is the Faraday constant, and ψ is the membrane potential:

$$\Delta G^{0'} = 2.3RT\Delta\text{pH} + Zf\psi$$

In actively respiring mitochondria, the observed change in pH across the membrane is 0.75−1.0 pH units and that in membrane potential is 0.15−2.0 V. Hence, the $\Delta G^{0'}$ calculated is roughly 200 kJ for the translocation of 10 H^+ across the membrane during transfer of electrons from NADH to O_2. The $\Delta G^{0'}$ can also be calculated from the difference in standard redox potentials of the redox pair. The calculated $\Delta G^{0'}$ for the redox pairs NADH and O_2 is 219 kJ/mol (Section 14.6), suggesting that the energy of the electron transfer reactions is efficiently captured in the electrochemical potential. The energy stored in the **proton and charge gradient** and the electrochemical gradient, also called the **protonmotive force**, drives the synthesis of ATP by the movement of protons down the electrochemical gradient through the ATP synthase by a mechanism discussed later in this section.

Coupling of ATP Synthesis and Electron Transport

The rate of ATP utilization regulates the rate of ATP synthesis in mitochondria, which in turn regulates the rate of electron transfer. The coupling of ATP synthesis to electron transport is achieved by the electrochemical gradient, as illustrated by the experiment shown in Figure 14.42. The rate of electron transport measured by the rate of O_2 consumption by a suspension of liver mitochondria increases only after addition of an electron donor (succinate in this experiment) and ADP (a phosphate acceptor) plus phosphate (P_i). Conversion of all the added ADP to ATP returns the rate to that observed prior to addition of ADP. Hence, the rate of electron transport, or O_2 uptake, is tightly coupled to ATP synthesis. Chemiosmosis readily explains this relationship, termed respiratory control. When the energy needs of the cell are low, ATP will accumulate and the proton gradient will not be used for ATP synthesis. The magnitude of the proton gradient increases until the energy required to pump protons across the membrane against the existing electrical gradient equals that released during transfer of electrons from NADH to O_2. At this point, electron transport ceases as equilibrium has been attained. In cells using ATP, ADP accumulates leading to stimulation of the ATP synthase. While ATP is synthesized, the magnitude of the proton gradient decreases as protons move through the ATP synthase to provide the energy for ATP synthesis. As a result, the proton back pressure on the electron transport chain will decrease. The increased rate of electron transport through the chain stimulates the oxidation of NADH, resulting in the formation of NAD^+. The increased concentrations of NAD^+ coupled with the increased concentrations of ADP in cells actively using ATP acts to stimulate reactions of the TCA cycle and fatty acid oxidation. In this way, the need for ATP in a cell regulates in a coordinated fashion the rate of electron flow through the electron transport chain and of the reactions of the TCA cycle and fatty acid oxidation.

P/O Ratios for Mitochondrial Electron Transport and Oxidative Phosphorylation

The P/O ratio (phosphate incorporated into ATP to atoms of O_2 utilized) is a measure of the number of ATP molecules formed during the transfer of two electrons through all or part of the electron transport chain. Classically, the P/O ratio had been thought to be a whole number, 3 for transfer of two electrons from NADH linked substrates to O_2, 2 for succinate to O_2, and 1 for reduced cytochrome *c* to O_2. These P/O ratios suggested that one ATP was produced during electron transfer through each of the proton-pumping complexes I, III, and IV. Questions about actual P/O ratios, however, have arisen with recent calculations that 10 protons are pumped across the mitochondrial membrane during the transfer of two electrons from NADH to O_2, whereas the synthesis of one ATP and its transport across the membrane requires four protons. These proton stoichiometries result in a calculated P/O ratio of 2.5. Indeed, recent experimental determinations of the P/O ratio gave values of approximately 2.5 with NADH-linked substrates and 1.5 with succinate.

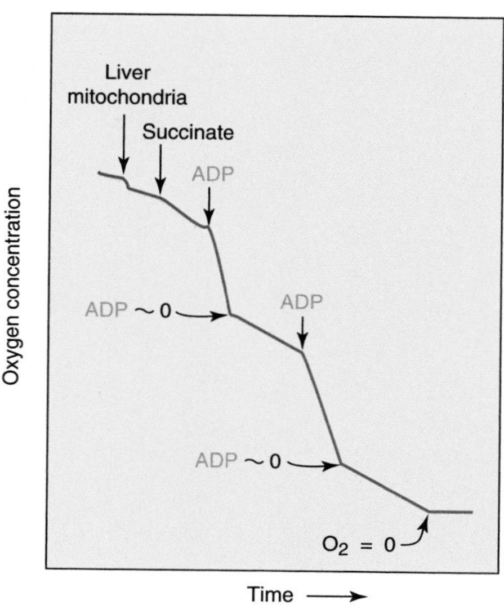

Figure 14.42 Demonstration of coupling of electron transport to oxidative phosphorylation in a suspension of liver mitochondria. In a medium containing Pi, addition of ADP stimulates rate of electron transfer measured as oxygen uptake. This is defined as respiratory control.

Effects of Uncouplers and Inhibitors of the Electron Transport–Oxidative Phosphorylation System

The coupling between electron transport and ATP synthesis can be abolished by chemical reagents, or **uncouplers**, such as 2,4-dinitrophenol or carbonylcyanide-p-trifluoromethoxyphenylhydrazone. After addition of an uncoupler to a suspension of tightly coupled liver mitochondria with a low rate of O_2 uptake, a rapid increase in the rate of O_2 consumption is observed (Figure. 14.43*a*). Because electron transport is uncoupled from ATP synthesis, electron transport may continue but without ATP synthesis. Uncouplers are hydrophobic

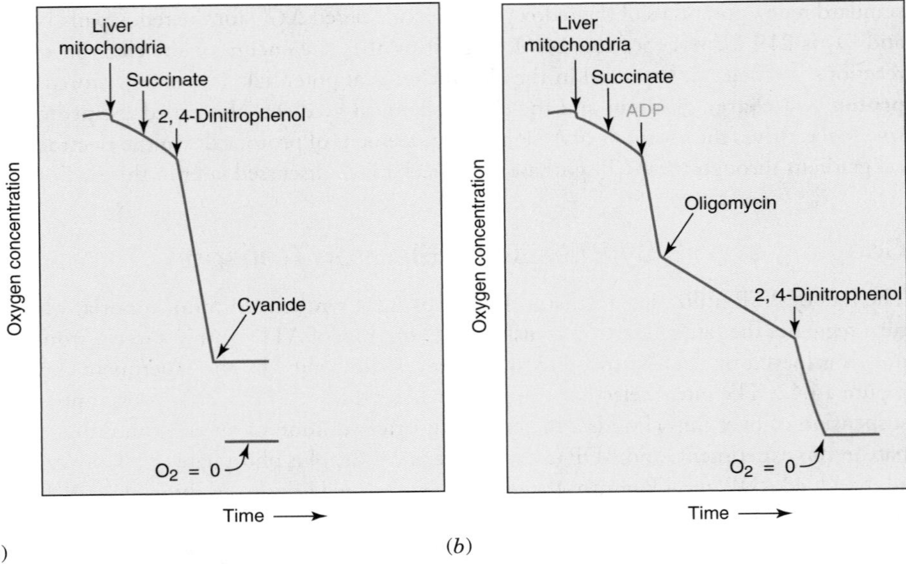

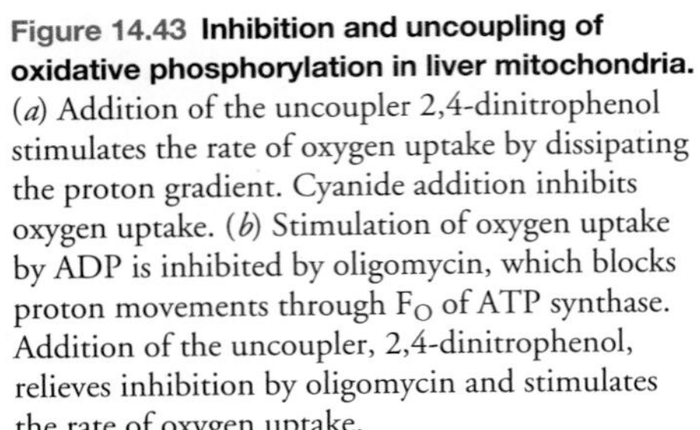

Figure 14.43 Inhibition and uncoupling of oxidative phosphorylation in liver mitochondria. (*a*) Addition of the uncoupler 2,4-dinitrophenol stimulates the rate of oxygen uptake by dissipating the proton gradient. Cyanide addition inhibits oxygen uptake. (*b*) Stimulation of oxygen uptake by ADP is inhibited by oligomycin, which blocks proton movements through F_O of ATP synthase. Addition of the uncoupler, 2,4-dinitrophenol, relieves inhibition by oligomycin and stimulates the rate of oxygen uptake.

weak acids that pick up a proton in the intermembrane space where a higher concentration of protons results from active electron transfer. These protonated uncouplers, being lipophilic, rapidly diffuse into the mitochondrial matrix where they lose their proton, because of the lower concentration of protons there. In this way the proton gradient can be completely dissipated as shown in Figure 14.44 and hence ATP synthesis is abolished.

As illustrated in Figure 14.43*b*, addition of **oligomycin**, an inhibitor of the ATP synthase, to liver mitochondria actively respiring in the presence of ADP results in an

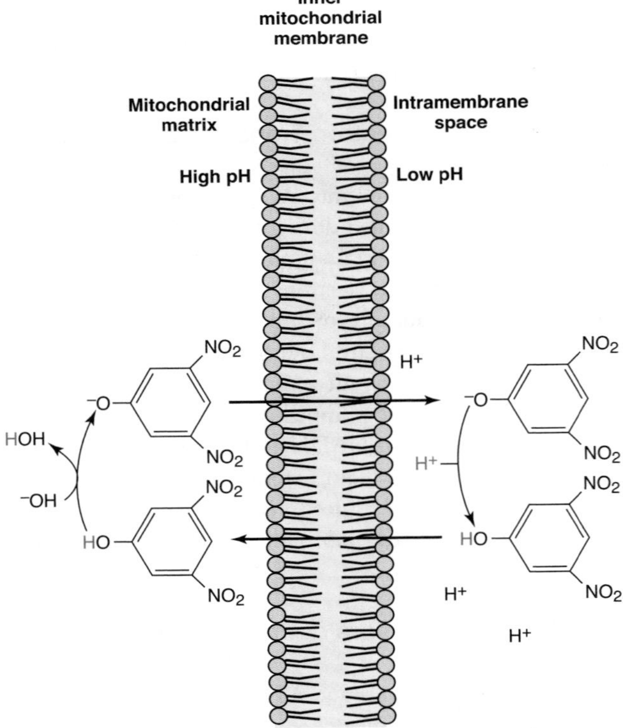

Figure 14.44 Action of the uncoupler, 2,4-dinitrophenol, a proton ionophore which equilibrates pH across the inner mitochondrial membrane. 2,4-Dinitrophenol is a weak acid that picks up a proton from the intermembrane space (P-side of membrane) that has a high proton concentration and carries it across the membrane to the matrix (N-side of membrane) where the proton dissociates due to the low proton concentration there.

inhibition of O_2 uptake. Oligomycin blocks the synthesis of ATP by preventing the movement of protons through the ATP synthase. Since ATP synthesis and electron flow are tightly coupled, the buildup of protons in the intermembrane space almost completely blocks the rate of electron transport as discussed for respiratory control. Subsequent addition of 2,4-dintirophenol that dissipates the proton gradient results in a rapid increase in the rate of O_2 uptake, because electron transport has been uncoupled from ATP synthesis.

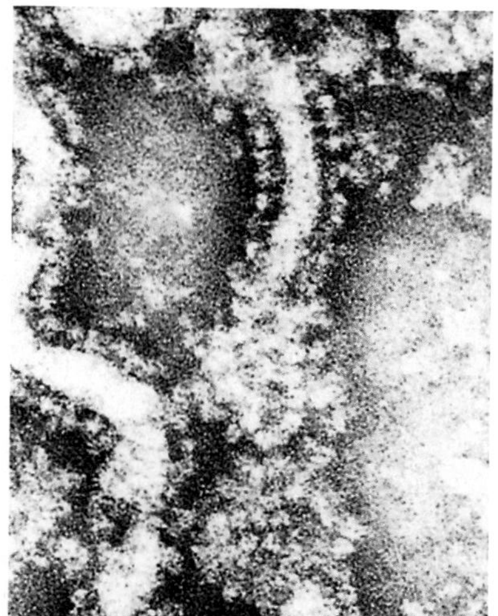

Figure 14.45 Electron micrograph of mitochondrial F_1.
Generously supplied by Dr. D. F. Parsons. From Parsons, D. F. *Science* 140:985, 1963. Reprinted with permission from AAAS.

ATP Synthase

ATP synthase, or complex V, located in the inner mitochondrial membrane of mammals, yeast, and fungi and in the cytoplasmic membrane of bacteria, catalyzes ATP synthesis using the energy of the proton gradient as protons flow through the ATP synthase. ATP synthase consists of two domains, $\mathbf{F_1}$, a peripheral complex first observed in electron micrographs as small particles bound to the inner mitochondrial membrane (Figure 14.45) and $\mathbf{F_0}$, an integral membrane protein complex. The F_1 domain contains the binding sites for ATP and ADP and catalyzes ATP synthesis (Figure 14.46). The F_0 domain provides a channel for the translocation of protons across the membrane. Removal of F_1 from the mitochondrial inner membrane by gentle agitation leaves an intact electron transport chain capable of electron transfer without formation of a proton gradient, since protons pumped across the membrane during electron transfer flow back into the matrix through the F_0 domain. Adding the F_1 domain back to the stripped membranes permits formation of a proton gradient as F_1 reconstitutes with F_0 and plugs the flow of protons through F_0. The entire ATP synthase can be isolated and when incorporated into artificial membrane vesicles, it will synthesize ATP when an electrochemical gradient is established across the membrane. The ATP synthase is a multicomponent complex of 480−500 kDa (Table 14.6 and Figure 14.46). The F_1 domain consists of five nonidentical subunits (α, β, γ, δ, and ε) with a subunit stoichiometry of α_3, β_3, γ, δ, and ε and a mass of 350−380 kDa. Binding sites for ATP and ADP are present on both α- and β-subunits. The catalytic sites are on the β-subunits, while the function of nucleotides bound to the α-subunits is unknown. The γ-subunit forms the central core of F_1, whereas the δ-subunit may be involved in the attachment of the F_1 domain to the membrane. The F_0 domain of the *E. coli* enzyme consists of three nonidentical hydrophobic subunits termed a, b, and c, which are present in the apparent stoichiometry of a_1, b_2, c_{9-12}. The c-subunits each contain an essential charged residue (aspartate-61 in *E. coli*) that is involved in proton pumping. Each c-subunit consists of two trans-membrane α-helices with aspartate residue-61 located in the middle. Mutation of this aspartate to asparagine abolishes proton pumping. Mutations of charged residues have also implicated the a-subunit of F_0 in proton movements, whereas the b-subunits appear to act to attach the F_1 domain to F_0. The F_0 domain found in mitochondria contains subunits homologous to subunits a, b, and c of the *E. coli* enzyme; however, additional subunits are also present.

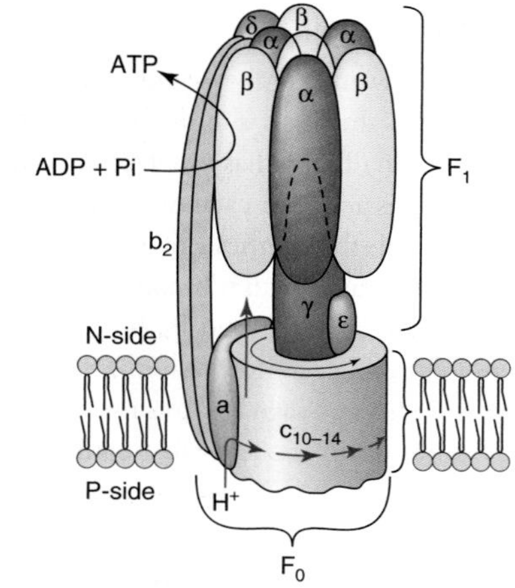

Figure 14.46 Model for mitochondrial F_1F_0–ATP synthase, a rotating molecular motor. ATP synthesis occurs on the β-subunits of F_1, whereas F_0 contains a proton channel. In F_0, the c-subunits in the membrane are attached to the shaft containing the γ and ε of F_1 and constitute a rotor. The two b-subunits of F_0 along with the α- and β-subunits and the δ-subunit constitute a stator (nonmoveable structural element). Protons flow through the a- and c-subunits of F_O causing the rotor to turn resulting in conformational changes in the β-subunits where ATP is synthesized.

TABLE 14.6 • Subunits of F_1F_0-ATP Synthase from *Escherichia coli*

Complex	*Protein Subunit*	*Mass (kDa)*	*Stoichiometry*
F_1	α	55	3
	β	52	3
	γ	30	1
	δ	15	1
	ϵ	5.6	1
F_0	a	30	1
	b	17	2
	c	8	9–12

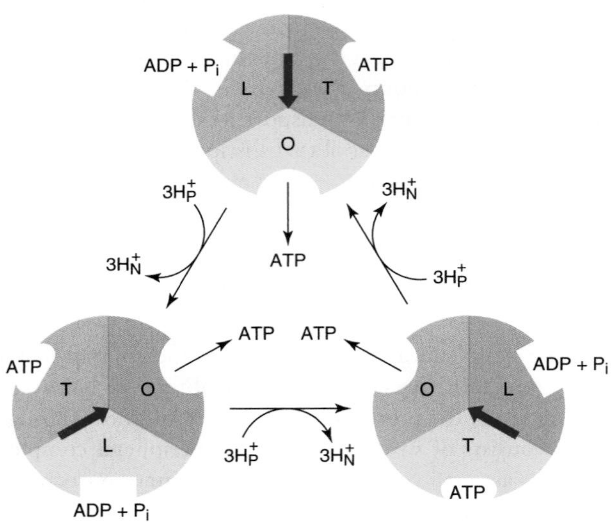

Figure 14.47 The binding change model for ATP synthesis by ATP synthase. Each β-subunit of F_1 has a non-identical adenine nucleotide binding site. At any time, one of these β-subunits is in the T (tight) conformation which binds ATP tightly, a second is in the L (loose) conformation which binds ATP and P_i loosely, and a third is in the O conformation (open) which does not bind nucleotides. The proton gradient causes rotation of the γ-subunit, the central shaft, which comes into contact with each β-subunit in succession producing a cooperative conformation change converting the T site to the O site releasing ATP, the L site to the T site promoting synthesis of ATP, and the O site to the L site binding ADP and P_i.

A CLOSER LOOK 14.3

Synthesis of ATP on F_1

The binding-charge mechanism suggests that the three β-subunits each adopt different conformations that change during catalysis with only one subunit acting as the catalytic site. As depicted in Figure 14.51, one subunit has an open conformation (O) with a low affinity for ligands and is empty. A second subunit has a loose conformation (L) with a low affinity for ligands and is inactive, while the third subunit has a tight conformation (T) that has a high affinity for ligands and is active in catalysis. According to this model, the synthesis of ATP occurs on the β-subunit in the T conformation. During catalysis, ADP and P_i, bind to the β-subunit in the L conformation. The energy provided by passage of protons through F_0 to F_1 results in the following conformational changes: The T site containing ATP changes to the O conformation with the release of ATP, the L site changes to the T conformation with ATP synthesis, and the O site changes to the L conformation binding ADP and P_i. According to this model, the energy released by electron transfer is conserved as a proton gradient that drives conformational changes in the ATP synthase resulting in the binding of substrates, the synthesis of ATP on the enzyme and the release of the product ATP.

Synthesis of ATP on F_1

The mechanism of ATP synthesis by F_1 was derived from isotope exchange experiments, which revealed that in the presence of stoichiometric amounts of ADP, ATP, and inorganic phosphate with isolated F_1, the reaction was essentially in equilibrium having a $\Delta G^{0'}$ close to zero. This exchange reaction

$$\text{Enz:ADP} + P_i \rightleftharpoons \text{Enz:ATP}$$

proceeds readily even in the absence of a proton gradient. The result indicated that the synthesis of ATP by F_1 does not require the input of energy; however, the movement of protons through the ATP synthase was required for release of ATP from their unique binding to the β-subunits of F_1. It was proposed that the energy released during the movement of protons across the membrane causes a conformational change in the ATP synthase, which results in the release of tightly bound ATP from one of the β-subunits, the binding of ADP and Pi to a second β-subunit in a loose conformation, and forcing the third β-subunit into the tight conformation where ATP synthesis occurs (Figure 14.47 and A Closer Look 14.3).

Mechanism of ATP Synthesis

Resolution of the crystal structure of F_1 has provided dramatic visualization of the conformations of the different β-subunits that provide evidence for the binding-change model described in A Closer Look 14.3. In this crystal structure, alternating α- and β-subunits form the knob of F_1 with the single γ-subunit forming a central shaft through the center of F_1 (Figure 14.48*a* and *b*). Each β-subunit had a different conformation depending on the presence of substrate. Thus F_1 crystallized in the presence of ADP and a non-hydrolyzable analog of ATP revealed binding of ATP analogue to one β-subunit, binding of ADP to a second β-subunit, and an empty third β-subunit (Figure 14.48*c*).

The model for ATP synthesis that has developed is that protons flow through the membrane by first binding to conserved acidic amino acid residues in subunit a of F_0. Protons then bind to a conserved amino acid residue present in the c-subunit causing the ring of c-subunits attached to γ- and ε-subunits to rotate. The movement of γ-subunit causes conformational changes in β-subunits as γ-subunit associates sequentially with each β-subunit in turn. The a- and b-subunits of F_0 plus δ-subunit of F_1 form the stator to hold α- and β-subunits in position, while γ- and c-subunits form the moving rotor. (See A Closer Look 14.4, p. 578, for experimental evidence for this proposed mechanism.)

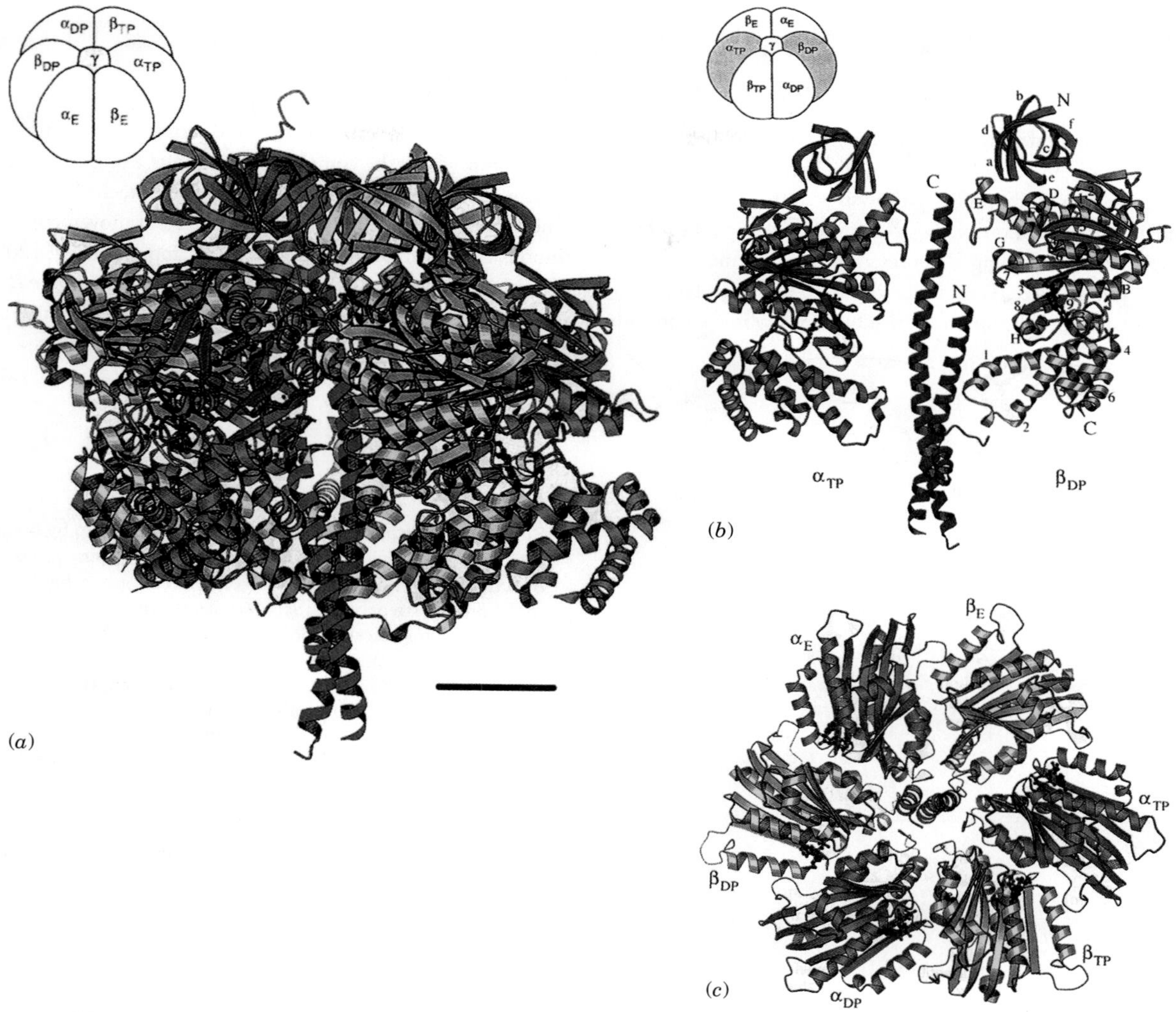

Figure 14.48 Mitochondrial ATP synthase complex. (*a*) Side view of F_1 complex structure deduced from the crystal structure. Three α- (red) and 3 β- (yellow) subunits alternate around a central shaft, the γ-subunit (blue). (*b*) Side view of F_1 subunit in which two α- and two β-subunits have been removed to reveal the central γ-subunit. Subunits are colored as indicated in caption (*a*). (*c*) Top view of F_1 complex shows alternating α- and β-subunits surrounding central γ-subunit.
Reproduced with permission from Abrahams, J. P., Leslie, A. G. W., Lutter, R., and Walker, J. E. *Nature* 370:621, 1994. Copyright © (1994 Macmillan Magazines Limited).

14.8 • MITOCHONDRIAL INNER MEMBRANE CONTAINS SUBSTRATE TRANSPORT SYSTEMS

Whereas the outer membrane presents little or no permeability barrier to substrate or nucleotide molecules of interest in energy metabolism, the inner membrane limits the types of substrates, intermediates, and nucleotides that can diffuse from cytosol into the matrix. Various transport systems have been described in mitochondria (Figure 14.49), some of which have been thoroughly characterized. These transport systems facilitate selective movement of various substrates and intermediates back and forth across the inner mitochondrial membrane. Through these transporters, various substrates can be accumulated in the matrix, since the transporters can move the substrate against a concentration gradient.

Transport of Adenine Nucleotides and Phosphate

Continued synthesis of ATP in the mitochondrial matrix requires that cytosolic ADP formed during energy-consuming reactions be transported back across the inner membrane into the matrix for conversion to ATP. Similarly, newly synthesized ATP must be transported back

A CLOSER LOOK 14.4

Experimental Evidence for Rotation of the γ- and c-Subunits by ATP Synthase

The binding-change model predicts that the γ-subunit should move in one direction during ATP synthesis and in the opposite direction during ATP hydrolysis, such that ATP hydrolysis results in formation of a proton gradient. The rotation of the γ-subunit in a single F_1 subunit was demonstrated by attaching a fluorescent actin polymer to the γ-subunit of an F_1 in which $\alpha_3\beta_3$ subunits were fixed to a microscope slide (see figure). Rotation of the fluorescent γ-subunit was observed on addition of ATP. Similar experiments were performed using the entire F_1/F_0 complex in which the complex of c-subunits along with the γ-subunit rotated as indicated by the fluorescent actin attached to one c subunit. Under both experimental conditions, the rotor movement was not continuous but occurred in discrete steps of approximately 120°, which is consistent with stepwise movement of γ-subunit from one β-subunit to another. ATP synthase is the smallest known molecular motor.

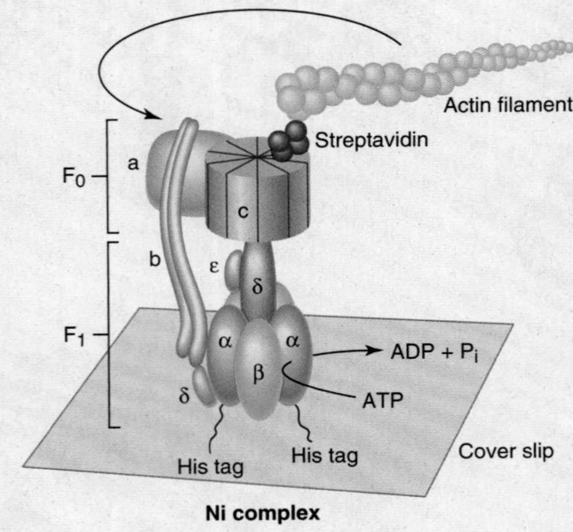

F_1 domain is attached to a nickel coated coverslip by histidine residues genetically engineered at the N-terminus of the α-subunits. Biotin, covalently attached to c-subunits, binds very tightly to the protein streptavidin that is covalently attached to an actin filament containing a fluorescent probe. Addition of ATP that is hydrolyzed by the ATPase of F_1 causes actin filament to rotate in one direction proving that c-subunit of F_0 rotates. Earlier experiments in which an actin filament was attached to γ-subunit provided evidence that γ-subunit can also rotate. Presumably, both γ- and c-subunits rotate as a unit. Redrawn from Sambongi, Y., Iko, Y., Tanabe, M., Omote, H., et al. *Science* 286:1722, 1999.

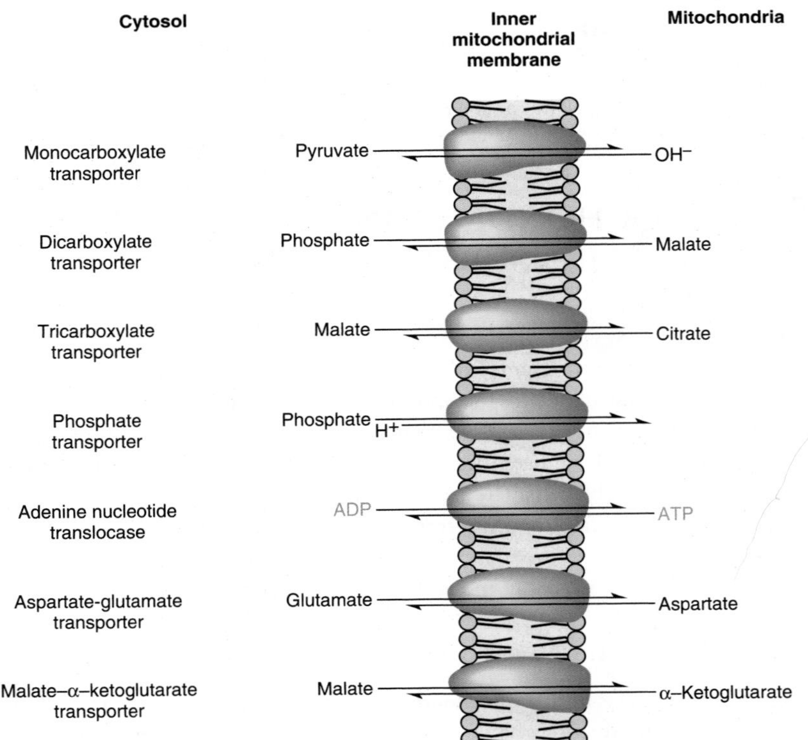

Figure 14.49 Mitochondrial metabolite transporters.

across the inner membrane into the cytosol to meet the energy needs of the cell. This exchange of the highly charged hydrophilic adenine nucleotides is catalyzed by a very specific **adenine nucleotide translocase** located in the inner membrane (Figure 14.50). The adenine nucleotide translocase, a homodimer of 30-kDa subunits, catalyzes a 1:1 exchange of ATP for ADP. The presence of one nucleotide-binding site on the transporter suggests that the enzyme alternately faces the matrix or the intermembrane space during the transport process. Newly synthesized ATP is bound to the translocase in the matrix, which then changes its conformation to face the cytosol where the ATP is released in exchange for an ADP. The translocase then changes conformation again to bring the nucleotide-binding site containing ADP back to face the matrix. The translocase favors outward movement of ATP and inward movement of ADP despite observations that both nucleotides bind equally well to the binding site. The explanation for this is that at pH 7, ADP has three negative charges whereas ATP has four. Hence, the exchange of one ATP for one ADP results in net outward movement of one negative charge, which is equivalent to import of one proton. The membrane potential established during electron transfer is positive outside, which would favor outward transport of more negatively charged ATP over that of ADP. The adenine nucleotide translocase is present in high concentrations, up to 14% of total protein, in the inner membrane. Hence, it is unlikely that transport of adenine nucleotides across the inner mitochondrial membrane is ever rate limiting for ATP synthesis.

A second transporter essential for oxidative phosphorylation is the **phosphate transporter**, which transports cytosolic phosphate plus a proton into the matrix (Figure 14.50). This symport also depends on the proton gradient as phosphate and protons are transported in a 1:1 ratio. Transport of ADP and phosphate requires a significant fraction of the energy present in the electrochemical gradient produced during electron transfer. Thus, the protonmotive force provides energy for ATP synthesis by ATP synthase as well as for uptake of the two required substrates.

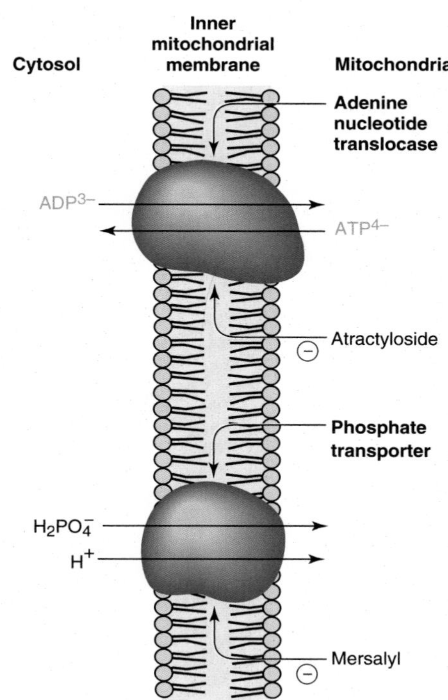

Figure 14.50 The adenine nucleotide translocase and phosphate transporter.

Substrate Shuttles Transport Reducing Equivalents across the Inner Mitochondrial Membrane

The nucleotides involved in cellular oxidation–reduction reactions (e.g., NAD^+, NADH, $NADP^+$, NADPH, FAD, and $FADH_2$) and CoA and its derivatives are not transported across the inner mitochondrial membrane. Thus, to transport reducing equivalents (e.g., protons and electrons) from cytosol to matrix or the reverse, substrate shuttle mechanisms are required.

Two substrate transport shuttles are shown in Figure 14.51. The malate–aspartate shuttle and α-glycerol–phosphate shuttle are employed in various tissues to translocate reducing equivalents from cytosol to matrix for oxidation to yield energy. Their operation requires that appropriate enzymes are localized on either side of the membrane and that appropriate transporters be present in the mitochondrial inner membrane.

In the **glycerol–phosphate shuttle**, two glycerol phosphate dehydrogenases, one in the cytosol and the other on the outer face of the inner mitochondrial membrane, are involved. NADH produced in the cytosol is used to reduce dihydroxyacetone phosphate to glycerol 3-phosphate by the cytosolic isozyme. The glycerol 3-phosphate in turn is oxidized by the mitochondrial isozyme, a flavoprotein, to produce dihydroxyacetone phosphate and $FADH_2$ that is oxidized by the electron transport chain.

The **malate–aspartate shuttle** operates on the same principle. NADH in the cytosol reduces oxaloacetate to malate, which enters mitochondria on the malate/α-ketoglutarate transporter. This malate is readily oxidized by mitochondrial malate dehydrogenase to oxaloacetate and NADH that is then oxidized by the electron transport chain. The oxaloacetate produced is converted to aspartate by mitochondrial aspartate aminotransferase and can then cross the membrane via the aspartate–glutamate transporter where cytosolic aspartate amino transferase converts it to oxaloacetate. The antiport transport of aspartate out of mitochondria in exchange for glutamate is driven by the membrane potential and is therefore irreversible.

Acetyl Units Are Transported as Citrate

The inner mitochondrial membrane does not have a transporter for acetyl CoA, but acetyl groups are transferred from the mitochondria to cytosol, where they are required

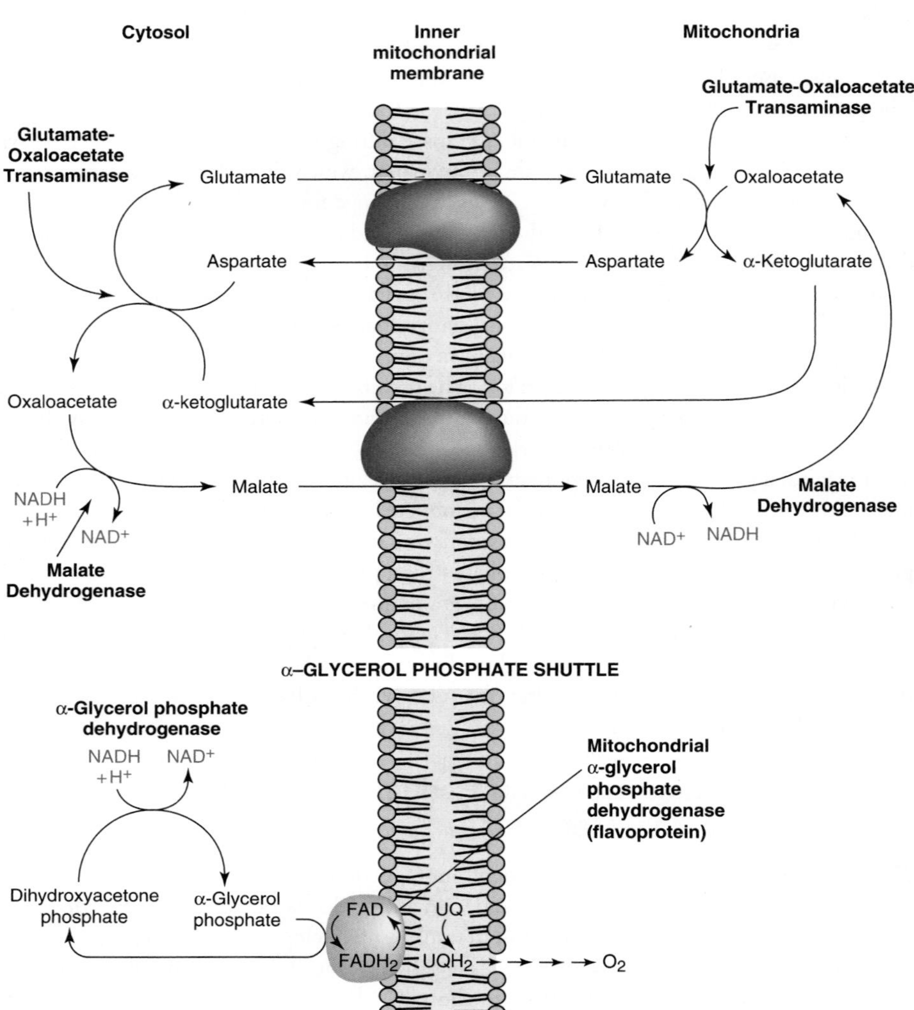

Figure 14.51 **Transport shuttles for reducing equivalents.**

for fatty acid or sterol biosynthesis (Figure 14.52). Intra-mitochondrial acetyl CoA is converted to citrate by citrate synthase of the TCA cycle. Citrate is then exported to cytosol by a tricarboxylate transporter in exchange for malate. Cytosolic citrate is cleaved to acetyl CoA and oxaloacetate at the expense of an ATP by **ATP-citrate lyase** (p. 684). Substrate shuttle mechanisms are involved in movement of appropriate substrates and

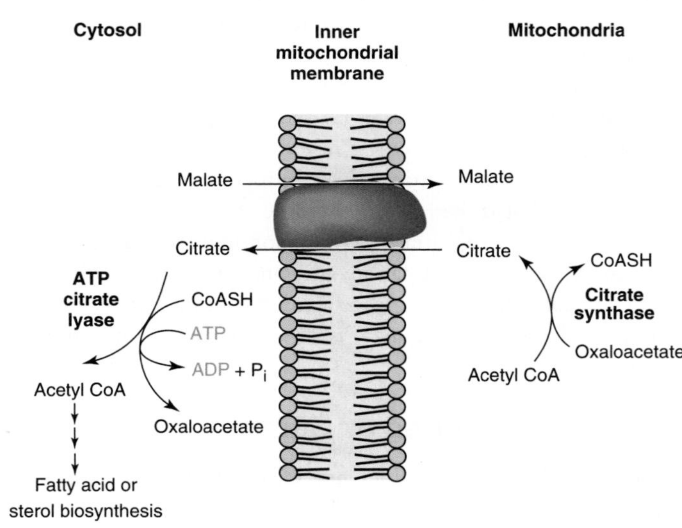

Figure 14.52 **Export of citrate generated in mitochondria to cytosol where it serves as a source of acetyl CoA for biosynthesis of fatty acids or sterols.**

intermediates in both directions across the inner membrane during periods of active gluconeogenesis (p. 619) and ureagenesis (p. 758) by the liver.

Mitochondria Have a Specific Calcium Transporter

In most mammalian tissues, mitochondria have a transport system for translocating Ca^{2+} across the inner membrane. The distribution/redistribution of cellular Ca^{2+} pools within cells is critical for muscle contraction, neural transmission, secretion, and hormone action. Distinct pools of Ca^{2+} have been detected in endoplasmic reticulum (or sarcoplasmic reticulum), mitochondria, nuclei, and Golgi. Some intracellular Ca^{2+} is bound to nucleotides, metabolites, or membrane ligands, while a portion is free in solution. Cytosolic Ca^{2+} concentration is about 10^{-7} M, whereas the extracellular concentration is at least four orders of magnitude greater. Ca^{2+} entry into mitochondria is by a uniporter in the inner membrane, which uses the energy of the electrochemical gradient (Figure 14.53). Confocal microscopy of living cells has provided convincing evidence that mitochondria take part in regulation of cytosolic Ca^{2+} concentration. Mitochondria are localized in close proximity to endoplasmic reticulum and sarcoplasmic reticulum. The binding of certain hormones to cell membranes results in release of inositol trisphosphate (IP_3) from phosphatidyl inositol, which releases Ca^{2+} from endoplasmic reticulum (p. 16). The resulting transient microdomains of high Ca^{2+} concentrations may be modulated by uptake into nearby mitochondria. In mitochondria, Ca^{2+} regulates the pyruvate dehydrogenase complex as well as isocitrate and α-ketoglutarate dehydrogenases. One consequence of uptake of high concentrations of Ca^{2+} into mitochondria is the opening of a pore on the outer membrane leading to release of cytochrome *c* into cytosol and activation of apoptosis.

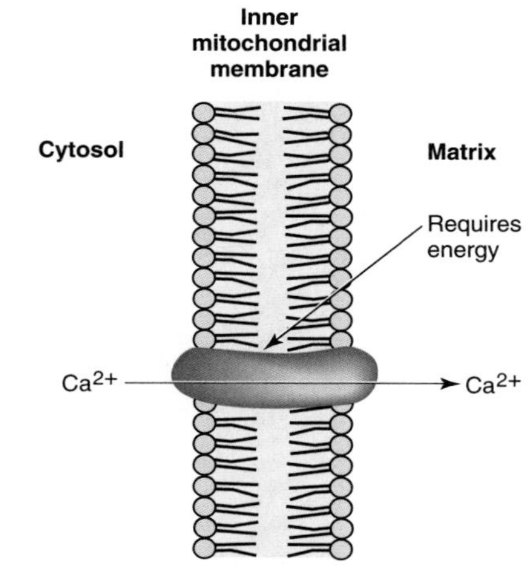

Figure 14.53 Mitochondrial calcium carrier.

Uncoupling Proteins

Brown adipose tissue plays a major role in nonshivering thermogenesis in newborns, in hibernating animals, and in experimental animals in diet-induced thermogenesis. The primary agent involved in cold-induced thermogenesis in brown fat is the uncoupling protein, UCP-1, which is localized exclusively in the inner mitochondrial membrane of brown adipose tissue. UCP-1 carries protons from the mitochondrial matrix and acts to uncouple ATP synthesis from electron transport (Figure 14.54). Thermogenesis results from activation of

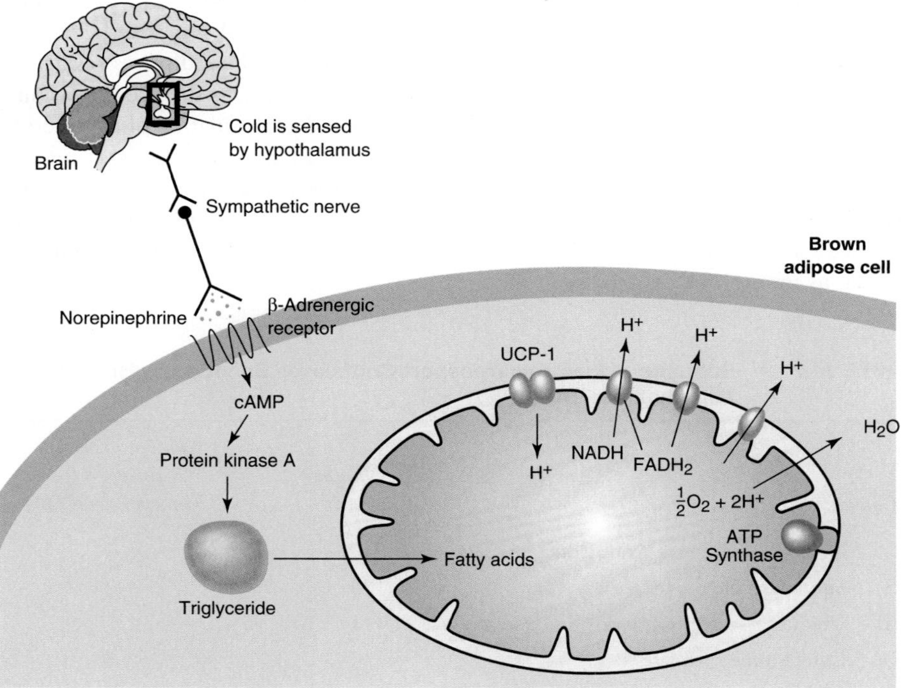

Figure 14.54 Activation of UCP-1 by cold adaptation. Cold stimulates release of norepinephrine from sympathetic nerve cells. The norepinephrine binds to the β-adrenergic receptor resulting in activation of a lipase with production of free fatty acids that activate proton-conducting protein, UCP-1.

sympathetic nerves by the brain responding to cold exposure with the release of norepinephrine that binds to β-adrenergic receptors on cell membranes of brown fat cells. This binding causes release of cAMP and the activation of protein kinase A, which stimulates lipolysis. The production of free fatty acids activates UCP-1, which transports protons back into the matrix (Figure 14.54). The stimulation of proton transport by free fatty acids is believed to result from release of a proton from the carboxyl group of the free fatty acid. UCP-1 is a member of the mitochondrial transporter family that includes the adenine nucleotide translocase and the phosphate transporter but with a specific pore for the transport of protons into the matrix. Chronic cold-induced stimulation of the β-adrenergic receptor by norepinephrine results in increased transcription of the UCP-1 gene, stimulation of mitochondrial biogenesis, and eventual hyperplasia of brown adipose tissue. In large mammals, such as dogs, cats, and primates who do not hibernate, including humans, discrete deposits of brown fat are present at birth, but become sparse during later development.

Four other uncoupling proteins, UCP-2, UCP-3, UCP-4, and UCP-5, with sequences similar to that of UCP-1 have been discovered in tissues other than brown adipose tissue. The presence of uncoupling proteins in tissues such as skeletal muscle has prompted investigations into the possible role of these proteins in regulation of energy expenditure and perhaps obesity. Development of pharmacological agents that affect uncoupling proteins has been suggested as a treatment for obesity. Recently, it has been suggested that uncoupling proteins may act to prevent the formation of reactive oxygen species in mitochondria.

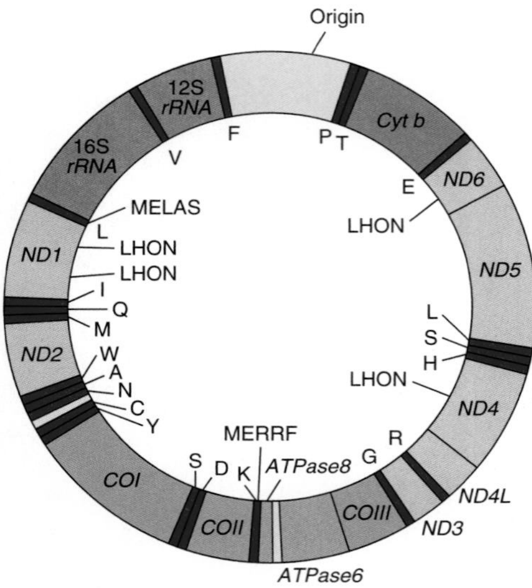

Figure 14.55 Map of genes on mitochondrial DNA. The indicated genes CO1, CO11, and CO111 encode subunits of cytochrome *c* oxidase, ND encode subunits of complex I, and ATPase encode subunits of ATP synthase. The dark red bands indicated by single letters are the genes for tRNAs. LHON indicates location of mutations causing Leber's Hereditary Optic Neuropathy. Mutations in the tRNA for leucine (L) cause MELAS (mitochondrial encephalopathy, lactic acidosis and strokelike activity), and those in the tRNA for lysine (K) cause MERRF (myoclonic epilepsy and ragged red fibers).

14.9 • MITOCHONDRIAL GENES AND DISEASES

Mitochondria contain their own genome, a circular double-stranded DNA that contains structural genes for 13 proteins of the electron transport chain including seven subunits of complex I (NADH:ubiquinone (oxidoreductase), one subunit (cytochrome *b*) of complex III (ubiquinol:cytochrome *c* oxidoreductase), three subunits of complex IV (cytochrome *c* oxidase), and two subunits of complex V (ATP synthase) (Table 14.7). Mitochondrial DNA (mtDNA) also contains genes encoding two ribosomal RNAs (rRNAs) and all the transfer RNAs (tRNAs) required for protein synthesis in mitochondria (Figure 14.55). Mitochondria, however, are not self-replicating organelles as over 90% of all mitochondrial proteins are encoded in nuclear DNA, synthesized in the cytosol, and then imported into mitochondria.

Mitochondrial defects have been implicated in several degenerative diseases of aging including Parkinson and Alzheimer diseases. Several diseases arise from point mutations in mtDNA involving either tRNAs or one of the structural genes. Other diseases arise from deletions of large portions of mtDNA. Generally, these diseases result from decreased activity of the electron transport chain, which leads to the accumulation of pyruvate and fatty acids with resultant lactate acidosis and accumulation of triglycerides. The rate of ATP synthesis is also decreased, resulting in muscle weakness and exercise intolerance. A hallmark of all mitochondrial diseases is their maternal inheritance, because essentially all mitochondria present in a fertilized ovum are derived from the egg. See Clin. Corrs. 14.4, 14.5, and 14.6 for a discussion of diseases resulting from mutations in mtDNA.

TABLE 14.7 • Subunits of Electron Transport Complexes Encoded by Human Mitochondrial DNA

Complex		*Total Number of Subunits*	*Number of Subunits Encoded by Mitochondrial DNA*
I	NADH–ubiquinone oxidoreductase	>40	7
II	Succinate dehydrogenase	4	0
III	Ubiquinol–cytochrome *c* oxidoreductase	11	1
IV	Cytochrome *c* oxidase	13	3
V	ATP synthase	12	2

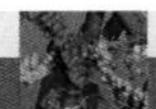

CLINICAL CORRELATION 14.4

Leber Hereditary Optic Neuropathy (OMIM 535000)

The first mitochondrial disease to be elucidated at the molecular level is maternally inherited Leber hereditary optic neuropathy (LHON), which affects the central nervous system, including the optic nerves, causing sudden-onset blindness in early adulthood due to the death of the optic nerve. In nearly all families, LHON results from single base changes in the mitochondrial genes encoding three subunits of complex I (ND1, ND4, and ND6), which lower activity of NADH: ubiquinone oxidoreductase (complex I).

The severity of mitochondrial diseases depends on the amount of mutated mtDNA present in a given cell or tissue. The presence of hundreds or thousands of mitochondria in each cell allows for considerable variation in the amount of mutant mtDNA in a tissue as a consequence of the random distribution of mutant mtDNA to the daughter cells at cell division. Patients with a lower percentage of mutant mtDNA develop sudden-onset blindness and other symptoms typical of LHON in early adulthood. Patients with a higher percentage of the mutant mtDNA in which a conserved alanine is replaced by a valine in ND6 develop a severe disease characterized by the early onset of generalized movement disorder, impaired speech, and mental retardation.

Chalmers, R. M., and Schapira, A. H. V. Clinical, biochemical and molecular genetic features of Leber hereditary optic neuropathy. *Biochim. Biophys. Acta* 1410:147, 1999.

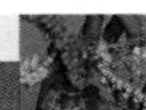

CLINICAL CORRELATION 14.5

Mitochondrial Myopathies from Mutations in Mitochondrial tRNA Genes

Point mutations in genes encoding mitochondrial tRNAs result in two of the most common mitochondrial diseases characterized by encephalomyopathy. A mutation in the tRNA gene for lysine causes **m**yoclonic **e**pilepsy and **r**agged **r**ed **f**ibers (MERRF) (OMIM 545000). Symptoms include myoclonus and ataxia with generalized seizures and myopathy. The skeletal muscles contain abnormally shaped mitochondria that contain paracrystalline structures giving an appearance of ragged red fibers (see figure) and decreased cytochrome *c* oxidase activity.

Mutation in the tRNA for leucine causes the common **m**itochondrial **e**ncephalopathy, **l**actic **a**cidosis and **s**troke-like activity (MELAS) (OMIM 540000). The skeletal muscles contain ragged red fibers but retain cytochrome *c* oxidase activity. The severity of symptoms varies with the percentage of mutated mtDNAs. Patients with >85% mutated DNA present with the more severe central nervous symptoms, those with 5%–30% mutated DNA often present with maternally inherited diabetes mellitus and deafness.

Example of paracrystalline inclusions in mitochondria from muscles of ocular myopathic patient (×36,000).
Courtesy of Dr. D. N. Landon, Institute of Neurology, University of London.

The biochemical consequence of both of these tRNA mutations is impaired mitochondrial protein synthesis leading to decreased activities of complex I and cytochrome *c* oxidase.

Wallace, D. C. Mitochondrial diseases in man and mouse. *Science* 283:1482, 1999.

Other mutations in mitochondrial genes lead to progressive muscular weakness, retinitis pigmentosa (loss of retinal response), hearing loss, and ataxia (uncoordinated muscular action), plus enlargement and deterioration of heart muscle. The deleterious effects of aging may also result from mutations that accumulate in mtDNA throughout the life of the individual, which are caused by DNA-damaging agents such as oxygen radicals.

14.10 • REACTIVE OXYGEN SPECIES (ROS)

Oxygen is essential to life. Most intracellular oxidations result in transfer of two electrons to appropriate acceptors such as NAD^+ or FAD, which are then oxidized by the electron transport chain. The terminal step in this chain is catalyzed by cytochrome *c* oxidase, which tightly binds O_2 to the binuclear center where stepwise reduction of O_2 occurs without release of intermediates in the oxidation process (see Section 14.6, p. 561). The electronic structure of O_2, however, favors its reduction by addition of one electron at a time leading

CLINICAL CORRELATION 14.6

Exercise Intolerance in Patients with Mutations in Cytochrome *b*

In 1993, a mutation in cytochrome *b* resulting in lowered activity of the cytochrome bc_1 complex was reported in a 25-year old man who presented with exercise intolerance and proximal weakness. The mutation substituted an aspartate residue for a conserved glycine at position 290. Subsequently, other patients with similar symptoms and lowered bc_1 complex activity were shown to have mutations in which glutamate was substituted for a conserved glycine at position 339, and serine for a conserved glycine at position 34. More recently, a patient with severe hypertrophic cardiomyopathy was shown to have a mutation in which a glutamate was substituted for a conserved glycine at position 166. The mutations of glycine to aspartate or glutamate were located in cytochrome *b* close to the Q_O site for ubiquinol oxidation, while the glycine to serine mutation was located near the Q_i site of ubiquinone reduction. All these mutations involved a guanine to adenine transition in the mtDNA suggesting that the mutation may have resulted from oxidative damage. Moreover, in each of the missense mutations a conserved glycine was replaced by a larger charged molecule, which altered the structure of cytochrome *b* resulting in lowered catalytic activity of the bc_1 complex. Nonsense mutations resulting in truncation of cytochrome *b* and mutations involving deletions of 4 to 24 base pairs of mtDNA have been identified. These nonsense and deletion mutations often lead to severe exercise intolerance, lactic acidosis in the resting state and occasionally myoglobinuria. In contrast to the majority of mutations in mtDNA, the mutations identified in the cytochrome *b* gene were not maternally inherited. Moreover, most have only been expressed in muscle tissues suggesting that these are somatic mutations, which occurred during germ-layer differentiation of myogenic stem cells.

Andreu, A. L., Hanna, M. G., Reichmann, H., Bruno, C., et al. Exercise intolerance due to mutations in the cytochrome *b* gene of mitochondrial DNA. *N. Engl. J. Med.* 341:1037, 1999.

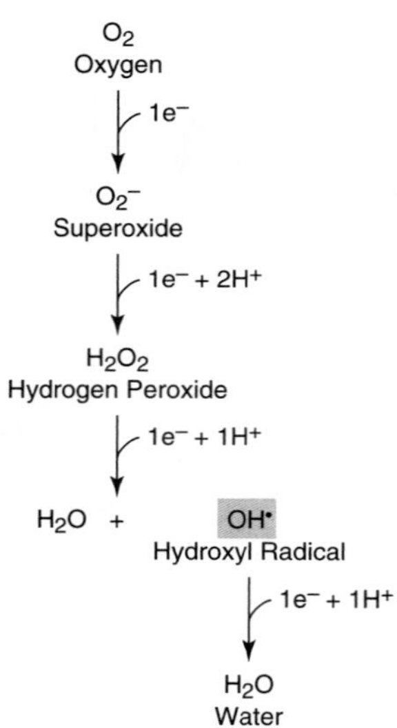

Figure 14.56 One electron steps in reduction of oxygen leading to formation of reactive oxygen species superoxide, hydrogen peroxide, and hydroxyl radical.

Fenton Reaction

$$Fe^{2+} + H_2O_2 \longrightarrow Fe^{3+} + OH^{\bullet} + OH^-$$

Haber–Weiss Reaction

$$O_2^- + H_2O_2 \xrightarrow{H^+} O_2 + H_2O + OH^{\bullet}$$

Figure 14.57 The Fenton and Haber–Weiss reactions for formation of toxic hydroxyl radical.

to the generation of oxygen radicals that can cause cellular damage. A radical is a molecule with a highly reactive unpaired electron in an outer orbital, which can initiate chain reactions by removal of an electron from another molecule to complete its own orbital. The stepwise transfer of electrons to O_2 results in the sequential formation of superoxide anions (O_2^-), hydrogen peroxide (H_2O_2), and finally **hydroxyl free radicals ($OH\cdot$)** (Figure 14.56). The hydroxyl radical is undoubtedly the most dangerous free radical as it is involved in reactions such as lipid peroxidation and generation of other toxic radicals. Hydrogen peroxide itself is not a free radical, but is converted by the Fenton or Haber–Weiss reactions to the hydroxyl radical in the presence of Fe^{2+} or Cu^+, which are prevalent in cells (Figure 14.57).

Production of Reactive Oxygen Species

While oxidative processes in cells generally result in transfer of electrons to O_2 to form water without release of intermediates, a small number of oxygen radicals are inevitably formed due to leakage in electron transfer reactions. The major intracellular source of oxygen radicals is the mitochondrial electron transport chain where superoxide is produced by transfer of one electron to O_2 from the stable semiquinone produced during reduction of ubiquinone by complexes I and II or during oxidation of ubiquinol by complex III (Figure 14.58). Superoxide can also be produced by transfer of an electron from a flavin such as FMN. The reactive oxygen species produced in mitochondria include superoxide, hydrogen peroxide, and hydroxyl radical. Toxic oxygen species are also produced in peroxisomes in which long-chain fatty acids and other compounds are oxidized by transfer of two electrons from $FADH_2$ to O_2 with formation of hydrogen peroxide, which is readily converted to hydroxyl radical (see Figure 14.56). The cytochrome P450 system localized in endoplasmic reticulum can also produce oxygen radicals.

An additional source of ROS in many tissues in the body including neutrophils is the membrane-bound NADPH-dependent oxidase system. Inflammation due to bacterial infection in neutrophils results in the activation of NADPH oxidase which produces superoxide in a process known as the respiratory burst. The conversion of superoxide to hydroxyl radical kills bacteria which are then engulfed by phagocytes (Figure 14.59). In an acute infection, production of oxygen radicals and killing of bacteria are efficient processes; however, in prolonged infections, phagocytes tend to die, releasing toxic oxygen radicals that affect surrounding cells.

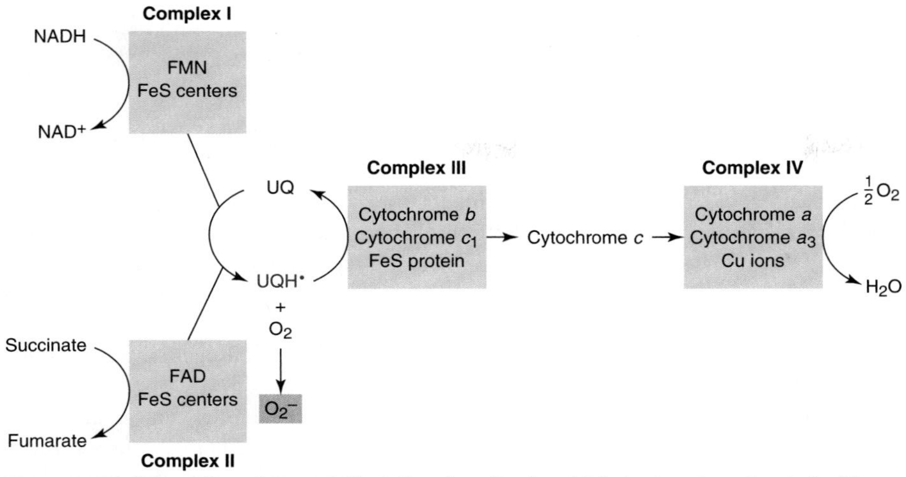

Figure 14.58 Generation of superoxide anions by mitochondrial electron transfer chain. The semiquinone formed during the two electron reduction of ubiquinone by the iron–sulfur centers of both complexes I and II can transfer an electron to oxygen to form the superoxide anion. By contrast, the binuclear center of cytochrome *c* oxidase prevents release of intermediates in the reduction of oxygen.

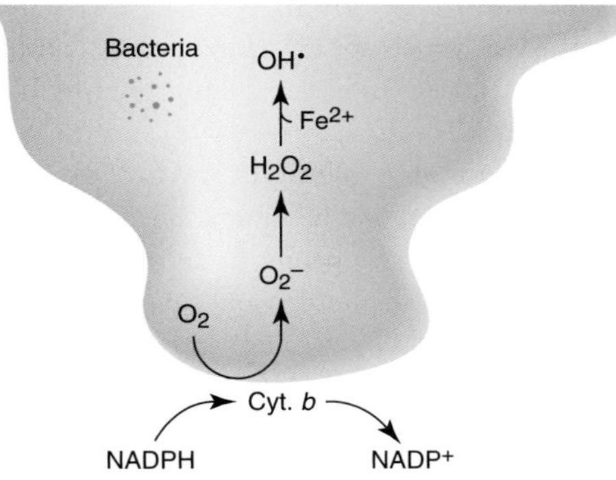

Figure 14.59 Respiratory burst in phagocytes. An electron transfer chain involving a unique cytochrome *b* transfers electrons from NADPH to oxygen with formation of superoxide anion. Superoxide is converted to the hydroxyl radical that kills bacteria subsequently engulfed by phagocytes.

In addition to its presence in phagocytes, different forms of NADPH oxidase have been reported in many tissues in the body, where they are the major non-mitochondrial source of ROS. The physiological role of these NADPH oxidases in cardiovascular tissues includes cellular processes such as signal transduction, cell proliferation, and apoptosis. Evidence suggests that excess NADPH oxidase can generate ROS contributing to pathological conditions such as atherosclcerosis, hypertension, and other vascular pathologies (Clin. Corr. 14.7).

Cosmic radiation, ingestion of chemicals, and drugs, as well as smog, can lead to formation of reactive oxygen species. Damage from reactive oxygen species often occurs during perfusion of tissues with solutions containing high concentrations of O_2 as happens in patients who have suffered an ischemic episode in which localized O_2 levels are lowered due to blockage of an artery and have then undergone thrombolytic or other procedures to remove the blockage (Clin. Corr. 14.8).

CLINICAL CORRELATION 14.7

NADPH Oxidase (NOX) in Health and Disease

The discovery and characterization of NADPH-dependent oxidase (NOX) in phagocytic cells had suggested that this tissue was the sole location of NOX in the body, where its physiological role was to produce large amounts of superoxide to kill microbes. Elucidation of the biochemical properties of phagocytic NOX also benefited from studies of patients with chronic granulomatous disease, which presents as an inability of affected individuals to combat infection resulting from mutations in NOX, which lead to loss of ability to produce adequate amounts of superoxide. The recent development of more sensitive methods to detect the reactive oxygen species, superoxide and hydrogen peroxide, indicated that NOX was more widely distributed in the body and that furthermore ROS might contribute to physiological processes in non-phagocytic tissues. Subsequently, seven different isoforms of NOX family have been identified in almost all tissues of the body. The role of NOX and superoxide production in nonphagocytic tissues has not been completely resolved as the low levels of these enzymes hinder direct measurements; however, NOX in all tissues is activated by various stimuli and may trigger redox-sensitive signaling pathways that are involved in endothelial activation, cell growth, and apoptosis. Recent studies have revealed the presence of three different isoforms of NOX in cardiovascular tissues; smooth muscle, endothelium, cardiomyocytes, and vascular adventitium. These isoforms of NOX are thought to contribute to maintenance of vascular tone, cell proliferation, angiogenesis, and apoptosis. In addition to the role of NOX in these pathways, the regulatory pathways that maintain vascular tone reflect a complex interplay between ROS and reactive nitrogen species (NO) produced by endothelial nitric oxide synthase.

Despite the beneficial roles of NOX in cell processes, excess ROS generated by NOX may result in numerous pathological conditions including endothelial cell dysfunction that contributes to atherosclerosis, hypertension, congestive heart failure, ischemic reperfusion injury, and the vascular problems associated with diabetes. Continued investigations of the role that NOX plays in normal physiological and in pathological processes will provide insights into the development of therapies needed to combat the chronic diseases that plague humans.

Nauseff, W. M. Biological roles for the NOX family NADPH oxidases. *J. Biol. Chem.* 283:16961, 2008.

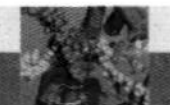

CLINICAL CORRELATION 14.8

Myocardial Reperfusion Injury

The occlusion of a major coronary artery during myocardial infarction results in ischemia or lowered oxygen supply to the affected area leading to damaged cells, the infarct. After an acute myocardial infarction, early reperfusion with appropriate therapy results in reduction of the size of the infarct and a better clinical outcome for the patient. Restoring blood flow to the ischemic tissue, however, may damage the heart in a process termed myocardial reperfusion injury. During reperfusion, the ischemic myocardium is subject to rapid biochemical changes, which may cause further damage to the myocardium. These changes include rapid restoration of electron transfer through the mitochondrial electron transfer chain with the concomitant generation of ROS, which are also generated by xanthine oxidase present in endothelial cells and several hours later by NADPH oxidase located in neutrophils. These increased levels of ROS cause damage to the heart by contributing to intracellular Ca^{2+}overload, rapid restoration of intracellular pH, and inflammation. The increase in intra-mitochondrial Ca^{2+} leads to the opening of the mitochondrial transition pore (MRP), which triggers apoptosis and subsequent cell death. ROS also damage the sarcoplasmic reticulum and cell membrane by lipid peroxidation, induce enzyme denaturation, and cause oxidative damage to DNA. The rapid increase in intracellular ATP levels that occurs when oxygen is introduced during reperfusion due to stimulation of mitochondrial electron transfer combined with the high intracellular levels of Ca^{2+} and restoration of physiologic pH may result in hypercontracture and the resultant death of cardiomyocytes.

The formation of ROS during myocardial reperfusion also results in lowered levels of nitric oxide (NO), an important signaling molecule, thus reducing its protective effects such as neutrophil accumulation, inactivation of superoxide radicals, and improvement of coronary blood flow. During reperfusion of the ischemic heart, the presence of the oxygen radical peroxynitrite ($ONOO^-$) produced from NO and superoxide, has been reported. Peroxynitrite may also contribute to the poor recovery of mechanical function in ischemic hearts.

Myocardial ischemia-reperfusion represents a clinical problem associated with thrombolysis, angioplasty, and coronary bypass surgery. Injuries to the myocardium due to ischemia-reperfusion include cardiac contractile dysfunction, arrhythmias, and irreversible myocyte damage. Ischemia may also arise during surgery especially during transplantation of tissues. Current investigations of possible interventions to prevent reperfusion injury are focused on better understanding of the factors involved in injury especially the opening of the mitochondrial PTP that results in cell death by apoptosis. Clinical trials are currently underway to test new hypotheses to reduce injury from reperfusion. The increased use of invasive procedures in clinical medicine indicates the importance of developing methods to protect against ischemic-reperfusion injury.

Yellon, D. M., and Hausenloy, D. J. Myocardial reperfusion injury. *N. Engl. J. Med.* 357:1121, 2007.

Damage Caused by Reactive Oxygen Species

Reactive oxygen species cause damage to all major classes of macromolecules in cells. The phospholipids of plasma and organelle membranes are subject to lipid peroxidation, a free-radical chain reaction initiated by removal of hydrogen from a polyunsaturated fatty acid by hydroxyl radical. The resulting lipid radicals then react with O_2 to form lipid peroxy radicals and lipid peroxide along with malondialdehyde, which is water soluble and can be detected in blood. The effect of lipid peroxidation in humans is exemplified by the brown spots commonly observed on hands of the elderly. These "age" spots contain the pigment lipofuscin, which is probably a mixture of cross-linked lipids and products of lipid peroxidation, which accumulate over the course of a lifetime. One significant consequence of lipid peroxidation is increased membrane permeability leading to an influx of Ca^{2+} and other ions with subsequent swelling of the cell. Similar increases in permeability of organelle membranes may result in maldistribution of ions and cause intracellular damage. For example, accumulation of excessive amounts of Ca^{2+} in mitochondria may trigger apoptosis.

Proline, histidine, arginine, cysteine, and methionine are susceptible to attack by hydroxyl radicals with subsequent fragmentation of proteins, cross-linking, and aggregation. Proteins damaged by oxygen radicals may be targeted for digestion by intracellular proteases.

The most important consequence of oxygen radicals is damage to mitochondrial and nuclear DNA resulting in mutations. The nonspecific binding of ferrous ions (Fe^{2+}) to DNA may result in localized formation of hydroxyl radicals that attack individual bases and cause strand breaks. Mitochondrial DNA is more susceptible to damage, since the electron transport chain is a major source of toxic oxygen radicals. In addition, nuclear DNA is protected from permanent damage by a protective coat of histones as well as by active and efficient mechanisms for DNA repair. Damage to mtDNA generally results in mutations that affect energy production. The symptoms in effected individuals are manifest in energy requiring processes such as muscle contraction. An example of the consequences of a somatic mutation in the mitochondrial gene for cytochrome *b* that may have been caused by oxygen radicals is presented in Clin. Corr. 14.6, page 584.

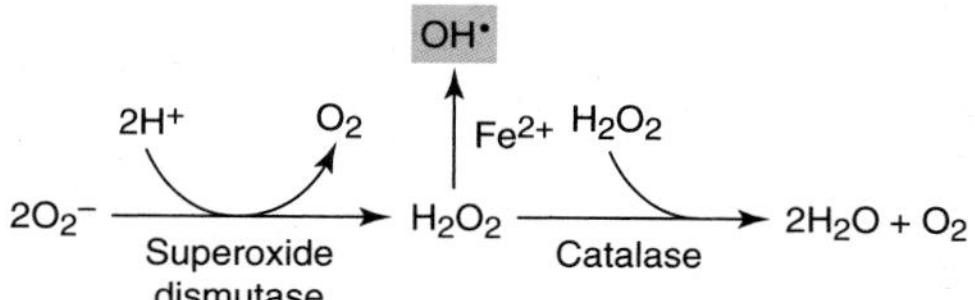

Figure 14.60 Superoxide dismutase and catalase protect cells by removing superoxide and hydrogen peroxide.

Cellular Defenses against Reactive Oxygen Species

Cells that live in an aerobic environment have developed multiple ways to remove reactive oxygen species and thus protect themselves against their deleterious effects. Mammals have three different isozymes of **superoxide dismutase** that catalyze conversion of superoxide to hydrogen peroxide (Figure 14.60). The cytosolic form of superoxide dismutase contains Cu/Zn at its active site, as does the extracellular enzyme; however, the mitochondrial enzyme contains Mn at its active site. Hydrogen peroxide is removed by catalase, a heme-containing enzyme present in highest concentration in peroxisomes and to a lesser extent in mitochondria and cytosol.

Glutathione peroxidase catalyzes reduction of both hydrogen peroxide and lipid peroxides (Figure 14.61). This selenium-containing enzyme uses the sulfhydryl groups of glutathione (GSH) as a hydrogen donor with formation of oxidized or disulfide form of glutathione (GSSG). **Glutathione reductase** converts the disulfide form back to the sulfhydryl form using NADPH produced in the pentose phosphate pathway as an electron donor. The importance of NADPH in maintaining glutathione levels and thus preventing oxidative damage in cells is manifest in the inherited disorder resulting in lowered levels of glucose 6-phosphate dehydrogenase (see Clin. Corr. 16.1, p. 648) in which oxidative stress results in acute hemolytic anemia. Glutathione levels in the liver are critical to prevent hepatic necrosis induced by the common pain killer acetaminophen (sold as Tylenol, Panadol, or Paracetomol) especially in young children.

Protection against reactive oxygen species may also be gained by ingestion of oxygen scavengers such as vitamins C and E and β-carotene. Recent evidence has indicated that substances present in green tea, berries (especially blueberries), and red wine may also protect against oxidative damage by ROS.

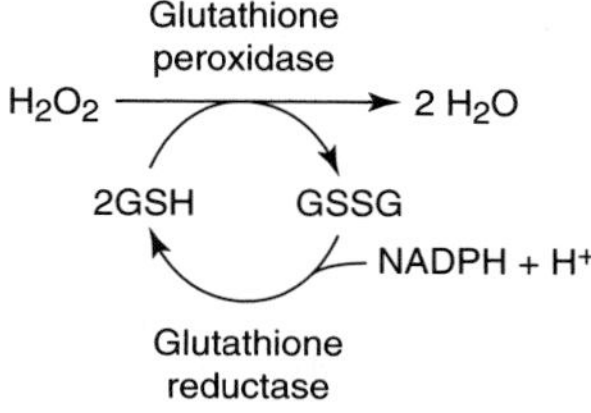

Figure 14.61 Glutathione peroxidase removes hydrogen peroxide as well as lipid peroxides. Electrons are transferred to hydrogen peroxide from the sulfhydryl groups of reduced glutathione (GSH) to form water and oxidized glutathione (GSSG). Glutathione reductase then reduces GSSG to GSH with NADPH as the reducing agent.

Bibliography

Energy-Producing and Energy-Utilizing Systems

Hanson, R. W. The role of ATP in metabolism. *Biochem. Educ.* 17: 86, 1989.

Sources and Fates of Acetyl Coenzyme A

Harris, R. A., Bowker-Kinley, M. M., Huang, B., and Wu, P. Regulation of the activity of the pyruvate dehydrogenase complex. *Adv. Enzyme Regulation* 42: 249, 2002.

Holness, M. J., and Sugden, M. C. Regulation of pyruvate dehydrogenase complex activity by reversible phosphorylation. *Biochem. Soc. Trans.* 31: 1143, 2003.

The Tricarboxylic Acid Cycle

Hajnoczky, G., Csordas, G., Krishnamurthy, R., and Szalai, G. Mitochondrial calcium signaling driven by the IP_3 receptor. *J. Bioenerg. Biomembr.* 32: 15, 2000.

Krebs, H. A. The history of the tricarboxylic acid cycle. *Perspect. Biol. Med.* 14: 154, 1970.

Ovadi, J., and Srere, P. A. Macromolecular compartmentation and channeling. *Int. Rev. Cytol.* 192: 255, 2000.

Owen, O. E., Kalhan, S. C., and Hanson, R. W. The key role of anaplerosis and cataplerosis for citric acid cycle function. *J. Biol. Chem.* 277: 30409, 2002.

Structure and Compartmentation of Mitochondrial Membranes

Scheffler, I.E. *Mitochondria.* New York: Wiley-Liss, 1999.

Electron Transport

Collman, J. P., Devaraj, N. K., Decreau, R. A., Yang, Y., et al. A cytochrome *c* oxidase model catalyzes oxygen to water reduction under rate-limiting electron flux. *Science* 315: 1565, 2007.

Hunte, C., Pasldottir, H., and Trumpower, B. L. Protonmotive pathways and mechanisms in the cytochrome bc_1 complex. *FEBS Letters* 545: 39, 2003.

Mills, D. A., and Ferguson-Miller, S. Understanding the mechanism of proton movement linked to oxygen reduction in cytochrome *c* oxidase: lessons from other proteins. *FEBS Letters* 545: 47, 2003.

Richter, O-M. H., and Ludwig, B. Cytochrome *c* oxidase: Structure, function, and physiology of a redox-driven molecular machine. *Rev. Physiol. Biochem. Pharmacol.* 147: 47, 2003.

Sazanov, L. A. Respiratory complex I: Mechanistic and structural insights provided by the crystal structure of the hydrophilic domain. *Biochemistry* 46: 2275, 2007.

Xia, D., Yu, C. A., Kim, H., Xia, J. Z., et al. Crystal structure of the cytochrome bc_1 complex from bovine heart mitochondria. *Science* 277: 60, 1997.

Yagi, T., and Matsuno-Yagi, A. The proton-translocating NADH-quinone oxidoreductase in the respiratory chain: The secret unlocked. *Biochemistry* 42: 2266, 2003.

Zhang, Z., Berry, E. A., Huang, L. S., and Kim, S. H. Mitochondrial cytochrome bc_1 complex. *Subcell. Biochem.* 35: 541, 2000.

Oxidative Phosphorylation

Boyer, P. D. A research journey with ATP Synthase. *J. Biol. Chem.* 277: 39045, 2002.

Junge, W., and Nelson, N. Nature's rotary electromotors. *Science* 308: 642, 2005.

Mitchell, P. Vectorial chemistry and the molecular mechanism of chemiosmotic coupling: Power transmission by proticity. *Biochem. Soc. Trans.* 4: 399, 1976.

Rich, P. Chemiosmotic coupling: the cost of living. *Nature* 421: 583, 2003.

Sambongi, Y., Iko, Y., Tanabe, M. Omote, H., et al. Mechanical rotation of the c subunit oligomer in ATP synthase (F_0F_1):Direct observation. *Science* 286: 1722, 1999.

Stock, D., Gibbons, C. Arechaga, I., Leslie, A. G., and Walker J. E. The rotary mechanism of ATP synthase. *Curr. Opin. Struc. Biol.* 10: 672, 2000.

Weber, J., and Senior, A. E. ATP synthesis driven by proton transport in F_1F_O–ATP synthase. *FEBS Letters* 545: 61, 2003.

Weber, J. ATP synthase: The structure of the stator stalk. *Trends in Biochemical Sciences* 32: 53, 2007.

Mitochondrial Transport Systems

Palmieri, F. The mitochondrial transporter family (SLC25): Physiological and pathological implications. *Pflugers Arch.* 447: 68909, 2003.

Pebay-Peyroula, E., Dahout-Gonzalez, C., Kahn, R., Trezeguet, V., et al. Structure of mitochondrial ADP/ATP carrier in complex with carboxyatractyloside. *Nature* 426: 39, 2003.

Rousset, S., Alves-Guerra, M.-C., Mozo, J., Miroux, B., et al. The biology of mitochondrial uncoupling proteins. *Diabetes* 53, Supp.1: S130, 2004.

Mitochondrial Genes and Mitochondrial Diseases

DiMauro, S., and Schon, E. A. Mitochondrial respiratory-chain diseases. *N. Engl. J. Med.* 348: 2656, 2003.

Schon, E. A., and Manfredi, G. Neuronal degeneration and mitochondrial dysfunction. *J. Clin. Invest.* 111: 303, 2003.

Wallace, D. C. Mitochondrial diseases in mice and man. *Science* 283: 1482, 1999.

Reactive Oxygen Species

Hansford, R., Tsuchiya, N., and Pepe, S. Mitochondria in heart ischemia and aging. *Biochem. Soc. Symp.* 66: 141, 1999.

Quinn, M. R., Ammons, M. D., and Deleo, F. R. The expanding role of NADPH oxidases in health and disease: No longer just agents of death and destruction. *Clin. Sci.* 111: 1, 2006.

Raha, S., and Robinson, B. H. Mitochondria, oxygen free radicals, disease, and ageing. *Trends in Biochem. Sci.* 25: 502, 2000.

Ray, R., and Shar, A. M. NADPH oxidase and endothelial cell function. *Clin. Sci.* 109: 217, 2005.

Sastre, J., Pallardo, F. V., and Vina, J. The role of mitochondrial oxidative stress in aging. *Free Radical Biology and Medicine* 35: 1, 2003

Turrens, J. F. Mitochondrial formation of reactive oxygen species. *J. Physiol.* 552: 335, 2003.

Key Terms

anabolism
catabolism
adenosine triphosphate
high-energy bonds
nicotinamide adenine dinucleotide
reducing equivalents
exergonic reactions
free energy
endergonic reactions
equilibrium constant
tricarboxylic acid cycle
flavin adenine dinucleotide
lipoamide
pyruvate dehydrogenase
thiamin pyrophosphate
electron transport chain
mitochondrial matrix
iron–sulfur centers
succinate dehydrogenase
biosynthetic processes
anaplerotic reactions
inner mitochondrial membrane
outer mitochondrial membrane
mitochondrial DNA
oxidation-reduction potential
redox couple
complex I: NADH–ubiquinone oxidoreductase
complex II: succinate–ubiquinone oxidoreductase
complex III: ubiquinol–cytochrome *c* oxidoreductase
complex IV: cytochrome *c* oxidase
cytochrome *b*, cytochrome *c*
ubiquinone–ubiquinol
flavin mononucleotide
glycerol 3-phosphate dehydrogenase
oxidative phosphorylation
protonmotive force
P/O ratios
uncouplers
F_1F_0
mitochondrial substrate transport
adenine nucleotide translocase
shuttles for reducing equivalents
mitochondrial diseases
hydroxyl free radicals
reactive oxygen species
ATP synthase
catalase
glutathione
glutathione peroxidase
superoxide dismutase

Questions • CAROL N. ANGSTADT

Multiple-Choice Questions

1. A bond may be *high energy* for any of the following reasons except
   A. products of its cleavage are more resonance stabilized than the original compound.
   B. the bond is unusually stable, requiring a large energy input to cleave it.
   C. electrostatic repulsion is relieved when the bond is cleaved.
   D. a cleavage product may be unstable, tautomerizing to a more stable form.
   E. the bond may be strained.

2. All of the following tricarboxylic acid cycle intermediates may be added or removed by other metabolic pathways except
   A. citrate.
   B. fumarate.
   C. isocitrate.
   D. α-ketoglutarate.
   E. oxaloacetate.

3. The inner mitochondrial membrane contains a transporter for
   A. NADH.
   B. acetyl CoA.
   C. GTP.
   D. ATP.
   E. NADPH.

4. During the transfer of electrons to O_2 via the mitochondrial electron transport chain,
   A. the energy released is used to translocate protons across the inner membrane.
   B. a proton gradient is generated with the matrix now being more positive than the intermembrane space.
   C. pumping of protons across the membrane occurs each time electrons are moved.
   D. no charge gradient develops because an OH^- moves each time a proton does.
   E. the energy is used directly in the addition of P_i to ADP to form ATP.

5. ATP synthase (also known as complex V) consists of two domains, F_1 and F_o.
   A. F_1 and F_o are both integral membrane protein complexes of the outer membrane.
   B. F_1 domain provides a channel for translocation of protons across the membrane.
   C. F_1 binds ATP but not ADP.
   D. F_1 domain catalyzes the synthesis of ATP.
   E. Only the F_o domain contains more than one subunit.

6. All of the following statements are correct except
   A. reactive oxygen species (oxygen radicals) result when there is a concerted addition of four electrons at a time to O_2.
   B. superoxide anion (O_2^-) and hydroxyl radical ($\cdot OH$) are two forms of reactive oxygen.
   C. superoxide dismutase is a naturally occurring enzyme that protects against damage by converting O_2^- to H_2O_2.
   D. reactive oxygen species damage phospholipids, proteins, and nucleic acids.
   E. glutathione protects against H_2O_2 by reducing it to water.

Questions 7 and 8: A child presented with severe neurological defects. Blood tests indicated elevated serum levels of lactate, pyruvate, and alanine. Cultures of skin fibroblasts showed deficient activity of the pyruvate–dehydrogenase complex. Further study might indicate which subunit of the complex is defective, although the metabolic effects are essentially the same regardless of which subunit is defective.

7. The active form of pyruvate dehydrogenase is favored by the influence of all of the following on pyruvate dehydrogenase kinase except
   A. low [Ca^{2+}].
   B. low acetyl CoA/CoASH.
   C. high [pyruvate].
   D. low $NADH/NAD^+$.

8. Suppose the specific defect were a mutant pyruvate dehydrogenase (the first catalytic subunit) with poor binding of its prosthetic group. In this type of defect, sometimes greatly increasing the dietary precursor of the prosthetic group is helpful. In this case, increasing which of the following might be helpful?
   A. Lipoic acid
   B. Niacin (for NAD)
   C. Pantothenic acid (for CoA)
   D. Riboflavin (for FAD)
   E. Thiamin (for TPP)

Questions 9 and 10: There are a number of mitochondrial diseases arising from mutations in the mitochondrial genes encoding for mitochondria proteins of mitochondrial tRNAs. Leber Hereditary Optic Neuropathy (LHON) arises from single base changes in the mitochondrial genes encoding three subunits of complex I, diminishing its activity. LHON affects the central nervous system, including the optic nerves, causing sudden-onset blindness in early adulthood.

9. Complex I
   A. transfers electrons directly from NADH to ubiquinone.
   B. can transfer electrons from the $FADH_2$ of succinate dehydrogenase as well as from NADH.
   C. does not contain FeS centers.
   D. transfers electrons coupled to the transport of four protons across the membrane.
   E. is considered a mobile electron carrier because it can travel along the outer face of the inner membrane.

10. An inability to reoxidize NADH because of the defect of complex I would
    A. inhibit isocitrate dehydrogenase and, therefore, inhibit the rate of the TCA cycle.
    B. force the oxaloacetate–malate equilibrium toward oxaloacetate.
    C. promote the α-glycerol phosphate shuttle for transporting reducing equivalents.
    D. cause NADH to freely diffuse from the mitochondria to the cytosol.
    E. force the succinate–fumarate equilibrium toward succinate.

Questions 11 and 12: Cyanide is a potent and rapidly acting poison. It binds to the heme of the cytochrome a_3 in cytochrome oxidase. Death occurs from tissue anoxia, especially in the central nervous system. An antidote to cyanide poisoning, which must be given rapidly, is the administration of a nitrite to convert oxyhemoglobin to methemoglobin. Methemoglobin competes with cytochrome a_3 for cyanide.

11. Cyanide
    A. only minimally inhibits the electron transport chain because cytochrome oxidase is a terminal component of the chain.
    B. inhibits mitochondrial respiration but energy production is unaffected.
    C. also binds the copper of cytochrome oxidase.
    D. binds to Fe^{3+} of cytochrome a_3.
    E. poisoning could also be reversed by increasing O_2 concentration.

12. If cyanide is added to tightly coupled mitochondria that are actively oxidizing succinate,
   A. subsequent addition of 2,4-dinitrophenol will cause ATP hydrolysis.
   B. subsequent addition of 2,4-dinitrophenol will restore succinate oxidation.
   C. electron flow will cease, but ATP synthesis will continue.
   D. electron flow will cease, but ATP synthesis can be restored by subsequent addition of 2,4-dinitrophenol.
   E. subsequent addition of 2,4-dinitrophenol and the phosphorylation inhibitor, oligomycin, will cause ATP hydrolysis.

Problems

13. For the reaction $A \rightleftharpoons B$, $\Delta G^{0'} = -29.7$ kJ/mol. At 37°C, $-2.303\ RT = -5.94$ kJ/mol. What is the equilibrium ratio of B/A?

14. Using pyruvate, labeled with ^{14}C in its keto group, via the pyruvate dehydrogenase reaction and the TCA cycle, where would the carbon label be at the end of one turn of the TCA cycle? Where would the carbon label be at the end of the second turn of the cycle?

Answers

1. **B** High energy does not refer to a high energy of formation (bond stability). A *high-energy* bond is so designated because it has a high free energy of hydrolysis. This could arise for reasons A, C, D, or E.
2. **C** A: Citrate is transported out of the mitochondria to be used as a source of cytoplasmic acetyl CoA. B: Fumarate is produced during phenylalanine and tyrosine degradation. D: α-Ketoglutarate can be formed from glutamate. E: Oxaloacetate is produced by pyruvate carboxylase and is used in gluconeogenesis. Clearly most of the tricarboxylic acid cycle intermediates play multiple roles in the body.
3. **D** ATP and ADP are transported in opposite directions. A and B: Reducing equivalents from NADH are shuttled across the membrane, as is the acetyl group of acetyl CoA, but NADH and acetyl CoA cannot cross. C: Of the nucleotides, only ATP and ADP are transported. E: Like NADH, NADPH does not cross the membrane.
4. **A** This is the direct utilization of the energy. B: Movement of protons is from the matrix to the intermembrane space so the latter becomes more positive. C: Only certain electron transfers have sufficient energy to pump electrons. D: No negative ion is transported with the proton so a charge gradient develops. E: The energy is required to release ATP from the subunit, not for the actual synthesis.
5. **D** This is correct. A: Both are associated with the inner mitochondrial membrane. B: This is the role of the F_o domain. C: Both bind and ATP synthesis occur. E: Changing conformations of different subunits is important in the actions of both domains.
6. **A** Four electron transfer to oxygen produces water. Oxygen radicals occur when electrons are added stepwise to O_2. B: H_2O_2 is also included as reactive oxygen because it can produce $\cdot OH$. C: The H_2O_2 is then degraded by catalase. D: All of these are subject to oxidation with deleterious effects. E: Glutathione peroxidase catalyzes this reaction.
7. **A** High Ca^{2+} favors the active dehydrogenase but by activating the phosphatase. B, C, and D: NADH and acetyl CoA activate pyruvate dehydrogenase kinase, thus inactivating pyruvate dehydrogenase. Pyruvate inhibits the kinase, favoring the active dehydrogenase.
8. **E** TPP is the cofactor for the first reaction which decarboxylate pyruvate. A and C: These are cofactors for the dihydrolipoyl transacetylase. B and D: These are cofactors for the dihydrolipoyl dehydrogenase.
9. **D** This contributes to the protonmotive force required for the synthesis of ATP. A and C: The transfer is not direct but proceeds through FMN and the FeS centers. B: Electrons are transferred from NADH to FMN linked to one of the subunits of complex I, but complex I cannot react with succinate dehydrogenase. E: This is cytochrome *c*.
10. **A** This is the key regulatory enzyme of the TCA cycle. B: Malate is the reduced form so it would be favored. C: The cytosolic enzyme uses NADH and transports reducing equivalents into the mitochondria. The mitochondrial enzyme is FAD linked. D: NADH does not cross the membrane. E: This reaction is $FADH_2$ linked.
11. **D** This is why methemoglobin is an effective antidote since it also has an Fe^{3+}. B: Respiration and energy production are coupled so inhibiting one inhibits the other as well. C: Copper is an important part of cytochrome oxidase but cyanide does not bind it. E: The problem is inability to react with O_2, not a lack of O_2.
12. **A** Cyanide inhibits electron transport at site III, blocking electron flow throughout the system. In coupled mitochondria, ATP synthesis ceases too. Addition of an uncoupler permits the mitochondrial ATPase (which is normally driven in the synthetic direction) to operate, and it catalyzes the favorable ATP hydrolysis reaction unless it is inhibited by a phosphorylation inhibitor such as oligomycin.
13. $\Delta G^{0'} = -RT \ln K = -2.303 RT \log K$. Substitution gives the $\log K = 5$. K then is 100,000 so the B/A = 100,000/1.
14. The labeled keto carbon of pyruvate becomes the labeled carboxyl carbon of acetyl CoA. After condensation with oxaloacetate, the first carboxyl group of citrate is labeled. This label is retained through subsequent reactions to succinate. However, succinate is a symmetrical compound to the enzyme so, in effect, both carboxyl groups of succinate are labeled. This means that the oxaloacetate regenerated is labeled in both carboxyl groups at the end of one turn (actually, half the molecules are labeled in one carboxyl and half in the other but this can't be distinguished experimentally). Note that CO_2 is not labeled. In the second turn, the same carboxyl labeled acetyl CoA is added but this time to labeled oxaloacetate. Both carboxyl groups of the oxaloacetate are released as CO_2 so it will be labeled, as will the regenerated oxaloacetate.

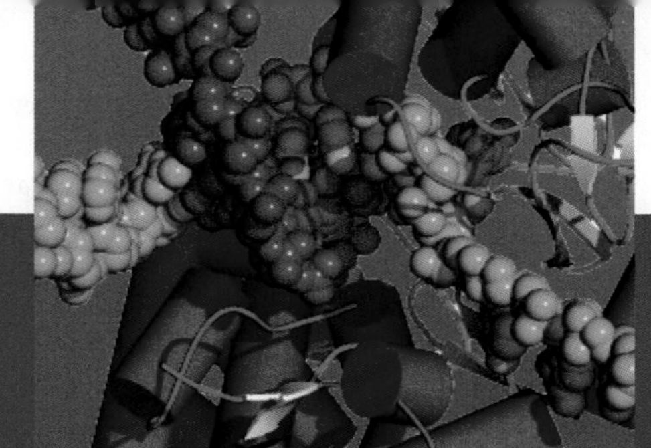

15

Carbohydrate Metabolism I: Major Metabolic Pathways and Their Control

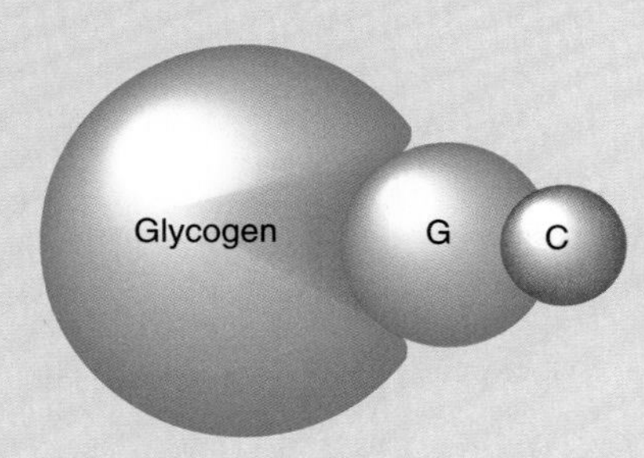

Robert A. Harris

Distinguished Professor Emeritus and Showalter Professor of Biochemistry Emeritus, Indiana University School of Medicine

Key Concepts

- Glucose is metabolized by all mammalian cells to form ATP. Anaerobic (absence of oxygen) glycolysis produces two molecules each of lactate and ATP from one molecule of glucose. Aerobic glycolysis (presence of oxygen) produces two molecules each of NADH and pyruvate. NADH must be reoxidized for glycolysis to continue.
- Glycolysis is regulated at three steps. 6-Phosphofructo-1-kinase catalyzes the committed step of glycolysis and is regulated both negatively and positively by allosteric effectors.
- Gluconeogenesis, the formation of glucose from non-carbohydrate substrates, is required to maintain blood glucose and involves those glycolytic enzymes that catalyze reversible reactions. The reactions of glycolysis that are irreversible are bypassed by specific reactions of gluconeogenesis. Some amino acids are glucogenic but fatty acids with an even number of carbons are not.
- Gluconeogenesis is inhibited by insulin and stimulated by glucagon, mediated by regulation of the phosphorylation states of regulatory enzymes. Long-term stimulatory effects of glucagon and inhibitory effects of insulin on gluconeogenesis are mediated by induction and repression of key enzymes of the glycolytic/gluconeogenic pathways.
- When glucose is plentiful, liver and skeletal muscle synthesize and store glycogen. Skeletal muscle utilizes stored glycogen for the synthesis of ATP during exercise and liver releases glucose

from glycogen when blood glucose levels are low. Muscle does not release free glucose.
- Glycogen synthase exists in a phosphorylated form that is dependent for activity on the presence of glucose 6-phosphate, a positive allosteric effector, and a nonphosphorylated form that is independent of glucose 6-phosphate.
- Glycogen phosphorylase, a key enzyme in degradation of glycogen, is activated by AMP and inhibited by glucose and ATP, and is also activated by a signaling cascade involving protein kinases.
- Glycogen storage diseases are caused by inherited defects in enzymes involved in glycogen degradation.

15.1 • INTRODUCTION

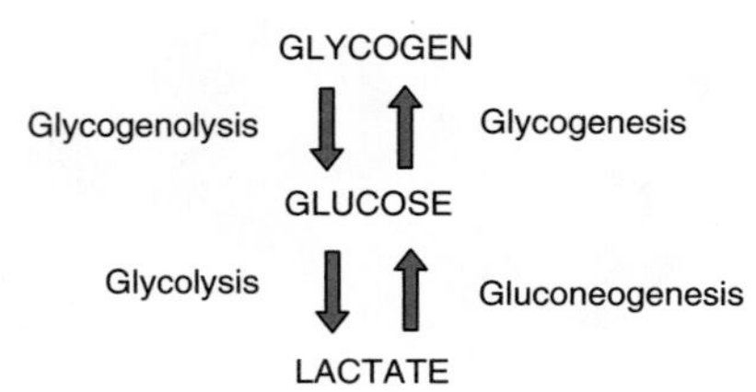

Figure 15.1 Relationship of glucose to major pathways of carbohydrate metabolism.

The major pathways of carbohydrate metabolism begin or end with glucose (Figure 15.1). This chapter describes the utilization of glucose as a source of energy, formation of glucose from non-carbohydrate precursors, storage of glucose in the form of glycogen, and release of glucose from glycogen. An understanding of the pathways and their regulation is necessary because of the important role played by glucose in the body. Glucose is the major form in which carbohydrate absorbed from the intestinal tract is presented to the cells of the body. It is the only fuel used to any significant extent by some specialized cells, and the major fuel of the brain. Therefore, blood glucose must be maintained at a level sufficient to meet these needs at all times. Several tissues of the body have evolved a coordinated working relationship that assures a continuous supply of this essential substrate for these cells and brain. On the other hand, glucose is toxic and can cause tissue damage when its concentration is not controlled within narrow limits. This occurs in diabetes and is a contributing factor to the development of atherosclerosis, hypertension, small-vessel disease, kidney disease, and blindness.

The discussion begins with glycolysis, a pathway used by all body cells to extract part of the chemical energy inherent in the glucose molecule. This pathway also converts glucose to pyruvate and sets the stage for complete oxidation of glucose to CO_2 and H_2O. Gluconeogenesis, the *de novo* synthesis of glucose, is conveniently discussed after glycolysis because it involves many of the same enzymes used in glycolysis, although the reactions being catalyzed are in the reverse direction. In contrast to glycolysis, which produces ATP, gluconeogenesis requires ATP and is therefore an energy-requiring process. Thus, some enzyme-catalyzed steps have to be different between glycolysis and gluconeogenesis. How regulation is exerted at key enzymes will be stressed throughout the chapter. This will be particularly true for glycogen synthesis (glycogenesis) and degradation (glycogenolysis). Many cells store glycogen for their own later needs. The liver is less selfish, storing glycogen mostly for maintenance of blood glucose to ensure that other tissues, especially the brain, have an adequate supply. Regulation of the synthesis and degradation of glycogen is a model for our understanding of how hormones work and how metabolic pathways are regulated. These topics contribute to our understanding of the diabetic condition, starvation, and how tissues of the body respond to stress, severe trauma and injury.

The chemistry and nomenclature of carbohydrates is presented in the Appendix.

15.2 • GLYCOLYSIS

Glycolysis Occurs in All Human Cells

The Embden–Meyerhof, or glycolytic, pathway represents an ancient process, possessed by all cells of the human body, in which anaerobic degradation of glucose to lactate occurs with release of energy as ATP. This is an example of *anaerobic fermentation,* a term used for metabolic pathways that organisms use to extract chemical energy from high-energy fuels in the absence of oxygen. For many tissues, glycolysis is only an emergency energy-yielding pathway, capable of producing 2 mol of ATP from 1 mol of glucose in the absence of oxygen (Figure 15.2). When oxygen supply to a tissue is shut off, ATP levels can still be maintained by glycolysis for at least a short period. Many examples could be given, but the capacity to use glycolysis as a source of energy is particularly important during natural birth of human babies. With the exception of the brain, circulation of blood decreases to most parts of the body of the baby during delivery. The brain is not normally deprived of oxygen during

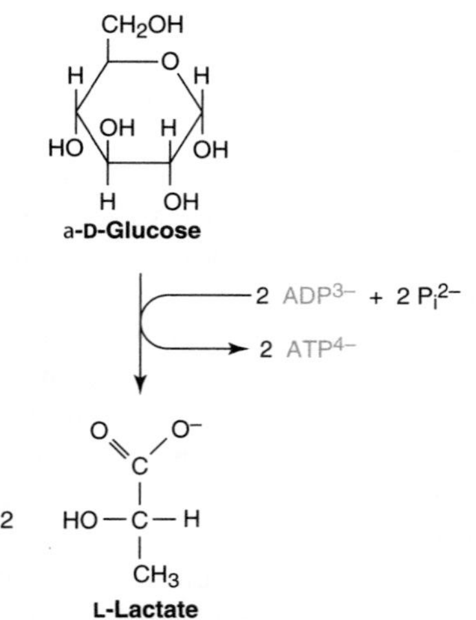

Figure 15.2 Overall balanced equation for the sum of reactions of glycolysis.

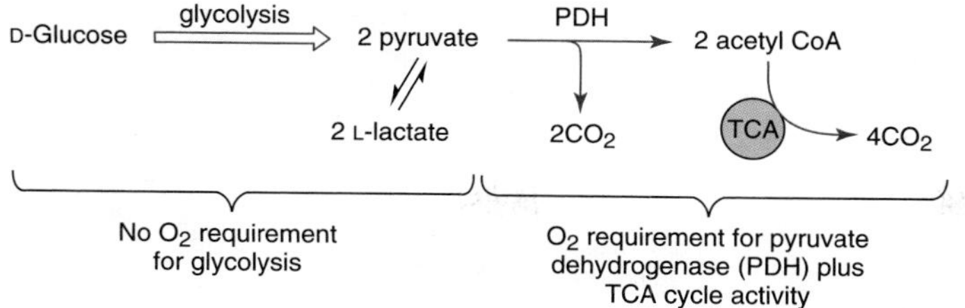

Figure 15.3 Glycolysis is a preparatory pathway for aerobic metabolism of glucose. PDH refers to the pyruvate dehydrogenase complex, TCA to the tricarboxylic acid cycle.

delivery, but other tissues have to depend on glycolysis for their supply of ATP until a normal oxygen supply is available. This conserves oxygen for use by the brain, illustrating one of many mechanisms that have evolved to assure survival of brain tissue in times of stress. Oxygen is not necessary for glycolysis; in fact, oxygen can indirectly suppress glycolysis by the Pasteur effect, which is considered later (p. 609). Nevertheless, glycolysis does occur in cells with an abundant supply of molecular oxygen. Provided cells also contain mitochondria, the end product of glycolysis in the presence of oxygen is pyruvate rather than lactate. Pyruvate is then completely oxidized to CO_2 and H_2O by the pyruvate dehydrogenase complex and enzymes of the TCA cycle housed within the mitochondria (Figure 15.3). Glycolysis therefore sets the stage for aerobic oxidation of carbohydrate. The overall process of glycolysis and mitochondrial oxidation of pyruvate to CO_2 and H_2O has the following sum equation:

$$\text{D-Glucose}(C_6H_{12}O_6) + 6O_2 + 32\ ADP^{3-} + 32\ P_i^{2-} \\ \rightarrow 6CO_2 + 6H_2O + 32\ ATP^{4-} + 32OH^-$$

Much more ATP is produced in complete oxidation of glucose to CO_2 and H_2O (32 ATP/glucose) than is produced in the conversion of glucose to lactate (2 ATP/glucose). This has important consequences to be considered in detail later. The importance of glycolysis as a preparatory pathway is best exemplified by the brain that has an absolute need for glucose. Pyruvate produced by glycolysis is oxidized to CO_2 in mitochondria. An adult human brain uses approximately 120 g of glucose each day to meet its need for ATP. In contrast, glycolysis with lactate as the end product is the major mechanism of ATP production in some other tissues. Red blood cells lack mitochondria and therefore cannot convert pyruvate to CO_2. The cornea, lens, and regions of the retina have a limited blood supply, and also lack mitochondria (because mitochondria would absorb and scatter light), and depend on glycolysis as the major mechanism for ATP production. Kidney medulla, testis, leukocytes, and white muscle fibers have relatively few mitochondria and are therefore almost totally dependent on glycolysis as a source of ATP. Tissues dependent primarily on glycolysis for ATP production consume about 40 g of glucose per day in a normal adult.

Starch is the storage form of glucose in plants and contains α-1,4-glycosidic linkages and α-1,6-glycosidic branches. Glycogen is the storage form of glucose in animal tissues and contains the same type of glycosidic linkages and branches. Exogenous glycogen refers to the glycogen we obtain from animal products; endogenous glycogen is that synthesized and stored in our tissues. Exogenous starch or glycogen is hydrolyzed to glucose in the intestinal tract, whereas stored glycogen endogenous to our tissues is converted to glucose or glucose 6-phosphate by enzymes within the cells. Disaccharides such as milk sugar (lactose) and grocery store sugar (sucrose) are important sources of glucose in our diet. Their hydrolysis by enzymes of the brush border of the intestinal tract is discussed on page 1050. Although glucose can be a source of energy for cells of the intestinal tract, these cells do not depend on glucose to any great extent; most of their energy requirement is met by glutamine catabolism (p. 756). Most of the glucose is absorbed by the cells of the intestinal tract into the portal vein blood and then the general circulation for use by other tissues. Liver is the first major tissue to have an opportunity to remove glucose from the portal vein blood. When blood glucose is high, the liver removes glucose for glycogenesis and glycolysis. When blood glucose is low, the liver supplies the blood with glucose by glycogenolysis and gluconeogenesis. The liver is also the first organ exposed to the blood arriving from the pancreas and therefore experiences the highest concentrations of glucagon and insulin. The effects of these important hormonal regulators of blood glucose levels will be discussed later.

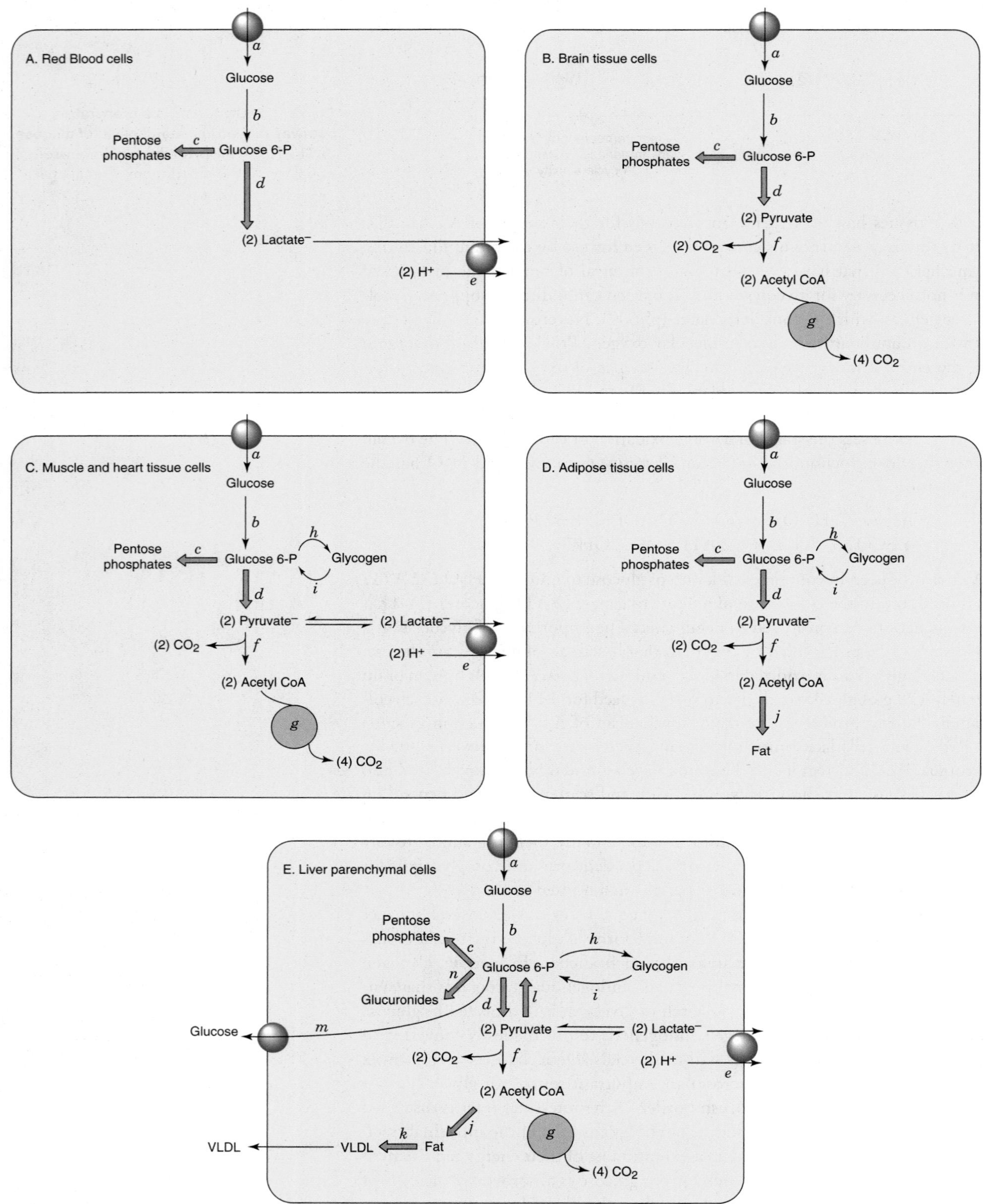

Figure 15.4 Overviews of the major ways in which glucose is metabolized within cells of selected tissues of the body. A. Red blood cells. B. Brain tissue cells. C. Muscle and heart tissue cells. D. Adipose tissue cells. E. Liver parenchymal cells. (*a*) Glucose transport into a cell by a glucose transporter (GLUT). (*b*) Glucose phosphorylation by hexokinase. (*c*) Pentose phosphate pathway. (*d*) Glycolysis. (*e*) Lactic acid transport out of the cell. (*f*) Pyruvate decarboxylation by pyruvate dehydrogenase. (*g*) TCA cycle. (*h*) Glycogenesis. (*i*) Glycogenolysis. (*j*) Lipogenesis. (*k*) Formation and release of very low density lipoproteins (VLDL). (*l*) Gluconeogenesis. (*m*) Hydrolysis of glucose 6-phosphate and release of glucose from the cell into the blood. (*n*) Formation of glucuronides (drug and bilirubin detoxification by conjugation) by the glucuronic acid pathway.

Glucose Is Metabolized Differently in Various Cells

Glucose is metabolized mainly by glycolysis in red blood cells (Figure 15.4A). Transport through the plasma membrane is catalyzed by GLUT1 (glucose transporter 1; p. 489). Since these cells lack mitochondria, the end product of glycolysis is lactic acid, which is released into the blood. Glucose used by the pentose phosphate pathway in red blood cells provides NADPH to keep glutathione in the reduced state, which has an important role in the destruction of organic peroxides and H_2O_2 (see Figure 14.61, p. 587). Peroxides cause irreversible damage to membranes, DNA, and other cellular components and must be removed to prevent cell damage and death.

The brain takes up glucose by facilitated transport in an insulin-independent manner by GLUT3 (glucose transporter 3) (Figure 15.4B). Glycolysis yields pyruvate, which is subsequently oxidized to CO_2 and H_2O by the pyruvate dehydrogenase complex and the TCA cycle. The pentose phosphate pathway is active in these cells, generating part of the NADPH needed for reductive synthesis and the maintenance of glutathione in the reduced state.

Muscle and heart cells readily utilize glucose (Figure 15.4C). Insulin stimulates transport of glucose into these cells by way of GLUT4 (glucose transporter 4). In the absence of insulin, GLUT4 exists in intracellular vesicles where it cannot facilitate glucose transport (Figure 15.5). Binding of insulin to its receptor on the plasma membrane initiates a signaling cascade that promotes translocation and fusion of GLUT4-containing vesicles with the plasma membrane, thereby placing GLUT4 where it can facilitate glucose transport. Glucose taken into muscle and heart cells can be utilized by glycolysis to give pyruvate, which is used by the pyruvate dehydrogenase complex and the TCA cycle to provide ATP. Muscle and heart synthesize significant quantities of glycogen, an important fuel that these tissues store for later use.

As in muscle, the uptake of glucose by adipose tissue is dependent upon and stimulated by insulin (Figures 15.4D and 15.5). Pyruvate, as in other cells, is generated by glycolysis and is oxidized by the pyruvate dehydrogenase complex to give acetyl CoA, which is used primarily for *de novo* fatty acid synthesis. Glycolysis also provides carbon for the synthesis of

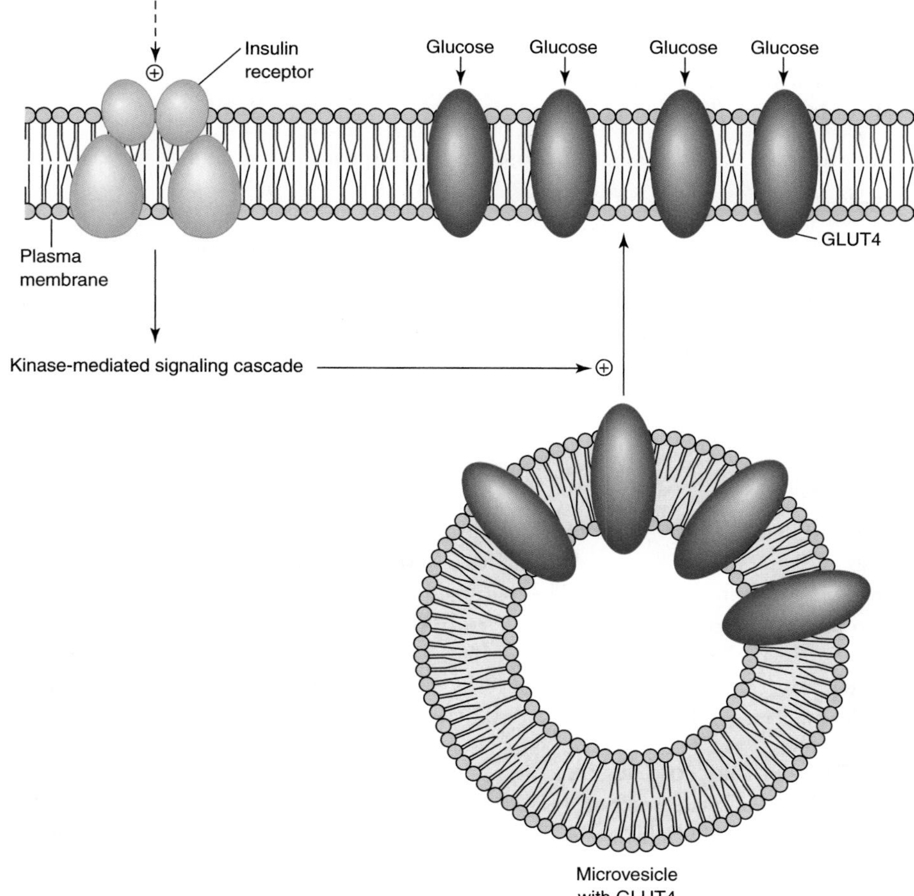

Figure 15.5 Insulin stimulates glucose uptake by adipose tissue and muscle by increasing the number of glucose transporters (GLUT4) in the plasma membrane.

glycerol 3-phosphate (not shown) required for synthesis of triacylglycerol (p. 687). Adipose tissue can carry out glycogenesis and glycogenolysis, but its capacity for these pro-cesses is very limited relative to that of muscle, heart, and liver.

Liver has the greatest number of ways to utilize glucose (Figure 15.4E, p. 594). Uptake of glucose occurs independently of insulin by means of GLUT2, a low-affinity, high-capacity glucose transporter. Glucose is used by the pentose phosphate pathway for the production of NADPH, which is needed for reductive synthesis (*de novo* synthesis of fatty acids and cholesterol), maintenance of reduced glutathione, and numerous reactions catalyzed by endoplasmic reticulum enzyme systems. A vital function of the pentose phosphate pathway is the provision of ribose phosphate for the synthesis of the sugar moiety of nucleotides such as ATP and those in DNA and RNA. Storage of glucose as glycogen is a particularly important feature of the liver. Glucose is also used in the glucuronic acid pathway, which is important in drug and bilirubin detoxification (pp. 654 and 797). The liver carries out glycolysis, the pyruvate produced being used as a source of acetyl CoA for complete oxidation by the tricarboxylic acid cycle and for fatty acid synthesis. Glycolysis also provides the carbon for the synthesis of the glycerol moiety of triacylglycerol, which is synthesized by the liver during production of very low-density lipoproteins (VLDL) (p. 724). The liver also converts three-carbon precursors (lactate, pyruvate, glycerol, and alanine) into glucose by the process of gluconeogenesis to meet the needs of other cells and the brain.

15.3 • GLYCOLYSIS PATHWAY

Glucose is combustible and burns in a test tube to yield heat and light, but of course, no ATP. Cells use about 30 steps to take glucose to CO_2 and H_2O, a seemingly inefficient process, since it can be done in a single step in a test tube. However, side reactions and some of the actual steps used by the cell to oxidize glucose to CO_2 and H_2O lead to the conservation of a significant amount of energy as ATP. In other words, cells produce ATP by the controlled "burning" of glucose, of which glycolysis represents only the first few steps, shown in Figure 15.6.

Glycolysis Occurs in Three Stages

Glycolysis has three major stages.

Priming stage (Figure 15.6*a*)

$$\text{D-Glucose} + 2\ \text{ATP}^{4-} \rightarrow \text{D-fructose 1,6-bisphosphate}^{4-} + 2\ \text{ADP}^{3-} + 2\ \text{H}^+$$

Splitting stage (Figure 15.6*b*)

$$\text{D-Fructose 1,6-bisphosphate}^{4-} \rightarrow 2\ \text{D-glyceraldehyde 3-phosphate}^{2-}$$

Oxidoreduction-phosphorylation stage (Figure 15.6*c*)

$$2\ \text{D-Glyceraldehyde 3-phosphate}^{2-} + 4\ \text{ADP}^{3-} + 2\ \text{P}_i^{2-} + 2\ \text{H}^+ \rightarrow$$
$$2\ \text{L-lactate}^{-} + 4\ \text{ATP}^{4-} + 2\ \text{H}_2\text{O}$$

Sum:

$$\text{D-Glucose} + 2\ \text{ADP}^{3-} + 2\ \text{P}_i^{2-} \rightarrow 2\ \text{L-lactate}^{-} + 2\ \text{ATP}^{4-} + 2\ \text{H}_2\text{O}$$

The priming stage involves input of two molecules of ATP to convert glucose into fructose 1,6-bisphosphate. ATP is "invested" rather than lost in this stage because it is subsequently regained plus more from later steps. The splitting stage "splits" fructose 1,6-bisphosphate into two molecules of glyceraldehyde 3-phosphate. In the oxidoreduction-phosphorylation stage two molecules of glyceraldehyde 3-phosphate are converted into two molecules of lactate with the production of four molecules of ATP. The overall process therefore generates two molecules of lactate and two molecules of ATP from one molecule of glucose.

Stage One: Priming of Glucose

Although the hexokinase reaction (Figure 15.6*a*) consumes ATP, it gets glycolysis off to a good start by trapping glucose as glucose 6-phosphate (G6P) within the cytosol where all the glycolytic enzymes are located. Phosphate esters are charged and hydrophilic and

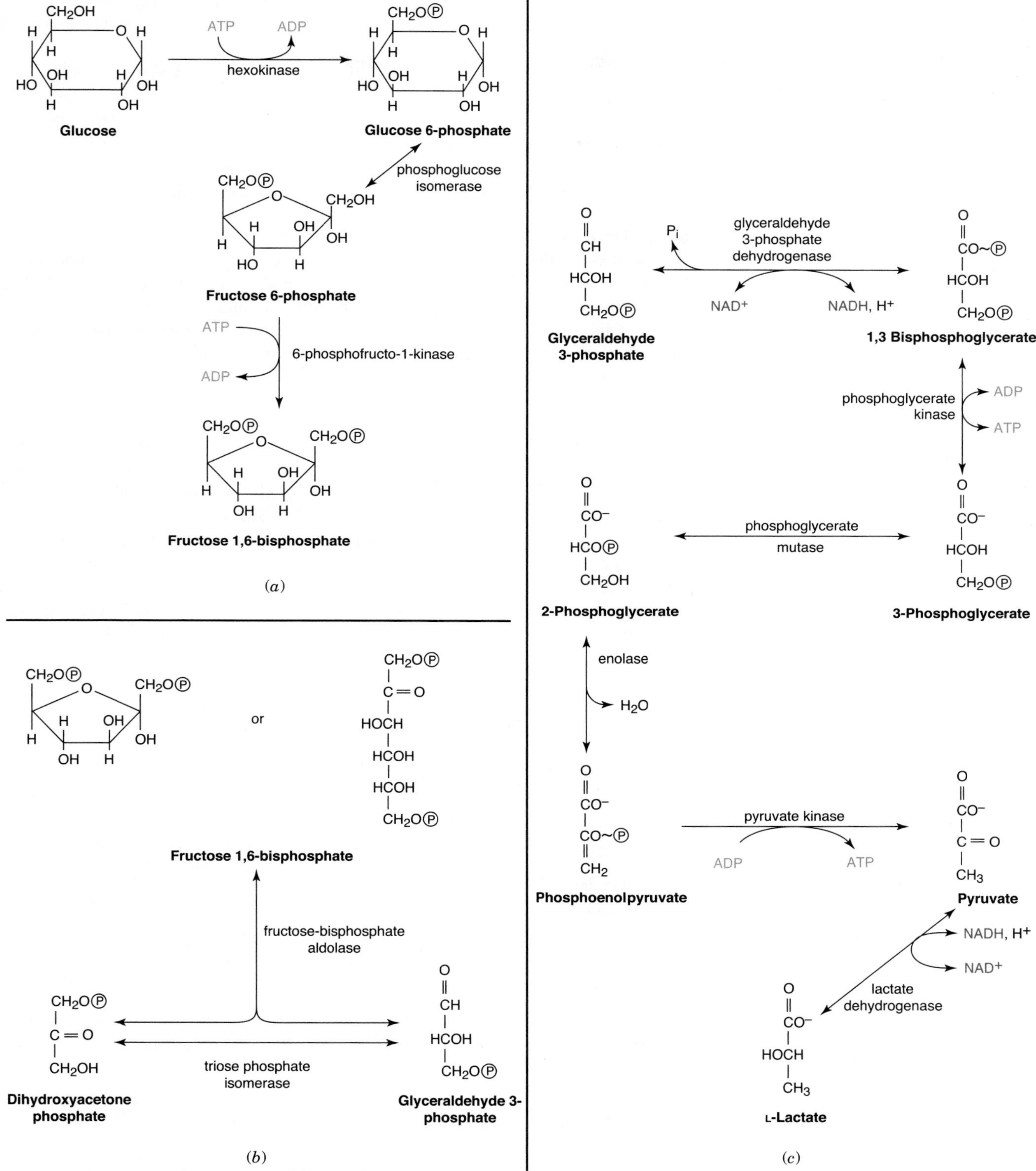

Figure 15.6 The glycolytic pathway, divided into three stages. The symbol P refers to phosphoryl group PO_3^{2-}; ~ indicates a high-energy phosphate bond. (*a*) Priming stage. (*b*) Splitting stage. (*c*) Oxidoreduction–phosphorylation stage.

therefore cannot cross cell membranes. The phosphorylation of glucose by ATP is a thermodynamically favorable, irreversible under cellular conditions, reaction. The reverse reaction cannot be used to synthesize ATP.

The next reaction catalyzed by phosphoglucose isomerase is readily reversible and is not subject to regulation.

6-Phosphofructo-1-kinase (or phosphofructokinase-1) catalyzes the ATP-dependent phosphorylation of fructose 6-phosphate (F6P) to fructose 1,6-bisphosphate (FBP). This enzyme is subject to regulation by several effectors and is often considered the key regulatory enzyme of glycolysis. The reaction is irreversible and uses the second ATP needed to "prime" glucose.

Stage Two: Splitting of a Phosphorylated Intermediate

Fructose 1,6-bisphosphate aldolase cleaves fructose 1,6-bisphosphate into a molecule each of dihydroxyacetone phosphate (DHAP) and glyceraldehyde 3-phosphate (GAP) (Figure 15.6*b*). This is a reversible reaction, corresponding to an aldol cleavage in one direction and an aldol condensation in the other. Triose phosphate isomerase catalyzes reversible interconversion of DHAP and GAP. With the transformation of DHAP into GAP, one molecule of glucose is converted into two molecules of GAP.

Stage Three: Oxidoreduction Reactions and ATP Synthesis

The reaction catalyzed by glyceraldehyde 3-phosphate dehydrogenase (Figure 15.6*c*) is of interest because of what is accomplished. An aldehyde (glyceraldehyde 3-phosphate) is oxidized to a carboxylic acid with the reduction of NAD^+ to NADH. The acid produced is 1,3-bisphosphoglycerate, a mixed anhydride of a carboxylic acid and phosphoric acid, which has a large negative free energy of hydrolysis, enabling it to participate in a subsequent reaction that forms ATP. The overall reaction can be visualized as the coupling of a very favorable exergonic reaction with an unfavorable endergonic reaction. The exergonic reaction is the one in which an aldehyde is oxidized to a carboxylic acid, which is then coupled with an endergonic half-reaction in which NAD^+ is reduced to NADH.

$$R{-}\overset{\displaystyle O}{\overset{\|}{C}}H + H_2O \longrightarrow R{-}\overset{\displaystyle O}{\overset{\|}{C}}OH + 2H^+ + 2e^-$$

$$NAD^+ + 2H^+ + 2e^- \longrightarrow NADH + H^+$$

The overall reaction (sum of the half-reactions) is quite exergonic.

$$R{-}\overset{\displaystyle O}{\overset{\|}{C}}H + NAD^+ + H_2O \longrightarrow R{-}\overset{\displaystyle O}{\overset{\|}{C}}OH + NADH + H^+, \quad \Delta G^{\circ\prime} = -10.3\ \text{kcal mol}^{-1}$$

A second endergonic component of the reaction is formation of a mixed anhydride between the carboxylic acid and phosphoric acid.

$$R{-}\overset{\displaystyle O}{\overset{\|}{C}}OH + HPO_4^{2-} \longrightarrow R{-}\overset{\displaystyle O}{\overset{\|}{C}}{-}OPO_3^{2-} + H_2O, \Delta G^{\circ\prime} = +\ 11.8\ \text{kcal mol}^{-1}$$

The overall reaction involves coupling of the endergonic and exergonic components with an overall standard free-energy change of +6.3 kJ/mol (+1.5 kcal/mol).

$$\text{Sum: } R{-}\overset{\displaystyle O}{\overset{\|}{C}}H + NAD^+ + HPO_4^{2-} \longrightarrow$$

$$R{-}\overset{\displaystyle O}{\overset{\|}{C}}OPO_3^{2-} + NADH + H^+, \Delta G^{\circ\prime} = +\ 1.5\ \text{kcal mol}^{-1}$$

This reaction is freely reversible in cells. The catalytic mechanism involves glyceraldehyde 3-phosphate reacting with a sulfhydryl group of a cysteine residue to generate a thiohemiacetal (Figure 15.7). An internal oxidation–reduction reaction occurs in which bound NAD^+ is reduced to NADH and the thiohemiacetal is oxidized to a high-energy thiol ester. The thiol ester reacts with P_i to form the mixed anhydride and regenerate the free sulfhydryl group. The mixed anhydride dissociates from the enzyme and exogenous

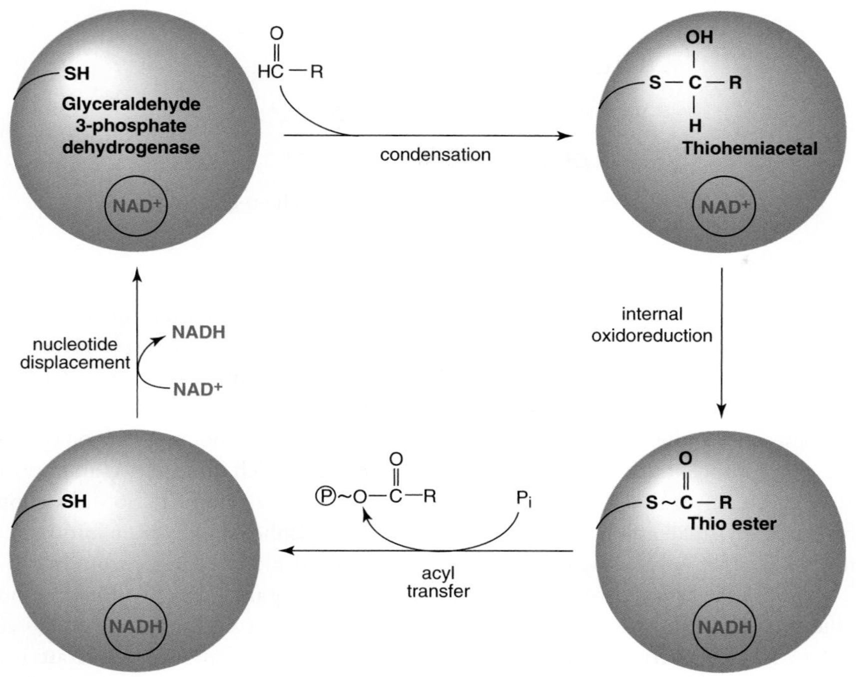

Figure 15.7 Catalytic mechanism of glyceraldehyde 3-phosphate dehydrogenase. Large circle represents enzyme; small circle binding site for NAD^+; RCOH, the aldehyde group of glyceraldehyde 3-phosphate; -SH, the sulfhydryl group of the cysteine residue located at the enzyme's active site; and ~, high-energy bonds in the thio ester and the mixed anhydride.

NAD^+ replaces the bound NADH. It should be noted that a free carboxylic acid group (–COOH) is not generated from the aldehyde group (–CHO) during the reaction. Instead, the enzyme generates a carboxyl group in the form of a high-energy thiol ester, which is converted by reaction with P_i into a mixed anhydride of carboxylic and phosphoric acids.

This reaction catalyzed by glyceraldehyde 3-phosphate dehydrogenase requires NAD^+ and produces NADH. Since the cytosol has only a limited amount of NAD^+, continuous glycolytic activity can occur only if NADH is oxidized back to NAD^+; otherwise glycolysis will stop for want of NAD^+. The options that cells have for regeneration of NAD^+ from NADH are described in a later section (p. 601).

In the next reaction phosphoglycerate kinase produces ATP from the high-energy compound 1,3-bisphosphoglycerate (Figure 15.6*c*, p. 597). This is the first site of ATP production in glycolysis. Since two ATP molecules were invested for each glucose molecule in the priming stage, and since two molecules of 1,3-bisphosphoglycerate are produced from each glucose, all of the "invested" ATP is recovered in this step. The glyceraldehyde 3-phosphate dehydrogenase-phosphoglycerate kinase system is an example of substrate-level phosphorylation in which a substrate participates in an enzyme-catalyzed reaction that yields ATP or GTP. Substrate-level phosphorylation stands in contrast to oxidative phosphorylation catalyzed by the mitochondrial electron transfer chain and ATP synthase (p. 572). Note that the combination of glyceraldehyde 3-phosphate dehydrogenase and phosphoglycerate kinase accomplishes the coupling of an oxidation (an aldehyde is oxidized to a carboxylic acid) to a phosphorylation (a mixed anhydride of a carboxylic acid and phosphoric acid is formed) without the involvement of a membrane system.

Phosphoglycerate mutase converts 3-phosphoglycerate to 2-phosphoglycerate. This is a freely reversible reaction in which 2,3-bisphosphoglycerate is an obligatory intermediate at the active site.

$$\begin{aligned} \text{E-phosphate} + \text{3-phosphoglycerate} &\rightleftharpoons \text{E} + \text{2,3-bisphosphoglycerate} \\ \text{E} + \text{2,3-bisphosphoglycerate} &\rightleftharpoons \text{E-phosphate} + \text{2-phosphoglycerate} \\ \text{Sum:} \quad \text{3-phosphoglycerate} &\rightleftharpoons \text{2-phosphoglycerate} \end{aligned}$$

Involvement of 2,3-bisphosphoglycerate in this reaction creates an absolute requirement for catalytic amounts of this compound in cells. This can be appreciated by noting that E-P cannot be generated without 2,3-bisphosphoglycerate and likewise that 2,3-bisphosphoglycerate cannot be generated without E-P. Cells solve this problem by synthesizing 2,3-bisphosphoglycerate from 1,3-bisphosphoglycerate with a 2,3-bisphosphoglycerate mutase.

$$\begin{array}{c} O \\ \| \\ COPO_3^{2-} \\ | \\ HCOH \\ | \\ CH_2OPO_3^{2-} \end{array} \longrightarrow \begin{array}{c} O \\ \| \\ CO^- \\ | \\ HCOPO_3^{2-} \\ | \\ CH_2OPO_3^{2-} \end{array} + H^+$$

1,3-Bisphospho-D-glycerate 2,3-Bisphospho-D-glycerate

This enzyme is bifunctional, serving as mutase for the formation of 2,3-bisphosphoglycerate and also as a phosphatase that hydrolyzes 2,3-bisphosphoglycerate to 3-phosphoglycerate and P_i. All cells contain the minute quantities of 2,3-bisphosphoglycerate necessary to produce the phosphorylated form (E-P) of newly synthesized phosphoglycerate mutase. In striking contrast to other cells of the body, red blood cells contain very high concentrations of 2, 3-bisphosphoglycerate, which functions as an important negative allosteric regulator of the binding of oxygen to hemoglobin (p. 365). In contrast to the extremely small amounts of glucose used for 2,3-bisphosphoglycerate in other cells, from 15% to 25% of glucose converted to lactate in red blood cells goes by way of the *BPG shunt* for the synthesis of 2,3-bisphosphoglycerate (Figure 15.8). It should be noted that the BPG shunt bypasses the PGK step. Therefore, no net production of ATP occurs when glucose is converted to lactate via the BPG shunt. 2,3-Bisphosphoglycerate is also hydrolyzed in red blood cells by multiple inositol polyphosphate phosphatases (MIP phosphatase) (Figure 15.8). By converting 2,3-BPG to 2-phosphoglycerate rather than 3-phosphglycerate, MIP phosphatase expands the BPG shunt to by-pass the reaction catalyzed by phosphoglycerate mutase. Exceptional sensitivity of MIP phosphatase to changes in pH suggest that the activity of this enzyme may set the concentration of 2,3-BPG, the most important allosteric effector of hemoglobin, in physiological and pathological conditions that affect the intracellular pH of red blood cells (A Closer Look 15.1).

Enolase eliminates water from 2-phosphoglycerate to form phosphoenolpyruvate (PEP) in the next reaction (see Figure 15.6*c*, p. 597). This reaction generates a high-energy phosphate from one of markedly lower energy level. $\Delta G^{0\prime}$ for the hydrolysis of phosphoenolpyruvate is -61.9 kJ/mol (-14.8 kcal/mol), while that for 2-phosphoglycerate is -17.6 kJ/mol (-4.2 kcal/mol). Pyruvate kinase (Figure 15.6*c*) accomplishes another substrate-level phosphorylation. This reaction is not reversible under intracellular conditions.

The last step of glycolysis is a freely reversible oxidoreduction reaction catalyzed by lactate dehydrogenase (Figure 15.6*c*). Pyruvate is reduced to give L-lactate and NADH is oxidized to NAD^+. The forward direction of this reaction is the only reaction of the body by which L-lactate can be produced. The reverse direction of this reaction is the only reaction of the body by which L-lactate can be utilized. Lactate dehydrogenase is therefore responsible for both the production and the utilization of L-lactate.

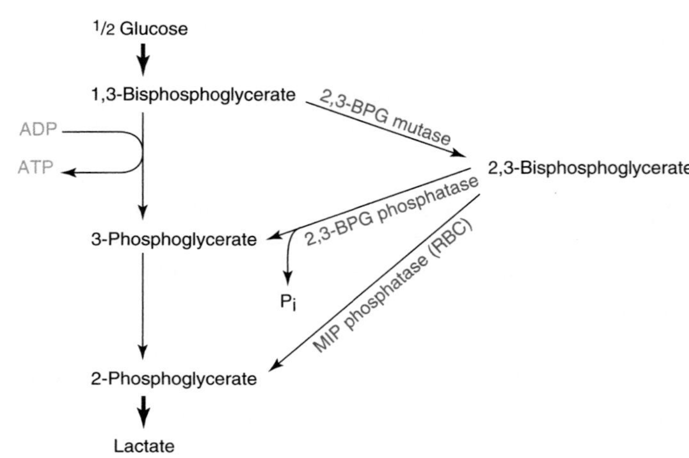

Figure 15.8 The reactions of the 2,3-bisphosphoglycerate (2,3-BPG) shunt are catalyzed by the bifunctional enzyme, 2,3-BPG mutase/phosphatase. 2,3-BPG is also hydrolyzed to 2-phoshoglycerate by multiple inositol polyphosphate phosphatase (MIP phosphatase), so named because multiple inositol polyphosphate was identified before 2,3-BPG to be a substrate for this enzyme.

A CLOSER LOOK 15.1

2,3-Bisphosphoglycerate and High Altitude

Delivery of oxygen to tissues depends on regulation of hemoglobin by allosteric effectors, the most important being 2,3-bisphosphoglycerate, which promotes oxygen release by binding to deoxyhemoglobin. Supply of oxygen to tissues in a variety of conditions is dependent upon regulation of the 2,3-bisphosphoglycerate concentration in red blood cells. Relatively small changes in blood pH and the opposite sensitivities of 6-phosphofructo-1-kinase and MIP phosphatase to pH changes appear to be most important factors. For example, in patients with metabolic acidosis, DPG levels are dramatically reduced in RBCs, most likely because a decrease in pH reduces 6-phosphofructo-1-kinase activity and increases MIP phosphatase activity. Conversely, in response to high altitude or anemia due to blood loss, DPG levels are dramatically increased in RBCs in these situations because an increase in pH increases 6-phosphofructo-1-kinase activity and decreases MIP phosphatase activity. Alkalosis is induced by blood loss and high altitude because hyperventilation due to hypoxia in these situations blows off CO_2 which lowers the blood H^+ concentration by forcing the reaction catalyzed by carbonic anhydrase to the left.

$$CO_2 + H_2O \leftrightharpoons H_2CO_3 \leftrightharpoons H^+ + HCO_3^-$$

Wallis, J. P., Wells, A. W., Whitehead, S., and Brewster, N. Recovery from postoperative anaemia. *Transfusion Med.* 15: 413, 2005; Cho, J., King, J. S., Qian, X., Harwood, A. J., and Shears, S. B. Dephosphorylation of 2,3-bisphosphoglycerate by MIPP expands the regulatory capacity of the Rapoport-Lubering glycolytic shunt. *Proc. Natl. Acad. Sci. USA* 105: 998, 2008.

ATP Yield and Balanced Equation for Anaerobic Glycolysis

Conversion of one molecule of glucose to two molecules of lactate results in the net formation of two molecules of ATP. Two molecules of ATP are used in the priming stage, but subsequent steps yield four molecules of ATP so that the overall net yield is two molecules of ATP.

$$\text{D-Glucose} + 2\ ADP^{3-} + 2\ P_i^{2-} \rightarrow 2\ \text{L-lactate}^- + 2\ ATP^{4-} + 2\ H_2O$$

Cells have only a limited amount of ADP and P_i. Flux through glycolysis is dependent on an adequate supply of these substrates. If the ATP is not utilized for performance of work, glycolysis stops for want of ADP and/or P_i. Consequently, the ATP generated has to be used in normal work-related processes for glycolysis to occur. Use of ATP for any work-related process is represented simply by

$$ATP^{4-} + H_2O \rightarrow ADP^{3-} + P_i^{2-} + H^+ + \text{"work"}$$

When the quantities are doubled and added to the previous equation for glycolysis, excluding the work accomplished since this is necessary for ATP turnover, the overall balanced equation becomes

$$\text{D-Glucose} \rightarrow 2\ \text{L-lactate}^- + 2\ H^+$$

This shows that anaerobic glycolysis generates acid that can create major problems for the body (described later in Clin. Corr. 15.5, p. 611) since intracellular pH must be maintained near neutrality for optimum enzyme activity.

NADH Generated by Glycolysis Has to Be Oxidized Back to NAD⁺: Role of Lactate Dehydrogenase and Substrate Shuttles

Anaerobic Glycolysis

NADH and NAD^+ do not appear in the balanced sum equation for anaerobic glycolysis because the generation of NADH and its utilization are coupled (balanced) in the pathway (see Figure 15.6*c*). Two molecules of NADH are generated by glyceraldehyde 3-phosphate dehydrogenase and two are utilized by lactate dehydrogenase. NAD^+ is available in only limited amounts and must be regenerated for glycolysis to continue unabated. The two reactions involved are

$$\text{D-Glyceraldehyde 3-phosphate} + NAD^+ + P_i \rightarrow \text{1,3-bisphospho-D-glycerate} + NADH + H^+$$

$$\text{Pyruvate} + NADH + H^+ \rightarrow \text{L-lactate} + NAD^+$$

The sum reaction is

$$\text{D-Glyceraldehyde 3-phosphate} + \text{pyruvate} + P_i \rightarrow \text{1,3-bisphosphoglycerate} + \text{L-lactate}$$

Perfect coupling of reducing equivalents by these reactions occurs under anaerobic conditions or in cells that lack mitochondria.

Mitochondrial Oxidation of NADH Produced by Glycolysis

When oxygen and mitochondria are present, the reducing equivalents of NADH generated by glyceraldehyde 3-phosphate dehydrogenase are shuttled into the mitochondria for oxidation, leaving pyruvate rather than lactate as the end product of glycolysis. The mitochondrial inner membrane is not permeable to NADH (p. 577) but the malate–aspartate shuttle and the glycerol–phosphate shuttle (see Figure 14.51, p. 580) transport reducing equivalents into the mitochondrial matrix space (p. 560). Liver makes greater use of the malate–aspartate shuttle, but some muscle cells are more dependent on the glycerol–phosphate shuttle. The shuttle systems move reducing equivalents from the cytosol into the mitochondria but will not shuttle reducing equivalents from the mitochondria into the cytosol. The sum of all reactions of the malate–aspartate shuttle reactions is simply

$$\text{NADH}_{\text{cytosol}} + \text{H}^+{}_{\text{cytosol}} + \text{NAD}^+{}_{\text{mito}} \rightarrow \text{NAD}^+{}_{\text{cytosol}} + \text{NADH}_{\text{mito}} + \text{H}^+{}_{\text{mito}}$$

The glycerol phosphate shuttle produces $FADH_2$ within the mitochondrial inner membrane (see Figure 14.40, p. 571). The active site of the mitochondrial glycerol 3-phosphate dehydrogenase is exposed on the cytosolic surface of the mitochondrial inner membrane. The sum of all reactions of the glycerol phosphate shuttle is

$$\text{NADH}_{\text{cytosol}} + \text{H}^+{}_{\text{cytosol}} + \text{FAD}_{\text{inner membrane}} \rightarrow \text{NAD}^+{}_{\text{cytosol}} + \text{FADH}_{2\ \text{inner membrane}}$$

The mitochondrial NADH formed by the malate–aspartate shuttle activity can be used by the mitochondrial electron transport chain for the production of 2.5 molecules of ATP by oxidative phosphorylation.

$$\text{NADH}_{\text{mito}} + \text{H}^+ + 0.5\ \text{O}_2 + 2.5\ \text{ADP} + 2.5\ \text{P}_i \rightarrow \text{NAD}^+{}_{\text{mito}} + 2.5\ \text{ATP} + \text{H}_2\text{O}$$

In contrast, the $FADH_2$ formed by the glycerol phosphate shuttle yields only 1.5 molecules of ATP.

$$\text{FADH}_{2\ \text{inner membrane}} + 0.5\ \text{O}_2 + 1.5\ \text{ADP} + 1.5\ \text{P}_i \rightarrow \text{FAD}_{\text{inner membrane}} + 1.5\ \text{ATP} + \text{H}_2\text{O}$$

Thus, the ATP yield from the oxidation of the NADH generated by glycolysis is 3 or 5, depending on the substrate shuttle used for its oxidation.

Shuttles Are Important in Other Oxidoreduction Pathways

Alcohol Oxidation

Alcohol (i.e., ethanol) is oxidized to acetaldehyde with production of NADH by alcohol dehydrogenase.

$$\underset{\text{Ethanol}}{\text{CH}_3\text{CH}_2\text{OH}} + \text{NAD}^+ \longrightarrow \underset{\text{Acetaldehyde}}{\text{CH}_3\overset{\overset{\text{O}}{\|}}{\text{C}}\text{H}} + \text{NADH} + \text{H}^+$$

This enzyme is located almost exclusively in the cytosol of hepatocytes. Acetaldehyde traverses the mitochondrial inner membrane for oxidation by an aldehyde dehydrogenase in the mitochondrial matrix space.

$$\underset{\text{Acetaldehyde}}{\text{CH}_3\overset{\overset{\text{O}}{\|}}{\text{C}}\text{H}} + \text{NAD}^+ + \text{H}_2\text{O} \longrightarrow \underset{\text{Acetate}}{\text{CH}_3\overset{\overset{\text{O}}{\|}}{\text{C}}\text{O}^-} + \text{NADH} + 2\text{H}^+$$

The NADH generated by the last step can be used directly by the mitochondrial electron transfer chain. However, the NADH generated by cytosolic alcohol dehydrogenase is oxidized back to NAD^+ by one of the substrate shuttles (see Figure 14.51, p. 580).

Thus, the capacity to oxidize alcohol is dependent on the ability of the liver to transport reducing equivalents from the cytosol into the mitochondria by these shuttle systems.

Glucuronide Formation

Water-soluble glucuronides of bilirubin and various drugs (p. 431) are eliminated in the urine and bile. For glucuronide formation, UDP-glucose (see p. 654 for structure) is oxidized to UDP-glucuronic acid (see p. 655 for structure) by

$$\text{UDP-D-glucose} + 2\,\text{NAD}^+ + \text{H}_2\text{O} \rightarrow \text{UDP-D-glucuronic acid} + 2\,\text{NADH} + 2\,\text{H}^+$$

Primarily in liver, this "activated" glucuronic acid is transferred to a nonpolar, acceptor molecule, such as bilirubin or a compound (R-OH) foreign to the body.

$$\text{UDP-D-glucuronic acid} + \text{R-OH} \rightarrow \text{R-O-glucuronic acid} + \text{UDP}$$

NADH generated by the first reaction is reoxidized by the substrate shuttles. Since ethanol oxidation and drug conjugation occurs in the liver, the two occurring together may overwhelm the capacity of the substrate shuttles. This explains the advice not to mix the intake of pharmacologically active compounds with alcohol (see Clin. Corr. 15.1).

Sulfhydryl Reagents and Fluoride Inhibit Glycolysis

Glyceraldehyde 3-phosphate dehydrogenase is inhibited by sulfhydryl reagents because of the catalytically important cysteine residue at its active site. During a catalytic cycle the sulfhydryl group reacts with glyceraldehyde 3-phosphate to form a thiohemiacetal (see Figure 15.7, p. 599). Sulfhydryl reagents, which are often mercury-containing compounds or alkylating compounds such as iodoacetate, prevent formation of the thiohemiacetal (Figure 15.9).

Fluoride is a potent inhibitor of enolase. Mg^{2+} and P_i form an ionic complex with fluoride that inhibits enolase by interfering with the binding of its substrate (Mg^{2+} 2-phosphoglycerate^{2-}).

Hyperglycemia Inhibits Glycolysis

Glyceraldehyde 3-phosphate dehydrogenase is vulnerable to inhibition in cells exposed to high glucose levels. This is because hyperglycemia promotes overproduction of reactive oxygen species which activate poly(ADP-ribose) polymerase (PARP). In a reaction that

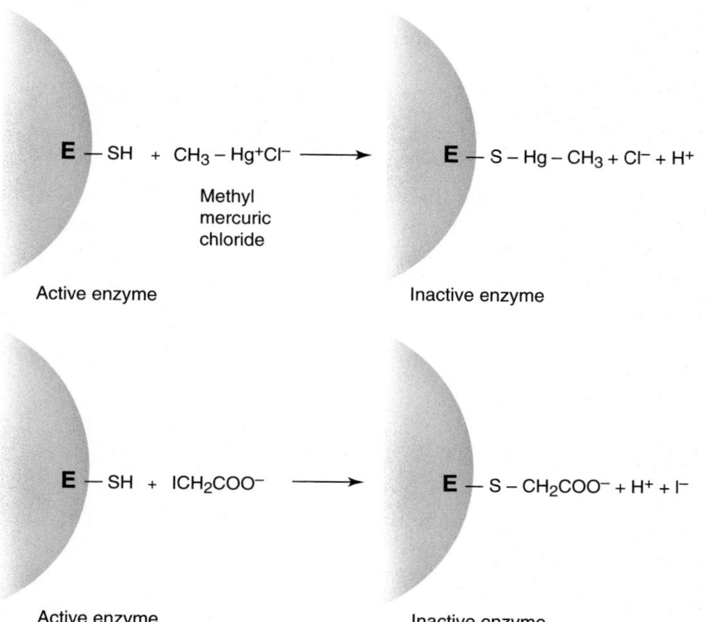

Figure 15.9 Mechanism for inactivation of glyceraldehyde 3-phosphate dehydrogenase by sulfhydryl reagents.

CLINICAL CORRELATION 15.1

Alcohol and Barbiturates

Acute alcohol intoxication increases sensitivity to the general depressant effects of barbiturates. Barbiturates and alcohol interact with the γ-aminobutyrate (GABA) activated chloride channel. Activation of this channel inhibits neuronal firing, which may explain the depressant effects of both compounds. This combination is very dangerous and normal prescription doses of barbiturates are potentially lethal when taken with ethanol. Furthermore, ethanol inhibits the metabolism of barbiturates, thereby prolonging the time barbiturates remain effective in the body. Hydroxylation of barbiturates by the NADPH-dependent cytochrome P^{450} system of the endoplasmic reticulum of the liver is inhibited by ethanol. This decreases formation of water-soluble derivatives of the barbiturates for elimination by the kidneys and bile. Blood levels of barbiturates remain high and cause increased CNS depression.

Surprisingly, the sober alcoholic is less sensitive to barbiturates. Chronic ethanol consumption apparently causes adaptive changes in the sensitivity to barbiturates (cross-tolerance) and induces cytochrome P450 of liver endoplasmic reticulum involved in drug hydroxylation reactions. Consequently, the sober alcoholic can metabolize barbiturates more rapidly. This sets up the following scenario: A sober alcoholic has trouble falling asleep, even after taking several sleeping pills, because his liver has increased capacity to hydroxylate the barbiturate contained in the pills. In frustration he consumes more pills and then alcohol. Sleep results, but may be followed by respiratory depression and death because the alcoholic, although less sensitive to barbiturates when sober, remains sensitive to the synergistic effect of alcohol.

Misra, P. S., Lefevre, A., Ishii, H., Rubin, E., and Lieber, C. S. Increase of ethanol, meprobamate, and pentobarbital metabolism after chronic ethanol administration in man and in rats. *Am. J. Med.* 51: 346, 1971; and Tabakoff, B., Cornell, N., Hoffman, P. L. Alcohol tolerance. *Ann. Emerg. Med.* 15:1005, 1986

A CLOSER LOOK 15.2

Glyceraldehyde 3-Phosphate Dehydrogenase and Hyperglycemia

When blood glucose levels are elevated, cells of the body that are unable to restrict the uptake of glucose are vulnerable to damage from too much intracellular glucose. The most susceptible cells are found in the retina, kidney, and neurons of peripheral nerves, early sites of damage in diabetes. Overproduction of the superoxide radical ($O_2^{-}\cdot$) by mitochondria is the primary cause of the damage, at least in part because of inactivation of glyceraldehyde 3-phophate dehydrogenase, an enzyme required for glycolysis. Oxidation of the overload of glucose in these cells results in excessive reduction of the mitochondrial electron transport chain. Transfer of a single electron from one of the overloaded carriers of the electron transport chain to oxygen produces the superoxide radical. The superoxide radical activates poly(ADP-ribose) polymerase, a DNA repair enzyme that ADP-ribosylates free sulfhydryl groups of many enzymes including the active site cysteine residue of glyceraldehyde 3-phosphate dehydrogenase. Partial inhibition of glycolysis by this mechanism may contribute to the damage induced by hyperglycemia in susceptible cells.

Brownlee, M. Banting Lecture 2004. The pathology of diabetic complications: A unifying mechanism. *Diabetes* 54: 1615, 2005.

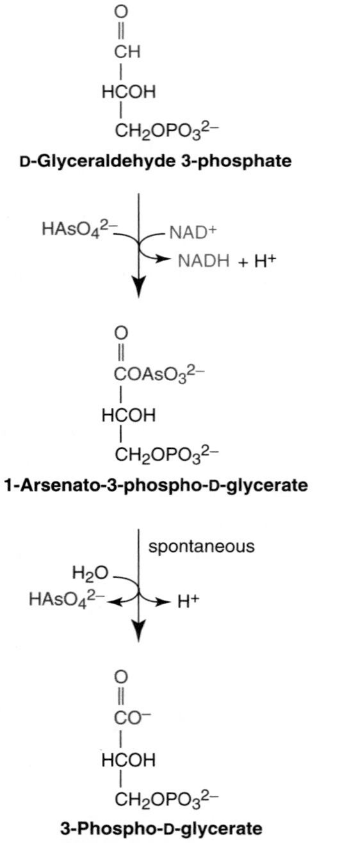

uses NAD^+ as substrate and produces nicotinamide, PARP catalyzes the poly(ADP-ribosyl) ation of cysteine residues of a number of proteins including the active site cysteine of glyceraldehyde 3-phosphate dehydrogenase (A Closer Look 15.2).

Arsenate Prevents Net Synthesis of ATP without Inhibiting Glycolysis

Pentavalent arsenic or arsenate prevents net synthesis of ATP by causing arsenolysis in the glyceraldehyde 3-phosphate dehydrogenase reaction (Figure 15.10). The structure of arsenate resembles that of P_i (also pentavalent) and readily substitutes for P_i in enzyme-catalyzed reactions. The mixed anhydride of arsenic acid and the carboxyl group of 3-phosphoglycerate is formed by glyceraldehyde 3-phosphate dehydrogenase. 1-Arsenato 3-phosphoglycerate is unstable, and undergoes spontaneous hydrolysis to 3-phosphoglycerate and inorganic arsenate. As a consequence, glycolysis continues unabated in the presence of arsenate, but 1,3-bisphosphoglycerate is not formed. ATP is therefore not synthesized by phosphoglycerate kinase. Thus net ATP synthesis does not occur when glycolysis takes place in the presence of arsenate, the ATP invested in the priming stage being balanced by the ATP generated in the pyruvate kinase step. Arsenolysis also interferes with ATP formation by oxidative phosphorylation, hence the toxicity of arsenate (Clin. Corr. 15.2).

O
||
CH
|
HCOH
|
$CH_2OPO_3^{2-}$
D-Glyceraldehyde 3-phosphate

$HAsO_4^{2-}$ → ; NAD^+ → $NADH + H^+$

O
||
$COAsO_3^{2-}$
|
HCOH
|
$CH_2OPO_3^{2-}$
1-Arsenato-3-phospho-D-glycerate

spontaneous; H_2O → ; → $HAsO_4^{2-}$ + H^+

O
||
CO^-
|
HCOH
|
$CH_2OPO_3^{2-}$
3-Phospho-D-glycerate

Figure 15.10 **Arsenate uncouples oxidation from phosphorylation at the glyceraldehyde 3-phosphate dehydrogenase reaction.**

15.4 • REGULATION OF GLYCOLYSIS

As in other complex pathways that involve multiple steps, flux through glycolysis is determined by the activities of several enzymes rather than a single *rate-limiting* enzyme. Quantitative information about the relative contribution of an enzyme to flux through a pathway under a particular set of conditions is best determined from the *control strength* for the enzyme. The effect that a small inhibition of the activity of an enzyme has upon flux through a pathway is used to determine an enzyme's control strength according to the equation

$$\text{Control Strength} = \frac{\text{Change in flux}}{\text{Change in enzyme activity}}$$

For the purpose of illustration, assume that inhibition of a particular enzyme by 10% has no effect on flux. The control strength of the enzyme is then zero (0 divided by 10). Now assume that inhibition of an enzyme by 10% results in 10% inhibition of flux. The control strength would be 1.0 (10 divided by 10), which means that flux is entirely dependent on the activity of the enzyme under these conditions. Now assume that inhibition of an enzyme by 10% results in only 5% inhibition of flux through the pathway. The control

CLINICAL CORRELATION 15.2

Arsenic Poisoning

Most forms of arsenic are toxic, the trivalent form (arsenite as AsO_2^-) being much more toxic than the pentavalent form (arsenate or $HAsO_4^{2-}$). Less ATP is produced whenever arsenate substitutes for P_i in biological reactions. Arsenate competes for P_i binding sites on enzymes, resulting in the formation of arsenate esters that are unstable. Arsenite works by a different mechanism, involving formation of a stable complex with enzyme bound lipoic acid (See illustration to right.).

For the most part arsenic poisoning is explained by inhibition of enzymes that require lipoic acid as a coenzyme. These include pyruvate dehydrogenase, α-ketoglutarate dehydrogenase, and branched-chain α-keto acid dehydrogenase. Chronic arsenic poisoning from well water contaminated with arsenical pesticides or through the efforts of a murderer is best diagnosed by determining the concentration of arsenic in the hair or fingernails of the victim. About 0.5 mg of arsenic is present in 1 kg of hair from a normal individual. The hair of a person chronically exposed to arsenic could have 100 times as much.

Hindmarsh, J. T., and McCurdy, R. F. Clinical and environmental aspects of arsenic toxicity. *CRC Crit. Rev. Clin. Lab.* Sci. 23: 315, 1986; and Mudur, G. Half of Bangladesh population at risk of arsenic poisoning. *British Medical Journal* 320: 822, 2000.

strength is now 0.5 (5 divided by 10), which means that half of control of flux is determined by this enzyme. The rest of the control is exerted by one or more other steps, since the control strength of steps of a linear pathway must sum to 1.0 by definition.

With the caveat that regulation of flux through glycolysis is dependent on the tissue and the nutritional and hormonal state, the enzymes of glycolysis with the greatest control strength are hexokinase, 6-phosphofructo-1-kinase, and pyruvate kinase (Figure 15.11). These enzymes are regulated by allosteric effectors and/or covalent modification.

A nonregulatory enzyme most likely catalyzes a *near-equilibrium reaction,* whereas a regulatory enzyme most likely catalyzes a *nonequilibrium reaction.* The activity of a nonregulatory enzyme readily brings its substrates and products to equilibrium concentrations. A regulatory enzyme is not active enough to equilibrate its substrates and products. Whether an enzyme-catalyzed reaction is near equilibrium or nonequilibrium can be determined by comparing the established equilibrium constant for the reaction with the mass-action ratio as it exists within a cell. The equilibrium constant for the reaction A + B → C + D is defined as

$$K_{eq} = \frac{[C][D]}{[A][B]}$$

where the brackets indicate the concentrations at equilibrium. The mass-action ratio is calculated in a similar manner, except that the steady-state (ss) concentrations of reactants and products within the cell are used.

$$\text{Mass-action ratio} = \frac{[C]_{ss}[D]_{ss}}{[A]_{ss}[B]_{ss}}$$

If the mass-action ratio is approximately equal to K_{eq}, the enzyme is said to catalyze a near-equilibrium reaction and is not considered likely to be regulated. When the mass-action ratio is considerably different from the K_{eq}, the enzyme is said to catalyze a nonequilibrium reaction and is usually regulated. Comparison of mass-action ratios and equilibrium constants for enzymes of glycolysis in liver indicates that many enzymes catalyze equilibrium reactions. The reactions of glucokinase (liver isoenzyme of hexokinase), 6-phosphofructo-1 -kinase, and pyruvate kinase in liver are far from equilibrium, suggesting that they are likely sites for regulation.

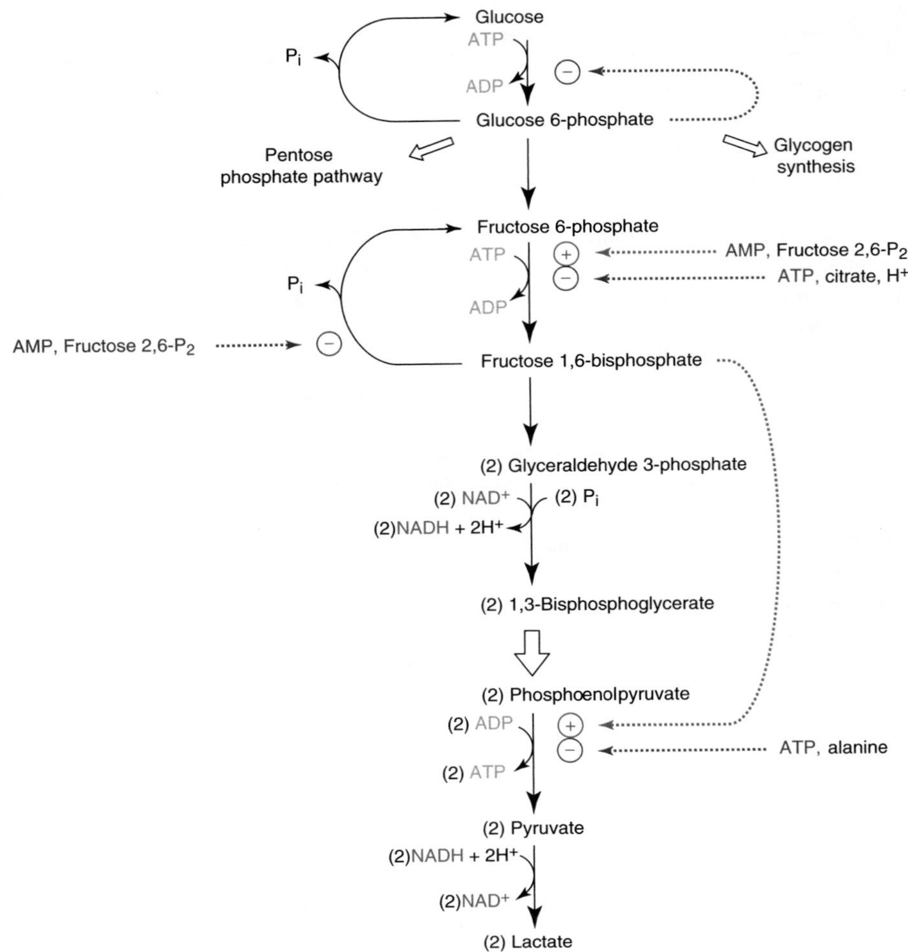

Figure 15.11 Important regulatory features of glycolysis. Because of tissue differences in isoenzyme expression, not all tissues of the body have all of the regulatory mechanisms shown here.

Hexokinase and Glucokinase Have Different Properties

Four different isoenzymes of hexokinase (I, II, III, and IV) are expressed in a tissue-specific manner in the body. The hexokinase isoenzymes found in most tissues (I, II, and III) have a low K_m for glucose (< 0.1 mM) relative to its concentration in blood (~ 5 mM) and are strongly inhibited by the product glucose 6-phosphate (G6P). The latter is important as it prevents hexokinase from tying up all of the inorganic phosphate of a cell in the form of phosphorylated hexoses (Clin. Corr. 15.3). Although the hexokinase reaction is not at equilibrium because of the inhibition imposed by G6P, the level at which hexokinase is expressed in cells can have a major impact on the rate of glycolysis in cells (A Closer Look 15.3).

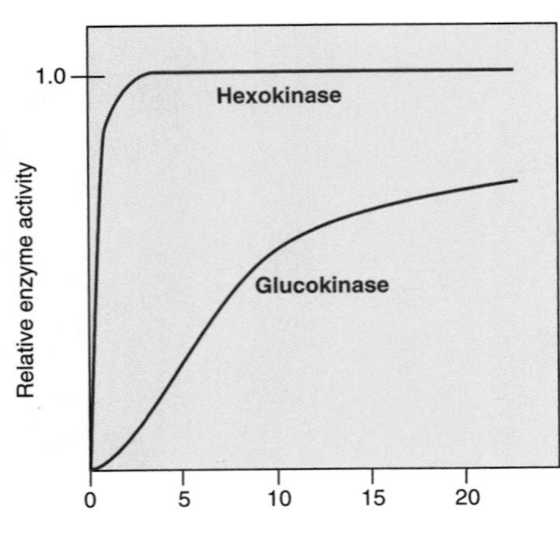

Figure 15.12 Comparison of the substrate saturation curves for hexokinase and glucokinase.

Liver parenchymal cells and β cells of pancreas are unique in that they contain hexokinase isoenzyme IV, usually called glucokinase, which has strikingly different kinetic properties. Glucokinase catalyzes ATP-dependent phosphorylation of glucose like other hexokinases, but its $S_{0.5}$ (substrate concentration that gives enzyme activity of one half maximum velocity) for glucose is considerably higher than the K_m for glucose of the other hexokinases (Figure 15.12). Furthermore, glucokinase is much less sensitive to product inhibition by G6P, and its glucose saturation curve is sigmoidal, which is indicative of cooperativity (p. 414). Because the Michaelis–Menten equation does not apply (p. 400), its kinetics are described by a $S_{0.5}$ value (concentration of substrate required to produce a rate one half of V_{max}) rather than a K_m value for glucose. Although not sensitive to inhibition by physiological concentrations of G6P, glucokinase is indirectly regulated by fructose 6-phosphate, which is just one step removed and in equilibrium with G6P. A special glucokinase inhibitory protein (GK-RP), which is located in the nucleus of liver cells, is responsible for this effect (Figure 15.13). GK-RP sequesters glucokinase as an inactive complex in the nucleus. Fructose 6-phosphate promotes binding of glucokinase to the regulatory protein,

CLINICAL CORRELATION 15.3

Fructose Intolerance (OMIM 229600)

Patients with hereditary fructose intolerance are deficient in the liver enzyme (aldolase B) that splits fructose 1-phosphate into dihydroxyacetone phosphate and glyceraldehyde. Three isoenzymes (A, B, and C) are expressed in mammals. Aldolase B is present in the largest amounts in liver. It acts on both fructose 1-phosphate and fructose 1,6-bisphosphate, but it has much greater affinity (lower *Km*) for fructose 1,6-bisphosphate. Consumption of fructose by an individual with an aldolase B deficiency results in accumulation of fructose 1-phosphate and depletion of P_i and ATP in the liver. The reactions involved are those of fructokinase and the enzymes of oxidative phosphorylation.

$$\text{Fructose} + \text{ATP} \rightarrow \text{fructose 1-phosphate} + \text{ADP}$$
$$\text{ADP} + P_i + \text{energy provided by electron transport chain'} \rightarrow \text{ATP}$$
$$\text{Net: } P_i + \text{fructose} \rightarrow \text{fructose 1-phosphate}$$

Tying up P_i as fructose 1-phosphate makes it impossible for liver mitochondria to generate ATP by oxidative phosphorylation. The ATP levels fall precipitously, making it impossible for the liver to carry out its normal functions. Cell damage results largely because of inability to maintain normal ion gradients by the ATP-dependent pumps. The cells swell and undergo osmotic lysis.

Although patients with fructose intolerance are particularly sensitive to fructose, humans in general have a limited capacity to handle this sugar. The capacity of the normal liver to phosphorylate fructose greatly exceeds its capacity to split fructose 1-phosphate. This means that fructose use by the liver is poorly controlled and that excessive fructose could deplete the liver of P_i and ATP. Fructose was tried briefly in hospitals as a substitute for glucose in patients being maintained by parenteral nutrition. The rationale was that fructose would be a better source of calories than glucose because its utilization is relatively independent of the insulin status of a patient. Delivery of large amounts of fructose by intravenous feeding was soon found to result in severe liver damage. Similar attempts have been made to substitute sorbitol and xylitol for glucose, but they also tend to deplete the liver of ATP and, like fructose, should not be used for parenteral nutrition.

Steinmann, B., Gitzelmann, R., and Van den Berghe, G. Disorders of fructose metabolism. In Scriver, C. R., Beaudet, A. L., Sly, W. S., and Valle, D. (Eds.), *The Metabolic and Molecular Bases of Inherited Disease,* 8th ed.. New York: McGraw Hill, 2001, p. 1489; Ali, M., Rellos, P., and Cox T.M. Hereditary fructose intolerance. *J. Med. Genetics* 35: 353, 1998; and Wong, D. Hereditary fructose intolerance. *Mol. Gen. Metab.* 85: 165, 2005.

A CLOSER LOOK 15.3

Hexokinase II and Cancer

As a rule, rapidly growing cancers metabolize glucose faster than normal cells, at least in part because overexpression of hexokinase II gives cancer cells greater enzymatic capacity for glucose phosphorylation. On top of this, tight binding of hexokinase II to the mitochondrial outer membrane provides the enzyme with first dibs for the ATP produced by oxidative phosphorylation. The exceptional capacity that cancer cells have for glucose metabolism is used to detect cancer by positron emission tomography (PET). ^{18}F-labeled 2-deoxyglucose is administered to individuals suspected of having cancer. Large amounts of ^{18}F-2-deoxyglucose-6-P accumulate in cancer cells that rapidly metabolize glucose. The absence of a hydroxyl group in the 2 position of the 2-deoxyglucose moiety prevents further metabolism of ^{18}F-2-deoxyglucose-6-P.

Mathupala, S. P., Ko, Y. H., and Pedersen, P. L. Hexokinase II: Cancer's double-edged sword acting as both facilitator and gatekeeper of malignancy when bound to mitochondria. *Oncogene* 25: 4777, 2006.

thereby inhibiting glucokinase. Fructose 6-phosphate in effect promotes translocation of glucokinase from the cytosol to the nucleus where glucokinase is completely inhibited by the inhibitory protein. This inhibitory effect of fructose 6-phosphate on glucokinase can be completely overcome by a large increase in glucose concentration. Glucose triggers dissociation of glucokinase from the regulatory protein, thereby promoting translocation of glucokinase from the nucleus to the cytosol. These special regulatory features of glucokinase (high $S_{0.5}$ for glucose, cooperativity with respect to glucose concentration, and glucose-stimulated translocation from the nucleus to the cytosol) contribute to the capacity of the liver to "buffer" blood glucose levels. Because GLUT2, the glucose transporter of liver cells, brings about rapid equilibration of glucose across the plasma membrane, the cellular concentration of glucose is the same as that in the blood. Since the $S_{0.5}$ of glucokinase for glucose ($\sim$ 7 mM) is greater than normal blood glucose concentrations ($\sim$ 5 mM) and glucose promotes translocation of glucokinase from the nucleus, any increase in portal blood glucose above normal leads to a dramatic increase in the rate of glucose phosphorylation by glucokinase in the liver (see Figures 15.12 and 15.13). Likewise, any decrease in glucose

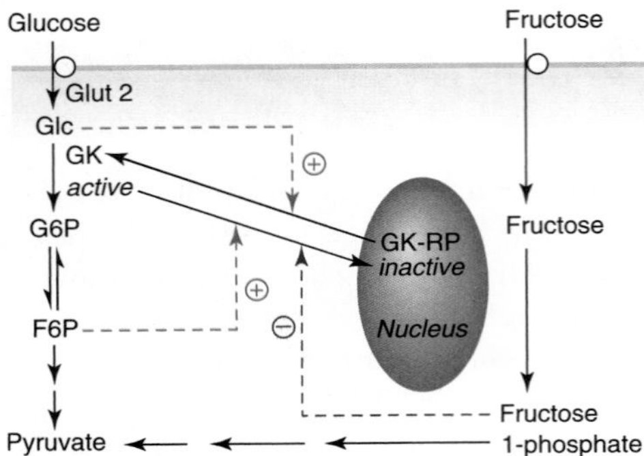

Figure 15.13 Glucokinase activity is regulated by translocation of the enzyme between the cytoplasm and the nucleus. Glucose increases glucokinase (GK) activity by promoting translocation of the enzyme to the cytoplasm. Fructose 6-phosphate decreases GK by stimulating translocation into the nucleus. Fructose 1-phosphate increases GK by inhibiting translocation into the nucleus. Binding of GK to regulatory protein (RP) in the nucleus completely inhibits GK activity.

concentration has the opposite effect. Thus, liver uses glucose at a significant rate only when blood glucose levels are greatly elevated and decrease its use of glucose when blood glucose levels are low. This buffering effect on blood glucose levels would not occur if glucokinase had the low K_m characteristic of other hexokinases, as it would then be completely saturated at physiological concentrations of glucose. On the other hand, a low K_m form of hexokinase is a good choice for tissues such as the brain in that it allows phosphorylation of glucose even when blood and tissue glucose concentrations are dangerously low.

The glucokinase reaction is not at equilibrium under normal intracellular conditions because of the rate restriction imposed by the high $S_{0.5}$ and inhibition by the regulatory protein. Another factor in opposition to the activity of glucokinase is glucose 6-phosphatase which like glucokinase has a high K_m (3 mM) for glucose 6-phosphate relative to its intracellular concentration (~ 0.2 mM). Thus, flux through this step is almost directly proportional to the intracellular concentration of glucose 6-phosphate. As shown in Figure 15.14, the combined action of glucokinase and glucose 6-phosphatase constitutes a futile cycle; that is, the sum of their reactions is hydrolysis of ATP to give ADP and P_i without the performance of any work. When blood glucose concentrations are about 5 mM, the activity of glucokinase is almost exactly balanced by the opposing activity of glucose 6-phosphatase. The result is that no net flux occurs in either direction. This futile cycling is wasteful of ATP but, combined with the process of gluconeogenesis (p. 619), contributes significantly to the "buffering" action of the liver on blood glucose levels. It also provides a mechanism for preventing glucokinase from tying up all of the P_i of the liver.

Fructose, a component of fruits, honey, vegetables, and high-fructose corn syrup used in popular carbonated beverages, promotes hepatic glucose utilization by an indirect mechanism. It is converted in liver to fructose 1-phosphate (Clin. Corr. 15.3, p. 607), which promotes dissociation of glucokinase from its regulatory protein (see Figure 15.13) and therefore translocation out of the nucleus. This action, which is capable of overriding the inhibition of glucokinase normally imposed by fructose 6-phosphate, may be a factor in the adverse effects sometimes associated with excessive dietary fructose consumption, e.g., increased hepatic carbohydrate utilization, lipogenesis, and hypertriacylglycerolemia.

Glucokinase is also an inducible enzyme. Induction and repression of synthesis of an enzyme are slow processes, usually requiring several hours before significant changes occur. Insulin promotes transcription of the glucokinase gene. An increase in blood glucose levels signals an increase in insulin release from the β cells of the pancreas, which promotes transcription and increases the amount of liver glucokinase enzyme protein. Thus, a person who has consumed a large carbohydrate-rich meal will have greater amounts of glucokinase than one who has not. The liver with glucokinase induced contributes more to the lowering of elevated blood glucose levels. The absence of insulin makes the liver of patients with diabetes mellitus deficient in glucokinase, in spite of high blood glucose levels, and decreases the ability of the liver to "buffer" blood glucose (Clin. Corr. 15.4). Defects in the gene-encoding glucokinase that alter its $S_{0.5}$ and/or V_{max} cause maturity-onset diabetes of the young (MODY), a form of type 2 diabetes mellitus.

6-Phosphofructo-1-kinase Is a Regulatory Enzyme of Glycolysis

6-Phosphofructo-1-kinase is an important regulatory site of glycolysis. It catalyzes the first committed step of glycolysis because the reaction catalyzed by phosphoglucose isomerase is reversible and cells make use of glucose 6-phosphate in the pentose phosphate pathway and

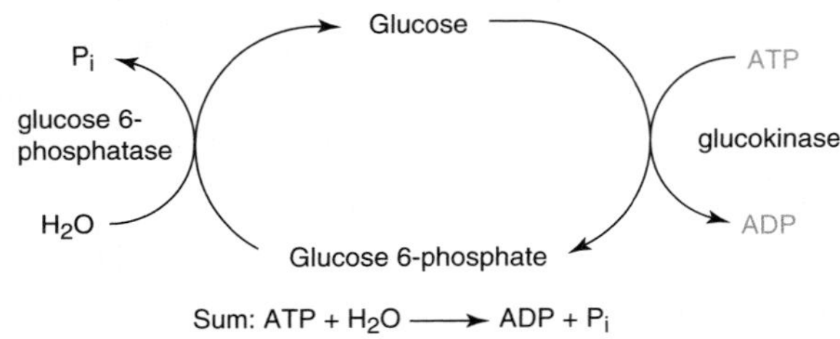

Figure 15.14 Phosphorylation of glucose followed by dephosphorylation constitutes a futile cycle in parenchymal cells of the liver.

CLINICAL CORRELATION 15.4

Diabetes Mellitus

Diabetes mellitus is a chronic disease characterized by derangements in carbohydrate, fat, and protein metabolism. Two major types are recognized clinically: type 1 (see Clin. Corr. 21.8, p. 868) and type 2 (see Clin. Corr. 21.6, p. 866).

In patients without fasting hyperglycemia, the oral glucose tolerance test can be used for diagnosis. It consists of determining the blood glucose level in the fasting state and at intervals of 30–60 min for 2 h or more after consuming a 100-g carbohydrate load. In normal individuals, blood glucose returns to normal levels within 2 h after ingestion of the carbohydrate. In diabetics, blood glucose reaches a higher level and remains elevated for longer periods of time, depending on the severity of the disease. However, many factors may contribute to an abnormal glucose tolerance test. The patient must have consumed a high carbohydrate diet for the preceding 3 days, presumably to allow for induction of enzymes of glucose-utilizing pathways, for example, glucokinase, fatty acid synthase, and acetyl-CoA carboxylase. Almost any infection (even a cold) and less well-defined stress, (presumably by effects on the sympathetic nervous system) can result in transient abnormalities of the glucose tolerance test. Because of these problems, fasting hyperglycemia (> 126 mg/dl) should probably be the *sine qua non* for the diagnosis of diabetes. Glucose uptake by insulin-sensitive tissues, that is, muscle and adipose, is decreased in the diabetic state. The diabetic patient either lacks insulin or has developed *insulin resistance* in these tissues. Resistance to insulin results from abnormality of the insulin receptor or in subsequent steps that mediate the metabolic effects of insulin. Liver parenchymal cells do not require insulin for glucose uptake. Without insulin, however, the liver has diminished capacity to remove glucose from the blood. This is explained in part by decreased glucokinase activity and the loss of insulin's action on key enzymes of glycogenesis and the glycolytic pathway.

Taylor, S. I. Insulin action, insulin resistance, and type 2 diabetes mellitus. In Scriver, C. R., Beaudet, A. L., Sly, W. S., and Valle, D. (Eds.), *The Metabolic and Molecular Bases of Inherited Disease,* 8th ed. New York: McGraw Hill, 2001, p. 1433; American Diabetes Association web site www.diabetes.org/home.jsp.

for glycogen synthesis. Citrate, ATP, and hydrogen ions (low pH) are the important negative allosteric effectors whereas AMP and fructose 2,6-bisphosphate are important positive allosteric effectors (see Figure 15.11, p. 606). These compounds signal the need for different rates of glycolysis in response to changes in (a) energy state of the cell (ATP and AMP), (b) internal environment of the cell (hydrogen ions), (c) availability of alternate fuels such as fatty acids and ketone bodies (citrate), and (d) the insulin/glucagon ratio in the blood (fructose 2,6-bisphosphate).

Regulation of 6-Phosphofructo-1-kinase by ATP and AMP

The Pasteur effect refers to the inhibition of glucose utilization and lactate accumulation that occurs when respiration (oxygen consumption) is initiated in anaerobic cells. It is readily understandable on a thermodynamic basis since complete oxidation of glucose to CO_2 and H_2O yields much more ATP than anaerobic glycolysis.

Glycolysis $\quad \text{D-Glucose} + 2\ \text{ADP}^{3-} + 2\ \text{P}_i^{2-} \rightarrow 2\ \text{L-lactate}^- + 2\ \text{ATP}^{4-}$

Complete Oxidation $\quad \text{D-Glucose} + 6\ \text{O}_2 + 32\ \text{ADP}^{3-} + 32\ \text{P}_i^{2-} + 32\ \text{H}^+ \rightarrow$

$$6\ \text{CO}_2 + 6\ \text{H}_2\text{O} + 32\ \text{ATP}^{4-}$$

Cells use ATP to provide the energy necessary for their inherent work processes. Since so much more ATP is produced from glucose in the presence of oxygen, much less glucose has to be consumed to meet the energy need. The Pasteur effect occurs in part because of ATP inhibition of glycolysis at the level of 6-phosphofructo-1-kinase. This can be readily rationalized since ATP is an inhibitor of 6-phosphofructo-1-kinase, and much more ATP is generated in the presence of oxygen than in its absence. Since 6-phosphofructo-1-kinase is severely inhibited at concentrations of ATP (2.5–6 mM) normally present in cells, the relatively small change in ATP concentration that occurs with oxygen versus without oxygen cannot account for large changes in flux through 6-phosphofructo-1-kinase. However, much greater changes occur in the concentration of AMP, a positive allosteric effector of 6-phosphofructo-1-kinase (see Figure 15.11, p. 606). The steady-state concentration of AMP when oxygen is introduced falls dramatically, decreasing 6-phosphofructo-1-kinase activity, suppressing glycolysis, and accounting in large part for the Pasteur effect. Levels of AMP automatically fall when ATP levels rise because the sum of adenine nucleotides in a cell, that is, ATP + ADP + AMP, is nearly constant under most physiological conditions and the amount of

ATP is always much greater than the amount of AMP. Furthermore, adenine nucleotides are maintained in equilibrium in the cytosol by adenylate kinase, which catalyzes the readily reversible reaction of 2ADP → ATP + AMP. The equilibrium constant (K'_{eq}) for this reaction is given by

$$K'_{eq} = \frac{[ATP][AMP]}{[ADP]^2}$$

Since this reaction operates near equilibrium under intracellular conditions, the concentration of AMP is given by

$$[AMP] = \frac{K'_{eq}[ADP]^2}{[ATP]}$$

Because intracellular [ATP] >> [ADP >> [AMP], a small decrease in [ATP] causes a substantially greater percentage increase in [ADP], and since [AMP] is related to the square of [ADP], an even greater percentage increase in [AMP]. Because of this relationship, a small decrease in [ATP] leads to a greater percentage increase in [AMP] than in the percentage decrease in [ATP]. This makes the [AMP] an excellent signal of the energy status of the cell and an important allosteric regulator of 6-phosphofructo-1-kinase activity. The effectiveness of 6-phosphofructo-1-kinase is influenced by [AMP] in yet another way. Fructose 1,6-bisphosphatase catalyzes an irreversible reaction that opposes the reaction of 6-phosphofructo-1-kinase.

$$\text{Fructose 1,6-bisphosphate} + H_2O \rightarrow \text{fructose 6-phosphate} + P_i$$

Fructose 1,6-bisphosphatase sits "cheek by jowl" with 6-phosphofructo-1-kinase in the cytosol of many cells. Together they catalyze a futile cycle (ATP + H_2O → ADP + P_i + "heat") and are therefore capable of decreasing each others "effectiveness." However, AMP inhibits fructose 1,6-bisphosphatase, which is opposite to the effect that AMP has on 6-phosphofructo-1-kinase. Thus, a small decrease in [ATP] triggers a large percentage increase in [AMP], which in turn signals a large increase in net conversion of fructose 6-phosphate into fructose 1,6-bisphosphate because of its effects on these two enzymes. This increases glycolytic flux by increasing the amount of fructose 1,6-bisphosphate available for the splitting stage.

The decrease in lactate that occurs due to the Pasteur effect is readily explained by decreased glycolytic flux. In addition, lactate dehydrogenase loses the competition with the substrate shuttle systems for NADH and the competition with the pyruvate dehydrogenase complex for pyruvate.

Regulation of 6-Phosphofructo-1-kinase by Intracellular pH

It would make sense if lactate, as the end product of glycolysis, would inhibit the most important regulatory enzyme of glycolysis. However, it is hydrogen ions rather than lactate that inhibit 6-phosphofructo-1-kinase (see Figure 15.11, p. 606). As shown in Figure 15.15, anaerobic glycolysis generates lactic acid, and the cell must dispose of it as such or suffer the negative consequences of acidification. This explains why excessive glycolysis lowers blood pH and leads to an emergency medical situation known as lactic acidosis (Clin. Corr.15.5). Plasma membranes contain a symport for lactate and hydrogen ions that allows transfer of lactic acid into the bloodstream. This defense mechanism prevents pH from getting so low that lactic-acid-producing tissues become pickled (Clin. Corr. 15.6). Transport of lactic acid out of a cell requires that blood be available to carry it away. When blood flow is inadequate, for example, in the heart during an attack of angina pectoris or in skeletal muscle during heavy exercise, hydrogen ions cannot escape from cells fast enough. Yet, the need for ATP, manifested by an increase in [AMP], may partially override inhibition of 6-phosphofructo-1-kinase by hydrogen ions. Unabated accumulation of hydrogen ions then causes pain, which, in the case of skeletal muscle, can be relieved by simply terminating the exercise. In the heart, rest or pharmacologic agents that increase blood flow or decrease the need for ATP within myocytes may be effective (Clin. Corr. 15.7).

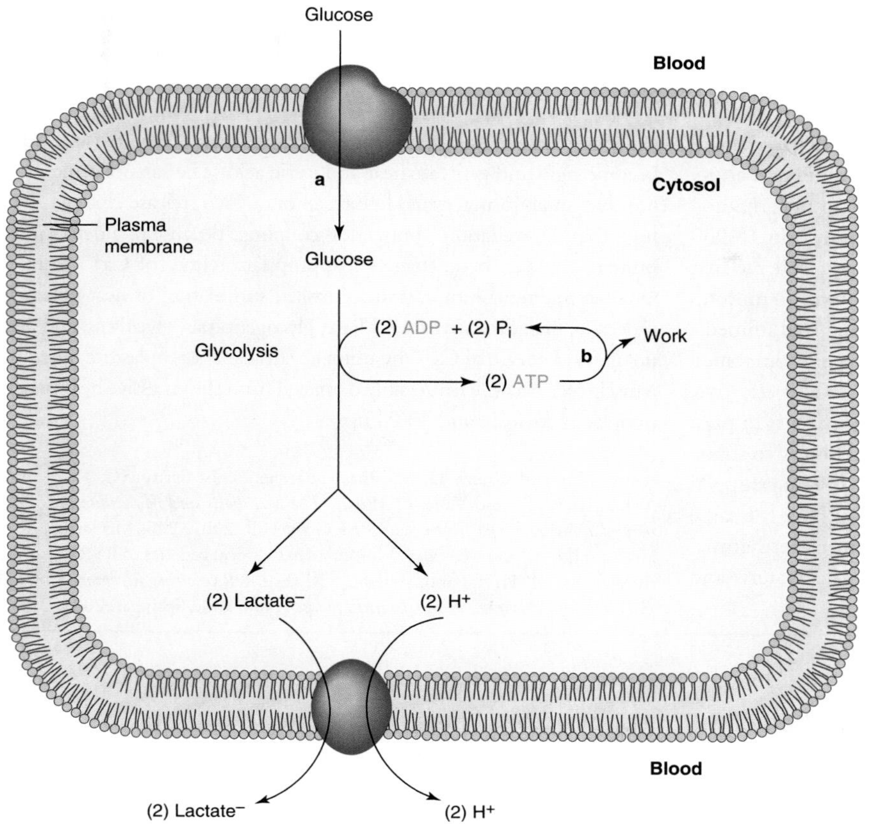

Figure 15.15 Unless lactate formed by glycolysis is transported out of the cell, the intracellular pH will decrease by the accumulation of intracellular lactic acid (equivalent to lactate$^-$ plus H^+ because lactic acid ionizes at intracellular pH). The low pH decreases 6-phosphofructo-1-kinase activity so that further lactic acid production by glycolysis is inhibited. (*a*) Glucose transport into the cell. (*b*) All work performances that convert ATP back to ADP and P_i. (*c*) lactate-H^+ symport (actual stoichiometry of one lactate$^-$ and one H^+ transported by the symport).

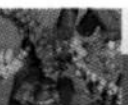

CLINICAL CORRELATION 15.5

Lactic Acidosis

This is characterized by elevated blood lactate levels, usually greater than 5 mM, with decreased blood pH and bicarbonate concentrations (see Clin. Corr. 1.2, p. 11). Lactic acidosis is the most frequent form of metabolic acidosis and can be the consequence of overproduction of lactate, underutilization of lactate, or both. Lactate production is normally balanced by lactate utilization so that lactate is usually not present in the blood at concentrations greater than 1.2 mM. All tissues can produce lactate by anaerobic glycolysis, but most tissues do not produce large quantities since they are well supplied with oxygen and mitochondria. However, all tissues respond with an increase in lactate generation when oxygenation is inadequate. A decrease in ATP because of less oxidative phosphorylation increases the activity of 6-phosphofructo-1-kinase. Thus, tissues have to rely on anaerobic glycolysis for ATP production under such conditions and overproduce lactic acid. A good example is muscle exercise, which can deplete the tissue of oxygen and cause an overproduction of lactic acid. Tissue hypoxia occurs, however, in all forms of shock, during convulsions, and in diseases involving circulatory and pulmonary failure.

The major fate of lactate in the body is either complete combustion to CO_2 and H_2O or conversion back to glucose by the process of gluconeogenesis. Both require oxygen. Decreased oxygen availability, therefore, increases lactate production and decreases lactate utilization. The latter can also be decreased by liver diseases, ethanol, and a number of other drugs. Phenformin, a drug that was once used to treat the hyperglycemia of type 2 diabetes, was well documented to induce lactic acidosis as a consequence of inhibition of complex I of the mitochondrial electron transport chain. Thiamin deficiency, common in poorly nourished alcoholics, causes lactic acidosis as a consequence of loss of activity of the pyruvate dehydrogenase complex. Thiamin should be administered immediately to patients with severe lactic acidosis.

Bicarbonate is usually administered in an attempt to control the acidosis associated with lactic acid accumulation. The key to successful treatment, however, is to find and eliminate the cause of the overproduction and/or underutilization of lactic acid and most often involves the restoration of circulation of oxygenated blood.

Newsholme, E. A., and Leech, A. R. *Biochemistry for the Medical Sciences.* New York, Wiley, 1983; Kruse, J. A., and Carlson, R. W. Lactate metabolism. *Crit. Care Clin.* 3: 725, 1985; Stacpoole, P. W. Lactic acidosis and other mitochondrial disorders. *Metabolism* 46:306, 1997; Ammari, A. N., and Schulze, K. F. Uses and abuses of sodium bicarbonate in the neonatal intensive care unit. *Curr. Opin. Pediatr.* 14: 151, 2002; and Klein, M., Weksler, N., and Gurman, G. M. Fatal metabolic acidosis caused by thiamine deficiency. *J. Emer. Med.* 26: 301, 2004.

CLINICAL CORRELATION 15.6

Pickled Pigs and Malignant Hyperthermia (OMIM 145600)

In malignant hyperthermia, a variety of agents, especially the general anesthetic halothane, produces a dramatic rise in body temperature, metabolic and respiratory acidosis, hyperkalemia, and muscle rigidity. This dominantly inherited abnormality occurs in about 1 in 15,000 children and 1 in 50,000–100,000 adults. Death may result the first time a susceptible person is anesthetized. Onset occurs within minutes of drug exposure, and the hyperthermia must be recognized immediately. Packing the patient in ice is effective and should be accompanied by measures to combat acidosis. The drug dantrolene is also effective.

A phenomenon similar to malignant hyperthermia occurs in pigs; it is called porcine stress syndrome. The pigs respond poorly to stress. This genetic disease usually manifests itself as the pig is being shipped to market. On exposure to halothane, susceptible pigs have the same response as that observed in humans with malignant hyperthermia. The meat of pigs that have died from the disease is pale, watery, and of very low pH (i.e., nearly pickled).

In response to halothane, the skeletal muscles of affected individuals become rigid and generate heat and lactic acid. The sarcoplasmic reticulum has an abnormal ryanodine receptor, a Ca^{2+} release channel that is important in excitation–contraction coupling. Because of a defect in this protein, the anesthetic triggers inappropriate release of Ca^{2+} from the sarcoplasmic reticulum and uncontrolled stimulation of heat-producing processes, including myosin ATPase, glycogenolysis, glycolysis, and cyclic uptake and release of Ca^{2+} by mitochondria and sarcoplasmic reticulum. Muscle cells become irreversibly damaged from the excessive heat production, lactic acidosis, and ATP loss.

Kalow, W., and Grant, D. M. Pharmacogenetics. In Scriver, C. R., Beaudet, A. L., Sly, W. S., and Valle, D. (Eds.), *The Metabolic and Molecular Bases of Inherited Disease,* 8th ed. New York: McGraw Hill, 2001, 225; McCarthy, T. V., Heffron, J. J. A., and Macrill, J. Molecular and clinical genetics of RYR1 disorders. In Wehrens, X. H. T., and Marks, A. R. (Eds.). *Ryanodine Receptors: Structure, Function and Dysfunction in Clinical Disease.,* New York: Springer, 2005, 219.

Regulation of 6-Phosphofructo-1-kinase by Citrate

Many tissues prefer to use fatty acids and ketone bodies as oxidizable fuels in place of glucose. Most such tissues can use glucose but prefer to oxidize fatty acids and ketone bodies. This helps preserve glucose for tissues, such as brain, that are absolutely dependent on it. Oxidation of both fatty acids and ketone bodies elevates levels of cytosolic citrate, which inhibits 6-phosphofructo-1-kinase (see Figure 15.11, p. 606) and decreases glucose utilization.

Hormonal Control of 6-Phosphofructo-1-kinase by cAMP and Fructose 2,6-bisphosphate

Fructose 2,6-bisphosphate (Figure 15.16), like AMP, is a positive allosteric effector of 6-phosphofructo-1-kinase and is a negative allosteric effector of fructose 1,6-bisphosphatase (see Figure 15.11, p. 806). Indeed, without its presence, glycolysis could not occur in liver because 6-phosphofructo-1-kinase would have insufficient activity and fructose 1,6-bisphosphatase would be too active for net conversion of fructose 6-phosphate to fructose 1,6-bisphosphate.

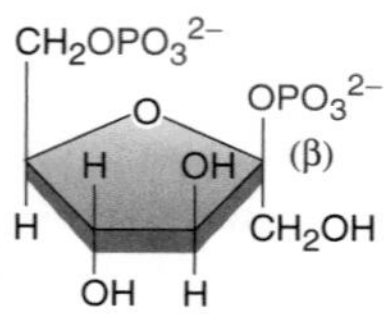

Figure 15.16 Structure of fructose 2,6-bisphosphate.

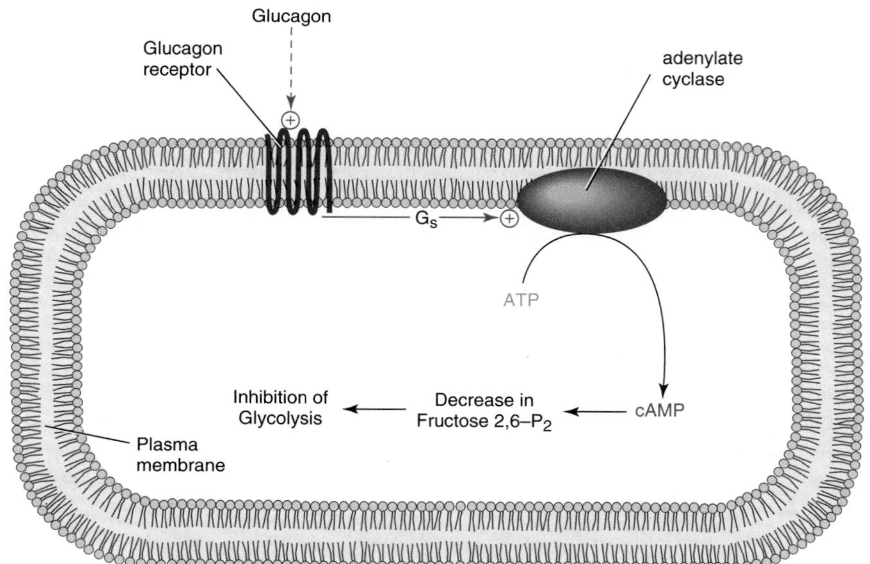

Figure 15.17 Mechanism by which glucagon inhibits hepatic glycolysis. Binding of glucagon to its membrane receptor activates adenylate cyclase (an intrinsic membrane protein) activity through the action of a stimulatory G protein (G_s; a peripheral membrane protein). The (+) symbol indicates activation.

CLINICAL CORRELATION 15.7

Angina Pectoris and Myocardial Infarction

Chest pain associated with reversible myocardial ischemia is termed angina pectoris (literally, strangling pain in the chest). The pain results from an imbalance between demand for and supply of blood flow to cardiac muscles and is most commonly caused by narrowing of the coronary arteries. The patient experiences a heavy squeezing pressure or ache substernally, often radiating to either the shoulder and arm, or occasionally to the jaw or neck. Attacks occur with exertion, last from 1 to 15 min, and are relieved by rest or by death. The coronary arteries involved are obstructed by atherosclerosis (i.e., lined with fatty deposits) or less commonly narrowed by spasm. Infarction occurs if the ischemia persists long enough to cause severe damage (necrosis) to the heart muscle. Commonly, a blood clot forms at the site of narrowing and completely obstructs the vessel. In infarction, tissue death occurs and the pain is longer lasting and often more severe.

Nitroglycerin and other nitrates are frequently prescribed to relieve the pain. They can be used prophylactically, enabling patients to participate in activities that would otherwise precipitate an attack. Nitroglycerin may work in part by causing dilation of the coronary arteries, improving oxygen delivery, and washing out lactic acid. Probably more important is the effect on the peripheral circulation. Breakdown of nitroglycerin produces nitric oxide (NO) (p. 447) that relaxes smooth muscle, causing venodilation throughout the body. This reduces arterial pressure and allows blood to accumulate in the veins. The result is decreased return of blood to the heart, and reduced volume of blood to be pumped, which reduces the energy requirement. The heart also empties itself against less pressure, which further spares energy. The overall effect is a lowering of the oxygen requirement of the heart, bringing it in line with the oxygen supply via the diseased coronary arteries. Other useful agents are calcium channel blockers, which are coronary vasodilators and β-adrenergic blockers. The β-blockers prevent the increase in myocardial oxygen consumption induced by sympathetic nervous system stimulation of the heart, as occurs with physical exertion.

The coronary artery bypass operation is used in severe cases of angina that cannot be controlled by medication. In this operation, veins are removed from the leg and interposed between the aorta and coronary arteries so as to bypass the portion affected by atherosclerosis and provide the affected tissue with a greater blood supply. Remarkable relief from angina can be achieved by this operation, with the patient being able to return to normal productive life in some cases.

Hugenholtz, P. G. Calcium antagonists for angina pectoris. *Ann. N.Y. Acad. Sci.* 522: 565, 1988; Feelishch, M. and Noack, E. A. Correlation between nitric oxide formation during degradation of organic nitrates and activation of guanylate cyclase. *Eur. J. Pharmacol.* 139: 19, 1987; Ignarro, L. J. Biological actions and properties of endothelium-derived nitric oxide formed and released from artery and vein. *Circ. Res.* 65:1, 1989; and Miller, M. R. and Megson, I. L. Recent developments in nitric oxide donor drugs. *British J. Pharm.* 151: 305, 2007.

Figure 15.17 summarizes the role of fructose-2, 6-bisphosphate in hormonal control of hepatic glycolysis. The mechanism uses cAMP (Figure 15.18) as the *second messenger* of hormone action. Glucagon is released from the α cells of pancreas and circulates in blood until it encounters glucagon receptors on the outer surface of liver plasma membrane (see p. 615 for more details). Binding of glucagon to its receptor triggers stimulation of adenylate cyclase through the second messenger cyclic AMP (cAMP), which results in a decrease in fructose 2,6-bisphosphate. This makes 6-phosphofructo-1-kinase less effective and fructose 1,6-bisphosphatase more effective, thereby severely restricting flux from fructose 6-phosphate to fructose 1,6-bisphosphate in glycolysis.

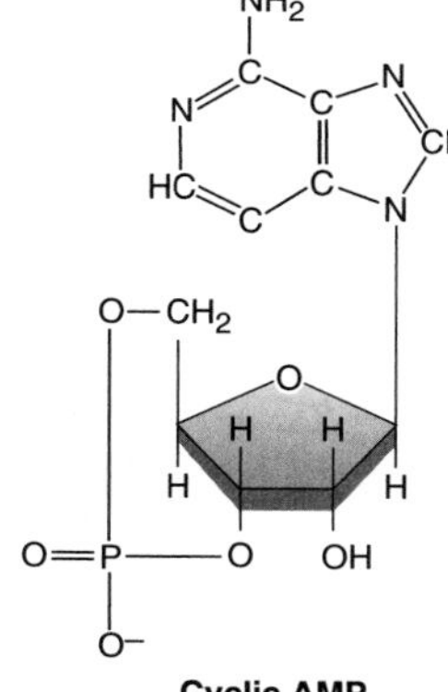

Figure 15.18 **Structure of cAMP.**

Fructose 2,6-bisphosphate is produced from F6P by the enzyme **6-phosphofructo-2-kinase** (Figure 15.19) as a side product rather than as an intermediate of glycolysis. This makes two phosphofructokinases for us to contend with, one (6-phosphofructo-1-kinase) produces an intermediate (fructose 1,6-bisphosphate) of glycolysis and the other (6-phosphofructo-2-kinase) produces a positive allosteric effector (fructose 2,6-bisphosphate) of the former enzyme. Fructose 2,6-bisphosphatase opposes 6-phosphofructo-2-kinase by converting fructose 2,6-bisphosphate to F6P by simple hydrolysis. These kinase and phosphatase activities that determine the amount of fructose 2,6-bisphosphate reside in the same protein, a bifunctional enzyme named 6-phosphofructo-2-kinase/fructose 2,6-bisphosphatase. cAMP

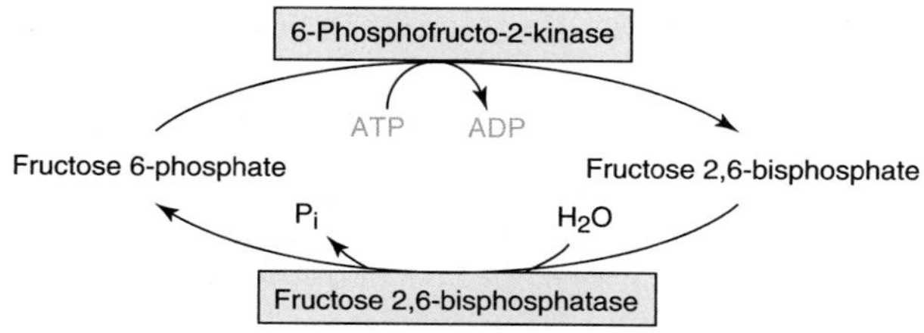

6-Phosphofructo-2-kinase

ATP ADP

Fructose 6-phosphate Fructose 2,6-bisphosphate

P_i H_2O

Fructose 2,6-bisphosphatase

Figure 15.19 **Reactions involved in the formation and degradation of fructose 2,6-bisphosphate.**

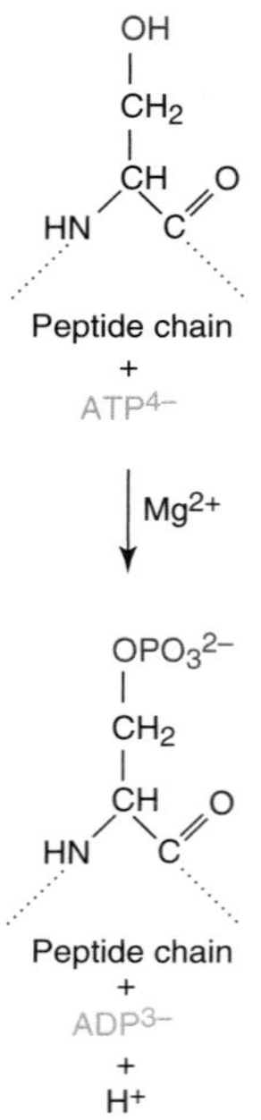

regulates fructose 2,6-bisphosphate levels in liver by activating the phosphatase and inactivating the kinase moieties of the bifunctional enzyme. This is achieved by cyclic-AMP-mediated activation of **protein kinase A**. Inactive protein kinase A consists of two regulatory and two catalytic subunits. Binding of cAMP to the regulatory subunits causes conformational changes that release and activate the catalytic subunits. The activated catalytic subunits then phosphorylate specific serine residues present in many enzymes (Figure 15.20).

Phosphorylation of an enzyme can conveniently be abbreviated as

$$\boxdot + \text{ATP} \rightarrow \odot - \text{P} + \text{ADP}$$

where ⊡ and ⊙-P are used to indicate dephosphorylated and phosphorylated enzymes, respectively. Phosphorylation causes a change in conformation that affects the active site of the enzyme. The activity of some enzymes is increased whereas that of others is decreased by phosphorylation. Many enzymes are regulated by this type of covalent modification that can be reversed by simply removing the phosphate. Regardless of whether phosphorylation or dephosphorylation activates the enzyme, the active enzyme is called the *a* form and the inactive enzyme the *b* form. Likewise, the action of a protein kinase is always opposed by that of a phosphoprotein phosphatase which catalyzes the reaction of

$$\odot - \text{P} + \text{H}_2\text{O} \rightarrow \boxdot + \text{P}_i$$

Putting these together creates a cyclic control system (Figure 15.21), such that the ratio of phosphorylated to dephosphorylated enzyme depends on the relative activities of the protein kinase and the phosphoprotein phosphatase, both of which are almost invariably also subject to regulation.

Figure 15.20 Enzymes subject to covalent modification are usually phosphorylated on specific serine residues. Tyrosine and threonine residues are also important sites of covalent modification by phosphorylation.

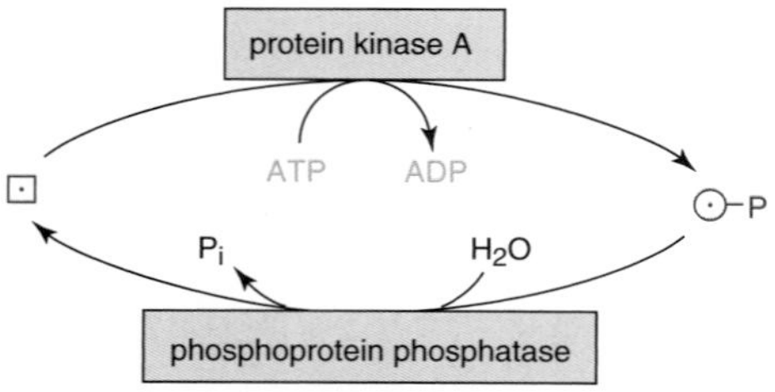

Figure 15.21 General model for regulation of enzymes by phosphorylation-dephosphorylation. The symbols ⊡ and ⊙-P indicate that different conformational and activity states of the enzyme are produced because of phosphorylation-dephosphorylation.

The Bifunctional Enzyme 6-Phosphofructo-2-kinase/Fructose 2,6-bisphosphatase Is Regulated by Phosphorylation

Usually an enzyme subject to phosphorylation is either activated or inactivated as a result of phosphorylation. The bifunctional nature of 6-phosphofructo-2-kinase/fructose 2,6-bisphosphatase makes it unique in this regard. A single phosphorylation of liver 6-phosphofructo-2-kinase/fructose 2,6-bisphosphatase inactivates the kinase enzyme activity but activates the phosphatase enzyme activity (Figure 15.22). Dephosphorylation has the opposite effects. This provides a very sensitive mechanism for setting the intracellular concentration of fructose 2,6-bisphosphate in liver cells in response to changes in blood levels of glucagon or epinephrine (Figure 15.23). Increased levels of glucagon or epinephrine, acting through plasma membrane glucagon receptors and β-adrenergic receptors, respectively, have the common effect of increasing intracellular levels of cAMP. This activates protein kinase A, which phosphorylates a serine residue of 6-phosphofructo-2-kinase/fructose 2,6-bisphosphatase (Figure 15.24), which inhibits the kinase activity and activates its phosphatase activity. The resulting decrease in fructose 2,6-bisphosphate makes 6-phosphofructo-1-kinase less effective and fructose 1,6-bisphosphatase more effective and thereby inhibits glycolysis at the level of the conversion of fructose 6-phosphate to fructose 1,6-bisphosphate. A decrease in glucagon or epinephrine in the blood results in the opposite effect. A phosphoprotein phosphatase removes phosphate from the bifunctional

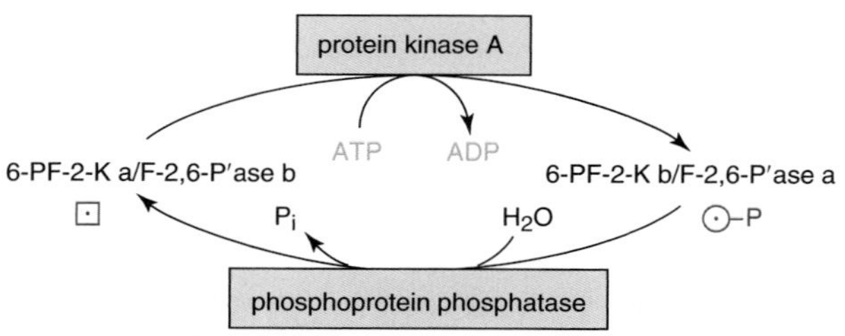

Figure 15.22 Mechanism for covalent modification of 6-phosphofructo-2-kinase/fructose 2,6-bisphosphatase. Name of the enzyme is abbreviated as 6-PF-2-K/F-2,6-P'ase. Letters a and b indicate the active and inactive forms of the enzymes, respectively.

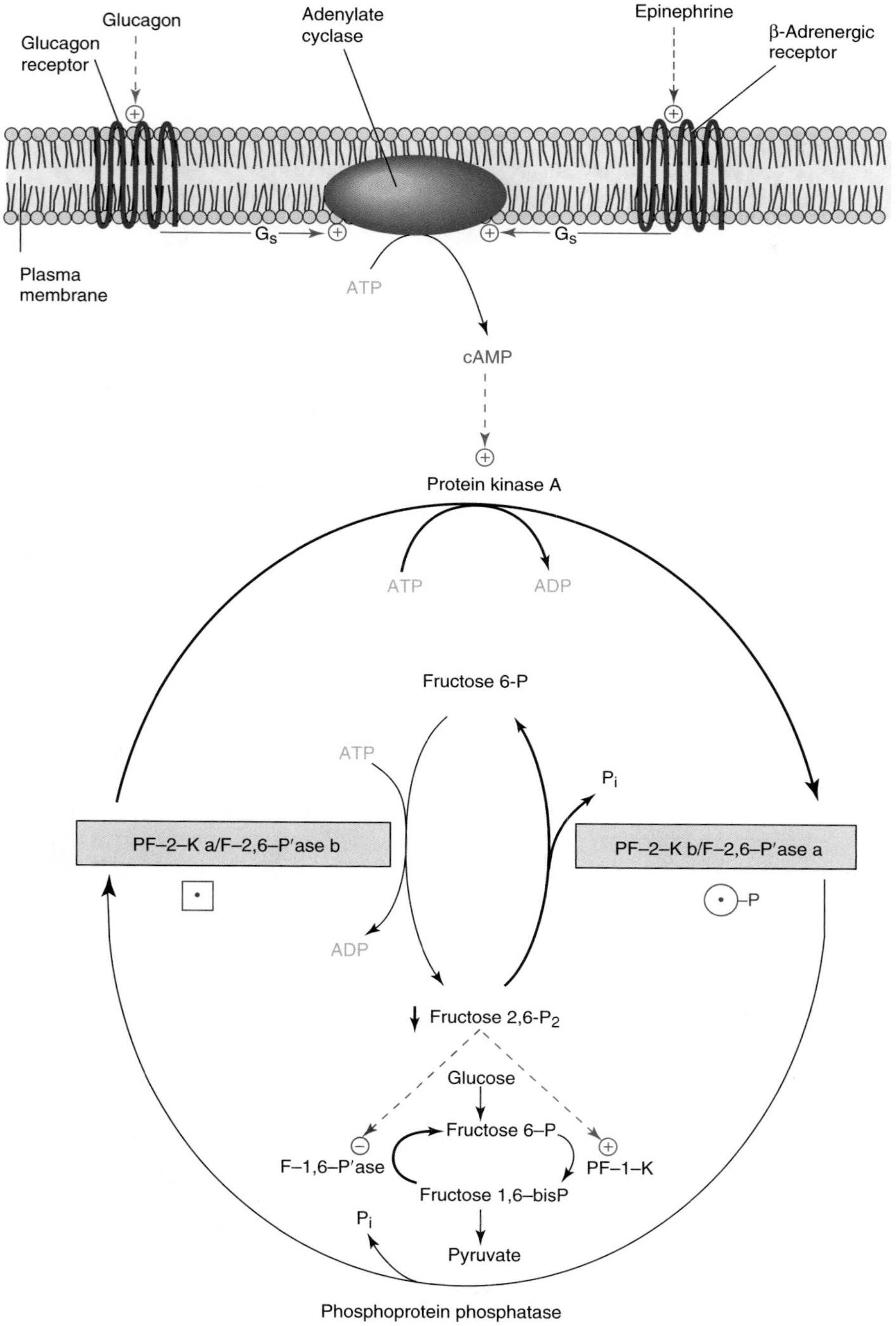

Figure 15.23 Mechanism of glucagon and epinephrine inhibition of hepatic glycolysis by cAMP-mediated decrease in fructose 2,6-bisphosphate.
See legend for Figure 15.19, p. 613. Heavy arrows indicate reactions that predominate in the presence of glucagon. Small arrow before fructose 2,6-bisphosphate indicates a decrease in its concentration.

enzyme to activate the kinase and inactivate the phosphatase. Fructose 2,6-bisphosphate accumulates to a higher steady-state concentration and increases the rate of glycolysis. Thus, glucagon and epinephrine are extracellular signals that stop liver from using glucose, whereas fructose 2,6-bisphosphate is an intracellular signal that promotes glucose utilization by this tissue.

Insulin opposes the actions of glucagon and epinephrine by means of a signaling cascade initiated by activation of the tyrosine kinase activity of its receptor (p. 910). Insulin also activates cAMP phosphodiesterase (lowers cAMP levels), inhibits protein kinase A, and activates phosphoprotein phosphatase, all of which oppose the effects of glucagon and epinephrine (Figure 15.25). Insulin therefore acts to stimulate the rate of glycolysis.

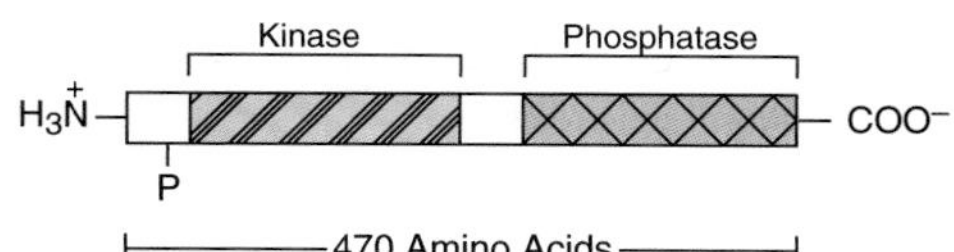

Figure 15.24 Schematic diagram of the primary structure of the liver isoenzyme of 6-phosphofructo-2-kinase/fructose-2,6-bisphosphatase. H_3N^+ and COO^- designate the N-terminal and C-terminal ends, respectively. Domain with kinase activity is located in the N-terminal half; domain with phosphatase activity in the C-terminal half. The letter P indicates the site (serine 32) phosphorylated by protein kinase A.

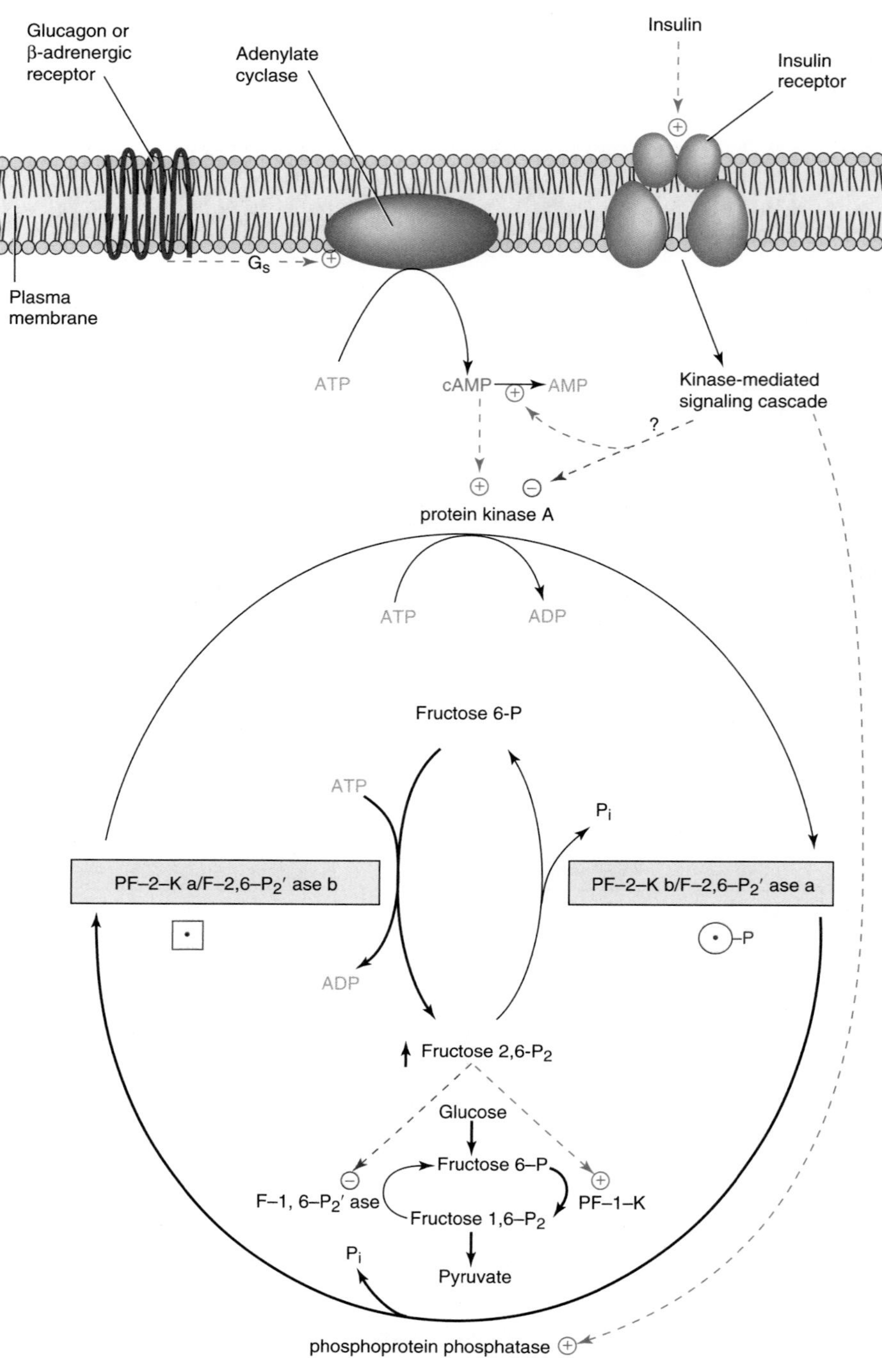

Figure 15.25 Mechanism for accelerated rates of hepatic glycolysis when the concentration of glucagon and epinephrine are low and that of insulin is high in the blood. See legends to Figures 15.17, p. 612, and 15.23, p. 615. The insulin receptor is an intrinsic protein of the plasma membrane. Small arrow before fructose 2,6-bisphosphate indicates an increase in concentration. cAMP is converted to AMP by cAMP phosphodiesterase.

Heart Contains a Different Isoenzyme of 6-Phosphofructo-2-kinase/Fructose 2,6-bisphosphate

An increase in blood level of epinephrine has a markedly different effect on glycolysis in heart than in liver. Although glycolysis is inhibited in liver to conserve glucose for use by other tissues, epinephrine stimulates glycolysis in heart to meet the increased need for ATP caused by an epinephrine-signaled increase in workload. As in liver, epinephrine acts on the heart by way of a β-adrenergic receptor on the plasma membrane, promoting formation of cAMP (Figure 15.26). This results in activation of protein kinase A that then phosphorylates 6-phosphofructo-2-kinase/fructose 2,6-bisphosphatase. However, in contrast to what

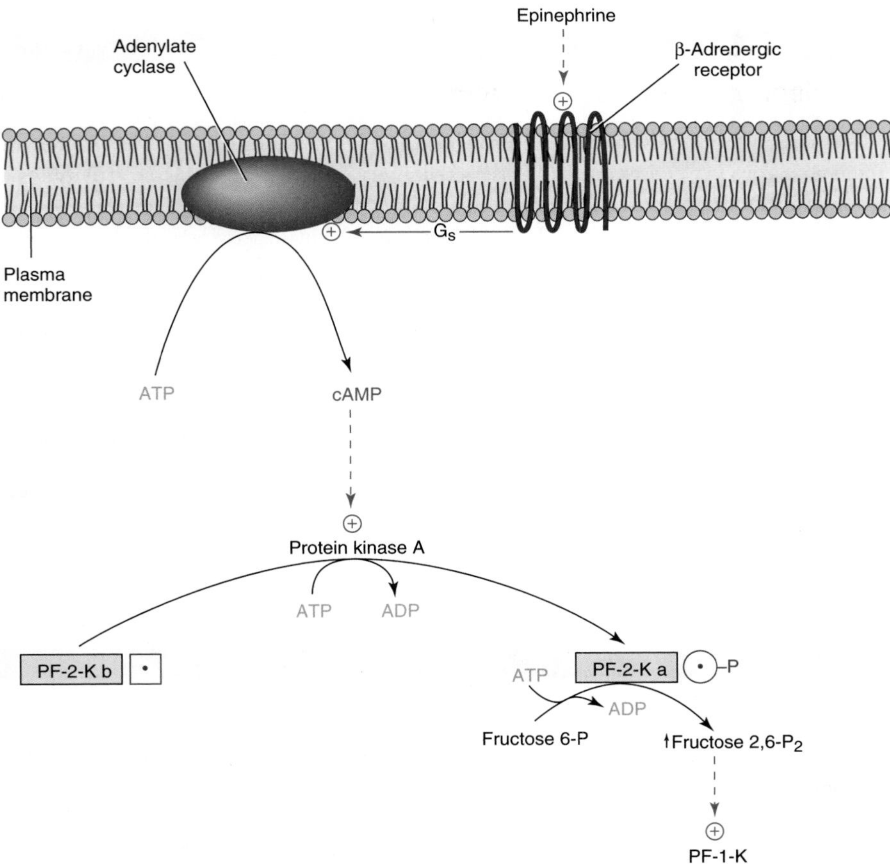

Figure 15.26 Mechanism for acceleration of glycolysis in the heart in response to epinephrine. See legends to Figures 15.17, p. 612, and 15.25, p. 616.

happens in liver, phosphorylation of the bifunctional enzyme in heart increases rather than decreases fructose 2,6-bisphosphate levels. This is because the isoenzyme of the bifunctional enzyme expressed in heart is different from the isoenzyme expressed in liver. Phosphorylation of the heart isoenzyme occurs at a site that activates rather than inhibits the kinase activity (Figure 15.27). Increased fructose 2,6-bisphosphate concentration then increases 6-phosphofructo-1-kinase activity and glycolytic flux.

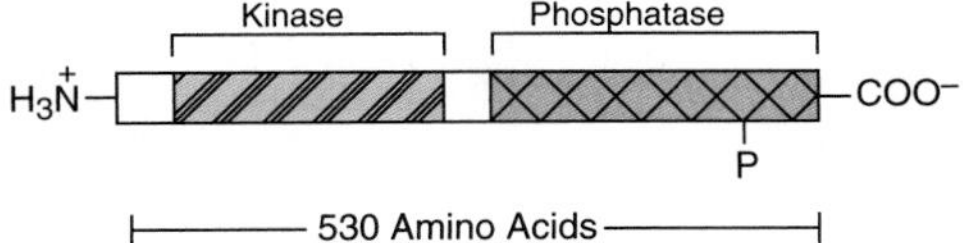

Figure 15.27 Schematic diagram of the primary structure of the heart isoenzyme of 6-phosphofructo-2-kinase/fructose-2,6-bisphosphatase. See legend to Figure 15.24, p. 615. The letter P indicates the site (serine 466) phosphorylated by protein kinase A.

Role of 6-Phosphofructo-2-kinase/Fructose 2,6-bisphosphate in Cancer

Many cancer cells express a special isoform of 6-phosphofructo-2-kinase/fructose 2,6-bisphosphatase in which the activity of the 2-kinase component greatly exceeds the activity of the 2-phosphatase component. This results in a high steady state concentration of fructose 2,6-bisphosphate which maximally stimulates of 6-phosphofructo-1-kinase activity. This is a major component of the mechanism responsible for the high rate of glycolysis characteristic of rapidly growing tumors (A Closer Look 15.4 and 15.5).

Pyruvate Kinase Is Also a Regulatory Enzyme of Glycolysis

Pyruvate kinase (Clin. Corr. 15.8) is drastically inhibited by physiological concentrations of ATP (see Figure 15.11, p. 606), so much so that its potential activity is probably never fully realized under physiological conditions. The liver isoenzyme is greatly activated by fructose 1,6-bisphosphate (FBP), and thereby linking regulation of pyruvate kinase to that of 6-phosphofructo-1-kinase. Thus, if conditions favor increased flux through 6-phosphofructo-1-kinase, the level of FBP increases and acts as a feed-forward activator of pyruvate kinase. The liver enzyme is also regulated by covalent modification by protein kinase A, being active in the dephosphorylated state and inactive in the phosphorylated state (Figure 15.28, p. 619). Concurrent inhibition of hepatic glycolysis and stimulation of hepatic gluconeogenesis by glucagon can be explained in part by inhibition of pyruvate kinase caused by activation of protein kinase A by cAMP. This aspect is explored more thoroughly in Section 15.5 in the discussion of gluconeogenesis.

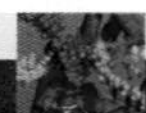

A CLOSER LOOK 15.4

6-Phosphofructo-2-kinase/Fructose-2,6-bisphosphatase and Cancer

Aggressive tumors often grow faster than new blood vessels can be formed from existing blood vessels. This can result in a core of oxygen-starved cells within tumors that could die for want of the ATP normally produced by oxidative phosphorylation. However, cancerous cells have an insidious ability to survive hypoxia, afforded by the exceptional capacity they have for the generation of ATP by glycolytic pathway. Expression of an hypoxia-inducible form of 6-phosphofructo-2-kinase/fructose-2,6-bisphosphatase is one of the major reasons for the unusual glycolytic activity of cancer cells. As in the liver, the kinase activity greatly exceeds the phosphatase activity of the nonphosphorylated bifunctional enzyme in cancerous cells. Remarkably, and in complete contrast to the liver, phosphorylation of the hypoxia-inducible form of the bifunctional enzyme by AMP-activated protein kinase (AMPK) causes an even greater increase in the kinase activity relative to the phosphatase activity. As a consequence, the increase in AMP that occurs in response to hypoxia activates AMPK which activates 2-kinase activity of the bifunctional enzyme which increases the production of fructose 2,6-bisphosphate which activates 6-phosphofructo-1-kinase which increases the rate of glycolysis which generates the ATP required to meet the energy needs of hypoxia cancerous cells.

Kim, S.-G., Manes, N. P., El-Maghrabi, M. R., and Lee, Y.-H. Crystal structure of the hypoxia-inducible form of 6-phosphofructo-2-kinase/fructose-2,6-bisphosphatase (PFKFB3): A possible new target for cancer therapy. *J. Biol. Chem.* 281: 2939, 2006.

A CLOSER LOOK 15.5

TIGAR and Cancer

TIGAR stands for TP53(tumor protein 53)-induced glycolysis and apoptosis regulator. As implied by its name, TIGAR regulates glycolysis and apoptosis and is induced by activation of the TP53 gene. The TP53 gene encodes the tumor-suppressor protein p53, a transcription factor that regulates the expression of several proteins, including TIGAR, that prevent the development of cancer. TIGAR is structurally related to the bisphosphatase domain of the bifunctional enzyme 6-phosphofructo-2-kinase/fructose-2,6-bisphosphatase. It lacks kinase activity but possesses phosphatase activity toward fructose 2,6-bisphosphate. Induction of TIGAR by p53 reduces fructose 2,6-biphosphate which reduces 6-phosphofructo-1-kinase activity which in turn reduces flux through the glycolytic pathway. Since glycolysis is important for the survival and growth of cancer cells (see A Closer Look 15.4, p. 618), up-regulation of TIGAR by p53 reduces the likelihood of tumor formation. Loss of this regulatory mechanism as a consequence of mutations in the TP53 gene is common in cancer.

Bensaad, K., Tsuruta, A., Selak, M. A., Vidal, M. N., et al. TIGAR, a p53-inducible regulator of glycolysis and apoptosis. *Cell* 126: 107, 2006; and Green, D. R. and Chipuk, J. E. p53 and metabolism: Inside the TIGAR. *Cell* 126: 30, 2006.

CLINICAL CORRELATION 15.8

Pyruvate Kinase Deficiency and Hemolytic Anemia (OMIM 266200)

Mature erythrocytes are absolutely dependent on glycolytic activity for ATP production. ATP is needed for the ion pumps, especially the Na^+, K^+- transporting-ATPase, which maintains the biconcave disk shape of erythrocytes that helps them slip through the capillaries as they deliver oxygen to the tissues. The cells swell and lyse without ATP. Anemia due to excessive erythrocyte destruction is referred to as hemolytic anemia. Pyruvate kinase deficiency is rare but is the most common genetic defect of glycolysis that causes hemolytic anemia. Most patients have 5% to 25% of normal erythrocyte pyruvate kinase level and flux through the glycolysis is severely restricted, resulting in markedly lower ATP concentrations. The glycolysis intermediates proximal to the pyruvate kinase step accumulate, whereas pyruvate and lactate concentrations decrease. 2,3-Bisphosphoglycerate levels are increased. As a result, anemia is better tolerated in some patients than would be predicted because the affinity of hemoglobin for oxygen is decreased by the increase in 2,3-bisphosphoglycerate. Normal ATP levels are observed in reticulocytes of patients. Although deficient in pyruvate kinase, these immature' red blood cells have mitochondria and can generate ATP by oxidative phosphorylation. Maturation of reticulocytes into erythrocytes results in the loss of mitochondria and complete dependence on glycolysis for ATP production. Thus, the mature cells are lost rapidly from circulation. Anemia results because they cannot be replaced rapidly enough by erythropoiesis.

Valentine, W. N. The Stratton lecture: Hemolytic anemia and inborn errors of metabolism. *Blood* 54: 549, 1979; Hirono, A., Kanno, H., Miwa, S., and Beutler, E. Pyruvate kinase deficiency and other enzymopathies of the erythrocyte. In Scriver, C. R., Beaudet, A. L., Sly, W. S., and Valle, D. (Eds.), *The Metabolic and Molecular Bases of Inherited Disease,* 8th ed. New York: McGraw Hill, 2001, 4637.

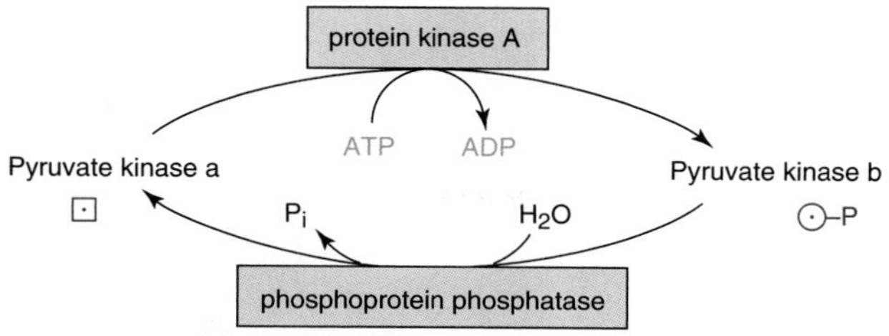

protein kinase A
ATP ADP
Pyruvate kinase a
Pyruvate kinase b
P_i H_2O
phosphoprotein phosphatase

Figure 15.28 Glucagon acts via cAMP to cause phosphorylation and inactivation of hepatic pyruvate kinase.

Pyruvate kinase, like glucokinase, is induced in liver by high carbohydrate intake and high insulin levels. This is one reason why liver of the well-fed individual has much greater capacity for utilizing carbohydrate than that of a fasting or diabetic person (see Clin. Corr. 15.4, p. 609).

Role of Pyruvate Kinase in Cancer

Cancer cells express a special isoform of pyruvate kinase called PKM2, a splicing variant of the muscle form of pyruvate kinase (PKM1). PKM2 is required for rapid growth of tumor cells. Although PKM1 and PKM2 are closely related, PKM1 cannot substitute for PKM2 in cancer cells. Why tumors require this specific pyruvate kinase isoform is an active area of research (A Closer Look 15.6).

A CLOSER LOOK 15.6

Pyruvate Kinase M2 and Cancer

All cancer cells express PKM2, an isoform of pyruvate kinase found in embryos but not in normal adult tissues. For reasons that are not understood, the exceptional capacity that cancer cells have for aerobic glycolysis depends on this particular isoform of pyruvate kinase. Aerobic glycolysis by cancer cells, discovered by Otto Warburg, the recipient of the 1931 Nobel Prize for Physiology and Medicine, refers to an unusually rapid rate of glycolysis and overproduction of lactate in the presence of oxygen. Noncancerous cells extract more ATP per glucose, and therefore use less glucose and produce little or no lactate when oxygen is present. This is because the rate of glycolysis is tightly coordinated with the rates of the TCA cycle and oxidative phosphorylation in normal cells. In cancer cells glycolysis exceeds the capacity for complete glucose oxidation, resulting in overproduction of lactic acid, in spite of the presence of oxygen. Warburg proposed that the replacement of respiration by fermentation is the primary cause of cancer. Subsequent research has shown this is not the case. Cancer is caused by mutations that activate proto-oncogenes and inhibit tumor suppressor genes. Nevertheless, altered expression of genes encoding PDKM2 and other enzymes that increase the capacity for glycolysis provide cancer cells with a survival and growth advantage over noncancer cells.

Christofk, H. R., Vander Heiden, M. G., Harris, M. H., Ramanathan, A., et al. The M2 splice isoform of pyruvate kinase is important for cancer metabolism and tumour growth. *Nature* 452: 230, 2008.

15.5 • GLUCONEOGENESIS

Glucose Synthesis Is Required for Survival

Net synthesis or formation of glucose from non-carbohydrate substrates is termed gluconeogenesis. Various amino acids, lactate, pyruvate, propionate, and glycerol are sources of carbon for the pathway (Figure 15.29). Glucose can also be produced from fructose. Glycogenolysis, that is, formation of glucose or glucose 6-phosphate from glycogen, should be differentiated from gluconeogenesis. Glycogenolysis refers to breakdown of glycogen to glucose and does not therefore correspond to *de novo* synthesis of glucose. Gluconeogenesis is essential for survival of humans and other animals as blood glucose levels have to be maintained to support metabolism of tissues that use glucose as their primary substrate (Clin. Corr. 15.9). These include brain, red blood cells, kidney medulla, lens, cornea, and testis. Gluconeogenesis enables the maintenance of blood glucose levels long after all dietary glucose has been absorbed and completely oxidized and glucose stored as glycogen has been used up.

Cori and Alanine Cycles

Gluconeogenesis is involved in two cycles that are critically important for maintaining blood glucose levels. The Cori cycle (or glucose–lactate cycle) and the alanine cycle (or glucose–alanine cycle) (Figure 15.30) depend on gluconeogenesis in liver followed by delivery of glucose and its use in a peripheral tissue. Both cycles provide for a continuous supply

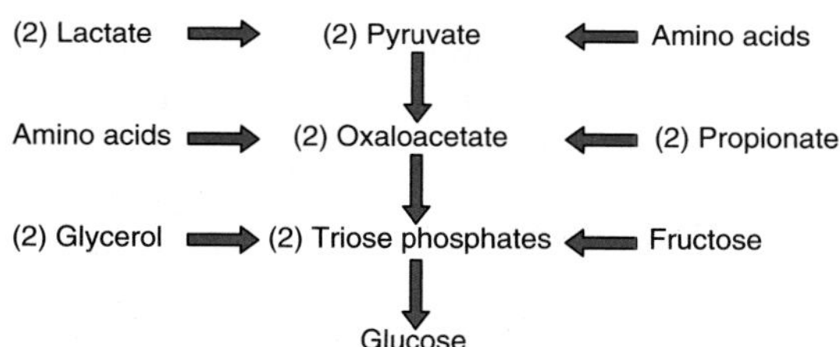

Figure 15.29 Abbreviated pathways of gluconeogenesis, illustrating the major substrate precursors for the process.

CLINICAL CORRELATION 15.9

Hypoglycemia and Premature Infants

Premature and small-for-gestational-age neonates are more susceptible to hypoglycemia than full-term, appropriate-for-gestational-age infants. Children in general are also more susceptible than adults, simply because they have larger brain/body weight ratios and the brain utilizes disproportionately greater amounts of glucose than the rest of the body. Newborn infants have a limited capacity for ketogenesis, apparently because the transport of long-chain fatty acids into their liver mitochondria is poorly developed. Since the use of ketone bodies by the brain is directly proportional to the circulating ketone body concentration, the neonate cannot spare glucose to any significant extent by using ketone bodies. Thus the neonate's brain is almost completely dependent on glucose obtained from liver glycogenolysis and gluconeogenesis.

The capacity for hepatic glucose synthesis from lactate and alanine is also limited in newborn infants, as the rate-limiting enzyme, phosphoenolpyruvate carboxykinase, is present in very low amounts during the first few hours after birth. Its induction to the level required to prevent hypoglycemia during the stress of fasting requires several hours. Premature and small-for-gestational-age infants also have smaller stores of liver glycogen. Fasting depletes their glycogen stores more rapidly, making them more dependent on gluconeogenesis than normal infants.

Newsholme, E. A. and Leech, A. R. *Biochemistry for the Medical Sciences.* New York, Wiley, 1983; Duvanel, C. B., Fawer, C.-L., Cotting, J., Hohlfeld, P., and Matthieu, J. M. Long-term effects of neonatal hypoglycemia on brain growth and psychomotor development in small-for-gestational-age preterm infants. *J. Pediatr.*134:492, 1999.

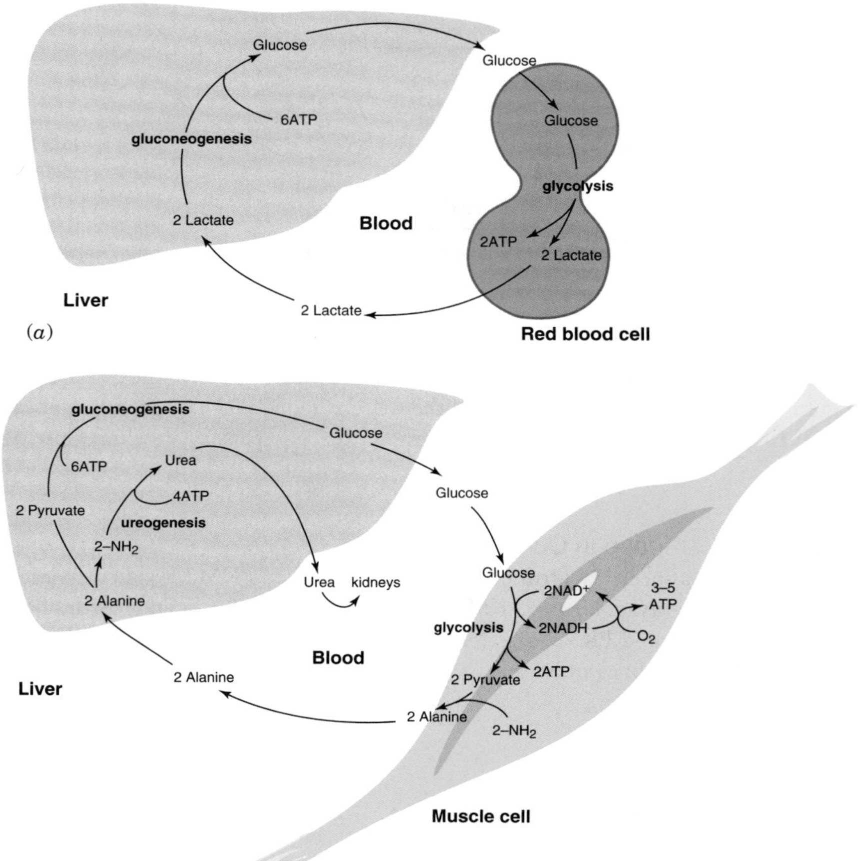

Figure 15.30 Relationship between gluconeogenesis in the liver and glycolysis in the rest of the body. (*a*) Cori cycle. (*b*) Alanine cycle.

of glucose to tissues that require it as their primary energy source. To participate in these cycles, peripheral tissues must release either alanine or lactate as the end product of glucose metabolism. A major difference between the cycles is that NADH generated by glycolysis in the alanine cycle is not used to reduce pyruvate to lactate; if it did, pyruvate would not be available for conversion to alanine by transamination with glutamate. In tissues that have mitochondria, reducing equivalents of NADH are transported into the mitochondria by the malate–aspartate shuttle or the glycerol–phosphate shuttle for the synthesis of ATP by oxidative phosphorylation.

$$\text{NADH} + \text{H}^+ + 0.5\ \text{O}_2 + 2.5\ \text{ADP} + 2.5\ \text{P}_\text{i} \rightarrow \text{NAD}^+ + 2.5\ \text{ATP} + \text{H}_2\text{O}$$

or

$$\text{FADH}_2 + 0.5\ \text{O}_2 + 1.5\ \text{ADP} + 1.5\ \text{P}_\text{i} \rightarrow \text{FAD} + 1.5\ \text{ATP} + \text{H}_2\text{O}$$

The consequence is that five to seven molecules of ATP can be formed per glucose molecule in peripheral tissues that participate in the alanine cycle. In the Cori cycle (Figure 15.30*a*), only two molecules of ATP per molecule of glucose are produced.

$$6\ \text{ATP}_{\text{liver}} + 2\ (\text{ADP} + \text{P}_\text{i})_{\text{red blood cells}} \rightarrow 6\ (\text{ADP} + \text{P}_\text{i})_{\text{liver}} + 2\ \text{ATP}_{\text{red blood cells}}$$

Six molecules of ATP are needed in liver for glucose synthesis. The alanine cycle (Figure 15.30*b*) transfers energy from liver to peripheral tissues and, because of the five to seven molecules of ATP produced per molecule of glucose, is energetically more efficient. However, the alanine cycle presents liver with amino nitrogen that must be disposed of as urea (p. 757). This requires four ATP molecules for every urea molecule produced and increases the amount of ATP required to 10 molecules per glucose molecule during the alanine cycle.

$$10\ \text{ATP}_{\text{liver}} + 5 - 7\ (\text{ADP} + \text{P}_i)_{\text{muscle}} + \text{O}_{2\ \text{muscle}} \rightarrow$$
$$10\ (\text{ADP} + \text{P}_i)_{\text{liver}} + 5 - 7\ \text{ATP}_{\text{muscle}}$$

This and the requirement in peripheral tissue for oxygen and mitochondria distinguish the alanine cycle from the Cori cycle.

Glucose Synthesis from Lactate

Gluconeogenesis from lactate is an ATP-requiring process.

$$2\ \text{L-Lactate}^- + 6\ \text{ATP}^{4-} + 6\ \text{H}_2\text{O} \rightarrow \text{glucose} + 6\ \text{ADP}^{3-} + 6\ \text{P}_i^{2-} + 4\ \text{H}^+$$

Many of the enzymes of glycolysis are used for gluconeogenesis from lactate. However, different reactions are necessary for the process because glycolysis produces two ATPs whereas gluconeogenesis requires six ATPs per molecule of glucose, and three steps of glycolysis are irreversible, namely the reactions catalyzed by glucokinase, 6-phosphofructo-1-kinase, and pyruvate kinase.

The initial step of lactate gluconeogenesis is the conversion of lactate to pyruvate by lactate dehydrogenase (Figure 15.31). The NADH generated is needed for a subsequent step

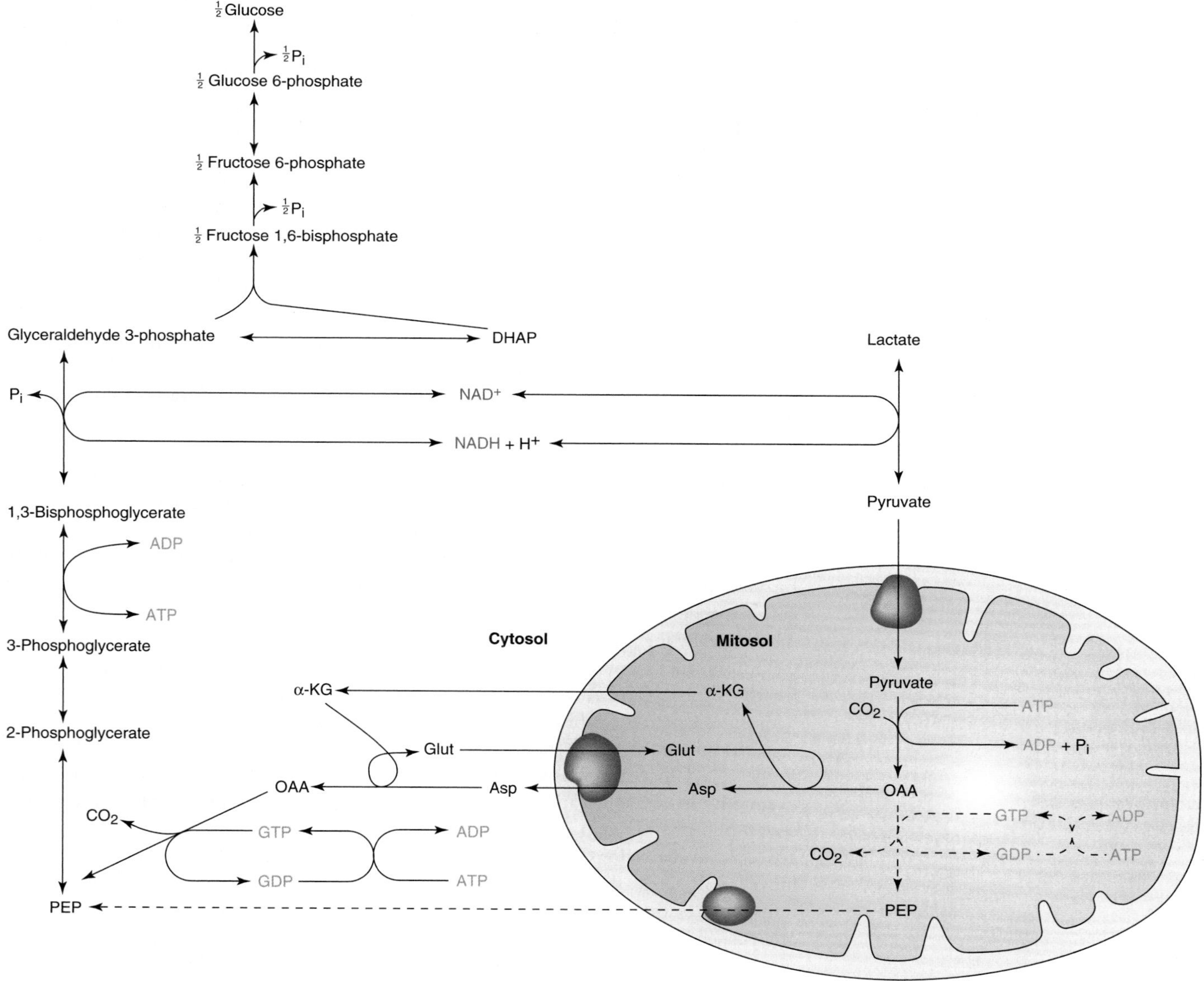

Figure 15.31 Pathway of gluconeogenesis from lactate. The involvement of the mitochondrion in the process is indicated. Dashed arrows refer to an alternative route, which employs mitochondrial PEP carboxykinase rather than the cytosolic isoenzyme. Abbreviations: OAA, oxaloacetate; α-KG, α-ketoglutarate; PEP, phosphoenolpyruvate; and DHAP, dihydroxyacetone phosphate.

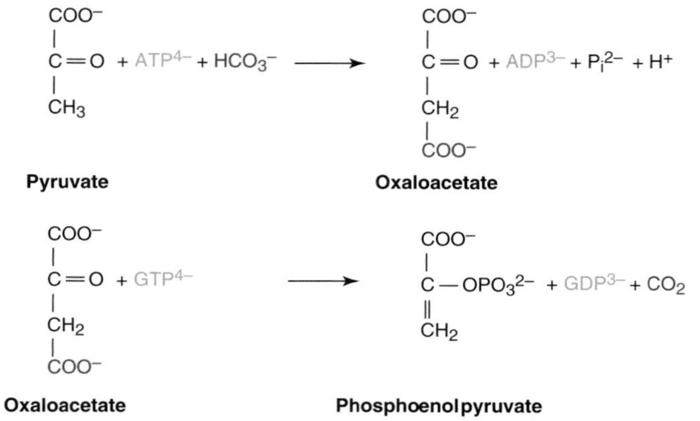

$$\text{Pyruvate } (CH_3\text{–}C(=O)\text{–}COO^-) + ATP^{4-} + HCO_3^- \longrightarrow \text{Oxaloacetate } (^-OOC\text{–}CH_2\text{–}C(=O)\text{–}COO^-) + ADP^{3-} + P_i^{2-} + H^+$$

$$\text{Oxaloacetate} + GTP^{4-} \longrightarrow \text{Phosphoenolpyruvate } (CH_2\text{=}C(OPO_3^{2-})\text{–}COO^-) + GDP^{3-} + CO_2$$

Figure 15.32 Energy-requiring steps involved in phosphoenolpyruvate formation from pyruvate. Reactions are catalyzed by pyruvate carboxylase and PEP carboxykinase, respectively.

in the pathway. Pyruvate cannot be converted to phosphoenolpyruvate (PEP) by pyruvate kinase because the reaction is irreversible under intracellular conditions. Pyruvate is converted into PEP by coupling of the reactions catalyzed by **pyruvate carboxylase**, which requires ATP, and PEP carboxykinase, which requires GTP (Figure 15.32). GTP is equivalent to an ATP through the action of nucleoside diphosphate kinase (GDP + ATP → GTP + ADP). The CO_2 generated by PEP carboxykinase and the HCO_3^- required by pyruvate carboxylase are linked by the reaction catalyzed by carbonic anhydrase ($CO_2 + H_2O \leftrightarrows H_2CO_3 \leftrightarrows H^+ + HCO_3^-$). Summing these reactions with the reactions of Figure 15.32 yields

$$\text{Pyruvate}^- + 2\ ATP^{4-} \rightarrow \text{phosphoenolpyruvate}^{3-} + 2\ ADP^{3-} + P_i^{2-} + 2\ H^+$$

Thus, conversion of pyruvate into PEP during gluconeogenesis costs the cell two molecules of ATP. This contrasts with the conversion of PEP to pyruvate during glycolysis that yields just one molecule of ATP.

The mitochondrial location of pyruvate carboxylase and the requirement for ATP make mitochondria mandatory for conversion of cytosolic pyruvate into cytosolic PEP (see Figure 15.31). However, because PEP carboxykinase is present in both the cytosol and mitochondrial matrix compartments, there are two routes that oxaloacetate takes to glucose. The first uses mitochondrial PEP carboxykinase, which converts oxaloacetate into PEP which then traverses the mitochondrial inner membrane. The second converts oxaloacetate into aspartate, which exits the mitochondrion by way of the glutamate–aspartate antiport. Aspartate contributes its amino group to α-ketoglutarate in the cytosol to produce oxaloacetate, which is used by cytosolic PEP carboxykinase to produce PEP.

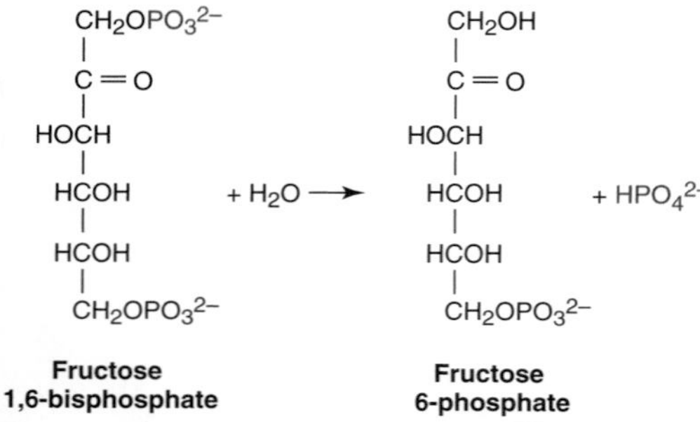

$$\text{Fructose 1,6-bisphosphate} + H_2O \longrightarrow \text{Fructose 6-phosphate} + HPO_4^{2-}$$

Figure 15.33 Reaction catalyzed by fructose 1,6-bisphosphatase.

Gluconeogenesis Uses Many Glycolytic Enzymes in the Reverse Direction

Enzymes of glycolysis operate in reverse to convert PEP to fructose 1,6-bisphosphate during gluconeogenesis. The generation of reducing equivalents (NADH) by lactate dehydrogenase is balanced by the use of reducing equivalents by glyceraldehyde 3-phosphate dehydrogenase. 6-Phosphofructo-1-kinase catalyzes an irreversible reaction and cannot convert FBP to fructose 6-phosphate. A way around this is provided by fructose 1,6-bisphosphatase, which hydrolyzes fructose 1,6-bisphosphate to F6P (Figure 15.33).

Phosphoglucose isomerase is freely reversible and functions in glycolysis and in gluconeogenesis. Glucose 6-phosphatase replaces glucokinase for the last step of gluconeogenesis (Figure 15.34), its function being to generate glucose for release into the blood stream. Glucose 6-phosphatase is an integral protein of the endoplasmic reticulum membrane with its active site available for G6P hydrolysis on the luminal surface (Figure 15.35). A transporter moves G6P across the endoplasmic reticulum membrane. A genetic defect in either the transporter or the phosphatase interferes with gluconeogenesis and results in accumulation of glycogen in liver, as discussed later for glycogen metabolism.

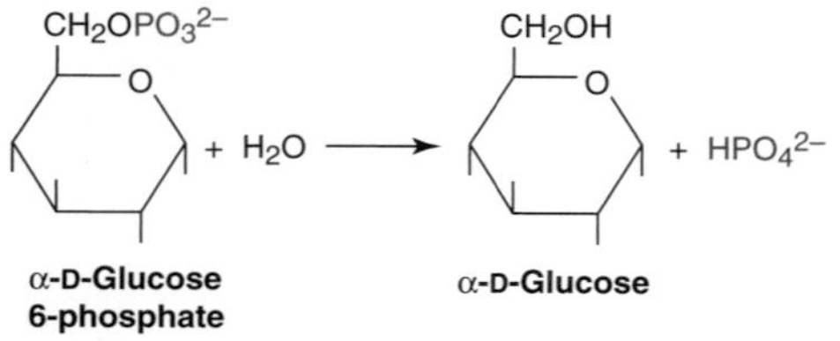

$$\alpha\text{-D-Glucose 6-phosphate} + H_2O \longrightarrow \alpha\text{-D-Glucose} + HPO_4^{2-}$$

Figure 15.34 Reaction catalyzed by glucose 6-phosphatase.

Glucose Is Synthesized from Most Amino Acids

All amino acids except leucine and lysine can supply carbon for net synthesis of glucose by gluconeogenesis (p. 619). If catabolism of an amino acid produces either pyruvate or oxaloacetate,

$$10\ ATP_{liver} + 5 - 7\ (ADP + P_i)_{muscle} + O_{2\ muscle} \rightarrow$$
$$10\ (ADP + P_i)_{liver} + 5 - 7\ ATP_{muscle}$$

This and the requirement in peripheral tissue for oxygen and mitochondria distinguish the alanine cycle from the Cori cycle.

Glucose Synthesis from Lactate

Gluconeogenesis from lactate is an ATP-requiring process.

$$2\ \text{L-Lactate}^- + 6\ ATP^{4-} + 6\ H_2O \rightarrow \text{glucose} + 6\ ADP^{3-} + 6\ P_i^{2-} + 4\ H^+$$

Many of the enzymes of glycolysis are used for gluconeogenesis from lactate. However, different reactions are necessary for the process because glycolysis produces two ATPs whereas gluconeogenesis requires six ATPs per molecule of glucose, and three steps of glycolysis are irreversible, namely the reactions catalyzed by glucokinase, 6-phosphofructo-1-kinase, and pyruvate kinase.

The initial step of lactate gluconeogenesis is the conversion of lactate to pyruvate by lactate dehydrogenase (Figure 15.31). The NADH generated is needed for a subsequent step

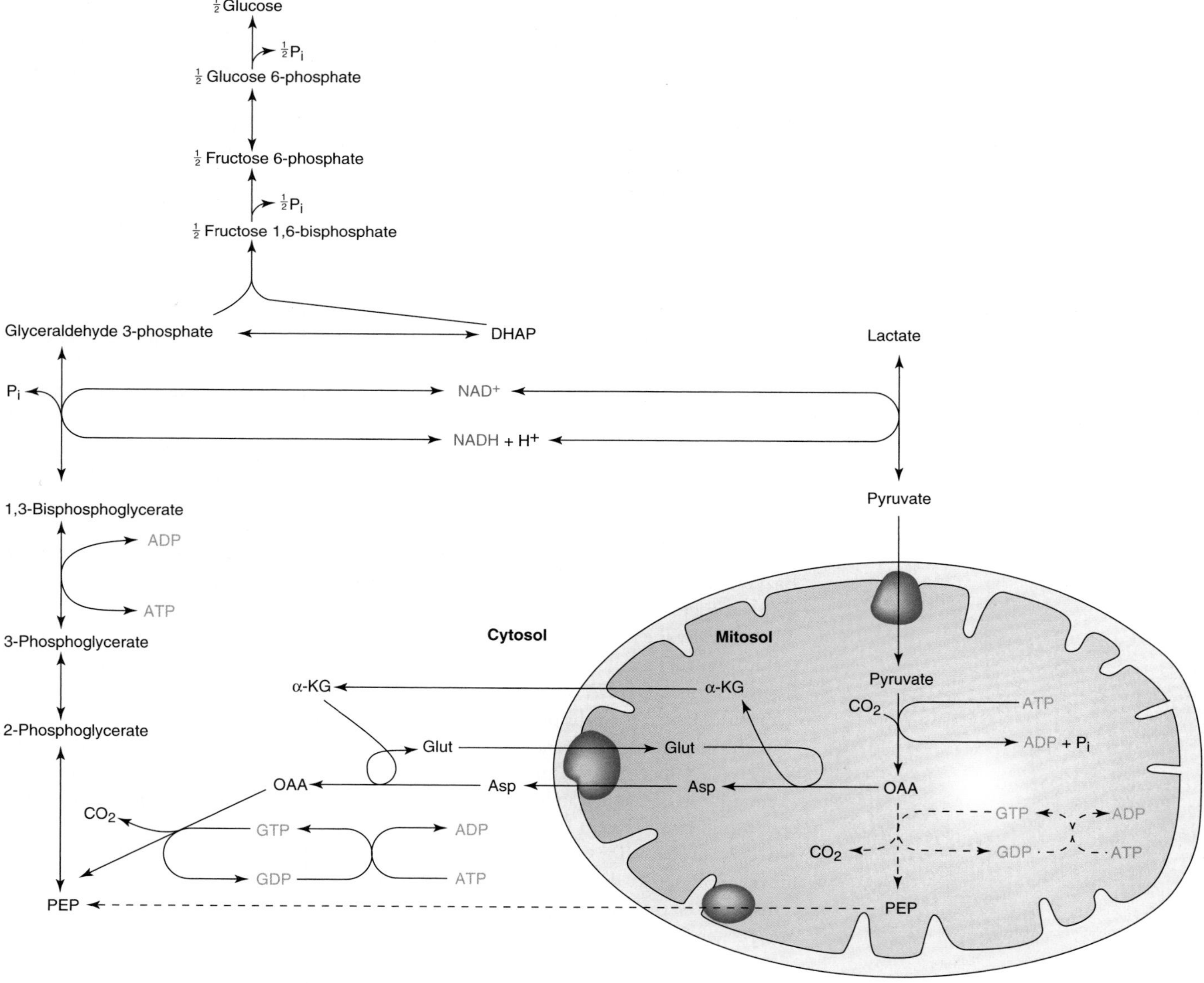

Figure 15.31 Pathway of gluconeogenesis from lactate. The involvement of the mitochondrion in the process is indicated. Dashed arrows refer to an alternative route, which employs mitochondrial PEP carboxykinase rather than the cytosolic isoenzyme. Abbreviations: OAA, oxaloacetate; α-KG, α-ketoglutarate; PEP, phosphoenolpyruvate; and DHAP, dihydroxyacetone phosphate.

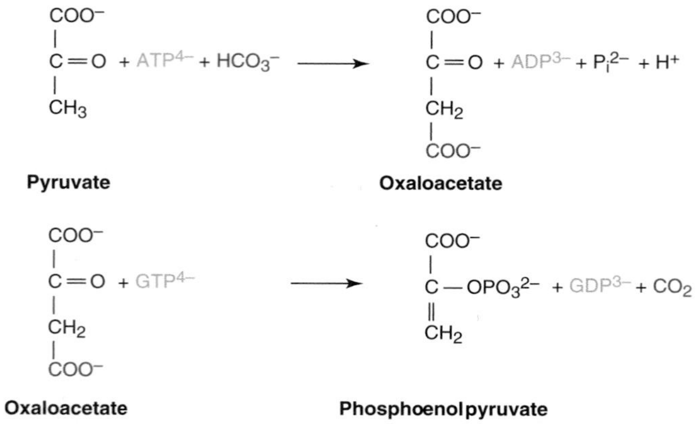

$$\underset{\text{Pyruvate}}{CH_3{-}CO{-}COO^-} + ATP^{4-} + HCO_3^- \longrightarrow \underset{\text{Oxaloacetate}}{^-OOC{-}CH_2{-}CO{-}COO^-} + ADP^{3-} + P_i^{2-} + H^+$$

$$\underset{\text{Oxaloacetate}}{^-OOC{-}CH_2{-}CO{-}COO^-} + GTP^{4-} \longrightarrow \underset{\text{Phosphoenolpyruvate}}{CH_2{=}C(OPO_3^{2-}){-}COO^-} + GDP^{3-} + CO_2$$

Figure 15.32 Energy-requiring steps involved in phosphoenolpyruvate formation from pyruvate. Reactions are catalyzed by pyruvate carboxylase and PEP carboxykinase, respectively.

in the pathway. Pyruvate cannot be converted to phosphoenolpyruvate (PEP) by pyruvate kinase because the reaction is irreversible under intracellular conditions. Pyruvate is converted into PEP by coupling of the reactions catalyzed by **pyruvate carboxylase**, which requires ATP, and PEP carboxykinase, which requires GTP (Figure 15.32). GTP is equivalent to an ATP through the action of nucleoside diphosphate kinase (GDP + ATP $\rightarrow$ GTP + ADP). The CO_2 generated by PEP carboxykinase and the HCO_3^- required by pyruvate carboxylase are linked by the reaction catalyzed by carbonic anhydrase ($CO_2 + H_2O \leftrightarrows H_2CO_3 \leftrightarrows H^+ + HCO_3^-$). Summing these reactions with the reactions of Figure 15.32 yields

$$\text{Pyruvate}^- + 2\ ATP^{4-} \rightarrow \text{phosphoenolpyruvate}^{3-} + 2\ ADP^{3-} + P_i^{2-} + 2\ H^+$$

Thus, conversion of pyruvate into PEP during gluconeogenesis costs the cell two molecules of ATP. This contrasts with the conversion of PEP to pyruvate during glycolysis that yields just one molecule of ATP.

The mitochondrial location of pyruvate carboxylase and the requirement for ATP make mitochondria mandatory for conversion of cytosolic pyruvate into cytosolic PEP (see Figure 15.31). However, because PEP carboxykinase is present in both the cytosol and mitochondrial matrix compartments, there are two routes that oxaloacetate takes to glucose. The first uses mitochondrial PEP carboxykinase, which converts oxaloacetate into PEP which then traverses the mitochondrial inner membrane. The second converts oxaloacetate into aspartate, which exits the mitochondrion by way of the glutamate–aspartate antiport. Aspartate contributes its amino group to α-ketoglutarate in the cytosol to produce oxaloacetate, which is used by cytosolic PEP carboxykinase to produce PEP.

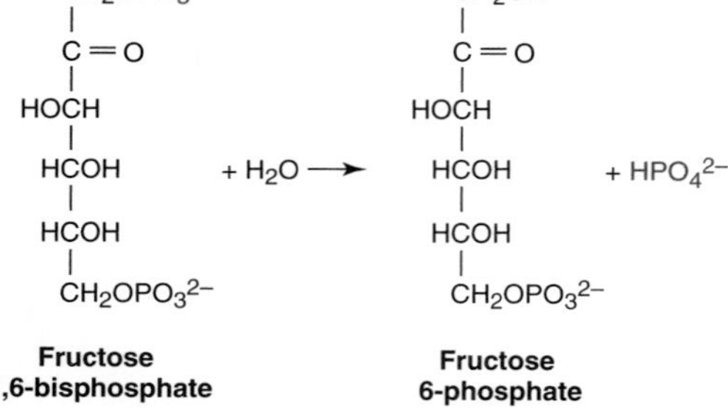

Figure 15.33 Reaction catalyzed by fructose 1,6-bisphosphatase.

Gluconeogenesis Uses Many Glycolytic Enzymes in the Reverse Direction

Enzymes of glycolysis operate in reverse to convert PEP to fructose 1,6-bisphosphate during gluconeogenesis. The generation of reducing equivalents (NADH) by lactate dehydrogenase is balanced by the use of reducing equivalents by glyceraldehyde 3-phosphate dehydrogenase. 6-Phosphofructo-1-kinase catalyzes an irreversible reaction and cannot convert FBP to fructose 6-phosphate. A way around this is provided by fructose 1,6-bisphosphatase, which hydrolyzes fructose 1,6-bisphosphate to F6P (Figure 15.33).

Phosphoglucose isomerase is freely reversible and functions in glycolysis and in gluconeogenesis. Glucose 6-phosphatase replaces glucokinase for the last step of gluconeogenesis (Figure 15.34), its function being to generate glucose for release into the blood stream. Glucose 6-phosphatase is an integral protein of the endoplasmic reticulum membrane with its active site available for G6P hydrolysis on the luminal surface (Figure 15.35). A transporter moves G6P across the endoplasmic reticulum membrane. A genetic defect in either the transporter or the phosphatase interferes with gluconeogenesis and results in accumulation of glycogen in liver, as discussed later for glycogen metabolism.

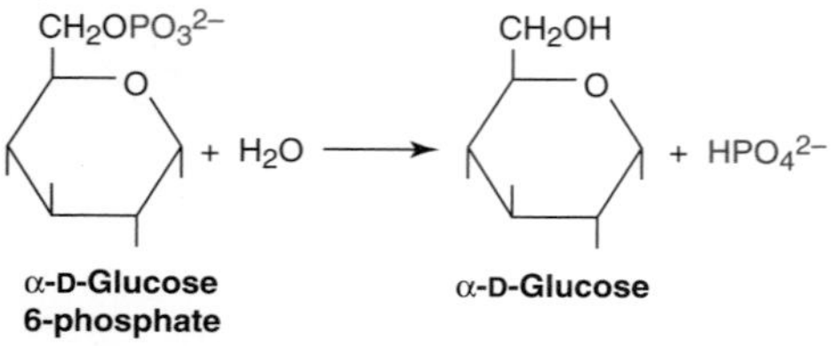

$$\underset{\alpha\text{-D-Glucose 6-phosphate}}{(CH_2OPO_3^{2-})} + H_2O \longrightarrow \underset{\alpha\text{-D-Glucose}}{(CH_2OH)} + HPO_4^{2-}$$

Figure 15.34 Reaction catalyzed by glucose 6-phosphatase.

Glucose Is Synthesized from Most Amino Acids

All amino acids except leucine and lysine can supply carbon for net synthesis of glucose by gluconeogenesis (p. 619). If catabolism of an amino acid produces either pyruvate or oxaloacetate,

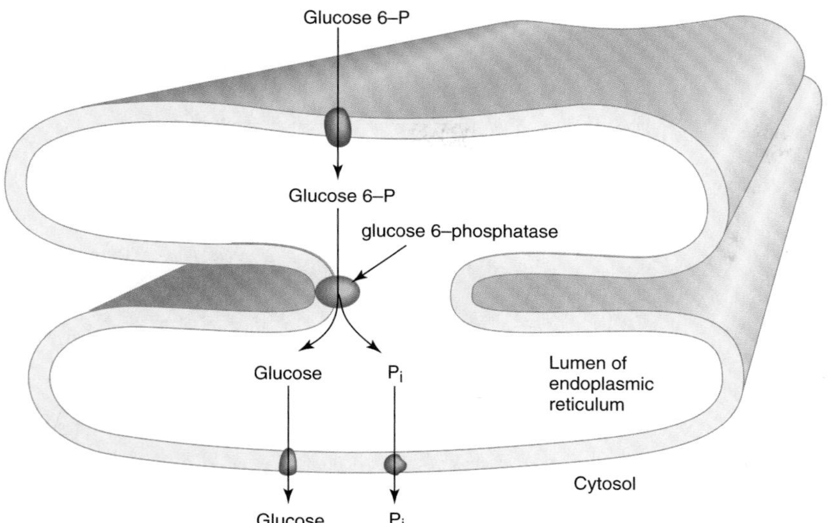

Figure 15.35 Glucose 6-phosphate is hydrolyzed by glucose 6-phosphatase located on the luminal surface of the endoplasmic reticulum. Three transporters are involved: One moves glucose 6-phosphate into the lumen, a second moves P_i back to the cytosol, and a third moves glucose back into the cytosol.

net glucose synthesis can occur from that amino acid. Oxaloacetate is an intermediate in gluconeogenesis and pyruvate is readily converted to oxaloacetate by pyruvate carboxylase (see Figure 15.31, p. 621). Catabolism of amino acids feeds carbon into the TCA cycle at several points. As long as net synthesis of a TCA cycle intermediate occurs, net synthesis of oxaloacetate will follow. Reactions that lead to net synthesis of TCA cycle intermediates are called anaplerotic reactions (anaplerosis) and support gluconeogenesis because they provide for net synthesis of oxaloacetate. Reactions catalyzed by pyruvate carboxylase and glutamate dehydrogenase are good examples of anaplerotic reactions.

$$\text{Pyruvate}^- + \text{ATP}^{4-} + \text{HCO}_3^- \rightarrow \text{oxaloacetate}^{2-} + \text{ADP}^{3-} + \text{P}_i^{2-} + \text{H}^+$$

$$\text{Glutamate}^- + \text{NAD(P)}^+ \rightarrow \alpha\text{-ketoglutarate}^{2-} + \text{NAD(P)H} + \text{NH}_4^+ + \text{H}^+$$

On the other hand, the reaction catalyzed by glutamate-oxaloacetate aminotransferase (α-ketoglutarate + aspartate $\rightarrow$ glutamate + oxaloacetate) is not anaplerotic because net synthesis of a TCA cycle intermediate is not accomplished; that is, generation of oxaloacetate from aspartate is counterbalanced by conversion of α-ketoglutarate to glutamate.

Since gluconeogenesis from amino acids imposes a nitrogen load on liver, a close relationship exists between urea synthesis and glucose synthesis from amino acids. In Figure 15.36, two alanine molecules are metabolized, one yielding NH_4^+ and the other aspartate, the primary substrates for the urea cycle. Aspartate leaves the mitochondrion and becomes part of the urea cycle after reacting with citrulline. The carbon of aspartate is released as fumarate that is converted to malate by cytosolic fumarase. This malate and another malate from mitochondria are converted to glucose by cytosolic enzymes of gluconeogenesis. A balance is achieved between reducing equivalents (NADH) generated and those required in the cytosol and the mitochondrial matrix. Summation of the reactions given in Figure 15.36 yields simply

$$2\ \text{Alanine} + 10\ \text{ATP} + \text{CO}_2 \rightarrow \text{Glucose} + \text{urea} + 10\ \text{ADP} + 10\ \text{P}_i$$

Leucine and lysine are the only amino acids that cannot function as carbon sources for net synthesis of glucose. They are ketogenic but not glucogenic. All other amino acids are glucogenic or both glucogenic and ketogenic (Table 15.1). Glucogenic amino acids give rise to net synthesis of either pyruvate or oxaloacetate, whereas amino acids that are both glucogenic and ketogenic also yield the ketone body acetoacetate, or acetyl CoA, which is readily converted into acetoacetate. Acetyl CoA is the end product of lysine metabolism, and acetoacetate and acetyl CoA are end products of leucine metabolism. No pathway exists that can convert acetoacetate or acetyl CoA into either pyruvate or oxaloacetate. Acetyl CoA cannot be used for net synthesis of glucose because the reaction of the pyruvate dehydrogenase complex is irreversible.

$$\text{Pyruvate} + \text{NAD}^+ + \text{CoASH} \rightarrow \text{acetyl CoA} + \text{NADH} + \text{CO}_2.$$

TABLE 15.1 • Glucogenic and Ketogenic Amino Acids

Glucogenic	*Ketogenic*	*Both*
Glycine	Leucine	Threonine
Serine	Lysine	Isoleucine
Valine		Phenylalanine
Histidine		Tyrosine
Arginine		Tryptophan
Cysteine		
Proline		
Hydroxyproline		
Alanine		
Glutamate		
Glutamine		
Aspartate		
Asparagine		
Methionine		

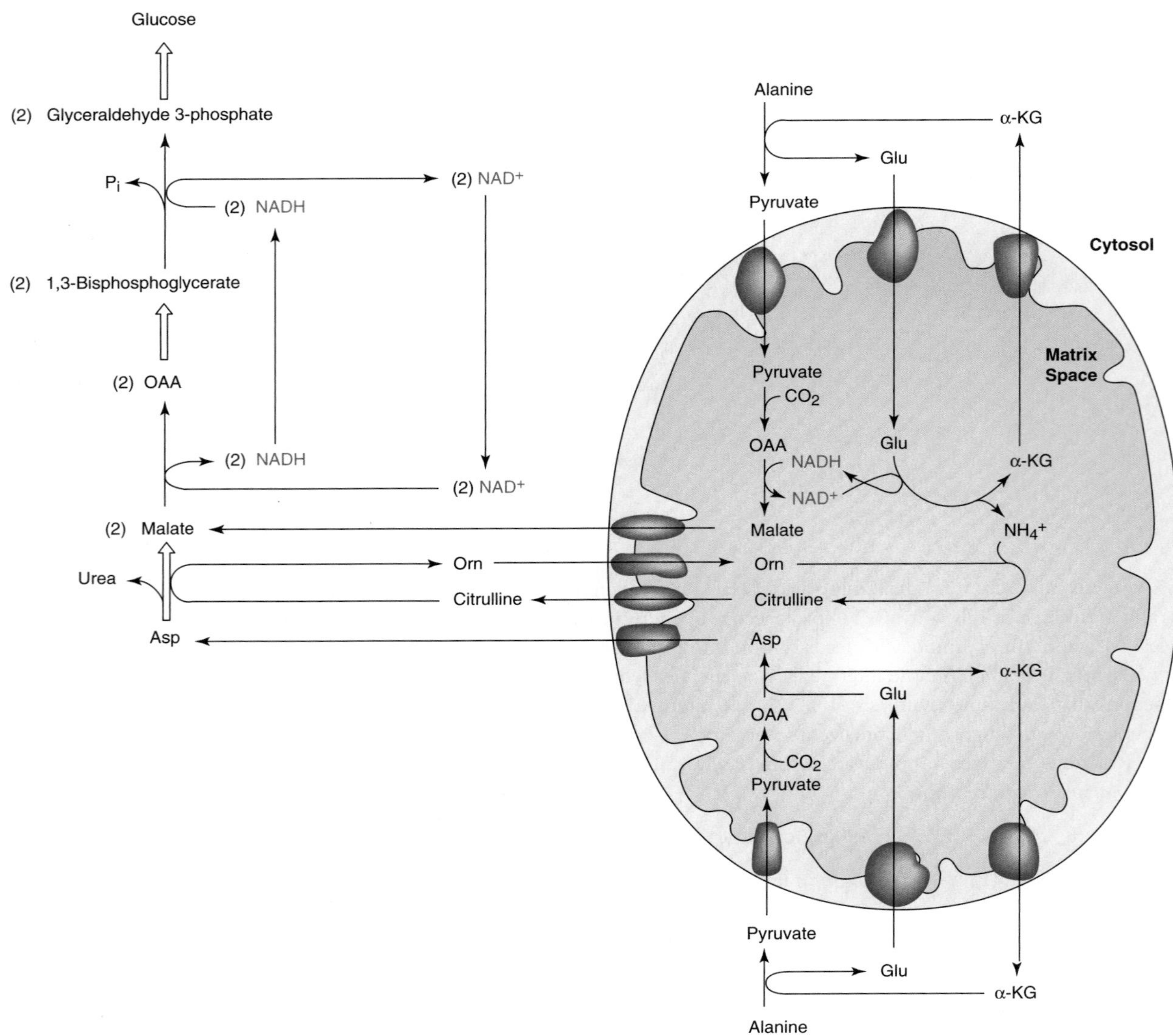

Figure 15.36 Pathway of gluconeogenesis from alanine and its relationship to urea synthesis. Abbreviations: OAA, oxaloacetate; Asp, aspartate; Orn, ornithine; Glu, glutamate; α-KG, α-ketoglutarate.

It could be argued that oxaloacetate is generated from acetyl CoA by the TCA cycle through this series of reactions.

$$\text{Acetyl CoA} \rightarrow \text{citrate} \rightarrow 2\ CO_2 + \text{oxaloacetate}$$

However, this fails to consider the oxaloacetate required for the formation of citrate from acetyl CoA.

$$\text{Acetyl CoA} + \text{oxaloacetate} \rightarrow \text{citrate} + \text{CoA}$$

Decarboxylation of citrate by the TCA cycle regenerates the oxaloacetate.

$$\text{Citrate} \rightarrow \rightarrow 2\ CO_2 + \text{oxaloacetate}$$

Summing these reactions gives the sum reaction for one turn of the TCA cycle.

$$\text{Acetyl CoA} \rightarrow 2CO_2 + \text{CoA}$$

Thus, net synthesis of a TCA cycle intermediate does not occur during complete oxidation of acetyl CoA by the TCA cycle. It is therefore impossible for animals to synthesize glucose from acetyl CoA.

Glucose Can Be Synthesized from Fatty Acids with an Odd but Not an Even Number of Carbons

Most fatty acids found in humans have straight chains with an even number of carbon atoms. Their catabolism by fatty acid oxidation followed by ketogenesis or complete oxidation to CO_2 is abbreviated in Figure 15.37. Since acetyl CoA and other intermediates of even-numbered fatty acid oxidation cannot be converted to oxaloacetate or any other intermediate of gluconeogenesis, it is impossible to synthesize glucose from fatty acids. An exception to this rule applies to fatty acids with methyl branches (e.g., phytanic acid, a breakdown product of chlorophyll; see discussion of Refsum disease, Clin. Corr. 17.6, p. 612), and fatty acids with an odd number of carbon atoms. Catabolism of such fatty acids yields propionyl CoA.

$$\text{Fatty acid with an odd number } (n) \text{ of carbon atoms} \rightarrow \frac{(n-3)}{2} \text{ acetyl CoA} + 1 \text{ propionyl CoA}$$

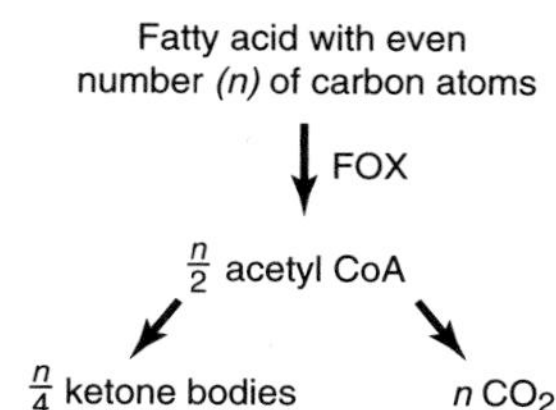

Figure 15.37 Overview of the catabolism of fatty acids to ketone bodies and CO_2. Abbreviation: FOX, fatty acid oxidation.

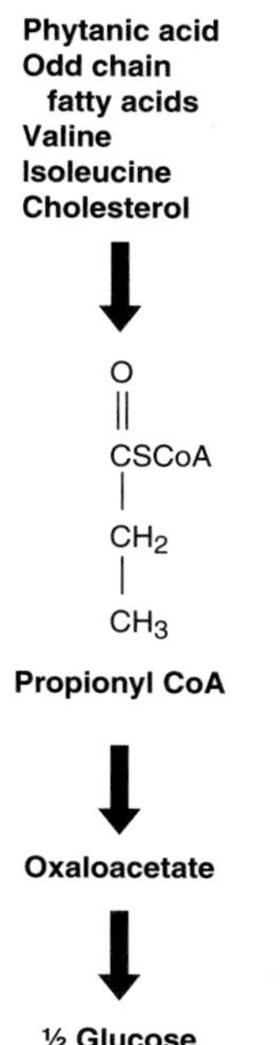

Figure 15.38 Pathway of gluconeogenesis from propionyl CoA. The large arrow refers to steps of the TCA cycle plus steps of gluconeogenesis.

Propionyl CoA is a good precursor for gluconeogenesis as it yields oxaloacetate (Figure 15.38) by an anaplerotic pathway (see Figure 19.53, page 79). Propionyl CoA is also produced during the catabolism of valine and isoleucine (see Figure 19.50, p. 777) and in the conversion of cholesterol into bile acids (see Figure 18.42, p. 728).

It is sometimes loosely stated that *fat cannot be converted into carbohydrate (glucose) by animals.* This is certainly true for fatty acids with an even number of carbon atoms. However, the term *fat* usually refers to triacylglycerols, which are composed of three O-acyl groups combined with one glycerol molecule. Hydrolysis of a triacylglycerol yields three fatty acids and glycerol, the latter compound being an excellent substrate for gluconeogenesis (Figure 15.39).

Phosphorylation of glycerol by glycerol kinase produces glycerol 3-phosphate, which is converted by glycerol 3-phosphate dehydrogenase into dihydroxyacetone phosphate, an intermediate of the gluconeogenic pathway (see Figure 15.31, p. 607). Depending on the nutritional state, the last stage of glycolysis can compete with the gluconeogenic pathway and convert dihydroxyacetone phosphate into lactate (or into pyruvate for subsequent complete oxidation to CO_2 and H_2O).

Glucose Is Also Synthesized from Fructose

Humans consume considerable quantities of fructose in the form of sucrose, which yields glucose and fructose on hydrolysis by sucrase in the intestine (p. 1050), and as a result of the widespread use of high-fructose corn syrup as a sweetener by the food industry. In the liver, fructose is phosphorylated by fructokinase (Figure 15.40), which yields fructose 1-phosphate rather than fructose 6-phosphate (see Clin. Corr. 15.3, p. 607). The same

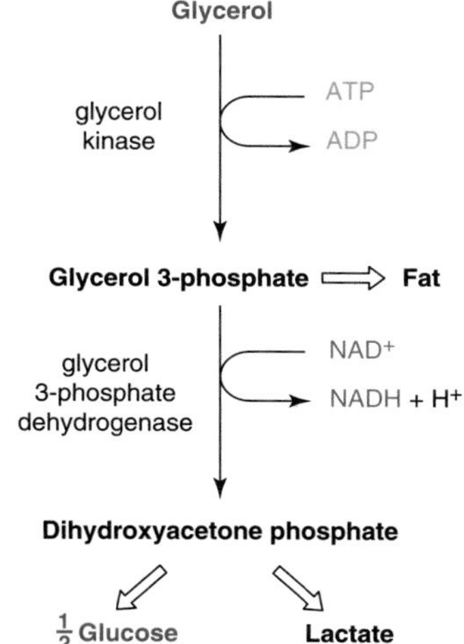

Figure 15.39 Pathway of gluconeogenesis from glycerol, along with competing pathways. Large arrows indicate steps of glycolysis and gluconeogenesis given in Figures15.5, p. 595, and 15.31, p. 621, respectively. The large arrow pointing to fat indicates triacylglycerol and glycerophospholipid synthesis.

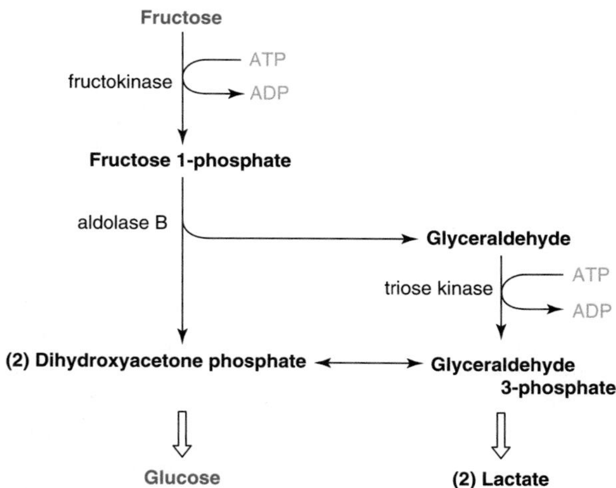

Figure 15.40 Pathway of glucose formation from fructose and the competing pathway of fructolysis. Large arrows indicate steps of the glycolysis and gluconeogenesis given in Figures15.6, p. 597, and 15.31, p, 621, respectively.

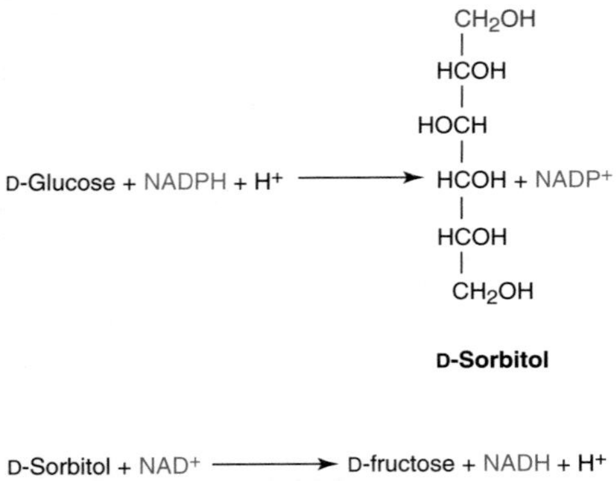

$$\text{D-Glucose} + \text{NADPH} + \text{H}^+ \longrightarrow \begin{array}{c} CH_2OH \\ | \\ HCOH \\ | \\ HOCH \\ | \\ HCOH \\ | \\ HCOH \\ | \\ CH_2OH \end{array} + \text{NADP}^+$$

D-Sorbitol

$$\text{D-Sorbitol} + \text{NAD}^+ \longrightarrow \text{D-fructose} + \text{NADH} + \text{H}^+$$

Figure 15.41 Pathway responsible for the formation of sorbitol and fructose from glucose.

aldolase that cleaves fructose 1,6-bisphosphate then cleaves fructose 1-phosphate to yield dihydroxyacetone phosphate and glyceraldehyde. The latter has several possible fates. It can be oxidized by an aldehyde dehydrogenase, reduced by an alcohol dehydrogenase, or phosphorylated by a kinase. The favored pathway is conversion to glyceraldehyde 3-phosphate by triose kinase. Two molecules of dihydroxyacetone phosphate obtainable from one molecule of fructose can be converted to glucose by enzymes of gluconeogenesis or, alternatively, into pyruvate or lactate by the last stage of glycolysis. In analogy to glycolysis, conversion of fructose into lactate is called fructolysis.

The major energy source of spermatozoa is fructose, formed from glucose by cells of seminal vesicles as shown in Figure 15.41. NADPH-dependent reduction of glucose to sorbitol is followed by an NAD^+-dependent oxidation of sorbitol to fructose. Fructose is secreted from seminal vesicles in a fluid that becomes part of semen. Although the fructose concentration in human semen can exceed 10 mM, tissues that are exposed to semen utilize fructose poorly, allowing this substrate to be conserved to meet the energy demands of spermatozoa in their search for ova. Spermatozoa contain mitochondria and thus can metabolize fructose completely to CO_2 and H_2O by the combination of fructolysis and TCA cycle activity.

Gluconeogenesis Requires Expenditure of ATP

Synthesis of glucose is costly in terms of ATP. Six molecules are required for synthesis of one molecule of glucose from lactate, 10 for synthesis of glucose from alanine. ATP for glucose synthesis is provided in large part by fatty acid oxidation. Metabolic conditions under which liver is required to synthesize glucose generally favor increased availability of fatty acids in the blood. These fatty acids are oxidized by liver mitochondria to ketone bodies with concurrent production of large amounts of ATP.

$$\text{Palmitate} + 7\ O_2 + 26\ \text{ADP} + 26\ P_i \rightarrow 4\ \text{Acetoacetate} + 26\ \text{ATP}$$

Gluconeogenesis Has Several Sites of Regulation

Regulation of gluconeogenesis occurs at multiple sites (Figure 15.42). Those enzymes that are used to bypass the irreversible steps of glycolysis are primarily involved in regulation of the pathway, that is, pyruvate carboxylase, PEP carboxykinase, fructose 1,6-bisphosphatase, and glucose 6-phosphatase. Regulation of hepatic gluconeogenesis is very similar to regulation of hepatic glycolysis. Inhibition of glycolysis at its chief regulatory sites or repression of synthesis of enzymes that bypass these sites (glucokinase, 6-phosphofructo-1-kinase, and pyruvate kinase) greatly increases effectiveness of the gluconeogenic enzymes. Activation of gluconeogenesis is therefore accomplished in large part by inhibition of glycolysis.

Fatty acid oxidation invariably occurs hand in hand with gluconeogenesis. It promotes glucose synthesis, in part by supplying the ATP needed and by increasing the steady-state concentrations of mitochondrial acetyl CoA and NADH. Both of these products of fatty acid oxidation are potent activators of pyruvate dehydrogenase kinase, which phosphorylates and inactivates the pyruvate dehydrogenase complex. This prevents pyruvate from being converted into acetyl CoA, thereby conserving pyruvate for glucose synthesis. Acetyl CoA is also a positive allosteric effector of pyruvate carboxylase, which along with inhibition of the pyruvate dehydrogenase complex, directs the carbon of pyruvate into oxaloacetate for the synthesis of glucose. The increase in oxaloacetate due to increased pyruvate carboxylase activity along with the increase in acetyl CoA from fatty acid oxidation also promotes greater synthesis of citrate, a negative allosteric effector of 6-phosphofructo-1-kinase. Inhibition of 6-phosphofructo-1-kinase decreases the fructose 1,6-bisphosphate concentration, an activator of pyruvate kinase. This decreases flux of PEP to pyruvate by pyruvate kinase and, by this means, increases effectiveness of the coupling of pyruvate carboxylase and PEP carboxykinase for the conversion of pyruvate to PEP. An increase in ATP levels with the consequent decrease in AMP levels favors gluconeogenesis by inhibition of 6-phosphofructo-1-kinase and pyruvate kinase and activation of fructose 1,6-bisphosphatase (see Figure 15.42 and p. 627). On the

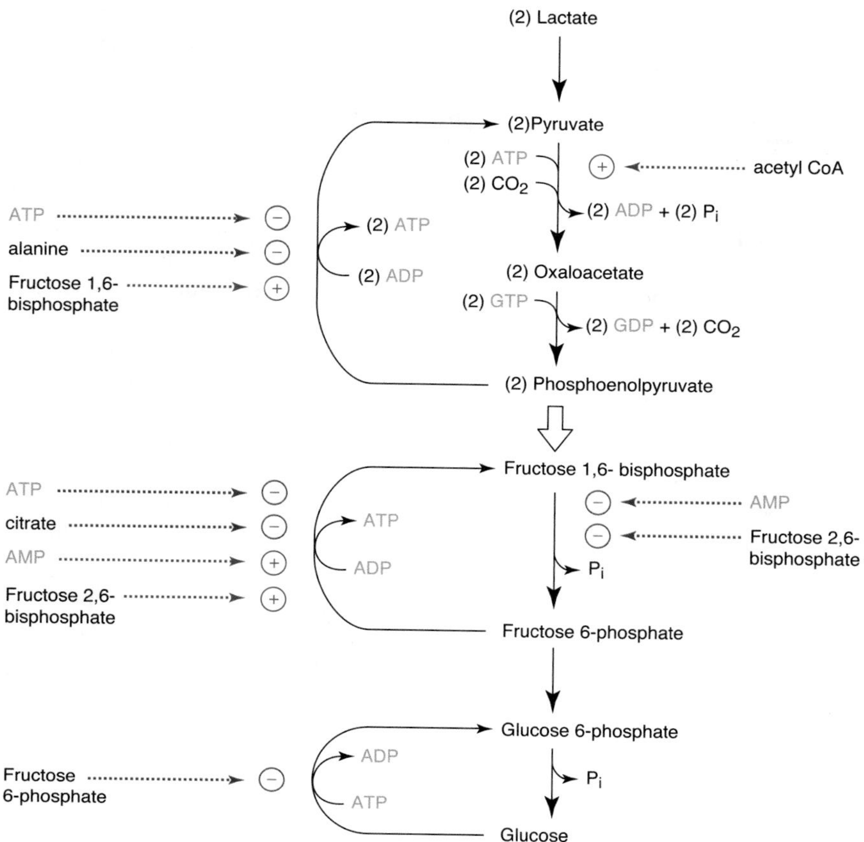

Figure 15.42 Important allosteric regulatory features of the gluconeogenesis.

other hand, a shortage of oxygen, a shortage of fatty acids for oxidation, or an inhibition or uncoupling of oxidative phosphorylation would cause the liver to turn from gluconeogenesis to glycolysis.

Hormonal Control of Gluconeogenesis Is Critical for Homeostasis

Hormonal control of gluconeogenesis is a matter of regulating the supply of fatty acids to liver as well as the activities of enzymes of glycolysis and gluconeogenesis. Catecholamines increase plasma fatty acids by promoting lipolysis in adipose tissue, an action opposed by insulin. The greater availability of fatty acids results in greater fatty acid oxidation by liver, which promotes glucose synthesis. Insulin produces the opposite effect by inhibiting lipolysis in the adipose tissue. Glucagon and insulin also regulate gluconeogenesis directly in the liver by influencing the state of phosphorylation of the hepatic enzymes that are subject to covalent modification. As discussed previously (see Figure 15.28, p. 619), pyruvate kinase is active when dephosphorylated and inactive when phosphorylated. Glucagon activates adenylate cyclase to produce cAMP, which activates protein kinase A, which phosphorylates and inactivates pyruvate kinase. Inactivation of this glycolytic enzyme stimulates gluconeogenesis by blocking the futile conversion of PEP to pyruvate. Glucagon also stimulates gluconeogenesis by decreasing the concentration of fructose 2,6-bisphosphate, an allosteric activator of 6-phosphofructo-1-kinase and an allosteric inhibitor of fructose 1,6-bisphosphatase. Glucagon through its second messenger cAMP lowers fructose 2,6-bisphosphate by stimulating phosphorylation of the bifunctional enzyme 6-phosphofructo-2-kinase/fructose 2,6-bisphosphatase which inactivates the kinase activity but activates the phosphatase activity. The glucagon-induced fall in fructose 2,6-bisphosphate levels leads to decrease in activity of 6-phosphofructo-1-kinase while fructose 1,6-bisphosphatase becomes more active (Figure 15.42). The overall effect is increased conversion of FBP to F6P and a corresponding increase in the rate of gluconeogenesis. An increase in fructose 6-phosphate also favors gluconeogenesis by inhibition of

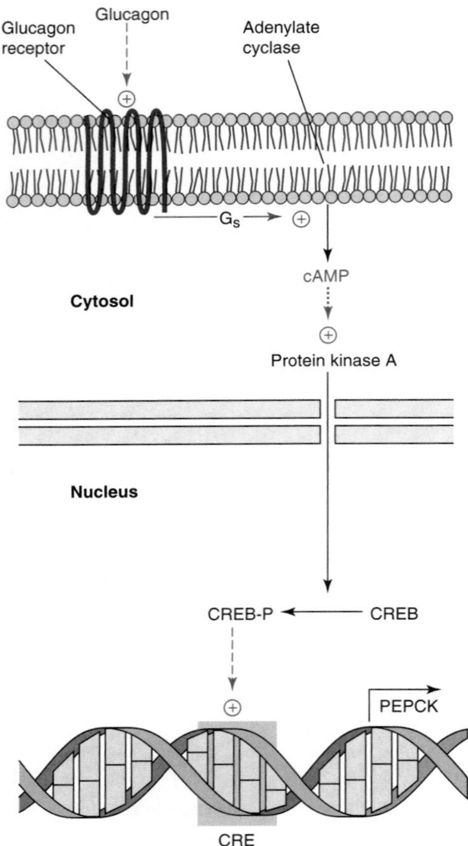

Figure 15.43 Glucagon promotes transcription of the gene for PEP carboxykinase. Abbreviations: PEPCK, PEP carboxykinase; CRE, cAMP-response element; CREB, cAMP-response element binding protein.

glucokinase via its inhibitory protein. Insulin effects are opposite to those of glucagon by signaling activation of cAMP phosphodiesterase, inhibition of protein kinase A, and activation of phosphoprotein phosphatase.

Glucagon and insulin also have long-term effects on hepatic glycolysis and gluconeogenesis by induction and repression of key enzymes of both pathways. A high glucagon/insulin ratio in blood increases the capacity for gluconeogenesis and decreases that for glycolysis in liver. A low glucagon/insulin ratio has the opposite effects. The glucagon/insulin ratio increases when gluconeogenesis is needed and decreases when glucose from the gastrointestinal track is plentiful. Glucagon signals induction of greater quantities of PEP carboxykinase, fructose-1,6-bisphosphatase, glucose 6-phosphatase, and various aminotransferases. A model for how this occurs is given in Figure 15.43. Binding of glucagon to its plasma membrane receptor increases cAMP which activates protein kinase A. Protein kinase A then phosphorylates a protein called the cAMP-response element binding protein (CREB), a transcription factor that when phosphorylated binds to a cAMP-response element (CRE), a cis acting element within the regulatory region of genes responsive to cAMP. Phosphorylated CREB promotes transcription of genes for key gluconeogenic enzymes such as PEP carboxykinase. By repression of gene transcription, glucagon decreases the amounts of glucokinase, 6-phosphofructo-1-kinase, and pyruvate kinase. Insulin opposes the action of glucagon through a kinase-mediated signaling cascade that culminates in the inactivation of a transcription factor for genes of key gluconeogenic enzymes (page 905; Figure 22.21). When glucose synthesis is not needed, synthesis of key gluconeogenic enzymes is decreased and synthesis of key glycolytic enzymes increases because of a decrease in the blood glucagon/insulin ratio.

Alcohol Oxidation Inhibits Gluconeogenesis

Oxidation of alcohol (ethanol) by the liver produces a large load of reducing equivalents in the form of NADH that have to be transported into the mitochondria by the malate–aspartate shuttle. Excess NADH in the cytosol interferes with gluconeogenesis (Clin. Corr. 15.10) because it forces the equilibrium of the reactions catalyzed by lactate dehydrogenase and malate dehydrogenase toward lactate and malate formation, respectively:

CLINICAL CORRELATION 15.10

Hypoglycemia and Alcohol Intoxication

Consumption of alcohol, especially by an undernourished person, can cause hypoglycemia. The same effect can result from drinking alcohol after strenuous exercise. In both cases the hypoglycemia results from the inhibitory effects of alcohol on hepatic gluconeogenesis and occurs under circumstances of hepatic glycogen depletion. The liver simply cannot handle the reducing equivalents provided by ethanol oxidation fast enough to prevent metabolic derangements. The extra reducing equivalents block the conversion of lactate to glucose and promote conversion of alanine into lactate with considerable lactate accumulation in the blood and lactic acidosis (see Clin. Corr. 15.5, p. 611) can develop, although it is usually mild.

Low doses of alcohol cause impaired motor and intellectual performance; high doses are depressant and can lead to stupor and anesthesia. Low blood sugar can contribute to these undesirable effects of alcohol. What is more, a patient may be thought to be inebriated when in fact she or he is suffering from hypoglycemia that may lead to irreversible damage to the central nervous system. Children are highly dependent on gluconeogenesis while fasting, and accidental ingestion of alcohol by a child can produce severe hypoglycemia (see Clin. Corr. 15.9, p. 620). Alcohol potentiates the hypoglycemic effect of insulin. It is therefore not uncommon for patients with diabetes to present at the emergency room with hypoglycemia due to the combination of self-administered insulin and alcohol consumption.

Krebs, H. A., Freedland, R. A., Hems, R., and Stubbs, M. Inhibition of hepatic gluconeogenesis by ethanol. *Biochem. J.* 112: 117, 1969; Service, F. J. Hypoglycemia. *Med. Clin. North America* 79:1, 1995; and Gibson, E. M., and Tingen, M. S. Nursing care for diabetic patients with alcohol induced hypoglycemia. *J. Emer. Nursing* 24: 165, 1998.

$$\underset{\text{Ethanol}}{CH_3CH_2OH} + NAD^+ \longrightarrow \underset{\text{Acetaldehyde}}{CH_3\overset{\overset{\displaystyle O}{\|}}{C}H} + NADH + H^+$$

$$\text{pyruvate} + NADH + H^+ \longrightarrow \text{lactate} + NAD^+$$

$$\text{Sum:} \quad \text{ethanol} + \text{pyruvate} \longrightarrow \text{acetaldehyde} + \text{lactate}$$

or

$$\text{oxaloacetate} + NADH + H^+ \longrightarrow \text{malate} + NAD^+$$

$$\text{Sum:} \quad \text{ethanol} + \text{oxaloacetate} \longrightarrow \text{acetaldehyde} + \text{malate}$$

This inhibits gluconeogenesis by limiting the availability of pyruvate and oxaloacetate for the reactions catalyzed by pyruvate carboxylase and PEP carboxykinase, respectively.

15.6 • GLYCOGENOLYSIS AND GLYCOGENESIS

Glycogen Is the Storage Form of Glucose

Glycogenolysis refers to the breakdown of glycogen to glucose or glucose-6-phosphate; glycogenesis to the synthesis of glycogen. These processes occur in almost every tissue but especially in muscle and liver. In the well-fed human, liver glycogen content can account for as much as 10% of wet weight of this organ. Muscle stores less glycogen when expressed on the same basis—a maximum of only 1% – 2% of its wet weight. However, most people have more muscle than liver, with total muscle glycogen adding up to about double the amount of liver glycogen.

Glycogen granules are abundant in liver of the well-fed animals (Figure 15.44) but are virtually absent from this organ after 24 h of fasting. Heavy exercise causes the same loss of glycogen granules in muscle fibers. These granules are clusters of individual glycogen molecules that have a mass up to 2×10^7 Da. Glycogen is composed of glucosyl residues, mostly linked together by α-1,4-glycosidic linkages (Figure 15.45). Branches arise from frequent α-1,6-glycosidic linkages. A limb of the glycogen "tree" (Figure 15.46) is characterized by branches at every fourth glucosyl residue within the more central core of the molecule and much less frequently in outer regions. Glycogen stands in contrast to proteins and nucleic acids because of this branching, but of course, it is a storage form of fuel and does not catalyze reactions or convey information within a cell.

Glycogen is stored in muscle and liver for quite different reasons. Muscle glycogen is a fuel reserved for the production of ATP within that tissue, whereas liver glycogen is a glucose reserve for the maintenance of blood glucose concentrations. Liver glycogen levels

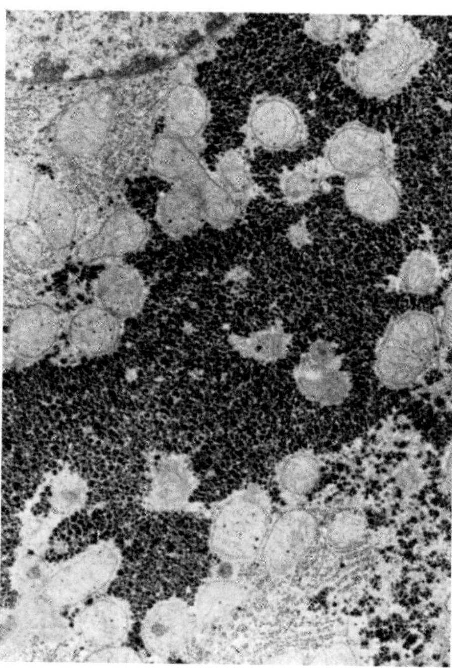

Figure 15.44 Electron micrograph showing glycogen granules (darkly stained material) in the liver of a fed rat.
Micrographs generously provided by Dr. Robert R. Cardell of the Department of Anatomy at the University of Cincinnati, Cincinnati, OH.

CH2OH CH2OH

α-1,4-Glycosidic linkage

(a)

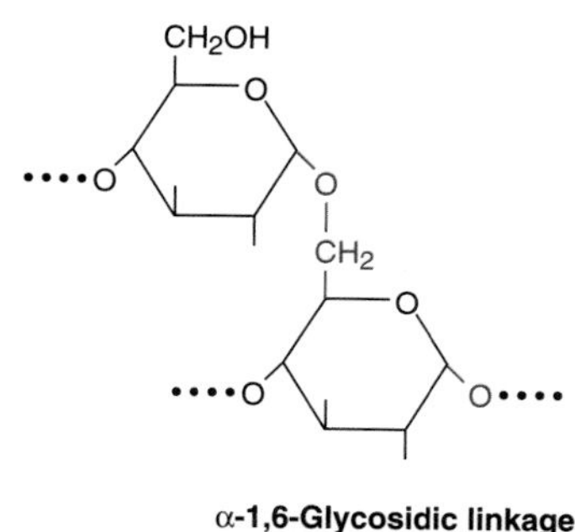

α-1,6-Glycosidic linkage

(b)

Figure 15.45 Two types of linkage between glucose molecules are present in glycogen.

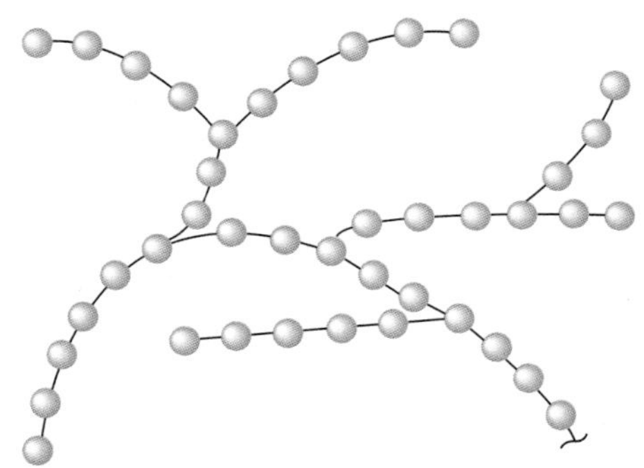

Figure 15.46 Branched structure of glycogen.

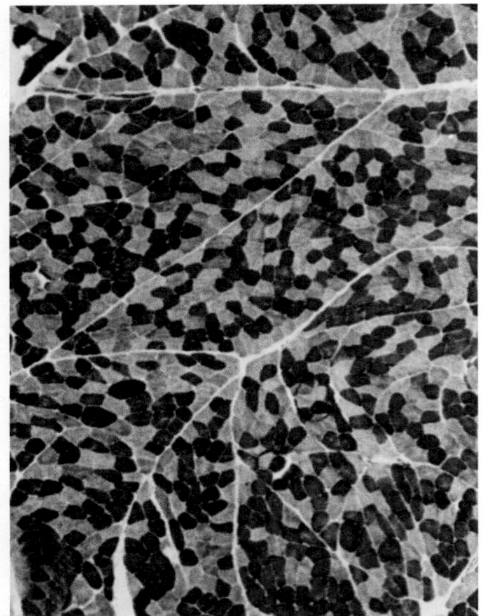

Figure 15.48 Cross section of human skeletal muscle showing red and white muscle fibers. Section was stained for NADH dehydrogenase activity. Red fibers are dark, and white fibers are light.
Picture generously provided by Dr. Michael H. Brooke of the Jerry Lewis Neuromuscular Research Center, St. Louis, MO.

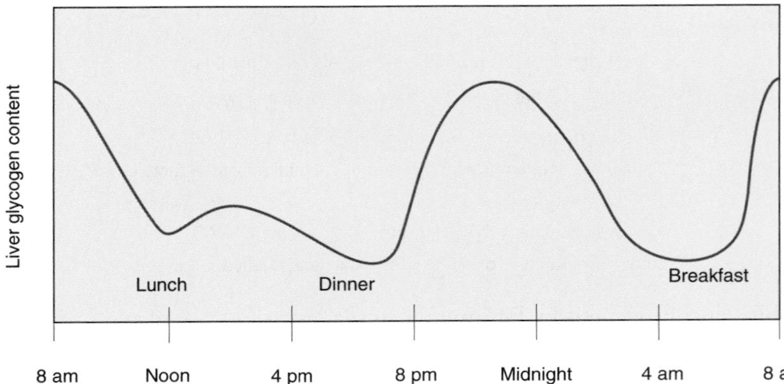

Figure 15.47 Variation of liver glycogen content between meals and during the nocturnal fast.

are high shortly after a meal and then decrease slowly as it is used to help maintain the blood glucose level (Figure 15.47) between meals and during the nocturnal fast. In humans and rats, the stored glycogen lasts between 12 and 24 h during fasting, depending greatly, of course, on whether the individual under consideration is caged or running wild.

Most of the muscle glycogen is consumed by that tissue without formation of free glucose. However, because of the way branched points are cleaved, about 8% of muscle glycogen is converted into free glucose within the tissue, most of which undergoes glycolysis in muscle. Since muscle lacks glucose 6-phosphatase and most of the free glucose formed is catabolized, muscle glycogen is not important for maintenance of blood glucose levels during fasting. In contrast, liver glycogen is a critical source of blood glucose in the postabsorptive state. On the other hand, muscle glycogenesis plays an important role in clearing glucose from the blood after a high carbohydrate meal. Liver glycogenesis contributes but is less important than glycogen synthesis in muscle.

Exercise mobilizes muscle glycogen for formation of ATP. Red muscle fibers have a rich blood flow, contain a large amount of myoglobin, and are packed with mitochondria. Glycogen mobilized within these cells is converted into pyruvate, which is then converted into CO_2 and H_2O by the TCA cycle. White muscle fibers contain less myoglobin and fewer mitochondria. Glycogenolysis within these cells only supplies substrate for glycolysis, with the end product being mostly lactate. White muscle fibers have greater capacity for glycogenolysis and glycolysis but can only function at full capacity for short periods of time. Breast and leg muscle of chicken are good examples of white and red muscles, respectively. Breast muscle of chicken is not continuously carrying out work, but enables the chicken to fly rapidly for short distances, as in fleeing from predators (or amorous roosters). It is designed for maximal activity over short periods. Most skeletal muscles of the human body are a mixture of red and white fibers, which provide for both rapid and sustained capacity for contraction. The distribution of white and red muscle fibers in cross sections of a human skeletal muscle is demonstrated by a special staining procedure in Figure 15.48.

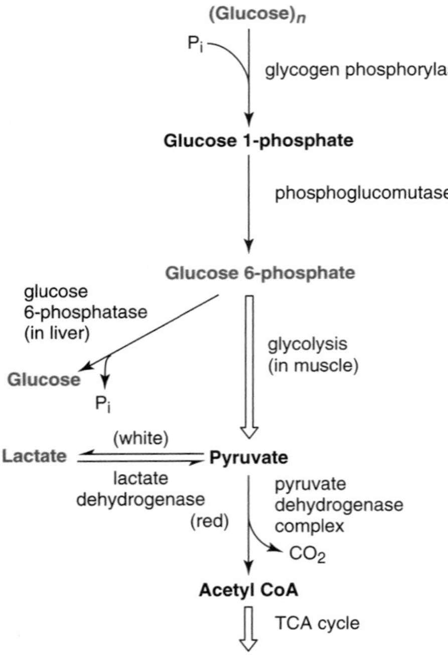

Figure 15.49 Glycogenolysis and the fate of glycogen degraded in liver versus its fate in peripheral tissues. White refers to white muscle, red to red muscle.

Glycogen Phosphorylase Initiates Glycogenolysis

Glycogen phosphorylase catalyzes phosphorolysis of glycogen, a reaction in which P_i is used to cleave α-1,4-glycosidic linkages to yield glucose 1-phosphate. This reaction always occurs at terminal, nonreducing ends of a glycogen molecule:

$$\text{Glycogen (partial structure)} \xrightarrow{HPO_3^{2-}} \alpha\text{-D-Glucose 1-phosphate} + \text{Glycogen}_{n-1}$$

This reaction is different from that of α-amylase, which uses water rather than inorganic phosphate (P_i) to cleave α-1,4-glycosidic bonds of glycogen and starch in the gut (p. 1050). Although a molecule of glycogen may contain up to 100,000 glucose residues, it is most convenient to abbreviate its structure as simply (glucose)*n* where *n* is the number of glucosyl residues in the molecule. The reaction can then be written as

$$(\text{Glucose})_n + \text{P}_i^{2-} \rightarrow (\text{glucose})_{n\text{-}1} + \alpha\text{-D-glucose 1-phosphate}^{2-}$$

The next step in glycogenolysis is catalyzed by phosphoglucomutase.

$$\text{Glucose 1-phosphate} \rightarrow \text{glucose 6-phosphate}$$

This is a near-equilibrium reaction used in both glycogen degradation and synthesis.

The next enzyme depends on the tissue (Figure 15.49). In liver, glucose 6-phosphate is hydrolyzed by glucose 6-phosphatase to free glucose.

$$\text{Glucose 6-phosphate}^{2-} + H_2O \rightarrow \text{glucose} + \text{P}_i^{2-}$$

Lack of either this enzyme or of the transporter for the translocation of G6P into the endoplasmic reticulum (see Figure 15.35, p. 623) results in type I glycogen storage disease (Clin. Corr. 15.11).

As a result of the action of glucose 6-phosphatase, the overall balanced equation for glycogenolysis in the liver adds up to be simply glycogen hydrolysis.

$$(\text{Glucose})_n + H_2O \rightarrow (\text{glucose})_{n\text{-}1} + \text{glucose}$$

Note that ATP is neither used nor produced in the process.

In peripheral tissues the G6P undergoes glycolysis, leading primarily to lactate in white muscle fibers and primarily to CO_2 in red muscle fibers (Figure 15.49). Since no ATP had to be invested in white muscle fibers to produce G6P from glycogen, the overall equation for glycogenolysis followed by glycolysis is

$$(\text{Glucose})_n + 3\,\text{ADP}^{3-} + 3\,\text{P}_i^{2-} + \text{H}^+ \rightarrow (\text{glucose})_{n\text{-}1} + 2\,\text{lactate}^- + 3\,\text{ATP}^{4-} + 2\,H_2O$$

Debranching Enzyme Is Required for Glycogenolysis

Glycogen phosphorylase is specific for α-1,4-glycosidic linkages. It stops attacking α-1,4-glucosidic linkages four glucosyl residues from an α-1,6-branch point. A glycogen molecule that has been degraded to the limit by phosphorylase because of the branches is called phosphorylase limit dextrin. The action of the debranching enzyme allows phosphorylase to continue to degrade glycogen. Debranching enzyme is a bifunctional enzyme that catalyzes two reactions necessary for debranching of glycogen. The first is a 4-α-D-glucanotransferase activity that removes a strand of three glucosyl residues from a four glucosyl residue branch of the glycogen molecule, leaving a single glucosyl residue in α-1,6-linkage (Figure 15.50). The three glucosyl residue strand remains covalently attached to the enzyme until transferred to a free 4-hydroxyl of a glucosyl residue at the end of the same or an adjacent glycogen molecule to produce a longer branch. The α-1,6-linkage of the single glucosyl residue is hydrolyzed by the second enzyme activity of debranching enzyme, which is its amylo-α-1,6-glucosidase activity.

CH₂OH — O — CH₂ ... Glycogen $\xrightarrow{H_2O}$ α-D-Glucose + Glycogen

The cooperative and repetitive action of phosphorylase and debranching enzyme results in complete breakdown of glycogen to glucose 1-phosphate and glucose. Glycogen storage diseases result when either of these enzymes is defective. An average molecule of glycogen yields about 12 molecules of glucose 1-phosphate by action of phosphorylase for every molecule of free glucose produced by debranching enzyme.

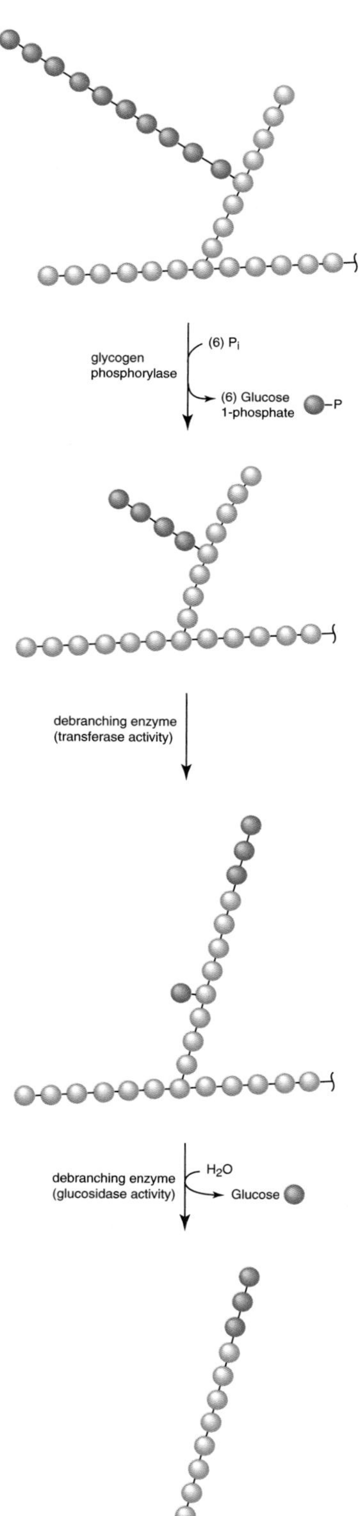

Figure 15.50 Concerted action of glycogen phosphorylase and glycogen debranching enzyme is required for glycogenolysis.

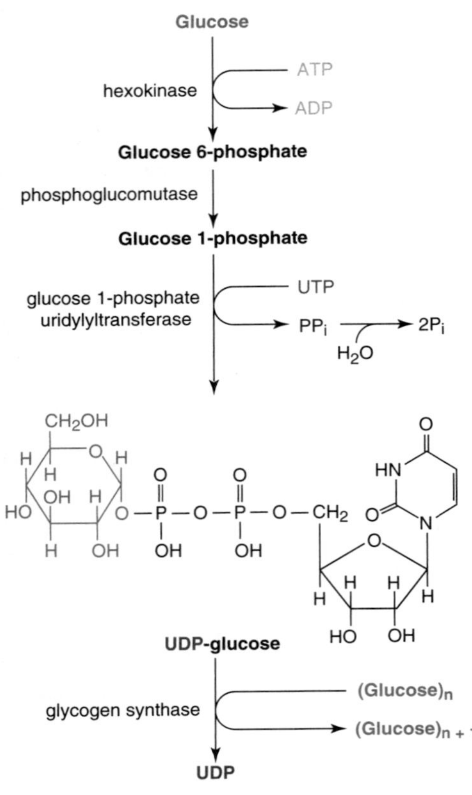

Figure 15.51 Pathway of glycogenesis.

There is another, albeit quantitatively less important, pathway for glycogen degradation that depends on glucosidases present in lysosomes. Glycogen that enters lysosomes during normal turnover of intracellular components has to be degraded. Failure to degrade glycogen taken up by lysosomes creates a major medical problem described in Clin. Corr. 15.11.

Glycogenesis Requires Unique Enzymes

The first reaction (Figure 15.51) is that of glucokinase in liver and hexokinase in peripheral tissues.

$$\text{Glucose} + \text{ATP} \rightarrow \text{glucose 6-phosphate} + \text{ADP}$$

Phosphoglucomutase then forms glucose 1-phosphate.

$$\text{Glucose 6-phosphate} \leftrightarrows \text{glucose 1-phosphate}$$

Glucose 1-phosphate uridylyltransferase then produces UDP-glucose.

$$\text{Glucose 1-phosphate} + \text{UTP} \rightarrow \text{UDP-glucose} + \text{PP}_i$$

The latter reaction generates UDP-glucose, an activated glucose molecule, from which glycogen can be synthesized. Formation of UDP-glucose is made energetically favorable and irreversible by hydrolysis of pyrophosphate to inorganic phosphate by pyrophosphatase.

$$\text{PP}_i^{4-} + H_2O \rightarrow 2\ \text{P}_i^{2-}$$

Glycogen synthase then transfers the activated glucosyl moiety of UDP-glucose to the carbon 4 of a glucosyl residue of the growing glycogen chain to form a new glycosidic bond at the hydroxyl group of carbon 1 of the activated sugar. The reducing end of glucose (carbon 1, an aldehyde that can reduce other compounds during its oxidation to a carboxylic acid) is always added to a nonreducing end (carbon 4 of a glucosyl residue) of the glycogen chain. According to this, each molecule of glycogen should have one free reducing end tucked away within its core. In fact it does not have a free reducing end because its one potentially free aldehyde group is covalently linked to a protein called glycogenin within its core (described on p. 634). The UDP formed as a product of the glycogen synthase reaction is converted back to UTP by nucleoside diphosphate kinase.

$$\text{UDP} + \text{ATP} \leftrightarrows \text{UTP} + \text{ADP}$$

Glycogen synthase cannot form the α-1,6 glycosidic linkages. Working alone, it would only produce amylose, a straight-chain polymer of glucose with α-1,4-glycosidic linkages. Once an amylose chain of at least 11 residues has been formed, a *branching enzyme* called 1,4-α-glucan branching enzyme removes a block of about 7 glucosyl residues from a growing chain and transfers it to another chain to produce an α-1,6 linkage (Figure 15.52). The new branch has to be introduced at least four glucosyl residues from the nearest branch point. Thus, the creation of the highly branched structure of glycogen requires the concerted actions of glycogen synthase and branching enzyme. The overall balanced equation for glycogen synthesis as outlined is

$$(\text{Glucose})_n + \text{glucose} + 2\ \text{ATP}^{4-} + H_2O \rightarrow (\text{glucose})_{n+1} + 2\ \text{ADP}^{3-} + 2\ \text{P}_i^{2-} + 2\ \text{H}^+$$

As already noted, the combination of glycogenolysis and glycolysis yields three molecules of ATP per glucosyl residue.

$$(\text{Glucose})_n + 3\ \text{ADP}^{3-} + 3\ \text{P}_i^{2-} + \text{H}^+ \rightarrow (\text{glucose})_{n\text{-}1} + 2\ \text{lactate}^- + 3\ \text{ATP}^{4-} + 2\ H_2O$$

Thus the combination of glycogenesis and glycogen degradation to lactate yields only one ATP.

$$\text{Glucose} + \text{ADP}^{3-} + \text{P}_i^{2-} \rightarrow 2\ \text{lactate}^- + \text{ATP}^{4-} + H_2O + \text{H}^+$$

Recall, however, that glycogen synthesis and degradation are normally carried out at different times in a cell. For example, white muscle fibers synthesize glycogen at rest when glucose is plentiful and less ATP is needed for muscle contraction. Glycogen is then used

CLINICAL CORRELATION 15.11

Glycogen Storage Diseases

There are a number of well-characterized glycogen storage diseases, all due to inherited defects in enzymes involved in the degradation of glycogen. The liver is usually the tissue most affected, but heart and muscle glycogen metabolism can also be defective.

Von Gierke Disease (OMIM 232200)

The most common glycogen storage disease, referred to as type I or von Gierke disease, arises from a deficiency of liver, intestinal mucosa, and kidney glucose 6-phosphatase. Thus, diagnosis by intestinal biopsy is possible. Patients with this disease may lack the glucose 6-phosphatase itself (type Ia) or the glucose 6-phosphate transporter (type Ib) (see Figure 15.37, p. 625). Glucose 6-phosphatase deficiency occurs in about 1 person in 200,000 and is transmitted as an autosomal recessive trait. It manifests as fasting hypoglycemia, lactic academia, hyperlipidemia, and hyperuricemia with gouty arthritis. The fasting hypoglycemia is a consequence of the glucose 6-phosphatase deficiency. The liver of these patients does release some glucose by the action of the glycogen debrancher enzyme. The lactic acidemia occurs because the liver cannot use lactate effectively for gluconeogenesis and because it inappropriately produces lactic acid in response to glucagon. This hormone should trigger glucose release without lactate production; however, the opposite occurs because of the lack of glucose 6-phosphatase. Hyperuricemia may result from increased purine degradation in the liver, hyperlipidemia because of increased availability of lactic acid for lipogenesis, and lipid mobilization from the adipose tissue caused by high catecholamine levels in response to hypoglycemia. These manifestations can be greatly diminished by providing carbohydrate throughout the day to prevent hypoglycemia. During sleep this can be done by infusion of carbohydrate into the gut by a nasogastric tube.

Pompe Disease (OMIM 232300)

Type II glycogen storage disease or Pompe disease is caused by the absence of α-1,4-glucosidase (or acid maltase) normally found in lysosomes. This leads to accumulation of glycogen mostly in lysosomes in virtually every tissue. This is somewhat surprising, but lysosomes take up glycogen granules and become defective with respect to other functions if they lack the capacity to hydrolyze the granules. Because other synthetic and degradative pathways of glycogen metabolism are intact, metabolic derangements such as those in von Gierke disease are not seen. Severe hypotonia, massive cardiomegaly, and cardiomyopathy occur and death results from heart failure, usually in the first months of life. An adult onset form of the disease causes severe muscle weakness, particularly of respiratory muscles (diaphragm), with death often resulting from respiratory insufficiency.

Cori Disease (OMIM 232400)

Type III glycogen storage disease or Cori disease is caused by deficiency of glycogen debranching enzyme. Glycogen accumulates because only the outer branches can be removed by phosphorylase. Hepatomegaly occurs, but diminishes with age. The clinical manifestations are similar to, but much milder than, those in von Gierke disease because gluconeogenesis is unaffected, and hypoglycemia and its complications are less severe.

McArdle Disease (OMIM 232600)

Type V glycogen storage disease or McArdle disease is caused by absence of muscle phosphorylase. Patients suffer from painful muscle cramps and are unable to perform strenuous exercise, presumably because muscle glycogen stores are not available to the exercising muscle. Thus, the normal increase in plasma lactate (released from the muscle) following exercise is absent. The muscles are probably damaged because of inadequate energy supply and glycogen accumulation. Release of creatine kinase, aldolase, and myoglobin from muscle is common; elevated levels of these proteins in the blood suggest a muscle disorder.

Chen, Y.-T. Glycogen storage diseases. In Scriver, C. R., Beaudet, A.R., Sly, W. S., and Valle, D. (Eds.), *The Metabolic and Molecular Bases of Inherited Disease,* 8th ed. New York: McGraw Hill, 2001, 1521.

Cori, G. T., and Cori, C. F. Glucose-6-phosphatase of the liver in glycogen storage disease. *J. Biol. Chem.* 199:661, 1952.

Raben, N., Plotz, P., and Byrne, B. J. Acid alpha-glucosidase deficiency (glycogenosis type II, Pompe Disease). *Curr. Mol. Med.* 2:145, 2002.

Van Hoff, F. and Hers, H. G. The subgroups of type III glycogenosis. *Eur. J. Biochem.* 2:265, 1967.

McArdle, B. Myopathy due to a defect in muscle glycogen breakdown. *Clin. Sci.* 10:13, 1951.

during periods of exertion. Although not a very efficient process, the storage of glucose in the form of glycogen provides cells with a fuel reserve that can be quickly mobilized.

Special Features of Glycogenolysis and Glycogenesis

Why Store Glucose as Glycogen?

The fact that glycogen is such a good fuel reserve makes it obvious why we synthesize and store glycogen in liver and muscle. But, why not store our excess glucose calories entirely as fat instead of glycogen? The answer is at least threefold: (1) Fatty acids cannot be released from fat nearly as rapidly as glucose can be released from glycogen, (2) fat cannot be used as a source of energy in the absence of oxygen, and (3) fat cannot be converted to glucose to maintain blood glucose levels required by the brain. Why not just pump glucose into cells and store it as free glucose until needed? Why waste ATP making a polymer out of glucose?

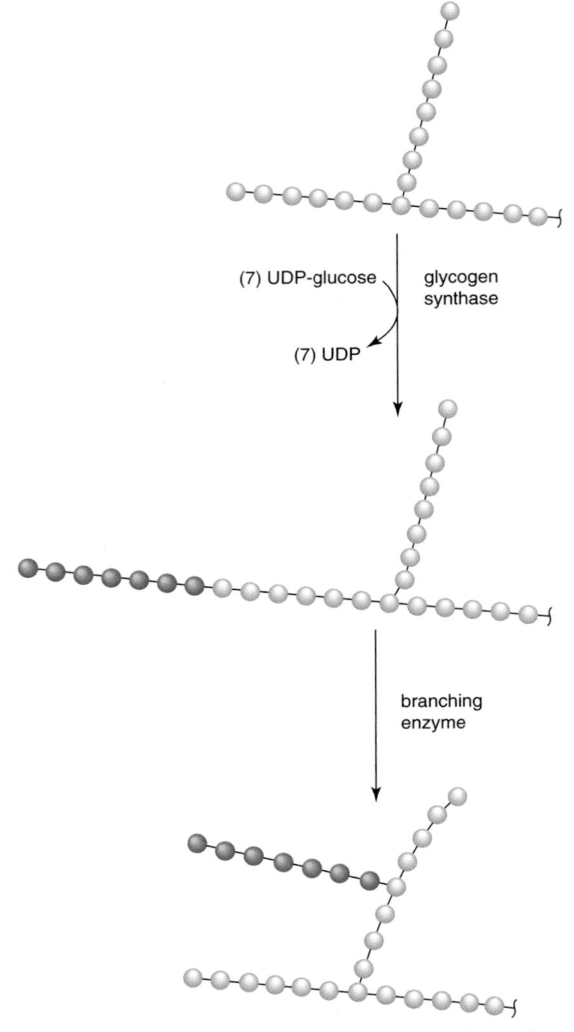

Figure 15.52 Concerted action of glycogen synthase and glycogen branching enzyme is required for glycogenesis. Action of glycogen-branching enzyme.

The problem is that glucose is osmotically active. It would cost ATP to "pump" glucose into a cell against a concentration gradient, and its concentration would have to reach about 400 mM in liver cells to match the *glucose reserve* provided by the usual liver glycogen content. Unless balanced by outward movement of some other osmotically active compound, accumulation of glucose would cause considerable uptake of water with osmotic lysis of the cell. Assuming the mass of a glycogen molecule is about 10^7 Da, 400 mM glucose is in effect stored at a glycogen concentration of only 0.01 μM. Storage of glucose as glycogen, therefore, creates no osmotic pressure problem for the cell.

Glycogenin Is Required as a Primer for Glycogen Synthesis

As in DNA synthesis, a primer is needed for glycogen synthesis. No template, however, is required. Glycogen itself is the usual primer, since glycogen synthesis adds glucosyl units to glycogen core molecules, which are almost invariably present in cells. The outer regions of glycogen molecules are removed and resynthesized more rapidly than the inner core. Glycogen within a cell is frequently sheared by the combined actions of glycogen phosphorylase and debranching enzyme but is seldom obliterated before one eats again and glycogen synthase and branching enzyme rebuild the molecule. This begs the question why glycogen is a branched molecule with only one real beginning (the original starting glucose residue) and many branches terminating in nonreducing glucosyl units. The answer is that this gives numerous sites of attack for glycogen phosphorylase on a mature glycogen molecule and the same number of sites to serve as primers for the addition of glucosyl units by glycogen synthase. If cells synthesized α amylose, an unbranched glucose polymer, there would only be one nonreducing end per molecule. This would make glycogen degradation and synthesis much slower. As it is, glycogen phosphorylase and glycogen synthase occur in tight association with glycogen granules and have ready access to a multitude of nonreducing sugars at the ends of the branches. This is not the case in Lafora disease, a juvenile-onset myoclonus epilepsy characterized by neuronal accumulation of insoluble poorly branched glycogen deposits called Lafora bodies (A Closer Look 15.7)

A primer is needed for glycogen synthesis because glycogen synthase cannot initiate glycogen synthesis with a single glucose molecule as the acceptor of an activated glucosyl residue from UDP-glucose. Glycogen synthase has a very low K_m for large glycogen molecules, and therefore readily adds glucosyl residues to make even larger glycogen molecules. However, the K_m gets larger and larger as the glycogen molecule gets smaller and smaller so that glucose, at its physiological concentration, cannot function as a primer. This led for some time to the notion that glycogen might be immortal; that is, some glycogen might have to be handed down from one cell generation to the next in order for glycogen to be synthesized. However, a polypeptide of 332 amino acids called glycogenin serves as a primer for glycogen synthesis. Glycogenin is a self-glucosylating enzyme that uses UDP-glucose to link glucose to one of its own tyrosine residues (Figure 15.53). Glycogenin forms a chain of glucosyl residues on itself by catalyzing the addition of seven more glucosyl residues with α-1,4 linkages. Glycosylated glycogenin then serves as a primer for synthesis of glycogen by glycogen synthase. Alas, glycogen is not immortal.

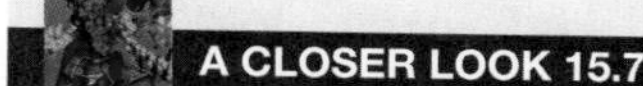

A CLOSER LOOK 15.7

Phosphorylated Glycogen and Lafora Disease

Lafora disease is due to a gene defect in laforin, a phosphatase that removes phosphate residues from phosphorylated glycogen. Surprisingly, prior to studies on the mechanism responsible for Lafora disease it was not appreciated that glycogen is phosphorylated. How this occurs and whether it serves a purpose is still not known. Glycogen isolated from mice with Lafora disease is less branched but more heavily phosphorylated than glycogen isolated from wild type mice. This suggests that the presence of phosphate groups on glycogen promotes the formation of Lafora bodies by inhibiting the branching enzyme.

Tagliabracci, V. S., Turnbill, J., Wang, W., Girard, J.-M., et al. Laforin is a glycogen phosphatase, deficiency of which leads to elevated phosphorylation of glycogen in vivo. *Proc. Natl. Acad. Sci. USA* 104:19262, 2007.

Glycogen Limits Its Own Synthesis

If glycogen synthase becomes more efficient as the glycogen molecule gets bigger, how is synthesis of this ball of sugar curtailed? Adipocytes have an almost unlimited capacity to pack away fat, but then adipocytes have nothing else to do. Muscle cells participate in mechanical activity and liver cells carry out many processes other than glycogen synthesis. Even in the face of excess glucose, there has to be a way to limit the intracellular accumulation of glycogen. Glycogen itself inhibits glycogen synthase by a mechanism discussed later.

Glycogen Synthesis and Degradation Are Highly Regulated

Glycogen synthase and glycogen phosphorylase are the regulatory enzymes of glycogen synthesis and degradation, respectively. Both catalyze nonequilibrium reactions, and both are subject to control by allosteric effectors and covalent modification.

Regulation of Glycogen Phosphorylase

Glycogen phosphorylase is activated by AMP and inhibited by glucose and ATP (Figure 15.54). Control by these allosteric effectors is integrated with a very elaborate control by covalent modification. Phosphorylase exists in an active *a* form and an inactive *b* form. These forms are interconverted by phosphorylase kinase and phosphoprotein phosphatase. A conformational change caused by phosphorylation creates a more active catalytic state. Nonphosphorylated phosphorylase, that is the *b* form, has low activity but is greatly stimulated by AMP. This effector has little activating effect on the already active phosphorylase *a*. Thus, regulation by covalent modification can be bypassed by an AMP-mediated mechanism.

Phosphorylase kinase phosphorylates and activates phosphorylase (Figure 15.54) and is itself subject to regulation by **phosphorylation-dephosphorylation**. Protein kinase A phosphorylates and activates phosphorylase kinase; phosphoprotein phosphatase dephosphorylates and inactivates it. Phosphorylase kinase is a complex (1.3 million Da) composed of four different subunits with four copies of each subunit in the complex ($\alpha_4\beta_4\gamma_4\delta_4$). Catalytic activity resides in the γ-subunit; α-, β-, and δ-subunits exert regulatory control. The α- and β-subunits are phosphorylated in the transition from the inactive *b* form to the active *a* form. Protein kinase A can only exert an effect on phosphorylase via phosphorylation and activation of phosphorylase kinase. Thus, a bicyclic system is required for activation of phosphorylase in response to cAMP-mediated signals.

The δ-subunit is the Ca^{2+}-binding regulatory protein calmodulin (p. 493). It is present in cells in its free form and bound tightly in enzyme complexes. It functions as a Ca^{2+} receptor in cells, responding to changes in intracellular Ca^{2+} concentration and affecting the activity of several enzyme systems. Binding of Ca^{2+} to calmodulin makes phosphorylase kinase more active. As shown in Figure 15.55, p. 637, Ca^{2+} is an activator of both phosphorylase

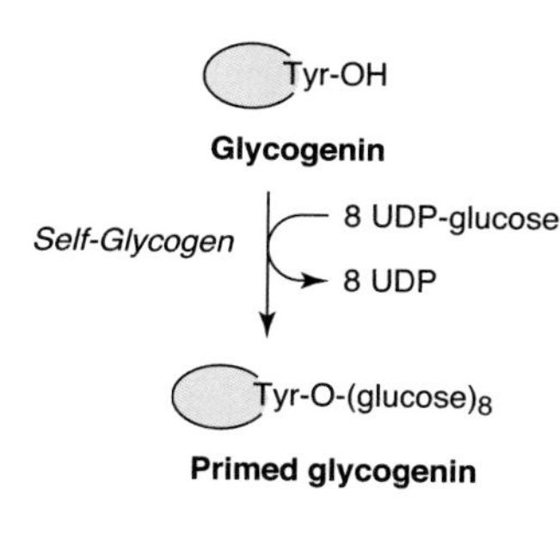

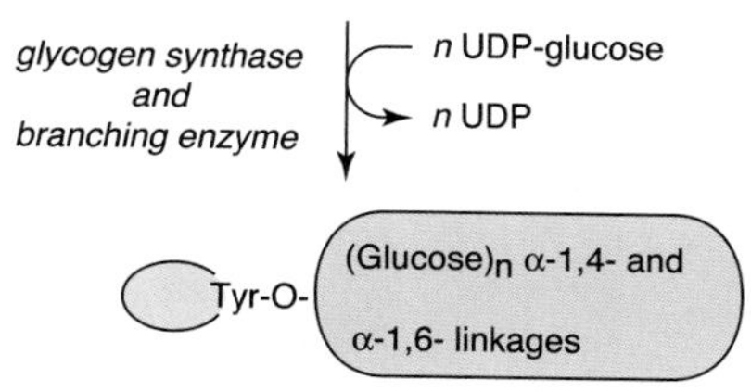

Figure 15.53 Glycogenin provides a primer for glycogen synthesis by glycogen synthase. Tyr designates a tyrosine residue of glycogenin.

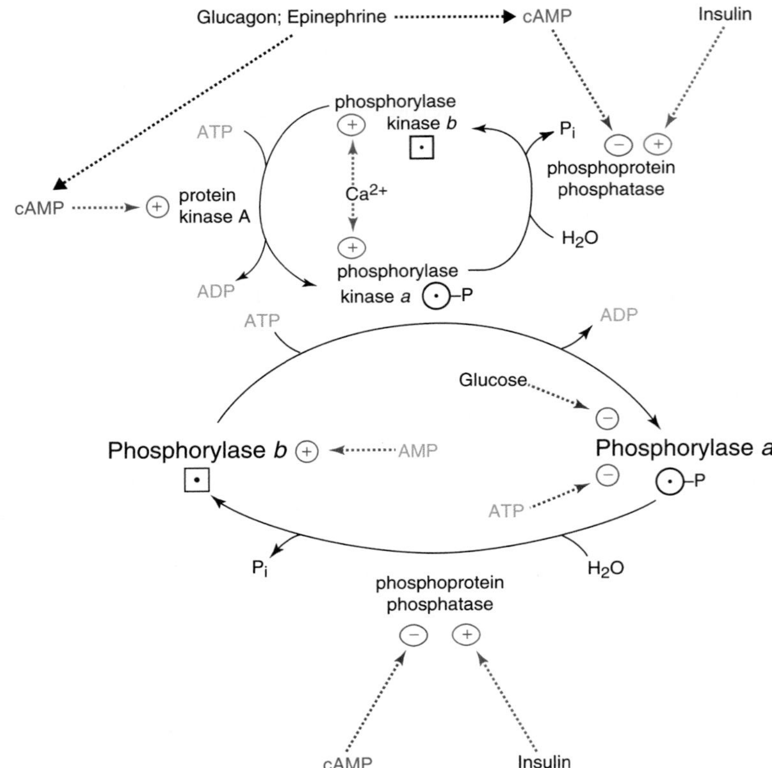

Figure 15.54 Regulation of glycogen phosphorylase by covalent modification and allosteric effectors. Phosphorylation converts glycogen phosphorylase and phosphorylase kinase from their inactive *b* forms to their active *a* forms.

kinase *a* and phosphorylase kinase *b*. Maximum activation of phosphorylase kinase requires phosphorylation of specific serine residues plus interaction of Ca^{2+} with calmodulin. This is one way by which Ca^{2+} functions as a second messenger of hormone action.

Activation of phosphorylase kinase by phosphorylation and Ca^{2+} has a substantial effect on the activity of glycogen phosphorylase. It is obvious, however, that inhibiting the phosphoprotein phosphatase that modulates both phosphorylase kinase and glycogen phosphorylase could achieve the same effect. Ultimate control of glycogen phosphorylase therefore involves reciprocal regulation of phosphoprotein phosphatase and phosphorylase kinase activities. Regulation of phosphoprotein phosphatase activity involves cAMP. Hormones that increase cAMP levels, such as glucagon and epinephrine, promote activation of glycogen phosphorylase by signaling activation of phosphorylase kinase and inactivation of phosphoprotein phosphatase. Insulin, though a kinase-mediated signal cascade, has the opposite effect on phosphorylase by promoting activation of phosphoprotein phosphatase activity.

Glycogen Phosphorylase Is Regulated by a Cascade That Greatly Amplifies a Small Signal into a Very Large Effect

The bicyclic control system for phosphorylation of glycogen phosphorylase amplifies enormously a very small initial signal. Activation of adenylate cyclase by one molecule of epinephrine causes formation of many molecules of cAMP. Each cAMP molecule activates a protein kinase A molecule, which in turn phosphorylates many molecules of phosphorylase kinase, which are activated, as well as many molecules of phosphoprotein phosphatase, which are inactivated. In turn phosphorylase kinase phosphorylates many molecules of glycogen phosphorylase, which in turn catalyzes phosphorolysis of many glycosidic bonds of glycogen. This is therefore an amplification system in which the signal provided by a few molecules of hormone is amplified into an enormous number of glucose 1-phosphate molecules. If each step represents an amplification factor of 100, then four steps would result in an amplification of 100 million! This system is so rapid that all of the stored glycogen of white muscle fibers could be converted to glucose 6-phosphate within seconds.

Regulation of Glycogen Synthase

Glycogen synthase has to be active for glycogenesis and inactive for glycogenolysis. The combination of the reactions of glycogen synthase, glycogen phosphorylase, glucose 1-phosphate uridylyltransferase, and nucleoside diphosphate kinase adds up to a futile cycle with the

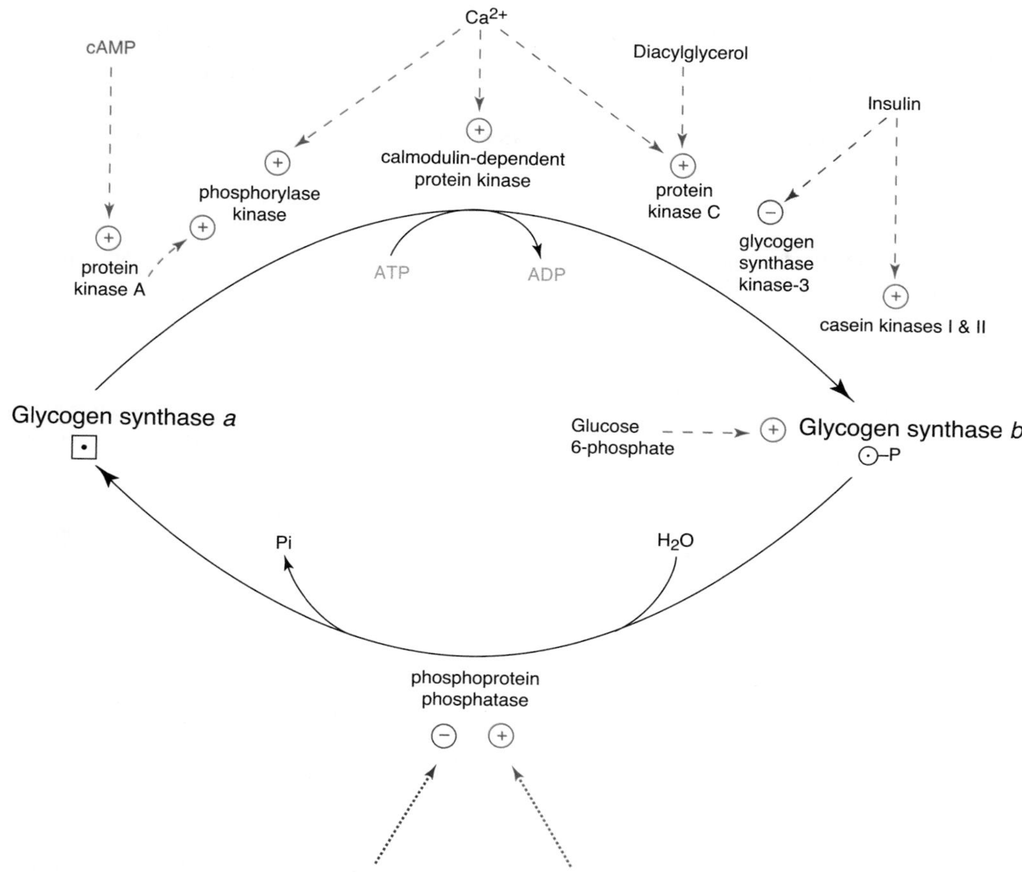

Figure 15.55 Regulation of glycogen synthase by covalent modification. Phosphorylation converts glycogen synthase from its active *a* form to its inactive *b* form.

overall equation ATP + $H_2O \rightarrow$ ADP + P_i. Hence, glycogen synthase needs to be inhibited when glycogen phosphorylase is active, and vice versa.

Whether G6P activates glycogen synthase depends on its phosphorylation state (Figure 15.55). Glycogen synthase exists in a phosphorylated "D" form that is dependent on the presence of G6P for activity and a nonphosphorylated "I" form which is independent of G6P. The D form corresponds to the *b*, or inactive, form of the enzyme, the I form to the *a*, or active, form of the enzyme. Phosphorylation of glycogen synthase is catalyzed by several protein kinases, which in turn are regulated by second messengers of hormone action, including cAMP, Ca^{2+}, diacylglycerol, and probably others yet to be identified. Each protein kinase identified in Figure 15.55 can phosphorylate and contribute to inactivation of glycogen synthase. Glycogen synthase is a homotetramer (α_4) of mass 85 kDa that can be phosphorylated on at least nine different serine residues. Eleven protein kinases have been identified that can phosphorylate glycogen synthase. This is in striking contrast to glycogen phosphorylase, which is regulated by phosphorylation at one site by one specific kinase.

Cyclic AMP induces opposite effects on the activities of glycogen synthase (Figure 15.55) and glycogen phosphorylase (Figure 15.54). An increase in cAMP signals activation of glycogen phosphorylase and inactivation of glycogen synthase via activation of protein kinase A and inhibition of phosphoprotein phosphatase. Ca^{2+} likewise influences the phosphorylation states of both enzymes and reciprocally regulates their activity by its effects on phosphorylase kinase. Two cAMP-independent, Ca^{2+}-activated protein kinases have been identified that may have physiological significance. One is a calmodulin-dependent protein kinase and the other a Ca^{2+}- and phospholipid-dependent protein kinase (protein kinase C). Both phosphorylate glycogen synthase, but neither can phosphorylate glycogen phosphorylase. Protein kinase C requires phospholipid, diacylglycerol, and Ca^{2+} for full activity. It is of considerable interest because tumor-promoting agents called phorbol esters mimic diacylglycerol as activators of its activity. Diacylglycerol is a second messenger of hormone action, acting via protein kinase C to regulate numerous cellular processes.

Glycogen synthase is also phosphorylated by glycogen synthase kinase-3, casein kinase I, and casein kinase II. These kinases are not subject to regulation by cAMP or Ca^{2+} but special regulatory mechanisms exist for them. An insulin-signaling cascade results in activation of protein kinase B, which inactivates glycogen synthase kinase-3 by phosphorylation, an action that allows activation of glycogen synthase to occur via dephosphorylation by phosphoprotein phosphatase.

The phosphoprotein phosphatase that converts glycogen synthase *b* to the *a* form (Figure 15.55) is regulated in a manner analogous to that of glycogen phosphorylase (Figure 15.54). Cyclic AMP promotes inactivation whereas insulin promotes activation of glycogen synthase through opposite effects upon phosphoprotein phosphatase activity.

In general, phosphoprotein phosphatases are present as catalytic subunits associated with regulatory subunits that control their activity, determine which substrate(s) can be dephosphorylated, and target association with specific structures within a cell. A regulatory protein important for glycogen metabolism is the G subunit or glycogen-binding protein. G subunit binds both glycogen and a phosphatase catalytic subunit (Figure 15.56) making the phosphatase 10 times more active toward glycogen synthase and glycogen phosphorylase. However, phosphorylation of G subunit by protein kinase A releases the phosphatase catalytic subunit, which is then less active. Interaction of the free catalytic subunit with yet another regulatory protein (called inhibitor 1) further inhibits phosphatase activity. Effective inhibition of the residual phosphatase activity requires phosphorylation of inhibitor 1 by protein kinase A, thereby creating yet another link to hormones that increase cAMP levels. Insulin has effects opposite to those of cAMP; that is, it promotes activation of the catalytic subunit of the phosphatase.

Effector Control of Glycogen Metabolism

Certain muscles mobilize their glycogen stores rapidly in response to anaerobic conditions without marked conversion of phosphorylase *b* to the *a* form or glycogen synthase *a* to the *b* form. Presumably this is accomplished by effector control in which ATP levels decrease, causing less inhibition of phosphorylase; glucose 6-phosphate levels decrease, causing less activation of glycogen synthase; and AMP levels increase, causing activation of phosphorylase. This enables muscle to keep working, for at least a short period, by using ATP produced by glycolysis of glucose 6-phosphate derived from glycogen.

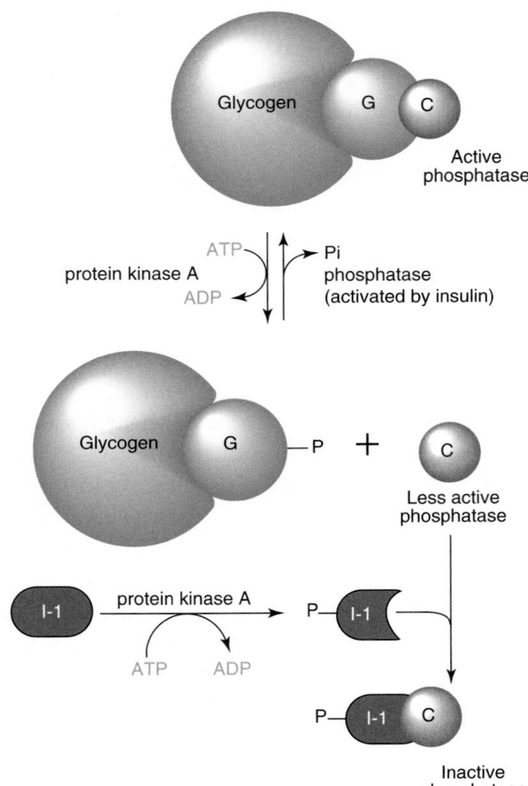

Figure 15.56 Mechanism for regulation of a phosphatase that binds to glycogen. The glycogen binding subunit G binds directly to glycogen; the phosphoprotein phosphatase catalytic subunit C binds to glycogen by the G subunit; and the phosphorylated inhibitor 1 (I-1) binds the free catalytic subunit. Protein kinase A inactivates the phosphatase by phosphorylating G subunit and I-1. Insulin signals activation of the phosphatase by promoting dephosphorylation of the G subunit and I-1 (not shown).

Proof that effector control can operate has also been obtained from a strain of mice deficient in muscle phosphorylase kinase. Phosphorylase *b* in muscle of such mice cannot be converted into phosphorylase *a*. Nevertheless, heavy exercise results in depletion of muscle glycogen because of stimulation of phosphorylase *b* by AMP.

Negative Feedback Control of Glycogen Synthesis by Glycogen

Glycogen exerts feedback control over its own formation. The portion of glycogen synthase in the active *a* form decreases as glycogen accumulates in a tissue. The mechanism is not well understood, but glycogen may make the *a* form a better substrate for a protein kinase or, alternatively, glycogen may inhibit dephosphorylation of glycogen synthase *b* by phosphoprotein phosphatase. Either mechanism would account for the shift in the steady state in favor of glycogen synthase *b* that occurs in response to glycogen accumulation.

Phosphorylase *a* Is a "Glucose Receptor" in Liver

A carbohydrate-rich meal increases blood and liver glucose, which increases glycogen synthesis in liver tissue. The mechanism involves glucose stimulation of insulin release from the pancreas and its effects on hepatic glycogen phosphorylase and glycogen synthase. However, hormone-independent mechanisms also appear to be important in liver (Figure 15.57). Direct inhibition of phosphorylase *a* by glucose is probably of importance. Binding of glucose to phosphorylase makes the *a* form of phosphorylase a better substrate for dephosphorylation by phosphoprotein phosphatase. Therefore, phosphorylase *a* in liver functions as an intracellular glucose receptor. Binding of glucose to phosphorylase *a* promotes its inactivation, thereby inhibiting glycogenolysis. This *negative-feedback* control of glycogenolysis by glucose would not necessarily promote glycogen synthesis. However, phosphorylase *a* inhibits dephosphorylation of glycogen synthase *b* by phosphoprotein phosphatase. This inhibition is lost once phosphorylase *a* has been converted to the *b* form (Figure 15.57). In other words, phosphoprotein phosphatase can turn its attention to glycogen synthase *b* only after dephosphorylation of phosphorylase *a*. Thus, as a result of interaction of glucose with phosphorylase *a*, glycogen is synthesized rather than degraded in liver. Phosphorylase *a* can serve this function of "glucose receptor" in liver because the concentration of glucose in liver reflects that in blood. This is not true for extrahepatic tissues. Liver cells have a very high-capacity transporter for glucose (GLUT2) and a high $S_{0.5}$ enzyme for glucose phosphorylation (glucokinase) whereas in extrahepatic tissues glucose

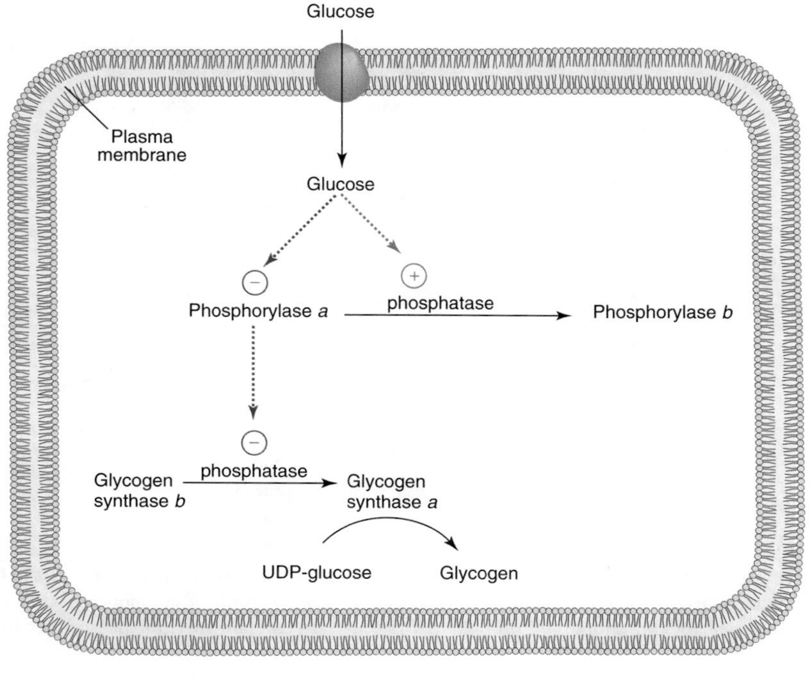

Figure 15.57 Overview of the mechanism for glucose stimulation of glycogenesis in the liver.

transport and phosphorylation systems maintain intracellular glucose at a concentration too low for phosphorylase *a* to function as a "glucose receptor."

Hormonal and Neural Control of Glycogen Synthesis and Degradation

Glucagon and Epinephrine Stimulate Glycogenolysis in Liver

Glucagon is released from α cells of pancreas in response to low blood glucose levels. Under such conditions, for example, during fasting, glucagon stimulates glycogenolysis so as to ensure that adequate blood glucose is available for glucose-dependent tissues (Figure 15.58). Binding of glucagon to its receptors on liver cells activates adenylate cyclase and triggers the cascades that activate glycogen phosphorylase and inactivate glycogen synthase (see Figures 15.54 and 15.55, respectively). It also inhibits glycolysis at the level of 6-phosphofructo-1-kinase and pyruvate kinase as shown in Figures 15.25, p. 616, and 15.28, p. 619, respectively. The net result of all these effects mediated by cAMP and covalent modification is a very rapid increase in normal blood glucose levels. Hypergly-cemia does not occur because less glucagon is released from the pancreas as blood glucose levels increase.

Epinephrine is released into blood from chromaffin cells of the adrenal medulla in response to stress. This "fright, flight or fight" hormone prepares the body for either combat or escape. Binding of epinephrine with β-adrenergic receptors on liver cells activates adenylate cyclase (Figure 15.58) and cAMP has the same effects as glucagon, that is, activation of glycogenolysis and inhibition of glycogenesis and glycolysis to maximize the release of glucose. Binding of epinephrine to α-adrenergic receptors on liver cells signals formation of inositol 1,4,5-trisphosphate (IP_3) and diacylglycerol (Figure 15.59). These are

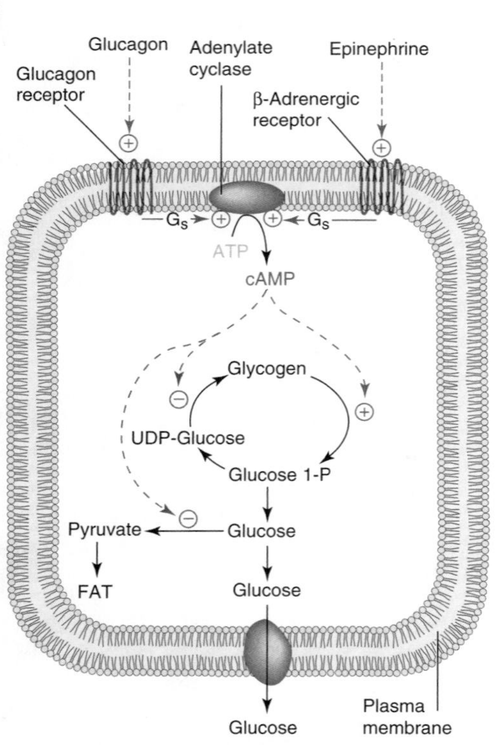

Figure 15.58 Cyclic AMP mediates stimulation of glycogenolysis in liver by glucagon and β agonists (epinephrine). See legends to Figures 15.17, p. 612, and 15.23, p. 615.

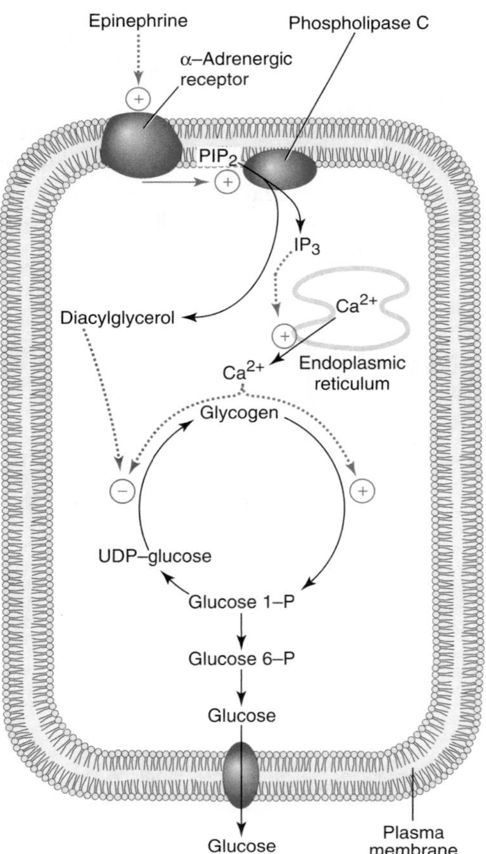

Figure 15.59 Inositol trisphosphate (IP_3) and Ca^{2+} mediate stimulation of glycogenolysis in liver by α agonists. The α-adrenergic receptor and glucose transporter are intrinsic components of the plasma membrane. Phosphatidylinositol 4,5-bisphosphate (PIP_2) is also a component of the plasma membrane.

second messengers, produced by the action of a phospholipase C on phosphatidylinositol 4,5-bisphosphate of the plasma membrane (Figure 15.60). IP_3 stimulates the release of Ca^{2+} from the endoplasmic reticulum, which activates phosphorylase kinase which in turn activates glycogen phosphorylase (see Figure 15.54, p. 636). In addition, Ca^{2+}-mediated activation of phosphorylase kinase and calmodulin-dependent protein kinase as well as diacylglycerol-mediated activation of protein kinase C may all contribute to inactivation of glycogen synthase (see Figure 15.55, p. 637).

An increased rate of glucose release into blood is a major consequence of epinephrine action on the liver. This makes glucose available to tissues that are called on to meet the challenge of the stressful situation that triggered the release of epinephrine from the adrenal medulla.

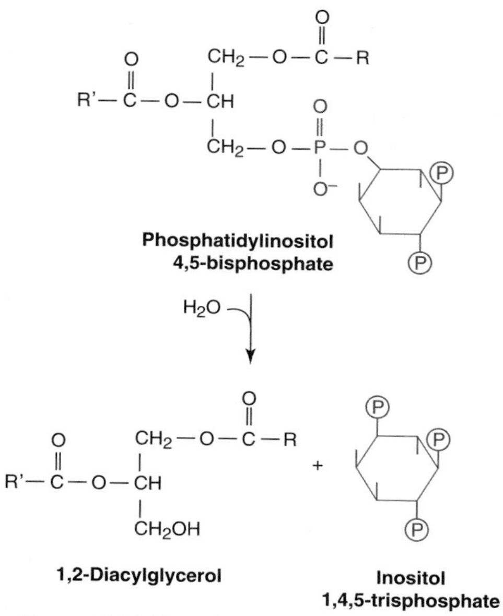

Phosphatidylinositol 4,5-bisphosphate

1,2-Diacylglycerol

Inositol 1,4,5-trisphosphate

Figure 15.60 Phospholipase C cleaves phosphatidylinositol 4,5-bisphosphate to 1,2-diacylglycerol and inositol 1,4,5-trisphosphate.

Epinephrine Stimulates Glycogenolysis in Heart and Skeletal Muscle

Epinephrine also stimulates glycogenolysis in heart and skeletal muscle (Figure 15.61). It binds to β-adrenergic receptors, which stimulates adenylate cyclase to produce cAMP which activates glycogen phosphorylase and inactivates glycogen synthase. Since these tissues lack glucose 6-phosphatase, this leads to stimulation of glycolysis rather than release of glucose into the blood. Thus, the effect of epinephrine in heart and skeletal muscle is to make more glucose 6-phosphate available for glycolysis. ATP generated by glycolysis can then meet the need for energy imposed on these muscles by the stress that triggered epinephrine release.

Neural Control of Glycogenolysis in Skeletal Muscle

Nervous excitation of muscle activity is mediated by changes in intracellular Ca^{2+} concentration (Figure 15.62). A nerve impulse causes membrane depolarization, which causes Ca^{2+}

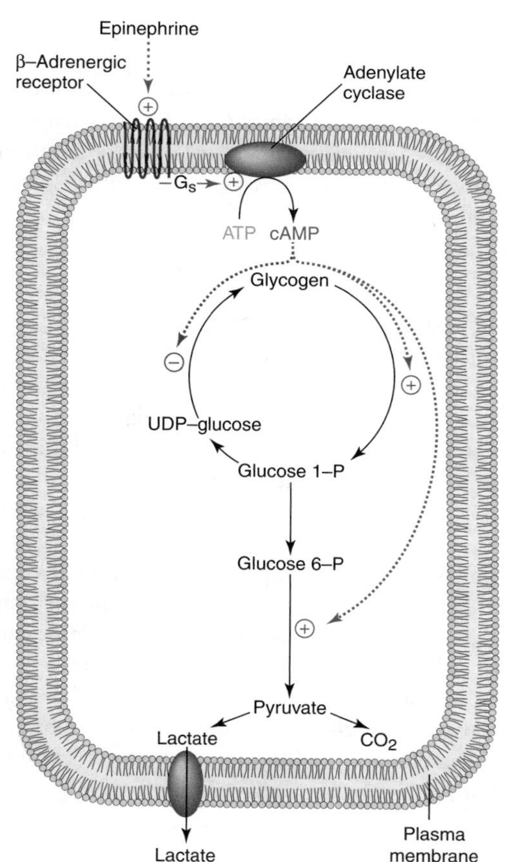

Figure 15.61 Cyclic AMP mediates stimulation of glycogenolysis in muscle by β agonists (epinephrine). The β-adrenergic receptor is an intrinsic component of the plasma membrane that stimulates adenylate cyclase by a stimulatory G protein (G_s).

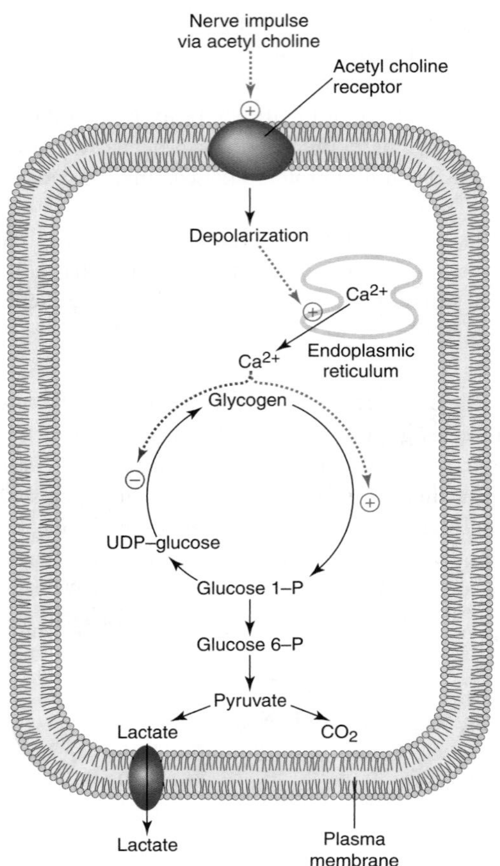

Figure 15.62 **Ca^{2+} mediates the stimulation of glycogenolysis in muscle by nervous excitation.**

release from the sarcoplasmic reticulum into the sarcoplasm of muscle cells. This triggers muscle contraction, whereas reaccumulation of Ca^{2+} by the sarcoplasmic reticulum causes relaxation. The change in Ca^{2+} concentration also activates phosphorylase kinase and glycogen phosphorylase and perhaps inactivates glycogen synthase. Thus, more glycogen is converted to glucose 6-phosphate so that more ATP is produced to meet the greater energy demand of muscle contraction.

Insulin Stimulates Glycogenesis in Muscle and Liver

An increase in blood glucose signals release of insulin from β cells of the pancreas. Insulin receptors on the plasma membranes of insulin-responsive cells respond to insulin binding through a signaling cascade that promotes glucose use (Figures 15.63 and 15.64). The pancreas responds to a decrease in blood glucose with less release of insulin and more release of glucagon. These hormones have opposite effects on glucose utilization by liver, thereby establishing the pancreas as a fine-tuning device that prevents dangerous fluctuations in blood glucose levels.

Insulin increases glucose utilization in part by promoting glycogenesis and inhibiting glycogenolysis in muscle and liver. Insulin stimulation of glucose transport is essential for these effects in muscle but not liver. Hepatocytes have a high-capacity, insulin-insensitive glucose transporter (GLUT2), whereas skeletal muscle cells and adipocytes have an insulin-sensitive glucose transporter (GLUT4) (p. 489). Insulin increases the number of glucose transporter 4 proteins associated with the plasma membrane by promoting their translocation from an intracellular pool (see Figure 15.5, p. 595). Insulin promotes glycogen accumulation in both tissues by activating glycogen synthase and inhibiting glycogen phosphorylase.

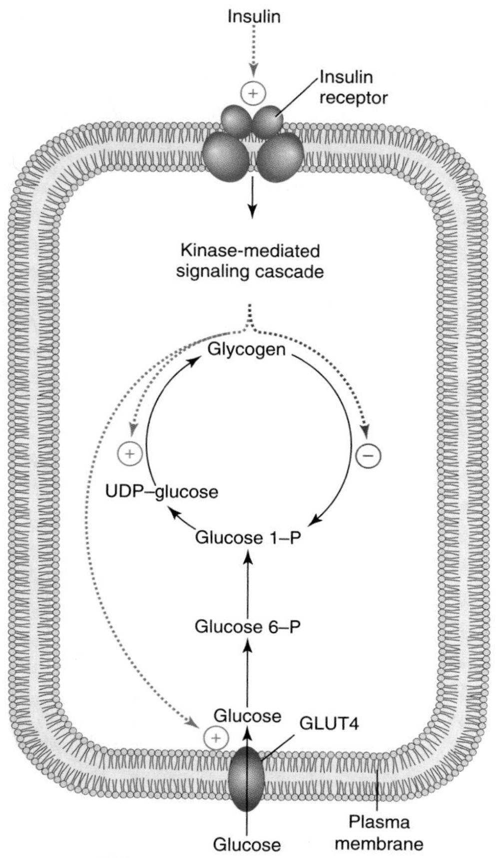

Figure 15.63 Insulin acts by a plasma membrane receptor to promote glycogenesis in muscle.

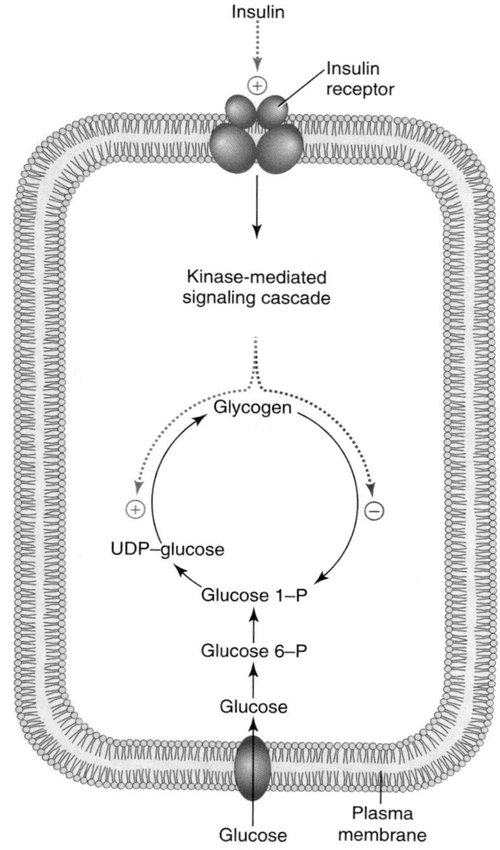

Figure 15.64 Insulin acts by a plasma membrane receptor to promote glycogenesis in liver.

Bibliography

Brosnan, J. T. Comments on metabolic needs for glucose and the role of gluconeogenesis. *E. J. Clin. Nutr.* 53: S107, 1999.

Chen, Y.-T. Glycogen storage diseases. In Scriver ,C. R., Beaudet, A.R., Sly, W. S., and Valle, D. (Eds.), *The Metabolic and Molecular Bases of Inherited Disease,* 8th ed. New York: McGraw Hill, 2001, 1521.

Cho, J., King, J. S., Qian, X., Harwood, A. J., and Shears, S. B. Dephosphorylation of 2,3-bisphosphoglycerate by MIPP expands the regulatory capacity of the Rapoport–Lubering glycolytic shunt. *Proc. Natl. Acad. Sci. USA* 105: 5998, 2008.

Croniger, C., Leahy, P., Reshef, L., and Hanson, R. W. C/EBP and the control of phosphoenolpyruvate carboxykinase gene transcription in liver. *J. Biol. Chem.* 273: 31629, 1998.

DePaoli-Roach, A. A., Park, I.-K., Cerovsky, V., Csortos, C., et al. Mechanisms of control of heart glycolysis. *Eur J. Biochem.* 258: 277, 1998.

Fell, D. *Understanding the Control of Metabolism.* London: Portland Press, 1997.

Gibson, D. M. and Harris, R. A. *Metabolic Regulation in Mammals.* London: Taylor & Frances, 2001.

Gould, G. W., and Holman, G. D. The glucose transporter family: Structure, function and tissue-specific expression. *Biochem. J.* 295: 329, 1993.

Gurney, A. L., Park, E. A., Liu, J., Giralt, M., McGrane, M. M., Patel, Y. M., Crawford, D. R., Nizielski, S. E., Savon, S., and Hanson, R. W. Metabolic regulation of gene transcription. *J. Nutr.* 124: 1533S, 1994.

Hanson, R. W., and Mehlman, M. A. (Eds.). *Gluconeogenesis, Its Regulation in Mammalian Species.* New York: Wiley, 1976.

Harper, E. T., and Harris, R. A. Glycolytic pathway. *Nature Encyclopedia of Life Sciences.* www.els.net. 2005.

Harris, R. A. Glycolysis overview. *Encylcopedia of Biological Chemistry,* Section of metabolism, vitamins, and hormones. Amsterdam: Elsevier/ Academic Press (2005).

Hunter, T. Protein kinases and phosphatases: The yin and yang of protein phosphorylation and signaling. *Cell* 80: 225, 1995.

King, M. W. Dr. King's Medical Biochemical Page. http://web.indstate.edu/thcme/mwking/. Accessed January 16, 2005.

King, M. W. Glycogen, starch, and sucrose. *Nature Encyclopedia of Life Sciences.* www.els.net. 2005.

Lalli, E., and Sassone-Corsi, P. Signal transduction and gene regulation: The nuclear response to cAMP. *J. Biol. Chem.* 269: 17359, 1994.

Lopaschuk, G. D. Glycolysis regulation. *Nature Encyclopedia of Life Sciences.* www.els.net. 2005.

Metzler, D. E. *Biochemistry, the Chemical Reactions of Living Cells.* San Diego, CA: Academic Press, 2001.

Newsholme, E. A., and Leech, A. R. *Biochemistry for the Medical Sciences.* New York: Wiley, 1983.

Newsholme, E. A., and Start, C. *Regulation in Metabolism.* New York: Wiley, 1973.

Pilkis, S. J., and El-Maghrabi, M. R. Hormonal regulation of hepatic gluconeogenesis and glycolysis. *Annu. Rev. Biochem.* 57: 755, 1988.

Pilkis, S. J., Claus, T. H., Kurland, I. J., and Lange, A. J. 6-Phosphofructo-2-kinase/fructose-2,6-bisphosphatase: A metabolic signaling enzyme. *Annu. Rev. Biochem.* 64: 799, 1995.

Price, T. B., Rothman, D. L., and Shulman R. G. NMR of glycogen in exercise. *Proc. Nutr. Soc.* 58: 851, 1999.

Roach, P. J., Skurat, A.V., and Harris, R. A. Regulation of glycogen metabolism: The endocrine pancreas and regulation of metabolism. In Jefferson, L. S., and Charrington, A. D. (Eds.), *Handbook of Physiology.* New York: Oxford, 2001, 609.

Scriver, C. R., Beaudet, A. L., Sly, W. S., and Valle, D. (Eds.). *The Metabolic and Molecular Bases of Inherited Disease,* 8th ed. New York: McGraw Hill, 2001.

Sitikov, A., Tang, P. M., Verin, A., and Zolnierowicz, S. Serine/threonine protein phosphatases in the control of cell function. *Adv. Enz. Reg.* 34: 199, 1994.

Smith, C., Marks, A. D., and Lieberman, M. *Basic Medical Biochemistry. A Clinical Approach.* Baltimore, Lippincott, Williams & Wilkins, 2005.

Steinmann, B., Gitzelmann, R., and Van den Berghe, G. Disorders of fructose metabolism. In Scriver ,C. R., Beaudet, A.R., Sly, W. S., and Valle, D. (Eds.), *The Metabolic and Molecular Bases of Inherited Disease,* 8th ed. New York: McGraw Hill, 2001, 1489.

Taylor, S. I. Insulin action, insulin resistance, and type 2 diabetes mellitus. In Scriver ,C. R., Beaudet, A.R., Sly, W. S., and Valle, D. (Eds.), *The Metabolic and Molecular Bases of Inherited Disease,* 8th ed. New York: McGraw Hill, 2001, 1433.

Van Schaftingen, E., and Geren, I. The glucose-6-phosphatase system. *Biochem. J.* 362: 513, 2002.

Van Schaftingen, E., Vandercammen, A., Detheux, M., and Davies, D. R. The regulatory protein of liver glucokinase. *Adv. Enz. Reg.* 32: 133, 1992.

Wallace, J. C., and Baritte, G. J. Gluconeogenesis. *Nature Encyclopedia of Life Sciences.* www.els.net. 2005.

Key Terms

anaerobic
glycolysis
ATP
lactate
Pasteur effect
insulin
acetyl CoA
GLUT 4
fructose 1,6-bisphosphate
oxido-reduction reactions
bifunctional enzyme
allosteric effectors
pyruvate kinase
substrate shuttles
ethanol
UDP-glucose
6-phosphofructo-1-kinase
glucokinase
hexokinase
fructose
glucose 6-phosphatase
diabetes
glycogen
induction and repression of enzymes
fructose 2,6-bisphosphate
fructose 2,6-bisphosphatase
citrate
6-phosphofructo-2-kinase
cyclic AMP
glucagon
covalent modification
protein kinase A
epinephrine
gluconeogenesis
Cori cycle
glycogenolysis
ketogenic amino acids
pyruvate carboxylase
phosphoenolpyruvate carboxykinase
glucogenic amino acids
fatty acid oxidation
glycogenesis
glycogen branching enzyme
glycogen phosphorylase
glycogen debranching enzyme
glycogen storage diseases
glycogen synthase
glycogenin
phosphorylase kinase
phosphorylation-dephosphorylation

Questions • CAROL N. ANGSTADT

Multiple-Choice Questions

1. Glucokinase
   - **A.** has a $S_{0.5}$ greater than the normal blood glucose concentration.
   - **B.** is found in muscle.
   - **C.** is inhibited by glucose 6-phosphate.
   - **D.** is also known as the GLUT-2 protein.
   - **E.** has glucose 6-phosphatase activity as well as kinase activity.

2. 6-Phosphofructo-1-kinase activity can be decreased by all of the following except
   - **A.** ATP at high concentrations.
   - **B.** citrate.
   - **C.** AMP.
   - **D.** low pH.
   - **E.** decreased concentration of fructose 2,6-bisphosphate.

3. In the Cori cycle,
   - **A.** only tissues with aerobic metabolism (i.e., mitochondria and O_2) are involved.
   - **B.** a three-carbon compound arising from glycolysis is converted to glucose at the expense of energy from fatty acid oxidation.
   - **C.** glucose is converted to pyruvate in anaerobic tissues, and this pyruvate returns to the liver, where it is converted to glucose.
   - **D.** the same amount of ATP is used in the liver to synthesize glucose as is released during glycolysis, leading to no net effect on whole body energy balance.
   - **E.** nitrogen from alanine must be converted to urea, increasing the amount of energy required to drive the process.

4. When blood glucagon rises, which of the following hepatic enzyme activities falls?
   - **A.** Adenyl cyclase
   - **B.** Protein kinase
   - **C.** 6-Phosphofructo-2-kinase
   - **D.** Fructose 1,6-bisphosphatase
   - **E.** Hexokinase

5. Glucose 6-phosphatase, which is deficient in Von Gierke disease, is necessary for the production of blood glucose from
   A. liver glycogen.
   B. fructose.
   C. amino acid carbon chains.
   D. lactose.
   E. all of the above.

6. UDP-glucose
   A. is formed by the reaction between UTP and glucose.
   B. is an intermediate in both the synthesis of glycogen and its degradation.
   C. formation is irreversible because it generates pyrophosphate.
   D. adds glucose units to the preexisting glycogen in α-1,6 linkages.
   E. is the substrate for branching enzyme.

Questions 7 and 8: Alcohol intoxication can lead to hypoglycemia especially if the alcohol is consumed by an undernourished individual or after strenuous exercise. In both cases the hypoglycemia results from the inhibitory effects of alcohol on hepatic gluconeogenesis and occurs under circumstances of hepatic glycogen depletion. Alcohol potentiates the hypoglycemic effect of insulin so diabetics who have self-administered insulin and then consumed alcohol are at increased risk.

7. Alcohol metabolism produces large amounts of NADH which inhibit gluconeogenesis by
   A. shifting the pyruvate–lactate equilibrium toward lactate.
   B. favoring the production of oxaloacetate from malate.
   C. preventing the movement of phosphoenolpyruvate from the mitochondria to the cytosol.
   D. inhibiting the electron transport chain.
   E. inhibiting the malate–aspartate shuttle.

8. Insulin promotes hypoglycemia by a variety of mechanisms including all of the following except
   A. inactivating a transcription factor for genes of key gluconeogenic enzymes.
   B. inhibiting lipolysis in adipose tissue, thus decreasing the energy supply for gluconeogenesis in the liver.
   C. decreasing levels of cAMP.
   D. increasing activation of phosphoprotein phosphatase.
   E. increasing the activity of cAMP-response element binding protein (CREB) and its binding to the cAMP-response element (CRE).

Questions 9 and 10: Malignant hyperthermia is a genetic abnormality in which exposure to certain agents, especially the widely used general anesthetic halothane, produces a dramatic rise in body temperature, acidosis, hyperkalemia, and muscle rigidity. Death is rapid if the condition is untreated and may occur the first time a susceptible person is anaesthetized. The defect causes an inappropriate release of Ca^{2+} from the sarcoplasmic reticulum of muscle. Many heat-producing processes are stimulated in an uncontrolled fashion by the release of Ca^{2+}, including glycolysis and glycogenolysis.

9. Ca^{2+} increases glycogenolysis by
   A. activating phosphorylase kinase b, even in the absence of cAMP.
   B. binding to phosphorylase b.
   C. activating phosphoprotein phosphatase.
   D. inhibiting phosphoprotein phosphatase.
   E. protecting cAMP from degradation.

10. Phosphorylation–dephosphorylation and allosteric activation of enzymes play roles in stimulating glycogen degradation. All of the following result in enzyme activation except
   A. phosphorylation of phosphorylase kinase.
   B. binding of AMP to phosphorylase b.
   C. phosphorylation of phosphorylase.
   D. phosphorylation of protein kinase A.
   E. dephosphorylation of glycogen synthase.

Questions 11 and 12: Patients with hereditary fructose intolerance are deficient in the liver form of the enzyme aldolase. Consumption of fructose leads to a depletion of ATP and P_i in the liver, which, in turn, leads to cell damage. Much of the cell damage can be attributed to the inability to maintain normal ion gradients by ATP-dependent pumps.

11. The first step in liver's metabolism of fructose is
   A. isomerization to glucose.
   B. phosphorylation to fructose 1,6-bisphosphate by ATP.
   C. phosphorylation to fructose 1-phosphate by ATP.
   D. phosphorylation to fructose 6-phosphate by ATP.
   E. cleavage by aldolase.

12. The products initially produced by aldolase action on the substrate formed from fructose are
   A. two molecules of dihydroxyacetone phosphate.
   B. two molecules of glyceraldehyde 3-phosphate.
   C. two molecules of lactate.
   D. dihydroxyacetone phosphate and glyceraldehyde 3-phosphate.
   E. dihydroxyacetone phosphate and glyceraldehyde.

Problems

13. If a cell is forced to metabolize glucose anaerobically, how much faster would glycolysis have to proceed to generate the same amount of ATP as it would get if it metabolized glucose aerobically?

14. The alanine cycle requires more ATP per glucose molecule formed than does the Cori cycle. Why is this?

Answers

1. **A** Blood glucose is ∼ 5 mM. $S_{0.5}$ of glucokinase is ∼ 7 mM. B: Glucokinase is hepatic, and unlike the muscle hexokinase, it is not inhibited by glucose 6-phosphate.
2. **C** AMP is an allosteric regulator that relieves inhibition by ATP. B and D are probably important physiological regulators in muscle, and E is critical in liver.
3. **B** The liver derives the energy required for gluconeogenesis from aerobic oxidation of fatty acids. A: The liver is an essential organ in the Cori cycle; it is aerobic. C: In anaerobic tissues the end product of glycolysis is lactate; in aerobic tissues it is pyruvate, but there the pyruvate would likely be oxidized aerobically. D: Gluconeogenesis requires six ATP per glucose synthesized; glycolysis yields two ATP per glucose metabolized. E: Alanine is not part of the Cori cycle.
4. **C** As blood glucagon rises, A is activated, producing cAMP; cAMP activates B, and B inactivates C. Low levels of fructose 2,6-bisphosphate increase the activity of D. E is not an important hepatic enzyme; its role is filled in liver by glucokinase.
5. **E** To get into the blood, glucose must be free, not phosphorylated. A: Glycogen is degraded to glucose 1-phosphate, which is converted to glucose 6-phosphate. B: Fructose is metabolized to dihydroxyacetone phosphate, which can either continue through glycolysis or reverse to glucose 6-phosphate, depending on the state of the cell. C and D: Amino acid carbon chains and lactate are substrates for gluconeogenesis.
6. **C** The hydrolysis of the pyrophosphate to inorganic phosphate makes the reaction irreversible. A: Glucose 1-phosphate reacts with UTP. B: It is an intermediate in synthesis but not degradation. D: It is added only in an α-1,4 linkage. E: Branching enzyme removes a chain of glucosyl units from the end of a long branch and moves it to another place.
7. **A** Lactate must be converted to pyruvate to start gluconeogenesis. B: NADH shifts oxaloacetate to malate. C and D: The NADH is not in the mitochondria. E: NADH reducing equivalents can be moved into the mitochondria by the shuttle but the large amounts generated by alcohol metabolism overwhelm the capacity of the shuttle and mitochondrial oxidation.
8. **E** CREB is activated by phosphorylation (and would lead to increased gluconeogenic enzymes) but insulin promotes dephosphorylation. A: This is true. B: Gluconeogenesis is dependent on a supply of fatty acids to liver to provide energy. C: This occurs by activating cAMP phosphodiesterase. D: This promotes dephosphorylation of key enzymes, for example, increasing glycogen synthesis and decreasing glycogenolysis.
9. **A** The γ-subunit of phosphorylase kinase is a calmodulin-type protein. Both a and b forms of the enzyme are activated by Ca^{2+}. B and E: These do not happen. C and D: Ca^{2+} does not affect the phosphatase.
10. **D** Protein kinase A catalyzes phosphorylations but is activated by binding cAMP. A and C: Both of these enzymes are phosphorylated in their a forms. B: Phosphorylase B is allosterically activated by binding AMP. E: The a form of the synthase is the nonphosphorylated form.
11. **C** The ADP formed is converted to ATP at the expense of P_i. Inability to further metabolize fructose 1-phosphate results in depletion of P_i. A: This does not happen. B: This would require two phosphorylations. D: This does not happen in liver. E: The fructose must be phosphorylated first.
12. **E** Glyceraldehyde can be converted to glyceraldehyde 3-phosphate so both products feed into the glycolytic pathway or gluconeogenesis.
13. Anaerobically there is a net of two moles of ATP/mol glucose. Aerobically the same net of 2 ATP is obtained plus 2 NADH because pyruvate is the product. Let us assume the cell uses the malate–aspartate shuttle where each NADH yields 2.5 ATP. Therefore, there is a net of 7 moles ATP/mol glucose. Each pyruvate is converted to AcCoA and the AcCoA is oxidized by the tricarboxylic acid cycle. Each mol of pyruvate then yields 12.5 moles ATP, or 25 moles ATP for the two pyruvates. This gives a total of 32 moles ATP/mol glucose aerobically. (See Chapter 14.) Therefore, glycolysis must proceed 16 times as rapidly under anaerobic conditions to generate the same amount of ATP as occurs aerobically.
14. Both cycles require the same amount of ATP to convert pyruvate to glucose (lactate to pyruvate and alanine to pyruvate). In the alanine cycle, the NADH produced is not used to reduce pyruvate to lactate and could be oxidized by the electron transport system to produce energy. But the alanine cycle presents the liver with amino nitrogen that must be disposed of as urea. This requires four ATP per urea molecule produced, and this ATP is not required in the Cori cycle.

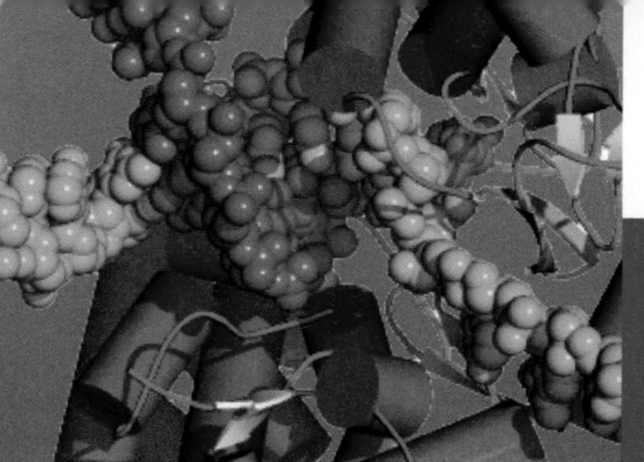

16

Carbohydrate Metabolism II: Special Pathways and Glycoconjugates

Repeat unit of chondroitin 4-sulfate

Nancy B. Schwartz

Professor, University of Chicago

Key Concepts

- Glucose phosphorylation mediated by hexokinases yields glucose-6-phosphate, which has a pivotal role as a common precursor for several metabolic pathways utilizing glucose.
- Glucose-6-phosphate metabolism through the pentose phosphate pathway conserves redox equivalents of glucose-6-phosphate as NADPH, whereas glycolysis promotes energy production.
- The pentose phosphate pathway supplies ribose-5-phosphate for nucleic acid synthesis.
- The pentose phosphate pathway degrades the sugar molecule one carbon at a time in two distinct phases.
- The first and rate-limiting step in the pentose phosphate pathway is catalyzed by glucose-6-phosphate dehydrogenase, which oxidizes glucose-6-phosphate to 6-phosphogluconate.
- Most sugar components of biomolecules are derived from glucose through a variety of chemical transformations and interconversions. Nucleotide-linked sugars are key to many sugar transformations as well as synthesis of complex polysaccharides.

- Oligo- or polysaccharides are linked to proteins via a limited number of *N*- or *O*-glycosyl bonds in glycoproteins and proteoglycans.
- *N*-glycosylation entails a dolichol-linked assembly pathway and a multi-compartment cellular processing pathway.
- Glycan structures modulate numerous molecular interactions like cell signaling, adhesion, and receptor activation.
- Genetic disorders of glycosylation cause a wide range of phenotypes, with examples from all clinical specialties. Many genetic diseases of complex carbohydrate metabolism result from deficiencies of glycosidases.

16.1 • PENTOSE PHOSPHATE PATHWAY

Pentose Phosphate Pathway Has Two Phases

The pentose phosphate pathway provides a means for degrading the carbon chain of a sugar molecule one carbon at a time. However, this pathway does not constitute a consecutive set of reactions that lead directly to CO_2, but occurs in two phases. In the first, hexose is decarboxylated to pentose via two oxidation reactions that form NADPH; in the second, by a series of transformations, six molecules of pentose undergo rearrangements to yield five molecules of hexose.

Glucose 6-Phosphate Oxidization Conserves Redox Equivalents as NADPH and Decarboxylation Supplies Pentose Phosphates

The first reaction, catalyzed by glucose 6-phosphate (G6P) dehydrogenase (Figure 16.1) is **dehydrogenation** of G6P to form **6-phosphoglucono-δ-lactone** and **NADPH**, and is a major regulatory site for this pathway. Special interest in this enzyme stems from the severe anemia that may result from absence of **G6P dehydrogenase** in erythrocytes or from the presence of one of many genetic variants of the enzyme (Clin. Corr.16.1). The lactone product of this reaction is a substrate for gluconolactonase, which ensures that the reaction goes to completion. The overall equilibrium of both reactions lies far in the direction of NADPH, maintaining a high NADPH/NADP$^+$ ratio within cells. A second dehydrogenation and **decarboxylation**, catalyzed by **6-phosphogluconate dehydrogenase**, produces the pentose phosphate, **ribulose 5-phosphate** and a second molecule of NADPH. Ribulose 5-phosphate is then **isomerized** to ribose 5-phosphate through an enediol intermediate.

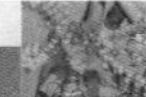

CLINICAL CORRELATION 16.1

Glucose 6-Phosphate Dehydrogenase Deficiency

Glucose-6-phosphate dehydrogenase (G6PD) deficiency is the most common human enzyme defect and may be present in as many as 400 million people worldwide. About 140 mutations have been described in this protein of 516 amino acids, accounting for a wide range of symptoms. The most frequent clinical manifestations of G6PD deficiency are neonatal jaundice and acute hemolytic anemia, which is usually triggered by an exogenous agent. When certain seemingly harmless drugs, such as antimalarials, antipyretics, or sulfa antibiotics, are administered to susceptible patients, an acute hemolytic anemia may result in 48–96 h. Susceptibility to drug-induced hemolytic disease is most often due to a deficiency of glucose 6-phosphate (G6P) dehydrogenase activity in erythrocytes, and was an early indication that X-linked genetic deficiencies of this enzyme exist. The enzyme is particularly important, since the pentose phosphate pathway is the major pathway of NADPH production in the red cell. Red blood cells with the relatively mild A-type G6P dehydrogenase deficiency can oxidize glucose at a normal rate when the demand for NADPH is normal. However, if the rate of NADPH utilization is increased, the cells cannot increase the activity of the pathway adequately. In addition, cells do not reduce enough NADP$^+$ to maintain glutathione in its reduced state, and hence to protect against lipid peroxidation. Reduced glutathione is necessary for the integrity of the erythrocyte membrane, thus rendering enzyme-deficient red cells more susceptible to hemolysis by a wide range of compounds. This deficiency illustrates the interplay of heredity and environment in the production of disease. The most effective management of G6PD deficiency is to prevent hemolysis by avoiding oxidative stress. Complete deficiency of G6PD is lethal.

Luzzatto, L., Mehta, A., and Vulliamy, T. Glucose 6-phosphate dehydrogenase deficiency. In Scriver ,C. R., Beaudet, A.R., Sly, W. S., and Valle, D. (Eds.). *The Metabolic and Molecular Bases of Inherited Disease,* 8th ed. New York: McGraw Hill, 2001, III: 4517; and Cappellini, M. D., and Fiorelli, G. *Lancet* 371:64, 2008.

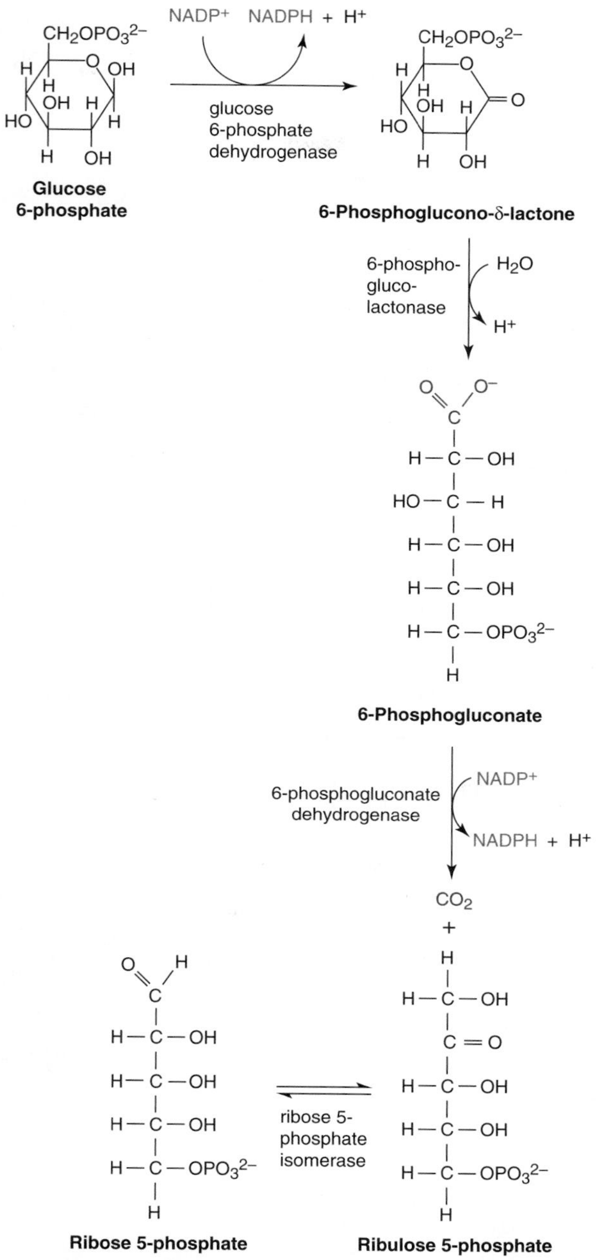

$NADP^+$ $NADPH + H^+$

glucose 6-phosphate dehydrogenase

Glucose 6-phosphate

6-Phosphoglucono-δ-lactone

6-phosphogluconolactonase

H_2O

H^+

6-Phosphogluconate

6-phosphogluconate dehydrogenase

$NADP^+$

$NADPH + H^+$

CO_2 +

ribose 5-phosphate isomerase

Ribose 5-phosphate

Ribulose 5-phosphate

Figure 16.1 **Oxidative phase of the pentose phosphate pathway: Formation of pentose phosphate and NADPH.**

Under certain metabolic conditions, the pentose phosphate pathway can end at this point, with utilization of NADPH for reductive biosynthetic reactions and ribose 5-phosphate as a precursor for nucleotide synthesis. The overall equation may be written as

$$\text{Glucose 6-phosphate} + 2\text{NADP}^+ + \text{H}_2\text{O} \rightleftharpoons \text{ribose 5-phosphate} + 2\text{NADPH} + 2\text{H}^+ + \text{CO}_2$$

Interconversions of Pentose Phosphates Lead to Intermediates of Glycolysis

If more NADPH is needed for **reductive biosynthesis** than ribose 5-phosphate for incorporation into nucleotides, a sugar **interconversion** system (Figure 16.2) forms triose, tetrose, hexose, and heptose sugars from the pentoses and provides a reversible link between the

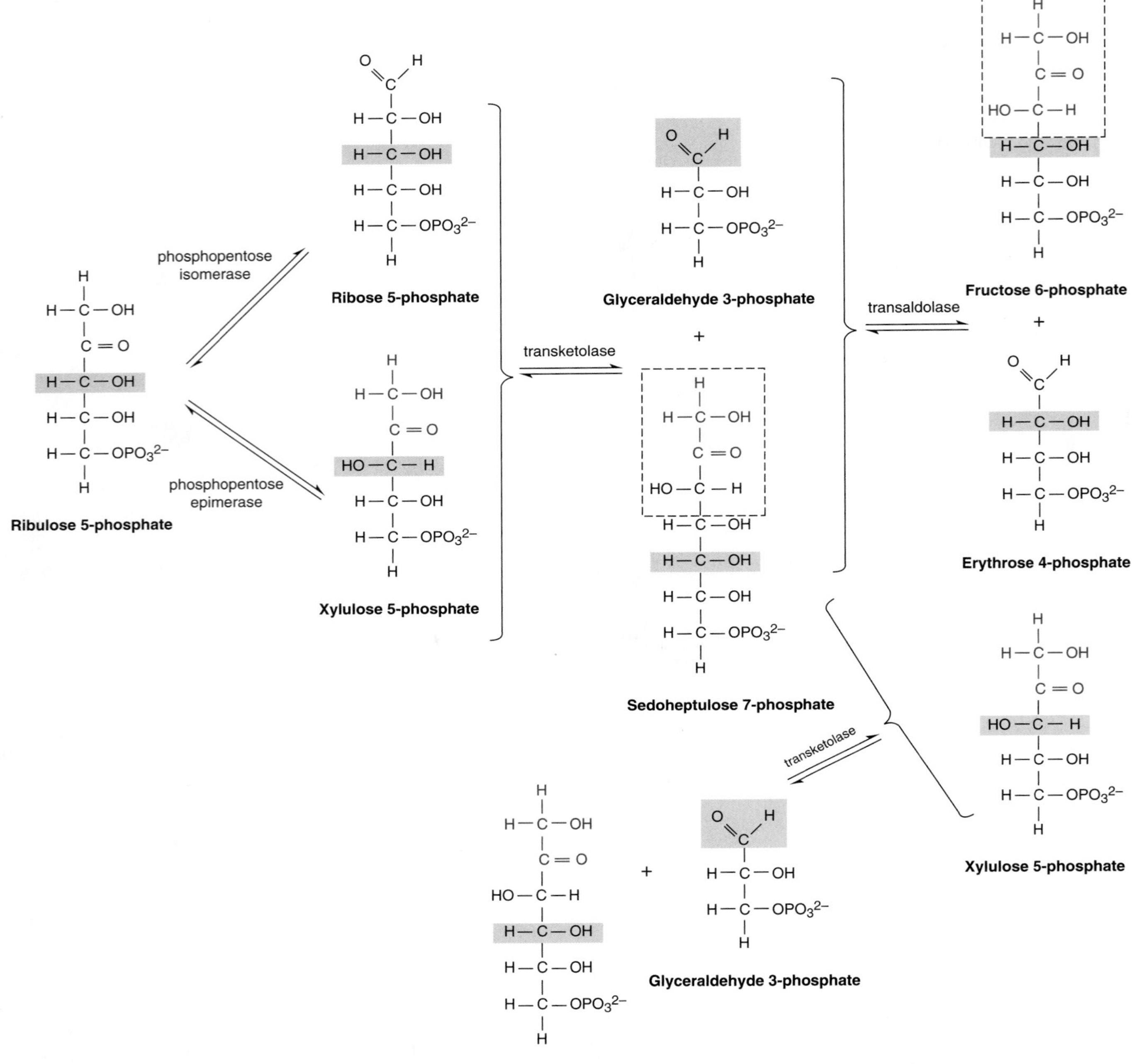

Figure 16.2 Nonoxidative reactions of the pentose phosphate pathway: Interconversions of pentose phosphates.

pentose phosphate pathway and glycolysis via common intermediates. **Xylulose 5-phosphate** is formed through isomerization of ribulose 5-phosphate by **phosphopentose epimerase**; ribulose 5-phosphate, ribose 5-phosphate, and xylulose 5-phosphate thus exist as an equilibrium mixture and can undergo transformations catalyzed by transketolase and transaldolase.

Transketolase requires **thiamin pyrophosphate** (TPP) and divalent cations, transfers a C_2 unit of *active glycolaldehyde* from xylulose 5-phosphate to ribose 5-phosphate, and produces **sedoheptulose** and **glyceraldehyde 3-phosphate**, an intermediate of glycolysis. Alterations in transketolase can lead to Wernicke–Korsakoff syndrome (Clin. Corr.16.2). **Transaldolase** transfers a C_3 unit (dihydroxyacetone) from sedoheptulose 7-phosphate to glyceraldehyde 3-phosphate forming **erythrose 4-phosphate**, and **fructose 6-phosphate**, another intermediate of glycolysis. Transketolase produces fructose 6-phosphate and glyceraldehyde 3-phosphate, from erythrose 4-phosphate and xylulose 5-phosphate. The sum of these reactions is

$$\text{2 Xylulose 5-phosphate} + \text{ribose 5-phosphate} \rightleftharpoons \text{2 fructose 6-phosphate} + \text{glyceraldehyde 3-phosphate}$$

Since xylulose 5-phosphate is derived from ribose 5-phosphate, the net reaction starting from ribose 5-phosphate is

$$\text{3 Ribose 5-phosphate} \rightleftharpoons \text{2 fructose 6-phosphate} + \text{glyceraldehyde 3-phosphate}$$

Thus, excess ribose 5-phosphate, produced in the first stage of the pentose phosphate pathway or from degradation of nucleic acids, is effectively converted to intermediates of glycolysis.

Glucose 6-Phosphate Can Be Completely Oxidized to CO_2

Complete oxidation of glucose 6-phosphate (G6P) to CO_2, with reduction of $NADP^+$ to NADPH, may also occur (Figure 16.3). G6P continually enters this pathway, and CO_2 and NADPH are produced in the first phase. A balanced equation includes the oxidation of six molecules of G6P to six of ribulose 5-phosphate and six of CO_2. This results in transfer of 12 pairs of electrons to $NADP^+$, the requisite amount for total oxidation of one glucose to six CO_2. Six molecules of ribulose 5-phosphate are then rearranged to generate five molecules of G6P. The overall equation then becomes

$$\text{6 Glucose 6-phosphate} + 12NADP^+ + 7H_2O \rightleftharpoons \text{5 glucose 6-phosphate} + 6CO_2 + 12NADPH + 12H^+ + P_i$$

And the net reaction is

$$\text{Glucose 6-phosphate} + 12NADP^+ + 7H_2O \rightleftharpoons 6CO_2 + 12NADPH + 12H^+ + P_i$$

Pentose Phosphate Pathway Serves as a NADPH Regenerating System and Supplier of Pentose Phosphates

The pentose phosphate pathway serves several purposes, including synthesis and degradation of sugars other than hexoses, particularly ribose 5-phosphate for synthesis of nucleotides, and

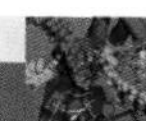

CLINICAL CORRELATION 16.2

Wernicke–Korsakoff Syndrome (OMIM 277730): Associated Anomalies in Transketolase Activity

Symptoms of Wernicke–Korsakoff syndrome become apparent after moderate stress that does not affect normal individuals, and an anomaly in transketolase has been noted. Cloning and sequencing of the transketolase gene appears to exclude a genetic defect. Rather, the dysfunction of transketolase may be related to a thiamin deficiency, since transketolase utilizes thiamin pyrophosphate as a cofactor. The syndrome presents as a mental disorder, with memory loss and partial paralysis, and can become manifest in alcoholics, whose diets may be vitamin deficient. The medical importance of the pentose phosphate pathway is also highlighted by deficiencies in transaldolase, which are linked to a spectrum of clinical diseases, including liver cirrhosis and male infertility.

Perl, A. The pathology of transaldolase deficiency. *Life* 59:365, 2007

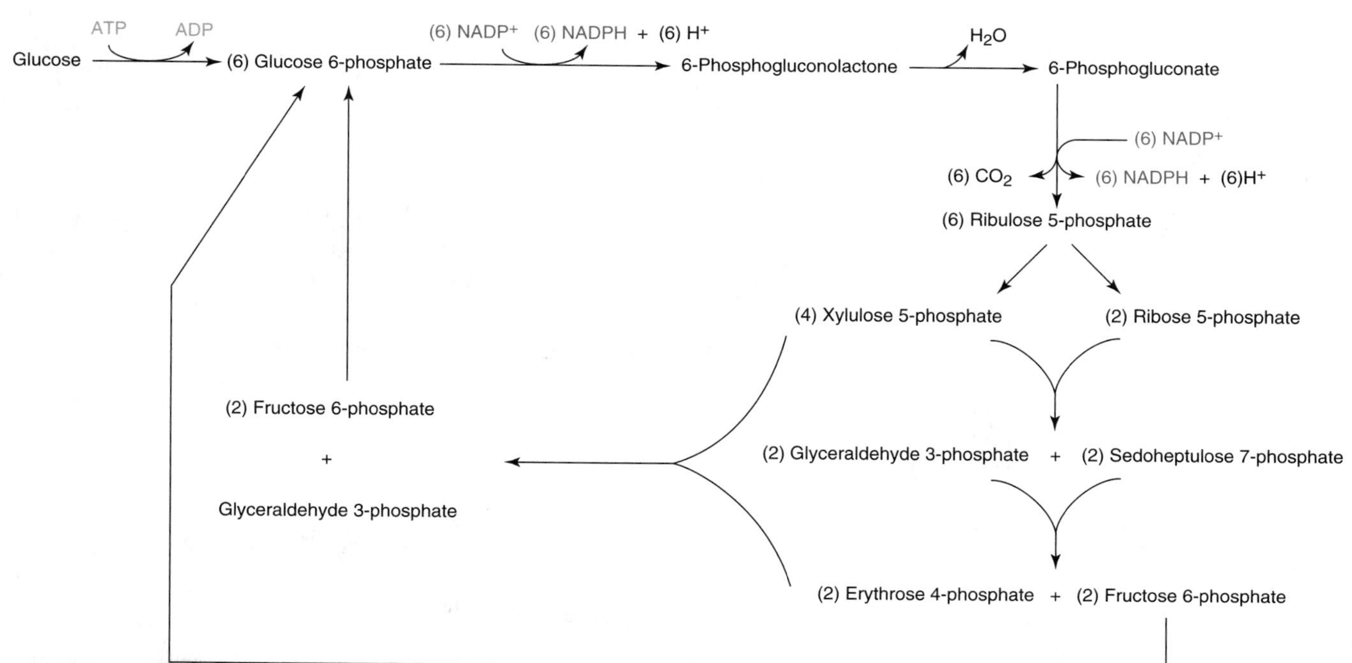

Figure 16.3 Pentose phosphate pathway.

synthesis of NADPH. The processing flow of G6P after entry into the pathway is determined largely by the needs of the cell for NADPH or sugar intermediates. When more NADPH than ribose 5-phosphate is required, the path leading to complete oxidation of G6P to CO_2 and resynthesis of G6P from ribulose 5-phosphate is favored. Alternatively, if NADPH demand is relatively low, conversion of G6P to ribulose 5-phosphate for nucleic acid synthesis or recycling to produce intermediates of the glycolytic pathway predominates.

The tissue distribution of the pentose phosphate pathway is consistent with its functions. NADPH, required to maintain reduced glutathione which protects the integrity of red blood cell membranes, is produced in erythrocytes, as well as in liver, mammary gland, testis, and adrenal cortex, which are sites of fatty acid or steroid synthesis. The balance between glucose entry into glycolysis or the pentose phosphate pathway depends on the metabolic requirements of the organ. 20%–30% of the CO_2 produced in the liver may arise from the pentose phosphate pathway. In mammalian striated muscle, which carries on little fatty acid or steroid synthesis, all catabolism of G6P proceeds via glycolysis and the TCA cycle with no direct oxidation of glucose 6-phosphate through the pentose phosphate pathway.

16.2 • SUGAR INTERCONVERSIONS AND NUCLEOTIDE-LINKED SUGAR FORMATION

Most monosaccharides found in biological compounds derive from glucose. The most common sugar **transformations** in mammalian systems are summarized in Figure 16.4.

Isomerization and Phosphorylation Are Common Reactions for Interconverting Carbohydrates

Formation of some sugars can occur directly, starting from glucose via modification reactions such as the **aldose-ketose isomerization** catalyzed by **phosphomannose isomerase**, which produces **mannose 6-phosphate**. Deficiency in this enzyme leads to one form of **c**ongenital **d**isorders of **g**lycosylation **s**yndrome (**CDGS**) (p. 662).

Phosphorylation and internal transfer of a phosphate group on the same sugar molecule are also common modifications. Glucose 1-phosphate, resulting from glycogenolysis, is converted to G6P by **phosphoglucomutase**. Galactose is phosphorylated to galactose 1-phosphate

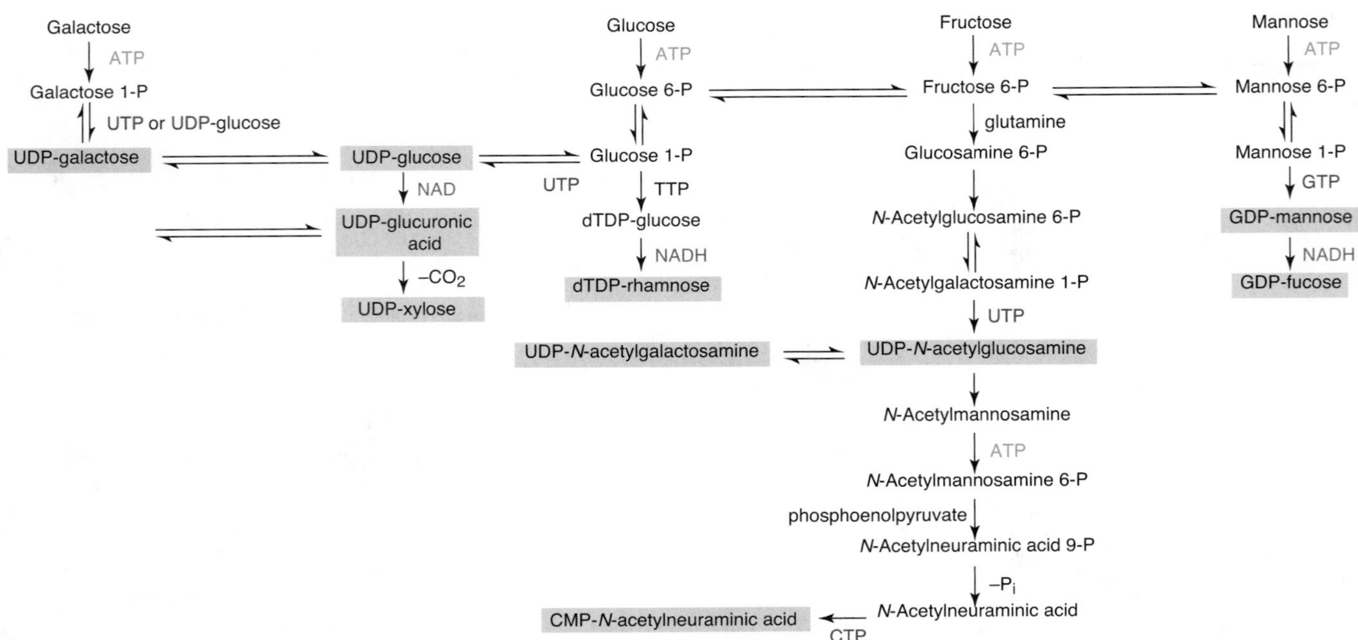

Figure 16.4 Pathways of formation of nucleotide-linked sugars and interconversion of some hexoses.

CLINICAL CORRELATION 16.3

Essential Fructosuria (OMIM 229800) and Fructose Intolerance (OMIM 229600): Deficiency of Fructokinase and Fructose 1-Phosphate Aldolase

Fructose may account for 30%–60% of the total carbohydrate intake of mammals and is predominantly metabolized by a fructose-specific pathway. Fructokinase is deficient in essential fructosuria. This disorder is a benign asymptomatic metabolic anomaly which appears to be inherited as an autosomal recessive trait. Following intake of fructose, blood and urinary fructose levels of affected individuals are unusually high; however, 90% of their fructose intake is eventually metabolized. In contrast, hereditary fructose intolerance is characterized by severe hypoglycemia, jaundice, hemorrhage, hepatomegaly, uricemia, and eventual kidney failure. Ingestion of fructose or prolonged ingestion by affected young children may lead to death. Fructose 1-phosphate aldolase may also be deficient, in which case fructose 1-phosphate accumulates intracellularly (see Clin. Corr. 15.3, p. 607).

Steinmann, B., Gitzelmann, R., and Vanden Berghe, G. Disorders of fructose metabolism. In Scriver ,C. R., Beaudet, A.R., Sly, W. S., and Valle, D. (Eds.). *The Metabolic and Molecular Bases of Inherited Disease,* 8th ed. New York: McGraw Hill, 2001, I:1489.

by **galactokinase** and mannose to mannose 6-phosphate by **mannokinase**. Free fructose, an important dietary constituent, is phosphorylated in the liver to fructose 1-phosphate by a special fructokinase. However, no mutase interconverts fructose 1-phosphate and fructose 6-phosphate, and phosphofructokinase cannot synthesize fructose 1,6-bisphosphate from fructose 1-phosphate. Rather, **fructose 1-phosphate aldolase** cleaves fructose 1-phosphate to **dihydroxyacetone phosphate** (DHAP), which enters the glycolytic pathway directly, and glyceraldehyde, which is first reduced to glycerol, phosphorylated, and then reoxidized to DHAP (Figure 15.39, p. 625). Lack of this aldolase leads to fructose intolerance (Clin. Corr. 16.3).

Nucleotide-Linked Sugars Are Intermediates in Many Sugar Transformations

Many sugar transformation reactions require conversion into **nucleotide-linked sugars**. A **pyrophosphorylase** joins hexose 1-phosphate and nucleoside triphosphate (NTP) to produce a nucleoside diphosphate (NDP)-sugar and pyrophosphate. **Pyrophosphatase** rapidly hydrolyzes the pyrophosphate, thereby driving the synthesis reaction. These reactions are summarized as follows:

$$\text{NTP} + \text{sugar 1-phosphate} + H_2O \rightleftharpoons \text{NDP-sugar} + PP_i$$
$$PP_i + H_2O \rightleftharpoons 2P_i$$

The net reaction is

$$\text{NTP} + \text{sugar 1-phosphate} + H_2O \rightleftharpoons \text{NDP-sugar} + 2P_i$$

For example, UDP-glucose is used in synthesis of glycogen and glycoproteins and is synthesized by **UDP-glucose pyrophosphorylase**.

HOCH$_2$ … H, H, OH, H, HO, H, OH … O—P(=O)(O$^-$)—O$^-$ + $^-$O—P(=O)(O$^-$)—O—P(=O)(O$^-$)—O—P(=O)(O$^-$)—O—Uridine ⇌

Glucose

HOCH$_2$ … H, H, OH, H, HO, H, OH … O—P(=O)(O$^-$)—O—P(=O)(O$^-$)—O—Uridine + PP$_i$

UDP-glucose

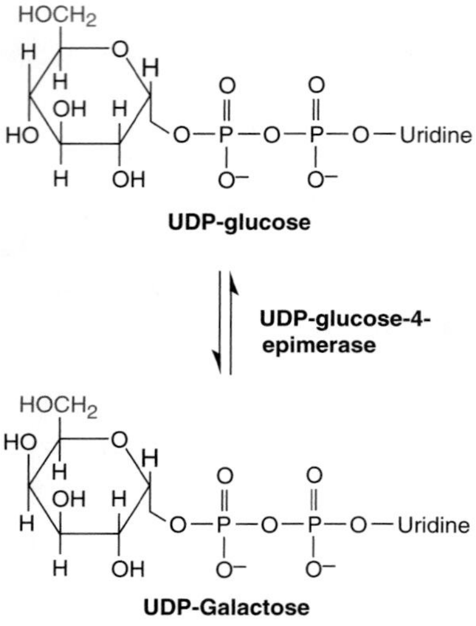

UDP-glucose

UDP-glucose-4-epimerase

UDP-Galactose

Figure 16.5 Conversion of glucose into galactose.

Nucleoside diphosphate-sugars are important constructs: They contain two phosphoryl bonds, each with a large negative ΔG of hydrolysis that underlies their value as glycosyl donors in further transformation and transfer reactions, and they confer substrate specificity in those reactions. UDP is usually the glucosyl carrier, whereas ADP, GDP, and CMP are carriers for other sugars. Many sugar transformation reactions occur only at the level of nucleotide-linked sugars (Figure 16.4, p. 652).

Epimerization Interconverts Nucleotide-Linked Glucose and Galactose

Interconversion of glucose and galactose in animal cells occurs by epimerization of UDP-glucose to UDP-galactose, catalyzed by **UDP-glucose-4-epimerase** (Figure. 16.5). UDP-galactose is also formed from free galactose, derived from hydrolysis of lactose in the intestinal tract. Galactose is phosphorylated by **galactokinase** and ATP to yield galactose 1-phosphate. Then **galactose 1-phosphate uridylyltransferase** forms UDP-galactose by galactose 1-phosphate displacing glucose 1-phosphate from UDP-glucose. These reactions are summarized as follows:

$$\text{Galactose} + \text{ATP} \rightleftharpoons \text{galactose 1-phosphate} + \text{ADP}$$

$$\text{UDP-glucose} + \text{galactose 1-phosphate} \rightleftharpoons \text{UDP-galactose} + \text{glucose 1-phosphate}$$

Dietary galactose can be transformed by a combination of these reactions into glucose 1-phosphate and metabolized as previously described, or the 4-epimerase can produce UDP-galactose needed for biosynthesis. A severe form of hereditary galactosemia results from absence of the uridylyltransferase (Clin. Corr. 16.4).

Synthesis of GDP-fucose (see Figure. 16.4) begins with the conversion of GDP-mannose to GDP-4-keto-6-deoxymannose by **GDP-mannose-4,6-dehydratase**, followed by epimerization to GDP-4-keto-6-deoxy-L-galactose, and finally reduction to GDP-fucose. These latter reactions are catalyzed by the bifunctional **GDP-4-keto-6-deoxymannose 3,5-epimerase-4-reductase** (FX protein), which is abundant in red blood cells. Epimerization of D-glucuronic acid to L-iduronic acid occurs after the former is incorporated into heparin or dermatan sulfate (p. 665).

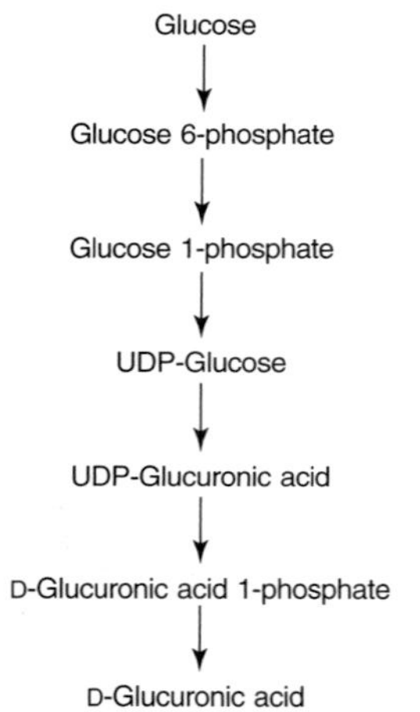

Figure 16.6 Biosynthesis of D-glucuronic acid from glucose.

Glucuronic Acid Is Formed by Oxidation of UDP-Glucose

The synthesis of glucuronic acid from glucose is summarized in Figure 16.6. A critical step is the oxidation of UDP-glucose by **UDP-glucose dehydrogenase** (Figure 16.7). Glucuronic acid is reduced by NADPH to L-gulonic acid (Figure 16.8). Gulonic acid can be oxidized to 3-ketogulonic acid and decarboxylated to L-xylulose. In humans, L-xylulose is the ketopentose excreted in essential pentosuria (Clin. Corr. 16.5). L-Xylulose is normally reduced to xylitol, reoxidized to D-xylulose and phosphorylated to xylulose 5-phosphate,

CLINICAL CORRELATION 16.4

Galactosemia: Inability to Transform Galactose into Glucose

Reactions of galactose are of particular interest because in humans they are subject to genetic defects that produce the hereditary disorder galactosemia. When a defect is present, individuals are unable to metabolize the galactose derived from lactose (milk sugar) to glucose metabolites, often resulting in cataract formation, growth failure, mental retardation, or eventual death from liver damage. These phenotypes may be due to a cellular deficiency of galactokinase (Type 2, GALK1 gene) (OMIM 230200), causing a relatively mild disorder characterized by early cataract formation; galactose 1-phosphate uridylyltransferase (Type 1, GALT gene) (OMIM 606999), resulting in severe disease; or UDP galactose-4-epimerase deficiency (Type 3, GALE gene) (OMIM 230350). Galactose is reduced to galactitol in a reaction similar to that of glucose to sorbitol. Galactitol initiates cataract formation in the lens and may play a role in central nervous system damage. Accumulation of galactose-1-phosphate is responsible for liver failure; the toxic effects of galactose metabolites disappear when galactose is removed from the diet.

Petry, K. G., and Reichardt, J. K. The fundamental importance of human galactose. *Trends Genet.* 14:98, 1998; Scriver, C. R., Beaudet, A. L., Valle, D., Sly, W. S., Vofelstrin, B., and Kinzler, K.W. Galactosemia Online Metabolic and Molecular Basis of Inherited Disease (http://www.ommbid.com/). New York: McGraw Hill, 2006.

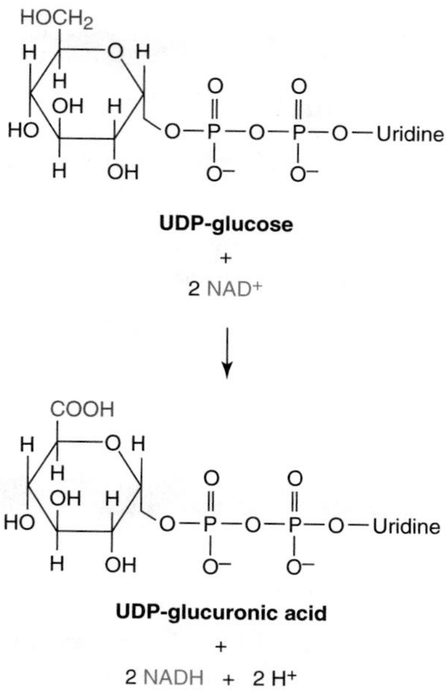

Figure 16.7 Formation of UDP-glucuronic acid from UDP-glucose.

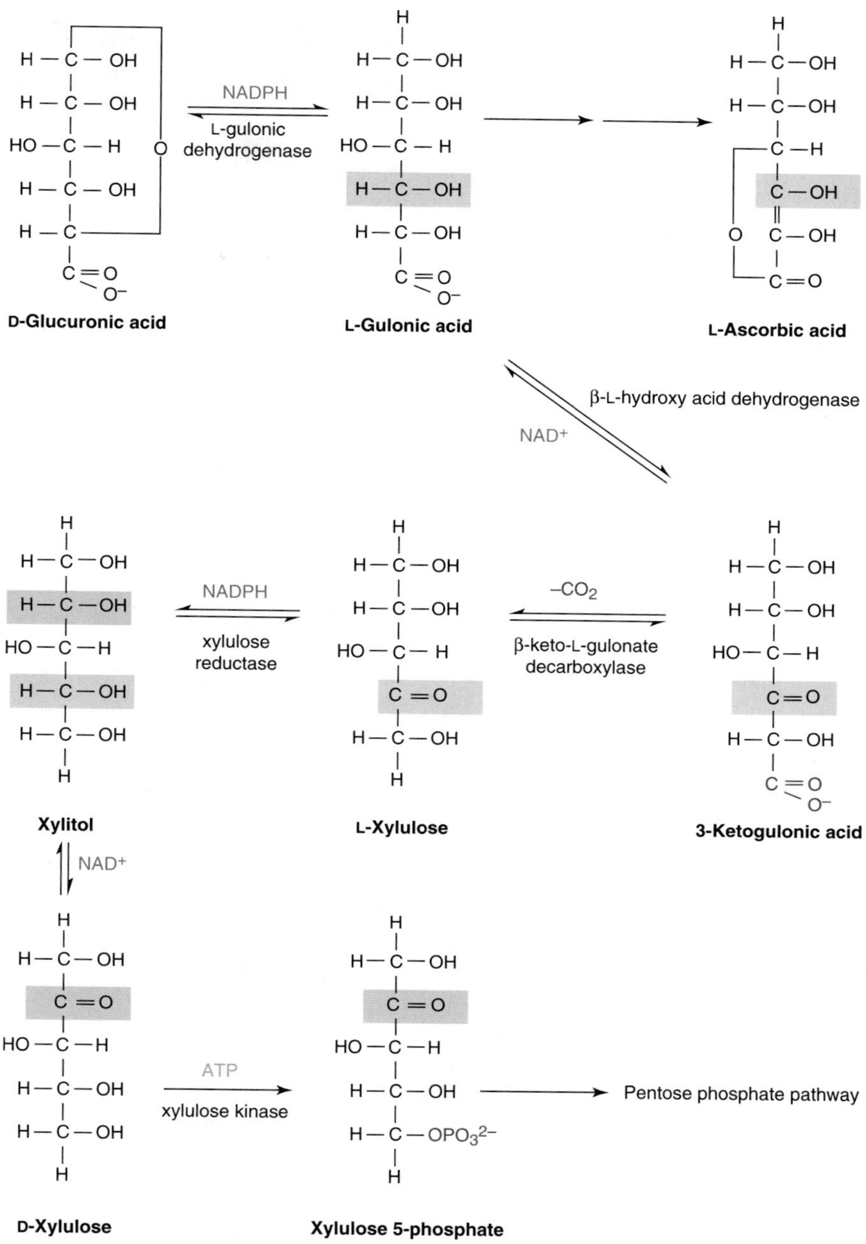

Figure 16.8 Glucuronic acid oxidation pathway.

which can enter the pentose phosphate pathway described previously. Glucuronic acid is also a precursor of **L-ascorbic acid** (Figure 16.8) in those animals that synthesize vitamin C (Clin. Corr. 16.6). Glucuronic acid also participates in detoxification by formation of glucuronide conjugates (Clin. Corr. 16.7). The glucuronic acid pathway operates in adipose tissue, and its activity is usually increased in tissue from starved or diabetic animals.

Decarboxylation, Oxidoreduction, and Transamidation of Sugars Yield Necessary Products

The only known **decarboxylation** of a nucleotide-linked sugar is the conversion of UDP-glucuronic acid to UDP-xylose, which is necessary for synthesis of proteoglycans (p. 664) and is a potent inhibitor of the production of UDP-glucuronic acid by UDP-glucose dehydrogenase (Figure 16.7), thus regulating the level of these nucleotide-linked sugar precursors by a sensitive feedback mechanism.

CLINICAL CORRELATION 16.5

Pentosuria (OMIM 260800): Deficiency of Xylitol Dehydrogenase; L-Xylulose Reductase

The glucuronic acid oxidation pathway does not appear to be essential for human carbohydrate metabolism, since individuals in whom the pathway is blocked suffer no ill effects. A metabolic variation, called idiopathic pentosuria, results from reduced activity of NADP-linked L-xylulose reductase that reduces xylulose to xylitol. Hence, affected individuals excrete large amounts of pentose into the urine, especially following intake of glucuronic acid.

Hiatt, H. Pentosuria. In Scriver,C. R., Beaudet, A. R., Sly, W. S., and Valle, D. (Eds.), *The Metabolic and Molecular Bases of Inherited Disease,* 8th ed. New York: McGraw Hill, 2001, I: 1589.

CLINICAL CORRELATION 16.6

Ascorbic Acid (Vitamin C) Is Derived from Glucuronic Acid

Glucuronic acid is reduced to L-gulonic acid and then converted through L-gulonolactone to L-ascorbic acid (vitamin C) in plants and most higher animals. Humans, other primates, and guinea pigs lack the enzyme that converts L-gulonolactone to L-ascorbic acid, and for them, ascorbic acid is a vitamin. The lack of vitamin C in the diet causes scurvy, a condition of connective tissues, which leads to multiple hemorrhages and decreased wound healing.

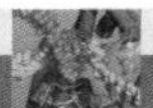

CLINICAL CORRELATION 16.7

Glucuronic Acid: Physiological Significance of Glucuronide Formation

The biological significance of glucuronic acid includes its conjugation with certain endogenous and exogenous substances to form glucuronides in a reaction catalyzed by UDP-glucuronyltransferase. Conjugation with glucuronic acid produces a strongly acidic compound that is more water soluble at physiological pH than its precursor, thereby enhancing its transport or excretion. Glucuronide formation is important in drug detoxification, steroid excretion, and bilirubin metabolism. Bilirubin is the major metabolic breakdown product of heme, the prosthetic group of hemoglobin. The central step in bilirubin excretion is conjugation with glucuronic acid by UDP-glucuronyltransferase. This enzyme may take from several days to two weeks after birth to become fully active in humans. Most cases of *physiological jaundice of the newborn* result from inability of neonatal liver to form bilirubin glucuronide at a rate comparable to that of bilirubin production. The mutant strain of Wistar Gunn rats has a deficiency of UDP-glucuronyltransferase, which results in hereditary hyperbilirubinemia. In humans, a similar defect occurs in congenital familial nonhemolytic jaundice (Crigler–Najjar syndrome); patients cannot efficiently conjugate foreign compounds with glucuronic acid.

Chowdhury, J. R., Wolkoff, W., Chowdhury, N. R., and Arias, I. W. Hereditary jaundice and disorders of bilirubin metabolism. In Scriver, C. R., Beaudet, A. R., Sly, W. S., and Valle, D. (Eds.), The *Metabolic and Molecular Bases of Inherited Disease,* 8th ed. New York: McGraw Hill, 2001, II: 3063.

Deoxyhexoses and **dideoxyhexoses** are also synthesized from precursor sugars attached to nucleoside diphosphates. L-rhamnose is synthesized from glucose by a series of oxidation–reduction reactions starting with dTDP-glucose and yielding dTDP-rhamnose; GDP-fucose is similarly synthesized from GDP-mannose as are various dideoxyhexoses.

Formation of amino sugars, major components of human complex oligo- and polysaccharides and constituents of antibiotics, occurs by **transamidation**; **glucosamine 6-phosphate** is formed from fructose 6-phosphate and glutamine by this mechanism (Figure 16.9).

Glucosamine 6-phosphate can be *N*-acetylated to ***N*-acetylglucosamine 6-phosphate**, followed by isomerization to *N*-acetylglucosamine 1-phosphate and subsequent formation of UDP-*N*-acetylglucosamine. The latter is a precursor of glycoprotein synthesis and of UDP-*N*-acetylgalactosamine, which is necessary for proteoglycan synthesis. Fructose 6-phosphate-glutamine transamidase is under negative feedback control by UDP-*N*-acetylglucosamine (see Figure 16.4, p. 652). This regulation is meaningful in certain tissues such as skin, in which this pathway accounts for up to 20% of the glucose flux.

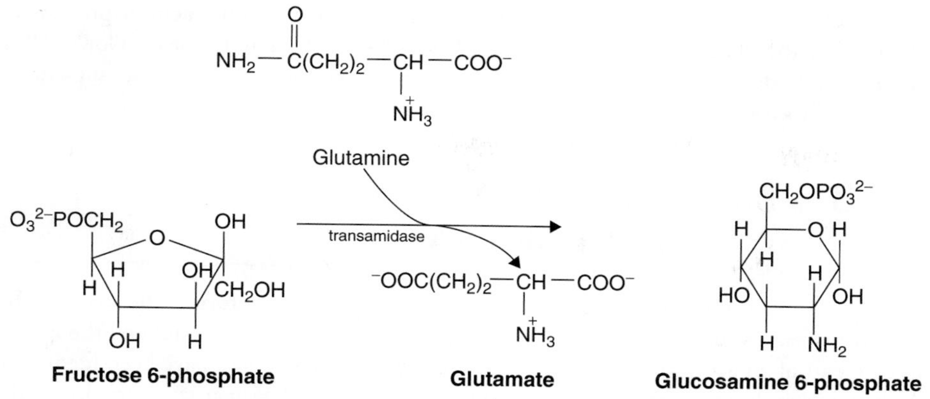

Glutamine

transamidase

Fructose 6-phosphate

Glutamate

Glucosamine 6-phosphate

Figure 16.9 Formation of amino sugars.

Sialic Acids Are Derived from *N*-Acetylglucosamine

Another product of UDP-*N*-acetylglucosamine is ***N*-acetylneuraminic acid**, one of a family of C_9 sugars called **sialic acids** (Figure 16.10). Epimerization of UDP-*N*-acetylglucosamine by a 2-epimerase produces *N*-acetylmannosamine. This reaction likely proceeds by *trans* elimination of UDP and formation of the unsaturated intermediate, 2-acetamidoglucal. In mammalian tissues *N*-acetylmannosamine is phosphorylated to *N*-acetylmannosamine 6-phosphate, which

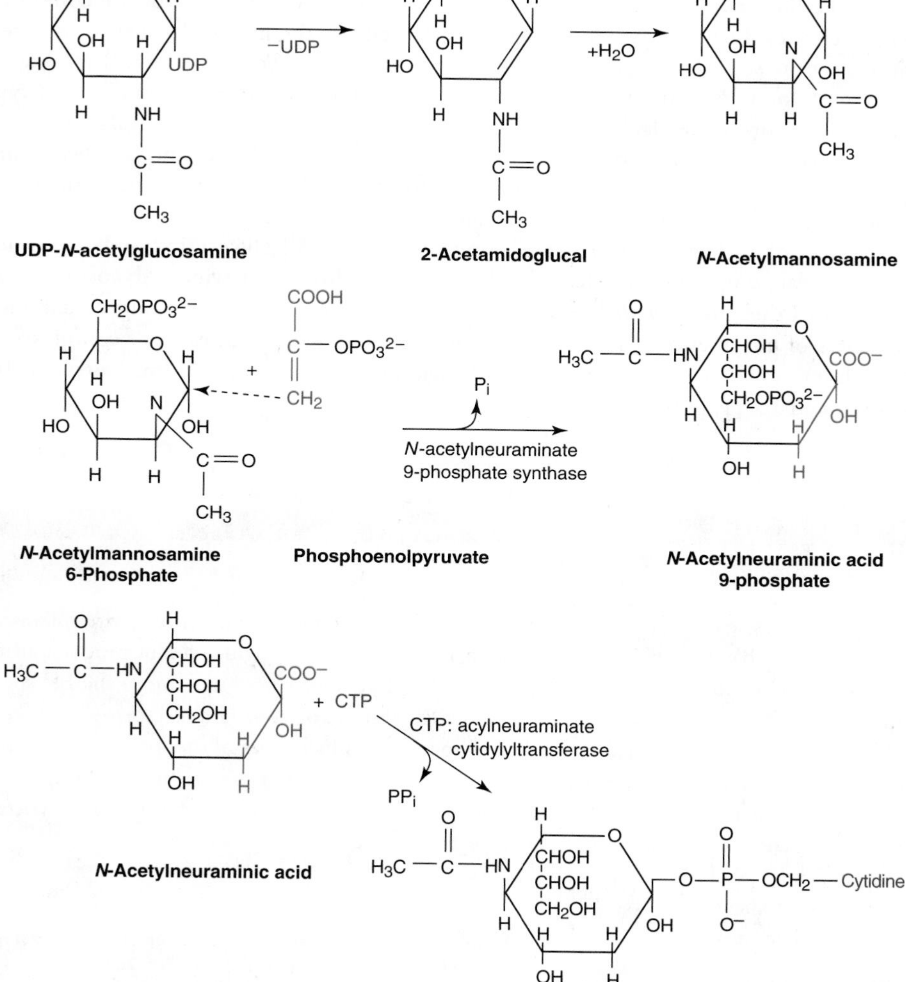

Figure 16.10 Biosynthesis of CMP-*N*-acetylneuraminic acid.

condenses with phosphoenolpyruvate to form ***N*-acetylneuraminic acid 9-phosphate**. Removal of the phosphate and formation of CMP-*N*-acetylneuraminic acid follows. All of these reactions occur in the cytosol, except the last which occurs in the nucleus with subsequent export of the CMP-*N*-acetylneuraminic acid to the cytoplasm.

16.3 • BIOSYNTHESIS OF COMPLEX POLYSACCHARIDES

Sugar moieties in polysaccharides are linked by glycosidic bonds formed by specific **glycosyltransferases** that transfer the glycosyl unit from a nucleotide derivative to the nonreducing end of an acceptor sugar. Over 180 glycosyltransferase genes have been identified. A given glycosyltransferase is specific for the sugar acceptor, the sugar transferred, and the linkage formed. A glycosyltransferase reaction is as follows:

$$\underset{\text{(donor)}}{\text{Nucleoside diphosphate-glycose}} + \underset{\text{(acceptor)}}{\text{glycose}} \xrightleftharpoons{\text{glycosyltransferase}} \underset{\text{(glycoside)}}{\text{glycosyl}_1\text{-}O\text{-glycose}_2} + \text{nucleoside diphosphate}$$

More than 40 types of glycosidic bonds have been identified in mammalian oligosaccharides and about 15 more in glycosaminoglycans. The multitude of linkages arises from the diversity of monosaccharides involved and from the formation of α and β linkages with each of the available hydroxyl groups on the acceptor saccharide. This suggests that oligosaccharides have the potential for great informational content; accordingly, the bioactivity of many molecules is determined by the nature of the composite sugar residues. For example, the antigenic specificity of the major blood types is determined by the sugar composition of cell surface molecules (Clin. Corr. 16.8). *N*-Acetylgalactosamine is the immunodeterminant of blood type A and galactose of blood type B. Removal of *N*-acetylgalactosamine from type A erythrocytes or of galactose from type B erythrocytes converts them to type O erythrocytes. Increasingly, other examples of sugars as determinants of specificity for cell surface receptor and lectin interactions, targeting of cells to certain tissues, and persistence in or clearance from the circulation of certain molecules, are being recognized. All glycosidic bonds identified in biological compounds are degraded by specific hydrolytic enzymes, **glycosidases**. In addition to their being valuable tools for the structural elucidation of oligosaccharides, this class of enzymes is of interest based on the many genetic diseases of complex carbohydrate metabolism that result from defects in glycosidases (see Clin. Corrs. 16.10, p. 663, and 16.11, p. 664).

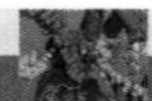

CLINICAL CORRELATION 16.8

Blood Group Substances

The surface of human erythrocytes is covered by a complex mosaic of specific antigenic determinants, many of which are complex polysaccharides. There are about 100 blood group determinants, belonging to 21 independent blood group systems. The most widely studied are those of the ABO blood group system and the closely related Lewis system. Genetic variation is achieved through specific glycosyltransferases responsible for synthesis of the heterosaccharide determinants. The H gene codes for a fucosyltransferase, which adds fucose to a peripheral galactose in the heterosaccharide precursor. The A allele encodes an *N*-acetylgalactosamine glycosyltransferase, the B allele a galactosyltransferase, and the O allele an inactive protein. The sugars transferred by the A and B enzymes are added to the H-specific oligosaccharide. The Lewis (Le) gene codes for another fucosyltransferase, which adds fucose to a peripheral *N*-acetylglucosamine residue in the precursor. Absence of the H gene product gives rise to the Le^a specific determinant, whereas absence of both H and Le enzymes is responsible for the Le^b specificity. Elucidation of the structures of these oligosaccharide determinants represents a milestone in carbohydrate chemistry. This knowledge is essential to blood transfusion practices and important for legal and historical purposes.

Yamamoto, F., Clausen, I., White, T., Mark, J., and Hakomori, S. Molecular genetic basis of the histo-blood group ABO system. *Nature* 345:229, 1990.

16.4 • GLYCOPROTEINS

Glycoproteins are defined as conjugated proteins, which contain one or more saccharides lacking a serial repeat unit and which are bound covalently to a protein. This definition excludes proteoglycans (p. 664). Glycoproteins in cell membranes have an important role in the behavior of cells and especially in the biological functions of the membrane. Glycoproteins are constituents of the mucus secreted by certain epithelial cells, serving to lubricate and protect tissues lining the respiratory, gastrointestinal, and female reproductive systems. Many secreted proteins are glycoproteins and include hormones such as follicle stimulating hormone, luteinizing hormone, and chorionic gonadotropin and plasma proteins, such as the orosomucoids, ceruloplasmin, plasminogen, prothrombin, and immunoglobulins.

Glycoproteins Contain Variable Amounts of Carbohydrate

The amount of carbohydrate in glycoproteins is highly variable. IgGs contain about 4% carbohydrate by weight, whereas glycophorin contains 60%, human ovarian cyst glycoprotein 70%, and human gastric glycoprotein 82%. The carbohydrate may be distributed fairly evenly along a polypeptide chain or concentrated in defined regions. In human glycophorin A the carbohydrate is restricted to the extracellular NH_2-terminal half of the polypeptide chain.

The carbohydrate components of glycoproteins usually contain less than 12–15 sugar residues. Some consist of a single sugar moiety, as in the submaxillary gland glycoprotein (one *N*-acetyl-α-D-galactosaminyl residue) and in some mammalian collagens (one α-D-galactosyl residue). In general, the sugar residues are of the D form, except for L-fucose, L-arabinose, and L-iduronic acid. Orthologous glycoproteins from different animal species often have identical primary protein structures, but different carbohydrate components. Heterogeneity of a given protein also occurs within an organism. For example, pancreatic ribonucleases A and B have identical primary structures and similar specificity toward substrates, but differ significantly in their carbohydrate compositions.

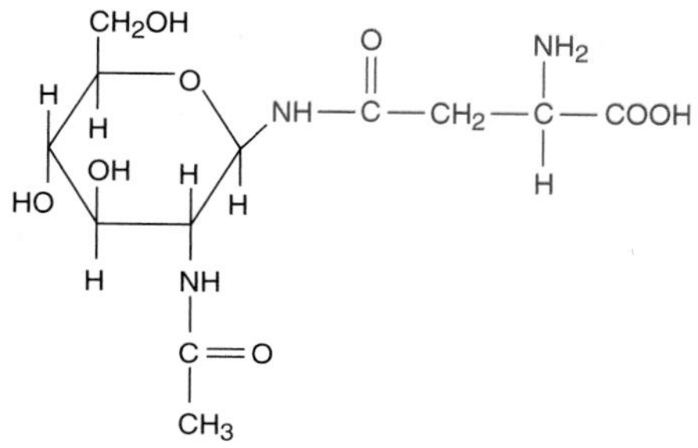

Carbohydrates Are Linked to Glycoproteins by *N*- or *O*-Glycosyl Bonds

Carbohydrate structures have been completely elucidated for only a limited number of glycoproteins. Microheterogeneity of glycoproteins, arising from incomplete synthesis or partial degradation, makes structural analyses extremely difficult. However, certain generalities have emerged. Covalent linkage of sugar to the protein is an essential feature of glycoprotein structure, and only a small number of bond types are found (p. 112). The three major types of glycopeptide bonds, as shown in Figure 16.11, are *N*-glycosyl to **asparagine** (Asn), *O*-glycosyl to **serine** (Ser) or **threonine** (Thr), and *O*-glycosyl to **5-hydroxylysine**, generally confined to the collagens. N- and O-linkages are present in many glycoproteins and bind two types of oligosaccharides (simple and complex). The simple class has a core structure of galactose(Gal)-linked β (1⟶3) to *N*-acetylgalactosamine (GalNAc) *O*-glycosidically linked to serine or threonine residues. L-Fucose (Fuc), sialic acid (NeuAc), and *N*-acetylgalactosamine are present at the nonreducing end. The general structure is as follows:

$$\text{GalNAc} \xrightarrow{1\to3} \underset{\substack{\uparrow 1\to2 \\ \text{Fuc}}}{\text{Gal}} \xrightarrow{1\to3} \underset{\substack{\uparrow 1\to2 \\ \text{NeuAc}}}{\text{GalNAc}} \longrightarrow O\text{-Ser/Thr}$$

The complex class is exemplified by the ABO and Lewis blood group substances (see Clin. Corr. 16.8, p. 658), in which the oligosaccharide is *N*-glycosidically linked to asparagine. These glycoproteins commonly contain a core structure consisting of mannose (Man) residues linked to *N*-acetylglucosamine (GlcNAc) in the structure.

$$(\text{Man})_n \xrightarrow[1\to4]{} \text{Man} \xrightarrow[1\to4]{} \text{GlcNAc} \xrightarrow[1\to4]{} \text{GlcNAc} \longrightarrow \text{Asn}$$

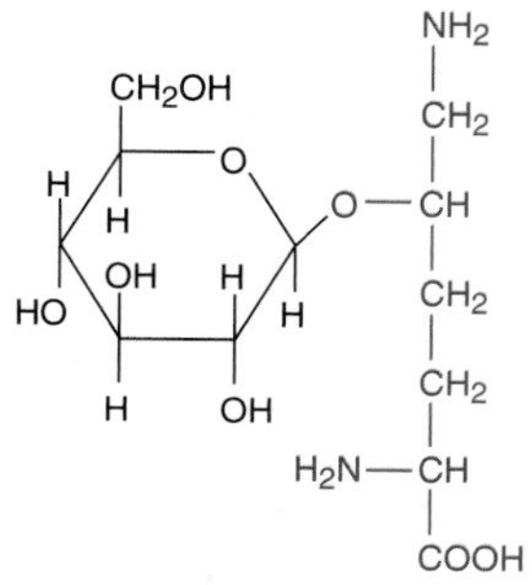

Type I ***N*-Glycosyl linkage to asparagine**

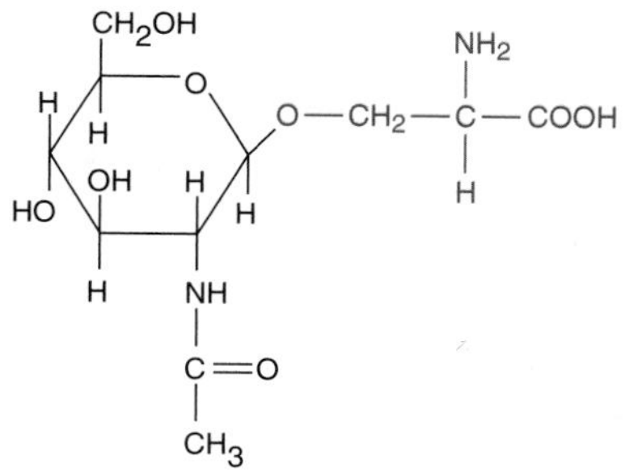

Type II ***O*-Glycosyl linkage to serine**

Type III ***O*-Glycosyl linkage to 5-hydroxylysine**

Figure 16.11 Structure of three major types of glycopeptide bond.

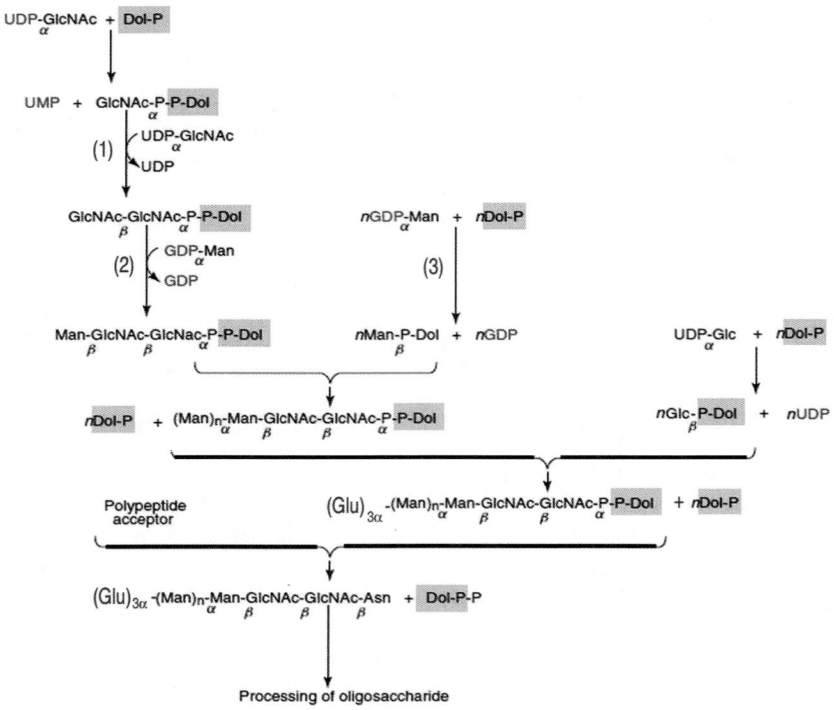

Figure 16.12 **Biosynthesis of the oligosaccharide core in asparagine-*N*-acetylglucosamine-linked glycoproteins. Dol, dolichol.**

Structural diversity of N-linked glycoproteins arises from assembly and processing of this core to produce a large repertoire of high-mannose, hybrid, or complex *N*-glycan subtypes.

Synthesis of *N*-Linked Glycoproteins Involves Dolichol Phosphate

Whereas synthesis of *O*-glycosidically linked glycoproteins involves sequential action of glycosyltransferases, synthesis of *N*-glycosidically linked glycoproteins involves a different and more complex mechanism (Figure 16.12). A common core is preassembled as a **lipid-linked oligosaccharide** on the cytoplasmic side of the ER and then "flipped" across the bilayer to the lumen of the ER and transferred as a unit to the polypeptide. During synthesis, the oligosaccharide intermediates are bound to **dolichol phosphate**.

$$(CH_2{=}\overset{\displaystyle CH_3}{\overset{|}{C}}-CH{=}CH)_n-CH_2-\overset{\displaystyle CH_3}{\overset{|}{C}H}-CH_2-CH_2O-PO_3H_2$$

Dolichol phosphate

Dolichols are polyprenols (C_{80}–C_{100}) that contain 16–20 isoprene units, in which the final unit is saturated. These lipids function in two ways during oligosaccharide synthesis. The first involves formation of *N*-acetylglucosaminyl pyrophosphoryldolichol from the UDP-linked and dolichol phosphate-linked sugars. The second involves *N*-acetylglucosamine, and the mannose is transferred directly from the nucleotide without formation of intermediates. In either case, the oligosaccharide is transferred in a final step from the dolichol pyrophosphate to an asparagine residue in the polypeptide chain.

After transfer to the polypeptide, the core structures are completed by glycosyltransferases without further participation of lipid intermediates (Figure 16.13). A series of early processing reactions, which are highly conserved among vertebrate species and cell types, occur largely in the ER and appear to be coupled with proper folding of the glycoprotein. Following the initial trimming and release from the ER, *N*-glycans undergo further

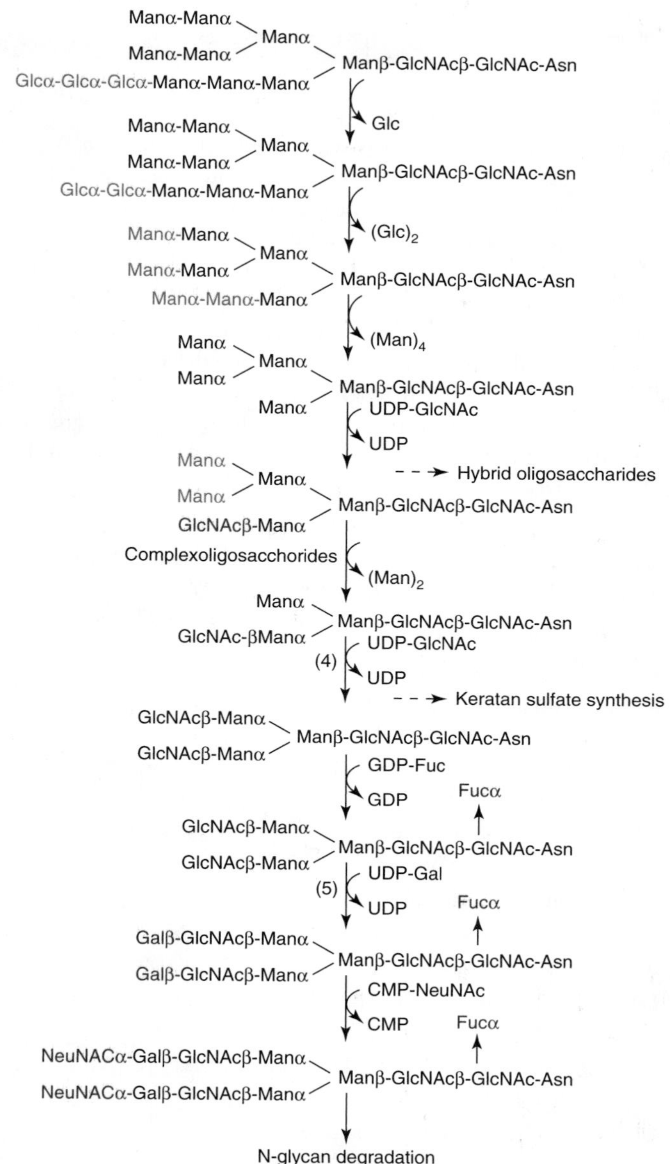

Figure 16.13 Processing pathway for N-linked oligosaccharides.

glycosidase and glycosyltransferase modifications, mostly in the Golgi. Several avenues exist in the processing pathway that determine the final diversity of glycan structure (i.e., high mannose, hybrid or complex subtypes) as well as the trafficking fate of glycoproteins. The Man_8 $GlcNAc_2$-Asn *N*-glycan on glycoproteins destined for the lysosomal compartment is modified by addition of a GlcNAc residue catalyzed by a **GlcNAc-phosphotransferase** and subsequent removal by **GlcNAc phosphodiester glycosidase**, exposing a Man-6-P residue. A defect in this process forms the basis of the **lysosomal storage disease** known as **I-cell disease** (see Clin. Corr. 6.8, p. 238).

As complex oligosaccharide structures require complicated and intricate pathways of synthesis, there has been a rapid expansion in the number of **c**ongenital **d**isorders of **g**lycosylation (**CDG**) identified as genetic diseases resulting from deficient or increased glycosylation. At present, 28 disorders have been identified, including 16 in protein *N*-glycosylation, 6 in protein *O*-glycosylation, 4 in both *O*- and *N*-glycosylation, and 2 in lipid glycosylation. A remarkably wide range of phenotypes, from mild to lethal and from organ-specific to multisystem, have been recorded (Clin. Corr. 16.9).

Catabolism of glycoconjugates may also produce abnormal phenotypes. Degradation of hetero-oligosaccharides is catalyzed by specific glycosidases. Exoglycosidases remove sugars

CLINICAL CORRELATION 16.9

Congenital Disorders of Glycosylation (CDGs)

The congenital disorders of glycosylation (CDGs) are clinically heterogeneous disorders of glycosylation. Using the glycosylation state of serum transferrin as a sensitive indicator, several types of CDGs have been identified. Type I is the most common, representing disorders of *N*-glycan assembly in the early mannose biosynthethic pathway. Numerous defective genes have been identified leading to human genetic disorders (see table). For example, mannosyltransferase I adds the first mannose residue to the lipid-linked oligosaccharide (see Figure 16.12, p. 660). Patients with deficits in this enzyme present with epilepsy, severe psychomotor retardation, dysmorphy, microcephaly, hypotonia, cardiomyop-athy, and nephrotic syndrome followed by early (1 week to 10 months) death.

CDG-II disorders involve enzymatic defects in *N*-glycan processing enzymes (see Figure 16.13 and table). Note the carbohydrate transformation enzymes (see Section 16.3, p. 658) CDG-1a (phosphomannomutase II) and CDG-1b (phosphomannose isomerase), which also cause glycosylation disorders. Other disorders may involve defects in *O*-glycosylation, such as Walker–Warburg syndrome, due sometimes to deficiencies of *O*-mannosyltransferase I; combined *N*- and *O*- glycosylation defects, such as CMP-sialic acid transporter deficiency and defects in lipid glycosylation like glycosylphosphatidylinositol deficiency. Little progress in therapy for any of these disorders has been made; only CDG-1b is efficiently treatable.

Enzymic Defects in *N*-linked Glycoprotein Synthesis

Disorder	*Defective protein*
CDG-Ia	Phosphomannomutase II
CDG-Ib	Phosphomannose isomerase
CDG-Ic	DoI-P-Glc:Man_9-$GlcNAc_2$-P-P-Dol glucosyltransferase (glucosyltransferase I)
CDG-Id	Dol-P-Man: Man_5-$GlcNAc_2$-P-P-Dol mannosyltransferase (mannosyltransferase VI)
CDG-Ie	GDP-Man: Dol-P-mannosyltransferase (Dol-P-Man synthase I)(3)
CDG-If	Lec35 (Man-P-DoI utilization 1)
CDG-Ig	Dol-P-Man: Man_7-$GlcNAc_2$-P-P-Dol mannosyltranferase (mannosyltransferase VIII)
CDG-Ih	Dol-P-Glc: Glc_1 -Man_9-$GlcNAc_2$-P-P-Dol glucosyltransferase (glucosyltransferase II)
CDG-Ii	GDP-Man: ManI-$GlcNAc_2$-P-P-Dol mannosyltransferase (mannosyltransferase II)
CDG-Ij	UDP-GlcNAc: Dol-P-GlcNAc-P transferase (1)
CDG-Ik	GDP-Man: $GlcNAc_2$-P-P-Dol mannosyltransferase (mannosyltransferase I) (2)
CDG-II	Dol-P-Man: Man_6-and Man_8-$GlcNAc_2$-P-P-Dol mannosyltransferase (mannosyltransferase VII-IX)
CDG- IIa	N-acetylglucosaminyltransferase II (4)
CDG-IIb	Glucosidase I
CDG-IIc	GDP-fucose transporter
CDG-IId	β-1, 4 galactosyltransferase (5)

Source: Jacken, J and Matthijs, G. Congenital disorders of glycosylation: A rapidly expanding disease family. *Ann. Rev. Gemonics Hum. Genetics* 8:261, 2007

sequentially from the nonreducing end, exposing the substrate for the next glycosidase. Absence of a particular glycosidase causes cessation of catabolism, resulting in accumulation of the proceeding product (Clin. Corr. 16.10). Endoglycosidases with broader specificity also exist, so that catabolism of glycoproteins results from the combined action of endo- and exoglycosidases. Many of the same *N*- or *O*-linked glycan chains are found on both

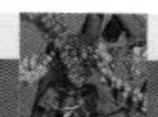

CLINICAL CORRELATION 16.10

Defects in Catabolism of Glycoproteins

Some human inborn errors of metabolism involve storage of glycolipids, glycopeptides, mucopolysaccharides, and hetero-oligosaccharides. These diseases are caused by defects in lysosomal glycosidase activity, which prevent the catabolism of oligosaccharides. They involve gradual accumulation in tissues and urine of compounds derived from incomplete degradation of the oligosaccharides, and may be accompanied by skeletal abnormalities, hepatosplenomegaly, cataracts or mental retardation. For example, a defect in catabolism of asparagine-*N*-acetylglucosamine-linked oligosaccharides leads to aspartylglycosylaminuria in which a deficiency of 4-L-aspartylglycosylamine amidohydrolase allows accumulation of aspartylglucosamine-linked structures (see table). Other disorders involve accumulation of oligosaccharides derived from both glycoproteins and glycolipids, which share common oligosaccharide structures (see table and Clin. Corr. 16.11).

Enzymic Defects in Degradation of Asn-GlcNAc Type Glycoproteins[a]

Disease	*Deficient Enzyme*[b]
Aspartylglycosylaminuria	4-L,-Aspartyilglycosylamine amidohydrolase (2)
β-Mannosidosis	β-Mannosidosis (7)
α-Mannosidosis	α-Mannosidosis (3)
GM_2 gangliosidosis variant O (Sandhoff-Jatzkewitz disease)	β-*N*-Acetylthexosarninidases (A and B) (4)
GM_1 gangliosidosis	β-Galactosidase (5)
Mucolipidosis I (sialidosis)	Sialidase (6)
Fucosidosis	α-Fucosidase (8)

[a]A typical Asn-GlcNAc oligosaccharide structure.

NeuAc —(6) α— Gal —(5) β— GlcNAc —(4) β— Man
NeuAc —α— Gal —β— GlcNAc —β— Man
Man —(3) α— Man —(7) β— GlcNAc —(2) β— GlcNAc — Asn
Man — β — GlcNAc (bisecting); GlcNAc —(8)— Fuc
NeuAc —α— Gal —β— GlcNAc —β— Man α
NeuAc —(6) α— Gal —(5) β— GlcNAc —(4) β— Man

[b]The numbers in parentheses refer to the enzymes that hydrolyze those bonds.

Aula, P., Janlanko, A., and Peltonen, L. Aspartylglucosaminuria. In Scriver, C. R., Beaudet, A. R., Sly, W. S., and Valle, D. (Eds.), *The Metabolic and Molecular Bases of Inherited Disease,* 8th ed. New York: McGraw Hill, 2001, III: 3507.

glycoproteins and glycolipids, hence certain enzyme defects may affect degradation of both types of glycoconjugates (Clin. Corr. 16.11).

Glycan Function

Glycosylation is one of the most common posttranslational modifications and nearly half of all known proteins in eukaryotes are glycosylated. The carbohydrate or **glycan** component of glycoproteins and glycolipids participate in many important biological processes including receptor activation, signal transduction, endocytosis, cell adhesion, and leukocyte trafficking. The cell surface glycans are the first molecules to be encountered and recognized by other cells, antibodies, viruses, and bacteria. For instance, the glycan repertoire provides oncofetal and stem cell biomarkers that reflect the binding specificity of various monoclonal antibodies; one of the barriers to antibody neutralization of HIV is the array of protective carbohydrates that cover the antigens on the virus surface. Although the diversity of glycan structures is large, specificity is provided by the glycosylation enzymes, glycosyltransferases and glycosidases. Thus, regulation of the biosynthetic processes of these abundant and diverse biopolymers direct the molecular mechanisms by which glycans contribute to development and disease.

CLINICAL CORRELATION 16.11

Glycolipid Disorders

A host of human genetic diseases arise from deficiencies in hydrolases which act predominantly on glycolipid substrates, resulting in accumulation of glycolipid and ganglioside products. The clinical symptoms associated with each of the glycoconjugates may vary greatly. However, because of the preponderance of lipids in the nervous system, such disorders often have associated neurodegeneration and severe mental and motor deterioration.

Enzymic Defects in Degradation of Glycolipids

Disease	*Deficiency Enzyme*
Tay-Sachs	β-Hexosaminidase A
Sandhoff	β-Hexosaminidases A and B
GM_1 gangliosidosis	β-Galactosidase
Sialidosis	Sialidase
Fabry	α-Galactosidase
Gaucher	β-Glucoceramidase
Krabbe	β-Galactoceramidase
Metachromatic leukodystrophy	Arylsulfatase A (cerebroside sulfatase)

Beutler, E., and Garabowski, G. Gaucher Disease. In Scriver, C. R., Beaudet, A. R., Sly, W. S., and Valle, D. (Eds.), *The Metabolic and Molecular Bases of Inherited Disease,* 8th ed. New York: McGraw Hill, 2001, III: 3635.

16.5 • PROTEOGLYCANS

Proteoglycans are complex macromolecules that may be composed of 95% or more carbohydrate, resembling polysaccharides more than proteins. Their carbohydrate chains are called **glycosaminoglycans** (or **mucopolysaccharides**) especially in reference to the storage diseases, the **mucopolysaccharidoses**, which result from an inability to degrade these molecules (see Clin. Corr. 16.14, p. 668).

There Are Six Classes of Proteoglycans

Proteoglycans consist of one or more of many different glycosaminoglycan chains linked covalently to a protein core. Six distinct classes are recognized: **chondroitin sulfate**, **dermatan sulfate**, **keratan sulfate**, **heparan sulfate**, **heparin**, and **hyaluronate**. Certain features are common to the different classes of glycosaminoglycans. These long, unbranched heteropolysaccharide chains are made up largely of disaccharide repeating units, consisting of a hexosamine and a uronic acid. Common substituents of glycosaminoglycans are sulfate groups, linked by ester bonds to certain monosaccharides or by amide bonds to the amino group of glucosamine. Only hyaluronate is not sulfated and is not covalently attached to protein. The carboxyls of uronic acids and the sulfate groups contribute to the highly charged nature of glycosaminoglycans. Their electrical charge and their macromolecular structure are important in their roles as lubricants and support elements in connective tissue. Glycosaminoglycans of proteoglycans are predominantly components of the extracellular matrices and cell surfaces where they participate in cell adhesion and signaling.

Hyaluronate Is a Copolymer of *N*-Acetylglucosamine and Glucuronic Acid

Hyaluronate differs from the other types of glycosaminoglycans. It is unsulfated, not covalently linked with protein, and produced by bacteria as well as animal cells. It is classified as a glycosaminoglycan because of its structural similarity to these polymers, consisting solely of repeating disaccharide units of *N*-acetylglucosamine and glucuronic acid (Figure 16.14). Although it has the least complex chemical structure of the glycosaminoglycans, the chains may reach 10^5–10^7 Da. The large mass, polyelectrolyte character, and large volume it occupies in solution contribute to the properties of hyaluronate as a lubricant and shock absorbent. It is ubiquitous in the extracellular matrix and found in synovial fluid, vitreous humor, and umbilical cord.

Chondroitin Sulfates Are the Most Abundant Glycosaminoglycans

The **chondroitin sulfates** are attached to specific serine residues in a protein core through a tetrasaccharide linkage module.

$$\text{GluUA} \xrightarrow[1\to3]{} \text{Gal} \xrightarrow[1\to3]{} \text{Gla} \xrightarrow[1\to4]{} \text{Xyl} \longrightarrow O\text{-Ser}$$

Each chain contains 30 to 50 disaccharide units (15–25 kDa), consisting of *N*-acetylgalactosamine and glucuronic acid, which are attached to the linkage module (Figure 16.14). The disaccharides can be sulfated at the C4 or C6 position of *N*-acetylgalactosamine. An average chondroitin sulfate proteoglycan molecule contains about 100 chondroitin sulfate chains attached to the protein core, giving a mass of $1.5–2 \times 10^6$ Da. Proteoglycan preparations are extremely heterogeneous, differing in number and distribution of polysaccharide chains, length of chondroitin sulfate chains, and degree of sulfation. Chondroitin sulfate proteoglycans may aggregate noncovalently with hyaluronate and are prominent components of cartilage, tendons, ligaments, and aorta, as well as brain, kidney, and lung.

Dermatan Sulfate Contains L-Iduronic Acid

Dermatan sulfate differs from chondroitin sulfate in that its predominant uronic acid is L-iduronic acid, although some D-glucuronic acid is also present. The epimerization of D-glucuronic acid to L-iduronic acid occurs after incorporation into the polymer chain and is coupled with the process of sulfation. The glycosidic linkages have the same position and configuration as in chondroitin sulfates, with average polysaccharide chains of $2–5 \times 10^4$ Da. Like heparin,

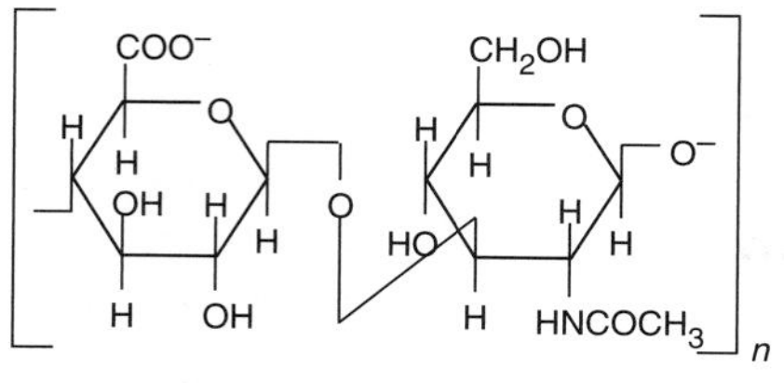

Repeat unit of hyaluronic acid

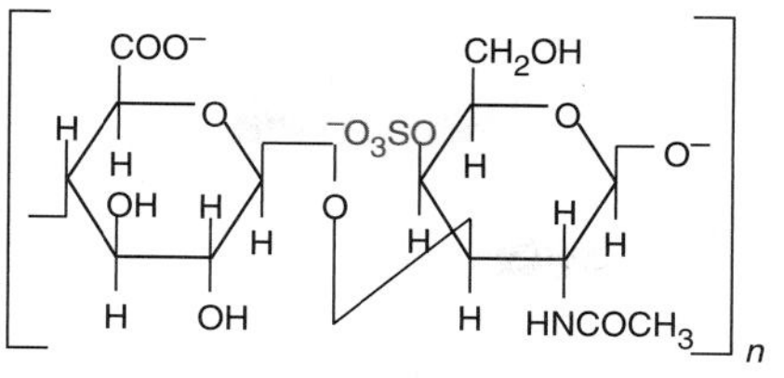

Repeat unit of chondroitin 4-sulfate

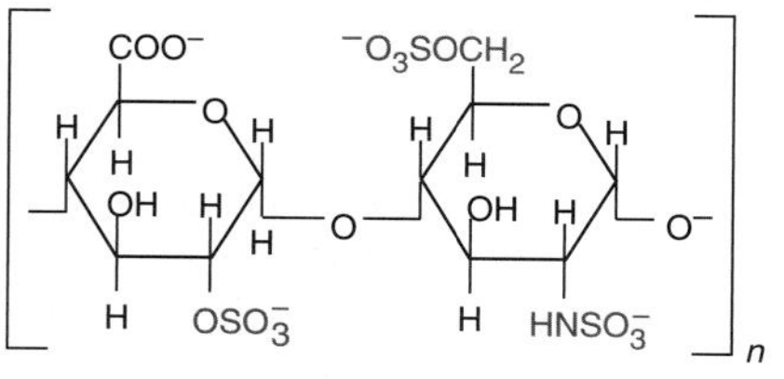

Repeat unit of heparin

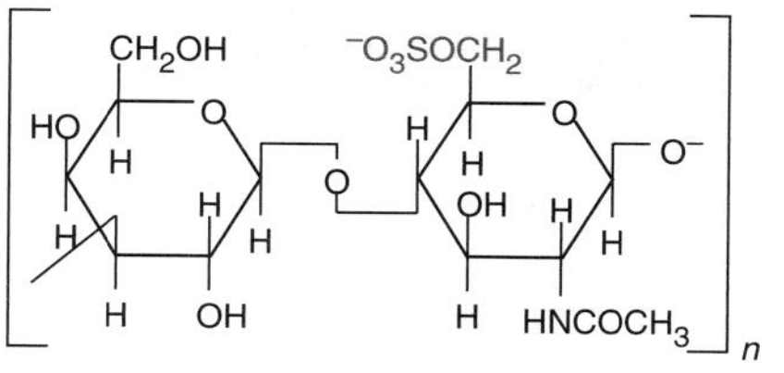

Repeat unit of keratan sulfate

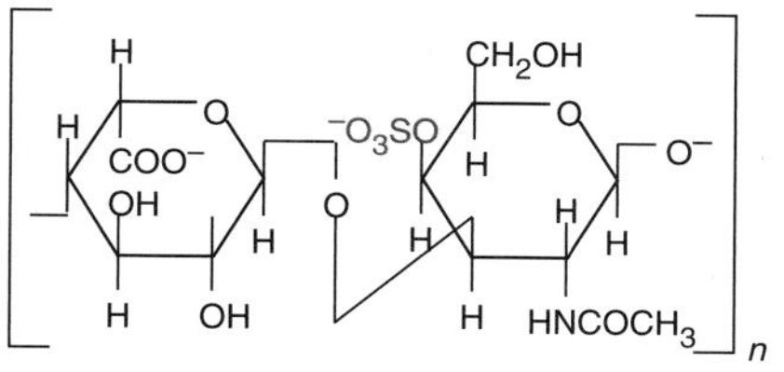

Repeat unit of dermatan sulfate

Figure 16.14 **Major repeat units of glycosaminoglycan chains.**

dermatan sulfate is antithrombic but has only minimal whole-blood anticoagulant and blood lipid-clearing activities. Dermatan sulfate is found in skin, blood vessels, and heart valves.

Heparin and Heparan Sulfate Differ from Other Glycosaminoglycans

Glucosamine and D-glucuronic acid or L-iduronic acid form the characteristic disaccharide repeat unit in **heparin** (Figure 16.14). In contrast to most other glycosaminoglycans, heparin contains β-glycosidic linkages. Almost all glucosamine residues contain sulfamide linkages, while a small number of glucosamine residues are *N*-acetylated. The sulfate content of heparin approaches 2.5 sulfate residues per disaccharide unit in preparations with the highest biological activity. In addition to *N*-sulfate and *O*-sulfate on C6 of glucosamine, heparin may contain sulfate on C3 of the hexosamine and C2 of the uronic acid. Unlike other glycosaminoglycans, heparin is an intracellular component of mast cells, and functions predominantly as an anticoagulant and lipid-clearing agent (Clin. Corr. 16.12).

Heparan sulfate contains a similar disaccharide repeat unit as heparin but has more *N*-acetyl groups, fewer *N*-sulfate groups, and a lower content of *O*-sulfate groups. Heparan sulfate as a component of proteoglycans may be extracellular or an integral and ubiquitous component of the cell surface in many tissues including blood vessel walls and brain, and may function as a coreceptor for certain growth factors.

Keratan Sulfate Exists in Two Forms

Keratan sulfate is composed principally of the disaccharide unit of *N*-acetylglucosamine and galactose, and contains no uronic acid (Figure 16.14). Sulfate content is variable, as ester sulfate on C6 of galactose and hexosamine. Two types of keratan sulfate differ in carbohydrate content and tissue distribution. Both also contain mannose, fucose, sialic acid, and *N*-acetylgalactosamine. Keratan sulfate I, from cornea, is linked to protein by an *N*-acetylglucosamine-asparaginyl bond, typical of glycoproteins. Keratan sulfate II, from cartilage, is linked through *N*-acetylgalactosamine to serine or threonine. Skeletal

CLINICAL CORRELATION 16.12

Heparin Is an Anticoagulant

Heparin is a naturally occurring sulfated glycosaminoglycan that is used to reduce the clotting tendency of patients. Both *in vivo* and *in vitro,* heparin prevents the activation of clotting factors by binding with an inhibitor of the coagulation process. The inhibitor is antithrombin III, a plasma protein inhibitor of serine proteases. In the absence of heparin, antithrombin III slowly (10–30 min) combines with several clotting factors, yielding complexes devoid of proteolytic activity; in the presence of heparin, inactive complexes are formed within a few seconds. Antithrombin III contains an arginine residue that combines with the active site serine of factors Xa and IXa; thus, the inhibition is stoichiometric. Heterozygous antithrombin III deficiency results in increased risk of thrombosis in the veins and resistance to the action of heparin.

Hirsh, J. Drug therapy: Heparin. *N. Engl. J. Med.* 324:1565, 1991.

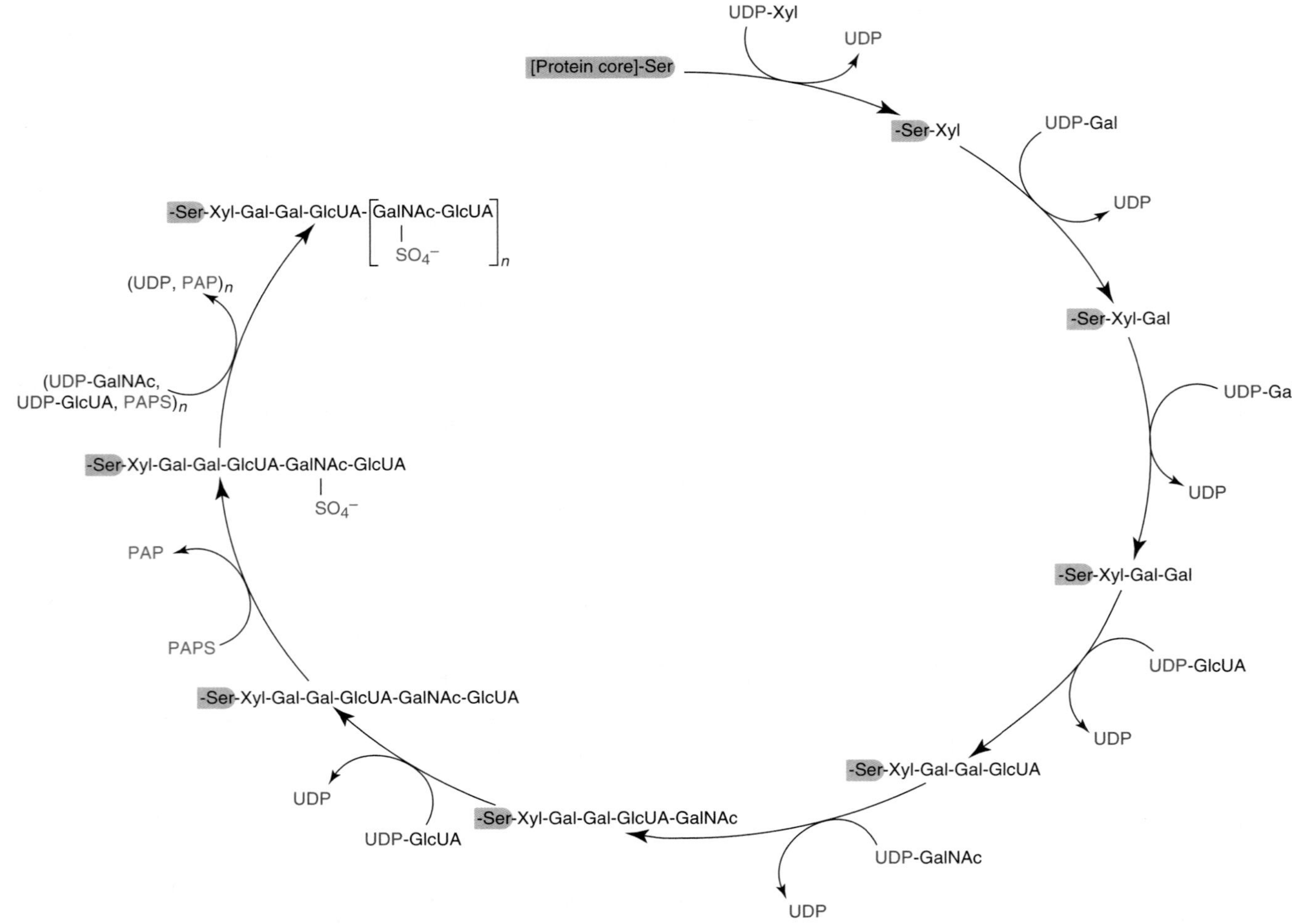

Figure 16.15 Synthesis of chondroitin sulfate proteoglycan. Xyl, xylose; Gal, galactose; GlcUA, glucuronic acid; GalNAc, *N*-acetylgalactosamine; PAPS, phosphoadenosine phosphosulfate.

keratan sulfates are often covalently attached to the same core protein as the chondroitin sulfate chains.

Biosynthesis of Chondroitin Sulfate Is Typical of Glycosaminoglycan Formation

The glycosaminoglycans are assembled by sequential action of glycosyltransferases, which transfer a monosaccharide from a nucleotide-linked derivative to an appropriate acceptor, either the nonreducing end of another sugar or a polypeptide. The biosynthesis of the chondroitin sulfates is most thoroughly understood (Figure 16.15).

Formation of the core protein is the first step, followed by the action of six glycosyltransferase reactions. Strict substrate specificity is required for completion of the unique tetrasaccharide linkage region. Polymerization then results from repeated *N*-acetylgalactosaminyltransferase and glucuronosyltransferase reactions to form the characteristic disaccharide units. Sulfation of *N*-acetylgalactosamine at the C4 or C6 position occurs with chain elongation. The sulfate donor, as in other biological systems, is **3′-phosphoadenosine 5′-phosphosulfate (PAPS)**, which is formed from ATP and sulfate in two steps catalyzed by the bifunctional PAPS synthetase (Figure 16.16). The importance of sulfation is highlighted by the preponderance of chondrodystrophic conditions in animals and humans caused by deficiencies in the sulfation process (Clin. Corr. 16.13).

Synthesis of other glycosaminoglycans requires additional transferase reactions specific for the appropriate sugars and linkages. Completion often involves *O*-sulfation,

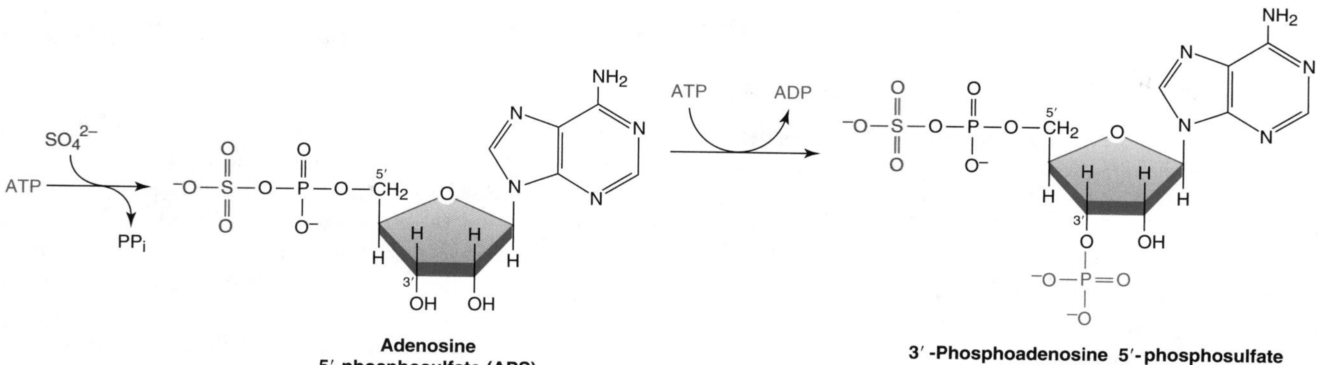

Adenosine 5′-phosphosulfate (APS)

3′-Phosphoadenosine 5′-phosphosulfate (PAPS)

Figure 16.16 Biosynthesis of 3′-phosphoadenosine 5′-phosphosulfate (PAPS).

epimerization, acetylation, or *N*-sulfation. Synthesis of both proteoglycans and glycoproteins is regulated by the same mechanism at the level of hexosamine synthesis. The fructose 6-phosphate-glutamine transamidase reaction (see Figure 16.4, p. 652) is subject to feedback inhibition by UDP-*N*-acetylglucosamine, which is in equilibrium with UDP-*N*-acetylgalactosamine. Similarly, the concentrations of UDP-xylose and UDP-glucuronic acid are stringently controlled by UDP-xylose inhibition of the UDP-glucose dehydrogenase-catalyzed conversion of UDP-glucose to UDP-glucuronic acid (see Figure 16.4). Since xylose is the first sugar added during synthesis of chondroitin sulfate, dermatan sulfate, heparin, and heparan sulfate, the earliest effect of decreased core protein synthesis would be accumulation of UDP-xylose, which aids in maintaining a balance between synthesis of protein and glycosaminoglycans.

Proteoglycans, like glycoproteins and glycolipids are degraded by the sequential action of proteases and glycosidases, **deacetylases** and **sulfatases**. Much of the information about metabolism and degradation of proteoglycans has been derived from the study of the **mucopolysaccharidoses** (Clin. Corr. 16.14). These genetic disorders are characterized by accumulation in tissues and excretion in urine of hetero-oligosaccharide products derived from incomplete breakdown of proteoglycans, due to a deficiency of one or more **lysosomal hydrolases**.

Although proteoglycans continue to be defined on the basis of the glycosaminoglycan chains they contain, new ones have been described largely on the basis of functional proper-

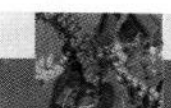

CLINICAL CORRELATION 16.13

Chondrodystrophies due to Sulfation Defects

Sulfation is an essential modification of glycosaminoglycans in the various proteoglycan families. The sulfation process involves transport of inorganic sulfate into the cell via plasma membrane transporters, activation by transformation into phosphoadenosylphosphosulfate (PAPS) via a two-step process catalyzed by PAPS synthetase in the cytosol, then either direct utilization by cytosolic sulfotransferases or transport of PAPS from the cytosol to the Golgi complex for utilization by a host of lumenal sulfotransferases. Three autosomal recessive disorders, diastrophic dysplasia (DTD), atelosteogenesis type II (AOII), and achondrogenesis type 1B (ACG-1B), result from mutations in the DTDST gene which encodes a sulfate transporter. Patients with DTD exhibit disproportionate short stature and generalized joint dysplasia, but usually have a normal lifespan; ACG-1B is characterized by extremely short extremities and trunk; AOII is a perinatally lethal chondrodysplasia. Genetic disorders due to defects in synthesis of PAPS by the bifunctional sulfurylase/kinase (PAPS synthetase) have been identified in both animals and humans. The brachymorphic mouse exhibits a severe growth disorder resulting in extremely short trunk and limbs and small skull. In humans, spondyloepimetaphyseal dysplasia (Pakistani type) is characterized by short and bowed lower limbs, enlarged knee joints, and early onset of degenerative joint disease. These phenotypes clearly highlight the importance of this posttranslational modification to the functioning of proteoglycans, especially in development and maintenance of the skeletal system.

Schwartz, N. B., and Domowicz, M. Chondrodysplasias due to proteoglycan defects. *Glycobiology* 12:57R, 2002; and Schwartz, N. B. Chondrodysplasias. *Encyclopedia of Endocrine Disorders,* 2004, 1:502.

CLINICAL CORRELATION 16.14

Mucopolysaccharidoses

Human genetic disorders characterized by excessive accumulation and excretion of the oligosaccharides of proteoglycans comprise the mucopolysaccharidoses. They result from deficiency of one or more lysosomal hydrolases that are responsible for the degradation of dermatan and/or heparan sulfate. Hurler syndrome and Sanfilippo syndrome are autosomal recessive conditions, whereas Hunter disease is X-linked. Both Hurler and Hunter syndromes are characterized by skeletal abnormalities and mental retardation, which in severe cases may result in early death. In contrast, the physical defects in the Sanfilippo syndrome are relatively mild, whereas the mental retardation is severe. Collectively, the incidence of the mucopolysaccharidoses is 1 per 30,000 births. Multiple sulfatase deficiency (MSD) is characterized by decreased activity of all known sulfatases. Recent evidence suggests that a co- or postranslational modification of a cysteine to 2-amino 3-oxopropionic acid is essential for sulfatase activity, and that a lack of this modification results in MSD. These disorders are amenable to prenatal diagnosis, since the pattern of metabolism exhibited by affected cells obtained from amniotic fluid is strikingly different from normal.

Enzyme Defects in the Mucopolysaccharidoses

Syndrome	*Accumulated Products*[a]	*Deficient Enzyme*[b]
Hunter	Heparan sulfate Dermatan sulfate	Iduronate sulfatase (1)
Hurler—Scheie	Heparan sulfate Dermatan sulfate	α-L-Iduronidase (2)
Maroteaux—Lamy	Dermatan sulfate	*N*-Acetylgalactosamine sulfatase (3)
Mucolipidosis VII	Heparan sulfate Dermatan sulfate	β-Glucuronidase (4)
Sanfilippo type A	Heparan sulfate	Heparan sulfamidase (6)
Sanfilippo type B	Heparan sulfate	*N*-Acetylglucosaminidase (9)
Sanfilippo type C	Heparan sulfate	Acetyl CoA: α-glucosaminide acetyltransferase
Sanfilippo type D	Heparan sulfate	*N*-Acetylglucosamine 6-sulfatase (8)
Morquio type A	Keratan/chondroitin sulfate	Galactose-6-sulfatase
Morquio type B	Keratan sulfate	β-Galactosidase

[a]Structures of dermatan sulfate and heparan sulfate.

Dermatan sulfate —IdUA $\overset{(2)}{\underset{\alpha}{—}}$ GalNAc $\underset{\beta}{—}$ GlcUA $\overset{(4)}{\underset{\beta}{—}}$ GalNAc $\underset{\beta}{—}$
IdUA: (1) OSO_3H; GalNAc: (3) OSO_3H; GalNAc: OSO_3H

Heparan sulfate —IdUA $\overset{(2)}{\underset{\alpha}{—}}$ GlcN $\overset{(7)}{\underset{\alpha}{—}}$ GlcUA $\underset{\beta}{—}$ GlcNAc $\overset{(9)}{\underset{\alpha}{—}}$
IdUA: (1) OSO_3H; GlcN: (6) OSO_3H; GlcNAc: (8) OSO_3H

[b]The numbers in parentheses refer to the enzymes that hydrolyze those bonds.

Neufeld, E., and Muenzer, J. Mucopolysaccharidoses. In Scriver, C. R., Beaudet, A. R., Sly, W. S., and Valle, D. (Eds.), *The Metabolic and Molecular Bases of Inherited Disease*, 8th ed. New York: McGraw Hill, 2001, III: 3421.

ties or location. **Aggrecan** and **versican** are the predominant extracellular species; **syndecan**, **CD44**, and **thrombomodulin** are integral membrane proteins; **neurocan**, **brevican**, **cerebrocan,** and **phosphacan** are largely restricted to the nervous system. Many proteoglycans (i.e., aggrecan, **syndecan**, and **betaglycan**) carry more than one type of glycosaminoglycan chain, and their sizes and relative amounts may change with development, age, or disease. Many genes encoding the proteoglycan core proteins and biosynthetic enzymes have been cloned, revealing that the relevant proteins belong to families of related origin and possible function.

Bibliography

Collins, P. M., and Ferrier, R. J. *Monosaccharides: Their Chemistry and Roles in Natural Products.* Chichester, UK: Wiley, 1995.

Dutton, G. J. (Ed.). *Glucuronic Acid, Free and Combined.* New York: Academic Press, 1966.

Horecker, B. L. *Pentose Metabolism in Bacteria.* New York: Wiley, 1962.

Kornfeld, R., and Kornfeld, S. Assembly of asn-linked oligosaccharides. *Annu. Rev. Biochem.* 54:631, 1985.

Kornfeld, S. Diseases of abnormal protein glycosylation: An emerging area. *J. Clin. Investigation* 101:1293, 1998.

Neufeld, E. F. Lysosomal storage diseases. *Annu. Rev. Biochem.* 60:257, 1991.

Ohtsubo, K., and Marth, J. D. Glycosylation in cellular mechanisms of health and disease. *Cell* 126:855, 2006.

Schwartz, N. B. Proteoglycans. In *Embryonic Encyclopedia of Life Sciences.* London: Nature Publishing Group, 2000. www.els.net.

Schwartz, N. B. Biosynthesis and regulation of expression of proteoglycans. *Front. Biosci.* 5:649, 2000.

Schwartz, N.B. PAPS synthetase. In *Encyclopedia of Molecular Medicine.* New York:Wiley, 2002, 1:284.

Scriver, C. R., Beaudet, A. L., and Valle, D., Sly, W. S. (Eds.). *The Molecular and Metabolic Bases of Inherited Disease,* 8th ed., New York: McGraw Hill, 2001.

Tozzi, M. G., Camici, M., Mascia, L., Sgarrella, F., and Ipata, P. L. Pentose phosphates in nucleoside interconversion and catabolism. *FEBS Journal* 273:1089, 2006.

Zhao, Y. Z., Takahashi, M., Gu, J.-G., Miyoshi, E., Matsumoto, A., Kitazume, S., and Taniguchi, N. Functional roles of *N*-glycans in cell signaling and cell adhesion in cancer. *Cancer Sci.* 99:1304, 2008.

Key Terms

pentose phosphate pathway
glucose-6-phosphate
glucose-6-phosphate dehydrogenase
dehydrogenation
NADPH
pyrophosphorylase
decarboxylation
dolichol phosphate
glycosyl transferase
oxidation
sialic acid
glycosidase
reduction
ribulose-5-phosphate
glycoprotein
transamidation
glucuronic acid
glycosamine glycan
glycosylation
chondroitin sulfate
proteoglycan
epimerization
dermatan sulfate
oligosaccharide
heparan sulfate
nucleotide sugar
heparin
carbohydrate
hyaluronate
mucopolysaccharide
aggrecan

Questions • CAROL N. ANGSTADT

Multiple-Choice Questions

1. All of the following interconversions of monosaccharides (or derivatives) require a nucleotide linked sugar intermediate except

A. galactose 1-phosphate to glucose 1-phosphate.
B. glucose 6-phosphate to mannose 6-phosphate.
C. glucose to glucuronic acid.
D. glucuronic acid to xylose.
E. glucosamine 6-phosphate to *N*-acetylneuraminic acid (a sialic acid).

2. All of the following are true about glucuronic acid except

A. it is a charged molecule at physiological pH.
B. as a UDP derivative, it can be decarboxylated to a component used in proteoglycan synthesis.
C. it is a precursor of ascorbic acid in humans.
D. its formation from glucose is under feedback control by a UDP-linked intermediate.
E. it can ultimately be converted to xylulose 5-phosphate and thus enter the pentose phosphate pathway.

3. $NADPH/NADP^+$ is maintained at a high level in cells primarily by

A. lactate dehydrogenase.
B. the combined actions of glucose 6-phosphate dehydrogenase and gluconolactonase.
C. the action of the electron transport chain.
D. exchange with $NAD^+/NADH$.
E. the combined actions of transketolase and transaldolase.

4. Fucose and sialic acid

A. are both derivatives of UDP-*N*-acetylglucosamine.
B. are the parts of the carbohydrate chain that are covalently linked to the protein.
C. can be found in the core structure of certain O-linked glycoproteins.
D. are transferred to a carbohydrate chain when it is attached to dolichol phosphate.
E. are the repeating unit of proteoglycans.

5. Glycosaminoglycans

A. are the carbohydrate portion of glycoproteins.
B. contain large segments of a repeating unit typically consisting of a hexosamine and a uronic acid.
C. always contain sulfate.
D. exist in only two forms.
E. are bound to protein by ionic interaction.

6. All of the following are true of proteoglycans except

A. specificity is determined, in part, by the action of glycosyltransferases.
B. synthesis is regulated, in part, by UDP-xylose inhibition of the conversion of UDP-glucose to UDP-glucuronic acid.
C. synthesis involves sulfation of carbohydrate residues by PAPS.
D. synthesis of core protein is balanced with synthesis of the polysaccharide moieties.
E. degradation is catalyzed in the cytosol by nonspecific glycosidases.

Questions 7 and 8: Galactosemia is an inability to transform galactose into glucose and can lead to such problems as cataract formation, growth failure, mental retardation, or eventual death from liver damage. Galactose is reduced to galactitol, which initiates cataract formation. Accumulation of galactose 1-phosphate, the most serious condition, leads to liver failure.

7. The most severe form of galactosemia
A. is a genetic deficiency of a uridylyltransferase that exchanges galactose 1-phosphate for glucose on UDP-glucose.
B. results from a deficiency of an epimerase.
C. is insignificant in infants but a major problem in later life.
D. is a defect in galactokinase.
E. would be expected to interfere with the use of fructose as well as galactose because the deficient enzyme is common to the metabolism of both sugars.

8. UDP-galactose
A. must be formed from galactose 1-phosphate.
B. is usually the first sugar linked to dolichol phosphate.
C. is used in the synthesis of chondroitin sulfate.
D. could not lead to the formation of sugar derivatives like glucuronic acid or xylose.
E. is the direct precursor of *N*-acetylgalactosamine.

Questions 9 and 10: A 6 month old infant presented with hypoglycemia, vomiting, diarrhea, protein-loss enteropathy, and hepatic fibrosis. Measurement of the glycosylation state of endogenous serum transferrin revealed Type Ib CDG (congenital disorders of glycosylation), which is a defect in phosphomannose isomerase activity.

9. The role of phosphomannose isomerase is the interconversion of
A. mannose 6-phosphate and mannose 1-phosphate.
B. glucose 6-phosphate and mannose 6-phosphate.
C. fructose 6-phosphate and mannose 6-phosphate.
D. fructose 1-phosphate and mannose 1-phosphate.
E. glucose 6-phosphate and mannose 1-phosphate.

10. In Type Ic CDG, a defect in an enzyme transferring a glucosyl residue to a high-mannose dolichol pyrophosphate precursor, the carbohydrate structure would be part of a(n)
A. *N*-linked glycoprotein.
B. *O*-linked glycoprotein.
C. proteoglycan.
D. glycosaminoglycan.
E. complex lipid.

Questions 11 and 12: An infant presented with multiple skeletal abnormalities, with the most prominent ones being a short trunk and short limbs. Urine was free of partially degraded oligosaccharides or oligosaccharides of proteoglycans. The blood sulfate concentration and extracellular acid hydrolase activity were normal.

11. The disorder described would most likely be caused by a defect in
A. the ability to generate mannose 6-phosphate.
B. lysosomal glycosidase activity.
C. the gene that codes for a sulfate transporter.
D. PAPS synthetase.
E. a sulfatase.

12. Sulfation is an important component of the synthesis of
A. most proteoglycans.
B. hyaluronate.
C. most glycoproteins.
D. conjugated bilirubin.
E. all of the above.

Problems

13. What is the role of transaldolase and transketolase in the metabolism of glucose?

14. Essential fructosuria is a defect in fructokinase whereas fructose intolerance is a defect in fructose 1-phosphate aldolase. Which of these two diseases leads to severe hypoglycemia after ingestion of fructose, and why?

Answers

1. B The glucose and mannose phosphates are both in equilibrium with fructose 6-phosphate by phosphohexose isomerases. A: Occurs via an epimerase at the UDP-galactose level. C and D: This oxidation of glucose is catalyzed by UDP-glucose dehydrogenase and the product can be decarboxylated to UDP-xylose. E: Again, an epimerization occurs on the nucleotide intermediate.

2. C Man does not make ascorbic acid. A: The charged acid group enhances water solubility, which is a major physiological role for glucuronic acid, for example, bilirubin metabolism. B and D: Decarboxylation of UDP-glucuronic acid gives UDP-xylose, which is a potent inhibitor of the oxidation of UDP-glucose to the acid. E: The reduction of *d*-glucuronic acid to *l*-gulonic acid leads to ascorbate as well as xylulose 5-phosphate for the pentose phosphate pathway.

3. B Although the glucose 6-phosphate dehydrogenase reaction, specific for NADP, is reversible, hydrolysis of the lactone assures that the overall equilibrium lies far in the direction of NADPH. A and C: These use NAD, not NADP. D: This does not occur. E: These enzymes are part of the pentose phosphate pathway but do not use NADP.

4. C Core structure also contains galactose and *N*-acetylgalactosamine. A: Only sialic acid does. Fucose comes from CDP-mannose. B: Usually found at the periphery of the carbohydrate. D: Core structure of *N*-linked carbohydrates contains mannose and *N*-acetylglucosamine. E: Repeating unit is hexosamine and uronic acids.

5. B This is a major distinction from glycoproteins, which, by definition, do not have a serial repeating unit. A: These are the carbohydrate of proteoglycans. C: Most do, but hyaluronate does not. D: There are at least six different classes. E: Carbohydrates are bound by covalent links.

6. E Degradation is lysosomal; deficiencies of one or more lysosomal hydrolases lead to accumulation of proteoglycans in the mucopolysaccharidoses. A: Strict substrate specificity of the enzymes is important in determining the type and quantity of proteoglycans synthesized. Formation of specific protein acceptors for the carbohydrate is also important. B and D: Both xylose and glucuronic acid levels are controlled by this; xylose is the first sugar added in the synthesis of four of the six types and would accumulate if core protein synthesis were decreased. C: This is necessary for the formation of most proteoglycans.

7. A B: The epimerase is normal. C: Galactose is an important sugar for infants. D: In this disease, the galactokinase is normal but is deficient in the mild form of the disease. E: Fructose metabolism does not use the uridylyltransferase that is deficient in galactosemia.

8. **C** The initial carbohydrate chain built on the serine –OH group contains two galactoses plus an *N*-acetylgalactosamine, supplied as the UDP-derivatives. A: It can also be formed by epimerization of UDP-glucose. B: This is an *N*-acetylglucosamine. D: Since it can be epimerized to UDP-glucose, it can lead to these products. E: This is one of the more complex interconversions that proceeds through nucleotide-bound derivatives of glucose and rhamnose.
9. **C** This is similar to the glucose 6-phosphate-fructose 6-phosphate interconversion. A: Mutases catalyze a 1-6 phosphate shift. B, D, and E: These conversions do not occur directly.
10. **A** A carbohydrate chain assembled on dolichol phosphate is characteristic of N-linked glycoproteins. B, C, and D: Synthesis of O-linked glycoproteins involves the sequential addition to the *N*-acetylgalactosamine linked to serine or threonine. E: Dolichol phosphate is a lipid but is eliminated when the carbohydrate chain is added to the protein.
11. **D** These structural abnormalities are associated with defects in sulfation of the proteoglycans. A: This would lead to an inability to target acid hydrolase to the lysosome and thus increase the extracellular concentration. B and E: These are lysosomal enzymes whose deficiency would lead to partially degraded oligosaccharides in urine. C: Lack of a transporter would lead to elevated serum levels of sulfate.
12. **A** B: Hyaluronate is one glycosaminoglycan which is not sulfated. C: Glycoproteins don't contain sulfate. D: Bilirubin is conjugated with glucuronic acid.
13. 3 Pentose phosphates (ribose 5-phosphatehosphate + 2 xylulose 5-phosphate) are converted to 2 fructose 6-phosphate and glyceraldehyde 3-phosphate by transaldolase and transketolase via a series of 2-carbon and 3-carbon transfers. These reactions are reversible.
14. Lack of fructokinase causes accumulation of fructose which is excreted. Lack of the aldolase results in increased fructose 1-phosphate, which sequesters cellular inorganic phosphate, inhibiting the cells ability to generate ATP.

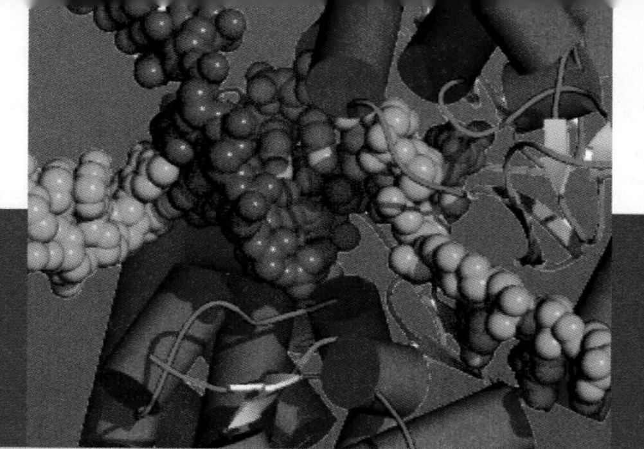

17

Lipid Metabolism I: Synthesis, Storage, and Utilization of Fatty Acids and Triacylglycerols

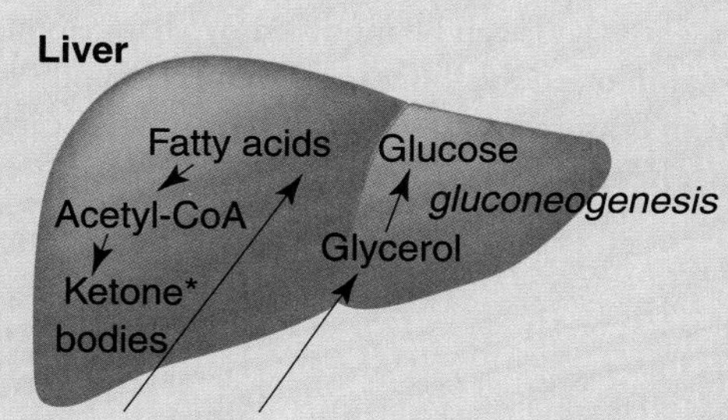

Martin D. Snider

Associate Professor, Case Western Reserve University School of Medicine

J. Denis McGarry (deceased)

Professor, University of Texas Southwestern Medical Center at Dallas

Richard W. Hanson

Leonard & Jean Skeggs Professor of Biochemistry, Case Western Reserve University School of Medicine

CLINICAL CORRELATIONS

Key Concepts

- Triacylglycerols are hydrophobic molecules that are the major fuel store in all organisms. In mammals, most triacylglycerol is found in adipose tissue.
- There is net deposition of fat as triacylglycerol in adipose tissue in the fed state, whereas adipose tissue breaks down the triacylglycerol to provide fuel for other tissues during fasting.
- Fatty acids are synthesized mainly in the liver by the sequential addition of two-carbon units, using acetyl CoA derived from dietary glucose and other fuels.
- Fatty acids can be modified by chain elongation and the addition of double bonds to produce a family of molecules. In mammals, some polyunsaturated fatty acids can be made only from essential fatty acid precursors in the diet.
- Most tissues can assemble triacylglycerols from acyl CoAs and glycerol 3-phosphate.
- Fatty acids are converted to acetyl CoA by β-oxidation and then used as fuel in many tissues.
- During prolonged fasting, the ketone bodies, acetoacetate and β-hydroxybutyrate are synthesized from acetyl CoA in the liver. The use of ketone bodies as fuel by the brain and other tissues is an important adaptation to prolonged fasting.
- The synthesis and breakdown of fatty acids is regulated to promote energy storage in the fed state and energy mobilization in the fasted state. In addition, the mobilization of fatty acids from stored triacylglycerol is regulated to meet cellular energy needs.

17.1 • INTRODUCTION

Lipids are **hydrophobic** molecules, meaning that they are not soluble in water and tend to self-associate in separate lipid phases. Most lipids contain or are derived from **fatty acids** (Figure 17.1). Lipids have many important biological functions. **Triacylglycerols** (Figure 17.2) are the major fuel store in the body. Other lipids, including **phospholipids**, **glycolipids**, and **cholesterol**, are crucial constituents of biological membranes; the unique surface active properties of these molecules allow them to form the membrane backbone, separating and defining aqueous compartments within cells. Surface active lipids have other important functions, including maintenance of alveolar integrity in the lungs and solubilization of nonpolar substances in body fluids. Finally, some lipids are important signaling molecules. Fatty acids, **steroid hormones**, and the **eicosanoids**, including **prostaglandins**, are important in communication between cells. Other lipids have signaling functions within cells.

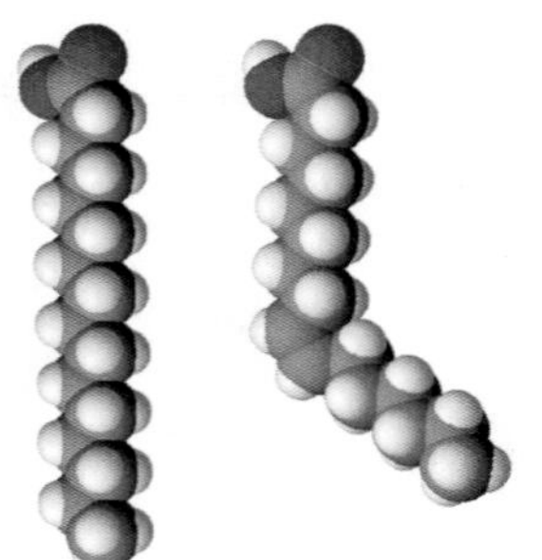

Figure 17.1 Long-chain fatty acids. Left: Palmitic (saturated). Right: Cis palmitoleic (unsaturated). Color scheme: H white, C gray, and O red.
Courtesy of Dr. Daniel Predecki, Shippensburg University, PA.

The metabolism of fatty acids and triacylglycerols is so important that imbalances and deficiencies in these processes can have serious pathological consequences in humans. Disease states related to dysfunctions in lipid metabolism include obesity, diabetes mellitus, hypertriglyceridemia, and ketoacidosis. There are also human genetic diseases that affect fatty acid metabolism, including Refsum disease and deficiencies in carnitine and in fatty acid oxidation.

The Appendix describes the nomenclature and chemistry of lipids, and a discussion of the digestion and absorption of fats is presented in Chapter 25.

17.2 • CHEMICAL NATURE OF FATTY ACIDS AND ACYLGLYCEROLS

Fatty Acids Are Alkyl Chains Terminating in a Carboxyl Group

Fatty acids (see Figure 17.1) consist of an **alkyl chain** with a terminal carboxyl group. The basic formula of a completely saturated species is $CH_3(CH_2)_nCOOH$. Nearly all fatty acids in humans have an even number of carbon atoms, although some organisms synthesize molecules with an odd number. Humans can use the latter for energy, but they are marginally incorporated into complex lipids. Unsaturated fatty acids occur commonly, with up to six double bonds per chain. If there is more than one double bond per molecule, they are separated by a **methylene** ($-CH_2-$) **group**. The double bonds are almost always in the cis configuration, which induces a kink in the fatty acid chain, preventing orderly packing of the chains in lipid phases. By inducing disorder, double bonds lower the melting temperature and increase the fluidity of the phases.

Figure 17.2 Triacylglycerol. 1-Palmityl 2,3-dioleoyl glycerol. See Figure 17.1 for color scheme.
Courtesy of Dr. Daniel Predecki, Shippensburg University, PA.

Most fatty acids in humans have 16–20 carbon atoms, but C_{14} molecules are found linked to proteins and very long fatty acids are found in complex lipids in the nervous system.

These include nervonic acid (22:1) and docosahexaenoic acid, a C_{22} acid that has six double bonds (Figure 17.3). Some fatty acids with an α-OH group are found as constituents of membrane lipids. **Branched-chain fatty acids**, which have methyl groups at one or more positions along the chain, are also found in some animals, including humans. These contribute specific physical properties to some secretory and structural molecules. For instance, large amounts of branched-chain fatty acids, particularly isovaleric acid (Figure 17.4), occur in lipids of echo-locating structures in marine mammals.

The most abundant fatty acids in humans are shown in Table 17.1. The symbols show the number of carbon atoms and, after the colon, the number of double bonds. Carbon atoms are numbered starting with the carboxyl carbon. The bond locations are shown in parentheses, using the number of the carbon atom on the carboxyl side of the bond.

$$CH_3—(CH_2)_7—CH{=}CH—(CH_2)_{13}—COOH$$

Nervonic acid

$$CH_3—(CH_2—CH{=}CH)_6—(CH_2)_2—COOH$$

All-*cis*-4,7,10,13,16,19,-docosahexaenoic acid

Figure 17.3 **Very long-chain fatty acids.**

Most Fatty Acids in Humans Occur as Triacylglycerols

Fatty acids are stored primarily as esters of glycerol (Figures 17.2 and 17.5). Most fatty acids in humans are found in **triacylglycerols**, in which all three hydroxyl groups of glycerol are esterified with fatty acids. Compounds with one (**monoacylglycerols**) or two (**diacylglycerols**) esterified fatty acids occur in relatively minor amounts, largely as metabolic intermediates in the synthesis and degradation of glycerol-containing lipids. The trivial names, mono-, di- and triglycerides, are also used for these compounds.

Most triacylglycerols have different fatty acids esterified to the three positions of glycerol. The distribution of fatty acids is influenced by many factors, including diet

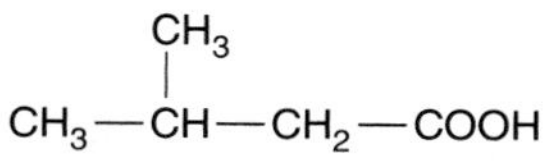

$$CH_3—CH(CH_3)—CH_2—COOH$$

Figure 17.4 **Isovaleric acid.**

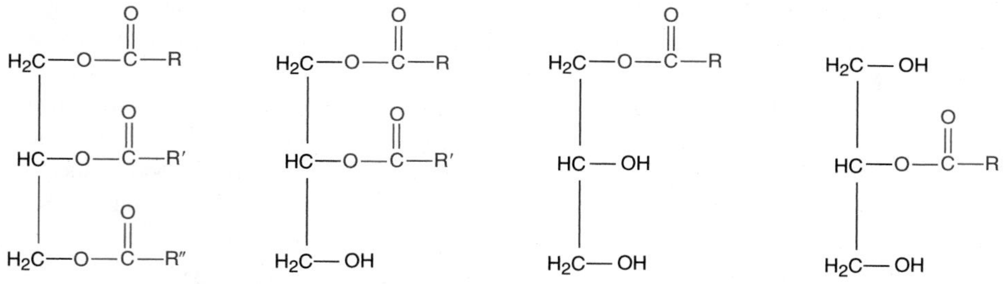

Triacylglycerol: $H_2C—O—C(=O)—R$ / $HC—O—C(=O)—R'$ / $H_2C—O—C(=O)—R''$

Diacylglycerol: $H_2C—O—C(=O)—R$ / $HC—O—C(=O)—R'$ / $H_2C—OH$

1–Monoacylglycerol: $H_2C—O—C(=O)—R$ / $HC—OH$ / $H_2C—OH$

2–Monoacylglycerol: $H_2C—OH$ / $HC—O—C(=O)—R$ / $H_2C—OH$

Figure 17.5 **Acylglycerols.**

TABLE 17.1 • Fatty Acids of Importance to Humans

Name	*Numerical Formula*	*Functions in Humans*
Formic acid	1	
Acetic acid	2:0	
Propionic acid	3:0	Produced by metabolism of odd-chain fatty acids as well as isoleucine, valine, and methionine
Butyric acid	4:0	Milk triacylglycerols contain short chain fatty acids
Myristic acid	14:0	Covalently linked to some proteins
Palmitic acid	16:0	Product of fatty acid synthase
Palmitoleic acid	16:1(9)	Fatty acids with 16–18 carbons comprise the bulk of the fatty acids in triacylglycerols and complex lipids
Stearic acid	18:0	
Oleic acid	18:1(9)	
Linoleic acid	18:2(9, 12)	Essential fatty acid
Linolenic acid	18:3(9, 12, 15)	Essential fatty acid
Arachidonic acid	20:4(5, 8, 11, 14)	Precursor of prostaglandins and other eicosanoids
Lignoceric acid	24:0	
Nervonic acid	24:1(15)	Enriched in sphingolipids

CLINICAL CORRELATION 17.1

Obesity

Obesity has become a worldwide epidemic. It has been estimated that in 2004, 25% of the population in the United States was obese, a figure that is projected to increase to 40% by 2020. Equally distressing figures have been noted for countries with rapidly growing economies such as China and India. The definition of obesity is arbitrary and is based on estimates of ideal body weight (IBW), that is, body weight that is associated with the lowest morbidity and mortality. Overweight is defined as weight up to 20% above IBW, and obesity is weight that is 20% (or greater) above IBW. Body Mass Index (BMI) is another commonly used measure of obesity. It is calculated by dividing weight (in kilograms) by height^2 (in meters squared). An individual with a BMI of 25 or greater is overweight, and of 30 or greater is considered obese. The anatomical location of fat deposition is a good indicator of the risk of morbidity and mortality from obesity; a central distribution of body fat (abdominal adipose tissue) is a greater health risk than a more peripheral distribution of fat (subcutaneous adipose tissue).

Obesity is part of what has been termed Metabolic Syndrome, a collection of symptoms that include central obesity, high blood pressure, high triacylglycerols and low levels of high density lipoprotein in the blood, as well as insulin resistance. As many as 47 million Americans have this syndrome. Individuals with Metabolic Syndrome are at high risk for diabetes and cardiovascular disease. The biochemical link between obesity and diabetes is discussed in Clin. Corr. 17.2. Because of these health risks, the control of obesity is a major public health goal.

The causes of obesity and the underlying reasons for its dramatic increase in human populations are complex. Endocrine disorders such as hypothyroidism or Cushing disease (an overproduction of corticosteroids) are rare, as are mutations in genes coding hormones involved in the control of food intake, such as leptin or its receptor. It seems more likely that the genetic disposition of an individual, together with the environment (food choices, level of exercise), result in obesity. A change in lifestyle is clearly a major element in the obesity epidemic. An increased consumption of foods high in fat and carbohydrate, together with a decrease in exercise is the major cause of weight gain and body fat deposition in populations throughout the world. Thus, obesity is not a single disorder but a heterogeneous group of conditions with multiple causes.

Unfortunately, obesity is a condition that is rarely curable; its treatment remains a persistent medical problem. The standard procedure of reducing caloric intake and increasing energy expenditure is ineffective in obese individuals since, to be effective, it must be continued indefinitely to ensure a reduced body weight. Markedly obese individuals often resort to surgical procedures to remove unwanted fat or reduce the stomach to prevent hunger and food absorption. Several pharmacological interventions have been used for the treatment of obesity. These include the use of compounds, such as β-phenylamine that selectively inhibits the reuptake of norepinephrine, serotonin, and dopamine. This compound reduces food intake and increases thermogenesis. Another approach is to prevent fat absorption using orlistat, which blocks pancreatic lipase, thereby decreasing triacylglycerol digestion. The thermogenic (increases the generation of heat) compound ephedrine has been shown to increase oxygen consumption by 10% in humans, when combined with caffeine. However, a major portion of the weight loss noted with this combination of drugs is due to decreased food intake. Potentially important new approaches to the treatment of obesity involve the development of compounds that regulate satiety. Gastrointestinal peptides such as cholecystokinin and glucagon-like peptide have long been known to reduce food intake. In addition, antagonists of the cannabinoid receptor in the brain reduce appetite. Small molecules that act by binding to these receptors are candidates for drugs that could be used to control appetite.

Hirsch, J., Salens, L. B., and Aronne, L. J. Obesity. In Becker, K. L. (Ed), *Principles and Practice of Endocrinology and Metabolism,* 3d ed. New York, Lippincott, Williams and Wilkins, 2001, 1239; and Freidmann, J. M. Obesity in the new millennium. *Nature* 404: 632, 2000.

and the anatomical location of the stored molecule. In humans, most fatty acids are either saturated or contain one double bond, with oleic acid (18:1) being the most common. Although triacylglycerols are readily catabolized, they are comparatively inert chemically. The highly unsaturated fatty acids in triacylglycerols are much more susceptible to nonenzymatic oxidation.

The Hydrophobicity of Triacylglycerols Is Important for Their Functions

Triacylglycerols and other complex lipids have limited solubility in water because the long hydrocarbon chains of the constituent fatty acids do not form hydrogen bonds. Consequently, they tend to associate with each other or with other hydrophobic groups, such as sterols and the side chains of hydrophobic amino acids. This insolubility in water is essential for the storage of triacylglycerols and for the assembly of biological membranes.

Triacylglycerols are a much more efficient form of energy storage than glycogen. On a weight basis, triacylglycerols yield nearly 2½ times more ATP on complete oxidation than

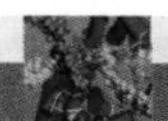

CLINICAL CORRELATION 17.2

The Key Role of Fatty Acid Metabolism in Type 2 Diabetes: A Tribute to J. Denis McGarry

The late Denis McGarry, a long-time contributor to this book, wrote a prescient article in 1992 entitled "What if Minkowski had been ageustic? An alternative angle on diabetes." Oskar Minkowski (1858–1931) and Josef von Mering were the first to recognize that the pancreas produces a substance that regulates the level of blood glucose; their research ultimately led to the isolation of insulin from β cells of the pancreas. The concept of diabetes as a disease of carbohydrate metabolism began with the ancient observations that diabetic urine is sweet. Folklore tells us that Dr. Minkowski put his finger into the urine of a diabetic patient, tasted it, and noted that it was as "sweet honey." To this day, diabetes is widely considered to be due to an imbalance in carbohydrate metabolism. Dr. McGarry raised the humorous supposition that if Minkowski had been ageustic (i.e., he could not taste), but could smell the acetone in the urine, diabetes would have been classified as a disease of lipid rather than carbohydrate metabolism. This is an important insight into Type 2 (noninsulin-dependent) diabetes, which is associated with obesity and resistance to insulin.

Because most of the human energy reserve is triacylglycerol this fuel is extremely important, particularly during fasting. After a meal, glucose is rapidly removed from the blood and either used for energy or stored as glycogen; within hours humans switch to the oxidation of fatty acids as the major source of energy, with ketone bodies becoming important during prolonged fasting. After 3 days of fasting, even the brain metabolizes ketone bodies. The metabolism of fat over the limited glucose supply is referred to as *fuel sparing*. Fatty acids play an important role in this process by inhibiting glucose utilization in skeletal muscle. In 1963, Phillip Randle and colleagues proposed what is now known as the Randle hypothesis, which states that fatty acid oxidation blocks glucose metabolism by inhibiting key regulatory enzymes involved in glucose utilization, including phosphofructokinase and pyruvate dehydrogenase complex (see Chapter 15) and inhibits glycogen synthesis in skeletal muscle. Fatty acids also markedly reduce the uptake of glucose from the blood by blocking the recruitment of GLUT 4 from the endoplasmic reticulum to the cell surface. This is due in part to the activation of specific isoforms of protein kinase C, as well as the transcription factor NF-κB, by acyl CoAs generated by fatty acid metabolism.

Although the fuel sparing effect of fatty acids is of clear survival advantage, there is also a disadvantage in obesity, which is frequently characterized by a high fatty acid concentration in the blood. This reduces glucose uptake and metabolism by skeletal muscle, resulting in an increase in the fasting levels of blood glucose and a concomitant increase in insulin secretion. The resulting condition is termed *insulin resistance* (elevated levels of blood glucose and insulin). During the early stages of insulin resistance, pancreatic β cells produce enough insulin to regulate blood glucose. However, prolonged overproduction of insulin can cause failure of the β cells, resulting in Type 2 diabetes. While this progression does not occur in all obese patients, insulin resistance evolves to diabetes in ~40% of patients over 5–10 years. Currently over 220 million people worldwide have Type 2 diabetes and the number is predicted to increase dramatically over the next 20 years. More alarmingly, Type 2 diabetes has replaced Type 1 diabetes as the predominant form of this disease among children. Prevention and treatment of Type 2 diabetes is clearly a major medical priority; in many cases insulin resistance is reversible with weight loss and increased exercise, particularly in the early stages.

In his article, Dennis McGarry noted that the re-esterification of fatty acids to triacylglycerol is a critical factor in regulating their concentration in the blood. This prediction is amply supported by the current, widespread clinical use of the antidiabetic thiazolidinedione drugs, which act at the nuclear receptor, PPARγ2. These drugs lower plasma-free fatty acids, in part by increasing the rate of fatty acid esterification to triacylglycerol in white adipose tissue, thereby removing fatty acids from the blood. Dr. McGarry was clearly ahead of his time in predicting the importance of the regulation of fatty acid metabolism in the control of diabetes.

McGarry, J. D. What if Minkowski had been ageustic? An alternative angle on diabetes. *Science* 285: 766, 1992; Roden, M. How FFA inhibits glucose utilization in human skeletal muscle. *News Physiol. Sci.* 19: 92, 2004; Randle, P. J. Regulatory interactions between lipids and carbohydrates: the glucose fatty acid cycle after 35 years. *Diabetes Metabol. Rev.* 14: 263, 1998; and Zimmet, P. Alberti, K. G. M. M., and Shaw, J. Global and social implications of the diabetes epidemic. *Nature* 414: 782, 2001.

glycogen. In addition, triacylglycerols are stored without associated water, whereas glycogen is hydrophilic and binds about twice its weight of water. Therefore, the energy yield per gram of stored triacylglycerol is about 4 times that of hydrated glycogen. Humans store much more fuel as triacylglycerol than as glycogen. The average 70-kg human stores about 250 g of glycogen in liver and muscle. This represents about 1000 kcal of energy, which is less than one day's energy requirement. By contrast, the same individual stores ~16 kg of triacylglycerol, which provides sufficient energy to survive several weeks of starvation. Unlike stores of glycogen and amino acids, which are limited, triacylglycerol stores, can greatly expand, depending on caloric intake. This makes triacylglycerol an excellent fuel store, but it can also lead to obesity and, in some cases, diabetes (Clin. Corrs. 17.1 and 17.2). Adipose tissue is a significant source of hormones that regulate appetite, and subcutaneous fat is also important in temperature regulation because it provides a layer of insulation.

17.3 • INTERORGAN TRANSPORT OF FATTY ACIDS AND THEIR PRIMARY PRODUCTS

In all mammals, the transport and storage of fatty acids is regulated by dietary status. Triacylglycerol is stored in the fed state, with net deposition in adipose tissue. During fasting, triacylglycerol in adipose tissue is hydrolyzed and the products are distributed throughout the body to be used for energy production. As fasting progresses, the liver converts fatty acids to the **ketone bodies, acetoacetate**, and ***β*-hydroxybutyrate**, which are released into the blood and become a major source of energy for many tissues. These processes are summarized in Figure 17.6.

The transport of lipids presents unique problems because these molecules (especially triacylglycerols, cholesterol and its esters) are insoluble in water. Consequently, lipids are transported through the bloodstream in plasma lipoproteins or bound to proteins. In addition, the transport of triacylglycerols across membranes usually involves breakdown to smaller constituents by lipases.

Lipid Transport in the Fed State

Triacylglycerols in the diet are digested in the stomach and small intestine by **gastric** and **pancreatic lipases** (p. 1052). The principal products are 2-monoacylglycerols and **free fatty acids**, which are absorbed by the epithelial cells that line the small intestine. These cells assemble the absorbed fatty acids and monoacylglycerols into triacylglycerols (p. 688),

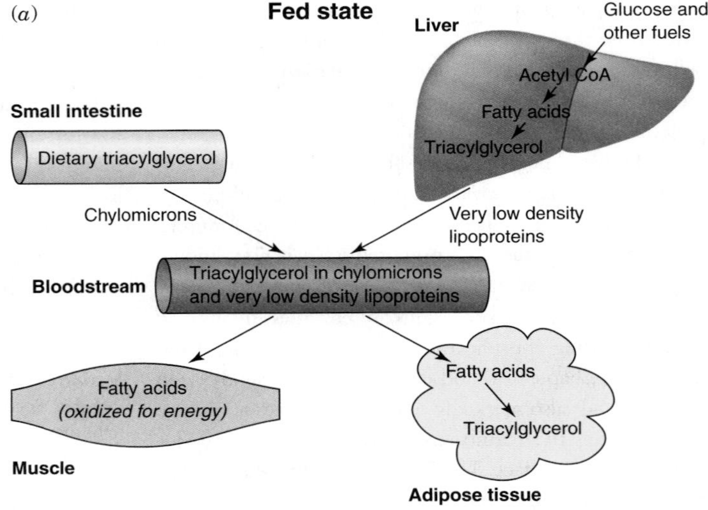

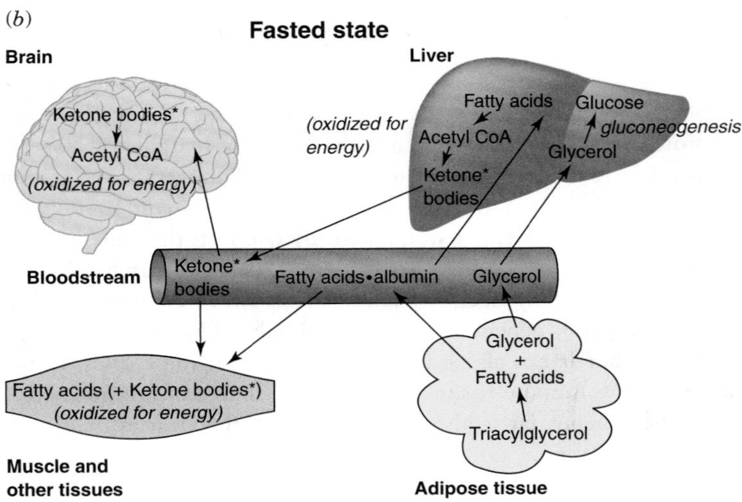

Figure 17.6 Interorgan transport of fatty acids. (*a*) In the fed state, there is net deposition of triacylglycerol in adipose tissue. The sources are dietary fat and fatty acids synthesized in the liver from excess carbohydrate and amino acids. (*b*) In the fasted state, triacylglycerols are hydrolyzed and free fatty acids and glycerol are released into the blood.
*During fasting, the liver synthesizes ketone bodies, which become a major fuel in the blood as fasting progresses.

which are then packaged into **chylomicrons**, a triacylglycerol-rich plasma lipoprotein (p. 1051). Chylomicrons are secreted into lymph and then circulate in the bloodstream.

The liver is another source of triacylglycerols in the fed state. Fatty acids are synthesized in this tissue from excess carbohydrate and amino acids. These fatty acids are assembled into triacylglycerols and packaged into **very low-density lipoprotein (VLDL)**, a second triacylglycerol-rich lipoprotein, which is secreted into the bloodstream (p. 724).

The triacylglycerols in chylomicrons and VLDL are hydrolyzed by **lipoprotein lipase**, located on the surface of capillary endothelial cells in tissues such as adipose tissue and skeletal muscle. The apoprotein ApoC-II, which is found in chylomicrons and VLDL, activates the process by binding the lipoproteins to the enzyme. The products of lipoprotein lipase action are free fatty acids and glycerol. The free fatty acids are utilized in the tissue where hydrolysis occurs. The glycerol is transported through the bloodstream and taken up by the liver where it is used in glycolysis or gluconeogenesis.

Lipoprotein lipase is present at high levels in adipose tissue, cardiac muscle, and skeletal muscle, allowing these tissues to utilize triacylglycerols from lipoproteins. In adipose tissue, the products of lipoprotein lipase are taken up and assembled into triacylglycerols, allowing net deposition of fuel (p. 688). In muscle, the fatty acid products of lipase action are taken up and used to generate energy, although some triacylglycerol synthesis occurs in this tissue.

Lipid Transport in the Fasted State

The triacylglycerols stored in adipose tissue are mobilized for use as fuel in the fasted state. This process is initiated by **hormone-sensitive lipase**, which is located within adipocytes. This enzyme is activated when it is phosphorylated by cAMP-dependent protein kinase A. Conversely, insulin inhibits the activity of this enzyme by inducing its dephosphorylation. The protein, **perilipin**, which coats the surface of fat droplets, is also important in this regulation. When perilipin is not phosphorylated, it blocks lipase access to the triacylglycerol. When perilipin is phosphorylated by protein kinase A, hormone-sensitive lipase translocates to the surface of the fat droplet and hydrolyzes triacylglycerols. The principal product of this enzyme is monoacylglycerol and free fatty acids. Other enzymes complete the hydrolysis. A summary of the regulation of triacylglycerol metabolism is presented in Table 17.2. This regulation results in the net mobilization of fatty acids during fasting and net deposition after a meal. The balance of triacylglycerol synthesis and hydrolysis helps to assure adequate energy stores and avoid obesity (see Clin. Corrs 17.1 and 17.2). Other lipases, including **adipose triglyceride lipase**, may also play a role in the degradation of triacylglycerols. These may contribute to slow unregulated **lipolysis**.

Other lipases rapidly complete the hydrolysis, releasing fatty acids and glycerol into the blood. The fatty acids are referred to as *free fatty acids* although they circulate bound to serum albumin. Each albumin molecule can bind ~10 fatty acids, so its binding capacity is very high. Free fatty acids bound to albumin, however, are a relatively small fraction of the total lipid in plasma when one considers the presence of the lipids in the plasma lipoproteins (p. 724). However, these free fatty acids turn over rapidly, so they represent a significant fraction of the lipid flux through the bloodstream.

The hydrolysis of triacylglycerols present in adipose tissue and in plasma lipoproteins also produces free glycerol. This glycerol is used by the liver, which contains high levels of glycerol kinase, an enzyme that synthesizes glycerol 3-phosphate from glycerol and ATP

TABLE 17.2 • Regulation of Triacylglycerol Metabolism

Protein	*Effect*	*Regulatory Agent*
	Triacylglycerol Mobilization	
Hormone-sensitive lipase and perilipin	Activation by PKA-mediated phosphorylation	Glucagon, epinephrine, ACTH
	Inhibition by dephosphorylation	Insulin
Lipoprotein lipase	Increased enzyme synthesis	Insulin
	Triacylglycerol Synthesis	
Phosphatidate phosphatase	Increased enzyme synthesis	Steroid hormones

(see Figure 15.39, p. 625). Glycerol is metabolized sparingly in other tissues due to low levels of this enzyme. Hepatic glycerol 3-phosphate dehydrogenase converts glycerol 3-phosphate to dihydroxyacetone phosphate, which enters the glycolytic pathway in the fed state. In the fasted state, it is converted to glucose via gluconeogenesis. During prolonged fasting, when much of the body's energy is derived from stored fat, the glycerol produced by the hydrolysis of triacylglycerol in adipose tissue is an important substrate for gluconeogenesis in the liver.

17.4 • SYNTHESIS OF FATTY ACIDS: LIPOGENESIS

Glucose Is the Major Precursor for Fatty Acid Synthesis

Dietary carbohydrate in excess of that needed for energy production and glycogen synthesis is converted to fatty acids in the liver during the fed state in mammals. Glucose provides the carbon for fatty acid synthesis (via acetyl CoA) as well as the reducing equivalents (NADPH) required for this process. Other substrates, such as amino acids, can also contribute to lipogenesis.

Pathway of Fatty Acid Biosynthesis

The synthesis of fatty acids occurs in the cytosol using acetyl CoA produced from glucose or from other precursors (i.e., the carbon skeletons of amino acids). The saturated C_{16} acid, palmitic acid, is synthesized first, and all other fatty acids are made by its modification. Fatty acids are synthesized by sequential addition of two-carbon units from acetyl CoA to the activated carboxyl end of a growing chain by **fatty acid synthase**. In bacteria, fatty acid synthase is a complex of several proteins, whereas in mammalian cells it is a single multifunctional protein.

Formation of Malonyl CoA Is the Commitment Step of Fatty Acid Synthesis

Synthesis of fatty acids from acetyl CoA requires the activated intermediate, malonyl CoA, which is made by carboxylation of acetyl CoA by acetyl-CoA carboxylase (Figure 17.7). The reaction requires ATP and bicarbonate as the source of CO_2. In the first step, CO_2 is linked to a biotin moiety on the enzyme, using energy derived from ATP hydrolysis; the CO_2 is then transferred to acetyl CoA. This reaction is similar to the carboxylation of pyruvate to oxaloacetate by pyruvate carboxylase (p. 622).

Acetyl-CoA carboxylase catalyzes the rate-limiting step in fatty acid synthesis. This enzyme exists in an inactive, protomeric state and, when activated, as linear polymers. The mammalian enzyme is activated by citrate or isocitrate. This represents feed-forward activation of fatty acid synthesis because citrate is exported from the mitochondria to generate cytosolic acetyl-CoA for fatty acid synthesis (see the section on the citrate cleavage pathway on p. 684). Acetyl-CoA carboxylase is also inhibited by long-chain acyl CoAs, resulting in feedback inhibition of fatty acid synthesis by the end product of the pathway. Phosphorylation of acetyl-CoA carboxylase by both cyclic-AMP-dependent protein kinase A and AMP-dependent protein kinases inhibit this enzyme. The importance of this regulation is discussed on page 702.

Reaction Sequence for the Synthesis of Palmitic Acid

The first step catalyzed in fatty acid synthesis is the transfer of the primer molecule, either an acetyl or butyryl group from CoA to a 4′-phosphopantetheine moiety on **acyl carrier protein** (ACP), a protein constituent of the multienzyme complex (Figure 17.8, Reaction 1). The phosphopantetheine unit is identical with that in CoA; both are derived from the

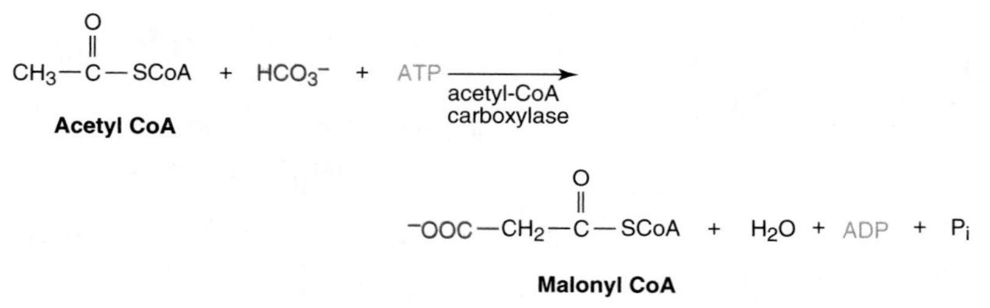

$$\underset{\textbf{Acetyl CoA}}{CH_3-\overset{O}{\overset{\|}{C}}-SCoA} + HCO_3^- + ATP \xrightarrow[\text{carboxylase}]{\text{acetyl-CoA}} \underset{\textbf{Malonyl CoA}}{^-OOC-CH_2-\overset{O}{\overset{\|}{C}}-SCoA} + H_2O + ADP + P_i$$

Figure 17.7 Acetyl-CoA carboxylase reaction.

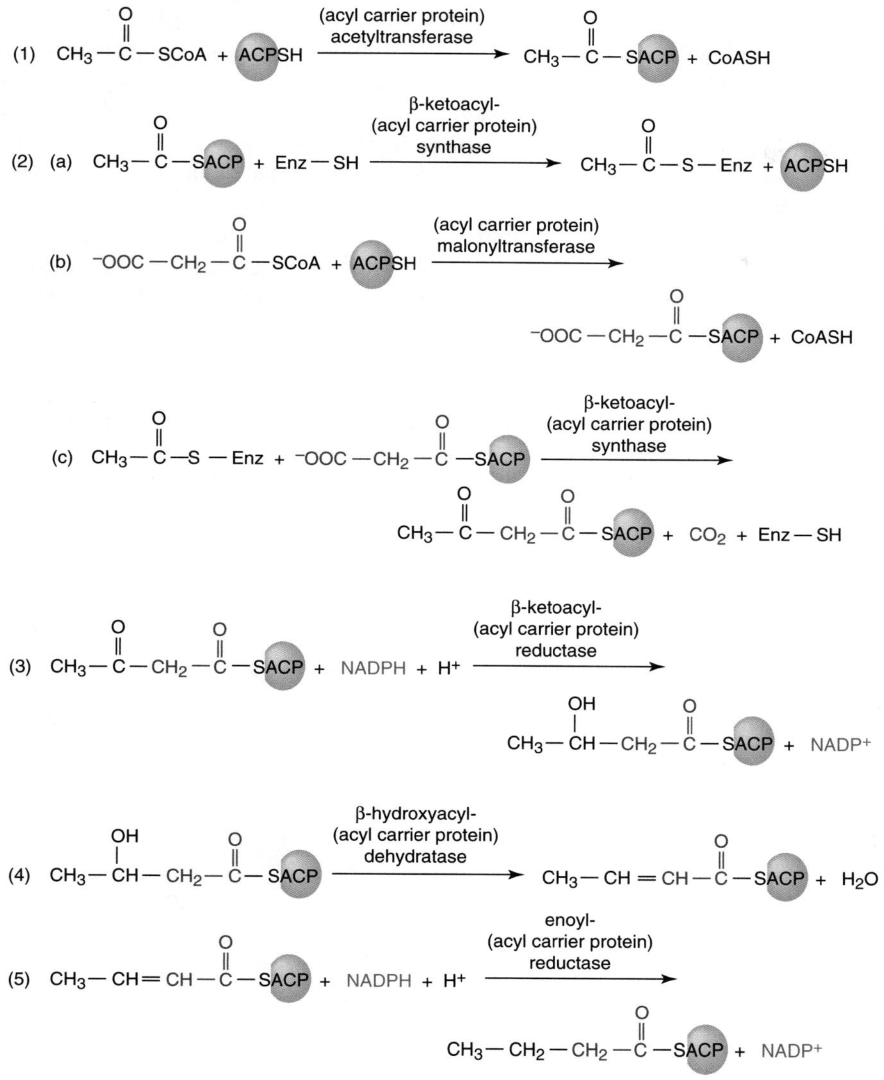

(1) $CH_3{-}C(=O){-}SCoA$ + ACPSH $\xrightarrow{\text{(acyl carrier protein) acetyltransferase}}$ $CH_3{-}C(=O){-}SACP$ + CoASH

(2) (a) $CH_3{-}C(=O){-}SACP$ + Enz—SH $\xrightarrow{\beta\text{-ketoacyl-(acyl carrier protein) synthase}}$ $CH_3{-}C(=O){-}S{-}Enz$ + ACPSH

(b) $^{-}OOC{-}CH_2{-}C(=O){-}SCoA$ + ACPSH $\xrightarrow{\text{(acyl carrier protein) malonyltransferase}}$ $^{-}OOC{-}CH_2{-}C(=O){-}SACP$ + CoASH

(c) $CH_3{-}C(=O){-}S{-}Enz$ + $^{-}OOC{-}CH_2{-}C(=O){-}SACP$ $\xrightarrow{\beta\text{-ketoacyl-(acyl carrier protein) synthase}}$ $CH_3{-}C(=O){-}CH_2{-}C(=O){-}SACP$ + CO_2 + Enz—SH

(3) $CH_3{-}C(=O){-}CH_2{-}C(=O){-}SACP$ + NADPH + H^+ $\xrightarrow{\beta\text{-ketoacyl-(acyl carrier protein) reductase}}$ $CH_3{-}CH(OH){-}CH_2{-}C(=O){-}SACP$ + $NADP^+$

(4) $CH_3{-}CH(OH){-}CH_2{-}C(=O){-}SACP$ $\xrightarrow{\beta\text{-hydroxyacyl-(acyl carrier protein) dehydratase}}$ $CH_3{-}CH{=}CH{-}C(=O){-}SACP$ + H_2O

(5) $CH_3{-}CH{=}CH{-}C(=O){-}SACP$ + NADPH + H^+ $\xrightarrow{\text{enoyl-(acyl carrier protein) reductase}}$ $CH_3{-}CH_2{-}CH_2{-}C(=O){-}SACP$ + $NADP^+$

Figure 17.8 Pathway of fatty acid synthesis.

vitamin, pantothenic acid. In bacteria ACP is a small protein, whereas in mammals it is a domain of fatty acid synthase. Six or seven two-carbon units are added to the enzyme complex (depending on whether acetyl CoA or butyryl CoA is the primer) in a repetitive sequence of reactions until the palmitate molecule is completed. The reaction sequence, starting with an acetyl CoA primer and leading to butyryl-ACP, is presented in Figure 17.8.

A round of fatty acid elongation is initiated by the addition of two carbon atoms to the chain in a three-step process. First the fatty acid chain is transferred from the 4′-phosphopantetheine moiety of ACP to a cysteine sulfhydryl group of β-ketoacyl-ACP synthase (Reaction 2a). The -SH group of ACP then accepts a malonyl unit from malonyl CoA (Reaction 2b). Then the condensation step links the growing acyl chain to C2 of the malonyl group, with the loss of CO_2 (Reaction 2c). This is the same CO_2 that was added by acetyl-CoA carboxylase, so the carbon atoms in palmitate are derived entirely from acetyl CoA. The acyl product is β-ketoacyl-ACP. This intermediate is reduced to β-hydroxyacyl-ACP using NADPH as electron donor (Reaction 3). β-Hydroxyacyl-ACP is dehydrated to an enoyl-ACP (Reaction 4) and then reduced to a saturated acyl CoA, using a second molecule of NADPH as reductant (Reaction 5). The growing fatty acid proceeds through six more reaction cycles to produce palmitoyl-ACP, which then is acted on by a thioesterase to produce free palmitic acid (Figure 17.9). The specificity of this enzyme determines the chain length of the fatty acid product. Note that at this stage, the sulfhydryl groups of ACP and synthase are both free, so that another cycle of fatty acid synthesis can begin. The released product is converted to palmitoyl CoA, preparing it for modification or incorporation into complex lipids.

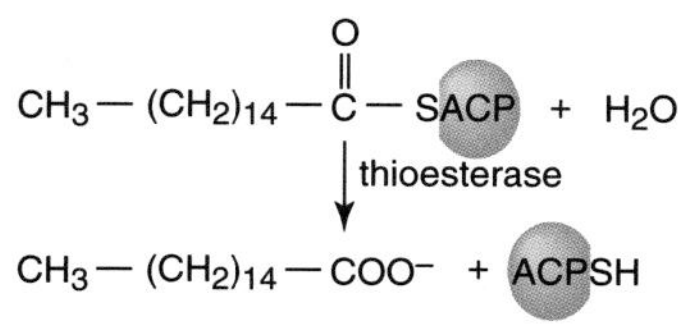

$CH_3{-}(CH_2)_{14}{-}C(=O){-}SACP$ + H_2O $\xrightarrow{\text{thioesterase}}$ $CH_3{-}(CH_2)_{14}{-}COO^-$ + ACPSH

Figure 17.9 Release of palmitic acid from fatty acid synthase by thioesterase.

Mammalian Fatty Acid Synthase Is a Multifunctional Polypeptide

The reactions outlined above are the basic sequence for fatty acid synthesis in all organisms. The enzyme complex, fatty acid synthase, catalyzes all these reactions, but its structure and properties vary considerably. In *Escherichia coli,* the complex is composed of separate enzymes. By contrast, mammalian fatty acid synthase is composed of two identical subunits each of which is a multienzyme polypeptide that contains all of the catalytic activities. There are also variations in the enzyme between mammalian species and tissues.

The growing fatty acid chain is always bound to the multifunctional protein and is transferred sequentially between the 4′-phosphopantetheine group of ACP, which is a domain of the protein, and the cysteine sulfhydryl group of β-ketoacyl-ACP synthase, during the condensation reaction (Reaction 2, Figures 17.8 (p. 681) and 17.10).

The levels of fatty acid synthase in tissues are controlled by the rate of its synthesis and degradation. As shown in Table 17.3, insulin increases the rate of fatty acid synthesis by stimulating transcription of the fatty acid synthase gene, thereby increasing enzyme levels

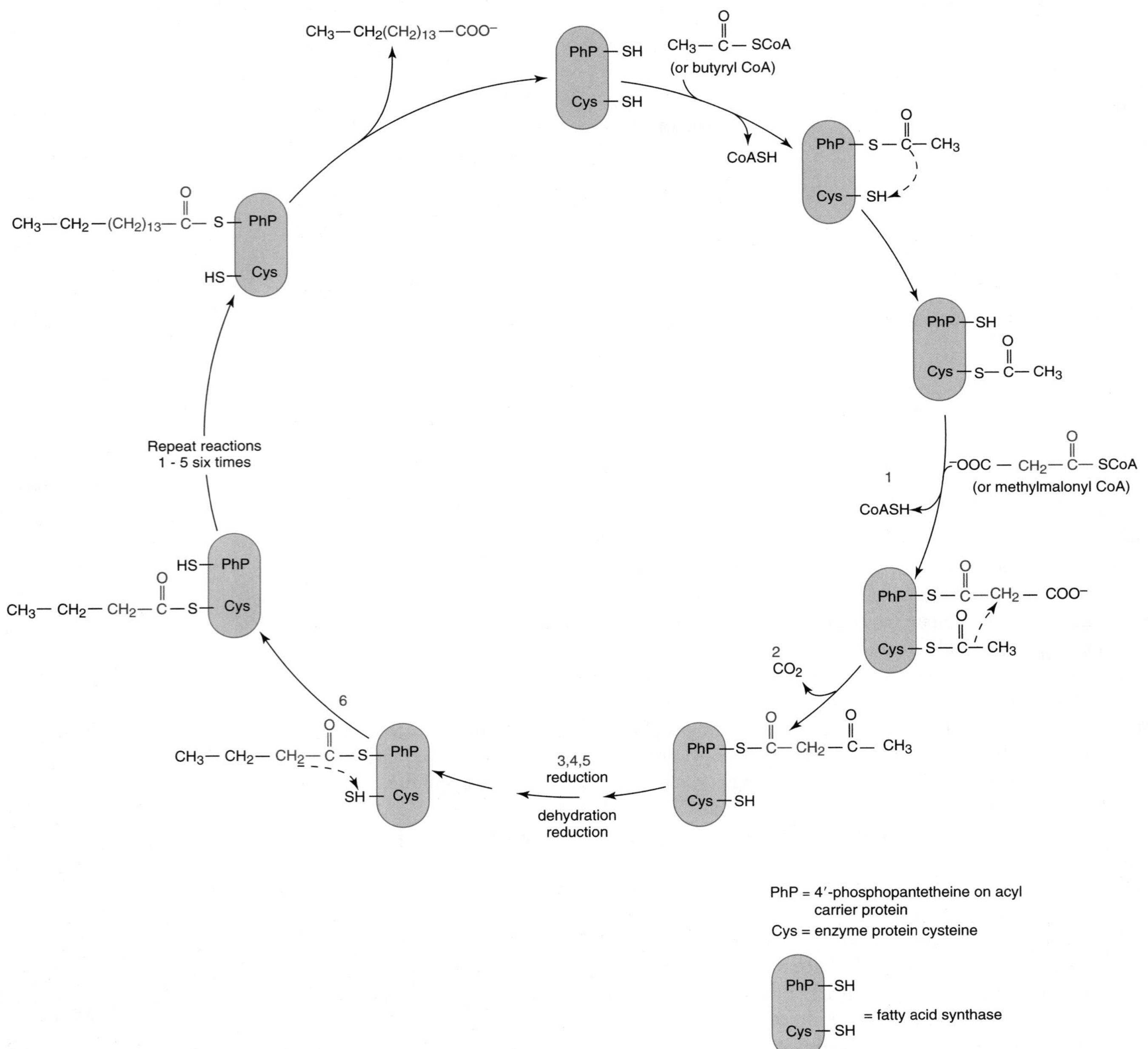

Figure 17.10 Proposed mechanism of fatty acid elongation by mammalian fatty acid synthase.

TABLE 17.3 • Regulation of Fatty Acid Synthesis

Enzyme		*Effect*	*Regulatory Agent*
		Palmitate Synthesis	
Acetyl-CoA carboxylase (ACC)	Short term	Allosteric activation	Citrate, isocitrate
		Allosteric inhibition	C16–C18 acyl CoAs
		Inhibition via PKA phosphorylation of ACC	Glucagon and epinephrine
		Inhibition via AMPK-mediated phosphorylation of ACC	AMP
		Activation by dephosphorylation of ACC	Insulin
	Long term	Increased enzyme synthesis	Thyroid hormone, high carbohydrate diet, insulin
		Decreased enzyme synthesis	Fasting, high fat diet, via low insulin and high glucagon. Polyunsaturated fatty acids (PUFAs) also inhibit synthesis
Fatty acid synthase	Long term	Increased enzyme synthesis	High carbohydrate diet via insulin
		Decreased enzyme synthesis	Fasting, high fat diet, via low insulin and high glucagon. PUFAs also inhibit synthesis
		Biosynthesis of Fatty Acids Other than Palmitate	
Fatty acid synthase		Increased synthesis of branched-chain fatty acids	Increased methylmalonyl CoA/malonyl CoA
		Synthesis of short-chain fatty acids	Expression of specific thioesterase in mammary gland
Stearoyl-CoA desaturase		Increased enzyme synthesis	Insulin, thyroid hormones, hydrocortisone, dietary cholesterol
		Decreased enzyme synthesis	Dietary PUFAs

in the liver and other tissues. The factors that are involved in the regulation of lipogenesis in mammals are presented on page 684.

Stoichiometry of Fatty Acid Synthesis

With acetyl CoA as the primer for palmitate synthesis, the overall reaction is

$$CH_3{-}\overset{\overset{\displaystyle O}{\|}}{C}{-}SCoA + 7\ {}^{-}OOC{-}CH_2{-}\overset{\overset{\displaystyle O}{\|}}{C}{-}SCoA + 14\ NADPH + 14\ H^+ \longrightarrow$$
$$CH_3{-}(CH_2)_{14}{-}COO^- + 7\ CO_2 + 14\ NADP^+ + 8\ CoASH + 6\ H_2O$$

To describe the overall conversion of acetyl CoA to palmitate, the ATP used in formation of malonyl CoA must be included.

$$7\ CH_3{-}\overset{\overset{\displaystyle O}{\|}}{C}{-}SCoA + 7\ CO_2 + 7\ ATP \longrightarrow 7\ {}^{-}OOC{-}CH_2{-}\overset{\overset{\displaystyle O}{\|}}{C}{-}SCoA + 7\ ADP + 7\ P_i$$

Then the stoichiometry for conversion of acetyl CoA to palmitate is

$$8\ CH_3{-}\overset{\overset{\displaystyle O}{\|}}{C}{-}SCoA + 7\ ATP + 14\ NADPH + 14\ H^+ \longrightarrow$$
$$CH_3{-}(CH_2)_{14}{-}\overset{\overset{\displaystyle O}{\|}}{C}{-}O^- + 8\ CoASH + 7\ ADP + 7\ P_i + 6\ H_2O + 14\ NADP^+$$

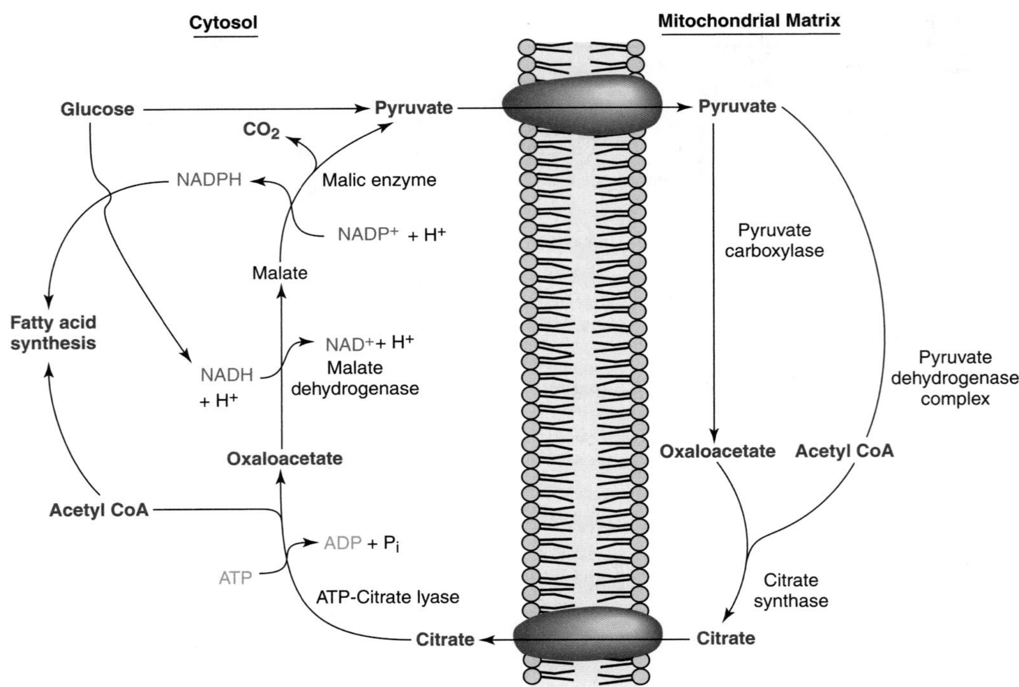

Figure 17.11 **Transfer of acetyl CoA from mitochondria to cytosol for fatty acid biosynthesis by the citrate cleavage pathway.**

The Citrate Cleavage Pathway Provides Acetyl CoA and NADPH for Lipogenesis in the Cytosol

Glucose breakdown in the liver via glycolysis results in the production of pyruvate, which is converted to acetyl CoA in the mitochondria by pyruvate dehydrogenase complex. However, the synthesis of fatty acids is a cytosolic process and acetyl CoA is not readily transported across the inner mitochondrial membrane. The **citrate cleavage pathway** overcomes this problem. Citrate, formed by citrate synthase in the tricarboxylic acid cycle, is transported across the mitochondrial inner membrane via the tricarboxylate transporter (see Figure 14.49, p. 578). The citrate is then cleaved in the cytosol by **ATP citrate lyase** (also called **citrate cleavage enzyme**) to form acetyl CoA and oxaloacetate (Figure 17.11).

$$\text{citrate} + \text{ATP} + \text{CoA} \rightarrow \text{acetyl CoA} + \text{oxaloacetate} + \text{ADP} + \text{P}_i$$

This reaction is not a reversal of citrate synthase, since it requires the hydrolysis of ATP. As mentioned previously, citrate has a second role in fatty acid synthesis as an allosteric activator of acetyl-CoA carboxylase, the rate-limiting enzyme in lipogenesis.

The oxaloacetate generated by citrate cleavage in the cytosol is not readily transported back into the mitochondria. It is instead reduced to malate by the cytosolic isoform of **NAD malate dehydrogenase**. The malate then undergoes oxidative decarboxylation to pyruvate, catalyzed by **NADP malate dehydrogenase** (also called **malic enzyme**). The pyruvate enters the mitochondria for further metabolism. The removal of citrate from the mitochondria (**cataplerosis**) must be accompanied by its replacement (**anaplerosis**) in order to maintain tricarboxylic acid cycle flux. This is achieved by the conversion of pyruvate to oxaloacetate by **pyruvate carboxylase** (p. 622), the major anaplerotic enzyme in mitochondria.

The conversion of oxaloacetate to pyruvate by the citrate cleavage pathway transfers a pair of electrons from NADH to generate NADPH that is used for fatty acid synthesis. The source of the NADH for this process is the glyceraldehyde-3-phosphate dehydrogenase reaction in the glycolytic pathway. The metabolism of glucose thus provides both the carbon atoms and the reducing equivalents required for lipogenesis.

The production of the reducing equivalents used in fatty acid synthesis can be summarized as follows: The citrate cleavage pathway transfers a pair of electrons from NADH to NADPH for each acetyl CoA transferred from mitochondria to cytosol. The transfer of the eight acetyl CoAs that are used for the synthesis of one molecule of palmitate supplies eight NADPHs. Since palmitate synthesis requires 14 NADPHs per mol, the other six NADPHs are supplied by the pentose phosphate pathway, which is also present in the cytosol. The actual stoichiometry of this process is more complicated because transport of citrate and other di- and tricarboxylic acids across the inner mitochondrial membrane occurs by one-for-one exchanges. The flow rates are probably controlled by a composite of the concentration gradients of several of these exchange systems.

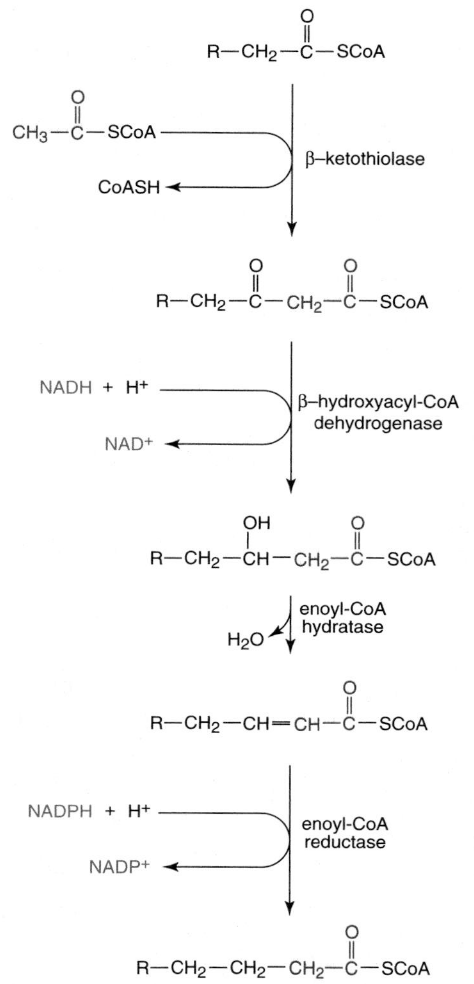

Modification of Fatty Acids

Humans can synthesize the other fatty acids they require from palmitate, except for some polyunsaturated fatty acids (see Table 17.1, p. 675). Fatty acids are modified as the CoA derivatives by three processes: elongation, desaturation, and hydroxylation.

Elongation Reactions

Fatty acid elongation in mammals occurs in either the endoplasmic reticulum or the mitochondria. These processes are slightly different in the two locations. In the endoplasmic reticulum, fatty acyl CoAs are elongated by reactions similar to the ones catalyzed by cytosolic fatty acid synthase; malonyl CoA is the source of two-carbon units and NADPH provides the reducing power. The preferred elongation substrate is palmitoyl CoA, which is converted almost exclusively to stearate (18:0) in most tissues. The brain, however, contains one or more additional elongation systems, which synthesize longer-chain acids (up to C_{24}) that are required for membrane lipids. These elongation systems also use malonyl CoA.

Fatty acid elongation in mitochondria uses acetyl CoA, and both NADH and NADPH as electron donors (Figure 17.12). This system operates by reversal of the pathway of fatty acid β-oxidation (p. 692) with the exception that NADPH-linked enoyl-CoA reductase (last step of elongation) replaces FAD-linked acyl-CoA dehydrogenase (first step in β-oxidation). This process has little activity with acyl CoA substrates of 16 carbons or longer; it serves primarily to elongate shorter molecules.

$$R{-}CH_2{-}\overset{O}{\overset{\|}{C}}{-}SCoA$$

$CH_3{-}\overset{O}{\overset{\|}{C}}{-}SCoA$ → CoASH (β-ketothiolase)

$$R{-}CH_2{-}\overset{O}{\overset{\|}{C}}{-}CH_2{-}\overset{O}{\overset{\|}{C}}{-}SCoA$$

NADH + H^+ → NAD^+ (β-hydroxyacyl-CoA dehydrogenase)

$$R{-}CH_2{-}\overset{OH}{\overset{|}{C}H}{-}CH_2{-}\overset{O}{\overset{\|}{C}}{-}SCoA$$

→ H_2O (enoyl-CoA hydratase)

$$R{-}CH_2{-}CH{=}CH{-}\overset{O}{\overset{\|}{C}}{-}SCoA$$

NADPH + H^+ → $NADP^+$ (enoyl-CoA reductase)

$$R{-}CH_2{-}CH_2{-}CH_2{-}\overset{O}{\overset{\|}{C}}{-}SCoA$$

Figure 17.12 Pathway of fatty acid elongation in mitochondria.

Formation of Monoenoic Acids by Stearoyl-CoA Desaturase

In vertebrates, **fatty acid desaturation** occurs in the endoplasmic reticulum, and the reactions and enzymes that introduce cis double bonds are significantly different from the acyl-CoA dehydrogenases of mitochondrial β-oxidation. Desaturation is carried out by **monooxygenases**, which have fatty acyl CoA, NADH and O_2 as substrates (p. 695). The three components of the system are the **desaturase enzyme, cytochrome b_5**, and **NADPH-cytochrome b_5 reductase**. The overall reaction is.

$$R{-}CH_2{-}CH_2{-}(CH_2)_7{-}\overset{O}{\overset{\|}{C}}{-}SCoA + NADPH + H^+ + O_2 \rightarrow$$
$$R{-}CH{=}CH{-}(CH_2)_7{-}\overset{O}{\overset{\|}{C}}{-}SCoA + NADP^+ + 2H_2O$$

The initial step in the formation of unsaturated fatty acids is the formation of the Δ^9 double bond by **stearoyl-CoA desaturase** in palmitic or stearic acid to produce palmitoleic or oleic acid, respectively. The synthesis of unsaturated fatty acids is important for regulating the fluidity of triacylglycerols and membrane phospholipids. It is also required for the synthesis of cholesterol esters in the liver and waxy secretions in the skin, which preferentially use newly synthesized, rather than dietary fatty acids. Expression of stearoyl-CoA desaturase is highly regulated by both dietary and hormonal mechanisms. Insulin, triiodothyronine, hydrocortisone, and dietary cholesterol increase gene transcription and thus the levels of stearoyl-CoA desaturase in the liver, whereas dietary polyunsaturated fatty acids have the opposite effect.

Formation and Modification of Polyunsaturated Fatty Acids

Polyunsaturated fatty acids, particularly arachidonic acid (20:4), are precursors of important signaling molecules: prostaglandins, thromboxanes, and leukotrienes (p. 737). Polyunsaturated fatty acids are also required for the synthesis of complex lipids, particularly in the nervous system. These fatty acids can also undergo nonenzymatic oxidation reactions, creating products that damage cellular constituents and may have pathological effects.

In mammals, double bonds can be added only to the proximal half of fatty acyl CoAs; they cannot be added beyond C9. Consequently, linoleic (18:2) and linolenic (18:3) are required polyunsaturated fatty acids in the diet (Figure 17.13). These **essential fatty acids**

$$CH_3{-}(CH_2)_3{-}(CH_3{-}CH{=}CH)_n{-}(CH_2)_m{-}COOH$$

Basic formula of the linoleic acid series

$$CH_3{-}(CH_2{-}CH{=}CH)_n{-}(CH_2)_m{-}COOH$$

Basic formula of the linolenic acid series

Figure 17.13 The linoleic and linolenic acid series.

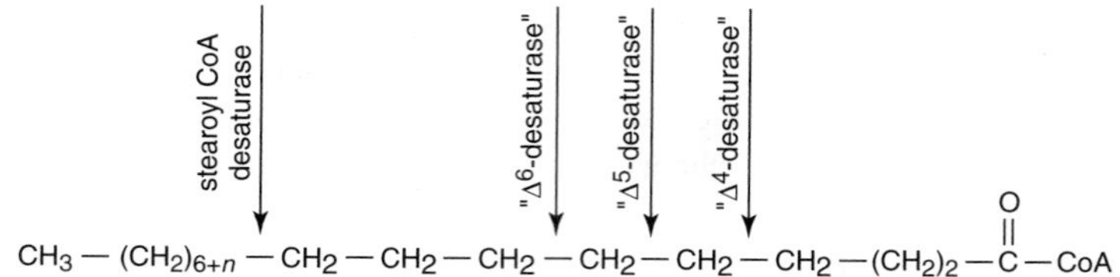

Figure 17.14 Positions in the fatty acid chain where desaturation can occur in the human. In mammals, there must be at least six carbons beyond the bond being desaturated.

are obtained in the diet from plants and cold-water fish. There are two series of polyunsaturated fatty acids. The distal double bond of linoleic acid is six carbons from the methyl group; it is referred to as n-6 or ω-6. A second series, n-3 or ω-3, has the distal double bond three carbons from the methyl group. One isomer of linolenic acid is part of this series.

The essential fatty acids are modified by elongation and desaturation to form the polyunsaturated fatty acids found in mammals. A variety of polyunsaturated fatty acids are synthesized by humans using desaturases that introduce double bonds at positions 4, 5, or 6 (Figure 17.14). These enzymes act only in the synthesis of polyunsaturated fatty acids because they can use only substrates with a double bond at position 9. This elongation and desaturation occurs in either order. Conversion of linolenic acid to all *cis*-4,7,10,13,16,19-docosahexaenoic acid in the brain is an example of such a reaction sequence.

$$\mathrm{CH_3{-}(CH_2{-}CH{=}CH)_3{-}CH_2{-}CH_2{-}CH_2{-}CH_2{-}CH_2{-}CH_2{-}CH_2{-}\overset{O}{\overset{\|}{C}}{-}CoA}$$

Linolenic acid

↓ "Δ^6-desaturase"

$$\mathrm{CH_3{-}(CH_2{-}CH{=}CH)_3{-}CH_2{-}CH{=}CH{-}CH_2{-}CH_2{-}CH_2{-}CH_2{-}\overset{O}{\overset{\|}{C}}{-}CoA}$$

↓ elongation

$$\mathrm{CH_3{-}(CH_2{-}CH{=}CH)_3{-}CH_2{-}CH{=}CH{-}CH_2{-}CH_2{-}CH_2{-}CH_2{-}CH_2{-}CH_2{-}\overset{O}{\overset{\|}{C}}{-}CoA}$$

↓ "Δ^5-desaturase"

$$\mathrm{CH_3{-}(CH_2{-}CH{=}CH)_3{-}CH_2{-}CH{=}CH{-}CH_2{-}CH{=}CH{-}CH_2{-}CH_2{-}CH_2{-}\overset{O}{\overset{\|}{C}}{-}CoA}$$

↓ elongation

$$\mathrm{CH_3{-}(CH_2{-}CH{=}CH)_3{-}CH_2{-}CH{=}CH{-}CH_2{-}CH{=}CH{-}CH_2{-}CH_2{-}CH_2{-}CH_2{-}CH_2{-}\overset{O}{\overset{\|}{C}}{-}CoA}$$

↓ "Δ^4-desaturase"

$$\mathrm{CH_3{-}(CH_2{-}CH{=}CH)_3{-}CH_2{-}CH{=}CH{-}CH_2{-}CH{=}CH{-}CH_2{-}CH{=}CH{-}CH_2{-}CH_2{-}\overset{O}{\overset{\|}{C}}{-}CoA}$$

All-*cis*-4,7,10,13,16,19-docosahexaenoic acid

Formation of Hydroxy Fatty Acids in Nerve Tissue

Two different processes produce **α-hydroxy fatty acids** in vertebrates. One occurs in the mitochondria of many tissues and acts on relatively short-chain fatty acids. The other has been demonstrated only in the nervous system, where it produces long-chain fatty acids hydroxylated on C2 that are required for the formation of some myelin lipids. The enzyme in the brain that catalyzes this reaction is a monooxygenase that requires O_2 and NADH or NADPH and preferentially uses C_{22} and C_{24} fatty acids. This process is closely coordinated with the synthesis of sphingolipids that contain hydroxylated fatty acids (see p. 729).

Fatty Acid Synthase Can Produce Fatty Acids Other than Palmitate

The principal fatty acids synthesized by humans for energy storage are palmitate and its modification products. However, smaller amounts of different fatty acids are synthesized for other purposes. Two examples are the production of the fatty acids shorter than palmitate in the **mammary gland** and the synthesis of branched-chain fatty acids in some secretory glands.

Milk contains fatty acids with chains that are shorter than palmitate. The relative amounts of the fatty acids produced by the mammary gland vary with species and with the physiological state of the animal. In ruminants, the pathway of fatty acid synthesis, which normally produces palmitate, is modified to synthesize fatty acids as short as C_4. This is accomplished by the expression of soluble **thioesterases**, which cleave the shorter chains from fatty acid synthase. Human milk contains no fatty acids with chains shorter than 10 carbons.

There are relatively few branched-chain fatty acids in vertebrates. Until recently, their metabolism has been studied mostly in bacteria such as Mycobacteria, where they are present in a greater variety and amount. Simple, branched-chain fatty acids are synthesized by tissues of vertebrates for specific purposes, such as the production of waxes in sebaceous glands and avian preen glands, and the elaboration of structures in echo-locating systems of porpoises.

Most of the branched-chain fatty acids in vertebrates are methylated derivatives of saturated, straight-chain acids that are synthesized by fatty acid synthase. When **methylmalonyl CoA** is used as a substrate instead of malonyl CoA, a methyl side chain is inserted into the fatty acid by the following reaction:

$$CH_3-(CH_2)_n-\overset{\overset{\displaystyle O}{\|}}{C}-SACP + HOOC-\overset{\overset{\displaystyle CH_3}{|}}{CH}-\overset{\overset{\displaystyle O}{\|}}{C}-SCoA \longrightarrow$$

$$CH_3-(CH_2)_n-\overset{\overset{\displaystyle O}{\|}}{C}-\overset{\overset{\displaystyle CH_3}{|}}{CH}-\overset{\overset{\displaystyle O}{\|}}{C}-SACP + CO_2 + CoA$$

Regular reduction steps then follow. These reactions occur in many tissues at a rate several orders of magnitude lower than the utilization of malonyl CoA in fatty acid synthesis. The proportion of branched-chain fatty acids that are synthesized is largely governed by the relative availability of the two precursors. An increase in branching can occur by decreasing the ratio of malonyl CoA to methylmalonyl CoA. A malonyl-CoA decarboxylase that is responsible for this decrease occurs in many tissues. It has been suggested that an increased concentration of methylmalonyl CoA, which occurs in vitamin B_{12} deficiency, can lead to excessive production of branched-chain fatty acids.

Fatty Acyl CoAs May Be Reduced to Fatty Alcohols

Many phospholipids contain fatty acid chains in ether linkage, rather than ester linkage. The synthetic precursors of these ether-linked chains are fatty alcohols (Figure 17.15) rather than fatty acids. These alcohols are formed in vertebrates by a two-step, NADPH-linked reduction of fatty acyl CoAs in the endoplasmic reticulum. In tissues that produce relatively large amounts of ether-containing lipids, the concurrent production of fatty acids and fatty alcohols is closely coordinated.

$$CH_3-(CH_2)_n-CH_2OH$$

Figure 17.15 Fatty alcohol.

17.5 • STORAGE OF FATTY ACIDS AS TRIACYLGLYCEROL

Most mammalian tissues convert fatty acids to triacylglycerols by a common sequence of reactions, but liver, adipose, and muscle tissue carry out this process to the greatest extent. Adipose tissue is a specialized organ for synthesis, storage, and hydrolysis of triacylglycerol and is the main site of long-term energy storage. Triacylglycerols are stored in the cytosol as liquid droplets surrounded by a monolayer of membrane lipids and the protein perilipin. These droplets are not a dead-end store; there is continuous synthesis and breakdown of triacylglycerol in adipose tissue. Some storage also occurs in skeletal and cardiac muscle, but only for local consumption. Triacylglycerol synthesis in liver is primarily for the production of

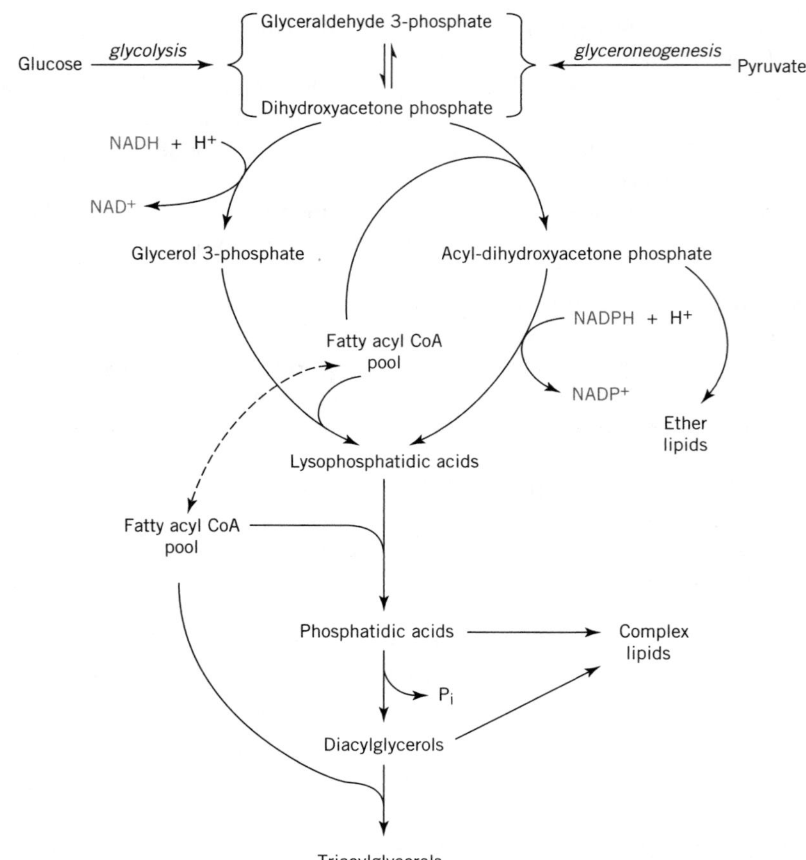

Figure 17.16 Pathways of triacylglycerol synthesis. Some of these intermediates are used as precursors for the synthesis of membrane lipids.

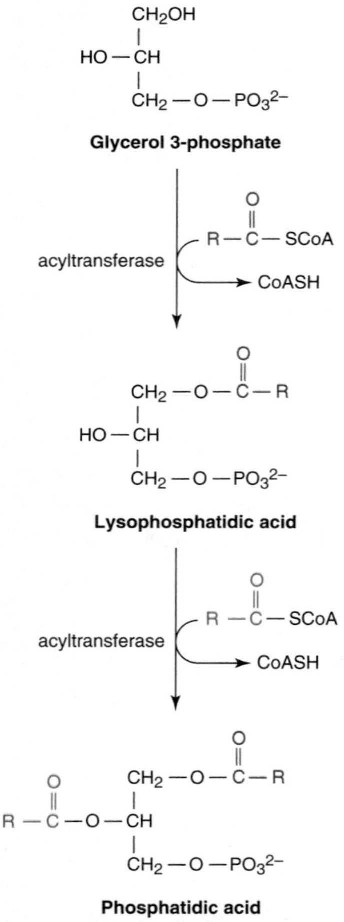

CH_2OH / $HO-CH$ / $CH_2-O-PO_3^{2-}$

Glycerol 3-phosphate

acyltransferase: $R-\overset{O}{\overset{\|}{C}}-SCoA \rightarrow CoASH$

$CH_2-O-\overset{O}{\overset{\|}{C}}-R$ / $HO-CH$ / $CH_2-O-PO_3^{2-}$

Lysophosphatidic acid

acyltransferase: $R-\overset{O}{\overset{\|}{C}}-SCoA \rightarrow CoASH$

$CH_2-O-\overset{O}{\overset{\|}{C}}-R$ / $R-\overset{O}{\overset{\|}{C}}-O-CH$ / $CH_2-O-PO_3^{2-}$

Phosphatidic acid

Figure 17.17 Synthesis of phosphatidic acid from glycerol 3-phosphate and fatty acyl CoAs.

plasma lipoproteins, rather than for energy storage. The fatty acids for this process may come from the diet, from adipose tissue via the blood, or from *de novo* synthesis, primarily from the catabolism of dietary glucose.

Triacylglycerols Are Synthesized from Fatty Acyl CoAs and Glycerol 3-Phosphate

Triacylglycerols are synthesized in most tissues from fatty acyl CoAs and a glycerol precursor, **glycerol 3-phosphate** (Figure 17.16). Glycerol 3-phosphate is derived from several sources. In most tissues, it is synthesized by the reduction of dihydroxyacetone phosphate. In the fed state, the dihydroxyacetone phosphate is derived from glucose via glycolysis; in the fasted state in adipose tissue and liver, glycerol 3-phosphate is derived from **glyceroneogenesis** (p. 689). In the liver, there is an additional source of glycerol 3-phosphate; glycerol can be phosphorylated by **glycerol kinase**, which is very active in this tissue.

Fatty acids are activated for further metabolism by conversion to their CoA esters in the following reactions:

$$R-\overset{O}{\overset{\|}{C}}-O^- + ATP + CoASH \xrightarrow[\text{synthetase}]{\text{acyl-CoA}} R-\overset{O}{\overset{\|}{C}}-SCoA + AMP + PP_i + H_2O$$

This two-step reaction has an acyl adenylate (fatty acyl-AMP) as an intermediate. The overall reaction is driven by the hydrolysis of the pyrophosphate product to two P_i.

The synthesis of triacylglycerols involves the formation of **phosphatidic acid**, which is formed by two sequential acylations of glycerol 3-phosphate to form lysophosphatidic and then phosphatidic acid (Figure 17.17). Phosphatidic acid is used for triacylglycerol synthesis by hydrolysis of the phosphate group by **phosphatidate phosphatase** to yield diacylglycerol, which is then acylated to triacylglycerol (Figure 17.18). Phosphatidic acid is also a key intermediate in the synthesis of other glycerolipids (p. 712). In an alternative pathway, dihydroxyacetone phosphate is acylated, reduced to lysophosphatidic acid, and then acylated a second time to form phosphatidic acid (see Figure 17.16). Although this is not a major pathway of triacylglycerol synthesis, it is important for the synthesis of membrane lipids with ether-linked alky chains.

Triacylglycerol synthesis follows a different pathway in epithelial cells of the small intestine. These cells take up 2-monoacylglycerols and free fatty acids from the gut, which are the major digestion products of dietary triacylglycerols by pancreatic lipase. An enzyme in the mucosal cells acylates these monoacylglycerols using acyl CoAs as substrates. The resulting diacylglycerols can then be acylated to form triacylglycerols, which are packaged into chylomicrons.

Analysis of human triacylglycerols shows that each position of glycerol is esterified with fatty acids of distinct composition. Saturated fatty acids are found preferentially at position 1 and unsaturated fatty acids at positions 2 and 3. Two main factors that determine the fatty acid composition at each position on glycerol are the specificity of the acyltransferases involved and the relative availability of different fatty acids in the fatty acyl CoA pool.

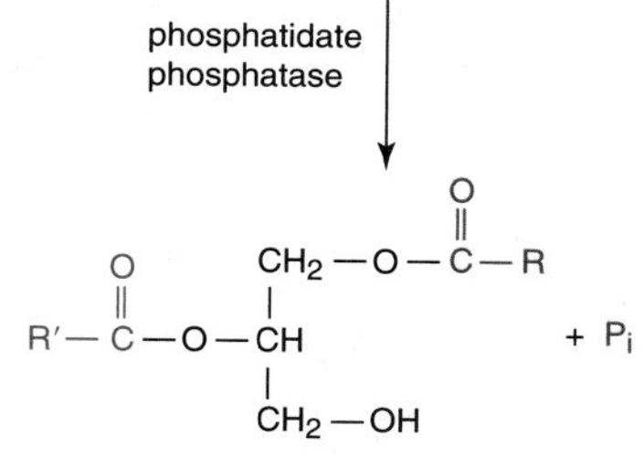

phosphatidate phosphatase

$$R'-C(=O)-O-CH(CH_2-O-C(=O)-R)(CH_2-OH) + P_i$$

Diacylglycerol

acyltransferase: $R''-C(=O)-SCoA$ → CoASH

$$R'-C(=O)-O-CH(CH_2-O-C(=O)-R)(CH_2-O-C(=O)-R'')$$

Triacylglycerol

Figure 17.18 Synthesis of triacylglycerol from phosphatidic acid.

Mobilization of Triacylglycerols Requires Hydrolysis

The first step in mobilizing stored fatty acids for energy production is hydrolysis of triacylglycerol. Several lipases catalyze this reaction and the sequence of hydrolysis of the three acyl chains is determined by the specificities of the lipases involved.

$$R'-C(=O)-CH(CH_2-O-C(=O)-R)(CH_2-O-C(=O)-R'') + 3\ H_2O \xrightarrow{\text{lipases}}$$

$$HO-CH(CH_2-OH)(CH_2-OH) + R-C(=O)-OH + R'-C(=O)-OH + R''-C(=O)-OH$$

Triacylglycerol Synthesis Occurs during Fasting as Part of a Triacylglycerol–Fatty Acid Cycle Involving Glyceroneogenesis

The release of free fatty acids from adipose tissue is a critical metabolic adaptation to fasting. The quantity of fatty acids released by adipose tissue, however, exceeds the amount used for energy by other tissues. As much as 60% of these fatty acids are redeposited in adipose tissue as triacylglycerols. Both liver and adipose tissue play a major role in this process (Clin. Corr. 17.3). In the fed state, the glycerol 3-phosphate for triacylglycerol synthesis is derived from glucose via glycolysis. During fasting, however, the entry of glucose into adipose tissue is limited because the insulin concentration is low and glucose is being used by other tissues. In this dietary state, glycerol 3-phosphate is synthesized in both the adipose tissue and the liver by glyceroneogenesis. As shown in Figure 17.19, substrates that enter the tricarboxylic acid cycle, such as pyruvate, glutamate, or aspartate, can support net glyceroneogenesis. This pathway is essentially an abbreviated version of gluconeogenesis, in which malate formed in the tricarboxylic acid cycle leaves the mitochondria and is converted to oxaloacetate in the cytosol. Phosphoenolpyruvate carboxykinase then converts oxaloacetate to phosphoenolpyruvate, which is converted to dihydroxyacetone phosphate and then to glycerol 3-phosphate, which is used for triacylglycerol synthesis. The key enzyme in this process is phosphoenolpyruvate carboxykinase; its activity is induced in liver and adipose tissue during fasting.

CLINICAL CORRELATION 17.3

Triacylglycerol/Fatty Acid Cycle

The triacylglycerol that is stored in adipose tissue is hydrolyzed to free fatty acids (FFA) during fasting to provide energy for tissues such as skeletal and cardiac muscle, and indirectly to the brain via ketone bodies. Hormones, most notably insulin, control this process. As the level of insulin falls during fasting, the rate of triacylglycerol hydrolysis (lipolysis) increases, resulting in FFA release from adipose tissue. A surprising aspect of this process is the fate of the FFA; in fasted humans up to 65% of this FFA is re-esterified to triacylglycerol in the liver and other peripheral tissues. In liver the triacylglycerol products are released into the blood as VLDL and sent back to the adipose tissue for deposition as triacylglycerol. This process has been termed the triacylglycerol/fatty acid cycle.

The synthesis of triacylglycerol in mammalian tissues requires glycerol 3-phosphate, which can be derived from dietary glucose via glycolysis in the fed state. During fasting, when low insulin inhibits glucose utilization, the glycerol 3-phosphate for the re-esterification of FFA is generated by glyceroneogenesis, an abbreviated version of gluconeogenesis. In this pathway, pyruvate—or compounds that can generate pyruvate, such as alanine or lactate—is converted to glycerol 3-phosphate via dihydroxyacetone phosphate (Figure 17.19). Recent studies have shown that glyceroneogenesis and not glycolysis is the predominant pathway for the synthesis of glycerol 3-phosphate, even during the fed state. The key enzyme in the pathway of glyceroneogenesis is phosphoenolpyruvate carboxykinase (PEPCK), which is very active in brown and white adipose tissue. If the expression of the PEPCK gene is ablated in the adipose tissue of mice, glyceroneogenesis is inhibited and triacylglycerol storage is reduced. Conversely, overexpression of the PEPCK gene in adipose tissue of transgenic mice increases the rate of glyceroneogenesis, resulting in obesity.

The metabolic logic of the triacylglycerol/fatty acid cycle most likely resides in the importance of fatty acids as a fuel during starvation. In order to ensure that there is sufficient FFA in the blood, more FFA is released from fat cells than is required; what is not used is reesterified to triacylglycerol and redeposited in adipose tissue, at a small energetic cost. The triacylglycerol/fatty acid cycle consumes 3%–6% of the energy in a molecule of triacylglycerol; it is apparently better to have the needed fuel available and pay for it energetically, than to run short at a critical time! The rate of FFA reesterification in the triacylglycerol/fatty acid cycle is a key factor in determining the steady-state concentration of FFA in the blood, a parameter that is directly involved in the etiology of Type 2 diabetes (see Clin. Corr. 17.2, p. 677). The thiazolidinediones, a class of antidiabetic drugs, induce the activity of PEPCK in adipose tissue and increase the rate of FFA reesterification to triacylglycerol via glyceroneogenesis in this tissue, supporting the important role of this process in maintaining lipid homeostasis.

Nye, C., Kalhan S. C., and Hanson, R. W. Reassessing the pathway of triglyceride synthesis in adipose tissue. *Trends in Endocrinol. and Metab.* 19: 2008; Reshef, L., Olswang, Y., Cassuto, H., Blum, B., et al. Glyceroneogenesis and the triglyceride/fatty acid cycle. *J. Biol. Chem.* 278: 30413, 2003; Jensen, M. D., Ekberg, K., and Landau, B. R. Lipid metabolism during fasting. *Am. J. Physiol.* 281: E789 2001; and Tordjman, J., Khazan, W., Antoine, B., Chauvet, G., et al. Regulation of glyceroneogenesis and phosphoenolpyruvate carboxykinase by fatty acids, retinoic acid and thiazolidinediones. *Biochimie* 85: 1213, 2003.

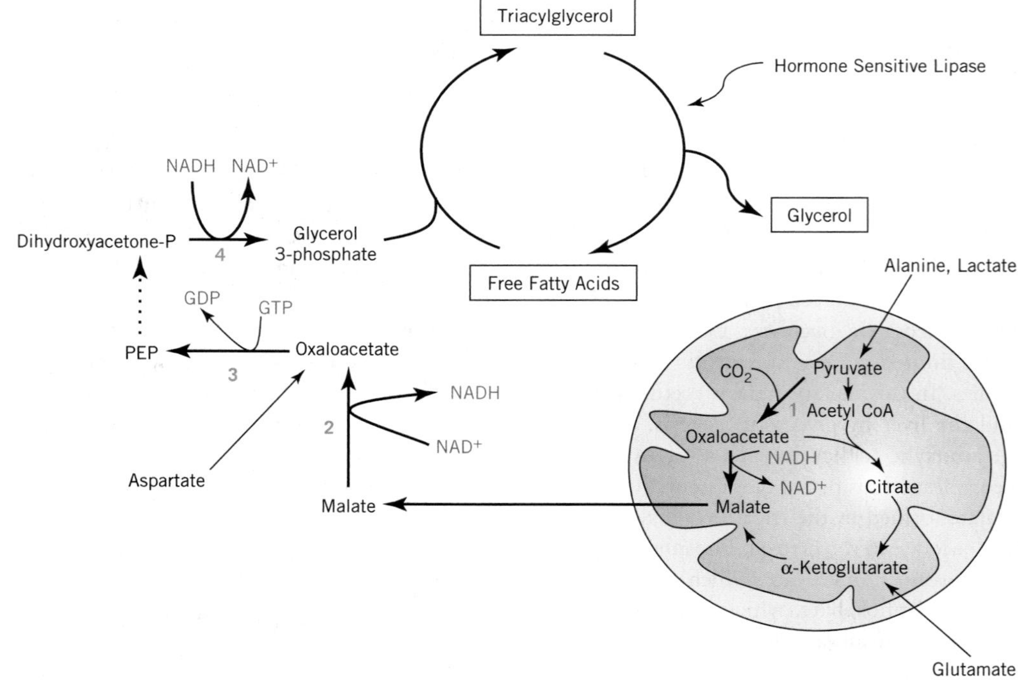

Figure 17.19 The pathway of glyceroneogenesis in the liver and adipose tissue. During fasting, this pathway, which is an abbreviated version of gluconeogenesis, produces glycerol 3-phosphate for triacylglycerol synthesis. The key enzymes of glyceroneogenesis are (1) pyruvate carboxylase, (2) the mitochondrial and cytosolic forms of NAD malate dehydrogenase, (3) phosphoenolpyruvate carboxykinase, and (4) glycerol 3-phosphate dehydrogenase.

17.6 • UTILIZATION OF FATTY ACIDS FOR ENERGY PRODUCTION

Fatty acids in the circulation are taken up by cells and used for energy production, primarily in mitochondria, in a process integrated with energy generation from other sources. Fatty acids are broken down to acetyl CoA in the mitochondria, with the production of NADH and $FADH_2$. These three products are then used in the mitochondrial matrix for energy production via the tricarboxylic acid cycle and oxidative phosphorylation.

The utilization of fatty acids for energy production varies considerably from tissue to tissue and depends to a significant extent on metabolic status, that is, fed or fasted, exercising or resting. Most tissues can use fatty acids as a fuel. Fatty acids are a major energy source in cardiac and skeletal muscle, but the brain does not oxidize fatty acids because they cannot cross the blood–brain barrier. Mammalian red blood cells cannot oxidize fatty acids because they lack mitochondria, the site of fatty acid oxidation. During fasting, the liver converts acetyl CoA, generated by fatty acid oxidation and the breakdown of amino acids, into **ketone bodies**, which become a major fuel after 2–3 days of fasting. Most tissues, including the brain, adapt to fasting by utilizing these ketone bodies.

β-Oxidation of Straight-Chain Fatty Acids Is a Major Energy-Producing Process

CoA esters of fatty acids are the substrates for oxidation. For the most part, the pathway of fatty acid oxidation is similar, but not identical to, a reversal of the process of fatty acid synthesis. That is, two-carbon fragments are removed sequentially from the carboxyl end of the fatty acid by enzymatic **oxidation, hydration**, and **oxidation** to form a β-keto acid, which is then split by **thiolysis**. The pathway is called **β-oxidation** because carbon 3 (the β-carbon) is oxidized twice prior to cleavage.

Fatty Acids Are Activated by Conversion to Fatty Acyl CoA

The first step in the utilization of a fatty acid is its activation to a fatty acyl CoA. This reaction is catalyzed by enzymes in the endoplasmic reticulum and outer mitochondrial membrane and is the same one used in the synthesis of triacylglycerols. The fatty acyl CoAs are released into the cytosol.

Carnitine Carries Acyl Groups across the Inner Mitochondrial Membrane

Most long chain fatty acyl CoAs are formed outside the mitochondria, but are oxidized in the mitochondrial matrix. The mitochondrial membrane is impermeable to CoA and its derivatives. Fatty acids are transported into the mitochondria using **carnitine (4-trimethylamino-3-hydroxybutyrate)** as a carrier. The process is outlined in Figure 17.20.

$$CH_3-(CH_2)_n-\overset{O}{\overset{\|}{C}}-SCoA + HO-\underset{}{CH}(CH_2-\overset{+}{N}-(CH_3)_3)-CH_2-COOH \xrightleftharpoons[\text{transferase}]{\text{carnitine palmitoyl-}} CH_3-(CH_2)_n-\overset{O}{\overset{\|}{C}}-O-CH(CH_2-\overset{+}{N}-(CH_3)_3)-CH_2-COOH + CoASH$$

The acyl group is transferred from CoA to carnitine by **carnitine palmitoyltransferase I (CPT I)** on the outer mitochondrial membrane. Acyl carnitine and free carnitine are then exchanged across the inner mitochondrial membrane by **carnitine-acylcarnitine translocase**. Finally, the fatty acyl group is transferred back to CoA by **carnitine palmitoyltransferase II (CPT II)** on the matrix side of the inner mitochondrial membrane. This process functions primarily in the transport of fatty acyl CoAs with 12–18 carbons. Genetic abnormalities in this system lead to serious pathological consequences (Clin. Corr. 17.4). By contrast, entry of shorter-chain fatty acids is independent of carnitine; they cross the inner mitochondrial membrane as free fatty acids and become activated to their CoA derivatives in the matrix. CPT I is an important site for the regulation of fatty acid oxidation because its rate controls the

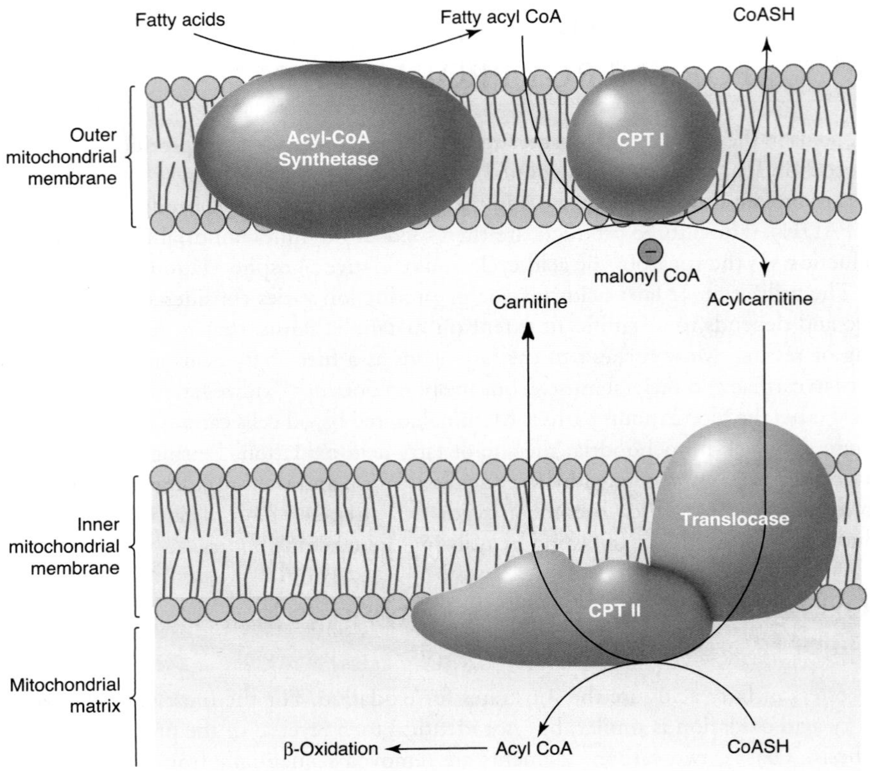

Figure 17.20 Fatty acids are transported into the mitochondria as acylcarnitines for oxidation. Inhibition of carnitine palmitoyltransferase I (CPT I) by malonyl CoA regulates the rate of fatty acid oxidation.

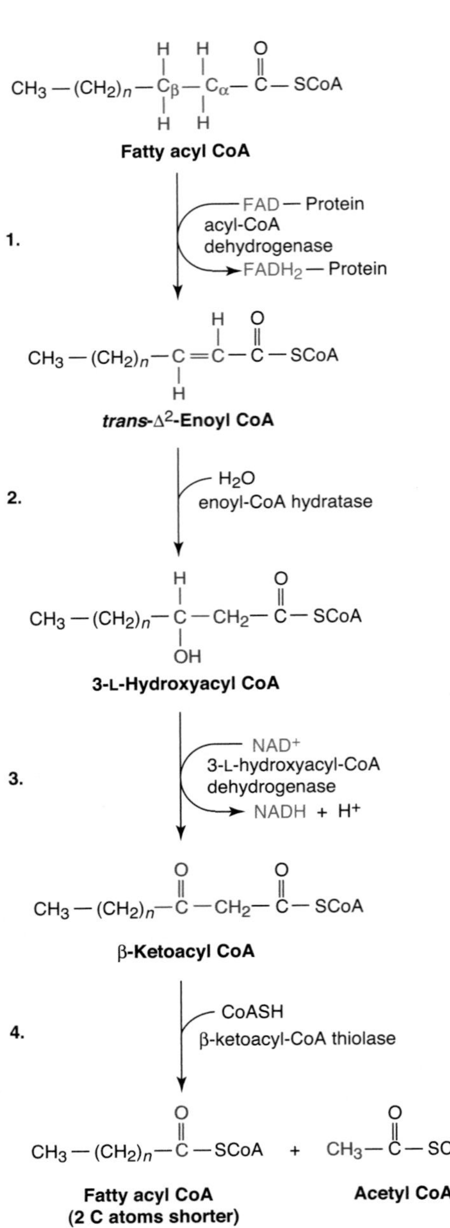

Figure 17.21 Pathway of fatty acid β-oxidation.

entry of fatty acids into the mitochondria and therefore determines the supply of substrate for β-oxidation in the mitochondrial matrix. **Malonyl CoA**, which is the product of acetyl-CoA carboxylase and a key intermediate in fatty acid synthesis (p. 680), is an inhibitor of CPT I.

β-Oxidation Is a Sequence of Four Reactions

β-oxidation is a series of four reactions that act on a fatty acyl Co A to produce an acetyl CoA and a new acyl CoA that is two carbon atoms shorter than the initial substrate (Figure 17.21). Once a fatty acyl CoA is formed at the inner surface of the inner mitochondrial membrane, it can be oxidized by acyl-CoA dehydrogenase, a flavoprotein that uses FAD as the electron acceptor (Reaction 1). The product is an enoyl CoA with a *trans* double bond between the C2 and C3 atoms and enzyme-bound $FADH_2$. As in the tricarboxylic acid cycle, the $FADH_2$ transfers its electrons to enzymes of oxidative phosphorylation, regenerating FAD.

The second step in β-oxidation is hydration of the trans-double bond to a **3-L-hydroxyacyl CoA**, which is oxidized to a **3-ketoacyl CoA** intermediate, with the generation of NADH in the third step. The final step is cleavage of the chain by ketothiolase, generating acetyl CoA and a fatty-acyl CoA that has been shortened by two carbon atoms. This shortened acyl CoA is ready for the next round of oxidation, starting with acyl-CoA dehydrogenase. In most tissues, the acetyl CoA will be used by the tricarboxylic acid cycle and the $FADH_2$ and NADH will be reoxidized by the oxidative phosphorylation pathway with the production of ATP.

Each of the four reactions shown in Figure 17.21 is catalyzed by several different enzymes that have specificity for substrates of different chain lengths. For example, at least four enzymes catalyze the first dehydrogenation step. These are very long-chain, long-chain, medium-chain, and short-chain acyl-CoA dehydrogenases (**VLCAD**, **LCAD**, **MCAD**, and **SCAD**). VLCAD, which oxidizes straight-chain acyl CoAs ranging from C12 to C24, differs from the other family members in that it is membrane associated. MCAD has broad chain length specificity but is most active with C6 and C8 substrates, whereas the order of preference for SCAD is C4 > C6 > C8. LCAD is involved in initiating the oxidation of branched chain fatty acids, for example, 2-methylpalmitoyl CoA.

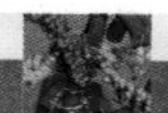

CLINICAL CORRELATION 17.4

Genetic Deficiencies in Carnitine Transport (OMIM 212140) or Carnitine Palmitoyltransferase (OMIM 600650)

Several diseases result from genetic abnormalities in the transport of long-chain fatty acids across the inner mitochondrial membrane. They stem from deficiencies in the level of carnitine or in the synthesis and transport of acylcarnitines. Mutations can affect carnitine palmitoyl transferases (CPT) or the mitochondrial carnitine-acylcarnitine translocase.

Two categories of carnitine deficiency, primary and secondary, are now recognized. Primary carnitine deficiency is caused by a defect in the high-affinity plasma membrane carnitine transporter in tissues such as muscle, kidney, heart, and fibroblasts (but apparently not in liver where a different transporter is operative). It results in extremely low levels of carnitine in affected tissues and in plasma (because of failure of the kidneys to reabsorb carnitine). The clinical symptoms of carnitine deficiency range from mild, recurrent muscle cramping to severe weakness and death. The very low carnitine level in heart and skeletal muscle seriously compromises long-chain fatty acid oxidation. Dietary carnitine therapy, which raises the plasma concentration of carnitine and forces its entry into tissues in a nonspecific manner, is frequently beneficial. Secondary carnitine deficiency is often associated with inherited defects in the β-oxidation pathway. These disorders frequently cause the accumulation of acylcarnitines, which are excreted in the urine (see Clin. Corr. 17.5, p. 694), thereby depleting the carnitine pool. These acylcarnitines may also impair the uptake of free carnitine by tissues.

There are several different CPT deficiencies. The most common form results from mutations in the CPT II gene that cause a partial loss of enzyme activity. Patients generally experience muscle weakness during prolonged exercise when the muscles rely heavily on fatty acids as an energy source. Myoglobinuria, due to breakdown of muscle tissue, is frequently observed. The disorder is usually referred to as the *muscular* form of CPT II activity deficiency. Mutations causing more severe loss of CPT II activity (90% or greater) can have serious consequences in early infancy. These are usually precipitated by periods of fasting and include hypoketotic hypoglycemia, hyperammonemia, cardiac malfunction, and sometimes death. Similar morbidity and mortality are associated with mutations in the gene for liver CPT I. To date, only a few patients with hepatic CPT I deficiency have been reported, possibly because the disease is frequently lethal. No defects in the muscle isoform of CPT I have been reported.

The first patient with carnitine-acylcarnitine translocase deficiency was described in 1992. Clinical features included intermittent hypoglycemic coma, hyperammonemia, muscle weakness, and cardiomyopathy. The condition proved fatal at age 3 years. Several additional cases with similar symptomatology have since been reported.

These disorders can be treated with a diet that is low in long-chain fatty acids and by avoiding fasting, to minimize conditions where tissues require fatty acid oxidation for energy. The diet can also be supplemented with medium-chain triacylglycerols, because these fatty acids enter the mitochondria by a carnitine-independent mechanism.

Stanley, C. A., Hale, D. E., Berry, G. T., Deleeno, S., et al. A deficiency of carnitine-acylcarnitine translocase in the inner mitochondrial membrane. *N. Engl. J. Med.* 327: 19, 1992; Roe, C. R., and Dong, J. Mitochondrial fatty acid oxidation disorders. In Scriver, C. R., Beaudet, A. L., Sly, W. S., and Valle, D. (Eds.), *The Metabolic and Molecular Bases of Inherited Disease,* 8th ed. New York: McGraw Hill, 2001, II:2297; and Bonnefont, J. P., Demaugre, F., Prip-Buus, C., Saudubray, J., M., et al. Carnitine palmitoyltransferase deficiencies. *Mol. Gen. Metab.* 68: 424, 1999.

A feature unique to the oxidation of long-chain fatty acids is that the enoyl-CoA hydratase, 3-hydroxyacyl-CoA dehydrogenase, and β-ketothiolase steps are all catalyzed by a membrane-bound complex of the three enzymes called **trifunctional protein**. This complex is distinct from the enzymes that catalyze the oxidation of medium and short chain acyl CoAs, all of which are soluble proteins in the mitochondrial matrix. Clin. Corr. 17.5 describes genetic deficiencies of acyl-CoA dehydrogenases.

Energy Yield from the β-Oxidation of Fatty Acids

Each round of β-oxidation produces one acetyl CoA, one $FADH_2$, and one NADH. In the oxidation of palmitoyl CoA, seven cleavages of carbon–carbon bonds take place, with the formation of two acetyl CoAs in the final cleavage. Thus, the β-oxidation of palmitate produces eight acetyl CoA, seven $FADH_2$, and seven NADH. On the basis of current estimates of the yields of ATP in oxidative phosphorylation (p. 572), each $FADH_2$ yields 1.5 ATPs, and each NADH yields 2.5 ATPs when oxidized by the electron transport chain. Therefore, the oxidation of the seven NADH and seven FADH produces 28 ATP. Oxidation of each acetyl CoA through the tricarboxylic acid cycle yields 10 ATP (p. 573), so the eight 2-carbon fragments from a palmitate molecule produce 80 ATP, for a total of 108 ATP. However, two ATP equivalents are used to activate palmitate to palmitoyl CoA (1 ATP is converted to 1 AMP + PP_i). Therefore, each palmitic acid yields 106 ATP/mol on complete oxidation. The importance of fatty acids in supplying the energy needs for human metabolism is discussed on page 849.

CLINICAL CORRELATION 17.5

Genetic Deficiencies in the Acyl-CoA Dehydrogenases

Deficiencies in the acyl-CoA dehydrogenases represent a recently discovered group of inherited disorders that affect the first reaction in the β-oxidation of fatty acids. Patients with mutations that affect enzymes with specificity for different chain length have been described. These include very long-chain acyl-CoA dehydrogenase (VLCAD), long-chain acyl-CoA dehydrogenase (LCAD), medium-chain acyl-CoA dehydrogenase (MCAD), and short-chain acyl-CoA dehydrogenase (SCAD). Patients with these autosomal recessive mutations share many of the same clinical features. The best characterized is MCAD deficiency which, though first recognized in 1982, is among the most common of all inborn errors of metabolism.

MCAD deficiency usually manifests itself within the first 2 years of life. Typical symptoms, which are seen after 12 h or more of fasting, include vomiting, lethargy, and frequently coma, accompanied by hypoketotic hypoglycemia and dicarboxylic aciduria. The absence of ketosis is due to the block in hepatic fatty acid oxidation, which also causes a slowing of gluconeogenesis. This blockade, coupled with impaired fatty acid oxidation in muscle, promotes glucose utilization, leading to profound hypoglycemia. Accumulation of medium-chain acyl CoAs in tissues forces their metabolism through alternative pathways including ω-oxidation and transesterification to glycine or carnitine. Excessive urinary excretion of the reaction products (medium-chain dicarboxylic acids together with medium-chain esters of glycine and carnitine) provide diagnostic clues to this disorder. Many cases previously diagnosed loosely as *Reye-like syndrome* or *sudden infant death syndrome* were in fact due to MCAD deficiency.

Patients with this disorder are treated by avoidance of prolonged fasting and consumption of a high-carbohydrate diet. Supplementing the diet with carnitine replenishes carnitine that is lost by the increased excretion of acylcarnitines. This is consistent with the fact that the metabolic complications of MCAD deficiency are seen only when tissues become heavily dependent on fatty acids as a source of energy.

Rinaldo, P., Matern, D., and Bennett, M. J. Fatty acid oxidation disorders. *Annu. Rev. Physiol.* 64: 477, 2002; and Wanders, R. J. A., Vreken, P., den Boer, M. E. J., et al. Disorders of mitochondrial fatty acyl CoA β-oxidation. *J. Inher. Metab. Dis.* 22: 442, 1999.

Comparison of Fatty Acid Synthesis and Oxidation

The pathways of fatty acid synthesis and oxidation are similar. However, like many paired anabolic and catabolic pathways, the synthesis and oxidation of fatty acids are not the reverse of each other. The critical differences between the two pathways are outlined in Table 17.4. They include different cellular locations, different cofactors (NADPH in synthesis and FAD and NAD^+ in oxidation) and the use of ATP to drive malonyl CoA formation for fatty acid synthesis. These differences permit both pathways to proceed in the forward direction because $\Delta G < 0$ for both. The differences also permit independent regulation that prevents futile cycling.

β-Oxidation of Some Fatty Acids Requires Additional Steps

The β-oxidation pathway oxidizes saturated fatty acids with even numbers of carbon atoms to acetyl CoA. However, other fatty acids in the diet, including those with cis double

TABLE 17.4 • Comparison of Schemes for Biosynthesis and β-Oxidation of Palmitate

Parameter	*Biosynthesis*	*β-Oxidation*
Subcellular localization	Primarily cytosolic	Primarily mitochondrial
Phosphopantetheine-containing acyl carrier	Acyl carrier protein	Coenzyme A
Small carbon fragment added or removed	C1 and C2 atoms of malonyl CoA after initial priming	Acetyl CoA
Oxidation–reduction coenzyme	NADPH	FAD when saturated chain dehydrogenated, NAD^+ when hydroxy acid dehydrogenated
Stereochemical configuration of β-hydroxy intermediates	D-β-Hydroxy	L-β-Hydroxy
Energy equivalents yielded or utilized in interconversion of palmitate and acetyl CoA	7 ATP + 14 NADPH = 49 ATP equiv	7 $FADH_2$ + 7 NADH − 2 ATP = 26 ATP equiv

bonds, branched chains, and odd numbers of carbon atoms, require additional steps for their complete oxidation. These steps allow these fatty acids to be used as fuels and prevent their accumulation. Other reactions catalyze the α- and ω-oxidation of fatty acids. **α-Oxidation** occurs at C2 instead of C3 as occurs in the β-oxidation whereas **ω-oxidation** occurs at the methyl end of the fatty acid molecule.

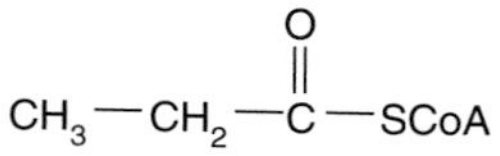

$$CH_3-CH_2-\overset{\overset{\displaystyle O}{\|}}{C}-SCoA$$

Figure 17.22 Propionyl CoA.

Oxidation of Odd-Chain Fatty Acids Produces Propionyl CoA

Fatty acids with an odd number of carbon atoms are oxidized by the β-oxidation pathway. The products of the final cleavage by thiolase are acetyl CoA and **propionyl CoA** (Figure 17.22). Propionyl CoA, which is also produced by the catabolism of isoleucine, valine, and methionine, is metabolized by carboxylation to methylmalonyl CoA and conversion to succinyl CoA (p. 779).

Oxidation of Unsaturated Fatty Acids Requires Additional Enzymes

Unsaturated fatty acids are utilized by β-oxidation, but additional reactions are required to deal with the cis double bonds. Metabolism begins with several rounds of β-oxidation, which generates intermediates that have cis double bonds near the carboxyl carbon. Double bonds beginning at odd- and even-numbered carbon atoms require different strategies. The oxidation of linoleoyl CoA (18:2) (Figure 17.23), illustrates this process. β-oxidation generates an enoyl CoA intermediate with a cis double bond between C3 and C4, instead of the intermediate with a trans bond between C2 and C3 that is required by enoyl-CoA hydratase. **Enoyl-CoA isomerase** changes the cis $\Delta3$ to a *trans*-$\Delta2$-enoyl CoA, which can then be metabolized by β-oxidation.

A second problem occurs when the cis double bond of the acyl CoA intermediate resides between C4 and C5. In this case, the action of acyl-CoA dehydrogenase gives rise to a *trans*-2, *cis*-4-enoyl CoA. This is acted on by 2,4-dienoyl-CoA reductase that produces a *trans*-3-enoyl CoA using reducing equivalents from NADPH. **Enoyl-CoA isomerase** then produces *trans*-2-enoyl CoA, which is a substrate for β-oxidation.

Some Fatty Acids Undergo α-Oxidation

As noted earlier, there are several mechanisms for **hydroxylation of fatty acids**. Some long-chain fatty acids are hydroxylated for sphingolipid synthesis and other shorter-chain fatty acids are hydroxylated on the α carbon in order to start their oxidation. The sequence is as follows:

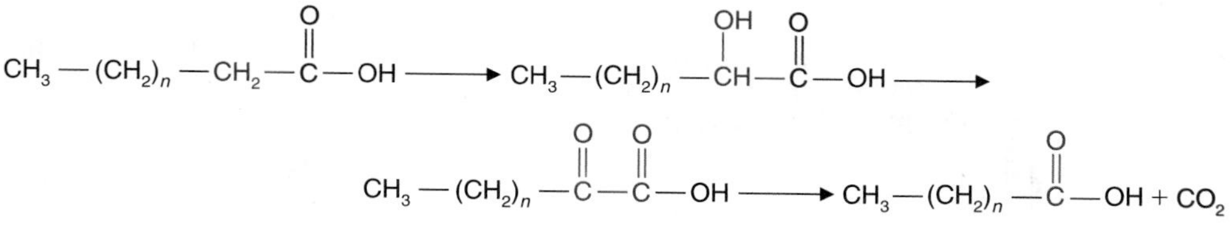

$$CH_3-(CH_2)_n-CH_2-\overset{\overset{\displaystyle O}{\|}}{C}-OH \longrightarrow CH_3-(CH_2)_n-\overset{\overset{\displaystyle OH}{|}}{CH}-\overset{\overset{\displaystyle O}{\|}}{C}-OH \longrightarrow$$

$$CH_3-(CH_2)_n-\overset{\overset{\displaystyle O}{\|}}{C}-\overset{\overset{\displaystyle O}{\|}}{C}-OH \longrightarrow CH_3-(CH_2)_n-\overset{\overset{\displaystyle O}{\|}}{C}-OH + CO_2$$

Some of these hydroxylations occur in the endoplasmic reticulum and mitochondria and involve monooxygenases (P450 family) that require O2 and NADH or NADPH. The α-hydroxylation of fatty acids also occurs in peroxisomes. This is particularly important for the metabolism of branched-chain fatty acids (Clin. Corr. 17.6, p. 697). A branched-chain fatty acid, such as phytanoyl CoA, which is derived from chlorophyll in the diet, is acted on by a hydroxylase in a reaction involving α–ketoglutarate, Fe^{2+}, and ascorbate, with the generation of 2-hydroxyphytanoyl CoA and formyl CoA. The latter is metabolized to CO_2 via formic acid. The 2-hydroxyphytanoyl CoA is then further metabolized to pristanic acid, which undergoes β-oxidation.

ω-Oxidation Gives Rise to Dicarboxylic Acids

ω-Oxidation is another minor pathway for fatty acid oxidation, which occurs in the endoplasmic reticulum of many tissues. In this pathway, hydroxylation takes place on the methyl carbon at the opposite end of the molecule from the carboxyl group or on

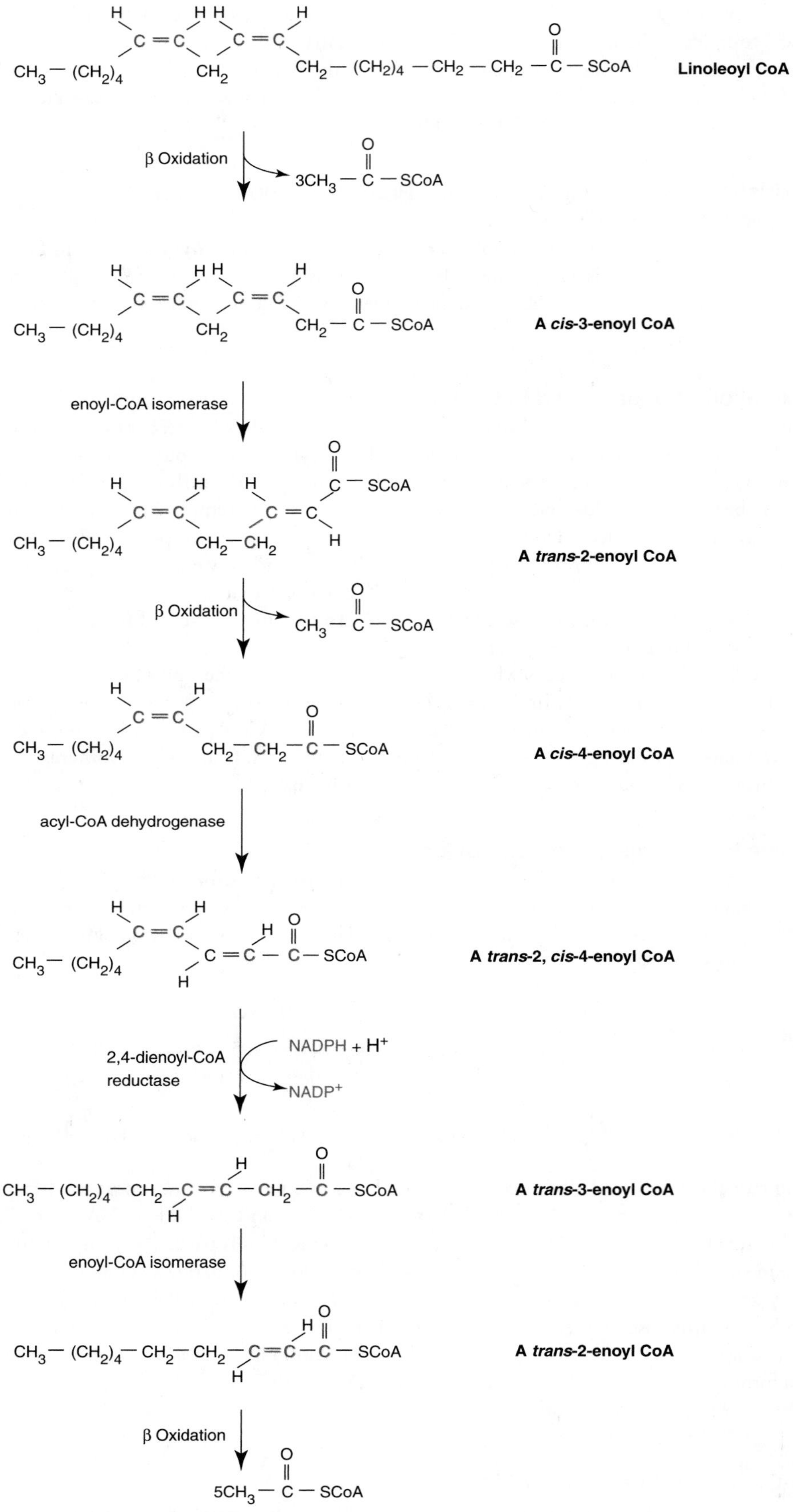

Figure 17.23 Oxidation of linoleoyl CoA.

CLINICAL CORRELATION 17.6

Refsum Disease (OMIM 266500)

Although α-oxidation of fatty acids is relatively minor in terms of total energy production, it is significant in the metabolism of branched-chain fatty acids in the diet. A principal example of these is phytanic acid, a metabolic product of phytol, which is a constituent of chlorophyll. Phytanic acid is a significant constituent of milk and animal fats. It cannot be oxidized by β-oxidation because of the presence of the 3-methyl group. It is metabolized by α-hydroxylation followed by dehydrogenation and decarboxylation. β-Oxidation can completely degrade the resulting molecule, producing three molecules of propionyl CoA, three molecules of acetyl CoA, and one molecule of isobutyryl CoA.

Patients with a rare genetic disease called Refsum disease lack the peroxisomal α-hydroxylating enzyme and accumulate large quantities of phytanic acid in their tissues and sera. This leads to serious neurological problems such as retinitis pigmentosa, peripheral neuropathy, cerebellar ataxia, and nerve deafness. The restriction of dietary dairy and meat products from ruminants results in lowering of plasma phytanic acid and regression of the neurological symptoms.

Wanders, R. J. A., van Grunsven, E. G., and Jansen, G. A. Lipid metabolism in peroxisomes: enzymology, functions and dysfunctions of the fatty acid α- and β-oxidation systems in humans. *Biochem. Soc. Trans.* 28: 141, 2000.

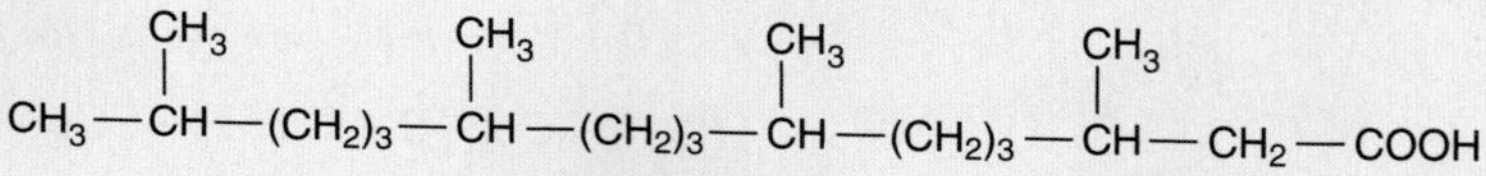

$$CH_3-\underset{\underset{CH_3}{|}}{CH}-(CH_2)_3-\underset{\underset{CH_3}{|}}{CH}-(CH_2)_3-\underset{\underset{CH_3}{|}}{CH}-(CH_2)_3-\underset{\underset{CH_3}{|}}{CH}-CH_2-COOH$$

Phytanic acid

the carbon next to the methyl end. It uses a monooxygenase, which requires O_2, and NADPH. Hydroxylated fatty acids can be further oxidized in the cytosol to **dicarboxylic acids** via the sequential action of cytosolic **alcohol** and **aldehyde dehydrogenases**. Medium-chain fatty acids are the principal substrates of this pathway. The overall reactions are

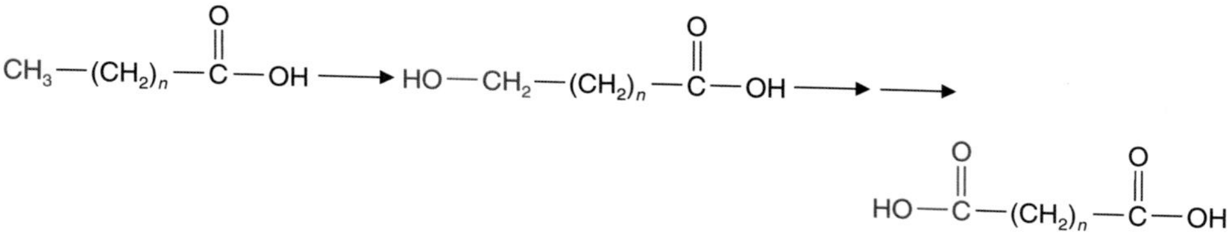

$$CH_3-(CH_2)_n-\overset{\overset{O}{\|}}{C}-OH \longrightarrow HO-CH_2-(CH_2)_n-\overset{\overset{O}{\|}}{C}-OH \longrightarrow \longrightarrow$$

$$HO-\overset{\overset{O}{\|}}{C}-(CH_2)_n-\overset{\overset{O}{\|}}{C}-OH$$

These dicarboxylic acids form CoA esters at either carboxyl group and then undergo β-oxidation to produce shorter-chain dicarboxylic acids such as adipic (C_6) and succinic (C_4) acids. This process also occurs primarily in peroxisomes.

Ketone Bodies Are Formed from Acetyl CoA

Ketone bodies are water-soluble products of lipid oxidation that are formed in liver and kidney mitochondria during prolonged fasting. The ketone bodies, **acetoacetic acid** and its reduction product **β-hydroxybutyric acid**, are made from acetyl CoA that is produced by fatty acid and amino acid catabolism (Figure 17.24). Ketone bodies are an important adaptation to prolonged fasting; they can be present at high concentrations in the blood (>3 mM) and are an important energy source for many tissues (see Clin. Corr. 17.3, p. 690).

HMG CoA Is an Intermediate in the Synthesis of Acetoacetate from Acetyl CoA

Ketone bodies are formed in the liver (and to a lesser extent in the kidney cortex during prolonged fasting). Their synthesis occurs in the mitochondrial matrix and begins with condensation of two acetyl CoA molecules to form acetoacetyl CoA, in a reaction that is the reverse of the final step of β-oxidation (Figure 17.25). The enzyme involved, **β-ketothiolase**, is an isozyme of the enzyme that functions in β-oxidation. HMG-CoA synthase catalyzes the condensation of acetoacetyl CoA with another molecule of acetyl CoA to form

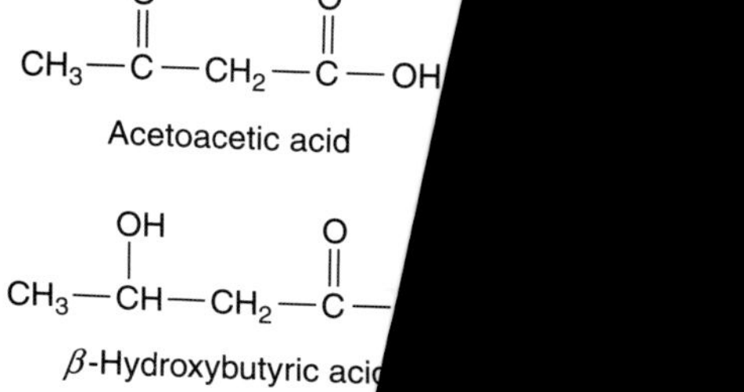

$$CH_3-\overset{\overset{O}{\|}}{C}-CH_2-\overset{\overset{O}{\|}}{C}-OH$$

Acetoacetic acid

$$CH_3-\overset{\overset{OH}{|}}{CH}-CH_2-\overset{\overset{O}{\|}}{C}-$$

β-Hydroxybutyric aci

Figure 17.24 Structures of keton

2 $H_3C-\overset{O}{\overset{\|}{C}}-SCoA$
Acetyl CoA

ketothiolase → CoA

$H_3C-\overset{O}{\overset{\|}{C}}-CH_2-\overset{O}{\overset{\|}{C}}-SCoA$
Acetoacetyl CoA

$H_3C-\overset{O}{\overset{\|}{C}}-SCoA$
Acetyl CoA

HMG-CoA synthase → CoA

$HOOC-CH_2-\overset{OH}{\underset{CH_3}{C}}-CH_2-\overset{O}{\overset{\|}{C}}-SCoA$
3-Hydroxy, 3-methyl-glutaryl CoA
(HMG CoA)

HMG-CoA lyase → $H_3C-\overset{O}{\overset{\|}{C}}-SCoA$
Acetyl CoA

$H_3C-\overset{O}{\overset{\|}{C}}-CH_2-COOH$
Acetoacetate

spontaneous → CO_2

NADH + H^+ → NAD^+
β-Hydroxybutyrate dehydrogenase

$H_3C-\overset{O}{\overset{\|}{C}}-CH_3$
Acetone

$H_3C-\overset{OH}{CH}-CH_2-COOH$
β-Hydroxybutyrate

...25 Ketone bodies are synthesized ... CoA in hepatic mitochondria.

...OH

...e bodies.

β-hydroxy-β-methylglutaryl coenzyme A (HMG CoA). HMG CoA lyase then cleaves HMG CoA to yield acetoacetic acid and acetyl CoA.

Acetoacetate Forms Both D-β-Hydroxybutyrate and Acetone

Some of the acetoacetate is reduced to D-β-hydroxybutyrate in mitochondria by **D-β-hydroxybutyrate dehydrogenase**. Note that the product of β-hydroxybutyrate dehydrogenase is D-β-hydroxybutyrate, whereas β-hydroxybutyryl CoA formed during β-oxidation is the L isomer. The extent of this reaction depends on the intramitochondrial NAD^+/NADH ratio. Because β-hydroxybutyrate dehydrogenase has high activity in the liver, the concentrations of its substrates and products are maintained close to equilibrium. Thus, the ratio of β-hydroxybutyrate to acetoacetate in the blood reflects the NAD^+/NADH ratio in liver mitochondria. During fasting, this ratio is relatively high because of the NADH generated by fatty acid oxidation, favoring β-hydroxybutyrate formation; in overnight fasted humans, the β-hydroxybutyrate/acetoacetate ratio is ~3:1. β-hydroxybutyrate and acetoacetate are released from liver and kidney for use by other tissues. β-Hydroxybutyrate also exports reducing equivalents from fatty acid oxidation.

Some acetoacetate undergoes spontaneous nonenzymatic decarboxylation to acetone:

$$CH_3-\overset{O}{\overset{\|}{C}}-CH_2-\overset{O}{\overset{\|}{C}}-O^- + H^+ \longrightarrow CH_3-\overset{O}{\overset{\|}{C}}-CH_3 + CO_2$$

Acetone formation is negligible under normal conditions, but at high concentrations of acetoacetate, which can occur in severe diabetic ketoacidosis (Clin. Corr. 17.7), acetone can reach levels high enough to be detectable in the breath.

HMG CoA is also an intermediate in cholesterol synthesis (p. 719). However, the HMG CoA used for ketone body and cholesterol synthesis is present in different metabolic pools. The HMG CoA used for **ketogenesis** is synthesized in hepatic (and renal) mitochondria by an isozyme of HMG-CoA synthase that is expressed at high levels during prolonged fasting. Moreover, HMG-CoA lyase, which converts HMG CoA to acetoacetate and acetyl CoA, is expressed only in hepatic (and renal) mitochondria. In contrast, HMG

CLINICAL CORRELATION 17.7

Ketone Bodies as a Fuel: The Atkins Diet

The current popularity of low-carbohydrate diets for weight loss underscores the importance of the metabolism of ketone bodies as fuels in humans. The best know of these diets was popularized by the late Dr. Robert Atkins in his book *Dr. Atkins' Diet Revolution,* which has sold more than 6 million copies. The Atkins' Diet is high in fat and protein and very low in carbohydrate (less that 20 g/day during the initial phase) and has been highly controversial among the medical establishment because of its high fat content. Individuals on the diet often lose a considerable amount of weight, despite the author's promise that dieters can "stop counting calories and measuring portions." Controlled clinical studies have shown that obese subjects lose more weight on a high-fat/low-carbohydrate diet than on an isocaloric diet with higher levels of carbohydrate. Surprisingly, a marked decrease was noted in the levels of triacylglycerol in the blood of individuals consuming the high-fat/low-carbohydrate diet. As an example, a 2-year trial that compared a low-carbohydrate/high-fat diet (an Atkins-type diet) against a low-fat/high-carbohydrate diet, a Mediterranean-style diet, was performed on 322 subjects, who had an average BMI of 31. Individuals on the high-fat/low-carbohydrate diet lost an average of 10.4 pounds compared with a loss of 6.4 pounds for the group on the high-carbohydrate/low-fat diet. In addition, there was a 20% reduction in the ratio of total cholesterol to HDL cholesterol for the high-fat group and only a 12% reduction in that ratio for the individuals on the high-carbohydrate diet. The authors concluded that the high-fat/low-carbohydrate diet was safe for long-term human consumption and that it provided metabolic benefits; they suggested that it be considered in weight loss programs for obese patients.

Central to the Atkins' Diet is the mobilization of fatty acids from adipose tissue and their conversion by the liver to ketone bodies (β-hydroxybutyrate, acetoacetate, and acetone). An assay of ketone bodies in the urine is the prescribed method of determining metabolic status while on the diet, since even small quantities of dietary carbohydrate will depress ketone body synthesis, largely by inhibiting lipolysis in adipose tissue. As the concentration of ketone bodies rise in the blood, a fraction is excreted in the urine and some via the breath. Could this account for the greater weight loss noted by individuals on the Atkins' Diet? For comparison, after 7 days of fasting, the daily urinary excretion rate of acetoacetate and β-hydroxybutyrate in humans is approximately 110 mmol/day; the excretion rate is even less early in starvation (60 mmol/day after 2 days of starvation). This excretion could contribute to the negative caloric balance and weight loss characteristic of the Atkins' Diet, although the energy loss does not exceed 100 kcal/day. It is most likely, however, that the high fat content of the Atkins Diet reduces appetite and thus food intake. In addition, in the absence of carbohydrate, the diet is monotonous and compliance is a major problem.

In general, ketosis develops when glucose oxidation is suppressed and fat catabolism is accelerated. There are two types of ketosis: the normal ketosis of fasting and the pathological hyperketonemia of diabetic ketoacidosis. No other fuel in the human blood can change so drastically as ketone bodies and still be compatible with life. After an overnight fast, the concentration of ketone bodies is approximately 0.05 mM, but this concentration can rise to 2 mM after 2 days of starvation and to 7 mM after 40 days, a 140-fold change in the concentration. In a seminal study, O. E. Owen and colleagues demonstrated that during prolonged starvation acetoacetate and β-hydroxybutyrate replaced glucose as the predominant fuel for the brain. This reduces the need to synthesize glucose from amino acids derived from muscle and liver protein. Muscle avidly consumes ketone bodies early in starvation but switches to fatty acid oxidation as starvation progresses, thereby sparing ketone bodies for metabolism by the brain. Thus, ketone bodies are a normal fuel for a variety of tissues and are part of a complex pattern of fuel metabolism that occurs during fasting in humans.

Owen, O. E., Morgan, A. P., Kemp, H. G., Sullivan, J. M., et al. *J. Clin. Invest.* 46: 1589, 1967; Feinman, R. D., and Fine, E. J. Thermodynamic and metabolic advantage of weight loss diets. *Metabolic Syndrome and Related Disorders* 1: 209, 2003; Samaha, F. F., Iqbal, N., Seshadri, P., Chicano, K. L., et al. A low-carbohydrate, as compared with a low-fat diet in severe obesity. *N. Engl. J. Med* 348: 21, 2003; and Shai, I., Schwarzfuchs, D., Henkin, Y., Shahar, D. R., et al. Weight loss with a low-carbohydrate, Mediterranean, or low-fat diet. *N. Engl. J. Med.* 359: 229, 2008.

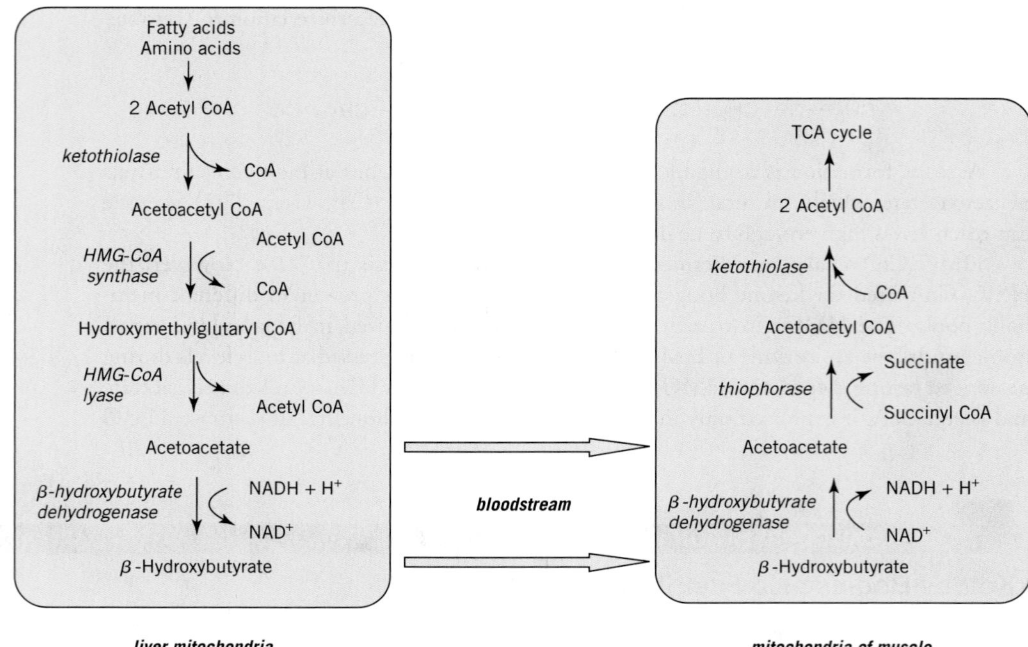

Figure 17.26 **Ketone body synthesis and utilization.**

CoA for cholesterol synthesis is made at low levels in the cytosol of many tissues by a cytosolic isozyme of HMG-CoA synthase.

Utilization of Ketone Bodies by Nonhepatic Tissues Requires Formation of Acetoacetyl CoA

Acetoacetate and β-hydroxybutyrate produced by the liver are excellent fuels for many nonhepatic tissues, including cardiac muscle, skeletal muscle, and brain, particularly when glucose is in short supply (prolonged fasting) or inefficiently used (insulin deficiency). Under these conditions, these tissues will oxidize free fatty acids, whose blood concentration rises as insulin levels fall. During prolonged fasting, ketone bodies replace glucose as a fuel, particularly in brain, which begins to use ketone bodies after 2–3 days of fasting. This reduces the requirement for glucose production by gluconeogenesis during a prolonged fast, thus "sparing" the muscle protein that contributes amino acids for gluconeogenesis (p. 619). Acetoacetate and β-hydroxybutyrate also serve as precursors for cerebral lipid synthesis during the neonatal period.

Ketone bodies are metabolized in the mitochondria of nonhepatic tissues. β-Hydroxybutyrate dehydrogenase converts β-hydroxybutyrate to acetoacetate by an NAD-linked oxidation in the mitochondrial matrix. Acetoacetate is then converted to its CoA derivative by **acetoacetate:succinyl–CoA transferase** (**thiophorase**), which is present in tissues that use ketone bodies but not in liver. Succinyl CoA serves as the source of the CoA. The reaction is depicted in Figure 17.26. β-Ketothiolase converts the acetoacetyl CoA into two acetyl CoAs, which enter the tricarboxylic acid cycle for energy production.

In summary, the pathways of ketone body synthesis and utilization share several steps. However, there are reactions that are unique to each pathway. The key enzymes of ketone body synthesis, HMG-CoA synthase and HMG-CoA lyase, are expressed in liver (and kidney cortex) but not in other tissues. The key enzyme of ketone body utilization, acetoacetate:succinyl–CoA transferase, is present in many tissues, but not in the liver. These differences ensure that ketone bodies are made in the liver and utilized in other tissues.

Peroxisomal Oxidation of Fatty Acids Serves Many Functions

Although the bulk of fatty acid oxidation occurs in mitochondria, a significant fraction also takes place in the **peroxisomes** of liver, kidney, and other tissues. Peroxisomes are a class of subcellular organelles with distinctive morphological and chemical characteristics

(p. 20). Peroxisomes in the liver contain the enzymes needed for β-oxidation. The mammalian **peroxisomal fatty acid oxidation** pathway is similar to that in plant glyoxysomes, but differs from mitochondrial β-oxidation in three important respects. First, the initial dehydrogenation is accomplished by an oxidase system that uses $\mathbf{O_2}$ and produces $\mathbf{H_2O_2}$ (Figure 17.27). The H_2O_2 is consumed by **catalase**. The remaining steps are the same as in the mitochondrial β-oxidation system. Second, the peroxisomal and mitochondrial enzymes differ in their specificity; the peroxisomal enzymes prefer fatty acids longer than eight carbons. Although rat liver mitochondria carry out the complete oxidation of fatty acyl CoAs to acetyl CoA, β-oxidation in liver peroxisomes will not proceed beyond octanoyl CoA (C_8). Thus, peroxisomes shorten long-chain fatty acids to a point where β-oxidation can be completed in the mitochondria. It is interesting to note that the thiazolidinediones, antidiabetic drugs that reduce peripheral insulin resistance and decrease triacylglycerol levels in patients, cause a marked increase in peroxisomes.

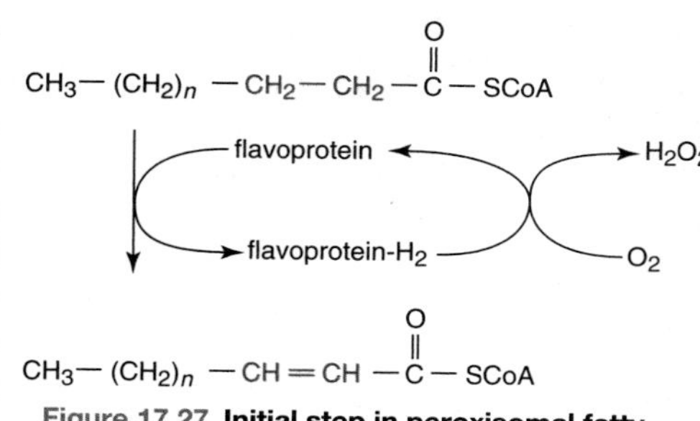

Figure 17.27 Initial step in peroxisomal fatty acid oxidation.

Other peroxisomal reactions include chain shortening of dicarboxylic acids, conversion of cholesterol into bile acids, and formation of ether lipids. Because of these diverse metabolic roles, it is not surprising that the congenital absence of functional peroxisomes, an inherited defect known as Zellweger syndrome, has such devastating effects (see Clin. Corr. 1.7, p. 21).

17.7 • REGULATION OF LIPID METABOLISM

Regulation in the Fed State

The metabolism of lipids in humans is controlled by the dietary status of the individual via a complex set of hormonal signals. After a meal that contains lipid, carbohydrate, and protein, the dietary lipid is deposited as triacylglycerol in adipose tissue. In addition, dietary carbohydrate and amino acids, in excess of that required for energy or for protein synthesis, are converted into fatty acids and deposited in adipose tissue as triacylglycerol. The major anabolic hormone, insulin, is required for both fatty acid synthesis and for the formation of triacylglycerol in adipose tissue. A summary is presented in Tables 17.2, p. 679, and 17.3, p. 683. Insulin acts at two levels; it induces the transcription of genes that code for critical enzymes in the pathways of lipid synthesis and storage (long-term regulation) and it controls processes such as glucose uptake and triacylglycerol hydrolysis (short-term regulation).

Insulin stimulates fatty acid synthesis by increasing the levels of key enzymes, including fatty acid synthase, NADP-malate dehydrogenase (malic enzyme), and acetyl-CoA carboxylase in the liver by inducing the transcription of their genes. Insulin also stimulates the synthesis of glucose-6-phosphate dehydrogenase and 6-phosphogluconate dehydrogenase, the two enzymes in the oxidative portion of the pentose pathway that generate some of the NADPH that is required for fatty acid synthesis. The short-term effect of insulin on hepatic fatty acid synthesis is exerted by activating a specific **phosphoprotein phosphatase**, which removes phosphate from acetyl-CoA carboxylase, thereby activating this enzyme. Increased flux through glycolysis is also important in providing acetyl CoA for fatty acid synthesis.

In adipose tissue, insulin is required in the fed state for glucose uptake via GLUT 4. The metabolism of this glucose via glycolysis provides glycerol 3-phosphate for the synthesis of triacylglycerol. Insulin also blocks the breakdown of triacylglycerol by inhibiting lipolysis, thereby preventing a futile cycle. As in the liver, insulin exerts its short-term effects by activating phosphoprotein phosphatases. This decreases the phosphorylation of key proteins, including hormone sensitive lipase and perilipin, leading to decreased breakdown of triacylglycerols.

Regulation in the Fasted State

Fasting results in a dramatic alteration in lipid metabolism, with cessation of lipid deposition and reliance on stored triacylglycerol. As the concentration of glucose in the blood decreases, there is a parallel decrease in the concentration of insulin in the circulation. There is also an increase in epinephrine and glucagon, which increases the level of hepatic cAMP and activates protein kinase A. In adipose tissue, there is increased phosphorylation of hormone sensitive lipase and perilipin, resulting in an increase in triacylglycerol breakdown and the release of free fatty acids and glycerol from this tissue (see Table 17.2, p. 679, for a summary of these controls).

In the liver, these hormonal changes lead to a decrease in fatty acid synthesis, due to reduction in the levels of the key enzymes (see Table 17.3, p. 683). There is also inhibition of the rate-limiting enzyme, acetyl-CoA carboxylase, due to cAMP-dependent phosphorylation of the enzyme. Glycolysis is also inhibited; this decreases the supply of acetyl CoA for lipogenesis. The liver begins to produce ketone bodies as fasting progresses, due to an increase in the rate of fatty acid oxidation caused by increased levels of fatty acids in the blood. During prolonged fasting, about half of the fatty acids that enter the liver are converted to ketone bodies and released into the blood for utilization by tissues such as the muscle, heart, and (after 2 days of fasting) the brain, thereby sparing the use of glucose.

Regulation of Fatty Acid Oxidation

The rate of fatty acid oxidation in mitochondria is controlled by regulating the entry of substrate into this organelle. The key enzyme is carnitine palmitoyltransferase I (CPT I), which synthesizes acyl carnitine from cytosolic acyl CoA (see Figure 17.20, p. 692). In the liver, acetyl-CoA carboxylase is activated in the fed state, because enzyme levels are increased, cyclic-AMP-dependent phosphorylation is low, and the enzyme is activated by citrate. The resulting high concentration of malonyl CoA stimulates fatty acid synthesis, but blocks fatty acid oxidation by inhibiting CPT I. This regulation prevents a futile cycle. Conversely, in the fasted state, the activity of acetyl-CoA carboxylase in the liver is low because enzyme levels are decreased, the enzyme is phosphorylated, and citrate is also decreased. CPT I is active and fatty acid oxidation occurs at a high rate under these conditions because of the low levels of malonyl CoA.

Fatty acid oxidation in muscle is also regulated by malonyl CoA, even though this tissue does not synthesize fatty acids. Muscle contains an isozyme of acetyl-CoA carboxylase, which produces malonyl CoA solely for the regulation of CPT I. The enzyme is activated by citrate and inhibited by phosphorylation. It is phosphorylated by both protein kinase A and an AMP-dependent kinase. Phosphorylation by the former enzyme allows fatty acid oxidation to be regulated by dietary state. In the fed state, the high concentration of insulin results in low levels of phosphorylation. Acetyl-CoA carboxylase produces malonyl CoA, which inhibits CPT I and blocks fatty acid oxidation. Conversely, in the fasted state, the high concentration of cAMP stimulates phosphorylation of acetyl-CoA carboxylase, resulting in its inhibition. As a consequence, CPT I facilitates fatty acid entry into the mitochondria for oxidation. This regulation promotes fatty acid oxidation in the fasted state and inhibits it in the fed state.

The second kinase, which is regulated by AMP, links the rate of fatty acid oxidation to the energy status of the muscle. In resting muscle, AMP levels are low. As a result, the AMP-dependent protein kinase is inactive, acetyl-CoA carboxylase is active, and the malonyl CoA that is generated inhibits CPT I and fatty acid oxidation. In exercising muscle, high levels of AMP activate the protein kinase. The resulting inhibition of acetyl-CoA carboxylase results in low levels of malonyl CoA and the activation of both CPT I and fatty acid oxidation. This regulation allows exercise to stimulate the rate of fatty oxidation through the increased production of AMP.

Fatty Acids as Regulatory Molecules

Fatty acids are themselves regulatory molecules in the liver, muscle, and adipose tissue. In muscle, long-chain acyl CoAs activate protein kinase C, as well as the transcription factors PPARγ and NFκB, which in turn inhibit insulin action by blocking the activation of intermediates in the insulin signaling pathway. These effects of fatty acids have dramatic consequences in the regulation of carbohydrate metabolism in muscle (Clin. Corrs. 17.2 and 17.8).

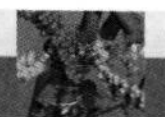

CLINICAL CORRELATION 17.8

Fatty Acids as Regulatory Molecules

Most textbooks of biochemistry deal extensively with the critical role of fatty acids in energy metabolism and thus stress the pathways of their synthesis, degradation, and modification. However, fatty acids also serve as regulatory molecules, which play a key role in the control of gene transcription, appetite, and hepatic glucose synthesis. Dietary fatty acids regulate gene expression by altering the activity of one of several **p**eroxisome **p**roliferator-**a**ctivated **r**eceptors (PPARs), which are transcription factors in the nuclear receptor family. The importance of the PPARs in the control of energy homeostasis is exemplified by the widespread clinical use of thiazolidinediones, which are PPARγ agonists, for the control of diabetes (see Clin. Corr. 17.1 for a detailed discussion of the role of these transcription factors). Fatty acids activate the PPARs and promote triacylglycerol deposition and glucose utilization by adipocytes. Long-chain fatty acids (LCFA) also act in the brain to regulate appetite via their CoA derivatives. Administration of oleic acid into rat brain ventricles suppresses appetite by lowering the expression of hypothalamic neuropeptides that inhibit appetite (neuropeptide Y and agouti-related protein); the elevated fatty acids in the hypothalamus signal a surfeit of energy. This effect is independent of leptin, a hormone produced in adipose tissue and a key inhibitor of appetite in mammals (see Clin. Corr. 21.1, p. 840). The administration of oleic acid into the brain also markedly decreases hepatic glucose production and the activity of glucose-6-phosphatase, a key enzyme in gluconeogenesis. This effect represents a brain–liver axis, since raising the concentration of oleic acid in the blood does not alter hepatic glucose output. The effect of oleic acid in the brain depends on the activation of K^+ channels by protein kinase C (PKC); several PKC isoforms are activated by fatty acids. Another site of regulation is the upper small intestine. There is evidence in both rodents and humans that fat activates an intestine–brain–liver axis, which signals satiety and controls the synthesis and release of glucose by the liver. The direct administration of fatty acids into the upper intestine increases LCFA CoA levels and suppresses hepatic glucose output. The importance of this axis is established by experiments showing that the effect of LCFA CoA is abolished when innervation of the gut via the vagus nerve is disrupted. Clearly, the view that fatty acids are only a fuel for energy metabolism and a constituent of complex lipids is an oversimplification. These molecules also have complex roles in regulating metabolic adaptations to nutrient availability.

Yki-Jarvinen, H. Thiazolidinediones. *N. Engl. J. Med.* 351: 1106, 2004; Wang, P. Y., Caspi, L., Lam, C. K., Chari, M., et al. Upper intestinal lipids trigger a gut-brain-liver axis to regulate glucose production. *Nature* 452: 1012, 2008; Jump, D. B., Botolin, D., Wang, Y., Xu, J., et al. Fatty acid regulation of hepatic gene transcription. *J. Nutr.* 135: 2503, 2005; and Lam, T. K., Schwartz, G. J., and Rossetti, L. Hypothalamic sensing of fatty acids. *Nat. Neurosci.* 8: 579, 2005.

Bibliography

Triacylglycerol Metabolism and Lipid Transport

Ducharme, N. A., and Bickel, P. E. Lipid droplets in lipogenesis and lipolysis. *Endocrinology* 149: 942, 2008.

Hanson, R. W., and Reshef, L. Glyceroneogenesis revisited. *Biochimie* 85: 1199, 2003.

Kane, J. P., and Havel, R. J. Disorders of the biogenesis and secretion of lipoproteins containing the B apolipoproteins. In Scriver, C. R., Beaudet, A.L., Sly, W. S., and Valle, D. (Eds.), *The Metabolic and Molecular Bases of Inherited Disease,* 8th ed. New York: McGraw Hill, 2001, 2: 2717.

Kurat, C. F., Natter, K., Petschnigg, J., Wolinski, H.,et al. Obese yeast: triglyceride lipolysis is functionally conserved from mammals to yeast. *J. Biol. Chem.* 281: 491, 2006.

Yeaman, S. J. Hormone-sensitive lipase—new roles for an old enzyme. *Biochem. J.* 379: 11, 2004.

Zimmermann, R., Strauss, J. G., Haemmerle, G., Schoiswohl, G., et al. Fat mobilization in adipose tissue is promoted by adipose triglyceride lipase. *Science* 306: 1383, 2004.

Fatty Acid Oxidation and Ketone Body Metabolism

Eaton, S., Bartlett, K., and Pourfarzam, M. Mammalian mitochondrial beta-oxidation. *Biochem. J.* 320(Pt. 2): 345, 1996.

Hashimoto, T. Peroxisomal β-oxidation enzymes. *Neurochem. Res.* 24: 551, 1999.

McGarry, J. D., and Brown, N. F. The mitochondrial carnitine palmitoyltransferase system: From concept to molecular analysis. *Eur. J. Biochem.* 244: 1, 1997.

Mitchell, G. A., and Fukao, T. Inborn errors of ketone body metabolism. In Scriver, C. R., Beaudet, A.L., Sly, W. S., and Valle, D. (Eds.), *The Metabolic and Molecular Bases of Inherited Disease,* 8th ed. New York: McGraw Hill, 2001, 2: 233.

Owen, O. E., and Hanson, R.W. Ketone bodies. *Encyclopedia of Endocrine Diseases* 3: 125, 2004.

Rinaldo, P., Matern, D., and Bennett, M. J. Fatty acid oxidation disorders. *Annu. Rev. Physiol.* 64: 477, 2002.

Wanders, R. J. A., Vreken, P., Den Boer, M. E. J., Wijburg, F. A., et al, Disorders of mitochondrial fatty acyl-CoA β-oxidation. *J. Inher. Metab. Dis.* 22: 442, 1999.

Fatty Acid Synthesis

Jump, D. B. The biochemistry of n-3 polyunsaturated fatty acids. *J. Biol. Chem.* 277: 8755, 2002.

Nakamura, M. T., and Nara, T. Y. Structure, function, and dietary regulation of delta6, delta5, and delta9 desaturases. *Annu. Rev. Nutr.* 24: 345, 2004.

Wakil, S. J. Fatty acid synthase, a proficient multifunctional enzyme. *Biochem.* 28: 4523, 1989.

Wakil, S. J., Stoops, J. K., and Joshi, V. C. Fatty acid synthesis and its regulation. *Annu. Rev. Biochem.* 52: 537, 1983.

Regulation of Lipid Metabolism in Health and Disease

Dowell, P., Hu, Z., and Lane, M. D. Monitoring energy balance: metabolites of fatty acid synthesis as hypothalamic sensors. *Annu. Rev. Biochem.* 74: 515, 2005.

Foster, D. W., and McGarry, J. D. Acute complications of diabetes: ketogenesis, hyperosmolar coma, lactic acidosis. In DeGroot, L. J. (Ed.), *Endocrinology*, 3d ed. Philadelphia: Saunders, 1995, 2: 1506.

Hardie, D. G., Carling, D., and Carlson, M. The AMP-activated/SNF1 protein kinase subfamily: Metabolic sensors of the eukaryotic cell? *Annu. Rev. Biochem.* 67: 821, 1998.

Hillgartner, F. B., Salati, L. M., and Goodridge, A. G. Physiological and molecular mechanisms involved in nutritional regulation of fatty acid synthesis. *Physiol. Rev.* 75: 47, 1995.

Itani, S. I., Ruderman, N. B., Schmieder, F., and Boden, G. Lipid-induced insulin resistance in human muscle is associated with changes in diacylglycerol, protein kinase C, and I kappa B-alpha. *Diabetes* 51: 2005, 2002.

Kim, K. H. Regulation of mammalian acetyl-coenzyme A carboxylase. *Annu. Rev. Nutr.* 17: 77, 1997.

Lowell, B. B., and Shulman, G. I. Mitochondrial dysfunction and Type 2 diabetes. *Science* 307: 384, 2005.

Ruderman, N. B., Saha, A. K., Vavvas, D., and Witters, L. A. Malonyl-CoA, fuel sensing, and insulin resistance. *Am. J. Physiol.* 276: E1, 1999.

Sampath, H., and Ntambi, J. M. Polyunsaturated fatty acid regulation of gene expression. *Nutr. Rev.* 62: 333, 2004.

Sul, H. S., and Wang, D. Nutritional and hormonal regulation of enzymes in fat synthesis: Studies of fatty acid synthase and mitochondrial glycerol-3-phosphate acyltransferase gene transcription. *Annu. Rev. Nutr.* 18: 331, 1998.

Willson, T. M., Lambert, M. H., and Kliewer, S. A. Peroxisome proliferator-activated receptor γ and metabolic disease. *Annu. Rev. Biochem.* 70: 341, 2001.

Key Terms

acylglycerols (mono-, di-, and tri-)
fatty acids (saturated, unsaturated, essential, branched chain)
chylomicron
free fatty acids
lipoprotein lipase
pancreatic lipase
very low-density lipoprotein (VLDL)
hormone-sensitive lipase
perilipin
adipose triglyceride lipase
fatty acid synthesis
lipogenesis
lipolysis
coenzyme A (CoA)
fatty acid synthase
malonyl CoA
acyl carrier protein
palmitic acid
ATP citrate lyase
citrate cleavage pathway
anaplerosis/cataplerosis
malic enzyme
fatty acid desaturation
fatty acid elongation
polyunsaturated fatty acids
glycerol 3-phosphate
glycerol kinase
glyceroneogenesis
phosphatidic acid
brown adipose tissue
white adipose tissue
insulin
fatty acid oxidation (β-oxidation)
carnitine/carnitine shuttle
propionyl CoA
ketone bodies
ketogenesis
peroxisomes

Questions • CAROL N. ANGSTADT

Multiple-Choice Questions

1. All of the following statements about acetyl-CoA carboxylase are correct except
   A. it catalyzes the rate-limiting step of fatty acid synthesis.
   B. it requires biotin.
   C. it is inhibited by cAMP-mediated phosphorylation.
   D. it is activated by palmitoyl CoA.
   E. its content in a cell responds to changes in fat content in the diet.

2. During the synthesis of palmitate in liver cells,
   A. the addition of malonyl CoA to fatty acid synthase elongates the growing chain by three carbon atoms.
   B. a β-keto residue on the 4′-phosphopantetheine moiety is reduced to a saturated residue by NADPH.
   C. palmitoyl CoA is released from the synthase.
   D. transfer of the growing chain from ACP to another -SH occurs prior to addition of the next malonyl CoA.
   E. the first compound to add to fatty acid synthase is malonyl CoA.

3. In humans, desaturation of fatty acids
   A. occurs primarily in mitochondria.
   B. is catalyzed by an enzyme system that uses NADPH and a cytochrome.
   C. introduces double bonds primarily of *trans* configuration.
   D. can occur only after palmitate has been elongated to stearic acid.
   E. introduces the first double bond at the methyl end of the molecule.

4. Lipoprotein lipase
   A. is an intracellular enzyme.
   B. is stimulated by cAMP-mediated phosphorylation.
   C. functions to mobilize stored triacylglycerols from adipose tissue.
   D. is stimulated by one of the apoproteins present in VLDL.
   E. produces free fatty acids and a monoacylglycerol.

5. α-Oxidation
   A. is important in the metabolism of branched chain fatty acids.
   B. metabolizes a fatty acid completely to acetyl CoA.
   C. produces hydrogen peroxide.
   D. prevents the fatty acid from producing energy.
   E. requires NADPH.

6. Another minor pathway of fatty acid oxidation is ω-oxidation, which results in a hydroxylation. ω-Oxidation
   A. occurs in mitochondria.
   B. introduces the –OH on the carbon adjacent to the carboxyl group.
   C. oxidizes primarily very long-chain fatty acids.
   D. oxidizes the terminal methyl group.
   E. produces dicarboxylic acids in the initial oxidation.

Questions 7 and 8: Following a severe cold which caused a loss of appetite, a 1-year-old boy was hospitalized with hypoglycemia, hyperammonemia, muscle weakness, and cardiac irregularities. These symptoms were consistent with a defect in the carnitine transport system. Dietary carnitine therapy was tried unsuccessfully, but a diet low in long-chain fatty acids and supplemented with medium-chain triacylglycerols was beneficial.

7. Carnitine transport of fatty acids from the cytosol to the mitochondria involves all of the following except
   A. hydrolysis of ATP.
   B. the exchange of acylcarnitine and free carnitine across the inner mitochondrial membrane.
   C. two carnitine palmitoyl transferases (CPT I and CPT II) located on different mitochondrial membranes.
   D. release of CoASH from fatty acyl CoA in the cytosol.
   E. consumption of mitochondrial CoASH.

8. The child was diagnosed with carnitine-acylcarnitine translocase deficiency. The dietary treatment was beneficial because
   A. the child could get all required energy from carbohydrate.
   B. the deficiency was in the peroxisomal system so carnitine would not be helpful.
   C. medium-chain fatty acids (8–10 carbons) enter the mitochondria before being converted to their CoA derivatives.
   D. medium-chain triacylglycerols contain mostly hydroxylated fatty acids.
   E. medium-chain fatty acids such as C_8 and C_{10} are readily converted into glucose by the liver.

Questions 9 and 10: Medium-chain acyl-CoA dehydrogenase deficiency (MCAD), a defect in β-oxidation, usually produces symptoms within the first 2 years of life after a period of fasting. Typical symptoms include vomiting, lethargy, and hypoketotic hypoglycemia. Excessive urinary secretion of medium-chain dicarboxylic acids and medium-chain esters of glycine and carnitine help to establish the diagnosis.

9. β-Oxidation of fatty acids
   A. generates ATP only if acetyl CoA is subsequently oxidized.
   B. is usually suppressed during starvation.
   C. uses only even-chain, saturated fatty acids as substrates.
   D. uses $NADP^+$.
   E. occurs by a repeated sequence of four reactions.

10. The lack of ketone bodies in the presence of low blood glucose in this case is unusual since ketone body concentrations usually increase with fasting-induced hypoglycemia. Ketone bodies
    A. are formed by removal of CoA from the corresponding intermediate of β-oxidation.
    B. are synthesized from cytoplasmic β-hydroxy-β-methyl glutaryl coenzyme A (HMG-CoA).
    C. are synthesized primarily in muscle tissue.
    D. include both β-hydroxybutyrate and acetoacetate, the ratio reflecting the intramitochondrial [NADH]/[NAD^+] ratio in liver.
    E. form when β-oxidation is interrupted.

Questions 11 and 12: One of the problems associated with obesity is the increased risk of Type 2 diabetes. High fatty acid concentration in the blood reduces glucose uptake and metabolism by skeletal muscle, increasing levels of blood glucose and insulin secretion (insulin resistance). Prolonged overproduction of insulin can cause failure of the β cells of the pancreas and Type 2 diabetes. This occurs in ~40% of obese individuals over 5–10 years. One way of regulating the concentration of fatty acids in blood is their reesterification into triacylglycerols. One type of antidiabetic drug (thiazolidinedione) acts on a nuclear receptor (PPARγ2) facilitating the rate of fatty acid esterification in white adipose tissue.

11. All of the following events are usually involved in the synthesis of triacylglycerols in adipose tissue except
    A. addition of a fatty acyl CoA to a diacylglycerol.
    B. addition of a fatty acyl CoA to a lysophosphatide.
    C. a reaction catalyzed by glycerol kinase.
    D. hydrolysis of phosphatidic acid by a phosphatase.
    E. reduction of dihydroxyacetone phosphate.

12. Glycerol-3-phosphate for triacylglycerol synthesis
    A. is always formed by reduction of dihydroxyacetone phosphate.
    B. can be formed in liver by glyceroneogenesis but not in adipose tissue.
    C. derives its carbons primarily from amino acids in the fed state.
    D. can be synthesized only in the presence of phosphoenolpyruvate carboxykinase.
    E. is derived primarily from glucose via glycolysis in the fed state.

Problems

13. How does β-oxidation of an unsaturated fatty acid deal with the naturally occurring *cis* double bond when the first step of β-oxidation produces a *trans* enoyl CoA intermediate?

14. How does oxidation of a 17-carbon fatty acid (from plants) lead to the production of propionyl CoA? Be specific in your answer.

Answers

1. **D** It is activated by citrate and inhibited by long-chain fatty acyl CoAs. C: Since cAMP increases at times when energy is needed, it is consistent that a process that uses energy would be inhibited. E: Long-term control is related to enzyme synthesis and responds appropriately to dietary changes.
2. **B** A: Splitting CO_2 from malonyl CoA is the driving force for the condensation reaction so the chain grows two carbon atoms at a time. C: In mammals, palmitate is released as the free acid; the conversion to the CoA ester is by a different enzyme. D: It is important to realize that only ACP binds the incoming malonyl CoA so it must be freed before another addition can be made. E: Acetyl CoA adds first to form the foundation for the rest of the chain.

3. **B** A: Desaturation occurs in the endoplasmic reticulum. C: Naturally occurring fatty acids are *cis*. D: Elongation and unsaturation can occur in any order. E: If this were true, we could make linoleic acid.
4. **D** ApoC-II binds the lipoprotein to the enzyme. A–C: These are characteristics of hormone-sensitive lipase. E: This is pancreatic lipase.
5. **A** The presence of a methyl group precludes the β-oxidation process at that point. B: Once past the methyl group, β-oxidation can proceed. C: The reaction is by a hydroxylase that adds an –OH. D: During the α-oxidation phases, energy is produced. E: This hydroxylase uses ascorbate as reducing agent.
6. **D** This is why it is called ω-oxidation. A: Occurs in endoplasmic reticulum. B: This is α-oxidation. C: Medium-chain fatty acids are preferred substrates. E: The hydroxyl group requires alcohol and aldehyde dehydrogenases to convert it to the acid.
7. **A** The exchanges of the acyl group between CoA and carnitine do not require additional energy. B: This is catalyzed by carnitine-acylcarnitine translocase. C: CPT I is on the outer membrane, and CPT II is on the matrix side of the inner membrane. D and E: Only the acyl portion of the molecule is transported; the CoASH molecules remain in their respective compartments.
8. **C** Because medium-chain fatty acids cross the mitochondrial membrane directly, they do not require the carnitine system. A: This is never true. B: Peroxisomal oxidation does not require carnitine, but this is not a peroxisomal system. D: Hydroxylated fatty acids are not a common constituent of triacylglycerols. E: Liver cannot synthesize glucose from fatty acids with an even number of carbon atoms.
9. **E** A and D: It is important to realize that β-oxidation, itself, generates $FADH_2$ and NADH, which can be reoxidized to generate ATP. B: Fatty acid oxidation is usually enhanced during fasting. C: β-Oxidation is a general process requiring only minor modifications to oxidize nearly any fatty acid in the cell.
10. **D** A and E: β-Oxidation proceeds to completion; ketone bodies are formed by a separate process. B and C: Ketone bodies are formed, but not used, in liver mitochondria; cytosolic HMG CoA is a precursor of cholesterol. Ketone bodies are not readily synthesized by muscle.
11. **C** This does not occur to any significant extent in adipose tissue. A, B, and D: The sequential addition of fatty acyl CoAs to glycerol 3-phosphate forms lysophosphatidic acid, then phosphatidic acid whose phosphate is removed before the addition of the third fatty acyl residue. E: This is the formation of α-glycerol phosphate in adipose.
12. **E** This is the primary pathway in liver and adipose tissue. A: Liver can form it from glycerol because it has glycerol kinase. B: Both tissues have this pathway. C and D: These are substrates for glyceroneogenesis in the fasted state and require phosphoenolpyruvate carboxykinase.
13. When several rounds of β-oxidation produce an intermediate with a *cis* double bond between C3 and C4, enoyl-CoA isomerase changes the *cis* Δ^3 to a *trans* Δ^2 enoyl CoA and β-oxidation continues. If another *cis* double bond is encountered between C4 and C5, first the normal dehydrogenase, then a reductase with NADPH, followed by the enoyl-CoA isomerase act on the molecule to produce a *trans* 2-enoyl CoA. β-Oxidation continues.
14. β-Oxidation proceeds normally, but the final thiolase cleavage yields acetyl CoA and propionyl CoA. Propionyl CoA is not a substrate for SCAD (short-chain acyl-CoA dehydrogenase) so β-oxidation terminates.

18

Lipid Metabolism II: Pathways of Metabolism of Special Lipids

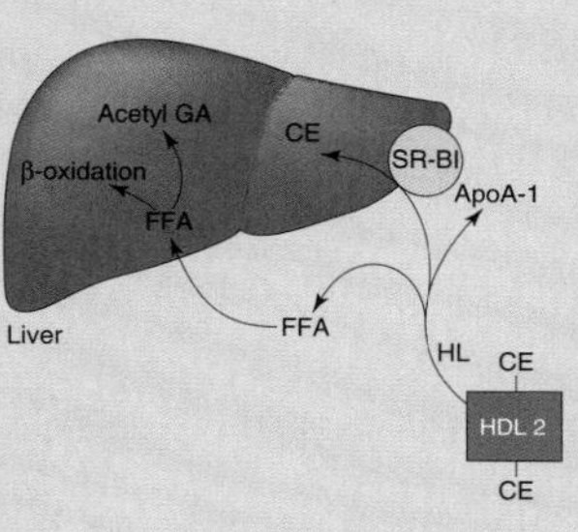

Robert H. Glew

Professor Emeritus, University of New Mexico

CLINICAL CORRELATIONS

Key Concepts

- Most cells synthesize the major glycerophospholipids and sphingolipids from dietary components and metabolic intermediates. Both are major components of cell membranes.
- Glycerophospholipids have other functions including as components of pulmonary surfactant. Phosphatidylcholine has a role in the HDL-dependent process that carries cholesterol from peripheral tissues to the liver, and inositol-containing phospholipids serve as second messengers and as attachment sites for proteins on the cell surface.
- Cholesterol is synthesized from acetyl CoA in a multistep process in which the synthesis of mevalonate, catalyzed by HMG-CoA reductase, is the rate-limiting, regulated step. The reductase is the target of the statin class of cholesterol-lowering drugs. Cholesterol is the precursor for the synthesis of bile acids and steroid hormones such as testosterone, estrogen, glucocorticoids, and mineralocorticoids.
- Plasma lipoproteins, including HDL, LDL, chylomicrons, VLDL, and a variety of accessory proteins, transport cholesterol and triacylglycerols in blood. Cholesteryl ester transferase has an important role in the process. LDL and LDL-receptors are the major regulatory elements that control the levels of cholesterol in plasma and tissues.
- There are several classes of sphingolipids, including sphingomyelin, cerebrosides, globosides, gangliosides, and sulfatides. Degradation of sphingolipids involves lysosomal enzymes: A genetic deficiency of one or more of these enzymes results in the accumulation of incompletely catabolized sphingolipids and mucopolysaccharides in tissues.
- Arachidonic acid is the precursor of lipid hormones that include prostaglandins, leukotrienes, and lipoxins that mediate various inflammatory phenomena. The key enzymes in these pathways are cyclooxygenase and lipoxygenases.

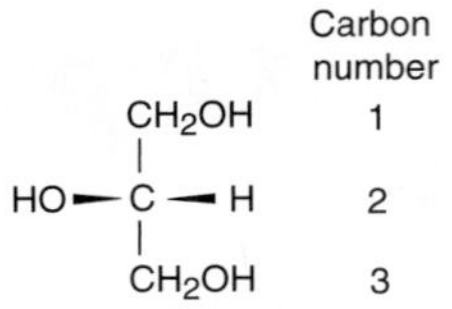

Figure 18.1 **Stereospecific numbering of glycerol.**

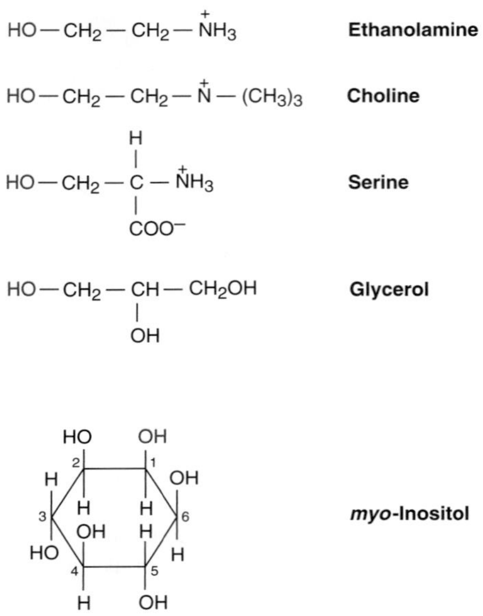

18.1 • INTRODUCTION

Lipid is a general term that describes substances that are relatively water insoluble and extractable by nonpolar solvents. Complex lipids of humans fall into two broad categories: (1) neutral, nonpolar lipids, such as triacylglycerols and cholesteryl esters, and (2) polar lipids, such as phospholipids and glycolipids. The polar lipids are amphipathic, containing a hydrophobic and hydrophilic region in the same molecule. The hydrophobic and hydrophilic regions in glycerophospholipids are bridged by a glycerol moiety and in sphingomyelin and glycosphingolipids by sphingosine. Triacylglycerol is found primarily in storage sites in adipose tissue, whereas polar lipids occur primarily in cellular membranes.

Complex lipids have many roles. Besides their function in the structure of membranes, some glycerophospholipids are required for the activity of membrane enzymes and inositol-containing phospholipids serve as precursors of signaling molecules Glycosphingolipids have a role in cell–cell recognition, phagocytosis, contact inhibition, and rejection of transplanted tissues and organs. Antigenic determinants of blood groups are primarily glycolipid in nature. Cholesterol is important in atherosclerosis and various sphingolipids accumulate in genetic disorders called sphingolipidoses.

The nomenclature and chemistry of lipids is presented in the Appendix.

Figure 18.2 **Structure of some common polar groups of phospholipids.**

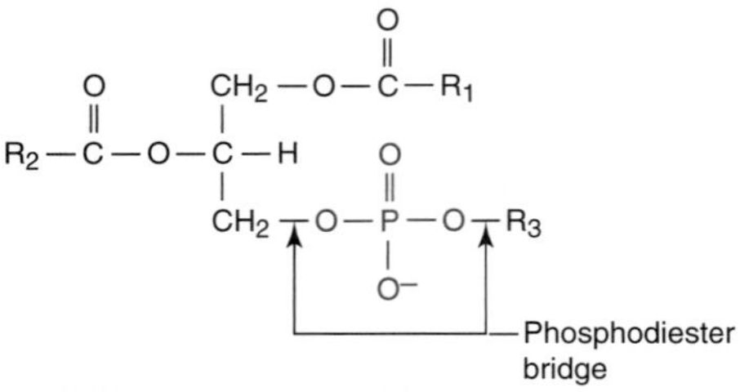

18.2 • PHOSPHOLIPIDS

Two major classes of acylglycerolipids are triacylglycerols and glycerophospholipids which have as their core the C_3 polyol, glycerol. Two primary alcohol groups of glycerol are not stereochemically identical, and in the case of phospholipids, it is usually the same hydroxyl group esterified to the phosphate residue. The stereospecific numbering system designates different hydroxyl groups. In this system, when the structure of glycerol is drawn in the Fischer projection with the C2 hydroxyl group projecting to the left of the page, the carbon atoms are numbered as shown in Figure 18.1. When the stereospecific numbering (sn) system is employed, the prefix *sn-* is used before the name of the compound. Glycerophospholipids usually contain a sn-glycerol 3-phosphate moiety. Although each contains the glycerol moiety as a fundamental structural element, neutral triacylglycerols and charged ionic phospholipids have very different physical properties and functions.

Phospholipids Contain Phosphatidic Acid Linked to a Base

Phospholipids are polar, ionic lipids composed of 1,2-diacylglycerol and a phosphodiester bridge that links the glycerol backbone to some base, usually a nitrogenous one, such as choline, serine, or ethanolamine (Figures 18.2 and 18.3). The most abundant phospholipids in human tissues are phosphatidylcholine (also called lecithin), phosphatidylethanolamine, and phosphatidylserine (Figure 18.4). At physiologic pH, phosphatidylcholine and phosphatidylethanolamine have no net charge and exist as dipolar zwitterions, whereas phosphatidylserine has a net charge of −1, causing it to be an acidic phospholipid. Phosphatidylethanolamine (PE) is related to phosphatidylcholine in that trimethylation of PE produces lecithin. Most phospholipids contain more than one kind of fatty acid per molecule, so that a given class of phospholipids from any tissue actually represents a family of molecular species. Phosphatidylcholine (PC) contains mostly palmitic acid (16:0) or stearic acid (18:0) in the sn-1 position and primarily unsaturated 18 carbon fatty acids oleic, linoleic, or α-linolenic in the sn-2 position. Phosphatidylethanolamine has the same saturated fatty acids as PC at the sn-1 position but contains more of the long-chain polyunsaturated fatty acids, namely, linoleic acid [18:2(9,12)], arachidonic acid [20:4(5,8,11,14)], and docosahexaenoic acid [22:6(4,7,10,13,16,19)], at the sn-2 position.

Figure 18.3 **Generalized structure of a phospholipid where R1 and R2 represent the aliphatic chains of fatty acids, and R3 represents a polar group.**

Phosphatidylinositol, an acidic phospholipid that occurs in mammalian membranes (Figure 18.5), is rather unusual because it often contains almost exclusively stearic acid (18:0) in the sn-1 position and arachidonic acid in the sn-2 position.

Another phospholipid composed of a polyol polar head group is phosphatidylglycerol (Figure 18.5), which occurs in relatively large amounts in mitochondrial membranes and pulmonary surfactant and is a precursor of cardiolipin. Phosphatidylglycerol and phosphatidylinositol both carry a formal charge of –1 at neutral pH and are therefore acidic lipids.

Cardiolipin, a very acidic (charge –2) phospholipid, is composed of two molecules of phosphatidic acid linked together covalently through a molecule of glycerol (Figure 18.6). It occurs primarily in the inner membrane of mitochondria of metabolically active tissues (e.g., heart muscle) and in bacterial membranes. Cardiolipin is present in the membrane of *Treponema palladium* and is the antigen detected in the Wasserman test for syphilis. Barth syndrome is a rare X-linked mitochondrial disorder caused by a defect in the TAZ gene, which encodes the protein taffazin which is required for cardiolipin synthesis. Patients with this inherited disorder exhibit cardiomyopathy, skeletal myopathy, and abnormal mitochondria.

Phospholipids mentioned so far contain only *O*-acyl residues attached to glycerol. *O*-(1-alkenyl) substituents occur at C1 of the sn-glycerol in phosphoglycerides in combination with an *O*-acyl residue esterified to the C2 position; compounds in this class are known as plasmalogens (Figure 18.7) or plasmenyl lipids. Relatively large amounts of ethanolamine plasmalogen (also called plasmenylethanolamine) occur in myelin with lesser amounts in heart muscle where **choline plasmalogen** is abundant. The alkenyl moiety is usually 16:0, 18:0 or 18:1(9).

An unusual phospholipid called **platelet activating factor** (**PAF**) (Figure 18.8) is a major mediator of hypersensitivity, acute inflammatory reactions, allergic responses, and anaphylactic shock. In hypersensitive individuals, cells of the polymorphonuclear (PMN) leukocyte family (basophils, neutrophils, and eosinophils), macrophages, and monocytes are coated with IgE molecules that are specific for a particular antigen (e.g., ragweed pollen and bee venom). Subsequent reexposure to the antigen and formation of antigen–IgE complexes on the surface of the aforementioned inflammatory cells provokes synthesis and release of PAF. Platelet activating factor contains an *O*-alkyl moiety at sn-1 and an acetyl residue instead of a long-chain fatty acid in position 2 of the glycerol moiety. PAF is not stored; it is synthesized and released when PMN cells are stimulated. Platelet aggregation, cardiovascular and pulmonary changes, edema, hypotension, and PMN cell chemotaxis are affected by PAF. Inactivation of PAF involves hydrolysis of the acetyl moiety followed by reacylation with a long-chain fatty acid to form an ether-type membrane phospholipid.

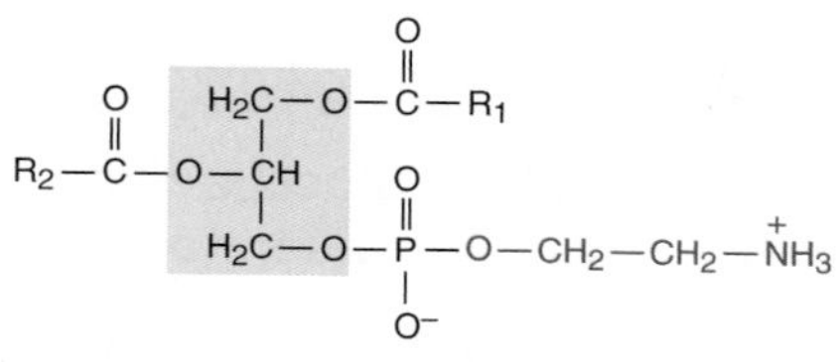

Phospholipids in Membranes Serve a Variety of Roles

Although present in body fluids such as plasma and bile, phospholipids are present in highest concentration in cellular membranes where they serve as structural and functional components. Nearly one-half the mass of the erythrocyte membrane is composed

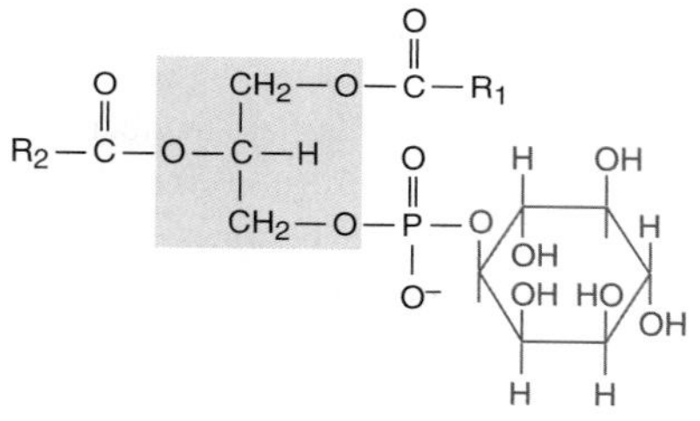

Phosphatidylethanolamine

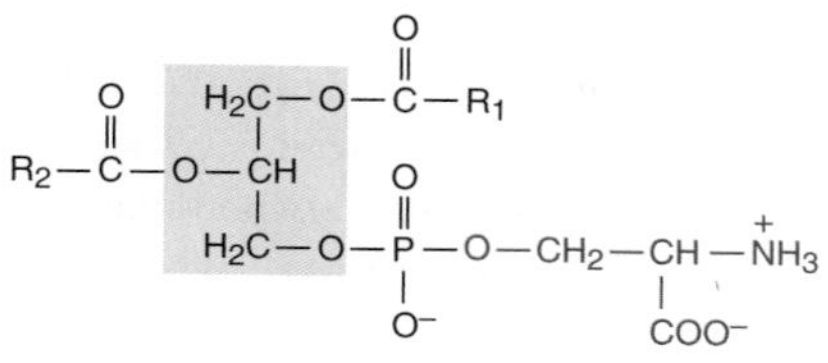

Phosphatidylserine

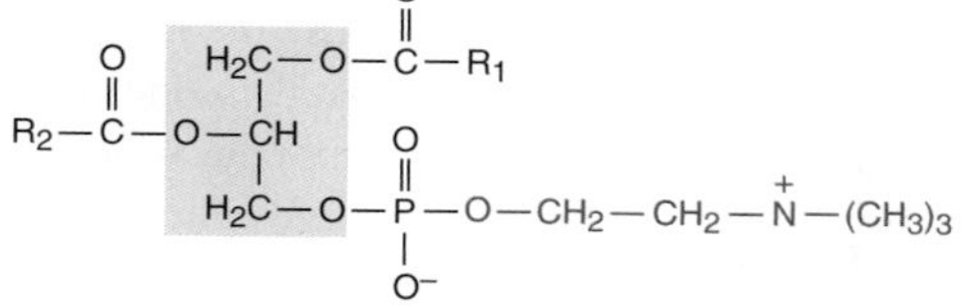

Phosphatidylcholine (lecithin)

Figure 18.4 Structures of some common phospholipids.

Phosphatidylinositol

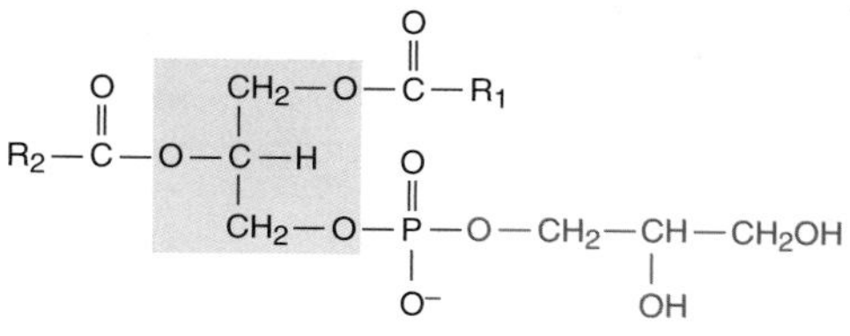

Phosphatidylglycerol

Figure 18.5 Structures of phosphatidylglycerol and phosphatidylinositol.

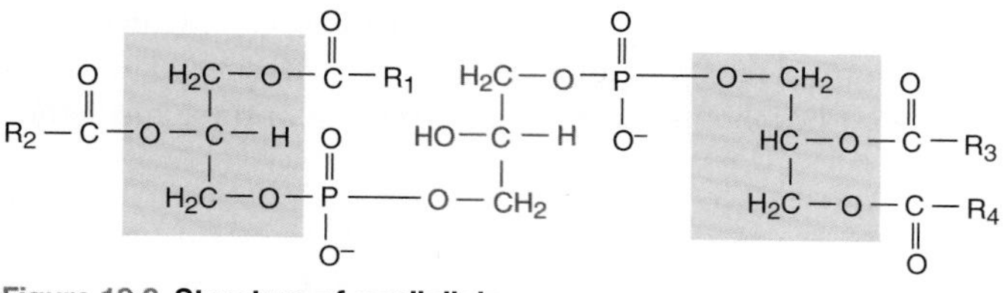

Figure 18.6 Structure of cardiolipin.

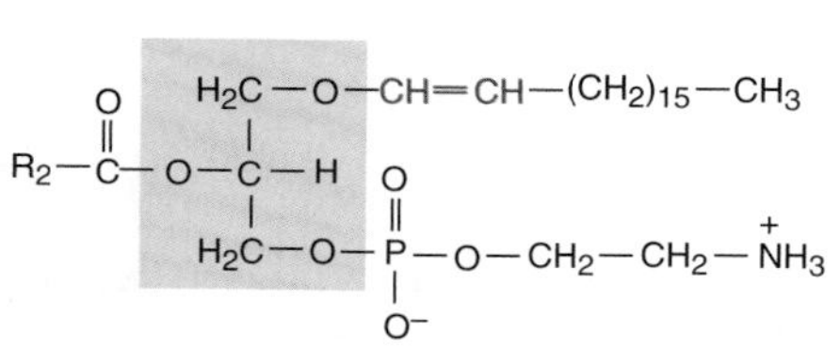

Figure 18.7 Structure of ethanolamine plasmalogen.

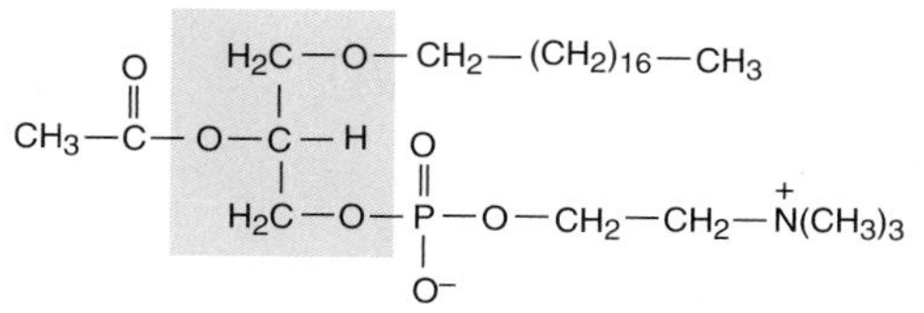

Figure 18.8 Structure of platelet activating factor (PAF).

CLINICAL CORRELATION 18.1

Clearance of Red Blood Cells: Role of Phosphatidylserine

The phospholipids of the plasma membrane of cells, including red blood cells, are distributed asymmetrically: The outer leaflet that faces the extracellular space contains the choline lipids phosphatidylcholine and sphingomyelin, whereas the inner leaflet facing the cytoplasm contains phosphatidylethanolamine and negatively charged phospholipids, including phosphatidylserine. Macrophages contain phosphatidylserine receptors that bind, internalize, and degrade phosphatidylserine-displaying cells. Normal aging of red cells is associated with the exposure of phosphatidylserine on the surface of the plasma membrane, which signals cells of the macrophage system to remove them from the circulation.

There is evidence that the anemia and shortened life span of red cells in certain pathologic conditions such as lead intoxication and uremia in patients with renal failure could be due to increased red cell phosphatidylserine exposure. An enzyme called semblase transports phosphatidylserine from the inner leaflet of the plasma membrane to the outer leaflet. Semblase is inactive in healthy cells; however, this calcium-sensitive enzyme becomes active when cells are exposed to stressors (e.g., osmotic shock, ATP depletion, exposure to reactive oxygen species) that increase the intracellular calcium concentration.

Kempe, S. K., Lang, P.A., Eisele, K., et al. Stimulation of erythrocyte phosphatidylserine exposure by lead ions. *Am. J. Cell Physiol.* 288:C396, 2005.

of various phospholipids (p. 458). They also activate certain enzymes; as an example, β-hydroxybutyrate dehydrogenase, in the inner membrane of mitochondria (p. 698), has an absolute requirement for phosphatidylcholine; phosphatidylserine and phosphatidylethanolamine cannot substitute. Phosphatidylserine apparently has a role in clearance of red cells (Clin. Corr. 18.1). Phosphatidylcholine also plays a central role in the process of reverse cholesterol transport.

Dipalmitoyllecithin Is Necessary for Normal Lung Function

Normal lung function depends on a constant supply of dipalmitoyllecithin in which the molecule contains palmitic acid (16:0) residues in the sn-1 and sn-2 positions. More than 80% of the phospholipid in the extracellular fluid layer that lines alveoli of normal lungs is dipalmitoyllecithin. This surfactant produced by type II epithelial cells prevents atelectasis at the end of the expiration phase of breathing (Figure 18.9). It decreases the surface tension of the fluid layer of the lung. Lecithin molecules that do not contain two residues of palmitic acid are not effective in the lowering of surface tension. Surfactant also contains phosphatidylglycerol, phosphatidylinositol, cholesterol, and 18- and 36-kDa proteins (designated surfactant proteins), which contribute significantly to the lowering of the surface tension. Surfactant proteins alter the molecular structure of the phospholipid film secreted by type II pneumocytes in such a way as to stabilize the film and maintain its flexibility. Most of the cholesterol in surfactant is derived from plasma lipoproteins, but the phospholipids are synthesized by type II cells. Before the 28th week of gestation, fetal lung synthesizes primarily sphingomyelin. Normally, at this time, glycogen that has been stored in epithelial type II cells is converted to fatty acids and then to dipalmitoyllecithin. During lung maturation there is a good correlation between the increase in intracellular lamellar inclusion bodies that represent phosphatidylcholine storage organelles, called lamellar bodies, and the decrease in glycogen content of these cells. At the 24th week of gestation, the type II granular pneumocytes appear in the alveolar epithelium and start to produce lamellar bodies. Their number increases until the 32nd week when surfactant appears in the lung and amniotic fluid. In the few weeks before term, screening tests on amniotic fluid can detect newborns that are at risk for respiratory distress syndrome (RDS) (Clin. Corr. 18.2). These are useful in timing elective deliveries, in determining if the mother should receive antenatal corticosteroid therapy to accelerate maturation of the fetal lung, or in applying preventive therapy to the newborn infant. Dexamethasone has been used in neonates with chronic lung disease (bronchopulmonary dysplasia). While corticosteroid therapy may be effective in some cases in improving lung function, in others it causes periventricular abnormalities in the brain. Exogenous human surfactant administered by means of intratracheal instillation is used widely to treat RDS.

Respiratory failure due to surfactant insufficiency also occurs in adults whose type II cells have been destroyed as an adverse side effect of immunosuppressive medications or chemotherapeutic drug (e.g., bleomycin).

The detergent properties of phospholipids, especially phosphatidylcholine, are important in bile to aid in solubilizing cholesterol. An impaired production and secretion of phospholipids into bile can lead to formation of cholesterol stones and bile pigment gallstones. Membrane phospholipids are a reservoir for lipid mediators that regulate many metabolic pathways and processes. Phospholipases catalyze the release of these mediators.

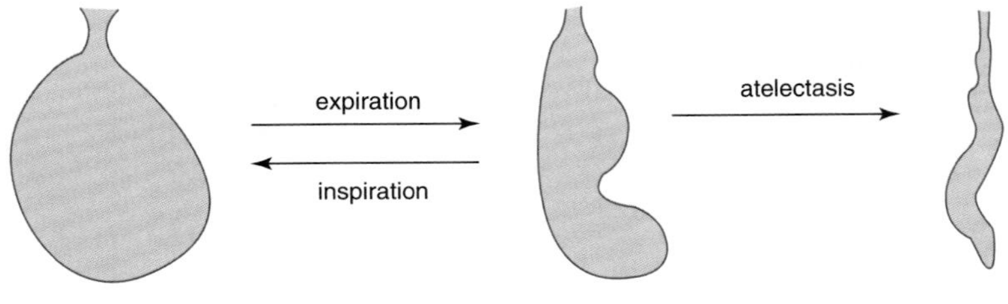

Figure 18.9 Role of surfactant in preventing atelectasis.

CLINICAL CORRELATION 18.2

Respiratory Distress Syndrome

Respiratory distress syndrome (RDS) is a major cause of neonatal morbidity and mortality in many countries. It accounts for about 15%–20% of all neonatal deaths in Western countries and somewhat less in developing countries. The disease affects only premature babies, and the incidence varies directly with the degree of prematurity. Premature babies develop RDS because of immaturity of their lungs from deficiency of pulmonary surfactant. The maturity of the fetal lung can be assessed from the lecithin/sphingomyelin (L/S) ratio in amniotic fluid. The mean L/S ratio in normal pregnancies increases gradually with gestation until about 31 or 32 weeks when the slope rises sharply. The ratio of 2.0 is characteristic of term birth, and it is achieved at the gestational age of about 34 weeks. At pulmonary maturity, the critical L/S ratio is 2.0 or greater. The risk of developing RDS when the L/S ratio is 1.5–1.9 is about 40%, and for less than 1.5 about 75%. Although the L/S ratio is still widely used to predict the risk of RDS, the results are unreliable if the amniotic fluid has been contaminated by blood or meconium. Determination of saturated palmitoyl phosphatidylcholine (SPC), phosphatidylglycerol, and phosphatidylinositol is also predictive of the risk of RDS. Replacement therapy using surfactant from human and animal lungs is effective in the prevention and treatment of RDS.

Merritt, T. A., Hallman, M., Bloom, B. T., et al., Prophylactic treatment of very premature infants with human surfactant. *N. Engl. J. Med.* 315:785, 1986; and Simon, N. V., Williams, G. H., Fairbrother, P. F., Elser, R. C., and Perkins, R. P. Prediction of fetal lung maturity by amniotic fluid fluorescence polarization, L/S ratio, and phosphatidylglycerol. *Obstet. Gynecol.* 57:295, 1981.

Phosphatidylinositol and phosphatidylcholine are sources of arachidonic acid for synthesis of prostaglandins, thromboxanes, leukotrienes, and related compounds.

Inositides Are Important in Membrane Function

Inositol-containing phospholipids (inositides), especially phosphatidylinositol 4,5-bisphosphate (PIP_2) (Figure 18.10), play a central role in signal transduction systems. When certain hormones bind to their receptors (p. 507), PIP_2 in the inner leaflet of the plasma membrane is hydrolysed by phosphoinositidase C (PIC) into inositol 1,4,5-trisphosphate (IP_3), which triggers release of Ca^{2+} from the endoplasmic reticulum, and 1,2-diacylglycerol (DAG), which increases the activity of protein kinase C (PKC) (Figure 18.11). Removal of the 5-phosphate of IP_3 abolishes the signal and the intracellular Ca^{2+} concentration declines. The 1,2-diacylglycerol is converted to phosphatidic acid by diacylglycerol kinase (Figure

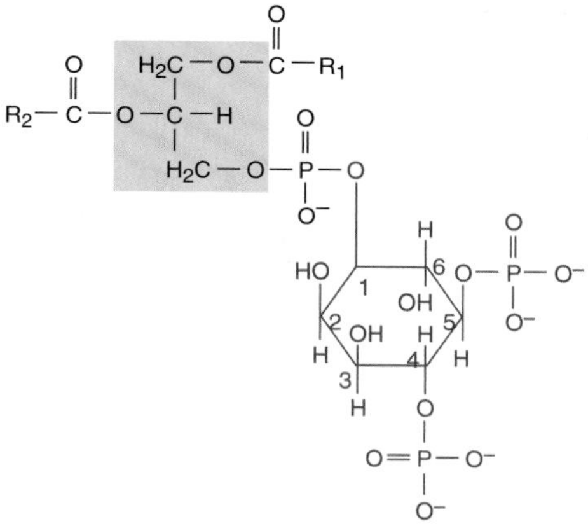

Figure 18.10 Structure of phosphatidylinositol 4,5-bisphosphate (PIP_2 or PtdIns $(4,5)P_2$).

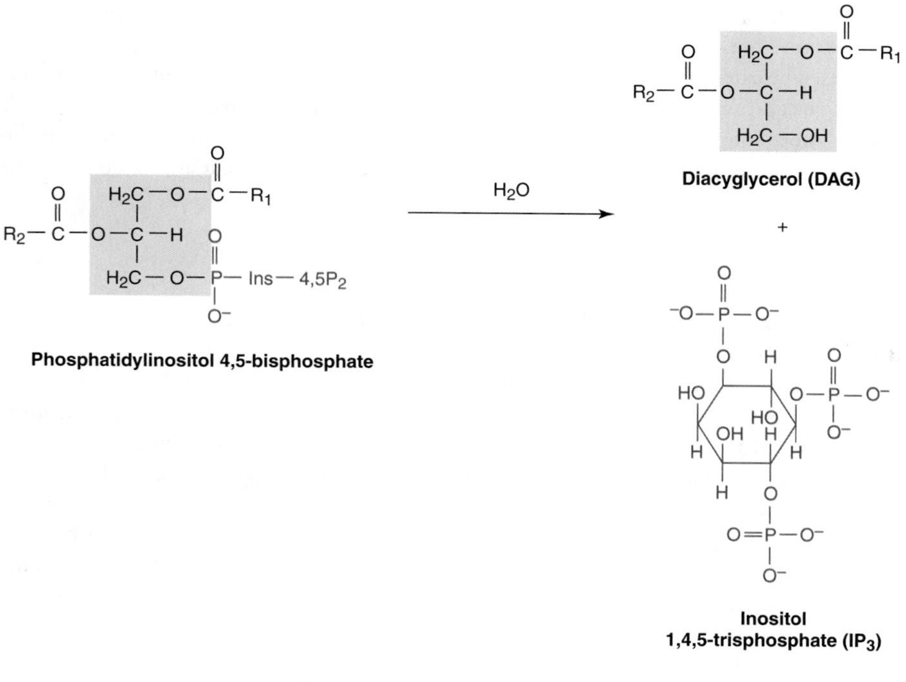

Figure 18.11 Generation of 1,2-diacylglycerol and inositol 1,4,5-trisphosphate by action of phospholipase C on phosphatidylinositol 4,5-bisphosphate.

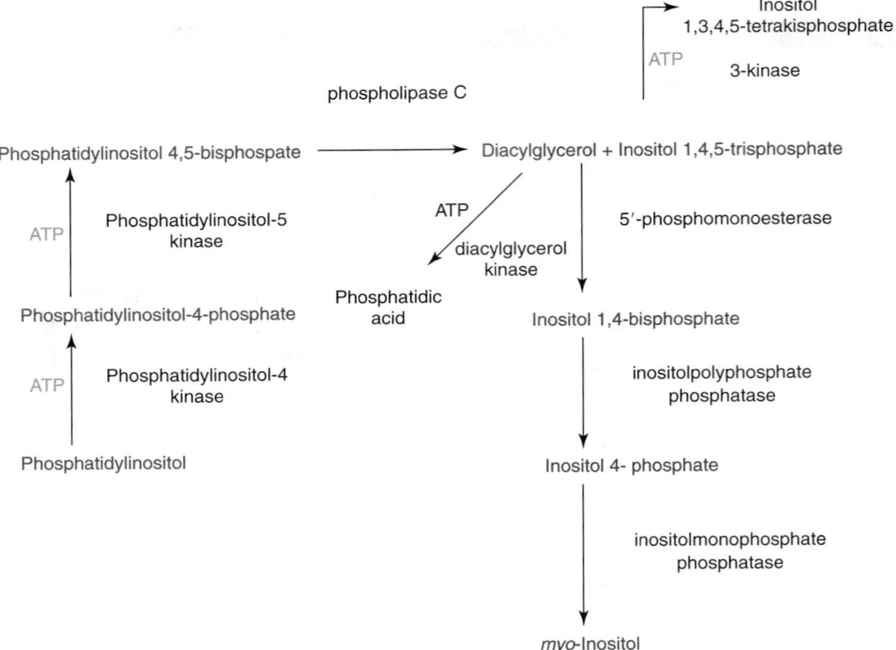

Figure 18.12 Pathways for the synthesis and removal of intracellular inositol 1,4,5-trisphosphate and diacylglycerol.

18.12). Phosphatidic acid, a product of phospholipase D action on phospholipids, has been implicated as a second messenger.

These pathways of inositol phosphate metabolism serve in: (1) removal and inactivation of IP_3, (2) conservation of inositol, and (3) synthesis of polyphosphates such as inositol pentakisphosphate ($InsP_5$) and inositol hexakisphosphate ($InsP_6$) whose functions have not been determined. IP_3 is metabolized by 5-phosphomonoesterase to inositol 1,4-bisphosphate and by a 3-kinase that forms inositol 1,3,4,5-tetrakisphosphate. A family of phosphatases converts Ins(1,4)P_2 to myo-inositol (Figure 18.12), which then enters the phospholipid pool.

Besides being a component of membranes and source of arachidonic acid for prostaglandin and leukotriene synthesis (p. 743), phosphatidylinositol serves to anchor (GPI anchor) certain glycoproteins to the external surface (p. 471). A major medical problem concerns trypanosomal parasites (e.g., *Trypanosoma brucei,* which causes sleeping sickness). This parasite has resisted immunological approaches to treatment by changing its surface antigens. The external surface of the plasma membrane is coated with a protein called variable surface glycoprotein (VSG) linked to the membrane through a phosphatidylinositol anchor. C-type phospholipase on the cell surface permits the shedding of the anchored protein allowing trypanosomes to discard surface antigens, thus changing their coat and escaping antibodies of the host's immune system.

Biosynthesis of Phospholipids

Phosphatidic Acid Is Synthesized from α-Glycerophosphate and Fatty Acyl CoA

l-α-Phosphatidic acid (commonly called phosphatidic acid) and 1,2-diacyl-sn-glycerol are intermediates common to the pathways of phospholipid and triacylglycerol synthesis (Figure 18.13; p. 713). All cells synthesize phospholipids to some degree (except mature erythrocytes), whereas triacylglycerol biosynthesis occurs only in liver, adipose tissue, and intestine. In most tissues, the pathway for phosphatidic acid synthesis begins with α-glycerol 3-phosphate (sn-glycerol 3-phosphate). The most general source of α-glycerol 3-phosphate, particularly in adipose tissue, is from reduction of the glycolytic intermediate, dihydroxyacetone phosphate by glycerol 3-phosphate dehydrogenase.

$$\text{Dihydroxyacetone phosphate} + \text{NADH} + \text{H}^+ \leftrightharpoons \text{glycerol 3-phosphate} + \text{NAD}^+$$

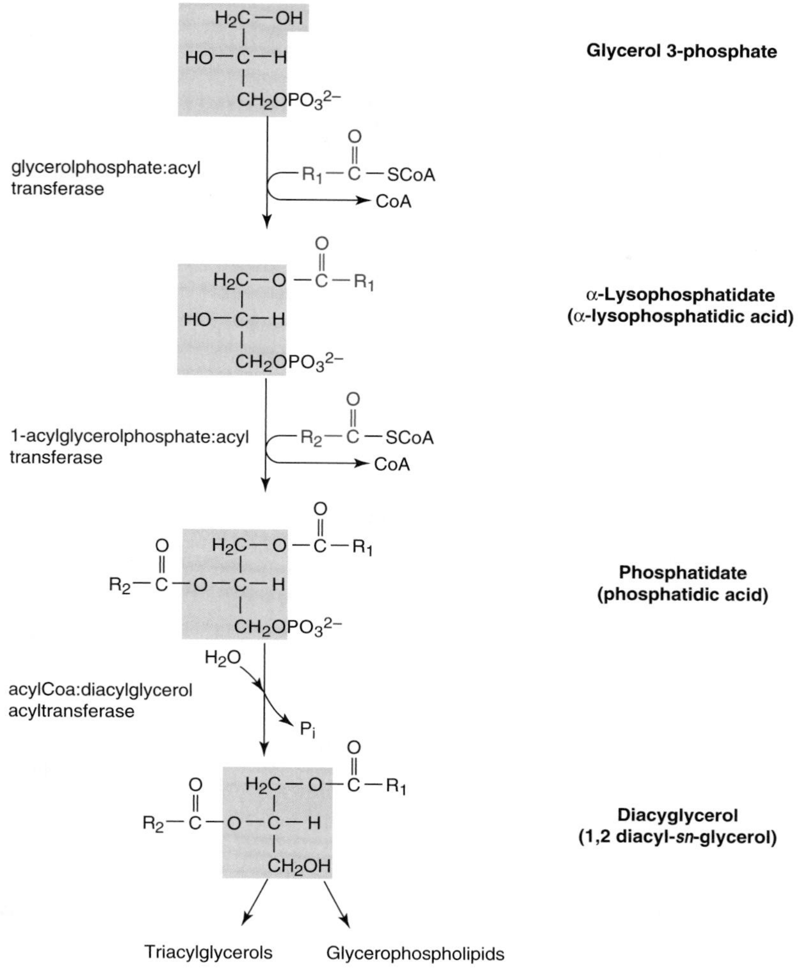

Glycerol 3-phosphate

glycerolphosphate:acyl transferase

α-Lysophosphatidate (α-lysophosphatidic acid)

1-acylglycerolphosphate:acyl transferase

Phosphatidate (phosphatidic acid)

acylCoa:diacylglycerol acyltransferase

Diacyglycerol (1,2 diacyl-*sn*-glycerol)

Triacylglycerols Glycerophospholipids

Figure 18.13 Phosphatidic acid biosynthesis from glycerol 3-phosphate and the role of phosphatidic acid phosphatase in synthesis of phospholipids and triacylglycerols. R_1 and R_2 represent long-chain fatty acids.

Liver and kidney derive glycerol 3-phosphate by means of the **glycerol kinase** reaction:

$$\text{Glycerol} + \text{ATP} \xrightarrow{Mg^{+2}} \text{glycerol 3-phosphate} + \text{ADP}$$

Phosphatidic acid synthesis starts with **glycerol 3-phosphate:acyltransferase** attaching predominantly saturated fatty acids or oleic acid to glycerol 3-phosphate to produce 1-acylglycerol phosphate or α-lysophosphatidic acid. 1-Acylglycerol phosphate:acyltransferase then acylates the sn-2 position, usually with an unsaturated fatty acid to produce phosphatidic acid (Figure 18.13). The donors of acyl groups are the CoA derivatives of the appropriate fatty acids.

The specificity of these acyltransferases does not always match the fatty acid asymmetry of the membrane phospholipids of a particular cell. Remodeling reactions, discussed in the following section, modify the composition at C1 and C2 of the glycerol phosphate backbone. Cytosolic **phosphatidic acid phosphatase** hydrolyzes phosphatidic acid generated on the endoplasmic reticulum, thereby yielding 1,2-diacyl-sn-glycerol that serves as the branch point in triacylglycerol and phospholipid synthesis (Figure 18.13).

Phospholipids Are Synthesized by Addition of a Base to Phosphatidic Acid

The major pathway for synthesis of phosphatidylcholine involves sequential conversion of choline to phosphocholine, **CDP-choline**, and phosphatidylcholine. Free choline, a dietary requirement for most mammals including humans, is phosphorylated by choline kinase (Figure 18.14). **Phosphocholine cytidylyltransferase** converts phosphocholine to CDP-choline. Inorganic pyrophosphate (PP_i) is a product of this reaction. The phosphocholine moiety is then transferred to C3 of 1,2-diacylglycerol by

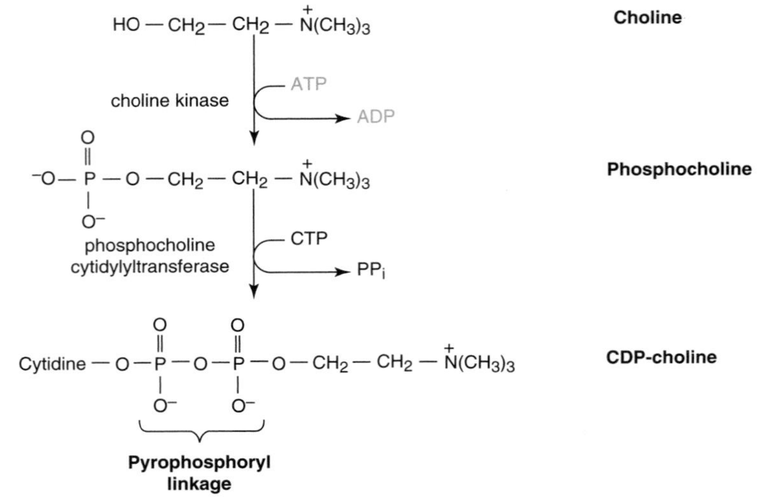

Figure 18.14 Biosynthesis of CDP-choline from choline.

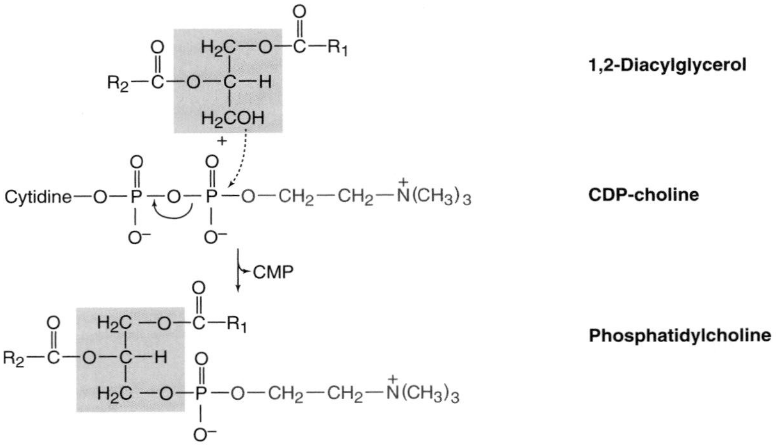

Figure 18.15 Choline phosphotransferase reaction.

choline phosphotransferase (Figure 18.15). This is the principal pathway for synthesis of dipalmitoyllecithin in lung.

The rate-limiting step for phosphatidylcholine synthesis is the cytidylyltransferase reaction (see Figure 18.14). This enzyme is regulated by translocation between cytosol and endoplasmic reticulum. The cytosolic form is inactive; binding of the enzyme to the ER membrane activates it. Translocation of cytidyl transferase from cytosol to endoplasmic reticulum is regulated by cAMP and fatty acyl CoA. Reversible phosphorylation of the enzyme by a cAMP-dependent protein kinase releases it from the membrane, rendering it inactive. Dephosphorylation causes it to bind to the ER membrane and become active. Fatty acyl CoAs promote its binding to the endoplasmic reticulum. In liver most of the phosphatidylcholine is formed by repeated methylation of phosphatidylethanolamine. Phosphatidylethanolamine *N*-methyltransferase of the ER transfers methyl groups in sequence from *S*-adenosylmethionine (AdoMet) (Figure 18.16).

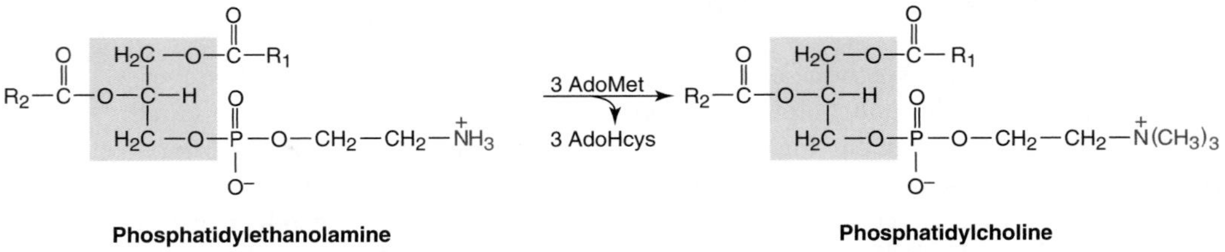

Figure 18.16 Biosynthesis of phosphatidylcholine from phosphatidylethanolamine and *S*-adenosylmethionine (AdoMet); *S*-adenosylhomocysteine (AdoCys).

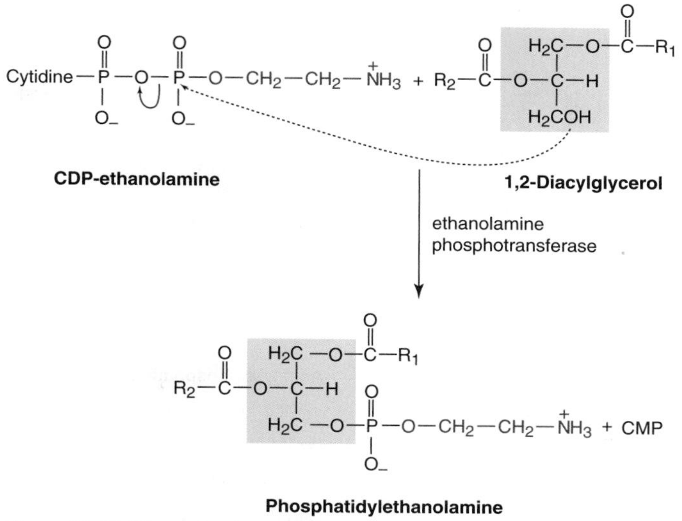

Figure 18.17 Biosynthesis of phosphatidylethanolamine from CDP-ethanolamine and diacylglycerol.

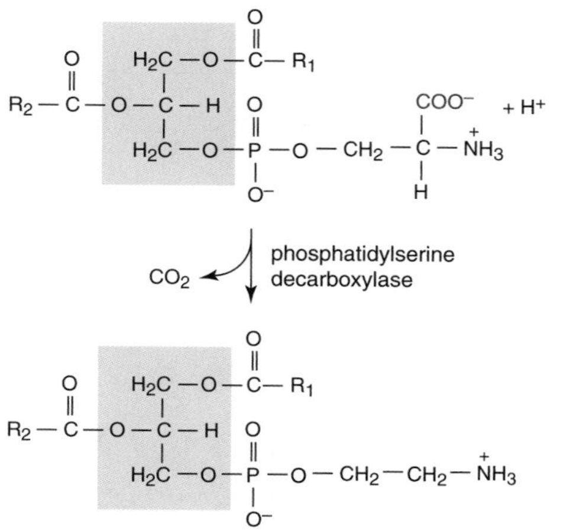

Figure 18.18 Formation of phosphatidylethanolamine by the decarboxylation of phosphatidylserine.

Phosphatidylethanolamine synthesis in liver and brain involves ethanolamine phosphotransferase of the endoplasmic reticulum (Figure 18.17). CDP-ethanolamine is formed by ethanolamine kinase

$$\text{Ethanolamine} + \text{ATP} \xrightleftharpoons{Mg^{2+}} \text{phosphoethanolamine} + \text{ADP}$$

and phosphoethanolamine cytidylyltransferase

$$\text{Phosphoethanolamine} + \text{CTP} \xrightleftharpoons{Mg^{2+}} \text{CDP-ethanolamine} + \text{PP}_i$$

Liver mitochondria also generate phosphatidylethanolamine by decarboxylation of phosphatidylserine, but this is considered a minor pathway (Figure 18.18).

The major source of phosphatidylserine in mammalian tissues is *base exchange* (Figure 18.19) in which the polar head group of phosphatidylethanolamine is exchanged for serine. Since there is no net change in the number or kind of bonds, this reaction is reversible and has no requirement for ATP or any other high-energy compound. Phosphatidylinositol is made via CDP-diacylglycerol and free myoinositol (Figure 18.20) by **phosphatidylinositol synthase** of the endoplasmic reticulum.

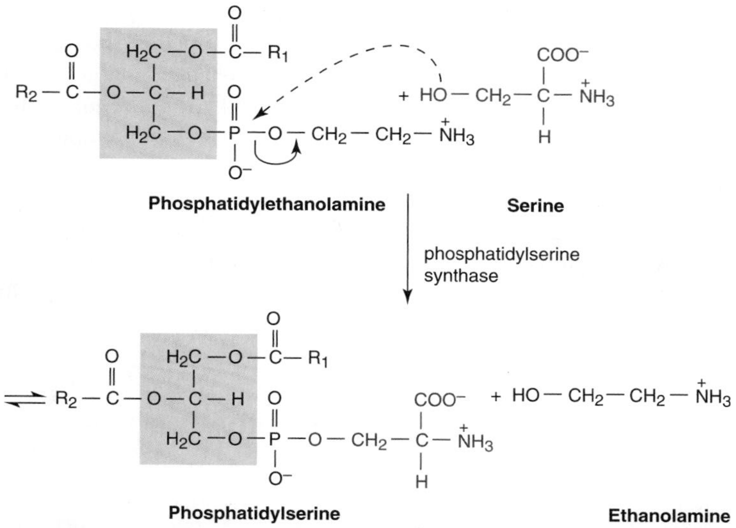

Figure 18.19 Biosynthesis of phosphatidylserine from serine and phosphatidylethanolamine by base exchange.

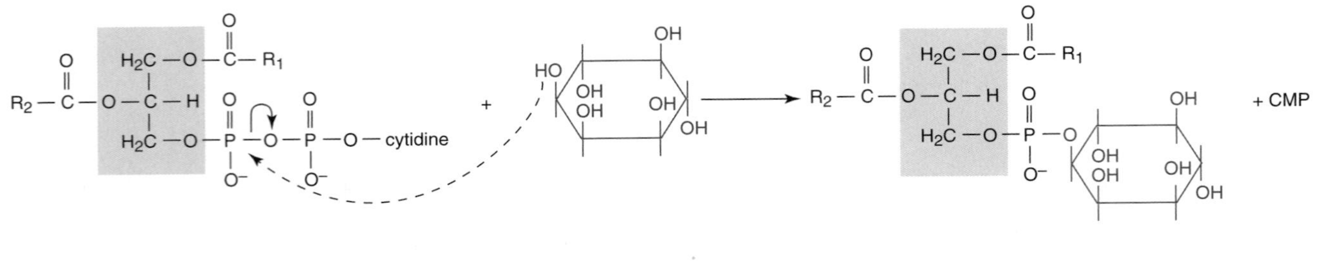

CDP-diacylglycerol + Inositol → Phosphatidylinositol + CMP

Figure 18.20 Biosynthesis of phosphatidylinositol.

2-Acyl lysophosphatide + $R_1—COO^-$ ← (phospholipase A_1, H_2O) Phospholipid → (phospholipase A_2, H_2O) 1-Acyl lysophosphatide + $R_2—COO^-$

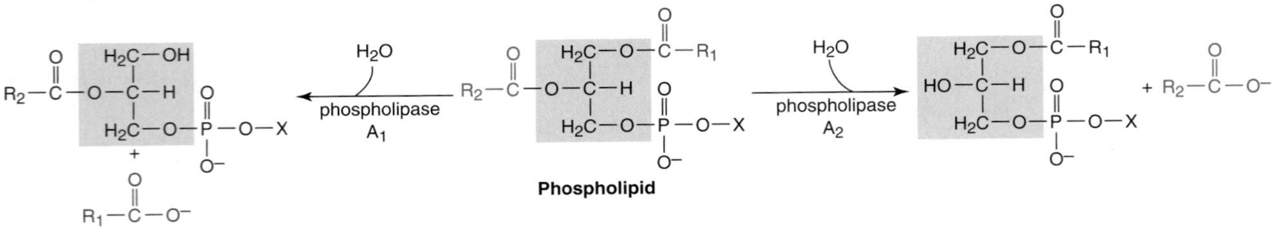

Figure 18.21 Reactions catalyzed by phospholipase A_1 and phospholipase A_2.

Asymmetric Distribution of Fatty Acids in Phospholipids Is due to Remodeling Reactions

Phospholipase A_1 and phospholipase A_2 occur in many tissues and function in the remodeling of specific phospholipid structures at the sn-1 and sn-2 positions. Most fatty acyl-CoA transferases and enzymes of phospholipid synthesis lack the specificity required to account for the distribution of fatty acids found in many tissue phospholipids. The fatty acids found in the sn-1 and sn-2 positions of the various phospholipids are often not the same ones transferred to the glycerol backbone in the initial acyl transferase reactions of phospholipid biosynthesis. Phospholipases A_1 and A_2 catalyze reactions indicated in Figure 18.21 where X represents the polar head group of a phospholipid. The phospholipid products are referred to as lysophosphatides or lysophospholipids.

If it becomes necessary for a cell to remove some undesired fatty acid, such as stearic acid from the sn-2 position of phosphatidylcholine, and replace it by a more unsaturated one like arachidonic acid, then this can be accomplished by the action of phospholipase A_2 followed by a reacylation reaction. Insertion of arachidonic acid into the 2 position of sn-2-lysophosphatidylcholine can be accomplished either by direct acylation from arachidonyl CoA by arachidonyl-CoA transacylase (Figure 18.22) or from some other arachidonyl-containing phospholipid by an exchange catalyzed by lysolecithin:lecithin acyltransferase (LLAT) (Figure 18.23). Since there is no change in either the number or nature of the bonds involved in products and reactants, ATP is not required. Reacylation of lysophosphatidylcholine from acyl CoA is the major route for remodeling of phosphatidylcholine.

Lysophospholipids, particularly sn-1-lysophosphatidylcholine, also serve as a source of fatty acid in remodeling reactions. Those involved in synthesis of dipalmitoyllecithin (surfactant) from 1-palmitoyl-2-oleoylphosphatidylcholine are presented in Figure 18.24. Note sn-1 = palmitoyl lysolecithin is the source of palmitic acid in the acyltransferase exchange reaction.

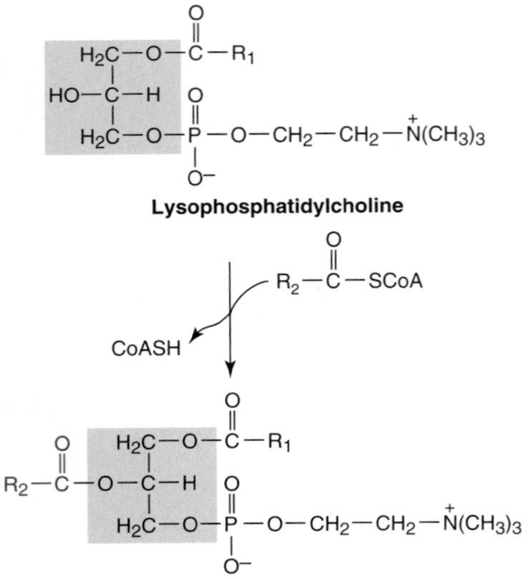

Lysophosphatidylcholine + $R_2—CO—SCoA$ → Phosphatidylcholine + CoASH

Figure 18.22 Synthesis of phosphatidylcholine by reacylation of lysophosphatidylcholine where $R_2—C(=O)—$ represents a rachidonic acid. This reaction is catalyzed by acyl-CoA:1-acylglycerol-3-phosphocholine *O*-acyltransferase.

Plasmalogens Are Synthesized from Fatty Alcohols

Ether glycerolipids are synthesized from DHAP, long-chain fatty acids and long-chain fatty alcohols as summarized in Figure 18.25. Acyldihydroxyacetone phosphate is formed by acyl-CoA: dihydroxyacetone phosphate acyltransferase (enzyme 1). The ether bond is

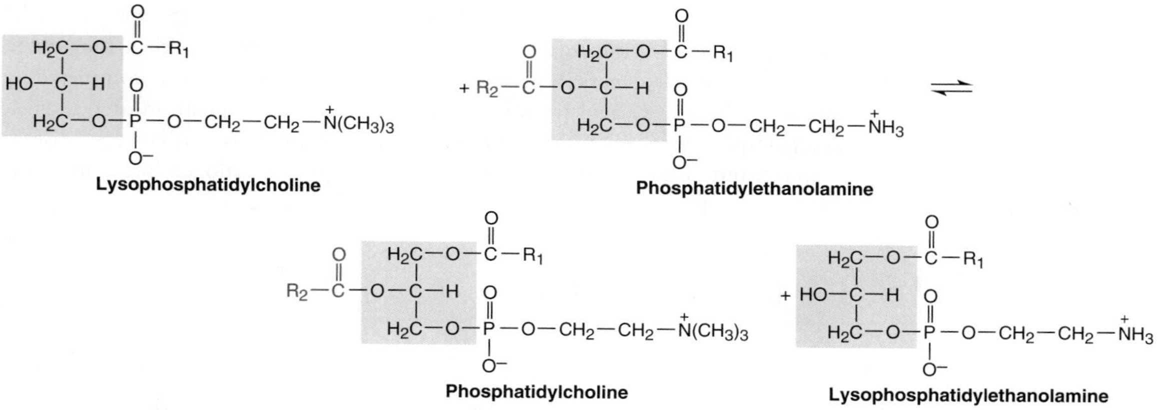

Lysophosphatidylcholine + Phosphatidylethanolamine ⇌ Phosphatidylcholine + Lysophosphatidylethanolamine

Figure 18.23 Formation of phosphatidylcholine by lysolecithin exchange, where $R_2-\overset{O}{\overset{\|}{C}}-$ represents arachidonic acid.

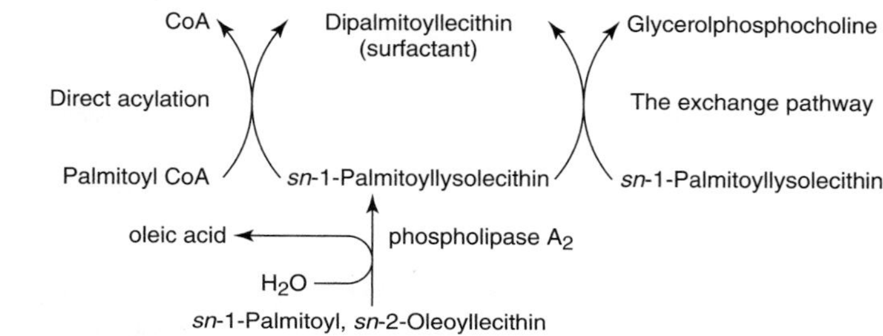

Figure 18.24 Two pathways for biosynthesis of dipalmitoyllecithin from *sn*-1 palmitoyllysolecithin.

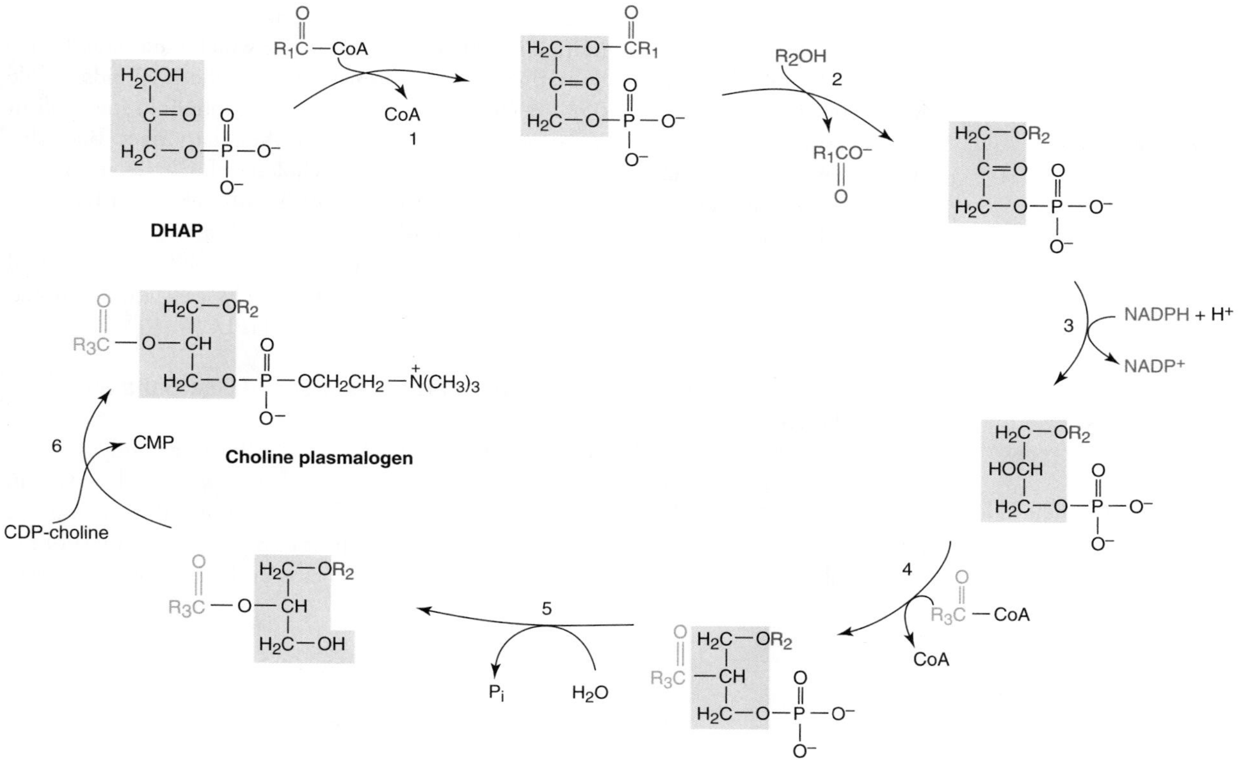

DHAP → (1: R_1COCoA → CoA) → (2: R_2OH → R_1COO^-) → (3: NADPH + H^+ → $NADP^+$) → (4: R_3CO–CoA → CoA) → (5: H_2O → P_i) → (6: CDP-choline → CMP) → Choline plasmalogen

Figure 18.25 Pathway of choline plasmalogen biosynthesis from DHAP. (1) Acyl CoA: dihydroxyacetone-phosphate acyltransferase. (2) Alkyldihydroxyacetone-phosphate synthase. (3) NADPH: alkyldihydroxyacetone-phosphate oxidoreductase. (4) Acyl CoA: 1-alkyl-2-lyso-sn-glycero-3-phosphate acyltransferase. (5) 1-alkyl-2-acyl-sn-glycerol-3-phosphate phosphohydrolase. (6) CDP-choline: 1-alkyl-2-acyl-sn-glycerol choline phosphotransferase. R_2OH is a long-chain fatty alcohol.

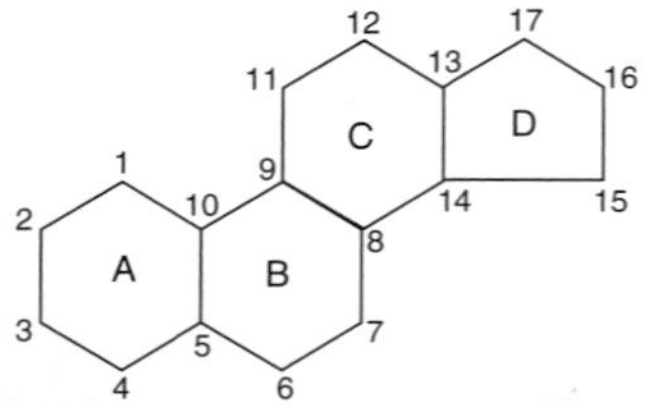

Figure 18.26 **The cyclopentenophenanthrene ring.**

introduced by alkyldihydroxyacetone phosphate synthase (Figure 18.25, enzyme 2) by exchange of the 1-*O*-acyl group of acyldihydroxyacetone phosphate with a long-chain fatty alcohol. The synthase occurs in peroxisomes. Plasmalogen synthesis is completed by transfer of a long-chain fatty acid from its CoA donor to the sn-2 position of 1-alkyl-2-lyso-sn-glycero-3-phosphate (Figure 18.25, Reaction 4). Patients with Zellweger disease lack peroxisomes and cannot synthesize adequate amounts of plasmalogen (see Clin. Corr. 1.7, p. 21).

18.3 • CHOLESTEROL

Cholesterol Is Widely Distributed in Free and Esterified Forms

Cholesterol is an alicyclic compound whose structure includes (1) the perhydrocyclopentenophenanthrene nucleus with its four fused rings, (2) a single hydroxyl group at C3, (3) an unsaturated center between C5 and C6, (4) an eight-membered branched hydrocarbon chain attached to the D ring at C17, and (5) a methyl group (designated C19) attached at position C-10, and another methyl group (designated C-18) attached at position C-13 (Figures 18.26 and 18.27).

Cholesterol has very low solubility in water; at 25°C, the limit of solubility is approximately 0.2 mg/100 dL, or 4.7 μM. The actual cholesterol concentration in plasma of healthy people is usually 150 to 200 mg/dL. This value is almost twice the normal concentration of blood glucose. Such a high concentration of cholesterol in blood is possible due to plasma lipoproteins (mainly LDL and VLDL) that contain large amounts of cholesterol (p. 111). Only about 30% of the total plasma cholesterol is free (unesterified); the rest is cholesteryl esters in which a long-chain fatty acid, usually linoleic acid, is esterified to the C3 hydroxyl of the A ring. This fatty acid residue increases the hydrophobicity of cholesterol (Figure 18.28).

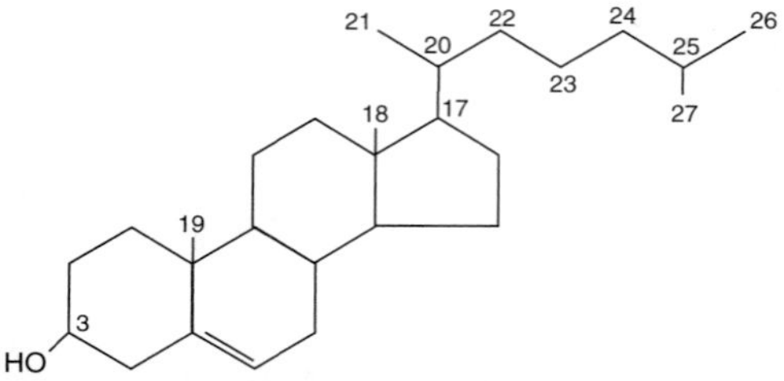

Figure 18.27 **Structure of cholesterol (5-cholesten-3 β-ol).**

Cholesterol, a ubiquitous and essential component of mammalian cell membranes, is abundant in bile (normal concentration is 390 mg/dL, only 4% of which is esterified). Free cholesterol is solubilized partially by the detergent property of the phospholipids in bile that are produced in liver (p. 1058). A chronic disturbance in phospholipid metabolism in liver can lead to deposition of cholesterol-rich gallstones in the biliary tract. Bile salts, which are metabolites of cholesterol, also aid in solubilizing cholesterol. Cholesterol in bile protects gallbladder membranes from potentially irritating or harmful effects of bile salts. Cholesterol is also a precursor of vitamin D (p. 1067).

Total cholesterol is estimated in the clinical laboratory by the Lieberman–Burchard reaction. The ratio of free and esterified cholesterol is determined by gas–liquid chromatography or reverse phase high-pressure liquid chromatography (HPLC).

Cholesterol Is a Membrane Component and Precursor of Bile Salts and Steroid Hormones

Cholesterol is derived from the diet or synthesized *de novo* in all cells of the body. It is the major sterol in humans and a component of all membranes. It is especially abundant in myelinated structures of brain and central nervous system but is present in small amounts in the outer membrane of mitochondria (p. 464). In contrast to plasma where most of the cholesterol is esterified, the cholesterol in cellular membranes is in the free form. The ring

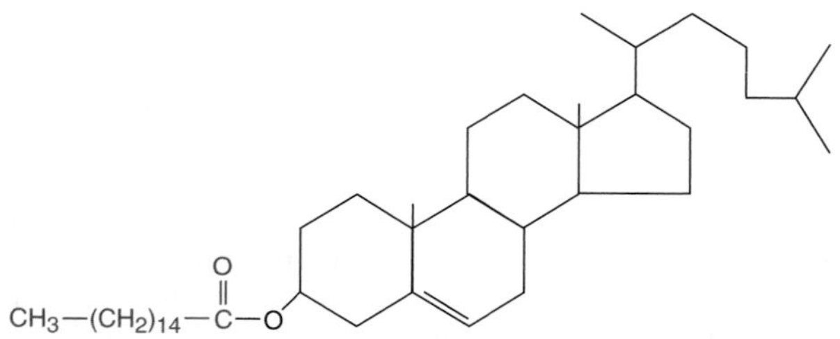

Figure 18.28 **Structure of cholesteryl (palmitoyl-) ester.**

structure of cholesterol cannot be metabolized to CO_2 and water in humans. Excretion of cholesterol is by way of the liver and gallbladder into the intestine in the form of bile acids. Cholesterol is the immediate precursor of bile acids synthesized in liver; they facilitate absorption of dietary triacylglycerols and fat-soluble vitamins (p. 1058).

Cholesterol is the precursor of various steroid hormones (p. 916) including progesterone, corticosteroids (corticosterone, cortisol, and cortisone), aldosterone, and the sex hormones, estrogen and testosterone. Although all steroid hormones are structurally related to and biochemically derived from cholesterol, they have widely different physiological properties. The hydrocarbon skeleton of cholesterol also occurs in plant sterols, for example in ergosterol (Figure 18.29), a precursor of vitamin D, which is converted in skin by ultraviolet irradiation to vitamin D_2 (p. 1067).

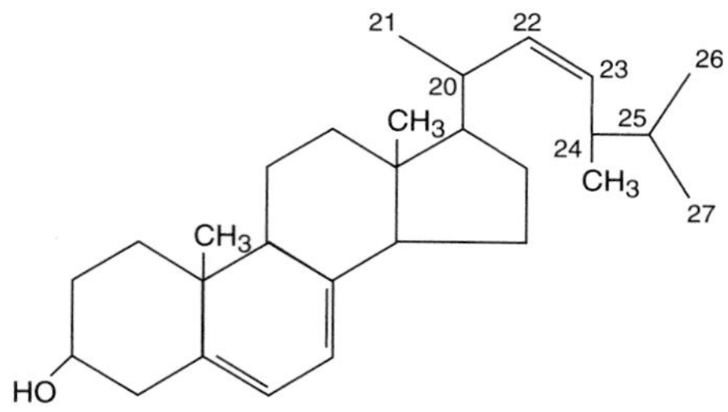

Figure 18.29 Structure of ergosterol.

Cholesterol Is Synthesized from Acetyl CoA

Although synthesis of cholesterol occurs in virtually all cells, the capacity is greatest in liver, intestine, adrenal cortex, and reproductive tissues, including ovaries, testes, and placenta. All carbon atoms of cholesterol are derived from acetate. Reducing power in the form of NADPH is provided mainly by glucose 6-phosphate dehydrogenase and 6-phosphogluconate dehydrogenase of the hexose monophosphate shunt pathway (p. 648). Cholesterol synthesis occurs in the cytosol and endoplasmic reticulum and is driven largely by hydrolysis of high-energy thioester bonds of acetyl CoA and phosphoanhydride bonds of ATP.

Mevalonic Acid Is a Key Intermediate

The first compound unique to cholesterol synthesis is mevalonic acid derived from acetyl CoA. There are several sources of acetyl CoA: (1) β-oxidation of fatty acids (p. 692), (2) oxidation of ketogenic amino acids such as leucine and isoleucine (p. 779), and (3) the pyruvate dehydrogenase reaction. Acetate is converted to acetyl CoA at the expense of ATP by acetokinase or acetate thiokinase.

$$\text{ATP} + \text{CH}_3\text{COO}^- + \text{CoASH} \longrightarrow \text{CH}_3\text{—}\overset{\text{O}}{\overset{\|}{\text{C}}}\text{—SCoA} + \text{AMP} + \text{PP}_i$$

As in the pathway that produces ketone bodies (p. 698), two molecules of acetyl CoA are condensed to acetoacetyl CoA by **acetoacetyl-CoA thiolase** (acetyl-CoA:acetyl-CoA acetyltransferase):

$$\text{CH}_3\text{—}\overset{\text{O}}{\overset{\|}{\text{C}}}\text{—SCoA} + \text{CH}_3\text{—}\overset{\text{O}}{\overset{\|}{\text{C}}}\text{—SCoA} \longrightarrow \text{CH}_3\text{—}\overset{\text{O}}{\overset{\|}{\text{C}}}\text{—CH}_2\text{—}\overset{\text{O}}{\overset{\|}{\text{C}}}\text{—SCoA} + \text{CoA—SH}$$

Formation of the carbon–carbon bond in acetoacetyl CoA is favored energetically by cleavage of a thioester bond and generation of coenzyme A.

A third molecule of acetyl CoA is used to form the branched-chain **3-hydroxy-3-methylglutaryl-CoA (HMG CoA)** (Figure 18.30) by **HMG-CoA synthase (3-hydroxy-3-methylutarylyl CoA: acetoacetyl-CoA lyase)**. Liver parenchymal cells contain a cytosolic form of HMG-CoA synthase, which is involved in cholesterol synthesis, and a mitochondrial form that functions in the synthesis of ketone bodies (p. 698). The enzyme catalyzes an aldol condensation between the methyl carbon of acetyl CoA and the β-carbonyl group of acetoacetyl CoA with hydrolysis of the thioester bond of acetyl CoA. The original

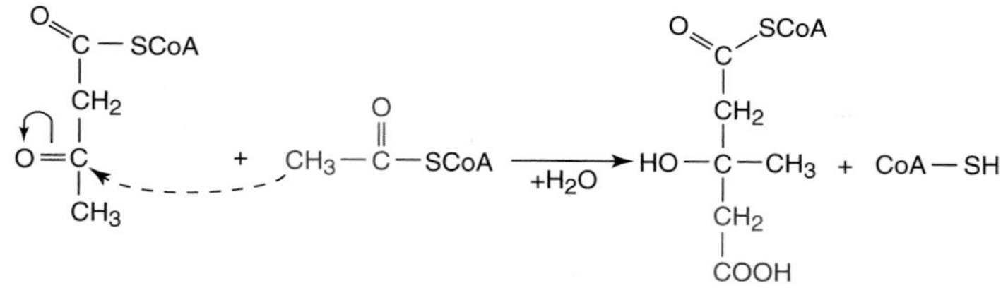

Acetoacetyl CoA + Acetyl CoA $\xrightarrow{+H_2O}$ HMG CoA + CoA—SH

Figure 18.30 HMG-CoA synthase reaction.

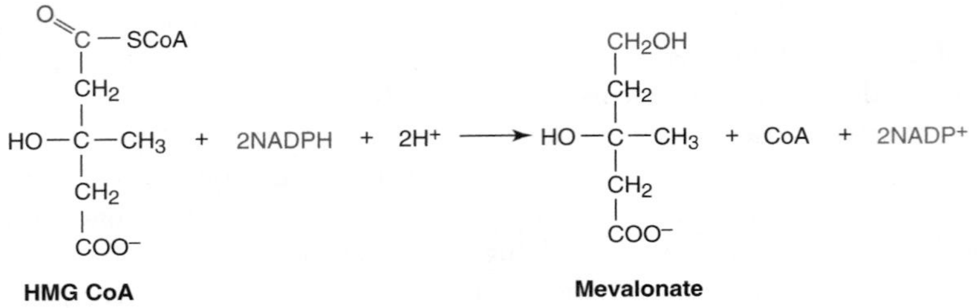

$$\text{HMG CoA} + 2\text{NADPH} + 2H^+ \longrightarrow \text{Mevalonate} + \text{CoA} + 2\text{NADP}^+$$

Figure 18.31 HMG-CoA reductase reaction.

thioester bond of acetoacetyl CoA remains intact. HMG CoA is also formed from oxidative degradation of the branched-chain amino acid leucine, through the intermediates 3-methylcrotonyl CoA and 3-methylglutaconyl CoA (p. 777).

Mevalonic acid is formed from HMG CoA by the endoplasmic reticulum enzyme **HMG-CoA reductase (mevalonate:NADP$^+$ oxidoreductase)** which has an absolute requirement for NADPH (Figure 18.31). The reduction consumes two molecules of NADPH, results in hydrolysis of the thioester bond of HMG CoA, and generates the primary alcohol group of mevalonate. This reduction is irreversible and produces (R)-(+) mevalonate. HMG-CoA reductase catalyzes the rate-limiting reaction in cholesterol biosynthesis. It is an intrinsic protein of the endoplasmic reticulum with its catalytic C-terminal domain extending into the cytosol. Phosphorylation of HMG-CoA reductase diminishes its catalytic activity (V_{max}) and enhances its degradation by increasing its susceptibility to proteolysis. Increased intracellular cholesterol stimulates phosphorylation of HMG-CoA reductase.

The central role of HMG-CoA reductase in cholesterol homeostasis is evidenced by the effectiveness of a family of drugs called statins used to lower plasma cholesterol levels. Statins (e.g., lovastatin, pravastatin, fluvastatin, cerivastatin, atorvastatin, simvastatin) inhibit HMG-CoA reductase activity, particularly in liver, and commonly decrease total plasma cholesterol and LDL-cholesterol by as much as 50%.

Mevalonic Acid Is Precursor of Farnesyl Pyrophosphate

Reactions that convert mevalonate to **farnesyl pyrophosphate** are summarized in Figure 18.32. The stepwise transfer of the terminal phosphate group from two molecules of ATP to mevalonate (A) to form 5-pyrophosphomevalonate (B) is catalyzed by mevalonate kinase (enzyme I) and phosphomevalonate kinase (enzyme II). Decarboxylation of 5-pyrophosphomevalonate by pyrophosphomevalonate decarboxylase generates Δ^3-isopentenyl pyrophosphate (D). In this ATP-dependent reaction, ADP,

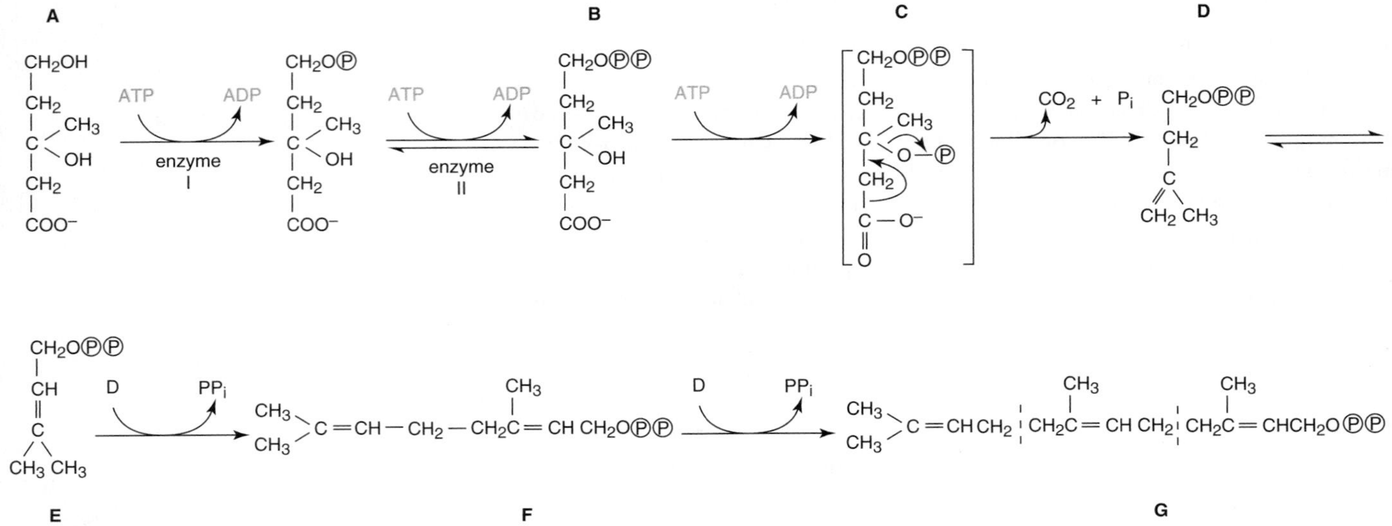

Figure 18.32 Formation of farnesyl-PP (F) from mevalonate (A). Dotted lines divide molecules into isoprenoid-derived units. D is 3-isopentenyl pyrophosphate.

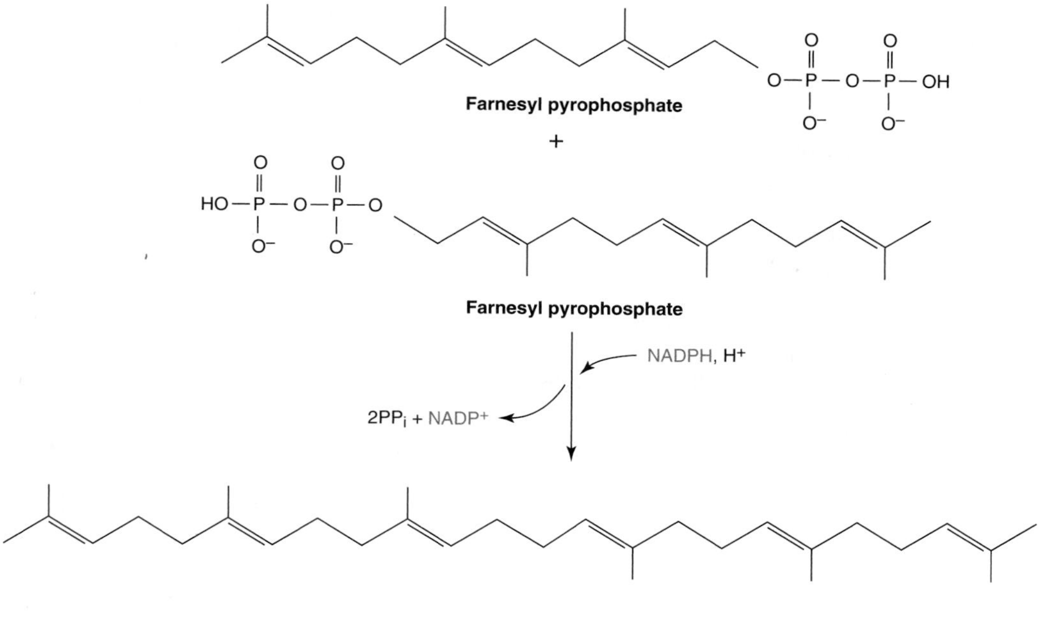

Farnesyl pyrophosphate

+

Farnesyl pyrophosphate

NADPH, H^+

$2PP_i$ + $NADP^+$

Squalene

Figure 18.33 **Formation of squalene from two molecules of farnesyl pyrophosphate.**

P_i, and CO_2 are produced. It is thought that decarboxylation–dehydration proceeds by way of the intermediate 3-phosphomevalonate 5-pyrophosphate (C). Isopentenyl pyrophosphate is converted to its allylic isomer 3,3-dimethylallyl pyrophosphate (E) by isopentenyl pyrophosphate isomerase in a reversible reaction. The condensation of 3,3-dimethylallyl pyrophosphate (E) and 3-isopentenyl pyrophosphate (D) generates geranyl pyrophosphate (F).

The stepwise condensation of three C_5 isopentenyl units to form the C_{15} farnesyl pyrophosphate (G) is catalyzed by a cytosolic prenyl transferase called **geranyltransferase.**

Cholesterol Is Formed from Farnesyl Pyrophosphate via Squalene

The last steps in cholesterol synthesis involve "head-to-head" fusion of two molecules of farnesyl pyrophosphate to form squalene and finally cyclization of squalene to yield cholesterol. Formation of the C_{30} carbon molecule of squalene (Figure 18.33) by squalene synthase of endoplasmic reticulum releases two pyrophosphate groups and requires NADPH. Several intermediates probably occur. By rotation about carbon–carbon bonds, the conformation of squalene indicated in Figure 18.34 can be obtained. Note that the overall shape of squalene resembles that of cholesterol and that squalene is devoid of oxygen atoms.

Cholesterol synthesis from squalene by squalene cyclase proceeds through the intermediate lanosterol, which contains the fused tetracyclic ring system and an eight-carbon side chain:

$$\text{Squalene} \longrightarrow \text{squalene 2,3-epoxide} \longrightarrow \text{lanosterol}$$

The ER enzyme that catalyzes this reaction is bifunctional and contains squalene epoxidase or monooxygenase and a cyclase (**lanosterol cyclase**) activity. The many carbon–carbon bonds formed during cyclization of squalene are generated in a concerted fashion as indicated in Figure 18.35. The OH group of lanosterol projects above the plane of the A ring; that is, it is in the β-orientation. In this reaction sequence, an OH group is added to C3, two methyl groups undergo shifts and a proton is eliminated.

Cyclization is initiated by epoxide formation between the future C2 and C3 of cholesterol, the epoxide being formed at the expense of NADPH.

$$\text{Squalene} + O_2 + \text{NADPH} + H^+ \longrightarrow \text{squalene 2,3-epoxide} + H_2O + \text{NADP}^+$$

The hydroxylation at C3 triggers the cyclization of squalene to lanosterol (Figure 18.35).

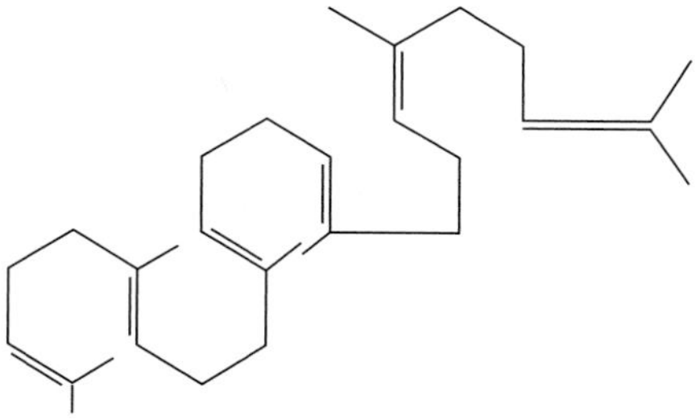

Figure 18.34 **Structure of squalene (C_{30}).**

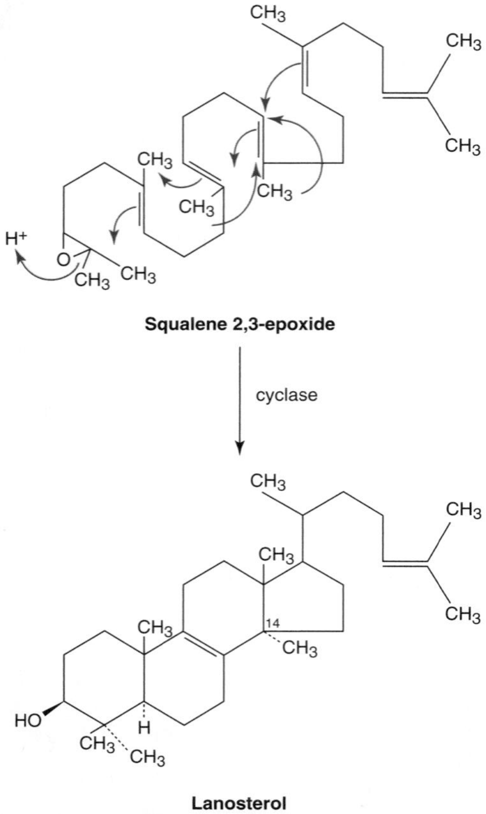

Figure 18.35 **Conversion of squalene 2,3-epoxide to lanosterol.**

Transformation of lanosterol to cholesterol involves many poorly understood steps and several enzymes. These reactions include (1) removal of the methyl group at C14, (2) removal of two methyl groups at C4. (3) migration of the double bond from C8 to C5, and (4) reduction of the double bond between C24 and C25 in the side chain (Figure 18.36).

Plasma Lipoproteins

Blood contains triacylglycerols, cholesterol, and cholesteryl esters at concentrations that far exceed their solubilities in water. Blood lipids are kept in solution, or at least thoroughly dispersed, by incorporation into macromolecular structures called lipoproteins. The plasma lipoproteins facilitate lipid metabolism and transfer of lipids between tissues. There are five major lipoprotein classes: **high-density lipoprotein** (**HDL**), **low-density lipoprotein** (**LDL**), **intermediate-density lipoproteins** (**IDL**s), **very low-density lipoproteins** (**VLDL**), and **chylomicrons**. Their physical characteristics are presented in Table 3.13, p. 108, and the lipid composition in Table 3.15, p. 111. All contain phospholipids and one or more proteins called apoproteins; there are 10 common apoproteins (Table 18.1).

Lipoproteins are spherical particles with the most hydrophobic lipids such as cholesteryl esters and triacylglycerols located in the core, sequestered away from water, whereas free (nonesterified) cholesterol, phospholipids, and the protein(s) are arrayed on the surface (see Figure 3.47, p. 110). The amphipatic phospholipids and proteins keep the otherwise highly insoluble lipids in solution. The apoproteins on the surface of the particles also serve as ligands for cell receptors, and cofactors for enzymes involved in lipoprotein metabolism.

Lipoproteins are vehicles by which cholesterol and its esters and triacylglycerols are transported in the body. For example, one function of LDL is to deliver cholesterol to peripheral tissues that require cholesterol for membrane formation or steroid hormone synthesis. The cholesterol carried to the liver from peripheral tissues and the cholesterol in chylomicrons regulates hepatic cholesterol synthesis (Figure 18.37, p. 724). In contrast, HDL, which is rich in cholesterol but poor in triacylglycerol, carries cholesterol from the periphery to the liver where it is excreted in bile as cholesterol or after conversion to bile salts. VLDL and chylomicrons transport triacylglycerols to be used for energy (e.g., muscle) or stored (e.g., fat cells).

Plasma lipoproteins are substrates for several enzymes in blood. Lipoprotein lipase is attached to the luminal surface of the vascular endothelium by binding to heparan sulfate proteoglycans and hydrolyses triacylglycerols of VLDL and chylomicrons. It is activated by apolipoprotein C-II (apoC-II) and by heparin released from mast cells and cells of the macrophage/reticuloendothelial system. The fatty acids released by lipoprotein lipase can then be taken up by cells and oxidized by β-oxidation, incorporated into phospholipids for membrane assembly or, in adipocytes, stored as triacylglycerols. These fatty acids are incorporated into milk fat in

TABLE 18.1 • Apoproteins of Human Plasma Lipoproteins

Apolipoprotein	*Molecular Weight*	*Plasma Concentration* ($g\ L^{-1}$)	*Lipoprotein Distribution*
ApoA-I	28,000	1.0–1.2	Chylomicrons, HDL
ApoA-II	17,000	0.3–0.5	Chylomicrons, HDL
ApoA-IV	46,000	0.15–0.16	Chylomicrons, HDL
ApoB-48	264,000	0.03–0.05	Chylomicrons
ApoB-100	512,000	0.7–1.0	VLDL, IDL, LDL
ApoC-I	7000	0.04–0.06	Chylomicrons, VLDL, HDL
ApoC-II	9000	0.03–0.05	Chylomicrons, VLDL, HDL
ApoC-III	9000	0.12–0.14	Chylomicrons, VLDL, HDL
ApoD	33,000	0.06–0.07	HDL
ApoE	38,000	0.03–0.05	Chylomicrons, VLDL, IDL, HDL

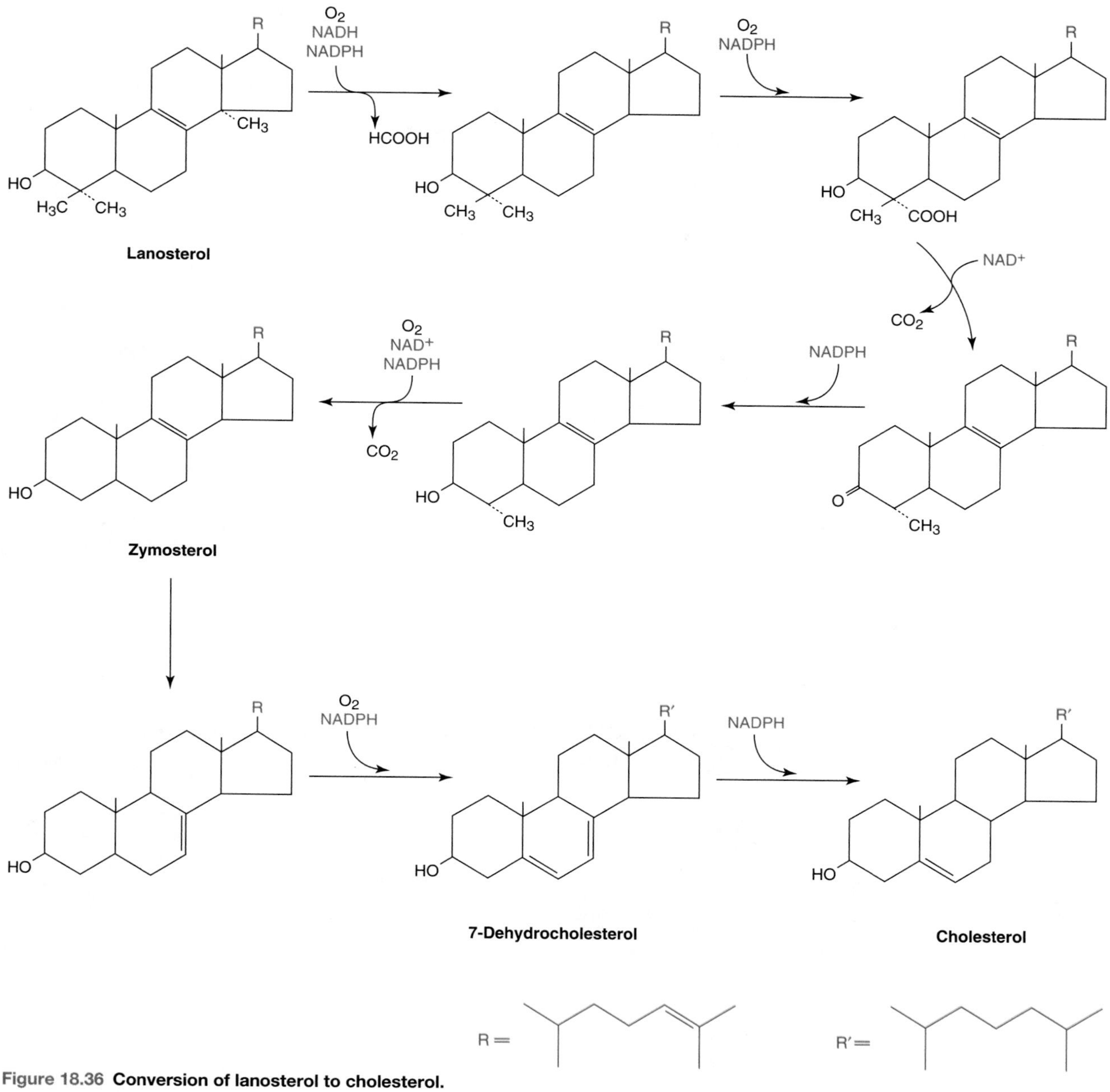

Figure 18.36 Conversion of lanosterol to cholesterol.

the mammary gland. Hydrolysis of triacylglycerols also results in the release of phospholipids, free cholesterol, and exchangeable apolipoproteins on the surface shell and their transfer to other circulating lipoproteins, especially HDL. As chylomicrons lose their triacylglycerol core they become the smaller chylomicron remnants that are rich in cholesteryl esters, bind specific receptors on hepatic membranes, and, are catabolized in the liver following endocytosis (Figure 18.37). Clearly, lipoproteins are dynamic in structure and composition.

Main sites of synthesis of the apoprotein components are liver and small intestine. For example, apoB-48 is produced in the intestine whereas apoB-100 is synthesized mainly in hepatocytes. ApoB-48 is 48% as long as apoB-100 and is formed from the same message as that for apoB-100. A stop codon introduced into the apoB mRNA in the intestine terminates translation 48% of the way along the mRNA molecule. ApoB-48 is synthesized during the course of fat digestion and is used in production of chylomicrons. ApoB-100 is used in the production of VLDL. The fatty acids of triacylglycerols of newly produced VLDL particles are synthesized from sugars, glucose in particular, after they have been oxidized to acetyl CoA, or from the acetate derived from oxidation of ethanol. VLDL acquires freshly synthesized

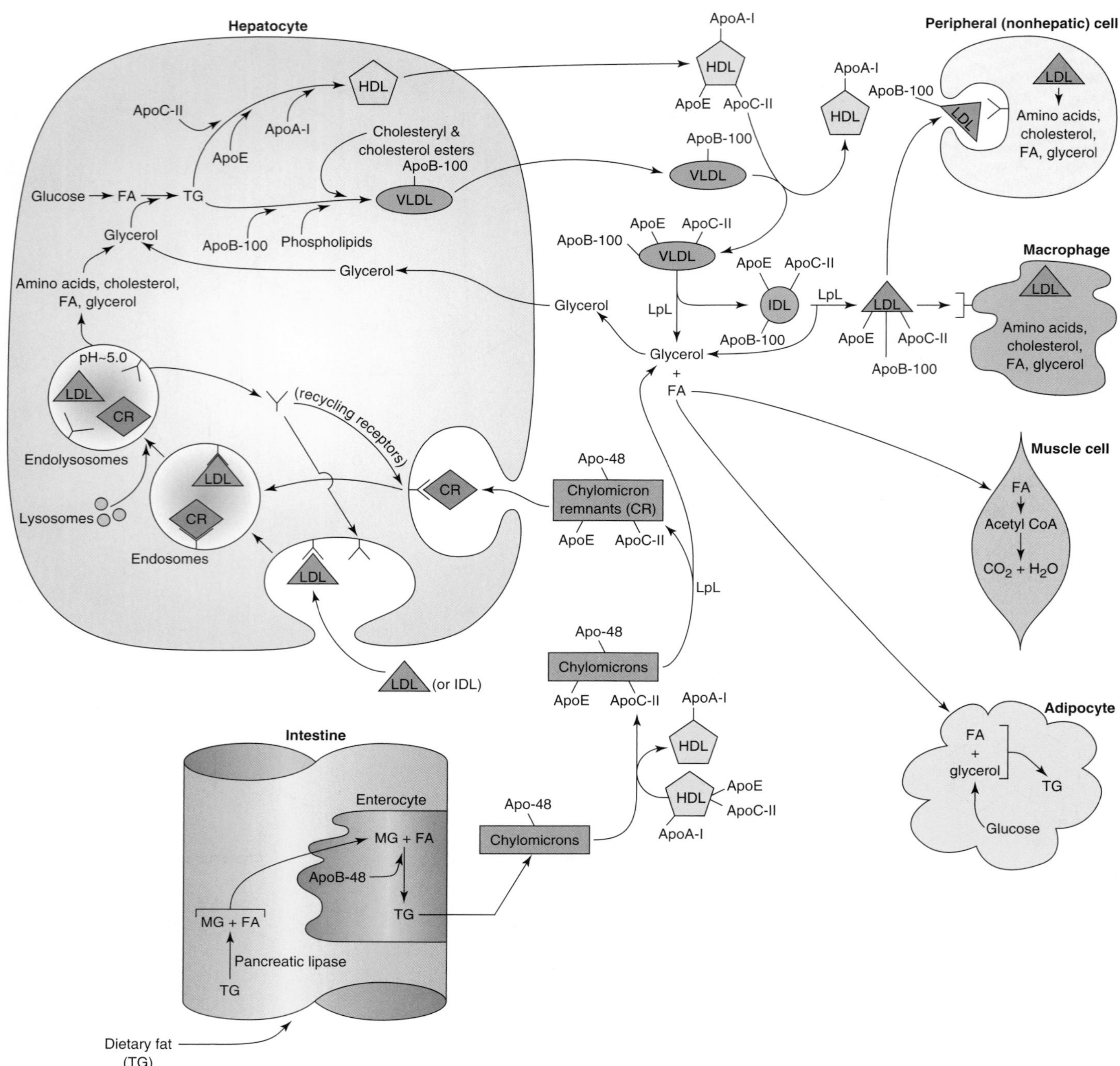

Figure 18.37 Organs and pathways involved in plasma lipoprotein metabolism. FA, fatty acid; TG, triacylglycerol (triglyceride); HDL, high density lipoprotein; LDL, low density lipoprotein; IDL, intermediate density lipoprotein; VLDL, very low-density lipoprotein; LpL, lipoprotein lipase; apo-, apoprotein.

cholesteryl esters in the liver, as well as from other lipoproteins, during its transport in the circulation. In the circulation, the net transfer of cholesteryl ester and triacylglycerols between HDL and VLDL and LDL is facilitated by **cholesteryl ester transfer protein** (**CETP**). CETP is associated with HDL in the plasma. CETP is synthesized and secreted by hepatocytes and adipocytes, and its expression is stimulated by diet-induced hypercholesterolemia. Lipoprotein lipase (LpL) converts VLDL into intermediate density lipoprotein (IDL), and then, as additional triacylglycerol is subjected to lipolysis, into LDL (Figure 18.37).

HDL is synthesized mainly in liver and to a lesser extent in the intestine and has the unique function of being a reservoir for apoE and apoC-II, which are activators of lipoprotein

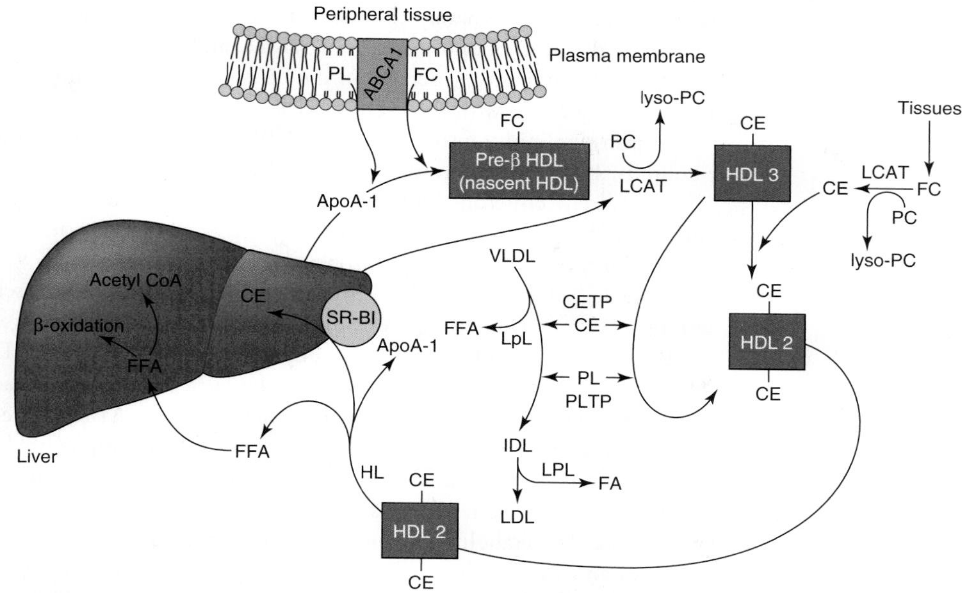

Figure 18.38 Reverse cholesterol transport showing the proteins and enzymes involved. PL, phospholipids; FC, free cholesterol; ABCA1, ATP-binding cassette 1; VLDL, very low-density lipoprotein; HDL, high-density lipoprotein; LDL-low-density lipoprotein; IDL, intermediate density lipoprotein; PC, phosphatidylcholine; lyso-PC, lysophosphatidylcholine; LCAT, lecithin cholesterol acyltransferase; CE, cholesteryl ester; CETP, cholesterol ester transfer protein; PLTP, phospholipid transfer protein; LpL, lipoprotein lipase; apoA-1, apolipoproteinA-1; FFA, free fatty acid; SR-BI, scavenger receptor class B type I; TG, triacylglycerol; HL, hepatic lipase. FA, fatty acid; TG, triacylglycerol (triglyceride); HDL, high-density lipoprotein; LDL, low-density lipoprotein; IDL, intermediate-density lipoprotein; VLDL, very low-density lipoprotein; LpL, lipoprotein lipase; apo-, apoprotein; Y, LDL receptor. Chylomicrons contain B-48, VLDL and LDL contains B-100.

lipase. It is secreted as apoprotein A-1 without phospholipids or triacylglycerol. HDL regulates exchange of apoproteins and lipids between various lipoproteins in the blood. HDL particles donate apoE and apoC-II to chylomicrons and VLDL. Once the triacylglycerols in chylomicrons and VLDL are extensively hydrolysed and these lipoproteins are transformed into LDL and chylomicron remnants, respectively, the apoE and apoC-II are returned to HDL (Figure 18.37).

HDL also participates in removal of excess cholesterol from cells and its transport to the liver for elimination as cholesterol and bile salts. This phenomenon is termed *reverse cholesterol transport* (Figure 18.38). It is free (nonesterified) cholesterol that exchanges readily between lipoproteins and the plasma membrane of cells. The transfer of free cholesterol from the plasma membrane to apoA-I or a lipid-poor species of HDL to produce what is called pre-β HDL or nascent HDL is mediated by a membrane transporter designated *ATP-binding cassette transporter* (ABCA1). ABCA1 also transfers phospholipids, along with free cholesterol, from the membrane to nascent HDL to produce HDL3. Absence of the ABCA1 transporter results in the HDL-deficiency disease called Tangier disease. Further addition of esterified cholesterol to HDL3 produces HDL2.

Cholesterol is esterified by **lecithin:cholesterol acyl transferase** (**LCAT**) associated with HDL. This freely reversible reaction (Figure 18.39) transfers the fatty acid in the

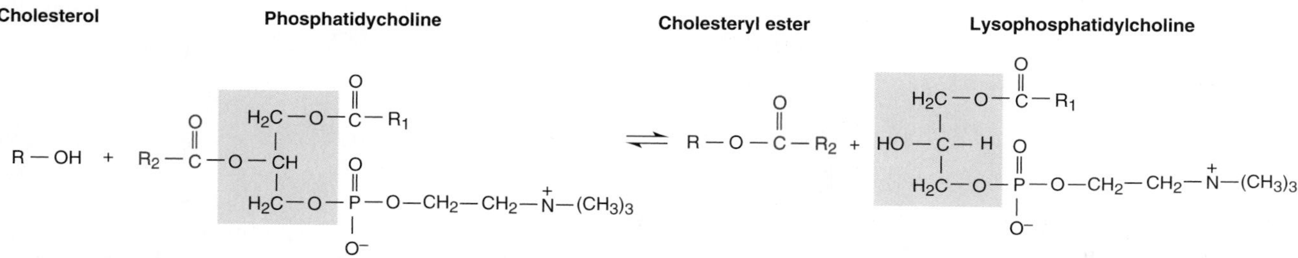

Figure 18.39 Lecithin:cholesterol acyltransferase (LCAT) reaction. R-OH = cholesterol.

sn-2 position of phosphatidylcholine to the 3-hydroxyl of cholesterol. LCAT is produced mainly by the liver, bound to HDL in plasma, and activated by the apoA-1 component of HDL. Cholesteryl ester generated in the LCAT reaction is transferred to VLDL and LDL by CETP associated with the HDL particle, and eventually taken up by the liver.

A phospholipid transfer protein (PLTP) catalyzes transfer of lipids, particularly phospholipids, between lipoproteins. As LpL-catalyzed triacylglycerol hydrolysis occurs and VLDL and chylomicrons both get smaller; PLTP removes excess phospholipid from the surface of these particles and transfers it to HDL. This provides substrate for the LCAT reaction in reverse cholesterol transport.

The degradation of HDL takes place in the liver following the selective uptake of cholesteryl esters mediated by a plasma membrane protein called *scavenger receptor-BI* (SR-BI). SR-BI is a multiligand receptor that binds not only HDL but VLDL and LDL as well. The uptake and degradation of HDL by the liver involve cell surface hepatic lipase that hydrolyzes the triacylglycerols of HDL particles. The apoA1 from the degradation of HDL is recycled for new HDL formation. An inverse relationship exists between plasma HDL concentration and the incidence of coronary artery disease, and a positive relationship exists between plasma cholesterol levels and coronary heart disease.

Liver cells metabolize chylomicron remnants by a similar mechanism; however, macrophages and many other cells have specific receptors that recognize chylomicron remnants and internalize them. These receptors recognize the apoE of chylomicron remnants. Some LDL is taken up via nonspecific scavenger receptors on certain cells, macrophages in particular.

Cholesterol Synthesis Is Regulated

Elevated plasma cholesterol predisposes to atherosclerotic vascular disease. In healthy individuals, plasma cholesterol levels are maintained within a relatively narrow concentration range largely by the liver which (1) expresses the majority of the body's LDL receptors, (2) is the major site for conversion of cholesterol to bile acids, and (3) has the highest level of HMG-CoA reductase activity. The cholesterol pool of the body is derived from dietary cholesterol and cholesterol synthesis primarily in liver and intestine. Cholesterol synthesis increases in liver and intestine when dietary cholesterol intake is reduced. This cholesterol is then transported from liver and intestine to peripheral tissues by VLDLs and chylomicrons, respectively.

The committed step and the rate-limiting reaction in cholesterol synthesis (Figure 18.40) is that of the HMG-CoA reductase which catalyses the step that produces mevalonic acid. Cholesterol exerts feedback inhibition on HMG-CoA reductase and promotes degradation of the enzyme by mechanisms that remain to be elucidated.

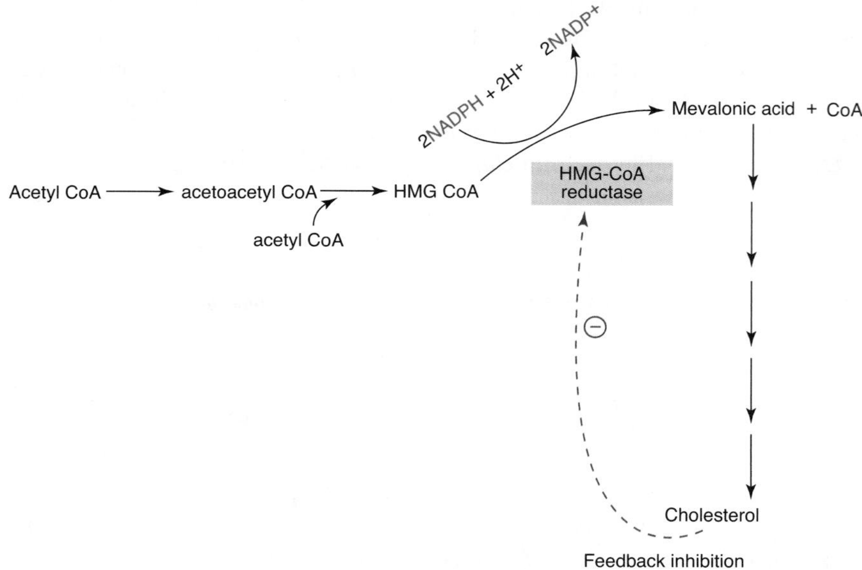

Figure 18.40 Summary of cholesterol synthesis indicating feedback inhibition of HMG-CoA reductase by cholesterol.

In a normal healthy adult on a low cholesterol diet, about 1300 mg of cholesterol is presented to the liver each day for disposal. This cholesterol comes from the diet and peripheral tissues and is disposed of by (1) excretion in bile of about 250 mg bile salts and about 550 mg of cholesterol, (2) storage as cholesteryl esters, and (3) incorporation into VLDL and secretion into the circulation. In the same adult, liver synthesizes about 800 mg of cholesterol per day to replace bile salts and cholesterol.

LDL is responsible for cholesterol delivery to and transport into peripheral tissues that require cholesterol for synthesis and maintenance of their membranes or by specialized cells for steroid hormone synthesis. The removal of LDL from circulation involves **LDL receptors** on the surface of hepatocytes and cells in the periphery.

About 75% of LDL catabolism occurs in liver via an LDL receptor-mediated process. The binding of LDL to hepatocytes and cells in the periphery is characterized by saturability, high affinity, and high degree of specificity. The receptor recognizes apolipoprotein E (apoE) and apolipoprotein B-100 (apoB-100) on LDL and VLDL. Binding occurs on the plasma membrane at pits coated with clathrin and leads to ligand-receptor complexes internalized as clathrin-coated vesicles. This is called receptor-mediated endocytosis. Intracellularly, the vesicles lose their clathrin and become endosomes that fuse with lysosomes that contain proteases and cholesteryl esterase. In this relatively acid (approximately pH 5.0) environment the receptor separates from LDL and returns to the cell surface while the cholesteryl esters are hydrolyzed by cholesteryl esterase to cholesterol and a long-chain fatty acid and the protein is degraded to amino acids that enter the cell's amino acid pool. The free cholesterol diffuses into the cytosol where it inhibits HMG-CoA reductase and suppresses synthesis of the enzyme. At the same time, **fatty acyl CoA:cholesterol acyltransferase** (**ACAT**) of the endoplasmic reticulum is activated by cholesterol, promoting formation of cholesteryl esters, principally cholesteryl oleate.

$$\text{Cholesterol} + \text{oleoyl CoA} \rightarrow \text{cholesteryl oleate} + \text{CoA}$$

Accumulation of intracellular cholesterol inhibits the synthesis and replenishment of LDL receptors by down-regulation of their expression, thereby blocking further uptake and accumulation of cholesterol and increasing the plasma cholesterol concentration. The mechanism of cholesterol regulation of HMG-CoA reductase involves a transcription factor, called the sterol regulatory element binding protein (SREBP), and two other proteins, designated COPII and InSig. Much of the LDL-cholesterol internalized by hepatocytes is metabolized to bile acids and secreted into bile or released directly as cholesterol into bile.

The LDL receptor is a single-chain glycoprotein; numerous mutations in its gene are associated with familial hypercholesterolemia. The receptor spans the plasma membrane once with the carboxyl terminus on the cytoplasmic face and the amino terminus, which binds apoB-100 and apoE-100, extending into the extracellular space.

The correlation between high plasma cholesterol level, particularly LDL cholesterol, and heart attacks and strokes has lead to dietary and therapeutic approaches to lower blood cholesterol (Clin. Corr. 18.3). Patients with familial (genetic) hypercholesterolemia suffer from accelerated atherosclerosis (Clin. Corr. 18.4). In most cases, called receptor negative, there is a lack of functional LDL receptors because the mutant alleles produce little or no LDL receptor protein. In other cases, the receptor is synthesized and transported normally to the cell surface, but an amino acid substitution or other alteration in the primary structure adversely affects LDL binding. As a result, there is little or no binding of LDL to the cell, cholesterol synthesis is not inhibited, and the blood cholesterol level increases. Some LDL-deficient patients synthesize the LDL receptor but have a defect in the transport mechanism that delivers the glycoprotein to the plasma membrane. Another group has LDL receptors that have a defect in the cytoplasmic carboxyl terminus and are unable to internalize the LDL–LDL receptor complex.

In specialized tissues such as the adrenal cortex and ovaries, cholesterol derived from LDL is the precursor for steroid hormones, such as cortisol and estradiol, respectively. In liver, cholesterol extracted from LDL and HDL is converted into bile salts that function in intestinal fat digestion.

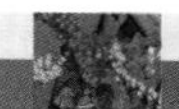

CLINICAL CORRELATION 18.3

Treatment of Hypercholesterolemia

Many authorities recommend screening asymptomatic individuals by measuring plasma cholesterol. A level less than 200 mg% is considered desirable; over 240 mg% requires lipoprotein analysis, especially determination of LDL cholesterol. Reduction of LDL cholesterol depends on dietary restriction of cholesterol to less than 300 mg/day, of calories to attain ideal body weight, and of total fat intake to less than 30% of total calories. Approximately two-thirds of the fat should be mono- or poly-unsaturated. The second line of therapy is with drugs. Cholestyramine and colestipol are bile salt-binding drugs that promote excretion of bile salts, increasing hepatic bile salt synthesis, and LDL uptake by the liver. Lovastatin inhibits HMG-CoA reductase, decreases endogenous synthesis of cholesterol and stimulates uptake of LDL via the LDL receptor. The combination of lovastatin and cholestyramine is sometimes used for severe hyperlipidemia.

Expert Panel: Evaluation and treatment of high blood cholesterol in adults. *Arch. Intern. Med.* 148: 36, 1988.

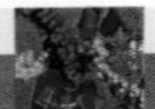

CLINICAL CORRELATION 18.4

Atherosclerosis

Atherosclerosis is the leading cause of death in Western industrialized countries. The risk of developing it is directly related to the plasma LDL cholesterol and inversely related to the HDL cholesterol level. This explains why the former is frequently called "bad" cholesterol and the latter "good" cholesterol, though chemically there is no difference. In atherosclerosis the arterial wall contains accumulated cholesteryl esters in cells derived from the monocyte-macrophage line, there is also smooth muscle cell proliferation and fibrosis. The earliest abnormality is migration of blood monocytes to the subendothelium of the artery. These cells then differentiate into macrophages and accumulate cholesteryl esters derived from plasma LDL. Some of the LDL may be taken up via pathways that do not require the LDL receptor. For instance, there are receptors that take up acetylated LDL or LDL complexed with dextran sulfate; however, this pathway is not regulated by cellular cholesterol content. Distortion of the subendothelium leads to platelet aggregation on the endothelial surface and release of platelet-derived mitogens such as platelet-derived growth factor (PDGF) which stimulates smooth muscle cell growth. Death of the foam cells leads to deposition of the cellular lipid and fibrosis. The resulting atherosclerotic plaque narrows the blood vessel and leads to thrombus formation, which precipitates myocardial infarction (heart attack).

Wick, G., Knoflach, M., and Xu, Q. B. Autoimmune and inflammatory mechanisms in atherosclerosis. *Annu. Rev. Immunol.* 22:361, 2004.

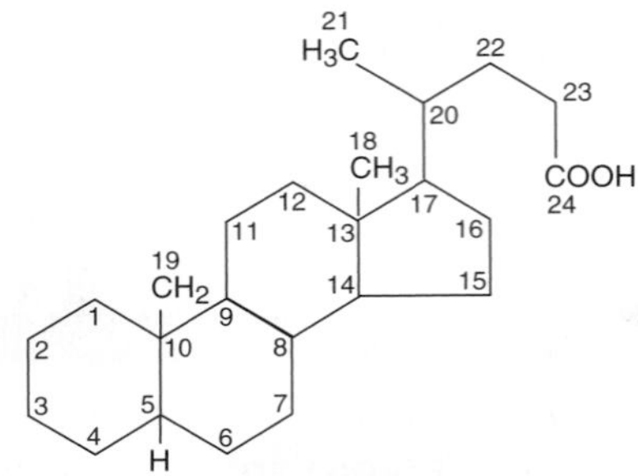

Figure 18.41 Structure of cholanic acid.

Cholesterol Is Excreted Primarily as Bile Acids

Bile acids are the products of cholesterol metabolism. Primary bile acids are synthesized in hepatocytes from cholesterol. The bile acids are derivatives of cholanic acid (Figure 18.41). Cholic acid and chenodeoxycholic acid (Figure 18.42) are C_{24} compounds containing three and two OH groups, respectively, and a C-5 side chain that ends in a carboxyl group that is ionized at pH 7.0 (hence the name bile salt). The carboxyl group is often conjugated via an amide bond to glycine ($NH_2—CH_2—COOH$) (Figure 18.43) or taurine ($NH_2—CH_2—CH_2—SO_3H$) to form glycocholic or taurocholic acid, respectively.

Microorganisms in the gut modify primary bile acids to secondary bile acids. Deoxycholic acid and lithocholic acid are derived from cholic acid and chenodeoxycholic acid,

Cholic acid

Deoxycholic acid

Chenodeoxycholic acid

Lithocholic acid

Figure 18.42 Structures of some common bile acids.

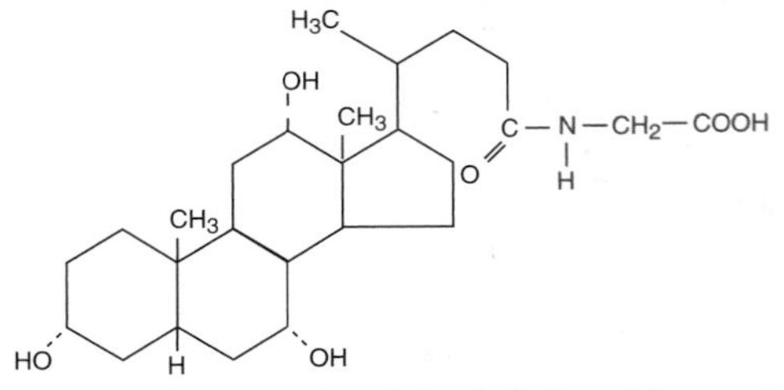

Figure 18.43 Structure of glycocholic acid, a conjugated bile acid.

respectively, by the removal of one OH group (see Figure 18.42). Transformation of cholesterol to bile acids requires (1) epimerization of the 3β-OH group, (2) reduction of the C5 double bond, (3) introduction of OH groups at C7 (chenodeoxycholic acid) or at C7 and C12 (cholic acid), and (4) conversion of the C27 side chain into a C24 carboxylic acid by elimination of a propyl equivalent.

Bile acids are secreted into bile, stored in the gallbladder, and then secreted into the small intestine. Liver production of bile acids is insufficient to meet the physiological needs, so the body relies on an enterohepatic circulation that carries the bile acids from the intestine back to the liver several times each day.

Bile acids and phospholipids solubilize cholesterol in bile, thereby preventing cholesterol from precipitating in the gall bladder. Bile acids in the gut act as emulsifying agents for dietary triacylglycerols, facilitating their hydrolysis by pancreatic lipase. Bile acids play a direct role in activating pancreatic lipase (p. 1051) and facilitate the absorption of fat-soluble vitamins, particularly vitamin D (p. 1067), from the intestine.

18.4 • SPHINGOLIPIDS

Synthesis of Sphingosine

Sphingolipids are complex lipids whose core structure is provided by the long-chain amino alcohol sphingosine (Figure 18.44) (4-sphingenine or trans-1,3-dihydroxy-2-amino-4-octadecene). Sphingosine has two asymmetric carbon atoms (C2 and C3); of the four possible optical isomers, naturally occurring sphingosine is the D-erythro form. The double bond has the *trans* configuration. The primary alcohol group at C1 is a nucleophilic center that is linked with sugars in glycosphingolipids. The amino group at C2 always bears a long-chain fatty acid (usually $C_{20}-C_{26}$) in amide linkage in ceramides. The secondary alcohol at C3 is always free. Note the structural similarity of this part of the sphingosine molecule to the glycerol moiety of the acylglycerols (Figure 18.44).

Sphingolipids occur in blood and the plasma membranes of nearly all cells. The highest concentrations are found in the white matter of the central nervous system.

Sphingosine is synthesized by way of sphinganine (dihydrosphingosine) from serine and palmitoyl CoA. Serine provides C1, C2, and the amino group of sphingosine and palmitic acid provides the remaining carbon atoms. The condensation of serine and palmitoyl CoA is catalyzed by **serine palmitoyltransferase**, a pyridoxal phosphate-dependent enzyme. The driving force for the reaction is provided by both cleavage of the reactive, high-energy, thioester bond of palmitoyl CoA and the release of CO_2 from serine (Figure 18.45). Reduction of the carbonyl group in 3-ketodihydrosphingosine by 3-ketosphinganine reductase to produce sphinganine (Figure 18.46) occurs at the expense of NADPH. Insertion of the double bond into sphinganine produces sphingosine.

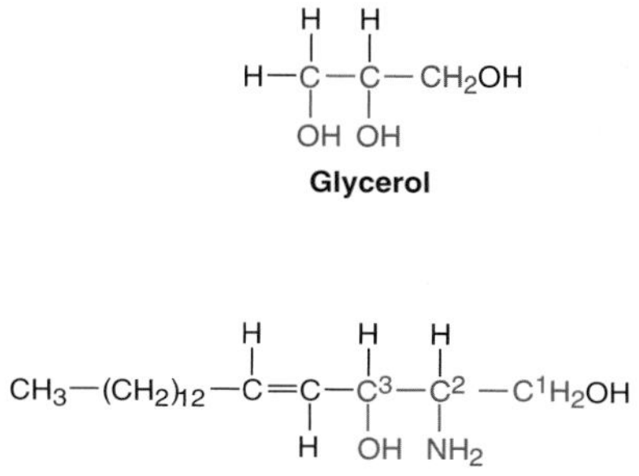

Figure 18.44 **Comparison of structures of glycerol and sphingosine (trans-1,3,dihydroxy-2-amino-4-octadecene).**

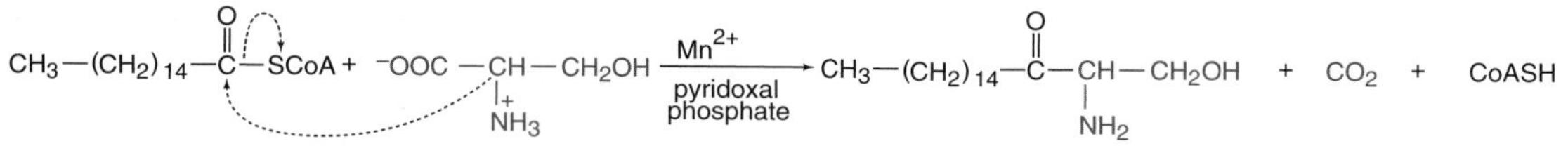

Palmitoyl CoA **L-Serine** **3-Ketodihydrosphingosine**

Figure 18.45 **Formation of 3-ketodihydrosphingosine from serine and palmitoyl CoA.**

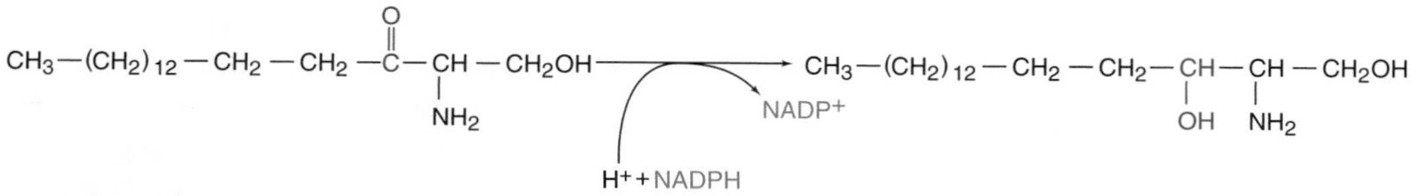

3-Ketodihydrosphingosine **Sphinganine (dihydrosphingosine)**

Figure 18.46 **Conversion of 3-ketodihydrosphingosine to sphinganine.**

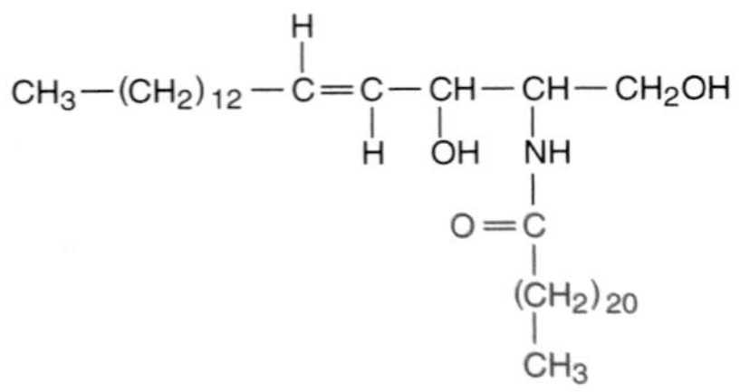

$$CH_3-(CH_2)_{12}-CH_2-CH_2-CH(OH)-CH(NH_2)-CH_2OH$$

Dihydrosphingosine

$$CH_3-(CH_2)_{20}-C(=O)-SCoA$$

Behenyl CoA

CoA

$$CH_3-(CH_2)_{12}-CH_2-CH_2-CH(OH)-CH(NH-C(=O)-(CH_2)_{20}-CH_3)-CH_2OH$$

Dihydroceramide

FAD

$FADH_2$

$$CH_3-(CH_2)_{12}-CH=CH-CH(OH)-CH(NH-C(=O)-(CH_2)_{20}-CH_3)-CH_2OH$$

Ceramide

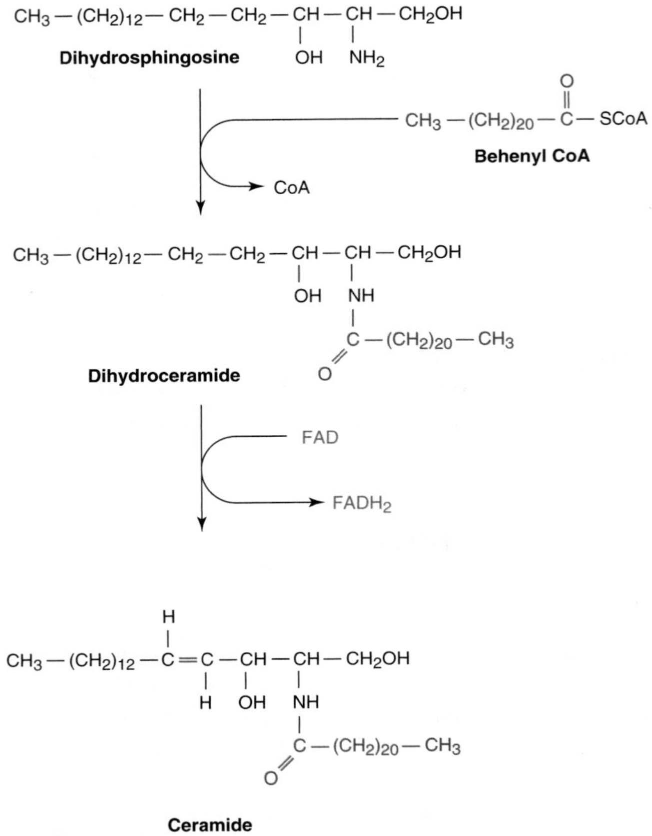

Figure 18.48 **Formation of ceramide from dihydrosphingosine.**

$$CH_3-(CH_2)_{12}-CH=CH-CH(OH)-CH(NH-C(=O)-(CH_2)_{20}-CH_3)-CH_2OH$$

Figure 18.47 **Structure of a ceramide (*N*-acylsphingosine).**

Ceramides Are Fatty Acid Amide Derivatives of Sphingosine

Sphingosine is the precursor of ceramide, a long-chain fatty acid amide derivative of sphingosine, which provides the core structure of sphingolipids. The fatty acid is attached to the 2-amino group of sphingosine through an amide bond (Figure 18.47). Most often the acyl group is behenic acid, a saturated 22-carbon fatty acid, but other long-chain acyl groups can be used. The two long-chain hydrocarbon domains in the ceramide molecule regions are responsible for the lipid character of sphingolipids.

Ceramide is synthesized from dihydrosphingosine (sphinganine) and a long-chain fatty acyl CoA by an endoplasmic reticulum enzyme. Dihydroceramide is an intermediate that is then desaturated at C4 and C5 (Figure 18.48). Ceramide is not a component of membrane lipids but rather is an intermediate in synthesis and catabolism of glycosphingolipids and sphingomyelin. Structures of prominent sphingolipids of humans are presented in Figure 18.49 in diagrammatic form.

Sphingomyelin Is a Phosphorus-Containing Sphingolipid

Sphingomyelin, a major component of membranes of nervous tissue, is a phospholipid. Since this ceramide phosphocholine contains one negative and one positive charge, it is neutral at physiological pH (Figure 18.50). The most common fatty acids in sphingomyelin are palmitic (16:0), stearic (18:0), lignoceric (24:0), and nervonic acid (24:1). The sphingomyelin of myelin contains predominantly lignoceric acid and nervonic acid, whereas that of gray matter contains largely stearic acid. Excessive accumulation of sphingomyelin occurs in Niemann–Pick disease. Conversion of ceramide to sphingomyelin by sphingomyelin synthase involves transfer of a phosphocholine moiety from phosphatidylcholine (lecithin) (Figure 18.51).

Sphingomyelin

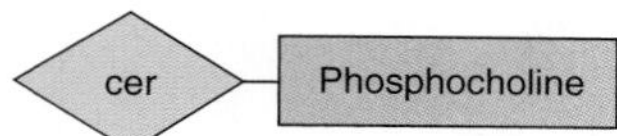

Neutral sphingolipids

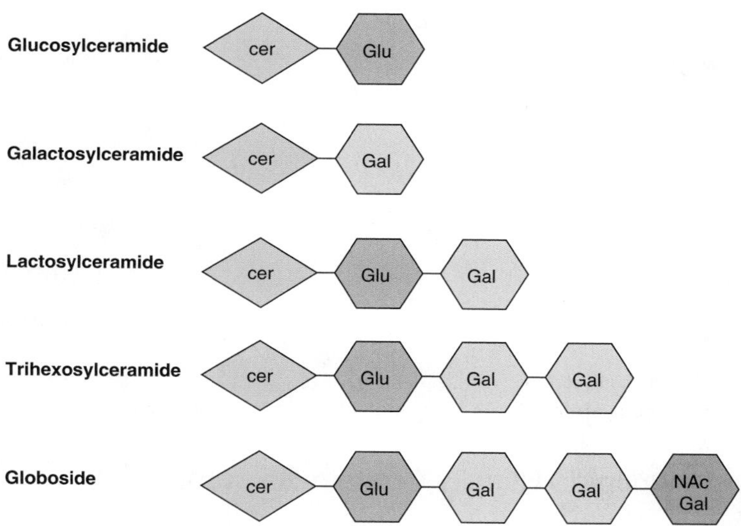

Acid sphingolipids

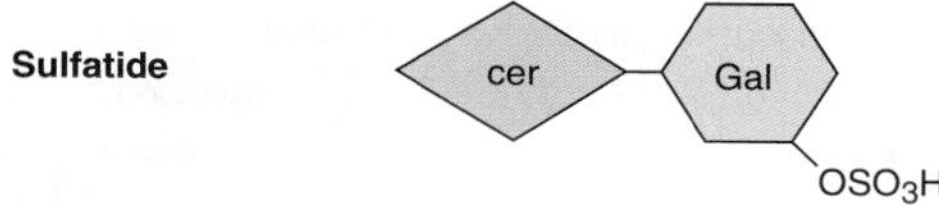

Gangliosides

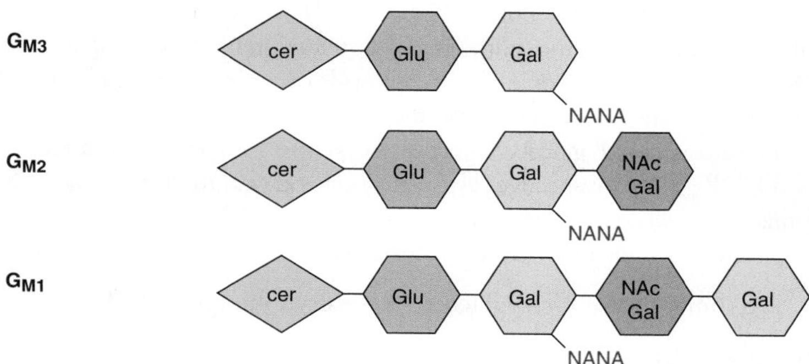

Figure 18.49 Structures of some common sphingolipids in diagrammatic form. Cer = ceramide, Glu = glucose, Gal = galactose, NAcGal = *N*-acetylgalactosamine, and NANA = *N*-acetylneuraminic acid (sialic acid).

$$CH_3-(CH_2)_{12}-\overset{H}{\overset{|}{C}}=\underset{\underset{H}{|}}{C}-\underset{\underset{OH}{|}}{CH}-\underset{\underset{\underset{\underset{O\!=\!C-(CH_2)_{16}-CH_3}{|}}{NH}}{|}}{CH}-CH_2-O-\overset{\overset{O}{\|}}{\underset{\underset{O^-}{|}}{P}}-O-CH_2-CH_2-\overset{+}{N}(CH_3)_3$$

Figure 18.50 Structure of sphingomyelin.

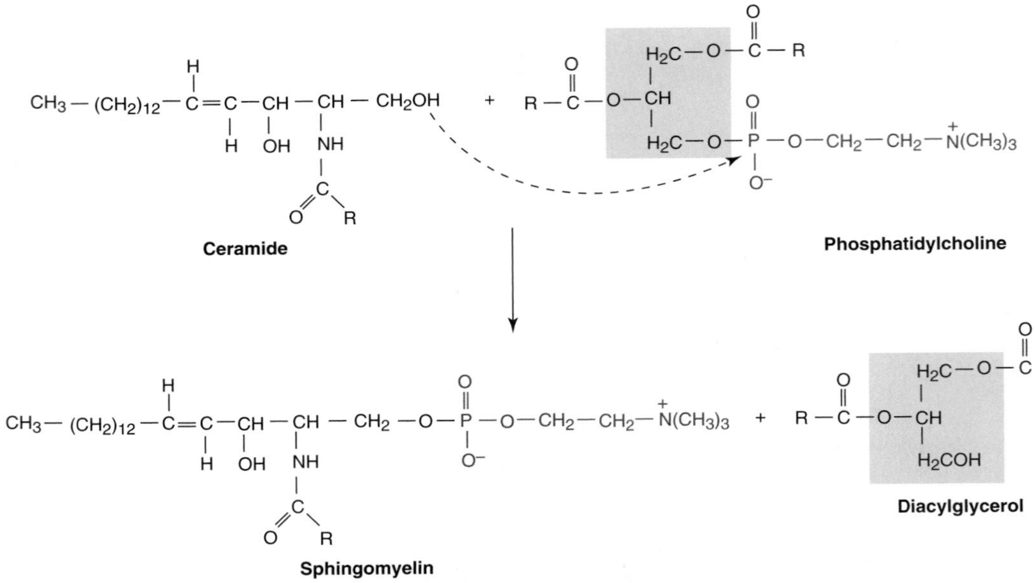

Ceramide + Phosphatidylcholine → Sphingomyelin + Diacylglycerol

Figure 18.51 Sphingomyelin synthesis from ceramide and phosphatidylcholine.

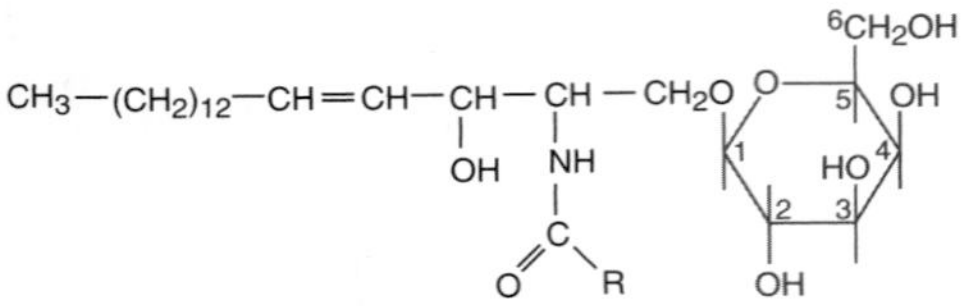

Figure 18.52 Structure of galactocerebroside (galactolipid).

Glycosphingolipids Usually Contain Galactose or Glucose

The principal glycosphingolipid classes are cerebrosides, sulfatides, globosides, and gangliosides. In glycolipids the polar head group attached to sphingosine is a sugar molecule.

Cerebrosides Are Glycosylceramides

Cerebrosides are ceramide monohexosides; the most common are **galactocerebroside** and **glucocerebroside**. Unless specified otherwise, the term *cerebroside* usually refers to galactocerebroside, also called *galactolipid.* In Figure 18.52 note that C1 of the monosaccharide unit is attached to C1 of the ceramide and the anomeric configuration of the sugar moiety in both galactocerebroside and glucocerebroside is β. Most galactocerebroside in healthy individuals is found in the brain. Moderate amounts of galactocerebroside accumulate in the white matter in Krabbe disease, or globoid leukodystrophy, a deficiency in lysosomal galactocerebrosidase.

Glucocerebroside (glucosylceramide) (Figure 18.53) is not normally a component of membranes. It is an intermediate in the synthesis and degradation of more complex glycosphingolipids. However, 100-fold increases in glucocerebroside occur in spleen and liver in the genetic lipid storage disorder called Gaucher disease, a deficiency of lysosomal glucocerebrosidase.

Galactocerebroside and glucocerebroside are synthesized from ceramide, and UDP-galactose and UDP-glucose, respectively. The glucosyl and galactosyl transferases that catalyze these reactions are associated with the endoplasmic reticulum (Figure 18.54). In some tissues, synthesis of glucocerebroside proceeds by glucosylation of sphingosine by glucosyltransferase.

$$\text{Sphingosine} + \text{UDP-glucose} \rightarrow \text{glucosylsphingosine} + \text{UDP}$$

followed by fatty acylation

$$\text{Glucosylsphingosine} + \text{stearoyl CoA} \rightarrow \text{glucocerebroside} + \text{CoASH}$$

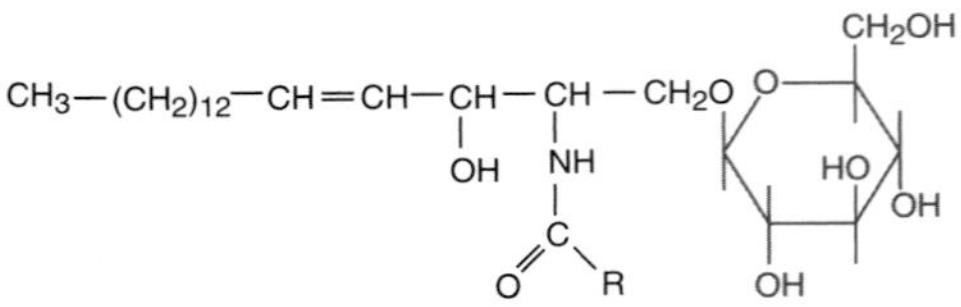

Figure 18.53 Structure of glucocerebroside.

Sulfatide Is a Sulfuric Acid Ester of Galactocerebroside

Sulfatide, or sulfogalactocerebroside, is a sulfuric acid ester of galactocerebroside. Galactocerebroside 3-sulfate is the major sulfolipid of brain, accounting for ~15% of lipids of white matter (Figure 18.55). It is synthesized from galactocerebroside and 3′-phosphoadenosine 5′-phosphosulfate (PAPS) by sulfotransferase.

$$\text{Galactocerebroside} + \text{PAPS} \rightarrow \text{PAP} + \text{galactocerebroside 3-sulfate}$$

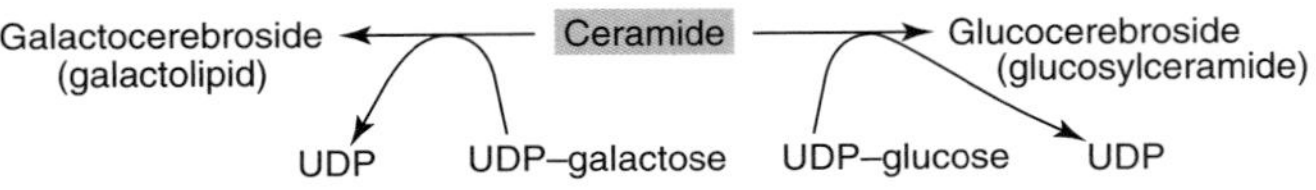

Figure 18.54 Synthesis of galacto- and glucocerebrosides.

The structure of PAPS, sometimes referred to as *activated sulfate,* is indicated in Figure 18.56. Large quantities of sulfatide accumulate in the central nervous system in metachromatic leukodystrophy due to a deficiency of lysosomal sulfatase.

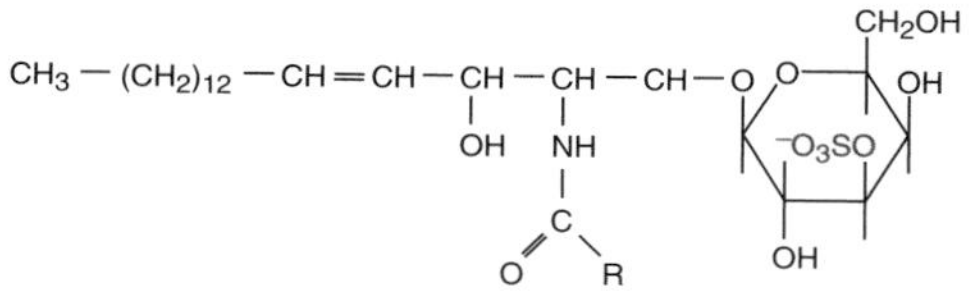

Figure 18.55 **Structure of galactocerebroside sulfate (sulfolipid).**

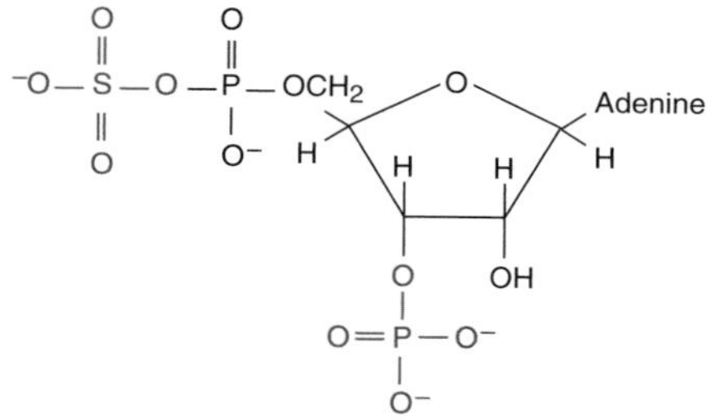

Figure 18.56 **Structure of PAPS (3′-phosphoadenosine-5′-phosphosulfate).**

Globosides Are Ceramide Oligosaccharides

Globosides are cerebrosides that contain two or more sugar residues, usually galactose, glucose, or *N*-acetylgalactosamine. The oligosaccharides are uncharged and contain no free amino groups. Lactosylceramide is present in the erythrocyte membrane (Figure 18.57). Another prominent globoside of the nervous system is ceramide trihexoside or ceramide galactosyllactoside: ceramide-β-glc(4 $\rightarrow$ 1)-β-gal-(4 $\rightarrow$ 1)-α-gal. Note that the terminal galactose residue of this globoside has the α-anomeric configuration. Ceramide trihexoside accumulates in kidneys of patients with Fabry disease, a deficiency of lysosomal α-galactosidase A.

Gangliosides Contain Sialic Acid

Gangliosides are sialic acid-containing glycosphingolipids highly concentrated in ganglion cells of the central nervous system, particularly in the nerve endings. The central nervous system is unique among human tissues because more than one-half of its sialic acid is in gangliosides, the remainder occurring in glycoproteins. Lesser amounts of gangliosides are present in the plasma membranes of cells of most extraneural tissues where they account for less than 10% of the total sialic acid.

Neuraminic acid (abbreviated Neu) is present in gangliosides, glycoproteins, and mucins. Its amino group occurs most often as the *N*-acetyl derivative, producing *N*-acetylneuraminic acid (NANA) or sialic acid (Figure 18.58). The OH group on C2 occurs most often in the α-anomeric configuration and is linked to the ceramide oligosaccharide through the OH group on position 2 of *N*-acetylneuraminic acid. Structures of some common gangliosides are indicated in Table 18.2. The principal gangliosides in brain are G_{M1}, G_{D1a}, G_{D1b}, and G_{T1b}. Nearly all of the gangliosides in the body are derived from glucosylceramide. In the naming of sialoglycosphingolipids, the letter G refers to ganglioside. The subscripts M, D, T, and Q indicate mono-, di-, tri-, and quatra (tetra)-sialic acid-containing gangliosides, and the subscripts 1, 2, and 3 designate the carbohydrate sequence that is attached to ceramide as follows: 1, Gal-GalNAc-Gal-Glc-ceramide; 2, GalNAc-Gal-Glc-ceramide; and 3, Gal-Glc-ceramide. In the Tay–Sachs ganglioside, the designation G_{M2} denotes the structure shown in Table 18.2.

A ganglioside on intestinal mucosal cells binds cholera toxin, an 84-kDa protein secreted by the pathogen *Vibrio cholerae.* The toxin stimulates secretion of chloride ions into the gut lumen, resulting in the copious diarrhea of cholera. The cholera toxin contains an A subunit (28 kDa) and five B subunits (~11,000 kDa each). After binding to the cell surface membrane through the B subunits, the A subunit enters the cell and acts as an ADP-ribosyltransferase that transfers ADP-ribose of NAD^+ onto the $G_{\alpha s}$ subunit of a G protein on the cytoplasmic side of the cell membrane (p. 522). This leads to activation of adenylate cyclase that stimulates secretion of chloride ion and produces diarrhea. The choleragenoid domain, as the B subunits are called, binds to the ganglioside G_{M1} (Table 18.2).

Gangliosides may also bind other toxins, such as tetanus toxin, and certain viruses, such as the influenza viruses. They may play an informational role in cell–cell interactions by providing specific recognition determinants on the cell surface.

Several lipid storage disorders involve accumulation of sialic acid-containing glycosphingolipids. The two most common gangliosidoses involve the gangliosides G_{M1} (G_{M1}

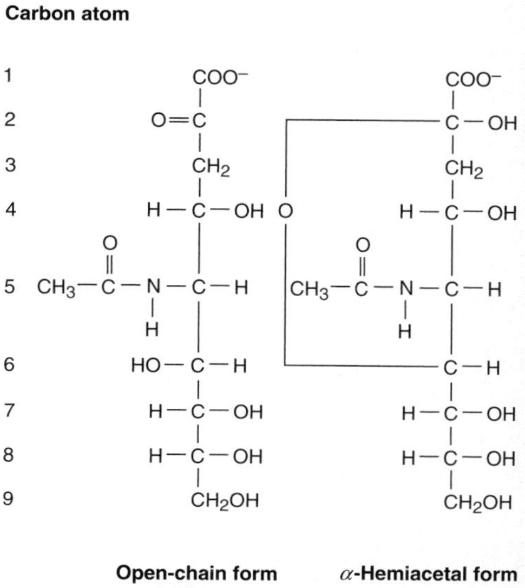

Open-chain form α-Hemiacetal form

Figure 18.58 **Structure of *N*-acetylneuraminic acid (NANA).**

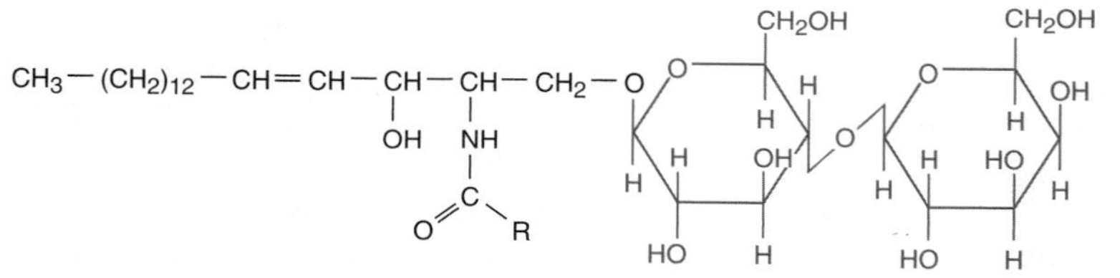

Figure 18.57 **Structure of ceramide-β-glc-(4→1)-β-gal (lactosylceramide).**

TABLE 18.2 • Structures of Some Common Gangliosides

Code Name	*Chemical Structure*
G_{M3}	Galβ → 4 Glcβ → Cer 3 ↑ αNANA
G_{M2}	GalNAcβ → 4 Galβ → 2Glcβ → Cer 3 ↑ αNANA
G_{M1}	Galβ → 3GalNAcβ → 4Galβ → 4Glcβ → Cer 3 ↑ αNANA
G_{D1a}	Galβ → 3GalNAcβ → 4Galβ → 4Glcβ → Cer 3 3 ↑ ↑ αNANA αNANA
G_{D1b}	Galβ → 3GalNAcβ → 4Galβ → 4Glcβ → Cer 3 ↑ αNANA8 ← αNANA
G_{T1a}	Galβ → 3GalNAcβ → 4Galβ → 4Glcβ → Cer 3 3 ↑ ↑ αNANA8 ← αNANA αNANA
G_{T1b}	Galβ → 3GalNAcβ → 4Galβ → 4Glcβ → Cer 3 3 ↑ ↑ αNANA αNANA8 ← αNANA
G_{Q1b}	Galβ → 3GalNAcβ → 4Galβ → 4Glcβ → Cer 3 3 ↑ ↑ αNANA8 ← αNANA αNANA8 ← αNANA

gangliosidosis) and G_{M2} (Tay–Sachs disease). G_{M1} gangliosidosis is an autosomal recessive metabolic disease characterized by impaired psychomotor function, mental retardation, hepatosplenomegaly, and death within the first few years of life. The massive cerebral and visceral accumulation of G_{M1} is due to a marked deficiency of β-galactosidase.

Sphingolipidoses Are Lysosomal Storage Diseases

Sphingolipids are degraded within lysosomes of phagocytic cells, particularly the histiocytes or macrophages of the reticuloendothelial system located primarily in liver, spleen, and bone marrow. Degradation begins with engulfment of membranes of white cells and erythrocytes that are rich in lactosylceramide (Cer-Glc-Gal) and hematoside (Cer-Glc-Gal-NANA). In the brain, the majority of cerebrosides are gangliosides and, particularly during the neonatal period, ganglioside turnover is extensive. Sphingolipid catabolism is summarized in Figure 18.59. This pathway requires enzymes that cleave specific bonds, including α- and β-galactosidases, a β-glucosidase, a neuraminidase, hexosaminidase, sphingomyelin-specific

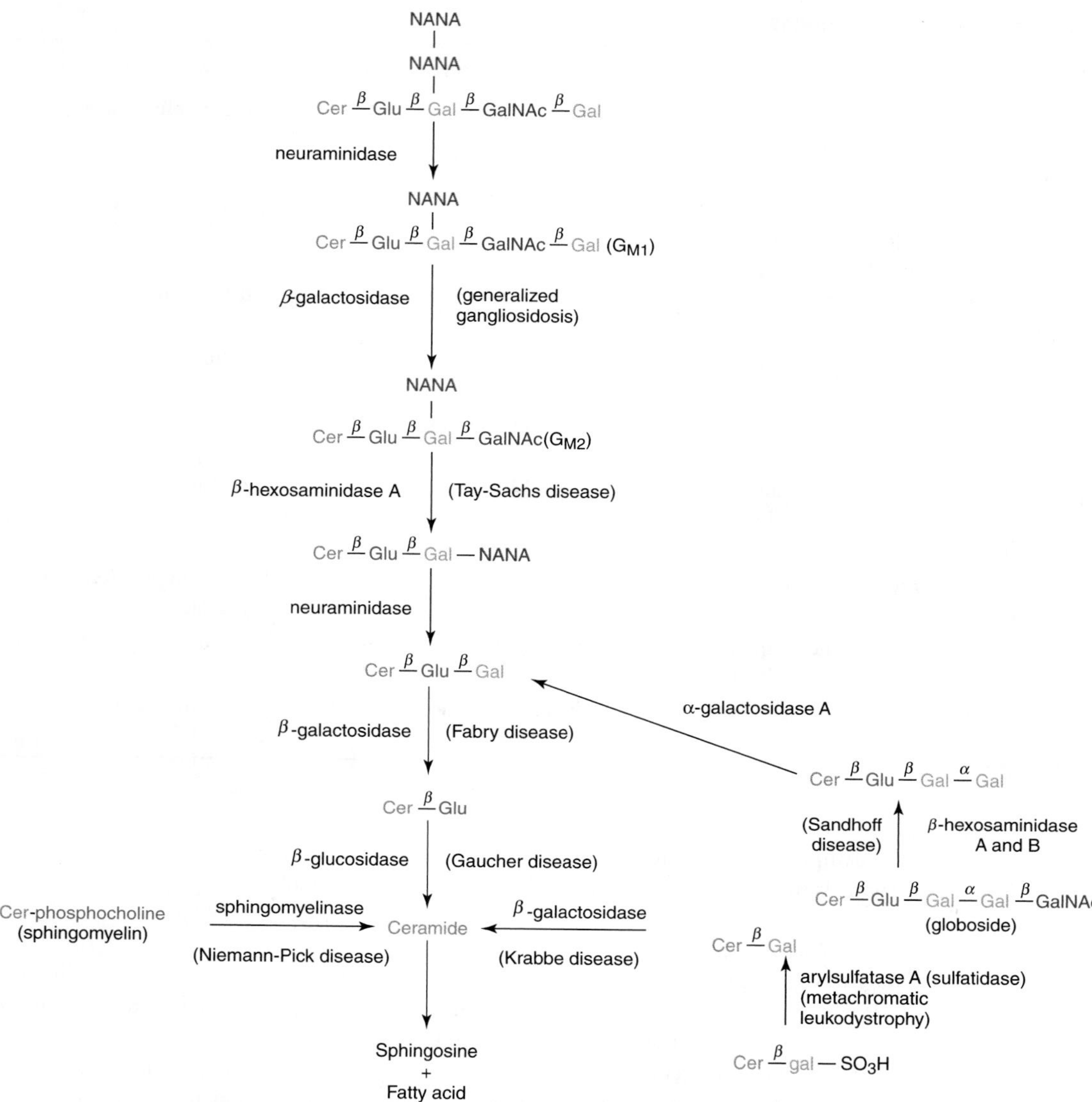

Figure 18.59 Summary of the pathways for catabolism of sphingolipids by lysosomal enzymes. The genetically determined enzyme deficiency diseases are indicated in the parentheses.

phosphodiesterase (sphingomyelinase), a sulfate esterase (sulfatase), and a ceramide-specific amidase (ceramidase). Important features of the catabolic pathway are (1) all the reactions occur within lysosomes; (2) the enzymes are hydrolases; (3) the pH optima of the enzymes is in the acid range, pH 3.5–5.5; (4) most of the enzymes are relatively stable and occur as isoenzymes, for example, hexosaminidase A (HexA) and hexosaminidase B (HexB); (5) the enzymes are glycoproteins and often occur firmly bound to the lysosomal membrane; and (6) the adjacent intermediates in the pathway differ by one sugar molecule, a sulfate group, or a fatty acid residue and are formed by the removal of one constituent at a time, such as sugars and sulfate, by irreversible reactions.

Sphingolipid catabolism normally functions smoothly, all of the glycosphingolipids and sphingomyelin being degraded to their constituents. However, when the activity of one enzyme in the pathway is markedly reduced due to a genetic error, then the substrate for that enzyme accumulates and is deposited within the lysosomes of the tissue in which catabolism of that sphingolipid normally occurs. An enzyme deficiency has been described for most reactions in Figure 18.59. These disorders, called sphingolipidoses, are summarized in Table 18.3.

Common features of lipid storage diseases are (1) usually only a single sphingolipid accumulates in the involved organs. (2) the ceramide portion is shared by the various

TABLE 18.3 • Sphingolipid Storage Diseases of Humans

Disorder	*Principal Signs and Symptoms*	*Principal Storage Substance*	*Enzyme Deficiency*
1. Tay–Sachs disease	Mental retardation, blindness, cherry red spot on macula, death between second and third year	Ganglioside G_{M2}	Hexosaminidase A
2. Gaucher disease	Liver and spleen enlargement, erosion of long bones and pelvis, mental retardation in infantile form only	Glucocerebroside	Glucocerebrosidase
3. Fabry disease	Skin rash, kidney failure, pain in lower extremities	Ceramide trihexoside	α-Galactosidase A
4. Niemann–Pick disease	Liver and spleen enlargement, mental retardation	Sphingomyelin	Sphingomyelinase
5. Globoid leukodystrophy (Krabbe disease)	Mental retardation, absence of myelin	Galactocerebroside	Galactocerebrosidase
6. Metachromatic leukodystrophy	Mental retardation, nerves stain yellowish brown with cresyl violet dye (metachromasia)	Sulfatide	Arylsulfatase A
7. Generalized gangliosidosis	Mental retardation, liver enlargement, skeletal involvement	Ganglioside G_{M1}	G_{M1} ganglioside: β-galactosidase
8. Sandhoff-Jatzkewitz disease	Same as 1; disease has more rapidly progressing course	G_{M2} ganglioside, globoside	Hexosaminidase A and B
9. Fucosidosis	Cerebral degeneration, muscle spasticity, thick skin	Pentahexosylfucoglycolipid	α-L-Fucosidase

storage lipids, (3) the rate of synthesis of the accumulating lipid is normal, (4) a catabolic enzyme is missing, and (5) the enzyme deficiency occurs in all tissues.

Diagnostic Assays for Sphingolipidoses

Diagnosis of a sphingolipidosis is made from biopsy of the involved organ, usually bone marrow, liver, or brain, on morphologic grounds based on the highly characteristic appearance of the storage lipid within lysosomes. Assay of enzyme activity confirms the diagnosis. Peripheral leukocytes, cultured skin fibroblasts, and chorionic villus that normally express the relevant enzyme are used for diagnostic purposes. In some cases (e.g., Tay–Sachs disease) serum or tears can be used for the diagnosis. Sphingolipidoses, for the most part, are autosomal recessive, the disease occurring only in homozygotes with a defect in both alleles. Enzyme assays can identify carriers or heterozygotes.

In Niemann–Pick disease, the deficient enzyme is sphingomyelinase (Figure 18.60). Sphingomyelin, C^{14}-labeled in the methyl groups of choline, is a useful substrate for

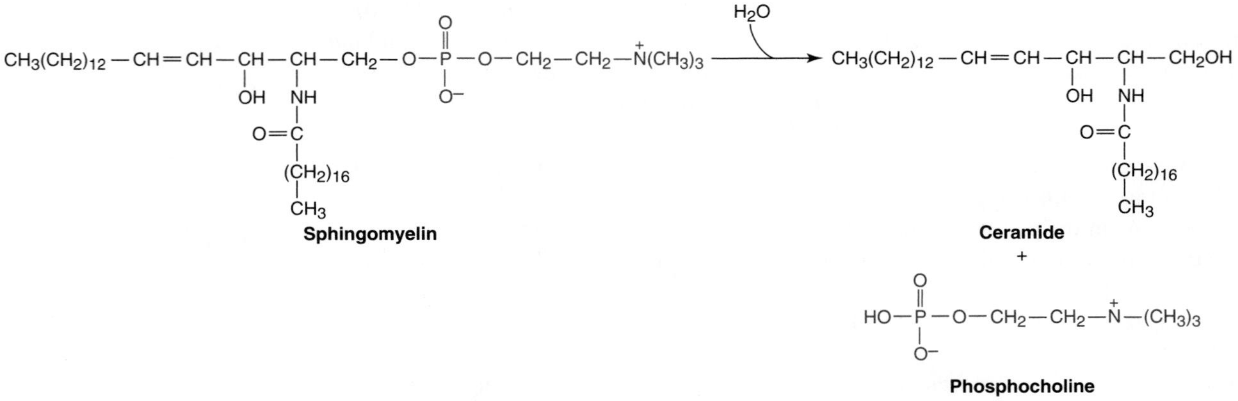

$$CH_3(CH_2)_{12}-CH{=}CH-\underset{OH}{CH}-\underset{NH-C(=O)-(CH_2)_{16}-CH_3}{CH}-CH_2-O-\overset{O}{\overset{\|}{P}}(O^-)-O-CH_2-CH_2-\overset{+}{N}(CH_3)_3 \xrightarrow{H_2O} CH_3(CH_2)_{12}-CH{=}CH-\underset{OH}{CH}-\underset{NH-C(=O)-(CH_2)_{16}-CH_3}{CH}-CH_2OH + HO-\overset{O}{\overset{\|}{P}}(O^-)-O-CH_2-CH_2-\overset{+}{N}-(CH_3)_3$$

Sphingomyelin → Ceramide + Phosphocholine

Figure 18.60 **Sphingomyelinase reaction.**

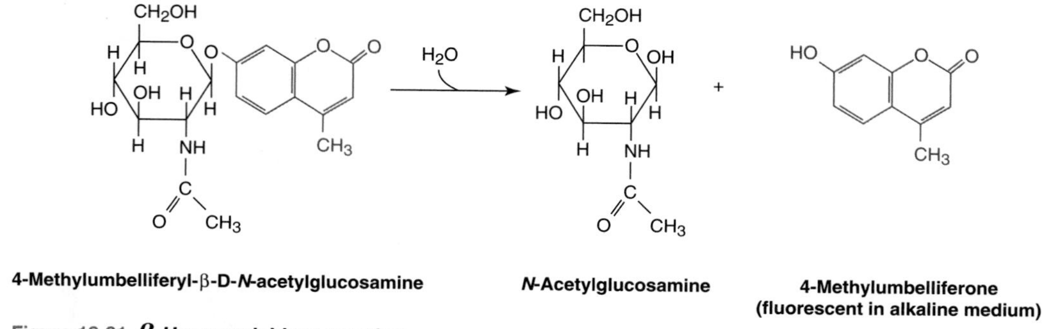

4-Methylumbelliferyl-β-D-*N*-acetylglucosamine $\xrightarrow{H_2O}$ *N*-Acetylglucosamine + 4-Methylumbelliferone (fluorescent in alkaline medium)

Figure 18.61 β-Hexosaminidase reaction.

determining sphingomyelinase activity. Extracts of white blood cells from healthy controls produce phosphocholine, which is water soluble. Extraction of the final incubation medium with an organic solvent such as chloroform will show radioactivity in the upper, aqueous phase. White blood cells from a patient with Niemann–Pick disease show little or no radioactivity (i.e., phosphocholine) in the aqueous phase.

Tay–Sachs disease, the most common form of G_{M2} gangliosidosis, is diagnosed by use of an artificial substrate. Ganglion cells of the cerebral cortex in this fatal disorder become swollen and many lysosomes are engorged with the acidic lipid, G_{M2} ganglioside. There is loss of ganglion cells, proliferations of glial cells, and demyelination of peripheral nerves. The pathognomonic finding is a cherry red spot on the macula caused by swelling and necrosis of ganglion cells in the eye. The substrate 4-methylumbelliferyl-β-*N*-acetylglucosamine is used to confirm the diagnosis. Hydrolysis of the substrate by hexosaminidase A, which is lacking in this disease, produces the intensely fluorescent 4-methylumbelliferone (Figure 18.61). Unfortunately, the diagnosis can be confused by the presence of hexosaminidase B, which is not deficient in this disease. The problem is usually resolved by taking advantage of the relative heat lability of hexosaminidase A and heat stability of hexosaminidase B. First, the total activity is assayed and then an aliquot of the tissue extract or serum specimen is heated at 55°C for 1 h and assayed. The difference between the two assays is a measure of hexosaminidase A activity and is used in making the diagnosis.

Once carriers of a lipid storage disease have been identified or if there has been a previously affected child in a family, the pregnancies at risk for these diseases can be monitored. All of the lipid storage disorders are transmitted as autosomal recessive genetic abnormalities. Fabry disease is X-chromosome linked. In the autosomal conditions, statistically one of four fetuses will be homozygous (or hemizygous in Fabry disease), two fetuses will be carriers, and one will be completely normal. Enzyme assays are used to detect affected fetuses and carriers *in utero*, using cultured fibroblasts obtained by amniocentesis as a source of enzyme.

Enzyme replacement therapy is available for Gaucher disease and Fabry disease and is being developed for several other sphingolipidoses. Prevention of the sphingolipidoses through genetic counseling based on enzyme assays and DNA analysis is available. A discussion of the diagnosis and therapy of Gaucher disease is presented in Clin. Corr. 18.5.

18.5 • PROSTAGLANDINS AND THROMBOXANES

Prostaglandins and Thromboxanes Are Derivatives of Monocarboxylic Acids

In mammalian cells, two major pathways of arachidonic acid metabolism produce important mediators of cellular and bodily functions: the cyclooxygenase and the lipoxygenase pathways. The substrate for both pathways is free arachidonic acid. The cyclooxygenase pathway leads to a series of compounds including prostaglandins and thromboxanes. Prostaglandins were discovered through their ability to promote contraction of intestinal and uterine muscle and lower blood pressure. Although the complexity of their structures and

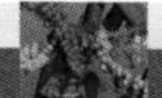

CLINICAL CORRELATION 18.5

Diagnosis of Gaucher Disease in an Adult

Gaucher disease (OMIM 230800) is an inherited disease of lipid catabolism in which glucocerebroside accumulates in macrophages of the reticuloendothelial system. Because of the large numbers of macrophages in the spleen, bone marrow, and, liver, hepatomegaly, splenomegaly and its sequelae (thrombocytopenia or anemia), and bone pain are the most common signs and symptoms of the disease.

Gaucher disease results from a deficiency of glucocerebrosidase. Some patients suffer severe neurologic deficits as infants, while others do not exhibit symptoms until adulthood. The diagnosis can be made by assaying leukocytes or fibroblasts for their ability to hydrolyze the β-glycosidic bond of artificial substrates (β-glucosidase activity) or of glucocerebroside (glucocerebrosidase activity). Gaucher disease has been treated with regular infusions of purified glucocerebrosidase and the long-term efficacy of the therapy appears good.

Brady, R. O., Kanfer, J. N., Bradley, R. M., and Shapiro, D. Demonstration of a deficiency of glucocerebroside-cleaving enzyme in Gaucher's disease. *J. Clin. Invest.* 45:1112, 1966.

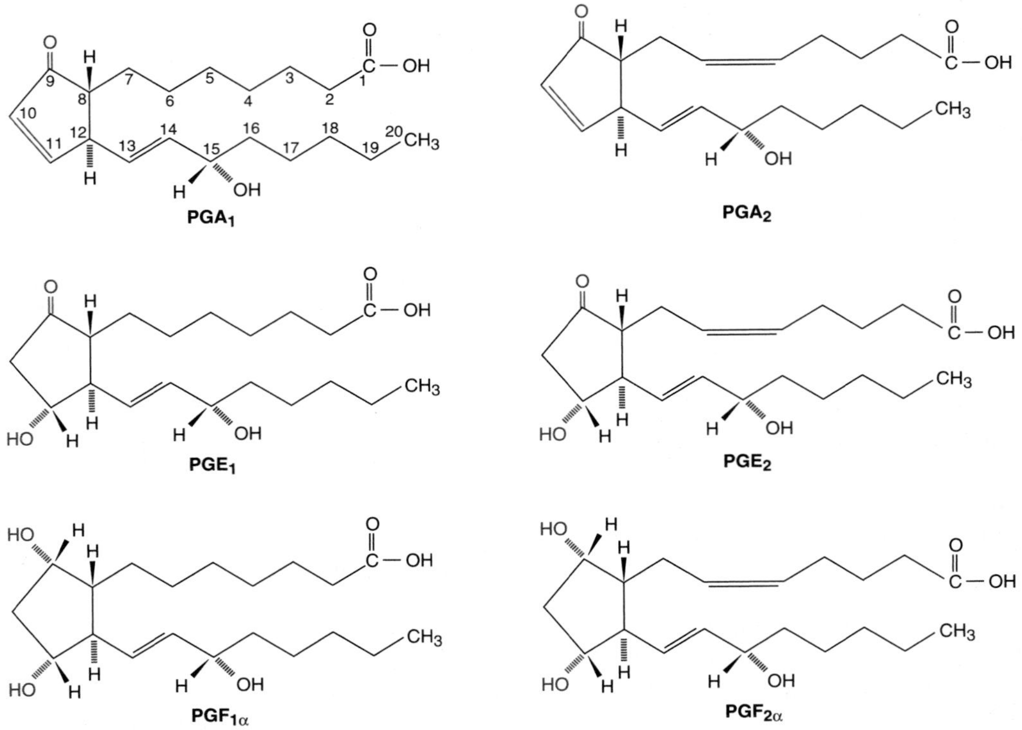

Figure 18.62 Structures of the major prostaglandins.

the diversity of their sometimes conflicting functions often create a sense of frustration, the potent pharmacological effects of the prostaglandins have made them important in human biology and medicine. With the exception of red blood cells, prostaglandins are produced and released by nearly all mammalian cells. Prostaglandins are not stored in cells but are released immediately after they are synthesized.

Structures of the more common prostaglandins A, E, and F are shown in Figure 18.62. All are structurally related to prostanoic acid (Figure 18.63). Note that the prostaglandins contain a multiplicity of functional groups; for example, PGE_2 contains a carboxyl group, a β-hydroxy-ketone ring, a secondary alkylic alcohol and two carbon–carbon double bonds. The three classes are distinguished by the functional groups of the cyclopentane ring: the E series contains a β-hydroxy-ketone ring, the F series are 1,3-diols, and the A series are α, β-unsaturated ketones. The subscript numerals 1, 2, and 3 refer to the number of double bonds in the side chains. The subscript α refers to the configuration of a C-9 OH group. An α-hydroxyl group projects *down* from the plane of the ring.

The most important dietary precursor of prostaglandins is linoleic acid [18:2(9,12)], an essential fatty acid. About 10 g of linoleic acid is ingested daily in adults. A small part of this intake is converted by elongation and desaturation in liver to arachidonic acid (eicosatetraenoic acid) and to some extent to dihomo-γ-linoleic acid. Since the total daily excretion of prostaglandins and their metabolites is only about 1 mg, formation of prostaglandins is a quantitatively unimportant pathway in the overall metabolism of fatty acids. However, the metabolism of prostaglandins is completely dependent on a regular and constant supply of linoleic acid.

Synthesis of Prostaglandins Involves a Cyclooxygenase

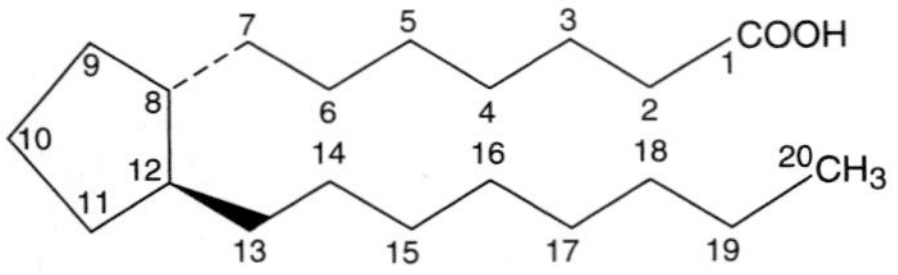

The immediate precursors of prostaglandins are C_{20} polyunsaturated fatty acids that contain three, four, or five carbon–carbon double bonds. Since arachidonic acid and most of its metabolites contain 20 carbons, they are referred to as eicosanoids. These fatty acids are cyclized and take up oxygen during their transformation into prostaglandins. The number of carbon–carbon double bonds in the side chain of a particular prostaglandin depends on the fatty acid precursor. Dihomo-γ-linolenic acid [20:3(8,11,14)] is the precursor of PGE_1

Figure 18.63 Structure of prostanoic acid.

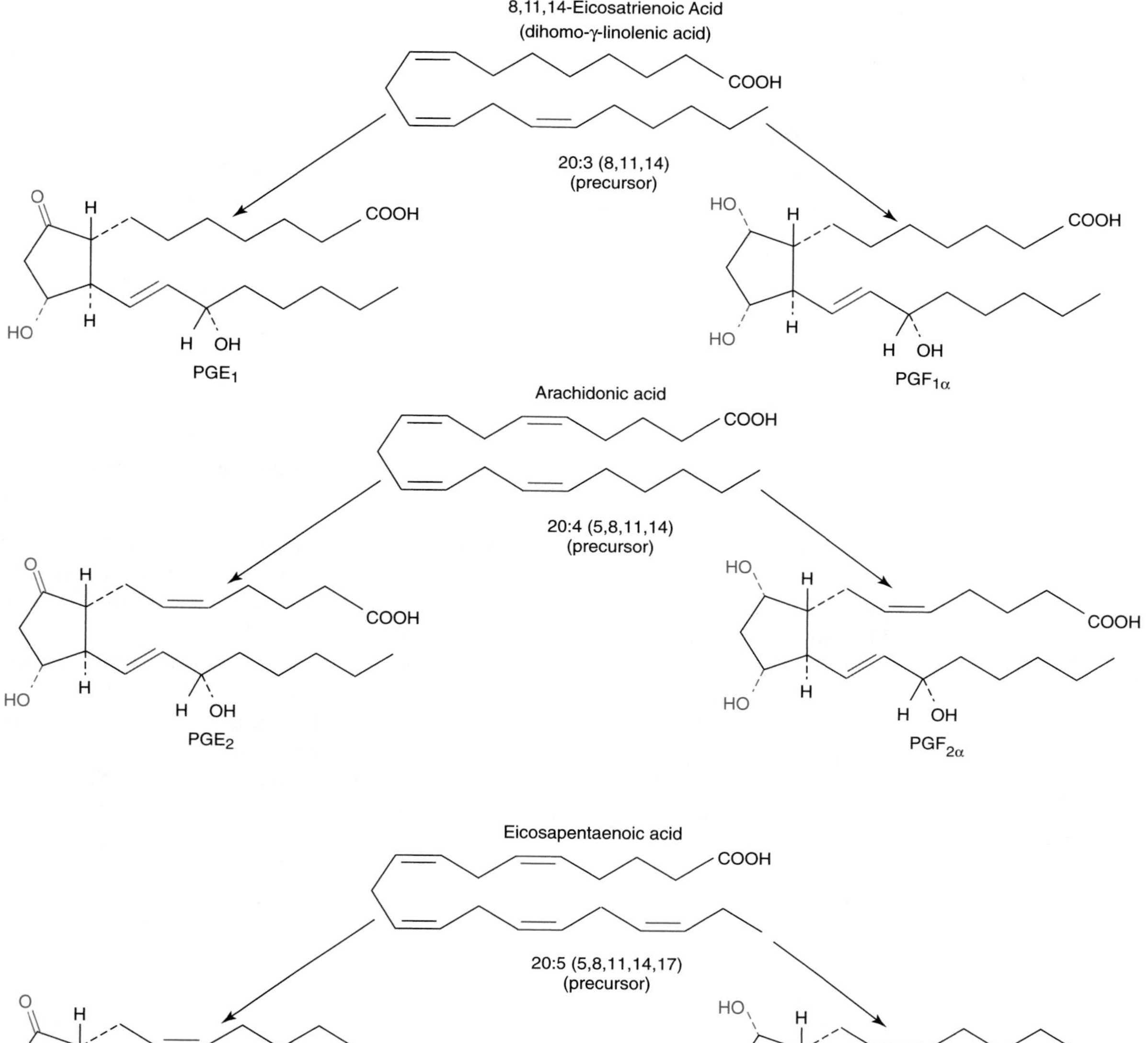

Figure 18.64 Synthesis of E and F prostaglandins from fatty acid precursors.

and $PGF_{1\alpha}$, arachidonic acid [20:4(5,8,11,14)] of PGE_2 and $PGF_{2\alpha}$, and eicosapentaenoic acid [20:4(5,8,11,14,17)] of PGE_3 and $PGF_{3\alpha}$ (Figure 18.64).

The 2-series prostaglandins derived from arachidonic acid are the principal prostaglandins in humans and are of the greatest significance biologically. Arachidonic acid is released from membrane phospholipids by the action of phospholipase A_2. This is the rate-limiting step in prostaglandin synthesis, and some agents that stimulate prostaglandin production do so by stimulating phospholipase A_2. Cholesteryl esters containing arachidonic acid also serve as a source of arachidonic acid. The key enzyme of prostaglandin biosynthesis is the bifunctional prostaglandin G/H synthase (PGS), which catalyzes oxidative cyclization of polyunsaturated fatty acids.

The cyclooxygenase (COX) component of PGS catalyzes cyclization of C8—C12 of arachidonic acid to form the 9,11-endoperoxide 15-hydroperoxide, PGG_2. This reaction requires two molecules of molecular oxygen (Figure 18.65) and involves stereospecific removal of the 13-pro-S-hydrogen of arachidonic acid. PGG_2 is then converted to

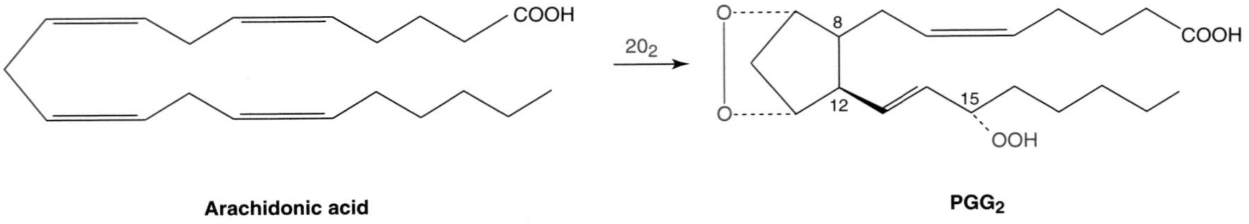

COOH
2O2
8
12
15
COOH
OOH
Arachidonic acid
PGG2

Figure 18.65 **Cyclooxygenase reaction.**

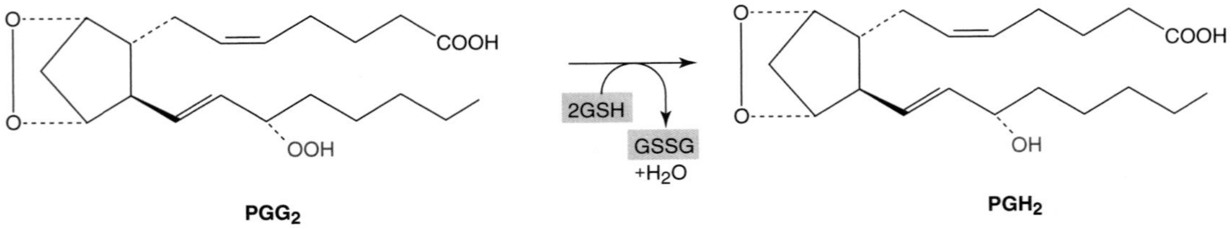

COOH
OOH
2GSH
GSSG
+H2O
COOH
OH
PGG2
PGH2

Figure 18.66 **Conversion of PGG_2 to PGH_2; PG hydroperoxidase (PGH synthase) reaction.**

prostaglandin H_2 (PGH_2) by the reduced glutathione-dependent peroxidase (PG hydroperoxidase) component of PGS (Figure 18.66). Reactions cyclizing polyunsaturated fatty acids occur in the membranes of the endoplasmic reticulum. Major pathways of prostaglandin biosynthesis are summarized in Figure 18.67. Depending on cell type, different enzymes mediate formation of prostaglandins of the D, E, and F series and of thromboxanes and prostacyclin (PGI_2). This produces some tissue specificity as to type and quantity of prostaglandin produced. In kidney and spleen, PGE_2 and $PGF_{2\alpha}$ are the major prostaglandins formed. In contrast, blood vessels produce mostly PGI_2 and $PGF_{2\alpha}$. In the heart PGE_2, $PGF_{2\alpha}$, and PGI_2 are formed in about equal amounts. Thromboxane A_2 (TXA_2) is the main prostaglandin endoperoxide formed in platelets.

Arachidonic acid
COOH
OH
Prostaglandin H2 (PGH2)
COOH
OH
OH
Prostaglandin I2
(prostacyclin, PGI2)
OH
COOH
OH
OH
Prostaglandin F2α
(PGF2α)
OH
COOH
OH
Prostaglandin D2
(PGD2)
COOH
OH
OH
Prostaglandin E2
(PGE2)

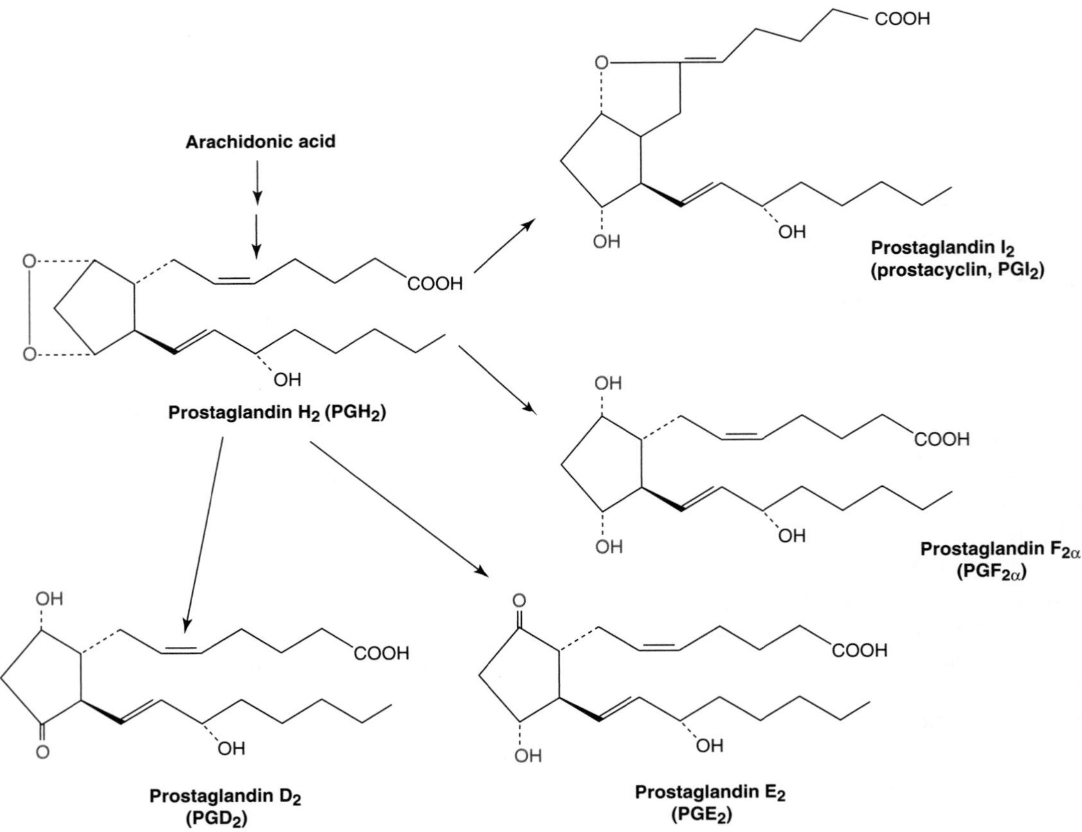

Figure 18.67 **Major routes of prostaglandin biosynthesis.**

There are two forms of cyclooxygenase (COX) or prostaglandin G/H synthase (PGS). COX-1, or PGS-1, is a constitutive enzyme of gastric mucosa, platelets, vascular endothelium, and kidney. COX-2, or PGS-2, is inducible and is produced in response to inflammation. The two prostaglandin synthases are homodimers. The two subunits each have PG hydroperoxidase and cyclooxygenase domains and these two catalytic centers are located close together. There is also a third domain, the membrane-binding domain, through which the enzyme is attached to the endoplasmic reticulum. COX-2 is mainly expressed in activated macrophages and monocytes that are stimulated by platelet-activating factor (PAF), interleukin-1, or bacterial lipopolysaccharide (LPS), and in smooth muscle cells, epithelial and endothelial cells, and neurons. PGS-2 induction is inhibited by glucocorticoids.

Prostaglandins have a very short half-life. They are rapidly taken up by cells soon after release through a receptor-mediated process, and inactivated by oxidation of the 15-hydroxy group or by β-oxidation from the carboxy end. The lungs have an important role in inactivating prostaglandins.

Thromboxanes are highly active metabolites of the PGG_2- and PGH_2-type prostaglandin endoperoxide in which the cyclopentane ring is replaced by a six-membered oxygen-containing (oxane) ring. The term **thromboxane** refers to their thrombus-forming potential. Thromboxane A_2 synthase of the endoplasmic reticulum is abundant in lung and platelets and converts endoperoxide PGH_2 to TXA_2. The half-life of TXA_2 is very short in water ($t_{1/2} \approx 1$ min) as it is transformed rapidly into inactive thromboxane B_2 (TXB_2) by the reaction shown in Figure 18.68.

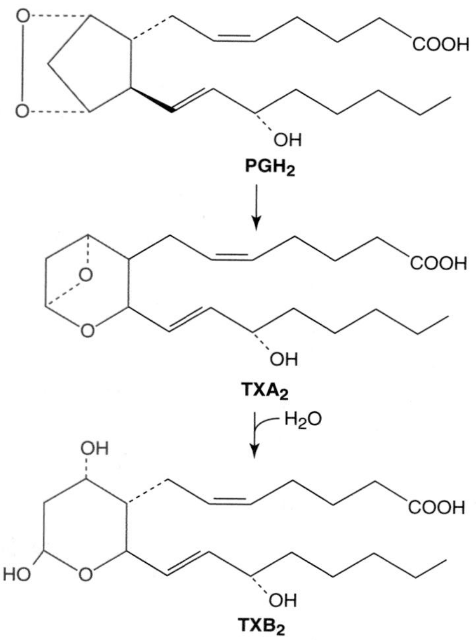

Figure 18.68 Synthesis of TXB_2 from PGH_2.

Prostaglandin Production Is Inhibited by Steroidal and Nonsteroidal Anti-inflammatory Agents

The **nonsteroidal anti-inflammatory drugs (NSAIDs)**, such as aspirin (acetylsalicylic acid), indomethacin, and phenylbutazone, block prostaglandin production by inhibiting cyclooxygenase. Aspirin inhibits by acetylating the enzyme. Other NSAIDs inhibit cyclooxygenase by binding noncovalently rather than by acetylating it; they are called *nonaspirin NSAIDS.* Aspirin is more potent against COX-1 than COX-2. Most NSAIDs inhibit COX-2 more than COX-1. These drugs have undesirable side effects; for example, aplastic anemia can result from phenylbutazone therapy. Steroidal anti-inflammatory drugs like hydrocortisone, prednisone, and betamethasone block prostaglandin release by inhibiting phospholipase A_2 activity and interfering with release of arachidonic acid (Figure 18.69). A family of proteins called anexins mediates corticosteroid inhibition of COX-2.

Control of the synthesis of prostaglandins is poorly understood. In general, prostaglandin release seems to be triggered by hormonal or neural excitation or by muscular activity. For example, histamine increases prostaglandin concentration in gastric perfusates. In addition, prostaglandins are released during labor and after cellular injury (e.g., platelets exposed to thrombin and lungs irritated by dust).

Prostaglandins Exhibit Many Physiological Effects

Prostaglandins are natural mediators of inflammation. Inflammatory reactions most often involve the joints (e.g., rheumatoid arthritis), skin (e.g., psoriasis), and eyes and are treated frequently with corticosteroids that inhibit prostaglandin synthesis. Administration of PGE_2 and PGE_1 induces redness and heat (due to arteriolar vasodilation), and swelling and edema resulting from increased capillary permeability characteristic of inflammation. PGE_2 generated in immune tissues (e.g., macrophages, mast cells, B cells) evokes chemotaxis of T cells. PGE_2 in amounts that do not cause pain, prior to administration of histamine and bradykinin, enhances the intensity and duration of pain caused by these two agents. Pyrogens (fever-inducing agents) activate the prostaglandin synthesis pathway with release of PGE_2 in the hypothalamus where body temperature is regulated. Aspirin is an antipyretic drug that acts by inhibiting cyclooxygenase.

Prostaglandins are synthesized in the uterus where they soften tissues and stimulate contraction to expel the fetus. Increased production of PGE_2 within the ovary follicle is

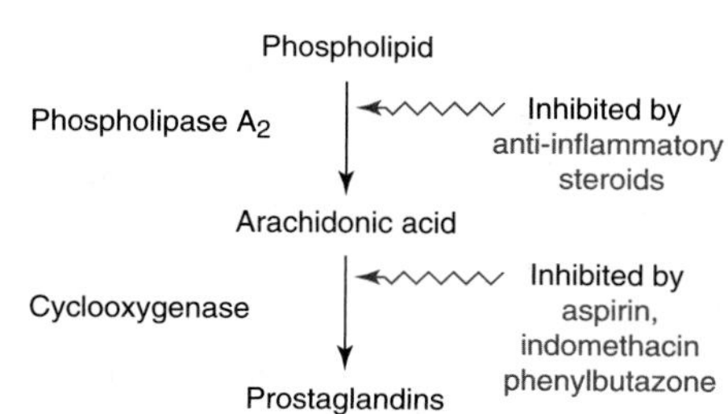

Figure 18.69 Site of action of inhibitors of prostaglandin synthesis.

essential for ovulation. The PGE agonist misoprostol has been used to induce parturition and to terminate unwanted pregnancies. There is also evidence that the PGE series of prostaglandins may be effective in treating infertility in males.

Synthetic prostaglandins are very effective in inhibiting gastric acid secretion in patients with peptic ulcer. They appear to inhibit cAMP formation in gastric mucosal cells and accelerate the healing of gastric ulcers. PGE, PGA, and PGI_2 are vasodilators that lower systemic arterial pressure, thereby increasing local blood flow and decreasing peripheral resistance. TXA_2 causes contraction of vascular smooth muscle and glomerular mesangium. In the fetus, PGE_2 maintains the patency of the ductus arteriosus prior to birth. If the ductus remains open after birth, closure can be hastened by the cyclooxygenase inhibitor indomethacin. In infants born with congenital abnormalities where the defect can be corrected surgically, infusion of prostaglandins maintains blood flow through the ductus until surgery is performed.

PGI_2 inhibits platelet aggregation, whereas PGE_2 and TXA_2 promote this clotting process. TXA_2 is produced by platelets and accounts for their aggregation when in contact with some foreign surface, collagen, or thrombin. Endothelial cells lining blood vessels release PGI_2, which may account for the lack of adherence of platelets to the healthy blood vessel wall. PGE_2 and PGD_2 dilate renal blood vessels and increase blood flow through the kidney. They also regulate sodium excretion and glomerular filtration rate.

18.6 • LIPOXYGENASE AND OXY-EICOSATETRAENOIC ACIDS

Cyclooxygenase directs polyunsaturated fatty acids into the prostaglandin pathway that has arachidonic acid as substrate. Lipoxygenase is a dioxygenase that also acts on arachidonic acid. Different lipoxygenases are specific for the double bond of arachidonic acid at which oxygen attack initially occurs (e.g., positions 5, 11, or 15). In humans, the most important leukotrienes are the 5-lipoxygenase products that mediate inflammatory disorders. Lipoxygenases occur widely in plants and fungi as well as in animals, but are absent from yeasts and most prokaryotes. They contain nonheme iron and are active when this iron is in the ferric state.

Monohydroperoxyeicosatetraenoic Acids Are Products of Lipoxygenase Action

Lipoxygenase adds a hydroperoxy group to arachidonic acid to produce monohydroperoxyeicosatetraenoic acids (HPETEs) (Figure 18.70). In contrast to the cyclooxygenase of prostaglandin endoperoxide synthase, which catalyzes the bis-dioxygenation of unsaturated fatty acids to endoperoxide, lipoxygenases catalyze the monodioxygenation of unsaturated fatty acids to allylic hydroperoxides. Hydroperoxy substitution of arachidonic acid by lipoxygenases may occur at position 5, 12, or 15. A 15-lipoxygenase (15-LOX) oxygenates arachidonic acid at carbon-15. 5-HPETE is the major product in basophils, polymorphonuclear (PMN) leukocytes, macrophages, mast cells, and any organ undergoing an inflammatory response; 12-HPETE predominates in platelets, pancreatic islet cells, vascular smooth muscle, and glomerular cells; and 15-HPETE is the principal product in reticulocytes, eosinophils, T-lymphocytes, and tracheal epithelial cells. The 5-, 12-, and 15-lipoxygenases occur mainly in the cytosol. Since the oxygenated carbon atom in HPETEs is asymmetric, there are two possible stereoisomers of the hydroperoxy acid, (*R*) or (*S*). For example, the stereo configuration is specified 12R-LOX or 12S-LOX. The three major HPETEs are of the (S) configuration. 5-LOX has a dioxygenase activity that converts arachidonic acid to 5-HPETE and a dehydrase activity that transforms 5-HPETE to LTA_4. 5-LOX activity is restricted to a few cell types, including B-lymphocytes, but not T lymphocytes. It is activated by an accessory protein called

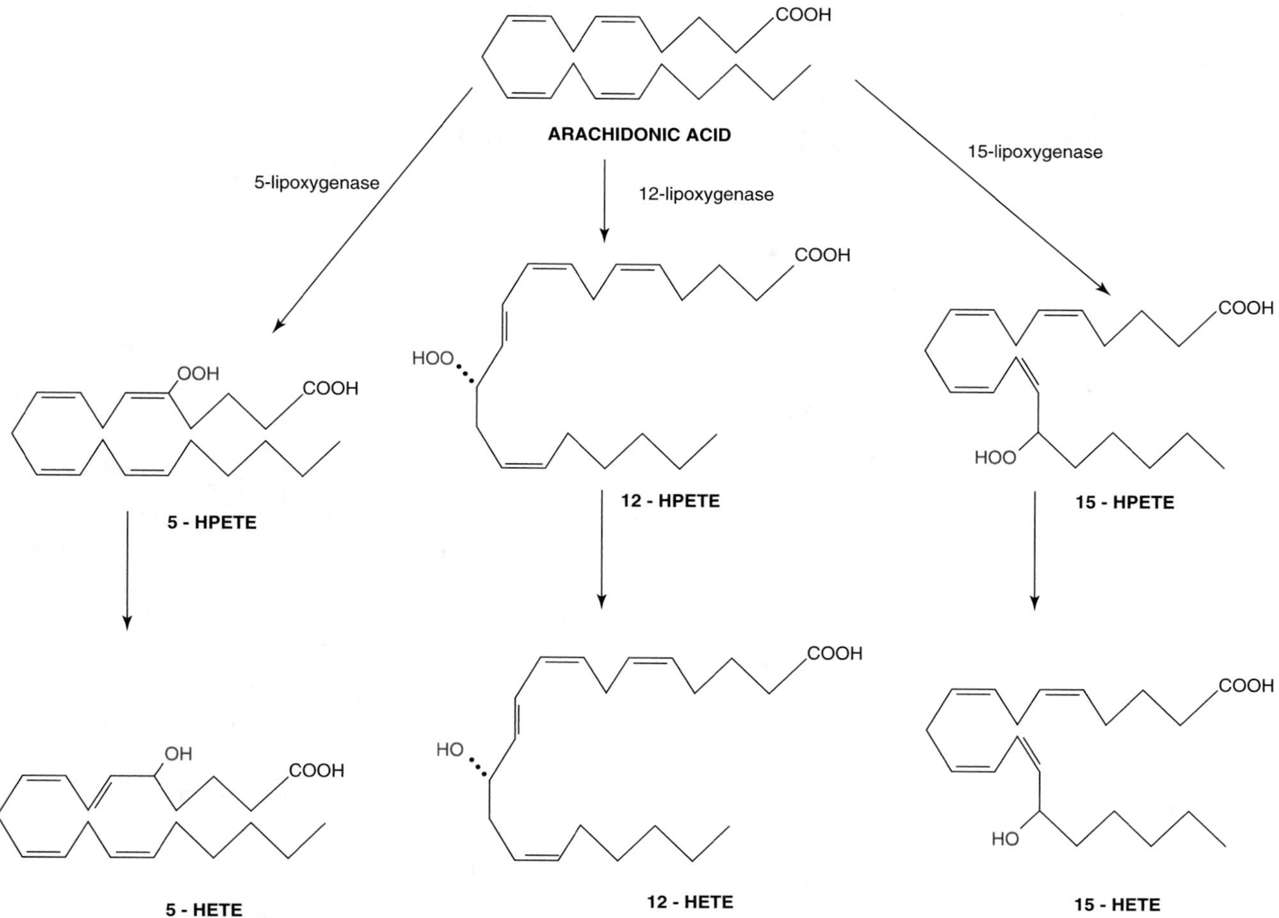

Figure 18.70 Lipoxygenase reaction and role of 5-hydroperoxyeicosatetraenoic acids (HPETEs) as precursors of hydroxyeicosatetraenoic acids (HETEs).

5-lipoxygenase activating protein (FLAP). In human leukocytes, FLAP is an arachidonic acid transfer protein that presents the fatty acid substrate to the 5-LOX located on the nuclear membrane.

Leukotrienes, Hydroxyeicosatetraenoic Acids, and Lipoxins Are Hormones Derived from HPETEs

HPETE-hydroperoxides are highly reactive, unstable intermediates that are converted either to the analogous alcohol (hydroxy fatty acid) by reduction of the peroxide moiety or to leukotrienes. The HPETEs are reduced either spontaneously or by the action of peroxidases to the corresponding hydroxyeicosatetraenoic acids (HETEs) (Figure 18.70). Leukotrienes contain at least three conjugated double bonds. Figure 18.71 shows how 5-HPETE is rearranged by LTA_4 synthase to the epoxide leukotriene A_4 (LTA_4), which is then converted by LTB_4 synthase (hydratase) to LTB_4 or LTC_4, emphasizing that 5-HPETE is an important branch point in the lipoxygenase pathway. Note that the subscript indicates the number of double bonds. Thus, while double-bond rearrangement may occur, the number of double bonds in the leukotriene product is the same as in the original arachidonic acid.

The conversion of LTA_4 to leukotrienes LTC_4, LTD_4, and LTE_4 requires the participation of reduced glutathione that opens the epoxide ring in LTA_4 to produce LTC_4 (Figure 18.71). Sequential removal of glutamic acid and glycine residues by specific dipeptidases yields the leukotrienes LTD_4 and LTE_4 (Figure 18.72).

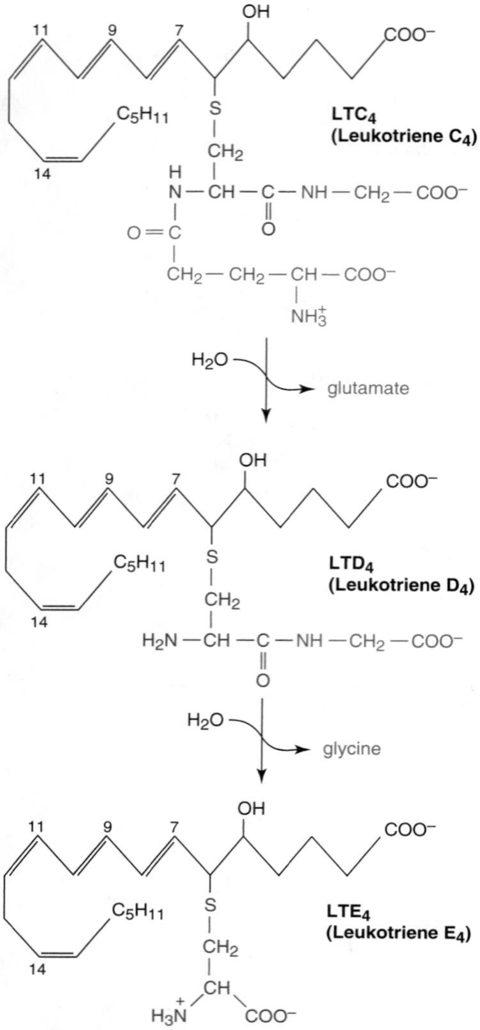

Figure 18.72 **Conversion of LTC_4 to LTD_4 and LTE_4.**

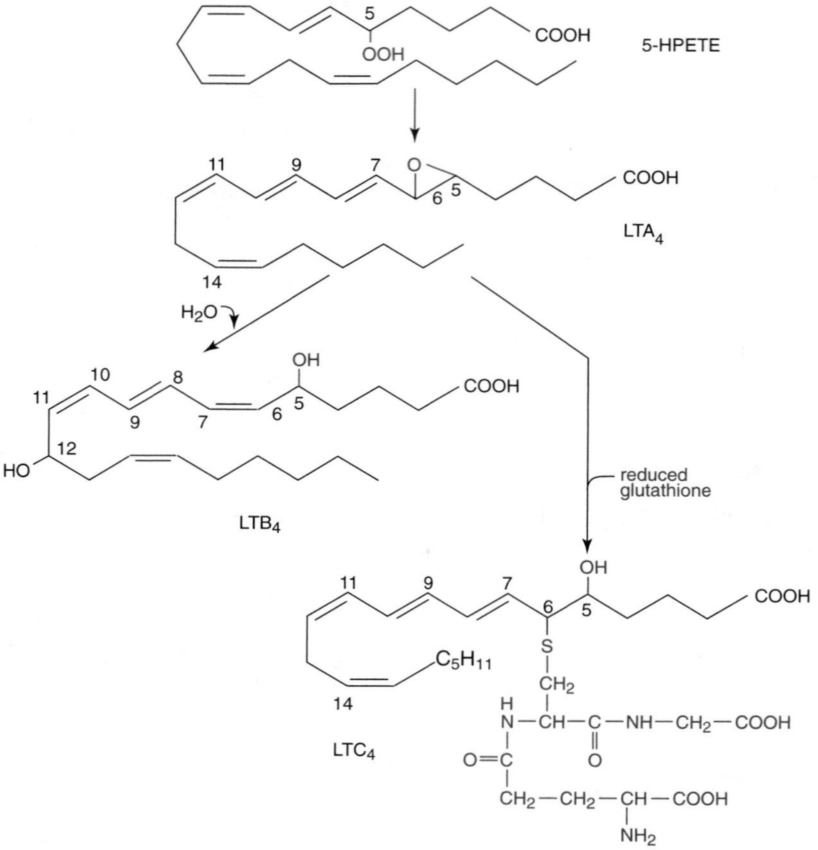

Figure 18.71 **Conversion of 5-HPETE to LTB_4 and LTC_4 through LTA_4 as intermediate.**

Lipoxins are another class of linear eicosanoids derived from arachidonic acid. They differ structurally from leukotrienes and HETEs in that they contain three hydroxyl groups and a conjugated tetraene system (Figure 18.73).

Leukotrienes and HETEs Affect Several Physiological Processes

Leukotrienes persist for up to 4 h in the body. Omega-oxidation of the methyl end followed by β-oxidation from this end inactivates LTB_4 and LTE_4. These reactions occur in mitochondria and peroxisomes. The thionyl peptides LTC_4, LTD_4, and LTE_4 constitute

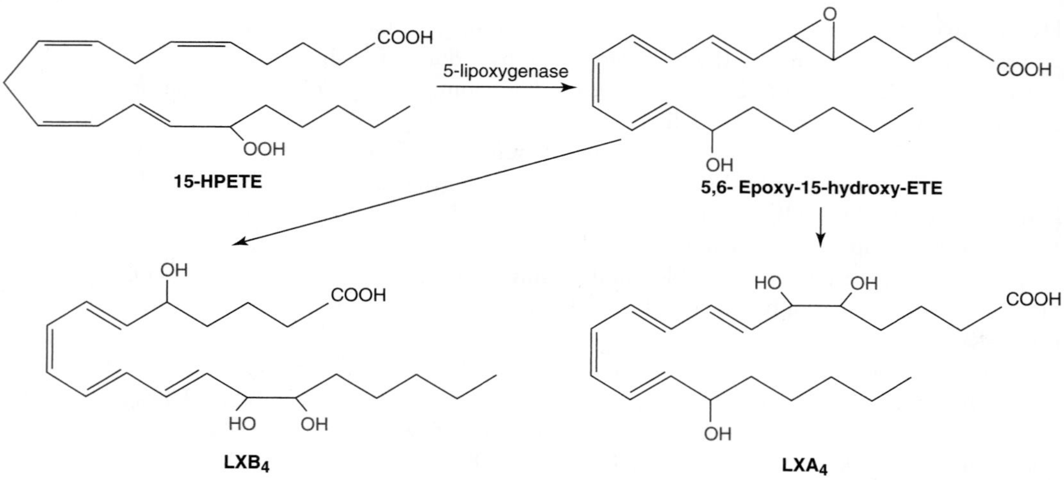

Figure 18.73 **Synthesis of Lipoxins.**

the **slow-reacting substance** of anaphylaxis (SRS-A). They cause slowly evolving but protracted contraction of smooth muscles in the airways and gastrointestinal tract. LTC_4 is rapidly converted to LTD_4 and then slowly converted to LTE_4. Enzymes in plasma catalyze these conversions. LTB_4 and the thionyl peptides LTC_4, LTD_4, and LTE_4 exert their actions through specific ligand–receptor interactions. In humans, activation of 5-LOX of leukocytes stimulates production of leukotrienes that provoke bronchoconstriction and inflammation. Current drugs for asthma include 5-LOX inhibitors and leukotriene receptor antagonists.

In general, HETEs (especially 5-HETE) and LTB_4 regulate neutrophil and eosinophil function: they mediate chemotaxis, stimulate adenylate cyclase, and induce polymorphonuclear (PMN) leukocyte cells to degranulate and release lysosomal hydrolytic enzymes. In contrast, LTC_4 and LTD_4 are humoral agents that promote smooth muscle contraction; constriction of pulmonary airways, trachea, and intestine; and increases in capillary permeability (edema). The HETEs appear to exert their effects by being incorporated into the phospholipids of membranes of target cells where the presence of fatty acyl chains containing a polar OH group may disturb lipid packing and thus the structure and function of the membrane. LTB_4 is immunosuppressive through inhibition of $CD4^+$ cells and proliferation of suppressor $CD8^+$ cells. LTB_4 also promotes neutrophil–endothelial cell adhesion.

Monohydroxy eicosatetraenoic acids of the lipoxygenase pathway are potent mediators of processes involved in allergy (hypersensitivity) and inflammation, secretion (e.g., insulin), cell movement, cell growth, and calcium fluxes. The initial allergic event, namely, the binding of IgE antibody to receptors on the surface of the mast cell, causes release of substances, including leukotrienes, referred to as mediators of immediate hypersensitivity. Lipoxygenase products are usually produced within minutes after the stimulus. The leukotrienes LTC_4, LTD_4, and LTE_4 are much more potent than histamine in contracting nonvascular smooth muscles of bronchi and intestine. LTD_4 increases the permeability of the microvasculature. LTB_4 stimulates migration (chemotaxis) of eosinophils and neutrophils, making them the principal mediators of PMN-leukocyte infiltration in inflammatory reactions.

Lipoxin A_4 (LXA4) and lipoxin B_4 (LXB4) have many physiologic functions, including inhibition of angiogenesis, promotion of clearance of pulmonary edema and protection against reperfusion injury.

Eicosatrienoic acids (e.g., dihomo-γ-linolenic acid) and eicosapentaenoic acid (see Figure 18.64, p. 739) also serve as lipoxygenase substrates. The amount of these 20 carbon polyunsaturated fatty acids with three and five double bonds in tissues is less than that of arachidonic acid, but special diets can increase their levels. The lipoxygenase products of these tri- and pentaeicosanoids are usually less active than LTA_4 or LTB_4. Since most Western diets contain about 10-fold more omega-6 fatty acids than omega-3 fatty acids, on a mass basis inflammatory prostaglandins, thromboxanes, leukotrienes, hydroxyl fatty acids, and lipoxins are produced in greater amounts than the less inflammatory products formed from omega-3 fatty acids such as eicosapentaenoic acid. Thus, a diet rich in omega-6 fatty acids shifts an individual into a state that is proinflammatory. It remains to be determined if fish oil diets rich in eicosapentaenoic acid are useful in the treatment of allergic and autoimmune diseases. Pharmaceutical research into therapeutic uses of lipoxygenase and cyclooxygenase inhibitors and inhibitors and agonists of leukotrienes in treatment of inflammatory diseases such as asthma, psoriasis, rheumatoid arthritis, and ulcerative colitis is very active.

Bibliography

Phospholipid Metabolism

Bogdanov, M., Mileykovskaya, E., and Dowhan, W. Lipids in the assembly of membrane proteins and organization of protein supercomplexes: Implications for lipid-linked disorders. *Subcell. Biochem.* 49: 197, 2008.

Brites, P, Waterham, H. R., and Wanders, R. J. Functions and biosynthesis of plasmalogens in health and disease. *Biochim. Biophys. Acta* 1636: 219, 2004.

Cohn, J. S., Wat, E., Kamili, A., and Tandy S. Dietary phospholipids, hepatic lipid metabolism and cardiovascular disease. *Curr. Opin. Lipidol.* 19: 257, 2008.

Fagone, P., and Jackowski, S. Membrane phospholipid synthesis and ER function. *J. Lipid Res.,* 50: S311, 2009.

Frasch, S. C., Henson, P. M, Nagaosa, K., Fessler, M. B., et al. Phospholipid flip-flop and phospholipids scramblase 1 (PLSCR1) co-localize to uropod rafts in formylated Met-Leu-Phe-stimulated neutrophils. *J. Biol. Chem.* 279: 17625, 2004.

Houtkooper R. H., and Vaz, F. M. Cardiolipin, the heart of mitochondrial metabolism. *Cell. Mol. Life Sci.* 65: 2493, 2008.

Kiessling, V., Wan, C., and Tamm, L. K. Domain Coupling in asymmetric lipid bilayers. *Biochim. Biophys. Acta* 1788: 64, 2008.

Lie, J., de Crom, R., van Gent, T., van Haperen, R., et al. Elevation of plasma phospholipids transfer protein increases the risk of atherosclerosis despite lower apolipoprotein B-containing lipoproteins. *J. Lipid Res.* 45: 805, 2004.

Okeley, N. M., and Gelb, M. H. A designed probe for acidic phospholipids reveals the unique enriched anionic character of the cytosolic face of the mammalian plasma membrane. *J. Biol. Chem.* 279: 21833, 2004.

Zwaal, R. F., Comfurius, P., and Bevers, E. M. Scott syndrome, a bleeding disorder caused by defective scrambling of membrane phospholipids. *Biochim. Biophys. Acta Mol. Cell Biol. Lipids* 1636: 119, 2004.

Lipoproteins

Clader, J. W. The discovery of ezetimibe: a view from outside the receptor. *J. Med. Chem.* 47: 1 2004.

Daniels, S. R.,and Greer, F. R.; Committee on Nutrition. Lipid screening and cardiovascular health in childhood. *Pediatrics* 122: 904, 2008.

De Beer, M. C., Castellani, L. W , Cai, L., Stromberg, A.J., et al. ApoA-II modulates the association of HDL with class B scavenger receptors SR-BI and CD36. *J. Lipid Res.* 45: 706, 2004.

Hoffman-Kuczynski, B., and Reo, N. V. Studies of myo-inositol and plasmalogen metabolism in rat brain. *Neurochem. Res.* 29: 843, 2004.

Navab, M., Reddy, S., Van Lenten, B. J., Anantharamaiah, G. M., and Fogelman, A. M. Role of dysfunctional HDL in atherosclerosis. *J. Lipid Res.* 2009.

Olofsson, S. O., Bostrom, P., Andersson, L., Rutberg, M., et al. Lipid droplets as dynamic organelles connecting storage and efflux of lipids. *Biochim. Biophys. Acta* 2009.

Venkatesh, P. K., Caskey, D., and Reddy, P. C. Therapies to increase high-density lipoprotein cholesterol and their effect on cardiovascular outcomes and regression of atherosclerosis. *Am. J. Med. Sci.* 336: 64, 2008.

Wang, M., and Briggs, M. R. HDL: The metabolism, function, and therapeutic importance. *Chem. Rev.* 104: 119, 2004.

Cholesterol Synthesis

Brown, M. S., and Goldstein, J. L. Cholesterol feedback: from Schoenheimer's bottle to Scap's MELADL. *J. Lipid Res.*, 50: S15, 2009.

Charlton-Menys, V., and Durrington, P. N. Human cholesterol metabolism and therapeutic molecules. *Exp. Physiol.* 93: 27, 2008.

Goldstein, J. L., and Brown, M. S. Regulation of the mevalonate pathway. *Nature* 343: 425, 1990.

Owen, C. G., Whincup, P. H., Kaye, S. J., Martin, R. M., et al. Does initial breastfeeding lead to lower blood cholesterol in adult life? A quantitative review of the evidence. *Am. J. Clin. Nutr.* 88: 305, 2008.

Tall, A. R. Cholesterol efflux pathways and other potential mechanisms involved in the athero-protective effect of high-density lipoproteins. *J. Intern. Med.* 263: 256, 2008.

Lung Surfactant

Abonyo, B. O., Gou, D., Wang, P., Narasaraju, T., et al. Syntaxin 2 and SNAP-23 are required for regulated surfactant secretion. *Biochem.* 43: 3499, 2004.

Dani, C., Martelli, E., Buonocore, G., Longini, M., et al. Influence of bilirubin on oxidative lung damage and surface tension properties of lung surfactant. *Pediatr. Res.* 53: S574A, 2003.

Halliday, H. L. Surfactant: Past, present, and future. *J. Perinatol.* 28: S47 2008.

Hallman, M. Lung surfactant, respiratory failure, and genes. *N. Engl. J. Med.* 350: 1278, 2004.

Rau, G. A, Dombrowsky, H., Gerbert, A., Thole, H. H., et al. Phophatidylcholine metabolism of rat trachea in relation to lung parenchyma and surfactant. *J. Appl. Physiol.* 95: 1145, 2003.

Schram, V., Anyan, W. R., and Hall, S. B. Non-cooperative effects of lung surfactant proteins on early adsorption to an air/water interface. *Biochim. Biophys. Acta* 1616: 165, 2003.

Shulenin, S., Nogee, L. M., Annilo, T., Wert, S. E., et al. ABCA3 gene mutations in newborns with fatal surfactant deficiency. *N. Engl. J. Med.* 350: 1296, 2004.

Prostaglandins, Thromboxanes, and Leukotrienes

Bos, C. L., Richel, D. J, Ritsema, T., Peppelenbosch, M. P., and Versteeg, H. H. Prostanoids and prostanoid receptors in signal transduction. *Int. J. Biochem. Cell Biol.* 36: 1187, 2004.

Cipollone, F., Rocca, B., and Patrono, C. Cyclooxygenase-2 expression and inhibition in atherothrombosis. *Atheroscler. Thromb.Vasc. Biol.* 24: 246, 2003.

De Caterina, R., and Zampolli, A. From asthma to atherosclerosis—5-lipoxygenase, leukotrienes, and inflammation. *N. Engl. J. Med.* 350: 4, 2004.

Dwyer, J. H., Allayee, H., Dwyer, K. M., Fan, J., et al. Arachidonate 5-lipoxygenase promoter genotype, dietary arachidonic acid, and atherosclerosis. *N. Engl. J. Med.* 350: 29, 2004.

Leslie, C. C. Regulation of the specific release of arachidonic acid by cytosolic phospholipase A2. *Prostaglandins, Leukotrienes, Essent. Fatty Acids* 70: 373, 2004.

Mayatepek, E., Okun, J. G., Meissner, T., Assmann, B., et al. Peroxidase activity, of cyclooxygenase-2 (COX-2) cross-links beta-amyloid (Abeta) and generates Abeta-COX-2 hetero-oligomers that are increased in Alzheimer disease. *J. Biol. Chem.* 279: 14673, 2004.

Takao, S. Lipid mediators in health and disease: Enzymes and receptors as therapeutic targets for the regulation of immunity and inflammation. *Annu. Rev. Pharm. Toxicol.* 49: 13, 2009.

Takezono, Y., Joh, T., Oshima, T., Suzuki, H., et al. Role of prostaglandins in maintaining gastric mucus-cell permeability against acid exposure. *J. Lab. Clin. Med.* 143: 52, 2004.

Bile Acids

Dawson, P. A., Haywood, J., Craddock, A. L., Wilson, M., et al. Targeted deletion of the ileal bile acid transporter eliminates enterohepatic cycling of bile acids in mice. *J. Biol. Chem.* 278: 33920, 2003.

De Fabiani, E., Mitro, N., Gilardi, F., Caruso, D., et al. Coordinated control of cholesterol catabolism to bile acids and of gluconeogenesis via a novel mechanism of transcription regulation linked to the fasted-to-fed cycle. *J. Biol. Chem.* 278: 39124, 2003.

Higuchi, H., and Gores, G. J. Bile acid regulation of hepatic physiology: IV: Bile acids and death receptors. *Am. J. Physiol. Gastrointest. Liver Physiol.* 284: G734, 2003.

Hofmann A. F., Hagey L. R. Bile acids: Chemistry, pathochemistry, biology, pathobiology, and therapeutics. *Cell. Mol. Life Sci.* 65: 2461 2008.

Knarreborg, A., Jensen, S. K., and Engberg, R. M. Pancreatic lipase activity as influenced by unconjugated bile acids and pH, measure in vitro and in vivo. *J. Nutr. Biochem.* 14: 259, 2003.

Lo, S. G., Petruzzelli, M., and Moschetta A. A translational view on the biliary lipid secretory network. *Biochim. Biophys. Acta* 1781: 79, 2008.

O'Byrne, J., Hunt, M. C., Rai, D. K., Saeki, M., and Alexson, S. E. The human bile acid-CoA: amino acid *N*-acyltransferase functions in the conjugation of fatty acids to glycine. *J. Biol. Chem.* 278: 34237, 2003.

Paumgartner, G., and Pusl, T. Medical treatment of cholestatic liver disease. *Clin. Liver Dis.* 12: 53, 2008.

Thomas, C., Auwerx, J., and Schoonjans, K. Bile acids and the membrane bile acid receptor TGR5-connecting nutrition and metabolism. *Thyroid.* 18: 167, 2008.

Sphingolipids and the Sphingolipidoses

Cox, T. M. Gaucher disease: Understanding the molecular pathogenesis of sphingolipidoses. *J. Inherit. Metab. Dis.* 24: S106, 2001.

Linke, T., Wilkening, G., Sadeghlar, F., Mozcall, H., et al. Interfacial regulation of acid ceramidase activity. Stimulation of ceramide degradation by lysosomal lipids and sphingolipid activator proteins. *J. Biol. Chem.* 276: 5760, 2001.

Platt, F. M., and Lachmann, R. H. Treating lysosomal storage disorders: Current practice and future prospects. *Biochim. Biophys. Acta* 2009.

Sandhoff, K., and Kolter, T. Biosynthesis and degradation of mammalian glycosphingolipids. *Philos. Trans. R. Soc. Lond. B. Biol. Sci.* 358: 847, 2003.

Watts, R. W. A historical perspective of the glycosphingolipids and sphingolipidoses. *Philos. Trans. R. Soc. Lond. B. Biol. Sci.* 358: 975, 2003.

Yu, R. K., Nakatani, Y., and Yanagisawa, M. Role of glycosphingolipid metabolism in the developing brain. *J. Lipid Res.* 2009.

Key Terms

phospholipids
cardiolipin
choline plasmalogen
platelet-activating factor
phosphatidylserine decarboxylase
phosphatidylinositol
phosphatidylinositol 4, 5-bisphosphate
CDP-choline
cytidyltransferase
glycerol 3-phosphate: acyltransferase
glycerol kinase
phosphatidic acid phosphatase
phosphatidylinositol synthase
cholesterol
acetoacetyl-CoA thiolase
HMG-CoA reductase
HMG-CoA synthase
farnesyl pyrophosphate
geranyltransferase
lanosterol cyclase
chylomicrons
high-density lipoprotein
low-density lipoproteins (LDL)
plasma lipoproteins
very low-density lipoproteins
cholesteryl ester transfer protein
lecithin:cholesterol acyl transferase
LDL receptors
CoA:cholesterol acyltransferase
bile acids
sphingosine
ceramide
serine palmitoyltransferase
sphingomyelin
galactocerebroside
glucocerebroside
sulfatide
gangliosides
globosides
neuraminic acid
prostaglandins
thromboxanes
cyclooxygenase
nonsteroidal anti-inflammatory drugs
lipoxygenase
leukotrienes
slow-reacting substance

Questions • CAROL N. ANGSTADT

Multiple-Choice Questions

1. Roles of various phospholipids include all of the following except
   A. cell–cell recognition.
   B. a surfactant function in lung.
   C. activation of certain membrane enzymes.
   D. signal transduction.
   E. mediator of hypersensitivity and acute inflammatory reactions.

2. CDP-X (where X is the appropriate alcohol) reacts with 1,2-diacylglycerol in the primary synthetic pathway for
   A. phosphatidylcholine.
   B. phosphatidylinositol.
   C. phosphatidylserine.
   D. all of the above.
   E. none of the above.

3. Plasmalogens differ from other membrane lipids in that they
   A. do not contain phosphorus.
   B. are found primarily in the inner mitochondrial membrane.
   C. have an alkene in ether linkage at the sn-1 position rather than a fatty acyl group.
   D. always contain a bound serine.
   E. contain a sugar molecule esterified to the phosphate.

4. Bile acids differ from their precursor cholesterol in that they
   A. are not amphipathic.
   B. contain an ionizable carboxyl group.
   C. contain less oxygen.
   D. are synthesized primarily in intestine.
   E. contain more double bonds.

5. A ganglioside may contain all of the following except
   A. a ceramide structure.
   B. glucose or galactose.
   C. phosphate.
   D. one or more sialic acids.
   E. sphingosine.

6. Phosphatidylinositols
   A. are neutral phospholipids.
   B. are found primarily in mitochondrial membrane.
   C. release Ca^{2+} from the endoplasmic reticulum.
   D. can serve to anchor glycoproteins to cell surfaces.
   E. are the major surfactants maintaining normal lung function.

Questions 7 and 8: Prostaglandins and leukotrienes are physiologically highly reactive compounds produced from polyunsaturated C20 acids like arachidonic acid. Both play some role in inflammatory processes. NSAIDs (nonsteroidal anti-inflammatory drugs) like aspirin are effective in inhibiting prostaglandin production but not leukotrienes.

7. Prostaglandin synthase, a bifunctional enzyme,
   A. catalyzes the rate-limiting step of prostaglandin synthesis.
   B. is inhibited by anti-inflammatory steroids.
   C. contains both a cyclooxygenase and a peroxidase component.
   D. produces PGG_2 as the end product.
   E. uses as substrate the pool of free arachidonic acid in the cell.

8. Leukotrienes
   A. synthesis begins with the addition of a hydroperoxy group to arachidonic acid.
   B. are inactivated by conversion to HPETEs.
   C. have an internal ring structure.
   D. are highly unstable, lasting only a few minutes.
   E. are converted to lipoxins.

Questions 9 and 10: Hypercholesterolemia is one of the risk factors for cardiovascular disease. Total cholesterol in blood is not as important in considering risk as distribution of cholesterol between LDL ("bad") and HDL ("good") particles. Efforts to reduce serum cholesterol are two-pronged: diet and drugs if reducing dietary cholesterol and saturated fat is not sufficient. The drugs of choice are statins, inhibitors of the rate-limiting enzyme of cholesterol biosynthesis. If necessary, bile acid-binding resins can be added.

9. In biosynthesis of cholesterol,
   A. 3-hydroxy-3-methyl glutaryl CoA (HMG CoA) is synthesized by mitochondrial HMG CoA synthase.
   B. HMG CoA reductase catalyzes the rate-limiting step.
   C. the conversion of mevalonic acid to farnesyl pyrophosphate proceeds via condensation of three molecules of mevalonic acid.
   D. condensation of two farnesyl pyrophosphates to form squalene is freely reversible.
   E. conversion of squalene to lanosterol is initiated by formation of the fused ring system, followed by addition of oxygen.

10. Cholesterol present in LDL (low-density lipoproteins)
   A. binds to a cell receptor and diffuses across the cell membrane.
   B. when it enters a cell, suppresses activity of ACAT (acyl CoA-cholesterol acyl transferase).
   C. once in the cell, is converted to cholesteryl esters by LCAT (lecithin-cholesterol acyl transferase).
   D. once it has accumulated in the cell, inhibits replenishment of LDL receptors.
   E. represents primarily cholesterol that is being removed from peripheral cells.

Questions 11 and 12: Sphingolipidoses (lipid storage diseases) are a group of diseases characterized by defects in lysosomal enzymes. Rate of synthesis of sphingolipids is normal, but un-degraded material accumulates in lysosomes. The severity of the particular disease depends, in part, on what tissues are most affected. Those affecting primarily brain and nervous tissue, like Tay–Sachs disease, are lethal within a few years of birth. Since the extent of enzyme deficiency is the same in all cells, easily accessible tissues like skin fibroblasts can be used to assay for enzyme deficiency. There are no treatments for most of the diseases so genetic counseling seems to be the only helpful approach. However, for Gaucher disease enzyme (glucocerebrosidase) replacement therapy is effective.

11. All of the following are true about degradation of sphingolipids except it
   A. occurs by hydrolytic enzymes contained in lysosomes.
   B. terminates at the level of ceramides.
   C. is a sequential, stepwise removal of constituents.
   D. may involve a sulfatase or a neuraminidase.
   E. is catalyzed by enzymes specific for a type of linkage rather than specific compound.

12. In Niemann–Pick disease, the deficient enzyme is sphingomyelinase. Sphingomyelins differ from other sphingolipids in that they are
   A. not based on a ceramide core.
   B. acidic rather than neutral at physiological pH.
   C. only types containing *N*-acetylneuraminic acid.
   D. only types that are phospholipids.
   E. not amphipathic.

Problems

13. Cells from a patient with familial hypercholesterolemia (FH) and cells from an individual without that disease were incubated with LDL particles containing radioactively labeled cholesterol. After incubation, the incubation medium was removed and the radioactivity of the cells measured. The cells were treated to remove any bound material and lysed, and internal cholesterol content was measured. Results are given below. What mutation of the gene for the LDL receptor protein could account for the results?

Cell Type	*Radioactivity of Cell*	*Cholesterol Content*
Normal	3000 cpm/mg cells	Low
FH	3000 cpm/mg cells	High

14. The distribution of fatty acids in many tissue phospholipids is different than that of the initially synthesized phospholipids. How does this redistribution occur?

Answers

1. **A** This function appears to be associated with complex glycosphingolipids. B: Especially dipalmitoyllecithin. C: For example, β-hydroxybutyrate dehydrogenase. D: Especially phosphatidylinositols. E: Platelet activity factor (PAF) does this.
2. **A** This is the main pathway for choline. B: Phosphatidylinositol is formed from CDP-diglyceride reacting with myoinositol. C: This is formed by base exchange.
3. **C** This linkage is formed by reaction with a long-chain alcohol. A: They are phospholipids. B: This is cardiolipin; plasmalogens are found in large amounts in myelin. D: Choline plasmalogen is the most abundant. E: They do not contain sugar; cerebrosides and gangliosides contain sugar but do not contain phosphate.
4. **B** This carboxyl group is often conjugated to glycine or taurine. A: Their role depends on being amphipathic. C: Hydroxylation is a major part of synthesis. D: Liver is site of synthesis. E: Primary bile acids are unsaturated.
5. **C** Glycosphingolipids do not contain phosphate. A and E: Ceramide, which is formed from sphingosine, is the base structure from which glycosphingolipids are formed. B: Glucose is usually the first sugar attached to the ceramide. D: By definition, gangliosides contain sialic acid.
6. **D** Glyceryl-phosphatidylinositols anchors have specific structural features. A: These are acidic, even more so as inositol is phosphorylated. B: Their functions reside at the plasma membrane. C: PIP_2 must be hydrolyzed to IP_3, which releases Ca^{2+}. E: This is dipalmitoyllecithin.
7. **C** Cyclooxygenase oxidizes arachidonic acid, and peroxidase converts PGG_2 to PGH_2. A and B: The release of the precursor fatty acid by phospholipase A_2 is the rate-limiting step and the one inhibited by

anti-inflammatory steroids. D: The peroxidase component converts the PGG_2 to PGH_2. E: Arachidonic acid is not free in the cell but is part of the membrane phospholipids or sometimes a cholesteryl ester.

8. A This is catalyzed by a lipoxygenase. B: HPETEs are intermediates in the synthesis. C: They are open chain derivatives of arachidonic acid. D: They can persist up to 4 h. E: Lipoxins are derived from the same intermediates (HPETEs) but are a different class of arachidonic acid derivatives.

9. B This enzyme is inhibited by cholesterol. A: Remember that cholesterol biosynthesis is cytosolic; mitochondrial biosynthesis of HMG CoA leads to ketone body formation. C: Mevalonic acid is decarboxylated (among other things) to produce the isoprene pyrophosphates which are the condensing units. D: Pyrophosphate is hydrolyzed which prevents reversal. E: The process is initiated by epoxide formation.

10. D This is one of the ways to prevent overload in the cell. A: LDL binds to the cell receptor and it is endocytosed and then degraded in lysosomes to release cholesterol. B: ACAT is activated to facilitate storage. C: LCAT is a plasma enzyme. E: The primary role of LDL is to deliver cholesterol to peripheral tissues; HDL removes cholesterol from peripheral tissues.

11. B Ceramides are hydrolyzed to sphingosine and fatty acid. D: Sulfogalactocerebroside contains sulfate and gangliosides contain one or more *N*-acetylneuraminic acids. E: Many sphingolipids share the same types of bonds, for example, a β-galactosidic bond, and one enzyme, for example, β-galactosidase, will hydrolyze it whenever it occurs.

12. D Sphingomyelins are not glycosphingolipids but phosphosphingolipids. A, B, and E: They are formed from ceramides, are amphipathic and are neutral. C is the definition of gangliosides.

13. FH cells have LDL receptors with normal binding properties as indicated by the bound radioactivity being the same as that of normal cells. They are unable to internalize the receptor–LDL complex so cholesterol synthesis is not inhibited as it is with normal cells. The mutation is most likely on the carboxy terminus of the protein that is involved in internalization. Make sure you understand why mutation in other regions would not lead to the observed results.

14. The acyltransferases in phosphatidic acid synthesis usually place a saturated fatty acid at sn-1 and unsaturated one at sn-2. Phospholipases A_1 and A_2 hydrolyze the fatty acids from these respective positions, generating either 2-acyl- or 1-acyl-lysophosphatide. The lysophosphatides can then be reacylated with the desired fatty acyl CoA.

19

Amino Acid and Heme Metabolism

Marguerite W. Coomes
Associate Professor, Howard University College of Medicine

Key Concepts

AMINO ACIDS

- Ammonia, used to synthesize amino acids and other nitrogenous compounds, is formed from nitrogen in the atmosphere by plants, bacteria, and the electrical charge from lightening.
- Humans synthesize *de novo* 11 amino acids; they are classified as *nonessential amino acids* in the diet. Metabolites of intermediary metabolism are precursors for synthesis of nonessential amino acids. Nine other amino acids are required in the diet and are classified as *essential amino acids.*
- Glutamate dehydrogenase catalyses incorporation of ammonia into an amino group, which can then be incorporated into other nonessential amino acids. Aminotransferases catalyze the

transfer of amino groups from most amino acids to an α-keto acid. Pyridoxal phosphate is the prosthetic group of aminotransferases.

- Amino group nitrogen is transported to the liver predominantly by glutamine and alanine where the urea cycle synthesizes urea from amino groups and ammonia for excretion. Glutamine is transported to kidneys for excretion of the amino group as ammonium ion or to intestines to initiate synthesis of arginine.
- Two amino acids are ketogenic, some are glucogenic, and some are both ketogenic and glucogenic. Many amino acids serve as precursors for secondary metabolites such as hormones and neurotransmitters.
- There are genetic errors in amino acid degradative and synthetic pathways.

HEME

- Heme, produced in virtually all mammalian tissues, consist of a ferrous ion and protoporphyrin IX. The organic portion of heme is derived totally from eight residues each of glycine and succinyl CoA.
- Four enzymatic steps are catalyzed by enzymes in mitochondria and four in the cytosol. Regulation of heme synthesis is at the δ-aminolevulinic acid synthase reaction, the first biosynthetic step.
- Derangements of porphyrin metabolism are known clinically as the porphyrias.
- Heme catabolism occurs in the endoplasmic reticulum of reticuloendothelial cells with the formation of bilirubin, CO, and iron. Bilirubin is transported to the liver on albumin where it is conjugated with glucuronic acid. Bilirubin glucuronide is excreted in bile into the intestines, where some is further metabolized by bacteria.

19.1 • INCORPORATION OF NITROGEN INTO AMINO ACIDS

Most Amino Acids Are Obtained from the Diet

The nitrogen found in macromolecules of the body come from the diet, after the nitrogen in the atmosphere has been made available (*fixed*) as ammonia by microorganisms and plants (Figure 19.1).

A healthy adult eating a varied and plentiful diet is generally in **nitrogen balance**, a state where the amount of nitrogen ingested each day is balanced by the amount excreted, resulting in no net change in total body nitrogen. In the well-fed condition, excreted nitrogen comes mostly from excess protein intake or from normal turnover. Protein turnover is defined as replacement of body protein by synthesis and degradation (p. 1103). Under some conditions, the body is either in negative or positive nitrogen balance. In negative nitrogen balance more nitrogen is excreted than ingested. This occurs in starvation and certain diseases. During starvation, carbon chains of amino acids from proteins are needed for gluconeogenesis; ammonia released from amino acids is excreted mostly as urea and is not reincorporated into protein. A diet deficient in essential amino acids (essential amino acids cannot be synthesized *de novo* by humans) also leads to a negative nitrogen balance, since body proteins are degraded and their essential amino acids reutilized, and the other amino acids liberated are metabolized. Negative nitrogen balance may also exist in senescence. Positive nitrogen balance occurs in growing children, who are increasing their body weight and incorporating more amino acids into proteins than they break down (A Closer Look 19.1). Cysteine, tyrosine, and arginine supplementation is sometimes required in premature infants and for people with liver disease. Positive nitrogen balance also occurs in pregnancy and during refeeding after starvation. Nonessential amino acids are synthesized from readily available intermediates (Table 19.1), and all amino acids are metabolized to intermediates of glycolysis, and the TCA cycle, and acetyl CoA (Figure 19.2).

There are many inherited metabolic diseases associated with amino acid metabolism. Information on these diseases may be found at www.ncbi.nlm.nih.gov/Omim. OMIM accession numbers associated with a disease or enzyme are indicated n the text.

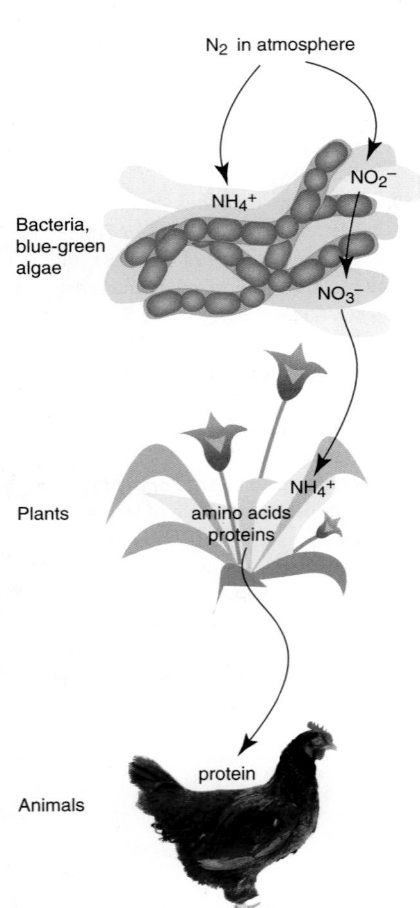

Figure 19.1 Outline of entry of atmospheric nitrogen into the animal diet. This occurs initially by reduction of nitrogen in the atmosphere to ammonia by nitrogen fixation or by reduction of nitrate to ammonia. Nitrate fixation and reduction is carried out by enzymes in microorganisms and plants. Source (photo, brown chicken): PhotoDisc, Inc.

Amino Groups Are Transferred from One Amino Acid to Form Another

Most amino acids used to synthesize protein or as precursors for amino acid derivatives are obtained from the diet or from protein turnover. When necessary, nonessential amino acids are synthesized from α-keto acid precursors by transfer of a preexisting amino group from another amino acid by aminotransferases, also called transaminases (Figure 19.3;

A CLOSER LOOK 19.1

Blood Urea Nitrogen and Measurement of Nitrogen Balance

Measurement of blood urea nitrogen (BUN) is a common test used to assess kidney function. It may be used in cases of kidney disease, or as part of a routine health screening. It is also useful to assess kidney function before administration of a wide range of drugs (such as allopurinol, indomethacin, thiazide diuretics, and tetracyclines) that have the potential for kidney damage. Creatinine (p. 789) may also be assessed along with BUN. Increases in BUN may be associated with heart failure, severe burns, dehydration, and obstructed urine flow. Excess protein intake also raises BUN. The normal range for the ratio of BUN to serum creatinine is 10:1 to 15:1.

Nitrogen balance is a measure of the difference between nitrogen intake (in the form of protein) and nitrogen excretion. During growth and tissue repair, especially after surgery and trauma, the body is in positive N balance; that is, intake is greater than loss. In fevers, fasting, and wasting diseases (cachexia), the loss is greater than the intake, the individual is in negative balance, and there is a net loss of protein from the body. A key relationship between protein intake and nitrogen output is the conversion of grams of protein in the diet into nitrogen equivalents. The average amino acid is considered to have a molecular weight of 120, of which 16% can be attributed (on average) to nitrogen. This allows for a comparison of protein nitrogen consumed with total urine nitrogen (TUN) output as follows:

$$N_2 \text{ balance} = (\text{protein intake (gm)} \times 0.16) - [(\text{total urine nitrogen (gm)}) + 3 \text{ gm})]$$

The 3-gm "fudge factor" represents the nitrogen lost in feces, skin cells, and nails.

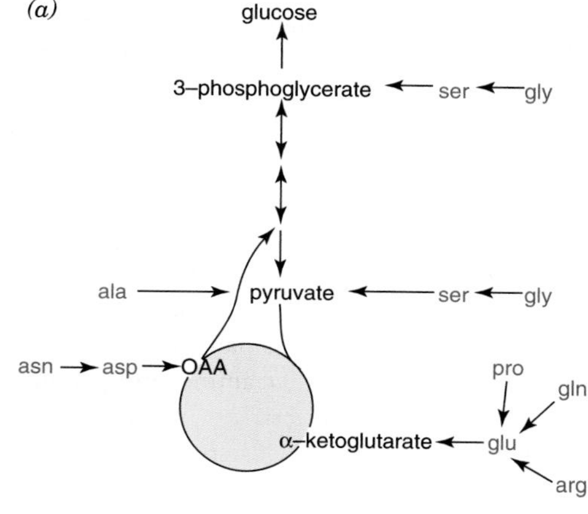

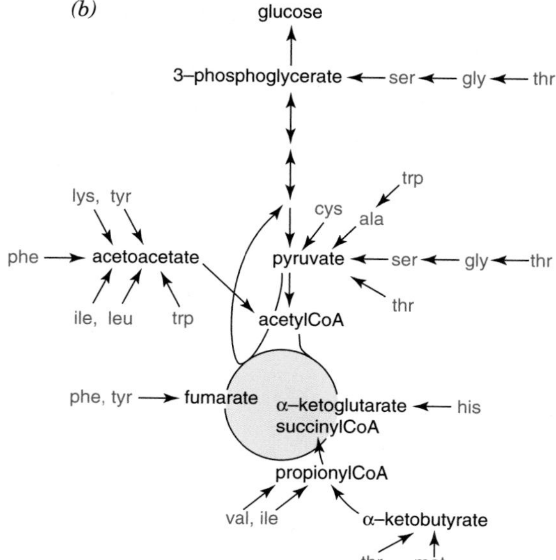

Figure 19.2 Metabolic fate of (*a*) nonessential amino acids and (*b*) essential amino acids plus cysteine and tyrosine.

TABLE 19.1 • Essential and Nonessential Amino Acids

Essential[a]	*Nonessential*
Histidine	Alanine
Isoleucine	Arginine
Leucine	Aspartate
Lysine	Asparagine
Methionine	Cysteine
Phenylalanine	Glutamate
Threonine	Glutamine
Tryptophan	Glycine
Valine	Proline
	Serine
	Tyrosine

[a] Arginine, cysteine, and tyrosine are considered to be "conditional essential"

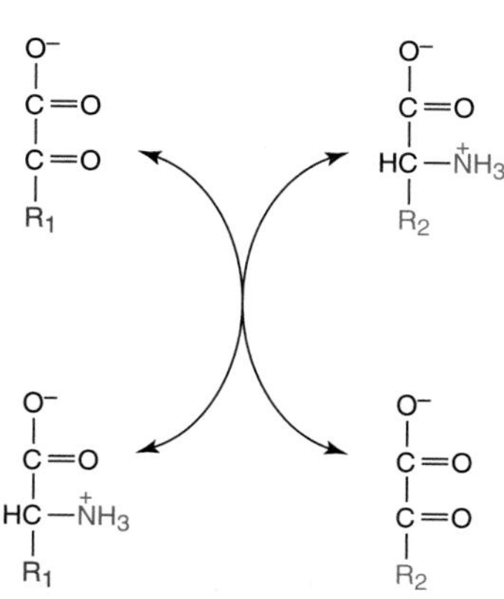

Figure 19.3 Aminotransferase reaction.

A CLOSER LOOK 19.2

Mechanism of Aminotransferases

The carbon in the Schiff base originates in the aldehyde group of pyridoxal phosphate (PLP), and the nitrogen in the lysine residue when PLP is covalently bound to the enzyme. When a substrate amino acid approaches the active site, its amino group displaces the lysine ε-amino group and a Schiff base linkage is formed with the amino group of the amino acid substrate (Figure 19.9). At this point the pyridoxal-phosphate-derived molecule is no longer covalently attached to the enzyme but is held in the active site only by ionic and hydrophobic interactions between it and the protein. The Schiff base linkage with the substrate amino acid substrate is in tautomeric equilibrium between an aldimine, $—CH{=}N—CHR_2$, and a ketimine, $—CH_2—N{=}CR_2$. Hydrolysis of the ketimine liberates an α-keto acid, leaving the amino group as part of the pyridoxamine structure. A reversal of the process is now possible; an α-keto acid reacts with the amine group, the double bond is shifted, and then hydrolysis liberates an amino acid. Pyridoxal phosphate now reforms its Schiff base with the "resting" enzyme (Figure 19.8). Many pyridoxal-phosphate-dependent reactions involve transamination, but the ability of the Schiff base to transfer electrons between different atoms allows the cofactor to participate when other groups, such as carboxyls, are to be eliminated. Figure 19.10 (p. 755) shows the reaction of a pyridoxal-phosphate-dependent decarboxylase and an α,β-elimination.

A Closer Look 19.2). Transfer of amino groups also occurs during degradation of amino acids. Figure 19.4 shows how the amino group of alanine is transferred to α-ketoglutarate to form glutamate. The pyruvate produced then provides carbons for gluconeogenesis or for energy production via the TCA cycle. This reaction is necessary since ammonia cannot enter the urea cycle directly from alanine but the amino group of glutamate can be utilized. The reverse reaction would occur if there were a need for alanine for protein synthesis that was not being met by dietary intake or protein turnover. Transamination involving essential amino acids is normally unidirectional since the body cannot synthesize the equivalent α-keto acid. Figure 19.5 (solid line) shows removal of the nitrogen from valine, an essential amino acid, by transamination, during catabolism. The resulting α-ketoisovalerate is further metabolized to succinyl CoA. Transamination is the most common reaction involving free amino acids, and only threonine and lysine do not participate in an aminotransferase reaction. An obligate amino and α-keto acid pair in all transamination reactions is glutamate and α-ketoglutarate. This means that amino group transfer between alanine and aspartate would have to occur via coupled reactions, with a glutamate intermediate (Figure 19.6). The equilibrium constant for aminotransferases is close to 1 so that the reactions are freely reversible. When nitrogen excretion is impaired and hyperammonemia occurs, as in liver failure, amino acids, especially branched-chain amino acids, can be replaced in the diet by α-keto acid analogs. The α-keto acids are transaminated to produce the corresponding amino acids. Figure19.5 (dotted line) shows valine formation after administration of α-ketoisovalerate as therapy for hyperammonemia.

Tissue distribution of some of the aminotransferases is used diagnostically by measuring the release of a specific enzyme during tissue damage; for instance, an increase in **glutamate aspartate aminotransferase** (**AST**, aspartate transaminase; previously SGOT, serum glutamate oxaloacetate transaminase) and **glutamate alanine aminotransferase** (**ALT**, alanine transaminase; previously SGPT, serum glutamate pyruvate transaminase) in plasma is a sign of liver damage.

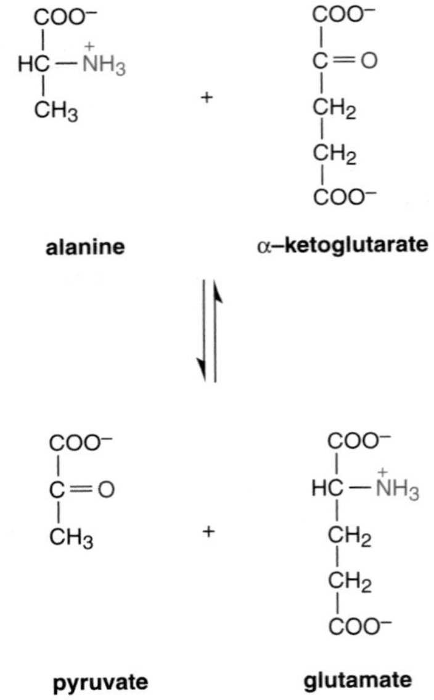

Figure 19.4 Alanine aminotransferase reaction.

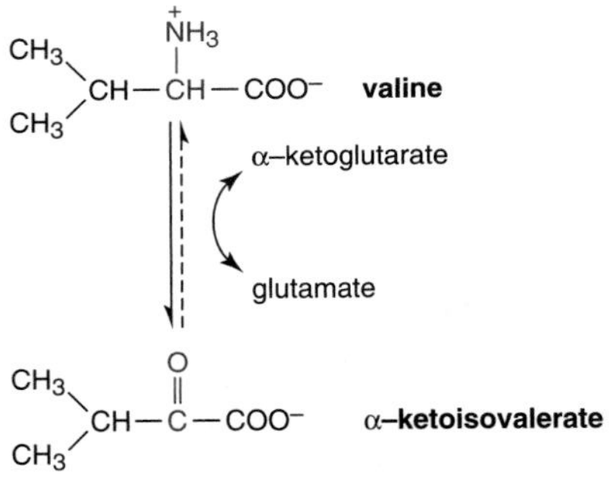

Figure 19.5 Transamination of valine. Valine can be formed from α-ketoisovalerate only when this compound is administered therapeutically.

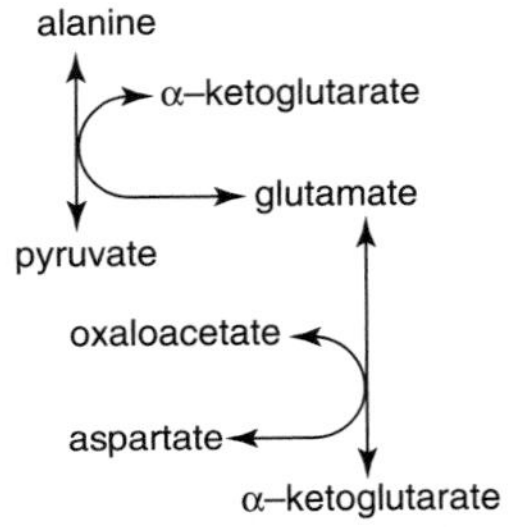

Figure 19.6 Coupled transamination reaction.

Pyridoxal Phosphate Is Cofactor for Aminotransferases

Transfer of amino groups occurs via enzyme-associated intermediates derived from **pyridoxal phosphate**, the functional form of vitamin B_6 (Figure 19.7). The active site of the "resting" aminotransferase contains pyridoxal phosphate covalently attached to a ε-amino group of a lysine residue of the enzyme (Figure 19.8). The complex is further stabilized by ionic and hydrophobic interactions. The linkage, —CH=N—, is called a Schiff base. The reaction occurs by a double-displacement (*ping-pong*) mechanism (Figure 19.9). The effective concentration of vitamin B_6 in the body may be decreased by administration of certain drugs, such as the antituberculosis drug, isoniazid, which forms a Schiff base with pyridoxal thus making it unavailable for catalysis. Pyridoxal phosphate plays an important role in many transferase reactions (Figure 19.10).

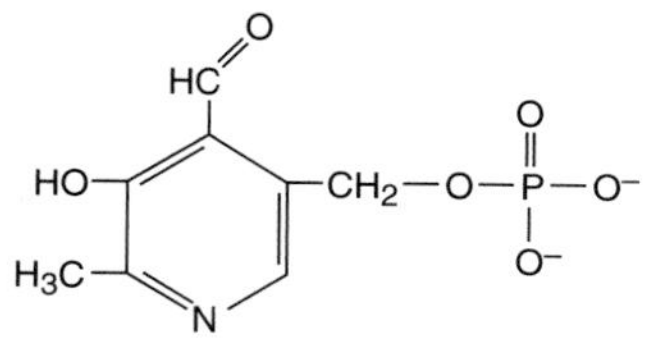

Figure 19.7 Pyridoxal phosphate.

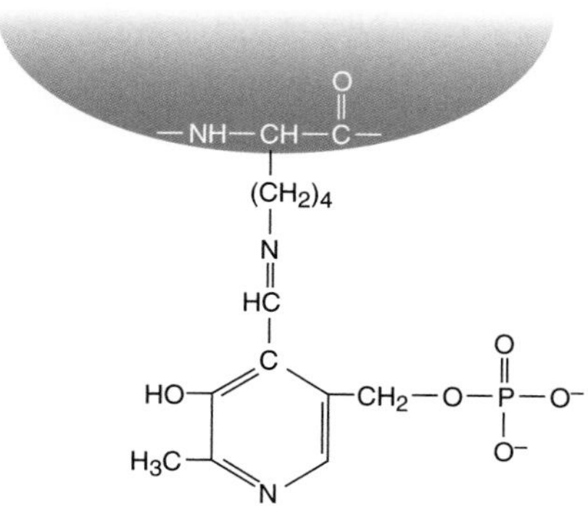

Figure 19.8 Pyridoxal phosphate in aldimine linkage to protein lysine residue.

Glutamate Dehydrogenase Incorporates and Produces Ammonia

In liver ammonia is incorporated into glutamate by glutamate dehydrogenase (Figure 19.11), which also catalyzes the reverse reaction. Glutamate always serves as one of the amino acids in transaminations and is thus the "gateway" between amino groups of most amino acids and free ammonia (Figure 19.12). NADPH is used in the synthetic reaction, whereas NAD^+ is used in liberation of ammonia, a degradative reaction. The enzyme produces ammonia from amino acids when these are needed as glucose precursors or for energy. Formation of NADH during the oxidative deamination reaction is a welcome bonus, since it can be reoxidized by the respiratory chain with formation of ATP. The reaction as shown is readily reversible in the test tube but it probably occurs more frequently in the direction of ammonia formation. The concentration of ammonia needed for production of glutamate is toxic and under normal conditions would rarely be attained except in the perivenous region of the liver.

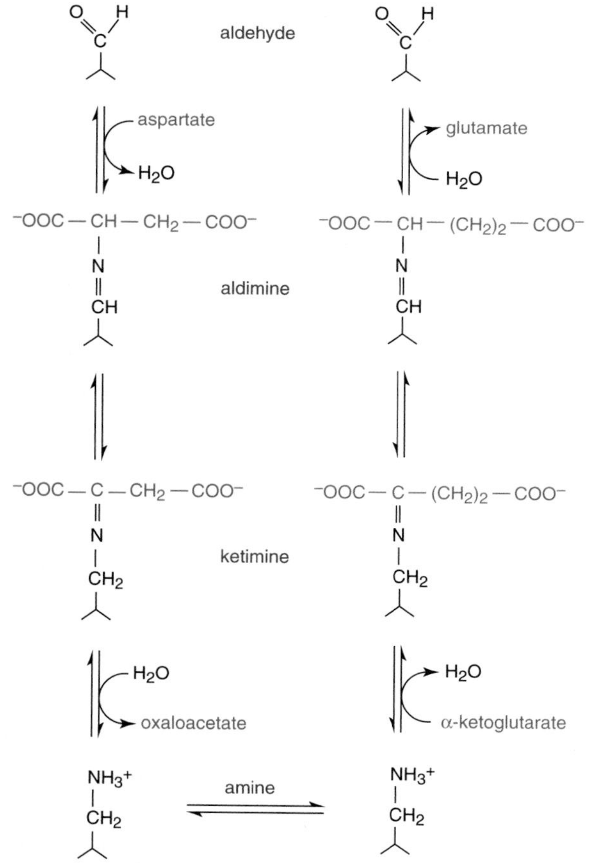

Figure 19.9 Different forms of pyridoxal phosphate during a transamination reaction.

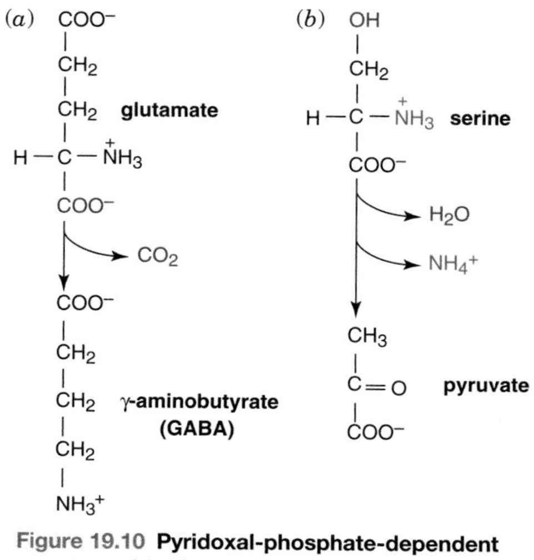

Figure 19.10 Pyridoxal-phosphate-dependent reactions. (*a*) Glutamate decarboxylase. (*b*) Serine dehydratase.

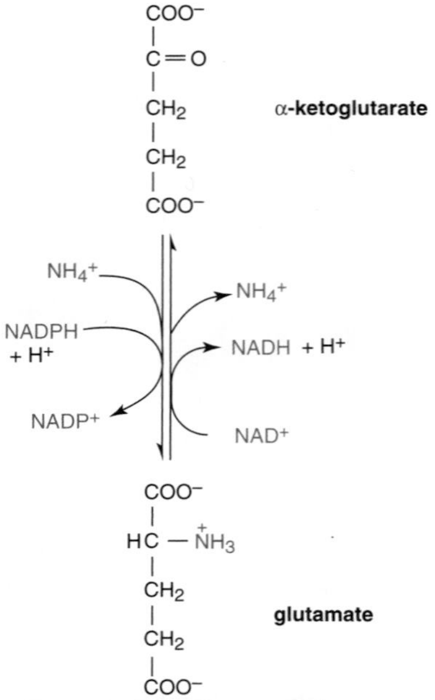

A major source of ammonia is bacterial metabolism in the intestinal lumen from where it is transported to the liver. The enzyme's dominant role in ammonia removal is emphasized by its location in liver mitochondria, where the initial reactions of the urea cycle occur.

Glutamate dehydrogenase is regulated allosterically by purine nucleotides. When there is need for oxidation of amino acids for energy, the activity is increased in the direction of glutamate degradation by ADP and GDP, which are indicative of a low cellular energy level. GTP and ATP, indicative of an ample energy level, are allosteric activators in the direction of glutamate synthesis (Figure 19.13).

Figure 19.11 Glutamate dehydrogenase reaction.

Free Ammonia Is Incorporated into and Produced from Glutamine

Free ammonia is toxic and is preferentially transported in blood in the form of amino or amide groups. The most abundant circulating amino acid is glutamine, an ammonia transporter. The amide group of glutamine is a nitrogen donor for several classes of molecules, including purine bases, and the amino group of cytosine. Glutamate and ammonia are substrates for glutamine synthetase (Figure 19.14). ATP is needed for activation of the γ-carboxyl group to make the reaction energetically favorable.

Congenital glutamine synthetase deficiency leads to severe brain malformations and death. Removal of the amide group is catalyzed by glutaminase (Figure 19.15) that has tissue-specific isozymes.

Amide Group of Asparagine Is Derived from Glutamine

The amide group of asparagine comes from that of glutamine (Figure 19.16), and not from free ammonia, as in the synthesis of glutamine. ATP is needed to activate the receptor carboxyl group. Asparagine is readily synthesized in most cells, but some leukemic cells

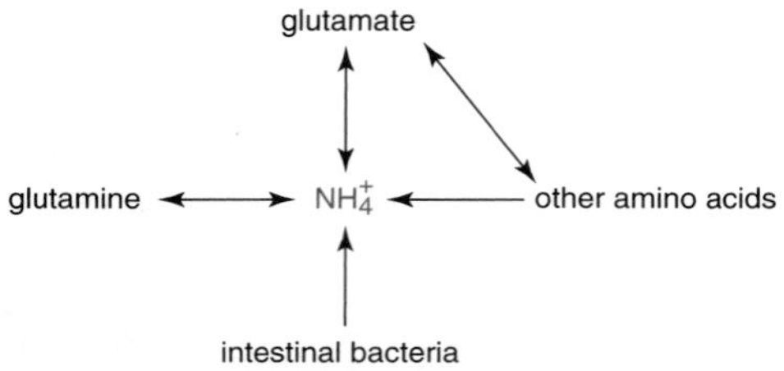

Figure 19.12 Role of glutamate in amino acid synthesis, degradation, and interconversion.

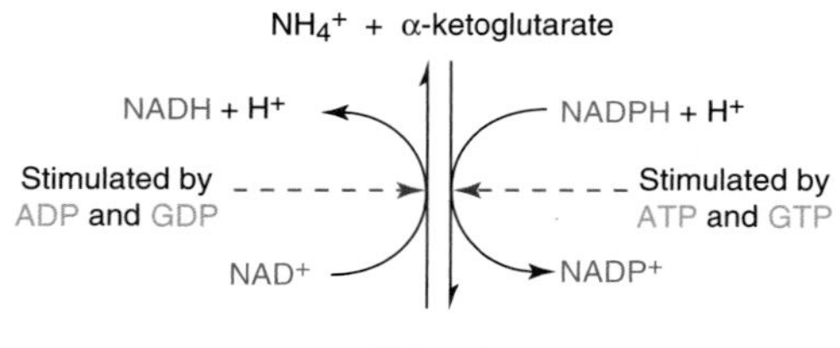

Figure 19.13 Allosteric regulation of glutamate dehydrogenase.

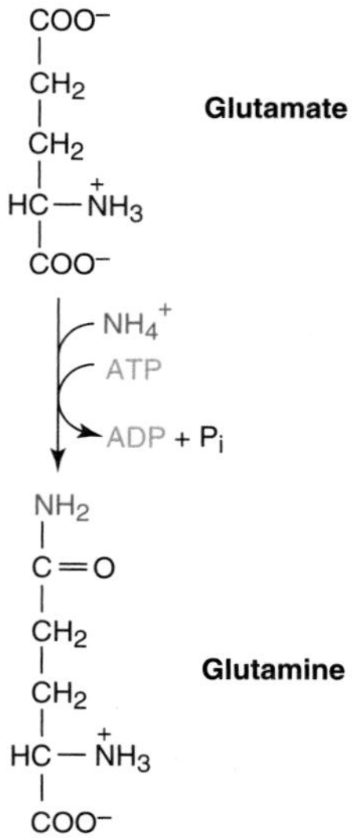

Figure 19.14 Reaction catalyzed by glutamine synthetase.

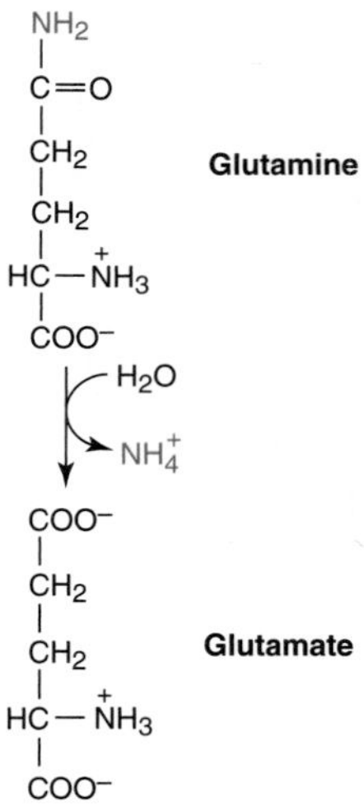

Figure 19.15 Reaction catalyzed by glutaminase.

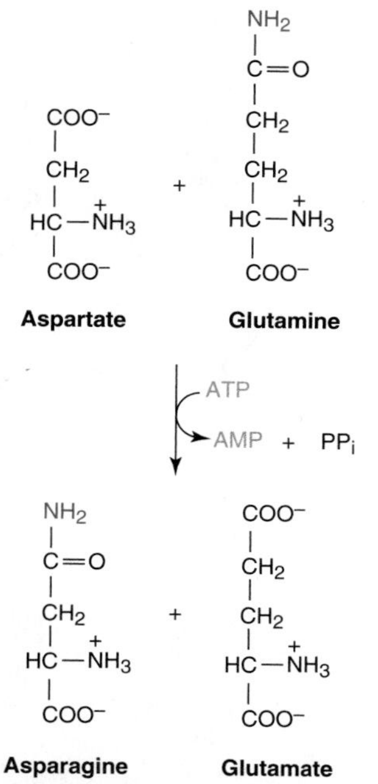

Aspartate + Glutamine → (ATP → AMP + PP_i) → Asparagine + Glutamate

Figure 19.16 Synthesis of asparagine.

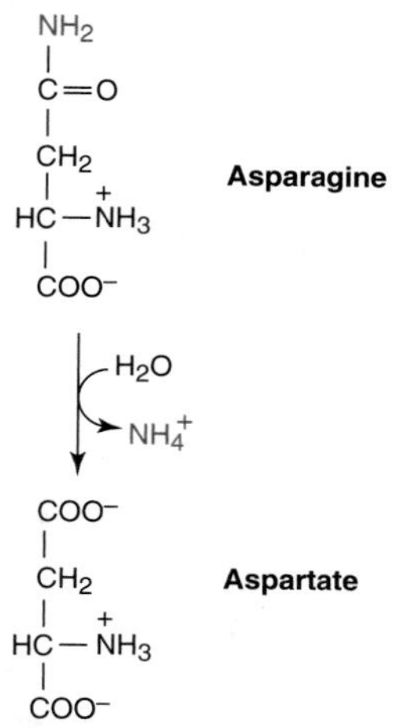

Asparagine → (H_2O → NH_4^+) → Aspartate

Figure 19.17 Reaction catalyzed by asparaginase.

seem to have lost this ability. A therapeutic approach that has been used for over 30 years in patients with asparagine synthetase-deficient leukemia is treatment with exogenous asparaginase to hydrolyze the blood-borne asparagine on which these cells rely (Figure 19.17).

Amino Acid Oxidases Remove Amino Groups

Many amino acids are substrates for **L-amino acid oxidase** (Figure 19.18). The significance of this reaction in metabolism is uncertain, but appears to be small. The enzyme contains flavin mononucleotide (FMN) and produces hydrogen peroxide. Catalase metabolizes the hydrogen peroxide to oxygen and water. The final products are an α-keto acid, ammonia, and water, the same products as for the glutamate dehydrogenase reaction. In the amino acid oxidase reaction, unlike the glutamate dehydrogenase reaction, there is no production of NADH, and therefore no production of ATP.

A D-amino acid oxidase occurs in human cells. Very little of the D-amino acid isomer is found in humans and the enzyme may degrade D-amino acids derived from intestinal bacteria.

19.2 • TRANSPORT OF NITROGEN TO LIVER AND KIDNEY

Protein Is Degraded Constantly

Cells die by necrosis and by programmed cell death, a process termed apoptosis (p. 1010), and their component molecules are metabolized. Individual proteins also undergo regular turnover under normal conditions (p. 248). Regulation of protein degradation is described in Chapter 6. The half-life of a protein can be an hour or less, such as for ornithine decarboxylase, phosphokinase C, and insulin, several months, such as for hemoglobin, collagen, and histones, or the life of the organism, such as for the crystallins of the lens. The majority, however, turn over every few days.

The majority of body protein, and consequently of amino acids, is in skeletal muscle. Under conditions of energy need, this protein is degraded and amino groups from the amino acids are transferred to glutamine and alanine. Some glutamine is transported directly to kidney or liver, but most glutamine is transported to the intestines where it is converted to alanine and ammonia. The alanine from the intestines, muscle, and other extrahepatic tissues is taken to liver. Urea is produced in liver and ammonia (from glutamine)

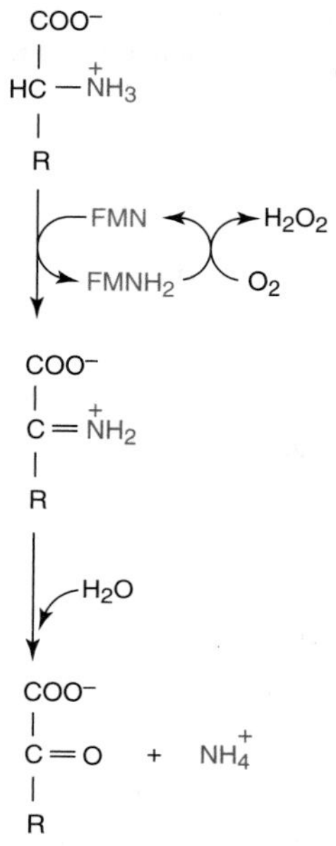

COO^-–HC–$\overset{+}{N}H_3$–R → (FMN → $FMNH_2$; O_2 → H_2O_2) → COO^-–C=$\overset{+}{N}H_2$–R → (H_2O) → COO^-–C=O–R + NH_4^+

Figure 19.18 Reaction of L-amino acid oxidase, a flavoprotein.

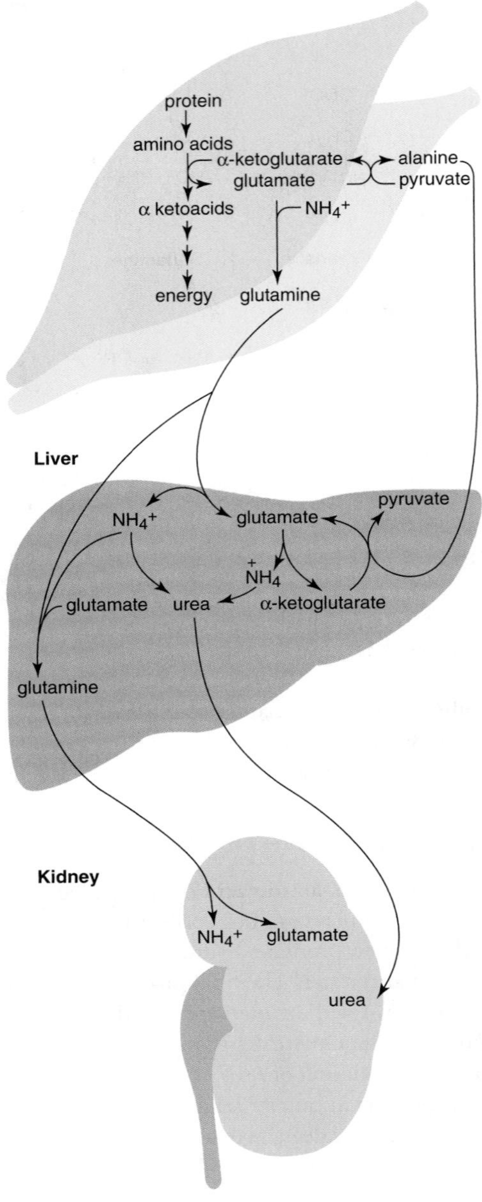

Figure 19.19 Major pathways of interorgan nitrogen transport following muscle proteolysis. Most glutamine goes directly to kidney

in kidneys (Figure 19.19). Carbon skeletons are used for energy or transported to the liver for gluconeogenesis. Muscle protein responds to conditions such as starvation, trauma, burns, and septicemia by undergoing massive degradation. Of the amino acids released, most important as a source of fuel are branched-chain amino acids (valine, leucine, and isoleucine) because steps in their degradation also produce large amounts of NADH and $FADH_2$. The first step in their degradation is transamination, which occurs almost exclusively in muscle. Protein, of course, is degraded throughout the body, but muscle is by far the greatest source of free amino acids for metabolism.

Ammonia Is Released in Liver and Kidney

Ammonia formed in muscle and other extrahepatic organs is transported through the blood circulation in nontoxic form as amino acids, predominantly glutamine and alanine. These two amino acids are the result of addition of ammonia to α-ketoglutarate and pyruvate.

The main destination of blood alanine is the liver (Figure 19.19), where ammonia is released by transamination and by action of glutamate dehydrogenase that releases ammonia, NADH, and α-ketoglutarate, a glucogenic intermediate. Under conditions of energy need, these products are very beneficial. A high circulating glucagon concentration, a signal for the liver to increase gluconeogenesis, increases amino acid uptake by this organ. Some glutamine is also transported to the liver, some is taken up by the kidneys and ammonia is released, protonated to ammonium ion, and excreted. Much of this amino acid is destined for the intestines, where it is converted to alanine and ammonia. Acidosis causes the body to shunt more glutamine to kidneys for excretion as ammonia by suppressing uptake in the liver. This helps to restore pH homeostasis by removing excess hydrogen ions and also by conserving bicarbonate that would otherwise be used in the liver for urea synthesis.

19.3 • UREA CYCLE

Nitrogen Atoms of Urea Come from Ammonia and Aspartate

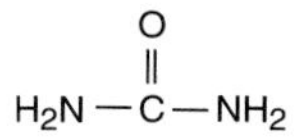

$$H_2N-\overset{\overset{\displaystyle O}{\|}}{C}-NH_2$$

Figure 19.20 Urea.

The urea cycle and the tricarboxylic acid (TCA) cycle were discovered by Sir Hans Krebs and coworkers; the former being described before the latter. In land-dwelling mammals, the urea cycle is the mechanism of choice for nitrogen excretion. The two nitrogens in each urea molecule (Figure 19.20) are derived from free ammonia and the amino group of aspartate respectively. The cycle starts and finishes with ornithine. Unlike the TCA cycle, where carbons of oxaloacetate at the start are different from those at the end, the carbons in the original and the final ornithine are the same. Ammonia (first nitrogen for urea) enters the cycle after condensation with bicarbonate to form carbamoyl phosphate (Figure 19.21), which reacts with ornithine to form **citrulline**. Aspartate (donor of the second urea nitrogen) and citrulline react to form argininosuccinate, which is then cleaved to arginine and fumarate. Arginine is hydrolyzed to urea and ornithine is regenerated. The liver has an intercellular glutamine cycle that serves to scavenge any unused ammonia (A Closer Look 19.3)

Synthesis of Urea Requires Five Enzymes

Carbamoyl phosphate synthetase I (CPSI) is technically not a part of the urea cycle, although it is essential for urea synthesis. Ammonia and bicarbonate are condensed, at the expense of two ATPs, to form carbamoyl phosphate. One ATP activates bicarbonate, and the other donates the phosphate group of carbamoyl phosphate (Figure 19.21). CPSI is found in the mitochondrial matrix, uses ammonia as nitrogen donor, and is absolutely dependent on ***N*-acetylglutamate** for activity (Figure 19.22). Carbamoyl phosphate synthase II (CPSII) is cytosolic, uses the amide group of glutamine, and is not affected by *N*-acetylglutamate. By having separate enzymes in different cellular compartments urea formation can be regulated independently without effect on pyrimidine biosynthesis (p. 821).

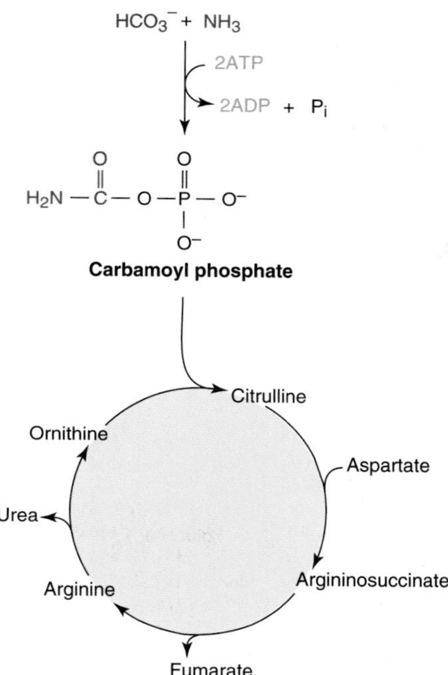

Figure 19.21 Synthesis of carbamoyl phosphate and entry into urea cycle.

Formation of citrulline is catalyzed by **ornithine transcarbamoylase** (Figure 19.23) in the mitochondrial matrix. Citrulline is transported out of the mitochondria into the cytosol where the other reactions of the cycle occur. Argininosuccinate production by argininosuccinate synthetase requires hydrolysis of ATP to AMP and PP_i. This is the equivalent of hydrolysis of two molecules of ATP since PP_i is irreversibly cleaved to 2 P_i. Although this step is very energy requiring, it apparently serves the purpose of maintaining the urea cycle

A CLOSER LOOK 19.3

Intercellular Glutamine Cycle

Liver contains glutamine synthetase and glutaminase but is neither a net consumer nor a net producer of glutamine. The two enzymes are present in parenchymal cells in different segments of the liver. Blood enters the liver lobules through sinusoids that originate in the portal vein. The cells in this periportal area contain glutaminase (and the urea cycle enzymes) and are in contact with blood coming from skeletal muscle. Blood flows along the sinusoids to the central vein of each lobule, from where it flows into the central vein and then to the vena cava and eventually to the kidney. The cells of the perivenous area represent 5% of parenchymal cells and contain glutamine synthetase. This *intercellular glutamine cycle* may be a mechanism for scavenging ammonia that has not been incorporated into urea. The enzymes of urea synthesis are found in the same periportal cells as glutaminase, whereas the uptake of glutamate and α-ketoglutarate for glutamine synthesis predominates in the perivenous region. The intercellular glutamine cycle makes it possible to control flux of ammonia to urea or glutamine, and then to excretion of ammonia by the kidney under different pH conditions (p. 876).

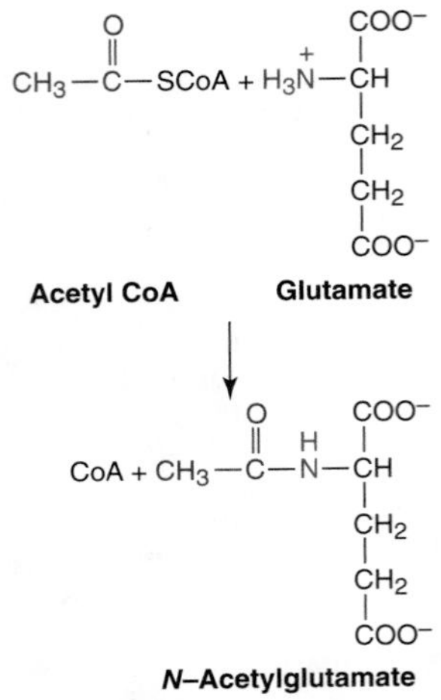

Acetyl CoA Glutamate

N-Acetylglutamate

Figure 19.22 Reaction catalyzed by *N*-acetylglutamate synthase. The enzyme is allosterically stimulated by arginine.

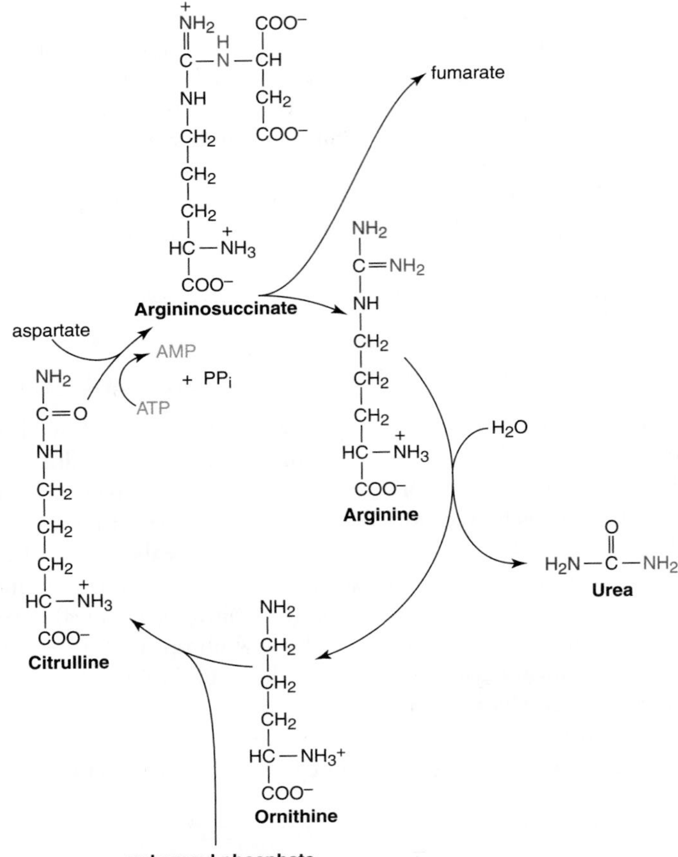

Figure 19.23 Urea cycle.

in the forward direction. Cleavage of argininosuccinate by argininosuccinate lyase produces fumarate and arginine. Arginine is cleaved by **arginase** to ornithine and urea. Ornithine reenters the mitochondrial matrix for another turn of the cycle. The inner mitochondrial membrane contains a citrulline/ornithine exchange transporter.

Fumarate released from the cleavage of argininosuccinate, as mentioned above, may be converted to malate in the cytosol and transported into mitochondria, where it be metabolized to oxaloacetate in the TCA cycle. Oxaloacetate may then be transaminated and transported to the cytosol where it can enter another turn of the urea. Thus the citric acid cycle and the urea cycle are linked.

About two-thirds of the oxaloacetate derived from fumarate is metabolized via oxaloacetate to aspartate or via phosphoenolpyruvate to glucose (Figure 19.24). The amount of fumarate used to form ATP is approximately equal to that required for the urea cycle and gluconeogenesis, meaning that the liver itself gains no net energy. Since humans cannot utilize urea, it is transported to the kidneys for filtration and excretion. Any urea that enters the intestinal tract is cleaved by urease-containing bacteria in the intestinal lumen and the ammonia is absorbed back into the body.

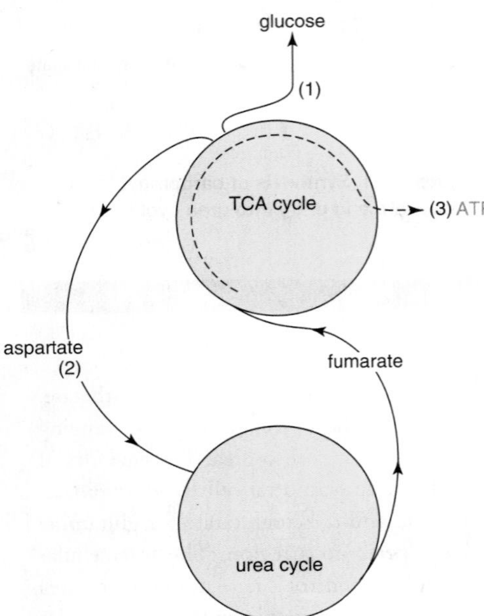

Figure 19.24 Fumarate from the urea cycle is a source of glucose (1), aspartate (2), or energy (3).

Urea Synthesis Is Regulated by an Allosteric Effector and by Enzyme Induction

CPSI requires the allosteric activator *N*-acetylglutamate (see Figure 19.22). This compound is synthesized from glutamate and acetyl CoA by *N*-acetylglutamate synthase (still sometimes called synthetase), that is in turn activated by arginine. Acetyl CoA, glutamate, and arginine are needed to supply intermediates or energy for the urea cycle, and the presence of *N*-acetylglutamate indicates that they are available. Tight regulation is desirable for a pathway that controls the plasma level of ammonia that is potentially toxic, and that is also highly energy dependent.

Induction of urea cycle enzymes (10–20 fold) occurs when delivery of ammonia or amino acids to liver rises. Concentration of intermediates also regulates activity through mass action. A high-protein (net-excess amino acids) diet and starvation (need to metabolize body protein in order to provide carbons for energy production) result in induction of urea cycle enzymes. A deficiency of essential amino acids also promotes excess protein degradation. *N*-acetylglutamate activation is short-term regulation and enzyme induction is long term.

Metabolic Disorders of Urea Synthesis Have Serious Consequences

Metabolic disorders that arise from abnormal function of enzymes of urea synthesis are potentially fatal and cause coma when ammonia concentrations become high. Loss of consciousness may be a consequence of ATP depletion. The major source of ATP is oxidative phosphorylation, which is linked to transfer of electrons from the TCA cycle down the electron transport chain (p. 561). A high concentration of ammonia sequesters α-ketoglutarate as glutamate, thus depleting the TCA cycle of important intermediates and reducing ATP production.

Urea cycle enzyme deficiencies are usually fatal in infants, but adults with a partial deficiency in one or other of the urea cycle enzymes have been found. Therapy for these deficiencies has a fourfold basis: (1) to limit protein intake and potential buildup of ammonia, (2) to remove excess ammonia, (3) to replace any intermediates missing from the urea cycle, and (4) to undergo a liver transplant. The first is accomplished by limiting ingestion of amino acids, replacing them if necessary with the equivalent α-keto acids to be transaminated *in vivo*. The bacterial source of ammonia in the intestines can be decreased by a compound that acidifies the colon, such as lactulose, a poorly absorbed synthetic disaccharide that is metabolized by colonic bacteria to acidic products. This promotes excretion of ammonia in feces as protonated ammonium ions. Antibiotics can also be administered to kill ammonia-producing bacteria. The second is achieved by compounds that bind covalently to amino acids and produce nitrogen-containing molecules that are excreted in urine. Figure 19.25 shows condensation of benzoate and glycine to form hippurate and of phenylacetate and glutamine to form phenylacetylglutamine. Phenylacetate is extremely unpalatable and is given as the precursor sodium phenylbutyrate. Both reactions require energy for activation of the carboxyl groups by addition of CoA. Clin. Corrs. 19.1 and 19.2 give examples of therapy for specific enzyme deficiencies, which often includes administration of urea cycle intermediates.

Gene therapy has been tried to correct urea cycle enzyme deficiencies. However, in 1999 this resulted in the death of a patient, who had only partial ornithine transcarbamoylase activity (p. 759). Gene therapy for urea cycle enzyme has subsequently been halted.

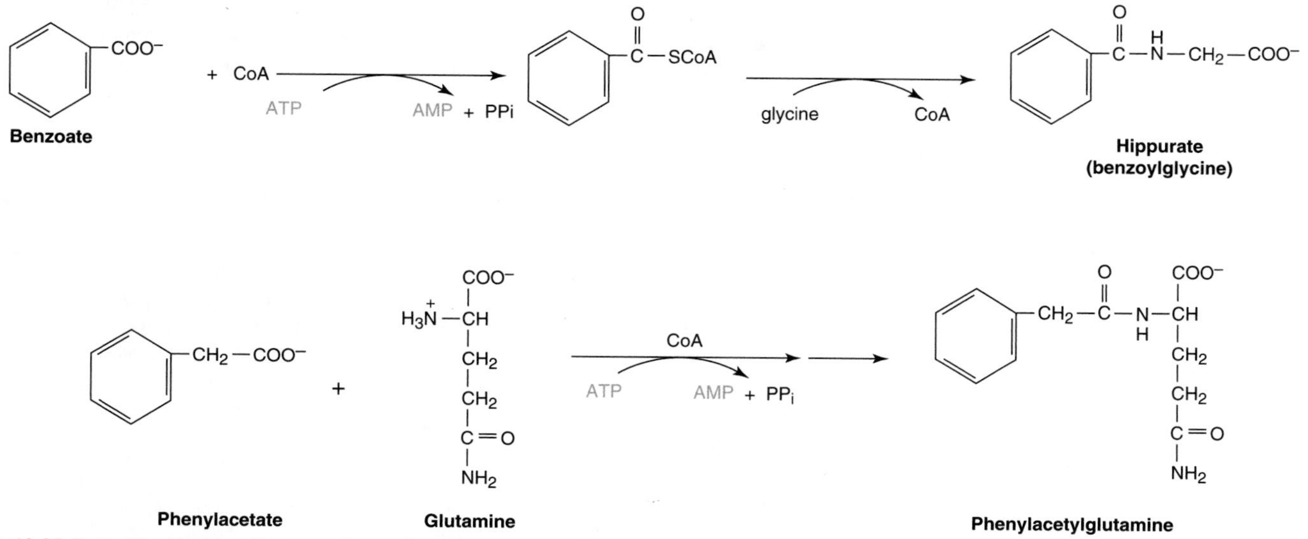

Benzoate + CoA → (ATP → AMP + PPi) → C_6H_5–C(=O)–SCoA → (glycine → CoA) → C_6H_5–C(=O)–NH–CH_2–COO^- **Hippurate (benzoylglycine)**

Phenylacetate (C_6H_5–CH_2–COO^-) + **Glutamine** ($H_3\overset{+}{N}$–CH(COO^-)–CH_2–CH_2–C(=O)–NH_2) → (CoA; ATP → AMP + PP_i) → **Phenylacetylglutamine** (C_6H_5–CH_2–C(=O)–NH–CH(COO^-)–CH_2–CH_2–C(=O)–NH_2)

Figure 19.25 Detoxification reactions as alternatives to the urea cycle.

CLINICAL CORRELATION 19.1

Carbamoyl Phosphate Synthetase and *N*-Acetylglutamate Synthetase Deficiencies

Deficiencies in these enzymes, as for deficiencies in enzymes of the urea cycle, result in hyperammonemia, encephalopathy, and respiratory alkalosis. Carbamoyl phosphate synthetase (CPSI) (OMIM 237300) deficiency exists in two forms. The first is neonatal and fatal, and the second is delayed onset that may manifest itself in childhood or later, frequently by seizures, vomiting, and abdominal pain. The mutations leading to this disease are autosomal recessive, and exist at many loci in the gene. CPSI is allosterically activated by *N*-acetylglutamate, and cases of apparent CPSI deficiency are caused by inactivity of *N*-acetylglutamate synthase (NAG) (OMIM 237310 and 608310). This latter enzyme is activated severalfold by arginine, and arginine therapy has been shown in some cases to ameliorate the inability to make adequate carbamoyl phosphate. NAG deficiency is so uncommon that no incidence data is available.

Caldovic, L., Morizono, H., and Tuchman, M. Mutations and polymorphisms in the human *N*-acetylglutamate synthase (NAGS) gene. *Hum. Mutat.* 28:754, 2007; Finckh, U., Kohlschutter, A., Schafer, H., Sperhake, K., et al. Prenatal diagnosis of carbamoyl phosphate synthetase I deficiency by identification of a missense mutation in CPS1. *Hum. Mutat.* 12:206–211, 1998; and Roth, K. S. *N*-Acetyl Synthetase Deficiency 2003, http://www.emedicine.com/ped/topic10.htm.

CLINICAL CORRELATION 19.2

Deficiencies of Urea Cycle Enzymes

Ornithine Transcarbamoylase Deficiencies

The most common deficiency of a urea cycle enzyme is lack of ornithine transcarbamoylase (OMIM 300461 and 311250). Mental retardation and death often result, but the occasional finding of normal development in treated patients suggests that the mental retardation is caused by excess ammonia before adequate therapy. Many different mutations exist. The gene for ornithine transcarbamoylase is on the X-chromosome, and males are generally more seriously affected than heterozygotic females, most of who show inactivation of one of the X-chromosomes and occasionally exhibit cellular mosaicism. In some cases mutations have been found in the coding region for the leader sequence of the protein, which prevents incorporation of the enzyme in the mitochondria. In addition to ammonia and amino acids appearing in the blood in increased amounts, orotic acid also increases, presumably because carbamoyl phosphate that cannot be used to form citrulline diffuses into the cytosol, where it condenses with aspartate, ultimately forming orotate and pyrimidines (p. 821).

Argininosuccinate Synthetase and Lyase Deficiency

The inability to condense citrulline with aspartate results in accumulation of citrulline in blood and its excretion in urine (citrullinemia). Inheritance is autosomal recessive and about 50% of cases are severe, due to hyperammonemia. There are heterogeneous mutations that cause this deficiency, and three distinct types. In Type I the enzyme usually has an altered Michaelis constant, and the enzyme is affected in both liver and kidney. In Type II the kidney is not affected, and the residual enzyme in liver is kinetically normal. Type III argininosuccinate synthetase (OMIM 215700, 603471, and 605814) deficiency is caused by lack of transcription of the gene. Impaired ability to split argininosuccinate (OMIM 207900 and 608310) to form arginine is also autosomal recessive, and due to a multiplicity of mutations. In patients with early-onset disease therapy with a low-protein diet and arginine supplementation has resulted in good outcomes.

Arginase Deficiency

Arginase (OMIM 207800, 608313, and 107830) deficiency is rare but causes many abnormalities in development and function of the central nervous system. Arginine accumulates and is then excreted. Precursors of arginine and products of arginine metabolism may also be excreted. In severe cases spastic paraplegia may result. Type I affects liver, but not kidney, brain, or intestinal tract. The enzyme involved in Type I argininemia is the enzyme found in cytosol and contributes to the production of urea. The enzyme identified as responsible for Type II argininemia is found in kidney mitochondrial matrix. The formation of nitric oxide from arginine and polyamines from ornithine are affected by a deficiency in this enzyme.

Mitochondrial Ornithine Transporter Deficiency (OMIM 603861 and 608157)

This disease is also called hyperornithemia, hyperammonemia, homocitrullinemia syndrome (HHH Syndrome.) The symptoms include mental retardation, cerebellar, ataxia, and episodic coma. The disease is caused by many different mutations in the gene, including nonsense, missense, and frameshift.

Hypoargininemia in Preterm Infants

Premature births occur before the burst of cortisol production that occurs late in pregnancy. Cortisol is an inducer of arginine-synthetic enzymes, and it is suggested that addition of cortisol to enteral and parenteral feeding of premature infants may improve survival and growth.

Yudkoff, M., Daikhin, Y., Nissim, I., Jawad, A. et al. In vivo nitrogen metabolism in ornithine transcarbamylase deficiency. *J. Clin. Invest.* 98:2167, 1996. Haberle, J., Pauli, S., Linnebank, M., Kleijer, W. J., et al. Structure of the human argininosuccinate synthetase gene and an improved system for molecular diagnostics in patients with classical and mild citrullinemia. *Hum. Genet.* 110:327, 2002. Iyer, R., Jenkinson, C. P., Vockley, J. G., Kern, R. M., et al. The human arginases and arginase deficiency. *J. Inherit. Metab. Dis.* 21:86, 1998. Camacho, J. A., Obie, C., Biery, B., Goodman, B. K., Hu, C. A., et al. Hyperornithinemia-hyperammoniemia-homocitrullinuria syndrome is caused by mutations in a gene encoding a mitochondrial ornithine transporter. *Nature Genet.* 22:151, 1999; and Salvi, S., Santorelli, F. M., Bertini, E., Boldrini, R., et al. Clinical and molecular findings in hyperornithinemia-hyperammonemia-homocitrullinuria syndrome. *Neurology* 57:911, 2001. Wu, G., Jaeger, L., Bazer, F., and Rhoads, J. Arginine deficiency in preterm infants: Biochemical mechanism nutritional implications. *Nutr. Biochem.* 15:442, 2004; and Berry, G., and Steiner, R. Long term management of patients with urea cycle disorders. *J. Pediatr.* 138:S56, 2001.

19.4 • BIOSYNTHESIS OF NONESSENTIAL AMINO ACIDS

Synthesis of glutamate, glutamine, aspartate, asparagine, and alanine has been described (p. 752).

Arginine: Arginine is synthesized by sequential reactions starting in the intestinal epithelial cells, and continuing in proximal tubules of the kidneys (Figure 19.26). In these cells arginase is not expressed. Any deficiencies in the enzymes needed for the urea cycle (except arginase) will also affect synthesis of arginine, hence the necessity of dietary arginine supplementation in urea cycle deficiencies.

Ornithine: The precursor of citrulline and arginine, and proline is synthesized from glutamate. Synthesis starts from glutamate by a reaction that uses ATP and NADH (Figure 19.27) and forms **glutamic semialdehyde**. This spontaneously cyclizes to form a Schiff base between the aldehyde and amino groups, which is then reduced by NADPH to proline. When semialdehyde is transaminated, it forms ornithine (Figure 19.28).

Serine: Synthesis of serine from 3-phosphoglycerate uses phosphorylated intermediates (Figure 19.29). Loss of the phosphate is the last step in the production of this amino acid. Serine plays a significant role in the central nervous system because of its role as a precursor of glycine and D-serine, which are neurotransmitters.

Serine is a precursor of the cofactor-like group in pyruvoyl enzymes (A Closer Look 19.4).Serine is converted reversibly to **glycine** in a reaction that requires pyridoxal

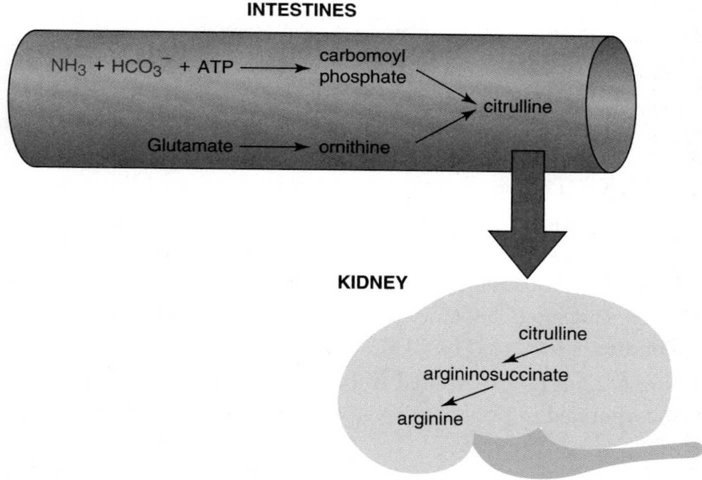

Figure 19.26 Synthesis of arginine in intestines and kidney.

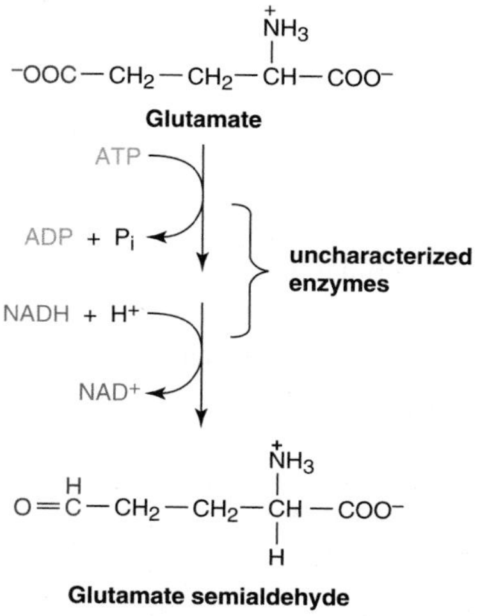

$^{-}OOC-CH_2-CH_2-CH(\overset{+}{N}H_3)-COO^-$

Glutamate

ATP → ADP + P_i

NADH + H^+ → NAD^+

uncharacterized enzymes

$O{=}C(H)-CH_2-CH_2-CH(\overset{+}{N}H_3)-COO^-$

Glutamate semialdehyde

Figure 19.27 Synthesis of glutamic semialdehyde.

$H_2N-CH_2-CH_2-CH_2-CH(\overset{+}{N}H_3)-COO^-$

Ornithine

α–ketoacid / α–amino acid — aminotransferase

$O{=}C(H)-CH_2-CH_2-CH(\overset{+}{N}H_3)-COO^-$

Glutamic semialdehyde

spontaneous

Δ[1]Pyrroline–5–carboxylate

NADPH + H^+ → $NADP^+$ — Pyrroline-5-carboxylate reductase

Proline

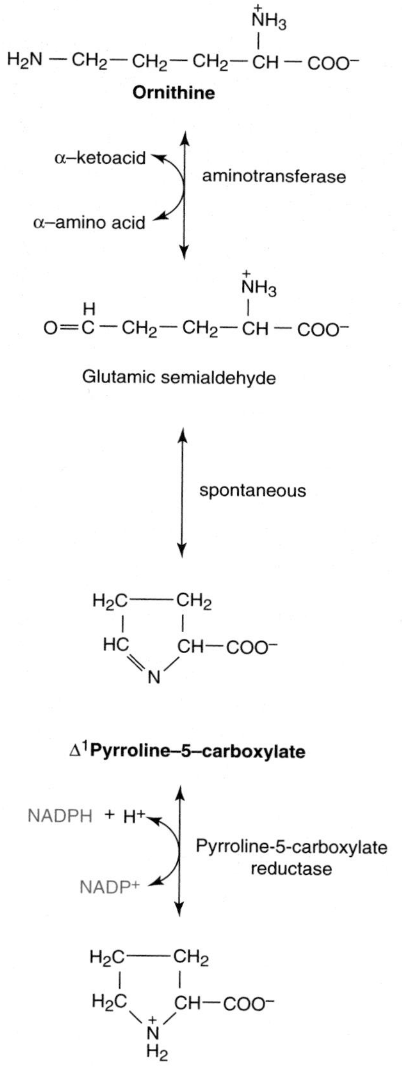

Figure 19.28 Synthesis of ornithine and proline from glutamic semialdehyde, a shared intermediate.

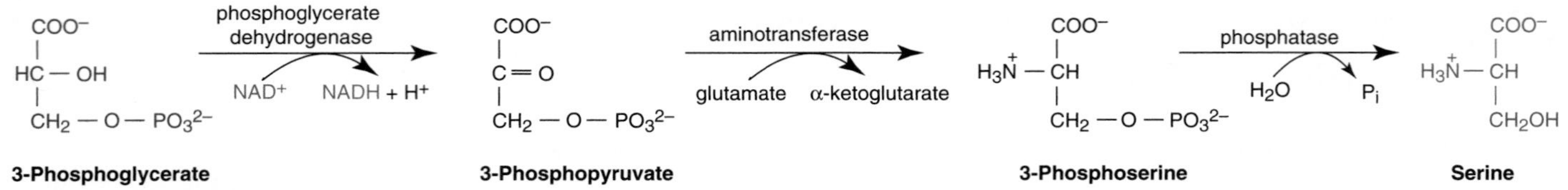

Figure 19.29 Synthesis of serine.

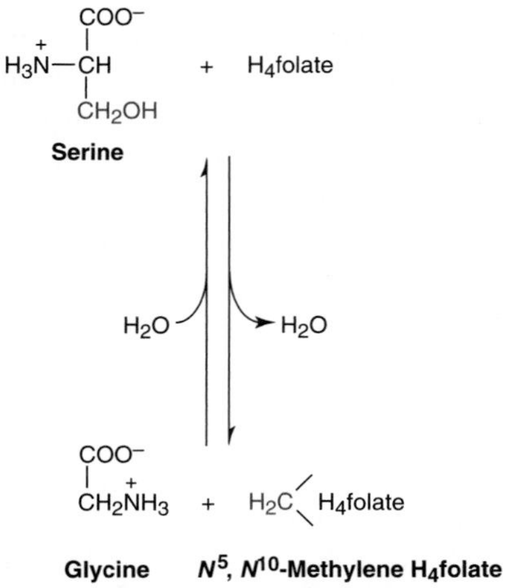

Figure 19.30 Formation of selenocysteinyl tRNA from seryl tRNA is via a phosphoseryl tRNA intermediate.

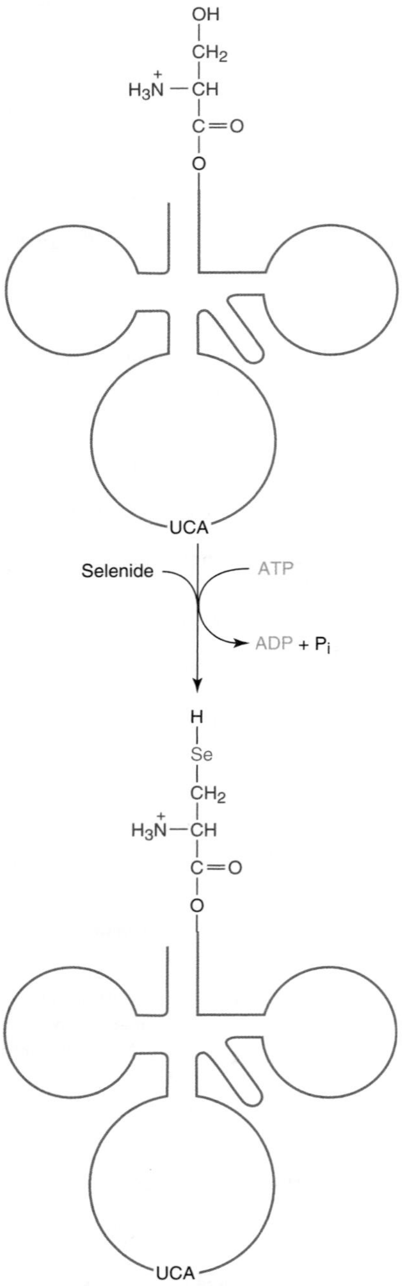

Figure 19.31 Glycine is the product of serine hydroxymethyltransferase.

phosphate and tetrahydrofolate (p. 1079). N^5,N^{10}-methylenetetrahydrofolate (N^5,N^{10}-THF) is produced (Figure 19.30). The demand for serine or glycine and the amount of N^5,N^{10}-THF available determine the direction of this reaction.

Selenocysteine: Serine is the precursor of an unusual but important amino acid, selenocysteine, which is present in some proteins, notably glutathione peroxidase (Figure 19.31). In mRNA for selenoproteins the codon UGA, which usually serves as a termination codon, codes for selenocysteine (Clin. Corr. 19.3). This amino acid is formed from serine after formation of the seryl-tRNA^{Ser} (p. 212).

Pyrrolysine: A 22nd DNA-encoded amino acid, pyrrolysine, was discovered recently. It also is incorporated in response to a stop codon and is found only in methanogenic bacteria.

19.5 • DEGRADATION OF AMINO ACIDS

Nonessential Amino Acids

Glutamate, Alanine, and Aspartate: Glutamate dehydrogenase, a reversible enzyme, deaminates glutamate. The amide groups of glutamine and asparagine are removed by the hydrolytic activity of glutaminase and asparaginase. Arginine is metabolized by arginase to ornithine. Transamination of alanine, aspartate, and glutamate yield pyruvate, oxaloacetate, and α-ketoglutarate respectively.

Glycine: A glycine cleavage complex degrades glycine to CO_2 and ammonia (Figure 19.32). This reaction is reversible in vitro, but not *in vivo,* as the K_m values for ammonia and N^5,N^{10}-THF are much higher than their respective physiological concentrations. The enzyme system for cleavage of glycine (glycine cleavage system) has several component enzymes and is confined to the mitochondria. Glycine encephalopathy, also called nonketotic hyperglycinemia (NKH), may be due to a defect in any one of these enzymes (Clin. Corr. 19.4, p. 766).

A CLOSER LOOK 19.4

Pyruvoyl Enzymes

A serine residue in some enzymes is modified to form a prosthetic group (see figure). In humans, the only example found so far is *S*-adenosylmethionine decarboxylase. The enzyme is synthesized in precursor form that is then cleaved autocatalytically between a glutamate and a serine residue to form two polypeptides. During cleavage the new N-terminal serine is converted into a pyruvoyl group that functions in decarboxylation by forming a Schiff base with the amino group of *S*-adenosylmethionine.

The figure shows the formation of *S*-adenosylmethionine decarboxylase with covalently bound pyruvoyl prosthetic group.

precursor protein

NH_3

β-subunit with glutamate carboxy terminus

α-subunit showing pyruvoyl group derived from serine

CLINICAL CORRELATION 19.3

Selenoproteins

Human selenoproteins include glutathione peroxidase-1, thioredoxin reductase, glutathione peroxidase-2, glutathione peroxidase-3, thyroxine deiodinase type 1, and mitochondrial capsule selenoprotein. Most of the known selenoproteins are members of the glutathione peroxidase or iodothyronine deiodinase families. Selenoprotein P (SEPP1) (OMIM 601484) is a major selenoprotein that is not a member of those families. It is an extracellular glycoprotein that is present in several isoforms and is the only selenoprotein known to contain multiple selenocysteine residues. It acts as a heparin-binding protein that appears to be associated with endothelial cells. Bifunctional SEPP1 serves to provide selenium to proliferating cells and also for glutathione-dependent reduction of phosphatidylcholine hydroperoxide in the extracellular space. The shift in usage of the UGA codon from a stop codon to a signal for selenocysteine incorporation is mediated by at least five different components. Two of these are cis sequences, one a region in the 3'UTR of the mRNA and the other the UGA codon itself. Three are trans-acting factors, a selenocysteine-specific translation elongation factor (eEFSec), a binding protein for the sequence in the 3'UTR (SBP2), and the selenocysteinyl-tRNA. Statin drugs used as therapy for hypercholesterolemia inhibit the cholesterol synthetic pathway, and therefore reduce synthesis of isopentyl groups. The myopathy and other side effects caused by statins resemble the symptoms of selenium deficiency. This suggests that the mechanism of tissue damage caused by statins may be inhibition of selenoprotein synthesis since selenocysteinyl-$tRNA^{[Ser]Sec}$ requires isopentylation for activity. Chronic inflammation has a pathologic role in many common diseases and is influenced by both genetic and environmental factors. SEPS1 (OMIM 607918) is associated with both TNF-alpha and IL1-beta and there is a direct mechanistic link between SEPS1 and the production of inflammatory cytokines that suggests that SEPS1 has a role in mediating inflammation.

Curran, J. E., Jowett, J. B. M., Elliott, K. S., Gao, Y., et al. Genetic variation in selenoprotein *S* influences inflammatory response. *Nature Genet.* 37:1234, 2005, and Kryukov, G. V., Castellano, S., Novoselov, S. V., Lobanov, A. V., et al. Characterization of mammalian selenoproteomes. *Science* 300:1439, 2003; and Moosmann, B., and Behl, C., Selenoprotein synthesis and side effects of statins. *Lancet* 363:2000, 2004.

Serine: Degradation of serine to 3-phosphoglycerate is similar to its synthesis except that degradation uses unphosphorylated intermediates, addition of a phosphate being the last step. The enzymes of these two pathways, however, are not the same (Figure 19.33). Alternatively serine may be metabolized to pyruvate from serine by **serine dehydratase** with loss of the amino group as NH_4^+ (Figure 19.34). Serine is also metabolized to glycine. Serine is one of the amino acids with various degradative pathways that are determined by physiological needs and conditions.

Ornithine and **Proline**: These two amino acids are degraded via glutamic semialdehyde (Clin. Corrs. 19.5 and 19.6) These reactions, together with the reactions that synthesize proline and ornithine, result in these three amino acids being readily interconvertible (see Figure 19.28, p. 763). Proline is converted back to the Schiff base intermediate, δ-1-pyrroline 5-carboxylate, which is in equilibrium with glutamic

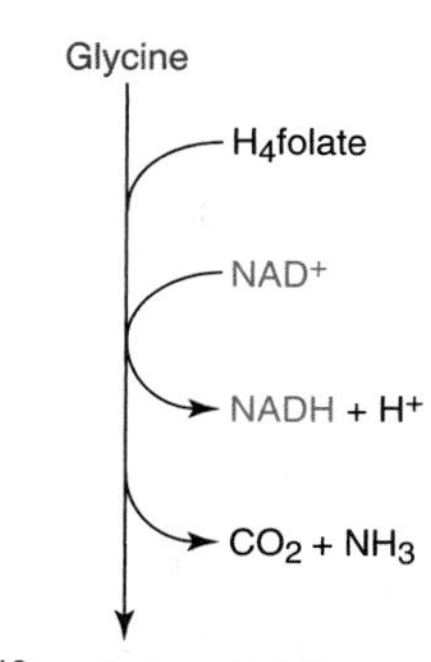

Figure 19.32 Glycine cleavage is pyridoxal-phosphate-dependent.

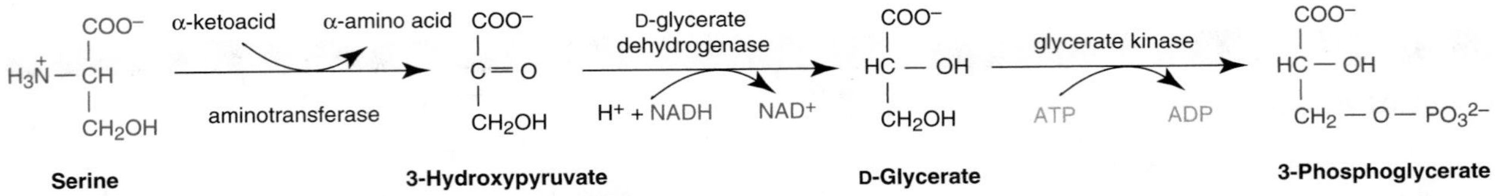

Figure 19.33 Metabolism of serine for gluconeogenesis.

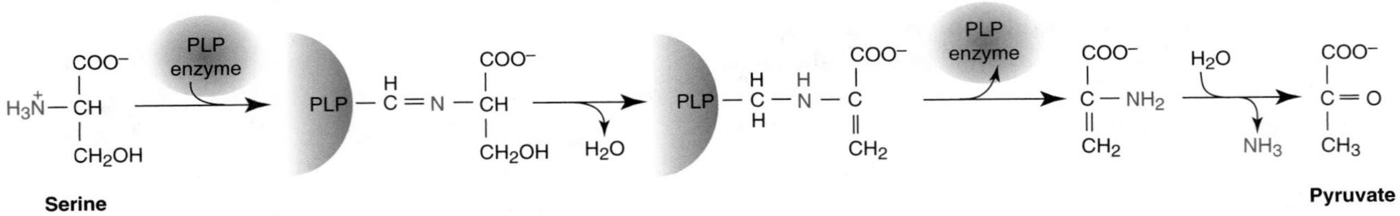

Figure 19.34 Reaction of serine dehydratase requires pyridoxal phosphate.

semialdehyde. The transaminase reaction in the ornithine synthetic pathway is freely reversible and forms glutamic semialdehyde from ornithine. Proline residues that have been post-translationally modified by hydroxylation to form 3-hydroxyproline or 4-hydroxyproline (Figure 19.35) produce glyoxalate and pyruvate, and 4-hydroxy-2-ketoglutarate, respectively.

CLINICAL CORRELATION 19.4

Nonketotic Hyperglycinemia: Glycine Encephalopathy

Neonatal nonketotic hyperglycinemia is characterized by severe mental deficiency and seizures. Many patients do not survive infancy. This very serious disease is to be distinguished from ketoacidosis in abnormalities of branched-chain amino acid metabolism in which the glycine level in the blood is also elevated. Deficiency of glycine cleavage complex has been demonstrated in homogenates of liver from several patients, and isotopic studies *in vivo* have confirmed that this enzyme is not active. Glycine encephalopathy (OMIM 605899) can present at about 6 mon of age or later. The late-onset disease is characterized by mild mental retardation and problems with the central nervous system. The severity of this disease suggests that glycine cleavage is of major importance in the catabolism of glycine. Glycine is a major inhibitory neurotransmitter, which probably explains some neurological complications of the disease. Treatment with ketamine or dextromethorphan, antagonist of the NMDA receptor, has been helpful in some cases.

Hamosh, A., and Johnston, M. V. Nonketotic hyperglycinemia. In Scriver, C. R., Beaudet, A. L., Sly, W. S., Valle, D. *The Metabolic and Molecular Bases of Inherited Disease,* 8th ed. New York: McGraw Hill, 2001. 2:2065.

CLINICAL CORRELATION 19.5

Proline Dehydrogenase Deficiency

A deficiency in proline dehydrogenase (oxidase) (PRODH1), which converts proline back to pyrroline-5-carboxylate, results in high concentrations of proline, glycine, and ornithine in serum and is generally benign, but is sometimes associated with seizures (OMIM 606810 and 237000). There are many allelic variants. The gene is expressed most strongly in human lung, skeletal muscle, and brain, but also in heart, kidney, liver, placenta, and pancreas. Recent studies have found a strong association between this enzyme deficiency and schizophrenia. PRODH2 deficiency causes an increase in hydroxyproline, a metabolite of collagen degradation in plasma and urine (the latter concentration can be up to 100-fold the normal value).

Baron, M. Genetics of schizophrenia and the new millennium: progress and pitfalls. *Am. J. Hum. Genet.* 68:299, 2001; Chakravarti, A. A compelling genetic hypothesis for a complex disease: PRODH2/DGCR6 variation leads to schizophrenia susceptibility. *Proc. Nat. Acad. Sci. USA* 99:4755, 2002; and Kamoun, P., Aral, and Saudubray, J., A new inherited metabolic disease: δ-Pyrroline-5-carboxyl synthetase deficiency. *Bull. Natl. Acad. Med.* 182:131, 1998.

CLINICAL CORRELATION 19.6

Deficiencies in the Glutamic Semialdehyde Pathway

Pyrroline-5-Carboxylate Synthetase

A deficiency in the enzyme that forms glutamic semialdehyde from glutamate by transamination, pyrroline-5-carboxylate synthetase (also found as synthase) (OMIM 138250) has serious symptoms. Hyperprolinemia, hypoornithinemia, and hyperammonemia lead to cataracts, mental retardation, joint lability, and skin hyperelasticity. This latter deficiency is called hyperprolinemia II. The elevated concentration of pyrroline-5-carboxylate in this disease inactivates pyridoxal phosphate, and this leads to neurological complications.

Ornithine Keto Acid Aminotransferase

Ornithine keto acid aminotransferase (OAT or OKT) (OMIM 258870) is a pyridoxal-phosphate-dependent reversible enzyme that converts glutamic semialdehyde to ornithine. A deficiency in the enzyme results in hyperornithemia and gyrate atrophy leading to night blindness. Eventual loss of muscle fibers may be linked to loss of creatine phosphate in muscle, since ornithine deficiency would lead to arginine deficiency. Arginine is necessary for synthesis of creatine. Some types of this enzyme deficiency respond to vitamin B_6 therapy. Protein restriction in the diet is advised.

Baumgartner, M. R., Hu, C. A., Almashanu, S., Steel, G., et al. Hyperammonemia with reduced ornithine, citrulline, arginine and proline: a new inborn error caused by a mutation in the gene encoding delta-1-pyrroline-5-carboxylate synthase. *Hum. Molec. Genet.* 9:2853, 2000; and Hu, C. A., Lin, W. W., Obie, C., and Valle, D. Molecular enzymology of mammalian delta-1-pyrroline-5-carboxylate synthase. *J. Biol. Chem.* 274:6754, 1999.

Essential Amino Acids

Threonine: Threonine is sometimes metabolized to pyruvate (Figure 19.36), but an intermediate in this pathway can undergo thiolysis with CoA to acetyl CoA and glycine. Thus, the α-carbon atom of threonine can contribute to the one-carbon pool through formation of glycine. In a more common pathway, serine dehydratase (p. 766) converts threonine to α-ketobutyrate. A complex similar to pyruvate dehydrogenase metabolizes this to propionyl CoA, which is converted to succinyl CoA (p. 776).

Phenylalanine: Phenylalanine and tyrosine are discussed together since tyrosine is produced by hydroxylation of phenylalanine and is the first product in phenylalanine degradation. Because of this, tyrosine is not usually considered to be essential, whereas phenylalanine is. Normally three-quarters of ingested phenylalanine is hydroxylated to tyrosine by **phenylalanine hydroxylase** (Figure 19.37) an irreversible liver enzyme that is **tetrahydrobiopterin** dependent (Figure 19.38). Biopterin resembles folic acid in containing a pteridine ring but is not a vitamin. It is synthesized from GTP.

Individuals with phenylketonuria (Clin. Corr. 19.7) are unable to convert phenylpyruvate to tyrosine. As a result phenylalanine may accumulate to approximately 20 times the normal level. Some of the excess is transaminated to phenylpyruvate that is in turn converted to phenyllactate and phenylacetate (Figure 19.39). These metabolites are formed in very small amounts in people without phenylketonuria.

Tyrosine: The metabolism of tyrosine starts with transamination by tyrosine aminotransferase to produce *p*-hydroxyphenylpyruvate (Figure 19.40). The enzyme is inducible by glucocorticoids and dietary tyrosine (Clin. Corr. 19.8). Further oxidation of *p*-hydroxyphenylpyruvate leads to homogentisate (Clin. Corr. 19.9). The aromatic ring is next cleaved by the iron-containing homogentisate oxidase to maleylacetoacetate. This is isomerized from *cis* to *trans* to give fumarylacetoacetate by maleylacetoacetate isomerase, an enzyme that seems to require glutathione for activity. Fumarate can be utilized in the TCA cycle for energy or for gluconeogenesis. Acetoacetate can be used as acetyl CoA for lipid synthesis or energy.

Methionine: Methionine is an essential amino acid and therefore no net synthesis occurs, although, as described later, it can be produced in a salvage pathway from **homocysteine**. Cysteine, however, is synthesized by transfer of the sulfur atom derived from methionine to the hydroxyl group of serine. As long as the supply of methionine is adequate, cysteine is nonessential. The disposition of individual atoms of methionine and cysteine is a prime example of how cells regulate pathways to fit their immediate needs for energy or for other purposes.

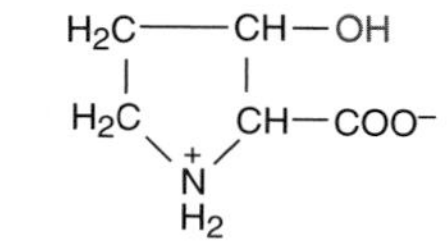

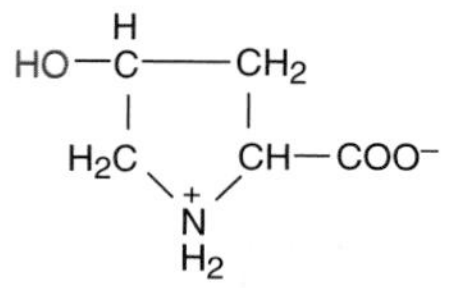

Figure 19.35 **Hydroxyprolines.**

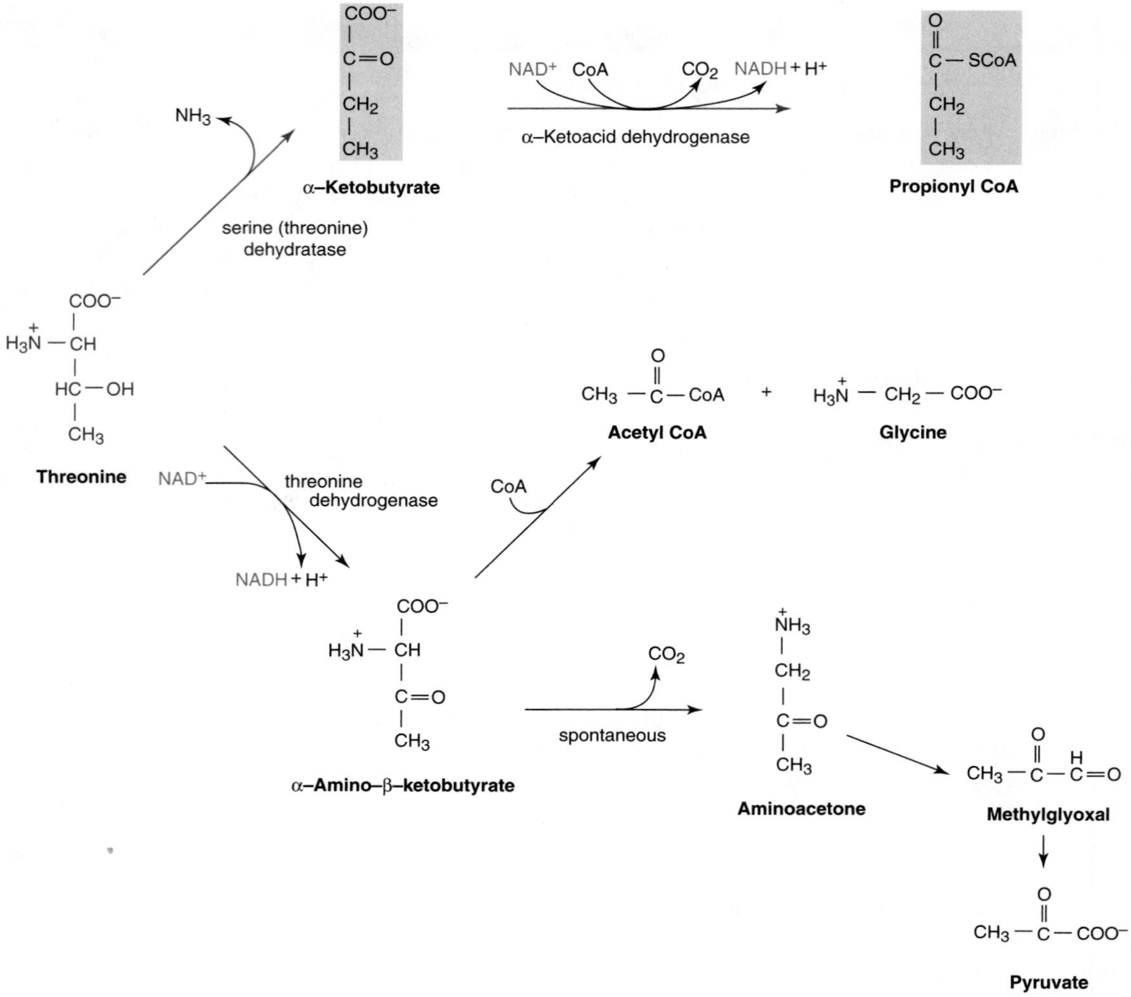

α–Ketobutyrate
NAD+ CoA CO2 NADH + H+
α–Ketoacid dehydrogenase
Propionyl CoA
NH3
serine (threonine) dehydratase
Threonine
Acetyl CoA
Glycine
NAD+
threonine dehydrogenase
NADH + H+
CoA
α–Amino–β–ketobutyrate
CO2
spontaneous
Aminoacetone
Methylglyoxal
Pyruvate

Figure 19.36 Outline of threonine metabolism. Major pathway is in color.

tetrahydrobiopterin dihydrobiopterin
O2 H2O
phenylalanine hydroxylase
Phenylalanine
Tyrosine

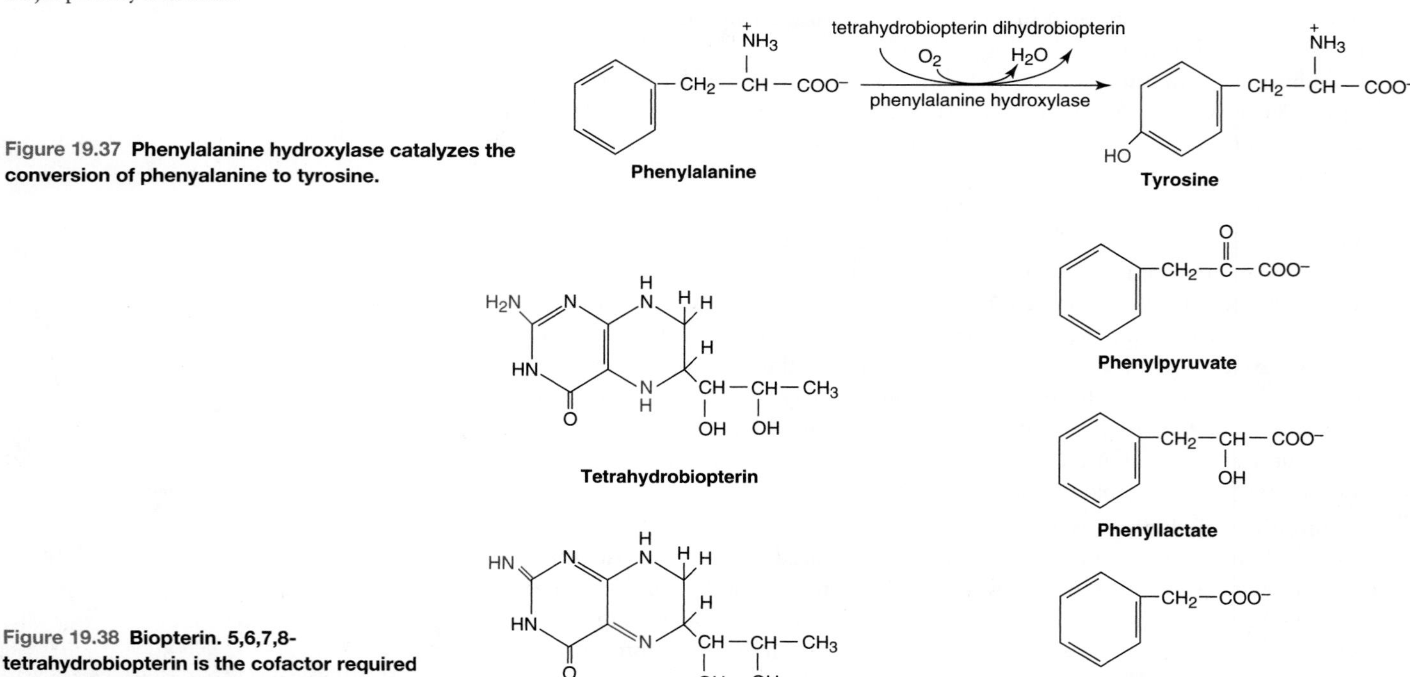

Figure 19.37 Phenylalanine hydroxylase catalyzes the conversion of phenyalanine to tyrosine.

Tetrahydrobiopterin
Dihydrobiopterin

Figure 19.38 Biopterin. 5,6,7,8-tetrahydrobiopterin is the cofactor required for the hydroxylation and is oxidized to 7,8-dihydrobiopterin. Dihydrobiopterin is reduced by dihydrobioterin reductase and NADH.

Phenylpyruvate
Phenyllactate
Phenylacetate

Figure 19.39 Minor products of phenylalanine metabolism.

CLINICAL CORRELATION 19.7

Phenylketonuria

Phenylketonuria (PKU) (OMIM 261600) is the most common disease caused by a deficiency of an enzyme of amino acid metabolism and has been widely studied. The name comes from the excretion of phenylpyruvic acid, a phenylketone, in the urine. Phenyllactate (Figure 19.39), the reduced form of phenylpyruvate, and phenylacetate are also excreted. The latter gives the urine a "mousey" odor. These three metabolites are found only in trace amounts in urine in the healthy person. The disease is autosomal recessive, and over 200 allelic variants have been described. The symptoms of mental retardation, possibly caused by the inhibition of pyruvate carboxylase in brain by high levels of phenylpyruvic acid, associated with this disease can be prevented by a diet low in phenylalanine. Another approach to treatment that has been tried is inclusion of the plant enzyme, phenylalanine ammonia lyase, in the diet. This is an unusually stable enzyme that can metabolize dietary phenylalanine in the digestive tract. Routine screening for PKU is required by governments in many parts of the world. Classical PKU is an autosomal recessive deficiency of phenylalanine hydroxylase. Untreated PKU almost always results in severe neurological symptoms and very low IQ. Babies born to mothers with untreated PKU also show these symptoms.

The characteristic light color of skin and eyes is due to under pigmentation because of tyrosine deficiency. Hypotyrosinemia may also lead to a decrease in catecholamine synthesis. Conventional treatment is by a synthetic diet low in phenylalanine, but including tyrosine, for about 4 to 5 years, followed by dietary protein restriction for several more years or for life. Aspartame, an artificial sweetener, is usually avoided, although it has been shown to increase blood phenylalanine only moderately. About 3% of infants with high levels of phenylalanine have normal hydroxylase but are defective in either synthesis of biopterin or reduction of the oxidized form of the coenzyme.

Platt, B. A., and Clarke, J. T. R. "Hypotyrosinemia" in phenylketonuria. *Molec. Genet. Metab.* 69:286, 2000; Koch, R., Burton, B., Hoganson, G., Peterson, R., et al. Phenylketonuria in adulthood: A collaborative study. *J. Inherit. Metab. Dis.* 25:333, 2002; and Hanley, W. B., Lee, A. W., Hanley, A. J. G., et al. A different approach to treatment of phenylketonuria: phenylalanine degradation with recombinant phenylalanine ammonia lyase. *Proc. Nat. Acad. Sci. USA* 96: 2339, 1999.

CLINICAL CORRELATION 19.8

Tyrosinemias

Tyrosinemia Type I

Tyrosinemia Type I (OMIM 276700) is a defect in fumarylacetoacetase (also called fumarylacetoacetate hydroxylase [FAH]). It is known as hepatorenal tyrosinemia because of the severe liver and kidney damage associated with the disease. The abnormal metabolites produced in this deficiency, succinyl acetone (SA) and succinyl acetoacetone, form Schiff bases with amino acids, primarily lysine. Resulting chromosomal damage is also resistant to repair because SA appears to inhibit DNA ligase I. There is high risk for hepatocarcinoma and chromosome damage. One of the common symptoms of tyrosinemia type I is acute hepatic porphyria. This results from inhibition of porphobilinogen synthetase by SA and is recognized by excretion of δ-aminolevulinic acid. The disease often leads to a need for liver transplantation, but 2-(2-nitro-4-trifluoromethylbenzoyl)-1, 3-cyclohexanedione (NTBC), a strong inhibitor of 4-hydroxyphenylpyruvate dioxygenase, has been shown to reverse the symptoms of tyrosinemia type I by diminishing formation of maleylacetoacetate and fumarylacetoacetate.

Tyrosinemia Type II

Tyrosinemia Type II (OMIM 276600) is an absence or deficiency of tyrosine aminotransferase leads to accumulation and excretion of tyrosine and metabolites. This disease is also known as oculocutaneous tyrosinemia and results in eye and skin lesions and mental retardation. The ulcerated skin lesions can be particularly severe on the soles of the feet. It is treated primarily by a diet low in phenylalanine and tyrosine. Both tyrosinemias types I and II are autosomal recessive and rare.

Tyrosinemia Type III

Tyrosinemia Type III (OMIM 276710) is caused by a deficiency in a recessively inherited disease, caused by a mutation in 4-hydroxyphenylpyruvate dioxygenase (HPD). This results in very high concentrations of tyrosine metabolites in urine. Patients have mild retardation, but no liver damage.

Aponte, J. L., Sega, G. A., Hauser, L. J., Dhar, M. S., et al. Point mutations in the murine fumarylacetoacetate hydrolase gene: Animal models for the human genetic disorder hereditary tyrosinemia type 1. *Proc. Nat. Acad. Sci. USA* 98:641, 2001; Holme, E., and Lindstedt, S. Tyrosinaemia type I and NTBC (2-(2-nitro-4-trifluoromethylbenzoyl)-1,3-cyclohexanedione). *J. Inherit. Metab. Dis.* 21:507, 1998; and Jorquera, R. and Tanguay, R. M. Fumarylacetoacetate, the metabolite accumulating in hereditary tyrosinemia, activates the ERK pathway and induces mitotic abnormalities and genomic instability. *Hum. Molec. Genet.* 10:1741, 2001.

Rettenmeier, R., Natt, E., Zentgraf, H., and Scherer, G. Isolation and characterization of the human tyrosine aminotransferase gene. *Nucleic Acids Res.* 18:3853, 1990; and Tomoeda, K., Awata, H., Matsuura, T., Matsuda, I., et al. Mutations in the 4-hydroxyphenylpyruvic acid dioxygenase gene are responsible for tyrosinemia type III and hawkinsinuria. *Molec. Genet. Metab.* 71:506, 2000.

When excess methionine is present its carbons can be used for energy or for gluconeogenesis, and the sulfur retained as the sulfhydryl of cysteine. Figure 19.41 shows the first step in methionine metabolism, catalyzed by methionine adenosyltransferase. All phosphates of ATP are removed, the product being ***S*-adenosylmethionine** (abbreviated **AdoMet**, or **SAM**). The sulfonium ion is highly reactive, and the methyl is a good leaving group. AdoMet as a methyl group donor will be described below. After a methyltransferase removes the methyl group, the resulting *S*-adenosylhomocysteine is cleaved to produce homocysteine (Figure 19.42). Note that homocysteine is one carbon longer than cysteine (Clin. Corr. 19.10). Although the carbons are destined for intermediary metabolism, the sulfur is conserved through transfer to serine to form cysteine. This requires the pyridoxal-phosphate-dependent cystathionine synthase and cystathionase. Since the bond to form cystathionine is made on one side of the sulfur and that cleaved is on the other side, the result is a transsulfuration. Homocysteine can also directly produce α-ketobutyrate and ammonia (Figure 19.43). α-Ketobutyrate is decarboxylated by a multienzyme complex resembling pyruvate dehydrogenase to yield propionyl CoA, which is then converted to succinyl CoA.

When cells need to resynthesize methionine (Figure 19.44), homocysteine methyltransferase catalyzes the transfer. This is one of two enzymes known to require vitamin B_{12} as cofactor (p. 1082). The methyl group comes from N^5-methyltetrahydrofolate. This is the only reaction known that uses this particular form of tetrahydrofolate as a methyl donor. A minor salvage pathway uses a methyl group from betaine instead of N^5-methyltetrahydrofolate.

Tyrosine

α-ketoglutarate → glutamate

p-Hydroxyphenylpyruvate

O_2 → CO_2 (*p*-hydroxyphenylpyruvate oxidase)

Homogentisate

O_2 (homogentisate oxidase)

Maleylacetoacetate

maleylacetoacetate isomerase

Fumarylacetoacetate

fumarylacetoacetate hydrolase

Fumarate + Acetoacetate

Figure 19.40 **Degradation of tyrosine.**

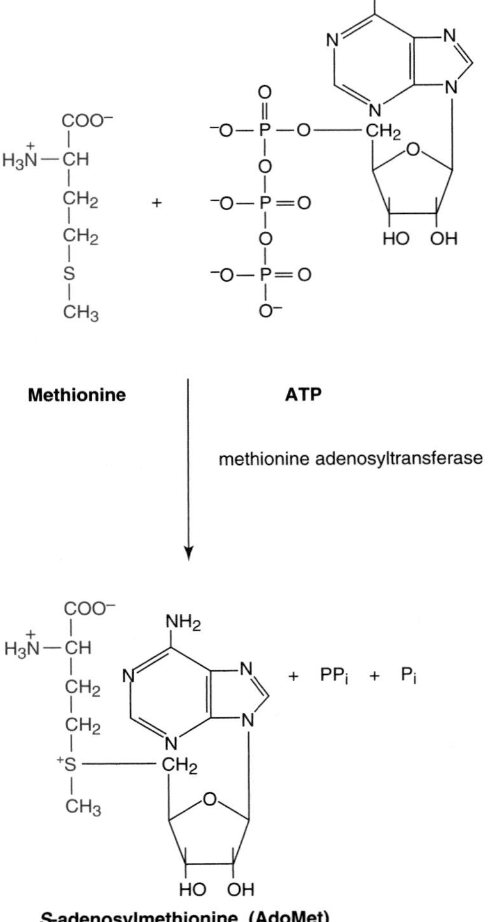

Figure 19.41 **Synthesis of *S*-adenosylmethionine.**

CLINICAL CORRELATION 19.9

Alkaptonuria (OMIM 203500)

The first identified "inborn error of metabolism" was alcaptonuria. Individuals deficient in homogentisate 1,2-dioxygenase (HGD) excrete almost all ingested tyrosine as the colorless homogentisic acid in their urine. Homogentisic acid autooxidizes to the corresponding quinone, which polymerizes to form an intensely dark color. Concern about the dark urine is the only consequence of this condition early in life. Homogentisate in body fluids is slowly oxidized to pigments that are deposited in bones, connective tissue, and elsewhere, a condition called ochronosis because of the ochre color of the deposits. This is thought to be responsible for the associated arthritis, especially of the spine. Nitisinone, a triketone herbicide that inhibits 4-hydroxphenylpyruvate dioxygenase, has been approved for therapy of alkaptonuria. Another inhibitor of this enzyme, 2(-2-nitro-4-trifluoromethylbenzoyl)-1, 3-cyclohexanedione (NTBC) has also been used as therapy. Dietary intervention is also recommended. The study of alkaptonuria by Archibald Garrod, who first indicated its autosomal recessive genetic basis, includes an unusual historical description of the iatrogenic suffering of the first patient treated for the condition, which is frequently benign.

Knox, A. E. Sir Archibald Garrod "inborn errors of metabolism." II. Alkaptonuria. *Am. J. Hum. Genet.* 10:95, 1958; and Phornphutkul, C., Introne, W. J., Perry, M. B., Bernardini, I., et al. Natural history of alkaptonuria. *N. Engl. J. Med.* 347:2111, 2002.

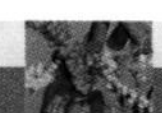

CLINICAL CORRELATION 19.10

Homocysteinemia and Homocysteinuria

Deficiency of cystathionine synthase or lack of pyridoxal phosphate causes homocysteinuria (OMIM 236200). Homocysteine (Hyc) accumulates, and its methylation leads to high levels of methionine (Met). Many minor products of Hyc and Met are formed and excreted. Other causes of increased homocysteine in plasma (homocysteinemia) are folate and vitamin B_{12} deficiency. These vitamin deficiencies prevent salvage of methionine by remethylation of homocysteine. Homocysteinemia (OMIM 603174) may lead to dislocation of the lens of the eye in early life, and other ocular abnormalities often occur. Osteoporosis may develop during childhood, and high homocysteine is a risk factor for osteoporotic fractures in the elderly. Mental retardation is frequently the first indication of high plasma homocysteine, and this has been linked to Alzheimer disease later in life. No mechanism has been firmly established to explain why accumulation of homocysteine should lead to some of the pathological changes. It has, however, been suggested that the associated atherosclerosis results from damage to blood vessels. Homocysteine injures the cells that line arteries and stimulates the growth of smooth muscle cells. Excess homocysteine can form homocysteine thiolactone, a highly reactive intermediate which thiolates free amino groups in low-density lipoproteins (LDL), and causes them to aggregate and be endocytosed by macrophages. The lipid deposits form atheromas. Homocysteine can have other effects, including lipid oxidation and platelet aggregation, which in turn lead to fibrosis and calcification of atherosclerotic plaques. Hyc can also disrupt normal blood-clotting mechanisms, increasing the risk of clots that can bring on a heart attack or stroke. Homocysteine may react with and block lysyl aldehyde groups on collagen and bind to fibrillin-1, producing Marfan-like symptoms. It increases DNA synthesis in vascular smooth muscle cells and in endothelial cells induces cyclins that promote quiescent cells to divide. About one quarter of patients with atherosclerosis who exhibit none of the other risk factors (such as smoking or oral contraceptive therapy) have been found to be deficient in cystathionine synthase activity. Attempts at treatment include restriction of methionine intake and feeding of betaine (or its precursor, choline). In some cases significant improvement has been obtained by feeding pyridoxine (vitamin B_6), suggesting that the deficiency may be caused by more than one type of gene mutation; one type may affect the K_m for pyridoxal phosphate and others may alter the K_m for other substrates, the V_{max}, or the amount of enzyme.

den Heijer, M., Koster, T., Blom, H. J., Bos, G. M. J., et al. Hyperhomocystinemia as a risk factor for deep-vein thrombosis. *N. Engl. J. Med.* 334: 759, 1996; Guttormsen, A. B., Ueland, P. M., Kruger, W. D., Kim, C. E., et al. Disposition of homocysteine in subjects heterozygous for homocystinuria due to cystathionine beta-synthase deficiency: Relationship between genotype and phenotype. *Am. J. Med. Genet.* 100: 204-213, 2001; and Seshadri, S., Beiser, A., Selhub, J., Jacques, P. F., et al. Plasma homocysteine as a risk factor for dementia and Alzheimer disease. *N. Eng. J. Med.* 346: 476, 2002.

Cysteine: Cysteine sulfinate is the major metabolite of cysteine (Figure 19.45). This is converted to sulfite and pyruvate, or to hypotaurine and taurine (secondary metabolites, p. 774; Clin. Corr. 19.11).

Tryptophan: Metabolism of tryptophan has many branch points. The major or oxidative pathway of tryptophan metabolism in humans (Figure 19.46, p. 773) starts with oxidation to *N*-formylkynurenine by a heme-containing enzyme, **tryptophan dioxygenase**. Tryptophan dioxygenase is induced in liver by glucocorticoids and glucagon.

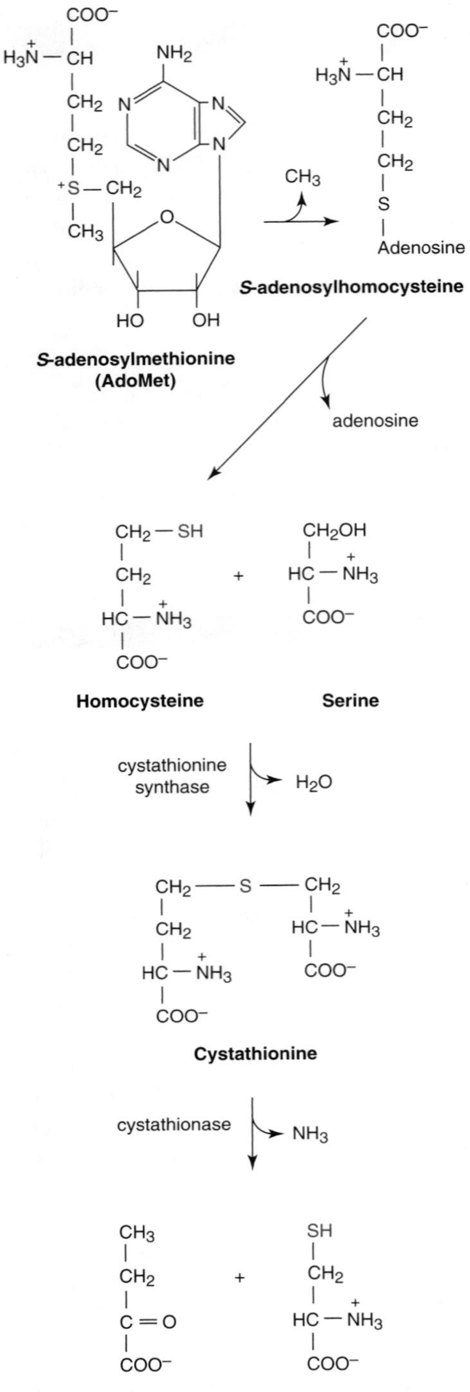

Figure 19.42 **Synthesis of cysteine from *S*-adenosylmethionine.**

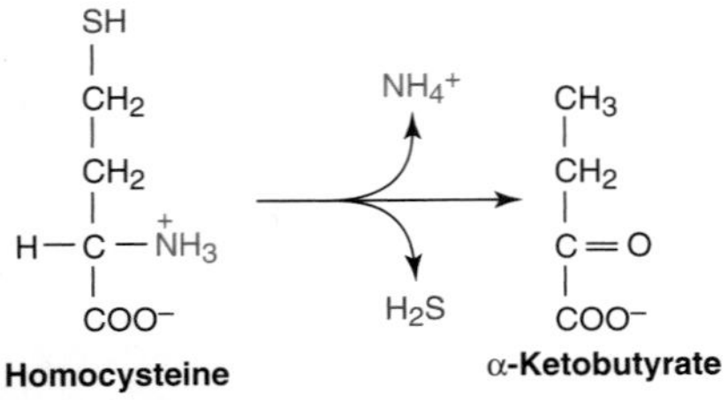

Figure 19.43 **Homocysteine desulfhydrase.**

Other tissues contain a similar enzyme called indolamine dioxygenase, which is less substrate specific. Formamidase then hydrolyzes formylkynurenine to formate and kynurenine. At this point the pathway begins to branch. The dominant pathway leads to 3-hydroxykynurenine, 3-hydroxyanthranilic acid and alanine, amino-carboxymuconic semialdehyde and, by decarboxylation, to aminomuconic semialdehyde. This can be further metabolized in several steps to glutaryl CoA (Clin. Corr. 19.12) and eventually acetoacetyl CoA or recyclized nonenzymatically to picolinic acid, which is excreted in the urine.

Tryptophan has been implicated in the postprandial "Thanksgiving nap" (A Closer Look 19.5, p. 775).

Many enzymes in this lengthy pathway are pyridoxal phosphate dependent. One of these is kynureninase. This enzyme is very sensitive to vitamin B_6 deficiency resulting in excess kynurenine and xanthurenate excretion, which gives urine a greenish-yellow color. This is a diagnostic symptom of vitamin B_6 deficiency.

Lysine: Lysine like leucine is entirely ketogenic. The carbon skeleton enters intermediary metabolism as acetoacetyl CoA. Lysine has an ε- and an α-amino group, which is transferred to α-ketoglutarate by a bifunctional enzyme with an intermediate called

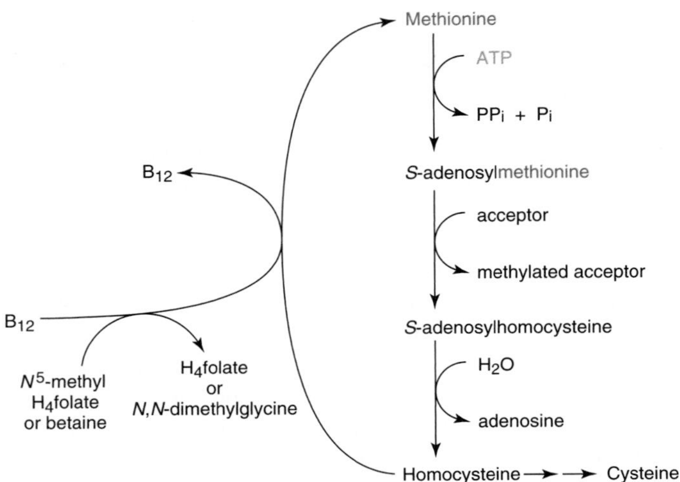

Figure 19.44 **Resynthesis of methionine.** The methyl group goes from folate to methionine via a cobalamin-dependent reaction.

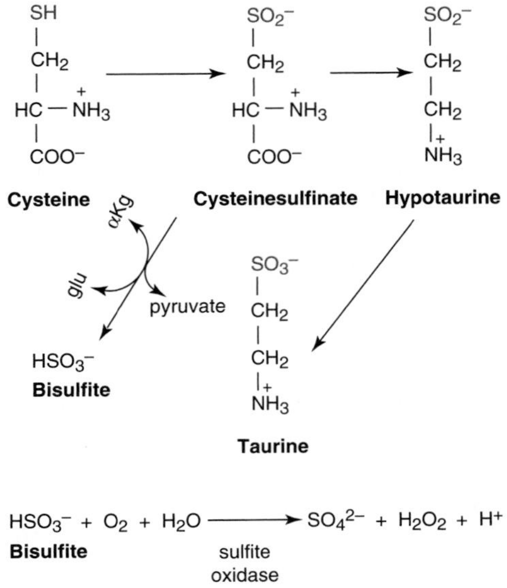

Figure 19.45 **Metabolism of cysteine.**

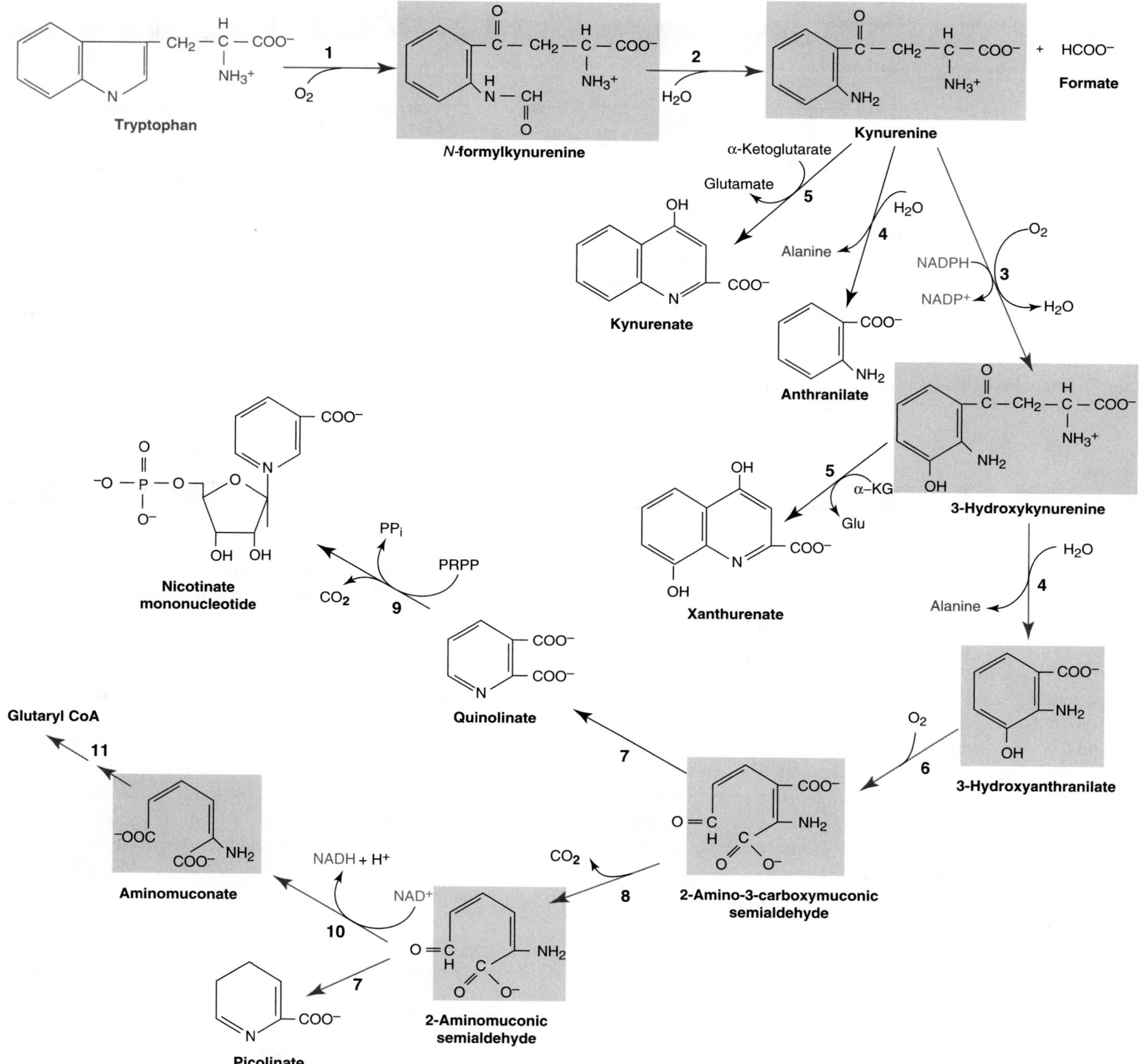

Figure 19.46 Metabolism of tryptophan. Major pathway is shown in colored boxes. Enzymes indicated by number are (1) tryptophan oxygenase, (2) kynurenine formamidase, (3) kynurenine hydroxylase, (4) kynureninase, (5) aminotransferase, (6) 3-hydroxyanthranilate oxidase, (7) spontaneous nonenzymatic reaction, (8) picolinate carboxylase, (9) quinolinate phosphoribosyltransferase, (10) aldehyde dehydrogenase, and (11) complex series of reactions.

saccharopine (Clin. Corr. 19.13, p. 776, and Figure 19.47). Glutamate and a semialdehyde compound are formed. The semialdehyde is then oxidized to a dicarboxylic amino acid, and a transamination of the α-amino group occurs in a pyridoxal-dependent manner. Further reactions lead to glutaryl CoA that is in turn metabolized to acetoacetyl CoA. A minor pathway leads to pipecolate (Figure 19.48, p. 776; A Closer Look 19.6, p. 778).

Histidine: Histidase releases free ammonia from histidine and leaves a compound with a double bond called urocanate (Figure 19.49, p. 776). Two other reactions lead

CLINICAL CORRELATION 19.11

Diseases Involving Cystine

Cystinuria (OMIM 220100), one of the most common of genetic disorders with a prevalence of about 1 in 7000, is a defect of membrane transport of cystine and basic amino acids (lysine, arginine, and ornithine) that results in their increased renal excretion. Extracellular sulfhydryl compounds are quickly oxidized to disulfides. Low solubility of cystine results in crystals and formation of renal calculi, a serious feature of this disease. Treatment is limited to removal of stones, prevention of precipitation by drinking large amounts of water or alkalinizing the urine to solubilize cystine or formation of soluble derivatives by conjugation with drugs. If these treatments do not work, penicillamine is used.

Much more serious is cystinosis (OMIM 219800) in which cystine accumulates in lysosomes. The stored cystine forms crystals in many cells, with a serious loss of function of the kidneys, usually causing renal failure within 10 years. The defect is believed to be in the cystine transporter of lysosomal membranes. Cystinosis is often treated by cysteamine therapy and, when serious enough, by kidney transplant.

Font-Llitjos, M., Jimenez-Vidal, M., Bisceglia, L., Di Perna, M., et al. New insights into cystinuria: 40 new mutations, genotype-phenotype correlation, and digenic inheritance causing partial phenotype. *J. Med. Genet.* 42:58, 2005; Gahl, W. A., Thoene, J. G., and Schneider, J. A. Cystinosis. *N. Engl. J. Med.* 347:111, 2002; Tsilou, E. T., Rubin, B. I., Reed, G., Caruso, R. C., et al. Nephropathic cystinosis: posterior segment manifestations and effects of cysteamine therapy. *Ophthalmology* 113:1002, 2006.

CLINICAL CORRELATION 19.12

Diseases of Glutaric Acid Metabolism

Glutaric Aciduria (OMIM 231670)

Glutaryl CoA is produced from the metabolism of tryptophan and lysine and is further metabolized by the mitochondrial enzyme glutaryl-CoA dehydrogenase to glutaconyl CoA. The earliest sign of a deficiency in this enzyme is microencephalic macrocephaly at birth with subdural hematoma and acute retinal hemorrhage. It may be diagnosed by increasing lysine intake, which makes the aciduria worse, and decreasing protein intake, which reduces the amount of glutaric acid excreted. Tissue degeneration and behavioral arrest follow in the first 2 years of life, often triggered by infection. Glutaric acid causes modulation of glutamatergic and GABAergic neurotransmission, and this may explain the mechanism of neural damage. Newborn screening by enzymatic assay of asymptomatic infants can identify this deficiency, and careful monitoring and treatment with a low lysine/tryptophan diet with carnitine supplementation during crises can prevent about two-thirds of dystonia and dyskinesis. It is believed that some cases of *postencephalitic* cerebral palsy result from glutaryl-CoA dehydrogenase deficiency.

Multiple Acyl-CoA Dehydrogenase Deficiency (MADD) (Glutaric Aciduria II) (OMIM 231680)

Many different acids are excreted due to deficiencies of these flavoprotein deficiencies. They include glutaric, isovaleric, and butyric acids. Mutations have been found in two of the subunits for the flavoprotein found in electron transport, and in flavoprotein dehydrogenase. Any one of these mutations can be responsible for the symptoms found in this disease. These mutations affect fatty acid degradation, metabolism of branched-chain amino acids and lysine, and degradation of choline, as well as metabolism of glutaric acid. Symptoms, which include hypoglycemia, appear in early childhood, and resemble those of poisoning caused by ingestions of unripe ackee fruit. Unripe ackee contains a substance called hypoglycin, which inhibits many acyl-CoA dehydrogenases.

Glutaryl-CoA Oxidase Deficiency (Glutaric Acidemia III) (OMIM 231690)

Excess glutaric acid in the urine usually indicates one of the two diseases just described. Some cases, however, have been observed where tests for these conditions have ruled them out as the cause of the aciduria. In these cases administration of lysine or pipecolic acid increased the excretion of glutaric acid and led to the discovery of a peroxisomal glutaryl-CoA oxidase deficiency. There are often no overt symptoms connected to this condition.

Kolker, S., Koeller, D., Okun, J., and Hoffman, G. Pathomechanisms of neurodegeneration in glutaryl CoA dehydrogenase deficiency. *Ann. Neurol.* 55:7, 2004; and Strauss, K. A., Puffenberger, E. G., Robinson, D. L., and Morton, D. H. Type I glutaric aciduria, Part 1: Natural history of 77 patients. *Am. J. Med. Genet.* 121C:38-52, 2003.

Olsen, R. K. J., Andresen, B. S., Christensen, E., Bross, P., et al. Clear relationship between ETF/ETFDH genotype and phenotype in patients with multiple acyl-CoA dehydrogenation deficiency. *Hum. Mutat.* 22:12, 2003; and Russell, A. P., Schrauwen, P., Somm, E., Gastaldi, G., et al. Decreased fatty acid beta-oxidation in riboflavin-responsive, multiple acyl-coenzyme A dehydrogenase-deficient patients is associated with an increase in uncoupling protein-3. *J. Clin. Endocr. Metab.* 88:5921, 2003.

Knerr, I., Zschocke, J., Trautmann, U., Dorland, L. at al. Glutaric aciduria type III: A distinctive non-disease? *J. Inherit. Metab. Dis.* 25:483, 2002.

to formation of forminoglutamate (FIGLU). The formimino group of FIGLU must then be transferred to tetrahydrofolate before the final product, glutamate, can be produced. When there is insufficient tetrahydrofolate available, this reaction decreases and FIGLU is excreted in urine. This is a diagnostic sign of folate deficiency if it occurs after a test dose of histidine is ingested (Clin. Corr.19.14).

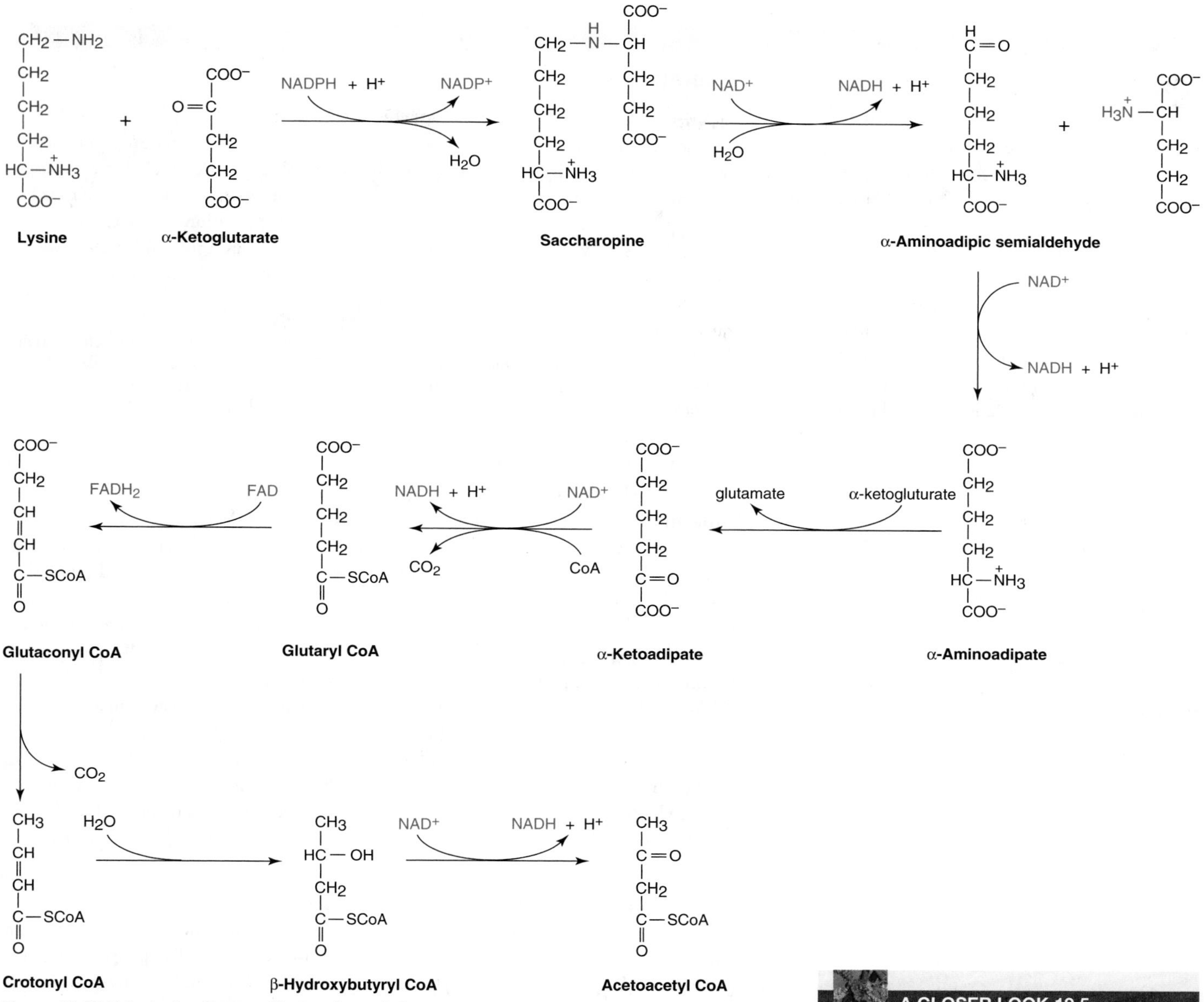

Figure 19.47 Principal pathway of lysine degradation.

Branched-Chain Amino Acids

Metabolism of the **branched-chain amino acids (BCAAs)** valine, isoleucine, and leucine is unusual, being initiated in muscle. NADH and $FADH_2$ are formed during their metabolism, making them an excellent source of energy. BCAA aminotransferase activity is higher in muscle than in liver. Although these amino acids produce different products, the initial steps in their metabolism are shared. BCAA aminotransferase exists in three isozymes with different tissue distributions. Some are found in cytosol and some in mitochondria (Figure 19.50, p. 777). Two of these isozymes metabolize all three BCAAs, and one is specific for leucine. Starvation induces the muscle BCAA aminotransferases. The resulting α-keto branched-chain acids are oxidatively decarboxylated by an inner mitochondrial membrane enzyme complex similar to the pyruvate dehydrogenase complex, producing NADH and CO_2. All three α-keto branched-chain acids are oxidized by the same enzyme. The more active form occurs in liver in the fed state and in muscle during starvation, reflecting the metabolism of dietary BCAAs by liver and of muscle BCAAs to provide energy during fasting. The resulting CoA compounds are one carbon shorter than the amino acids and are next acted on by an enzyme that resembles the first dehydrogenase of fatty acid β-oxidation (Clin. Corr. 19.15, p. 778).

A CLOSER LOOK 19.5

Tryptophan, Carbohydrates, and Sleep

Tryptophan availability is reduced when other amino acids compete with it for transport through the blood–brain barrier. Elevated plasma concentrations of other amino acids, after a high-protein meal, diminish transport of tryptophan and induce wakefulness. The sleep-inducing effect of carbohydrates is due to decreased plasma amino acid levels, since carbohydrate stimulates release of insulin, and insulin causes removal of amino acids from plasma and uptake into muscle. This alleviates competition and increases the amount of tryptophan that can enter the brain.

CLINICAL CORRELATION 19.13

Hyperlysinemia and Lysinuric Protein Intolerance

The enzyme α-aminoadipic semialdehyde synthase is deficient in a small number of patients who excrete lysine and smaller amounts of saccharopine. This has led to the discovery that the enzyme has both lysine-α-ketoglutarate reductase and saccharopine dehydrogenase activities. In saccharopinuria some of the reductase activity is retained. Hyperlysinemia (OMIM 238700) can cause lax ligaments and muscles, convulsions, and anemia. Lysine is an inhibitor of arginase, and hyperammonemia may also occur. More serious is familial lysinuric protein intolerance (OMIM 222700) due to failure to transport dibasic amino acids across intestinal mucosa and renal tubular epithelium. Plasma lysine, arginine, and ornithine are decreased to one-third or one-half of normal. Patients may be undiagnosed as infants, but develop gastrointestinal symptoms and hyperammonemia after being changed to a meat-containing diet. Citrulline excretion is enhanced. This is thought to arise from deficiency of urea cycle intermediates ornithine and arginine in liver, limiting the capacity of the cycle. Consistent with this view, oral supplementation with citrulline prevents hyperammonemia. Other features are thin hair, muscle wasting, and osteoporosis, which may reflect protein malnutrition due to lysine and arginine deficiency.

McManus, D. T., Moore, R., Hill, C. M., Rodgers, C., et al. Necropsy findings in lysinuric protein intolerance. *J. Clin. Path.* 49:345, 1996; and Sperandeo, M. P., Andria, G., and Sebastio, G. Lysinuric protein intolerance: update and extended mutation analysis of the SLC7A7 gene. *Hum. Mutat.* 29:14, 2008.

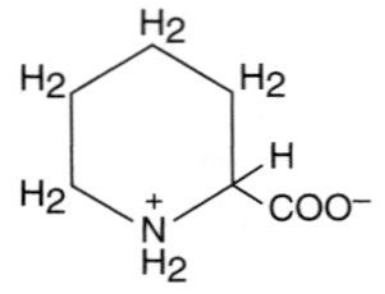

Pipecolate

Figure 19.48 Pipecolate, a minor product of lysine metabolism.

Valine and **isoleucine**: These two amino acids continue down a common pathway, with addition of water across the double bond to form a hydroxylated intermediate (Figure 19.51). The hydroxyl group on the isoleucine derivative is oxidized by NAD^+ followed by thiolysis to give acetyl CoA and propionyl CoA. The valine derivative loses CoA and is then oxidized by NAD^+ to methylmalonate semialdehyde, which is converted to propionyl CoA.

Leucine: At this point, the metabolism of leucine diverges from that of the other two BCAAs. β-Methylcrotonyl CoA is carboxylated, hydroxylated, and cleaved to acetoacetate and acetyl CoA (Figure 19.52). One intermediate is β-hydroxy-β-methylglutaryl CoA, an intermediate in cytoplasmic sterol synthesis (p. 719). Since BCAA degradation occurs in mitochondria the two pools do not mix. Leucine also has a minor alternative pathway (not shown), which results in excretion of 3-hydroxyvaleric acid and can be utilized in the case of blockage in the leucine degradative pathway.

Propionyl CoA Is Metabolized to Succinyl CoA

An end product of isoleucine, valine, threonine, and methionine metabolism, odd-chain fatty acid oxidation, and degradation of the side chain of cholesterol is **propionyl CoA**. The first step in conversion of propionyl CoA to succinyl CoA is catalyzed by **propionyl-**

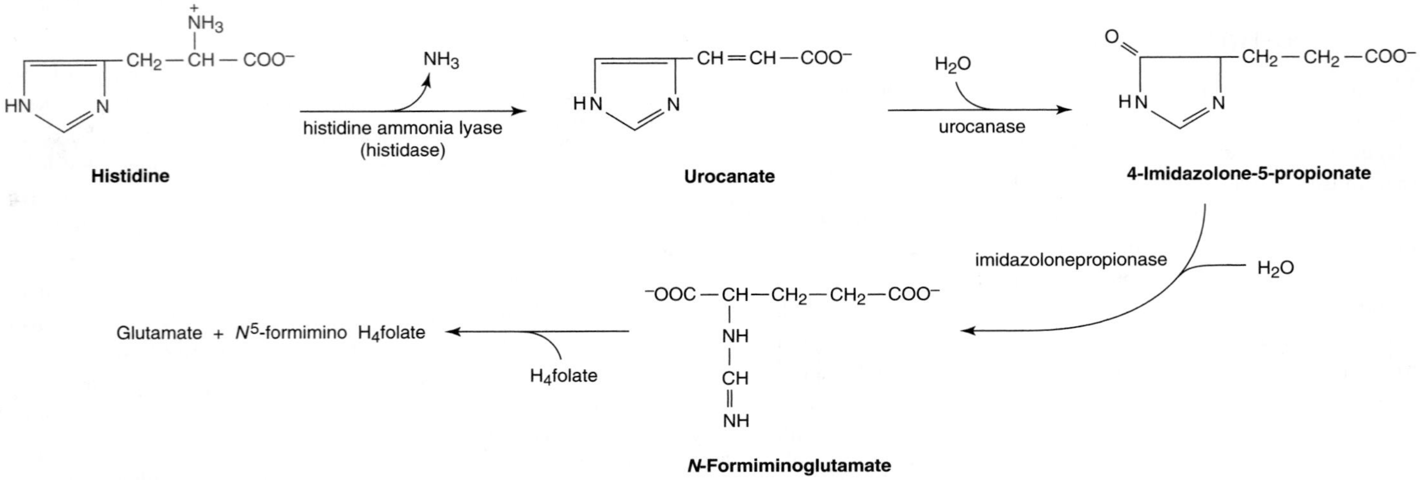

Figure 19.49 Degradation of histidine.

CLINICAL CORRELATION 19.14

Histidinemia and Formiminotransferase Deficiency

Histidinemia (OMIM 235800) is due to histidine ammonia-lyase (histidase) deficiency. A convenient assay for this enzyme uses skin, which produces urocanate as a constituent of sweat; urocanase and other enzymes of histidine catabolism in liver do not occur in skin. Histidase deficiency can be confirmed from a skin biopsy. Incidence of the disorder is high, about 1 in 10,000 newborns screened. Most reported cases of histidinemia show normal mental development. Restriction of dietary histidine normalizes the biochemical abnormalities but is not usually required. Glutamate formiminotransferase (OMIM 229100) deficiency is the second most common error of folate metabolism. In mild cases the only symptom is excretion of formiminoglutamate (FIGLU), but in severe cases mental retardation may occur. The difference between lack of activity of this enzyme and folic acid deficiency may be established by measuring folic acid in plasma.

Hilton, J. F., Christensen, K. E., Watkins, D., Raby, B. A., et al. The molecular basis of glutamate formiminotransferase deficiency. *Hum. Mutat.* 22:67, 2003; and Levy, H. L., Taylor, R. G., and McInnes, R. R. Disorders of histidine metabolism. In Scriver, C. R., Beaudet, A. L., Sly, W. S., Valle, D. (Eds.), *The Metabolic and Molecular Bases of Inherited Disease,* 8th ed. New York: McGrawHill, 2001, 2:1807.

CoA carboxylase that contains a biotin covalently linked to the ε-amino group of a lysine residue (Figure 19.53) and yields D-methylmalonyl CoA. A racemase converts this to a mixture of D- and L-methylmalonyl CoA. Methylmalonyl mutase, which requires 5′-deoxyadenosylcobalamin (a derivative of vitamin B_{12}), converts the L-isomer to succinyl CoA. This is the second enzyme known to be dependent on vitamin B_{12} (p. 1082). The reaction is very unusual, removing a methyl side chain and inserting it as a methylene group into the backbone of the compound (Clin. Corr. 19.16, p. 780).

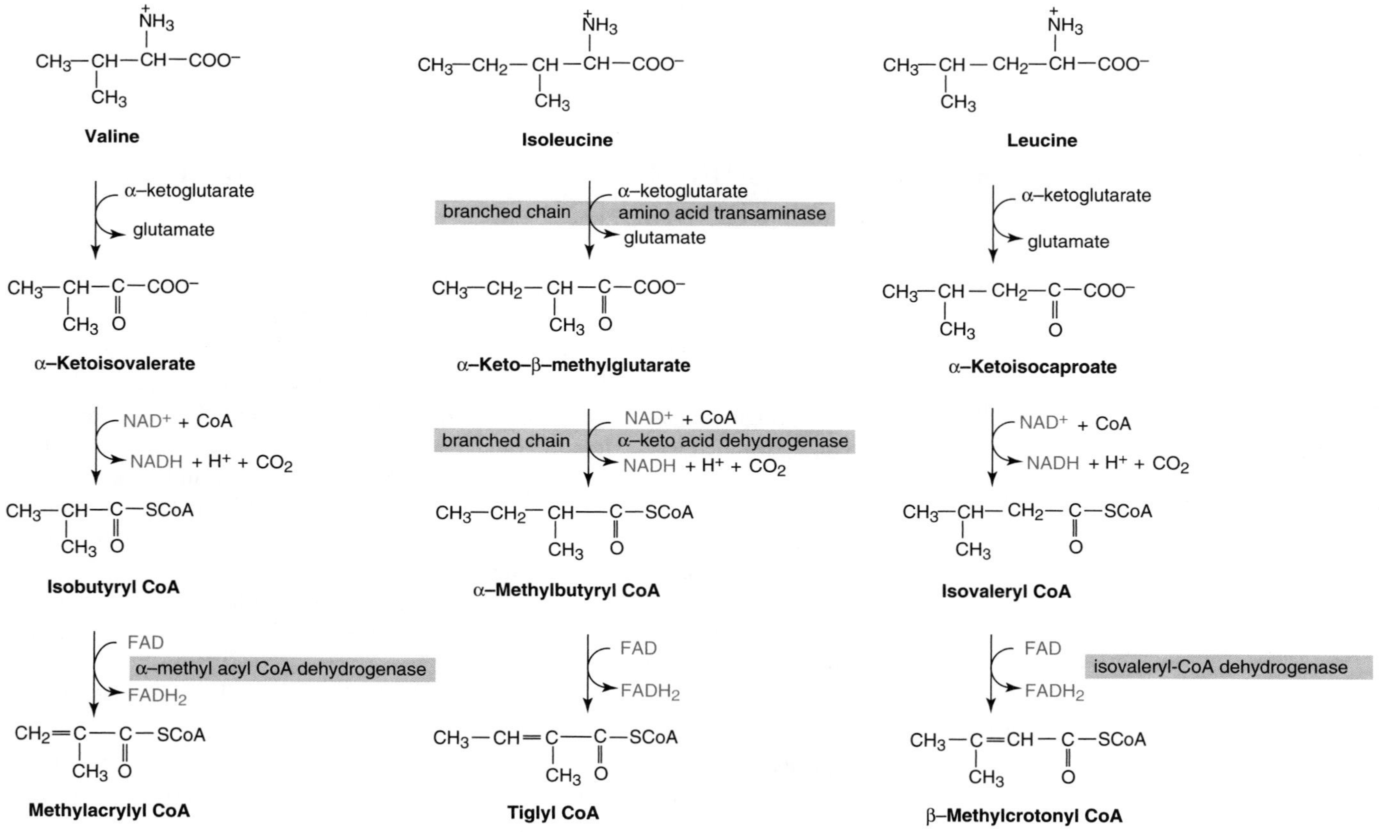

Valine | Isoleucine | Leucine

↓ α-ketoglutarate → glutamate (branched chain amino acid transaminase)

α-Ketoisovalerate | α-Keto-β-methylglutarate | α-Ketoisocaproate

↓ NAD^+ + CoA → NADH + H^+ + CO_2 (branched chain α-keto acid dehydrogenase)

Isobutyryl CoA | α-Methylbutyryl CoA | Isovaleryl CoA

↓ FAD → $FADH_2$ (α-methyl acyl CoA dehydrogenase; isovaleryl-CoA dehydrogenase)

Methylacrylyl CoA | Tiglyl CoA | β-Methylcrotonyl CoA

Figure 19.50 Common reactions in degradation of branched-chain amino acids.

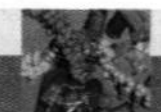

CLINICAL CORRELATION 19.15

Maple Syrup Urine Disease (OMIM 248600) and Other Diseases in the Branched-Chain Amino Acid Degradative Pathways

Enzyme deficiency in catabolism of branched-chain amino acids is not common and results in acidosis in newborns or young children. Very rare instances have been reported of hypervalinemia, isovaleric acidemia, and hyperleucine-isoleucinemia. It has been suggested that these conditions indicate existence of specific aminotransferases for valine, leucine, and isoleucine. Alternatively, mutation could alter the specificity of a single enzyme. The most common abnormality is deficiency of branched-chain keto acid–dehydrogenase complex which contains four catalytic components as well as regulatory proteins. Some of the mutations appear to involve protein misfolding, and trimethylamine-amine oxide, a chemical chaperone, has been found to improve dehydrogenase activity. There are several variations, but all patients excrete the branched-chain α-keto acids, corresponding hydroxy acids, and other side products; an unidentified product imparts the characteristic odor as indicated by the name maple syrup urine disease. Some cases respond to high doses of thiamin, and these are caused by a mutation in the thiamin-binding site on one of the proteins. Many show serious mental retardation, ketoacidosis, and short life span. Severity of the disease appears to be linked to the identity of the protein carrying the mutation. Dietary treatment to reduce the branched-chain ketoacidemia is effective in some cases. Deficiency of enzymes in later reactions of branched-chain amino acids include a blockage of oxidation of isovaleryl CoA with accumulation of isovalerate (which gives urine a smell of sweaty feet: OMIM 607036), β-methylcrotonyl-CoA carboxylase deficiency (in which urine smells like that of a cat: OMIM 609010), deficiency of β-hydroxy-β-methylglutaryl-CoA lyase (OMIM 246450), and deficiency of β-ketothiolase that splits β-methylacetoacetyl CoA (with no defect in acetoacetate cleavage: OMIM 203750). In the latter condition, development is normal and symptoms appear to be related only to episodes of ketoacidosis.

2-Methyl-3-hydroxybutyryl-CoA dehydrogenase deficiency results in excretion of the substrate and glycylglycine. At about the age of 1 year motor and mental deterioration occurs and without dietary treatment the mental retardation becomes severe.

Baumgartner, M. R., Almashanu, S., Suormala, T., Obie, C., et al. The molecular basis of human 3-methylcrotonyl-CoA carboxylase deficiency. *J. Clin. Invest.* 107: 495, 2001; Chuang, D. T., and Shih, V. E. Maple syrup urine disease (branched-chain ketoaciduria). In Scriver, C. R., Beaudet, A. L., Sly, W. S., Valle, D. (Eds.), *The Metabolic and Molecular Bases of Inherited Disease,* 8th ed. New York: McGraw Hill, 2001, 2: 1971; Gallardo, M. E., Desviat, L. R., Rodriguez, J. M., Esparza-Gordillo, J., et al. Kinetic and expression analyses of seven novel mutations in mitochondrial acetoacetyl-CoA thiolase (T2):identification of a K_m mutant and an analysis of the mutational sites in the structure. *Molec. Genet. Metab.* 90: 370, 2007; and Song, J. L., and Chuang, D., T. Natural osmolyte trimethylamine *N*-oxide corrects assembly defects of mutant branched-chain alpha-ketoacid decarboxylase in maple syrup urine disease. *J. Biol. Chem.* 276: 40241, 2001.

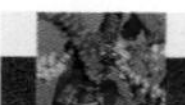

A CLOSER LOOK 19.6

Lysine and Pipecolate

A minor pathway of lysine metabolism starts with removal of the α-amino group and goes via the cyclic compound pipecolate (see Figure 19.48, p. 776) to join the major pathway at the level of the semialdehyde intermediate. This does not replace the major pathway even in a deficiency of enzymes in the early part of the pathway (see Figure 19.47, p. 775).

19.6 • IMPORTANT METABOLITES DERIVED FROM AMINO ACIDS

Glutamate: Glutamate is a component of glutathione, which is discussed at the end of this chapter (p. 789). It is also a precursor of **γ-amino butyric acid (GABA)**, a neurotransmitter (see Figure 19.10, p. 755), and of proline and ornithine (p. 763).

Serine: Serine is itself a head group for phospholipids. Ethanolamine, choline, and betaine (Figure 19.54, p. 780) are derivatives of serine. It should be noted that choline is now classified as a vitamin. Ethanolamine (decarboxylated serine) and choline (*N*-trimethylethanolamine) are components of phospholipids, and betaine (an oxidized derivative of choline) is a methyl donor in a minor pathway leading to methionine salvage (p. 767). Metabolism of choline and **carnitine** (see Figure 19.68, p. 788) leads to trimethylamine, which is normally oxidized for excretion. Loss of the enzyme for the further metabolism of trimethylamine leads to a condition called fish malodor syndrome. Serine also provides the carbon skeleton of cysteine, the sulfur being transferred from homocysteine (see Figure 19.42, p. 772).

A serine residue in some enzymes is metabolized in pyruvoyl enzymes (see A Closer Look 19.4, p. 765).

Arginine: Arginine is a precursor for nitric oxide (p. 447) in brain; agmatine (Figure 19.55, p. 780), a compound with properties of a neurotransmitter, is derived by decarboxylation of arginine and may have antihypertensive properties.

Glycine: Glycine is the precursor of glyoxalate, which can be transaminated back to glycine or oxidized to oxalate (Figure 19.56, p. 780). Excessive production of oxalate

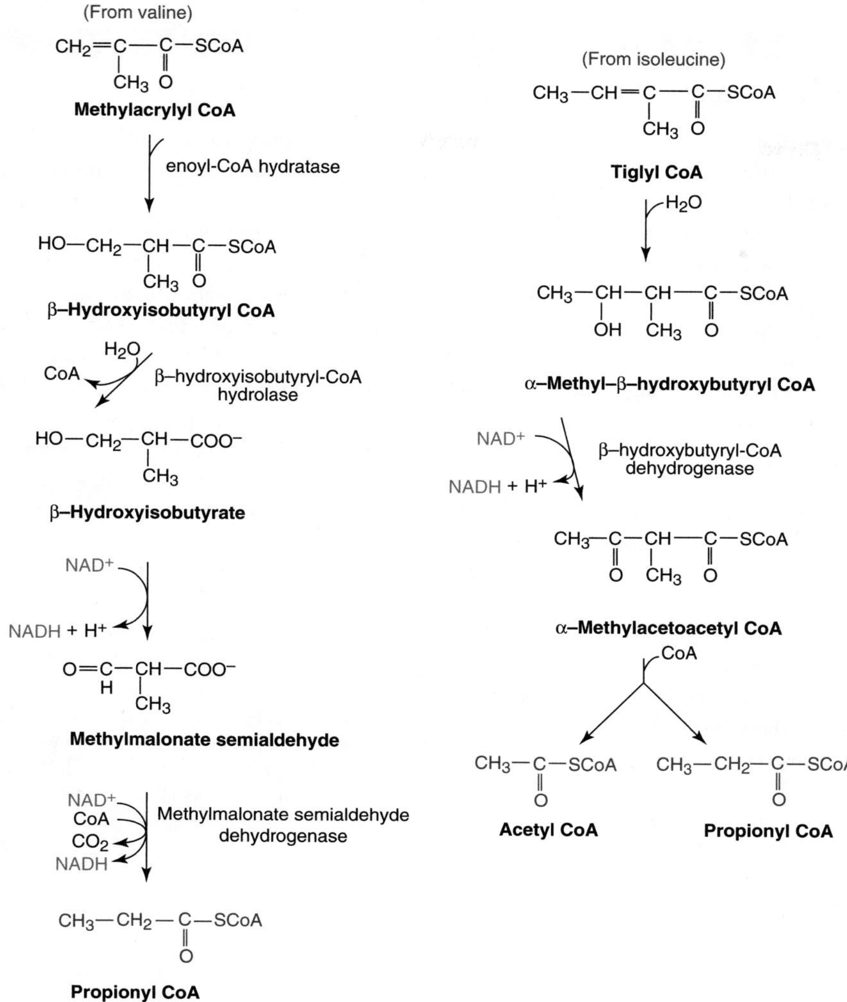

Figure 19.51 Terminal reactions in degradation of valine and isoleucine.

$CH_3-CH_2-C(=O)-SCoA$

Propionyl CoA

propionyl CoA carboxylase (CO_2)

$^-OOC-CH(CH_3)-C(=O)-SCoA$

D–Methylmalonyl-CoA

methylmalonyl-CoA racemase

$^-OOC-CH(CH_3)-C(=O)-SCoA$

L–Methylmalonyl-CoA

methylmalonyl-CoA mutase

$^-OOC-CH_2-CH_2-C(=O)-SCoA$

Succinyl CoA

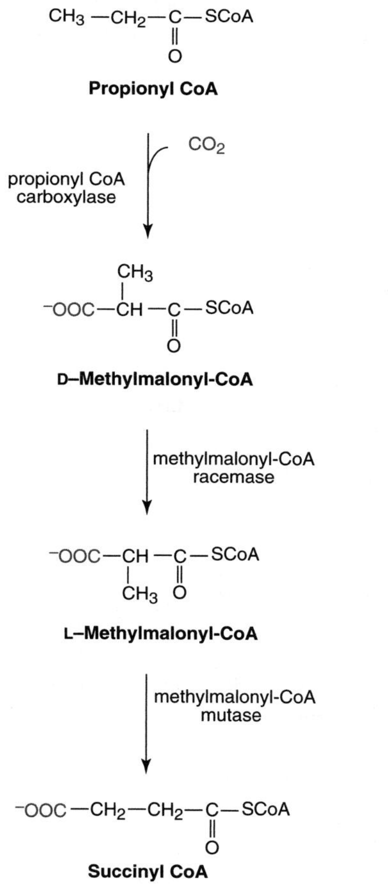

Figure 19.53 Interconversion of propionyl CoA, methylmalonyl CoA, and succinyl CoA. The mutase requires 5′-deoxyadenosylcobalamin for activity.

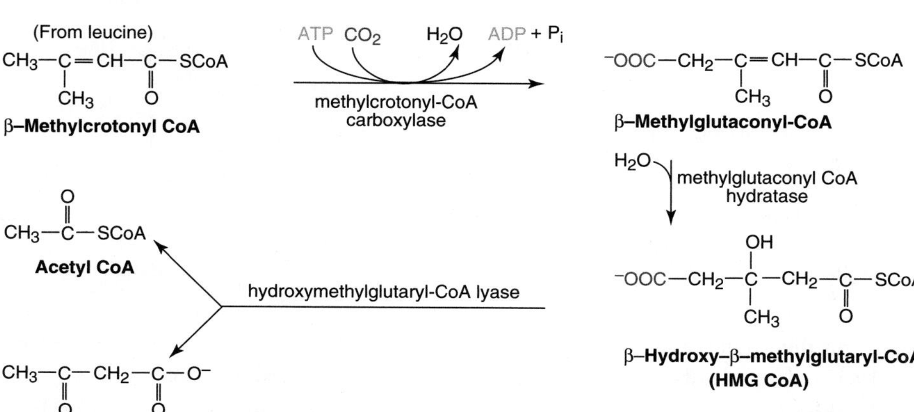

Figure 19.52 Terminal reactions of leucine degradation.

CLINICAL CORRELATION 19.16

Propionic Acidemia and Methylmalonic Aciduria

Deficiency of any of the three enzymes shown in Figure 19.53 contributes to ketoacidosis. Propionate is formed in the degradation of valine, isoleucine, methionine, threonine, the side chain of cholesterol, and odd-chain fatty acids. The amino acids appear to be the main precursors since decreasing or eliminating dietary protein immediately minimizes acidosis. A defect in propionyl-CoA carboxylase (OMIM 606054) results in accumulation of propionate, which is diverted to alternative pathways including incorporation into fatty acids in place of the first acetyl group to form odd-chain fatty acids. Its hallmark diagnostic criteria are hyperglycinemia and hyperglycinuria that are accompanied by symptoms such as vomiting, lethargy, and ketosis. A low-protein diet can ameliorate these symptoms. In one case, large amounts of biotin were reported to produce beneficial effects, suggesting that more than one defect decreases propionyl-CoA carboxylase activity. Possibilities are a lack of intestinal biotinidase that liberates biotin from ingested food for absorption (OMIM 253260) or a lack of biotin holocarboxylase that incorporates biotin into biotin-dependent enzymes (OMIM 253270). Acidosis in children may be caused by high levels of methylmalonate, which is normally undetectable in blood. Liver taken at autopsy or cultured fibroblasts showed in some cases deficiency of methylmalonyl-CoA mutase. Some samples were unable to convert methylmalonyl CoA to succinyl CoA under any conditions, but other samples carried out the conversion when 5′-adenosylcobalamin was added. Clearly, those with an active site defect cannot metabolize methylmalonate, but those with defects in handling vitamin B_{12} respond to massive doses of the vitamin. Other cases of methylmalonic aciduria have a more fundamental inability to use vitamin B_{12} that leads to methylcobalamin deficiency (coenzyme of methionine salvage) and 5′-adenosylcobalamin deficiency (coenzyme of methylmalonyl CoA isomerization). Methylmalonic aciduria (OMIM 251000) can be anywhere from benign to fatal, often causing mental retardation and kidney failure. An interesting case of false accusation of murder is that of Patricia Stalling, who was jailed for causing the death of her infant son by feeding him ethylene glycol. While in jail, she gave birth again, and the baby was diagnosed with methylmalonic academia. Reexamination of the first infant's blood showed that the compound that had originally been identified as ethylene glycol was in fact propionic acid.

Desviat, L. R., Clavero, S., Perez-Cerda, C., Navarrete, R., et al. New splicing mutations in propionic acidemia. *J. Hum. Genet.* 51:992, 2006; Fenton, W. A., Gravel, R. A., and Rosenblatt, D. S. Disorders of propionate and methylmalonate metabolism. In Scriver, C. R., Beaudet, A. L., Sly, W. S., Valle, D. (Eds.), *The Metabolic and Molecular Bases of Inherited Disease,* 8th ed. New York: McGraw Hill, 2001, 2:2176; and Shoemaker, J. D., Lynch, R. E., Hoffmann, J. W., and Sly, W. S. Misidentification of propionic acid as ethylene glycol in a patient with methylmalonic acidemia. *J. Pediat.* 120:417, 1992.

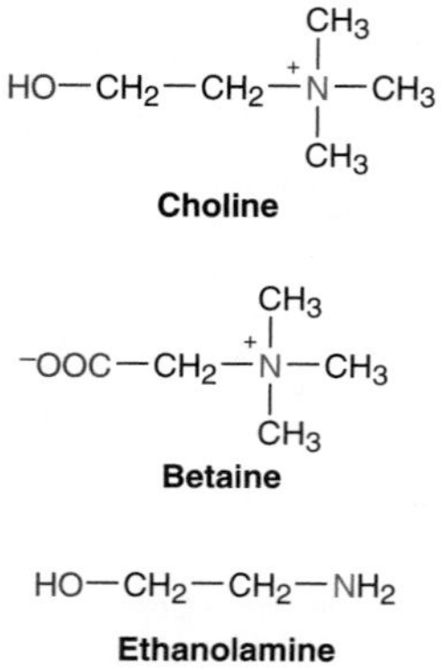

$HO{-}CH_2{-}CH_2{-}\overset{+}{N}(CH_3)_3$

Choline

$^{-}OOC{-}CH_2{-}\overset{+}{N}(CH_3)_3$

Betaine

$HO{-}CH_2{-}CH_2{-}NH_2$

Ethanolamine

Figure 19.54 Choline and related compounds.

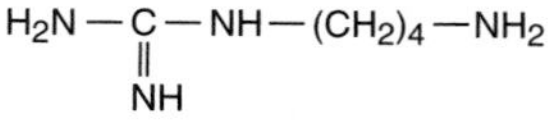

$H_2N{-}C({=}NH){-}NH{-}(CH_2)_4{-}NH_2$

Agmatine

Figure 19.55 Agmatine, a product of arginine metabolism.

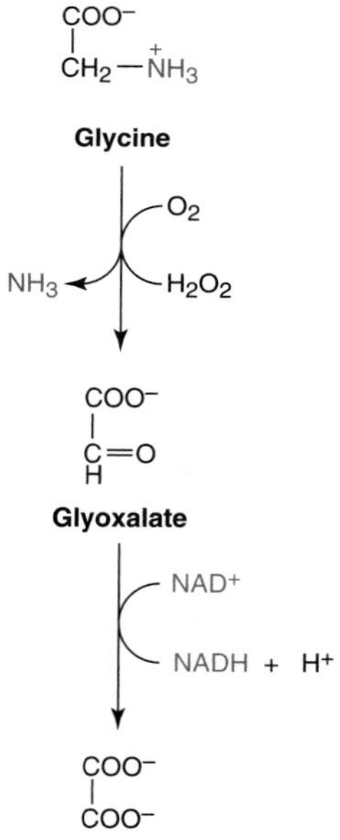

$COO^{-}{-}CH_2{-}\overset{+}{N}H_3$ **Glycine**

$\xrightarrow{O_2 \rightarrow H_2O_2,\ NH_3}$

$COO^{-}{-}CHO$ **Glyoxalate**

$\xrightarrow{NAD^+ \rightarrow NADH + H^+}$

$COO^{-}{-}COO^{-}$ **Oxalate**

Figure 19.56 Oxidation of glycine.

forms the insoluble calcium oxalate salt, which may lead to kidney stones (Clin. Corr. 19.17). The role of glycine as a neurotransmitter is described on page 942.

Methionine: The vast majority of methyltransferase reactions utilize ***S*-adenosylmethionine** (see Figure 19.41, p. 770, and Clin. Corr. 19.18). Methyl group transfer from AdoMet (also called SAM, for *S*-adenosylmethionine), a methyl acceptor, is

CLINICAL CORRELATION 19.17

Primary Hyperoxaluria

Idiopathic hyperoxaluria is caused by excess oxalate production. About 50% of the oxalate comes from the diet and may be related to changes in the ability of the intestines to absorb oxalate. Primary hyperoxaluria is a rare disease and a result of enzyme deficiency. It has serious consequences. More than 50% of people with this genetic defect have renal failure by age 15, and 80% by age 30. The amount of excess serum oxalate is so great that even dialysis (the normal treatment for high oxalic acid) is ineffective. Calcium oxalate stones in the kidney are the cause of the damage, and the oxalate buildup is due to the deficiency in either one of two enzymes. Type I primary hyperoxaluria (OMIM 259900) results from a loss of activity of alanine: glyoxylate aminotransferase. This enzyme is liver specific and found in peroxisomes. Its function is the conversion of glyoxylate to glycine. The aminotransferase is pyridoxal-phosphate-dependent, and the disease sometimes responds to vitamin B_6 therapy. Children with kidney stones are usually screened for this enzyme deficiency because, when identified, any younger siblings can start therapy as early as possible to avoid eventual kidney damage. Early liver transplantation is sometimes successful in preventing further damage. Later symptoms are caused by blockage of the circulatory system by oxalate and include uremia, Raynaud syndrome, spasms of large arteries, gangrene, and vision-associated problems.

Type II primary hyperoxaluria (OMIM 260000), is characterized by a loss of D-glyceric dehydrogenase. This enzyme catalyzes the conversion of glyoxalate to oxalate. It is less serious than type I.

The clinical outcome of both diseases is related to the degree of loss of enzyme activity.

Latta, K., and Brodehl, J. Primary hyperoxaluria type I. *Eur. J. Pediat.* 149:518–522, 1990; and Milliner, D. S., Eickholt, J. T., Bergstralh, E. J., Wilson, D. M., and Smith, L. H. Results of long-term treatment with orthophosphate and pyridoxine in patients with primary hyperoxaluria. *N. Engl. J. Med.* 331:1553, 1994.

irreversible. An example is shown in Figure 19.57. A butylamino group derived from AdoMet is used for posttranslational modification of a specific lysine residue in eIF-4D, an initiation factor that promotes formation of the first peptide bond in eukaryotic protein synthesis. The group is first added to putrescine to make spermidine, and then cleaved to form deoxyhypusine. It is then hydroxylated, and the modified residue that results is called hypusine (Figure 19.58). Use of the iron chelator desferrioxamine, a strong inhibitor of hypusine synthesis, has been shown to potentiate apoptosis in cancer cells and shows promise as an anticancer agent.

Cysteine: Cysteine is metabolized in several ways as determined by the needs of the cell. The major metabolite is cysteine sulfinate (see Figure 19.45, p. 772). This is converted to bisulfite and pyruvate or to hypotaurine and taurine. Taurine is an abundant intracellular free amino acid. It appears to play a necessary role in brain development. It forms conjugates with bile acids (p. 1058) and may enhance bile flow and increase cholesterol clearance by the liver. Taurine may also play a role as an antioxidant in salvaging toxic intermediates, in immunity, in regulating intracellular calcium, in controlling hypertension, and, because of its abundance, in osmoregulation.

Sulfite produced from cysteine metabolism is oxidized to sulfate (see Figure 19.45, p. 772) and used in formation of 3′-phosphoadenosine-5′-phosphosulfate (PAPS), the source of sulfate groups for addition to biological molecules (Figure 19.59).

Another reaction of cysteine metabolism catalyzed by cystathionase moves the sulfur from one cysteine to another cysteine (Figure 19.60) to form thiocysteine. Thiosulfate is formed from cysteine as shown in Figure 19.61. An enzyme called rhodanese (named for the deep red color of thiocyanate) can incorporate a sulfur from thiosulfate or thiocysteine into other molecules such as cyanide ion (Figure 19.62). Cyanide-related compounds occur in some plant material, notably those of the genus *Brassica*.

Tyrosine: Most tyrosine not incorporated into proteins is metabolized to acetoacetate and fumarate but some is used for the synthesis of the catecholamines: **dopamine**, **norepinephrine**, and **epinephrine**. The eventual metabolic fate of tyrosine carbons is determined by the first step in each pathway. Catecholamine synthesis (Figure 19.63) starts with tyrosine hydroxylase, which, like phenylalanine and tryptophan hydroxylase, is dependent on tetrahydrobiopterin. All three are affected by biopterin deficiency or a defect in dihydrobiopterin reductase (see Figure 19.38,

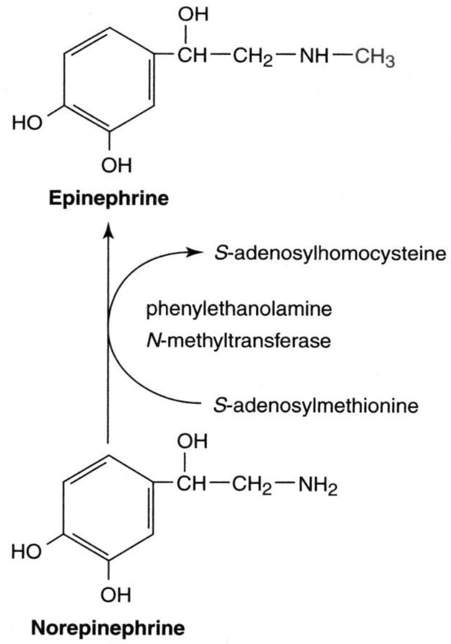

Epinephrine

S-adenosylhomocysteine

phenylethanolamine *N*-methyltransferase

S-adenosylmethionine

Norepinephrine

Figure 19.57 *S*-Adenosylmethionine is the methyl donor used in the conversion of norepinephrine to epinephrine.

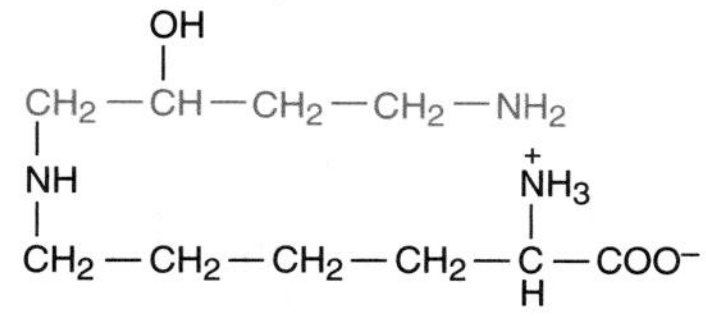

Figure 19.58 Hypusine.

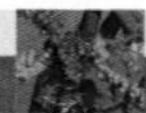

CLINICAL CORRELATION 19.18

Methionine Adenosyltransferase (MAT) Deficiency (OMIM 250850)

This inborn error of metabolism is probably more prevalent than reported, since it is generally symptom free and therefore remains undiagnosed. Only a severe loss of enzyme activity results in symptoms that include growth retardation, digestive problems, and anorexia. Frequent manifestations of the condition are halitosis or other unpleasant body odors due to buildup of dimethylsulfide.

MAT deficiency can cause hypermethioninemia (over 1500 μM; normal is 35 μM), and is differentiated from that caused by cystathionine synthase deficiency, tyrosinemia-associated liver failure, and liver disease by liver biopsy. At extremely high levels of methionine it can be transaminated, and the transaminated metabolites are found in blood. Treatment is a low methionine diet, but a severe lack of AdoMet that can result from MAT inactivity is associated with neurological symptoms and must be treated with AdoMet therapy.

The disease is believed to be inherited in an autosomal recessive manner, but heterozygotes exhibiting truncation, missense, and splicing mutations have been found. The defect may be an example of a dominant negative mutant where one defective subunit of the dimeric or tetrameric enzyme can cause inactivation of the other subunits.

Chamberlin, M. E., Ubagai, T., Mudd, S. H., Thomas, J., et al. Methionine adenosyltransferase I/III deficiency: novel mutations and clinical variations. *Am J Hum Genet.* 66(2): 347, 2000; and Ubagai, T., Lei, K.-J., Huang, S., Mudd, S. H., et al. Molecular mechanisms of an inborn error of methionine pathway: methionine adenosyltransferase deficiency. *J. Clin. Invest.* 96: 1943, 1995.

p. 768). Tyrosine hydroxylase produces **dihydroxyphenylalanine** (dioxophenylalanine, **DOPA**). DOPA decarboxylase, with pyridoxal phosphate as cofactor, forms dopamine, the active neurotransmitter. Because dopamine is metabolized primarily by monoamine oxidase B in humans, and Parkinson disease (Clin. Corr. 19.19, p. 785) is characterized by a reduction in the brain dopamine concentration, selective inactivation of MAO-B has been shown to be effective in increasing the dopamine concentration and thus treat the disease.

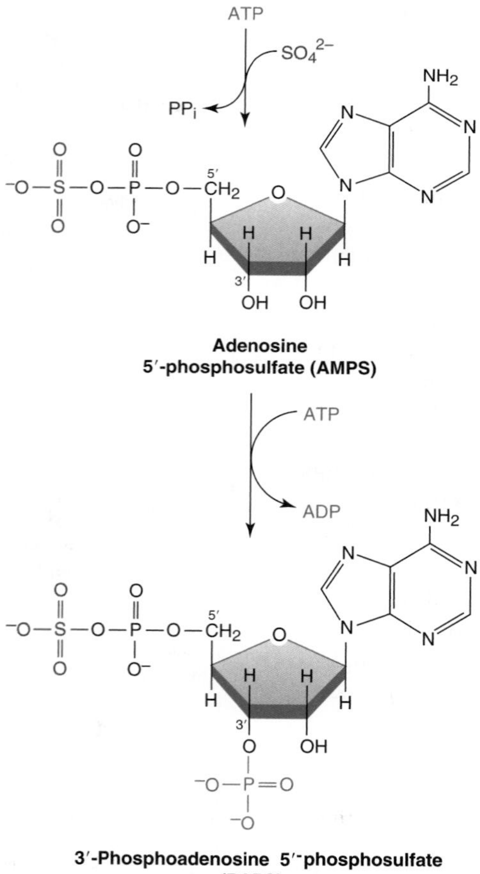

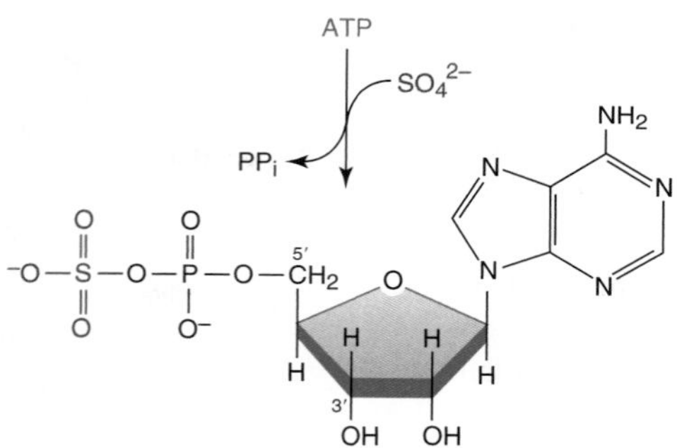

Figure 19.59 **Synthesis of PAPS.**

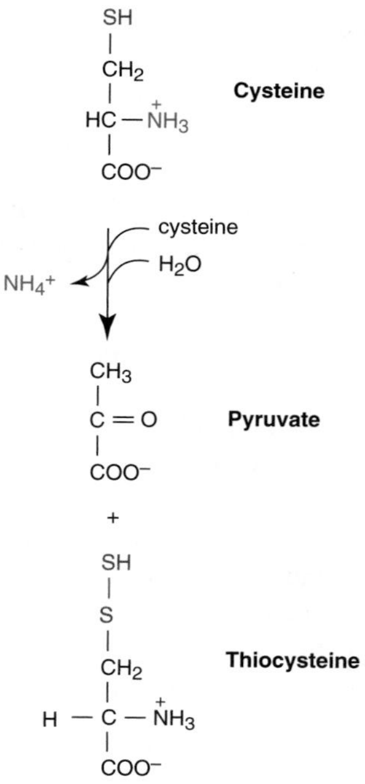

Cysteine

cysteine, H_2O, NH_4^+

Pyruvate + Thiocysteine

Figure 19.60 **Synthesis of thiocysteine.**

In the substantia nigra and some other parts of the brain, this is the end of this pathway. The adrenal medulla converts dopamine to norepinephrine and epinephrine (adrenaline). The methyl group of epinephrine is derived from *S*-adenosylmethionine (see Figure 19.41, p. 770).

Tyrosine in brain cells regulates norepinephrine formation. Estrogens decrease tyrosine concentration and increase tyrosine aminotransferase activity, diverting tyrosine into the catabolic pathway. Estrogen sulfate competes for the pyridoxal phosphate site on DOPA decarboxylase. These effects combined may help explain mood variations during the menstrual cycle. Tyrosine is therapeutic in some cases of depression and stress. Its transport appears to be reduced in skin fibroblasts from schizophrenic patients, indicating other roles for tyrosine derivatives in mental disorders. Catecholamines are metabolized by monoamine oxidase and catecholamine *o*-methyltransferase. Major metabolites are shown in Figure 19.64. Absence of these metabolites in urine is diagnostic of a deficiency in synthesis of catecholamines. Lack of synthesis of serotonin is indicated by lack of 5-hydroxyindole-3-acetic acid in urine. Biopterin deficiency is described in Clin. Corr. 20.

Tyrosine is also required for synthesis of **melanin**, thyroid hormone, and quinoproteins.

Conversion of tyrosine to melanin requires tyrosinase, a copper-containing protein (Figure19.65*a*). The reaction produces dopaquinone. During melanogenesis, following exposure to UVB light, tyrosinase and tyrosinase-related protein, which may function

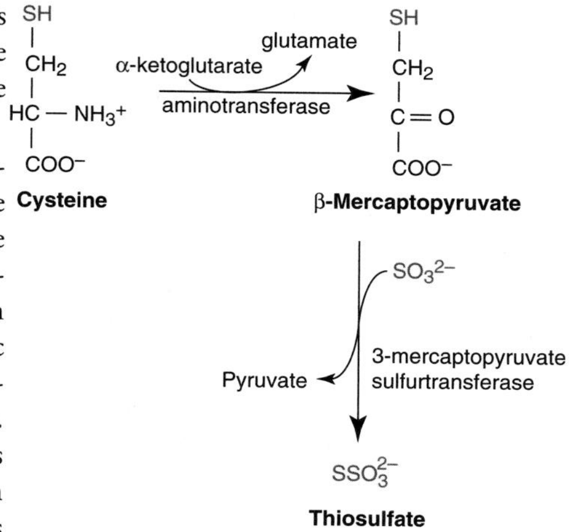

Figure 19.61 Formation of thiosulfate.

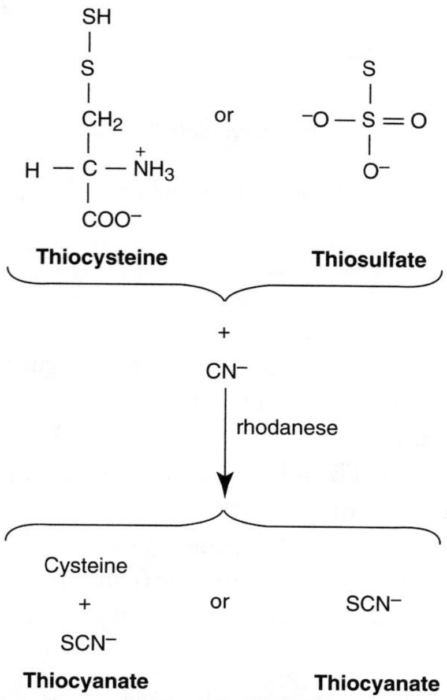

Figure 19.62 Detoxification of cyanide by products of cysteine metabolism.

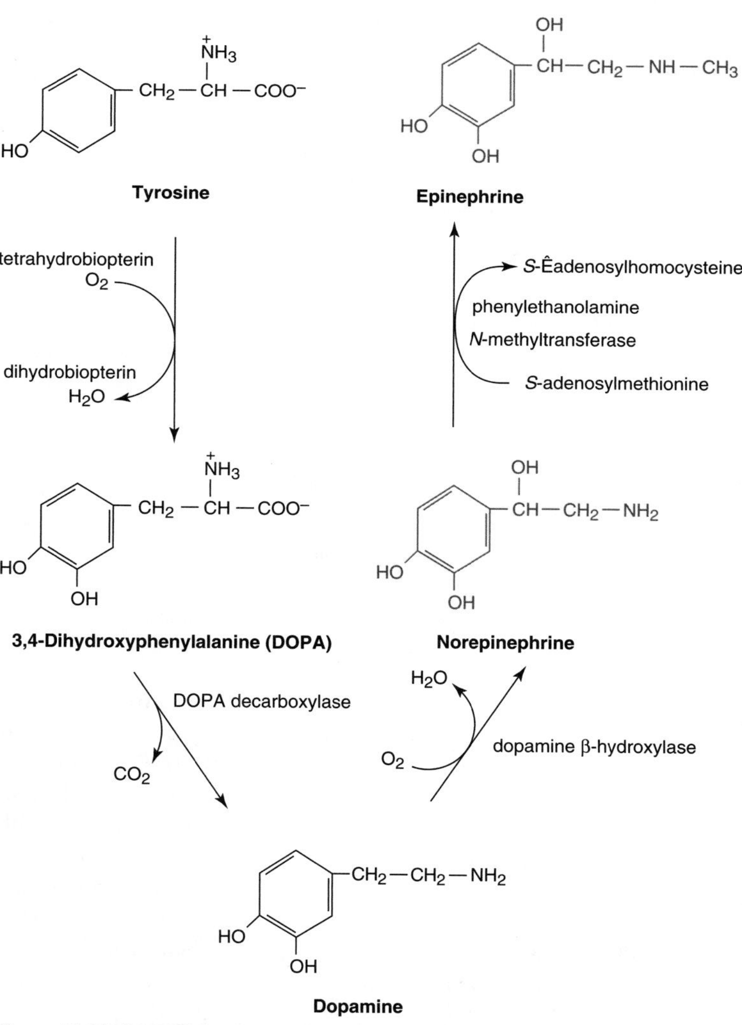

Figure 19.63 Catecholamine synthesis. DOPA decarboxylase is also called aromatic L-amino acid decarboxylase.

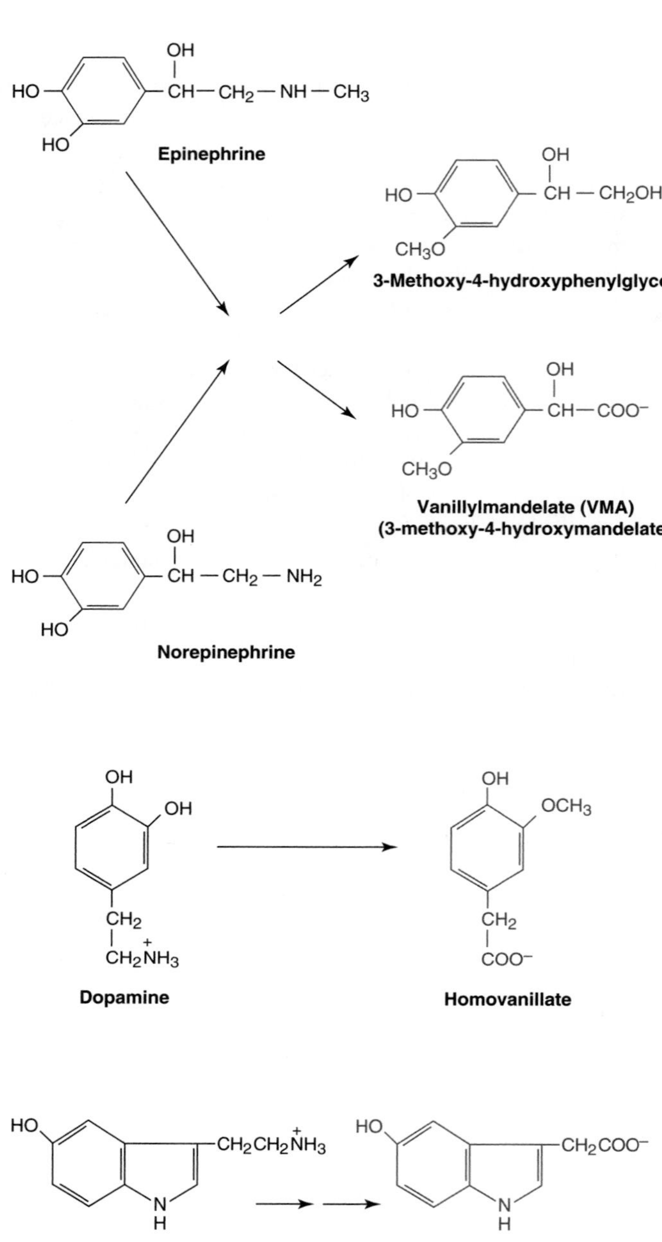

Figure 19.64 Major urinary excretion products of epinephrine, norepinephrine, dopamine, and serotonin.

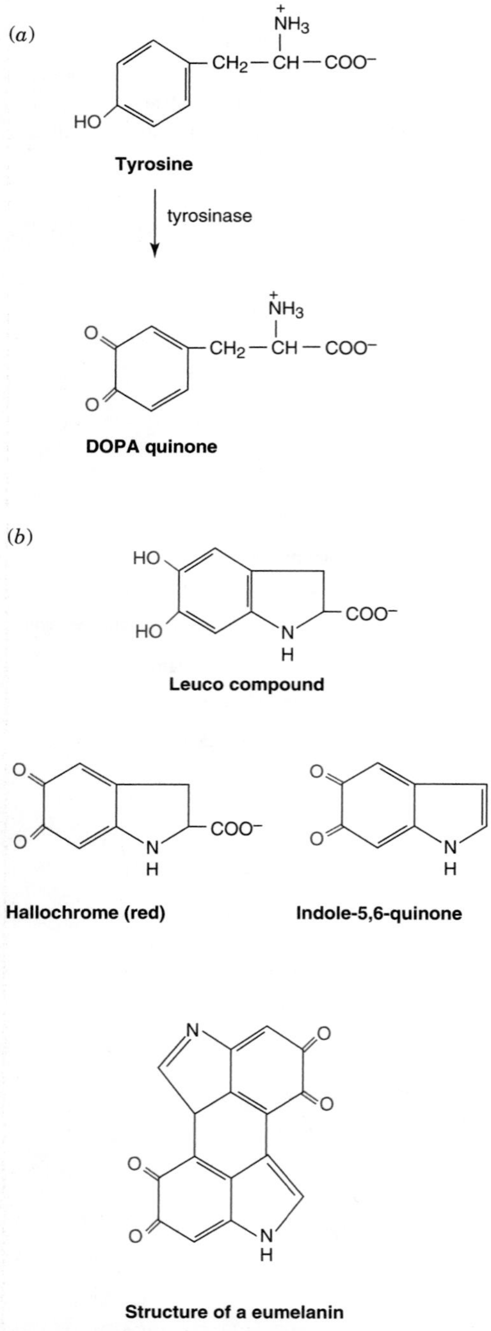

(*a*)

Tyrosine

tyrosinase

DOPA quinone

(*b*)

Leuco compound

Hallochrome (red)

Indole-5,6-quinone

Structure of a eumelanin

Figure 19.65 Tyrosinase and intermediates in melanin formation. (*a*) Tyrosinase. (*b*) Some intermediates in melanin synthesis and an example of the family of black eumelanins.

in posttranslational modification of tyrosinase, are induced. A lack of tyrosinase activity produces albinism (Clin. Corr. 19.21). There are various types of melanin (Figure 19.65*b*). All are aromatic quinones in which the conjugated bond system gives rise to color. The dark pigment that is usually called melanin is eumelanin, from the Greek for "good melanin." Other melanins are yellow or colorless. The role of tyrosine residues of thyroglobulin in thyroid hormone synthesis is presented on page 896.

Some proteins use a modified tyrosine residue as a prosthetic group in oxidation–reduction reactions. The only example reported in humans is topaquinone (trihydroxyphenylalanylquinone) or dihydroxyphenylalanine (TOPA), which is present in some plasma amine oxidases (Figure 19.66).

Tryptophan: Tryptophan is the precursor of approximately 50% of the body's pyridine nucleotides. The rest is obtained from the diet. The branch point leading to

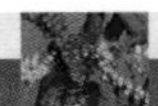

CLINICAL CORRELATION 19.19

Parkinson Disease

Idiopathic Parkinson disease (PD) (OMIM 168600 and 168601) generally manifests itself in people over the age of 60 years but occasionally earlier. Tremors may develop that gradually interfere with motor function and muscular rigidity of various muscle groups. Prominent symptoms include a loss of dopaminergic neurons in the substantia nigra (SN) and the presence of Lewy bodies (see the following) in cells in other areas of the brain. Of neurological diseases only Alzheimer occurs more frequently.

Parkinsonism is made up of diseases that mimic the symptoms of PD. These conditions may or may not exhibit Lewy bodies. PD can be definitively diagnosed only after postmortem pathological examination, with a finding of degeneration of cells in certain small nuclei of the cells of substantia nigra. These cells normally produce dopamine as a neurotransmitter, the amount released being proportional to the number of surviving cells.

Many causes have been suggested for PD. Among them are mutations of the alpha-synuclein (SNCA) gene. These mutations may cause deficiencies in vesicle movement, and consequent accumulation of dopamine in cytoplasm. Free dopamine in cytosol is subject to metabolism and excretion. Other possibilities for the etiology of PD include disruption in the ubiquitin system and interaction with factors leading to cell death. Some mutations appear to be mitochondrial and some present as X linked. In most cases of PD Lewy bodies, accumulated bits of alpha-synuclein protein, inside the nuclei of neurons, are seen in various parts of the brain.

A dramatic outbreak of parkinsonism occurred in young adult drug addicts using a derivative of pyridine (methylphenyltetrahydropyridine [MPTP]). It (or a contaminant produced during its manufacture) appears to be directly toxic to dopamine-producing cells of the substantia nigra. Symptomatic relief from PD, often dramatic, is obtained by administering DOPA, the precursor of dopamine. Clinical problems developed when DOPA (L-DOPA, levo-DOPA) was used for treatment of many people who have Parkinson disease. Side effects included nausea, vomiting, hypotension, cardiac arrhythmias, and various central nervous system symptoms. These were explained as effects of dopamine produced outside the central nervous system. Administration of DOPA analogs, such as carbidopa, that inhibit DOPA decarboxylase in peripheral organs but are unable to cross the blood–brain barrier has been effective in decreasing side effects and increasing effectiveness of the DOPA. The interactions of the many brain neurotransmitters are very complex, cell degeneration continues after treatment, and elucidation of the major biochemical abnormality has not yet led to complete control of the disease. Limited trials with injection of the brain glial cell line-derived neurotrophic factor (GDNF) into the putamen have been found to have a beneficial effect. A recent discovery suggests that loss of norepinephrine-producing cells in the locus coeruleus may contribute to symptoms of PD.

Feany, M. B. New genetic insights into Parkinson disease. *N. Engl. J. Med.* 351:1937, 2004; Lotharius, J., and Brundin, P. Impaired dopamine storage resulting from alpha-synuclein mutations may contribute to the pathogenesis of Parkinson disease. *Hum. Molec. Genet.* 11:2395, 2002; and Warner, T. T., and Schapira, A. H. V. Genetic and environmental factors in the cause of Parkinson disease. *Ann. Neurol.* 53, Suppl. 3:S16, 2003.

nicotinate mononucleotide (see Figure 19.46, p. 773) is at amino-carboxymuconic semialdehyde. Picolinate carboxylase forms 2-aminomuconic semialdehyde; this enzyme has a low K_m and is easily saturated with substrate. Since picolinate carboxylase has low activity in liver, some amino-carboxymuconic semialdehyde is spontaneously cyclized to quinolinic acid. Phosphoribosyl pyrophosphate provides a ribonucleotide moiety, and the final step is a decarboxylation leading to nicotinate mononucleotide. Note that the nicotinic acid ring is synthesized as a part of a nucleotide. Because kynurenine hydroxylase is inhibited by estrogen, women are more susceptible to pellagra, the disease produced by niacin deficiency (from the Italian pelle, "skin," and agra, "rough").

Several intermediates in the tryptophan degradative pathway are neuroactive, including L-kynurenine, kynurenate and quinolinate, and some of their derivatives. They function frequently by binding to the *N*-methyl D-aspartate (NMDA) receptor.

Serotonin and melatonin are tryptophan derivatives. Serotonin (5-hydroxytryptamine) results from hydroxylation of tryptophan by a tetrahydrobiopterin-dependent enzyme (Clin. Corr. 19.22) and decarboxylation by a pyridoxal-phosphate-containing enzyme (Figure 19.67*a*). It is a neurotransmitter in brain and causes contraction of smooth muscle of arterioles and bronchioles. It occurs widely in the body and may have other physiological roles. Melatonin, a sleep-inducing molecule, is *N*-acetyl-5-methoxytryptamine (Figure 19.67*b*). The acetyltransferase needed for its synthesis is present in pineal gland and retina. Melatonin is involved in regulation of circadian rhythm, being synthesized mostly at night. It appears to function by inhibiting

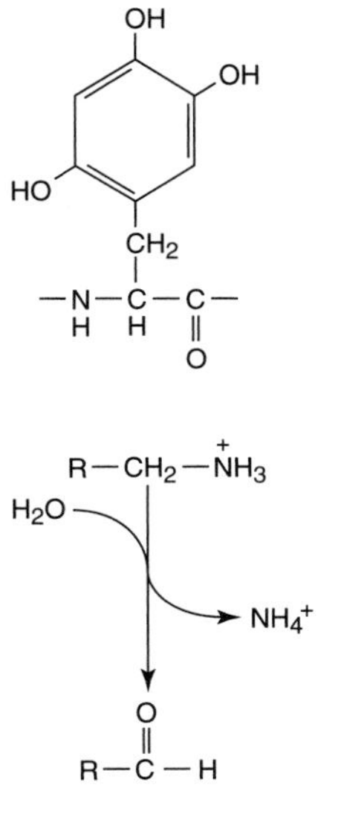

(*a*)

(*b*)

Figure 19.66 *(a)* Trihydroxyphenylalanine (TOPA). *(b)* Amine oxidase reaction.

CLINICAL CORRELATION 19.20

Tetrahydrobiopterin

A general test for biopterin deficiency is measurement of pterin metabolites in the urine.

An indirect test for biopterin deficiency measures loss of urinary metabolites resulting from the activity of the three aromatic amino acid hydroxylases and is diagnosed by a reduction in the urinary metabolites of these pathways (Figure 19.65). Studies with cultured skin fibroblasts can allow for determination of the source of the deficiency. Possible enzyme deficiencies are GTP cyclohydrolase (OMIM 233910), 6-pyruvoyltatrahydrobiopterin synthase (OMIM 261640), sepiapterin reductase (OMIM 182125), and dihydropteridine reductase (OMIM 261630). The first two of these enzymes are involved in the synthesis of biopterin, and the other two in its regeneration. Symptoms include red hair, psychomotor retardation, and progressive neurological deterioration. Some cases of biopterin deficiency respond well to supplementation of the diet with biopterin, and some are very difficult to treat. Neurotransmitter precursors (L-DOPA and hydroxytryptophan) can be administered. A compound called quinonoid dihydrobiopterin is also synthesized by enzymes that are encoded by different genes from those needed for synthesis of tetrahydrobiopterin. Addition of a methyl group from N^5,N^{10}-tetrahydrofolate to quinonoid dihydrobiopterin forms dihydrobiopterin. Supplementation with folinic acid can therefore sometimes be a successful treatment for biopterin deficiency.

In some cases of vitiligo it has been noted that there is a low level of catalase in the affected cells, leading to a buildup of hydrogen peroxide. The peroxide inactivates enzymes involved in regeneration of tetrahydrobiopterin, and therefore to reduction in amounts of tyrosine, the precursor of melanin.

The oxygenase domain of each nitric oxide synthase isoform also contains tetrahydrobiopterin. Unlike other enzymes where tetrahydrobiopterin is used as a source of reducing equivalents and is recycled by dihydrobiopterin reductase, in this case tetrahydrobiopterin activates heme-bound O_2 by donating a single electron. A combination of NO depletion and loss of synthesis of catecholamines and serotonin have been blamed on tetrahydrobiopterin deficiency.

Blau, N., Bonafe, L., and Thony, B. Tetrahydrobiopterin deficiencies without hyperphenylalininemis diagnosis and genetics of dopa-repensive distonia and sepiapterin reductase deficiency. *Mol. Genet. Metab.* 74: 172, 2001; Schallreuter, K. U., Moore, J., Wood, J. M., Beazley, W. D., et al. Epidermal H_2O_2 accumulation alters tetrahydrobiopterin (6BH4) recycling in vitiligo: identification of a general mechanism in regulation of all 6BH4-dependent processes? *J. Invest. Dermatol.* 116: 167, 200; and Shintaku, H. Disorders of tetrahydrobiopterin metabolism and their treatment. *Curr. Drug Metab.* 3: 123, 2002.

CLINICAL CORRELATION 19.21

Albinism (OMIM 203100)

Skin and hair color are controlled by numerous genetic loci in humans and exist in infinite variation. There are many cases in which the skin has little or no pigment. The chemical basis for classical oculocutaneous albinism (OCA1) results from a lack of tyrosinase ("tyrosinase-negative"). Lack of pigment in the skin makes albinos sensitive to sunlight, increasing the incidence of skin cancer in addition to burns; lack of pigment in the eyes may contribute to photophobia. In OCA1B the enzyme is partially missing. Other forms of albinism can be attributed to a deficiency in tyrosinase-related protein (TRP) or membrane-associated transporter protein (MATP). The skin of albinos with OCA1 has been found to contain melanocytes, the cells where melanin synthesis occurs. It is suggested that any residual tyrosinase in these is not available to metabolize tyrosine.

King, R. A., Hearing, V. J., Creel, D. J., and Oetting, W. S. Albinism. In: Scriver, C. R., Beaudet, A. L., Sly, W. S., Valle, D. (Eds.), *The Metabolic and Molecular Bases of Inherited Disease,* 8th ed. New York: McGraw Hill, 2001, 2:5587; and Witkop, C. J., Jr., Quevedo, W. C., Jr., Fitzpatrick, T. B., and King, R. A. Albinism. In Scriver, C. R., Beaudet, A. L., Sly, W. S., Valle, D. (Eds.), *The Metabolic and Molecular Bases of Inherited Disease,* 6th ed. New York: McGraw Hill, 1989, 2:2905.

synthesis and secretion of other neurotransmitters such as dopamine and GABA (see A Closer Look 19.5, p. 775).

Lysine: Medium- and long-chain fatty acids are transported into mitochondria for β-oxidation as carnitine conjugates (p. 691). **Carnitine** is synthesized from lysine residues in certain proteins. The first step is trimethylation of the ε-amino group of the lysine side chain, with AdoMet as the methyl donor (Figure 19.68). Free trimethyllysine is released by hydrolysis of these proteins and is converted in four steps to carnitine.

Histidine: Histamine (Figure 19.69), released from cells as part of an allergic response, is produced from histidine by histidine decarboxylase. Histamine has many physiological roles, including dilation and constriction by interacting with different types of

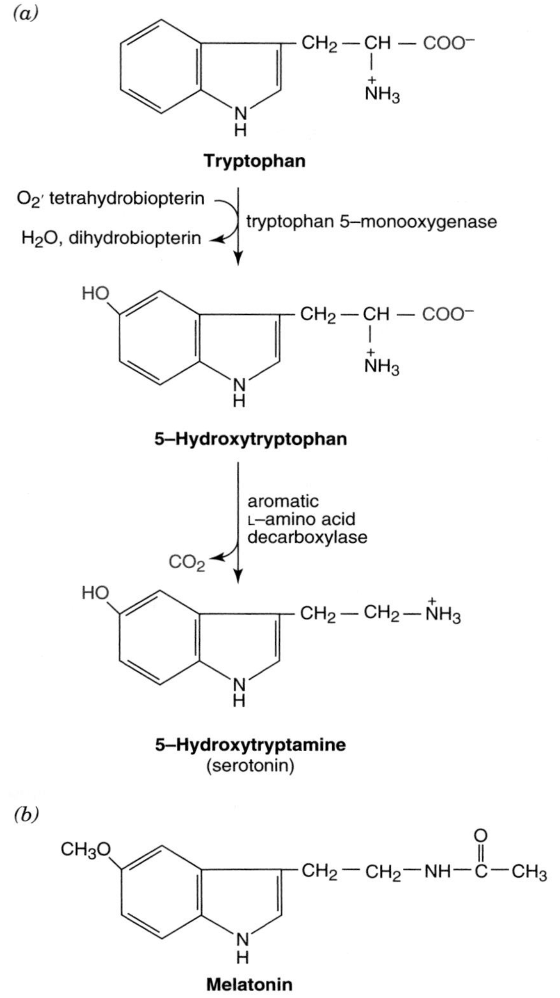

(a)

Tryptophan

O_2, tetrahydrobiopterin → H_2O, dihydrobiopterin; tryptophan 5-monooxygenase

5-Hydroxytryptophan

aromatic L-amino acid decarboxylase; CO_2

5-Hydroxytryptamine (serotonin)

(b)

Melatonin

Figure 19.67 ***(a)* Synthesis of serotonin (5-hydroxytryptamine). *(b)* Structure of melatonin.**

receptors on endothelium of blood vessels. An overreaction to histamine can lead to asthma and other allergic reactions.

Metabolites Made from More than One Amino Acid

Carnosine and **Anserine**: Carnosine (Figure19.70) is a dipeptide of the amino acids histidine and β-alanine. It is highly concentrated in muscle and brain tissues. Several roles have been proposed for this dipeptide, but only its role as an antioxidant has been well documented. Anserine (Figure19.70) is a dipeptide of β-alanine and *N*-methylhistidine. It is also an antioxidant.

Polyamines: Ornithine is a precursor of putrescine, the foundation molecule of the polyamines, highly cationic molecules that interact with DNA. Ornithine decarboxylase (Figure 19.71) is regulated by phosphorylation at several sites, presumably in response to specific hormones, growth factors, or cell cycle regulatory signals. It can also be induced, often the first easily measurable sign that cell division is imminent, since polyamines must be synthesized before mitosis can occur. Propylamine derived from AdoMet, leaving methylthioadenosine, is added to putrescine to form spermidine and spermine. Putrescine is formed by decarboxylation of ornithine and, with propylamine, forms spermidine. Addition of another propylamine gives spermine (Figure 19.72). The methylthioadenosine that remains can be used to regenerate methionine. Much of the polyamine needed by the body is provided by microflora in the gut or from the diet and is carried by the enterohepatic circulation. Meat has a high content

CLINICAL CORRELATION 19.22

Tryptophan Hydroxylase Deficiency

Unipolar major depression is associated with a low level in serotonin in the synaptic cleft (p. 943). In order to prevent reuptake into the presynaptic neuron, where it is degraded, a class of drugs called selective serotonin reuptake inhibitors (SSRIs), such as fluoxetine (Prozac), have been developed. A significant number of patients with unipolar depression do not respond to therapy with SSRIs. In some of these cases the defect is not in increased degradation of serotonin but in the loss of activity of the first enzyme in the synthetic pathway for this neurotransmitter, tryptophan hydroxylase. The two isozymes, tryptophan hydroxylases (TPH) 1 (OMIM 191060) and 2 (OMIM 607478), are coded on different chromosomes. TPH 1 is expressed in raphe neurons, pinealocytes, mast cells, mononuclear leukocytes, β cells of the pancreas, and intestinal and pancreatic enterochromaffin cells. TPH 1 controls several neural processes, including mammary homeostasis and liver regeneration. Mutations in TPH 1 have been associated with unipolar depression and suicidal ideation. The other isozyme, TPH2, is known as neuronal TPH, and mutations in this enzyme are related to unipolar major depression. Bipolar disorder does not seem to be associated with loss of either isozyme.

Bellivier, F., Chaste, P., and Malafosse, A. Association between the TPH gene A218C polymorphism and suicidal behavior: a meta-analysis. *Am. J. Med. Genet.* 124B:87, 2004; Walther, D. J.; Peter, J. U., Bashammakh, S., Hortnagl, H., et al. Synthesis of serotonin by a second tryptophan hydroxylase isoform. *Science* 299:76, 2003; Zhang, X., Beaulieu, J. M., Sotnikova, T. D., Gainetdinov, R. R., and Caron, M. G. Tryptophan hydroxylase-2 controls brain serotonin synthesis. *Science* 305:217, 2004; and Zhang, X., Gainetdinov, R. R., Beaulieu, J.-M., Sotnikova, T. D., et al. Loss-of-function mutation in tryptophan hydroxylase-2 identified in unipolar major depression. *Neuron* 45:11, 2005.

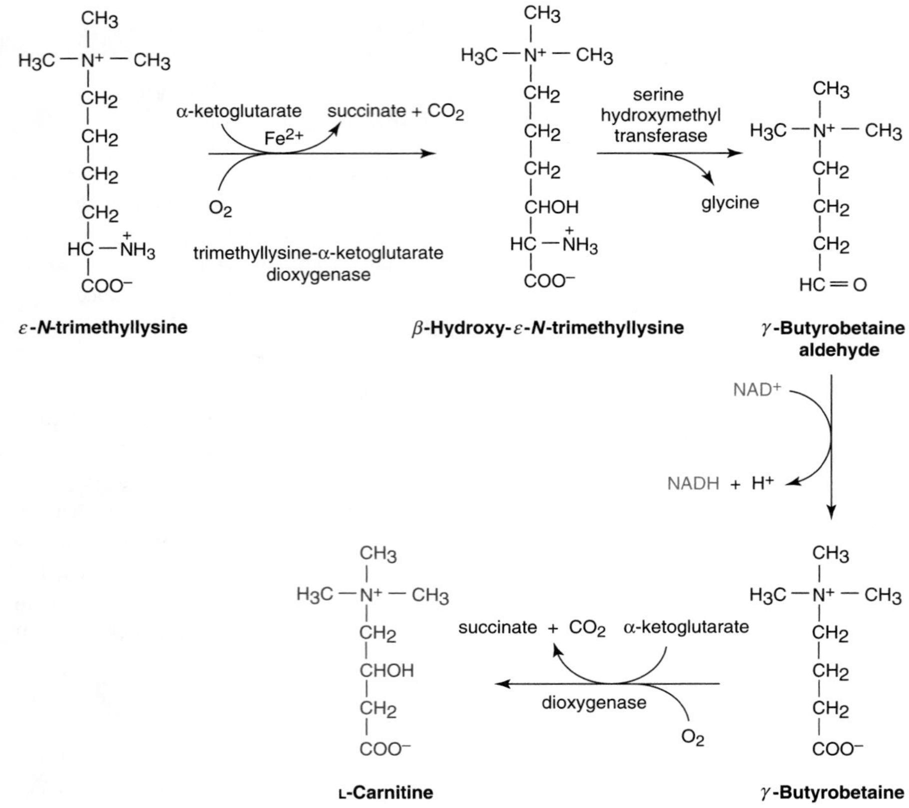

Figure 19.68 Biosynthesis of carnitine.

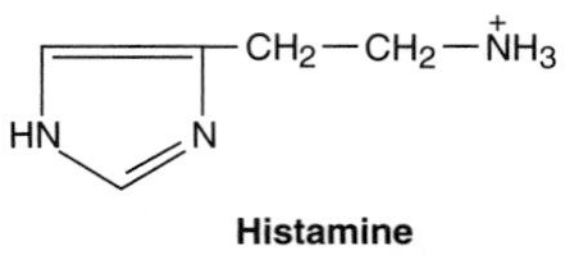

Figure 19.69 Histamine.

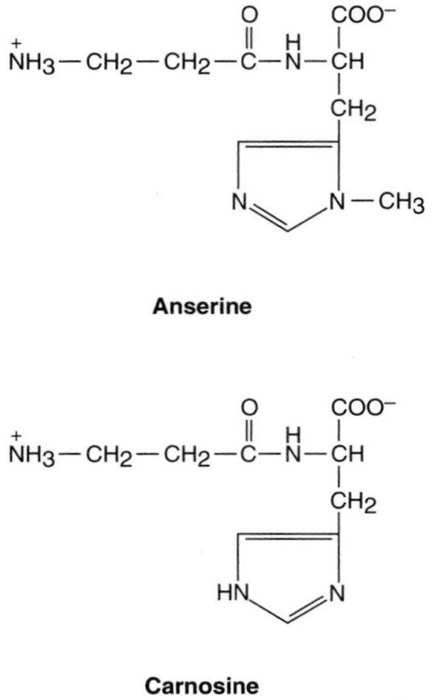

Figure 19.70 Anserine and carnosine.

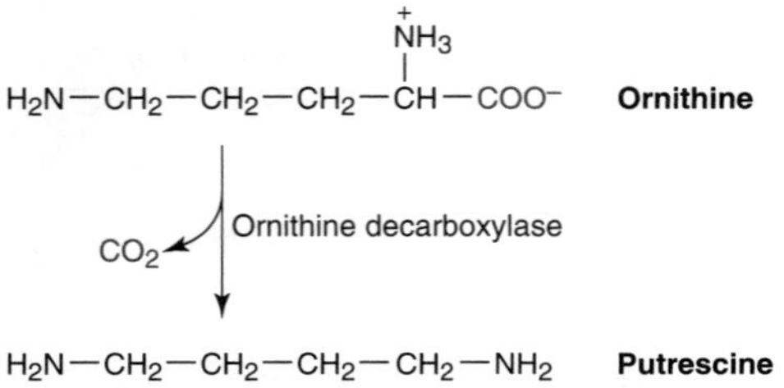

Figure 19.71 Decarboxylation of ornithine to putrescine. Putrescine is converted to spermidine and spermine.

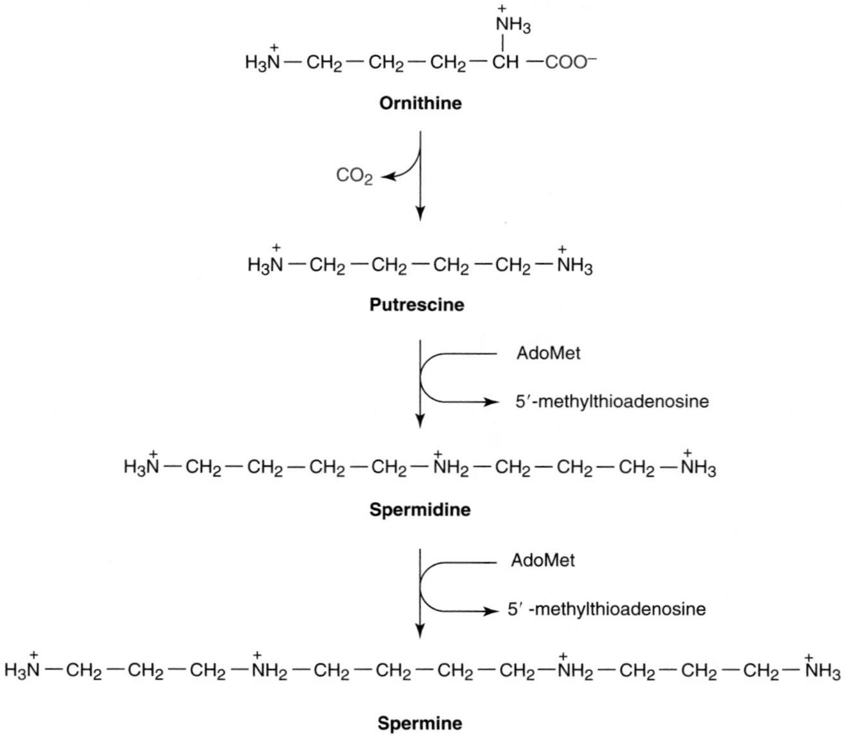

Figure 19.72 Polyamine synthesis.

of putrescine, but other foods contain more spermidine and spermine. Inhibition of polyamine synthesis is being investigated as a therapy for cancer.

Creatine: Storage of high-energy phosphate, particularly in cardiac and skeletal muscle, occurs by transfer of the phosphate group from ATP to creatine (p. 979). Creatine is synthesized by transfer of the guanidinium group of arginine to glycine, followed by addition of a methyl group from AdoMet (Figure 19.73). The amount of creatine in the body is related to muscle mass, and a certain percentage of this undergoes turnover each day. About 1% to 2% of preexisting creatine phosphate is cyclized nonenzymatically to creatinine (Figure 19.74) and excreted in urine, and new creatine is synthesized to replace it. The amount of creatinine excreted by an individual is therefore constant from day to day. When a 24-h urine sample is requested, the amount of creatinine in the sample can be used to determine whether the sample truly represents a whole day's urinary output.

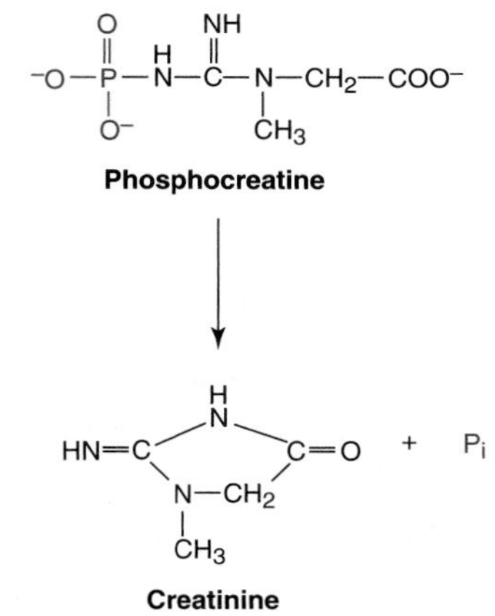

Phosphocreatine → Creatinine + P_i

Figure 19.74 Spontaneous reaction forming creatinine.

Glutathione

The tripeptide γ-glutamylcysteinylglycine, glutathione, has several important functions. It is a reductant, is conjugated with drugs to make them more water soluble (p. 431), is involved in transport of amino acids across cell membranes, is part of some leukotriene structures (p. 743), is a cofactor for some enzymatic reactions, and participates in the rearrangement of protein disulfide bonds. Glutathione as reductant is very important in maintaining stability of erythrocyte membranes. Its sulfhydryl group can be used to reduce peroxides formed during oxygen transport (p. 583). The resulting oxidized form consists of two molecules of GSH joined by a disulfide bond. This is reduced by glutathione reductase to two molecules of GSH at the expense of NADPH (Figure 19.75). The usual steady-state

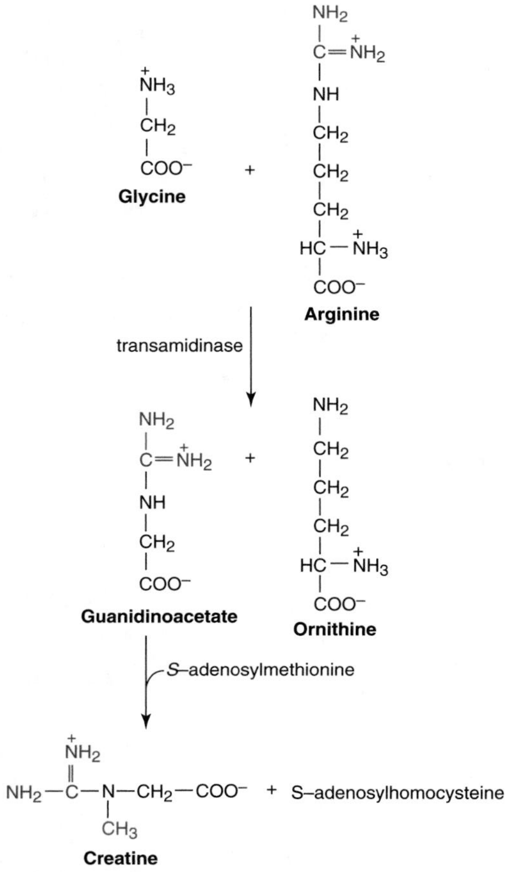

Glycine + Arginine —(transamidinase)→ Guanidinoacetate + Ornithine

Guanidinoacetate —(S-adenosylmethionine)→ Creatine + S-adenosylhomocysteine

Figure 19.73 Synthesis of creatine.

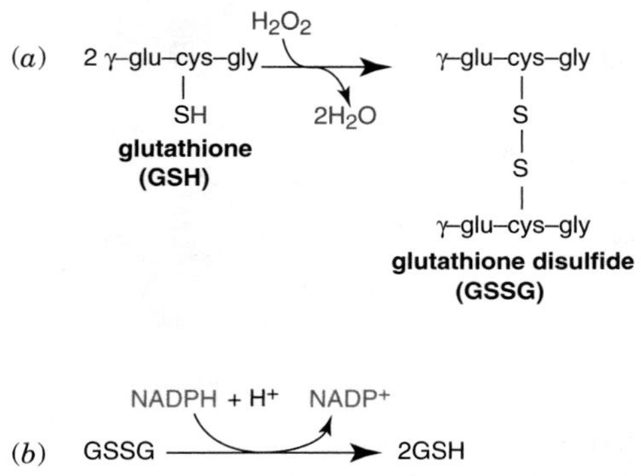

(*a*) 2 γ-glu–cys–gly (SH) [glutathione (GSH)] + H_2O_2 → γ-glu–cys–gly–S–S–γ-glu–cys–gly [glutathione disulfide (GSSG)] + $2H_2O$

(*b*) GSSG + NADPH + H^+ → 2GSH + $NADP^+$

Figure 19.75 (*a*) Scavenging of peroxide by glutathione peroxidase. (*b*) Regeneration of reduced glutathione by glutathione reductase.

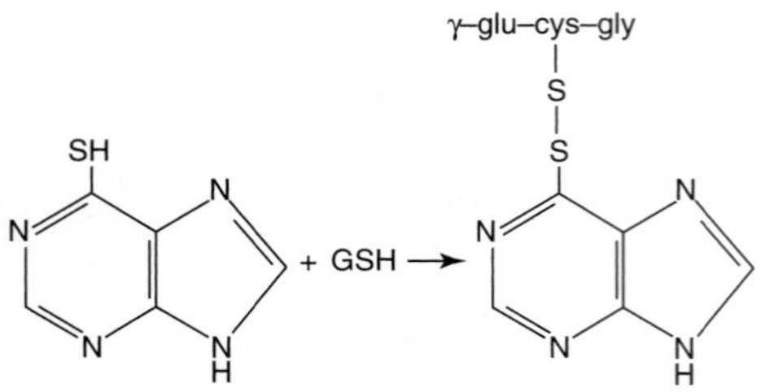

6–thiopurine + GSH → 6–thiopurine–glutathione

Figure 19.76 **Conjugation of a drug by glutathione transferase.**

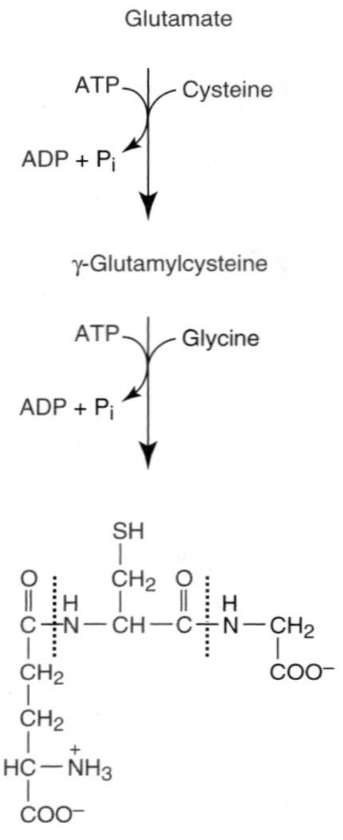

Glutamate → (ATP, Cysteine → ADP + P_i) → γ-Glutamylcysteine → (ATP, Glycine → ADP + P_i) → Glutathione

Glutathione (γ-glutamylcysteinylglycine)

Figure 19.77 **Synthesis of glutathione.**

ratio of GSH to GSSG in erythrocytes is 100:1. Conjugation of drugs such as 6-thiopurine with glutathione renders them more polar for excretion (Figure 19.76).

Glutathione is synthesized by formation of the dipeptide γ-glutamylcysteine and the subsequent addition of glycine. Both reactions require activation of carboxyl groups by ATP (Figure 19.77). Synthesis of glutathione is largely regulated by cysteine availability.

The γ-Glutamyl Cycle Transports Amino Acids

There are several mechanisms for transport of amino acids across cell membranes. Many are symport or antiport mechanisms (p. 486) coupled to sodium transport. The γ-glutamyl cycle for transmembrane transport of amino acids functions in kidneys and some other tissues but is of particular importance in renal epithelial cells. It is more energy-requiring than other mechanisms, but is rapid and has high capacity. The cycle involves several enzymes. The γ-glutamyl transpeptidase, located in the plasma membrane, adds glutamate from GSH to an extracellular amino acid. The resulting γ-glutamyl amino acid is transported into the cell by an amino acid transporter, where the γ-glutamyl amino acid is hydrolyzed to liberate the amino acid (Figure 19.78) and 5-oxoproline. Cysteinylglycine formed in the transpeptidase reaction is cleaved to its component amino acids. To regenerate GSH, glutamate is reformed from oxoproline in an ATP-requiring reaction, and GSH is resynthesized from its three component parts. Three ATPs are used in the regeneration of glutathione, one in formation of glutamate from oxoproline and two in formation of the peptide bonds.

Glutathione Concentration Affects Response to Toxins

When the body encounters toxic conditions such as peroxide formation, ionizing radiation, alkylating agents, or other reactive intermediates, it is beneficial to increase the level of GSH. Cysteine and methionine have been administered as GSH precursors, but they have the disadvantage of being precursors of an energy-expensive pathway to GSH. A more promising approach is administration of a soluble diester of GSH, such as γ-(α-ethyl) glutamylcysteinylethylglycinate. Very premature infants have a very low concentration of cysteine because of low cystathionase activity in liver. This keeps the GSH concentration low and makes them more susceptible to oxidative damage, especially from hydroperoxides formed in the eye after hyperbaric oxygen treatment. Under certain circumstances, such as rendering tumor cells more sensitive to radiation or parasites more sensitive to drugs, it is desirable to lower GSH levels. This can be achieved by administration of the glutamate analog buthionine sulfoximine (Figure 19.79) as a competitive inhibitor of GSH synthesis.

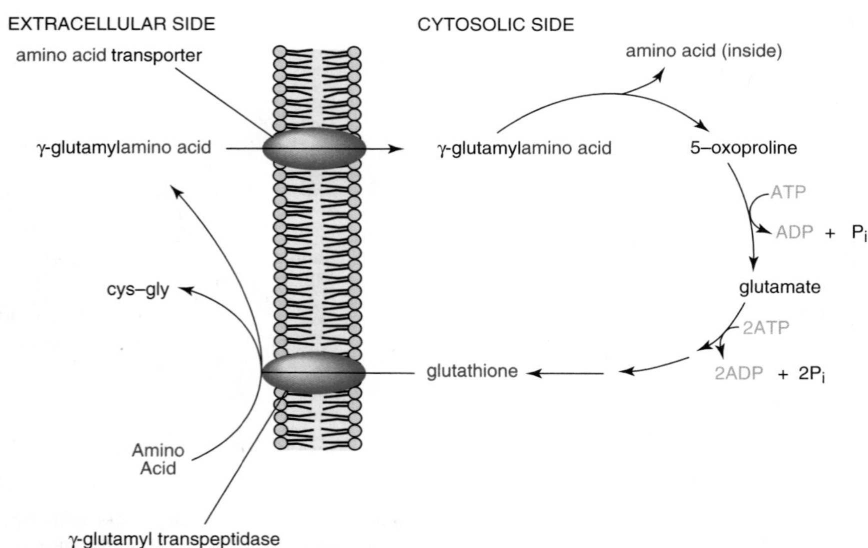

Figure 19.78 **γ-Glutamyl cycle for transporting amino acids.**

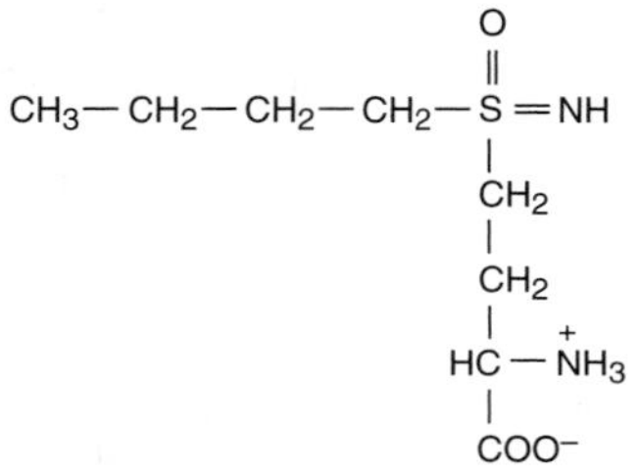

$CH_3-CH_2-CH_2-CH_2-S(=O)(=NH)-CH_2-CH_2-CH(\overset{+}{N}H_3)-COO^-$

Figure 19.79 **Buthionine sulfoximine.**

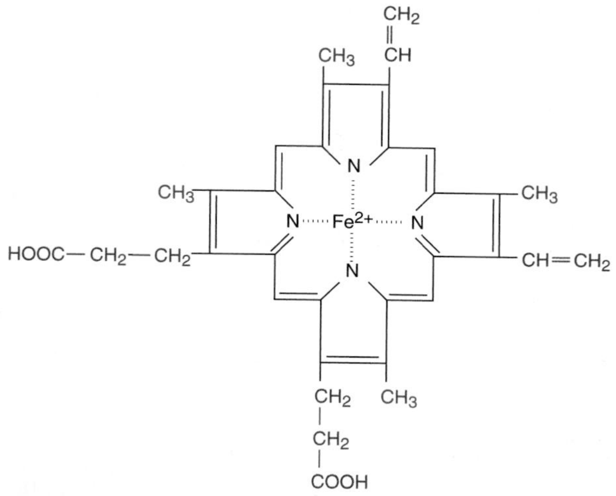

Figure 19.80 Structure of heme.

19.7 • HEME BIOSYNTHESIS

Heme is produced in virtually all mammalian tissues. Synthesis is most pronounced in the bone marrow and liver because of the requirements for incorporation into hemoglobin and the cytochromes, respectively. As depicted in Figure 19.80, heme is a largely planar molecule. Heme is one of the most stable of compounds, reflecting its strong resonance features.

Figure 19.81 depicts the pathway for heme biosynthesis. Porphyrinogens are unstable and can be readily oxidized, especially in the presence of light, by nonenzymatic means to their stable porphyrin products. In the latter cases, resonance between pyrrole groups is established by oxidation of the four methylene bridges. Figure 19.82, p. 793, depicts the enzymatic conversion of protoporphyrinogen to protoporphyrin by this oxidation mechanism. This is the only enzyme-catalyzed oxidation of a porphyrinogen in humans; all other porphyrinogen–porphyrin conversions are nonenzymatic and catalyzed by light.

Enzymes Involved in Heme Biosynthesis

δ-Aminolevulinic Acid Synthase: δ-Aminolevulinic acid (ALA) synthase controls the rate-limiting step of heme synthesis in all tissues studied. The synthesis of the enzyme occurs in the cytosol, being directed by mRNA derived from the nucleus. The enzyme is transferred into the matrix of the mitochondrion, where it interacts with succinyl CoA, an intermediate of the tricarboxylic acid cycle. In the cytosol, each subunit is in the unfolded state, the only form that can pass into mitochondria directed by a basic N-terminal signaling sequence. An ATP-dependent cytosolic molecule, known as a chaperone protein, maintains it in the unfolded extended state. Thereafter, the N-terminal sequence is cleaved by a metal-dependent protease in the mitochondrial matrix, to yield an ALA synthase subunit of 65 kDa. Within the matrix another oligomeric chaperon protein catalyzes the correct folding in a second ATP-dependent pro-cess (Figure 19.83, p. 793). The ALA synthase has a short biological half-life (~60 min). Both the synthesis and the activity of the enzyme are subject to regulation by a variety of substances; 50% inhibition of activity occurs in the presence of 5 mM hemin, and virtually complete inhibition at a 20 mM concentration. The enzymatic reaction involves the condensation of glycine with succinyl CoA to produce δ-aminolevulinic acid. The reaction has an absolute requirement for pyridoxal phosphate. Two isoenzymes exist for ALA synthase; only the mRNA of the erythrocytic form contains an iron response element (IRE). Mutations in the erythrocytic form give rise to sideroblastic anemia, where excess iron is found in the mitochondria of developing red blood cells, but heme synthesis is defective. The condition can be X linked or acquired.

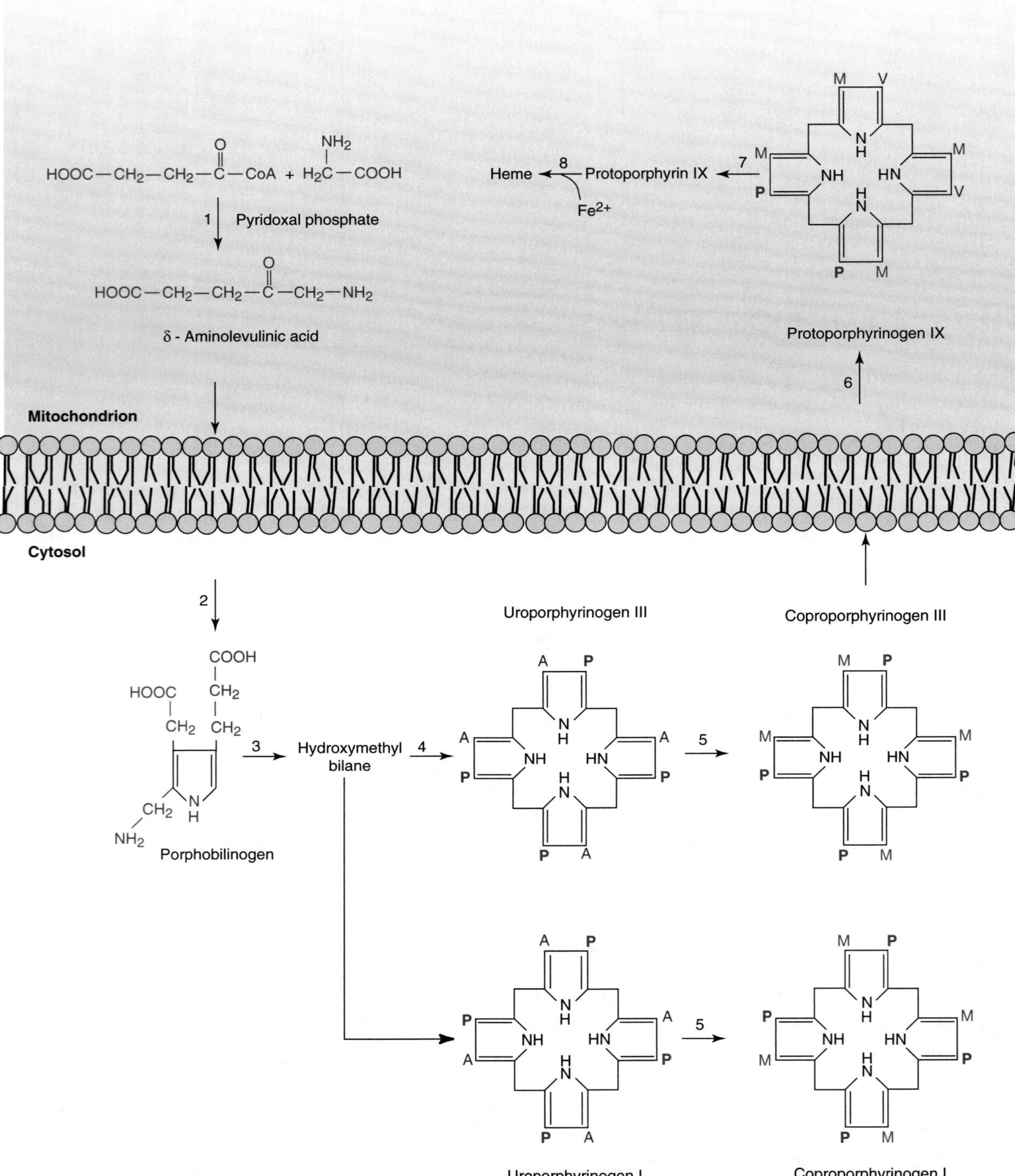

Figure 19.81 Pathway for heme biosynthesis. Numbers indicate enzymes involved in each step as follows: 1. ALA synthase, 2. ALA dehydratase (porphobilinogen synthase), 3. porphobilinogen deaminase (hydroxymethylbilane synthase), 4. uroporphyrinogen III synthase, 5. uroporphyrinogen decarboxylase, 6. coproporphyrinogen III oxidase, 7. protoporphyrinogen IX oxidase, and 8. ferrochelatase. Pyrrole ligands are indicated as follows: P = propionic, A = acetic, M = methyl, and V = vinyl.

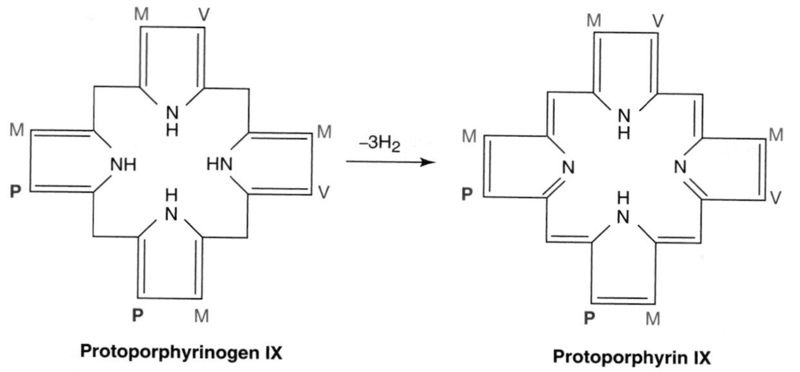

Protoporphyrinogen IX $\xrightarrow{-3H_2}$ **Protoporphyrin IX**

Figure 19.82 Action of protoporphyrinogen IX oxidase, an example of the conversion of a porphyrinogen to a porphyrin.

Aminolevulinic Acid Dehydratase: The second enzyme in the pathway, aminolevulinic acid dehydratase (porphobilinogen synthase), is cytosolic and consists of eight subunits, of which only four interact with the substrate. This protein interacts with the substrate to form a Schiff base, but in this case the ε-amino group of a lysine residue binds to the carbonyl carbon of the substrate molecule (Figure 19.84). Two molecules of ALA condense asymmetrically to form porphobilinogen, a five member heterocyclic ring structure with three side chains, two of which are acetic acid and propionic acid. The ALA dehydratase is a zinc-containing enzyme and is very sensitive to inhibition by heavy metals, particularly lead. A characteristic of **lead poisoning** is the elevation of ALA in the absence of an elevation of porphobilinogen.

Porphobilinogen Deaminase and Uroporphyrinogen III Synthase: The synthesis of the porphyrin ring is a complicated process. A sulfhydryl group on porphobilinogen deaminase (hydroxymethylbilane synthase) (Clin. Corr. 19.23) forms a thioether bond with a porphobilinogen residue through a deamination reaction. Thereafter, five additional porphobilinogen residues are deaminated successively to form a linear hexapyrrole adduct with the enzyme. The adduct is cleaved hydrolytically to form both an enzyme–dipyrrolomethane complex and the linear tetrapyrrole, hydroxymethylbilane. The enzyme–dipyrrolomethane complex is then ready for another cycle to generate another tetrapyrrole. Thus, dipyrrolomethane is the covalently attached cofactor for the enzyme. Porphobilinogen deaminase has no ring-closing function; hydroxymethylbilane closes spontaneously in an enzyme-independent step to form uroporphyrinogen I, a structure consisting of four linked pyrrole rings, if no additional factors are present. However, the deaminase is closely associated with a second protein, **uroporphyrinogen III synthase**, which directs the synthesis of the III isomer. The formation of the latter involves an intermediate where the rings are linked by only one atom (a spiro structure) generated from hydroxymethylbilane; this allows the inversion of one of the pyrrole groups (Figure 19.85, p. 795). In the absence of the uroporphyrinogen III synthase, uroporphyrinogen I is synthesized slowly; in its presence, the III isomer is synthesized rapidly. The uroporphyrinogens have two different substituents for each pyrrole group. Going clockwise around the ring, the substituents can be arranged as ABABABAB (where A and B are different substituents), forming a type I porphyrinogen, or the arrangement can be ABABABBA, forming a type III porphyrinogen. In principle, two other arrangements can occur to form porphyrinogens II and IV, and these can be synthesized chemically; however, they do not occur naturally. Later in the heme synthetic pathway are **protoporphyrinogen** and **protoporphyrin** with three types of substituents, and the classification becomes more complicated; type IX is the only form that is synthesized naturally. A rare recessively inherited disease, erythropoietic porphyria, associated with marked cutaneous light sensitization, is due to an abnormality of reticulocyte uroporphyrinogen III synthase. Here large amounts of the type I isomers of uroporphyrinogen and coproporphyrinogen are synthesized in the bone marrow.

Uroporphyrinogen Decarboxylase: The reactions occurring on the side groups attached to the tetrapyrrole ring involve the colorless intermediates known as **porphyrinogens**. Though exhibiting resonance features within each pyrrole ring, these do not demonstrate resonance between the pyrrole groups. Hence, the porphyrinogens are

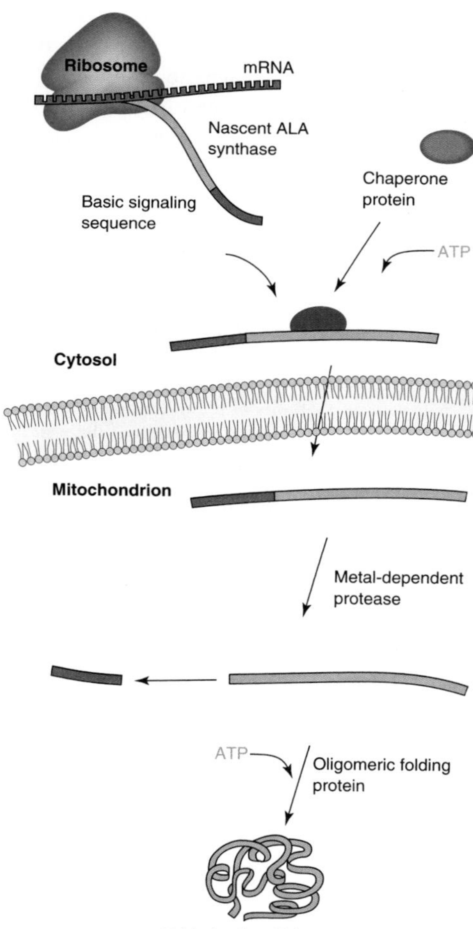

Figure 19.83 Synthesis of δ-aminolevulinic acid (ALA) synthase.

CLINICAL CORRELATION 19.23

Acute Intermittent Porphyria

Acute intermittent porphyria (AIP) (OMIM 176000) is caused by mutation in the gene encoding hydroxymethylbilane synthase (HMBS), also referred to as porphobilinogen deaminase (PBGD). First attacks usually occur after puberty and are frequently precipitated by drugs such as barbiturates, by alcohol, and by infection. This is due to induction of hepatic 5-aminolevulinate synthase. Women are more prone to AIP. AIP is characterized by increased urinary excretion of the HMBS precursors, delta-aminolevulinic acid (ALA) and porphobilinogen (PBG). It is inherited as an autosomal dominant trait, so even heterozygotes may exhibit symptoms. Only about 10% to 20% of AIP gene carriers become symptomatic during their lifetime. Prevention consists of notification of family members so that they may avoid the precipitating factors.

Symptoms include marked weakness in the arms and legs, a slightly rapid heart rate, and moderate hypertension. There may have been have been earlier episodes of severe abdominal pain without a diagnosis. High levels of porphobilinogen are often noted in urine.

Puy, H., Deybach, J. C., Lamoril, J., Robreau, A. M., et al. Molecular epidemiology and diagnosis of PBG deaminase gene defects in acute intermittent porphyria. *Am. J. Hum. Genet.* 60:1373, 1997; and Solis, C., Martinez-Bermejo, A., Naidich, T. P., Kaufmann, W. E., et al. Acute intermittent porphyria: Studies of the severe homozygous dominant disease provides insights into the neurologic attacks in acute porphyrias. *Arch. Neurol.* 61:1764, 2004.

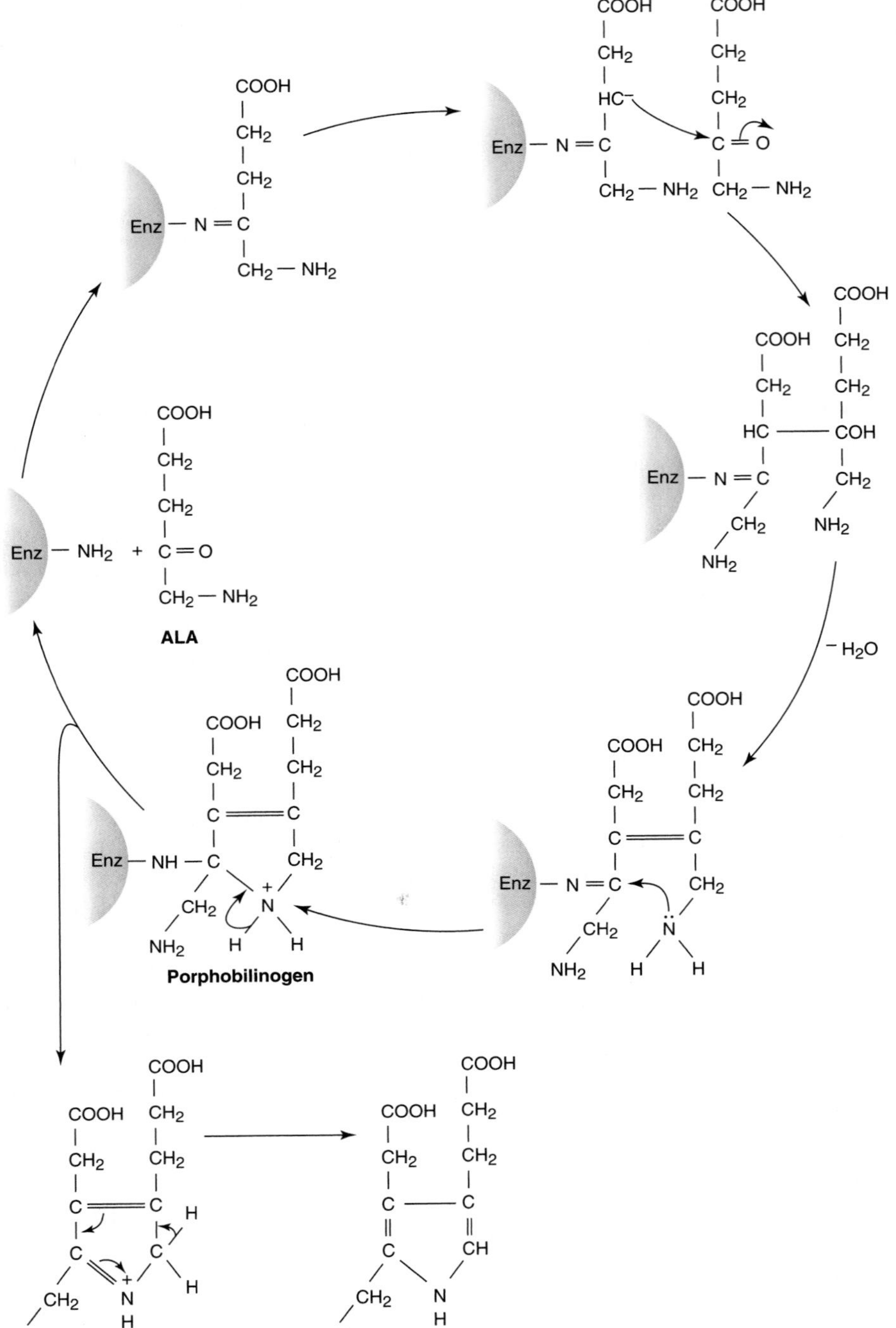

Figure 19.84 **Synthesis of porphobilinogen.**

unstable and can be readily oxidized, especially in the presence of light, by nonenzymatic means to their stable **porphyrin** products. In the latter cases, resonance between pyrrole groups is established by oxidation of the four methylene bridges. Figure 19.82 (p. 793) depicts the enzymatic conversion of protoporphyrinogen to protoporphyrin by this oxidation mechanism. This is the only enzyme-catalyzed oxidation of a porphyrinogen in humans; all other porphyrinogen–porphyrin conversions are nonenzymatic and catalyzed by light.

Uroporphyrinogen decarboxylase acts on the side chains of uroporphyrinogens to form coproporphyrinogens, where acetic acid groups have been decarboxylated to leave methyl groups. The protein catalyzes the conversion of both I and III isomers of

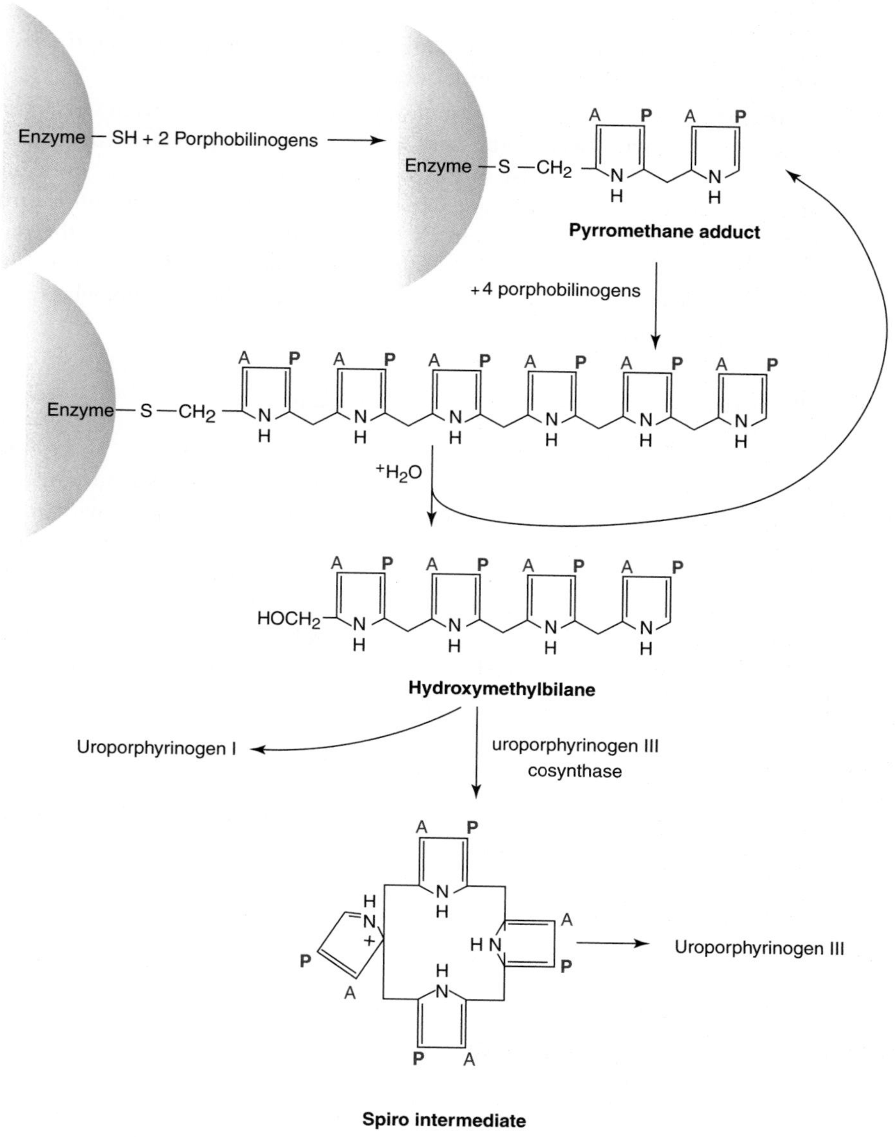

Figure 19.85 **Synthesis of uroporphyrinogens I and III.** Enzyme in blue is uroporphyrinogen I synthase.

uroporphyrinogen to the respective coproporphyrinogen isomers. Uroporphyrinogen decarboxylase is inhibited by iron salts. Clinically the most common cause of porphyrin derangement is associated with patients who have a single gene abnormality for uroporphyrinogen decarboxylase, leading to 50% depression of the enzyme's activity. This disease, which shows cutaneous manifestations primarily with sensitivity to light, is known as **porphyria cutanea tarda**. The condition is not expressed unless patients either take drugs that cause an increase in porphyrin synthesis or drink large amounts of alcohol. Alcohol-induced cirrhosis leads to the accumulation of iron, which then acts to inhibit further the activity of uroporphyrinogen decarboxylase. The treatment for this condition is phlebotomy.

Coproporphyrinogen Oxidase: Coproporphyrinogen oxidase, a mitochondrial enzyme, is specific for type III coproporphyrinogen. It does not act on the type I isomer. Coproporphyrinogen III enters mitochondria and it is converted to protoporphyrinogen IX. A dominant hereditary disease associated with a deficiency of this enzyme leads to a form of hereditary hepatic porphyria, known as **hereditary coproporphyria**.

Protoporphyrinogen Oxidase: Another mitochondrial enzyme, protoporphyrinogen oxidase, oxidizes the methylene bridges and generates protoporphyrin IX, which, in contrast to the other heme precursors, is very water insoluble. Excess amounts of protoporphyrin IX that are not converted to heme are excreted by the biliary system into the intestinal tract. An autosomal dominant disease, **variegate porphyria**, is due to a deficiency of protoporphyrinogen oxidase.

Ferrochelatase: The last enzyme of the pathway is ferrochelatase, which inserts ferrous iron into protoporphyrin IX. Ascorbic acid and cysteine are required as reductants for its activity. The protein is sensitive to the effects of heavy metals, especially lead and, of course, to iron deprivation. In these latter instances, zinc instead of iron is incorporated to form a zinc–protoporphyrin IX complex. In contrast to heme, the zinc–protoporphyrin IX complex is brilliantly fluorescent and easily detectable in small amounts. Prokaryotic ferrochelatase does not contain a prosthetic group, whereas the mammalian enzyme contains a Fe_2S_2 group.

ALA Synthase Catalyzes Rate-Limiting Step of Heme Biosynthesis

ALA synthase controls the rate-limiting step of heme synthesis in all tissues. Succinyl CoA and glycine are substrates for a variety of reactions. The modulation of the activity of ALA synthase determines the quantity of the substrates that will be shunted into heme biosynthesis. Heme and hematin, a derivative of oxidized heme, act both as a repressor of the synthesis of ALA synthase and as an inhibitor of its activity. Since heme resembles neither the substrates nor the product of the enzyme's action, it is probable that it acts at an allosteric site. Almost 100 different drugs and metabolites can cause induction of ALA synthase, some as much as 40-fold. The effect of pharmacological agents has led to the important clinical

TABLE 19.2 • Derangements in Porphyrin Metabolism

Enzyme	*Substrate*	*Prevalence*	*Indicators and Symptoms*	*OMIM**	*Porphyria*
ALA synthase	Glycine, Succinyl CoA			125290	Contributes to porphyrias by induction
ALA dehydratase (porphobilinogen synthase)	D-Amino-levulinic acid	Rare	Mild to severe neuropathy; intermittent fever, hypertension	125270	Ala-dehydratase deficiency
PBG deaminase (hydroxymethylbilane synthase)	Porphobilinogen	Among most prevalent	Mild to severe neuropathy; intermittent fever, hypertension and gastric problems	176000	Acute intermittent porphyria
Uroporphyrinogen III synthase	Hydroxymethyl-bilane	Very rare	Very severe; mutilating; evidence in infancy Pink, red, or violet urine staining diapers	606938	Congenital or hereditary erythropoietic porphyria
Uroporphyrinogen III decarboxylase	Uroporphyrinogen III	Among most prevalent	Mild to severe dermapathy; siderosis; liver	176100	Porphyria cutanea tarda
Coproporphyrinogen III oxidase	Coproporphyrinogen III	Among most prevalent	Intermittent fever; hypertension; liver	121300	Coproporphyria
Protoporphyrinogen oxidase	Protoporphyrinogen IX	Among most prevalent	Intermittent fever; hypertension; skin and neurological involvement	176200	Variegate porphyria
Ferrochelatase	Protoporphyrin IX	Quite rare	Mild to severe dermopathy; intermittent fever; hypertension; liver damage; gallstones	177000	Erythropoietic protoporphyria
No particular enzyme		Among most prevalent	Dermatopathy in some cases but not in others; usually neuropathy; variable intermittent fever; hypertension		Intoxication porphyria
Ferrochelatase; ALA dehydratase		Frequent but declining	All tissues		Lead poisoning

*OMIM: Online Mendelian Inheritance in Man. http://www.ncbi.nlm.nih.gov/Omim/Entry#/.

feature where some patients with certain kinds of porphyria have had exacerbations of their condition following the inappropriate administration of certain drugs (e.g., barbiturates). ALA dehydratase is also inhibited by heme; but this is of little physiological consequence, since the activity of ALA dehydratase is about 80-fold greater than that of ALA synthase, and thus heme-inhibitory effects are reflected first in the activity of ALA synthase.

Glucose or one of its proximal metabolites inhibits heme biosynthesis by a mechanism involving inactivation of transcription factors. This is of clinical relevance, since some patients manifest their porphyric state for the first time when placed on a very low caloric (and therefore glucose) intake. Other regulators of porphyrin metabolism include certain steroids. Steroid hormones (e.g., oral contraceptive pills) with a double bond in ring A between C4 and C5 atoms can be reduced by two different reductases. The product of 5α-reduction has little effect on heme biosynthesis; however, the product of 5β-reduction serves as a stimulus for the synthesis of ALA synthase.

Porphyrias

Porphyrias are a family of diseases of great interest because they have revealed that the regulation of heme biosynthesis is complicated. The clinical presentations of the different porphyrias provide a fascinating exposition of biochemical regulatory abnormalities and their relationship to pathophysiological processes. Table 19.2 lists some characteristics of the different porphyrias.

19.8 • HEME CATABOLISM

Catabolism of heme-containing proteins presents two requirements to the mammalian host: (1) a means of processing the hydrophobic products of porphyrin ring cleavage and (2) retention and mobilization of the iron atom so that it may be reutilized. Red blood cells have a life span of approximately 120 days. Senescent cells are recognized by changes in their membranes and engulfed by the reticuloendothelial system at extravascular sites. The globin chains denature, releasing heme into the cytoplasm. The globin is degraded to its constituent amino acids, which are reutilized for general metabolic needs.

Figure 19.86 depicts the events of heme catabolism. Heme is degraded primarily by an endoplasmic reticulum enzyme system in reticuloendothelial cells that requires molecular oxygen and NADPH. **Heme oxygenase** has two isomers; type I is substrate inducible and type II is constitutive. The enzyme catalyzes cleavage of the α-methene bridge, which joins the two pyrrole residues that contain vinyl substituents. The α-methene carbon is converted quantitatively to carbon monoxide. This is the only endogenous source of **carbon monoxide** in humans. Most of the carbon monoxide is released via the respiratory tract, and measurement of CO in respired air is a good measure of heme destruction. The oxygen present in the carbon monoxide and in the newly derivatized lactam rings of biliverdin come directly from molecular oxygen. The stoichiometry of the reaction requires 3 mol of oxygen for each ring cleavage. Heme oxygenase will only use heme as a substrate, with the iron possibly participating in the cleavage mechanism. Thus free protoporphyrin IX is not a substrate. The linear tetrapyrrole **biliverdin IX** is formed by heme oxygenase. Biliverdin IX is reduced by **biliverdin reductase** to bilirubin IX. The products of heme oxygenase action are found to be cytoprotective (Clin. Corr. 19.24).

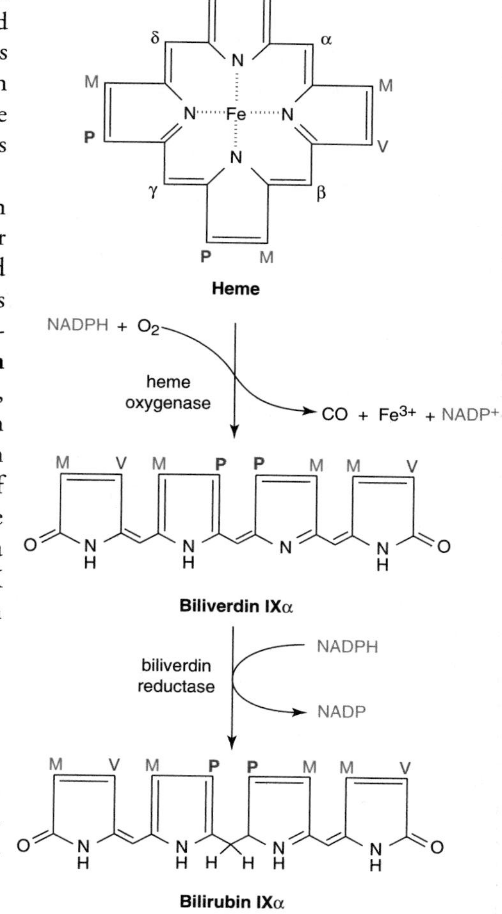

Figure 19.86 Formation of bilirubin from heme Greek letters indicate labeling of methene carbon atoms in heme.

Bilirubin Is Conjugated to Form Bilirubin Diglucuronide in Liver

Bilirubin is derived from senescent red cells but also from turnover of other heme-containing proteins, such as the cytochromes. Studies with labeled glycine as a precursor have revealed that an early-labeled bilirubin, with a peak within 1–3 h, appears very quickly after a pulsed administration of the labeled precursor. A larger amount of bilirubin appears much later at about 120 days, reflecting the turnover of heme in red blood cells. Early-labeled bilirubin can be divided into two parts: an early-early part, which reflects turnover of heme proteins, such as the cytochromes, in the liver, and a late-early part, which is a result

CLINICAL CORRELATION 19.24

Cytoprotective Role of Heme Oxygenase

Carbon monoxide (CO) and biliverdin are not merely passive products of heme oxygenase action. Biliverdin is an antioxidant and as such performs a significant role when heme oxygenase is induced by stress. CO, like nitric oxide (NO) which it structurally resembles, acts on smooth muscle, and thus as a vasodilator has been shown to have a protective effect, for instance in stroke victims. Like NO, CO acts apparently through cyclic GMP. The interaction with NO is complex; at times CO is complementary to and at other settings antagonistic to NO. Simplistically, CO appears generally to function protectively whereas NO, depending on the circumstance, can either be protective or cause cell damage. Bilirubin has been shown to inhibit the expression of inducible NO-synthase.

Baranano, D. E., and Snyder, S. Neural roles for heme oxygenase: Contrasts to nitric oxide. *Proc. Natl. Acad. Sci. USA* 98:10996, 2001; Baranano, D. E., Rao, M., Ferris, C. D., et al. Biliverdin reductase: A major physiologic cytoprotectant. *Proc. Natl. Acad. Sci. USA* 99:16095, 2002; Chen, Y. M., Yet, S. F., and Perrella, M. A. Role of heme oxygenase in the regulation of blood pressure and cardiac function. *Exp. Biol. Med.* 228:447, 2003; and Wang, W. W., Smith, D. L., and Zucker, S. D. Bilirubin inhibits iNOS expression and NO production in response to endotoxin in rats. *Hepatology* 40:424, 2004.

CLINICAL CORRELATION 19.25

Neonatal Isoimmune Hemolysis (OMIM 111689)

Rh-negative women pregnant with Rh-positive fetuses develop antibodies to Rh factors. These antibodies cross the placenta to hemolyze fetal red blood cells. Usually this is not of clinical relevance until about the third Rh-positive pregnancy, in which the mother has had antigenic challenges from earlier babies. Antenatal studies reveal rising maternal levels of IgG antibodies against Rh-positive red blood cells, indicating that the fetus is Rh-positive. DNA testing can be used to determine RhD (Rhesus blood group, D antigen) type in chorionic villus samples or amniotic cells. A safer method for diagnosis of RhD status has recently been developed. PCR technology can be used to amplify DNA from fetal cells in the mother's bloodstream, obviating the need for more invasive procedures. Damage from Rhesus is initiated at birth. Placental transfer of fetal bilirubin occurs with excretion through the maternal liver. Because hepatic enzymes of bilirubin metabolism are poorly expressed in the newborn, infants may not be able to excrete the large amounts of bilirubin that can be generated from red cell breakdown. At birth these infants usually appear normal; however, the unconjugated bilirubin in the umbilical cord blood is elevated up to 4 mg/dL, due to the hemolysis initiated by maternal antibodies. During the next 2 days the serum bilirubin rises, reflecting continuing isoimmune hemolysis, leading to jaundice, hepatosplenomegaly, ascites, and edema. If untreated, signs of central nervous system damage can occur, with the appearance of lethargy, hypotonia, spasticity, and respiratory difficulty, constituting the syndrome of kernicterus.

Treatment involves exchange transfusion with whole blood, which is serologically compatible with both the infant's blood and the maternal serum. This is necessary to prevent hemolysis of the transfused cells. Additional treatment includes external phototherapy, which facilitates the breakdown of bilirubin. The problem can be prevented by treating Rh-negative mothers with anti-Rh globulin. These antibodies recognize the fetal red cells, block the Rh antigens, and cause them to be destroyed without stimulating an immune response in the mothers.

Innan, H. A two-locus gene conversion model with selection and its application to the human RHCE and RHD genes. *Proc. Nat. Acad. Sci. USA*100:8793, 2003; Lo, Y. M. D., Hjelm, N. M., Fidler, C., Sargent, I. L., et al. Prenatal diagnosis of fetal RhD status by molecular analysis of maternal plasma. *N. Engl. J. Med.* 339:1734, 1998.

of incomplete erythropoesis. The latter is a measurement of ineffective erythropoiesis and can be very pronounced in disease states such as pernicious anemia and the thalassemias. Bilirubin is produced in cells of the reticulo–endothelial system, including phagocytes, the Kupffer cells of the liver, and cells in the spleen and bone marrow (Clin. Corr. 19.25). Bilirubin is poorly soluble in aqueous solutions at physiological pH values. When transported in plasma, it is bound to serum albumin with an association constant greater than 10^6. Albumin contains one high-affinity site and another with a lesser affinity. However, bilirubin toxicity **(kernicterus)**, which is manifested by the transfer of bilirubin to membrane lipids, suggests that the weak affinity of the second site does not allow it to serve effectively in the transport of bilirubin. Bilirubin is released from albumin and taken up by liver cells using a transporter mechanism. Once in the cell, bilirubin is bound to two major proteins, cytosolic Y protein (glutathione S-transferase B, also called ligandin) and cytosolic Z protein (also known as fatty-acid-binding protein [FABP]). The binding of bilirubin to these proteins prevents the return flow out of the cell. Ligandin has been purified and

CLINICAL CORRELATION 19.26

Bilirubin UDP-Glucuronosyltransferase Deficiency

Three forms of mutations can lead to bilirubin UDP-glucuronosyltransferase deficiency (OMIM 191740). All three are in the UGT1A1 gene (UDP-glucuronosyltransferase 1A1), and all three have different outcomes. The mutation that leads to Crigler–Najjar type I deficiency produces a situation where very low levels of bilirubin diglucuronides are found in blood, and, unlike the other types, type I does not respond to phenobarbital therapy. Crigler–Najjar syndrome type II that results in high levels of free bilirubin and the third type of transferase deficiency, Gilbert syndrome (intermediate amounts of bilirubin), are both responsive to phenobarbital therapy. Over 40 separate mutations have been described, and the levels of bilirubin in blood vary from patient to patient. In some cases very little free bilirubin is found, and this is due to a low rate of bilirubin production. The effectiveness of phenobarbital treatment is based on the ability of this drug to induce synthesis of UDP-glucuronosyltransferase.

Chowdhury, J. R., Bakker, C., et al. The genetic basis of the reduced expression of bilirubin UDP-glucuronosyltransferase 1 in Gilbert syndrome. *N. Engl. J. Med.* 333:1171, 1995; Burchell, B, and Hume, R. Molecular genetic basis of Gilbert syndrome. *J. Gastroenterol. Hepatol.* 14:960, 1999; Gantla, S., Bakker, C. T. M., Deocharan, B., Thummala, N. R., et al. Splice-site mutations: A novel genetic mechanism of Crigler–Najjar syndrome type 1. *Am. J. Hum. Genet.* 62:585, 1998; and Gong, Q.-H., Cho, J. W., Huang, T., Potter, C., et al. Thirteen UDP-glucuronosyltransferase genes are encoded at the human UGT1 gene complex locus. *Pharmacogenetics* 11:357, 2001.

characterized as having two subunits (22 kDa and 27 kDa). The stoichiometry of binding is one bilirubin molecule per complete ligandin molecule.

Once in the hepatocyte the propionyl side chains of bilirubin are conjugated to form a diglucuronide (Clin. Corr. 19.26 and Figure 19.87). The reaction utilizes uridine diphosphoglucuronate derived from the oxidation of uridine diphosphoglucose. In normal bile the diglucuronide is the major form of excreted bilirubin, with only small amounts of the monoglucuronide or other glycosidic adducts. **Bilirubin diglucuronide** is much more water soluble than free bilirubin, and thus the action of the transferase facilitates the excretion of the bilirubin into bile. Bilirubin diglucuronide is poorly absorbed by the intestinal mucosa. The glucuronide residues are released in the terminal ileum and large intestine by bacterial hydrolases; the released bilirubin is reduced to the colorless linear tetrapyrroles

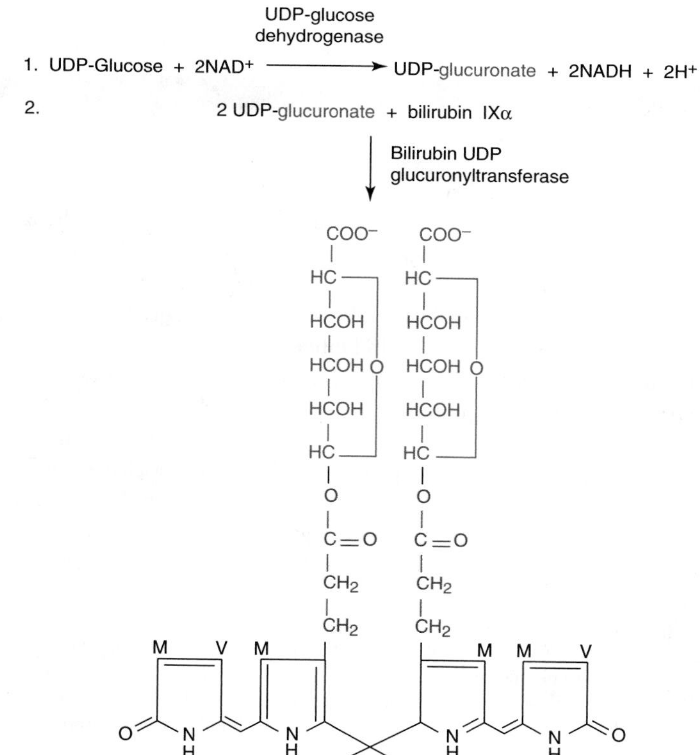

Figure 19.87 Biosynthesis of bilirubin diglucuronide.

A CLOSER LOOK 19.7

Urochrome

Urochrome was the original name given to the pigment thought to give urine its color. It is now known that several related pigments are responsible, so the original name has been dropped. Urine is yellow because of **urobilins**, specifically D-urobilin, L-urobilin, L-stercobilin, and possibly others. Urobilins are composed of four modified pyrrole rings connected by methylene bridges.

known as **urobilinogens**, which become oxidized to colored products known as **urobilins**, which are excreted in the feces. A small fraction of urobilinogen is reabsorbed by the terminal ileum and large intestine to be removed by hepatic cells and resecreted in bile. When urobilinogen is reabsorbed in large amounts in certain disease states, the kidney serves as a major excretory site (A Closer Look 19.7).

Plasma bilirubin concentrations are 0.3–1 mg/dL in the normal state, and this is almost all in the unconjugated state (Clin. Corr. 19.27). Conjugated bilirubin is expressed as **direct bilirubin** because it can be coupled readily with diazonium salts to yield azo dyes in the direct **van den Bergh reaction**. Unconjugated bilirubin is bound noncovalently to albumin, and it does not react until released by the addition of an organic solvent such as ethanol. This is the indirect van den Bergh reaction, and this measures the **indirect bilirubin** or the unconjugated bilirubin. Unconjugated bilirubin binds so tightly to serum albumin and lipid that it does not diffuse freely in plasma and therefore does not appear in urine. It has a high affinity for membrane lipids, which leads to the impairment of cell membrane function, especially in the nervous system. In contrast, conjugated bilirubin is relatively water soluble, and elevations of conjugated bilirubin lead to high urinary concentrations with the characteristic deep yellow-brown color. The deposition of conjugated and unconjugated bilirubin in skin and the sclera gives the yellow to yellow-green color seen in patients with jaundice.

A third form of plasma bilirubin occurs only with hepatocellular disease in which a fraction of the bilirubin binds so tightly that it is not released from serum albumin by the usual techniques and is thought to be linked covalently to the protein. In some cases, up to 90% of total bilirubin can be in this covalently bound form.

The normal liver has a very large capacity to conjugate and excrete the bilirubin that is delivered. Therefore, hyperbilirubinemia due to excess heme destruction (see Clin. Corr. 19.25), as in hemolytic diseases, rarely leads to bilirubin levels that exceed 5 mg/dL, except in situations in which functional derangement of the liver is present (see Clin. Corr.19.26). Thus, marked elevation of unconjugated bilirubin reflects primarily a variety of hepatic diseases, including those inherited and those acquired.

Increase in conjugated bilirubin in plasma is attributable to liver and/or biliary tract disease. In simple uncomplicated biliary tract obstruction, the major component is the

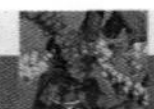

CLINICAL CORRELATION 19.27

Elevation of Serum-Conjugated Bilirubin

Elevations of serum conjugated bilirubin are attributable to liver and/or biliary tract disease. In simple uncomplicated biliary tract obstruction, the major component of the elevated serum bilirubin is the diglucuronide form, which is released by the liver into the vascular compartment. Biliary tract disease may be extrahepatic or intrahepatic, the latter involving the canaliculi and biliary ductules.

Hydrophobic compounds have an affinity for membranes and thus enter cells. To remove these compounds, ATP-dependent membrane channels pump them into the circulation where they can be bound by serum albumin and transferred to the liver. These pumps were initially described in the setting of cancer chemotherapy where the agents are commonly hydrophobic and used in high concentrations. One kind of resistance to therapy was found to be due to the increased activity of the pumps in tumor cells, reducing intracellular drug concentrations. Accordingly, these pumps have been called MRPs (multidrug resistance proteins) (OMIM 601107). At least six kinds are known. The property of hydrophobicity reduces the need of a specific pump for each compound since capture from a dilute aqueous medium is a minor issue, in contrast to hydrophilic compounds where channels have to be very specific. The MRPs serve also to transfer physiological hydrophobic compounds such as steroids from the adrenal gland and bilirubin from hepatocytes into bile. Dubin–Johnson syndrome is an autosomal recessive disease involving a defect in the biliary secretory mechanism of the liver. Excretion from the hepatocyte to the canaliculi relies upon MRP 2. In Dubin–Johnson syndrome mutations occur in this protein (known also as cMOAT, **c**analicular **m**ultispecific **o**rganic **a**nion **t**ransporter). Excretion through the biliary tract of a variety (but not all) of organic anions is affected. Retention of melanin-like pigment in the liver in this disorder leads to a characteristic gray-black color of this organ. A second heritable disorder associated with elevated levels of serum-conjugated bilirubin is Rotor syndrome. In this poorly defined disease no hepatic pigmentation occurs.

Iyanagi, E. Y., and Accoucheur, S. Biochemical and molecular disorders of bilirubin metabolism. *Biochim. Biophys. Acta* 1407:173, 1998; and Tsugi, H., Konig, J., Rost, D., et al. Exon–intron organization of the human multidrug-resistance protein 2 (MRP 2) gene mutated in Dubin–Johnson syndrome. *Gastroenterology* 117:653, 1999.

diglucuronide form, which is released by the liver into the vascular compartment. Biliary tract disease may be extrahepatic or intrahepatic, the latter involving the canaliculi and biliary ductules (see Clin. Corr. 19.27).

Intravascular Hemolysis Requires Scavenging of Iron

In certain diseases, destruction of red blood cells occurs in the intravascular compartment rather than in the extravascular reticuloendothelial cells. In the former case, the appearance of free hemoglobin and heme in plasma could lead potentially to their excretion through the kidney with substantial loss of iron. To prevent this, specific plasma proteins are involved in scavenging mechanisms. Transferrin binds free iron and thus permits its reutilization. Free hemoglobin, after oxygenation in the pulmonary capillaries, dissociates into α,β-dimers, which are bound to a family of circulating plasma proteins, the **haptoglobins**, which have a high affinity for the oxyhemoglobin dimer. Since deoxyhemoglobin does not dissociate into dimers in physiological settings, it is not bound by haptoglobin. Two α,β-oxyhemoglobin dimers are bound by a haptoglobin molecule.

Haptoglobins are α_2-globulins, synthesized in the liver. They consist of two pairs of polypeptide chains (α being the lighter and β the heavier). The α- and β-chains are derived from a single polypeptide that is cleaved to form the two different chains. The β-chains are glycopeptides of 39 kDa and are invariant in structure; α-chains are of several kinds. The haptoglobin chains are joined by disulfide bonds between the α- and β-chains and between the two α-chains.

The haptoglobin–hemoglobin complex is too large to be filtered through the renal glomerulus. Free hemoglobin (appearing in renal tubules and in urine) will occur during intravascular hemolysis only when the binding capacity of circulating haptoglobin has been exceeded. Haptoglobin delivers hemoglobin to the reticuloendothelial cells. The heme in free hemoglobin is relatively resistant to the action of heme oxygenase, whereas the heme residues in an α,β-dimer of hemoglobin bound to haptoglobin are very susceptible.

Measurement of serum haptoglobin is used clinically as an indication of the degree of intravascular hemolysis. Patients who have significant intravascular hemolysis have little or no plasma haptoglobin because of the removal of haptoglobin–hemoglobin complexes by the reticuloendothelial system. Haptoglobin levels can also be low in severe extravascular hemolysis, in which the large load of hemoglobin in the reticuloendothelial system leads to the transfer of free hemoglobin into plasma.

Free heme and hematin (with iron in the ferric state) appearing in plasma are bound by a β-globulin, **hemopexin** that binds one heme residue per molecule. Hemopexin transfers heme to liver, where further metabolism by heme oxygenase occurs. Normal plasma hemopexin contains very little bound heme, whereas in intravascular hemolysis, it is almost completely saturated by heme and is cleared with a half-life of about 7 h. In the latter excess heme binds to albumin, with newly synthesized hemopexin serving as a mediator for the transfer of the heme from albumin to the liver. Hemopexin also binds free protoporphyrin.

Bibliography

General

An excellent source of information on metabolic disorders of amino acid metabolism is Online Mendelian Inheritance in Man (OMIM) developed by Johns Hopkins University and accessible at www.ncbi.nlm.nih.gov/Omim/.

Brosnan, J. T. Amino acids, then and now—A reflection on Sir Hans Krebs, contribution to nitrogen metabolism. *IUBMB Life.* 52: 265, 2001.

Carey, F.A. Amino acids are carboxylic acids that contain an amine function. In Carey, F. A. *Organic Chemistry,* 6th ed., New York: McGraw Hill, 2006, 1149.

Nitrogen Transport and Urea Cycle

Brosnan, J. T. Interorgan amino acid transport and its regulation. *J. Nutr.* 133:2068S, 2003.

Brosnan, J. T., and Brosnan, M. E. Hepatic glutaminase—A special role in urea synthesis? *Nutrition* 18: 455, 2002.

Brosnan, J. T, Brosnan , M. E, and Nissim, I. Application of 15N to the study of hepatic nitrogen metabolism. *Adv. Enzyme Regul.* 42: 305, 2002.

Curis, E., Nicolis, I., Moinard, C., Osowska, S, et al. Almost all about citrulline in mammals. *Amino Acids* 29: 177, 2005.

Individual Amino Acids

Brosnan, J. Glutamate, at the interface between amino acid and carbohydrate metabolism. *J. Nutr.* 130: 988S, 2000.

Brosnan, J. T., and Brosnan, M. E. Branched-chain amino acids: enzyme and substrate regulation. *J. Nutr.*136: 207S, 2006.

Brosnan, J. T., and Brosnan, M. E. The sulfur-containing amino acids: An overview. *J .Nutr.*136: 1636S, 2006.

House. J. D., Hall, B. N., and Brosnan, J. T. Threonine metabolism in isolated rat hepatocytes. *Am. J. Physiol. Endocrinol. Metab.* 281: E1300, 2001.

Morris, S. M. Enzymes of arginine metabolism. *J. Nutr.* 134: 2743S, 2004.

Ratnam, S., Maclean, K. N., Jacobs, R. L., Brosnan, M. E., et al. Hormonal regulation of cystathionine beta-synthase expression in liver. *J. Biol. Chem.* 277: 42912, 2002.

Reeds, P. J., Burrin, D. G., Stoll, B., and Jahoor, F. Intestinal Glutamate Metabolism. *J. Nutr.*130: 978S, 2000.

Watford, M., and Reeds, P. Glutamate metabolism in the gut. *Forum Nutr.* 56: 81, 2003.

Selenoproteins

Chen, J., and Berry, M. Selenium and selenoproteins in the brain and brain disease. *J. Neurochem.* 86: 1, 2003.

Driscoll, D., and Copeland, P. Mechanism and regulation of selenoprotein synthesis. *Annu. Rev. Nutr.* 23: 17, 2003.

Namy, O., Rousset, J., Napthine, S., and Brierly, I. Reprogrammed genetic decoding in cellular gene expression. *Mol. Cell* 13: 157, 2004.

Tetrahydrobiopterin

Thony, B., Auerbach, G., and Blau, N. Tetrahydrobiopterin biosynthesis, regeneration and functions. *Biochem. J.* 347: 1, 2000.

Secondary Metabolites

Bouckenooghe, T., Remacle, C., and Reusens, B. Is taurine a functional nutrient? *Curr. Opin. Clin. Nutr.* 9: 728, 2006.

Brosnan, M. E., Edison, E. E., da Silva, R., and Brosnan, J. T. New insights into creatine function and synthesis. *Adv. Enzyme Regul.* 47: 252, 2007.

Caraglia, M., Tagliaferri, P., Budillon, A., Abbruzzese, A. Post-translational modifications of eukaryotic initiation factor-5A (eIF-5A) as a new target for anti-cancer therapy. *Adv. Exp. Med. Biol.* 472: 187, 1999.

Reddy, V. P., Garrett, M. R., Perry, G., and Smith, M. A. Carnosine: A versatile antioxidant and antiglycating agent. *Sci. Aging. Knowledge. Environ.* 2005: 12, 2005.

Schuller-Levis, G. B., and Park, E. Taurine: New implications for an old amino acid. *FEMS Microbiol. Lett.* 226: 195, 2003.

Wallace, H. M., and Fraser, A. V. Inhibitors of polyamine metabolism: Review article. *Amino Acids* 26: 353, 2004.

Wu, G., Fang, Y., Yang, S., Lupton, J., and Turner, N. Glutathione metabolism and its implications for health. *J. Nutr.* 134: 489, 2004.

Metabolic Diseases of Amino Acid Metabolism

Finkelstein, J. D. Inborn errors of sulfur-containing amino acid metabolism. *J. Nutr.* 136: 1750S, 2006.

Finkelstein, J. D. and Martin, J. J. Homocysteine. *Int. J. Biochem. Cell Biol.* 32: 385, 2000.

Hamosh, A., and Johnston, M. V. Nonketotic hyperglycinemia. In Scriver, C. R., Beaudet, A. L., Sly, W. S., Valle, D. (Eds.), *The Metabolic and Molecular Bases of Inherited Disease*, 8th ed. New York: McGraw Hill, 2001, 2: 2065.

Kelly, A., and Stanley, C. Disorders of glutamate metabolism. *Ment. Retard. Develop. Disord. Res. Rev.* 7: 287, 2001.

Mitchell, S.C., and Smith, R.L. Trimethylaminuria: The fish malodor syndrome. *Drug Metab Dispos.* 29: 517, 2001.

Schulze, A. Creatine deficiency syndromes. *Mol. Cell Biochem.* 244: 143, 2003.

Heme Synthesis

Davids, L. M., Corngall, A. V., and Meissner, P. N. Mitochondrial targeting of human protoporphyrinogen oxidase. *Cell Biol. Int.* 30: 416, 2006.

Ferreira, G. C. Ferrochelatase binds the iron-responsive element present in the erythroid δ-aminolevulinate synthase mRNA. *Biochem. Biophys. Res. Commun.* 214: 875, 1995.

Hamza, I. Intracellular trafficking of porphyrins. *ACS Chem. Biol.* 1: 627, 2006.

Hunter, G. A., Sampson, M. P., and Ferreira, G. C. Metal ion inhibition of ferrochelatase. *J. Biol. Chem.* 283: 23685, 2008.

Hunter, G. A., Zhang, J., and Ferreira, G. C. Transient kinetic studies support refinements to the chemical and kinetic mechanisms of aminolevulinate synthase. *J. Biol. Chem.* 282: 23025, 2007.

Jaffe, E. K., Martins, J., Li, J., Kervinen, J., and Dunbrack, R. L. The molecular mechanism of lead inhibition of human porphobilinogen synthase. *J. Biol. Chem.* 276: 1531, 2001.

Martini, C. N., Romero, D. G., Yanes, L. L., and Vila, M. del C. Induction of 5-aminolevulinate synthase by activators of steroid biosynthesis. *Life Sci.* 81: 19, 2007.

Schubert, H. L., Phillips, J. D., Heroux, A., and Hill, C. P. Structure and mechanistic implications of a uroporphyrinogen III synthase-product complex. *Biochemistry* 47: 8648, 2008.

Zhao, Y., Wang, L., Shen, H. B., Wang, Z. X., Wei, O. Y., and Chen, E. Association between delta-aminolevulinic acid dehydratase (ALAD) polymorphism and blood lead levels: A meta-regression analysis. J. *Toxicology Environ. Health* 70: 1986, 2007.

Heme Degradation

Almolki, A., Guenegou, A., Golda, S., Boyer, L., Benallaoua, M., et al. J. Heme oxygenase-1 prevents airway mucus hypersecretion induced by cigarette smoke in rodents and humans. *Am. J. Pathol.* 173: 981, 2008.

Converso, D. P., Taille, C., Carreras, M. C., Jaitovich, A., et al. HO-1 is located in liver mitochondria and modulates mitochondrial heme content and metabolism. *FASEB J.* 20: 1236, 2006.

Gohya, T., Sato, M., Zhang, X., and Migita, C. T. Variation of the oxidation state of verdoheme in the heme oxygenase reaction. *Biochem. Biophys. Res. Commun.* 376: 293, 2008.

Kamisako, T., Kobayashi, Y., Ishihara, T., Higushi, K., et al. Recent advances in bilirubin metabolism research: The molecular mechanism of hepatocyte bilirubin transport and its clinical relevance. *J. Gastroenterol.* 35: 659, 2000.

Tudor, C., Lerner-Marmarosh, N., Engelborghs, Y., Gibbs, P. E., and Maines, M. D. Biliverdin reductase is a transporter of haem into the nucleus and is essential for regulation HO-1 gene expression by haematin. *Biochem. J.* 413: 405, 2008.

Heme-Related Diseases

Anderson, K. E., Sassa, S., Bishop, D. F., and Desnick, R. J. Disorders of heme biosynthesis: X-linked sideroblastic anemia and the porphyrias. In Scriver, C. R., Beaudet, A. L., Sly, W. S., Valle, D. (Eds.), *The Metabolic and Molecular Bases of Inherited Disease*, 8th ed. New York: McGraw Hill, 2001, 2: 2991.

Beutler, E., Bothwell, T. M., Charlton, R. W., and Motulsky, A. G. Hereditary hemachromatosis. In Scriver, C. R., Beaudet, A. L., Sly, W. S., Valle, D. (Eds.), *The Metabolic and Molecular Bases of Inherited Disease*, 8th ed. New York: McGraw Hill, 2001, 2: 3127.

Chowdhury, J. R., Wolkoff, A. W., Chowdhury, N. R., and Arias, I. M. Hereditary jaundice and disorders of bilirubin metabolism. In Scriver, C. R., Beaudet, A. L., Sly, W. S., Valle, D. (Eds.), *The Metabolic and Molecular Bases of Inherited Disease*, 8th ed. New York: McGraw Hill, 2001, 2: 3063.

Holme, S. A., Worwood, M., Anstey, A. V., Elder, G. H., and Badminton, M. N. Erythropoiiesis and iron metabolism in dominant erythropoietic protoporphyria. *Blood* 110: 4108, 2007.

Kappas, A., Sassa, S., Galbraith, R. A., and Nordmann, V. The porphyrias. In Scriver, C. R., Beaudet, A. L., Sly, W. S., Valle, D. (Eds.), *The Metabolic and Molecular Bases of Inherited Disease*, 7th ed. New York: McGraw Hill, 1995, 2: 2103.

Otterbein, L. E., Soares, M. P., Yamashita, K., and Buch, F. H. Heme oxygenase-1: Unleashing the protective properties of heme. *Trends Immunol.* 24: 449, 2003.

Perrella, M. A., and Yet, S. F. Role of heme oxygenase-1 in cardiovascular function. *Curr. Pharm. Des.* 9: 2479, 2003.

Schertner, H. A., and Vitek, L. Gilbert syndrome, UGT1A1* allele, and cardiovascular disease risk: possible protective affects and therapeutic applications of bilirubin. *Atherosclerosis* 198: 1, 2008.

Weiss, G., and Goodnough, L. T. Anemia of chronic disease *N. Engl. J. Med.* 352: 1011, 2005.

Yokooji, T., Murakami, T., Yumoto, R., Nagai, J., and Takano, M. Function of multidrug resistance-associated protein 2 in acute hepatic failure rats. *Eur. J. Pharmacol.* 546: 152, 2006.

Key Terms

nitrogen balance
aminotransferases
glutamate alanine aminotransferase
glutamate aspartate aminotransferase
glutamate dehydrogenase
pyridoxal phosphate
amino acid oxidase
asparaginase
glutaminase
glutamine synthetase
urea cycle
arginase
citrulline
N-acetylglutamate
ornithine transcarbamoylase
carbamoyl phosphate synthetase I
glutamic semialdehyde
serine dehydratase
phenylalanine hydroxylase
tetrahydrobiopterin
homocysteine
S-adenosylmethionine
propionyl-CoA carboxylase
tryptophan dioxygenase
dihydroxyphenylalanine
dopamine
epinephrine
melanin
tyrosine hydroxylase
3′-phosphoadenosine-5′-phosphosulfate
carnitine
serotonin
histamine
polyamines
spermine
glutathione
heme
porphyrinogens
porphyrin
protoporphyrin
δ-aminolevulinic acid synthase
aminolevulinic acid dehydratase
porphobilinogen deaminase
uroporphyrinogen III synthase
ferrochelatase
biliverdin reductase
heme oxygenase
bilirubin diglucuronide
urobilins

Questions • CAROL N. ANGSTADT

Multiple-Choice Questions

1. Aminotransferases
- **A.** usually require α-ketoglutarate or glutamine as one of the reacting pair.
- **B.** catalyze reactions that result in a net use or production of amino acids.
- **C.** catalyze irreversible reactions.
- **D.** require pyridoxal phosphate as an essential cofactor for the reaction.
- **E.** are not able to catalyze transamination reactions with essential amino acids.

2. The production of ammonia in the reaction catalyzed by glutamate dehydrogenase
- **A.** requires the participation of NADH or NADPH.
- **B.** proceeds through a Schiff base intermediate.
- **C.** may be reversed to consume ammonia if it is present in excess.
- **D.** is favored by high levels of ATP or GTP.
- **E.** would be inhibited when gluconeogenesis is active.

3. *S*-Adenosylmethionine
- **A.** contains a charged sulfur that carries the methyl group to be transferred.
- **B.** yields α-ketobutyrate in the reaction in which the methyl is transferred.
- **C.** donates a methyl group in a freely reversible reaction.
- **D.** generates H_2S by transsulfuration.
- **E.** provides the carbons for the formation of cysteine.

4. Glutathione does all of the following except
- **A.** participate in the transport of amino acids across some cell membranes.
- **B.** scavenge peroxides.
- **C.** form conjugates with some drugs to increase water solubility.
- **D.** decreases the stability of erythrocyte membranes.
- **E.** acts as a cofactor for some enzymes.

5. Heme oxygenase
- **A.** can oxidize the methene bridge between any two pyrrole rings of heme.
- **B.** requires molecular oxygen.
- **C.** produces bilirubin.
- **D.** produces carbon dioxide.
- **E.** can use either heme or protoporphyrin IX as substrate.

6. Conjugated bilirubin is
- **A.** transported in blood bound to serum albumin.
- **B.** deficient in Crigler–Najjar syndrome, a deficiency of a UDP-glucuronosyltransferase.
- **C.** reduced in serum in biliary tract obstruction.
- **D.** the form of bilirubin most elevated in hepatic (liver) disease.
- **E.** less soluble in aqueous solution than the unconjugated form.

Questions 7 and 8: Defects in the metabolism of the branched-chain amino acids are rare but serious. The most common one is called maple syrup urine disease (named from the smell of the urine), which is a deficiency of branched-chain keto acid dehydrogenase complex. The disease is characterized by mental retardation and a short life span.

7. All of the following are true about the branched-chain amino acids except they
   A. are essential in the diet.
   B. differ in that one is glucogenic, one is ketogenic and one is classified as both.
   C. are catabolized in a manner that bears a resemblance to β-oxidation of fatty acids.
   D. are oxidized by a dehydrogenase complex to branched-chain acyl CoAs one carbon shorter than the parent compound.
   E. are metabolized initially in the liver.

8. Valine and isoleucine give rise to propionyl CoA, a precursor of succinyl CoA. A disease related to a defect in this conversion is methylmalonic aciduria. Some patients respond to megadoses of vitamin B_{12}. Which of the following statements about the conversion of propionyl CoA to succinyl CoA is/are correct?
   A. The first step in the conversion is a biotin-dependent carboxylation.
   B. Some methylmalonic aciduria patients respond to B_{12} because the defect in the mutase converting malonyl CoA to succinyl CoA is poor binding of the cofactor.
   C. The same pathway of propionyl CoA to succinyl CoA is part of the metabolism of odd-chain fatty acids.
   D. All of the above.
   E. None of the above.

Questions 9 and 10: Hyperammonemia caused by deficiencies of the enzymes involved in carbamoyl phosphate synthesis or any of the enzymes of the urea cycle is a very serious condition. Untreated, the result is early death or mental retardation and other developmental abnormalities. Ornithine transcarbamoylase deficiency is the most common error in the cycle. Treatment aims to relieve the hyperammonemia and sometimes there is supplementation with arginine.

9. In the formation of urea from ammonia, all of the following are correct except
   A. aspartate supplies one of the nitrogens found in urea.
   B. this is an energy-expensive process, utilizing several ATPs.
   C. the rate of the cycle fluctuates with the diet.
   D. fumarate is produced.
   E. ornithine transcarbamoylase catalyzes the rate-limiting step.

10. Carbamoyl phosphate synthetase I
   A. is a flavoprotein.
   B. is controlled primarily by feedback inhibition.
   C. is unresponsive to changes in arginine.
   D. requires *N*-acetyl glutamate as an allosteric effector.
   E. requires ATP as an allosteric effector.

Questions 11 and 12: A woman appears in an emergency room in an agitated state with severe abdominal pain and marked weakness in all her limbs. Usual laboratory tests are normal and no specific abnormality can be found. She is sedated with phenobarbital (a barbiturate) and observed over the next few hours. Her condition rapidly deteriorates and further study finds an elevated level of porphobilinogen in her urine. A diagnosis of acute intermittent porphyria is made. Enzyme abnormalities in this disease are a marked increase in ALA synthase and a 50% reduction in porphobilinogen deaminase.

11. Aminolevulinic acid synthase
   A. requires NAD for activity.
   B. is allosterically activated by heme.
   C. synthesis is inhibited by steroids.
   D. is synthesized in mitochondria.
   E. synthesis can be induced by a variety of drugs.

12. Normally, porphobilinogen is an intermediate in the pathway to heme biosynthesis. There are many other intermediates. Uroporphyrinogen III
   A. is synthesized rapidly from porphobilinogen in the presence of a cosynthase.
   B. does not contain a tetrapyrrole ring.
   C. is formed from coproporphyrinogen III by decarboxylation.
   D. is converted directly to protoporphyrinogen IX.
   E. formation is the primary control step in heme synthesis.

Problems

13. An inability to generate tetrahydrobiopterin would have what specific effects on the metabolism of phenylalanine, tyrosine, and tryptophan?

14. Ornithine, an intermediate of the urea cycle, and proline, found in proteins, are both synthesized from glutamate. How does this occur?

Answers

1. **D** The mechanism of action begins with the formation of a Schiff base with pyridoxal phosphate. A: Most mammalian aminotransferases use glutamate or α-ketoglutarate. B: One amino acid is converted into another amino acid; there is neither net gain nor net loss. C: The reactions are freely reversible. E: Only lysine and threonine do not have aminotransferases.
2. **C** This is an important mechanism for reducing toxic ammonia concentrations. A: This would favor ammonia consumption. B: The cofactor is a pyridine nucleotide, not pyridoxal phosphate. D: These are inhibitory. E: Since part of the role is to provide amino acid carbon chains for gluconeogenesis, this would be active.
3. **A** The reactive, positively charged sulfur reverts to a neutral thioether when the methyl group is transferred to an acceptor. B: The product, *S*-adenosylhomocysteine, is hydrolyzed to homocysteine. C: Transmethylations from AdoMet are irreversible. D: Transsulfuration refers to the combined action of cystathionine synthase and cystathionase transferring methionine's sulfur to serine to yield cysteine. E: Methionine provides only the sulfur; carbons are from serine.
4. **D** Most of the functions of glutathione listed are dependent on the sulfhydryl group (-SH). It increases membrane stability. Glutathione reductase helps to maintain the ratio of GSH to GSSG at about 100:1.

5. **B** Oxygenases usually use O_2. A: The enzyme is specific for the methene between the two rings containing the vinyl groups (α-methene bridge). C and D: The products are biliverdin and CO; the measurement of CO in the breath is an index of heme degradation. E: Iron is necessary for activity.
6. **B** Conjugated bilirubin is bilirubin diglucuronide. A: Unconjugated bilirubin is transported bound to albumin. C: Liver conjugates bilirubin, but if the bile duct is obstructed, it won't be excreted. D: Liver is responsible for the conjugation, which would be impaired in hepatic damage. E: The purpose of conjugation is to increase water solubility.
7. **E** A: Aminotransferase, the first enzyme, is much higher in muscle than in liver. B, C, and D: Their catabolism is similar but the end products are different because of the differences in the branching. The similarity to β-oxidation comes in steps like oxidation to an α,β-unsaturated CoA, hydration of the double bond, and oxidation of an hydroxyl to a carbonyl.
8. **D** A: This is a typical carboxylation reaction like pyruvate carboxylase. B: Megadoses of a cofactor if the defect is in binding has been used in a number of diseases. C: The final cleavage of odd-chain fatty acids produces propionyl CoA.
9. **E** Carbamoyl phosphate synthetase I catalyzes the rate-limiting step. A, B, and D: One of the nitrogen atoms is supplied as aspartate, with its carbon atoms being released as fumarate. C: The level of CPSI and the synthesis of *N*-acetylglutamate increase as protein in the diet increases.
10. **D** The primary control is by the allosteric effector, *N*-acetylglutamate. B: This is an activation, not an inhibition. C: Synthesis of the effector, and therefore activity of CPSI, is increased in the presence of arginine. E: ATP is a substrate.
11. **E** The enzyme is induced in response to need as well as by drugs and metabolites. Giving the porphyria patient phenobarbital increased her already high ALA synthase. A: The mechanism involves a Schiff base with glycine; therefore the coenzyme is pyridoxal phosphate. B: Heme both allosterically inhibits and suppresses synthesis of the enzyme. C: One reduction product of catabolic steroids stimulates the synthesis. D: The gene for this enzyme is on nuclear DNA.
12. **A** In the absence of cosynthase, uroporphyrinogen I is synthesized slowly. B: The tetrahydropyrrole ring has formed by this point. C and D: The decarboxylation goes from uroporphyrinogen to coproporphyrinogen, which is then converted to protoporphyrinogen. E: The synthesis of aminolevulinic acid is the rate-limiting step.
13. Tetrahydrobiopterin is a necessary component of phenylalanine, tyrosine and tryptophan hydroxylases. Its deficiency would inhibit normal degradation of both phenylalanine and tyrosine because their degradative pathways begin with the respective hydroxylases. Catecholamine formation (norepinephrine and epinephrine) begins with the formation of DOPA from tyrosine via tyrosine hydroxylase so catecholamine synthesis would be inhibited. The initial step in the conversion of tryptophan to serotonin is catalyzed by tryptophan hydroxylase.
14. Glutamate, using ATP and NADH, can be reduced to glutamic semialdehyde. The semialdehyde spontaneously cyclizes to form a Schiff base between the aldehyde and amino groups which is then reduced by NADPH to proline. If, instead, the semialdehyde undergoes transamination, ornithine is produced.

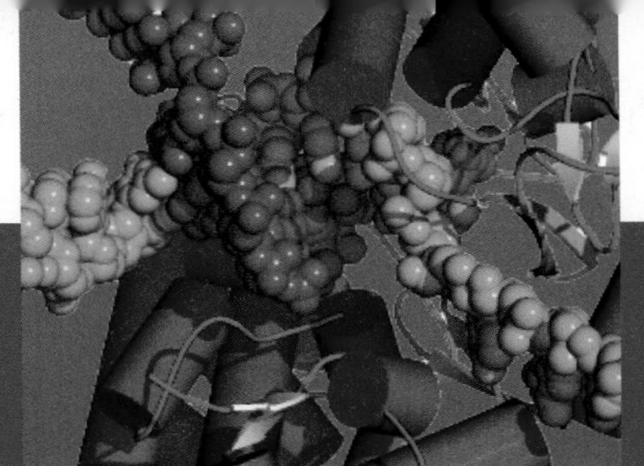

20

Purine and Pyrimidine Nucleotide Metabolism

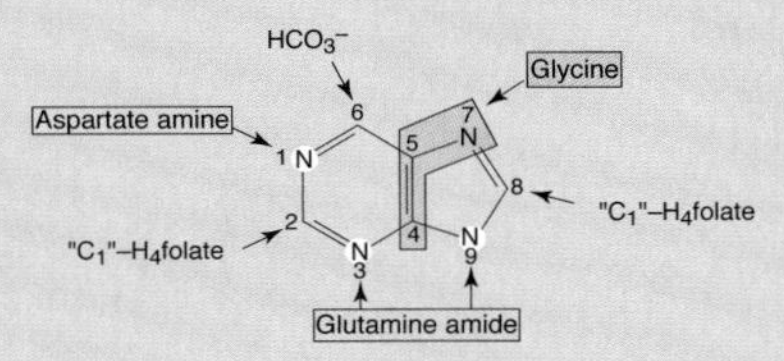

Joseph G. Cory

Professor, Brody School of Medicine, East Carolina University

Ann H. Cory

Research Associate, Brody School of Medicine, East Carolina University

CLINICAL CORRELATIONS

Key Concepts

- Purine and pyrimidine nucleotides are required for many normal cellular functions including DNA and RNA synthesis. Nucleotides in the cell are supplied by *de novo* synthesis and by salvage of preformed nucleobases or nucleosides.
- Synthesis of both purine and pyrimidine nucleotides requires specific amino acids, tetrahydrofolate, and 5-phosphoribosyl-1-pyrophosphate (PRPP). IMP is a precursor for both AMP and GMP synthesis. Synthesis is tightly regulated by allosteric control of the committed steps in the individual pathways.
- Deoxyribonucleotides are formed by reduction of ribonucleoside 5′-diphosphates. Deoxythymidylate synthesis requires N^5, N^{10}-methylene H_4folate

- PRPP is required for the salvage pathway. Uric acid is the end product of purine degradation in man, and β-aminoisobutyric acid of thymidine nucleotide.
- Nucleotide kinases convert nucleoside 5′-monophosphates to nucleoside 5′-diphosphates and nucleoside 5′-diphosphates to nucleoside 5′-triphosphates.
- Defects or alterations in nucleotide metabolism lead to defined clinical problems, such as gout, Lesch–Nyhan, and immunodeficiency. Inhibitors of specific steps in the synthesis of the nucleotides are effective anti-tumor drugs.
- NAD, FAD, and coenzyme A, critical coenzymes in metabolism, are synthesized by mammalian cells.

20.1 • INTRODUCTION

There are major differences between nucleotide metabolism in bacteria and mammalian cells, and even differences between humans and animals. The following discussion is limited to nucleotide metabolism in mammalian cells and where appropriate to nucleotide metabolism in humans.

Purine and pyrimidine nucleotides participate in many critical cellular functions. The cellular levels of the purine and pyrimidine nucleotides are maintained by *de novo* synthetic pathways and "salvage" reactions. Amino acids, CO_2, "Carbon-1"-tetrahydrofolate, and ribose-5-phosphate serve as sources for carbon, nitrogen, and oxygen atoms.

Intracellular concentrations of nucleotides are finely controlled by allosterically regulated enzymes in the pathways. Nucleotide end products of the pathways serve as effectors and regulate key steps in the pathways. 2′-Deoxyribonucleotides required for DNA replication are generated directly from ribonucleotides, and their production is also carefully regulated by nucleoside 5′-triphosphate nucleotides acting as positive and negative effectors. In addition to the regulation of nucleotide metabolism via allosteric regulation, concentrations of key enzymes in their metabolic pathways are altered during the cell cycle with many of the increases in enzyme activity occurring especially during late G_1/early S phase just preceding DNA replication.

The importance of both the *de novo* synthesis and the salvage pathways is shown by the fact there are clinical diseases or syndromes due to defects in either pathway. These conditions include gout (defect in *de novo* purine nucleotide synthesis), **Lesch–Nyhan syndrome** (defect in purine nucleobase salvage), orotic aciduria (defect in *de novo* pyrimidine nucleotide synthesis), and immunodeficiency diseases (defects in purine nucleoside degradation). Because nucleotide synthesis is required for DNA replication and RNA synthesis in dividing cells, drugs that block *de novo* pathways of nucleotide synthesis have been successfully used as antitumor and antiviral agents.

The structures, chemistry, and properties of nucleobases, nucleosides, and nucleotides are presented on pages 1132–1133 and in the Appendix.

20.2 • METABOLIC FUNCTIONS OF NUCLEOTIDES

Nucleotides and their derivatives play critical and diverse roles in cellular metabolism. Many different nucleotides are present in mammalian cells. Some, such as ATP and NAD are present in millimolar concentrations while others such as dATP and cyclic AMP are orders of magnitude lower in concentration. The functions of nucleotides are summarized in Table 20.1 with some examples given.

Distribution of Nucleotides Varies with Cell Type

The principal purine and pyrimidine compounds found in cells are the 5′-nucleotide derivatives with ATP being present in the highest concentration. The distribution of the various nucleotides in cells varies with cell type. In red blood cells, adenine nucleotides far exceed the concentrations of guanine, cytosine, and uracil nucleotides; in

TABLE 20.1 • Functions of Nucleotides

Function	*Selected Examples*
1. Energy metabolism	ATP (muscle contraction; active transport; ion gradients; phosphate donor)
2. Monomeric units of nucleic acids	NTPs and dNTPs (substrates for RNA and DNA)
3. Physiological mediators	Adenosine (coronary blood flow); ADP (platelet aggregation); cAMP and cGMP (second messengers); signal transduction via GTP-binding proteins
4. Precursor function	GTP (mRNA capping); tetrahydrobiopterin, (hydroxylation of aromatic amino acids)
5. Components of coenzymes	NAD, FAD, FMN, and coenzyme A
6. Activated intermediates	UDP-glucose (glycogen); CDP-choline (phospholipids); SAM (methylation); PAPS (sulfation)
7. Allosteric effectors	ATP (negative effector of PFK-I); AMP (positive effector of phosphorylase b); dATP (negative effector of ribonucleotide reductase)

other tissues, such as liver, there is a complete spectrum of nucleotides, which also include NAD^+, NADH, UDP-glucose, and UDP-glucuronic acid. In normally functioning cells, nucleoside 5′-triphosphates predominate, whereas in hypoxic cells the concentrations of nucleoside 5′-monophosphates and nucleoside 5′-diphosphates are greatly increased. Free nucleobases, nucleosides, nucleoside 2′- and 3′-monophosphates, and modified bases are found in the cytosol represent degradation products of endogenous or exogenous nucleotides or nucleic acids.

The concentrations of ribonucleotides in cells are in great excess over the concentrations of 2′-deoxyribonucleotides. For example, the concentration of ATP in Ehrlich tumor cells is 3600 pmol per 10^6 cells compared to dATP concentrations of 4 pmol per 10^6 cells. However, at the time of DNA replication the concentrations of dATP and other deoxyribonucleoside 5′-triphosphates are markedly increased to meet the substrate requirements for DNA replication. In normal cells, the total concentration of nucleotides is essentially constant. Thus, the total concentration of AMP plus ADP plus ATP remains constant, but there can be major changes in the individual concentration of each such that the ratio of ATP/(ATP + ADP + AMP) is altered depending on the energy state of the cell. The same is true for NAD^+ and NADH. The total concentration of NAD^+ plus NADH is normally fixed within rather narrow concentration limits. Consequently, when it is stated that the NADH level is increased, it follows that the concentration of NAD^+ is correspondingly decreased in that cell. The basis for this fixed concentration of nucleotides is that *de novo* synthesis and salvage pathways for nucleotides, nucleosides, and nucleobases are very rigidly controlled under normal conditions.

20.3 • 5′-PHOSPHORIBOSYL-1-PYROPHOSPHATE AND GLUTAMINE IN *DE NOVO* SYNTHESIS OF NUCLEOTIDES

5′-Phosphoribosyl-1-pyrophosphate

Ribose 5-phosphate generated from the pentose phosphate pathway or from the phosphorylysis of nucleosides supplies the **5′-phosphoribosyl-1-pyrophosphate (PRPP)** required for the *de novo* and salvage pathways of nucleotides. The reaction catalyzed by PRPP synthetase is shown in Figure 20.1. This reaction is normally under tight regulation. The properties of PRPP synthetase are listed in Table 20.2. There is a rare but severe clinical

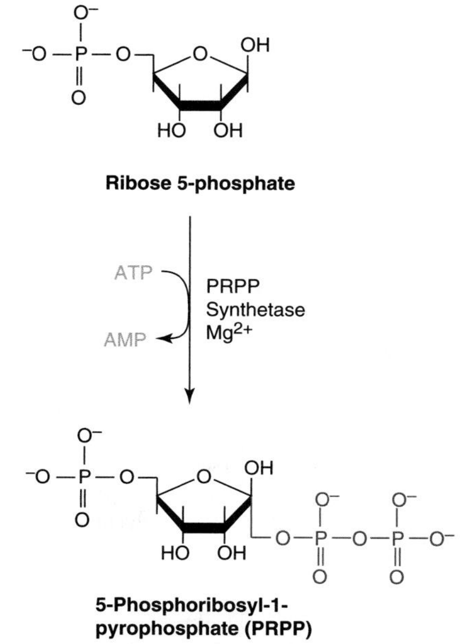

Figure 20.1 **Synthesis of PRPP.**

CLINICAL CORRELATION 20.1

Loss-of-Function Mutations in Phosphoribosylpyrophosphate Synthetase1 (PRPS1): Arts Syndrome (OMIM 301835)

Arts syndrome is a very rare genetic disorder that is caused by mutations in phosphoribosyl pyrophosphate synthetase (PRPS1) that results in loss of PPRS1 activity. It is X linked, and the child suffers from mental retardation, delayed motor development, and optic atrophy. In the patients studied, at least two different missense mutations have been identified. The clinical problems are likely the result of the lack of purine nucleotide synthesis. Hypoxanthine is undetectable in the urine and there is reduced uric acid in serum. Arts syndrome represents the completely opposite case in which PPRS1 activity is elevated leading to over production of purine nucleotides and the clinical condition of gout.

De Brouwer, A. P., Williams, K. I., Duley, J. A., Kuilenburg, A. B., et al. Arts syndrome is caused by loss-of-function mutations in PRPS1. *Am. J. Hum. Genet.* 81:507, 2007; Iizasa, T. Increased activity of PRPP synthetase. *Nippon Rinshi* 4:694, 2008; and Kim, H. J., Shy, M. E., Krajewski, K. M., Huang, M., et al. Mutations in PRPS1, which encodes the phosphoribosylpyrophosphate synthetase enzyme critical for nucleotide biosynthesis, causes hereditary peripheral neuropathy with hearing loss and optical neuropathy. *Am J. Hum. Genet.* 81:522, 2007.

TABLE 20.2 • Properties of 5-Phosphoribosyl-1-pyrophosphate Synthetase

1. Ribose supplied by glucose 6-phosphate via the pentose phosphate pathway.
2. Absolute requirement for inorganic phosphate; the v vs [Pi] curve is sigmoidal.
3. Inhibited by bis, 2,-3 DPG and other nucleotides.
4. ADP is a competitive inhibitor to ATP.
5. 2,3-Bisphosphoglycerate is a competitive inhibitor to ribose 5-phosphate.

TABLE 20.3 • Reactions and Pathways Requiring 5-Phosphoribosyl-1-pyrophosphate

1. *De novo* synthesis of purine nucleotides	PRPP + glutamine ⟶ 5-phosphoribosylamine + glutamate + PPi
2. Salvage of purine bases	PRPP + hypoxanthine(guanine) ⟶ IMP (GMP) + PPi PRPP + adenine ⟶ AMP + PPi
3. *De novo* synthesis of pyrimidine nucleotides	PRPP + orotate ⟶ OMP + PPi
4. Salvage of pyrimidine bases	PRPP + uracil ⟶ UMP + PPi
5. NAD synthesis	PRPP + nicotinate ⟶ nicotinate mononucleotide + PPi PRPP + nicotinamide ⟶ nicotinamide mononucleotide + PPi PRPP +quinolate ⟶ nicotinate mononucleotide + PPi

condition that has been documented in a Dutch family in which there is a loss of function of 5-phosphoribosylpyrophosphate synthetase (PRS1) (Clin. Corr. 20.1). On the other hand there is the involvement of excess PRPP levels contributing to gout (Clin. Corr. 20.2). PRPP levels are critical not only for the *de novo* syntheses of nucleotides but also for the salvage of nucleobases and synthesis of NAD. Table 20.3 lists the reactions and pathways requiring PRPP.

Glutamine

The amino acid, glutamine, while not considered as an essential amino acid, is a critical substrate in five specific reactions involved in the *de novo* synthesis of nucleotides. These reactions are summarized in Table 20.4. A limiting supply of glutamine in serum or cells could greatly influence the rate of nucleotide synthesis if the concentrations of glutamine were less than saturating. This in turn, could have severe consequences on the ability of a cell to synthesize RNA and DNA at the time of cell replication.

The overall *de novo* syntheses of purine and pyrimidine nucleotides are highly regulated at **PRPP amidotransferase** and **carbamoyl phosphate synthetase (CPS II)**, respectively. The regulation of CTP synthetase is critical for maintaining the appropriate cellular ratio of UTP to CTP in cells.

TABLE 20.4 • Reactions Requiring Glutamine for Nucleotide Synthesis

1. Purine nucleotide synthesis
   (a) glutamine PRPP amidotransferase
   (b) 5′-phosphoribosylformylglycinamide synthetase
   (c) GMP synthetase
2. Pyrimidine nucleotide synthesis
   (a) carbamonyl phosphate synthetase II
   (b) CTP synthetase

20.4 • SYNTHESIS OF PURINE NUCLEOTIDES

The *de novo* synthesis pathways for purine and pyrimidine nucleotides in mammalian cells utilize amino acids as carbon and nitrogen donors. Although all of the enzymes of the purine nucleotide pathway are cytosolic, the enzymes of pyrimidine nucleotide synthesis pathway are located in the cytosol and mitochondria. Both types of nucleotides derive the ribose 5-phosphate from PRPP and both pathways are very finely regulated. The energy from the hydrolysis of ATP is used to drive several reactions. Overall, the *de novo* pathways are expensive in terms of moles of ATP utilized per mole of nucleotide synthesized.

All enzymes involved in synthesis of purine nucleotides are present in the cytosol. However, not all cells (e.g., red cells) are capable of *de novo* purine nucleotide synthesis. In the *de*

CLINICAL CORRELATION 20.2

Gout

Gout is characterized by elevated uric acid concentrations in blood and urine due to a variety of metabolic abnormalities that include the overproduction of purine nucleotides or the decreased excretion of uric acid. It appears to be a growing health problem, related to lifestyle and the aging population. Many of the clinical symptoms associated with elevated concentrations of uric acid arise because of the very poor solubility of uric acid in the aqueous environment. Sodium urate crystals deposit in joints of the extremities and in renal interstitial tissue. These events tend to trigger the sequelae. Hyperuricemia from overproduction of uric acid via the *de novo* pathway can be distinguished from hyperuricemia that results from kidney disease or excessive cell death (e.g., increased degradation of nucleic acids from radiation therapy or cancer chemotherapy). Feeding of ^{15}N-glycine to a patient who is an overproducer results in uric acid excreted in urine that is enriched in ^{15}N at the N7 of uric acid. In contrast, a patient who is not an overproducer will not have enrichment of ^{15}N in the excreted uric acid.

Studies of patients with gout have shown that multiple and heterogenous metabolic defects can cause overproduction of uric acid. In some cases, biochemical defects have not been unequivocally defined. Examples of biochemical defects that result in increased purine nucleotide synthesis include the following: (1) Increased PRPP synthetase (OMIM 300661) activity leads to increased intracellular concentrations of PRPP, which serve as a positive effector of glutamine-PRPP amidotransferase leading to increased flux through the *de novo* pathway since activity of the rate-limiting step is markedly increased. (2) Partial decrease in HGPRTase activity (OMIM 308000) has two effects that lead to increased *de novo* pathway for purine nucleotide synthesis. Since there is decreased salvage of hypoxanthine and guanine, PRPP is not consumed by the HGPRtase. PRPP not consumed by the HGPRTase reaction can serve to activate glutamine-PRPP amidotransferase activity. With decreased salvage of hypoxanthine and guanine, IMP and GMP are not formed via this pathway so that regulation of the PRPP amidotransferase step by IMP and GMP as negative effectors is compromised. (3) Glucose 6-phosphatase deficiency (von Gierke disease, Type I glycogen storage disease, (OMIM 232200)) is frequently associated with hyperuricemia and gout as well since loss of glucose 6-phosphatase activity results in more glucose 6-phosphate being shunted to the pentose phosphate pathway. As a result, more ribose 5-phosphate is generated and the intracellular levels of PRPP are increased stimulating PRPP amidotransferase activity. Increased levels of cellular PRPP in each of these cases appears to be sufficient to attenuate the regulation of purine nucleotide synthesis by IMP, AMP, and GMP at the PRPP amidotransferase step.

These examples demonstrate that factors that increase the rate-limiting step in *de novo* synthesis of purine nucleotide synthesis lead to increased generation of uric acid. Under normal conditions, increased purine nucleotide synthesis occurs only to meet the cellular needs for purine nucleotides for RNA and DNA synthesis.

There are different approaches for the treatment of gout. These include colchicine (antimitotic agent), uricosuric drugs (to enhance renal excretion of uric acid), and allopurinol. Allopurinol and its metabolite, alloxanthine, are effective inhibitors of xanthine oxidoreductase and will cause a decrease in uric acid levels. In overproducers who have only a partial deficiency of HGPRTase activity, allopurinol treatment inhibits xanthine oxidoreductase, causing increases in the concentrations of hypoxanthine and xanthine which can be salvaged via HGPRTase to form IMP and XMP. These reactions consume PRPP and generate purine nucleotides that inhibit PRPP amidotransferase. The overall effect is that allopurinol treatment decreases both uric acid formation and *de novo* synthesis of purine nucleotides in these individuals with partial HGPRTase activity. However, there is a population of patients who do not respond to conventional allopurinol treatment and are considered treatment resistant. Studies are in progress to treat these patients with the use of recombinant polyethylene glycol-uricase (PEG-uricase), which enzymatically converts uric acid to allantoin, a product that is more water soluble than uric acid and readily excreted.

Becker, M.A. Hyperuricemia and gout. In Scriver, C. R., Beaudet, A. L. Sly, W. S., and Valle, D. (Eds.), *The Metabolic and Molecular Bases of Inherited Disease,* 8th ed. New York: McGraw Hill, 2001, vol 2, Chap. 106, 2513–2535; Iiassa, T., Increased activity of PRPP synthetase. *Nippon Rinsho* 66:6948, 2008; Sundy, J. S, Ganson, J. L., Kelly, J. S., Scarlet, J. L., et al. Pharmokinetics and pharmacodynamics of intravenous PEGylated recombinant mammalian urate oxidase. *Arthritis and Rheumatism* 56:1021, 2007.

novo pathway, a series of reactions leads to synthesis of IMP, which serves as the common precursor for both adenosine 5′-monophosphate (AMP) and guanosine 5′-monophosphate (GMP). This pathway is highly regulated by AMP, GMP, and IMP; although IMP is the end product, it is not normally found to any extent under aerobic conditions.

Formation of IMP

Formation of IMP is shown in Figure 20.2. Several points should be emphasized about this pathway: 5′-phosphoribosyl-1-pyrophosphate (PRPP) is synthesized from ribose 5-phosphate that is generated primarily by the pentose phosphate pathway; the equivalent of 6 mol of ATP are utilized per mole of IMP synthesized; formation of 5-phosphoribosylamine (the first step) is the committed and regulated step. In formation of 5-phosphoribosylamine, the N–C bond is formed that will ultimately be the *N*-glycosidic bond of the purine nucleotide; there are no known regulated

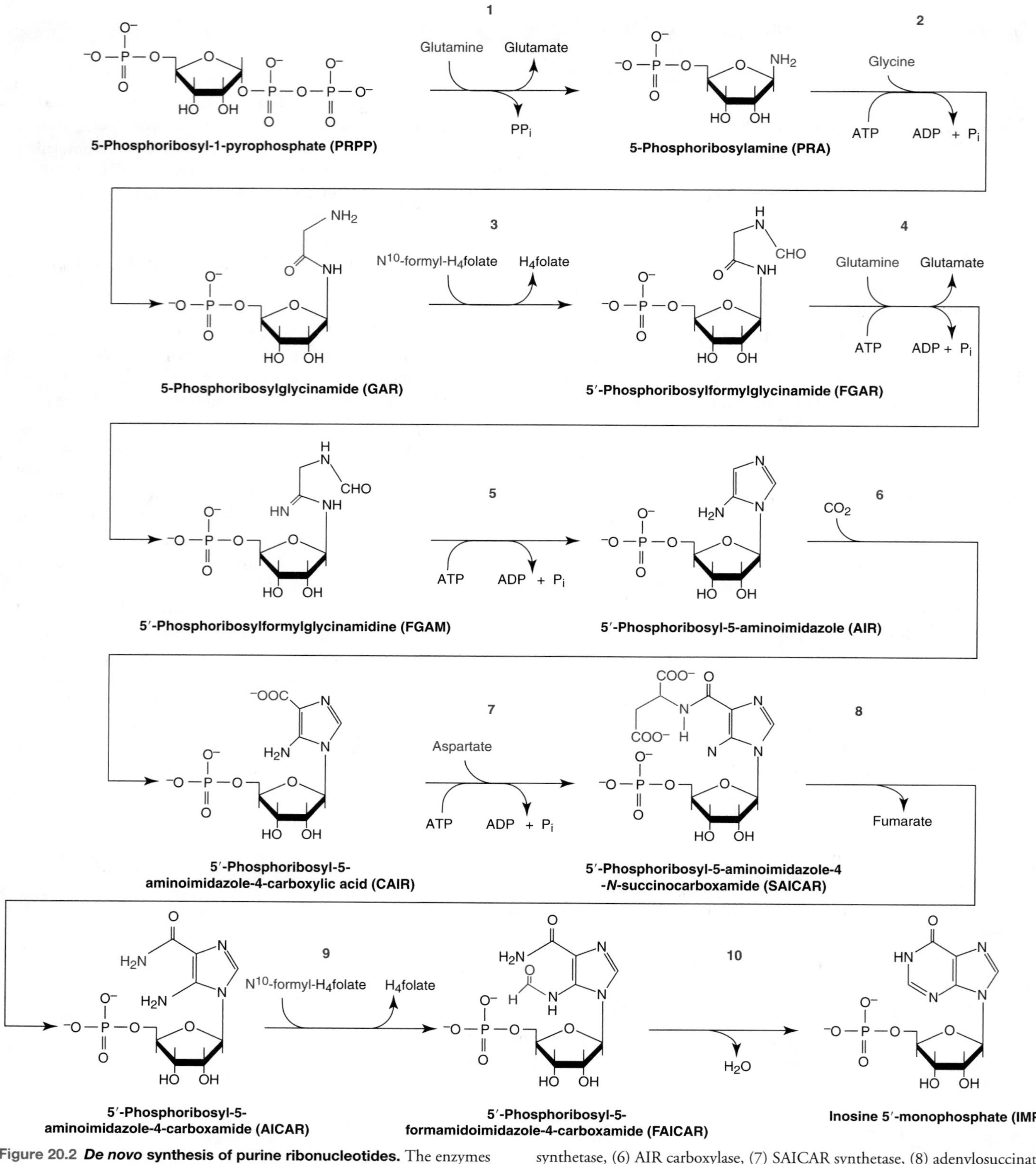

Figure 20.2 *De novo* synthesis of purine ribonucleotides. The enzymes catalyzing the reactions are (1) glutamine PRPP amidotransferase, (2) GAR synthetase, (3) GAR transformylase, (4) FGAM synthetase, (5) AIR synthetase, (6) AIR carboxylase, (7) SAICAR synthetase, (8) adenylosuccinate lyase, (9) AICAR transformylase, and (10) IMP cyclohydrolase.

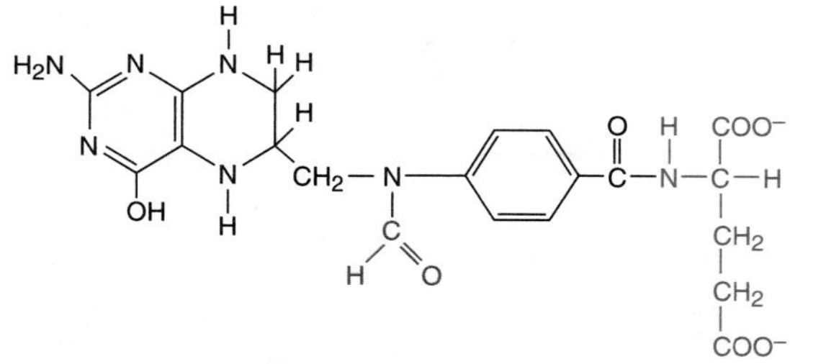

Figure 20.3 **Structure of N^{10}-formyl H_4folate.**

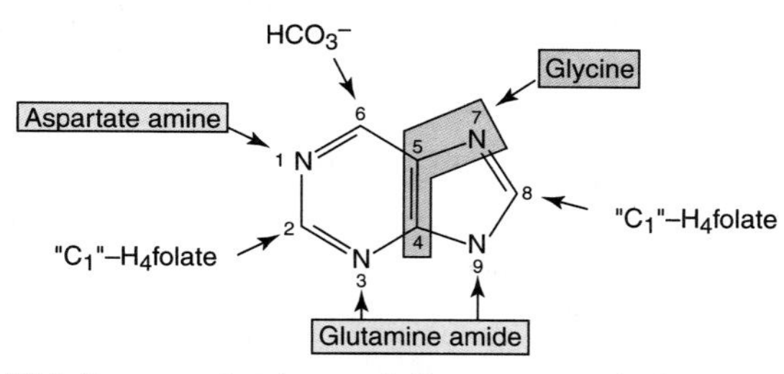

Figure 20.4 **Sources of carbon and nitrogen atoms in the purine ring.** C4, C5, and N7 from glycine; N3 and N9 from glutamine; C2 and C8 from C_1"-H_4folate; N1 from aspartate; and C6 from CO_2.

steps between the formation of 5-phosphoribosylamine and IMP; and tetrahydrofolate serves as a C_1-carrier (N^{10}-formyl H_4folate, Figure 20.3) in this pathway. It is important to point out that in Reactions 3 and 9, tetrahydrofolate is only a one-carbon carrier and is regenerated. Phosphoribosyl-5-aminoimidazole carboxylase, which catalyzes the reaction in which CO_2 is used to introduce carbon-6 of the ring, is not a biotin-dependent carboxylase.

The enzyme activities catalyzing several steps the pathway reside on separate domains of **multifunctional proteins**. (1) Activities of 5′-phosphoribosylglycinamide synthetase (step 2), 5′-phosphoribosylglycinamide transformylase (step 3), and 5′-phosphoribosylaminoimidazole synthetase (step 5), form part of a trifunctional protein. (2) 5′-Phosphoribosylaminoimidazole carboxylase (step 6) and 5′-phosphoribosyl-4-(*N*-succinocarboxamide)-5-aminoimidazole synthetase (step 7) activities are on the same bifunctional protein. (3) 5′-Phosphoribosyl-4-carboxamide-5-aminoimidazole transformylase (step 9) and IMP cyclohydrolase (step 10) activities are present on a bifunctional protein.

To summarize, *de novo* synthesis of purine nucleotides requires amino acids as carbon and nitrogen donors, CO_2 as a carbon source, and C_1-units transferred via H_4-folate. The contributions of these sources to the purine ring are shown in Figure 20.4. Several amino acids including serine, glycine, tryptophan and histidine can provide C_1-units to H_4folate (p. 1079), and therefore they may contribute to carbon-2 and carbon-8 of the purine ring.

IMP Is Common Precursor for AMP and GMP

IMP, the first ribonucleotide formed in the *de novo* pathway, is the common precursor for AMP and GMP synthesis (Figure 20.5). AMP and GMP are converted to ATP and GTP respectively, by nucleoside 5′-monophosphate kinases and nucleoside 5′-diphosphate kinases, which are not rate limiting. Conversion of IMP to AMP and GMP is regulated such that the appropriate cellular ratios of adenine and guanine nucleotides are maintained. Formation of GMP from IMP requires ATP as an energy source whereas formation of AMP from IMP requires GTP as the energy source. This reciprocal relationship ensures that when there is sufficient ATP in the cell, GMP will be synthesized from IMP and, when there is sufficient GTP, AMP will be synthesized from IMP.

Purine Nucleotide Synthesis Is Highly Regulated

The committed step of a metabolic pathway is generally the site of major regulation. In *de novo* purine nucleotide synthesis, formation of 5-phosphoribosylamine from glutamine and 5-phosphoribosyl-1-pyrophosphate (PRPP) is the committed step in IMP formation. The enzyme catalyzing this reaction, glutamine PRPP-amidotransferase, is rate limiting and is regulated allosterically by the end products of the pathway. IMP, GMP, and AMP serve as negative effectors of PRPP amido tranferase, while PRPP is a positive effector. Glutamine PRPP amidotransferase is a monomer of 135 kDa. In the presence of IMP, AMP, or GMP the enzyme forms a dimer that is much less active while

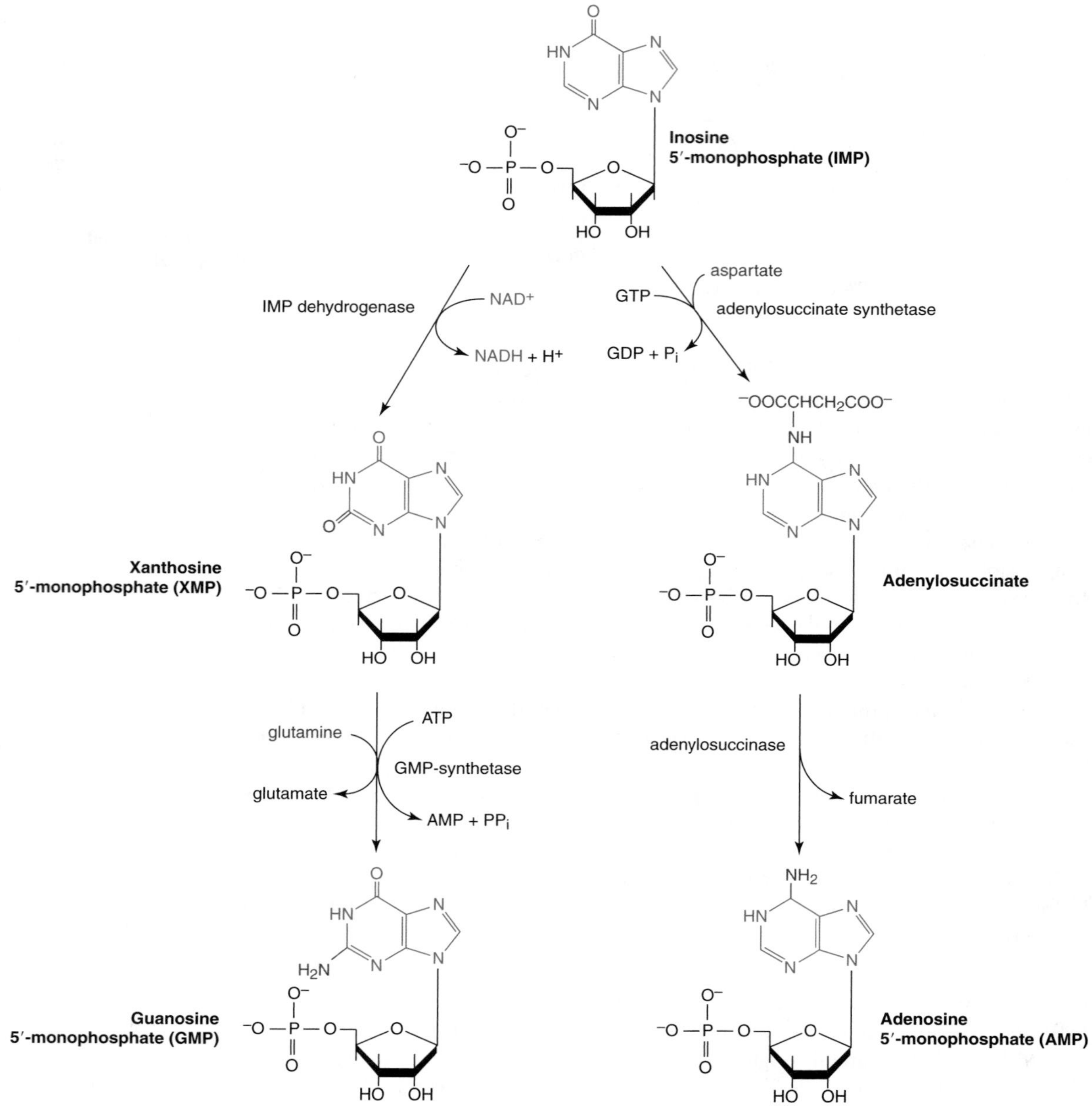

Figure 20.5 Formation of AMP and GMP from IMP branch point.

the presence of PRPP shifts the equilibrium in favor of the active monomeric form of the enzyme (Figure 20.6).

The enzyme from human tissues has distinct nucleotide binding sites. One site specifically binds oxypurine nucleotides (IMP, XMP, and GMP) whereas the other site specifically binds aminopurine nucleotides (AMP). When AMP and GMP (or IMP) are present simultaneously, the enzyme activity is inhibited synergistically by the combination of an oxy and amino purine nucleotide. Glutamine PRPP amidotransferase displays hyperbolic kinetics with respect to glutamine as the substrate and sigmoidal kinetics with respect to PRPP (Figure 20.7). Since the intracellular concentration of glutamine normally varies relatively little and is close to the K_m of the enzyme, the glutamine concentration is thought to have little effect in regulating IMP synthesis. However, in cancer patients who are treated with asparaginase (which also possesses glutaminase activity), the decreased concentration of glutamine as a result of asparaginase treatment may have an effect on the *de novo* pathway by limiting the availability of glutamine. The intracellular concentration of PRPP varies widely and can be 10 to 100 times less than the K_m for PRPP. Therefore, the

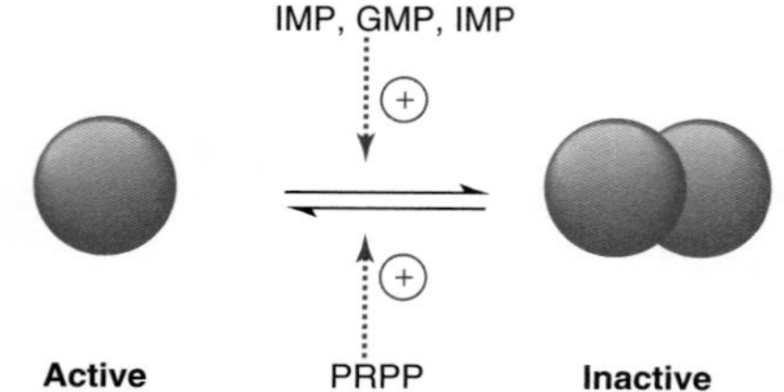

Figure 20.6 Effects of allosteric modulators on molecular forms of glutamine PRPP amidotransferase.

concentration of PRPP plays an important role in regulating synthesis of purine nucleotides under normal conditions.

Between formation of 5-phosphoribosylamine and IMP, there are no known regulated steps although alterations in the tetrahydrofolate levels will have an effect on the *de novo* synthesis of purine nucleotides. Methotrexate, a drug used extensively in the treatment of cancer, is cytotoxic because it markedly alters the folate pools. There is regulation at the branch point of IMP to AMP and IMP to GMP. From IMP to GMP, IMP dehydrogenase (IMPDH) is the rate-limiting enzyme and it is regulated by GMP acting as a competitive inhibitor with respect to IMP of IMPDH. There are two distinct genes for IMPDH with IMPDH-I serving as the constitutive form of the enzyme while IMPDH-II is related to cell growth and proliferation. Adenylosuccinate synthetase is rate limiting in conversion of IMP to AMP with AMP acting as a competitive inhibitor.

Since the ATP to GTP ratio in various cell types is relatively constant, it would suggest that additional control mechanisms function at the IMP *branch point.* Total cellular concentration of adenine nucleotides (ATP + ADP + AMP) in most cells is four to six times that of guanine nucleotides (GTP + GDP + GMP). The overall regulation of purine nucleotide synthesis is summarized in Figure 20.8. Defects in the metabolic pathway that lead to loss of regulation of purine nucleotide synthesis result in overproduction of purine nucleotides and the end product, uric acid. This results in a relatively common clinical condition known as gout (see Clin. Corr. 20.2, p. 811).

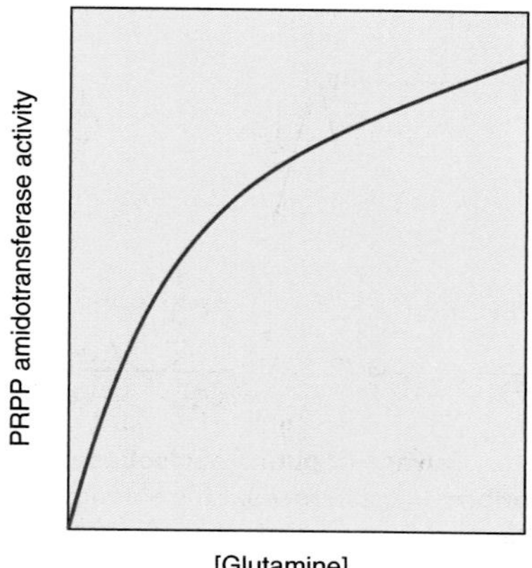

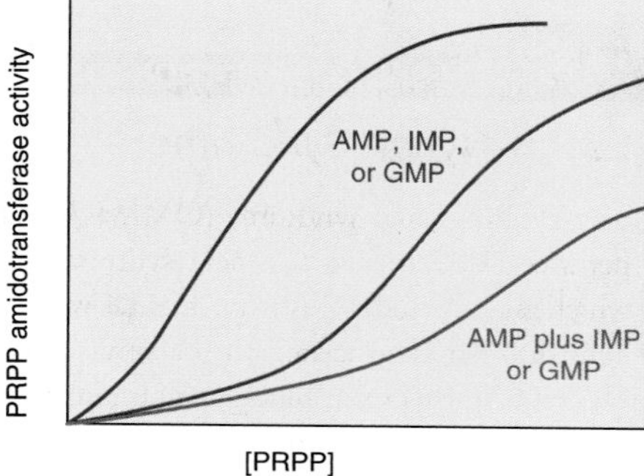

Figure 20.7 Glutamine PRPP amidotransferase activity as a function of glutamine or PRPP concentrations.

Purine Bases and Nucleosides Are Salvaged to Reform Nucleotides

The efficiency of normal metabolism is shown by the presence of two distinct "salvage pathways". One pathway utilizes the nucleobases, hypoxanthine, guanine, and adenine as substrates whereas the other pathway utilizes preformed nucleosides as the substrates. Each pathway is specific with respect to the nucleobase or nucleoside that is salvaged. The salvage of nucleobases requires the activity of phosphoribosyl transferases that utilize PRPP as the ribose phosphate donor. There are two distinct phosphoribosyl transferases. **Hypoxanthine-guanine phosphoribosyl transferase (HGPRTase)** catalyzes the reactions.

$$\text{Hypoxanthine} + \text{PRPP} \rightleftharpoons \text{IMP} + \text{PPi}$$

and

$$\text{Guanine} + \text{PRPP} \rightleftharpoons \text{GMP} + \text{PPi}$$

and **adenine phosphoribosyl transferase (APRTase)** catalyzes

$$\text{Adenine} + \text{PRPP} \rightleftharpoons \text{AMP} + \text{PPi}.$$

These two enzymes do not overlap in substrate utilization. The reactions are regulated by their end products. IMP and GMP are competitive inhibitors of HGPRTase with respect to PRPP; AMP is a competitive inhibitor with respect to PRPP of APRTase. In this way, salvage of purine bases is regulated; this has the overall effect of regulating the *de novo* pathway.

Hypoxanthine and guanine for salvage arise from degradation of endogenous or exogenous purine nucleotides. Adenine utilized in the APRTase reaction is generated mainly from synthesis of polyamines (p. 787). For each molecule of spermine synthesized, two molecules of 5′-methylthioadenosine are released that are then degraded to 5-methylthioribose-1-phosphate and adenine by the 5′-methylthioadenosine phosphorylase-catalyzed reaction. Adenine is salvaged through the APRTase reaction to form AMP.

Generation of AMP and GMP through these phosphoribosyltransferase reactions is very effective in decreasing the *de novo* pathway at the PRPP amidotransferase step. First, PRPP (a positive effector) is consumed in the reaction, decreasing the rate of formation of 5-phosphoribosylamine, and second, the products, AMP, IMP and GMP serve as negative effectors of PRPP amidotransferase (Figure 20.9).

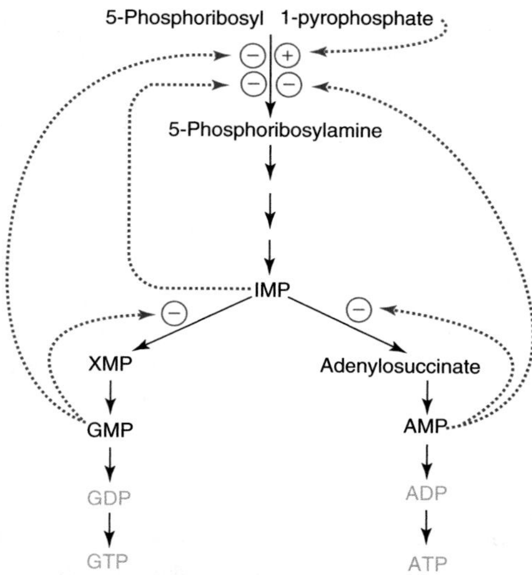

Figure 20.8 Regulation of purine nucleotide synthesis.

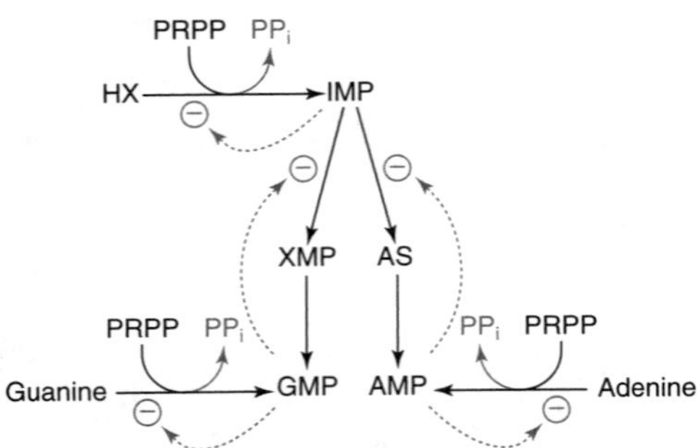

Figure 20.9 Salvage of purine nucleobases via phosphoribosyl transferases. Effects of products on AMP and GMP synthesis from IMP. The dashed lines represent sites of regulation.

HGPRTase activity is markedly depressed in the Lesch–Nyhan syndrome (Clin. Corr. 20.3), which is characterized clinically by hyperuricemia, mental retardation, and self-mutilation. In contrast to HGPRTase deficiency, there are no neurological problems such as mental retardation or self-mutilation associated with APRTase deficiency. APRTase-deficient individuals show increased excretion of adenine, 8-hydroxyadenine (8-HA) and 2,8-dihydroxyadenine (2,8-DHA). 8-HA and 2,8-DHA are generated from adenine by the action of xanthine oxidoreductase (xanthine dehydrogenase). The accumulation of 2,8-DHA can lead to the formation of calculi consisting of 2,8-DHA (which are sometimes misinterpreted as sodium urate calculi). Allopurinol treatment can be used to decrease the amount of 2,8-DHA formed and the associated nephrotoxicity.

Nucleosides such as adenosine are salvaged by adenosine kinase, a 5′-phosphotransferase that utilizes ATP as the phosphate donor. The substrate specificity of the 5′-phosphotransferases varies with the particular nucleoside kinase.

CLINICAL CORRELATION 20.3

Lesch–Nyhan Syndrome

The Lesch–Nyhan syndrome (OMIM 300322) is characterized by hyperuricemia, excessive uric acid synthesis, and neurological problems, which may include spasticity, mental retardation, and self-mutilation. This disorder is associated with a very severe or complete deficiency of HGPRTase (hypoxanthine-guanine phosphoribosylsyltranferase). HGPRTase is on the X chromosome; hence, the deficiency is virtually limited to males. Exceptions have been reported. In a detailed study of the available patients, it was observed that if HGPRTase activity were less than 2% of normal, mental retardation was present, but if the activity were less than 0.2% of normal, the unique self-mutilation phenomenon was present. The defect also leads to excretion of elevated levels of hypoxanthine and xanthine.

There are more than 200 mutations in the HGPRTase gene from patients with the Lesch–Nyhan syndrome. These mutations have led to the loss of HGPRTase protein, loss of HGPRTase activity, K_m mutants, and HGPRTase protein with a short half-life. HGPRTase catalyzes the reactions in which hypoxanthine, xanthine, and guanine are converted to IMP, GMP, and XMP with PRPP as the ribose 5-phosphate donor. The hyperuricemia and excessive uric acid production that occur in patients with the Lesch–Nyhan syndrome are easily explained by the lack of HGPRTase activity. As a consequence, hypoxanthine and guanine are not salvaged, leading to increased intracellular pools of PRPP and decreased levels of IMP or GMP. Both of these factors promote *de novo* synthesis of purine nucleotides without regard for appropriate regulation of this pathway at PRPP amidotransferase, the rate-limiting step. It is not understood why a severe defect in this salvage pathway leads to neurological problems. Adenine phosphoribosyltransferase (APRTase) activity in these patients is normal or slightly elevated. With APTRase present, the cellular needs for purine nucleotides could be met by conversion of AMP to IMP followed by the conversion of IMP to GMP if the cell's *de novo* pathway were not functioning. The normal tissue distribution of HGPRTase activity perhaps could explain the neurological symptoms. The brain (frontal lobe, basal ganglia, and cerebellum) has from 10 to 20 times the level of enzyme activity found in liver, spleen, or kidney and from 4 to 8 times that of erythrocytes. Individuals who have primary gout with excessive uric acid formation and hyperuricemia do not display neurological problems so that it is argued that products of purine degradation (hypoxanthine, guanine, xanthine, and uric acid) should not be toxic to the central nervous system (CNS). However, it is possible that these metabolites are toxic to the developing CNS or that lack of enzyme leads to an imbalance in the concentrations of adenine and guanine nucleotides at critical times during development. If IMP dehydrogenase activity in brain were extremely low, lack of HGPRTase could lead to decreased amounts of intracellular GTP due to decreased salvage of guanine. Since GTP is a precursor of tetrahydrobiopterin, a required cofactor in the biosynthesis of neurotransmitters and nitric oxide, and is required in other functions such as signal transduction via G proteins and protein synthesis, low concentrations of GTP during development could be the triggering factor in the observed neurological manifestations.

Some data indicate that patients with Lesch–Nyhan disease have impairment with dopamine function. This could be related to the role of tetrahydrobiopterin in the hydroxylation of tyrosine leading to dopamine synthesis. However, at this time there is no unequivocal explanation for why the loss of HGPRTase activity leads to the unusual neurological symptoms seen in the patients.

Because of the lack of a curative treatment for Lesch–Nyhan, there have been various treatments tried to reduce the self-injury aspect of the disease. These treatments have included therapy with levodopa, *S*-adenosylmethionine, and even injections with botulinum toxin A. There is no cure. Treatment with allopurinol serves only to decrease the amount of uric acid formed; this relieves some of the problems caused by sodium urate deposits. However, since the Lesch–Nyhan patient has a marked reduction in HGPRTase activity, hypoxanthine and guanine are not salvaged, PRPP is not consumed, and consequently *de novo* synthesis of purine nucleotides is not regulated. This is in contrast to patients with gout who will have decreased uric acid levels and decreased *de novo* synthesis in response to allopurinol treatment. There is no successful treatment for the neurological problems. These patients usually die from kidney failure.

Iiassa, T. Increased activity of PRPP synthetase. *Nippon Rinsho* 66:694, 2008; and Nyhan, W. L. Lesch–Nyhan disease. *Nucleosides, Nucleotides Nucleic Acids* 6: 559, 2008.

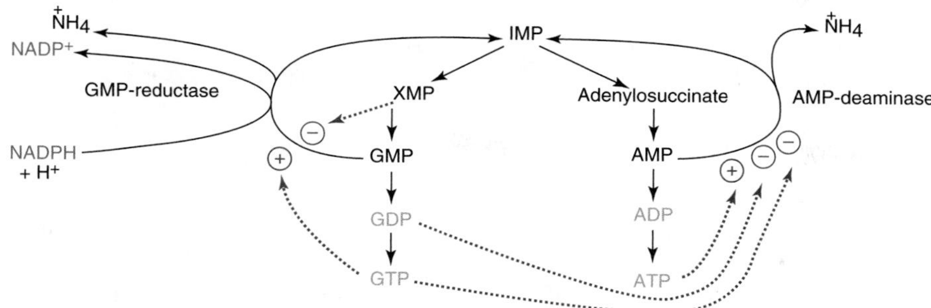

Figure 20.10 Interconversions of purine nucleotides.

Overall, these salvage reactions conserve energy and permit cells to form nucleotides from the free nucleobases or nucleosides. Erythrocytes, for example, do not have glutamine PRPP amidotransferase and hence cannot synthesize 5-phosphoribosylamine, the first unique metabolite in the pathway of purine nucleotide synthesis. Therefore, they must depend on purine phosphoribosyl transferases and 5′-phosphotransferase (adenosine kinase) to replenish their nucleotide pools.

Purine Nucleotides Are Interconverted to Balance Cellular Levels of Adenine and Guanine Nucleotides

De novo synthesis of purine nucleotides is under very fine control by the regulation at glutamine PRPP amidotransferase and at the branch points, IMP to AMP (adenylosuccinate synthetase) and IMP to GMP (IMP dehydrogenase). Additional enzymes present in mammalian cells allow for interconversions of adenine and guanine nucleotides to maintain the appropriate balance of cellular concentrations of these purine nucleotides. These interconversions occur by indirect steps since there is no direct one-step pathway for conversion of GMP to AMP or AMP to GMP. In each case, AMP or GMP is converted to IMP (Figure 20.10). These reactions are catalyzed by separate enzymes, each of which is under separate regulation. Reductive deamination of GMP to IMP is catalyzed by GMP reductase. GTP activates this step whereas xanthosine 5′-monophosphate (XMP) is a strong competitive inhibitor of the reaction. GTP, although not required by the enzyme, increases enzyme activity by lowering the K_m with respect to GMP and by increasing V_{max}.

AMP deaminase (5′-AMP aminohydrolase) catalyzes deamination of AMP to IMP and is activated by K^+ and ATP and inhibited by P_i, GDP, and GTP. In the absence of K^+, the velocity-versus-AMP concentration curve is sigmoidal. The presence of K^+ is not required for maximum activity but serves as a positive allosteric effector reducing the apparent K_m for AMP.

The net effect of these reactions is that cells can interconvert adenine and guanine nucleotides to meet cellular needs, while maintaining control over these reactions.

20.5 • GTP IS THE PRECURSOR OF TETRAHYDROBIOPTERIN

GTP is the direct precursor for **tetrahydrobiopterin** synthesis (Figure 20.11). Reactions from GTP to tetrahydrobiopterin are catalyzed by GTP cyclohydrolase I, 6-pyruvoyl-tetrahydropterin synthase, and sepiapterin reductase. GTP cyclohydrolase I catalyzes the rate-limiting step. Many cell types synthesize tetrahydrobiopterin, which is a required cofactor in hydroxylation reactions involving phenylalanine, tyrosine, and tryptophan (p. 767). In addition, tetrahydrobiopterin is involved in the generation of nitric oxide by nitric oxide synthases. Inhibitors of IMP dehydrogenase cause a marked reduction in cellular levels of tetrahydrobiopterin, demonstrating the importance of GTP as the precursor of tetrahydrobiopterin and of IMP dehydrogenase as the rate-limiting enzyme in GTP formation.

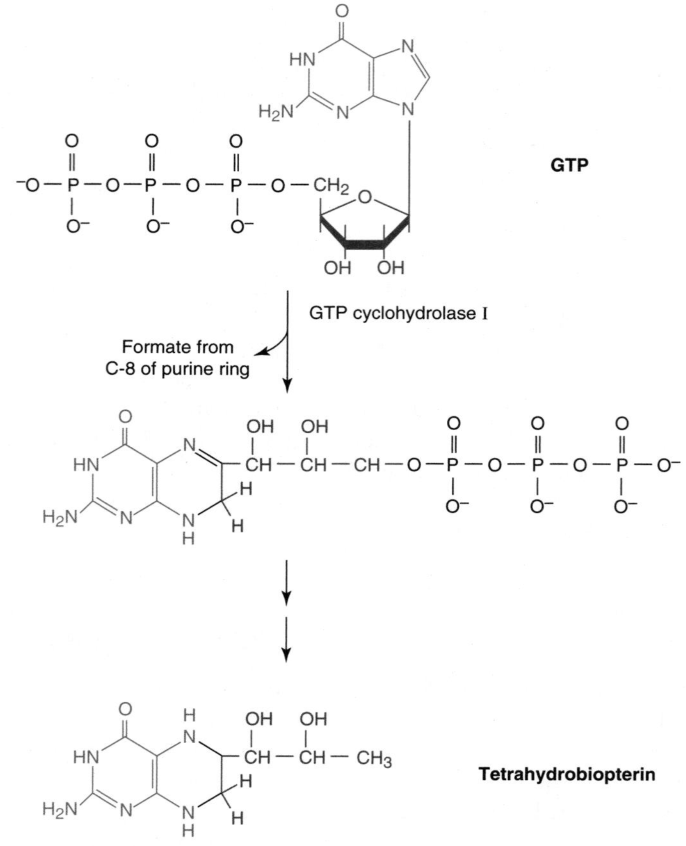

GTP

GTP cyclohydrolase I

Formate from C-8 of purine ring

Tetrahydrobiopterin

Figure 20.11 Synthesis of tetrahydrobiopterin from GTP.

The degradation of purine nucleotides, nucleosides, and nucleobases follow a common pathway that leads to formation of **uric acid** (Figure 20.12). The enzymes involved in degradation of nucleic acids, nucleotides and nucleosides vary in specificity. **Nucleases** are specific toward either RNA or DNA and also toward the bases and position of cleavage site at the 3′, 5′-phosphodiester bonds. Nucleotidases range from those with relatively high specificity, such as 5′-AMP nucleotidase to those with broad specificity, such as the acid and alkaline phosphatases, which will hydrolyze any of the 3′- or 5′-nucleotides (Clin. Corr. 20.4). AMP deaminase is specific for AMP. Adenosine deaminase is less specific, since adenosine, 2′-deoxyadenosine, and many other 6-aminopurine nucleosides are deaminated by this enzyme.

Purine nucleoside phosphorylase catalyzes the reversible reactions.

$$\text{Inosine} + \text{Pi} \rightleftharpoons \text{hypoxanthine} + \text{ribose 1-P}$$

or

$$\text{Guanosine} + \text{Pi} \rightleftharpoons \text{guanine} + \text{ribose 1-P}$$

or

$$\text{Xanthosine} + \text{Pi} \rightleftharpoons \text{xanthine} + \text{ribose 1-P}$$

with 2′-deoxyinosine and 2′-deoxyguanosine also being natural substrates for purine nucleoside phosphorylase. This is important since removal of deoxyguanosine prevents uncontrolled accumulation of dGTP, which is toxic to cells at high concentrations. Although the equilibrium constants for reactions catalyzed by purine nucleoside phosphorylase favor the direction of nucleoside synthesis, cellular concentrations of free purine base and ribose-1-phosphate are too low to support nucleoside synthesis under normal conditions. Therefore, the main function of the enzyme under cellular conditions is the degradative rather than synthetic pathway. Adenosine deaminase deficiency is associated with severe combined immunodeficiency whereas purine nucleoside phosphorylase deficiency leads to a defective T-cell immunity but normal B-cell immunity (Clin. Corr. 20.5).

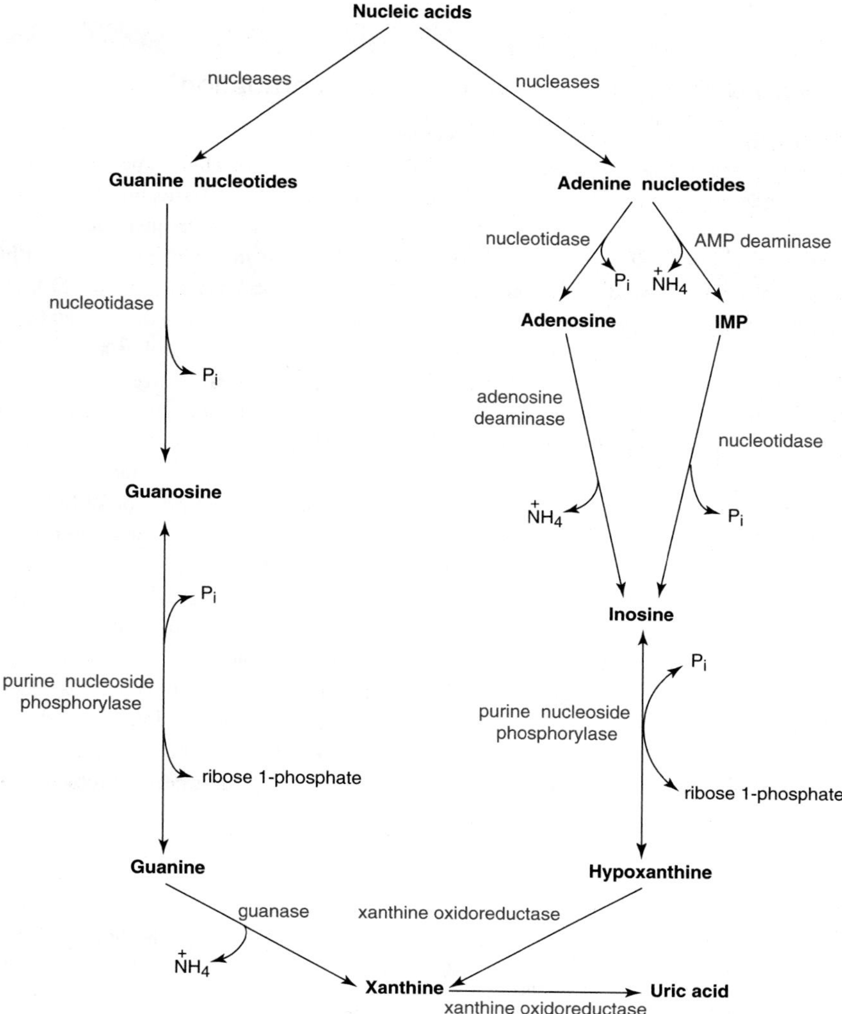

Figure 20.12 Degradation of purine nucleotides.

CLINICAL CORRELATION 20.4

Increased Cytosolic 5′-Nucleotidase Activity

Four patients were identified in whom 5′-nucleotidase activity in fibroblast lysates as measured using 5′-AMP or 5′-UMP as substrate was increased to 6- to 10- fold in these patients compared to the controls. These four unrelated children had problems associated with developmental delay, seizures, ataxia, infections, severe language deficit, and an unusual behavioral phenotype characterized by hyperactivity, short attention span, and poor social interaction. Since it was possible that the increased cellular 5′-nucleotidase activity decreased nucleotide pools and these patients were anemic, they were treated with oral uridine. Remarkably, all four patients treated with uridine improved dramatically in all aspects of the physical and clinical behaviors. The concentration of uridine needed to treat these children was much higher than the concentration needed to treat the children with orotic aciduria. How the increased 5′-nucleotidase activity leads to the clinical symptoms is not known. It is possible that the increased cellular adenosine level that results from the increased enzyme activity is involved in the problems as adenosine has metabolic interactions even at low concentrations (adenosine receptors, potassium ion channels, etc.).

These findings again point to the fact that either overproduction or blockage of nucleotide synthesis or decreased or increased degradation of nucleotides or nucleosides can have major consequences on the well-being of the individual.

Careddu, M. G., Alegruni, S. Pesi, R., Camici, M., et al. Knockdown of cytolosic 5′-nucletidease II(cN-II) reveals that its activity is essential for survival in astrocytoma cells. *Biochem.Biophys Acta* 1783:1529, 2008; Chiarelli, L. R, Fermo, E, Zanella, G., Valentini, G., et al. Hereditary erythrocyte pyrimidine 5′-nucleotidase deficiency: A biochemical, genetic and clinical overview. *Hemetology* 11:67, 2006; and Page, T., Yu, A., Fontanesi, J., and Nyhan, W. L. Developmental disorder associated with increased cellular nucleotidase activity. *Proc. Natl. Acad. Sci. USA* 94:11601, 1997.

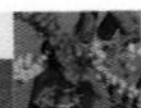
CLINICAL CORRELATION 20.5

Immunodeficiency Diseases Associated with Defects in Purine Nucleoside Degradation

Distinct immunodeficiency diseases are associated with defects in adenosine deaminase (ADA) and purine nucleoside phosphorylase (PNP), respectively. These enzymes are involved in the degradative pathways leading to formation of uric acid as the end product. Natural substrates for adenosine deaminase are adenosine and deoxyadenosine whereas natural substrates for purine nucleoside phosphorylase are inosine, guanosine, deoxyinosine and deoxyguanosine. A deficiency in ADA is associated with a severe combined immunodeficiency involving both T-cell and B-cell functions. PNP deficiency is associated with an immunodeficiency involving T-cell functions with little or no effect on B-cell function. In neither case is the specific mechanism(s) by which the lack of these enzymes leads to immune dysfunction known. The deficiency in ADA activity arises from mutations in various exons that cause either missense or nonsense effects. In ADA-deficient patients, intracellular concentrations of dATP and *S*-adenosylhomocysteine are greatly increased. Several hypotheses have been put forth to explain the biochemical consequences of a lack of ADA. (1) High concentrations of dATP inhibit ribonucleotide reductase activity and as a consequence inhibit DNA synthesis. (2) deoxyadenosine inactivates *S*-adenosyl homocysteine hydrolase leading to decreased *S*-adenosylmethionine that is required for methylation of bases in RNA and DNA. (3) Increased concentrations of adenosine result in increased cAMP levels. It is possible that each of these mechanisms contributes to the overall effect of immune dysfunction. There is not, however, a suitable explanation for the specificity of the effects on only T-cells and B-cells.

Treatments of children with ADA deficiency have included (1) blood transfusions, (2) bone marrow transplantation, (3) enzyme replacement therapy with ADA-polyethylene glycol conjugate (ADA-PEG), and most recently (4) gene therapy. Each of these treatments has disadvantages. Blood transfusions produce problems of iron overload and safety of the blood source. Bone marrow transplantation, while curative, requires a suitably matched donor. Enzyme replacement therapy with adenosine deaminase covalently attached to polyethylene glycol (ADA-PEG) has been the most successful to date, but the treatment requires constant monitoring of ADA levels, frequent injections of ADA-PEG and there is considerable cost involved for the ADA-PEG. Studies using erythrocytes from SCID patients treated with ADA-PEG have shown that as the serum level of ADA increased, there was a corresponding increase in adenosylhomocysteine hydrolase activity and a decrease in dATP levels to essentially zero. Concurrently, there were the reappearances of circulating T and B lymphocytes. Gene therapy presents the hope for the future. There are indications in gene therapy trials that the ADA gene has been successfully transfected into stem cells of ADA-deficient children. There is a second type of SCID not related to adenosine deaminase deficiency. This type of SCID is caused by a mutant form of the gamma C interleukin receptor that blocks the development of T lymphocytes and natural killer cells. Gene therapy to correct this defect was essentially successful in treating the immune dysfunction. However, two of the nine children who had received the gene therapy for this defect, developed a rare form of leukemia. It is believed that the genetic disruptions caused by the retroviral vector used in the gene transfer process led to the leukemia. Gene therapy trials have proceeded with extreme caution following this unexpected problem.

Gaspar, H. B., Bjorkegren, E., Parsley, K., Gilmour, K. C., et al. Successful reconstitution of immunity in ADA-SCID by stem cell gene therapy following cessation of PEG-ADA and use of mild preconditioning. *Mol. Therapy* 14:505, 2006; Hershfield, M. Adenosine deaminase deficiency. In Pagon, R. A. (Ed.), *GENERevіews*. Retrieved from NCBI bookshelf database, http://www.bcbi/nlm.nih.gov/bookshelf; Hoogerbrugge, P. M., von Beusechem, V. W., Kaptein, L. C., Einerhard, M. P. and Valerio, D. Gene therapy for adenosine deaminase deficiency. *Brit. Med. Bull.* 51:72, 1995; Hershfield, M. S. Adenosine deaminase deficiency: Clinical expression, molecular basis, and therapy. *Seminars in Hematology* 35:291, 1998; and Russell, C. S. and Clarke, L. A. Recombinant proteins for genetic disease. *Clinical Genetics* 55:389, 1999.

20.6 • URIC ACID IS THE END PRODUCT OF PURINE DEGRADATION IN HUMANS

As shown in Figure 20.12, adenine nucleotides are degraded to hypoxanthine whereas guanine nucleotides are metabolized to xanthine. Hypoxanthine and xanthine are oxidized by xanthine oxidoreductase which has both dehydrogenase and oxidase activities. Xanthine dehydrogenase activity requires NAD as the electron acceptor whereas the oxidase activity utilizes molecular oxygen with the generation of hydrogen peroxide as product. Xanthine oxidoreductase, which contains FAD, Fe, and Mo, can exist in either the dehydrogenase or oxidase form. Uric acid is the unique end product of purine nucleotide degradation in man. The reactions are shown in Figure 20.13.

Since uric acid is not very soluble in aqueous medium, in gout (see Clin. Corr. 20.2, p. 811) there are elevated levels of uric acid that can result in deposition of sodium urate crystals primarily in joints. Hyperuricemia is a clinical condition characterized by excess levels of uric acid in the blood and accompanied by increased levels of uric acid in urine (hyperuricuria). Since uric acid is the unique end product of purine degradation in

CLINICAL CORRELATION 20.6

Tumor Lysis Syndrome (TLS)

Cancer patients with large tumor burdens who undergo radiation therapy or chemotherapy treatments also show increased serum and urine concentrations of uric acid. The source of this increased uric acid is not due to increased purine nucleotide synthesis but rather from the destruction of the tumor cells by radiation or cytoxic drugs that in turn release degraded nucleic acids and cellular nucleotides that are further metabolized to uric acid by the normal pathway. Many of the cancer treatment protocols include allopurinol as one of the drugs for the sole purpose of limiting the buildup of uric acid in the patients.

Tumor lysis syndrome (TLS) represents a group of metabolic consequences that can result in response to the cancer treatment. Concurrent with the increase in uric acid formation as a result of xanthine oxidase activity, the patients can suffer from hypercalcemia, hyperkalemia, hyperphosphatemia, and acute renal failure. Rasburicase, a recombinant urate oxidase, has been used successfully, especially in children, to treat the hyperuricemia that is associated with TLS. Rasburicase functions to catalyze the degradation of uric acid, to allantoin, which is a more water-soluble product than uric acid and much more readily excreted. Rasburicase, although useful in the treatment of TLS, is contraindicated in patients who have a deficiency in glucose 6-phosphate dehydrogenase (G6PDH). The urate oxidase reaction generates hydrogen peroxide as a product which in the absence G6PDH activity, leads to red cell lysis due to the lack of reduced glutathione in cells. Although the use of uricase has certain pharmacological advantages over allopurinol, it has the disadvantage that it is very expensive and probably not readily available to all.

Bessmertny, D., Robitaille, L. M., and Cairo, M. S. Rasburicase: A new approach for preventing and/or treating tumor lysis syndrome. *Curr. Pharm. Design* 11:4177, 2005; Jeha, S., and Pui, C. H. Recombinant urate oxidase (rasburicase) in prophylaxis and treatment of tumor lysis syndrome. *Contributions Nephrology* 147:69, 2005; Wang, L. Y., Shee, L. Y., Chang, H., Jou, S. T., et al. Recombinant urate oxidase (Rasburicase) for prevention and treatment of tumor lysis syndrome in patients with hematologic malignancies. *Acta Haematol.* 115:35, 2006; and Ribeiro, R. C., and Pui, C. H. Recombinant urate oxidase for prevention of hyperuricemia and tumor lysis syndrome in lymphoid malignancies. *Clin. Lymphoma* 3: 252, 2003.

humans, excess levels of uric acid indicate some metabolic situation which may or may not be serious. There are several instances in which the cause of the hyperuricemia/hyperuricuria can be defined as a metabolic defect related to the overproduction of purine nucleotides and other situations in which there are not defined metabolic alterations (see Clin. Corr. 20.1, p. 810). Cancer patients undergoing chemotherapy or radiation treatment become hyperuricemic due to cellular destruction and the release of uric acid (Clin. Corr. 20.6).

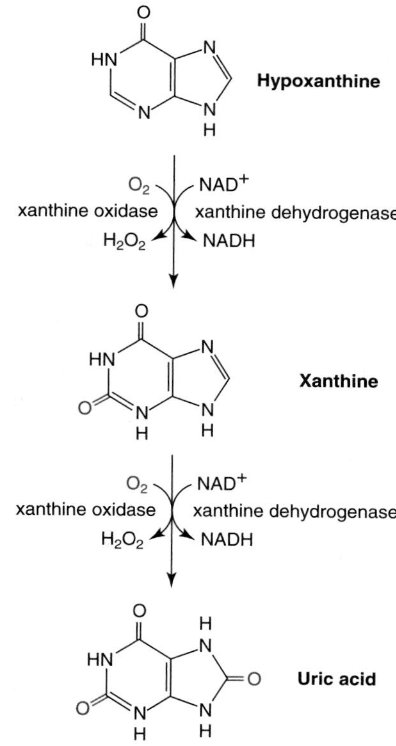

Figure 20.13 Reactions catalyzed by xanthine oxidoreductase (XOR). XOR contains both xanthine dehydrogenase and oxidase activities. The dehydrogenase and oxidase forms of the enzyme are interconvertible.

20.7 • METABOLISM OF PYRIMIDINE NUCLEOTIDES

The *de novo* synthesis of the pyrimidine ring in mammalian cells utilizes amino acids as carbon and nitrogen donors in addition to CO_2. **Uridine 5′-monophosphate** (UMP) is synthesized in a metabolic pathway. Energy from ATP hydrolysis (or equivalent) is required to drive several steps in the pathway.

Synthesis of Pyrimidine Nucleotides

In contrast to *de novo* purine nucleotide synthesis, not all enzymes for *de novo* synthesis of pyrimidine nucleotides are cytosolic. Reactions leading to formation of UMP are shown in Figure 20.14. Important aspects of the pathway should be noted. The pyrimidine ring is formed first and then ribose 5-phosphate is added with PRPP as the ribose 5-phosphate donor. The enzyme-catalyzing formation of cytosolic carbamoyl phosphate, carbamoyl phosphate synthetase II (CPS II), is cytosolic and is distinctly different from mitochondrial carbamoyl phosphate synthetase I (CPS I) in mitochondria and which functions as part of the urea cycle. Formation of *N*-carbamoylaspartate is the committed step in pyrimidine nucleotide synthesis but formation of cytosolic carbamoyl phosphate (CPS II) is the regulated step. Formation of orotate from dihydroorotate is catalyzed by mitochondrial **dihydroorotate dehydrogenase (DHODH)**. DHODH is located on the outer surface of the inner mitochondrial membrane. This location is the functional connection between the respiratory electron transport chain

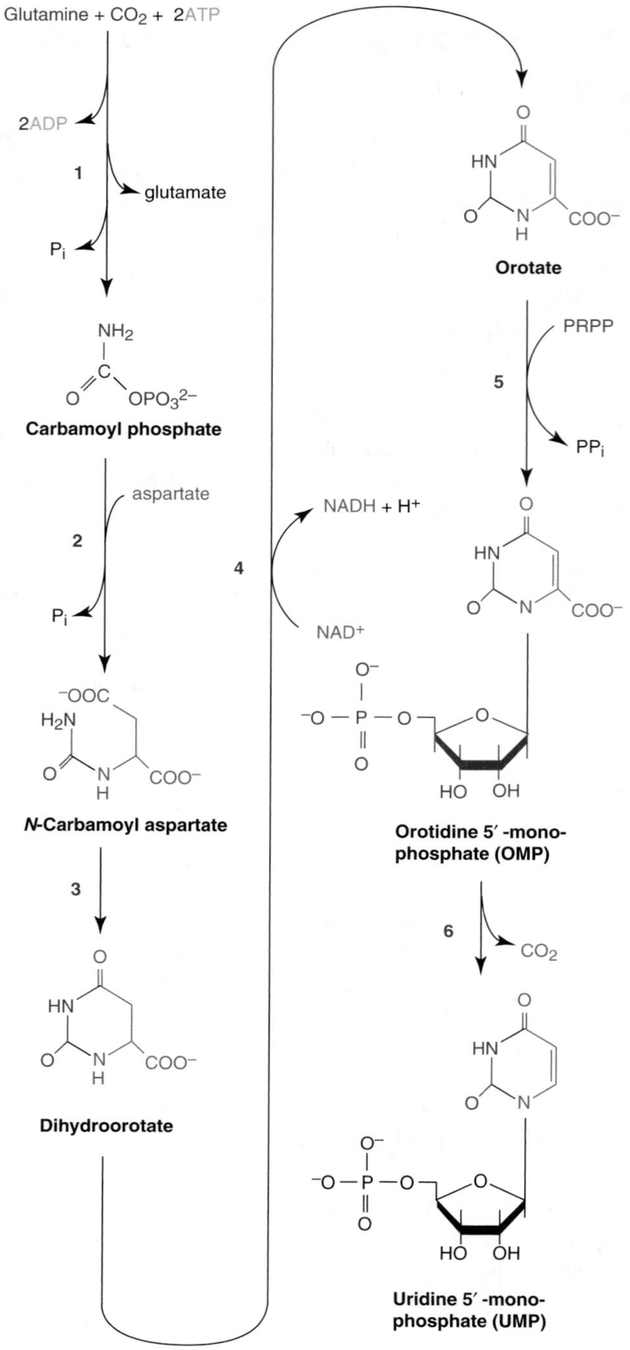

Figure 20.14 *De novo* synthesis of pyrimidine nucleotides. Enzyme activities catalyzing the reactions are (1) carbamoyl phosphate synthetase II, (2) aspartate carbamoyltransferase, (3) dihydroorotase, (4) dihydroorotate dehydrogenase, (5) orotate phosphoribosyltransferase, and (6) OMP decarboxylase. The activities of 1, 2, and 3 are on a trifunctional protein (CAD); the activities of 5 and 6 are on a bifunctional protein (UMP synthase).

through ubiquinone and dihydroorotate metabolism. The relationship of this connection is borne out by the fact that pyrimidine nucleotide synthesis is inhibited under conditions in which mitochondrial respiration is reduced. The other activities of the pathway are found in the cytosol on multifunctional proteins. The activities of CPS II, aspartate carbamoyl transferase and dihydroorotase are found on a trifunctional protein (CAD), whereas orotate phosphoribosyltransferase and OMP decarboxylase activities are found on a bifunctional protein (UMP synthase). A defect in this bifunctional protein that affects either phosphoribosyl transferase activity or decarboxylase activity leads to a rare clinical condition known as hereditary orotic aciduria (Clin. Corr. 20.7). The immunosuppressive drug, leflunomide, which is used in the treatment of rheumatoid arthritis, inhibits *de novo* synthesis of pyrimidine nucleotides specifically at dihydroorotate dehydrogenase.

Nucleotide kinases convert UMP to UTP (Figure 20.15). UTP is the direct substrate for CTP synthetase. CTP synthetase catalyzes formation of CTP from UTP with glutamine

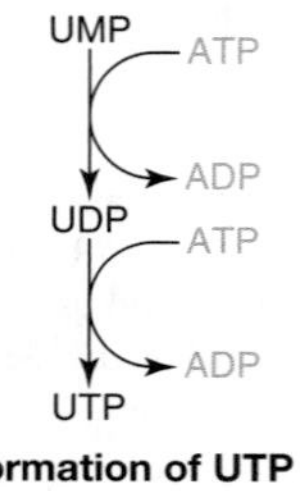

Figure 20.15 Formation of UTP from UMP.

CLINICAL CORRELATION 20.7

Hereditary Orotic Aciduria

Hereditary orotic aciduria (OMIM 258900) results from a defect in *de novo* synthesis of pyrimidine nucleotides. This genetic disease is characterized by severe anemia, growth retardation, and high levels of orotic acid excretion. The biochemical basis for orotic aciduria is a defect in one or both of the activities (orotate phosphoribosyl transferase or orotidine decarboxylase) associated with UMP synthase, the bifunctional protein. It is a very rare disease, but the understanding of the metabolic basis for this disease has led to successful treatment of the disorder. Patients are fed uridine, which leads not only to reversal of the anemia but also to decreased formation of orotic acid. Uridine taken up by cells is salvaged by uridine phosphotransferase to UMP that is sequentially converted to UDP and then to UTP. UTP formed from exogenous uridine in turn inhibits carbamoyl phosphate synthetase II, the major regulated step in the *de novo* pathway. As a result, orotic acid synthesis via the *de novo* pathway is markedly decreased to essentially normal levels. Since UTP is also the substrate for CTP synthesis, uridine treatment serves to replenish both the UTP and CTP cellular pools. In effect, then, exogenous uridine bypasses the defective UMP synthase and supplies cells with UTP and CTP required for nucleic acid synthesis and other cellular functions. The success of treatment of hereditary orotic aciduria with uridine provides *in vivo* data that show the importance of the step catalyzed by carbamoyl phosphate synthase II as the site of regulation of pyrimidine nucleotide synthesis in humans.

Suchi, M., Mizuno, H., Kawai, Y., Tonboi, T., et al. Molecular cloning of the human UMP synthase gene and characterization of a point mutation in two hereditary orotic aciduria families. *Am. J. Hum. Genet.* 60: 525, 1997; Webster, D. R., Becroft, D. M. O., van Gennip, A. H., and Van Kuilenburg, A. B. P. Hereditary orotic aciduria and other disorders of pyrimidine metabolism. In Scriver, C. R., Beaudet, A. L., Sly, W. S. and Valle, D.(Eds.), *The Metabolic and Molecular Bases of Inherited Disease,* 8th ed. New York: McGraw Hill, 2001, vol. 2, Chap. 113, 2663.

as the amino group donor (Figure 20.16). CTP synthetase displays homotropic sigmoidal kinetics with respect to UTP whereas CTP, the product, is a negative effector of the reaction as shown in Figure 20.17. The regulation of CTP synthetase in this way is critical as it allows cells to maintain an appropriate ratio of UTP and CTP for cellular functions and RNA synthesis.

To summarize, *de novo* synthesis of pyrimidine nucleotides requires aspartate as a carbon and nitrogen donor, glutamine as a nitrogen donor, and CO_2 as a carbon donor (Figure 20.18). Five of the six reactions in the pathway take place in the cytosol of the cell, whereas the other reaction occurs in mitochondria. The cytosolic enzyme activities reside on multifunctional proteins. UTP is the direct precursor of CTP.

Genetic disorders of pyrimidine nucleotide metabolism as a result of defects in either the synthetic or degradative pathways that lead to clinical disorders indicate that the substrate or product is critical for normal cellular function. Interestingly, these defects present in a variety of clinical situations, ranging from anemia to neurological conditions.

Pyrimidine Nucleotide Synthesis Is Regulated at the Level of Carbamoyl Phosphate Synthetase II

Regulation of pyrimidine nucleotide synthesis in mammalian cells occurs at the carbamoyl phosphate synthetase II step. CPS II is a cytosolic enzyme and is distinct from mitochondrial CPS I, which utilizes ammonia as the amino donor instead of glutamine and which requires activation by *N*-acetyl glutamate. CPS II is inhibited by UTP, an end product of the pathway and is activated by PRPP. The K_i for UTP (at CPSII) and the K_a for PRPP (at CPSII) are in the range of values that would allow intracellular levels of UTP and PRPP to have an effect on the control of pyrimidine nucleotide synthesis. CPS II is the only source of carbamoyl phosphate in extrahepatic tissues. However, in liver, under stressed conditions in which there is excess ammonia, CPS I generates carbamoyl phosphate in mitochondria, which moves into the cytosol and serves as a substrate for orotic acid synthesis. This pathway serves to detoxify excess ammonia. Elevated levels of orotic acid are excreted during ammonia toxicity in humans. This points to carbamoyl phosphate synthetase II as being the major regulated activity of pyrimidine nucleotide metabolism. However, the lack of sufficient PRPP leads to the accumulation of orotic acid rather than to excess pyrimidine nucleotide synthesis.

UMP does not inhibit carbamoyl phosphate synthetase II but competes with OMP to inhibit OMP decarboxylase (Figure 20.19). As discussed earlier, conversion of UTP to CTP is also regulated so that cells maintain a balance between uridine and cytidine nucleotides in the cell.

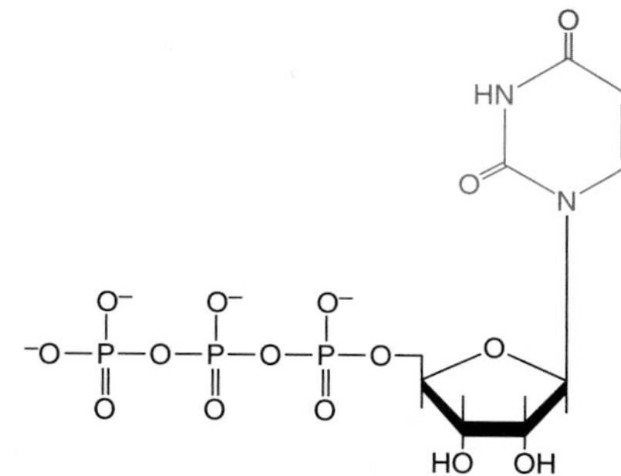

Uridine 5′-triphosphate (UTP)

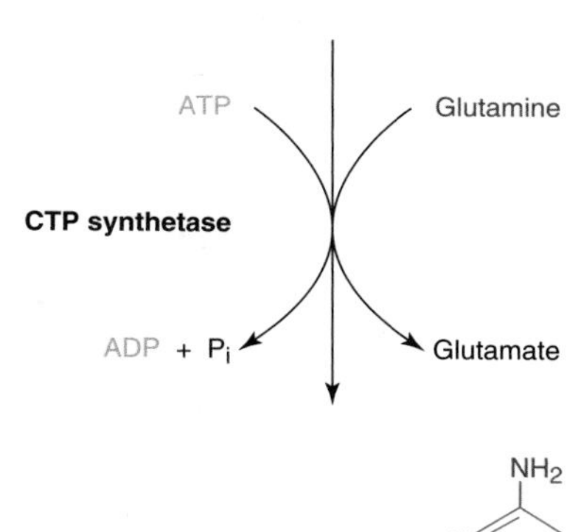

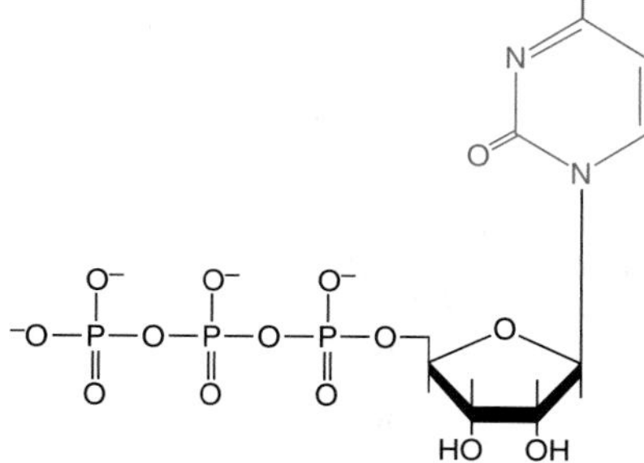

Cytidine 5′-triphosphate (CTP)

Figure 20.16 Formation of cytidine 5′-triphosphate from uridine 5′-triphosphate CTP from UTP.

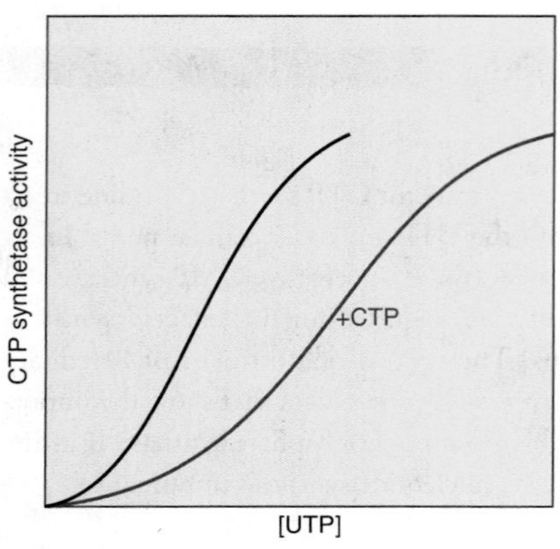

Figure 20.17 Regulation of CTP synthetase.

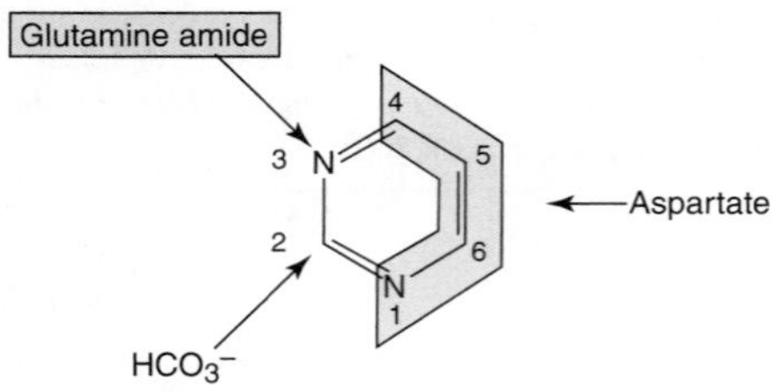

Figure 20.18 Sources of carbon and nitrogen atoms in pyrimidines. C4, C5, C6, and N1 from aspartate; N3 from glutamine; and C2 from CO_2.

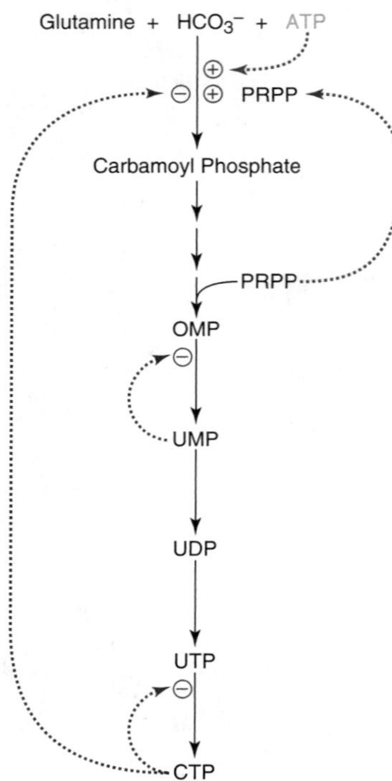

Figure 20.19 Regulation of pyrimidine nucleotide synthesis. Solid arrows represent enzyme catalyzed reactions, and dashed arrows inhibition by products of the reactions.

Pyrimidine Bases Are Salvaged to Reform Nucleotides

Pyrimidines are salvaged by conversion to nucleotides by pyrimidine phosphoribosyltransferase. The general reaction is

$$\text{Pyrimidine} + \text{PRPP} \rightleftharpoons \text{pyrimidine nucleoside 5}'\text{-monophosphate} + \text{PP}_i$$

The enzyme from human erythrocytes utilizes orotate, uracil, and thymine as substrates but not cytosine. These reactions divert the pyrimidine bases from the degradative pathway to nucleotide formation. As a pyrimidine base becomes available to cells, there are competing reactions that will either result in degradation and excretion of the products or reutilization of the bases for nucleotide synthesis. For example, when normal liver is presented with uracil, it is rapidly degraded to β-alanine, whereas in proliferating tumor cells uracil would be converted to UMP. This is the result of the availability of PRPP, enzyme levels and metabolic state of the cells.

20.8 • DEOXYRIBONUCLEOTIDE FORMATION

The concentrations of the 2′-deoxyribonucleoside 5-triphosphates (dNTPs) are extremely low in nonproliferating cells. However, the cellular levels of dNTPs rapidly expand during DNA replication (S phase of the cell cycle) and repair (following DNA damage) due to increased ribonucleotide reductase activity. The relative and absolute concentrations of dNTPs during these times are critical for determining the fidelity of DNA replication and repair. Unbalanced levels of the individual dNTPs can lead to a wide range of genetic disturbances or ultimately cell death.

Deoxyribonucleotides Are Formed by Reduction of Ribonucleoside 5′-Diphosphates

Nucleoside 5′-diphosphate reductase (ribonucleotide reductase) catalyzes the reaction in which ribonucleoside 5′-diphosphates are converted to the corresponding 2′-deoxyribonucleoside 5′-diphosphates. The reaction is controlled by the amount of enzyme present in cells and by a very finely regulated allosteric control mechanism. The reaction is summarized in Figure 20.20.

Mammalian ribonucleotide reductase consists of two nonidentical protein subunits (R1 and R2). The larger subunit (R1) has at least two different effector-binding sites and the smaller subunit (R2) contains a nonheme iron and a stable tyrosyl free radical. The two subunits are encoded by genes on separate chromosomes. The ratios of the mRNAs for these subunits, and consequently the proteins, do not remain constant as cells transit the cell cycle. Recently, a homolog of R2 (referred to as p53R2) was identified. The p53R2 contains a tyrosyl free radical and complements R1 to generate ribonucleotide reductase activity. The exact role of p53R2 has yet to been defined, but there are developing data that indicate that p53R2 is involved in repair of DNA damage and mitochondrial DNA replication. To complete the catalytic cycle, small molecular weight proteins, **thioredoxin** or glutaredoxin, along with NADPH are required to regenerate the free sulfhydryl groups on thioredoxin or glutaredoxin.

The activity of ribonucleotide reductase is under very strict allosteric control. Although reduction of each substrate requires a specific positive effector (nucleoside 5′-triphosphate),

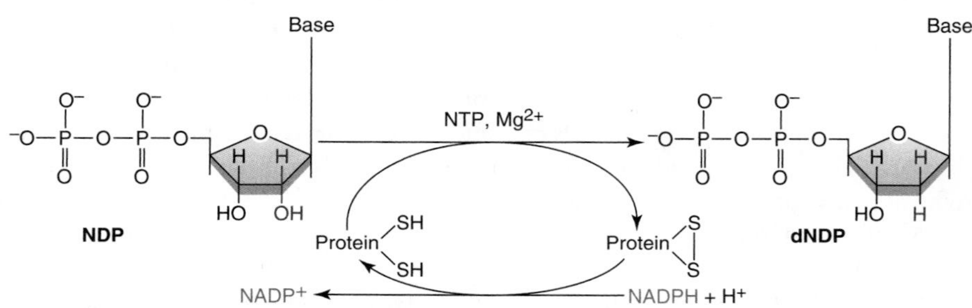

Figure 20.20 *De novo* synthesis of 2′-deoxyribonucleotides from ribonucleotides.

TABLE 20.5 • Nucleoside 5′-Triphosphates as Regulators of Ribonucleotide Reductase Activity

Substrate	*Major Positive Effector*	*Major Negative Effector*
CDP	ATP	dATP, dGTP, dTTP[a]
UDP	ATP	dATP, dGTP, dTTP[a]
ADP	dGTP	dATP
GDP	dTTP	dATP

[a]In decreasing order of effectiveness.

the dNTP products can serve as potent negative effectors of the enzyme. The effects of nucleoside 5′-triphosphates as regulators of ribonucleotide reductase activity are summarized in Table 20.5. 2′-DeoxyATP is a potent inhibitor of the reduction of all four substrates, CDP, UDP, GDP, and ADP; dGTP inhibits reduction of CDP, UDP, and GDP, and dTTP inhibits its reduction of CDP, UDP, and ADP. Thus, dGTP and dTTP serve as either positive or negative effectors of ribonucleotide reductase activity depending on the substrate. This means that while dGTP is the required positive activator for ADP reduction, it also serves as an effective inhibitor of CDP and UDP reductions; dTTP is the positive effector of GDP reduction and serves as an inhibitor of CDP and UDP reductions. Effective inhibition of ribonucleotide reductase by dATP, dGTP, or dTTP explains why high concentrations of exogenous 2′-deoxyadenosine, 2′-deoxyguanosine, and thymidine are toxic to a variety of mammalian cells due to the intracellular accumulation of dATP, dGTP, and dTTP.

Ribonucleotide reductase is uniquely responsible for catalyzing the rate-limiting reactions in which 2′-deoxyribonucleoside 5′-triphosphates are synthesized *de novo* from ribonucleoside 5′-diphosphates for DNA replication and repair as summarized in Figure 20.21. Inhibitors of ribonucleotide reductase are potent inhibitors of DNA synthesis and hence, of cell replication.

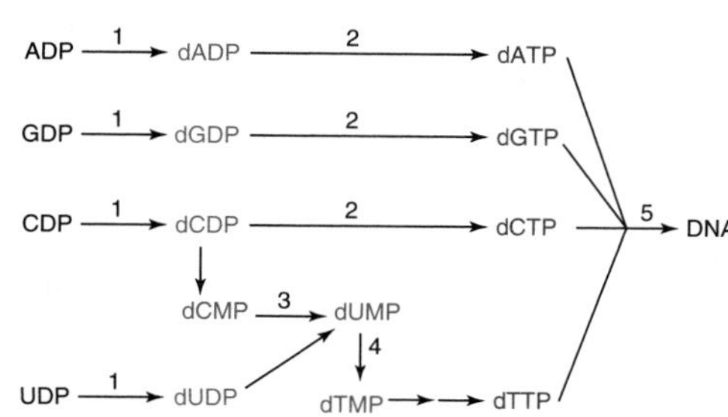

Figure 20.21 Role of ribonucleotide reductase in DNA synthesis. The enzymes catalyzing the reactions are (1) ribonucleotide reductase, (2) nucleoside 5′-diphosphate kinase, (3) deoxycytidylate deaminase, (4) thymidylate synthase, and (5) DNA polymerase.

Deoxythymidylate Synthesis Requires N^5, N^{10}-Methylene H_4Folate

2′-Deoxythymidine 5′-monophosphate (dTMP) is formed directly from 2′-deoxyuridine 5′-monophosphate (dUMP) in a unique reaction. **Thymidylate synthase** catalyzes the transfer of a one-carbon unit from N^5, N^{10}-methylene H_4folate (Figure 20.22) to dUMP, which is simultaneously reduced to a methyl group. The reaction is presented in Figure 20.23. In this reaction, N^5, N^{10}-methylene H_4folate serves as a one-carbon donor and as a reducing agent. This is the only reaction in which H_4folate, acting as a one-carbon carrier, is oxidized to H_2folate. There are no known regulatory mechanisms for this reaction.

The substrate for this reaction can come from two different pathways.

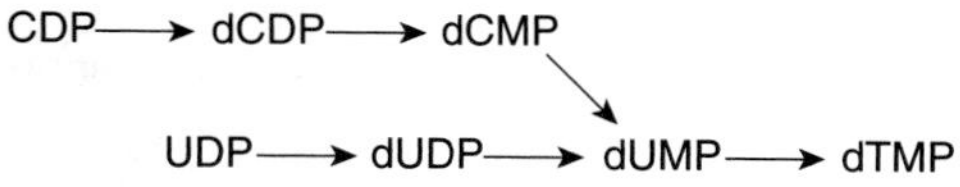

CDP ⟶ dCDP ⟶ dCMP ↘

UDP ⟶ dUDP ⟶ dUMP ⟶ dTMP

In both pathways, dCDP or dUDP, are generated in the reactions catalyzed by ribonucleotide reductase. In one pathway, dUMP is generated from dUDP whereas in the other pathway, dCMP is deaminated to dUMP. Labeling studies indicate that the major pathway for formation of dUMP involves deamination of dCMP by dCMP deaminase, an enzyme that is subject to allosteric regulation by dCTP acting as a positive effector and dTTP acting as a negative effector (Figure 20.24). This regulation of dCMP deaminase by dCTP and dTTP allows cells to maintain the correct ratio of dCTP and dTTP for DNA synthesis.

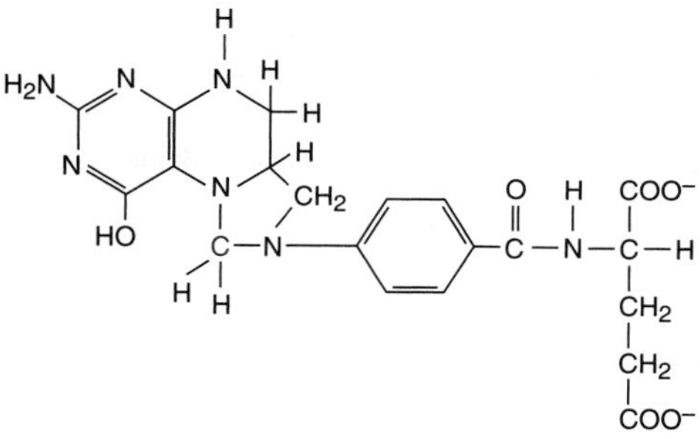

Figure 20.22 Structure of N^5,N^{10}-methylene H_4folate.

Pyrimidine Interconversions: Deoxyribopyrimidine Nucleosides and Nucleotides

Metabolic pathways for interconversions of AMP and GMP (see Figure 20.10, p. 817) are regulated to maintain appropriate intracellular levels of adenine and guanine nucleotides.

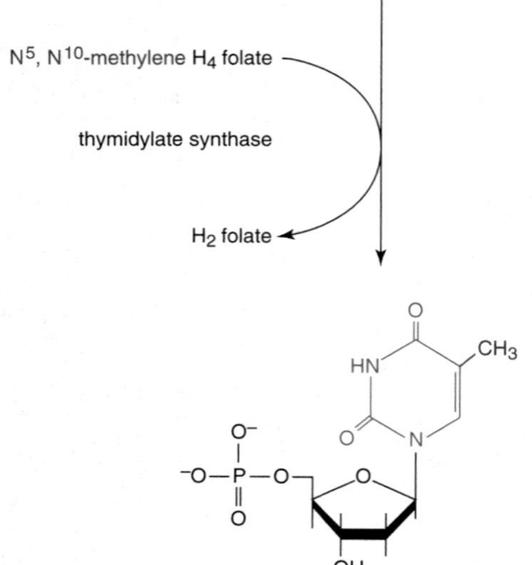

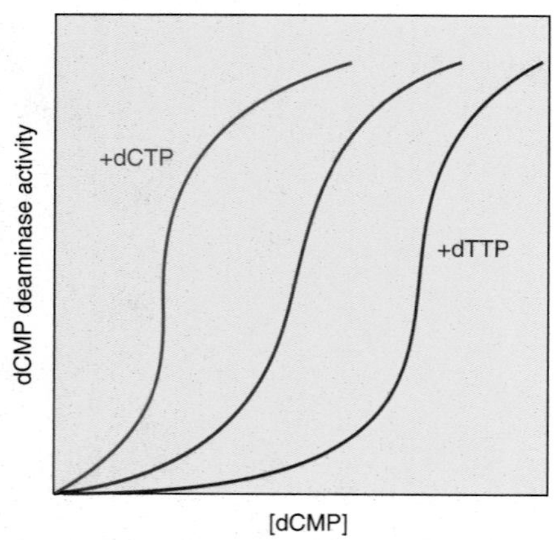

Figure 20.23 Synthesis of 2′-deoxythymidine monophosphate.

Figure 20.24 Regulation of dCMP deaminase.

Pathways also exist for interconversions of pyrimidine nucleotides and are of particular importance for pyrimidine deoxyribonucleosides and deoxyribonucleotides as summarized in Figure 20.25. Note that dCTP and dTTP are major positive and negative effectors of the interconversions and salvage of deoxyribonucleosides. There are specific deoxyribonucleoside kinases in mammalian cells that catalyze the committed step in the formation of 2′-deoxyribonucleoside 5′-monophosphates. These include the kinases for thymidine, deoxyguanosine, and deoxycytidine. The monophosphates are converted to the triphosphates by relatively nonspecific nucleotide kinases. Of clinical importance is the fact that these kinases play an important role in converting exogenous analogs to chemotherapeutic dNTPs as antitumor or antiviral agents.

20.9 • DEGRADATION OF PYRIMIDINE NUCLEOTIDES

Turnover of nucleic acids results in release of pyrimidine and purine nucleotides. Degradation of pyrimidine nucleotides follows the pathways shown in Figure 20.26. Pyrimidine nucleotides are converted to nucleosides by nonspecific phosphatases. Cytidine and 2′-deoxycytidine are deaminated to uridine and 2′-deoxyuridine, respectively, by pyrimidine nucleoside deaminase. Uridine phosphorylase catalyzes phosphorolysis of uridine, deoxyuridine, and deoxythymidine to uracil and thymine.

Uracil and thymine are degraded further by analogous reactions, although the final products are distinctly different as shown in Figure 20.27. Uracil is degraded to β-alanine, NH_4^+, and CO_2. None of these products is unique to uracil degradation, and consequently the turnover of cytosine or uracil nucleotides from RNA or DNA cannot be estimated from the end products of this pathway. Thymine degradation on the other hand, yields **β-aminoisobutyric acid**, NH_4^+, and CO_2. β-Aminoisobutyric acid is excreted in urine of humans, and it originates exclusively from degradation of thymine from DNA. Thus, its excretion can be used to estimate the turnover of DNA or deoxythymidine nucleotides by measurement of β-aminoisobutyric acid in urine. Increased levels of β-aminoisobutyric acid are excreted in cancer patients undergoing chemotherapy or radiation therapy in which large numbers of cells are killed and DNA is degraded.

Enzymes catalyzing the degradation of uracil and thymine (dihydropyrimidine dehydrogenase, dihydropyrimidinase, and uriedopropionase) do not show a preference for either uracil or thymine as substrates or their reaction products.

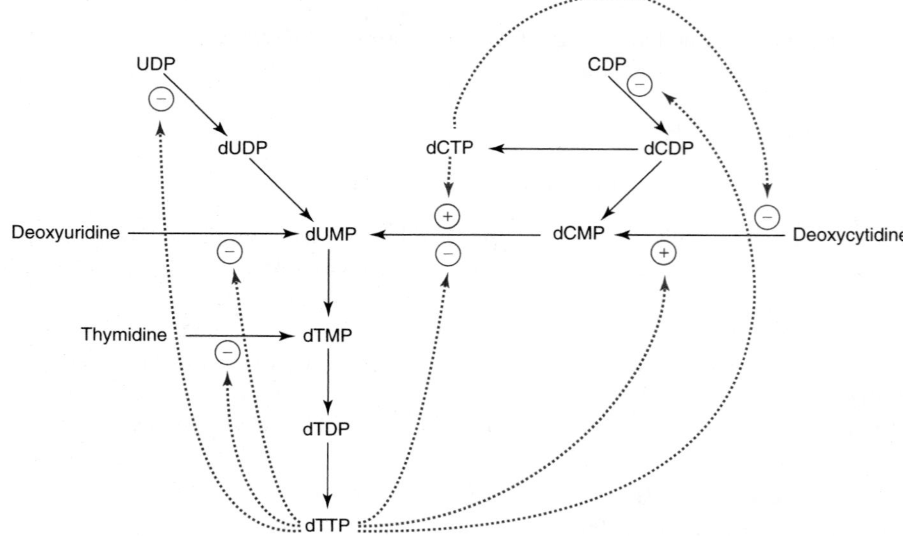

Figure 20.25 Interconversions of pyrimidine nucleotides with emphasis on deoxyribonucleotide metabolism. The solid arrows indicate enzyme-catalyzed reactions; the dashed lines represent positions of negative control points.

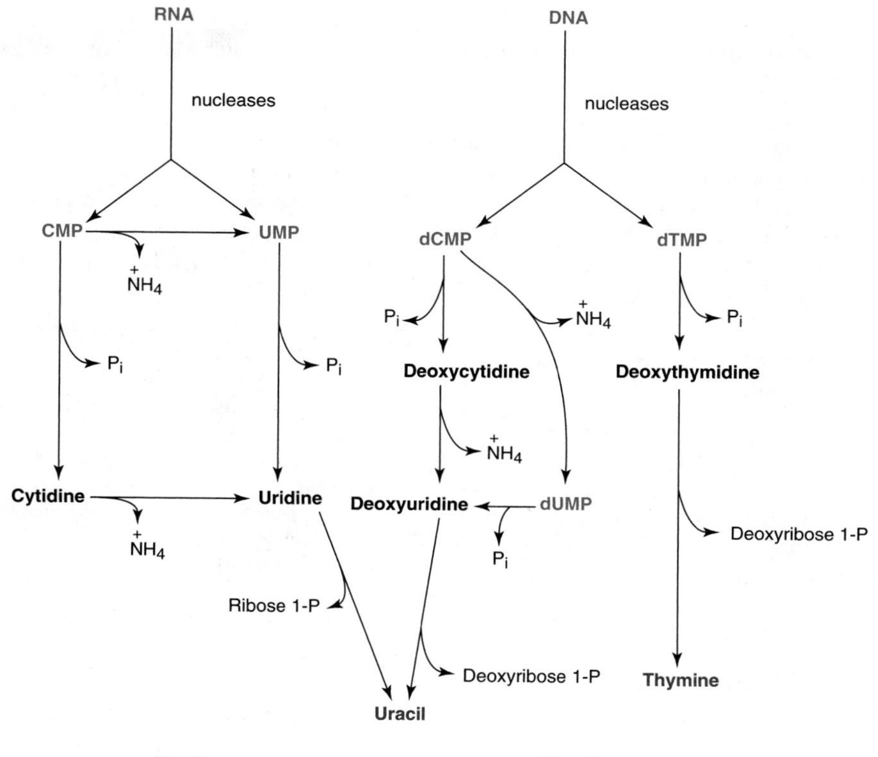

Figure 20.26 **Pathways for degradation of pyrimidine nucleotides.**

Uracil

NADPH + H+

NADP+

(dihydropyrimidine dehydrogenase)

Dihydrouracil

(dihydropyrimidinase)

β-Ureidopropionate

$CO_2 + NH_4^+$

(ureidopropionase)

$H_3\overset{+}{N}CH_2$ | CH_2COO^-

β-Alanine

Thymine

NADPH +

NADP+

Dihydrothymine

β-Ureidoisobutyrate

$CO_2 + NH_4^+$

$H_3\overset{+}{N}CH_2$ | CH_3CHCOO^-

β-Aminoisobutyrate

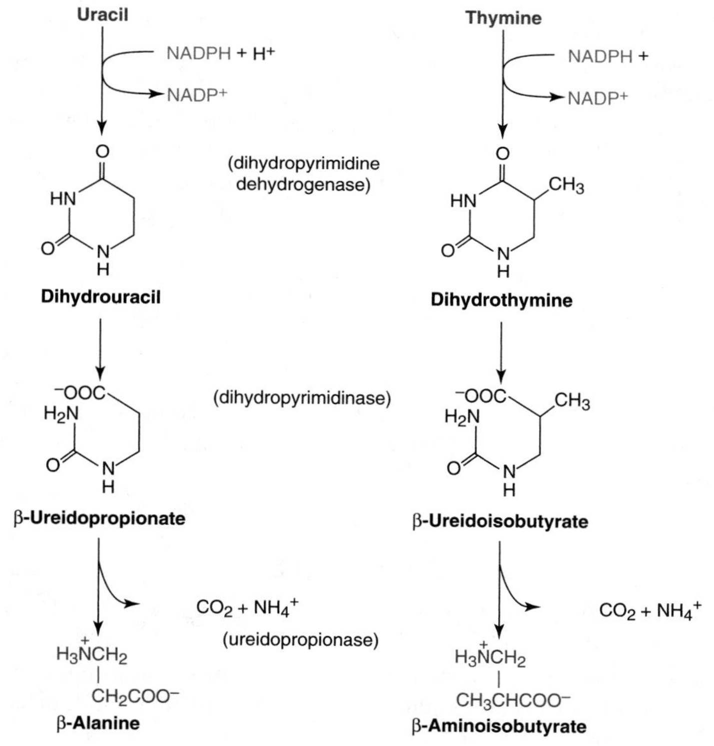

Figure 20.27 **Degradation of uracil and thymine to end products.**

20.10 • NUCLEOSIDE AND NUCLEOTIDE KINASES

De novo synthesis of both purines and pyrimidine nucleotides yields nucleoside monophosphates. Likewise, the salvage of nucleobases by phosphoribosyl transferases or nucleosides by nucleoside kinases also yields nucleoside 5′-monophosphates. This is particularly important in cells such as erythrocytes that cannot form nucleotides *de novo*. Nucleotide kinases convert nucleoside 5′-monophosphates to nucleoside 5′-diphosphates and nucleoside

CLINICAL CORRELATION 20.8

Mitochondrial Neurogastrointestinal Encephalopathy Syndrome (MNGIE)

MNGIE (OMIM 603041) is characterized by external disorders of the eye muscles, gastrointestinal problems, cachexia, and peripheral neuropathy. Multiple mutations have been reported for ECGF1 (platelet-derived Endothelial Growth Factor) gene that has been implicated. It is a progressive disease. Biochemically, the condition arises from a defect in the gene for cytosolic thymidine phosphorylase activity. Thymidine phosphorylase catalyzes the reversible phosphorylsis of dThd or dUrd to thymine or uracil, respectively. Although the reaction is thermodynamically reversible, only the catabolic direction is physiologically relevant because the concentrations of products would not favor the reverse reaction. As a result of decreased thymidine phosphorylase activity, the systemic levels of thymidine and deoxyuridine are greatly increased because they are not further degraded. Consequently, these nucleosides are salvaged and lead to increased dTTP and/or dUTP in mitochondria and are incorporated into mitochondrial DNA during mitochondrial DNA replication. As a result of the abnormal mitochondrial DNA, clinical manifestations are observed but the exact basis for these cannot be explained. This is an interesting case and another example of an effect at the nuclear DNA level that has major consequences on mitochondrial DNA synthesis and cellular function.

Although it is generally accepted that MNGIE is due to the loss of thymidine phosphorylase, it has been questioned whether MNGIE results from the loss of 5′-thymine phosphorylase activity as a result of their studies.

Carod-Artal, F. J., Herrerro, M. D., Lara, M. C., Lopez-Gallardo, E., et al. Cognitive dysfunction and hypogonadotropic hypogonadoism in a Brazilian patient with mitochondrial neurogastrointestnal encephalolyopathy and a novel ECGF1 mutation. *Eur. J. Neuro.*14:581, 2007; Lara, M. C., Valentino, M. L., Torres-Torrontras, J., Hirano, J. and Marti. R., Mitochondrial neurogastrointestinal encephalomyopathy (MMNGIE): Biochemical features and therapeutic approaches. *Biosci Rev.* 27: 15, 2007; and Valentino, M. L., Marti, R., Tadesse, S., Lppez, L. C., et al. Thymidine and deoxyuridine accumulate in tissues of patients with mitochondrial neurogastrointestinal encephalomyopathy (MNGIE). *FEBS Lett.* 581: 3410, 2007.

5′-diphosphates to nucleoside 5′-triphosphates. These are important reactions since most reactions in which nucleotides function require nucleoside 5′-triphosphates (primarily) or nucleoside 5′-diphosphates. Some of these nucleoside kinases show a high degree of specificity, especially deoxynucleoside kinases (Clin. Corr. 20.8), with respect to the base and sugar moieties while others are much less specific. There is also some substrate specificity with the nucleotide kinases. Mammalian cells contain, in high concentration, nucleoside diphosphate kinase that is relatively nonspecific for either phosphate donor or phosphate acceptor in terms of purine or pyrimidine base or the sugar. This reaction is as follows:

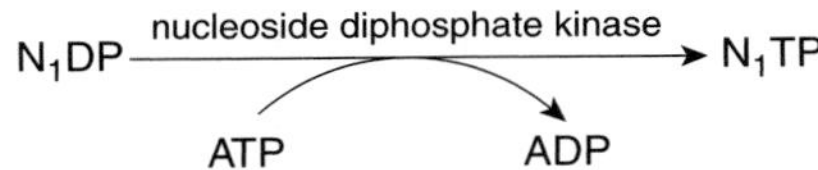

Since ATP is present in the highest concentration in most cells and is most readily regenerated by glycolysis or oxidative phosphorylation, it is probably the major phosphate donor for these reactions.

20.11 • NUCLEOTIDE METABOLIZING ENZYMES AS A FUNCTION OF THE CELL CYCLE

The strict regulation of nucleotide synthesis requires that mechanisms must be available in cells to meet the requirements for ribonucleotide and deoxyribonucleotide precursors at the time of increased RNA synthesis and DNA replication. To meet these needs, cells increase levels of specific enzymes involved with nucleotide formation during very specific periods of the cell cycle (p. 157).

Enzymes involved in purine nucleotide synthesis and interconversions that are elevated during the S phase of the cell include PRPP amidotransferase and IMP dehydrogenase. Enzymes involved in pyrimidine nucleotide synthesis that are elevated during S phase include aspartate carbamoyltransferase, dihydroorotase, dihydroorotate dehydrogenase, orotate phosphoribosyltransferase, and CTP synthetase. Many enzymes involved in synthesis and interconversions of deoxyribonucleotides are also elevated during the S phase. Included in these enzymes are ribonucleotide reductase, thymidine kinase, dCMP deaminase, thymidylate synthase, and dTMP kinase.

The deoxyribonucleotide pool is extremely small (less than 1 μM) in resting cells. As a result of the increase in ribonucleotide reductase activity, concentrations of deoxyribonucleotides reach 10–20 μM during DNA synthesis. However, this concentration would sustain DNA synthesis for only minutes, whereas complete DNA replication requires hours. Consequently, levels of ribonucleotide reductase activity must not only increase but must also be sustained during S-phase order to provide the necessary substrates for DNA synthesis. Pools of dNTPs must also be maintained for DNA repair. There is evidence that ribonucleotide reductase activity also increases in response to DNA damage. Since growing tissues such as regenerating liver, embryonic tissues, and intestinal mucosal cells are geared toward DNA replication and RNA synthesis, these tissues will also show elevated levels of the key enzymes involved with purine and pyrimidine nucleotide synthesis and interconversions with complementary decreases in amount for enzymes that catalyze reactions in which these precursors are degraded. These changes in enzyme levels actually reflect the proportion of the cells that are in S-phase in that tissue.

20.12 • NUCLEOTIDE COENZYME SYNTHESIS

The nucleotide coenzymes NAD and FAD, and coenzyme A have common features in that they each have a vitamin moiety as part of their structure and AMP is part of the structure although not directly involved in the reaction in which each of these functions. The structure and synthesis of NAD and the pathway of its synthesis in mammalian cells is presented in Figure 20.28.

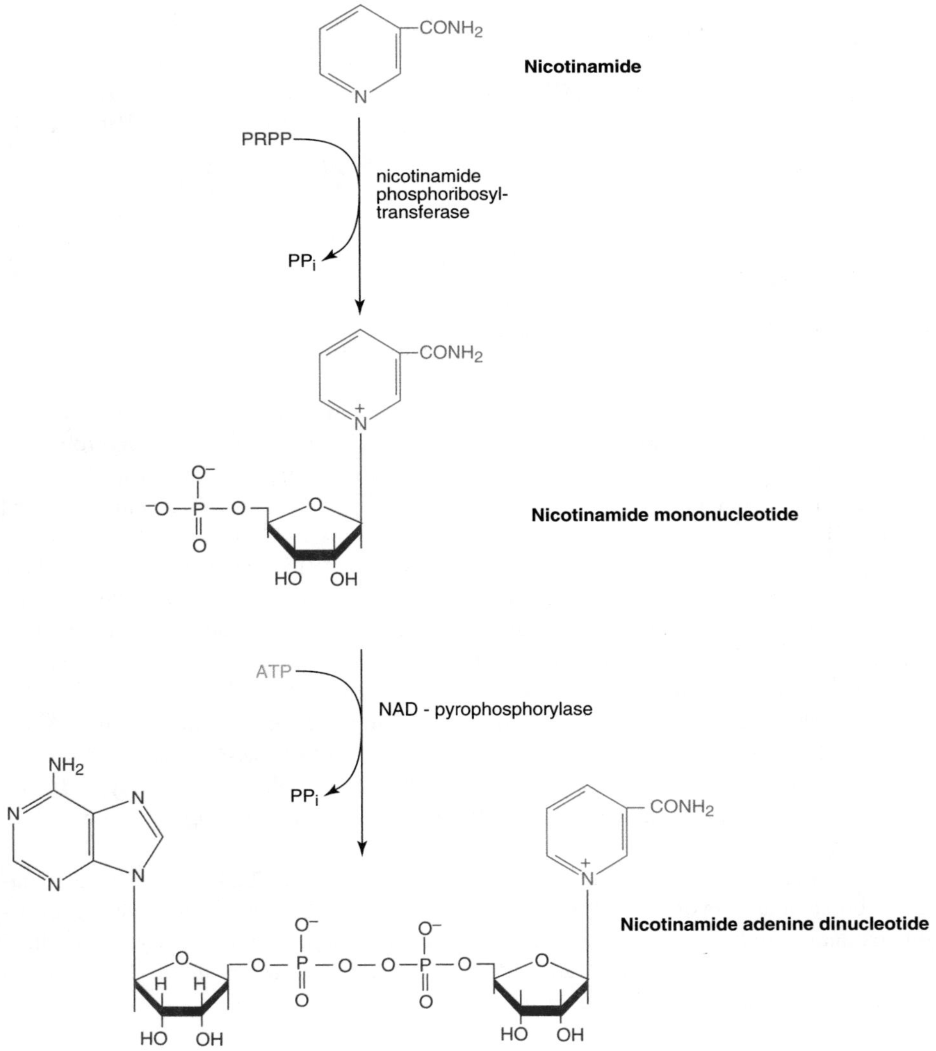

Nicotinamide

PRPP → PP_i (nicotinamide phosphoribosyltransferase)

Nicotinamide mononucleotide

ATP → PP_i (NAD - pyrophosphorylase)

Nicotinamide adenine dinucleotide

Figure 20.28 Pathway for nicotinamide adenine nucleotide synthesis.

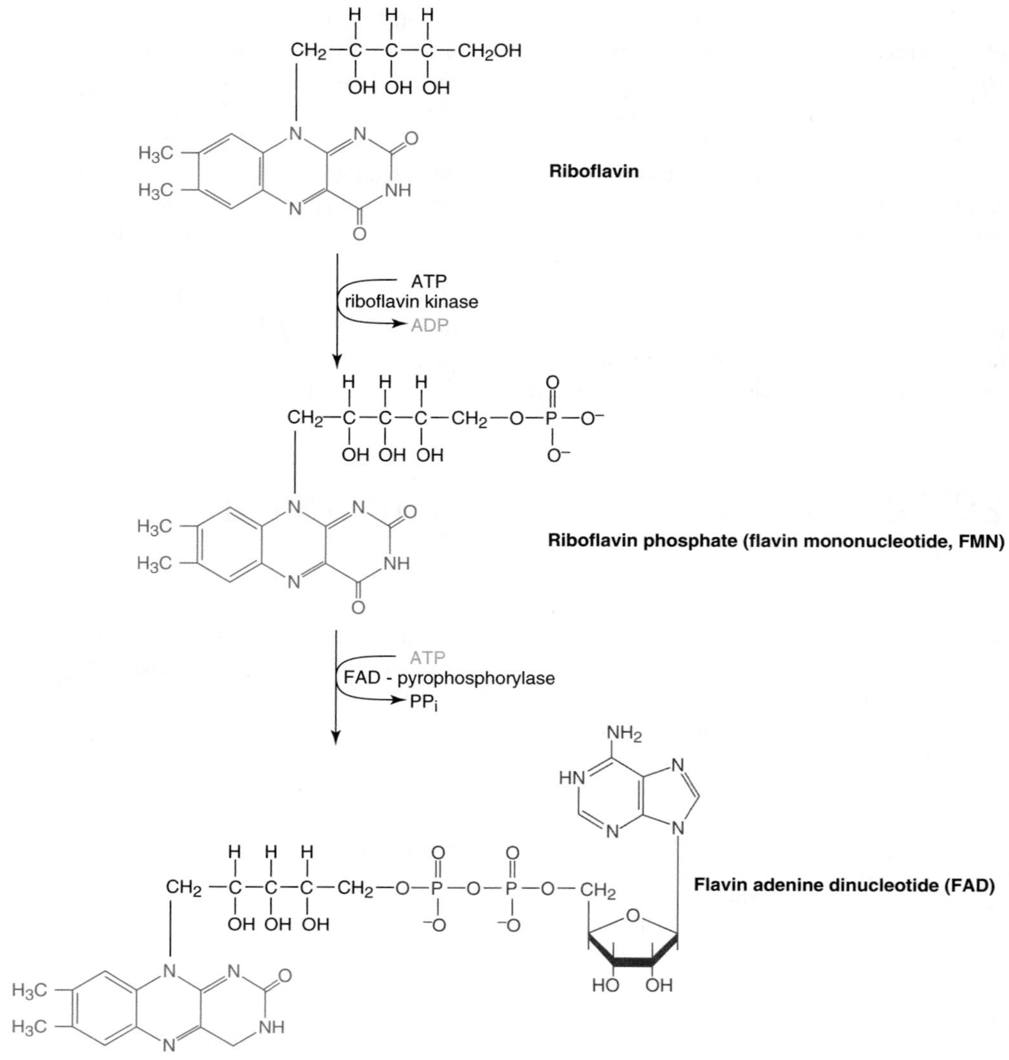

Figure 20.29 Synthesis of flavin adenine dinucleotide.

NAD synthesis requires niacin. NAD can be synthesized by three different pathways starting with tryptophan, nicotinamide, or nicotinic acid, respectively. When tryptophan is in excess of the amount required for protein or serotonin synthesis, it can be used for NAD synthesis. This is not a likely situation for most individuals and therefore niacin is required in the diet.

Synthesis of NAD by any of the pathways requires PRPP as the ribose 5-phosphate donor. Nicotinamide adenine dinucleotide (NADP) is synthesized by the phosphorylation of NAD. NAD, NADH, NADP, and NADPH have specific roles in cellular metabolism with respect to energy production. In these roles, they are involved in reversible oxidation–reduction reactions. However, there are important pathways in which NAD serves as a substrate and is consumed in the reaction. These pathways include ADP-ribose transferases (poly ADP polymerases [PARPs]), ADP ribose synthases, and sirtuins (type III protein lysine deacetylases). NAD utilization through these pathways is linked to processes that control gene expression, calcium mobilization, neuroprotection, death, and aging.

The structures of FMN and FAD are presented in Figure 20.29. They have the vitamin riboflavin as part of their structures. FAD and FMN have important roles in reversible oxidation–reduction reactions. Depending on the reaction and/or enzyme in which it participates, the flavin nucleotides can serve as either a coenzyme in enzyme-catalyzed reactions or as prosthetic groups in proteins.

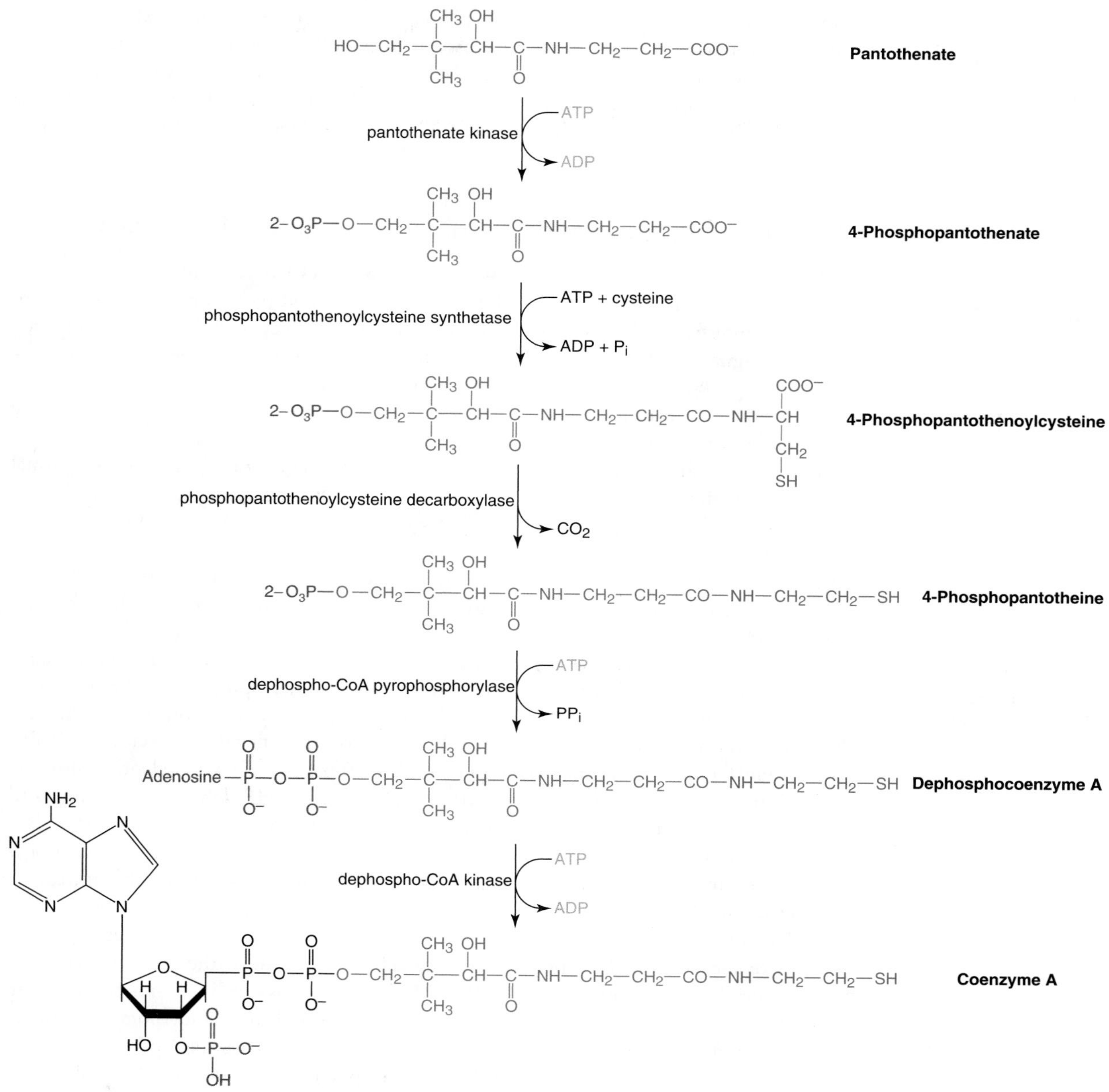

Pantothenate

pantothenate kinase ($ATP \rightarrow ADP$)

4-Phosphopantothenate

phosphopantothenoylcysteine synthetase (ATP + cysteine $\rightarrow ADP + P_i$)

4-Phosphopantothenoylcysteine

phosphopantothenoylcysteine decarboxylase ($\rightarrow CO_2$)

4-Phosphopantotheine

dephospho-CoA pyrophosphorylase ($ATP \rightarrow PP_i$)

Dephosphocoenzyme A

dephospho-CoA kinase ($ATP \rightarrow ADP$)

Coenzyme A

Figure 20.30 Synthesis of coenzyme A.

The pathway of synthesis and structure of pantothenic acid are presented in Figure 20.30. This coenzyme is easily recognized for its central and critical role in intermediary metabolism as a component of acetyl CoA. However, acetyl CoA also serves an important function in the acetylation of the amino termini of proteins and the ε amino groups of lysine in proteins. These posttranslational modifications serve to alter the biological properties of these modified proteins and play important roles in cellular function.

20.13 • CHEMOTHERAPEUTIC AGENTS THAT INTERFERE WITH PURINE AND PYRIMIDINE NUCLEOTIDE METABOLISM

De novo synthesis of purine and pyrimidine nucleotides is critical for normal cell replication, maintenance, and function. Regulation of these pathways is important since disease states have been identified that arise from defects in the regulatory enzymes. Synthetic

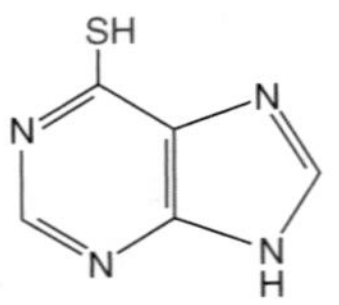

6-Mercaptopurine

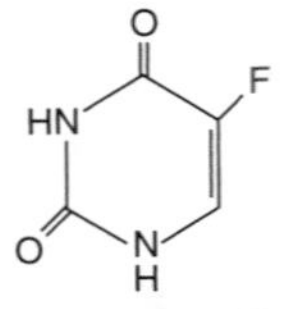

5-Fluorouracil

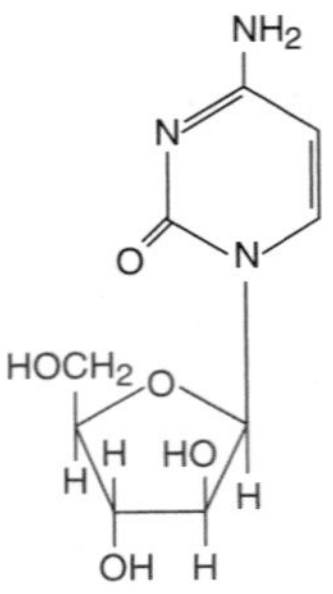

Cytosine arabinoside

Figure 20.31 Structures of antimetabolites: 6-mercaptopurine, 5-fluorouracil, and cytosine arabinoside.

compounds and natural products from plants, bacteria, or fungi that are structural analogs of the nucleobases or nucleosides used in metabolic reactions have been shown to be cytotoxic. These compounds are relatively specific inhibitors of enzymes involved in nucleotide synthesis or interconversions and have proven to be useful in therapy of diverse clinical problems. They are generally classified as antimetabolites, antifolates, glutamine antagonists. and other agents.

Inhibitors of Purine and Pyrimidine Nucleotide Metabolism

Antimetabolites Are Structural Analogs of Bases or Nucleosides

Antimetabolites are usually structural analogs of purine and pyrimidine bases or nucleosides that interfere with very specific metabolic reactions. They include 6-mercaptopurine and **6-thioguanine** used in the treatment of acute leukemia, azathioprine for immunosuppression in patients with organ transplants, allopurinol for treatment of hyperuricemia, and acyclovir for treatment of herpesvirus infection. The detailed understanding of purine nucleotide metabolism aided in the development of these compounds as drugs. Conversely, study of the mechanism of action of these drugs has led to a better understanding of normal nucleotide metabolism in humans.

Three antimetabolites will be specifically discussed to show (1) the importance of *de novo* synthetic pathways in normal cell metabolism, (2) that regulation of these pathways occurs *in vivo,* (3) the requirement for metabolic activation of the drugs utilizing the cellular salvage enzymes, and (4) that inactivation of these compounds greatly influences their usefulness.

6-Mercaptopurine (6-MP) (Figure 20.31) is a useful antitumor drug in humans. Its cytotoxic activity depends on the formation of 6-mercaptopurine ribonucleotide by the tumor cells. Utilizing PRPP and HGPRTase, 6-mercaptopurine ribonucleoside 5′-monophosphate is formed in cells and serves as a negative effector of PRPP-amidotransferase, the committed step in the *de novo* pathway. This nucleotide also acts as an inhibitor of the conversion of IMP to GMP at the IMP-dehydrogenase step and of IMP to AMP at the adenylosuccinate synthetase step. Since 6-mercaptopurine is a substrate for xanthine oxidoreductase and is oxidized to 6-thiouric acid, allopurinol is usually administered to inhibit degradation of 6-MP and to potentiate the antitumor properties of 6-MP.

5-Fluorouracil (Fura) (Figure 20.31) is an analog of uracil. 5-Fluorouracil is not the active species as it must be converted by cellular enzymes to the active metabolites, 5-fluorouridine 5′-triphosphate (FUTP) and 5-fluoro-2′-deoxyuridine 5′-monophosphate (FdUMP). FUTP is efficiently incorporated into RNA and, once incorporated, inhibits maturation of 45 S precursor rRNA into the 28S and 18S RNA species and alters splicing of pre-mRNA into functional mRNA. FdUMP is a potent and specific inhibitor of thymidylate synthase. In the presence of H_4folate, FdUMP and thymidylate synthase, a ternary complex is formed with the covalent binding of FdUMP to thymidylate synthase, which results in irreversible inhibition of thymidylate synthase. This inhibits dTMP synthesis and leads to “thymineless death” for cells.

Cytosine arabinoside (AraC) (Figure 20.31) is used in treatment of several forms of human cancer. It must be metabolized by cellular enzymes to cytosine arabinoside 5′-triphosphate (araCTP) to exert its cytotoxic effects. AraCTP competes with dCTP in the DNA polymerase reaction, and araCMP is incorporated into DNA. This inhibits synthesis of the growing DNA strand. Clinically, the efficacy of araC as an antileukemic drug correlates with the concentration of araCTP that is achieved in the leukemic cells; this in turn determines the amount of araCMP incorporated into DNA. Formation of araCMP by deoxycytidine kinase appears to be the rate-limiting step in the activation to araCTP. Ara-C is inactivated by deamination to araU.

Antifolates Inhibit Reactions Involving Tetrahydrofolate

Folate analogs, depending on their specific structure, interfere with metabolic steps in which tetrahydrofolate is involved either as a substrate or product. **Methotrexate (MTX)**, a synthetic structural analog of folic acid, interferes with formation of H_2folate and H_4folate

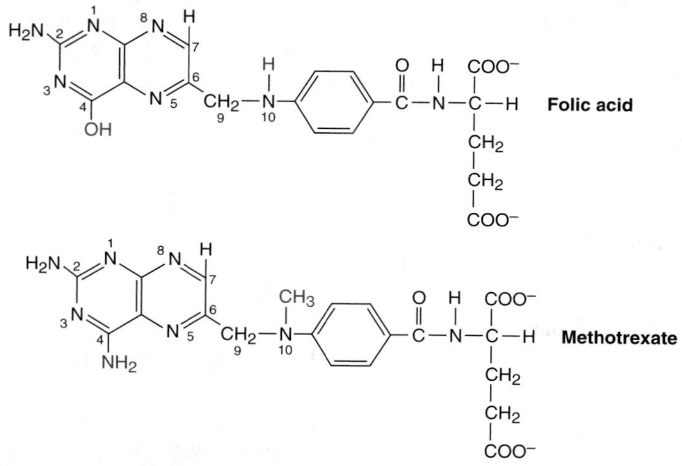

Figure 20.32 Comparison of the structures of folic acid and methotrexate.

from folate by specifically inhibiting H_2folate reductase (DHFR). MTX is used as an antitumor agent in the treatment of human cancers. The structures of folic acid and MTX are compared in Figure 20.32. MTX and folate differ only at C-4 where an amino group replaces a hydroxyl group and at N-10 where a methyl group replaces a hydrogen atom. MTX specifically inhibits H_2folate reductase with a K_i in the range of 0.1 nM. The reactions inhibited by MTX are shown in Figure 20.33.

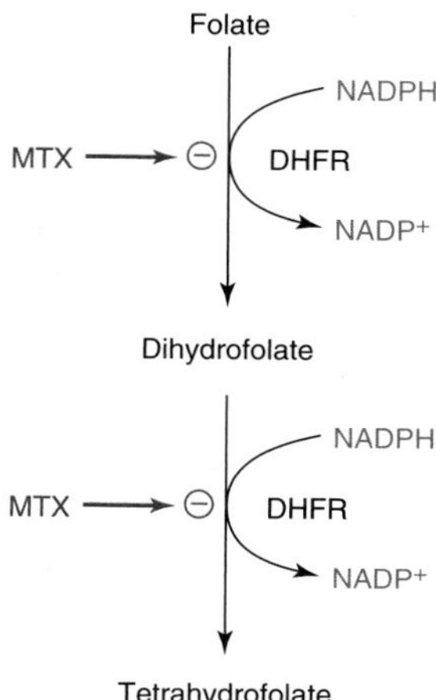

Figure 20.33 Sites of inhibition by methotrexate.

MTX at very low concentrations is cytotoxic to mammalian cells in culture. The inhibition of dihydrofolate reductase by MTX results in lowering the intracellular pools of both ribonucleoside 5′-triphosphates and 2′-deoxyribonucleoside 5′-triphosphates. The effects can be prevented, at least partially, by addition of deoxythymidine and hypoxanthine to the culture medium. Reversal of the MTX effects by thymidine and hypoxanthine indicates that MTX causes depletion of both deoxythymidine and purine nucleotides in cells. Figure 20.34 shows the relationship between H_4-folate, *de novo* purine nucleotide synthesis, and dTMP formation. Note that in the thymidylate synthase reaction, H_2-folate is generated. Unless it were readily reduced back to H_4folate via dihydrofolate reductase, cells would not be capable of *de novo* synthesis of purine nucleotides or thymidylate synthesis due to depletion of H_4 folate pools.

In treatment of human leukemias, normal cells can be rescued from the toxic effects of high-dose MTX by N^5-formyl-H_4 folate (leucovorin). This increases the clinical efficacy of MTX in the treatment of cancer. MTX is also being successfully used at very low doses in the treatment of rheumatoid arthritis (RA) although the molecular basis by which MTX acts in this disorder is not known. It is not clear that the effects of MTX treatment in RA are related to its effects on nucleotide metabolism. However, it is interesting to note that leflunomide, another drug being used to treat RA, has as one of its defined sites of action the inhibition of mitochondrial dihydroorotate dehydrogenase.

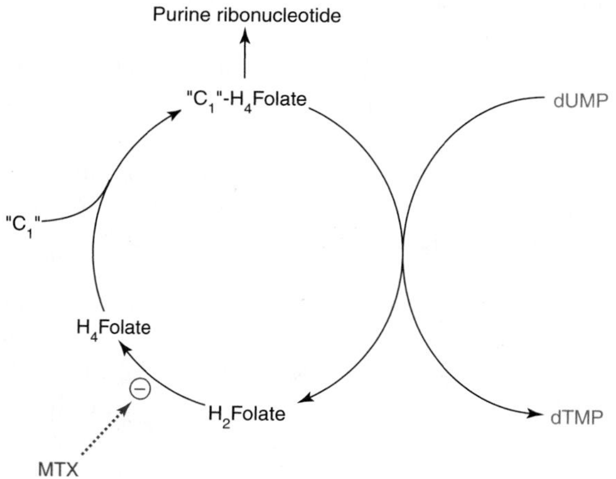

Figure 20.34 Relationship between H_4folate, *de novo* purine nucleotide synthesis, and dTMP synthesis.

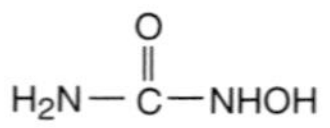

Hydroxyurea

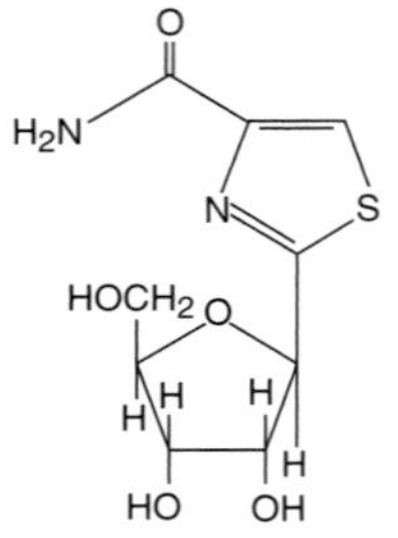

Tiazofurin

Figure 20.35 **Structures of hydroxyurea and tiazofurin.**

Newer antifolates have been synthesized that are relatively specific inhibitors of either thymidylate synthase or *de novo* purine nucleotide synthesis, but not both. These compounds are in clinical trials for the treatment of cancer.

Glutamine Antagonists Inhibit Enzymes That Utilize Glutamine as Nitrogen Donor

Many reactions in mammalian cells require glutamine as the amino group donor. In contrast, bacteria primarily utilize ammonia as the amino donor in a similar reaction. These amidation reactions are critical in *de novo* synthesis of purine nucleotide (N-3 and N-9), synthesis of GMP from IMP, formation of cytosolic carbamoyl phosphate, synthesis of CTP from UTP (see Table 20.3, p. 810), and synthesis of NAD^+.

Compounds that inhibit these reactions are referred to as glutamine antagonists. **Azaserine** and **DON** (diazo-5-oxo-L-norleucine), which were first isolated from cultures of *Streptomyces,* are very effective inhibitors of enzymes that utilize glutamine as the amino donor. Since azaserine and DON inactivate many of the glutamine-dependent enzymes involved in nucleotide metabolism, these compounds proved to be too toxic for use clinically as antitumor agents.

Other Agents That Inhibit Cell Growth by Interfering with Nucleotide Metabolism

Tumor cells treated with hydroxyurea (Figure 20.35) show specific inhibition of DNA synthesis with little or no inhibition of RNA or protein synthesis. Hydroxyurea inhibits ribonucleotide reductase activity by destroying the tyrosyl free radical on the small subunit of ribonucleotide reductase (R2). This results in the inhibition of the reductions of CDP, UDP, GDP, and ADP to the corresponding 2′-deoxyribonucleoside 5′-diphosphates. Toxiciy to cells results from depletion of 2′-deoxyribonucleoside 5′-triphosphates that are required for DNA replication. Clinical use of hydroxyurea as an antitumor agent is limited because of its rapid rate of clearance and the high drug concentration required for effective inhibition. However, hydroxyurea has recently been utilized in the treatment of sickle-cell anemia in both adults and children. By a mechanism not completely understood, hydroxyurea treatment of sickle cell patients results in the re-expression of the fetal (γ) hemoglobin gene resulting in the increased fetal hemoglobin in the red cells. The increased concentration of fetal hemoglobin in the red cells decreases HbS precipitation and, as a result, decreases the frequency of sickle cell crises in the patients during hypoxia. It does not appear that the effect of hydroxyurea on sickle-cell erythrocytes is directly related to inhibition of ribonucleotide reductase.

Tiazofurin (Figure 20.35) is not an active drug, but must be converted by cellular enzymes to the NAD^+ analog, tiazofurin adenine dinucleotide (TAD), which is the active agent. TAD inhibits IMP dehydrogenase, the rate-limiting enzyme in GTP synthesis, with a K_I of 0.1 μM. As a result of IMP dehydrogenase inhibition, the concentration of GTP is markedly depressed with a corresponding decrease in dGTP. Although there are many dehydrogenases that utilize NAD^+ as a substrate, IMP dehydrogenase is most affected, possibly because it catalyzes a rate-limiting step in a critical pathway and is quantitatively limiting in concentration.

These clinically useful drugs serve as examples in which the knowledge of basic biochemical pathways and mechanisms lead to generation of effective drugs.

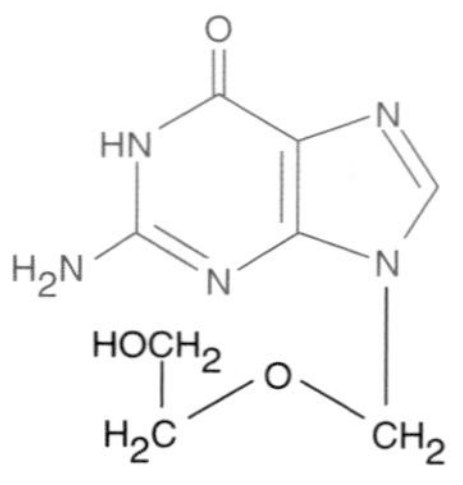

Acycloguanosine

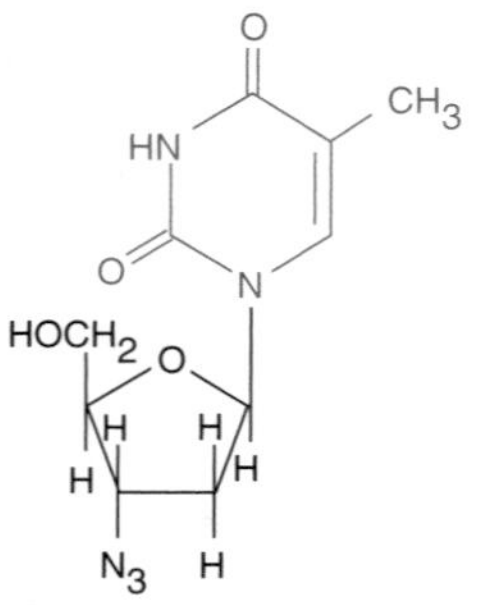

AZT

Figure 20.36 **Structures of the antiviral agents, acyclovir and AZT.**

Purine and Pyrimidine Analogs as Antiviral Agents

Infections caused by herpes virus (HSV) and the human immunodefciency virus (HIV) infections, present major clinical problems. Two antimetabolites used in the control or treatment (but not cure) of HSV and HIV infections are Acyclovir (acycloguanosine), a purine analog, and 3′-azido-3′-deoxythymidine (AZT), a pyrimidine analog (Figure 20.36). Both drugs must be phosphorylated to yield the active drug. Acycloguanosine is converted to the monophosphate by a specific HSV-thymidine kinase (encoded by the HSV genome), which is present only in the virally infected cells. The host cellular thymidine kinases cannot utilize acyclovir as a substrate. Acycloguanosine monophosphate is then phosphorylated by the cellular enzymes to the di- and triphosphate forms. Acycloguanosine triphosphate serves as a substrate for the HSV-specific DNA polymerase and is incorporated into the growing viral DNA chain. This causes chain termination. The specificity of acycloguanosine and its high therapeutic index result from the fact that only HSV-infected cells can form the acycloguanosine monophosphate. Unfortunately, resistant strains of HSV have evolved.

AZT is phosphorylated by cellular kinases to AZT-triphosphate, which blocks HIV replication by inhibiting HIV-DNA polymerase (an RNA-dependent polymerase). The selectivity of AZT for HIV-infected versus normal cells occurs because DNA polymerase from HIV is at least 100-fold more sensitive to AZT-triphosphate than is host cell DNA-dependent DNA polymerase. Resistance to AZT has been observed.

These two antiviral agents demonstrate the diversity of responses required for selectivity. In one case, enzyme activity encoded specifically by the viral genome is mandatory for activation of the drug (acycloguanosine); in the second example, although cellular enzymes activate AZT, the viral gene product (HIV-DNA polymerase) is the selective target.

Biochemical Basis for Responses to Chemotherapeutic Agents

There are two aspects to consider with respect to a patient's response to cancer chemotherapy. In the first case there is the development of, or selection of, drug-resistant populations as a result of the drug treatment. In the second case preexisting genetic polymorphisms cause altered drug metabolism such that at the standard dose, the patient shows increased or even severe toxicity to the chemotherapeutic drug or even a reduced response to the drug.

Failure of chemotherapy in treatment of human cancer is often related to development of, or selection of, tumor cell populations that are resistant to the cytotoxic effects of the particular drug. Tumors contain very heterogenous populations of cells and in many instances drug-resistant cells are already present. On therapy, drug-sensitive cells are killed off and resistant cell populations become enriched. In some cases, drug treatments produce genetic alterations that result in the drug-resistant phenotype. Resistance to drugs can be categorized as *specific drug resistance* or *multidrug resistance.*

Biochemical and molecular mechanisms that account for drug resistance have been determined for many drugs. For example, resistance to methotrexate can develop because of several different alterations. These include (1) a defect in or loss of the transporter for N^5-formyl-H_4folate and N^5-methyl-H_4folate that results in decreased cellular uptake of MTX; (2) amplification of the dihydrofolate reductase gene that results in large increases in the amount of dihydrofolate reductase, the target enzyme; (3) alterations in the dihydrofolate reductase gene that result in a mutant dihydrofolate reductase that is less sensitive to inhibition by MTX; and (4) decreased levels of folylpolyglutamate synthetase, which results in lower levels of polyglutamylated MTX, the trapped form of MTX. MTX-resistant populations could have any one or a combination of these alterations. The net result of any of these resistance mechanisms is to decrease the ability of MTX to inhibit dihydrofolate reductase at clinically achievable MTX concentrations. Other specific drug resistance mechanisms could be described for compounds such as cytosine arabinoside, 5-fluorouracil, hydroxyurea, and other drugs.

In multiple drug resistance (MDR), drug-resistant cells are cross-resistant to a variety of seemingly unrelated antitumor agents, such as the vinca alkaloids, adriamycin, actinomycin D, and etoposide. All of these are natural products or are derived from natural products and they are not chemically related in structure. They have different mechanisms of action as antitumor agents but appear to act on some nuclear event.

Multidrug-resistant tumor cells express high levels (compared to the drug-sensitive tumor cells) of membrane transport proteins (P-glycoproteins and multidrug resistance-associated proteins [MRPs], p. 494) that transport drugs out of the cell. These are ATP-dependent pumps and effectively reduce the cellular concentration of drug to below its cytotoxic concentration.

Development of drug-resistant tumor cells presents major clinical problems. However, study of the mechanisms of drug resistance has greatly aided in our understanding of cancer cells and how best to design chemotherapeutic protocols to treat the various forms of cancer.

Many different examples can be given in which genetic polymorphisms play a large role in the metabolism or disposition of various classes of drugs involving distinct gene products. As a result, there will be variability in the responses of patients to a specific drug regimen. As one example, patients who inherit nonfunctional alleles for thiopurine *S*-methyltransferase (TPMT), an enzyme involved in the metabolism of 6-mercaptopurine (see Figure 20.31, p. 832) have very adverse effects to the standard dose of 6-mercaptopurine. These patients can be successfully treated by reducing the conventional dose of 6-MP and still achieve the necessary clinical response with decreased morbidity. Pharmacogenomics, a relatively new field that is rapidly developing, will allow the identification of many additional candidate genes such as MDR1 with the use of molecular techniques that will allow the physician to move toward personalized medicine so that drug doses for patients can be optimized before therapy has begun rather than later in the course of treatment. The use of pharmcogenomics will apply not only for the defined treatment of cancer but also for many other diseases as well.

Bibliography

An, S., Kumar, R., Sheets, E. D., and Benkovic, S. J. Reversible compartmentalization of *de vovo* purine biosynthetic complexes in living cells. *Science* 320:103, 2008.

Belenky, P., Bogan, K. L., and Brenner, C. NAD metabolism in health and disease. *Trends Biochem. Sci.* 32:12, 2006.

Bertino, J. R., and Hait, W. Principles of cancer therapy. In Goldman, L. and Ausiello, D. A. (Eds.), *Textbook of Medicine,* 22d ed. Maryland Heights, MO: Saunders, 2004, Chap, 191, 1137.

Cory, A. H., and Cory, J. G. Critical roles of glutamine as nitrogen donors in purine and pyrimidine nucleotide synthesis: Asparaginase treatment in childhood acute lymphoblastic leukemia. *In Vivo* 20:587, 2006.

De Brouwer, A. P., Williams, K. I., Duley, J. A., Kuilenburg, A. B., et al. Arts syndrome is caused by loss-of-function mutations in PRPS1. *Am. J. Hum. Genet.* 81:507, 2007.

Evans, W. E., and Relling, M. V. Moving towards individualized medicine with pharmacogenomics. *Nature* 429:464, 2004.

Hara, N., Yamada, K., Shibita, T., Osago, H., et al. Elevation of cellular NAD levels by nicotinic acid levels by nicotinic acid and involvement of nicotinic acid phosphoribosyltransferase in human cells. *Biol. Chem.* 282:24574, 2007.

Loffler, M., Fairbanks, D. L., Elke, Z., Marinaki, A. M., and Simmons H. C. Pyrimidine pathways in health and disease. *Trends Biochem. Sci.* 11:430, 2005.

Munch-Petersen, B., and Piskur, J. Deoxynucleoside kinases and their potential role in deoxynucleoside cytotoxicity. In Peters, G. J. (Ed.), *Cancer Drug Discovery and Development: Deoxynucleoside Analogs in Cancer Therapy.* Totowa, NJ: Humana Press, 2006, 53.

Sauna, Z. E., Kimichi-Sarfaty, C., Ambdkur, S. V., and Gottesman, M. M., Silent polymorphisms speak: How they affect pharmacogenomics and the treatment of cancer. *Cancer. Res.* 67:9609, 2007.

Scriver, C. R., Beaudet, A. L., Sly, W. S., and Valle, D. (Eds.). *The Metabolic and Molecular Bases of Inherited Disease,* 8th ed. New York: McGraw Hill, 2004, vol. 2, Chaps. 106–113.

Xu, X., Williams, J. W., Gong, H., Finnegan, A., and Chong, A. S. Two activities of the immunosuppressive metabolite of leflunomide, A771726. Inhibition of pyrimidine nucleotide synthesis and protein tyrosine phosphorylation. *Biochem. Pharmacol.* 52:527, 1996.

Zalkin, H., and Dixon, J. E. De novo purine nucleotide biosynthesis. Prog. *Nucleic Acid Res.* 42:259, 1992.

Key Terms

5-phosphoribosyl-1-pyrophosphate
carbamoyl phosphate synthetase
guanosine 5′-monophosphate
glycine
multifunctional protein
nucleoside 5′-diphosphate kinase
nucleoside 5′-monophosphate kinase
PRPP amidotransferase
adenine phosphoribosyl transferase
hypoxanthine-guanine phosphoribosyl transferase
Lesch–Nyhan syndrome
N^{10}-formyltetrahydrofolate
xanthine 5′-monophosphate
xanthine oxidoreductase
adenosine 5′-monophosphate
adenosine kinase
AMP deaminase
tetrahydrobiopterin
inosine 5′-monophosphate
nucleases
uric acid
dihydroorotate dehydrogenase
orotic acid
uridine 5′-monophosphate
aspartate
β-alanine
nucleoside 5′-diphosphate reductase
thioredoxin reductase
thymidylate synthase
cytosine 5′-monophosphate
β-aminoisobutyric acid
NAD synthesis
FAD synthesis
6-thioguanine
6-mercaptopurine
glutamine
methotrexate
5-fluorouracil
azaserine
hydroxyurea

Questions • CAROL N. ANGSTADT

Multiple-Choice Questions

1. The two purine nucleotides found in RNA
   A. are formed in a branched pathway from a common intermediate.
   B. are formed in a sequential pathway.
   C. must come from exogenous sources.
   D. are formed by oxidation of the deoxy forms.
   E. are synthesized from nonpurine precursors by totally separate pathways.

2. The type of enzyme known as a phosphoribosyltransferase is involved in all of the following except
   A. salvage of pyrimidine bases.
   B. the *de novo* synthesis of pyrimidine nucleotides.
   C. the *de novo* synthesis of purine nucleotides.
   D. salvage of purine bases.

3. Deoxyribonucleotides
   - **A.** cannot be synthesized so they must be supplied preformed in the diet.
   - **B.** are synthesized *de novo* using PRPP.
   - **C.** are synthesized from ribonucleotides by an enzyme system involving thioredoxin.
   - **D.** are synthesized from ribonucleotides by nucleotide kinases.
   - **E.** can be formed only by salvaging free bases.

4. The conversion of nucleoside 5′-monophosphates to nucleoside 5′-triphosphates
   - **A.** is catalyzed by nucleoside kinases.
   - **B.** is a direct equilibrium reaction.
   - **C.** utilizes a relatively specific nucleotide kinase and a relatively nonspecific nucleoside diphosphate kinase.
   - **D.** generally uses GTP as a phosphate donor.
   - **E.** occurs only during the S phase of the cell cycle.

5. The synthesis of the coenzymes NAD, FAD, and coenzyme A have in common
   - **A.** the same vitamin.
   - **B.** PRPP.
   - **C.** AMP.
   - **D.** ATP.
   - **E.** a nucleotidase.

6. Tetrahydrobiopterin
   - **A.** is synthesized from GTP.
   - **B.** is an inhibitor of IMP dehydrogenase.
   - **C.** is synthesized from ATP.
   - **D.** is a degradation product of guanine.
   - **E.** requires the addition of glutamine to a purine nucleotide.

Questions 7 and 8: Hereditary orotic aciduria is characterized by severe anemia, growth retardation, and high levels of orotic acid excretion. The defect may be in orotate phosphoribosyl transferase, orotidine decarboxylase, or both. The preferred treatment for this disease is dietary uridine, which reverses the anemia and decreases the formation of orotic acid.

7. Elements involved in the effectiveness of the dietary treatment include
   - **A.** conversion of exogenous uridine to UMP by uridine phosphotransferase.
   - **B.** UTP from exogenous uridine providing substrate for synthesis of CTP.
   - **C.** inhibition of carbamoyl phosphate synthetase II by UTP.
   - **D.** all of the above.
   - **E.** none of the above.

8. In the *de novo* synthesis of pyrimidine nucleotides
   - **A.** reactions take place exclusively in the cytosol.
   - **B.** a free base is formed as an intermediate.
   - **C.** PRPP is required in the rate-limiting step.
   - **D.** UMP and CMP are formed from a common intermediate.
   - **E.** UMP inhibition of OMP-decarboxylase is the major control of the process.

Questions 9 and 10: Gout is a disease characterized by hyperuricemia from an overproduction of purine nucleotides via the *de novo* pathway. The specific cause of Lesch–Nyhan syndrome is a severe deficiency of HGPRTase. Allopurinol is used in the treatment of gout to reduce the production of uric acid. In Lesch–Nyhan syndrome, the decrease in uric acid is balanced by an increase in xanthine plus hypoxanthine in blood. In the other forms of gout, the decrease in uric acid is greater than the increase in xanthine plus hypoxanthine.

9. The explanation for this difference in the two forms of gout is
   - **A.** it is an experimental artifact and the decrease in uric acid and increase in xanthine plus hypoxanthine in non-Lesch–Nyhan gout is the same.
   - **B.** allopurinol is less effective in non-Lesch–Nyhan gout.
   - **C.** there is an increased excretion of xanthine and hypoxanthine in non-Lesch–Nyhan gout.
   - **D.** PRPP levels are reduced in Lesch–Nyhan.
   - **E.** in non-Lesch–Nyhan gout hypoxanthine and xanthine are salvaged to IMP and XMP and inhibit PRPP amidotransferase.

10. Which of the following is/are aspects of the overall regulation of *de novo* purine nucleotide synthesis?
    - **A.** AMP, GMP, and IMP shift PRPP amido transferase from a small form to a large form.
    - **B.** PRPP amidotransferase shows hyperbolic kinetics with PRPP.
    - **C.** AMP inhibits the conversion of IMP to GMP.
    - **D.** Change in glutamine concentration is a major regulator.
    - **E.** Direct interconversion of AMP to GMP maintains balance of the two.

Question 11 and 12: There are two distinct immunodeficiency diseases that lead to the formation of uric acid as the end product. Mutation in genes for adenosine deaminase (ADA) leads to severe combined immunodeficiency (SCID) in which both T-cells and B-cells are affected. Defects in purine nucleoside phosphorylase (PNP) affect only T-cells. These two enzymes are in the pathways for degradation of nucleic acids. Gene therapy has had some success in treating ADA deficiency.

11. In nucleic acid degradation, all of the following are correct except
    - **A.** there are nucleases that are specific for either DNA or RNA.
    - **B.** nucleotidases convert nucleotides to nucleosides.
    - **C.** the conversion of a nucleoside to a free base is an example of a hydrolysis.
    - **D.** because of the presence of deaminases, hypoxanthine rather than adenine is formed.
    - **E.** both DNA and RNA degradation lead to uric acid.

12. The best estimate of the turnover of DNA comes from a measurement in urine of
    - **A.** uric acid.
    - **B.** NH_4^+ and CO_2
    - **C.** β-alanine.
    - **D.** β-aminoisobutyrate.
    - **E.** cytidine.

Problems

13. If a cell capable of *de novo* synthesis of purine nucleotides has adequate AMP but is deficient in GMP, how would the cell regulate synthesis to increase [GMP]? If both AMP and GMP were present in appropriate concentrations, what would happen?

14. Inhibitors of purine and pyrimidine nucleotide metabolism take many forms. 5-Fluorouracil is an antimetabolite, methotrexate is an antifolate, and azaserine is an antagonist. How do these three types differ in their action?

Answers

1. **A** GMP and AMP are both formed from the first purine nucleotide, IMP, in a branched pathway. B: The pyrimidine nucleotides UMP and CTP are formed in a sequential pathway from orotic acid. C: Humans are capable of synthesizing purine nucleotides. D: Deoxy forms are formed by reduction of the ribose forms. E: IMP is the common precursor.
2. **C** In purine nucleotide synthesis, the purine ring is built up stepwise on ribose-5-phosphate and not transferred to it. A, B, and D: Phosphoribosyl transferases are important salvage enzymes for both purines and pyrimidines and are also part of the synthesis of pyrimidines since OPRT catalyzes the conversion of orotate to OMP.
3. **C** Deoxyribonucleotides are synthesized from the ribonucleoside diphosphates by nucleoside diphosphate reductase that uses thioredoxin as the direct hydrogen-electron donor. A, B, and E: There is a synthetic mechanism as just described but it is not a *de novo* pathway. D: Nucleotide kinases are enzymes that add phosphate to a base or nucleotide.
4. **C** These two enzymes are important in interconverting the nucleotide forms. A: These convert nucleosides to nucleoside monophosphates. B: Two steps are required. D: ATP is present in highest concentration and the phosphate donor. E: Occurs during the S phase, but this is a general reaction for the cell.
5. **D** ATP is used both as a source of energy and part of it is incorporated into the product. A: They are different vitamins. B: This is necessary for the formation of NAD but not the other two. C: AMP is part of the final structure but must be incorporated as ATP. E: The removal of the phosphate is associated with the use of ATP as energy, not a simple hydrolysis.
6. **A** This begins with a cyclohydrolase acting on GTP. B: Such an inhibitor would reduce the level of tetrahydrobiopterin by inhibiting production of GTP. C: It is not a derivative of ATP. D: GTP is not converted to guanine. E: Glutamine is not an adduct.
7. **D** It is common for an exogenous agent to require conversion to an active form; in this case the uridine is salvaged to the monophosphate and ultimately to the triphosphate. The cell is deficient in UTP and CTP because the conversion of orotic acid is blocked so the exogenous uridine provides a bypass around the block. Orotic acid formation is decreased since UTP inhibits carbamoyl phosphate synthetase II, the control enzyme.
8. **B** This is in contrast to purine *de novo* synthesis. A: One enzyme is mitochondrial. C: PRPP is required to convert orotate to OMP but this is not rate limiting. D: OMP to UMP to CTP is a sequential process. E: This does occur, but the rate-limiting step is that catalyzed by CPS II.
9. **E** Not only is uric acid production directly inhibited, *de novo* synthesis is as well, thus reducing production of xanthine and hypoxanthine. In Lesch–Nyhan, the only effect is the direct inhibition of xanthine oxidase. A: It is a real effect. B: Actually, it is more effective because of the dual roles. C: This does not happen. D: PRPP levels are very high because of the lack of the salvage of bases, leading to improper or lack of regulation of pathways.
10. **A** This is a mechanism of inhibition since the large form of the enzyme is inactive. B: PRPP amidotransferase shows sigmoidal kinetics with respect to PRPP so large shifts in concentration of PRPP have the potential for altering velocity. C: AMP inhibits the conversion of IMP to itself. D: Glutamine concentration is relatively constant. E: This does not happen.
11. **C** The product is ribose-1-phosphate rather than the free sugar catalyzed by PNP. A: They can also show specificity toward the bases and positions of cleavage. B: A straight hydrolysis. D: AMP deaminase and adenosine deaminase remove the 6-NH_2 as NH_3. The IMP or inosine formed is eventually converted to hypoxanthine. E: This is called severe combined immunodeficiency.
12. **D** This is generated exclusively from the degradation of thymine from DNA. A: Uric acid is formed from purine nucleotides from any source. B: These are generated from many things. C: This comes from more than pyrimidines. E: Cytidine is converted to uridine which is metabolized to β-alanine.
13. AMP would partially inhibit *de novo* synthesis of IMP by its allosteric inhibition of PRPP amidotransferase. The IMP formed would be directed toward GMP because AMP is an inhibitor of its own synthesis from IMP. If both AMP and GMP are adequate, the synergistic effect of the two on PRPP amidotransferase would severely inhibit *de novo* synthesis.
14. 5-Fluorouracil can be converted by the same enzyme as uracil to the mono- and triphosphate nucleotides and incorporated into RNA, disrupting many functions as well as inhibiting thymidylate synthase. That is, it is a structural analog of the normal compound. Antifolates inhibit dihydrofolate reductase so preventing the reformation of tetrahydrofolate for its multiple functions. Azaserine, as an antagonist, is an inhibitor of many enzymes that use glutamine as an amino donor.

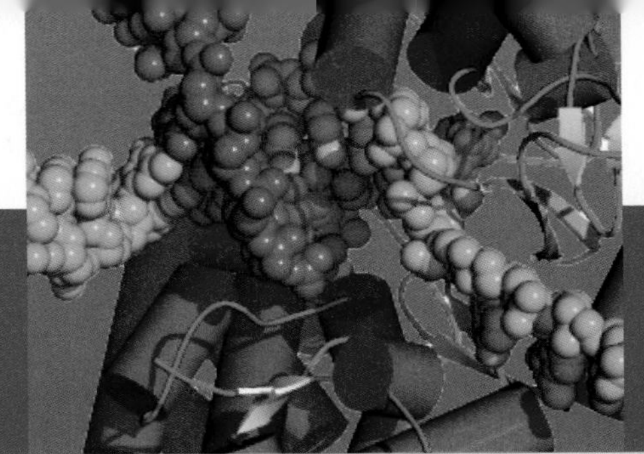

21

Metabolic Interrelationships

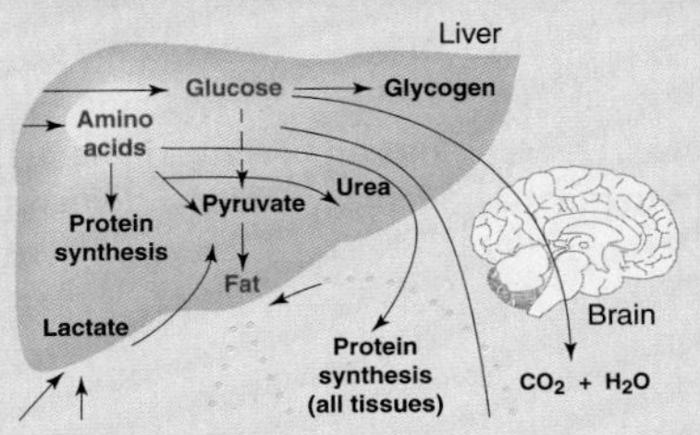

Robert A. Harris
Distinguished Professor Emeritus and Showalter Professor of Biochemistry Emeritus, Indiana University School of Medicine

David W. Crabb
John B. Hickam Professor of Medicine and Chair, Indiana University School of Medicine

CLINICAL CORRELATIONS

Key Concepts

- Metabolic pathways are controlled in different nutritional and disease states to maintain sources of energy and amino acids in the blood for all tissues.
- Pathways that remove excess fuels from the blood (glycogenesis, glycolysis, fatty acids synthesis, and lipogenesis) are active in the fed state.
- Pathways that maintain adequate levels of fuels in the blood (glycogenolysis, gluconeogenesis, lipolysis, proteolysis, and ketogenesis) are active in the starved state.
- Pathways are controlled by substrate availability, allosteric effectors, covalent modification, and induction or repression of key enzymes.
- The changes in metabolism that accompany common disease states are variations on the themes that function in the fed and fasted states.

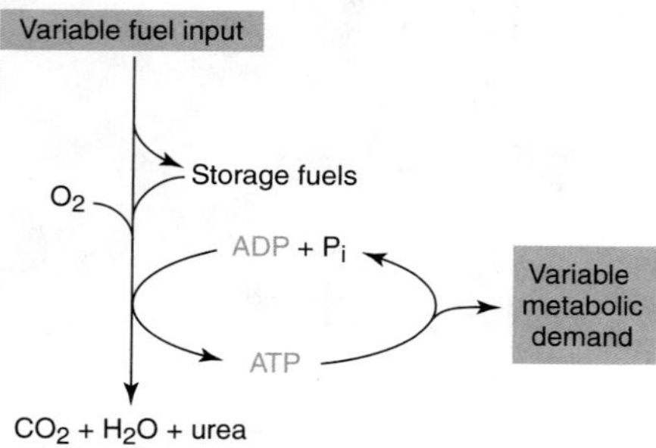

Figure 21.1 Humans can use a variable fuel input to meet a variable metabolic demand.

21.1 • INTRODUCTION

The interdependence of metabolic processes of the major tissues of the body will be stressed in this chapter. Not all of the major metabolic pathways operate in every tissue at any given time. Given the nutritional and hormonal status of a patient, it is necessary to know qualitatively which pathways are functional and how they relate to one another.

The metabolic processes of concern are glycogenesis, glycogenolysis, gluconeogenesis, glycolysis, fatty acid synthesis, lipogenesis, lipolysis, glyceroneogenesis, fatty acid oxidation, glutaminolysis, tricarboxylic acid (TCA) cycle activity, ketogenesis, amino acid oxidation, protein synthesis, proteolysis, and urea synthesis. It is important to know (1) which tissues are most active in these various processes, (2) when these processes are most active, and (3) how these processes are controlled and coordinated in different metabolic and disease states.

The best way to gain an understanding of the interrelationships of the pathways is to learn the changes in metabolism during the starve–feed cycle (Figure 21.1). This cycle allows a variable fuel and nitrogen intake to meet variable metabolic and anabolic needs. Feed refers to the intake of meals (the variable fuel input) after which the fuel is stored (as glycogen and triacylglycerol) to meet metabolic needs of fasting. An ATP cycle functions within the starve–feed cycle (Figure 21.1). Cells of the body die without the provision of continuous supply of energy for ATP synthesis to meet their needs.

Humans can consume food at a rate far greater than their basal caloric requirements; this allows them to survive from meal to meal. Unfortunately, an almost unlimited capacity to consume food is matched by an almost unlimited capacity to store it as triacylglycerol. Obesity is the consequence of excess food consumption and the commonest form of malnutrition in affluent countries (Clin. Corr. 21.1). Other forms of malnutrition are more prevalent in developing countries (Clin. Corrs. 21.2 and 21.3). The regulation of food

CLINICAL CORRELATION 21.1

Obesity and the Metabolic Syndrome

Obesity is the most common nutritional problem in the United States, in fact a majority of the U.S. population is either obese or overweight, and the problem may be even greater for children. It is a risk factor in development of diabetes mellitus, hypertension, endometrial carcinoma, osteoarthritis, cirrhosis, gallstones, and cardiovascular diseases. Clinically, the quartet of obesity, insulin resistance, dyslipidemia, and hypertension is called either syndrome X or the metabolic syndrome and contributes greatly to the high rate of cardiovascular death in Western countries. Obesity is easy to explain: An obese person has eaten more calories than he or she expended. The accumulation of massive amounts of body fat is not otherwise possible. For unknown reasons, the neural control of caloric intake to balance energy expenditure is abnormal. Rarely, obesity is secondary to a correctable disorder, such as hypothyroidism or Cushing syndrome. The latter is the result of increased secretion of glucocorticoids, which cause fat deposition in the face and trunk, with wasting of the limbs, and glucose intolerance. These effects are due to increased protein breakdown in muscle and conversion of the amino acids to glucose and fat. Less commonly, tumors, vascular accidents, or maldevelopment of the nervous system hunger control centers in the hypothalamus cause obesity. However, the rapid increase in the prevalence of obesity cannot be explained by genetic or biochemical mechanisms and must reflect cultural changes in food preparation and consumption and reduced physical activity.

The obese (*ob/ob*) mouse was discovered in the 1950s, and the defective gene cloned in 1994. This *ob* gene encodes a 146 amino-acid-secreted protein (alternatively called OB protein or leptin, for its slimming effect) that is produced in adipocytes and detectable in blood. *ob/ob* mice have a nonsense mutation in the gene and produce no leptin. Injection of leptin into them causes increased energy expenditure and reduced eating, with marked weight loss. This effect on appetite is mimicked by intracerebroventricular injection. Leptin also reduced appetite and weight of normal mice. Obese humans do not generally have defective *ob* genes, and in fact tend to have high blood levels of leptin. This suggests that their nervous system is insensitive to leptin, analogous to the insulin resistance seen in many diabetic patients.

In the most common type of obesity, the number of adipocytes does not increase, but rather they get larger as they become engorged with triacylglycerols. If obesity develops before puberty, there may also be an increase in the number of adipocytes. In this case, the hyperplasia (increase in cell number) and hypertrophy (increase in cell size) contribute to the magnitude of the obesity. Obesity in men tends to be centered on the intraabdominal (so called visceral) fat, while in women it is more likely to be on the hips. The male pattern, characterized by a high waist-to- hip circumference ratio, is more predictive of premature coronary heart disease. Moreover, visceral adipose tissue seems more resistant to suppression of lipolysis by insulin, and thus the release of fatty acids into the portal vein may contribute to the relative inability to suppress hepatic glucose production in obesity and type 2 diabetes (see Clin. Corrs. 15.4, p. 594, and 21.6, p. 848).

The only effective treatment of obesity is reduction in the ingestion or increase in the use of calories. Practically speaking, this means dieting, since even vigorous exercise such as running only consumes 10 kcal per minute of exercise. Thus, an hour-long run (perhaps 5–6 mi) uses the energy present in about two candy bars. However, exercise programs can be useful to help motivate individuals to remain on their diets. There is interest in low-carbohydrate, high-fat, and protein diets, sometime referred to as the Atkins diet. This diet reduces carbohydrate intake to a low enough level to induce ketonemia. As weight loss progresses, carbohydrate is reintroduced until weight stabilizes. Several studies have shown that this strategy is more effective in inducing weight loss than standard calorie restriction and better at reducing triglyceride levels without raising LDL cholesterol. Beneficial effects on weight management, triglyceride levels, cholesterol levels, and blood pressure have also been achieved with high-protein, low-carbohydrate diets.

Unfortunately for individuals trying to lose weight, the body compensates for decreased energy intake with reduced formation of triiodothyronine and a decrease in the basal metabolic rate. Thus, there is a biochemical basis for the universal complaint that it is far easier to gain than to lose weight. What is more, about 95% of people who lose a significant amount of weight regain it within 1 year. Most patients who have successfully lost weight long term engage in regular exercise and weigh themselves daily (perhaps compensating for inadequate neural response to satiety signals).

Jequier, E., and Tappy, L. Regulation of body weight in humans. *Physiol. Rev.* 79:451, 1999; Roberts, K., Dunn, K., Jean, S. K., and Lardinois, C. K. Syndrome X: Medical nutrition therapy. *Nutr. Rev.* 58:154, 2000; and Kumanyika, S. K., Obarzanek, E., Stettler, N., Bell, R., et al. Population-based prevention of obesity: The need for comprehensive promotion of healthful eating, physical activity, and energy balance. American Heart Association Council on Epidemiology and Prevention, Interdisciplinary Committee for Prevention (Formerly the Expert Panel on Population and Prevention Science). *Circulation* 118:428, 2008.

CLINICAL CORRELATION 21.2

Protein–Calorie Malnutrition

Protein–calorie malnutrition is the most important and widespread nutritional problem among young children in the developing world. The clinical syndrome, called kwashiorkor, occurs mainly in children 1–3 years of age and is precipitated by weaning an infant from breast milk onto a starchy, protein-poor diet. The name originated in Ghana, and means "the sickness of the older child when the next baby is born." It is a bit of a misnomer since the major deficiency is protein, not calories. It is a consequence of feeding the child a diet adequate in calories but deficient in protein. It may become manifest when protein requirements are increased by infection, for example, malaria, helminth infestation, or gastroenteritis. It is characterized by poor growth, low plasma protein and amino acid levels, muscle wasting, edema, diarrhea, and increased susceptibility to infection. The presence of subcutaneous fat clearly differentiates it from simple starvation. Fat stores are maintained by the high carbohydrate intake and resulting high insulin levels. In fact, the high insulin level interferes with the adaptations that occur in starvation. Fat is not mobilized as an energy source, ketogenesis does not take place, and there is no transfer of amino acids from skeletal muscle to the liver, kidneys, heart, and immune cells. Lack of dietary amino acids results in diminished protein synthesis in all tissues. The liver becomes enlarged and infiltrated with fat, reflecting the need for hepatic protein synthesis for the formation and release of VLDL. In addition, protein malnutrition impairs the function of the gut, and leads to malabsorption of carbohydrate, protein, and vitamins, which accelerates the disease. The consequences depend on when in development the deficiency occurs. Children with low weight for height are called "wasted," but can make a good recovery when properly fed. Those with low height for weight are called "stunted", and never regain full height or cognitive potential.

Protein–calorie malnutrition is a problem for the elderly when they become sick. The requirement for energy and food intake of healthy elderly may decline with age. There is the risk that insufficient intake of protein and of nutrients such as iron, calcium, and vitamins will occur. Such deficiency may accelerate loss of lean body mass and strength (leading to falls), anemia, loss of bone strength (resulting in breaking of the hip during falls), and rarely, vitamin deficiency states. Chronic illnesses that are more common in the elderly frequently impair appetite, food intake, or nutrient assimilation. As a result, elderly patients are more often found to suffer from protein–calorie malnutrition than younger adults.

Protein–calorie malnutrition is common in patients with liver cirrhosis. Low serum albumin levels as a consequence of reduced hepatic synthesis are prognostic of poor survival rates. Dietary deficiency of protein coupled with loss of control of the oxidation of branched-chain amino acids (BCAA: leucine, isoleucine, valine) causes low serum BCAA levels in patients with severe liver cirrhosis. In some countries, such patients are given BCAA supplements in an effort to raise serum albumin levels without worsening blood ammonia levels. This is not standard practice in the United States for want of convincing clinical evidence of the effectiveness of this intervention.

Omran, M.L., and Morley, J. E., Assessment of protein energy malnutrition in older persons, Part I: History, examination, body composition, and screening tools. *Nutrition* 16:50, 2000; Omran, M. L., and Morley, J. E. Assessment of protein energy malnutrition in older persons, Part II: Laboratory evaluation. *Nutrition* 16:131, 2000; and Corish, C. A., and Kennedy, N. P. Protein-energy under nutrition in hospital in-patients. *Br. J. Nutr.* 83:575, 2000.

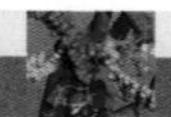

CLINICAL CORRELATION 21.3

Starvation

Starvation leads to a syndrome known as marasmus. *Marasmus* is a word of Greek origin meaning "to waste." It is most frequent in children under 1 year of age. In developing countries, early weaning of infants from breast milk is a common cause. This may result from pregnancies in rapid succession, the desire of the mother to return to work, or the use of overdiluted artificial formulas (to make the expensive formulas last longer). This practice leads to insufficient intake of calories as well as protein. Likewise, diarrhea and malabsorption can develop if safe water and sterile procedures are not used in preparation of formula.

In contrast to kwashiorkor (see Clin. Corr. 21.2, p. 841), subcutaneous fat, hepatomegaly, and fatty liver are absent in marasmus because fat is mobilized for energy and muscle temporarily provides amino acids to the liver for synthesis of glucose and hepatic proteins. Low insulin levels allow the liver to oxidize fatty acids and to produce ketone bodies for other tissues. Ultimately, energy and protein reserves are exhausted, and the child starves to death. The immediate cause of death is often pneumonia, because the child is too weak to cough. Adults can suffer from marasmus because of diseases that prevent swallowing (cancer of the throat or esophagus), reduce the ability of the gut to absorb nutrients (celiac disease or Crohn disease, short-bowel syndrome from surgical resection of the gut), or interfere with access to food (strokes or dementia). Thousands of the residents confined to the Warsaw Ghetto during World War II died from starvation. Studies by Jewish physicians confined to the ghetto showed that death, when not complicated by a severe infection, occurred when substrates for gluconeogenesis could no longer be supplied by the breakdown of muscle.

A related and very common disorder is cancer cachexia, which is due to anorexia (loss of appetite, and therefore starvation) and to body wasting. The latter differs from simple starvation in that skeletal muscle is not spared and both muscle and fat is used to supply energy needs (see Clin. Corr. 21.9, p. 869).

Studies with experimental animals suggest females are more resistant to the adverse effects of starvation than males. Physical examination of survivors of the POW camps and orphanages after World War II suggest this is also true for humans. Females are usually smaller and have a lower basal metabolic rate than males, and therefore need less food per day than males. Furthermore, the evolutionary selection pressures for survival have surely been greater for females than males during the many periods of famine that have occurred over time. Since for most species many females can mate with one male, males are not needed during pregnancy and lactation, and males compete with offspring for food, survival of males is less important than survival of females for maintenance of the species during long periods of food shortage. Genes may therefore have been selected that confer females with greater resistance to starvation. Support for this notion comes from the finding that mitochondria isolated from female experimental animals are more highly differentiated, as evidenced by more enzymatic machinery for oxidative phosphorylation, than mitochondria isolated from males.

Uvin, P. The state of world hunger. *Nutr. Rev.* 52:151, 1994; Body, J. J. The syndrome of anorexia-cachexia. *Curr. Opin. Oncol.* 11:255, 1999; and Wolfe, R. R. The underappreciated role of muscle in health and disease. *Am. J. Clin. Nutr.* 84:475, 2006.

consumption is complex and not well understood. One important factor is leptin, a protein synthesized and secreted into the blood by adipocytes that regulates energy expenditure and appetite through its effects on the hypothalamus (see Clin. Corr. 21.1). The tight control needed is indicated by the calculation that eating two extra pats of butter (~100 cal) per day over caloric expenditures results in a weight gain of 10 lb per year. A weight gain of 10 lb may not sound excessive, but multiplied by 10 years it equals severe obesity!

21.2 • STARVE–FEED CYCLE

In the Well-Fed State the Diet Supplies the Energy Requirements

Figure 21.2 shows the fate of glucose, amino acids, and fat obtained from food. Glucose passes from the intestinal epithelial cells via the portal vein to the liver. Amino acids are partially metabolized in the gut before being released into portal blood. Chylomicrons containing triacylglycerol are secreted by the intestinal epithelial cells into lymphatics. The lymphatics empty into the thoracic duct, which delivers the chylomicrons to the subclavian vein and thence to the rest of the body.

In the liver, dietary glucose can be converted into glycogen by glycogenesis or into pyruvate and lactate by glycolysis or can be used in the pentose phosphate pathway for the generation of NADPH for synthetic processes. Pyruvate can be oxidized to acetyl CoA, which, in turn, can be converted into triacylglycerol or oxidized to CO_2 and water by the

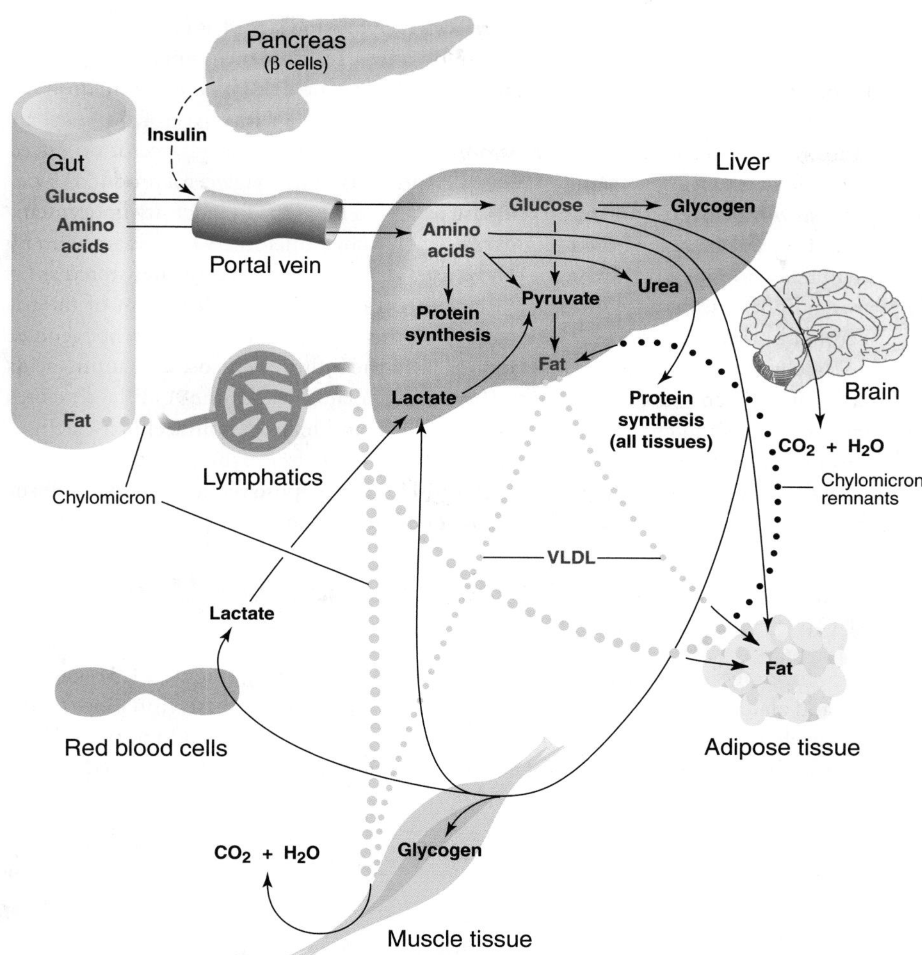

Figure 21.2 Disposition of glucose, amino acids, and fat by various tissues in the well-fed state. Chylomicrons (large yellow dots) are converted to chylomicron remnants (black dots) by the action of lipoprotein lipase in the adipose tissue and skeletal muscle. VLDL particles (small yellow dots) are converted to LDL particles (not shown) by lipoprotein lipase in these same tissues.

TCA cycle. Much of the dietary glucose passes through the liver to reach other organs, including brain, which is almost completely dependent on glucose for ATP generation, red blood cells and renal medulla, which can only carry out glycolysis, and adipose tissue, which primarily converts it into the glycerol moiety of triacylglycerol. Muscle also uses glucose, converting it to glycogen or using it in glycolysis and the TCA cycle. Lactate and pyruvate produced by glycolysis in other tissues are taken up by the liver and oxidized to CO_2 or converted to triacylglycerol. In the well-fed state, the liver uses glucose and does not engage in gluconeogenesis. Thus, the Cori cycle, conversion of glucose to lactate in peripheral tissues followed by conversion of lactate to glucose in liver, is interrupted in this state.

The intestinal cells use some dietary amino acids as an energy source but transport most of them into the portal blood for distribution. Liver removes some absorbed amino acids from the portal blood (Figure 21.2), but most pass through. This is especially important for essential amino acids, which are needed by all cells for protein synthesis. Liver metabolizes amino acids, but the K_m values of the enzymes involved are high, meaning that the amino acids have to be present in high concentration before significant catabolism can occur. In contrast, the low K_m values of tRNA charging enzymes ensures that protein synthesis can occur when amino acids are present. Excess amino acids can be oxidized completely to CO_2, urea, and water, or the intermediates generated can be used for lipogenesis. Amino acids that escape the liver are used for protein synthesis or energy in other tissues.

Dietary triacylglycerol reaches the bloodstream as chylomicrons (p. 1057), which in turn are acted on by lipoprotein lipase attached to the surface of endothelial cells in the lumen of the capillaries of various tissues, but particularly adipose tissue (Figure 21.2). This lipase hydrolyzes a large portion but not all of the triacylglycerol present in chylomicrons. The released fatty acids are taken up by the adipocytes, reesterified with glycerol 3-phosphate (derived from glucose via glycolysis) to form triacylglycerols, and stored as fat droplets within the adipocytes. Chylomicron remnants remaining after digestion by lipoprotein lipase are cleared from the blood by the liver. Triacylglycerols present in remnants are hydrolyzed by a lysosomal lipase. The released fatty acids are reesterified with glycerol 3-phosphate (derived from free glycerol and glucose) to form triacylglycerols. Triacylglycerol produced from dietary fat in this manner along with smaller amounts of triacylglycerol produced by *de novo* synthesis from glucose and amino acids are packaged into very low-density lipoproteins (VLDL) and secreted into the blood (Figure 21.2). Like chylomicrons, VLDL is acted on by lipoprotein lipase to give fatty acids that can be used to form triacylglycerol and stored as fat droplets in adipocytes. Because of the high fat content of the human diet, most of the triacylglycerol of human adipose tissue originates from the diet rather than *de novo* lipogenesis.

The β-cells of the pancreas are very responsive to the influx of glucose and amino acids in the fed state. When glucose enters the β-cell, its oxidation raises the ATP level, closes ATP-sensitive potassium channels, depolarizes the cell, and increases intracellular calcium, leading to insulin release. β-cells release insulin during and after eating, which is essential for the metabolism of these nutrients by liver, muscle, and adipose tissue. The role of insulin in the starve–feed cycle is discussed in more detail in Section 21.3, p. 852.

In the Early Fasting State Hepatic Glycogenolysis Maintains Blood Glucose

Hepatic glycogenolysis maintains blood glucose during early fasting (Figure 21.3). Lactate, pyruvate, and alanine are diverted from oxidation and fatty acid synthesis into glucose formation, completing the Cori cycle. The alanine cycle, in which carbon and nitrogen return to the liver in the form of alanine, also becomes important (see Figure 15.32, p. 622).

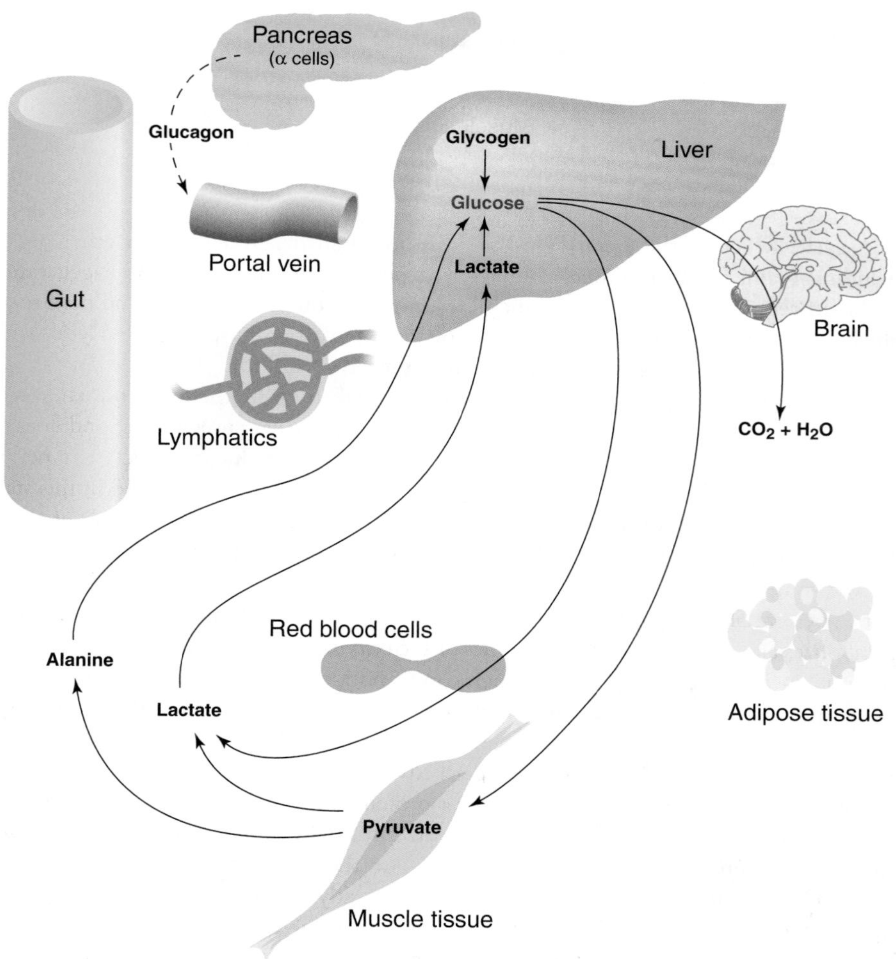

Figure 21.3 Metabolic interrelationships of major tissues in early fasting state.

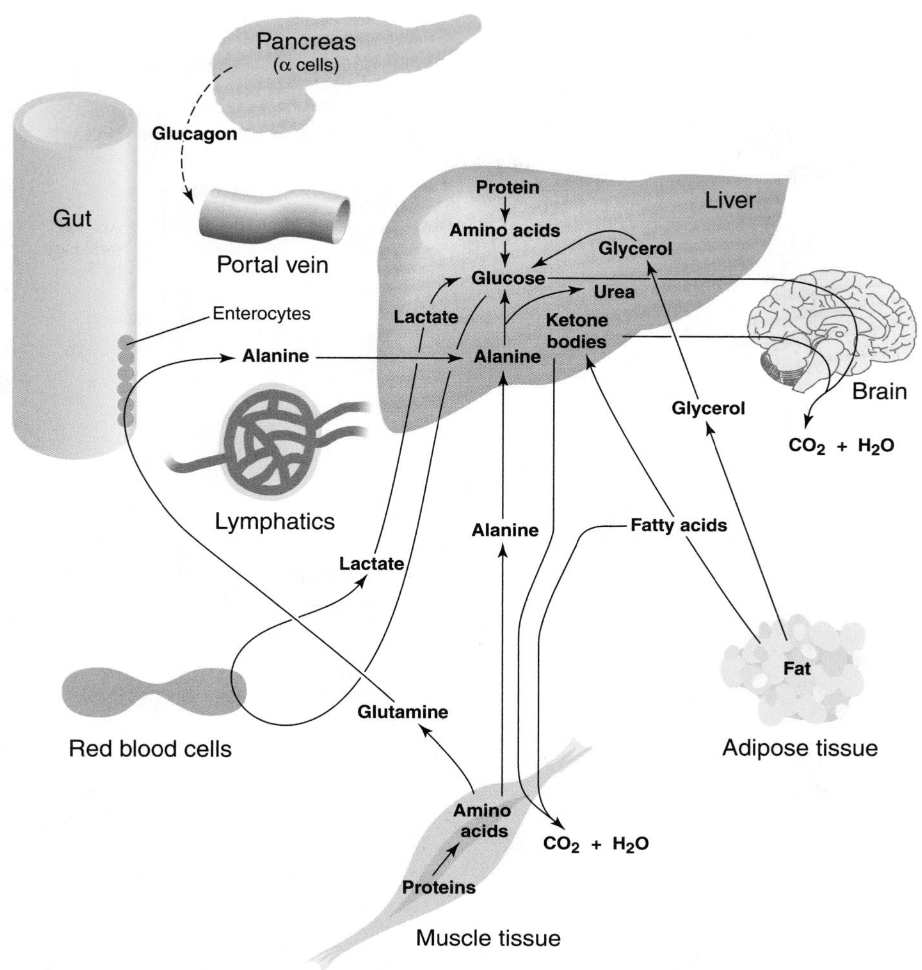

Figure 21.4 Metabolic interrelationships of major tissues in fasting state.

The Fasting State Requires Gluconeogenesis from Amino Acids and Glycerol

Since no dietary fuel enters from the gut and little glycogen is left in the liver after 10–12 h of fasting state (Figure 21.4), the body is dependent on hepatic gluconeogenesis, primarily from lactate, glycerol, and alanine. The Cori and alanine cycles play important roles, but do not provide carbon for net synthesis of glucose. This is because glucose formed from lactate and alanine by the liver merely replaces that which was converted to lactate and alanine by peripheral tissues. In effect, these cycles transfer energy from fatty acid oxidation in the liver to peripheral tissues that cannot oxidize triacylglycerol. The brain oxidizes glucose completely to CO_2 and water. Hence net glucose synthesis from some other source of carbon is mandatory in fasting. Fatty acids cannot be used for the synthesis of glucose because no pathway exists for the conversion of acetyl CoA, the two-carbon compound produced by fatty acid oxidation, into glucose. Glycerol, a by-product of lipolysis in adipose tissue, is an important substrate for glucose synthesis. However, protein, especially from skeletal muscle, supplies most of the carbon needed for net glucose synthesis. Proteins are hydrolyzed within muscle cells, and most amino acids are partially metabolized. Of the amino acids, alanine and glutamine are released in the largest amounts. The other amino acids are for the most part metabolized to intermediates (pyruvate and α-ketoglutarate), which can yield alanine and glutamine. Branched-chain amino acids are a major source of nitrogen for the production of alanine and glutamine in muscle. Branched-chain α-keto acids produced by transamination are partially released into the blood for uptake by the liver, which produces glucose from the α-keto acid of valine, ketone bodies from that of leucine, and both glucose and ketone bodies from that of isoleucine.

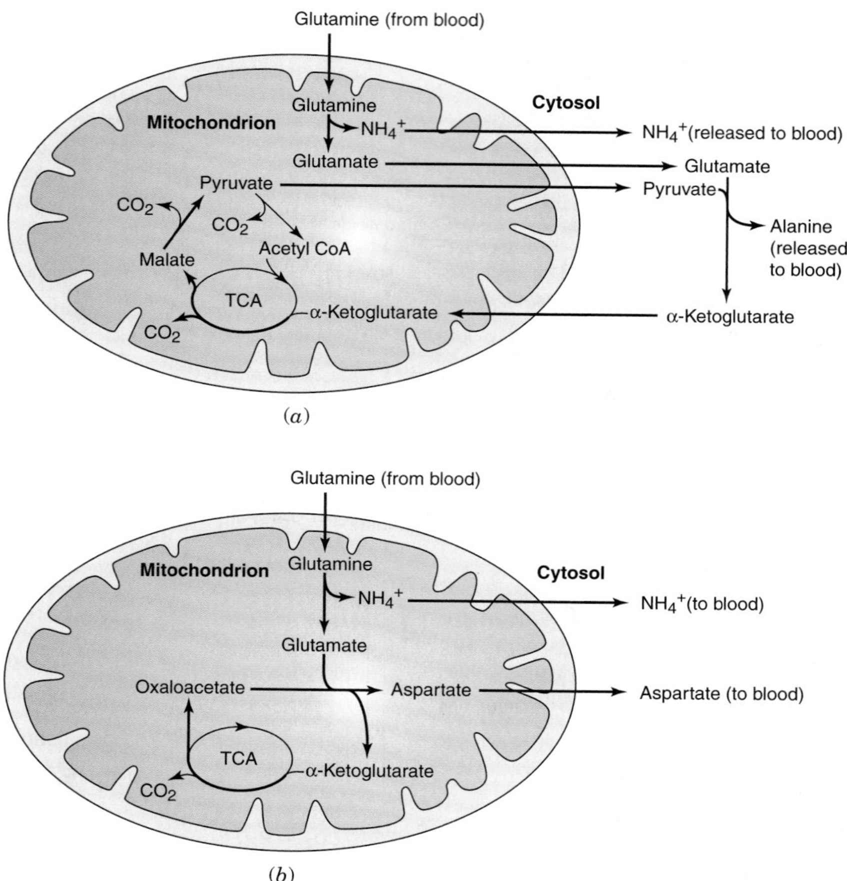

Figure 21.5 Glutamine catabolism by rapidly dividing cells. (*a*) Enterocytes. (*b*) Lymphocytes. Redrawn from Duée, P.-H., Darcy-Vrillon, B., Blachier, F., and Morel, M.-T. Fuel selection in intestinal cells. *Proc. Nutr. Soc.* 54:83, 1995.

Part of the glutamine released from muscle is used by intestinal epithelium, lymphocytes, and macrophages. Glutamine is an important fuel for enterocytes and lymphocytes, which divide rapidly and thus require it for synthesis of pyrimidines and purines. In this process, glutamine is converted to glutamate, which is transaminated with pyruvate to form α-ketoglutarate and alanine. In the TCA cycle, the α-ketoglutarate is converted to malate, which is then converted to pyruvate by malic enzyme, generating the pyruvate needed for alanine formation by enterocytes. This pathway is called glutaminolysis because glutamine is only partially oxidized (Figure 21.5a). Glutaminolysis, with aspartate as the major end product, is also used by lymphocytes and macrophages to meet a large portion of their energy needs (Figure 21.5*b*).

Synthesis of glucose in the liver during fasting is closely linked to synthesis of urea. Most amino acids can give up their amino nitrogen by transamination with α-ketoglutarate, forming glutamate and a new α-keto acid, which often can be utilized for glucose synthesis. Glutamate provides the two forms of nitrogen required for urea synthesis: ammonia from oxidative deamination by glutamate dehydrogenase, and aspartate from transamination of oxaloacetate by aspartate aminotransferase. An important source of ammonia and precursors of ornithine such as citrulline is the gut mucosa (described in Section 21.4, p. 863).

Because of low blood insulin levels during fasting, lipolysis is greatly activated in adipose tissue, raising the blood level of fatty acids, which are used in preference to glucose by many tissues. In heart and muscle, oxidation of fatty acids inhibits glycolysis and pyruvate oxidation. In liver, fatty acid oxidation provides most of the ATP needed for gluconeogenesis. Very little acetyl CoA generated by fatty acid oxidation in liver is oxidized completely. Rather, it is converted into ketone bodies by liver mitochondria. Ketone bodies (acetoacetate and β-hydroxybutyrate) are released into the blood and are a source of energy for many tissues. Like fatty acids, they are preferred to glucose by many tissues. Fatty acids are not oxidized by the brain as they cross the blood–brain barrier poorly. Once their blood concentration is high enough, ketone bodies enter the brain and serve as an alternative fuel. They are unable, however, to replace completely the brain's need for glucose. Ketone bodies may also suppress proteolysis and branched-chain amino acid oxidation in muscle and decrease alanine release. This decreases muscle wasting and reduces the amount of glucose synthesized in liver. As long as high ketone body levels are maintained by hepatic fatty acid oxidation, there is less need for glucose, less need for gluconeogenic amino acids, and less need for breaking down precious muscle tissue.

This may be because insulin levels remain high enough to suppress partially muscle proteolysis as long as glucose levels remain high enough to stimulate some release of insulin from the pancreas. The working relationship between the liver, muscle, and adipose tissue in providing glucose for the brain are shown in Figure 21.4, p. 845. Liver synthesizes the glucose, muscle and gut supply the substrate (alanine), and adipose tissue supplies the ATP (via fatty acid oxidation in the liver) needed for hepatic gluconeogenesis. This cooperation among major tissues is dependent on the appropriate blood hormone levels. Glucose levels are lower in fasting, reducing the secretion of insulin but favoring release of glucagon from the pancreas and epinephrine from the adrenal medulla. Fasting also reduces formation of triiodothyronine, the active form of thyroid hormone, from thyroxine. This reduces the daily basal energy requirements by up to 25%. This response is useful for survival, but makes weight loss more difficult than weight gain (see Clin. Corr. 21.1).

In the Early Refed State Glycogen Is Formed by the Indirect Pathway

Triacylglycerol is metabolized as described for the well-fed state. In contrast, the liver extracts glucose poorly and, in fact, remains in the gluconeogenic mode for a few hours after feeding. Rather than providing blood glucose, however, hepatic gluconeogenesis provides glucose 6-phosphate for glycogenesis. This means that liver glycogen is not entirely repleted after a fast by direct synthesis from blood glucose. Rather, glucose is catabolized in peripheral tissues to lactate, which is converted in the liver to glycogen by gluconeogenesis, that is, indirectly:

$$\begin{array}{c} \text{Glycogen} \\ \uparrow \\ \text{glucose} \xrightarrow{\textit{direct}} \text{glucose 6-phosphate} \xleftarrow{\textit{indirect}} \text{lactate} \end{array}$$

Gluconeogenesis from amino acids entering from the gut also plays an important role in reestablishing liver glycogen levels by the indirect pathway. After the rate of glucose synthesis declines, liver glycogen is sustained by direct synthesis from blood glucose.

Important Interorgan Metabolic Interactions

The intestinal epithelium converts glutamine to citrulline (Figure 21.6); it is the only tissue that expresses an ATP-dependent glutamate reductase necessary for this conversion.

$$\text{Glutamate} + \text{NADPH} + \text{H}^+ + \text{ATP} \rightarrow \text{glutamate semialdehyde} + \text{NADP}^+ + \text{ADP} + \text{Pi}$$

The gut also converts dietary arginine to citrulline. Citrulline released from the gut passes through the liver and is converted by the kidney to arginine, which can be converted to creatine or released into the blood. This pathway can be viewed as a way to reduce the release of arginine and ammonia in the portal blood, both of which stimulate ureagenesis. This may be particularly important under conditions of low protein intake. In addition, citrulline is a better source of arginine in conditions of small bowel disease (such as small bowel resection or celiac disease), but may not be utilized normally in renal failure. The liver uses arginine to generate ornithine, which expands the capacity of the urea cycle during periods of increased protein intake. The liver irreversibly converts ornithine into glutamate:

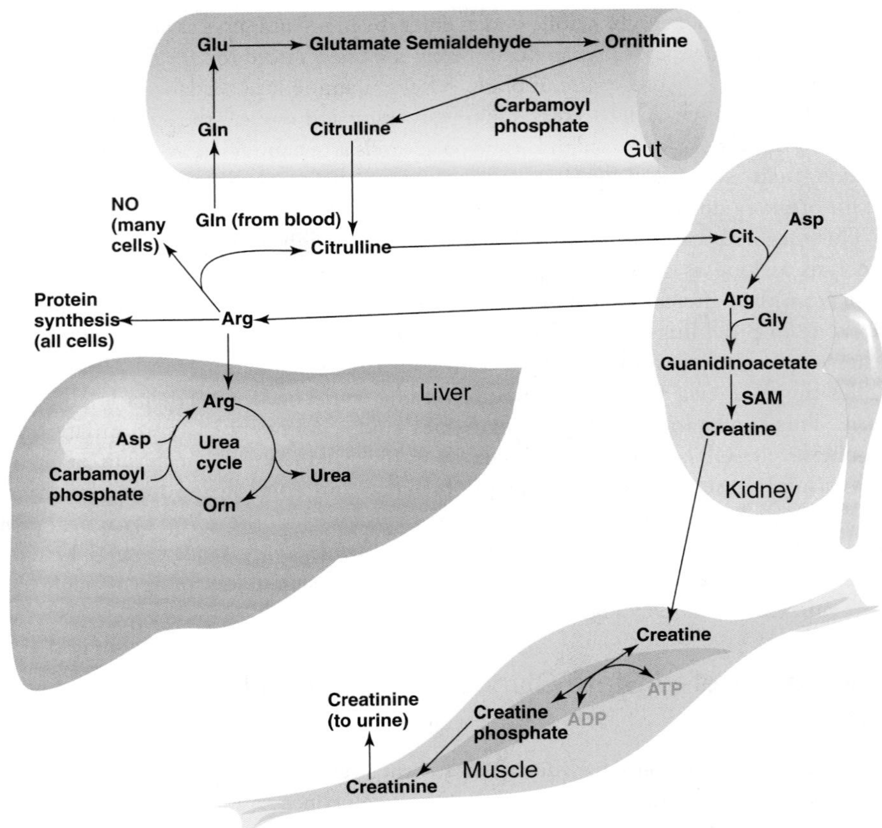

Figure 21.6 Gut and kidney function together in synthesis of arginine from glutamine. Abbreviations: Cit, citrulline; Arg, arginine; Asp, aspartate; Gln, glutamine; Glu, glutamate; NO, nitric oxide; Gly, glycine; Orn, ornithine; and SAM, *S*-adenosylmethionine.

$$\text{Ornithine} \xrightarrow{\text{Transamination}} \text{Glutamate semialdehyde} \xrightarrow{\text{Oxidation}} \text{Glutamate}$$

Depletion of ornithine in this way inhibits urea synthesis. Replenishment of ornithine is completely dependent on a source of arginine. Thus, urea synthesis in the liver is dependent on citrulline and arginine produced by the gut and kidney, respectively. Arginine is also used by many cells for the production of nitric oxide (NO) (Figure 21.6; see p. 848). The arginine generated from citrulline in the kidney is used to synthesize creatine (Figure 21.6). Glycine transamidinase (GTA) generates guanidinoacetate from arginine and glycine (p. 789). GTA is found predominantly in renal cortex, pancreas, and liver. After methylation using the methyl donor *S*-adenosylmethionine (SAM), creatine is formed. This is quantitatively the most important use of SAM in the body. One to two grams of creatine are synthesized per day. Creatine then circulates to other tissues, and accumulates in muscle, where it serves as a high-energy reservoir when phosphorylated to creatine phosphate. This undergoes nonenzymatic conversion to creatinine, which is released to the bloodstream and removed by the kidney. Excretion of creatinine is used clinically both as a measure of muscle mass and of renal function.

Glutathione (GSH) is important in detoxification of endogenously generated peroxides and exogenous chemical compounds (p. 789). Liver is a major synthesizer of GSH from glutamate, cysteine, and glycine (see Figure 19.77, p. 790). Synthesis is limited by the availability of cysteine. Plasma cysteine is not taken up well by liver, which uses dietary methionine to form cysteine via the cystathionine pathway (p. 770). Hepatic GSH is released to the bloodstream and bile. Kidney removes a substantial amount of plasma GSH. Enterocytes may be able to take up biliary-excreted GSH from the intestinal lumen. Release to plasma is the same in fed and fasting states, providing a stable source of this compound and its constituent amino acids, especially cysteine, for most tissues of the body.

Carnitine is derived from lysyl residues of various proteins, which are *N*-methylated using SAM to form trimethyllysyl-residues (Figure 21.7 and p. 849). Trimethyllysine is released when the proteins are degraded. It is hydroxylated and then cleaved, releasing glycine and γ-butyrobetaine aldehyde. The latter is oxidized to γ-butyrobetaine and then

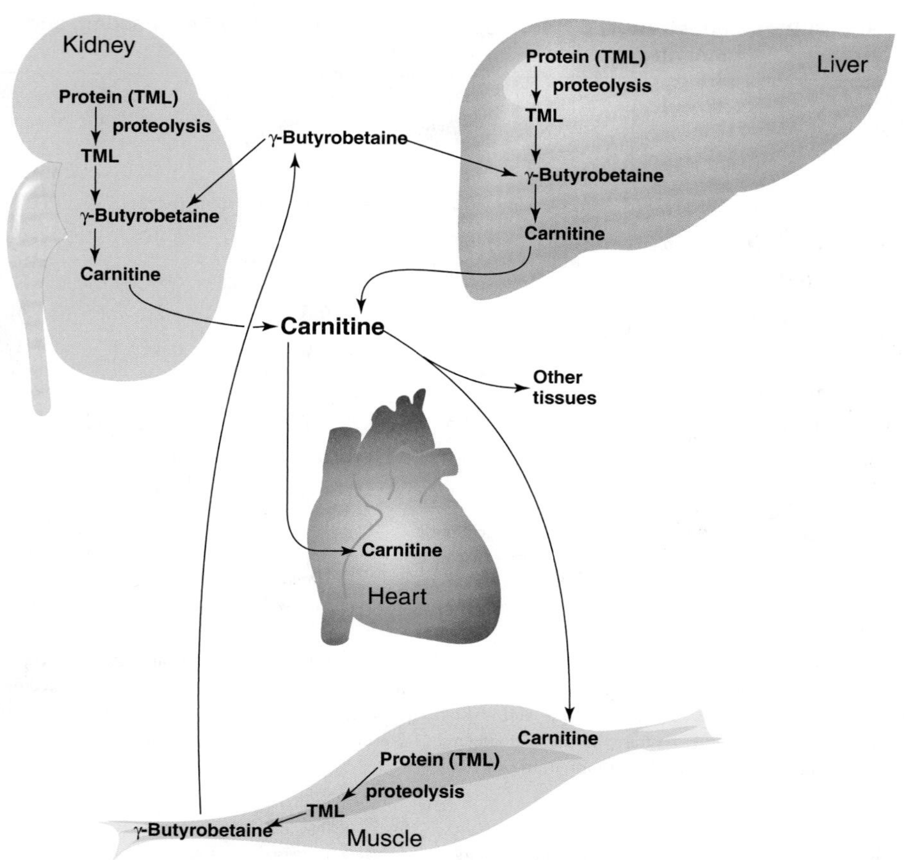

Figure 21.7 Kidney and liver provide carnitine for other tissues. Abbreviations: (TML), trimethyllysyl residues in protein molecules; TML, free trimethyllysine.

hydroxylated to form carnitine. Kidney and to a lesser extent liver carry out the complete pathway and supply other tissues, especially muscle and heart, with carnitine. Skeletal muscle can form γ-butyrobetaine, but must release it for conversion to carnitine by liver or kidney.

Energy Requirements, Reserves, and Caloric Homeostasis

The average person leading a sedentary life consumes 180–280 g of carbohydrate, 70–100 g of protein, and 70–100 g of fat daily. This meets a daily energy requirement of 1600–2400 kcal. As shown in Table 21.1, the energy reserves of an average-sized person are considerable. They are called on between small meals and overnight to maintain blood glucose. Although glycogen can be mobilized rapidly, glycogen reserves are small relative to fat reserves (Table 21.1). Fat stores are called on only during more prolonged fasting. Those of obese subjects can amount to 80 kg, adding another 585,000 kcal to their energy reserves. Protein is listed in Table 21.1 as a reserve of energy because it can be used to provide amino acids for oxidation. Recall that protein is not dispensable like triacylglycerol and glycogen and is utilized by the body for energy more reluctantly. The constant availability of fuels in the blood is termed caloric homeostasis, which, as illustrated in Table 21.2, means that regardless of whether a person is well-fed, fasting, or starving to death, the blood contains fuels in amounts that on combustion in cells of the body, provide a comparable amount of ATP. Note that blood glucose concentrations are controlled within very tight limits, whereas fatty acid and ketone body concentrations in the blood can vary by one or two orders of magnitude, respectively. Glucose is carefully regulated because of the absolute need of the brain for this substrate. If the blood glucose level falls too low (< 1.5 mM), coma and death will follow shortly unless the glucose concentration is restored. On the other hand, hyperglycemia must be avoided because glucose will be lost in the urine, resulting in dehydration and sometimes hyperosmolar, hyperglycemic coma (Clin. Corr. 21.4). Chronic hyperglycemia results in glycation of many proteins and endothelial dysfunction, which contributes to the complications of diabetes (Clin. Corr. 21.5). The changes in insulin/glucagon ratio shown in Table 21.2 are crucial for the maintenance of caloric homeostasis. Simply stated,

TABLE 21.1 • Energy Reserves of Humans[a]

		Fuel Reserves	
Stored Fuel	*Tissue*	*(g)*	*(kcal)*
Glycogen	Liver	70	280
Glycogen	Muscle	120	480
Glucose	Body fluids	20	80
Fat	Adipose	15,000	135,000
Protein	Muscle	6,000	24,000

[a]Data are for a normal subject weighing 70 kg. Carbohydrate supplies 4 kcal/g; fat, 9 kcal/g; protein, 4 kal/g.

TABLE 21.2 • Substrate and Hormone Levels in Blood of Well-Fed, Fasting, and Starving Humans[a]

Hormone or Substrate (units)	*Very Well Fed*	*Postabsorptive 12 hours*	*Fasted 3 days*	*Starved 5 weeks*
Insulin (μU/mL)	40	15	8	6
Glucagon (pg/mL)	80	100	150	120
Insulin/glucagon ratio (μU/pg)	0.50	0.15	0.05	0.05
Glucose (mM)	6.1	4.8	3.8	3.6
Fatty acids (mM)	0.14	0.6	1.2	1.4
Acetoacetate (mM)	0.04	0.05	0.4	1.3
β-Hydroxybutyrate (mM)	0.03	0.10	1.4	6.0
Lactate (mM)	2.5	0.7	0.7	0.6
Pyruvate (mM)	0.25	0.06	0.04	0.03
Alanine (mM)	0.8	0.3	0.3	0.1
ATP equivalents (mM)	262	235	301	428

Source: From Ruderman, N. B., Aoki, T. T., and Cahill, G. F. Jr. Gluconeogenesis and its disorders in man. In: R. W. Hanson and M. A. Mehlman (Eds.), *Gluconeogenesis, Its Regulation in Mammalian Species.* New York: Wiley, 1976, p. 515.

[a]Data are for normal-weight subjects except for the 5-week starvation values, which are from obese subjects undergoing therapeutic starvation. ATP equivalents were calculated on the basis of the ATP yield expected on complete oxidation of each substrate to CO_2 and H_2O: 32 molecules of ATP for each molecule of glucose; 106 for the average fatty acid (palmitate); 19 for acetoacetate; 21.5 for β-hydroxybutyrate; 15 for lactate; 12.5 for pyruvate; and 13 (corrected for urea formation) for alanine.

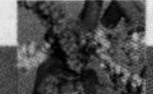
CLINICAL CORRELATION 21.4

Hyperglycemic, Hyperosmolar Coma

Patients with type 2 diabetes (see Clin. Corrs. 15.4, p. 594, and 21.6, p. 848) sometimes develop a condition called hyperglycemic, hyperosmolar coma. This is particularly common in the elderly and may occur in individuals under metabolic stress who were not previously recognized as having diabetes. Hyperglycemia, perhaps worsened by failure to take insulin or hypoglycemic drugs, an infection, or a coincidental medical problem such as a heart attack, leads to urinary losses of water, glucose, and electrolytes (sodium, chloride, and potassium). This osmotic diuresis reduces the circulating blood volume, a physiological stress that releases hormones that worsen insulin resistance and hyperglycemia. In addition, elderly patients may be less able to sense thirst or to obtain fluids. Over the course of several days, they can become extremely hyperglycemic (glucose >1000 mg/dL), dehydrated, and comatose. Ketoacidosis does not develop in these patients, possibly because free fatty acids are not always elevated or because adequate insulin concentrations exist in the portal blood to inhibit ketogenesis (although not high enough to inhibit gluconeogenesis). Therapy is aimed at restoring water and electrolyte balance and correcting the hyperglycemia with insulin, and identifying the underlying cause. Mortality from this syndrome is considerably higher than that of diabetic ketoacidosis.

Arieff, A. I., and Carroll, H. J. Nonketotic hyperosmolar coma with hyperglycemia. Clinical features, pathophysiology, renal function, acid-base balance, plasma-cerebrospinal fluid equilibria, and the effects of therapy in 37 cases. *Medicine* 51:73, 1972; Gaglia, J. L., Wyckoff, J., and Abrahamson, M. J. Acute hyperglycemia crisis in the elderly. *Med. Clin. North Am.* 88:1063, 2004; and Trence, D. L., and Hirsch, I. B. Hyperglycemic crises in diabetes mellitus type 2. *Endocrinol. Metab. Clin. North Am.* 30:817, 2001.

well-fed individuals have high insulin/glucagon ratios that favor storage of glycogen and triacylglycerol, while starving individuals have low insulin/glucagon ratios that stimulate glycogenolysis, lipolysis, ketogenesis, proteolysis, and gluconeogenesis.

The Five Phases of Glucose Homeostasis

Figure 21.8 summarizes the work of Cahill and his colleagues on obese patients undergoing long-term starvation for weight loss. It illustrates the effects of starvation on glucose homeostasis, and it is divided arbitrarily into five phases. Phase I is the well-fed state, in which glucose is provided by dietary carbohydrate. Once this supply is exhausted, hepatic

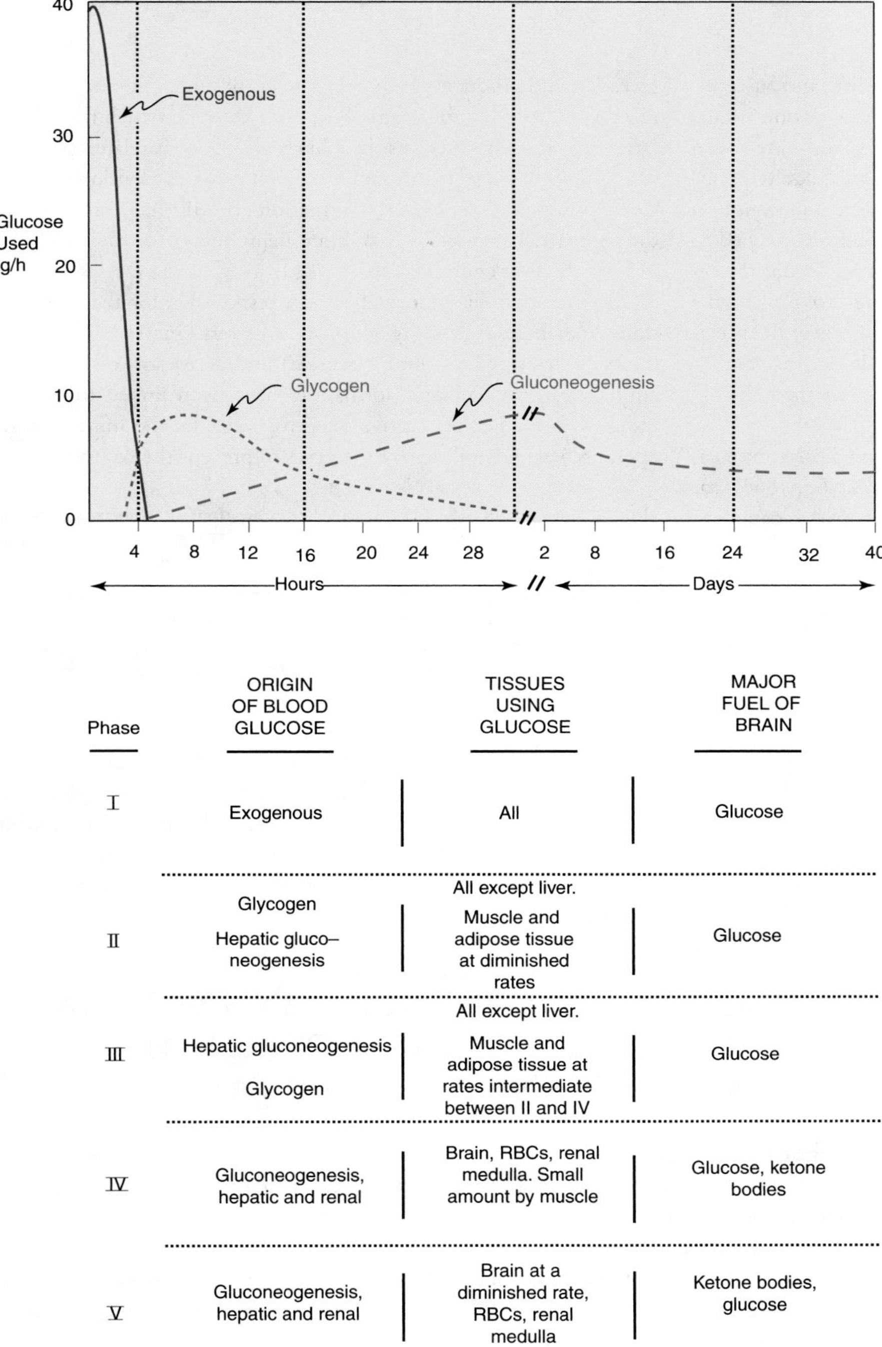

Phase	ORIGIN OF BLOOD GLUCOSE	TISSUES USING GLUCOSE	MAJOR FUEL OF BRAIN
I	Exogenous	All	Glucose
II	Glycogen Hepatic gluco–neogenesis	All except liver. Muscle and adipose tissue at diminished rates	Glucose
III	Hepatic gluconeogenesis Glycogen	All except liver. Muscle and adipose tissue at rates intermediate between II and IV	Glucose
IV	Gluconeogenesis, hepatic and renal	Brain, RBCs, renal medulla. Small amount by muscle	Glucose, ketone bodies
V	Gluconeogenesis, hepatic and renal	Brain at a diminished rate, RBCs, renal medulla	Ketone bodies, glucose

Figure 21.8 The five phases of glucose homeostasis.
Reproduced with permission from Ruderman, N. B., Aoki, T. T., and Cahill, G. F., Jr. Gluconeogenesis and its disorders in man, in R. W. Hanson, and M. A. Mehlman (Eds.), *Gluconeogenesis, Its Regulation in Mammalian Species*. New York: Wiley, 1976, 515.

glycogenolysis maintains blood glucose levels during phase II. As this supply of glucose starts to dwindle, hepatic gluconeogenesis from lactate, glycerol, and alanine becomes increasingly important until, in phase III, gluconeogenesis is the major source of blood glucose. These changes occur within 20 h or so of fasting, depending on how well-fed the individual was prior to the fast, how much hepatic glycogen was present, and the sort of physical activity occurring during the fast. Several days of fasting lead to phase IV, when dependence on gluconeogenesis decreases. By now, ketone bodies have accumulated to high enough concentrations for them to enter the brain and meet some of its energy needs. Renal gluconeogenesis also becomes significant in this phase. Phase V occurs after very prolonged starvation of extremely obese individuals and is characterized by even less dependence on gluconeogenesis. In this phase, the energy needs of almost every tissue are largely met by fatty acid or ketone body oxidation.

CLINICAL CORRELATION 21.5

Hyperglycemia and Protein Glycation

Glycation of enzymes changes their activity, solubility, and susceptibility to degradation. In the case of hemoglobin A, glycation occurs by a nonenzymatic reaction between glucose and the amino-terminal valine of the β-chain. A Schiff base forms between the aldehyde carbon of glucose and the free amino group of this valine, followed by a rearrangement of the molecule to give a 1-deoxyfructose attached to the valine. The reaction is favored by high glucose levels and the resulting protein, called hemoglobin A_{1c}, is a good index of how high a person's average blood glucose concentration has been over the previous several weeks. The concentration of hemoglobin A_{1c} increases in uncontrolled diabetes and is low in patients who control their glucose level closely.

Glycation of proteins may contribute to the medical complications of diabetes (see Clin. Corrs. 15.4, p. 594, and 21.6, p. 866), for example, coronary heart disease, retinopathy, nephropathy, cataracts, and neuropathy. Increased glycation of lens proteins may contribute to the development of diabetic cataracts. Collagen, laminin, vitronectin, and other matrix proteins can become glycated, with alterations in self-assembly and binding of other matrix molecules. Advanced glycation end products (AGEs) present in the circulation have specific receptors (receptors for AGEs [RAGEs]), through which they stimulate inflammatory reactions. It is likely that RAGE-related phenomena underlie the medical complications of diabetes mentioned above. Compounds which inhibit the formation of AGEs (e.g., aminoguanidine) or which block the RAGE are being tested for their ability to prevent diabetic complications.

Other metabolic abnormalities are responsible for the complications of diabetes, including activation of stress kinases (via increased levels of diacylglycerol and ceramide) and activation of the hexosamine pathway. These abnormalities have been linked to glucose-induced formation of reactive oxygen species by the mitochondria, opening new possibilities for therapies to prevent the complications.

He, Z., and King, G. L. Microvascular complications of diabetes. *Endocrinol Metab. Clin. N. Am.* 33:215, 2004; Brownlee, M., The pathobiology of diabetic complications. A unifying mechanism. *Diabetes* 54:1615, 2005; and Yan, S. F., Ramasamy, R., and Schmidt, A. M. Mechanisms of disease: Advanced glycation end products and their receptor in inflammation and diabetes complications. *Nat. Clin. Pract. Endocrinol. Metab.* 4:285, 2008.

As long as ketone body concentrations are high and glucose levels are maintained, proteolysis will be somewhat restricted, perhaps by small amounts of insulin still produced by the pancreas, and conservation of muscle proteins and enzymes will occur. This continues until practically all of the fat is used up and ketone body levels fall. After all of the fat is gone, the body has to use muscle protein to maintain blood glucose. Before it's gone, you're gone (see Clin. Corr. 21.3, p. 842).

21.3 • MECHANISMS INVOLVED IN SWITCHING LIVER METABOLISM BETWEEN THE WELL-FED AND STARVED STATES

The liver of a well-fed person actively synthesizes glycogen and triacylglycerol; such a liver is glycogenic, glycolytic, and lipogenic. In contrast, the liver of the fasting person is glycogenolytic, gluconeogenic, ketogenic, and proteolytic. The strategy employed is to store calories when food is available and mobilize them when the rest of the body is in need. The liver is moved between these metabolic extremes by a variety of regulatory mechanisms: substrate supply, allosteric effectors, covalent modification, and **induction–repression** of enzymes.

Substrate Availability Controls Many Metabolic Pathways

This mechanism of control is often ignored. However, the concentration of fatty acids in blood entering the liver is a major determinant of the rate of ketogenesis. Glucose synthesis by the liver is affected by the rate at which gluconeogenic substrates flow to the liver. Delivery of amino acids to the liver in diabetes, because of accelerated and uncontrolled proteolysis, stimulates gluconeogenesis and exacerbates hyperglycemia. On the other hand, failure to supply the liver adequately with glucogenic substrate explains some types of **hypoglycemia,** such as that observed during pregnancy or advanced starvation. Urea synthesis is also regulated by substrate supply. Amino acid metabolism in the intestine provides a substantial fraction of the ammonia used by the liver for urea production. The intestine releases citrulline as discussed, the metabolic precursor of ornithine. A larger ornithine pool permits increased urea synthesis after a high protein meal. In protein deficiency, the rate of urea formation declines.

We can conclude that substrate supply is a major determinant of the rate at which virtually every metabolic process of the body operates. However, variations in substrate supply are not sufficient to account for the marked changes in metabolism that must occur in the starve–feed cycle. Finer-tuning of the pathways is required.

Allosteric Effectors Regulate Key Enzymes

Figures 21.9 and 21.10 summarize the effects of allosteric effectors in the liver in well-fed and starved states, respectively. As shown in Figure 21.9, glucose activates glucokinase (indirectly by promoting its translocation from the nucleus to the cytoplasm [p. 606]), thereby promoting phosphorylation of glucose.

Glucose also inactivates glycogen phosphorylase and activates glycogen synthase indirectly, thereby preventing degradation and promoting synthesis of glycogen. Fructose 2,6-bisphosphate stimulates 6-phosphofructo-1-kinase and inhibits fructose 1,6-bisphosphatase, thereby stimulating glycolysis and inhibiting gluconeogenesis. Fructose 1,6-bisphosphate activates pyruvate kinase, thereby stimulating glycolysis, and pyruvate activates the pyruvate dehydrogenase complex (indirectly by inhibition of pyruvate dehydrogenase kinase (p. 551)). Citrate activates **acetyl-CoA carboxylase**, thereby stimulating fatty acid synthesis and malonyl CoA inhibits **carnitine palmitoyltransferase I**, thereby inhibiting fatty acid oxidation.

As shown in Figure 21.10, acetyl CoA stimulates gluconeogenesis in the fasted state by activating pyruvate carboxylase and inhibiting the pyruvate dehydrogenase complex (indirectly through stimulation of pyruvate dehydrogenase kinase (p. 551)). Long-chain acyl CoA

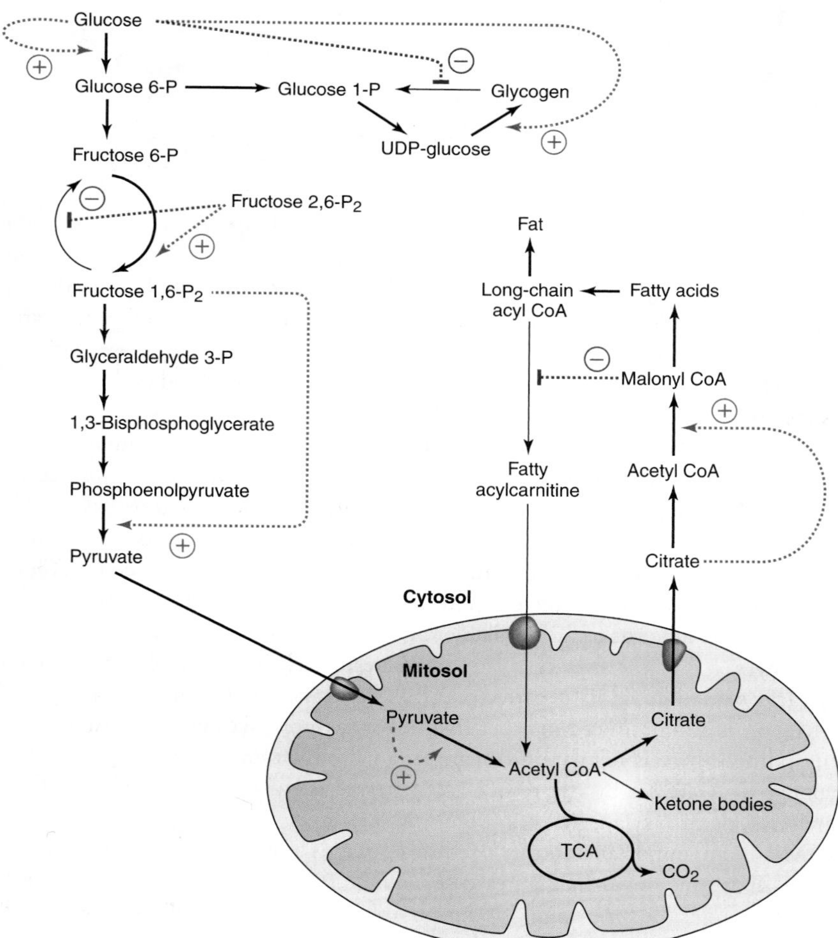

Figure 21.9 **Control of hepatic metabolism by allosteric effectors in the well-fed state.**

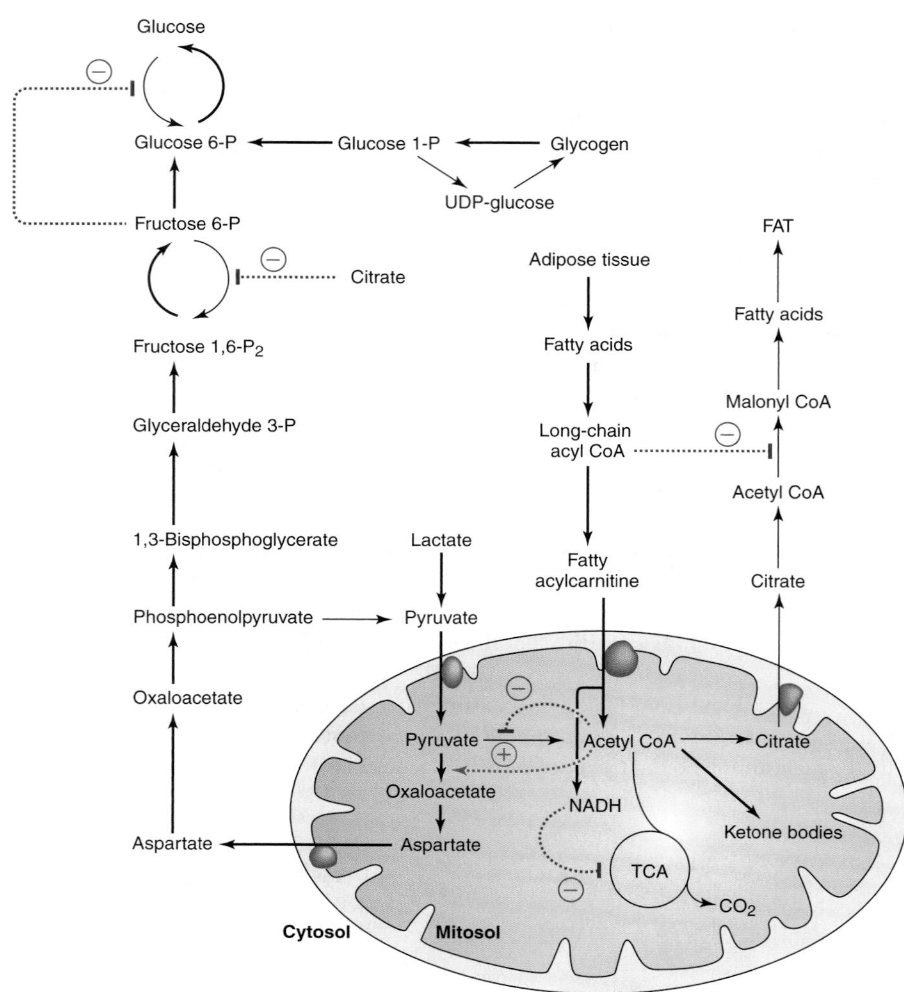

Figure 21.10 **Control of hepatic metabolism by allosteric effectors in the fasting state.**

esters inhibit acetyl-CoA carboxylase, which lowers the level of malonyl CoA and increases carnitine palmitoyltransferase I activity and fatty acid oxidation. Fructose 6-phosphate inhibits glucokinase (indirectly by promoting its translocation from the cytoplasm to the nucleus [p. 606]). Citrate, which is increased in concentration as a consequence of greater fatty acid oxidation, inhibits 6-phosphofructo-1-kinase and 6-phosphofructo-2-kinase (not shown), and NADH produced by fatty acid oxidation inhibits tricarboxylic acid cycle.

Allosteric effectors also regulate flux through metabolic pathways in nonhepatic tissues. Citrate for example serves as a sensor of excess fuel availability in a number of tissues. As negative effector for 6-phosphofructo-1-kinase and a positive effector for acetyl-CoA carboxylase, citrate regulates flux through both glycolysis and fatty acid oxidation. The latter effect is indirect, involving citrate activation of acetyl-CoA carboxylase, which increases the level of malonyl CoA, a negative effector of carnitine palmitoyltransferase I. Since glucose and fatty acid catabolism increase citrate levels and both can be inhibited by citrate, cells can sense the amount of fuel available for catabolism by their citrate level.

Malonyl CoA in the liver is an intermediate in fatty acid synthesis and a regulator of fatty acid oxidation via its negative effect on carnitine palmitoyltransferase I (p. 691). Malonyl CoA is also produced in other tissues, for example, skeletal muscle and heart, but its only purpose here is regulation of carnitine palmitoyltransferase I. Steady-state levels of malonyl CoA are set by the relative activities of acetyl-CoA carboxylase (ACC) and malonyl-CoA decarboxylase (MDC) (Figure 21.11).

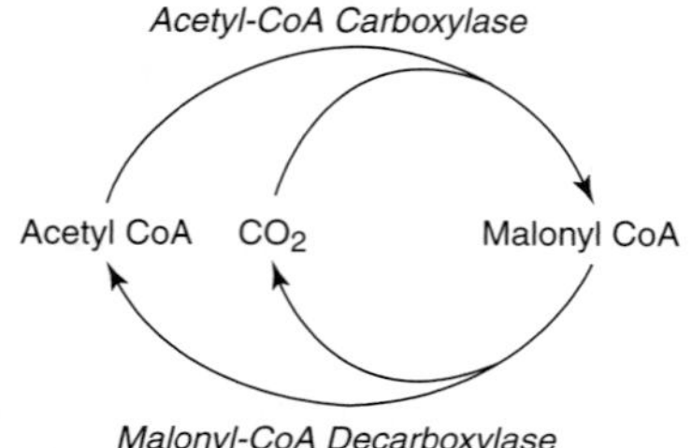

Figure 21.11 **Relative activities of acetyl-CoA carboxylase and malonyl-CoA decarboxylase determine the concentration of malonyl CoA.**

cAMP is an important allosteric effector that is greatly increased in liver during starvation. It is not shown in Figures 21.9 and 21.10 because it is part of the signaling mechanism responsible for bringing about phosphorylation of enzymes subject to covalent modification.

AMP is another important allosteric effector not shown in Figures 21.9 or 21.10. Its concentration is maintained very low in both the fed and the starved state. The high ATP concentration usually present in cells keeps AMP low by driving the reaction catalyzed by adenylate kinase to the right by mass action: ATP + AMP $\rightleftharpoons$ 2ADP. However, conditions that induce an energy deficit in cells, for example, hypoxia, excessive demand for energy, or muscle contraction during exercise, cause the reaction catalyzed by adenylate kinase to shift to the left as ATP decreases. The resulting increase in AMP causes activation of glycogen phosphorylase and 6-phosphofructo-1-kinase and inhibition of fructose-1,6-bisphosphatase by allosteric mechanisms (p. 413), thereby increasing glycogenolysis and glycolysis which combine to increase the rate of ATP production. The increase in AMP also activates **AMP-activated protein kinase (AMPK)**, which also promotes the recovery of normal ATP levels by the phosphorylation of a number of regulatory enzymes.

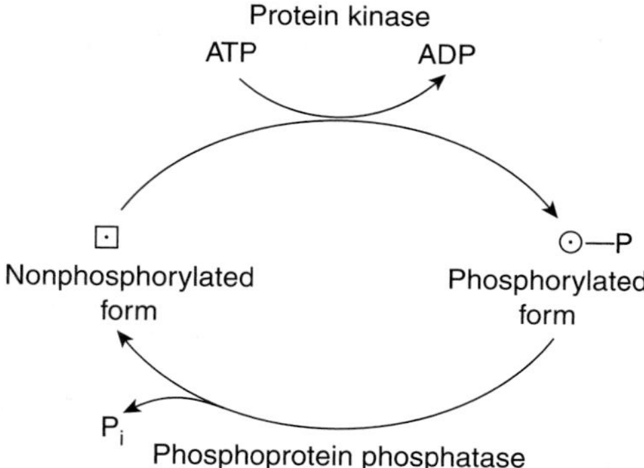

Figure 21.12 Regulation of the activity of key enzymes by covalent modification. The symbols ⊡ and ⊙-P designate the nonphosphorylated and phosphorylated states of enzymes, respectively.

Covalent Modification Regulates Key Enzymes

The activities of many enzymes are modified by covalent modification, especially by phosphorylation of serine and threonine residues (Figure 21.12 and p. 413). Some important points about regulation by this type of control are (1) some enzymes undergo phosphorylation on one or more serine or threonine residues by protein kinases that are also subject to regulation; (2) dephosphorylation of enzymes is carried out by phosphoprotein phosphatases that are also subject to regulation; (3) phosphorylation status affects the catalytic activities of the enzymes; (4) some enzymes are active in the dephosphorylated state, others in the phosphorylated state; (5) cAMP signals the phosphorylation of many enzymes by activating **protein kinase A** (cAMP-dependent protein kinase) (Figure 21.13); (6) glucagon and α-adrenergic agonists (epinephrine) activate protein kinase A by increasing cAMP levels (Figure 21.13); (7) AMP also signals the phosphorylation of many enzymes by activating AMPK (AMP-activated protein kinase) (Figure 21.14); (8) stress (extra work) imposed on a cell that causes energy deprivation results in an increase in AMP concentration and activation of AMPK (Figure 21.14); (9) insulin (p. 615) opposes the action of protein kinase A and AMPK by activating phosphoprotein phosphatases; (10) metabolic enzymes are dephosphorylated in the fed state because the insulin/glucagon ratio is high and both cAMP and AMP are low; (11) metabolic enzymes are phosphorylated in the fasted state because the insulin/glucagon ratio is low and cAMP is increased (see Figure 21.13); and (12) metabolic enzymes are phosphorylated in energy-deprived states because AMP levels are increased (see Figure 21.14).

The hepatic enzymes subject to covalent modification are all relatively dephosphorylated in well-fed animals (Figure 21.15). Insulin is high but glucagon is low in the blood, resulting in low cAMP levels in the liver. A resulting low protein kinase A activity and high phosphoprotein phosphatase activity induce the dephosphorylated state of the enzymes

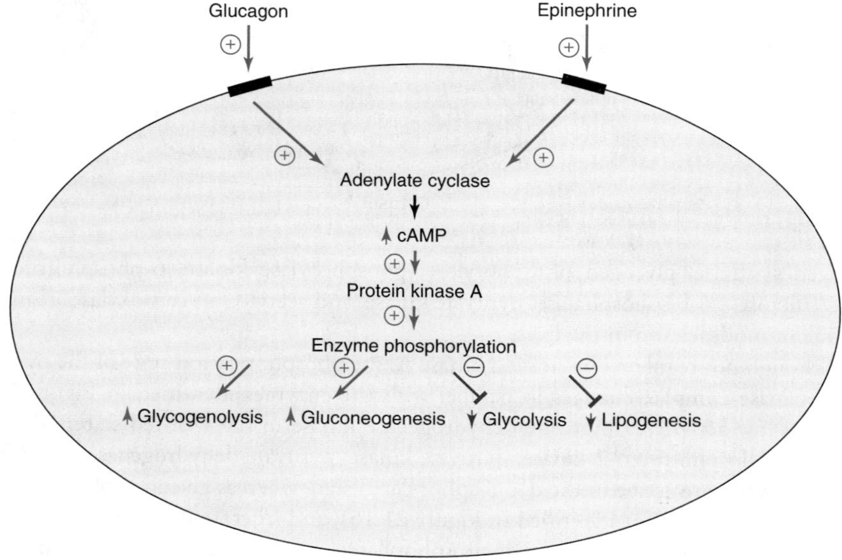

Figure 21.13 Glucagon and epinephrine stimulate glycogenolysis and gluconeogenesis and inhibit glycolysis and lipogenesis in liver.

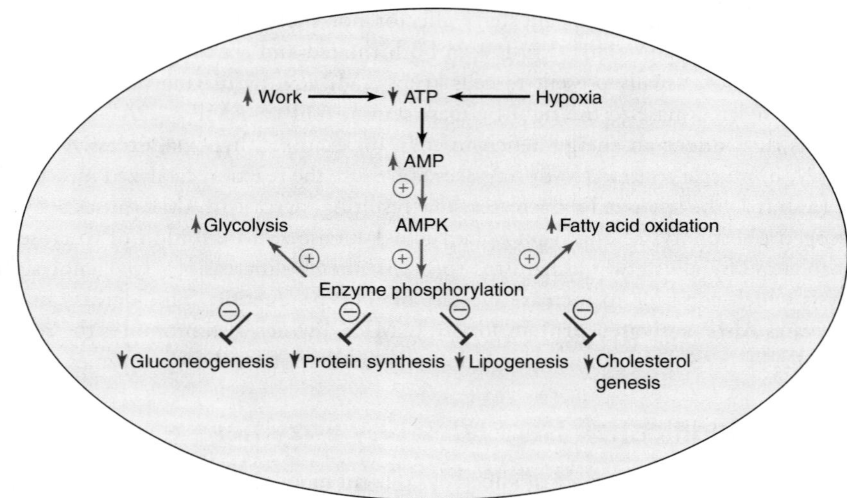

Figure 21.14 Activation of AMPK shuts down ATP-requiring processes and stimulates ATP-producing processes.

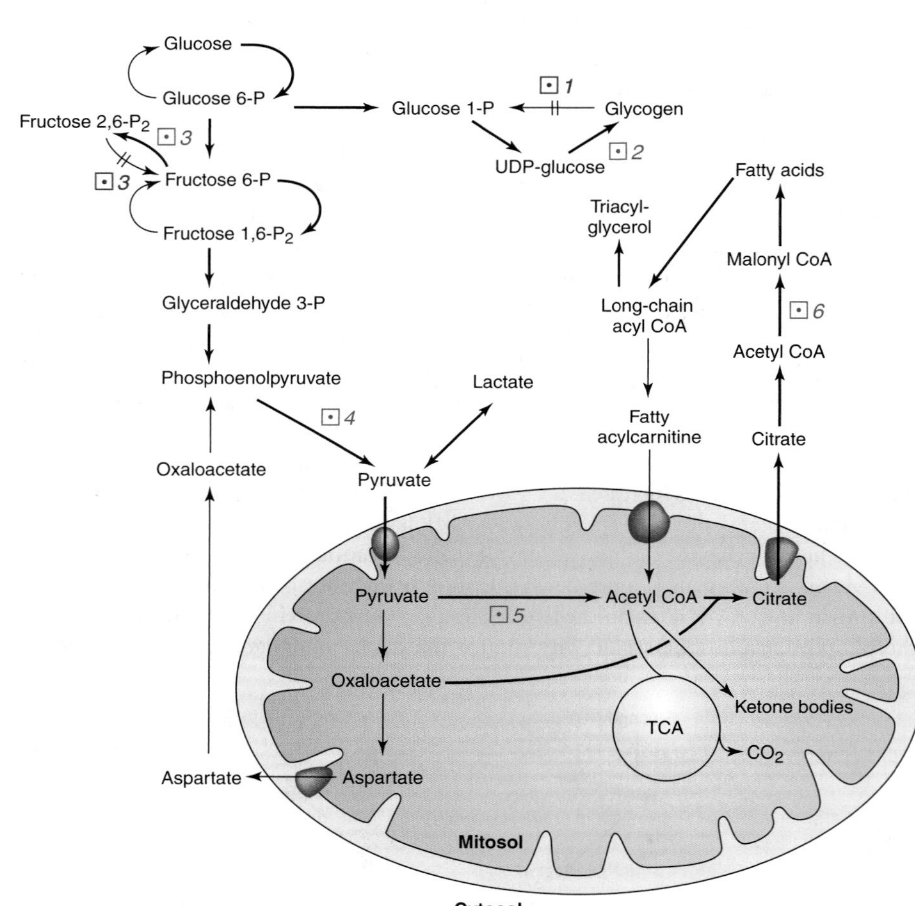

Figure 21.15 Control of hepatic metabolism by covalent modification in the well-fed state. Dephosphorylated mode is indicated by ⊡. The enzymes subject to covalent modification are (1) glycogen phosphorylase, (2) glycogen synthase, (3) 6-phosphofructo-2-kinase/fructose- 2,6-bisphosphatase (bifunctional enzyme), (4) pyruvate kinase, (5) pyruvate dehydrogenase; and (6) acetyl-CoA carboxylase.

(glycogen synthase, glycogen phosphorylase, phosphorylase kinase, 6-phosphofructo-2-kinase/fructose 2,6-bisphosphatase, pyruvate kinase, and acetyl-CoA carboxylase) regulated by covalent modification in the liver.

Although not regulated by protein kinase A, the phosphorylation state of the pyruvate dehydrogenase complex changes in parallel with the enzymes identified in Figure 21.15 because of low activity of pyruvate dehydrogenase kinase in the well-fed state. Glycogen synthase, 6-phosphofructo-2-kinase, pyruvate kinase, pyruvate dehydrogenase, and acetyl-CoA carboxylase are active in the dephosphorylated state whereas glycogen phosphorylase, phosphorylase kinase (not identified in Figure 21.15), and fructose 2,6-bisphosphatase are all inactive. As a consequence of the dephosphorylated state of these enzymes, glycogenesis,

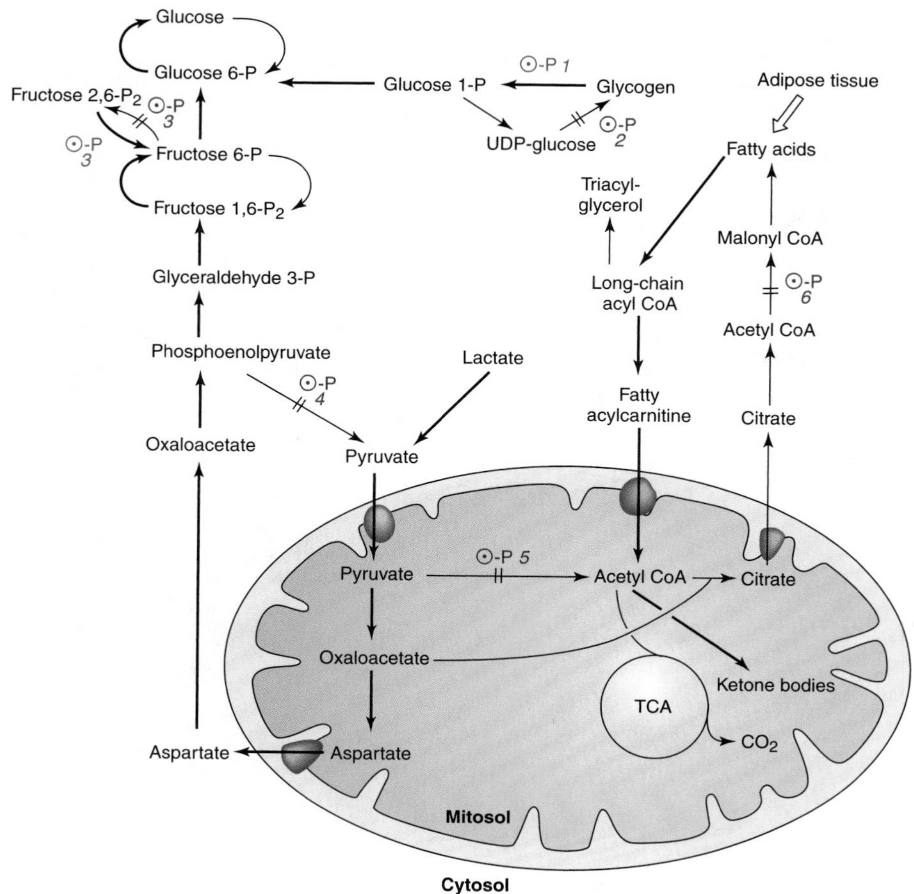

Figure 21.16 Control of hepatic metabolism by covalent modification in the fasting state. Phosphorylated mode is indicated by ⊙ -P. Numbers refer to the same enzymes as in Figure 21.15.

glycolysis, and lipogenesis are greatly favored in the liver of the well-fed animal whereas the opposing pathways (glycogenolysis, gluconeogenesis, and ketogenesis) are inhibited.

As shown in Figure 21.16, the hepatic enzymes subject to covalent modification are all relatively phosphorylated in the fasting animal. Insulin level in the blood is low but glucagon is high, resulting in high cAMP levels in the liver. This activates protein kinase A and inactivates phosphoprotein phosphatase. The net effect is a greater degree of phosphorylation of the regulatory enzymes than in the well-fed state. Three enzymes (glycogen phosphorylase, phosphorylase kinase, and fructose 2,6-bisphosphatase) are activated as a consequence of phosphorylation. All other enzymes subject to covalent modification are inactivated. As a result, glycogenolysis, gluconeogenesis, and ketogenesis dominate and glycogenesis, glycolysis, and lipogenesis are shut down.

Metabolic enzymes are also regulated through phosphorylation by AMP-activated protein kinase (AMPK) as summarized in Figures 21.14, p. 856, and 21.17. AMPK is activated by an increase in AMP concentration, which is set by the energy status of cells. Under conditions of high-energy demand that decreases ATP and therefore increases AMP, AMPK turns off anabolic pathways that consume ATP and turns on catabolic pathways that generate ATP. As shown in Figure 21.17, AMPK inhibits fatty acid synthesis by phosphorylating acetyl-CoA carboxylase, inhibits triacylglycerol synthesis by phosphorylating glycerol-3-phosphate acyltransferase, inhibits cholesterol synthesis by phosphorylating 3-hydroxy-3-methylglutaryl-CoA reductase, and inhibits glycogen synthesis by phosphorylating glycogen synthase. AMPK also inhibits protein synthesis (not shown) by phosphorylating components of the mTOR (mammalian target of rapamycin) pathway that activate mRNA translation. The strategy is to minimize use of ATP by all pathways not immediately essential for cell survival. At the same time, AMPK promotes ATP generation by fatty acid oxidation by lowering the concentration of malonyl CoA, a potent allosteric inhibitor of carnitine palmitoyl-transferase 1 (p. 691). This is achieved by phosphorylation-mediated inactivation of acetyl-CoA carboxylase and activation of malonyl-CoA decarboxylase by AMPK.

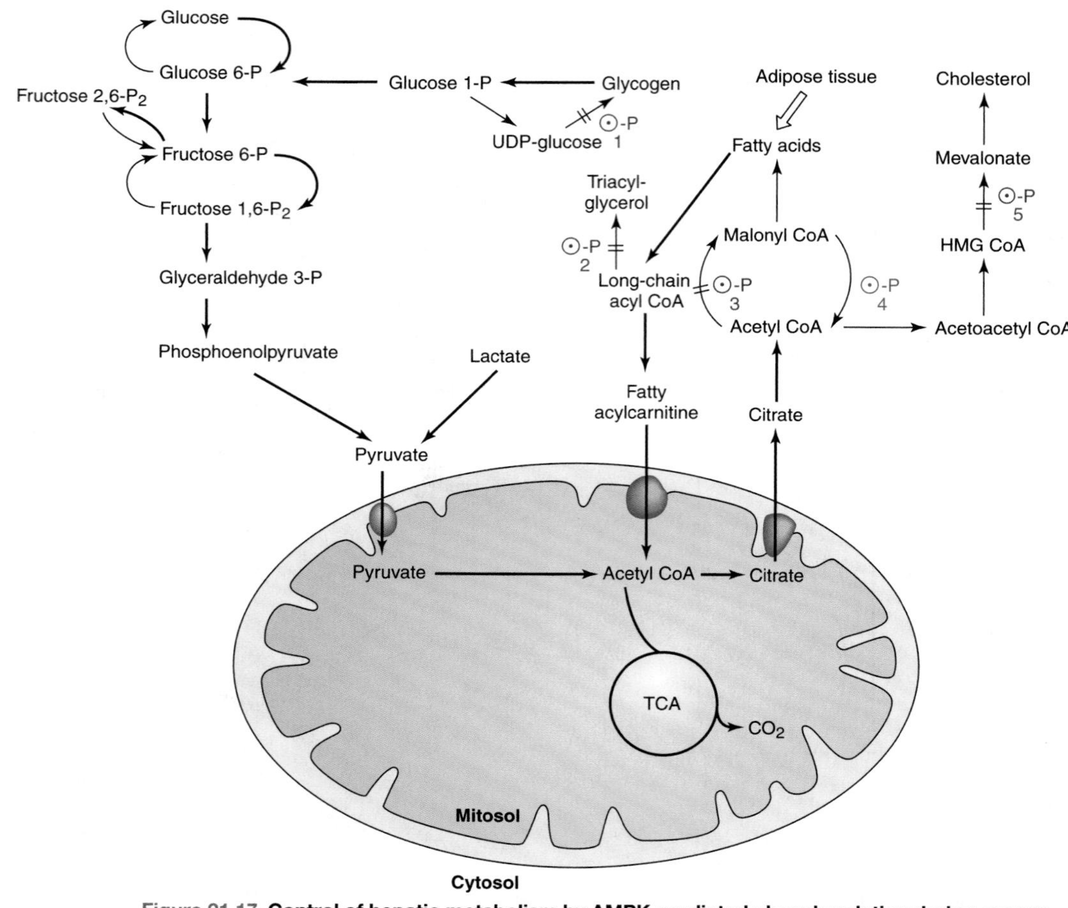

Figure 21.17 Control of hepatic metabolism by AMPK-mediated phosphorylation during energy deprivation. Phosphorylated mode is indicated by ⊙-P. The enzymes phosphorylated by AMPK are (1) glycogen synthase, (2) glycerol-3-phosphate acyltransferase, (3) acetyl-CoA carboxylase, (4) malonyl-CoA decarboxylase, and (5) 3-hydroxy-3-methylglutaryl-CoA (HMG CoA) reductase.

Adipose tissue responds almost as dramatically as liver to the starve–feed cycle. Pyruvate kinase, the pyruvate dehydrogenase complex, acetyl-CoA carboxylase, and hormone-sensitive lipase (not found in liver) are dephosphorylated in adipose tissue in the well-fed state. The first three enzymes are active while hormone-sensitive lipase is inactive in this state. A high insulin level in the blood and a low cAMP concentration in adipose tissue are important determinants of the phosphorylation state of these enzymes, which favors lipogenesis in the well-fed state. During fasting, a decrease in the insulin level and an increase in epinephrine shut down lipogenesis and activate lipolysis because of phosphorylation of these enzymes. In this manner, adipose tissue is transformed from a fat storage tissue into a source of fatty acids for oxidation in other tissues and glycerol for gluconeogenesis in the liver.

Covalent modification of enzymes in skeletal muscle is also important in the starve–feed cycle. Glycogen synthase, glycogen phosphorylase, pyruvate dehydrogenase complex, acetyl-CoA carboxylase, and malonyl-CoA decarboxylase are dephosphorylated in the fed state. This, along with insulin-mediated stimulation of glucose uptake by glucose transporter 4 (GLUT4) (p. 858), favors glucose uptake, glycogen synthesis, and complete oxidation by skeletal muscle. The increase in malonyl-CoA favored by the combination of an active acetyl-CoA carboxylase and an inactive malonyl-CoA decarboxylase limits fatty acid oxidation at the level of carnitine palmitoyltransferase 1. During fasting, conservation of glucose, lactate, alanine, and pyruvate is crucial for survival. Tissues of the body that can use alternative fuels invariably shut down their use of glucose and three-carbon compounds that can be used for the synthesis of glucose. Increased availability of fatty acids and enzymatic activity for oxidation spares glucose in the starved state. The latter is due to decreased malonyl CoA levels and therefore less inhibition of carnitine palmitoyl-transferase 1 induced by phosphorylation-mediated inactivation of acetyl-CoA carboxylase and activation of malonyl-CoA decarboxylase. Regulation of the utilization of glucose by fatty acid catabolism has been named the glucose–fatty acid cycle. Inactivation of the pyruvate dehydrogenase complex in skeletal muscle by phosphorylation is the key to conservation of glucose and three carbon compounds for hepatic gluconeogenesis during fasting. This is mediated by pyruvate dehydrogenase kinase (p. 551), which is induced to a higher level of expression and stimulated to greater activity by its allosteric effectors acetyl CoA and NADH produced by fatty acid oxidation.

Exercise induces profound effects on metabolic pathways in skeletal muscle. The energy demand of muscle contraction increases AMP and activates AMPK. AMPK stimulates the transport of vesicles bearing GLUT4 to the plasma membrane for greater glucose uptake and catabolism for ATP production. AMPK-mediated phosphorylations also decrease malonyl CoA by inactivating acetyl-CoA carboxylase and activating malonyl-CoA decarboxylase (see Figures 21.14, p. 856, and 21.17). Less malonyl CoA results in greater carnitine palmitoyltransferase 1 activity and fatty acid oxidation to help meet muscle's need for ATP for contraction.

Covalent modification, like allosteric effectors and substrate supply, is a short-term regulatory mechanism, operating on a minute-to-minute basis. On a longer time scale, enzyme activities are controlled at the level of expression, most frequently by the rate of gene transcription.

Changes in the Amounts of Key Enzymes Provide Long-Term Adaptation

Whereas allosteric effectors and covalent modification affect either the K_m or V_{max} of an enzyme, enzyme activity is also regulated by the rate of its synthesis or degradation, and thus the quantity of enzyme in a cell. For example, in a person maintained in a well-fed or overfed condition, the liver has increased amounts of enzymes involved in triacylglycerol synthesis (Figure 21.18). Many enzymes are induced by an increase in the insulin/glucagon ratio and glucose in the blood. These include glucokinase, 6-phospho-1-fructokinase, and pyruvate kinase for faster rates of glycolysis; glucose 6-phosphate dehydrogenase, 6-phosphogluconate

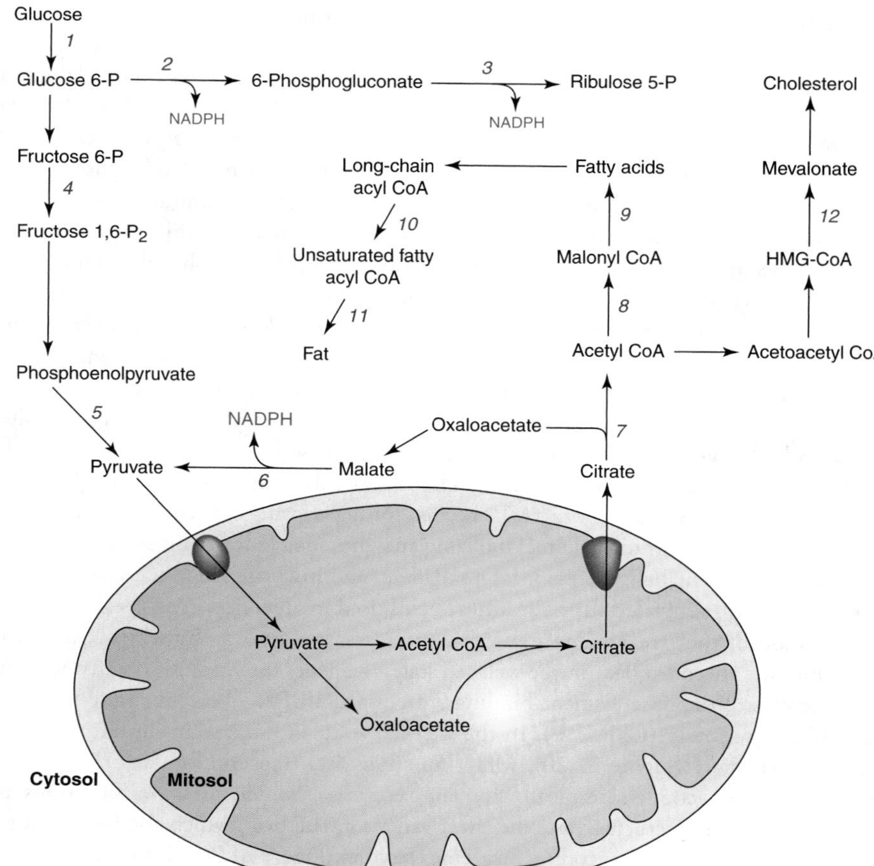

Figure 21.18 Hepatic enzymes induced in the well-fed state. The inducible enzymes are numbered (1) glucokinase, (2) glucose 6-phosphate dehydrogenase, (3) 6-phosphogluconate dehydrogenase, (4) 6-phosphofructo-1-kinase, (5) pyruvate kinase, (6) malic enzyme, (7) citrate cleavage enzyme, (8) acetyl-CoA carboxylase, (9) fatty acid synthase, (10) Δ^9-desaturase, (11) glycerol-3-phosphate acyltransferase, and (12) 3-hydroxy-3-methylglutaryl-CoA (HMG-CoA) reductase.

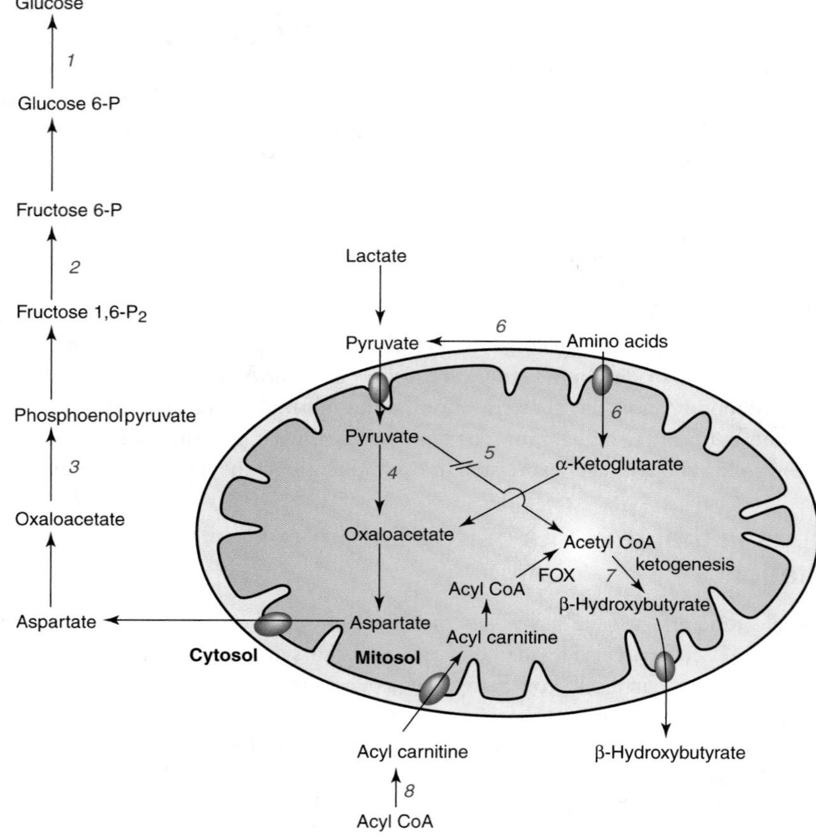

Figure 21.19 Hepatic enzymes induced in the fasted state. The inducible enzymes are numbered (1) glucose 6-phosphatase, (2) fructose 1,6-bisphosphatase, (3) phosphoenolpyruvate carboxykinase, (4) pyruvate carboxylase, (5) pyruvate dehydrogenase kinase, (6) various aminotransferases, (7) mitochondrial 3-hydroxy-3-methylglutaryl-CoA synthase, and (8) carnitine palmitoyltransferase I. Parallel lines intersecting arrow from pyruvate to acetyl CoA denote inhibition (due to phosphorylation) of the pyruvate dehydrogenase complex due to induction of pyruvate dehydrogenase kinase. Abbreviation: FOX, fatty acid oxidation.

dehydrogenase, and malic enzyme to provide greater quantities of the NADPH required for synthesis of fatty acids and cholesterol; citrate cleavage enzyme, acetyl-CoA carboxylase, fatty acid synthase, and Δ^9-desaturase for fatty acid synthesis; 3-hydroxy-3-methylglutaryl-CoA reductase for cholesterol synthesis; and glycerol 3-phosphate acyltransferase for triacylglycerol and phospholipid synthesis. At the same time, phosphoenolpyruvate carboxykinase, pyruvate dehydrogenase kinase, pyruvate carboxylase, fructose 1,6-bisphosphatase, glucose 6-phosphatase, and some aminotransferases are decreased in amount.

In fasting, the lipogenic enzymes decrease dramatically in quantity, while those that favor gluconeogenesis (glucose-6-phosphatase, fructose 1,6-bisphosphatase, phosphoenolpyruvate carboxykinase, pyruvate carboxylase, and various amino transferases) are remarkably induced (Figure 21.19). Starvation also induces pyruvate dehydrogenase kinase, responsible for phosphorylation and inactivation of the pyruvate dehydrogenase complex, which prevents conversion of pyruvate to acetyl CoA, thereby conserving lactate, pyruvate, and the carbon of some amino acids for glucose synthesis. Induction of carnitine palmitoyltransferase I and mitochondrial 3-hydroxy-3-methylglutaryl-CoA synthase likewise increases the capacity of the liver for fatty acid oxidation and ketogenesis. This is particularly important because fatty acid oxidation is the primary source of the ATP needed for hepatic glucose synthesis. Enzymes of the urea cycle and other amino acid metabolizing enzymes such as liver glutaminase, tyrosine aminotransferase, serine dehydratase, proline oxidase, and histidase are induced for disposal of nitrogen, as urea, generated from the amino acids used in gluconeogenesis.

Regulation of the rate of gene transcription is the primary way by which changes in enzyme amounts are controlled. In the well-fed state, the genes for synthesis of fatty acids are controlled by sterol-response-element-binding proteins (**SREBP**) and carbohydrate-response-element-binding protein (**ChREBP**). In the well-fed state an increase in insulin signals an increase in SREBP-1c (Figure 21.20), which functions as a transcription factor to increase transcription of the genes encoding the lipogenic enzymes. Under the same conditions, an increase in glucose indirectly increases the effectiveness of ChREBP, which also functions as a transcription factor for genes encoding lipogenic enzymes (Figure 21.20). ChREBP is subject to regulation by phosphorylation/dephosphorylation, with the phosphorylated form being

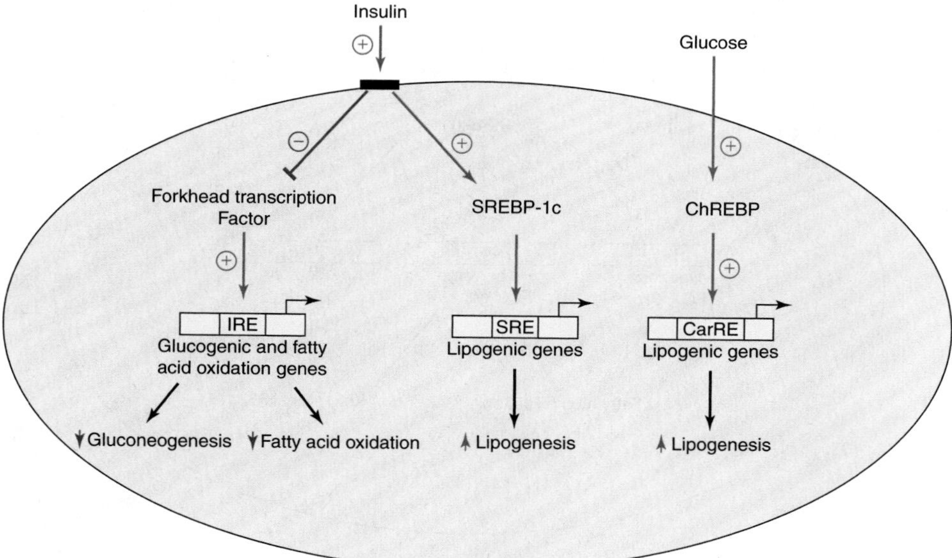

Figure 21.20 Regulation of gene transcription in liver by insulin and glucose. Insulin inactivates forkhead transcription factor by a signaling pathway involving activation of the insulin receptor, phosphorylation of insulin receptor substrate (IRS), activation of phosphatidylinositol-3-kinase (PI3K), and activation of protein kinase B (see Figure 13.33, p. 536). Abbreviations: IRE, insulin responsive element; SRE, SREBP-1c responsive element; and CarRE, carbohydrate responsive element.

inactive and sequestered outside of the nucleus. Activation of a phosphoprotein phosphatase by a metabolite of glucose results in dephosphorylation of ChREBP, which is transported into the nucleus where it increases transcription of genes encoding the lipogenic enzymes. Glucagon acts in opposition to insulin and glucose by signaling activation via cAMP of protein kinase A, which phosphorylates ChREBP, rendering it ineffective. Protein kinase A also phosphorylates **c**AMP-**r**esponse-**e**lement(CRE)-**b**inding (CREB) protein, a transcription factor that inhibits transcription of the genes encoding the lipogenic enzymes (Figure 21.21).

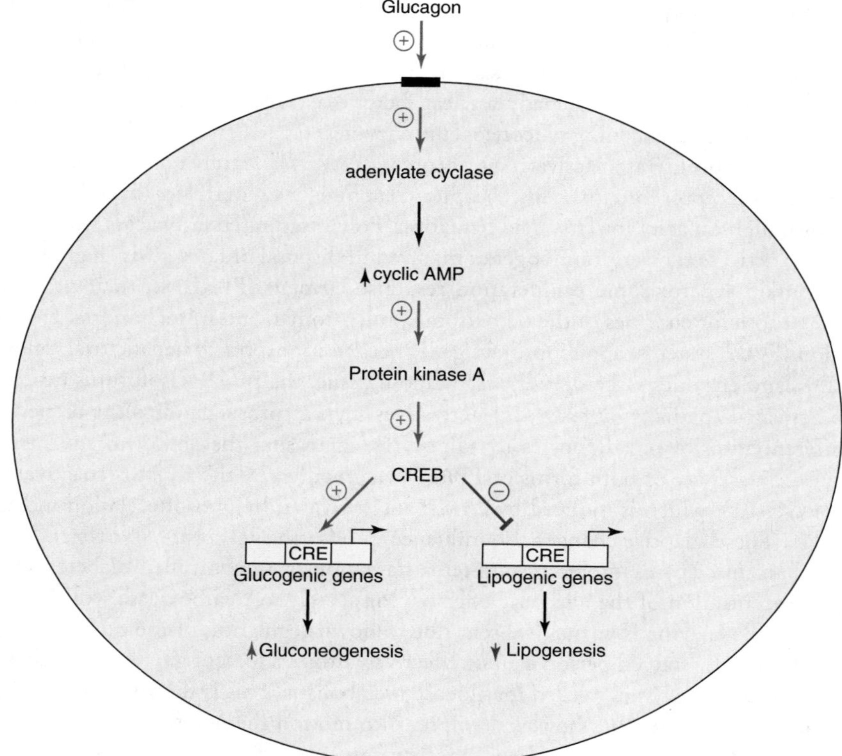

Figure 21.21 Regulation of gene transcription in liver by glucagon. Glucagon activates protein kinase A by the pathway given in Figure 21.13. Abbreviation: CRE, CREB responsive element.

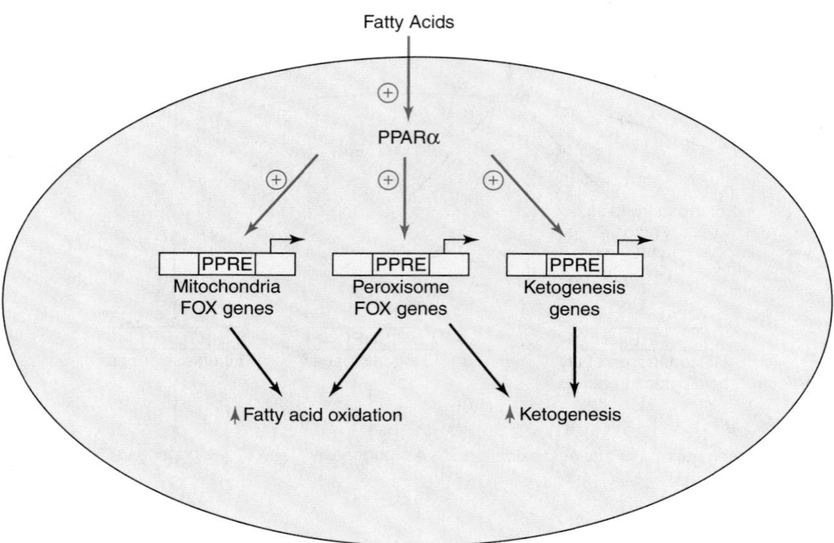

Figure 21.22 PPARα activation by fatty acids promotes transcription of fatty acid oxidation (FOX) and ketogenesis genes. Abbreviation: PPRE, PPAR responsive element

A second form of SREBP designated SREBP-2, regulates cholesterol synthesis (p. 727). When levels of cholesterol fall in cells, SREBP, which is anchored in the endoplasmic reticulum, moves to the Golgi apparatus where it is cleaved by a protease, releasing the N-terminal fragment (SREBP-2) that acts as a transcriptional activator of genes encoding for cytosolic 3-hydroxy-3-methylglutaryl-CoA synthase, 3-hydroxy-3-methylglutaryl-CoA reductase, squalene synthase, and the LDL receptor.

Glucagon promotes the transcription of genes encoding gluconeogenic enzymes via activation of adenylate cyclase, protein kinase A, and the transcription factor CREB (Figure 21.21). Insulin opposes this action of glucagon. Several mechanisms are involved but one of the most important involves inhibition of the activity of forkhead transcription factors that are required for the transcription of the genes encoding insulin response elements (IRE) and gluconeogenic enzymes (see Figure 21.20). Energy deficiency leads to inhibition of fat, cholesterol, and glucose synthesis by liver cells. Activation of AMPK by AMP both reduces the transcription of SREBP and inhibits its transcriptional activity and therefore inhibits both fat and cholesterol synthesis (Figure 21.20). Activation of AMPK also inhibits the transcriptional activity of **h**epatic **n**uclear **f**actor **4**α (HNF4α), which is required for transcription of genes encoding gluconeogenic enzymes.

Peroxisome **p**roliferator-**a**ctivated **r**eceptor α (PPARα), a member of the nuclear receptor family, is a receptor for fatty acids, is expressed at high levels in tissues (liver, kidney, and heart) with high capacity for fatty acid oxidation. Polyunsaturated fatty acids stimulate the receptor to activate transcription of genes involved in disposal of fatty acids (Figure 21.22) which contain a **p**eroxisome **p**roliferation **r**esponse **e**lement (PPRE) in their promoters, including those for enzymes of the peroxisomal, microsomal, and mitochondrial fatty acid oxidation (FOX) systems, apolipoprotein genes needed for export of hepatic triacylglycerol as VLDL, and enzymes of ketogenesis. In adipose tissue, the PPARγ isoform is expressed. When activated (perhaps by fatty acid derivatives such as prostaglandins), it orchestrates the differentiation of preadipocytes to adipocytes, increasing the ability to store triacylglycerol. The activity of both forms of PPAR is augmented by the PPARγ-coactivator 1-alpha (PGC-1α), which is induced by cAMP and activated by phosphorylation mediated by AMPK. These adaptive changes also influence the effectiveness of the short-term regulatory mechanisms. For example, in long-term starvation or uncontrolled diabetes, changing the concentration of the allosteric effectors for acetyl-CoA carboxylase will have little effect when the enzyme is virtually absent due to down-regulation of the expression of its gene. A chronically starved person cannot effectively utilize a load of glucose because of the absence of the key enzymes needed for glucose metabolism. This is the glucose intolerance of starvation. A glucose load, however, will set into motion the necessary adaptations and reestablishment of short-term regulatory mechanisms.

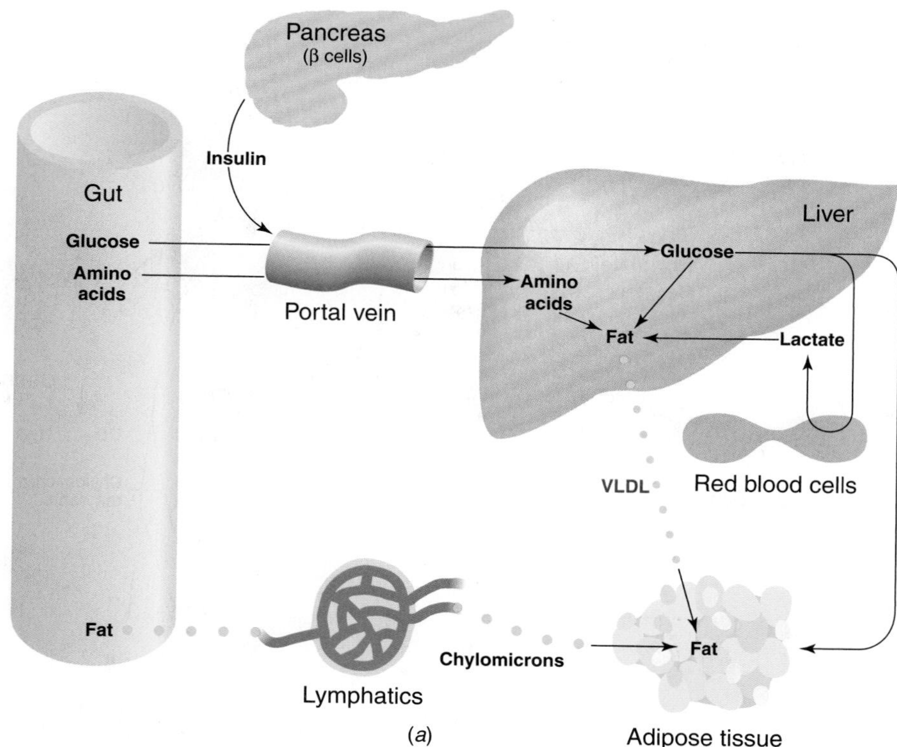

Figure 21.23 **Metabolic interrelationships of tissues in various nutritional, hormonal, and disease states: Obesity.**

21.4 • INTERRELATIONSHIP OF TISSUES IN NUTRITIONAL AND HORMONAL STATES

Many changes that occur in the different nutritional and hormonal states are variations on the starve–feed cycle. Some examples are given in Figure 21.23. Others are obvious, for example, the rapid growth of a child, when amino acids are directed away from catabolism and into protein synthesis. The changes that occur in some physiologically important situations, however, are rather subtle and poorly understood. For instance, in aging there appears to be a decreased sensitivity of the major tissues of the body to hormones, with decreased ability of tissues to respond normally during the starve–feed cycle. Whether this is a contributing factor to or a consequence of the aging process is not known.

Obesity

Figure 21.23 illustrates the metabolic interrelationships prevailing in an obese person. Body fat originates primarily from the diet. Only small amounts are synthesized in the liver and transported to the adipose tissue or synthesized in adipocytes. Obesity results from overeating. It follows from being too long in the well-fed state because of the amount of food consumed. The fasting phase of the starve–feed cycle is too short to use up the fat stored by the body during the feeding phase of the cycle (see Clin. Corr. 21.1, p. 840).

Obesity is epidemic in the affluent countries of the world. Enticing food is plentiful and cheap. Obesity needs to be aggressively treated to prevent progression to the metabolic syndrome, type 2 diabetes mellitus, and cardiovascular disease. Unfortunately, there is no effective intervention other than dieting. Metabolic syndrome, strongly associated with the development of early atherosclerosis, refers to a constellation of medical problems, including abdominal obesity, elevated blood pressure, high blood lipids, and insulin resistance.

Obesity frequently causes insulin resistance. In some patients, the number or affinity of insulin receptors is reduced, while others have normal insulin binding, but abnormal postreceptor responses, such as the activation of glucose transport. Generally, the quantity of body fat is proportional to the degree of insulin resistance. Some of the resistance is due to peptides (TNFα and resistin) produced by adipocytes that are known to oppose the

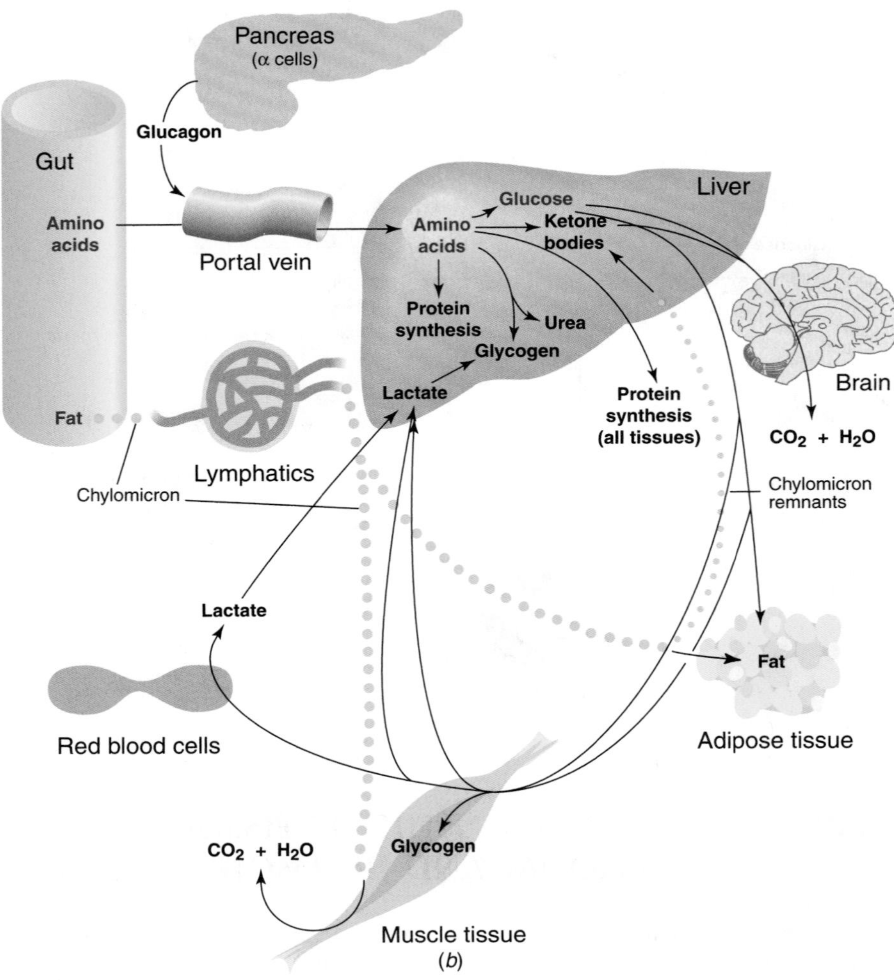

Figure 21.24 **Metabolic interrelationships of tissues in dieting.**

action of insulin. An additional adipokine, adiponectin, is reduced in obesity, which may also contribute to insulin resistance. Plasma insulin levels are often greatly elevated in obese individuals, a harbinger for the development of type 2 diabetes mellitus.

Dieting

To lose weight requires a negative energy balance, which means fewer calories must be consumed than are expended each day. Consuming less food with the same macronutrient composition has little effect on the starve–feed cycle, other than the length of time in the well-fed state. The responsibilities of tissues will stay the same in the well-fed state (see Figure 21.2, p. 843), except that less glycogen and triacylglycerol will be stored and the switch to the function of tissues in the fasted state (see Figure 21.4, p. 845) will occur sooner after meals. Another way to lose weight, long advocated by most health agencies, is to cut calories by specifically reducing the amount of fat consumed. Again, the well-fed state of tissues will stay the same except for a reduction in stored triacylglycerol. The switch to the fasting state will occur sooner after meals, unless the decrease in fat intake has been compensated by an increase in carbohydrate intake, a common problem for dieters. Another way to lose weight, originally advocated by Robert Atkins, is to decrease specifically the amount of carbohydrate consumed. The Atkins controlled-carbohydrate diet and variations thereof, frequently called the very low-carbohydrate ketogenic diet, produces interesting effects on tissues. On this diet, intake of protein is high, fat is moderate, and carbohydrate is extremely low (< 50 g/day; $< 10\%$ of a 2000 kcal/day diet). The metabolic functions of tissues in the fed state for an individual on such a diet are summarized in Figure 21.24. The fasting state is little changed from that of other diets, but the almost complete lack of dietary carbohydrate requires that the liver remain gluconeogenic as well as ketogenic in the fed

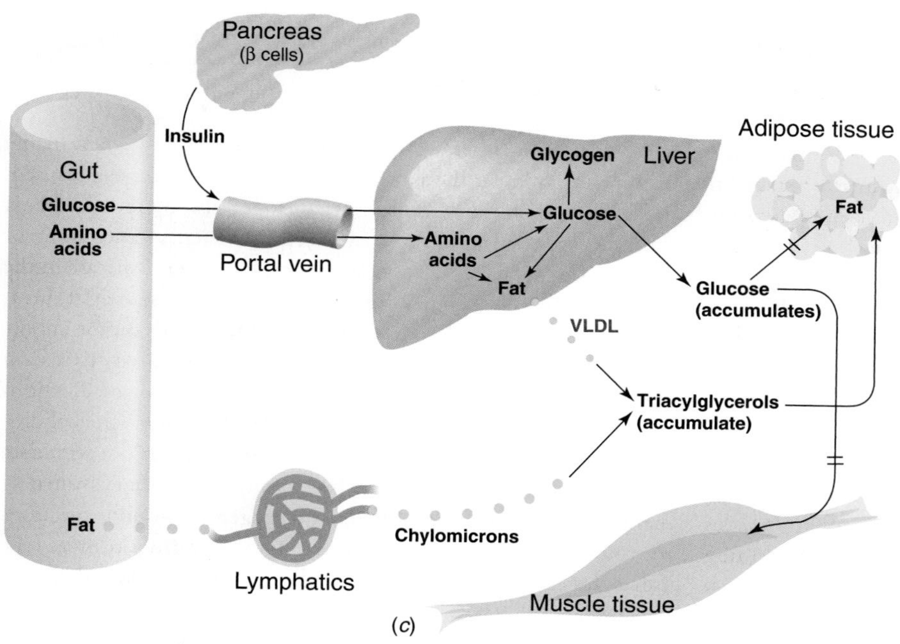

Figure 21.25 Metabolic interrelationships of tissues in type 2 diabetes mellitus.

state. Little rise in blood glucose and a less than normal rise in insulin occur in response to meals. Amino acids present in excess beyond that needed for protein synthesis are converted to liver glycogen, blood glucose, and ketone bodies. Large amounts of amino acids coming from the gut minimize the need for release of amino acids from peripheral tissues for hepatic gluconeogenesis. Fatty acids delivered to the liver in chylomicron remnants are primarily converted to ketone bodies to provide ATP for gluconeogenesis. Glucose and ketone bodies are therefore produced in both the fed and fasted states. Although it would seem that overproduction and/or underutilization of ketone bodies could cause ketoacidosis, as it does in type 1 diabetes (p. 865), the need for a supply of energy in the form of ketone bodies in peripheral tissues largely balances the production of ketone bodies by the liver. The low-carbohydrate diet has been tested in controlled trials, which have been relatively short term (6–12 mon): Weight loss is somewhat more rapid than achieved with low-fat, low-calorie diets, but the extent of weight loss, compliance with the diet, and tendency to regain weight are similar. Of interest, this diet didn't increase blood cholesterol or LDL levels, and improved VLDL and HDL levels better than low-fat, low-calorie diets.

Type 2 Diabetes Mellitus

Figure 21.25 shows the metabolic interrelationships characteristic of a person with type 2 diabetes mellitus. These individuals are resistant to insulin and have insufficient production of insulin to overcome the resistance (Clin. Corr. 21.6). The majority of patients are obese, and although their insulin levels often are high, they are not as high as those of nondiabetic but similarly obese individuals. Hence, β-cell failure and insulin resistance are components of this form of diabetes. Although the body still produces insulin in type 2 diabetes, it is not enough to control glucose production by the liver or promote uptake of glucose by the skeletal muscle. Hyperglycemia results for both reasons. The normal increase in fructose 2,6-bisphosphate and down-regulation of phosphoenolpyruvate carboxykinase does not occur in these patients. Translocation of intracellular vesicles bearing GLUT4 to the plasma membrane in response to insulin is decreased in the skeletal muscle and adipose tissue of these patients. Ketoacidosis rarely develops, perhaps because enough insulin is present to prevent uncontrolled release of fatty acids from adipocytes and fatty acids reaching the liver or synthesized *de novo* are directed into triacylglycerol. Hypertriacylglycerolemia is characteristic and usually results from an increase in VLDL without hyperchylomicronemia. This is most likely explained by hepatic synthesis of fatty acids and diversion of fatty acids reaching the liver into triacylglycerol and VLDL. Concurrent lipogenesis and gluconeogenesis should never occur but result in this disease from a state of mixed insulin resistance of

CLINICAL CORRELATION 21.6

Diabetes Mellitus, Type 2

Type 2 diabetes mellitus accounts for 80%–90% of the diagnosed cases of diabetes (see Clin. Corr. 15.4, p. 609). It usually occurs in middle-aged to older obese people and is characterized by hyperglycemia, often with hypertriglyceridemia and other features of the metabolic syndrome (see Clin. Corr. 21.1, p. 840). The ketoacidosis characteristic of type 1 diabetes is usually not observed, although some patients can develop transient episodes of ketoacidosis. They tend to develop many of the same complications as type 1 diabetes patients (see Clin. Corr. 21.8, p. 868), that is, nerve, eye, kidney, and coronary artery disease. Increased levels of VLDL are probably the result of increased hepatic triacylglycerol synthesis stimulated by hyperglycemia and hyperinsulinemia. Insulin is present at normal to elevated levels in this form of the disease. Obesity often precedes the development of type 2 diabetes and is the major contributing factor. Obese patients are usually hyperinsulinemic and have high levels of free fatty acids, which impair insulin action. Recent data implicate increased levels of tumor necrosis factor α (TNFα) and resistin and reduced secretion of adiponectin by adipocytes of obese individuals as a cause of insulin resistance. The greater the adipose tissue mass, the greater the production of TNFα and resistin, which acts to impair insulin receptor function. The higher the basal level of plasma insulin, the fewer receptors present on the plasma membranes of target cells, and there are defects within insulin-responsive cells at sites beyond the receptor, for example, the ability of insulin to recruit glucose transporters (GLUT4) from intracellular sites to the plasma membrane. As a consequence, insulin levels remain high, but glucose levels are poorly controlled. Even though the insulin level is high, it is not as high as in a person who is obese but not diabetic. In other words, there is a relative deficiency in the insulin supply from the pancreatic β cells. This disease is caused, therefore, not only by insulin resistance but also by impaired beta cell function resulting in relative insulin deficiency. Diet alone or bariatric surgery to induce weight loss can control the disease in the obese diabetic. If the patient can be motivated to lose weight, insulin receptors will increase in number, and the postreceptor abnormalities will improve, which will increase tissue sensitivity to insulin and glucose tolerance. There are now a multitude of medications that either sensitize peripheral tissue to insulin action (thiazolidinediones), reduce hepatic gluconeogenesis (metformin), or stimulate insulin secretion from beta cells (sulfonylureas), but ultimately many type 2 diabetes patients require exogenous insulin injections to control blood sugar. The most recent class of drugs to treat type 2 diabetes are based on the incretin effect, that is, the stimulation of beta cell growth and release of insulin caused by the gastrointestinal hormone incretin (also called glucagon-like peptide 1 or GLP-1). GLP-1 and its analogs only augment glucose-stimulated insulin release which reduces the risk of hypoglycemia. GLP-1 has a short plasma half-life because of degradation by dipeptidyl peptidase-4 (DPP-4). GLP-1 analogs resistant to degradation, and inhibitors of DPP-4, which increase endogenous levels of GLP-1, are entering clinical practice.

Kahn, B. B., and Flier, J. S. Obesity and insulin resistance. *J. Clin. Invest.* 106:473, 2000; Trayhurn, P., and Wood, I. S. Adipokines: Inflammation and the pleiotropic role of white adipose tissue. *Brit. J. Nutr.* 92:347, 2004; and Wajchenberg, B. L. Beta cell failure in diabetes and preservation by clinical treatment. *Endocr. Rev.* 28:187, 2007.

the insulin signaling pathways controlling these processes. A defect in the insulin-signaling pathway controlling gluconeogenesis prevents suppression of hepatic glucose production (via PI_3 kinase, p. 535) in the face of high insulin levels. A more responsive insulin-signaling pathway controlling fatty acid synthesis and esterification (via SREBP-1c, p. 727) leads to overproduction of triacylglycerol.

Diet, exercise, and weight control are the first choices for treating type 2 diabetes. When blood glucose cannot be controlled by this means, several prescription medications (**metformin,** glipizide, or rosiglitazone) are available. In spite of insulin resistance, exogenous insulin is the most effective medication and often must be administered to control the blood glucose level of these patients. Tight control by intensive therapy is desirable, but it increases the risk of life-threatening hypoglycemia (Clin. Corr. 21.7).

Type 1 Diabetes Mellitus

Figure 21.26 shows the metabolic interrelationships in type 1 diabetes mellitus (Clin. Corr. 21.8). In contrast to type 2, there is complete absence of insulin production by the pancreas in this disease. Since the insulin/glucagon ratio cannot increase, the liver is always gluconeogenic and ketogenic and cannot properly buffer blood glucose levels. Indeed, since hepatic gluconeogenesis is continuous, the liver contributes to hyperglycemia in the well-fed state. In muscle and adipose tissue, GLUT4 remains sequestered within cells. Accelerated gluconeogenesis, fueled by uncontrolled proteolysis in skeletal muscle, maintains the hyperglycemia even in the starved state. Uncontrolled lipolysis in adipose tissue increases

CLINICAL CORRELATION 21.7

Hypoglycemia and Diabetes

The microvascular and macrovascular complications that increase morbidity and mortality in type 1 and 2 diabetic patients can be reduced by tight control of blood glucose concentration. Tight control requires intensive therapy, with either insulin or a combination of prescription medications that maintains an average blood glucose comparable to that of nondiabetic individuals (no more than 120 mg/dL). Although not appropriate for every person with diabetes, for example, the young who are still developing, tight control requires frequent self-monitoring of blood glucose concentration, resolute compliance, and a good understanding of the effects of physical activity and diet. Since available technology for insulin delivery cannot match the normal pattern of insulin secretion from the pancreas, intensive control therapy greatly increases the likelihood of life-threatening hypoglycemic episodes. Indeed, the threat of hypoglycemia is a major impediment against maintaining tight control of blood glucose. Severe hypoglycemia can result in neurological dysfunction, coma, convulsions, cardiac arrhythmia, and sudden death. In fact, in a recent clinical trial (ACCORD, Action to Control Cardiovascular Risk in Diabetes), extremely tight control of blood glucose was concluded to be inadvisable in very high-risk type 2 diabetic patients.

Hypoglycemia triggers a series of counterregulatory responses that are designed to prevent blood glucose from falling to dangerously low concentrations. These occur sequentially, beginning in most individuals when blood glucose levels fall below a threshold of 65 mg/dL. Suppression of insulin secretion from the pancreas occurs first. This is effective in normal individuals but not in patients with diabetes using insulin. As hypoglycemia worsens, counterregulatory hormones are released, first glucagon from the alpha cells of the pancreas, followed by epinephrine from the adrenal medulla and growth hormone from the anterior pituitary, and finally cortisol from the adrenal cortex. Glucagon and epinephrine stimulate hepatic glycogenolysis and gluconeogenesis; growth hormone stimulates adipose tissue lipolysis, which provides substrate (glycerol) and energy (fatty acids for the synthesis of ATP) required for hepatic gluconeogenesis; and cortisol promotes transcription of genes encoding enzymes (e.g., phosphoenolpyruvate carboxykinase) required for hepatic gluconeogenesis.

Hypoglycemia is detected by glucose-sensing neurons of the ventromedial hypothalamus. The reduced availability of glucose reduces glucose catabolism in these neurons, resulting in lower ATP and higher AMP concentrations. Subsequent activation of AMP-activated protein kinase triggers nerve activity that initiates release of counterregulatory hormones. The increase in blood level of epinephrine induces symptoms (tachycardia, palpitations, unsteadiness, and sweating) that patients recognize as warning signs of hypoglycemia. Immediate consumption of carbohydrate, for example, Life Saver candies, or the injection of glucagon (1 mg intramuscularly) usually prevents progression to severe hypoglycemia. Unfortunately, for reasons that are not understood, repeated episodes of hypoglycemia in diabetic patients often induce "hypoglycemia unawareness," a condition in which release of the counterregulatory hormones and the symptoms of distress no longer occur in response to mild hypoglycemia. Tight control of blood glucose is not advisable in such patients.

Davis, A., and Alonso, M. D. Hypoglycemia as a barrier to glycemic control. *J. of Diabetes and Its Complications* 18:60, 2004; McCrimmons, R. J., Shaw, M., Fan, X., Cheng, H., et al. Key role for AMP-activated kinase in the ventromedial hypothalamus in regulating counterregulatory hormone response to acute hypoglycemia. *Diabetes* 57:444, 2008; and Cryer, P. E. Preventing hypoglycemia: What is the appropriate glucose alert value? *Diabetologia* 52:35, 2009.

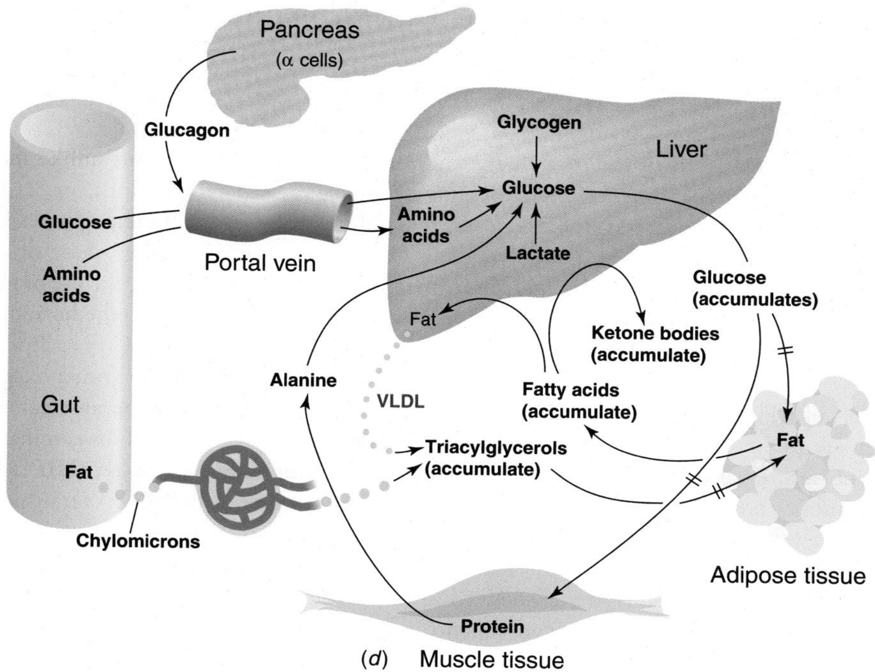

Figure 21.26 Metabolic interrelationships of tissues in type 1 diabetes mellitus.

CLINICAL CORRELATION 21.8

Diabetes Mellitus, Type 1

Type 1 diabetes mellitus usually appears in childhood or in the teens, but it is not limited to these patients (see Clin. Corr. 15.4, p. 609). Insulin secretion is very low because of defective beta cell function, the result of an autoimmune process. Clinical trials of immunosuppression to prevent complete destruction of the islets are underway. Untreated, type 1 diabetes is characterized by hyperglycemia, hypertriglyceridemia (chylomicrons and VLDL), and episodes of severe ketoacidosis. Thus, there is severe derangement of carbohydrate, lipid, and protein metabolism. The hyperglycemia results from the inability of the insulin-dependent tissues to take up glucose and from accelerated hepatic gluconeogenesis from amino acids derived from muscle protein. The ketoacidosis results from increased lipolysis in adipose tissue and accelerated fatty acid oxidation in liver. Hyperchylomicronemia results from low lipoprotein lipase activity in adipose tissue capillaries, an enzyme dependent on insulin for its synthesis.

Although insulin does not cure type 1 diabetes, it markedly alters the clinical course. Insulin promotes glucose uptake and inhibits gluconeogenesis, lipolysis, and proteolysis. It is a difficult job to adjust the insulin dose to variable dietary intake and physical activity, the other major determinant of glucose disposal by muscle. Tight control of blood sugar requires several injections of insulin per day and close blood sugar monitoring by the patient, but it has now been proven to reduce the microvascular complications of diabetes (renal and retinal disease). Increasingly, patients with type 1 diabetes and renal failure are being treated by combined transplantation of a kidney and pancreas to provide an endogenous source of insulin. Islet cell transplantation remains largely experimental.

Atkinson, M. A., and Maclaren, N. K. The pathogenesis of insulin dependent diabetes mellitus. *N. Engl. J. Med.* 331:1428, 1994; Clark, C. M., and Lee, D. A. Prevention and treatment of the complications of diabetes mellitus. *N. Engl. J. Med.* 332:1210, 1994; Luppi, P., and Trucco, M. Immunological models of type 1 diabetes. *Horm. Res.* 52:1, 1999; and Kukreja, A., and Maclaren, N. K. Autoimmunity and diabetes. *J. Clin. Endocrinol. Metab.* 84:4371, 1999.

plasma fatty acid levels and ketone body production by the liver. Ketoacidosis develops due to accumulation of ketone bodies and hydrogen ions. Fatty acid oxidation and ketogenesis cannot completely dispose of fatty acids taken up by liver, and the excess is esterified and directed into VLDL synthesis. Hypertriacylglycerolemia results because VLDL and chylomicrons cannot be cleared from the blood by lipoprotein lipase, whose expression is dependent upon insulin. Thus, in these patients every tissue plays the catabolic role that it was designed to play in starvation, in spite of delivery of adequate or even excess fuel from the gut. In effect, patients with type 1 diabetes are stuck in the starved state without benefit of the brake usually imposed during starvation by low but continuous production of insulin by the pancreas. This leads to severe wasting of body tissues and ultimately in death unless insulin is administered.

Exogenous insulin is the only effective medication in these patients. As noted above for type 2 diabetes, tight control by intensive therapy is desirable, but it increases the risk of hypoglycemia (see Clin. Corr. 21.7, p. 867).

Cancer

Tumors are composed of cancer cells that have to be fed like all cells, but unlike most normal tissues, tumors function independently of the starve–feed cycle (Clin. Corr. 21.9). Their demand for glucose as a source of energy and amino acids for protein synthesis is incessant. They usually prefer glucose and they rarely adapt in the fasting phase of the starve–feed cycle to use fatty acids and ketone bodies in order to minimize their use of glucose for the benefit of the rest of the body. Most tumors are unresponsive to the hormonal changes that alter the metabolic processes in normal tissues. They establish a Cori cycle with the liver but can still completely oxidize substantial amounts of glucose provided oxygen is available (Figure 21.27). Cells in the core of a tumor are often hypoxic because cancers often outgrow the development of blood vessels that supply oxygen. Lack of oxygen in any cell, normal or cancerous, leads to an increase in **hypoxia-inducible factor 1α (HIF-1α)**, a remarkably potent transcription factor for activation of genes encoding glucose transporters, the enzymes of glycolysis, and one of the kinases (pyruvate dehydrogenase kinase 1) responsible for phosphorylation and inactivation of the pyruvate dehydrogenase complex. HIF-1α also becomes constitutively active in some cancer cells because of mutations that activate certain oncogenes. As a consequence of the action of HIF-1α, most cancerous tumors have exceptional capacity for generation of ATP by glycolysis. Glycolysis is not

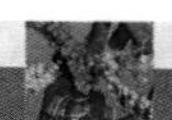

CLINICAL CORRELATION 21.9

Cancer Cachexia

Unexplained weight loss may be a sign of malignancy, and weight loss is common in advanced cancer. This is not completely accounted for by decreased appetite and food intake. The weight loss is largely from skeletal muscle and adipose tissue, with relative sparing of visceral protein (i.e., liver, kidney, and heart). Although tumors commonly have high rates of glycolysis, the energy requirement of the tumor probably does not explain weight loss as this can occur with even small tumors. Skeletal muscle protein degradation is mediated by lysosomes, cytosolic proteases, and the ubiquitin–proteosome pathway; the latter is considered the major pathway of protein degradation in infection, injury, and cancer. Endocrine abnormalities have been identified in cancer patients. Patients tend to be insulin-resistant, have higher cortisol levels, and have a higher basal metabolic rate. Some tumors synthesize and secrete biologically active peptides such as ACTH, nerve growth factor, and insulin-like growth factors, which can modify energy metabolism. The host response to a tumor, by analogy to chronic infection, includes release of interleukin-1 (IL-1), interleukin 6 (IL-6), interferon γ (IFN γ), and tumor necrosis factor α (TNFα) by immune cells. TNFα is also called cachexin because it produces wasting. TNFα and IL-1 may act in a paracrine fashion, as their plasma levels are not increased. IL-6 has been detected in cachectic patients' blood at increased levels. These cytokines stimulate fever, proteolysis, lipolysis, and secretion of acute phase reactants by the liver and can elevate resting energy expenditure by inducing uncoupling proteins. More recent studies have identified proteolysis-inducing factor (PIF) and lipid mobilizing factor (LMF) as products of tumors which stimulate skeletal protein catabolism and wasting of adipose tissue, respectively. Similar pathways can be activated in other wasting illness such as chronic inflammatory states and AIDS.

Tisdale, M. J. Cancer anorexia and cachexia. *Nutrition* 17:438, 2001; Wray, C. J., Mammen, J. M., and Hasselgren, P. O. Catabolic response to stress and potential benefits of nutrition support. *Nutrition* 18:971, 2002; and Melstrom, L. G., Melstrom, K. A., Jr, Ding, X. Z., and Adrian, T. E. Mechanisms of skeletal muscle degradation and its therapy in cancer cachexia. *Histol. Histopathol.* 22:805, 2007.

an efficient process compared to complete oxidation of glucose, but efficient utilization of body resources is not characteristic of a cancer, and shutting off mitochondrial oxidative processes at the level of the pyruvate dehydrogenase complex may reduce production of toxic reactive oxygen species by the electron transport chain. Exceptional capacity for ATP generation by glycolysis enables cancer cells to survive and grow as they spread and metastasize into regions of low oxygen tension.

Aerobic and Anaerobic Exercise

Aerobic exercise is exemplified by long-distance running, anaerobic exercise by sprinting or weight lifting. During anaerobic exercise, there is very little interorgan cooperation. The blood vessels within the muscles are compressed during peak contraction; thus their cells are isolated from the rest of the body and largely rely on their own glycogen and

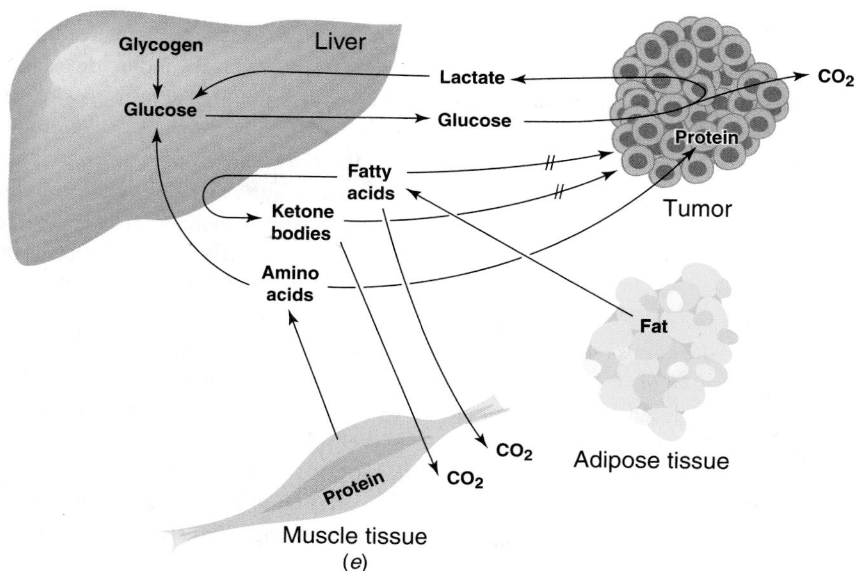

Figure 21.27 Metabolic interrelationships of tissues in cancer.

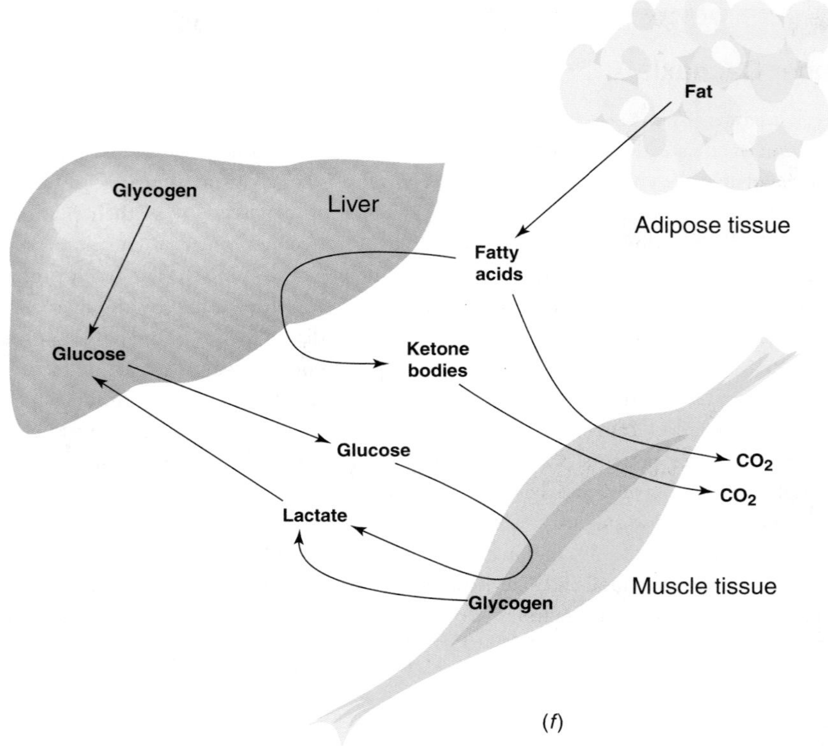

Figure 21.28 Metabolic interrelationships of tissues in exercise.

phosphocreatine. Phosphocreatine is a source of high-energy phosphate for ATP synthesis (see Figure 21.6, p. 866) until glycogenolysis and glycolysis are stimulated. For want of oxygen, glycolysis becomes the primary source of ATP. During moderate aerobic exercise (Figure 21.28), much energy is derived from glycolysis of muscle glycogen, which is the basis for carbohydrate loading. Muscle glycogen content can be increased by exhaustive exercise that depletes glycogen, followed by rest and a high-carbohydrate diet. Glucose uptake also increases due to insulin-independent translocation of GLUT4 to the plasma membrane, a process involving activation of AMPK. A decrease in ATP due to the demand of muscle contraction results in an increase in AMP, which allosterically activates glycogen phosphorylase, 6-phosphofructo-1-kinase, and AMPK. Increased glucose uptake, glycogen degradation, and glycolysis help provide the ATP needed for muscle contraction. There is also stimulation of branched-chain amino acid oxidation, ammonium production, and alanine release from the exercising muscle.

A well-fed individual does not store enough glycogen to meet the energy needs of running long distances. The respiratory quotient, the ratio of liters of carbon dioxide exhaled to oxygen consumed, falls during distance running. This indicates the progressive switch from glycogen to fatty acid oxidation during a race. Lipolysis gradually increases as glucose stores are exhausted, and, as in the fasted state, muscles oxidize fatty acids in preference to glucose. The increase in AMP due to the demand for ATP also activates AMPK, which phosphorylates and inactivates acetyl-CoA carboxylase. This, along with an increase in long chain acyl CoA esters, which are negative allosteric effectors of acetyl-CoA carboxylase, reduces synthesis of malonyl CoA. AMPK also activates malonyl-CoA decarboxylase, removing malonyl-CoA. This allows greater carnitine palmitoyltransferase I activity and fatty acid oxidation to provide ATP for the contracting muscle. Interestingly, exercise also has effects on the liver mediated by activation of AMPK. Phosphorylation of acetyl-CoA carboxylase, malonyl-CoA decarboxylase, and glycerol-3-phosphate acyltransferase by AMPK directs fatty acids into oxidation and away from esterification to triacylglycerol (p. 688). Unlike fasting, however, there is little increase in blood ketone body concentration during exercise because hepatic ketone body production is balanced by the muscle oxidation of ketone bodies for energy.

In contrast, blood lactate levels can increase greatly during exhaustive exercise, particularly when anaerobic glycolysis is stimulated. Lactate accumulates in the blood in this situ-

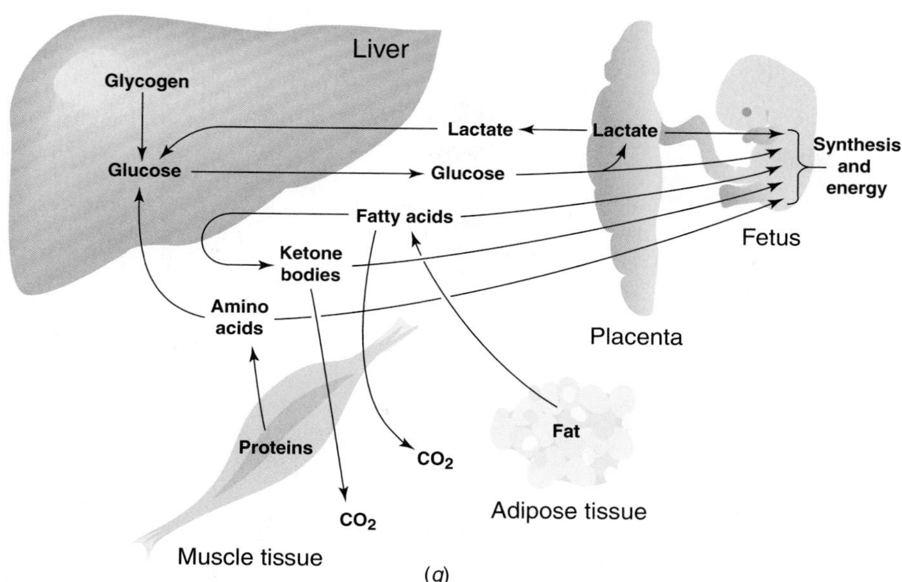

Figure 21.29 **Metabolic interrelationships of tissues in pregnancy.**

ation because the rate of lactate production by muscle exceeds the rate of lactate utilization for glucose synthesis by the liver (Figure 21.28). Normally the brain does not use blood lactate as a fuel because blood lactate concentration (normally about 1 mM) is much lower than the K_m of the lactate transport system across the blood brain barrier. However, when blood lactate levels increase into the range of 10–20 mM during exhaustive exercise, lactate penetrates the blood brain barrier to become an important fuel for the brain. That this helps clear both lactate and acid from the blood can be seen from the balanced equation for complete oxidation of lactate: lactate$^-$ + H^+ + 3 $O_2 \rightarrow$ 3 CO_2 + 3 H_2O. It is also more efficient since direct utilization of lactate by the brain avoids the energy requirement for the conversion of lactate to glucose in the liver (6 mol of ATP per mole of glucose).

Pregnancy

The fetus is a nutrient-requiring organism (Figure 21.29). It mainly uses glucose for energy, but may also use amino acids, lactate, fatty acids, and ketone bodies. Lactate produced in the placenta by glycolysis is partly directed to the fetus, and the rest enters the maternal circulation to establish a Cori cycle with the liver. Maternal LDL cholesterol is an important precursor of placental steroids (estradiol and progesterone). During pregnancy, the starve–feed cycle is perturbed. The placenta secretes placental lactogen and two steroid hormones, estradiol and progesterone. Placental lactogen stimulates lipolysis in adipose tissue, and the steroid hormones induce insulin resistance. After meals, pregnant women enter the starved state more rapidly because of increased consumption of glucose and amino acids by the fetus. Plasma glucose, amino acids, and insulin levels fall rapidly, and glucagon and placental lactogen levels rise and stimulate lipolysis and ketogenesis. Consumption of glucose and amino acids by the fetus may be great enough to cause **maternal hypoglycemia.** In the fed state, pregnant women have increased levels of insulin and glucose and demonstrate resistance to exogenous insulin. These swings of plasma hormones and fuels are even more exaggerated in pregnant diabetic women and make control of their blood glucose difficult. This is important, as hyperglycemia adversely affects fetal development.

Lactation

In late pregnancy placental (progesterone) and maternal (prolactin) hormones induce lipoprotein lipase in the mammary gland and promote the development of milk-secreting cells and ducts. During lactation (Figure 21.30), the breast uses glucose for lactose and triacylglycerol synthesis and as its major energy source. Amino acids are taken up for protein synthesis, and chylomicrons and VLDL provide fatty acids for triacylglycerol synthesis. If these compounds are not supplied by the diet, proteolysis, gluconeogenesis, and lipolysis must supply them, eventually resulting in maternal malnutrition and poor-quality milk. The

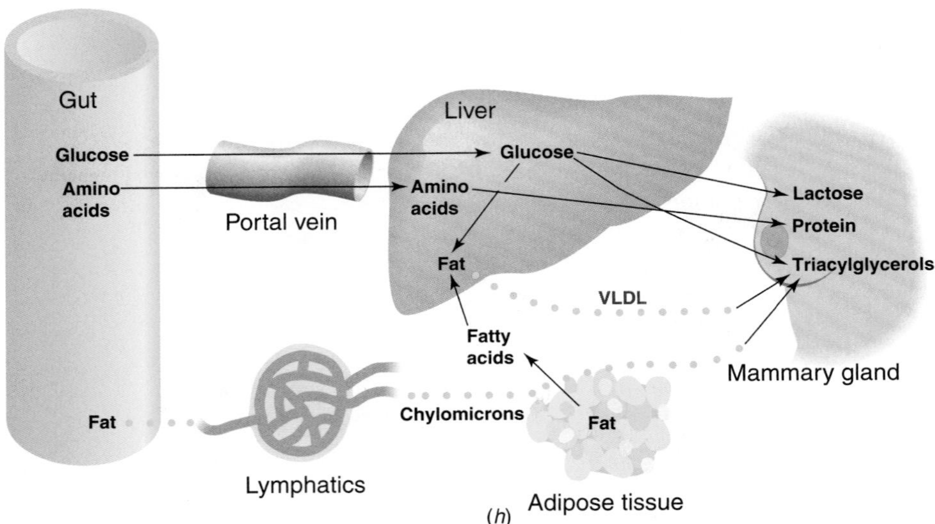

Figure 21.30 **Metabolic interrelationships of tissues in lactation.**

lactating breast secretes **parathyroid hormone-related protein (PTHrP,** p. 891), which mimics the effects of **parathyroid hormone (PTH)** in stimulating absorption of calcium and phosphorus from the gut and bone. The release of PTH from the parathyroid and PTHrP from the breast is under the control of the calcium-sensing receptor, a G protein coupled receptor that senses extracellular calcium and thus may serve to coordinate release of calcium mobilizing hormone with plasma calcium levels.

Stress and Injury

Physiological stresses include injury, surgery, renal failure, burns, and infections (Figure 21.31). Characteristically, blood cortisol, glucagon, catecholamines, and growth hormone levels increase and there is resistance to insulin. Basal metabolic rate and blood

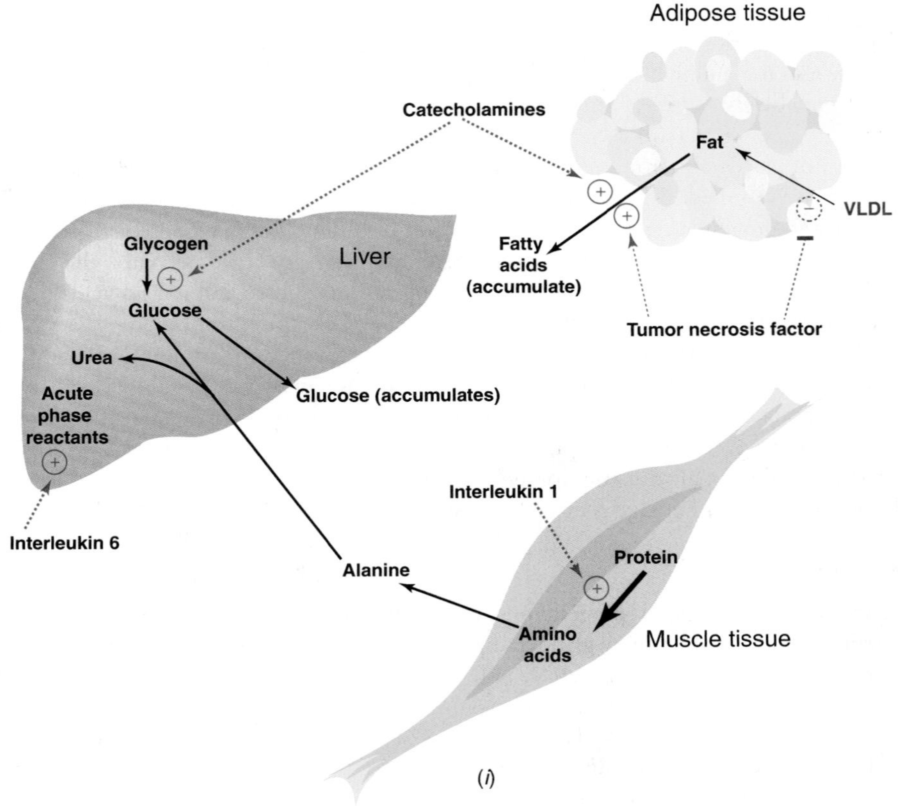

Figure 21.31 **Metabolic interrelationships of tissues in stress and injury.**

glucose and free fatty acid levels are elevated. However, for poorly understood reasons, ketogenesis is not accelerated, as it is in fasting where it provides a good fuel (ketone bodies) that can substitute for glucose in many tissues, especially the brain, thereby reducing the need for gluconeogenesis and conserving body protein. The muscle glutamine and branched-chain amino acid pools are reduced, resulting in reduced protein synthesis and increased protein breakdown. It can be very difficult to reverse this protein breakdown, despite intravenous administration of solutions containing amino acids, glucose, and triacylglycerol. However, solutions used to feed patients intravenously lack glutamine, tyrosine, and cysteine because of stability and solubility constraints. Supplementation of these amino acids, perhaps by the use of more stable dipeptides, may help to reverse the catabolic state better than now possible. In fact, there is growing recognition that enteral feeding (via tubes in the stomach or intestine) is superior to intravenous feeding in reducing or reversing catabolism.

Negative nitrogen balance of injured or infected patients is mediated by **interleukin 1, interleukin 6,** and **tumor necrosis factor α** (TNFα) produced by monocytes and lymphocytes (Clin. Corr. 21.9, p. 869). These cytokines induce fever and other metabolic changes. Interleukin 1 activates proteolysis in skeletal muscle. Interleukin 6 stimulates the liver synthesis of acute phase reactants, such as fibrinogen, complement proteins, some clotting factors, and $\alpha 2$ macroglobulin, which may defend against injury and infection. TNFα suppresses adipocyte triacylglycerol synthesis, inhibits lipoprotein lipase, stimulates lipolysis, inhibits release of insulin, and promotes insulin resistance. These cytokines may be responsible for the wasting seen in chronic infections. Sick patients in intensive care units are increasingly recognized as developing an acute myopathy that may virtually paralyze them. This has been attributed to the use of paralyzing drugs (to assist in use of ventilators), malnutrition, and increased cytokines and sympathetic nervous system activity.

Liver Disease

Advanced **liver disease** is associated with major metabolic derangements (Figure 21.32), especially for amino acids. In patients with **cirrhosis,** the liver cannot convert ammonia into urea and glutamine rapidly enough, and the blood ammonia level rises. Shunting of blood around the liver and interference with the intercellular glutamine cycle (p. 759) contribute to the problem. Ammonia arises from the action of glutaminase, glutamate dehydrogenase, and adenosine deaminase in intestine and liver and from the intestinal lumen, where bacterial **urease** splits urea into ammonia and carbon dioxide. Ammonia levels rise particularly after upper GI bleeding (i.e., bleeding from the esophagus,

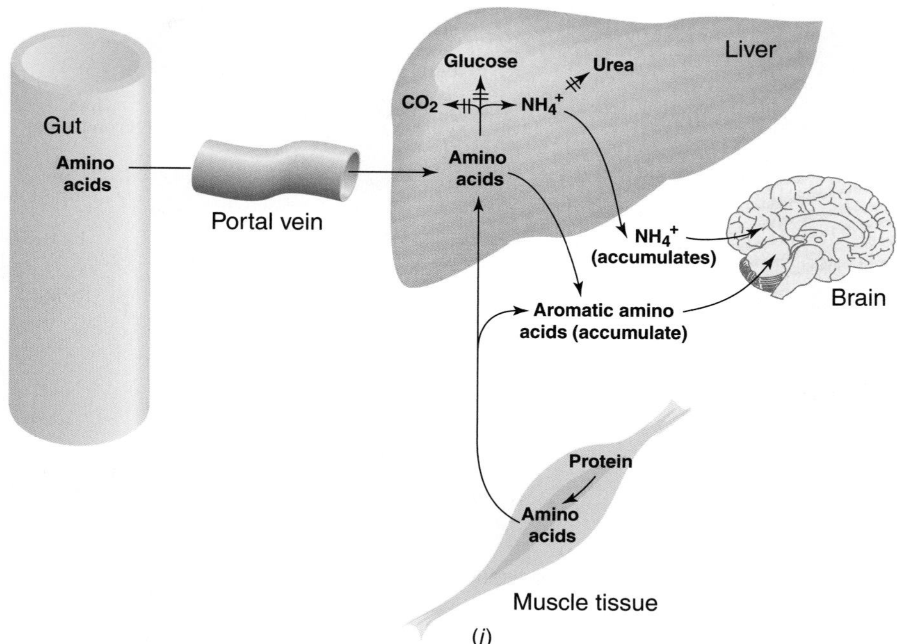

Figure 21.32 Metabolic interrelationships of tissues in liver disease.

stomach, or duodenum). In the past this was simply ascribed to the high protein content of blood, but recently it has been suggested that the unusual amino acid composition of hemoglobin is to blame. Hemoglobin totally lacks isoleucine, and thus absorption of amino acids derived from hemoglobin results in a reduction in plasma isoleucine, with impairment of protein synthesis and thus increased net rates of protein degradation and ammonia production. Ammonia toxicity for the central nervous system leads to coma that sometimes occurs in patients with liver failure. In advanced liver disease, plasma branched-chain amino acids are reduced while aromatic amino acids are elevated, resulting in a reduced Fischer ratio, defined as the molar ratio of the branched-chain amino acids to the aromatic amino acids. These two groups of amino acids are transported into the brain by the same carrier system. Because of less competition by the branched-chain amino acids, increased brain uptake of aromatic amino acids may increase synthesis of neurotransmitters such as serotonin, which may be responsible for some of the neurological abnormalities of liver disease. The liver is a major source of insulin-like growth factor I (IGF-1). Patients with cirrhosis suffer muscle wasting because of deficient IGF-1 synthesis in response to growth hormone. They also often have insulin resistance and may have diabetes mellitus. Finally, in outright liver failure, patients sometimes die of hypoglycemia because the liver is unable to maintain the blood glucose level by gluconeogenesis.

Renal Disease

In chronic renal disease, levels of amino acids normally metabolized by kidney (glutamine, glycine, proline, and citrulline) increase and nitrogen end products, for example, urea, uric acid, and creatinine, also accumulate (Figure 21.33). This is worsened by high dietary protein intake or accelerated proteolysis. Since gut bacteria can split urea into ammonia and liver uses ammonia and α-keto acids to form nonessential amino acids, a diet high in carbohydrate and an amino acid intake limited as much as possible to essential amino acids ensures that the liver synthesizes nonessential amino acids from TCA cycle intermediates. This type of diet therapy may delay the need for dialysis, but this therapy has largely been supplanted by earlier institution of dialysis. An additional abnormality in dialysis patients is carnitine deficiency, resulting from reduced intake of dietary carnitine (in meat) and

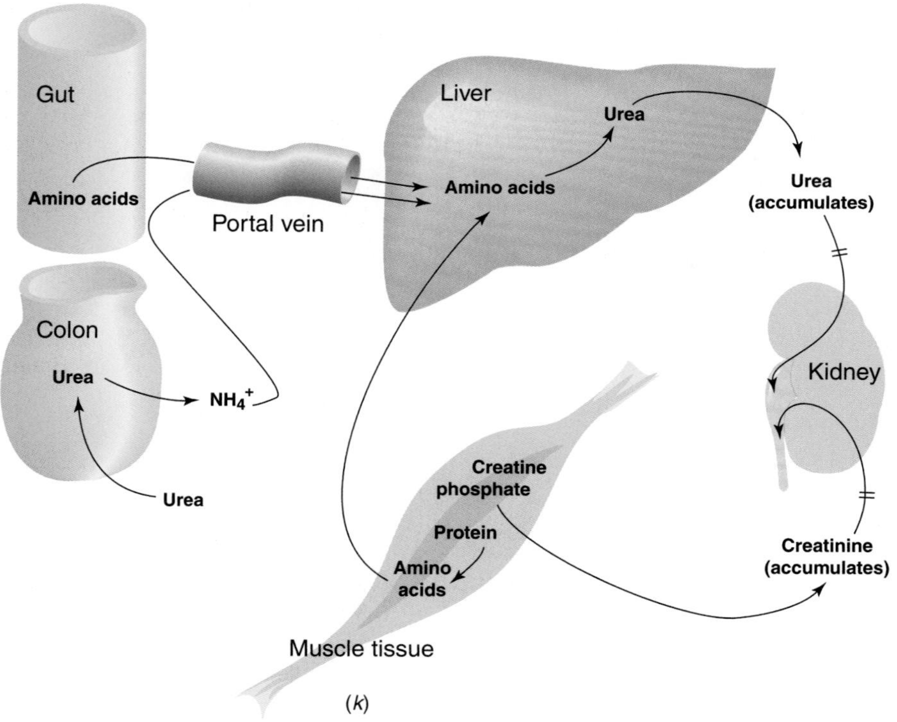

Figure 21.33 **Metabolic interrelationships of tissues in kidney failure.**

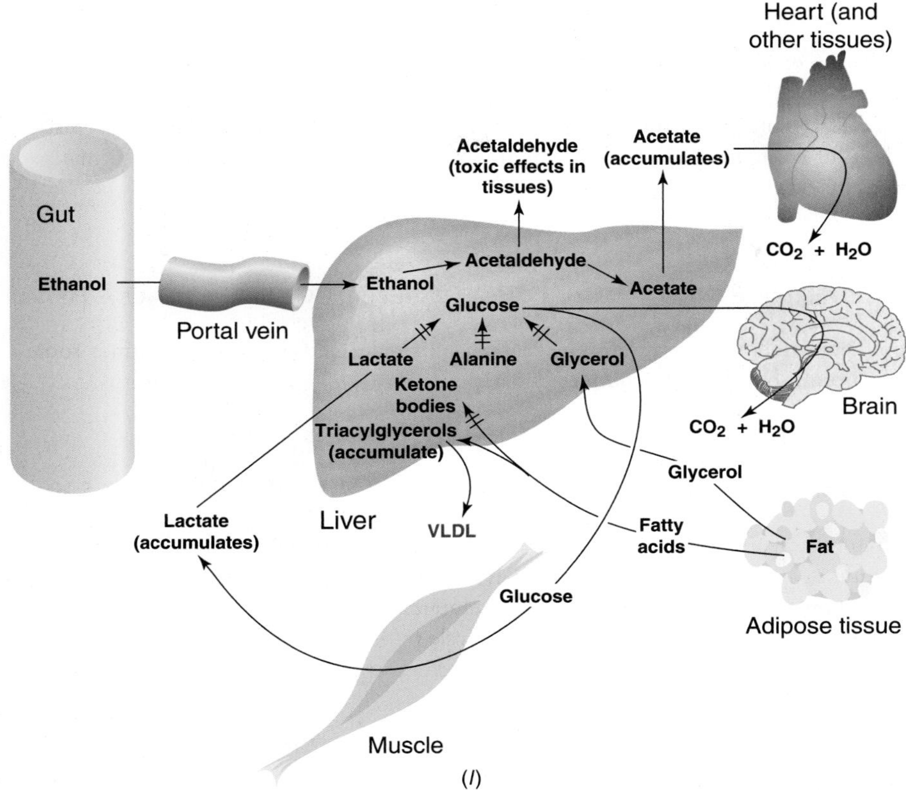

Figure 21.34 **Metabolic interrelationships of tissues in consumption of alcohol.**

reduced functional renal mass. Carnitine may also be lost from the bloodstream during hemodialysis. This may lead to cardiac and skeletal myopathy due to reduced ability of these tissues to oxidize fatty acids.

Consumption of Alcohol

The liver is primarily responsible for the first steps of **ethanol** catabolism.

$$\text{Ethanol } (CH_3CH_2OH) + NAD^+ \rightarrow \text{acetaldehyde } (CH_3CHO) + NADH + H^+$$
$$\text{Acetaldehyde } (CH_3CHO) + NAD^+ + H_2O \rightarrow \text{acetate } (CH_3COO^-) + NADH + 2H^+$$

The first, catalyzed by **alcohol dehydrogenases,** generates NADH in the cytosol; the second, catalyzed by aldehyde dehydrogenase, also generates NADH, but in the mitochondrial matrix space. Liver disposes of the NADH generated through the mitochondrial electron transport chain. Intake of even moderate amounts of ethanol generates too much NADH. Enzymes involved in gluconeogenesis (lactate dehydrogenase and malate dehydrogenase) and fatty acid oxidation (β-hydroxyacyl-CoA dehydrogenase) require NAD^+ as a substrate. Thus, these pathways are inhibited by alcohol intake (Figure 21.34), and fasting hypoglycemia and the accumulation of hepatic triacylglycerols (fatty liver) may develop. Lactate may accumulate from inhibition of the conversion of lactate to glucose but rarely causes overt metabolic acidosis.

Liver mitochondria have a limited capacity to oxidize **acetate** to CO_2 because the TCA cycle is inhibited by high NADH and ATP levels during ethanol oxidation. Much of the acetate derived from ethanol escapes the liver to the blood. Virtually every other tissue can oxidize it to CO_2 by way of the TCA cycle.

Acetaldehyde can also escape from the liver and readily forms covalent bonds with functional groups of biologically important compounds. Formation of acetaldehyde adducts with proteins in liver and blood of animals and humans drinking alcohol has been demonstrated. Such adducts may provide a marker for past drinking activity of an individual, just as hemoglobin A_{1c} is an index of blood glucose control in diabetic patients.

Acid–Base Balance

Regulation of acid–base balance, like that of nitrogen excretion, is shared between the liver and kidney. Although complete catabolism of most amino acids yields neutral products (CO_2, H_2O, and urea), the oxidation of positively charged arginine, lysine, and histidine and sulfur-containing amino acids methionine and cysteine result in net formation of protons (acid). For example,

$$\text{Arginine}^{+} + 5.5\ O_2 \longrightarrow 4\ CO_2 + 2\ \text{urea}\ (NH_2\overset{\overset{\displaystyle O}{\|}}{C}NH_2) + 3\ H_2O + H^{+}$$

Complete catabolism of negatively charged glutamate and aspartate consumes some of these protons but does not completely compensate.

$$\text{Glutamate}^{-} + 4.5\,O_2 + H^{+} \longrightarrow 4.5\ CO_2 + 0.5\ \text{Urea}\ (NH_2\overset{\overset{\displaystyle O}{\|}}{C}NH_2) + 3.5\ H_2O$$

Therefore, for acid–base balance, the excess protons must be matched by an equivalent amount of base. In the kidney glutamine is readily taken up, deaminated by glutaminase to give glutamate, oxidatively deaminated by glutamate dehydrogenase to give α-ketoglutarate, and converted by enzymes of the TCA cycle to malate, which is converted to glucose.

Summation of all steps reveals net production of glucose, and more importantly the production of ammonium and bicarbonate ions.

$$\text{Glutamine} + 1.5\ O_2 + 3\ H_2O \rightarrow 0.5\ \text{glucose} + 2\ HCO_3^{\ -} + 2\ NH_4^{\ +}$$

The ammonium ions are excreted into the glomerular filtrate destined to become urine while the bicarbonate ions enter the blood to neutralize protons.

$$HCO_3^{\ -} + H^{+} \rightleftharpoons H_2CO_3 \rightleftharpoons CO_2 + H_2O$$

CO_2 is released in the lungs, thereby effectively eliminating excess protons (acid) produced by amino acid oxidation. In metabolic acidosis (Figure 21.35), more acid than normal is produced in the body because some metabolic process is out of control, for example, lactic acid formation by anaerobic glycolysis or β-hydroxybutyric acid production by ketogenesis.

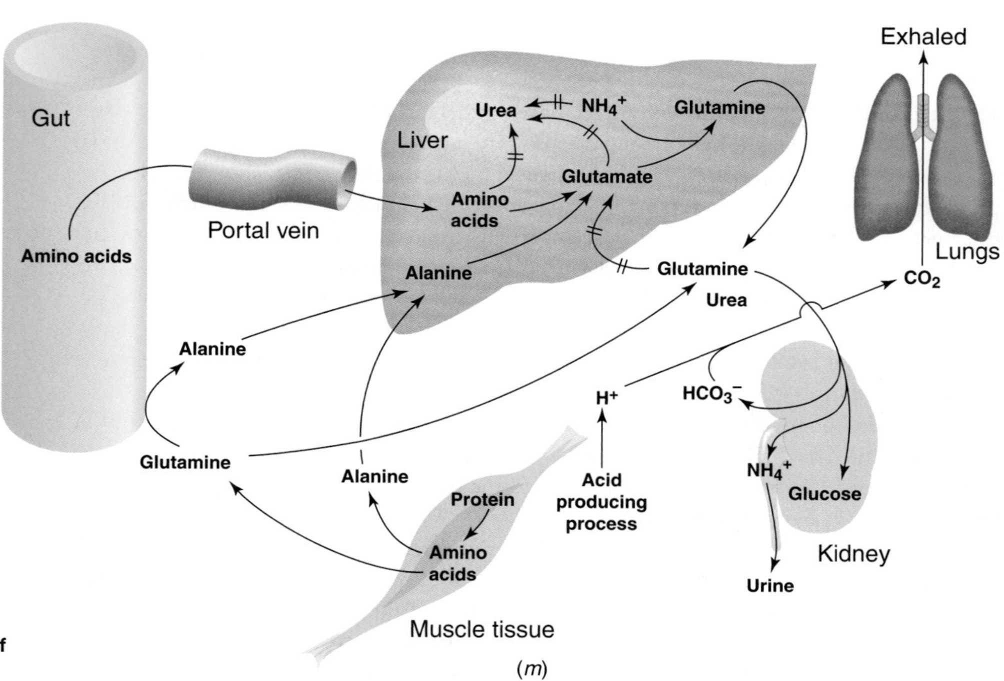

Figure 21.35 Metabolic interrelationships of tissues in acidosis.

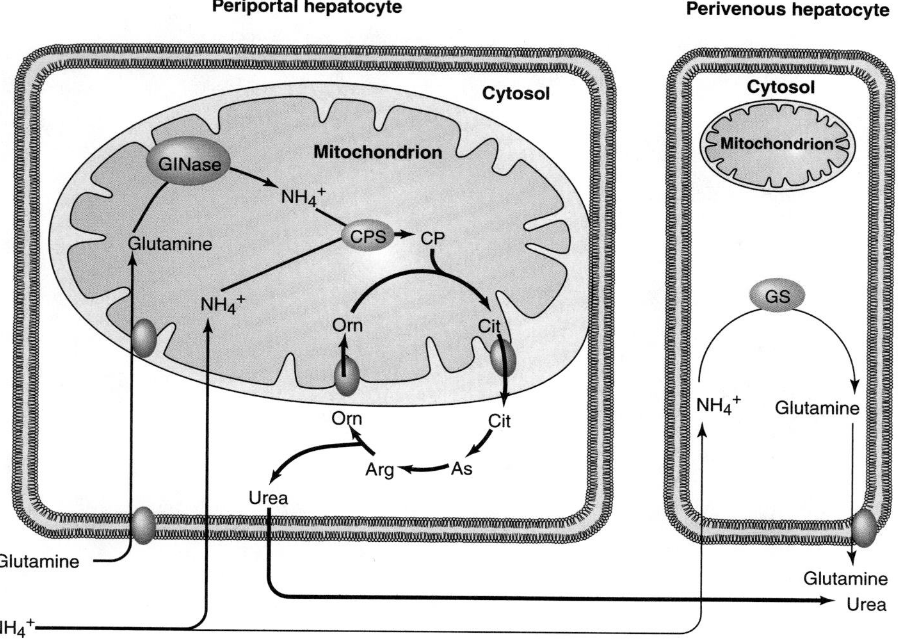

Figure 21.36 Intercellular glutamine cycle of the liver. Abbreviations: GlNase = Glutaminase; GS = Glutamine synthetase; CPS = carbamoyl phosphate synthetase I; CP = carbamoyl phosphate; Cit = citrulline; AS = argininosuccinate; Arg = arginine; and Orn = ornithine.
Redrawn from Häussinger, D. Hepatic glutamine transport and metabolism. *Advances in Enzymology and Related Areas of Molecular Biology* 72:43, 1998.

In these conditions, renal glutaminase, glutamate dehydrogenase, phosphoenolpyruvate carboxykinase, and the mitochondrial glutamine transporter are induced to promote glucose synthesis from glutamine by the reactions given above. The consequence is increased urinary excretion of ammonium ions and greater generation of bicarbonate ion for neutralization of acid. The liver adapts in metabolic acidosis to synthesize less urea, making more glutamine available to the kidney. The opposite occurs in alkalosis. Urea synthesis increases in liver while glucose synthesis, ammonium ion excretion, and bicarbonate generation by kidney decrease.

Liver regulates the fate of glutamine by an intercellular cycle involving periportal hepatocytes near the hepatic arteriole and portal venule and perivenous scavenger hepatocytes located near the central venule (Figure 21.36). Blood enters the liver by way of the hepatic artery and portal vein and leaves by way of the central vein. Glutaminase and urea cycle enzymes are concentrated in the periportal hepatocytes while glutamine synthetase is found exclusively in perivenous scavenger hepatocytes (p. 756). Glutamine entering the periportal cells is hydrolyzed to ammonium ion for urea synthesis; hence, the bulk of glutamine and ammonium nitrogen that enters the liver leaves as urea. The ammonium ion that escapes conversion to urea is converted to glutamine by glutamine synthetase in the perivenous hepatocytes.

Glutamine is released to circulate in the blood before reentering the glutamine cycle in the periportal hepatocytes. Thus, in liver, release of ammonium ion by glutaminase for urea synthesis and its consumption in the synthesis of glutamine are important in maintaining low blood ammonium levels. In acidosis, some of the blood glutamine escapes hydrolysis in the liver because glutamine uptake by hepatocytes and glutaminase activity are partially inhibited by a fall in blood pH. Carbamoyl phosphate synthetase I of periportal hepatocytes is also less active when the blood pH decreases, thereby limiting urea synthesis. This permits perivenous cells to convert more ammonium ion to glutamine, and makes more glutamine available for production of bicarbonate ion by the kidneys.

Colon

The small intestine uses glutamine as its major energy source, but the colon uses the short-chain fatty acids butyrate, propionate, isobutyrate, and acetate (Figure 21.37) derived from bacterial fermentation of unabsorbed dietary components, predominantly carbohydrates such as fiber and pectin, in the lumen. Since they would otherwise be lost in stool, their

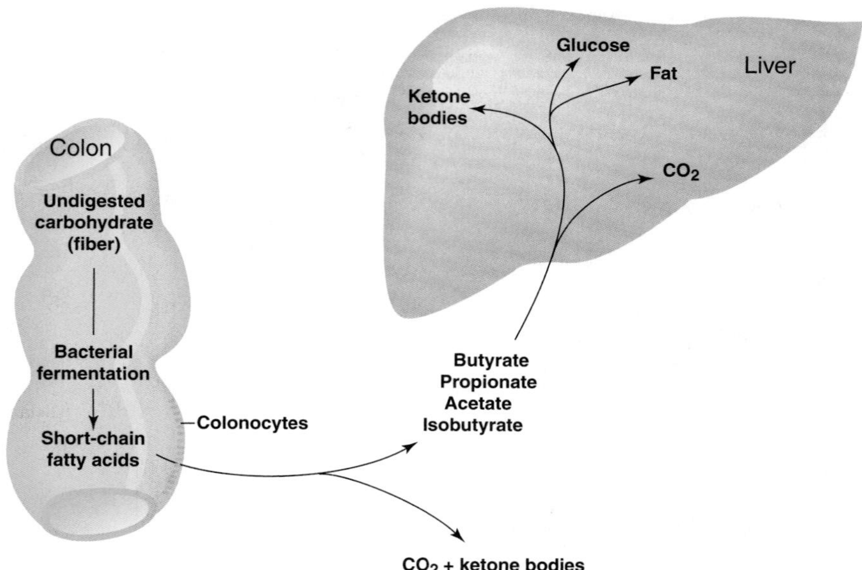

Figure 21.37 Bacterial fermentation generates fuel for colonocytes.

use by colonocytes is a way to salvage additional energy from dietary sources. Short-chain fatty acids produced in excess of the energy needs of colonocytes pass into the portal blood for use by the liver. Interestingly, colonocytes produce ketone bodies from butyrate and release them into the portal blood for use as fuel by extrahepatic tissues. When surgery is performed that bypasses the colon, for example, an ileostomy, some patients develop diversion colitis. In some cases, enemas that contain short-chain fatty acids have healed the colitis.

Bibliography

Amiel, S. A. Organ fuel selection: Brain. *Proc. Nutr. Soc.* 54:151, 1995.

Barthel, A., and Schmoll, D. Novel concepts in insulin regulation of hepatic gluconeogenesis. *Am. J. Physiol. Endocrinol. Metab.* 285:E685, 2003.

Brosnan, J. T. Interorgan amino acid transport and its regulation. *J. Nutr.* 133:2068S, 2003.

Cahill, G. F. Fuel metabolism in starvation. *Annu. Rev. Nutr.* 26:1, 2006.

Calder, P. C. Fuel utilization by cells of the immune system. *Proc. Nutr. Soc.* 54:65, 1995.

Clarke, S. D. Polyunsaturated fatty acid regulation of gene transcription: A molecular mechanism to improve the metabolic syndrome. *J. Nutr.* 131:1129, 2001.

Crabb, D. W., Matsumoto, M., Chang, D., and You, M. Overview of the role of alcohol dehydrogenase and aldehyde dehydrogenase and their variants in the genesis of alcohol-related pathology. *Proc. Nutr. Soc.* 63:49, 2004.

Curthoys, N. P., and Watford, M. Regulation of glutaminase activity and glutamine metabolism. *Ann. Rev. Nutr.* 15:133, 1995.

De Feo, P., Di Loreto, C., Lucidi, P., Murdolo, G., et al. Metabolic response to exercise. *J. Endocrinol. Invest.* 26:851, 2003.

Du, X., Matsumura, T., Edelstein, D., Rossetti, L., et al. Inhibition of GAPDH activity by poly(ADP-ribose) polymerase activates three major pathways of hyperglycemic damage in endothelial cells. *J. Clin. Invest.* 112:1049, 2003.

Duée, P.-H., Darcy-Vrillon, B., Blachier, F., and Morel, M.-T. Fuel selection in intestinal cells. *Proc. Nutr. Soc.* 54:83, 1995.

Elia, M. General integration and regulation of metabolism at the organ level. *Proc. Nutr. Soc.* 54:213, 1995.

Evans, R. M., Barish, G. D., and Wang, Y. X. PPARs and the complex journey to obesity. *Nat. Med.* 10:355, 2004.

Feinman, R.D., and Makowskie, M. Metabolic syndrome and low-carbohydrate ketogenic diets in the medical school biochemistry curriculum. *Metabolic Syndrome and Related Disorders* 1:189, 2003.

Frayn, K. N., Humphreys, S. M., and Coppack, S. W. Fuel selection in white adipose tissue. *Proc. Nutr. Soc.* 54:177, 1995.

Gibson, D. M., and Harris, R. A. *Metabolic Regulation in Mammals.* New York: Taylor & Francis, 2002.

Grimble, R. F. Fasting in healthy individuals and adaption to undernutrition during chronic disease. *Curr. Opin. Clin. Nutr. Metab Care* 1:369, 1998.

Grimble, R. F. Nutritional modulation of immune function. *Proc. Nutr. Soc.* 60:389, 2001.

Halperin, M. L., and Rolleston, F. S. *Clinical Detective Stories. A Problem-Based Approach to Clinical Cases in Energy and Acid-Base Metabolism.* London: Portland, 1993.

Handschin, C., and Spiegelman, B. M. Peroxisome proliferator-activated receptor coactivator 1 coactivators, energy homeostasis, and metabolism. *Endocrine Reviews* 27:728, 2006.

Hardie, D. G. AMP-activated protein kinase: The guardian of cardiac energy status. *J. Clin. Invest.* 114:465, 2004.

Harris, R. A., Huang, B., and Wu, P. Control of pyruvate dehydrogenase kinase gene expression. *Adv. Enz. Regul.* 41:269, 2001.

Häussinger, D. Hepatic glutamine transport and metabolism. *Advances in Enzymology and Related Areas of Molecular Biology* 72:43, 1998.

Henriksson, J. Muscle fuel selection: Effect of exercise and training. *Proc. Nutr. Soc.* 54:125, 1995.

Holness, M. J., and Sugden, M. C. Regulation of pyruvate dehydrogenase complex activity by reversible phosphorylation. *Biochem. Soc. Trans.* 31:1143, 2003.

Homko, C. J., Sivan, E., Reece, E. A., and Boden, G. Fuel metabolism during pregnancy. *Semin. Reprod. Endocrinol.* 17:119, 1999.

Horton, J. D., Goldstein, J. L, and Brown, M. S. SBREPs: Activators of the complete program of cholesterol and fatty acid synthesis in the liver. *J. Clin. Invest.* 109:1125, 2002.

Kersten, S., Seydoux, J., Peters, J. M., Gonzalez, F. J., et al. Peroxisome proliferator-activated receptor alpha mediates the adaptive response to fasting. *J. Clin. Invest.* 103:1489, 1999.

Kim, J.-W., Tchernyshyov, I., Semenza, G. L., and Dang, C. V. HIF-1-mediated expression of pyruvate dehydrogenase kinase: A metabolic switch required for cellular adaptation to hypoxia. *Cell Metabolism* 3:177, 2006.

Krebs, H. A. Some aspects of the regulation of fuel supply in omnivorous animals. *Adv. Enzyme Regul.* 10:387, 1972.

Krebs, H. A., Williamson, D. H., Bates, M. W., Page, M. A., and Hawkins, R. A. The role of ketone bodies in caloric homeostasis. *Adv. Enzyme Regul.* 9:387, 1971.

Kurkland, I. J., and Pilkis, S. J. Indirect and direct routes of hepatic glycogen synthesis. *FASEB J.* 3:2277, 1989.

Large, V., Peroni, O., Letexier, D., Ray, H., and Beylot, M. Metabolism of lipids in human white adipocytes. *Diabetes Metab.* 30:294, 2004.

Lecker, S. H., Solomon, V., Mitch, W. E., and Goldberg, A. L. Muscle protein breakdown and the critical role of the ubiquitin-proteasome pathway in normal and disease states. *J. Nutr.* 129:227S, 1999.

MacDonald, I. A., and Webber, J. Feeding, fasting and starvation: Factors affecting fuel utilization. *Proc. Nutr. Soc.* 54:267, 1995.

McGarry, J. D., and Brown, N.F. The mitochondrial carnitine-palmitoyl-CoA transferase system. *Eur. J. Biochem.* 244:1, 1997.

Newsholme, E. A., and Leech, A. R. *Biochemistry for the Medical Sciences.* New York: Wiley, 1983.

Newsholme, P. Why is L-glutamine metabolism important to cells of the immune system in health, postinjury, surgery, or infection? *J. Nutr.*131:2515S, 2001.

Nosadini, R., Avogaro, A., Doria, A., Fioretto, P., et al. Ketone body metabolism: A physiological and clinical overview. *Diabetes Metab. Rev.* 5:299, 1989.

Olde Damink, S. W. M., Dejong, C. H. C., Deutz, N. E. P., van Berlo, C. L. H., and Soeters, P. B. Upper gastrointestinal bleeding: An ammoniagenic and catabolic event due to the total absence of isoleucine in the haemoglobin molecule. *Medical Hypotheses* 52:515, 1999.

Ookhtens, M., and Kaplowitz, N. Role of the liver in interorgan homeostasis of glutathione and cyst(e)ine. *Seminars in Liver Disease* 18:313, 1998.

Pedersen, O. The impact of obesity on the pathogenesis of noninsulin-dependent diabetes mellitus: A review of current hypotheses. *Diabetes/Metabolism Rev.* 5:495, 1989.

Pilkis, S. J., Claus, T. H., Kurland, I. J., and Lange, A. J. 6-Phosphofructo-2-kinase/fructose-2,6-bisphosphatase: A metabolic signaling enzyme. *Ann. Rev. Biochem.* 64:799, 1995.

Quistorff, B., Secher, N. H., and Van Lieshout, J. J. Lactate fuels the human brain during exercise. *FASEB J.* 22:3443, 2008.

Randle, P. J. Metabolic fuel selection: general integration at the whole-body level. *Proc. Nutr. Soc.* 54:317, 1995.

Rider, M. H., Bertrand, L., Vertommen, D., Michels, P. A., et al. 6-phosphofructo-2-kinase/fructose-2,6-bisphosphatase: head-to-head with a bifunctional enzyme that controls glycolysis. *Biochem. J.* 381:561, 2004.

Roach, P. J. Glycogen and its metabolism. *Curr. Mol. Med.* 2:101, 2002.

Ruderman, N. B., Park, H., Kaushik, V. K., Dean, D., et al. AMPK as a metabolic switch in rat muscle, liver and adipose tissue after exercise. *Acta Physiol. Scand.* 178:435, 2003.

Ruderman, N. B., Saha, A. K., Vavvas, D., and Witters, L. A. Malonyl CoA, fuel sensing, and insulin resistance. *Am. J. Physiol. Endocrinol. Metab.* 276:E1, 1999.

Shulman, G. I., and Landau, B. R. Pathways of glycogen repletion. *Physiological Rev.* 72:1019, 1992.

Speigelman, B. M., and Heinrich, R. Biological control through regulated transcriptional coactivators. *Cell* 119:157, 2004.

Steppan, C. M., Bailey, S. T., Bhat, S., Brown, E. J., et al. The hormone resistin links obesity to diabetes. *Nature* 409:307, 2001.

Stubbs, M., Bashford, C. L., and Griffiths, J. R. Understanding the tumor metabolic phenotype in the genomic era. *Curr. Mol. Med.* 3:485, 2003.

Taylor, S. I. Diabetes mellitus. In Scriver, C. R., Beaudet, A. R., Sly, W. S., and Valle, D. (Eds.), *The Metabolic and Molecular Bases of Inherited Disease,* 8th ed. New York: McGraw Hill, 2001.

Uyeda, K., and Repa, J. J. Carbohydrate response element binding protein, ChREBP, a transcription factor coupling hepatic glucose utilization and lipid synthesis. *Cell Metabolism* 4:107, 2006.

Wannemacher, R.W., Jr., Pace, J. G., Beall, F. A., Dinterman, R. E., Petrella, V. J., and Neurfeld, H. A. Role of the liver in the regulation of ketone body production during sepsis. *J. Clin. Invest.* 64:1565, 1979.

Williamson, D. H., and Lund, P. Substrate selection and oxygen uptake by the lactating mammary gland. *Proc. Nutr. Soc.* 54:165, 1995.

Zick, Y. Molecular basis of insulin action. *Novartis Found. Symp.* 262:36, 2004.

Key Terms

metabolic interrelationships
metabolic pathways
starve–feed cycle
glycogenesis
glycolysis
6-phosphofructo-1-kinase
fatty acid synthesis
lipogenesis
Cori cycle
gluconeogenesis
glycogenolysis
glutamine
glutaminolysis
ketone bodies
lipolysis
proteolysis
caloric homeostasis
hyperglycemia
glucose homeostasis
glycation
starvation
induction and repression
ketogenesis
allosteric effectors
glucokinase
acetyl-CoA carboxylase
carnitine palmitoyltransferase I
malonyl CoA
AMP-activated protein kinase
cAMP
covalent modification
protein kinase A
fructose 1,6-bisphosphatase
metabolic fuels
sterol response-element-binding protein
peroxisome proliferator-activated receptor α
PPARγ-coactivator 1 α
obesity
type 2 diabetes mellitus
type 1diabetes mellitus
cancer
aerobic exercise
anaerobic exercise
pregnancy
lactation
stress and injury
acid-base balance

Questions • CAROL N. ANGSTADT

Multiple-Choice Questions

1. Since the K_m of aminotransferases for amino acids is much higher than that of aminoacyl-tRNA synthetases,
   A. at low amino acid concentrations, protein synthesis takes precedence over amino acid catabolism.
   B. liver cannot accumulate amino acids.
   C. amino acids will undergo transamination as rapidly as they are delivered to the liver.
   D. any amino acids in excess of immediate needs for energy must be converted to protein.
   E. amino acids can be catabolized only if they are present in the diet.

2. Carnitine
   A. is formed in all cells for their own use.
   B. is synthesized directly from free lysine.
   C. formation requires that lysyl residues in protein be methylated by *S*-adenosylmethionine.
   D. is important in the detoxification of peroxides.
   E. is cleaved to γ-butyrobetaine.

3. All of the following represent control of a metabolic process by substrate availability except
   A. increased urea synthesis after a high protein meal.
   B. rate of ketogenesis.
   C. hypoglycemia of advanced starvation.
   D. response of glycolysis to fructose 2,6-bisphosphate.
   E. gluconeogenesis in diabetes.

4. Conversion of hepatic enzymes from nonphosphorylated to their phosphorylated form
   A. always activates the enzyme.
   B. is more likely to occur in the fasted than in the well-fed state.
   C. is signaled by insulin.
   D. is always catalyzed by a cAMP-dependent protein kinase.
   E. usually occurs at threonine residues of the protein.

5. Long-term adaptation to changes in the nutritional state can be regulated by the change in synthesis of enzymes, regulated by the rate of gene transcription. All of the following statements concerning transcription are correct except
   A. insulin increases a sterol-response-element-binding protein (SREBP), which increases transcription of lipogenic genes.
   B. glucose increases a carbohydrate-response-element-binding protein (ChREBP), which decreases the transcription of lipogenic genes.
   C. glucagon promotes the transcription of genes encoding gluconeogenic enzymes via activation of adenylate cyclase.
   D. a SRBEP regulates cholesterol synthesis.
   E. energy deficiency activates AMPK and reduces the ability of SREBP to transcribe genes for fat and cholesterol synthesis.

6. Arginine and ornithine are intermediates in the utilization of ammonia and in other processes. All of the following statements are correct except
   A. ornithine for the urea cycle is synthesized from glutamate in the kidney.
   B. citrulline is a precursor for arginine synthesis by the kidney.
   C. kidney uses arginine in the synthesis of creatine for distribution to muscle.
   D. arginine is the source of nitric oxide.
   E. creatinine cleared by the kidney is generated from creatine phosphate in muscle.

Questions 7 and 8: Protein and calorie malnutrition are important nutritional problems, especially among children. Both often occur in developing countries when the child is weaned from breast milk. Protein malnutrition, kwashiorkor, occurs when the child is fed a diet adequate in calories (mostly carbohydrate) but deficient in protein. Inadequate caloric intake is called marasmus. In kwashiorkor, insulin levels are high and there is subcutaneous fat. Children with low weight for height can make a good recovery when properly fed although those with the reverse situation do not. Children with marasmus lack subcutaneous fat.

7. Adipose tissue responds to low insulin:glucagon ratio by
   A. dephosphorylating the interconvertible enzymes.
   B. stimulating the deposition of fat.
   C. increasing the amount of pyruvate kinase.
   D. activation of hormone-sensitive lipase.
   E. releasing glutamine.

8. Which of the following would favor gluconeogenesis in the fasted state?
   A. Fructose 1,6-bisphosphate stimulation of pyruvate kinase
   B. Acetyl CoA activation of pyruvate carboxylase
   C. Citrate activation of acetyl-CoA carboxylase
   D. Malonyl CoA inhibition of carnitine palmitoyltransferase I
   E. Fructose 2,6-bisphosphate stimulation of 6-phosphofructo-1-kinase

Questions 9 and 10: Advanced liver disease, cirrhosis, leads to a number of metabolic problems, especially of amino acids. Patients suffer from muscle wasting, with an impairment of protein synthesis and an increased rate of protein degradation and ammonia production. The liver cannot convert ammonia into urea and glutamine rapidly enough and blood ammonia level rises; this may lead to coma. Patients also often have insulin resistance. With outright liver failure, patients sometimes die of hypoglycemia because liver is unable to maintain the blood glucose level by gluconeogenesis.

9. Normally the liver helps to maintain low blood ammonia levels by the glutamine cycle of which glutaminase is an important component. Glutaminase
   A. is found in the same liver compartment as the urea cycle enzymes.
   B. requires ATP for the reaction it catalyzes.
   C. is the sole source of ammonia for urea synthesis.
   D. is more active in liver in acidosis.
   E. provides the product that combines with ornithine in the urea cycle.

10. Muscle proteolysis releases branched-chain amino acids which
    A. can also be synthesized from alanine.
    B. can be catabolized by muscle but not liver.
    C. are the main amino acids metabolized by intestine.
    D. are the amino acids released in largest amounts by muscle.
    E. are a major source of nitrogen for alanine and glutamine produced in muscle.

Questions 11 and 12: Diabetes mellitus is a disease in which glucose metabolism is impaired. Type 2 occurs primarily in middle-aged to older obese individuals whereas type 1 usually appears in childhood or in the teens. Insulin's ability to control blood glucose levels is affected either because of lack (or very low levels) of insulin or inadequate function of insulin (or both).

11. In type 2 (noninsulin-dependent) diabetes mellitus,
   A. hypertriglyceridemia does not occur.
   B. ketoacidosis in the untreated state is always present.
   C. β-cells of the pancreas are no longer able to make any insulin.
   D. may be accompanied by high levels of insulin in the blood.
   E. severe weight loss always occurs.

12. Insulin normally does all of the following except
   A. recruit glucose transporters 4 (GLUT4) from intracellular sites to the plasma membrane.
   B. activates protein kinase A and AMP-dependent protein kinase (AMPK).
   C. activates phosphoprotein phosphatases.
   D. signals an increase in sterol-response-element-binding proteins (SREBP-1).
   E. inhibits ketogenesis at levels lower than required to inhibit gluconeogenesis.

Problems

13. What metabolic and hormonal changes account for decreased gluconeogenesis in phase IV (2 to 24 days of starvation) of glucose homeostasis in humans?

14. How does the activation of AMP-dependent protein kinase (AMPK) during aerobic exercise foster the switch to fatty acid oxidation for long-distance running?

Answers

1. **A** High K_m means a reaction will proceed slowly at low concentrations, whereas a low K_m means the reaction can be rapid under the same circumstances. Protein synthesis requires only that all amino acids be present.
2. **C** These trimethyllysines are released when protein is hydrolyzed. A: Only liver and kidney have the complete synthetic pathway. B: Lysine must first be present in cellular protein. D: This is glutathione. E: This is a precursor.
3. **D** Fructose 2,6-bisphosphate is an allosteric effector (activates the kinase and inhibits the phosphatase) of the enzyme controlling glycolysis. A: After a high-protein meal, the intestine produces ammonia and precursors of ornithine for urea synthesis. B: Ketogenesis is dependent on the availability of fatty acids. C: This represents lack of gluconeogenic substrates. E: Amino acids from excess muscle breakdown provide gluconeogenic substrates to the liver.
4. **B** In the well-fed state, insulin:glucagon is high and cAMP levels are low. A: Some enzymes are active when phosphorylated; for others the reverse is true. C: Glucagon signals the phosphorylation of hepatic enzymes by elevating cAMP. D: This is the most common, though not only, mechanism of phosphorylation. E: The most common site for phosphorylation in metabolic enzymes is serine.
5. **B** ChREBP increases the transcription of lipogenic genes. A: This occurs in the well-fed state. C: Glucagon is released in response to low blood glucose so would be expected to increase gluconeogenic enzymes. D: The cytosolic HMGCoA synthase is increased which leads to cholesterol synthesis. E: These activities are energy requiring so need to be inhibited when energy is deficient.
6. **A** Kidney lacks the enzyme needed to convert glutamate to glutamate semialdehyde. B: This is true in both kidney and liver. C: The reaction requires *S*-adenosylmethionine. E: Creatinine is thus a measure of both muscle mass and renal function.
7. **D** Phosphorylation activates hormone-sensitive lipase to mobilize fat. A: Low insulin:glucagon means high cAMP and high activity of cAMP-dependent protein kinase and protein phosphorylation. C: cAMP works by stimulating covalent modification of enzymes. E: This occurs in muscle.
8. **B** Pyruvate carboxylase is a key gluconeogenic enzyme. A and E: Stimulation of these enzymes stimulates glycolysis. C and D: Malonyl CoA inhibits transport of fatty acids into mitochondria for β-oxidation, a necessary source of energy for gluconeogenesis.
9. **A** This is the periportal hepatocytes. B: Glutamine synthetase requires ATP; glutaminase catalyzes hydrolysis of glutamine to glutamate and ammonia. C: Glutamate dehydrogenase and bacterial urease in the intestine also provide ammonia. D: Less flux through liver glutaminase during acidosis permits glutamine to escape liver for use by the kidney. E: Ammonia reacts with bicarbonate to form carbamoyl phosphate, which combines with ornithine.
10. **E** Transamination of branched-chain amino acids transfers the nitrogen to alanine or glutamine. A: Branched-chain amino acids are essential amino acids and cannot be synthesized from other amino acids. B: Muscle transaminates them, and then the branched-chain α-ketoacids are catabolized by liver. C: Intestine metabolizes primarily glutamine. D: Alanine and glutamine are.
11. **D** The problem is insulin resistance, not complete failure to produce insulin. A: Hypertriglyceridemia is characteristic. B: Ketoacidosis is common only in the insulin-dependent type. C: See correct answer. E: Most patients are obese and remain so.
12. **B** These enzymes are active in their phosphorylated states and insulin promotes dephosphorylation. A: This promotes uptake of glucose. D: This increases transcription of genes encoding lipogenic enzymes. E: This may account in part for the lack of ketoacidosis in type 2 diabetics.
13. In phase IV, most tissues are using primarily fatty acids and ketone bodies. Ketone bodies are now sufficiently high that they can enter the brain and reduce the glucose requirement. Low insulin:glucagon stimulates lipolysis and gluconeogenesis. Increased fatty acid oxidation increases acetyl CoA and NADH. High NADH allosterically inhibits the tricarboxylic acid cycle and the accumulating acetyl CoA is converted to ketone bodies. Low insulin:glucagon also decreases catabolism of muscle protein. Decreased gluconeogenesis conserves protein for a longer period of time.
14. AMP is increased as ATP is used for contraction, so AMPK is activated. AMPK phosphorylates and activates enzymes that foster glycogenolysis and glycolysis and glycogen is used until it is exhausted. AMPK also phosphorylates and inactivates acetyl-CoA carboxylase and activates malonyl-CoA decarboxylase reducing malonyl CoA. This allows greater carnitine palmitoyltransferase I activity to move long-chain acyl CoA esters into the mitochondria for fatty acid oxidation.

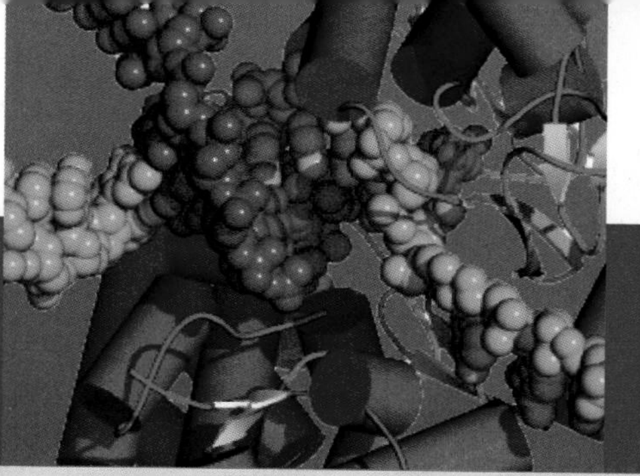

22

Biochemistry of Hormones

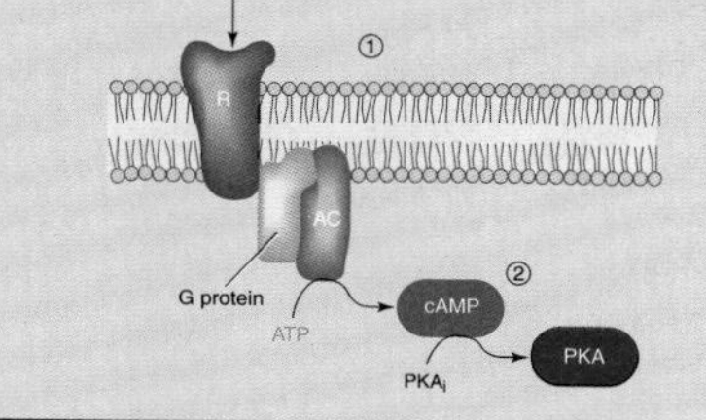

Thomas J. Schmidt

Professor, Carver College of Medicine, University of Iowa

Key Concepts

- The hormonal cascade refers to (1) synthesis and secretion by hypothalamic neurons of specific releasing hormones, (2) releasing hormones stimulate the synthesis and secretion of tropic hormones by specific cells of the pituitary anterior lobe, and (3) tropic hormones increase the synthesis and secretion of specific hormones by target endocrine glands.
- Some genes for hormones code for large proteins that serve as precursors for a number of smaller proteins with distinctive hormonal activities. Other genes code for multiple copies of the same hormone.
- Norepinephrine and epinephrine are synthesized from tyrosine in the adrenal medulla. Thyroid hormone is synthesized by incorporation of iodine into tyrosine residues of thyroglobulin stored in the lumen of thyroid gland follicles.
- Protein hormones transmit their signals to target cells by binding to specific, high affinity membrane receptors; this leads to an increase in intracellular second messengers including cyclic AMP, cyclic GMP, inositol 1,4,5, trisphosphate, diacylglycerol, and phosphatidylinositol 3,4,5-trisphosphate.
- The female ovarian cycle is controlled by the pulsatile and cyclic secretion of gonadotropin-releasing hormone from the hypothalamus, which stimulates synthesis and secretion of both follicle stimulating and luteinizing hormone in the anterior pituitary. These hormones stimulate ovarian synthesis of estradiol and progesterone.
- Many second messengers activate specific protein kinases. Binding of insulin to its receptor activates auto-tyrosine kinase.
- Steroid hormones are derived from cholesterol and are synthesized in the adrenal cortex (aldosterone, cortisol, androgens), the Leydig cells of the male testes (testosterone, estrogen), and the female ovaries (estrogen, progesterone, androgens).
- Steroid hormone binding to specific plasma proteins protects them from degradation.

- Intracellular receptors for steroid and thyroid hormones, as well as vitamin D are members of the same superfamily and function as ligand-activated transcription factors, increasing gene transcription by binding to hormone response elements.
- Receptor proteins of this superfamily have three major functional domains: a carboxy-terminal ligand-binding domain; a DNA-binding domain; and a highly variable amino-terminal domain containing an antigenic region and a region that modulates transcriptional activation.

22.1 • INTRODUCTION

The hormones that will be discussed in this chapter fall into three major categories: **peptide and protein hormones**, hormones derived from the amino acid tyrosine (**thyroid hormones** and the **catecholamine hormones**), and **steroid hormones**. Taken collectively, these hormones regulate the growth, differentiation, and function of a wide variety of target cells that express receptors specific for these hormones. The peptide hormones and the catecholamine hormones interact with **cell surface receptors** (Figure 22.1) and transmit their signals through **second messengers** that are generated intracellularly. The binding of insulin to its cell surface receptor activates an intrinsic tyrosine kinase activity. Steroid hormones are derived from cholesterol and include glucocorticoid hormones, mineralocorticoid hormones, and the sex hormones. Steroid hormones diffuse freely across the plasma membrane and bind to **intracellular receptors** that function as **ligand-activated transcription factors**. The intracellular receptors for all of the steroid hormones and for nonsteroid hormones, including thyroid hormone, the active metabolite of vitamin D3, and retinoic acid, belong to the **steroid receptor super family** and share sequence homologies. The three major domains of steroid receptor are a C-terminal ligand-binding domain, a DNA-binding domain and an N-terminal immunogenic domain.

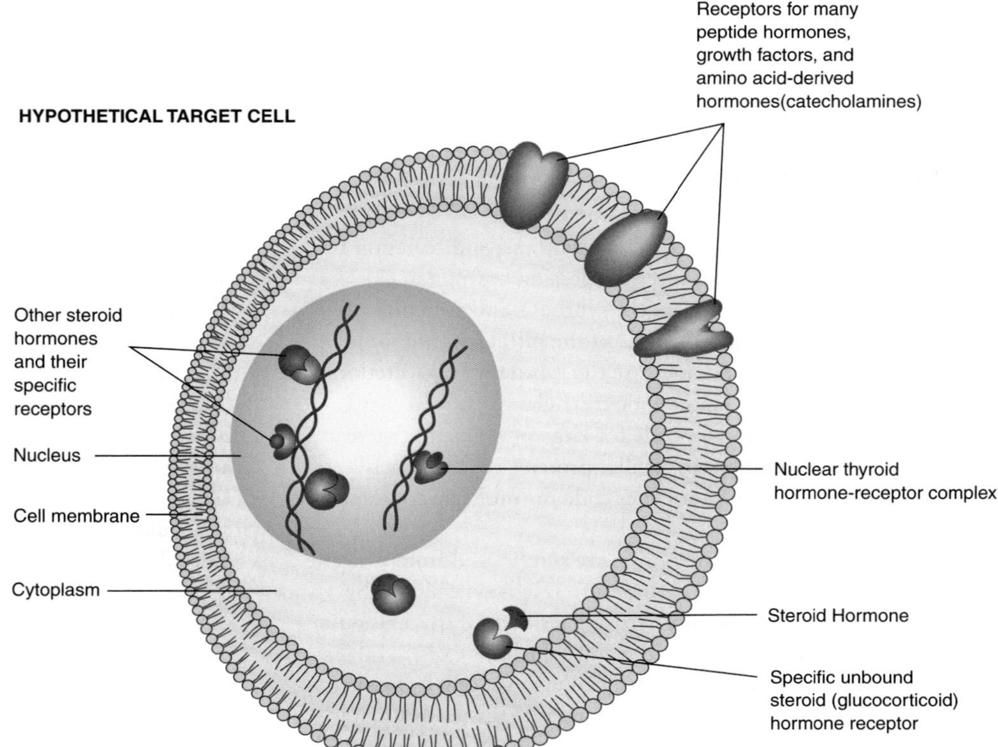

Figure 22.1 **Diagram showing the different locations of classes of hormone receptors expressed by a target cell.**

22.2 • HORMONES AND THE HORMONAL CASCADE SYSTEM

The definition of a hormone has expanded over the past several decades. Hormones secreted by endocrine glands were considered for many years to represent all of the physiologically relevant hormones. The term **hormone** now refers to any substance in an organism that carries a "signal" to generate some alteration at the cellular level. **Endocrine hormones** are synthesized in one tissue or gland and travel through the general circulation to reach distant target cells that express cognate receptors. **Paracrine hormones** are secreted by a cell and travel a relatively short distance to interact with cognate receptors on a neighboring cell. **Autocrine hormones** are produced by a cell that is also the target for that hormone (neighboring cells may also be targets). Endocrine hormones are frequently more stable than autocrine hormones that exert their effects over very short distances.

Hormonal Cascade Systems Amplify Specific Signals

Before we focus on details concerning each hormone, we need to look more broadly at the organization of the endocrine system and the hormonal hierarchy. For many hormonal systems in higher animals the signal pathway originates in the brain and culminates within the target cell. Figure 22.2 outlines the sequence of events in this cascade. A stimulus may originate in the external environment or within the organism and can be transmitted as action potentials, chemical signals, or both. In many cases, such signals are forwarded to the limbic system and subsequently to the hypothalamus, the anterior pituitary, and the target gland that secretes the ultimate hormone that then affects various target cells, frequently in proportion to the number of cognate receptors expressed by those cells. This may be a true **cascade** in the sense that increasing amounts of hormone are generated at successive levels (hypothalamus, anterior pituitary, and target gland) and in the sense that the half-lives ($t_{1/2}$) of blood-borne hormones tend to become longer the further they are down the cascade.

Consider a specific hormone secreted via one of these cascades. An environmental stress such as change in temperature, noise, or trauma results in a signal to the hippocampal structure

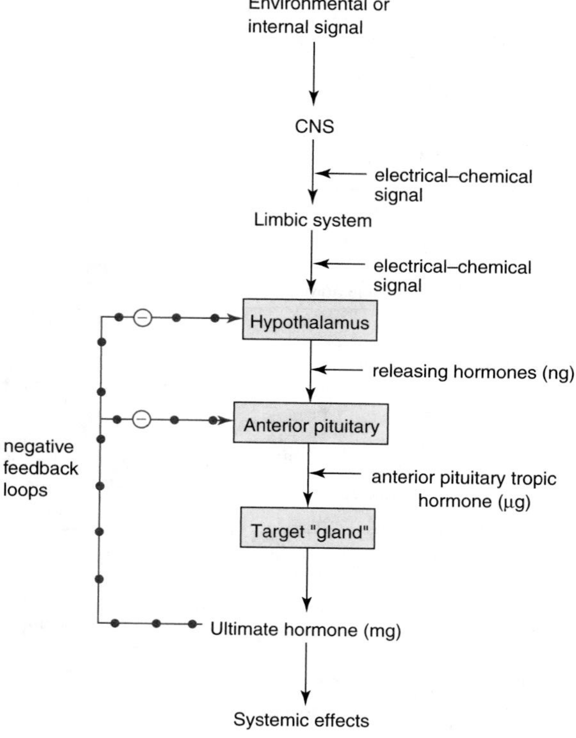

Figure 22.2 Hormonal cascade of signals from CNS to ultimate hormone. The target gland is the last hormone-producing tissue in the cascade, which is stimulated by an appropriate anterior pituitary hormone. Examples are thyroid gland, adrenal cortex, ovary, and testes. Ultimate hormone feeds back negatively on sites producing intermediate hormones in the cascade. Amounts (nanogram [ng], microgram [µg], and milligram [mg]) represent approximate quantities of hormone released.
Redrawn from Norman, A. W., and Litwack, G. *Hormones.* New York: Academic Press, 1987, 38.

from the limbic system for release of nanogram amounts of a hypothalamic-releasing hormone, corticotropin-releasing hormone (CRH), which has a $t_{1/2}$ in the bloodstream of several minutes. CRH travels down a closed portal system to the **anterior pituitary**, where it binds its cognate receptor in the membrane of corticotropic cells and initiates intracellular events that result in release of adrenocorticotropic hormone (ACTH) and β-lipotropin. ACTH is released in microgram amounts and has a longer $t_{1/2}$ than CRH. ACTH circulates until it binds to its cognate receptors expressed on the membranes of cells in the zona fasciculata of the adrenal cortex (target gland). Here it increases the synthesis and release of the steroid hormone cortisol in milligram amounts. Circulating cortisol has a longer $t_{1/2}$ than ACTH. Cortisol then interacts with target cells throughout the body that express intracellular glucocorticoid receptors. The ultimate hormone, cortisol, feeds back negatively on cells of the anterior pituitary and hypothalamus and reduces its own rate of synthesis and secretion from the adrenal cortex. At the level of the **target cell**, cortisol–receptor complexes mediate specific transcriptional responses, which collectively produce the systemic effects of cortisol. Other systems operate through similar cascades although through different specific **releasing hormones**, **anterior pituitary tropic hormones**, and ultimate hormones. Clearly, the number of target cells affected depends on their expression of receptor for ultimate hormones.

A distant but related system involves the **posterior pituitary hormones**, oxytocin and vasopressin (antidiuretic hormone), which are stored in and secreted by the posterior pituitary but are synthesized in neuronal cell bodies in the hypothalamus. This system is represented in Figure 22.3, which is an extended version of Figure 22.2. The posterior pituitary system branches to the right from the hypothalamus. Oxytocin and vasopressin are synthesized in separate cell bodies of hypothalamic neurons. Synthesis of vasopressin occurs largely in the supraoptic nucleus and synthesis of oxytocin occurs largely in the paraventricular nucleus. Their release from the posterior pituitary is independent and occurs in response to separate stimuli.

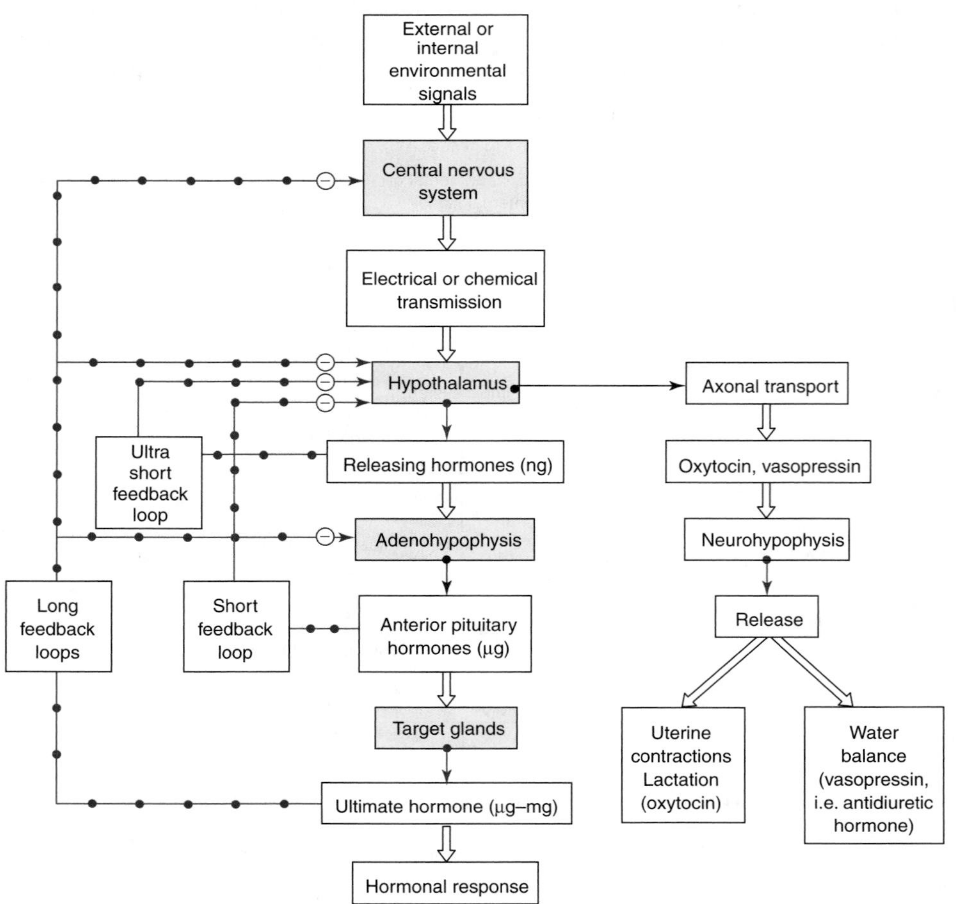

Figure 22.3 Many hormonal systems involve hypothalamus. Cascade of hormonal responses starting with an external or internal signal. This signal is transmitted first to the CNS and may involve the limbic system, including the hippocampus and amygdala. These components innervate the hypothalamus in a specific region, which responds by secreting (nanogram amounts) a specific releasing hormone. Releasing hormones are transported down a closed portal system to the anterior pituitary, where they cause secretion of microgram amounts of specific anterior pituitary hormones. These access the general circulation through fenestrated local capillaries and trigger release of an ultimate hormone in microgram to milligram daily amounts. The ultimate hormone generates its response by binding to receptors in target tissues. Overall, this system is an amplifying cascade. Consequently, the organism is in intimate association with the external environment. Solid arrows indicate a secretory process. Long arrows studded with open or closed circles indicate negative feedback pathways (ultrashort, short and long feedback loops).
Redrawn from Norman, A. W., and Litwack, G. *Hormones.* New York: Academic Press, 1987, 102.

TABLE 22.1 • Hypothalamic Releasing Hormones[a]

Releasing Hormone	*Number of Amino Acids in Structure*	*Anterior Pituitary Hormone Released or Inhibited*
Thyrotropin-releasing hormone (TRH)	3	Thyrotropin (TSH)
Gonadotropin-releasing hormone (GnRH)	10	Luteinizing and follicle-stimulating hormones (LH and FSH) from the same cell type; leukotriene C_4 (LTC_4) can also release LH and FSH by a different mechanism
Gonadotropin release-inhibiting factor (GnRIF)	Not established	
Corticotropin-releasing hormone (CRH)	41	ACTH, β-lipotropin (β-LPH), and some β-endorphin
Arginine vasopressin (AVP)	9	Stimulates CRH action on ACTH release
Angiotensin II (AII)	8	Stimulates CRH action on ACTH release; releases ACTH weakly
Growth hormone releasing hormone (GHRH)	49	Growth hormone (GH) release
Somatostation (growth hormone release-inhibiting hormone, GHIH)	14	GH release inhibited
Hypothalamic gastrin-releasing peptide		Inhibits release of GH and PRL
Prolactin-releasing factor (PRF)	Not established	Releases prolactin (PRL)
Prolactin release-inhibiting factor (PIF)		Evidence that a new peptide may inhibit PRL release; dopamine also inhibits PRL release and may be a secondary PIF: oxytocin may inhibit PRL release

[a] Melanocyte-stimulating hormone (MSH) is a major product of the pars intermedia (Figure 23.5) in the rat and is under the control of aminergic neurons. Humans may also secrete α-MSH from pars intermedia-like cells, although this structure is anatomically indistinct in the human.

Highly specific signals dictate release of polypeptide hormones along the cascade. Thus, **aminergic neurons** that secrete dopamine and/or serotonin relay to neurons involved in the synthesis and secretion of **releasing hormones** from the hypothalamus. Releasing hormones are summarized in Table 22.1. Aminergic neurons respond to various types of internal or external signals. Their activities account for the **pulsatile release** of hormones such as gonadotropin-releasing hormone (GnRH), and the **rhythmic cyclic release** of hormones like cortisol.

A prominent feature of the hormonal cascade (Figure 22.3) is the **negative feedback** that operates when sufficiently high levels of an ultimate hormone have been secreted. In long-loop feedback, the final hormone binds a cognate receptor in/on cells of the anterior pituitary, hypothalamus, and CNS to prevent further synthesis/secretion of releasing hormones. Short-loop feedback is exemplified by the pituitary tropic hormone that feeds back negatively on the hypothalamus and operates through a cognate receptor. In ultrashort feedback loops the hypothalamic releasing factor feeds back on the hypothalamus to inhibit its own further secretion.

Major Polypeptide Hormones and Their Actions

Since cellular communication is so specific, it is not surprising that there are numerous hormones in the body and new hormones continue to be discovered. Table 22.2 presents some major polypeptide hormones and their actions and shows that many hormones cause release of other hormones. This is particularly the case for hormonal cascade systems like that presented in Figures 22.2 and 22.3.

Polypeptide Hormones of Anterior Pituitary

Polypeptide hormones of the anterior pituitary are shown in Figure 22.4 along with their release-controlling hormones from the hypothalamus. The major hormones are growth hormone (GH), thyrotropin or thyroid-stimulating hormone (TSH), adrenocorticotropic hormone (ACTH), β-lipotropin (β-LTH), β-endorphin (from pars intermedia-like cells), α-MSH (from pars intermedia-like cells), β-MSH (from pars intermedia-like cells),

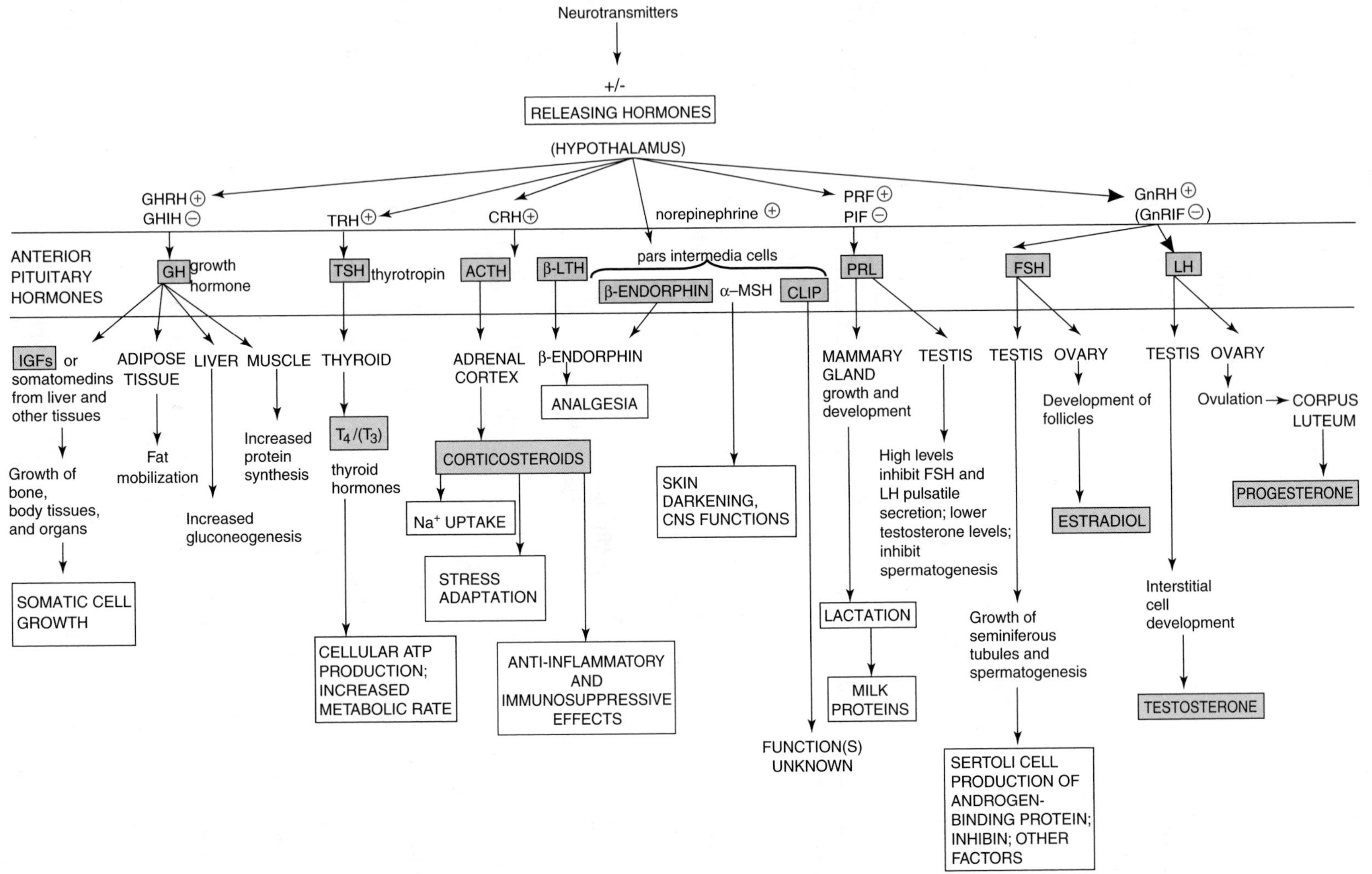

Figure 22.4 Overview of anterior pituitary hormones with hypothalamic releasing hormones and their actions.

TABLE 22.2 • Important Polypeptide Hormones in the Body and Their Actions[a]

Source	*Hormone*	*Action*
Hypothalamus	Thyrotropin-releasing hormone (TRH)	Acts on thyrotrope to release TSH
	Gonadotropin-releasing hormone (GnRH)	Acts on gonadotrope to release LH and FSH from the same cell
	Growth hormone-releasing hormone or somatocrinin (GRH)	Acts on somatotrope to release GH
	Growth hormone release inhibiting hormone or somatostatin (GIH)	Acts on somatotrope to prevent release of GH
	Corticotropin-releasing hormone (CRH)	Acts on corticotrope to release ACTH and β-lipotropin
	Angiotensin II and vasopressin stimulate CRH action in releasing ACTH	
	Prolactin-releasing factor (PRF) (not well established)	Acts on lactotrope to release PRL
	Prolactin release-inhibiting factor (PIF) (not well established; may be a peptide hormone under control of dopamine or may be dopamine itself)	Acts on lactotrope to inhibit release of PRL
Anterior pituitary	Thyrotropin (TSH)	Acts on thyroid follicle cells to bring about release of $T_4(T_3)$
	Luteinizing hormone (LH) (human chorionic gonadotropin, hCG, is a similar hormone from the placenta)	Acts on Leydig cells of testes to increase testosterone synthesis and release; acts on corpus luteum of ovary to increase progesterone production and release
	Follicle-stimulating hormone (FSH)	Acts on Sertoli cells of seminiferous tubule to increase secretion of androgen-binding protein (ABP) and increase estradiol production from testosterone; acts on ovarian follicles to stimulate maturation of ovum and production of estradiol
	Growth hormone (GH)	Acts on a variety of cells to produce IGFs (or somatomedins), cell growth, and bone growth
	Adrenocorticotropic hormone (ACTH)	Acts on cells in the adrenal cortex to increase cortisol production and secretion
	β-Endorphin	Acts on cells and neurons to produce analgesic and other effects
	Prolactin (PRL)	Acts on mammary gland to cause differentiation of secretory cells (with other hormones) and to stimulate synthesis of components of milk
	Melanocyte-stimulating hormone (MSH)	Acts on skin cells to cause the dispersion of melanin (skin darkening)
Ultimate gland hormones	Insulin-like growth factors (IGF)	Respond to GH and produce growth effects by stimulating cell mitosis
	Thyroid hormone (T_4/T_3) (amino acid-derived hormone)	Responds to TSH and stimulates oxidation in many cells
	Opioid peptides	May arise as breakdown products of γ-lipotropin or β-endorphin or from specific gene products; can respond to CRH or dopamine and may produce analgesia and other effects

(continued)

TABLE 22.2 • (continued)

Source	*Hormone*	*Action*
Ovarian granulosa cells; testicular Sertoli cells.	Inhibin	Stimulates steroidogenesis in ovaries and testes; regulates secretion of FSH from anterior pituitary. Second form of inhibin (activin) may stimulate FSH secretion
Intermediate lobe of pituitary gland	Corticotropin-like intermediary peptide (CLIP)	Derives from intermediate pituitary by degradation of ACTH; may be an endogenous modulator of pancreatic exocrine function.
Peptide hormones responding to other signals than anterior pituitary hormones	Arginine vasopressin (AVP; antidiuretic hormone, ADH)	Responds to increased activity in osmoreceptor, which senses extracellular $[Na^+]$; increases water reabsorption from distal kidney tubule
	Oxytocin	Responds to suckling reflex and estradiol; causes milk "let down" or ejection in lactating female, involved in uterine contractions of labor; luteolytic factor produced by corpus luteum; decreases steroid synthesis in testis
β Cells of pancreas respond to glucose and other blood constituents	Insulin	Increases tissue utilization of glucose
α Cells of pancreas respond to low levels of glucose and falling serum calcium	Glucagon	Decreases tissue utilization of glucose to elevate blood glucose
Derived from angiotensinogen by actions of renin and converting enzyme	Angiotensin II and III (AII and AIII)	Renin initially responds to decreased blood volume or decreased $[Na^+]$ in the macula densa of the kidney. AII/AIII stimulate outer layer of adrenal cortex to synthesize and release aldosterone
Released from heart atria in response to hypovolemia; regulated by other hormones	Atrial natriuretic factor (ANF) or atriopeptin	Acts on adrenal cortex cells to decrease aldosterone release; has other effects also
Generated from plasma, gut, or other tissues	Bradykinin	Modulates extensive vasodilation resulting in hypotension
Hypothalamus and intestinal mucosa	Neurotensin	Effects on gut; may have neurotransmitter actions
Hypothalamus, CNS, and intestine	Substance P	Pain transmitter, increases smooth muscle contractions of the GI tract
Nerves and endocrine cells of gut	Bombesin (mammalian equivalent is gastrin-releasing peptide)	Increases gastric acid secretion
	Cholecystokinin (CCK)	Stimulates gallbladder contraction and bile flow; increases secretion of pancreatic enzymes
Stomach antrum	Gastrin	Increases secretion of gastric acid and pepsin
Duodenum at pH values below 4.5	Secretin	Stimulates pancreatic acinar cells to release bicarbonate and water to elevate duodenal pH
Hypothalamus and GI tract	Vasointestinal peptide (VIP)	Acts as a neurotransmitter in peripheral autonomic nervous system; relaxes vascular smooth muscles; increases secretion of water and electrolytes from pancreas and gut
Kidney	Erythropoietin	Acts on bone marrow for terminal differentiation and initiates hemoglobin synthesis
Ovarian corpus luteum	Relaxin	Inhibits myometrial contractions; relaxes pelvic ligaments and increases dilation of cervix.
	Human placental lactogen (hPL)	Acts like PRL and GH.

TABLE 22.2 • (continued)

Source	*Hormone*	*Action*
Salivary gland	Epidermal growth factor	Mitogenic; stimulates proliferation of various epidermal and epithelial cell types
Thymus	Thymopoietin (α-thymosin)	Stimulates phagocytes; stimulates differentiation of precursors into immune competent T cells
Parafollicular C cells of thyroid gland	Calcitonin (CT)	Lowers serum calcium
Parathyroid glands	Parathyroid hormone (PTH)	Stimulates bone resorption; stimulates phosphate excretion by kidney; raises serum calcium levels
Endothelial cells of blood vessels	Endothelin	Vasoconstriction

Source: Part of this table is reproduced from Norman, A. W. and Litwack, G. Hormones. Orlando, FL: Academic Press, 1987.

[a] This is only a partial list of polypeptide hormones in humans. TSH, thyroid-stimulating hormone or thyrotropin; LH, luteinizing hormone; FSH, follicle-stimulating hormone; GH, growth hormone; ACTH, adrenocorticotropic hormone; PRL, prolactin; T_4, thyroid hormone (also T_3); IGF, insulin-like growth factor. For the releasing hormones and for some hormones in other categories, the abbreviation may contain "H" at the end when the hormone has been well characterized, and "F" in place of H to refer to "Factor" when the hormone has not been well characterized. Names of hormones may contain "tropic" or "trophic" endings; tropic is mainly used here. Tropic refers to a hormone generating a change, whereas trophic refers to growth promotion. Both terms can refer to the same hormone at different stages of development. Many of these hormones have effects in addition to those listed here.

corticotropin-like intermediary peptide (CLIP; from pars intermedia-like cells), prolactin (PRL), follicle-stimulating hormone (FSH), and luteinizing hormone (LH). All of these hormones are single polypeptide chains, except TSH, FSH, and LH, all of which are dimers that share a similar or identical α-subunit. Since the intermediate lobe in humans is rudimentary, the circulating levels of free α- and β-MSH are relatively low. It is of interest, particularly in the human, that MSH receptors recognize and are activated by ACTH, because the first 13 amino acids of ACTH contain the α-MSH sequence. For this reason, ACTH may be an important contributing factor to skin pigmentation and may exceed the importance of MSH, especially in conditions where the circulating level of ACTH is high. The clinical consequences of hypopituitarism are presented in Clin. Corr. 22.1.

CLINICAL CORRELATION 22.1

Hypopituitarism

The hypothalamus is connected to the anterior pituitary by a delicate stalk that contains the portal system through which releasing hormones, secreted from the hypothalamus, gain access to the anterior pituitary cells. Located in the cell membranes of these cells are specific receptors for releasing hormones. In most cases, different cells express different releasing hormone receptors. The connection between the hypothalamus and anterior pituitary can be disrupted by trauma or tumors. Trauma can occur in an automobile accident or other local damaging events that may result in severing of the stalk and preventing the releasing hormones from reaching their target anterior pituitary cells. When this happens, the anterior pituitary cells no longer receive the appropriate signals for the release of anterior pituitary hormones. *Panhypopituitarism* is a term used to describe this overall deficiency of anterior pituitary hormones. In the case of tumors of the pituitary gland, all of the anterior pituitary hormones may not be shut off to the same degree, or the secretion of some may disappear sooner than others. Thus the symptoms of hypopituitarism sometimes develop slowly. In any case, if hypopituitarism occurs, this condition may result in a life-threatening situation in which the clinician must determine the extent of loss of each of the pituitary hormones, especially ACTH. Posterior pituitary hormones—oxytocin and vasopressin—may also be lost, resulting in excessive urination (vasopressin deficiency) which must be addressed. Panhypopituitarism can also result in an increased sensitivity to the actions of insulin because of the decreased secretion of the insulin antagonists, growth hormone and cortisol, leading to hypoglycemia. The usual therapy involves administration of the end organ hormones, such as thyroid hormone, cortisol, sex hormones, and progestin; with female patients it is also necessary to maintain the ovarian cycle. These hormones can be easily administered in oral form. Growth hormone deficiency is not a problem in the adult but would be a serious problem in a growing child. The patient suffering from panhypopituitarism must learn to anticipate needed increases of cortisol in the face of stressful situations. Fortunately, these patients are usually maintained in reasonably good condition.

Marshall, J. C., and Barkan, A. L. Disorders of the hypothalamus and anterior pituitary. In Kelley, W. N. (Ed.), *Internal Medicine.* New York: Lippincott, 1989, 2159; and Robinson, A. G. Disorders of the posterior pituitary. In Kelley, W. N. (Ed.), *Internal Medicine.* New York: Lippincott, 1989, 2172.

22.3 • SYNTHESIS OF POLYPEPTIDE AND AMINO-ACID-DERIVED HORMONES

Polypeptide Hormones: Gene Coding

Genes for polypeptide hormones contain the coding sequence for the hormone and the control elements upstream of the structural gene. In some cases, more than one hormone is encoded in a gene. For example, proopiomelanocortin generates at least nine peptide hormones from a single gene product. As is seen with many other protein hormones, both antidiuretic hormone (ADH, vasopressin) and oxytocin are synthesized as preprohormones. The prohormones that are generated contain segments called **neurophysin**, which are cleaved during transport to the posterior pituitary. When secretion occurs equal amounts of the hormone and its neurophysin enter the circulation. These neurophysins have no known physiological function.

Proopiomelanocortin Is Precursor for Numerous Hormones

Proopiomelanocortin is a precursor for several hormones including: ACTH, β-lipotropin, and γ-lipotropin, γ-MSH, α-MSH, CLIP, and β-endorphin, and potentially β-MSH and enkephalins (Figure 22.5). All of these are not expressed simultaneously in a single cell type but are produced in separate cells depending on their content of specific proteases, metabolic controls, and regulators. Thus, while proopiomelanocortin is expressed in both corticotropes of the anterior pituitary and pars intermedia cells, the stimuli and products are different (Table 22.3). The pars intermedia is a discrete anatomical structure located between the anterior and posterior pituitary (Figure 22.6) in some species including the rat. In the human, however, pars intermedia are not a discrete anatomical structure, although some residual pars intermedia-like cells may be present in the equivalent location.

Genes of Polypeptide Hormones May Encode Additional Peptides

Other genes that encode more than one peptide are those for vasopressin and oxytocin and their accompanying neurophysins. Vasopressin, neurophysin II, and a glycoprotein of unknown function are released from the vasopressin precursor. A similar situation exists for

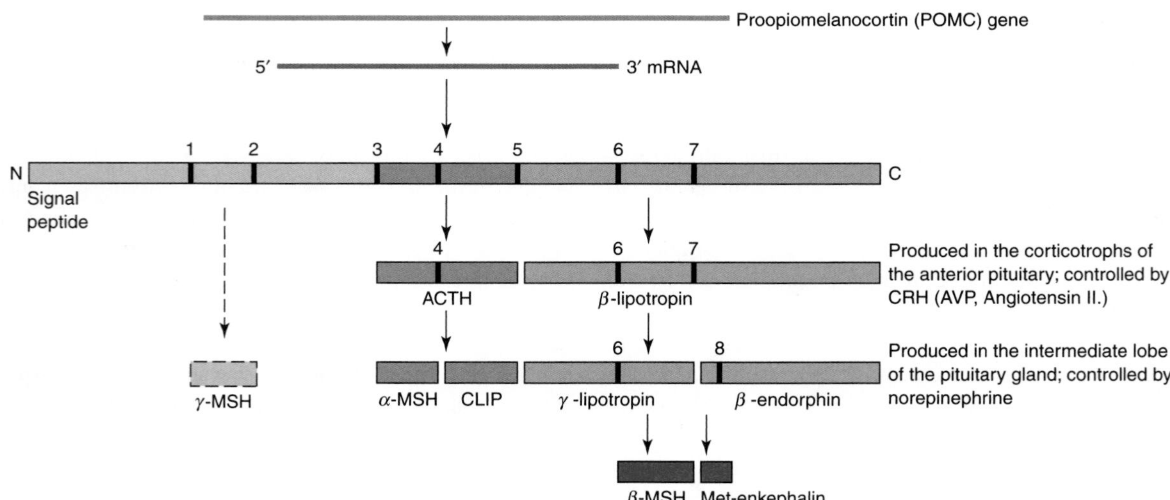

Figure 22.5 Proopiomelanocortin is a polypeptide encoded by one gene. The dark vertical bars represent proteolytic cleavage sites for specific enzymes. The cleavage sites are Arg-Lys, Lys-Arg, or Lys-Lys. Some specificity also may be conferred by neighboring amino acid residues. In the anterior pituitary, enzymes cleave at sites 3 and 5, releasing the major products, ACTH and β-lipotropin. In the pars intermedia, especially in vertebrates below humans, these products are further cleaved at major sites 4, 6, and 7 to release α-MSH, CLIP, γ-lipotropin, and β-endorphin. Some β-lipotropin may be further degraded to form β-endorphin. The anterior pituitary is under the positive control of the CRH and its stimulators, arginine vasopressin (AVP), and Angiotensin II. AVP by itself does not release ACTH but enhances the action of CRH in this process. The intermediary pituitary is under the positive control of norepinephrine. β-Endorphin also contains a pentapeptide, enkephalin, which potentially could be released at some point (hydrolysis at 8).

TABLE 22.3 • Summary of Stimuli and Products of Proopiomelanocortin[a]

Cell type	Corticotroph	Pars intermedia
Stimulus	CRH (+) (Cortisol (−))	Dopamine (−) Norepinephrine (+)
Auxiliary stimulus	AVP, AII	
Major products	ACTH, β-lipotropin (β-endorphin)	α-MSH, CLIP, γ-lipotropin, β-endorphin

[a]CRH, corticotropin-releasing hormone; AVP, arginine vasopressin; AII, angiotensin II; ACTH, adrenocorticotropin; α-MSH, α melanocyte-stimulating hormone; CLIP, corticotropin-like intermediary peptide.

Note: Although there are *pars intermedia cells* in the human pituitary gland, they do not represent a distinct lobe.

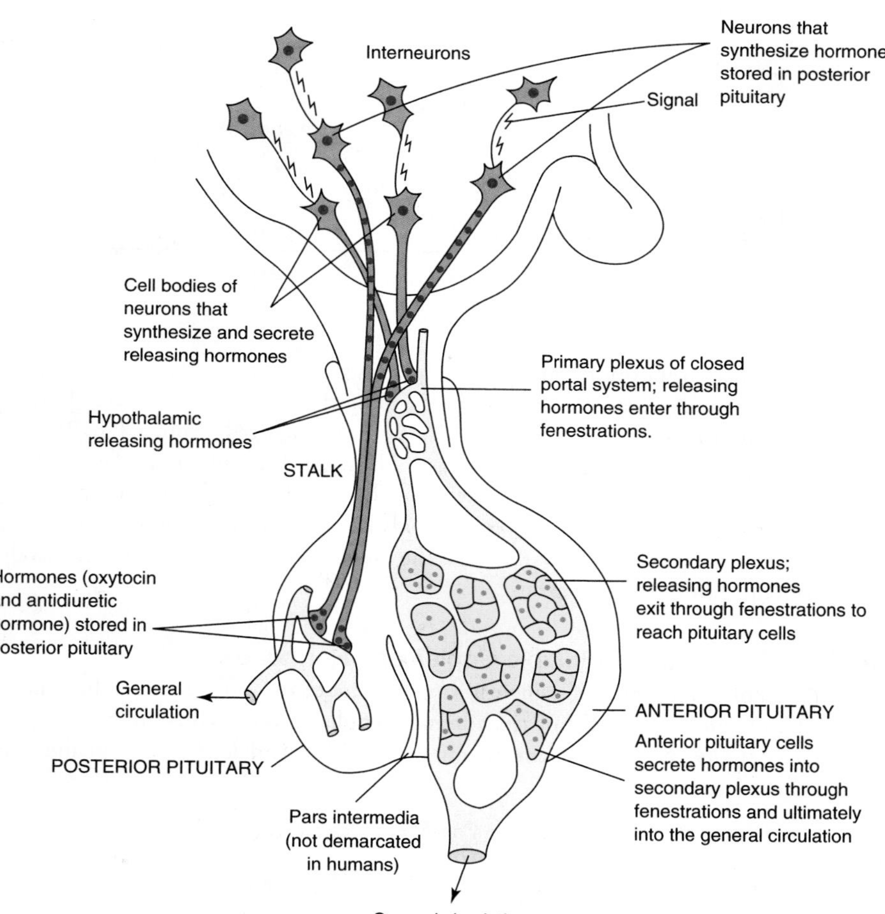

Figure 22.6 Anatomical relationship between hypothalamus and pituitary gland. The major vascular network is a primary plexus which releasing hormones enter through fenestrations. The secondary plexus is in the anterior pituitary where the releasing hormones are transported out of fenestrations to interact with the anterior pituitary target cells. Hypothalamic releasing hormones cause secretion of the anterior pituitary hormones, which enter the general circulation. Adapted from Norman, A. W., and Litwack, G. *Hormones.* New York: Academic Press, 1987, 104.

oxytocin and neurophysin I, except that no glycoprotein is released (Figure 22.7). **Vasopressin** and **neurophysin II** are co-released in response to stimuli from baroreceptors and osmoreceptors, which sense a fall in blood pressure or a rise in extracellular sodium ion concentration, respectively. **Oxytocin** and **neurophysin I** are coreleased by the suckling response in lactating females or as part of a conditioned reflex, such as a mother hearing her infant cry. Oxytocin is well known for its milk "let-down" action in the lactating female. Although oxytocin from the mother is probably not involved in initiating labor in humans, it may help maintain labor. Oxytocin from the fetus could be involved in the onset of labor.

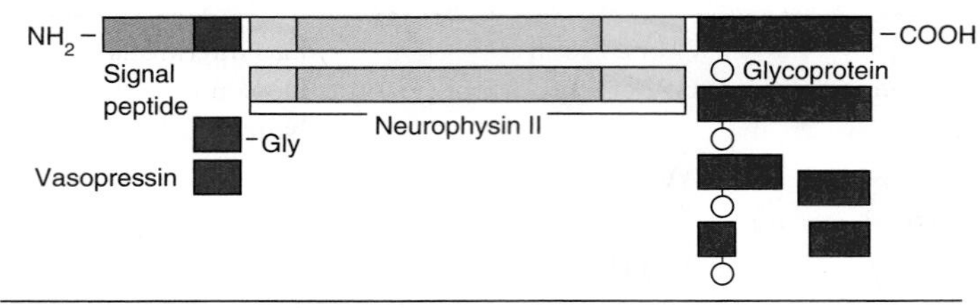

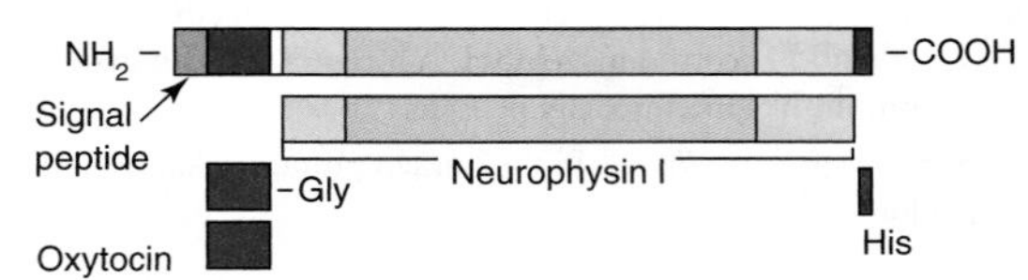

Figure 22.7 Preprovasopressin and preprooxytocin. Proteolytic maturation proceeds from top to bottom for each precursor. The organization of the gene translation products is similar except that a glycopeptide is included on the vasopressin precursor in the C-terminal region. Orange bars of the neurophysin represent conserved amino acid regions; gray bars represent variable C- and N-termini.
Redrawn with permission from Richter, D. VP and OT are expressed as polyproteins. *Trends Biochem. Sci.* 8:278, 1983.

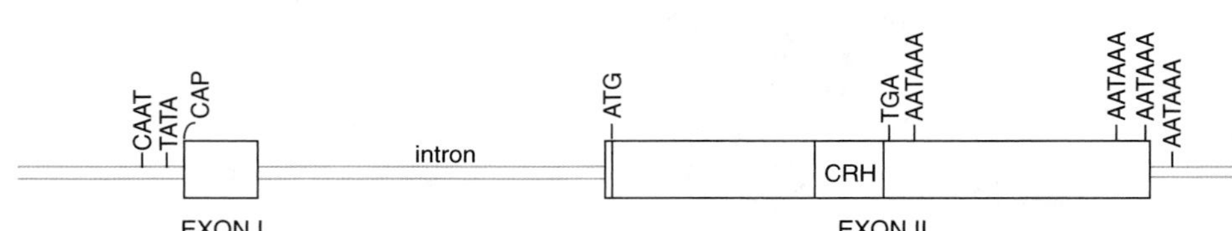

Figure 22.8 Nucleic acid sequence of rat proCRH genes. Schematic representation of the rat proCRH gene. Exons are shown as blocks, and the intron by a double red line. The TATA and CAAT sequence, putative cap site, translation initiation ATG, translation terminator TGA, and poly(A) addition signals (AATAAA) are indicated. The location of the CRH peptide is indicated by CRH.
Redrawn from Thompson, R. D., Seasholz, A. F., and Herbert, E. *Molec. Endocrinol.* 1:363, 1987.

Other polypeptide hormones are encoded by a single gene that does not encode another protein or hormone. An example is the gene encoding the decapeptide GnRH. This gene appears to reside to the left of a gene for the GnRH-associated peptide (GAP), which may be capable of inhibiting prolactin release. Thus, GnRH and the prolactin release-inhibiting factor (GAP) appear to be cosecreted by the same hypothalamic cells, but are not co-encoded. Many genes for hormones encode only one copy of the hormone and this may be the more common situation. An example is shown in Figure 22.8. Information encoding CRH is contained in the second axon.

One Gene Can Code for Multiple Copies of a Hormone

An example of multiple copies of a single hormone encoded on a single gene is the enkephalins secreted by chromaffin cells of the adrenal medulla. **Enkephalins** are pentapeptides with opioid activity; methionine-enkephalin (Met-ENK) and leucine-enkephalin (Leu-ENK) have the structures

Tyr-Gly-Gly-Phe-Met (Met-ENK)

Tyr-Gly-Gly-Phe-Leu (Leu-ENK)

A model of the enkephalin precursor which encodes several Met-ENK (M) molecules and a molecule of Leu-ENK (L) is presented in Figure 22.9. The processing sites for release of

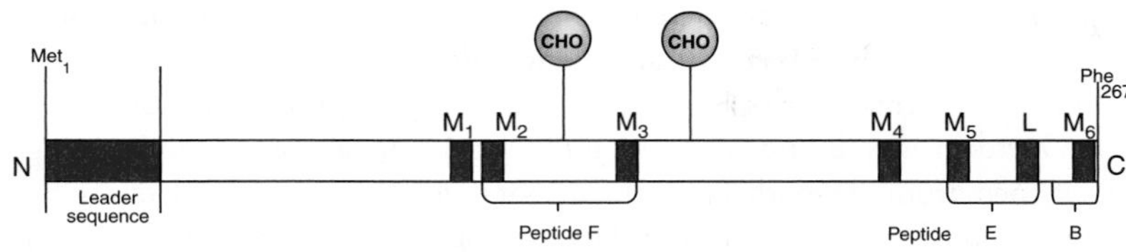

Figure 22.9 Model of enkephalin precursor. Distribution of Met-enkephalin sequences (M_1–M_6) and Leu-enkephalin (L) sequences within the precursor of bovine adrenal medulla. CHO, potential carbohydrate attachment sites.
Redrawn from Comb, M., Seeburg, P. H., Adelman, J., Eiden, L., and Herbert, E. *Nature* 295:663, 1982.

enkephalin molecules from the protein precursor contain Lys-Arg, Arg-Arg, and Lys-Lys bonds. Another example is the gene for the tripeptide hormone TRH. The TRH peptide sequence is present six times within the human TRH pre-prohormone.

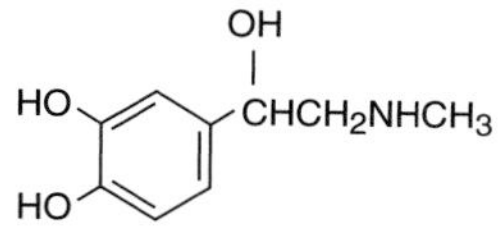

OH

HO

$CHCH_2NHCH_3$

HO

Figure 22.10 Structure of catecholamine hormone epinephrine.

Amino-Acid-Derived Hormones

Epinephrine Is Synthesized from Phenylalanine/Tyrosine

Epinephrine (Figure 22.10) is synthesized from phenylalanine/tyrosine in the adrenal medulla (p. 895). This catecholamine hormone is secreted along with some norepinephrine, enkephalins, and dopamine-β-hydroxylase by medullary chromaffin cells. Its secretion is signaled by the neuronal response to stress, which is transmitted by way of preganglionic acetylcholinergic neurons (Figure 22.11*a*). This signal increases intracellular Ca^{2+}, which in turn stimulates exocytosis and release of hormone stored in the **chromaffin granules** (Figure 22.11*b*). Once secreted, epinephrine and norepinephrine mediate their specific effects by interacting with receptors located on the plasma membranes of target cells. These receptors are generically grouped as α and β; each group has several subtypes. Epinephrine has a greater affinity for β-receptors than α-receptors, while norepinephrine acts primarily via α receptors.

In contast to the catecholamine hormones, the steroid hormones including aldosterone, cortisol, and dehydroepiandrosterone are synthesized and secreted by cells in the adrenal cortex (p. 885). Like epinephrine, the secretion of cortisol by the adrenal cortex is increased in response to stress. This secreted cortisol perfuses the adrenal medulla where it induces **phenylethanolamine *N*-methyltransferase (PNMT)** that converts norepinephrine to epinephrine. Thus, in biochemical terms, the stress response at the level of the adrenal cortex ensures the production of epinephrine from the adrenal medulla (Figure 22.12).

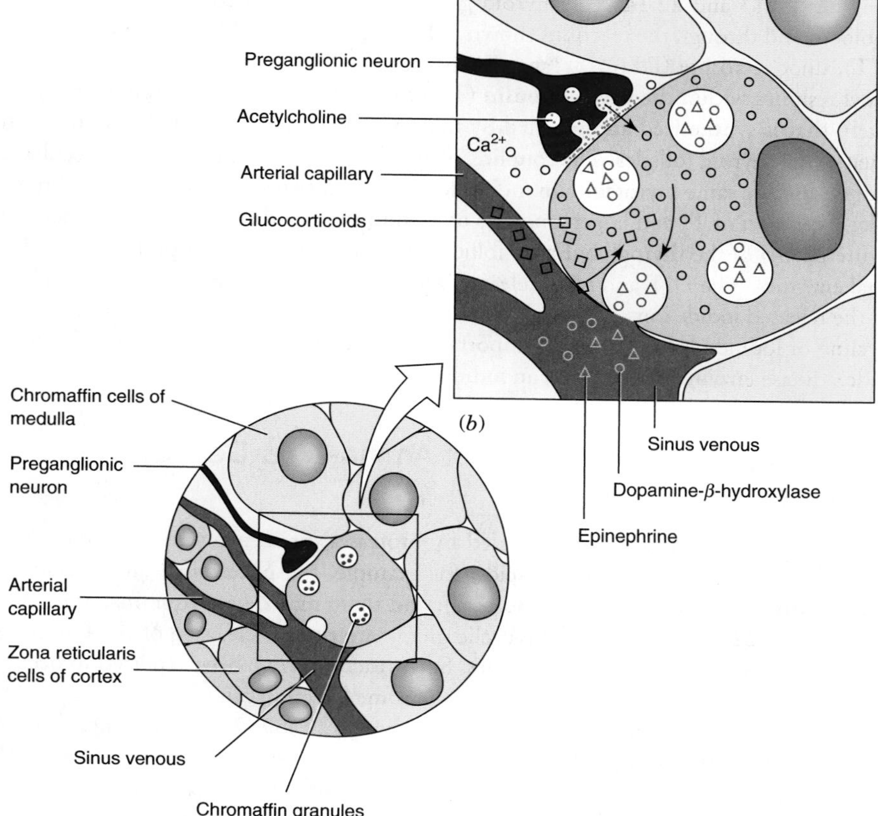

Figure 22.11 Relationship of adrenal medulla chromaffin cells to preganglionic neuron innervation and the structural elements involved in synthesis of epinephrine and discharge of catecholamines in response to acetylcholine. (*a*) Functional relationship between cortex and medulla for control of synthesis of adrenal catecholamines. Glucocorticoids that induce the PNMT enzyme reach the chromaffin cells from capillaries shown in (*b*). Discharge of catecholamines from storage granules in chromaffin cells after release of acetylcholine from nerve fiber stimulation. Calcium enters the cells, causing the fusion of granules and plasma membranes and exocytosis of the contents.
Redrawn from Krieger, D. T., and Hughes, J. C. (Eds.). *Neuroendocrinology.* Sunderland, MA: Sinauer Associates, 1980.

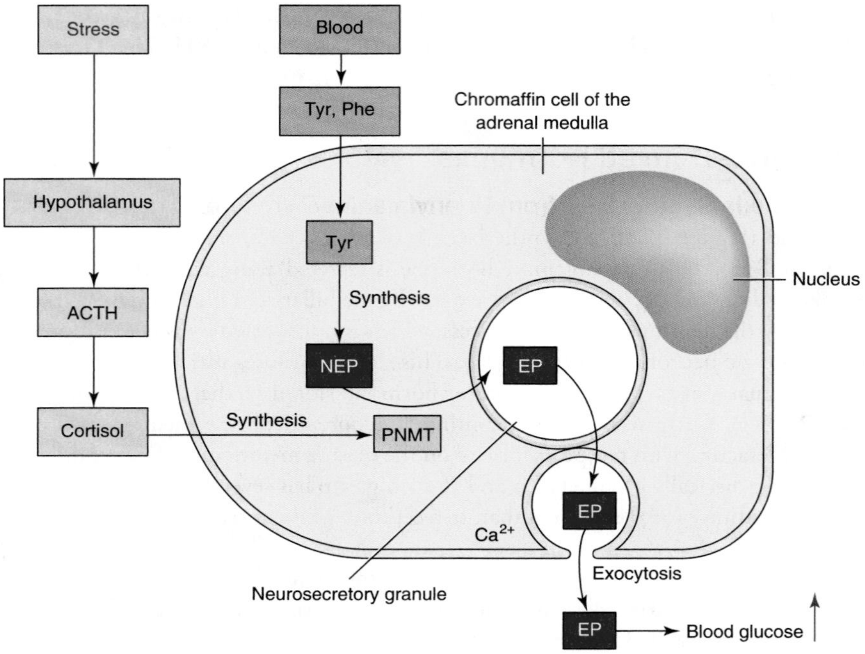

Figure 22.12 Biosynthesis, packaging, and release of epinephrine in adrenal medulla chromaffin cell. PNMT, phenylethanolamine *N*-methyltransferase; EP, epinephrine; and NEP, norepinephrine. Neurosecretory granules contain epinephrine, dopamine β-hydroxylase, ATP, Met or Leu-enkephalin, and larger enkephalin-containing peptides or norepinephrine in place of epinephrine. Epinephrine and norepinephrine are stored in different chromaffin granules.
Adapted from Norman, A. W., and Litwack, G. *Hormones.* New York: Academic Press, 1987, 464.

Synthesis of Thyroid Hormone Requires Incorporation of Iodine into Tyrosines of Thyroglobulin

An outline of the biosynthesis and secretion of thyroid hormone, **tetraiodo-L-thyronine (T_4)**, or **thyroxine**, and its more active metabolite, **triiodo-L-thyronine (T_3)**, is presented in Figures 22.13 and 22.14. The thyroid gland is specialized to concentrate iodide from the blood, and through the reactions shown in Figures 22.13 and 22.14, monoiodotyrosine (MIT), diiodotyrosine (DIT), T_4, and T_3 and reverse T3 are produced via iodination of tyrosyl residues within the **thyroglobulin** (TG) molecule. **Thyroglobulin** is a large glycoprotein that is synthesized and secreted by the thyroid epithelial cells and is stored in the lumen of the thyroid follicles. The coupling of an MIT and DIT or two DIT molecules can occur within the same thyroglobulin molecule or between two adjacent thyroglobulin molecules. Secretion of T_3 and larger amounts of T_4 into the bloodstream requires endocytosis (Figure 22.14) of the thyroglobulin by follicular epithelial cells and its proteolysis by lysosomal enzymes. The DIT and MIT released within the epithelial cell are then deiodinated and the released iodide ions are recycled and reutilized for thyroid hormone synthesis. This recycling of iodine ions is extremely important, and mutations resulting in inactivation of the deiodinase enzyme can result in an iodide deficiency.

Inactivation and Degradation of Amino-Acid-Derived Hormones

Most polypeptide hormones are degraded by proteases, presumably in lysosomes. Some hormones contain modified amino acids; for example, the N-terminal amino acid may be **cycloglutamic acid** (**pyroglutamic acid**), and there may be a C-terminal amino acid amide (Table 22.4). Breakage of the cyclic glutamate ring or cleavage of the C-terminal amide inactivates many of these hormones. Such reactions have been reported to occur in blood and may account for the short half-life of some hormones in plasma.

Some hormones contain cystine disulfide bonds (Table 22.5), and these may be degraded by **cystine aminopeptidase** and **glutathione transhydrogenase** (Figure 22.15). Alternatively, the peptide may undergo partial proteolysis to shorter peptides, some of which may have hormonal actions. Maturation or processing of **prohormones** into mature hormones involves selective proteolysis (see Figure 22.5, p. 892).

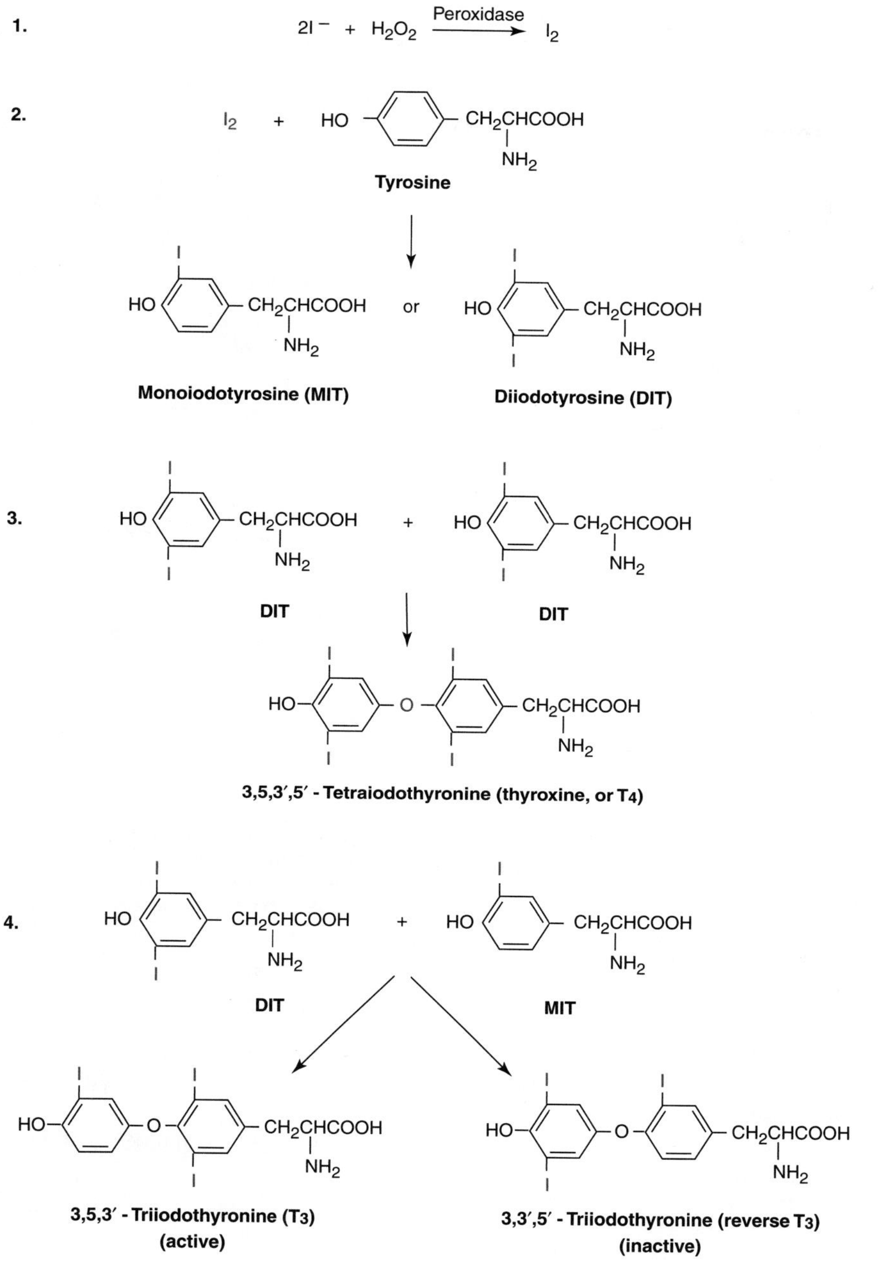

1. $2I^- + H_2O_2 \xrightarrow{\text{Peroxidase}} I_2$

2. I_2 + HO–(C₆H₄)–CH₂CHCOOH(NH₂) Tyrosine → Monoiodotyrosine (MIT) or Diiodotyrosine (DIT)

Tyrosine

Monoiodotyrosine (MIT) **Diiodotyrosine (DIT)**

3. **DIT** + **DIT** →

3,5,3′,5′ - Tetraiodothyronine (thyroxine, or T4)

4. **DIT** + **MIT** →

3,5,3′ - Triiodothyronine (T3) (active) **3,3′,5′ - Triiodothyronine (reverse T3) (inactive)**

Figure 22.13 Synthesis and structures of thyroid hormones T_4 and T_3 and reverse T_3. Step 1, oxidation of iodide; step 2, iodination of tyrosine residues; step 3, coupling of DIT to DIT; and step 4, coupling of DIT to MIT (coupling may be intramolecular or intermolecular).

22.4 • PROTEIN HORMONE SIGNALING

Overview of Signaling

Membrane Receptors

In the cascade system displayed in Figures 22.2 and 22.3 (pp. 885 and 886), hormones emanate from one source, cause hormonal release from the next step, and so on, down the cascade. Correct responses follow each specific stimulus. Polypeptide hormones generally bind to membrane receptors expressed specifically on target cells. The receptor recognizes structural features of the hormone with an affinity constant for the interaction of 10^9–10^{11}/M. This activates or inactivates an effector protein in or on the membrane (p. 474). Some receptors undergo **internalization** to the cell interior and others open a membrane ion channel (p. 478).

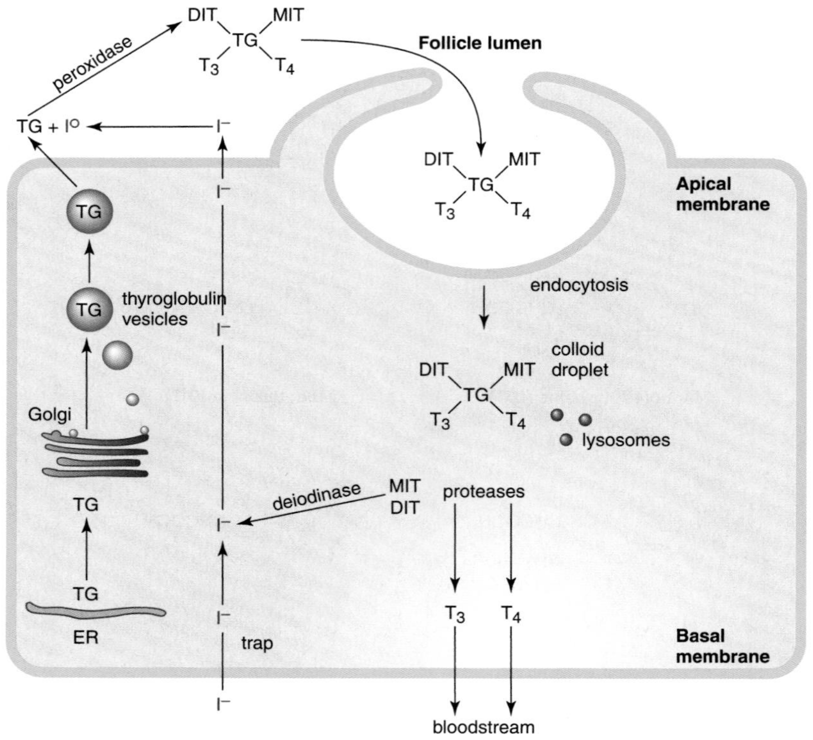

Figure 22.14 Cellular mechanisms for T_3 and T_4 release into bloodstream. Iodide trapping by basal membrane concentrates iodide approximately 30-fold. Secretion requires endocytosis of thyroglobulin and subsequent proteolysis. DIT and MIT are deiodinated and the released iodide ions are reutilized for hormone synthesis. Redrawn from Berne, R. M., and Levy, M. L. (Eds.). *Physiology,* 2d ed. New York: Mosby, 1990, 938.

TABLE 22.4 • Hypothalamic Releasing Hormones Containing an N-Terminal Pyroglutamate,[a] a C-Terminal Amino Acid Amide, or Both

Hormone	*Sequence*
Thyrotropin-releasing hormone (TRH)	*pGlu*-H-*Pro*-NH_2
Gonadotropin-releasing hormone (GnRH)	*pGlu*-HWSYGLRP-*Gly*-NH_2
Corticotropin-releasing hormone (CRH)	SQEPPISLDLTFHLLREVLEMTKADQLAQQAHSNRKL-LDI-*Ala*-NH_2
Growth hormone-releasing hormone (GRH)	YADAIFTNSYRKVLGQLSARKLLQDIMSRQQGESNQE-RGARAR-*Leu*-NH_2

[a]The pyroglutamate structure is

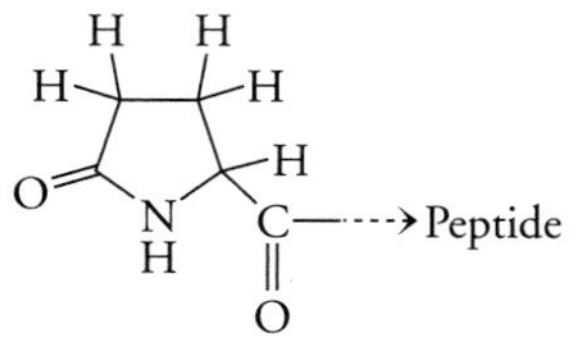

[b]Single-letter abbreviations used for amino acids: Ala, A; Arg, R; Asn, N; Asp, D; Cys, C; Glu, E; Gln, Q; Gly; G; His, H; Ile, I; Leu, L; Lys, K; Met, M; Phe, F; Pro, P; Ser, S; Thr, T; Trp, W; Tyr, Y; Val, V.

TABLE 22.5 • Examples of Hormones Containing a Cystine Disulfide Bridge Structure

Hormone	*Sequence*[a]
Somatostatin (GHIH)	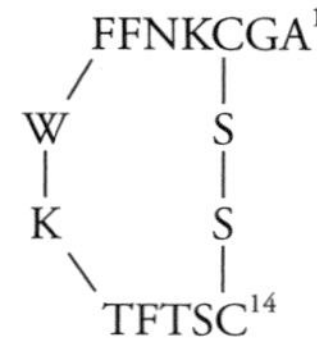 FFNKCGA[1] W S K S TFTSC[14]
Oxytocin	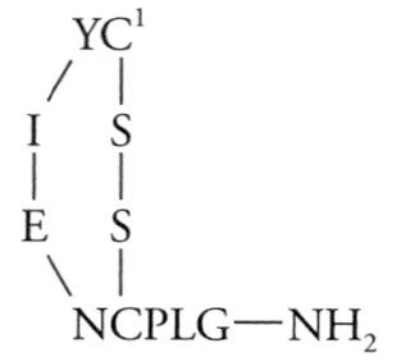 YC[1] I S E S NCPLG—NH_2
Arginine vasopressin	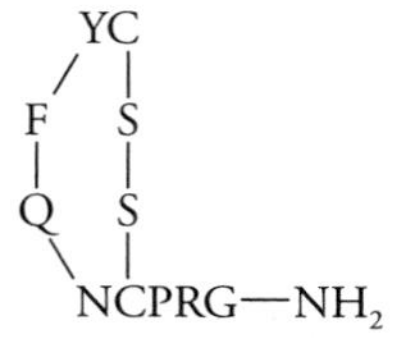 YC F S Q S NCPRG—NH_2

[a]Letters refer to single-letter amino acid abbreviations (see Table 23.4).

Intracellular Signal Cascade: Second Messengers

After binding to their cognate membrane receptors, many peptide and protein hormones transmit their signal intracellularly via second messengers, which transmit and amplify the hormonal signal (p. 509). Some hormones transmit their signal by increasing the intracellular concentration of one second messenger, while others increase the concentration of several second messengers, either simultaneously or sequentially. Second messengers include cyclic AMP (cAMP), cyclic GMP (cGMP), inositol trisphosphate (IP_3), diacylglycerol

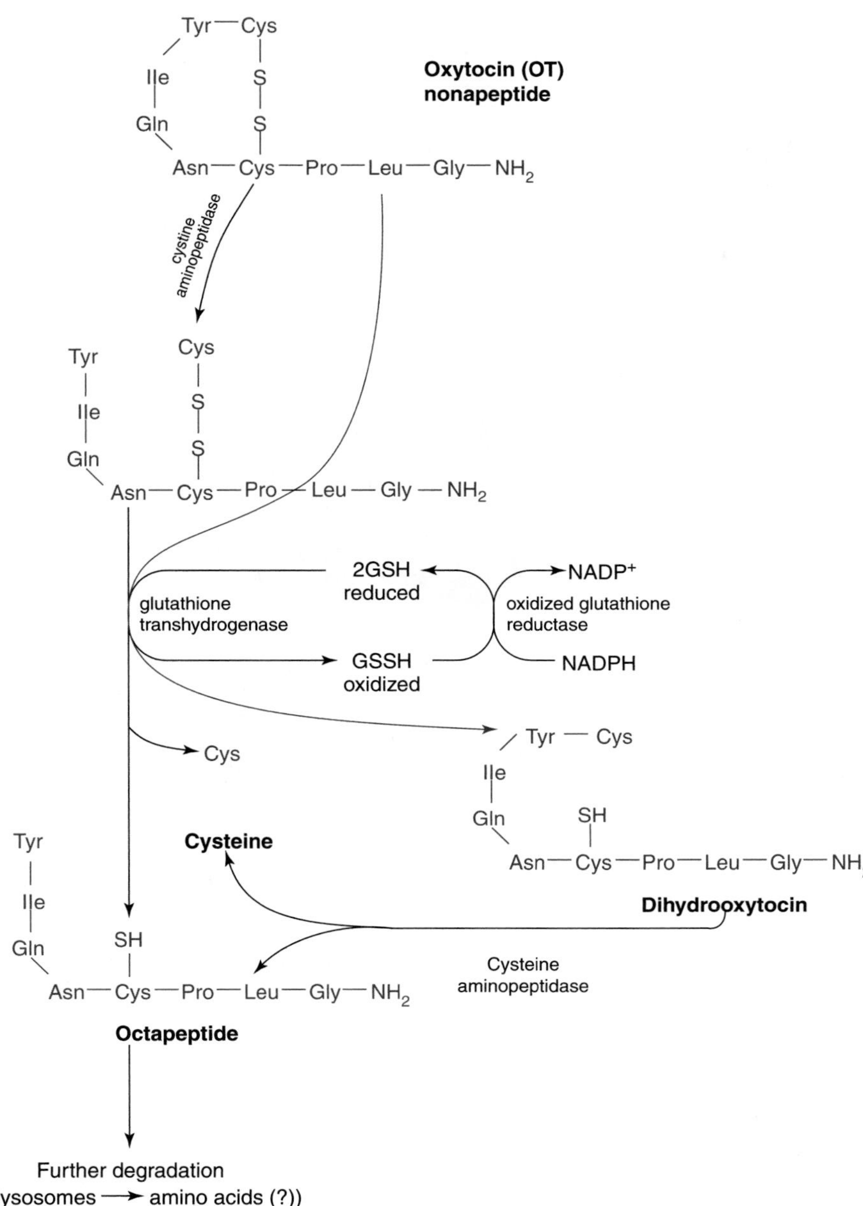

Figure 22.15 Degradation of posterior pituitary hormones. Oxytocin transhydrogenase is similar to degrading enzymes for insulin; presumably, these enzymes also degrade vasopressin. Redrawn from Norman, A. W., and Litwack, G. *Hormones*. New York: Academic Press, 1987, 167.

(DG), and phosphatidylinositol 3,4,5-trisphosphate (PIP_3). Different hormones bind to receptors that activate either a stimulatory or an inhibitory G protein subunit (G_s or G_i, respectively) and result in activation or inhibition of an effector enzyme and, thus, an increase or decrease in the corresponding intracellular second messenger. The intracellular second messengers activate specific kinases that initiate a cascade of phosphorylation/dephosphorylation reactions and result in activation of some, and inactivation of other, enzymes (p. 510). Stimulation of adenylate cyclase by G protein-coupled receptors generates cAMP, which activates **protein kinase A**, whereas stimulation of guanylate cyclase by different G protein-coupled receptors generates cGMP, which activates **protein kinase G**. Stimulation of phospholipase C with generation of DG and IP_3 results in the mobilization of Ca^{2+}stores and activation of **protein kinase C**.

An example of a hormone that transduces a signal via the generation of a second messenger is depicted in Figure 22.16. Thyrotropin-releasing hormone synthesized by hypothalamic neurons reaches the thyrotropes in the anterior pituitary and stimulates them to synthesize and secrete thyroid-hormone-stimulating hormone (TSH). TSH binds to its G

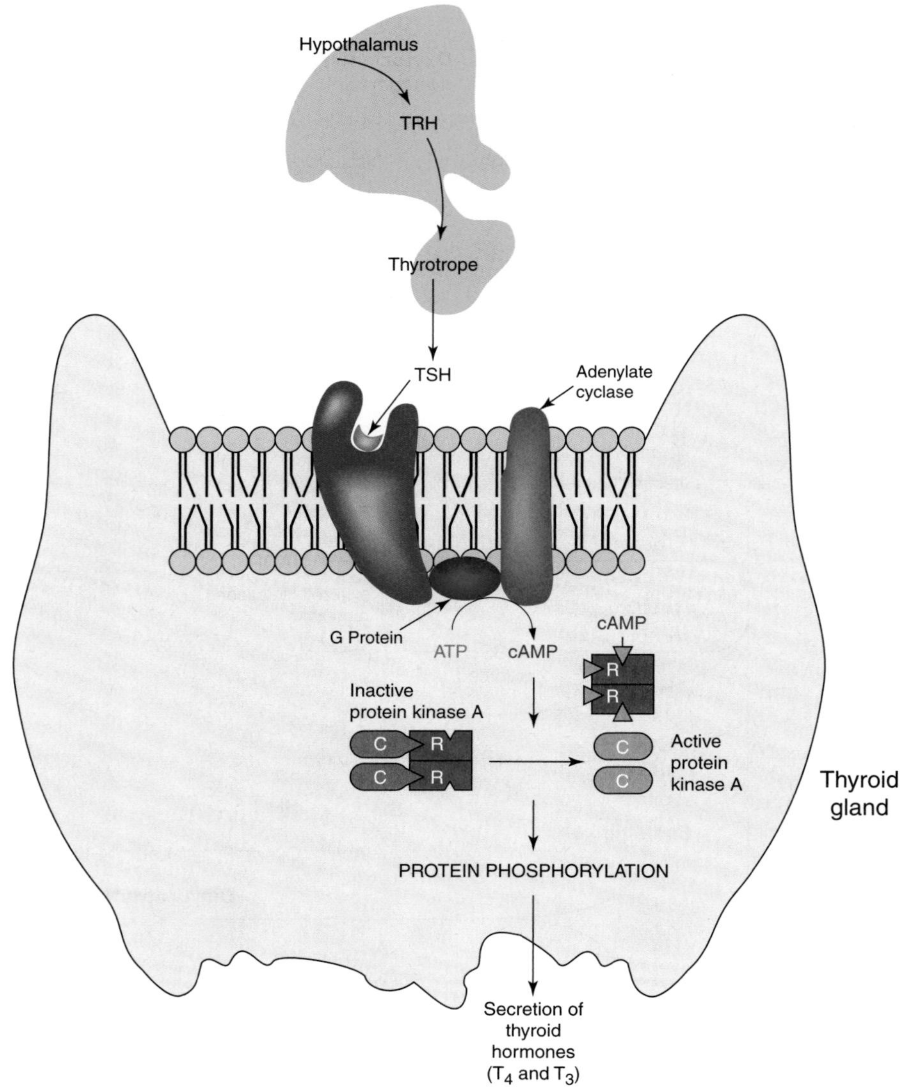

Figure 22.16 Effect of TSH on secretion of thyroid hormone. TSH stimulates all steps in synthesis and secretion of T_3 and T_4. These are mediated by its binding to TSH receptors located on basal membrane of thyroid epithelial cells, elevation of cAMP levels, and subsequent cascade of phosphorylation reactions.

protein-coupled membrane receptors in the thyroid gland and activates adenylate cyclase with generation of cAMP. cAMP in turn binds to the regulatory subunits in the inactive form of protein kinase A leading to their dissociation from the catalytic subunits, which are fully active (p. 510) and initiate a cascade of protein phosphorylations that results in secretion of thyroid hormone.

Amplification occurs at each step of this signal transduction pathway. For example, activation of one molecule of adenylate cyclase may result in the generation of ~100 molecules of cAMP and the ultimate phosphorylation of ~10,000 enzyme molecules. The effects of cAMP are terminated when it is hydrolyzed by **phosphodiesterase**. Since phosphodiesterase is also modulated by hormones via a G protein, the level of cAMP is actually under dual regulation. Two different hormones can have antagonistic effects as one stimulates adenylate cyclase and the other stimulates phosphodiesterase. Hormonal activation of protein kinase A can also alter the rate of transcription of genes (p. 529). After activation by cAMP, the catalytic subunit of PKA diffuses into the nucleus where it catalyzes phosphorylation of a serine residue in **CREB (cAMP-response element binding protein)**, a ubiquitously expressed transcription factor. The activated CREB then binds as a dimer to the conserved consensus **cAMP response element (CRE)**. A conserved palindromic CRE has been identified in the promoter of various genes regulated by cAMP. Two other transcription factors, CREM (CRE modulator) and ATF-1 (Activating Transcription

Factor), are also phosphorylated by protein kinase A. While CREB and ATF-1 stimulate transcription, some isoforms of CREM inhibit CRE activity. Thus hormonal activation of a protein kinase can increase or decrease gene transcription.

Cyclic Hormonal Systems

The **diurnal variation** in the secretion of cortisol from the adrenal cortex is regulated by the sleep/wake transition whereas secretion of **melatonin** from the pineal gland is dictated by daylight and darkness. The female ovarian cycle also operates on a cyclic basis dictated by the central nervous system. These are all examples of **chronotropic control** of hormone secretion.

Melatonin and Serotonin Synthesis Are Controlled by Light/Dark Cycles

In the release of melatonin from the pineal gland (Figure 22.17*a*, p. 902), the internal signal is provided by norepinephrine released by an adrenergic neuron. Control is exerted by light entering the eyes, which inhibits the pineal gland and hence the release of melatonin. Norepinephrine released in the dark stimulates cAMP formation through a β receptor in the pinealocyte membrane. Increased activity of PKA enhances synthesis of ***N*-acetyltransferase** and conversion of **serotonin**, synthesized from tryptophan (p. 785), to ***N*-acetylserotonin**, and this is the rate-limiting step that dictates the circadian rhythm of melatonin. **Hydroxyindole-*O*-methyltransferase (HIOMT)** then converts *N*-acetylserotonin to **melatonin** (Figure 22.17*b*), which is secreted during the dark hours. Relatively small doses of melatonin can induce sleep and basically reset the daily rhythm. This physiological response could benefit workers whose shifts alternate between daylight and nighttime hours. Melatonin is also a potent antioxidant and may provide some protection against damaging oxygen free radicals. Although melatonin inhibits reproductive functions in animals that breed during specific seasons, there is no proof that it influences human reproductive functions.

Ovarian Cycle Is Controlled by Pulsatile and Cyclic Secretion of Gonadotropin-Releasing Hormone

GnRH is secreted from hypothalamic neuroendocrine cells in pulses about 1 h apart in response to norepinephrinergic neurons in both adult males and females. In females the frequency of pulses, and therefore the total amount of GnRH secreted during a 24-h period, changes over the course of the monthly menstrual cycle. Figure 22.18 (p. 903) summarizes this important role of pulsatile GnRH secretion in terms of FSH and LH secretion from the female anterior pituitary. Entry of GnRH into the portal system is through fenestrations in the blood vessels to reach the **gonadotropes** located in the anterior pituitary. Here GnRH binds to its membrane receptors and mediates its effects via the phosphatidylinositol second messenger system (p. 534), with release of FSH and LH from the same gonadotrope. Clin. Corr. 22.2 describes how the premature secretion of large amounts of GnRH can result in precocious puberty in a young child. **FSH** operating through protein kinase A via cAMP elevation, stimulates synthesis and secretion of 17β-estradiol, and matures the ovarian follicle and ovum. **Inhibin**, a disulfide-linked dimeric glycoprotein hormone, is also synthesized and secreted by the granulosa cells of the ovarian follicle. These hormones are feedback inhibitors of FSH production by gonadotropes. **Activins** are dimeric proteins closely related to the inhibins. These hormones are produced by the same tissue that secretes inhibins, but stimulate, rather than inhibit, FSH secretions by the gonadotropes. When a follicle reaches maturity, a surge of LH and prostaglandin $F_{2\alpha}$ triggers ovulation. The residual follicle under primary control by LH (Figure 22.18) becomes the functional corpus luteum. LH binds to its cognate receptors in the corpus luteum and through stimulation of protein kinase A increases synthesis of progesterone. **Estradiol** and **progesterone** bind to specific intracellular receptors in the uterine endometrium and promote thickening of the wall, vascularization, and increased secretory activity in preparation for implantation of the fertilized egg. Estradiol is synthesized in large amounts prior to production of progesterone and induces expression of progesterone receptors. This induction of progesterone receptors primes the uterus for subsequent stimulation by progesterone.

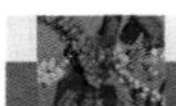

CLINICAL CORRELATION 22.2

Precocious Puberty

Children with brain tumors or other hypothalamic lesions may undergo what is called precocious puberty. In this endocrine disorder, sexual maturation occurs at a very early age due to premature secretion of large amounts of GnRH. As an extreme example, the youngest recorded mother to give birth to a full-term, healthy infant by cesarean section was only 5 years, 8 mon of age. Of course such pregnancies are in fact the result of sexual abuse of a child with true precocious puberty.

In young boys the testes usually enlarge under gonadotropin stimulation before any other signs of puberty are seen. In young girls an increased growth rate, breast development, an increase in the size of ovaries and uterus, and changes in the vaginal mucosa are the usual presenting features. The rapid growth in these young girls is associated with a rise in estrogen synthesis and secretion, and this leads to increased growth hormone secretion. Spermatogenesis in males and ovulation in females may occur, and fertility is certainly possible.

Three major drugs that have been used successfully to treat this disorder include medroxyprogesterone, cyproterone acetate, and superactive GnRH agonists, which are the most effective. Medroxyprogesterone inhibits gonadotropin secretion and also functions as a competitive inhibitor of a specific enzyme involved in steroidogenesis. Cyproterone acetate has antiandrogenic (androgen receptor antagonist), antigonadotropic, and progestational properties. GnRH agonists are synthetic analogues of the amino acid sequence of the endogenous decapeptide. Somewhat paradoxically, when given chronically, these agents suppress pulsatile LH and FSH release, steroid production by the gonads, and gametogenesis in both males and females.

Kaplan, S. L., and Grumbach, M. M. Pathogenesis of sexual precocity. In Grumbach, M. M., Sizonenko, P. C., and Aubert, M. L. (Eds.), *Control of the Onset of Puberty.* Baltimore: Williams & Wilkins, 1990, 620; and Belchetz, P. E., Plant, T. M., Nakai, Y., Keogh, E. J. and Knobil, E. Hypophysial response to continuous and intermittent delivery of hypothalamic gonadotropin-releasing hormone. *Science* 202:631, 1978.

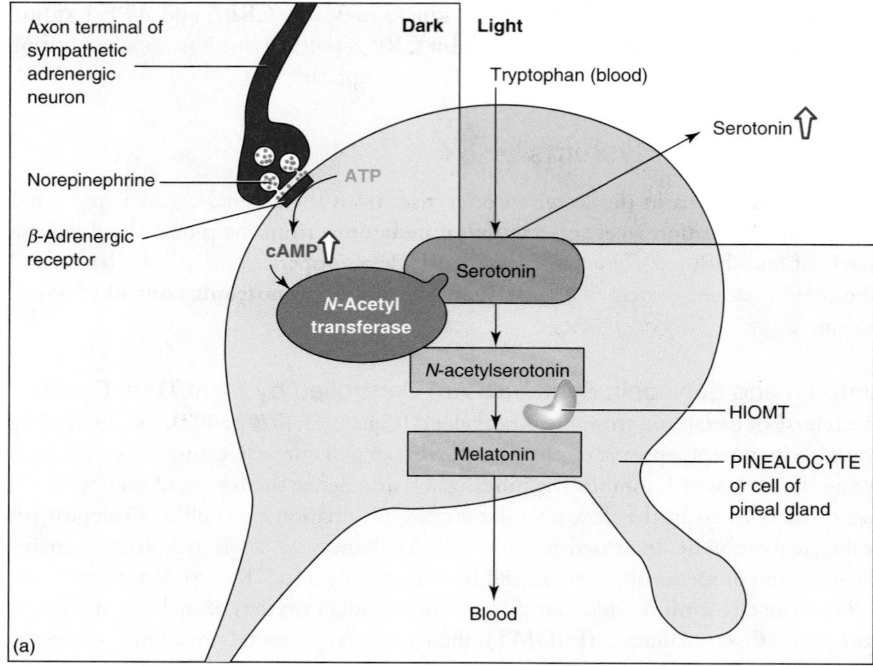

$H_2N-CH_2-CH_2-$ (indole, 5-HO)

Serotonin
(5-hydroxy-L-tryptamine, 5HT)

↓ *N*-acetyltransferase (Rate limiting step)

$H_3C-C(=O)-HN-CH_2-CH_2-$ (indole, 5-HO)

N-Acetylserotonin

↓ hydroxyindole-O-methyltransferase (HIOMT)

$H_3C-C(=O)-HN-CH_2-CH_2-$ (indole, 5-H_3C-O)

Melatonin (Secreted during dark hours)

(b)

Figure 22.17 Biosynthesis of Melatonin.
(*a*) Synthesis of melatonin in pinealocytes.
(*b*) Rate-limiting step in biosynthesis of melatonin.
HIOMT, hydroxyindole-*O*-methyl transferase.
Redrawn from Norman, A. W., and Litwack, G. *Hormones*. New York: Academic Press, 1987, 710.

Absence of Fertilization

If fertilization does not occur, the corpus luteum involutes or degenerates because of diminished LH supply, and progesterone and estrogen levels fall sharply. The hormonal stimuli for a thickened and vascularized uterine endometrial wall are thus lost and menstruation occurs as a consequence of cellular necrosis. The fall in blood steroid levels releases the feedback inhibition on the gonadotropes and hypothalamus and the cycle starts again. The time course for the ovarian menstrual cycle in humans is shown in Figure 22.19 (p. 904).The first monthly cycle occurs at the time of puberty when GnRH secretion begins to increase (day 1 of Figure 22.19). GnRH is released in a pulsatile fashion, causing the gonadotrope to release FSH and LH, and the blood concentrations of these hormones gradually increase in

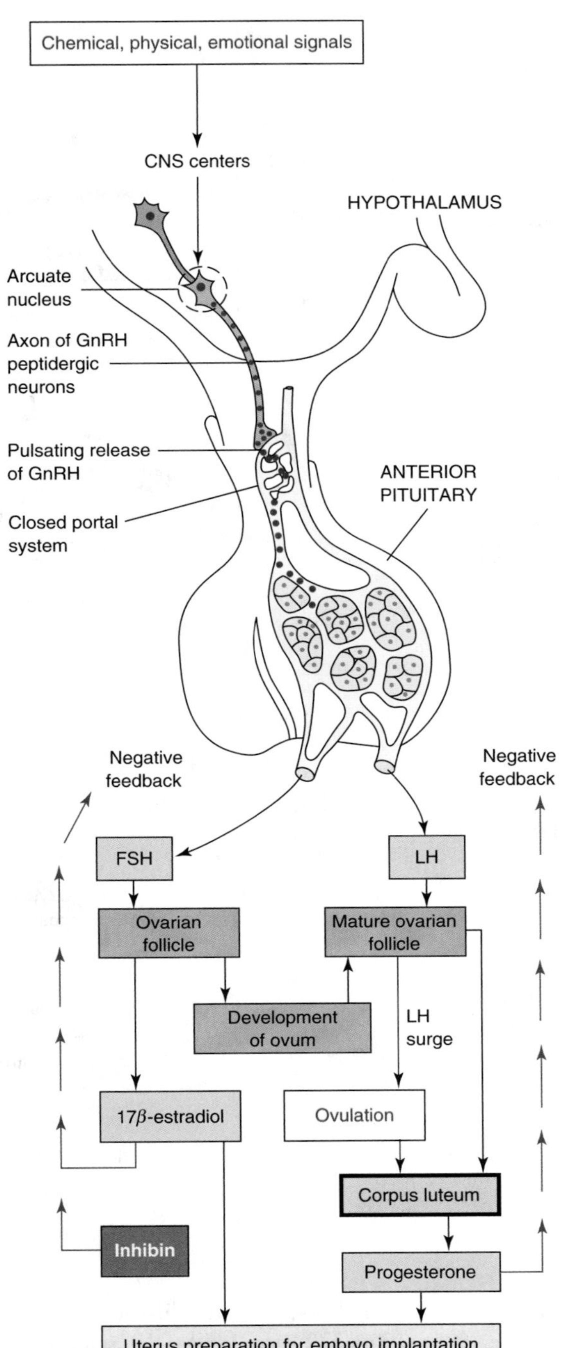

Figure 22.18 Ovarian cycle in terms of generation of hypothalamic releasing hormone, pituitary gonadotropic hormones, and sex hormones. At puberty, several centers in CNS coordinate with hypothalamus so that GnRH is released in a pulsatile fashion. This causes release of LH and FSH, which affects the ovarian follicle, ovulation, and corpus luteum. Inhibin B selectively inhibits FSH secretion. Products of the follicle and corpus luteum, respectively, are β-estradiol and progesterone; GnRH, gonadotropin-releasing hormone; FSH, follicle-stimulating hormone; and LH, luteinizing hormone.

subsequent days. Under the stimulation of FSH, the follicle begins to mature (lower section of Figure 22.19) and 17β-estradiol (E_2) is produced, causing the uterine endometrium to thicken. Under the continued action of FSH, the follicle matures, and high concentrations of estradiol are produced (around day 13 of the cycle). This elevated level of estradiol now mediates **positive feedback** (rather than negative feedback mediated by lower estradiol levels) resulting in a surge of LH and a decreased release of FSH from the gonadotropes. The FSH response is smaller because this gonadotropic hormone stimulates the ovarian production of **inhibin B** that inhibits FSH, but not LH, secretion. The high midcycle peak of LH is referred to as the *LH spike.* Ovulation then occurs at about day 14 (midcycle) through the effects of high LH concentration and other factors, such as prostaglandin F_{2a} (PGF_{2a}). After ovulation, LH promotes differentiation at the ruptured follicle (Figure 22.19, bottom) into a corpus luteum that produces high levels of progesterone that further thicken the uterine

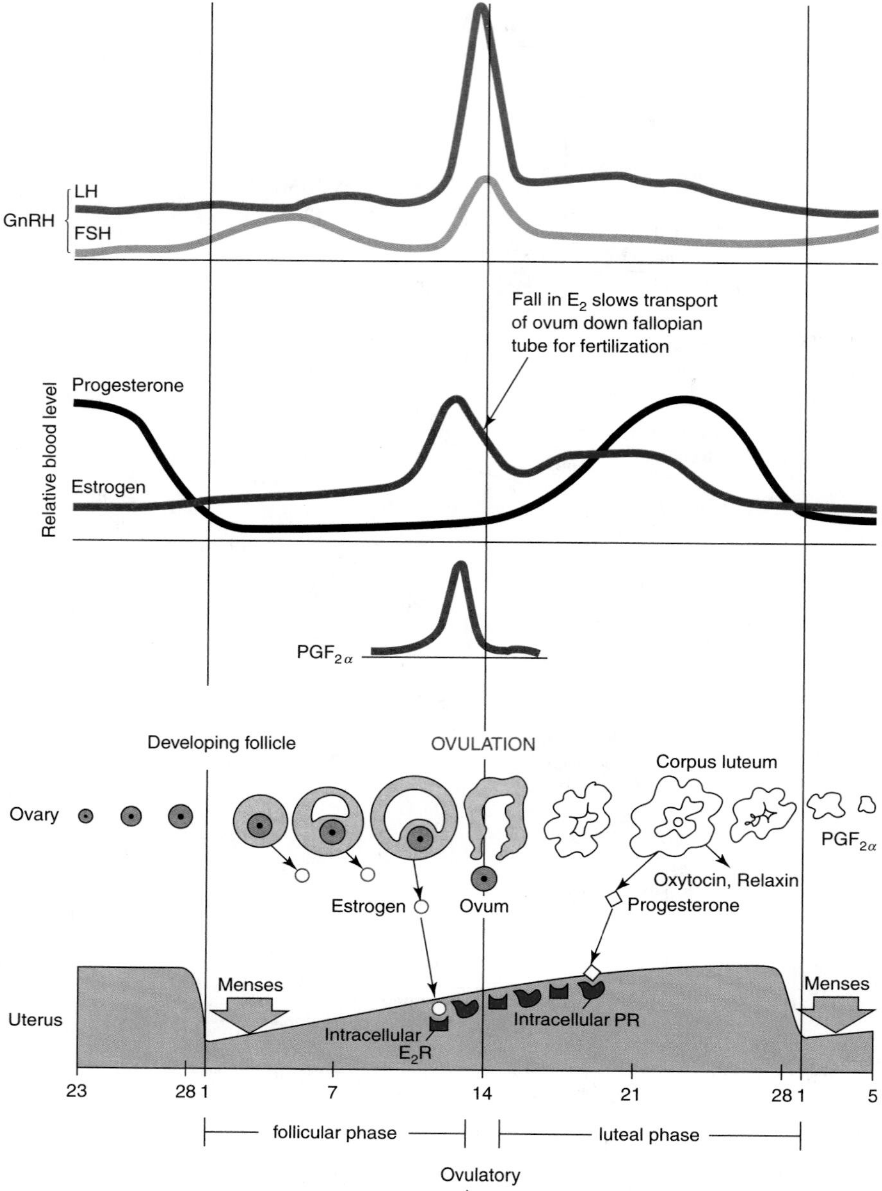

Figure 22.19 Ovarian cycle. In the upper diagram, relative blood levels of GnRH, LH, FSH, progesterone, estrogen, and $PGF_{2\alpha}$ are shown. In the lower diagram, events in ovarian follicle, corpus luteum, and uterine endometrium are diagrammed. Abbreviations: GnRH, gonadotropin-releasing hormone; LH, luteinizing hormone; FSH, follicle-stimulating hormone; $PGF_{2\alpha}$, prostaglandin $F2_{2\alpha 1}$; E_2, estradiol; E_2R, intracellular estrogen receptor; and PR, intracellular progesterone receptor.

endometrial wall. The corpus luteum also secretes **inhibin A**, which along with estradiol and progesterone, suppresses FSH and LH secretion by the anterior pituitary during the luteal phase of the cycle. In the absence of fertilization, the corpus luteum remains functional for about 2 weeks. It then involutes or regresses due to the decline in LH levels and an age-related reduction in LH sensitivity. With the death of the corpus luteum, there is a profound decline in estradiol and progesterone levels. The endometrial wall can no longer be maintained and menstruation occurs, followed by the start of another menstrual cycle.

Fertilization

If fertilization occurs as shown in Figure 22.20, the corpus luteum remains viable because of production of **chorionic gonadotropin** (CG), which resembles and acts like LH, from the trophoblast cells. The secretion of CG reaches a peak about 80 days after the last menstrual period. It then declines very rapidly and is produced by the placenta at a relatively low level for the remainder of pregnancy. Once CG levels fall, the corpus luteum begins to involute, and by about 12 weeks of pregnancy, the placenta takes over production and secretion of progesterone and estrogens (primarily estriol). From the seventh month onward estrogen

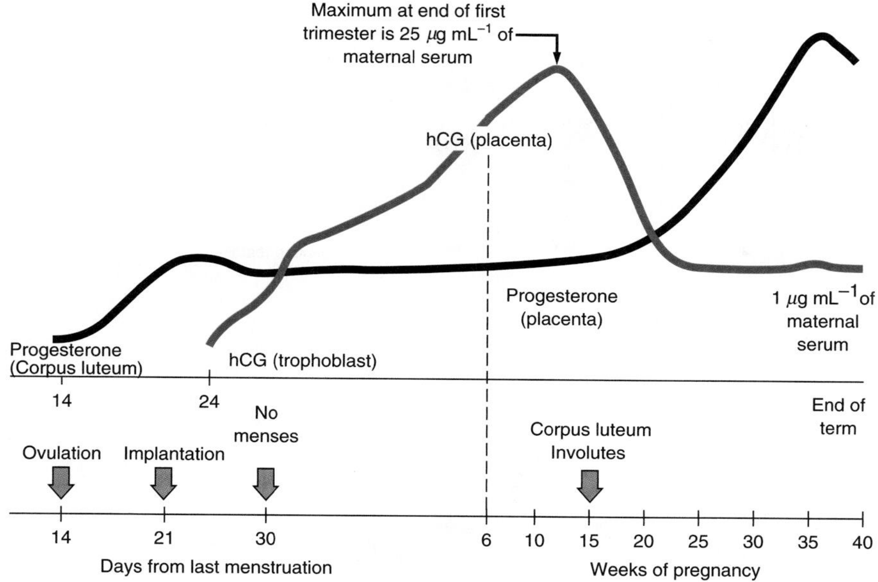

Figure 22.20 Effect of fertilization on ovarian cycle in terms of secretion of progesterone and human chorionic gonadotropin (hCG).

secretion continues to increase while progesterone secretion remains constant or may even decrease slightly. The estrogen/progesterone ratio increases toward the end of pregnancy and may be partly responsible for increased uterine contractions. Oxytocin from the posterior pituitary contributes to these uterine contractions. The fetal membranes release prostaglandins ($PGF_{2\alpha}$) at the time of parturition, which increases the intensity of uterine contractions. Finally, the fetal adrenal cortex secretes cortisol, which stimulates fetal lung maturation by inducing the synthesis of surfactant-associated proteins.

22.5 • MEMBRANE HORMONE RECEPTORS

Some Hormone-Receptor Interactions Involve Multiple Hormone Subunits

Thyrotropin (TSH), **luteinizing hormone (LH)**, and **follicle-stimulating hormone (FSH)** each contain an α- and a β-subunit. The α-subunits for all three hormones are similar or identical. The specificity of receptor recognition is imparted by the β-subunit, whose structure is unique for each hormone. A model of the interaction of LH with its receptor is shown in Figure 22.21. The LH receptor recognizes both subunits of the hormone, but the β-subunit is specifically recognized by the receptor to elicit a hormonal response. The TSH–receptor complex stimulates adenylate cyclase and the phosphatidylinositol pathway. The preferred model is one in which there is a single receptor whose interaction with hormone activates both the adenylate cyclase and the phospholipid second messenger systems, as shown in Figure 22.22 (p. 906).

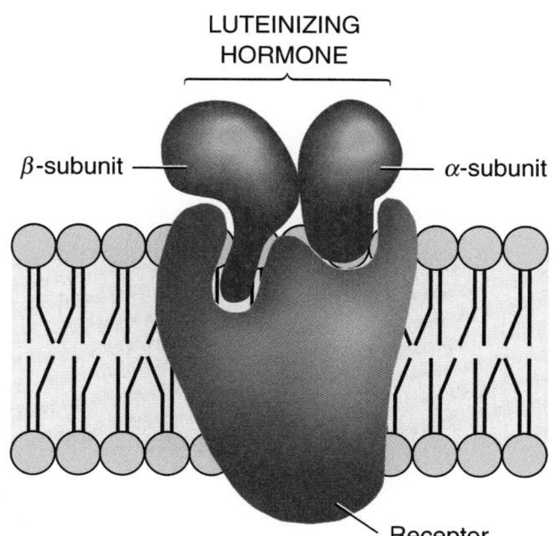

Figure 22.21 The interaction of α- and β-subunits of LH with LH receptor of rat Leydig cells. Both α- and β-subunits participate in LH receptor binding.
Adapted from Alonoso-Whipple, C., Couet, M. L., Doss, R. Koziarz, J., et al. *Endocrinology* 123:1854, 1988.

β-Adrenergic Receptor

Structures of receptors are conveniently discussed in terms of functional domains. For membrane receptors there are **ligand-binding domains**, **transmembrane domains**. and **intracellular domains**; this last domain may have intrinsic protein kinase activity. Specific **immunological domains** contain primary epitopes of antigenic regions. The β-adrenergic receptors (β_1 and β_2) recognize the catecholamines norepinephrine and epinephrine, and hormone binding stimulates adenylate cyclase. The subtypes differ in their affinity for norepinephrine and for synthetic antagonists. β_1 receptors bind norepinephrine with a higher affinity than epinephrine, whereas the reverse is the case for the β_2 receptors. Isoproterenol, an analog of epinephrine that is a β-receptor stimulator, has a greater affinity for both

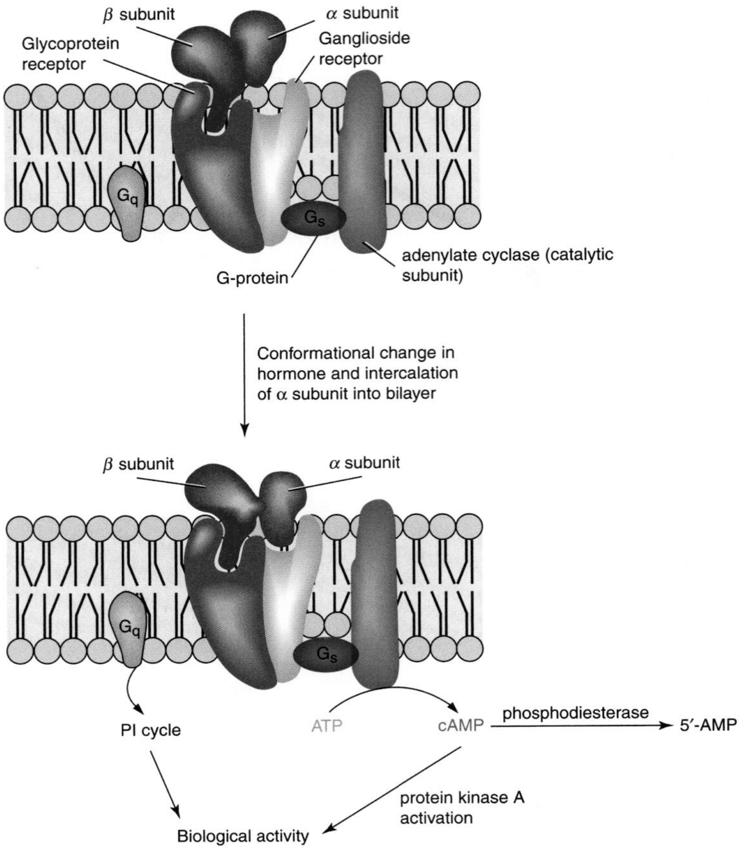

Figure 22.22 Model of TSH receptor. Receptor is composed of glycoprotein and ganglioside component. After TSH β-subunit interacts with receptor, hormone changes its conformation and α-subunit interacts with other membrane components. β-subunit of TSH may carry primary determinants recognized by glycoprotein receptor component. It is suggested that the TSH signal to adenylate cyclase is via the ganglioside; the glycoprotein component appears more directly linked to phospholipid signal system. PI, phosphatidylinositol; G_s, G protein linked to activation of adenylate cyclase; G_q, G protein linked to PI cycle.
Adapted with modifications from Kohn, L. D., et al. In Litwack, G. (Ed.), *Biochemical Actions of Hormones*. New York: Academic Press, 1985, 466.

receptors than either norepinephrine or epinephrine. Figure 22.23 shows the amino acid sequence for the β_2-adrenergic receptor (see p. 79 for single-letter abbreviations of amino acids). The N-terminal segment extends from α-helix I into the extracellular space, and there are seven membrane-spanning domains. The β_1 receptor shows extensive homology with the β_2 receptor. Intracellular loops join helices I and II, III and IV, and V and VI. The long chain extended from VII is the intracellular C-terminal region and contains phosphorylation sites (serine and threonine residues), which are important for receptor desensitization. Phosphorylation results in binding of an inhibitory protein, called **β arrestin**, which blocks the receptor's ability to activate G (p. 522). Extracellular loops join helices II and III, IV and V, and VI and VII, but mutational analysis suggests that these loops do not participate in ligand binding. Ligand binding may occur in a pocket formed by the clustering of these membrane-spanning α-helices, as shown from a top view in Figure 22.24, p. 908. Transmembrane domain VI may function in stimulation of adenylate cyclase. Substitution of a specific cysteine residue in this domain produces a mutant with normal ligand-binding properties but decreased ability to stimulate adenylate cyclase.

Internalization of Receptors

Many types of cell membrane hormone–receptor complexes are internalized by **endocytosis** (Figure 22.25, p. 908). For endocytosis to occur, the polypeptide–receptor complex enters **coated pits**, which are indentations of the plasma membrane into the cytoplasm. These coated pits pinch off from the membrane to form **coated vesicles**, which shed their coats, fuse with each other, and form vesicles called **receptosomes**. The receptors and ligands on the inside of these receptosomes have different fates. Receptors may be recycled to the cell surface following fusion with the Golgi apparatus. Alternatively, the coated vesicles fuse with lysosomes, which contain proteolytic enzymes that degrade both receptor and hormone. Some hormone–receptor complexes dissociate in the lysosome and only the hormone is degraded; the receptor is then returned intact to the plasma membrane. The

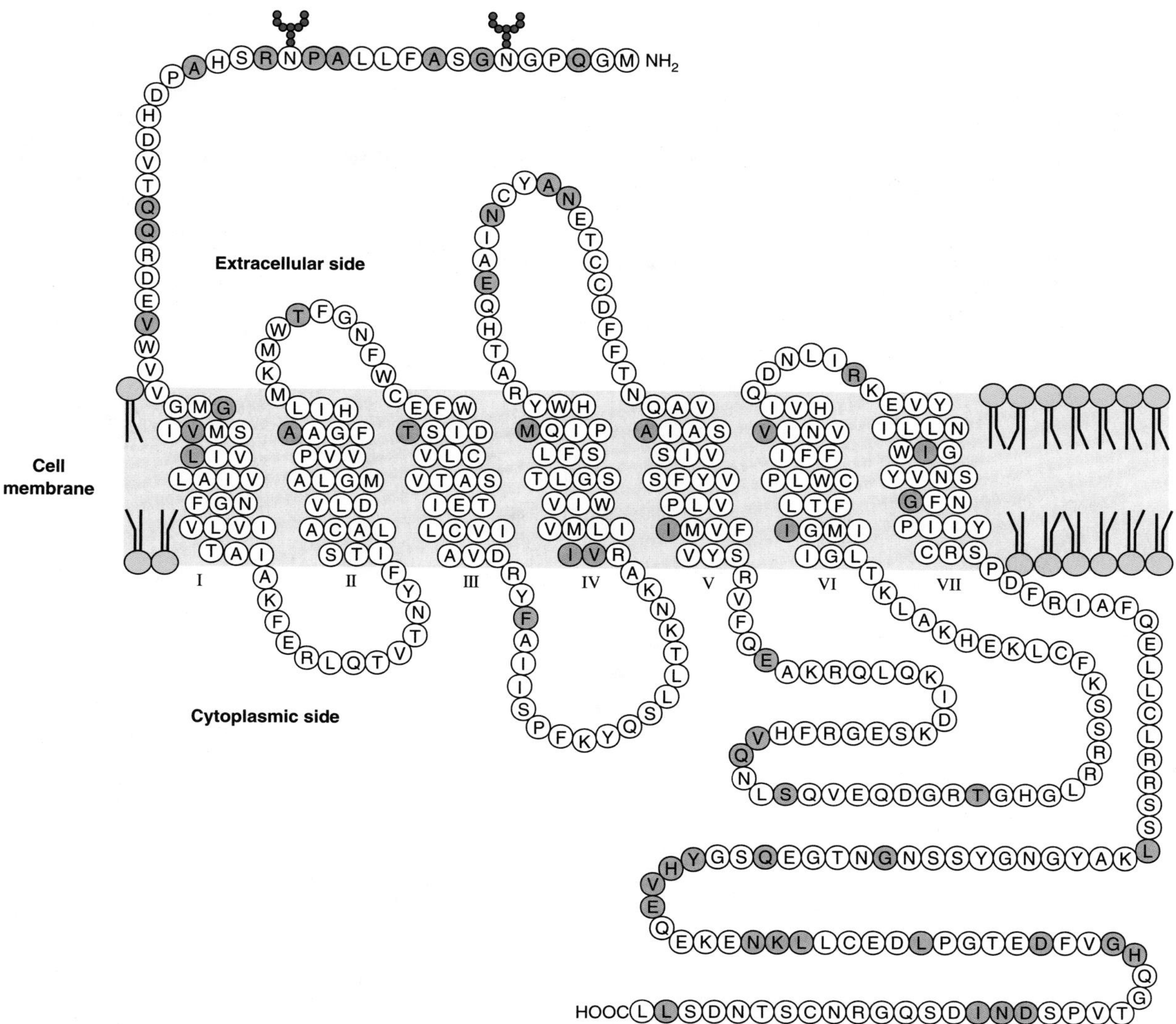

Figure 22.23 Proposed model for insertion of β_2-adrenergic receptor (AR) in cell membrane. The model is based on hydropathicity analysis of human β_2-AR. Standard one-letter codes for amino acid residues are used. Hydrophobic domains are represented as transmembrane helices. Pink circles with black letters indicate residues in the human sequence that differ from those in hamster. Also noted are the potential sites of N-linked glycosylation. Redrawn from Kobilka, B. K., Dixon, R. A., Frielle, T., Doblman, H.G., et al. *Proc. Natl. Acad. Sci. USA* 84:46, 1987.

receptor may also be concentrated in coated pits in the absence of exogenous ligand and cycle in and out of the cell in a constitutive, non-ligand-dependent manner.

Clathrin Directs Internalization of Hormone–Receptor Complexes from Plasma Membrane

The major protein component of a coated vesicle is **clathrin**, a nonglycosylated protein (180 kDa) whose amino acid sequence is highly conserved. The coated vesicle contains 70% clathrin, 5% polypeptides of about 35 kDa, and 25% polypeptides of 50–100 kDa. Coated vesicles have a lattice-like surface structure composed of hexagons and pentagons (Figure 22.26). Three clathrin molecules on the cytoplasmic side of the plasma membrane generate each polyhedral vertex and two clathrin molecules contribute to each edge. A 200-nm-diameter coated vesicle contains about 1000 clathrin molecules, which form flexible lattice structures or cages that act as scaffolds for vesicular budding. Completion of the budding process results in entry of the mature clathrin-coated vesicle into the cell by endocytosis.

Endocytosis introduces an intact receptor or ligand to the cell interior in cases where the nucleus may contain a receptor-binding site or a ligand-binding site. For example, growth factors bind to a cell membrane receptor but trigger events that lead to mitosis. Signal transmission may occur by an effect on a specific cytoplasmic protein (transcription factor), which translocates into the nucleus. Internalization of an intact ligand could allow

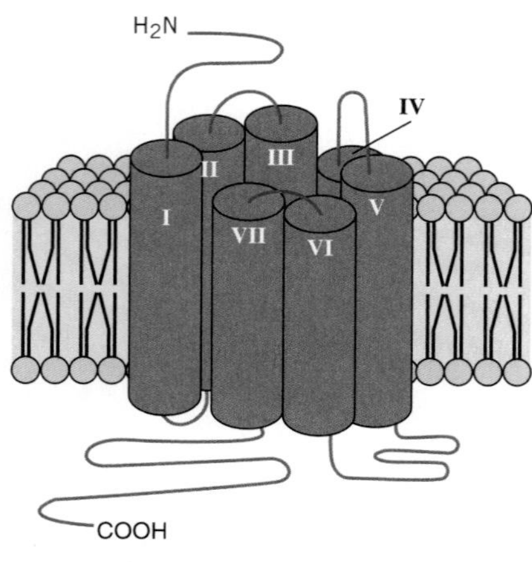

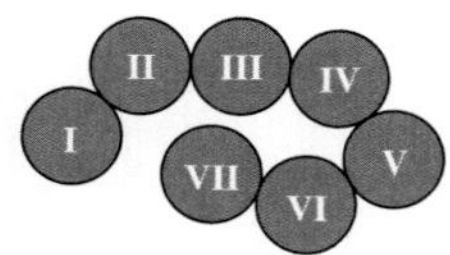

Figure 22.24 Proposed arrangement of the β-adrenergic receptor in membrane. Lower portion of figure is a view from above the plane of the plasma membrane. It is proposed that helices IV, VI, and VII form a ligand-binding pocket, with helix VII most centrally located.
Adapted from Frielle, T., Daniel, K. W., Caron, M. G., and Lefkowitz, R. J. *Proc. Natl. Acad. Sci. USA* 85:9494, 1988.

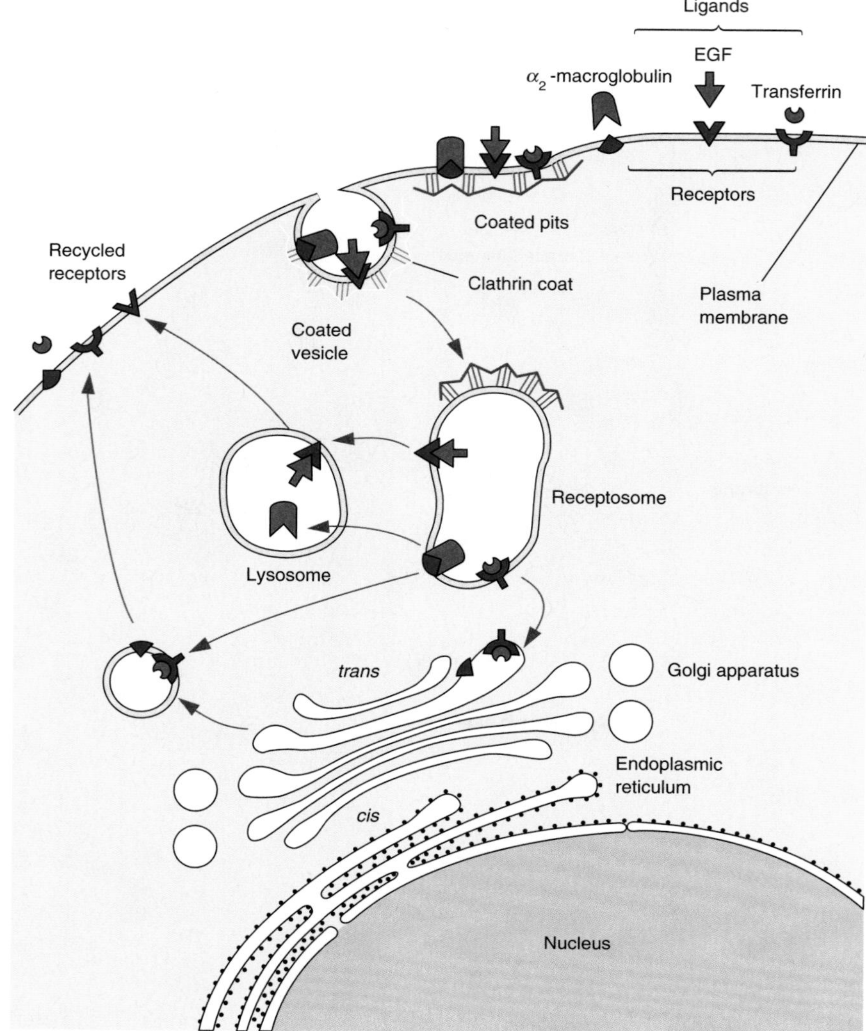

Figure 22.25 A diagrammatic summary of endocytosis in cells. The morphological elements of the pathway of endocytosis are not drawn to scale. The ligands shown are EGF, transferrin, and β_2-macroglobulin. For EGF, both ligand and receptor are delivered to lysosomes; for transferrin, both the ligand and receptor recycle to the surface; and for β_2-macroglobulin, ligand is delivered to lysosomes but the receptor recycles back to cell surface via Golgi apparatus.
Adapted from Pastan, I., and Willingham, M. C. (Eds.). *Endocytosis.* New York: Plenum Press 1985, 3.

Distal clathrin heavy chain
Proximal clathrin heavy chain
Clathrin light chain
Terminal domain

(*a*) Coated vesicle (*b*) Triskelion (*c*) Intermediate assembly

Figure 22.26 Structure and assembly of a coated vesicle. (*a*) A typical coated vesicle is 40 nm in diameter and surrounded by a fibrous network of proteins. One clathrin triskelion is centered on each of 36 vertices of the coat. (*b*) Detail of a clathrin triskelion. Each of three clathrin heavy chains is bent into a proximal arm and a distal arm. A clathrin light chain is attached to each heavy chain, most likely near the center. (*c*) An intermediate in the assembly of a coated vesicle.
Crowther, R. A., and Pearse, B. M. F. J. *Cell. Biol.* 91:790, 1981.
Redrawn from Nathke, I. S., Heuser, J., Lupas, A., Stock, J., et al. *Cell* 68:899, 1992.
Redrawn from Darnell, J., Lodish, H., and Baltimore, D. *Molecular Cell Biology.* New York: Scientific American, 1986, 647.

it to interact with a nuclear receptor. Although such mechanisms are speculative, they constitute a rationale for the participation of endocytosis in signal transmission. Endocytosis renders a cell less responsive to hormone, since it reduces the number of cell surface receptors. Internalization of receptors by endocytosis thus leads to receptor **down-regulation** and a decrease in hormone sensitivity.

22.6 • INTRACELLULAR HORMONE CASCADES: PROTEIN KINASES

Many hormones that bind to cell membrane receptors transmit their signals through second messengers that activate specific protein kinases; cAMP activates protein kinase A, DAG activates protein kinase C, and cGMP activates protein kinase G (p. 522). Other less prevalent systems involve hydrolysis of membrane phosphatidylcholine or sphingomyelin. Some hormones such as TSH use more than one intracellular second messenger system to transmit their signal.

Specific proteins are phosphorylated by protein kinase A and others are phosphorylated by protein kinase C. Both PKA and PKC phosphorylate on **serine** or **threonine** residues. Protein kinases that phosphorylate **tyrosine** residues are found in the cytoplasmic domains of some membrane receptors, especially growth factor receptors. The insulin and IGF-1 receptors possess tyrosine kinase domains and activities. The location of these domains of some receptor tyrosine kinases with respect to the plasma membrane is presented in Figure 22.27. The protein kinase domains are similar in amino acid sequence, suggesting that they have all evolved from a common primordial gene. Several **tyrosine specific kinases** that function as transmembrane receptors are shown in Figure 22.27. The α (extracellular) and β (transmembrane) polypeptides of the insulin receptor are encoded by a single gene,

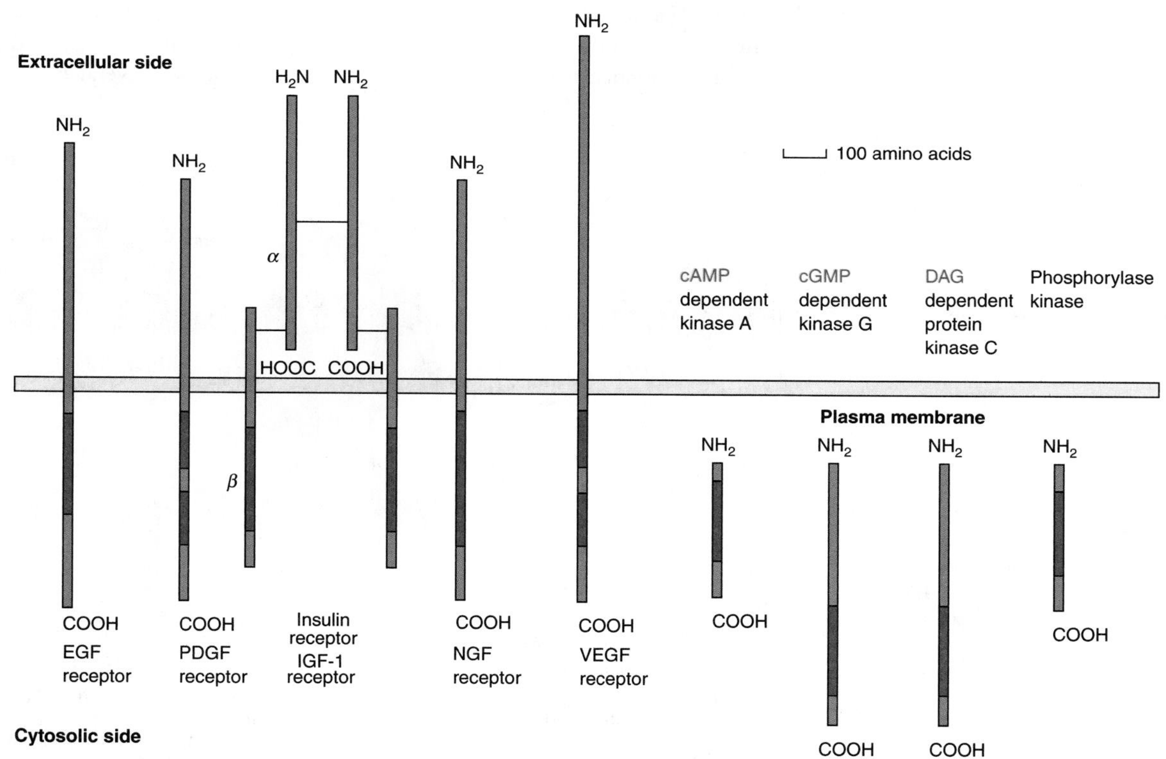

Figure 22.27 Protein kinases of plasma membrane or cytosol showing size and location of catalytic domains. In each case, the catalytic domain (red region) is about 250 amino acid residues long. Abbreviations: EGF, epidermal growth factor; PDGF, platelet derived growth factor; NGF, nerve growth factor; and VEGF, vascular endothelial growth factor.

Redrawn from Alberts, B., Bray, D., Lewis, J., Raff, M., Roberts, K., and Watson, J. D. *Molecular Biology of the Cell,* 3d ed. New York: Garland, 1994, 760.

which produces a precursor protein that is cleaved into the α and β disulfide-linked chains. In general, proteins regulated by phosphorylation-dephosphorylation usually have multiple phosphorylation sites and may be phosphorylated by more than one type of protein kinase.

Insulin Receptor: Transduction through Tyrosine Kinase

The α subunits of the **insulin receptor** are located extracellularly and are the insulin-binding sites (Figure 22.27). Ligand binding induces **autophosphorylation** of tyrosine residues located in the cytoplasmic portions of β-subunits. This autophosphorylation facilitates binding of cytoplasmic substrate proteins, such as **insulin receptor substrate-1 (IRS-1)**. IRS-1 when phosphorylated acts as a docking protein for proteins that mediate insulin action. It is not clear whether protein phosphorylation is the only mechanism for all actions of insulin. The net responses to this hormone include short-term metabolic effects, such as a rapid increase in the uptake of glucose, and longer-term effects on cellular differentiation and growth. Although the insulin receptor becomes auto-phosphorylated on tyrosines and phosphorylates tyrosines of IRS-1, other mediators are phosphorylated predominantly on serine and threonine residues, as indicated in Figure 22.28. Insulin stimulates phosphorylation of some proteins and dephosphorylation of others. Activation or inhibition of specific enzymes suggests that separate signal transduction pathways may originate from the insulin receptor. An insulin second messenger may be released at the cell membrane to account for the short-term metabolic effects of insulin. This may be a glycoinositol derivative that stimulates phosphoprotein phosphatase. A detailed diagram of the signal transduction pathways for insulin is presented Figure 22.29. Many proteins that are activated or inactivated subsequent to insulin binding have now been identified. This diagram shows that binding of insulin to the α-subunit of the receptor promotes autophosphorylation of a β-subunit on several tyrosine residues. The activated β-subunit then catalyzes the phosphorylation of cellular proteins such as members of the IRS family, including Shc and Cbl (members of IRS family) (Figure 22.29). On tyrosine phosphorylation, these proteins interact with other signaling molecules through their SH2 (Src-homolog-2) domains, which bind to a distinct amino acid sequence neighboring a phosphotyrosine residue. Several pathways that are

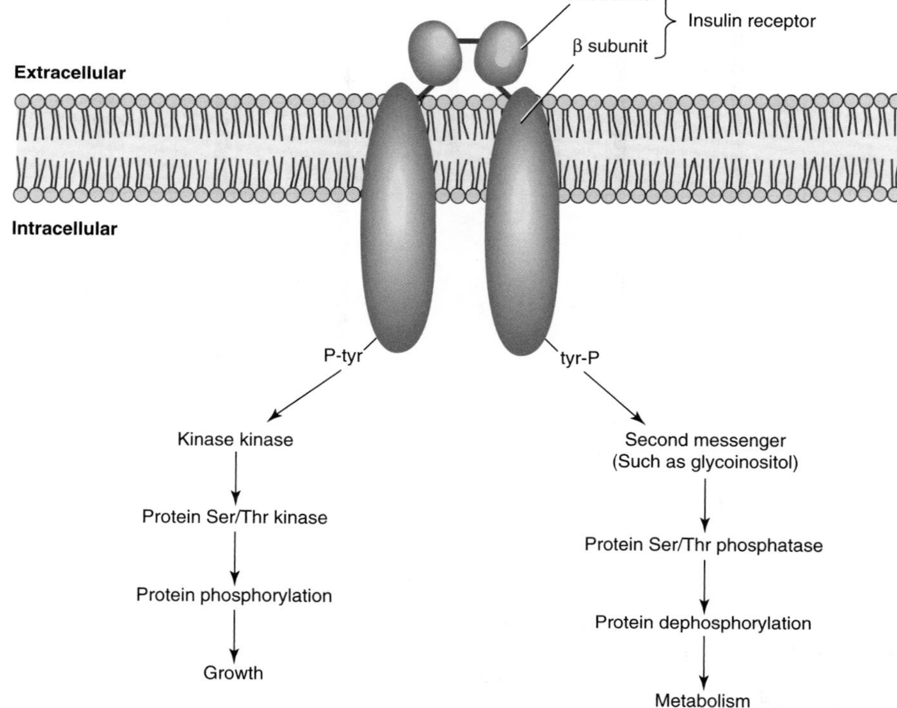

Figure 22.28 Hypothetical model depicting two pathways to explain paradoxical effects of insulin on protein phosphorylation. Insulin simultaneously produces increases in the serine/threonine phosphorylation of some proteins and decreases in others. This paradoxical effect may result from the activation of both kinases and phosphatases. Model explains (1) the generation of a soluble second messenger that directly or indirectly activates serine/threonine phosphatase, and (2) the stimulation of a cascade of protein kinases, resulting in the phosphorylation of cellular proteins.
Redrawn from Saltiel, A. R. *FASEB J.* 8:1034, 1994.

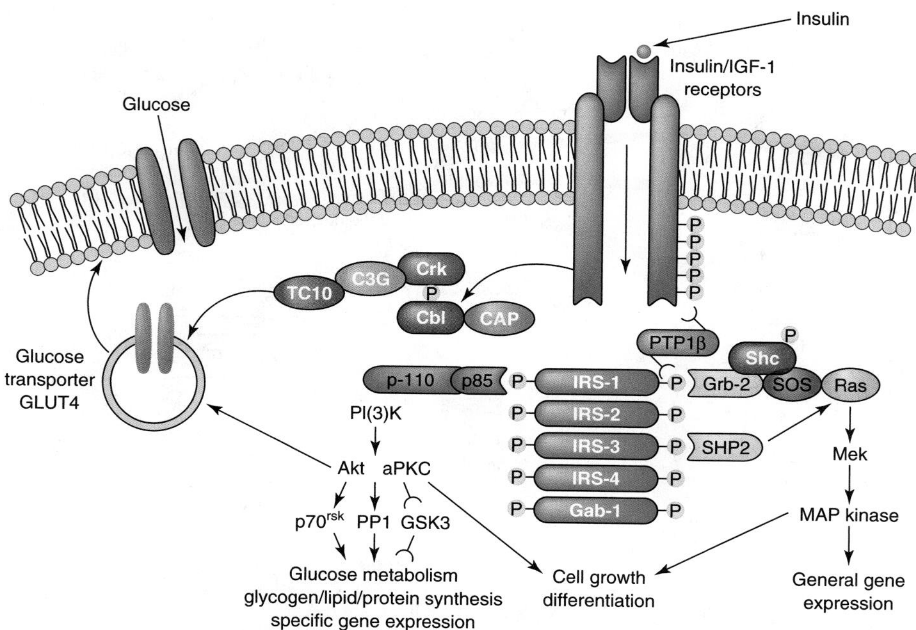

Figure 22.29 Hypothetical scheme for signal transduction in insulin action. The insulin receptor undergoes tyrosine autophosphorylation and kinase activation on hormone binding. It phosphorylates intracellular substrates including IRS-1, Shc, and Cbl, which associate with SH2-containing proteins like p85 and Grb2. Formation of the IRS-1/p85 complex activates PI (3-) kinase; the IRS-1/Grb2 complex activates MAP kinase. Abbreviations: IRS-1, insulin receptor substrate-1; SH, src homology; MAP kinase, mitogen-activated protein kinase; MeK, MAP kinase kinase; GPI, glycosyl-phosphatidylinositol; PLC, phospholipase; and SOS, son of sevenless. Redrawn from: Saltiel, A. R., and Kahn, C. R. Insulin signaling and the regulation of glucose and lipid metabolism. *Nature* 414:799, 2001.

activated include PI_3 kinase (phosphatidylinositol 3′-kinase), Ras (small GTP binding protein), the MAP kinase (mitogen-activated protein kinase), and TC10 (small GTP binding protein). Once activated via an exchange of GTP for GDP, TC10 promotes translocation of the glucose transporter GLUT4 vesicles to the plasma membrane, perhaps by stabilizing microfilaments. These pathways act in a concerted fashion to coordinate the regulation of vesicle trafficking (incorporation of GLUT4 into plasma membrane), protein synthesis, enzyme activation and inactivation, and gene expression. The net result of these pathways is regulation of glucose, lipid, and protein metabolism as well as cell growth and differentiation. The importance of the activity of the insulin receptor kinase in the overall signal transduction pathway of insulin is emphasized in Clin. Corr. 22.3.

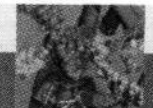

CLINICAL CORRELATION 22.3

Decreased Insulin Receptor Kinase Activity in Gestational Diabetes Mellitus

During pregnancy, an important maternal metabolic adaptation is a decrease in insulin sensitivity. This adaptation helps provide adequate glucose for the developing fetus. However, in some pregnant women glucose intolerance develops. Gestational diabetes mellitus (GDM) is a serious pregnancy associated disorder affecting up to 14% of all pregnant women. This disease is characterized by an additional decrease in insulin sensitivity and an inability to compensate with increased insulin secretion. Although both pregnancy-induced insulin resistance and GDM are generally reversible after pregnancy, approximately 30%–50% of women with a history of GDM go on to develop type 2 diabetes later in life, particularly if they are obese. Although the cellular mechanisms responsible for the insulin resistance in GDM are not fully understood, the resistance to insulin-mediated glucose transport appears to be greater in skeletal muscle from GDM subjects than in women who are pregnant but do not have GDM. Recent data indicate that defects in insulin action, rather than a decrease in insulin receptor binding affinity, may contribute to the pathogenesis of GDM. More specifically, skeletal muscle cells of GDM subjects appear to over-express plasma cell membrane glycoprotein-1 (PC-1), which has been reported to inhibit the tyrosine kinase activity of the insulin receptor by directly interacting with the α-subunits and blocking the insulin-induced conformational change. Additionally, excessive phosphorylation of serine/threonine residues located within muscle insulin receptors appears to down-regulate tyrosine kinase activity in GDM. There also appears to be a decreased expression and phosphorylation (tyrosine residues) of the insulin receptor substrate-1 (IRS-1; see Figure. 22.39, p. 920) in skeletal muscle cells from GDM subjects. These postreceptor defects in insulin signaling may thus contribute to the pathogenesis of GDM and the increased risk for type 2 diabetes later in life.

Shao, J., Catalono, P. M., Hiroshi, Y., Ruyter, I., et al. Decreased insulin receptor tyrosine kinase activity and plasma cell membrane glycoprotein-1 overexpression in skeletal muscle from obese women with gestational diabetes mellitus (GDM). *Diabetes* 49(4):603, 2000; and Maddux, B. A., and Goldfine, I.D. Membrane glycoprotein PC-1 inhibition of insulin receptor function occurs via direct interaction with the receptor alpha-subunit. *Diabetes* 49(1):13, 2000.

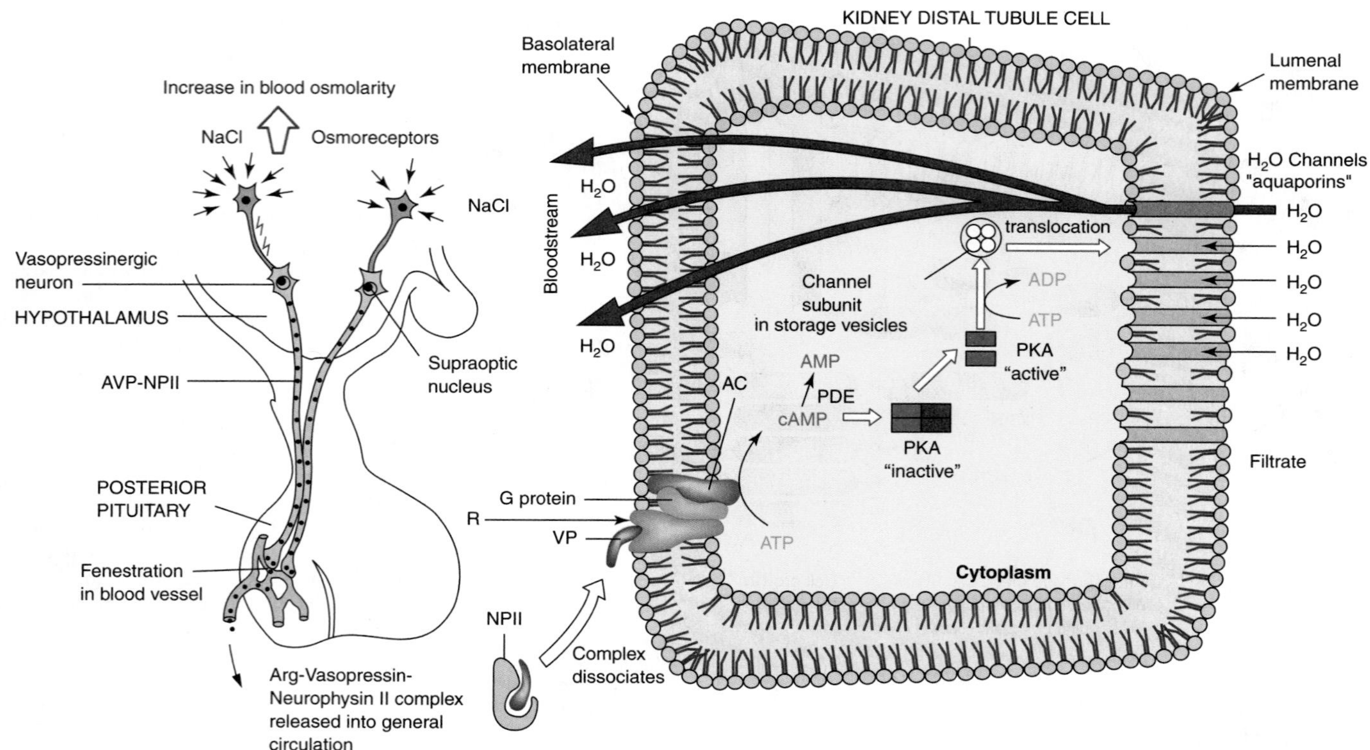

Figure 22.30 Secretion and action of arginine vasopressin in distal kidney tubules. Release of arginine vasopressin (AVP or VP) from the posterior pituitary is triggered by osmoreceptors or baroreceptors (not shown). This signal is transmitted down a vasopressinergic neuron and promotes the release of a VP–neurophysin complex from the posterior pituitary gland, where the hormone is normally stored. The neurophysin bound to the VP eventually dissociates, and the posterior pituitary hormone binds to cognate membrane receptors on the kidney distal tubule cell. Through this G protein-coupled receptor, adenylate cyclase is stimulated to increase levels of cAMP from ATP. Cyclic-AMP-dependent protein kinase A is then activated and it phosphorylates various proteins including subunits of the aquaporin channels. The phosphorylated subunits aggregate and functional water channels (aquaporins) are inserted in the lumenal plasma membrane, thus increasing the reabsorption of water. Abbreviations: NPII, neurophysin II; VP, vasopressin; R, receptor; AC, adenylate cyclase; and PDE, phosphodiesterase.
Redrawn in part from Dousa, T. P., and Valtin, H. Cellular actions of vasopressin in the mammalian kidney. *Kidney Int.* 10:45, 1975.

Activity of Vasopressin: Protein Kinase A

Arginine vasopressin (AVP), the antidiuretic hormone, causes increased water reabsorption from the urine in the distal kidney. A mechanism of action of this system is shown in Figure 22.30. Neurons that synthesize and secrete AVP (vasopressinergic neurons) release this hormone in response to stimuli from **baroreceptors** that respond to a fall in blood pressure or from **osmoreceptors** that respond to an increase in extracellular salt concentration. VP binds to its cognate membrane receptors in the distal kidney, anterior pituitary, hepatocytes, and perhaps other cell types. In the kidney, AVP binding to its G protein-coupled receptor stimulates adenylate cyclase activity and activates protein kinase A, which phosphorylates subunits that aggregate to form specific water channels, or aquaporins (p. 479). Water crosses the kidney cell to the basolateral side and then enters the general circulation, where it dilutes the salt concentration. Specific mutations in the intracellular and extracellular loop sequences of the aquaporin channels result in loss of function and in development of nephrogenic diabetes insipidus, which is characterized by increased thirst and production of a large volume of urine. Other examples of hormones that activate protein kinase A are listed in Table 22.6.

Gonadotropin-Releasing Hormone (GnRH): Protein Kinase C

Table 22.7 lists polypeptide hormones that stimulate the phosphatidylinositol pathway and activate **protein kinase C** (p. 535). Although AVP activates protein kinase A in

TABLE 22.6 • Examples of Hormones that Operate through the Protein Kinase A Pathway

Hormone	*Location of Action*
CRH	Corticotrope of anterior pituitary
TSH	Thyroid follicle
LH	Leydig cell of testis Mature follicle at ovulation and corpus luteum
FSH	Sertoli cell of seminiferous tubule and ovarian follicle
ACTH	Inner layer of cells of adrenal cortex
Opioid peptides	Some in CNS function on inhibitory pathway through G_i
AVP	Kidney distal tubular cell
PGI_2 (prostacyclin)	Blood platelet membrane
Norepinephrine/ epinephrine	β-Receptor: expressed in various tissues and cell types

TABLE 22.7 • Examples of Polypeptide Hormones that Stimulate the Phosphatidylinositol Pathway and Activate Protein Kinase C

Hormone	*Location of Action*
TRH	Thyrotrope of the anterior pituitary releasing TSH
GnRH	Gonadotrope of the anterior pituitary releasing LH and FSH
AVP	Corticotrope of the anterior pituitary; assists CRH in releasing ACTH: hepatocyte; causes increase in intracellular Ca^{2+}
TSH	Thyroid follicle: releasing thyroid hormones; causes increase in phosphatidylinositol cycle as well as increase in protein kinase A activity
Angiotensin II/III	Zona glomerulosa cell of adrenal cortex: releases aldosterone
Epinephrine	Smooth muscle cells that express α_1-receptors

renal cells, this hormone activates protein kinase C in other target cells. **GnRH** mediates its effects through activation of protein kinase C and the action of this hypothalamic releasing hormone is summarized in Figure 22.31. An aminergic nerve fiber most likely stimulates the appropriate neurons to secrete GnRH, which then enters the closed portal system, which connects the hypothalamus and anterior pituitary through fenestrations. GnRH binds to cognate membrane receptors in the gonadotrope (enlarged view in Figure 22.31) and activates phospholipase C. Hydrolysis of PIP_2 forms diacylglycerol (DAG) and IP_3, and DAG activates protein kinase C, which phosphorylates specific proteins. IP_3 binds to a receptor on the membrane of the endoplasmic reticulum and releases Ca^{2+} stored therein. The released Ca^{2+} also activates protein kinase C, which ultimately results in the secretion of LH and FSH from the same cell. There are at least 11 isoforms of protein kinase C, nine being activated by DAG. The isoforms of protein kinase C normally exist as monomeric proteins in the cytoplasm. The free enzyme is folded in such a way that the binding site for substrate proteins is blocked. When DAG binds to the enzyme it translocates it to the inner surface of the plasma membrane, where it can also bind to acidic phospholipids. The increased Ca^{2+} concentration also increases the binding of the enzyme to the membrane. When protein kinase C is bound, the protein unfolds and exposes binding sites for protein substrates. This results in phosphorylation of the bound proteins that then mediate a variety of intracellular effects.

Activity of Atrial Natriuretic Factor (ANF): Protein Kinase G

The receptor for atrial natriuretic factor (ANF) is a transmembrane protein whose cytoplasmic C-terminal domain has guanylate cyclase activity and whose extracellular

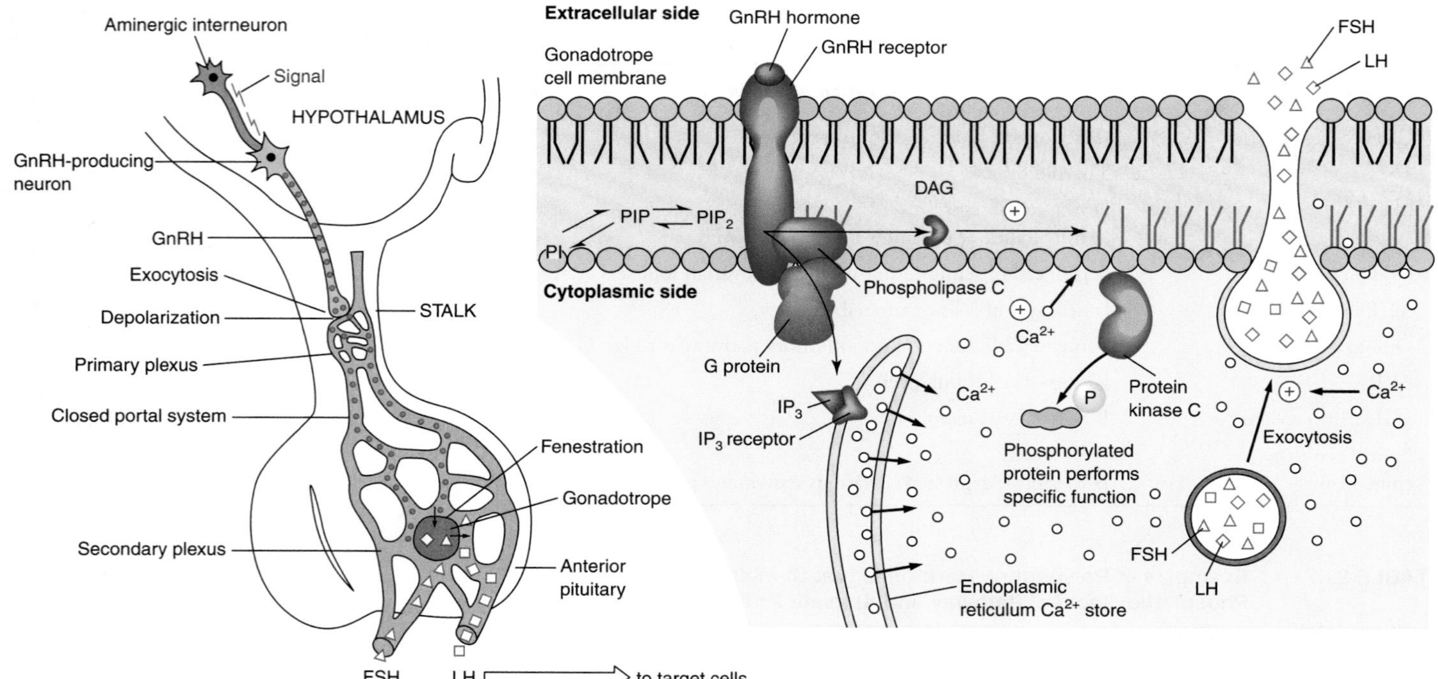

Figure 22.31 Regulation of secretion of LH and FSH by protein kinase C. A general mode of action of GnRH to release the gonadotropins from the gonadotropes of the anterior pituitary is presented. Abbreviations: GnRH, gonadotropin-releasing hormone; FSH, follicle-stimulating hormone; LH, luteinizing hormone; and DAG, diacylglycerol.

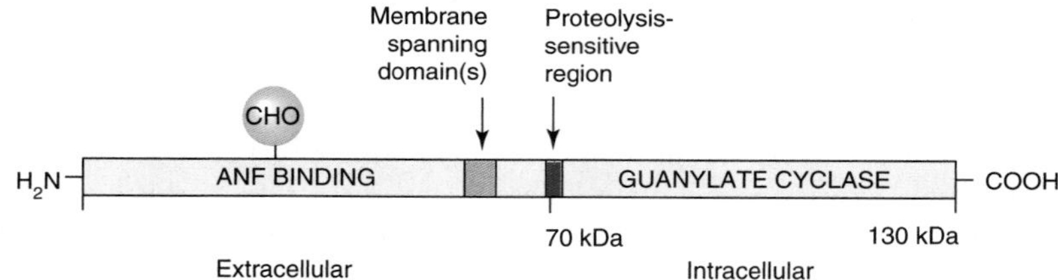

Figure 22.32 Functional domains of ANF–R$_1$ receptor. Model shows an ANF-binding domain, a membrane spanning domain(s), a proteolysis-sensitive region, a guanylate cyclase domain, glucosylation site (CHO), and amino and carboxyl terminals of the receptor.
Redrawn from Liu, B., Meloche, S, McNicoll, N., Lord, C., and DeLéan, A. *Biochemistry* 28:5599, 1989.

N-terminal domain binds ANF (Figure 22.32). A model for signal transduction by ANF is presented in Figure 22.33. ANF is a member of a family of peptides (Figure 22.34). This hormone is secreted by cardiac myocytes in response to signals such as blood volume expansion, high salt intake, increased right atrial pressure, and increased heart rate. Its secretion is stimulated by activators of cardiac protein kinase C and decreased by activators of protein kinase A. These opposing actions may be mediated by α- and β-adrenergic receptors, respectively. An overview of the secretion of ANF and its general effects is shown in Figure 22.35, p. 916. ANF is secreted as a dimer but only the monomeric form binds the receptor. ANF increases the glomerular filtration rate, leading to increased urine volume and excretion of sodium ion. Renin and aldosterone secretion are reduced, and the vasoconstriction produced by angiotensin II is inhibited, causing relaxation of the renal vessels and of other vascular beds and large arteries. ANF mediates these effects via its membrane receptor, whose intracellular domain has guanylate cyclase activity (see Figure 22.33, p. 915). The cGMP that is produced activates protein kinase G, which then phosphorylates cellular proteins involved in this pathway. Many analogs of ANF bind to receptors in the kidney but fail to elicit a physiological response. This suggests that these receptors may serve as specific peripheral storage-clearance binding sites for ANF and modulate its plasma levels.

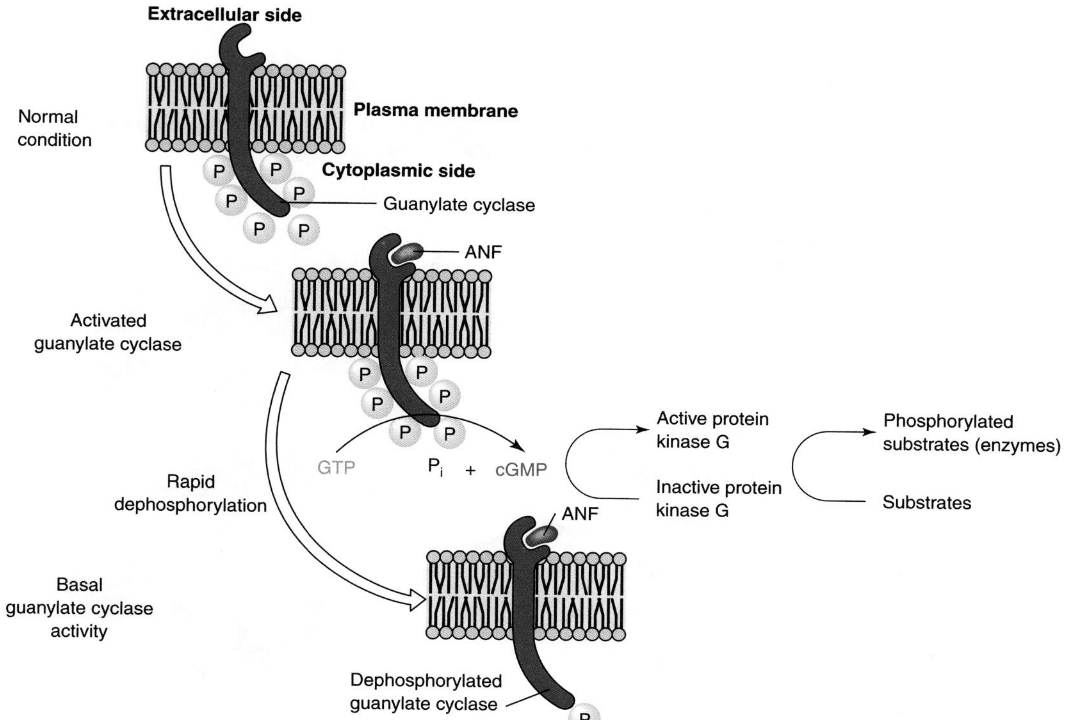

Figure 22.33 Model for signal transduction by ANF receptor. The guanylate cyclase domain is in a highly phosphorylated state under normal conditions. Binding of hormone markedly enhances enzyme activity and dephosphorylation of the guanylate cyclase domain.
Redrawn from Schultz, S., Chinkers, M., and Garbers, D. L. *FASEB J.* 3:2026, 1989.

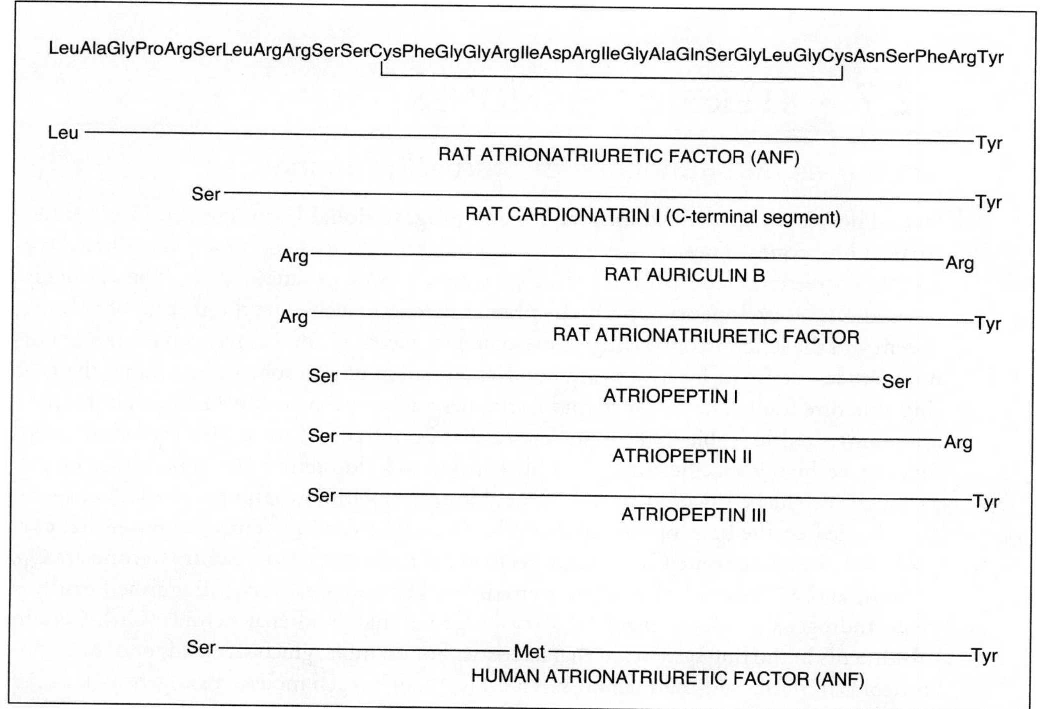

Figure 22.34 Atrial natriuretic peptides. These ANF derived peptides relax vascular smooth muscle and produce vasodilation and natriuresis as well as other effects discussed in the text.
Adapted from Cantin, M., and Genest, J. The heart and the atrial natriuretic factor. *Endocrine Rev.* 6:107, 1985.

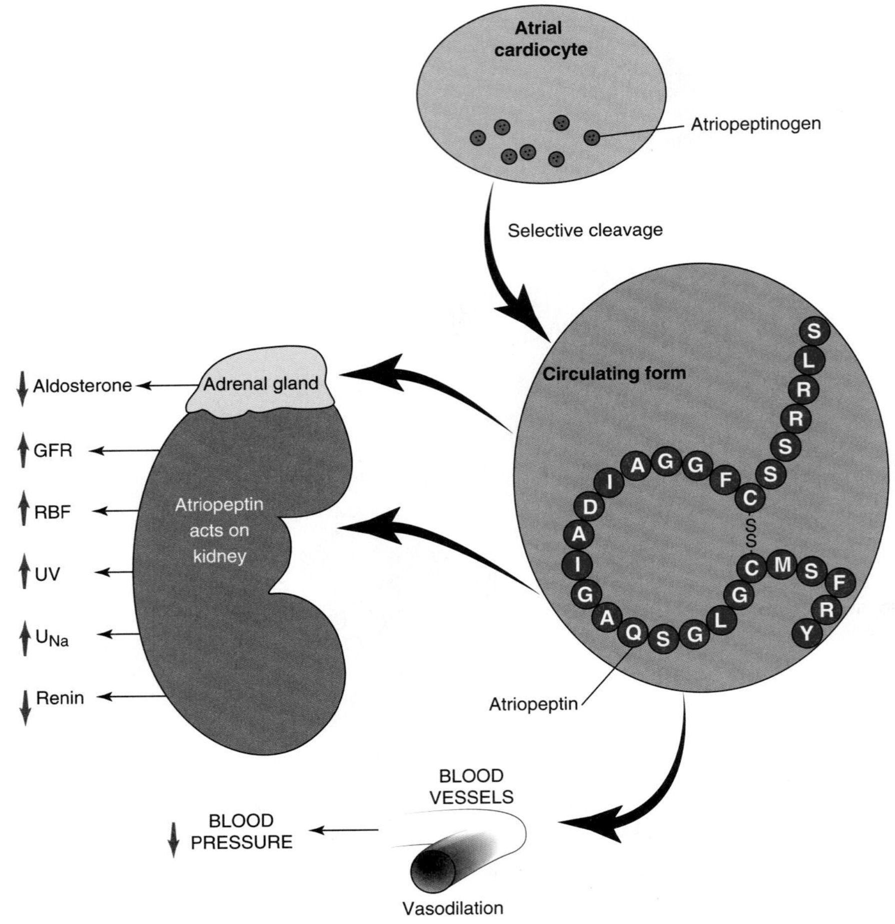

Figure 22.35 Schematic diagram of atrial natriuretic factor–atriopeptin hormonal system. Prohormone is stored in perinuclear granules in atrial cardiocytes. An elevated vascular volume results in cleavage of atriopeptinogen and release of atriopeptin, which increases the glomerular filtration rate (GFR), renal blood flow (RBF), urine volume (UV), and sodium excretion (U_{Na}), and decreases plasma renin activity and the secretion of aldosterone and arginine vasopressin. Vasodilation lowers blood pressure (BP). In contrast, diminution of vascular volume suppresses circulating levels of atriopeptin.
Redrawn from Needleman, P., and Greenwald, J. E. Atriopeptin: a cardiac hormone intimately involved in fluid, electrolyte, and blood pressure homeostasis. *N. Engl. J. Med.* 314:828, 1986.

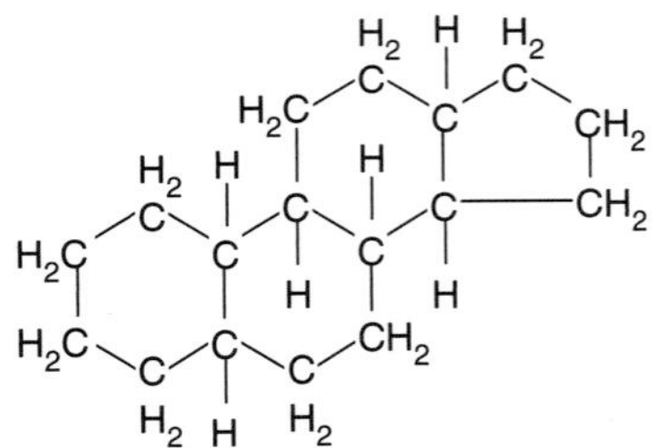

Cyclopentanoperhydrophenanthrene nucle

Numbering system of carbons

Figure 22.36 The steroid nucleus.

22.7 • STEROID HORMONES

Structures and Functions of Steroid Hormones

Steroid hormones are divided into the **sex** and **progestational hormones**, and the **adrenocortical hormones**. They are synthesized in the gonads (ovaries and testes) and adrenal cortex from cholesterol (p. 719) through Δ^5-pregnenolone as an intermediate. Their structure is based on the **cyclopentanoperhydrophenanthrene** nucleus; the numbering of this ring system and the lettering of the rings is presented in Figure 22.36. Conversion of steroid hormones to less active or inactive forms involves alteration of ring substituents rather than the ring structure itself. The major steroid hormones and their important functions in humans are summarized in Table 22.8. Many are similar in overall structure, although their receptors can be highly specific. Receptors for cortisol and aldosterone can bind either ligand, although the glucocorticoid receptor binds aldosterone with low affinity. Steroid hormones are classified on the basis of the number of carbons they contain. Thus **progesterone**, **cortisol**, and **aldosterone** are C_{21} steroids; **testosterone** and **dehydroepiandrosterone** are C_{19} steroids; and **17β-estradiol** is a C_{18} steroid. Sex hormones can be distinguished easily as being androgens (C_{19}), estrogens (C_{18}), or progestational or adrenal steroids (C_{21}). Certain substituents in the ring system are characteristic. For example, glucocorticoids and mineralocorticoids (typically aldosterone) possess a C11 OH or oxygen moiety. Estrogens lack a C19 methyl group and their A ring contains three double bonds. Many steroid receptors recognize primarily the A ring of their specific hormone. For instance, the estrogen receptor can distinguish the A ring of estradiol, which is stretched out of the plane of the B–C–D rings, from the A rings in other steroids, which are coplanar with the B–C–D rings. This relationship between the A ring and the B–C–D rings is illustrated in Figure 22.37, p. 918.

TABLE 22.8 • Major Steroid Hormones of Humans

Hormone	*Structure*	*Secretion from*	*Secretion Signal*	*Functions*
Progesterone	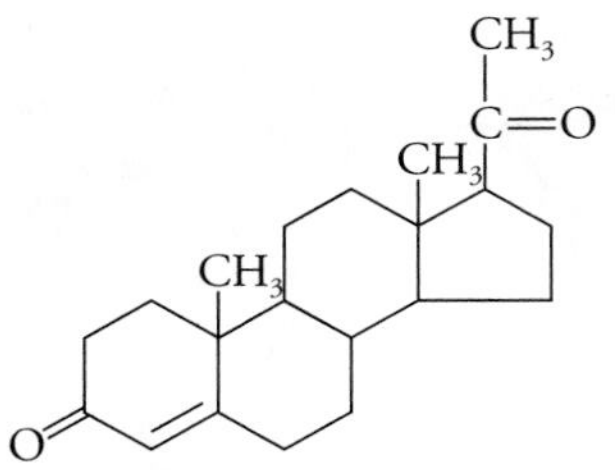 CH_3, C=O, CH_3, CH_3, O	Corpus luteum	LH	Maintains (with estradiol) the uterine endometrium for implantation of fertilized oocyt differentiation of mammary glands
17β-Estradiol	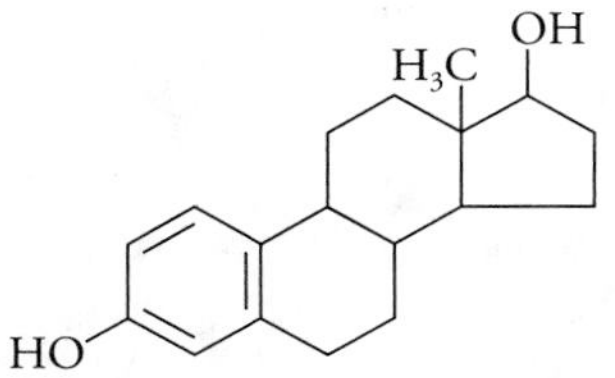 OH, H_3C, HO	Ovarian follicle; corpus luteum; (Sertoli cell)	FSH	Female: regulates gonadotropin secretion in ovarian cycle; maintains (with progesterone) uterine endometrium; growth of mammary gland. Male: negative feedback inhibitor of Leydig cell synthesis of testosterone
Testosterone	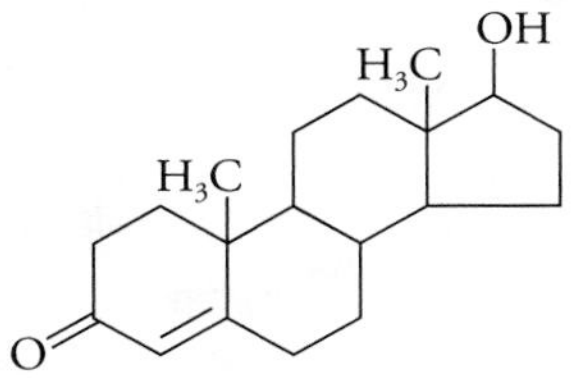 OH, H_3C, H_3C, O	Leydig cells of testis; (adrenal gland); ovary	LH	Male: required for spermatogenesis; converted to more potent androgen, dihydrotestosterone, in some target tissues like prostate gland; secondary sex characteristics (in some tissues testosterone is active hormone)
Dehydroepian drosterone	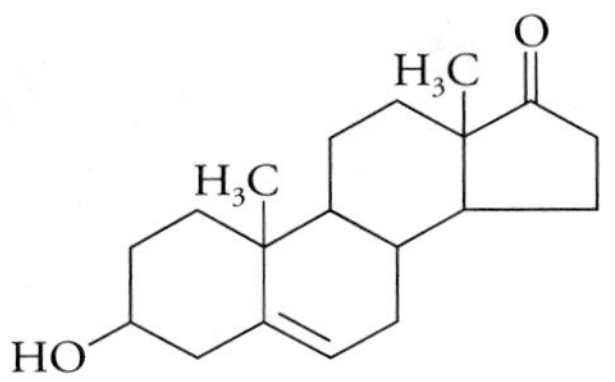 O, H_3C, H_3C, HO	Reticularis cells of adrenal cortex	ACTH	Various protective effects of adrenal cortex (anticancer, antiaging); weak androgen; can be converted to estrogen; no receptor yet isolated.
Cortisol	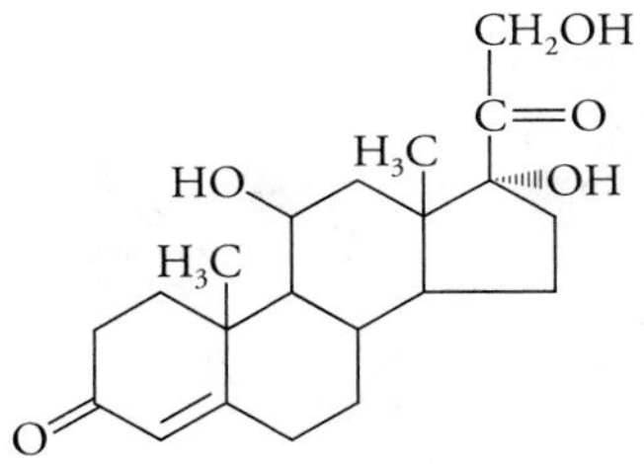 CH_2OH, C=O, H_3C, OH, HO, H_3C, O	Fasciculata cells of adrenal cortex	ACTH	Stress adaptation of adrenal cortex through various cellular phenotypic expressions; regulates protein, carbohydrate, and lipid metabolism; Immunosuppressive effects.
Aldosterone	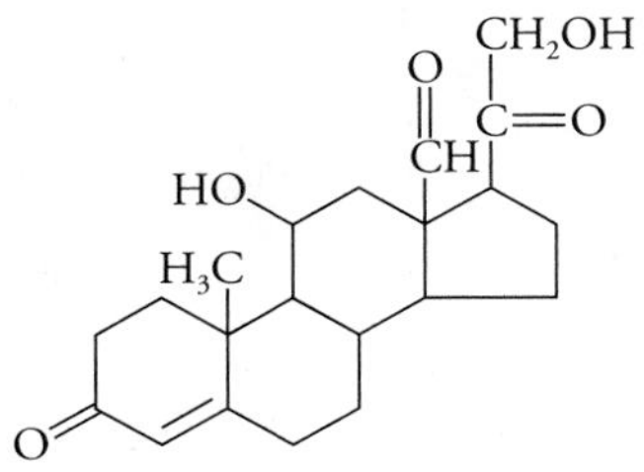 CH_2OH, O, C=O, CH, HO, H_3C, O	Glomerulosa cells of adrenal cortex	Angiotensin II/III	Causes sodium ion reabsorption in kidney via conductance channel; controls salt and water balance; raises blood pressure by increasing fluid volume.
1,25-Dihydroxy-vitamin D_3	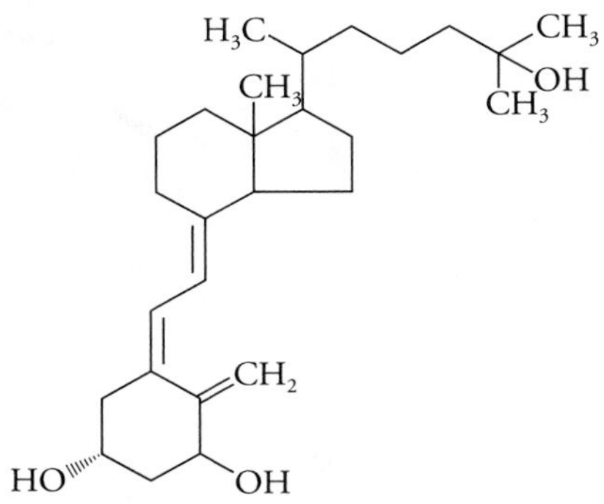 H_3C, CH_3, OH, CH_3, CH_3, CH_2, HO, OH	Vitamin D arises in skin cells after exposure to UV light; successive hydroxylations in liver and kidney yield active form of hormone	PTH (stimulates kidney proximal tubule hydroxylation system)	Facilitates Ca^{2+} and phosphate absorption by intestinal epithelial cells; induces intracellular calcium-binding protein

[a]LH, luteinizing hormone; FSH, follicle-stimulating hormone; ACTH, adrenocorticotropic hormone; PTH, parathyroid hormone.

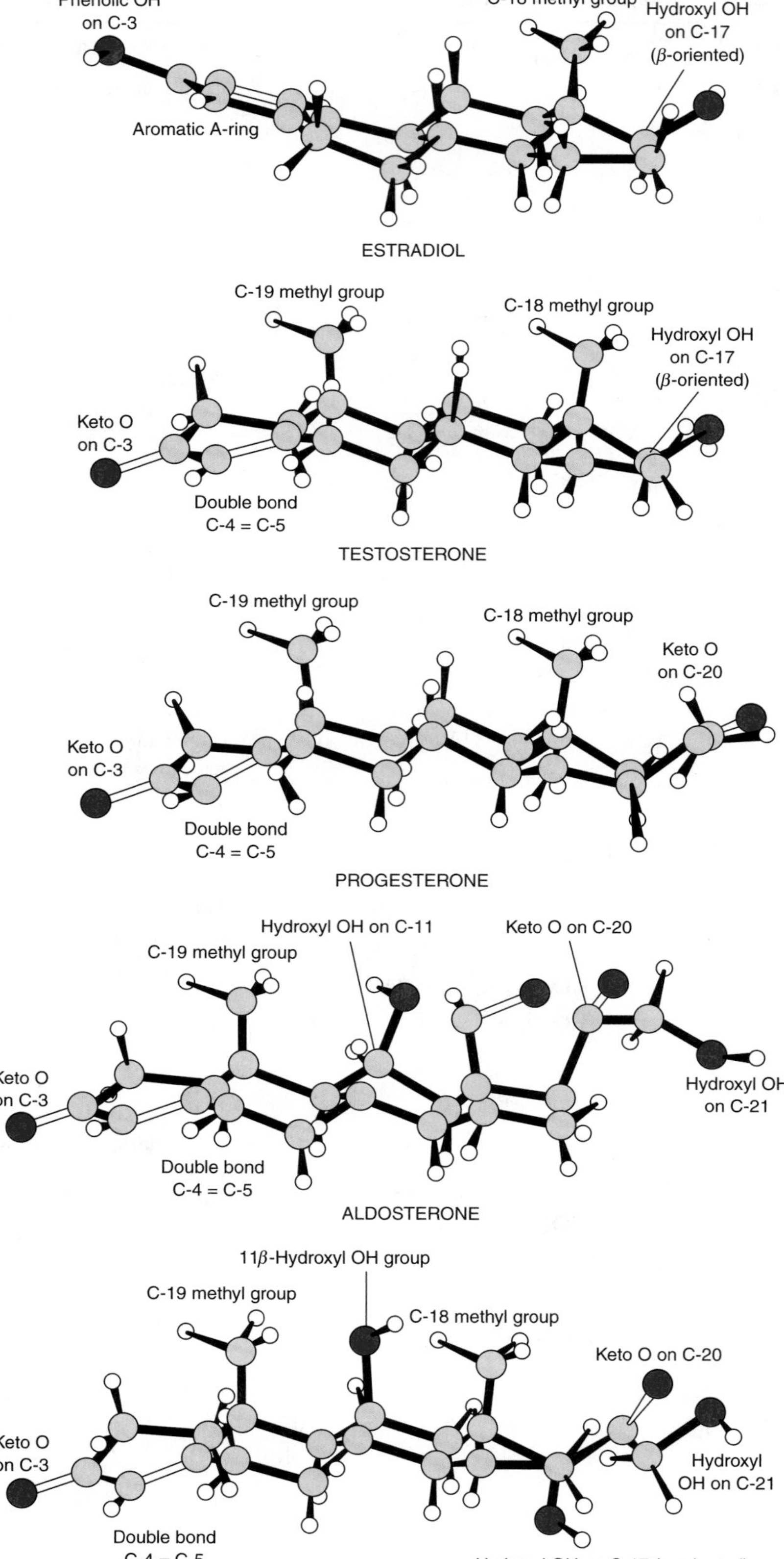

Figure 22.37 "Ball-and-stick" representations of some steroid hormones determined by X-ray crystallographic methods. Details of each structure are labeled. In aldosterone the acetal grouping is $R{-}CH\langle^{OR_1}_{OR_2}$ and the hemiketal grouping is $\begin{matrix} R_1 & & OR_3 \\ & C & \\ R_2 & & OH \end{matrix}$ where R_1, R_2, and R_3 refer to different substituents.

Reprinted with permission from Glusker, J. P. In Litwack, G. (Ed.). *Biochemical Actions of Hormones.* New York: Academic Press, 1979, 6:121.

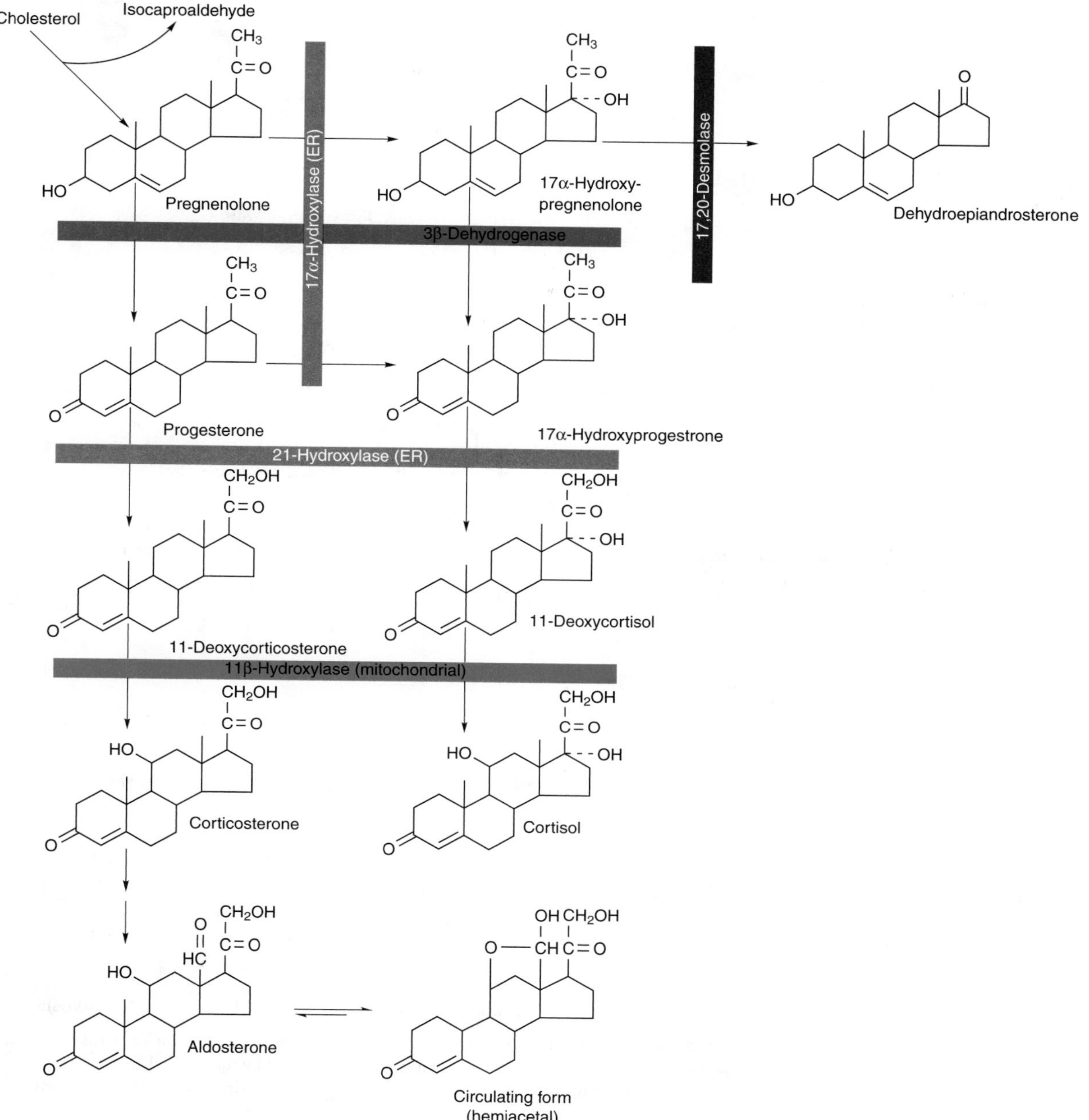

Figure 22.38 Conversion of cholesterol to adrenal cortical hormones. Not all intermediates are included and only enzymes of clinical significance are shown. ER, endoplasmic reticulum. Redrawn based on figure in Porterfield, S. P., ed. *Endocrine Physiology*. New York: Mosby, 1997, 136.

Biosynthesis of Steroid Hormones

Pathways for conversion of cholesterol to the adrenal cortical steroid hormones are presented in Figure 22.38. Cholesterol undergoes side chain cleavage to form Δ^5-pregnenolone and isocaproaldehyde. **Δ^5-Pregnenolone** is a required precursor in the synthesis of all steroid hormones. Pregnenolone is converted directly to progesterone by **3β-ol dehydrogenase** and **$\Delta^{4,5}$-isomerase**. The dehydrogenase converts the 3-OH group of pregnenolone to a 3-keto group and the isomerase moves the double bond from the B ring to the A ring to produce progesterone. In the ovarian corpus luteum the bulk of steroid synthesis stops at this point. Conversion of pregnenolone to **aldosterone** in the adrenal zona glomerulosa cells requires the endoplasmic reticulum 21-hydroxylase, and 11β-hydroxylase and 18-hydroxylase located in mitochondria. To form cortisol, primarily in adrenal zona fasciculata cells, **17-hydroxylase** and **21-hydroxylase** in the endoplasmic reticulum are

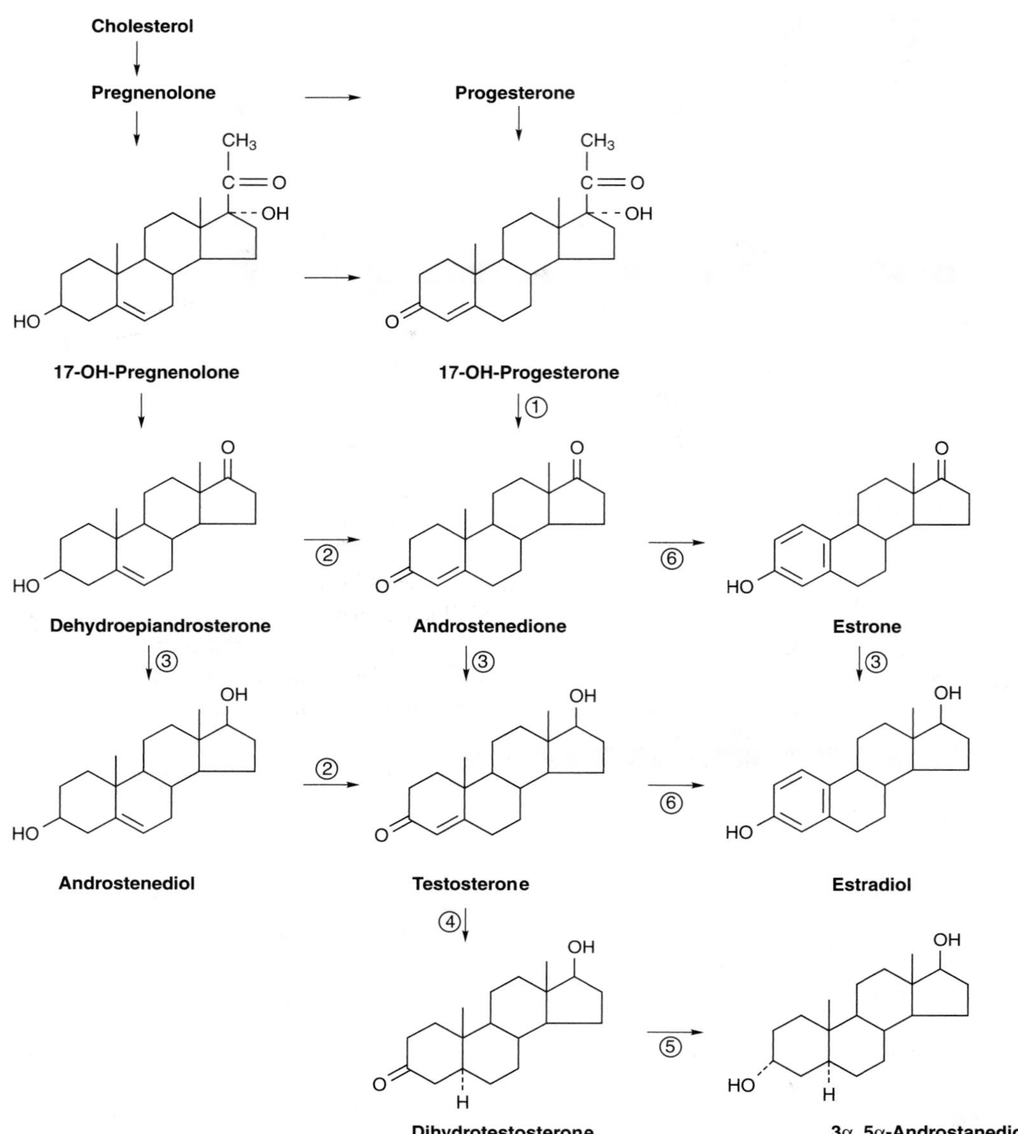

Figure 22.39 Conversion of cholesterol to sex hormones. Testosterone is the major steroid secreted by the testis. Estradiol and progesterone are the major steroids secreted by the ovary. Enzymes are (1)17,20-desmolase, (2) dehydrogenase and isomerase, (3) 17β-OH dehydrogenase, (4) 5α-reductase, (5) 3α-reductase, and (6) aromatase.

required along with mitochondrial **11β-hydroxylase**. The endoplasmic reticulum hydroxylases are cytochrome P450 (CYP) enzymes (p. 426). Δ^5-Pregnenolone is converted to **dehydroepiandrosterone** in the adrenal zona reticularis cells by 17α-hydroxylase localized in the endoplasmic reticulum to form 17α-hydroxypregnenolone and then by a side chain cleavage system to form dehydroepiandrosterone. Cholesterol is converted to the sex steroids by way of Δ^5-pregnenolone and progesterone (Figure 22.39). **Progesterone** is converted to testosterone by the action of cytoplasmic enzymes and 17-dehydrogenase. **Testosterone** is a major secretory product in the Leydig cells of the testis and is converted to dihydrotestosterone in some, but not all, androgen target cells before binding with high affinity to the intracellular androgen receptor. This reduction requires the activity of **5α-reductase** located in the endoplasmic reticulum and nucleus. Pregnenolone can enter an alternative pathway to form dehydroepiandrosterone as described above. This weak androgen can then be converted to androstenedione and then to testosterone. Estradiol is formed from testosterone by the action of the **aromatase** system.

The endoplasmic reticulum hydroxylases involved in steroid hormone synthesis use molecular oxygen (O_2) to introduce one oxygen atom into the steroidal substrate (as an

OH), while the second atom is reduced to water. Electrons generated from NADH or NADPH through a flavoprotein are transferred to ferredoxin or a similar nonheme protein. Note that there is movement of intermediates in and out of the mitochondrial compartment during biosynthesis of steroids. Once the specific steroids are synthesized, they diffuse through the plasma membrane and enter the general circulation, where they often bind to transport proteins. Unlike peptide hormones, steroids are not stored in secretory vesicles.

Metabolism of Steroid Hormones

Steroid hormones are metabolized slowly because of their binding to plasma proteins, which protects them from degradation. In humans, cortisol binds extensively to serum transcortin and has a plasma half-life of about 60–70 min. In contrast, aldosterone, which is not extensively bound to plasma proteins, has a half-life of only about 20 min. The liver is the principal site for steroid metabolism and a large number of metabolites are produced for each steroid. In general, the enzymatic reactions involved tend to decrease biological activity and increase solubility in water, thus facilitating their excretion in urine. Conjugation (p. 438) also increases the water solubility of steroid metabolites; glucuronides and sulfates are the most common conjugates. Estimates of steroid hormone secretion are often based on urinary metabolite levels. A number of factors influence the hepatic metabolism of steroids and hence their elimination from the circulation. For example, age influences hepatic steroid metabolism and the clearance of some steroids is slower (lower metabolic clearance rate) in infants and the elderly. The rates of steroid metabolism are also increased in patients with hyperthyroidism and decreased in patients with hypothyroidism and in different types of liver disease.

An example of steroid metabolism, more specifically testosterone metabolism, is presented in Figure 22.40. This does not necessarily inactivate this androgen. Testosterone

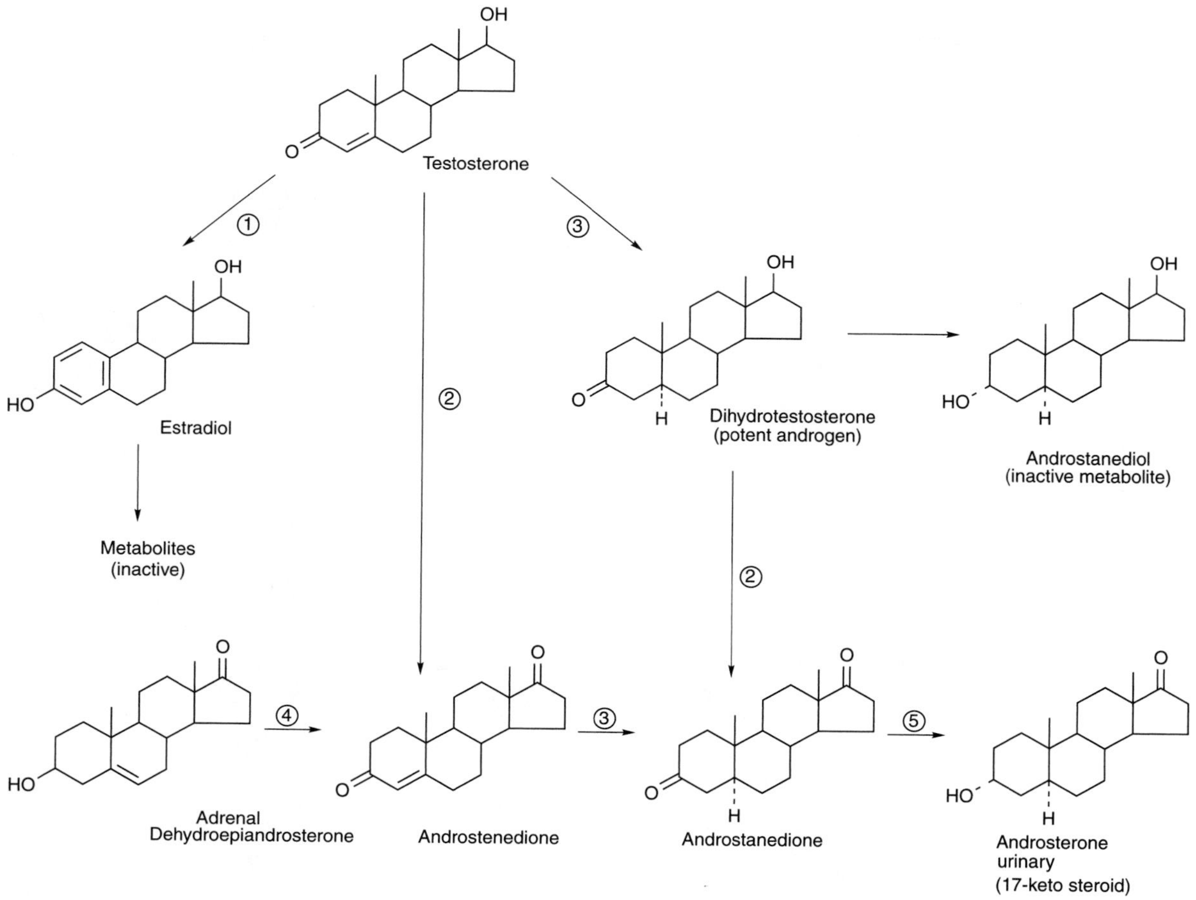

Figure 22.40 Metabolism of testosterone to generate active steroids or inactive 17-ketosteroids. Enzymes indicated (1) aromatase, (2) 17β-hydroxysteroid dehydrogenase, (3) 5β reductase; (4) 3β-hydroxysteroid dehydrogenase isomerase; and (5) 3β-hydroxysteroid dehydrogenase.

is converted to the active steroid, estradiol, by **aromatase**. Reduction of testosterone by 5α-reductase in some target cells generates **dihydrotestosterone**, which is a more potent androgen. Androsterone, the major inactive metabolic end product produced in this pathway, is conjugated and excreted in the urine. The major adrenal androgen, dehydroepiandrosterone (DHEA), is a known source of urinary 17-ketosteroids. Two other testosterone metabolites appear in urine in relatively small amounts. These are androstanediol, which is formed by reduction of the 3-keto group of dihydrotestosterone, and estrogen metabolites, which are formed once testosterone has been converted into estradiol.

Regulation of Steroid Hormone Synthesis

Regulation of steroid hormone biosynthesis is mediated by increased intracellular concentrations of cAMP and Ca^{2+}, although generation of IP_3 may also be involved (Figure 22.41). cAMP exerts rapid (seconds to minutes) and slow (hours) effects on steroid synthesis. The rapid effect is the mobilization and delivery of cholesterol to the inner membrane of mitochondria, where it is metabolized to pregnenolone by the cholesterol side chain cleavage enzyme (p. 434). In contrast, the slower effects involve increased transcription of genes for steroidogenic enzymes that are responsible for maintaining optimal long-term steroid production. The 30-kDa phosphoprotein, **steroidogenic acute regulatory (StAR)** protein, facilitates translocation of cholesterol from the outer to the inner mitochondrial membranes. In humans, StAR mRNA is specifically expressed in testes, ovaries, and adrenals, which are known sites of steroidogenesis. Patients with lipoid congenital adrenal hyperplasia (LCAH), an inherited disease in which adrenal and gonadal steroidogenesis is significantly impaired and lipoidal deposits occur, express truncated and nonfunctional StAR proteins. This suggests that the StAR protein is the hormone-induced protein that mediates acute regulation of steroid hormone biosynthesis.

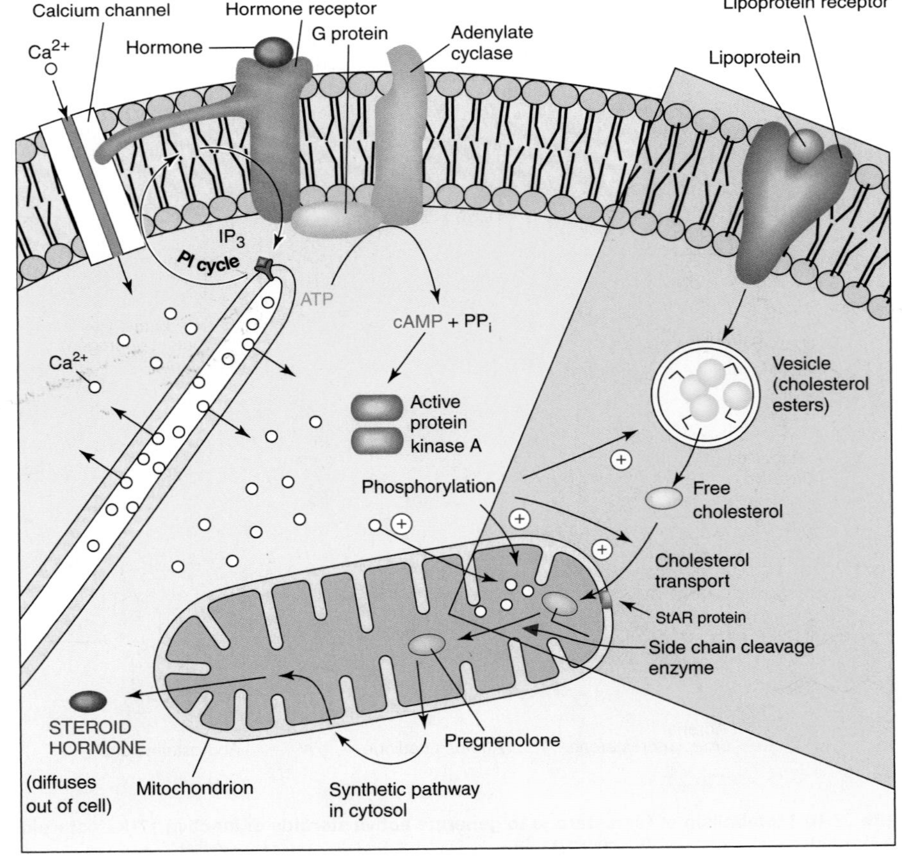

Figure 22.41 Overview of hormonal stimulation of steroid hormone biosynthesis. Nature of the hormone (top of figure) depends on the cell type and receptor (ACTH for cortisol synthesis, FSH for estradiol synthesis, LH for testosterone synthesis, etc., as given in Table 22.1). This activates adenylate cyclase via a stimulatory G protein or may stimulate a calcium channel directly or indirectly by activating the phosphatidylinositol cycle (PI cycle). If the PI cycle is stimulated, IP_3 augments cytosol Ca^{2+} levels from the ER store. cAMP activates protein kinase A, which, via phosphorylation, increases hydrolysis of cholesteryl esters from the vesicles to free cholesterol and increase cholesterol transport into the mitochondrion. The elevated Ca^{2+} levels and protein phosphorylation and induction of the StAR protein, result in increased side chain cleavage and steroid biosynthesis. These reactions overcome the rate-limiting steps (availability of cholesterol from cholesterol esters stored in vesicles, transport of cholesterol to inner mitochondrial membrane, and side chain cleavage reaction) in steroid biosynthesis, and more steroid is synthesized and secreted.

TABLE 22.9 • Hormones that Directly Stimulate Synthesis and Release of Steroid Hormones

Steroid Hormone	*Steroid-Producing Cell or Structure*	*Signal*[a]	*Second Messenger*	*Signal System*
Cortisol	Adrenal zona fasciculata	ACTH	cAMP, PI cycle, Ca^{2+}	Hypothalamic-pituitary cascade
Aldosterone	Adrenal zona glomerulosa	Angiotensin II/III	PI cycle, Ca^{2+}	Renin-angiotensin system
Testosterone	Leydig cell	LH	cAMP	Hypothalamic-pituitary cascade
17β-Estradiol	Ovarian follicle	FSH	cAMP	Hypothalamic-pituitary-ovarian cycle
Progesterone	Corpus luteum	LH	cAMP	Hypothalamic-pituitary-ovarian cycle
1,25$(OH)_2$ D_3	Kidney	PTH	cAMP	Sunlight, parathyroid glands, plasma Ca^{2+} level

[a]ACTH, adrenocorticotropic hormone; LH, luteinizing hormone; FSH, follicle-stimulating hormone; PI, phosphatidylinositol; PTH, parathyroid hormone.

Polypeptide hormones that stimulate biosynthesis and secretion of steroid hormones are summarized in Table 22.9. They operate through cognate cell membrane receptors. Where both the cAMP and the phosphatidylinositol cycle are involved, it is not clear whether one second messenger predominates. For the synthesis and secretion of aldosterone, probably several components (i.e., acetylcholine muscarinic receptor, atriopeptin receptor) and their second messengers are involved in addition to those listed in Table 22.9. Figure 22.42 summarizes the effects of ACTH on cortisol biosynthesis and secretion.

Aldosterone

Figure 22.43 shows reactions leading to secretion of aldosterone in a cell located in the adrenal zona glomerulosa. This outermost zone of the adrenal cortex cannot synthesize cortisol because it does not express the 17α-hydroxylase enzyme. A major driving force for aldosterone biosynthesis is **angiotensin II**, which is generated by the **renin–angiotensin system** shown in Figure 22.44. The signal for aldosterone secretion is generated under conditions when blood Na^+ concentration and blood pressure (blood volume) need to be increased. The N-terminal decapeptide of a plasma **α_2-globulin (angiotensinogen)** is cleaved by the proteolytic enzyme **renin**. This decapeptide is the inactive angiotensin I, which is then converted to the active octapeptide, angiotensin II, by **angiotensin-converting enzyme (ACE)**, which is found on the surface of pulmonary and renal endothelial cells. Angiotensin II is then converted to the heptapeptide, angiotensin III, by an aminopeptidase. Both angiotensins II and III bind to the angiotensin receptor (see Figure 22.43), which activates phospholipase C to generate IP_3 and DAG. The IP_3 then triggers release of Ca^{2+} from the endoplasmic reticulum. In addition, the plasma membrane Ca^{2+} channel is opened by the angiotensin–receptor complex. These events lead to a greatly increased level of cytoplasmic free Ca^{2+}, which together with DAG stimulates protein kinase C. **Acetylcholine** released through neuronal stress signals produces similar effects on calcium levels and protein kinase C activity, effects that are mediated by the muscarinic acetylcholine receptor (p. 515). Phosphorylation of enzymes that stimulate the rate-limiting steps in aldosterone synthesis lead to increased synthesis of aldosterone. In renal distal tubule cells, this adrenal steroid binds to its intracellular receptor and enhances expression of genes for proteins, including subunits for membrane Na^+ channels, which increase absorption of Na^+ from the glomerular filtrate. Although ACTH is not considered a major regulator of aldosterone synthesis, it promotes maximal secretion of this mineralocorticoid hormone. In contrast, ACTH is the major stimulator of the synthesis and secretion of cortisol by cells in the zona fasciculata (see Figure 22.42). Cortisol also accounts for substantial reabsorption of Na^+ in the kidney, probably through glucocorticoid stimulation of the Na^+/H^+ antiport transporter in the luminal membrane of renal epithelial cells.

Physiological conditions opposite to those that activate the formation of angiotensin I and II generate **atrial natriuretic factor (ANF)**, or atriopeptin, from the heart atria (see Figures 22.43 and 22.44). An increase in blood volume results in increased stretching of the atria and increased synthesis and secretion of ANF. In the zona glomerulosa cell, ANF

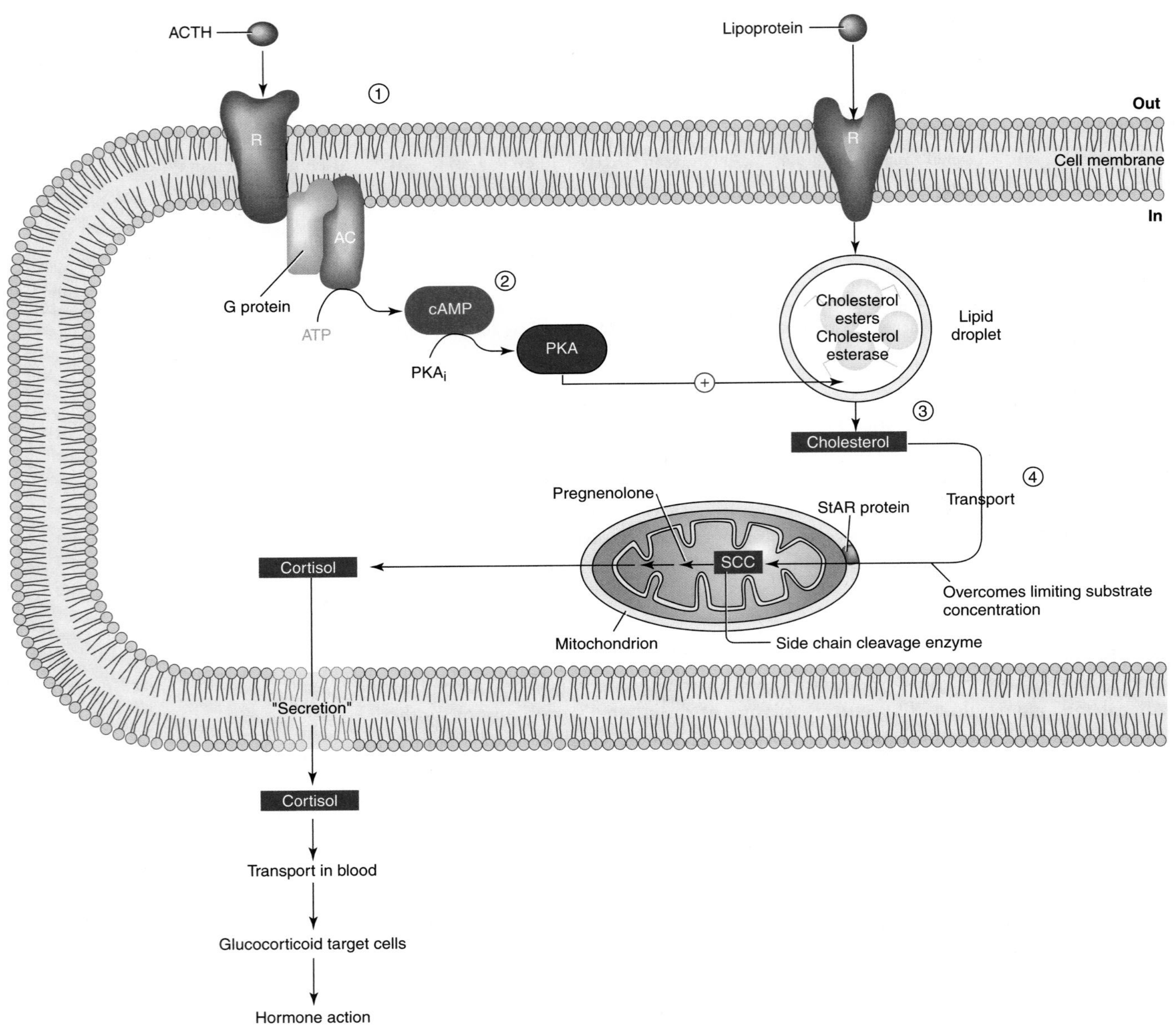

Figure 22.42 Action of ACTH on adrenal fasciculata cells to enhance production and secretion of cortisol. AC, adenylate cyclase; cAMP, cyclic AMP; PKA, protein kinase A; SCC, side chain cleavage system of enzymes. StAR (steroidogenic acute regulatory) protein is a cholesterol transporter functioning between the outer and inner mitochondrial membranes.

activates its receptor guanylate cyclase activity to increase cGMP levels and inhibit synthesis and secretion of aldosterone and formation of cAMP by adenylate cyclase.

Estradiol

Hormonal control of synthesis and secretion of **17β-estradiol** is shown in Figure 22.45. At puberty in females the secretion of GnRH, which is under control from higher brain centers, increases significantly. During adult life this hypothalamic-releasing hormone stimulates the release of FSH and LH from the anterior pituitary gonadotropes (p. 901). The FSH stimulates synthesis and secretion of 17β-estradiol in the ovary. This steroid exerts negative feedback on the pituitary **gonadotropes**, to suppress further secretion of FSH, and on the hypothalamic cells that secrete GnRH. Near ovarian midcycle, however, 17β-estradiol exerts a positive effect on the gonadotropes, (Figure 22.45), which causes very high levels of LH to be released (LH spike), as well as increased secretion of FSH (see Figure 22.19, p. 904). This LH spike is essential for ovulation to occur (p. 904). After

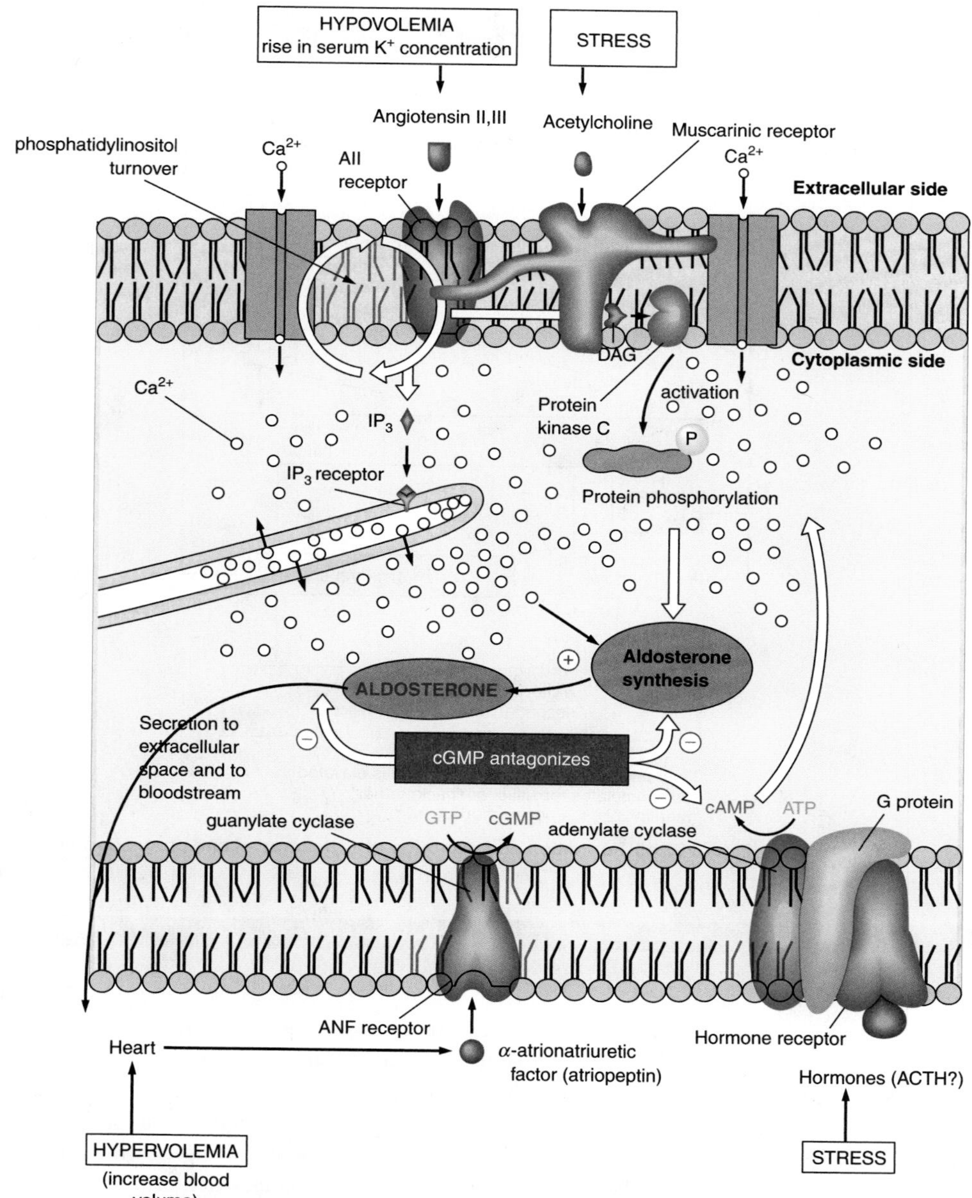

Figure 22.43 Reactions leading to secretion of aldosterone in adrenal zona glomerulosa cell. Abbreviations: cGMP, cyclic GMP; ANF, atrial natriuretic factor; see Figure 22.42, p. 924, for additional abbreviations.

ovulation has occurred, a functional corpus luteum (CL) forms from the ruptured follicle and synthesizes progesterone and some estradiol. Progesterone, however, inhibits continued LH synthesis and release. Eventually the corpus luteum dies (involutes), owing to a fall in the plasma LH level. The blood levels of progesterone and estradiol fall significantly, resulting in menstruation and a decline in their negative effects on the anterior pituitary and hypothalamus. Once the negative feedback exerted by estradiol and progesterone has been eliminated, a new menstrual cycle is initiated. Clin. Corr. 22.4 describes how oral contraceptives interrupt this sequence.

In males LH acts principally on Leydig cells to stimulate synthesis of testosterone. FSH acts on Sertoli cells to stimulate the conversion of testosterone secreted by the Leydig cells, into 17β-estradiol, which is required for **spermatogenesis.** FSH also stimulates Sertoli cells to secrete an androgen-binding protein that binds testosterone and estradiol and carries them into the lumen of the seminiferous tubules, where they are required by the maturing sperm. Testosterone itself exerts negative feedback that decreases secretion of GnRH. Sertoli cells also secrete the glycoprotein hormone **inhibin B**, the same hormone secreted by ovarian granulosa cells, which selectively inhibits FSH secretion. These hormonal negative feedback loops

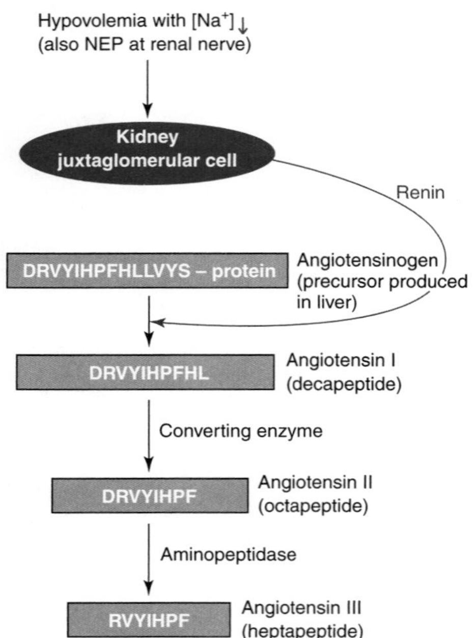

Figure 22.44 Renin–angiotensin system. See p. 79 for amino acid abbreviations. NEP, norepinephrine.

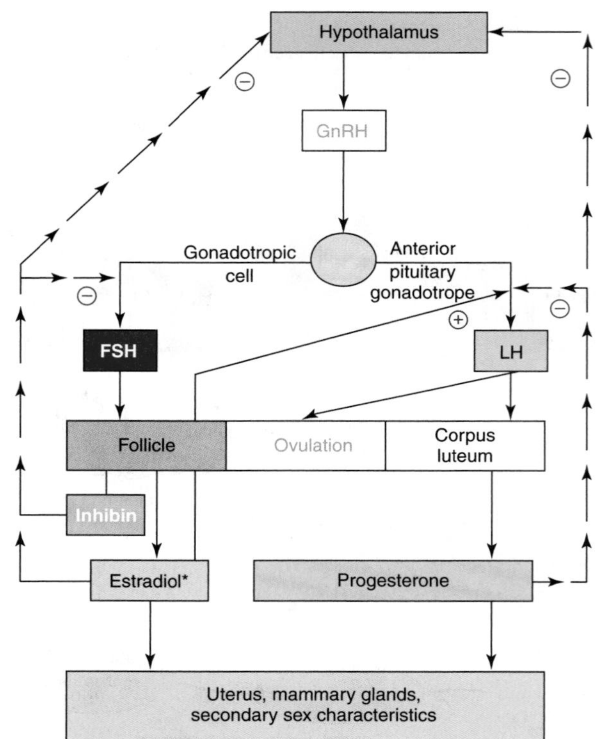

Figure 22.45 Formation and secretion of 17β-estradiol and progesterone.

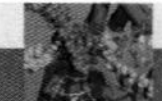

CLINICAL CORRELATION 22.4

Oral Contraception

Many forms of oral contraception are based on the fact that estrogens and progesterone can inhibit pituitary secretion of FSH and LH, thereby preventing maturation of the ovarian follicle and ovulation. Most oral contraceptives are a combination of a synthetic estrogen and a progesterone-like substance or progestin. This combination of steroids blocks the LH surge required for ovulation and results in thickening and increased vascularization of the uterine endometrium. Pills that lack any steroids (placebos) are usually inserted in the regimen at about the 28th day, causing blood levels of estrogens and progestins to fall dramatically and menstruation to occur. When the combination oral contraceptives are resumed, the blood levels of estrogen and progestin increase again and the uterine endometrium thickens. This creates a false cycling because of the occurrence of menstruation at the expected time point in the cycle. Some oral contraceptives, such as the minipill, contain only a progestin. Although these pills do not block the LH surge required for ovulation, they are still effective at blocking fertility because they mediate other contraceptive effects. For example, these progestins can affect the composition of the cervical mucus, thus preventing passage of sperm through the cervix. They also inhibit the estrogen-induced proliferation of the endometrium and thus prevent implantation. Other hormonal methods of contraception include surgical implantation of progestin-containing silicone capsules under the skin. In this situation the progestin is released slowly, thus providing contraception for up to 3–5 years.

Zatuchni, G. I. Female contraception. In Becker, K. L. (Ed.), *Principles and Practice of Endocrinology and Metabolism.* New York: Lippincott, 1990, 861; and Shoupe, D., and Mishell, D. R. Norplant: Subdermal implant system for long term contraception. *Am. J. Obstet. Gyn.* 160:1286, 1988.

thus reduce testosterone and estradiol secretion and inhibit the process of spermatogenesis. The biosynthesis of testosterone from cholesterol is summarized in Figure 22.39, p. 920.

Vitamin D_3

The active form of vitamin D, called **calcitriol**, is referred to as a **secosteroid**, which is a steroid in which one of the rings has been opened. As shown in Figure 22.46, 7-dehydrocholesterol is activated in the skin by sunlight to generate **vitamin D_3 (cholecalciferol)**. This then is

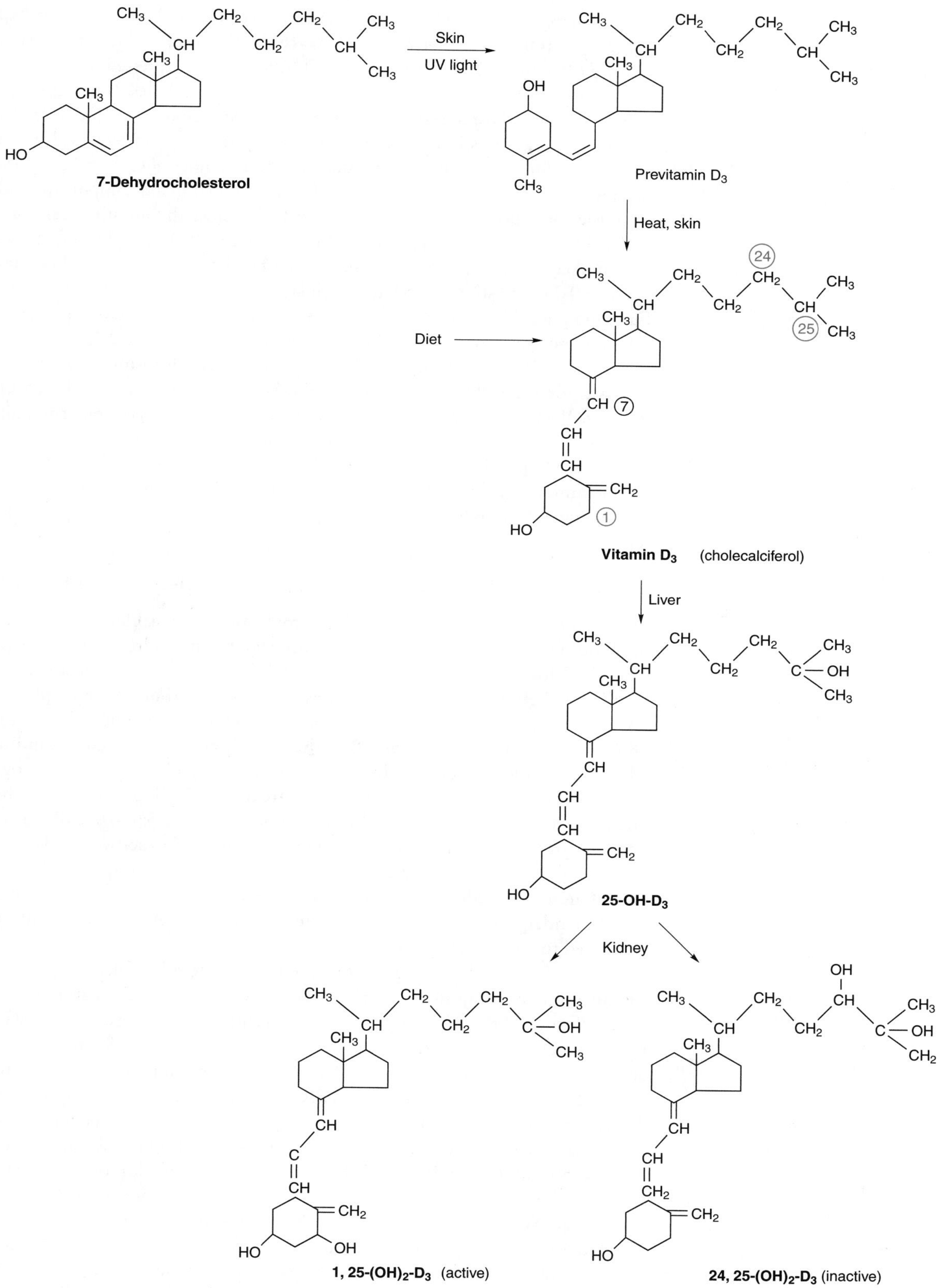

Figure 22.46 Structure of 1α,25-dihydroxycholecalciferol (calcitriol), the biologically active form of vitamin D.

hydroxylated in the liver to generate **25-hydroxy vitamin D_3 (25-hydroxycholecalciferol)**. In the kidney this is further hydroxylated to either **1α,25-vitamin D_3 (1,25$(OH)_2D_3$) (1α, 25-dihydroxychol-ecalciferol)**, which is the active form of the hormone, or **24,25-vitamin D_3 (24, 25 $(OH)_2$ D_3) (24,25-dihydroxycholecaliciferol)**, which is an inactive metabolite of the hormone. Nuclear 1,25$(OH)_2D_3$ receptors are expressed in target cells including intestinal epithelial cells, bone cells, and cells of kidney tubules. Binding of 1,25$(OH)_2D_3$ induces receptor phosphorylation, and binding of the hormone–receptor complex to vitamin D_3 -response elements in the DNA. This binding results in an increased rate of transcription of vitamin D_3-responsive genes that encode a number of Ca^{2+}-binding proteins called **calbindins**, Ca^{2+}-ATPase and other ATPases, and membrane components and facilitators of vesicle formation. Calbindins may ferry Ca^{2+} across the intestinal cell, or may simply buffer the cytoplasm against high Ca^{2+} levels (Figure 22.47). 1,25$(OH)_2$ D_3 also increases the number of Ca^{2+} pump molecules in the basolateral membrane. The major effect of 1,25$(OH)_2D_3$ is to stimulate the transcellular (across intestinal epithelial cells) absorption of Ca^{2+} and phosphate from the intestinal lumen against a concentration gradient, and thus it plays an important role in bone mineralization. However, elevated 1,25$(OH)_2D_3$ levels can actually stimulate bone resorption by osteoclasts, which surprisingly do not express nuclear receptors for 1,25$(OH)_2D_3$. Osteoblasts do express these receptors and 1,25$(OH)_2D_3$ may stimulate them to secrete a paracrine factor that increases the recruitment and differentiation of precursors into active osteoclasts, which can then mediate bone resorption. Vitamin D may also be responsible for autocrine and paracrine effects that are important in the regulation of immune responses. In the kidney 1,25$(OH)_2D_3$ weakly stimulates Ca^{2+} reabsorption by increasing the number of Ca^{2+} pumps.

Transport of Steroid Hormones: Plasma-Binding Proteins

Four major plasma proteins bind the steroid hormones in blood. They are corticosteroid-binding globulin, sex-hormone-binding globulin, androgen-binding protein, and albumin. Most of the circulating cortisol (75%–80%) binds to a specific **corticosteroid-binding α2-globulin (CBG)** known as **transcortin**. About 15% of plasma cortisol is bound to albumin with a much lower affinity. Thus even though albumin is present at 1000-fold the concentration of CBG, the cortisol occupies the CBG-binding sites first. The concentration of transcortin increases during pregnancy and after estrogen administration. During stress, when cortisol levels are high, the CBG binding sites become saturated first and the excess cortisol binds to albumin. Only 5%–10% of plasma cortisol is normally free (unbound). It is this free cortisol that diffuses across the plasma membrane, binds to glucocorticoid receptors, and mediates a biological response. About 50%–70% of the circulating aldosterone binds with low affinity to albumin and transcortin, and the remainder is unbound. Thus aldosterone has a lower plasma half-life (20 min) as compared to cortisol (70 min).

About 65% of circulating testosterone is bound to a liver-derived glycoprotein called **sex hormone-binding globulin (SHBG)**. Only 1%–2% of circulating testosterone is in the free form and the rest is bound to albumin and other proteins. The SHBG-bound fraction thus serves as circulating reservoir of testosterone. Approximately 60% of estrogens are transported bound to SHBG, 20% are bound to albumin, and 20% are in the free form. However, estradiol binds to SHBG with a much lower affinity than testosterone. Thus, estradiol bound to SHBG dissociates very rapidly, and it is taken up by target tissues. Adult males have about one-half as much circulating SHBG as females, so free testosterone is about 20 times greater in males than in females. In addition, the total (bound plus unbound) concentration of testosterone is about 40 times greater in males. Testosterone itself lowers SHBG levels and increases the amount of free testosterone in blood, whereas 17β-estradiol and thyroid hormone raise SHBG levels in blood. These effects have important ramifications in pregnancy and in other endocrine conditions.

Androgen-binding protein (ABP) is produced by the Sertoli cells in response to testosterone and FSH. ABP is also called **testosterone-estrogen-binding globulin (TeBG)** and this protein helps to maintain high androgen levels within the testis and seminal fluids. These high local androgen levels are important for the development and maturation

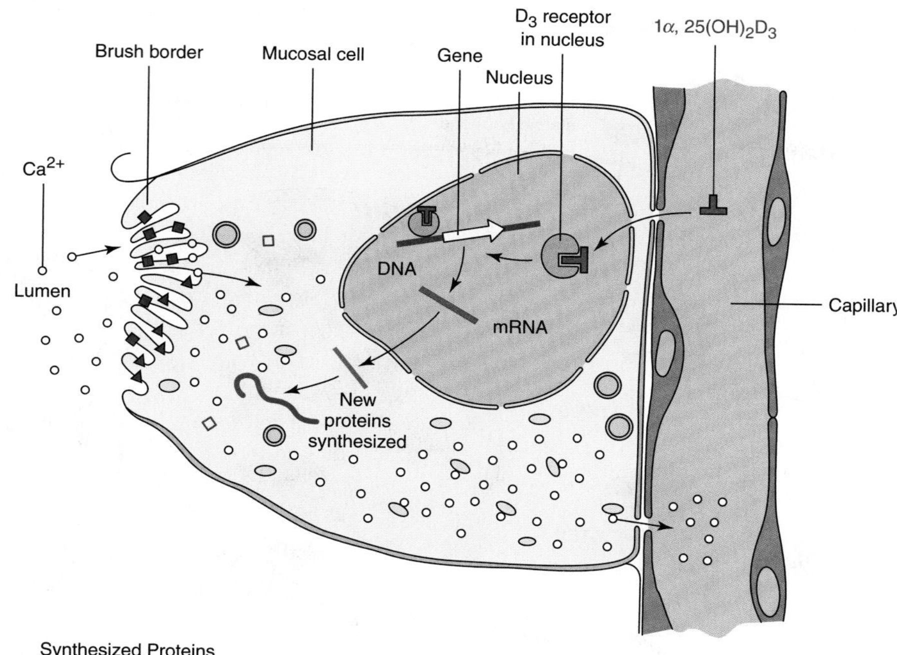

Figure 22.47 Schematic model of action of 1,25(OH)$_2$D$_3$ in intestinal mucosal cell in stimulating calcium absorption from lumen. Redrawn from Nemere, I., and Norman, A. W. *Biochim. Biophys. Acta* 694:307, 1982.

of sperm. Progesterone binds primarily to transcortin and albumin. Because this steroid has a relatively low affinity for these plasma proteins, its circulating half-life is only about 5 min.

22.8 • STEROID HORMONE RECEPTORS

Steroid Hormones Bind Intracellular Receptor Proteins

Receptors for steroids and receptors for nonsteroid hormones (i.e., thyroid hormone, retinoic acid, and Vitamin D_3) are located intracellularly. The unbound glucocorticoid receptor and possibly the aldosterone receptor appear to reside in the cytoplasm, whereas the rest are located within the nucleus, presumably in association with chromatin. In Figure 22.48, step 1 shows a steroid hormone dissociating from a plasma transport protein. The free steroid diffuses into the cell through the lipid bilayer (step 2). Cortisol binds to its receptor with a binding constant of 10^9/M, compared to a binding constant of about 10^7/M for CBG. The unbound receptor, in this case glucocorticoid receptor, is a complex (~300 kDa) that contains other associated proteins, including a dimer of a 90-kDa **heat shock protein** which masks the DNA-binding domain of the receptor (Figure 22.49), and heat shock protein Hsp56, which is an immunophilin that binds several potent immunosuppressive drugs. Binding of the steroidal ligand (step 3) causes a conformational change (activation) in the receptor protein causing release of the associated proteins, including the dimer of Hsp 90, and exposure of positively charged amino acid residues located within the DNA-binding domain (step 4). The ligand–receptor complex translocates to the nucleus (step 5), binds to DNA, frequently as a homodimer, and searches the DNA for specific, high-affinity acceptor sites. Once the ligand–receptor complex has bound to specific hormone response elements (HRE) (step 6) in the DNA, it regulates gene transcription and frequently enhances gene transcription. New mRNAs are translocated to the cytoplasm and direct the synthesis of proteins (step 7) that alter the metabolism and functioning of

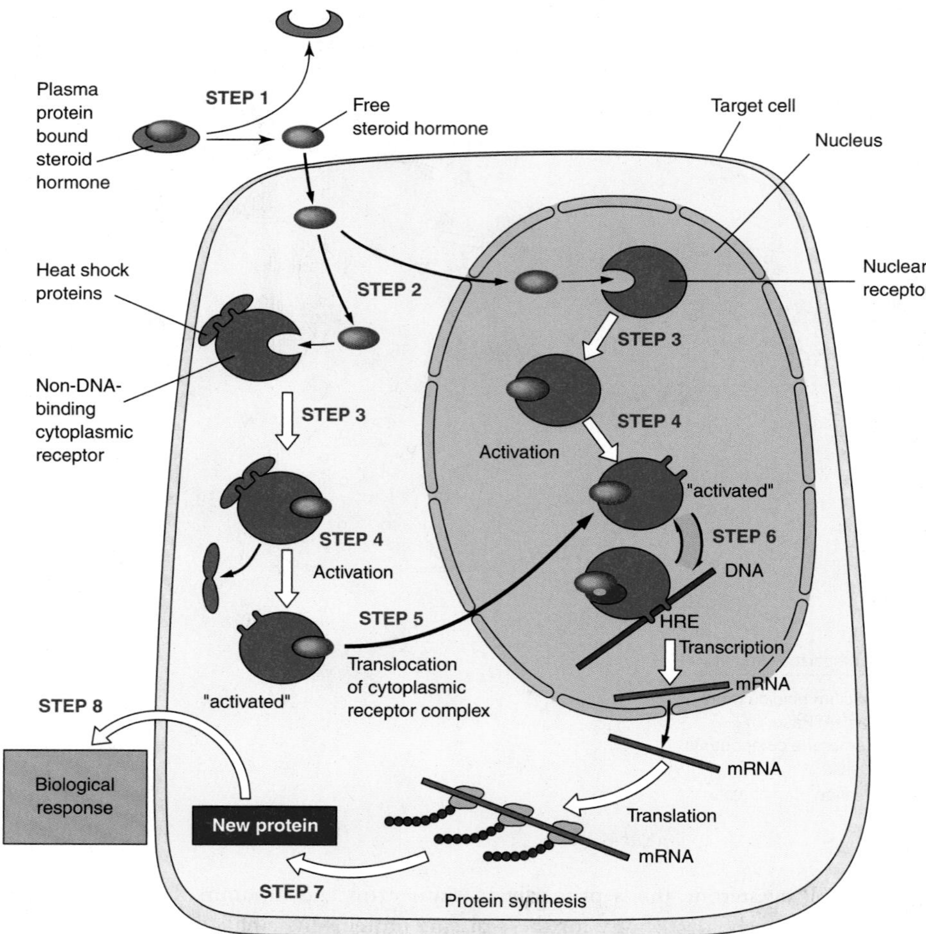

Figure 22.48 Model of steroid hormone action. Step 1: Dissociation of free hormone from circulating transport protein; step 2: diffusion of free ligand into cytosol or nucleus; step 3: binding of ligand to cytoplasmic or nuclear receptor; step 4: activation of cytosolic or nuclear hormone–receptor complex to DNA-binding form; step 5: translocation of activated cytosolic hormone-receptor complex into nucleus; step 6: binding of activated hormone receptor complexes to specific response elements within the DNA; step 7: synthesis of new proteins encoded by hormone-responsive genes; and step 8: alteration in phenotype or metabolic activity of target cell mediated by specifically induced proteins.

the target cell (step 8). In some situations steroid–receptor complexes repress, rather than induce, specific gene transcription.

The unbound steroid hormone receptors for estradiol, progesterone, androgens, and the secosteroid vitamin D_3 (see Figure 22.48) are located in the nucleus. Once inside the nucleus the hormone binds to its specific receptor, induces the dissociation of receptor-associated proteins, and regulates gene transcription. The transcriptional effects of 17β-estradiol in humans and rodents are mediated by the α and β forms of the **estrogen receptor** (**ER**α and **ER**β, respectively). Both bind 17β-estradiol with the same high affinity and bind to estrogen response elements within the DNA. The ERα and ERβ are highly homologous. ERα contains two distinct transcriptional activation domains, AF-1 and AF-2. ERβ contains an AF-2 domain and a repressor domain. ERβ is an inhibitor of ERα transcriptional activity at subsaturating hormone levels and decreases overall cellular sensitivity to 17β-estradiol. Results from knockout mice suggest that these two forms of the ER may have two different biological roles. When ERα and ERβ are coexpressed in neurons, they trigger different intracellular signals and distinct metabolic responses. ERβ may be relevant for neural cell differentiation, whereas ERα may be implicated in synaptic plasticity, both during development and in mature neurons. ERα and Erβ may also play different roles in the uterus. The uterus contains a variety of cell types that undergo continuous synchronized changes in proliferation rates and differentiation in response to estrogens and progesterone. ERα and ERβ levels in uterine tissue vary during the menstrual cycle, the highest level for both occurs during the proliferative phase. 17β-Estradiol regulates the expression of the gene for the **progesterone receptor (PR)** in uterine and breast tissue. ERα induces PR in the uterine stroma and glandular epithelial cells. In contrast, ERβ down-regulates PR in the luminal epithelium. Thus, the differential expression of ERα and ERβ within specific cell types dictates what type of response will be elicited by 17β-estradiol.

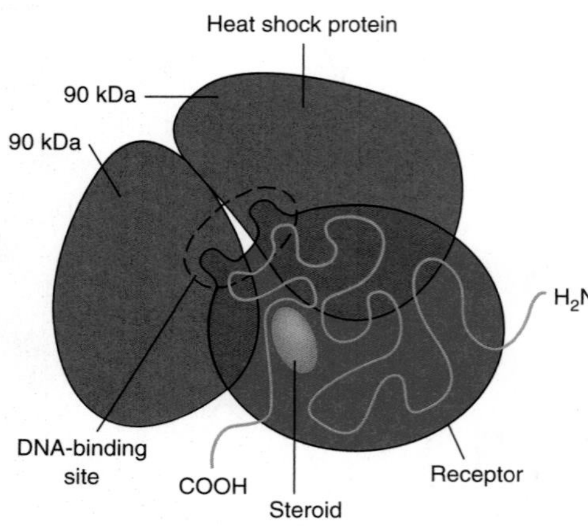

Figure 22.49 Hypothetical model of a non-DNA-binding form of a steroid receptor. This form of the receptor cannot bind to DNA because the DNA-binding site is blocked by the 90 kDa hsp proteins or by some other constituent. Mass of this complex is approximately 300 kDa.

TABLE 22.10 • Steroid Hormone Receptor Responsive DNA Elements: Consensus Acceptor Sites

Element	*DNA Sequence*[a]
POSITIVE	
Glucocorticoid response element (GRE)	5′-GGTACAnnnTGTTCT-3′
Mineralocorticoid response element (MRE)	
Progesterone response element (PRE)	
Androgen response element (ARE)	
Estrogen response element (ERE)	5′-AGGTCAnnnTCACT-3′
NEGATIVE	
Glucocorticoid response element	5′-ATYACNnnnTGATCW-3′

Source: Data are summarized from work of Beato, M. *Cell* 56:355, 1989.
[a]n, any nucleotide; Y, a purine; W, a pyrimidine

Consensus DNA sequences have been defined for specific **hormone response elements (HREs)** (Table 22.10). Receptors for glucocorticoids, mineralocorticoids, progesterone, and androgen bind to the same HRE. Thus, in a given target cell, the type of receptor expressed determines the hormone sensitivity. Sex hormone and progesterone receptors are expressed in only a few cell types, whereas the glucocorticoid receptor is present in a wide variety of cells. Where aldosterone and cortisol receptors are coexpressed, only one receptor form may predominate. However, tissues such as the kidney and colon are targets for aldosterone and express relatively high levels of both mineralocorticoid and glucocorticoid receptors. They also express **11β-hydroxysteroid dehydrogenase** type 2 (Clin. Corr. 22.5) that converts cortisol, which binds to the mineralocorticoid receptor with high affinity, to cortisone, which binds poorly to this receptor. This inactivation of cortisol facilitates binding of aldosterone to mineralocorticoid receptors in these tissues. In certain tissues, mineralocorticoid receptors may mediate the effects of low, circulating levels of cortisol, which are approximately 100-fold higher than the circulating aldosterone levels. Thus, the mineralocorticoid and glucocorticoid receptors may regulate the expression of an overlapping gene network in various target tissues. Note that the estrogen–receptor complex recognizes a unique response element (Table 22.10). Glucocorticoids inhibit transcription of the **proopiomelanocortin (POMC)** gene and thus regulate the rate of secretion of ACTH and hence of cortisol. **Negative glucocorticoid response elements (nGREs)** mediate repression of the POMC gene and of other genes.

Numerous models have been proposed to explain how steroid–receptor complexes function as positive or negative regulators of gene transcription. Binding of a steroid–receptor homodimer to a hormone response element (HRE) may allow it to interact synergistically with a positive transcription factor and hence induce gene transcription. Alternatively, binding of the homodimer to a HRE may sterically block the binding of a positive transcription factor, thus causing repression of gene transcription. Direct protein–protein interactions between a steroid receptor and a positive transcription factor could, at least theoretically, prevent either from binding to DNA and result in a decreased rate of transcription.

Complex physiological responses to steroid hormones may involve induction or inhibition of gene transcription. For example, glucocorticoids exert major anti-inflammatory actions and they have been utilized for decades for this very purpose. They suppress the immune system in several ways. Glucocorticoids inhibit immune responses by inhibiting **prostaglandin** production. Glucocorticoid–receptor complexes induce **annexin-I** (or **lipocortin**), a 40-kD protein, which inhibits membrane phospholipase A_2, and hence the release of arachidonic acid for prostaglandin synthesis. Glucocorticoids also inhibit expression of **cyclooxygenase (COX)** (p. 741) that generates prostaglandins and related compounds. COX1 is constitutively expressed and produces prostaglandins under noninflammatory conditions. COX2 is induced in inflammatory cells and suppression of its synthesis by glucocorticoids accounts for a major part of their anti-inflammatory effects.

CLINICAL CORRELATION 22.5

Apparent Mineralocorticoid Excess Syndrome

Some patients (usually children) exhibit the hypertension, hypokalemia, and suppression of the renin–angiotensin–aldosterone system that would be expected if they were hypersecreting aldosterone. Since assays of plasma and urine may fail to identify excess mineralocorticoids, these patients are said to have the *apparent mineralocorticoid excess (AME)* syndrome. This syndrome is an autosomal recessive disease caused by a deficiency in 11β-hydroxysteroid dehydrogenase type 2 (11β-HSD2). Since the plasma levels of cortisol are about 100-times higher than the plasma levels for aldosterone, cortisol saturates the renal mineralocorticoid receptor, causing sodium retention and suppression of the renin–angiotensin–aldosterone axis. Although this syndrome can result from a congenital defect in renal 11β-HSD2, it can also be acquired by ingesting excessive amounts of licorice. The major components of licorice are glycyrrhizic acid and its hydrolytic product, glycyrrhetinic acid (GE), which is a potent inhibitor of 11β-HSD2. By blocking the activity of this inactivating enzyme, GE facilitates the binding of cortisol to renal mineralocorticoid receptors and hence induces the hypertension and hypokalemia characteristic of the AME syndrome.

Edwards, C. R. W. Primary mineralocorticoid excess syndromes. In DeGroot, L. J. (Ed.), *Endocrinology*. Philadelphia: Saunders, 1995, 1775; and Shackleton, C. H. L., and Stewart, P. M. The hypertension of apparent mineralocorticoid excess syndrome. In Biglieri, E. G., and Melby, J. C. (Eds.), *Endocrine Hypertension*. New York: Raven, 1990, 155; and Mumoli, N., and Cei, M. Licorice-induced hypovolemia. *Int. J. Cardiol.* 124(3):e42, 2008.

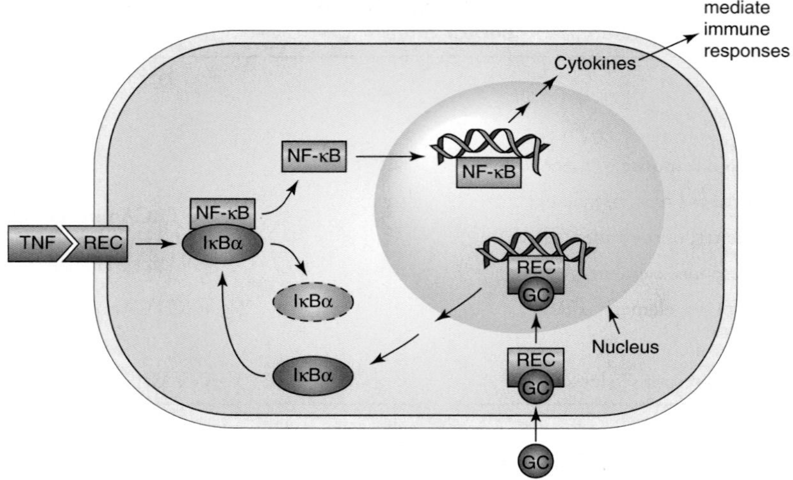

Figure 22.50 Action of glucocorticoids in suppressing immune and inflammatory responses mediated by cytokines. Abbreviations: REC, glucocorticoid receptor; GC, glucocorticoid hormones; TNF, tumor necrosis factor; and NF-κB, nuclear factor kappa B. Redrawn from Marx, J. *Science* 270:232, 1995.

Glucocorticoids also interfere with the transcription factor **nuclear factor kappa B (NF-κB)**, which in unstimulated immune cells is retained in the cytoplasm complexed with either **IκB**α or the structurally related **IκB**β (I for "inhibitor"). Stimulation of immune cells by any one of several immune signals, for example tumor necrosis factor, causes phosphorylation of **IκB**α on serines 32 and 36, which leads to ubiquitination and subsequent degradation via proteasomes. This degradation then releases NF-κB, which has been trapped in the cytoplasm in an inactive form, from this complex and it migrates into the nucleus. Once in the nucleus it induces genes for cytokines, such as interferons and interleukins, which activate immune cells, as well as cell adhesion molecules that draw immune cells into inflammatory sites. Glucocorticoids suppress this immune cell activation (Figure. 22.50) by inducing IκBα gene transcription and ensuring that NF-κB is retained in the cytoplasm in its inactive form under conditions where it should migrate into the nucleus and induce gene transcription.

Some members of the steroid receptor super family mediate gene silencing. Silencer elements in DNA, like enhancer elements, function independently of their position and orientation. The silencer for a gene consists of modules that independently repress gene activity. Unliganded **thyroid hormone receptors (T_3R)** and **retinoic acid receptors (RAR)** bind specific silencer elements and repress gene transcription. After ligand binding, these receptors lose this silencing activity and function as transactivators.

Dimerization is a prerequisite for efficient DNA binding and transcriptional activation by most steroid receptors, and is mediated by their ligand-binding domains. The dimerization region of this domain may form a leucine zipper-like structure or a helix-turn-helix motif (p. 322), which is necessary for dimerization in other transcription factors. Most steroid receptors form homodimers. However, **retinoid X receptors** (RXRs) form heterodimers with the retinoic acid receptor, the thyroid hormone receptor, or other members of this receptor superfamily. A model for the stabilization of the transcriptional preinitiation complex by an RXR/RAR heterodimer is presented in Figure 22.51.

Steroid Receptor Domains

Glucocorticoid receptors contain three major **functional domains** (Figure 22.52). Located at the C-terminus, the steroid-binding domain has 30%–60% homology with the **ligand-binding domains** of other steroid receptors (Clin. Corr. 22.6). This domain may also be involved in the binding of a dimer of the 90-kDa heat shock protein, which (1) maintains this domain in optimal conformation for steroid binding and (2) prevents unliganded receptor from binding to DNA. To the left of this domain are a transcription activation region and a nuclear localization signal, which appear to provide recognition for the nuclear pore. At about the center is the **DNA-binding domain**, which shows 60%–95% homology among steroid receptors and contains two zinc fingers that recognize specific HREs and stabilize binding to these DNA sequences. The N-terminal

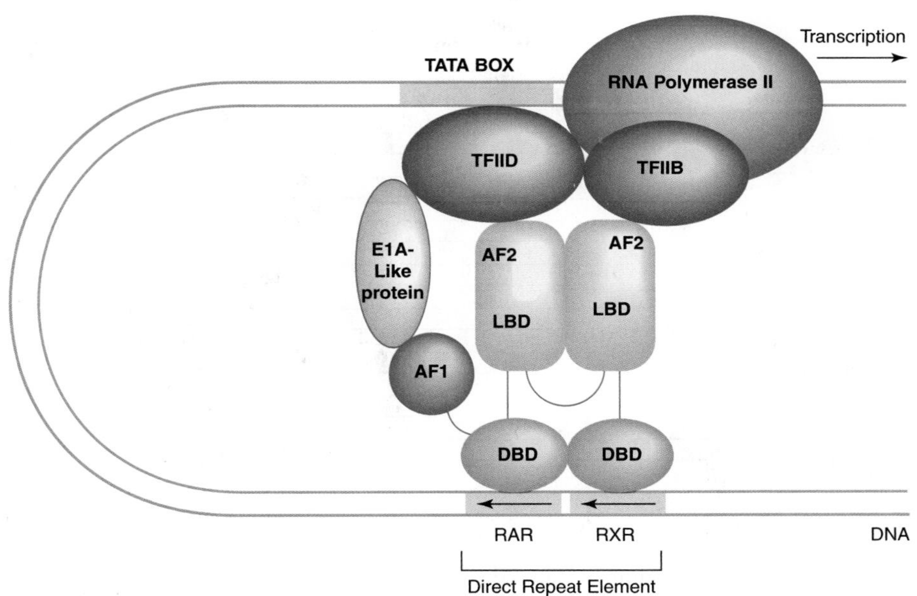

Figure 22.51 Model for stabilization of preinitiation complex by an RXR/RAR heterodimer. Abbreviations: TF, transcription factor; LBD, ligand-binding domain; DBD, DNA-binding domain; AF1, activation function located in amino terminal region of receptor, which may provide contact with cell specific proteins; AF2, activation function located within ligand-binding domain, which interacts directly with transcriptional machinery; and E1A, adenovirus oncoprotein that functions as a tumor suppressor.

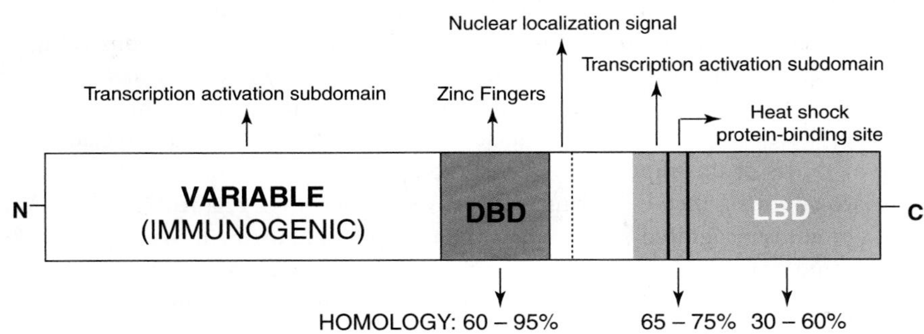

Figure 22.52 Major functional domains of steroid receptor proteins. Abbreviations: DNA, DNA-binding domain; and LBD, ligand-binding domain.

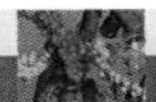

CLINICAL CORRELATION 22.6

Mineralocorticoid Receptor Mutation Results in Hypertension and Toxemia of Pregnancy

The underlying causes of hypertension, especially the hypertension associated with the toxemia of pregnancy, which is called eclampsia, include the renin–angiotensin system and the aldosterone receptor. Hypertension occurs in about 6% of pregnancies and in some cases is associated with a mutation in the mineralocorticoid receptor in which a serine residue at position 810 is replaced by a leucine residue (referred to as S810L mutation). The mutated serine is in the hormone-binding domain of the receptor and is conserved in all mineralocorticoid receptors across many species. Interestingly, the S810L mutant receptor binds progesterone with the same high affinity as aldosterone. In the normal wild-type receptor, progesterone binds with low affinity and functions as an antagonist. However, in the mutant form of the receptor progesterone functions as an agonist and induces reabsorption of sodium ions in the kidney. Since plasma progesterone levels increase significantly during pregnancy (see Figure 22.20, p. 905), the mutated receptor is continuously saturated with this steroid. In subjects under age 35 carrying the mutated receptor, the systolic to diastolic blood pressure ratio is 167/110, as compared with 126/78 (normal range) in noncarriers. Spironolactone, which functions as an aldosterone antagonist when bound to the normal wild-type receptor, functions as an agonist when bound to the mutated receptor. Therefore, spironolactone should not be used to treat hypertensive patients bearing the S810L mutation. This mutated receptor could be an important factor contributing to the early development of heart failure in seriously hypertensive patients.

Geller, D. S., Farhi, A., Pinkerton, N., Fradley, M., et al. Activating mineralocorticoid receptor mutation in hypertension exacerbated by pregnancy. *Science* 289:119, 2000.

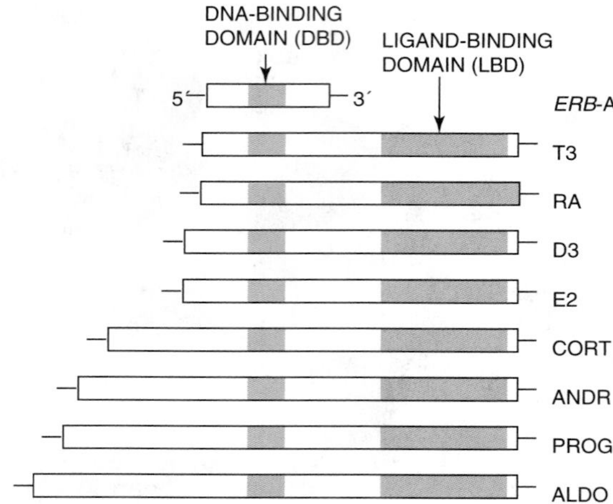

Figure 22.53 Steroid receptor gene superfamily. Abbreviations: T3, triiodothyronine; RA, retinoic acid; D3, dihydroxy vitamin D3; E2, estradiol; CORT, cortisol; ANDR, androgen; PROG, progesterone; ALDO, aldosterone. Figure shows roughly the relative sizes of the genes for these receptors.

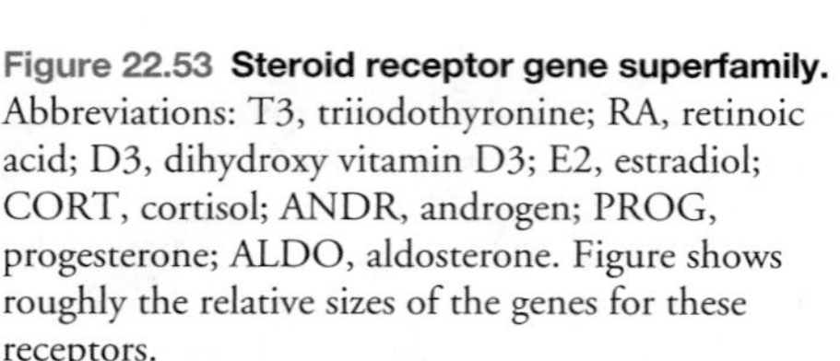

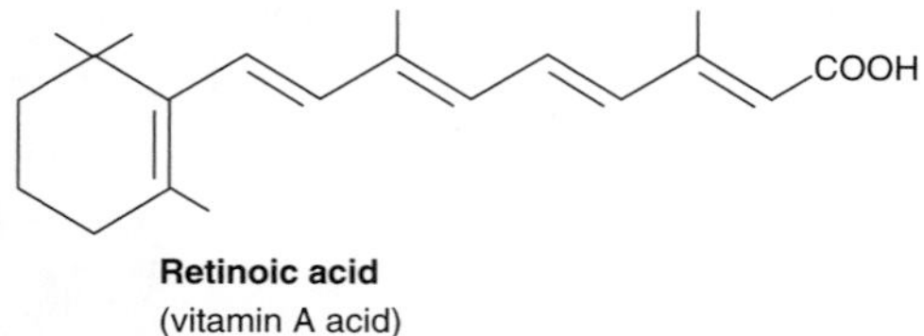

Figure 22.54 Structures of retinoic acid (Vitamin A acid) and 3,5,3′-triiodothyronine.

domain is highly variable and contains the principal **antigenic region** and a region that modulates transcriptional activation. These features are common to all members of the steroid receptor superfamily (Figure 22.53). The ancestor of these receptor genes is the gene for v-erbA or c-erbA, which is an oncogene product that binds to DNA but has no ligand-binding domain. The DNA-binding domains of some of these receptors are homologous enough that more than one receptor binds to a common responsive element (see Table 22.10). The **aryl hydrocarbon receptor** (Ah) may also be a member of this family. It binds carcinogens with an affinity that parallels their carcinogenic potency and translocates these compounds into the nucleus. Thyroid hormone and retinoic acid receptors are members of this superfamily of receptors. Their ligands are not steroids, but do contain two or one six-membered rings, respectively, as shown in Figure 22.54. The A ring of most steroids is recognized by the appropriate receptor and inserts into the pocket of the ligand-binding domain. Again, all of these receptors form homodimers or heterodimers, function as ligand-activated transcription factors, and modulate (induce or repress) the expression of specific genes.

Orphan Receptors

Many related receptors were originally called **orphan receptors**, as no physiological ligands or activators were known for them. They are found in almost all animal species. Examples of orphan receptors for which the ligand has now been identified include BXR (**b**enzoate **X** **r**eceptor), **RXR** (**r**etinoid **X** **r**eceptor), **PPAR** (**p**eroxisome-**p**roliferator-**a**ctivated **r**eceptor), **CAR**β (**c**onstitutive **a**ndrostane **r**eceptor), PXR (**p**regnane **X** **r**eceptor); **SXR** (**s**teroid and **x**enobiotic **r**eceptor), and **FXR** (**f**arnesoid **X** **r**eceptor). PXR, SXR, and CARβ are highly expressed in the liver, respond to specific steroid ligands, and require heterodimerization with RXR for DNA binding. They and their respective ligands may be important physiologically and have an impact on specific human diseases. For example, the human SXR can be activated by a diverse group of steroid agonists and antagonists. The activation induces transcription of several genes encoding key degrading enzymes and may facilitate the detoxification and removal of various endogenous hormones, dietary steroids, drugs, and xenobiotic compounds with biological activity. In patients receiving steroid therapy or women taking oral contraceptives, certain drugs (i.e., rifampicin) that bind to SXR can cause rapid depletion of the administered steroids by this increased steroid catabolism.

Down-regulation of Steroid Receptor by Ligand

Many hormone receptors are down-regulated when the cell has been exposed to a certain concentration of the cognate hormone. In the case of intracellular hormone receptors, down-regulation generally means a ligand-induced decrease in the half-life of the receptor protein and a decreased receptor gene expression with a decrease in the concentration of receptor molecules. The promoter of a receptor gene may thus have a negative response element and binding of the steroid–receptor complex to that element will repress transcription of the receptor gene. Down-regulation of receptors by their own ligands plays an important physiological role, because it desensitizes the target cell and therefore prevents overstimulation when circulating hormone levels are elevated.

Although down-regulation of steroid receptors by their cognate hormones appears to be the most common form of autoregulation, it is not detected in all target cells. Glucocorticoid-mediated up-regulation of its own receptor levels has been reported in a number of responsive cells. In theory this **homologous up-regulation** could increase hormonal responsiveness. The ability of the estrogen receptor to increase the concentration of progesterone receptors in key target tissues is an example of **heterologous up-regulation**.

Nuclear Hormone Receptors, Coactivators, and Corepressors

Coactivators and corepressors are cofactors that increase or decrease the transcriptional functions of most nuclear steroid–receptor complexes. Coactivators, such as the p160 family of coactivators, steroid receptor coactivator 1, transcriptional intermediary factor 2 (TIF2), and GR-interacting protein (GRIP1) all increase the amount of induced gene product with a saturating concentration of a steroid hormone. Conversely, corepressors such as nuclear receptor corepressor (NcoR) and the silencing mediator of retinoid and thyroid hormone receptor (SMRT) decrease the amount of gene product. Binding of the cognate ligand to one of these receptors acts like a “molecular switch”, causing dissociation of corepressors from the receptor and association of coactivators. The sites of interaction in steroid and nuclear receptors for both coactivators and corepressors have been shown to be present in the ligand-binding domain, and these two binding sites may overlap.

Although corepressors do not bind to ligand-bound nuclear receptors such as the thyroid hormone receptor, they do appear to interact with nuclear ligand-bound steroid receptors. This interaction may provide a mechanism of differentiating between activated complexes of various steroid receptors (androgens, glucocorticoid, mineralocorticoid, and progestin) in a cell-specific manner. Although each of theses receptors binds its ligand specifically, once activated they bind to the same hormone response element (see Table 22.10). Interactions between cell-specific corepressors and DNA-bound receptors can restore some of the specificity that appears to be lost on binding to a common HRE.

Nongenomic Steroid Effects

Not all effects of steroid hormones are mediated at the level of gene transcription. Many steroid hormones, including aldosterone, 17β-estradiol, progesterone, glucocorticoids, and androgens, exert rapid (within minutes) stimulatory effects on the activities of a wide variety of signal transduction molecules (protein kinase C, diacylglycerol, and IP_3) and pathways. These **nongenomic effects** appear to be initiated at the plasma membrane rather than in the nucleus of a target cell. These non-transcriptional responses may be mediated either by a subpopulation of conventional nuclear receptors localized in the cell membrane or by distinct **membrane receptors** that are unrelated to classical intracellular steroid receptors. Some arguments in favor of novel membrane receptors include (1) rapid steroid responses appear to be mediated through receptors with pharmacological properties very different from those of classic intracellular receptors. (2) steroids exert rapid nongenomic effects in cells or tissues that do not express classic intracellular receptors, and (3) these rapid steroid effects are not blocked by classic receptor antagonists. The cloning and functional reexpression of unrelated proteins transmitting rapid steroid effects has yet to be accomplished. Drugs that specifically affect nongenomic steroid action may find applications in various clinical areas including cardiovascular and central nervous system disorders, electrolyte homeostasis, and infertility.

Bibliography

Alberts, B., Johnson, A., Lewis, L., Raff, M., et al. (Eds.). *Molecular Biology of the Cell*, 5th ed. New York: Garland Science, 2008.

Beato, M. Gene regulation by steroid hormones. *Cell* 56:335, 1989.

Blumberg, B., and Evans, R. M. Orphan nuclear receptors B new ligands and new possibilities. *Genes. Dev.* 12:3149, 1998.

Carson-Jurica, M. A., Schrader, W. T., and O'Malley, B. W. Steroid receptor family: structure and functions. *Endocrine Rev.* 11:201, 1990.

Chen, J. D. Steroid/nuclear receptor coactivators. In Litwack, G. (Ed.), *Vitamins and Hormones*. New York: Academic Press, 2000, 391.

Cuatrecasas, P. Hormone receptors, membrane phospholipids, and protein kinases. *Harvey Lectures Series* 80:89, 1986.

DeGroot, L. J. (Ed.). *Endocrinology*. Philadelphia: Saunders, 1995.

Dostert, A., and Heinzel, T. Negative glucocorticoid response elements and their role in glucocorticoid action. *Curr. Pharm. Des.* 10:2807, 2004.

Evans, R. M. The steroid and thyroid hormone receptor superfamily. *Science* 240:889, 1988.

Falkenstein, E., Tillmann, H.-C., Christ, M., Feuring, M., and Wehling, M. Multiple actions of steroid hormones: A focus on rapid, nongenomic effects. *Pharmacol. Rev.* 52:513, 2000.

Fritsche, L., Weigert, C., Häring, H. U., and Lehmann, R. How insulin receptor substrate proteins regulate the metabolic capacity of the liver implications for health and disease. *Curr. Med. Chem.* 15:1316, 2008.

Green, S., Kumar, V., Theulaz, I., Wahli, W., and Chambon, P. The N-terminal DNA-binding "zinc-finger" of the estrogen and glucocorticoid receptors determines target gene specificity. *EMBO J.* 7:3037, 1988.

Gustafsson, J. A., Carlstedt-Duke, J., Poellinger, L., Okret, S., et al. Biochemistry, molecular biology, and physiology of the glucocorticoid receptor. *Endocrine Rev.* 8:185, 1987.

Kakizaki, S., Yamazaki, Y., Takizawa, D., and Negishi, M. New insights on the xenobiotic-sensing nuclear receptors in liver diseases—CAR and PXR. *Curr. Drug Metab.* 9:614, 2008.

Krieger, D. T., and Hughes, J. C. (Eds.). *Neuroendocrinology*. Sunderland, MA: Sinauer Associates, 1980.

Larsen, P. R., Kronenberg, H. M., Melmed, S., and Polonsky, K. S. (Eds.). *Williams Textbook of Endocrinology*, 10th ed. Philadelphia: Saunders, 2003.

Levy, M., Koeppen, B., and Stanton, B. (Eds.). *Berne and Levy Principles of Physiology*, 4th ed. Amsterdam: Elsevier, 2008.

Litwack, G. (Ed.). *Vitamins and Hormones*, Vol. 50. Orlando, FLA: Academic Press, 1995.

Litwack, G. (Ed.). *Biochemical Actions of Hormones*, Vols. 1–14. New York: Academic Press, 1973–1987.

Mester, J., and Baulieu, E. E. Nuclear receptor superfamily. In DeGroot, L. J. et al. (Eds.). *Endocrinology*. 3d ed. Philadelphia: Saunders, 1995, 93.

Norman, A. W., and Litwack, G. *Hormones*. Orlando, FL: Academic Press, 1987.

O'Malley, B. W., Tsai, S. Y., Bagchi, M., Weigel, N. L., et al. Molecular mechanism of action of a steroid hormone receptor. *Recent Prog. Horm. Res.* 47:1, 1991.

Richter, D. Molecular events in expression of vasopressin and oxytocin and their cognate receptors. *Am. J. Physiol.* 255:F207, 1988.

Saltiel, A. R. The paradoxical regulation of protein phosphorylation in insulin action. *FASEB J.* 8:1034, 1994.

Schmidt, T. J., and Meyer, A. S. Autoregulation of corticosteroid receptors: How, when, where and why? *Receptor* 4:229, 1994.

Spiegel, A. M., Shenker, A., and Weinstein, L. S. Receptor-effector coupling by G proteins: Implication for normal and abnormal signal-transduction. *Endocr. Rev.* 13:536, 1992.

Stahn, C., and Buttgereit, F. Genomic and nongenomic effects of glucocorticoids. *Natl. Clin. Pract. Rheumatol.* 4:525, 2008.

Struthers, A. D. (Ed.). *Atrial Natriuretic Factor*. Boston: Blackwell Scientific, 1990.

Wahli, W., and Martinez, E. Superfamily of steroid nuclear receptors-positive and negative regulators of gene expression. *FASEB J.* 5:2243, 1991.

Key Terms

second messengers
ligand-activated transcription factors
proopiomelanocortin
oxytocin
phenylethanolamine *N*-methyltransferase
thyroxine
protein kinase A
protein kinase G
protein kinase C
phosphodiesterase
cAMP response element
N-acetyltransferase
N-acetylserotonin
hydroxyindole-*O*-methyltransferase
inhibin
estradiol
progesterone
ligand-binding domains
tyrosine-specific kinases
insulin receptor
arginine vasopressin
cyclopentanoperhydrophenanthrene
cortisol
aldosterone
testosterone
dehydroepiandrosterone
5α-reductase
aromatase
dihydrotestosterone
angiotensin converting enzyme
atrial natriuretic factor
1α, 25-vitamin D_3
calbindins
heat shock protein
11β-hydroxysteroid dehydrogenase
cyclooxygenase
nuclear factor kappa B
DNA-binding domain
nongenomic effects

Questions • CAROL N. ANGSTADT

Multiple-Choice Questions

1. If a single gene contains information for synthesis of more than one hormone molecule,
   A. all the hormones are produced by any tissue that expresses the gene.
   B. all hormone molecules are identical.
   C. cleavage sites in the gene product are typically pairs of basic amino acids.
   D. all peptides of the gene product have well-defined biological activity.
   E. hormones all have similar function.

2. In the interaction of a hormone with its receptor, all of the following are true except
   A. more than one polypeptide chain of the hormone may be necessary.
   B. more than one second messenger may be generated.
   C. an array of transmembrane helices may form the binding site for the hormone.
   D. receptors have a greater affinity for hormones than for synthetic agonists or antagonists.
   E. hormones released from their receptor after endocytosis could theoretically interact with a nuclear receptor.

3. Some hormone–receptor complexes are internalized by endocytosis. This process may involve
   A. binding of hormone–receptor complex to a clathrin coated pit.
   B. recycling of receptor to cell surface.
   C. degradation of receptor and hormone in lysosomes.
   D. formation of a receptosome.
   E. all of the above.

4. Epinephrine
   A. mediates its effects by binding to cytoplasmic receptors.
   B. is synthesized in the adrenal cortex.
   C. is synthesized from norepinephrine by a methyl transferase (PNMT).
   D. leads to the formation of cGMP.
   E. produces DAG and IP_3.

5. Binding of insulin to its receptor
   A. occurs on the β-subunit.
   B. induces autophosphorylation.
   C. reduces binding of cytosolic substrate proteins.
   D. leads only to phosphorylation of proteins.
   E. does not lead to release of a second messenger.

6. Glucocorticoid receptors are in the cytoplasm. All of the following statements about the process by which the hormone influences transcription are correct except
   A. the hormone must be in the free state to cross the cell membrane.
   B. cytoplasmic receptors may be associated with heat shock proteins.
   C. the receptor–hormone complex is not activated/transformed until it is translocated to the nucleus.
   D. in the nucleus, the activated/transformed receptor–hormone complex searches for specific sequences on DNA called HREs (hormone receptor elements).
   E. the activated receptor–hormone complex may either activate or repress transcription of specific genes (only one activity per gene).

Questions 7 and 8: Aldosterone bound to its receptor promotes sodium reabsorption in the distal nephron of the kidney. Elevated sodium in the blood leads to hypertension which can be a serious problem in pregnancy. In some cases, hypertension that appears in early pregnancy and increases with time has been shown to be caused by a mutation in the mineralocorticoid receptor. The mutation allows progesterone to bind with the same affinity as aldosterone and thus act as an agonist. Because of the high levels of progesterone during pregnancy, the mutated receptor remains saturated and blood pressure can become dangerously high. Spironolactone, which acts as an antagonist of aldosterone with a normal receptor, acts as an agonist with the mutated receptor and should not be used to treat this kind of hypertension.

7. All of the following are normal events leading to secretion of aldosterone from the adrenal gland except
   A. renin is released by the kidney in hypovolemia.
   B. angiotensinogen binds to membrane receptors.
   C. the PI cycle is activated producing IP_3 and DAG.
   D. Ca^{2+} levels in the cell rise.
   E. aldosterone is secreted into the blood.

8. Once ovulation occurs, the pathway followed differs when the egg is fertilized and when it is not. Which of the following statements about this process is/are correct?
   A. FSH, via cAMP as a second messenger, stimulates the follicle to release 17β-estradiol.
   B. Blood levels of progesterone fall as pregnancy progresses as the corpus luteum dies.
   C. Inhibin produced by the follicle prevents release of LH.
   D. The primary influence for the corpus luteum to produce progesterone and estradiol is FSH.
   E. All of the above.

Questions 9 and 10: Hypopituitarism may result from trauma, such as an automobile accident severing the stalk connecting the hypothalamus and anterior pituitary, or from tumors of the pituitary gland. In trauma, usually all of the releasing hormones from hypothalamus fail to reach the anterior pituitary. With a tumor of the gland, some or all of the pituitary hormones may be shut off. Posterior pituitary hormones may also be lost. Hypopituitarism can be life threatening. Usual therapy is administration of end organ hormones in oral form.

9. If the stalk between the hypothalamus and anterior pituitary is severed, the pituitary would fail to cause the ultimate release of all of the following hormones except
   A. ACTH.
   B. estradiol.
   C. oxytocin.
   D. testosterone.
   E. thyroxine.

10. In hypopituitarism the ovarian cycle would be affected because
    A. FSH and LH are synthesized in the anterior pituitary.
    B. gonadotropin-releasing hormone (GnRH) from the hypothalamus stimulates the release of FSH and LH.
    C. inhibins would be secreted in much larger than normal amounts.
    D. the corpus luteum would be maintained.
    E. GnRH would bind to receptors to activate protein kinase A.

Questions 11 and 12: Manic-depression may be caused by overactivity of certain central nervous system cells, perhaps caused by an overstimulation of the phosphatidylinositol (PI) cycle. Lithium has been used in the treatment of this condition as it slows the PI cycle despite continued stimulation and cells may become less sensitive to these stimuli. Li^+ may have two functions: inhibition of the phosphatase that dephosphorylates IP_3 and direct interference with the function of G proteins.

11. The PI cycle begins with activation of phospholipase C; this initiates a sequence of events including all of the following except
A. activation of IP_3 by action of a phosphatase.
B. increase in intracellular Ca^{+2} concentration.
C. release of diacylglycerol (DAG).
D. activation of protein kinase C.
E. phosphorylation of certain cytoplasmic proteins.

12. Which of the following statements concerning G proteins is correct?
A. G proteins bind the appropriate hormone at the cell surface.
B. G proteins interact with cytoplasmic receptors.
C. G proteins are second messengers.
D. G protein causes the regulatory subunits of the protein kinase to dissociate from the catalytic subunits.
E. activated G protein may either activate or inhibit the production of second messenger.

Problems

13. How is the iodine present in thyroid hormones conserved by the body?

14. What is the relationship between 7-dehydrocholesterol and 1α,25-dihydroxycholecalciferol?

Answers

1. C One or more trypsin-like proteases catalyze the reaction. A: The POMC gene product is cleaved differently in different parts of the anterior pituitary. B: Multiple copies of a single hormone may occur, but not necessarily. D: Some fragments have no known function. E: ACTH and β-endorphin, for example, hardly have similar functions.
2. D β-receptors bind isoproterenol more tightly than their hormones. A and B: These are true of glycoprotein hormones. C: This appears to be true for β_1-receptor. E: This is possible, but entirely speculative; there are currently no known examples.
3. E A and D always happen. B or C happens sometimes.
4. C This is induced by cortisol under stress. A: Its receptor is on the plasma membrane. B: This is where steroid hormones are made. Epinephrine is made in the adrenal medulla. D and E: Epinephrine stimulates protein kinase A and leads to cAMP.
5. B This occurs on tyrosine residues of the β-subunit. A: Binding is to the α-subunit. C: Autophosphorylation facilitates binding. D: Some proteins are dephosphorylated. E: A second messenger may account for short-term metabolic effects.
6. C Dissociation of the heat shock protein from the receptor–hormone complex in the cytosol activates the complex. A: Steroid hormones travel bound to plasma proteins, but some is always free. D: These are consensus sequences in DNA. E: Activation is more common, but glucocorticoids repress transcription of the proopiomelanocortin gene.
7. B Angiotensinogen is cleaved by renin to angiotensin I, which must further be cleaved by converting enzyme to active angiotensin II. A: This is a major signal. C and D: These lead to increased Ca^{2+} and activation of protein kinase C.
8. A Activation of cAMP-dependent protein kinase stimulates the synthesis and secretion of estradiol. B: This is what happens in the absence of fertilization. In pregnancy, the corpus luteum eventually dies but the placenta produces high levels of progesterone. C: Inhibin controls FSH release. D: LH controls the corpus luteum.
9. C Oxytocin is released from the posterior pituitary. A, B, D, and E all require releasing hormones from the hypothalamus for the anterior pituitary to release them.
10. B FSH and LH are released from the anterior pituitary. A: See previous statement. C: Inhibins are formed and secreted by the ovarian follicle. D: FSH and LH are needed to stimulate the ultimate formation of the corpus luteum. E: GnRH receptors are in the anterior pituitary and stimulate the PI second messenger system.
11. A IP_3 is the active form. Action of phosphatase renders IP_3 inactive. B: IP_3 causes release of Ca^{+2}. C and D: This is a second messenger and it activates protein kinase C. E: This is what protein kinase C does.
12. E There are both stimulatory and inhibitory G protein subunits; which is activated depends on the hormone–receptor complex. A: G protein interacts with receptors, not with the hormone. B: This is a plasma membrane phenomenon. C: They lead to the production of the second messenger. D: That is the function of the second messenger.
13. Various levels of iodinated tyrosyl residues are present in thyroglobulin, which is taken into cells by endocytosis. Hydrolysis by lysosomal proteases releases the hormones T_3 and T_4, and also mono- and diiodothyronines. The last two are deiodinated and the iodide ion trapped and reused for thyroid hormone synthesis.
14. 1α,25-dihydroxycholecalciferol ($1,25(OH)_2D_3$) is the active from of vitamin D. Ultraviolet light acting on the skin converts 7-dehydrocholesterol to cholecalciferol. This compound must be hydroxylated in liver to 25-hydroxycholecalciferol and subsequently in kidney to yield the active $1,25(OH)_2D_3$.

PART V

PHYSIOLOGICAL PROCESSES

23

Molecular Cell Biology

Thomas E. Smith

Professor, Howard University College of Medicine

Key Concepts

NERVOUS TISSUE: METABOLISM AND FUNCTION

- Nerve tissue primarily uses glucose as an energy source. The membrane potential of nerve cells is maintained at –70 mV through the action of the Na^+, K^+ ion pump.
- Nerve impulses are propagated through a depolarization process involving voltage-gated ion channels leading to localized depolarization. Transmission from neuron to neuron proceeds through synapses that are electrically or chemically coupled.
- Chemically coupled synapses operate by secreting into the spaces between neurons neurotransmitters that bind to receptors on the postsynaptic side. This initiates an electrical response.
- Actions of neurotransmitters can be either excitatory or inhibitory. Effects are rapid and transient and are terminated by reuptake, metabolism, and/or diffusion of the neurotransmitter.

THE EYE: METABOLISM AND VISION

- The eye is an extension of the nervous system and derives most of its energy from glucose metabolism. Eye structures through which light must pass before reaching the retina have very few mitochondria or other pigmented subcellular particles.
- The corneal epithelium is permeable to atmospheric oxygen and is protected from active oxygen species by operation of the NADPH glutathione reductase system.

- The lens is an active metabolizing tissue with no blood supply and no subcellular structures. Osmotic balance is maintained by Na^+/K^+ ATPase and redox balance by a glutathione reductase system.
- The retina is a vascular tissue that contains the photo (visual) receptors (rods and cones). It has no blood vessels in the area of greatest visual acuity, but the rods and cones contain a blood supply and mitochondria.
- Vision is a four-event process: involving photons of light, a light-induced isomerization, a signal transduction, and a mental interpretation of the objects in question—"in the mind's eye."
- The proteins of the visual pigments contain 11-*cis* retinal, a derivative of β-carotene (vitamin A_1). Isomerization of 11-*cis*-retinal to all-*trans*-retinal forces a conformation change of the protein of the visual pigments, leading to activation of transducin, a G-protein.
- The α-transducin–GTP subunit complex activates a phosphodiesterase that hydrolyzes cGMP resulting in closure of Na^+ channels, hyperpolarization of the membrane, and generation of an electric impulse. Hydrolysis of GTP in the α-transducin-GTP complex allows return of the membrane to the resting state.
- The same processes occur in the cones for color vision. There are three different pigments sensitive to red, green, or blue light found in different cones. Color vision results from graded stimulation of cones containing specific pigments and interpretation of these stimuli by the brain—"in the mind's eye."

MOLECULAR MOTORS AND ASSOCIATED PROTEINS

- Muscles contract when myosin and actin are pulled past each other shortening the sarcomere, the contractile unit. Myosin's actions are driven by the different conformations it takes when bound to ATP, ADP, and during the myosin catalyzed conversion of ATP to ADP.
- Signal transduction for contraction goes from electrical to chemical to mechanical. The initial stimulus originates at the neuromuscular junction with release of acetylcholine and its binding to receptors. This results in release into the sarcomere of Ca^{+2} that binds troponin-C. Troponin-C–Ca^{2+} complexes cause conformation changes in tropomyosin that initiates the contractile process.
- In skeletal muscle the *power stroke* for contraction occurs upon release of phosphate following ATP hydrolysis. The concentration of ATP in muscle is maintained by metabolism, and by actions of creatine phosphokinase and adenylate kinase.
- Smooth muscle contraction occurs more slowly. Ca^{2+} influx and the phosphorylation of myosin by a specific Ca^{2+}-calmodulin-activated kinase initiates contraction. Control is influenced by hormones that affect Ca^{2+} influx.
- Several classes of *unconventional myosins* are involved in cellular functions such as membrane–membrane interaction, melanosome transport, and movement of hairs in the inner ear.
- Kinesins are microtubule-based molecular motors that generate motion by conformational changes associated with ATP binding and hydrolysis. A major function of kinesins is to effect movement of intracellular cargo.
- The two classes of dynein molecular motors are axonemal and cytoplasmic. Dynein movement along tubules is in the opposite direction of that of kinesins and they are cargo-moving motors. Their movement is load dependent; the heavier the load of the cargo, the shorter the individual steps.

MECHANISM OF BLOOD COAGULATION

- The process of blood coagulation begins at the site of injury to the blood vessel with formation of multienzyme complexes. There are two general pathways: extrinsic and intrinsic. They merge at the point of activation of factor X, an enzyme that converts prothrombin to thrombin.
- The extrinsic pathway plays a major role in initiating the process and the intrinsic pathway plays a major role in amplifying the process.
- Thrombin catalyzes the conversion of fibrinogen to fibrin, the major clot-forming protein. Formation of some of the active multienzyme complexes requires protein cofactors and Ca^{2+}, which facilitates formation of the complexes by binding to γ-carboxyglutamate residues in the proteins.
- Catalytic action of the proteases is inhibited by interaction with specific protein inhibitors in the blood.
- The clot is dissolved through action of plasmin formed by activation of plasminogen by tissue plasminogen activator, t-PA.

23.1 • NERVOUS TISSUE: METABOLISM AND FUNCTION

Essential Concepts

Approximately 2.4% of an adult's body weight is nerve tissue, 83% of which is the brain. The nervous system provides the communications network between the senses, the environment, and all parts of the body. The brain is the command center. It must function all the time, and as a consequence of this demand, it requires a large amount of energy. Under normal conditions, the brain derives its energy from glucose metabolism. Ketone bodies can cross the **blood–brain barrier** and are metabolized by brain tissue, especially during **starvation**, but they cannot replace the requirement for glucose. The human brain uses approximately 103 to 120 grams of glucose/day. For a 1.4-kg brain, this corresponds to an average rate of utilization of approximately 0.3 micromole/min/g of tissue and represents a capacity for ATP production through the tricarboxylic acid (TCA) cycle alone of approximately

6.2 micromoles/min/g of tissue. The TCA cycle is multifunctional, and all of the carbon that flows through it is not used for ATP production. Also, all glucose metabolized by nerve tissue is not metabolized by the TCA cycle. The TCA cycle, however, does function at near maximum capacity and generates most of the ATP used by nerve tissue. Much of the energy used by nerve tissue is to maintain ionic gradients across plasma membranes, to affect various storage and transport processes, and to synthesize neurotransmitters and other cellular components. Glycolysis functions at approximately 20% capacity, and in some parts of the nervous system is the predominant source of ATP production.

The central nervous system contains a higher level of lipids than any other tissue in the body except adipose tissue. Many of these lipids are specialized and complex lipids, and they play important roles in many aspects of brain metabolism, but generally not as a major source for energy production.

There is a relatively rapid **turnover rate** of brain proteins compared to that of other tissues such as muscle and liver although neuronal cells generally do not divide after they have differentiated. Nerve tissue can be injured in many ways. and there is experimental evidence suggesting that regeneration of neuronal tissues may occur faster than originally thought.

Cells of the nervous system responsible for collecting and transmitting messages are highly specialized **neurons** (Figure 23.1). Each neuron consists of a cell body, **dendrites** that are short antenna-like protrusions that receive signals from other cells, and an **axon** that extends from the cell body (presynaptic neuron) to other neurons (postsynaptic neurons), cells to which it transmits signals. The central nervous system (CNS) is highly integrated where individual neurons can receive inhibitory and excitatory stimuli from a variety of different sources. A normal adult has between 10^{11} and 10^{13} neurons, and communication between them is by electrical and chemical signals. Stem cells found in brain can produce three major types of cells: neurons, **astrocytes**, and oligodendrocytes. Astrocytes are glial cells that are associated with greater than 90% of the blood–brain-barrier endothelium and help to determine blood–brain-barrier morphology and function. Astrocytes send out processes at the external surfaces of the CNS that are linked, with other cell types, to form anatomical complexes that provide sealed barriers at the level of the capillaries and isolate the CNS from the external environment. These tight junctions prevent the passive entry into the brain of water-soluble molecules and form what is commonly known as the **blood–brain barrier** (Figure 23.2). In general, therefore, water-soluble compounds enter the brain only if there are specific membrane transporters for them (Clin. Corr.23.1). Oligodendrocytes surround the axons like insulators and facilitate efficient signal conduction. There are other types of glial cells, each of which has a specialized function, but only astrocytes appear to be directly associated with a biochemical function related to metabolic neuronal activity (p. 950). Some glial cells, astrocytes specifically, have been shown to use intermediary metabolites other than glucose as a source of energy. Astrocytes in culture can oxidize fatty acids; apparently by β-oxidation, but also yielding ketone bodies, specifically from the last four carbons (ω-end) of the chain, as major products. Ketone bodies can be used by other brain cells as a source of energy.

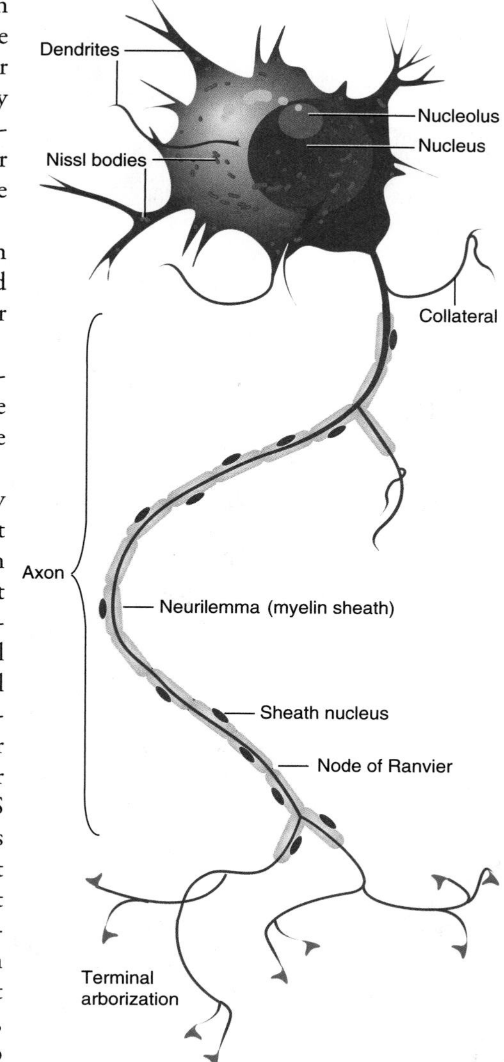

Figure 23.1 A motor nerve cell and investing membranes.

CLINICAL CORRELATION 23.1

The Blood–Brain Barrier and Defects in Glucose Transport

De Vivo disease is an inherited autosomal dominant disease caused by a deficiency of glucose in brain due to defective GLUT1, a glucose transporter responsible for getting glucose across the blood–brain barrier (p. 489) (OMIM 138140). The symptoms of De Vivo disease, also referred to as GLUT1 syndrome, are expressed early in life. Deficiency of brain glucose may result in infantile seizures, delayed development, microcephaly, and other mental disorders. Infantile seizures may appear at 1–4 months of age. The frequency of seizures may decrease during an illness that causes the infant to become somewhat ketotic. A ketogenic diet is used to help control the seizures. There appears to be a critical time during developmental stages when glucose and/or some of its metabolites are essential for activities other than energy production.

De Vivo, D. C., Trifiletti, R. R., Jacobson, R. I., Ronen, G. M., et al. Defective glucose transport across the blood-brain barrier as a cause of persistent hypoglycorrhachia, seizures, and developmental delay. *N. Engl. J. Med.* 325:703, 1991; Wang, D., Pascual, J. M., Yang, H., Engelstad, K., et al. Glut-1 deficiency syndrome: clinical, genetic, and therapeutic aspects. *Annals Neurology* 57:111, 2005; Pascual, J. M., Wang, D., Hinton, V., Engelstad, K., et al. Brain glucose supply and the syndrome of infantile neuroglycopenia. *Arch. Neurol.* 64:507, 2007.

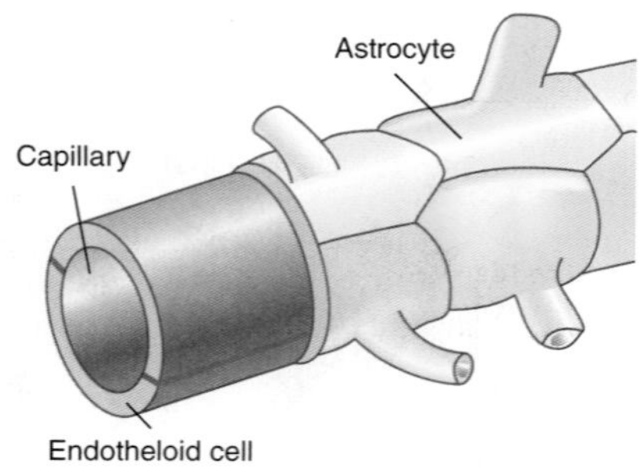

(*a*)

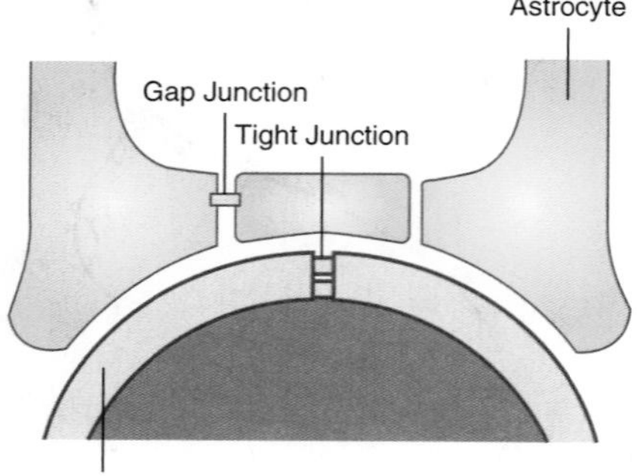

(*b*)

Figure 23.2 Diagrammatic representation of the structure of the blood–brain barrier. (*a*) Spatial view showing astrocyte end feet surrounding a cerebral capillary. (*b*) A cross-sectional view demonstrating tight junctions between astrocyte feet and that astrocyte feet cover tight junctions of the endothelial cells of capillaries.
Redrawn from Deetjen, P., and Speckmann, E.-J. *Physiology*. Munich. Urban & Schwarzenburg, 1994.

TABLE 23.1 • Some Neurotransmitters Found in Nervous Tissue

EXCITATORY
Acetylcholine
Aspartate
Dopamine
Histamine
Norepinephrine
Epinephrine
Glutamate
5-Hydroxytryptamine
INHIBITORY
4-Aminobutyrate
Glycine
Taurine

ATP and Transmembrane Electrical Potential in Neurons

Neurons have a resting transmembrane potential of approximately −70 mV. This transmembrane potential is due to the following factors: The membrane is essentially completely permeable to K^+, only very slightly permeable to Na^+, and impermeable to intracellular anions; the membrane also contains a **Na^+/K^+-exchanging ATPase** that pumps Na^+ out of the cell by an antiport mechanism where K^+ is moved into the cell. The action of the pump (Na^+/K^+ ATPase) and differential permeability of the membrane to Na^+ and K^+ establish ionic gradients across the membrane where Na^+ is higher on the outside and K^+ is higher on the inside of the cell. There is the tendency for K^+ to flow out of the cell down its chemical concentration gradient; this would leave the inside negative due to the intracellular concentrations of anions that cannot cross the membrane. As K^+ tends to flow out of the cell down its concentration gradient, negative charges of the anions tend to pull K^+ back into the cell. These two opposing forces (plus the very slight permeability to Na^+) achieve an equilibrium at a transmembrane potential of −70 mV; that is, the chemical concentration gradient that moves K^+ out of the cell is balanced by the negative charge inside the cell that operates to pull K^+ back into the cell.

Generation of an action potential in a neuron requires the initiation of an event that will effect depolarization of the membrane such that the transmembrane potential reaches +20 to +30 mV. This is accomplished by opening Na^+ channels and permitting enough Na^+ to flow into the cell to reverse the membrane's polarity by 90 to 100 mV. Thus, when Na^+ channels open under the influence of electrical stimulation generating an action potential, sodium ions move down their chemical concentration gradient into neurons through voltage-gated channels. This is followed by movement of potassium ions out of neurons. The Na^+/K^+-exchanging ATPase eventually reestablishes the ionic gradients and restores the resting state.

An electrical impulse is generated and propagated down the axon by a process of depolarization of voltage-gated channels during which time Na^+ enters the cell. Channel proteins undergo a charge-dependent conformational change that triggers their opening when the electrical potential across the membrane reaches a threshold voltage of +20 to +30 mV. When the membrane becomes depolarized, Na^+, whose concentration is higher outside of the cell than inside, flows into the cell and K^+, whose concentration is higher inside the cell, flows out of the cell; both move down their respective concentration gradients. The channels are open in a defined region of the cell membrane for fractions of milliseconds (Figure 23.3). The localized depolarization (voltage changes due to Na^+ influx) causes a conformational change in the neighboring proteins of **voltage-gated ion channels** (p. 481). These adjacent channels open momentarily in response to the localized depolarization and allow the process to continue down the axon. Since there is a finite recovery time greater than the resident time of Na^+ within that region during which time the signal for reopening diminishes, charge propagation proceeds in one direction. This recovery time for channels to reopen is related to the time required to reestablish a membrane potential below the excitation threshold value (by action of the Na^+/K^+ exchanging ATPase) and protein conformational changes required to close and reset the channel gate. Thus, progressive depolarization and repolarization along the length of the axon allows electrical impulses to be propagated undiminished in amplitude and direction. A cartoon of how a sensor for a voltage-gated potassium channel may appear and operate in a membrane is shown in Figure 23.3*c*. When the inside of the cell becomes more positive due to a depolarization event, the channel protein undergoes a conformation change, opens the gate, and K^+ is permitted to flow out of the cell down its concentration gradient. This helps to minimize the extent to which that localized region of the membrane stays above the excitation threshold voltage.

Voltage-gated ion channels of primary concern are Na^+, K^+, and Ca^{2+}. They show varying degrees of overall structural diversity, but a high degree of conservation within the pore-forming regions. Each has a selectivity filter that determines which ion will be permitted to flow through its pore (Figure 23.4, p. 944). The selectivity filter uses chemical features of the protein's structural components to mimic the hydration sphere of the ions leading to dehydration; rehydration occurs as the ions exit the filter.

Neuron–Neuron Interaction Occurs through Synapses

Neuron–neuron interaction occurs through **electrical synapses** or through **chemical synapses**. Electrical synapses permit the more rapid transfer of signals from cell to cell. Chemical synapses allow for a greater degree of chemical versatility in neuron–neuron communication. Pathological events that affect proper functioning of chemical synapses are also more accessible to therapeutic drug interventions.

Chemical synapses are of two types: those where the neurotransmitter binds directly to an ion channel (neuron–neuron) and causes it to open or to close and those where the neurotransmitter binds to a receptor that leads to generation of a second messenger (smooth muscles, for example) that reacts with ion channels causing them to open or to close. Note that propagation of an impulse down an axon occurs by the operation of voltage-gated channels and that the chemical transmission of an impulse across the synapse is by ligand-gated channels.

Chemical neurotransmitters are found in presynaptic axon termini and so are enzymes for their syntheses. Electrical or physiological stimulation of the presynaptic axon results in their release. Mechanisms exist within the synaptic junction for rapid termination of their action. Direct application of appropriate neurotransmitters to the postsynaptic terminal mimics in that neuron the action of nervous stimulation. Drugs that modify the metabolism of neurotransmitters have physiological effects consistent with *in vivo* modification of actions of the neurotransmitters.

Chemical neurotransmitters are of two general types: excitatory and inhibitory. **Excitatory neurotransmitters** include acetylcholine and the catecholamines. **Inhibitory neurotransmitters** include **γ-aminobutyric acid** (**GABA** or 4-aminobutyric acid), glycine, and taurine (Table 23.1). Glycine acts predominantly in the spinal cord and the brain stem; GABA acts predominantly in other parts of the brain. **Strychnine** (Figure 23.5), a highly poisonous alkaloid obtained from *Nux vomica* and related plants of the genus *Strychnos,* binds to **glycine receptors**. It has been used in very small doses as a CNS stimulant. Can you propose how it works? **GABA receptors** also react with a variety of pharmacologically significant agents such as **benzodiazepines (Valium)** (Figure 23.6) and barbiturates. There is little structural similarity between GABA and benzodiazepines.

Genes for some of the receptors have been cloned, including those for **nicotinic/acetylcholine**, glycine, glutamate, and GABA and X-ray crystallographic structures of some have been determined.

A model of the GABA receptor is shown in Figure 23.7. This receptor has an $\alpha_2\beta_2\gamma_2$ composition. This model shows areas where GABA binds and where some pharmacological agents such as benzodiazepines (Bz), an antagonist, bind. The subunits of the receptor form channels through which small negative ions (Cl^-) flow into the cell under normal stimulation by GABA making the inside of the cell more negative. Thus, inhibitory neurotransmitters make depolarization more difficult because enough Na^+ must flow into the neuron to overcome the increased negative charges of excess Cl^- as well as create a polarization value above the threshold potential.

Chloride ion (Cl^-) channels differ significantly from cation channels (Na^+, K^+, and Ca^{2+}). The Cl^- channel is a two-barrel antiparallel structure in which the ion selectivity filter is formed mostly by the charged polarized ends of helices. These are shown in Figure 23.8*a*, p. 945, in a 2-D linear conformation. A 3-D conformational arrangement with a Cl^- within the filter is shown in Figure 23.8*b*. It is assumed that this selectivity filter does not use more positively charged amino acid residues because the stronger ionic interaction would impede Cl^- movement through the channel.

Excitatory neurotransmitters are released after stimulation of the neuron, traverse the synapse, and bind to specific receptors on the postsynaptic junction to elicit a response in its targeted postsynaptic cell. They cause depolarization of the membrane by opening channels permitting the influx of Na^+. They interact with ligand-gated receptors (p. 478). When no ligand is bound, the receptor is at rest and the ion channel is closed. Binding of ligand causes conformational changes in the membrane helices that open the ion channel. The channel becomes desensitized and closes as the ligand begins to dissociate, and the channel returns to its resting state.

(a)

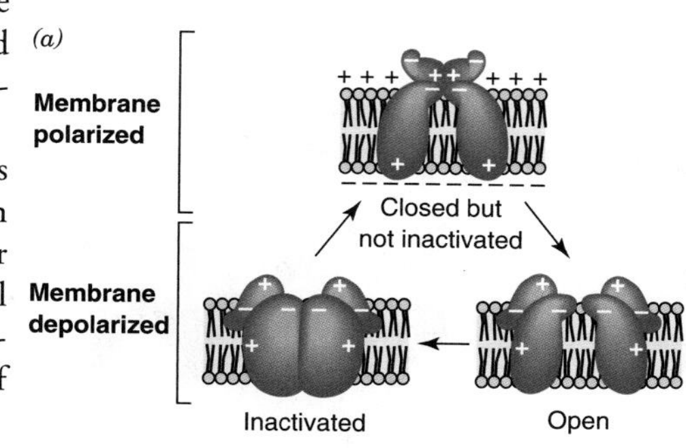

(b)

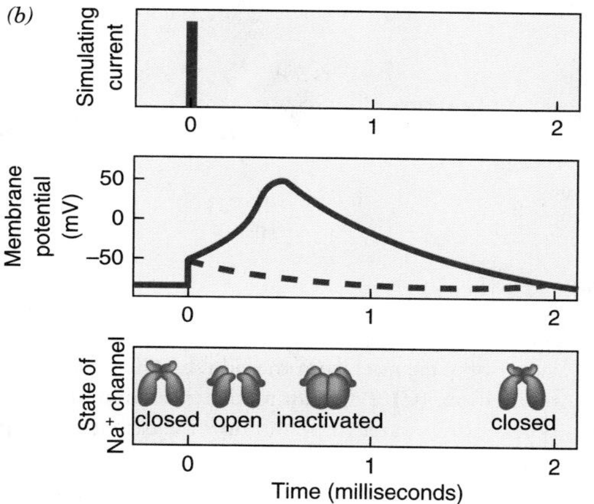

(c)

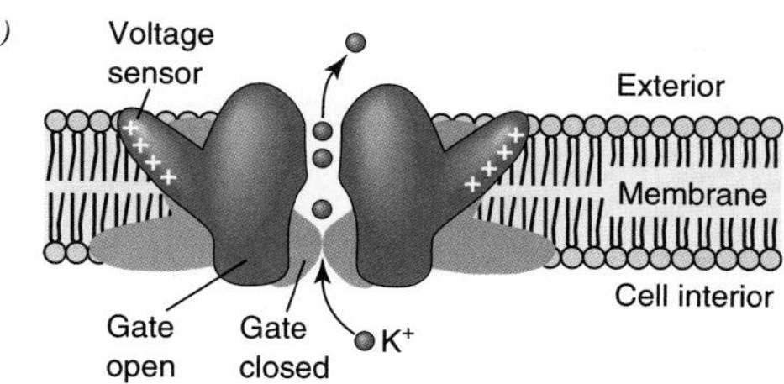

Figure 23.3 Opening and closing of Na^+ channels. Schematic (*a* and *b*) of Na^+ channels opening and closing during nerve impulse transmission. (*c*) Schematic representation of a voltage sensor. The lighter color shows the position of the sensor in its closed state. Positive ions cause charge–charge repulsion as they approach the sensor resulting in a conformation change in the sensor that opens the channel.
Redrawn from Alberts, B., Bray, D., Lewis, J., Raff, M., Roberts, K., and Watson, J. *Molecular Biology of the Cell,* 2d ed. New York: Garland, 1989, 1071.
Redrawn from Sigworth, F. J. Structural biology: Life's transistors. *Nature* 423: 21, 2003.

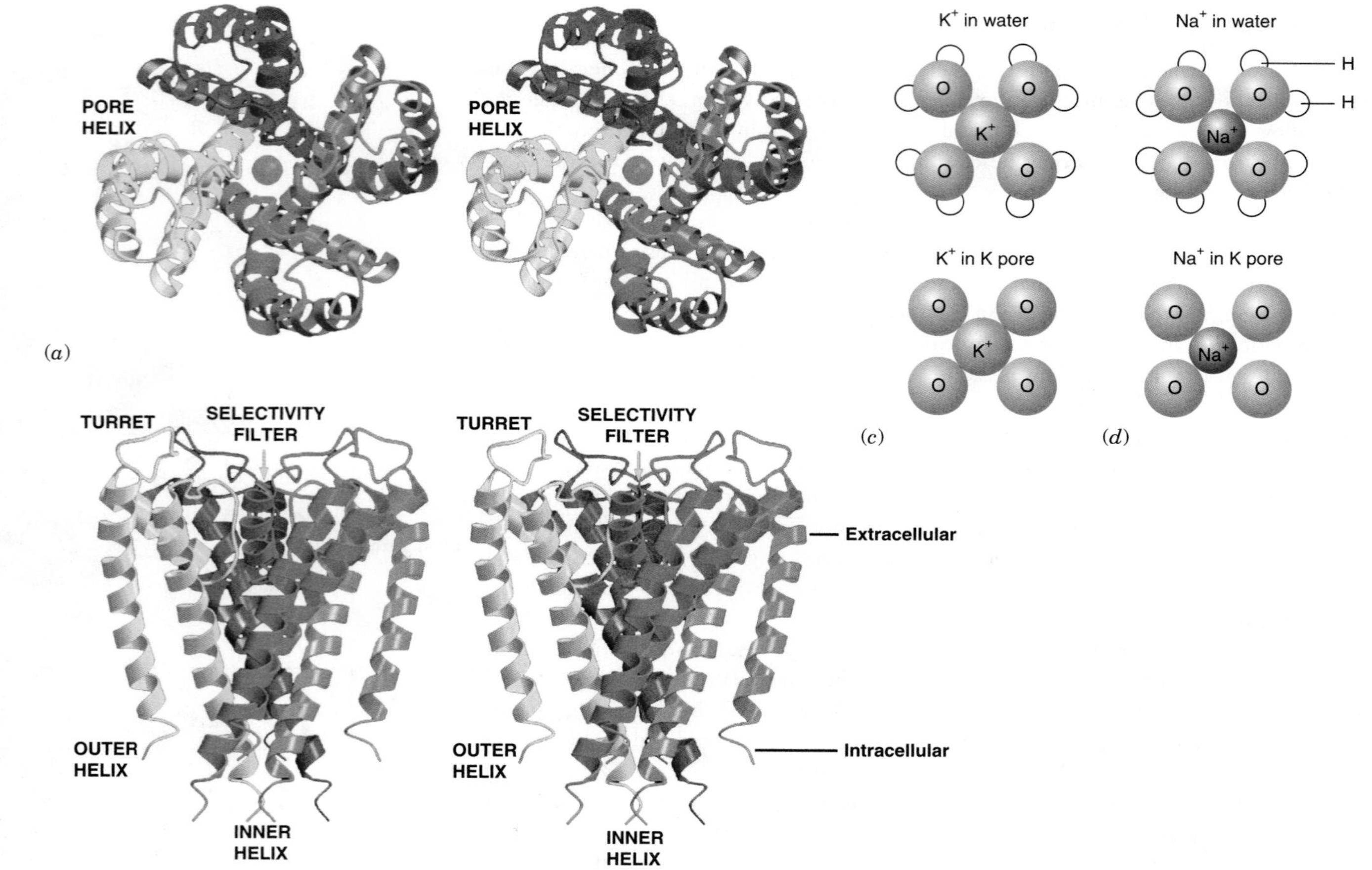

PORE HELIX
PORE HELIX
(a)
TURRET
SELECTIVITY FILTER
TURRET
SELECTIVITY FILTER
Extracellular
Intracellular
OUTER HELIX
OUTER HELIX
INNER HELIX
INNER HELIX
(b)
K^+ in water
Na^+ in water
H
H
K^+ in K pore
Na^+ in K pore
(c)
(d)

Figure 23.4 Structure of the potassium channel. (A) A stereo view from the top of the channel showing a potassium ion inside. (B) A stereo view of the channel from the side showing its extracellular and intracellular orientation. (C) A diagrammatic representation of the hydration sphere of potassium in water (top) and how the carbonyl groups of helices within the channel (bottom) mimic its aqueous hydration sphere. (D) A diagrammatic representation of sodium in water (top) and how the carbonyl groups within the channel cannot mimic the aqueous hydration sphere. Note that there are spaces within the channel that are too big to permit carbonyl groups within the channel to satisfy the hydration requirements of sodium ions.
From Doyle, D. A., Cabral, J. M, Pfuetzner, R. A., Kuo, A, et al. *Science* 280:69, 1998. Reprinted with permission from AAAS.
Modifications of a diagram from Lodish, H., Berk, A., Kaiser, C. A., Kreiger, M., et al. *Molecular Cell Biology*. New York: Freeman,1995.

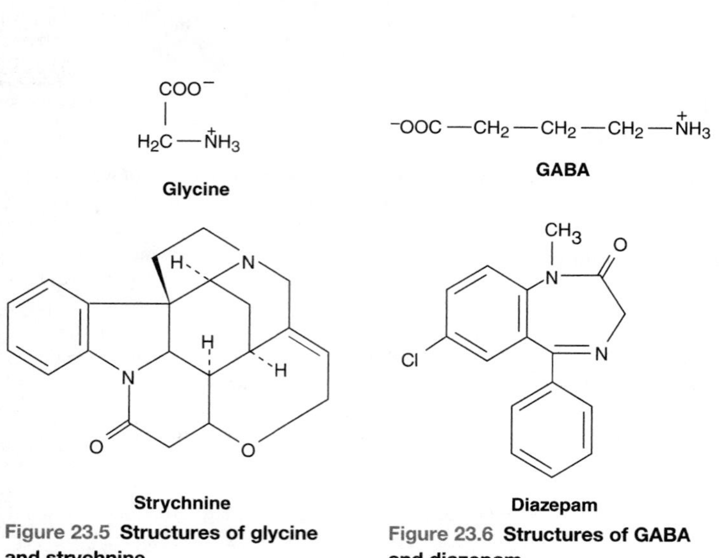

Figure 23.5 Structures of glycine and strychnine.

Figure 23.6 Structures of GABA and diazepam.

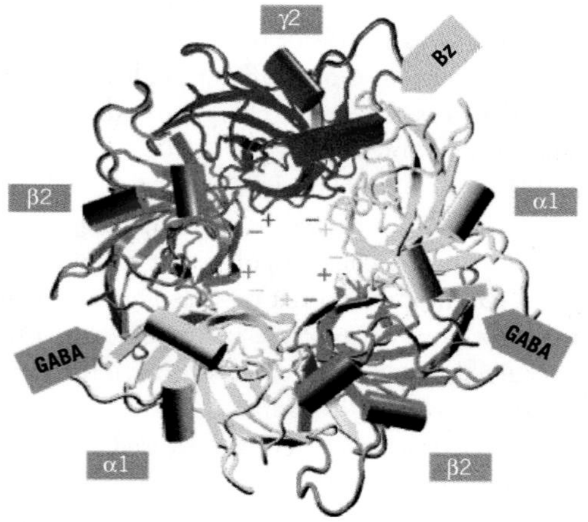

γ2
Bz
β2
α1
GABA
GABA
α1
β2

Figure 23.7 Model of the crystal structure of the GABA receptor; top view. The GABA receptor has an $\alpha_2\beta_2\gamma_2$ structure and forms an ion channel for transport of negative ions, Cl^-. GABA binding sites are labeled and so is the site where benzodiazepam (Bz) binds.
Reproduced with permission from Ernst, M., Brauchart, D., Boresch, S., and Sieghart W. *Neuroscience* 119:933, 2003. Copyright (2003) Elsevier.

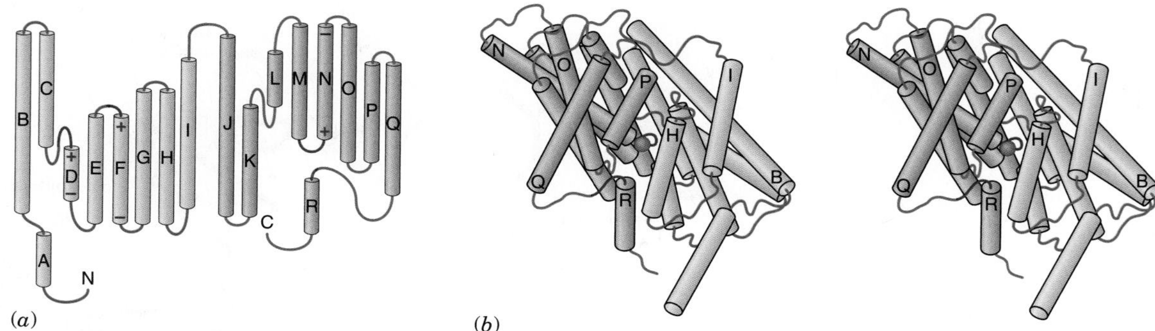

Figure 23.8 Structural model of a ClC chloride channel. This is essentially a two-barrel channel, one green and one blue in this diagram. (*a*) Linear arrangement of the helices showing the positive (+) and negative (−) ends. (*b*) Stereo view of the channel with a Cl^- (colored ball) between the two barrels. Cl^- interacts with the positive (+) ends of the helices.

Reproduced with permission from Dutzler, R., Campbell, E. B., Cadene, M., Chait, B. T., and MacKinnon, R. X-ray structure of a ClC chloride channel at 3.0 Å reveals the molecular basis of anion selectivity. *Nature* 415:287, 2002. Copyright (2002) *Nature*.

Synthesis, Storage, and Release of Neurotransmitters

Nonpeptide neurotransmitters may be synthesized in almost any part of the neuron, in the cytoplasm near the nucleus, or in the axon. Most nonpeptide neurotransmitters are amino acids or derivatives of amino acids.

Released neurotransmitters travel rapidly across the **synaptic junction** (which is about 20 nm across) and bind to receptors on the postsynaptic side. On binding, they initiate the process of electrical impulse propagation in the postsynaptic neuron as described above. Storage and release of neurotransmitters are multistep processes, many of which have been defined. Some neurons appear to contain more than one chemical type of neurotransmitter. The physiological significance of this observation is not clear. Release of neurotransmitters is a quantal event. A nerve impulse reaching the presynaptic terminal causes release of transmitters from a fixed number of **synaptic vesicles**. An important step in the process involves attachment of synaptic vesicles to the presynaptic membrane and **exocytosis** of their content into the synaptic cleft.

Storage of neurotransmitters occurs in large or small vesicles in the presynaptic terminal. Small vesicles predominate and exist in two pools; free and attached to cytoskeletal proteins, mainly actin. They contain nonpeptide small-molecule-type transmitters. A schematic diagram of a small synaptic vesicle is shown in Figure 23.9*a*. Large vesicles may contain both nonpeptide- and peptide-type neurotransmitters. Some may also contain enzymes for synthesis of norepinephrine from dopamine. A list of some proteins in synaptic vesicles is presented in Table 23.2. Figure 23.10, p. 947, shows schematically how some of the proteins may be arranged on the synaptic vesicle and how they may interact with the plasma membrane of the presynaptic neuron.

Synapsin has a major role in determining whether the small synaptic vesicles are free and available for binding to the presynaptic membrane (Figure 23.11, p. 947) or whether they are attached to cytoskeletal proteins. There is a family of synapsin proteins, encoded by two genes that differ primarily in the C-terminal end (see Figure 23.9*b*). Synapsins constitute about 9% of the total protein of the synaptic vesicle membrane. All can be phosphorylated near their *N*-termini by either **cAMP-dependent protein kinase** and/or **calcium-calmodulin (CaM) kinase I**. Synapsins Ia and Ib can also be phosphorylated by **CaM kinase II** near their C-termini, a region that is missing in synapsins IIa and IIb.

Nerve stimulation leads to the entry of Ca^{2+} into the presynaptic neuron (see Clin. Corr. 23.2, p. 948), where it binds to calmodulin and activates CaM kinases I and CaM kinases II (producing Ca^{2+}-CaM kinases I and II), which phosphorylate synapsin. This causes release of synaptic vesicles from the cytoskeletal protein matrix and/or prevents free phosphorylated vesicles from binding to those matrices. The result is an increase in the free pool of synaptic vesicles. **Calcium-calmodulin** (p. 493) can also bind synapsin directly and competitively block its interaction with actin and presumably other cytoskeletal proteins. Calcium-calmodulin and the phosphorylation of synapsin, therefore, regulate the number of synaptic vesicles that are in the free state.

TABLE 23.2 • Some Synaptic Vesicle Proteins

Synapsin	Ia
	Ib
	IIa
	IIb
Synaptophysin	
Synaptotagmin	
Syntaxin	
Synaptobrevin/VAMP	
Rab3 and rabphilin	
SV-2	
Vacuolar proton pump	

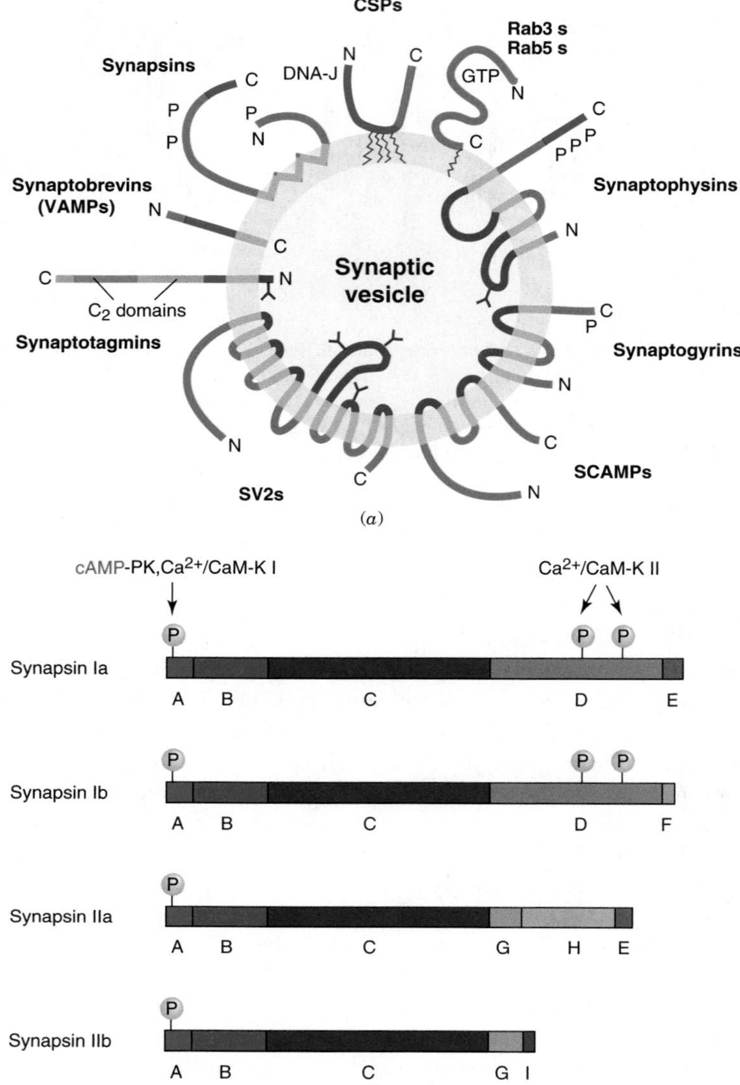

Figure 23.9 Synaptic vesicle. (*a*) A schematic drawing of the relative arrangement of proteins of the synaptic vesicle (SV). RAB proteins are attached by isoprenyl groups and cysteine string proteins (CSPs) by palmitoyl chains to SVs. The N- and C-termini of proteins are marked by N and C, respectively. Phosphorylation sites are indicated by P. (*b*) Structural arrangement of the synapsin family of proteins.
Redrawn from Sudhof, T. C. *Nature* 375:645, 1995.
Redrawn from the work of T. J. Chilcote, T. J., Siow, Y. L. Schaeffer, E., Greengard, P., and Thiel G. *J. Neurochem.* 63:1568, 1994.

Following is a brief summary of some properties of other synaptic vesicle proteins:

Synaptophysin is an integral membrane protein of synaptic vesicles that is structurally similar to gap junction proteins. It may be involved in formation of a channel from the synaptic vesicle through the presynaptic membrane to permit passage of neurotransmitters into the synaptic cleft.

Synaptotagmin is also an integral membrane protein of synaptic vesicles. It interacts in a Ca^{2+}-dependent manner with specific proteins localized on the presynaptic membrane. It is probably involved in docking of synaptic vesicles to the membrane.

Syntaxin is an integral protein of the plasma membrane of presynaptic neurons. It binds synaptotagmin and mediates its interaction with Ca^{2+} channels at the site of release of the neurotransmitters. It also appears to have a role in exocytosis.

Synaptobrevin/VAMP (vesicle associated membrane protein) consists of a family of two small proteins of 18 and 17 kDa that are anchored in the cytoplasmic side of the membrane through a single C-terminal domain. It appears to be involved in **vesicle transport** and/or **exocytosis**. VAMPs appear to be involved in the release of synaptic vesicles from the plasma membrane of the presynaptic neuron. **Tetanus toxin** and **botulinum toxin**, a zinc-dependent endoprotease, bind and cleave VAMPs, resulting in a slow and irreversible inhibition of transmitter release.

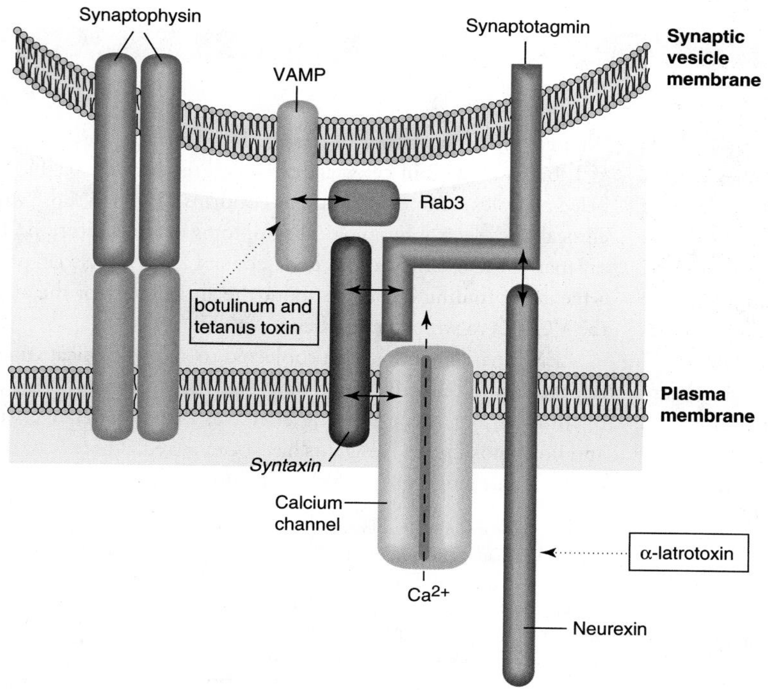

Figure 23.10 Schematic diagram showing how some synaptic vesicle proteins may interact with plasma membrane proteins. Redrawn from Bennett, M. K., and Scheller, R. H. *Proc. Natl. Acad. Sci. USA* 90:2559, 1993.

Rab3 belongs to the large rab family of **GTP-binding proteins**. Rab3 is specific for synaptic vesicles and is involved in docking and fusion of vesicles for exocytosis. Rab3 is anchored to the membrane through a polyprenyl side chain near its C-terminal.

SV-2 is a large glycoprotein with 12 transmembrane domains. Its function is unclear.

Vacuolar proton pump (p. 491) functions to transport neurotransmitters back into the synaptic vesicles after their reformation and release from the presynaptic membrane.

CSPs are cysteine string proteins and members of a family of chaperones (p. 231) that function during latter stages of Ca^{2+}-regulated exocytosis.

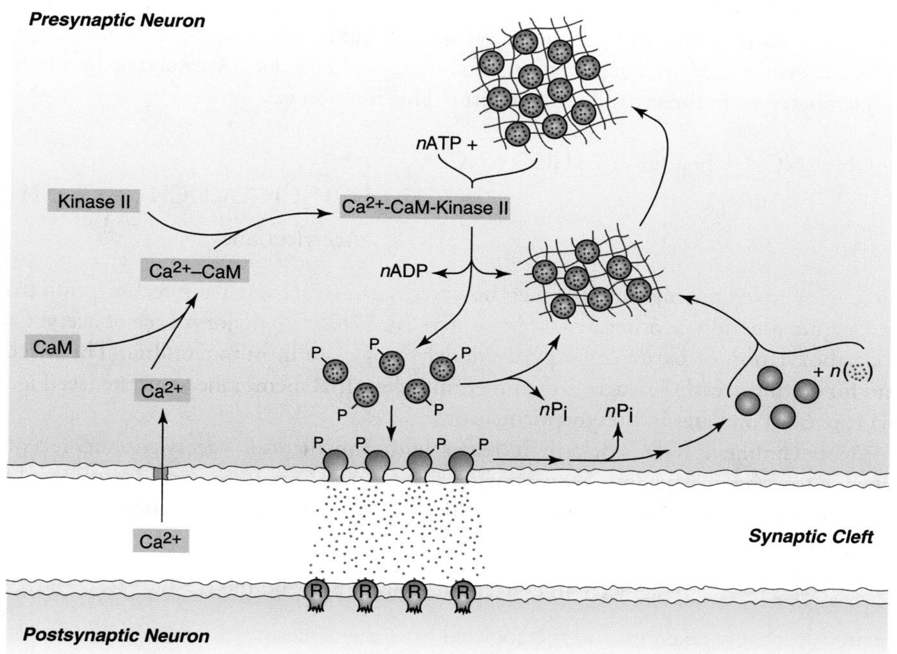

Figure 23.11 A model of the mechanism of regulation of synaptic vesicles function by calcium ions and calmodulin kinase II. Green circles within the skeletal mesh represent bound, nonphosphorylated synaptic vesicles. Green circles with P attached have been freed from the mesh (bound pool) after phosphorylation of synapsin. These phosphorylated synaptic vesicles can now interact with the presynaptic membrane and release neurotransmitters into the synaptic cleft. Receptors on the postsynaptic membrane (red) will bind to some of those neurotransmitters. The recovery of synaptic vesicles and repackaging with neurotransmitters is also illustrated schematically.

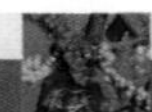

CLINICAL CORRELATION 23.2

Lambert–Eaton Myasthenic Syndrome

Lambert–Eaton myasthenic syndrome (LEMS) (OMIM 600003) is an autoimmune disease in which the body raises antibodies against voltage-gated calcium channels (VGCC) located on presynaptic nerve termini. On depolarization of presynaptic neurons, these calcium channels open permitting the influx of Ca^{2+}. This increase in Ca^{2+} concentration initiates events of the synapsin cycle resulting in release of neurotransmitters into synaptic junctions. When autoantibodies against VGCC react with neurons at neuromuscular junctions, Ca^{2+} cannot enter and the amount of acetylcholine released into the synaptic junction is diminished. Since action potentials to muscles may not be induced, the effect mimics that of classic myasthenia gravis.

LEMS has been observed in conjunction with other conditions such as small-cell lung cancer. Some patients manifest subacute cerebellar degeneration (SCD). Plasma exchange (removal of antibodies) and immunosuppressive treatments have been effective for LEMS, but the latter treatment is less effective on SCD.

Diagnostic assays for LEMS depend on the detection of antibodies against VGCC in serum. There are at least four subtypes of VGCC, T, L, N, and P. The P-subtype may be the one responsible for initiating neurotransmitter release at the neuromuscular junction in mammals. A peptide toxin produced by a cone snail (Conus magnus) binds to P-type VGCC in cerebella extracts. This small peptide, labeled with ^{125}I, binds VGCC in cerebella extracts. Precipitation of this radiolabeled complex with sera of patients confirms LEMS in those who have clinical and electrophysiological symptoms of the condition. This assay may prove useful not only in detecting LEMS but also in providing a means of finding out more about the antigenicity of the area(s) on the VGCCs to which antibodies are raised.

LEMS patients have been subjected to several clinical trials in attempts to find effective treatments, but there is insufficient data to quantify any of them as being effective. Plasma exchange, steroids, and immunosuppressive agents have been tested. There is no cure and clinical trials looking for effective therapy continue.

http://www.ninds.nih.gov/disorders/lambert_eaton/lambert_eaton.htm; Goldstein, J. M., Waxman, S. G., Vollmer, T. L., Lang, B., et al. Subacute cerebellar degeneration and Lambert–Eaton myasthenic syndrome associated with antibodies to voltage-gated calcium channels: differential effect of immunosuppressive therapy on central and peripheral defects. *J. Neurol. Neurosurgery, Psych.* 57:1138, 1994; Motomura, M., Johnston, I., Lang, B., Vincent, A., and Newsom-Davis, J. An improved diagnostic assay for Lambert–Eaton myasthenic syndrome. *J. Neurol. Neurosurgery, Psych.* 58:85, 1995; and Maddison, P., and Newsom-Davis, J. Treatment for Lambert–Eaton myasthenic syndrome (Cochrane Review). In *Cochrane Library*, Issue 4. Chichester, UK: Wiley, 2004.

Termination of Signals at Synaptic Junctions

Neurotransmitter action may be terminated by reuptake into the presynaptic neurons, metabolism, or uptake into other cell types. One or more of these mechanisms function for inactivation of neurotransmitters. Here are examples of some pathways involved in synthesis and degradation of some representative neurotransmitters.

Acetylcholine

Reactions involving acetylcholine at the synapse are summarized in Figure 23.12. Acetylcholine is synthesized by condensation of choline and acetyl CoA catalyzed by **choline acetyltransferase** in the cytosol of the neuron This reaction is

$$\underset{\text{Choline}}{(CH_3)_3\overset{+}{N}CH_2CH_2OH} + CH_3CO{-}SCoA \longrightarrow \underset{\text{Acetylcholine}}{(CH_3)_3\overset{+}{N}CH_2CH_2OCOCH_3} + CoASH$$

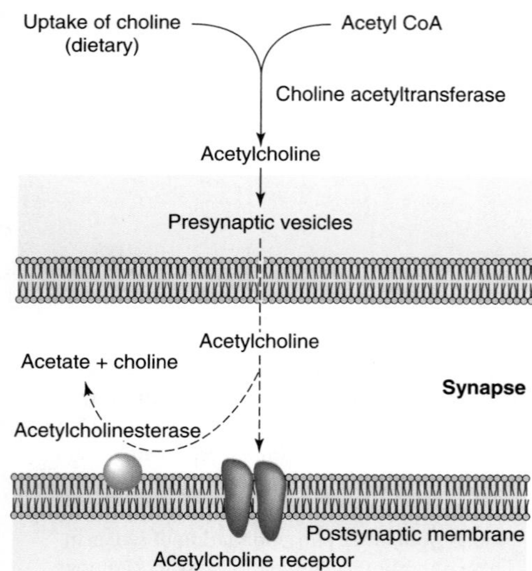

Figure 23.12 Summary of reactions of the neurotransmitter acetylcholine at the synapse. Abbreviations: AcCoA, acetylcoenzyme A, and AcChEase, acetylcholinesterase.

Choline is derived mainly from the diet; however, some may come from reabsorption from the synaptic junction or other metabolic sources (p. 778). The major source of acetyl CoA is decarboxylation of pyruvate by pyruvate dehydrogenase in mitochondria. The mechanism for getting acetyl CoA across the inner mitochondrial membrane as citrate (see Figure 17.11, p. 684) operates in presynaptic neurons.

Acetylcholine is released, and it reacts with the nicotinic–acetylcholine receptor embedded in the postsynaptic membrane (Clin. Corr. 23.3). Its action is terminated by the action of **acetylcholinesterase**, which hydrolyzes it to acetate and choline (A Closer Look 23.1).

$$\text{Acetylcholine} + H_2O \rightarrow \text{acetate} + \text{choline}$$

Acetate is probably taken up and metabolized by other tissues.

CLINICAL CORRELATION 23.3

Myasthenia Gravis: A Neuromuscular Disorder

Myasthenia gravis (OMIM 254200) is an acquired autoimmune disease characterized by muscle weakness due to decreased neuromuscular signal transmission. The neurotransmitter involved is acetylcholine. The sera of more than 90% of patients with myasthenia gravis have antibodies to the nicotinic–acetylcholine receptor (AChR) located on the postsynaptic membrane of the neuromuscular junction. Antibodies against the AChR interact with and inhibit either its ability to bind acetylcholine or its ability to undergo conformational changes necessary to affect ion transport. The number of functional AChRs is reduced in patients with the disease. Experimental models of myasthenia gravis have been generated either by immunizing animals with the AChR or by injecting them with antibodies against it.

Events that trigger the onset of the disease are unknown. Some environmental antigens have epitopes resembling those on the AChR. A rat monoclonal IgM antibody prepared against AChRs reacts with two proteins obtained from the intestinal bacterium *E. coli.* Both are membrane proteins of 38 and 55 kDa, the smaller of which is located in the outer membrane. This does not suggest that exposure to *E. coli* proteins is likely to trigger the disease. The sera of both normal individuals and myasthenia gravis patients have antibodies against a large number of *E. coli* proteins. Some environmental antigens from other sources also react with antibodies against AChRs.

The thymus gland, which is involved in antibody production, is implicated in this disease. Antibodies have been found in thymus glands of myasthenia gravis patients that react with AChRs and with environmental antigens. The relationship between environmental antigens, thymus antibodies against AChRs, and onset of myasthenia gravis is not clear.

Myasthenia gravis patients may receive one or a combination of several therapies. Pyridostigmine bromide, a reversible inhibitor of acetylcholine esterase (AChE) that does not cross the blood–brain barrier, has been used. The inhibition of AChE within the synapse increases the half-time for acetylcholine hydrolysis. This leads to an increase in the concentration of acetylcholine, stimulation of more AChR, and increase in signal transmission. Other treatments include immunosuppressant drugs, steroids, and surgical removal of the thymus gland to decrease the rate of production of antibodies. Future treatment may include the use of anti-idiotype antibodies to the AChR antibodies, and/or the use of small nonantigenic peptides that compete with AChR epitopes for binding to the AChR antibodies.

http://www.ninds.nih.gov/disorders/myasthenia_gravis/; Stefansson, K., Dieperink, M. E., Richman, D. P., Gomez, C. M., and Marton, L. S. *N. Engl. J. Med.* 312:221, 1985; Drachman, D. B. (Ed.). Myasthenia gravis: Biology and treatment. *Ann. N.Y. Acad. Sci.* 505:1, 1987; and Steinman, L., and Mantegazza, R. *FASEB J.* 4:2726, 1990.

A CLOSER LOOK 23.1

Structure of Acetylcholine Esterase

An X-ray crystallographic structure of acetylcholine esterase is shown in the following figure. Its mechanism of action is similar to that of serine proteases (p. 343). It contains a **catalytic triad**, but the amino acids in that triad, from N- to C-termini, are in reverse order to those of serine proteases like trypsin and chymotrypsin; in addition, glutamate replaces aspartate. The serine that forms the covalent intermediate during catalysis is visible at the bottom of a hydrophobic channel. Glutamate, one of the other residues of the catalytic triad, is also visible within the channel of that structure.

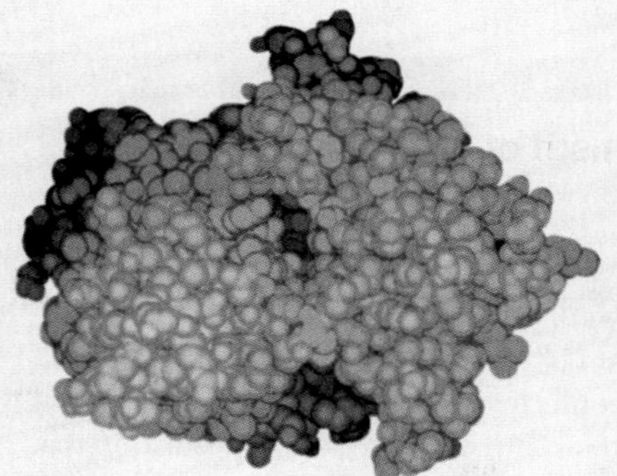

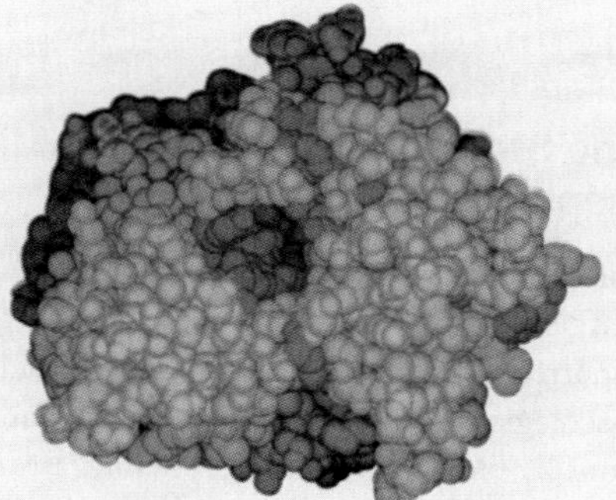

Space-filling stereo view of acetylcholinesterase looking down into the active site. Aromatic residues are in green, Ser^{200} is red, Glu^{199} is cyan, and other residues are gray.

From Sussman, J. L., Harel, M., Frolow, F., Oefner, C., et al. *Science* 253:872, 1991. Reproduced with permission from AAAS.

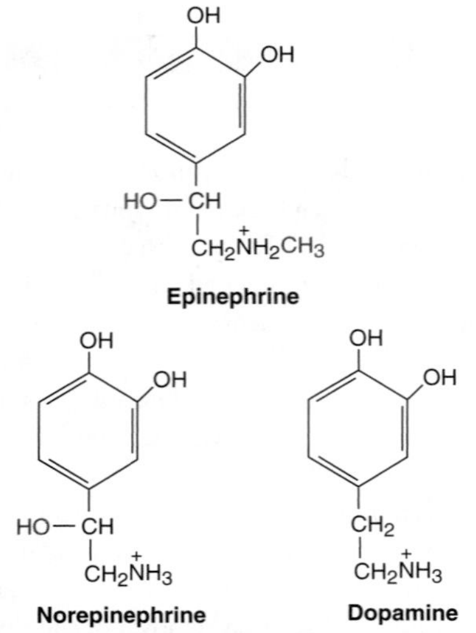

Epinephrine

Norepinephrine

Dopamine

Figure 23.13 Catecholamine neurotransmitters.

Catecholamines

The **catecholamine neurotransmitters** are **dopamine** (3,4-dihydroxyphenylethylamine), **norepinephrine**, and **epinephrine** (Figure 23.13 and p. 781). Their action is terminated by reuptake into the presynaptic neuron by specific transporter proteins (A Closer Look 23.2). Cocaine, for example, binds specifically to the **dopamine transporter** and blocks dopamine reuptake. In the presence of cocaine, dopamine remains in the synaptic cleft for a prolonged period and stimulates excessively its receptors on the postsynaptic neuron. After reuptake, catecholamine neurotransmitters may be either repackaged into synaptic vesicles or metabolized by **catechol-*O*-methyltransferase** that catalyzes the transfer of a methyl group from *S*-adenosylmethionine to one of the phenolic OH groups, and **monoamine oxidase** (Figure 23.14) that catalyzes the oxidative deamination of these amines to aldehydes and ammonium ions. They are substrates for monoamine oxidase whether or not they have been altered by catechol-*O*-methyltransferase. The end product of dopamine metabolism is **homovanillic acid**, and that of epinephrine and norepinephrine is **3-methoxy-4-hydroxymandelic acid**.

5-Hydroxytryptamine (Serotonin)

Serotonin is derived from tryptophan (p. 785). Like dopamine, its action is terminated by **reuptake** by a specific transporter. Some types of depression are associated with low brain levels of serotonin; antidepressants such as Paxil (paroxetine hydrochloride), Prozac (fluoxetine hydrochloride), and Zoloft (sertraline hydrochloride) specifically inhibit serotonin reuptake. Once inside the presynaptic neuron, serotonin may be either repackaged in synaptic vesicles or oxidatively deaminated to the corresponding acetaldehyde by monoamine oxidase (Figure 23.15, p. 952). The aldehyde is further oxidized to 5-hydroxyindole-3-acetate by an aldehyde dehydrogenase.

γ-Aminobutyrate (GABA)

γ-Aminobutyrate, an inhibitory neurotransmitter, is synthesized and degraded through reactions commonly known as the **GABA shunt**. In brain tissue, GABA and glutamate, an excitatory neurotransmitter, may share common routes of metabolism (Figure 23.16, p. 952) where they are taken up by astrocytes and converted to glutamine, which is then transported back into presynaptic neurons. In excitatory neurons, glutamine is converted to glutamate and repackaged in synaptic vesicles. In inhibitory neurons, glutamine is converted to glutamate and then to GABA, which is packaged in synaptic vesicles.

It has been suggested that brain levels of GABA in some epileptic patients may be low. **Valproic acid** (2-propylpentanoic acid) can cross the blood–brain barrier and apparently increases brain levels of GABA. The mechanism by which it does so is not clear. Valproic

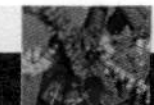

A CLOSER LOOK 23.2

LeuT–Desimpramine Structure: A Clue to the Mechanism of Neuronal Reuptake of Dopamine, Epinephrine, and Serotonin

Serotonin and catecholamine neurotransmitters are among those whose actions are stopped at the synapse by reuptake into the presynaptic neuron by transporters that are specific for each individual neurotransmitter. Pharmacological agents and some drugs of abuse generally effect their actions by blocking reuptake of neurotransmitters. The specifics of how these agents work is poorly understood because of a lack of mechanistic information at the molecular level.

Transporters for serotonin (SERT), norepinephrine (NET), and dopamine (DAT) are members of a family of proteins that have a high degree of homology with the bacterial leucine transporter (LeuT) whose structure has been determined at 2.9-Å resolution in a complex with desimpramine, a tricyclic inhibitor of LeuT's action. Desimpramine binds at a different site than leucine, and its binding at the inner end of the extracellular cavity of the transporter prevents conformational changes that would open the gate to permit leucine transport. This allows the speculation that inhibitors of SERT, NET, and DAT may function in a similar manner. Structural and other information are available in the cited article.

Zhou, Z., Zhen, J. Karpowich, J. K., Goetz, R. M., Law, C. J., Reith, M. E., and Wang, D.-N. LeuT–desipramine structure reveals how antidepressants block neurotransmitter reuptake. *Science* 317:1390, 2007.

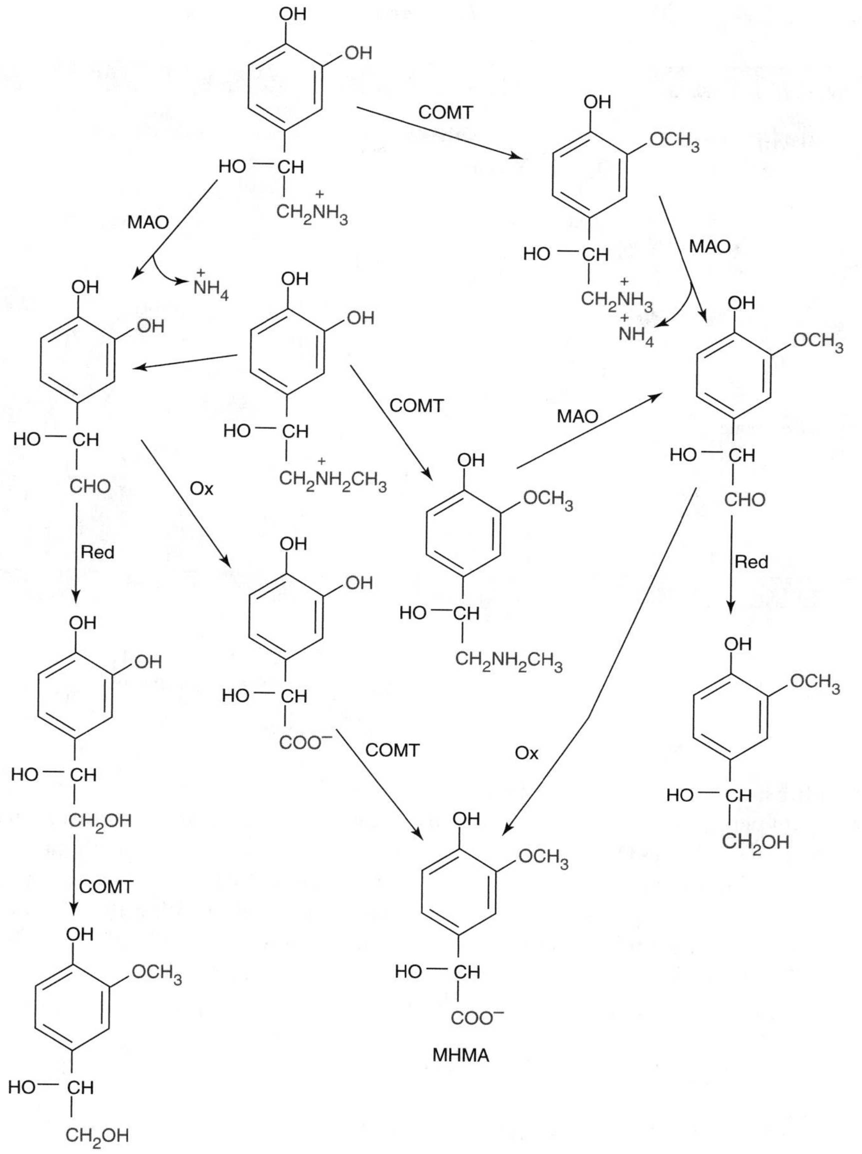

Figure 23.14 Reactions of catecholamine degradation. Abbreviations: COMT, catechol-*O*-methyl transferase (requires *S*-adenosylmethionine); MAO, monoamine oxidase; Ox, oxidation; and Red, reduction. The end product of epinephrine and norepinephrine metabolism is 3-methoxy-4-hydroxymandelic acid (MHMA).

acid is metabolized primarily in the liver by glucuronidation and urinary excretion of the glucuronides or by mitochondrial β-oxidation and oxidation by endoplasmic reticulum enzymes.

Neuropeptides Are Derived from Precursor Proteins

Peptide neurotransmitters are synthesized as larger proteins. Proteolysis of these larger proteins releases the neuropeptide molecules. Synthesis of the larger proteins takes place in the cell body and not the axon. The neuropeptides travel down the axon to the presynaptic region by one of two mechanisms: **fast axonal transport** at a rate of about 400 mm/day or **slow axonal transport** at a rate of 1–5 mm/day. Since axons may vary in length from 1 mm to 1 m in length, theoretically the transit time could vary from 150 ms to 200 days. It is unlikely that the latter transit time occurs under normal physiological conditions, and the upper limit is probably hours rather than days. There is experimental evidence that suggest that the faster transit time prevails. Kinesins and myosins, molecular motor proteins (p. 980), facilitate this transport process.

Neuropeptides mediate sensory and emotional responses such as those associated with hunger, thirst, sex, pleasure, pain, and so on. Included in this category are **enkephalins**,

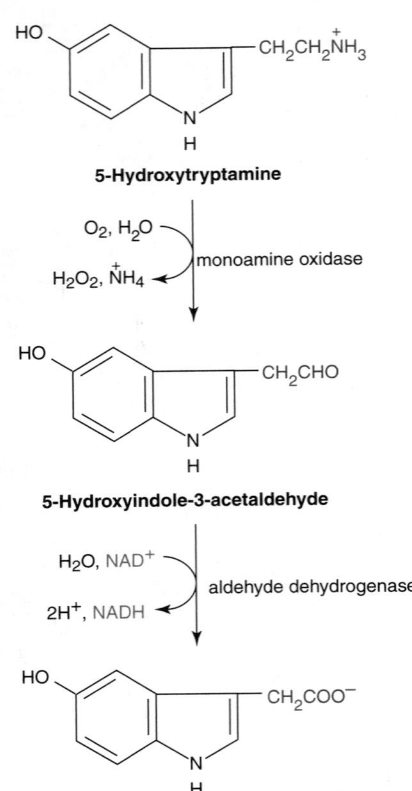

Figure 23.15 Degradation of 5-hydroxytryptamine (serotonin).

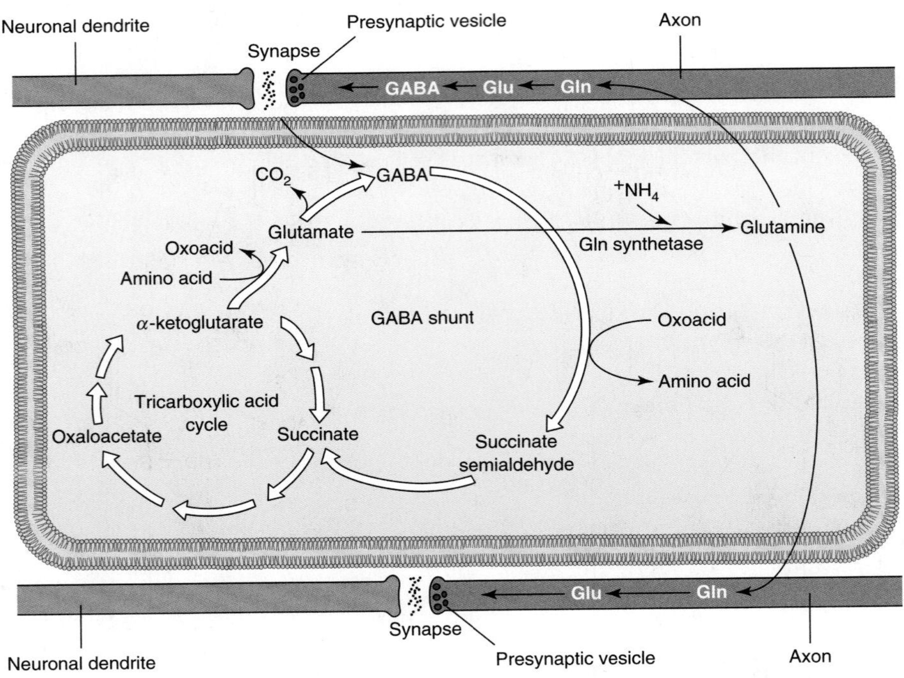

Figure 23.16 Involvement of astrocytes in the metabolism of GABA and glutamate.

endorphins, and **substance P**. Substance P is an excitatory neurotransmitter that has a role in pain perception. It is among a class of neuropeptides called **neurokinins**. Its receptor, **NK-1** (**n**euro**k**inin-**1**) is a G-type protein consisting of seven transmembrane helical elements. Endorphins and opioids bind to receptors that also have seven transmembrane helical elements. Endorphins and enkephalins have roles in eliminating the sensation of pain. Some of the peptides found in brain tissue are shown in Table 23.3. Note that Met-enkephalin is derived from the *N*-terminal region of β-endorphin. The *N*-terminal or both the *N*- and C-terminal amino acids of many of the neuropeptide transmitters are modified.

TABLE 23.3 • Peptides Found in Brain Tissue[a]

Peptide	*Structure*
β-Endorphin	Y G G F M T S E K S Q T P L V T L F K N A I I K N A Y K K G E
Met-enkephalin	Y G G F M
Leu-enkephalin	Y G G F L
Somatostatin	A G C K N F F W | | C S T F T K
Luteinizing hormone-releasing hormone	p-E H W S Y G L R P G NH_2
Thyrotropin-releasing hormone	p-E H P-NH_2
Substance P	R P K P E E F F G L M-NH_2
Neurotensin	p-E L Y E N K P R R P Y I L
Angiotensin I	D R V Y I H P F H L
Angiotensin II	D R V Y I H P F
Vasoactive intestinal peptide	H S D A V F T D N Y T R L R- K E M A V K K Y L N S I L N-NH_2

[a]Peptides with "p" preceding the structure indicate that the N-terminal is pyroglutamate. Those with NH_2 at the end indicate that the C-terminal is an amide.

23.2 • THE EYE: METABOLISM AND VISION

The eye, our window to the outside world, allows us to view the beauties of nature. A view through any window or any camera lens is clearest when unobstructed. The eye has evolved in such a way that a similar objective has been achieved. It is composed of live tissues that require continuous nourishment obtained through use of conventional metabolic pathways appropriate to their unique needs. Pigmented structures such as cytochromes of mitochondria are either not present in some structures or arranged and distributed so as not to interfere with the visual process. In addition, the brain has devised an enormously efficient filtering system that makes objects within the eye invisible that may otherwise appear to lead to visual distortion. A schematic diagram of a cross section of the eye is shown in Figure 23.17.

Light enters the eye; passes progressively through the **cornea**, the anterior chamber that contains aqueous humor, the lens, the vitreous body that contains vitreous humor; and finally focuses on the **retina** that contains the visual sensing apparatus. Tears bathe the exterior of the cornea while the interior is bathed by the aqueous humor, an isoosmotic fluid that contains salts, albumin, globulin, glucose, and other constituents. The aqueous humor brings nutrients to the cornea and to the lens, and it removes end products of metabolism from them. The vitreous humor is a gelatinous mass that helps maintain the shape of the eye while allowing it to remain somewhat pliable.

Cornea Derives ATP from Aerobic Metabolism

The eye is an extension of the nervous system and, like other tissues of the central nervous system, its major metabolic fuel is glucose. The cornea is not a homogeneous tissue. It consists of (1) an anterior corneal epithelium, (2) Bowman's membrane, (3) stroma (substantia propria) that is made up primarily of type 1 collagen and constitutes about 90% of the corneal thickness, (4) Descemet's membrane, and (5) endothelium (posterior corneal epithelium). The cornea is a clear tissue that, like the lens, diffracts light. Its clearness is due in part to the arrangement of collagen molecules of the stroma. The cornea is permeable to water and oxygen. The water content of the corneal stroma must be controlled for it to maintain clarity, and this is done by an ATP-driven water pump. Another reason for its clarity is the lack of blood vesicles in the epithelial layer. There is a large amount of the protein **VEGFR-3 (vascular endothelial growth factor receptor-3)** on the anterior epithelial layer of the cornea. VEGFR-3 prevents growth of blood vessels by binding to or neutralizing growth factors that are produced to stimulate growth of blood vessels.

The cornea obtains its ATP from aerobic glucose metabolism; glycolysis, and the TCA cycle. Lactate does not accumulate to any significant extent because of efficient use of pyruvate by oxidative metabolism. About 30% of the glucose it metabolizes is by glycolysis, and about 65% by the hexose monophosphate pathway. On a relative weight basis, the cornea has the highest activity of the hexose monophosphate pathway of any mammalian tissue. It also has a high activity of **glutathione reductase** that requires NADPH, a product

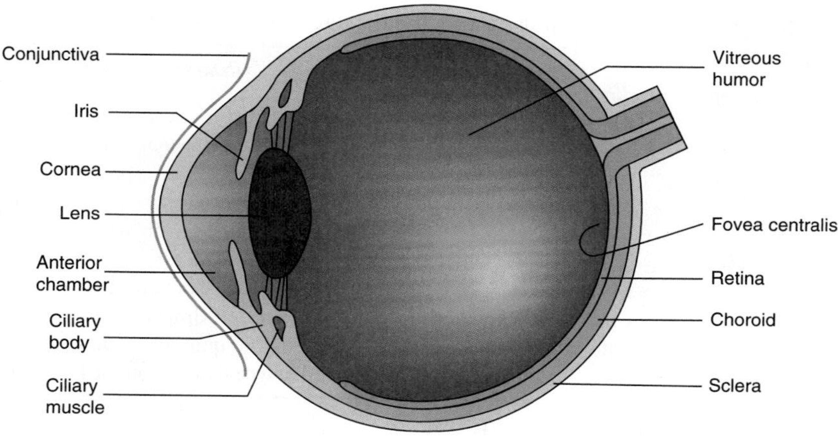

Figure 23.17 Schematic of a horizontal section of the left eye.

of the hexose monophosphate pathway. Corneal epithelium is permeable to atmospheric oxygen. Reactions of oxygen can lead to formation of various **active oxygen species** that are harmful to tissues, in some cases by oxidizing protein sulfhydryl groups to disulfides and by lipid peroxidation, mostly of cellular medium-chain lipids of six carbons or more (p. 583). Reduced glutathione (GSH) is used to reduce those disulfide bonds and lipid peroxides back to their original native states while GSH itself is converted to oxidized glutathione (GSSG). GSSG may also be formed directly by active oxygen species. Glutathione reductase uses NADPH to reduce GSSG to 2GSH.

$$\text{GSSG} + \text{NADPH} + \text{H}^+ \xrightleftharpoons{\text{GSH reductase}} 2\text{GSH} + \text{NADP}^+$$

The pentose phosphate pathway and glutathione reductase help protect the cornea by effectively neutralizing active oxygen species.

Some lipids that are subjected to peroxidation may spontaneously form active aldehydes that react with other tissue components and lead to various pathological conditions. The cornea also contains an isoform of aldehyde dehydrogenase (ALDH3A1), a member of a superfamily of enzymes that uses either NAD^+ or $NADP^+$ to inactivate these active aldehydes by oxidizing them to their corresponding acids.

Lens Consists Mostly of Water and Protein

The **lens** is bathed on one side by aqueous humor and supported on the other side by vitreous humor. It has no blood supply, but is metabolically active. It obtains nutrients from and eliminates waste into the aqueous humor. The lens is mostly water and proteins. The majority of vertebrate lens proteins are **α-, β-, and γ-crystallins** (Table 23.4). There are also albuminoids, enzymes, and membrane proteins that are synthesized in an epithelial layer around the edge of the lens. Other animals have different crystallins, some of which are enzymes that probably function as such in other tissues. The most important physical requirement of these proteins is that they maintain a clear crystalline state. They are sensitive to changes in oxidation–reduction, osmolarity, excessively increased concentrations of metabolites, and UV irradiation. Structural integrity of the lens is maintained: for osmotic balance by the Na^+/K^+ exchanging ATPase (p. 491); for redox-state balance by glutathione reductase and for growth and maintenance by protein synthesis and other metabolic processes that take place mostly in cells on the periphery of the lens.

The primary role of most lens proteins is to function as crystallins, but many are expressed in other tissues and serve other roles such as enzymes and/or have other functional roles. **α- and β-Crystallins** are **small heat shock proteins (sHSP)** or **chaperones** that function to help maintain lens proteins in their native, unaggregated states. Their highest expression is in eye lens, but they also occur in other tissues such as skeletal and cardiac muscle where they are involved in filament assembly. Mutations in crystallins, therefore, not only predispose an individual to cataract formation, but also to possible muscle weakness and heart failure.

TABLE 23.4 • Eye Lens Crystallins and Their Relationships with Other Proteins

Crystallin	*Distribution*	*[Related] or Identical*
α	All vertebrates	Small heat shock proteins (αB) [*Schistosoma mansoni* antigen]
β	All vertebrates	[*Myxococcus xanthus* protein S]
γ	(embryonic γ not in birds)	[*Physarum polycephatum* spherulin 3a]
Taxon-specific enzyme crystallins		
δ	Most birds, reptiles	Argininosuccinate lyase (δ2)
ε	Crocodiles, some birds	Lactate dehydrogenase B
ξ	Guinea pig, camel, llama	NADPH: quinone oxidoreductase
η	Elephant shrew	Aldehyde dehydrogenase I

Source: Wistow, G. *Trends Biochem. Sci.* 18: 301, 1993.

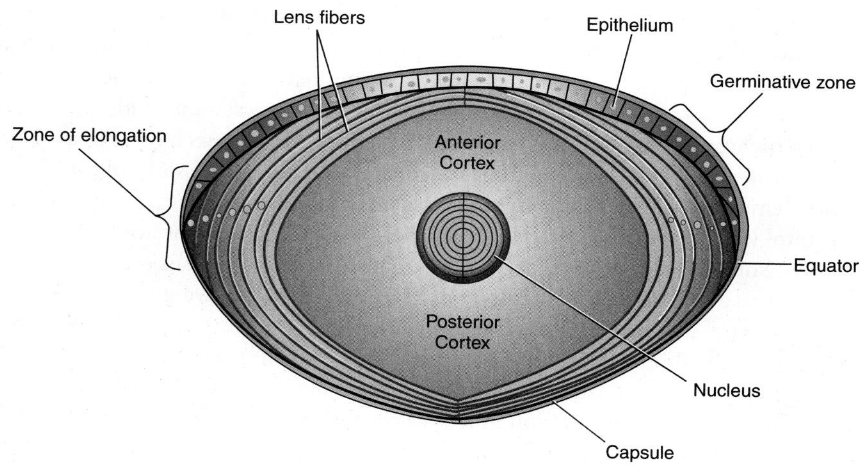

Figure 23.18 Schematic representation of a meridional section of a mammalian lens.

Energy for these processes comes from the metabolism of glucose. About 85% of glucose used by lens tissue is metabolized by glycolysis and about 3% by the TCA cycle, presumably by cells located at the periphery. Most of the remaining portion of glucose metabolized by lens goes through the pentose phosphate pathway.

The central area of the lens, the nucleus or core, consists of lens cells that were present at birth. The lens grows from the periphery (Figure 23.18) and in humans increases in weight and thickness with age and becomes less elastic. This leads to a loss of near vision (Table 23.5), a normal condition referred to as **presbyopia.** On average, the lens may increase threefold in size and approximately 12-fold in thickness from birth to about age 80.

Cataract, the only known disease of the lens, is an opacity of lenses brought about by many different conditions. The two most common types of cataracts are (1) **senile cataracts** in which changes in the architectural arrangement of the lens crystallins and other lens proteins are age related and due to such changes as breakdown of the protein molecules starting at the C-terminal ends, deamidation, and racemization of aspartyl residues and (2) **diabetic cataracts** which result from loss of control of osmolarity of the lens due to increased activity of **aldose reductase** and **polyol (aldose) dehydrogenase** of the polyol metabolic pathway. When glucose concentration in the lens is high, aldose reductase converts some of it to **sorbitol** (Figure 23.19) that may be converted to

TABLE 23.5 • Changes in Focal Distance with Age

Age (years)	*Focal Distance (in.)*
10	2.8
20	4.4
35	9.8
45	26.2
70	240.0

Source: Adapted from Koretz, J. F. and Handelman, G. H. *Sci. Am., 92,* July 1988.

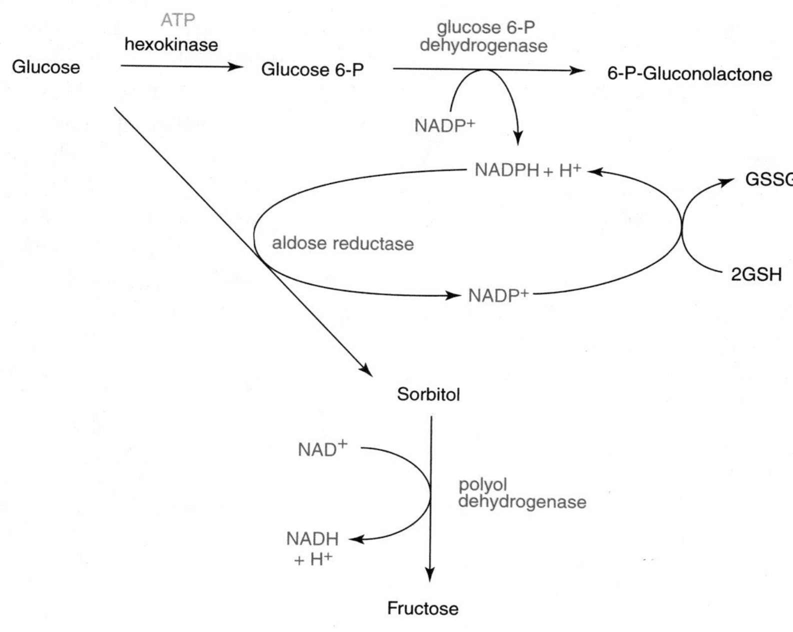

Figure 23.19 Metabolic interrelationships of lens metabolism.

fructose by polyol dehydrogenase. In human lens, the ratio of activities of these two enzymes favors sorbitol accumulation, especially; since sorbitol is not used by other pathways and it diffuses out of the lens rather slowly. Accumulation of sorbitol increases osmolarity of the lens, affects the structural organization of the crystallins, and enhances the rate of protein aggregation and denaturation. Areas where this occurs have increased light-scattering properties, which is the definition of cataracts. Normally, sorbitol formation is not a problem because the K_m of aldose reductase for glucose is about 200 mM and very little sorbitol would be formed. In diabetics, where circulating concentration of glucose is high, activity of this enzyme can be significant. Cataracts affect millions of people per year throughout the world, and there are no known cures or preventative measures, especially for the senile type. The most common remedy is lens replacement; a routine operation in many countries. A side effect of cataract and surgical treatment for it can be glaucoma; but this is rare. A third cause of cataracts, especially among young people, is due to inherited mutations in crystallins that function in lens as chaperones. When there are mutations in chaperones that interfere with their function, misfolded proteins can occur and result in cataract formation.

Retina Derives ATP from Anaerobic Glycolysis

The **retina**, like the lens, depends heavily on anaerobic glycolysis for ATP production. Unlike the lens, the retina is a vascular tissue. In its center is the macula, and in the center of the macula is the **fovea centralis**, an avascular concave area that contains only cones. This is the area of greatest visual acuity (Clin. Corr. 23.4). Mitochondria are present in retinal rods and cones but not in the outer segments where visual pigments are located.

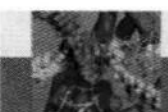

CLINICAL CORRELATION 23.4

Macula Degeneration and Loss of Vision

Many diseases of the eye affect vision, but not all have clear, direct biochemical origins. The most serious eye diseases are those that result in blindness. Glaucoma is the most common and is often associated with diabetes mellitus, the biochemistry of which is fairly well known. Glaucoma is treatable and blindness does not have to be a result.

Macula degeneration leads to blindness and there is no cure. The macula is a circular area of the retina, the center of which is the fovea centralis, the area of greatest visual acuity, which contains the highest number of cones. Age-related macula degeneration (AMD) may be among the leading causes of blindness of people over the age of 50 and is of two types: dry and wet. The dry form develops gradually over time, whereas the wet form develops rapidly and can lead to blindness within days. It occurs when blood vessels invade and/or rupture under the macula leading to a rapid loss of vision.

One current therapy under assessment for neovascular age-related macular degeneration is intravitreal ranibizumab. Ranibizumab is a humanized monoclonal antibody fragment that inhibits VEGF (vascular endothelial growth factor) (p. 909). In 2006, a Phase 3 trial showed that this treatment is somewhat effective. Another therapy involves use of siRNA to inhibit expression of the gene for VEGF. VEGF promotes excessive blood vesicle growth behind the retina. These blood vesicles leak and cause blindness. siRNA treatment has been quite effective in treating AMD. This is the first siRNA therapy to be given to patients in clinical trials. Thus, advances in basic science and medical research continue to reveal better ways of treating and possibly preventing onset of AMD.

Rupture of blood vessels that obscure macula details and result in rapid onset of blindness may be temporary. Several cases of sudden visual loss associated with sexual activity, but not with a sexually transmitted disease, have been reported. Vision was lost in one eye apparently during, but most often was reported a few days after engaging in, highly stimulatory sexual activity. Blindness was due to rupture of blood vessels in the macula area. Most patients were reluctant to discuss with their ophthalmologist what they were doing when sight loss was first observed. Four patients recovered with restoration of vision on reabsorption of blood. In one case, blood was trapped between the vitreous humor and the retinal surface directly in front of the fovea. The hemorrhage cleared only slightly during the next month, but visual acuity did not improve. That patient did not return for a follow-up examination, but there was no indication during the initial examination that the condition was permanent. Since most of the victims of this phenomenon were over the age of 39, it may be a worry more to professors than to students. It also may give a new meaning to the phrase, "love is blind."

http://www.molvis.org/molvis/v9/a31; Friberg, T. R., Braunstein, R. A., and Bressler, N. M. *Arch. Ophthalmol.* 113: 738, 1995; Reich, S. J., Fosnot, J., Kuroki, A., Tang W., et al. Small interfering RNA (siRNA) targeting effectively inhibits ocular neovascularization in a mouse model. *Molecular Vision* 9: 210, 2003; and Jager, R. D., Mieler, W. F., and Miller, J. W. Age-related macular degeneration: Review. *N. Engl. J. Med.* 358: 2606, 2008.

Visual Transduction Involves Photochemical, Biochemical, and Electrical Events

Figure 23.20*a* shows an electron micrograph and schematic of the retinal membrane. Light enters the eye, and when it reaches the retinal membrane, it passes the optic nerve fibers, the ganglion neurons, the bipolar neurons, and the nuclei of rods and cones before it reaches the outer segment of the rods and cones where the signal transduction process begins. The tips of the rods and cones penetrate the **pigmented epithelial** layer of the retina. The pigmented

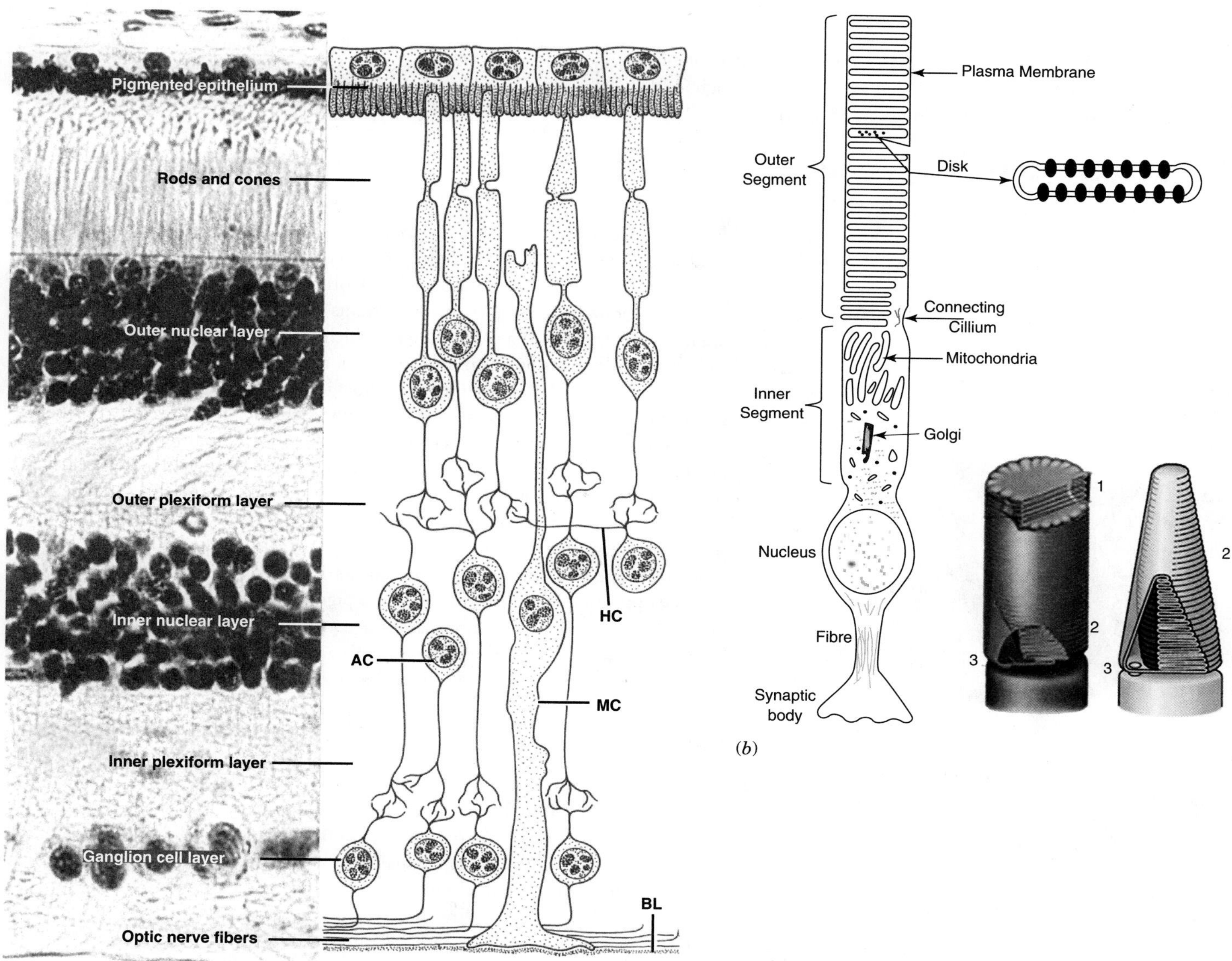

Figure 23.20 Layers of the human retinal membrane. (*a*) Electron micrograph and schematic representation of cells of the human retina. Tips of rods and cones are imbedded in the pigmented epithelium of the outermost layer. Rods and cones form synaptic junctions with many bipolar neurons that form synapses with cells in the ganglion layer. Cells in this layer send axons through the optic nerve to the brain. Synaptic interactions of rods and cones with many neurons are important for the integration of information. Abbreviations: HC, horizontal cells; AC, amacrine cell; MC, Muller cell; and BL, basal lamina. (*b*) More detailed structural features of rods and cones showing the relationships of disks to plasma membrane. Dots on the disk in the rod and in the enlarged diagram represent rhodopsin (Figure 23.21) that is imbedded in the disk membrane. The same relationship holds for cones and color pigments. Numbers next to the blue (rod) and tan (cone) inserts indicate structural areas of disk arrangements. Number 1 identifies disk in rods that are free floating and not directly attached to the plasma membrane, number 2 shows folding of outer cell membrane that is continuous with disk in both rods and cones, and number 3 is the connecting cilium of both rods and cones. Reproduced with permission from R. G. Kessel and R. H. Kardon. *Tissues and Organs: A Text-Atlas of Scanning Electron Microscopy.* New York: W. H. Freeman and Company, p. 87, 1979. Copyright (1979) R. G. Kessel and R. H. Kardon.
Redrawn and modified from Young, R. *Sci. Am.* 223: 81, 1970.

CLINICAL CORRELATION 23.5

Niemann–Pick Disease and Retinitis Pigmentosa

There are central nervous system disorders associated with the Niemann–Pick group of diseases that can become evident by ocular changes. Some of these are observed as abnormal macula with gray discoloration and granular pigmentation or granule opacities about the fovea.

Acute type I Niemann–Pick disease (OMIM 257220), lipidosis with sphingomyelinase deficiency and primary sphingomyelin storage, may show a cherry red spot in the retina in up to 50% of these patients. Macula halo describes the crystalloid opacities seen in some patients with subacute type I disease. These halos are approximately one-half the disk diameter at their outer edge and are scattered throughout the various layers of the retina. They do not interfere with vision.

An 11-year-old girl with type II disease had more extensive ocular involvement. There was sphingomyelin storage in the keratocytes of the cornea, the lens, the retinal ganglion cells, the pigmented epithelium, the corneal tract, and the fibrous astrocytes of the optic nerve.

Thus, retinitis pigmentosa may also be a secondary effect of the abnormal biochemistry associated with Niemann–Pick disease.

Spence, M. W., and Callahan, J. W. Sphingomyelin-cholesterol lipidoses: The Niemann–Pick group of diseases. In Schriver, C. R., Beaudet, A. L., Sly, W., and Valle, D. (Eds.), *The Metabolic Bases of Inherited Disease.* New York: McGraw Hill, 1989, 1656.

epithelial layer of the retina is involved in the *trans*- to *cis*-retinal recycling phase of the visual cycle, and it also absorbs excess light and prevents reflections back into rods and cones where it may cause distortion or blurring of images (Clin. Corr. 23.5). The choroid lies behind the retina and contains blood vesicles that supply nutrients to the retina.

The eye may be compared to a video camera, which collects images, converts them into electrical pulses, records them on magnetic tape, and allows their visualization by decoding the taped information. The eye focuses on an image by projecting that image onto the retina. A series of events begins, the first of which is photochemical, followed by biochemical events that amplify the signal, and finally electrical impulses are sent to the brain where the image is reconstructed—"in the mind's eye." In effecting this process, the initial event is transformed from a physical event to a chemical event through a series of biochemical reactions, to an electrical event, to a conscious acknowledgment of the presence of an object in the environment outside of the body.

Photons (light) are absorbed by photoreceptors in the **outer segments** of **rods** or **cones**, where they cause isomerization of the visual pigment, **retinal**, from the 11-*cis*-form to the all-*trans*-form. This isomerization causes a conformational change in the protein moiety of the complex and affects the resting membrane potential of the cell, resulting in an electrical signal being transmitted by way of the optic nerve to the brain.

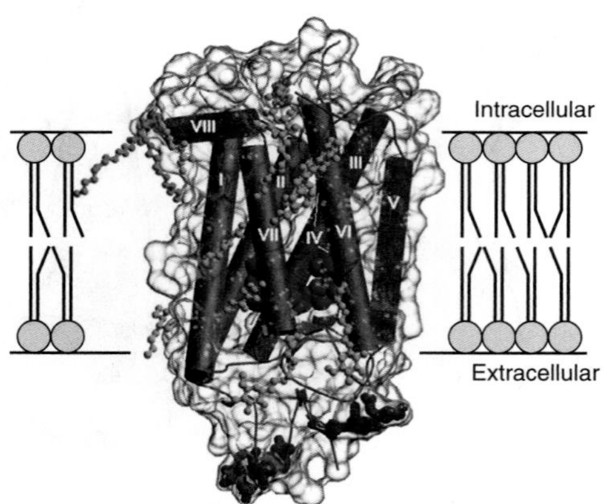

Figure 23.21 Crystal structure of bovine rhodopsin at 2.8-Å resolution. Rhodopsin is a transmembrane protein. The width of the disk membrane into which it is imbedded is approximately equivalent to the length of its helices (blue rods). The intracellular side of the membrane approximately transects helix VIII. β-Strands are shown as blue arrows. Structures in green on the intracellular side are two palmitoyl groups oriented such that the hydrophobic groups can interact with hydrophobic regions of the membrane. The blue ball-and-stick structures at the bottom (extracellular side) of the molecule are carbohydrates. The yellow structures located near the hydrophobic surface of the protein are nonylglucoside and heptanol molecules. Reproduced with permission from Teller, D. C., Okada, T., Behnke, C. A., Palczewski, K., and Stenkamp, R. E. *Biochem.* 40: 7761, 2001. Copyright, (2001) American Chemical Society. Figure generously supplied by Dr. K. Palczewski and Dr. C. A. Behnke.

Rods and Cones Are Photoreceptor Cells

Photoreceptor cells of the eye are the **rods** and the **cones** (Figure 23.20*b*, p. 957). Each type has flattened disks that contain a photoreceptor pigment that consists of a protein and a prosthetic group, 11-*cis*-retinal. The pigment is **rhodopsin** in rod cells, and they are red (long), green (medium), or blue (short) pigments in cone cells. Rhodopsin, as are other photoreceptor pigments, is a transmembrane protein containing **11-*cis*-retinal**, the protein moiety of which is called **opsin**. The three proteins that form the red, green, and blue pigments of cone cells are different from each other and from the opsin of rhodopsin; but the prosthetic group for all visual pigments is the same.

Rhodopsin ($\sim$40 kDa) contains seven transmembrane α-helices. An 11-*cis*-retinal molecule is attached through a protonated Schiff base to the ε-amino group of Lys^{296} on the seventh helix and lies about midway between the two faces of the membrane (Figure 23.21; Clin. Corr. 23.6, p. 960).

A schematic of the formation of 11-*cis*-retinal from **β-carotene** and rhodopsin formation from opsin and 11-*cis*-retinal are shown in Figure 23.22. 11-*Cis*-retinal is derived from **vitamin A** and/or β-carotene of the diet. Cleavage of β-carotene yields two molecules of **all-*trans*-retinol**, which is isomerized by an enzyme in the pigmented epithelial cell layer of the retina to **11-*cis*-retinol**. Oxidation of 11-*cis*-retinol to 11-*cis*-retinal and its binding to opsin occur in the rod outer segment.

$\frac{1}{2}$ β-Carotene

O_2 β-Carotene-15, 15'-dioxygenase

2NADH (NADPH) + $2H^+$
Retinal reductase
$2NAD^+$ ($NADP^+$)

CH_2OH

All-*trans*-retinol (Vitamin A_1)

Δ^{11}-*cis*-Retinol

(a)

Δ^{11}-*cis*-Retinal

CHO 15

+

Lysine side chain

$H_2N-(CH_2)_4-$opsin $\overset{H^+}{\rightleftharpoons}$ $R-\underset{15}{\overset{H}{C}}=\underset{H}{N}-(CH_2)_4-$opsin + H_2O

cis-Retinal

Rhodopsin (Protonated Schiff base)

(b)

Figure 23.22 Formation of 11-*cis*-retinal and rhodopsin from β-carotene.

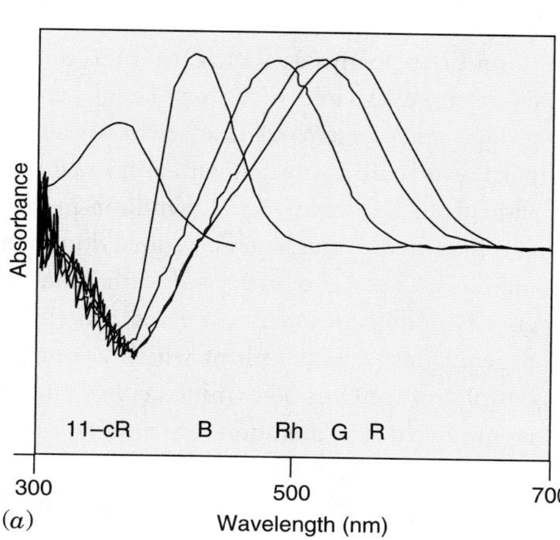

Blue

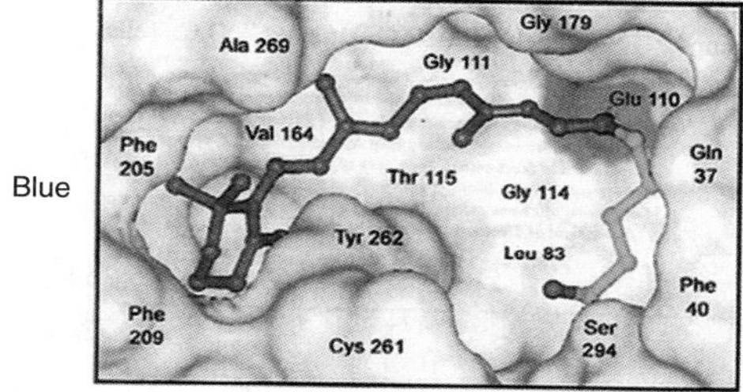

Green

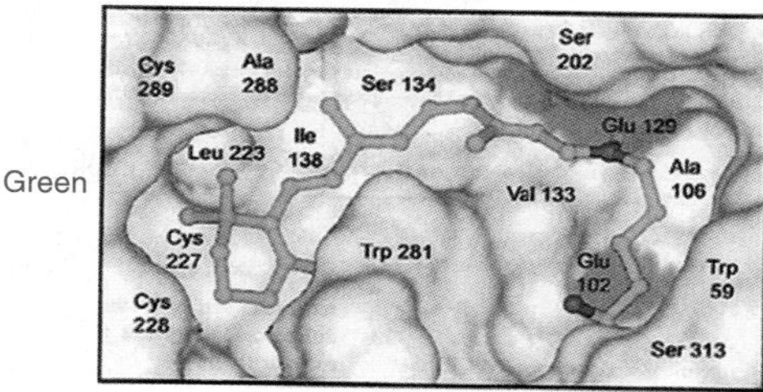

Red

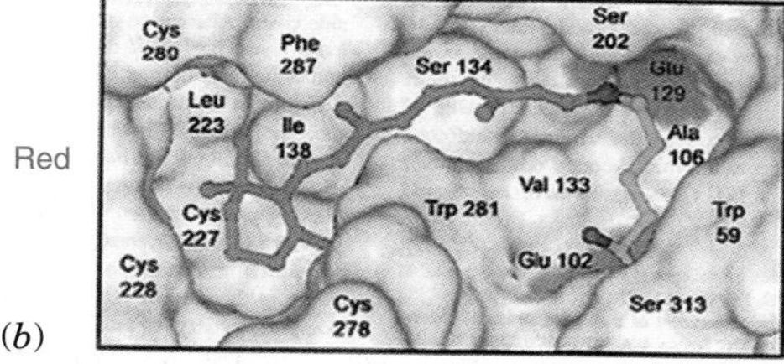

(*b*)

Figure 23.23 Absorption spectra of 11-*cis*-retinal and the four visual pigments.
(*a*) Absorbance is relative and represents difference spectra obtained by subtracting the spectra for the recombinant apoproteins. The spectrum for 11-*cis*-retinal (11cR) is in the absence of protein. Other abbreviations: B, blue pigment; Rh, rhodopsin; G, green; and R, red. (*b*) Some of the amino acids in the vicinity of 11-*cis*-retinal in the blue, green, and red pigments.
Reproduced with permission from Stenkamp, R. E., Filipek, S., Driessen, C. A. G. G., Teller, D. C., and Palczewski, K. Crystal structure of rhodopsin: a template for cone visual pigments and other G protein-coupled receptors. *Biochim. Biophys. Acta* 1565: 168, 2002. Copyright (2002) Elsevier.
Modified from Nathans, J. *Cell* 78: 357, 1994.

The absorption spectra of 11-*cis*-retinal and the four visual pigments are shown in Figure 23.23*a*. The wavelength of maximum absorption of 11-*cis*-retinal is different when it is bound to opsin (λ_{max} = 498 nm) or to the protein moieties of the other visual pigments. Spectral differences are due to subtle differences in the chemical environments in which the 11-*cis*-retinal resides (Figure 23.23*b*). Absorption bands for the pigments reflect spectral areas of their light sensitivity.

CLINICAL CORRELATION 23.6

Retinitis Pigmentosa from a Mutation in the Gene for Peripherin

Retinitis pigmentosa (RP) (OMIM 268000) is a slowly progressive condition associated with loss of night and peripheral vision. It is a group of heterogeneous diseases of variable clinical and genetic origins; several are associated with abnormal lipid metabolism. This disease affects approximately 1.5 million people throughout the world. It is inherited through an autosomal dominant, recessive, or X-linked mode. RP has been associated with mutations in the protein moiety of rhodopsin and in a related protein, peripherin/RDS (retinal degeneration slow), both of which are integral membrane proteins. Peripherin contains 344 amino acid residues and is located in the rim region of the disc membrane. Structural models of these two proteins are shown in the following figure. Filled circles and other notations in the figure mark residues or regions that have been correlated with RP or other retinal degenerations.

A *de novo* mutation in exon 1 of the gene coding for peripherin resulted in the onset of RP. It caused a C-to-T transition in the first nucleotide of codon 46. This resulted in changing an arginine to a stop codon (R46X). The pedigree of this family is shown in the adjacent figure. Neither parent had the mutation and genetic typing analysis (20 different short tandem repeat polymorphisms) showed that the probability that the proband's parents are not his actual biological parents is less than one in 10 billion, establishing with near certainty that the mutation is *de novo.*

This R46X mutation has been observed in another unrelated patient. This demonstrates the importance of DNA analysis to establish the genetic basis for RP, as opposed to that associated with other abnormal metabolic conditions.

Shastry, B. S. Retinitis pigmentosa and related disorders: phenotypes of rhodopsin and peripherin/RDS mutations. *Am. J. Med. Genetics* 52: 467, 1994; and Lam, B. L., Vandenburgh, K., Sheffield, V. C., and Stone, E. M. Retinitis pigmentosa associated with a dominant mutation in codon 46 of the peripherin/RDS gene (Arg[46]stop). *Am. J Ophthalmol.* 119: 65, 1995.

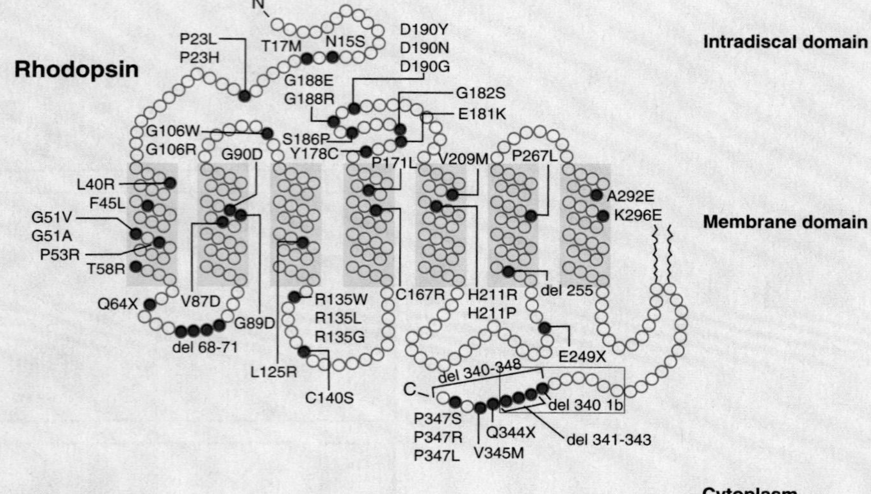

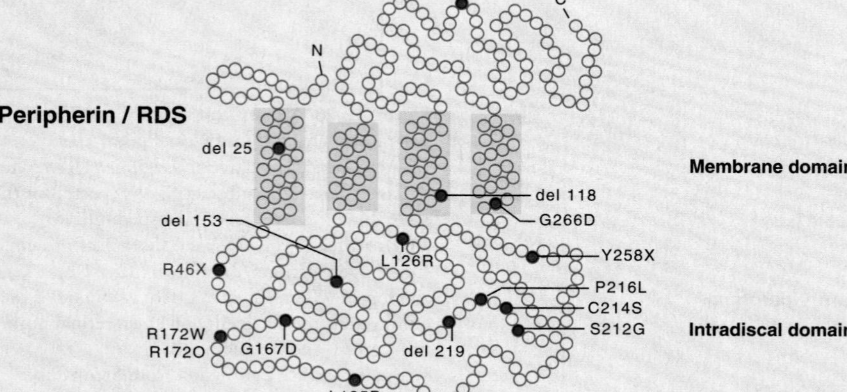

Schematic representation of structural models for rhodopsin (top) and peripherin/RDS (Retinal Degeneration Slow) (bottom). The location of mutations in amino acid residues that segregate with RP or other retinal degenerations are shown as solid circles. From Lam et al. (in Clin. Corr. 23.6 references).

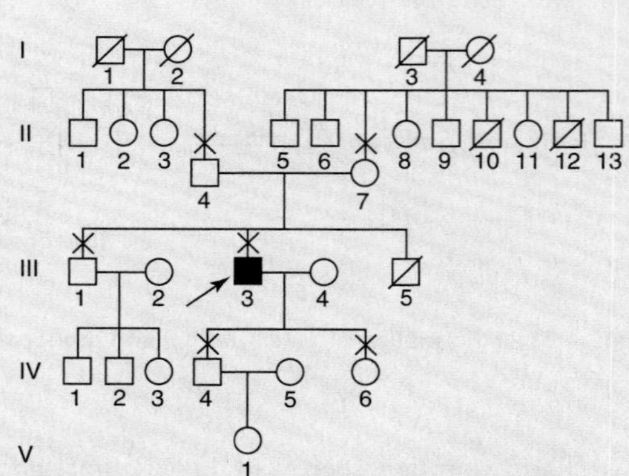

Pedigree of family. Males are squares; females are circles. Solid square indicates the proband. A slashed-through symbol indicates deceased. From Lam et al. (in Clin. Corr. 23.6 references).

The magnitude of change in the electrical potential of photoreceptor cells following exposure to a light pulse differs from that of neurons during depolarization. The **resting potential** of rod cell membrane is approximately −30 mV compared to −70 mV for neurons. Excitation of rod cells by a light pulse causes **hyperpolarization** of the membrane; from about -30 mV to about -35 mV (Figure 23.24). It takes hundreds of milliseconds for the potential to reach its maximum state of hyperpolarization, during which time a number of biochemical events occur.

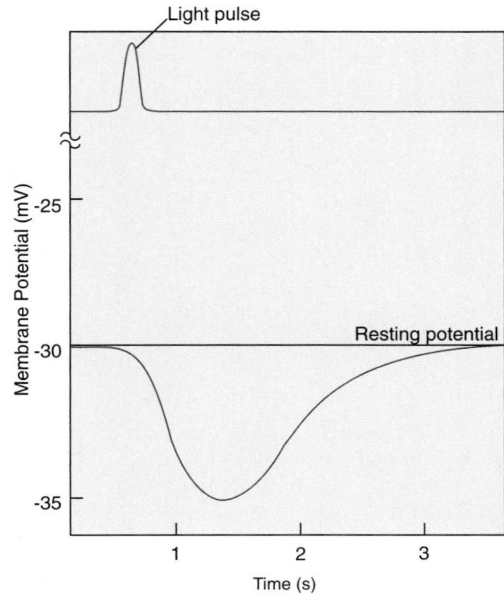

Figure 23.24 Changes in the potential of a rod cell membrane after a light pulse.
Redrawn from Darnell, J., Lodish, H., and Baltimore, D. *Molecular Cell Biology*. New York: Scientific American, 1986, 763.

Absorption of photons of light and isomerization of 11-*cis*-retinal in rhodopsin, are rapid, requiring only picoseconds. Following this, rhodopsin undergoes a series of conformation changes to accommodate the structural change of the 11-*cis*- to all-*trans*-retinal (Figure 23.25). One of these species, **metarhodopsin II**, has a half-life of approximately 1 min and is *active rhodopsin* (R*) that is involved in the more important biochemical phase of the visual cycle (A Closer Look 23.3, p. 964).

Structural changes are all kinetic events; that is, one does not have to be completed before the other begins, and metarhodopsin II will have begun to form within hundredths of microseconds of the initial event. Finally, dissociation of metarhodopsin to opsin and all-*trans*-retinal occurs. All-*trans*-retinal is enzymatically converted to all-*trans*-retinol by **all-*trans*-retinol dehydrogenase** in the rod outer segment. All-*trans*-retinol is transported into the pigmented epithelium where a specific isomerase converts it to 11-*cis*-retinol that is then oxidized to the aldehyde and transported back into the rod outer segment. After the aldehyde is transported back into the rod outer segment, it can recombine with opsin to form rhodopsin and the cycle can begin again. A diagrammatic representation of these events is shown in Figure 23.26 (see also Clin. Corr. 23.7, p. 965). Similar events occur in the cones with the three proteins of the red, green, and blue pigments.

There are three interconnected biochemical cycles of events that occur in the conversion of light energy to nerve impulses (Figure 23.27). These cycles describe the reactions of rhodopsin, **transducin**, and **phosphodiesterase**, respectively. The net result is a hyperpolarization of the plasma membrane of the rod (or cone) cells from −30 mV to approximately −35 mV. Maintenance of a steady-state potential of the plasma membrane of rod and cone cells at −30 mV is an important function of this series of reactions.

Rod cells of a fully dark-adapted human eye can detect a flash of light that emits as few as five photons. The rod is a specialized type of neuron in that the signal generated does not depend on an all-or-none event. The signal may be graded in intensity and reflect the extent that the mV potential changes from its steady-state value of −30 mV as well as the number of photoreceptor cells stimulated. The **steady-state potential** of −30mV is maintained at this more positive value because **Na^+ channels** of the photoreceptor cells are **ligand-gated**, and a fraction of them responsible for the influx of Na^+ are maintained in an open state. The ligand responsible for keeping some of the Na^+ channels open is **cyclic GMP (cGMP)**, which binds to them in a concentration-dependent, kinetically dynamic manner. Biochemical events that affect the concentration of cGMP within rod and cone cells also affect the number of Na^+ channels open, the concentration of Na^+ within the cells, and hence, the membrane potential (Figure 23.27).

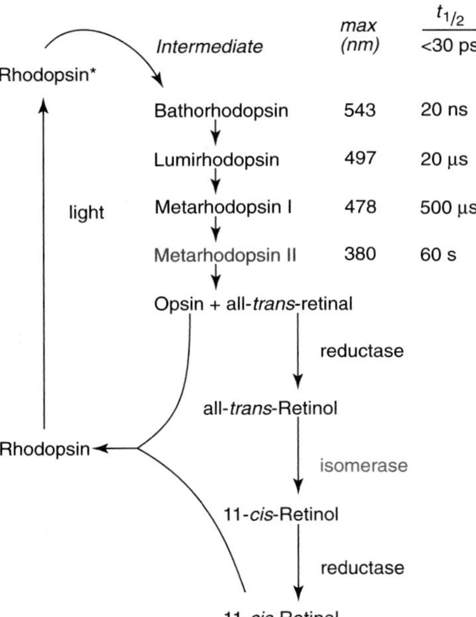

Figure 23.25 Conformation changes that rhodopsin undergoes after photoactivation that leads to Metarhodopsin II, active rhodopsin.

Active rhodopsin (R*, namely, **metarhodopsin II**) forms a complex with transducin. **Transducin** is a classic trimeric **G protein** and functions in a manner very similar to that described on page 522. In the R*-transducin complex (R*-$T_{\alpha,\beta,\gamma}$), transducin undergoes a conformational change that facilitates exchange of its bound GDP with GTP. When this occurs, the α-subunit (T_α) dissociates from the β-,γ-subunits. T_α activates **phosphodiesterase (PDE)**, which hydrolyzes cGMP to 5′-GMP. This decreased concentration of cGMP decreases the amount of cGMP available to bind to Na^+ channels and, therefore, decreases the number of open Na^+ channels. The influx of Na^+decreases, but the efflux of Na^+ is not affected by these events and Na^+ concentration inside the cells decreases causing the membrane potential to become more negative, that is, hyperpolarized.

The diagram of Figure 23.27 shows in cartoon form two such channels in the plasma membrane, one of which has cGMP bound to it and is open. The other does not have cGMP bound to it and is closed. By this mechanism, the concentration of Na^+ in the cell is directly linked to the concentration of cGMP and, thus, also the membrane potential.

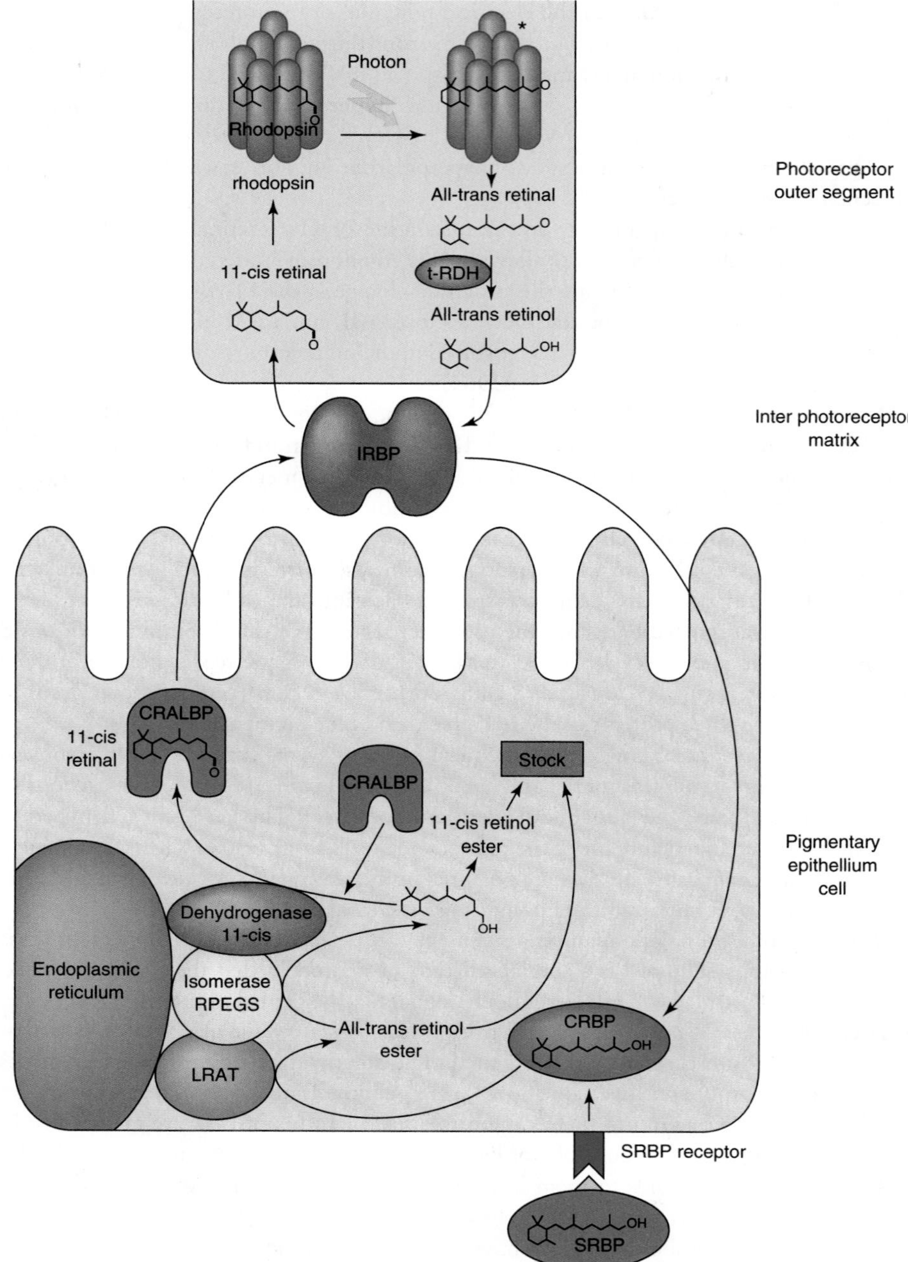

Figure 23.26 Transport and metabolism of 11-*cis*- and all-trans-retinal into the pigmented epithelium.
Redrawn from Perrault, I., Rozet, J.-M., Gerber, S. , Ghazi, I., Leowski, C., et al. Leber congenital amaurosis. *Mol. Gen. & Metab.* 68: 200, 1999.

PDE in rod cells is a **heterotetrameric protein** consisting of one each α and β catalytic subunits and two γ regulatory subunits. T_α-GTP forms a complex with the γ-subunits of PDE effecting dissociation of the catalytically active α,β-dimeric PDE subunit complex. T_α-GTP has GTPase activity that hydrolyzes bound GTP to GDP and inorganic phosphate (P_i). This results in dissociation of T_α-GDP from the regulatory γ-subunits of PDE permitting them to reassociate with the catalytic subunits to reestablish the α,β,γ_2 structure of PDE—its inactive form. The same reactions occur in cone cells, but the catalytic subunits of cone cells PDE differ from those of rods in that cone cells PDE consists of two α catalytic subunits (α_2) instead of an α and a β catalytic subunit ($\alpha\beta$).

cGMP concentration is regulated by intracellular Ca^{2+}. Ca^{2+} enters rod cells in the dark through sodium channels increasing its concentration to the 500-nM range. At

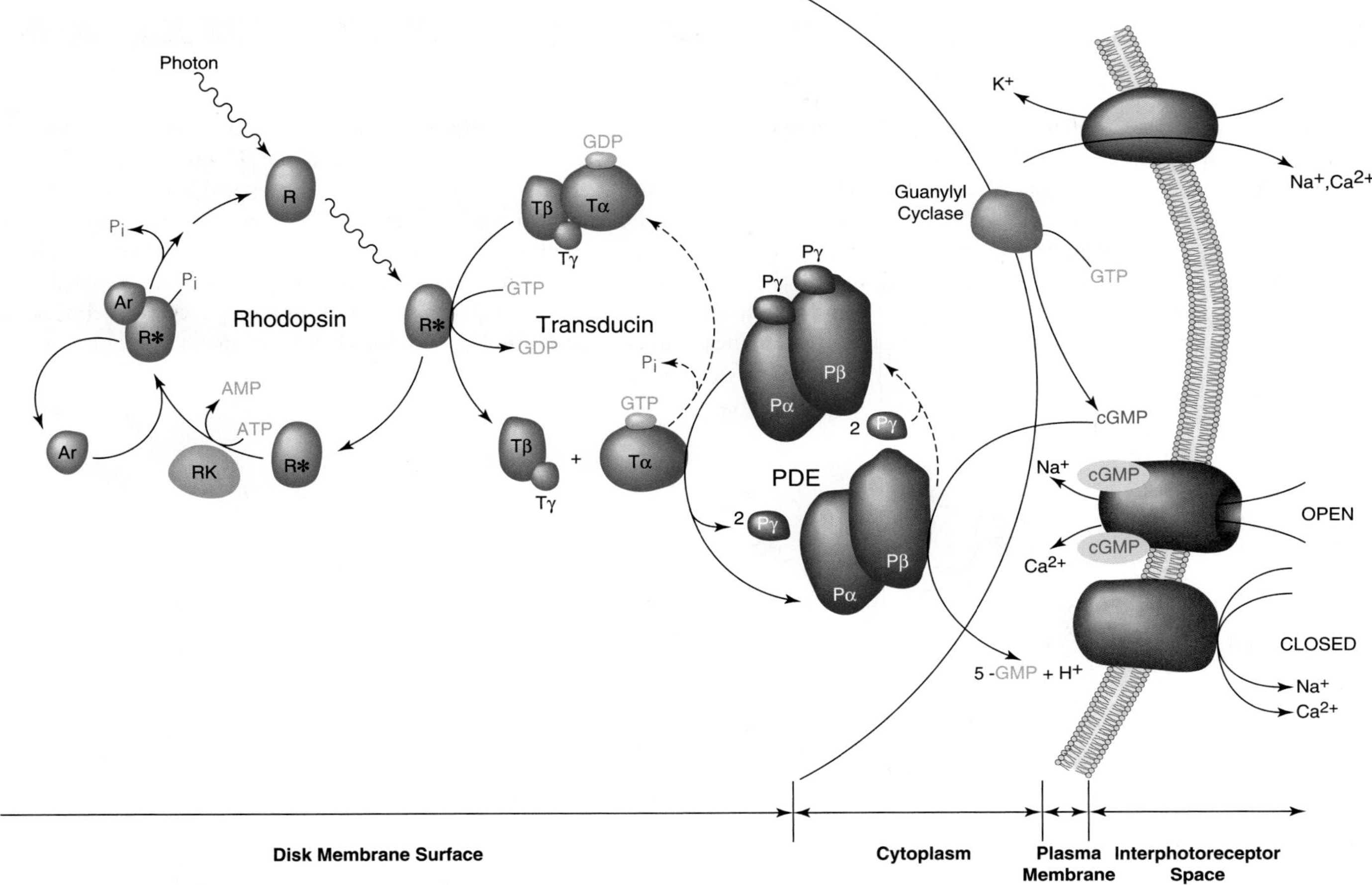

Figure 23.27 Cascade of biochemical reactions involved in the visual cycle.
Redrawn from Farber, D. B. *Investigative Ophthalmology & Vis. Sci.* 36: 263, 1995.

these concentrations, activity of **guanylate cyclase** is low. When sodium channels are closed, Ca^{2+} entry is inhibited also, but efflux of Ca^{2+} by the sodium/calcium–potassium exchanger is unchanged (top complex of the plasma membrane in Figure 23.27). This decreases the intracellular Ca^{2+} concentration resulting in activation of guanylate cyclase and increased production of cGMP from GTP. Resynthesis of cGMP and the hydrolysis of GTP of the T_α-GTP complex play important roles in stopping reactions of the visual cycle.

Inactivation of activated rhodopsin R* is also important in stopping this cascade of events. Activated rhodopsin R* is phosphorylated by an ATP-dependent **rhodopsin kinase** (Figure 23.27). The R*-P_i has high binding affinity for the cytosolic protein, **arrestin**. The arrestin-R*-P_i complex is no longer capable of interacting with transducin. The kinetics of binding of arrestin to the activated-phosphorylated rhodopsin is sufficiently rapid *in vivo* to stop the cascade of reactions. When rhodopsin is regenerated, the cycle can be initiated again by photons of light.

Mutations can occur in any of the major proteins involved in the visual cycle, some of which are listed in Table 23.6, and result in visual impairment (Clin. Corr. 23.7).

Color Vision Originates in Cones

Even though photographic artists, such as the late Ansel Adams, make the world look beautiful in black and white, the intervention of colors in the spectrum of life's pictures brings another degree of beauty to the wonders of nature and the beauty of life, even the ability to make a distinction between tissues from histological staining. The ability of humans to distinguish colors resides within a relatively small portion of the visual system, the cones. The number of cones within the human eye (6 to 7 million) is few compared with the number

A CLOSER LOOK 23.3

Conformation Changes during Formation of "Active Rhodopsin"

The conformation changes leading to formation of "active rhodopsin" are stepwise events with some of the species being very short lived as evident by the data shown in Figure 23.25, p. 961. The exact structures for these intermediates are not known, but it is clear that they result from changes in the neighboring group environment of retinal after its isomerization. Panel *a* of the following figure is a schematic from the 3D structure of rhodopsin showing the amino acid side chains surrounding 11-*cis*-retinal as viewed from the cytoplasmic side. When 11-*cis*-retinal undergoes isomerization to the all-*trans*-retinal at the position within the red circle of panel *b*, the β-ionone ring of all-*trans*-retinal reaches Ala[169] in helix IV. Ala[169] is not within the range of interactions shown in panel *a*. The intermediate species listed in Figure 23.25 represent some of the conformations rhodopsin undergoes before it finally dissociates to opsin and all-*trans*-retinal. At 37°C, activated rhodopsin decays in slightly more than 1 ms through several intermediates to **metarhodopsin II**, which has a half-life of approximately 1 min.

(*a*)

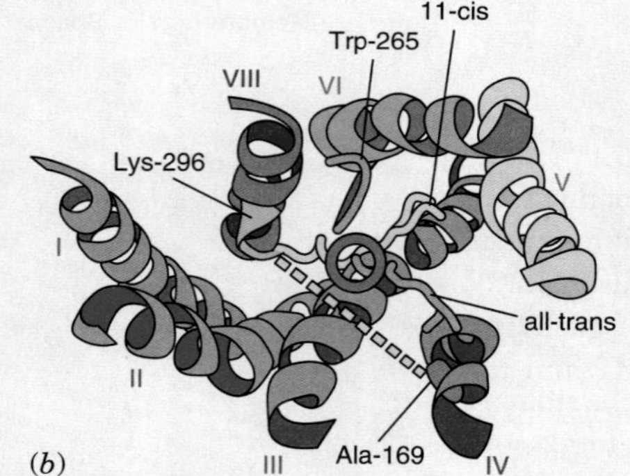

(*b*)

Schematics of 11-*cis*-retinal and all-*trans*-retinal in their protein environments. (*a*) Amino acid side chains surrounding 11-*cis*-retinal. When 11-*cis*-retinal isomerizes to the all-*trans* form (*b*), the β-ionone ring of all-*trans*-retinal interacts with Ala[169] of helix IV.
(*a*) From Palczewski, K., Kumasaka, T., Hori, T., Behnke, C. A., et al. *Science* 289:739, 2000. (*b*) From Bourne, H. R., and Meng, E. C. *Science* 289:733, 2000. Reprinted with permission from AAAS.

TABLE 23.6 • Major Proteins Involved in the Phototransduction Cascade

Protein	*Relation to Membrane*	*Molecular Mass (kDa)*	*Concentration in Cytoplasm (μM)*
Rhodopsin	Intrinsic	39	—
Transducin ($\alpha + \beta + \gamma$)	Peripheral or soluble	80	500
Phosphodiesterase	Peripheral	200	150
Rhodopsin kinase	Soluble	65	5
Arrestin	Soluble	48	500
Guanylate cyclase	Attached to cytoskeleton	112	?
cGMP-activated channel	Intrinsic	66	?

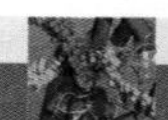

CLINICAL CORRELATION 23.7

Leber Congenital Amaurosis: Retinal Dystrophy Leading to Blindness

Leber congenital amaurosis (LCA) is one of the more serious conditions leading to congenital blindness. Leber first described LCA in 1869. Clinical heterogeneity in LCA was recognized early but essentially ignored. Genetic heterogeneity has been accepted since 1963 and three genes have been shown to account for about 27 % of the dystrophy. (1) retGC1 (aka GUCY2D) on chromosome 17p13 codes for the photoreceptor specific guanylate cyclase. (2) CRX on chromosome 19q13.3 codes for a homeobox protein essential for photoreceptor maintenance and outer rod/cone segment biogenesis. RPE65 on chromosome 1p31 codes for a specific retinal pigment epithelium (RPE) protein involved in vitamin A metabolism, specifically, retinol *trans/cis* isomerase. Structural representation of the human RPE65 gene and positions of mutations that have been identified in LCA is shown.

A canine model of LCA has been treated by gene replacement therapy in which a normal RPE65 gene was delivered intraocularly to one eye of each test animal and vision was restored in that eye to near normal. The function of this enzyme in vision is evident from Figures 23.25 and 23.26. Gene therapy to restore vision of humans affected by this mutation is now in progress.

LCA has been linked to mutations in the gene encoding another protein, the photoreceptor-specific gene Aryl hydrocarbon interacting protein like AIPL 1 . The protein, AIPL 1, enhances protein farnesylation. Retinal proteins known to be farnesylated are cGMP phosphodiesterase, transducin, and rhodopsin kinase. Mutationally defective AIPL 1 affects cGMP phosphodiesterase localization and amount in rods and cones prior to their degeneration and loss of vision.

Perrault, I., Rozet, J.-M., Gerber, S., Ghazi, I., Leowski, C., et al. Leber congenital amaurosis. *Molecular Genetics and Metabo.* 68: 200, 1999; Acland, G. M., Aguirre, G. D., Ray, J., Zhang, Q., et al. J. Gene therapy restores vision in a canine model of childhood blindness. *Nature Genetics* 28: 92, 2001; Maguire, A. M. , Simonelli, F., Pierce, E. A., Pugh, E. N., Jr., et al. Safety and efficacy of gene transfer for Leber congenital amaurosis. *N. Engl. J. Med.* 358: 2240, 2008; and Ramamurthy. V., Niemi, G. A., Reh, T. A., and Hurley, J. B. Leber congenital amaurosis linked to AIPL1: A mouse model reveals destabilization of cGMP phosphodiesterase. *Proc. Natl. Acad. Sci. USA* 101: 13897, 2004.

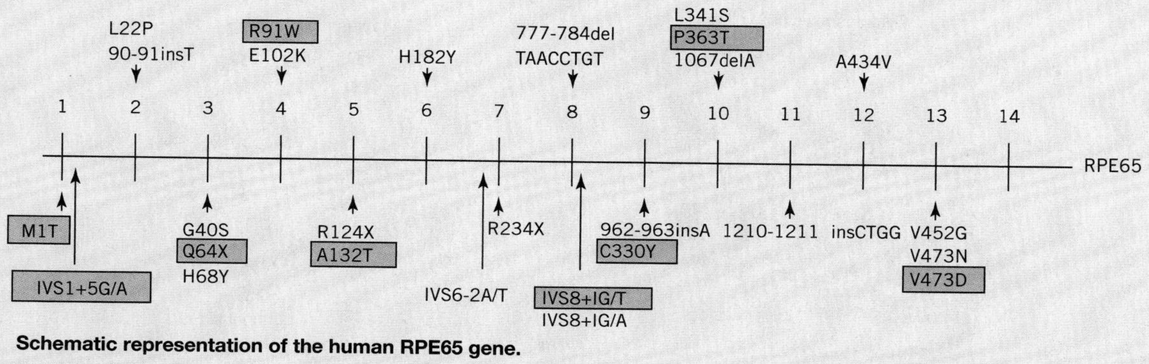

Schematic representation of the human RPE65 gene.

of rods (120 million). Some animals (e.g., dogs) have even fewer cones, and other animals (e.g., birds) have many more.

The mechanism by which light stimulates cone cells is the same as it is for rod cells. The initial event in all cases is the photo-induced isomerization of 11-*cis*-retinal. There are three types of cone cells, defined by whether they contain the blue, green, or red visual pigments (see Figure 23.23*b*, p. 959). They are also referred to in medical literature as short (S), medium (M), or long (L) wavelength-specific opsins, respectively. Color discrimination by cone cells is an inherent property of the proteins of the visual pigments to which the 11-*cis*-retinal is attached. The 11-*cis*-retinal is attached to each of the proteins through a protonated Schiff base as in rhodopsin. The absorption spectrum produced by the conjugated double-bond system of 11-*cis*-retinal is influenced by its chemical environment (see Figure 23.23*b*). When 11-*cis*-retinal is bound to different visual proteins, amino acid residues in the local areas around the protonated base and the conjugated π-bond system influence the energy level and give different absorption spectra with absorption maxima that are characteristic of the different color pigments: approximately 420 nm for blue, 535 nm for green, and 565 nm for red (A Closer Look 23.4).

Normally, only one type of visual pigment occurs per cell. During fetal development, however, multiple pigments may occur in the same cell. During development the blue pigment seems to appear first, followed by green and then red pigments. The number of cones containing these multiple pigments decreases after birth and is

A CLOSER LOOK 23.4

Sequence Homology of Visual Pigments

Significant homology exists among the amino acid sequences of the visual pigments (figure follows). Open circles represent amino acids that are the same, and closed circles those that are different. A string of closed circles at either end represents an extension of the chain of one protein relative to the other. The red and the green pigments show the greatest degree of homology, about 96% identity, whereas the degree of homology between different pairs of the others is between 40% and 45%.

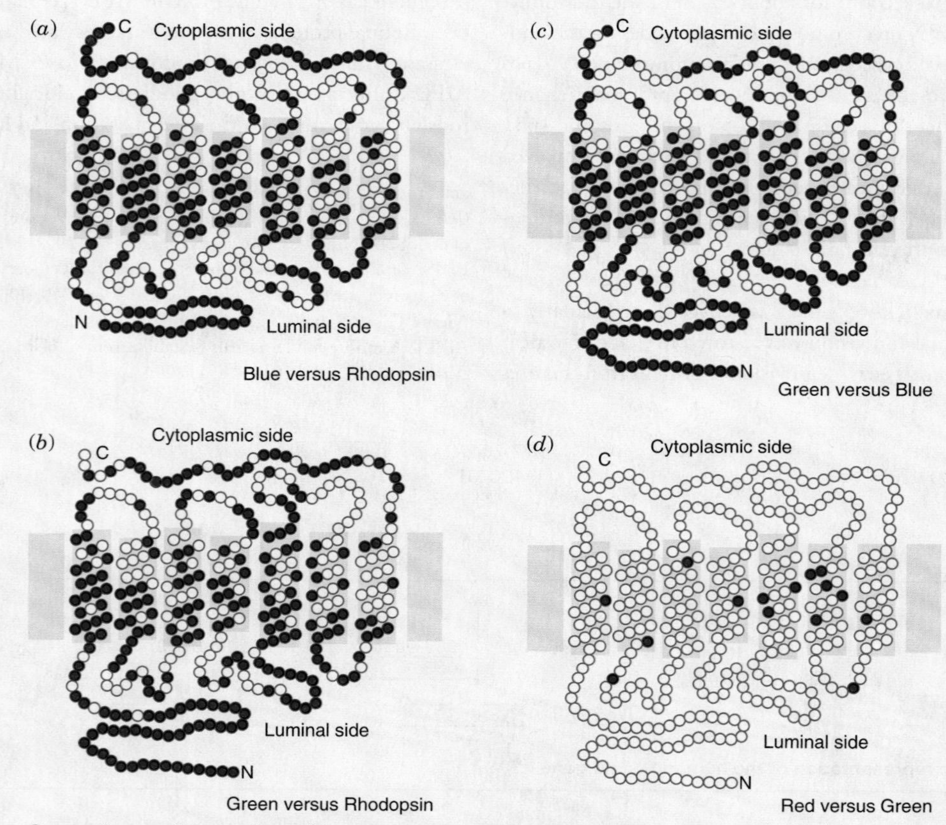

Comparisons of the amino acid sequences of the human visual pigments. Each filled dot indicates an amino acid difference.
Adapted from Nathans, J. *Annu. Rev. Neurosci.* 10:163, 1987.

insignificantly small or nonexistent in adults. Cones containing the blue pigment constitutes only about 2% of the total number of cones, and they are outside of the fovea. Cones containing the green and red pigments are approximately 32% and 64% of the total, respectively.

Color Vision Is Trichromatic

Genes encoding the visual pigments have been mapped to specific chromosomes. The rhodopsin gene resides on **chromosome 3**, the gene encoding the blue pigment resides on **chromosome 7**, and genes for the red and green pigments are on the **X chromosome**. In spite of the great similarity among the red and green pigments, they are distinctly different proteins. Individuals have been identified with inherited variations that affect one but not both pigments simultaneously. There may be more than one gene for the green pigment, but it appears that only one is expressed. Colors other than those reflected by the absorption maxima of the visual pigments are distinguished by graded stimulation of the different cones and comparative analysis by the brain (A Closer Look 23.5 and 23.6).

A CLOSER LOOK 23.5

Diagnosis of John Dalton's Color Blindness

John Dalton (1766–1844), who developed the atomic theory of chemistry, was color blind. He first described his visual problem in a lecture in 1794 and published his observation in 1798. He thought his color blindness was due to the vitreous humor being tinted blue, selectively absorbing longer wavelengths of light. He instructed that after his death his eyes be examined to determine whether his theory was correct. An autopsy revealed that the vitreous humor was *perfectly pellucid*, or normal. Using DNA analysis on Dalton's preserved eyes, obtained from the British Museum, it has now been shown that Dalton was missing the green pigment. Thus, instead of having trichromatic vision, he was dichromatic with a vision type known as **deuteranopia** or "Daltonism" since his publication was the first recognized account of red-green color-blindness. The type color blindness of one who is missing the red pigment is called **protanopia**.

23.3 • MOLECULAR MOTORS AND ASSOCIATED PROTEINS

There are three families of molecular motors: myosin, kinesin, and dynein. Myosin is an actin-based motor, and the other two are microtubule-based motors. Actin-based motors include those involved in muscle contraction, and the latter two include those involved in intracellular trafficking—moving cargo or arranging intracellular structures. There are *unconventional myosins* (actin-based motors) that also move cargo. Motors involved in trafficking function as single units and do not form fibers as those involved in muscle contraction do. This includes also cargo-carrying actin-based myosins. Fuel for all of these motors is ATP. The mechanism for ATP use in myosin- and kinesin-based motors to affect their function is similar and well known. Dynein is structurally very different from myosin and kinesin, and the mechanisms of its load-dependent cargo movement and multiple ATP use are currently not well defined. Protein conformation changes on binding ATP, its hydrolysis, and release of products are common to the mechanics of action of all of them.

A CLOSER LOOK 23.6

Other Differences between Rods and Cones

Sensitivity and response time of rods differs from those of cones. Absorption of a single photon by photoreceptors in rods generates a current of approximately 1–3 picoAmps ($1–3 \times 10^{-12}$), whereas the same event in the cones generates a current of approximately 10 femtoAmps (10×10^{-15}), about 1/100th of the rod response. The response time of cones, however, is about four times faster than that of rods. Thus, the cones are better suited for discerning rapidly changing events and the rods are better suited for low-light visual sensitivity.

Muscle Contraction

There are three types of muscles in the human body: skeletal, cardiac, and smooth. Skeletal muscles are striated in appearance—they have discrete bands and filaments—and the individual controls their actions voluntarily. Cardiac muscles are striated also, but their contractile units are not arranged in a linear manner as are skeletal muscles and their contraction is not voluntary. Smooth muscles are not striated, their thick and thin filaments do not show a visibly discrete organizational pattern, and they, like cardiac muscle, are not under voluntary control. Common to the contraction of each of these classes of muscle is the *sliding filament model,* that is, movement of the actin filament over the myosin filament and use of ATP as the energy source to effect contraction.

Skeletal Muscle: Structural Organization of Its Components

A schematic diagram of the structural organization of skeletal muscle is shown in Figure 23.28. Skeletal muscle consists of bundles of fibers (panel *c*). Each bundle is called a fasciculus (panel *b*). The fibers are made up of myofibrils (panel *d*), and each myofibril is a tandem series of muscle cells or units called **sarcomeres**. The muscle cell is multinucleated and is not capable of division. Most muscle cells survive for the life of the animal, but they can be replaced when lost or lengthened by fusion of **myoblast cells**.

A muscle cell is shown diagrammatically in Figure 23.29. Note that the myofibrils are surrounded by a membranous structure called the **sarcoplasmic reticulum**. At discrete intervals along the fasciculi and connected to the terminal cisterna of the sarcoplasmic reticulum are transverse tubules. These are connected to the external plasma membrane that surrounds the entire structure. Nuclei and mitochondria lie just inside the plasma membrane.

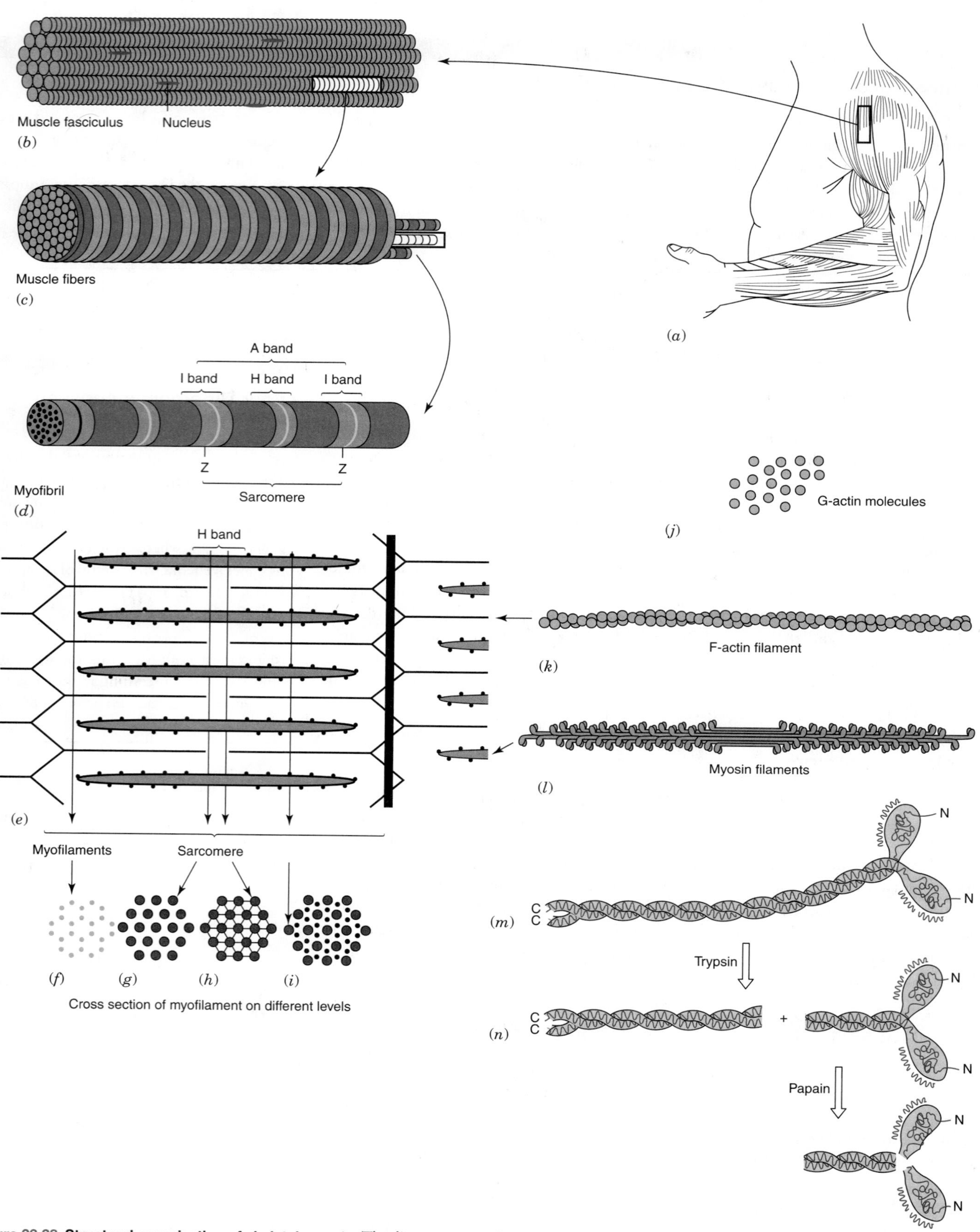

Figure 23.28 Structural organization of skeletal muscle. The diagram shows schematically the aggregation of G-actin to F-actin and the relative positions of cleavage of myosin by trypsin and papain.
Redrawn from Bloom, W. D., and Fawcett, D. W. *Textbook of Histology,* 10th ed. Philadelphia: Saunders, 1975. Sections *m* and *n* redrawn from Alberts, B., Bray, D., Lewis, J., Raff, M., Roberts, K., and Watson, J. *Molecular Biology of the Cell,* 2nd ed. New York: Garland, 1983.

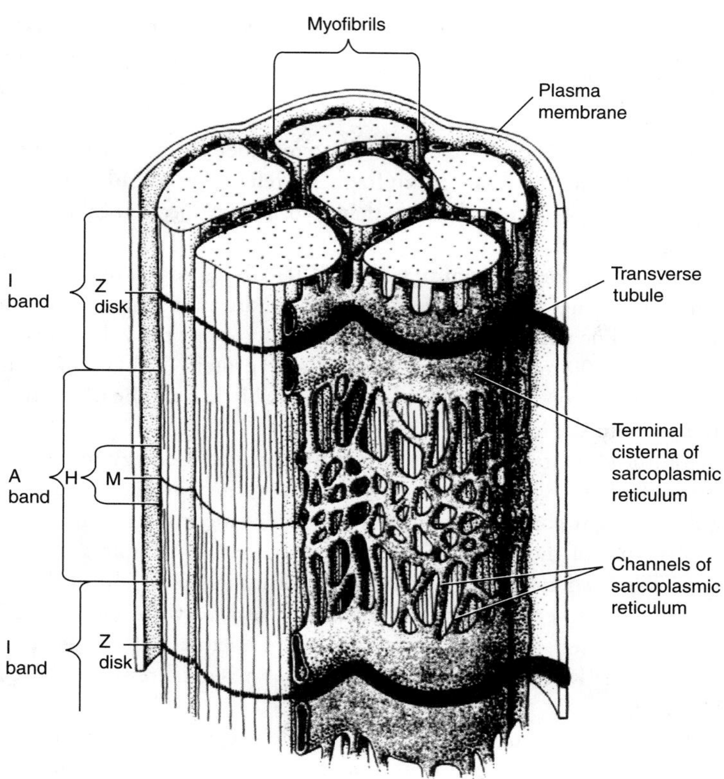

Figure 23.29 A schematic representation of a bundle of six myofibrils. The lumen of the transverse tubules connects with the extracellular medium and enters the fibers at the Z disk. Reprinted with permission from Darnell, J., Lodish, H., and Baltimore, D. *Molecular Cell Biology*. New York: Scientific American, 1986, 827.

The single contractile unit, the sarcomere, consists of all structural components between Z bands (or Z disk) (see Figures 23.28*d* and 23.29). Bands in the sarcomere are due to the arrangement of specific proteins (see Figure 23.28*e*). Two types of fibers are apparent: long thick fibers with protrusions on both ends are anchored at the center of the sarcomere, and long thin fibers that are attached to proteins within the **Z disk. I (isotropic) bands** extend for a short distance on both sides of the Z disk (two adjacent sarcomeres) and contain only **thin filaments**. The **H band** is in the center of the sarcomere and contains thick or heavy filaments but no thin filaments. In the middle of the H band, there is a somewhat diffuse **M band** which contains additional proteins that assist in anchoring fibers of the **heavy filaments** (Figure 23.28*h*) to the center of the sarcomere. The **A (anisotropic) band** contains both thin and heavy filaments and is located between the inner edges of the I bands. When a muscle contracts, the H and I bands shorten, but the distance between the Z disk and near edge of the H band remains constant. The distance between the innermost edges of I bands on both ends of the sarcomere also remains constant indicating that the length of thin filaments and of thick filaments does not change during contraction. Contraction, therefore, results when these filaments "slide" past each other.

Sarcomeres consist of many different proteins, eight of which are listed in Table 23.7. The two most abundant are myosin and actin. Approximately 60%–70% of the muscle protein is myosin and 20%–25% is actin. Thick filaments consist mostly of myosin and thin filaments consist mostly of actin. Thin filaments are formed from end-to-end polymerization of globular actin subunit monomers. After monomer polymerization, thin filaments are referred to as F-actin, fibrous actin (Figure 23.28*k*). Tropomyosin and troponin are associated with F-actin and have major roles in regulation of contraction. Other proteins listed are mostly associated with formation and/or maintenance of structural integrity of the contractile unit.

TABLE 23.7 • Approximate Molecular Masses (kDa) of Some Skeletal Muscle Proteins

Myosin	500
Heavy chain	200
Light chain	20
Actin monomer (G-actin)	42
Tropomyosin	70
Troponin	76
Tn-C subunit	18
Tn-I subunit	23
Tn-T subunit	37
α-Actinin	200
C-protein	150
β-Actinin	60
M-protein	100

Myosin Forms Thick Filaments of Muscle

Myosin, a long fibrous molecule with two globular heads on one end, consists of two heavy chains of about 230 kDa each. Bound near each head group is a dissimilar pair of light chains, each of which is approximately 20 kDa. The light chains are "calmodulin-like" proteins and are involved in various aspects of regulation of the process, not all of

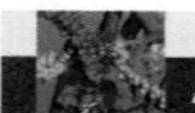

A CLOSER LOOK 23.7

Experimental Evidence for Hinge Regions and for the Section of the Molecule That Promotes Aggregation

Structural analyses show that the carboxyl end of each myosin chain is located in the tail section, where the two heavy chains are held together in a coil–coil arrangement (Figure 23.28*m*). *In vitro* experiments show that **trypsin** cleaves the tail at about one-third of its length from the head to produce **heavy meromyosin** (the head group and a short tail) and **light meromyosin** (the remainder of the tail section). Only light meromyosin can aggregate under physiological conditions *in vitro*, suggesting that effecting aggregation is one of the roles of this part of the tail section in heavy-chain formation *in vivo.*

which have been clearly defined. They bind to IQ motifs (distinctive amino acid sequence patterns) in myosin. These IQ motifs are located near the head groups and have the consensus sequence IQXXXRGXXXR (Ile-Gln-X-X-X-Arg-Gly-X-X-X-Arg, where X is any amino acid).

The head section can be cleaved from the remainder of the tail section by **papain** (Figure 23.28*n*), resulting in a head-group fragment referred to as **Subfragment 1** or **S-1**. Action of papain and trypsin demonstrates that myosin has at least two hinge regions near the head–tail junction (A Closer Look 23.7).

Heavy chains are formed by **symmetrical tail-to-tail aggregation** of myosin centered around M lines of H zones in sarcomeres. Tail sections are aligned in a parallel manner on both sides of the M line with the head groups pointing toward the Z line. Each thick filament contains about 400 molecules of myosin. The C protein (Table 23.7) is involved in their assembly. The M protein is also involved, presumably to hold the tail sections together and to anchor them to the M line.

The myosin head contains **ATPase** activity that provides energy for contraction and **actin binding**. The S-1 fragment also contains binding sites for the **essential light chain** and the **regulatory light chain** that bind at IQ motifs as mentioned. A model of the three dimensional structure of the myosin S-1 fragment is shown in Figure 23.30. The actin-binding region is located at the lower right-hand corner in the region of the visible cleft. The S-1 fragment has several structural domain-like regions with masses of 25, 50, and 20 kDa colored green, red, and blue, respectively. The essential light chain (ELC) and the regulatory light chain (RLC) are shown in yellow and magenta, respectively (Clin. Corrs. 23.8 and 23.9).

The ATP binding site, just above the visible cleft, is also an open cleft about 13 Å deep and 13 Å wide. It is separated from the actin-binding site by approximately 35 Å. Myosin binds to actin in a stereospecific manner. The ELC and the RLC are associated with a single long helix that connects the head region with the tail section. There is room for flexibility between the ELC and the connecting single helix. The conformation of myosin–ATP complex has an affinity for actin that is 1/10,000th that of the conformation of myosin that contains ADP or no ADP. Thus, the transduction of chemical energy to mechanical work depends on protein conformational changes that occur on binding of ATP, its hydrolysis, and product dissociation—initially inorganic phosphate followed by ADP (A Closer Look 23.8).

Actin, Tropomyosin, and Troponin Are Thin-Filament Proteins

Actin is the major protein of thin filaments and constitutes about 20%–25% of muscle protein. It is synthesized as a 42-kDa globular protein called G-actin, which aggregates to form fibrous actin or F-actin (see Figure 23.28*k*). More than 90% of its amino acid sequence is conserved among a variety of species (Table 23.8, p. 972). For these examples, differences are observed in about seven positions. Sequences of more than 30 actin isotypes,

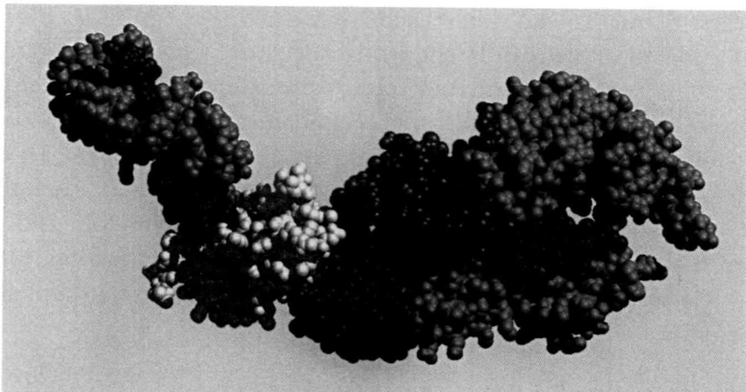

Figure 23.30 A space-filling model of the three-dimensional structure of the S-1 fragment of myosin. 25−, 50−, and 20-kDa domains of the heavy chain are green, red, and blue, respectively. Essential and regulatory light chains are yellow and magenta, respectively.
From Rayment, I., Rypniewski, W. R., Schmidt-Base, K., Smith, R., et al. *Science* 261: 50, 1993. Reproduced with permission from AAAS.

CLINICAL CORRELATION 23.8

Voltage-Gated Ion Channelopathies

There are three important types of voltage-gated cation channels: Na^+, Ca^{2+}, and K^+. Each is a heterogeneous protein consisting of various numbers of α- and β-subunits. In membranes, they are arranged in a more-or-less circular manner with a funnel-like channel formed through the middle of α-subunits. Roles of β-subunits are still being elucidated, but they appear to help stabilize and/or regulate activity of α-subunits.

Voltage-gated channels from nerve and muscle tissue show high sequence homology in many transmembrane domains, but are less conserved in the intracellular connecting loops. A common effect of mutations in Na^+ channels is muscle weakness or paralysis. Some inherited Na^+ voltage-gated ion channelopathies are listed following. Each of these is reported to result from a single amino acid change in the α-subunit and is transmitted as autosomal dominant.

It has been suggested that if the membrane potential is slightly more positive (i.e., changes from –70 to –60 mV) the myofiber can reach the threshold more easily and the muscle becomes hyperexcitable. If the membrane potential becomes even more positive (i.e., up to –40 mV) the fiber cannot fire an action potential leading to paralysis.

Catterall, W. A. Structure and function of voltage-gated ion channels. *Annu. Rev. Biochem.* 64: 493, 1995; Hoffmann, E. P. Voltage-gated ion channelopathies: Inherited disorders caused by abnormal sodium, chloride, and calcium regulation in skeletal muscle. *Annu. Rev. Med.* 46: 431, 1995; and Abraham, M. R., Jahangir, A., Alekseev, A. E., and Terzic, A. Channelopathies of inwardly rectifying potassium channels. *FASEB J.* 13: 1901, 1999.

Disorder	*Unique Clinical Feature*
Hyperkalemic periodic paralysis	Induced by rest after exercise, or the intake of K^+
Paramyotonia congenita	Cold-induced myotonia
Sodium channel myotonia	Constant myotonia

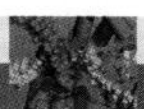

CLINICAL CORRELATION 23.9

Familial Hypertropic Cardiomyopathies and Mutations in Muscle Proteins

Hypertropic cardiomyopathies are characterized by enlargement/thickening of the left and/or right ventricle. Arrhythmias and premature death can result from these conditions. Familiar hypertrophic cardiomyopathy results from mutations involving genes coding for β-myosin heavy chain, ventricular myosin regulatory light chain, cardiac troponin T, cardiac troponin I, α-tropomyosin, and cardiac myosin-binding protein C. Some of these genes express isoforms that may be expressed only in cardiac tissue. Thus, some of the effects are not applicable to other types of muscle. Mutations can alter structure and function of the sarcomere and lead to changes in cardiac function. Alteration in cardiac function may include decrease in force generated from myosin–actin interaction, faulty anchoring of myosin (protein C) within the sarcomere, and/or interference with any of the functions of tropomyosin and troponin subunits.

Bonne, C., Carrier, L., Pascal, R., Hainque, B., and Schwartz, K. Familiar hypertropic cardiomyopathy: from mutations to functional defects. *Circ. Res.* 83: 580, 1998.

the longest consisting of 375 amino acid residues, reveal that no more than 32 residues have been substituted (Clin. Corr. 23.10). A significant number of these substitutions occur at the *N*-terminal where essentially all actin molecules are posttranslationally modified. The *N*-terminal amino acid residue may be acetylated and hydrolyzed one or two times before the final acetylated product is achieved.

A crystal structure of G-actin is shown in Figure 23.31. Actin has two distinct domains of approximately equal size; however, one has been designated large (left) and the other small (right). Each of them consists of two subdomains. The *N*-terminal and the C-terminal residues are located within subdomain 1 of the small domain.

The molecule has polarity, and aggregation to **F-actin** can occur from either end. *In vitro* kinetic data indicate that the preferred direction of aggregation is by addition of monomers to the large end of the molecule and that the rate of addition to this end is diffusion controlled, that is, at a rate as fast as the monomer can diffuse to that end. Each G-actin contains a specific binding site between the two major domains for ATP and a

A CLOSER LOOK 23.8

Homology among Myosin from Different Sources

cDNA analyses of myosins from many different species and types of muscle indicate that there is a very high degree of homology among myosins from different sources, particularly within the head region. There is somewhat less sequence homology within the tail region, but functional homology exists to an extraordinarily high degree regardless of length, which ranges from about 86 to about 150 nm for different species. The myosin head contains nearly one-half (about 839 to about 850) of the amino acid residues in the entire molecule in mammals (Clin. Corr. 23.8, p. 971).

TABLE 23.8 • Summary of the Amino Acid Differences Among Chicken Gizzard Smooth Muscle Actin, Skeletal Muscle Actin, and Bovine Cardiac Actin

	Residue Number						
Actin Type	*1*	*2*	*3*	*17*	*89*	*298*	*357*
Skeletal muscle[a]	Asp	Glu	Asp	Val	Thr	Met	Thr
Cardiac muscle[b]		Asp	Glu			Leu	Ser
Smooth muscle[c]	Absent		Glu	Cys	Ser	Leu	Ser

Source: Adapted from Vandekerckhove, J. and Weber, K. *FEBS Lett.* 102:219, 1979.
[a]From rabbit, bovine, and chicken skeletal muscle.
[b]From bovine heart.
[c]From chicken gizzard.

divalent metal ion, Mg^{2+}, but Ca^{2+} competes with Mg^{2+} for the same binding site. The **G-actin–ATP–Mg2^{+} complex** aggregates faster to form **F-actin**. Subdomains 1 and 2 of G-actin molecules in F-actin are to the outside where myosin binding sites are located. The conformational form of F-actin is helical. F-Actin may be viewed as either (1) a single-start, left-handed single-stranded helix with rotation of the monomers through an approximate 166° with a rise of 27.5 Å or (2) a two-start, right-handed double-stranded helix with a half pitch of 350–380 Å.

β-Actinin binds to F-actin and assists in limiting the length of the thin filaments. **α-Actinin**, a homodimer of 90 to110 kDa subunits, binds adjacent actin monomers of F-actin at positions 86–117 of one and 350–375 of the other and strengthens the fiber. It also helps to anchor the actin filament to the Z disk. Other major proteins associated with the thin filament are **tropomyosin** and **troponin**.

Tropomyosin is a rod-shaped protein consisting of two dissimilar subunits, each of about 35 kDa. It forms aggregates in a head-to-tail configuration. It interacts in a flexible manner with the thin filament throughout its entire length. It fits within the groove of the helical assembly of the actin monomers of F-actin. Each tropomyosin molecule interacts with about seven actin monomers between subdomain one and three. Tropomyosin helps to

CLINICAL CORRELATION 23.10

Dilated Cardiomyopathy and Mutations in Actin

Thin walls of the heart and inability to pump blood effectively are characteristics of dilated cardiomyopathy. A genetic cause of this condition can be defective actin. Although the amino acid sequence of actin is highly conserved, there are several isoforms expressed by different genes; five of six are expressed in skeletal and cardiac myocytes. In adult cardiac myocytes, approximately 80% of actin results from expression of specific cardiac isoforms. Mutations in the invariant regions of these cardiac isoforms can result in serious health effects. An inherited single amino acid substitution (see following), Arg312His in one patient and Glu361Gly in another patient, resulted in dilated cardiomyopathy, a condition in which cardiac transplant is the only definitive treatment for end-stage disease.

Comparison of the positions of those substitutions with the structure of actin shown in Figure 23.31 reveals that they occur in two different functional regions of actin.

Olson, T. M., Michels, V. V., Thibodeau, S. N., Tai, Y-S., and Keating, M. T. Actin mutations in dilated cardiomyopathy, a heritable form of heart failure. *Science* 280: 750, 1998.

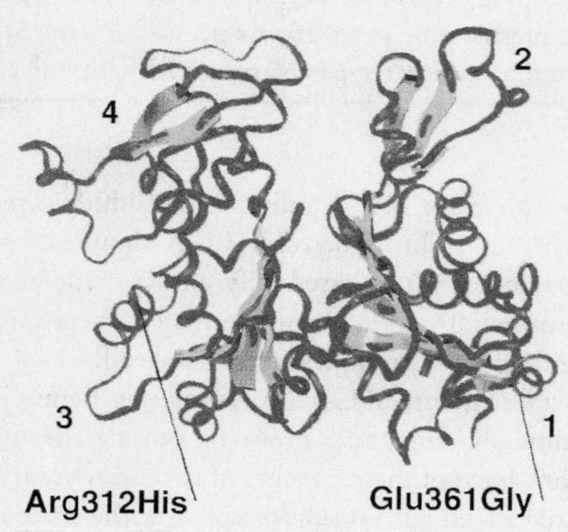

Schematic representation of the cardiac actin monomer and location of idiopathic dilated cardiomyopathic mutations.
From Olson, et al. (in Clin. Corr. 23.10 reference).

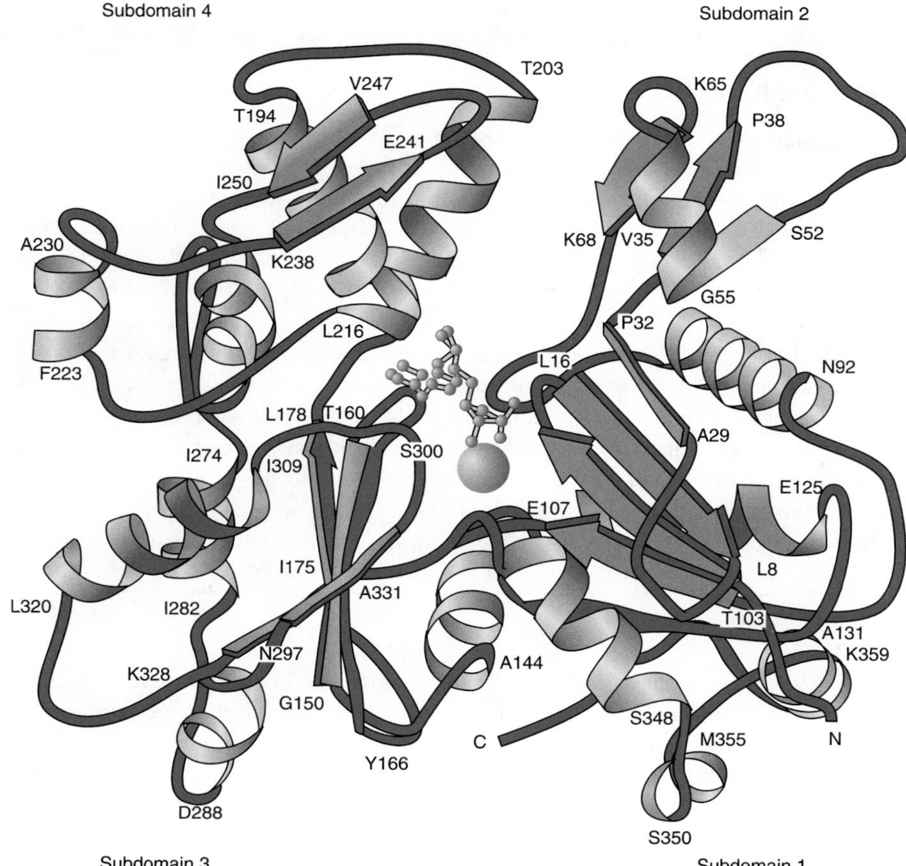

Figure 23.31 Crystal structure of G-actin showing secondary structural elements. ADP and a divalent metal ion are shown in the cleft between the two large domains.
Reproduced with permission from Lorenz, M., Popp, D., and Holmes, K. C. *J. Mol. Biol.* 234: 826, 1993 by permission of the publisher Copyright (1993) Academic Press Limited, London.

stabilize the thin filament and to transmit signals for conformational change to other components of the thin filament when Ca^{2+} binds to troponin that is attached to tropomyosin.

Troponin consists of three dissimilar subunits designated TnC, TnI, and TnT with molecular mass of about 18 kDa, 23 kDa, and 37 kDa, respectively. The TnT subunit binds to tropomyosin. The TnI subunit inhibits binding of actin to myosin. The TnC subunit is a calmodulin like protein that binds Ca^{2+}.

The three-dimensional structure of TnC shows it to be dumbbell shaped and very similar to calmodulin. A structural model of the calcium-saturated TnC–TnI complex is shown in Figure 23.32. The TnI subunit fits around the central region of TnC as a helical

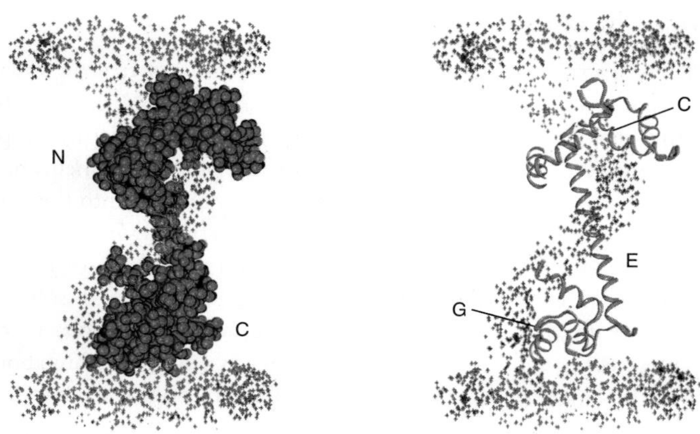

Figure 23.32 Best-fit model for the $4Ca^{2+}$–TnC–TnI complex. A model for the $4Ca^{2+}$–TnC–TnI complex based on neutron-scattering studies with deuterium labeling and contrast variation (Olah, C. C., and Trewhella, J., *Biochemistry* 33: 12800, 1994). (Right) A view showing the spiral path of TnI (green crosses) winding around the $4Ca^{2+}$-TnC that is represented by an α-carbon backbone trace (red ribbon). C, E, and G helices are labeled. (Left) The same view with $4Ca^{2+}$-TnC represented as a CPK model.
Photograph generously supplied by Dr. J. Trewhella. The publisher recognizes that the U. S. government retains a nonexclusive, royalty-free license to publish or reproduce the published form of this contribution or to allow others to do so for U. S. government purposes.

CLINICAL CORRELATION 23.11

Troponin Subunits as Markers for Myocardial Infarction

Troponin contains three subunits (TnT, TnI, and TnC) and each is expressed by more than one gene. Two genes code for skeletal muscle TnI, one for fast and one for slow skeletal muscle, and one gene codes for cardiac muscle TnI. The genes that code for TnT have that same distribution pattern. The gene for the cardiac form of TnI appears to be specific for heart tissue. Two genes encode TnC, but neither gene appears to be expressed only in cardiac tissue.

The cardiac form of TnI in humans is about 31 amino acids longer than the skeletal muscle form. Serum levels of TnI increase within 4 h of an acute myocardial infarction and remain high for about seven days in about 68% of patients tested. Almost 25% of one group of patients showed a slight increase in the cardiac form of TnI after acute skeletal muscle injury indicating that it would not be a good and sensitive test for myocardial infarction.

Two isoforms of cardiac TnT, TnT_1 and TnT_2, are present in adult human cardiac tissue and two more isoforms are present in fetal heart tissue. These isoforms may result from alternative splicing of mRNA. Serum levels of TnT_2 increase within four hours of acute myocardial infarction and remain high for up to fourteen days. TnT_2 in serum is 100% sensitive and 95% specific for detection of myocardial infarction. TnT_2 assay is used to detect acute myocardial infarction. Myocardial infarcts are either undiagnosed or misdiagnosed in hospital patients admitted for other causes or in 5 million or more people who go to doctors for episodes of chest pain. This test is sufficiently specific to diagnose myocardial incidents in these patients and to help direct doctors to proper treatment for them.

Anderson, P. A. W., Malouf, N. N., Oakeley, A. E., Pagani, E. D., and Allen, P. D. Troponin T isoform expression in humans. *Circulation Res.* 69: 1226, 1991; and Ottlinger, M. E., and Sacks, D. B. *Clin. Lab. News,* 20: 33, 1994.

coil and forms caps over it at each end. The cap regions of TnI are in close contact with TnC when TnC is fully saturated with Ca^{2+}. TnC contains four divalent metal ion-binding sites. Two are in the C-terminal region, have high affinity for calcium ions (K_D of about 10^{-7} M), and are presumed to be always occupied by divalent metal ions (Ca^{2+} or Mg^{2+}) since the concentrations of these ions in resting cells are within the same order of magnitude as the K_D. Under resting conditions, TnI has a conformation that prevents proper orientation of tropomyosin and inhibits myosin binding to actin, thus, preventing contraction. On excitation, the calcium ion concentration increases to about 10^{-5}M, high enough to bind to sites within the *N*-terminal region of TnC. TnI now binds preferentially to TnC in a capped structural conformation as shown in Figure 23.32. This permits movement of tropomyosin, and myosin-binding sites on actin become exposed. The nature of the interaction of tropomyosin with actin permits it the flexibility to alter its conformation as a function of calcium ion concentration and to assist in blockage of the myosin binding sites on actin at low Ca^{2+} concentration (Clin. Corr. 23.11).

Figure 23.28*i*, p. 968, shows schematically a cross section of a sarcomere and the relative arrangement of thin and thick filaments. Six thin filaments surround each thick filament. The arrangement and flexibility of myosin head groups make it possible for each thick filament to interact with multiple thin filaments. When cross-bridges form between thick and thin filaments, they do so in patterns consistent with that shown in the colorized electron micrograph of Figure 23.33. This shows a two-dimensional view of thick filaments interacting with two thin filaments.

Skeletal Muscle Contraction

Skeletal muscle contraction is initiated by excitation of a motor neuron and propagation of an action potential to a neuromuscular junction (Figure 23.34). Note from an earlier section of this chapter on nerve transmission (p. 974) that when the impulse reaches the neuromuscular junction, Ca^{2+} channels open on the presynaptic neuron, Ca^{2+} enters those channels and initiates the process for release of acetylcholine into the neuromuscular junction (see Clin. Corr. 23.1, p. 941). Acetylcholine, on entering into the neuromuscular junction, binds to its (ligand-gated) receptors on the muscular side of the junction and initiates an impulse that travels along the sarcolemma, through the T-tubules, to sacs of the sarcoplasmic reticulum. Ca^{2+} is released into the cytosol (sarcoplasm) of the sarcomere resulting in an increase in its concentration of about 100-fold, from about 10^{-7} M to about 10^{-5} M. At this higher concentration, Ca^{2+} binds troponin; specifically, it binds to

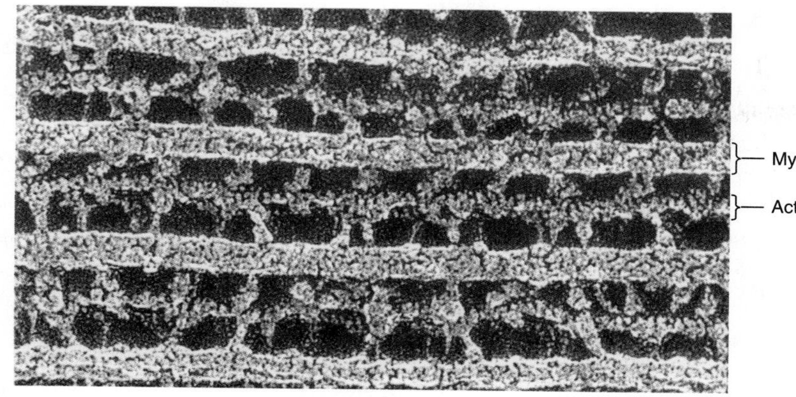

Figure 23.33 A colorized electron micrograph of actin–myosin cross-bridges in a striated insect flight muscle.
Reproduced with permission from Darnell, J., Lodish, H., and Baltimore, D. *Molecular Cell Biology*. New York: Scientific American Books, 1986.

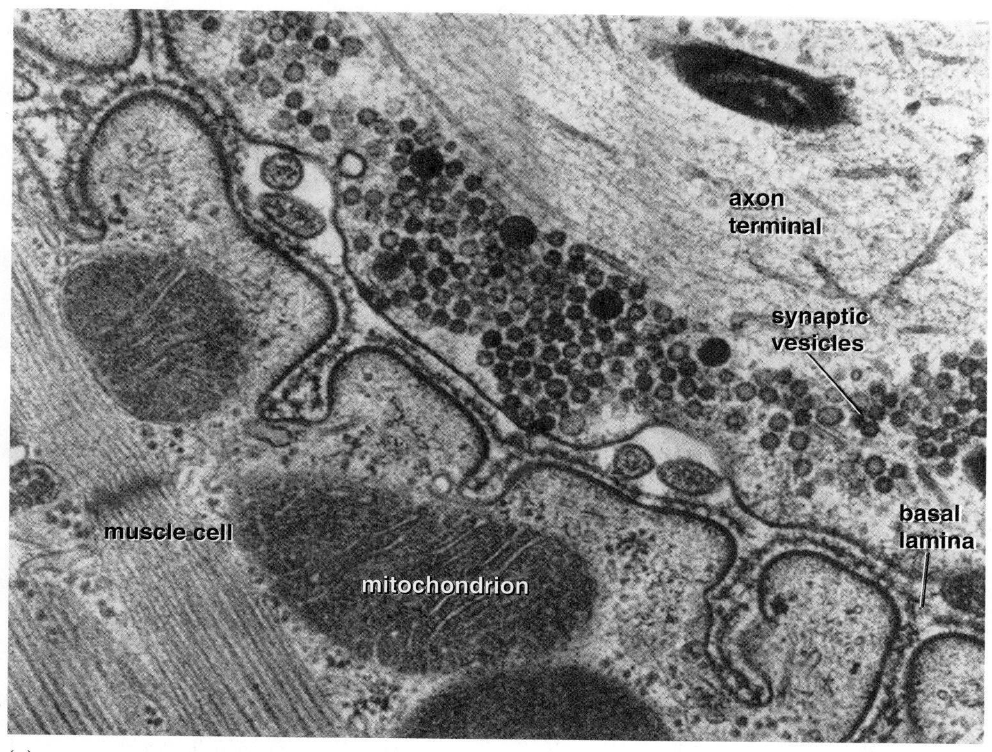

(*a*)

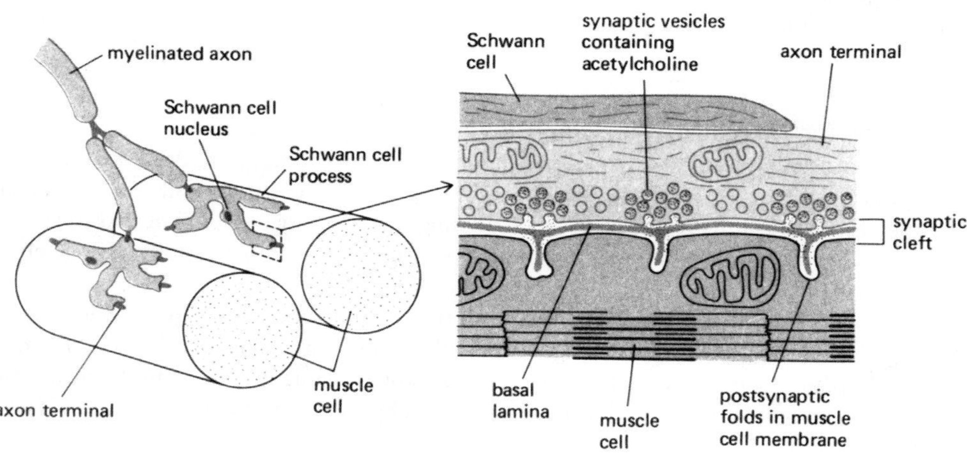

(*b*)

Figure 23.34 Neuromuscular junction.
(*a*) Electron micrograph of a neuromuscular junction. (*b*) Schematic diagram of the neuromuscular junction shown in (*a*). Reproduced with permission from Alberts, B., Bray, D., Lewis, J., Raff, M., Roberts, K., and Watson, J. *Molecular Biology of the Cell.* New York: Garland , 1983.

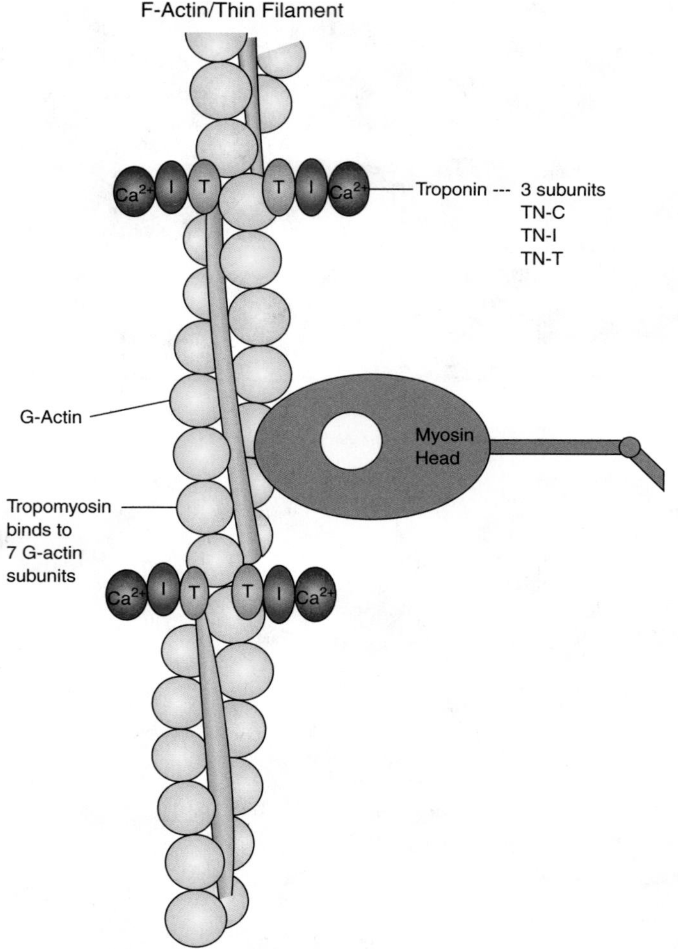

Figure 23.35 **Diagram showing the relationship between the three subunits of troponin, tropomyosin, the actin filament, and a myosin head unit.** TnC and TnI subunits are shown in Figure 23.32. These subunits interact with tropomyosin through TnT.

TnC subunits of the heterotrimeric troponin molecules (Figure 23.35). Binding of Ca^{2+} to TnC induces a conformation change that also affects other subunits of troponin. There is a change in TnI conformation and a conformational change is eventually transmitted through the TnT subunit of troponin to tropomyosin. Tropomyosin lies in the grove of the helical arrangement of the F-actin filament. Tropomyosin is forced into a rotational conformation that exposes myosin-binding sites on the F-actin filament permitting actin–myosin interaction.

Model for Skeletal Muscle Contraction: The Power Stroke

Myosin head groups undergo conformational changes on binding of ATP, hydrolysis of ATP, and release of products. ATP binding leads to closure of the active site cleft and opening of the cleft in the region of the actin binding site. Hydrolysis of ATP results in closure of the cleft in the actin-binding region that opens again on or for release of inorganic phosphate (A Closer Look 23.9).

Conformational changes in myosin that occur in conjunction with closing and opening of the cleft in the actin-binding region with release of inorganic phosphate (P_i) during ATP hydrolysis are associated with the **power stroke** (Figure 23.36). Actin filaments can be moved as much as 10 nm during this process. The open conformation of myosin is now in a position to bind ATP and start the cycle over again.

Individual myosin units functioning in an asynchronous manner, possibly like changes in the position of hands on a rope in the game of tug-of-war can maintain force or

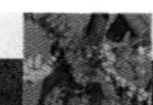

A CLOSER LOOK 23.9

Evidence for Conformation Changes on P_i and ADP Release

This conformational change is evident by the movement of two cysteine-containing helices. The distance between Cys^{697} and Cys^{707} changes from about 19 Å to about 2 Å. Experimentally cross-linking of these two cysteines in the closed position traps ADP within its binding site. A stereo view of myosin showing the reactive cysteine pocket (cysteine residues that can react chemically with externally added cysteine reactive agents) is shown in the figure.

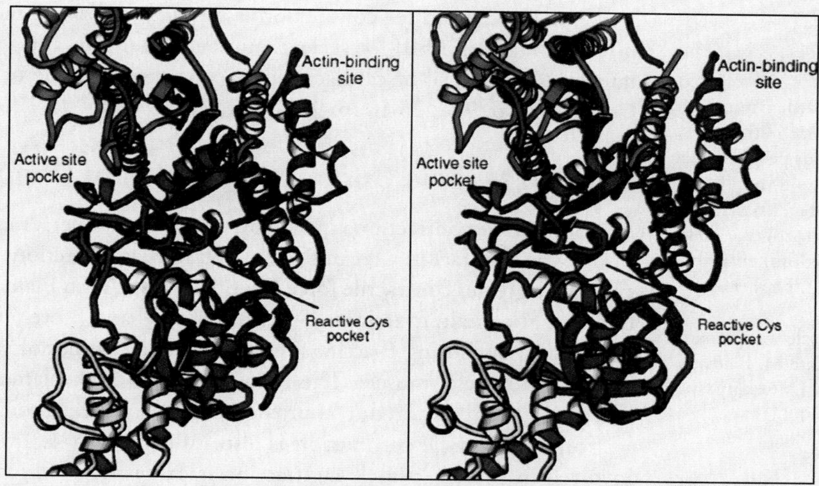

Stereo view of myosin showing the pocket that contains the mobile reactive cysteine residues.
Reproduced with permission from Rayment, I., and. Holden, H. M. The three dimensional structure of a molecular motor. *Trends Biol. Sci.* 19:129, 1994. Copyright (1994) Elsevier.

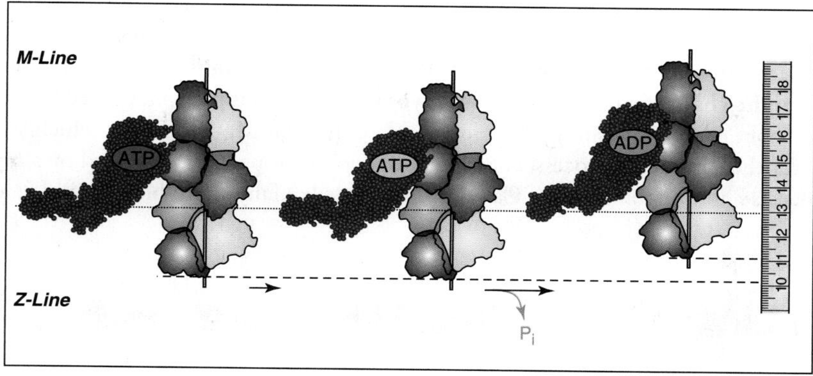

Figure 23.36 A model of actin–myosin interaction. A schematic diagram showing how conformation changes in the myosin head group and arm can pull actin filaments in a direction away from the Z disk during hydrolysis of ATP and release of inorganic phosphate.
Reproduced with permission from Rayment, I., and.Holden, H. M. The three dimensional structure of a molecular motor. *Trends Biol. Sci.* 19: 129, 1994. Copyright (1994) Elsevier.

create maximum force. Thus, when some myosin head groups bind with high affinity, others have low affinity, and this binding, pulling, release, binding action of myosin shortens the sarcomere in an appropriately forceful manner.

Contraction is stopped by action of the Ca^{2+}-transporting ATPase (Ca^{2+}-ATPase) (see Figure 12.55, p. 992) that rapidly pumps Ca^{2+} from the sarcomere back into the sarcoplasmic reticulum where it is sequestered by calsequestrin.

Cardiac Muscle: Structure and Contraction

Cardiac muscle is striated but in a somewhat less pronounced way than skeletal muscle. A diagram of cardiac muscle structure is shown in Figure 23.37. Cardiac muscle is under control of the *autonomic nervous system* and its contraction is involuntary. The sarcomeres of cardiac muscle are not arranged in a linear manner but are branched and held together

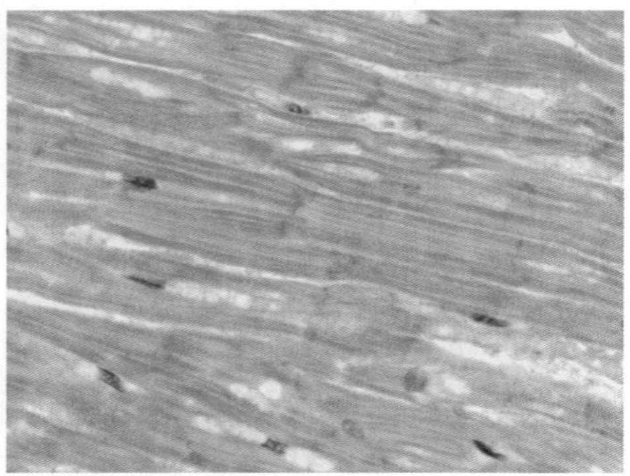

Figure 23.37 Cardiac muscle. Cardiac muscle cells are striated and connected to each other by intercalated discs; the dark red bands perpendicular to the fibers. Intercalated discs contain three types of membrane junctions – fascia adherens, macula adherens (desmosomes), and gap junctions. These membranes form a network (syncytium) ensuring that all cells work together as a unit. Darker areas are nuclei.
Photograph of human cardiac muscle was prepared and generously provided by Dr. Kate M. Baldwin, Department of Anatomy, Howard University College of Medicine, Washington, DC.

tightly by adheren junctions. These junctions are assembled from the association of a transmembrane protein, cadherin, a polymerized rodlike protein that crosses the membranes of adjacent cells and interacts with another protein, catenin, that binds to actin and α-actinin in adjacent sarcomeres.

Cardiac muscle contracts in a rhythmic manner originating from an impulse generated within the heart itself. Impulse transmission is electronic through gap junctions, which permit faster signaling between sarcomeres. There is a longer time period between each contractile cycle than usually occurs in skeletal muscle, but this can be altered by hormones or other agents.

The mechanism for contraction is the *sliding filament* model involving myosin–actin interaction as described for skeletal muscle. Other ions and ion channels play very important roles in control of cardiac muscle contraction. Some of these are discussed in Clin. Corrs. 23.9, p. 971, 23.10, p. 972, 23.11, p. 974, and 23.12, p. 978.

Smooth Muscle Contraction: Calcium Regulation

Smooth muscle contraction is generally slower than skeletal muscle contraction; but smooth muscle can contract to a greater extent and in more directions than skeletal muscle and can generate a greater contractile force (Figure 23.38). Actin filaments are attached to the dense bodies that contain α-actinin. α-Actinin is a Z-band protein in skeletal muscle that is also involved in anchoring F-actin. Presumably dense bodies of smooth muscle function as Z lines do in skeletal muscle. The ratio of thin to thick filaments is approximately 15:1 in smooth muscle compared to approximately 6:1 in skeletal muscle.

Smooth muscle contraction is also influenced by hormones. Among other proteins involved in smooth muscle contraction is **caldesmon**, which serves a function similar to troponin in skeletal muscle. Calcium-calmodulin binds to caldesmon, affects its release from actin, and permits contraction. Smooth muscle does not have a well-defined sarcoplasmic reticulum, and Ca^{2+} that is required for contraction comes from internal storage or from outside of the cell.

Calcium ions play another important role in smooth muscle contraction. A mechanism for calcium regulation of smooth muscle contraction is shown in Figure 23.39. Some key elements of this mechanism are (1) release of Ca^{2+} from intracellular stores or increased flux across the plasma membrane, (2) formation of a Ca^{2+}–CaM complex is dependent on the concentration of intracellular Ca^{2+}, (3) A Ca^{2+}–calmodulin (CaM) complex activates **myosin light-chain kinase (MLCK)**, (4) myosin light chain is phosphorylated by MLCK, (5) a phosphorylated myosin light chain stimulates myosin Mg^{2+}-ATPase, which supplies energy for the contractile process, and (6) contraction is stopped or decreased by a **myosin phosphatase** and/or transport of Ca^{2+} out of the cell. Many more biochemical steps are

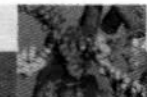

CLINICAL CORRELATION 23.12

Ion Channels and Cardiac Muscle Disease

Voltage-gated ion channels in cardiac muscles, like those of other tissues, require a finite recovery time after excitation. The heart contracts and relaxes on a continuous basis and in a rhythmic manner that cannot be altered significantly without causing problems such as arrhythmias, fibrillation, and possibly death. The recovery time between contractions of cardiac muscle is measured on electrocardiograms as the QT interval. Initiating the excitation phase is the opening of Na^+ channels for a finite period. Na^+ moves into the cell causing depolarization. K^+ channels then open to permit K^+ to move out of the cell. Both ions flow down their respective chemical concentration gradients. Opening of the K^+ channels is important for shutting off the action potential. Inherited conditions referred to as long QT syndrome or LQTS, have been linked to genes that affect K^+ channels (KVLQT1, HERG, mink) and Na^+ channels (SCN5A). Defects in these channels in cardiac muscle can cause sudden death; particularly in young people who are physically active, have never had an electrocardiogram, and have no knowledge that they have LQTS. Prevalence of this condition is approximately 1 in 10,000.

Balser, J. R. Structure and function of the cardiac sodium channels. *Cardiovasc. Res.* 42: 327, 1999; Ackerman, M. J., Schroeder, J. J., Berry, R., Schaid, D. J., et al. A novel mutation in KVLQT1 is the molecular basis of inherited long QT syndrome in a near-drowning patient's family. *Pediatr. Res.* 44: 148, 1998; and Barinaga, M. Tracking down mutations that can stop the heart. *Science* 281: 32, 1998.

involved in the regulation of smooth muscle contraction, steps that can be regulated by hormones and other agents such as NO and cGMP. These diverse interactions give smooth muscles the ability to develop various degrees of tension and to retain it for prolonged periods.

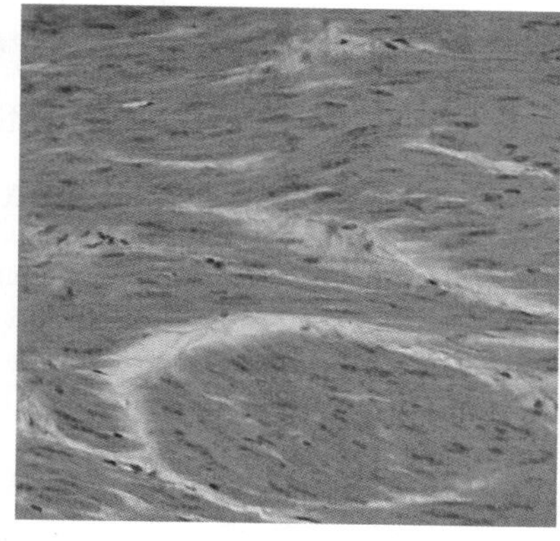

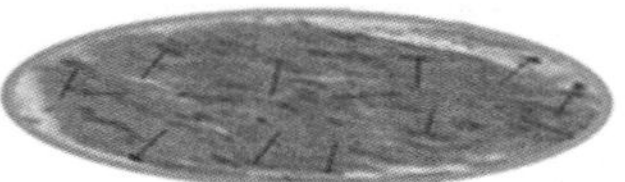

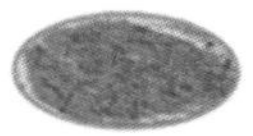

Figure 23.38 Smooth muscle. (*a*) Histological stain of smooth muscle. (*b*) A diagrammatic representation of how smooth muscle can contract in essentially all directions.
From http://en.wikipedia.org/wiki/Smooth_muscle.
From http://www.health.howstuffworks.com/muscle4.htm.

Energy Reservoirs for Muscle Contraction

In normal muscle, the concentration of ATP remains fairly constant even during strenuous activity because of increased metabolic activity and the actions of **creatine phosphokinase** and **adenylate kinase**. Creatine phosphokinase catalyzes transfer of phosphate from phosphocreatine to ADP in an energetically favored manner.

$$\text{Phosphocreatine} + \text{ADP} \rightleftharpoons \text{ATP} + \text{Creatine}$$

If the metabolic activity is insufficient to keep up with the need for ATP, the creatine phosphokinase helps to maintain cellular levels relatively constant. Adenylate kinase provides additional ATP by catalyzing the reaction

$$2\text{ADP} \rightleftharpoons \text{ATP} + \text{AMP}$$

The most obvious pathological consequence of ATP depletion is development of a state of **rigor**. Effects of ATP depletion are (1) intracellular Ca^{2+} concentration is no longer controlled and (2) myosin will exist exclusively in the myosin–ADP complex and bound to actin, a condition called **rigor mortis**. Recall that ATP is required for dissociation of the actin–myosin complex.

Other Classes of Myosins and Molecular Motors

Eighteen classes of myosin have been identified in the human genome. They are generally written as Myosin- followed by [Roman numbers I-XVIII]. Skeletal, cardiac, and smooth muscle myosins are all in Class II and have gene designations MYH1–MYH8 for either skeletal or cardiac, whereas smooth muscle is designated MYH11. There are also some nonmuscle myosins in Class II.

Unconventional Myosins and Their Functions

Analyses of the human genome suggest that there are more than 20–30 myosins and/or myosin-like genes. Functions for some of their products are known and are listed here because they may be implicated in some disease states (Clin. Corr. 23.13).

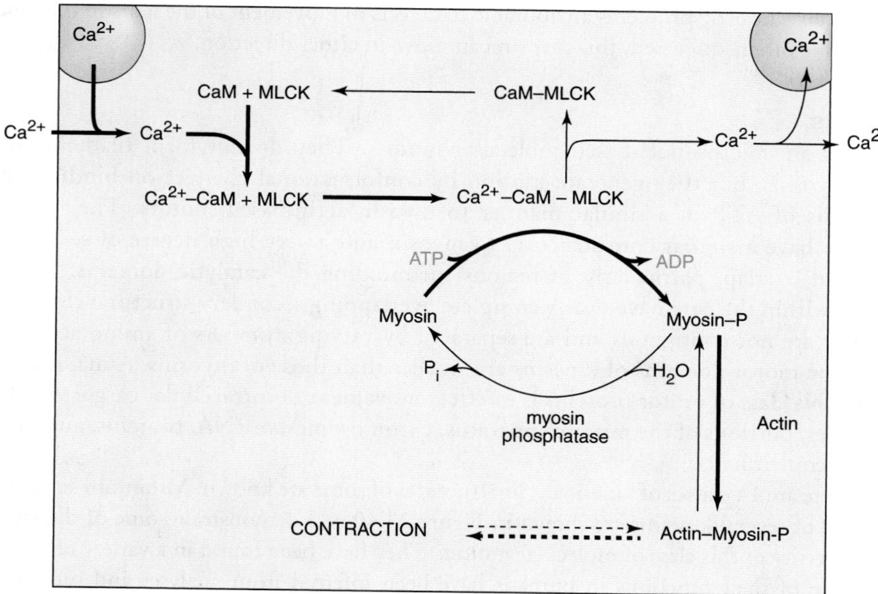

Figure 23.39 A schematic representation of the mechanism of regulation of smooth muscle contraction. Heavy arrows show the pathway for tension development, and light arrows show the pathway for release of tension. Mg^{2+}–ATPase activity is highest in the actin–myosin–P complex. Abbreviations: CaM, calmodulin, and MLCK, myosin light chain kinase.
Adapted from. Kramm, K. E., and Stull, J. T. *Ann. Rev. Pharmacol. Toxicol.* 25: 593, 1985.

CLINICAL CORRELATION 23.13

Mutations Affecting Pigmentation: Is There a Molecular Motor Connection?

Elejalde syndrome is characterized by silvery hair and severe dysfunction of the central nervous system. Large granules of melanin are unevenly distributed in the hair shaft, and abnormal melanocytes and melanosomes may be present in fibroblasts. These observations suggest that mechanisms for their movement and distribution might be defective.

Chediak–Higashi syndrome is characterized by enlarged lysosomes, melanosomes, and other cytoplasmic granules. Eyes of humans and animals with this condition are hypopigmented. Intraocular melanin granules vary in size and can become extremely large. Some of the manifestations of this condition suggest similarity in mechanism with Elejalde syndrome.

Elejalde, B. R., Holguin, J., Valencia, A., Gilbert, E. F., et al. Mutations affecting pigmentation in humans: Neuroectodermal melanolysosomal disease. *Am. J. Med. Genet.* 3: 65, 1979; Duran-McKinster, C., Rodriguez-Jurado, R., Ridaura, C., de la Luz Orozco-Covarrubias, M., et al. Elejalde syndrome: A melanolysosomal neurocutaneous syndrome: clinical and morphological findings in 7 patients. *Arch. Dermatol.* 136: 120, 2000; and Mottonen, M., Lanning, M., Baumann, P., and Saarinen-Pihkala, U. M. Chediak–Higashi syndrome: Four cases from northern Finland. *Acta Paediatr.* 92: 1047, 2003.

Myosin I is a low-molecular-mass motor protein that is associated with cell membranes. It contains a tail section rich in basic amino acid residues and can associate with anionic phospholipids. There is a tail section domain rich in glycine, proline, and alanine (GPA domain) that facilitates binding to actin. The tail also contains an src homology 3 (SH3) domain that mediates protein–protein interactions. SH3 domains select peptides sharing the consensus motif LXXRPLXΨP, where Ψ is an aliphatic residue. This class of myosins appears to be involved in membrane–membrane interactions and to function in brush boarders of the small intestines. They may also be involved in transport/movement of Golgi-derived vesicles. They have several IQ motifs in the neck region near the head suggesting interactions with calmodulin-like molecules and regulation by them.

Myosin V is involved in organelle transport such as synaptic vesicles, melanosomes, vacuoles, and mRNA. It has an extended neck region and a tail with globular ends that permit dimerization but not filament formation. Light chains of myosin V are of the EF hand family and bind to an altered repeat IQ motif (IQXXXRGXXXR). Its cargo binds to the tail region. It is found in all brain neurons and is required for melanosome transport. Mutations in myosin V are associated with neurological malfunctions and depigmentation, especially of hair. It shows processivity in its movement like some kinesins and takes steps of about 36 nm along an actin fiber. Its rate of movement is limited by ADP dissociation and ATP binding, which causes the trailing head of myosin V to dissociate from actin.

Myosin VI is expressed in most cells and tissues. The neck region is extended and contains one calmodulin (IQ motif) binding region per head. It is a dimeric molecule with a coiled tail that has two globular ends (one for each monomer). Defects in the gene for this protein cause hearing problems attributable to defects in movement of the hairs in the inner ear. Unlike others discussed, this myosin can move in either direction.

Kinesin-1

Kinesin-2

Kinesin-14

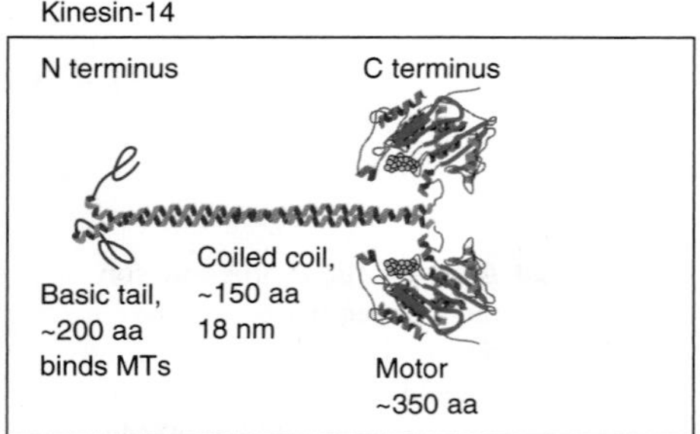

Figure 23.40 Kinesin structures. Representative structures of Kinesin-1, Kinesin-2, and Kinesin-14. All structures were taken from the Kinesin Home Page and associated links, http://www.proweb.org/kinesin/index.html.

Kinesins

Kinesins are microtubule-based molecular motors. They do not form filaments like some myosins, but they generate motion by conformational changes on binding and hydrolysis of ATP in a similar manner to myosin actin-based motors. Their motor domains have a similar core structure to myosin and a very high degree of secondary structural overlap, particularly in regions surrounding the catalytic domains. Several regions within the core have nearly complete overlapping secondary structural elements, but they are not contiguous and are separated by varying numbers of amino acid residues. The motor domains of kinesins are smaller than those of myosins. A major function of this class of motor proteins is to effect movement of intracellular cargo: vesicles, organelles, portions of the mitotic apparatus, chromosomes, mRNA, proteins, and other cellular constituents.

There are 14 classes of kinesins. The structures of some are known. Minimum structural elements of several kinesins are shown in Figure 23.40 and demonstrate some of the structural diversity of this class of molecular motors. They have been found in a variety of species and some of their functions in humans have been inferred from analyses and mutations induced in them in other species of animals. In most cases, kinesins move their cargo toward the plus end of microtubules.

Kinesin-1 is a motor for fast axonal transport (p. 951). Among the cargo it transports are vesicles containing membrane proteins such as those which make ion channels along the axon during development, synaptic vesicle proteins, and those with which they interact at the axon termini such as syntaxin and synaptotagmin. This kinesin is expressed ubiquitously in human neural tissue. Mutations in kinesin-1 may lead to neuronal defects in humans that are reflected by a variety of symptoms.

Kinesin-2 is involved with membrane-associated movements in axons, axonemes, and melanophores. It has been shown to be involved in dispersion of melanosomes in fish melanophores. It has not been detected in humans, but there are human diseases associated with the lack of dispersion and/or movement of melanin containing granules, a defect that may be related more to defects in myosin-V in humans.

Kinesin-4 is associated with maintenance of spindle bipolarity and movement of chromosomes to the metaphore plate.

Kinesin-5 is associated with the mitotic apparatus of dividing cells and is involved in centrosomes and spindle pole body separation.

Kinesin-6 is present in the spindle midbody during telophase and is presumed to mediate the sliding of spindle fibers in the late stages of anaphase that are needed for spindle elongation and separation of chromosomes.

Kinesin-7 is also chromosome associated and provides a direct link between the chromosomes and microtubules of the spindle.

Kinesin-13 is also a chromosomal kinesin that is probably involved in mitotic chromosome movement.

Kinesin-14 is among the minus-end motors like the Ncd mitotic kinesin in *Drosophila* that functions in early stages of mitosis. It is localized to spindles in oocytes.

Kinesins-1 and -2 are related more to material discussed in this chapter, whereas the others are associated with more general aspects of cell division and movement of various components associated with that process.

Dynein

There are two classes of dynein motors: axonemal that effect flagella and cilia movement and cytoplasmic that effect distribution and organization of cytoplasmic structures. These functions include protein sorting and movement; chromosome organization during various stages of its function; distribution and/or redistribution of organelles such as endosomes, lysosomes, and others; and retrograde axonal transport—transportation of cargo in the opposite direction of most kinesins.

The structure of dynein is much more complex than that of the other two classes of motors. Dynein has a seven-member planar ring structure that is overall approximately 10 times the molecular mass of kinesins. A diagrammatic representation of its structure is shown in Figure 23.41.

ATP binds to an AAA motif in domain 1. Its binding and hydrolysis induce conformational changes that are transmitted through domains 2–4 to the stalk that interacts with microtubules and causes step movements of 24–32 nm for an unloaded dynein. Dynein movement responds to load in a downshift, gear-like manner and, under heavy load, takes steps of approximately 8 nm. Down shifting appears to be associated with conformation changes in several of its other domains and the availability of ATP. Under heavy-load conditions, ATP also appears to bind AAA motifs in domain 3. AAA motifs are conserved regions of 220–230 amino acid residues that exist in a family of proteins that participate in a number of diverse cellular activities that depend on energy from ATP hydrolysis to effect their functions, which may include proteolysis, protein folding and unfolding, metal ion metabolism, and other activities in addition to those associated with dynein. The motif name AAA refers to **A**TPase **A**ssociated with diverse cellular **A**ctivities.

In summary, note that dyneins as motors are (1) structurally more complex than myosins or kinesins, (2) generally involved in retrograde movement of cellular material, and (3) involved in various other aspects of structural organization, and (4) dyneins step movement along the microtubule is load dependent.

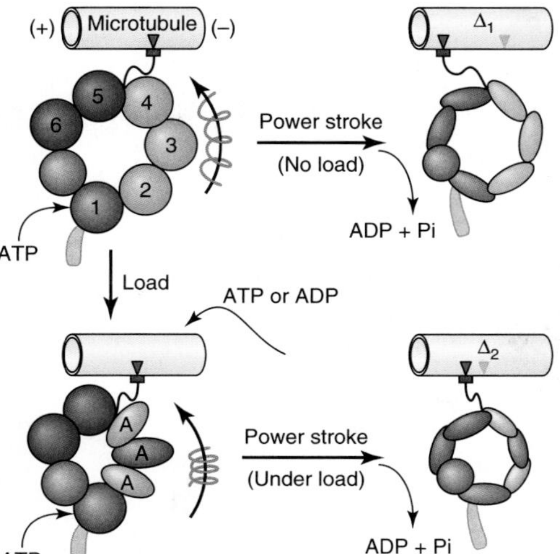

Figure 23.41 Dynein functions as a molecular motor.
Redrawn from Mallik, R., Carter, B. C., Lex, S. A., King, S. J., and Gross, S. P. Cytoplasmic dynein functions as a gear in response to load. *Nature* 427: 649, 2004.

23.4 • MECHANISM OF BLOOD COAGULATION

Biochemical Processes of Hemostasis

Blood circulation occurs in a very specialized type of closed system in which the volume of circulating fluid is maintained within a constant range. Multiple functions of the system make the transfer of solutes across its boundaries necessary. In order to maintain a state of hemostasis, no loss of blood, leaks that occur from various types of insults must be repaired.

Hemostasis requires that the process of clot formation (**procoagulation,** designated as **Phase 1**) is in balance with processes for stopping clot formation (**anticoagulation**, **Phase 2**) and for clot dissolution (**fibrinolysis, Phase 3**). Procoagulation leads to production of fibrin from fibrinogen and aggregation of fibrin into an insoluble network, or clot that covers the ruptured area and prevents further loss of blood. Concomitantly, aggregation of blood platelets occurs at the site of injury. Platelet aggregation forms a physical plug to help stop the leak. Platelets also undergo morphological changes and release some chemicals that aid in other aspects of the overall process, such as vasoconstriction to reduce blood flow to the area, and enzymes that aid directly in clot formation. Shown schematically in the following diagram are steps involved in the initial phase of this process.

INJURY
↓

EXTRINSIC PATHWAY	PLATELET AGGREGATION
↓	↓ ↑
THROMBIN FORMATION	RELEASE of FACTORS for COAGULATION and FORMATION of PLUG
↓ ↑	
INTRINSIC PATHWAY	

The dual arrows are intended not to imply that those steps are reversible but to indicate that the processes they join are mutually facilitative.

The next phase of hemostasis, the anticoagulation phase, begins early in the process, as soon as it is kinetically feasible, and prevents excessive clot formation. Finally, the process of clot dissolution, or fibrinolysis, occurs when the injured vessel is repaired sufficiently well to prevent further bleeding. Many proteins involved in blood coagulation contain **epidermal growth factor(EGF)** –like domains that may act directly or indirectly to facilitate re-growth of damaged areas in blood vessels. These processes are dynamic, kinetic, and mass-action driven. One phase does not end before another phase begins. Blood clotting is a dynamic process of signal amplification and modulation.

Some major proteins involved in this process are listed in Table 23.9, not necessarily in order of appearance. All are important and others are likely to be added. It is important to emphasize that structural anomalies, primarily mutations in critical regions of any of these proteins may affect adversely the clotting and clot dissolution processes and lead to disease states.

Clot formation involves two pathways: **intrinsic** or **contact factor pathway** and **extrinsic** or **tissue factor pathway**, both of which lead to the activation of Factor X. From this point on, there is a single pathway for clot formation. Historically, the term *intrinsic pathway* came from the observation that blood clotting could occur spontaneously when blood was placed in clean glass test tubes, leading to the idea that all components of the clotting process were intrinsic to circulating blood. Glass contains **anionic surfaces** that form nucleation points that initiate the process. In mammals, anionic surfaces are exposed on rupture of the **endothelial lining** of blood vessels, as are binding and activation sites for specific protein factors that initiate clotting in the intrinsic pathway. Similarly, *extrinsic* came from the observation that there was a factor extrinsic to circulating blood that facilitates blood clotting. This factor was identified as **Factor III**, or tissue factor, an integral membrane protein that becomes exposed on rupture of blood vessels. Whether intrinsic or extrinsic, blood coagulation is initiated and continued on the membrane at the site of injury. There is a great degree of necessity and interdependence on both traditionally designated pathways for affecting the coagulation process.

Throughout this section, coagulation factors in their nonactivated form are referred to by the letter "F" followed by the traditional Roman number designation for that factor,

TABLE 23.9 • Some Proteins Involved in Blood Coagulation, Control, and Clot Dissolution

Factor	*Name*	*Pathway*	*Characteristic*	*Concentration*[a]
I	Fibrinogen	Both		9.1
II	Prothrombin	Both	Contains *N*-terminal Gla residues	1.4
III	Tissue factor	Extrinsic	Transmembrane protein	—
IV	Calcium ions	Both		—
V	Proaccelerin	Both	Protein cofactor	0.03[b]
VII	Proconvertin	Extrinsic	Endopeptidase with Gla residues	0.01[c]
VIII	Antihemophilic factor	Intrinsic	Protein cofactor	0.0003[b]
IX	Christmas factor	Intrinsic	Endopeptidase with Gla residues	0.089
X	Stuart factor	Both	Endopeptidase with Gla residues	0.136
XI	Thromboplastin antecedent	Intrinsic	Endopeptidase	0.031
XII	Hageman factor	Intrinsic	Endopeptidase	0.375
XIII	Proglutamidase	Both	Transpeptidase	0.031[b]
	α_2-Antiplasmin		Plasmin inhibitor	0.953
	Antithrombin III	Both	Thrombin inhibitor	3.0
	Heparin Co-II	Both	Thrombin inhibitor	1.364
	HMWK[d]	Intrinsic	Receptor protein	0.636
	α_2-Macroglobulin		Proteinase inhibitor	2.9
	Plasminogen		Zymogen/clot dissolution	2.4
	Prekallikrein	Intrinsic	Zymogen/activator of factor XII	0.581
	Protein C	(Both)	Endopeptidase with Gla residues	0.065
	Protein C inhibitor		Protein C inhibitor	0.070
	Protein S	(Both)	Cofactor with Gla residues	0.030
	Protein Z	Both	Gla-containing cofactor for protein Z inhibitor	0.08
	Protein Z inhibitor	Both	Inhibitor of factor Xa	0.018
	TFPI[e]		Tissue factor pathway inhibitor	0.003

[a]Concentrations are approximate and shown as micromolar.
[b]These values approximate solution concentrations since some are complexed with other proteins in platelets.
[c]This factor probably circulates as both FVII and FVIIa.
[d]HMWK is high-molecular-weight kininogen.
[e]TFPI is tissue factor pathway inhibitor, formerly known as lipoprotein-associated coagulation factor (LAC1).

for example, **FVII** for Factor VII. The letter "a" will indicate activated forms, for example, FVIIa for the activated form of Factor VII.

Procoagulation Phase of Hemostasis (Phase 1)

The extrinsic pathway initiates formation of a blood clot. This has been demonstrated from kinetic analyses of the appearances and concentrations of various proteins and zymogens involved in the blood coagulation cascade. The extrinsic pathway is a one-step process that activates the key enzyme, FX to FXa, responsible for prothrombin activation. Thrombin, which results from prothrombin during the activation process, is a key enzyme for clot formation. Not only is it of prime importance for fibrin formation, but also it activates many other factors involved in other phases of hemostasis. A schematic of the reactions involved in the procoagulation phase, including both extrinsic and intrinsic pathways, is shown in Figure 23.42.

Extrinsic Pathway and Initiation of Coagulation

Tissue Factor (TF), or FIII, is the membrane receptor that initiates the procoagulation phase. TF is a transmembrane protein of 263 amino acids that is exposed on rupture of the endothelium lining of blood vessels and forms the receptor to which FVII binds. Residues 1–219 are on the extracellular side of the membrane and are exposed upon injury. FVII is a **γ-carboxyglutamyl** (Gla)-containing protein that binds to tissue factor only in the presence of Ca^{2+}. The resulting complex (FVII–Ca^{2+}–TF) is the initial enzyme complex of the extrinsic pathway, and its action initiates blood clotting by activating FX to FXa, the catalytic part of the enzyme complex (FXa:FV) responsible for thrombin formation from prothrombin. **TF** and **FVII** are unique to the extrinsic pathway and are essentially all of

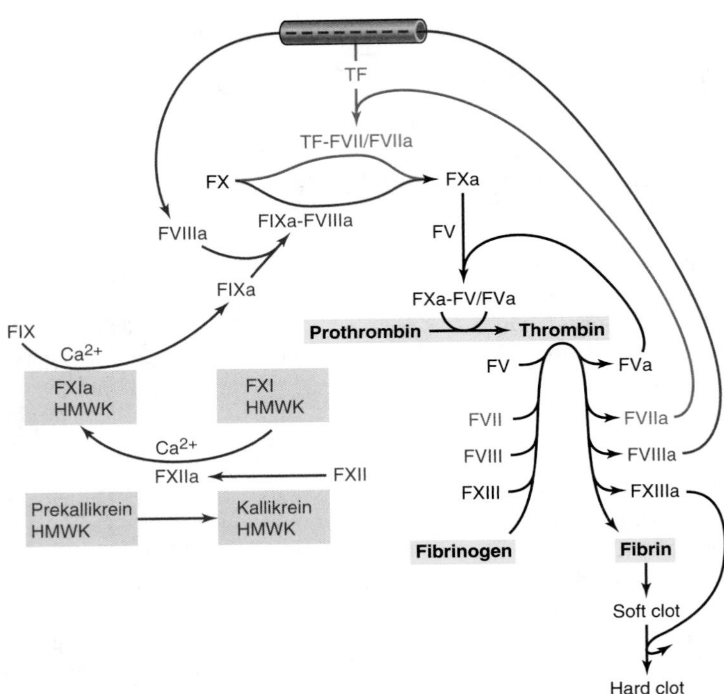

Figure 23.42 Diagrammatic representation of phase one of the blood coagulation scheme. Reactions of the extrinsic pathway are shown in red, those of the intrinsic pathway are shown in blue, and those common to both pathways are shown in black. Major end products of the scheme are highlighted in yellow. The inverted U in the lower right of the figure shows five proteins that are subjected to proteolytic cleavage by thrombin. The inactive forms are on the left, and the activated forms are on the right of the inverted U. Membrane-bound FXII activates prekallikrein to kallikrein, which activates XII to XIIa. These reactions occur in a cyclic manner. Abbreviation: HMWK, high-molecular-weight kininogen. Activated factors are designated with an "a."

its major components. The zymogen form of FVII is activated initially through protein–protein interaction resulting from its binding to TF. Additional FVII is activated by thrombin through specific proteolytic cleavage. FVIIa has a long half-life in circulating blood, and a small amount normally exists. This has no adverse effects since FVIIa is not catalytically active until it forms a complex with TF.

Thrombin Formation

FXa and FV form a complex sometimes referred to as prothrombinase that catalyzes formation of thrombin from prothrombin by a proteolytic cleavage reaction. The relatively small amount of thrombin formed in this initiation phase (Figure 23.42) catalyzes the activation of FV, FVII, FVIII, and FXIII (all of the components on the left side of the inverted loop of Figure 23.42), in addition to its ultimate functional goal of converting fibrinogen to fibrin. Thrombin converts fibrinogen to fibrin by cleavage of four highly negatively charged peptides from the center domain of fibrinogen. Those negatively charged peptides of fibrinogen prevent it from aggregating. Once they are removed by thrombin, the resulting fibrin can aggregate into a network forming a soft clot.

Reactions of the Intrinsic Pathway

Also shown in Figure 23.42 are reactions of the intrinsic pathway. Injury to the endothelial lining of blood vessels exposes anionic membrane surfaces. The zymogen, **FXII**, binds directly to some of these anionic surfaces and undergoes a conformational change that increases its catalytic activity 10^4- to 10^5-fold. **Prekallikrein** and **FXI** are also zymogens and circulate in blood as complexes with **high-molecular-weight kininogen (HMWK)**, either as a FXI–HMWK complex or a prekallikrein–HMWK complex. FXI and prekallikrein bind, through their interactions with HMWK, to anionic sites that become exposed on injury to membrane surfaces. This brings those zymogens to the site of injury and in direct proximity to FXII. The membrane-bound activated form of FXII activates prekallikrein to **kallikrein**, which then permanently activates FXII by specific proteolytic cleavage to give FXIIa. The kallikrein-activated XIIa activates more prekallikrein to kallikrein, which then activates more XII to XIIa in a cyclic autocatalytic manner.

FXI, in the HMWK–membrane-bound complex, is activated by FXIIa by proteolytic cleavage to give FXIa. FXIa activates **FIX** to FIXa (Clin. Corr. 23.14), which in the presence of **FVIIIa**, also activates **FX** to FXa.

CLINICAL CORRELATION 23.14

Intrinsic Pathway Defects: Prekallikrein Deficiency

Components of the intrinsic pathway include Factor XII (Hageman Factor), Factor XI, prekallikrein (Fletcher Factor), and High-Molecular-Weight Kininogen. Inherited disorders in each appear to be autosomal recessive and associated with an increase in activated partial thromboplastin time (APTT). Factor XI deficiency is directly associated with a clinical bleeding disorder.

In prekallikrein (Fletcher Factor) deficiency, autocorrection after prolongation of the preincubation phase of the APTT test occurs. This is explained by activation of Factor XII by an autocatalytic mechanism. The reaction is very slow in prekallikrein deficiency since the rapid reciprocal autoactivation between Factor XII and prekallikrein cannot take place. Prekallikrein deficiency may be due to a decrease in the amount of the protein synthesized, to a genetic alteration in the protein itself that interferes with its ability to be activated or its ability to activate Factor XII. A lack of knowledge of the structure of the gene or the protein will not permit an explanation of the mechanism of this deficiency. Specific deficiencies of the intrinsic pathway, however, can be localized to a specific factor if the appropriate tests are performed. These may include measurement of the amount of each factor in plasma and an APTT test performed with and without prolonged preincubation time. In a 9-year-old girl with prolonged APTT, the functional level of prekallikrein was less than 1/50th of the minimum normal value. An immunological test showed an antigen level of 20%–25% suggesting that she was synthesizing a dysfunctional molecule.

Coleman, R. W., Rao, A. K., and Rubin, R. N. Fletcher factor deficiency in a 9-year-old girl: Mechanisms of the contact pathway of blood coagulation. *Amer. J. Hematol.* 48:273, 1995.

The intrinsic pathway is essentially a four-step cascade started by the contact activation of FXII and the autocatalytic activation of FXII and kallikrein to give FXIIa (step 1). FXIIa activates FXI (step 2). In step 3, FXIa activates FIX; and in step 4, FIXa, in the presence of FVIIIa, activates FX. If each activated enzyme molecule catalyzed the activation of 100 molecules of the next enzyme of the cascade, an amplification factor of 1×10^6 of the intrinsic pathway would be achieved. As shown in the diagram and as stated, several feedback loops accelerate the overall process and produce a fibrin clot in a rapid and efficient manner. During this time, **FXIII**, a transglutamidase (often referred to as transglutaminase) that has been activated also by thrombin (lower right corner of Figure 23.42), is actively catalyzing the formation of cross-links between fibrin monomers of the soft clot to form a hard clot. This is the overall process of clot formation.

Formation of Platelet Plug

Thrombin mediates clumping of platelets at the site of injury. Endothelial cells contain a **thrombin receptor**, a member of the seven-transmembrane-helix family of receptors. This receptor is exposed on injury and is activated by α-thrombin. Aggregation of platelets is facilitated by binding to this activated receptor. In addition to forming a physical plug, platelets undergo a morphological change and release ADP, serotonin, some types of phospholipids, and proteins that aid in coagulation and tissue repair (Figure 23.43). They also release a glycoprotein, von Willebrand factor (vWF) that concentrates in the area of the injury and forms links between the exposed receptor and platelets.

Platelet aggregation becomes autocatalytic on release of ADP and **thromboxane A_2**. Another protein released from platelets is **FIV**, a **heparin-binding protein**, which prevents premature formation of heparin–antithrombin III complexes and it attracts cells with anti-inflammatory activity to the site of injury. About 20% of FV and one form of FXIII, the transglutamidase, are present in platelets. Intact normal vascular endothelium does not bind platelets because (1) receptors and other elements are not exposed, (2) activators such as ADP are rapidly degraded or are not present in blood in sufficient concentration to be effective, and (3) the endothelium secretes prostacyclin (PGI_2), a potent inhibitor of platelet aggregation.

Some Properties of Proteins Involved in Clot Formation

Tissue Factor: (TF, or FIII), (Figure 23.44*a*) is a transmembrane protein of 263 amino acids. Residues 243–263 are located on the cytosolic side of the membrane. Residues 220–242 are hydrophobic residues and represent the transmembrane sequence. Residues 1–219 are on the outside of the membrane, are exposed after injury, and form the receptor for FVII binding and formation of the initial complex of the extrinsic pathway. This domain is glycosylated

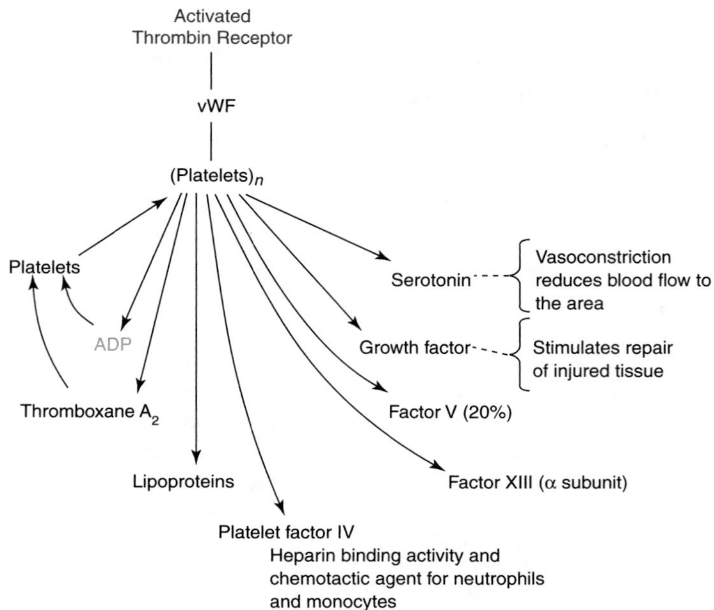

Figure 23.43 Action of platelets in blood coagulation.

and contains four cysteine residues. A stereo representation of a section of it highlighting some of the amino acid residues involved in **FVII** binding is shown in Figure 23.44*b*.

Factor VII: A 3-D ribbon structural representation of FVIIa is shown in Figure 23.45. The regions for TF interaction, Ca^{2+} binding, and the substrate-binding pocket are highlighted (Clin. Corr. 23.17).

Factor X: A stereo view of FXa is shown in Figure 23.46. Both the extrinsic and the intrinsic pathways lead to formation of FXa from FX. FXa is the catalytic component of the FXa:FVa complex.

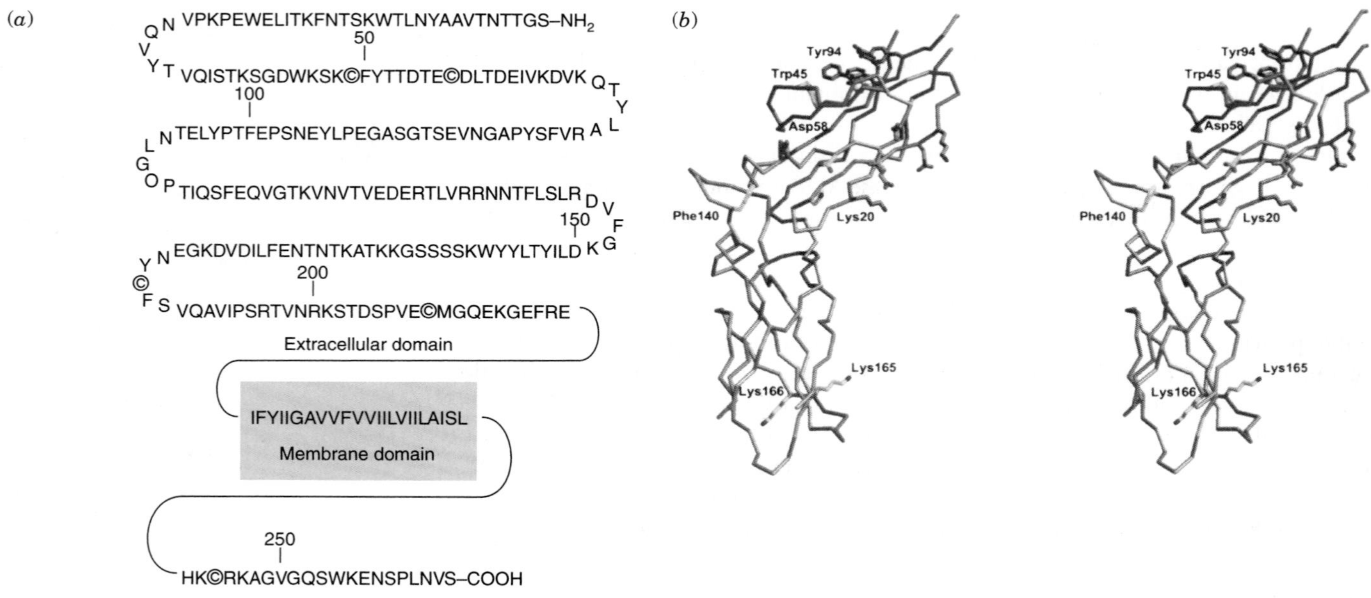

Figure 23.44 Tissue Factor. (*a*) Amino acid sequence of human tissue factor derived from its cDNA sequence. (*b*) A stereo representation of the carbon chain of the extracellular domain of tissue factor. Residues important for binding of Factor VII are shown in yellow. Clusters of aromatic and charged residues are shown in light blue.

Redrawn from Spicer, E. K., Horton, R., Bloem, L., Bach, R., et al. *Proc. Natl. Acad. Sci. USA* 84: 5148, 1987.
Reproduced with permission from Muller, Y. A., Ultsch, M. H., Kelley, R. F., and deVos, A. M. *Biochem.* 33: 10864, 1994. Copyright (1994) American Chemical Society. Figure generously supplied by deVos, A. M., and Muller, Y.A., Genenteck, Inc., San Francisco, CA.

Factor V: FV does not have catalytic activity, but it is a protein cofactor for FXa. In its inactivated form it has some activity as a cofactor but is much more active after activation to FVa. FV is a 330-kDa protein that is activated by proteolytic cleavage at Arg^{709} and Arg^{1545}. FVa is a heterodimer consisting of an *N*-terminal domain (105 kDa) and a C-terminal domain (74 kDa) that are held together noncovalently by Ca^{2+} (see Figure 23.51, p. 989). The substrate for the FXa:FVa complex is prothrombin and the complex is sometimes referred to as prothrombinase.

Thrombin: Thrombin circulates in plasma as prothrombin. Prothrombin is a 72 kDa protein (Figure 23.47) that contains ten **γ-carboxyglutamate (Gla) residues** in its *N*-terminal region. Binding of Ca^{2+} to these residues neutralizes negative charges and facilitates binding of prothrombin to membrane surfaces and to the prothrombinase complex (FXa:FVa) at the site of injury. Prothrombin is activated by two proteolytic cleavages on the carboxyl side of arginine residues; first at position 320 and then at position 284. The active thrombin molecule (α-thrombin) consists of two chains, 6 kDa and 31 kDa, that are covalently linked by a disulfide bond. A stereo view of the active α-thrombin molecule is shown in Figure 23.48. Regions involved in some of its functions are highlighted. Thrombin activates many blood coagulation factors as shown in Figure 23.42, p. 984. Its major substrate for clot formation, however, is fibrinogen.

Fibrinogen/Fibrin: Fibrinogen, a large molecule of approximately 340 kDa consists of two sets of tripeptide units with α,β,γ structure (Figure 23.49) that are linked together at their *N*-terminal regions by disulfide bonds. Fibrinogen has three globular domains, one on each end and one in the middle joined by rodlike domains. Short segments of the free *N*-terminal regions project out from the central globular domain. The *N*-terminal regions of α-, α'-, β-, and β'-subunits are highly negatively charged and through charge–charge repulsion prevent

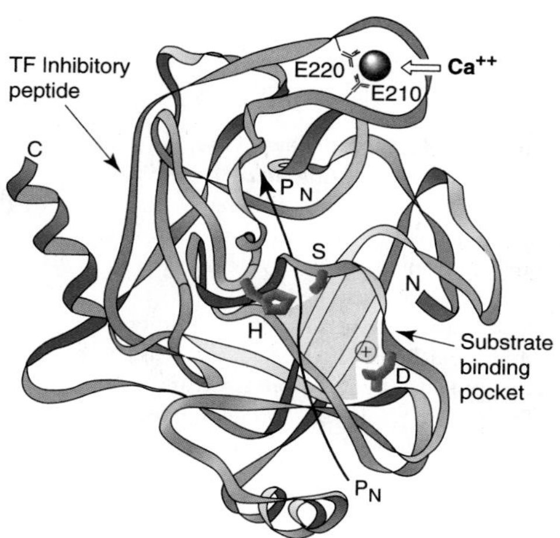

Figure 23.45 Ribbon structural representation of the protease domain of Factor VIIa. The dark ribbon labeled "TF Inhibitory peptide" represents a section involved in binding to Tissue Factor. The catalytic triad is shown in the substrate-binding pocket as H, S, and D for His-193, Ser-344, and Asp-338, respectively. The arrow labeled P_N–P'_N lies in the putative extended substrate binding region. Reproduced with permission from Sabharwal, A. K., Birktoft, J. J., Gorka, J., Wildgoose P., et al. J. *Biol. Chem.* 270: 15523, 1995.

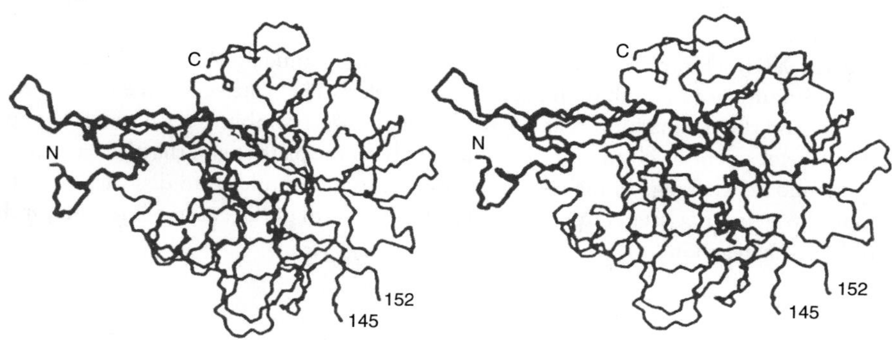

Figure 23.46 Stereo view of the CN-backbone structure of Factor Xa. The EGF-like domain is in bold.
Reproduced with permission from Padmanabhan, K., Padmanabhan, K. P., Tulinsky, A., Park, C. H., et al. *J. Mol. Biol.* 232: 947, 1993.

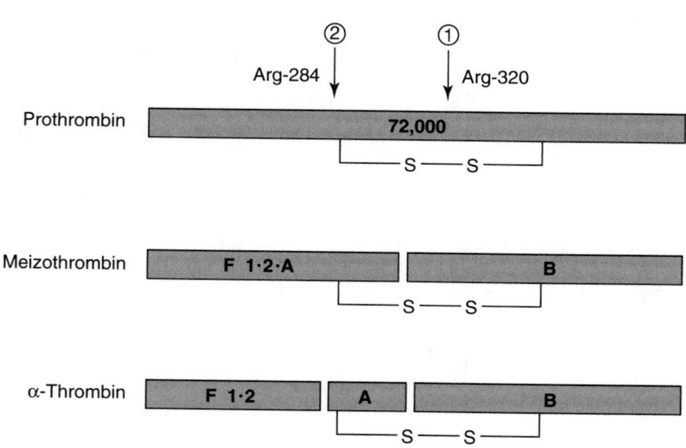

Figure 23.47 Schematic diagram of prothrombin activation.

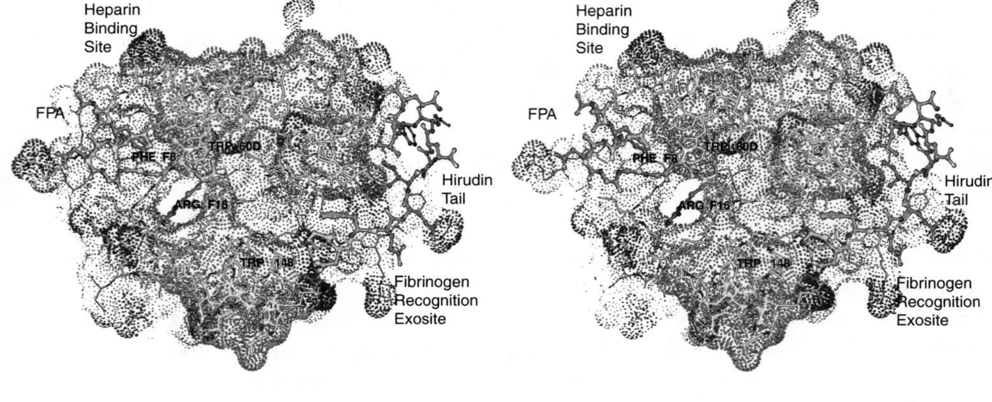

Figure 23.48 Stereo view of the active-site cleft of human α-thrombin. Dark blue, basic amino acids; red, acid; and light blue, neutral. The active-site cleft is oriented from left to right. Site of heparin binding is also illustrated.
Reproduced with permission from Stubbs, M. T., and Bode, W. The clot thickens: clues provided by thrombin structure. *Trends Biol. Sci.* 20: 23, 1995, Copyright (1995) Elsevier.

aggregation of fibrinogen. Thrombin cleaves these *N*-terminal peptides and allows the resulting fibrin molecules to aggregate and to form a **"soft" clot**. **FXIIIa** stabilizes the soft clot by catalyzing formation of isopeptide linkages between the δ-carboxylamide of glutamine residues of one fibrin molecule and the ε-amino group of lysine residues of another fibrin molecule (Figure 23.50), that is, exchanges the amide nitrogen of glutamine for the ε-amino group of lysine. Ammonia is released in this reaction and a **hard clot** is formed in which individual fibrin molecules throughout the clot are interconnected by covalent linkages.

Factor VIII: FVIII circulates in plasma noncovalently bound to **von Willebrand Factor (vWF)**. FVIII, a 285-kDa protein, is activated by thrombin cleavage at Arg^{372}, Arg^{740}, Arg^{1648}, and Arg^{1689}. This last cleavage releases FVIIIa from vWF. FVIIIa is a heterotrimer (Figure 23.51) consisting of N-terminal peptides of 40 kDa (A_2) and 50 kDa (A_1), and a C-terminal peptide of 74 kDa (A_3). FVIIIa also contains a Ca^{2+} bridge between the *N*- and C-terminal domains. Classic hemophilia A results from a deficiency in FVIII (Clin. Corrs. 23.15 and 23.16).

Factor XIII: Thrombin also activates FXIII, transglutamidase (Figure 23.52, p. 989). **Protransglutamidase** exists in both plasma and platelets. The platelet enzyme is an α_2 configuration, and the plasma form is an $\alpha_2\beta_2$ configuration. Thrombin activates FXIII by specific cleavage of a peptide bond in the α-subunit of both the platelet and the plasma forms of transglutamidase. Cleavage in the α-subunit of the plasma form leads to dissociation of the β-subunit, which is not catalytically active. The platelet form of the enzyme is released at the site of fibrin aggregation.

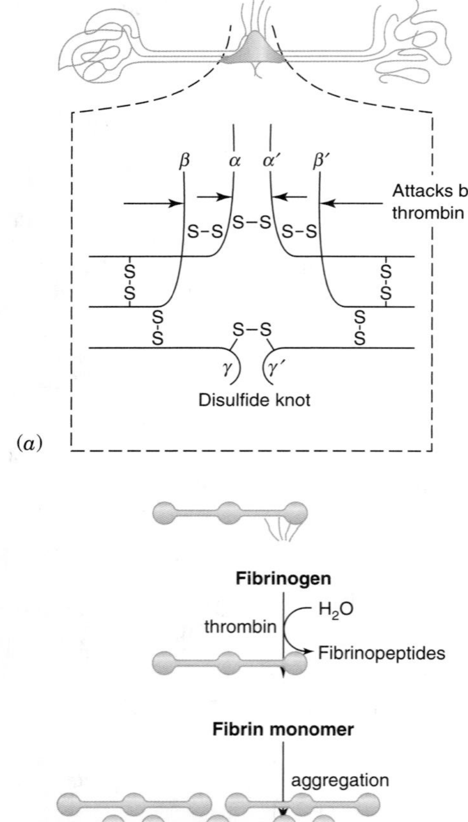

Figure 23.49 Diagrammatic representation of the fibrinogen molecule and its conversion to the soft clot of fibrin.

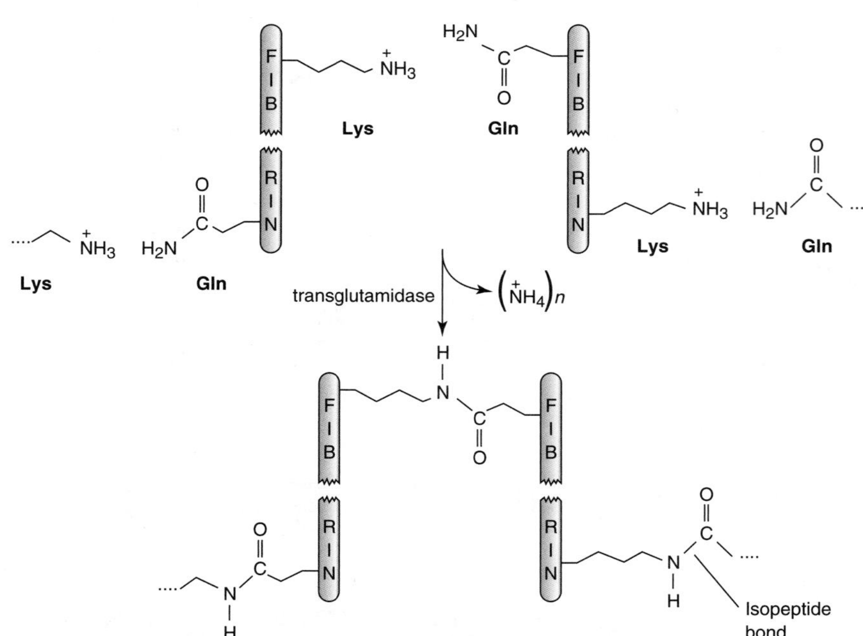

Figure 23.50 Reactions catalyzed by transglutamidase.

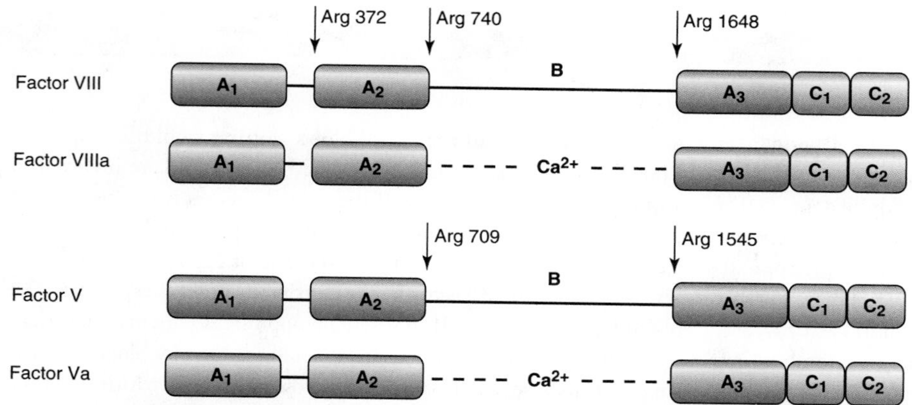

Figure 23.51 Organizational structure of Factors VIII and V. Positions for thrombin cleavage are shown. Structural domains are represented by the letters A and C.
Redrawn from Kalafatis, M., Swords, N. A., Rand, M. D., and Mann, K. G. *Biochim. Biophys. Acta* 1227: 113, 1994.

High-Molecular-Weight Kininogen (HMWK), Prekallikrein, and **Factor XI:** Two of the proteins involved in the intrinsic pathway, prekallikrein and FXI, circulate in blood as complexes with HMWK. Figure 23.53 shows a schematic diagram of the functional regions of HMWK. The binding site on HMWK for prekallikrein consists of approximately 31 amino acid residues, relative positions of which are shown. FXI binds to approximately 58 amino acid residues (not shown) that overlap the 31 amino acid residues to which prekallikrein binds. A molecule of HMWK can bind to one or the other of these proteins, but not to both simultaneously. Bradykinin, a vasodilator, is released from HMWK by the action of kallikrein. Prekallikrein is a 619-amino-acid protein. Prekallikrein is converted to kallikrein by FXIIa, which cleaves a peptide bond between Arg^{371} and Ile^{372}. Kallikrein contains two chains covalently linked by a single disulfide bond. The C-terminal domain of 248 amino acids contains the catalytic site.

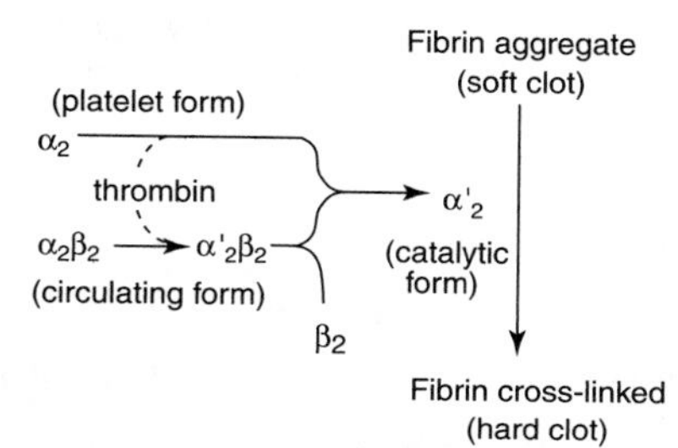

Figure 23.52 Activation of transglutamidase by thrombin.

Anticoagulation Phase of Hemostasis (Phase 2)

Inhibition of the hydrolyases involved in blood coagulation is a kinetic process that begins almost as soon as coagulation itself. Initially, formation of inhibitor complexes is slow because concentrations of the enzymes with which the inhibitors interact are low. As activation of the zymogens proceeds, inhibition increases and becomes more prominent. These reactions, and destruction of protein cofactors, eventually stop the coagulation process completely. In general, **protease–inhibitor complexes** do not dissociate readily and they are removed intact from blood by the liver.

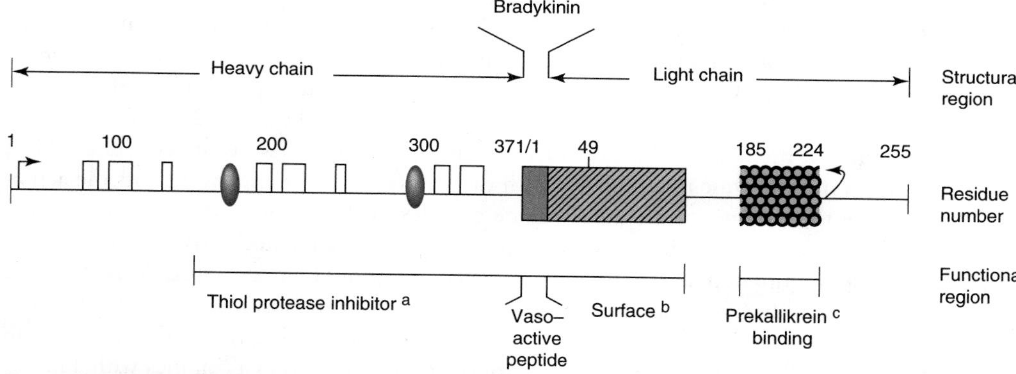

Figure 23.53 Schematic diagram of the functional regions of human high-molecular-weight kininogen (HMWK). Bradykinin is derived from near the middle of HMWK by proteolysis. The resulting two chains are held together by disulfide bonds, horizontal arrows.
Redrawn from Tait, J. F., and Fujikawa, K. *J. Biol. Chem.* 261: 15396, 1986.

CLINICAL CORRELATION 23.15

Classic Hemophilia

Hemophilia is an inherited disorder characterized by a permanent tendency for hemorrhages, spontaneous or traumatic, due to a defective blood-clotting system. Classic hemophilia, hemophilia A (OMIM 306700), is an X-linked recessive disorder characterized by a deficiency of Factor VIII. It affects about 1 in 10,000 males. Of the approximately 25,000 hemophiliacs in the United States, more than 80% are of the A type. Hemophilia B is due to a dysfunction in Factor IX.

Some hemophilia A patients may have a normal prothrombin time if the concentration of tissue factor is high. One possible explanation for this is that Factor V in human plasma is much lower in concentration than Factor X. Activation of an amount of Factor X to Xa in excess of that required to bind all of Factor Va would initiate blood clotting by the extrinsic pathway and give a normal prothrombin time. The intrinsic pathway would not function normally due to the deficiency in Factor VIII. Without the two pathways operating in concert, the overall process of blood clotting would be impaired. Both Factor Xa and thrombin activate Factor V and are involved in a number of other reactions. If the overall process is not accelerated at its onset by intervention of the intrinsic pathway, due to kinetics of the interaction of thrombin and Factor Xa with the normally low concentration of Factor V, the clotting disorder is expressed. The blood level of Factor VIII in severe hemophilia A patients is less than 5% of normal. These patients are generally treated by blood transfusion with its associated dangers: the possibility of hepatitis or HIV/AIDS and a 6% possibility of patients making autoantibodies. Treatment of hemophiliacs has been made much safer as a result of cloning and expression of the gene for Factor VIII. The pure recombinant protein is administered to patients with minimum danger.

Nemerson, Y. *Blood* 71: 1, 1988.

CLINICAL CORRELATION 23.16

Use of Recombinant Factor VIIa to Control Bleeding

Factor VIIa (FVIIa) is a major component of the extrinsic pathway for blood coagulation. It can be present in circulating blood without causing harmful effects because it is inactive until it forms a complex with Tissue Factor (Factor III). A recombinant form of FVIIa (rFVIIa) was initially developed as a treatment for hemophilia patients who had acquired inhibitors associated with their treatment. It has been commercialized and approved for treatment of various traumatic bleeding problems, particularly some types of surgery and bleeding strokes. It has also been demonstrated that rFVIIa given along with phospholipids enhances thrombin generation and improves its effectiveness.

Aggarwal, A., Malkovska, V., Catlett, J. P., and Alcorn, K. Recombinant activated factor VII (rFVIIa) as salvage treatment for intractable hemorrhage. *Thrombosis J.* 2: 9, 2004; Butenas, S., Brummel, K. E., Paradis, S. G., and Mann, K. G. Influence of Factor VIIa and phospholipids on coagulation in "acquired" hemophilia. *Arterioscler Thromb Vasc. Biol.* 23: 123, 2003.

Inhibition of the Extrinsic Pathway

Inhibition of the extrinsic pathway, that is, the TF–FVIIa–Ca^{2+}–FXa complex, is unique and involves specific interaction with **Tissue Factor Pathway Inhibitor (TFPI)**, formerly known as **lipoprotein-associated coagulation inhibitor (LACI)** and as **anticonvertin**. TFPI is a 32-kDa protein that contains three tandem domains (Figure 23.54). Each domain is a functionally homologous protease inhibitor (sometimes referred to as Kunitz domains) that resembles other individual protease inhibitors such as the bovine **pancreatic trypsin inhibitor**. TFPI inhibits the extrinsic pathway by interacting specifically with the TF–FVIIa–Ca^{2+}–FXa complex. First, domain 1 binds to FXa and domain 2 binds to FVIIa of the complex. Binding of TFPI to FVIIa does not occur unless FXa is present. Thus, TFPI is truly a multienzyme inhibitor in which each of its separate domains inhibits the action of one of the enzymes of the multienzyme complex of the extrinsic pathway. Second, the TFPI–FXa complex mediates internalization of FVIIa by an endocytosis mechanism. The C-terminal (third domain) of TFPI appears to be necessary for endocytosis. Most of FVIIa is degraded within the cells but a small amount returns to the surface as intact protein and is apparently one of the sources of circulating FVIIa. As previously stated, FVIIa has no detrimental effects since it is active as a protease only when it is in a complex with TF.

Protease inhibitors of the **serine protease inhibitor (Serpin)** family of proteins in blood interact with and inhibit other enzymes of the blood coagulation system. There is a similarity in tertiary structure among them with a common core domain of about 350 amino acids.

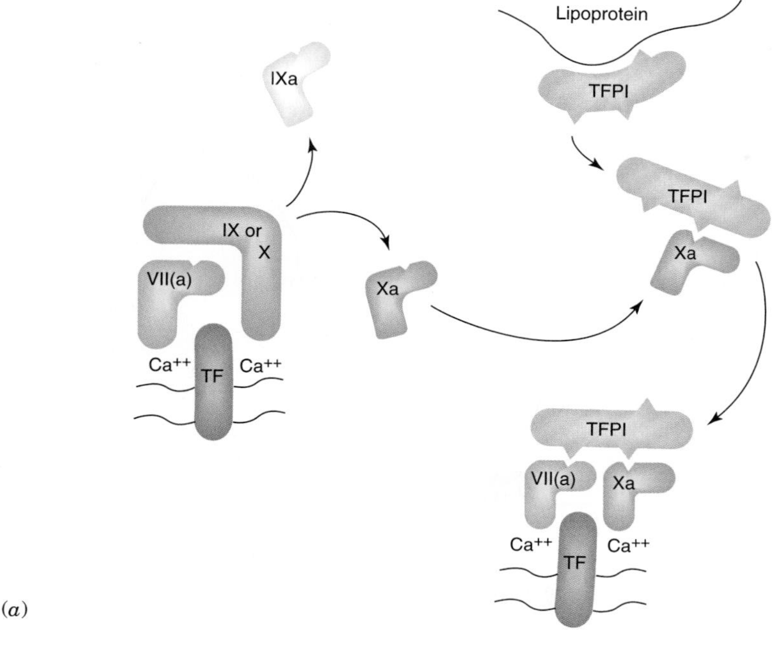

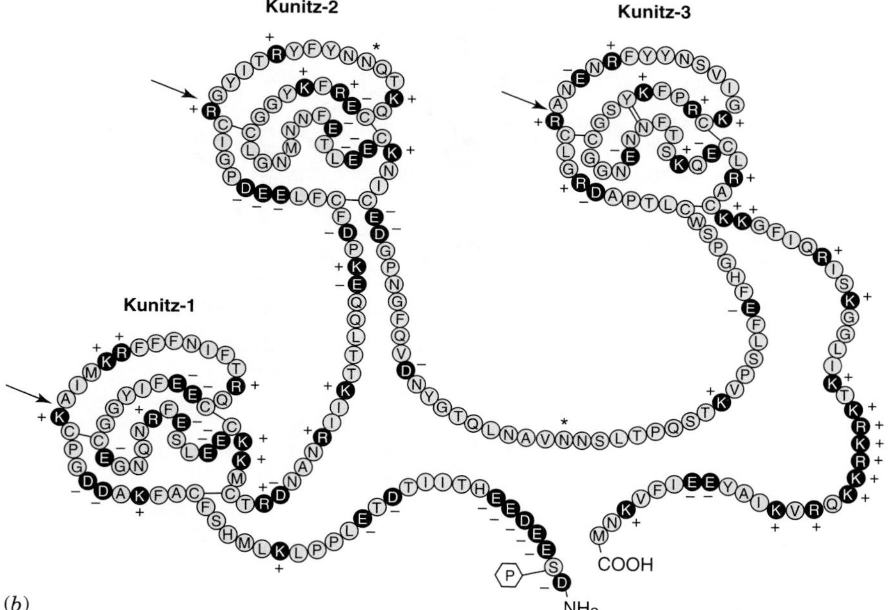

Figure 23.54 Mechanism of inhibition of the extrinsic pathway. (*a*) TFPI is tissue factor pathway inhibitor. A schematic of its secondary structure is shown in (*b*). Kunitz domain 1 inhibits Factor VIIa and Kunitz domain 2 inhibits Factor Xa. Domain 3 is necessary for endocytosis of the complex. Arrows indicate the presumed location of the active-site inhibitor region for each domain. Reproduced with permission from Brooze, G. J., Girard, T. J., and Novotny, W. F. *Biochem.* 29: 7539, 1990. Copyright (1990) American Chemical Society.

Antithrombin III (AT3) is a serpin that inhibits several hydrolases of the blood coagulation system, but most specifically thrombin and FXa. AT3 inhibits thrombin and FXa more efficiently as complexes with different oligosaccharide groups of heparin. Heparin is a highly sulfated oligosaccharide of the glucosaminoglycan type. It exists as a mixture of oligosaccharides spanning a range of molecular sizes. The interaction of heparin in the different inhibitory complexes is size dependent. At least 18 saccharide units are required for the effective formation of the thrombin inhibitor complex.

$$\text{Thrombin} + \text{Heparin}_{(>18)} + \text{AT3} \rightarrow \text{AT3:Heparin}_{(>18)}\text{:Thrombin}_{(\text{inactive})}$$

AT3 also forms a complex with a specific pentasaccharide form of heparin. The structure of this pentasaccharide is shown in Figure 23.55. Shown in Figure 23.56 is a structural model of AT3 bound to this heparin pentasaccharide. This complex inhibits FXa.

$$\text{FXa} + \text{Heparin}_{(5)} + \text{AT3} \rightarrow \text{AT3:Heparin}_{(5)}\text{:FX}_{\text{inh}}$$

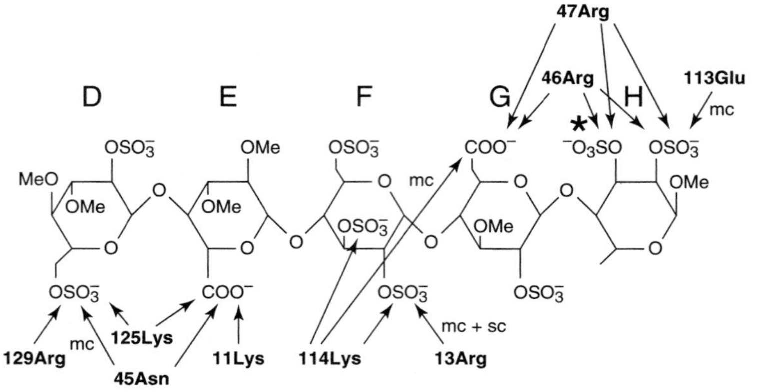

Figure 23.55 Chemical structure of the heparin pentasaccharide and positions of its interactions with specific residues of antithrombin.
Redrawn from Whisstock, J. C., Pike, R. N., Jin, L., Skinner, R., et al. *J. Mol. Biol.* 301: 1287, 2000.

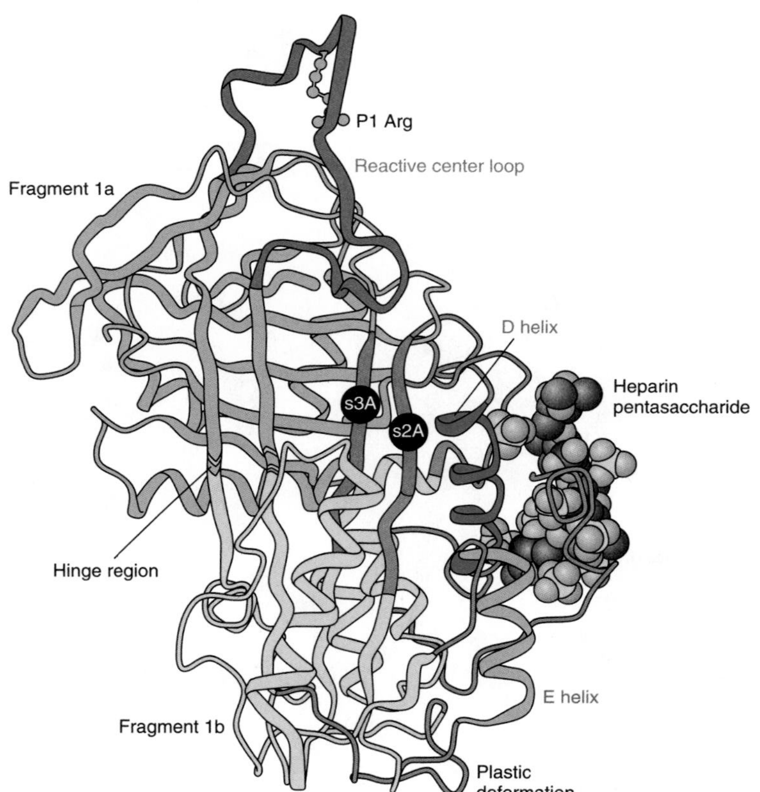

Figure 23.56 Fragment of antithrombin to which the heparin pentasaccharide binds.
Fragment of antithrombin in the presence of the heparin pentasaccharide.
Reproduced with permission from Whisstock, J. C., Pike, R. N., Jin, L., Skinner, R., et al. *J. Mol. Biol.* 301:1287, 2000.

Although AT3 inhibits thrombin and FXa in the absence of heparin, heparin enhances inhibition of thrombin by a factor of approximately 9,000 and inhibition of FXa by a factor of approximately 17,000.

There is another pathway for inhibition of FXa. Blood contains a 62-kDa Gla-containing glycoprotein, **Protein Z (PZ)**. Protein Z is a protein cofactor that, in the presence of Ca^{2+}, interacts with membrane and forms a complex with another plasma protein, **Protein Z-dependent protease inhibitor (ZPI)**, a 72-kDa protein. This complex inhibits FXa.

$$\mathrm{FXa} + \mathrm{Ca}^{2+} + \mathrm{PZ} + \mathrm{ZPI} \rightarrow (\mathrm{Ca}^{2+}:\mathrm{PZ}):\mathrm{ZPI}:\mathrm{FX}_{\mathrm{inh}}$$

Protein Z-dependent protease inhibitor (ZPI) in the absence of PZ will inhibit FXIa, and this inhibition is enhanced by heparin. ZPI with or without PZ shows no measurable ability to inhibit other proteases, including thrombin, FVIIa, FIXa, and Protein C.

Inactivation of FVa and FVIIIa

Protein C (PC), a Gla-containing protein, is activated in a membrane-bound complex of thrombin, **thrombomodulin**, and calcium ions. Protein C requires the presence of another

CLINICAL CORRELATION 23.17

Thrombosis: Defects of the Protein C Pathway and Increased Levels of Coagulation Factors

Four major proteins are involved in the action of Protein C in regulating blood coagulation: Protein C itself; Protein S, a cofactor for Protein C; Factor Va; and Factor VIIIa. The latter two are substrates for catalytic action of the Proteins C–S complex. Mutations in any of them can result in venous thrombosis with various degrees of severity.

De novo mutations have been identified in patients with type I Protein C deficiency. One was a missence mutation, a transition of T to C, which changed amino acid residue 270 from serine to proline (Ser270Pro) resulting in decreased activity. The gene for Protein C is on chromosome 2 and has nine exons and eight introns. Another was a *de novo* 5-bp deletion (underlined below) located at the exon VI-intron f junction resulting in a read through of sections of the intron.

Exon VI ◇ Intron f

Normal sequence:

CAC CCC GCA*G* ◇ *G*TGAGAAGCCCCAATAT- - - -

His Pro Ala

Mutated sequence:

CAC CCC GCAGGA GCC CCC AAT AT- - - -

His Pro Ala Gly Ala Pro Asn- - - - - -

The sequence normally translated is in bold type. The degree of clinical severity of thrombotic events depends on the extent to which the gene inherited from the other parent is normal and expressed.

Resistance to activated Protein C because of single-point mutations in its substrates, Factor Va and Factor VIIIa, can occur and prevent or retard their inactivation (proteolysis) by Protein C. The most commonly identified cause is a single-point mutation in the gene for Factor V results in R506Q, also known as FV-Leiden.

A third cause of Protein C-related thrombosis is a defect in Protein S. Fewer specific details are available that permit a definition of the mechanism of the interaction between Protein C and Protein S, and likewise of the mutations that affect its function. It is quite clear, however, that Protein S deficiency also leads to thrombotic events. Venous thrombosis occurs in almost one-half of patients at some stage of their lives if they have deficiencies in functional amounts of Protein S.

Patients with elevated levels of FVIII, FIX, or FXI appear to be predisposed to venous thrombosis. Risk for thrombosis from these factors is low compared to the Protein-C-associated condition described above, but multiple elevations of these factors increase the risk. The biochemical cause for this is not clear, but hereditary and undetected gene mutations are leading suspects.

Gandrille, S., Jude, B., Alhenc-Gelas, M., Emmerich J., and Aiach, M. First *de novo* mutations in the protein C gene of two patients with type I deficiency: A missence mutation and a splice site deletion. *Blood* 84: 2566, 1994; Reitsma, P. H., Bernardi, F., Doig, R. G., Gandrille, S., et al. Protein C deficiency: A database of mutations, 1995 update. *Thrombosis and Haemostasis* 73: 876, 1995; and Crowther, M. A., and Kelton, J. G. Congential thrombophilic states associated with venous thrombosis: A qualitative overview and proposed classification system. *Ann. Internal Med.* 138: 128, 2003.

protein cofactor, **Protein S (PS)**, a 75-kDa Gla-containing protein. The PC:PS complex inhibits coagulation by inactivating Factors Va and VIIIa. Inactivation of FVa and FVIIIa occurs by cleavage of peptide bonds at specific arginine residues. Deficiencies and/or mutations in Protein S or Protein C can lead to **thrombotic diseases** (Clin. Corr. 23.17). Disease states are also possible if there are mutations of FV or FVIII that do not affect their ability to function as cofactors but do affect the cleavage sites for the PC:PS complex inactivation reactions.

$$V_a \text{ and } VIII_a \xrightarrow{PC:PS} V_{inh} \text{ and } VIII_{inh}$$

Thrombomodulin is an integral glycoprotein of endothelial cell membranes (A Closer Look 23.10). It is a receptor for thrombin. In the thrombin–thrombomodulin–Ca^{2+} complex, thrombin has a decreased affinity for fibrinogen and an increased affinity for Protein C. Thus, thrombin's role switches from that of procoagulation to anticoagulation.

Thrombin (Figure 23.57) may exist *in vitro* in two conformations: one has high specificity for conversion of fibrinogen to fibrin, and the other conformation has low specificity for fibrinogen conversion but high specificity for thrombomodulin binding and for proteolytic activation of Protein C. These forms are referred to as *fast* and *slow* forms, respectively. This type of dynamic feedback mechanism is important for stopping the clotting process at its point of origin.

There is also a specific inhibitor for Protein C. **Protein C inhibitor (PCI)** has been found in plasma, platelets, and megakaryocytes. ADP, epinephrine, thrombin, and other molecules that stimulate platelet activity cause release of about 30% of PCI from platelets on stimulation. Inactivation of activated Protein C (APC) by PCI occurs on membrane surfaces like most of the other reactions discussed.

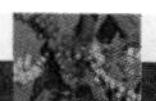

A CLOSER LOOK 23.10

Thrombomodulin

Thrombomodulin contains 560 amino acids and shows sequence homology with the low-density lipoprotein receptor, but very little with tissue factor. There is a great deal of similarity in functional domains between tissue factor and thrombomodulin, each of which is a receptor and activator for a protease.

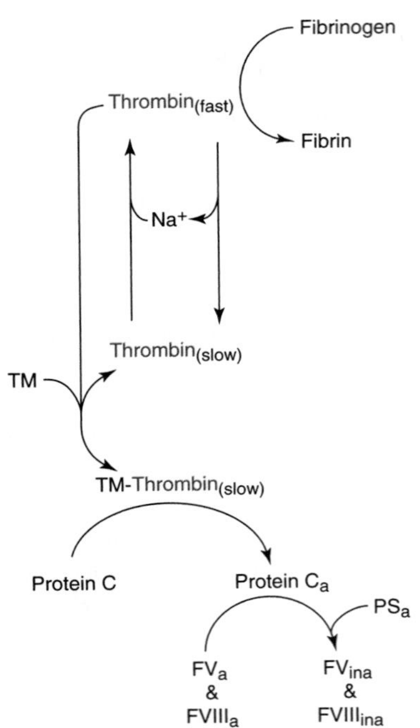

Figure 23.57 Conformational forms of thrombin induced by its interaction with thrombomodulin. Thrombin has high affinity for fibrinogen in the absence of thrombomodulin, but changes to high affinity for Protein C in the presence (complex) with thrombomodulin.

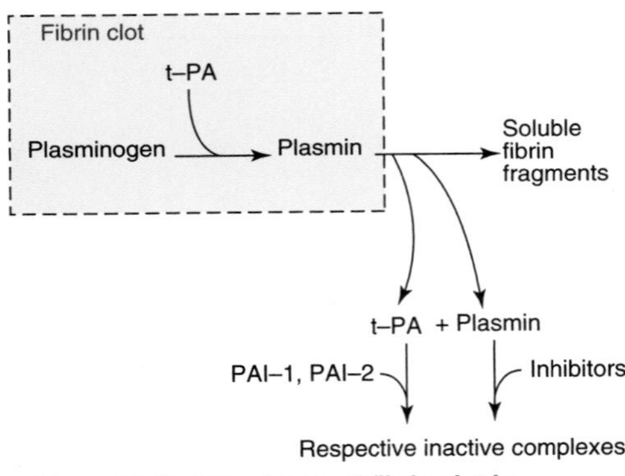

Figure 23.58 Dissolution of fibrin clot by plasmin. Plasminogen is activated to plasmin by t-PA within the matrix of the clot. These reactions are shown within the box in this figure.

Fibrinolysis Phase of Hemostasis (Phase 3)

Fibrinolysis Requires Plasminogen and Tissue Plasminogen Activator (t-PA) to Produce Plasmin

Reactions of fibrinolysis are shown in Figure 23.58. Lysis of a fibrin clot occurs through action of the enzyme plasmin that is formed from plasminogen through the action of **tissue plasminogen activator (t-PA** or **TPA)**. Plasminogen has high affinity for fibrin clots. It diffuses or becomes immersed within the clot and forms complexes with fibrin throughout various regions of the fibrin network. t-PA also binds to fibrin clots (A Closer Look 23.11) and activates plasminogen to plasmin by specific bond cleavage. Plasmin hydrolyzes the fibrin clot producing soluble peptides that are taken out of circulation by the liver and destroyed.

Protein inhibitors regulate activity of t-PA. Four immunologically distinct types of inhibitors of t-PA have been identified, two of which react rapidly with t-PA and are specific for it. They are **plasminogen activator-inhibitor type 1 (PAI-1)** and **plasminogen activator-inhibitor type 2 (PAI-2)**. The human PAI-2 contains 415 amino acid residues.

Starting and stopping blood coagulation follow essentially the same type process, interactions of proteins, formation of multienzyme complexes, and proteolysis. Both are unidirectional and the only mechanism for replenishing the proteins is by resynthesis.

Role of Gla Residues in Blood Coagulation Factors

Posttranslational modification of many proteins involved in blood coagulation results in formation of γ-carboxyglutamyl (Gla) residues that transforms them into excellent chelators of calcium ions. Ca^{2+} forms complexes with Gla residues of those factors and induces conformational and electronic states that facilitate their interaction with membrane receptors. Ca^{2+} also binds at sites other than Gla residues producing protein conformational changes that enhance catalytic activity. Evidence for this second effect of Ca^{2+} comes from the observation that activation of at least one of the enzymes leads to cleavage and elimination of the *N*-terminal region containing the Gla residues, but calcium ions are still required for its effective participation in blood coagulation.

Schematic representations of structures of five of the **Gla-containing proteins** listed in Table 23.9 are shown in Figure 23.59. Gla residues are located in the *N*-terminal region of the molecules followed by a structural area that resembles epidermal growth factor that may play a role in facilitating the healing process. Activation of zymogens involved in blood coagulation generally occurs at peptide bonds located between cysteine residues that have formed disulfide bonds (A Closer Look 23.12). Prothrombin is the only zymogen whose activation occurs by cleavage of a bond in the primary sequence outside of a bridging disulfide bond, and dissociation of this moiety containing the Gla peptide is facilitated (see Figure 23.47, p. 987).

Role of Vitamin K in Protein Carboxylase Reactions

Modification of prothrombin, **Protein C**, **Protein S**, **Protein Z**, and Factors VII, IX, and X, to form Gla residues occurs during their synthesis by a carboxylase located on the luminal side of the rough endoplasmic reticulum. **Vitamin K** (phytonadione, the "koagulation"

A CLOSER LOOK 23.11

Structural Arrangement of Tissue Plasminogen Activator

t-PA is a 72-kDa protein that contains a growth factor domain near its *N*-terminus, two adjacent **Kringle domains** that interact with fibrin, and a protease domain close to its C-terminus. Kringle domains are conserved sequences that fold into large loops stabilized by disulfide bonds. They are important for protein–protein interactions that occur with several blood coagulation factors. t-PA is activated by cleavage between an Arg-Ile bond to give a molecule with a heavy and a light chain. The light chain contains the serine protease activity.

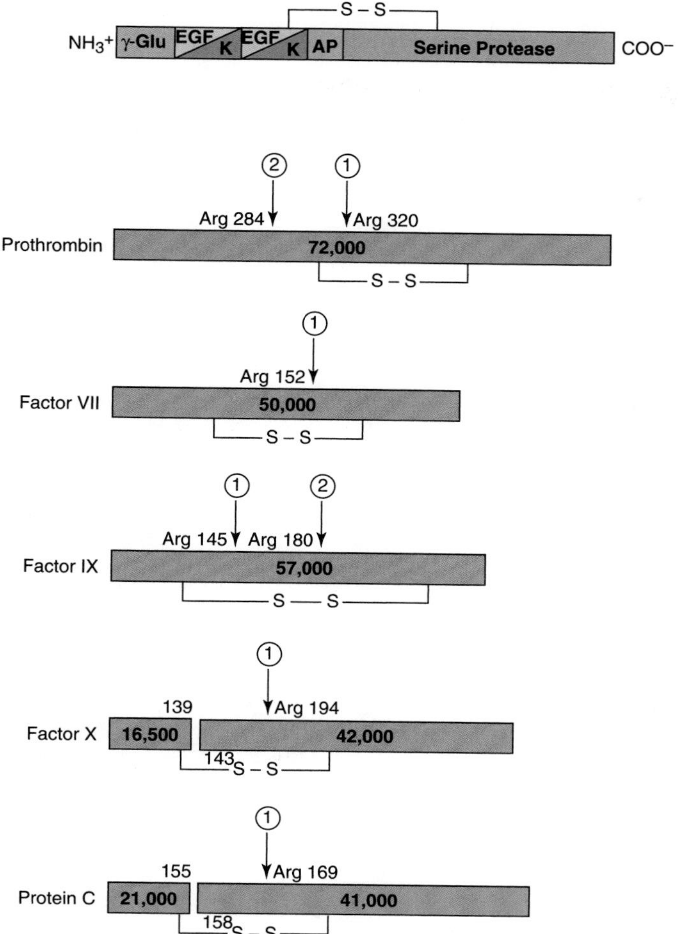

Figure 23.59 ***(a)*** **General structure of the γ-carboxyglutamyl containing proteins. *(b)* Structural organization of the zymogens and their cleavage sites for activation.**

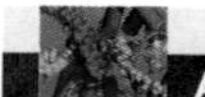

A CLOSER LOOK 23.12

Activation of Zymogens Involved in Blood Coagulation

Activation may or may not result in loss of small peptides located between disulfide bonds. Factor VII is activated by cleavage of a single Arg^{152}–Ile^{153} bond. Factor IX is activated by cleavages at Arg^{145} and Arg^{180} with the release of an approximate 11-kDa peptide. Factor X consists of two chains connected by a disulfide bridge. It is activated by cleavage of its heavy chain at Arg^{194}-Ile^{195}. The Gla residues are located in the light chain. Protein C also consists of a heavy and a light chain connected by a disulfide bond. It is activated by cleavage of an Arg-Ile bond at position 169.

vitamin) is an essential cofactor for this carboxylase. The dihydroquinone, or reduced form of vitamin K (Figure 23.60), is oxidized to the epoxide form by O_2 (A Closer Look 23.13). The **epoxide** is converted back to the **dihydroquinone** by enzymes that require as cofactors dithiols like **thioredoxin**. Analogs of vitamin K inhibit the dihydroquinone reductases and result in conversion of all available vitamin K to the epoxide form that is not functional in the carboxylation reaction. The overall carboxylation reaction is

$$\text{HOOC–CH}_2\text{–CH}_2\text{–CH(NH–R}_1\text{)–C(=O)–R}_2 \xrightarrow[\text{Protein carboxylase}]{\text{vitamin K(H}_2\text{)} + O_2 + CO_2} (\text{HOOC})_2\text{CH–CH}_2\text{–CH(NH–R}_1\text{)–C(=O)–R}_2 + \text{vitamin K(O)} + H_3O^+$$

Structures of two analogs, **dicoumarol** and **warfarin**, that interfere with the action of vitamin K are shown in Figure 23.60. In animals treated with high doses of these compounds, prothrombin, Protein C, Protein S, Factors VII, IX, and X, and other Gla-containing proteins are not posttranslationally modified, are ineffective in binding Ca^{2+}, and cannot participate in blood coagulation. Since dicoumarol and warfarin affect only synthesis of Gla-containing factors, they have no effect on blood coagulation in the test tube.

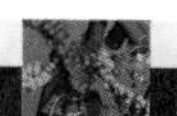

A CLOSER LOOK 23.13

Possible Mechanism of Vitamin K Action in Protein Carboxylation

A plausible mechanism involves the addition of molecular oxygen to the C1 position of dihydro-vitamin K and its subsequent rearrangement to an alkoxide with a pKa of ~20. This intermediate serves as a strong base and abstracts a proton from the γ-methylene carbon of glutamate yielding a carbanion to which CO_2 can add by a nucleophilic mechanism (Figure 23.60, p. 996).

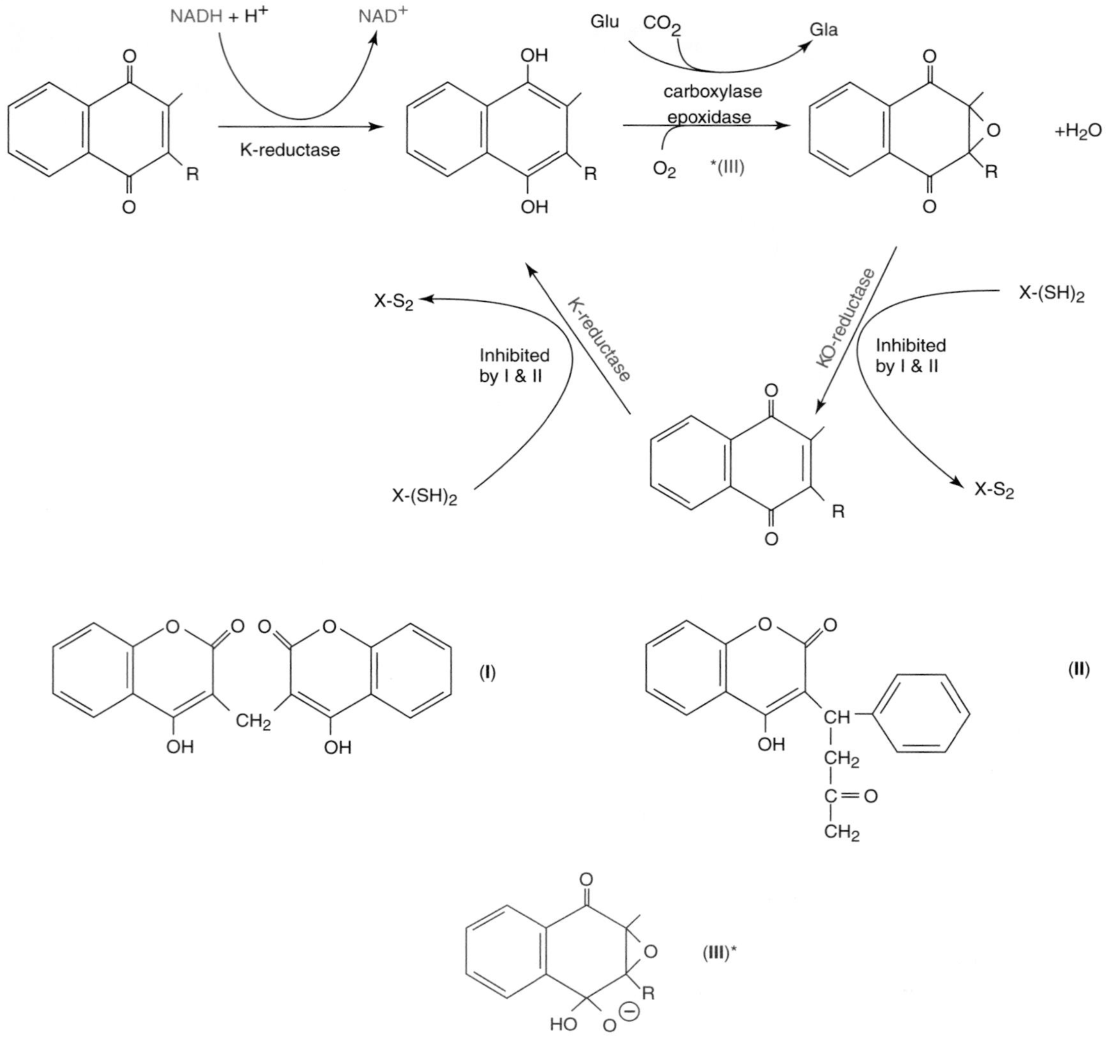

NADH + H^+ NAD$^+$ K-reductase; Glu CO_2 Gla carboxylase epoxidase O_2 *(III) $+H_2O$; K-reductase; KO-reductase; X-S_2; X-$(SH)_2$; Inhibited by I & II; (I); (II); (III)*

pKa-20

Figure 23.60 The vitamin K cycle as it functions in protein γ-glutamyl carboxylation reactions. X-(SH)2 and X-S2 represent the reduced and oxidized forms, respectively, of a thioredoxin. The NADH-dependent and the dithiol-dependent vitamin K reductases are different enzymes. The dithiol-dependent K and KO reductases are inhibited by dicoumarol (I) and warfarin (II). (*Possible alkoxide intermediate (III), see SIB 13.)
Redrawn and modified from Vermeer, C. *Biochem. J.* 266: 625, 1990.

Control of Synthesis of Gla-Proteins

N-terminal Gla-containing peptides that are released from prothrombin on activation are removed from circulation by the liver. They stimulate *de novo* synthesis of Gla-containing proteins required for blood coagulation (Figure 23.61). The proteins are synthesized even in the absence of vitamin K or in the presence of antagonists of vitamin K, but they do not contain Gla residues and are ineffective in blood coagulation. In addition, they are not secreted into the circulation. Some may remain in the liver and some are degraded. When vitamin K is restored or is added in high enough concentrations to overcome effects of antagonists, the preformed proteins are carboxylated and secreted into the circulation.

Activation of blood coagulation is a one-way process. The use of prothrombin Gla-containing peptides to signal the liver to synthesize more of these proteins maintains their concentrations in blood at effective levels. Monitoring of patients on long-term therapy with vitamin K antagonists is necessary to assure that production of the Gla-containing proteins is not shut down completely. When all of these processes function properly, hemostasis is achieved.

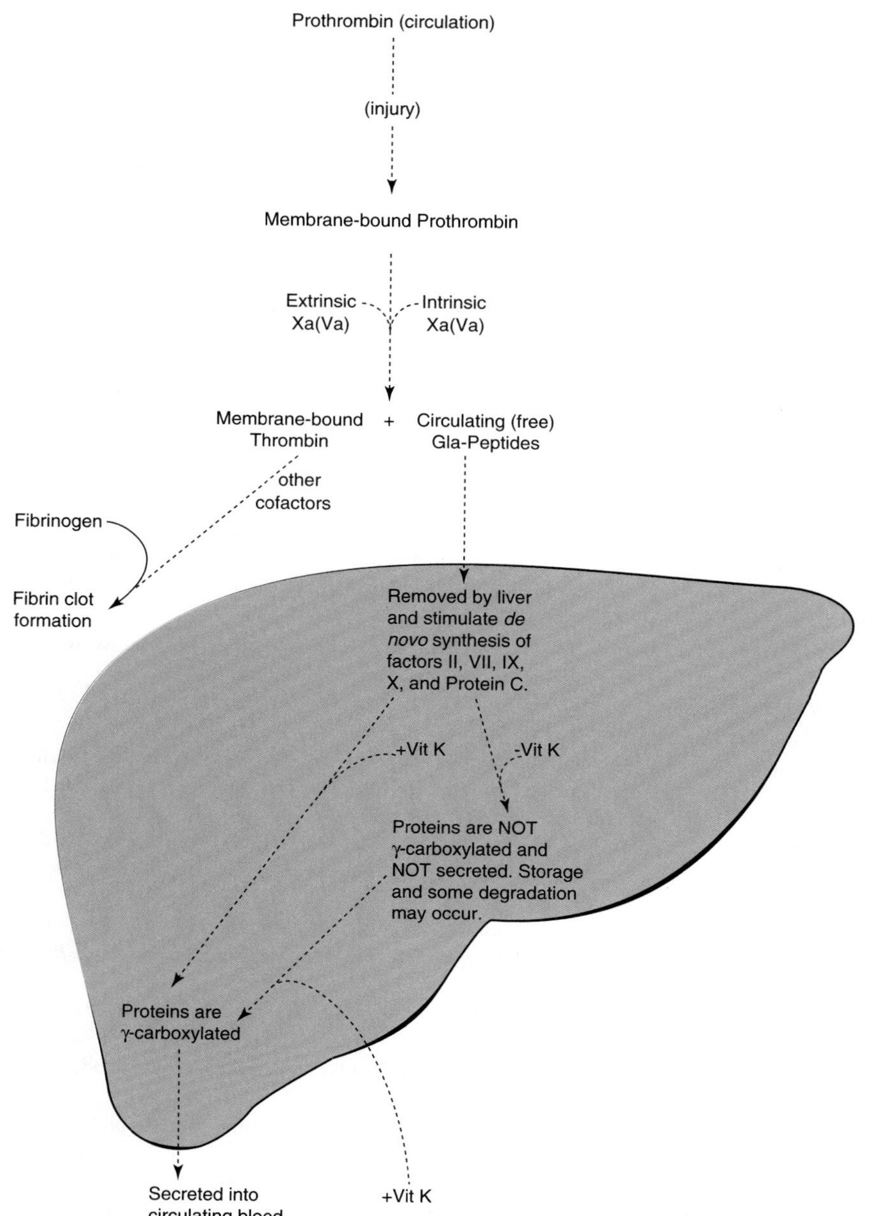

Figure 23.61 Role of Gla-peptides in the regulation of *de novo* synthesis of coagulation factors.

Bibliography

Nerve

Angulo, M. C., Le Meur, K., Kozlov, A. S, Charpak, S., and Audinat, E. GABA, a forgotten gliotransmitter. *Prog Neurobiol.* 86: 297, 2008.

Auestad, N., Korsak, R. A., Morrow, J. W., and Edmond, J. Fatty acid oxidation and ketogenesis by astrocytes in primary culture. *J. Neurochem.* 56: 1376, 1991.

Chapman, E. R., How does synaptotagmin trigger neurotransmitter release? *Annu. Rev. Biochem.* 77: 615, 2008.

Evans, G. J. O., and Morgan, A. Phosphorylation-dependent interaction of the synaptic vesicle proteins cysteine string protein and synaptotagamin I. *Biochem. J.* 364: 343, 2002.

Greengard, P., Valtorta, F., Czernik, A. J., and Benfenati, F. Synaptic vesicle phosphoproteins and regulation of synaptic function. *Science* 259: 780, 1993.

Jlang, Y., Lee, A., Chen, J., Cadene, M., Chalt, B. T., and MacKinnon, R. Crystal structure and mechanism of a calcium-gated potassium channel. *Nature* 417: 515, 2002.

Kuner, T., Seeburg, P. H., and Guy, H. R. A common architecture for K^+ channels and ionotropic glutamate receptors? *Trends in Neurosciences* 26: 27, 2003.

Pleribone, V. A., Shupllakov, O., Brodin, L., Hilfiker-Rothenfluh, S., et al. Distinct pools of synaptic vesicles in neurotransmitter release. *Nature* 375: 493, 1995.

Reid, C.A., Berkovic, S. F., and, Petrou S. Mechanisms of human inherited epilepsies. *Prog Neurobiol.* 87: 41, 2009.

Roth-Deri, I., Green-Sadan, T., and Yadid, G. Beta-endorphin and drug-induced reward and reinforcement. *Prog Neurobiol.* 86: 1, 2008.

Sheng, M., and Hoogenraad, C. C. The postsynaptic architecture of excitatory synapses: A more quantitative view. *Annu. Review Biochem.* 76: 823, 2007.

Sudhof, T. The synaptic vesicle cycle: A cascade of protein–protein interactions. *Nature* 375: 645, 1995.

Tarsa, L., and Goda, Y. Synaptophysin regulates activity-dependent formation in cultured hippocampal neurons. *PNAS* 99: 1012, 2002.

Wang, D. D., and Bordey A. The astrocyte odyssey. *Prog. Neurobiol.* 86: 342, 2008.

Vision

Cursiefen, C., Chen, L., Saint-Geniez, M., Hamrah, P., et al. Nonvascular VEGF receptor 3 expression by corneal epithelium maintains avascularity and vision. *Proc. Natl. Acad. Sci. USA* 103: 11405, 2006.

Filipek, S., Teller, D. C., Palczewski, K., and Stenkamp, R. The crystallographic model of rhodopsin and its use in studies of other G protein-coupled receptors. *Annu. Rev. Biophy. and Biomolec. Struct.* 32: 375, 2003.

Jager, R. D., Mieler, W. F., and. Miller, J. W. Age-related macular degeneration. *N. Engl. J. Med.* 358: 2606, 2008.

Nathans, J., Davenport, C. M., Maumenee, I. H., Lewis, R. A., et al. Molecular genetics of human blue cone monochromacy. *Science* 245: 831, 1989.

Nathans, J. Molecular biology of visual pigments. *Ann. Rev. Neurosci.* 10: 163, 1987.

Pappa, A., Estey, T., Manzer, R., Brown, D., and Vasiliou, V. Human aldehyde dehydrogenase 3A1 (ALDH3A1): biochemical characterization and immunohistochemical localization in the cornea. *Biochem. J.* 376: 615, 2003.

Parry, J. W. L., Poopalasundaram, S., Bowman, J. K., and Hunt, D. M. A novel amino acid substitution is responsible for spectral tuning in a rodent violet-sensitive visual pigment. *Biochemistry* 43: 8014, 2004.

Reich, S. J., Fosnot, J., Kuroki, A., Tang, W., et al. Small interfering RNA (siRNA) targeting *VEGF* effectively inhibits ocular neovascularization in a mouse model. *Molecular Vision* 9: 210, 2003.

Riley, M. V., and Winkler, B. S. Strong Pasteur effect in rabbit corneal endothelium preserves fluid transport under anaerobic conditions. *J. Physiology* 426: 81, 1990.

Stenkamp, R. E., Filipek, S., Driessen, C. A., Teller, D. C., and Palczewski, K. Crystal structure of rhodopsin: A template for cone visual pigments and other G protein-coupled receptors. *Biochim. Biophys. Acta* 1565: 168, 2002.

Winger, J. A, Derbyshire, E. R., Lamers1, M. H., Marletta, M. A., and Kuriyan, J. The crystal structure of the catalytic domain of a eukaryotic guanylate cyclase. *BMC Structural Biology* 8: 42, 2008.

Zhang, Q., ZulWqar, F., Xiao, X., Riazuddin, S.A., et al. Severe retinitis pigmentosa mapped to 4p15 and associated with a novel mutation in the PROM1 gene. *Hum. Genet.* 122: 293, 2007.

Zigler, J. S., Jr., and Goosey, J. Aging of protein molecules: Lens crystallins as a model system. *Trends Biochem. Sci.* 7: 133, 1981.

Molecular Motors

Anderson, P. A. W., Malouf, N. N., Oakley, A. E., Pagani, E. D., and Allen, P. D. Troponin T isoform expression in humans: A comparison among normal and failing adult heart, fetal heart, and adult and fetal skeletal muscle. *Circulation Res.* 69: 1226, 1991.

Bassel-Duby, R., and Olson, E. N. Signaling pathways in skeletal muscle remodeling. *Annu. Rev. Biochem.* 75: 19, 2006.

Bonnevier, J., Fassler, R., Somlyo, A. P., Somlyo, A. V., and Arner, A. Modulation of Ca^{2+} sensitivity by cyclic nucleotides in smooth muscle from protein kinase G-deficient mice. *J. Biol. Chem.* 279: 5146, 2004.

Braunwald, E. Biomarkers in Heart Failure. *N. Engl. J. Med.* 358: 2148, 2008.

Denz, C. R., Narshi, A., Zajdel, R. W., and Dube, D. K. Expression of a novel cardiac-specific tropomyosin isoform in humans. *Biochem. Biophys. Research. Commun.* 320: 1291, 2004.

dos Remedios, C. G., and Moens, P. D. J. Actin and the actomyosin interface: A review. *Biochim. Biophys. Acta* 1228: 99, 1995.

Geeves, M. A., and Holmes, K. C. Structural mechanism of muscle contraction. *Annu. Rev. Biochem.* 68: 687, 1999.

Goode, B. L., Michael, J., and Eck, M. J. Mechanism and function of formins in the control of actin assembly *Annu. Rev. Biochem.* 76: 593, 2007.

Huxley, H. E. The mechanism of muscular contraction. *Science* 164: 1356, 1969.

Mallik, R., Carter, B. C., Lex, S. A., King, S. J., and Gross, S. P. Cytoplasmic dynein functions as a gear in response to load. *Nature* 427: 649, 2004.

Peacock, W. F., De Marco T., Fonarow G. C., Diercks, D., et al. Cardiac troponin and outcome in acute heart failure. *N. Engl. J. Med.* 358: 2117, 2008.

Terrak, M., Wu, G., Stafford, W. F., Lu, R. C., and Dominguez, R. Two distinct myosin light chain structures are induced by specific variations within the bound IQ motifs: Functional implications. *EMBO J.* 22: 362, 2003.

Vale, R. D., and Milligan, R. A. The way things move: Looking under the hood of molecular motor proteins. *Science* 288: 88, 2000.

Blood Coagulation

Aggarwal, A., Malkovska, V., Catlett, J. P., and Alcorn, K. Recombinant activated factor VII (rVII) as salvage treatment for intractable hemorrage. *Thrombosis J.* 2: 9, 2004.

Brummel, K. E., Butenas, S., and Mann, K. G. An integrated study of fibrinogen during blood coagulation. *J. Biol. Chem.* 274: 22862, 1999.

Butenas, S., Brummel, K. E., Branda, R. F., Sara G., et al. Mechanism of factor VIIa-dependent coagulation in hemophilia blood. *Blood* 99: 923, 2002.

Colombatti, A., and Bonaldo, P. The superfamily of proteins with von Willebrand factor type A-like domains: One theme common to components of extracellular matrix, hemostasis, cellular adhesion, and defense mechanisms. *Blood* 77: 2305, 1991.

Crowther, M. A., and Kelton, J. G. Congenital thrombophilic states associated with venous thrombosis: A qualitative overview and proposed classification system. *Ann. Intern. Med.* 138: 128, 2003.

Davie, E. W. A brief historical review of the waterfall/cascade of blood coagulation. *J. Biol. Chem.* 278: 50819, 2003.

Fressinaud, E., Mazurier, C., and Meyer, D. Molecular genetics of type 2 von Willebrand disease. *Int. J. Hematol.* 75: 9, 2002.

Furie, B., and Furie, B. C. Mechanisms of thrombus formation. *N. Engl. J. Med.* 359: 938, 2008.

Han, X., Huang, Z. F., Fiehler, R., and Broze, G. J., Jr. The protein Z-dependent protease inhibitor is a serpin. *Biochem.* 38: 11073, 1999.

Huai, Q., Mazar, A. P., Kuo, A., Parry, G. C., et al. Activator in complex with its receptor structure of human urokinase plasminogen. *Science* 311: 656, 2006.

Iakhiaev, A., Pendurthi, U. R., Voigt, J., Ezban, M., and Rao, L. V. M. Catabolism of factor VIIa bound to tissue factor fibroblasts in the presence and absence of tissue factor pathway inhibitor. *J. Biol. Chem.* 274: 36995, 1999.

Kalafatis, M., Sworde, N. A., Rand, M. D., and Mann, K. G., Membrane-dependent reactions in blood coagulation: Role of the vitamin K-dependent enzyme complexes. *Biochim. Biophys. Acta* 1227: 113, 1994.

Kuliopulus, A., Hubbard, B. R., Lam, Z., Koski, I. J., et al. Dioxygen transfer during vitamin K-dependent carboxylase catalysis. *Biochem.* 31: 7722, 1992.

Monroe, D. M., Hoffman, M., and Roberts, H. R. Platelets and thrombin generation. *Arterioscler. Thromb. Vasc. Biol.* 22: 381, 2002.

Orfeo, T., Brummel-Ziedins, K. E., Gissel, M., Butenas, S., and Mann, K. G. The nature of the stable blood clot procoagulant activities. *J. Biol. Chem.* 283: 9776, 2008.

Orfeo, T., Butenas, S., Brummel-Ziedins, K. E., and Mann, K. G. The tissue factor requirement in blood coagulation. *J. Biol. Chem.* 280: 42887, 2005.

Palston, P. A., and Gettings, P. G. W. A database of recombinant wild-type and mutant serpins. *Thrombosis and Haemostasis* 72: 166, 1994.

Patnaik, M. M., and Moll, S. Inherited antithrombin deficiency: a review. *Haemophilia* 14: 1229, 2008.

Roberts, H. R., Monroe, D. M., and White, G. C. The use of recombinant factor VIIa in the treatment of bleeding disorders. *Blood* 104: 3858, 2004.

Shriver, Z., Sundaram, M., Venkataraman, G., Fareed, J., et al. Cleavage of the antithrombin III binding site in heparin by heparinases and its implication in the generation of low molecular weight heparin. *Proc. Natl. Acad. Sci. USA* 97: 10365, 2000.

Zhang, B., and Ginsburg, D. Familial multiple coagulation factor deficiencies: new biologic insight from rare genetic bleeding disorder. *Thrombosis and Haemostasis* 2: 1564, 2004.

Key Terms

NERVOUS TISSUE

blood–brain barrier
voltage-gated channels
neurotransmitter
calcium–calmodulin kinase
acetylcholinesterase
choline acetyltransferase
catechol-*O*-methyltransferase
monoamine oxidase
GABA shunt
endorphins
enkephalins
neurokinins

THE EYE

glutathione reductase
vascular endothelial growth factor receptor-3
α- and β-crystallins
aldose reductase
polyol (aldose) dehydrogenase
small heat shock proteins
11-*cis*-retinol
all-*trans*-retinol
all-*trans*-retinol dehydrogenase
cyclic GMP (cGMP)
transducin
phosphodiesterase (PDE)
arrestin
guanylate cyclase
rhodopsin kinase

MOLECULAR MOTORS

IQ motifs
kinesin
tropomyosin
troponin subunits
power stroke
cadherin and catenin
caldesmon
adenylate kinase
creatine phosphokinase
myosin light-chain kinase
myosin phosphatase
unconventional myosins
AAA motif
dynein

BLOOD COAGULATION

anticoagulation
fibrinolysis
hemostasis
procoagulation
epidermal growth factor
extrinsic pathway
intrinsic pathway
transglutamidase
thrombomodulin
protein carboxylase
prostacyclin (PGI_2)

Questions • CAROL N. ANGSTADT

Multiple-Choice Questions

1. There are both excitatory and inhibitory neurotransmitters. All of the following are excitatory except
   A. acetylcholine.
   B. γ-aminobutyric acid (GABA).
   C. dopamine.
   D. epinephrine.
   E. serotonin.

2. Which of the following classes of molecular motors can be involved in both contraction and trafficking?
   A. Dyneins
   B. Kinesins
   C. Myosins
   D. All of the above

3. Platelet aggregation
   A. is initiated at the site of an injury by conversion of fibrinogen to fibrin.
   B. is inhibited in uninjured blood vessels by prostacyclin of intact vascular endothelium.
   C. causes morphological changes and a release of the vasodilator serotonin.
   D. is inhibited by the release of ADP and thromboxane A_2.
   E. is inhibited by von Willebrand factor (vWF).

4. In the formation of a blood clot,
   A. proteolysis of γ-carboxyglutamate residues from fibrinogen to form fibrin is required.
   B. the clot is stabilized by the cross-linking of fibrin molecules by the action of Factor XIII, transglutaminase.
   C. thrombin's only role is in activation of Factor VII.
   D. tissue factor, Factor III or TF, must be inactivated for the clotting process to begin.
   E. the role of calcium is primarily to bind fibrin molecules together to form the clot.

5. Lysis of a fibrin clot
   A. is in equilibrium with formation of the clot.
   B. begins when plasmin binds to the clot.
   C. requires the hydrolysis of plasminogen into heavy and light chains.
   D. is regulated by the action of protein inhibitors on plasminogen.
   E. requires the conversion of plasminogen to plasmin by t-PA.

6. The eye relies heavily on the metabolism of glucose but different parts of the eye use glucose differently. Which of the following statements is/are correct?
   A. The retina uses glucose primarily by aerobic glycolysis and the TCA cycle.
   B. The cornea uses over half its glucose via the hexose monophosphate pathway.
   C. The lens has the glycolytic pathway but not the hexose monophosphate pathway.
   D. The role of glutathione reductase is to inactivate active oxygen species.
   E. All of the above.

Questions 7 and 8: Two autoimmune diseases that lead to muscle weakness because of interference with neuromuscular signal transmission are Lambert–Eaton myasthenia syndrome (LEMS) and myasthenia gravis. In LEMS, the body raises antibodies against voltage-gated calcium channels. In myasthenia gravis, the antibodies are to the nicotinic–acetylcholine receptor (AchR). In both cases, the amount of acetylcholine released in the synaptic juction is diminished.

7. In the propagation of a nerve impulse by an electrical signal,
   A. the electrical potential across the membrane maintained by the Na^+/K^+-exchanging ATPase becomes more negative.
   B. local depolarization of the membrane causes protein conformational changes in ion channels that allow Na^+ and K^+ to move down their concentration gradients.
   C. charge propagation is bidirectional along the axon.
   D. voltage-gated ion channels have a finite recovery time so the amplitude of the impulse changes as it moves along the axon.
   E. astrocytes function as antenna-like protrusions and receive signals from other cells.

8. The nerve impulse that initiates muscular contraction
   A. begins with binding of acetylcholine to receptors in the sarcoplasmic reticulum.
   B. causes both plasma membrane and transverse tubules to undergo hyperpolarization.
   C. causes opening of calcium channels, leading to an increase in Ca^{2+} within the sarcomere.
   D. prevents Na^+ from entering the sarcomere.
   E. prevents Ca^{+2} from binding to troponin C.

Questions 9 and 10: Genes coding for proteins of visual pigments are located on different chromosomes. The rhodopsin gene is on the third chromosome; blue pigment gene on the seventh; and genes for the red and green pigments on the X chromosome. Females have two X chromosomes so color vision abnormalities are rare, but males have only one X chromosome. About 8 % of males have abnormal color vision that affects either red or green perception. At least one individual is known with blue color blindness.

9. The cones of the retina
   A. are responsible for color vision.
   B. are much more numerous than the rods.
   C. have red, blue, and green light-sensitive pigments that differ because of small differences in the retinal prosthetic group.
   D. do not use transducin in signal transduction.
   E. are better suited for discerning rapidly changing visual events because a single photon of light generates a stronger current than it does in the rods.

10. Which of the following statements about rhodopsin is true?
   A. Rhodopsin is the primary photoreceptor of both rods and cones.
   B. The prosthetic group of rhodopsin is all-*trans*-retinol derived from β-carotene.
   C. Conversion of rhodopsin to activated rhodopsin, R*, by a light pulse requires depolarization of the cell.
   D. Rhodopsin is located in the cytosol of the cell.
   E. Absorption of a photon of light by rhodopsin causes an isomerization of 11-*cis*-retinal to all-*trans*-retinal.

Questions 11 and 12: Ion channels in cardiac muscle, as in other tissues, require a finite recovery time after excitation. This is measured on an electrocardiogram as the QT interval. Initiating the excitation phase is the opening of Na^+ channels, causing depolarization. Then K^+ channels open to permit K^+ to move out of the cell, which is a key element in shutting off the action potential. Inherited defects in these channels lead to a condition called long QT syndrome (LQTS). LQTS can cause sudden death, especially in physically active young people, who have not previously had any symptoms of cardiac irregularities.

11. All of the following statements about actin and myosin are true except
   A. the globular head section of myosin has domains for binding ATP and actin.
   B. actin is the major protein of the thick filament.
   C. binding of ATP to the actin–myosin complex promotes dissociation of actin and myosin.
   D. F-actin is stabilized when tropomyosin is bound to it.
   E. binding of Ca^{2+} to the calmodulin-like subunit of troponin induces conformational changes that permit myosin to bind to actin.

12. When a muscle contracts, the
   A. transverse tubules shorten, drawing the myofibrils and sarcoplasmic reticulum closer.
   B. thin filaments and thick filaments of the sarcomere shorten.
   C. light chains dissociate from the heavy chains of myosin.
   D. H bands and I bands of the sarcomere shorten because thin filaments and thick filaments slide past each other.
   E. cross-linking of proteins in the heavy filaments increases.

Problems

13. In the presence of warfarin, an analog of vitamin K, several proteins of the blood coagulation pathway are ineffective because they cannot bind Ca^{2+} efficiently. Why?

14. Organophosphate compounds are irreversible inhibitors of acetylcholinesterase. What effect does an organophosphate inhibitor have on the transmission of nerve impulses?

Answers

1. **B** GABA stimulates its receptor to form channels through which Cl^- flows, making the inside of the cell more negative.
2. **C** Myosins are involved in contraction and "unconventional myosins" in trafficking. A and B: These are both microtubule-based motors involved in moving cargo.
3. **B** The yin-yang nature of PGI_2 and TXA_2 help to control platelet aggregation until there is a need for it. A: Initiation is by contact with an activated receptor at the site of injury. Clot formation requires activation of various enzymes. C: Serotonin is a vasoconstrictor. D: TXA_2 facilitates aggregation. E: vWF forms a link between the receptor and platelets, promoting aggregation.
4. **B** Cross-linking occurs between a glutamine and a lysine. A and E: γ-Carboxyglutamate residues are on various enzymes; they bind calcium and facilitate the interaction of these proteins with membranes that form the sites for initiation of reaction. C: Thrombin activates Factors V, VII, VIII, XIII and fibrinogen. D: TF, Factor III, is the primary receptor for initiation of the clotting process.
5. **E** The clot is solubilized by plasmin. A: Both formation and lysis of clots are unidirectional. B: Both plasminogen and t-PA bind to the clot. C and D: Both of these refer to t-PA.
6. **B** The NADPH generated, especially in the cornea, is necessary to convert glutathione back to its reduced form. A: The retina depends mostly on anaerobic glycolysis. C: Around 10% of glucose is used by the hexose monophosphate pathway in lens. D: Reduced glutathione (GSH) inactivates these species; glutathione reductase is necessary to reduce glutathione back to GSH.
7. **B** This is the mechanism for impulse propagation. A: The potential becomes less negative. C: It is unidirectional. D: Voltage-gated channels do have a finite recovery time so the amplitude remains constant. E: This describes dendrites. Astrocytes are glial cells involved in processes that insulate neurons from their external environment.
8. **C** Ca^{2+} enters with Na^+. A: Acetylcholine receptors are on the plasma membrane. B: The impulse results in depolarization of both of these structures. D. Both Ca^{+2} and Na^{+2} enter the sarcomere when the channels open. E: Binding of Ca^{+2} to TnC initiates contraction.
9. **A** Rods are responsible for low light vision. C: All three pigments have 11-*cis*-retinal; differences in the proteins are responsible for the different spectra. D: Biochemical events are believed to be the same in rods and cones. E: Cones are better suited for rapid events because their response rate is about four times faster than rods, even though their sensitivity to light is much less.
10. **E** This causes the conformational change of the protein that affects the resting membrane potential and initiates the rest of the events. A: Cones have the same prosthetic group but different proteins, so rhodopsin is in rods only. B: This is the precursor of the prosthetic group 11-*cis*-retinal. C: Isomerization of the prosthetic group leads to hyperpolarization. D: Rhodopsin is a transmembrane protein.
11. **B** Myosin forms the thick filament. Actin is in the thin filament. A: These are both important in myosin's role. C: The role of ATP in contraction is to favor dissociation, not formation, of the actin–myosin complex. D and E: Tropomyosin, troponin, and actin are the three major proteins of the filament.
12. **D** This occurs because of association–dissociation of actin and myosin. A: Depolarization in transverse tubules may be involved in transmission of the signal but not directly in the contractile process. B: Filaments do not change length, but slide past each other. C: This is not physiological. E: Cross-linking occurs in the H band of the sarcomere but does not change during the contractile process.
13. Proteins affected are those with γ-carboxyglutamyl (Gla) residues, which are excellent calcium chelators. Formation of Gla is a post-translational modification catalyzed by a carboxylase whose essential cofactor is vitamin K. Warfarin prevents reduction of vitamin K epoxide formed during carboxylation back to the dihydroquinone form.
14. Acetylcholine is an excitatory neurotransmitter, causing channels to open, Na^+ to enter, and a depolarization of the membrane. Hydrolysis of acetylcholine allows repolarization. The inhibitor keeps the channel open and prevents repolarization, and thus continued transmission of nerve impulses.

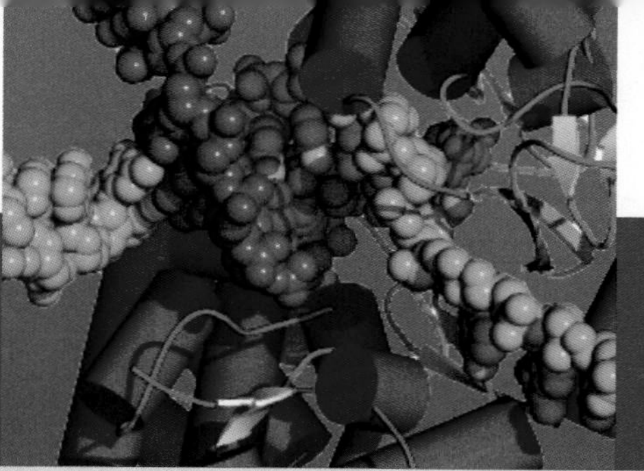

24

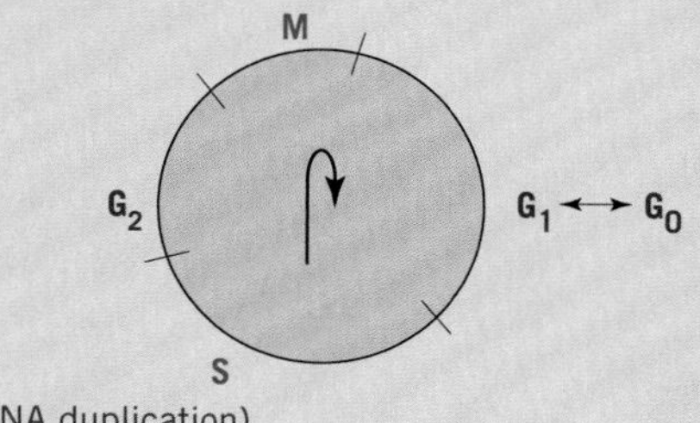

Cell Cycle, Programmed Cell Death and Cancer

Richard M. Schultz

Professor, Stritch School of Medicine, Loyola University of Chicago

Key Concepts

CELL DIVISION CYCLE

- The cell division cycle is divided into highly regulated biochemically distinct phases that are controlled by cyclin-dependent kinases (Cdks). Cdks are regulated by synthesis and degradation of cyclins, phosphorylation of the Cdks, and inhibition.
- Many transcription factors (tfs) are involved in control of cell division. Phosphorylation of retinoblastoma sensitivity protein (Rb) leads to release of E2F (a tf) and entry into S phase. p53 (a tf) halts the cell cycle by increasing the expression of a Cdk inhibitor.
- Mitogen signal pathways activate Ras, which triggers a MAPK kinase cascade that results in expression of Myc to increase the transcription of G_1/S-cyclins.

PROGRAMMED CELL DEATH (APOPTOSIS)

- Cell division is balanced by apoptotic cell death to maintain homeostasis. The death receptor pathway (extrinsic pathway) and the mitochondrial pathway (intrinsic pathway) promote cell death with the activation of a cascade of cytoplasmic proteases (caspases) enzymes.
- The death receptor pathway signal is transmitted by activation of pro-caspase 8.
- The intrinsic pathway involves release of cytochrome c when the mitochondrial proteins Bax and Bak permeabilize the membrane. Other proteins prevent Bax and Bak from acting. Released cytochrome c activates pro-caspase 9.

CANCER

- Malignant tumor cells are characterized by unregulated cell division, resistance to apoptosis, immortality, and ability to metastasize and induce angiogenesis.
- Multiple mutations are required in protooncogenes and tumor suppressor genes that in combination generate a cancer. Cells within a tumor are heterogenic, but are all progeny of a single precursor cell.
- Carcinogenesis is initiated by a mutagenic event and propagated by tumor promoters. Diet, environment, and cultural factors are major determinants in the formation of cancers.
- Genomic, transcriptomic, and proteomic analyses characterize the abnormal biochemical pathways present in cancer, leading to individualized diagnosis and treatment specific to that cancer.

24.1 • INTRODUCTION

At the interface of cell biology and biochemistry are the biochemical pathways that regulate cell division and cell death. These pathways involve proteins that transmit signals, often through their action as enzymes but also through their ability as signaling proteins to bind to other proteins and transmit a signal through induced-conformation changes in the associated proteins. Enzymatic activities commonly present in signal pathways are those of kinases, phosphatases, proteases, and ubiquitin ligases. While the signals are composed of recognizable enzymatic reactions or protein-induced conformational changes, the pathways regulating cell division and cell death are complex. Rather than a simple linear progression of signals from one molecule to the next constituting a pathway, these pathways can be depicted as a network of parallel signaling pathways joined at nodes or hubs that can cross-talk. The inputs are interpreted by these networks in a manner dependent on cell type and on the strength and time course of the extrinsic and intrinsic signals received by the cell, to give an output that causes a cell either to divide, to die, to senesce, or to differentiate. We have only a preliminary understanding of how the networks operate to generate a specific outcome. For many inputs, the complexity of the network makes the outcomes impossible to predict in the absence of computer algorithms to model the multiple node interactions that differ in different cell types and environmental conditions. In this chapter, these pathways will be described in a traditional linear manner to promote a basic understanding of the primary pathways that are the basis of these complex signaling networks. An insight into the complexity of these pathways may be discerned in the discussion of the Ras downstream kinases and their regulation of both cell division and cell death (p. 1010).

In order to maintain cellular homeostasis in a mature organism, an equal number of cells are born and die within a set unit of time. Thus cell division must be balanced by cell death and both cell division and cell death pathways are equally active. Aberrations in these pathways often lead to cancer, and the biochemical basis of cancer will be considered in the last section of this chapter.

24.2 • CELL DIVISION CYCLE

A model of the cell cycle divides the steps of cell division into four phases (Figure 24.1). In the **S phase** the chromosomal DNA is duplicated, and in the **M phase** mitosis occurs leading to the separation of chromosomes, cellular organelles, and cytoplasm of the parent cell into two progeny cells. The gap phases, $\mathbf{G_1}$ and $\mathbf{G_2}$, separate the S phase and M phase. In a simpler cell cycle model, cell division is divided into mitosis (M) and an **interphase** (I) that combines G_1, S, and G_2 phases. Although the time span of the M phase is shorter than the other phases, it is particularly complex. In a fibroblast cell that divides once every 24 h, the M phase spans about 1 h while the S phase takes 10–12 h, and the other two phases (G_1, G_2) occupy the remaining 11–13 h. The observable steps in M phase are presented in Table 24.1. While the processes

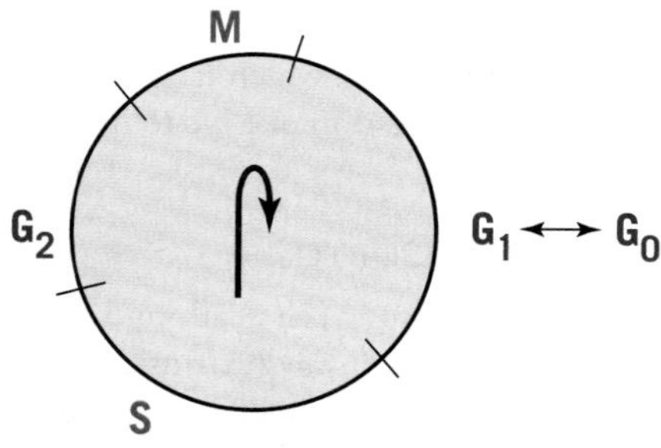

Figure 24.1 Phases of the cell cycle. Mitosis occurs in M phase, and DNA synthesis in S phase. Interphase consists of G_1, S, and G_2. G_0 phase is in equilibrium with G_1 phase. Cells in G_0 are either quiescent or senescent.

TABLE 24.1 • Microscopically Observable Steps in the M Phase

1. DNA condenses into chromosomes (46 chromosomes in human cells).
2. Centrosomes separate and mitotic spindle forms from the centrosomes.
3. Nuclear membrane disintegrates.
4. Spindle microtubule connects chromosomes.
5. Chromosomes align along equator of cell.
6. Sister chromatids separate and group around the two centrosomes, creating two sets of chromosomes.
7. Nuclear membrane forms around each set of chromosomes.
8. Chromosomes decondense.
9. Plasma membrane pinches parental cell into two progeny cell (cytokinesis).

that occur in M phase are the most dramatic, essential biochemical processes occur in each of the phases and each phase must be completed before the cell progresses to the next phase. Each phase is under close regulation by biochemical checkpoints that comprise stop or go directives that regulate proceeding to the next biochemical step in the cell division pathway.

An additional phase, in equilibrium with the G_1 phase, is the G_0 phase in which the cell is either in a quiescent or a senescent state. A **quiescent cell**, while not participating in the cell cycle, can be induced to reenter the cell cycle by a mitotic stimulus, such as an increase in concentration of a growth factor in its external environment. In contrast, a **senescent cell** cannot reenter the cell cycle even in the presence of mitotic growth factors. Different cell types vary in their frequency of cell division and thus in the amount of quiescent time spent in G_0. Although fibroblasts and epithelial cells spend very little or no time in G_0, adult liver cells divide about once a year and adult brain cells almost never divide. Thus, adult liver and brain cells spend most of their time in G_0. In G_0 the critical proteins of the cell cycle pathway including cyclin-dependent kinases are absent.

Regulation of Cell Cycle

The cell cycle is regulated by the activity of **cyclin-dependent kinases (Cdks)**, which catalyze phosphorylation of serine and threonine side chains in substrate proteins. The phosphorylated proteins then perform critical functions in the cell cycle pathways. The activities of the Cdks are closely regulated by other protein kinases and phosphatases, by their association with a **cyclin** protein, and by the presence or absence of Cdk inhibitor proteins.

In human cells different cyclin-Cdk heterodimers regulate different steps or phases of the cell cycle. In some cases a particular Cdk activity is required in more than one phase or at an interphase (Figure 24.2). In each case, a Cdk-specific cyclin protein is first synthesized to bind to its Cdk, forming a Cdk-cyclin heterodimer. Then the proper activating phosphates are

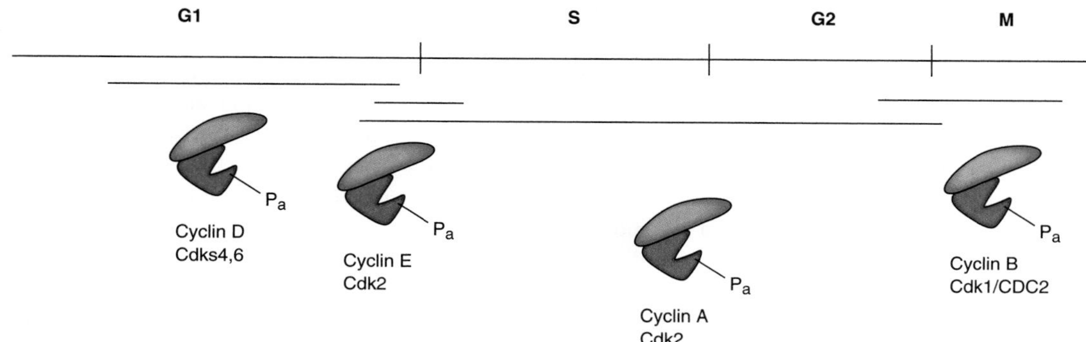

Figure 24.2 Cyclin and cyclin-dependent kinase complexes of the human cell cycle. Top line shows the cell cycle phases; intermediate lines are time frame for activity of the indicated cyclin/Cdk isoform. Both Cdk4 and Cdk6 bind to cyclin D in late G_1. Cyclin E and cyclin A associate with Cdk2 at different time points in S and G_2. Cdk1 is also known as CDC2. Multiple isoforms of some of the cyclins, such as D2 and D3, exist. P_a is an activating phosphate added by a CAK (see text).
Redrawn based on diagram in Morgan, D. O. *Annu. Rev. Cell. Dev. Biol.* 13: 261, 1997.

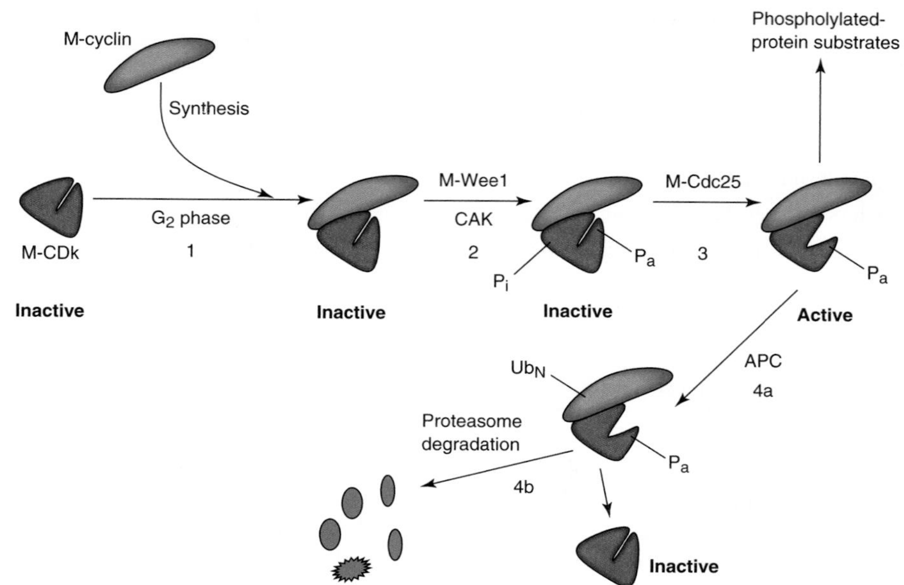

Figure 24.3 Regulation of M phase cyclin-dependent kinase (Cdk) activity. M-cyclin (green) is synthesized in the G_2 phase and combines with M Cdk (red) to form an inactive heterodimer. Phosphorylation by Wee1 kinase at an inhibitory site (Pi) maintains M–Cdk/M–cyclin heterodimer in an inactivated state. Entry into M phase occurs on phosphorylation by CAK at an activation site (P_a) and removal of the inhibitory phosphate by Cdc25 phosphatase. Activated M-Cdk phosphorylates protein substrates that promote M phase. Loss of M-Cdk activity occurs when M-cyclin is polyubiquitinated (Ub_N) by the APC ubiquitin ligase complex and degraded in a proteasome.

added by a **cyclin-dependent activating kinase** (**CAK**) and inhibitory phosphates removed by a **CDC25 phosphatase** to generate the Cdk activity that regulates the cell cycle step. At the end of the cell cycle step, the cyclin protein of the cyclin–Cdk complex is polyubiquitinated by an ubiquitin ligase and the polyubiquitinated cyclin degraded, turning off its associated Cdk activity and allowing the cell cycle pathway to proceed to the next stage of the pathway. Thus for the M phase in human cells, the **M-phase Cdk** (**Cdk1**) binds **M cyclin** (**cyclin B**). Cdk activity is then regulated by both its inhibitory and activating kinases and activating phosphatase (Figure 24.3).

Wee1 is a G_2/M-phase kinase that phosphorylates Cdk1 at an inhibitory site (P_i), preventing Cdk1 activity and entry into M phase. Wee1 obtained its name in yeast experiments in which knockout of the *wee1* gene results in "tiny" or "wee" cells. Thus Wee1 appears in yeast to be a terminal signal of a kinase cascade that prevents a cell from proceeding into M phase when the cell is too small to divide, and the absence of this checkpoint leads to tiny progeny cells. In mammalian cells, signals that prevent a cell from proceeding into M phase are generated by incomplete DNA replication and DNA damage. When the cell is ready to proceed into the M phase, the Wee1 stop signal is released by removal of the inhibitory phosphate group by the phosphatase, Cdc25. In addition to removal of inhibitory phosphates, the Cdk1/cyclin is activated by an M-Cdk activating kinase, M-CAK (**C**dk **a**ctivating **k**inase), that places a phosphate on an activation site residue (P_a) in the Cdk1 molecule. At the end of M phase, ubiquitylation of cyclin B terminates the M-Cdk/cyclin activity (p. 248). The polyubiquitinated cyclin B is degraded by the cellular proteasomes. The ubiquitin system that regulates cyclin B is named the **anaphase-promoting complex** (**APC**), so called because loss of cyclin B and its associated Cdk1 kinase activity is required for initiation of sister chromatid separation in the anaphase stage of mitosis.

While a single Cdk/cyclin regulates the M phase in mammalian cells, at least four different Cdk/cyclin heterodimers regulate progression through the G_1 and S phases (see Figure 24.2). Each of the G_1/S-Cdks is regulated by specific CAK and Wee1-like kinase isoforms that act by mechanisms similar to those that regulate Cdk1. Each Cdk also has a Cdc25 phosphatase isoform to remove inhibitory phosphate groups. A G1/S-Cdk/cyclin is required in G1 for the cell to proceed from G1 into S phase. Several Cdks are active at the G_1/S interface. One of the G1/Cdk/cyclin isoforms is active through the S and G_2 phases. Each Cdk loses its activity on ubiquitylation and degradation of its cyclin (Figure 24.4). The ubiquitin ligase complex that regulates the G1/S- and S-phase cyclins is known as the **SCF complex**.

Rb Regulation of E2F Activity and G_1/S Transition

A cell cycle critical substrate of the G_1/S Cdks is **Rb protein** (**r**etino**b**lastoma sensitivity **protein**), named after its deficiency in the cancer retinoblastoma. Rb protein is unphosphorylated

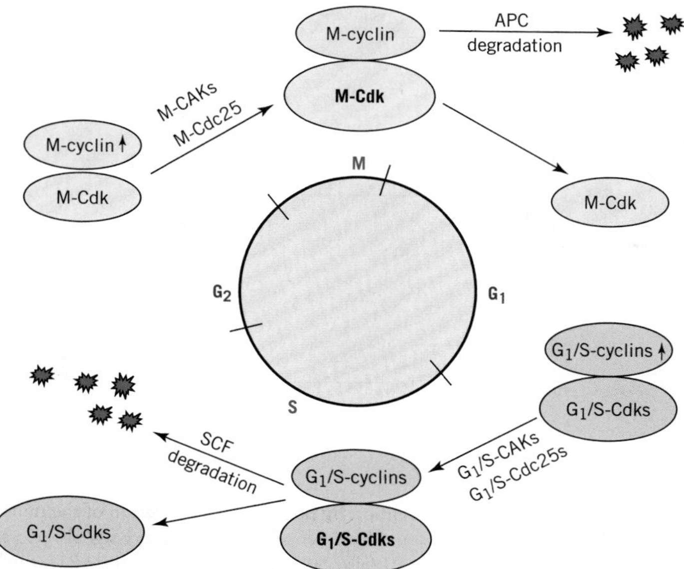

Figure 24.4 Reactions leading to activation and inactivation of the cyclin dependent kinases (Cdks). Binding to their cyclin and the action of their appropriate Cdk-activating kinase (CAK) and CDC25 phosphatase activates the Cdks. Cdk activities are lost on degradation of their cyclin following its ubiquitination by APC in the M phase and SCF in the G_1-S phase. The Cdks are not degraded while the cell remains in the cell cycle, but are degraded when the cell enters G_0.

in the G_1 and G_0 phases, and unphosphorylated Rb binds and sequesters the transcription factor **E2F** (Figure 24.5). On phosphorylation by the G_1 and G_1/S-Cdks, Rb changes conformation and releases E2F. The E2F binds to regulatory elements in target genes to increase transcription of the gene products required for S phase such as DNA polymerase, dihydrofolate reductase, thymidine kinase, and the G_1/S phase cyclin proteins. The G_1/S cyclin proteins increase the activity of G_1/S Cdks. Generation of phosphorylated Rb by the G_1/S-Cdks and release of the E2F transcription factor is a key requirement for entry into S phase from G_1. Unphosphorylated Rb is the active form of Rb, as it acts to sequester and inhibit E2F. The phosphorylated form of Rb that no longer acts to sequester E2F is the inactive form of Rb.

p53 Regulation of the Cdk Inhibitor p21

p53 is a transcription factor that has been described as "the guardian of the genome." On damage to the cellular DNA, p53 stops the cell cycle to give the cell time to repair its DNA. If the DNA damage is irreparable, p53 induces apoptotic cell death. The loss of p53 function results in cell division with damaged DNA, a property of cancer cells.

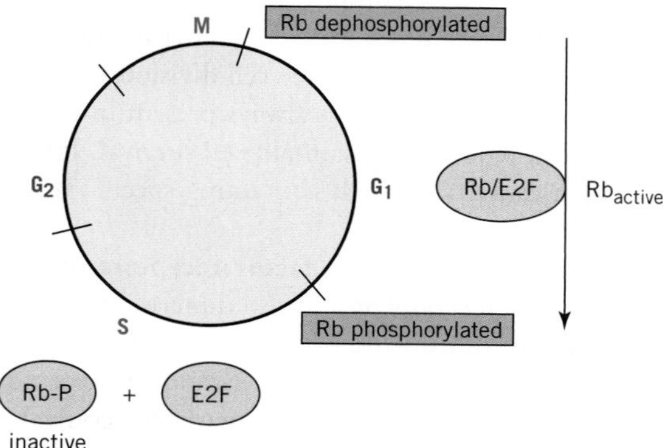

Figure 24.5 Role of Rb in cell cycle regulation. Active Rb is present in G_1, where it binds and inhibits the transcription factor E2F. On phosphorylation by the G_1-S Cdks, Rb releases E2F, which increases expression of genes required for S phase. Phospho-Rb is dephosphorylated by a phosphatase on entering G_1 phase from M phase.

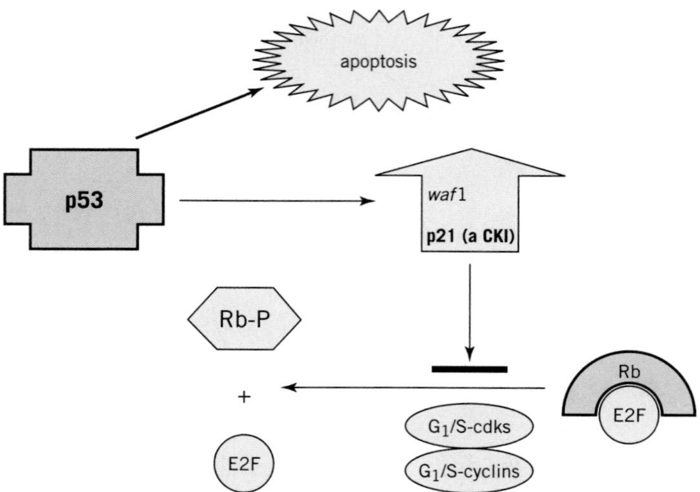

Figure 24.6 Regulation of the G_1-to-S transition by p53. The concentration of the transcription factor p53 is increased on DNA damage, by stress, and by other cell cycle regulatory signals. p53 stops the cell cycle by increasing the transcription of the *waf1* (also called *cip1*) gene to express p21, which is a Cdk inhibitor (CKI) and binds to G_1-S Cdks to inhibit phosphorylation of Rb. Higher concentrations of p53 than required for p21 synthesis initiate apoptosis.

Kinases monitor DNA integrity and are activated on DNA damage (p. 164). The activated kinases phosphorylate p53 directly or activate other kinases that phosphorylate p53. Phosphorylated p53 is resistant to degradation, and its concentration in the cell increases. p53 is a transcription factor protein and higher levels increase the transcription of the p53 target gene *waf1* (also referred to as *cip1*). The product of the *waf1* gene is the protein **p21**, a Cdk inhibitor. **Cdk inhibitor** proteins are called **CKI**s, and p21 protein is the key CKI increased by p53. The increase in p21 inhibits the G_1- and G_1/S-Cdks and their phosphorylation of Rb. Unphosphorylated Rb retains bound E2F, preventing cell entry into S phase of the cell cycle (Figure 24.6). With DNA damage, this mechanism (DNA-kinase → p53 → p21 ⊣ G_1-Cdk) gives the cell time to repair its DNA before proceeding into S phase with replication of the cellular DNA. In over 50% of cancers, the p53 genes are mutated or lost. In the 50% of cancers that contain normal p53 genes, the pathways that regulate p53 are often defective. These defects in p53 genes and in p53 regulation permit the propagation of cancer cells with damaged and unstable chromatin.

Growth Factor Signal Transduction Pathway

The most closely regulated step of the mammalian cell cycle is entry into the S phase, which commits the cell to either completing a cell division cycle or, in the absence of the completion of cell division, to die. Entry into S phase normally requires an extracellular signal such as that given by mitogen factors, commonly called **growth factors**. Growth factors are proteins secreted by cells to either induce their own division (**autocrine mechanism**) or secreted by neighboring or distant cells to induce cell division (**paracrine mechanism**). A low concentration of growth factor proteins is always present in the extracellular environment of mammalian cells and required to maintain cell survival. Higher concentrations of growth factor proteins then cause target cells containing specific receptors for the growth factor proteins to divide.

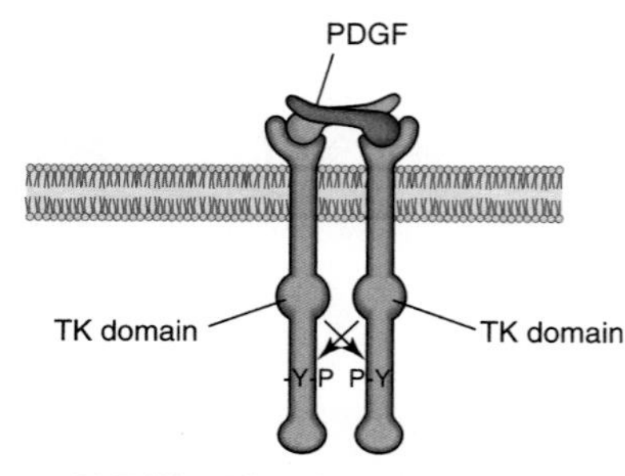

Figure 24.7 Phosphorylation of tyrosine residues within cytoplasmic domains of PDGF receptor. Binding of PDGF as a dimer causes its membrane receptor to also dimerize. The tyrosine kinase activity in the cytoplasmic region of one receptor protein then phosphorylates tyrosine residues (Y) in the cytoplasmic region of the neighboring receptor protein.

Growth factor proteins bind to **growth factor receptors** on cellular plasma membranes. This binding causes the receptor proteins to dimerize or polymerize in the plasma membrane, an essential step for transmission of the growth factor signal. The growth factor proteins utilize various mechanisms for oligomerization of receptor proteins. Platelet-derived growth factor (PDGF) forms a homodimer with two equivalent receptor binding sites that bind and join two receptor proteins together to form a receptor dimer in the plasma membrane (Figure 24.7). Other growth factors associate with cofactor proteins that facilitate their binding and simultaneously induce oligomerization of the receptor proteins in the plasma membrane.

Receptor proteins pass through the plasma membrane to connect a growth factor receptor domain on the extracellular surface and a tyrosine kinase catalytic domain on the cytoplasmic side of the plasma membrane. Unlike the serine/threonine kinases (e.g., PKC, PKA, and the Cdks) that place a phosphate group on side chains of serine and threonine residues, tyrosine kinases place a phosphate group on the phenolic side chain of a tyrosine residue. The initial tyrosines phosphorylated on receptor activation are within the cytoplasmic sequence of the receptor protein. The kinase activity of one receptor protein in the oligomer acts on a neighboring receptor protein of the receptor dimer or oligomer complex, and vice versa, to autophosphorylate tyrosines of the partner receptor protein (Figure 24.7). The phosphotyrosines are binding sites for cytoplasmic signaling proteins.

Some growth factor receptors lack tyrosine kinase activity, and on binding growth factor and oligomerization in the membrane, the receptor proteins bind an independent cytoplasmic protein with a tyrosine kinase activity such as **Src**. The addition of the Src protein gives these tyrosine-kinase-deficient receptors a tyrosine kinase activity.

Cellular proteins of the growth factor signaling pathway typically contain **SH2** (**S**rc **h**omology domain **2**) and **SH3** domains, which contain binding sites that lead to the protein–protein interactions that transmit the signal. The SH2 domain in signaling proteins binds to a phosphorylated tyrosine site, and the SH3 domain binds to a region in target proteins having a polyproline helix secondary structure (Figure 24.8). Different P-Ys in the receptor protein sequence are specific to the binding of a particular signal molecule. Thus, each of the signaling molecule SH2 domains has a secondary site specificity that differentiates between the amino acid sequences surrounding the P-Ys to find its P-Y binding site in the receptor protein. Accordingly, the SH2 domain of the signal molecule PI3K has specificity not only for phosphotyrosine but for the phosphotyrosine at tyrosine-740 in the PDGF receptor protein sequence, and PLCγ for the phosphotyrosine residue at position 1021 in the

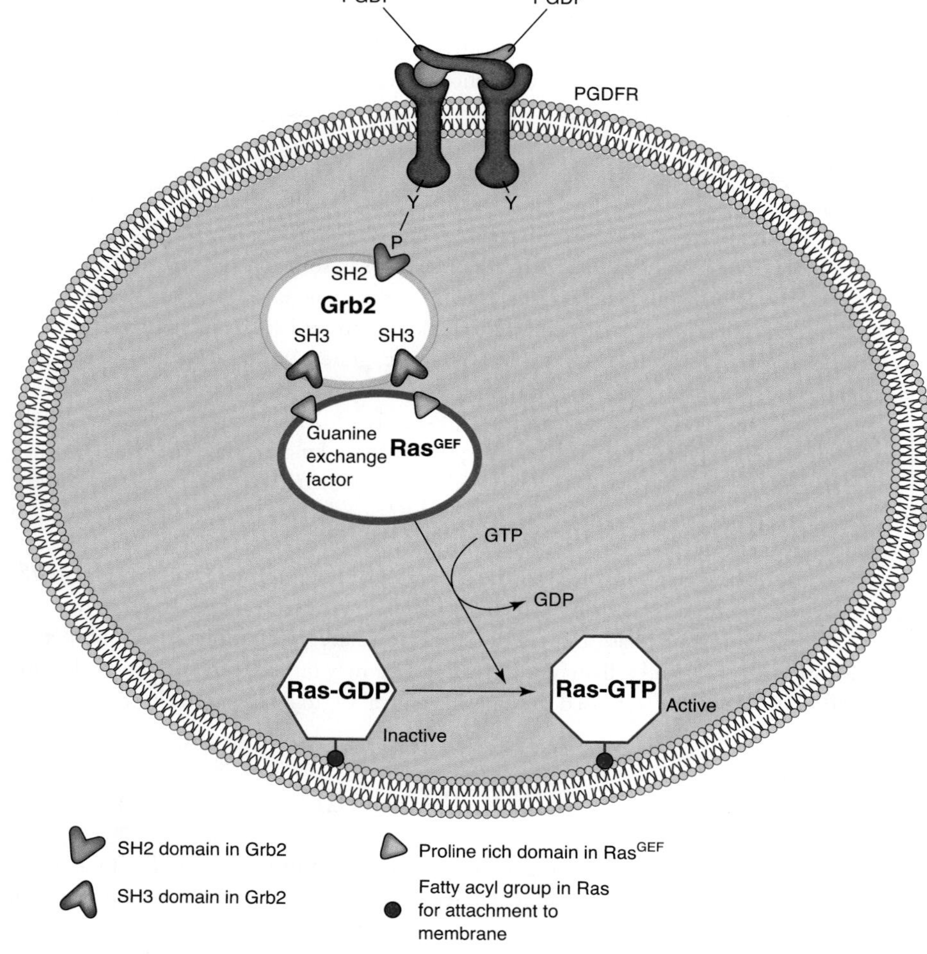

Figure 24.8 Transmission of signal from PDGF receptor to Ras. Plasma membrane containing PDGF receptor becomes autophosphorylated on tyrosine residues. The SH2 domain of Grb-2 binds to phospho-tyrosine residue 416 in the PDGF receptor, and its SH3 domains bind to polyproline helix region in GEF^{Ras} (Ras guanine exchange factor), which promotes exchange of GDP for GTP in Ras. Ras-GTP then transmits its downstream signals.

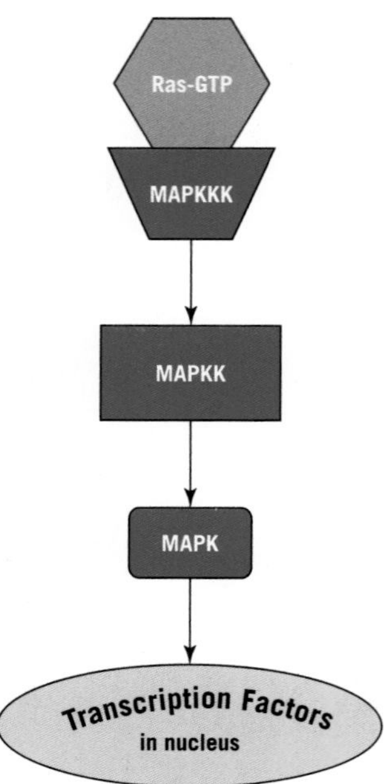

Figure 24.9 GTP-Ras activates a kinase cascade. Ras-GTP activates a kinase cascade resulting in the phosphorylation and activation of the terminal kinase, MAPK. Activated MAPK enters the nucleus and phosphorylates transcription factors that regulate the expression of proteins involved in S phase. MAPK is mitogen-activated protein kinase. MAPKK is MAP kinase kinase, and MAPKKK is MAP kinase kinase kinase.

receptor sequence. On binding to the receptor protein, PI3K and PLC become substrates for the receptor tyrosine kinase. Phosphorylation of tyrosines in the signal molecules activates the signaling protein, and it may dissociate from the receptor to transmit its signal at another site in the cell.

The signal for cell division is transmitted to **Grb2**, which binds to the phosphorylated tyrosine residue 716 in the PDGF receptor. Grb2 has no catalytic activity but is an adaptor protein that transmits its signal through conformational change and protein–protein associations. Grb2 contains both **SH2** and **SH3** domains (Figure 24.8). Binding of Grb2 to the PDGF receptor through its SH2 domain induces a conformation change that opens its SH3 domains to bind a polyproline helix region in the **GEF^{Ras}** protein (**g**uanine-nucleotide **e**xchange **f**actor). Conformationally activated GEF^{Ras} then makes active the signaling protein **Ras** by promoting the exchange of its bound GDP for GTP in the Ras nucleotide binding site (Figure 24.8). GTP-Ras is the active form of Ras that gives a downstream signal by binding to a kinase that becomes activated on GTP-Ras association. The kinase then activates a downstream kinase cascade. Ras is also a GTPase enzyme that inactivates itself by catalyzing the hydrolysis of its bound GTP to GDP, converting the signaling active GTP-Ras to an inactive GDP-Ras.

Activating *ras* mutations resulting in permanently active GTP-Ras occur in approximately 30% of human cancers. The activating mutations are in amino acid residues 12, 13, and 61 of Ras, amino acid residues essential for the Ras GTPase catalytic activity. The oncogenic GTP-Ras conformation cannot deactivate and continuously transmits a downstream signal to enter cell division.

In the normal Ras, the Ras GTPase deactivating activity is increased 100- to 1000-fold by the binding of **GAP^{Ras}** (**G**TPase **a**ctivating **p**rotein). However, the binding of GAP^{Ras} to the GTPase inert oncogenic Ras has no effect, as no GTPase activity is possible due to the catalytic residue mutations in Ras.

Ras also needs to associate with the plasma membrane to be activated. Posttranslational modifications of Ras are required for membrane association. Modifications include removal of the four amino acids from the C-terminal end, methylation of the new C-terminal carboxylic acid group, and addition of a farnesyl fatty acid group to a cysteine side chain near the new COOH-terminal end that acts to anchor the Ras protein to the cytoplasmic surface of the plasma membrane. In some Ras isoforms a second fatty acyl group is added, also near the COOH-terminal end.

Three different Ras genes exist in humans, Ha-*ras,* N-*ras,* and K-*ras,* which produce homologous proteins of 21 kDa. The amino acid sequences of the first 85 residues in each of the isoforms are identical and the next 80 residues show 85% homology, but the C-terminal ends differ more dramatically. While there are differences in some of the downstream signals between the Ras isoforms, in general their signals overlap. These isoforms are characteristically expressed in different cell types, and mutations of residues 12, 13, and 61 constitutively activate all isoforms. Wild-type *ras* genes are **protooncogenes**. Protooncogenes are changed to **oncogenes** by an activating mutation that acts to maintain a cancer cell.

The kinase activated by association with GTP-Ras is a MAP kinase kinase kinase (MAPKKK) (Figure 24.9). The MAPKKK phosphorylates a MAP kinase kinase (MAPKK), which phosphorylates a MAP kinase (MAPK). Activated (phosphorylated) **MAPK** translocates to the nucleus where it phosphorylates and activates transcription factors such as **Jun** and **Fos**. An important target gene of Jun or Fos is the gene for the transcription factor **Myc**. Myc increases the expression of many genes involved in the S phase including those of the G_1- and G_1/S-cyclins and the transcription factor E2F (p. 189). Increasing the levels of the transcription factors Myc and E2F initiate S phase and cell division. However, if conditions are not proper for completion of the S phase and cell division, Myc and E2F may instead initiate a process leading to cell death in the cell that has passed through the critical G_1 to S transition checkpoint.

24.3 • APOPTOSIS: PROGRAMMED CELL DEATH

Apoptosis is the Greek word for "falling leaves," a natural autumn event, and cellular apoptosis is the natural biochemical process of cell death. In an adult human, it is estimated that 3 billion cells are born each minute and therefore 3 billion cells per minute must undergo apoptotic death to maintain homeostasis. Apoptosis is required to restructure tissues and

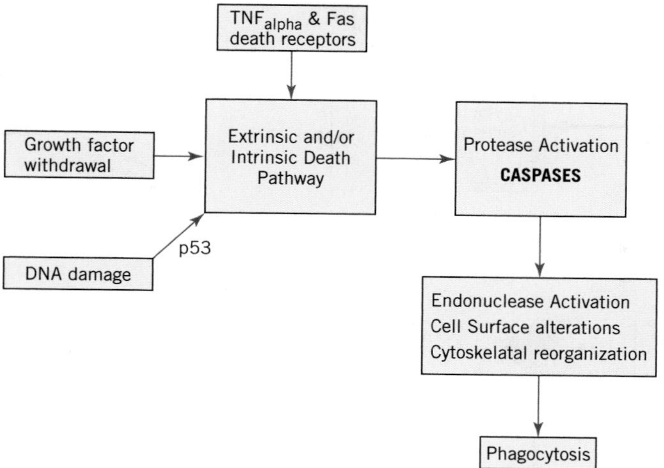

Figure 24.10 Outline of pathway for programmed cell death.

organs in embryogenesis and development. Apoptosis also eliminates immunoresponsive T-cells and B-cells at the end of their response to infection, and regulates wound healing. Apoptotic death differs from **necrotic** cell death. In necrotic cell death, lysis of the cell membrane occurs and releases the cell's contents into the extracellular space generating an inflammatory response, as often occurs in bacterial or viral infections and in trauma. In contrast, apoptosis is often unobservable and does not induce an inflammatory or immunological response. It can be initiated by a variety of signals, including those from damaged DNA, entrance of a cell into the S phase under improper conditions, lack of proper contacts of a cell with extracellular matrix, lack of necessary growth factors in the cellular extracellular environment, or the presence of death signal proteins in the environment of a cell. These signals activate cytoplasmic protease enzymes called **caspases**. The caspases hydrolyze specific peptide bonds in target proteins that on activation by a caspase promote cell death by either a gain or loss of function. Apoptotic death is characterized by caspase-generated changes in the plasma membrane, cytoskeleton, and nuclear DNA. The plasma membrane becomes blebbed and sheds membranous particles that encapsulate the intracellular content. Endonucleases are activated that degrade the chromosomal DNA into nucleosome-size fragments ($\sim$170 base pairs in length). The remnants of the cell are engulfed and digested by phagocytic cells such as macrophages (Figure 24.10).

The caspase proteases are activated by precursor caspases in a protease cascade mechanism similar to the protease cascades of blood coagulation and the protease complement reactions of the immunological response. Each caspase is activated by cleavage of a peptide bond by an upstream caspase in the cascade pathway. In turn, the activated caspases act on protein substrates that generate the changes characteristic of apoptotic death.

Major Pathways of Apoptosis

Death Receptor Pathway

The two major pathways that activate the caspase cascade and promote cell death are the **death receptor pathway (extrinsic pathway)** and the **mitochondrial pathway (intrinsic pathway)** (Figure 24.11). The receptor pathway is initiated by binding of a death ligand in the extracellular environment to its receptor in the plasma membrane of a target cell. **Death ligand** binding promotes binding of intracellular adaptor proteins to the cytoplasmic region of the receptor protein in the plasma membrane. Such an adaptor protein is FADD (the DD in the acronym refers to a protein-binding domain present in many proteins of this pathway called the **d**eath **d**omain). The aggregates formed by the adaptor proteins through DD interactions at the cytoplasmic side of the receptor include the caspase 8 precursor protein, pro-caspase 8. The pro-caspase 8 within the aggregate acquires an auto-proteolytic activity, and cleaves peptide bonds in its sequence that change pro-caspase 8 to an active caspase 8. Caspase 8 leaves the aggregate to initiate the protease caspase cascade (Figure 24.11). The cytoplasmic receptor aggregate containing multiple adaptor proteins such as FADD and pro-caspase 8 molecules is referred to as a **DISC** (**d**eath **i**nducing **s**ignaling **c**omplex).

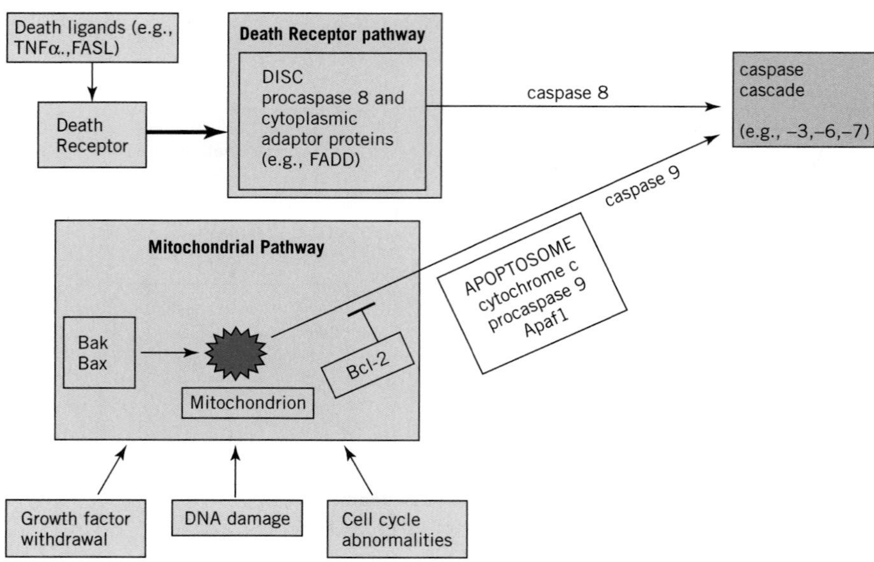

Figure 24.11 Death receptor and mitochondrial pathways activate caspase cascade. The death receptor (extrinsic) and mitochondrial (intrinsic) pathways converge on the activation of caspases. Death ligands activate death receptor to bind adaptor proteins and pro-caspase 8, which is activated to caspase 8. Caspase 8 then activates the caspase cascade. In mitochondrial pathway, Bak and Bax oligomerize to permeabilize the outer mitochondrial membrane to allow cytochrome c to pass into the cytoplasm where it forms an apoptosome with multiple molecules of Apaf1 and pro-caspase 9. Activated caspase 9 then activates caspase cascade. Bcl-2 inhibits cytochrome c passage from mitochondria. Activators of the mitochondrial and receptor pathways (i.e., TNFα and DNA damage) are shown.

Mitochondrial Pathways

The intrinsic, or mitochondrial, pathway is regulated by the relative concentrations of **BH domain** proteins in the outer mitochondrial membrane (A Closer Look 2.1, p. 1013). Cytochrome c is the major mitochondrial signaling protein, and apoptosis is activated by the passing of **cytochrome c** through the outer mitochondrial membrane after permeabilization of the membrane by the pro-apoptotic BH domain proteins Bak and Bax (Figure 24.12). The membrane permeabilization to cytochrome c is generated by hetero- or homo-oligomers of **Bax** and/or of **Bak** BH proteins formed in the outer membrane. In opposition, the anti-apoptotic BH domain proteins **Bcl-2** and **Bcl-X$_L$** prevent Bax and Bak self-association by binding to Bak and Bax and sequestering them from associating with each other. Whether cell death occurs then depends on the relative concentrations of pro-apoptotic Bak and Bax and anti-apoptotic proteins such as Bcl-2 and Bcl-X$_L$ in the outer mitochondrial membrane. If the concentration of Bcl-2 and Bcl-X$_L$ proteins with empty binding sites is greater than the concentration of Bak and Bax, aggregates of Bak/Bax will not form and the cell survives. However, if the relative concentrations of Bak and Bax proteins are greater than Bcl-2 and Bcl-X$_L$ binding capacity, these proteins will self-aggregate to induce mitochondrial

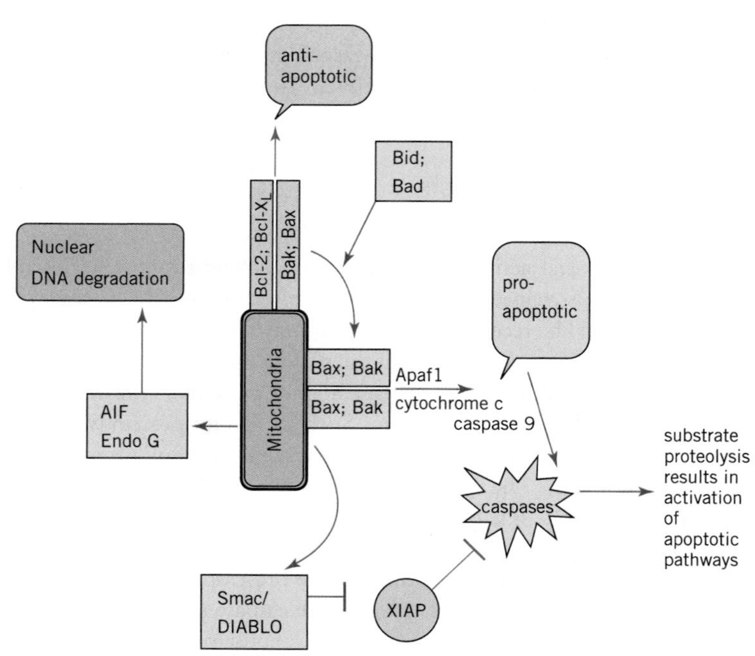

Figure 24.12 Regulation of apoptosis by cytochrome c and other mitochondrial proteins. Oligomerization of Bak and/or Bax in the outer mitochondrial membrane permeabilizes the membrane for transfer of cytochrome c from mitochondria to cytoplasm. Bcl-2 and Bcl-X$_L$ inhibit membrane permeabilization by binding with Bax and Bak to prevent Bax and Bak self-association. Facilitator proteins, such as Bad and Bid, promote apoptosis by either promoting the dissociation of Bak and Bax from Bcl-2 and Bcl-X$_L$ or promoting an active conformation of the Bax/Bak oligomers. In the cytoplasm, cytochrome c binds in apoptosome complex, which leads to the activation of pro-caspase 9 to active caspase 9. Smac/DIABLO can also enter cytoplasm and activate the caspases by inhibiting XIAP. XIAP is an inhibitor of caspase activity. AIF and EndoG can pass from mitochondria into nucleus to promote degradation of the DNA. Blue lines show anti-apoptotic and red lines pro-apoptotic effects.

A CLOSER LOOK 24.1

The Bcl-2 Protein Family

The Bcl-2 protein family has pro- or anti-apoptotic roles in the mitochondrial membrane. Members of the family are characterized by their containing one or more **B**cl-2-**h**omology (**BH**) domains. These domains function to bind other proteins of the family through BH-domain interactions. The anti-apoptotic Bcl-2 and Bcl-X_L proteins contain four BH-type domains numbered BH1, BH2, BH3, and BH4. The Bax and Bak proteins that self-polymerize to permeabilize the mitochondrial membrane contain three BH-type domains, BH1, BH2, and BH3. The facilitator BH proteins contain only the BH3 domain and are referred to as BH3-only proteins (see figure).

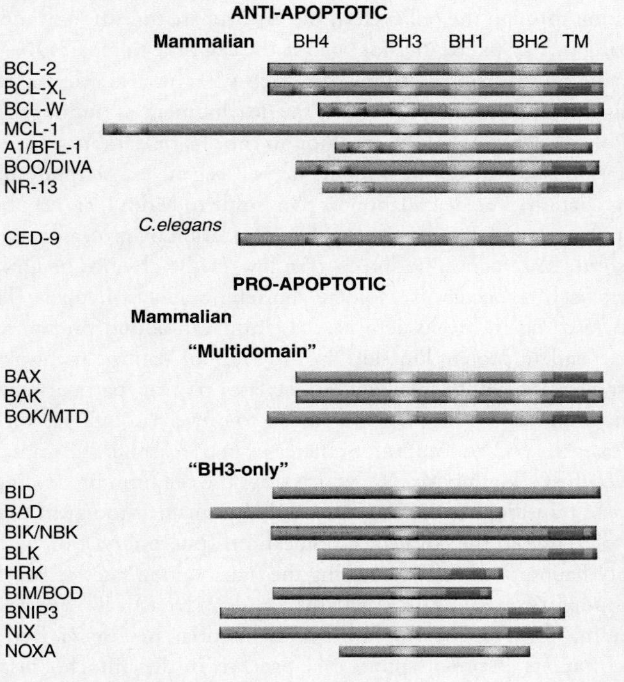

Schematic diagram showing the BH domains in anti- and pro-apoptotic mitochondrial membrane proteins. Anti-apoptotic proteins that bind and sequester pro-apoptotic proteins contain four different domains (top group): BH1, BH2, BH3, and BH4 (homologues include Bcl-X_L, Bcl-w, A1/Bfl-1, Mcl-1, Boo/Diva, and NR-13). Pro-apoptotic proteins that permeabilize the outer mitochondrial membrane contain three domains (middle group): BH1, BH2, and BH3 domains, but no BH4 domain (homologues are Bak, Bax, and Box/Mtd). Pro-apoptotic facilitator proteins (bottom group) only contain a BH3 domain (homologues include Bid, Bik, Puma, Noxa, Blk, Hrk, Bnip3, and Bim). TM is a transmembrane domain present in many of the BH domain proteins. Different groups of homologues are expressed in different cell types.
Redrawn from Scorrano, L., and Korsmeyer, S. J. *Biochem. Biophys. Res. Commun.* 304:437, 2003

outer membrane permeability and cytochrome c will pass from the mitochondria to the cytoplasm to initiate caspase activation and cell death. In addition to Bax and Bak, other BH domain proteins in the mitochondrial membrane act as facilitator proteins to promote dissociation of Bak or Bax from Bcl-2 or Bcl-X_L by competing for the binding sites in Bcl-2 or Bcl-X_L. Other BH domain facilitator proteins in the plasma membrane bind directly to the Bax/Bak oligomers to promote membrane permeabilization by the Bax/Bak oligomer.

There are at least 24 BH domain mitochondrial membrane proteins in humans, divided into the following three functional classes: (1) anti-apoptotic proteins such as Bcl-2 and Bcl-X_L, (2) pro-apoptotic membrane permeabilization proteins Bax and Bak, and (3) pro-apoptotic facilitator proteins such as Bid, Bad, Bim, PUMA, and Noxa. The proteins in all the functional classes contain **BH domains** (**B**cl-2 **H**omology **domains**), which are the binding sites for self-association and for heteroassociation with the other BH domain proteins (A Closer Look 2.1, p. 1013). Which of the Bcl-2 isoforms are expressed by a cell depend on the cell type.

In the cytoplasm the cytochrome c binds into a protein aggregate containing multiple molecules of the adaptor protein **Apafl** (**a**poptotic **p**rotease-**a**ctivating **f**actor **1**) and **pro-caspase 9**. This cytoplasmic complex is called an **apoptosome** and is functionally similar to a DISC (p. 1011) of the death receptor pathway. The addition of the cytochrome c to the apoptosome causes an autocatalytic hydrolysis of specific peptide bonds in the sequence of pro-caspase 9, resulting in the active caspase 9. **Caspase 9** leaves the apoptosome and activates the caspase cascade to execute apoptotic death (Figure 24.12).

Proteins from mitochondria other than cytochrome c pass through the permeable mitochondria membrane to initiate cytochrome-c-independent apoptosis under certain conditions (Figure 24.12). The protein **Smac/DIABLO** passes into the cytoplasm and inhibits **XIAP** (**X**-linked **i**nhibitor of **ap**optosis), a caspase protease inhibitor protein that is normally present to limit basal caspase activity generated in the absence of a pro-apoptotic-activating signal. By inhibiting the inhibitor of caspases (XIAP), Smac/DIABLO activates the caspase cascade. The mitochondrial proteins **AIF** and **endoG** are released from the mitochondria and enter the nucleus to activate DNA degradation in cellular apoptosis.

p53 Induction of Apoptosis

Cells that have damaged DNA are stopped from completing cell division by p53-dependent mechanisms. p53 can act to increase levels of the CKI p21 (p. 1008) to inhibit Cdk-promoted progression through the cell cycle at the G_1-S-phase transition or the G_2-M-phase transition. Increasing p21 stops cell division, giving the cell time to repair DNA. If the DNA damage is irreparable in a reasonable time span, then p53 initiates cellular apoptosis. The elimination of cells with DNA damage prevents the development of tumor cells, and the loss of p53 regulation of cellular apoptosis is common in cancers (p. 1016).

p53 concentration is not determined by the rate of its gene transcription but by the rate of p53 protein degradation. The degradation of p53 is due to Mdm2, an E3 ubiquitin ligase. Mdm2-catalyzed polyubiquitylation promotes p53 degradation by proteasomes. Mdm2 regulation keeps the steady-state concentration of p53 low in the absence of DNA damage or p53-inducing stress such as occurs in telomere shortening and in hypoxia (Figure 24.13). DNA-damage-generated chain breaks activate the chromatin-bound protein kinases ATM, ATR, and DNA-dependent protein kinase (DNA-PK). DNA damage pathways also activate the lysine acetyl-transferase p300/CBP, which acetylates p53 on particular lysine residues. Acetylation acts in conjunction with phosphorylation to prevent ubiquitylation by Mdm2 of p53 and thus increases the p53 concentration. Increases in p53 lead to the transcription of the p21 gene to produce the Cdk inhibitor p21, which stops the cell from proceeding through the G_1 to S and G_2 to M transitions (p. 1004). If the DNA damage is too great to repair, higher concentrations of p53, due to the covalent modifications (phosphorylation and acetylation), result in p53-induced apoptosis by (1) increasing the transcription rates of Bax and several of the facilitator pro-apoptotic proteins such as PUMA and (2) by activation of cytoplasmic Bax causing the cytoplasmic form to associate in the mitochondrial membrane. Both mechanisms increase the concentrations of pro-apoptotic BH proteins in the mitochondrial membrane, resulting in cellular apoptosis.

p53 levels are also regulated by the transcription factors Myc and E2F in the S phase of the cell cycle (p. 1004). While these transcription factors promote the expression of proteins critical for the S phase in cell division, they also initiate cell death when conditions are unfavorable for cell division. This emphasizes that the commitment to enter the S phase is

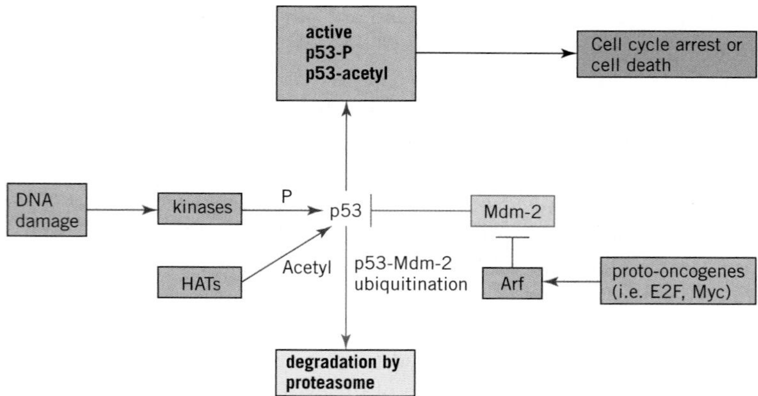

Figure 24.13 Regulation of p53 activity. Under normal conditions p53 concentrations are kept low by Mdm2, which promotes ubiquitination of p53 and its degradation by proteasomes. Phosphorylation and acetylation of appropriate residues by kinase and HAT enzymes prevents degradation of p53 and also increases its transcriptional activity. Certain protooncogene transcription factor proteins, such as E2F and Myc, here act as tumor suppressor proteins. Under favorable conditions E2F and Myc promote the expression of genes required in cell division, but under abnormal conditions increase Arf expression. Arf binds to Mdm2 to prevent its interaction with p53 and thereby increases p53 concentration. The Arf-induced mechanism may be a cellular safety valve to stop a cell from dividing when certain negative signals or conditions arise after the cell has initiated the cell division process. Increased p53 activity leads to cell cycle arrest, or apoptosis. Red lines show steps leading to activation of p53 and blue lines to inactivation of p53.

a critical irreversible decision for the cell that results in either cell division or death, depending on the circumstances. Under conditions unfavorable for cell division in the S phase, the elevated levels of E2F and Myc increase the transcription of the gene for the protein Arf. Arf binds to Mdm2 and prevents its association with p53 (Figure 24.13). Thus Myc and E2F through Arf increase p53 concentration and cellular apoptosis. The E2F- or Myc-initiated cell death pathway is not functional in the p53-deficient cancer cells.

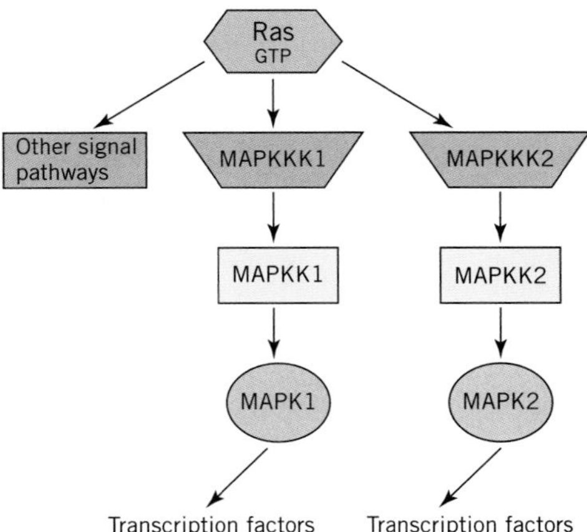

Figure 24.14 Multiple pathways may be regulated by Ras. Ras is the hub of multiple pathways that regulate not only cell division but also cell apoptosis, differentiation, and other important cellular functions.
MAPKKK is a MAPK kinase kinase and MAPKK is a MAPK kinase. MAPKs on activation by a MAPKK can localize to the nucleus to phosphorylate and thereby regulate transcription factors. The kinases may also phosphorylate proteins other than transcription factors. Alternative targets of the kinases include BH domain pro- and anti-apoptotic proteins and p53.

MAPK Pathways Regulate Both Cell Death and Cell Survival

Like Myc and E2F, MAPKs have dual roles in regulating both cell division and cell death. The MAPK pathways are activated by a mitogen growth factor signal and transmit the signal to transcription factors in the nucleus (see Figure 24.9, p. 1010). In the primary signal pathway transmitting the cell division signal, the MAPKKK is Raf, followed by the MAPKK MEK, and the MAPK ERK. In actuality, there are multiple MAPK pathways downstream from Ras, so that Ras is a focal point for a network of pathways (Figure 24.14). Which of these Ras downstream pathways are activated, the level of activation and the duration of the MAPK pathway signal depend on cell type and on cellular context. Although the Ras upstream signal can activate multiple MAPK cascades, other signaling molecules can separately activate individual Ras downstream MAPK pathways. Thus, particular MAPK pathways are activated by UV light, radiation, cytokines, or the stress molecules of inflammation, in the absence of Ras activation. How these MAPK pathways then regulate cell division, differentiation, or cellular apoptosis depends on the integration and interpretation by the cell of the intensity, timing, and duration of all the MAPK pathway signals, some of which may be negative and others positive for a particular process or cellular outcome. Thus, activation of the MAPK **c-Jun N-terminal kinase (JNK)** commonly promotes apoptosis, but more rarely and only in particular cell types, JNK promotes cell survival. JNK promotes cell

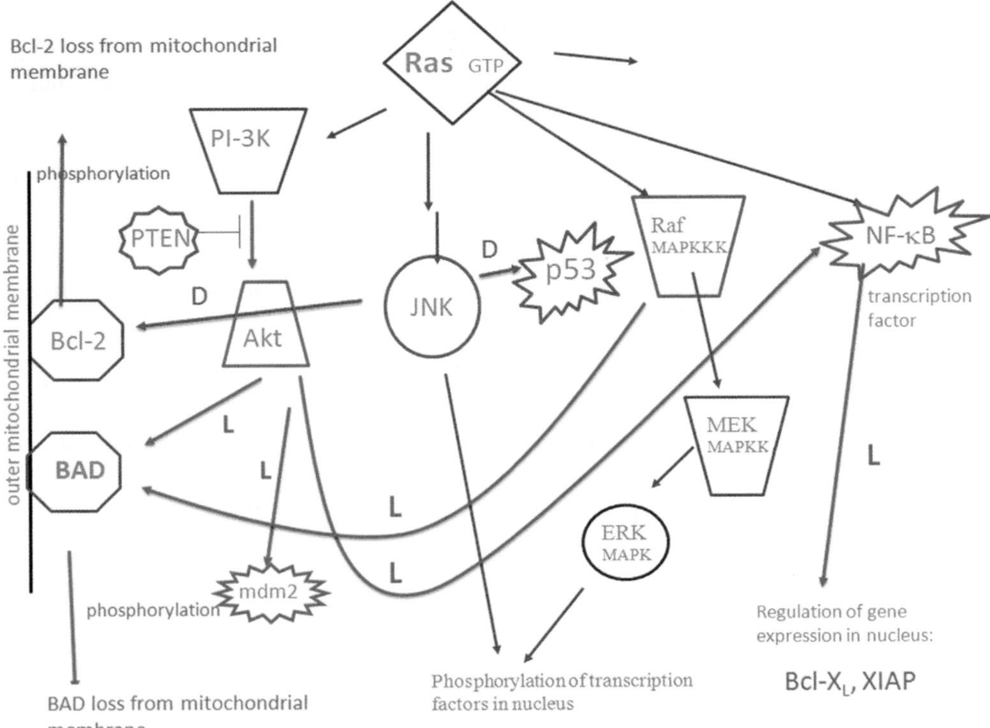

Figure 24.15 Ras downstream effectors regulation of apoptosis. Abbreviations: **L**, Live (pathway promotes cellular survival); **D**, Die (pathway promotes apoptotic cellular death). Raf, MEK, ERK, JNK, and Akt are protein kinases. PI3K is the signaling molecule 1-phosphatidylinositol-3-kinase upstream from Akt. Activation of Akt by PI3K is inhibited by the PTEN phosphatase. By inhibiting the Akt survival pathway, PTEN is an important tumor suppressor activity that is lost in many cancer cells. NF-κB is a transcription fact that promotes cell survival by increasing expression of Bcl-X and XIAP. Under certain conditions and in particular cell types, NF-κB can promote apoptosis rather than survival. Bcl-2 and BAD are phosphorylated by kinases that cause them to leave the mitochondrial membrane. Symbols: *circles*, MAPKs; *trapezoids*, other protein kinases; *stars*, transcription factors and other molecular types such as ubiquitin ligases and phosphatase; and *octagons*, mitochondrial membrane proteins. Red arrows show death pathways and green arrows survival pathways.

death by phosphorylating anti-apoptotic Bcl-2 to promote its dissociation from the mitochondrial membrane (Figure 24.15). JNK can also phosphorylate p53, making it resistant to Mdm-2 ubiquitylation. In contrast, the MAPK ERK, the MAPKKK Raf, and the kinase Akt, promote cell survival by phosphorylating BH domain pro-apoptotic proteins, which decreases their mitochondrial membrane concentrations and activities (Figure 24.15). The kinase Akt also activates the transcription factor NFκB, which increases the transcription of genes that promote cell survival. Thus, different kinases downstream from Ras transmit opposing survival and death signals; the actual outcome of the MAPK signaling depends on the presence or absence of other extracellular signals, the physiological status of the cell, and the cell type.

24.4 • CANCER

Oncogenes and Tumor Suppressor Genes

The promotion of the cell division cycle and resistance to apoptosis are critical properties of a cancer cell. Genes that encode proteins that promote cell division or promote resistance to apoptosis and cell survival are **protooncogenes** (Table 24.2). Activating mutations in a protooncogene convert it to an **oncogene**. Activating mutations can also occur in the protooncogene noncoding regions when the mutation leads to overexpression of the protooncogene product. A single activated allele of a protooncogene is sufficient to promote an oncogenic effect in a cell. Thus activating mutations in protooncogenes are autosomal dominant for promotion of a cancer.

Protooncogene products promote cell division or cell survival and include genes for growth factors, growth factor receptors, Grb-like adaptor molecules, Src-like tyrosine kinases, kinases of MAPK cascades, Cdks, cyclins, CAKs, Cdc25s, and transcription factors such as Jun, Fos, Myc, and E2F that increase expression of cell division cycle proteins. Anti-apoptotic or survival-promoting protooncogenes are Bcl-2 and Mdm-2, whose expressions are elevated in many cancers. The most prominent oncogene in human cancers is activated *ras,* which is found in approximately 30% of human cancers.

In contrast to the effect of oncogenes, products of **tumor suppressor genes** suppress cell division or promote apoptosis. For tumor suppressor activities, the loss of its

TABLE 24.2 • Examples of Protooncogene Products That Regulate Cell Division and Apoptosis

A. Activators of Cell Division	
Growth factors	PDGF (*sisI*), FGF, and other mitogen factors
Growth factor receptors	PDGFR and other growth factor receptors
Tyrosine kinase cytoplasmic proteins	Src family of proteins
Adaptor proteins	Grb-2
Ras GTP-binding proteins	K-Ras, H-Ras, N-Ras
Kinases and kinase cofactors	Cyclins, Cdk, CAK, and MAPKs
Phosphatases	Cdc25
Transcription factors	Myc, Jun, Fos, and E2F
B. Activators of Apoptotic Resistance (Survival)	
BH domain anti-apoptotic proteins	Bcl-2, Bcl-X_L
Promotes degradation of p52	Mdm2
Caspase inhibitors	XIAP
Anti-apoptotic kinases	Akt, Raf (MAPKKK)
Transcription factor	NFκB

activity promotes a cancer. p53, which induces apoptosis with DNA damage, is the most prominent of the tumor suppressor proteins. The p53 gene is inactive or lost in approximately 50% of human cancers and p53 pathways are dysregulated in the great majority of the other 50%. Rb protein is a prominent tumor suppressor, and its loss releases E2F to continuously induce entrance into S phase from G_1. CKIs, such as p21, are tumor suppressors as they act to inhibit cell cycle progression on DNA damage. Pro-apoptotic BH domain proteins such as Bak, Bax, and Bad are tumor suppressors (Clin. Corr. 24.1). Both alleles of a tumor suppressor gene must be inactivated to eliminate their tumor suppressor activity in a cell. In contrast to protooncogenes, inactivating mutations in tumor repressor genes must therefore occur in both alleles for the loss of the suppressor activity. Hence, inactivating mutations in both suppressor gene alleles are most often required for the inactivations to effectively contribute to a cancer phenotype.

Properties of Cancer Cells

Two major properties acquired by a cancer cell have been emphasized, namely, unregulated cell division and resistance to apoptosis. Malignant cancer cells also exhibit cellular **immortality**, the ability to induce angiogenesis, and the ability to metastasize. Genes that

CLINICAL CORRELATION 24.1

Oncogenic DNA Viruses

Oncogenic DNA viruses produce viral proteins that act to promote survival of their host cell during viral replication and to induce the expression of S-phase enzymes from host cell genes to replicate its viral DNA. After viral replication is complete, the virus kills the host cell by lysing the host cell plasma membrane (lytic cycle). In rare instances during a viral infection, the viral genes that promote entry into S phase or promote resistance to apoptosis become incorporated into their host cell DNA by a recombinational event. This recombination must occur in a host cell that also survives the viral infection, in order to be carcinogenic to the host organism. If the viral gene within the host DNA is then overexpressed in an unregulated manner, it is a viral oncogene and can initiate a carcinogenic process leading to a cancer. DNA viruses known to cause cancers are human papillomavirus (HPV), Epstein–Barr virus (EBV), Kaposi's sarcoma-associated herpevirus (KSHV), and Hepatitis B virus (HBV) (see following table).

Papillomavirus is a causative factor in over 95% of uterine cervical cancers. The viral EBV E7 protein binds to the cellular-Rb-binding pocket and displaces E2F from Rb, resulting in the E2F-dependent up-regulation of S-phase proteins. The viral oncoprotein E6 binds to cellular p53 and promotes its degradation, eliminating p53-dependent cell cycle regulatory and apoptotic mechanisms. Two vaccines are currently approved for prevention of infection by strains of papillomavirus commonly associated with female cervical cancers. Other viral oncoproteins similarly interfere with molecules within cellular pathways to promote unregulated cell division, cell survival, and cellular immortality; all properties of cancer cells (see following table).

Human DNA Viruses[1]

DNA Virus Types	*Cancer Types*	*Viral Oncogene Proteins and their Actions*[2]
Human papillomavirus (HPV), certain subtypes	cervical, anogenital cancers	E7 inhibits Rb, E6 inhibits p53
Epstein-Barr virus (EBV)	lymphoproliferative disease, Burkitt lymphoma, nasopharyngeal cancers, Hodgkin lymphoma, non-Hodgkin lymphoma in immunosuppressed patients	LMP-1 increases telomerase, increases c-Myc, decreases Bax, decreases caspases, increases c-IAP, and increases Bcl-2
Kaposi sarcoma-associated herpevirus (KSHV)	Kaposi sarcoma, AIDS related primary effusion lymphoma, multicentric Castleman's disease	LANA-I increases telomerase and decreases p53. EBNA2 increases NFκB activity to promote survival
Hepatitis B virus (HBV)	Hepatocellular carcinoma	HBV X disrupts mitotic spindle, decreases DNA repair, dysregulates centrosome synthesis and function, and decreases p53

[1]Table based on review by D.E. de Oliveira, *Cancer Letters* 247:182 (2007).
[2]Effects of viral oncoproteins are cell type specific.

De Oliveira, D. E. DNA viruses in human cancer: an integrated overview on fundamental mechanisms of viral carcinogenesis. *Cancer Letters* 247:182, 2007; and Stanley, M. Prevention strategies against the human papillomavirus: The effectiveness of vaccination. *Gynecol. Oncol.* 107:S19, 2007.

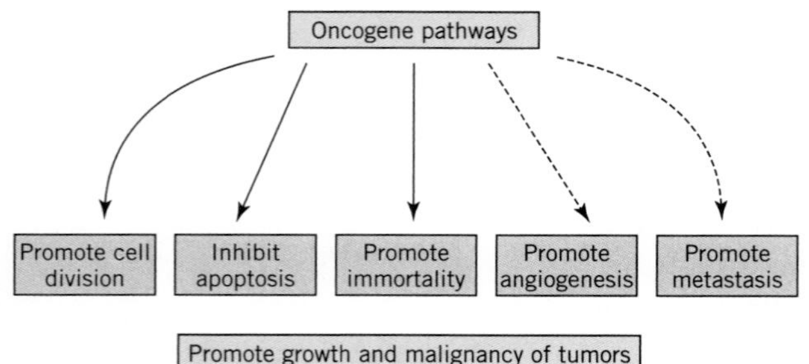

Figure 24.16 Oncogene pathways. Oncogene pathways promote cell division, inhibit apoptosis, promote cellular immortality, promote tumor cell metastasis, and promote angiogenesis. Dotted line indicates involvement of non-transformed neighboring cells to that of the tumor cell.

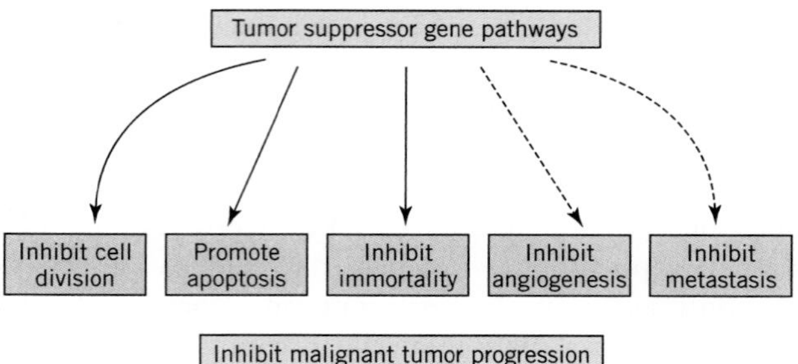

Figure 24.17 Tumor suppressor pathways. Tumor suppressor pathways inhibit cell division, promote apoptosis, inhibit cellular immortality, inhibit cancer cell metastasis, and inhibit angiogenesis. Dotted line indicates involvement of non-transformed neighboring cells to that of the tumor cell.

contribute to these properties may also be considered protooncogenes, and those which inhibit these properties tumor suppressor genes (Figures 24.16 and 24.17).

Cellular **immortality** is the ability of a cell to produce infinite generations of progeny. Mortal cells, such as normal human epithelial cells in cell culture, divide to produce 20–40 generations and then enter senescence. Cells in culture derived from an immortal cell continue to divide indefinitely.

Angiogenesis refers to the growth of new blood vessels. This is required to provide a growing tumor mass with nutrients and oxygen for survival and continued growth. The malignant tumor cell also has the ability to metastasize. The **metastasis** of cancer cells to grow tumors at multiple sites is what usually kills patients with cancer. In blood-borne metastasis, tumor cells dissociate from their contacts with other cells within the tumor mass, pass through the basement membrane and the surrounding extracellular matrix into capillaries, travel through the blood to a target tissue, extravasate out of the blood vessels, and proliferate in the new site to form a secondary tumor.

Immortality of Cancer Cells

Cellular immortality primarily depends on the ability of most cancer cells to maintain expression of their telomerase activity (p. 156). Telomerase maintains the length of the telomer regions at the ends of chromatids. A terminal nucleotide segment is lost each time chromatids are duplicated, and if telomer extension by telomerase is lacking, progeny cells enter senescence to protect against DNA damage. Whereas most normal cells lose telomerase activity with increasing age of the organism and repeated generations of cell division, 90% of cancer cells retain the ability to express telomerase and acquire the ability to indefinitely proliferate.

Metastasis of Cancer Cells

The major barrier to metastasis is the basement membrane surrounding the primary tumor. Tumor cell invasion through the basement membrane can be divided into steps. (1) The tumor cell laminin receptors bind to laminin protein in the basement membrane. (2) Transmission by the activated laminin receptors of intracellular signals leads to the secretion of proteases into the extracellular matrix (ECM). (3) The tumor cell moves through the holes in the basement membrane created by the protease activity. These three steps are reiterated multiple times as the cell invades through the basement membrane. Many tumor cells also express urokinase plasminogen activator receptor (uPAR) in their plasma membranes to bind secreted urokinase plasminogen activator (uPA) and give their outer membrane a plasminogen activator activity.

Step 3 requires the tumor cell to have a **cellular motility**. Development of motility in a normally stationary cell involves biochemical and structural changes in the cellular cytoskeleton and membrane.

In step 2 the tumor or neighboring cells must secrete enzymes that degrade the basement membrane and extracellular matrix to produce a path for tumor cell migration. Basement membrane (or *basal lamina*) consists of an ordered collagen type IV matrix, proteoglycans, the large protein laminin, and other proteins (p. 367). Metastatic tumor

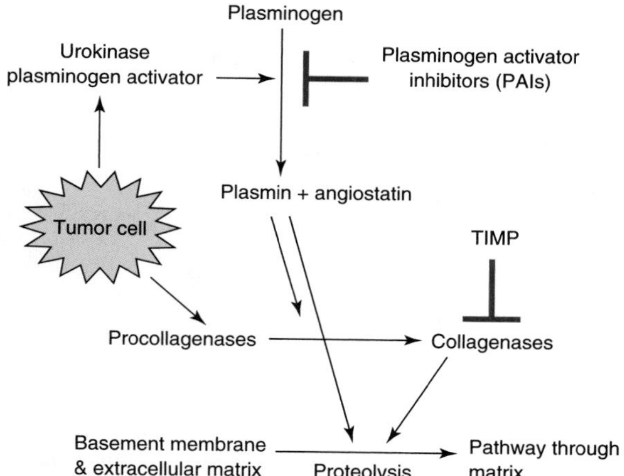

Figure 24.18 Secreted proteases generate pathway through basement membrane in metastasis. Tumor or normal neighboring cells secrete proteases **u**rokinase **p**lasminogen **a**ctivator (uPA) and type IV procollagenases (members of the group of **m**atrix **m**etallo**p**roteinases [MMPs]) and decrease secretions of protease inhibitor proteins TIMP (**t**issue **i**nhibitor of **m**etallo**p**roteinases) and PAIs (**p**lasminogen **a**ctivator **i**nhibitors). uPA cleaves plasminogen into plasmin and angiostatin (an angiogenesis inhibitor, discussed in next section). Plasmin activates procollagenase to collagenase and also acts on other basement membrane proteins to promote their degradation.

cells express an elevated concentration of laminin receptors in their plasma membrane, which on binding to basement membrane laminin (step 1) induces signals that initiate cellular processes related to metastasis. Malignant cells either produce the proteases **urokinase plasminogen activator (uPA)** and **type IV procollagenase** or induce neighboring stroma cells to produce and secrete these proteases. The type IV collagenases are members of the matrix metalloprotease (MMP) class of enzymes containing a catalytically essential Zn metal atom. Two gene products, MMP2 and MMP9, are type IV procollagenases. Activation of **type IV collagenase** from its procollagenase form is catalyzed by uPA-generated **plasmin** results in the type IV collagenase proteolytic cleavage of the type IV collagen in the *basal lamina.* The uPA generates the plasmin protease from plasminogen (Figure 24.18). The **plasminogen** is synthesized in the liver and ubiquitously distributed in extracellular matrix, where it is remains inert till activated by a plasminogen activator (uPA) protein. In addition to plasmin-catalyzed activation of type IV collagenase, the plasmin protease degrades other proteins of the *basal lamina* to allow the tumor cells to invade through the membrane barrier (Figure 24.18).

Under normal conditions, the turnover of extracellular matrix is regulated by the balance of protease and protease inhibitor activities. Two plasminogen activator inhibitors are known, **PAI-1** and **PAI-2** (**p**lasminogen **a**ctivator **i**nhibitor 1 and 2), and several collagenase inhibitors or **TIMPs** (**t**issue **i**nhibitor of **m**etallo**p**roteinases) (Figure 24.18). Cancer cells or neighboring cells increase the protease activity in their extracellular environment by both increasing cellular secretions of proteases and decreasing their synthesis and secretions of the protease inhibitors (the PAIs and TIMPs).

Angiogenesis Induced by Cancer Cells

Angiogenesis is a normal process during embryonic and fetal development and in wound healing and is induced by cancer cells. Tumor cells secrete vascular endothelial cell chemoattractant proteins to promote angiogenesis. Angiogenic proteins are **VEGF** (**v**ascular **e**ndothelial cell **g**rowth **f**actor) and **FGF** (**f**ibroblast **g**rowth **f**actor). Their secretions by tumor cells induce blood vessel endothelial cells to migrate and proliferate to form new blood vessels into the tumor (Figure 24.19). The mechanism of migration and invasion by the endothelial cells mimics the mechanism utilized by tumor cells in metastasis. On receiving the chemoattractant signal, the endothelial cells increase their secretion of plasminogen activator and MMPs, including type IV procollagenase, and decrease their secretion of PAIs and TIMPs. This gives them the ability to pass through basement membrane and ECM.

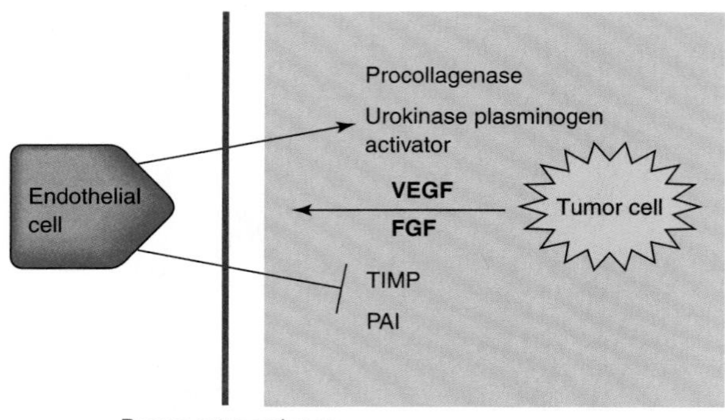

Figure 24.19 Induction of invasion and growth of blood vessel endothelial cells by tumor cell in angiogenesis. Tumor cells secrete endothelial mitogenic and chemotactic factors VEGF and/or FGF, which promote endothelial cells proliferation and migration toward the tumor and formation of blood vessels. In their migration through the extracellular matrix toward the tumor, the endothelial cells use a similar mechanism to that of the tumor cell in metastasis. Endothelial cells increase their secretion of uPA and procollagenases and decrease their secretion of proteases inhibitors PAI and TIMP.

The plasminogen activator uPA cleaves a peptide bond that divides the plasminogen molecule into two halves. The C-terminal segment has the plasmin protease activity, which facilitates the migration of vascular endothelial cells toward the tumor. The N-terminal sequence of plasminogen, produced by plasminogen activator proteolysis, surprisingly has an anti-angiogenic activity and is called **angiostatin**. The angiostatin generated by plasminogen cleavage at the primary tumor site has insufficient activity to inhibit angiogenesis at the primary site where the pro-angiogenic processes predominate. However, the angiostatin has a long lifetime and travels through blood to distant sites to inhibit angiogenesis at secondary metastatic sites. This anti-angiogenic activity is overcome when and if the metastatic tumors develop, but initially this anti-angiogenic activity slows the development of the metastatic tumors at the secondary sites. Other anti-angiogenic proteins are also generated by proteolysis of ECM proteins near the primary tumor. One of these is **endostatin**, which is produced from the C-terminal end of type XVIII collagen. Endostatin, angiostatin, and other anti-angiogenic proteins from extracellular matrix introduced at high concentrations into mice have been shown to cure cancers in mice. These observations have stimulated the investigation of anti-angiogenic drugs for treatment of human cancers.

Multiple Mutations Are Required to Form a Cancer

A single mutation is not sufficient to transform a normal cell into a cancer cell unless it occurs on a background of previous oncogenic mutations. The evidence for a requirement of multiple mutations comes from many observations. A cancer typically forms many years after an initial mutagenic event, indicating that time is required for the accumulation of additional multiple mutations. The increasing incidence of cancer with age also shows the need to accumulate mutations over time in carcinogenesis. More direct is the documentation of mutations in the progressive stages in the development of a cancer. Thus in cervical cancer, a region of the cervix initially shows epithelial cells with a higher-than-normal proliferative rate. These abnormal cells may proceed to form a benign adenoma. A minority of the adenomas progress through additional stages that form a malignant tumor. Analysis of cells from each of the progressive stages shows the acquisition of new activating mutations in protooncogenes and inactivations of tumor suppressor genes that in combination generate the malignant cancer cell.

Experiments with primary human epithelial cells show that a minimum of four changes will generate a transformed epithelial cell. In these experiments, different oncogenes and inhibitors of tumor suppressor gene products were introduced by transfecting normal epithelial cells with gene expression plasmids. A minimum of three particular genes that affect four biochemical activities could transform the cells. Required for

CLINICAL CORRELATION 24.2

Molecularly Targeted Anticancer Drug

Imatinib (formerly known as STI-571 and marketed as Gleevec in the United States) is the first successful drug targeted to a specific oncogene product. It inhibits the tyrosine kinase activity of the chimeric Bcr-Abl tyrosine kinase protein expressed in chronic myelogenous leukemia (CML) and the c-Kit growth factor receptor overexpressed in many gastrointestinal stromal tumors. It has been approved by the U.S. Food and Drug Administration for treatment of these cancers. Tyrosine kinases transfer phosphate from ATP to protein substrates in growth factor signaling pathways. Imatinib binds to the ATP binding site of the Abl and Kit tyrosine kinases, inhibiting the normal binding of the ATP substrate (see figure).

CML arises by a translocation between chromosome 9 and 22 in a pluripotent hematopoietic stem cell to produce aberrant chromosomes in which a part of chromosome 22 is translocated to chromosome 9 and a part of 9 to 22. The smaller of the two abnormal chromosomes, $22q^-$, is easily recognizable in leukemia cells and is called the Philadelphia chromosome. At the translocation site, the reading frame of the 5′ end of the Bcr gene is fused with the 3′ sequences of the Abl gene to produce the chimeric protein Bcr-Abl. Abl is a Src-like tyrosine kinase (p. 1009). The gene fusion results in a highly expressed and active Abl tyrosine kinase activity, which is critical for initiating and maintaining the leukemia.

The progression of CML is divided into stages; the initial chronic stage characterized by the expansion of terminally differentiated neutrophils typically spans a 5-year period. The later blastic stage is characterized by excessive numbers of undifferentiated lymphoid progenitor cells and carries a poor prognosis. The Bcr-Abl oncogene is expressed at all stages, but cells in the blast stage contain additional molecular and cytogenetic changes. The initial chronic stage has the highest response in treatment with imatinib (complete or nearly complete cytogenetic remissions of Philadelphia chromosome containing cells in 80%–90% of patients). Blast stage CML often shows partial remission (cytogenetic response rate of 16%–24% reported). It is hoped that better responses in blastic-phase disease will be achieved in combinations with other drugs. There are reports of the development of resistance to imatinib in the blastic stage by the appearance of point mutations in Bcr-Abl that decrease its affinity for imatinib. Other mutations can lead to imatinib resistance by causing an amplified expression of the Bcr-Abl fusion protein.

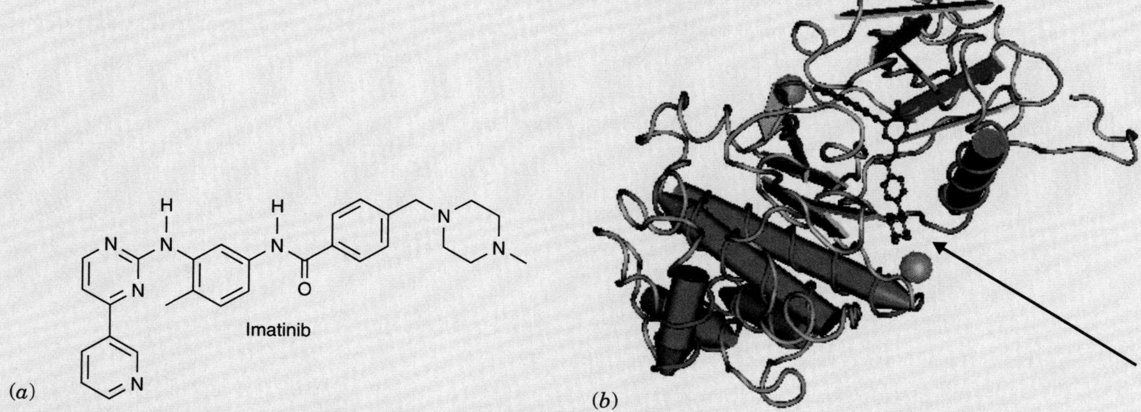

Binding of Imatinib into ATP binding site in tyrosine kinase domains of Abl. (a) Structure of Imatinib (STI-571). (b) Structure of the Abl tyrosine kinase domain with Imatinib in ATP binding site. Imatinib shown with stick-and-ball model (see arrow).
MMDB structure 16291 visualized with CN3D software. Citation for CN3D is Chen, J., Anderson, J. B., DeWeese-Scott, C., Fedorova, N. D., et al. *Nucleic Acids Res.* 31:474, 2003; and citation for X-ray structure is Nagar, B., Bornmann, W. G., Pellicena, P., Schindler, T., et al. *Cancer Res.* 62:4236, 2002.

Goldman, J. M., and Melo, J. V. Chronic myeloid leukemia: Advances in biology and new approaches to treatment. *N. Engl. J. Med.* 349:1451, 2003; and Kantarjian, H. M., Cortes, J. E., O'Brien, S., Giles, F., et al. Imatinib mesylate therapy in newly diagnosed patients with Philadelphia chromosome-positive chronic myelogenous leukemia: high incidence of early complete and major cytogenetic responses. *Blood* 101:97, 2003.

transformation were a gene that overexpressed telomerase, a *ras* oncogene, and the gene for the SV40 virus T antigen protein. The T antigen protein binds and inhibits both Rb and p53 in the infected epithelial cells. Rb suppression, p53 suppression, Ras activation, and telomerase overexpression were all essential for cellular transformation (Clin. Corr. 24.2.).

CLINICAL CORRELATION 24.3

Environmental Cause of Human Cancers

Environmental and cultural factors are the predominant causes of human cancer. If these factors can be regulated, the incidence of cancers may be reduced by greater than 75%. Cigarette smoking is responsible for approximately 30% of U.S. cancer deaths. It is estimated that there are 1,100,000 worldwide lung cancer deaths per year of which 85% are caused by tobacco. While cigarette consumption rates have recently begun to decrease, the time lag in cancer development will result in dramatic increases in smoking-induced cancers over the next two or three decades to 2 million per annum.

It is estimated that diet can prevent the development of approximately 30% of U.S. cancers (estimates vary from 10% to 70%). These numbers are primarily based on epidemiological data of the incidence of cancer types in different geographical locations, cultures, and environments (see table). On migration from one culture or location to another, the incidence of a cancer type often approaches that of the population in the new culture as the individuals or their offspring assimilate. Also changes in diet or lifestyle over time within a country show the importance of these factors in the incidence of cancers.

The quarter of the U.S. population with the highest dietary fruit and vegetable content has a 30% to 40% lower incidence for many cancer types. However, the constituents of these foods that inhibit the formation of these cancers are not yet clear. Inadequate folic acid levels in the U.S. diet have been implicated as a risk factor for colon and breast cancers. Folic acid intake may be particularly critical in individuals with a polymorphism in their gene for methylenetetrahydrofolate reductase, an enzyme involved in folic acid metabolism, which results in decreased activity in the enzyme. Folate deficiency causes excessive incorporation of uracil into DNA and chromosome breaks. Folate deficiency is common in alcoholics and may be the reason for an increase in the incidence of certain cancers with alcohol. Preadolescent diet and lifestyle affect the onset of menarche, and the early onset of menarche and length of time between menarche and menopause are associated with increased risk of breast cancer in women. Obesity is inversely correlated with risk of breast cancer in premenopausal women and positively correlated in postmenopausal. High animal fat in the diet has been associated with increased risk of colon cancer, but the data are inconsistent. Rather then total animal fat in the diet, the data may suggest that it is the ratio of polyunsaturated to saturated fat that correlates with colorectal cancer risk. Colon cancer incidence has also been inversely correlated with lack of activity or exercise. Environmental and dietary studies are difficult as large population must be studied over a generation of time, and multiple variables must be controlled. However, advances in our knowledge and control of these dietary and environmental factors have an enormous potential for decreasing the incidence of cancer.

The effect of environment and lifestyle on cancer incidence is interpreted through a genetic background. An individual's susceptibility to environmental factors is influenced by her or his genetic polymorphisms and acquired mutations that control metabolic, regulatory, and hormonal pathways.

Variation in the Incidence of Cancer Type in a Two Location Comparison[a]

	Location Comparison	*Incidence in Location 1*	*Incidence in Location 2*	*Fold Difference in incidence*
Lung	New Orleans (blacks)–Madras (India)	110	5.8	19
Breast	Hawaii (Hawaiians)–Israel (non-Jews)	94	14	7
Prostate	Atlanta (blacks)–China (Tianjin)	91	1.3	70
Uterine cervix	Brazil (Racife)–Israel (non-Jews)	83	3.0	18
Liver	China (Shanghai)–Canada (Nova Scotia)	34	0.7	49
Colon	United States (Connecticut, whites)–India (Madras)	34	1.8	19
Melanoma	Australia (Queensland)–Japan (Osaka)	31	0.2	155

[a]Incidence in number of new cases per year per 100,000 people, adjusted for age variation with the specific population. Data were taken Alberts, B., Johnson, A., Lewis, J., Raff, M., Roberts, K., and Watson, J. D. *Molecular Biology of the Cell,* 4th ed. New York: Garland, 2002, Table 24-2, which was adapted from DeVita, V. T., Hellman, S., and Rosenberg, S. A. (Eds.). *Cancer: Principles and Practice of Oncology,* 4th ed. Philadelphia: Lippincott, 1993; based on data from C. Muir et al. *Cancer Incidence in Five Continents,* Vol. 5, Lyon: International Agency for Research on Cancer, 1987. General References: Shibuya, K., Mathers, C. D., Boschi-Pinto, C., Lopez, A. D., and Murray, C. J. L. Global and regional estimates of cancer mortality and incidence by site: II. Results for the global burden of disease 2000. *BMC Cancer* 2:37, 2002. Pisani, P., Parkin, D. M., Bray, F., and Ferlay, J. Estimates of the worldwide mortality from 25 cancers in 1990. *Int. J. Cancer* 93:18, 1999.

Willett, W. C. Diet and cancer. *Oncologist* 5:393, 2000; Bingham, S., and Riboli, E. Diet and cancer: the European prospective investigation into cancer and nutrition. *Nature Reviews Cancer* 4:206, 2004; and Ames, B. N., and Wakimoto, P. Are vitamin and mineral deficiencies a major cancer risk? *Nature Reviews Cancer* 2:694, 2002.

Genetic and Biochemical Heterogeneity of Cancers

Cells in a tumor contain mutations in multiple protooncogenes and tumor suppressor genes. Although specific oncogenes and inactivated tumor suppressor genes are commonly present in all the cells of a tumor, other mutations occur that are different among the cells within a tumor mass. In addition, epigenetic marks may differ in the cells within the tumor, resulting in differences in protooncogene and tumor suppressor gene expressions among the population of tumor cells. This heterogeneity is demonstrated experimentally by showing that cells from different regions of the same tumor have differences in metastatic ability. The heterogeneity was generated in the cells over time with successive cell divisions. However, genetic markers can trace each of the cells in a tumor to a single parental cell, in which the initial mutational events occurred that led to a cancer cell. Thus while the tumor cell population is genetically and epigenetically heterogenic, all the cells within a tumor are derived from a single precursor cell.

Mutagens and Promoters Cause Cancers

Most cancers are caused by environmental factors (Clin. Corr. 24.3), which act as mutagens or as promoters in carcinogenesis. A promoter acts as a growth factor after the mutagenic event to increase the number of cells containing an initial mutagenic defect. It expands clones of mutant cells to give a higher probability of additional mutations in these cells. Natural factors such as estrogen act as tumor promoters on epithelial breast and ovarian cells, and women with high levels of estrogen over time have a higher incidence of breast and ovarian cancers. Diet can inhibit carcinogenesis by providing components that inhibit mutagens and tumor promoters. Alternatively factors within the diet can act as tumor repressors by providing cofactors for antioxidant enzymes and enzymes involved in DNA repair (Clin. Corr. 24.3).

Biochemical Analysis of Individual Cancers

Each cancer has a different pattern of activated oncogenes and inactivated tumor suppressor genes (Figure 24.20). Although certain gene mutations are commonly found in particular cancer types, individual cancers have genetic differences from other cancers even within the same cancer type. Because of the differences in activated oncogenes and inactivated tumor suppressor genes in individual cancers, the biochemical characterization of an individual cancer may be critical for its accurate prognosis and efficacious treatment. Gene microarray and proteomic techniques can precisely determine the abnormal biochemical patterns due to the oncogene activations and tumor suppressor losses within a particular cancer (Figure 24.21). Future pharmacological treatments will be based on the exact molecular characterization of a cancer to which will be designed selective treatments against the particular biochemical activities that sustain the cancer.

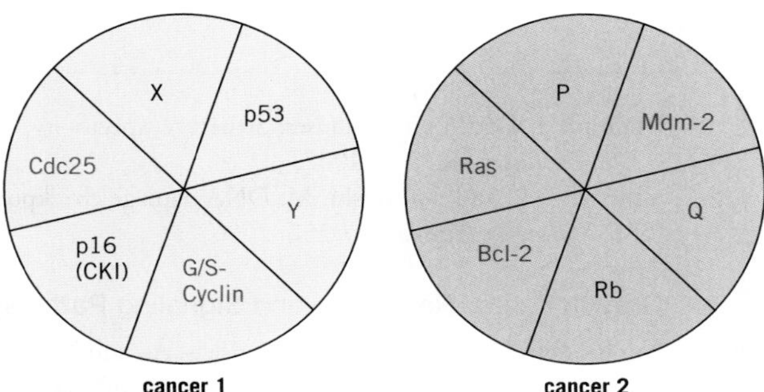

Figure 24.20 Each cancer has a different pattern of activated protooncogenes and deactivated tumor suppressor genes. Oncogenes are colored red and deactivated tumor suppressor genes black. X, Y, P, and Q are unidentified oncogenes or deactivated tumor repressor genes in the particular cancer.

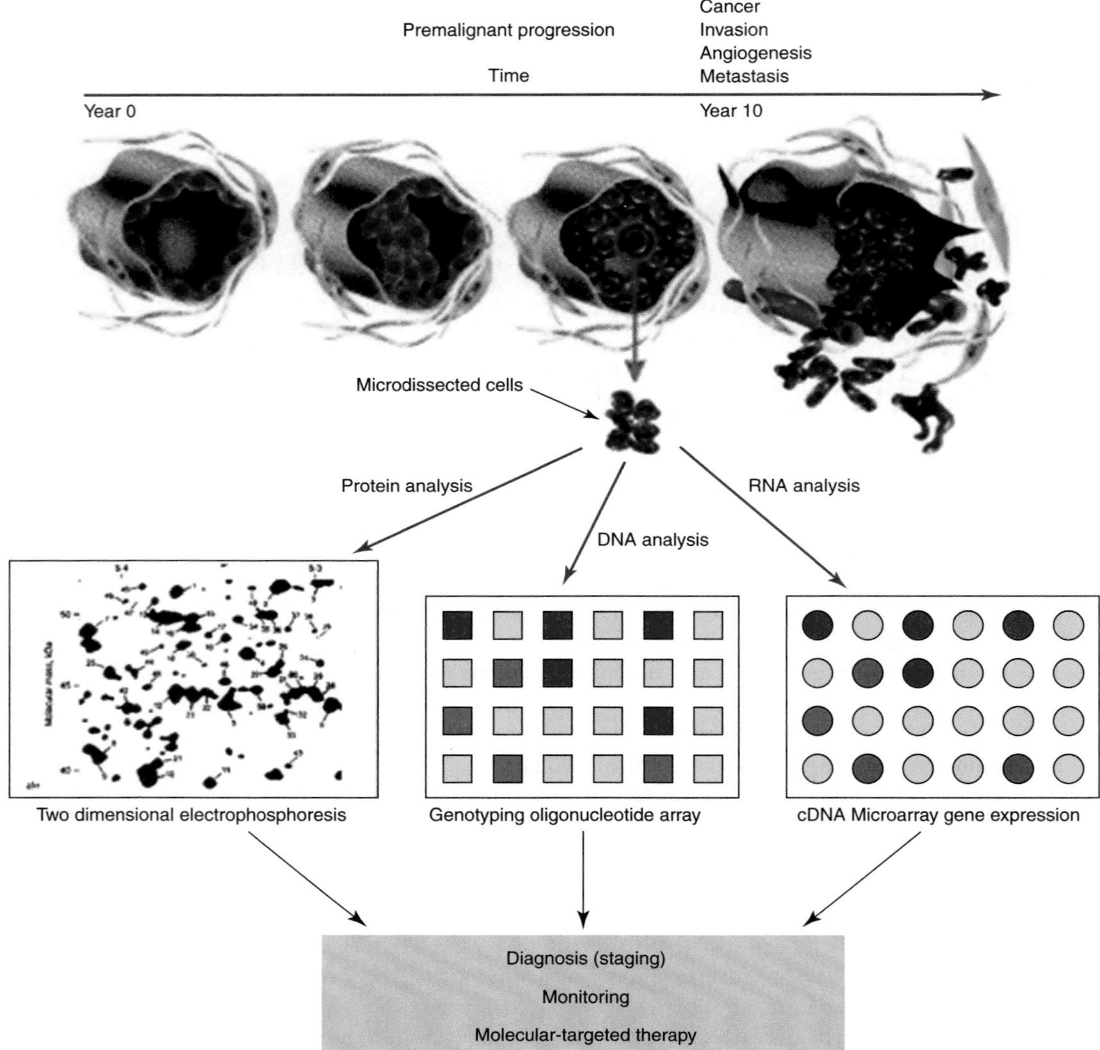

Figure 24.21 **Molecular analysis of laser microdissected cells characterizes the specific tumor.** Different stages in development of a malignant cancer may be analyzed for their genetic and epigenetic expression patterns with microdissected cells extracted for molecular analysis (top). Molecular analyses are by two-dimensional gel electrophoresis to determine expressed proteins (p. 124), genomic analysis, and cDNA analysis for expressed mRNAs. Staging, monitoring, and treatment of cancer are then based on biochemical and genetic analysis of the particular cancer. Biochemical characterization of an individual cancer will lead to a personalized treatment specific to the cancer.
Reproduced and modified with permission from Simone, N. L., Bonner, R. F., Gillespie, J. W., Emmert-Buck, M. R., and Liotta., L. A. Laser-capture microdissection: opening the microscopic frontier to molecular analysis. *Trends Genet.* 14:272, 1998. Copyright (1998) Elsevier.

Bibliography

Cell Cycle

Bloom, J., and Cross, F. R. Multiple levels of cyclin specificity in cell-cycle control. *Nature Reviews Molecular Cell Biology* 8:149, 2007.

Giacinti, C., and Giordana, A. Rb and cell cycle progression. *Oncogene* 25:5220, 2006.

Giono, L. E., and Manfredi, J. J. The p53 tumor suppressor participates in multiple cell cycle checkpoints. *J. Cellular Physiology* 209:13, 2006.

Nakayama, K. I., and Nakayama, K. Ubiquitin ligases: Cell-cycle control and cancer. *Nature Reviews Cancer* 6:369, 2006.

Nurse, P. A long twentieth century of the cell cycle and beyond. *Cell* 100:71, 2000.

Rudolph, J. Cdc25 phosphatases: Structure, specificity, and mechanism. *Biochemistry* 46:3595, 2007.

Shimada, M., and Nakanishi, M. DNA damage checkpoints and cancer. *J. Molecular Histology* 37:253, 2006.

Growth Factor Receptors and Signaling Pathway

Fambrough, D., McClure, K., Kazlauskas, A., and Lander, E. S. Diverse signaling pathways activated by growth factor receptors induce broadly overlapping, rather than independent, sets of genes. *Cell* 97:727, 1999.

Perkins, N. D. Integrating cell-signalling pathways with NF-κB and IKK function. *Nature Reviews Molecular Cell Biology* 8:49, 2007.

Raman, M., Chen, W., and Cobb, M. H. Differential regulation and properties of MAPKs. *Oncogene* 26:3100, 2007.

Schlessinger, J., and Lemmon, M. A. SH2 and PTB domains in tyrosine kinase signaling. *Sci. STKE* 2003:re12 (2003).

Turjanski, A. G., Vaqué, J. P., and Gutkind, J. S. MAP kinases and the control of nuclear events. *Oncogene* 26:3240, 2007.

Weng, G., Bhalla, U. S., and Iyengar, R. Complexity in biological signaling systems. *Science* 284:92, 1999.

Apoptosis

Cheng, E. H.-Y. A., Wei, M. C., Weiler, S., Flavell, R. A., et al. Mitochondria and Apoptosis. *Science* 281:1309, 1998.

Chipuk, J.E., and Green, D.R. How do BCL-2 proteins induce mitochondrial outer membrane permeabilization? *Trends in Cell Biology* 18:157, 2008.

Saelens, X., Festjens, N., Vande Walle, L., van Gurp, M., et al. Toxic proteins released from mitochondria in cell death. *Oncogene* 23:2861, 2004.

Scorrano, L., and Korsmeyer, S. J. Mechanisms of cytochrome c release by preapoptotic BCL-2 family members. *Biochem. Biophys. Res. Commun.* 304:437, 2003.

Thompson, C. B. Apoptosis in the pathogenesis and treatment of disease. *Science* 267:1456, 1995.

Yu, X., Acehan, D., Ménétret, J.-F., Booth, C. R., et al. A structure of the human apoptosome at 12.8 Å resolution provides insights into this cell death platform. *Structure* 13:1, 2005.

Youle, R. J., and Strasser, A. The BCL-2 protein family: opposing activities that mediate cell death. *Nature Rev. Molecular Cell Biol.* 9:47, 2008.

Zinkel, S., Gross, A., and Yang, E. BCL2 family in DNA damage and cell cycle control. *Cell Death and Differentiation* 13:1351, 2006.

Cancer

Cancer Biology and Molecularly Based Cancer Therapy

Ames, B. N., Gold, L. S., and Willett, W. C. The causes and prevention of cancer. *Proc. Natl. Acad. Sci. USA* 92:5258, 1995.

Ciardiello, F., and Tortora, G. EGFR antagonists in cancer treatment. *N. Engl. J. Med.* 358:1160, 2008.

Chabner, B. A., and Roberts, T. G., Jr. Chemotherapy and the war on cancer. *Nature Reviews Cancer* 5:65, 2005.

Croce, C. M. Oncogenes and cancer. *N. Engl. J. Med.* 358:502, 2008.

Fröhling, S., and Döhner, H. Chromosomal abnormalities in cancer. *N. Engl. J. Med.* 359:722, 2008.

Hahn, W.C., and Weinberg, R.A. Rules for making human tumor cells. *N. Engl. J. Med.* 347:1593, 2002.

Halazonetis, T. D., Gorgoulis, V. G., and Bartek, J. An oncogene-induced DNA damage model for cancer development. *Science* 319:1352, 2008.

Luo, J.I., and Elledge, S. J. Deconstructing oncogenesis. *Nature* 453:995, 2008.

Schatton, T., Murphy, G. F., Frank, N. Y., Yamaura, K., et al. Identification of cells initiating human melanomas. *Nature* 451:345, 2008.

Sharma, S. V., and Settleman, J. Oncogene addiction: Setting the stage for molecularly targeted cancer therapy. *Genes and Development* 21:3214, 2007.

Vogelstein, B., and Kinzler, K. W. (Eds.). *The Genetic Basis of Human Cancer*, 2d ed. New York, McGraw Hill, 2002.

Molecular-Based Analysis of Cancer

Acharya, C. R., Hsu, D. S., Anders ,C. K., Anguiano, A., et al. Gene expression signatures, clinicopathological features, and individualized therapy in breast cancer. *JAMA* 299:1574, 2008.

Bild, A. H., Yao, G., Chang, J. T., Wang Q., et al. Oncogenic pathway signatures in human cancers as a guide to targeted therapies. *Nature* 439:353, 2006.

Chanrion, M., Negre, V., Fontaine, H., Salvetat, N., et al. A gene expression signature that can predict the recurrence of tamoxifen-treated primary breast cancer. *Clin. Cancer Res.* 14:1744, 2008.

García-Escudero, R., and Paramio, J. M. Gene expression profiling as a tool for basic analysis and clinical application of human cancer. *Mol. Carcinog.* 47:573, 2008.

Liu, R., Wang, X., Chen, G. Y., Dalerba, P., et al. The prognostic role of a gene signature from tumorigenic breast-cancer cells. *N. Engl. J. Med.* 356:217, 2007.

Nagaraja, G. M., Othman, M., Fox, B. P., Alsaber, R., et al. Gene expression signatures and biomarkers of noninvasive and invasive breast cancer cells: Comprehensive profiles by representational difference analysis, microarrays and proteomics. *Oncogene* 25:2328, 2006.

Pao, W., Miller, V., Zakowski, M., Doherty, J., et al. EGF receptor gene mutations are common in lung cancers from "never smokers" and are associated with sensitivity of tumors to gefitinib and erlobinib. *Proc. Natl. Acad. Sci. USA* 101:13306, 2004.

Angiogenesis in Cancer

Abdollahi, A., Schwager, C., Kleeff, J., Esposito, I., et al. Transcriptional network governing the angiogenic switch in human pancreatic cancer. *Proc. Natl. Acad. Sci., USA* 104, 12890, 2007.

Boehm, T., Folkman, J., Browder, T., and O'Reilly, M. S. Antiangiogenic therapy of experimental cancer does not induce acquired drug resistance. *Nature* 390:404, 1997.

Carmeliet, P. Angiogenesis in health and disease. *Nature Medicine* 9:653, 2003.

Metastasis in Cancer

Cairns, R. A., Khokha, R., and Hill, R. P. Molecular mechanisms of tumor invasion and metastasis: An integrated view. *Current Molecular Medicine* 3:659, 2003.

Chang, C., and Werb, Z. The many faces of metalloproteases: cell growth, invasion, angiogenesis and metastasis. *Trends in Cell Biol.* 11:S37, 2001.

Nguyen, D. X., and Massagué, J. Genetic determinants in cancer metastasis. *Nature Reviews Genetics* 8:341, 2007.

Signaling Networks in Cancer

Downward, J. Targeting Ras signalling pathways in cancer therapy. *Nature Reviews Cancer* 3:11, 2003.

Finkel, T., and Gutkind, J. S. (Eds.). *Signal Transduction and Human Disease.* New York: Wiley-Liss, 2003.

Hornberg, J. J., Binder, B., Bruggeman, F. J. Schoeberl, B., et al. Control of MAPK signalling: From complexity to what really matters. *Oncogene* 24:5533, 2005.

Irish, J. M., Hovland, R., Krutzik, P. O., Perez, O. D., et al. Single cell profiling of potentiated phospho-protein networks in cancer cells. *Cell* 118:217, 2004.

Roberts, P. J., and Der, C. J. Targeting the Raf-MEK-ERK mitogen-activated protein kinase cascade for the treatment of cancer. *Oncogene* 26:3291, 2007.

p53 in Cancer

Aylon, Y., and Oren, M. Living with p53, dying of p53. *Cell* 130:597, 2007.

Oren, M. Decision making by p53: Life, death and cancer. *Cell Death and Differentiation* 10:431, 2003.

Van Dyke, T. p53 and tumor suppression. *N. Engl. J. Med.* 356:79, 2007.

Vousden, K. H., and Lane, D. P. p53 in health and disease. *Nature Reviews Mol. Cell Biol.* 8:275, 2007.

Key Terms

Cell Cycle

G1, G2, S, and M phases
quiescent
senescent
CDC25 phosphatase
Cdk inhibitor
Cdk-activating kinase
cyclin
cyclin-dependent kinase
Weel kinase
Anaphase-promoting complex
E2F
SCF complex
p21
p53
retinoblastoma sensitivity protein

Growth Factors

autocrine and paracrine mechanisms
growth factor receptors
tyrosine kinase
Grb2
GEF^{Ras}
Ras
GAP^{Ras}
MAPK cascade
Jun/Fos

Apoptosis

caspase cascade
cytochrome c
extrinsic (death receptor) pathway
intrinsic (mitochondrial) pathway
Bax, Bak
Bcl-2, Bcl-X_L
apoptosome
DISC
inhibitor of apoptosis
Smac/DIABLO
Mdm-2
Arf

Cancer

oncogene
protooncogene
tumor suppressor gene
angiogenesis
immortality
metastasis
matrix metalloproteinase
type IV collagenase
urokinase plasminogen activator
VEGF
carcinogenesis

Questions • CAROL N. ANGSTADT

Multiple-Choice Questions

1. In the cell cycle,
   A. M phase is both the most complex and the longest phase.
   B. there is a G_0 phase in equilibrium with the G_1 phase.
   C. quiescent cells cannot be induced to reenter the cell cycle.
   D. microtubule spindles form during the S phase.
   E. the order of phases is G_1, G_2, S, and M.

2. Cyclin-dependent kinases (Cdks)
   A. occur only in the M phase.
   B. are always inactivated by phosphorylation.
   C. typically phosphorylate proteins on tyrosine residues.
   D. in addition to binding cyclin, require other modifications for activity.
   E. that phosphorylate Rb (retinoblastoma sensitivity) protein inhibit the synthesis of S-phase proteins.

3. Elements of growth factor pathway regulation of the cell cycle include all of the following except
   A. dimerization or polymerization of receptor proteins in the plasma membrane.
   B. binding of ATP to a tyrosine kinase domain of growth factor receptor proteins.
   C. nonspecific binding of SH2 domains to any phosphorylated tyrosine.
   D. conformational changes of proteins as part of signal transmission.
   E. location of the growth factor receptor proteins in the plasma membrane.

4. Ras protein of the signal pathway
   A. is active in the GDP-Ras conformation.
   B. transmits a signal by initiating a kinase cascade.
   C. from a mutated gene in many cancers that has inactivated GTPase activity so it inhibits the downstream signal to divide.
   D. is formed from an oncogene.
   E. becomes inactive as a signaling protein by phosphorylating transcription factors.

5. Apoptotic death of a cell may include all of the following except
   A. lysis of the cell membrane with inflammation.
   B. activation of enzymes called caspases.
   C. phagocytosis.
   D. binding of a ligand in the extracellular environment to a plasma membrane receptor.
   E. transfer of toxic molecules from the mitochondria to cytoplasm through protein-induced outer membrane permeabilization.

6. Malignant cancer cells have all of the following properties except
   A. unregulated cell division.
   B. inhibition of angiogenesis.
   C. resistance to apoptosis.
   D. cellular immortality.
   E. ability to metastasize.

7. The property of cellular immortality refers to a cell's ability to
   A. grow new blood vessels.
   B. invade through the basement membrane of its tissue.
   C. resist apoptosis.
   D. continue to divide without limit to the number of cell generations.
   E. grow tumors at multiple sites.

8. Telomerase
   A. activity generally increases as an individual ages.
   B. catalyzes the resynthesis of telomer regions.
   C. activity is high during all stages of the carcinogenesis process.
   D. plays a role in apoptosis but not immortality.
   E. when present at a high level, leads to cell senescence.

Questions 9 and 10. Human papillomavirus (HPV) is a causative factor in over 95% of cervical cancers. Other viruses such as Epstein–Barr, Kaposi's sarcoma-associated herpes virus, and hepatitis B virus also are known to cause cancer. For this to happen, viral genes that promote entry into S phase or that promote resistance to apoptosis must become incorporated into the host cell DNA in a host cell that also survives the viral infection. If the viral gene within the host DNA is then overexpressed in an unregulated manner, it is a viral oncogene.

9. The Epstein–Barr protein binds to the cellular Rb-binding pocket and displaces E2F from Rb. The effect would be to
A. require that Rb be phosphorylated.
B. stall the cell cycle in the G_1 phase.
C. increase the transcription of gene products for S-phase proteins.
D. inhibit the action of cyclin-dependent kinases (Cdks).
E. inhibit apoptosis.

10. A viral protein could bind to cellular P53 and promote its degradation. P53, which is an inducer of apoptosis,
A. is activated by kinases located in chromatin.
B. increases the transcription of several pro-apoptotic genes for Bcl-2 proteins.
C. concentration is normally low because it is subject to ubiquitination.
D. concentration may be increased by acetylation.
E. all of the above.

Questions 11 and 12: Environmental and cultural factors are the predominant causes of human cancer. Cancers from environmental damage may not appear until years after exposure to carcinogens (for example, lung cancer from smoking). It takes mutations of numerous protooncogenes and tumor suppressor genes to convert a protooncogene into an oncogene or to render a tumor suppressor gene less active.

11. Protooncogenes
A. produce protein products that promote cell division.
B. produce protein products that promote apoptosis.
C. are converted to oncogenes by a mutation that inhibits their expression.
D. require both alleles to be mutated to show a pro-proliferative effect.
E. are constitutively activated.

12. Tumor suppressor genes
A. have the same effect as protooncogenes.
B. contribute to cancer if only one allele is inactivated by mutation.
C. produce protein products that promote apoptosis.
D. produce products that promote cancer.
E. all of the above.

Problems

13. What is the role of type IV collagenase in metastasis of a tumor cell?

14. What is the difference between the two pathways that activate the caspase cascade?

Answers

1. **B** A: M phase is the shortest although it is highly complex. C: Senescent cells cannot be induced to reenter the cell cycle. D: This occurs in the M phase. E: G_1 and G_2 separate the M and S phases.
2. **D** For example, M-Cdk-cyclin heterodimer must be correctly phosphorylated. A: Other phases also have Cdks. B: There are both activation and inhibitory phosphorylation sites. C: Serine and threonine residues are the phosphorylation sites. E: Phosphorylation of Rb causes release of E2F and thus promotes synthesis.
3. **C** Each SH2 domain has a secondary specificity that differentiates between phosphorylated tyrosines. A: This seems to be critical to transmission of the signal. B: Many receptors operate this way. Those that don't cause dimerization by binding a protein with a tyrosine kinase domain. D: This is a common occurrence.
4. **B** GTP-Ras activates MAPKKK, which phosphorylates the next kinase, and so on. A: GTP-Ras is the active form. C: Inhibiting GTPase activity prevents the conversion to GDP-Ras so the signal remains active. D: The Ras gene is a protooncogene and requires an activating mutation to become an oncogene. E: The transcription factors are at the end of the chain so don't react directly with Ras.
5. **A** This occurs in necrotic cell death. B: These proteases degrade cell protein. C: This happens to the remainder of the cell after multiple digestions. D: This is the extrinsic or death receptor pathway. E: This is the intrinsic pathway with pro-apoptotic BH domain proteins acting to permeabilize the outer mitochondrial membrane.
6. **B** Tumor cells must have the ability to induce the growth of blood vessels in order to grow. A, C, D, and E are all properties that allow tumors to grow.
7. **D** This may occur because such cells have high levels of telomerase. A: This allows a cancer cell to grow a tumor mass. B and E: These are part of metastasis. C: Apoptosis is a separate property from immortality.
8. **B** A terminal nucleotide segment is lost each time chromatids are duplicated in cell division and must be replaced. A, and E: Telomerase decreases on cell aging, leading to senescence. C: There is believed to be a low level at early and intermediate stages, leading to chromosomal instability and mutations. D: This is believed to be a major factor leading to immortality.
9. **C** By removing the normal control of releasing and sequestering E2F, the virus is promoting the entry into the S phase. A: This is the usual mechanism for releasing E2F which is bypassed. B: This promotes entry into S phase. D: These are the enzymes that usually phosphorylate Rb. E: This is not involved.
10. **E** A: DNA chain breaks and changes in chromatin structure initiate this. B: There is also a nontranscriptional method; both lead to the intrinsic pathway. C: Binding Mdm-2 catalyzes polyubiquitination. D: Acetylation prevents binding of Mdm-2.
11. **A** This is part of the definition. B: The products promote resistance to apoptosis. C: Oncogenes result in higher gene activity. D: One is sufficient. E. This occurs on mutation.
12. **C** Also products that suppress the cell cycle. A: These effects are opposite to those of protooncogenes. B: Both alleles need to be inactivated. D: Their normal products promote apoptosis which is opposite to what cancer cells require.
13. Type IV collagen is part of the basement membrane. Tumor cells typically promote the secretion of a plasminogen activator and a type IV procollagenase. This leads to formation of plasmin, which cleaves proteins of the extracellular matrix and activates procollagenases to collagenases. Type IV collagen is degraded. These two activities acting together provide a path through which the tumor cell passes to the blood.
14. The death receptor (extrinsic) pathway is initiated by binding of a death ligand in the extracellular environment to its plasma membrane receptor in the target cell. The mitochondrial (intrinsic) pathway occurs by BH domain proteins in the outer mitochondrial membrane altering the membrane to allow cytochrome c to pass. In both cases, after the initial stages, protein aggregates form; this ultimately leads to activation of a procaspase, which initiates the caspase cascade.

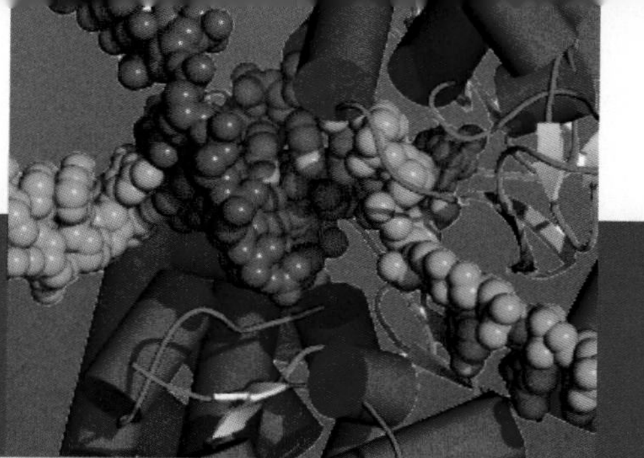

25

Digestion and Absorption of Basic Nutritional Constituents

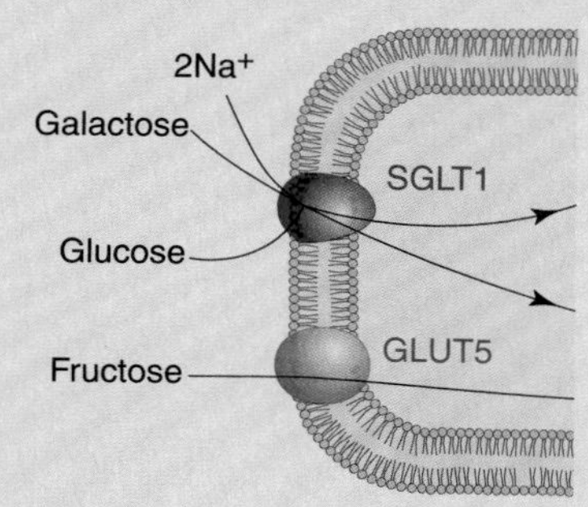

Ulrich Hopfer

Professor, Case Western Reserve University School of Medicine

CLINICAL CORRELATIONS

Key Concepts

- Dietary proteins are polymers that are broken down to amino acids or di- and tripeptides for absorption. Dietary carbohydrates consist mostly of dimers and polymers that need to be digested to monosaccharides for absorption. Dietary fats consist mainly of fatty acids esterified to glycerol (triacylglycerols) that require biological detergents (bile acids) for their dispersion and digestion to monoglycerides and free fatty acids.
- Digestion of nutrients requires a range of conditions, some very harsh (e.g., pH < 1), and large amounts of an array of digestive enzymes that sequentially degrade nutrient molecules to smaller pieces. Soluble digestive enzymes are secreted mostly as inactive precursors by salivary glands, the stomach, and the pancreas.
- Terminal digestion of peptides and carbohydrates to absorbable molecules is accomplished by ectoenzymes on the brush border surface of enterocytes.
- Monosaccharides, free amino acids, and di- and tripeptides are absorbed predominantly by co-transport with sodium, a positive charge, or a proton. Free fatty acids are also absorbed actively.
- Active electrolyte secretion, electrolyte absorption, and secondary active transport are all powered by Na^+/K^+-exchanging ATPase in the contraluminal plasma region of epithelial cells.
- Absorbed lipids are resynthesized within cells to triacylglycerols and released into lymph as small globules consisting of neutral lipids, phospholipids, and specific lipoproteins.

- Bile acids are biological detergents necessary to facilitate lipid digestion and absorption by dispersion of lipids into micelles and increasing transport of free fatty acids and monoglycerides through an unstirred layer over the lining enterocytes. Bile acids synthesized by the liver are secreted as conjugates with taurine and glycine.
- Bile acids are mostly retained by the body through an enterohepatic cycle of active reabsorption in the ileum, transport through blood to the liver, and resecretion by the liver.

25.1 • INTRODUCTION

Nutrient Types

Three types of biochemicals serve as nutrients for humans: proteins, carbohydrates, and fats. Many different foods can satisfy the nutritional needs of humans, even though they differ in the ratio of different types of nutrients and of digestible-to-nondigestible material. Unprocessed plant products are especially rich in fibrous material that cannot be digested by human enzymes or easily degraded by intestinal bacteria. These fibers are mostly polymers of carbohydrates, such as cellulose (β-1,4-glucan) or pectins (mixtures of methyl esters of polygalacturonic acid, polygalactose, and polyarabinose). High-fiber diets enjoy a certain popularity nowadays because they add bulk to stool and may have a preventive effect on development of colonic cancer.

Table 25.1 describes average contributions of different food classes to the diet of North Americans. Intake by individuals may substantially deviate from the average, as food consumption depends mainly on availability and individual tastes. The ability to utilize a wide variety of food reflects great adaptability and digestive reserve capacity of the gastrointestinal tract.

Knowledge of the nature of proteins and carbohydrates in the diet is important from a clinical point of view. Certain proteins and carbohydrates, although good nutrients for most humans, cannot be properly digested by some individuals and produce gastrointestinal disease. In this case, omission of the offending food constituent eliminates the problems. For example, many humans, particularly non-Caucasians, lose the ability to digest the common milk disaccharide lactose after childhood and ingested milk produces diarrhea, gas, and bloating (lactose intolerance). Another common example is provided by gluten, one of the protein fractions of wheat. One of its digestion products produces an immune and inflammatory response in some people, which can lead to reduced intestinal mucosal surface, reduced digestive/absorptive capacity, and diarrhea (gluten enteropathy or celiac disease).

Several Gastrointestinal Organs Contribute to Food Digestion

Secretion of digestive fluids and digestion of food were among the earliest biochemical events to be investigated at the beginning of the era of modern science. The discovery of gastric HCl production goes back to the American physician William Beaumont (1785–1853). In 1822 he treated a patient with a stomach wound. The patient recovered from the wound, but retained a gastric fistula (abnormal opening through the skin) so that Beaumont could obtain and study gastric juice during and after meals. Chemical analysis revealed, to the surprise of chemists and biologists, the presence of the inorganic acid HCl. This discovery established the principle of unique secretions by specialized glands. Soon thereafter, in 1836, Theodor Schwann, a German anatomist and physiologist (1810–1882), noticed that gastric juice would degrade albumen in the presence of dilute acid. He recognized that this

TABLE 25.1 • Contribution of Major Food Groups to Daily Nutrient Supplies in the United States

Type of Nutrient	*Total Daily Consumption* (*g*)	*Dairy Products, Except Butter* (%)	*Meat, Poultry, Fish* (%)	*Eggs* (%)	*Fruits, Nuts, Vegetables* (%)	*Flour, Cereal* (%)	*Sugar, Sweeteners* (%)	*Fats Oils* (%)
Protein	100	22	42	6	12	18	0	0
Carbohydrate	381	7	0.1	0.1	19	36	37	0
Fat	155	13	35	3	4	1	0	42

process involved a new chemical principle, namely, enzymatic digestion, and coined the word *pepsin* from the Greek *pepsis,* meaning "digestion."

The bulk of ingested nutrients consists of large polymers that have to be broken down to smaller units, mostly monomers, before they can be absorbed and made available to all cells of the body. The complete process from food intake to absorption of nutrients into the blood includes (Figure 25.1) (1) mechanical homogenization of food and mixing of

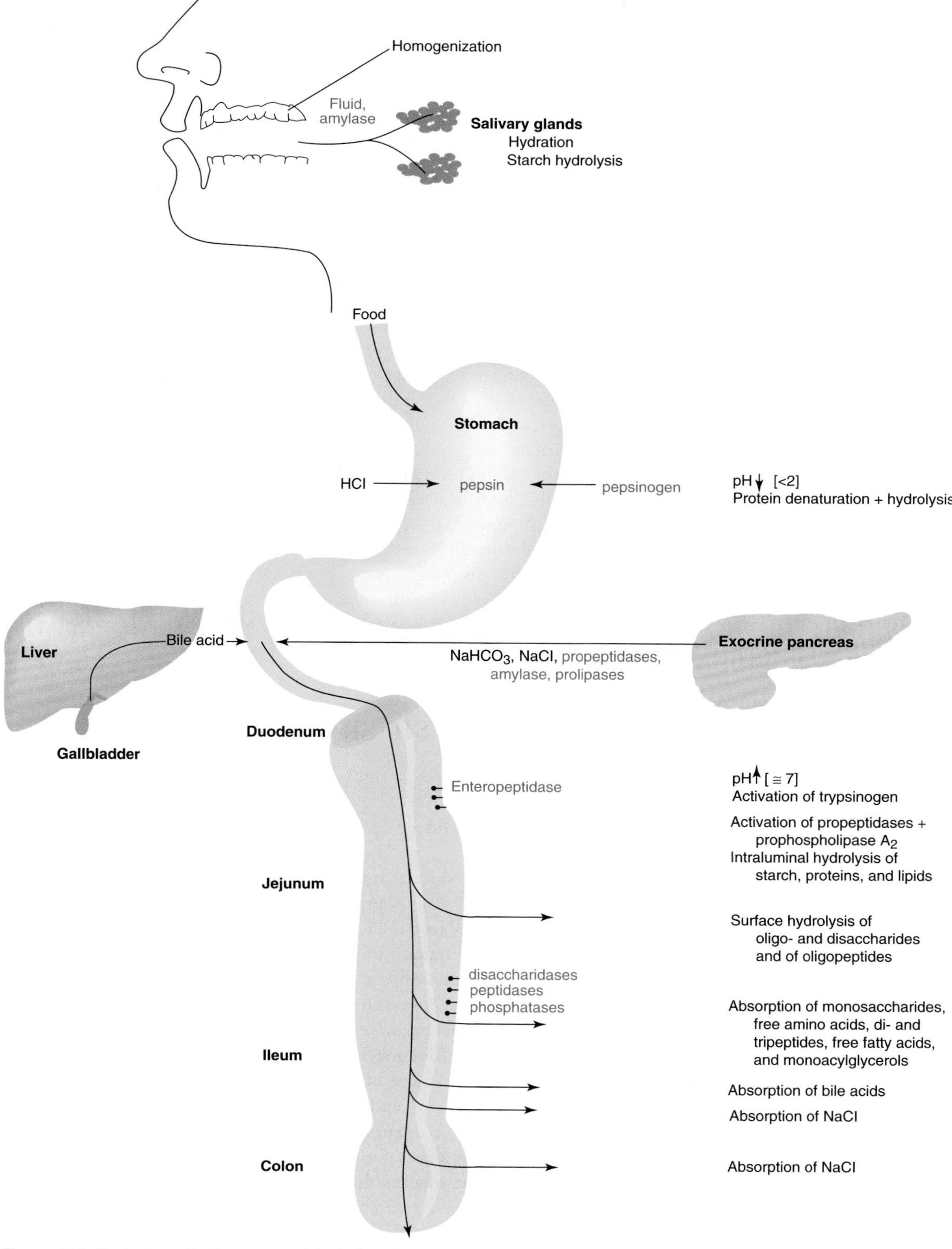

Figure 25.1 Gastrointestinal organs and their functions.

TABLE 25.2 • Functions of Specialized Glands and Tissues in Digestion and Absorption

Organ	*Major Function in Digestion and Absorption*
Salivary glands	Elaboration of fluid and digestive enzymes
Stomach	Elaboration of HCl and digestive enzymes
Pancreas	Elaboration of $NaHCO_3$ and enzymes for intraluminal digestion
Liver	Elaboration of bile acids
Gallbladder	Storage and concentration of bile
Small intestine	Terminal digestion of food, absorption of nutrients and electrolytes
Large intestine	Absorption of electrolytes

ingested solids with fluids secreted by glands of the gastrointestinal tract; (2) secretion of digestive enzymes that hydrolyze macromolecules to oligomers, dimers, or monomers; (3) secretion of electrolytes, acid, or base to provide an appropriate environment for optimal enzymatic digestion; (4) secretion of bile acids as detergents to solubilize lipids and facilitate their absorption; (5) hydrolysis of nutrient oligomers and dimers by enzymes on the intestinal surface; and (6) absorption of nutrient molecules and electrolytes from the intestinal lumen by intestinal epithelial cells. To accomplish these functions, the gastrointestinal tract contains specialized glands and surface epithelia (Table 25.2).

The pancreas and small intestine are essential for digestion and absorption of all basic nutrients. Fortunately, both organs have large reserve capacities. Thus, maldigestion due to pancreatic failure generally becomes a clinical problem only when the pancreatic secretion rate of digestive enzymes drops to below 1/10th of the normal rate. Secretion of bile by liver is important for efficient lipid digestion and absorption, which depend on bile acids. In contrast, gastric digestion of food is nonessential for adequate nutrition, and loss of this function can be compensated for by the pancreas and the small intestine. Yet normal gastric digestion greatly increases the smoothness and efficiency of the total digestive process. The stomach aids in the digestion through its reservoir function, its churning ability, and initiation of protein and lipid hydrolysis, which, although minor, is important for stimulation of pancreatic and gallbladder output. Peptides, amino acids, and fatty acids liberated in the stomach stimulate the coordinated release of pancreatic juice and bile into the lumen of the small intestine, thereby ensuring efficient digestion of food.

25.2 • GENERAL CONSIDERATIONS

Different Sites of Digestion

The first steps in the breakdown of food are catalyzed by soluble enzymes and occur within the lumen of stomach and small intestine. Digestive enzymes are secreted by the salivary glands, stomach, and pancreas; the pancreas makes the largest and essential contributions. Secreted enzymes amount to at least 30 g of protein per day in a healthy adult. Pancreatic enzymes together with bile are poured into the lumen of the second (descending) part of the duodenum, so that the bulk of the intraluminal digestion occurs distal to this site. Whereas lipid digestion completely relies on soluble lipases, a substantial portion of carbohydrates and peptides remains as dimers and oligomers even after exhaustive luminal digestion and depends for further breakdown on the digestive enzymes of intestinal epithelial cells (enterocytes).

The luminal plasma membrane of enterocytes is enlarged by a regular array of microvilli that look like a brush and have led to its name of brush border. This brush border membrane is covered on its outer surface with many di- and oligosaccharidases, amino- and dipeptidases, and esterases (Table 25.3). Many of these enzymes protrude up to 100 Å into the lumen, attached to the plasma membrane by an anchoring polypeptide. This arrangement provides effective surface digestion generating small molecules that can be absorbed by enterocytes. Carbohydrates have to be broken down to monosaccharides for absorption, whereas amino acids as well as di- and tripeptides generated in the process of luminal and surface digestion

TABLE 25.3 • Digestive Enzymes of Small Intestinal Surface

Enzyme (Common Name)	*Substrate*
Maltase	Maltose
Sucrase/isomaltase	Sucrose/α-limit dextrin
Glucoamylase	Amylose
Trehalase	Trehalose
β-Glucosidase	Glucosylceramide
Lactase	Lactose
Endopeptidase	Protein (cleavage at internal hydrophobic amino acids)
Aminopeptidase A	Oligopeptide with acidic NH_2 terminus
Aminopeptidase N	Oligopeptide with neutral NH_2 terminus
Dipeptidyl aminopeptidase IV	Oligopeptide with X-Pro or X-Ala at NH_2 terminus
Leucine aminopeptidase	Peptides with neutral amino acid at NH_2 terminus
γ-Glutamyltransferase	Glutathione + amino acid
Enteropeptidase (enterokinase)	Trypsinogen
Alkaline phosphatase	Organic phosphates

can be absorbed by enterocytes. For the peptides, intracellular breakdown to free amino acids is accomplished by cytoplasmic peptidases, thus completing protein digestion.

An interesting question is how the surface enzymes, which are themselves proteins, escape digestion by soluble proteases. It appears that their heavy glycosylation affords some protection as it prevents access of proteases to relevant peptide bonds.

Digestive Enzymes Are Secreted as Proenzymes

Soluble digestive enzymes are produced and secreted by specialized cells of salivary glands and pancreas (acinar cells) and gastric mucosa (chief cells). The secretion is termed exocrine because of its direction toward the lumen (Figure 25.2). Proteins destined to become soluble digestive enzymes are synthesized on polysomes of the rough endoplasmic reticulum (p. 231) and transported via the Golgi complex to storage vesicles in the apical cytoplasm. The storage vesicles (zymogen granules) have a diameter of about 1 μm and are bounded by a typical cellular membrane. Most digestive enzymes are produced and stored as inactive proenzymes (zymogens) (Figure 25.3; p. 1036). When an appropriate stimulus for secretion is received, zymogen granules move to the luminal plasma membrane and fuse with the plasma membrane to release their contents into the lumen (exocytosis). Activation of digestive proenzymes occurs only after they have been released from the cells.

Secretion Is Regulated by Many Secretagogues

Secretion of enzymes is regulated and coordinated with that of electrolytes and acid or base (sodium bicarbonate). Transport of electrolytes from blood into the ductal system of glands draws water into this compartment, and the resulting fluid solubilizes enzymes and flushes them out of glands into the alimentary canal. Regulation of secretion is through secretagogues that interact with receptors on the surface of exocrine cells (Table 25.4).

TABLE 25.4 • Physiological Secretagogues

Organ	*Secretion*	*Secretagogue*
Salivary gland	NaCl, amylase	Acetylcholine, catecholamines?
Stomach	HCl, pepsinogen	Acetylcholine, histamine, gastrin
Pancreas-acini	NaCl, digestive enzymes	Acetylcholine, cholecystokinin, secretin
Pancreas-duct	$NaHCO_3$, NaCl	Secretin
Small intestine	NaCl	Acetylcholine, serotonin, vasoactive intestinal peptide (VIP), guanylin

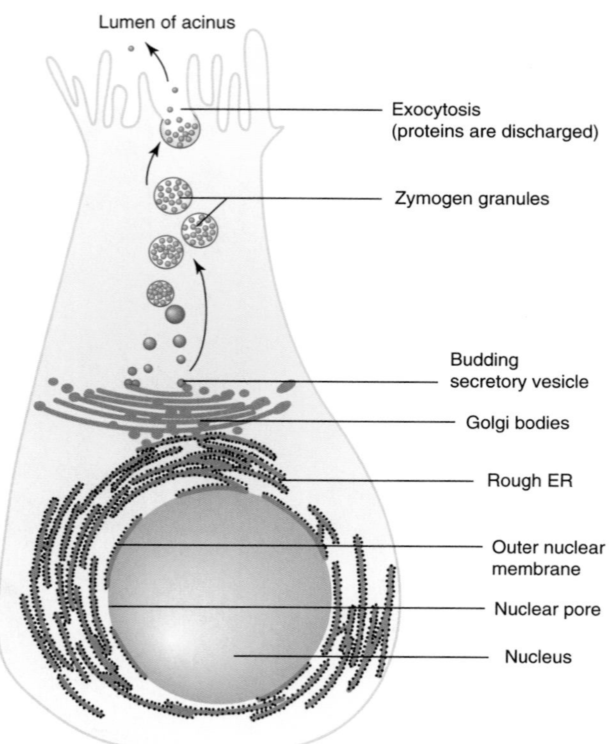

Figure 25.2 Exocrine secretion of digestive enzymes. Redrawn with permission from Jamieson, J. D. Membrane and secretion. In G. Weissmann and R. Claiborne (Eds.), *Cell Membranes: Biochemistry, Cell Biology, and Pathology*. New York: HP Publishing Co., 1975. Figure by B. Tagawa.

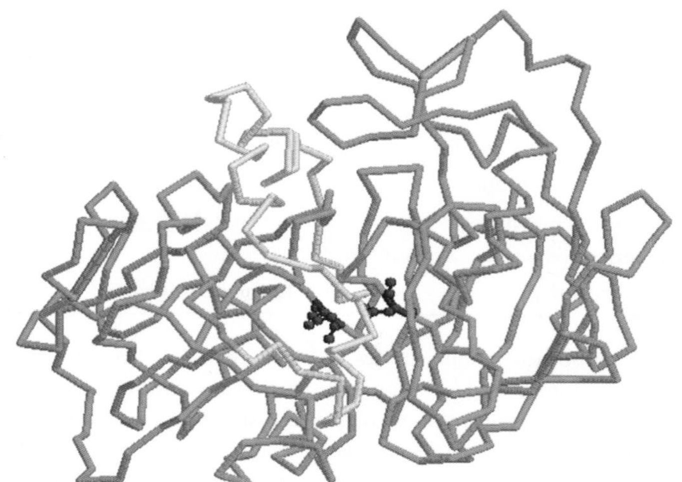

Figure 25.3 Model of porcine pepsinogen: NH2-terminus (grey) occupies the central groove. For activation, the NH2-terminus is cleaved and removed. This exposes the aspartic acid residues (red) at the active site. Model 3PSG from Protein Data Bank and illustrated with Protein Explorer.

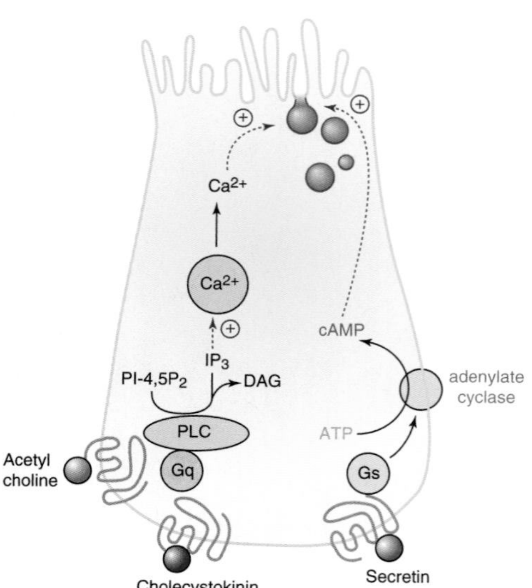

Figure 25.4 Cellular regulation of exocrine secretion in pancreas. Abbreviations: PI-4,5P_2, phosphatidylinositol-4,5-bisphosphate; DG, diacylglycerol; IP_3, inositol-1,4,5-trisphosphate; and PLC, phospholipase C.
Adapted from Gardner, J. D. *Annu. Rev. Physiol.* 41:63, 1979. Copyright (1979) by Annual Reviews, Inc.

Neurotransmitters, hormones, pharmacological agents, and certain bacterial toxins can be secretagogues. Secretagogue–receptor interactions are cell specific, and different glands usually possess different sets of receptors. Binding of secretagogues to receptors activates intracellular signaling events that lead to release of the contents of zymogen granules into the lumen. Major signaling pathways for secretion are (Figure 25.4) (1) activation of phosphatidylinositol-specific phospholipase C with liberation of inositol 1,4,5-trisphosphate and diacylglycerol (p. 534), in turn, triggering Ca^{2+} release from the endoplasmic reticulum into the cytosol and activation of protein kinase C, respectively, and (2) activation of adenylate or guanylate cyclase, resulting in elevated cAMP or cGMP levels, respectively.

Acetylcholine (Figure 25.5) is the major neurotransmitter for stimulating enzyme and electrolyte secretion throughout the gastrointestinal tract. It mediates regulation of salivary and gastric secretions by the autonomic nervous system during the early phase of digestion (cephalic phase of gastric secretion) and of pancreatic and intestinal secretions by the enteric nervous system during subsequent digestion. The acetylcholine receptor of exocrine cells is of the muscarinic type; that is, it can be stimulated by muscarinic acid and blocked by atropine (Figure 25.6). Atropine used to be given by dentists to "dry up" the mouth for dental work.

The biogenic amines **histamine** and 5-hydroxytryptamine (**serotonin**) are secretagogues that function mainly in a paracrine fashion, that is, affecting neighboring cells. Histamine is produced by specialized regulatory cells in the gastric mucosa (enterochromaffin-like or ECL cells). 5-Hydroxytryptamine is produced by many specialized epithelial cells that are diffusely distributed throughout the lining epithelium of the stomach and the intestines. The amines are stored in vesicles and released when cells receive appropriate stimuli. Since 5-hydroxytryptamine-producing cells release most of their product toward the blood side, they are also called epithelial endocrine cells.

Histamine (Figure 25.7) is a potent stimulator of HCl secretion. It interacts with the H_2 receptor on the contraluminal plasma membrane of gastric HCl-producing (parietal) cells. Antagonists at the H_2 receptor (H_2 blockers) are used medically as antacids.

TABLE 25.5 • Secretory Intestinal Neuropeptides and Hormones (Human)

Vasoactive intestinal peptide (VIP)

His-Ala-Asp-Gly-Val-Phe-Thr-Ser-Asp-Phe-Ser-Lys-Leu-Leu-Gly-Gln-Leu-Ser-Ala-Lys
|
[c]NH_2-Met-Leu-Ser-Glu-Leu-Tyr -Lys

Secretin

His-Ser-Asp-Gly-Thr-Phe-Thr-Ser-Glu-Leu-Ser-Arg-Leu-Arg-Glu-Gly-Ala-Arg-Leu-Gln
|
[c]NH_2-Val-Leu-Gly-Gln-Leu-Leu-Arg

Guanylin

Pro-Gly-Thr-Cys-Glu-Ile-Cys-Ala-Tyr-Ala-Ala-Cys-Thr-Gly-Cys

Gastrin G-34-II[a] **G-17-II**

[b]Glp-Leu-Gly-Pro-Gln-Gly-Pro-Pro-His-Leu-Val-Ala-Asp-Pro-Ser-Lys-Lys-Gln
|
[c]NH_2-Phe-Asp-Met-Trp-Gly-Tyr(SO_3H)-Ala-(Glu)$_5$-Leu-Trp-Pro-Gly

Cholecystokinin

Lys-Ala-Pro-Ser-Gly-Arg-Met-Ser-Ile-Val-Lys-Asn-Leu-Gln-Asn-Leu-Asp-Pro
|
[c]NH_2-Phe-Asp-Met-Trp-Gly-Met-Tyr(SO_3H)-Asp-Arg-Asp-Ser-Ile-Arg-His-Ser

Source: Yanaihara, C. In: B. B. Rauner, G. M. Makhlouf, and S. G. Schultz (Eds.), *Handbook of Physiology: Section 6: Alimentary Canal Vol II: Neural and Endocrine Biology.* Bethesda, MD: American Physiological Society, 1989, pp. 95–62; UniProtKB/Swiss-Prot entry P01282

[a]Gastrin I is not sulfated.

[b]Glp denotes pyrrolidino carboxylic acid, derived from Glu through internal amide formation.

[c]NH_2 denotes amide of carboxy-terminal amino acid.

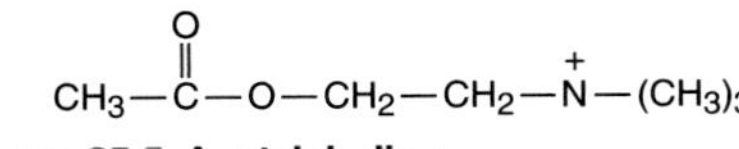

Figure 25.5 **Acetylcholine.**

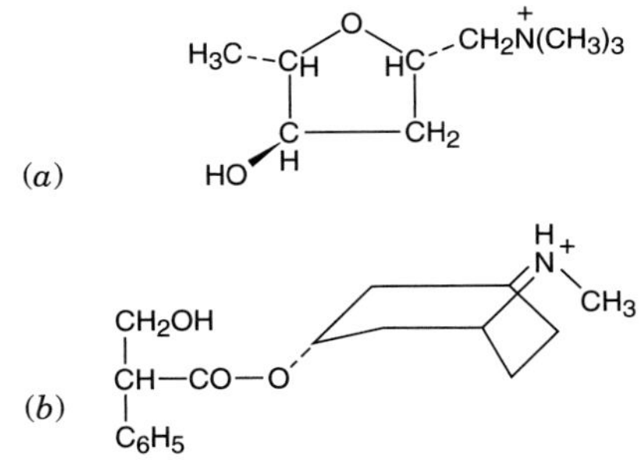

Figure 25.6 **(a) L(+)-Muscarine. (b) Atropine.**

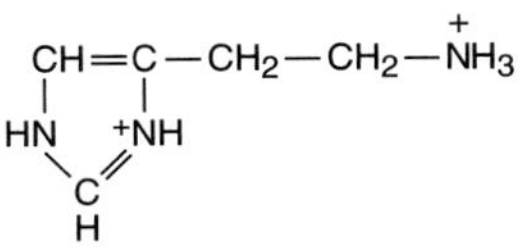

Figure 25.7 **Histamine.**

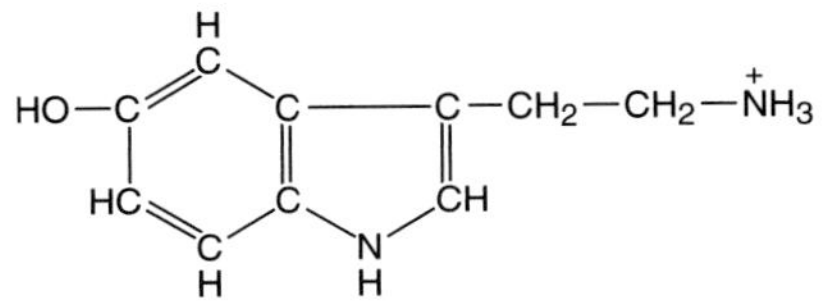

Figure 25.8 **5-OH-Tryptamine (serotonin).**

5-Hydroxytryptamine (Figure 25.8) stimulates intestinal NaCl secretion, but also functions as neurotransmitter, for example, initiating the sensation of nausea.

A third class of secretagogues consists of peptides, some of which function predominantly as hormones, others as neurotransmitters, and again others as both (Table 25.5). Many of these peptides are stabilized against digestion and chemical degradation by formation of an amide at the carboxyl-terminus (A Closer Look 25.1). The intestinal peptide hormones are produced by a number of different epithelial endocrine cells, of which many

A CLOSER LOOK 25.1

Amidation of Peptide Hormones and Neuropeptides

Most peptide hormones or neuropeptides, including most gastrointestinal ones (Table 25.5, p. 1035), are stabilized by carboxyl-terminal α-amidation. This amide group is usually required for full biological activity. Amidation is generated by a single enzyme with two different functions, namely, peptidyl-glycine α-amidating mono-oxygenase and peptidyl-α-hydroxyglycine α-amidating lyase activities The enzyme is targeted within cells to secretory granules where it acts on glycine extended prohormones (peptidyl-glycine) that are produced from larger precursors by proteolytic cleavage. The enzyme converts the carboxy terminal glycine to glyoxalate while leaving behind the amino group of glycine as amide of the finished hormone peptide (see figure).

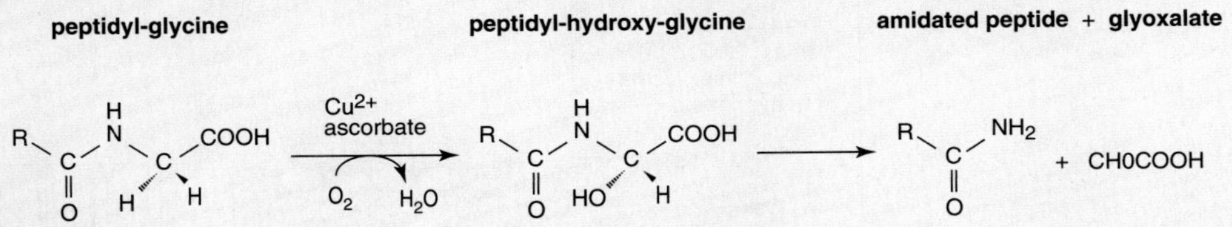

peptidyl-glycine → (Cu^{2+}, ascorbate; O_2 → H_2O) → **peptidyl-hydroxy-glycine** → **amidated peptide + glyoxalate** (+ CHOCOOH)

Eipper, B. A., Milgram, S. L., Husten, E. J., Yun, H. Y., and Mains, R. E. Peptidylglycine alpha-amidating monooxygenase: A multifunctional protein with catalytic, processing, and routing domains. *Protein Sci.* 2:489, 1993: and Takahashi, K., Satani, M., Gao, Y., and Nogucki, M. The reaction mechanism of peptidylglycine α-hydroxylating monooxygenase. *Internation Congress Series* 1233:235, 2002.

also produce 5-hydroxytryptamine and store both products in small intracellular vesicles. Among the many peptide hormones, gastrin, cholecystokinin (pancreozymin), secretin, and guanylin have particular importance for digestive enzyme and fluid secretion. **Gastrin** occurs naturally in two forms: "big" gastrin is a peptide of 34 amino acids (G-34), which is cleaved to yield "little" gastrin comprising the 17 residues at the carboxyl terminus (G-17). The functional portion of gastrin resides mainly in the carboxyl-terminal pentapeptide, and synthetic pentagastrin can be administered intravenously to assess the potential for gastric HCl and pepsin secretion in a patient. Gastrin and cholecystokinin have a sulfated tyrosine, which considerably enhances the potency of each hormone.

Cholecystokinin and **pancreozymin** denote the same peptide, but *cholecystokinin* is the preferred term. The peptide stimulates gallbladder contraction (hence cholecystokinin) and secretion of pancreatic enzymes (hence pancreozymin). Cholecystokinin is released into the bloodstream by epithelial endocrine cells of the small-intestine lining, particularly in the duodenum, in response to ingestion of food and contact with cholecystokinin releasing factors secreted into the lumen by pancreas and intestinal cells. The releasing factors are themselves proteins and thus substrates for pancreatic proteases (p. 1044). The state of their degradation serves as feedback for adequate pancreatic enzyme delivery to the intestine as lack of digestion of the releasing factors stimulates further cholecystokinin and hence pancreatic enzyme secretion. Cholecystokinin and gastrin are thought to be evolutionarily related because they share an identical amino acid sequence at the carboxyl terminus.

Secretin, a polypeptide of 27 amino acids, is released by another type of endocrine cell in the small intestine. Its release is stimulated particularly by luminal pH less than 5. Its major biological activity is stimulation of secretion of pancreatic juice rich in $NaHCO_3$. This is essential for neutralization of gastric HCl in the duodenum. Secretin also enhances pancreatic enzyme release, acting synergistically with cholecystokinin.

Guanylin is a peptide that is produced by goblet (mucin-producing) cells and released mainly into the lumen. It stimulates NaCl secretion by binding to a brush border receptor/guanylate cyclase and thereby elevating cytosolic cGMP levels. The peptide was discovered because it activates the same receptor/guanylate cyclase as a heat-stable *E. coli* enterotoxin, one of the factors responsible for traveler's diarrhea.

Among neuropeptides, **vasoactive intestinal peptide (VIP)** is a potent and physiologically important secretagogue for NaCl and fluid in the intestines and the pancreas. Enterocytes possess receptors for VIP and release of this neuropeptide by enteric nerves regulates intestinal electrolyte and fluid secretion.

25.3 • EPITHELIAL TRANSPORT

Solute Transport May Be Transcellular or Paracellular

The barrier properties of epithelia are determined by the plasma membranes of the epithelial cells and their intercellular tight junctional complexes (Figure 25.9). Tight junctions extend in a beltlike manner around the perimeter of each epithelial cell and connect neighboring cells. They constitute part of the barrier between the two extracellular spaces on either side of the epithelium, that is, the gastrointestinal lumen and the intercellular (interstitial) space on the blood or serosal side. Tight junctions mark the boundary between luminal and contraluminal regions of the plasma membrane of epithelial cells.

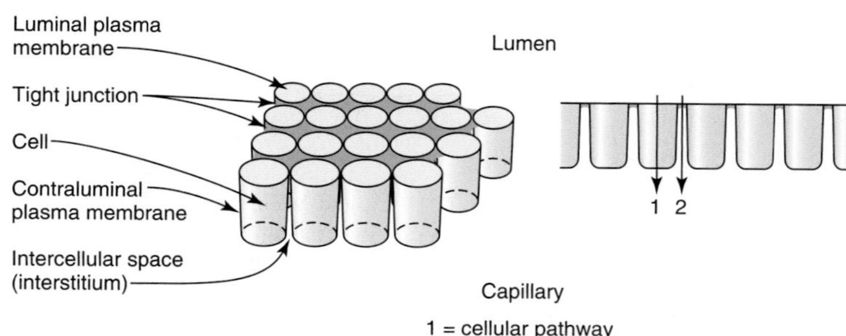

Figure 25.9 Pathways for transport across epithelia.

TABLE 25.6 • Characteristic Differences between Luminal and Contraluminal Plasma Membranes of Enterocytes

Parameter	*Luminal*	*Contraluminal*
Morphological appearance	Microvilli arranged as brush border	Few microvilli
Enzymes	Di- and oligosaccharidases Aminopeptidase Dipeptidases γ-Glutamyltransferase Alkaline phosphates Guanylate cyclase C	Na^+/K^+ exchanging ATPase Adenylate cyclase
Transporters/channels	Na^+-glucose cotransporter (SGLT1, SLC5A1)	Facilitative glucose transporter (GLUT2, SLC2A2)
	Facilitative fructose transporter (GLUT5, SLC2A5)	Facilitative neutral amino acid transporter (SLC3A2/SLC7A8)
	Na^+-neutral amino acid cotransporter (SLC6A19)	
	Apical Na^+-bile acid cotransporter (ASBT, SLC10A2)	Sodium-potassium-2-chloride cotransporter (NKCC 1, SLC12A2)
	H^+-peptide cotransporter (PEPT1, SLC15A1))	
	Cystic Fibrosis Transmembrane Regulatory protein (CFTR, ABCC7)	

Names in parentheses are original abbreviations and corresponding new classifications according to SLC (solute carrier) or ABC (ATP Binding Casette) nomenclature; see http://www.bioparadigms.org/slc/menu.asp and http://nutrigene.4t.com/humanabc.htm.

Two parallel pathways can be distinguished for solute transport across epithelial cell layers, one through the cells (transcellular) and the other through the tight junctions between cells (paracellular) (Figure 25.9). Nutrient or electrolyte flux through either pathway can influence flux through the other one by changing driving forces. The transcellular route in turn consists mainly of two barriers in series, formed by the luminal and contraluminal plasma membranes.

Different regions of the gastrointestinal tract differ with respect to permeability characteristics of the tight junction. In regions with steep solute concentration gradients across the epithelium, such as stomach and distal colon, tight junctions have low permeability (or high resistance) to Na^+ and other ions, whereas they possess high permeability in regions with nutrient absorption or NaCl secretion as both processes require paracellular solute flux (p. 1036).

A major function of intestinal epithelial cells is active absorption of nutrients, electrolytes, and vitamins. The cellular basis for this vectorial solute movement lies in the different complement of transporters in luminal and contraluminal membranes. The small-intestinal epithelial cells provide an example of the differentiation and specialization of the two types of membrane in terms of morphological appearance, chemical composition, and complement of enzymes and transporters (Table 25.6). The luminal membrane is in contact with the nutrients in chyme (semifluid mass of partially digested food) and is specialized for terminal digestion of nutrients through hydrolytic enzymes on its external surface and for nutrient absorption through transporters or channels for monosaccharides, amino acids, peptides, fatty acids, cholesterol, and electrolytes. In contrast, the contraluminal plasma membrane is in contact with the intercellular fluid and therewith indirectly with capillaries and lymph vessels. It has properties similar to the plasma membrane of most cells and possesses receptors for hormonal or neuronal regulation of cellular functions, a Na^+/K^+-exchanging ATPase for removal of Na^+ from the cell, and transporters for entry of nutrients needed for cell maintenance. It also contains the transporters necessary for exit of absorbed nutrients so that the digested food becomes available to all cells of the body. Many of the nutrient transporters in the contraluminal membrane are bidirectional and serve to release nutrients into blood or lymph after a meal and provide nutrients to enterocytes in between meals.

NaCl Absorption Depends on Na^+/K^+-exchanging ATPase, Membrane Transporters, and Channels

Transport of Na^+ plays a crucial role not only for epithelial NaCl absorption or secretion but also for energization of nutrient uptake. The **Na^+/K^+-exchanging ATPase** (p. 491)

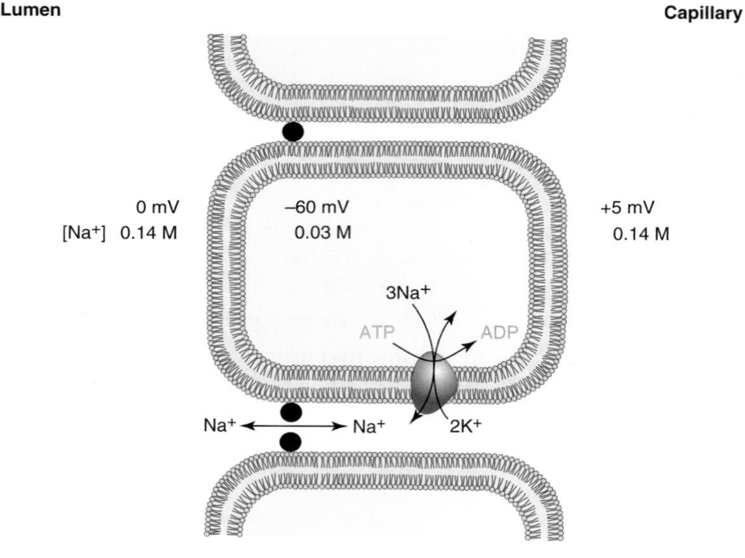

Figure 25.10 **Na^+ concentrations and electrical potentials in enterocytes.**

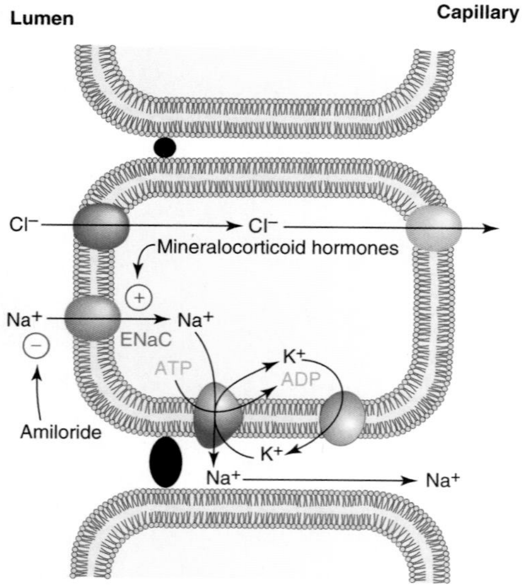

Figure 25.11 **Model for electrogenic NaCl absorption in lower large intestine.** Na^+ absorption proceeds via luminal epithelial sodium channel (ENaC) and Na^+, K^+-exchanging ATPase. The nature of Cl^- channels is not known.

is the dominant mechanism for transduction of chemical energy in the form of ATP into osmotic energy of a concentration (chemical) or a combined concentration and electrical (electrochemical) ion gradient across the plasma membrane. In gastrointestinal epithelial cells this enzyme is located exclusively in the contraluminal membrane (Figure 25.10). The ATPase maintains the high K^+ and low Na^+ concentrations in the cytosol and is responsible, in conjunction with a K^+ channel in the same membrane, for an electrical potential of about −60 mV of the cytosol relative to the extracellular solution. Transepithelial NaCl movements result from the combined actions of the Na^+/K^+-exchanging ATPase and additional passive transporters in the plasma membrane, which allow the entry of Na^+ or Cl^- into the cell. NaCl absorption results from Na^+ entry from the lumen and its extrusion by the Na^+/K^+-exchanging ATPase across the contraluminal membrane. Epithelial cells of the lower portion of the large intestine possess a luminal Na^+ channel (Epithelial Na Channel, ENaC, or SCNN1. *Note:* Acronyms used to identify the channels and transporters are being changed. Thus, where appropriate, both the older and newer acronym will be given). The channel allows the uncoupled entry of Na^+ down its electrochemical gradient (Figure 25.11). This Na^+ flux is electrogenic, that is, it is associated with an electrical current that, in turn, changes any electrical potential. It can be inhibited by the diuretic drug amiloride at micromolar concentrations (Figure 25.12). This transport system, and hence NaCl absorption, is regulated by mineralocorticoid hormones of the adrenal cortex.

Epithelial cells of the small intestine have a transporter in their brush border membrane, which catalyzes an electrically neutral Na^+/H^+ exchange; Na^+/H^+ transporter-3 (NHE3, or SLC9A3) is the predominant isoform in the intestine (Figure 25.13). This transporter is not affected by low concentrations of amiloride and not regulated by mineralocorticoids. The Na^+/H^+ exchange sets up a H^+ gradient that secondarily drives Cl^- absorption through a specific Cl^-/HCO_3^- transporter in the luminal plasma membrane, as illustrated in Figure 25.13. The intestinal Cl^-/HCO_3^- transporter is coded for by the "down-regulated in adenoma" (DRA) gene (SLC26A3) (Clin. Corr. 25.1). The need for two types of NaCl absorption mechanisms arises from the different physiological functions of upper and lower intestine, which require different regulation. The upper intestine absorbs the bulk of NaCl derived from diet and secretions of the exocrine glands, whereas the lower intestine scavenges remaining NaCl depending on overall NaCl balance of the body.

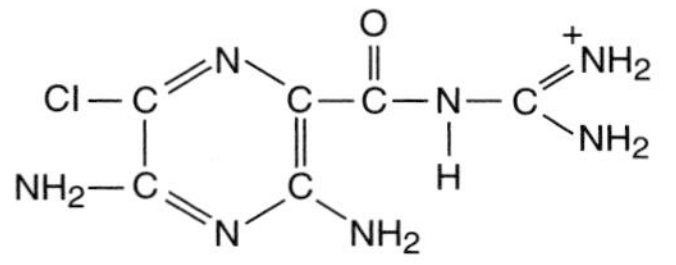

Figure 25.12 Amiloride.

NaCl Secretion Depends on Na^+/K^+-exchanging ATPase, Membrane Transporters, and Channels

Glands and intestinal crypts secrete each day a large amount of electrolytes and fluids into the gastrointestinal lumen (equivalent to at least a third of the extracellular fluid in humans). This is an energy-consuming process with active transport of electrolytes and passive flow of water. The water movement is due to osmotic forces exerted by secreted electrolytes. In other words, any primary secreted fluid has a greater osmolarity than plasma or cytosol and is said to be *hypertonic*. However, most gastrointestinal epithelia have a very high permeability to water due to the presence of water channels, aquaporins (p. 479), in the plasma membranes so that osmotic equilibrium is rapidly approached. As a result, the secretions actually emerging from the acini of glands or intestinal crypts are essentially *isotonic* (same osmolarity as plasma). Ionic compositions of gastrointestinal secretions are presented in Figure 25.14.

Major secreted ions are Na^+ and Cl^-, and net NaCl secretion is an energy-consuming process that depends on Na^+/K^+-exchanging ATPase located in the contraluminal plasma membrane (Figure 25.15). Hence, the secretion mechanism provides insight into energy coupling. Experimental evidence for the involvement of Na^+/K^+-exchanging ATPase in both NaCl secretion and absorption is provided by the finding that specific inhibitors of this enzyme, so-called cardiac glycosides, abolish both salt secretion and absorption. How can the Na^+/K^+-exchanging ATPase power Na^+ movement from the capillary side to the lumen when it actually extrudes Na^+ from the cell toward the capillary side? This paradox is resolved by an electrical coupling of Cl^- secretion across the luminal plasma membrane and Na^+ movements via the paracellular route, illustrated in Figure 25.15. Cl^- secretion depends on coupled uptake of two Cl^- ions with Na^+ and K^+ across the contraluminal membrane, extrusion of Na^+ and K^+ across the contraluminal membrane, and luminal exit

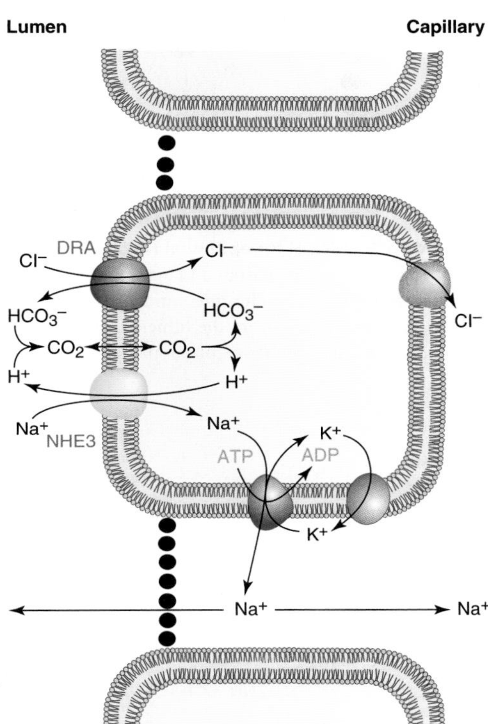

Figure 25.13 Model for electrically neutral NaCl absorption in the small intestine. Na^+ absorption proceeds via sodium/proton exchanger 3 (NHE3) and Na^+, K^+-exchanging ATPase. Cl^- absorption proceeds via 'down-regulated in adenoma" protein (DRA). The nature of the contraluminal Cl^- channel is not known.

CLINICAL CORRELATION 25.1

Familial Chloridorrhea Causes Metabolic Alkalosis

Loss-of-function mutations in the human DRA (down-regulated in adenoma) gene (SLC26A3, OMIM 126650) causes familial chloridorrhea. Patients with this disease have moderate diarrhea, generate acidic stool, and suffer from metabolic alkalosis (alkaline plasma at normal carbon dioxide concentration). The normal DRA gene product is a Na^+-independent Cl^-/HCO_3^- exchanger in epithelial cells of the lower ileum and colon. The exchanger absorbs Cl^- from stool in exchange for bicarbonate. Bicarbonate secretion is important for neutralization of protons secreted in exchange for Na^+ via sodium/proton exchanger NHE3 (SLC9A3). Chronic HCl loss with the stool leads to a corresponding metabolic alkalosis of the body fluids. In addition, the extra electrolytes in stool are responsible for its greater fluid contents, i.e., diarrhea, due to osmotic effects.

Mount, D. B. and Romero, M. F. The SLC26 gene family of multifunctional anion exchangers. *Pflugers Arch.* 447:710, 2004.

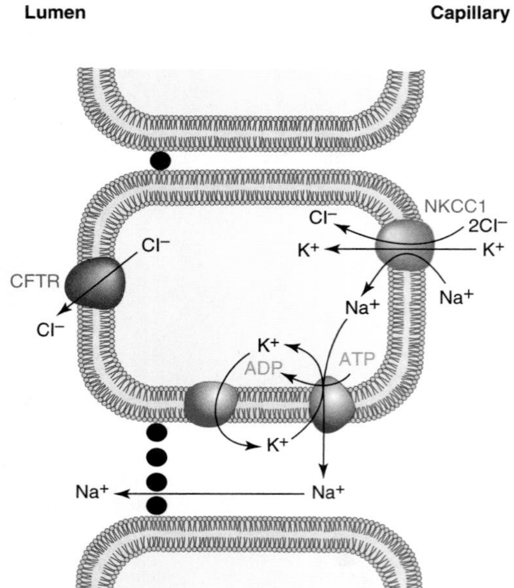

Figure 25.15 **Model for epithelial NaCl secretion.** Cl^- is concentrated in cytosol via sodium-potassium-two-chloride-cotransporter (NKCC1) and released into the lumen via Cystic Fibrosis Transmembrane Regulator protein (CFTR).

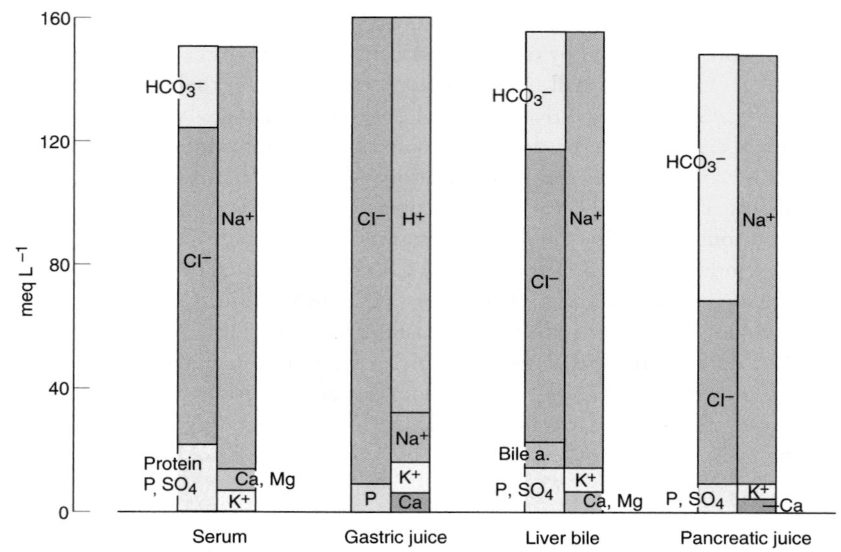

Figure 25.14 **Ionic composition of gastrointestinal secretions.** Serum included for comparison. Note high H^+ concentration in gastric juice (pH = 1) and high HCO_3^- concentration in pancreatic juice. Abbreviations: P, organic and inorganic phosphate; SO_4, inorganic and organic sulfate; Ca, calcium; Mg, magnesium; and bile a., bile acids.
Adapted from *Biological Handbooks. Blood and Other Body Fluids.* Federation of American Societies for Experimental Biology, Bethesda, MD.1961.

of Cl^- via channels. The uptake is mediated by a $Na^+/K^+/\ 2Cl^-$-cotransporter (NKCC1 or SLC12A2), which is pharmacologically characterized by inhibition by the diuretic furosemide (Figure 25.16), and utilizes energy of the Na^+ gradient to accumulate Cl^- within the cytosol above its electrochemical equilibrium. As the Na^+/K^+-exchanging ATPase generates and maintains the Na^+ gradient across the plasma membrane, it indirectly also powers Cl^- uptake from capillaries into cells and its movement into the lumen. The exit of Cl^- via luminal channels is associated with loss of a negative charge; this, in turn, sets up an electrical potential that draws Na^+ into the lumen via the paracellular route. The predominant luminal Cl^- channel in pancreatic ducts and the intestines is the cystic fibrosis transmembrane regulatory (CFTR) protein, and malfunctions of this channel lead to reduced amounts of secretions in the human disease cystic fibrosis (Clin. Corrs. 25.2 and 25.3).

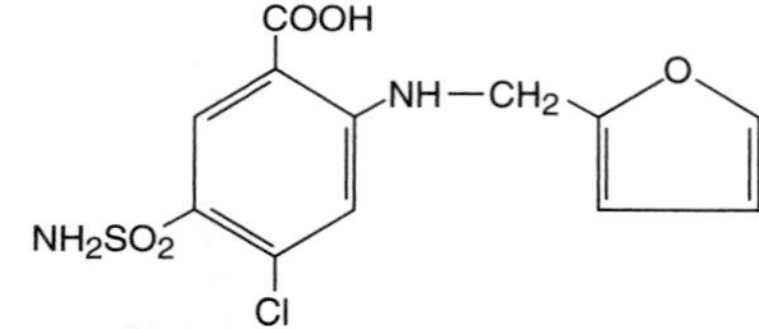

Figure 25.16 **Furosemide.**

Pancreatic acinar cells secrete a fluid rich in Na^+ and Cl^-, which provides the vehicle for transport of digestive enzymes from the acini to the lumen of the duodenum. This fluid is modified in the ducts by the additional secretion of $NaHCO_3$ (Figure 25.17). Bicarbonate concentrations can reach up to 120 mM in humans.

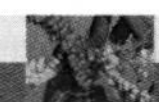

CLINICAL CORRELATION 25.2

Cystic Fibrosis of the Pancreas

Cystic fibrosis (OMIM 602421) transmembrane regulatory (CFTR) protein is a member of the ABC transporter family (subfamily ABCC7) and the predominant Cl^- channel in the luminal plasma membrane of epithelial cells in tissues affected in cystic fibrosis (airways, pancreatic duct, intestine, vas deferens, sweat gland ducts) (see Clin. Corr. 12.5, p. 496). The channel is normally closed but opens when phosphorylated by protein kinase A and ATP is present. Cl^- flux through CFTR depends on the existing electrochemical Cl^- gradient, which is set up by other cellular transporters and is different in secretory and absorptive cells. In most tissues, CFTR controls secretion of NaCl and fluid. (See Clin. Corr. 25.3 for activation of the CFTR Cl^- channel.) However, in sweat gland duct cells, which are absorptive, CFTR mediates efficient reabsorption of Cl^-, which is initially secreted in sweat gland acini. A defect in CFTR explains both excessive loss of Cl^- in sweat (sweat test for cystic fibrosis) and insufficient NaCl and fluid secretion in lungs, pancreas, and intestine. The gastrointestinal symptoms in cystic fibrosis patients (maldigestion, meconium ileus, and fecal impaction) stem from the diminished fluid secretion and the resulting partial or total blockage of pancreatic ducts and the intestine. The decrease in pancreatic juice delivered to the intestine explains maldigestion, whereas retention of digestive enzymes within the pancreas produces autodigestion, inflammation, scarring, and cyst formation.

CLINICAL CORRELATION 25.3

Bacterial Toxigenic Diarrheas and Electrolyte Replacement Therapy

Voluminous, life-threatening intestinal electrolyte and fluid secretion (diarrhea) occurs in patients with cholera, an intestinal infection by Vibrio cholerae. Certain strains of *E. coli* also cause (traveler's) diarrhea that can be serious in infants. The secretory state is a result of enterotoxins produced by the bacteria. The mechanisms of action of some of these enterotoxins are well understood at the biochemical level. Cholera toxin activates adenylate cyclase by causing ADP-ribosylation of the $G_{\alpha s}$ protein resulting in constitutive stimulation of the cyclase (p. 527). Elevated cAMP levels in turn activate protein kinase A and protein phosphorylation, which opens the luminal CFTR Cl^- channel in secretory cells and inhibits the Na^+/H^+ exchanger (NHE3) in absorptive cells. The net result is gross NaCl secretion. *E. coli* produces a heat-stable toxin that binds to guanylate cyclase C, which has an extracellular binding domain and intracellular catalytic domain. Binding of *E. coli* heat-stable toxin or of the peptide guanylin produced physiologically by intestinal goblet cells, activates guanylate cyclase resulting in increased cGMP levels. Elevated cGMP, similarly as elevated cAMP, inhibits NaCl reabsorption and stimulates Cl^- secretion.

Modern, oral treatment of cholera takes advantage of the presence of Na^+-glucose cotransport in the intestine, which is not altered by elevated cAMP and remains fully active in this disease. In this case, the presence of glucose allows uptake of Na^+ to replenish body NaCl. Composition of solution for oral treatment of cholera patients is glucose 110 mM, Na^+ 99 mM, Cl^- 74 mM, HCO_3^- 29 mM, and K^+ 4 mM. The major advantages of this form of therapy are its low cost and ease of administration when compared with intravenous fluid therapy.

The composition of sport drinks for electrolyte replacement is based on the same principle, namely more rapid sodium absorption in the presence of glucose.

Carpenter, C. C. J. In Field, M., Fordtran, J. S., and Schultz, S. G (Eds.), *Secretory Diarrhea.* Bethesda, MD: American Physiological Society, 1980, 67.

Ion Concentration Gradients and Electrical Potentials Energize Transport of Nutrients

Many solutes are absorbed across the intestinal epithelium against a concentration gradient. Energy for this active transport is directly derived from a Na^+ or H^+ concentration gradient or the electrical potential across the luminal membrane, and only indirectly from ATP hydrolysis. Intestinal glucose transport is an example of such uphill solute transport, driven in this case by an electrochemical Na^+ gradient (Figure 25.18).

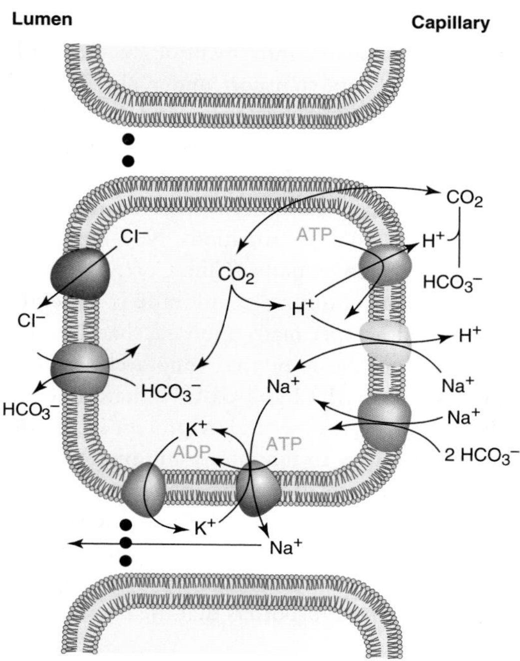

Figure 25.17 Model for $NaHCO_3$ secretion by pancreatic duct cells. Luminal Cl^- efflux and Na^+ flux are similar to those in NaCl secretion (see Figure 25.15). *Note:* Three mechanisms exist for bicarbonate influx into the cell (or its equivalent proton secretion) at the contraluminal membrane: (1) Na^+/H^+ exchange, (2) H^+-ATPase, and (3) Na^+-$2HCO_3^-$-co-transport.

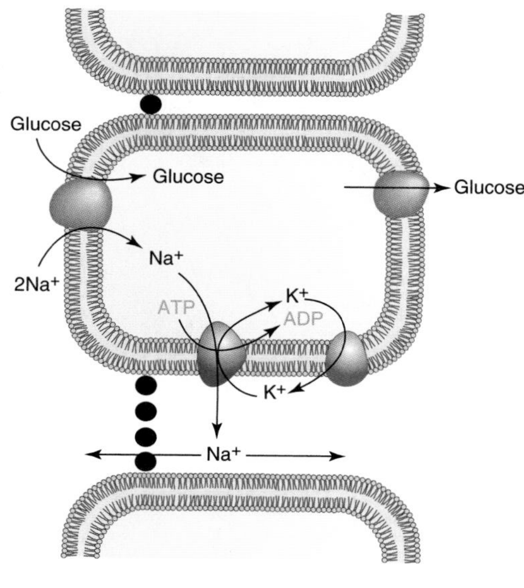

Glucose concentration profile normal to epithelial plane

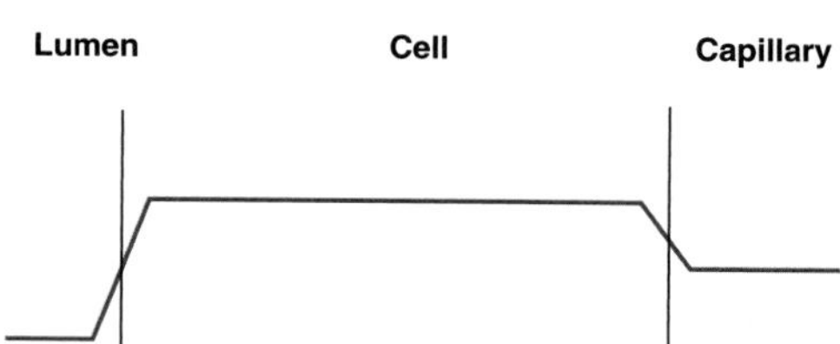

Figure 25.18 Model for epithelial glucose absorption. Note the indirect role of Na^+, K^+-exchanging ATPase.

$$\begin{array}{lrcl}
2X & 3Na^{+}_{cell} + 2K^{+}_{interstitium} + ATP_{cell} & \xrightarrow{Na^{+}, K^{+}-ATPase} & 3Na^{+}_{interstitium} + 2K^{+}_{cell} + ADP_{cell} + P_{cell} \\
4X & K^{+}_{cell} & \xrightleftharpoons{K^{+}\ channel} & K^{+}_{interstitium} \\
3X & 2Na^{+}_{lumen} + Glc_{lumen} & \xrightleftharpoons{SGLT1} & 2Na^{+}_{cell} + Glc_{cell} \\
3X & Glc_{cell} & \xrightleftharpoons{GLUT2} & Glc_{interstitium} \\
6X & Na^{+}_{interstitium} & \xrightleftharpoons{Tight\ junction} & Na^{+}_{lumen} \\
\hline
Sum & 3Glc_{lumen} + 2ATP_{cell} & \longrightarrow & 3Glc_{interstitium} + 2ADP_{cell} + 2P_{cell}
\end{array}$$

Figure 25.19 Transepithelial glucose transport as translocation reactions across plasma membranes and the tight junction. Transporters SGLT1 (sodium glucose transporter 1) and GLUT2 (glucose transporter 2) mediating Na^+-glucose cotransport and facilitative glucose transport, respectively. Numbers in the left column indicate the minimal turnover of individual reactions to balance the overall reaction.

Serum glucose is about 5 mM so that complete extraction of glucose from chyme can only be achieved by active transport across the epithelium. This vectorial process is the result of several separate membrane events (Figure 25.19). (1) Primary active, ATP-dependent Na^+ efflux at the contraluminal membrane establishes an electrochemical Na^+ gradient across the contraluminal membrane. (2) K^+-efflux via channels in the contraluminal membrane contributes further to the membrane potential. (3) **Secondary active cotransport** of glucose and Na^+ drives glucose uptake from the lumen into the cell. (4) Facilitative glucose movement downhill into the interstitial and capillary space completes the transepithelial glucose absorption. This scenario is possible because of the presence of a cotransporter for glucose and Na^+ in the luminal membrane (**sodium glucose transporter 1** [SGLT1 or SLC5A1]) (p. 489), a facilitative transporter for glucose in the contraluminal membrane (GLUT2 or SLC2A2), and a leaky "tight junction" that allows the electrochemical Na^+ gradient across the contraluminal membrane to spread over the entire plasma membrane.

SGLT1 facilitates a tightly coupled movement of Na^+ and D-glucose (or structurally similar sugars) with a stoichiometry of two Na^+ ions and one glucose molecule. While the cotransporter *per se* facilitates coupled movement of glucose and Na^+ across the membrane equally well in either direction, transport under physiological conditions is from lumen to cell because of the lower Na^+ concentration and the negative potential in the cell. As a result, glucose is concentrated in the cell. Thus, downhill Na^+ movement normally supports concentrative glucose transport. Concentration ratios of up to 20-fold between intracellular and extracellular glucose have been observed in vitro under conditions of blocked efflux of cellular glucose. In some situations, Na^+ uptake via this route is physiologically more important than glucose uptake (Clin. Corr. 25.3).

GLUT2 (p. 489) in the contraluminal membrane is a member of the facilitative **glu**cose **t**ransporter family and accepts many monosaccharides, including glucose. The direction of net flux is determined solely by the monosaccharide concentration gradient. The two glucose transport systems SGLT1 and GLUT2 share glucose as substrate, but differ considerably in amino acid sequence, secondary protein structure, Na^+ as cosubstrate, specificity for other sugars, sensitivity to inhibitors, and biological regulation. Since both SGLT1 and GLUT2 are not inherently directional, active transepithelial glucose transport relies on the Na^+/K^+-exchanging ATPase to continually move Na^+ out of the cell and maintain the electrochemical Na^+ gradient. One advantage of this arrangement for supplying energy to nutrient absorption is that the Na^+/K^+-exchanging ATPase can energize transport of many different nutrient transporters that use Na^+ as cosubstrate.

Gastric Parietal Cells Secrete HCl

The parietal (oxyntic) cells of gastric glands secrete HCl into the gastric lumen. Luminal H^+ concentrations of up to 0.14 M (pH 0.8) have been observed (see Figure 25.14, p. 1040). At plasma pH = 7.4, the parietal cell transports protons against a concentration gradient of $10^{6.6}$. The free energy required for HCl secretion under these conditions is minimally 38.1 kJ/mol (9.1 kcal/mol) of HCl. This active HCl secretion is achieved through a combination of primary active K^+/H^+ exchange by a K^+/H^+-exchanging ATPase (ATP4) (or gastric proton pump), Cl^- and K^+ channels in the luminal membrane, and Cl^-/HCO_3^- exchange in the contraluminal membrane. The K^+/H^+-exchanging ATPase is unique to the parietal cell and translocates from intracellular vesicles to the luminal membrane for periods of active HCl secretion. This enzyme couples the hydrolysis of ATP to an electrically neutral obligatory exchange of K^+ for H^+, secreting H^+ and moving K^+ into the cell. The stoichiometry appears to be 1 mol of transported H^+ and K^+ for each mole of ATP.

$$\text{ATP}_{\text{cell}} + \text{H}^+_{\text{cell}} + \text{K}^+_{\text{lumen}} \rightarrow \text{ADP}_{\text{cell}} + \text{P}_{\text{i,cell}} + \text{H}^+_{\text{lumen}} + \text{K}^+_{\text{cell}}$$

As this ATPase generates a very acidic solution, protein reagents that are activated by acid can become specific inhibitors of this enzyme. Figure 25.20 shows the mechanism of a proton pump inhibitor that is sold as antacid.

In the steady state, HCl is elaborated by H^+/K^+-exchanging ATPase only if the luminal membrane is permeable to K^+ and Cl^- and the contraluminal membrane catalyzes an exchange of Cl^- for HCO_3^- (Figure 25.21). This exchange is essential to replenish the cell's Cl^- and prevent accumulation of base within the cell. Thus, under steady-state conditions, secretion of HCl into the gastric lumen is coupled to movement of HCO_3^- into plasma.

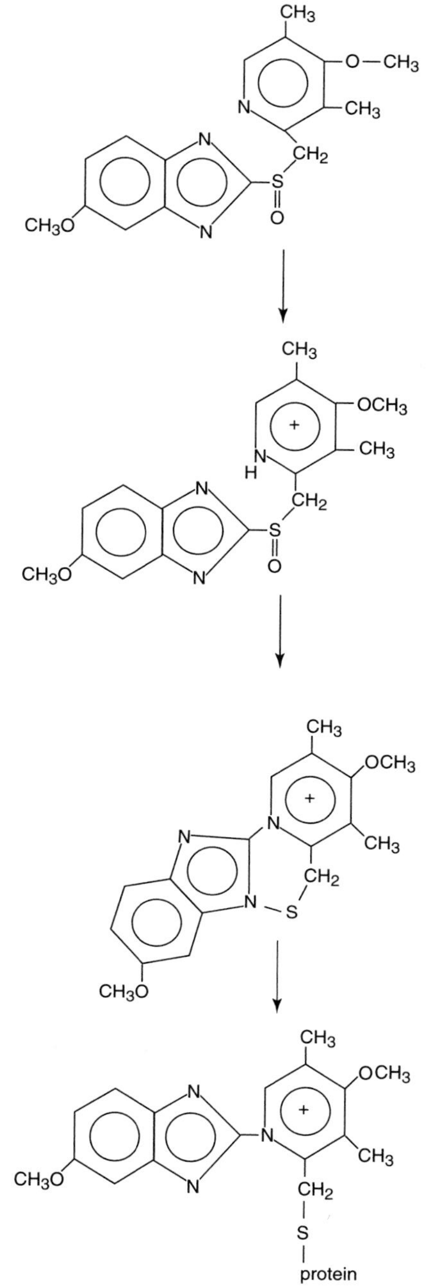

Figure 25.20 Omeprazole, an inhibitor of K^+/H^+-exchanging ATPase. This drug accumulates in acidic compartments (pKa ~4) and is converted to a sulfenamide that reacts with cysteine SH groups.
Redrawn from Sachs, G. The gastric K^+/H^+-ATPase. In Johnson, L. R. (Ed.), *Physiology of the Gastrointestinal Tract.* New York: Raven, 1994, 1133.

25.4 • DIGESTION AND ABSORPTION OF PROTEINS

Peptidases Assure Efficient Protein Digestion

The total daily protein load to be digested consists of about 70–100 g of dietary proteins and 35–200 g of endogenous proteins from digestive enzymes and sloughed-off cells. Digestion and absorption of proteins are very efficient processes in healthy humans, since only about 1–2 g of nitrogen are lost through feces each day, which is equivalent to 6–12 g of protein.

Except for a short period after birth, oligo- and polypeptides are not absorbed intact in appreciable quantities by the intestine. Proteins are hydrolyzed by a whole range of peptidases, each with specificity for peptide bonds between different amino acids. **Endopeptidases** (proteases) attack internal bonds and liberate large peptide fragments, while exopeptidases cleave off one amino acid at a time from either the carboxyl (carboxypeptidases) or the amino terminus (aminopeptidases). Endopeptidases are important for an initial breakdown of long polypeptides into smaller products, which can then be attacked

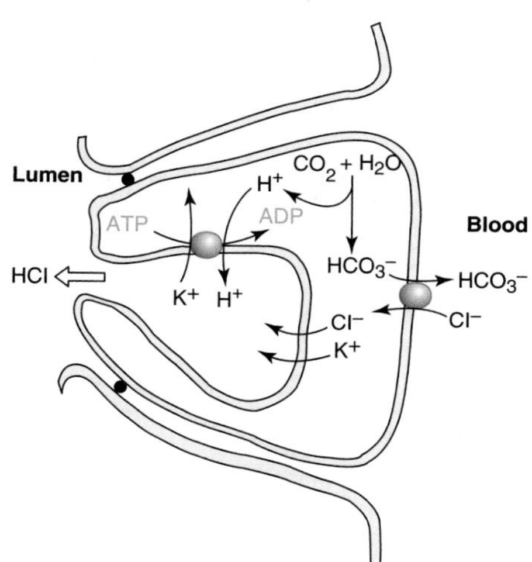

Figure 25.21 Model for secretion of hydrochloric acid.

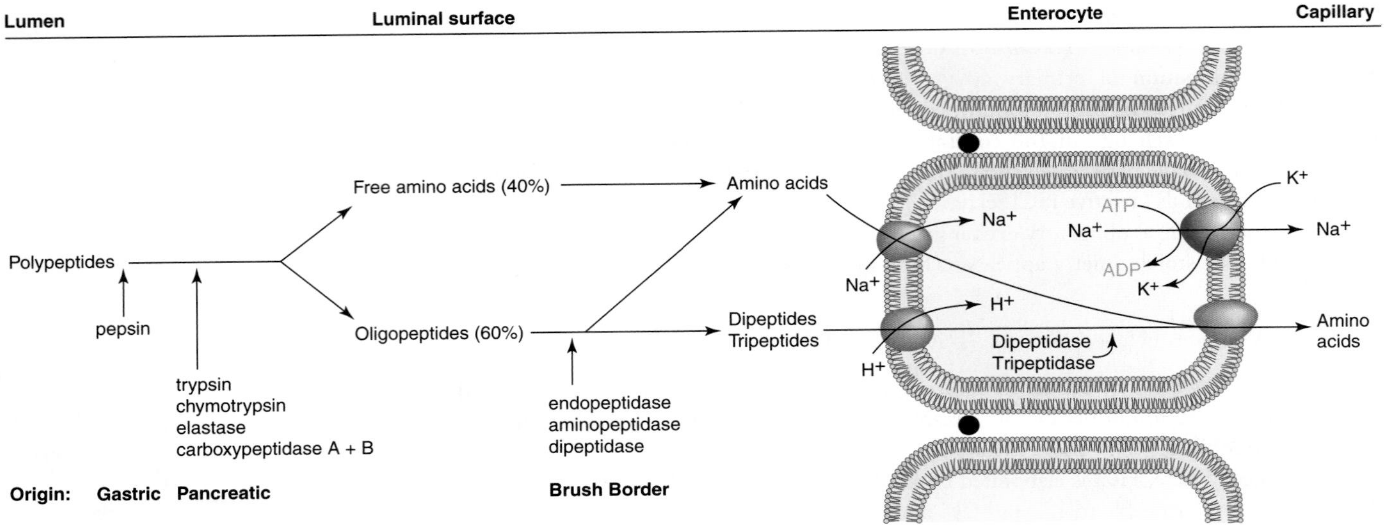

Figure 25.22 Digestion and absorption of proteins.

more efficiently by exopeptidases. The final products are free amino acids and di- and tripeptides, which are absorbed by epithelial cells (Figure 25.22).

Protein digestion can be divided into a gastric, a pancreatic, and an intestinal phase, depending on the source of peptidases.

Pepsins Catalyze Gastric Digestion of Protein

Gastric juice contains HCl at a pH of less than 2 and proteases of the pepsin family. The acid serves to kill off microorganisms and to denature proteins. Denaturation makes proteins more susceptible to hydrolysis by proteases. Pepsins are unusual enzymes in that they are stable and active at acid pH. The catalytic mechanism depends on the carboxylic groups from two aspartic acid residues at the active site (see Figure 25.3, p. 1034). Pepsin A, the major gastric protease, prefers peptide bonds formed by the amino group of hydrophobic and aromatic acids (Phe, Tyr, Trp, and Leu) (Table 25.7).

TABLE 25.7 • Gastric and Pancreatic Peptidases

Enzyme	*Proenzyme*	*Activator*	*Cleavage Point*	*R*
Carboxyl Proteases				
Pepsin A	Pepsinogen A	Autoactivation, pepsin	$-CO{\downarrow}NHCH(R)CO{\downarrow}NHCH(R')CO-$	Tyr, Phe, Trp, Leu
Serine Proteases				
Trypsin	Trypsinogen	Enteropeptidase, trypsin	$-CO-NHCH(R)CO{\downarrow}NHCH(R')CO-$	Arg, Lys
Chymotrypsin	Chymotrypsinogen	Trypsin	$-CO-NHCH(R)CO{\downarrow}NHCH(R')CO-$	Tyr, Trp, Phe, Met, Leu
Elastase	Proelastase	Trypsin	$-CO-NHCH(R)CO{\downarrow}NHCH(R')CO-$	Ala, Gly, Ser
Zinc Peptidases				
Carboxypeptidase A	Procarboxypeptidase A	Trypsin	$-CO{\downarrow}NHCH(R)COO^-$	Val, Leu, Ile, Ala
Carboxypeptidase B	Procarboxypeptidase B	Trypsin	$-CO{\downarrow}NHCH(R)CO^-$	Arg, Lys

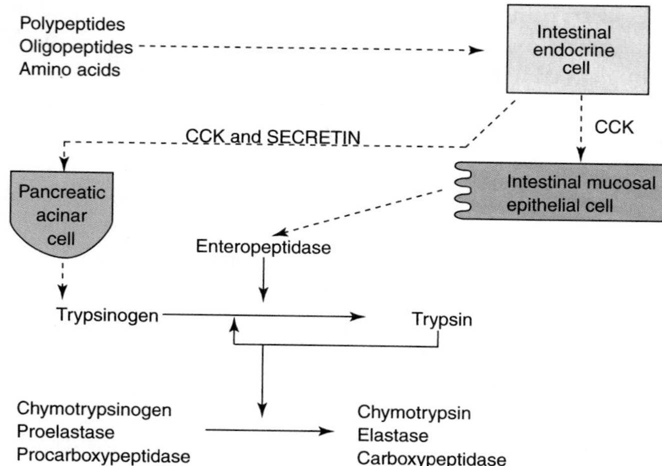

Figure 25.23 Secretion and activation of pancreatic enzymes. Abbreviation: CCK, cholecystokinin. Reproduced with permission from Freeman, H. J., and Kim, Y. S. *Annu. Rev. Med.* 29:102, 1978. Copyright (1978) Annual Reviews, Inc.

Active pepsin is generated from the proenzyme **pepsinogen** by removal of 46 amino acids from the NH_2-terminus. Cleavage between residues 46 and 47 occurs as an intramolecular reaction (autoactivation) below pH 5 or by the action of pepsin. The liberated peptide remains bound to pepsin and acts as pepsin inhibitor above pH 2. This inhibition is released by a drop of the pH below 2 or by further degradation of the peptide by pepsin. Thus, pepsinogen is activated at an exponential rate once the pH becomes acidic. Pepsin cuts proteins mainly into large peptide fragments, but also produces some smaller peptides and free amino acids. These latter products are important for initiation of the pancreatic phase of protein digestion.

Pancreatic Zymogens Are Activated in Small Intestine

Pancreatic juice is rich in proenzymes that are activated after they reach the lumen of the small intestine (Figure 25.23). **Enteropeptidase** (or enterokinase), a protease produced by duodenal epithelial cells, activates pancreatic trypsinogen to trypsin by scission of a hexapeptide from the NH_2-terminus. Trypsin activates more trypsinogen to trypsin and also acts on the other proenzymes, thus liberating the endopeptidases chymotrypsin and elastase and the carboxypeptidases A and B. Pancreatic juice normally contains a 6-kDa trypsin inhibitor protein that inhibits any trypsin formed prematurely within the pancreatic cells or pancreatic ducts (Clin. Corr. 25.4).

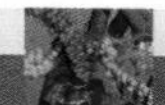

CLINICAL CORRELATION 25.4

Trypsin and Pancreatic Autodigestion

Trypsinogen actually refers to a family of proteins (e.g., coded by the genes TRYP 1–7). Premature activation of any of the trypsinogens within the pancreas results in activation of other members of the trypsin family and other types of digestive enzymes. The resulting active enzymes autodigest the pancreatic tissue itself, leading to an acute, painful inflammation of this organ. Clinically, the autodigestion can be detected by increases in serum levels of pancreatic amylase. To prevent premature activation of trypsinogens, pancreatic secretions normally contain a pancreatic secretory trypsin inhibitor (SPINK1). Rare autosomal-dominant mutations N29I and R122H of trypsin-1 (PRSS1 or cationic trypsinogen) and the variant N34S of the pancreatic secretory trypsin inhibitor are associated with childhood or youth onset of acute recurrent pancreatitis. Presumably, these mutations are associated with a greater tendency for autoactivation of trypsinogen and lesser inhibitory activity of the pancreatic secretory trypsin inhibitor, respectively. Digestive enzymes also activate within the pancreas when the ducts are clogged, e.g. when the major excretory duct that is shared with the common bile duct, is blocked by a gall stone (gall stone pancreatitis). Similarly, acinar cell injury, as suggested for chronic alcohol abuse, results in autoactivation of trypsinogen and pancreatitis (alcoholic pancreatitis).

Keim, V. Role of genetic disorders in acute recurrent pancreatitis. *World J. Gastroenterology* 14:1011, 2008; and Apte, M. V., Pirola, R. C., and Wilson, J. S. Molecular mechanisms of alcoholic pancreatitis. *Digestive Diseases* 23:232, 2005.

Specificities of **trypsin**, **chymotrypsin**, and **elastase** are presented in Table 25.7. The enzymes are active only at neutral pH and depend on pancreatic $NaHCO_3$ for neutralization of gastric HCl. Their mechanism of catalysis involves an essential serine residue (p. 343) and is thus similar to serine esterases, such as acetylcholine esterase. The serine proteases and esterases are inhibited by reagents that chemically modify serine in the active site. An example of such an inhibitor is diisopropylphosphofluoridate, which was developed for chemical warfare with acetylcholine esterase as intended target (p. 948).

Peptides generated from ingested proteins are further degraded within the small intestinal lumen by pancreatic carboxypeptidases A and B, which use Zn^{2+} as part of the catalytic mechanism. The combined action of pancreatic proteases and peptidases results in formation of free amino acids and small peptides of 2–8 residues; peptides account for about 60% of the amino nitrogen at this point.

Brush Border and Cytosolic Peptidases Digest Small Peptides

Since pancreatic juice does not contain appreciable aminopeptidase activity, final digestion of di- and oligopeptides depends on enzymes of the small intestine. The luminal surface of epithelial cells is particularly rich in endopeptidase and aminopeptidase activity, and also contains **dipeptidases** (see Table 25.3, p. 1033). Their action at the brush border surface produces free amino acids and di- and tripeptides, which are absorbed via specific amino acid or peptide transport systems. Transported di- and tripeptides are generally hydrolyzed within the intestinal epithelial cell before they leave the cell. This explains why practically only free amino acids are found in the portal blood after a meal. The virtual absence of peptides used to be taken as evidence that luminal protein digestion proceeded all the way to free amino acids before absorption occurred (Clin. Corr. 25.5). However, it is now established that a large portion of dietary amino nitrogen is absorbed in the form of small peptides with subsequent intracellular hydrolysis. Exceptions are di- and tripeptides that contain proline, hydroxyproline, or unusual amino acids, such as β-alanine in carnosine (β-alanylhistidine) or anserine (β-alanyl 1-methylhistidine). These peptides are absorbed and released intact into portal blood. Although unusual, β-alanine is nevertheless part of a normal diet as it is found, for example, in chicken meat.

Amino Acid and Di- and Tripeptide Transporters

The small intestine has a high capacity for absorbing free amino acids and di- and tripeptides. Most L-amino acids can be transported across the epithelium against a concentration gradient, although the need for concentrative transport *in vivo* is not obvious as luminal

CLINICAL CORRELATION 25.5

Gluten Enteropathy

Gluten enteropathy, or celiac disease, is due to intolerance of a protein fraction, termed gluten. Gluten is present in the cereals wheat, barley, and rye, but not rice and maize. The frequency of this disease in Caucasians is estimated as 1 in 250, that is, so prevalent that gluten-free diets can often be ordered. The disease appears to be caused by an enzyme deficiency in a yet to be identified brush border peptidase that results in incomplete digestion of gluten. Specific remaining peptides of 7 to 18 amino acid residues have been recognized, whereby peptides with the motifs of PSQQ or QQQP appear to be toxic to the mucosa. Such toxic peptides are produced at much higher concentrations during the digestion of gluten in celiac patients than in unaffected individuals. The presence of toxic peptides results in an increased permeability of the mucosal barrier and T-cell mediated immune and inflammatory responses to gluten-derived and other peptides that reach the immune cells. The inflammatory response, in turn, reduces the digestive surface area of the small intestinal mucosa and capacity of brush border digestion. Symptoms range from minor gastrointestinal complaints (bloating, diarrhea) to severe malnutrition. They are usually a consequence of reduced small intestinal brush border digestion and bacterial degradation of the remaining nutrients in the colon. Treatment consists of either avoidance of gluten-containing diets or, still experimental, the use of animal intestinal extracts to more completely digest gluten peptides.

Cornell, H. J. and Stelmasiak, T. A unified hypothesis of coeliac disease with implications for management of patients. *Amino Acids* 33:43, 2007.

TABLE 25.8 • Amino Acid Transporters in the Small Intestines

Transporter	*Other name(s)*	*Substrate Specificity*	*Mechanism*	*Disease Due to Loss of Function*
		Luminal Membrane		
SLC1A1	EAAT3	Asp, Glu	Secondary active transport: cotransport with Na^+, countertransport with K^+	Dicarboxylic aminoaciduria
SLC3A1/SLC7A9	rBAT/$b^{0,+}$AT	Lys, Arg, Cys-Cys	Facilitative, concentrating: cotransport with positive charge, countertransport with neutral amino acids	Cystinuria
SLC6A19	B^0AT1, NBB	Phe, Tyr, Met, Val, Leu, Ile	Secondary active transport: cotransport with Na^+	Hartnup disease, neutral aminoaciduria
SLC1A5	ASCT2, ASC	Ala, Ser, Cys	Exchanging: cotransport with Na^+	
SLC6A20	IMINO	Pro, hydroxy-Pro	Secondary active transport: cotransport with Na^+, cotransprot with Cl^-(in some species)	
SLC6A6	BETA	β-Ala, taurine	Secondary active transport: cotransport with Na^+, cotransprot with Cl^- (in some species)	
SLC36A1	PAT1	Gly, Ala, Pro, GABA, D-Ala	Secondary active transport: cotransport with H^+	
SLC15A1	PEPT1	Dipeptides, tripeptides, penicillin	Secondary active transport: cotransport with H^+	
		Contraluminal Membrane		
SLC3A2/SLC7A7	y^+L, y^+LAT 1	Lys, Arg	Exchanging: countertransport with Na^+ plus neutral amino acids	Lysinuric protein intolerance
SLC3A2/SLC7A8	L, LAT 2	Small and large neutral amino acids	Facilitative	
SLC16A10	TAT 1	Phe, Tyr, Trp	Facilitative	

concentrations are usually higher than the plasma levels of 0.1–0.2 mM. The uptake into cells is mediated by several different transporters in the luminal membrane, while the release into blood is mediated by several additional, different transporters in the contraluminal membrane (Table 25.8). Remarkably, several loss-of-function mutations in amino acid transporters were discovered because small intestine and renal proximal tubules share transporter types and renal loss of any particular amino acid transporter produces an easily measured result, namely, amino acid excretion in urine (amino aciduria). Similarly, importance of di- and tripeptide absorption for nutrition was discovered when a loss-of-function mutation in the major neutral amino transporter (neutral aminoaciduria) was not accompanied by an expected deficiency in the corresponding amino acids, suggesting compensation by at least one additional transporter type (Clin. Corr. 25.6).

The mechanism for active absorption of neutral L-amino acids appears to be similar to that discussed for D-glucose (see Figure 25.18, p. 1041). An Na^+-dependent cotransporter from the SLC6 family (also termed NBB for **N**eutral amino acid **B**rush **B**order or B^0AT1) and Na^+-independent, facilitative transporter (SLC3A2/SLC7A8, or L for **L**eucine preferring) have been characterized in the luminal and contraluminal membranes, respectively. The brush border transport of amino acids other than the neutral ones is energized in more complicated ways. For example, acidic amino acids can be concentrated by cotransport with 2 Na^+ ions and countertransport with one K^+ ion (SLC1A1) whereas basic amino acids rely on cotransport of a positive charge and the cell-inside negative potential (SLC3A1/SLC7A9). The amino acids alanine, serine, and cysteine are substrates for several transporters, one of which appears to catalyze an obligatory amino acid/amino acid exchange (SLC1A5).

Molecular analysis has revealed the existence of heteromeric amino acid transporters in which a "heavy" subunit (SLC3 family) directs cellular trafficking of the complex to either luminal or contraluminal membrane and a "light" subunit (SLC7 family) determines amino acid specificity and transport mechanism (Figure 25.24). For example, SLC3A1

TABLE 25.9 • Dietary Carbohydrates

Carbohydrate	*Typical Source*	*Structure*
Fructose	Fruit, honey	α-Fru
Glucose	Fruit, honey, grape	β-Glc
Amylopectin	Potatoes, rice, corn, bread	α-Glc(1 → 4)$_n$ Glc with α-Glc(1 → 6) branches
Amylose	Potatoes, rice, corn, bread	α-Glc(1 → 4)$_n$ Glc
Sucrose	Table sugar, desserts	α-Glc(1 → 2)β-Fru
Trehalose	Young mushrooms	α-Glc(1 → 1)α-Glc
Lactose	Milk, milk products	β-Gal(1 → 4)Glc
Raffinose	Leguminous seeds	α-Gal(1 → 6)α-Glc (1 → 2)β-Fru

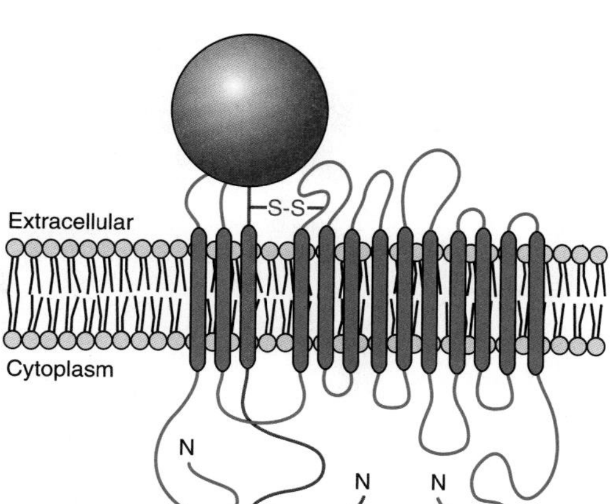

Figure 25.24 Model for heteromeric amino acid transporter. The heavy subunit with 1 transmembrane domaine (dark gray) is linked by a disulfide bridge to the light subunit with 12 transmembrane domains (light gray).
Redrawn from Chillaron, J., Roca, R., Valencia, A., Zorzano, A., and Palacin, M. Heteromeric amino acid transporters: biochemistry, genetics, and physiology. *Am. J. Physiol. Renal Physiol.* 281:F995, 2001.

directs the complex to the luminal membrane while SLC3A2 determines contraluminal location. Loss-of-function mutations of either subunit lead to a transport defect. The ability to form different combinations of heavy and light subunits in different tissues explains the large variability observed for the properties and specificity of amino acid transport.

Di- and tri-peptides are cotransported with H^+ and thus are energized through the proton electrochemical gradient across the luminal membrane (PEPT1, or SLC15A1). This H^+ electrochemical gradient is established by Na^+/H^+ exchange and is thus indirectly powered by Na^+/K^+-exchanging ATPase. The dipeptide transporter also accepts β-lactam antibiotics (aminopenicillins) and is important for absorption of orally administered antibiotics of this class.

CLINICAL CORRELATION 25.6

Neutral Amino Aciduria: Hartnup Disease

Hartnup disease (OMIM 234500) is an autosomal recessive defect of the Na^+-coupled transporter SCL6A19 that mediates active absorption of neutral amino acids from the lumen of the small intestine and the proximal tubule. Therefore, all patients have neutral amino aciduria. The disease was first described in the family that gave the disease its name as an association of neutral amino aciduria with a pellagra-type skin rash and episodes of cerebellar ataxia. The latter two symptoms are expressions of tryptophan deficiency in plasma and bacterial degradation products of tryptophan in the intestine, respectively. Pellagra-like features (p. 785), point to a decreased availability of tryptophan for conversion to nicotinamide. The clinical symptoms are variable and depend on the amount of protein in the diet. The variability indicates the existence of compensatory mechanisms for neutral amino acid absorption and subsequent studies revealed the existence of intestinal and renal transporters for di- or tripeptides, PEPT1 and PEPT2, respectively. PEPT1 (SLC15A1) is the major intestinal transporter for absorption of small peptide products of digestion.

Bröer, S. Amino acid transport across mammalian intestinal and renal epithelia. *Physiol. Rev.* 88: 249, 2008: and http://www.emedicine.com/derm/topic713.htm.

25.5 • DIGESTION AND ABSORPTION OF CARBOHYDRATES

Disaccharides and Polysaccharides Require Hydrolysis

Dietary carbohydrates provide a major portion of the daily caloric requirement. They consist of mono-, di-, and polysaccharides (Table 25.9). Major carbohydrates in Western diets are sucrose, starch, and lactose. Monosaccharides, such as glucose and fructose, are absorbed directly. Disaccharides require enzymes of the small intestinal surface for hydrolysis into monosaccharides, whereas polysaccharides depend on pancreatic amylase and intestinal surface enzymes for digestion (Figure 25.25).

Starch, a major nutrient, is a plant storage polysaccharide of over100 kDa. It is a mixture of linear chains of glucose molecules linked by α-1,4-glucosidic bonds (**amylose**) and of branched chains with branch points of α-1,6-glucosidic bonds (amylopectin). The ratio of branch points to 1,4-glucosidic bonds is about 1:20. Glycogen is the animal storage polysaccharide and similar in structure to amylopectin, except that the number of branch points is greater in glycogen.

Hydrated starch and glycogen are digested by the endosaccharidase ***α*-amylase** of saliva and pancreatic juice (Figure 25.26). Hydration of the polysaccharides occurs during heating and is essential for efficient digestion. Amylase is specific for internal α-1,4-glucosidic bonds; α-1,6 bonds are not attacked nor are α-1,4 bonds of glucose units that serve as branch points. Pancreatic amylase is secreted in large excess relative to starch intake and is more important for digestion than the salivary enzyme. The products are mainly the disaccharide **maltose**, the trisaccharide maltotriose, and the so-called α-limit dextrins that contain on average eight glucose units with one or more α-1,6-glucosidic bonds.

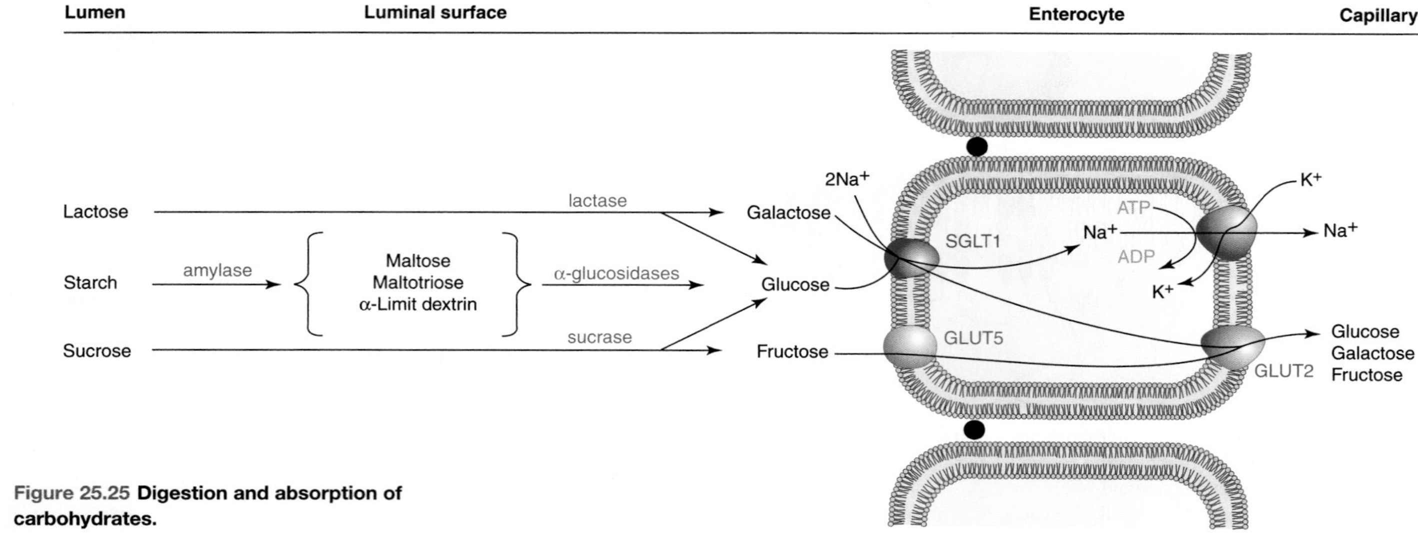

Figure 25.25 Digestion and absorption of carbohydrates.

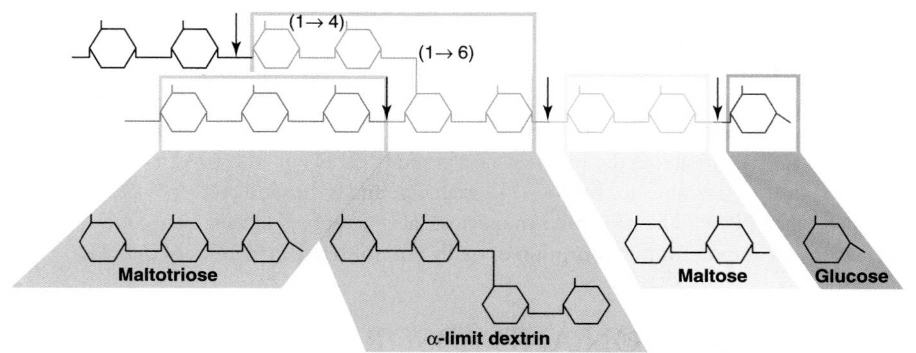

Figure 25.26 Digestion of amylopectin by salivary and pancreatic α-amylase.

TABLE 25.10 • Saccharidases of the Surface Membrane of Small Intestine

Enzyme	*Specificity*	*Natural Substrate*	*Product*
exo-1,4-α-Glucosidase (glucoamylase)	α-(1 → 4)Glucose	Amylose	Glucose
Oligo-1,6-glucosidase (isomaltase)	α-(1 → 6)Glucose	Isomaltose, α-dextrin	Glucose
α-Glucosidase (maltase)	α-(1 → 4)Glucose	Maltose, maltotriose	Glucose
Sucrose-α-glucosidase (sucrase)	α-Glucose	Sucrose	Glucose, fructose
α,α-Trehalase	α-(1 → 1)Glucose	Trehalose	Glucose
β-Glucosidase	β-Glucose	Glucosylceramide	Glucose, ceramide
β-Galactosidase (lactase)	β-Galactose	Lactose	Glucose, galactose

Final hydrolysis of di- and oligosaccharides to monosaccharides is carried out by enzymes on the luminal surface of small intestinal epithelial cells (Table 25.10). The surface oligosaccharidases are exoenzymes that cleave off one monosaccharide at a time from the nonreducing end. The capacity of α-glucosidases is normally much greater than needed for completion of digestion of starch. Similarly, there is usually excess capacity for **sucrose** (table sugar) hydrolysis. In contrast, **β-galactosidase** (**lactase**) for hydrolysis of **lactose**, the major milk carbohydrate, can be rate limiting in humans (Clin. Corr. 25.7).

Di-, oligo-, and polysaccharides not hydrolyzed by α-amylase and/or intestinal surface enzymes cannot be absorbed; therefore they reach the lower tract of the intestine, which from the lower ileum on contains bacteria. Bacteria can utilize many of these remaining carbohydrates because they possess many more types of saccharidases than humans. Monosaccharides released by bacterial enzymes are predominantly metabolized anaerobically by the bacteria themselves, resulting in degradation products such as short-chain fatty acids, lactate, hydrogen gas (H_2), methane (CH_4), and carbon dioxide (CO_2). In excess, these compounds can cause fluid secretion, increased intestinal motility, and cramps, either because of increased intraluminal osmotic pressure and distension of the gut or because of a direct irritant effect by the bacterial degradation products on the intestinal mucosa.

The flatulence that follows ingestion of leguminous seeds (beans, peas, and soy) is caused by oligosaccharides that cannot be hydrolyzed by human intestinal enzymes. The seeds contain modified sucrose to which one or more galactose moieties are linked. The glycosidic bonds of galactose are in the α configuration, which can be split only by bacterial enzymes. The simplest sugar of this family is raffinose. Trehalose is a disaccharide that occurs in young mushrooms and requires a special **disaccharidase, trehalase,** for digestion (see Table 25.9, p. 1048).

Monosaccharide Transporters

The major monosaccharides produced by digestion of di- and polysaccharide are D-glucose, D-galactose, and D-fructose. At least two monosaccharide transporters catalyze monosaccharide uptake from the intestinal lumen into the lining epithelial cells: (1) a Na^+-monosaccharide cotransporter (SGLT1) that mediates active uptake of D-glucose and D-galactose into cells and (2) a Na^+-independent, facilitative monosaccharide transporter with specificity for D-fructose (GLUT5) (p. 489). Another facilitative monosaccharide transporter (GLUT2) accepts all three monosaccharides and is present in the contraluminal plasma membrane. The physiological role of GLUT2 is to facilitate exit of monosaccharides from cells into the interstitial and capillary compartments, thereby completing the absorption process (Figure 25.25, p. 1050). GLUT2 is a member of the widely distributed GLUT family of transporters and present in tissues such as intestine, liver, and kidney that absorb or produce glucose and release it into blood. Properties of intestinal SGLT1 and of GLUT2 are compared in Table 25.11.

Disaccharidase Deficiency

Intestinal disaccharidase deficiencies are encountered relatively frequently in humans. Deficiency can be present in one or several enzymes for a variety of reasons (genetic defect, physiological decline with age, or the result of injuries to the mucosa). Lactase is the most commonly deficient enzyme. Absolute or relative deficiency is experienced as milk intolerance. Consequences of a lack of lactose hydrolysis in the upper small intestine are inability to absorb lactose, which then becomes available for bacterial fermentation in the lower small intestine. Bacterial fermentation produces gas (distension of gut and flatulence) and osmotically active solutes that draw water into the intestinal lumen (diarrhea). Lactose in yogurt has already been hydrolyzed during the fermentation process of making yogurt. Thus, individuals with lactase deficiency can usually tolerate yogurt better than unfermented dairy products. Lactase is commercially available to pretreat milk so that its lactose is hydrolyzed.

Swallow, D. M. Genetics of lactase persistence and lactose intolerance. *Annu. Rev. Genetics* 37:197, 2003; Solomons, N. W. Fermentation, fermented foods and lactose intolerance. *Eur. J. Clin. Nutrition* 56:S50, 2002; and Lactose intolerance: http://digestive.niddk.nih.gov/ddiseases/pubs/lactoseintolerance/and http://www.lactose.net.

25.6 • DIGESTION AND ABSORPTION OF LIPIDS

Digestion of Lipids Requires Overcoming Their Limited Water Solubility

An adult human ingests about 60–150 g of lipid per day. Triacylglycerols constitute more than 90% of this intake. The rest is made up of phospholipids, cholesterol, cholesteryl esters, and free fatty acids. In addition, 1–2 g of cholesterol and 7–22 g of phosphatidylcholine (lecithin) are secreted each day by the liver and reach the small intestine with bile.

The poor water solubility of lipids presents problems for digestion because substrates are not easily accessible to the digestive enzymes in the aqueous phase. In addition, most products of lipid digestion are themselves lipids with poor water solubility so they tend to form aggregates that hinder effective absorption. These problems are overcome by (1) generating/secreting surfactive molecules that increase the interfacial area between aqueous

TABLE 25.11 • Characteristics of Glucose Transporters in the Plasma Membranes of Enterocytes

Characteristic	*Luminal*	*Contraluminal*
Designation	SLC5A1 (SGLT1)	SLC2A2 (GLUT2)
Subunit H mass(kDa)	75	57
Effect of Na^+	Cotransport with Na^+	None
Good substrates	D-Glc, D-Gal, α-methyl-D-Glc	D-Glc, D-Gal, D-Man, 2-deoxy-D-Glc, D-Fru

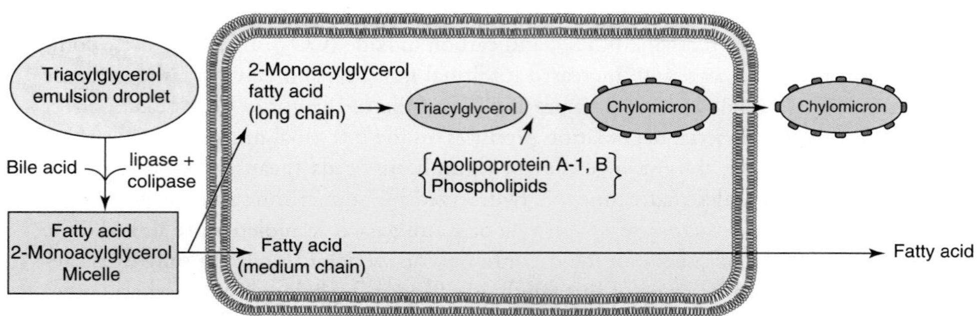

Figure 25.27 Digestion and absorption of lipids.

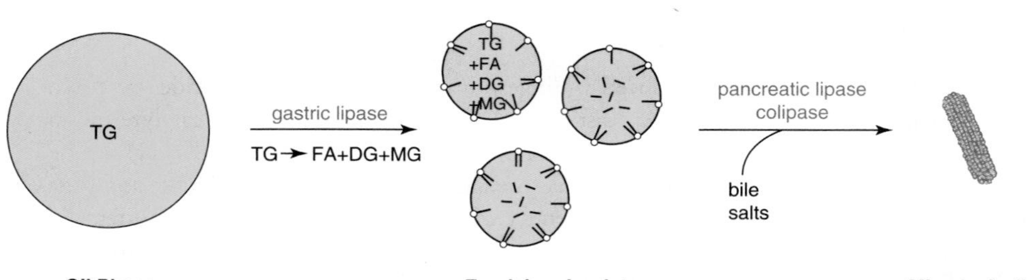

Figure 25.28 Changes in physical state during triacylglycerol digestion. Abbreviations: TG, triacylglycerol; DG, diacylglycerol; MG, monoacylglycerol; and FA, fatty acid.

and lipid phases and (2) "solubilization" of lipids with detergents. Thus changes in the physical state of lipids are intimately connected to chemical changes during digestion and absorption.

At least five phases of lipid digestion can be distinguished (Figure 25.27): (1) hydrolysis of triacylglycerols to free fatty acids and monoacylglycerols within the lumen of the gastrointestinal tract; (2) solubilization of lipids by detergents (bile acids) and transport from the intestinal lumen toward the surface of the lining epithelial cells; (3) uptake of free fatty acids and monoacylglycerols into the epithelial cell and synthesis to triacylglycerols; (4) packaging of newly synthesized triacylglycerols into special lipid-rich globules, called **chylomicrons**; and (5) exocytosis of chylomicrons from the intestinal epithelial cells into lymph.

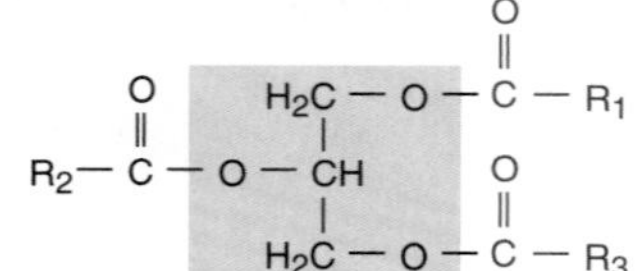

Triacylglycerol

HO—C—R_1
+
R_2—C—O—CH
H_2C—OH
H_2C—OH
+
HO—C—R_3

Fatty acids and monoacylglycerol

R = hydrocarbon chain

Figure 25.29 Mechanism of action of lipase.

Lipids Are Digested by Gastric and Pancreatic Lipases

Hydrolysis of triacylglycerols is initiated in the stomach by lingual and gastric **lipases**. Gastric digestion can account for up to 30% of total triacylglycerol hydrolysis. However, the rate of hydrolysis is slow because ingested triacylglycerols form lipid droplets with a limited interfacial area to which lipases can adsorb. Nevertheless, some lipase molecules adsorb and hydrolyze triacylglycerols into fatty acids and diacylglycerols (Figure 25.28), which converts a water-immiscible compound to products with both polar and nonpolar groups. Such surfactive products spontaneously adsorb to water–lipid interfaces and confer a hydrophilic surface to lipid droplets, thereby allowing an increase in interfacial area. At constant volume of the lipid phase, any increase in interfacial area causes dispersion of the lipid phase into smaller droplets (emulsification) and provides more sites for adsorption of lipase molecules. The dispersion of lipids into smaller droplets is also aided by the peristaltic and churning movements of the stomach. Gastric chyme that is released into the duodenum generally contains only **emulsion droplets** of less than two millimeter in diameter.

Pancreatic lipase is the major enzyme for triacylglycerol hydrolysis (Figure 25.29). It hydrolyzes esters in the α-position of glycerol and prefers long-chain fatty acids (longer than 10 carbon atoms). The products are free fatty acids and β-monoacylglycerols. Hydrolysis occurs at the water–lipid interface of emulsion droplets or bile acid **micelles** (p. 465). However, bile acids, which are present in the intestinal lumen, inhibit purified lipase, indi-

CLINICAL CORRELATION 25.8

Pharmacological Interventions to Prevent Fat Absorption and Obesity

Obesity is a major problem in modern society as food is generally very abundant. Therefore, interest in weight reduction is widespread. Two commercial products exploit the understanding of intestinal lipid absorption for this purpose. Olestra® is a commercial lipid produced by esterification of natural fatty acids with sucrose instead of glycerol (Figure *a*). With six to eight fatty acids covalently linked to sucrose the compound tastes like natural lipids, however, it cannot be hydrolyzed and is excreted unchanged.

Pancreatic lipase is the major enzyme that hydrolyzes dietary triacylglycerols to absorbable fatty acids and glycerol. Orlistat® (Figure *b*) is a non-hydrolyzable analog of a triacylglycerol and a powerful inhibitor of pancreatic lipase. Ingestion of Orlistat® slows down lipid digestion and hence absorption. Benefits derive from some lipid excretion, but also from release of a hormone (peptide YY) when lipids reach the terminal small intestine and colon. Peptide YY is postulated to increase the sensation of satiety and slow down the transit time of food through the gastrointestinal tract.

Olestra is a registered trademark of Proctor and Gamble, and Orlistat of Roche, Basel, Switzerland.

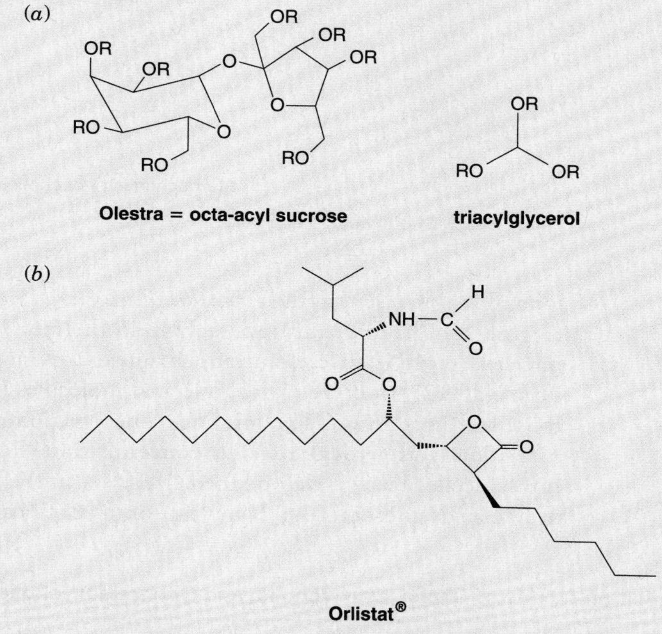

(*a*)

Olestra = octa-acyl sucrose **triacylglycerol**

(*b*)

Orlistat®

Thomson, A. B. R., et al. In Christophe, A. B., and DeVriese, S. (Eds), *Fat Digestion and Absorption.* Champaign, IL: AOCS Press, 2000, 383; Golay, A. In Christophe, A. B., and DeVriese, S. (Eds), *Fat Digestion and Absorption.* Champaign, IL: AOCS Press, 2000, 420; and http://en.wikipedia.org/wiki/Orlistat; and http://en.wikipedia.org/wiki/Olestra.

cating a more complex situation *in vivo.* Pancreatic juice contains a small protein (12 kDa) that binds to lipase and the micellar surface and prevents inhibition of lipase by bile acids. This protein has been termed **colipase.** It is secreted as procolipase and depends on tryptic removal of an NH_2-terminal decapeptide for activation. Clin. Corr. 25.8 describes two commercial strategies to reduce lipid absorption as a means to decrease obesity.

Pancreatic juice also contains an unspecific lipid esterase, which acts on cholesteryl esters, monoacylglycerols, or other lipid esters, such as esters of vitamin A with carboxylic acids. In contrast to triacylglycerol lipase, this esterase requires bile acids for activity.

Phospholipids are hydrolyzed by specific phospholipases. Pancreatic juice is especially rich in prophospholipase A_2 (Figure 25.30). Like other pancreatic proenzymes, it is activated by trypsin. Phospholipase A_2 requires bile acids for activity.

Bile Acid Micelles Solubilize Lipids during Digestion

Bile acids are biological detergents synthesized by liver and secreted as conjugates of glycine or taurine with the bile into the duodenum. At physiological pH, they are

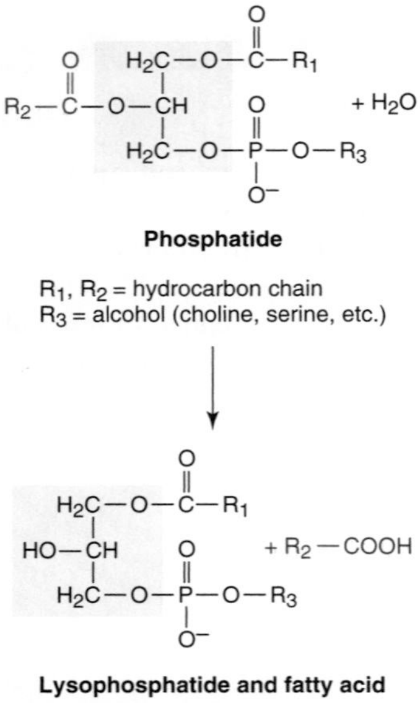

Phosphatide

R_1, R_2 = hydrocarbon chain
R_3 = alcohol (choline, serine, etc.)

Lysophosphatide and fatty acid

Figure 25.30 Mechanism of action of phospholipase A_2.

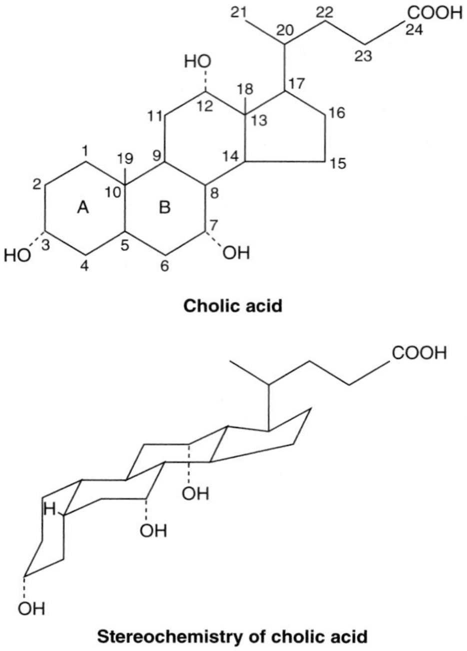

Cholic acid

Stereochemistry of cholic acid

Figure 25.31 Cholic acid, a bile acid.

ionized (anions) so that the terms *bile acids* and *bile salts* are often used interchangeably (Figure 25.31). Bile acids reversibly form thermodynamically stable aggregates, called micelles (p. 465), at concentrations above 2–5 mM and at pH values above the pK (Table 25.12). In other words, bile acid molecules in micelles are in equilibrium with those free in solution. The minimal concentration of a bile acid necessary for micelle formation is its **critical micellar concentration** (Figure 25.32). As equilibrium structure, micelles reach a well-defined size (≤ 4 nm) that depends on the concentrations of bile acids and other lipids but not on mechanical dispersion forces. The difference

TABLE 25.12 • Effect of Conjugation on Acidity of Cholic, Deoxycholic, and Chenodeoxycholic (Chenic) Acids

Bile Acid	*Ionized Group*	pK_a
Unconjugated bile acids	—COO^- of cholestanoic acid	≃ 5
Glycoconjugates	—COO^- of glycine	≃ 3.7
Tauroconjugates	—SO_3^- of taurine	≃ 1.5

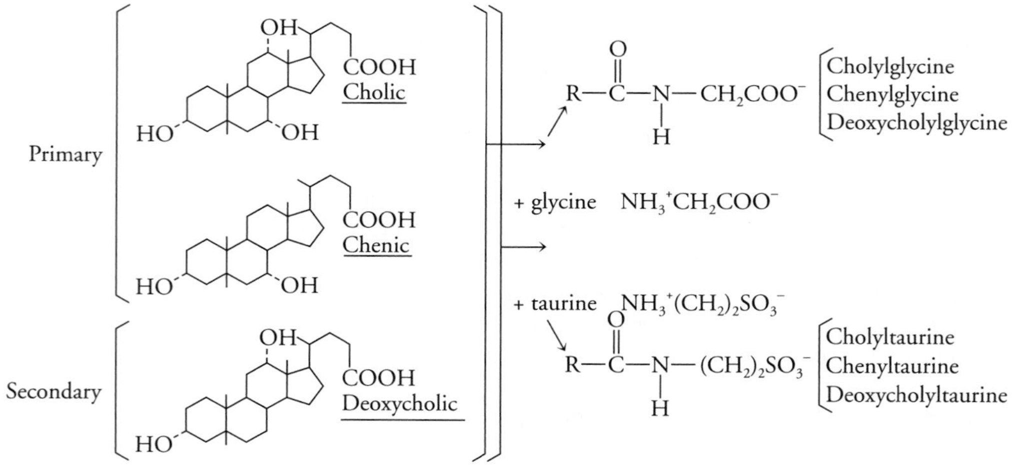

Primary: Cholic, Chenic

Secondary: Deoxycholic

+ glycine $NH_3^+CH_2COO^-$ → $R—C(=O)—NH—CH_2COO^-$: Cholylglycine, Chenylglycine, Deoxycholylglycine

+ taurine $NH_3^+(CH_2)_2SO_3^-$ → $R—C(=O)—NH—(CH_2)_2SO_3^-$: Cholyltaurine, Chenyltaurine, Deoxycholyltaurine

Source: Reproduced with permission from Hofmann, A. F. *Handbook of Physiology* 5:2508, 1968.

between micelles and emulsion droplets becomes obvious on ultracentrifugation of duodenal content: Emulsion droplets collect at the top as oily layer while micelles stay in the clear or slightly turbid subphase.

Major driving forces for micelle formation are the sequestering of apolar, hydrophobic groups away from water and the interaction of polar groups with water molecules. Bile acids have a fused ring system, which is hydrophobic on one side and hydrophilic on the other, and, in addition, a highly polar head group. The geometry of polar and apolar regions in bile acids is very different from that of ionized fatty acids (soap) or phospholipids, and hence their micelles have different geometries. For example, fatty acid or phospholipid micelles are spherical whereas pure bile acid micelles form "sandwich "structures" (Figure 25.33). Micelles, including the ones formed by bile acids, can solubilize other lipids, such as phospholipids and fatty acids, and form "mixed" micelles. Liver actually secretes bile acids together with phospholipids (mainly phosphatidyl choline) and cholesterol. The structure of these mixed micelles of bile acids and phospholipids is that of a rod, in which phospholipids are arranged radially along the length of the rod with the polar groups oriented outward and bile acids acting as wedge between the phospholipids head groups, also with their polar groups facing outward. The rods are capped by the polar groups of bile acids and phospholipids, and they become longer with increasing ratio of phospholipids to bile acids (Figure 25.34). Within the mixed phospholipid-bile acid micelles, other water-insoluble lipids, such as cholesterol, can be accommodated and thereby "solubilized." There are limits to the amounts of phospholipids and cholesterol that can be solubilized by physiological concentrations of bile acids; excess phospholipids form small unilamellar vesicles (20–60 nm in diameter), which can accommodate also some cholesterol. Excess cholesterol that cannot be solubilized by micelles or vesicles has a tendency to come out of solution by crystallization (Clin. Corr. 25.9).

During digestion of triacylglycerol, free fatty acids and monoacylglycerols are released at the surface of fat emulsion droplets and micelles. In contrast to triacylglycerols, which are water insoluble, free fatty acids and monoacylglycerols are slightly water soluble, and molecules at the surface equilibrate with those in solution and in bile acid micelles. Thus the products of triacylglycerol hydrolysis are continually transferred from emulsion droplets to the micelles (see Figure 25.28, p. 1052).

Micelles are the major vehicle for transferring lipids from the lumen to the mucosal surface where absorption occurs. Because the fluid layer near the cell surface is poorly mixed, the major mechanism for solute flux across this <u>unstirred fluid layer</u> is diffusion

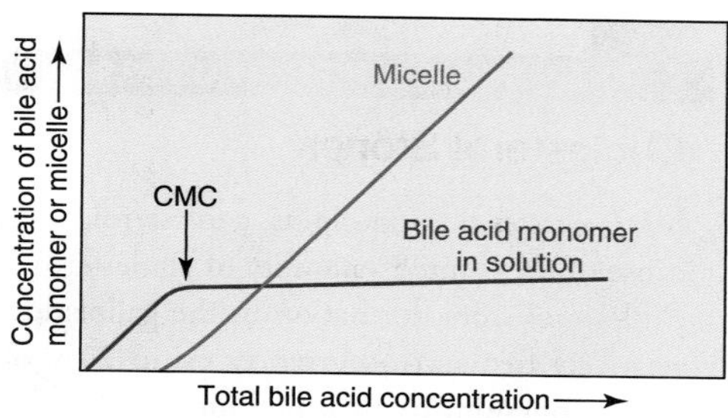

Figure 25.32 Solubility properties of bile acids in aqueous solutions. Abbreviation: CMC, critical micellar concentration.

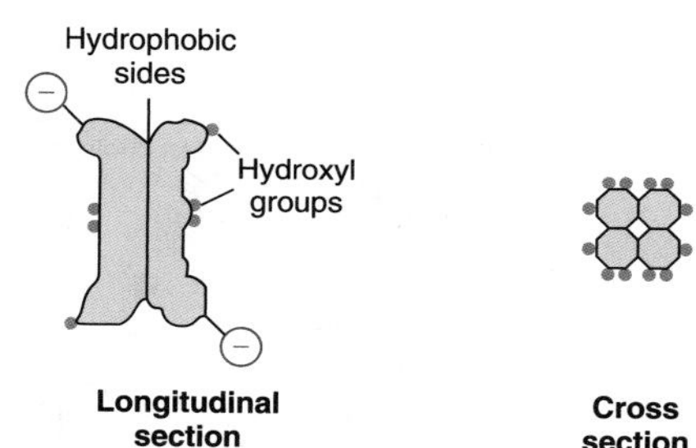

Figure 25.33 Diagrammatic representation of a Na^+ cholate micelle. Adapted from Small, D. M. *Biochim. Biophys. Acta* 176:178, 1969.

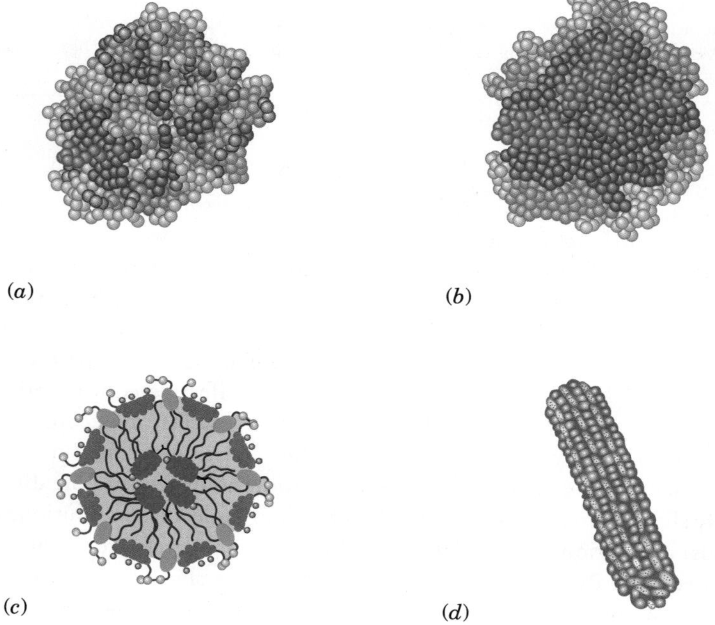

Figure 25.34 Proposed structure of mixed bile acid–phosphatidyl choline micelles. (*a*) View from outside. (*b*) View from interior of an equilibrium structure of mixed micelle from molecular dynamics simulations. (*c*) Diagram of cross section. (*d*) Diagram of a rod-shaped micelle. Note that phosphatidyl choline molecules are arranged radially with bile acids wedged in between and dominating top and bottom (radial shell model). Rod length grows with increasing phospholipid to bile acid ratio. Mixed micelles can accommodate fatty acids, monoacylglycerol, and cholesterol. Colors: Light blue = phospholipid head groups; dark blue = phospholipid tails; red = hydroxyl groups of bile acids; pink = ionic group of bile acid; purple = hydrophobic portion of bile acid; and green = cholesterol.
Adapted with permission from Marrink, S. J., and Mark, A. E. *Biochem.* 41:5375, 2002. Copyright (2002) American Chemical Society.

CLINICAL CORRELATION 25.9

Cholesterol Stones

Liver secretes phospholipids, cholesterol, and bile acids into bile. Because of the limited solubility of cholesterol, its secretion can lead to cholesterol stone formation in the gallbladder. Stone formation is a relatively frequent occurrence; up to 20% of North Americans develop stones during their lifetime. Cholesterol is practically insoluble in aqueous solutions. However, it can be incorporated into mixed phospholipid–bile acid micelles and thereby "solubilized" (see figure). Liver can produce bile that is supersaturated with respect to cholesterol. Excess cholesterol tends to come out of solution and crystallize. Supersaturated bile is considered lithogenic, that is, stone forming. Crystal formation usually occurs in the gallbladder, rather than the hepatic bile ducts, because contact time between bile and any crystallization nuclei is greater in the gallbladder. In addition, bile is concentrated here by absorption of electrolytes and water. The bile salts chenodeoxycholate (= chenate, Table 25.12) and its stereoisomer ursodeoxycholate (7-hydroxy group in β position) are available for oral use to dissolve gallstones. Their ingestion increases the bile acid pool and secretion rates by the liver, thereby reducing the cholesterol concentration of bile and allowing cholesterol from gall stones to be solubilized.

The tendency to secrete bile supersaturated with respect to cholesterol is inherited, occurs more frequently in females than in males, and is associated with obesity. Supersaturation also appears to be a function of the size and nature of bile acid pool and secretion rate.

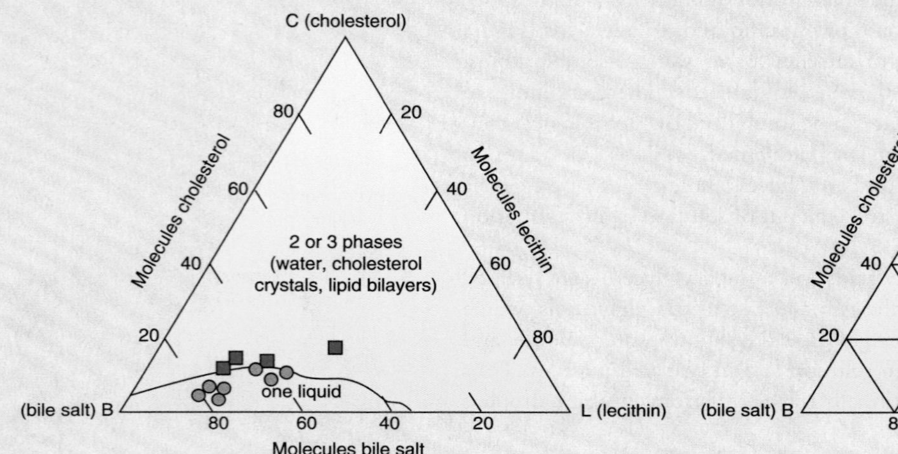

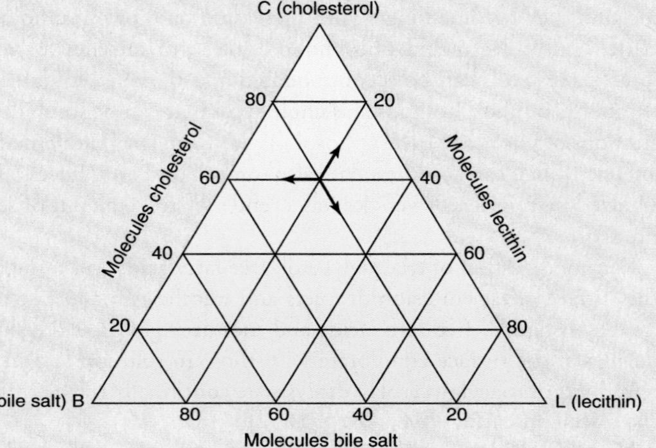

Diagram of the physical states of mixtures of 90% water and 10% lipid. The 10% lipid is made up of bile acids, phosphatidylcholine (lecithin), and cholesterol, and the triangle represents all possible ratios of the three lipid constituents. Each point within the triangle corresponds to a particular composition of the three components, which can be read off the graph as indicated; each point on one of the sides corresponds to a particular composition of just two components. The left triangle contains the composition of gallbladder bile samples from patients without stones (red o) and with cholesterol stones (blue □). Lithogenic bile has a composition that falls outside the "one liquid" area in the lower left corner.
Redrawn from Hofmann, A. F., and Small, D. M. *Annu. Rev. Med.* 18:362, 1967. Copyright 1967 by Annual Reviews, Inc.

Schoenfield, L. J., and Lachin, J. M. Chenodiol (chenodeoxycholic acid) for dissolution of gallstones: The National Cooperative Gallstone Study. A controlled trial of safety and efficacy. *Ann. Intern. Med.* 95:257, 1981; and Portincasa, P., Moschetta, A., van Erpecum, K. J., Calamita G. Margari, A., et al. Pathways of cholesterol crystallization in model bile and native bile. *Digestive Liver Disease* 35:118, 2003.

down the concentration gradient. The delivery rate of solutes at the cell surface by diffusion is proportional to the concentration difference between luminal bulk phase and cell surface. Diffusion through an unstirred layer becomes a problem for sparingly soluble or insoluble nutrients, in that reasonable concentration gradients and delivery rates cannot be achieved. Bile acid micelles overcome this problem for lipids by increasing their effective concentration in the unstirred layer. The increase in rate of transport is nearly proportional to the increase in effective concentration and can be 1000-fold over that of individually solubilized fatty acids, in accordance with the different water solubility of fatty acids as micelles and as individual molecules. This relationship between flux and effective concentration holds because the diffusion constant is only slightly smaller for micelles than for lipid molecules in solution. In absence of bile acids, absorption of triacylglycerols does not completely stop, but efficiency is drastically reduced. Residual absorption depends on the slight water solubility of free fatty acids and monoacylglycerols. Unabsorbed lipids reach the lower intestine where a small part can be metabolized by bacteria. The bulk of unabsorbed lipids, however, is excreted with the stool (steatorrhea).

Micelles also transport cholesterol and the lipid-soluble vitamins A, D, E, and K through the unstirred fluid layers. Bile acid secretion is essential for their absorption.

Most Absorbed Lipids Are Incorporated into Chylomicrons

Uptake of lipids by intestinal epithelial cells occurs by diffusion through the plasma membrane. In addition, long-chain fatty acid uptake is enhanced by a transporter (FATP4, or SLC27A4). Absorption is virtually complete for fatty acids and monoacylglycerols, which are slightly water soluble. It is less efficient for water-insoluble lipids. For example, only 30%–40% of the dietary cholesterol is usually absorbed (Clin. Corr. 25.10).

Within the absorbing epithelial cells, the fate of fatty acids depends on chain length (Figure 25.27, page 1052). Fatty acids of short and medium chain length ($\leq$ 10 carbon atoms) pass into portal blood without modification. Long-chain fatty acids ($>$ 12 carbon atoms) or their monoacylglycerols become bound to a cytosolic fatty acid-binding protein (intestinal FAB, FABP2) and are transported to the endoplasmic reticulum, where they are converted into triacylglycerols. Glycerol for this process is derived from the absorbed 2-monoacylglycerols and, to a minor degree, from glucose. Cholesterol is esterified by cholesterol acyltransferase. The newly synthesized triacylglycerols and cholesteryl esters form lipid globules into which phospholipids and **apolipoproteins** adsorb. The globules are called chylomicrons because they can grow up to several micrometers in diameter and leave the intestine through lymph vessels (chyle is milky lymph; the name is derived from the Greek *chylos,* which means "juice"). Chylomicrons are synthesized within the lumen of the endoplasmic reticulum from which they migrate through the Golgi and then in vesicles to the contraluminal membrane. They are released into the intercellular space by fusion of these vesicles with the plasma membrane. Interestingly, chylomicrons do not enter the capillary space and the portal vein; instead they travel through the intestinal lymph vessels (or lacteals) and the thoracic duct to the systemic venous system. The

CLINICAL CORRELATION 25.10

Cholesterol Absorption

Intestinal absorption of dietary and biliary cholesterol occurs via an incompletely understood transport system across the brush border membrane and inside enterocytes. The proteins scavenger receptor B-I (SCARB1) and CD36 appear to mediate brush border uptake of cholesterol, while the Niemann–Pick C1-like protein (NPC1L1) appears to serve to conduct cholesterol to intracellular sites. Knockout of NPC1L1 reduces cholesterol absorption by $>$ 70%. The drug ezetimibe (structure shown here) is approved by the Federal Drug Administration for adjunct therapy with statins to lower serum cholesterol levels. This drug is taken up in the intestine and resecreted after esterification with glucuronic acid in the liver. The glucuronated form is a potent inhibitor of intestinal cholesterol uptake.

A portion of cholesterol taken up by cells is actually returned to the lumen mediated by an ABC transporter (p. 494) consisting of two half-transporters (ABCG5 and ABCG8). The export of sterols by the ABC transporters is particularly important for rejecting plant sterols, and normally, plant sterols are not found in serum. Loss-of-function mutations in either of the half-transporters are associated with increased uptake and plasma levels of the plant sterol sitosterol (phytosterolemia or sitosterolemia).

Structure of ezetimibe [(3*R*,4*S*)-1-(4-fluorophenyl)-3-((3*S*)-3-(4-fluorophenyl)-3-hydroxypropyl)-4-(4-hydroxyphenyl)-2-azetidinone].

Ezetimibe is sold in the U.S.A. as prescription drug ZETIA®, which is a registered trademark of MSP Singapore Company, LLC, and marketed by Merck/Schering-Plough Pharmaceuticals.

Wang, D. Q. Regulation of intestinal cholesterol absorption. *Annu. Rev. Physiol.* 69:221, 2007; and Hui, D. Y, Labonté, E. D., and Howles, P. N. Development and physiological regulation of intestinal lipid absorption. III. Intestinal transporters and cholesterol absorption. *Am. J. Physiol. Gastrointest. Liver Physiol.* 294:G839, 2008.

CLINICAL CORRELATION 25.11

A-β-Lipoproteinemia

Apolipoprotein B (apoB) is a key component of lipoproteins: A 48-kDa splice variant is used by intestinal epithelial cells in the assembly of chylomicrons, while a 100-kDa variant is important for the assembly of very low density lipoproteins (VLDLs) by the liver. ApoB serves as acceptor for newly synthesized triglycerides that are transferred by microsomal triglyceride transfer protein. Mutations in the gene for this latter enzyme are the basis for a-β-lipoproteinemia characterized by the absence of liver and intestinal lipoproteins from plasma. Serum cholesterol is extremely low in this condition. A-β-lipoproteinemia is associated with severe malabsorption of triacylglycerol and lipid-soluble vitamins (especially tocopherol and vitamin E) and accumulation of apoB in enterocytes and hepatocytes.

Fisher, E. A., and Ginsberg, H. N. Complexity in the secretory pathway: The assembly of apolipoprotein B-containing lipoproteins. *J. Biol. Chem.* 277:17377, 2002; and Hussain, M. M., Iqbal, J., Anwar, K., Rava, P., and Dai, K. Microsomal triglyceride transfer protein: a multifunctional protein. *Frontiers in Bioscience* 8: s500, 2003.

intestinal apolipoproteins are designated A-1 and B48 (Clin. Corr. 25.11); they are different from those of the liver with similar function (p. 724).

While dietary medium-chain fatty acids reach the liver directly with the portal blood, the long-chain fatty acids first reach adipose tissue and muscle via the systemic circulation before coming into contact with the liver. Fat and muscle cells take up large amounts of dietary lipids for storage or metabolism. A bypass of the liver may have evolved to protect this organ from lipid overload after a meal.

The differential handling of medium- and long-chain fatty acids by intestinal cells can be exploited to provide the liver with high-caloric nutrients in the form of fatty acids. Short- and medium-chain fatty acids smell and taste rancid and are not very palatable; however, triacylglycerols that contain these fatty acids are quite palatable and can be used as part of the diet. Short-chain fatty acids are produced physiologically from residual carbohydrates, particularly in the colon, by bacteria.

25.7 • BILE ACID METABOLISM

Bile Acid Chemistry and Synthesis

Bile acids are synthesized in liver cells (hepatocytes) from cholesterol, secreted into bile together with phospholipids, and modified by bacterial enzymes in the intestinal lumen. Primary bile acids synthesized by the liver are **cholic** and **chenodeoxycholic** (chenic) acid. **Secondary bile acids** are derived from primary bile acids by bacterial reduction in position 7 of the ring structure, resulting in **deoxycholate** and **lithocholate**, respectively (see Figure 18.42, p. 728 for structures).

Primary and secondary bile acids are reabsorbed by the intestine (lower ileum) into portal blood, taken up by liver cells, and then resecreted into bile. In liver cells, primary as well as secondary bile acids are linked to either glycine or taurine via an isopeptide bond. These glyco- and tauroconjugates constitute the forms that are secreted into bile. Conjugation is important for converting the mildly acidic carboxyl group to more polar and acidic ones, which is expressed as lower pK value and implies ionization over a wider pH range (Table 25.12). The conjugation is partially reversed within the intestinal lumen by hydrolysis of the isopeptide bond.

Bile Acid Transport

The total amount of conjugated and unconjugated bile acids secreted per day is 20–30 g for an adult. However, the body maintains a pool of only 3–5 g. A small pool is advantageous because bile acids become toxic at high concentrations due to their detergent properties, for example, through their ability to lyse cells. Therefore, to achieve the observed secretion rates, bile acids are reabsorbed by ileal enterocytes, recirculated to the liver and resecreted 4 to 10 times per day. The secretion and reuptake is referred to as the enterohepatic circulation (Figure 25.35). Reabsorption of bile acids is quite efficient since only about 0.8 g of bile acids each day is excreted with the feces. Serum levels of bile acids normally vary with the rate of reabsorption and therefore are highest during a meal. Cholate, deoxycholate, chenodeoxycholate, and their conjugates continuously participate in the enterohepatic circulation. In contrast, most of the lithocholic acid produced by bacterial enzymes is sulfated during the next passage through the liver. The sulfate ester of lithocholic acid is not reabsorbed and therefore excreted in the feces. The transporters mediating enterohepatic circulation of bile acids are shown in Figure 25.36. Ileal absorption of bile acids is mediated by secondary active transport via a luminal Na^+ bile acid cotransport system (apical sodium-dependent bile acid transporter, ASBT, or SLC10A2) with a stoichiometry of 2:1 for Na^+ to bile acid. Bile acids are moved from ileal enterocytes into blood predominantly by exchange for another anion via a specific anion exchanger consisting of two different gene products (OSTα–OSTβ). Uptake of bile acids from blood by liver cells is predominantly by secondary active Na^+-bile acid cotransport (Na^+ taurocholate co-transporting polypeptide, NTCP, or SLC10A1). Unconjugated bile acids can also be taken up from blood by a family of Na^+-independent, organic anion transporters (SLC01A2, SLC01B1, SLC01B3, formerly SLC21 family). In contrast,

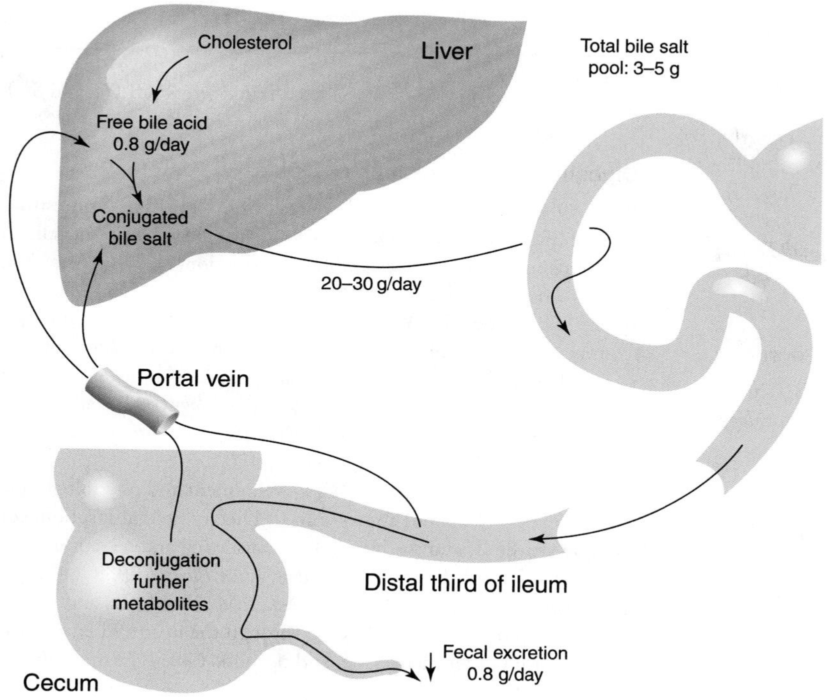

Figure 25.35 Enterohepatic circulation of bile acids. Redrawn from Clark, M. L., and Harries, J. T. In McColl. I., and Sladen, G. E. (Eds.), *Intestinal Absorption in Man.* New York: Academic Press, 1975, 195.

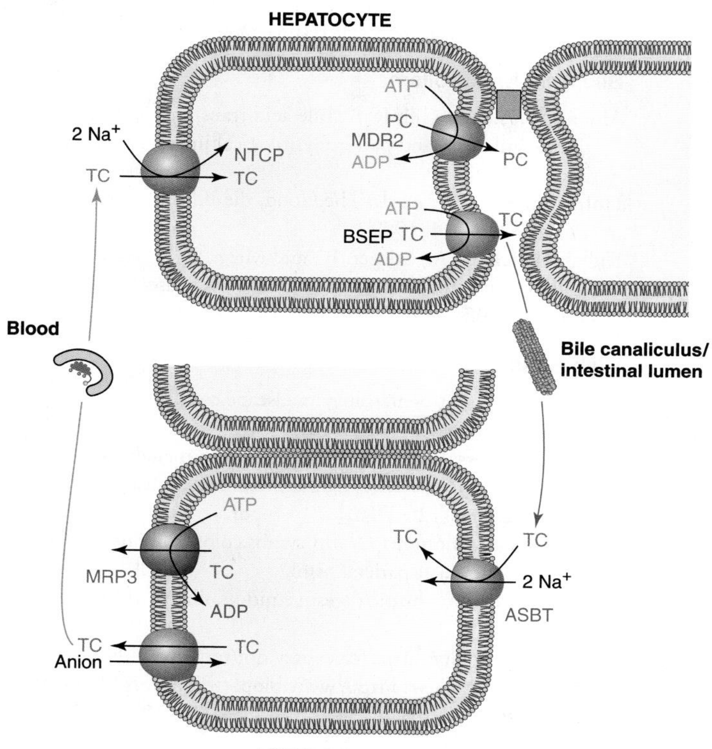

Figure 25.36 Transporters for taurocholate (TC) and phosphatidyl choline (PC) during enterohepatic circulation.

secretion of bile acids by liver cells across the canalicular plasma membrane into bile is by primary active transport (**Bile Salt Export Pump, BSEP**, or ABCB11, another member of the ABC transporters; p. 494). Phospholipids, which are secreted simultaneously with bile acids, are transported by the pump MDR2 (ABCB4). To achieve reasonable total bile acid concentrations within cells and in plasma while maintaining low free concentrations to prevent detergent action, bile acids are bound to specific binding proteins in the cytosol of hepatocytes and enterocytes. Albumin functions as such a binding protein for bile acids in plasma.

Bibliography

General

Hediger, M. A., Romero, M. F., Peng, J. B., Rolfs, A., et al. The ABCs of solute carriers: Physiological, pathological and therapeutic implications of human membrane transport proteins. Introduction. *Pflugers Arch.* 447:465. 2004.

Johnson, L. R. (Ed.-in-chief). *Physiology of the Gastrointestinal Tract,* vols. 1 and 2. 2d and 3d eds. New York: Raven, 1987 and 1994; 4th ed. Amsterdam and Boston: Elsevier, 2006.

Epithelial Electrolyte Transport and Gastric Acid Secretion

Dorwart, M. R., Shcheynikov, N., Yang, D., and Muallem S. The solute carrier 26 family of proteins in epithelial ion transport. *Physiology* (Bethesda) 23:104, 2008.

Geibel, J, P. Secretion and absorption by colonic crypts. *Annu. Rev. Physiol.* 67:471, 2005.

Rao, M.C. Oral rehydration therapy: new explanations for an old remedy. *Annu. Rev. Physiol.*66:385, 2004.

Van Itallie, C. M., and Anderson, J. M. Claudins and epithelial paracellular transport. *Annu. Rev. Physiol.* 68:403, 2006.

Yao, X., and Forte, J. G. Cell biology of acid secretion by the parietal cell. *Annu. Rev. Physiol.* 65:103, 2003.

Zachos, N, C., Tse, M., and Donowitz, M. Molecular physiology of intestinal Na^+/H^+ exchange. *Annu. Rev. Physiol.* 67:411, 2005.

Digestion and Absorption of Proteins

Bröer, S. Amino acid transport across mammalian intestinal and renal epithelia. *Physiol. Rev.* 88:249, 2008.

Bröer, S. Apical transporters for neutral amino acids: physiology and pathophysiology. *Physiology* (Bethesda) 23:95, 2008.

Richter, C., Tanaka, T., and Yada, R. Y. Mechanism of activation of the gastric aspartic proteinases: Pepsinogen, progastricsin and prochymosin. *Biochem. J.* 335:481, 1998.

Whitcomb, D. C., and Lowe, M. E. Human pancreatic digestive enzymes. *Dig Dis. Sci.* 52:1, 2007.

Digestion and Absorption of Carbohydrates

Cristofaro, E., Mottu, F., and Wuhrmann, J. J. Involvement of the raffinose family of oligosaccharides in flatulence. In Sipple, H. L., and McNutt, K. W. (Eds.). *Sugars in nutrition.* New York: Academic Press, 1974, 314.

Robayo-Torres, C. C., Quezada-Calvillo, R., and Nichols B. L. Disaccharide digestion: clinical and molecular aspects. *Clin. Gastroenterol. Hepatol.* 4:276, 2006.

Van-Loo, J., Cummings, J., Delzenne, N., Englyst, H., et al. Functional food properties of non-digestible oligosaccharides: a consensus report from the ENDO project. *Brit. J. Nutr.* 81:121, 1999.

Wong, J. M., and Jenkins D. J. Carbohydrate digestibility and metabolic effects. *J. Nutr.* 137:2539S, 2007.

Wood, I. S., and Trayhurn, P. Glucose transporters (GLUT and SGLT): expanded families of sugar transport proteins. *Brit. J. Nutr.* 89:3, 2003.

Digestion and Absorption of Lipids

Black, D. D. Development and physiological regulation of intestinal lipid absorption. I. Development of intestinal lipid absorption: cellular events in chylomicron assembly and secretion. *Am. J. Physiol. Gastrointest. Liver Physiol.* 293:G519, 2007.

Bonen, A., Chabowski, A., Luiken, J. J., and Glatz J. F. Is membrane transport of FFA mediated by lipid, protein, or both? Mechanisms and regulation of protein-mediated cellular fatty acid uptake: molecular, biochemical, and physiological evidence. *Physiology* (Bethesda) 22:15. 2007.

Christophe, A. B., and DeVriese, S. (Eds.). *Fat Digestion and Absorption.* Champaign, IL: AOCS Press, 2000.

Mansbach, C. M., and Gorelick F. Development and physiological regulation of intestinal lipid absorption. II. Dietary lipid absorption, complex lipid synthesis, and the intracellular packaging and secretion of chylomicrons. *Am. J. Physiol. Gastrointest. Liver Physiol.* 293:G645, 2007.

Thomson, A. B. R., Schoeller, C., Keelan, M., Smith, L., and Clandinin, M.T. Lipid Absorption: Passing through the unstirred layers near the brush border membrane, and beyond. *Can. J. Physiol. Pharmacol.* 71:531, 1993.

Topping, D. L., and Clifton, P. M. Short-chain fatty acids and human colonic function: Roles of resistant starch and nonstarch polysaccharides. *Physiological Reviews* 81:1031, 2001.

Bile Acid Metabolism

Alrefai, W. A., and Gill, R. K. Bile acid transporters: Structure, function, regulation, and pathophysiological implications. *Pharm. Res.* 24:1803, 2007.

Hofmann, A. F. Bile Acids: The Good, the Bad, and the Ugly. *News Physiol Sci.* 14:24. 1999.

Kullak-Ublick, G. A., Stieger, B., and Meier. P. J. Enterohepatic bile salt transporters in normal physiology and liver disease. *Gastroenterology* 126:322, 2004.

Web Sites

General: http://gastroenterology.medscape.com.

General: http://www.nutrition.gov/.

ABC transporters: http://nutrigene.4t.com/humanabc.htm.

Aminoaciduria: http://cnserver0.nkf.med.ualberta.ca/cn/Schrier/Volume2/chapt12/ADK2_12_4-6.pdf.

Gastric acid secretion: http://arbl.cvmbs.colostate.edu/hbooks/pathphys/digestion/stomach/parietal.html.

Gluten enteropathy: http://digestive.niddk.nih.gov/ddiseases/pubs/celiac/index.htm.

Peptidase specificity: http://ca.expasy.org/tools/peptidecutter/.

Superfamily of carriers: http://www.bioparadigms.org/slc/menu.asp.

Key Terms

gastrin
serotonin
cholecystokinin
guanylin
secretin
vasoactive intestinal peptide
K^+/H^+-exchanging ATPase
secondary active transport
facilitative glucose transporter
sodium-monosaccharide cotransporter
primary active transport
endopeptidase
pepsinogen, pepsin
zymogen
chymotrypsinogen, chymotrypsin
enteropeptidase
trypsinogen, trypsin
dipeptidase
amylase
amylopectin
amylose
sucrose
lactase
lactose
maltose
disaccharide
facilitative fructose transporter

sucrase/isomaltase
chylomicron
emulsion
lipase
bile acid, bile salt
colipase
micelle, mixed micelle
taurine
conjugated bile acid
critical micellar concentration
unstirred fluid layer
apolipoprotein
cholic acid
deoxycholic acid
lithocholic acid
primary bile acid
secondary bile acid
bile salt export pump

Questions • CAROL N. ANGSTADT

Multiple-Choice Questions

1. Active forms of most enzymes that digest food may normally be found in all of the following except
   A. in soluble form in the lumen of the stomach.
   B. in the saliva.
   C. attached to the luminal surface of the plasma membrane of intestinal epithelial cells.
   D. dissolved in the cytoplasm of intestinal epithelial cells.
   E. in zymogen granules of pancreatic exocrine cells.

2. Histamine is a potent secretagogue of
   A. amylase by the salivary glands.
   B. HCl by the stomach.
   C. gastrin by the stomach.
   D. hydrolytic enzymes by the pancreas.
   E. $NaHCO_3$ by the pancreas.

3. The contraluminal plasma membranes of small intestinal epithelial cells contain
   A. aminopeptidases.
   B. Na^+/K^+- exchanging ATPase.
   C. disaccharidases.
   D. GLUT5.
   E. Na^+-monosaccharide transport (SGLT1).

4. The enterohepatic circulation of bile acids
   A. necessitates participation of secondary active Na^+-bile acid cotransport.
   B. results in the reabsorption of only the primary bile acids.
   C. maintains the body pool of bile acids at 20–30 g.
   D. maintains serum levels of bile acids at a constant value.
   E. uses the same transport mechanism to secrete bile acids from the liver to bile as from the enterocyte to the blood.

5. Epithelial cells of the lower ileum express a Cl^-/HCO_3^- exchange coded for by the DRA gene. These cells
   A. mediate an electrogenic exchange of 2 luminal Na^+ for 1 cytosolic H^+.
   B. absorb Cl^- into the cell in exchange for HCO_3^- moving into the lumen.
   C. prevent a metabolic acidosis due to loss of HCl.
   D. mediate Na^+ movement out of the cell as Cl^- moves into the cell.
   E. mediate Na^+ movement into the cell via a Na^+ channel.

6. Peptidases involved in protein digestion
   A. are all endopeptidases.
   B. must have a neutral pH.
   C. are synthesized in the stomach and pancreas as proenzymes.
   D. each have a different activator.
   E. all have an essential serine at the active site.

Questions 7 and 8: A young woman finds that every time she eats dairy products she feels highly uncomfortable. Her gut becomes distended, and she has gas and, frequently, diarrhea. A friend suggested that she try yogurt to get calcium, and she is able to tolerate that. These symptoms do not appear when she eats food other than dairy products. Like many adults, she is deficient in an enzyme required for carbohydrate digestion.

7. The most likely enzyme in which she is deficient is
   A. α-amylase.
   B. β-galactosidase (lactase).
   C. α-glucosidase (maltase).
   D. sucrose-α-glucosidase (sucrase).
   E. α,α-trehalase.

8. Monosaccharides are absorbed from the intestine
   A. by a Na^+-dependent cotransporter for glucose and galactose.
   B. by a Na^+-independent facilitated transport for fructose.
   C. by a Na^+-independent transporter (GLUT2) across the contraluminal membrane.
   D. against a concentration gradient if the transporter is Na^+ dependent.
   E. all of the above.

Questions 9 and 10: A woman comes to the emergency room with severe abdominal pain in the right upper quadrant as well as severe pain in her back. The pain began several hours after she consumed a meal of fried chicken and cheese-coated french fries. The symptoms indicated gallstones, and this was confirmed by ultrasound. Although surgery might be necessary in the future, conservative treatment was tried first. She was instructed to limit fried foods and high-fat dairy products. She was also given chenodeoxycholate to take orally to try to dissolve the gallstones.

9. Cholesterol stones
   A. usually form during passage of bile through the hepatic bile duct.
   B. occur when the mixed phospholipid-bile acid micelles are very high in phospholipid.
   C. can be dissolved by excess bile acid micelles.
   D. rarely occur because cholesterol is not a normal part of bile.
   E. are a necessary part of lipid digestion.

10. In the metabolism of bile acids,
    A. the liver synthesizes cholic and deoxycholic acids, which are primary bile acids.
    B. secondary bile acids are produced by conjugation of primary acids to glycine or taurine.
    C. 7-dehydroxylation of bile acids by intestinal bacteria produces secondary bile acids, which have similar detergent and physiological properties as primary bile acids.
    D. daily bile acid secretion by the liver is approximately equal to daily bile acid synthesis.
    E. conjugation reduces the polarity of bile acids, enhancing interaction with lipids.

Questions 11 and 12: Hartnup disease is a genetic defect in an amino acid transport system. The specific defect is in the neutral amino acid transporter in both intestinal and renal epithelial cells. Clinical symptoms of the disease result from deficiencies of essential amino acids and nicotinamide (because of a deficiency specifically of tryptophan).

11. In addition to tryptophan, which of the following amino acids is likely to be deficient in Hartnup disease?
A. Aspartate
B. Leucine
C. Lysine
D. Proline
E. All of the above

12. Hartnup disease patients are able to get some of the benefit of the protein they consume because
A. only the neutral amino acid carrier is defective.
B. di- and tripeptides from protein digestion are absorbed by a different carrier (PepT1).
C. their endo- and exopeptidases are normal.
D. all of the above.
E. none of the above.

Problems

13. Using known endo- and exopeptidases, suggest a pathway for the complete degradation of the following peptide:

His-Ser-Lys-Ala-Trp-Ile-Asp-Cys-Pro-Arg-His-His-Ala

14. Certain tissues effect Cl^- secretion via a Cl^- channel (CFTR-cystic fibrosis transmembrane regulatory protein). Cholera toxin abnormally opens the channel leading to a loss of NaCl. A relatively inexpensive way to treat cholera is to administer fluids high in Na^+ and glucose. Why is this effective?

Answers

1. **E** Zymogen granules contain inactive proenzymes or zymogens, which are not activated until after release from the cell (amylase from the pancreas and salivary glands is an exception).
2. **B** Its binding to H_2 receptors of the stomach causes HCl secretion. A: Acetylcholine is the secretagogue. C: Gastrin itself is a secretagogue. D: Acetylcholine and cholecystokinin are the secretagogues. E. Secretin is the secretagogue.
3. **B** Only the contraluminal surface contains the Na^+, K^+-ATPase. All other activities are associated with the luminal surface.
4. **A** There are two different transporters from lumen to enterocyte and blood to hepatocyte. B: Both primary and secondary bile acids (except lithocholic acid) are reabsorbed. C: The body pool is very low (3–5 g); high concentrations are toxic. D: Levels fluctuate and are highest during a meal. E: Secretion from liver to bile is a primary active transport.
5. **B** The proton gradient generated by an electrically neutral Na^+/H^+ exchange drives this. A: The Na^+/H^+ exchange via the expressed NHE3 transporter is electrically neutral. C: Constant loss of HCl leads to a metabolic alkalosis. D: The direction of Na^+ movement is into the cell with subsequent removal by the Na^+/K^+-ATPase. E: This occurs in the large intestine, not here.
6. **C** Pepsinogen in the stomach and trypsinogen plus others in the pancreas are proenzymes. A: Carboxy- and aminopeptidases are exopeptidases. B: Pepsin is active at pH 2 and lower. D: Trypsin activates chymotrypsinogen, proelastase, and procarboxypeptidase. E: Trypsin, chymotrypsin, and elastase do, but carboxypeptidase has a different mechanism.
7. **B** Dairy products contain lactose. Undigested lactose is fermented and the products produce the symptoms. Yogurt fermentation hydrolyzes lactose. A, C, and D: Deficiency of any of these would cause problems with most carbohydrates. E: Trehalose is found in mushrooms, not dairy products.
8. **E** All of these play an important role in absorbing the monosaccharides from digestion. D: This is especially important for the uptake of most of dietary glucose.
9. **C** Stones occur when bile is supersaturated (> 1:1 ratio of cholesterol/phospholipid). The ingested bile salts increase the bile salt pool and hence bile salt and phospholipid secretion; this lowers the cholesterol/phospholipid ratio. A: Stones usually form in the gallbladder. B: The problem is too little phospholipid relative to cholesterol. D: Actually, stones are relatively common. Cholesterol is a normal component of bile. E: Bile salts are necessary for lipid digestion but stones are not.
10. **C** Primary bile acids (cholic and chenodeoxycholic acids) are synthesized in liver. In intestine they may be reduced by bacteria to form the secondary bile acids, deoxycholate and lithocholate. D: Only a small fraction of bile acids escapes reuptake; this must be replaced by synthesis. Both primary and secondary bile acids are reabsorbed and recirculated (enterohepatic circulation). E: Both are conjugated to glycine or taurine, increasing their polarity.
11. **B** Trp shares a carrier with tyr, phe, val, leu, ile, and met. A: Asp is acidic. C: Lysine is basic. D: Proline is an imino acid. All of these use separate carriers.
12. **D** The body has at least seven transporters for the various classes of amino acids (see question 11). It also has a carrier for di- and tripeptides. C: Since we cannot absorb intact proteins, the presence of the endo- and exopeptidases is essential to generate smaller fragments.
13. Trypsin cleavage gives (a)His-Ser-Lys + (b)Ala-Trp-Ile-Met-Cys-Gly-Pro-Arg + (c) His-His-Ala. Further degradation of (a) is accomplished by elastase and dipeptidase. Further degradation of (b) would start with chymotrypsin and use dipeptidases, tripeptidase, and carboxypeptidase B. To degrade (c) carboxypeptidase A and dipeptidase would be enough. The point is that several peptidases with varying specificities are required.
14. Cholera toxin activates adenylate cyclase leading to elevated cAMP levels and active protein kinase A. In addition to opening the CFTR Cl^{-1} channel, the Na^+/H^+ exchanger is also inhibited and there is a gross NaCl excretion. The Na^+-glucose cotransport is not affected. The presence of glucose allows uptake of Na^+ to replenish body NaCl.

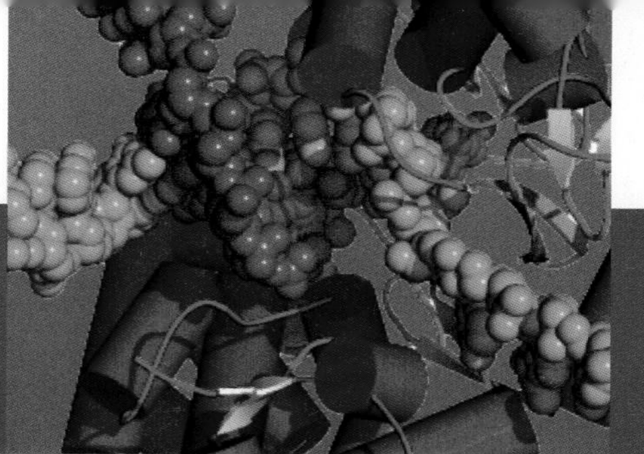

26

Vitamins and Minerals: Requirements and Function

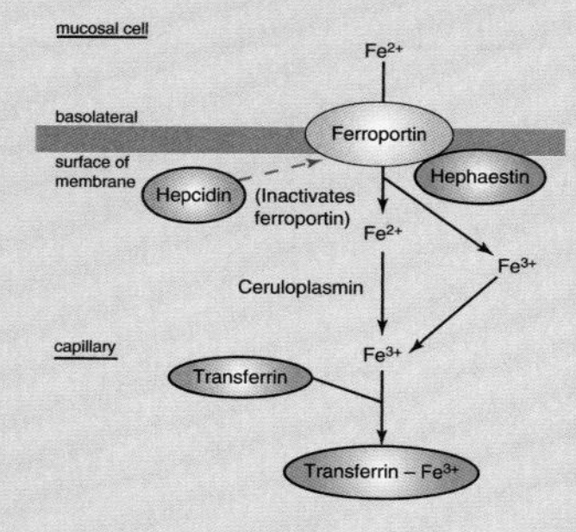

Stephen G. Chaney

Professor, School of Medicine, University of North Carolina at Chapel Hill

CLINICAL CORRELATIONS

Key Concepts

- Dietary Reference Intakes (DRIs) are quantitative estimates of nutrient intake to be used for planning and assessing diets for healthy people. There is often a fine line between micronutrient sufficiency and toxicity.
- Vitamin A can occur in several forms and can act as an antioxidant, a glycosyl donor, a hormone, or an essential component of the visual cycle.
- Vitamin D regulates cell growth and differentiation, important metabolic processes, and immune function in addition to its role in calcium homeostasis.
- Vitamin E occurs in multiple forms and affects gene expression through cellular redox signaling pathways in addition to its role as an antioxidant.

- Vitamin K is essential for the biological activity of a number of calcium-dependent enzymes, especially those involved in blood clotting and bone metabolism.
- The biological activity and deficiency symptoms of B vitamins are best understood on the basis of their conversion to coenzymes required in key metabolic processes.
- Vitamin C is an antioxidant, is a cofactor for some mixed function oxidases, and aids in iron absorption.
- Both dietary calcium and bone serve as sources of calcium to maintain serum calcium levels required for the activity of some enzymes, blood coagulation, muscle contractility, and nerve function.
- Magnesium is required for neuromuscular transmission and for many enzyme activities, especially those utilizing an ATP–Mg^{++} complex.
- Iron deficiency causes anemia and decreased immunocompetence.

26.1 • INTRODUCTION

Micronutrients play a vital role in human metabolism, as they are involved in almost every biochemical reaction pathway. However, the science of nutrition is concerned not only with the biochemistry of the nutrients but with also whether they are present in adequate amounts in the diet. The American diet is undoubtedly the best it has ever been. Our current food supply provides an abundant variety of foods all year long, and deficiency diseases have become medical curiosities. However, our diet is far from optimal. The old adage is that we can get everything we need from a balanced diet. Unfortunately, many Americans do not consume a balanced diet. Foods of high caloric density and low nutrient density (often referred to as empty calories or junk food) are abundant and popular, and our nutritional status suffers because of these food choices. Obviously then, neither alarm nor complacency is justified. We need to know how to evaluate the adequacy of our diet.

26.2 • ASSESSMENT OF MALNUTRITION

There are three increasingly stringent criteria for measuring **malnutrition**.

1. Dietary intake studies, which are usually based on a 24-h recall, are the least stringent. Twenty-four-h recalls tend to overestimate the number of people with deficient diets. In addition, poor dietary intake alone is usually not a problem in this country unless the situation is compounded by increased need.

2. Biochemical assays, either direct or indirect, are a more useful indicator of nutritional status. At their best, they indicate **subclinical nutritional deficiencies**, which can be treated before deficiency diseases develop. However, all biochemical assays are not equally valid, an unfortunate fact that is not sufficiently recognized. Changes in biochemical parameters due to stress need to be interpreted with caution. The distribution of many nutrients in the body changes dramatically in a stress situation such as illness, injury, and pregnancy. A drop in level of a nutrient in one tissue compartment (usually blood) need not signal a deficiency or an increased requirement. It could simply reflect a normal metabolic adjustment to stress.

3. The most stringent criterion is the appearance of **clinical symptoms**. However, it is desirable to intervene long before symptoms became apparent.

The question remains: When should dietary surveys or biochemical assays be interpreted to indicate the need for nutritional intervention? Dietary surveys are seldom a valid indication of general malnutrition unless the average intake for a population group falls significantly below the Estimated Average Requirement (EAR) for one or more nutrients. However, by looking at the percentage of people within a population group who have suboptimal intake, it is possible to identify high-risk population groups that should be monitored more closely. Biochemical assays can definitely identify subclinical cases of malnutrition where nutritional intervention is desirable provided (1) the assay has been shown to be reliable, (2) the deficiency can be verified by a second assay, and (3) there is no unusual stress situation that may alter micronutrient distribution. In assessing nutritional status, it is important to be aware of those population groups at risk, the most reliable biochemical assays for monitoring nutritional status, and the symptoms of deficiencies.

26.3 • DIETARY REFERENCE INTAKES

Dietary Reference Intakes (DRIs) are quantitative estimates of nutrient intakes to be used for planning and assessing diets for healthy people and refer to either RDAs or AIs, depending on the nutrient (Figure 26.1). In assessing quantitative standards for nutrient intake the Food and Nutrition Board of the National Research Council considers the amount of nutrients and food components required for preventing deficiency diseases and, where the data are definitive, for promoting optimal health. The first step is determining the **Estimated Average Requirement (EAR)**, the amount of nutrient estimated to meet the nutrient requirement of half of the healthy individuals in an age and gender group. The **Recommended Dietary Allowance (RDA)** is normally set at two standard deviations above the EAR and is assumed to be the dietary intake amount that is sufficient to meet the nutrient requirement of nearly all (97%–98%) healthy individuals in a group. If a nutrient is considered essential but the experimental data are inadequate for determining an EAR, an **Adequate Intake (AI)** is set rather than an RDA. The AI is believed to cover the needs of all individuals in a group, but uncertainty of the data prevent being able to specify with certainty the percentage of individuals covered by this intake. AIs are often based on approximations of nutrient intake by a group of individuals. For example, AIs for young infants are often based on the daily mean nutrient intake supplied by human milk for healthy, full-term infants who are exclusively breast-fed. Finally, for most nutrients the Food and Nutrition Board sets a **Tolerable Upper Intake Level (UL)**. The UL is defined as the highest level of daily nutrient intake that is likely to pose no risks of adverse health effects to almost all individuals in the general population. The RDAs, AIs, and ULs are designed to be of use in planning and evaluating diets for individuals. The EAR is designed to be used in setting goals for nutrient intake and assessing the prevalence of inadequate intake in a population group.

These determinations are relatively easy to make for those nutrients associated with dramatic deficiency diseases, for example, vitamin C and scurvy. Measures that are more indirect must be used in other instances, such as tissue saturation or extrapolation from animal studies. The Food and Nutrition Board normally meets every 6 to 10 years to consider currently available information and update their recommendations.

DRIs serve as a useful general guide in evaluating adequacy of individual diets. However, the DRIs have several limitations.

1. DRIs are designed to meet the needs of healthy people and do not take into account special needs arising from infections, metabolic disorders, or chronic diseases.
2. Since present knowledge of nutritional needs is incomplete, there may be unrecognized nutritional needs. To provide for these needs, the DRIs should be met from as varied a selection of foods as possible. No single food can be considered complete, even if it meets the DRI for all known nutrients. This is important, especially in light of the current practice of fortifying foods of otherwise low nutritional value.
3. As currently formulated, DRIs may not define the "optimal" level of any nutrient, since optimal amounts are difficult to define. Because of information suggesting that optimal intake of certain micronutrients may reduce heart disease and cancer risk, the DRIs for these nutrients have recently been increased slightly; however, some experts feel that the current DRIs may not be sufficient to promote optimal health.

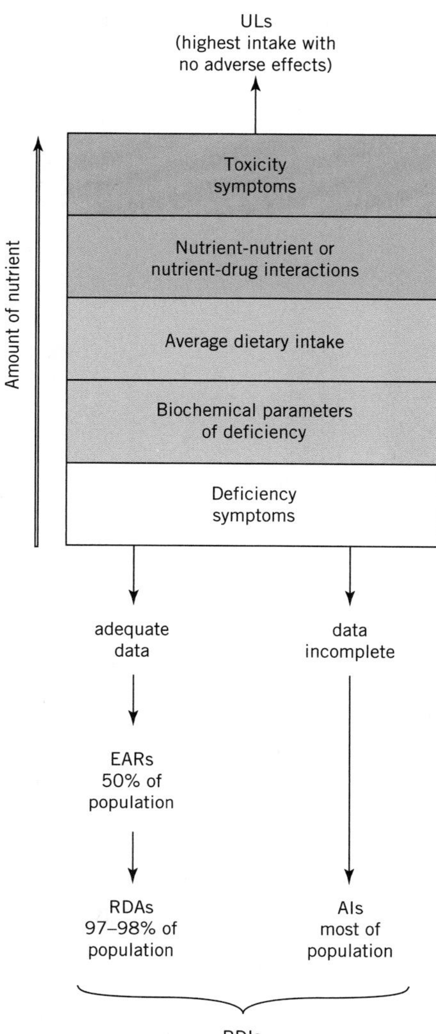

Figure 26.1 Dietary reference intakes. A schematic representation of the relationship between EAR (Estimated Average Requirement, the amount of nutrient estimated to meet the needs of 50% of a population group), RDA (Recommended Dietary Allowance; the amount of nutrient estimated to meet the needs of 97-98% of a population group), AI (Adequate Intake, the amount of nutrient estimated to meet the needs of most of a population group), DRI (Dietary Reference Intake, either RDA or AI, depending on the nutrient), and UL (Tolerable Upper Intake Level, highest level of nutrient intake that is likely to pose no risk of adverse health effects to almost all individuals in the general population).

26.4 • FAT-SOLUBLE VITAMINS

Vitamin A Is Derived from Plant Carotenoids

The active forms of vitamin A are retinol, retinal (retinaldehyde), and retinoic acid. Their precursors are synthesized by plants as the carotenoids (Figure 26.2), some of which are cleaved to retinol and stored in the liver as retinol palmitate. Liver, egg yolk, butter, and whole milk are good sources of retinol. Dark green and yellow vegetables are generally good

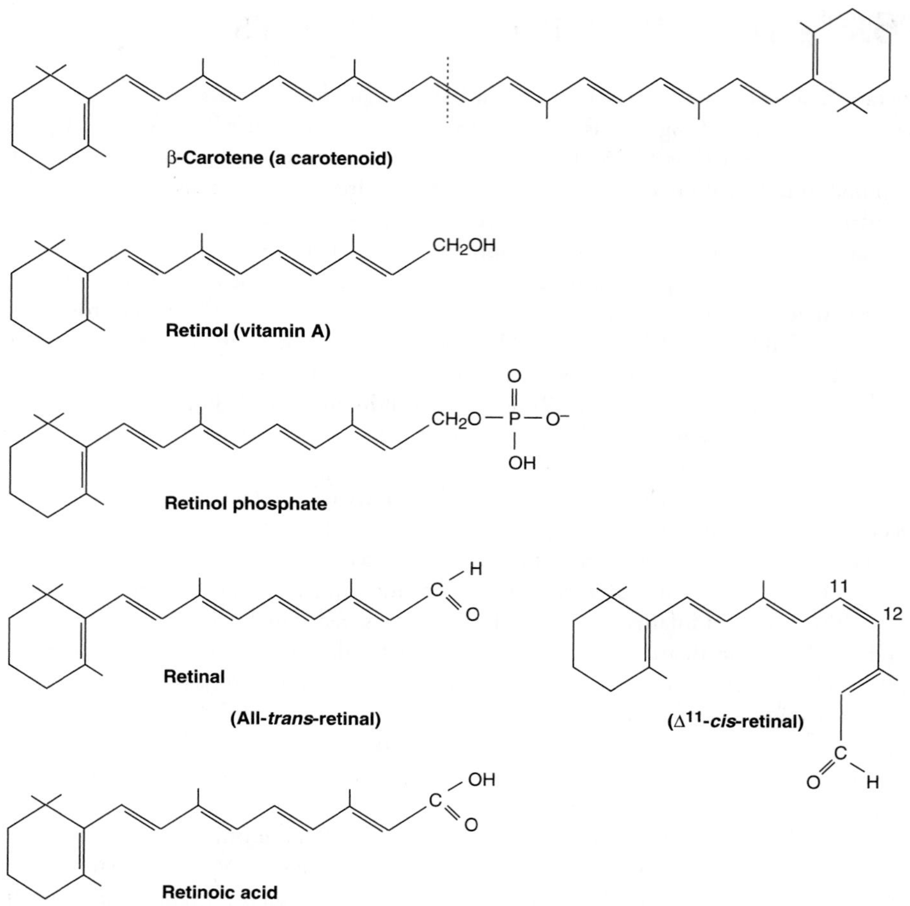

β-Carotene (a carotenoid)

Retinol (vitamin A)

Retinol phosphate

Retinal

(All-*trans*-retinal)

(Δ^{11}-*cis*-retinal)

Retinoic acid

Figure 26.2 Structures of vitamin A and related compounds.

sources of the carotenoids. Conversion of carotenoids to retinol is rarely 100%, so the vitamin A potency of various foods is expressed in terms of milligrams per day of retinol activity equivalents (1 RAE is equal to 1 μg retinol, 12 μg of ***β*-carotene**, and 24 μg of α-carotene or β-cryptoxanthin). The carotenoids are major sources of vitamin A in the American diet because they can be cleaved to retinol and converted to other vitamin A metabolites in the body (Figure 26.2). Carotenoids are also thought to function as antioxidants, although they may have additional metabolic functions.

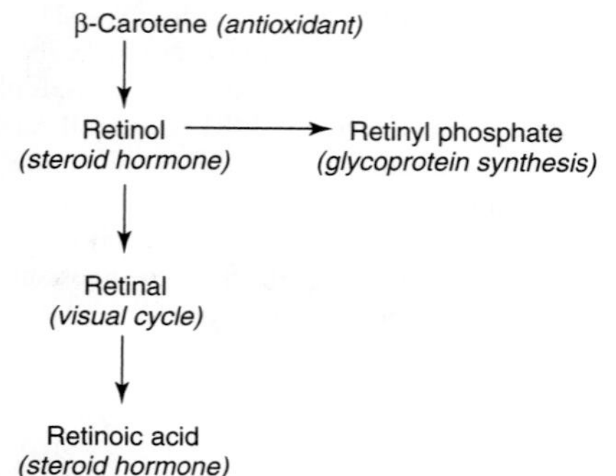

Figure 26.3 Vitamin A metabolism and function.

Only in recent years has the biochemistry of vitamin A become well understood (Figure 26.3). β-Carotene and some other carotenoids have an important role as antioxidants. At the low oxygen tensions prevalent in the body, β-carotene is a very effective antioxidant and may reduce the risk of those cancers initiated by free radicals and other strong oxidants. Several epidemiologic studies suggested that adequate dietary β-carotene may be important in reducing the risk of lung cancer, especially in people who smoke. However, supplemental β-carotene did not provide any detectable benefit and may have actually increased cancer risk for smokers in several multicenter prospective studies. This illustrates the danger of making dietary recommendations on epidemiologic studies alone.

Retinol is converted to **retinyl phosphate**, which appears to serve as a glycosyl donor in the synthesis of some glycoproteins and mucopolysaccharides in much the same manner as **dolichol phosphate** (p. 660). It is essential for the synthesis of glycoproteins needed for normal growth regulation and for mucus secretion. Retinoic acid binds to **retinoic acid receptors** (**RARs**) and retinoid X receptors (RXRs), which then bind to DNA and modulate the synthesis of proteins involved in the regulation of cell growth and differentiation. Thus, it can be considered to act like a steroid hormone in regulating growth and differentiation.

In the Δ^{11}-*cis*-retinal form, **vitamin A** becomes reversibly associated with visual proteins (the **opsins**). When light strikes the retina, a number of complex biochemical changes take place, resulting in generation of a nerve impulse, conversion of the *cis*-retinal to the all-*trans* form, and its dissociation from the visual protein (p. 962). Regeneration of functional visual pigments requires isomerization back to the Δ^{11}-*cis* form (Figure 26.4). In addition to the direct role of vitamin A in the visual cycle, clinical studies suggest that the carotenoids lutein and zeaxanthin reduce the risk of macular degeneration.

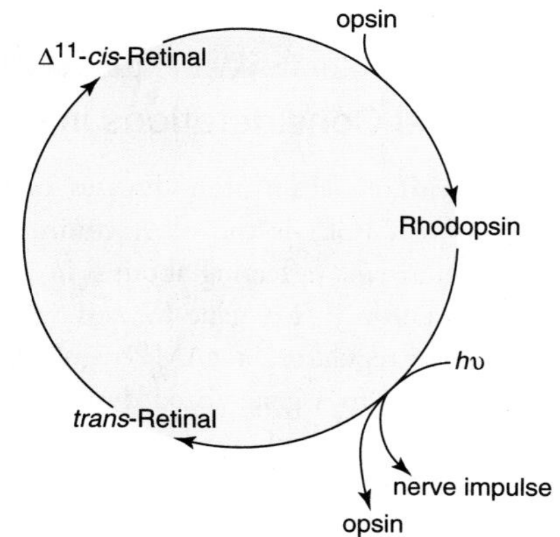

Figure 26.4 Role of vitamin A in vision.

On the basis of what is known about the biochemical mechanisms of vitamin A action, its biological effects are easy to understand. For example, retinyl phosphate is required for the synthesis of glycoproteins (an important component of mucus), and lack of mucus secretion leads to a drying of epithelial tissues. Retinol and/or retinoic acid down-regulate the synthesis of **keratin**, and excess keratin synthesis leaves a horny keratinized surface in place of the normal moist and pliable epithelium. Thus, vitamin A is required for maintenance of healthy epithelial tissue. In addition, retinol and/or retinoic acid are required for the synthesis of the iron transport protein transferrin. Thus, vitamin A deficiency can lead to anemia from impaired transport of iron.

Vitamin A-deficient animals are more susceptible to infections and cancer. Decreased resistance to infection may be due to keratinization of mucosal cells lining the respiratory, gastrointestinal, and genitourinary tracts. Fissures readily develop in the mucosal membranes, allowing microorganisms to enter. Vitamin A deficiency may also impair the immune system. The protective effect of vitamin A against many forms of cancer may result from the antioxidant potential of carotenoids and the effects of retinol and retinoic acid in regulating cell growth.

Since vitamin A is stored in the liver, deficiency can develop only over prolonged periods of inadequate intake. Mild vitamin A deficiency is characterized by follicular hyperkeratosis (rough keratinized skin resembling goosebumps), **anemia** (biochemically equivalent to iron deficiency anemia, but in the presence of adequate iron intake), and increased susceptibility to infection and cancer. **Night blindness** is an early symptom of deficiency. Severe deficiency leads to progressive keratinization of the cornea, known as xerophthalmia in its most advanced stages. Infection usually sets in, with resulting hemorrhaging of the eye and permanent loss of vision.

For most people, (unless they happen to eat liver) dark green and yellow vegetables are the most important dietary sources of vitamin A. Unfortunately, these foods are most often missing from the American diet. Dietary surveys indicate that 40%–60% of the population consumes less than two-thirds of the RDA for vitamin A. Clinical symptoms of vitamin A deficiency are rare in the general population, but are a fairly common consequence of severe liver damage or diseases that cause fat malabsorption (Clin. Corr. 26.1).

Vitamin A accumulates in the liver. Excess intake over prolonged periods can be toxic. Doses of 25,000–50,000 g/day of vitamin A over months or years are toxic for many children and adults. The usual symptoms include bone pain, scaly dermatitis, enlargement of liver and spleen, nausea, and diarrhea. It is virtually impossible to ingest toxic amounts of vitamin A from normal foods unless one eats polar bear liver (6000 μg/serving) regularly. Most instances of vitamin A toxicity are caused by massive doses of vitamin A supplements. Fortunately, this practice is relatively rare because of increased public awareness of vitamin A toxicity.

Vitamin D Synthesis Requires Sunlight

Technically, **vitamin D** should be considered a pro-hormone rather than a vitamin. **Cholecalciferol (D_3)** is produced in skin by UV irradiation of 7-dehyrocholesterol (p. 719 for a discussion of the synthesis of vitamin D). Thus, as long as the body is exposed to adequate sunlight, there is little or no dietary requirement for vitamin D. The best dietary sources of vitamin D_3 are saltwater fish (especially salmon, sardines, and herring), liver, and egg yolk. Milk, butter, and other foods are routinely fortified with ergocalciferol (D_2) prepared by

CLINICAL CORRELATION 26.1

Nutritional Considerations in Cystic Fibrosis

Patients with malabsorption diseases often develop malnutrition. Cystic fibrosis (CF) is the most common lethal inherited disease among Caucasians (affecting about 1 in 3500 newborns) and results from a mutation in the gene for the cystic fibrosis transmembrane conductance regulator, a cAMP-regulated chloride channel. This mutation leads to a generalized dysfunction of the exocrine glands that causes formation of viscid mucus, which progressively plugs their ducts. Obstruction of the bronchi and bronchioles leads to pulmonary infections, which are usually the direct cause of death. In many cases, however, the exocrine cells of the pancreas are also affected, leading to a deficiency of pancreatic enzymes and sometimes a partial obstruction of the common bile duct.

Deficiency of pancreatic lipase and bile salts leads to severe malabsorption of fat and fat-soluble vitamins. Calcium tends to form insoluble salts with the long-chain fatty acids, which accumulate in the intestine. Starches and proteins are also trapped in the fatty bolus of partially digested foods. This physical entrapment, along with the deficiencies of pancreatic amylase and pancreatic proteases, can lead to severe protein–calorie malnutrition, which increase the risk of death in patients with cystic fibrosis. Excessive mucus secretion on the luminal surface of the intestine may also interfere with the absorption of several micronutrients, including iron.

Fortunately, microsphere preparations of pancreatic enzymes are now available that greatly alleviate many of these malabsorption problems. With these preparations, protein and carbohydrate absorption are returned to near normal. Fat absorption is improved greatly but not normalized, since deficiencies of bile salts and excess mucus secretion persist. Because dietary fat is a major source of calories, these patients have difficulty obtaining sufficient calories from a normal diet. This is complicated by increased protein and energy needs because of the chronic infections often seen in these patients. Thus, the current recommendations are for energy intakes ranging from 110%–200% of the RDA to combat the poor growth and increased susceptibility to infection. The recommendation is for high-energy high-protein diets without any restriction of dietary fat (50% carbohydrate, 15% protein, and 35% fat). If caloric intake from the normal diet is inadequate, dietary supplements or enteral feeding may be used. The dietary supplements usually contain easily digested carbohydrates and milk protein mixtures. Medium-chain triglycerides are sometimes used as a partial fat replacement since they can be absorbed directly through the intestinal mucosa in the absence of bile salts and pancreatic lipase.

Since some fat malabsorption is present, deficiencies of fat-soluble vitamins often occur. Children aged 2–8 years need a standard adult multiple-vitamin preparation containing 400 IU of vitamin D and 5000 IU of vitamin A/day. Older children, adolescents, and adults need a standard multivitamin at a dose of 1–2/day. If serum vitamin A or vitamin E levels become low, water-miscible vitamin preparations should be used. Vitamin K deficiency has not been adequately studied, but supplementation is usually recommended, especially when on antibiotics or if cholestatic liver disease is present. Iron deficiency is common but iron supplementation is not usually recommended because of concern that higher iron levels in the blood might encourage systemic bacterial infections. Calcium levels in the blood are usually normal. However, since calcium absorption is probably suboptimal, it is important to ensure that the diet provides at least RDA levels of calcium.

Borowitz, D., Baker, R. B., and Stallings, V. Consensus report on nutrition for pediatric patients with cystic fibrosis. *J. Pedatr. Gasteroenterol. Nutr.* 35:246, 2002; and Yankaskas, J. R., Marshall, B. C., Sufian, B., Simon, R. H., and Rodman, D. Cystic fibrosis adult care. Consensus conference report. *Chest* 125:1S, 2004; Stallings, V. A., Stark, L. J., Robinson, K. A., Feranchak, A. P., et al. Evidence-based practice recommendations for nutrition-related management of children and adults with cystic fibrosis and pancreatic insufficiency: Results of a systematic review. J. *Am. Diet. Assoc.* 108:832, 2008.

irradiating ergosterol from yeast (Figure 26.5). Vitamin D potency is measured in terms of micrograms cholecalciferol (1 μg cholecalciferol, or ergocalciferol, = 40 IU). There has been some recent controversy about whether ergocalciferol is as biologically potent as cholecalciferol, but that has not yet been resolved.

Both cholecalciferol and ergocalciferol are metabolized in liver where **25-hydroxycholecalciferol [25-(OH)D]** is formed (Figure 26.5). This is the major circulating derivative of vitamin D, and is converted into the biologically active **1-α,25-dihydroxycholecalciferol (calcitriol)** by the enzyme 25-hydroxyvitamin D-1 α-hydroxylase (Figure 26.5). It used to be thought that this reaction occurred exclusively in the proximal convoluted tubules of kidney (Clin. Corr. 26.2, p. 1070). The kidney does appear to be the major source of 1-α,25-dihydroxycholecalciferol [1, 25-$(OH)_2$D] in the blood. However, it is now recognized that many other tissues including colon, prostate, breast, brain, β-islet cells of the pancreas, vascular smooth muscle cells and macrophages can also produce 1,25-$(OH)_2$D. These cells also have vitamin D receptors, so the 1,25-$(OH)_2$D that they produce acts in a paracrine fashion, making little if any contribution to circulating 1,25-$(OH)_2$D levels in the bloodstream.

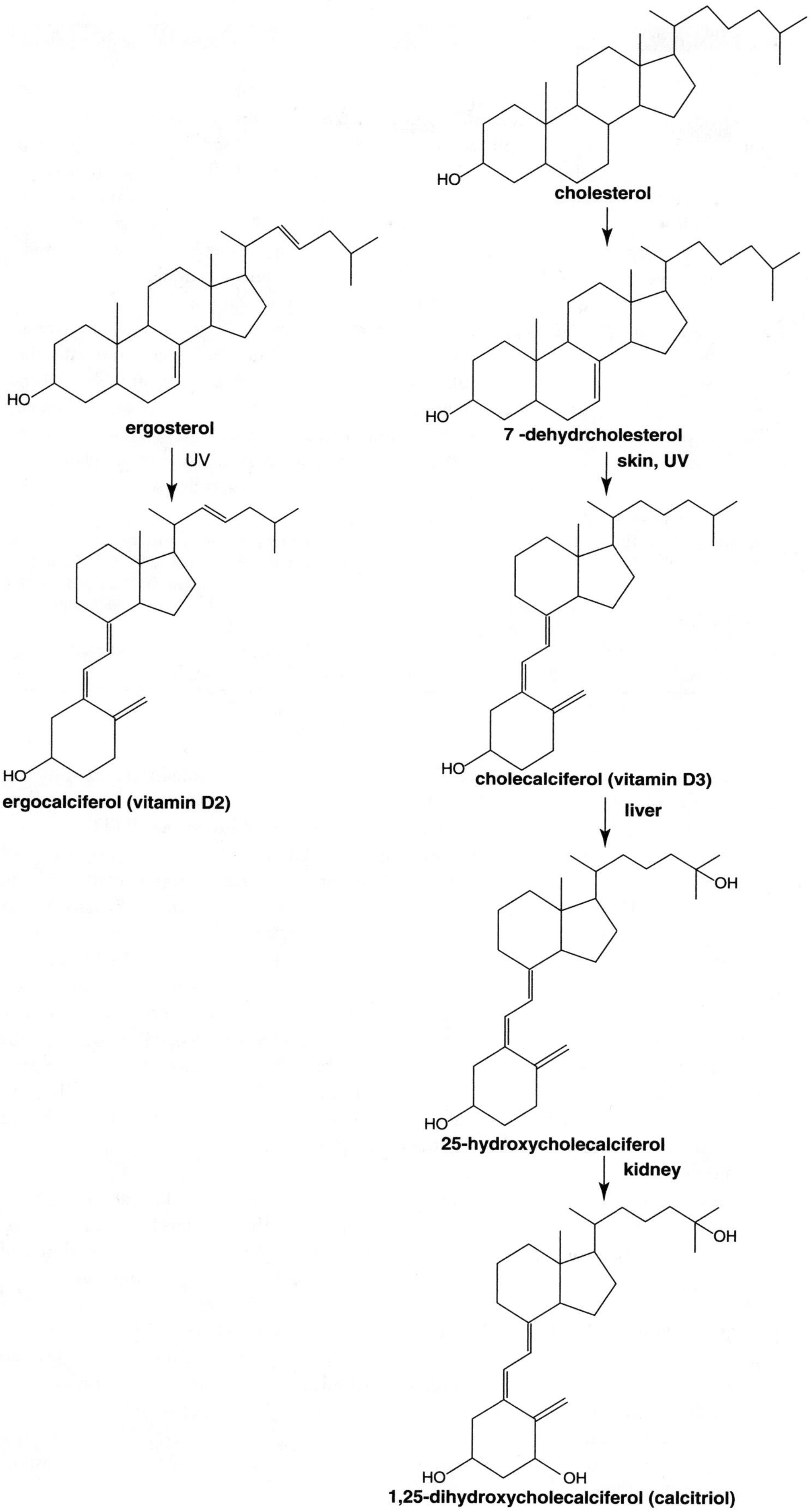

HO
cholesterol

HO
ergosterol
UV

HO
7 -dehydrcholesterol
skin, UV

HO
ergocalciferol (vitamin D2)

HO
cholecalciferol (vitamin D3)
liver

OH
HO
25-hydroxycholecalciferol
kidney

OH
HO OH
1,25-dihydroxycholecalciferol (calcitriol)

Figure 26.5 **Structures of vitamin D_3 and related metabolites.**

CLINICAL CORRELATION 26.2

Renal Osteodystrophy

In chronic renal failure, a complicated chain of events leads to renal osteodystrophy. The renal failure results in an inability to produce 1,25-$(OH)_2$D, and thus bone calcium becomes the only important source of serum calcium. In the later stages, the situation is complicated by increased renal retention of phosphate and the resulting hyperphosphatemia. The serum phosphate levels are often high enough to cause metastatic calcification (i.e., calcification of soft tissue), which tends to lower serum calcium levels further (the solubility product of calcium phosphate in the serum is very low and a high serum level of one component necessarily causes a decreased concentration of the other). The hyperphosphatemia and hypocalcemia stimulate parathyroid hormone secretion, which further accelerates the rate of bone loss. The result is bone loss and metastatic calcification. Administration of high doses of vitamin D or its active metabolites would not be sufficient since the combination of hyperphosphatemia and hypercalcemia would only lead to more extensive metastatic calcification. The readjustment of serum calcium levels by high calcium diets and/or vitamin D supplementation must be accompanied by phosphate reduction therapies. It is difficult to reduce dietary phosphate sufficiently because most protein sources are also high in phosphate. In this regard, vegetable proteins are a better choice than animal proteins because a significant portion of the phosphate in vegetable protein is in the form of phytates and are unavailable for absorption. It is also important to avoid processed, fast, and convenience foods because phosphate is also added to those foods. For example, sodium phosphate is often added to processed meats to keep the meat from drying out. Because it is so difficult to achieve sufficient phosphate restriction in the diet, phosphate binders are often used to make dietary phosphate unavailable for absorption. Calcium acetate and a cationic polymer called Sevelamer hydrochloride are the most commonly used phosphate binders at present. Orally administered 1,25-$(OH)_2$D is effective at stimulating calcium absorption in the mucosa but does not enter the peripheral circulation in significant amounts. Thus, in severe hyperparathyroidism intravenous 1,25-$(OH)_2$D may be necessary. Research is in progress with calciumimetic agents that bind to a calcium sensor located on the extracellular membrane of the parathyroid gland and decrease parathyroid hormone production and release.

Hoyland, J. A., and Picton, M. L. Cellular mechanisms of osteodystrophy. *Kidney Intl.* 56, Suppl. 73:508, 1999; and Slatopolsky, E., Gonzalez, E., and Marin, K., Pathogenesis and treatment of renal osteodystrophy. *Blood Purif.* 21:318, 2003; Kestenbaum, B. Phosphate metabolism in the setting of chronic kidney disease. *Seminars in Dialysis* 20:286, 2007; and Uribarri, J. Phosphorous homeostasis in normal health and in chronic kidney disease patients with special emphasis on dietary phosphorous intake. *Seminars in Dialysis* 20:295, 2007.

Traditionally, vitamin D has been thought to primarily play a role in maintaining calcium homeostasis (Figure 26.6). 1,25-$(OH)_2$D production by the kidney is increased when calcium levels are low, and it acts in concert with **parathyroid hormone** (**PTH**), which is also produced in response to low serum calcium. High PTH levels stimulate production of 1,25-(OH)D whereas low PTH amounts cause formation of 24,25-$(OH)_2$D by the kidney. The 1,25-$(OH)_2$ D acts as a typical steroid hormone in intestinal mucosal cells, where it induces synthesis of a calcium transport protein, TRPV5, and a calcium-binding protein, calbindin, that are required for calcium transport. In bone 1,25-$(OH)_2$D and PTH act synergistically to promote bone resorption (demineralization) by stimulating osteoblast formation and activity. Finally, PTH and 1,25-$(OH)_2$D inhibit calcium excretion in the kidney by stimulating calcium reabsorption in the distal renal tubules. 24,25-$(OH)_2$D was thought to be inactive, but recent studies with knockout mice lacking the 24-hydroxylase enzyme have shown that 24,25-$(OH)_2$D plays an essential but ill-defined role in bone metabolism. Calcitonin is produced when serum calcium levels are high (usually right after a meal) and lowers serum calcium levels by blocking bone resorption and stimulating calcium excretion by the kidney.

The response of calcium metabolism to several different physiological situations is summarized in Figure 26.6. The response to low serum calcium levels is characterized by elevation of PTH and 1,25-$(OH)_2$D, which act to enhance calcium absorption and bone resorption and to inhibit calcium excretion (Figure 26.6*a*). High serum calcium levels block production of PTH. The low PTH levels cause 25-(OH)D to be converted to 24,25-$(OH)_2$D instead of 1, 25-$(OH)_2$D. In the absence of PTH and 1,25-$(OH)_2$D bone resorption is inhibited and calcium excretion is enhanced. High serum calcium levels also stimulate production of **calcitonin**, which contributes to the inhibition of bone resorption and the increase in calcium excretion. Finally, the high levels of both serum calcium and phosphate increase the rate of bone mineralization (Figure 26.6*b*). Thus, bone is a very important reservoir of the calcium and phosphate needed to maintain homeostasis of serum levels. When dietary vitamin D and calcium are adequate, no net loss of bone

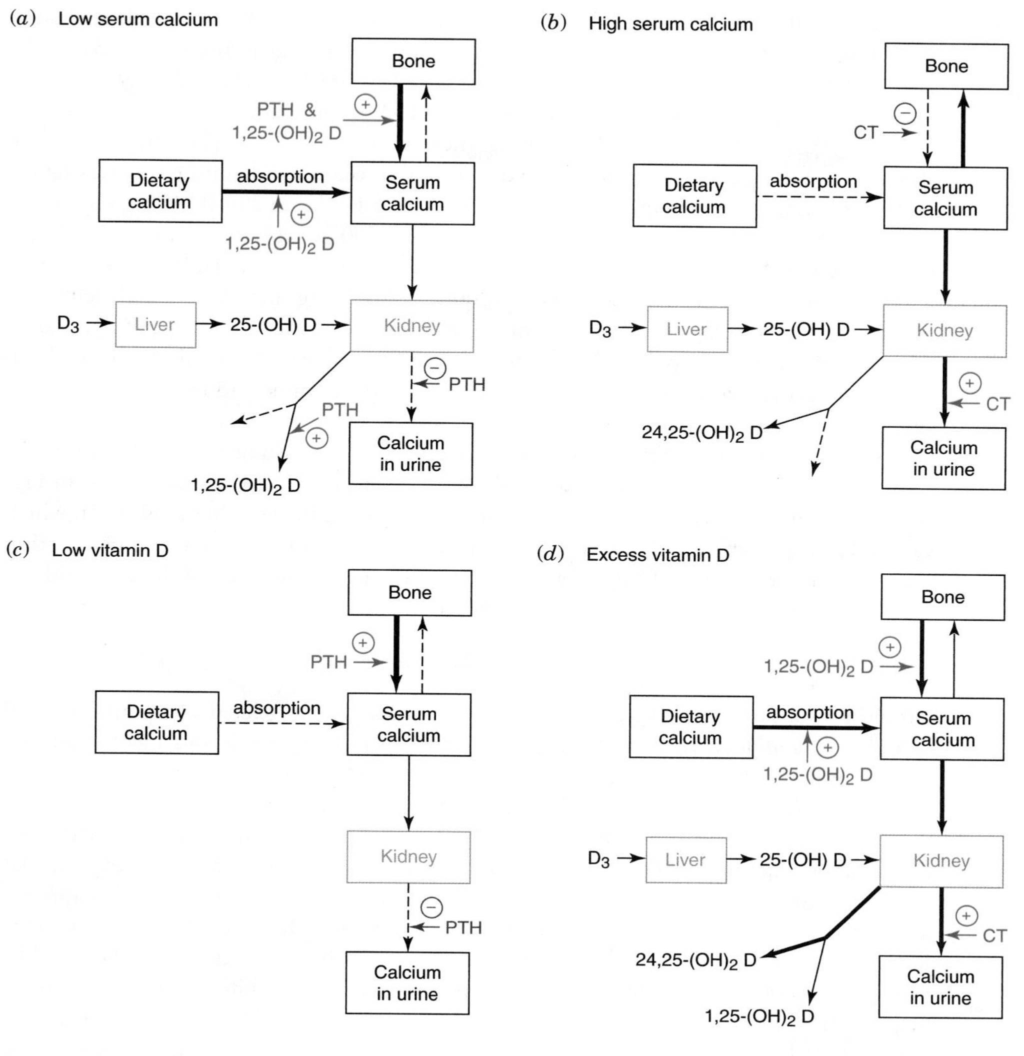

Figure 26.6 Vitamin D and calcium homeostasis.
The dominant pathways of calcium metabolism under each set of metabolic conditions are shown with heavy arrows. Effect of various hormones is shown by red arrows for stimulation or blue arrows for repression. Abbreviations: PTH, parathyroid hormone; CT, calcitonin; D, cholecalciferol; 25-(OH) D, 25-hydroxycholecalciferol; 1, 25-$(OH)_2$D, 1-α, 25-dihydroxycholecalciferol; and 24, 25-$(OH)_2$D, 24, 25-dihydroxycholecalciferol.

calcium occurs. However, when dietary calcium is low, PTH and 1,25-$(OH)_2$D cause net demineralization of bone to maintain normal serum calcium levels. Vitamin D deficiency also causes net demineralization of bone due to elevation of PTH (Figure 26.6*c*).

The most commonly recognized symptoms of vitamin D deficiency are rickets in young children and osteomalacia in adults. Rickets is characterized by continued formation of osteoid matrix and cartilage that are improperly mineralized resulting in soft, pliable bones. In adults, demineralization of preexisting bone causes it to become softer and more susceptible to fracture. **Osteomalacia** is easily distinguishable from the more common osteoporosis, by the fact that the osteoid matrix remains intact in the former but not in the latter. Because of fortification of dairy products with vitamin D, rickets and osteomalacia are very rare and are most often seen in low-income groups, the elderly (who often also have minimal exposure to sunlight), strict vegetarians (especially if their diet is also low in calcium and high in fiber), and chronic alcoholics.

However, we now know that vitamin D is involved in much more than regulation of calcium homeostasis. Receptors for 1,25-$(OH)_2$D occur in many tissues and most of those tissues have the ability to produce 1,25-$(OH)_2$D in a paracrine fashion from 25-(OH)D. It now appears that vitamin D also plays an important role in regulating cell proliferation, the functioning of the adaptive and innate immune system, secretion of insulin by the pancreatic β-cells, regulation of blood pressure and normal neuromuscular function. Recent research suggests that inadequate vitamin D intake may increase the risk of certain types of cancer (particularly breast, colon, and prostate), hypertension, and autoimmune diseases (particularly multiple sclerosis, rheumatoid arthritis, Crohn disease, and type I diabetes).

TABLE 26.1 • Recommended Serum Levels of 25-Hydroxy Vitamin D

Serum Levels of 25-Hydroxy Vitamin D	
Toxicity	> 150 ng/ml (150 nmol/liter)
Sufficiency	≥ 30 ng/ml (75 nmol/liter)
Relative insufficiency	21-29 ng/ml (52 to 72 nmol/liter)
Deficiency	≤ 20 ng/ml (50 nmol/liter)

Because many tissues can make 1,25-$(OH)_2$D from 25-(OH)D, blood levels of vitamin 25-(OH)D are now considered the best indicators of vitamin D sufficiency (Table 26.1). Most experts now define vitamin D deficiency as 25-(OH)D levels of 20 ng/ml or less, vitamin D insufficiency as 25-(OH)D levels of 21–29 ng/ml, and vitamin D sufficiency as 25-(OH)D levels >30 ng/ml. Based on these standards, a large percentage of North Americans and Europeans are vitamin D deficient and recommendations for increasing the RDIs for vitamin D are being considered. The current RDIs are 200 IU (5 μg/day) up to age 50, 400 IU (10 μg/day) from age 51 to 70, and 600 IU (15 μg/day) over 71. Many experts are now recommending that without adequate sun exposure the RDIs for children and adults be increased to at least 800 to 1000 IU (20–25 μg/day). Vitamin D deficiency can also result from fat malabsorption or severe liver and kidney disease (see Clin. Corrs. 26.1, p. 1068, and 26.2, p. 1070). Certain drugs also interfere with vitamin D metabolism. For example, corticosteroids stimulate the conversion of vitamin D to inactive metabolites and cause bone demineralization when used for long periods.

Vitamin D can be toxic in high doses. The tolerable upper intake level (UL) for adults is 2000 IU/day (50 μg/day). The mechanism of vitamin D toxicity is summarized in Figure 26.6*d*. Enhanced calcium absorption and bone resorption cause hypercalcemia, which can lead to metastatic calcification. The enhanced bone resorption causes demineralization similar to that of vitamin D deficiency. Finally, the high serum calcium leads directly to hypercalciuria, which predisposes to formation of renal stones.

Vitamin E Is a Mixture of Tocopherols and Tocotrienols

Vitamin E occurs in the diet as a mixture of several closely related compounds, called tocopherols and **tocotrienols** (Figure 26.7). All of the tocopherols and tocotrienols are important naturally occurring antioxidants. Due to their lipophilic character, they accumulate in circulating lipoproteins, cellular membranes, and fat deposits, where they act as scavengers for free radicals, protecting unsaturated fatty acids (especially in membranes) from peroxidation reactions. **α-Tocopherol** is the most potent scavenger of reactive oxygen species, but γ-tocopherol is a more potent scavenger of reactive nitrogen species. **γ-Tocopherol** also appears to inactivate fat-soluble electrophilic mutagens, thus complementing glutathione, which inactivates electrophilic mutagens in the aqueous compartments of the cell. The tocopherols appear to play a role in cellular respiration, either by stabilizing ubiquinone or by helping transfer electrons to ubiquinone (p. 563). Tocopherols and tocotrienols also prevent oxidation of LDL, which may be important in reducing the risk of cardiovascular disease since the oxidized form of LDL is atherogenic. Although many of the biological properties of the tocopherols are the result of their antioxidant potential, some of their benefits appear to be due to effects on enzyme activity or transcription. For example, they appear to enhance heme synthesis by increasing the levels of δ-aminolevulinic acid (ALA) synthase and ALA dehydratase. Recent studies have shown that vitamin E is required for maintaining normal immune function, particularly in the elderly, and may be important in preventing macular degeneration and cognitive decline. Finally, neurological symptoms have been reported following prolonged vitamin E deficiency associated with malabsorption diseases.

Tocopherols

Tocotrienols

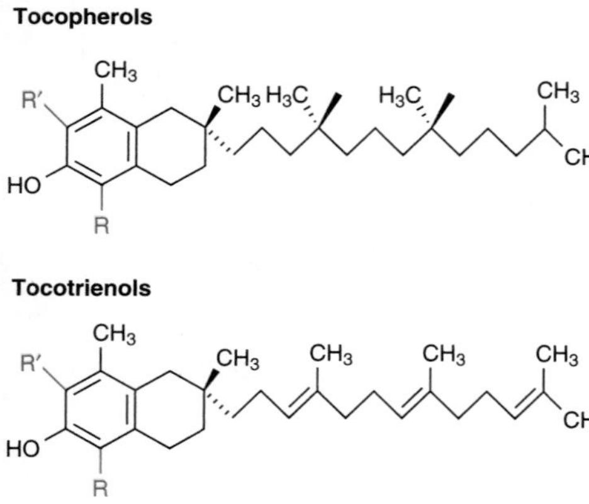

Naturally occurring homologs	R	R′
α	CH_3	CH_3
β	CH_3	H
γ	H	CH_3
δ	H	H

Figure 26.7 The structures of tocopherols and tocotrienols.

Setting the recommended intake levels of vitamin E has been hampered by the difficulty of producing severe vitamin E deficiency in humans. It is generally assumed that the vitamin E content of the American diet is sufficient, since no major vitamin E deficiency diseases have been found. However, vitamin E requirements increase as intake of polyunsaturated fatty acids increases. While the recent emphasis on high polyunsaturated fat diets to reduce serum cholesterol may be of benefit in controlling heart disease, the propensity of polyunsaturated fats to form free radicals on exposure to oxygen may lead to increased cancer risk. It is prudent to increase vitamin E intake for diets rich in polyunsaturated fats.

The controversy about the relationship between vitamin E and cardiovascular disease risk illustrates the difficulty in characterizing the role of vitamins in promoting optimal health, as opposed to preventing deficiency diseases. On one hand relationships between nutrient status and disease risk suggested by biochemical and epidemiologic studies are often not borne out by large-scale intervention trials. On the other hand intervention trials are not generally designed to identify the high-risk population groups that would benefit most from optimal nutrient intake. For example, vitamin E prevents the oxidation of LDL particles to a more atherogenic form, so it was logical that vitamin E supplementation might decrease the risk of atherosclerosis. Epidemiologic studies suggested that people consuming 100 mg/day of vitamin E had reduced risk of myocardial infarction. However, large, randomized, double-blind, placebo-controlled intervention trials with supplemental α-tocopherol failed to show any significant reduction in mortality from cardiovascular disease. This could indicate that α-tocopherol can prevent atherosclerosis in the early stages, but is ineffective in the more advanced clinical conditions represented in the clinical trials. Alternatively, it could indicate that γ-tocopherol, or some other tocopherol or tocotrienol found in food, is more effective than α-tocopherol in preventing atherosclerosis, since high levels of supplemental α-tocopherol are known to interfere with utilization of other forms of vitamin E. More importantly, most large-scale intervention trials performed to date have ignored the effects of genetic polymorphisms on disease prevention. (The effect of polymorphisms on nutritional status is discussed in more detail in Section 26.13.) For example, the haptoglobin 2-2 polymorphism is associated with increased production of oxygen free radicals and lower endogenous serum levels of vitamins E and C. Recent randomized, double-blind, placebo controlled clinical trials show an association between the haptoglobin 2-2 polymorphism and preventive effects of vitamin E supplementation against myocardial infarction and cardiovascular death.

Because the α-tocopherol transfer protein in the liver specifically binds to the natural *RRR*-α-tocopherol, it is retained by the body 4–6 times longer than the synthetic all rac or d, l form of α-tocopherol. The specificity of this protein for α-tocopherol also explains why high intakes of α-tocopherol interfere with the utilization of γ-tocopherol. Vitamin E appears to be the least toxic of the fat-soluble vitamins. The UL for vitamin E has been set at 1000 mg/day, primarily because high levels of vitamin could potentiate the effects of blood-thinning medications such as dicumarol.

Vitamin K Is a Quinone Derivative

Vitamin K is found naturally as K_1 (**phytylmenaquinone**) in green vegetables and K_2 (multiprenylmenaquinone), which is synthesized by intestinal bacteria (Figure 26.8). The body converts synthetic menaquinone (Menadione) and a number of water-soluble analogs to a biologically active form of vitamin K.

Vitamin K is required for conversion of glutamic acid residues to **γ-carboxyglutamic acid** residues in several precursor proteins (Figure 26.9). The γ-carboxyglutamic acid residues are good chelators and allow the proteins to bind Ca^{2+}, which is required for their biological activity. In the carboxylase reaction, the active hydroquinone form of vitamin K is converted

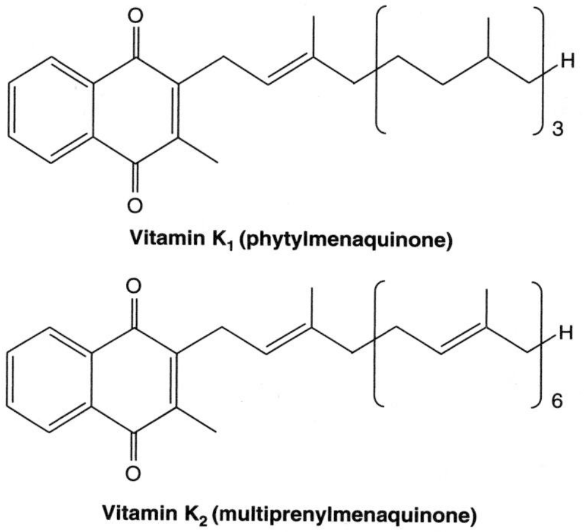

Figure 26.8 The structures of vitamin K_1 (phytylmenaquinone) and K_2 (multiprenylmenaquinone).

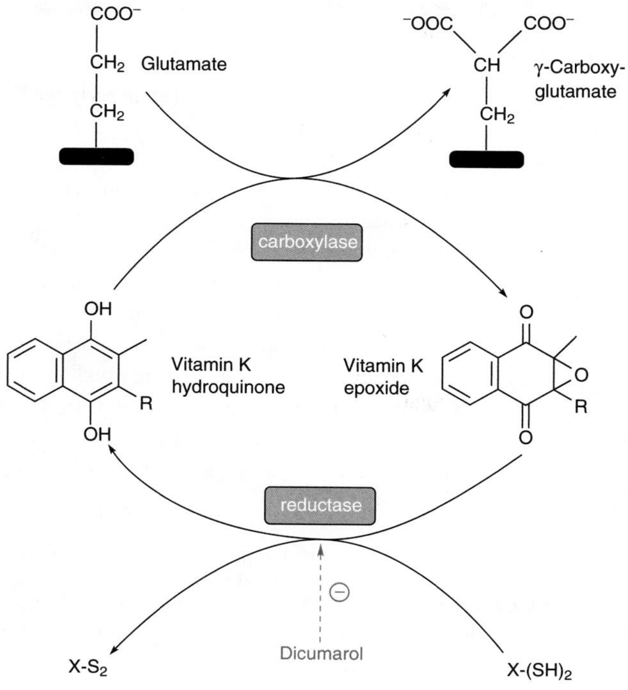

Figure 26.9 Function of vitamin K. Vitamin K is required for the conversion of glutamic acid residues to γ-carboxyglutamic acid residues by the enzyme vitamin K-dependent carboxylase.
In the process, the hydroquinone form of vitamin K is converted to the inactive 2, 3-epoxide form. Conversion of the 2,3-epoxide back to the active hydroquinone requires a dithiol-dependent reductase that is inhibited by dicumarol.

to an inactive 2,3-epoxide form (Figure 26.9). Regeneration of active vitamin K requires vitamin K epoxide reductase, which is inhibited by coumarin-type drugs like **dicumarol**. There are seven proteins involved in blood coagulation that require vitamin K-dependent activation, so vitamin K is essential for blood clotting. The mechanism of activation has been most clearly delineated for prothrombin (p. 964). The γ-carboxyglutamic acid residues allow prothrombin to bind Ca^{2+}, and the prothrombin–Ca^{2+} complex binds to the negatively charged phospholipid surfaces of the platelets and endothelial cells at the site of the injury, where the proteolytic conversion of prothrombin to thrombin occurs.

Vitamin K is also essential for the synthesis of γ-carboxyglutamic acid residues in three proteins found in bone. For example, **osteocalcin** accounts for 15%–20% of the noncollagen protein in the bone, and the presence of γ-carboxyglutamic acid residues are required for its binding to hydroxyapatite crystals in the bone. The physiological role of osteocalcin and the other γ-carboxylated bone proteins is not clear, but they do appear to be required for normal bone mineralization. Under carboxylated osteocalcin is associated with low bone density and increased risk of fracture. A protein called Gas6, which is a ligand for several receptor protein kinases and is involved in cell cycle regulation, has been shown to require vitamin K-dependent γ-carboxylation for activity. The physiological significance of this observation is not known in most tissues. However, carboxylation of Gas6 and a protein called matrix Gla protein (MGP) appears to be important in preventing vascular calcification.

Vitamin K_1 preferentially accumulates in the liver, where the clotting factors are formed. Vitamin K_2 preferentially accumulates in the peripheral tissues, and higher vitamin K intakes appear to be required for saturation of peripheral tissues with vitamin K_2. The most easily detectable symptom in humans of vitamin K deficiency is increased coagulation time, which reflects the requirement of vitamin K_1 for normal blood coagulation. Because of the higher requirements of vitamin K_2 in peripheral tissues, the RDI for vitamin K has recently been increased. Since vitamin K is synthesized by intestinal bacteria, deficiencies have long been assumed to be rare. However, intestinally synthesized vitamin K may not be efficiently absorbed and marginal vitamin K deficiencies, especially deficiencies that could interfere with bone mineralization, may be more common than originally thought. The most common deficiency occurs in newborn infants (Clin. Corr. 26.3), especially those of mothers on anticonvulsant therapy (Clin. Corr. 26.4). Vitamin K deficiency also occurs in

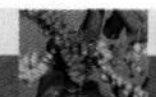

CLINICAL CORRELATION 26.3

Nutritional Considerations in Newborn Infants

Newborn infants are at special nutritional risk because of very rapid growth and needs for many nutrients are high. Some micronutrients (such as vitamins E and K) do not cross the placental membrane well, and tissue stores are low in the newborn. The gastrointestinal (GI) tract may not be fully developed, leading to malabsorption problems (particularly with respect to the fat-soluble vitamins). The GI tract is also sterile at birth and the intestinal flora that normally provide significant amounts of certain vitamins (especially vitamin K) take several days to become established. If the infant is born prematurely, the nutritional risk is slightly greater, since the GI tract will be less well developed and the tissue stores will be less.

The most serious nutritional complication appears to be hemorrhagic disease. Newborns, especially premature infants, have low tissue stores of vitamin K and lack the intestinal flora necessary to synthesize the vitamin. Breast milk is a relatively poor source of vitamin K. Approximately 1 out of 400 live births shows some signs of hemorrhagic disease, which can be prevented by 0.5–1 mg of the vitamin given at birth.

Most newborn infants have sufficient reserves of iron to last 3–4 months. Since cow's milk and breast milk contain little iron, supplementation with iron is usually initiated at an early age by the introduction of iron-fortified cereal. Vitamin D levels are also low in breast milk and supplementation with 200 IU/day of vitamin D is usually recommended. When infants must be maintained on assisted ventilation with high oxygen concentrations, supplemental vitamin E may reduce the risk of bronchopulmonary dysplasia and retrolental fibroplasia, potential complications of oxygen therapy. The anemia of prematurity may respond to supplemental folate and vitamin B_{12}.

In summary, supplemental vitamin K is given at birth to prevent hemorrhagic disease. Breast-fed infants are usually provided with supplemental vitamin D, with iron being introduced along with solid foods. Bottle-fed infants are provided with supplemental iron. If infants must be maintained on oxygen, supplemental vitamin E may be beneficial.

Mueller, D. P. R. Vitamin E therapy in retinopathy of prematurity. *Eye* 6:221, 1992; Morin, K. H. Current thoughts on healthy term infant nutrition, MCN. *Am. J. Matern. Child Nurs.* 29:312, 2004; and Collier, S., Fulhan, J., and Duggan, C. Nutrition for the pediatric office: Update on vitamins, infant feeding and food allergies. *Curr. Opin. Pediatr.* 16:314, 2004.

CLINICAL CORRELATION 26.4

Anticonvulsant Drugs and Vitamin Requirements

Anticonvulsant drugs such as phenobarbital or diphenylhydantoin (DPH) are an excellent example of drug–nutrient interactions that are of concern to physicians. Metabolic bone disease appears to be the most significant side effect of prolonged anticonvulsant therapy. Whereas children and adults who take these drugs seldom develop rickets or severe osteomalacia, up to 65% of those on long-term therapy have abnormally low serum calcium and phosphorus, abnormally high serum alkaline phosphatase, and some bone loss. Vitamin D supplements appear to correct both the hypocalcemia and osteopenia. Anticonvulsants also tend to increase needs for vitamin K, leading to an increased incidence of hemorrhagic disease in infants born to mothers on anticonvulsants. Anticonvulsants appear to increase the need for folic acid and B_6. Low serum folate levels occur in 75% of patients on anticonvulsants, and megaloblastic anemia may occur in as many as 50% without supplementation. This is of particular concern for women of childbearing age. When infants are exposed to antiepileptic drugs in utero, the risk of congenital malformations, particularly neural tube defects, is doubled compared to the general population. The recommended daily allowance of folic acid is 600 μg/day for pregnant women, and the need may be greater for patients on anticonvulsants. Since folates may speed up the metabolism of some anticonvulsants, it is important that excess folic acid not be given.

Rivery, M. D., and Schottelius, D. D. Phenytoin-Folic Acid: A review. *Drug Intelligence and Clinical Pharmacy* 18:292, 1984; Tjellesen, L. Metabolism and action of vitamin D in epileptic patients on anticonvulsant treatment and healthy adults. *Danish Medical Bulletin* 41:139, 1994; and Yerby, M. S., Management issues for women with epilepsy Neural tube defects and folic acid supplementation. *Neurology* 61, Suppl. 2:S23, 2003.

patients with **obstructive jaundice** and other diseases leading to severe fat malabsorption (see Clin. Corr. 26.1, p. 1068) and patients on long-term antibiotic therapy (which may destroy vitamin K-synthesizing organisms in the intestine). Deficiency is sometimes seen in the elderly, who are prone to fat malabsorption.

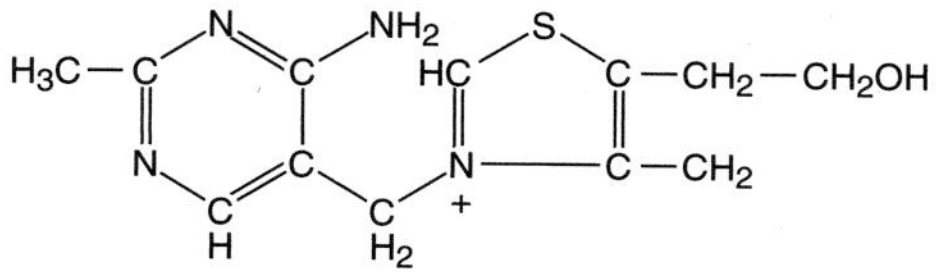

Thiamin

Figure 26.10 Structure of thiamin.

26.5 • WATER-SOLUBLE VITAMINS

Water-soluble vitamins differ from fat-soluble vitamins in several respects. Most are readily excreted once their concentration surpasses the renal threshold so toxicities are rare. Their metabolic stores are labile and depletion can often occur in a matter of weeks or months, so deficiencies occur relatively quickly on an inadequate diet. Since water-soluble vitamins are coenzymes for many common biochemical reactions, it is often possible to assay vitamin status by measuring one or more enzyme activities in isolated red blood cells. These assays are especially useful if one measures the endogenous activity and the stimulation of that activity by addition of the active coenzyme derived from the vitamin.

Most water-soluble vitamins are converted to coenzymes that are used in pathways for energy generation or hematopoiesis. Deficiencies of the energy-releasing vitamins produce a number of overlapping symptoms and show up first in rapidly growing tissues. Typical symptoms include dermatitis, glossitis (swelling and reddening of the tongue), cheilitis at the corners of the lips, and diarrhea. In many cases, nervous tissue is also involved due to its high-energy demand or specific effects of the vitamin. Common neurological symptoms include peripheral neuropathy (tingling of nerves at the extremities), depression, mental confusion, lack of motor coordination, and malaise. Demyelination and degeneration of nervous tissues may also occur. These deficiency symptoms are so common and overlapping that they can be considered as properties of the energy-releasing vitamins as a class, rather than being specific for any one.

26.6 • ENERGY-RELEASING WATER-SOLUBLE VITAMINS

Thiamin Forms the Coenzyme Thiamin Pyrophosphate

Thiamin (Figure 26.10) is rapidly converted to **thiamin pyrophosphate (TPP)** (also referred to as thiamin diphosphate [TDP]), a coenzyme required for the pyruvate dehydrogenase and **α-ketoglutarate dehydrogenase** reactions (Figure 26.11), and thiamin triphosphate, which is localized in peripheral nerve membranes and appears to function in transmission

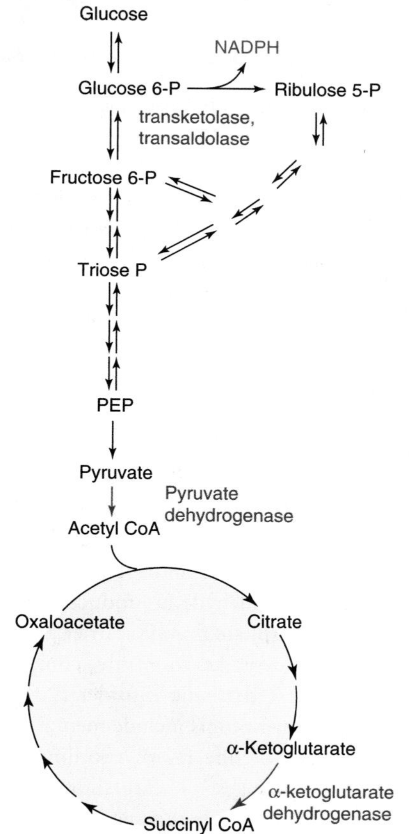

Figure 26.11 Summary of important reactions involving thiamin pyrophosphate.
Reactions involving thiamin pyrophosphate are indicated in red.

of nerve impulses. Loss of pyruvate dehydrogenase activity compromises cellular energy generation and causes an accumulation of pyruvate and lactate. Loss of α-ketoglutarate dehydrogenase activity causes reduced oxidative decarboxylation of α-keto acids (p. 554). Symptoms of thiamin deficiency involving neural tissue may result from the direct role of thiamin triphosphate in nerve transmission or from the accumulation of pyruvate and lactate in neural tissue. Thiamin pyrophosphate is also required for the transketolase and transaldolase reactions of the pentose phosphate pathway. Red blood cell transketolase is commonly used for measuring thiamin status in the body.

Symptoms of mild **thiamin deficiency** include loss of appetite, constipation, nausea, mental depression, peripheral neuropathy, irritability, and fatigue. These symptoms are most often seen in the elderly and low-income groups on restricted diets. Mental confusion, ataxia (unsteady gait while walking and general inability to achieve fine control of motor functions), and ophthalmoplegia (loss of eye coordination) are symptoms of moderately severe thiamin deficiency. This set of symptoms is referred to as Wernicke–Korsakoff syndrome, and is most commonly seen in chronic alcoholics (Clin. Corr. 26.5). Severe thiamin deficiency is known as beriberi. Dry beriberi is characterized by advanced neuromuscular symptoms, including muscular atrophy and weakness. When this is coupled with edema, the disease is called wet beriberi. Both forms of beriberi can be associated with an unusual type of heart failure characterized by high cardiac output. Beriberi occurs primarily in populations that rely exclusively on polished rice for food, although cardiac failure is sometimes seen in alcoholics as well. Coffee and tea contain substances that destroy thiamin, but this is not a problem with normal consumption. Routine enrichment of cereals has assured that most Americans have an adequate intake of thiamin on a mixed diet.

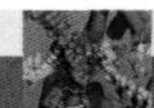

CLINICAL CORRELATION 26.5

Nutritional Considerations in Alcoholics

Chronic alcoholics run considerable risk of neurologic symptoms associated with thiamin or pyridoxine deficiencies and hematological problems associated with folate or pyridoxine deficiencies. The deficiencies are not necessarily due to poor diet alone, although it is often a strong contributing factor. Alcohol causes pathological alterations of the gastrointestinal tract that directly interfere with absorption of certain nutrients. The severe liver damage associated with chronic alcoholism appears to interfere with storage and activation of nutrients and vitamins.

Up to 40% of hospitalized alcoholics have megaloblastic erythropoiesis due to folate deficiency. Alcohol interferes with folate absorption and alcoholic cirrhosis impairs folate storage. Another 30% of hospitalized alcoholics have sideroblastic anemia or identifiable sideroblasts in erythroid marrow cells characteristic of pyridoxine deficiency. Some alcoholics develop a peripheral neuropathy that responds to pyridoxine supplementation. This problem may result from impaired activation or increased degradation of pyridoxine. In particular, acetaldehyde (a product of alcohol metabolism) displaces pyridoxal phosphate from its carrier protein in the plasma, causing it to be rapidly degraded to inactive compounds and excreted.

The most dramatic disorder is the Wernicke–Korsakoff syndrome. The symptoms include mental disturbances, ataxia (unsteady gait and lack of fine motor coordination), and uncoordinated eye movements. Congestive heart failure similar to that seen with beriberi may also occur. While this syndrome may only account for 1%–3% of alcohol-related neurologic disorders, the response to supplemental thiamin is dramatic. The thiamin deficiency probably arises from impaired absorption, although alcoholic cirrhosis may also affect storage of thiamin in the liver.

Deficiencies of most water-soluble vitamins can occur and cases of alcoholic scurvy and pellagra are occasionally reported. Chronic ethanol consumption causes redistribution of vitamin A stores in the body. Liver stores are rapidly depleted while levels of vitamin A in the serum and other tissues may be normal or slightly elevated. Apparently, ethanol causes increased mobilization of vitamin A from the liver and increased catabolism of liver vitamin A to inactive metabolites by the hepatic cytochrome P450 system. Alcoholic patients have decreased bone density and an increased incidence of osteoporosis. This probably relates to a defect in the 25-hydroxylation step in the liver and an increased rate of metabolism of vitamin D to inactive products by the cytochrome P450 system. Alcoholics generally have decreased serum levels of zinc, calcium, and magnesium due to poor dietary intake and increased urinary losses. Iron-deficiency anemia is very rare unless there is gastrointestinal bleeding or chronic infection. In fact, excess iron is a more common problem with alcoholics. Many alcoholic beverages contain relatively high iron levels, and alcohol may enhance iron absorption.

Lieber, C. S. Alcohol, liver and nutrition. *J. Am. College Nutr.* 10:602, 1991; and Markowitz, J. S., McRae, A. L., and Sonne, S. C. Oral nutritional supplementation for the alcoholic patient: A brief overview. *Ann. Clin. Psychiatry* 12:153, 2000.

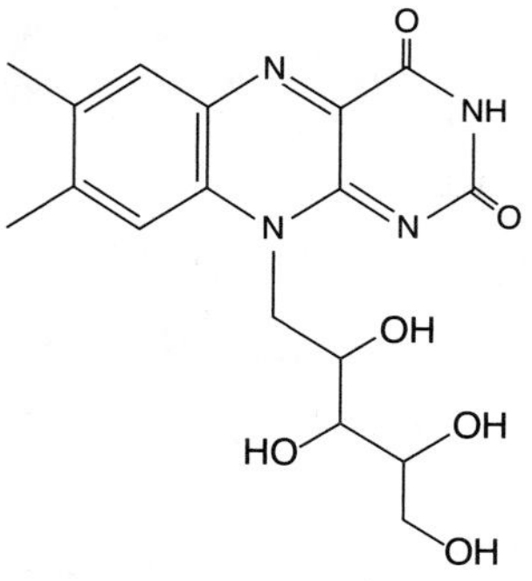

riboflavin

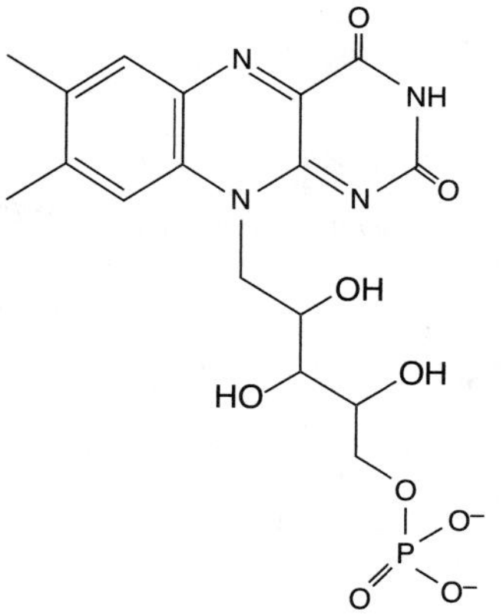

flavin mononucleotide (FMN)

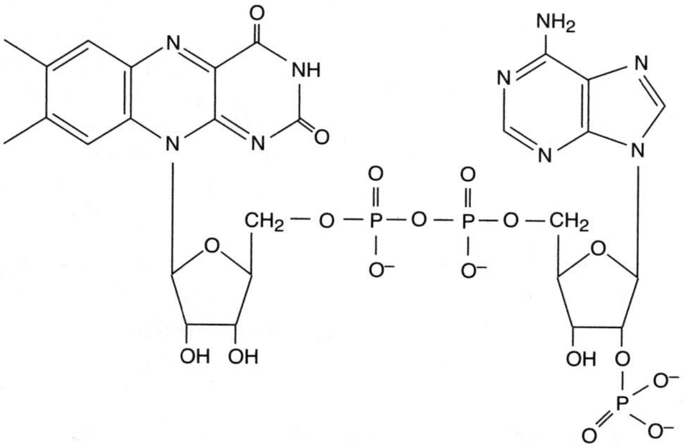

flavinadenine dinucleotide (FAD)

Figure 26.12 Structures of riboflavin, flavin mononucleotide (FMN) and flavin adenine dinucleotide (FAD).

Riboflavin Forms the Coenzymes FAD and FMN

Riboflavin (Figure 26.12) is the precursor of **flavin adenine dinucleotide** (**FAD**) and **flavin mononucleotide** (**FMN**), both of which are coenzymes in a wide variety of redox reactions essential for energy production and cellular respiration. Riboflavin is also required for iron mobilization, and riboflavin deficiency can contribute to anemia when iron intake is low. Characteristic symptoms of riboflavin deficiency are angular cheilitis, glossitis, and scaly dermatitis (especially around the nasolabial folds and scrotal areas). The best enzyme for assaying riboflavin status is erythrocyte glutathione reductase. Foods rich in riboflavin include milk, meat, eggs, and cereal products. Riboflavin deficiencies are quite rare in this country, and are usually seen in chronic alcoholics. Hypothyroidism slows the conversion of riboflavin to FMN and FAD, but it is not known whether this affects riboflavin requirements.

Niacin Forms the Coenzymes NAD and NADP

Niacin (Figure 26.13) is not a vitamin in the strictest sense of the word, since niacin can be synthesized from tryptophan (Figure 26.14). However, conversion of tryptophan to niacin is relatively inefficient (60 mg of tryptophan is required for the production of 1 mg of niacin) and occurs only after all of the body requirements for tryptophan (protein synthesis, production of serotonin and melatonin, and energy production) have been met. Since synthesis of niacin requires pyridoxine, riboflavin and iron (Figure 26.14), it is also very inefficient on a marginal diet. Dietary niacin (nicotinic acid) and niacinamide (nicotinamide) are both converted to the ubiquitous oxidation–reduction coenzymes NAD and NADP (Figure 26.14). These coenzymes are electron acceptors or hydrogen donors in many redox reactions and cellular respiration. NAD is also required for the poly-ADP-ribose polymerase reaction, which is part of the cellular DNA damage recognition system and regulates DNA replication, DNA repair, and cell cycle progression.

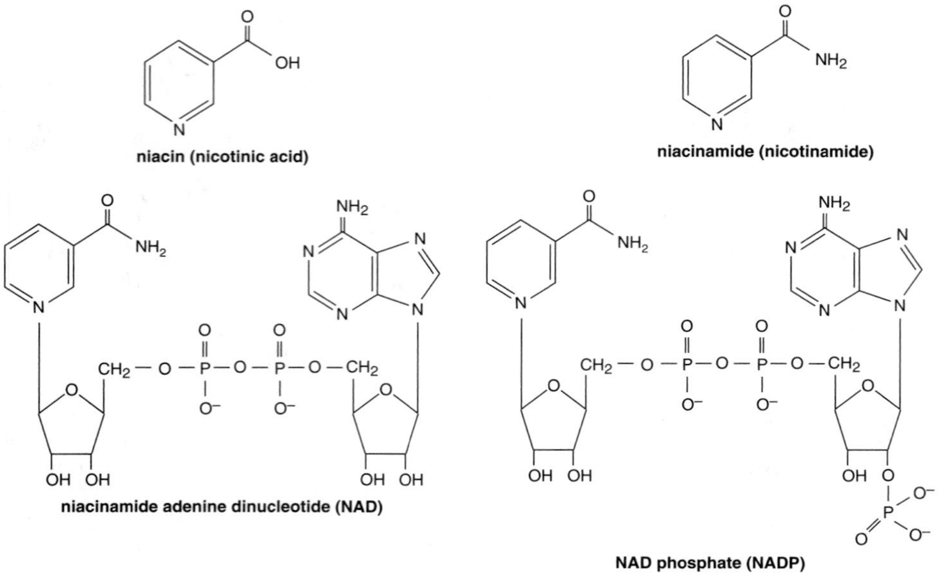

niacin (nicotinic acid)

niacinamide (nicotinamide)

niacinamide adenine dinucleotide (NAD)

NAD phosphate (NADP)

Figure 26.13 **Structures of niacin and related metabolites**

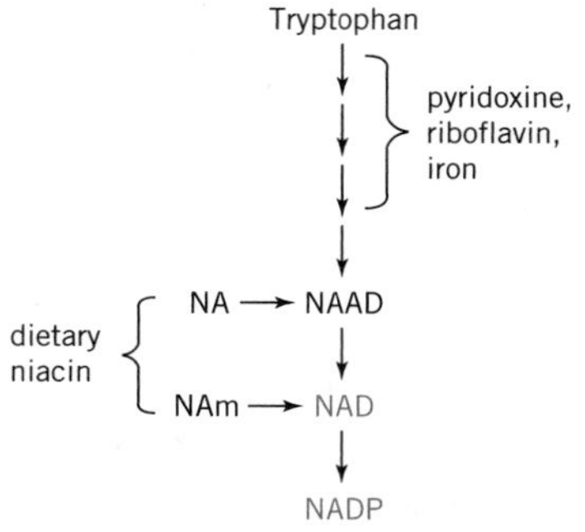

Tryptophan

pyridoxine, riboflavin, iron

dietary niacin

NA → NAAD

NAm → NAD

NADP

Figure 26.14 **Formation of NAD and NADP.** Pathways for formation of NAD and NADP. Abbreviations: NA = nicotinic acid; NAm = nicotinamide; and NAAD = nicotinic acid adenine dinucleotide.

Borderline **niacin deficiency** results in a glossitis (redness) of the tongue, somewhat similar to riboflavin deficiency. Pronounced deficiency leads to pellagra, which is characterized by the three Ds: dermatitis, diarrhea, and dementia. The dermatitis is usually seen only in skin areas exposed to sunlight and is symmetric. The neurologic symptoms are associated with actual degeneration of nervous tissue. Because of food fortification, pellagra is a medical curiosity in the developed world, being primarily seen in alcoholics, patients with severe malabsorption, and elderly on very restricted diets. Pregnancy, lactation, and chronic illness lead to increased needs for niacin, but a varied diet usually provides sufficient amounts. The richest sources of niacin are meats, peanuts and other legumes, and enriched cereals.

Nicotinic acid in pharmacologic doses (1.4–4 g/day) is used to lower LDL cholesterol and triglycerides, and increase HDL cholesterol levels. The decrease in LDL cholesterol and triglycerides appears to be due to a direct, noncompetitive inhibition by niacin of liver diacylglycerol acyltransferase-2, a key enzyme in triglyceride synthesis. The decreased hepatic triglyceride synthesis results in intracellular apo B degradation and decreased secretion of VLDL particles. The increased HDL levels appear to be due to an effect of niacin on a hepatic apo AI receptor, which inhibits removal of HDL particles from the circulation. Side effects of niacin therapy include flushing of the skin, hyperuricemia, and elevation of liver enzymes. The flushing of the skin can be avoided by using niacinamide or slow-release preparations of nicotinic acid, but careful monitoring of the patient for hepatic alterations is still required.

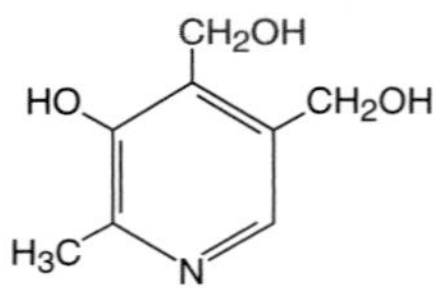

Pyridoxine

Pyridoxamine

Pyridoxal

Pyridoxal phosphate

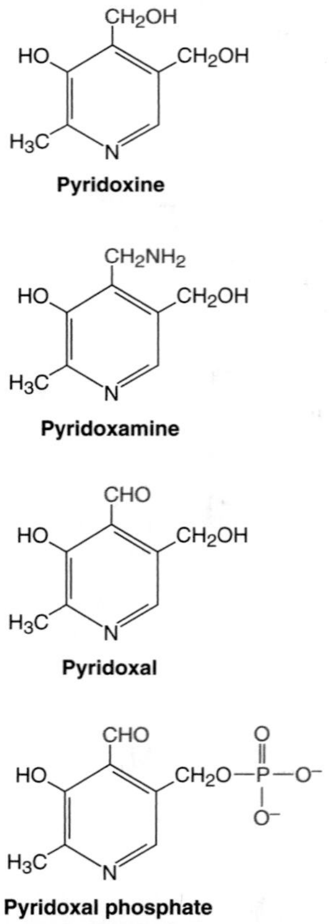

Figure 26.15 **Structures of vitamin B_6.**

Pyridoxine (Vitamin B_6) Forms the Coenzyme Pyridoxal Phosphate

Pyridoxine, **pyridoxamine**, and **pyridoxal**, are naturally occurring forms of vitamin B_6 (Figure 26.15). They are efficiently converted to pyridoxal phosphate, which is required for the synthesis, catabolism, and interconversion of amino acids. **Pyridoxal phosphate** is essential for the transaminase reactions that allow the interconversion of amino acids and their entry into energy generating pathways (p. 754) and can, therefore, be considered an energy-releasing vitamin. Some symptoms of severe deficiency are similar to those of the other energy-releasing vitamins. It is also required for the synthesis of the neurotransmitters serotonin, norepinephrine, epinephrine, and γ-aminobutyrate (GABA) and for synthesis of the sphingolipids necessary for myelin formation. This may explain the irritability, nervousness, and depression seen with mild deficiency and the peripheral neuropathy and convulsions observed with severe deficiency. It is required for synthesis of δ-aminolevulinic acid, a precursor of heme, and B_6 deficiencies can cause sideroblastic microcytic anemia. Pyridoxal phosphate is covalently linked to a lysine residue of glycogen phosphorylase, which stabilizes the enzyme. This may explain the decreased glucose tolerance associated with deficiency, although B_6 appears to have some direct effects on the glucocorticoid receptor as well. Pyridoxal phosphate is also required for the conversion of homocysteine to cysteine. Hyperhomocysteinemia is associated with increased risk for cardiovascular disease and cognitive decline in the elderly (see Section 26.7, p. 1079).

The dietary requirement for B_6 is roughly proportional to the protein content of the diet. The requirement is increased during pregnancy and lactation, and it may increase with age. Vitamin B_6 is widespread in foods, but meat, vegetables, whole-grain cereals, and egg yolks are among the richest sources. It had been assumed that the average American diet was adequate in B_6 and it was, therefore, not routinely added to flour and other fortified foods. However, recent nutritional surveys have found that a significant fraction of the US population consumes less than the recommended intake for B_6.

Pantothenic Acid and Biotin Form Coenzymes Involved in Energy Metabolism

Pantothenic acid is a component of **coenzyme A (CoA)** and the phosphopantotheine moiety of fatty acid synthase and is required for metabolism of fat, protein, and carbohydrate via the citric acid cycle and for fatty acid and cholesterol synthesis. More than 70 enzymes have been described to date that utilize CoA or its derivatives. Therefore, one would expect pantothenic acid deficiency to be a serious concern in humans. This is not the case since (1) pantothenic acid is very widespread in natural foods, probably reflecting its widespread metabolic role and (2) when pantothenic acid deficiency does occur, it is usually associated with multiple nutrient deficiencies, thus making it difficult to discern symptoms specific to the pantothenic acid deficiency.

Biotin is covalently bound to an ϵ-amino group of a lysine residue in pyruvate carboxylase, acetyl-CoA carboxylase, propionyl-CoA carboxylase, and β-methylcrotonyl-CoA carboxylase. Biotin occurs in peanuts, chocolate, and eggs and is synthesized by intestinal bacteria. However, the biotin synthesized by intestinal bacteria may not be present in a location or a form that can contribute significantly to absorbed biotin.

α-Lipoic Acid Plays Multiple Roles in the Body

α-Lipoic acid is an essential coenzyme for the pyruvate dehydrogenase, α-ketoglutarate dehydrogenase and branched-chain α-keto acid dehydrogenase reactions, so it plays a major role in energy generation via the citric acid cycle. In cells it is reduced to dihydrolipoic acid, which is a potent antioxidant. Finally, α-lipoic acid increases the activity of adenylate kinase, PPAR-α, and PPAR-γ by unknown mechanisms. The effects on PPAR-α and PPAR-γ may explain clinical studies showing that α-lipoic acid supplementation appears to improve glucose utilization and decrease insulin resistance in patients with metabolic syndrome and type 2 diabetes.

26.7 • HEMATOPOIETIC WATER-SOLUBLE VITAMINS

Folic Acid Functions as Tetrahydrofolate in One-Carbon Metabolism

The simplest form of **folic acid** is pteroylmonoglutamic acid. Folic acid in the diet usually occurs as **polyglutamate derivatives** with from 2 to 7 glutamic acid residues joined in γ-peptide linkages (Figure 26.16). An enzyme called **folylpoly-γ-glutamate carboxylase II (glutamate conjugase)** removes the extra glutamate in the intestine. The monoglutaminated folic acid is then taken up by the reduced folate carrier (RFC) in intestinal mucosal cells. The monoglutaminated folic acid in the blood is taken into cells via the folate receptor (FR) α and

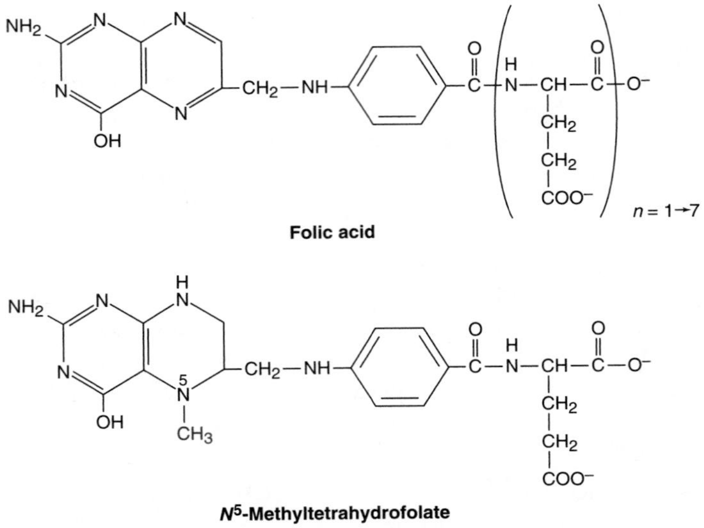

Folic acid

N^5-Methyltetrahydrofolate

Figure 26.16 Structure of folic acid and N^5-methyltetrahydrofolate.

the polyglutamate tail is added by folylpoly-γ-glutamate synthetase. The polygluamated folic acid is then reduced to **tetrahydrofolate** polyglutamate by **dihydrofolate reductase (DHFR)** (Figure 26.17). Inside cells tetrahydrofolates are present primarily as polyglutamates, which appear to be the biologically active form. Tetrahydrofolate polyglutamate is also stored in liver.

Tetrahydrofolate is a one-carbon carrier that facilitates interconversion of methenyl, formyl, formimino, methylene, and methyl groups (Figure 26.18). This occurs at the expense of pyridine nucleotide reduction or oxidation and while the carbon moiety is attached to THF. The interconversions permit use of a carbon that is removed in one oxidation state from a molecule for addition in a different oxidation state to a different molecule.

The various one-carbon tetrahydrofolate derivatives are used in the synthesis of serine, glycine, purines, and dTMP (see Figure 26.17). N^5-methyl tetrahydrofolate is also required for the B_{12}-dependent conversion of homocysteine to methionine. Homocysteine is a nonessential amino acid that is toxic when it accumulates within the cell. Methionine is a precursor to *S*-adenosyl methionine which is required for cellular methylation reactions,

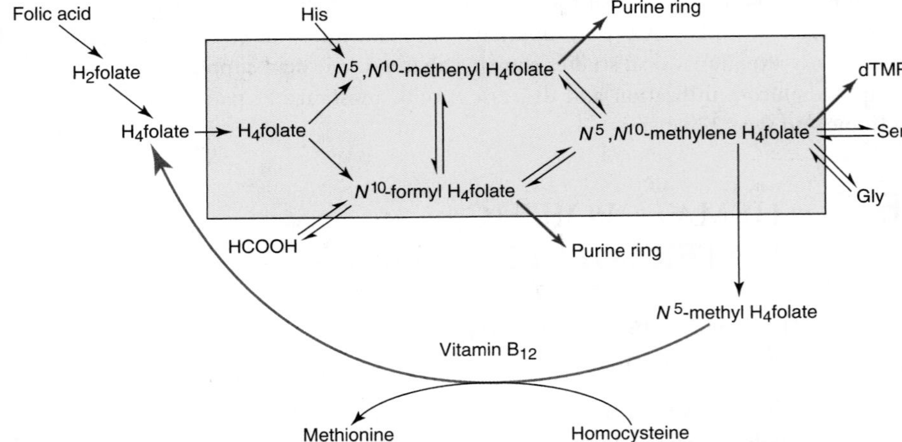

Figure 26.17 Metabolic roles of folic acid and vitamin B_{12} in one-carbon metabolism.
Metabolic interconversions of folic acid and its derivatives are indicated with black arrows. Pathways relying exclusively on folate are shown with red arrows. The important B_{12}-dependent reaction converting N^5-methyl tetrahydrofolate (H_4folate) back to H_4folate is shown with a blue arrow. The box encloses the 'pool' of one-carbon derivatives of H_4folate.

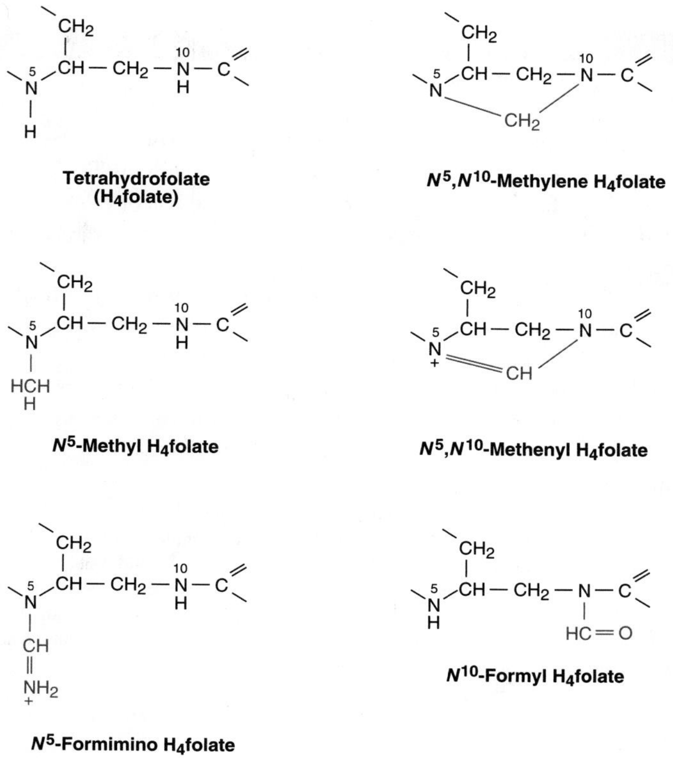

Tetrahydrofolate (H$_4$folate)

N^5,N^{10}-Methylene H$_4$folate

N^5-Methyl H$_4$folate

N^5,N^{10}-Methenyl H$_4$folate

N^5-Formimino H$_4$folate

N^{10}-Formyl H$_4$folate

Figure 26.18 Active center of tetrahydrofolate. N5 is the site of attachment of methyl and formimino groups; N10 is the site for formyl; methylene and methenyl groups form bridges between N5 and N10.

including choline biosynthesis and methylation of DNA and histones. Thus, adequate folate status is not only required for DNA synthesis and cellular proliferation, but is also required for normal regulation of gene expression.

Folate deficiency inhibits DNA synthesis by decreasing the availability of purines and dTMP. This leads to arrest of cells in S phase (p. 1004), a characteristic megaloblastic change in size and shape of the nuclei and a slower maturation of red blood cells, causing production of abnormally large macrocytic red blood cells with fragile membranes. Thus a macrocytic anemia associated with megaloblastic changes in the bone marrow is characteristic of folate deficiency. Folate deficiency in pregnant women increases the risk of birth defects, especially neural tube defects (Clin. Corr. 26.6), either due to effects on cell division or gene regulation during development. In addition, hyperhomocysteinemia is fairly common in the elderly population and may be due to inadequate intake and/or decreased utilization of folate, vitamin B_6, and vitamin B_{12}. Hyperhomocysteinemia is associated with increased risk for cardiovascular disease and cognitive decline in the elderly and usually responds to supplementation with folic acid, vitamin B_6, and vitamin B_{12}. Finally, folate deficiency appears to be associated with several forms of cancer, especially colon and cervical cancer.

Folate deficiency is caused by inadequate intake, increased need, impaired absorption, increased demand, and impaired metabolism. Some dietary surveys suggest that inadequate intake may be more common than previously supposed. As with most other vitamins, inadequate intake is probably not sufficient to trigger symptoms of folate deficiency in the absence of increased requirements or decreased utilization. For example, gene polymorphisms that increase the need for folate may be common (Clin. Corr. 26.6). Increased need also occurs during pregnancy and lactation. By the third trimester the folic acid requirement has almost doubled. In the United States, 20%–25% of otherwise normal pregnant women have low serum folate, but actual megaloblastic anemia is rare and is usually seen only after multiple pregnancies. However, inadequate folate levels during the early stages of pregnancy increase the risk for neural tube defects, a type of birth defect. Normal diets seldom supply the 600 μg/day of folate needed during pregnancy. Thus, enriched grain products are now fortified with folic acid at a concentration of 1.4 μ/g of product. This level of fortification was designed to increase average intake of folic acid by 100 μ/day. Since fortification, the incidence of neural

CLINICAL CORRELATION 26.6

Gene Polymorphisms and Folic Acid Requirement

Folic acid supplementation lowers the risk of neural tube defects and decreases serum homocysteine levels; this may lower the risk of heart disease. These data lead to an increase in the RDA for folic acid and to fortification of grain products with folic acid. Yet even on a marginal diet, not all adults have elevated homocysteine levels and not all mothers give birth to babies with neural tube defects. What determines these individual responses to inadequate folate intake? There is a common genetic polymorphism in the gene for 5,10-methylenetetrahydrofolate reductase (MTHFR) that produces the 5-methyltetrahydrofolate required for the conversion of homocysteine to methionine (see Figure 26.17, p. 1080). A C677T (OMIM 607093) single nucleotide polymorphism (SNP) results in a substitution of valine for alanine that lowers specific activity and reduces stability of the enzyme. Approximately 12% of the Caucasians and Asians are homozygous (T/T) and 50% are heterozygous (C/T) for this polymorphism. Plasma folate concentrations are significantly lower and plasma homocysteine levels are significantly higher in T/T individuals consuming diets low in folate. When coupled with low folate intake, the T/T genotype may account for 15% of neural tube defects. In addition, older individuals with the T/T genotype and low folate intake appear to be at increased risk of colon cancer. An active investigation of genetic polymorphisms in the other genes involved in folate metabolism is underway. Polymorphisms have been described in a number of other genes involved in tetrahydrofolate one-carbon metabolism, but none of them have been definitively shown to increase the risk of neural tube defects to date. However, absorption of folate by the intestine may be lower in mothers with a history of neural tube defect pregnancies than in control mothers. The genetics of this effect has not yet been determined.

Bailey, L. B., and Gregory, J. F. Polymorphisms of methylenetetrahydrofolate reductase and other enzymes: Metabolic significance, risks, and impact on folate requirement. *J. Nutr.* 129:919, 1999; Fang, J. Y., and Xiao, S. D. Folic acid, polymorphism of methyl-group metabolism genes, and DNA methylation in relation to GI carcinogenesis. *J. Gastroenterol.* 38:821, 2003; and Beaudin, A. E., and Stover, P. J. Folate-mediated one-carbon metabolism and neural tube defects: Balancing genome synthesis and gene expression. *Birth Defects Res.*, Part C, 81:183, 2007.

tube defects has decreased by 19%. Addition of an extra 200 μ/day would offer much greater protection against neural tube defects and hyperhomocysteinemia, but this level of folic acid supplementation could mask the symptoms of vitamin B_{12} deficiency as described below. Thus, most physicians routinely recommend supplementation for women during the child-bearing years and for the elderly. Folate deficiency is also common in alcoholics (see Clin. Corr. 26.5, p. 1076) and in those with malabsorption diseases.

Anticonvulsants and oral contraceptives may interfere with folate absorption, and anticonvulsants appear to increase catabolism of folates (see Clin. Corr. 26.4, p. 1075). Long-term use of these drugs can lead to folate deficiency unless adequate supplementation is provided. For example, 20% of patients using oral contraceptives develop megaloblastic changes in the cervicovaginal epithelium, and 20%–30% show low serum folate.

Vitamin B_{12} (Cobalamin) Contains Cobalt in a Tetrapyrrole Ring

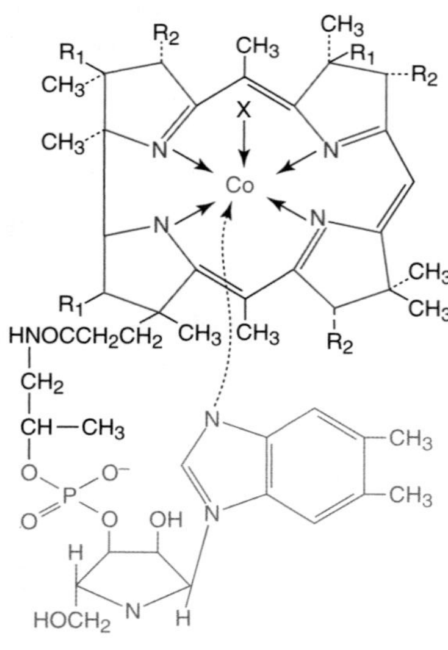

Figure 26.19 Structure of vitamin B_{12} (cobalamine).

Pernicious anemia, a megaloblastic anemia associated with neurological deterioration caused by progressive demyelination of nervous tissue, was invariably fatal until 1926 when liver extracts were shown to be curative. Subsequent work showed the need for both an **extrinsic factor** present in liver and an **intrinsic factor** produced by the body; **vitamin B_{12}** was the extrinsic factor. Vitamin B_{12} contains **cobalt** in a coordination state of six, coordinated in four positions by a tetrapyrrole (corrin) ring, in one position by a benzimidazole nitrogen, and in the sixth position by one of several different ligands (Figure 26.19). The crystalline forms of B_{12} used in supplementation are usually hydroxycobalamin or cyanocobalamin. B_{12} in foods usually occurs bound to protein in the methyl or 5′-deoxyadenosyl form. To be utilized, the B_{12} must be released from the protein by acid hydrolysis in the stomach or trypsin digestion in the intestine. It then combines with the intrinsic factor, a protein secreted by the stomach; this carries it to the ileum for absorption.

Vitamin B_{12} participates in only two chemical reactions in humans. The methyl derivative of B_{12} is required for **methionine synthase**, in which homocysteine is methylated to methionine. The 5-deoxyadenosyl derivative is required for **methylmalonyl-CoA mutase**, which converts methylmalonyl CoA to succinyl CoA, a key reaction in catabolism of valine, isoleucine, methionine, threonine, odd chain fatty acids, thymine, and the side chain of cholesterol. As might be expected, B_{12} deficiency causes the accumulation of both homocysteine and methylmalonic acid.

The megaloblastic anemia associated with B_{12} deficiency is thought to reflect the effect of B_{12} on folate metabolism. B_{12}- dependent methionine synthesis (homocysteine + N^5-methyl THF –> methionine + THF) is the only pathway by which N^5-methyltetrahydrofolate can return to the tetrahydrofolate pool (see Figure 26.17, p. 1080). Thus, in B_{12} deficiency essentially all of the folate becomes "trapped" as the N^5-methyl derivative, causing a buildup of N^5-methyltetrahydrofolate and a deficiency of the tetrahydrofolate derivatives needed for purine and dTMP biosynthesis. By replenishing the tetrahydrofolate pool, large amounts of supplemental folate can overcome the megaloblastic anemia, but not the neurological problems. This is the crux of the current debate on the optimal levels for folate fortification of foods. It is the megaloblastic anemia that usually brings the patient into the doctor's office. Thus, by masking the anemia, routine fortification of foods with high levels of folate could prevent detection of B_{12} deficiency until the neurological damage had become irreversible.

It has been proposed that the demyelination associated with B_{12} deficiency is caused by methylmalonyl CoA accumulation in two ways. (1) Methylmalonyl CoA is a competitive inhibitor of malonyl CoA in fatty acid biosynthesis. Because the myelin sheath is continually turning over, any severe inhibition of fatty acid biosynthesis will lead to its degeneration. (2) Methylmalonyl CoA can substitute for malonyl CoA in fatty acid synthesis leading to synthesis of branched-chain fatty acids, which may disrupt membrane structure. However, the neurologic symptoms of B_{12} deficiency cannot be fully explained by either mechanism, since accumulation of both methylmalonic acid and homocysteine is required for demyelination. Recent studies have shown that B_{12} deficiency is associated with increased expression of tumor necrosis factor α (TNF-α) and nerve growth factor (NGF) and decreased expression of epidermal growth factor (EGF) and interleukin 6 (IL-6) in cerebrospinal fluid, but the mechanism of these changes and their effect on neural function remain unknown.

Vitamin B_{12} is widespread in foods of animal origin, especially meats. Since the liver stores up to a 6-year supply of vitamin B_{12}, deficiencies of B_{12} are rare except in older people who produce insufficient intrinsic factor and/or HCl in the stomach (Clin. Corr. 26.7), patients with severe malabsorption diseases (see Clin. Corr. 26.1, p. 1068) and long-term vegetarians.

CLINICAL CORRELATION 26.7

Nutritional Needs of Elderly Persons

If current trends continue, one out of five Americans will be over the age of 65 by the year 2030. With this projected aging of the American population, there has been increased interest in defining the nutritional needs of the elderly. Recent research shows altered needs of elderly persons for several essential nutrients. For example, the absorption and utilization of vitamin B_6 decreases with age. Dietary surveys have consistently shown that B_6 is a problem nutrient for many Americans, and the elderly are no exception. Many older Americans get less than 50% of the RDA for B_6 from their diet. Vitamin B_{12} deficiency is also more prevalent in the elderly. Many older adults develop atrophic gastritis (decreased acid production in the stomach) and decreased production of intrinsic factor, which lead to poor absorption of B_{12}. The blood level of homocysteine, a possible risk factor for atherosclerosis, dementia, and Alzheimer disease, is often elevated in the elderly. Homocysteine is a by-product of DNA methylation and is normally metabolized to methionine or cysteine in reactions requiring folic acid, B_{12} and B_6 (see Figure 26.17, p. 1080). Simple supplementation with those B vitamins is generally sufficient to normalize homocysteine levels. Vitamin D can be a problem as well. Many elderly do not spend much time in the sunlight, and the conversion of 7-dehydrocholesterol to vitamin D in the skin and 25-(OH)D to 1,25-$(OH)_2$D in the kidney decreases with age. These factors lead to significant deficiencies of 1,25-$(OH)_2$D in the elderly, which can cause a negative calcium balance. These changes may contribute to osteoporosis.

There is some evidence for increased need for chromium and zinc as well. Many elderly appear to have difficulty converting dietary chromium to the biologically active chromodulin. Chromium deficiency could contribute to type 2 diabetes. Similarly, most elderly consume between one-half and two-thirds the RDA for zinc, and conditions such as atrophic gastritis can interfere with zinc absorption. Symptoms of zinc deficiency include loss of taste acuity, dermatitis, and a weakened immune system. All of these symptoms are common in the elderly population, and zinc deficiency may contribute.

Not all of the news is bad, however. Vitamin A absorption increases with age and its clearance by the liver decreases, so vitamin A remains in the circulation for a longer time. Not only does the need for vitamin A decrease as we age, but the elderly also need to be particularly careful to avoid vitamin A toxicity. Although this does not restrict their choice of foods or multivitamin supplements, they should generally avoid separate vitamin A supplements.

Johnson, K. A., Bernard, M. A., and Funderburg, K. Vitamin nutrition in older adults. *Clin. Geriatr. Med.* 18:773, 2002; Johnson, M. A., and Kimlin, M. G. Vitamin D, aging, and the 2005 Dietary Guidelines for Americans. *Nutr. Rev.* 64:410, 2006; and Baker, H. Nutrition in the elderly: An overview. *Geriatrics* 62:28, 2007.

26.8 • OTHER WATER-SOLUBLE VITAMINS

Ascorbic Acid Functions in Reduction and Hydroxylation Reactions

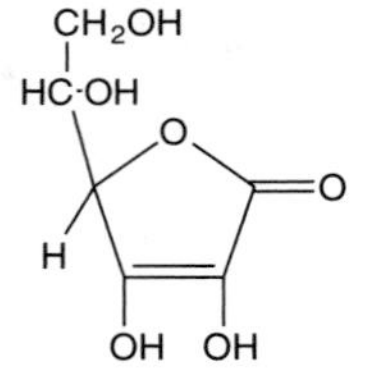

Figure 26.20 **Structure of vitamin C (ascorbic acid).**

Vitamin C or ascorbic acid (Figure 26.20) is a cofactor for mixed function oxidases involved in the hydroxylation of lysine and proline, synthesis of carnitine, and synthesis of norepinephrine. Hydroxylation of lysine and proline is required for protocollagen to cross-link properly into normal collagen fibrils. Vitamin C is required for maintenance of normal connective tissue and for wound healing. It is also necessary for bone formation, since the organic matrix of bone tissue consists largely of collagen. Collagen is also a component of the ground substance surrounding capillary walls, so vitamin C deficiency causes capillary fragility. Carnitine is required for transport of long-chain fatty acids into the mitochondria (p. 691), and decreased carnitine levels may be responsible for the fatigue associated with vitamin C deficiency. Since vitamin C is concentrated in the adrenal gland, it may be required for hydroxylation reactions in synthesis of some corticosteroids especially in periods of stress. Vitamin C also appears to modulate signal transduction pathways and gene expression affecting vascular endothelial cells. Finally, ascorbic acid also acts as a nonenzymatic reducing agent. For example, it aids in **absorption of iron** by reducing it to the ferrous state in the stomach. It spares vitamin A, vitamin E, and some B vitamins by protecting them from oxidation. It enhances utilization of folic acid, either by aiding conversion of folate to tetrahydrofolate or formation of polyglutamate derivatives of tetrahydrofolate.

Symptoms of mild **vitamin C deficiency** include capillary fragility, which leads to easy bruising and formation of petechiae (small, pinpoint hemorrhages in skin) and decreased immunocompetence. Scurvy, a more severe form of deficiency, is associated with decreased wound healing, osteoporosis, hemorrhaging, and anemia. Osteoporosis results from the inability to maintain the collagenous organic matrix of the bone, which leads to demineralization. Anemia results from extensive hemorrhaging coupled with defects in iron absorption and folate metabolism.

Vitamin C is readily absorbed so that deficiency invariably results from poor diet and/or increased need. In severe stress or trauma there is a rapid drop in serum vitamin C levels, and most of the body's supply of vitamin C is mobilized to the adrenals and/or the traumatized area. It is not clear whether this represents an increased demand for vitamin C or merely a normal redistribution to those areas where it is needed most. Nor is it clear whether the lowered serum levels of vitamin C impair its functions in other tissues in the body. The current consensus seems to be that the lowered serum vitamin C levels indicate an increased demand, but there is little agreement as to how much.

Smoking causes lower serum levels of vitamin C. In fact, the RDAs for smokers are 110–125 mg of vitamin C/day versus 75–90 mg/day for nonsmoking adults. Aspirin appears to block uptake of vitamin C by white blood cells. Oral contraceptives and corticosteroids also lower serum levels of vitamin C. The possibility of marginal vitamin C deficiency should be considered with any patient using these drugs over a long period, especially if dietary intake of vitamin C is suboptimal.

The use of megadoses of vitamin C to prevent and cure the common cold has generated considerable controversy. While vitamin C supplementation does not appear to prevent the common cold, it may moderate the symptoms or shorten the duration. It has been suggested that vitamin C is required for normal leukocyte function or that it decreases histamine levels. Although megadoses of vitamin C are probably no more harmful than the widely used over-the-counter cold medications, some potential side effects of high vitamin C intake should be considered. For example, oxalate is a major metabolite of ascorbic acid. Thus, high ascorbate intakes could theoretically lead to the formation of oxalate kidney stones in predisposed individuals. However, most studies have shown that excess vitamin C is primarily excreted as ascorbate rather than oxalate. Pregnant mothers taking megadoses of vitamin C may give birth to infants with abnormally high vitamin C requirements, but this is easily treated. The UL for vitamin C has been set at 2000 mg/day because higher levels can cause diarrhea in some individuals.

Choline and Carnitine Perform Several Functions

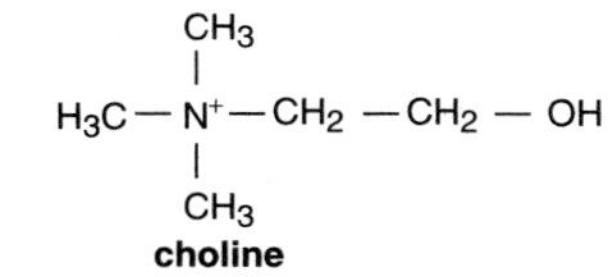

Choline and carnitine have traditionally been considered as nonessential because they can be synthesized *de novo.* However, choline has recently been reclassified as essential and carnitine as conditionally essential. Choline (Figure 26.21) is required for synthesis and release of **acetylcholine**, an important neurotransmitter involved in memory storage, motor control, and other functions. It is also a precursor for synthesis of the phospholipids **phosphatidylcholine** (lecithin) and **sphingomyelin**, which are important for membrane function, intracellular signaling, and hepatic export of very low-density lipoproteins. Phosphatidylcholine is also important in removal of cholesterol from tissues, as it is a substrate for lecithin-cholesterol acyltransferase in reverse cholesterol transport (p. 725). Finally, choline is a precursor for the methyl donor betaine.

$$H_3C-\overset{\displaystyle CH_3}{\underset{\displaystyle CH_3}{\overset{|}{\underset{|}{N^+}}}}-CH_2-CH_2-OH$$

choline

Figure 26.21 **Structure of choline.**

Studies in rodents indicate that choline deficiency increases the risk of liver cancer and memory deficits in aged animals, but these effects have not been demonstrated in humans. Both choline and betaine supplementation appear to lower serum homocysteine levels in humans. However, current data are insufficient to draw firm conclusions about whether choline and/or betaine supplementation have any effects on cardiovascular risk.

A significant portion of the daily choline requirement can be met by the endogenous conversion of phosphatidylethanolamine to phosphatidylcholine by the liver enzyme phosphatidylethanolamine *N*-methyltransferase (PEMT). Because choline can be synthesized *de novo* and is abundant in food, choline deficiency is very rare. Hepatic complications (fatty livers and elevated serum alanine aminotransferase) that respond to choline supplementation have been observed in patients on total parenteral nutrition solutions devoid of choline, with small intestinal bypass, and with liver cirrhosis. Choline is critical during fetal development because it affects DNA methylation, which influences neural cell precursor cell proliferation and apoptosis. Fortunately, PEMT expression is increased severalfold during pregnancy.

Carnitine (Figure 26.22) is required for transport of fatty acids across the mitochondrial membrane, so it is essential for normal fatty acid metabolism (p. 691). In muscle an enzyme called carnitine acyltransferase uses carnitine to convert acetyl CoA to acetylcarnitine, releasing free coenzyme A in the process. This is important because the supply of mitochondrial coenzyme A is very limited and acetyl CoA synthesis by pyruvate dehydrogenase during high intensity exercise is much more rapid than the citric acid cycle can utilize This would quickly deplete the coenzyme A supply and shut down the pyruvate dehydrogenase reaction in the absence of the carnitine acyltransferase reaction. Thus, in exercising muscle, carnitine is essential for both fatty acid and carbohydrate metabolism.

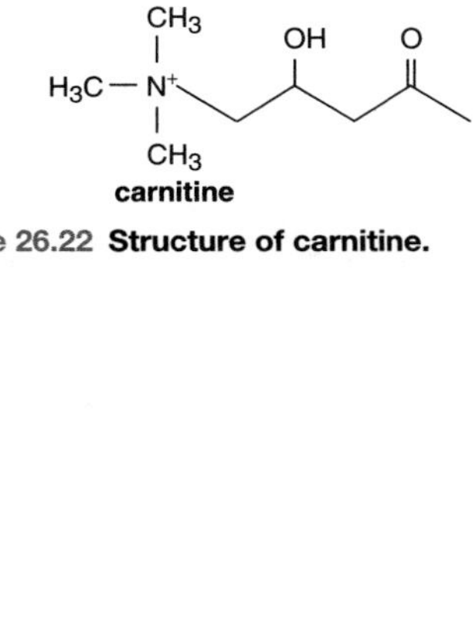

carnitine

Figure 26.22 **Structure of carnitine.**

Because carnitine can be synthesized *de novo,* it is nonessential for normal healthy adults. However, it is considered conditionally essential overall because human genetic disorders of carnitine metabolism have been described and some of them respond to carnitine supplementation. Carnitine is a popular dietary supplement for athletes. However, for supplemental carnitine to have any effect on muscle carnitine levels it must be administered along with sufficient carbohydrate to significantly increase serum insulin levels. Most carnitine supplements on the market do not have any effect on muscle carnitine levels.

26.9 • MACROMINERALS

Calcium Has Many Physiological Roles

Calcium is the most abundant mineral in the body. Most is in bone, but the small amount of Ca^{2+} outside of bone functions in a variety of essential processes. It is required for many enzymes; it mediates some hormonal responses; and it is essential for blood coagulation, muscle contractility, and normal neuromuscular irritability. In fact, only a relatively narrow range of serum Ca^{2+} levels is compatible with life. Since maintenance of constant serum levels is so vital, an elaborate homeostatic control system has evolved (see Figure 26.6, p. 1071). Low serum Ca^{2+} stimulates formation of 1,25-dihydroxycholecalciferol, which enhances intestinal Ca^{2+} absorption and, with parathyroid hormone, stimulates bone resorption. Thus, long-term dietary Ca^{2+} insufficiency almost always results in net loss of calcium from the bones.

Dietary Ca^{2+} requirements, however, vary considerably from individual to individual due to the existence of other factors that affect availability of Ca^{2+}. For example, vitamin D is required for optimal utilization of calcium, while excess dietary protein may cause more rapid excretion of Ca^{2+}. Exercise facilitates calcium utilization for bone formation. Calcium balance studies carried out on Peruvian Indians, who have extensive exposure to sunlight, get extensive exercise, and subsist on low-protein vegetarian diets, indicate a need for only 300–400 mg Ca^{2+}/day. However, similar studies carried out in this country consistently show higher requirements and the RDA has been set at 1000–1300 mg/day.

Symptoms of **Ca^{2+} deficiency** resemble those of vitamin D deficiency, but other symptoms such as muscle cramps are possible with marginal deficiencies. A significant portion of low-income children and adult females in this country do not consume adequate Ca^{2+}. This is of particular concern because these are the population groups with particularly high needs for Ca^{2+}. For this reason, the U.S. Congress has established the WIC (Women and Infant Children) program to assure adequate protein, Ca^{2+}, and iron for indigent families with pregnant/lactating mothers or with young infants.

Dietary surveys show that 34%–47% of the population over 60 years of age consumes less than the EAR for Ca^{2+}. This is the group most at risk of developing osteoporosis, characterized by loss of bone organic matrix and progressive demineralization. Causes of osteoporosis are multifactorial and largely unknown, but it is likely that part of the problem has to do with Ca^{2+} metabolism (Clin. Corr. 26.8). Recent studies suggest that inadequate intake of Ca^{2+} may result in elevated blood pressure. This is of great concern because most low-sodium diets (which are recommended for patients with high blood pressure) severely limit dairy products, the main source of Ca^{2+} for Americans.

Magnesium Is Required by Many Enzymes

Magnesium is required for many enzyme activities, particularly those utilizing an ATP–Mg^{2+} complex, and for neuromuscular transmission. Mg^{2+} content is significantly reduced during processing of food items, and recent dietary surveys have shown that the average

CLINICAL CORRELATION 26.8

Diet and Osteoporosis

There is strong consensus that the years from age 10 to 35, when bone density is reaching its maximum, are the most important for reducing the risk of osteoporosis. The maximum bone density obtained during these years depends on both calcium intake and exercise, and dense bones are less likely to become seriously depleted of calcium following menopause. Unfortunately, most American women consume far too little calcium during these years. The RDA for calcium is 1300 mg/day (4 or more glasses of milk/day) for women from age 11 to 18, 1000 mg/day (3 or more glasses of milk/day) for women from 19 to 50, and 1200 mg/day (4 glasses of milk/day) for women over 50. Some experts think that calcium requirements for postmenopausal women should be even higher. In 1994, an NIH consensus panel on osteoporosis recommended that postmenopausal women consume up to 1500 mg of calcium/day. Unfortunately, the median calcium intake for women 19 and older is only about 500 mg/day, and with the recent concern about the fat content of dairy products, calcium intakes appear to be decreasing rather than increasing. Thus, it is clear that increased calcium intake should be encouraged in this group. Even with drug therapy to prevent osteoporosis, calcium intake should not be ignored. Recent studies have shown that calcium intakes in the range of 1000–1500 mg/day make drug or estrogen therapy more effective at preserving bone mass.

Although most of the focus is on calcium intake, we need to remember that bones are not made of calcium alone. If the diet is deficient in other nutrients, the utilization of calcium for bone formation will be impaired. Vitamin C is needed to form the bone matrix, and magnesium and phosphorus are important components of bone structure. Vitamin K and a variety of trace minerals, including copper, zinc, manganese, and boron, are important for bone formation. Thus, calcium supplements may not be optimally utilized if the overall diet is inadequate. Vitamin D is required for absorption and utilization of calcium. It deserves special mention as it may be a problem for the elderly (see Clin. Corr. 28.7, p. 1083), and some experts feel that the current recommendation for vitamin D intake in adults may be too low. Finally, an adequate exercise program is just as important as drug therapy and an adequate diet for preventing the loss of bone density.

Heaney, R. P. Calcium in the prevention and treatment of osteoporosis. *J. Internal Med.* 231:169, 1992; National Institutes of Health. Optimal Calcium Intake. NIH Consensus Statement. 12 (November 4), 1994; and Murphy, N. M. and Carroll, P. The effect of physical activity and its interaction with nutrition on bone health. *Proc. Nutr. Soc.* 62:829, 2003.

Mg^{2+} intake in Western countries is often below the EAR. Deficiency occurs in alcoholism, with use of certain diuretics and in metabolic acidosis. The main symptoms of **Mg^{2+} deficiency** are weakness, tremors, and cardiac arrhythmia. Supplemental Mg^{2+} may help prevent formation of calcium oxalate stones in the kidney. Mg^{2+} supplementation has also been shown to lower blood pressure in several clinical studies, and there is an inverse effect between dietary Mg^{2+} intake and the risk of stroke.

26.10 • TRACE MINERALS

Iron Deficiency Causes Anemia and Decreased Immunocompetence

Iron is a component of heme in hemoglobin and myoglobin, which are essential for O_2 transport; the cytochromes, which are involved in mitochondrial electron transport; the **P_{450} enzymes**, which are involved in hydroxylation reactions; and the lysosomal enzyme myeloperoxidase, which is required for killing of bacteria and other pathogens by neutrophils. **Nonheme iron proteins** such as ribonucleotide reductase are also involved in a number of redox reactions. Thus, iron is required for O_2 transport, energy metabolism, cell proliferation, and immune defense against pathogens. Total body iron in human adults is 3–4 g. Two-thirds of that is incorporated into the heme moiety of erythrocytes. Erythrocytes normally live only 120 days; this means that erythrocytes containing approximately 20 mg of iron are destroyed by the reticuloendothelial system each day. Fortunately, almost all of that iron is reutilized. The only net loss of iron for males and postmenopausal women is shedding of intestinal and skin cells, which amounts to about 1–2 mg/day. Blood loss due to menstruation and disease and increase of blood volume in children increases the need for iron in those population groups. Assuming a 10%–15% efficiency of absorption, an RDA of 8 mg/day for normal adult males and 18 mg/day for menstruating females has been set. For pregnant females it is 27 mg/day. Although 8 mg/day of iron can easily be obtained from a normal diet, 18 mg is marginal at best and 27 mg can almost never be obtained. The best dietary sources are meats, dried legumes, dried fruits, and enriched cereal products.

While iron is absolutely essential for life, it is also extremely toxic. **Free iron** can generate dangerous free radicals through the Fenton reaction, and free iron in the circulation can support the growth of microbial pathogens and increase the risk of systemic infections. Thus, iron is sequestered in the cell by **ferritin** and in the bloodstream by **transferrin**. **Apoferritin** (the term given to ferritin before it binds iron) is a complex of 24 subunits with a capacity for binding 4500 iron atoms. Each apoferritin molecule is a mixture of two very similar subunits. Apoferritin in every tissue contains some of both subunits, but the H form predominates in nucleated blood cells and heart, while the L subunit predominates in the liver and spleen. Transferrin is a single polypeptide with two binding sites. Once cellular iron exceeds the binding capacity of apoferritin the excess iron is deposited on the outside of ferritin in an amorphous mixture of iron hydroxide, iron phosphate, and proteins called hemosiderin in the liver, heart, pancreas, and pituitary, leading to organ dysfunction.

Because the loss of iron (tissue shedding and blood loss) occurs in an unregulated fashion, regulation of iron homeostasis occurs almost entirely at the level of iron uptake and delivery to the bloodstream by the intestine. The uptake of iron by the small intestine is summarized in Figure 26.23. Heme iron is taken up most efficiently by the intestine, but the mechanism of heme iron uptake is not clear at present. The digestion of nonheme iron proteins in the intestinal lumen releases iron in the +3 state. For the remainder of iron metabolism its oxidation state is critical. Iron is transported across cell membranes in the +2 state and is stored or transported in the +3 state. Thus, **ferrireductases** and **ferroxidases** play an important role in iron metabolism. In the intestine Fe^{3+} is reduced to Fe^{2+} by the ferrireductase duodenal cytochrome b (Dcytb). Ascorbate is the primary source of reducing equivalents for Dcytb, so the absorption of nonheme iron is significantly enhanced when foods containing vitamin C are consumed at the same time.

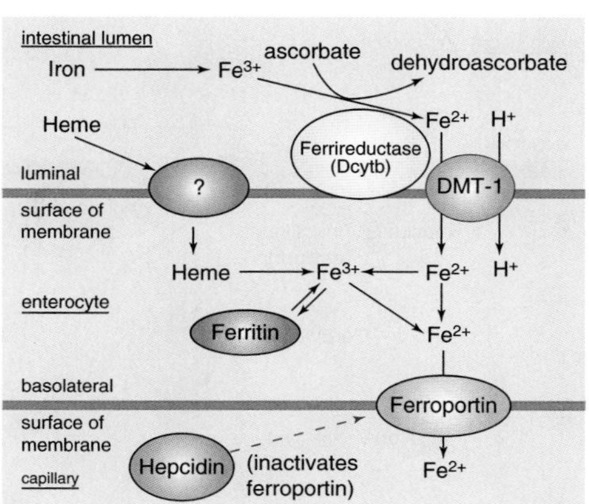

Figure 26.23 Uptake and export of iron by the intestine. Abbreviations: Dcytb, intestinal ferrireductase duodenal cytochrome b; and DMT-1, divalent metal transporter 1.

Fe^{2+} is transported into the intestinal mucosal cells by a transporter called **divalent metal transporter (DMT-1)**. As the name suggests, DMT-1 can also transport several other

TABLE 26.2 • Regulation of Key Enzymes Involved in Iron Homeostasis

Protein	*Regulatory Mechanism*	*Function*
Proteins whose expression increases when iron is scarce		
Intestinal DMT-1	IRE in 3′ UTR	Intestinal absorption of iron
Ferroportin	Hepcidin-dependent degradation when iron high	Export of iron from intestine
Ceruloplasmin	?	Conversion of Fe^{2+} to Fe^{3+}
Transferrin	IRE in 3′ UTR	Transport of iron in the blood
Transferrin receptor 1	IRE in 3′ UTR	Uptake of iron by cells
Proteins whose expression increases when iron is abundant		
Apoferritin	IRE in 5′ UTR	Intracellular storage and decreased intestinal transport
Transferrin receptor 2	IRE in 5′ UTR	Stimulates increased expression of hepcidin
Hepcidin	Expression stimulated by HJV, HFE and TfR2	Facilitates ferroportin degradation
Aminolevulinic acid synthase	IRE in 5′ UTR	Porphyrin biosynthesis

transition metals that occur naturally in the +2 state such as zinc, copper, and manganese. A large excess of any one of these essential trace metals in the diet can create a deficiency of the others. Iron transport by DMT-1 requires co-transport of protons, so it is most active in the upper portion of the duodenum where stomach acid empties into the intestine. This is why antacids and H2 histamine blockers interfere with iron absorption.

Because of the cytotoxicity of free iron, most of it is sequestered by binding to ferritin once it enters the intestinal mucosal cell. In addition to its role in protecting the cell from the cytotoxic effects of iron, the sequestration of iron by ferritin in the intestinal mucosal cell reduces the net delivery of iron to the circulation, and the sequestration of iron by ferritin in the liver removes iron from the circulation under conditions of iron excess. The H subunit of ferritin has the ferroxidase activity required for binding of iron by the ferritin complex. Release of iron into the circulation requires another ferrireductase to reduce it to the +2 state and a transporter called ferroportin. Levels of ferroportin are controlled by a peptide produced by the liver called **hepcidin**, which binds to ferroportin triggering its tyrosine phosphorylation, internalization, and ubiquitin-mediated degradation in proteasomes. When iron is in excess, hepcidin levels are increased, leading to down-regulation of ferroportin and decreasing iron transport into the circulation by the intestinal mucosal cells, by the liver and by macrophages in the reticuloendothelial system.

The key enzymes in this transport process are coordinately regulated to maintain iron homeostasis (Table 26.2). When iron levels are low, DMT-1 expression is up-regulated and ferritin expression is down-regulated. In addition hepcidin expression is down-regulated; this results in stabilization and accumulation of ferroportin. These effects combine to increase uptake, decrease sequestration, and increase export of iron by intestinal mucosal cells. In contrast, when iron levels are high, DMT-1 expression is down-regulated and both ferritin and hepcidin expression are up-regulated. This decreases export of iron by intestinal mucosal cells and increases retention of iron stores by the liver.

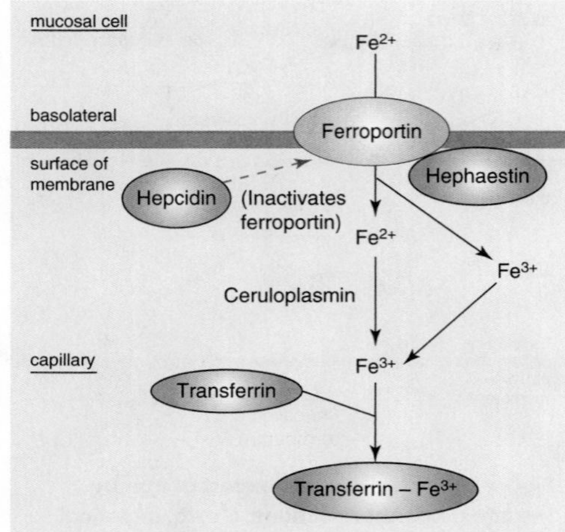

Figure 26.24 Transport of iron in the circulation.

Transport of iron in the bloodstream is summarized in Figure 26.24. Once Fe^{2+} enters the circulation it is oxidized to the +3 state by two ferroxidases called **hephaestin** and **ceruloplasmin**. Both hephaestin and ceruloplasmin are involved in oxidation of Fe^{2+} to Fe^{3+} in the intestine, but hepatocytes and macrophages appear to use ceruloplasmin exclusively. Ceruloplasmin and hephaestin are copper-containing enzymes; this is thought to be the reason that anemia is one symptom of copper deficiency (Clin. Corr. 26.9). Fe^{3+} is sequestered and transported through the bloodstream by transferrin. Transferrin levels are increased under conditions of iron deficiency and decreased under conditions of iron excess, but transferrin levels are generally in excess so this effect is much less important than the regulation of the other proteins involved in maintaining iron homeostasis.

CLINICAL CORRELATION 26.9

Ceruloplasmin and Iron Metabolism

A deficiency, but not the absence of ceruloplasmin, a copper-containing protein, is associated with Wilson disease because the ATP7B copper transporter, which is defective in Wilson disease (OMIM 27900), is essential for copper delivery to ceruloplasmin and copper-deficient ceruloplasmin is unstable. Because there was no evidence for significant impairment of mobilization of iron in Wilson disease, it was formerly thought that the ferroxidase activity of ceruloplasmin was not physiologically important. However, a very rare genetic defect in ceruloplasmin biosynthesis, where the protein was virtually absent in serum, leads to a marked elevation of liver-iron content and serum ferritin levels. These patients develop diabetes mellitus, retinal degeneration, and central nervous system changes. The diabetes and central nervous system findings are associated with increased iron in the pancreas and brain, respectively.

Iron-deficiency anemia is not associated with ceruloplasmin deficiency because the intestine has a second ferroxidase called hephaestin. However, both ceruloplasmin and hephaestin are copper-containing enzymes, so copper deficiency can lead to an iron-deficiency anemia. In addition, transcription of the ceruloplasmin gene in iron deficiency increases fourfold. Thus, in contrast to earlier considerations, it appears that ceruloplasmin has a significant role in iron metabolism.

Harris, E. D. The iron-copper connection: The link to ceruloplasmin grows stronger. *Nutr. Rev.* 53:226, 1995; Mukhopadhyay, C. K., Mazumder, B., and Fox, P. L. Role of hypoxia-inducible factor-1 in transcriptional activation of ceruloplasmin by iron deficiency. *J. Biol. Chem.* 275:21048, 2000; and Van Eden, M. E., and Aust S. D. Intact human ceruloplasmin is required for the incorporation of iron into human ferritin. *Arch. Biochem. Biophys.* 381:119, 2000.

As shown in Figure 26.25, transferrin is taken up by cells by binding to transferrin receptor 1 (TfR1). The transferrin–receptor complex clusters in clathrin-coated pits and is taken up by endocytosis. As the interior of the endosome is acidified, Fe^{3+} is released from transferrin and reduced by a ferrireductase called Streap3; the resulting Fe^{2+} is transported into the cytosol by DMT-1. In this case the activity of DMT-1 is optimal because of the low pH of the endosome. At the end of this process the transferrin receptor is recycled to the cell surface. Expression of TfR1 is increased under conditions of iron deficiency and decreased under conditions of iron excess. There is a homologous protein called transferrin receptor 2 (TfR2), but it appears to serve primarily as an iron sensor and its level is increased under conditions of iron excess. Cytosolic Fe^{2+} is transported into the mitochondria by a transporter called mitoferrin and inserted into protoporphyrin IX to form heme by the enzyme ferrochelatase (p. 796).

Regulation of key enzymes involved in iron homeostasis is summarized in Table 26.2. Several of these enzymes are regulated at a translational level by **iron responsive elements (IREs)** and **iron responsive proteins (IRPs)** (Figure 26.26). Iron responsive elements are stem-loop structures in either the 3′- or the 5′- untranslated regions of mRNAs for proteins involved in iron homeostasis. Iron-responsive proteins 1 and 2 (IRP1 and IRP2) are proteins that bind to the IREs. When the IRE is located in the 3′ untranslated region, such as with transferrin receptor mRNA, binding of an IRP stabilizes the mRNA and enhances its

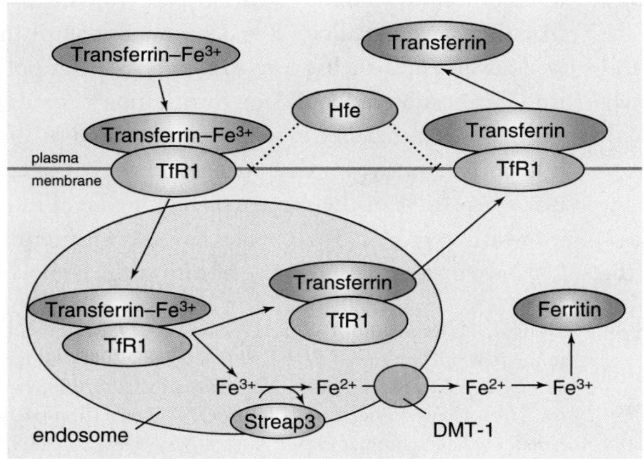

Figure 26.25 **Cellular uptake of transferrin**

Control of Transferrin Receptor Synthesis

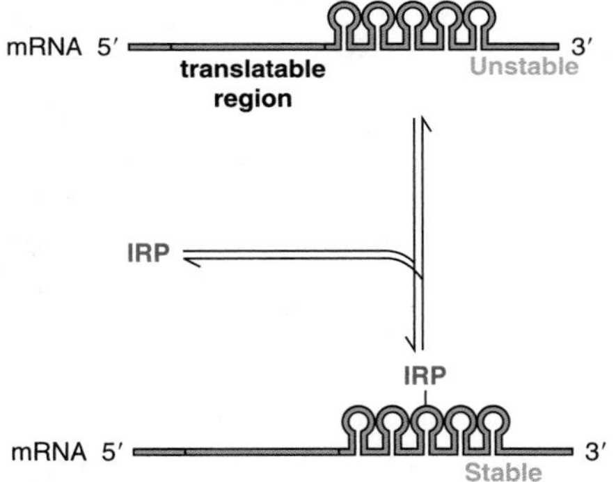

Control of Apoferritin Synthesis

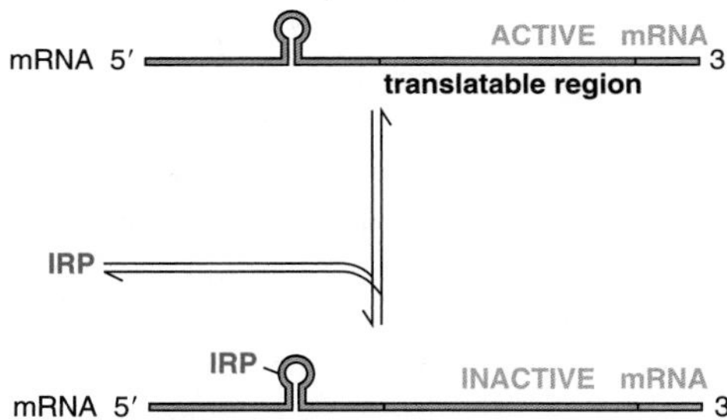

Figure 26.26 **Control of transferrin and apoferritin synthesis by iron responsive elements (IREs) and iron responsive proteins (IRPs).** A: Control of transferrin receptor synthesis; B: Control of apoferritin synthesis

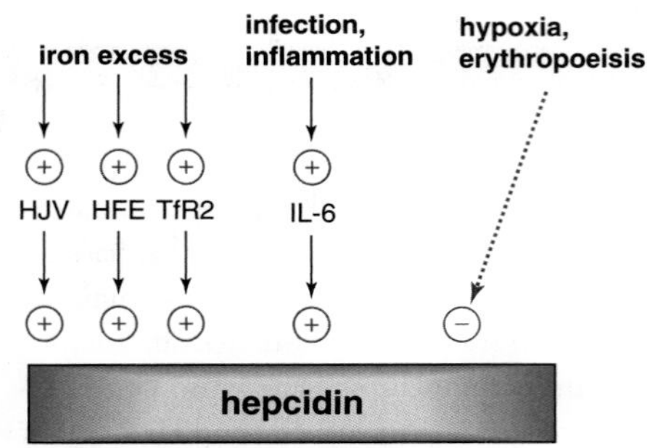

Figure 26.27 Regulation of hepcidin synthesis by the liver. HJV, hemojuvelin; HFE, histocompatibility factor E; TfR2, transferrin receptor 2; IL-6, interleukin 6.

translation. When the IRE is located in the 5′ untranslated region, such as with apoferritin, binding of an IRP interferes with ribosome loading and results in inhibition of translation.

IRP1 has a very interesting mechanism of regulation in response to iron levels. When iron is abundant, it contains an iron–sulfur cluster and has aconitase activity, but no IRE-binding activity. When iron is scarce, it loses the iron–sulfur cluster and undergoes a conformational change that results in loss of aconitase activity and gain of IRE-binding activity. The net result of these effects is that expression of enzymes like the transferrin receptor, which has IREs in the 3′ untranslated region of its mRNA, is increased when iron is scarce, and expression of enzymes like apoferritin, which has an IRE in the 5′ untranslated region of its mRNA, is decreased when iron is scarce. IRP2 is regulated in a more classical manner. IRP2 levels are increased when iron is scarce and decreased when iron is abundant, but the mechanism of this regulation is not clear.

Synthesis of hepcidin by the liver is another key control point for iron homeostasis (Figure 26.27). Hepcidin synthesis is stimulated by the proteins hemojuvelin (HJV), transferrin receptor-2 (TfR2) and the major histocompatibility class I-like molecule HFE. Synthesis of these three proteins is, in turn, up-regulated when iron is abundant. Mutations in any of the three can lead to an iron overload disease called hemochromatosis (Clin. Corr. 26.10) because hepcidin is not available to downregulate ferroportin levels under conditions of iron excess. Iron deficiency anemia can lead to hypoxia and increased erythropoiesis, and both of those lead to decreased hepcidin expression which up-regulates ferroportin levels and increases iron availability. Finally, infection and inflammation result in increased hepcidin levels via the effects of cytokines such as interleukin 6 (IL-6). This results in sequestration of iron in the tissues and reduces the risk of systemic infections.

The best-known symptom of **iron deficiency** is a microcytic hypochromic **anemia** (Clin. Corr. 26.11). Iron deficiency is also associated with decreased immunocompetence. Dietary surveys indicate that 95% or more of children and menstruating females do not obtain adequate dietary iron. Biochemical measurements reveal a 10%–25% incidence of

CLINICAL CORRELATION 26.10

Hemochromatosis

Hemochromatosis is primarily a genetic disease of iron overload. Patients with a predisposition to hemochromatosis develop iron deposits in the liver, heart, and endocrine tissue even though they have normal intakes of dietary iron. Eventually, those iron deposits can lead to cirrhosis, cardiomyopathy, diabetes, and other endocrine disorders. In most cases the iron overload is secondary to decreased hepcidin expression, which leads to an inability to appropriately downregulate ferroportin expression when iron is in excess. The regulation of hepcidin expression is summarized in Figure 26.27, p. 1090. The most common form of hemochromatosis (OMIM 235200) is caused by a homozygous Cys282Tyr mutation in HFE. Approximately 9% of the U.S. population is heterozygous and 0.25% is homozygous for this mutation. Hemochromatosis caused by this genetic defect is relatively mild, with onset at middle age, and penetrance is incomplete, with many patients being affected to only a modest degree.

As might be expected mutations in most of the other genes involved in regulation of hepcidin (see Figure 26.27, p. 1090) can also cause hemochomatosis, but those mutations are much rarer than the HFE polymorphism. Hemochromatosis caused by a homozygous deletion of TfR2 is somewhat more severe than the homozygous HFE mutation. Homozygous mutations in hemjuvelin or hepcidin cause a very severe form of hemochromatosis, called juvenile hemochromatosis. Untreated patients with juvenile hemochromatosis usually develop iron overload and damage to their liver and other organs in their teens. Finally, mutations in ferroportin are of two types. Loss-of-function mutations lead to anemia, whereas missense mutations that lead to an inability of ferroportin to interact with hepcidin lead to hemochromatosis.

Treatment for hemochromatosis consists of regular phlebotomy, which is effective at preventing the symptoms of hemochromatosis if started early enough. People with hereditary hemochromatosis are also generally advised to avoid foods and supplements containing high levels of iron or vitamin C. Unfortunately, many people don't realize that they have hereditary hemochromatosis until the symptoms are well advanced. This has lead to a public health policy debate about iron fortification of foods. Iron fortification of foods was instituted to prevent iron deficiency in young children and pregnant women, and it has been effective for that purpose. However, in countries like Sweden where 42% of the average dietary intake of iron is derived from iron-fortified foods, 5% of males have elevated serum iron values and 2% have iron stores indicative of early-stage hemochromatosis.

Dunn, L. L., Rahmanto, Y. S., and Richardson, D. R. Iron uptake and metabolism in the new millennium. *Trends Cell Biol.* 17:93, 2006; Andrews, N. C. Forging a field: The golden age of iron biology, *Blood* 112:219, 2008; and Allen, K. J., Gurrin, L. C., Constantine, C. C., Osburne, N. J., et al. Iron-overload disease in HFE hereditary hemochromatosis. *N. Engl. J. Med.* 358:221, 2008.

CLINICAL CORRELATION 26.11

Clinical Tests for Iron-Deficiency Anemia and Hemochromatosis

There are a number of clinical tests that can be used to determine iron status. Anemia in general leads to a decrease in hemoglobin (normal = 12.1–15.1 gm/dL for females and 13.8–17.2 gm/dL for males) and hematocrit (% of red blood cells found in whole blood; normal = 36.1%–44.3% for females and 40.7%–50.3% for males). Iron-deficiency anemia is characterized as a microcytic hypochromic anemia, meaning that the red cells are smaller and paler than normal due to their decreased hemoglobin content. Small amounts of ferritin end up in the circulation due to normal cellular turnover (normal = 12–150 ng/mL for females and 12–300 ng/mL for males), and serum ferritin levels are proportional to cellular ferritin levels. Serum ferritin is almost nonexistent in iron deficiency and increases with cellular iron load. Serum iron (normal = 60–170 mcg/dl) and total iron-binding capacity (TIBC) of serum transferrin (240–450 mcg/dl) are often measured and used to calculate percentage saturation of transferrin (normal = 20%–50%), a very sensitive indicator of iron status.

Iron-deficiency anemia is generally diagnosed on the basis of low hemoglobin and hematocrit levels couple with a microcytic, hypochromic morphology of the red blood cells. Serum iron, serum ferritin, and TIBC may be used as confirming tests. Serum iron and TIBC are most often used to diagnose hemochromatosis. In hemochromatosis, serum iron is high, TIBC is low or normal, and transferrin saturation is high.

Serum transferrin levels can also be measured, but are most frequently used to measure a patient's liver function or nutritional status. Because it is made in the liver, transferrin will be low with liver disease. Transferrin levels also drop when there is not enough protein in the diet, so this test can be used to monitor nutritional state.

iron deficiency anemia in this same group. Iron-deficiency anemia is also a problem with the elderly (see Clin. Corr. 26.7, p 1083) due to poor dietary intake and increased frequency of achlorhydria, which decreases the intestinal uptake of iron by DMT-1.

Because iron-deficiency anemia is widespread, government programs of nutritional intervention such as the WIC program have emphasized iron-rich foods. However, since recent studies have suggested that excess iron intake may increase the risk of cardiovascular disease, iron supplementation and the consumption of iron-fortified foods may be inappropriate for adult men and postmenopausal women. **Excess iron** can lead to the rare condition **hemochromatosis** in which iron deposits are found in abnormally high levels in many tissues and cause liver, pancreatic, and cardiac dysfunction as well as pigmentation of the skin (Clin. Corr. 26.10). Hemochromatosis is also occasionally seen in chronic hemolytic anemias that require frequent blood transfusions such as β-thalassemia and in liver disease.

Iodine Is Incorporated into Thyroid Hormones

Dietary **iodine** is efficiently absorbed and transported to the **thyroid gland**, where it is stored and used for synthesis of triiodothyronine and thyroxine. These hormones function in regulating the **basal metabolic rate** of adults and the growth and development of children. Adequate maternal thyroid levels are particularly important for fetal brain development. Saltwater fish are the best natural food sources of iodine, and in the past population groups living in inland areas suffered from the endemic deficiency disease goiter, an enlargement (sometimes massive) of the thyroid gland. Since iodine has been routinely added to table salt, goiter has become relatively rare. However, in some inland areas, mild forms of goiter still occur in up to 5% of the population.

Zinc Is Required by Many Proteins

Zinc is part of the catalytic center of over 300 metalloenzymes, including RNA and DNA polymerases, alkaline phosphatase, and carbonic anhydrase. In addition, it forms zinc fingers (Zn^{2+} coordinated to four amino acid side chains), which provide structural stability to another 300–700 proteins. Zinc fingers facilitate binding of proteins to DNA and are common motifs in transcription factors and nuclear hormone receptors. They are also important for protein–protein interactions and are found in many signal transduction proteins. Zinc also binds to the metal response element (MRE) of MRE-binding transcription factor-1 (MTF-1) and controls gene expression in a manner analogous to the effect of iron on binding of IRPs to IREs. Finally, relatively large amounts of loosely bound zinc are also found in vesicular structures including synaptic vesicles of nerve terminals and β-cells of the islets of Langerhans, suggesting that zinc plays a much broader physiological role than might be suggested by its presence in zinc metalloproteins.

Intracellular zinc levels are controlled in part by a group of 14 transporters called ZIPs which transport zinc into cells and 10 transporters called ZnTs which control the efflux of zinc from the cytosol to either intracellular vesicles or extracellular space. Most of the intracellular zinc is tightly bound to cysteine residues in metallothioneins and related proteins. When these cysteines are oxidized, free zinc is released into the cytosol. Thus, intracellular levels of free zinc are closely linked to the redox state of the cell and may be part of the redox signaling pathway.

Zinc deficiency in children is usually marked by poor growth and impairment of sexual development. In both children and adults zinc deficiency results in poor wound healing and dermatitis. Zinc is present in gustin, a salivary polypeptide that appears to be necessary for normal development of taste buds, so zinc deficiency leads to decreased taste acuity. Zinc is required for cytokine production by monocytes and T-cells. Thus, zinc deficiency is associated with impaired immune function. Zinc is required for the activity of porphobilinogen synthase. In lead poisoning, lead replaces the zinc, which leads to anemia and accumulation of γ-aminolevulinic acid (p. 793).

Dietary surveys indicate that zinc intake may be marginal for many individuals, and zinc supplementation has been shown to improve immune status in the elderly. Severe zinc deficiency is seen primarily in alcoholics (especially if they have cirrhosis), patients with chronic renal disease or severe malabsorption diseases, and occasionally in people after long-term parenteral nutrition (TPN). The most characteristic early symptom of zinc-deficient patients on TPN is dermatitis. Zinc is occasionally used therapeutically to promote wound healing and may be of some use in treating gastric ulcers.

Copper Is a Cofactor for Important Enzymes

Important **copper**-containing enzymes include ceruloplasmin and hephaestin (oxidize iron to facilitate its binding to transferrin), cytochrome c oxidase (electron transport), dopamine *ß*-hydroxylase (norepinephrine synthesis), lysyl oxidase (collagen cross-linking), superoxide dismutase (disproportionation of superoxide), tyrosinase (pigment formation), peptidylglycine α-amidating monoxygenase (neurotransmitter metabolism) and C_{18}, Δ^9 desaturase (addition of double bonds to long-chain fatty acids). The C_{18},Δ^9 desaturase is responsible for converting stearic acid (a C_{18} saturated fatty acid) to oleic acid (a C_{18} monounsaturated fatty acid). This may explain why dietary stearic acid does not raise blood cholesterol like the other saturated fatty acids. Symptoms of **copper deficiency** include anemia, hypercholesterolemia, demineralization of bones, leukopenia, fragility of large arteries, and dem yelination of neural tissue. Anemia may reflect the reduced ceruloplasmin and hephaestin activity. Bone demineralization and blood vessel fragility can be directly traced to defects in collagen and elastin formation. Hypercholesterolemia may be related to an increase in the ratio of saturated to monounsaturated fatty acids of the C_{18} series due to reduced activity of the C_{18},Δ^9 desaturase.

Cellular uptake of copper is catalyzed by a high-affinity **copper transporter** called **CTR1**. In mice genetic inactivation of CTR1 is embryonic lethal. Copper efflux from cells is catalyzed by two copper transporting ATPases, **ATP-7A** and **ATP-7B**. ATP-7A is found in most tissues except liver and is essential for exit of copper from intestinal cells. ATP-7B is found in greatest abundance in the liver and brain and is responsible for exit of copper from those tissues. Intracellular copper concentrations regulate the cellular location of both ATP-7A and ATP-7B. When copper levels are low, both ATP-7A and ATP-7B are localized primarily within the *trans* Golgi network. However, when copper levels are high, ATP-7A is transferred to the basolateral surface and plasma membrane of intestinal mucosal cells for transport of copper into the bloodstream and ATP7B is transferred to the bile canaliculus for excretion of copper into the bile. Copper deficiency is relatively rare and is usually seen only because of excess zinc intake (zinc and copper compete for absorption) and in Menkes syndrome, a relatively rare X-linked hereditary disease associated with

CLINICAL CORRELATION 26.12

Diseases of Copper Metabolism

Menkes disease (OMIM 309400) is an X-linked disorder characterized by global copper deficiency. It is caused by mutations in the ATP7A copper transporter that interfere with the ability of intestinal mucosal cells to transport copper into the bloodstream. The symptoms of Menkes disease are mental retardation, growth retardation, hypothermia, loose skin and joints, hypopigmentation, and kinky hair and are caused by the inability to load copper-dependent enzymes with copper. Patients with severe loss of ATP7A activity develop symptoms by 2–3 months and seldom live beyond 3 years. Treatment consists of administration of a copper–histidine complex and is only modestly successful.

Wilson disease is an autosomal recessive disease characterized by copper overload, particularly in the liver and the brain. It is caused by mutations in the ATP7B copper transporter that prevent the liver and neurological tissue from getting rid of excess copper. The accumulation of copper in the liver leads to cirrhosis, chronic hepatitis, and eventually liver failure. The accumulation of copper in the brain leads to parkinsonian symptoms, seizures and psychiatric symptoms. Copper also accumulates as a characteristic gold-brown ring, called the Kayser–Fleischer ring, around the periphery of the cornea. Treatment of Wilson disease consists of limiting copper-rich foods and increasing dietary zinc intake to reduce dietary copper absorption and use of copper chelating agents such as penicillamine and trientine to increase copper excretion from body. These treatments are very effective if begun early enough.

Ala, A., Walker, A. P., Ashkan, K., Dooley, J. S., and Schilsky, M. L. Wilson disease. *Lancet* 369:397, 2007; de Bie, P., Muller, P., Wijmenga, C., and Klamp, L. W. J. Molecular pathogenesis of Wilson and Menkes disease: Correlation of mutations with molecular defects and disease phenotypes. *J. Med. Genet.* 44:673, 2007.

a defect in the ATP-7A copper transporter. Wilson disease, an autosomal recessive disease that leads to copper overload, is associated with a defect in the ATP-7B copper transporter (Clin. Corr. 26.12).

Chromium Is a Component of Chromodulin

Chromium is a component of the low-molecular-weight protein **chromodulin**, which potentiates the effects of insulin by facilitating insulin binding to its receptor and receptor kinase signaling. The chief symptom of chromium deficiency is impaired glucose tolerance, a result of decreased insulin effectiveness. Chromium deficiency appears to be rare in healthy adults. However, diabetes causes increased urinary loss of chromium, which can lead to chromium deficiency over time. Supplementation with chromium appears to improve glycemic control in patients with type 2 diabetes.

Selenium Is Found in Selenoproteins

Selenium is incorporated into approximately 25 selenoproteins in humans including **glutathione peroxidase**, phospholipid-hydroperoxide, **thioredoxin reductase**, iodothyronine deiodinase, selenoprotein P, sperm capsule selenoprotein GPx4, and muscle selenoprotein W. These proteins contain one or more selenocysteine residues, which are incorporated during translation (p. 212). The incorporation of selenocysteine into protein requires a unique selenocystyl-tRNA that binds to UGA codons in mRNAs that also have a stem-loop structure called a Sec-insertion sequence (SECIS) in the 3′-untranslated region. The selenocysteine is synthesized directly on the tRNA from selenide, ATP, and seryl-tRNA.

Glutathione peroxidase destroys peroxides in the cytosol (p. 789), which complements the effect of vitamin E since vitamin E is limited primarily to the membrane. Phospholipid-hydroperoxide glutathione peroxidase catalyzes reductive destruction of phospholipid and cholesterol ester hydroperoxides in oxidized low-density lipoproteins. Iodothyronine deiodinase catalyzes the conversion of thyroxine (T_4) to the active thyroid hormone 3,3′,5-triiodothyronine (T_3). Selenoprotein P is an extracellular protein that delivers selenium to extrahepatic tissues. Selenoprotein GPx4 is important for sperm motility, and selenoprotein W appears to be essential for muscle metabolism. Selenium is one of the few nutrients not removed by the milling of flour and is usually thought to be present in adequate amounts in the diet. Selenium levels are very low in the soil in certain parts of the country, however; and foods raised in these regions are low in selenium. Fortunately, this effect is minimized by the current food distribution system, which assures that the foods marketed in any one area are derived from a variety of different geographical regions. Clinical studies have suggested that supplementation with selenium may reduce the risk of lung, breast, and bladder cancers.

Manganese, Molybdenum, Fluoride, and Boron Are Essential Trace Elements

Manganese is a component of arginase, glutamine synthetase, Mn superoxide dismutase, and phosphoenolpyruvate decarboxylase and activates a number of other enzymes. **Molybdenum** is present in xanthine oxidase (p. 816). **Fluoride** strengthens bones and teeth and is usually added to drinking water. **Boron** appears to be important role in bone formation, nerve function, and immune response.

26.11 • THE AMERICAN DIET: FACT AND FALLACY

Much has been said about the supposed deterioration of the American diet. Americans are eating much more processed food than did their ancestors. These foods have a higher caloric density and a lower nutrient density than the foods they replace. However, they are almost uniformly enriched with iron, thiamin, riboflavin, niacin, and low levels of folic acid. In many cases they are even fortified (usually as much for sales promotion as for nutritional reasons) with as many as 11–15 vitamins and minerals. Unfortunately, it is not practical to replace all of the nutrients lost during processing, especially the trace minerals and phytonutrients such as the carotenoids. Imitation foods present a special problem, as they are usually incomplete in more subtle ways. For example, imitation cheese and milkshakes are widely sold in this country. They usually contain the protein and calcium one would expect of the food they replace, but often do not contain the riboflavin which one would obtain from them. Fast food meals tend to be high in calories and fat, and low in certain vitamins and trace minerals. For example, the standard fast food meal provides over 50% of the calories the average adult needs for the entire day, while providing $< 5\%$ of the vitamin A and $< 30\%$ of biotin, folic acid, and pantothenic acid. Unfortunately, much of the controversy in recent years has centered on whether these trends are good or bad. This simply obscures the issue at hand. Clearly, it is possible to obtain a balanced diet that includes some processed, imitation, and fast foods if one compensates by selecting foods for the other meals that are low in caloric density and rich in nutrients. Without such compensation the balanced diet becomes a myth.

26.12 • ASSESSMENT OF NUTRITIONAL STATUS IN CLINICAL PRACTICE

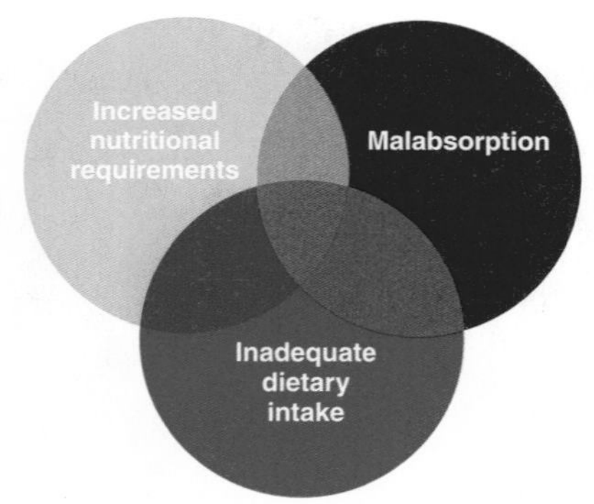

Figure 26.28 Factors affecting individual nutritional status. Schematic representation of three important risk factors in determining nutritional status. A person in the periphery would have very low risk of any nutritional deficiency, whereas those in the green, orange, purple, or center areas would be much more likely to experience some symptoms of nutritional deficiencies.

It might seem that the process of evaluating the **nutritional status** of an individual patient would be an overwhelming task after surveying the major micronutrients and their biochemical roles. There are three factors that can contribute to nutritional deficiencies: poor diet, **malabsorption**, and increased nutrient need. Only when two or three components overlap in the same person (Figure 26.28) does the risk of symptomatic deficiency become significant. For example, infants and young children have increased needs for iron, calcium, and protein. Dietary surveys show that many of them consume diets inadequate in iron and some consume diets that are low in calcium. Protein is seldom a problem unless the children are being raised as strict vegetarians. Thus, the chief nutritional concerns for most children are iron and calcium. Teenagers tend to consume diets low in calcium, magnesium, vitamin A, vitamin B_6, and vitamin C. Of these, their needs are particularly high for calcium and magnesium during the teenage years, so these are the nutrients of greatest concern. Young women are likely to consume diets low in iron, calcium, magnesium, vitamin B_6, folic acid, and zinc, and all of these nutrients are needed in greater amounts during pregnancy and lactation. Adult women often consume diets low in calcium, yet they may have a particularly high need for calcium to prevent rapid bone loss. Finally, the elderly have unique nutritional needs (see Clin. Corr. 26.7, p. 1083) and tend to have poor nutrient intake due to restricted income, loss of appetite, and loss of the ability to prepare a wide variety of foods. They are also more prone to suffer from malabsorption problems and to use multiple prescription drugs that increase nutrient needs (Table 26.3).

Illness and metabolic stress often cause increased demand or decreased utilization of certain nutrients. For example, diseases leading to fat malabsorption cause a particular problem with absorption of calcium and the fat-soluble vitamins. Other malabsorption diseases can result in deficiencies of many nutrients depending on the particular disease. Liver and kidney disease can prevent hydroxylation of vitamin D and storage or utilization of many other nutrients including vitamin A, vitamin B_{12}, and folic acid. Severe illness and trauma increase the need for calories, protein, and possibly vitamin C and certain B vitamins. Long-term use of many drugs in the treatment of chronic disease can affect the need for certain micronutrients. Some of these are listed in Table 26.3.

Who then is at a nutritional risk? Obviously, the answer depends on many factors. Nutritional counseling is an important part of treatment for infants, young children, and pregnant/lactating females. A brief analysis of a dietary history and further nutritional counseling are important when dealing with high-risk patients.

TABLE 26.3 • Drug–Nutrient Interactions

Drug	*Potential Nutrient Deficiencies*
Alcohol	Thiamin Folic acid Vitamin B_6
Anticonvulsants	Vitamin D Folic acid Vitamin K
Cholestyramine	Fat-soluble vitamins Iron
Corticosteroids	Vitamin D and calcium Zinc Potassium
Diuretics	Potassium Zinc
Isoniazid	Vitamin B_6
Oral contraceptives and estrogens	Vitamin B_6 Folic acid and B_{12}

26.13 • NUTRIGENOMICS—THE FUTURE OF NUTRITION

Rare gene defects that affect nutrient uptake and utilization (e.g., Wilson disease, Menkes disease, Vitamin D-resistant rickets, and phenylketonuria) have been known for years. However, more recently attention is being focused on common genetic polymorphisms that have more subtle effects on nutrient status and disease risk. The best characterized of these genetic polymorphisms at present are those affecting folate status in pregnant women and risk of giving birth to babies with neural tube defects (see Clin. Corr. 26.6, p. 1083) and those causing hemochromatosis (see Clin. Corr. 26.10. p. 1090).

The term ***nutrigenomics*** encompasses three distinct areas of nutrient–gene interactions. (1) **Nutritional genetics** describes how the response to nutrients in the diet is influenced by individual genetic differences. Genes involved in the metabolism or utilization of most nutrients are being systematically screened for commonly occurring polymorphisms. Most of these polymorphisms have no affect on enzyme activity and/or nutrient requirements. However, other examples of polymorphisms that affect nutrient status and disease risk have been identified and many more are likely to be discovered in the future. (2) **Nutritional epigenetics** describes nutrient induced changes in DNA methylation, histone post-translational modifications and other chromatin alterations. This is an important property of those nutrients (folate, vitamin B_{12}, choline, and methionine) that are involved in cellular methylation reactions, but may involve other nutrients as well. (3) **Nutritional transcriptomics** describes the effect of nutrients on gene expression. This is an important property of fat-soluble vitamins (vitamins A and D) that bind to nuclear receptors and directly affect gene expression, but it also appears to be a property of several of the antioxidant vitamins that affect redox signaling pathways which regulate gene expression.

Nutrigenomics has the potential to transform both clinical and public health nutritional practice. It can lead to genome-based nutrient and food guidelines for disease prevention, individualized nutritional recommendations for disease prevention and treatment, and better targeted public health nutrition interventions that maximize benefit and minimize risk. This is particularly important when we consider what types of nutritional interventions decrease the risk of multifactorial diseases like cardiovascular diseases, obesity, type 2 diabetes, and cancer. Most of the current large-scale intervention studies have focused on the effects of various nutrients on the risk of those diseases in the population as a whole. Those studies are likely to become obsolete in the future as we increasingly focus on nutrients that affect those diseases in genetically defined subpopulations (see the discussion of vitamin E supplementation and cardiovascular risk in Section 26.4, p. 1072).

Bibliography

Dietary Reference Intakes

Food and Nutrition Board, Institute of Medicine of the National Academy of Sciences. www.iom.edu/board.asp?id=3788.

Vitamin A

Fields, A. L., Soprano, D. L., and Soprano, K. J. Retinoids in biological control and cancer. *J. Cell. Biochem.* 102:886, 2007.

Soprano, D. R., Qin, P., and Soprano, K. J. Retinoic acid receptors and cancers. *Annu. Rev. Nutr.* 24:201, 2004.

Thurnham, D. I., and Northrop-Clews, C. A. Optimal nutrition: Vitamin A and the carotenoids. *Proc. Nutr. Soc.* 58:449, 1999.

Vitamin D

DeLuca, H. F. Overview of general physiologic features and functions of vitamin D. *Am. J. Clin. Nutr.* 80:1689S, 2004.

Holick, M. F. Vitamin D deficiency. *N. Engl. J. Med.* 357:266, 2007.

Supplement: Vitamin D and health in the 21st century: an update. *Am. J. Clin. Nutr.* 483S, 2008.

Vitamin E

Brizeluis-Flohe, R., and Traber, M. G. Vitamin E: Function and metabolism. *FASEB J.* 13:1145, 1999.

Hensley, K., Benaksas, E. J., Bolli, R., Comp, P., et al. New perspectives on vitamin E: α-Tocopherol and carboxyethylhydroxychroman metabolites in biology and medicine. *Free Rad. Biol. Med.* 36:1, 2004.

Zingg, J. M., Azzi, A., and Meydani, M. Genetic polymorphisms as determinants for disease-preventative effects of vitamin E. *Nut. Rev.* 66:406, 2008.

Vitamin K

Bugel, S. Vitamin K and bone health. *Proc. Nutr. Soc.* 62:839, 2003.

Danzinger, J. Vitamin K-dependent proteins, warfarin and vascular calcification. *Clin. J. Am. Soc. Nephrol.* 3:1504, 2008.

Nelsestuen, G. L., Shah, A. M., and Harvey, S. B. Vitamin K-dependent proteins. *Vitam. Horm.* 58:355, 2000.

Thiamin

Bates, C. J. Thiamin. In Bowman, B. A., and Russell, R. M. (Eds.), *Present Knowledge in Nutrition,* 8th ed. Washington, DC: ILSI Press, 2001, 184.

Kluger R., and Tittmann, K. Thiamin diphosphate catalysis: Enzymatic and non-enzymatic covalent intermediates. *Chem. Rev.* 108:1797, 2008.

Singleton, C. K., and Martin, P. R. Molecular mechanisms of thiamine utilization. *Curr. Molec. Med.* 1:197, 2001.

Riboflavin

Powers, H. J. Riboflavin (vitamin B-2) and health. *Am. J. Clin. Nutr.* 77:1352, 2003.

Niacin

Carlson, L. A. Niaspan, the prolonged release preparation of nicotinic acid (niacin), the broad-spectrum lipid drug. *Int. J. Clin. Pract.* 58:706, 2004.

Kamanna, V S., and Kashyap, M. L. Mechanism of action of niacin. *Am. J. Cardiol.* 101, Suppl.:20B, 2008.

Meyers, C. D., and Kashyap, M. L. Management of metabolic syndrome-nicotinic acid. *Endocrinol. Metab. Clin. North Am.* 33:557, 2004.

Suave, A. A. NAD and vitamin B_3: From metabolism to therapies. J. *Pharmacol. Exper. Therap.* 324:883, 2008.

Pyridoxine

Spinneker, A., Sola, A., Lemmen, V., Costillo, M. J., et al. Vitamin B6 status, deficiency and its consequences: An overview. *Nutr. Hosp.* 22:7, 2007.

Pantothenic Acid

Miller, J. A., Rogers, L. M. and Rucker, R. R. Pantothenic acid. In Bowman, B. A., and Russell, R. M. (Eds.), *Present Knowledge in Nutrition,* 8th ed. Washington, DC: ILSI Press, 2001, 253.

Biotin

McMahon, R. J. Biotin in metabolism and molecular biology. *Annu. Rev. Nutr.* 22:221, 2002.

Rodriguez-Melendez, R., and Zempleni, J. Regulation of gene expression by biotin. *J. Nutr. Biochem.* 14:680, 2003.

α-Lipoic Acid

Pershadsingh, H. A. α-Lipoic acid: Physiologic mechanisms and indications for the treatment of metabolic syndrome. *Expert Opin. Investig. Drugs* 16:291, 2007.

Folic Acid

Bailey, L.B. Folate and B_{12} recommended intakes and status in the United States. *Nutr. Rev.* 62:S14, 2004.

Beaudin, A. E., and Stover, P. J. Folate-mediated one-carbon metabolism and neural tube defects: Balancing genome synthesis and gene expression. *Birth Defects Res.* (Part C) 81:183, 2007.

Landgren, F., Israelsson, B., Lindgren, A., Hultsberg, B., et al. Plasma homocysteine in acute myocardial infarction: Homocysteine-lowering effects of folic acid. *J. Int. Med.* 237:381, 1995.

McNulty, H., Cuskelly, G. J., and Wood, M. Response of red blood cell folate to intervention: Implications for folate recommendations for the prevention of neural tube defects. *Am. J. Clin. Nutr.* 71, Suppl.:13085, 2000.

Stover, P. J. Physiology of folate and vitamin B_{12} in health and disease. *Nutr. Rev.* 62:S3, 2004.

Zanibbi, G.A. Homocysteine and cognitive function in the elderly. *CMAJ* 171:897, 2004.

Vitamin B_{12}

Banerjee, R., and Ragsdale, S. W. The many faces of vitamin B_{12}: Catalysis by cobalamin-dependent enzymes. *Annu. Rev. Biochem.* 72:209, 2003.

Scalabrino, G., and Peracchi M. New insights into the pathophysiology of cobalamin deficiency. *Trends Mol. Med.* 12:247, 2006.

Vitamin C

Bsoul, S. A., and Terezhalmy, G. T. Vitamin C in health and disease. *J. Contemp. Den. Pract.* 15:1, 2004.

Li, Y., and Schellhorn. H. E. New developments and novel therapeutic perspectives for vitamin *C. J. Nutr.* 137:2171, 2007.

Villacorta, L., Azzi, A. and Zingg, J. M. Regulatory roles of vitamins E and C on extracellular matrix components of the vascular system. *Molec. Aspects Med.* 28:507, 2007.

Choline

Zeisel, S. H. Choline: An essential nutrient for humans. *Nutrition* 16:669, 2000.

Zeisel, S. H. Choline: Critical role during fetal development and dietary requirements in adults. *Annu. Rev. Nutr.* 26:229, 2006.

Zeisel, S. H. Choline: needed for normal memory development. *J. Am. Coll. Nutr.* 19, 5 Suppl.:528S, 2000.

Carnitine

Stephens, F. B., Constantin-Teodosiu, D., and Greenhaff, P. L. New insights into the role of carnitine in the regulation of fuel metabolism in skeletal muscle. *J. Physiol.* 581:431, 2007.

Calcium

Flynn, A. The role of dietary calcium in bone health. *Proc. Nutr. Soc.* 62:851, 2003.

Hatton, D. C., and McCarron, D. A. Dietary calcium and blood pressure in experimental models of hypertension. *Hypertension* 23:513, 1994.

Meacham, S., Grayscott, D., Chen, J. J., and Bergman, C. Review of the dietary reference intake for calcium: Where do we go from here? *Crit. Rev. Food Sci. Nutr.* 48:378, 2008.

Magnesium

Champagne, C. M. Magnesium in hypertension, cardiovascular disease, metabolic syndrome and other conditions: A review. *Nutr. Clin. Pract.* 23:142, 2007.

Gums, J. G. Magnesium in cardiovascular and other disorders. *Am. J. Health Syst. Pharm.* 61:1569, 2004.

Saris, N. E., Mervaala, E., Karppanen, H., Khawaja, J.A., and Lewenstam, A., Magnesium. An update on physiological, clinical and analytical aspects. *Clin. Chim. Acta.* 294:1, 2000.

Vormann, J. Magnesium: nutrition and metabolism. *Mol. Aspects Med.* 24:27, 2003.

Iron

Andrews, N. C., and Schmidt, P. J. Iron homeostasis. *Annu. Rev. Physiol.* 69:69, 2007.

Andrews, N. C. Forging a field: The golden age of iron biology. *Blood* 112:219, 2008.

Dunn, L. L., Rahmanto, Y. S., and Richardson, D. R. Iron uptake and metabolism in the new millennium. *Trends Cell Biol.* 17:93, 2006.

Galy, B., Ferring-Appel, D., Kaden, S., Grone, H. J. and Hentze, M. W. Iron regulatory proteins are essential for intestinal function and control key iron absorption molecules in the duodenum. *Cell Metab.* 7:79, 2008.

Ganz, T., and Nemeth, E. Regulation of iron acquisition and iron distribution in mammals. *Biochim. Biophys. Acta* 1763:690, 2006.

Shaw, G. C., Cope, J. J., Li, L., Corson, K., et al. Mitoferrin is essential for erythroid iron assimilation. *Nature* 440:96, 2006.

Walden, W. E., Selezeva, A. I., Dupuy, J., Volbeda, A., et al. Structure of dual function iron regulatory protein 1 complexed with ferritin IRE-RNA. *Science* 314:1903, 2006.

Iodine

Pearce, E. N. National trends in thyroid nutrition: Is everyone getting enough? *Thyroid* 17:823, 2007.

Zimmermann, M. B. Iodine requirements and the risks and benefits of correcting iodine deficiency in populations. *J. Trace Elements Med. Biol.* 22:81, 2008.

Zinc

Devirgilis, C., Zalewski, P. D., Perozzi, G., and Murgia, C. Zinc fluxes and zinc transporters in chronic diseases. *Mutat. Res.* 622:84, 2007.

Krezel, A., Hao, Q., and Maret, W. The zinc/thiolate redox biochemistry of metallothionein and the control of zinc fluctuation in cell signaling. *Arch. Biochem. Biophys.* 463:188, 2007.

Prasad, A. S. Zinc deficiency: its characteristic and treatment. *Met. Ions. Biol. Syst.* 41:103, 2004.

Copper

Arredondo, M., and Nunez, M. T. Iron and copper metabolism. *Molec. Aspects Med.* 26:313, 2005.

de Bie, P., Muller, P., Wijmenga, C., and Klamp, L.W.J. Molecular pathogenesis of Wilson and Menkes disease: Correlation of mutations with molecular defects and disease phenotypes. *J. Med. Genet.* 44:673, 2007.

Schumann, K., Classen, H. G., Dieter, H. H., Konig, J., et al. Hohehheim consensus workshop: Copper. *Eur. J. Clin. Nutr.* 56:469, 2002.

Chromium

Anderson, R. A. Chromium in the prevention and control of diabetes. *Diabetes Metab.* 26:22, 2000.

Vincent, J. B. Recent advances in the nutritional biochemistry of trivalent chromium. *Proc. Nutr. Soc.* 63:41, 2004.

Vincent, J. B. The biochemistry of chromium. *J. Nutr.* 130:715, 2000.

Selenium

Brown, K. M., and Arthur, J. R. Selenium, selenoproteins and human health: a review. *Public Health Nutr.* 4:593, 2001.

Hesketh, J. Nutrigenomics and selenium: Gene expression patterns, physiological targets and genetics. *Annu. Rev. Nutr.* 28:157, 2008.

Other Trace Minerals

Aschner, J. L., and Aschner, M. Nutritional aspects of manganese homeostasis. *Molec. Ed.* 26:353, 2005.

Nielsen, F. H. Is boron nutritionally relevant? *Nutr. Rev.* 66:183, 2008.

Nielson, F. H. Boron, manganese, molybdenum, and other trace elements. In Bowman, B. A., and Russell, R. M. (Eds.), *Present Knowledge in Nutrition,* 8th ed. Washington, DC: ILSI Press, 2001, 253.

Turnland, J. R. Molybdenum metabolism and requirements in humans. *Met. Ions Biol. Syst.* 39:727, 2002.

Dietary Surveys

Block, G. Dietary guidelines and the results of food consumption surveys. *Am. J. Clin. Nutr.* 53:3565,1991.

Kritchevsy, D, Dietary guidelines. The rationale for intervention. *Cancer* 72:1011, 1993.

Nutrigenomics

Ross, S. A. Nutritional genomic approaches to cancer prevention research. *Exp. Oncol.* 29:250–256, 2007.

Stover, P. J., and Caudill, M. A. Genetic and epigenetic contributions to human nutrition and health: Managing genome-diet interactions. J. *Am. Diet. Assoc.* 108:1480, 2008.

Key Terms

Dietary Reference Intake
Recommended Dietary Allowance
vitamin A
retinol
retinoic acid
carotenoids
vitamin D
cholecalciferol
calcitonin
vitamin E
tocopherol
vitamin K
thiamin
thiamin pyrophosphate
riboflavin
niacin
pyridoxal phosphate
pantothenic acid
coenzyme A
biotin
α-lipoic acid
tetrahydrofolate
dihydrofolate reductase
vitamin B_{12}
cobalamin
vitamin C
ascorbic acid
choline
carnitine
calcium
magnesium
iron
ferritin
ceruloplasmin
iron responsive elements
iodine
zinc
copper
chromium
chromodulin
selenium
selenoproteins
manganese
molybdenum
fluoride
boron
nutrigenomics
nutritional epigenetics
nutritional transcriptomics

Questions • CAROL N. ANGSTADT

Multiple-Choice Questions

1. The Estimated Average Requirement (EAR) of a nutrient is
   A. the same as the RDA of that nutrient.
   B. an amount that should meet the requirement of half of the healthy individuals of a particular group.
   C. based on the observed nutrient intake of a particular group.
   D. the highest level of nutrient deemed to pose no risk or adverse health effects on the particular population.
   E. two standard deviations higher than the RDA.

2. The effects of vitamin A may include all of the following except
   A. prevention of anemia.
   B. serving as an antioxidant.
   C. cell differentiation.
   D. the visual cycle.
   E. induction of certain cancers.

3. Ascorbic acid may be associated with all of the following except
   A. iron absorption.
   B. bone formation.
   C. acute liver disease when taken in high doses.
   D. wound healing
   E. participation in hydroxylation reactions.

4. Choline, recently reclassified as essential,
   A. cannot be synthesized *de novo.*
   B. affects DNA methylation so is critical during fetal development.
   C. deficiency has never been demonstrated in humans because it is abundant in the diet.
   D. is necessary to maintain the coenzyme A supply in mitochondria.
   E. supplementation raises serum homocysteine.

5. Copper is an essential element. All of the following are correct about copper except
   A. it is a component of several oxidase enzymes.
   B. it is involved in the conversion of stearic acid to oleic acid.
   C. copper deficiency may lead to anemia.
   D. cellular uptake and efflux of copper are catalyzed by different transporters.
   E. copper deficiency occurs in tandem with a zinc deficiency.

6. Alcohol impairs both absorption and storage of folate. Megaloblastic erythropoiesis occurs because cells are arrested in the S phase since DNA synthesis is inhibited. DNA synthesis is inhibited in folate deficiency because tetrahydrofolate is required
   A. in the synthesis of purine nucleotides and dTMP.
   B. in the conversion of homocysteine to methionine.
   C. for the utilization of Vitamin B_{12}.
   D. all of the above.
   E. none of the above.

Questions 7 and 8: Cystic fibrosis is a generalized dysfunction of the exocrine glands leading to viscid mucus, which plugs various ducts. Pulmonary infections are common and are usually the direct cause of death. Cystic fibrosis patients, however, also have severe malabsorption problems because pancreatic enzymes are deficient and there may be a partial obstruction of the common bile duct. Malabsorption of fat, fat-soluble vitamins, and calcium is the most common, but not only, problem. Patients have increased protein and energy needs because of chronic infections.

7. Serum calcium levels are usually normal in spite of suboptimal calcium absorption and Vitamin D deficiency. Serum calcium is being maintained
   A. by low parathyroid hormone (PTH) levels inhibiting calcium excretion.
   B. by an increase in calcitonin.
   C. by increased bone resorption stimulated by elevated PTH.
   D. because PTH stimulates calcium absorption from the stomach.
   E. by lack of 1,25-dihydroxy Vitamin D which prevents bone from taking calcium from blood.

8. Cystic fibrosis patients are frequently on antibiotics for infections. Antibiotics exacerbate the fat malabsorption problem for obtaining
   A. vitamin A.
   B. vitamin C.
   C. vitamin D.
   D. vitamin E.
   E. vitamin K.

Questions 9 and 10: Hemochromatosis is a genetic disease of iron overload in which iron deposits in liver, heart, and endocrine tissue even though dietary intake of iron is normal. In most cases, the iron overload is secondary to a decrease in expression of hepcidin because of the mutation of the gene. This leads to an inability to regulate iron release to the circulation when iron is in excess.

9. Normally when iron is in excess, hepcidin
A. increases the formation of ferritin.
B. inactivates ferroportin.
C. prevents the reduction of Fe^{+3} to Fe^{+2}.
D. transports the iron across the cell membrane.
E. catalyzes the oxidation of Fe^{+2} to Fe^{+3}.

10. In the intestinal absorption of iron,
A. the presence of a reductant like ascorbate enhances the availability of iron.
B. the regulation of uptake occurs between the lumen and the mucosal cells.
C. the amount of apoferritin synthesized in the mucosal cell is directly related to the need for iron by the host.
D. iron bound tightly to a ligand, such as phytate, is more readily absorbed than free iron.
E. low pH in the stomach inhibits absorption by favoring Fe^{+2}.

Questions 11 and 12: Nutritional needs of the elderly are of interest because of ageing of the American population. Atrophic gastritis results in decreased acid production in the stomach. Absorption problems and decreased ability to convert some vitamins and minerals to their active forms contribute to nutritional deficiencies in the elderly. Chromium is not abundant in the diet and many elderly have difficulty in converting it to its active form, chromodulin.

11. The major symptom of chromium deficiency is impaired glucose tolerance. Chromodulin
A. increases the number of insulin receptors.
B. facilitates insulin binding to cell receptors.
C. protects cell receptors by salvaging peroxides.
D. facilitates the absorption of copper and zinc.
E. increases the metabolism of homocysteine.

12. Atrophic gastritis can interfere with zinc absorption. Zinc normally plays a role in all of the following except
A. growth in children.
B. wound healing.
C. taste acuity.
D. prevention of goiter.
E. immune function.

Problems

13. The neurological disorders seen in vitamin B_{12} deficiency are caused by progressive demyelination of nervous tissue. How does lack of B_{12} interfere with formation of the myelin sheath?

14. What is the chemical reaction in which vitamin K participates? How is this reaction involved in blood coagulation and bone formation?

Answers

1. B This is the first step in defining nutritional guidelines. A and E: The RDA is set at two standard deviations above the EAR. C: This is called the Adequate Intake (AI). D: This is the Total Upper Intake Level (UL).

2. E May have protective effects against some cancers. A. Retinol and/or retinoic acid is required for synthesis of transferrin. B. Various carotenoids are antioxidants. C. Retinol and retinoic acid may function like steroid hormones. D. Retinol cycles between the Δ11-*cis* and all-*trans* forms.

3. C Excess ascorbate is readily excreted. A: Ascorbic acid aids in iron absorption by reducing iron. B: Ascorbic acid is essential for collagen synthesis, which is critical in bone formation. D and E: Ascorbic acid is required for the hydroxylation of lysine and proline residues in protocollagen and, therefore, is required for wound healing.

4. B Choline can be a source of methyl groups (after conversion to betaine). A: It can be synthesized by methylation of phosphatidylethanolamine to phosphatidylcholine. C: Such conditions as liver cirrhosis could lead to a deficiency. D: This is carnitine. E: It lowers homocysteine.

5. E Zinc and copper compete for absorption so high zinc leads to low copper absorption. A: For example, cytochrome c oxidase and superoxide dismutase. B: Copper is part of C18, Δ^9 desaturase. C: Ceruloplasmin and hephaestin activity needed for iron absorption are copper containing. D: CTR1 is involved in uptake and ATP7A and ATP7B in efflux.

6. A *De novo* synthesis of purine nucleotides and dUMP to dTMP conversion require tetrahydrofolate. These components are required for DNA synthesis. B: Tetrahydrofolate is required in this process, but the process is not involved in DNA synthesis. C: Vitamin B_{12} is required for the release of tetrahydrofolate from N^5-methyl tetrahydrofolate, but this is not part of DNA synthesis.

7. C PTH promotes bone resorption and inhibits calcium excretion. A: PTH is elevated when vitamin D is low. B: Elevated calcitonin is a response to high serum calcium. D: Calcium is absorbed from the intestines, not the stomach. E: 1,25-dihydroxy vitamin D promotes bone resorption bone's does not necessarily affect bone's ability to take up calcium from blood.

8. E Some of our vitamin K is obtained from bacterial synthesis in the intestine, which is wiped out by antibiotics. A, C, and D: All of these require adequate fat absorption but come from foods not intestinal bacteria. B: Vitamin C is water soluble.

9. B This down-regulation of ferroportin decreases the transport of iron into the circulation. A: Ferritin sequesters iron in cells. C and E: Hepcidin is not an enzyme and is not involved in oxidation or reduction of iron. D: Hepcidin is not a transporter but binds to ferroportin and triggers its destruction.

10. A Ascorbate facilitates reduction to the ferrous state and, therefore, dissociation from ligands and absorption. B: Substantial iron enters the mucosal cell regardless of need, but the amount transferred to the capillary beds is controlled. C: Iron bound to apoferritin is trapped in mucosal cells and not transferred to the host. D: Iron must dissociate from ligands for absorption. This is why iron is spinach is not a good source of iron. E: Oxidation to Fe^{+3} is favored by higher pH.

11. B This would facilitate the uptake of glucose. A: Does not do this. C: This is the role of selenium. D: Metallothionein is the protein that influences copper and zinc absorption. E: Folate, B_6, and B_{12} are necessary for homocysteine catabolism.

12. **D** This is a role for iodine. A: RNA and DNA polymerases are zinc proteins. B: This is shown to be true. C: Zinc is part of a salivary polypeptide involved in development of taste buds. E: Zinc is required for cytokine production by monocytes and T-cells.
13. One of the reactions in which vitamin B_{12} participates is conversion of methylmalonyl CoA to succinyl CoA (a step in the catabolism of valine and isoleucine). Methylmalonyl CoA is a competitive inhibitor of malonyl CoA in fatty acid biosynthesis, necessary for the maintenance of the myelin sheath. Secondly, methylmalonyl CoA can be used in fatty acid synthesis leading to formation of branched-chain fatty acids, which might disrupt normal membrane structure.
14. Vitamin K is necessary for carboxylation of specific glutamic acid residues in certain proteins to form γ-carboxyglutamic acid residues. In blood coagulation, this step is required for the conversion of preprothrombin to prothrombin. In bone formation, this is required to form the calcium binding residues of the protein osteocalcin.

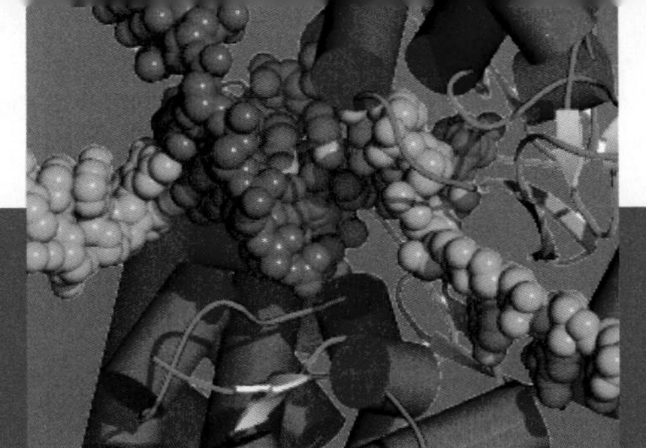

27

Macronutrients: Metabolic Effects and Health Implications

Stephen G. Chaney
Professor, School of Medicine, University of North Carolina at Chapel Hill

Key Concepts

- Energy balance is the relationship between food energy consumed and energy expended. Nitrogen balance is the relationship between intake of nitrogen and excretion of nitrogen.
- Essential amino acids must be present in the diet. Protein requirements are increased during growth, trauma, and illness.
- Protein-energy malnutrition can present in several forms depending on the age of the patient and the precipitating conditions, but the common feature is compromised immune function leading to decreased resistance to infection.
- Obesity is associated with insulin resistance and has significant health implications.
- The most common forms of carbohydrate intolerance are diabetes mellitus and lactase insufficiency.
- The amount and type of fats in the diet may have long-term health implications.
- The optimal diet composition varies from individual to individual.

27.1 • INTRODUCTION

The study of human nutrition can be divided into three areas: undernutrition, overnutrition, and ideal nutrition. The primary concern in this country is not with undernutrition because nutritional-deficiency diseases are now quite rare. However, overnutrition is a particularly serious problem in developed countries. The current estimate is that over 34% of the U.S. population is obese and another 32% is overweight, and obesity is associated with increased risk for a number of serious health consequences. Along with the concern about rising rates of obesity, there is increasing interest in the optimal macronutrient composition. Are there ideal ratios of carbohydrate, fat, and protein to reduce the risk of obesity and obesity-related diseases? Finally, there is the burgeoning area of nutrigenomics. What role does genetics play in obesity and the ideal diet composition for each of us as individuals? This is probably the most exciting area of nutrition today.

27.2 • ENERGY METABOLISM

Energy Content of Food Is Measured Primarily in Kilocalories

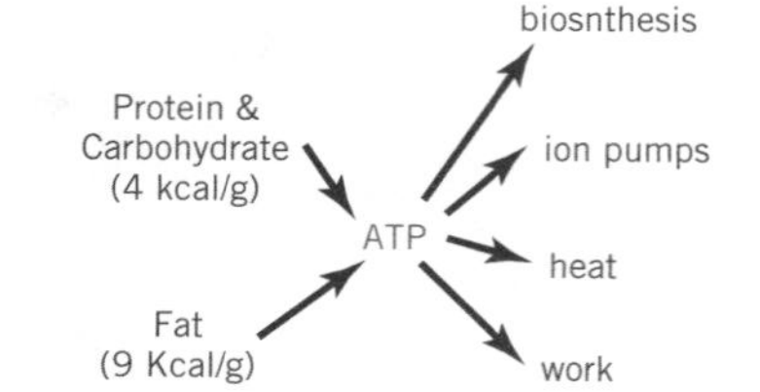

Figure 27.1 Metabolic fate of the foods we eat.

Much of the food we eat is converted to ATP and other high-energy compounds, which are utilized to drive biosynthetic pathways, generate nerve impulses, and power muscle contraction (Figure 27.1). The energy content of foods is generally described in terms of **calories**. Technically speaking, this refers to kilocalories of heat energy released by combustion of that food in the body. This is complicated by the fact that the international standard for energy measurements is the kilojoule (kJ). Since the public is likely to be counting calories rather than kilojoules in the foreseeable future, calories will be used in this chapter, with conversion to kilojoules where appropriate. Caloric values of protein, fat, carbohydrate, and alcohol are roughly 4, 9, 4, and 7 calories/g (16.7, 37.7, 16.7. and 29.3 kJ/g), respectively. Given these values and the amount and composition of the food, it is simple to calculate the caloric content (input) of the foods we eat. Calculating caloric content of foods is not a major problem in this country. Millions of Americans are able to do it with ease. The problem lies in balancing caloric input with caloric output. Where do these calories go?

Energy Expenditure Is Influenced by Four Factors

The four principal factors that affect an individual's energy expenditure are listed in Table 27.1. The effects of surface area are thought to be simply related to the rate of heat loss by the body; the greater the surface area, the greater the rate of heat loss. Although it may seem surprising, a lean individual actually has a greater surface area, and thus a greater energy requirement, than an obese individual of the same weight. Age may reflect two factors: growth and lean muscle mass. In infants and children, more energy expenditure is required for rapid growth, which is reflected in a higher basal metabolic rate (rate of energy utilization in resting state). In adults (even lean adults), muscle tissue is gradually replaced by fat and water during the aging process, resulting in a 2% decrease in **basal metabolic rate (BMR)** per decade of adult life. Women tend to have a lower BMR than men because of a smaller percentage of lean muscle mass and the effects of female hormones on metabolism. The effect of activity levels on energy requirements is obvious. However, most overemphasize the immediate, as opposed to the long-term, effects of exercise. For example, one would need to jog for over an hour to burn up the calories present in one piece of apple pie.

TABLE 27.1 • Factors that Influence Energy Expenditure

Factors
Surface area
Age
Sex
Activity level

Regular **exercise** increases basal metabolic rate, allowing calories to burn up more rapidly 24 h a day. A regular exercise program should be designed to increase lean muscle mass and should be repeated 3–5 days a week but need not be aerobic exercise to have an effect on basal metabolic rate. For an elderly or infirm individual, even daily walking may help to increase basal metabolic rate slightly.

Hormone levels are also important, since thyroxine, sex hormones, growth hormone, and, to a lesser extent, epinephrine and cortisol increase BMR. The effects of epinephrine and cortisol probably explain in part why severe stress and major trauma significantly increase energy requirements. Finally, energy intake itself has an inverse relationship to expenditure in that during periods of starvation or semistarvation BMR can decrease up to 50%. This is of great survival value in cases of genuine starvation, but not much help to the person who wishes to lose weight on a calorie-restricted diet.

27.3 • PROTEIN METABOLISM

Dietary Protein Serves Many Roles Including Energy Production

Protein carries a certain mystique as a body-building food. Although it is an essential structural component of all cells, it is also important for maintaining essential secretions such as digestive enzymes and peptide or protein hormones. Protein is also needed for synthesis of plasma proteins, which are essential for maintaining osmotic balance, transporting substances through the blood, and maintaining immunity. However, the average North American adult consumes far more protein than needed to carry out these essential functions. Excess protein is treated as a source of energy, with glucogenic amino acids being converted to glucose and ketogenic amino acids to fatty acids and keto acids. Both kinds of **amino acids** are eventually converted to triacylglycerol in adipose tissue if fat and carbohydrate supplies are already adequate to meet energy requirements. Thus for most of us the only body building obtained from high-protein diets is in adipose tissue.

It has been popular to say that the body has no storage depot for protein, and thus adequate dietary protein must be supplied with every meal. However, this is not quite accurate. Although there is no separate class of storage protein, there is a certain percentage of body protein that undergoes a constant process of breakdown and resynthesis. In the fasting state, the breakdown of this protein is enhanced, and the resulting amino acids are utilized for glucose production, synthesis of nonprotein nitrogenous compounds, and the essential secretory and plasma proteins mentioned above. Even in the fed state, some of these amino acids are utilized for energy production and as biosynthetic precursors. Thus, the turnover of body protein is a normal process and an essential feature of what is called nitrogen balance.

Nitrogen Balance Relates Intake to Excretion of Nitrogen

Nitrogen balance (Figure 27.2) is the relationship between intake of nitrogen, primarily in the form of protein, and excretion of nitrogen, chiefly in the form of undigested protein in the feces and urea and ammonia in urine. A normal adult is in nitrogen equilibrium, with losses just balanced by intake. Negative nitrogen balance results from inadequate intake of protein, since amino acids utilized for energy and biosynthetic reactions are not replaced. It also occurs in injury when there is net destruction of tissue and in major trauma or illness when the body's adaptive response causes increased protein catabolism. Positive nitrogen balance occurs when there is a net increase in body protein, such as in growing children, pregnant women, or convalescing adults.

Essential Amino Acids Must Be Present in the Diet

Several other factors must be considered in addition to the amount of protein in the diet. One is the complement of **essential amino acids** ingested. Essential amino acids are amino acids that cannot be synthesized by the body (Table 27.2). If just one of these essential amino acids is missing from the diet, the body cannot synthesize new protein to replace that lost due to normal turnover, and a negative nitrogen balance results (Figure 27.2). Obviously, the complement of essential amino acids in dietary protein determines how well it can be used by the body.

Most **animal proteins** contain all essential amino acids in about the quantities needed by the human body. **Vegetable proteins**, on the other hand, often lack one or more essential amino acids and may, in some cases, be more difficult to digest. Even so, **vegetarian diets** can provide adequate protein provided enough extra protein is consumed to provide sufficient quantities of the essential amino acids and/or two or more different proteins are consumed together, which complement each other in amino acid content. For example, if corn

TABLE 27.2 • Essential Amino Acids

Essential Amino Acids
Histidine
Isoleucine
Leucine
Lysine
Methionine
Phenylalanine
Threonine
Tryptophan
Valine

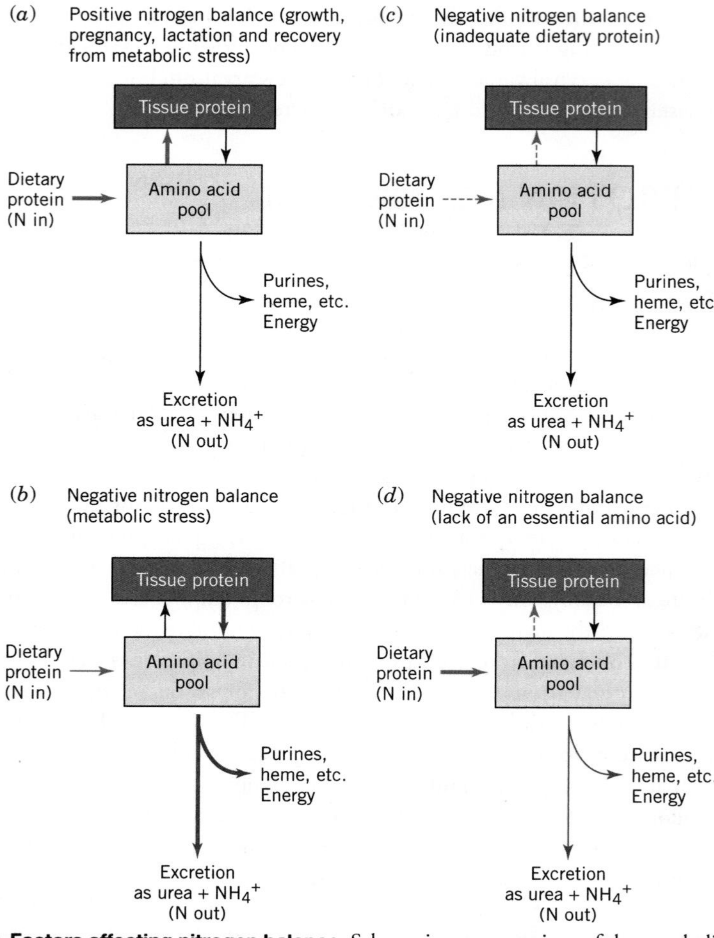

Figure 27.2 Factors affecting nitrogen balance. Schematic representations of the metabolic interrelationship involved in determining nitrogen balance. (*a*) Positive nitrogen balance (growth, pregnancy, lactation, and recovery from metabolic stress). (*b*) Negative nitrogen balance (metabolic stress). (*c*) Negative nitrogen balance (inadequate dietary protein). (*d*) Negative nitrogen balance (lack of an essential amino acid). Each figure represents the nitrogen balance resulting from a particular set of metabolic conditions. The dominant pathways in each situation are indicated by heavy red arrows.

(which is deficient in lysine) is combined with legumes (deficient in methionine but rich in lysine), the efficiency of utilization of the two vegetable proteins approaches that of animal protein. The adequacy of vegetarian diets with respect to protein and calories for children is discussed in Clin. Corr. 27.1; the need for high quality protein in the low protein diets used for treatment of renal disease is discussed in Clin. Corr. 27.2.

Protein Sparing Is Related to Dietary Content of Carbohydrate and Fat

Another factor that determines protein requirement is dietary intake of fat and carbohydrate. If they are present in insufficient quantities, some dietary protein must be used for energy generation, and it becomes unavailable for building and replacing tissue. Thus, as energy (calorie) content of the diet from carbohydrate and fat increases, the need for protein decreases. This is referred to as **protein sparing**. Carbohydrate is somewhat more efficient at protein sparing than fat presumably because it can be used as an energy source by almost all tissues, whereas fat cannot.

Normal Adult Protein Requirements

Assuming adequate calorie intake and a 75% efficiency of utilization, which is typical of mixed protein in the average American diet, the **recommended protein intake** is 0.8 g/kg

CLINICAL CORRELATION 27.1

Vegetarian Diets and Protein-Energy Requirements for Children

One of the most important problems of a purely vegetarian diet (as opposed to a lacto-ovo vegetarian diet) is the difficulty in obtaining sufficient calories and protein. Potential caloric deficit results because the caloric densities of fruits and vegetables are much less than the meats they replace (30–50 cal/100 g versus 150–300 cal/100 g; 7–12 kJ/100 mg versus 36–72 kJ/100 g). The protein problem is threefold. (1) Most plant products contain much less protein (1–2 g of protein/100-g serving versus 15–20-g/100-g serving). (2) Most plant protein is of low biological value. (3) Some plant proteins are not completely digested. Actually, well-designed vegetarian diets usually provide enough calories and protein for the average adult. In fact, the reduced caloric intake may well be of benefit because strict vegetarians tend to be lighter than their nonvegetarian counterparts.

However, whereas an adult male may require about 0.8 g/kg of protein and 40 cal/kg (9.6 kJ/kg) of body weight, a young child may require 2–3 times that amount. Similarly, a pregnant woman needs an additional 10 g of protein and 300 cal/day (72 kJ/day), and a lactating woman an extra 15 g of protein and 500 cal (120 kJ)/day. Thus both young children and pregnant and lactating women run a risk of protein-energy malnutrition. Children of vegetarian mothers generally have a lower birth weight than children of mothers consuming a mixed diet. Similarly, vegetarian children generally have a slower rate of growth through the first 5 years, but generally catch up by age of 10.

Sufficient calories and protein for these high-risk groups can be provided if the diet is adequately planned. Three principles should be followed to design a calorie–protein-sufficient vegetarian diet for young children.(1) Whenever possible, include eggs and milk; they are excellent sources of calories and high-quality protein. (2) Include liberal amounts of those vegetable foods with high-caloric density, including nuts, grains, dried beans, and dried fruits. (3) Include liberal amounts of high-protein vegetable foods that have complementary amino acid compositions. It used to be thought that these complementary proteins must be present in the same meal. Recent animal studies, however, suggest that a meal low in (but not devoid of) an essential amino acid may be supplemented by adding the limiting amino acid at a subsequent meal.

Messina, V., and Mangels, A. R. Considerations in planning vegan diets: Children. *J. Am. Diet. Assoc.* 101:661, 2001; Mangels, A. R., and Messina, V. Considerations in planning vegan diets: Infants. *J. Am. Diet. Assoc.* 101:670, 2001; and Dunham, L., and Kollar, L. M. Vegetarian eating for children and adolescents. *J. Ped. Health Care* 20:27, 2006.

CLINICAL CORRELATION 27.2

Dietary Protein Intake and Renal Disease

Chronic renal failure is characterized by accumulation of the end products of protein catabolism, mainly urea. Some degree of dietary protein restriction is usually necessary because these toxic end products are responsible for many of the symptoms associated with renal failure. The amount of protein restriction depends on the severity of the disease. It is easy to maintain patients in nitrogen balance for prolonged periods on diets containing as little as 40 g of protein/day if the diet is sufficient in calories. Diets containing less than 40 g/day pose problems. Protein turnover continues and a balance found between providing enough protein to avoid negative nitrogen balance, but little enough to avoid buildup of waste products.

The strategy employed in such diets is to (1) provide a physiologically adequate amount of protein, primarily of high biological value, and (2) provide the bulk of the daily caloric requirement as carbohydrates and fats. The goal is to provide just enough essential amino acids to maintain positive nitrogen balance. In turn, the body should be able to synthesize the nonessential amino acids from other nitrogen-containing metabolites. Enough carbohydrate and fat is provided so that essentially all dietary protein can be spared from energy metabolism. With this type of diet, it is possible to maintain a patient on 20 g of protein/day for considerable periods. Because of the difficulty in maintaining nitrogen equilibrium at such low-protein intakes, the patient's protein status should be monitored. This can be done by measuring serum albumin and transferrin.

Unfortunately, such diets are extremely monotonous and difficult to follow. A typical 20 g of protein diet consists of (1) one egg plus 3/4 cup milk or one additional egg or 1 oz of meat, (2) one-half lb of deglutenized (low-protein) wheat bread; all other breads and cereals must be avoided, and this includes almost all baked goods. (3) a limited amount of low-protein, low-potassium fruits and vegetables, and (4) sugars and fats to make up the rest of the needed calories; however, cakes, pies, and cookies should be avoided.

In contrast, hemodialysis causes a net protein catabolic state that can lead to decreased muscle mass and increase the risk of morbidity and mortality. Thus, patients with chronic renal failure who are on dialysis often have increased protein requirements. Recent studies have shown that either intravenous or dietary protein supplementation during dialysis can help restore normal protein homeostasis. Similarly, patients with acute renal failure caused by sepsis, shock, trauma, or burn injury often have increased protein requirements because of the catabolism associated with these conditions (see Clin. Corr. 27.3, p. 1106).

Beto, J. A., and Bansal, V. K. Medical nutrition therapy in chronic kidney failure. *J. Am. Diet. Assoc.* 104:404, 2004; Pupim, L. B., Flakoll, P. J., and Ikizler, T. A. Protein homeostasis in chronic hemodialysis patients. *Curr. Opin. Clin. Nutr. Metab. Care* 7:89, 2004; and Btaiche, I. F., Mohammad, R. A., Alaniz, C., and Mueller, B. A. Amino acid requirements in critically ill patients with acute kidney injury treated with continuous renal replacement therapy. *Pharmacotherapy* 28:600, 2008.

body wt/day. This amounts to about 58 g protein/day for a 72-kg (160-lb) man and about 44 g/day for a 55-kg (120-lb) woman. These recommendations need to be increased on a vegetarian diet if overall efficiency of utilization is less than 75%.

Protein Requirements Are Increased during Growth and Illness

Because dietary protein is essential for synthesis of new body tissue, as well as for maintenance and repair, the need for protein increases markedly during periods of rapid growth as occurs during pregnancy, infancy, childhood, and adolescence. Once growth requirements have been considered, age does not seem to have much effect on protein requirements. If anything, the protein requirement may decrease slightly with age. However, older people need and generally consume fewer calories, so high-quality protein should provide a larger percentage of their total calories. Some older people may have special protein requirements due to malabsorption problems.

Illness, major trauma, and surgery cause a major **catabolic response**. Energy and protein needs in these situations are very large, and the body responds by increasing production of glucocorticoids, epinephrine, and cytokines. Breakdown of body protein is greatly accelerated and a negative nitrogen balance results unless protein intake is increased (Figure 27.2). Although the increased protein requirement is of little significance in short-term illness, it can be vitally important in the recovery of hospitalized patients as discussed in the next section (Clin. Corr. 27.3).

CLINICAL CORRELATION 27.3

Providing Adequate Protein and Calories for Hospitalized Patients

The normal metabolic response to infection, trauma, and surgery is a complex and carefully balanced catabolic state. Glucocorticoids, interleukin-6 (IL-6), and other cytokines are released, greatly accelerating the rates of lipolysis, proteolysis, and gluconeogenesis. The net result is an increased supply of fatty acids and glucose to meet the increased energy demands of such major stress. The high serum glucose causes an increase in circulating insulin levels, which is more than counterbalanced by increased levels of cytokines and glucocorticoids. Skeletal muscle uses very little of the blood glucose, but relies on free fatty acids and its own catabolized protein as a primary source of energy. It continues to export amino acids, primarily alanine, for use elsewhere in the body, resulting in a very rapid depletion of body protein stores.

A highly catabolic hospitalized patient may require 35–45 kcal/kg/day (8.4–10.8 kJ/kg/day) and 2–3 g of protein/kg/day. A patient with severe burns may require even more. Several options are available to provide the postoperative patient with sufficient calories and protein to ensure optimal recovery. When the patient is unable to ingest enough food, it may be adequate to supplement the diet with high-calorie–high-protein preparations, which are usually mixtures of homogenized cornstarch, egg, milk protein, and flavorings. When the patient is unable to ingest solid food or to digest complex mixtures of foods adequately, elemental diets are usually administered via a nasogastric tube. Elemental diets consist of small peptides or amino acids, glucose and dextrins, some fat, vitamins, and electrolytes. These diets are sufficient to meet most of the short-term caloric and protein needs of a moderately catabolic patient.

When a patient is severely catabolic or unable to digest and absorb foods normally, parenteral (intravenous) nutrition is necessary. The least invasive method is to use a peripheral, slow-flow vein as for any other IV infusion. The main limitation of this method is hypertonicity. However, a solution of 5% glucose and 4.25% amino acids can be used safely. This solution will usually provide enough protein to maintain positive nitrogen balance, but will rarely provide enough calories for long-term maintenance of a severely catabolic patient.

The most aggressive nutritional therapy is total parenteral nutrition. Usually an indwelling catheter is inserted into a large fast-flow vessel such as the superior vena cava, so that the very hypertonic infusion fluid can be diluted rapidly. This allows solutions of up to 60% glucose and 4.25% amino acids to be used, providing sufficient protein and most of the calories for long-term maintenance. Intravenous infusions of lipid are often added to boost calories and provide essential fatty acids. Any of these methods can prevent or minimize the negative nitrogen balance associated with surgery and trauma. The choice of method depends on the patient's condition. As a general rule it is preferable to use the least invasive technique.

Wilmore, D. W. Metabolic response to severe surgical illness: An overview. *World J. Surg.* 24:705, 2000; Wolfe, R. R., and Martini, W. Z. Changes in intermediary metabolism in severe surgical illness. *World J. Surg.* 24:639, 2000; and Biffl, W. L., Moore, E. E., and Haenel, J. B. Nutrition support of the trauma patient. *Nutrition* 18:960, 2002.

27.4 • PROTEIN-ENERGY MALNUTRITION

The most common form of malnutrition in the world is **protein-energy malnutrition** (PEM). In developing countries, inadequate intake of protein and energy is all too common, especially in infants and young children. While the symptoms vary widely from case to case, it is common to classify most cases as either **marasmus** or **kwashiorkor**. Marasmus is caused by inadequate intake of both protein and energy. Kwashiorkor is caused by inadequate intake of protein with adequate energy intake. Often the diets that lead to marasmus and kwashiorkor are similar, with kwashiorkor being precipitated by conditions of increased protein demand such as infection. Marasmic infants will have a thin, wasted appearance and will be small for their age. Children are permanently stunted in physical and mental development if PEM continues long enough. The patient with kwashiorkor often has a deceptively plump appearance due to edema. Other telltale symptoms associated with kwashiorkor are dry brittle hair, diarrhea, dermatitis of various forms, and retarded growth. The most devastating result of both conditions is reduced ability to fight off infection. These individuals have a reduced number of T lymphocytes (and thus diminished cell-mediated immune response) as well as defects in generation of phagocytic cells, and production of immunoglobulins, interferon, and other components of the immune system. Many die from secondary infections rather than from starvation.

The most common form of PEM seen in the United States occurs in the hospital setting. A typical course of events is as follows: The patient has not been eating well for several weeks or months prior to entering the hospital due to chronic or debilitating illness. He or she enters the hospital with major trauma, severe infection, or for major surgery, all of which cause a large negative nitrogen balance. This is often compounded by difficulties in feeding the patient or by the need for fasting in preparation for surgery or diagnostic tests. The net result is PEM as reflected by low levels of serum albumin and other serum proteins or by decreased cellular immunity. Hospitalized patients with demonstrable PEM have delayed wound healing, decreased resistance to infection, increased mortality, and increased length of hospitalization. Most major hospitals have programs that monitor the nutritional status of their patients and will intervene where necessary to maintain a positive nitrogen and energy balance (see Clin. Corr. 27.3).

27.5 • EXCESS PROTEIN-ENERGY INTAKE

Much has been said in recent years about the large amount of protein that the average American consumes. Certainly most consume far more than needed to maintain nitrogen balance. An average American currently consumes about 99 g of protein, of which 68% is from animal sources. However, a healthy adult can consume this amount of protein with no apparent harm. Concern has been raised about possible effects of high protein intake on calcium requirements. Some studies suggest that high protein intake increases urinary loss of calcium and may accelerate bone demineralization associated with aging. However, this issue is far from settled.

Obesity Has Dietary and Genetic Components

Perhaps the more serious and frequent nutritional problem in this country is excessive energy consumption. In fact, obesity has been described as an epidemic in the United States and much of the developed world. **Obesity** is defined in terms of body mass index (BMI. weight in kilograms divided by height in meters squared). Ideal weight is usually considered to be a BMI of 24.9 or less. People with a BMI of 25 to 29 are considered overweight; people with a BMI of 30 to 40 are considered obese; and people with BMIs over 40 are considered morbidly obese. An inexact way of thinking about it is that a 5-foot, 4-inch person is generally obese if she or he is 30 pounds or more overweight.

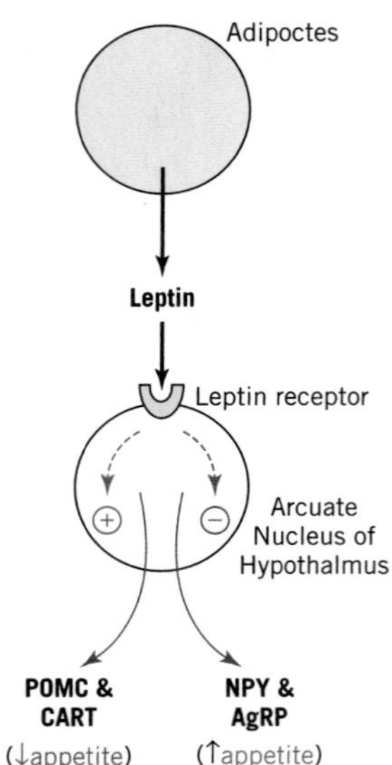

Figure 27.3 The leptin appetite suppression pathway. Schematic representation of the leptin appetite suppression pathway. Adipocytes produce leptin, which binds to its receptor in the arcuate nucleus of the hypothalamus and stimulates neurons that produce the appetite-suppressing hormones POMC (pro-opiomelanocortin) and CART (cocaine- and amphetamine-regulated transcript) and blocks neurons that produce the appetite-stimulating neuropeptides NPY (neuropeptide Y) and AgRP (agouti-related protein). Normally, the amount of leptin increases as the fat stores in adipocytes increase.

Obesity has a major genetic component, which is evident from its strong familial inheritance pattern and studies on monozygotic twins. Many experts think that the contribution of genetic factors to obesity is in the range of 30%–70%. However, to truly understand the role of genetics in obesity we need to think beyond classical Mendelian genetics. Most people tend to think of genetics in terms of rare genetic defects that lead directly to a disease such as phenylketonuria (p. 767) or cystic fibrosis (p. 494). In the case of obesity, it is more fruitful to think of genetics in terms of three levels of genetic influences: monogenic obesity, polygenic predisposition to obesity, and monogenic resistance to obesity.

Monogenic obesity refers to single genetic defects that are strongly linked to obesity irrespective of environmental and behavioral influences. These gene defects obey classical Mendelian inheritance and are exceedingly rare in the general population. For example, recent research has shown that fat cells (adipocytes) produce a hormone called **leptin** that suppresses appetite (Figure 27.3; see Clin. Corr. 17.8, p. 703). The leptin appetite suppression pathway was initially considered a very promising target for pharmacological intervention because a genetically obese strain of mice (ob/ob) was shown to be unable to produce leptin and administration of leptin to those mice resulted in weight loss. However, it turns out that most overweight individuals overproduce leptin, and defects in both the leptin gene and the leptin receptor gene are very rare in the human population.

In the case of polygenic predisposition to obesity, there are common polymorphisms in certain genes that increase the risk of obesity, but only in those individuals who consume more calories than they expend for extended periods of time. Some examples of these polymorphisms will be discussed in the section on nutrigenetics (see Section 27.10). In many cases the predisposition to obesity is relatively weak, so obesity only becomes likely in individuals who consume excess calories and have polymorphisms contributing to obesity in two or more genes. However, because there are many genes in this category and the polymorphisms that contribute to obesity are very common, polygenic predisposition to obesity is common. Finally, there are genetic polymorphisms that predispose to leanness even in individuals consuming excess calories over an extended period of time. Unfortunately, those polymorphisms are relatively rare in the general population.

In summary, even though genetics may exert a 30%–70% influence on obesity, most genetic traits that influence obesity do not cause it directly; they only predispose to obesity when calories consumed exceed calories expended for a prolonged period of time. Furthermore, genetic backgrounds that predispose to obesity when excess calories consumed are the most common genotypes in the general population. Genotypes that allow individuals to consume excess calories without gaining weight are relatively rare. In this light, it is easy to understand why diet and lifestyle play such a critical role in determining the incidence of obesity. The latest report of the National Center for Health Statistics shows 32.7% of Americans are overweight, 34% are obese. and 6% are morbidly obese. That means that over two-thirds of the U.S. population is now either overweight or obese. The prevalence of obesity has increased by 50% in adults and has more than doubled in children in the past 20 years. This clearly reflects recent lifestyle changes in this country because genetics does not change in a matter of a few years. The environmental and behavioral changes that have lead to this "epidemic" of obesity are clearly complex but include the increased availability of calorie-dense foods, an increase in portion size, and the more sedentary lifestyle of the U.S. population.

Obesity, Insulin Resistance, Metabolic Syndrome, and Type 2 Diabetes

Obesity is strongly associated with **type 2 diabetes**. Not only are 80% of people with type 2 diabetes overweight, but the incidence of type 2 diabetes has increased in parallel with the incidence of obesity over the last 20 years or more. However, not everyone who is obese has type 2 diabetes. In fact, there is a gradual, but fairly predictable, transition from simple obesity with no observable metabolic changes through **insulin resistance** with its many associated metabolic changes to type 2 diabetes. Why is this? Obesity is obviously associated with an increased number and/or size of adipose tissue cells. However, it is important to understand that adipocytes are endocrine-producing cells, not just repositories of fat stores.

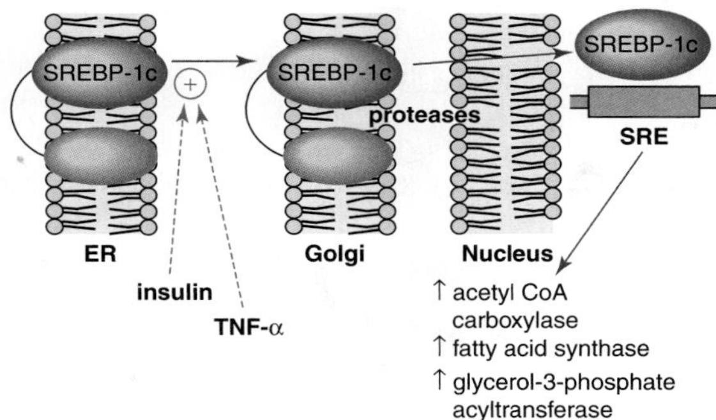

Figure 27.4 **Effect of TNFα on the expression of enzymes involved in fatty acid and triglyceride synthesis in the liver.** Adipokines such as TNFα mimic the effect of insulin on fatty acid and triglyceride synthesis in the liver by stimulating the movement of the transcription factor SREBP-1c (sterol regulatory element binding protein-1c) from the endoplasmic reticulum to the Golgi, where its membrane bound portion is cleaved by proteases. This allows SREBP-1c to diffuse into the nucleus where it binds to the sterol regulatory element (SRE) and increases the expression of acetyl CoA carboxylase, fatty acid synthase and glycerol-3-phosphate acy transferase.

When engorged with fat, adipocytes cells overproduce hormones such as leptin and **resistin** and **cytokines** such as **tumor necrosis factor alpha (TNFα)**. Although it is not clear at this point whether it is TNFα or other adipokines that exert their effects on muscle and liver metabolism, TNFα does exert a strong paracrine effect on adipose tissue. It stimulates hormone-sensitive lipase, which results in elevated levels of free fatty acids in the circulation, and inhibits lipoprotein lipase, which decreases the clearance rate of triacylglycerol-rich VLDL particles from the circulation.

The increased influx of the circulating free fatty acids into the liver increases the levels of fatty acyl CoA available for triacylglycerol synthesis and acetyl CoA available for fatty acid synthesis. At the same time TNFα activates the transcription factor SREBP-1c which, in turn, increases expression of key enzymes involved in fatty acid and triacylglycerol biosynthesis (Figure 27.4). The net result is increased production of triacylglycerol-rich VLDL particles by the liver. This, coupled with the decreased clearance of VLDL particles just described, leads to elevated levels of triglycerides (triacylglycerol-rich VLDL particles) in the circulation. TNFα also decreases lecithin: cholesterol acyltransferase (LCAT) expression and activity, ATP-binding cassette (ABCA1 and ABCG1) expression and apo A-I and apo A-IV expression, all of which are thought to contribute to a decline in HDL levels associated with obesity. Thus, obesity is often associated with a **dyslidemia** characterized by elevated triglycerides and decreased HDL levels. This dyslipidemia is highly associated with increased risk of cardiovascular disease.

The circulating free fatty acids associated with obesity also appear to be responsible for insulin resistance of the muscle and the liver. Fatty acids stimulate **PKC-θ** catalyzed serine phosphorylation of the insulin receptor 1 and 2 substrates, which in turn interferes with insulin stimulation of the PKB signaling pathway (Figure 27.5). In muscle, this interferes with the ability of insulin to stimulate translocation of the GLUT4 transporter to the membrane. In liver, it interferes with the ability of insulin to down-regulate gluconeogenesis. In addition, fatty acids are competitive inhibitors of glucose uptake by both the GLUT4 transporter in muscle and the GLUT2 transporter in liver. The decreased uptake of glucose by both muscle and liver and the increased production of glucose by the liver tend to lead to hyperglycemia.

In the early stages of obesity the pancreas compensates for the insulin resistance by overproducing insulin so that glucose homeostasis is maintained at normal or near normal levels. However, the pancreas cannot maintain this overproduction of insulin forever. The elevated levels of free fatty acids and/or cytokines lead to a gradual loss of the ability of the pancreas to overproduce insulin, a process called decompensation. Once the pancreas is no longer able to produce enough insulin to compensate for insulin resistance, hyperglycemia ensues. Since hyperglycemia is the defining criterion for diabetes, it is at this point that the patient is classified at having type 2 diabetes. **Type 2 diabetes** differs from type 1 diabetes

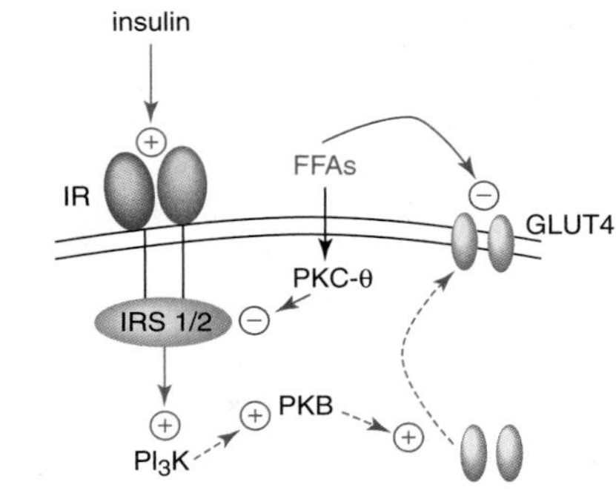

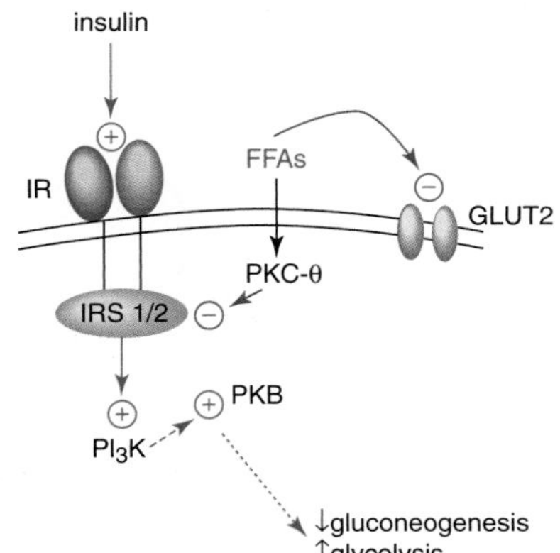

Figure 27.5 **Mechanisms involved in insulin resistance in muscle and liver.** Free fatty acids (FFA) stimulate PKC-θ (protein kinase C-θ), which serine phosphorylates and inactivates IRS1/2 (insulin receptor substrate 1 and 2). This interferes with the PI_3K (phosphoinositide-3 kinase)–PKB (protein kinase B) signaling pathway which normally causes transfer of the GLUT4 transporter to the cell surface in muscle and decreases gluconeogenesis and increases glycolysis in the liver. Free fatty acids also competitively inhibit transport of glucose by both the GLUT4 and GLUT2 transporters. + symbols denote positive regulation and – symbols denote negative regulation. The dotted lines denote responses to insulin that are not occurring because of insulin resistance.

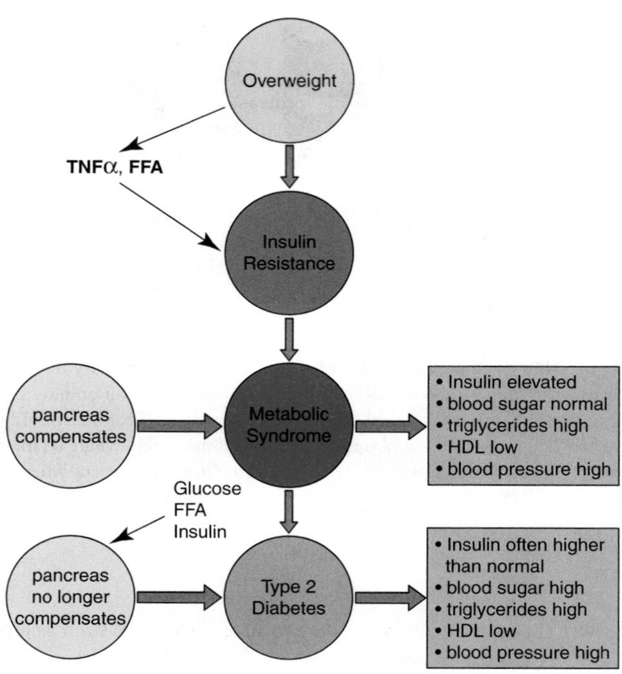

Figure 27.6 Schematic of the metabolic response to obesity over time.

in several fundamental aspects. Type 1 diabetes is caused by an inability of the pancreas to produce insulin, whereas type 2 diabetes is caused by an obesity-associated insulin resistance. In fact insulin levels in a type 2 diabetic are often elevated or near normal; they are just no longer sufficient to overcome the insulin resistance.

There is a relatively predictable sequence of metabolic changes associated with obesity (Figure 27.6). As described, one of the earliest changes is an overproduction and decreased clearance of VLDL particles, leading to a dyslipidemia characterized by elevated triacylglycerol-rich VLDL particles and decreased HDL levels. Insulin resistance is also a relatively early metabolic change associated with obesity, but the expected hyperglycemia often does not occur for many years because of the ability of the pancreas to compensate by overproducing insulin. However, the higher-than-normal levels of insulin required to maintain glucose homeostasis are not entirely benign. Insulin-signaling pathways lead to increased cell proliferation in many cells, and obesity appears to be associated with increased risk of several types of cancer. This hyperinsulinemia also appears to stimulate the sympathetic nervous system, leading to sodium and water retention and vasoconstriction, which increase blood pressure. Finally, if the obesity is maintained long enough, the ability of the pancreas to overproduce insulin is lost and type 2 diabetes ensues.

Thus, the insulin resistance associated with obesity increases the risk of developing type 2 diabetes, heart disease, hypertension, and several types of cancer. Since increased risk of some of those diseases may occur long before the patient becomes diabetic, there has been a movement to label the condition that exists between the onset of insulin resistance and the development of type 2 diabetes as **metabolic syndrome**. The World Health Organization defines metabolic syndrome as two or more of the following conditions: abdominal obesity, dyslipidemia (as defined by an elevation of triacylglycerol-rich VLDL and a decrease in HDL), hypertension, insulin resistance, modest increases in fasting blood sugar, a prothrombic state, or a pro-inflammatory state (usually defined by increase in C-reactive protein). However, because all of these conditions do not occur simultaneously in a given individual, the term *metabolic syndrome* is not universally accepted.

Obesity Has Significant Health Implications

Obesity is a primary risk factor in coronary heart disease, hypertension, and diabetes mellitus. Obesity is also associated with inflammatory diseases, some forms of cancer, bone and joint disorders, and breathing disorders. This is nutritionally significant because all of these metabolic changes are reversible. Quite often reduction to ideal body weight is the single most important aim of nutritional therapy. When the individual is at ideal body weight, the composition of the diet becomes a less important consideration in maintaining normal serum lipid and glucose concentrations.

As mentioned, obesity can lead to increased retention of both sodium and water. As the fat stores are metabolized, they produce water (which is denser than fat), and the water may be largely retained. In fact, some individuals may observe short-term weight gain when they diet, even though the diet is working perfectly well in terms of breaking down their adipose tissue. This metabolic fact of life can be psychologically devastating to dieters, who expect quick results for all their sacrifice.

27.6 • CARBOHYDRATES

The chief metabolic role of **dietary carbohydrates** is for energy production. Any carbohydrate in excess of that needed for energy is converted to glycogen and triacylglycerol for storage. The body can adapt to a wide range of dietary carbohydrate levels (see Clin. Corr 27.7, p. 1118). Diets high in carbohydrate result in higher steady-state levels of glucokinase and some of the enzymes involved in the pentose phosphate pathway and triacylglycerol synthesis. Diets low in carbohydrate result in higher steady-state levels of some of the enzymes involved in gluconeogenesis, fatty acid oxidation, and amino acid catabolism. Glycogen stores are also affected by the carbohydrate content of the diet (Clin. Corr. 27.4).

The most common form of **carbohydrate intolerance** is diabetes mellitus, caused by either subnormal insulin production or insulin resistance. This causes intolerance to glucose and sugars that are readily converted to glucose. Dietary treatment of diabetes is discussed in

CLINICAL CORRELATION 27.4

Carbohydrate Loading and Athletic Endurance

The practice of carbohydrate loading dates to observations made in the early 1960s that endurance during vigorous exercise was limited primarily by muscle glycogen stores. Of course, glycogen is not the sole energy source for muscle. Free fatty acids increase in blood during vigorous exercise and are utilized by muscle along with its glycogen stores. Once glycogen has been exhausted, however, muscle cannot rely entirely on free fatty acids without tiring rapidly, probably because muscle becomes increasingly hypoxic during vigorous exercise. Although glycogen is utilized equally well aerobically or anaerobically, fatty acids can only be utilized aerobically. Under anaerobic conditions, fatty acids cannot provide ATP rapidly enough to serve as the sole energy source.

The practice of carbohydrate loading to increase glycogen stores was devised for track and other endurance athletes. The original carbohydrate loading regimen consisted of a 3–4-day period of heavy exercise while on a low-carbohydrate diet, followed by 1–2 days of light exercise while on a high-carbohydrate diet. The initial low-carbohydrate–high-energy demand period caused a depletion of muscle glycogen stores. The subsequent change to a high-carbohydrate diet resulted in the production of higher-than-normal levels of insulin and growth hormone and glycogen stores reached almost twice the normal amounts. This practice did increase endurance significantly. In one study, test subjects on a high-fat and high-protein diet had less than 1.6 g of glycogen per 100 g of muscle and could perform a standardized workload for only 60 min. When the same subjects then consumed a high-carbohydrate diet for 3 days, their glycogen stores increased to 4 g per 100 g of muscle and the same workload could be performed for up to 4 h.

Although the technique clearly worked, the athletes often felt lethargic and irritable during the low-carbohydrate phase of the regimen, and the high-fat diet ran counter to current health recommendations. Recent studies indicate that regular consumption of a high-complex carbohydrate low-fat diet during training increases glycogen stores without sudden dietary changes. Current recommendations are for endurance athletes to consume a high-carbohydrate diet (with emphasis on complex carbohydrates) during training. Then carbohydrate intake is increased further (to 70% of calories) and exercise tapered off during the 2–3 days just prior to an athletic event. This increases muscle glycogen stores to levels comparable to the previously described carbohydrate-loading regimen.

Lambert, E. V., and Goedecke, J. H. The role of dietary micronutrients in optimizing endurance performance. *Curr. Sports Med. Rep.* 2:194, 2003; Hargreaves, M., Hawley, J. A., and Jeukendrup, A. Pre-exercise carbohydrate and fat ingestion: Effects on metabolism and performance. *J. Sports Sci.* 22:31, 2004; and Burke, L. M., Kiens, B., and Ivey, J. L., Carbohydrates and fat for training and recovery. *J. Sports Sci.* 22:15, 2004.

CLINICAL CORRELATION 27.5

High-Carbohydrate versus High-Fat Diet for Diabetics

For years the American Diabetes Association has recommended diets that were low in fat and high in complex carbohydrates and fiber for diabetics. The logic of such a recommendation seemed to be inescapable. Diabetics are prone to hyperlipidemia with attendant risk of heart disease, and low-fat diets seemed likely to reduce the risk of hyperlipidemia and heart disease. In addition, numerous clinical studies had suggested that the high-fiber content of these diets improved control of blood sugar. This recommendation has proved to be controversial and illustrates the difficulties in making dietary recommendations for population groups rather than individuals. There is significant variation in how individual diabetics respond to these diets. Some diabetic patients show poorer control (as evidenced by higher blood glucose levels, elevated VLDL and/or LDL levels, and reduced HDL levels) on the high-carbohydrate–high-fiber diets than on diets high in monounsaturated fatty acids. However, diets high in monounsaturated fatty acids tend to have higher caloric density and may be inappropriate for overweight individuals with type 2 diabetes. Thus, a single diet may not be equally appropriate for all diabetics. Even the glycemic index concept (see Table 27.2, p. 1103) may also turn out to be difficult to apply to the diabetic population as a whole, because of individual variation. In 2004, the American Diabetes Association abandoned the concept of a single diabetic diet. Instead their recommendations focus on achievement of glucose, lipid, and blood pressure goals, with weight reduction and dietary recommendations based on individual preferences and what works best to achieve metabolic control in the individual. However, that doesn't necessarily imply that any weight-reduction diet is satisfactory. In 2006 the Diabetes and Nutrition Study Group of the European Association for the Study of Diabetes made a very specific set of evidenced-based recommendations for the all diets used for the treatment and prevention of diabetes. Their grade A recommendations are that such diets should (1) reduce energy intake and increase energy expenditure among those who are overweight and to prevent weight regain once weight loss has been achieved, (2) reduce saturated fats and *trans*-unsaturated fatty acids to below 10% of total energy (below 8% if LDL-cholesterol levels are elevated), (3) restrict dietary cholesterol to less than 300 mg/day (especially if LDL-cholesterol is elevated), (4) include naturally occurring carbohydrate-rich foods that are high in dietary fiber and low in glycemic index (vegetables, legumes, fruits and whole grains) with a total dietary fiber intake of 40g/day, and (5) restrict salt to less than 6 g/day. Their position is that a wide range of carbohydrate, protein, and fat intakes are suitable for a diabetic as long as the foods chosen meet these criteria.

Klein, S., Sheard, N. F., Pi-Sunyer, X., Daly, A., et al. Weight management through lifestyle modification for the prevention and management of type 2 diabetes: rationale and strategies. A statement of the American Diabetes Association for the Study of Obesity and the American Society for Clinical Nutrition. *Am. J. Clin. Nutr.* 80:257, 2004; Grundy, S. M., Hansen, B., Smith, S. C., Cleeman, J. I., and Kahn, R. A. Clinical management of metabolic syndrome. Report of the American Heart Association/National Heart, Lung, and Blood Institute/American Diabetes Association Conference on Scientific Issues Related to Management. *Circulation* 109:551, 2004; and Mann, J. I. Nutrition recommendations for the treatment and prevention of type 2 diabetes and the metabolic syndrome: An evidenced based review. *Nutr. Rev.* 64:422, 2006.

Clin. Corr. 27.5. **Lactase insufficiency** (p. 1050) is also a common disorder of carbohydrate metabolism affecting over 30 million people in the United States alone. It is most prevalent among blacks, Asians, and Hispanics. Without intestinal lactase, dietary lactose is not significantly hydrolyzed or absorbed. It remains in the intestine where it acts osmotically to draw water into the gut, and it is converted to lactic acid and CO_2 by intestinal bacteria. The result is bloating, flatulence, and diarrhea, all of which can be avoided simply by eliminating milk and milk products from the diet.

27.7 • FATS

Triacylglycerols, or fats, are directly utilized by many tissues as an energy source, and phospholipids are important constituents of membranes. Excess dietary fat can be stored only as triacylglycerol in adipose tissue. As with carbohydrate, the body adapts to a wide range of fat intakes. However, problems develop at the extremes (either high or low) of fat consumption. At the low end, **essential-fatty-acid** (**EFA**) deficiency may become a problem. The fatty acids linoleic and linolenic cannot be made by the body and thus are essential components of the diet. They are needed for maintaining the function and integrity of membrane structure, for fat metabolism and transport, and for synthesis of **prostaglandins** and related compounds. The most characteristic symptom of essential-fatty-acid deficiency is a scaly dermatitis. EFA deficiency is very rare in the United States, occurring primarily in low-birth-weight infants fed on artificial formulas lacking EFA and in hospitalized patients maintained on total parenteral nutrition for long periods. At the other extreme, there is concern that excess dietary fat causes elevation of serum lipids and thus an increased risk of heart disease. Recent studies suggest that high fat intakes are associated with increased risk of colon, breast, and prostate cancer, but it is not clear whether the cancer risk is associated with fat intake *per se,* or with the excess calories and obesity associated with a high-fat diet. Animal studies suggest that polyunsaturated fatty acids of the ω-6 series may be more tumorigenic than other unsaturated fatty acids. The reason for this is not known, but it has been suggested that prostaglandins derived from the ω-6 fatty acids may stimulate tumor progression.

27.8 • FIBER

Dietary fiber comprises those components of food that cannot be broken down by human digestive enzymes. It is incorrect, however, to assume that fiber is indigestible since some fibers are, in fact, at least partially broken down by intestinal bacteria. Our current understanding of the metabolic roles of dietary fiber is based on three important observations: (1) there are several different types of dietary fiber. (2) they each have different chemical and physical properties, and (3) they each have different effects on human metabolism, which can be understood, in part, from their unique properties.

The major types of fiber and their properties are summarized in Table 27.3. **Cellulose** and most **hemicelluloses** increase stool bulk, decrease transit time, and are associated with the effects of fiber on regularity. They decrease intracolonic pressure and appear to play a beneficial role with respect to diverticular diseases. By diluting out potential carcinogens and speeding their transit through the colon, they may also play a role in reducing the risk of colon cancer. **Lignins** have bulk-enhancing properties and they adsorb organic substances such as cholesterol to lower plasma cholesterol concentration. Mucilaginous fibers, such as **pectin** and **gums**, tend to form viscous gels in the stomach and intestine and slow the rate of gastric emptying, thus slowing the rate of absorption of many nutrients. Their most important clinical role is to slow the rate at which carbohydrates are digested and absorbed. Thus, both the rise in blood sugar and the rise in insulin levels are significantly decreased if these fibers are ingested with carbohydrate-containing foods. Water-soluble fibers (pectins, gums, and some hemicelluloses, and storage polysaccharides) also help to lower serum cholesterol levels in most people. Whether this is due to their effect on insulin levels (insulin stimulates cholesterol synthesis and export) or to other metabolic effects (perhaps caused by end products of partial bacterial digestion) is not known. Vegetables, wheat, and most grain fibers are the best sources of the water-insoluble cellulose, hemicellulose, and lignin. Fruits, oats, and legumes are the best sources of the water-soluble fibers. Obviously, a balanced diet should include food sources of both soluble and insoluble fiber.

TABLE 27.3 • Major Types of Fiber and Their Properties

Type of Fiber	*Major Source in Diet*	*Chemical Properties*	*Physiological Effects*
Cellulose	Unrefined cereals Bran Whole wheat	Nondigestible Water-insoluble Absorbs water	Increases stool bulk Decreases intestinal transit time Decreases intracolonic pressure
Hemicellulose	Unrefined cereals Some fruits and vegetables Whole wheat	Partially digestible Usually water-insoluble Absorbs water	Increases stool bulk Decreases intestinal transit time Decreases intracolonic pressure
Lignin	Woody parts of vegetables	Nondigestible Water-insoluble Absorbs organic substances	Increases stool bulk Bind cholesterol Bind carcinogens
Pectin	Fruits	Digestible Water-soluble Mucilaginous	Decreases rate of gastric emptying Decreases rate of sugar uptake Decreases serum cholesterol
Gums	Dried beans Oats	Digestible Water-soluble Mucilaginous	Decreases rate of gastric emptying Decreases rate of sugar uptake Decreases serum cholesterol

27.9 • COMPOSITION OF MACRONUTRIENTS IN THE DIET

Since there are relatively few instances of macronutrient deficiencies in the American diet, much of the interest in recent years has focused on whether there is an ideal diet composition consistent with good health.

Composition of the Diet Affects Serum Cholesterol

With respect to heart disease, the current discussion centers around two key issues. (1) Can **serum cholesterol** and triacylglycerol concentration be controlled by diet? (2) Does lowering serum cholesterol and triacylglycerol levels protect against heart disease? The controversies around dietary control of cholesterol levels illustrate perfectly the trap one falls into by trying to look too closely at each individual component of the diet instead of at the diet as a whole. For example, at least four dietary components have an effect on serum cholesterol: cholesterol itself, **polyunsaturated fatty acids** (**PUFA**), **saturated fatty acids** (**SFA**), and fiber. It would seem that the more cholesterol one eats, the higher the serum cholesterol would be. However, cholesterol synthesis is tightly regulated, and decreases in dietary cholesterol have relatively little effect on serum cholesterol levels (p. 724). One can obtain a more significant reduction in cholesterol and triacylglycerol levels by increasing the **ratio of PUFA/SFA** in the diet. Finally, some plant fibers, especially the water-soluble fibers, appear to decrease cholesterol levels significantly.

While the effects of various lipids in the diet can be dramatic, the biochemistry of their action is still uncertain. Saturated fats inhibit receptor-mediated uptake of LDL, but the mechanism is complex. Palmitic acid (saturated, C_{16}) raises serum cholesterol levels whereas stearic acid (saturated, C_{18}) has no effect. Polyunsaturated fatty acids lower both LDL and HDL cholesterol levels, whereas oleic acid (monounsaturated, C_{18}) appears to lower LDL without affecting HDL levels. The ω-3 and ω-6 polyunsaturated fatty acids have slightly different effects on lipid profiles (Clin. Corr. 27.6). However, these

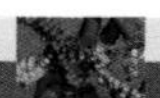

CLINICAL CORRELATION 27.6

Polyunsaturated Fatty Acids and Risk Factors for Heart Disease

Because reduction of elevated serum cholesterol levels can reduce risk of heart disease, there is considerable interest in the effects of diet on serum cholesterol levels and other risk factors for heart disease. One important dietary factor regulating serum cholesterol levels is the ratio of polyunsaturated fats (PUFAs) to saturated fats (SFAs) in the diet. Furthermore, recent research shows that different types of polyunsaturated fatty acids have different effects on lipid metabolism and on other risk factors for heart disease. Essential polyunsaturated fatty acids can be classified as either ω-6 or ω-3. Clinical studies have shown that the ω-6 PUFAs (chief dietary source is linoleic acid from plants and vegetable oils) primarily decrease serum cholesterol levels, with only modest effects on serum triacylglycerol levels. The ω-3 PUFAs (chief dietary source is eicosapentaenoic acid (EPA) and docosahexaenoic acid (DHA) from certain ocean fish and fish oils) cause only modest decreases in serum cholesterol levels but significantly lower serum triacylglycerol levels. The mechanisms behind these effects on serum lipid levels are unknown.

In addition, the ω-3 PUFAs have other effects that may decrease the risk of heart disease; they decrease platelet aggregation, inflammation, and arrhythmia and increase endothelial relaxation. In the case of platelet aggregation, the mechanism is clear. Arachidonic acid (ω-6 family) is a precursor of thromboxane A_2 (TXA_2), which is a potent pro-aggregating agent, and prostaglandin I_2 (PGI_2), which is a weak anti-aggregating agent (p. 742). The ω-3 PUFAs are converted to thromboxane A_3 (TXA_3), which is only weakly pro-aggregating, and prostaglandin I_3 (PGI_3), which is strongly anti-aggregating. Thus, the balance between pro-aggregation and anti-aggregation is shifted toward a more anti-aggregating condition as the ω-3 PUFAs displace ω-6 PUFAs as precursors to the thromboxanes and prostaglandins. The ω-3 PUFAs also have been shown to reduce cardiac arrhythmia and promote plaque stabilization. Several clinical studies have shown that diets rich in ω-3 fatty acids significantly decrease the risk of sudden cardiac death in patients who have previously had a myocardial infarction. Because of these studies, both the American and European Heart Associations have incorporated recommendations for ω-3 PUFAs into their treatment and prevention guidelines for myocardial infarction.

Marchioli, R., Schweiger, C., Tavazzi, L., and Valagussa, F. Efficacy of n-3 polyunsaturated fatty acids after myocardial infarction: Results of the GISSI-Prevenzione Trial. Gruppo Italiano per lo Studio della Sopravvivenza nell'Infarcto Miocardico. *Lipids* 36, Suppl.: S119, 2001; Marchioli, R., Barzi, F., Bomba, E., Chieffo, C., et al. Early protection against sudden death by n-3 polyunsaurated fatty acids after myocardial infarction: Time-course analysis of the results of the Gruppo Italiano per lo Studio della Sopravivenza nell'Infarcto Miocardico (GISSI)-Prevenzione. *Circulation* 105:1897, 2002; Kris-Etherton, T., Harris, W. H., and Appel, L. J. Omega-3 fatty acids and cardiovascular disease. New recommendations from the American Heart Association. Arterioscler. *Thromb. Vasc. Biol.* 23:151, 2003; and Biscione, F., Pignalberi, C., Totteri, A., Messina, F., and Altamura, G. Cardiovascular effects of omega-3 fatty acids. *Curr. Vascular Pharmacol.* 5:163, 2007.

complexities do not significantly affect dietary recommendations. Most foods high in saturated fats contain both palmitic and stearic acid, and are atherogenic. Since oleic acid lowers LDL levels, olive oil, and possibly peanut oil, may be as beneficial as polyunsaturated oils.

There is little disagreement with respect to these data. The question is: What can be done with the information? Much of the disagreement arises from the tendency to look at each dietary factor in isolation. For example, it is debatable whether it is worthwhile placing a patient on a highly restrictive 300-mg cholesterol diet (1 egg contains about 213 mg of cholesterol) if his serum cholesterol is lowered by only 5%–10%. Likewise, changing the PUFA/SFA ratio from 0.3 (the current value) to 1.0 would either require a radical change in the diet by elimination of foods containing saturated fat (largely meats and fats) or an addition of large amounts of rather unpalatable polyunsaturated fats to the diet. For many Americans, this would be unrealistic. Fiber is another good example. One could expect, at the most, a 5% decrease in serum cholesterol by adding a reasonable amount of fiber to the diet. Very few people would eat the 10 apples/day needed to lower serum cholesterol by 15%. Are we to conclude then that any dietary means of controlling serum cholesterol levels is useless? This would be the case only if each element of the diet is examined in isolation. For example, recent studies have shown that vegetarians, who have lower cholesterol intakes plus higher PUFA/SFA ratios and higher fiber intakes, may average 25%–30% lower cholesterol levels than their nonvegetarian counterparts. Perhaps, more to the point, diet modifications acceptable to the average American have been shown to cause a 10%–15% decrease in cholesterol levels in long-term studies, and a recent study called OmniHeart has shown that as long as saturated fat, cholesterol, fiber, sodium, calcium, magnesium, and potassium were at appropriate levels, a wide range of macronutrient composition was compatible with a heart-healthy diet.

Carbohydrates, Glycemic Index, and Glycemic Load

Much of the nutritional dispute in the area of carbohydrate intake centers around the effect of carbohydrate on blood glucose and triacylglycerol levels. The old paradigm that simple sugars raise blood sugar and triacylglycerol levels to a greater extent than complex carbohydrates is an oversimplification. The effect of the carbohydrates in a particular food is determined by the rate of digestion and absorption of the carbohydrate and by other components of the food. In particular soluble fiber, protein, and fat all blunt the effect of carbohydrates on blood glucose levels. Thus, the concept of **glycemic index** was developed to describe better the effects of carbohydrates on blood glucose. Glycemic index is determined empirically and is defined as the effect of 50 g of carbohydrate in a particular food on blood glucose levels compared to 50 g of glucose. In general, pastries, refined cereals, rice, and starchy vegetables have high glycemic indices, whereas nonstarchy vegetables, fruits, legumes, and nuts have low glycemic indices (Table 27.4). However, because the carbohydrate content of foods varies widely, even glycemic indices can be misleading. For example, carrots have a much higher glycemic index than ice cream (Table 27.4). Thus, the term ***glycemic load*** has recently been introduced. Glycemic load is the glycemic index times the amount of carbohydrate in a standard serving size of that food. As might be expected carrots have a much lower glycemic load than ice cream.

Nutritional Protein Requirements Are Achieved by Mixing Vegetable and Animal Proteins

Epidemiologic data and animal studies suggest that consumption of animal protein is associated with increased incidence of heart disease and various forms of cancer. One could assume that it is probably not the animal protein itself that is involved, but the associated fat and cholesterol. What sort of protein should we consume? Although the present diet may not be optimal, a strictly vegetarian diet may not be acceptable to many Americans. Perhaps a middle road is best. Clearly, there are no known health dangers associated with a mixed diet that is lower in animal protein than the current American standard.

TABLE 27.4 • Glycemic Index[a] of Selected Foods

Food	Glycemic Index
Grain and cereal products	
Bread (white)	69 ± 5
Bread (whole wheat)	72 ± 6
Rice (white)	72 ± 9
Sponge cake	46 ± 6
Breakfast cereals	
All bran	51 ± 5
Cornflakes	80 ± 6
Oatmeal	49 ± 8
Shredded wheat	67 ± 10
Vegetables	
Sweet corn	59 ± 11
Frozen peas	51 ± 6
Dairy products	
Ice cream	36 ± 8
Milk (whole)	34 ± 6
Yoghurt	36 ± 4
Root vegetables	
Beets	64 ± 16
Carrots	92 ± 20
Potato (white)	70 ± 6
Potato (sweet)	48 ± 6
Dried legumes	
Beans (kidney)	29 ± 8
Beans (soy)	15 ± 5
Peas (blackeye)	33 ± 4
Fruits	
Apple (Golden Delicious)	39 ± 3
Banana	62 ± 9
Orange	40 ± 3
Sugars	
Fructose	20 ± 5
Glucose	100
Honey	87 ± 8
Sucrose	59 ± 10

Source: Data from Jenkins,D. A., et al. Glycemic index of foods: A physiological basis for carbohydrate exchange. *Am. J. Clin. Nutr.* 34:362, 1981.

[a]Glycemic index is defined as the area under the blood glucose response curve for each food expressed as a percentage of the area after taking the same amount of carbohydrate as glucose (mean: 5–10 individuals).

Fiber from Varied Sources Is Desirable

Because of current knowledge about effects of fiber on human metabolism, most suggestions for a prudent diet recommend an increase in dietary fiber. The current fiber content of the American diet is about 14–15 g/day. Most experts feel that an increase to at least 25–30 g would be safe and beneficial. Since different types of fibers have different physiological roles, the increase in fiber intake should come from a wide variety of sources, including fresh fruits, vegetables, and legumes as well as the more popular cereal fibers (which are primarily cellulose and hemicellulose).

Dietary Recommendations

Several private and government groups have made specific recommendations with respect to the **ideal dietary composition** for the American public. This movement was spearheaded by the Senate Select Committee on Human Nutrition, which first published its *Dietary Goals for the United States* in 1977. This Committee recommended that the American public reduce consumption of total calories, total fat, saturated fat, cholesterol, simple sugars, and salt to "ideal" goals more compatible with good health (Figure 27.7). In recent years the U.S. Department of Agriculture (USDA), the American Heart Association, the American Diabetes Association, the National Research Council, and the Surgeon General have published similar recommendations, and the USDA has used these recommendations to design revised recommendations for its **Food Guide Pyramid** (Figure 27.8). How valid is the scientific basis of the recommendations for a **prudent diet**? Is there evidence that it will improve the health of the general public? How much does individual variability affect these recommendations? These remain controversial questions.

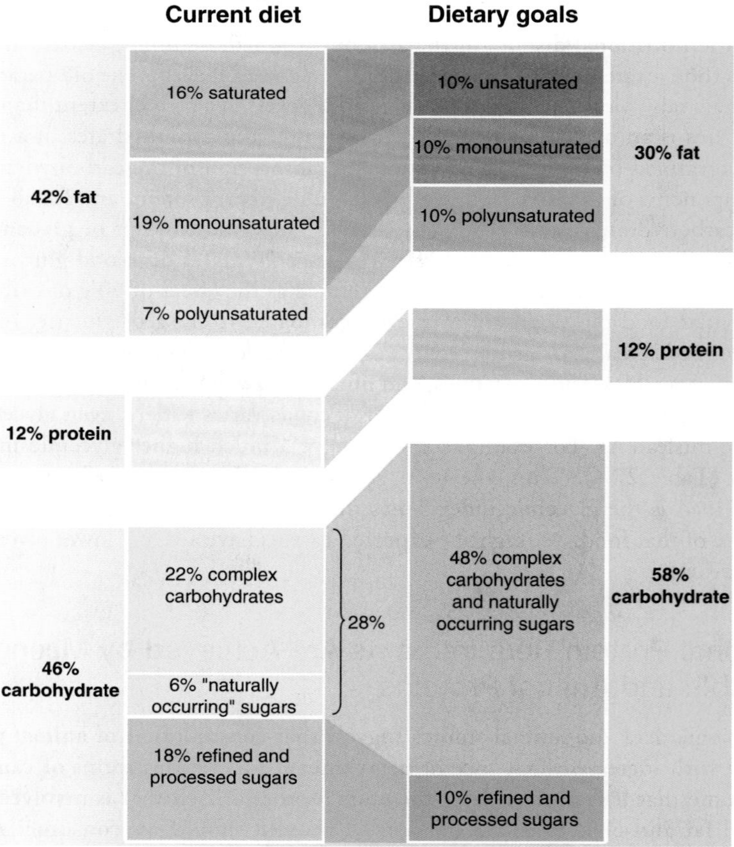

Figure 27.7 U.S. dietary goals. Graphical comparison of the composition of the current U.S. diet and the dietary goals for the U.S. population suggested by the Senate Select Committee on Human Nutrition. From *Dietary Goals for the United States,* 2d ed. Washington, DC: U.S. Government Printing Office, 1977.

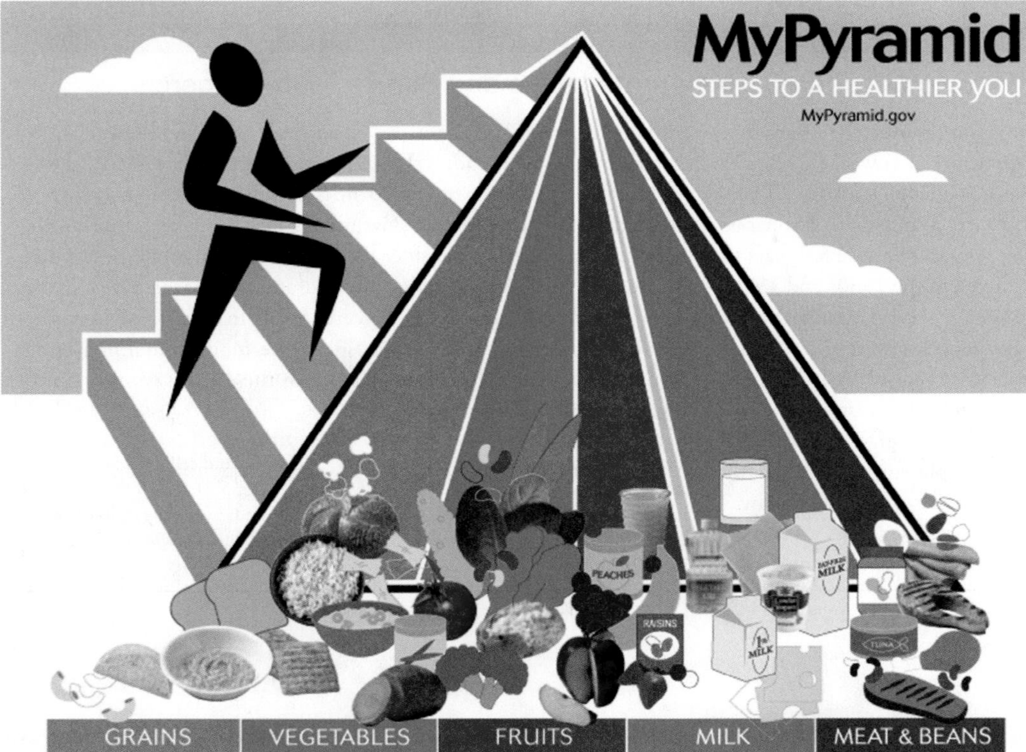

Figure 27.8 USDA food pyramid. Graphical representation of USDA recommendations for a balanced diet.www.mypyramid.gov.

The recent debate about diet composition and weight loss illustrates the complexity of these considerations. The debate was fueled by Dr. Atkins *Diet Revolution* and *New Diet Revolution* books claiming that weight loss was more effective with low-carbohydrate diets, and that fat, even saturated fat, did not adversely affect serum cholesterol levels. In fact, short-term studies seemed to confirm that weight loss was more rapid and control of blood sugar and improvement of lipid parameters were better on low-carbohydrate diets. There have now been a number of well-controlled clinical trials comparing low-fat (typically high carbohydrate, moderate-protein, and low-fat), low carbohydrate (typically low-carbohydrate, moderate-protein, and high-fat), high-protein (typically moderate-carbohydrate, high-protein, and moderate fat) and Mediterranean (typically moderate-carbohydrate, moderate-protein and moderate-fat with carbohydrates coming primarily from vegetables, protein coming primarily from poultry and fish, and fat coming primarily from olive oil) diets. Systematic analyses of these studies have concluded that weight loss is somewhat greater at 6 months on the low-fat and high-protein diets, but that there is very little difference in net weight loss on any of these diets at 1 year or more. Triglyceride levels (triacylglycerol-rich VLDL) and HDL levels appeared to improve slightly more on the low-carbohydrate diets, whereas total and LDL levels improved slightly more on the low-fat diets and blood sugar control was slightly better on the Mediterranean diet. However, all of the differences between the diets were very small, and there was a huge individual variation in response to the diets.

In evaluating the results of these trials, it is important to realize that the best of the studies were performed under controlled conditions using diets designed by trained dietitians. Thus, the high-carbohydrate diets generally incorporated a high percentage of low glycemic load carbohydrates and even the high-fat diets were low in saturated fats and cholesterol. This is important because the type of carbohydrates and fats in the diet appear to be just as important as their amounts. High-carbohydrate diets that are low in glycemic load appear to be just as effective in achieving weight loss and control of blood sugar as diets that are low in carbohydrate. Similarly, diets containing heart-healthy monounsaturated and ω-3 polyunsaturated fats are just as effective in achieving weight loss and lowering triglyceride levels and are better at lowering total and LDL cholesterol than diets containing saturated fats.

CLINICAL CORRELATION 27.7

Metabolic Adaptation: Relationship between Carbohydrate Intake and Serum Triacylglycerols

In evaluating the nutrition literature, it is important to be aware that most clinical trials are of rather short duration (2–6 weeks), whereas some metabolic adaptations may take considerably longer. Thus, even apparently well-designed clinical studies may lead to erroneous conclusions that will be repeated in the popular literature for years to come. For example, several studies conducted in the 1960s and 1970s tried to assess the effects of carbohydrate intake on serum triacylglycerol levels. Typically, young college-age males were given a diet in which up to 50% of their fat calories were replaced with sucrose or other simple sugars for a period of 2–3 weeks. In most cases serum triacylglycerol levels increased markedly (up to 50%). This led to the tentative conclusion that high intake of simple sugars, particularly sucrose, could increase the risk of heart disease, a notion that was popularized by nutritional best sellers such as *Sugar Blues* and *Sweet and Dangerous.* Unfortunately, while the original conclusions were promoted in the lay press, the experiments themselves were questioned. Subsequent studies showed that if these trials were continued for longer periods (3–6 months), the triacylglycerol levels usually normalized. The nature of this slow metabolic adaptation is unknown. It is also important to consider the type of carbohydrate in the diet. For many Americans, a high-carbohydrate diet means a diet that is high in simple sugars. Triacylglycerol levels in these individuals respond dramatically to diets that substitute foods containing either fat or complex carbohydrates and fiber for these foods contain simple sugars as a carbohydrate source.

Leahy, P., Croniger, C., and Hanson, R. W. Molecular and cellular adaptations to carbohydrate and fat intake. *Eur. J. Clin. Nutr.* 53, Suppl. 1:S6, 1999; and Parks, E. J., and Hellerstein, M. K. Carbohydrate-induced hypertriacylglycerolemia: Historical perspective and review of biological mechanisms. *Am. J. Clin. Nutr.* 71:412, 2000.

Finally, it is important to remember that dietary recommendations are for populations, not for individuals. The diet that works best for achieving weight control, blood sugar control, and healthy lipoprotein patterns is determined by an individual's genetic makeup (see Clin. Corr. 27.7, p. 1118, and Clin. Corr. 27.5, p. 1112).

27.10 • NUTRIGENETICS AND DIET COMPOSITION

In the past dietary recommendations have been made for the population as a whole without taking into account the influence of genetic background on whether or not the recommendations were advisable for any given individual. Moreover, it has often been difficult to make general dietary recommendations because of significant individual variation in the response to dietary interventions. However, as we learn more about the genetics underlying this individual variation, it may soon be possible to individualize dietary recommendations on the basis of the individual's genetic makeup. For example, it has been difficult in the past to draw firm conclusions about how the ratio of polyunsaturated fats and saturated fats (PUFA/SFA ratio) in the diet affects cardiovascular risk factors because of conflicting results from clinical trials conducted with different population groups. However, when differences in genetic background of these population groups are taken into account, the picture becomes much clearer.

For example, there is an A/G **SNP** (**single nucleotide polymorphism**) at the –75 position in the promoter region of the apoAI gene that affects the response of LDL-cholesterol to the relative amounts of polyunsaturated and saturated fats in the diet. An increase in the PUFA/SFA ratio results in a decrease in LDL-cholesterol in G/G homozygotes. However, the same increase in PUFA/SFA ratio results in an increase in LDL-cholesterol levels in G/A heterozygotes. Similarly, there is a SNP in the coding region of the PPARα gene that results in a L162V polymorphism that affects the response of serum triglycerides (triacylglycerol-rich VLDL) to the relative amounts of polyunsaturated and saturated fats in the diet. An increase in polyunsaturated fat intake leads to a decrease in triglyceride levels in V162 heterozygotes, but not in L162 homozygotes. Finally, there are two G/A SNPs in the promoter region of the TNFα gene that affect the response of HDL levels to the relative amounts of polyunsaturated and saturated fats in the diet. An increase in polyunsaturated fat intake results in increased HDL levels in –238 G/A heterozygotes and decreased HDL levels in –238 G/G monozygotes. In contrast, an increase in polyunsaturated fat intake results in decreased HDL levels in –308 G/A heterozygotes and does not affect HDL levels in –308 G/G homozygotes.

Nutrigenics holds the promise of allowing a truly individualized approach to making dietary recommendations to reduce disease risk in the future. However, because obesity and the diseases associated with obesity are polygenic diseases, this will not be a simple task. The examples given illustrate the difficulties that lie ahead. The relationships between fatty acid intake and LDL-cholesterol, triglycerides, and HDL are determined by at least four distinct SNPs that we know about, and possibly many more that we don't yet know about. It is easy to imagine that in a single individual, an increase in dietary polyunsaturated fats could lead to a decrease in LDL cholesterol, an increase in triglycerides and a decrease in HDL. What would the individualized recommendation regarding polyunsaturated fat be for that individual?

Bibliography

Protein Energy Malnutrition in Hospitalized Patients

Biffl, W. L., Moore, E. E., and Haenel, J. B. Nutrition support of the trauma patient. *Nutrition* 18:960, 2002.

Wilmore, D. W. Metabolic response to severe surgical illness: An overview. *World J. Surg.* 24:705 2000.

Wolfe, R. R., and Martini, W. Z. Changes in intermediary metabolism in severe surgical illness. *World J. Surg.* 24:639, 2000.

Genetic Predisposition to Obesity

Evans, R. M., Barish, G. D., and Wang, Y. X. PPARs and the complex journey to obesity. *Nat. Med.* 10:355, 2004.

Hainer, V., Zamrazilova, H., Spalova, J., Hainerova, I., et al. Role of hereditary factors in weight loss and its maintenance. *Physiol. Res.* 57:S1–S15, 2008.

Ochoa, M. C., Marti, A., Azcona, C., Cheuca, M., et al. Gene-gene interactions between PPAR gamma 2 and ADR beta 3 increases obesity risk in children and adolescents. *Int. J. Obes. Relat. Metab. Disord.* 28, Suppl. 3:S37, 2004.

Pihlajamaki, J., Vanhala, M., Vanhala, P., and Laakso, M. The Pro12Ala polymorphism of the PPAR gamma 2 gene regulates weight from birth to adulthood. *Obes. Res.* 12:187, 2004.

Obesity, Insulin Resistance, Metabolic Syndrome, and Type 2 Diabetes

American Diabetes Association: Diagnosis and Classification of Diabetes Mellitus. *Diabetes Care* 28, Suppl. 1:S37–S42, 2005.

Boden, G. Fatty acid: Induced inflammation and insulin resistance in skeletal muscle and liver. *Cur. Diabetes Rep.* 6:177–181, 2006.

Hotamisligii, G. S. The Role of TNFα and TNF receptors in obesity and insulin resistance. *J. Intern. Med.* 245:621, 1999.

Kahn, R., Buse, J., Ferrannini, E., and Stern, M. The metabolic syndrome: Time for a critical appraisal. *Diabetes Care* 28:2289, 2005.

Le Roith, R., and Zick, Y. Recent advances in our understanding of insulin action and insulin resistance. *Diabetes Care* 24:588, 2001.

Oda, E. The metabolic syndrome as a concept of adipose tissue disease. *Hypertens. Res.* 31:1283–1291, 2008.

Schinner, S., Scherbaum, W. A., Bornstein, S. R., and Barthel, A. Molecular mechanisms of insulin resistance. *Diabetic Med.* 22:674, 2005.

Diet and Cardiovascular Disease

Holub, D. J., and Holub, B. J. Omega-3 fatty acids from fish oils and cardiovascular disease. *Mol. Cell Biochem.* 263:217, 2004.

Krauss, R. M., Eckel, R. H., Howard, B., Appel, L. J., et al. AHA Dietary Guidelines Revision 2000: A statement for healthcare professionals from the nutrition committee of the American Heart Association. *Circulation* 102:2284, 2000.

Kris-Etherton, P., M., Harris, W. S., and Appel, L. J. Omega-3 fatty acids and cardiovascular disease. New recommendations from the American Heart Association. *Arterioscler. Thromb. Vasc. Biol.* 23:151, 2003.

Mozaffarian, D., Rimm, E. B., and Herrington, D. M. Dietary fats, carbohydrate, and the progression of atherosclerosis in postmenopausal women. *Am. J. Clin. Nutr.* 80:1175, 2004.

Swain, J., F., McCarron, P., B., Hamilton, E. F., Sacks, F. M., and Appel, L. J. Characteristics of the diet patterns tested in the optimal macronutrient intake trial to prevent heart disease (OmniHeart):Options for a heart-healthy diet. J. *Am. Dietetic Assoc.* 108:257, 2008.

Tanasecu, M., Cho, E., Manson, J. E., and Hu, F. B. Dietary fat and cholesterol and the risk of cardiovascular disease among women with type 2 diabetes. *Am. J. Clin. Nutr.* 79:999, 2004.

Third Report of the Expert Panel on Detection, Evaluation and Treatment of High Blood Cholesterol in Adults (Adult Treatment Panel III), www.nhlbi.nih.gov/guidelines/cholesterol.

Dietary Fiber and Health

Higgins, J. A., Resistant starch: Metabolic effects and potential health benefits. *J. AOAC Int.* 87:761, 2004.

Kendall, C. W., Emam, A., Augustin, L. S., and Jenkins, D. J. Resistant starches and health. *J. AOAC Int.* 87:769, 2004.

Shankar, S., and Lanza, E. Dietary fiber and cancer prevention. *Hematol. Oncol. Clinics of North Amer.* 5:25, 1991.

Glycemic Index and Glycemic Load

Brand-Miller, J. C. Glycemic load and chronic disease. *Nutr. Rev.* 61:S49, 2003.

Brand-Miller, J. C., Thomas, M., Swan, V., Ahmad, Z. I., et al. Physiological validation of the concept of glycemic load in lean young adults. *J. Nutr.* 133:2728, 2003.

Schulze, M. B. Glycemic index, glycemic load, and dietary fiber intake and the incidence of type 2 diabetes in younger and middle-aged women. *Am. J. Clin. Nutr.* 80:348, 2004.

Dietary Recommendations

Anderson, J. W., Randles, K. M., Kendall, C. W., and Jenkins, D. J. Carbohydrate and fiber recommendations for individuals with diabetes: A quantitative assessment and meta-analysis of the evidence. *J. Am. Coll. Nutr.* 23:5, 2004.

Assman, G., Sacks, F., Awad, A., Ascherio, A., et al. 2000 consensus statement on dietary fat, the Mediterranean diet, and lifelong good health. *Am. J. Med.* 113 (9B):5S, 2002.

Dansinger, M. L., Gleason, J. A., Griffith, J. L., and Schaefer, E. J. Comparison of the Atkins, Ornish, Weight Watchers, and Zone diets for weight loss and heart disease reduction. *J. Am. Med. Assoc.* 293:43, 2005.

Dubnov-Raz, G., and Berry, E. M. The dietary treatment of obesity. *Endocrinol. Metab. Clin. N. Am.* 37:873, 2008.

Food and Nutrition Board of the National Academy of Sciences. *Effects of Diet Composition on Weight Control, Blood Glucose Levels, and Blood Lipid Levels: Towards Healthful Diets.* Washington, DC: U.S. Government Printing Office, 1980.

Gardner, C. D., Kiazand, A., Alhassan, S., Kim, S., et al. Comparison of the Atkins, Zone, Ornish, and LEARN diets for change in weight and related risk factors among overweight premenopausal women. *J. Am. Med. Assoc.* 297:969, 2007.

Gerhard, G. T., Ahmann, A., Meeuws, K., McMurray, M. P., et al. Effects of a low-fat diet compared with those of high-monounsaturated fat diet on body weight, plasma lipids and lipoproteins, and glycemic control in type 2 diabetes. *Am. J. Clin. Nutr.* 80:668, 2004.

Hession, M., Rolland, C., Kulkarni, U., Wise, A., and Broom, J. Systematic review of randomized controlled trials of low-carbohydrate vs. low-fat/low-calorie diets in the management of obesity and its comorbidities. *Obes. Rev.* [epub ahead of print, PMID: 18700873], 2008.

Meckling, K. A., O'Sullivan, C, and Saari, D. Comparison of a low-fat diet to a low-carbohydrate diet on weight loss, body composition, and risk factors for diabetes and cardiovascular disease in free-living, overweight men and women. *J. Clin. Endocrin. Metab.* 89:2717, 2004.

National Research Council. *Diet, Nutrition and Cancer.* Washington, DC: National Academy Press, 1982.

Nordmann, A. J., Nordmann, A., Briel, M., Keller, U., et al. Effects of low-carbohydrate vs low-fat diets on weight loss and cardiovascular risk factors: A meta-analysis of randomized controlled trials. *Arch. Int. Med.* 166:285, 2006.

Opperman, A. M., Venter, C. S., Oosthuizen, W., Thompson, R. L., and Vorster, H. H. Meta-analysis of the health effects of using the glycemic index in meal-planning. *Brit. J. Nutr.* 92:367, 2004.

Pelkman, C. L., Fishell, V. K., Maddox, D. H., Pearson, T. A., et al. Effects of moderate fat (from monounsaturated fat) and low-fat weight-loss diets on the serum lipid profile in overweight and obese men and women. *Am. J. Clin. Nutr.* 79:204, 2004.

Senate Select Committee on Human Nutrition. *Dietary goals for the United States,* 2d ed., Stock No. 052-070-04376-8. Washington, DC: U.S. Government Printing Office. 1977.

Shai, I., Schwarzfuchs, D., Henkin, Y., Shahar, D. R., et al. Weight loss with a low-carbohydrate, Mediterranean, or low-fat diet. *N. Engl. J. Med.* 359:229, 2008.

Stern, L., Iqbal, N., Seshadri, P., Chicano, K. L., et al. The effects of low-carbohydrate versus conventional weight loss diets in severely obese adults: One year follow-up of a randomized trial. *Ann. Intern. Med.* 18:140:778, 2004.

Truby, H., Baic, S., deLooy, A., Fox, K.R., et al. Randomized controlled trial of four commercial weight loss programs in the UK: Initial findings from the BBC "diet trials". *Brit. Med. J.* 332:1309, 2006.

Truswell, A. S. Evolution of dietary recommendations, goals, and guidelines. *Am. J. Clin. Nutr.* 45:1060, 1987.

U.S. Department of Agriculture, The Food Guide Pyramid, www.mypyramid.gov.

U.S. Department of Agriculture. *Nutrition and Your Health, Dietary Guidelines for Americans,* Stock No. 017-001-00416-2. Washington, DC: U.S. Government Printing Office, 1980.

Nutrigenetics and Diet Composition

Cecil, J. E., Watt, P., Palmer, C. N., and Hetherington, M. Energy balance and food intake: The role of PPARα gene polymorphisms. *Physiol. Behav.* 88:227, 2006.

Dahlman, I., and Arner, P. Obesity and polymorphisms in genes regulating human adipose tissue. *Int. J. Obesity* 31:1629, 2007.

Duval, C., Muller, M., and Kersten, S. PPARα and dyslipidemia. *Biochem. Biophys. Acta* 1771:961, 2007.

Fontaine-Bisson, B., Wolever, T. M. S., Chaisson, J. L., Robasa-Lhoret, R., et al. Genetic polymorphisms of tumor necrosis factor-α modify the association between dietary polyunsaturated fatty acids and fasting HDL-cholesterol and apo A-I concentrations. *Am. J. Clin. Nutr.* 86:768, 2007.

Hainer, V., Zamrazilova, H., Spalova, J., Hainerova, I., et al. Role of hereditary factors in weight loss and its maintenance. *Physiol. Rev.* 57. Suppl. 1:S1–S15, 2008.

Lovegrove, J. A., and Gitau, R. Nutrigenetics and CVD: what does the future hold? *Proc. Nutr. Soc.* 67:206, 2008.

Low, Y. L., and Tai, E. S. Understanding diet-gene interactions: Lessons from studying nutrigenomics and cardiovascular disease. *Mutat. Res.* 622:7, 2007.

Masson, L. F., McNeill, G., and Avenell, A. Genetic variation and the lipid response to dietary intervention: a systematic review. *Am. J. Clin. Nutr.* 77:1098, 2003

Mutch, D. M., Wahli, W., and Williamson, G. Nutrigenomics and nutrigenetics: The emerging faces of nutrition. *FASEB J.* 19:1602, 2005.

Key Terms

basal metabolic rate
nitrogen balance
essential amino acids
protein sparing
protein energy malnutrition
marasmus
kwashiorkor
obesity
insulin resistance
metabolic syndrome
type 2 diabetes
adipokines
lactase insufficiency
essential fatty acids
dietary fiber
glycemic index
glycemic load
monounsaturated fats
ω-6 polyunsaturated fats
ω-3 polyunsaturated fats
nutrigenetics

Questions • CAROL N. ANGSTADT

Multiple-Choice Questions

1. Protein energy malnutrition (PEM)
A. is always manifested by a thin, wasted appearance.
B. causes problems only when there is an inadequate consumption of calories.
C. causes problems only when there is an inadequate consumption of protein.
D. leads to a reduced ability to fight infection.
E. occurs only in infants.

2. Basal metabolic rate
A. is not influenced by energy intake.
B. increases in response to starvation.
C. may decrease up to 50 % during periods of starvation.
D. increases in direct proportion to daily energy expenditure.
E. is not responsive to changes in hormone levels.

3. Obesity
A. usually does not have adverse metabolic consequences.
B. causes metabolic changes that are usually irreversible.
C. frequently leads to elevated serum levels of fatty acids, cholesterol, and triglycerides.
D. is frequently associated with an increased sensitivity of insulin receptors.
E. is caused solely by high caloric consumption.

4. Dietary fat
A. is usually present, although there is no specific need for it.
B. if present in excess, can be stored as either glycogen or adipose tissue triacylglycerol.
C. should include linoleic and linolenic acids.
D. should increase on an endurance training program to increase the body's energy stores.
E. if present in excess, does not usually lead to health problems.

5. Which of the following statements about dietary fiber is/are correct?
A. Water-soluble fiber helps to lower serum cholesterol in most people.
B. Mucilaginous fiber slows the rate of digestion and absorption of carbohydrates.
C. Insoluble fiber increases stool bulk and decreases transit time.
D. All of the above.
E. None of the above.

6. Nutrigenetics indicates that
A. dietary recommendations for the population as a whole may not be advisable for a given individual.
B. a single nucleotide polymorphism (SNP) in a given gene may alter the response to a dietary change.
C. homozygotes for an SNP may have a different response to a dietary change than heterozygotes for that SNP.
D. SNPs may occur in the promoter region or in the coding region of a gene.
E. all of the above.

Questions 7 and 8: A young man suffered third-degree burns over much of his body and is hospitalized in a severe catabolic state. An individual in this state requires about 40 kcal kg per day and 2 g protein kg per day to be in positive caloric and nitrogen balance. This young man weighs 140 lb (64 kg). Total Parenteral Nutrition (TPN) is started with a solution containing 20 % glucose and 4.25 % amino acids (the form in which protein is supplied).

7. If 3000 g of solution is infused per day,
A. the patient would not be getting sufficient protein.
B. the calories supplied would be inadequate.
C. both protein and calories would be adequate to meet requirements.
D. this is too much protein being infused.

8. Sometimes a lipid solution is also infused in a patient on TPN. In the case of this young man, the purpose of the lipid solution would be to
A. supply additional calories to meet caloric needs.
B. supply essential fatty acids.
C. improve the palatability of the mixture.
D. provide fiber.
E. assure an adequate supply of cholesterol for membrane building.

Questions 9 and 10: For many years, the American Diabetic Association recommended a diet high in complex carbohydrates and fiber and low in fat for diabetics. It was later found that some individuals did not do as well on such a diet as on one high in monounsaturated fatty acids. Since 1994, the ADA has abandoned the concept of a single diabetic diet and now recommends a focus on achieving glucose, lipid, and blood pressure goals with weight reduction if necessary.

9. Which of the following statements is/are correct?
A. A high-carbohydrate–high-fiber diet often results in significant weight reduction because it has a lower caloric density than a diet high in fat.
B. A diet high in monounsaturated fatty acids would be most appropriate for an overweight diabetic.
C. The goal for lipids is to reduce all lipoprotein levels in the blood.
D. Obesity aggravates diabetes because it inhibits the production of insulin by the pancreas.
E. All of the above.

10. For diabetics,
A. the only carbohydrate that must be eliminated in the diet is sucrose.
B. fiber increases the rate at which carbohydrate is digested and absorbed.
C. not all carbohydrate foods raise blood glucose levels at the same rate because the glycemic index of all foods is not the same.
D. who are normally in good control, stress will have no effect on their blood sugar levels.
E. a vegetarian diet is the only appropriate choice.

Questions 11 and 12: Recent studies have confirmed that reducing elevated serum cholesterol levels can reduce the risk of heart disease. Obesity is another risk factor. Epidemiological studies suggest that consumption of animal protein is associated with an increased incidence of heart disease, although it is probably the fat and cholesterol in animal foods rather than the animal protein itself. Most Americans would probably not accept a strictly vegetarian diet.

11. All of the following have been shown to have some effect, although it may be small, on lowering serum cholesterol except
    A. restricting dietary cholesterol.
    B. increasing the ratio of polyunsaturated to saturated fatty acids.
    C. replacing oleic acid with palmitic acid.
    D. increasing the amount of soluble fiber.
    E. changing to a vegetarian diet.

12. A complete replacement of animal protein in the diet by vegetable protein
    A. would be expected to have no effect at all on the overall diet.
    B. would reduce the total amount of food consumed for the same number of calories.
    C. might reduce the total amount of iron and vitamin B_{12} available.
    D. would be satisfactory regardless of the nature of the vegetable protein used.
    E. could not satisfy protein requirements.

Problems

13. Calculate the number of grams each of carbohydrate, lipid, and protein a person on a 2300-kcal diet should consume to meet the guidelines established by the Senate Select Committee on Human Nutrition. Assuming the individual weighs 180 lb and the protein is from mixed animal/vegetable sources with a 75% efficiency of utilization, does the amount of protein you calculated meet the recommended amount of protein?

14. A 120-lb woman is consuming a diet with adequate total calories and 44 g of protein per day. The protein is exclusively from vegetable sources, primarily corn based. What would be her state of nitrogen balance?

Answers

1. **D** They have reduced numbers of T lymphocytes and reduced production of immunoglobulins and other components of the immune system. A: This is true of marasmus but not kwashiorkor. B and C: Marasmus is a deficiency of both protein and calories whereas kwashiorkor is a deficiency of protein with adequate calories. E: Hospitalized adults can also suffer from PEM.
2. **C** This is part of the survival mechanism in starvation. A and B: BMR decreases when energy intake decreases. D: BMR as defined is independent of energy expenditure. Only when the exercise is repeated on a daily basis so that lean muscle mass is increased does BMR also increase. E: Many hormones increase BMR.
3. **C** Probably because of an increased number and/or size of adipose cells, which are less sensitive to insulin. A: Obesity has multiple effects. B: Most changes can be reversed if weight is lost. D: Sensitivity is decreased perhaps by interfering with autophosphorylation of the receptor. E: Inadequate exercise and a genetic component also play a role.
4. **C** A and C: Linoleic and linolenic acids are essential fatty acids and so must be present in the diet. B and D: Excess carbohydrate can be stored as fat but the reverse is not true. D: Carbohydrate loading has been shown to increase endurance. E: High-fat diets are associated with many health risks.
5. **D** These each illustrate the different properties and roles of the common kinds of fiber.
6. **E** They are all correct. A: The genetic makeup of the individual matters. B and C: For example, A/G SNP in the promoter of the apoA1 gene results in a decrease in LDL-cholesterol in response to increased PUFA/SFA in G/G homozygotes but an increase in G/A heterozygotes. D: The SNP is in the promoter of the apoA1 gene but in the coding region of the PPARα gene.
7. **C** This amount of solution would supply 128 g of protein and 2912 kcal (2400 from glucose and 512 from amino acids), both enough to meet the stated requirements.
8. **B** Patients on TPN need to have essential fatty acids supplied. A: The original solution supplies adequate calories, although additional calories should not hurt this young man. C: The patient is not tasting this mixture, so palatability is irrelevant. D: Fiber is supplied by complex carbohydrate sources. E: The body can make its own cholesterol.
9. **A** Since carbohydrate has less than half the caloric density of fat and fiber provides no calories, such diets have low caloric density. B: High fat would tend to increase weight. C: The goal is to reduce the LDL and VLDL but not HDL. D: The problem is that factors released by adipose cells invoke insulin resistance.
10. **C** Foods vary widely in their glycemic index. Bread and rice raise blood glucose levels more rapidly than does sucrose. A: See C. B: One of the benefits of fiber is that it decreases the rate of carbohydrate absorption. D: Stress raises blood glucose for everyone because of release of epinephrine and glucocorticoids. E: An appropriately designed vegetarian diet is perfectly acceptable but is certainly not the only choice.
11. **C** Oleic acid lowers cholesterol, but palmitic acid increases it. A, B, and D: These all have positive effects in lowering cholesterol. E: Vegetarians tend to have lower cholesterol (the diets actually incorporate A, B, and D).
12. **C** A and C: This would reduce the amount of fat, especially saturated fat, but could also reduce the amount of necessary nutrients that come primarily from animal sources. B: The protein content of vegetables is quite low, so much larger amounts of vegetables would have to be consumed. D and E: It is possible to satisfy requirements for all of the essential amino acids completely if vegetables with complementary amino acid patterns, in proper amounts, are consumed.
13. According to the SSCHN guidelines a 2300-kcal diet should consist of 333.5 g of carbohydrate (no more than 57.5 g as simple sugar); 69 g of protein and 76.7 g of fat (no more than 25.6 g of saturated fat). A 180-lb man weighs 81.8 kg times 0.8 g /kg = 65.5 g. of protein per day to meet requirements; 12 % of 2300 kcal supplies sufficient protein.
14. 44 g protein/120 lb is 0.8 g protein/kg/day. However, this is inadequate because pure vegetable protein is less than 75 % efficient in utilization. Also, heavy reliance on one protein (corn is deficient in lysine) would likely lead to a deficiency of one or more essential amino acids. Therefore, the woman in this problem is in negative nitrogen balance.

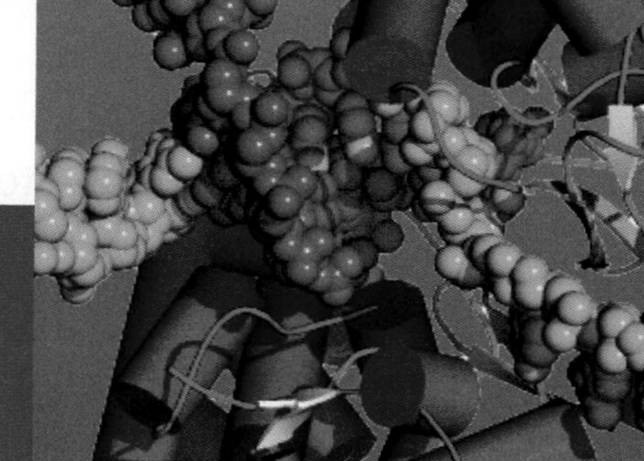

Appendix

Review of Organic Chemistry

Carol N. Angstadt

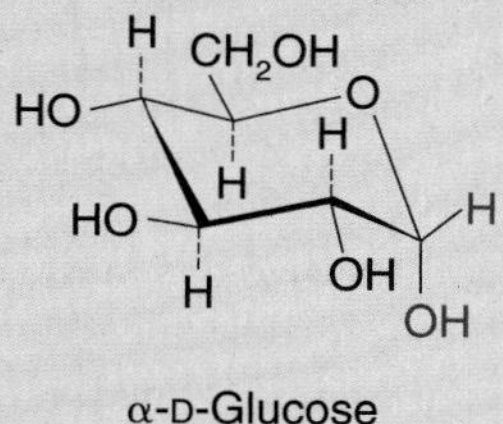

α-D-Glucose

FUNCTIONAL GROUPS

Alcohols

The general formula of **alcohols** is R—OH, where R equals an alkyl or aryl group. They are classified as *primary, secondary,* or *tertiary,* according to whether the hydroxyl (OH)-bearing carbon is bonded to no carbon or one, two, or three other carbon atoms:

$$-\underset{|}{\overset{|}{C}}-\underset{H}{\overset{H}{C}}-OH \qquad -\underset{|}{\overset{|}{C}}-\underset{H}{\overset{-C-}{C}}-OH \qquad -\underset{|}{\overset{|}{C}}-\underset{-C-}{\overset{-C-}{C}}-OH$$

Primary Secondary Tertiary

Aldehydes and Ketones

Aldehydes and **ketones** contain a carbonyl group:

$$-\overset{O}{\overset{\|}{C}}-$$

Aldehydes are

$$R-\overset{O}{\overset{\|}{C}}-H$$

and a *ketone* has two groups (alkyl and/or aryl) at the carbonyl group

$$R-\overset{O}{\overset{\|}{C}}-R'$$

Acids and Acid Anhydrides

Carboxylic acids contain the functional group

$$-\overset{O}{\overset{\|}{C}}-OH$$

(—COOH). Dicarboxylic and tricarboxylic acids contain two or three carboxyl groups. A carboxylic acid dissociates in water to a negatively charged carboxylate ion:

$$R-\overset{O}{\overset{\|}{C}}-OH \longrightarrow R-\overset{O}{\overset{\|}{C}}-O^- + H^+$$

Carboxylic acid Carboxylate ion

Names of carboxylic acids usually end in -ic and the carboxylate ion in -ate. **Acid anhydrides** are formed when two molecules of acid react with loss of a molecule of water. An acid anhydride may form between two organic acids, two inorganic acids, or an organic and an inorganic acid:

$$R-\overset{O}{\overset{\|}{C}}-O-\overset{O}{\overset{\|}{C}}-R \qquad HO-\underset{OH}{\overset{O}{\overset{\|}{P}}}-O-\underset{OH}{\overset{O}{\overset{\|}{P}}}-OH \qquad R-\overset{O}{\overset{\|}{C}}-O-\underset{OH}{\overset{O}{\overset{\|}{P}}}-OH$$

Organic anhydride Inorganic anhydride Organic–inorganic anhydride

Esters

Esters form in the reaction between a carboxylic acid and an alcohol:

$$R-COOH + R'-OH \longrightarrow R-\overset{O}{\overset{\|}{C}}-OR'$$

Esters may form between an inorganic acid and an organic alcohol, for example, glucose 6-phosphate.

Hemiacetals, Acetals, and Lactones

A reaction between an aldehyde and an alcohol gives a **hemiacetal**, which may react with another molecule of alcohol to form an **acetal**:

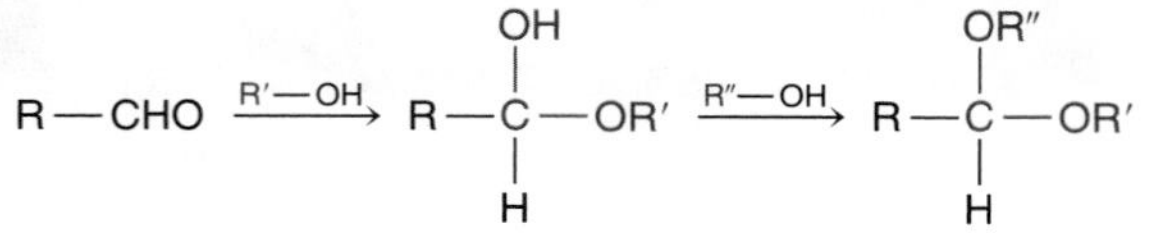

$$\text{R—CHO} \xrightarrow{\text{R'—OH}} \text{R—C(OH)(H)—OR'} \xrightarrow{\text{R''—OH}} \text{R—C(OR'')(H)—OR'}$$

Hemiacetal Acetal

Lactones are cyclic esters formed when an acid and an alcohol group on the same molecule react and usually require that a five- or six-membered ring be formed.

Unsaturated Compounds

Unsaturated compounds are those containing one or more carbon—carbon multiple bonds, for example, a double bond: —C=C—.

Amines and Amides

Amines, R—NH_2, are organic derivatives of NH_3 and are classified as *primary, secondary,* or *tertiary,* depending on the number of alkyl groups (R) bonded to the nitrogen. When a fourth substituent is bonded to the nitrogen, the species is positively charged and called a *quaternary ammonium ion*:

R—NH_2	R—N(H)—R′	R—N(R″)—R′	R—N^+(H(or R‴))(R″)—R′
Primary amine	Secondary amine	Tertiary amine	Quaternary ammonium ion

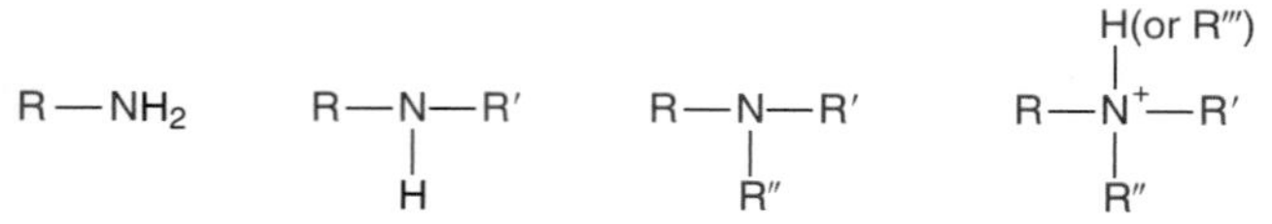

Amides contain the functional group

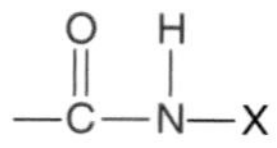

$$\text{—C(=O)—N(H)—X}$$

where X can be H (simple) or R (*N* substituted). The carbonyl group is from an acid, and the *N* is from an amine. If both functional groups are from amino acids, the amide bond is referred to as a **peptide bond**.

TYPES OF REACTIONS

Nucleophilic Substitutions at an Acyl Carbon

If the acyl carbon is on a carboxylic group, the leaving group is water. Nucleophilic substitution on carboxylic acids usually requires a catalyst or conversion to a more reactive intermediate; biologically this occurs via enzyme catalysis. X—H may be an alcohol (R—OH), ammonia, amine (R—NH_2), or another acyl compound. Types of nucleophilic substitutions include *esterification, peptide bond* formation, and *acid anhydride* formation.

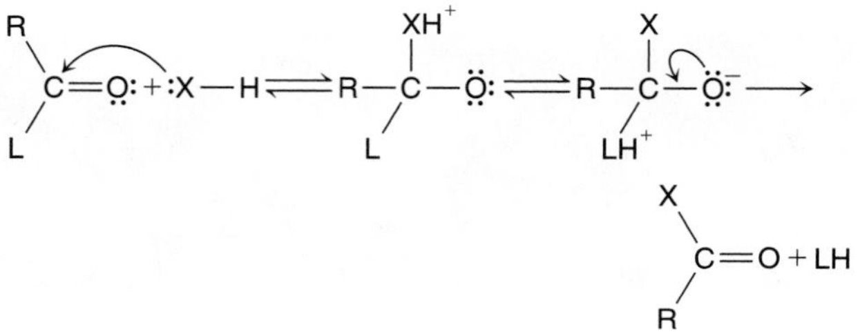

$$\text{R(L)C=O} + \text{:X—H} \rightleftharpoons \text{R—C(XH}^+\text{)(L)—O:}^- \rightleftharpoons \text{R—C(X)(LH}^+\text{)—O:}^- \longrightarrow \text{X(R)C=O} + \text{LH}$$

New compound Leaving group

Hydrolysis and Phosphorolysis Reactions

Hydrolysis is the cleavage of a bond by water:

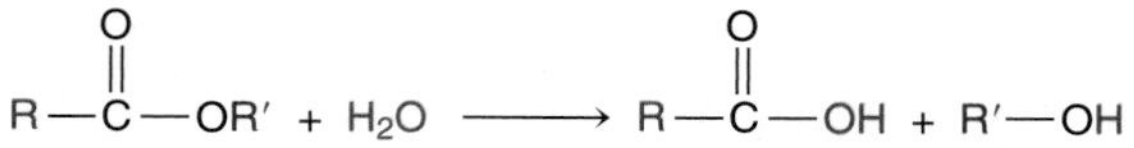

$$\text{R—C(=O)—OR'} + H_2O \longrightarrow \text{R—C(=O)—OH} + \text{R'—OH}$$

Hydrolysis is often catalyzed by either acid or base. *Phosphorolysis* is the cleavage of a bond by inorganic phosphate:

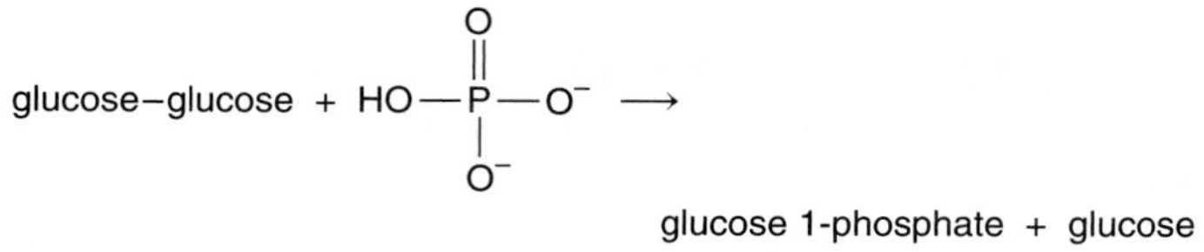

$$\text{glucose–glucose} + \text{HO—P(=O)(O}^-\text{)—O}^- \longrightarrow \text{glucose 1-phosphate} + \text{glucose}$$

Oxidation–Reduction Reactions

Oxidation is the loss of electrons; **reduction** is the gain of electrons. Examples of oxidation are as follows:

1. Fe^{2+} + acceptor → Fe^{3+} + acceptor · e^-
2. S(ubstrate) + O_2 + DH_2 → S—OH + H_2O + D
3. S—H_2 + acceptor → S + acceptor · H_2

Some of the group changes that occur on oxidation-reduction are:

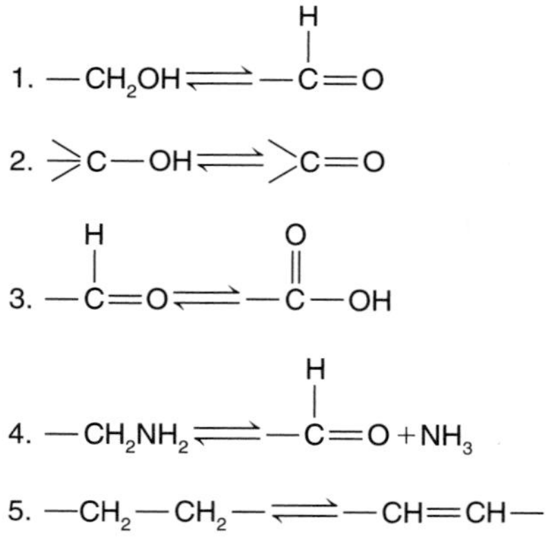

1. —CH_2OH ⇌ —C(H)=O
2. >C—OH ⇌ >C=O
3. —C(H)=O ⇌ —C(=O)—OH
4. —CH_2NH_2 ⇌ —C(H)=O + NH_3
5. —CH_2—CH_2— ⇌ —CH=CH—

STEREOCHEMISTRY

Stereoisomers are compounds with the same molecular formulas and order of attachment of constituent atoms but with different arrangements of these atoms in space.

Enantiomers are stereoisomers in which one isomer is the mirror image of the other and requires the presence of a chiral atom.

A chiral carbon (also called an asymmetric carbon) is one that is attached to four different groups:

```
    B          B
    |          |
 D—C—A      A—C—D
    :          :
    E          E
```

Enantiomers will be distinguished from each other by the designations *R* and *S* or D and L. The maximum number of stereoisomers possible is 2^n, where *n* is the number of chiral carbon atoms. A molecule with more than one chiral center will be an achiral molecule if it has a point or plane of symmetry.

Diastereomers are stereoisomers that are not mirror images of each other and need not contain chiral atoms. **Epimers** are diastereomers that contain more than one chiral carbon and differ in configuration about *only one* asymmetric carbon.

Anomers are a special form of carbohydrate epimers in which the difference is specifically about the anomeric carbon (see p. 1126. Diastereomers can also occur with molecules in which there is restricted rotation about carbon—carbon bonds. Double bonds exhibit ***cis–trans* isomerism**. The double bond is in the *cis* configuration if the two end groups of the longest contiguous chain containing the double bond are on the same side and is *trans* if the two ends of the longest chain are on opposite sides. Fused ring systems, such as those found in steroids (see p. 1131), also exhibit *cis–trans* isomerism.

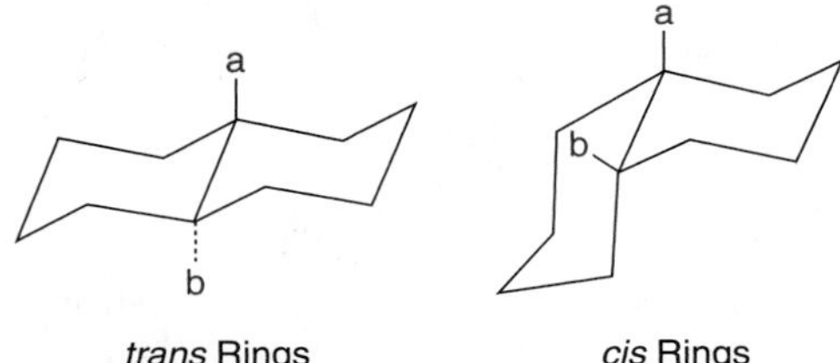

trans Rings   *cis* Rings

TYPES OF FORCES INVOLVED IN MACROMOLECULAR STRUCTURES

A **hydrogen** bond is a dipole–dipole attraction between a hydrogen atom attached to an electronegative atom and a non-bonding electron pair on another electronegative atom:

$$:\ddot{X}—H.......:\ddot{X}—H$$
$$\delta^-\ \delta^+ \qquad \delta^-\ \delta^+$$

Hydrogen bonds of importance in macromolecular structures occur between two nitrogen atoms, two oxygen atoms, or an oxygen and a nitrogen atom.

A **hydrophobic interaction** is the association of nonpolar groups in a polar medium. *Vander Waals* forces consist of dipole and induced dipole interactions between two nonpolar groups. A nonpolar residue dissolved in water induces a highly ordered, thermodynamically unfavorable, solvation shell. Interaction of nonpolar residues with each other, with the exclusion of water, increases the entropy of the system and is thermodynamically favorable.

Ionic (electrostatic) interactions between charged groups can be attractive if the charges are of opposite signs or repulsive if they are of the same sign. The strength of an electrostatic interaction in the interior of a protein molecule may be high. Most charged groups on the surface of a protein molecule interact with water rather than with each other. A **disulfide bond** (S—S) is a covalent bond formed by the oxidation of two sulfhydryl (SH) groups.

CARBOHYDRATES

Carbohydrates are polyhydroxy aldehydes or ketones or their derivatives. **Monosaccharides** (simple sugars) are those carbohydrates that cannot be hydrolyzed into simpler compounds. The generic name of a monosaccharide includes the type of function, a Greek prefix indicating the number of carbon atoms, and the ending -ose; for example, *aldohexose* is a six-carbon aldehyde and *ketopentose* a five-carbon ketone. Monosaccharides may react with each other to form larger molecules. With fewer than eight monosaccharides, either a Greek prefix indicating the number or the general term **oligosaccharide** may be used. **Polysaccharide** refers to a polymer with more than eight monosaccharides. Oligo- and polysaccharides may be either homologous or mixed.

Most *monosaccharides* are asymmetric, an important consideration since enzymes usually work on only one isomeric form. The simplest carbohydrates are glyceraldehyde and dihydroxyacetone whose structures, shown as Fischer projections, are as follows:

```
H—C=O            H—C=O           CH2OH
  |                |               |
H—C—OH          HO—C—H            C=O
  |                |               |
  CH2OH            CH2OH          CH2OH
```

D-Glyceraldehyde   L-Glyceraldehyde   Dihydroxyacetone

D-Glyceraldehyde may also be written as follows:

```
        H
       /
      C=O
      |
  H ◄—C—► OH
      |
     CH2OH
```

In the Cahn–Ingold–Prelog system, the designations are (*R*) (*rectus*; right) and (*S*) (*sinister*, left).

The configuration of monosaccharides is determined by the stereochemistry at the asymmetric carbon furthest from the carbonyl carbon (number 1 for an aldehyde; lowest possible number for a ketone). Based on the *position* of the OH on the highest number asymmetric carbon, a monosaccharide is D if the OH projects to the *right* and L if it projects to the *left*. The D and L monosaccharides with the same name are **enantiomers**, and the substituents on all asymmetric carbon atoms are reversed as in

```
 H—C=O           H—C=O
   |               |
 H—C—OH         HO—C—H
   |               |
HO—C—H           H—C—OH
   |               |
 H—C—OH         HO—C—H
   |               |
 H—C—OH         HO—C—H
   |               |
   CH2OH           CH2OH
```

D-Glucose   L-Glucose

Epimers (e.g., glucose and mannose) are stereoisomers that differ in the configuration about *only one* asymmetric carbon. The relationship of OH groups to *each other* determines the specific monosaccharide. Three aldohexoses and three pentoses of importance are

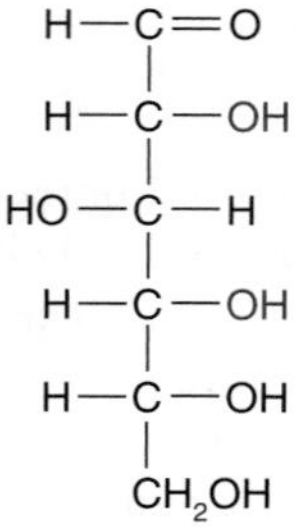

D-Glucose 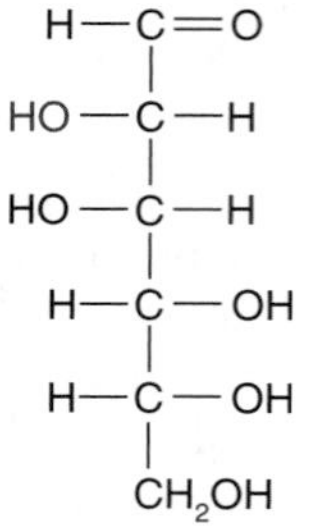 D-Mannose 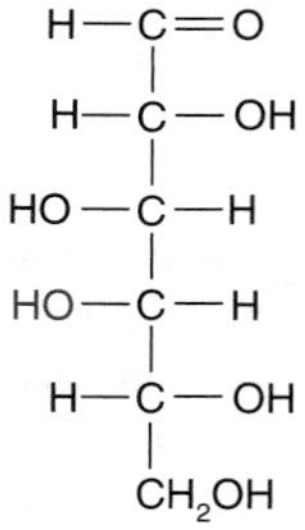 D-Galactose

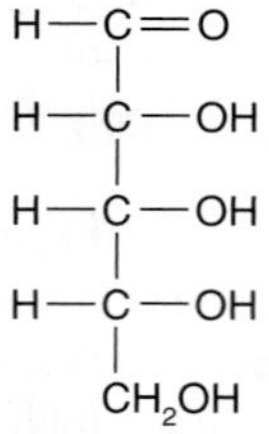

D-Ribose 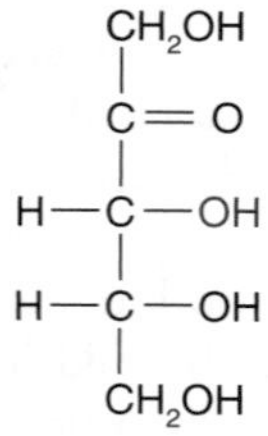 D-Ribulose 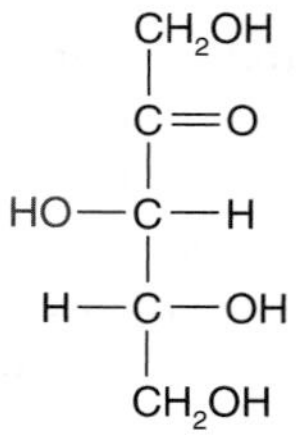 D-Xylulose

Fructose, a ketohexose, differs from glucose only on carbon atoms 1 and 2:

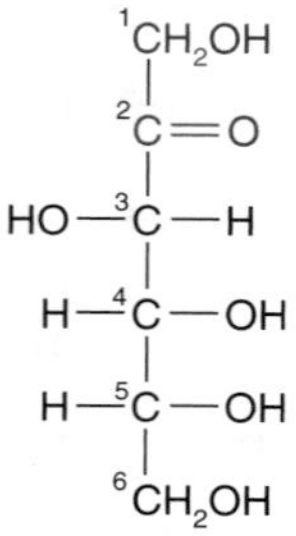

C_5 (pentose) and C_6 (hexose) monosaccharides form **cyclic hemiacetals** or **hemiketals** in solution. A new asymmetric carbon is generated so two isomeric forms are possible:

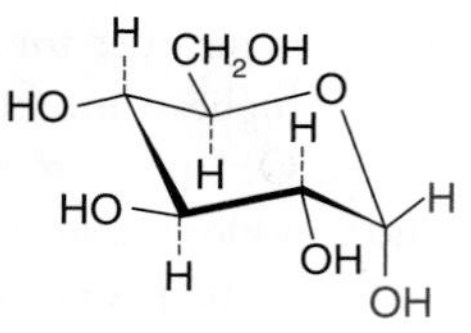

α-D-Glucose 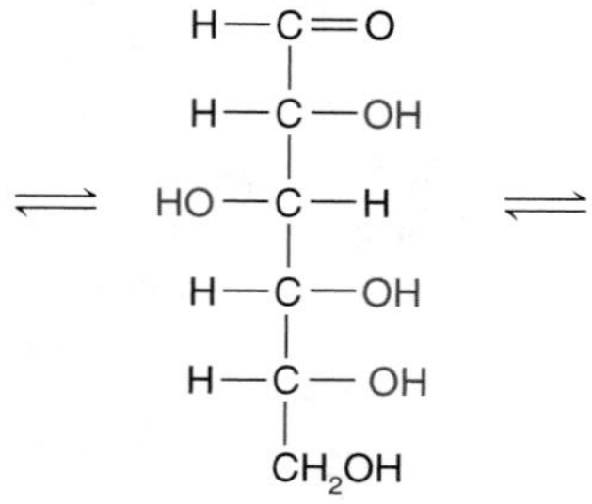 ⇌ D-Glucose 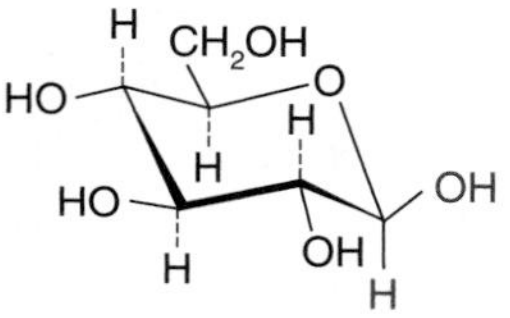 ⇌ β-D-Glucose

Both five-membered (furanose) and six-membered (pyranose) ring structures are possible, although pyranose rings are more common. A furanose ring is written as follows:

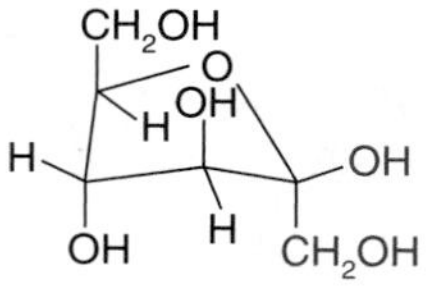

β-D-Fructose

The isomer is designated **α** if the OH group and the CH_2OH group on the two carbon atoms linked by the oxygen are *trans* to each other and **β** if they are *cis*. The hemiacetal or hemiketal forms may also be written as modified *Fischer projection formulas*: **α** if OH on the acetal or ketal carbon projects to the same side as the ring and **β** if on the opposite side:

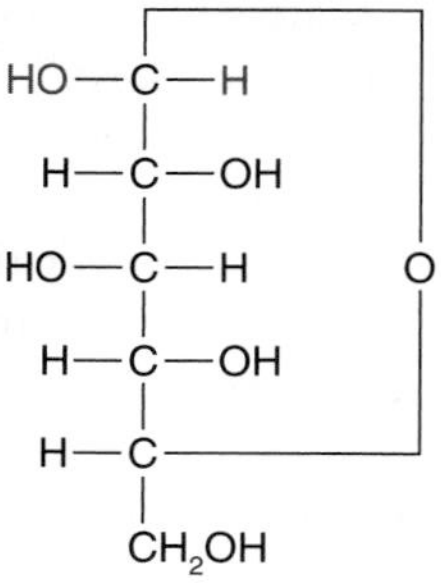

β-D-Glucose 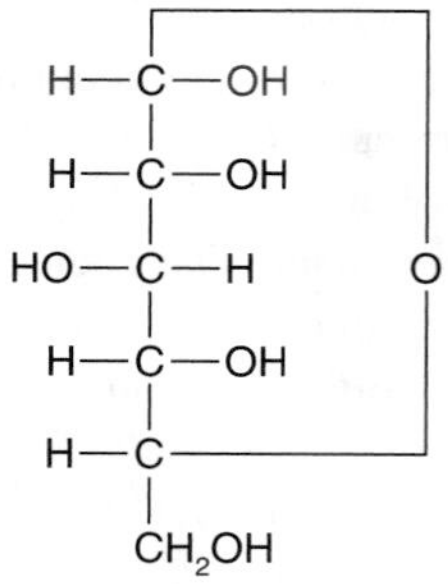 α-D-Glucose

Haworth formulas are used most commonly:

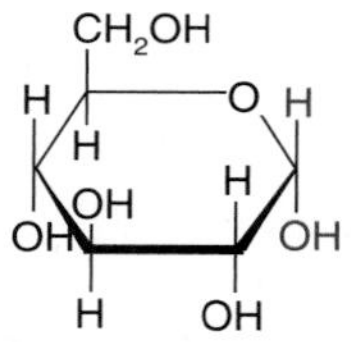

α-D-Glucose 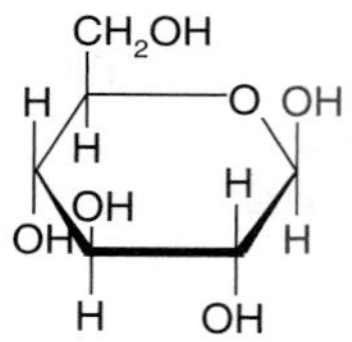 β-D-Glucose

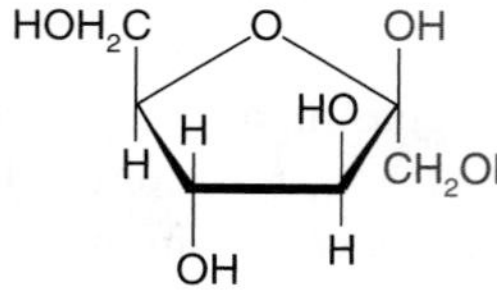

β-D-Fructose

The ring is perpendicular to the plane of the paper with the oxygen written to the back (upper) right, C1 to the right, and substituents above or below the plane of the ring. The OH at the acetal or ketal carbon is below in the **α** isomer and above in the **β**. Anything written to the right in the Fischer projection is written down in the Haworth formula.

The **α** and **β** forms of the same monosaccharide are special forms of epimers called *anomers*, differing only in the configuration about the anomeric (acetal or ketal) carbon. Monosaccharides exist in solution primarily as a mixture of the hemiacetals (or hemiketals) but react chemically as aldehydes or ketones. **Mutarotation** is the equilibration of **α** and **β** forms through the free aldehyde or ketone. Substitution of the H of the anomeric OH prevents mutarotation and fixes the configuration in either the **α** or **β** form.

Monosaccharide Derivatives

A **deoxymonosaccharide** is one in which an OH has been replaced by H. In biological systems, this occurs at C2 unless otherwise indicated. An **amino monosaccharide** is one in which an OH has been replaced by NH_2, again at C2 unless otherwise specified. The amino group of an amino sugar may be *acetylated*:

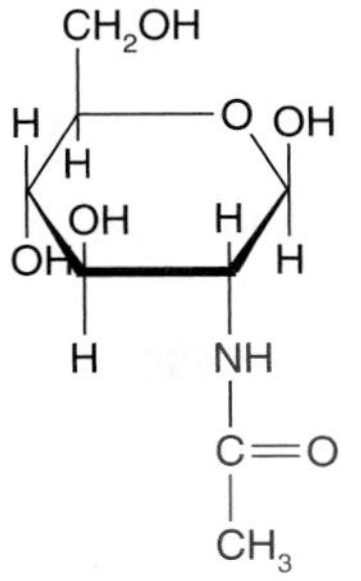

β-*N*-Acetylglucosamine

An aldehyde is reduced to a primary and a ketone to a secondary **monosaccharide alcohol (alditol)**. Alcohols are named with the base name of the sugar plus the ending *-itol* or with a trivial name (glucitol = sorbitol). Monosaccharides that differ around only two of the first three carbon atoms yield the same alditol. D-Glyceraldehyde and dihydroxyacetone give glycerol:

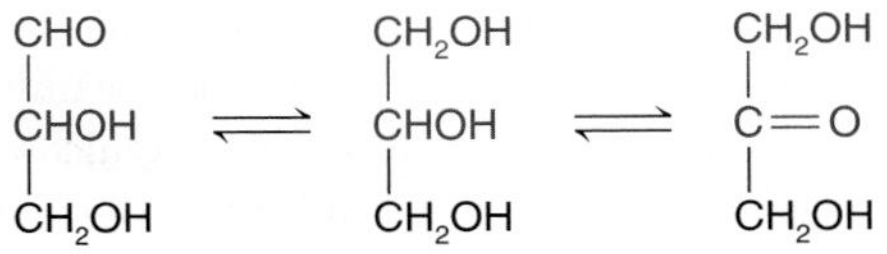

$$\begin{array}{c} CHO \\ | \\ CHOH \\ | \\ CH_2OH \end{array} \rightleftharpoons \begin{array}{c} CH_2OH \\ | \\ CHOH \\ | \\ CH_2OH \end{array} \rightleftharpoons \begin{array}{c} CH_2OH \\ | \\ C{=}O \\ | \\ CH_2OH \end{array}$$

D-Glucose and D-fructose give D-sorbitol; D-fructose and D-mannose give D-mannitol. Oxidation of the terminal CH_2OH, but not of the CHO, yields a **-uronic acid**, a *monosaccharide acid*:

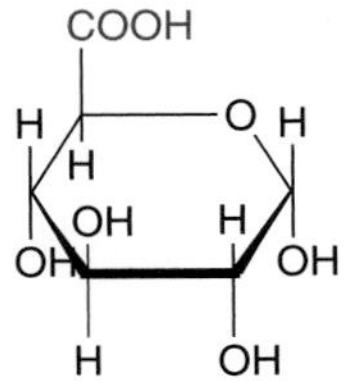

D-Glucuronic acid

Oxidation of the CHO, but not the CH_2OH, gives an **-onic acid**:

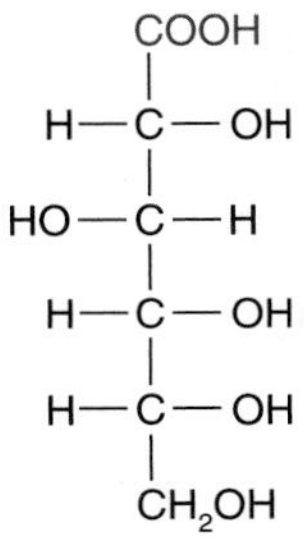

D-Gluconic acid

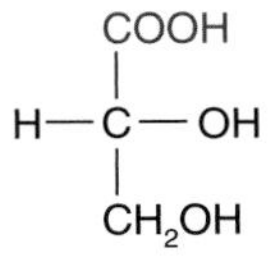

D-Glyceric acid

Oxidation of both the CHO and CH_2OH gives an **-aric acid**:

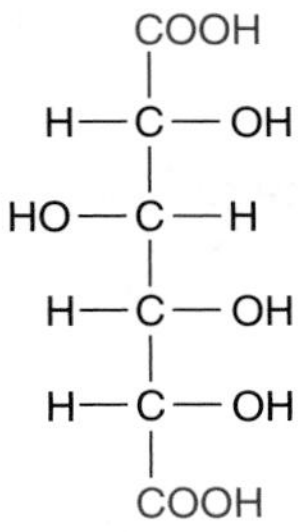

D-Glucaric acid

Ketones do not form acids. Both -onic and -uronic acids can react with an OH in the same molecule to form a **lactone** (see p. 1124):

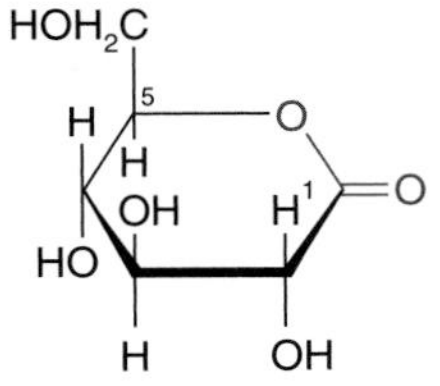

D-Glucono-5-lactone

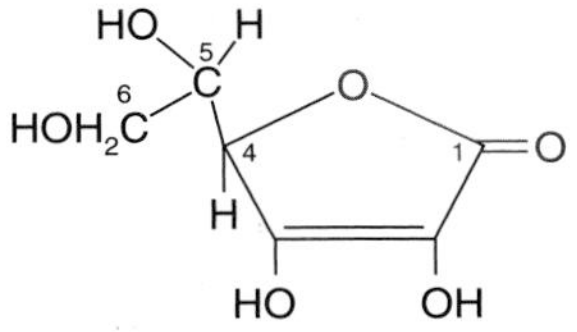

L-Ascorbic acid (derivative of L-gulose)

Reactions of Monosaccharides

The most common *esters* of monosaccharides are phosphate esters at carbon atoms 1 and/or 6:

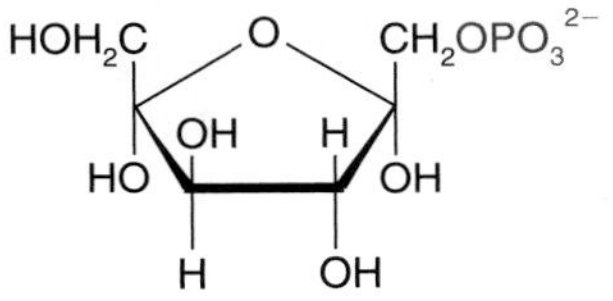

Fructose 1-Phosphate

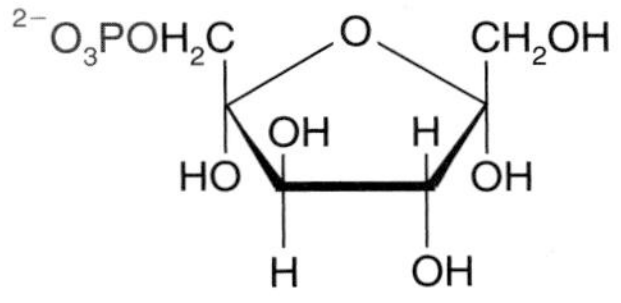

Fructose 6-Phosphate

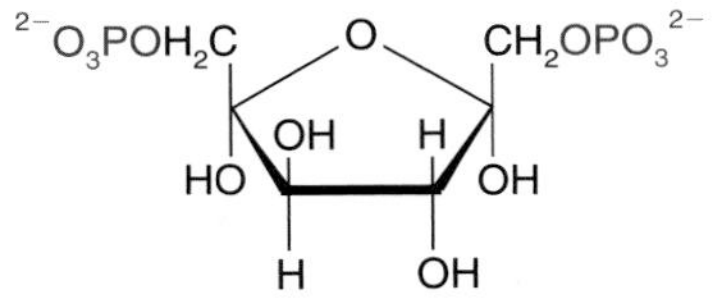

Fructose 1, 6 Bisphosphate

To be a **reducing sugar**, mutarotation must be possible. In alkali, enediols form that may migrate to 2,3 and 3,4 positions:

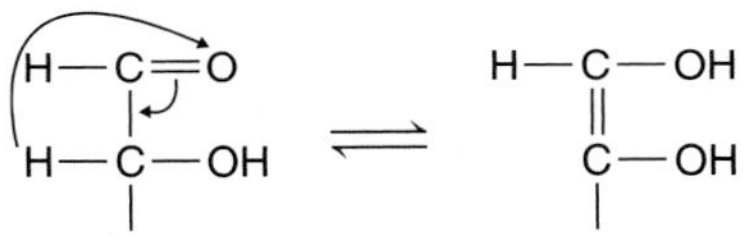

Enediols may be oxidized by O_2, Cu^{2+}, Ag^+, and Hg^{2+}. Reducing ability is more important in the laboratory than physiologically. A hemiacetal or hemiketal may react with the OH of another monosaccharide to form a disaccharide (*acetal: glycoside*) (see below):

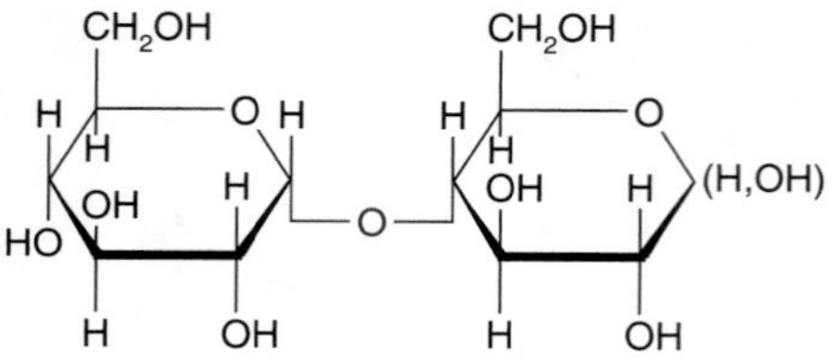

(May be either α or β)

α-1,4-Glycosidic linkage

One monosaccharide still has a free anomeric carbon and can react further. Reaction of the anomeric OH may be with any OH on the other monosaccharide, including the anomeric one. The anomeric OH that has reacted is fixed as either **α** or **β** and cannot mutarotate or reduce. If the glycosidic bond is not between two anomeric carbon atoms, one of the units will still be free to mutarotate and reduce.

Oligo- and Polysaccharides

Disaccharides have two monosaccharides, either the same or different, in glycosidic linkage. If the glycosidic linkage is between the two anomeric carbon atoms, the disaccharide is nonreducing:

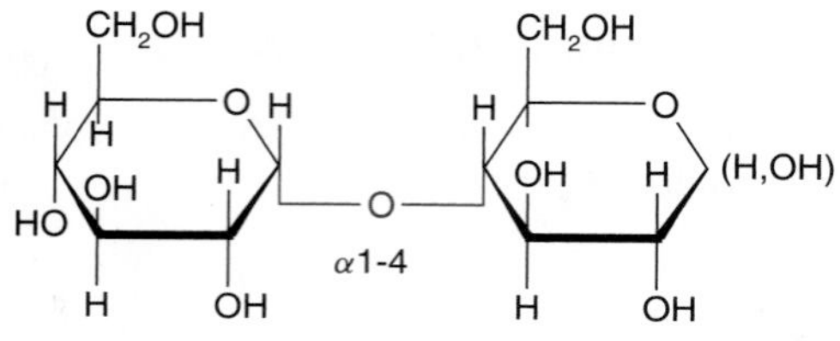

α1-4

Maltose

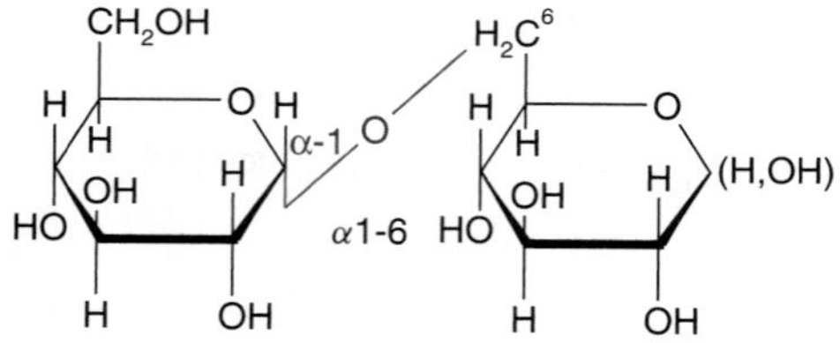

α-1, α1-6

Isomaltose

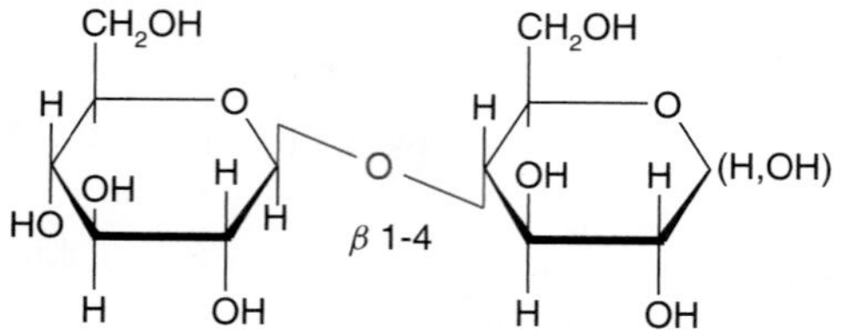

β 1-4

Cellobiose

β 1-4

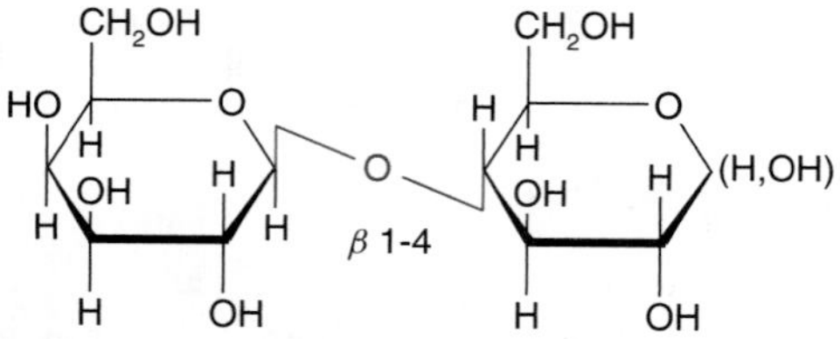

Lactose

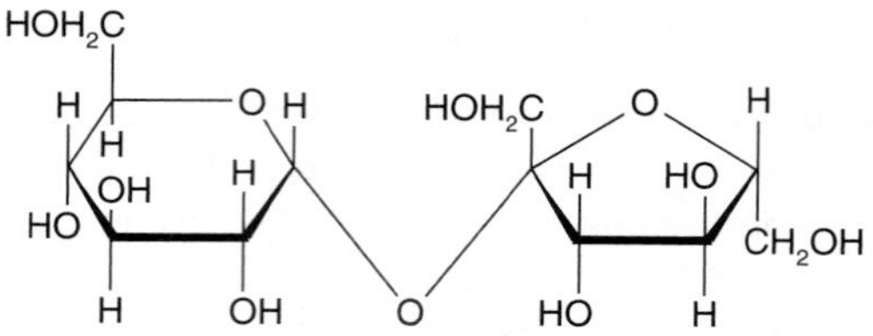

α1, β2

Sucrose

Maltose = 4-*O*-(**α**-D-glucopyranosyl)D-glucopyranose; reducing
Isomaltose = 6-*O*-(**α**-D-glucopyranosyl)D-glucopyranose; reducing
Cellobiose = 4-*O*-(**β**-D-glucopyranosyl)D-glucopyranose; reducing
Lactose = 4-*O*-(**β**-D-galactopyranosyl)D-glucopyranose; reducing
Sucrose = **α**-D-glucopyranosyl-**β**-D-fructofuranoside; nonreducing

As many as thousands of monosaccharides, either the same or different, may be joined by glycosidic bonds to form *polysaccharides.* The anomeric carbon of one unit is usually joined to C4 or C6 of the next unit. The ends of a polysaccharide are not identical (reducing end = free anomeric carbon; nonreducing = anomeric carbon linked to next unit; branched polysaccharide = more than one nonreducing end). The most common carbohydrates are homopolymers of glucose; for example, starch, glycogen, and cellulose. Plant starch is a mixture of **amylose**, a linear polymer of maltose units, and **amylopectin**, branches of repeating maltose units (glucose–glucose in **α**-1,4 linkages) joined via isomaltose linkages. **Glycogen**, the storage form of carbohydrate in animals, is similar to amylopectin, but the branches are shorter and occur more frequently. **Cellulose**, in plant cell walls, is a linear polymer of repeating cellobioses (glucose–glucose in **β**-1,4 linkages). Heteropolysaccharides contain more than one kind of monosaccharide.

Mucopolysaccharides contain amino sugars, free and acetylated, uronic acids, sulfate esters, and sialic acids in addition to the simple monosaccharides. ***N*-Acetylneuraminic acid**, a sialic acid, is

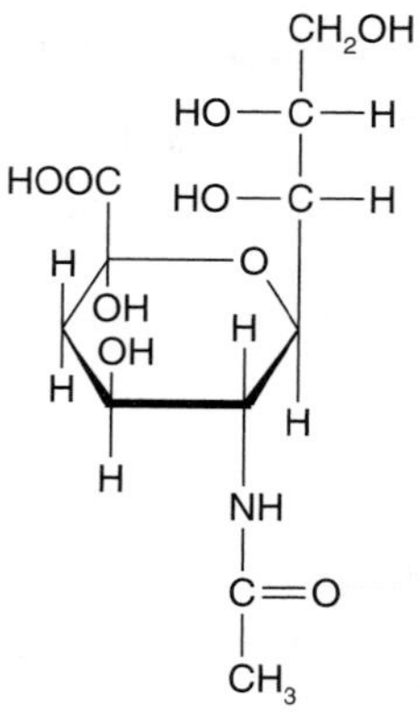

LIPIDS

Lipids are a diverse group of chemicals related primarily because they are insoluble in water, soluble in nonpolar solvents, and found in animal and plant tissues.

Saponifiable lipids yield salts of fatty acids upon alkaline hydrolysis. *Acylglycerols* = glycerol + fatty acid(s); *phosphoacylglycerols* = glycerol + fatty acids + HPO_4^{2-} + alcohol; *sphingolipids* = sphingosine + fatty acid + polar group (phosphorylalcohol or carbohydrate); *waxes* = long-chain alcohol + fatty acid. *Nonsaponifiable lipids* (*terpenes, steroids, prostaglandins,* and related compounds) are not usually subject to hydrolysis. *Amphipathic* lipids have both a polar "head" group and a nonpolar "tail." Amphipathic molecules can stabilize emulsions and are responsible for the lipid bilayer structure of membranes.

Fatty acids are monocarboxylic acids with a short (<6 carbon atoms), medium (8–14 carbon atoms), or long (>14 carbon atoms) aliphatic chain. Biologically important ones are usually linear molecules with an even number of carbon atoms (16–20). Fatty acids are numbered using either arabic numbers (COOH is 1) or the Greek alphabet (COOH is not given a symbol; adjacent carbon atoms are α, β, γ, etc.). **Saturated fatty acids** have the general formula $CH_3(CH_2)_nCOOH$. (*Palmitic acid* = C_{16}; *stearic acid* = C_{18}.) They tend to be extended chains and solid at room temperature unless the chain is short. Both trivial and systematic (prefix indicating number of carbon atoms + *anoic acid*) names are used. $CH_3(CH_2)_{14}COOH$ = palmitic acid or hexadecanoic acid.

Unsaturated fatty acids have one or more double bonds. Most naturally occurring fatty acids have *cis* double bonds and are usually liquid at room temperature. Fatty acids with *trans* double bonds tend to have higher melting points. A double bond is indicated by Δ^n, where n is the number of the first carbon of the bond. *Palmitoleic* = Δ^9-hexadecenoic acid; *oleic* = Δ^9 -octadecenoic acid; *linoleic* = $\Delta^{9,12}$-octadecadienoic acid; *linolenic* = $\Delta^{9,12,15}$-octadecatrienoic acid; and *arachidonic* = $\Delta^{5,8,11,14}$-eicosatetraenoic acid. Since fatty acids are elongated *in vivo* from the carboxyl end, biochemists use alternate terminology to assign these fatty acids to families: omega (ω) minus x (or $n - x$), where x is the number of carbon atoms from the methyl end where a double bond is first encountered. *Palmitoleic* and *oleic* are $\omega - 9$ acids, *linoleic* and *arachidonic* are $\omega - 6$ acids, and *linolenic* is an $\omega - 3$ acid. Addition of carbon atoms does not change the family to which an unsaturated fatty acid belongs.

Since the pK values of fatty acids are about 4–5, in physiological solutions, they exist primarily in the ionized form, called salts or "soaps." Long-chain fatty acids are insoluble in water, but soaps form micelles. Fatty acids form esters with alcohols and thioesters with CoA.

Biochemically significant reactions of unsaturated fatty acids are:

1. *Reduction* **—CH=CH— + XH_2 → —CH_2CH_2— + X**
2. *Addition of water* **—CH=CH— + H_2O→ —CH(OH)—CH_2—**
3. *Oxidation* **R—CH=CH—R' → R—CHO + R'—CHO**

Prostaglandins, thromboxanes, and *leukotrienes* are derivatives of C_{20}, polyunsaturated fatty acids, especially arachidonic acid. **Prostaglandins** have the general structure:

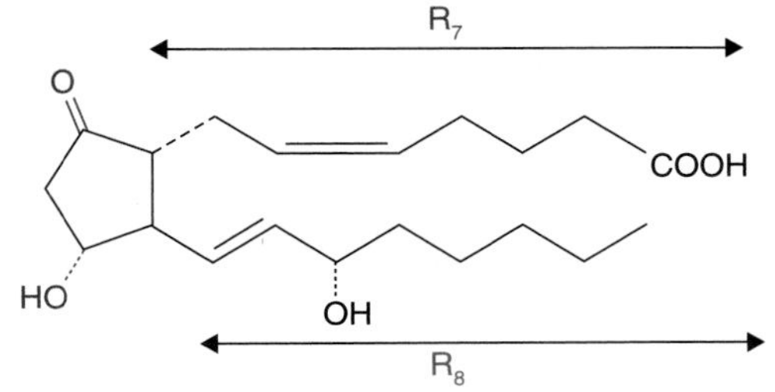

R7 COOH HO OH R8

PGE$_2$

The series differ from each other in the substituents on the ring and whether C15 contains an OH or O · OH group. The subscript indicates the number of double bonds in the side chains. Substituents indicated by — (β) are above the plane of the ring; . . . (α) below:

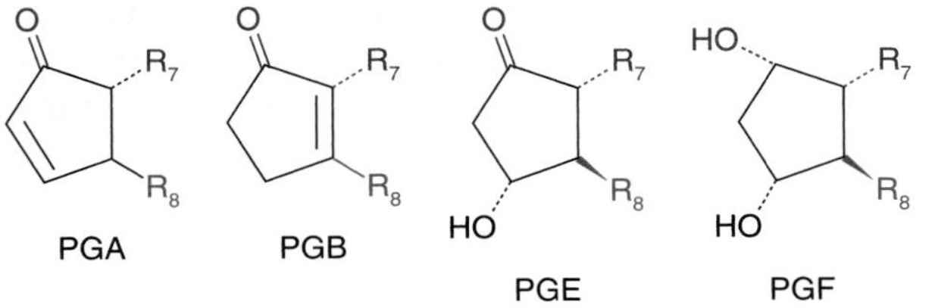

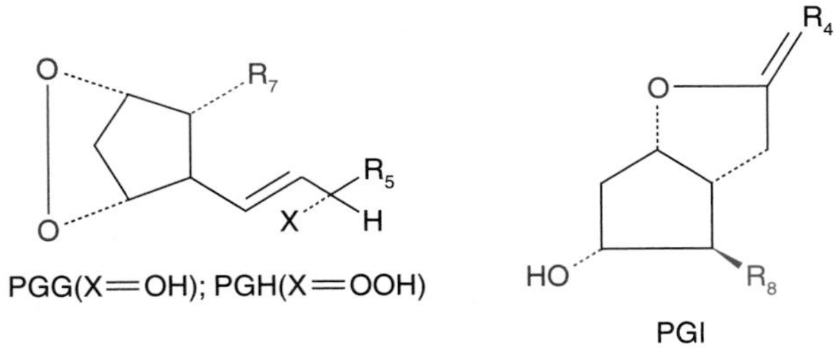

PGA PGB PGE PGF

PGG(X=OH); PGH(X=OOH)

PGI

Thromboxanes have an oxygen incorporated to form a six-membered ring:

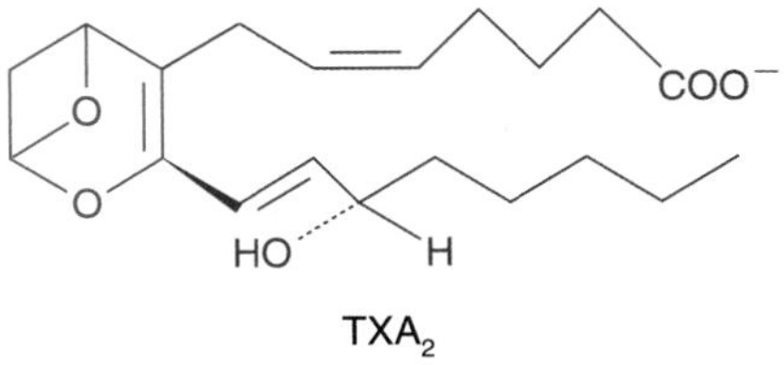

TXA$_2$

Leukotrienes are substituted derivatives of arachidonic acid in which no internal ring has formed; R is variable:

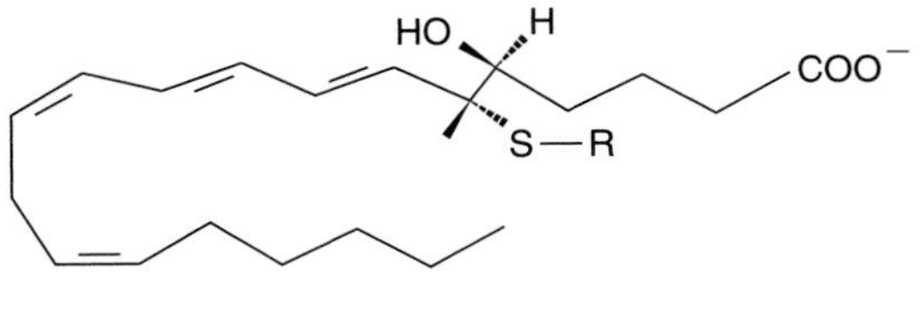

Leukotriene C, D, or E

Acylglycerols are compounds in which one or more of the three OH groups of glycerol is esterified. In **triacylglycerols** (triglycerides) all three OH groups are esterified to fatty acids. At least two of the three substituent groups are usually different. If R_1 is not equal to R_3, the molecule is asymmetric and of the L configuration:

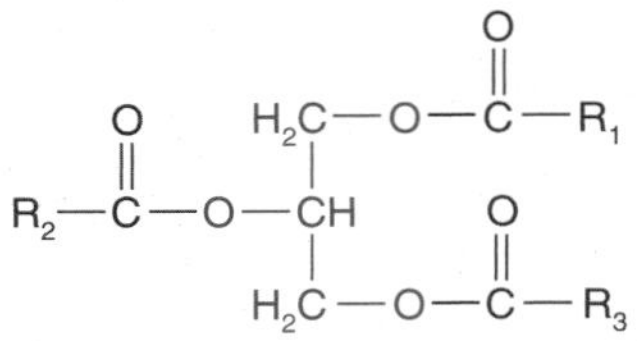

$$R_2-\overset{O}{\overset{\|}{C}}-O-CH\begin{cases}H_2C-O-\overset{O}{\overset{\|}{C}}-R_1\\ H_2C-O-\overset{O}{\overset{\|}{C}}-R_3\end{cases}$$

The properties of the triacylglycerols are determined by those of the fatty acids they contain; *oils* are liquids at room temperature (preponderance of short-chain and/or *cis*-unsaturated fatty acids), and *fats*, solid (preponderance of long-chain, saturated, and/or *trans*-unsaturated).

Triacylglycerols are hydrophobic and do not form stable micelles. They may be hydrolyzed to glycerol and three fatty acids by strong alkali or enzymes (lipases). *Mono-* [usually with the fatty acid in the **β**(2) position] and *diacylglycerols* also exist in small amounts as metabolic intermediates. Mono- and diacylglycerols are slightly more polar than triacylglycerols. *Phosphoacylglycerols* are derivatives of L-**α**-glycerolphosphate (L-glycerol 3-phosphate):

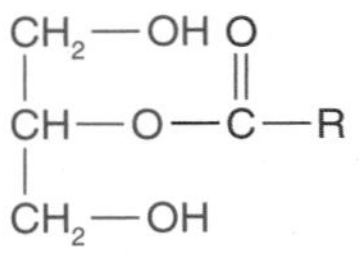

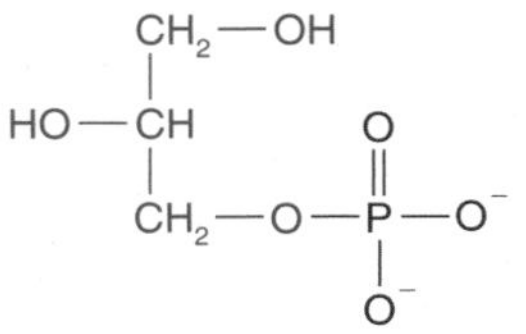

$$CH_2OH-CH(O-\overset{O}{\overset{\|}{C}}-R)-CH_2OH \qquad CH_2OH-\underset{HO}{CH}-CH_2-O-\overset{O}{\overset{\|}{P}}(O^-)-O^-$$

The parent compound, **phosphatidic acid** (two OH groups of L-**α**-glycerolphosphate esterified to fatty acids), has its phosphate esterified to an alcohol (XOH) to form several series of phosphoacylglycerols. These are amphipathic molecules, but the net charge at pH 7.4 depends on the nature of X—OH.

X—OH	*Phosphoacylglycerol*
$HO-CH_2-CH_2-\overset{+}{N}-(CH_3)_3$	Phosphatidylcholines (lecithins)
$HO-CH_2-CH_2-\overset{+}{N}H_3$	Phosphatidylethanolamines (cephalins)
$HO-CH_2-CH(\overset{+}{N}H_3)-COO^-$	Phosphatidylserines
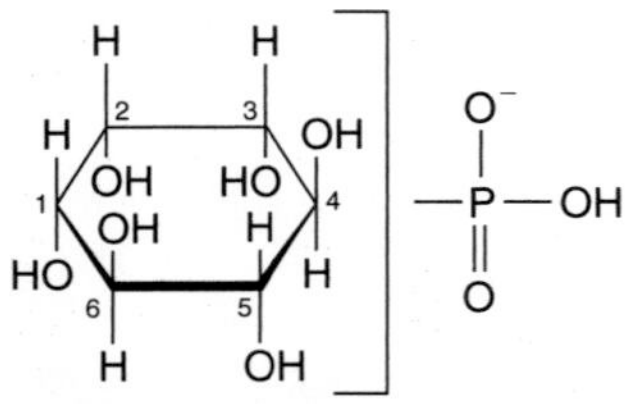 Inositol (ring positions 1–6) $[\;]-\overset{O^-}{\overset{\vert}{P}}(=O)-OH$	Phosphatidylinositols phospate on 4, or 4 and 5

In **plasmalogens**, the OH on C1 is in *ether*, rather than ester, linkage to an alkyl group. If *one* fatty acid (usually **β**) has been hydrolyzed from a phosphoacylglycerol, the compound is a *lyso-compound*; for example, lysophosphatidylcholine (lysolecithin):

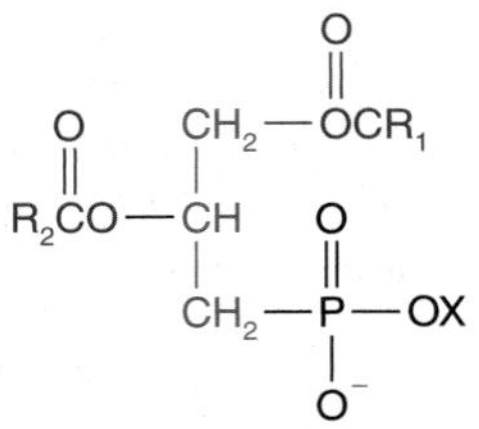

$$CH_2-O\overset{O}{\overset{\|}{C}}R_1,\; R_2\overset{O}{\overset{\|}{C}}O-CH,\; CH_2-\overset{O}{\overset{\|}{P}}(O^-)-OX$$

A phosphoacylglycerol

$$CH_2-O\overset{O}{\overset{\|}{C}}R_1,\; HO-CH,\; CH_2-\overset{O}{\overset{\|}{P}}(O^-)-OX$$

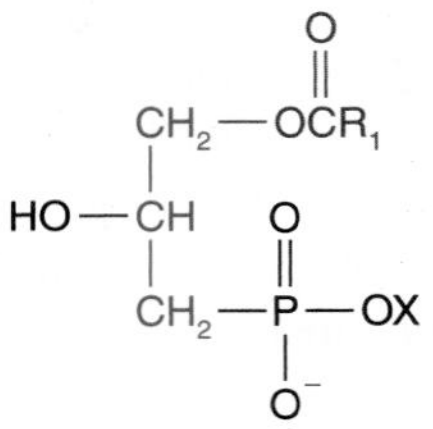

A lyso-compound

Sphingolipids are complex lipids based on the C_{18}, unsaturated alcohol, sphingosine. In *ceramides*, a long-chain fatty acid is in amide linkage to sphingosine:

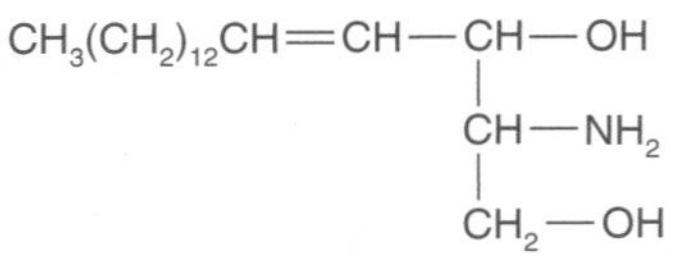

$$CH_3(CH_2)_{12}CH=CH-CH(OH)-CH(NH_2)-CH_2-OH$$

Sphingosine

$$CH_3(CH_2)_{12}CH=CH-CH(OH)-CH(N(H)-\overset{O}{\overset{\|}{C}}-R)-CH_2-OH$$

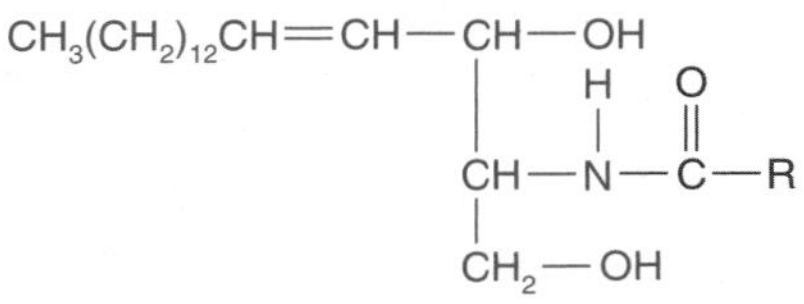

A ceramide

Sphingomyelins, the most common sphingolipids, are a family of compounds in which the primary OH group of a ceramide is esterified to phosphorylcholine (phosphoryl-ethanolamine):

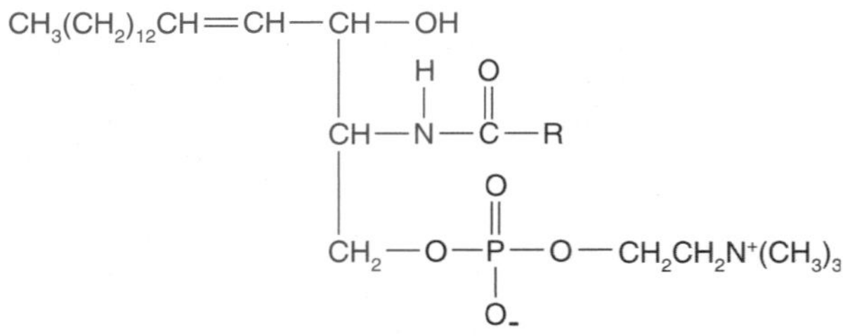

$$CH_3(CH_2)_{12}CH=CH-CH(OH)-CH(N(H)-\overset{O}{\overset{\|}{C}}-R)-CH_2-O-\overset{O}{\overset{\|}{P}}(O^-)-O-CH_2CH_2N^+(CH_3)_3$$

They are amphipathic molecules, existing as zwitterions at pH 7.4 and the only sphingolipids that contain phosphorus. *Glycosphingolipids* do not contain phosphorus but contain carbohydrate in glycosidic linkage to the primary alcohol of a ceramide. They are amphipathic and either neutral or acidic if the carbohydrate moiety contains an acidic group. **Cerebrosides** have a single glucose or galactose linked to a ceramide. *Sulfatides* are galactosylceramides esterified with sulfate at C3 of the galactose:

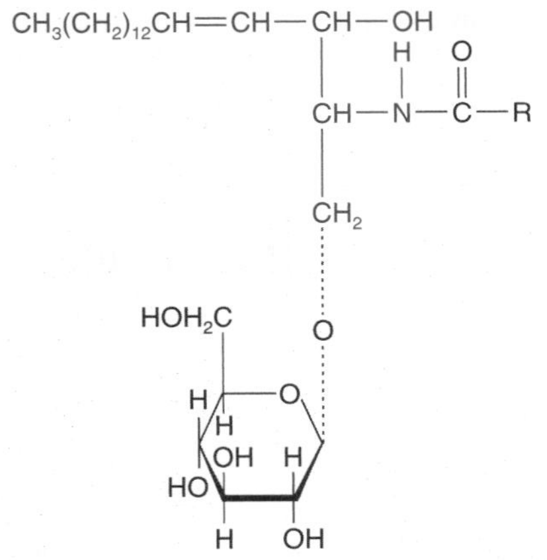

Glucosylceramide (glucocerebroside)

Globosides (*ceramide oligosaccharides*) are ceramides with two or more neutral monosaccharides, whereas in **gangliosides** the oligosaccharide contains one or more sialic acids.

Steriods are derivatives of cyclopentanoperhydrophenanthrene. The steroid nucleus is a rather rigid, essentially planar structure with substituents above the plane of the rings designated β (solid line) and those below called α (dotted line):

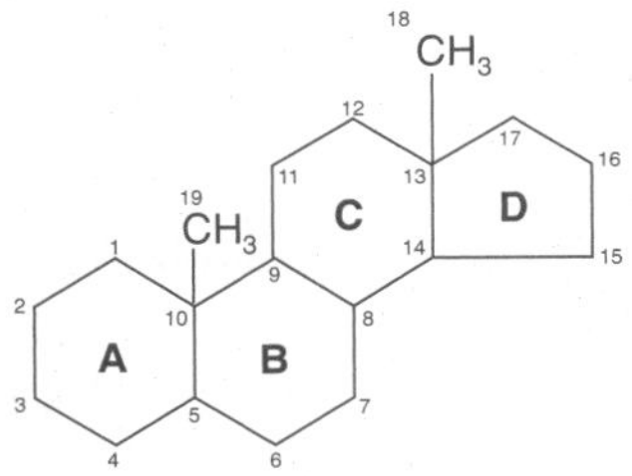

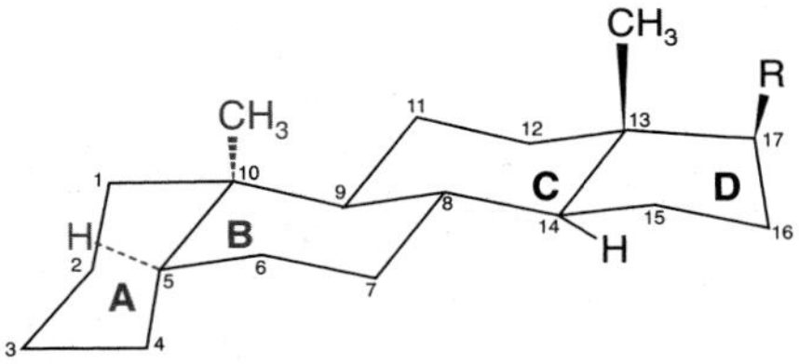

A and B rings—*cis*;
the others—*trans*

Most steroids in humans have methyl groups at positions 10 and 13 and frequently a side chain at position 17. *Sterols* contain one or more OH groups, free or esterified to a fatty acid. Most steroids are nonpolar. In a liposome or cell membrane, **cholesterol** orients with the OH toward any polar groups; cholesterol esters do not. **Bile acids** (e.g., cholic acid) have a polar side chain and so are amphipathic:

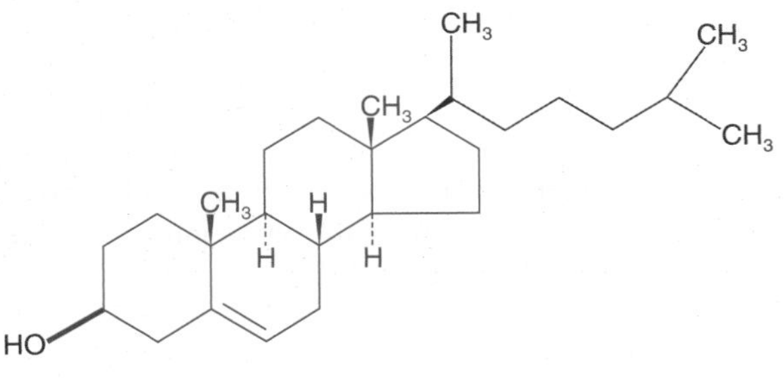

Cholesterol

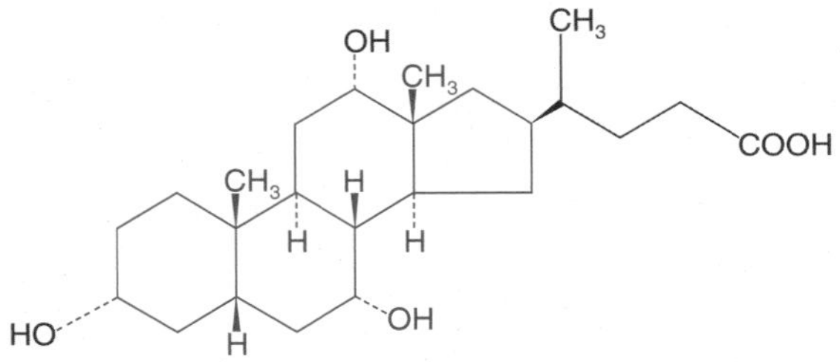

Cholic acid

Steroid hormones are oxygenated steroids of C_{18}, C_{19}, or C_{21} *Estrogens* have C_{18}, an aromatic ring A, and no methyl at C10. *Androgens* have C_{19} and no side chain at C17. *Glucocorticoids* and *mineralocorticoids* have C_{21} including a C_2, oxygenated side chain at C17. *Vitamin* D_3 (*cholecalciferol*) is not a sterol but is derived from 7-dehydrocholesterol in humans:

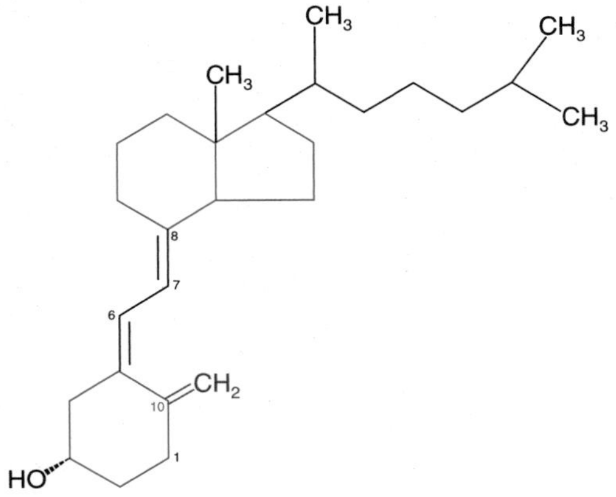

Cholecalciferol

Terpenes are polymers of two or more isoprene units. **Isoprene** is

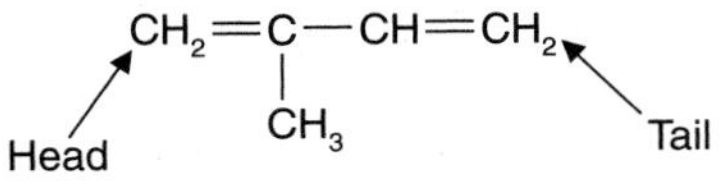

$$CH_2{=}C(CH_3){-}CH{=}CH_2$$

Head   Tail

Terpenes may be linear or cyclic, with the isoprenes usually linked head to tail and most double bonds *trans* (but may be *cis* as in vitamin A). *Squalene*, the precursor of cholesterol, is a linear terpene of six isoprene units. Fat-soluble *vitamins* (A, D, E, and K) contain isoprene units:

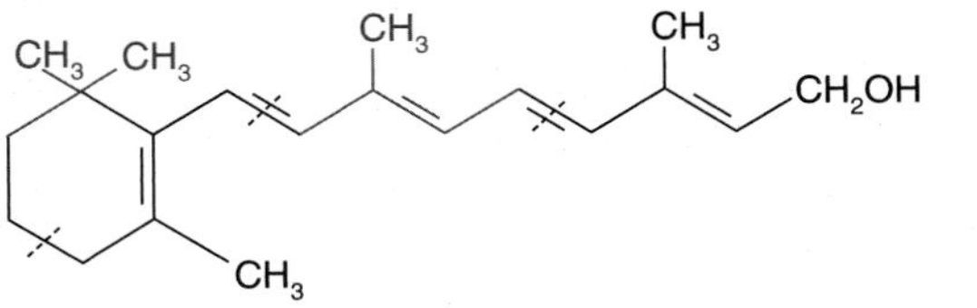

Vitamin A

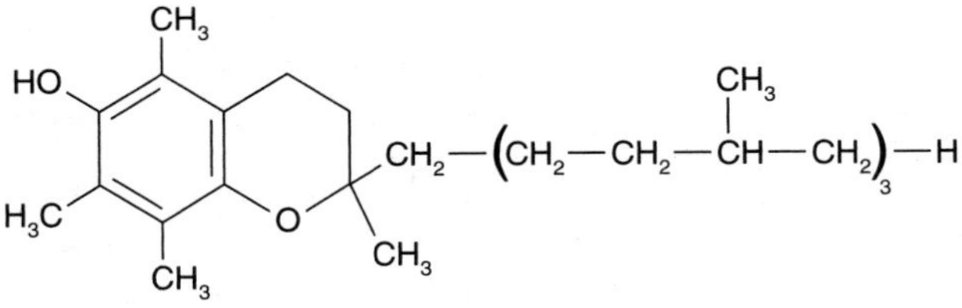

Vitamin E (α-tocopherol)

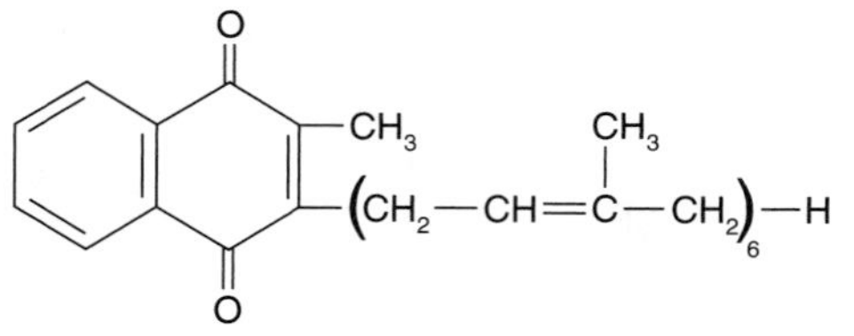

Vitamin K_2

AMINO ACIDS

Amino acids contain an *amino* (NH_2) and a *carboxylic acid* (COOH) group. Biologically important amino acids are usually **α**-amino acids with the formula

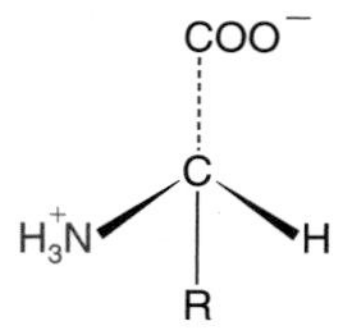

L-α-Amino acid

The amino group, has an unshared pair of electrons, is basic, with a pK_a of about 9.5, and exists primarily as $—NH_3^+$ at pH values near neutrality. The carboxylic acid group ($pK \sim 2.3$) exists primarily as a carboxylate ion. If R is anything but H, the molecule is asymmetric with most naturally occurring ones of the L configuration (same relative configuration as L-glyceraldehyde: see p. 1125).

The *polarity* of amino acids is influenced by their side chains (R groups) (see p. 78 for complete structures). *Nonpolar* amino acids include those with large, aliphatic, aromatic, or undissociated sulfur groups (aliphatic = Ala, Ile, Leu, Val; aromatic = Phe, Trp; sulfur = Cys, Met). *Intermediate* polarity amino acids include Gly, Pro, Ser, Thr, and Tyr (undissociated).

Amino acids with ionizable side chains are *polar.* The pK values of the side groups of arginine, lysine, glutamate, and aspartate are such that these are nearly always charged at physiological pH, whereas the side groups of histidine ($pK = 6.0$) and cysteine ($pK = 8.3$) exist as both charged and uncharged species at pH 7.4 (acidic = Glu, Asp, Cys; basic = Lys, Arg, His). Although undissociated cysteine is nonpolar, cysteine in dissociated form is polar. Asparagine and glutamine are polar even though their side chains are not charged (ionized).

All amino acids are at least *dibasic acids* because of the presence of both the **α**-amino and **α**-carboxyl groups, the ionic state being a function of pH. The presence of another ionizable group will give a tribasic acid as shown for cysteine.

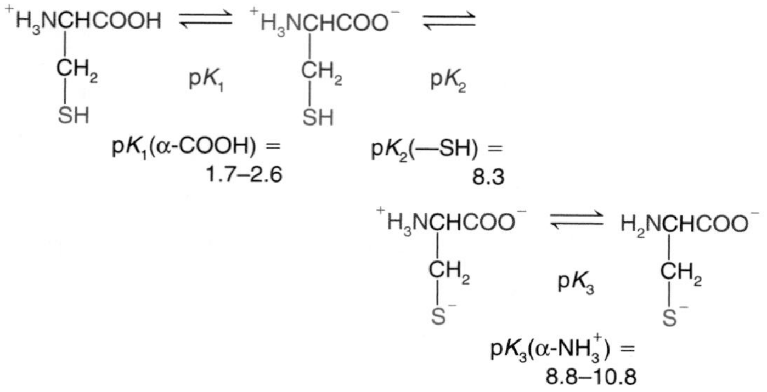

$$^{+}H_3NCH(CH_2SH)COOH \underset{pK_1}{\rightleftharpoons} {}^{+}H_3NCH(CH_2SH)COO^- \underset{pK_2}{\rightleftharpoons}$$

$pK_1(\alpha\text{-COOH}) = 1.7–2.6$ $\qquad pK_2(—SH) = 8.3$

$$^{+}H_3NCH(CH_2S^-)COO^- \underset{pK_3}{\rightleftharpoons} H_2NCH(CH_2S^-)COO^-$$

$pK_3(\alpha\text{-}NH_3^+) = 8.8–10.8$

The **zwitterionic form** is the form in which the *net* charge is zero. The *isoelectric point* is the average of the two pK values involved in the formation of the zwitterionic form. In the above example this would be the average of $pK_1 + pK_2$.

PURINES AND PYRIMIDINES

Purines and pyrimidines, often called *bases,* are nitrogen-containing heterocyclic compounds with the structures

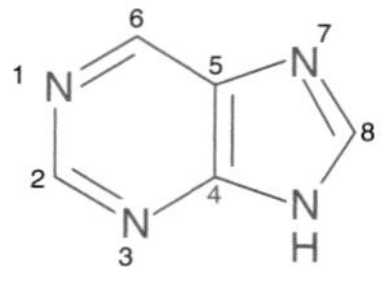

Purine

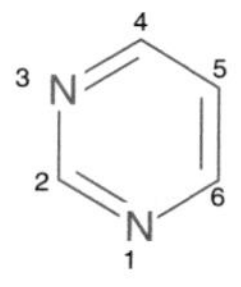

Pyrimidine

Major bases found in nucleic acids and as cellular nucleotides are the following:

Purines	*Pyrimidines*
Adenine: 6-amino	Cytosine: 2-oxy, 4-amino
Guanine: 2-amino, 6-oxy	Uracil: 2,4-dioxy Thymine: 2,4-dioxy, 5-methyl
Other important bases found primarily as intermediates of synthesis and/or degradation are	
Hypoxanthine: 6-oxy	Orotic acid: 2,4-dioxy, 6-carboxy
Xanthine: 2,6-dioxy	

Oxygenated purines and pyrimidines exist as *tautomeric* structures with the keto form predominating and involved in hydrogen bonding between bases in nucleic acids:

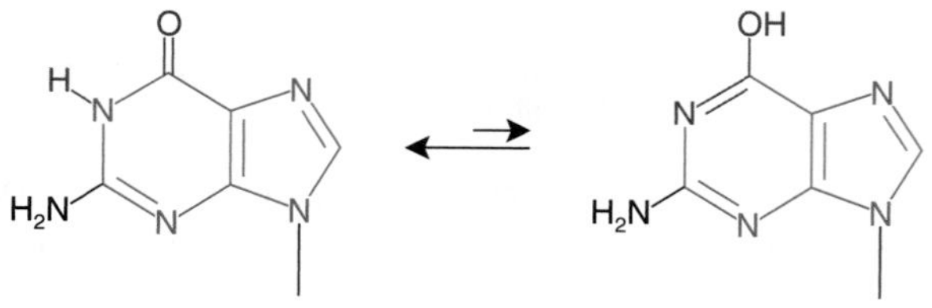

Keto $\longleftrightarrow$ *Enol*

Nucleosides have either **β**-D-ribose or **β**-D-2-deoxyribose in an *N*-glycosidic linkage between C1 of the sugar and N9 (purine) or N1 (pyrimidine).

Nucleotides have one or more phosphate groups esterified to the sugar. Phosphates, if more than one is present, are usually attached to each other via phosphoanhydride bonds. Monophosphates may be designated as either the base monophosphate or as an *-ylic acid* (AMP: adenylic acid):

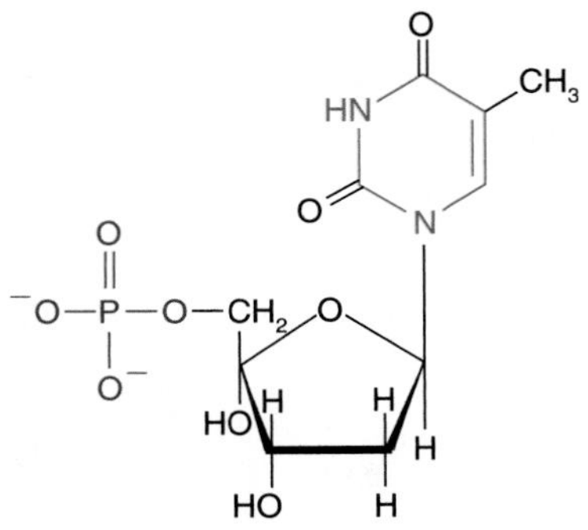

By conventional rules of *nomenclature*, the atoms of the base are numbered 1–9 in purines or 1–6 in pyrimidines and the carbon atoms of the sugar 1′–5′. A nucleoside with an unmodified name indicates that the sugar is ribose and the phosphate(s) is/are attached at C-5′ of the sugar. Deoxy forms are indicated by the prefix d (dAMP = deoxyadenylic acid). If the phosphate is esterified at any position other than 5′, it must be so designated [3′-AMP; 3′-5′-AMP (cyclic AMP = cAMP)]. The nucleosides and nucleotides (ribose form) are named as follows:

Base	*Nucleoside*	*Nucleotide*
Adenine	Adenosine	AMP, ADP, ATP
Guanine	Guanosine	GMP, GDP, GTP
Hypoxanthine	Inosine	IMP
Xanthine	Xanthosine	XMP
Cytosine	Cytidine	CMP, CDP, CTP
Uracil	Uridine	UMP, UDP, UTP
Thymine	dThymidine	dTMP, dTTP
Orotic acid	Orotidine	OMP

Minor (modified) bases and nucleosides also exist in nucleic acids. *Methylated* bases have a methyl group on an amino group (*N*-methyl guanine), a ring atom (1-methyl adenine), or on an OH group of the sugar (2′-*O*-methyl adenosine). *Dihydrouracil* has the 5–6 double bond saturated. In *pseudouridine*, the ribose is attached to C5 rather than to N1.

In **polynucleotides** (*nucleic acids*), the mononucleotides are joined by phosphodiester bonds between the 3′-OH of one sugar (ribose or deoxyribose) and the 5′-OH of the next (see p. 30 for the structure).

Glossary

Francis Vella

30-nm fiber: Fiber (30-nm diameter) of chromatin which has a solenoid structure consisting of six nucleosomes and their linker DNA and associated histone H1 per turn, stacked on top of each other.

5′ cap: 7-Methylguanosine linked to the 5′-terminal nucleotide of an mRNA via a 5′-5′-triphosphate bridge. It is formed during processing of pre-mRNA in the nucleus, inhibits degradation and enhances translation of the mRNA.

A site: Binding site on both ribosomal subunits that holds the tRNA that carries the amino acid to be added to the growing peptide chain during polypeptide synthesis. Also called aminoacyl site.

ABC transporter protein: Integral plasma membrane protein that consists of two domains (each of six transmembrane segments) and two loops (each containing an ATP-binding site) that uses the energy of ATP to transport a variety of hydrophobic natural products and synthetic drugs out of the cytoplasm. Also called ATP-binding cassette protein.

absorption spectrum: Pattern of absorption of incident light observed with a spectroscope or spectrometer.

absorption: Passage across intestinal cell membranes of the products of digestion.

abzyme: Catalytic antibody, natural or one that is raised against an antigen that has a transition state structure for a reaction. Also called catalytic antibody.

acceptor stem: In tRNA, the part in which the 5′- and 3′-ends are base paired. The 3′-end terminates in an unpaired CCA sequence with the A being the site of attachment of a cognate aminoacyl residue.

acid anhydride: See anhydride.

acid-base catalysis: Catalysis of a reaction that is aided by transfer of a proton from an acid or abstraction of a proton by a base.

acidosis: Abnormal physiological state in which an increase in proton concentration lowers the pH of plasma below 7.35.

acrosome: Organelle at the tip of a spermatozoon that is rich in enzymes for digestion of the protective coat of an oocyte at fertilization.

actin: Globular protein that forms the microfilaments of cytoskeleton and the thin filaments of skeletal and cardiac muscle. Its interaction with the heads of thick filaments produces muscle contraction.

activation energy: Energy required for a specific chemical reaction to occur.

active site: Region usually in a cleft on an enzyme surface that binds the appropriate substrate and catalyzes its conversion to product.

active transport: Passage of solutes across a cell membrane that is effected with energy expenditure from ATP.

activin: Glycoprotein hormone secreted by the gonads, that stimulates secretion of follicle-stimulating hormone. It consists of two B_A or B_B chains in inhibins A and B respectively, or of a B_A and a B_B chain, joined by a disulfide bond.

acyl carrier protein (ACP): In prokaryotes, the protein with an attached phosphopantetheine moiety that is the central component of fatty acid synthase complex. In eukaryotes, it is present as a domain of the multicatalytic fatty acid synthase protein.

adaptin: Major multimeric protein of clathrin-coated vesicles that binds on the cytoplasmic portion of various integral membrane proteins to mediate formation of the clathrin coat. Also called adaptor protein.

adenylate cyclase system: Pathway of signal transduction in which a hormone binding to its cognate serpentine receptor activates a trimeric G protein to stimulate or inhibit adenylate cyclase, thereby affecting production of cyclic AMP as a second messenger.

adenylate cyclase: Enzyme that is an integral protein of the plasma membrane. Its active site on the inner membrane surface converts ATP to cyclic AMP. Also called adenylyl cyclase.

adenylyl cyclase: See adenylate cyclase.

adherens junction: Part of a plasma membrane that contains cadherins, is involved in cell–cell adhesion and is linked on the inner surface to actin microfilaments via catenins and other proteins.

adipose tissue: Loose connective tissue dominated by adipocytes whose major component is neutral fat (i.e., triacylglycerol). Two varieties are recognized, white and brown.

A-DNA: Double-helical DNA that has 11 base pairs per right-handed turn, a diameter of ~26 Å, and forms in pure dehydrated DNA. It is also the common form of double-stranded RNA and of RNA–DNA heteroduplexes.

adrenal gland: One of two small endocrine glands each situated above a kidney. It consists of an outer cortex that secretes steroid hormones, and a central medulla that secretes catecholamines, particularly adrenaline. Also called suprarenal gland.

adrenaline: *N*-methyl noradrenaline, the major hormone derived from tyrosine in the adrenal medulla and some neurons. It is secreted in response to stress and increases glycogenolysis (in muscle and liver), gluconeogenesis (in liver), glycolysis (in muscle), and lipolysis (in adipocytes). Also called epinephrine.

adrenocortical hormone: Any glucocorticoid, mineralocorticoid or sex hormone secreted by the adrenal cortex.

adrenocorticotropic hormone (ACTH): Polypeptide hormone secreted by the anterior pituitary in response to corticotrophin-releasing hormone of the hypothalamus. It stimulates production and secretion of glucocorticoids by the zona fasciculata. Also called corticotropin, cortin.

aerobic metabolism: Complete breakdown of glucose and fatty acids in muscle in the presence of adequate amounts of oxygen into carbon dioxide and water via acetyl CoA. The oxygen-requiring part releases large amounts of energy as ATP and occurs within mitochondria.

aerobic: Occurring only in the presence of oxygen.

agglutination: Aggregation of erythrocytes produced by interaction between antigens on their surface and specific antibodies in plasma.

agglutinin: Antibody in plasma that reacts with antigens on the surface of erythrocytes when donor and recipient differ in their blood group.

agglutinogen: Antigen on the surface of erythrocytes whose presence and structure are genetically determined.

aggrecan: Proteoglycan rich in chondroitin sulfate and keratan sulfate attached to a large protein that is present in cartilage.

agonist: Hormone, neurotransmitter drug, or other compound that binds to a specific receptor and elicits a functional response.

alanine cycle: See glucose-alanine cycle.

aldimine: Compound formed by condensation of a primary amine with an aldehyde. Also called Schiff base.

aldosterone: Mineralocorticoid produced by zona glomerulosa and secreted in response to angiotensin II. It enhances reabsorption of Na^+ and water by renal tubules. Excessive secretion produces aldosteronism.

alkalosis: Abnormal physiological state in which a deficiency of protons or an excess of bicarbonate ions raises the pH of plasma above 7.45.

alkaptonuria: Autosomal recessive condition in which homogentisate, produced during catabolism of phenylalanine and tyrosine, accumulates because of a deficiency of homogentisate oxidase activity.

allosteric effector: See allosteric modulator.

allosteric site: Site on a regulatory enzyme that is remote from the active site and, when bound to an appropriate effector or modulator, increases or decreases the affinity of the enzyme for its substrate.

α + β-protein: Protein in which α-helical and β-strand regions are present in a nonrandom fashion.

α/β-barrel: Protein structure in which alternating α helices and β strands form a barrel of parallel β-strands, surrounded by a barrel of α-helices.

α/β protein: Protein in which α-helical and β-strands are interspersed or alternate with each other.

A-helix: Common form of secondary structure of polypeptides; spiral conformation that results from maximal hydrogen bonding between components of peptide bonds of adjacent turns of the spiral, each turn being 3.6 residues long.

all-α protein: Protein that consists solely or mostly of α-helical segments.

all-β protein: Protein that consists solely or mostly of β-strands.

allostery: Property of an enzyme or other protein whereby non-covalent binding of a ligand at one site induces a conformational change that affects binding of a ligand at another site.

alternative RNA splicing: Deletion of certain exons or incorporation of alternative exons during splicing of pre-mRNA, thus forming more than one mRNA and generating more than one protein from a primary transcript. Also called differential RNA splicing.

amino terminus: See N-terminus.

aminoacyl site: Site on a ribosome that binds aminoacyl-tRNA during polypeptide synthesis. Also called A site.

aminoacyl-tRNA: tRNA carrying an amino acid to be added to the growing peptide chain during polypeptide synthesis.

aminotransferase: Enzyme that catalyzes transfer of an amino group from an α-amino to an α-keto acid, yielding a new α-keto acid and a new α-amino acid and using pyridoxal phosphate as coenzyme. Also called transaminase.

amphipathic: Having both a hydrophilic region and a hydrophobic region.

amphoteric: Capable of serving as an acid (by donating protons) or as a base (by accepting protons).

amyloid: Insoluble protein highly resistant to proteolysis that is present in the brain in a variety of neurological diseases. It consists of aggregates of fibrils composed of subunits rich in β-sheet structure. The β-amyloid present in plaques in Alzheimer disease consists of a polypeptide (39 to 43 amino acid residues) derived from β-amyloid precursor protein.

anabolism: Phase of intermediary metabolism in which energy-utilizing metabolic reactions are used for synthesis of macromolecules and other biomolecules from simpler precursors.

anaerobic metabolism: Breakdown of glucose with production of lactate that occurs in vigorous exercise when muscle lacks adequate oxygen.

anaphase: Stage in mitosis in which the paired chromatids separate and move to the opposite ends of the spindle apparatus.

anaplerotic reaction: Reaction that replenishes the supply of intermediates in a metabolic pathway or cycle.

androgen: Steroid hormone secreted mainly by interstitial cells of the testis, and in small amounts by the adrenal cortex in both genders. It stimulates development and maintenance of the reproductive system and secondary sex characteristics in males. Oversecretion leads to virilization in either gender.

angiogenesis: Formation of new blood vessels or their development and growth from existing ones.

angiotensin I: N-terminal decapeptide released by renin acting on the plasma protein angiotensinogen.

angiotensin II: N-terminal octapeptide released in the lung capillaries by angiotensin-converting enzyme acting on angiotensin I. It produces arteriolar constriction and a rise in blood pressure and promotes thirst and release of antidiuretic hormone and aldosterone.

angiotensin-converting enzyme (ACE): Transmembrane glycoprotein present in the lung capillaries and plasma. It is a zinc-containing peptidase that cleaves off the C-terminal dipeptide of angiotensin I to produce angiotensin II.

angiotensinogen: Plasma glycoprotein secreted by the liver, the precursor of angiotensin I.

angstrom (Å): Å unit of length, 10^{-10} m.

anhydride: Product of condensation, with loss of water, of two molecules of an acid. Also called acid anhydride.

ankyrin: Large globular protein that is associated with the channel for antiport of Cl^- and HCO_3^- in the membrane, and with spectrin of the cytoskeleton, of erythrocytes.

anomer: Either of the stereoisomers formed by cyclization of an aldose or a ketose, differing only in the configuration around the carbonyl carbon.

anoxia: State of tissue deprivation of oxygen.

ANP: See atrial natriuretic peptide.

antagonist: Agent that binds to a receptor and inhibits binding of the specific agonist for the receptor.

anterior pituitary hormone: Hormone secreted by the anterior pituitary gland, that is, one of the glycoproteins luteinizing hormone, follicle-stimulating hormone, and thyroid-stimulating hormone, the simple proteins growth hormone and prolactin, and the polypeptide adrenocorticotrophic hormone.

anterograde transport: Transport toward the periphery of an axon.

antibody-based immunity: Type of immunity mediated by circulating antibodies secreted by B cells in response to an antigen. Also called humoral immunity.

anticodon arm: In tRNA, the stem-and-loop portion that contains the anticodon region.

anticodon: Three bases on a tRNA molecule that interact with a complementary codon on an mRNA molecule.

antidiuretic hormone (ADH): Nonapeptide hormone produced in the hypothalamus but secreted by the posterior pituitary. It causes reabsorption of water by causing recruitment of an aquaporin into the apical membranes of renal tubules and collecting ducts, and raises blood pressure. Deficient secretion causes diabetes insipidus. Also called vasopressin.

antigen: Substance or cluster of chemical groups capable of inducing production of specific antibodies by B cells.

antigenic determinant: Portion of an antigen that is responsible for eliciting production of an antibody and that interacts with the antibody.

antigen-presenting cell (APC): Cell that is essential for initiating a normal immune response by processing antigens and displaying them on its surface bound to MHC proteins.

antiparallel: Relative orientation of two β-strands (N to C and C to N, respectively) within the same or on two polypeptides, or within two strands of DNA (5′ to 3′, and 3′ to 5′, respectively).

antiport: Cotransport of two different solutes in opposite directions across a membrane.

antisense RNA: RNA, synthetic or transcribed from the noncoding strand of genetic DNA by genetic engineering, that is complementary to and inhibits the RNA transcript of that gene.

AP endonuclease: See apurinic/apyrimidinic endonuclease.

AP site: See apurinic site, apyrimidinic site.

APC protein: Tumor suppressor protein that promotes degradation of β-catenin. It is absent or nonfunctional in adenomatous polyposis coli.

apoenzyme: Catalytically inactive protein component of an enzyme freed from any required cofactor, coenzyme or prosthetic group.

apolipoprotein (apoLp): Protein component of a plasma lipoprotein freed from any associated lipids.

apoprotein: Protein component of a conjugated protein freed from any non-amino acid moiety.

apoptosis: Form of cell death in which the cell shrinks and condenses, its cytoskeleton collapses, the nuclear membrane disintegrates, and the DNA becomes fragmented. The cell membrane becomes altered and evokes rapid phagocytosis without loss of cell contents into the surrounding medium or induction of an inflammatory response. The process is genetically programmed and depends on activation of a caspase cascade. Also called programmed cell death.

apurinic site (AP site): Site on a strand of DNA from which a purine base is missing.

apurinic/apyrimidinic endonuclease (AP endonuclease): Endonuclease that removes a segment of DNA that contains an apurinic or an apyrimidinic site during base excision repair.

apyrimidinic site (AP site): Site on a strand of DNA from which a pyrimidine base is missing.

aquaporin: Integral membrane homotetrameric protein that forms a hydrophilic channel for rapid movement of water down a concentration gradient. Each subunit contains six transmembrane segments.

arrestin: Protein that binds to the phosphorylated intracellular segment of a serpentine receptor and prevents its interaction with a trimeric G protein.

asymmetric carbon atom: See chiral carbon atom.

ATM protein: Large protein kinase that is mutated in ataxia telangiectasia and is required to activate p53 in response to double-strand DNA breaks induced by ionizing radiation or oxidative damage.

ATP synthase: See F-type ATPase.

ATP-binding cassette protein: See ABC transporter protein.

atrial natriuretic peptide (ANP): Peptide hormone secreted by cardiocytes of the atrium when stretched by increased venous return. It promotes reduction of blood pressure and of venous return, and renal water loss. Its receptor contains guanylate cyclase activity in its intracellular segment. Also called atriopeptin.

atriopeptin: See atrial natriuretic peptide.

attenuator: Sequence within the leader region of nascent mRNA that is involved in regulating expression of certain genes for enzymes of amino acid synthesis in prokaryotes by terminating transcription in the presence of an adequate supply of the particular amino acid.

autoantibody: Antibody that interacts with antigens on the surface of a person's own cells and tissues.

autocatalysis: Process in which the product of a reaction is the catalyst for the same reaction (e.g., activation of pepsinogen by pepsin).

autocrine hormone: Biomolecule that acts on the cell that secretes it or on neighboring cells of the same type.

autoimmune disease: Disorder (such as myasthenia gravis) in which the immune system behaves abnormally by responding to antigens that are normal components of an individual's cells or tissues.

autolysis: Self-digestion of a proteolytic enzyme or the breakdown of a cell or tissue by its own degradative enzymes.

autophosphorylation: Phosphorylation of a protein kinase by another molecule of the same enzyme.

axonal transport: Movement of organelles and other particles along microtubules in the axon of a nerve cell, being anterograde (away from the cell body) and effected by kinesin or retrograde (toward the cell body) and effected by dynein.

B cell: Type of lymphocyte derived from the bone marrow that is capable of maturing and differentiating into a plasma cell that produces antibodies or into a B cell which retains a memory for a previous immune response to an antigen.

backbone: Chain of atoms (N-C_α-C) formed by the amino acid residues in a polypeptide or of atoms ($C_{4'}$-$C_{3'}$-O-P-O-$C_{5'}$) formed by the repeating sugar phosphate units in a nucleic acid.

basal lamina: Thin layer of extracellular matrix beneath epithelial sheets or tubes or surrounding individual muscle, adipose tissue, or Schwann cells that separates them from the underlying connective tissue.

basal metabolic rate (BMR): Number of kilocalories that a normal, resting, fasting individual requires to maintain that state during a period of time.

base pairing: Formation of a maximum number of hydrogen bonds between a pair of complementary bases (a purine and a pyrimidine) on two strands of a nucleic acid or within one strand folded back on itself.

base-excision repair: Correction of an abnormal base in a strand of DNA by removal of that base by a DNA glycosylase, removal of the remaining deoxyribosephosphate unit and possibly neighboring nucleotides by an AP endonuclease, and filling in of the gap by DNA polymerase and DNA ligase.

basement membrane: Layer of filaments and fibers that attach an epithelium to the connective tissue beneath it.

basic leucine zipper (bZIP): Protein motif in which a DNA-binding basic α-helix is the N-terminal extension of a leucine zipper. It is commonly present in transcription factors.

basophil: White blood cell with an affinity for basic stains. It contains granules of vasoactive amines and is important in the local response to infection.

B-DNA: Double-helical DNA that has 10.5 base pairs per right-handed turn and a diameter of 20 Å and is the most stable conformation under physiological conditions.

Bence Jones protein: Free light chain of an immunoglobulin produced in excess in myelomatosis. It is detectable in plasma and/or urine.

benign: Tumor that is noninvasive and usually grows slowly.

β–α–β loop: Protein structural motif consisting of two parallel β-strands separated by an α-helix.

β-amyloid precursor protein (β-APP): Single-pass integral glycoprotein of plasma membrane. The isoform that contains 695 residues is expressed in neurons; those which contain 751 and 770 are expressed mainly in non-neuronal tissues. Cleavage of β-APP 695 within the transmembrane segment by α-secretase releases β-amyloid, accumulation of which forms amyloid plaques that injure nerve cells and are associated with Alzheimer disease.

β-amyloid: See amyloid.

β-barrel: Supersecondary structure of proteins that consists of an antiparallel β-sheet rolled into a cylinder.

β- bend: See reverse turn.

β-conformation: See β-strand.

β-hairpin: Protein motif in which two antiparallel β-strands are connected by a turn or short loop.

***B* oxidation:** Metabolism of fatty acids by oxidation of the β-carbon (or C3) with release of acetyl CoA. This process occurs in the mitochondria of eukaryotes.

β-pleated sheet: See β-sheet.

$\beta\beta$-sheet: Common protein motif in which β-strands in different regions of a polypeptide or on two polypeptides, run beside each other in parallel or antiparallel direction and are held together by hydrogen bonds between the nitrogen and oxygen atoms of their peptide bonds. Also called β-pleated sheet.

β-strand: Common secondary structure in which a segment of a polypeptide assumes an extended zigzag arrangement. Also called β-conformation.

bilayer: Double layer of amphipathic lipid molecules arranged with their polar ends oriented toward the solvent surface and their nonpolar regions associated in the center.

bile pigment: Either of the linear tetrapyrroles, bilirubin and biliverdin, that are produced by catabolism of heme. They are excreted in bile as the bisglucuronides.

bile salt: Na or K salt of a glycine or taurine conjugate of a bile acid (e.g., cholic acid) derived from cholesterol. It is responsible for emulsification of dietary lipids and their digestion by pancreatic lipase.

bile: Alkaline secretion produced by the liver and contains bile pigments, bile salts, cholesterol and phospholipids, which become concentrated during storage in the gall bladder. It is released into the duodenum and is essential for emulsification and digestion of dietary lipid.

binding site: Region, usually a crease or pocket, on a protein surface in which a ligand binds.

biocytin: ϵ-*N*-biotin lysine, present as a prosthetic group in the holoprotein of biotin-dependent carboxylases.

bioenergetics: Analysis of energy production and utilization during metabolism.

biomolecule: Organic compound normally present in, or formed by, living organisms.

biopolymer: Biological macromolecule formed by covalent linking of similar or identical small units, end to end, to form a long chain. Examples are polysaccharides, polypeptides, and polynucleotides.

biotransformation: Chemical reactions responsible for conversion of drugs and other xenobiotics into metabolites that are more water soluble for excretion. Consists of Phase I and/or Phase II reactions.

blood clot: Network of fibrin fibers and trapped blood cells that is formed when blood is shed or a blood vessel is damaged. When it forms within a blood vessel, it is called a thrombus.

body mass index (BMI): Measure of adiposity obtained from the weight in kilograms divided by the square of the height in meters. An index up to 25 is considered normal.

Bohr effect: Decreased affinity of hemoglobin for oxygen when the carbon dioxide and H^+ levels in blood are increased.

bond energy: Energy that is required to break or that is released in making a chemical bond. Also called bond strength.

bond strength: See bond energy.

botulinum toxin: Mixture of neurotoxic zinc metalloproteases formed by Clostridium botulinum, the causative agent of botulism (severe food poisoning). It is a potent inhibitor of acetylcholine release from cholinergic neurons.

bound ribosome: Ribosome that is attached to the outer surface of the endoplasmic reticulum and that synthesizes proteins destined for the endoplasmic reticulum lumen, Golgi complex, lysosomes, secretory vesicles or plasma membrane.

branched-chain ketonuria: See maple syrup urine disease.

BRCA: Either of two genes (BRCA 1 or BRCA 2) that when mutated, greatly increase the risk of development of breast cancer. Both genes encode large proteins that are involved in DNA binding and protein–protein interactions.

brown fat: Specialized form of adipose tissue rich in mitochondria in which fuel oxidation is used for thermogenesis rather than for ATP synthesis. It is the site of nonshivering thermogenesis.

brush border: Dense covering of microvilli that are formed by the plasma membrane on the apical surface of cells of the intestinal mucosa and renal tubules.

buffer: Mixture of nearly equal amounts of a weak acid and its salt that stabilizes the pH of a solution by releasing or removing hydrogen ions.

cadherin: Single-pass intrinsic glycoprotein of adherens junctions of the plasma membrane. It mediates Ca^{2+}-dependent cell–cell adhesion. Several types (E - in epithelia, N - in nerve and muscle, P - in placenta, and epidermis, and VE - in vascular endothelium) exist.

calbindin: Calcium-binding protein of absorptive cells of the small intestine whose synthesis is dependent on vitamin D.

calcitonin: Polypeptide hormone that contains 32 residues, is secreted by parafollicular or C cells of the thyroid gland in response to an increase in circulating Ca^{2+} concentration and increases the rate of bone calcification and of Ca^{2+} loss from the kidney.

calcitriol: 1,25-dihydroxycholecalciferol, the active form of vitamin D_3.

calmodulin (CaM): Dumbbell-shaped protein present in cytosol that contains 148 residues and four Ca^{2+}-binding sites. It activates a variety of Ca^{2+}-calmodulin-dependent protein kinases (CaM kinases).

calorie (cal): Amount of energy required to raise the temperature of a gram of water from 14.5 to 15.5°C, being equal to 4.184 J. In nutrition the energy released by combustion of a food in the body is measured in kilocalories (1000 cal).

CaM kinase: Ca^{2+}-calmodulin-dependent protein kinase.

cap-binding complex (CBC): Complex of proteins, including initiation factor 4F, that binds the 5′cap of mRNA.

capsid: Protein shell composed of subunits (called capsomeres) that encloses a viral genome.

capsomere: Protein subunit of a capsid.

carbaminohemoglobin: Hemoglobin with carbon dioxide molecules bound to its N-terminal amino groups.

carbohydrate loading: Practice common among athletes of exercising intensively for a period (to deplete muscle glycogen) followed by consumption of a high-carbohydrate diet (to promote larger-than-normal glycogen stores).

carboxyl terminus: End of a polypeptide chain that contains an α carboxyl group that is free or, less frequently, aminated. Also called C-terminus.

carcinoma: Most common form of cancer in humans, arising from epithelial cells.

cardiac muscle: Specialized form of striated muscle that is characteristic of the heart. It is rich in mitochondria that are in contact with fibrils that branch and interdigitate. The membranes at the ends that abut each other are extensively folded to form intercalated disks, whereas those along the sides fuse over considerable distances and form gap junctions that permit the muscle to function like a syncytium. Also called myocardium.

cardiomyopathy: Progressive disease characterized by damage to cardiac muscle tissue.

carnitine shuttle: Transport of fatty acyl CoA across the inner mitochondrial membrane by formation of a fatty acyl-carnitine that is exchanged for free carnitine in its passage through an acylcarnitine/carnitine transporter.

carnitine: 4-*N*-trimethyl-3-hydroxybutyrate, derived metabolically from ϵ-amino methylated lysine residues of certain proteins. It is present especially in muscle and liver, and participates in transfer of fatty acyl CoA across the inner mitochondrial membrane.

carotene: Yellow-orange unsaturated tetraterpene hydrocarbon present in carrots and green and orange leafy vegetables. It is the precursor of retinol (vitamin A), retinaldehyde, and retinoic acid in the body.

cartilage: Flexible type of connective tissue that lacks blood vessels and nerves, and consists of chondrocytes and a matrix composed of collagen fibrils and the proteoglycan aggrecan that is rich in chondroitin sulfate that they secrete. It is present mostly at the ends of long bones in the articulating surfaces of joints, and in the intervertebral disk. During development it is the precursor of the long bones.

cascade: Process that consists of sequential activation of an increasing number of components of a signaling system so as to amplify that signal.

caspase cascade: Amplifying proteolytic cascade in the cytosol in which a procaspase is activated by an initiator of apoptosis and then activates another procaspase and so on until apoptosis is produced.

caspase: Cytosolic proteolytic enzyme that contains a cysteine residue in its active site, cleaves its substrates on the carboxyl-terminal side of an aspartate residue, is activated from a procaspase, and is involved in apoptosis.

catabolism: Phase of intermediary metabolism in which nutrient molecules are degraded with release of energy.

catalytic antibody: See abzyme.

catalytic triad: Histidine residue hydrogen-bonded with a serine and aspartate residues in the active site of chymotrypsin and of other serine proteases.

catechol: Aromatic ortho-diol, the characteristic chemical group present in catecholamines.

catecholamine: Catechol derivative of tyrosine such as dopamine, noradrenaline, and adrenaline that are major elements in the body's response to stress.

catenin: Either of two proteins (α or β) that mediate linking of cadherin in adherens junctions to the actin microfilaments of cytoskeleton.

cathepsin: Proteolytic enzyme found in lysosomes.

CCAAT box: Consensus sequence CCAAT present in promoter elements in transcription promoters of many eukaryotic genes.

CDC gene: See cell-division-cycle gene.

cell adhesion molecule (CAM): Integral protein of the plasma membrane that mediates binding of cell to cell or to extracellular matrix in a Ca^{2+}-dependent (e.g., a cadherin) or Ca^{2+}-independent (e.g., neural CAM) fashion.

cell coat: See glycocalyx.

cell junction: Specialized region of a plasma membrane at which cell to cell or extracellular matrix adhesion occurs.

cell-division-cycle gene (cdc gene): Gene that controls one or more steps in the cell cycle.

cell-mediated immunity: Type of immunity mediated by T cells in lymphoid tissue and in the circulation that functions in defense against microorganisms (including viruses within cells) and foreign tissue transplants.

centriole: Intracellular organelle that consists of nine groups of short microtubules with three in each group, that organizes the microtubules of the spindle apparatus during cell division.

centromere: Region where daughter chromatids remain connected after a chromosome has replicated, and where fibers of the spindle apparatus become attached.

centrosome: Region of the cytoplasm that contains a pair of centrioles oriented at right angles to each other.

ceramide: Metabolic precursor of the sphingolipids consisting of a fatty acid in amide linkage with the C2 amino group of sphingosine.

cerebroside: Glycosphingolipid that contains a glucose or galactose moiety linked to C1 of ceramide.

cerebrospinal fluid (CSF): Fluid that bathes the internal and external surfaces of the central nervous system. It is secreted and reabsorbed via aquaporins in membranes of cells of the choroid plexus.

channeling: Direct transfer of the product of an enzyme reaction to the active site of another enzyme for which it is the substrate.

chaotropic agent: Ion or small organic molecule that increases the solubility of nonpolar substances in water. It is useful for dissolving membranes, solubilizing particular proteins, and denaturing proteins and nucleic acids.

chaperone: Protein that helps other proteins avoid misfolding and production of inactive or aggregated species. Also called heat shock protein, molecular chaperone.

charge repulsion: Mutual repulsion of chemical groups with like charge.

chemiosmosis: Mechanism by which certain membranes in cells use the energy of electron transfer to pump protons across the membrane and to harness the energy stored in the H^+ gradient to drive cellular work, including ATP synthesis.

chief cell: Cell in the gastric mucosa that secretes pepsinogen. Also called peptic cell.

chiral atom: Carbon atom with four different substituents that can exist in two configurations making it optically active. Also called asymmetric carbon atom, chiral center.

chloride shift: Exchange of plasma Cl^- for HCO_3^- produced by dissociation of carbonic acid within erythrocytes.

cholecystokinin (CCK): Polypeptide hormone that contains 39 residues, is secreted by the duodenal mucosa, and causes contraction and emptying of the gall bladder and secretion of enzymes by the pancreas. Also called pancreozymin.

cholelithiasis: Presence or formation of biliary calculi or gall stones.

choline: *N*-trimethylethanolamine, a vitamin component of acetylcholine, phosphatidylcholine, plasmalogens, and platelet-activating factor.

chondroitin sulfate: Predominant glycosaminoglycan of cartilage. It consists of the repeating unit glucuronic acid β1-3*N*-acetylgalactosamine, and forms part of the proteoglycan aggrecan.

chorionic gonadotropin (CG): Protein hormone homologous with growth hormone and prolactin. It is secreted by the placenta, has most of the actions of growth hormone, and maintains the corpus luteum during the first third of human pregnancy. Also called chorionic somatomammotropin, placental lactogen.

chorionic somatomammotropin: See chorionic gonadotropin.

chorionic villus sampling: Technique of obtaining a sample of the fetal portion of the placenta for determining the presence of genetic abnormalities in the fetus while it is still in the womb.

chromaffin cell: Cell found in clusters in the adrenal medulla and elsewhere, that contains granules for biosynthesis, storage and, secretion of the catecholamines.

chromatid: Complete copy of the double-stranded DNA of a chromosome with its associated nucleoproteins.

chromatin: Complex of DNA, histones, and other proteins in the nucleus of a eukaryotic cell.

chromosome condensation: Process by which, prior to the M phase of the cell cycle, chromatin becomes packed into chromosomes.

chromosome decondensation: Process by which chromosomes present in the M phase of the cell cycle become less compact and form chromatin when that phase is passed.

chromosome walking: Technique for identifying and sequencing long parts of a DNA strand such as a chromosome by breaking these into smaller fragments and cloning them. A cloned fragment that has already been identified is used as a probe to select clones that contain adjacent sequences which are then used to select clones adjacent to them. In this way a walk is made along a chromosome from the starting fragment.

chromosome: DNA strand that contains a portion of the genome along with associated proteins. In humans there are 23 pairs of chromosomes.

chronotropic: Denoting the rate and strength of rhythmic movements such as the heart beat.

chylomicron: Particle present in plasma during the absorption of dietary lipid. It consists of triacylglycerols, phospholipids, cholesterol esters, fat-soluble vitamins, and apolipoprotein B48. It is synthesized and secreted by cells of the intestinal mucosa and enters venous blood via the lymphatic duct.

chyme: Acidic, semifluid mixture of ingested food and digestive secretions that enters the small intestine from the stomach during the early phases of digestion.

cirrhosis: Disorder of the liver in which degeneration of the hepatocytes is accompanied by their replacement by fibrous connective tissue.

cis-acting regulatory element: Sequence of DNA that regulates expression of a gene on the same chromosome, e.g. a promoter or an enhancer.

cis–Golgi network: Interconnected cisternae and tubules of the Golgi complex that are closest to the endoplasmic reticulum and receive protein- and lipid-containing vesicles from that reticulum. It is the site of phosphorylation of oligosaccharides on proteins destined for lysosomes.

cisterna: Flattened membrane-bound compartment of the Golgi complex and endoplasmic reticulum.

clathrin: Protein composed of three large and three small polypeptides arranged into a three-legged structure (triskelion) that can assemble into a basket-like framework on the inner surface of a portion of the membrane to form a structure which can be released into the cytoplasm as a coated vesicle.

clathrin-coated vesicle: Type of vesicle that mediates transport from the plasma membrane or Golgi complex. Its outer surface is covered by a framework of clathrin triskelions bound to transmembrane proteins via adaptins.

cloning vector: Agent such as a plasmid or modified virus that is used to transfer DNA in genetic engineering.

cloning: Construction of a recombinant DNA molecule, its insertion into a vector and then into a cell, and the production of multiple copies by expression in the progeny of that cell.

coactivator: Protein that binds to a DNA-bound protein or protein complex thereby activating transcription of an adjacent gene. Also an ion or biomolecule that with one or more participates in activating an allosteric enzyme.

coated vesicle: Small membrane-enclosed particle with a meshwork of clathrin or other proteins on its outer surface. It may be formed at the plasma membrane, endoplasmic reticulum, or components of the Golgi complex.

coding strand: Strand of genetic DNA that has the same sequence as RNA transcribed from that region; it is complementary to the template strand from which RNA is transcribed.

codon: Triplet of purine and/or pyrimidine bases in an mRNA molecule that specifies the insertion of a particular amino acid residue during polypeptide synthesis.

coenzyme A (CoA): Nucleotide coenzyme that contains pantetheine as a constituent. Its terminal sulfhydryl group functions as an acyl carrier in many reactions.

coenzyme B12: 5′-deoxyadenosylcobalamine, derived by binding of a 5′-deoxyadenosyl group to the cobalt atoms of cobalamine (vitamin B_{12}), and the coenzyme in conversion of methylmalonyl CoA to succinyl CoA; also methylcobalamine, derived by binding of a methyl group to the cobalt of cobalamine, the coenzyme in conversion of homocysteine to methionine.

coenzyme Q (Q): Ubiquinone, the benzoquinone isoprenoid that functions as a lipophilic electron carrier from complexes I and II to complex III and cytochrome c in the mitochondrial electron transport system.

coenzyme: Complex nonprotein organic cofactor required for activity of some enzymes. Most vitamins are precursors of coenzymes.

cofactor: Ion or molecule that on binding to the catalytic site of an apoenzyme renders it active.

cognate: Two biomolecules that normally interact (e.g., a receptor and its proper ligand).

coiled-coil: Structure present in some fibrous proteins (e.g., keratins, tropomyosin, myosin) formed by two polypeptides that contain α-helical regions that consist of repeats of seven residues (abcdefg) in which "a" and "d" are hydrophobic or in which "g" is a leucine residue (leucine zipper proteins).

compartmentation: Subdivision of the enzymes of a eukaryotic cell into functionally distinct, membrane-enclosed spaces or organelles.

complement: System of plasma proteins secreted by the liver that interact on exposure to activated antibodies or to the surface of certain pathogens, to promote cell lysis, phagocytosis and an inflammatory response.

complementary DNA (cDNA): DNA produced by retroviral reverse transcriptase from an RNA template.

complex oligosaccharide: Oligosaccharide of an N-linked glycoprotein that, by trimming and further modification in the Golgi complex, retains three of the mannose residues introduced in the endoplasmic reticulum. One or more trisaccharide units terminating in a sialic acid residue are added to the end of the oligosaccharide.

complex protein: See conjugated protein.

condensation reaction: Covalent linking of two molecules through loss (usually) of water.

configuration: Arrangement in space of atoms in an organic molecule that results from the presence of one or more chiral centers or of one or more double bonds. It can only be changed by breaking or reforming of covalent bonds.

conformation: Arrangement in space of an organic molecule whose substituent groups can assume different positions around single bonds.

conjugated protein: Protein that contains one or more tightly associated non-amino acid moieties. Examples include metalloprotein, phosphoprotein, hemoprotein, glycoprotein, lipoprotein, nucleoprotein, and flavoprotein. Also called complex protein.

conjugation: Addition, usually in the liver, of a moiety of glucuronic acid, acetate, sulfate, glycine, glutamine, or glutathione, to increase the polarity and excretability of nonpolar metabolites or foreign compounds introduced into the body. Also called Phase II reaction.

connectin: See titin.

connective tissue: Supporting tissue found between other tissues. It contains fibroblasts, chondrocytes, osteoblasts, osteoclasts, adipocytes, or smooth muscle cells and the extracellular material that they produce.

connexin: Integral protein that spans the plasma membrane four times. A cluster of six connexins constitute a connexon.

connexon: Channel in the plasma membrane that consists of six connexin units. When affixed to a connexon on an adjacent cell, a gap junction is formed that permits transit of material between both cells.

consensus sequence: Sequence of nucleotides or of amino acid residues that occurs most frequently within a related region of two or more nucleic acids or polypeptides respectively.

conservative substitution: Replacement in a polypeptide of a residue by another of similar characteristics (e.g., serine by threonine).

constitutive protein: Protein present in relatively constant amounts in the cells that produce it.

cooperative binding: Change in affinity (increase or positive cooperativity, decrease or negative cooperativity) for binding of a ligand at one site by prior binding of a molecule of the same (homotropic effect) or of a different (heterotropic effect) ligand at another site of a multimeric protein.

cooperativity: Interaction by which a conformational change induced in one subunit is transmitted to all others in a multimeric protein.

corepressor: Protein that binds to a DNA-bound protein or protein complex thereby blocking transcription of a gene.

Cori cycle: See glucose–lactate cycle.

corpus luteum: Mass of yellow follicle cells that develops in the ovary after ovulation. It secretes progesterone.

corticoliberin: See corticotropin-releasing hormone.

corticosteroid: Steroid hormone synthesized from cholesterol and secreted by the adrenal cortex. It may be a glucocorticoid, a mineralocorticoid or a sex hormone.

corticotropin: See adrenocorticotropic hormone.

corticotropin-releasing hormone (CRH): Polypeptide hormone that contains 41 residues that is secreted from the hypothalamus and causes release of adrenocorticotropic hormone (ACTH) by the anterior pituitary. Also called corticoliberin.

cortin: See adrenocorticotropic hormone.

cotransport: See symport.

coupled reaction: Linkage of an exergonic with an endergonic reaction that transfers energy from one to the other.

covalent catalysis: Catalysis that occurs when part or all of a substrate molecule forms a covalent bond with a component of the enzyme's active site before transfer to a second substrate.

cristae: Infoldings of the inner mitochondrial membrane.

cross-talk: Interaction of pathways for hormonal signal transduction, for example, through effects on a common component, common second messenger, or similar patterns of phosphorylation of target proteins.

C-terminal residue: Residue in a polypeptide that contains an α-carboxyl group that is free or occasionally aminated.

C-terminus: End containing the C-terminal residue in a polypeptide.

cyanosis: Bluish discoloration of skin that results from presence of deoxygenated blood in capillaries near the body surface.

cyclin: Protein whose concentration fluctuates during the cell cycle and that activates a cyclin-dependent protein kinase thereby regulating progression through the cycle.

cyclin-dependent kinase (cdk): Protein kinase whose activity is modulated by the concentration of a cyclin.

cysteamine: Decarboxylated cysteine, a component (with pantothenic acid) of pantetheine in coenzyme A.

cytochrome: Heme protein involved in an electron transport system in the endoplasmic reticulum or inner mitochondrial membrane.

cytokine receptor: Plasma membrane receptor for a cytokine that is stably associated with a cytosolic tyrosine protein kinase and functions by modulation of specific gene activity.

cytokine: Small protein secreted by cells of the immune system that mediates local cell–cell communication.

cytoskeleton: Network of microtubules, microfilaments and intermediate filaments present in the cytoplasm.

D arm: In tRNA the stem and loop portion that contains two or three dihydrouracil bases.

dalton (Da): Unit of atomic mass, one twelfth the mass of a ^{12}C atom.

***de novo* pathway:** Pathway for synthesis of a biomolecule (e.g., a nucleotide) from preformed precursors.

deletion: Loss of a fragment of a chromosome through breakage or unequal crossing over, or through mutational loss of one or more contiguous nucleotides from a gene.

denaturation: Loss of higher-order structure in a protein or nucleic acid as a result of change in pH, salt concentration or temperature.

depolarization: Change in transmembrane electrical potential from a negative value toward a more positive value.

detoxification: Removal of a harmful substance. Often refers to hydroxylation reactions catalyzed by cytochrome P450 enzymes of the hepatic smooth endoplasmic reticulum that render lipophilic drugs or other substances more water soluble for excretion.

diabetes insipidus: Condition of excessive thirst and water intake and urinary excretion of a dilute urine, produced by decreased secretion of, or resistance to, antidiuretic hormone (vasopressin).

differential RNA splicing: See alternative RNA splicing.

differentiation: During development, the gradual appearance of specific cellular characteristics caused by differential gene activation and repression.

dinucleotide fold: See nucleotide-binding fold.

diploid: Having two sets of chromosomes, one set coming from each parent, as in somatic cells.

dipolar ion: See zwitterion.

disulfide bond: Covalent bond formed between sulfhydryl groups of two cysteine residues within the same polypeptide (intrachain) or between two polypeptides (interchain).

DNA fingerprinting: Technique of breaking a piece of DNA containing a variable number of repeat sequences with a restriction enzyme, and the resulting fragments being separated by electrophoresis on a suitable support. The pattern of bands is characteristic of the DNA donor. (Also called genetic profiling).

DNA glycosylase: Enzyme that hydrolyses the N-glycosidic bond between a base and the DNA backbone as the first step in base-excision repair.

DNA helicase: Enzyme that utilizes the chemical energy of ATP to catalyse strand separation in DNA before replication.

DNA library: Collection of bacterial or phage clones that contain cDNA that represents the mRNA of an organism, specific cell type, or tissue.

DNA ligase: Enzyme that seals a nick (creates a phosphodiester bond between adjacent sugars) in the backbone of DNA.

DNA melting: Unwinding and separation of the two strands of DNA on heating, the melting temperature being determined by the base composition of the DNA.

DNA methylation: Enzymatic transfer of the methyl group of S-adenosylmethionine to adenine or cytosine in DNA that keeps genes in an inactive state.

DNA microarray: Compact arrangement of numerous short DNA sequences, derived by chemical synthesis or from PCR fragments immobilized on a solid that can be easily probed by hybridization with other nucleic acids that have been fluorescently labeled.

DNA probe: Natural or synthetic labelled oligonucleotide that is used to detect a gene of interest by hybridization to a complementary sequence.

dolichol: Long-chain polymer of isoprene present in endoplasmic membranes that, in the diphosphate form, is used to carry activated sugars in the biosynthesis of glycoproteins and some polysaccharides.

domain: Segment of a polypeptide or polynucleotide chain that may fold and function independently from the rest of the molecule.

dominant: Gene (or allele) that affects the phenotype when it is received from only one parent (i.e., in the heterozygous form). It also describes a character or trait due to such a gene.

down-regulation: Decrease in transcription of a DNA genetic sequence in response to an external variable.

dynamin: Cytosolic protein with GTPase activity that binds to the neck of a clathrin-coated pit and participates in separating it from the membrane as an intracellular vesicle.

dynein: Motor protein unrelated to the kinesins, composed of two or three heavy chains that contain a motor domain and a variable number of light chains. Cytoplasmic dyneins contain two heavy chains, are probably present in all eukaryotic cells, and are important in vesicle trafficking. Axonemal dyneins contain two or three heavy chains and are responsible for the sliding movement that drives the beating of cilia and flagella.

dynorphin: Polypeptide that has the same N-terminal pentapeptide as Leu-enkephalin. It is a powerful analgesic and is present in hypothalamus and brain stem and in the duodenum.

dystrophin: Cytoskeletal protein of cardiac and skeletal muscle that consists of a fibrous antiparallel homodimer of a large polypeptide whose central region consists of 24 repeats of a sequence characteristic of β-spectrin. It links the plasmalemma to actin filaments. It is absent in Duchenne muscular dystrophy and reduced or altered in structure in Becker muscular dystrophy.

E site: See exit site.

effector enzyme: Membrane-associated enzyme that responds to an extracellular signal transmitted via a membrane receptor by producing an intracellular messenger.

electrogenic transport: Transport across a plasma membrane of ionic solutes by a process that produces a change in the membrane potential.

emulsification: Physical breaking up of ingested lipids in the intestinal tract, by mixing with bile salts and other emulsifying agents, to form smaller particles that are accessible to lipolytic digestive enzymes.

endergonic: Requiring energy input to occur; a nonspontaneous process.

endocytosis: Entry of extracellular material by invagination of the cell plasma membrane to form vesicles called endosomes

endoplasmic reticulum (ER): Extensive network of membrane channels in the cell cytoplasm that is studded with (rough ER) or lacks (smooth ER) ribosomes and functions in protein synthesis and lipid metabolism and in intracellular transport, storage, packaging, or secretion.

endorphin: Neuromodulator and analgesic peptide that includes in its structure the pentapeptide of Leu-enkephalin or Met-enkephalin, is produced mostly in the central nervous system and gastrointestinal tract, and binds to an opiate receptor. Also called opioid peptide.

endosome: Small intracellular vesicle formed by endocytosis that transfers its contents to a lysosome for degradation.

enterohepatic circulation: Secretion of bile salts by the liver followed by their absorption by the intestinal mucosa for return to the liver via the portal vein.

enthalpy (*H*): In thermodynamics the heat content of a system.

entropy (*S*): In thermodynamics the degree of disorder or randomness in a system.

enzyme cascade: Series of proenzymes or weakly active enzymes, in which activation of the first causes that of the second and so on. Activation may be by proteolytic cleavage (e.g., in blood clotting) or by phosphorylation (e.g., in the insulin receptor pathway) and results in amplification of the original activation signal.

enzyme-linked receptor: Integral protein receptor of the plasma membrane that has an intracellular catalytic domain (e.g., a protein kinase or protein phosphatase, guanylate cyclase) or becomes associated with a cytosolic enzyme, and is activated on ligand binding.

enzymopathy: Disorder produced by lack of an enzyme activity.

epigenetic: Describing a change in a characteristic of a cell or organism that is not the result of a change in the nucleotide sequence of the DNA.

epinephrine: See adrenaline.

epitope: Region on the surface of an antigen that elicits secretion of a specific antibody and also binds to it. Also called antigenic determinant.

ER retention signal: Sequence Lys-Asp-Glu-Leu (KDEL) at the C-terminal end of proteins that are retained and function within the endoplasmic reticulum.

ER signal sequence: N-terminal hydrophobic sequence that directs polypeptides destined for lysosomes, the endoplasmic reticulum, Golgi complex, secretory vesicles, or plasma membrane to enter the endoplasmic reticulum where it is cleaved off by a signal peptidase soon after its entry.

ER-resident protein: Protein that functions within the endoplasmic reticulum being retained there via the sequence Lys-Asp-Glu-Leu (KDEL) at the C-terminal end.

erythropoietin: Glycoprotein hormone secreted mostly by kidneys on exposure to low oxygen concentration that stimulates erythropoiesis.

essential fatty acid: Polyunsaturated fatty acid that is required but cannot be synthesized by an animal.

euchromatin: Decondensed form of chromatin that is available for transcription.

eukaryote: Organism that consists of one or more cells that each contain a nucleus and other membrane-bounded organelles.

excinuclease: Complex enzyme system that excises the abnormal segment of the affected DNA strand during nucleotide excision repair. Also called excision-repair endonuclease.

Excision-repair endonuclease: See excinuclease.

exergonic: Accompanied by release of energy; a spontaneous process.

exit site: Site, mainly on the large subunit of a ribosome, from which the uncharged tRNA is released after an elongation step in protein synthesis. Also called E site.

exocytosis: Release or secretion of material contained within an intracellular vesicle by fusion of its membrane with the plasma membrane.

exon: Segment of coding DNA of a eukaryotic gene that is represented in mRNA and the amino acid sequence of its translated protein, or in mature tRNA or rRNA. Also called expressed sequence.

expressed sequence: See exon.

expression vector: Plasmid or virus into which a foreign DNA gene sequence and its regulatory elements have been introduced and which can produce that gene product in a host cell.

extracellular matrix (ECM): Secreted intercellular component of connective tissue, consisting mostly of glycoproteins and proteoglycans.

extreme obesity: See morbid obesity.

extrinsic protein: See peripheral protein.

facilitated diffusion: Rapid permeation of solutes, based on their concentration gradient across a cell membrane, by interaction with specific membrane transport proteins. Also called passive transport.

familial hypercholesterolemia: Presence of abnormally high concentrations of cholesterol in LDL in plasma, the result of a lack of functional LDL receptors, that is transmitted as an autosomal dominant trait.

feedback inhibition: Slowing down or stoppage, in a metabolic pathway, of an early reaction by the product of a later reaction

feedforward activation: Activation, in a metabolic pathway, of a later reaction by a product of an earlier reaction.

FeS center: An iron–sulfur cluster.

fibril-associated collagen: Collagen, such as type IX and type XII, that forms short triple-helical and nonhelical domains and binds to the side of collagen type I or type II fibrils, respectively.

fibrillar collagen: Collagen, such as type I, II, III, V, or XI, that forms fibrils and occurs in connective tissue.

fibrillin: Large glycoprotein component of the microfibrils present in the extracellular matrix of many tissues, that is secreted by fibroblasts and forms a scaffold for deposition of elastin. Hereditary abnormalities in structure of the protein are associated with Marfan syndrome.

fibroblast growth factor (FGF): Family of proteins classed as acidic (from brain or retina) or basic (from brain, retina, or cartilage). They are mitogenic and angiogenic, act as growth factors in tissue culture, and exert their effects through receptors that are tyrosine kinases. Mutation in one form of FGF receptor is responsible for achondroplasia (the commonest form of dwarfism).

fibroblast: Type of cell common in loose connective tissue, that secretes collagen and other macromolecules of the extracellular matrix and readily migrates to and proliferates in wounded tissue.

first law of thermodynamics: Law that requires that in all processes the total energy of the universe remains constant.

flagellum: Extension of the cell membrane of a spermatozoon that forms an organ of locomotion driven by dyneins. It consists of two single microtubules surrounded by a ring of nine doublet microtubules.

flippase: Protein that facilitates translocation of a membrane lipid from one membrane monolayer to the other.

flux: Flow of material through a metabolic pathway or of a solute across a membrane.

frameshift: Change in the coding frame of codons during protein synthesis as a result of insertion or deletion of one or more nucleotides into the DNA sequence of an exon.

free energy of activation: Energy required to raise a substrate from the ground state to the transition state at which it will react. Also called activation energy.

free ribosome: Ribosome in the cytoplasm that is unattached to any membrane and is used for synthesis of proteins destined for the cytosol, mitochondria, peroxisomes, or nucleus.

F-type ATPase: H^+-transporting F_0F_1–ATP synthase of the inner mitochondrial membrane that catalyses synthesis of ATP during transport of electrons from metabolic substrates to oxygen. Also called ATP synthase or complex V.

futile cycle: Two opposing sets of enzyme-catalyzed reactions that result in release of energy as heat by the net hydrolysis of ATP.

G protein: GTP-binding protein with inherent GTPase activity. It is active with bound GTP but not with GDP, may be monomeric (e.g., ras) or heterotrimeric (containing α-, β-, and γ-subunits), and plays an important role in intracellular signaling pathways.

G protein-linked receptor: Serpentine receptor of the plasma membrane that on binding of a ligand activates a trimeric G protein.

G_0 phase: Nondividing state of a cell that has exited the cell cycle at the restriction point during the G_1 phase.

G_1 phase: Phase in the cell cycle that consists of the growth portion in interphase before DNA replication begins.

gated ion channel: Ion channel that opens and closes to allow alteration of the cell membrane potential. It may be voltage-gated, mechanically gated, or ligand-gated (extracellularly or intracellularly).

gene amplification: Selective synthesis of extra copies of a gene or genes. This is normal for some organisms and occurs for some oncogenes in certain cancers.

gene cloning: Formation by a bacterium that carries a foreign gene, introduced into it via a recombinant vector, of a clone of cells that contain the replicated gene.

gene knockout: Deletion or inactivation of a gene in an experimental animal by genetic-engineering methods.

gene: Segment of chromosomal DNA that encodes a functional polypeptide or RNA molecule, including exons, introns, and regulatory sequences.

genomic DNA library: Collection of bacterial or phage clones that contain overlapping parts of the genome of an organism.

genomics: Study of DNA sequences and properties of genomes.

genotype: Genetic constitution or complement of an individual.

germ-line cell: Cell of the lineage that forms the gametes (sperm or ova).

gestational diabetes: Condition in which previously normal woman exhibits an abnormally high level of blood sugar during a pregnancy

glial cell: See neuroglia cell.

glicentin: Polypeptide derived from preproglucagon in special cells of the lower intestinal tract. It consists of glucagon extended by additional residues at either end and by glicentin-related polypeptide at the N-terminal end. Its activity is similar to that of glucagon.

globin: Apoprotein of myoglobin or of hemoglobin.

globular protein: Protein whose tertiary structure makes it compact and spheroidal.

gluconeogenesis: Metabolic pathway for synthesis of glucose from noncarbohydrate precursors in the liver.

glucose tolerance: Ability of the body to utilize glucose as ascertained by the nature of the blood glucose curve following administration of a test amount of glucose. It is decreased most commonly in diabetes mellitus and in conditions associated with liver damage.

glucose–alanine cycle: Interorgan metabolic conversion of alanine released from skeletal muscles during prolonged fasting or starvation, into glucose that is produced and secreted into the blood stream by the liver. Also called alanine cycle or Randle cycle.

glucose–lactate cycle: Interorgan conversion of lactate released from skeletal muscle after exertion, into glucose produced and secreted into the blood stream by the liver. Also called lactate cycle or Cori cycle.

glyceroneogenesis: An abbreviated version of gluconeogenesis in which the glycerol 3-phosphate that is required for triglyceride synthesis is generated from noncarbohydrate precursors such as lactate and amino acids.

glycocalyx: Fuzzy coat on the external surface of cells, consisting of the carbohydrate component of glycoproteins and glycolipids of the cell membrane and of secreted glycoproteins and proteoglycans that adhere to that surface. Also called cell coat.

glycoform: Glycoprotein that differs only in the location and/or structure of the carbohydrate component from another.

glycolipid: Lipid that contains a monosaccharide (e.g., in glucosyl cerebroside) or an oligosaccharide (e.g., in a ganglioside) component.

glycolysis: Metabolic pathway by which a molecule of glucose is catalyzed to two of pyruvate with net production of two ATP and two of NADH.

glycoprotein: Conjugated protein that contains covalently bound carbohydrate.

glycosuria: Presence of abnormally high concentration of glucose (most commonly) or of other sugars (e.g., galactose, fructose, or pentose) in urine.

Golgi complex: Series of stacked membranous cisternae (cis, medial, and trans) in the cytoplasm of eukaryotic cells in which proteins transferred from the endoplasmic reticulum are modified and sorted, and also the site of synthesis of glycosaminoglycans of the extracellular matrix.

Golgi stack: Pile, usually of three cisternae, between the cis- and the trans-Golgi networks, the site of processing of N-linked oligosaccharides and of *O*-glycosylation, and also production of proteoglycans.

gonadotrope: Cell in the anterior pituitary gland that produces the gonadotropins, that is, the glycoproteins luteinizing hormone and follicle-stimulating hormone.

granulocyte: Leukocyte that contains granules visible under light microscopy and may be basophilic, eosinophilic, or neutrophilic on appropriate staining.

growth factor: Protein that promotes synthesis of proteins and other cell constituents and produces an increased cell mass but not cell number. It is frequently used to include mitogens and survival factors.

guide RNA: Small RNA that is involved in editing of pre-mRNA and of pre-rRNA (when it is call snoRNA).

H chain: See heavy chain.

half-life ($t^{1/2}$): Time required for decay or disappearance of one-half of an amount of a given component in a system. Also called half-time.

half-time: See half-life.

haploid: Having one set of chromosomes as in germ cells.

hb S: Hemoglobin that contains the mutant β globin that causes erythrocyte sickling.

heat shock protein: See chaperone protein.

heavy chain: Larger of the two types of polypeptide in a protein. Also called H chain.

helix cap: Structure formed when the side chain of residues such as asparagine or glutamine that flank an α-helix fold back to form hydrogen bonds with peptide bond components of the first or of the last turn of the helix.

helper T cell: Type of lymphocyte that by its secretion and other activities helps trigger and coordinate cell-mediated and antibody-mediated immunity.

hemoglobinopathy: Hereditary condition that manifests itself as a structural variant of a hemoglobin (e.g. sickle cell anemia), a thalassemia, or persistence of fetal hemoglobin.

heptad repeat: Seven-residue sequence (abcdefg) that is repeated in tandem in a polypeptide, in which "d" and "g" are hydrophobic or in which "g" is leucine (as in a leucine zipper). Such heptads allow the polypeptides to dimerize as coiled-coils.

heterochromatin: Condensed form of chromatin that is not transcribed and forms a dense region, visible with a light microscope, in resting cells.

heterogeneous nuclear ribonucleoprotein (hnRNP): Complex of hnRNA present in the nucleus of eukaryotic cells.

heterogeneous nuclear RNA (hnRNA): RNA in the nucleus that is usually the primary transcript of genes for polypeptides and contains one or more intron sequences. Also called pre-mRNA.

heteropolysaccharide: Polysaccharide that consists of more than one type of monosaccharide (e.g., a glycosaminoglycan).

heterozygous: Having two different alleles at corresponding sites on a chromosome pair.

high-mannose oligosaccharide: Oligosaccharide of an N-linked glycoprotein that retains all or most of the nine mannose residues transferred onto it in the endoplasmic reticulum and undergoes little or no further modification in the Golgi complex.

histone core: Protein core of a nucleosome, consisting of two copies of each of histones H2A, H2B, H3, and H4.

HLA protein: Human leucocyte antigen or major histocompatibility complex protein.

homeobox: Conserved DNA sequence (~100 bp long) that encodes a protein domain (homeodomain) that regulates differentiation during development of the organism.

homeodomain: Protein domain (~60 residues long) that forms a helix–loop–helix structural motif and is encoded by a homeobox.

homeostasis: Maintenance of a relatively constant internal environment within the body.

homeotic gene: Gene involved in controlling the overall body plan by controlling the fate of groups of cells during development of the organism.

homologous chromosome: One of a pair of chromosomes in somatic cells that are similar in length, position of centromere, and location of alleles for the same traits and are derived from the father and the mother respectively. In germ cells, only one member of a pair is carried by each gamete.

homologous: Macromolecules that share similarity because of a common evolutionary origin, and more specifically in the primary structure of polypeptides or nucleic acids.

homopolysaccharide: Polysaccharide that consists of one type of monosaccharide (e.g., starch or glycogen).

homozygous: Having the same alleles at the corresponding sites on a chromosome pair.

hormone response element: DNA sequence to which a nuclear receptor for a hormone of the steroid family binds to and alters transcription of an adjacent DNA genetic sequence that is regulated by that hormone.

humoral immunity: See antibody-based immunity.

hyaluronidase: Enzyme that hydrolyses the glycosidic bonds of hyaluronic acid, is the spreading factor secreted by some bacteria, and is present at the acrosomal cap of spermatozoa.

hybridoma: Cell line obtained by fusing antibody-secreting lymphocytes with lymphoma cells, and used for production of monoclonal antibodies.

hydrogen bond: Weak electrostatic bond in which a hydrogen atom in covalent linkage with an electronegative atom is partially shared with another electronegative atom.

hydrolase: Enzyme that catalyses a hydrolytic cleavage reaction.

hydrophilic: Capable of associating freely with water molecules.

hyperammonemia: Presence of abnormally high concentrations of ammonia in blood usually associated with neurological and other abnormalities. It results from underutilization of ammonia in urea synthesis as a result of liver disease, of inborn errors of urea synthesis, or of organic acidemias.

hyperchromic shift: Increase in absorption of ultraviolet light by a DNA solution as the DNA melts on raising the temperature.

hyperglycemia: Elevated plasma glucose concentration relative to that during fasting. It occurs physiologically within 2–3 h after a meal, but when higher than normal, it is usually symptomatic of diabetes mellitus.

hyperplasia: Abnormal enlargement of an organ or tissue from increase in number of cells.

hypertrophy: Abnormal enlargement of an organ or tissue from increase in the size of its cells.

hypervariable region: Any of the three short loop segments in the variable region of the light or heavy chains of immunoglobulins that are poorly conserved and form part of the antigen-binding site. Also, a region in a polypeptide that shows great variability in sequence in different species.

hypervariable residue: Residue in a hypervariable region.

hypoglycin: Toxic non-proteogenic α-amino acid present in the unripe ackee fruit. Its deaminated and decarboxylated derivative is a potent inhibitor of β-oxidation of short-chain fatty acids and produces Jamaican vomiting sickness.

hypopituitarism: Condition of partial or total loss of function of the pituitary gland that leads to decreased secretion of one or more of the hormones produced there.

IF protein: See intermediate filament protein.

Ig domain: See immunoglobulin domain.

Ig superfamily: Family of proteins that contain one or more domains like those which characterize the immunoglobulins. It includes Ca^{2+}-dependent cell adhesion molecules, T and B cell receptors, MHC proteins, and transmembrane IgM.

immunoglobulin domain: Domain (~100 residues) that forms a sandwich consisting of a three- and a four- stranded antiparallel β sheet and is characteristic of the variable and constant regions of L and H chains of immunoglobulins. Also called immunoglobulin fold or Ig-domain.

immunoglobulin fold: See immunoglobulin domain.

induced fit: Change in conformation induced in an enzyme by binding to its substrate, also the change in conformation induced in any macromolecule by ligand binding.

inducer: Compound that promotes the activity of a specific gene; also a stimulus that provokes a physiological response.

inducible protein: Protein that is produced in amounts that depend on conditions in the environment.

inflammation: Nonspecific local defense mechanism that is characterized by swelling, warmth, redness and, pain in a part of the body.

inhibin: Glycoprotein hormone secreted by the gonads that inhibits secretion of follicle-stimulating hormone. It consists of a glycosylated α chain linked to a β_A chain (inhibin A) or to a β_B chain (inhibin B) by a disulfide bond.

inhibitory G protein (Gi): Heterotrimeric G protein that inhibits adenylate cyclase or regulates ion channels.

inhibitory neurotransmitter: Neurotransmitter (e.g., GABA, glycine) that opens transmitter-gated channels for Cl^- or K^+ in the target cell plasma membrane and makes formation of an action potential more difficult.

initiation codon: Nucleotide triplet AUG that codes for methionine as the first amino acid residue during polypeptide synthesis in eukaryotes and for *N*-formylmethionine in prokaryotes, mitochondria and chloroplasts.

innate immunity: Antigen-nonspecific mechanisms involved in the early phase of the response to a pathogen. It is not increased by repeated presence of the pathogen and includes phagocytic cells, the complement proteins and secretion of cytokines.

insertion mutation: Mutation that results from addition of one or more extra nucleotides into coding DNA.

integral protein: Protein that traverses a cell membrane one or more times and can only be released by disruption of the membrane. Also called transmembrane protein and intrinsic protein.

integrin: Heterodimeric intrinsic protein of plasma membrane that links a cell to proteins of the extracellular matrix (e.g., collagen and laminin) or to other cells including white blood cells and platelets.

intercalating agent: Dye or other compound that contains an aromatic ring or other planar structure that can fit between two successive bases of DNA and may cause insertion or deletion mutations.

interchain: Interaction or linkage between parts of two strands of a biopolymer.

interconvertible enzyme: Enzyme whose activity is modulated by a reversible posttranslational modification (e.g.. phosphorylation).

interferon (IFN): Protein released by cells infected by a virus or in response to other inducing agents. IFNα is secreted by leukocytes, IFNβ by fibroblasts, and IFNγ by macrophages and B cells. IFNs have nonspecific antiviral and anticancer activity.

intermediate filament (IF): Rope-like fiber (10 nm diameter) of the cytoskeleton that consists of one of a variety of IF proteins depending on cell type. The IFs of the nuclear lamina are composed of lamin proteins.

intermediate filament protein: Protein component of intermediate filaments of the cytoskeleton or nuclear lamina.

interphase: Major portion of the cell cycle in which the chromosomes are uncoiled and form chromatin between one M phase and the next.

intervening sequence: See intron.

intrachain: Interaction or linkage between parts within a strand of a biopolymer.

intrinsic factor: Glycoprotein secreted by parietal cells of the gastric mucosa that binds and facilitates absorption of cobalamin (vitamin B_{12}). Its deficiency, as from gastric atrophy, results in pernicious anemia.

intrinsic protein: See integral protein.

intron: Segment of noncoding DNA in a eukaryotic gene that is transcribed but is then excised from the primary transcript. Also called intervening sequence.

invariant residue: Amino acid residue that occurs in the same position in a homologous polypeptide derived from a different species.

ion channel: Integral membrane protein that provides regulated passage for a specific ion or ions across a membrane.

ionophore: Compound (often an antibiotic) that increases ion flux across a plasma membrane by forming an ion channel or functioning as a mobile ion carrier in the membrane.

isoaccepting tRNA: tRNA species that accepts the same amino acid as another tRNA species but has a different anticodon.

isoenzyme: Protein that can catalyze the same reaction as one or more different proteins from the same species. Also called isozyme.

isomerase: Enzyme that catalyses transfer of a group within a molecule to yield an isomeric form.

isopeptide bond: Amide bond formed between an amino group of an amino acid and the carboxyl group of another where either or both occupy a position other than α (C-2).

isoprenoid: Lipid (e.g.. a steroid or lipid-soluble vitamin) that is structurally derived from isoprene units.

isoschizomer: Restriction endonuclease that cleaves at the same nucleotide sequence as another such enzyme.

isozyme: See isoenzyme.

jaundice: Yellowing of skin and conjunctiva due to elevated plasma and tissue bilirubin concentration caused by overproduction or under excretion of this bile pigment.

junk DNA: Regions of DNA in a genome for which no function has yet been discovered.

KDEL: C-terminal tetrapeptide (Lys-Asp-Glu-Leu) signal for retention of proteins resident in the endoplasmic reticulum.

keratin: Fibrous protein that is a coiled-coil consisting of a type I (acidic) and a type II (neutral or basic) polypeptide cross-linked by disulfide bonds. They form a family of intermediate filament proteins synthesized by keratinocytes in epithelial tissues. Over 20 types occur in human epithelia, and at least 10 are specialized for hair and nails.

ketimine: Product of the condensation of a primary amine with a ketone. Also called Schiff base.

Keto acid: Organic acid such as pyruvic acid or oxaloacetic acid that contains a keto group adjoining the carboxylic group (an α-keto acid) or elsewhere.

ketogenesis: Production of acetoacetate and its derivatives, β-hydroxybutyrate and acetone, within the mitochondrial matrix in liver from the acetyl CoA, derived mainly from the increased catabolism of fatty acids that occurs during prolonged fasting.

ketone bodies: Acetoacetate and its derivatives, β-hydroxybutyrate and acetone, which are made by ketogenesis in the liver and become an important fuel for non-hepatic tissues during prolonged fasting.

killer T cell: Type of lymphocyte involved in cell-mediated immunity that secretes pore-forming proteins (perforans) that kill the target cells.

kilobase pair (kb): Unit of length of DNA, 1000 base pairs.

kinesin: Cytosolic motor protein that associates with an organelle and uses the energy of ATP to propel it along a microtubule.

kinetochore: Complex protein structure that links each sister chromatid to the spindle microtubules during metaphase.

knockin: Functional DNA genetic sequence introduced into germ cells of an organism by genetic-engineering technology.

knockout: Deleted or inactivated DNA genetic sequence produced in germ cells of an organism by genetic-engineering technology.

L chain: See light chain.

lactase deficiency: See lactose intolerance.

lactate cycle: See glucose–lactate cycle.

lacteal: Lymph vessel that extends into the core of a villus in the small intestinal mucosa into which chylomicrons are passed during absorption of ingested lipid.

lactic acidosis: Condition in which blood pH is lowered by an excess of lactic acid. It results from severe exercise, shock, hypoxia, cardiovascular insufficiency, intoxication by certain drugs, or as part of several inborn errors of metabolism.

lactose intolerance: Condition characterized by acid diarrhea and flatulence produced by the inability to digest lactose and resulting from intestinal lactase deficiency. Also called lactase deficiency.

lagging strand: DNA strand that is synthesized discontinuously and away from the replication fork during replication.

lamin: Protein component of intermediate filaments of the nuclear lamina. It may be isoprenylated for attachment to the inner surface of the nuclear membrane.

laminin: Flexible major component of basal lamina that consists of three long polypeptides (α, β, and γ) held together by disulfide bonds in the shape of a cross. It binds to the surface of cells and to other constituents of the basal lamina. Mutations in some forms of laminin are associated with epidermolysis bullosa.

leader sequence: Sequence of nucleotides, which may be coding or noncoding, that is present upstream of the initiation codon in mRNA and may have regulatory or targeting function. Also signal sequence of proteins synthesized on bound ribosomes.

leading strand: DNA strand that is synthesized continuously in the direction of the replicating fork during replication.

lectin: Highly specific carbohydrate-binding protein of plasma membranes, lumen of trans-Golgi network, or blood plasma.

leucine zipper protein: Protein that dimerizes by coiled-coil formation via interaction between leucine zippers that may be immediately preceded by a region rich in basic residues that form an α helix or a helix–loop–helix fold.

leucine zipper: Structural motif present in many DNA-binding proteins that dimerize via coiled-coil formation by interaction between the side chains of leucine residues in tandem heptad repeats within α-helical regions.

leukotriene: Eicosanoid produced by leukocytes, macrophages, and platelets from arachidonic acid, that contains four double bonds of which three are conjugated, and may be linked to glutathione, cysteinylglycine, or cysteine through their sulfur atom. Leukotrienes activate and attract leukocytes, increase capillary permeability, and are involved in inflammatory or immediate hypersensitivity reactions.

ligand: Small molecule that binds specifically and noncovalently to a larger one.

ligase: Enzyme that catalyzes bond formation between two substrates which is coupled with hydrolysis of ATP or another energy-rich compound. Also called synthetase.

light chain: Smaller of two types of polypeptide in a protein. Also called L chain.

lipid-anchored protein: See lipid-linked protein.

lipid-linked protein: Peripheral protein attached to a membrane by a fatty acyl (myristoyl or palmitoyl), isoprenoid (farnesyl or geranylgeranyl), or glycosylphosphatidylinositol group. Also called lipid- anchored protein.

lipogenesis: The *de novo* synthesis of fatty acids from sugars (mainly glucose and fructose), amino acids, and ethanol.

lipoic acid (lipoate): 6,8-Dithiooctanoic acid, that may exist in oxidized or reduced form and is joined in amide linkage to the side chain of a lysine residue in the dehydrogenase complexes specific for pyruvate, α-ketoglutarate or branched-chain α-keto acids.

lipolysis: The hydrolysis of triglycerides that generates glycerol and fatty acids.

lipophilic: Hydrophobic, soluble in lipid.

lipoprotein: Multimolecular complex of protein, phospholipid, and cholesterol that encloses a core composed of triacylglycerols, cholesterol esters, or other lipids.

liposome: Artificial vesicle that consists of one or more closed concentric phospholipid bilayers that enclose some of the suspending aqueous medium in the central compartment.

lipoxin: Eicosanoid produced in leukocytes from arachidonic acid that contains four conjugated double bonds and functions in an autocrine or paracrine manner.

locus: Particular site on a chromosome where a gene for a particular trait is located.

lyase: Enzyme that catalyzes elimination of a group from a molecule with formation of a double bond.

lymphocyte: Spherical leukocyte with a large nucleus and sparse cytoplasm, the predominant cell in lymphoid tissue. It is either a B cell and responsible for antibody-based immunity or a T cell and responsible for cellular immunity.

lymphokine: Cytokine secreted by activated lymphocytes.

lysosome: Cytoplasmic membrane-bounded organelle rich in hydrolytic enzymes for degrading and recycling unneeded cell components.

lysozyme: Enzyme of egg white and human body fluids that, by hydrolyzing polysaccharides in the cell wall of some bacteria, has antibiotic properties.

M phase: Portion of the cell cycle in which the nucleus and cytoplasm divide to form two daughter cells.

M protein: Abnormal protein present in plasma in myelomatosis. Also called myeloma protein or paraprotein.

macromolecule: Large polymeric molecule characteristic of living matter, formed by joining together of small subunits. Polysaccharides, polypeptides, and nucleic acids are macromolecules.

macrophage: Mononuclear actively phagocytic cell that migrates into tissues and ingests particulate material such as microorganisms and dead cells. It is rich in lysozyme and lysosomes and an oxidative system that is microbicidal when activated.

major histocompatibility complex (MHC) protein: Integral membrane glycoprotein of antigen-presenting cells that carries on its extracellular region a peptide derived by intracellular digestion of a protein antigen during synthesis of the MHC protein. Also called human leukocyte antigen (HLA).

malignant: Quality of a tumor or cell that grows rapidly and is invasive.

mast cell: Cell of connective tissue that on stimulation initiates an inflammatory response by releasing histamine, serotonin, heparin and other constituents.

matrix: Aqueous fluid of the cytoplasm, nucleus, or other cell compartment.

M-cyclin: Cyclin that regulates progression of the cell cycle from G_2 phase to M phase.

MDR protein: See multidrug resistance protein.

mechanism-based inhibitor: See suicide inhibitor.

melatonin: *N*-acetyl-5-methoxytryptamine, synthesized in the pineal gland in amounts that vary in a daily cycle and is involved in regulation of several functions that also exhibit a daily cycle.

membrane potential: Difference in charge across the plasma membrane of a cell (positive outside, negative inside) caused by unequal distribution of ions between the extracellular fluid and the cytoplasm.

memory B cell: Long-lived non-proliferating lymphocyte that retains a memory of a previous immune response to an antigen.

metabolic syndrome: See syndrome X.

metabolic turnover: Continuous breakdown and synthesis of biomolecules within living cells.

metabolite: Substance that participates as an intermediate in metabolism.

metal ion catalysis: Catalysis that occurs when an enzyme-bound metal ion participates in labilizing a substrate.

metaphase: Stage in cell division when the nuclear membrane disintegrates, chromosomes line up at the equatorial plan, and microtubules growing from their centromeres attach to those of the mitotic spindle.

micelle: Spherical particle made up of amphipathic compounds such as bile acids, fatty acids, and monoacylglycerols, that form a hydrophilic surface and a hydrophobic interior that contains virtually no water.

microarray: Glass slide or other support on which is mounted a large collection of known, short DNA sequences that is used to monitor the expression of many genes simultaneously.

microfilament: Fibrous (5 to 9 nm in diameter) component of the cytoskeleton and of the contractile part of skeletal and cardiac muscle. It consists of actin and functions in cell structure and in movement. Also called actin filament.

microsome: Small membranous vesicle produced by fragmentation, mostly of the endoplasmic reticulum, on homogenization of eukaryotic cells.

microtubule: Long, hollow cylindrical structure (25 nm diameter) of the cytoskeleton. It is composed of tubulin and is present in cilia, flagella, centrioles, and spindle fibers.

microvillus: Small finger-like extension of the exposed cell membrane of an epithelial cell in the intestinal mucosa or renal tubules that enhances absorption. A dense covering of microvilli constitutes a brush border.

mineralocorticoid: Steroid hormone (e.g. aldosterone) secreted from the adrenal cortex that affects mineral metabolism by enhancing renal absorption of Na^+ and excretion of K^+ and H^+.

mismatch repair: Enzymatic repair of a DNA strand that contains a base that cannot pair properly with that of the complementary strand.

missense mutation: Mutation in which substitution of a nucleotide that is part of a coding triplet changes the codon to one for another amino acid.

mitochondrial DNA (mtDNA): Each mitochondrion contains several copies of closed-circular double-stranded DNA molecule. In humans each mtDNA molecule contains about 16,600 base pairs and codes for 13 polypeptides of the electron transport chain, 2 for mitochondrial rRNA, and 22 for mitochondrial tRNA species.

mitochondrial signal sequence: Usually an N- terminal sequence of residues that can form an amphipathic helix that directs polypeptides synthesized on free ribosomes for entry across the mitochondrial membranes and is removed by a signal peptidase soon after its entry.

mitogen: Protein (e.g., platelet-derived growth factor) that stimulates cell division primarily by acting at the restriction point of the cell cycle.

modular protein: Polypeptide that contains two or more modules with specific binding properties that are also found in other polypeptides alone or in various combinations.

modulator: Molecule that binds the allosteric site of an enzyme and increases or decreases its activity. Also called allosteric effector.

module: Autonomously folding unit or domain of a polypeptide that usually functions as a specific binding site.

molar solution: Aqueous solution that contains one mole of solute in a total volume of 1000 mL.

mole: Amount of a substance that is its molecular weight in grams.

molecular chaperone: See chaperone.

monoclonal antibody: Antibody produced under laboratory conditions by cloned hybridoma cells which are therefore genetically identical and produce antibody against the same antigenic epitope.

monokine: Cytokine released by monocytes.

monotopic protein: Integral membrane protein that spans the membrane once. Also called single-pass protein.

motif: Supersecondary structure or fold (e.g., β-α-β, zinc finger) that is present in different polypeptides. Also a set of amino acid residues present in different polypeptides or of nucleotides in nucleic acids, that is associated with a particular function (e.g., KDEL in ER-resident proteins, TATA in some promotor sequences).

motor domain: Globular head that has ATPase activity in motor proteins such as myosin.

motor protein: Cytosolic dimeric protein that consists of two globular heads with ATPase activity and a coiled-coil body, that propels itself along a filament or microtubule.

mucus: Viscous secretion composed of water and glycoproteins called mucins that are produced by cells of the mucosa of respiratory, gastrointestinal, and urogenital tracts, and lubricates and forms a protective barrier as a sticky trap for foreign particles and microorganisms.

multicatalytic polypeptide: Polypeptide that contains two or more catalytic sites (e.g., mammalian fatty acid synthase).

multidrug resistance protein (MDR protein): ABC transporter protein that transports hydrophobic natural products or synthetic drugs out of the cytoplasm across the plasma membrane.

multienzyme complex: Cluster of catalytic proteins (enzymes) that are isolated together and catalyze several related metabolic reactions (e.g., pyruvate dehydrogenase complex).

multipass protein: Integral protein that spans a membrane two or more times. Also called polytopic protein.

muscarinic receptor: Serpentine receptor of plasma membrane that binds acetylcholine or the alkaloid muscarine and activates a trimeric G protein to modulate adenylate cyclase, K^+ channels or phospholipase C, thereby transmitting a nerve impulse.

mutagen: Physical or chemical agent that induces a change in DNA that converts one allele into another. Mutagens are frequently carcinogenic.

myelin: Specialized cell membrane that ensheathes an axon to form a myelinated nerve fiber in peripheral nerves (where it is produced by Schwann cells) and in the central nervous system (where it is produced by oligodendrocytes). It is rich in glycolipid and specific proteins.

myelomatosis: Condition in which a cancerous B cell proliferates to produce one or more tumors that secrete a single type of immunoglobulin or of its constituent polypeptides. The abnormal protein (M protein or paraprotein) is detectable in plasma and frequently in urine (Bence Jones protein).

myoblast: Undifferentiated mononucleated cell that develops into a fiber of a skeletal muscle cell.

myofibril: Contractile filament that consists of actin filaments, myosin and associated proteins within a muscle cell.

nebulin: Large fibrous polypeptide that is associated with the thin filament of muscle sarcomeres and stretches from the Z line to the M line. It is made up almost entirely of repeating actin-binding motifs.

necrosis: Death of cells or tissues from injury or disease while part of the body. The cells swell, burst, and spill their contents into their environment and may induce an inflammatory response.

negative feedback: Metabolic regulation in which a late product of a pathway inhibits an enzyme that functions early in that pathway. Also called feedback inhibition.

N-end rule: Rule stating that the half-life of a cytoplasmic protein is determined by the identity of its N-terminal residue. Also called Varshavsky rule.

network-forming collagen: Type of collagen (e.g., type IV) that forms a meshwork-like structure in the basal lamina of epithelia.

neurofilament: Type of cytoskeletal intermediate filament present in nerve cells and axons, being a heterodimer of neurofilament proteins NF-L, NF-M, or NF-H.

neuroglia: Supporting nonneural cells (microglia, oligodendrocytes, or astrocytes) of the central nervous system. These cells provide trophic substances to neurons, maintain a balance between glutamate and GABA, and are responsible for myelin formation. Also called glial cells.

neuromodulator: Neuropeptide secreted, usually with one or more neurotransmitters, that modulates the sensitivity of another neuron to specific neurotransmitters.

neuropeptide: Peptide secreted at synapses or elsewhere by a neuron to exert an effect on neighboring cells.

neurosecretory cell: Neural cell (e.g., in hypothalamus or neuronal plexuses of gastrointestinal tract) in which the axon ends against a blood vessel or sinus into which it secretes a hormone rather than transmitting an electrical impulse.

nicotinic receptor: Multimeric integral protein of plasma membrane that binds acetylcholine or the alkaloid nicotine and causes depolarization, thereby transmitting a nerve impulse. It consists of α, β, γ, and δ type of subunits.

N-linked glycoprotein: Glycoprotein that contains a high-mannose or complex oligosaccharide linked to the side-chain amide group of an asparagine residue.

N-linked glycosylation: Enzymatic transfer to the side-chain amide of an asparagine residue in a polypeptide of an oligosaccharide that has been synthesized in the ER lumen on dolichol present in the ER membrane.

nonconservative substitution: Replacement in a polypeptide of a residue by another of different characteristics (e.g., glutamate by valine).

nonsense codon: Nucleotide triplet that does not code for any amino acid but signals termination of translation of a mRNA. Also called stop codon.

nonshivering thermogenesis: Production of heat by oxidation of fatty acids in mitochondria of brown adipose tissue without synthesis of ATP.

Northern blotting: Technique in which RNA of a particular base sequence is identified, after electrophoretic separation from a complex mixture and transfer to a blotting surface, by hybridization with a labeled complementary nucleic acid probe.

N-terminus: End of a peptide or polypeptide that contains the (usually) free amino group of the first residue. Also called amino terminus.

nuclear export receptor: Soluble protein of the nuclear matrix that binds to the export signal of macromolecules for transfer from the nucleus into the cytoplasm. The receptor also binds to components of the nuclear pore complex and guides transport out of the nucleus across the nuclear pore complex.

nuclear export signal: Part of a macromolecule (e.g., ribosomal subunit, tRNA, or mRNA) that is bound by a nuclear export receptor as a prerequisite for passage across a nuclear pore complex into the cytoplasm.

nuclear import receptor: Soluble cytosolic protein that binds the nuclear localization signal of a protein to be transferred from cytoplasm into the nucleus. It also binds components of the nuclear pore complex and guides transport inward across the nuclear pore complex.

nuclear import signal: See nuclear localization signal.

nuclear lamina: Network of intermediate filaments composed of lamins that is anchored to the inner surface of the nuclear membrane and provides attachment sites for chromosomes and the nuclear pore complex.

nuclear localization signal: One or two basic oligopeptide regions which may be situated anywhere in the sequence of a nuclear protein and which signal its transfer from cytoplasm into the nucleus via the nuclear pore complex by an energy-dependent process. Also called nuclear import signal.

nuclear pore complex: Large structure composed of proteins (nucleoporins) in the nuclear membrane, that permits selective movement of proteins and nucleic acids between the nuclear matrix and the cytoplasm.

nuclear receptor superfamily: Family of intracellular receptors for hydrophobic hormones, (thyroid hormones, steroid hormones, retinoids, 1,25-dihydroxyvitamin D_3, and other signalling molecules. They contain zinc-finger motifs that bind at specific response elements in DNA and act as transcription factors. For some family members (orphan receptors), the ligand has not been characterized.

nuclease protection assay: Method for detection and quantitation of particular RNA species in a complex mixture of cellular RNA. A small antisense probe for the target RNA is used for hybridization, followed by nuclease digestion of unhybridized material, inhibition of the nuclease, and precipitation of the hybridized material.

nucleation: Formation of an initial structure during folding of a polypeptide, during renaturation of DNA or RNA, or during assembly of a polymeric structure.

nucleoid: Dense region that contains the DNA in prokaryotic cells. Also the cluster of DNA within a mitochondrion or chloroplast.

nucleolus: Structure within the nucleus where rRNA is transcribed from decondensed chromatin that contains highly amplified genes for rRNA and where ribosomal subunits are assembled from rRNA and specific proteins imported from the cytoplasm.

nucleoporin: Any protein component of the nuclear pore complex.

nucleosome: Fundamental structural subunit of chromatin, formed by wrapping of a segment of DNA twice around a histone core plus linker DNA, and its attached histone H1 that joins the adjacent nucleosomes.

nucleotide coenzyme: Coenzyme that resembles (e.g., FMN) or contains a nucleotide as a component (e.g., FAD, NAD, NADP, or CoA).

nucleotide excision repair: Correction of a lesion that distorts the helical structure of a DNA strand by removal of the abnormal segment by an excinulease and filling up of the gap by DNA polymerase and DNA ligase.

nucleotide sugar: Nucleoside diphosphate sugar, e.g., UDP-galactose, that is a substrate for a glycosyl transferase.

nucleotide-binding fold: Protein structural motif present in many enzymes that bind ATP or nucleotide coenzymes (e.g., NAD). It consists of a β–α–β–α–β fold. One fold is present in ATP-binding proteins, two in those which bind dinucleotide coenzymes. In each case the β-strands form a parallel sheet. Also called dinucleotide fold and Rossmann fold.

Okazaki fragment: DNA (100–200 nucleotides long in eukaryotic cells) produced on the lagging strand by extension of a primer RNA sequence.

oligomer: Short polymer consisting of amino acids, monosaccharides, or nucleotides. Also a protein that consists of more than one subunit.

oligopeptide: Peptide that contains up to 10 amino acid residues (approximately).

O-linked glycoprotein: Glycoprotein that contains a monosaccharide or an oligosaccharide bound to the side-chain hydroxyl group usually of serine or threonine.

O-linked glycosylation: Addition of a monosaccharide or stepwise addition of an oligosaccharide to a hydroxyl group, usually of a serine or threonine residue, in a polypeptide.

oncogene: Mutant version of DNA genetic sequence (a protooncogene), which by interfering with normal cell growth or division, results in uncontrolled cell proliferation and formation of a benign or malignant cancer.

oncogenesis: Conversion of a normal cell into a benign or malignant cancer cell, one in which cell proliferation is uncontrolled.

open-reading frame (ORF): Section of genomic DNA that contains about 100 or more nucleotide triplets that code for amino acids, beginning with an initiation codon and ending with a termination codon, that potentially codes for a polypeptide.

ordered reaction: Reaction in which binding of substrates to, and release of products from, an enzyme follow a particular sequence.

organelle: Intracellular structure that may be membrane bounded (e.g., a nucleus) or not (e.g., ribosomes) and performs a specific function or group of functions.

organic aciduria: Presence of abnormal amounts of an organic acid in urine (e.g., one or more amino acids, methylmalonic acid, orotic acid, or uric acid).

organogenesis: Formation of an organ during embryonic and fetal development as a result of programmed expression or repression of specific genes.

origin of replication complex (ORC): Multisubunit protein complex that binds at origins of replication in chromosomes throughout the cell cycle and recruits additional regulatory proteins during G_1 phase.

orphan receptor: Member of the nuclear receptor superfamily for which the ligand has not been identified.

osmotic pressure: Pressure generated by flow of solvent across a semipermeable membrane from a more concentrated solution on one side to a less concentrated solution on the other.

osteoblast: Bone-forming cell that produces and secretes collagen and other material of the bone matrix.

osteoclast: Macrophage-like cell that erodes bone by digesting and dissolving the bone matrix.

osteomalacia: Condition in adults characterized by softening of bones from lack of mineralization and excessive excretion of calcium and phosphate. It results from dietary deficiency or malabsorption of vitamin D.

oxidant: Reactant that accepts one or more electrons and becomes reduced in a redox reaction.

oxidative phosphorylation (oxphos): Phosphorylation of ADP to ATP that is coupled to electron transfer from metabolic fuels and dependent on the electron transfer chain and formation of a proton gradient across the inner mitochondrial membrane (respiration-linked phosphorylation).

oxidoreductase: Enzyme that catalyzes an oxidation–reduction reaction using NAD, NADP, FMN, or FAD as coenzyme.

oxyntic cell: Type of cell in the gastric mucosa that secretes hydrochloric acid and intrinsic factor. Also called parietal cell.

P site: See peptidyl site.

p53: Gene regulatory protein (53 kDa) that is activated by DNA damage and involved in cell cycle control, apoptosis, and maintaining genetic stability. It is mutated in about half of human cancers.

palindrome: Segment of double-stranded DNA in which the base sequences in both strands are inverted repeats that have the same sequence when read in the same chemical direction.

pantetheine: *N*-pantothenylcysteamine, derived from pantothenic acid and decarboxylated cysteine. A structural component of coenzyme A.

paracrine hormone: Hormone that is secreted by a cell into the extracellular space and acts on neighboring cells.

parathyroid gland: One of several (usually four) small glands embedded in the posterior surface of the thyroid gland that secrete parathyroid hormone that increases osteoclastic activity and raises calcium and decreases phosphate concentrations in plasma.

parietal cell: See oxyntic cell.

passive transport: See facilitated diffusion.

pathway: Particular sequence of metabolic reactions.

pentose phosphate pathway: Metabolic pathway for interconversion of hexose and pentose phosphates that generates NADPH and ribose-5-phosphate for biosynthetic processes

peptidyl site: Site on a ribosome that binds peptidyl-tRNA during protein synthesis. Also called P site.

peptidyl transferase: Catalytic activity that transfers the growing peptide chain from the peptidyl-tRNA to the amino group of the incoming aminoacyl-tRNA during polypeptide synthesis. The activity resides in the 23S rRNA of the large subunit of bacterial ribosomes.

peptidyl-tRNA: tRNA carrying the growing peptide chain during polypeptide synthesis.

peripheral protein: Protein that is associated with a membrane surface but does not span the membrane and can be released without disruption of the membrane. Also called extrinsic protein.

perlecan: Proteoglycan of the basal lamina that consists of heparin sulfate attached to a very large polypeptide.

peroxisomal signal sequence: Sequence Ser-Lys-Leu (SKL) at the C-terminus of many polypeptides synthesized on free ribosomes that targets them for import into peroxisomes.

PEST sequence: Sequence motif rich in proline, glutamate, serine, and threonine residues, which targets a polypeptide for rapid degradation within a cell.

phagocyte: Cell (e.g., a macrophage or neutrophil) specialized for engulfing and ingesting particles and microorganisms.

phagocytosis: Endocytosis of large particles and microorganisms by macrophages and neutrophils.

phagosome: Large membrane-bounded vesicle that contains material ingested by phagocytic cells. On fusing with a lysosome, the material is degraded or, if nondigestible, forms a residual body.

phase 1 reaction: Metabolic reaction (e.g., oxidation, reduction, or hydrolysis) in biotransformation that introduces or unmasks functional groups that make the product more water soluble.

phase II reaction: Metabolic reaction (e.g., acylation, sulfation, or conjugation with glucuronate) that usually follows a phase I reaction in biotransformation and inactivates the product and makes it water soluble for excretion.

phenotype: Observable characteristics of an individual determined by interaction of the environment with the genotype.

phosphagen: Compound such as creatine phosphate that stores high-energy phosphate which is used to form ATP when required.

phosphoinositide pathway: Sequence of reactions of signal transduction in which a hormone binding to a serpentine receptor activates a trimeric G protein to stimulate hydrolysis of phosphatidylinositol-4,5-bisphosphate (PIP_2) into the second messengers inositol-1,4,5-trisphosphate (IP_3) and diacylglycerol (DAG).

ping-pong reaction: Group-transfer reaction in which a functional group of a substrate is transferred onto an enzyme with release of a product, and the group is then transferred onto a second substrate with release of second product. Also called double displacement reaction.

plaque: Deposit of insoluble material in a tissue, for example, the fatty degeneration in the middle coat of an artery in atherosclerosis and the β-amyloid fibrils in the brain in Alzheimer disease.

plasma cell: Fully differentiated B lymphocyte that secretes antibodies. It arises from bone marrow in adults and from the liver in the fetus.

plasma membrane: Outer bounding membrane of a cell.

plasma: Protein-rich fluid portion of whole blood that has been prevented from clotting.

platelet: Small cell fragment consisting of cytoplasm derived by fragmentation from megakaryocytes in the bone marrow, present in large numbers in blood and important for initiating clotting of blood on damage to a blood vessel.

platelet-derived growth factor (PDGF): Protein that consists of two homologous polypeptides joined by disulfide bonds, is secreted by platelets and several other tissues, and is mitogenic for mesenchymal and glial cells. It exerts its effect by binding to a receptor tyrosine kinase.

point mutation: Replacement of a complementary base pair in DNA by a different base pair. Also called base-pair substitution.

polar bond: Covalent bond in which electrons are unequally shared.

polarity: Nonuniform distribution of electrons or charge in a molecule, also the distinction between the 5′ and 3′ ends of a nucleic acid.

polyA tail: Polyadenylate (up to 250 nucleotides long) at the 3′ end of mRNA that is added during pre-mRNA processing.

polyclonal antibody: Mixture of antibodies produced by different B cells in response to an antigen, each recognizing a different part of the antigen.

polyhormone: Polypeptide that is cleaved into more than one copy of a peptide hormone or more than one peptide hormone.

polyprotein: Polypeptide that is cleaved into two or more distinct polypeptides.

polysome: Cluster of ribosomes on an mRNA molecule each in the process of translating the mRNA. Also called polyribosome.

polytopic protein: See multipass protein.

porphyria: Condition characterized by accumulation in the urine of porphyrins and/or of their intermediates in the heme synthesis pathway. It may be secondary to hepatic cirrhosis or to lead poisoning or primary and result from hereditary increase or decrease in activity of an enzyme in this pathway.

positive feedback: Intracellular signaling process in which the output increases its own production.

postabsorptive state: Physiological condition of the body 3–4 h after a meal until the mobilization of metabolic reserves commences.

posttranscriptional processing: Covalent modification of a primary RNA transcript into a functional molecule of mRNA, tRNA or rRNA.

posttranslational modification: Covalent modification of one or more residues of a polypeptide after its synthesis; also limited proteolysis of the polypeptide. Also called posttranslational processing.

posttranslational processing: See posttranslational modification.

precursor mRNA (pre-mRNA): Primary transcript of a gene that encodes a polypeptide.

prenylation: Enzymatic transfer of a prenyl moiety (i.e., geranyl, farnesyl, or geranylgeranyl) to a cysteine residue within the C-terminal region of a polypeptide.

preproprotein: Proprotein with an attached N-terminal signal peptide. The precursor of a secreted proprotein.

primary structure: Sequence of amino acids in a polypeptide or of nucleotides in a nucleic acid.

primary transcript: RNA as transcribed from a gene for a polypeptide or for functional RNA, before any posttranscriptional processing has occurred.

primase: Enzyme that synthesizes an RNA that is a primer for synthesis of DNA by DNA polymerase.

primer: Oligosaccharide or oligoribonucleotide on to which other monomers can be added by an appropriate enzyme.

primosome: Complex consisting of primase and DNA helicase that synthesizes an RNA primer that initiates DNA replication.

prion: Protein that can exist in two 3-dimensional conformations, a normal one that is noninfectious and an abnormal one that is infectious. The latter can then induce the former in a host organism to assume the infectious form.

probe: Labeled polynucleotide with sequence complementary to that of a gene or other piece of DNA that is used to hybridize and detect that DNA.

processive enzyme: Polymerase that remains bound to its growing product through many additions of monomeric units.

proenzyme: See zymogen.

prokaryote: Single-celled organism that lacks a membrane-bounded nucleus.

promoter: Specific segment of DNA adjacent to a gene at which RNA polymerase binds to initiate transcription of that gene; also a compound that promotes growth of a tumor.

propeptide: Peptide segment present at the N-terminus or within a polypeptide that must be removed to produce one or more functional polypeptides.

properdin: Plasma glycoprotein of the alternative pathway of the complement system that enhances binding of components of that system to bacterial cell walls without dependence on antibody binding.

prophase: Opening stage of cell division when chromosomes appear, the nuclear membrane starts to disintegrate and the mitotic spindle starts to form.

proprotein: Polypeptide that contains a propeptide segment that when removed, produces one or more functional polypeptides.

proteasome: Cylindrical ATP-dependent proteolytic protein complex that degrades cell proteins that are marked for degradation by ubiquitination or other means.

protein family: Group of proteins that resemble each other in primary structure and in their overall conformation.

protein targeting: Mechanism by which newly synthesized polypeptides are selected and directed to their proper final location.

protein translocator: Intrinsic protein that mediates transport of a protein across an organelle's membrane (e.g., TIM or TOM complexes).

proteogenic: α-Amino acid that is used in protein biosynthesis.

proteoglycan: Macromolecule consisting of one or more glycosaminoglycan chains attached to a core protein, the former making up the largest component.

proteomics: Study of the protein complement of a cell, tissue, organ, or organism under different conditions.

protofilament: End-to-end assemblage of protein subunits that associates laterally with similar structures so as to form fibrils such as cytoskeletal components.

proton pump: Ion channel, specific for rapid transport of protons across a membrane, that is coupled to hydrolysis or synthesis of ATP.

pseudogene: Segment of DNA that is homologous in sequence to a particular gene but contains several termination codons and, if transcribed, does not produce a functional product.

P-type ATPase: ATPase of plasma membrane of animal cells whose function is changed by reversible phosphorylation of an aspartate residue (e.g., Na^+-K^+-ATPase, Ca^{2+}-ATPase and H^+/K^+-ATPase) (of stomach). It usually consists of two α and two β transmembrane subunits.

purine nucleotide cycle: Metabolic cycle of muscle in which AMP is deaminated to IMP and is then reformed by transfer of the amino group of aspartate with production of fumarate.

Q cycle: Model proposed for the cyclic flow of electrons through complex III of the electron transport chain, cytochrome C and coenzyme Q.

quaternary structure: Overall structure of a protein that results from interaction between protein subunits.

ragged red fiber: Muscle fiber with enlarged and abnormal mitochondria that contain highly organized inclusions and appear red on special staining. Such fibers are usually associated with mitochondrial disease.

Ran: Monomeric G protein with GTPase activity that is required for active transport of macromolecules across a nuclear pore complex.

random reaction: Reaction in which there is no preference for the sequence of binding of substrates or release of products.

Ras: Monomeric G protein with GTPase activity that is attached to the inner surface of the plasma membrane and participates in transduction of signals for growth factor receptors along a variety of pathways. A hyperactive mutant is present in about 25% of human cancers.

rate-determining step: See rate-limiting step.

rate-limiting step: Slowest reaction in a metabolic pathway or the slowest step in an enzyme-catalyzed reaction. Also called rate-determining step.

reading frame: Sequence in mRNA of contiguous coding triplets that are preceded by a start codon and end in a stop codon.

receptor tyrosine kinase: Integral protein of plasma membrane that is a hormone receptor and has a tyrosine kinase domain in its intracellular portion.

receptosome: Cytoplasmic vesicle derived by receptor-mediated endocytosis from coated pits on the cell surface following ligand binding to the receptor.

recessive: Gene that affects the phenotype only when it is received from both parents, i.e. in a homozygote. It also describes a trait or character due to such a gene.

recombinant DNA: DNA molecule that consists of DNA sequences joined together from different sources.

redox reaction: Reaction in which oxidation of one reactant is coupled to reduction of a second reactant.

reductant: Reactant that donates one or more electrons and becomes oxidized in a redox reaction.

regulator gene: Gene that codes for a protein that determines the expression of another gene or group of genes.

relaxin: Protein hormone related in structure to insulin-like growth factors. It is secreted by the corpus luteum mainly during pregnancy, relaxes pelvic ligaments, and softens the cervix.

releasing hormone: Peptide hormone secreted by specific neurosecretory cells of the hypothalamus that causes secretion of a hormone of the anterior pituitary.

renin: Proteolytic enzyme secreted by cells of the juxtaglomerular apparatus when blood flow through the kidney decreases. It converts angiotensinogen to angiotensin I.

rennin: Proteolytic enzyme secreted by the stomach of young mammals. It converts the caseinogen of milk into casein, which coagulates and therefore takes longer to leave the stomach.

replisome: Multiprotein complex that contains DNA polymerase and synthesizes the leading and lagging strands of DNA at a replication fork.

repressor: Protein that prevents transcription of a gene or group of genes by binding to the promoter or regulatory sequence.

residue: Monomeric unit in a polypeptide, polysaccharide, or polynucleotide.

respiration-linked phosphorylation: Phosphorylation of ADP to ATP that is dependent on flow of electrons along the electron transfer chain and generation of a proton gradient across the inner mitochondrial membrane.

restriction fragment: Fragment that results from cleavage of DNA by a restriction endonuclease.

restriction-fragment-length polymorphism (RFLP): Inherited variation in length of certain DNA restriction fragments that arises from a change in the nucleotide sequence that destroys or creates a cleavage site for a restriction endonuclease.

restriction point: Time during G_1 phase of the cell cycle when a cell proceeds to the S phase or exits into the G_o phase.

restriction site: Specific sequence of DNA that is recognized and cleaved by a restriction endonuclease.

retinoid X receptor: A nuclear receptor that is activated by the retinoid 9-cis retinoic acid.

retrograde transport: Transport away from the periphery of an axon.

retrovirus: Virus that has an RNA genome and is transcribed into DNA by its own reverse transcriptase for incorporation into the host cell genome.

reverse genetics: Study of gene function that starts from the DNA that encodes a polypeptide and then creates mutants of that gene.

reverse transcriptase: Retroviral zinc-containing enzyme that catalyses RNA-directed DNA synthesis, RNA-directed degradation, and DNA-directed DNA synthesis. It usually consists of two homologous subunits.

reverse turn: Usually a tetrapeptide segment in a polypeptide at which the chain abruptly changes its direction. Also called β-bend.

ribosomal protein: Protein of the small or the large ribosomal subunit. Generally located on the subunit surface filling gaps and crevices of the folded rRNA which the proteins stabilize.

ribozyme: Catalytic RNA.

Rieske iron–sulfur protein: Transmembrane protein component of complex III of the mitochondrial electron transport chain. It contains a 2Fe2S center in which one of the Fe atoms is coordinated to two histidine residues and the other is coordinated to two cysteine residues.

Rossmann fold: See nucleotide-binding fold.

S phase: Phase of the cell cycle in which DNA and chromosome replication occur.

salvage pathway: Metabolic pathway in which an intermediate (e.g., a purine or pyrimidine) in the degradation of a biomolecule (e.g.. a nucleotide) is reused in synthesis of that biomolecule.

sarcoplasmic reticulum (SR): Network of modified endoplasmic reticulum that surrounds striated and cardiac muscle fibrils, stores Ca^{2+}, and releases it during excitation to trigger contraction. With T tubules, it forms the sarcotubular system.

sarcotubular system: System composed of sarcoplasmic reticulum and T tubules in striated muscle.

satellite DNA: DNA that has a distinctive composition, consists of highly repetitive sequences, and is mostly associated with the centromeres. It forms bands (satellites) separate from those of other DNA on density gradient centrifugation.

saturated fatty acid: Fatty acid that contains a saturated alkyl chain.

Schiff base: See aldimine.

second law of thermodynamics: Law that requires that in any physical or chemical process, the entropy of the universe tends to increase.

second messenger: Ion or biomolecule released or produced at the cell plasma membrane in response to binding of a signal molecule (i.e., the first messenger) to its receptor located therein.

secondary structure: Pattern of regular local folding of a polymeric structure; in proteins formation of α-helices and/or β-sheets.

secretase: Proteolytic enzyme that cleaves certain integral membrane proteins (e.g., β-APP) within the transmembrane segment causing release of the extracellular portion of these proteins.

selectin: Integral carbohydrate-binding protein (or lectin) of leukocytes (L-), platelets (P-), and endothelial cells (E-) that mediates short-lasting Ca^{2+}-dependent cell–cell interactions in the blood stream.

sequential reaction: Reaction in which all substrates must be bound to the enzyme, in random or in specific order, before the reaction can proceed.

serpentine receptor: Integral protein that spans the plasma membrane seven times and binds a hormone, neurotransmitter, or other ligand to activate adenylate cyclase, ion channels, or phospholipase C.

serum: Fluid part of whole blood that is formed naturally when the blood has been allowed to clot. It is plasma that lacks fibrinogen and other proteins involved in clot formation.

sex hormone-binding globulin: Glycoprotein of the plasma globulin group that is secreted by the liver and transports testosterone and estradiol.

sex-linked: Gene that is located on a sex-determining chromosome, usually an X chromosome. Also the trait or characteristic due to such a gene.

sickle-cell anemia: Condition characterized by severe hemolytic anemia and circulating sickle-shaped erythrocytes. It is caused by homozygosity for the β-globin allele in which the normal glutamate in position 6 is replaced by valine.

sickle-cell trait: Usually benign condition associated with the heterozygous state for the hbS allele.

signal patch: Group of nonadjacent residues in a polypeptide that form a specific secondary or tertiary structure as a signal for import into mitochondria or export from the nucleus.

signal recognition particle (SRP): Ribonucleoprotein complex consisting of 7SL-RNA (300 nucleotides long) and six protein subunits that binds an ER signal sequence and directs the ribosome and partially synthesized polypeptide for binding to the surface of the rough ER.

signal sequence: Peptide sequence in a polypeptide that targets it to a specific destination in a cell.

silent mutation: Change in the sequence of a gene that produces no change in the sequence of the encoded protein or RNA.

simple protein: Protein that consists only of α-amino acids.

single nucleotide polymorphism: Change in a single nucleotide in DNA that distinguishes an individual from another of that species.

site-directed mutagenesis: Technique for creation of a specific alteration in a DNA genetic sequence

small-interfering RNA: Species of RNA, 21–26 nucleotides long, that inhibits gene expression by binding to, and leading to the destruction of, complementary mRNA

small nuclear ribonucleoprotein (snRNP): Complex of snRNA and several protein subunits. It forms the core of a spliceosome.

small nuclear RNA (snRNA): RNA component of a small nuclear ribonucleoprotein that is required for RNA splicing.

small nucleolar RNA (snoRNA): Small RNA present in the nucleolus that base pairs with parts of precursor rRNA and specifies methylation of ribose moieties or isomerization of uracil nucleotides to pseudouridine. Also called guide RNA.

SNARE: see synaptosome-associated protein receptor.

Southern blotting: Technique in which one or more DNA fragments in a complex mixture are transferred to a blotting surface and detected by hybridization with a complementary labeled nucleic acid probe.

spectrin: Major cytoskeletal protein of erythrocytes. It consists of two very large homologous polypeptides that each contains numerous copies of a repeat of ~106 residues that forms a triple helix.

spindle apparatus: Array of microtubules formed during cell division at the end of prophase. It radiates from a centriole at each pole to the centromeres of the duplicated chromosomes. It serves to move the sister chromosomes apart. Also called mitotic spindle.

spleen: Lymphoid organ in the abdomen that is perfused by blood and is important for lymphocyte production and for removal of effete, damaged, or abnormal erythrocytes and platelets.

spliceosome: Complex of snRNA and proteins that interacts with the ends of an intron in pre-mRNA causing release of the intron and joining of the freed ends of the adjacent exons.

Src homology 2 (SH2) domain: Module (~100 residues long) present in Src and other proteins that binds oligopeptide sequences that contain phosphotyrosine and are present in many intracellular signalling proteins.

Src homology 3 (SH3) domain: Module (50–75 residues long) present in Src and other proteins that binds a proline-rich oligopeptide sequence that is present in many intracellular signaling proteins.

Src: Family of protein tyrosine kinases anchored to the inner surface of the plasma membrane and associated with the intracellular domain of certain receptors that lack intrinsic catalytic activity. Each contains two important homology domains called SH2 and SH3 domains.

SRP receptor: Heterodimeric integral protein of the rough ER that binds the signal recognition particle bound to the ER signal sequence of a polypeptide being synthesized on a bound ribosome.

stem cell: Self-regenerating cell in a multicellular organism that can differentiate, if its source is an embryo, into any of the embryo's specialized tissues and, if its source is an adult, into the specialized tissue of its origin.

steroidogenic acute regulatory protein: Protein of mitochondrial matrix especially of steroid hormone-producing cells that regulates transfer of cholesterol into that organelle and regulates production of such hormones.

stimulatory G protein (Gs): Heterotrimeric G protein that on activation stimulates adenylate cyclase.

stop codon: See nonsense codon.

structural gene: Gene that codes for a polypeptide or a functional RNA.

substitution mutation: Replacement of one base pair by another in the coding part of a gene. It may be a missense or a nonsense mutation.

substrate-level phosphorylation: Transfer of a high-energy phosphate group from a metabolic intermediate directly to ADP or other nucleoside 5′-diphosphate, in contrast to oxidative phosphorylation.

suicide inhibitor: Molecule devised to inactivate a specific enzyme irreversibly but only after it has become activated by undergoing some chemical change at the active site. Also called suicide substrate, mechanism-based inhibitor.

suicide substrate: See suicide inhibitor.

suppressor T cell: Type of lymphocyte that inhibits activation of antibody synthesis and secretion by B cells.

surfactant: Secretion that coats the surface of pulmonary alveoli, preventing their collapse through lowering their surface tension. It contains phospholipids (mainly bispalmitoyl-phosphatidylcholine) and several specific proteins.

symport: Transport of two different solutes in the same direction across a membrane by a transport protein. Also called cotransport.

synaptosome-associated protein receptor (SNARE): Family of complementary transmembrane proteins involved in vesicle transport in the secretion and the endocytosis pathways. They function in pairs, v- on vesicle and t- on target membrane, and guide the vesicles to their destination. In nerve endings they mediate fusion of synaptic vesicles with presynaptic plasma membrane.

syndrome: Discrete set of signs and symptoms that occur together and indicate a particular disease.

synthase: Enzyme that catalyzes a condensation reaction without need for a nucleoside 5′-triphosphate as an energy source.

synthetase: Enzyme that catalyzes a condensation reaction and requires a nucleoside 5′-triphosphate as an energy source.

T cell: Type of lymphocyte derived from the thymus that is responsible for cell-mediated immunity and regulation of the immune response. It includes regulatory (helper or suppressor) and killer (cytotoxic) T cells.

T system: In skeletal and cardiac muscle a system of tubules formed by invagination of the sarcolemma that transversely contacts the myofibrils and as part of the sarcotubular system transmits the excitation signal to the sarcomeres thus ensuring speedy and synchronous contraction.

TATA box: Conserved DNA sequence rich in A and T that is present in promoter elements of many eukaryotic genes and occurs 25–30 nucleotides upstream of the transcription start site.

telomerase: Ribonucleoprotein complex whose RNA is a template for synthesis of the repetitive DNA of telomeres by the protein component which functions as a reverse transcriptase.

telophase: Final stage in cell division in which the spindle disappears, a nuclear membrane forms around each set of chromosomes which condense to form chromatin, and two daughter cells are formed by constriction of a cleavage furrow that consists mainly of actin filaments.

template strand: Strand of DNA that serves as a template for transcription of pre-mRNA; it is complementary to the sequence of the pre-mRNA.

template: Strand of DNA or RNA whose nucleotide sequence determines the sequence of the complementary strand or, in the case of mRNA, the sequence of the encoded polypeptide.

teratogen: Agent that causes congenital malformations.

tertiary structure: Three-dimensional form of a polymeric chain, especially a polypeptide or RNA molecule, that results from interaction between distant parts.

thymosin: Polypeptide hormone secreted by the thymus that is essential for development and differentiation of T cells.

thyroglobulin: Glycoprotein produced in epithelial cells of the thyroid gland and stored in thyroid vesicles where some of its tyrosine residues are modified to form thyroxine and triiodothyronine which are then released by proteolysis of the protein.

thyroid-stimulating hormone: Glycoprotein hormone secreted by the anterior pituitary gland that stimulates the thyroid gland to secrete thyroxine and triiodothyronine.

TIM complex: Multiprotein complex of inner mitochondrial membrane that translocates proteins into the matrix and in some cases mediates their insertion into that membrane.

titin: Very large fibrous protein that is partly associated with the thick filament of muscle sarcomeres and stretches from the Z disc to the M line. It contains a long series of Ig-like domains that act as a spring to protect the sarcomere against overstretching. Also called connectin.

TOM complex: Multiprotein complex of the outer mitochondrial membrane that mediates translocation of proteins across this membrane.

trans-acting regulatory element: Sequence of DNA that regulates expression of one or more genes on a variety of chromosomes.

transaminase: See aminotransferase.

transferase: Enzyme that catalyzes transfer of a functional group from one molecule to another.

transformation: Conversion of a normal cell into a cancerous cell; also the alteration of a cell's phenotype by incorporation of foreign DNA.

transgenic: Describing an organism into whose germ line one or more genes from another organism have been introduced by recombinant DNA technology

trans-Golgi network: Interconnected cisternae and tubules of the Golgi complex that are farthest away from the ER and release vesicles that contain proteins and lipids destined for lysosomes, secretory vesicles, or the plasma membrane.

translocation: Change in position during translation of mRNA on a ribosome of the peptidyl-tRNA and the deacylated tRNA; also transport of a protein across a membrane or to a different site in a cell, or of a portion of a chromosome onto a nonhomologous chromosome.

transmembrane protein: See integral protein.

transverse (T) tubule: In skeletal and cardiac muscle a tubular extension of the sarcolemma and a component of the sarcotubular system.

triplet: Trinucleotide in an exon in DNA, in the coding region of mRNA or in the anticodon region of tRNA, that specifies a particular residue in a polypeptide.

triplet repeat: Stretch of DNA in which the same three nucleotides, for example, (CAG), are repeated in tandem fashion. Expansion in the repeat number is associated with several human genetic disorders.

triskelion: See clathrin.

tumor-suppressor gene: Gene whose product prevents cell proliferation; its inactivation by mutation or deletion enhances susceptibility to cancer.

ubiquitin pathway: Sequence of reactions that require a ubiquitin-activating enzyme, a ubiquitin-conjugating enzyme, and a ubiquitin ligase to form a ubiquitinated or a polyubiquitinated target polypeptide. The former changes the activity of the polypeptide while the latter directs it to degradation by proteasomes.

ubiquitin: Intracellular, monomeric 76-residue protein that is highly conserved and ubiquitous in eukaryotic organisms. When covalently linked through its C-terminal residue to lysine side chains of other proteins, it targets them to degradation by proteasomes.

ubiquitination: Addition to the side chain of a lysine residue in a polypeptide of a ubiquitin monomer or of a series of such monomers joined together by isopeptide linkage.

uncoupling agent: Compound such as 2,4-dinitrophenol that dissociates phosphorylation of ADP from electron transport in mitochondria and dissipates the energy as heat.

uniport: Transport of one specific solute across a membrane by a transport protein.

up-regulation: Increase in transcription of a DNA genetic sequence in response to an external variable.

V(D)J joining: Bringing together during differentiation of a bone marrow stem cell into a mature B cell of V and J, or of V, D, and J segments to form a functional gene for the light chain and the heavy chain respectively of an immunoglobulin.

V(D)J recombinase: Enzyme complex that mediates V(D)J joining.

variable residue: Position of a residue that is occupied by different amino acids when the polypeptide is derived from different species. Also called hypervariable residue.

Varshavsky rule: See N-end rule.

vasopressin: See antidiuretic hormone.

vegan: Someone who eats only plant material.

V-type ATPase: ATPase of lysosomal, endosomal, and secretory vesicle membranes that functions to create an acidic pH in those compartments. It is similar in structure to F-type ATPases.

wobble: Relatively nonspecific pairing between the base at the 3′-end of a codon in mRNA with that at the 5′-end of the anticodon in tRNA.

Z-DNA: Double-helical DNA that contains 12 base pairs in a left-handed turn, has a diameter of ~18 A° and a zigzag appearance and is formed by sequences in which pyrimidine or purine bases alternate.

zinc finger: Protein motif common in DNA-binding proteins that contains four residues of cysteine or two each of cysteine and histidine that are coordinated to a zinc atom.

zwitterion: Molecule that contains charged groups of opposite polarity. Also called dipolar ion.

zymogen: Catalytically inactive protein that becomes an active enzyme on limited proteolytic cleavage. Also called proenzyme.

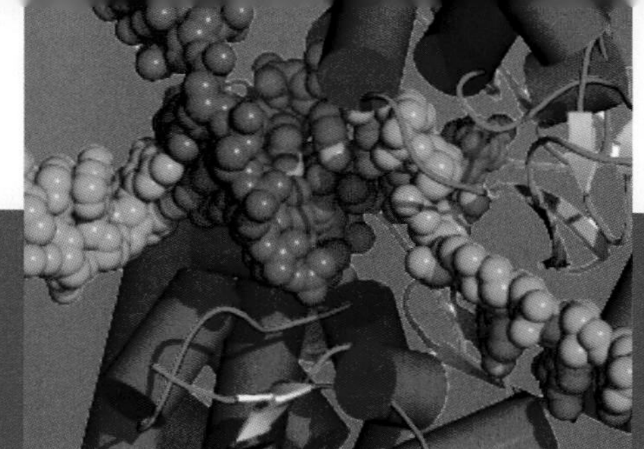

Index

G